ELECTRONICS DICTIONARY

BOOKS by JOHN MARKUS

ELECTRONIC CIRCUITS MANUAL

ELECTRONICS DICTIONARY

ELECTRONICS FOR COMMUNICATION ENGINEERS

ELECTRONICS FOR ENGINEERS

ELECTRONICS MANUAL FOR RADIO ENGINEERS

ELECTRONICS AND NUCLEONICS DICTIONARY

GUIDEBOOK OF ELECTRONIC CIRCUITS

HANDBOOK OF ELECTRONIC CONTROL CIRCUITS

HANDBOOK OF INDUSTRIAL ELECTRONIC CIRCUITS

HANDBOOK OF INDUSTRIAL ELECTRONIC CONTROL CIRCUITS

HOW TO GET AHEAD IN THE TELEVISION AND RADIO SERVICING BUSINESS

SOURCEBOOK OF ELECTRONIC CIRCUITS

TELEVISION AND RADIO REPAIRING

WHAT ELECTRONICS DOES

ELECTRONICS DICTIONARY

Accurate, easy-to-understand, and up-to-date definitions for 17,090 terms used in solid-state electronics, computers, television, radio, medical electronics, industrial electronics, satellite communication, and military electronics

FOURTH EDITION

JOHN MARKUS

Consultant, McGraw-Hill Book Company
Senior Member, Institute of Electrical and Electronics Engineers

McGraw-Hill Book Company

New York St. Louis San Francisco Auckland
Bogotá Düsseldorf Johannesburg London Madrid
Mexico Montreal New Delhi Panama Paris
São Paulo Singapore Sydney Tokyo Toronto

Library of Congress Cataloging in Publication Data

Markus, John, date.
Electronics dictionary.

Third ed. published under title: Electronics and nucleonics dictionary.
1. Electronics—Dictionaries. 2. Nuclear engineering—Dictionaries. I. Title.
TK7804.M35 1978 621.381'03 77-13876
ISBN 0-07-040431-3

1234567890 KPKP 7654321098

The editors for this book were Harold B. Crawford, Alice Goehring, and Beatrice Carson, the designer was Naomi Auerbach, and the production supervisor was Teresa F. Leaden. It was set in Baskerville by University Graphics, Inc.

Printed and bound by The Kingsport Press.

Contents

Preface

The *Electronics Dictionary* is intended to serve as a guide for engineers, technical writers, advertising copywriters, technicians, students, and secretaries. It is the fourth edition of a book that was launched in 1945 with the collaboration of the late Nelson M. Cooke. Each edition has presented an updated picture of the continually changing and expanding language of electronics.

To keep pace with the growth of the electronics industry, this edition had to be increased in size, even though many obsolete terms were dropped. The dictionary contains a total of 17,090 new and updated entries, supplemented by 1159 illustrations that were carefully chosen to clarify the more complex definitions. Thousands of new terms cover recent developments in digital watches, lasers, microcomputers, pocket calculators, satellite communication, space electronics, video games, and a host of other new developments of the past decade. Interesting new terms include ambisonics, angle jamming, autoranging, bubble memory, bucket counter, cache memory, codar, coercimometer, cruise missile, dibit, difar, domain-tip memory, EFTS, etalon, ferpic, handshaking, MIRV, pretersonics, quadraphonic, slot mask, sonography, unbundling, VFET, video game, and voiceprint. Nucleonics terms generally have been dropped in order to make room for more of the strictly electronics terms.

Completely new in this edition is an Electronics Style Manual, which is a convenient and quick reference to the abbreviating, hyphenating, and spelling rules that were followed in this dictionary. The style manual was developed because the three previous editions were official style guides for many commercial and government organizations. Used in conjunction with the dictionary, this style manual can eliminate time-wasting hours of arguments and greatly reduce manuscript editing and composition resetting costs. In addition, the style manual will apply to new terms expected to evolve during the life of this edition.

Many new terms go through an evolutionary phase between first usage and eventual adoption as part of the language of electronics. Thus a compound term starts out as two words, next takes on a hyphen, and then becomes one word, as in *push button, push-button,* and *pushbutton.* To help trace this evolution of both old and new terms, more than 50,000 citations were mounted on cards coded as to source, and file drawers were filled with article tearsheets on which possible new terms and examples of their uses were circled. After being alphabetically sequenced, this source material was analyzed to determine the word forms and meanings that best represent current usage. For example, research for the third edition indicated that *programming* and its related forms would follow the normal trend toward simplified spelling, so only one m was used in spelling such terms. How wrong that assumption was! *Programming* has two m's in this edition, and we're shopping for a new crystal ball.

Another example of language in transition is the abbreviation for operational amplifier. Today it is seen in many different forms, including Op Amp, Op-Amp, OP AMP, op amp, op-amp, and Op. Amp. The most logical form, based on historical precedent, is opamp. This was accordingly selected for the fourth edition, even though rarely used today. Will it click?

For abbreviations, company ads and catalogs and recent periodicals and books were studied. Actual usage counts were made of the various forms seen for each abbreviation, with maximum weight being given to publications that have a consistent style. This analysis showed a strong trend to all capital letters for abbreviations of common words and phrases, so practically all the abbreviations were changed to capital letters in the fourth edition.

There are two innovations in this edition: (1) a change to the internationally adopted SI abbreviations for units of measure, and (2) the addition of metric equivalents or changeover to metric units.

All terms are alphabetized letter by letter, ignoring spaces and hyphens, to give a consistent sequence in which a desired word is always where you expect it to be. Entries starting with a Greek letter are alphabetized as if the letter were spelled out.

A definition is given only once, to keep down the size of the dictionary. Synonyms are listed in their own alphabetic order, each followed by the preferred term, in italics, under which the complete definition is given.

Trademarks are capitalized. The correct use of trademarks will avoid unpleasant correspondence with corporate lawyers.

Thanks are again extended to The Institute of Electrical and Electronics Engineers and other professional organizations for permission to use in earlier editions their official definitions and illustrations, many of which were transferred unchanged to this new fourth edition. Credit for illustrations goes also to Bell Telephone Laboratories, Central Scientific Company, General Electric Company, NBS, Philco Corporation, RCA, United States Navy, Westinghouse Electric Corporation, and other electronics firms; to *IEEE Spectrum* and other IEEE publications; to *Electronics, Microwave Journal, MicroWaves, Popular Science, Radio-Electronics, QST, 73 Magazine,* and other industry publications;

and to the authors of many McGraw-Hill books. Thanks are also extended to the many engineers in the San Francisco Bay area who answered hundreds of questions about current usage and contributed to definitions of new terms.

Finally, to Joan Fife goes full credit for typing new words and definitions exactly as dictated, for her constructive help in establishing styles that represent current usage, and for her many helpful discussions, which contributed so much to the accuracy and usefulness of the definitions.

A dictionary is a growing thing, never quite complete and never perfect, no matter how much time is spent in its compilation. I accordingly welcome all suggestions for additions and corrections, which should be addressed in care of McGraw-Hill Book Company.

JOHN MARKUS

ELECTRONICS DICTIONARY

A

a Abbreviation for *atto-*.

A 1. Abbreviation for *ampere*. 2. Symbol for *anode*. 3. Symbol for *argon*.

Å Abbreviation for *angstrom*.

A− [A minus] The negative terminal of the filament voltage source for an electron tube.

A+ [A plus] The positive terminal of the filament voltage source for an electron tube.

abacus An instrument for performing arithmetic calculations manually by sliding markers on rods. Similar to biquinary notation used in some computers.

A battery The battery that supplies power for filaments or heaters of electron tubes in battery-operated equipment.

abbreviated dialing A telephone exchange service in which a connection is established with fewer than the normal number of digits. Widely used for emergency calls; for example, in Sunnyvale, California, dialing 911 connects one to the radio dispatcher for police, fire, and ambulance service.

ABC 1. Abbreviation for *automatic bass compensation*. 2. Abbreviation for *automatic brightness control*.

aberration An image defect that occurs when an optical lens or mirror does not bring all light rays to the same focus or when an electron lens does not bring the electron beam to the same sharp focus at all points on the screen of a cathode-ray tube.

ABM Abbreviation for *antiballistic missile*.

abnormal glow discharge A glow discharge characterized by an increase in the voltage drop as the current increases. It occurs when the current is increased beyond the point at which the cathode of the gas tube is completely covered with glow.

abnormal propagation Radio wave propagation in which unstable atmospheric and/or ionospheric conditions interfere with communication.

abnormal reflection A sharply defined reflection of radio waves from an ionized layer of the ionosphere, occurring at frequencies higher than the critical or penetration frequency of the layer. Also called sporadic reflection.

abort Failure of a guided missile or aerospace vehicle to accomplish its objective for any reason other than enemy action.

abrasion resistance The ability of a material to withstand mechanical wear, such as that produced by movement of a contact, brush, or wiper arm.

abrupt junction A junction in which the transition from P- to N-type material is effectively discontinuous in a single-crystal semiconductor.

abscissa The horizontal distance from a point on a graph to the zero reference line. The units of this distance are indicated on a scale at the bottom or top of the graph.

absolute address An address assigned to a particular storage location during the design of a computer. Also called specific address.

absolute altimeter An altimeter that registers the absolute altitude of an aircraft above the earth or sea over which the aircraft is flying. The frequency-modulated altimeter and the radar altimeter are the most common examples in current use.

absolute altitude The height or altitude of an aircraft above the surface or terrain over which it is flying.

absolute code A code that indicates the exact location where an item of data is to be found or stored in a computer.

absolute cutoff frequency The lowest frequency at which a waveguide will propagate energy without attenuation.

absolute delay The predetermined time interval between the transmission of two synchronized radio, radar, or loran signals from the same station or different stations.

absolute drift The amount of inherent unbalance in a magnetic amplifier, measured in terms of the watts, amperes, or ampere-turns of input signal required for rebalancing.

absolute error The magnitude of an error, disregarding its algebraic sign or direction.

absolute pressure Pressure with respect to a vacuum.

absolute temperature scale A temperature scale in which zero is the absolute zero of temperature, −273.16°C or −459.69°F. The most commonly used scale is the kelvin scale, which uses Celsius (centigrade) degrees; here absolute zero is 0 K, water freezes at 273.16 K and boils at 373.16 K. The less-used Rankine scale is based on Fahrenheit degrees; here water freezes at 491.69°R and boils at 671.69°R.

absolute unit A unit defined in terms of fundamental units of mass, length, time, and charge, such as the centimeter-gram-second electromagnetic and electrostatic units and the meter-kilogram-second-ampere electromagnetic units.

absolute value The numerical value of a number without regard to sign. Vertical lines on each side of a symbol specify that its absolute value is intended. Thus the absolute value of Z is written $|Z|$.

absolute-value converter A converter that changes an AC input signal to a unidirectional output signal while preserving instantaneous waveform amplitudes.

absolute-value device A computing element that produces an output equal to the magnitude of the input signal but always of one polarity.

absolute zero The lowest temperature that can exist, corresponding to a complete absence of molecular motion. Absolute zero is approximately −273.16°C or −459.69°F.

absorbed dose *Dose.*

absorbed dose rate The dose per unit of time, measured in rads per unit time.

absorber A material or device that takes up and dissipates radiated energy. It may be used to shield an object from that energy, prevent reflection of the energy, determine the nature of the radiation, or selectively transmit one or more components of the radiation. Examples are acoustic absorbers and microwave absorbers.

absorptance The ratio of the radiant energy absorbed in a body of material to the incident radiant energy.

absorptiometer An instrument that determines the concentration of substances by their absorption of nearly monochromatic radiation at a wavelength selected by filters or by a simple radiation-dispersing system.

absorption The dissipation of energy by radiation passing through a medium. Thus some electromagnetic energy is lost when radio waves travel through the atmosphere. Acoustic energy is lost when sound waves pass through an object. The kinetic energy of a nuclear particle is reduced when it passes through a body of matter. In another nuclear example of absorption, a particle is absorbed by a nucleus in the medium, with a different type of particle sometimes being emitted as a result.

absorption band A region of the absorption spectrum of a material in which the amount of absorption passes through a maximum.

absorption circuit A series resonant circuit used to absorb power at an unwanted signal frequency. The circuit provides a low impedance to ground at this frequency.

absorption coefficient The fraction of the intensity of a radiation that is absorbed by a unit thickness of a particular substance.

absorption cross section The sum of the cross sections for all neutron reactions with an atom except elastic and inelastic collisions.

absorption current The component of dielectric current that is proportional to the rate of accumulation of electric charges within the dielectric.

absorption discontinuity A discontinuity in the absorption coefficient of a substance for a particular type of radiation.

absorption edge The wavelength that corresponds to an absorption discontinuity.

absorption fading Gradual changes in the strength of a received radio signal, caused primarily by slow changes in absorption by the atmosphere along the signal path.

absorption frequency meter *Absorption wavemeter.*

absorption line A dark line corresponding to a peak in the absorption spectrum of a gas or a vapor.

absorption loss 1. That part of the transmission loss which is converted into heat when radiated energy is transmitted or reflected by a material. 2. Power loss in a transmission circuit caused by coupling to an adjacent circuit.

absorption mesh A filter used in a waveguide to absorb electromagnetic energy at undesired frequencies.

absorption modulation A system of amplitude modulation in which a variable-impedance device is inserted in or coupled to the output circuit of the transmitter, to absorb carrier power in accordance with the intelligence to be transmitted. In one system the modulator tubes control the absorption of the transmission line directly by means of stub connections, to achieve the same result. Also called loss modulation.

absorption peak Abnormally high attenuation at a particular frequency as a result of absorption loss.

absorption spectrophotometer A spectrophotometer in which absorption of radiation by a sample at a given wavelength is used to identify the unknown material.

absorption spectroscopy Spectroscopy that involves measurement of the energies and wavelengths of radiation absorbed by atoms and molecules of matter under various conditions.

absorption spectrum The spectrum obtained when continuous radiation is passed through an absorbing medium before it enters a spectroscope. The resulting recorded spectrum shows dark lines at wavelengths corresponding to maximum absorption.

absorption trap A parallel-tuned circuit used to absorb and thereby attenuate interfering signals.

absorption wavemeter A wavemeter that consists of a calibrated tuned circuit and a resonance indicator. When the wavemeter is lightly coupled to a signal source and tuned to resonance, maximum energy is absorbed from the source. The

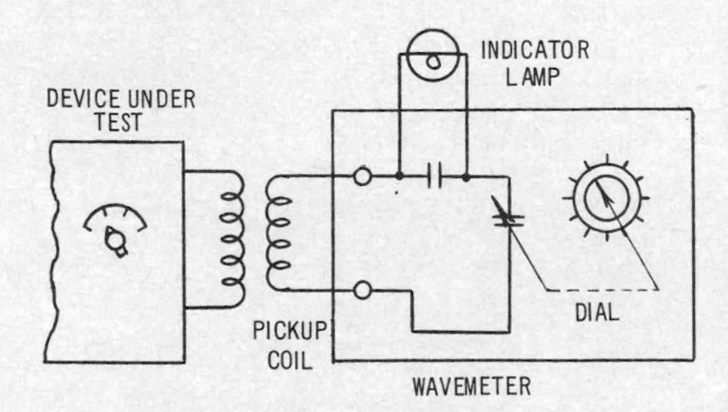

Absorption wavemeter using lamp to indicate resonance.

unknown wavelength or frequency is then read on the calibrated tuning dial. With waveguides, a cavity-type resonant circuit is used. Also called absorption frequency meter. When a vacuum-tube oscillator is a part of the resonance indicator, the instrument is usually called a grid-dip meter.

absorptive attenuator A waveguide section that contains dissipative material which gives a desired transmission loss.

absorptivity A measure of the portion of incident radiation or sound energy absorbed by a material.

AB test A method of comparing two sound systems by switching inputs so the same recording is heard in rapid succession over one system and then the other.

abundance ratio The ratio of the number of atoms of one isotope to the number of atoms of another isotope of the same element in a given sample.

AC Abbreviation for *alternating current.*

AC adapter A small power supply that plugs into an AC power outlet and delivers the DC voltage required by a portable calculator, tape recorder, or other portable battery-operated device.

AC bias An AC signal that is applied to a magnetic tape recording head along with the signal being recorded, to improve frequency response and minimize distortion and noise. The bias frequency must be several times the highest frequency value recorded.

accelerated life test Operation of a device, circuit, or system above maximum ratings to produce premature failure. Used to estimate normal operating life.

accelerating chamber An evacuated glass, metal, or ceramic envelope in which charged particles are accelerated.

accelerating electrode An electrode used in cathode-ray tubes and other electron tubes to increase the velocity of the electrons that constitute the space current or form a beam. Such an electrode is operated at a high positive potential with respect to the cathode.

accelerating tube A tubular accelerating chamber. It may be toroidal, as in a betatron, or in the form of a long cylinder, as in a linear accelerator.

acceleration The rate at which the velocity of a body changes.

acceleration space The region just outside the output aperture of the electron gun in an electron tube, in which electrons are accelerated to a desired higher velocity.

accelerator Any machine that accelerates charged particles to high velocities so they have high kinetic energy. It can be used for electrons, protons, deuterons, and ions. Also called particle accelerator. Examples include the betatron, cyclotron, linear accelerator, synchrocyclotron, synchrotron, and Van de Graaff electrostatic generator. Also called atom smasher.

accelerograph An accelerometer that records the acceleration of a point on the earth during an earthquake or records any other type of acceleration.

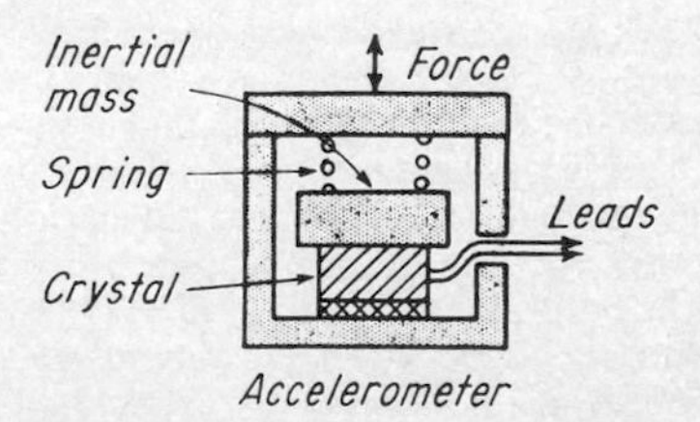

Accelerometer using piezoelectric crystal having constant-pressure loading produced by spring and inertial mass. Upward or downward acceleration changes this pressure, giving output voltage proportional to acceleration.

accelerometer A device that measures the acceleration of a moving body and translates it into a corresponding electrical quantity.

accentuation *Preemphasis.*
accentuator A circuit that provides preemphasis of certain audio frequencies.
acceptable quality level [abbreviated AQL] The percentage of defects that will be accepted a predetermined percentage of the time by a sampling plan during inspection or test of a product.
acceptable reliability level [abbreviated ARL] The percentage of failures allowed per thousand operating hours for acceptance of production parts or equipment. It is a measure of the reliability that will be accepted a predetermined percentage of the time by a reliability sampling plan.
acceptance angle The solid angle within which all received light reaches the light-sensitive area of a phototube, photodiode, or other light-sensitive device in its housing.
acceptance sampling plan A plan that specifies the sample sizes for incoming inspection and the test criteria for acceptance, rejection, or taking of another sample.
acceptance test A test that determines conformance of a product to design specifications, as a basis for acceptance.
acceptor An impurity element that increases the number of holes in a semiconductor crystal like germanium and silicon. Current flow is then essentially due to transfer of holes. Because these holes are equivalent to positive charges, the resulting alloy is called a P-type semiconductor. Aluminum, gallium, and indium are examples of acceptors.
acceptor circuit A series resonant circuit that has a low impedance at the frequency to which it is tuned and a higher impedance at all other frequencies. Used in series with a signal path to pass the desired frequency.
acceptor level An intermediate level close to the normal band in the energy-level diagram of an extrinsic semiconductor. It is empty at absolute zero. At other temperatures some electrons corresponding to the normal band can acquire energies corresponding to this intermediate level.
accessory A part, subassembly, or assembly that contributes to the effectiveness of a piece of equipment without changing its basic function. An accessory may be used for testing, adjusting, calibrating, recording, or other purposes.
access time 1. *Read time.* 2. *Write time.*
accidental coincidence Coincidence caused by the chance occurrence of unrelated counts in separate radiation detectors. Also called chance coincidence and random coincidence.
accidental coincidence correction The correction made in coincidence counting to offset the chance occurrence of unrelated signals within the resolving time of the apparatus.
accidental jamming Jamming due to transmission of radio or radar signals by friendly equipment.

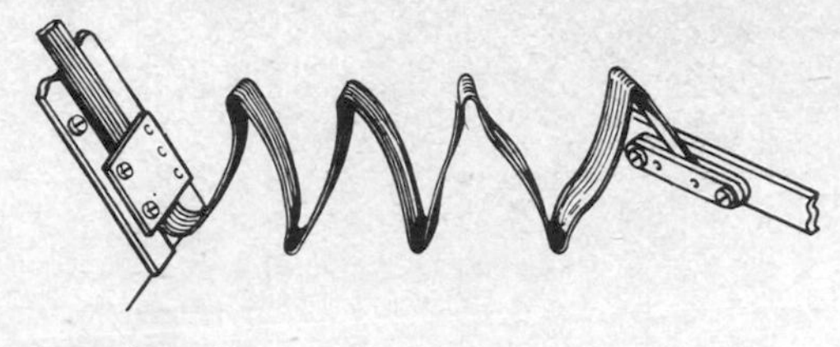

Accordion cable, as used with stress relief clamps to prevent tension on soldered connections.

accordion cable A flat multiconductor cable that is prefolded into a zigzag shape and used to make connections to movable equipment like a chassis mounted on pullout slides.
AC coupling A coupling arrangement that will not pass direct current or a DC component of a signal.
accumulating stimulus A current that is increased gradually, so it is less effective than if suddenly increased to final intensity. Used in electrobiology.
accumulator 1. A computer device that stores a number and, on receipt of another number, adds it to the number already stored and stores the sum. In another version, stored integers can be increased by unity or an arbitrary integer. An accumulator can be reset to either zero or an arbitrary integer. Also called counter. 2. British term for *storage battery.*
accuracy 1. The quality of being free from errors. 2. The extent to which the indications of an instrument approach the true values of the quantities measured.
AC/DC An abbreviation used to indicate that a receiver or other equipment will operate from either an AC or DC power line.
AC/DC receiver A radio receiver that operates from either an AC or DC power line. Also called universal receiver.
AC dump The removal of all AC power from a computer intentionally, accidentally, or conditionally. It usually results in the removal of all power.
AC erase Use of alternating current to energize an erasing head.
AC erasing head A magnetic head that uses alternating current to produce the gradually decreasing magnetic field necessary for erasing recorded signals.
acetate *Cellulose acetate.*
acetate base A transparent backing film for magnetic recording tape and motion-picture film, made from cellulose acetate. Also called safety base.
acetate disk A mechanical recording disk, either solid or laminated, made of various acetate and cellulose nitrate compounds.
acetate tape A magnetic recording tape that has an acetate base.
AC fan-out The fan-out limit of a logic circuit under high-speed conditions. Parasitic capacitances can reduce the permissible number of fan-

outs to almost half that for DC conditions.

AC generator A rotating electric machine that converts mechanical power into AC electric power.

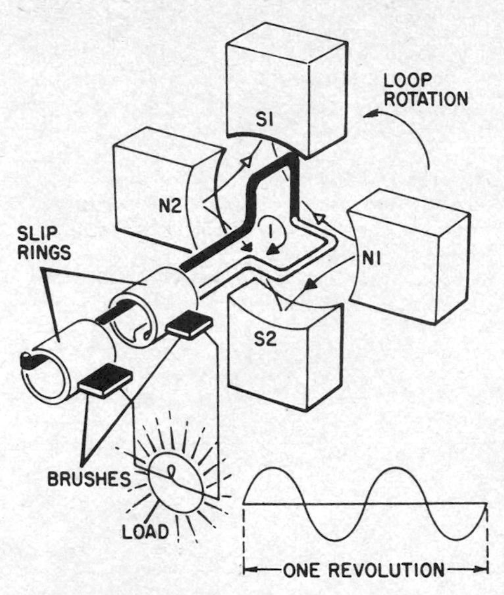

AC generator having four poles, simplified to show operating principle.

achromatic 1. Without color. 2. Capable of transmitting light without breaking it up into constituent colors.

achromatic antenna An antenna whose characteristics are uniform in a specified frequency band.

achromatic color A shade of gray.

achromatic lens A lens combination that gives correction for chromatic aberration. It is usually a convex lens of crown glass and a concave lens of flint glass; one lens corrects for the errors of the other. The combination brings all colors of light rays nearer to the same focus point.

achromatic locus An area on a chromaticity diagram that contains all points that represent acceptable reference white standards. Also called achromatic region.

achromatic point A point on a chromaticity diagram that represents an acceptable reference white standard.

achromatic region *Achromatic locus.*

achromatic stimulus A visual stimulus that gives the sensation of white light and thus has no hue.

AC Josephson effect *Josephson effect.*

aclinic line *Isoclinic line.*

acorn tube A UHF vacuum tube that resembles an acorn in shape and size. Leads come out directly through the sides of the tube. Small electrodes give low interelectrode capacitances, and close electrode spacings give low electron transit time.

acoubuoy An acoustic listening device similar to a sonobuoy, used on land to form an electronic fence that will pick up sounds of enemy movements and transmit them to orbiting aircraft or land stations.

acoustic Containing, producing, arising from, actuated by, related to, or associated with sound. The adjective acoustic is used (rather than acoustical) when the term being qualified designates something that has the properties, dimensions, or physical characteristics associated with sound waves.

acoustic absorption coefficient *Sound absorption coefficient.*

acoustic absorption loss Energy lost by conversion into heat or other forms when sound passes through or is reflected by a medium.

acoustic absorptivity *Sound absorption coefficient.*

acoustical Containing, producing, arising from, actuated by, related to, or associated with sound. The adjective acoustical (rather than acoustic) is used when the term being qualified does not explicitly designate something that has the properties, dimensions, or physical characteristics associated with sound waves.

acoustical attenuation constant The real part of the acoustical propagation constant. The commonly used unit is the neper per section or per unit distance.

acoustical intelligence Intelligence derived from sounds made by enemy sources.

acoustical ohm A unit of acoustic resistance, acoustic reactance, or acoustic impedance. The magnitude is 1 acoustical ohm when a sound pressure of 1 dyn/cm^2 (1 μbar) produces a volume velocity of 1 cm^3/s. Also called acoustic ohm.

acoustical phase constant The imaginary part of the acoustical propagation constant. The commonly used unit is the radian per section or per unit distance.

acoustical propagation constant A rating for a sound medium. It is the natural logarithm of the complex ratio of particle velocities, volume velocities, or pressures at two points in the path of a sound wave. The ratio is determined by dividing the value at the point nearer the sound source by the value at the more remote point. The real part of this constant is the acoustical attenuation constant, and the imaginary part is the acoustical phase constant.

acoustical reciprocity theorem A theorem that applies to an acoustic system. The theorem states that a simple sound source at point A in a region will produce the same sound pressure at another point B as would have been produced at A had the source been located at B.

acoustic amplifier An amplifier that increases the strength of a bulk or surface acoustic wave by an interaction involving energy transfer from traveling electric fields generated by acoustic waves in or on a piezoelectric semiconductor. The resulting charge carriers lose velocity during bunching if

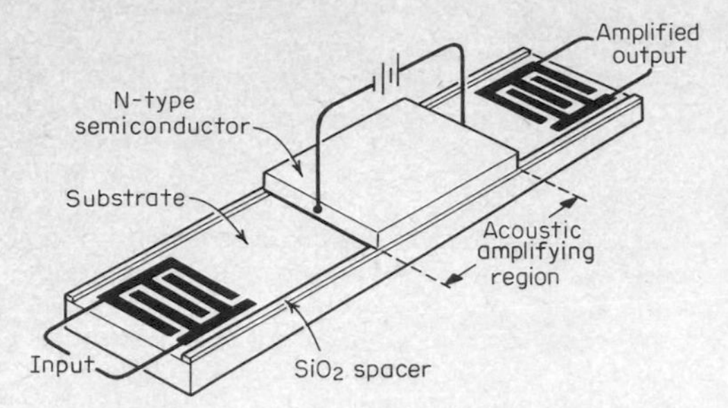

Acoustic amplifier using interdigital transducers at input and output.

the drift field is optimized for maximum amplification, and the excess kinetic energy is transferred to the acoustic wave. Also called acoustoelectric amplifier.

acoustic burglar alarm A burglar alarm that is responsive to sounds produced by an intruder. Microphones concealed in the rooms to be protected are connected to audio amplifiers that trip an alarm when sounds exceed a predetermined normal level. Also called acoustic intrusion detector.

acoustic clarifier A system of cones loosely attached to the baffle of a loudspeaker, to vibrate and absorb energy during sudden loud sounds in order to suppress these sounds.

acoustic compensator A device that matches acoustical path lengths in binaural or stereophonic audio equipment.

acoustic compliance The reciprocal of acoustic stiffness.

acoustic convolver *Convolver.*

acoustic coupler A device used between the modem of a computer terminal and a standard telephone line to permit transmission of digital data in either direction without making direct connections. When the handset is placed in the coupler, a loudspeaker converts modem output pulses to sounds for the handset microphone. Similarly, a microphone in the coupler converts computer return tone data to audio signals for amplification to the correct level for the modem.

acoustic delay line A device capable of transmitting and delaying sound pulses by recirculating them in a liquid or solid medium. For computers the pulses are usually in binary form. Also called acoustic storage and sonic delay line.

acoustic depth finder *Fathometer.*

acoustic dispersion The separation of a complex sound wave into its frequency components. It is usually caused by variation of the wave velocity of the medium with frequency. The rate of change of the velocity with frequency is a measure of the dispersion.

acoustic dissipation element An element that dissipates some or all of the acoustic energy reaching it.

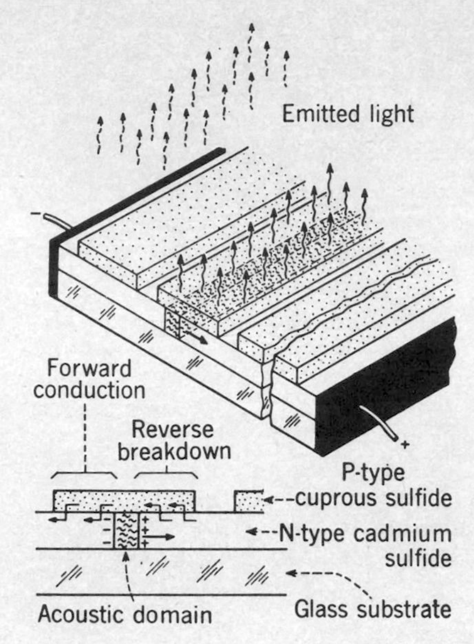

Acoustic domain moving from left to right in cadmium sulfide produces electroluminescent light.

acoustic domain A concentration of crystal lattice vibrations traveling at the speed of sound, used to generate light from an array of PN junctions.

acoustic Doppler effect The change heard in the frequency of a sound when there is relative motion between source and observer. The observed frequency increases as the distance decreases.

acoustic feedback The feedback of sound waves from a loudspeaker to a preceding part of an audio system, such as to the microphone, to aid or reinforce the input. When feedback is excessive, a howling sound is heard from the loudspeaker. Also called howling.

acoustic filter A sound-absorbing device that selectively suppresses certain audio frequencies.

acoustic generator A transducer that converts electric, mechanical, or other forms of energy into sound. Buzzers, headphones, and loudspeakers are examples.

acoustic hologram A three-dimensional image produced on a cathode-ray screen by using single-frequency sound waves to produce a phase interference pattern that is converted to a visible image by a laser light source and appropriate optics or by scanning microphones. Medical applications include detection of breast tumors.

acoustic holography Holography in which single-frequency sound waves are used to produce a three-dimensional image of an object in water, human flesh, or any other media capable of transmitting sound waves. The hologram is viewed on an oscilloscope or television screen.

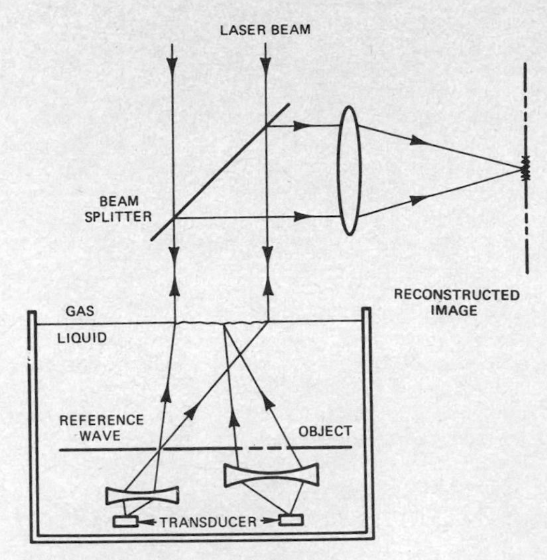

Acoustic-hologram image of underwater object is produced by laser optics from sound-diffraction pattern on liquid surface.

acoustic homing Homing on sound sources. Used in torpedoes to home on sounds made by the propellers of an enemy ship or submarine.

acoustic horn *Horn.*

acoustic imaging The production of real-time images of the internal structure of a metallic or biological object that is opaque to light. In a Bragg-diffraction version, the object is immersed in water and irradiated by plane waves of ultrasound. The resulting scattered waves are used to generate a Bragg-diffracted laser beam that produces an optical image. Also called ultrasonic imaging.

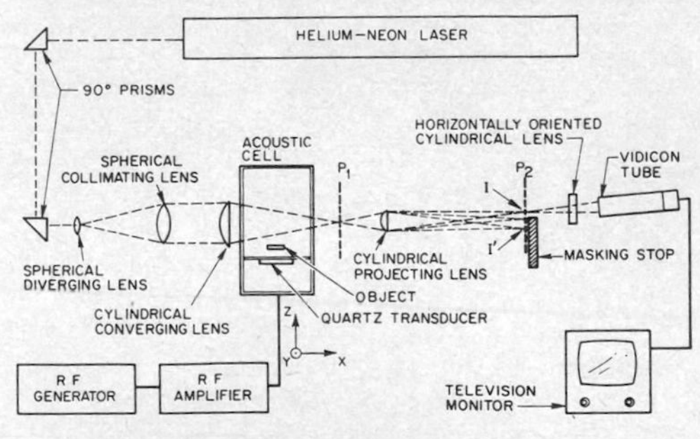

Acoustic imaging for object immersed in acoustic cell.

acoustic impedance The sound pressure on a unit area of surface divided by the sound flux through that surface, expressed in acoustical ohms. The real component of acoustic impedance is acoustic resistance, and the imaginary component is acoustic reactance. The two types of acoustic reactance are acoustic compliance and acoustic inertance.

acoustic inertance *Acoustic mass.*

acoustic interferometer An instrument that measures the velocity of sound waves in a liquid or gas. Variations of sound pressure are observed in the medium between a sound source and a reflector as the reflector is moved or the frequency is varied. Interference between direct and reflected waves produces standing waves that are related to the velocity of sound in the medium.

acoustic intrusion detector *Acoustic burglar alarm.*

acoustic jamming Generation of sound waves that interfere with enemy ground or underwater listening or acoustic homing devices.

acoustic labyrinth A loudspeaker baffle that consists of a long absorbent-walled duct folded into the volume of a cabinet, with a loudspeaker mounted at one end. The other end is open to the air in front of or underneath the cabinet. Used to reinforce bass response and prevent cavity resonance.

acoustic lens An array of obstacles that refracts sound waves in the same way that an optical lens refracts light waves. The dimensions of the obstacles are small compared to the wavelengths of the sounds being focused.

acoustic mass The quantity that when multiplied by 2π times the frequency gives the acoustic reactance associated with the kinetic energy of a medium. The unit is the gram per centimeter to the fourth power. Also called acoustic inertance.

acoustic memory A computer memory that uses an acoustic delay line in which a train of pulses travels through a medium like mercury or quartz.

acoustic methanometer An instrument that detects methane concentration in mines, based on the fact that sound travels much faster in methane gas than in air.

acoustic microscope A microscope that uses acoustic holography techniques to produce on a

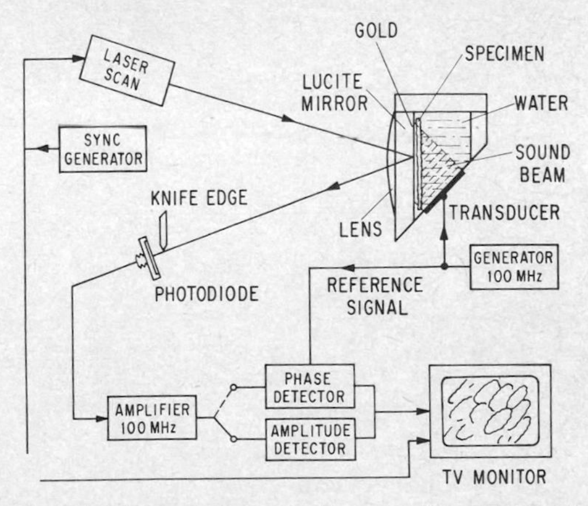

Acoustic-microscope setup used to magnify specimen of onion skin 400 times, showing individual cells on screen of monitor.

television monitor screen a greatly magnified image of an object immersed in water.

acoustic mine An underwater mine that is detonated by sound waves, such as from a ship's propeller or engines. Also called sonic mine.

acoustic mirage The distortion of a sound wavefront by a large temperature gradient in air or water, creating the illusion of two sound sources.

acoustic mode A type of thermal vibration of a crystal lattice in which neighboring points in the lattice move almost in unison.

acoustic ocean-current meter An instrument that measures current flow in rivers and oceans by transmitting acoustic pulses in opposite directions parallel to the flow and measuring the difference in pulse travel times between transmitter-receiver pairs.

acoustic ohm *Acoustical ohm.*

acoustic pickup A pickup that transforms phonograph-record groove modulations directly into sound, as in early phonographs. The phonograph needle is mechanically linked to a flexible diaphragm. Also called sound box and mechanical reproducer.

acoustic position reference system An acoustic system used in offshore oil drilling to provide continuous information on ship position with respect to an ocean-floor acoustic beacon transmitting an ultrasonic signal to three hydrophones on the bottom of the drilling ship.

acoustic radar Use of sound waves with radar techniques for remote probing of the lower atmosphere, up to heights of about 1500 m, for measuring wind speed and direction, humidity, temperature inversions, and turbulence.

acoustic radiation pressure The steady-state unidirectional pressure exerted on a surface by a sound wave.

acoustic radiator A vibrating surface that produces sound waves, such as a loudspeaker cone and a headphone diaphragm.

acoustic radiometer An instrument that measures sound intensity by determining the unidirectional steady-state pressure caused by the reflection or absorption of a sound wave at a boundary.

acoustic reactance The imaginary component of acoustic impedance. The unit is the acoustical ohm.

acoustic reflection coefficient *Sound reflection coefficient.*

acoustic reflectivity *Sound reflection coefficient.*

acoustic refraction A bending of sound waves when they pass obliquely from one medium to another in which the velocity of sound is different, as from warm water to cool water in the ocean or from warm air to cool air.

acoustic regeneration *Acoustic feedback.*

acoustic resistance The real component of acoustic impedance. The unit is the acoustical ohm.

acoustic resonator A resonator in the form of an enclosure that exhibits resonance at a particular frequency of acoustic energy.

acoustics 1. The science that deals with the production, transmission, and effects of sound, including its absorption, reflection, refraction, diffraction, and interference. 2. The properties of a room or location which control reflections of sound waves and therefore determine the character of sounds heard in that location.

acoustic scattering The irregular and diffuse reflection, refraction, or diffraction of sound in many directions.

acoustic shock Dizziness, physical pain, and sometimes nausea produced by a sudden loud sound.

acoustic sounding 1. Use of sound waves to determine water depth, by measuring the time required for a sound pulse to travel from the surface to the bottom and back. 2. Use of sound waves to study the lower atmosphere, as with acoustic radar.

acoustic spectrograph A spectrograph used with sound waves of various frequencies to study the transmission and reflection properties of thermal layers and marine life in the ocean.

acoustic stiffness The quantity that when divided by 2π times the frequency gives the acoustic reactance associated with the potential energy of a sound medium. The unit is the dyne per centimeter to the fifth power. The reciprocal of acoustic stiffness is acoustic compliance.

acoustic storage *Acoustic delay line.*

acoustic strain gage An instrument used for measuring structural strains; it consists of a length of fine wire mounted so its tension varies with strain. The wire is plucked with an electromagnetic device, and the resulting frequency of vibration is measured to determine the amount of strain.

acoustic surface wave *Surface acoustic wave.*

acoustic surveillance Use of sound pickup, amplifying, recording, and/or transmitting equipment to obtain intelligence from enemy sound sources.

acoustic theodolite An instrument that uses sound waves to provide a continuous vertical profile of ocean currents at a specific location.

acoustic transmission coefficient *Sound transmission coefficient.*

acoustic transmission system An assembly of elements adapted for the transmission of sound.

acoustic transmittivity *Sound transmission coefficient.*

acoustic treatment The use of sound-absorbing materials to give a room a desired degree of freedom from echo and reverberation.

acoustic velocity *Velocity of sound.*

acoustic wave An elastic nonelectromagnetic wave that has a frequency which may be up into the gigahertz range. One type is a surface acoustic wave, which travels on a surface that is an interface between two media (such as between a piezoelectric crystal and air). The other type is a bulk or volume acoustic wave, which travels through the material (as in a quartz delay line). Also called elastic wave.

acoustic-wave amplifier An amplifier in which the charge carriers in a semiconductor are coupled to an acoustic wave that is propagated in a piezoelectric material, to produce amplification.

acoustic-wave filter A filter that separates sound waves of different frequencies.

acoustic well logging Use of sound waves to determine depth and other properties of a borehole or liquid level in a well.

acoustoelectric amplifier *Acoustic amplifier.*

acoustoelectric effect The development of a DC voltage in a semiconductor or metal by an acoustic wave traveling parallel to the surface of the material. Also called electroacoustic effect.

acoustoelectronics *Pretersonics.*

acoustooptical cell An electric-to-optical transducer in which an acoustic or ultrasonic electric input signal modulates or otherwise acts on a beam of light.

acoustooptical filter An optical filter that is tuned across the visible spectrum by acoustic waves in the frequency range of 40 to 68 MHz.

acoustooptical material A material in which the refractive index or some other optical property can be changed by an acoustic wave.

acoustooptical modulator An arrangement for modulating a beam of light by passing it through an acoustic wave in a light-transparent solid or gaseous medium. The resulting amplitude modulation of the light beam is generally produced by diffraction, with the frequency of the acoustic wave determining the amount of diffraction or bending of the light beam.

acoustooptics The science that deals with interactions between acoustic waves and light.

AC power supply A power supply that provides one or more AC output voltages, such as an AC generator, dynamotor, inverter, or transformer.

acquisition The process of locating and following a radar target, satellite, or space probe, to obtain gun or missile firing data, telemetry data, or orbital information.

acquisition and tracking radar A radar set that locks onto a strong signal, tracks the object emitting the signal, and feeds position data directly and continuously to gun or missile control systems.

acquisition laser In an optical guidance system, a laser that radiates over a relatively large solid angle, such as 10°, for picking up the target during search or chase. When the target comes within range, a narrow-beam tracking laser takes over.

acquisition radar A radar set that detects an approaching target and feeds approximate position data to a fire-control or missile-guidance radar, which takes over the function of tracking the target.

AC receiver A radio receiver that operates only from an AC power line.

AC resistance *High-frequency resistance.*

acrylic resin A glasslike thermoplastic resin made by polymerizing esters of acrylic or methacrylic acid. Widely used for transparent parts. Trademark names include Lucite and Plexiglas.

AC tacho-generator An AC generator whose output voltage and output frequency are proportional to rotational speed.

actinic Pertaining to electromagnetic radiation capable of initiating photochemical reactions, as in photography or the fading of pigments. Photographic equipment for space vehicles requires protection from all such radiation, far beyond the wavelength ranges of ultraviolet and visible light.

actinides The series of heavy radioactive elements that have atomic numbers 89 through 103.

actinium [symbol Ac] A radioactive element. Atomic number is 89.

actinium series The series of nuclides that results from the decay of ^{235}U, including actinium A, B, C, C′, C″, D, K, and X. Mass numbers of all members are given by $4n + 3$, where n is an integer. The sequence is also known as the $4n + 3$ series or the actinouranium series.

actinoelectric Having photoconductivity.

actinometer An instrument that measures the intensity of radiation, such as by determining the amount of fluorescence produced by that radiation.

actinon [symbol An] The common name for 3.92s ^{219}Em, a member of the actinium series. Actinon is an isotope of radon.

actinouranium [symbol AcU] A common name for uranium isotope ^{235}U, the natural parent of the actinium series.

actinouranium series *Actinium series.*

action current A brief and very small electric current flowing in a nerve during a nervous impulse.

action potential The instantaneous value of the voltage between excited and resting portions of an excitable living structure.

action spectrum A graph of the action of incident light on a process or material, as a function of wavelength.

action spike The greatest in magnitude and briefest in duration of the characteristic negative waves in an action potential.

activated water Water that has ions, atoms, radicals, or molecules which are temporarily in a chemically reactive state because of exposure to ionizing radiation.

activation 1. The process of treating the cathode or target of an electron tube to create or increase its emission. Also called sensitization. 2. The process of inducing radioactivity by bombardment with neutrons or other types of radiation. 3. The process of adding liquid to a manufactured cell or battery to make it operative.

activation analysis A method of chemical analysis in which the material being analyzed is bombarded with nuclear particles, and the resulting characteristic radionuclides are detected.

activation cross section The cross section for the formation of a specified radionuclide, generally by a neutron-induced reaction.

activation detector A material that measures neutron flux or neutron density by the radioactivity induced in the material as a result of neutron capture.

activation energy The excess energy required for a particular nuclear process. An example is the energy needed by an electron to reach the conduction band in a semiconductor.

activation time The time interval from the moment activation is initiated to the moment the desired operating voltage is obtained in a cell or battery.

activator 1. An impurity atom that increases the luminescence of a solid material, such as copper in zinc sulfide and thallium in potassium chloride. 2. An impurity atom used to activate the target of a camera tube. Also called sensitizer.

active 1. Contributing to signal energy, as in transistors, electron tubes, repeaters, and other amplifying devices and systems. 2. *Radioactive.*

active air defense Air defense concerned with combatant action taken to prevent or interfere with a hostile attack by aircraft or guided missiles. It includes electronic countermeasures, air-to-air guided missiles, and surface-to-air guided missiles.

active area The portion of the rectifying junction of a metallic rectifier that carries forward current.

active chaff An expendable battery-powered jammer, usually supported by parachute or balloon, dropped by aircraft to saturate enemy radars or produce delayed false returns when triggered by enemy radars.

active communication satellite A communication satellite that amplifies a received signal before transmitting it back to earth.

active component A component capable of controlling voltages or currents, to produce gain or switching action in a circuit. Examples include diodes, ferromagnetic cores, saturable reactors, electron tubes, and transistors. Also called active device and active element.

active deposit A radioactive decay product deposited on a surface.

active device *Active component.*

active electric network An electric network or circuit that contains one or more sources of energy.

active electronic countermeasures Electronic countermeasures that involve actions of such nature that their use in jamming or otherwise disrupting enemy radio, radar, or sonar transmissions is detectable by the enemy.

active element *Active component.*

active filter A filter that uses an amplifier with conventional passive filter elements to provide a desired fixed or tunable pass or rejection characteristic.

active homing Homing in which the missile contains both the source of energy for illuminating the target and the receiver for energy reflected from the target.

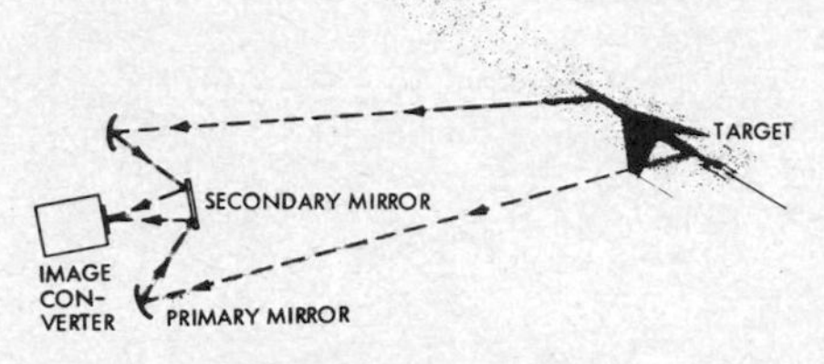

Active infrared detection system.

active infrared detection An infrared detection system in which a beam of infrared rays is transmitted toward possible targets, and rays reflected from a target are detected.

active jamming Intentional radiation or reradiation of electromagnetic energy in such a way as to impair use of a specific band of frequencies.

active line A horizontal line that carries picture information in television, as opposed to retrace lines which are blanked out during horizontal and vertical retrace.

active location A navigation system in which a navigation satellite interrogates an appropriately equipped craft, and the craft responds. When used for automated navigation, the satellite subsequently transmits position data.

active logic Logic that incorporates active components which provide such functions as level restoration, pulse shaping, pulse inversion, and power gain.

active material 1. A fluorescent material used in screens for cathode-ray tubes. Examples include calcium tungstate, zinc phosphate, and zinc silicate. 2. The lead oxide or other energy-storing material used in the plates of a storage battery. 3. Fissionable material, such as plutonium, uranium enriched in the isotope 233 or 235, and any other material capable of releasing substantial quantities of nuclear energy.

active medium A medium in which at some wavelength stimulated emission of light is more probable than absorption. The medium must have at least one quantum transition for which the higher state or energy level is more densely populated than the lower state.

active network A network whose output is dependent on a source of power other than that associated with the input signal.

active product A radioactive decay product of a radionuclide.

active region The region in which amplifying, rectifying, light-emitting, or other dynamic action occurs in a semiconductor device.

active satellite A satellite that transmits a signal. It may also have equipment for receiving and regenerating signals before retransmitting, as in a communication satellite.

active sonar Underwater sonar equipment that generates bursts of ultrasonic sound and picks up echoes reflected from submarines, fish, and other objects within range.

active substrate A semiconductor or ferrite material in which active elements are formed; it is also a mechanical support for the other elements of a semiconductor device or integrated circuit.

active tracking system A tracking system that requires a transponder, responder, or transmitter on board the vehicle to repeat, transmit, or retransmit information to the tracking equipment.

active transducer A transducer that has a power source whose output is controlled by the electric or other actuating signal.

active water homing Homing on the trail of radioactive sea water left by a nuclear-powered submarine.

activity 1. The intensity of a radioactive source. It can be expressed as the number of atoms disintegrating in unit time or the number of scintillations or other effects observed per unit time. One unit of activity is the curie, equal to 3.7×10^{10} disintegrations per second. 2. A measure of the amplitude of vibration of a crystal unit, generally expressed as the rectified grid current of the oscillator circuit in which the crystal is used. 3. A computer transaction that results in use or modification of information in the master file. 4. Short form of *radioactivity*.

activity curve A graph that shows how the activity of a radioactive source varies with time.

activity dip A decrease in the value of crystal activity, other than a band-break, that occurs over a small temperature interval. It is usually the result of loose coupling to other modes of vibration or variations in the mounting system.

activity ratio The fraction of the total records in a computer master file that is updated or otherwise processed in a given period.

AC transducer A transducer that needs an AC voltage source for proper operation.

AC transmission A mode of television transmission in which a fixed setting of the controls makes any instantaneous value of signal correspond to the same value of brightness for only a short time.

actual frequency The measured frequency of a crystal-controlled oscillator, as distinguished from the nominal frequency value that is marked on the crystal unit.

actuating signal The reference input minus the primary feedback in a control system.

actuating transfer function The transfer function that relates a feedback control loop actuating signal to the corresponding loop input signal.

actuator A mechanism that produces a desired form of motion.

AcU Symbol for *actinouranium.*

acute angle An angle numerically smaller than a right angle, hence less than 90°.

AC voltage *Alternating voltage.*

AC welder A welding machine that uses alternating current for welding purposes.

acyclic Following no regularly repeated cycles of variations.

acyclic machine *Homopolar generator.*

A/D Abbreviation for analog-to-digital, usually used in connection with converters.

adapter A device that makes electric or mechanical connections between items not originally intended for use together.

adaptive array An antenna array in which each element is independently phased according to information received from incoming signals. Used in automatic beam steering and focusing, retrodirective steering, and adaptive radar arrays. Also called self-phased array.

adaptive control system A process control system that monitors its own performance and automatically adjusts its parameters for optimum performance.

adaptive device A nonlinear device which has a memory capability, so that its resistance or other parameter depends on past operations and on changes in input.

adaptive filter A filter circuit that automatically adjusts itself to respond to fixed-waveform signals which occur in a random manner and are completely buried in noise. The filter does this without prior knowledge of the existence or shape of the waveform. Used for extending the detection range of sonar, radar, and electromagnetic reconnaissance systems. Also called adaptive waveform recognition.

adaptive quantizer A quantizer in which step size is matched to signal variance by comparing each new signal segment with the previous step size as stored in a small memory.

adaptive radar A radar system capable of adjusting its own performance parameters automatically to meet specific situations.

adaptive system A system that can change itself to meet new requirements. This involves identification of a change, comparison with previous conditions, and initiation of the corrective action required to restore optimum performance.

adaptive waveform recognition *Adaptive filter.*

ADAR [Advanced Design Array Radar] A radar system that uses two antennas and a data-processing center to locate and identify enemy intercontinental ballistic missiles.

ADC Abbreviation for *analog-to-digital converter.*

Adcock antenna A directional antenna that consists of two vertical wires spaced one half-wavelength apart or less, connected in phase opposition to give a figure-eight radiation pattern.

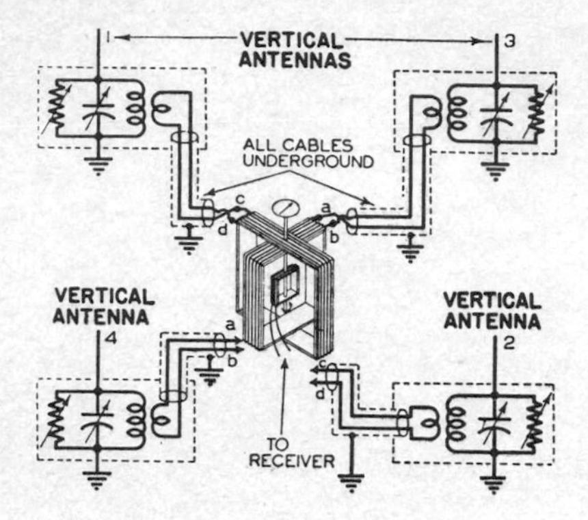

Adcock direction finder.

Adcock direction finder A radio direction finder that uses one or more pairs of Adcock antennas.

Adcock radio range An A-N radio range that uses Adcock antennas arranged at the four corners of a square on the ground. The vertical antennas at one set of opposite corners transmit the letter A in international Morse code, and the other two antennas transmit the letter N.

adconductor cathode A cathode in which adsorbed alkali metal atoms provide electron emission in a glow or arc discharge. Used in gas lasers.

A/D converter Abbreviation for *analog-to-digital converter.*

A/D-D/A An abbreviation which indicates that a converter will convert in both directions, from analog-to-digital and digital-to-analog.

add-and-subtract relay A stepping relay that can be actuated in either direction.

addend The number or quantity to be added to an existing quantity called the augend.

adder 1. A circuit in which two or more signals are combined to give an output signal amplitude that is proportional to the sum of the input signal amplitudes. In a color television receiver, the adder combines the chrominance and luminance signals. 2. A computer device that can form the sum of two or more numbers or quantities. Also called full adder.

adder-subtracter A device whose output is equal or proportional to the sum or difference of quantities represented by its two inputs, as determined by a control signal.

additive color system A system that adds two colors to form a third.

additive primaries Sources of color or light that, by additive mixture in varying proportions, can be made to match a large range of colors. The three additive primaries used in television are red, green, and blue.

additive printed circuit Printed wiring in which the conductor pattern of copper is deposited on the board. The process eliminates waste of copper etched away in conventional printed wiring.

additron A form of beam-switching tube used as a binary adder in digital computers.

address A numerical or other expression that designates a particular location in a storage device or other source or destination of information in a computer.

address computation A computation that produces or modifies the address part of a computer instruction.

address format The arrangement of the address parts of a computer instruction.

addressing system The procedure used to label storage locations on a magnetic drum or other storage of a computer.

address register A register that stores an address in a computer.

ADF Abbreviation for *automatic direction finder.*

adhesion *Bond strength.*

adiabatic Occurring without change in heat content.

adiabatic damping A reduction in the size of an accelerator beam as its energy is increased.

adiabatic demagnetization A technique used to obtain temperatures within thousandths of a degree of absolute zero. A strong magnetic field is applied to a precooled paramagnetic salt to align its atomic magnets. The material is then thermally insulated from its surroundings, and the magnetic field is removed. The resulting disorientation of the atomic magnets absorbs heat energy, thereby reducing the temperature.

adiactinic Not transmitting photochemically active rays.

adion An ion that has been adsorbed on a surface and cannot move out of it.

A display A radarscope display in which targets appear as vertical deflections from a line that rep-

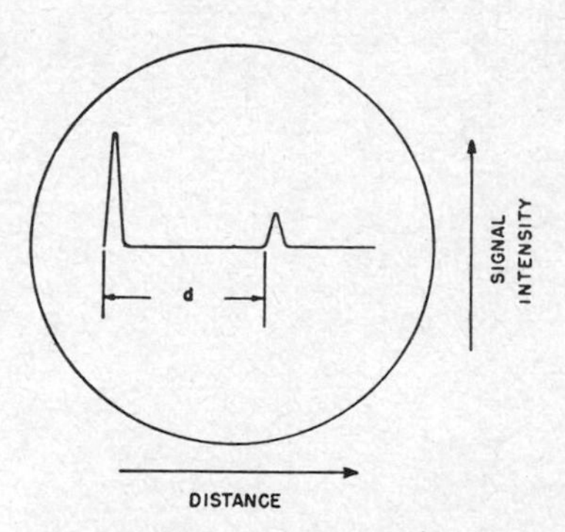

A display, with transmitted pulse at left. Distance d to echo pulse gives range of target.

resents a time base. Target distance is proportional to the horizontal position of the deflection from one end of the time base, and target echo signal intensity is proportional to the amplitude of the vertical deflection. On some scopes the display is rotated 90°, so the time base is vertical and the signal pips increase from left to right horizontally.

adjacent channel The channel immediately above or below the channel under consideration.

adjacent-channel attenuation *Selectance.*

adjacent-channel interference Interference caused by a transmitter operating in an adjacent channel. It is recognized as a peculiar garbled sound heard along with the desired program when the sidebands of the adjacent-channel transmitter beat with the carrier signal of the desired station. Also called monkey chatter, sideband interference, and sideband splash.

adjacent-channel selectivity The ability of a receiver to reject signals on channels adjacent to that of the desired station.

adjacent sound carrier The RF carrier which carries the sound modulation for the television channel immediately below that to which the receiver is tuned.

adjacent video carrier The RF carrier that carries the picture modulation for the television channel immediately above the channel to which the receiver is tuned.

adjustable resistor A wirewound resistor that has a sliding contact whose position can be changed by loosening a locking screw. Extra sliders can be added if desired. Used chiefly as a voltage divider.

adjustable short A waveguide section in which a movable wiper acts as a variable short-circuit that changes the reactance for tuning or other purposes.

adjustable voltage divider A wirewound resistor that has one or more movable terminals which can be slid along the length of the exposed resistance wire until the desired voltage values are obtained.

adjusted decibel *Decibels adjusted.*

admittance [symbol Y] A measure of how readily alternating current will flow in a circuit. Admittance is the reciprocal of impedance and is expressed in siemens. The real part of admittance is conductance, and the imaginary part is susceptance.

admittance meter A null-type instrument for measuring complex impedance and admittance in coaxial systems. It may also be used for measuring standing-wave ratios and reflection coefficients.

ADP Abbreviation for *automatic data processing.*

ADP crystal Abbreviation for *ammonium dihydrogen phosphate crystal.*

ADP microphone A crystal microphone that uses an ammonium dihydrogen phosphate crystal which has piezoelectric properties.

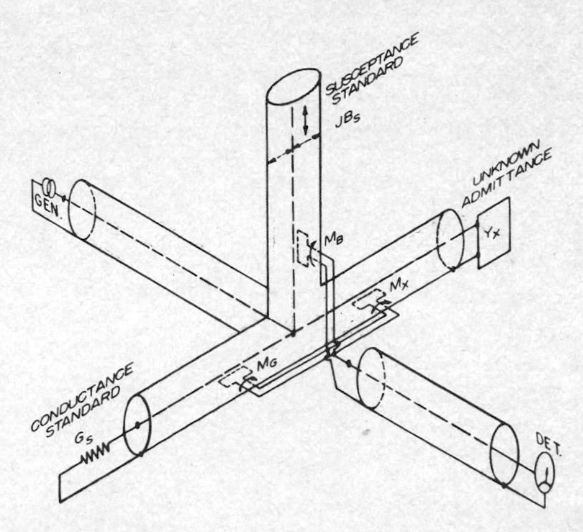

Admittance meter for frequency range of 20 MHz to 1.5 GHz. Three identical loops M in parallel, magnetically coupled to three coaxial lines, drive null detector.

Advance Trademark of Driver-Harris Co. for an alloy of copper and nickel, used in the construction of electric instruments.

advance ball A rounded support, often sapphire, that rides ahead of or alongside the cutting stylus when a mechanical recording such as a phonograph record is being made. The ball maintains a uniform depth of cut regardless of irregularities in the surface of the disk.

advanced integrated landing system [abbreviated AILS] An all-weather instrument landing system that provides precise three-dimensional guidance for aircraft during approach, flareout, landing, and rollout, as well as ground-based radar monitoring.

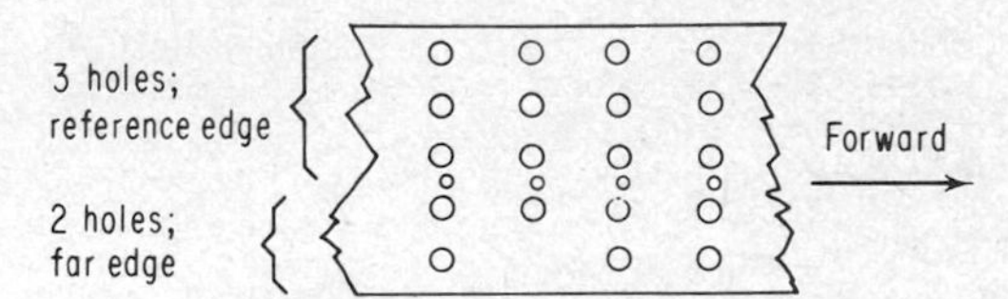

Advance feed tape having five levels of data.

advance feed tape Perforated paper tape that has the leading edges of the feed holes in line with the leading edges of the intelligence holes.

aelotropic *Anisotropic.*

aeolight A glow discharge lamp that uses a cold cathode and a mixture of inert gases. The intensity of illumination varies with the applied signal voltage. Used to produce a modulated light for motion-picture sound recording.

aerial 1. Pertaining to, existing in, or moving through the air or atmosphere. 2. British term for *antenna.*

aerial monitoring Monitoring of ground radioactivity with low-flying aircraft carrying radiation-detection equipment. Used in uranium exploration and in surveying areas in which radioactive fallout has occurred.

aerial torpedo A missile guided through the air to its target by remote control, as a glide bomb.

aeroballistic missile A wingless vehicle for flight at hypersonic speeds within the earth's atmosphere. After ballistic flight to apogee, the vehicle descends partly ballistically and partly through aerodynamic lift to an altitude of about 60,000 ft (18,300 m), then makes a ballistic dive to the surface. A slow continuous roll is imparted to the wingless vehicle during the aerodynamic portion of flight, to distribute frictional heat evenly over the airframe.

aerobiology The study of the distribution of living organisms suspended in the atmosphere.

aerodynamic heating Heating produced by air friction and air compression at high speeds.

aerodynamic missile A missile that uses aerodynamic forces to maintain its flight path. It generally will have a winged configuration.

aerodynamics The science that deals with the forces acting on bodies moving through air.

aerograph A meteorograph carried aloft by a balloon, kite, or airplane. Also called aerometeorograph.

aeromagnetic Pertaining to the magnetic field of the earth as surveyed from the air.

aerometeorograph *Aerograph.*

aerometric Pertaining to measurement of the properties of the atmosphere.

aeronautical mile A unit of length equal to 6076.11549 ft (1852 m) or about 1.15 mi, the same length as a nautical mile. Also called air mile.

aeronomy The study of the upper atmosphere, where physical and chemical reactions due to solar radiation take place.

aeropause A region of indeterminate limits in the upper atmosphere, considered the boundary between the atmosphere and outer space.

aerophare *Radio beacon.*

aerophysics Physics as related to the design, construction, and operation of vehicles that move rapidly through the atmosphere of the earth.

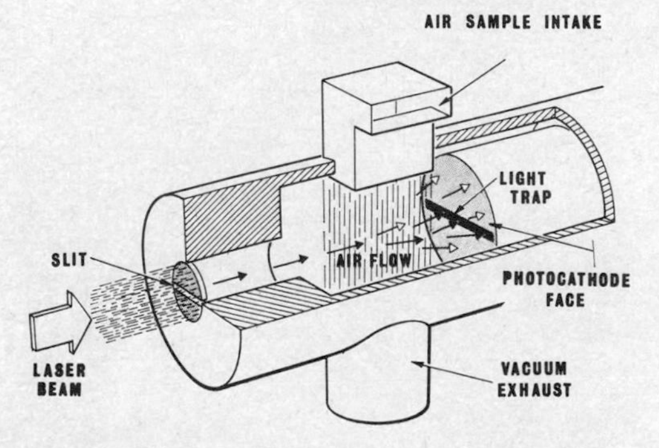

Aerosol monitor in which laser serves as light source for photoelectric pickup.

aerosol monitor An instrument capable of detecting particles smaller than 0.3 μm, to check air contamination in clean rooms. Particles in the monitored air cause scattering of light from a laser beam, with scattered light being converted to a proportional pulse by a multiplier phototube.

aerospace Pertaining to the earth's atmosphere and the space beyond.

aerospace electronics The field of electronics as applied to aircraft and spacecraft.

aerosphere The sphere that encompasses the atmosphere of the earth.

aerothermodynamic border The altitude, at about 100 mi (160 km), above which the atmosphere is so rarefied that the motion of an object through it at high speeds generates no significant surface heat.

aerothermodynamics The study of the problems connected with aerodynamic heating.

AEW radar Abbreviation for *airborne early-warning radar.*

AF Abbreviation for *audio frequency.*

AFC Abbreviation for *automatic frequency control.*

AF noise Any electric disturbance, in the audio-frequency range, that is introduced from a source extraneous to the signal.

AFSK Abbreviation for *audio frequency-shift keying.*

afterbody A section of launch vehicle that has been discarded and trails a satellite or missile. An afterbody will generally burn up during reentry.

afterglow 1. *Phosphorescence.* 2. *Persistence.*

afterpulse A spurious pulse induced in a multiplier phototube by a previous pulse.

AGC Abbreviation for *automatic gain control.*

agglomeration The union of small particles suspended in a fluid medium into larger aggregates by the action of sound or ultrasonic waves.

aggregate recoil The ejection, from the surface of a sample, of a cluster of atoms attached to an atom that is recoiling as the result of alpha-particle emission.

aging 1. Allowing a permanent magnet, capacitor, meter, or other device to remain in storage for a period of time, sometimes with voltage applied, until the characteristics of the device become essentially constant. 2. Changes in the characteristics of a device during use. 3. The process of slowing down neutrons.

agravic Pertaining to zero gravity.

Ah Abbreviation for *ampere-hour.*

aided tracking A radar antenna control system in which a constant rate of motion of the tracking mechanism is maintained by a DC motor and selsyn system so an equivalent constant rate of movement of a target in bearing, elevation, distance, or any combination of these variables can be followed. An operator adjusts the rate of motion from time to time with a potentiometer in the DC motor circuit, as required to compensate for target speed and course changes.

AIEE Abbreviation for *American Institute of Electrical Engineers.* (Now IEEE.)

AILS Abbreviation for *advanced integrated landing system.*
AIP American Institute of Physics.
AI radar Abbreviation for *airborne intercept radar.*
airborne beacon *Radar safety beacon.*
airborne collision avoidance *Collision avoidance system.*
airborne early-warning radar [abbreviated AEW radar] An early-warning radar carried by aircraft. The radar signals are relayed from the aircraft to surface stations, or their significance is reported by radio.
airborne electromagnetic prospecting Prospecting for metallic ore bodies by measuring the magnetic field of the earth with airborne instruments.
airborne intercept radar [abbreviated AI radar] Airborne radar used for detecting and tracking other aircraft at night or in clouds. It may also include computers that provide fire-control data. Also called aircraft intercept radar.
airborne magnetometer A magnetometer used in an airplane to measure variations in the magnetic field of the earth as produced by underground ore bodies or submerged submarines. The submarine-detecting version is also known as a magnetic airborne detector.
airborne moving-target indicator [abbreviated AMTI] A moving-target indicator system for airborne radar operating close to the ground, where moving targets are obscured by ground clutter, and both the ground and the target are moving with respect to the radar in the airplane.
airborne radar A self-contained radar installed in aircraft. It may provide information about ground landmarks, ships at sea, shoreline contours, other aircraft, storm clouds, or weather fronts.
airborne warning and control system [abbreviated AWACS] An Air Force program that uses airborne radar to detect low-flying enemy aircraft and high-altitude attacking bombers and to provide associated command and control functions for strategic interceptor forces.
air capacitor A capacitor that has only air as the dielectric material between its plates.
air cell A cell in which depolarization at the positive electrode is accomplished chemically by reduction of the oxygen in the air.
air cleaner *Precipitator.*
air column The air space within a horn or acoustic chamber for a loudspeaker.
air conduction The process by which sound is conducted to the inner ear through the air in the outer ear canal.
air controller *Aircraft controller.*
air-cooled tube An electron tube in which the generated heat is dissipated to the surrounding air directly, through metal heat-radiating fins, or with the aid of channels or chimneys that increase air flow.
air-core coil A coil wound on a fiber, plastic, or other nonmagnetic form, with no iron in its vicinity.
air-core transformer A transformer that has two or more coils wound on a fiber or other nonmagnetic form and no iron in its magnetic circuits. Usually designed for use as an RF transformer, IF transformer, antenna coil, or oscillator coil.
air count Measurement of radioactivity in a standard volume of air.
aircraft A vehicle that travels through the air when given lift by its own buoyancy or by dynamic reaction of air particles with its surfaces.
aircraft controller A person who controls the movements of aircraft, including guided missiles, by radar, radio communication, and other electronic devices. Also called air controller.
aircraft dB rating A rating in decibels assigned to each type of aircraft to indicate its approximate radar cross section or echo area. Used primarily with a radar coverage indicator.
aircraft flutter Sudden, flickering changes in the contrast of a television picture, caused by reflection of the television signal from an aircraft flying somewhere over the direct path between transmitter and receiver. The reflected signal alternately reinforces and cancels the normal signal at the receiving antenna. Also called airplane flutter.
aircraft instrument A mechanical, electric, or electronic device used aboard an aircraft for indicating engine performance, aircraft performance, or navigation data.
aircraft intercept radar *Airborne intercept radar.*
air defense All types of defense against enemy airborne vehicles, including missiles.
air defense controller An aircraft controller responsible for controlling and vectoring friendly aircraft during air defense and coordinating the operations of antiaircraft artillery.
air dose The x-ray dose in roentgens at a point in free air, including only the radiation of the primary beam and that scattered from surrounding air.
air equivalent A thickness of material that has the same stopping power as air for nuclear particles. Applied chiefly to materials used for walls and electrodes of ionization chambers.
air-equivalent ionization chamber *Air-wall ionization chamber.*
airframe The complete structure of an aircraft or guided missile, including the framework and skin but not the engines.
air gap 1. A short gap or equivalent filler of nonmagnetic material across the core of a choke, transformer, or other magnetic device. The gap prevents the core from being saturated by direct current or permits required mechanical movement of coils or an armature. 2. A spark gap that consists of two conducting electrodes separated by air.
air-gap crystal unit A crystal unit in which the electrodes are separate metallic plates rigidly spaced apart by an amount slightly greater than

the thickness of the quartz plate.

airglow The visible light that appears in the upper atmosphere. Caused by recombination of molecules and atoms acted on by solar radiation. In the daytime, airglow keeps the sky light even at heights of 50 mi (80 km); at night it limits photographic exposures at astronomical observatories.

air interception Visual or radar contact by a friendly aircraft with an unidentified aircraft. When broadcast-controlled, the interceptor is given the area of interception by a surface or air station and effects interception without further control. When close-controlled, the interceptor is continuously controlled by a surface or air station.

air intercept radar An airborne radar that searches for, acquires, and tracks a target to provide data needed for control of an air-to-air guided missile.

air log A distance-measuring device used in certain guided missiles to control range.

air mile *Aeronautical mile.*

air monitor A device for detecting and measuring airborne radioactivity for warning and control purposes.

air navigation aid A radar beacon, radio range, or other system, instrument, or device used in air navigation.

airplane A heavier-than-air craft supported by the dynamic reaction of air flowing over fixed or rotating plane surfaces, including piston-driven and jet airplanes, gliders, helicopters, gyroplanes, and winged guided missiles.

airplane dial A round radio receiver dial that has a pointer rotating over a scale to indicate the frequency of the station to which the receiver is tuned.

airplane flutter *Aircraft flutter.*

air pollution detector A detector that measures gaseous and solid-particle pollutants in the atmosphere. Detection techniques involve use of a laser

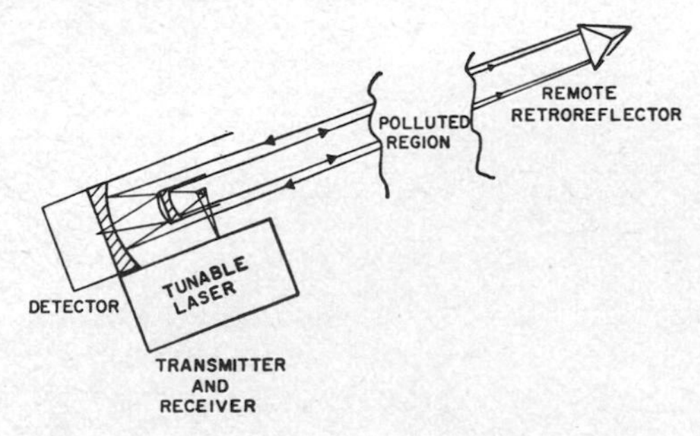

Air pollution detector using resonance absorption technique.

beam with Raman backscattering, resonance backscattering, or resonance absorption techniques.

airport A defined area on land or water, including any buildings and installations, normally used for the takeoff and landing of aircraft.

air-portable Readily carried in aircraft, with only minor or no dismantling and reassembly.

airport surface detection equipment [abbreviated ASDE] A short-range ground radar that shows the positions of all aircraft and vehicles on the surface of an airport at night and in fog. Runways, taxiways, and ramps also show clearly on the radar screen. Also called taxi radar.

airport surveillance radar [abbreviated ASR] A radar located on or near an airport to provide an indication of the bearing and distance of each aircraft within the terminal area. It is used by itself for air traffic control and with precision approach radar to form a ground-controlled approach system.

air-position indicator [abbreviated API] An airborne computing system that presents a continuous indication of aircraft position based on aircraft heading, airspeed, and elapsed time. Position is indicated in latitude and longitude values or other coordinates. True heading and air mileage flown are also shown.

air route surveillance radar [abbreviated ARSR] A long-range (approximately 150-mi or 240-km) radar used by the Federal Aviation Agency to control air traffic between terminals.

air sampler An instrument for obtaining a sample of air that can be taken to a laboratory for measurement of airborne radioactivity.

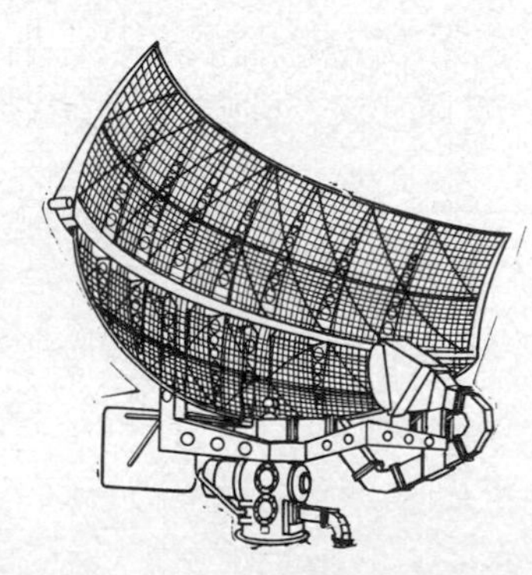

Air search radar designed for continuous rotation while tilting up and down, for long-range detection of aircraft.

air search radar Radar used to detect aircraft targets more than 50 mi (80 km) away and give range and bearing of each while maintaining complete 360° azimuth search.

air shower A grouping of cosmic-ray particles in the atmosphere.

air sounding Measuring atmospheric pressure, humidity, and other characteristics of the atmosphere with instruments carried aloft.

airspace The space above the earth, or above a specified part of the earth, in which there is sufficient air for the operation of aircraft.

air-spaced coax Coaxial cable in which the conductor is centered by beads, spirally wound plastic, or other dielectric material equivalent to air.

airspeed The speed of an aircraft, measured along its longitudinal axis relative to the air through which it moves.

airspeed computer A computer that determines true airspeed from indicated airspeed, temperature, and pressure data.

airspeed indicator A flight instrument that shows the approximate speed of an aircraft relative to the air through which it flies.

air surveillance Systematic observation of airspace by electronic, visual, or other means, for identifying and tracking friendly and enemy aircraft and missiles.

air-to-air guided missile A guided missile that fires at an airborne target from an airborne aircraft. Examples include Falcon, Sidewinder, and Sparrow.

air-to-surface guided missile A guided missile that fires at a surface vessel from an airborne aircraft.

air traffic control [abbreviated ATC] A service that monitors and controls the flow of air traffic.

air traffic control radar beacon system [abbreviated ATCRBS] A modern version of IFF, used worldwide to track aircraft for air traffic control purposes. All interrogation is at 1.03 GHz and all replies at 1.09 GHz, as established by the International Civil Aviation Organization (a United Nations agency). The reply pulses from an aircraft identify it and give its altitude. Also called secondary surveillance radar.

air-wall ionization chamber An ionization chamber in which the materials of the wall and electrodes are selected to produce ionization essentially equivalent to that in a free-air ionization chamber. Also called air-equivalent ionization chamber.

air warning system A system for warning of hostile aircraft approaching a defended area. An air warning system may include radar and communication facilities.

airwaves Slang for radio waves used in radio and television broadcasting.

alarm display A computer-generated display produced on a radarscope to alert the operator to a condition needing attention.

alarm signal The international radiotelegraph alarm signal, transmitted on 500 kHz as twelve 4-s dashes 1 s apart, to actuate automatic devices which sound an alarm to indicate that a distress message is about to be broadcast.

albedo 1. The reflection factor of a surface for neutrons. 2. A neutron field close to the earth, produced by interaction between primary cosmic rays and the upper atmosphere of the earth. 3. The ratio of the amount of light or other radiation reflected from an object to the total radiation falling on it.

albedometer An instrument that measures the albedo of an object or surface.

Alford loop A multielement antenna that has approximately equal in-phase currents uniformly distributed along each of its peripheral elements. The radiation pattern is very nearly circular in the plane of polarization.

Alford slotted tubular antenna A horizontally polarized antenna that consists of a metal cylinder which has a full-length slot. Currents flow in horizontal circles, simulating the operation of a vertical stack of in-phase loop antennas.

Alfven wave An electromagnetic wave that propagates along lines of force in magnetized plasma.

algebraic language The conventional method of writing the symbols, parentheses, and other signs of formulas and mathematical expressions. Many scientific calculators accept algebraic language directly for keyboarding of problems.

ALGOL [ALGOrithmic Language] An internationally accepted arithmetic language by which numerical procedures may be precisely presented to a computer in a standard form.

algorithm A set of well-defined rules for the solution of a problem in a finite number of steps.

algorithm translation A specific computational method for obtaining a translation from one computer language to another.

alias An alternate designation for a given data element or a given point in a computer program.

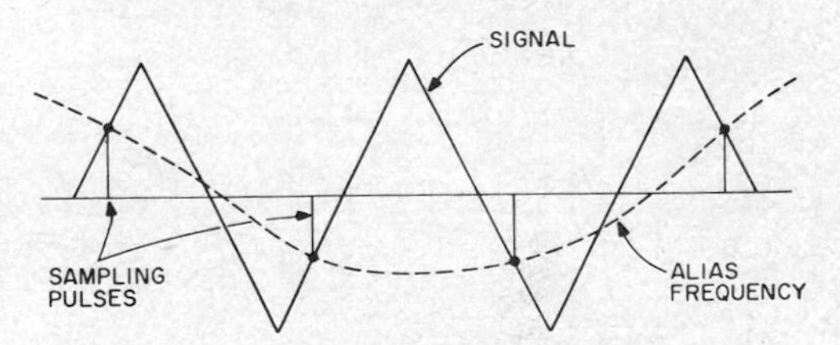

Alias frequency caused by low sampling rate.

alias frequency An erroneous lower frequency obtained when a periodic signal is sampled at a rate less than twice per cycle.

aliasing Sampling of a time function at too low a rate, so high-frequency components can impersonate low frequencies.

Alice 1. [ALaska Integrated Communications Exchange] A network of radio stations, generally using scatter propagation equipment, that links early-warning radar stations. Also called White Alice.

align To adjust two or more sections of a circuit or system so their functions are properly synchronized. Trimmers, padders, or variable inductances in tuned circuits are adjusted to give a desired response for fixed-tuned equipment or to provide tracking for tunable equipment.

aligned-grid tube A multigrid vacuum tube in which at least two of the grids are aligned one behind the other to give such effects as beam formation or noise suppression.

aligning plug The plug in the center of the base of an octal, loktal, or other tube; it has a single vertical projecting rib that prevents the tube from being inserted incorrectly into its socket.

aligning tool A small screwdriver, socket wrench, or special tool constructed partly or entirely of nonmagnetic materials, used to align tuned circuits.

alignment The process of aligning.

alignment chart *Nomograph.*

alignment pin A pin that ensures the correct mating of a tube in its socket or the correct mating of connectors.

alive *Energized.*

alkali metal An alkali-producing metal like lithium, cesium, or sodium that has photoelectric characteristics. Used in phototubes and camera tubes.

alkaline cell A primary cell that uses an alkaline electrolyte, usually potassium hydroxide, and delivers about 1.5 V at much higher current rates than the common carbon-zinc cell. In newer versions a number of cylindrical cells are mounted in a flat two-terminal plastic package that slips into a battery compartment, much as a tape cassette does. Also called alkaline-manganese cell.

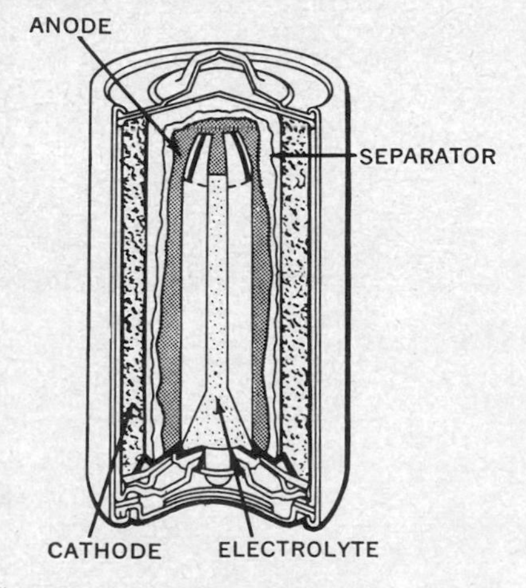

Alkaline cell providing high current drain as required for tape recorders, electric shavers, and portable television sets.

alkaline-manganese cell *Alkaline cell.*

alkaline storage battery A storage battery in which the electrolyte consists of an alkaline solution, usually potassium hydroxide.

all-channel tuning The ability of a television set to receive UHF channels 14–83 as well as VHF channels 2–13.

Allen screw A screw that has a hexagonal hole in the head.

Allen wrench A wrench made from a straight or bent hexagonal rod, used to turn an Allen screw.

alligator clip A long, narrow spring clip with meshing jaws, used with test leads to make temporary connections quickly.

Alligator clip for test lead.

allobar A form of an element that differs in isotopic composition and hence has a different atomic weight than the natural form.

allocate To assign storage locations to the main routines and subroutines in a computer, thereby fixing the absolute values of any symbolic addresses.

allochromatic Having photoelectric properties due to microscopic particles occurring naturally in a crystal or resulting from exposure to certain forms of radiation.

allochromy Fluorescence in which the wavelength of the emitted light differs from that of the absorbed light.

allowed band A band that contains a group of energy levels which electrons may occupy in a given material, such as a conduction or valence band.

allowed transition The most probable type of transition between two states of a quantum-mechanical system.

alloy A material that has metallic properties and consists of two or more elements, of which at least one is a metal.

alloy-diffused transistor A high-frequency transistor in which a single alloy contact containing both N- and P-type impurities is divided during processing. Extra impurities are added to one alloy part to form an emitter region, and a base region is formed in the other part by diffusion from within the crystal.

alloy diode A semiconductor diode that uses an alloy PN junction to give very low leakage along with high conductance.

alloy film A thin film of an alloy such as nickel-chromium, used for resistors in thin-film circuits.

alloying The process of making semiconductor junctions by melting an acceptor or donor on the surface of a semiconductor and letting it recrystallize.

alloy junction A junction produced by alloying one or more impurity metals to a semiconductor. A small button of impurity metal is placed at each desired location on the semiconductor wafer, heated to its melting point, and cooled rapidly. The impurity metal alloys with the semiconductor material to form a P or N region, depending on the impurity used. Also called fused junction.

alloy-junction photocell A photodiode in which an alloy junction is produced by alloying an indium disk into a thin wafer of N-type germanium.

alloy-junction transistor A junction transistor made by placing pellets of a P-type impurity such as indium above and below an N-type wafer of

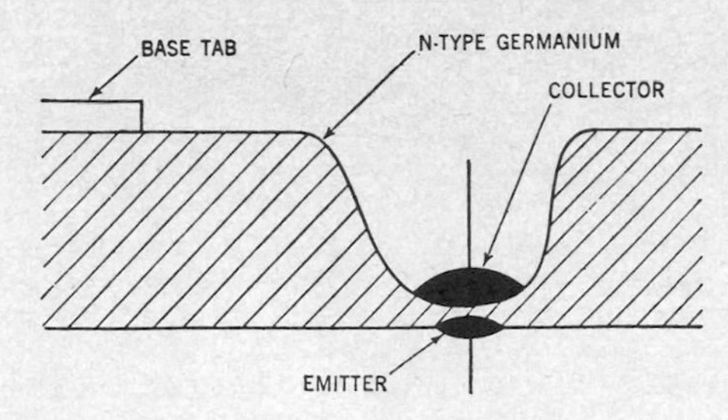

Alloy-junction transistor construction.

germanium, then heating until the impurity alloys with the germanium to give a PNP transistor.

all-pass network A network that introduces phase shift or delay without introducing appreciable attenuation at any frequency.

all-wave antenna A radio receiving antenna that responds reasonably well to a wide range of frequencies, including the shortwave and broadcast bands.

all-wave receiver A radio receiver that can be tuned to all bands used for communication and entertainment broadcasting, including FM stations.

all-weather landing system An instrument landing system that has optimum operational capability in low and zero visibility.

alnico [ALuminum NIckel CObalt] An alloy that consists chiefly of iron, aluminum, nickel, and cobalt and has high retentivity. Used to make permanent magnets, as required in loudspeakers, magnetrons, and other devices needing strong magnetic fields. Not a trademark. Usually used with an alloy number; thus, alnico V gives stronger permanent magnets than earlier alnico alloys.

alpha [Greek letter α] 1. Symbol for the current amplification factor of a transistor in a grounded-base circuit. It is the ratio of an incremental change in collector current to an incremental change in emitter current when collector voltage is held constant. 2. The ratio of radiative capture to fission cross section for a fissionable element. 3. An alphabetic character, as differentiated from numerical. 4. Symbol for *attenuation constant.*

alphabetic coding A system of abbreviations used in preparing information for input to a computer, so information may be handled in letters and words as well as numbers.

alpha chamber *Alpha counter tube.*

alpha counter An instrument that counts alpha particles; it comprises an alpha counter tube, amplifier, pulse height discriminator, scaler, and recorder.

alpha counter tube An electron tube consisting chiefly of a chamber for detecting alpha particles. It is usually operated in the nonmultiplying or proportional region. Pulse height selection is used to discriminate against pulses due to beta or gamma rays. Also called alpha chamber.

alpha cutoff The high frequency at which the alpha of a transistor drops 3 dB from its low-frequency value. The current amplification at alpha cutoff is thus 0.7 of the alpha rating for the transistor.

alpha decay The radioactive transformation that occurs when an alpha particle is emitted by a nuclide. The decay product is a new nuclide that has a mass number four units smaller and an atomic number two units smaller than the original nuclide.

alpha disintegration energy The disintegration energy of an alpha disintegration process, equal to the sum of the kinetic energy of the alpha particle and the kinetic energy of recoil of the product atom.

alpha emitter A radionuclide that undergoes transformation by alpha-particle emission.

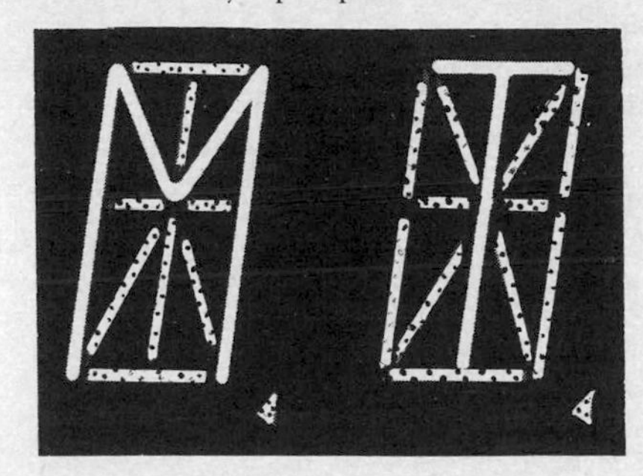
Alphameric display using 14 segments can form all numbers from 0 to 9, all capital letters of alphabet, and most special symbols found on standard typewriter.

alphameric Having letters of the alphabet, special symbols, and numerals, for data processing and display terminals. Also called alphanumeric.

alphameric readout A display of up to several hundred letters and numbers that may be generated by a cathode-ray tube, light-emitting diodes, or other character-generating methods.

alpha-neutron reaction The capture of a bombarding alpha particle by the target nucleus, producing a compound nucleus that ejects a neutron and thus changes into a new nucleus.

alphanumeric *Alphameric.*

alpha particle A positively charged particle that has two protons and two neutrons. It is emitted from certain radioactive elements or isotopes, has high ionizing power but little penetrating ability, and can damage living tissue. An alpha particle is identical in all measured properties with the nucleus of a helium atom.

alpha-particle binding energy The energy required to remove an alpha particle from a nucleus.

alpha-particle spectrum A line chart that shows the distribution, with respect to energy or momentum, of the alpha particles emitted by a radionuclide.

alpha-proton reaction The capture of a bombarding alpha particle by the target nucleus, producing a compound nucleus that ejects a proton and thus changes into a new nucleus.

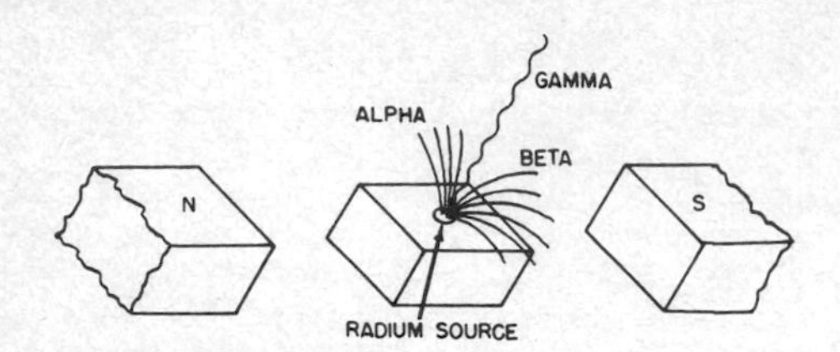

Alpha radiation is bent by magnetic field.

alpha radiation Alpha particles emerging from radioactive atoms.

alpha radiator A radioactive substance that emits alpha particles.

alpha ratio The ratio of the capture cross section of a fissionable nucleus to the fission cross section.

alpha ray A stream of alpha particles. It is only slightly deflected by a magnetic field.

alpha-ray spectrometer An instrument that determines the energy distribution of alpha particles emitted by a radionuclide.

alpha-ray spectrum The spectrum produced by separating alpha particles according to their speed.

alpha-ray vacuum gage An ionization gage in which the ionization is produced by alpha particles emitted by a radioactive source instead of by electrons emitted from a hot filament. Used chiefly for pressures above 10^{-3} mmHg, where filament life of the conventional ionization gage is seriously shortened by the positive-ion bombardment and chemical reaction with the residual gas.

alpha rhythm *Alpha wave.*

alpha spectrometer A spectrometer that measures the energies of alpha particles, generally by passing the particles through a uniform, known magnetic field. The energies of the particles can then be calculated from the curvature of the tracks.

alpha uranium An allotropic modification of uranium metal that is stable below approximately 660°C.

alpha wave A brain-wave current that has a frequency of 9 to 14 Hz. Also called alpha rhythm.

AlSb Symbol for *aluminum antimonide.*

alternating current [abbreviated AC] An electric current that is continually varying in value and reversing its direction of flow at regular intervals, usually in a sinusoidal manner. Each repetition, from zero to a maximum in one direction and then to a maximum in the other direction and back to zero, is called a cycle. The number of cycles occurring in 1 s is called the frequency in hertz. The average value of an alternating current is zero. For alternating-current terms, see AC entries.

alternating gradient A magnetic field in which successive magnets have gradients of opposite sign, so the field increases with radius in one magnet and decreases with radius in the next. Used in synchrotrons and cyclotrons.

alternating-gradient accelerator A high-energy accelerator that uses an alternating gradient.

alternating-gradient synchrotron A proton synchrotron that uses an alternating gradient for focusing, to generate beams of protons which have extremely high energy (above 25 GeV).

alternating quantity A periodic quantity whose average value is zero over a complete cycle.

alternating voltage The voltage generated by an alternator or developed across a resistance or impedance through which alternating current is flowing. This voltage is continually varying in value and reversing its direction at regular intervals. Also called AC voltage.

alternation Half an AC cycle, consisting of a complete rise and fall of voltage or current in one direction. There are 120 alternations per second in 60-Hz AC power.

alternator A machine that generates an alternating voltage when its armature or field is rotated by a motor, an engine, or other means. The output frequency is directly proportional to the speed at which the generator is driven. Also called synchronous generator.

alternator transmitter A radio transmitter that utilizes power generated by an RF alternator.

altigraph A recording altimeter.

altimeter An instrument used in air navigation to indicate altitude above sea level, above ground level at the point of measurement, or above ground level at some other point for which the altimeter was calibrated. An ordinary pressure altimeter uses an aneroid barometer to measure changes in barometric pressure with altitude. An absolute altimeter determines altitude above ground or water by measuring the time it takes for a radio or radar wave to travel straight down and be reflected back to the aircraft.

altimetric flareout An aircraft descent path in which the rate of descent is reduced as the touchdown point is approached. Also called exponential flareout.

altitude The height of an aircraft or other object above a given level, as above the sea or ground. Also called elevation.

altitude chamber A chamber in which air pressure, humidity, and temperature can be controlled to simulate conditions at various altitudes for test purposes.

altitude hole The small circle in the center of an airborne PPI radar presentation of ground terrain, corresponding to the time required for the radar signal to travel from the aircraft to the ground and back.

altitude signal The radar signal returned to an airborne radar set by the ground or water surface directly beneath the aircraft.

ALU Abbreviation for *arithmetic-logic unit.*

alumina A ceramic used for insulators in electron tubes because it is easily degassed. It can withstand temperatures up to 1400°C continuously and has low dielectric loss over a wide frequency range. Also called aluminum oxide.

alumina passivation Passivation of a semiconductor region by depositing a layer of alumina on silicon oxide previously applied by the planar process.

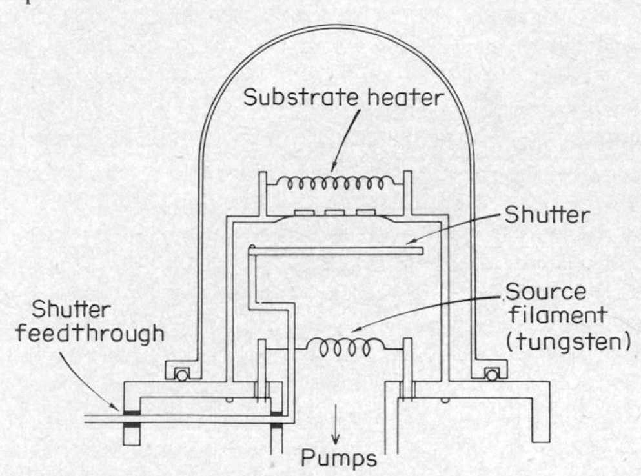

Aluminization of substrate (mounted above shutter) in vacuum by evaporation of aluminum placed over heated source filament.

aluminization Evaporation of a thin film of aluminum onto a substrate for a semiconductor device or integrated circuit, to provide conductive patterns corresponding to openings in a mask over the substrate.

aluminized screen A television picture tube screen that has a thin coating of aluminum on the back of its phosphor layer. Electrons in the beam readily penetrate the coating and activate the phosphors to produce an image. The aluminum reflects outward the light that would otherwise go back inside the tube, thereby improving the brilliance and contrast of the image. Also called metal-backed screen, metallized screen, and mirror-backed screen.

aluminum [symbol Al] A lightweight silvery white metal whose atomic number is 13. Widely used in electronic equipment for capacitor foil, tuning capacitor plates, shields, and equipment housings.

aluminum antimonide [symbol AlSb] A semiconductor that has a forbidden band gap of 2.2 eV and a maximum operating temperature of 500°C when used in a transistor.

aluminum electrolytic capacitor An electrolytic capacitor that uses plain or etched aluminum foil for both electrodes.

aluminum oxide *Alumina.*

alundum A form of aluminum oxide used in some electron tubes to insulate the heater from the cathode.

AM Abbreviation for *amplitude modulation.*

A/m Abbreviation for *ampere per meter.*

amateur A person holding a license, issued by the Federal Communications Commission or by corresponding authorities in other countries, which authorizes that person to operate a licensed amateur radio station for pleasure and service only, not for profit. Also called ham (slang).

amateur band A band of frequencies assigned exclusively to amateur operators.

ambience The acoustic environment of a stereo high-fidelity system, as determined by sound reflections from walls and furnishings, sound absorption by drapes and carpets, and loudspeaker positioning to approximate the acoustics of the concert hall or studio in which the original musical program was broadcast or recorded.

ambient condition The condition of the surrounding medium.

ambient fuze A proximity fuze that is activated by some characteristic of the environment in which the target is normally found, rather than by the presence of the target.

ambient light Normal room light, such as that reaching a television picture tube screen from the outside.

ambient-light filter A filter used in front of a television picture tube screen to reduce the amount of ambient light reaching the screen and to minimize reflections of light from the glass face of the tube. The filter, generally a dull color, can be incorporated into the glass faceplate of the tube or the safety-glass window, or it can be a separate sheet of plastic.

ambient noise Noise associated with a given environment, made up of more or less continuous sounds from near and far sources. It is generally undesirable.

ambient temperature The temperature of the immediately surrounding medium. If the device is giving off heat, the air in its vicinity will be heated, and the ambient temperature will then be higher than room temperature.

ambiguity 1. The condition in which navigation coordinates define more than one point, direction, line of position, or surface of position. 2. The condition in which a synchro or servomechanism seeks more than one null position.

ambipolar photoconductivity Photoconductivity in which the probability of capture of photoionized carriers in a semiconductor is about the same for either N- or P-type material.

ambisonics The reproduction of sound as it would be heard at a given point in the concert hall, including reverberations and other ambience characteristics of the hall.

American Institute of Electrical Engineers [abbreviated AIEE] A nonprofit professional organization of engineers and scientists established for the advancement of the theory and practice of electrical engineering. Now a part of IEEE.

American Morse code A dot-dash code used in wire telegraphy. It has a different spacing method from the international Morse code used in radio and entirely different letter codes.

American National Standards Institute [abbreviated ANSI] An organization supported by U.S. industry to establish uniform standards that permit maximum interchangeability of products. Formerly known as United States of America Standards Institute and American Standards Association.

American Standard Code for Information Interchange [abbreviated ASCII] A standard code that has seven channels and an eighth channel for parity. Developed to simplify the interconnection of computers, communication circuits, and digital output equipment.

American Standards Association [abbreviated ASA] Former name of American National Standards Institute, Inc. [abbreviated ANSI].

American wire gage [abbreviated AWG]. A gage used chiefly for specifying nonferrous wire and sheet metal. Sizes range from 0000 for the largest (0.46 in or 1.17 cm) to 0 (0.325 in or 0.8255 cm) and 1 (0.289 in or 0.654 cm) to 50 (0.001 in or 0.00254 cm). Also called Brown and Sharpe gage.

americium [symbol Am] An unstable transuranic radioactive element produced artificially by bombarding plutonium with helium ions. Atomic number is 95.

AM/FM An abbreviation which indicates that a radio receiver or other equipment will handle either AM or FM signals.

AML Abbreviation for *automatic modulation limiting.*

AMM Abbreviation for *antimissile missile.*

ammeter An instrument that measures current flow. Its scale may be calibrated in amperes or smaller units. An ammeter that indicates milliampere values is often called a milliammeter. An ammeter that indicates microampere values is often called a microammeter.

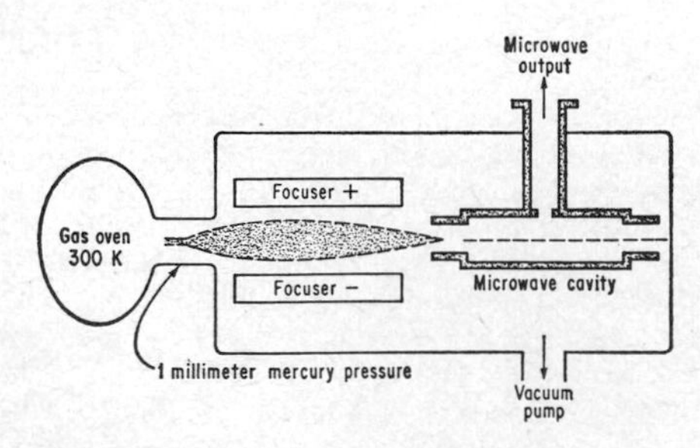

Ammonia-beam maser.

ammonia-beam maser A gas maser that uses ammonia as its paramagnetic material.

ammonia maser clock A gas maser that utilizes the transition of high-energy ammonia molecules to generate a stable microwave output signal for use as a time standard.

ammonium dihydrogen phosphate crystal [abbreviated ADP crystal] A piezoelectric crystal used in sonar transducers.

amorphous film A magnetically ordered metallic film that can be deposited on a semiconductor chip or on almost any other material without need for a crystal substrate, for magnetic bubble memories.

amorphous laser *Glass laser.*

amorphous semiconductor A noncrystalline semiconductor that has unique optical and electrical properties applicable to luminescent display panels, fast electric switching, and memory devices.

ampacity Current-carrying capacity in amperes. Used as a rating for power cables.

amperage The amount of current in amperes.

ampere [abbreviated A] The practical unit of electric current. A voltage of 1 V will send a current of 1 A through a resistance of 1 Ω.

ampere-hour [abbreviated Ah] A unit of quantity of electricity. Multiplying current in amperes by time of flow in hours gives ampere-hours. Used chiefly to indicate the amount of energy that a storage battery can deliver before it needs recharging or that a primary battery can deliver before it needs replacing.

ampere-hour efficiency The efficiency of a storage battery, equal to the ratio of ampere-hour output to ampere-hour input required for recharging.

ampere-hour meter A meter that measures current drawn per unit of time, integrated for reading in ampere-hours.

ampere per meter [abbreviated A/m] The SI unit of magnetic field strength.

Ampere's law The magnetic intensity at any point near a current-carrying conductor can be computed on the assumption that each infinitesimal length of the conductor contributes an amount which is directly proportional to that length and to the current it carries, inversely proportional to distance, and directly proportional to the sine of the angle.

Ampere's rule The magnetic field surrounding a conductor will have a counterclockwise direction when the electron flow is away from the observer.

ampere-turn The SI unit of magnetomotive force, equal to coil turns multiplied by coil current in amperes.

ampere-turn amplification The ratio of the change in output ampere-turns of a magnetic amplifier to the change in control ampere-turns.

amplidyne [AMPLIfier DYNE] A rotating magnetic amplifier that consists of a combination DC motor and generator which has special windings and brush connections to give power amplification, so small changes in power input to the field coils produce large changes in power output. Widely used as a power amplifier in servosystems.

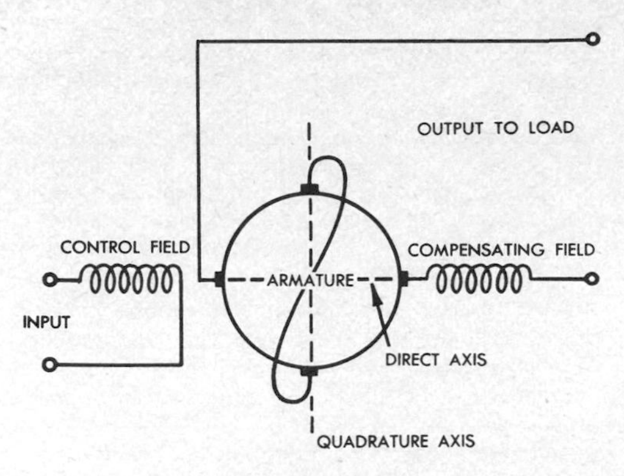

Amplidyne connections.

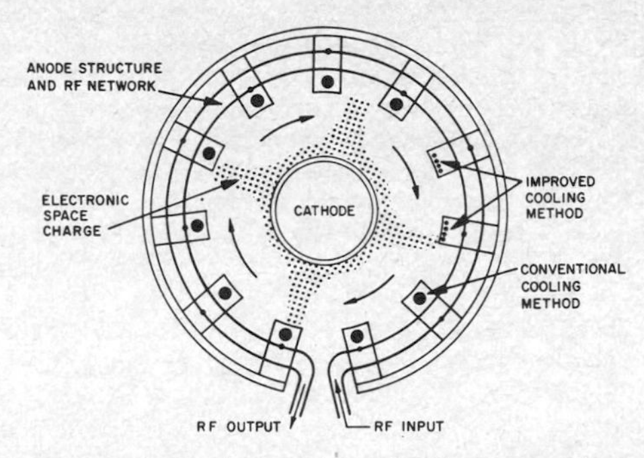

Amplitron interaction region, showing space-charge spokes (shaded areas) that rotate about center cathode.

amplification The process of increasing the strength (current, voltage, or power) of a signal.

amplification factor [symbol μ] The ratio of the change in anode voltage of an electron tube to a change in control-electrode voltage that produces the same change in anode current when other tube voltages and currents are held constant.

amplified AGC An automatic gain-control circuit in which the control voltage is amplified before being applied to the tube whose gain is to be controlled in accordance with the strength of the incoming signal.

amplifier A device that uses an electron tube, transistor, magnetic unit, or other amplification-producing component to increase the strength of a signal without appreciably altering its characteristic waveform. An amplifier transfers power to

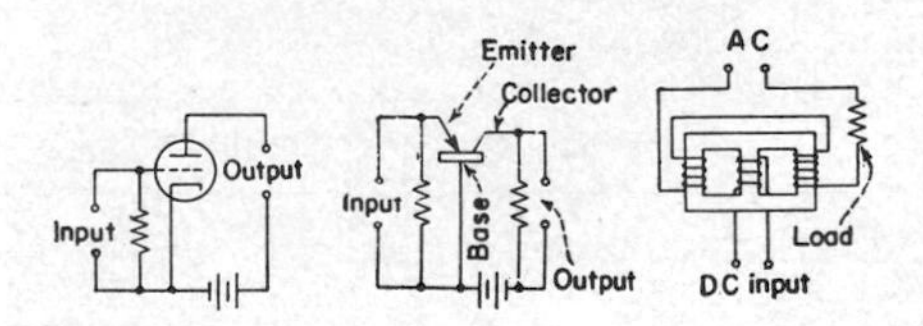

Amplifiers. Left: electron tube; center: transistor; right: magnetic.

the signal from an external source, whereas a transformer changes signal voltage or current without adding power.

amplifier noise Unwanted signals existing in a completely isolated amplifier when there is no input signal.

amplify To increase in magnitude.

amplitron A crossed-field device that provides efficient amplification of microwave signals by using a space-charge rotor which has very little mass and rotates at extremely high speed. The corresponding oscillator version is a magnetron.

amplitude The value of a varying quantity at a specified instant.

amplitude balance control *Sensitivity-time control.*

amplitude discriminator *Pulse-height discriminator.*

amplitude distortion *Frequency distortion.*

amplitude fading Fading in which all frequency components of a modulated carrier signal are uniformly attenuated in amplitude.

amplitude-frequency distortion *Frequency distortion.*

amplitude-frequency response A graph that shows how the gain or loss of a device or system

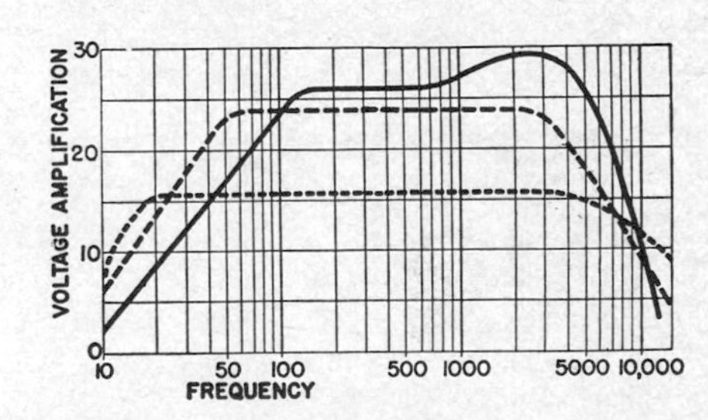

Amplitude-frequency response curves for three different AF amplifiers.

varies with frequency. Also called frequency characteristic, frequency response, response, response characteristic, and sine-wave response.

amplitude gate A gate circuit that transmits only the portions of an input wave lying between two amplitude values.

amplitude limiter *Limiter.*

amplitude-modulated transmitter A transmitter that transmits an amplitude-modulated wave.

amplitude-modulated wave A sinusoidal wave whose envelope contains a component similar to the waveform of the signal to be transmitted.

amplitude modulation [abbreviated AM] Modulation in which the amplitude of the carrier-frequency current is varied above and below its normal value in accordance with the audio, picture, or other intelligence signal to be transmitted.

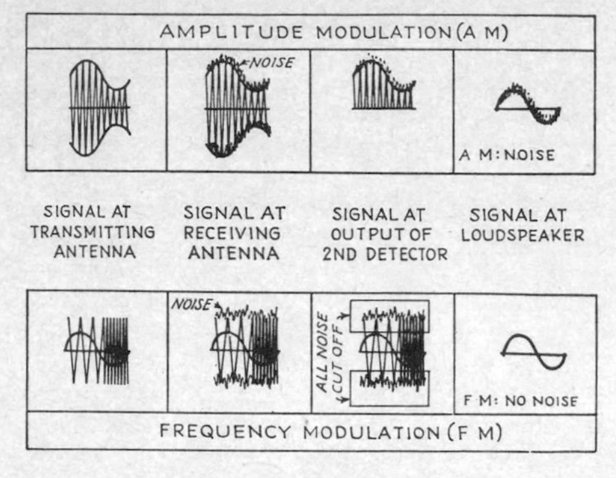

Amplitude modulation compared with frequency modulation in noise-suppressing action.

amplitude-modulation noise level The noise level produced by undesired amplitude variations of an RF signal in the absence of any intended modulation.

amplitude-modulation rejection The ability of an FM radio receiver to reject amplitude-modulated RF interference from man-made sources or electric storms.

amplitude noise Noiselike variations in the amplitude of the signal reflected from a radar target, caused by changes in target aspect.

amplitude range The ratio between the upper and lower limits of audio signal amplitudes that contain all significant energy contributions.

amplitude selection The selection of the portion of a waveform that lies above or below a given value or between two given values.

amplitude selector *Pulse-height selector.*

amplitude separator A circuit used to isolate the portion of a waveform that is above or below a given value or between two given values.

amplitude-shift keying [abbreviated ASK] Data transmission in which the signals from computers or data terminals produce a number of different amplitude levels of a sine-wave carrier.

AM/SSB *SSB/AM.*

AMTI *Airborne moving-target indicator.*

AMVER [Automated Merchant VEssel Report] A U.S. Coast Guard service that stores in a computer the positions of merchant vessels and aircraft, for locating quickly the vessels nearest the location from which an SOS is transmitted by a craft in distress.

An Symbol for *actinon.*

anacoustic zone The zone of silence in space, starting at about 100 mi (160 km) altitude, where the distance between air molecules is greater than the wavelength of sound and sound waves can no longer be propagated.

analog The representation of numerical quantities by physical variables such as translation, rotation, voltage, and resistance.

analog channel A channel on which the information transmitted can have any value between the channel limits. A voice channel is an analog channel.

analog comparator A comparator that produces a high digital output signal (binary 1) when the sum of two analog voltages is positive, or a low output signal (binary 0) when their sum is negative.

analog computer A computer that solves problems by setting up equivalent electric circuits and making measurements as the variables are changed in accordance with the corresponding physical phenomena. An analog computer gives approximate solutions, whereas a digital computer gives exact solutions.

analog data Data represented in a continuous form, as contrasted with digital data that have discrete values.

analog device A control device that operates with variables represented by continuously measured voltages or other quantities.

analog-digital computer *Hybrid computer.*

analog multiplier A device that accepts two or more inputs in analog form and produces an output proportional to the product of the input quantities.

analog panel meter The conventional meter in which the value being measured is indicated by a pointer moving over a calibrated scale, as contrasted to digital panel meters that show the exact value on a numerical display.

analog signal A control signal whose magnitude represents information content.

analog telemetering Telemetering in which some characteristic of the transmitted signal is proportional to a quantity being measured.

analog-to-digital converter [abbreviated A/D converter or ADC] A converter in which analog input signals are changed to essentially proportional digital signals. Also called digitizer.

analog-to-frequency converter A converter in which an analog input in some form other than frequency is converted to a proportional change in frequency.

analyst A person skilled in the definition of problems and the development of algorithms for their solution on a computer.

analytic inertial navigation Inertial navigation in which outputs of accelerometers that have inertia-maintained orientations are converted to geographic navigational data by automatic computers.

analyzer 1. A Nicol prism or other device for detecting and testing polarized light. 2. A computer routine used to analyze a program written for the same or a different computer. 3. A test instrument used to check performance or locate trouble in electronic equipment by measuring and displaying the voltages, currents, and/or other parameters at one or more test points simultaneously with meters, digital displays, cathode-ray displays, or other indicating means. Examples include logic, pulse, and set analyzers. 4. A test instrument used to display and identify frequencies or other parameters of interest in an input signal. A spectrum analyzer is an example.

anamorphic Having a difference in optical magnification along two mutually perpendicular axes.

ancillary equipment Auxiliary equipment.

AND A logic operator in which the output is a logic 1 if all the inputs are a logic 1.

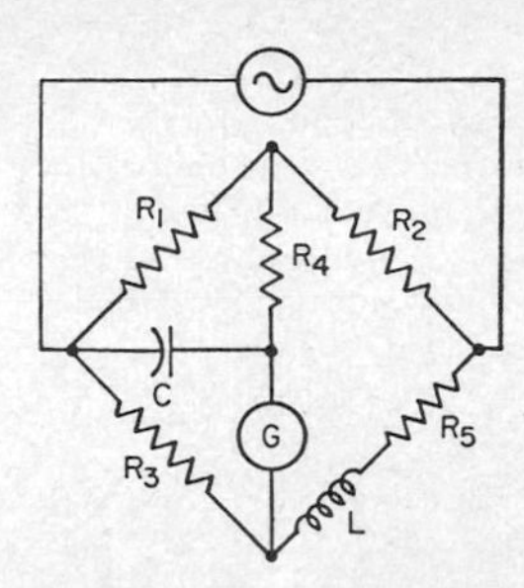

Anderson-bridge circuit.

Anderson bridge A six-branch modification of the Maxwell-Wien bridge, used to measure self-inductance in terms of capacitance and resistance. Bridge balance is independent of frequency.

AND gate A multiple-input gate circuit whose output is energized only when every input is energized simultaneously in a specified manner. Used in digital computers.

AND NOT A logic operator that is equivalent to the EXCEPT operator.

AND/OR gate A gate that produces an output for one of several possible combinations of inputs. It combines the characteristics of AND and OR gates.

anechoic room 1. A room in which the floor, ceiling, and all walls are lined with a sound-absorbing material to reduce reflections of sound to a minimum. Also called dead room and free-field room. 2. A room completely lined with a material that absorbs radio waves at a particular frequency or over a range of frequencies. Used principally at microwave frequencies, such as for measuring radar beam cross sections.

anelectrotonus Reduced excitability of a nerve or muscle that is near the anode during passage of direct current through living tissue. See also catelectrotonus.

anemograph An instrument that records wind velocity.

anemometer An instrument that measures wind velocity.

aneroid An evacuated chamber that has flexible metal walls, used to convert changes in atmospheric pressure into corresponding mechanical movements. When atmospheric pressure increases, the chamber becomes smaller.

aneroid altimeter An aircraft altimeter in which an aneroid is used to sense changes in atmospheric pressure and actuate an indicator calibrated in altitude above sea level or some other reference level. Atmospheric pressure decreases with altitude.

aneroid barometer A barometer that uses an aneroid to drive a pointer which indicates atmospheric pressure.

angel 1. A radar echo that comes from an invisible and sometimes unknown origin. High-flying birds and swarms of insects are some causes. Unusual atmospheric conditions can also cause radar signal reflections that result in target indications on a radar screen when the sky is perfectly clear. 2. A corner reflector or other metallic reflecting material suspended from a balloon or kite to simulate a radar target and thereby confuse the enemy.

angiocardiogram An x-ray of the blood vessels of the heart, made after injecting a radiopaque material into the bloodstream.

angiogram An x-ray of blood vessels, made after injecting a radiopaque material into the bloodstream.

angle The measure of the progression of a sine wave in time or space from a chosen instant or position, or the corresponding amount through which the rotating vector of the wave has progressed.

angle diversity Diversity reception in which beyond-the-horizon tropospheric scatter signals are received at slightly different angles, equivalent to paths through different scatter volumes in the troposphere.

angle jamming An electronic countermeasures technique in which a jamming pulse is transmitted that is similar to the returning echo pulse of an enemy scanning fire-control radar, except it has the phase angle of the modulation information changed.

angle modulation Modulation in which the angle of a sine-wave carrier is the characteristic varied from its normal value. Phase modulation and frequency modulation are particular forms of angle modulation.

angle noise In radar, tracking error caused by variations in echo arrival area as the target changes its aspect.

angle of elevation The angle between the horizontal plane and the line ascending to the object.

angle of incidence The angle between a wave or

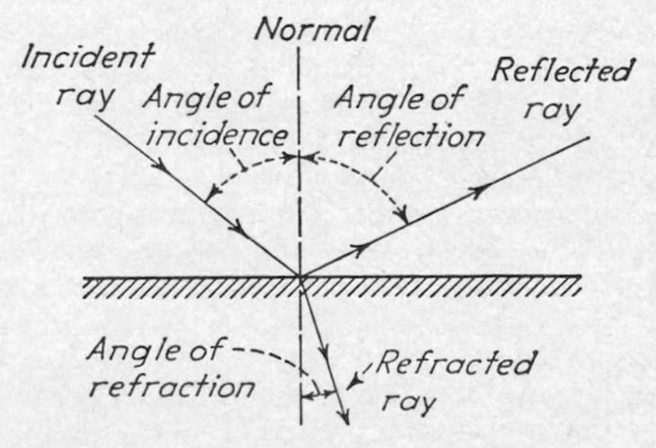

Angle of incidence, angle of reflection, and angle of refraction for light ray or electromagnetic wave.

beam arriving at a surface and the perpendicular to the surface at the point of arrival.

angle of reflection The angle between a wave or beam leaving a surface and the perpendicular to the surface.

angle of refraction The angle between a refracted wave or beam and the perpendicular to the refracting surface.

angstrom [abbreviated Å] A unit of wavelength of light and other radiation. One angstrom is 10^{-8} cm or 10^{-4} μm, so 1 μm is 10,000 Å. The range of visible light is between about 4000 and 7500 Å (0.4 and 0.75 μm). Ultraviolet radiation is below 0.4 μm, and infrared radiation is above 0.75 μm. The term micron, as a contraction of micrometer, was abrogated by international agreement in 1967.

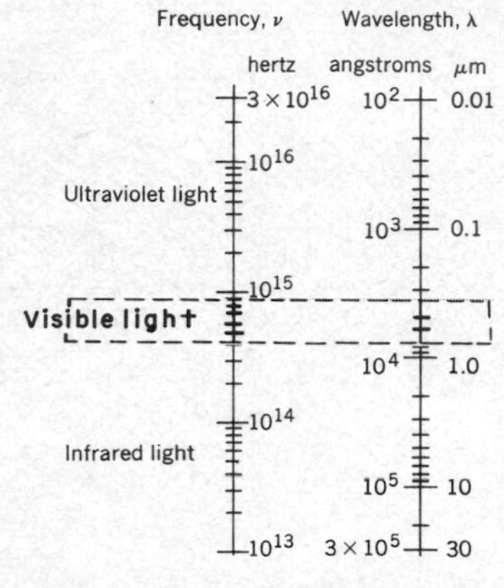

Angstrom nomograph, giving equivalent wavelength and frequency values covering range of lasers.

angular acceleration The rate of change of angular velocity about a rotational axis.

angular accelerometer An accelerometer that measures the rate of change of angle between two objects under observation.

angular deviation loss The ratio of the response of a microphone or loudspeaker on its principal axis to the response at a specified angle from the principal axis, expressed in decibels.

angular deviation sensitivity The ratio of the change in a course-indicator reading to the actual angular change in the course of an aircraft or ship.

angular distance Distance expressed in radians or degrees. It is equal to the distance in wavelengths multiplied by 2π rad or 360°.

angular distribution The relative number of particles coming off at various angles from a nuclear reaction or event, expressed with respect to some direction, such as that of the incident beam.

angular frequency The frequency expressed in radians per second. It is equal to the frequency in hertz multiplied by 2π.

angular momentum The angular velocity of a body multiplied by its moment of inertia.

angular-momentum quantum number A quantum number that determines the total angular momentum of a molecule exclusive of nuclear spin.

angular-position pickup An instrument that translates variations in angular position into an electrical quantity, such as a change in inductance, resistance, or voltage.

angular resolution The ability of a radar to distinguish between two targets solely by the measurement of angles. It is generally expressed as the minimum angle by which targets must be spaced to be separately distinguishable.

angular velocity [symbol ω] 1. The speed of a rotating object measured in radians per second. It is equal to revolutions per second multiplied by 2π. 2. The rate of change of phase of an alternating quantity. It is equal to the frequency in hertz multiplied by 2π.

angular width *Course width.*

anhysteresis Magnetization with a unidirectional field upon which is superposed an alternating field of gradually decreasing amplitude.

anion A negative ion.

anisotropic Having different properties in different directions. Also called aelotropic.

annihilation 1. A process in which a particle and its antiparticle meet and convert spontaneously into one or more photons; it is the inverse of pair production. The most common example is the annihilation of an electron and a positron, the rest masses of which are converted into photons. 2. The conversion of rest mass into electromagnetic radiation.

annihilation radiation Electromagnetic radiation produced by the union, and consequent annihilation, of a positron and an electron. Each such annihilation usually produces two photons, which have the same properties as gamma rays.

AN nomenclature system A joint Army-Navy-Air Force code system for designating communication and electronic equipment. The prefix AN indicates that the designation was assigned under this system.

annular transistor A transistor in which the characteristic semiconductor channels are in concentric circles around the emitter.

annunciator An electric remote signaling device, such as a buzzer or signal lamp.

anode 1. The positive terminal of a cell or battery. Electron flow is toward the anode through the connected load; current flow is away from the anode to the load. 2. [symbol A] The positive terminal of an electron tube. Electrons flow through the tube to its anode, and from there to the positive terminal of the connected voltage source. Also called plate. 3. In a semiconductor diode, the terminal toward which forward current flows from the external circuit. This anode terminal is normally marked negative. Current flow is from the anode to the cathode inside the diode, so electron flow is from the cathode to the anode inside the diode, and electrons flow away from the anode to the external circuit.

anode bend The curved portion at the bottom of the anode-current/grid-voltage characteristic for an electron tube.

anode-bend detector A detector that uses a triode vacuum tube operating over the lower curved portion of its anode-current/grid-voltage characteristic. Positive swings of the carrier signal are then amplified much more than negative swings, giving the effect of rectification as required for detection.

anode breakdown voltage The anode voltage necessary to cause conduction across the main gap of a gas tube when the starter gap is not conducting and all other tube elements are at cathode potential.

anode bypass capacitor A capacitor connected between anode and ground in an electron-tube circuit to bypass high-frequency currents and keep them out of the load. Also called plate bypass capacitor.

anode characteristic A graph plotted to show how the anode current of an electron tube is affected by changes in anode voltage.

anode circuit A circuit that includes the anode voltage source and all other parts connected between the cathode and anode of an electron tube. Also called plate circuit.

anode current The electron current flowing through an electron tube from the cathode to the anode. Also called plate current.

anode dark space A narrow dark zone next to the surface of the anode in a gas tube.

anode detection Detection in which rectification of RF signals takes place in the anode circuit of an electron tube. The grid bias is made sufficiently negative to bring anode current nearly to cutoff for no signal, so average anode current follows changes in signal amplitude. Also called plate detection.

anode dissipation Power dissipated as heat in the anode of an electron tube because of bombardment by electrons and ions.

anode efficiency The ratio of the AC load circuit power to the DC anode power input for an electron tube. Also called plate efficiency.

anode fall The voltage between the positive column and the anode of a gas tube. It may be positive, zero, or negative.

anode follower A tube circuit with heavy feedback from anode to grid, such that the output voltage is nearly equal and opposite to the input voltage. The input impedance is then very high.

anode glow A narrow bright zone on the anode side of the positive column in a gas tube.

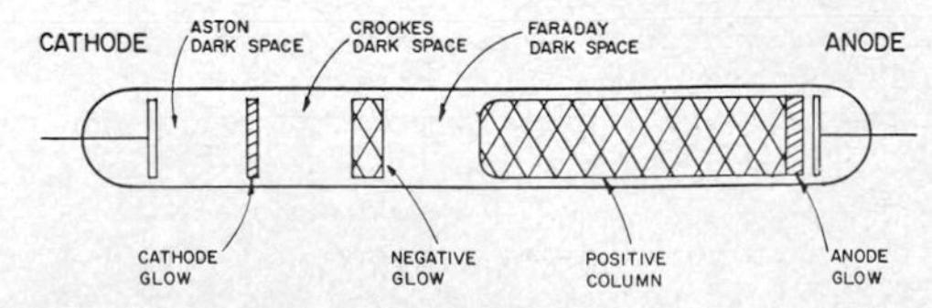

Anode glow in gas discharge tube.

anode input power The product of the direct anode voltage applied to the tubes in the last radio stage of a transmitter and the total direct current flowing to the anodes of these tubes, measured without modulation. Also called plate input power.

anode keying Keying of a radiotelegraph transmitter by interrupting the anode supply circuit. Also called plate keying.

anode load impedance The total impedance between the anode and cathode of an electron tube, exclusive of the electron stream. Also called plate load impedance.

anode modulation Modulation produced by introducing the modulating signal into the anode circuit of any tube in which the carrier is present. Also called plate modulation.

anode neutralization Neutralization in which a portion of the anode-cathode AC voltage is shifted 180° and applied to the grid-cathode circuit through a neutralizing capacitor. Also called plate neutralization.

anode power input The DC power delivered to the anode of an electron tube by the power supply. It is the product of average anode voltage and average anode current. Also called plate power input.

anode pulse modulation Modulation produced in an amplifier or oscillator by application of externally generated pulses to the anode circuit. Also called plate pulse modulation.

anode pulsing An RF oscillator circuit arrangement in which the anode voltage is normally reduced to such a low value that no anode current flows and no oscillations occur. A pulse equal to the full anode voltage is then introduced in series with the anode. Oscillations begin and last for the duration of the pulse. This circuit requires a modulator capable of supplying full anode power.

anode rays Positive ions that come from the anode of an electron tube. They are generally due to impurities in the metal of the anode.

anode region The positive column, anode glow, and anode dark space in a gas tube.

anode resistance The resistance value obtained when a small change in the anode voltage of an electron tube is divided by the resulting small change in anode current.

anode saturation The condition in which the anode current of an electron tube cannot be further increased by increasing the anode voltage. The electrons are then being drawn to the anode at the same rate as they are emitted from the cathode. Also called current saturation, plate saturation, saturation, and voltage saturation.

anode sheath A layer of electrons that surrounds the anode of a gas tube when the anode current is high.

anode shield A shield that partially surrounds the anode in a mercury-arc rectifier. It protects the anode from excessive ionization or radiation.

anode sputtering The emission of fine particles from the anode of an electron tube as a result of electron bombardment.

anode strap A metallic connector used between selected anode segments of a multicavity magnetron, principally for mode separation.

anode supply The direct voltage source used in an electron-tube circuit to place the anode at a high positive potential with respect to the cathode. Also called plate supply.

anode terminal The semiconductor-diode terminal that is positive with respect to the other terminal when the diode is biased in the forward direction.

anode voltage The direct voltage that exists between anode and cathode of an electron tube. Also called plate voltage and high tension (British).

anode voltage drop The voltage that exists between anode and cathode in a cold-cathode gas tube after conduction has been established in the main gap.

anodized dielectric film An insulating film produced on a conducting surface by anodizing. Used for producing thin-film capacitors, trimming resistor values, and passivation in the manufacture of integrated circuits.

anodizing An electrolytic process for producing a protective or decorative film on certain metals, chiefly aluminum and magnesium.

anomalous propagation Freak propagation of VHF radio waves beyond the horizon, apparently caused by temperature inversion in the lower atmosphere.

anomaly A deviation beyond normal variations, such as the change produced in the magnetic field of the earth by a submarine under water.

anomaly finder A computer-controlled data-plotting system used on ships to measure and record seismic, gravity, magnetic, and other geophysical data and water depth, time, course, and speed. Used in locating new underwater oil and mineral deposits.

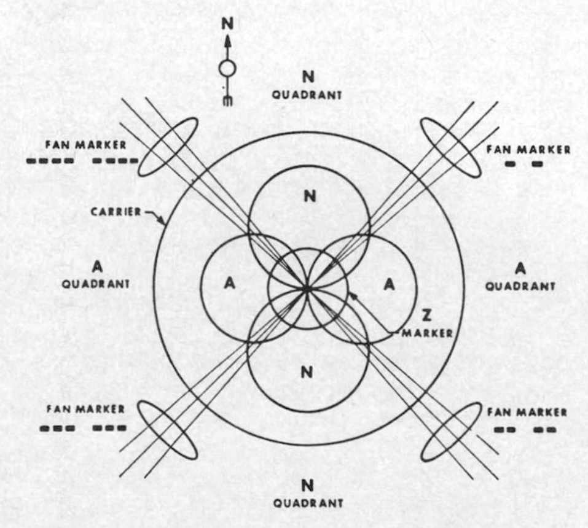

A-N radio range courses, radiation patterns, and fan marker codes.

A-N radio range A radio range that provides four radial lines of position for aircraft guidance, each line identified aurally as a continuous tone resulting from the interlocking of equal-amplitude A and N international Morse code letters. The sense of deviation from one of these lines is indicated by deterioration of the steady tone into audible A or N code signals. The two types of A-N radio ranges are the Adcock and the loop.

ANSI Abbreviation for *American National Standards Institute.*

antenna A device that radiates or receives radio waves. British term is aerial.

antenna array An arrangement of two to several thousand individual radiating elements, appropriately spaced and energized to give desired directional characteristics. Also called array and beam antenna.

antenna beam width The angle in degrees between two opposite half-power points of an antenna beam.

antenna coil The first coil in a receiver, through which antenna current flows. When this coil is inductively coupled to a secondary coil, the combination of two coils should be called an antenna transformer or RF transformer.

antenna counterpoise *Counterpoise.*

antenna coupler An RF transformer, tuned line, or other device that transfers energy efficiently from a transmitter to a transmission line or from a transmission line to a receiver.

antenna cross section A microwave receiving-antenna rating, expressed as the area, perpendicular to incident radiation, that intercepts an amount of energy equal to that delivered to a receiver by the antenna.

antenna crosstalk A measure of undesired power transfer through space from one antenna to another. It is the ratio of the power received by one antenna to the power transmitted by the other, usually expressed in decibels.

antenna current The RF current that flows in a transmitting antenna. It is generally measured when there is no modulation.

antenna drive A motor or other device that rotates or positions an antenna, such as for tracking a radar target.

antenna duplexer A circuit that permits two transmitters to transmit simultaneously from the same antenna without interaction.

antenna effect 1. An undesired output signal that results from a directional array acting as a nondirectional antenna in an electronic navigation system. Also called height effect. 2. A spurious effect caused by the capacitance of a loop antenna to ground.

antenna eliminator A device that plugs into a wall electric outlet and provides connections through capacitors to the powerline wires, to serve as an antenna connection for a radio receiver. Widely used before the development of efficient built-in loop antennas.

antenna feed The device that supplies energy to a transmitting antenna. It may be a transmission line for direct feed, or a dipole or horn for indirect feed to a reflector.

antenna field gain A Federal Communications Commission rating for a transmitting antenna. It is the effective free-space field intensity in millivolts per meter that is produced in the horizontal plane at a distance of 1 mi (1.609 km) by an antenna input power of 1 kW, divided by 137.6 mV/m (the value for a half-wave dipole).

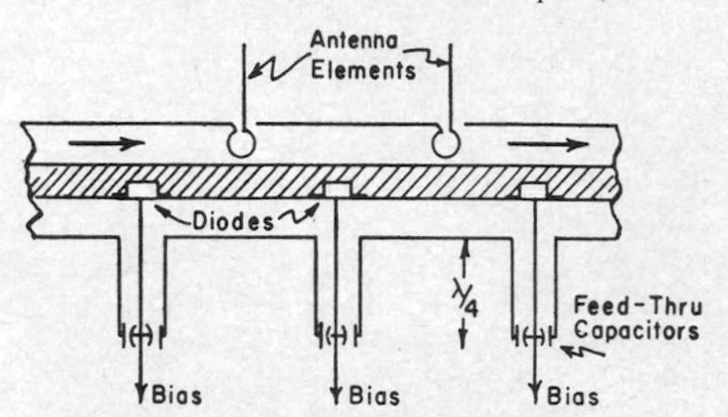

Antennafier having array-type structure with built-in amplifying diodes.

antennafier An antenna and amplifier in a single structure. In one version, each element of the antenna contains amplifying diodes.

antenna gain The effectiveness of a directional antenna as compared to a standard nondirectional antenna. It is usually expressed as the ratio in decibels of standard antenna input power to directional antenna input power that will produce the same field strength in the desired direction. For a receiving antenna, the ratio of signal power values produced at the receiver input terminals is used. The more directional an antenna is, the higher its gain.

antenna height above average terrain A Federal Communications Commission rating for transmitting antennas. It is the average of the antenna heights above the terrain from 2 to 10 mi (3.2 to 16 km) from the antenna for the eight directions, spaced evenly for each 45° of azimuth, starting with true north. The averages for each direction are averaged to get the final value.

antenna lens An arrangement of shaped metal vanes or dielectric material in front of a microwave antenna to concentrate the beam of transmitted or received radio waves.

antenna loading Use of lumped reactances to tune an antenna.

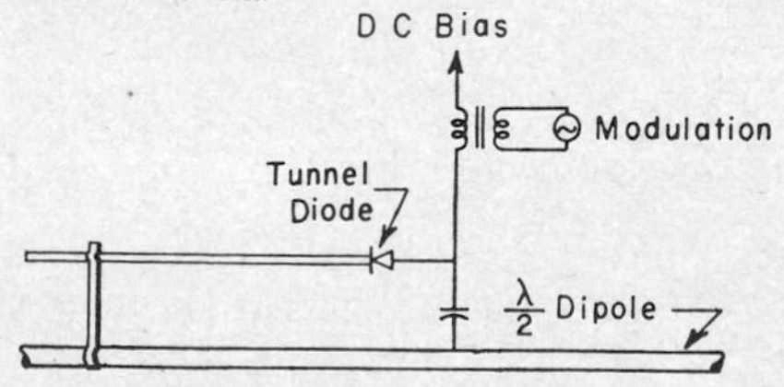

Antennamitter using built-in tunnel-diode oscillator.

antennamitter An antenna combined with a tunnel-diode oscillator to serve as a low-power transmitter. The modulation signal is applied to the DC bias line to the antennamitter.

antenna power A transmitter rating equal to the square of the antenna current multiplied by the antenna resistance at the point where the current is measured.

antenna power gain A transmitting antenna rating equal to the square of the antenna gain, expressed in decibels.

antenna relay A relay used in radio communication stations to switch an antenna from a receiver to a transmitter and vice versa.

antenna resistance A transmitting antenna rating that expresses the total resistance of the antenna system at the operating frequency. The antenna resistance in ohms is equal to the power in watts supplied to the entire antenna circuit divided by the square of the effective antenna current in amperes measured at the point where power is supplied to the antenna. Components of antenna resistance include radiation resistance, ground resistance, RF resistance of conductors in the antenna circuit, and the equivalent resistance due to corona, eddy currents, insulator leakage, and dielectric power loss.

antenna resonant frequency The frequency or frequencies at which an antenna appears to be a pure resistance.

antenna series capacitor A capacitor used in series with an antenna to shorten the electrical length of the antenna.

Antennaverter using conical spiral broadband antenna and untuned-diode mixer.

antennaverter A receiving antenna combined with a converter as a single unit, feeding directly into the IF amplifier of the receiver.

antiaircraft missile A guided missile launched from the surface against an airborne target.

antiballistic missile [abbreviated ABM] A defensive guided missile that intercepts and destroys a ballistic missile.

antibaryon An antiparticle of a baryon.

antiblocking contact A semiconductor contact that maintains a linear characteristic under all operating conditions and hence does not block current flow.

antibonding orbital An orbital electron whose energy increases monotonically as the two atoms to which it belongs move closer, so it does not lead to closer binding of the molecule.

anticapacitance switch A switch that has low capacitance between its terminals when open.

anticipation mode A mode of storing binary digits in a cathode-ray memory tube, wherein one type of digit is represented by a continuous line of excitation on the screen and the other type by gaps in the line.

anticlutter circuit A radar circuit that attenuates undesired reflections, to permit detection of targets otherwise obscured by such reflections.

anticlutter gain control *Sensitivity-time control.*

anticoincidence Occurrence of a count in a specified detector unaccompanied, simultaneously or within an assignable time interval, by a count in one or more other specified detectors.

anticoincidence circuit A circuit that produces a specified output pulse when one of two inputs receives a pulse and the other receives no pulse within an assigned time interval.

anticoincidence counter An arrangement of counters and associated circuits that will record a count only if an ionizing particle passes through certain counters but not the others.

anticoincidence counting The recording, from one or more counters of a coincidence circuit, of all counts except coincidences.

anticollision radar A radar set that gives warning of possible collisions during movements of ships or aircraft.

anticyclotron A type of traveling-wave tube.

antideuteron An antiparticle that consists of one antielectron and one antiproton.

antielectron *Positron.*

antifading antenna An antenna that confines radiation mainly to small angles of elevation, to minimize radiation of the sky waves which cause fading.

antiferroelectricity A dielectric phenomenon in which neighboring lines of spontaneously polarized ions are aligned in antiparallel directions.

antiferromagnetic material A material in which spontaneous magnetic polarization occurs in equivalent sublattices. The polarization in one sublattice is aligned antiparallel to the other.

antiferromagnetism A form of magnetism, occurring primarily at low temperatures, in which interaction between elementary atomic magnets causes adjacent magnets to try to have their magnetic directions oppose each other. If the atomic magnets are equal in strength, this leads to cancellation of the magnetic moments at very low temperature and a diminishing magnetic susceptibility with decreasing temperature.

antigravity A hypothetical condition in which the gravitational field of the earth is somehow canceled.

antihunt circuit A stabilizing circuit used in a closed-loop feedback system to prevent self-oscillations.

antihunt device An electric or mechanical device used in positioning systems to prevent hunting or oscillation of the load around an ordered position. It usually involves some form of feedback.

antihunt transformer A transformer used as a stabilizing network in a DC feedback system. Its primary winding is in series with the load. Its secondary winding provides a voltage, proportional to the derivative of the primary current, that is fed back into some other part of the loop to prevent self-oscillation.

antijamming A device, method, system, or technique that reduces or eliminates jamming.

antilambda A particle of antimatter corresponding to a lambda particle.

antilog Abbreviation for *antilogarithm.*

antilogarithm [abbreviated antilog] The number corresponding to a given logarithm. Example: If the logarithm of 563.2 is 2.75066, then 563.2 is the antilogarithm of 2.75066.

antimagnetic Made of alloys that will not remain magnetized.

antimatter A form of matter in which protons, electrons, and other particles have charges opposite those with which they are normally associated.

antimicrophonic Not affected by vibrations or sound waves from a loudspeaker to the extent that feedback and howling occur. Also called nonmicrophonic.

antimissile missile [abbreviated AMM] A guided missile that intercepts and destroys a missile in flight.

antimony [symbol Sb] A metallic element. Atomic number is 51.

antineutrino A particle with near-zero mass, no electric charge, and a spin opposite that of a neutrino.

antineutron A hypothetical particle that has the same mass and the same zero charge as a neutron and is capable of combining with a neutron to give complete conversion of both particles into mesons.

antinode A point, line, or surface in a standing-wave system at which some characteristic of the wave has maximum amplitude. Also called loop.

antinoise microphone A microphone that has characteristics which discriminate against undesired noise, such as a close-talking microphone, lip microphone, or throat microphone.

antinucleon A particle that has the same mass as a nucleon but an opposite charge or magnetic moment.

antiparticle A particle identical to the original particle in mass and spin but with opposite electric, magnetic, and other properties. When a particle interacts with its antiparticle, annihilation may occur.

antiplugging relay A relay used in some control systems to prevent plugging.

antiproton An elementary particle that differs from a proton only in that its charge is negative. Also called negative proton.

antiradar coating A coating applied to an aircraft to reduce reflection of radio waves and thereby minimize detection by radar.

antireflection coating A thin film applied to an optical surface to reduce reflectance and increase transmittance.

antiresonance *Parallel resonance.*

antiresonant circuit *Parallel resonant circuit.*

antisatellite missile A missile that destroys an orbiting satellite.

antisidetone circuit A circuit that has a balancing network for reducing sidetones in a telephone set.

antistatic coating A metallic or other coating applied to reduce or prevent buildup of static electricity. Antistatic materials are sometimes mixed with plastics and other materials during manufacture.

antistickoff voltage A small voltage applied to the rotor winding of the coarse synchro control transformer in a two-speed control system to eliminate the possibility of ambiguous behavior.

antistiction oscillator An oscillator or multivibrator that generates a small AC voltage to combat stiction near the null position in a direct-writing pen recorder. Also called dither injector and keep-alive oscillator.

antisubmarine barrier A line formed by a series of static devices or mobile units arranged to detect, deny passage to, or destroy hostile submarines.

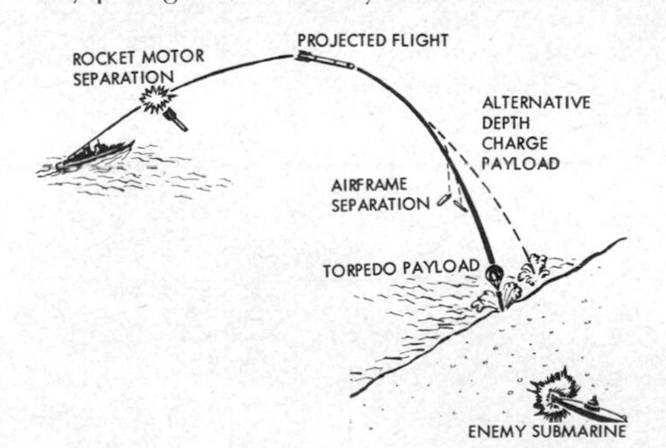

Antisubmarine rocket launched from destroyer.

antisubmarine rocket [abbreviated ASROC] A rocket-propelled homing torpedo that is launched from a ship and travels through the air for most of the distance to a submerged enemy submarine. Alternatively, it may carry a nuclear depth charge.

antisubmarine torpedo [abbreviated ASTOR] A submarine-launched wire-guided torpedo used against enemy submarines and surface ships. It may carry a nuclear warhead.

antisubmarine warfare [abbreviated ASW] General term for all techniques, weapons, and equipment used for detection, surveillance, and destruction of enemy submarines.

antitank guided missile A surface-to-surface guided missile whose flight path to an enemy tank is controlled by command signals produced by an automatic computer from data obtained by optical sighting, with the control signals being fed to the missile over multiple-wire command links. The missile may also contain an infrared homing device for final range correction.

antivoice-operated transmission [abbreviated ANTIVOX] Use of a voice-actuated circuit to prevent operation of a transmitter when an associated receiver is in use.

ANTIVOX Abbreviation for *antivoice-operated transmission.*

anti-xi-zero The thirty-fourth and last of the theoretically predicted elementary particles of matter, discovered in 1963. It is the antiparticle of xi-zero and has no electric charge.

APC Abbreviation for *automatic phase control.*

aperiodic Not responsive to any particular frequency.

aperiodic antenna An antenna that has essentially constant impedance over a wide range of frequencies. Examples are terminated rhombic antennas and terminated wave antennas. Also called untuned antenna.

aperiodic circuit A circuit in which it is impossible to produce free oscillation.

aperiodic damping Damping so great that a disturbed system or instrument comes to a position of rest without passing through this position. The point of change between aperiodic and periodic damping is called critical damping.

aperture An opening through which electrons, light, radio waves, or other radiation can pass. The aperture in the electron gun of a cathode-ray tube determines the size of the electron beam. The aperture in a television camera is the effective diameter of the lens that controls the amount of light entering the camera tube. The dimensions of the horn mouth or parabolic reflector determine the aperture of a microwave antenna.

aperture antenna An antenna in which the beam width is determined by the dimensions of a horn, lens, or reflector.

aperture card A punched card that has an opening which holds one or more frames of microfilm.

aperture distortion Attenuation of the high-frequency components of a television picture signal caused by the finite cross-sectional area of the scanning beam in the camera. The beam then covers several mosaic globules in the camera simultaneously, causing loss of picture detail.

aperture grille *Slot mask.*

aperture illumination The strength distribution of an electromagnetic wave in an aperture.

aperture mask *Shadow mask.*

aperture time The time required for a sampling circuit to change to its hold mode after receipt of a control command. Aperture time should generally be less than 50 ns, to prevent error in hold level if the signal changes during this time.

apex step A flexible step or corrugation at the center of a loudspeaker cone to modify high-frequency response.

aphelion The point most distant from the sun on the orbit of a planet or spacecraft.

API Abbreviation for *air-position indicator.*

APL [A Programming Language] A powerful, versatile computer programming language that was developed initially for mathematical problems. It now has additional features that make it particularly useful for interactive business and economics problems. In general, a character in APL is equivalent to an entire statement in most other programming languages.

apochromatic lens A lens that has been corrected for chromatic aberration for three colors.

apogee 1. The point in an elliptical orbit of an earth satellite farthest from the earth. 2. The point in the trajectory of a ballistic missile farthest from the earth.

apogee rocket A rocket that is attached to a satellite or spacecraft and which fires when the craft is at apogee to establish a new orbit farther from the earth or to escape from earth orbit.

Apollo A spacecraft designed for manned earth orbital flights and manned or unmanned lunar landings. For the latter, Apollo carried lunar excursion modules and was launched into space by Saturn boosters.

apostilb [abbreviated asb] A unit of luminance equal to 1×10^{-4} lambert. Use of the SI unit of luminance, the candela per square meter, is preferred.

apparent horizon The visible line of demarcation between land or sea and sky.

apparent power The power value obtained in an AC circuit by multiplying the effective values of voltage and current. The result is expressed in voltamperes and must be multiplied by the power factor to secure the average or true power in watts.

apparent precession The relative angular movement of the spinning axis of a gyroscope in relation to a line on the earth, resulting from the rotation of the earth.

Applegate diagram A diagram that illustrates the behavior of the electrons in a velocity-modulation tube. The distances of electrons from the buncher in the drift space are plotted as vertical coordinates against time on the horizontal axis. Close spacing of these vertical lines indicates bunching of electrons.

Appleton layer *F layer.*

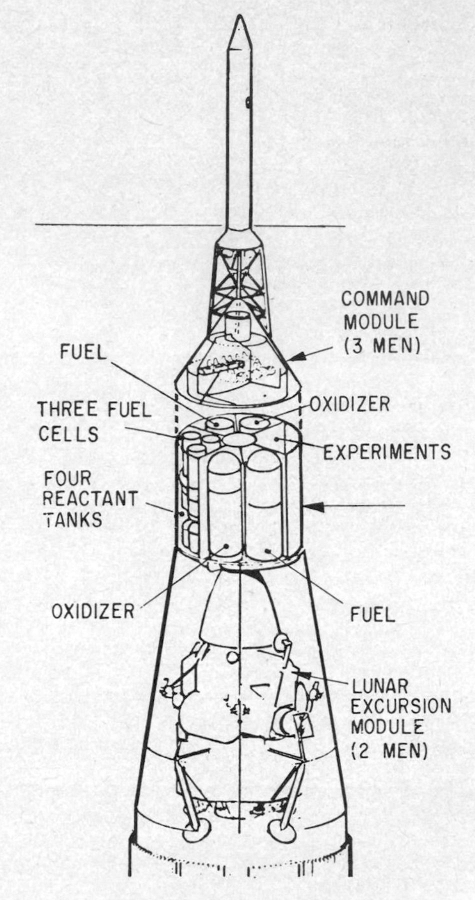

Apollo spacecraft with lunar excursion module.

appliance A piece of equipment that draws electric or other energy and produces a desired work-saving or other result, such as an electric heater, radio, or electronic range.

applicator An appropriately shaped metal electrode that establishes an alternating electric field in material to be heated by dielectric heating.

applied research Research directed toward practical use of knowledge gained by basic research.

applied shock Excitation that produces shock motion within a system.

approach control radar Any radar set or system used in a ground-controlled approach system, such as an airport surveillance radar, a precision approach radar, or a combination of both.

approach path The portion of the flight path, in the immediate vicinity of a landing area, that terminates at the touchdown point.

APT 1. [Automatically Programmed Tools] A computer language developed primarily for programming numerically controlled machine tools.

It is written in convenient Englishlike computer language, using such practical instructions as line, sphere, and tangent. 2. Abbreviation for *automatic picture transmission.*

AQL Abbreviation for *acceptable quality level.*

Aquadag Trademark of Acheson Colloids Co. for their brand of colloidal graphite in water. Widely used to produce a conductive coating on the inside surface of the glass envelope for cathode-ray picture tubes, where it collects secondary electrons emitted by the fluorescent screen. Also used on the outside of some picture tubes, where it serves as the final capacitor of the high-voltage filter circuit.

A quadrant One of the two quadrants in which the A signal of an A-N radio range is heard.

arbiter A computer unit that determines the priority sequence in which two or more processor inputs are connected to a single functional unit such as a multiplier or memory.

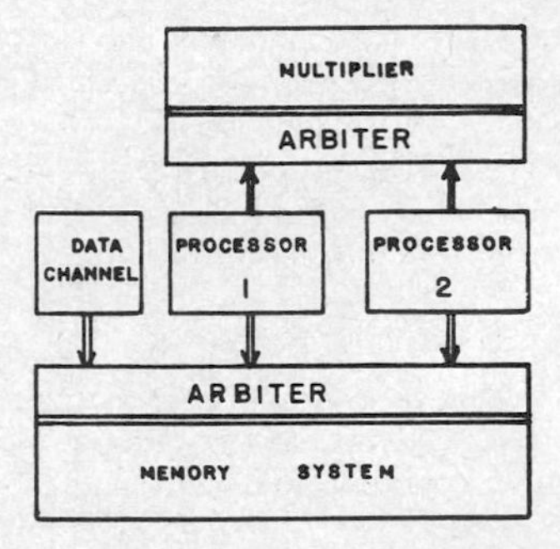

Arbiters used between multiple processors and two functional units of computer.

arbitrary constant A constant to which various values may be assigned by decision alone, with these values being unaffected by any of the variables in an equation.

arbitrary function generator A function generator in which a mask that has the desired waveform outline is inserted in a photoelectric scanning system, as in the photoformer.

arc *Electric arc.*

arcback The flow of a principal electron stream in the reverse direction in a mercury-vapor rectifier tube because of formation of a cathode spot on an anode. This results in failure of the rectifying action. Also called backfire. The action corresponds to reverse emission in electron tubes.

arc baffle A baffle used in mercury-pool tubes to prevent mercury from splashing the anode and causing an electric arc. Also called splash baffle.

arc cathode A cathode in which electron emission is self-sustaining at a low voltage drop, approximately equal to the ionization potential of the gas.

arc converter A form of oscillator that uses an electric arc as the generator of alternating or pulsating current. Used in some induction-heating generators.

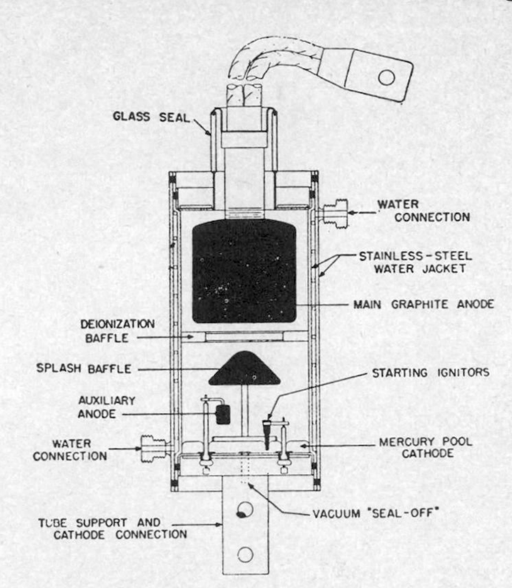

Arc baffle is mushroom-shaped cap, also called splash baffle, over mercury-pool cathode of ignitron.

arc discharge A discharge between electrodes in a gas or vapor, characterized by a relatively low voltage drop and a high current density.

arc-discharge tube A discharge tube in which a high-current arc discharge passes through the gas between the electrodes, generally for the purpose of producing an intense flash of light, as in a strobotron.

arc drop The voltage drop between the anode and cathode of a gas rectifier tube during conduction.

arc-drop loss The product of the instantaneous values of tube voltage drop and current averaged over a complete cycle of operation of a gas tube.

arcing The production of an arc, as at the brushes of a motor or the contacts of a switch.

arc lamp An electric lamp in which the light is produced by an arc made when current flows through ionized gas between two electrodes.

arc resistance The time required for a given electric current to render the surface of a material conductive because of carbonization by the arc flame.

arc spectrum The spectrum of light produced by vaporizing an element in an electric arc.

arc spraying Spraying with metal that has been melted by an electric arc.

arcthrough In multielectrode gas tubes, the loss of control that results from the flow of a principal electron stream in the normal direction during a scheduled nonconducting period.

arc transmitter A radio transmitter that uses an arc to generate RF carrier signals.

arc welding A fusion welding process in which

welding heat is obtained from an electric arc struck between an electrode and the metal being welded or between two separate electrodes, as in atomic hydrogen welding.

area control radar A radar set or system for air traffic control over a relatively large area, to provide a smooth flow of air traffic to the approach control radar.

area monitor A device that detects and/or measures radiation levels at a given location for warning or control purposes.

area monitoring Routine monitoring of levels of ionizing radiation in an area in which radiation hazards are present or suspected.

Arecibo Ionospheric Observatory An observatory in Puerto Rico that has one of the world's most powerful scanning radar installations, used primarily for detailed studies of the ionosphere.

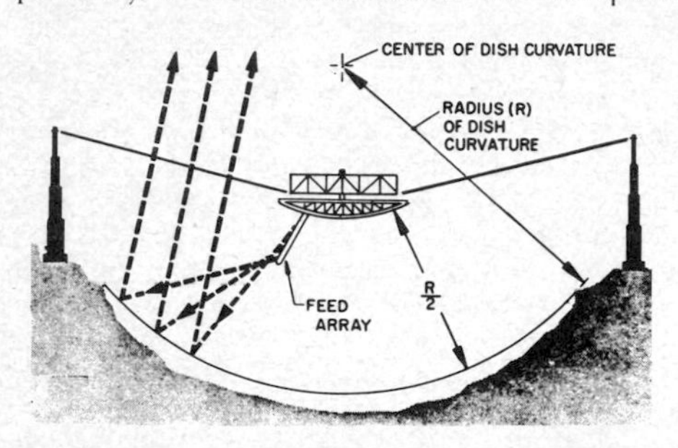

Arecibo Ionospheric Observatory, showing how feed array is tilted to scan up to 20° off from vertical in all directions.

The antenna-feed array for the 430-MHz 2.5-MW transmitter is suspended over a 1000-ft-diameter (1609-km) reflector built into a natural mountain bowl.

argon [symbol A] An inert gas element that gives a purple glow when ionized. Atomic number is 18. Sometimes used in electron tubes, electric lamps, and neon signs.

argon glow lamp A glow lamp that contains argon gas. It produces a pale blue-violet light. Wattages of standard sizes range from ¼ to 2 W.

argon laser A gas laser that uses ionized argon to produce strong radiation at 0.488 μm and infrared radiation.

arithmetic circuit A computer circuit that provides an arithmetic operation.

arithmetic-logic unit [abbreviated ALU] The section of a computer that performs all arithmetic and logic operations.

arithmetic mean *Mean.*

arithmetic operation A digital computer operation in which numerical quantities are added, subtracted, multiplied, divided, or compared.

arithmetic shift A computer operation in which a quantity is multiplied or divided by a power of the number base. Thus, binary 1011 represents decimal 11, and therefore two arithmetic shifts to the left is binary 101100, which represents decimal 44.

ARL Abbreviation for *acceptable reliability level.*

arm *Branch.*

armature The moving portion of a magnetic circuit, such as the rotating section of a generator or motor, the movable iron part of a relay, or the spring-mounted iron part of a vibrator or buzzer.

armature chatter Undesired vibration of the armature of a relay in an AC circuit.

armature contact *Movable contact.*

armature reaction Interaction between the magnetic flux produced by armature current and that of the main magnetic field in an electric motor or generator. The resulting distortion of the main magnetic field affects the speed of motors and the voltage regulation of generators.

armature relay The most common type of relay; it has a pivoted iron armature that is attracted by a coil and thereby completes the magnetic circuit which passes through the coil. The relay contacts are mounted on the armature or actuated by it through mechanical linkage.

armature travel The total distance traveled during relay operation by a point on the armature that is nearest the pole-face center.

armature voltage control Control of electric motor speed by changing the voltage applied to its armature windings.

arming The act of changing an explosive fuze from a safe condition to a state of readiness for initiation. A fuze may be armed manually, by acceleration, by rotation, by a clock mechanism, or by air travel.

arming signal A radio signal for arming the warhead of a missile.

armored cable A cable with a sheath of metal, primarily for mechanical protection.

Armstrong frequency-modulation system A phase-shift modulation system developed by E. H. Armstrong. A low-frequency carrier is modulated at a low level, and the signal is passed through several frequency-multiplier amplifier stages to obtain the desired high carrier frequency and high level. The low-level modulation is produced by a balanced modulator that has a quadrature carrier reinserted in its output.

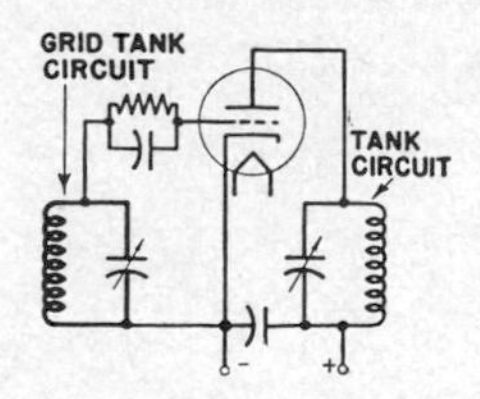

Armstrong-oscillator circuit.

Armstrong oscillator A tuned-grid, tuned-anode oscillator circuit developed by E. H. Armstrong. A parallel resonant circuit is the required inductive anode load when tuned slightly above the resonant frequency of the grid tank circuit or crystal. The interelectrode capacitance of the oscillator tube serves as the feedback path.

array *Antenna array.*

array radar A radar that has a fixed array of many thousands of antennas, each with an electronically variable phase shifter which permits steering the beam over wide angles without antenna motion. Also called phased-array radar.

array sonar A sonar system that incorporates digital display and magnetic storage techniques, variable transmitter frequency for best penetration of sea water, and other features that increase detection range and capability for mapping ocean-bottom terrain at the great depths in which nuclear submarines operate while carrying Polaris missiles.

arrester *Lightning arrester.*

arrhythmia monitor An instrument that provides continuous monitoring of electrocardiograph waveforms of patients in hospital intensive coronary care wards, with provisions for detecting transient abnormal patterns and sounding an alarm or otherwise alerting medical personnel to the need for action.

ARRL American Radio Relay League, an organization of licensed amateur radio operators.

arsenic [symbol As] An element. Atomic number is 33.

ARSR Abbreviation for *air route surveillance radar.*

Artemis A fixed underwater submarine surveillance system that involves multiple listening posts, each weighing many tons, moored in deep water beyond the continental shelf. Hydrophone arrays attached to the mooring cable at various levels monitor sounds in each thermocline interval, to prevent enemy submarines from escaping detection by hiding under a thermocline.

arteriogram An instrument that records arterial pulsations.

arteriometer An instrument that measures arterial pulsations.

articulation The percentage of speech units understood correctly by a listener in a communication system. Also called intelligibility. Articulation generally applies to unrelated words, as in code messages. Intelligibility is customarily used for regular messages where the context aids the listener's perception.

artificial antenna *Dummy antenna.*

artificial delay line *Delay line.*

artificial dielectric A three-dimensional arrangement of metallic conductors, which are usually small compared to a wavelength. The resulting medium acts as a dielectric to electromagnetic waves.

artificial ear A device that presents an acoustic impedance to an earphone equivalent to the impedance presented by the average human ear. It is equipped with a microphone for measuring the sound pressures developed by the earphone.

artificial earth satellite A man-made earth satellite, as distinguished from the moon.

artificial echo A radar echo signal generated by artificial means for test purposes, as by an echo box.

artificial gravity A simulated gravity established within a space vehicle, as by rotating a cabin about an axis of a spacecraft to produce centrifugal force.

artificial hand A device that duplicates the impedance of a human body. It is connected between an appliance housing and ground for testing electric shock hazards.

artificial horizon A device that indicates the attitude of an aircraft with respect to the true horizon. Also called flight indicator and gyro horizon.

artificial intelligence The capability of a computer or other device to perform functions normally associated with human intelligence, such as reasoning, learning, and self-improvement.

artificial ionization The introduction of an artificial reflecting or scattering layer into the upper atmosphere to improve beyond-the-horizon radio communication.

artificial language A language specifically designed for ease of communication in a particular area of endeavor, in contrast to a natural language that has evolved through long usage.

artificial line A network that simulates the electrical characteristic of a transmission line.

artificial load A dissipative but essentially nonradiating device that has the impedance characteristics of an antenna, transmission line, or other practical load.

artificial mouth A small loudspeaker that simulates voice sounds coming from the center of the lips of a person, for such applications as testing microphones and telephone receivers.

artificial radioactive element A radioactive element produced from another element or an isotope of the same element by bombarding the element with protons, neutrons, deuterons, gamma rays, or other particles or by exposing it to radiation.

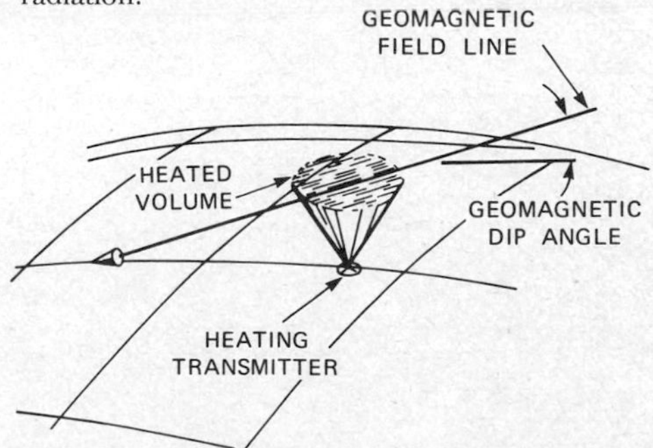

Artificial radio aurora created by heating electrons of ionosphere with powerful vertically aimed transmitter.

artificial radio aurora Modification of the ionosphere by high-power high-frequency radio transmitters to improve scatter and auroral long-distance communication. Also called radio aurora.

artificial satellite *Satellite.*

artificial voice A small loudspeaker mounted in a baffle that simulates the acoustical constants of the human head. Used for calibrating and testing close-talking microphones.

ASA Abbreviation for *American Standards Association.*

asb Abbreviation for *apostilb.*

asbestos A noninflammable fibrous mineral used in electronics for heat-insulating and fireproofing, as in a line-cord resistor.

ASCII Abbreviation for *American Standard Code for Information Interchange.*

A scope 1. A radarscope that produces an A display. 2. *Waveform monitor.*

ASDE Abbreviation for *airport surface detection equipment.*

asdic [Anti-Submarine Detection Investigation Committee] British term for sonar and underwater listening devices.

asfir [Active Swept Frequency Interferometer Radar] A dual-radar air surveillance system that gives high-precision angle and range information for pinpointing target locations by trigonometric techniques. A master radar transmits signals to a slave radar at the target location. The slave radar reflects the signals back to the master, where the elapsed travel time is converted into distance. Use of a second slave radar permits measurement of both range and altitude.

A signal A dot-dash signal heard in either a bisignal zone or an A quadrant of a radio range.

ASK Abbreviation for *amplitude-shift keying.*

aspect ratio The ratio of the frame width to the frame height in television. It is 4 to 3 in the United States and Great Britain.

ASR Abbreviation for *airport surveillance radar.*

ASROC Abbreviation for *antisubmarine rocket.*

assemble To prepare a machine-language program from a symbolic-language program by substituting absolute operation codes and addresses for symbolic operation codes and addresses.

assembler A computer program that assembles instructions in usable form by translating item by item, without changing program length. In contrast, a compiler will replace certain items with a series of instructions, to give an expanded program. Also called assembly program and assembly routine.

assembly A number of parts of subassemblies joined together to perform a specific function.

assembly language A computer language that has a one-to-one relationship between machine instructions and the program supplied by the user.

assembly machine A machine used for inserting components automatically in printed-wiring boards, such as an in-line assembly machine or a single-station assembly machine.

assembly program *Assembler.*

assembly routine *Assembler.*

assigned frequency The center of the frequency band assigned to a station.

assigned frequency band The frequency band whose center coincides with the frequency assigned to a station. The width of the band is the necessary bandwidth plus twice the absolute value of the frequency tolerance.

associated corpuscular emission The full complement of secondary charged particles associated with an x- or gamma-ray beam in its passage through air.

associative memory *Content-addressable memory.*

associative processor A digital computer that consists of a content-addressable memory and means for searching rapidly changing random digital data stored within, at speeds up to 1000 times faster than conventional digital computers. One application is for air traffic control systems.

astable circuit A circuit that alternates automatically and continuously between two unstable states at a frequency dependent on circuit constants. It can be readily synchronized at the frequency of any repetitive input signal. Blocking oscillators and certain multivibrators are examples.

astable multivibrator A multivibrator in which each tube or transistor alternately conducts and is cut off for intervals of time determined by circuit constants, without use of external triggers. Also called free-running multivibrator.

astatic Without orientation or directional characteristics; having no tendency to change position.

astatic galvanometer A sensitive galvanometer that consists of two small magnetized needles mounted parallel to each other inside the galvanometer coil. The needles reduce errors caused by the earth's magnetic field.

astatic microphone *Omnidirectional microphone.*

astatic wattmeter An electrodynamic wattmeter that is insensitive to uniform external magnetic fields.

astatine [symbol At] A radioactive element. Atomic number is 85.

A station The loran station whose signal is always transmitted more than half a repetition period before the signal from the slave or B station of the pair. Also called master station.

astigmatism 1. A type of spherical aberration in which light rays from a single point of an object do not converge at the corresponding point in the image. 2. An electron-beam tube defect in which electrons in the beam come to a focus in different axial planes as the beam is deflected, so the spot on the screen is distorted in shape and the image is blurred.

ASTM index A card index published by the American Society for Testing Materials, in which the x-ray powder diffraction lines obtained from a large number of substances are recorded and classified in terms of the three most intense lines. Other information that might assist in the x-ray identification of an unknown substance is also included.

Aston dark space The dark space next to the cathode of a glow discharge tube, in which the emitted electrons do not have enough velocity to excite the gas.

Aston mass spectrograph A spectrograph that uses successive electric and magnetic fields to focus rays of a constant charge-mass ratio at a focal line.

Aston rule The atomic weights of isotopes are approximately integers, and deviations of the atomic weights of the elements from integers are due to the presence of several isotopes with differing weights.

ASTOR Abbreviation for *antisubmarine torpedo.*

astrionics The science of electronics as applied to space travel.

astrobiology The study of living organisms on planets other than the earth.

astrocompass A compass that gives directions with respect to certain stars.

astrodome A transparent plastic dome that projects from the upper surface of an aircraft, used for celestial navigation.

astrodynamics Dynamics as applied to the motions of bodies in space, including artificial satellites and deep-space probes, and the forces acting on these moving bodies.

astrometer An instrument that determines the brightness of stars.

astron A system that involves use of a cylindrical sheath of rotating electrons, moving at speeds approaching that of light, to produce plasma pinch. An electron gun with several million electronvolts of energy shoots the plasma beam into a vacuum chamber that contains deuterium, to produce controlled thermonuclear reactions.

astronautics The science of designing, building, and operating space vehicles.

astrophysics Physics as applied to stellar astronomy.

A supply The A battery, transformer filament winding, or other voltage source that supplies power for heating the cathode of an electron tube.

ASW Abbreviation for *antisubmarine warfare.*

asymmetrical Not symmetrical.

asymmetrical distortion Distortion that affects two-condition modulation, in which the intervals corresponding to one of the two significant conditions have longer or shorter durations than the original signal.

asymmetrical-sideband transmission *Vestigial-sideband transmission.*

asynchronous Not synchronous.

asynchronous computer A computer in which the performance of any operation starts as a result of a signal that the previous operation has been completed, rather than on signal from a master clock.

asynchronous control A method of control in which the time allotted for performing an operation depends on the time actually required for the operation, rather than on a predetermined fraction of a fixed machine cycle.

asynchronous device A device in which the speed of operation is not related to any frequency in the system to which it is connected.

asynchronous logic Logic in which computer operations occur independently of time, in a sequential manner.

asynchronous machine An AC machine whose speed is not proportional to the frequency of the power line.

asynchronous multiplex A multiplex transmission system in which two or more transmitters occupy a common channel without provision for preventing simultaneous operation.

asynchronous transmission Data transmission in which each character contains its own start and stop pulses and there is no control over the time between characters.

ATC Abbreviation for *air traffic control.*

ATCRBS Abbreviation for *air traffic control radar beacon system.*

AT-cut crystal A quartz crystal slab cut at a 35° angle to the Z axis of the mother crystal. Temperature variations have little effect on its natural vibration frequency.

athermanous Opaque to infrared radiation.

Atlantic Missile Range A 6000-mi (9600-km) range for testing missiles and space vehicles. Instrument stations are at Cape Canaveral, Jupiter Inlet (Florida), Grand Bahama, Eleuthera, San Salvador, Mayaguana, Grand Turk, Dominican Republic, Mayaguez, Antigua, St. Lucia, Fernando de Noronha, and Ascension.

Atlas An Air Force surface-to-surface intercontinental ballistic missile. The Atlas vehicle has also been used for space launches. Newer missile versions have all-inertial guidance with nuclear warheads and are deployed in hardened and dispersed configurations.

atm Abbreviation for *atmosphere.*

atmosphere [abbreviated atm] The mixture of gases, chiefly oxygen and nitrogen, that surrounds the earth for a distance of about 100 mi (160 km) from the surface. Atmospheric pressure at sea level is about 14.7 lb/in^2 (a pressure of 760 mmHg at 0°C); this value is often used as a unit of pressure called 1 atmosphere.

atmospheric absorption The attenuating effect of gases and moisture in the earth's atmosphere on the propagation of microwaves.

atmospheric duct An atmospheric layer that conducts radio waves in the same manner as a waveguide under certain conditions of temperature and humidity, to give signal transmission far outside the usual reception area.

atmospheric interference Static caused by natural electric disturbances such as lightning and northern lights. Heard as crackling and hissing noises in radios. Also called atmospherics and sferics.

atmospheric noise Noise heard during radio reception because of atmosphere interference.

atmospheric radio wave A radio wave that reaches its destination after reflection from the upper ionized layers of the atmosphere.

atmospheric radio window A portion of the electromagnetic spectrum in which radio waves pass readily through the atmosphere of the earth. The major window extends from the longest radio waves (approaching zero frequency) up to about 19 GHz. Small windows for millimeter waves occur at 35 and 94 GHz. Also called window.

atmospheric refraction The bending of the path of electromagnetic radiation from a distant point as the radiation passes obliquely through varying air densities.

atmospherics *Atmospheric interference.*

atom The smallest particle into which an element can be divided and still retain the chemical properties of that element.

atomic Pertaining to atoms

atomic absorption coefficient The fractional decrease in intensity of a beam of photons or particles, per number of atoms per unit area of an element. It is equal to the linear absorption coefficient divided by the number of atoms per unit volume, or to the mass absorption coefficient divided by the number of atoms per unit mass. For the absorption of photons it is equal to the atomic cross section.

atomic absorption spectrophotometer A spectrophotometer for determining concentrations of different metallic elements in solutions or solid samples by measuring absorption of energy in the sample at different wavelengths.

atomic battery *Nuclear battery.*

atomic beam A stream of atoms, which may or may not be ionized.

atomic-beam frequency standard A frequency standard that provides one or more precise frequencies derived from an element such as cesium.

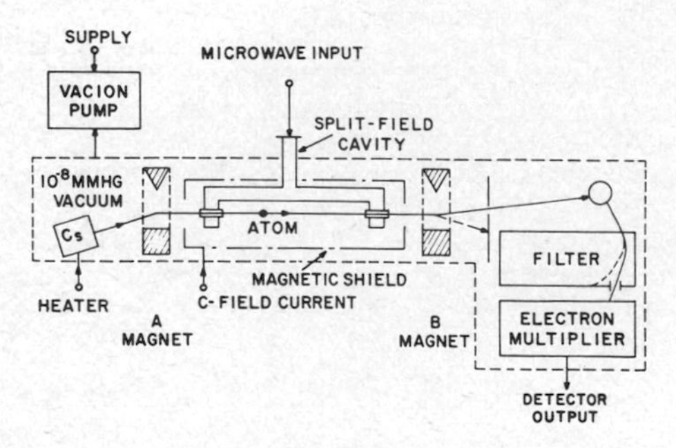

Atomic-beam frequency standard.

Frequency determination is accomplished by measuring the nuclear magnetic moment of the element by resonance absorption techniques. When the element is acted on by an applied frequency at its atomic spectrum line frequency, the element absorbs a quantum of energy, indicating coincidence between the two frequencies. The applied frequency thus becomes a standardized frequency and may be utilized at its fundamental or harmonics for calibration of other equipment.

atomic charge The electric charge of an ion. It is equal to the number of electrons the atom has gained or lost in its ionization multiplied by the charge on one electron.

atomic clock A highly accurate source of frequency or time that depends on the unchanging nuclear resonance of atoms of elements such as cesium when subjected to an RF electromagnetic field. Accuracy of 1 part in 10 million is achievable. Another type of atomic clock utilizes the ammonia molecule in a gas maser.

atomic cocktail A solution of radioisotopes that is swallowed or otherwise internally administered for treatment of some types of cancer.

atomic core *Atomic kernel.*

atomic cross section The cross section of an atom for a particular process. For the absorption of photons it is equal to the atomic absorption coefficient.

atomic disintegration Conversion of the nucleus of an atom of one element into that of some other element.

atomic energy level The energy corresponding to one of the stationary states of an isolated atom.

atomic frequency The natural vibration frequency of an atom.

atomic frequency standard A frequency standard that consists of a quantum-mechanical resonator which is frequency-locked to an electronic frequency converter if the resonator is passive or phase-locked if the resonator is active. The three types are atomic-beam, molecular-beam, and gas-cell frequency standards.

atomic fusion *Nuclear fusion.*

atomic gas laser A gas laser in which electrons and ions are accelerated between electrodes by an electric field. Collisions of these fast-moving electrons excite atoms and ions to higher energy levels, and laser action occurs in the gas during the subsequent decay back to lower energy levels. Neutral noble gases are used, consisting chiefly of argon, helium, krypton, and xenon. The normal wavelength range is 0.2 to 3 μm. The most common example is the helium-neon laser.

atomic hydrogen maser A maser in which dissociated hydrogen atoms from an electric discharge source are formed into a beam that undergoes selective magnetic processing. Hydrogen atoms in selected energy states then enter a microwave cavity where the atoms interact and radiate, giving up useful microwave energy at 1.42 GHz.

Can be used as an atomic clock.

atomic hydrogen welding An AC arc-welding process in which welding heat is obtained from an arc between two suitable electrodes in an atmosphere of hydrogen. The arc changes atomic hydrogen into molecular hydrogen in a reaction that produces the intense heat required for fusing metals that have high melting temperatures.

atomic kernel An atom that has lost the valence electrons in its outermost shell. Also called atomic core.

atomic lamp A lamp in which radioactivity is the source of energy. In one version a zinc sulfide phosphor is activated by a radioisotope of krypton gas that has a half-life of 10 years.

atomic mass The mass of a neutral atom of a nuclide, usually expressed in atomic mass units. Also called nuclidic mass. Formerly called isotopic mass.

atomic mass conversion factor The experimentally determined ratio of the atomic weight unit to the atomic mass unit. Also called mass conversion factor.

atomic mass unit [abbreviated u] A unit of mass (unified) defined as one-twelfth the mass of one atom of the ^{12}C nuclide.

atomic migration The transfer of the valence bond of one atom to another atom in the same molecule.

atomic moisture meter An instrument that measures the moisture content of coal instantaneously and continuously by bombarding the coal with neutrons and measuring the neutrons which bounce back to a detector tube after striking hydrogen atoms of water in coal.

atomic nucleus *Nucleus.*

atomic number [symbol Z] The number of elementary positive charges in the nucleus of an atom. It is a different number for each element, starting with 1 for hydrogen and going up beyond 103. For a neutral atom, the atomic number is also the number of electrons outside the nucleus of the atom. In the symbol of a nuclide, the atomic number is given as a subscript preceding the element symbol; thus in $^{59}_{26}Fe_{33}$ the atomic number for iron is 26, the neutron number is 33, and the mass number is 59.

atomic photoelectric effect *Photoionization.*

atomic scattering factor A factor that represents the efficiency with which x-rays are scattered by a given atom, as compared to the scattering by a single electron under the same conditions.

atomic second The International Unit of time adopted in 1967, defined as the duration of 9,192,631,770 periods of the radiation corresponding to the transition between two specific hyperfine levels of the atom of cesium 133.

atomic spectrum The spectrum of radiation emitted by an excited atom because of changes within the atom. An atomic spectrum is characterized by more or less sharply defined lines at certain frequencies, representing quanta of energy.

atomic stopping power The average energy loss suffered by a particle per atom of the material through which it passes. Also called stopping cross section.

atomic time Time based on atomic resonance.

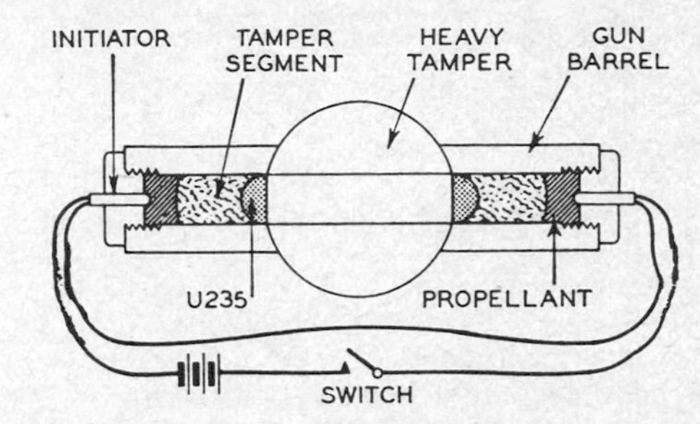

Atomic weapon in which two masses of ^{235}U are driven toward each other by explosive charges in gun barrel, to initiate fission reaction.

atomic weapon A bomb, shell, guided missile, or the like in which the warhead consists of nuclear-fissionable radioactive material such as uranium 235 or plutonium 239.

atomic weight The relative weight of a neutral atom of an element, based on an atomic weight of 16 for the neutral oxygen atom. On this basis, hydrogen has an atomic weight of 1.0078.

atomic weight unit [abbreviated awu] A unit of weight equal to exactly one-sixteenth of the average weight of ^{16}O, ^{17}O, ^{18}O, the three kinds of naturally occurring oxygen atoms. One atomic weight unit is equal to 1.660×10^{-24} g or 1.000272 atomic mass units.

atom smasher *Accelerator.*

ATP Abbreviation for *automatic telephone payment.*

atran [Automatic Terrain Recognition And Navigation] A map-matching system for missile navigation. One version uses a tape or film recording of radar or other data for terrain covered in the flight, previously obtained by reconnaissance. A continuous comparison of desired position with actual position is made to provide appropriate corrections.

ATR tube [Anti-Transmit-Receive tube] A gas-filled RF switching tube used in radar to isolate the transmitter from the antenna during the interval for pulse reception.

attack plotter An instrument used with navigation and sonar to show, on the screen of a cathode-ray tube, a plot of the course of the ship making an antisubmarine attack, the path of each sonar search beam from the ship, the position of the underwater target when each sound contact is made, the course of the target as successive target positions appear, and the range and bearing of the target.

attenuation A reduction in energy. Attenuation occurs naturally during wave travel through lines, waveguides, spaces, or a medium like water. It also occurs naturally when nuclear radiation passes through a medium but may be produced intentionally by inserting an attenuator in a circuit or placing an absorbing device in the path of the radiation. The amount of attenuation is generally expressed in decibels or decibels per unit length.

attenuation coefficient The fraction of a beam of light that is removed by scatter or absorption in passing through unit thickness of a material.

attenuation constant [symbol α] A rating for a line or medium through which a plane wave of a given frequency is being transmitted. It is the real part of the propagation constant and is the relative rate of decrease of amplitude of a field component (or of voltage or current) in the direction of propagation, in nepers per unit length. The imaginary part of the propagation constant is the phase constant.

attenuation distortion A deviation from uniform amplification or attenuation over a required frequency range.

attenuation equalizer An equalizer used to make the total transmission loss of a line or circuit essentially the same for all frequencies in the range being transmitted.

attenuation factor 1. The ratio of incident intensity to transmitted intensity for radiation passing through a layer of material. 2. The ratio of input current to output current for a transmission line or network.

attenuation ratio The magnitude of the propagation ratio.

attenuator An arrangement of fixed and variable resistive elements used to reduce the strength of an RF or AF signal a desired adjustable amount without introducing appreciable distortion. It is designed so the impedance of the attenuator will match that of the circuit in which it is connected,

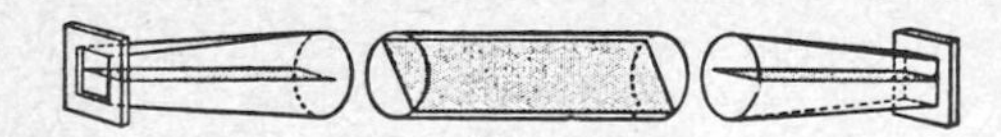

Attenuator for insertion in rectangular waveguide, consisting of rotatable resistive film mounted between two fixed resistive films. When all three films are in same plane, attenuation is zero.

regardless of the amount of attenuation introduced. The corresponding nonadjustable device is usually called a pad.

attenuator tube A gas-filled RF switching tube in which a gas discharge, initiated and regulated independently of RF power, is used to control this power by reflection or absorption.

attitude The position of an aircraft as determined by the inclination of its axes to some frame of reference, such as the earth.

attitude control A control system or mechanism, as an automatic pilot, that puts or keeps an aircraft or missile in a desired attitude.

attitude gyro A gyro used in attitude control of aircraft.

attitude indicator An indicator that shows the roll and pitch angles of an aircraft in relation to the earth.

attitude sensor A sensor that provides a reference signal for keeping a spacecraft in a desired attitude with respect to a given direction, such as toward the sun or earth.

atto- [abbreviated a] A prefix representing 10^{-18}, which is 0.000 000 000 000 000 001, or one-millionth of a millionth of a millionth.

audibility A measure of the strength of a specified sound as compared with the strength of a sound that can just be heard. Usually expressed in decibels.

audibility limit A threshold of hearing. The lower limit is the minimum effective sound pressure that can be heard at a specified frequency. The upper limit is the minimum effective sound pressure that causes pain in the ear.

audibility threshold The lower limit of audibility. For a specified signal it is the minimum effective sound pressure capable of producing an auditory sensation in a specified fraction of the trials. The characteristics of the signal, the manner in which it is presented to the listener, the point at which the sound pressure is measured, and the ambient noise all affect the value. Usually expressed in decibels above 0.0002 μbar or above 1 μbar.

audible Capable of being heard by the average human ear. The approximate range of human hearing is between 15 and 20,000 Hz, but actual limits vary greatly with different individuals.

audible Doppler enhancer A Doppler circuit that serves as a third detector for the received pulses. It separates the envelope from the pulses, thus giving an audio signal that can be amplified and fed to a loudspeaker.

audio [Latin for "I hear"] 1. Pertaining to signals, equipment, or phenomena that involve frequencies in the range of human hearing. 2. Slang for *sound.*

audio amplifier *Audio-frequency amplifier.*

audio frequency [abbreviated AF] A frequency that can be detected as a sound by the human ear. The range of audio frequencies extends approximately from 15 to 20,000 Hz. Also called sonic frequency and sound frequency.

audio-frequency amplifier An amplifier that has one or more electron-tube or transistor amplifier stages for amplifying an AF signal. In a superheterodyne receiver it follows the second detector and amplifies the AF signal after demodulation. Also used separately to amplify the AF output of a microphone, phonograph, magnetic tape recorder, or other AF signal source. Also called audio amplifier.

audio-frequency harmonic distortion Distortion in which integral multiples of a single AF input signal are generated by the amplifier.

audio-frequency oscillator An oscillator circuit that uses an electron tube, transistor, or other nonrotating device to produce an AF alternating current. Also called audio oscillator.

audio-frequency peak limiter A circuit used in an AF system to cut off signal peaks that exceed a predetermined value. Also called audio peak limiter.

audio frequency-shift keying [abbreviated AFSK] Radioteletype keying in which the RF carrier is transmitted continuously and pulses are transmitted by frequency-shifted tone modulation. Commonly used audio tones are 2.125 kHz for mark and 2.975 kHz for space.

audio frequency-shift modulation A facsimile system in which picture tones are represented by audio frequencies. In one example a 1.5-kHz tone represents black, a 2.3-kHz tone represents white, and frequencies in between represent shades of gray.

audio-frequency signal generator A signal generator that can be set to generate a sinusoidal AF signal voltage at any desired frequency in the audio spectrum. Also called audio signal generator.

audio-frequency transformer An iron-core transformer used for coupling between AF circuits. Also called audio transformer.

audiogram A graph that shows hearing loss, percent hearing loss, or percent hearing as a function of frequency.

audio level meter An instrument that measures AF power with reference to a predetermined level. Its scale is usually calibrated in decibels.

audiology 1. The science of hearing. 2. The branch of medicine dealing with causes and treatment of defective hearing.

audio masking *Masking.*

audiometer An instrument that measures hearing ability. In one form it consists of an audio oscillator that has variable calibrated output and is capable of generating a wide range of audio tone frequencies. Recorded speech sounds may also be used.

audiometry The study of hearing ability by audiometers.

audion The original three-element vacuum tube invented by Dr. Lee de Forest.

audio oscillator *Audio-frequency oscillator.*

audio peak limiter *Audio-frequency peak limiter.*

audiophile A person interested in listening to broadcasts and recordings that are reproduced with high fidelity.

audio response A form of computer output in which prerecorded spoken syllables, words, or messages are selected and put together by a computer as the appropriate verbal response to a keyboarded inquiry on a time-shared on-line information system.

audio response unit A magnetic recording system that provides voice response to an inquiry made from a typewriter or telephone-type terminal connected to a computer by a data-transmission line. The appropriate audio response is selected by the computer from spoken words previously recorded on a magnetic disk or other storage device. Applications include automatic stock-price-quotation service from any telephone when the query code for a particular stock is punched or dialed.

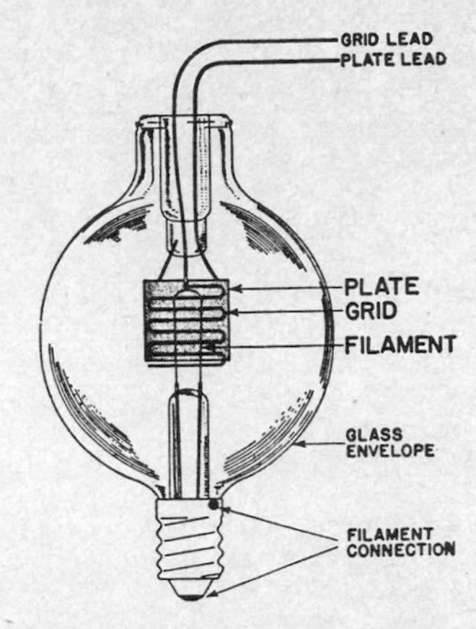

Audion-tube construction.

audio signal An electric signal that has an audio frequency.

audio signal generator *Audio-frequency signal generator.*

audio spectrum The continuous range of frequencies extending from the lowest to the highest audio frequency (from about 15 to 20,000 Hz).

audio subcarrier A subcarrier whose frequency lies within the audio range.

audio taper A taper commonly used in volume and tone controls, in which the resistance increases slowly at the beginning of shaft rotation and increases much faster as the shaft or knob is rotated toward the limit of its clockwise rotation. Used to compensate for the fact that the frequency range of the human ear is less at low volume levels.

audio transformer *Audio-frequency transformer.*

audiovisual Involving both sight and sound.

audition A preliminary studio test of a performer, act, or complete program for a television or radio show.

auditory perspective Three-dimensional realism of sound, as produced by an actual orchestra or a stereophonic sound system.

auditory sensation area The region enclosed by curves defining the thresholds of feeling and audibility as functions of frequency.

audit trail A data-processing system feature that permits tracing the flow of data step by step from input to output. Used primarily for locating the cause of an error in data processing.

augend The number to which a new quantity, called the addend, is to be added.

Auger coefficient The ratio of Auger yield to fluorescence yield, or the ratio of the number of Auger electrons to the number of x-ray photons emitted from a large number of similarly excited atoms.

Auger effect A nonradiative transition of an atom from an excited energy state to a lower energy state, accompanied by the emission of an electron.

Auger electron An electron ejected from an atom by a photon in the Auger effect. It has a kinetic energy equal to the difference between the energy of the x-ray photon of the corresponding radiative transition and the binding energy of the ejected electron.

Auger shower *Extensive shower.*

Auger yield The ratio of the number of Auger electrons emitted to the number of events that result in an electron vacancy in the inner shell of an atom.

aural Pertaining to the ear or the sense of hearing.

aural center frequency The average frequency of a carrier modulated by an AF signal or the carrier frequency without modulation.

aural harmonic A harmonic generated in the human ear.

aural masking *Masking.*

aural null The condition of weakest sound when tuning or otherwise adjusting a circuit that has an audio output.

aural-null direction finder A radio direction finder that consists of a radio receiver and rotatable loop antenna. When the loop is rotated to give an aural null, the plane of the loop is at right angles to the direction of the transmitted signal.

aural radio range A radio range station that provides lines of position by aural identification or comparison of signals at the output of a receiver, as in the A-N radio range.

aural signal 1. A signal that can be heard. 2. The sound portion of a television signal; the picture portion is called the visual signal.

aural transmitter The radio equipment used to transmit the aural portion of a television program. The aural and visual transmitters together are called a television transmitter.

aurora The sporadic visible emission from the upper atmosphere over middle and high latitudes.

aurora gating Operator-controlled gating of a radar receiver to eliminate undesired radar returns from aurora.

auroral absorption Absorption of radio waves by abnormal particle radiation from the sun.

auroral reflection Reflection of radio waves back to earth by a rapidly fluctuating ionized layer in the upper atmosphere of the polar regions, usually during magnetic storms accompanied by auroral displays.

auroral storm *Ionospheric storm.*

autoabstract A collection of significant words, phrases, or sentences selected from a document by a computer, to serve as an abstract.

autoalarm *Automatic alarm receiver.*

autocall A signal system that generates preset combinations of gong or tone signals for paging people in buildings or other locations.

autocode *Automatic code.*

autocoder The computer section which converts a symbolic input code language into the machine language used in that particular computer.

autocorrelation A mathematical technique that detects cyclic activity in a complex signal.

autocorrelation function A mathematical quantity defined as the time average of the product of a function of time and a delayed version of that function of time.

autocorrelator A correlator in which the input signal is delayed, then multiplied by the undelayed signal. The product is then smoothed in a low-pass filter to give an approximate computation of the autocorrelation function. It can be used to detect a weak periodic signal hidden in noise, if the chosen time delay is equal to the period of the signal. The autocorrelator may also be applied to the detection of nonperiodic signals.

autodyne circuit A circuit in which the same tube elements serve as oscillator and detector simultaneously. The output frequency is equal to the difference between the frequencies of the received signal and the oscillator signal, just as in a conventional superheterodyne circuit.

autoelectric emission The emission of electrons from a cold cathode under the influence of an intense electric field.

autoencode To select key words from a document by a machine method in order to develop search patterns for information retrieval.

autofluoroscope A scintillation camera in which the entire area of nuclide distribution in the human body is sensed at one time by a matrix of hundreds of sodium iodide crystals; each crystal emits light in proportion to its incident gamma radiation. Phototubes convert the light outputs to signals that produce the desired complete image on an oscilloscope screen.

autoindexing *Automatic indexing.*

automatic Having a self-acting mechanism that performs a required act at a predetermined time or in response to certain conditions.

automatic aiming *Automatic tracking.*

automatic alarm receiver A complete receiving, selecting, and warning device capable of being actuated automatically by intercepted RF waves that form the international automatic alarm signal. Also called autoalarm.

automatic-alarm-signal keying device A device that keys automatically the radiotelegraph transmitter on board a vessel, to transmit the international automatic alarm signal.

automatic approach and landing A control mode in which the speed and flight path of an aircraft are automatically controlled for approach, flareout, and landing.

automatic back bias Use of one or more automatic gain control loops in a radar receiver to prevent overloading by strong radar echoes or jamming signals.

automatic background control *Automatic brightness control.*

automatic bass compensation [abbreviated ABC] A circuit used in some radio receivers and audio amplifiers to make bass notes sound more natural at low volume-control settings. The circuit usually consists of a resistor and capacitor in series,

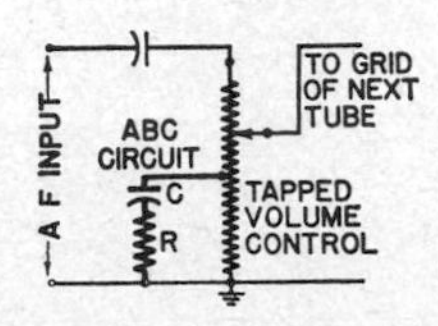

Automatic bass compensation.

connected between ground and a tap on the volume control. This circuit automatically compensates for the poor response of the human ear to weak low-frequency sounds.

automatic beam steering Antenna steering that involves the use of an adaptive array, with servo-controlled phase shifters in each antenna element, to create a highly directional receiving-antenna pattern that can be aimed electronically.

automatic brightness control [abbreviated ABC] A circuit used in a television receiver to keep the average brightness of the reproduced image essentially constant. Its action is like that of an automatic volume control circuit in a sound receiver. Also called automatic background control.

automatic carrier landing system A combination radio and radar system used in landing aircraft on carriers during adverse weather conditions. Radar is used to locate the aircraft and determine its exact position at each instant with respect to the landing deck. A computer converts the radar data into course-correction data that is fed to the approaching aircraft by a radio transmitter. If the approach is incorrect, the system automatically waves off the aircraft.

automatic celestial navigation Automatic navigation that uses equipment capable of locking on to a star.

automatic chart-line follower A device that automatically derives error signals proportional to the deviation of the track of a vehicle from a predetermined course line drawn on a chart. Also called automatic track follower.

automatic check A check performed by equipment built into a computer specifically for that purpose, and automatically accomplished each time the pertinent operation is performed.

automatic chroma control *Automatic color control.*

automatic chrominance control *Automatic color control.*

automatic code A code that allows a computer to translate or convert a symbolic language into a machine language. Also called autocode.

automatic coding Any method or technique by which a computer performs a significant portion of the coding for a problem.

automatic color control A circuit used in a color television receiver to keep color intensity levels essentially constant despite variations in the strength of the received color signal. Control is usually achieved by varying the gain of the chrominance bandpass amplifier. Also called automatic chroma control and automatic chrominance control.

automatic color purifier *Automatic degausser.*

automatic computer A computer that performs long sequences of operations in accordance with a predesigned program, without human intervention.

automatic contrast control A circuit that varies the gain of the RF and video IF amplifiers in such a way that the contrast of the television picture is maintained at a constant average level. Control is achieved by varying the bias on one or more variable-mu tubes. The manual contrast control determines the average level and the automatic contrast control maintains this average, despite variations in signal strength as different stations are tuned in.

automatic control Control in which regulating and switching operations are performed automatically in response to predetermined conditions. Also called automatic regulation.

automatic controller An instrument that continuously measures the value of a variable quantity or condition, then automatically acts on the controlled equipment to correct any deviation from a desired preset value. Also called controller.

automatic control system A control system that has one or more automatic controllers connected in closed loops with one or more processes. Also called regulating system.

automatic crossover A circuit used in some regulated power supplies to change automatically from voltage to current regulation in response to changes in load resistance and voltage- and current-level settings.

automatic cutout A device, usually operated by centrifugal force or an electromagnet, that automatically shorts part of a circuit at a particular time. Used on some induction motors to cut out the starting winding when operating speed is reached.

automatic data processing [abbreviated ADP] The processing of digital information by electronic or electric equipment. Includes the use of electric accounting machines (EAM) and electronic data processing (EDP).

automatic data-processing system An interacting assembly of procedures, processes, methods, personnel, and automatic data-processing equipment, used for data processing.

automatic degausser An arrangement of degaussing coils mounted around a color television picture tube, combined with a special circuit that energizes these coils only while the set is warming up after being turned on. The coils demagnetize any parts of the receiver that have been affected

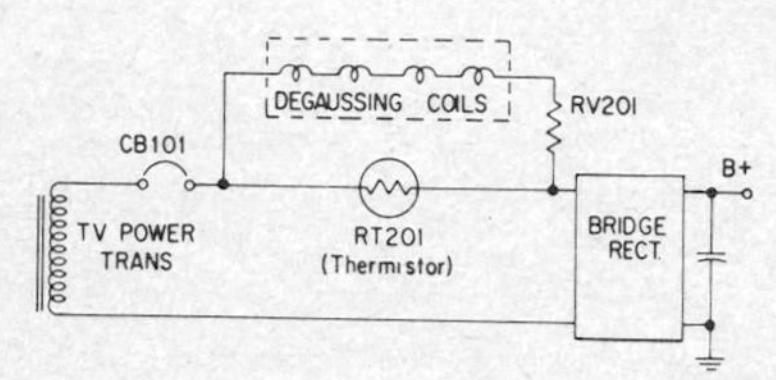

Automatic degausser in power supply of color television receiver. Varistor in series with degaussing coils, and thermistor shunting coils, act together to cut off current to coils by time that receiver tubes have warmed up.

by the earth's magnetic field or by the field of any nearby home appliance. Automatic degaussing permits a color television receiver to be moved around a home without readjusting purity controls. Also called automatic color purifier.

automatic degaussing control system A system that automatically changes the degaussing current for a ship to compensate for changes in the ship's heading. A synchro transmitter installed in the gyrocompass equipment of the ship provides a signal that is rectified and amplified for use in controlling the degaussing current source.

automatic dialer A device that enables a person to automatically dial a telephone number which has up to 14 digits by pressing a single button or by inserting a magnetically coded card. A prestored number usually can be changed by setting a control and dialing the desired new number once.

automatic direction finder [abbreviated ADF] A direction finder that automatically and continuously indicates the direction of arrival of a radio signal. The directional antenna is automatically aimed in the direction of minimum signal from the radio station being used for navigation guidance, by means of electronic circuits used alone or in conjunction with a motor drive for the antenna. Modern versions are generally used with an additional fixed antenna to provide a unidirectional bearing indication that tells whether the aircraft or ship is traveling toward or away from the radio station. Also called automatic loop radio compass, automatic radio compass, automatic radio direction finder, and radio compass.

automatic exchange A telephone, teletypewriter, or data-transmission exchange in which communication between subscribers is effected, without the intervention of an operator, by devices set in operation by the originating subscriber's instrument.

automatic fine-tuning control A circuit used in a color television receiver to keep the frequency of the oscillator in the tuner correct for best color picture by compensating for drift and incorrect tuning. Eliminates the need for careful manual fine-tuning each time a station is changed.

automatic flight control A control system that includes an automatic pilot and the additional equipment needed to maintain automatically a desired aircraft altitude and heading.

automatic focusing Electrostatic focusing in which the focusing anode of a television picture tube is internally connected through a resistor to the cathode so that no external focusing voltage is required.

automatic frequency control [abbreviated AFC] 1. A circuit that maintains the frequency of an oscillator within specified limits, as in a transmitter. 2. A circuit that keeps a superheterodyne receiver tuned accurately to a given frequency by controlling its local oscillator, as in an FM receiver. 3. A circuit used in radar superheterodyne receivers to vary the local oscillator frequency, to compensate for changes in the frequency of the received echo signal. 4. A circuit used in television receivers to make the frequency of a sweep oscillator correspond to the frequency of the synchronizing pulses in the received signal.

automatic gain control [abbreviated AGC] A control circuit that automatically changes the gain (amplification) of a receiver or other piece of equipment so the desired output signal remains essentially constant despite variations in input signal strength.

automatic indexing Selection of key words from a document by computer, for use as index entries. Generally based on statistics of frequency, location, and other features of selected words, individually or in context. Also called autoindexing.

automatic keyer A device that sends a predetermined signal or message in code, such as the series of 12 dashes which actuates an automatic alarm received on a ship.

automatic light control Automatic adjustment of illumination reaching a film, television camera, or other imaging device as a function of scene brightness.

automatic loop radio compass *Automatic direction finder.*

automatic message-switching center A location at which an incoming message is automatically directed to one or more outgoing circuits according to intelligence contained in the message.

automatic modulation control A transmitter circuit that reduces the gain for excessively strong audio input signals without affecting the strength of normal signals. This permits higher average modulation without overmodulation, equivalent to an increase in carrier-frequency power output.

automatic modulation limiting [abbreviated AML] A circuit that prevents overmodulation in some citizens band radio transmitters. This circuit reduces the gain of one or more audio amplifier stages when the voice signal becomes stronger, to keep the modulation level below 100%.

automatic noise limiter *Noise limiter.*

automatic pattern recognition *Pattern recognition.*

automatic peak limiter *Limiter.*

automatic pedestal control A process by which the pedestal height in a received television signal is automatically adjusted as a function of input signal strength or some other specified parameter.

automatic phase control [abbreviated APC] 1. A circuit used in color television receivers to reinsert a 3.58-MHz carrier signal with exactly the correct phase and frequency by synchronizing it with the transmitted color-burst signal. 2. An automatic frequency-control circuit in which the difference between two frequency sources is fed to a phase detector that produces the required control signal.

automatic picture control A multiple-contact switch used in some color television receivers to disconnect one or more of the regular controls and make connections to corresponding preset controls. Pushing one button thus corrects for accidental misadjustment of controls.

automatic picture transmission [abbreviated APT] A slow-scan television system used in weather satellites; it is capable of transmitting conventional television pictures of clouds in the daytime and infrared pictures of clouds at night. Each image is stored for about 200 s in a vidicon while being scanned for transmission to earth.

automatic pilot *Autopilot.*

automatic programming Any technique whereby a digital computer itself transforms programming from a form that is easy for a human being to produce into a form that is efficient for the computer to carry out.

automatic radio compass *Automatic direction finder.*

automatic radio direction finder *Automatic direction finder.*

automatic ranging *Autoranging.*

automatic record changer An electric phonograph that automatically plays a number of records one after another.

automatic regulation *Automatic control.*

automatic routine A computer routine that is executed independently of manual operations, but only if certain conditions occur within a program or record.

automatic-scanning receiver A receiver that automatically sweeps back and forth through a preselected frequency range. It may be set to plot signal occupancy in the range or stop when a signal is found.

automatic-search jammer A search receiver and jamming transmitter used together to search for and automatically jam enemy signals that have specified radiation characteristics.

automatic secure voice communication A worldwide military network that uses wire lines and radio links to provide cryptographically secure voice communication by voice-digitizing techniques.

automatic selectivity control A circuit that makes a receiver less selective when the received signal is strong and more selective when the signal is weak. The reduction in the anode-cathode resistance of a tube when it is handling a strong signal is used to damp a tuned circuit and thereby make the receiver less selective.

automatic sensitivity control A circuit that maintains receiver sensitivity at a predetermined level.

automatic sequencing Ability of a computer to perform successive operations without human intervention.

automatic shutoff A switch incorporated into some tape recorders to stop the machine when the tape breaks or runs out.

automatic stop Automatic halting of a computer processing operation as the result of an error detected by built-in checking devices.

automatic telephone payment [abbreviated ATP] Payment of bills or transfer of funds between accounts via an ordinary Touch-Tone telephone. The user calls the computer conventionally, and then enters an identification number, transaction code, and dollar amount for routing to the user's bank by the computer.

automatic test equipment Test equipment that makes two or more tests in sequence without manual intervention; it usually stops when the first out-of-tolerance value is detected.

automatic tint control A circuit used in color television receivers to maintain correct flesh tones when a station changes cameras or switches to commercials, by correcting phase errors before the chroma signal is demodulated.

automatic toll ticketing A telephone system that automatically records telephone numbers and call times under control of dial pulses from the calling telephone and prints toll tickets for billing use.

automatic track follower *Automatic chart-line follower.*

automatic tracking Tracking in which a servomechanism keeps the radar beam trained on the target. The servomechanism is actuated by circuits that respond to some characteristic of the echo signal from the target. Also called automatic aiming and autotrack.

automatic track-shift A system used with multiple-track magnetic tape recorders to index the tape head, after one track is played, to the correct position for the start of the next track.

automatic tuning system An electric, mechanical, or electromechanical system that tunes a radio receiver or transmitter automatically to a predetermined frequency when a button or lever is

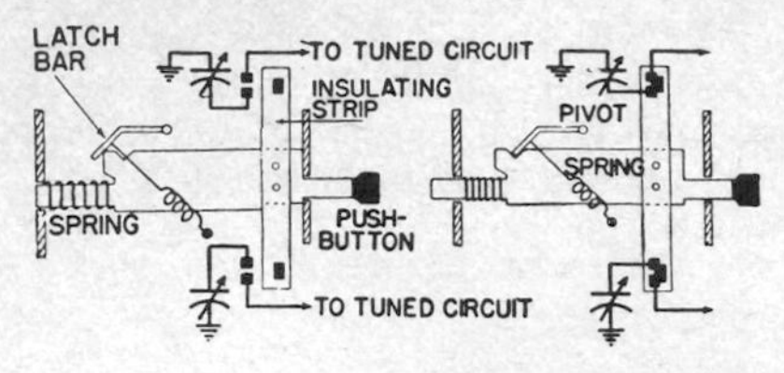

Automatic tuning system using latching switches controlled by pushbuttons.

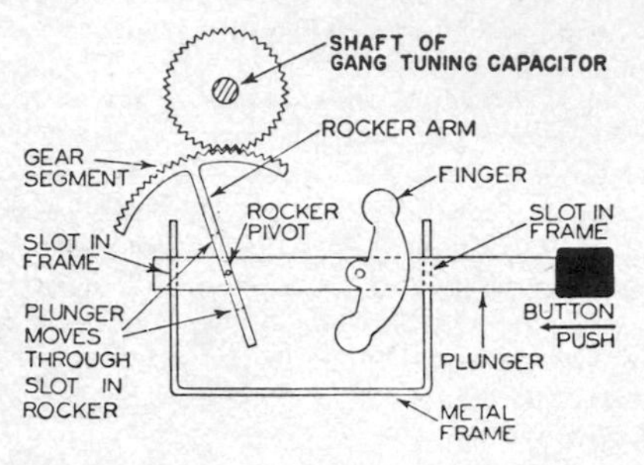

Automatic tuning system using mechanical gear-and-finger arrangement. Widely used in auto radios.

pressed, a knob is turned, or a telephone-type dial is operated.

automatic typewriter An electric typewriter that produces a punched paper tape or magnetic recording simultaneously with the conventional typed copy, for subsequent automatic retyping at high speed. Generally includes provisions for error correction and changing of personalized or other variable information during each retyping.

automatic voltage regulator *Voltage regulator.*

automatic volume compressor *Volume compressor.*

automatic volume control [abbreviated AVC] An automatic gain control that keeps the output volume of a radio receiver essentially constant despite variations in input signal strength during fading or when tuning from station to

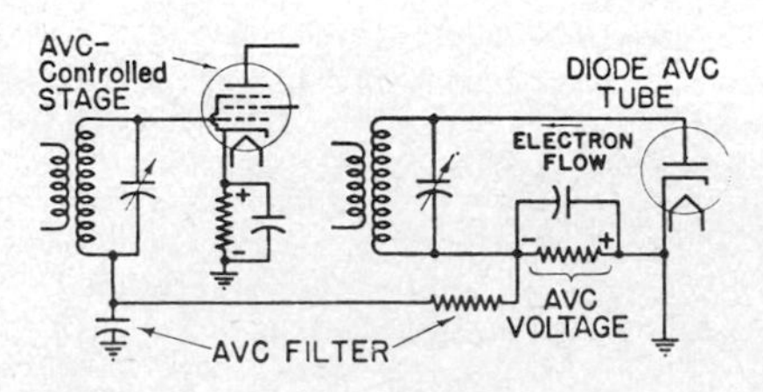

Automatic volume control circuit for radio receiver, in which diode detector provides AVC voltage.

station. A DC voltage proportional to audio output signal strength is obtained from the second detector and used to change the bias of one or more preceding RF and IF amplifier stages.

automatic volume expander [abbreviated AVE] *Volume expander.*

automation Continuous automatic operation in which control functions are performed by mechanisms instead of people.

automaton A device that automatically performs a predetermined sequence of operations, responds to encoded instructions, or otherwise simulates human actions.

autonavigator A navigation system that includes means for coupling the outputs of inertial and other navigation sensors to the control system of a vehicle.

autonetics The branch of electronics that deals with automatic guidance and control systems.

autopilot An arrangement of gyroscopes combined with amplifiers and servomotors to detect deviations in the flight of an aircraft and apply the required corrections directly to the controls. Also called automatic pilot.

autopilot coupler A coupling system that links the output of the navigation system receiver to the automatic pilot in an aircraft.

autopolarity Automatic interchanging of connections to a digital meter when polarity is wrong. A minus sign appears ahead of the value on the digital display if the reading is negative.

auto radio A radio receiver that can be installed in an automobile and operated from the automobile's storage battery.

autoradiograph A photographic record of radiation from radioactive material in an object, made by placing the object against photographic film. Also called radioautograph (deprecated).

autoranging Automatic switching of a multirange meter from its lowest range to the next higher range, with the switching process repeated until a range is reached for which the full-scale value is not exceeded. Automatic downranging is also provided in some multimeters, such as for choosing the resistance range that gives highest accuracy for a particular measurement. Also called automatic ranging.

autoregistration A technique used in the manufacture of semiconductor devices, in which previously applied metal contacts also act as masks during ion implantation of impurity atoms.

autoregulation induction heater An induction heater in which a desired control is effected by the change in characteristics of a magnetic charge as it is heated at or near its Curie point.

autosevocom [AUTOmatic SEcure VOice COMmunication] A defense communication network accessory that will make voice communication cryptographically secure by using a combination of wideband and narrow-band voice-digitizing techniques.

autotrack *Automatic tracking.*

autotracking Single-control operation of two or more DC power supplies. When the master supply, which generally has the highest positive output voltage, is turned on or off or changed in voltage, the interconnected slave supplies are similarly switched or changed proportionally in output voltages.

autotransductor British term for a magnetic amplifier in which the same windings serve as both control windings and power windings.

autotransformer A power transformer that has one continuous tapped winding. Part of the winding serves as the primary, and all of it serves as the secondary, or vice versa.

auxiliary anode An anode located adjacent to the pool cathode in an ignitron. It helps the main anode to maintain a cathode spot.

auxiliary relay A relay that assists another relay or device in the performance of a function.

auxiliary routine A routine that assists in the operation of a computer and in debugging other routines.

auxiliary storage A storage that supplements the main storage of a computer, like magnetic tape, magnetic disks, or a magnetic drum.

auxiliary transmitter A transmitter maintained only for transmitting the regular programs of a station in case the main transmitter fails.

avalanche breakdown Nondestructive breakdown in a semiconductor diode when the electric field across the barrier region is strong enough so current carriers collide with valence electrons to produce ionization and cumulative multiplication of carriers. Often confused with zener breakdown, in which the electric field across the barrier region becomes high enough to produce a form of field emission that suddenly increases the number of carriers in this region.

avalanche diode A semiconductor breakdown diode, usually made of silicon, in which avalanche breakdown occurs across the entire PN junction and voltage drop is then essentially constant and independent of current. The two most important types are IMPATT and TRAPATT diodes.

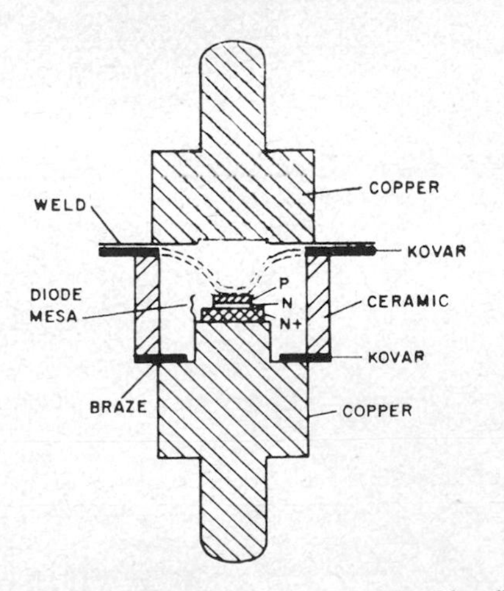

Avalanche diode using silicon mesa, capable of handling peaks up to 400 W as pulsed microwave source.

Breakdown characteristics are considerably sharper than for zener diodes.

avalanche effect 1. The cumulative process in which an electron or other charged particle accelerated by a strong electric field collides with and ionizes gas molecules, thereby releasing new electrons that in turn have more collisions. The discharge is thus self-maintained. Also called cascade, cumulative ionization, Townsend avalanche, and Townsend ionization. 2. Cumulative multiplication of carriers in a semiconductor because of avalanche breakdown.

avalanche-induced migration A technique of forming interconnections in a field-programmable logic array by applying appropriate voltages for shorting selected base-emitter junctions.

avalanche ionization Ionization in which a single ion creates a large number of additional ions through successive ionizing collisions in a gas.

avalanche noise Noise produced when a PN junction diode is operated at the onset of avalanche breakdown.

avalanche oscillator An oscillator that uses an avalanche diode as a negative resistance to achieve

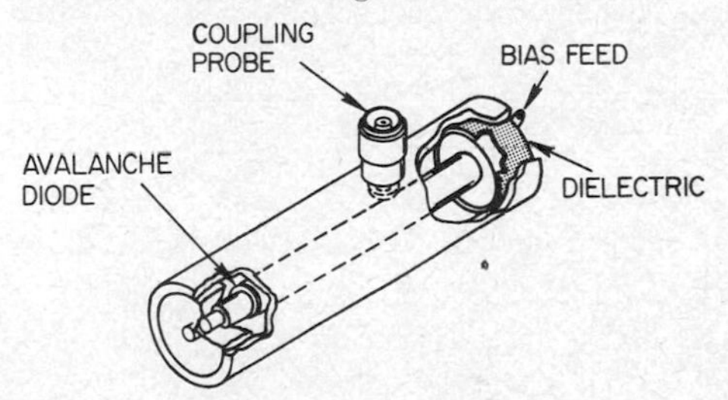

Avalanche oscillator using coaxial cavity.

one-step conversion from DC to microwave outputs in the gigahertz range. The diode is mounted in a coaxial cavity or a waveguide structure.

avalanche photodiode A photodiode operated in the avalanche breakdown region to achieve internal photocurrent multiplication, thereby providing rapid light-controlled switching operation. It can have good infrared response, as required for detection of modulated light from lasers and light-emitting diodes.

avalanche rectifier An avalanche diode designed specifically to act as a rectifier in power supplies.

avalanche silicon controlled rectifier A silicon controlled rectifier in which avalanche breakdown occurs across the entire PN junction.

avalanche silicon controlled rectifier A silicon controlled rectifier in which avalanche breakdown occurs across the entire PN junction.

avalanche transistor A transistor that utilizes avalanche breakdown to produce chain generation of charge-carrying hole-electron pairs. It can be operated to provide very high common-emitter current gain or very short switch-on time.

AVC Abbreviation for *automatic volume control.*
AVE Abbreviation for *automatic volume expander.*
average calculating operation A computer calculating operation that is longer than an addition and shorter than a multiplication. Often taken as the mean of nine additions and one multiplication.
average life *Mean life.*
average noise factor The ratio of the total delivered noise power of a linear system to the portion of the noise produced by the input termination. Also called average noise figure.
average noise figure *Average noise factor.*
average pulse amplitude The average of the instantaneous amplitudes taken over the duration of a pulse.
average speech power The average of instantaneous speech power values during a given time interval.
average value The average of many instantaneous amplitude values taken at equal intervals of time during half a cycle of alternating current. For a sine wave, the average value is 0.637 times the peak value.
averaging A means for improving the precision of measurement of a given quantity by averaging a number of measured values. The chief drawback is the danger of smoothing out real perturbations of the variable, along with the fluctuations due to noise and other factors.
averaging multiplier A multiplier whose output is a discrete function of the input variables sampled at specific intervals and averaged. Also called sampling multiplier.
avionics [AVIation electrONICS] The field of airborne electronics.
AWACS Abbreviation for *airborne warning and control system.*
AWG Abbreviation for *American wire gage.*
A wind Magnetic tape wound on the reel with the dull oxide-coated side of the tape toward the inside. This wind is now almost universally used. Recorder design determines whether A or B wind tape is required.
awu Abbreviation for *atomic weight unit.*

Axial leads on resistor.

axial lead A wire lead that comes out from the end along the axis of a resistor, capacitor, or other component.
axiotron An axially controlled magnetron, in which the magnetic field produced by the filament current controls the anode current.
Ayrton-Perry winding Winding of two wires in parallel but opposite directions to give better cancellation of magnetic fields than is obtained with a single winding. Used in some wirewound resistors.

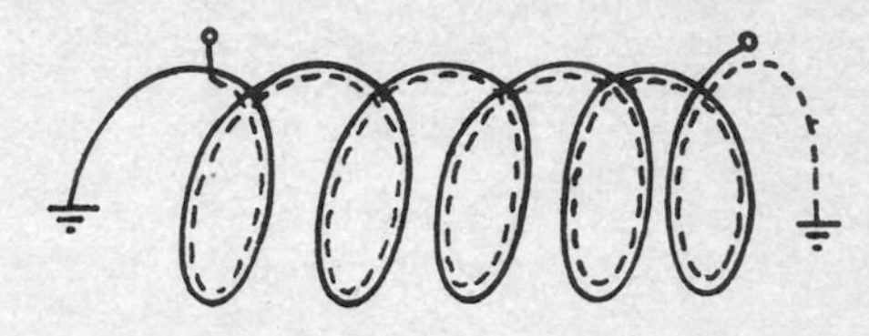

Ayrton-Perry winding as used in some resistors.

Ayrton shunt A shunt that increases the range of a galvanometer without changing the damping. Also called universal shunt.
azel indicator [AZimuth-ELevation indicator] *Expanded plan-position indicator.*
azel scope A cathode-ray tube on which both azimuth and elevation of a target are presented at the same time. Used in precision approach systems such as GCA.
azimuth *Bearing.*
azimuthally varying field A cyclotron magnetic field produced by adding spiral ridges to the magnet poles, to provide vertical focusing through the alternating-gradient principle. This permits use of a magnetic field which increases with radius in such a manner that the orbits are isochronous and rotation frequency is constant.
azimuth blanking 1. Automatic blanking of a radar transmitter beam as the antenna scans a predetermined horizontal sector of its scanning region, such as for preventing interference with television receivers in a city close to the search radar site. 2. Automatic blanking of a radar PPI display for a selected sector of the horizontal region scanned by the antenna.
azimuth deception An electronic countermeasure technique used to prevent an enemy tracking radar from obtaining accurate azimuth information. The incident radar signal is generally amplified and retransmitted with distortion to create a false echo.
azimuth gating Brightening of a selected sector of a radar PPI display, usually by applying a step waveform to the automatic-gain-control circuit.
azimuth marker A radar receiver circuit that produces a bright radial line on a PPI display at an angle which can be adjusted by a control dial so the line passes through a target indication on the screen.
azimuth rate computer A computer that calculates the rate of change of horizontal angular measurements from a base line.
azimuth resolution The minimum azimuth angle at which two targets that have the same range can be individually distinguished by a given radar set.
azimuth-stabilized PPI A plan-position indicator that is stabilized by a gyrocompass so either true or magnetic north is at the top of the screen, regardless of equipment orientation. In a north-stabilized PPI, the reference direction is magnetic north.

azimuth vs. amplitude A counter-countermeasure radar that has a PPI display which shows strobe lines at angles corresponding to locations of jamming aircraft.

azon [AZimuth ONly] A bomb that has movable control surfaces in the tail which can be adjusted by radio to control the bomb in azimuth.

azran [AZimuth and RANge] A radar or other system that locates a target by polar coordinates, which give angle and distance to the target.

azusa A guidance or range instrumentation system in which directive antennas and phase-comparison (coherent carrier) techniques provide angle determination, and multichannel subcarriers provide range measurements by means of time delay. The equipment includes an elaborate ground antenna array, a transmitting and receiving station, and a missile-born transponder.

B

b 1. Abbreviation for *barn*. 2. Abbreviation for *bit*.

B 1. Symbol for *base*. Used on transistor circuit diagrams. 2. Abbreviation for *bel*.

B− [B minus] The negative terminal of a B battery or other anode voltage source for an electron tube, or the anode circuit terminal to which this negative source terminal should be connected.

B+ [B plus] The positive terminal of a B battery or other anode voltage source for an electron tube, or the anode circuit terminal to which this positive source terminal should be connected.

B Symbol for *magnetic flux density*.

babble The aggregate crosstalk from a large number of channels.

babs [Blind Approach Beacon System] A pulse-type ground-based navigation beacon used for runway approach at airports. When interrogated by an aircraft making an instrument approach, the beacon sends out signals that produce range and runway position information on the L-scan cathode-ray indicator in the aircraft.

back contact A normally closed stationary contact on a relay. Its circuit is opened when the relay is energized. Also called break contact.

back diode *Backward diode*.

back echo An echo signal produced on a radar screen by one of the minor back lobes of a search radar beam.

backed stamper A thin metal stamper attached to a solid backing material for use in molding phonograph records.

back electromotive force *Counterelectromotive force*.

back emission *Reverse emission*.

backfire *Arcback*.

background Ever-present effects in physical apparatus above which a phenomenon must manifest itself to be measured or observed.

background count A count caused by ionizing radiation coming from sources other than that being measured.

background monitor An ionization chamber or other radiation counter used to measure prevailing background counts.

background music 1. Music selected to create a stimulating or relaxing environment in an office, factory, or other location. 2. Music selected to accompany dialogue or action other than singing or dancing on a radio or television program or movie.

background noise Undesired noise heard along with desired signals or sound.

background processing Automatic processing of lower-priority programs by a computer when there are no higher-priority programs on hand.

background radiation The radiation of people's natural environment, including that which comes from cosmic rays, the naturally radioactive elements of the earth, and within a person's body.

background response The response caused by ionizing radiation that comes from sources other than those to be measured by a radiation detector.

background return *Clutter*.

backheating Heating of a magnetron cathode by the high-velocity electrons that return to the cathode surface. Once the tube is in operation, backheating is often sufficient to keep the cathode at emitting temperature without heater current.

backlash The difference between the values obtained for a parameter when a control dial is set to an indicated value from opposite directions.

back lobe A radiation pattern lobe that is directed away from the intended direction.

backout Reversal of the countdown sequence during a missile launching.

backplane A wiring board, usually constructed as a printed circuit, that is used in microcomputers and minicomputers to provide the required connections between logic, memory, and input/output modules.

backplate The electrode to which the stored charge image of a camera tube is capacitively coupled.

back porch The portion of a composite picture signal that follows the horizontal sync pulse and extends to the trailing edge of the corresponding blanking pulse. The color burst, if present, is not considered part of the back porch.

back-porch effect The continuation of collector current in a transistor for a short time after the input signal has dropped to zero. The effect is due to storage of minority carriers in the base region. The effect also occurs in junction diodes.

back resistance The contact resistance that opposes the inverse current of a contact rectifier.

backscattering 1. Deprecated term for radar echoes, which may include both clutter and desired echoes from a target. 2. Deflection of radiation or nuclear particles by scattering processes through angles greater than 90° with respect to the original direction of travel. 3. Undesired radiation of energy to the rear by a directional antenna.

backscattering coefficient The ratio of reflected power to incident power for a plane wave. Also called echoing area.

backscattering thickness gage A radioactive thickness gage in which the radiation source and radiation meter are mounted on the same side of the material being measured. This gives a measurement of the reflected or backscattered radiation. Used chiefly for measuring or controlling the thickness or density of a coating being applied to a moving sheet of material.

backscatter radar An over-the-horizon radar system designed primarily to detect an enemy missile attack from far beyond the horizon, by detecting radio energy reflected back from the ionosphere disturbance created by the missiles.

back-shunt keying A method of keying a transmitter in which the RF energy is fed to the antenna when the telegraph key is closed and to an artificial load when the key is open.

backspace To move the printing or display position backward one space along the reading line.

backsputtering Sputtering used for plasma cleaning of samples, accomplished by appropriate DC and RF biasing of the target samples during bombardment with both electrons and ions.

backstop The part of a relay that limits the movement of the armature away from the core.

back-to-back connection A method of connecting a pair of tubes so that each operates on half of an AC cycle, thereby permitting control of alternating current. The anode of one tube is connected to the cathode of the other, and vice versa. Used chiefly for thyratrons and ignitrons. Transistors are similarly connected in parallel in opposite directions to control current in either direction without causing rectification.

back-to-back repeater A repeater in which the output of the receiver is connected directly to the input of the transmitter.

backup An item kept available as a replacement for an item that fails to perform satisfactorily.

backward diode A semiconductor diode, similar to a tunnel diode, in which maximum current flows in the reverse direction (when the N-type region of the junction is positive). Also called back diode.

backward read Transfer of data from a magnetic-tape unit to computer storage while the tape is moving in a reverse direction. Normally used during external sort phases, to reduce rewind time.

backward wave A wave traveling opposite to the normal direction, such as a wave whose group velocity is opposite to the direction of electron-stream motion in a traveling-wave tube, or the reflected wave in a mismatched transmission line.

backward-wave magnetron A magnetron oscillator in which the electron beam travels in a direction opposite to the flow of RF energy. Characterized by high power output, and it can be voltage-tuned over a wide band.

backward-wave oscillator [abbreviated BWO] An oscillator that uses a special vacuum tube in which electrons are bunched by an RF magnetic field as they flow from cathode to anode. This bunching action produces a backward wave that

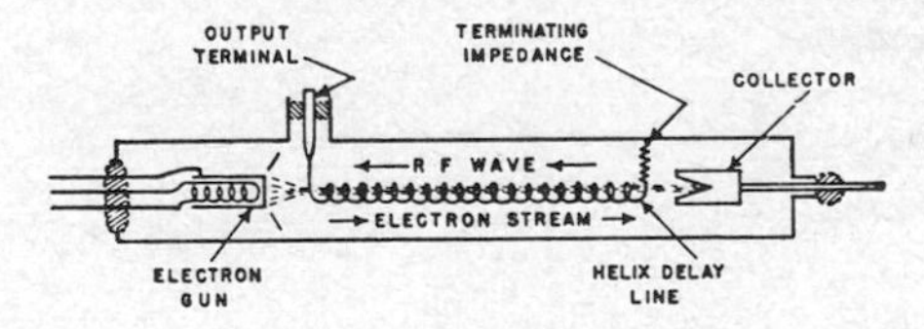

Backward-wave oscillator construction.

becomes larger as it progresses toward the electron-gun end of the tube. The magnetic field is produced by a folded-line structure centered on the axis of the electron beam. The output signal is taken from the gun end of this folded line.

backward-wave tube A traveling-wave tube in which the electrons travel in a direction opposite to that in which the wave is propagated.

back wave *Spacing wave.*

baffle 1. A cabinet or partition used with a loudspeaker to increase the effective length of the air path from the front to the rear of the moving diaphragm. By reducing interaction between sound waves produced simultaneously by the two surfaces of the diaphragm, a baffle improves the fidelity of reproduction. 2. An auxiliary member that is placed in the arc path of a gas tube and which has no external connection. It may be used for controlling the flow of mercury particles or deionizing the mercury vapor following conduction.

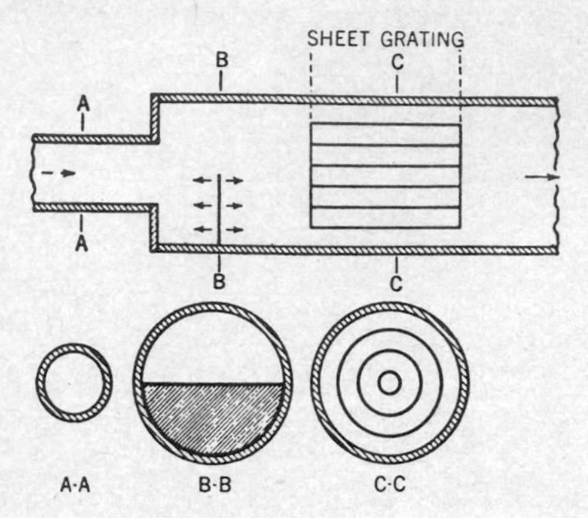

Baffle plate at B-B, used with concentric-cylinder sheet grating at C-C to convert transverse magnetic wave entering from left to transverse electric wave leaving at right.

baffle plate A metal plate inserted in a waveguide to reduce the cross-sectional area for wave-conversion purposes.

bail A loop of heavy wire snap-fitted around two or more parts of a connector or other device to hold the parts together.

Bakelite Trademark of Bakelite Corp. for their plastics and resins. Originally it applied only to their phenolic compound, widely used in radio parts because of its excellent insulating qualities.

bakeout Heating of an object to facilitate removal of absorbed gases.

balance The condition whereby both speakers in a stereo system produce the same average sound levels.

balance control A control used in a stereo sound system to vary the volume of one loudspeaker system relative to the other while maintaining their combined volume essentially constant.

balanced amplifier An amplifier circuit in which there are two identical signal branches connected to operate in phase opposition, with input and output connections each balanced to ground. A push-pull amplifier is an example.

balanced bridge A bridge adjusted for zero output, occurring when bridge branch currents are equal.

balanced circuit A circuit whose two sides are electrically alike and symmetrical with respect to a common reference point, usually ground.

balanced converter *Balun.*

balanced currents Currents flowing in the two conductors of a balanced line that, at every point along the line, are equal in magnitude and opposite in direction. Also called push-pull currents.

balanced detector A detector used in FM receivers. In one form the audio output is the rectified difference between the voltages produced across two resonant circuits, one circuit being tuned slightly above the carrier frequency and the other slightly below.

balanced input A two-terminal input circuit that has the same inpedance between each terminal and ground.

balanced line 1. A transmission line that consists of two conductors which are capable of being operated in such a way that the voltages of the two conductors at any transverse plane are equal in magnitude and opposite in polarity with respect to ground. The currents in the two conductors are then equal in magnitude and opposite in direction. 2. A production line for which the time cycles of the operators are made approximately equal so the work flows at a desired steady rate from one operator to the next.

balanced-line logic element A logic circuit that contains two matched tunnel diodes (as in a locked pair), performs majority logic, and uses a multiphase AC clock to direct the flow of information.

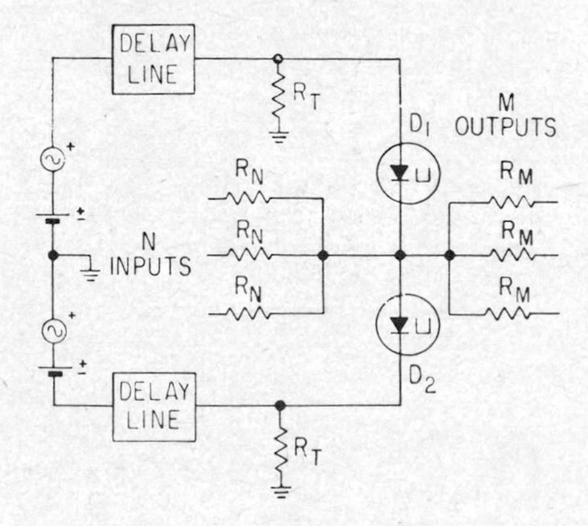

Balanced-line logic element using tunnel diodes.

The output pulse has a flatter top than does the locked pair, thus improving the timing tolerance.The DC bias and superimposed AC voltage are applied to the diodes through two delay lines.

balanced-line system A system that consists of generator, balanced line, and load adjusted so the voltages of the two conductors at each transverse plane are equal in magnitude and opposite in polarity with respect to ground.

balanced low-pass filter A low-pass filter designed to be used with a balanced line.

balanced magnetic amplifier A magnetic amplifier formed by mixing the outputs of two identical single-ended magnetic amplifiers in such a way that the output polarity can be reversed. The bias for each single-ended amplifier can be chosen to give class A, B, or C operation for the balanced amplifier.

balanced method A measuring method in which the reading is taken at zero or in the absence of a signal, as with a balanced bridge.

balanced mixer A mixer circuit used in superheterodyne receivers to reduce leakthrough and suppress even-order harmonics of one input, usually the local oscillator. Balance is achieved with a bridge or ring configuration of diodes, a balanced pair of transistors, or a variety of tube circuits.

balanced modulator A modulator in which the carrier and modulating signal are introduced in such a way that the output contains the two sidebands without the carrier. Used in color television transmitters to apply the I and Q signals to the subcarriers, as well as in suppressed-carrier communication transmitters.

balanced network A network that has equal impedances in opposite branches.

balanced oscillator An oscillator in which the impedance centers of the tank circuits are at ground potential and the voltages between either end and their centers are equal in magnitude and opposite in phase. A push-pull oscillator is an example.

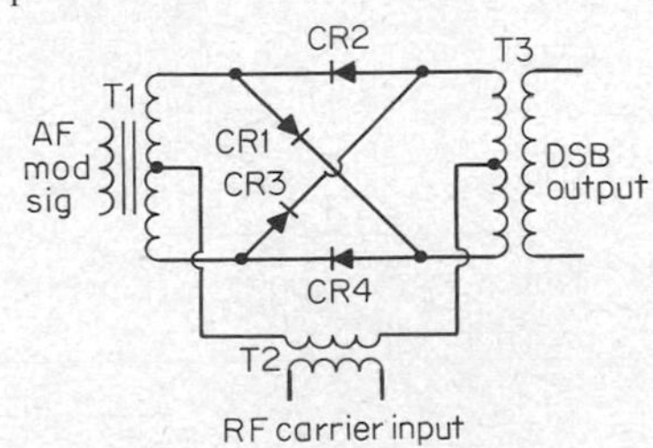

Balanced ring modulator using four diodes to suppress carrier.

balanced ring modulator A modulator that uses tubes or diodes to suppress the carrier signal while providing double-sideband output.

balanced termination A two-terminal load in which both terminals present the same impedance to ground.

balanced transmission line *Twin-line.*

balanced voltages Voltages that are equal in magnitude and opposite in polarity with respect to ground. Also called push-pull voltages.

ballast An iron-core inductance connected in series with a fluorescent lamp or other arc discharge lamp to provide the required high starting voltage and limit the operating current.

ballast lamp A lamp that increases in resistance when current increases, to maintain a nearly constant current.

ballast resistor A resistor that increases in resistance when current increases, thereby maintaining essentially constant current despite variations in line voltage. Also called barretter (British).

ballast tube A ballast resistor mounted in an evacuated glass or metal envelope like that of a vacuum tube, to reduce radiation of heat from the resistance element and thereby improve the voltage-regulating action.

ball bonding A type of thermocompression bonding in which a flame is used to cut a wire as it leaves a capillary feed tube. The molten end of the wire solidifies as a ball, and this ball is pressed against the bonding pad on the integrated circuit.

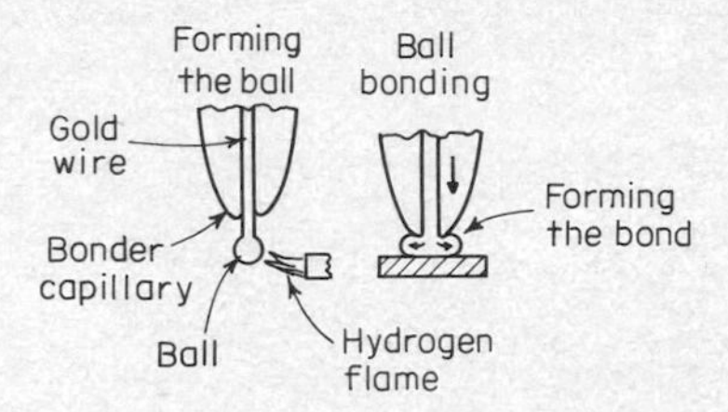

Ball bonding with gold wire.

Ball bonds must be made one at a time, whereas wedge bonding can be used to interconnect several components simultaneously.

ballistic galvanometer An instrument that measures the total quantity of electricity in a transient current, such as the discharge current of a capacitor.

ballistic missile A missile that is guided during powered flight in the upward part of its trajectory but becomes a free-falling body in the latter stages of its flight toward its target. The German V-2 was an example.

ballistic-missile early-warning system [abbreviated BMEWS] An electronic system that provides detection and early warning of attack by enemy intercontinental ballistic missiles. One example is a network of three long-range-radar bases at Thule (Greenland), Clear (Alaska), and in the British Isles, to provide warning of missile attack across the polar region.

ballistic trajectory The trajectory followed by a body being acted upon only by gravitational forces and the resistance of the medium through which it passes.

ballistocardiograph An instrument that measures cardiac performance by recording the recoil movements of the body which result from contractions of heart muscles.

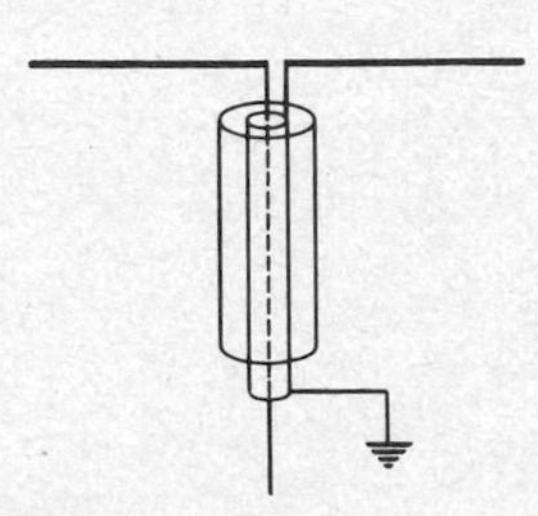

Balun as used to feed balanced dipole with unbalanced coaxial cable.

balun [BALanced to UNbalanced] A device that matches an unbalanced coaxial transmission line or system to a balanced two-wire line or system. It is a quarter-wavelength cylindrical sleeve placed over the end of a coaxial cable feed to an antenna, and isolates the outer conductor of the cable from ground. Also called balanced converter, bazooka (slang), and line-balance converter.

banana jack A jack that accepts a banana plug. Generally designed for panel mounting.

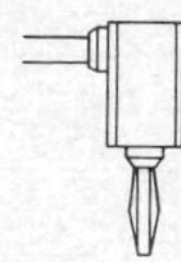

Banana plug with right-angle lead.

banana plug A plug that has a spring-metal tip which somewhat resembles a banana, used on test leads or as terminals for plug-in components.

Band	Frequency		Wavelength	
ELF	Below 300	Hz	Above 1000	km
ILF	300–3000	Hz	100–1000	km
VLF	3–30	kHz	10–100	km
LF	30–300	kHz	1–10	km
MF	300–3000	kHz	100–1000	m
HF	3–30	MHz	10–100	m
VHF	30–300	MHz	1–10	m
UHF	300–3000	MHz	10–100	cm
SHF	3–30	GHz	1–10	cm
EHF	30–300	GHz	1–10	mm
	300–3000	GHz	0.1–1	mm
P	225–390	MHz	133.3–76.9	cm
L	390–1550	MHz	76.9–19.37	cm
S	1.55–5.2	GHz	19.37–5.77	cm
C	3.9–6.2	GHz	7.69–4.84	cm
X	5.2–10.9	GHz	5.77–2.75	cm
K	10.9–36	GHz	27.5–8.34	mm
Q	36–46	GHz	8.34–6.52	mm
V	46–56	GHz	6.52–5.36	mm

Band limits for radio spectrum. Lower frequency limit of band is excluded, and upper limit is included. Bands designated by single letters were used unofficially in World War II. Frequency limits for subdivisions of K, L, and X bands are given in entries for these bands.

band 1. A range of frequencies between two definite limits. 2. One track, or a group of tracks, on a magnetic disk or drum in a computer.

band-edge energy The energy of the edge of the conduction band or valence band in a solid. It is the minimum energy needed by an electron to move freely in a semiconductor or the maximum energy it may have as a valence electron.

band-elimination filter A filter that attenuates alternating currents whose frequencies are between given upper and lower cutoff values while transmitting frequencies above and below this band. It is the opposite of a bandpass filter. The band rejected is generally much wider than that suppressed by a trap. Also called band-rejection filter, bandstop filter, and rejector circuit.

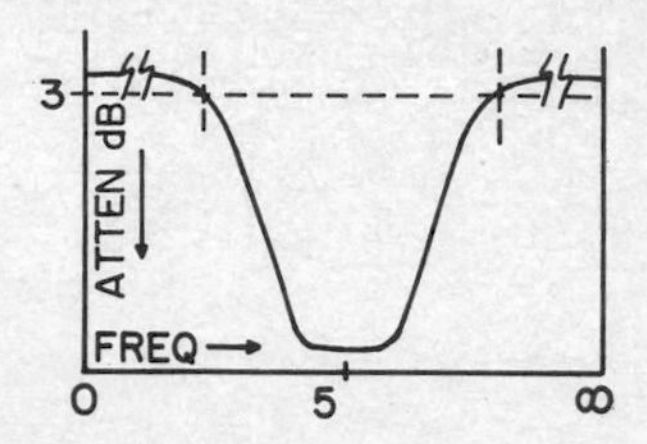

Band-elimination filter response curve for center frequency of 5 kHz.

band-ignitor tube A glow-discharge tube in which conduction is initiated by applying a high voltage between the cathode and a metal band wrapped around the envelope.

band-limited function A function whose Fourier transform is very small or vanishes outside some finite interval.

bandpass amplifier An amplifier that passes a definite band of frequencies with essentially uniform response.

bandpass filter A filter that transmits alternating currents whose frequencies are between given upper and lower cutoff values, while substantially attenuating all frequencies outside this band.

bandpass response A response characteristic in which a definite band of frequencies is transmitted with essentially uniform response. In IF transformers, this response is obtained by tuning the primary and secondary resonant circuits to slightly different frequencies. The response curve then usually has two humps. Also called double-hump response and flat-top response.

band pressure level The effective sound pressure level of the sound energy contained within a specified frequency band.

band-rejection filter *Band-elimination filter.*

band selector A switch that selects any one of the bands in which a receiver, signal generator, or transmitter is designed to operate. It usually has two or more sections, to make the required changes in all tuning circuits simultaneously. Also called band switch.

B and S gage Abbreviation for *Brown and Sharpe gage.*

band spectrum A spectrum that has the appearance of bands rather than separate lines. Usually applied to the spectra of molecules, even though the bands of most molecules have now been resolved into their separate lines. In nuclear physics, band spectra are useful in determining nuclear spin, nuclear statistics, and isotopic abundances.

bandspread tuning control A separate tuning control on some receivers to spread stations in a single band of frequencies over an entire tuning dial. It controls small variable capacitors that are

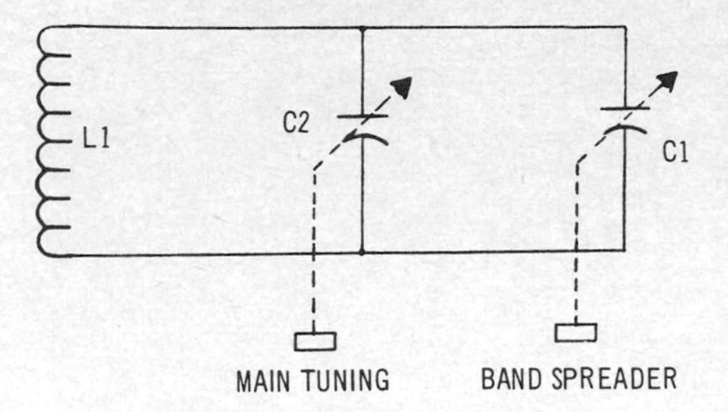

Bandspread tuning control.

connected in parallel with each main tuning capacitor.

bandstop filter *Band-elimination filter.*

band switch *Band selector.*

bandwidth [abbreviated BW] 1. The width of a band of frequencies used for a particular purpose. Thus the bandwidth of a television station is 6 MHz. 2. The range of frequencies within which a performance characteristic of a device is above specified limits. For filters, attenuators, and amplifiers these limits are generally taken to be 3 dB below the average level. Half-power points are also used as limits.

bang-bang control A missile or pilotless airplane control system in which corrective action is always applied to the full extent of servomotion, to give two-position or ON/OFF control.

bank 1. A number of similar devices connected together for use as a single device. A bank of resistors is an example. 2. A planned accumulation of subassemblies stored at a point on a production line to permit reasonable fluctuations in line speed before and after the bank.

banked winding A method of winding RF coils: single turns are wound one over the other in a flat outward spiral. The entire coil consists of many such spirals side by side, giving a multilayer coil without going back to the starting point. This construction reduces the distributed capacitance of the coil.

bantam tube A tube that has a small glass envelope on a standard octal base, identified by the letters GT following the type number of the equivalent standard octal tube.

bar A unit of pressure equal to 1×10^6 dyn/cm^2 or 1×10^5 N/m^2 (slightly less than 1 atmosphere). The microbar, equal to 1×10^{-6} bar, was the unit of pressure formerly used in acoustics. The newton per square meter (N/m^2) is now the SI unit of pressure.

bar code reader An optical reader that reads combinations of printed bars which represent numerical characters.

bar generator A signal generator that delivers pulses uniformly spaced in time and synchronized to produce a stationary bar pattern on a television screen. A color-bar generator produces these bars in different colors on the screen of a color television set.

BARITT diode [BARrier Injection Transit-Time diode] A microwave diode in which the carriers that traverse the drift region are generated by minority carrier injection from a forward-biased junction instead of being extracted from the plasma of an avalanche region. Can be used as an oscillator in the gigahertz range. One version has a thin slice of N-type silicon between platinum silicide Schottky barrier contacts.

barium [symbol Ba] An element used in cathode coatings of electron tubes. Atomic number is 56.

barium fuel cell A fuel cell in which barium is used with either oxygen or chlorine to convert chemical energy into electric energy.

barium titanate A ceramic that has piezoelectric properties and is capable of withstanding much higher temperatures than Rochelle salt crystals. Used in crystal pickups and sonar transducers.

barium titanate microphone A crystal microphone that uses a barium titanate ceramic bar which has piezoelectric properties.

Barkhausen effect The succession of abrupt changes in magnetization that occur when the magnetizing force acting on a piece of iron or other magnetic material is varied.

Barkhausen-Kurz oscillator A retarding-field-type oscillator in which the frequency of oscillation depends solely on the electron transit time within the tube. Electrons oscillate about a highly positive grid before reaching the less positive anode. Also called Barkhausen oscillator.

Barkhausen magnet A permanent magnet mounted on the horizontal output tube of a television receiver to reduce Barkhausen oscillations.

Barkhausen oscillation An undesired oscillation in the horizontal output tube of a television receiver; it causes one or more ragged dark vertical lines on the left side of the picture.

Barkhausen oscillator *Barkhausen-Kurz oscillator.*

bar magnet A bar of hard steel that has been strongly magnetized and holds its magnetism, thereby serving as a permanent magnet.

barn [abbreviated b] A unit of nuclear cross section, equal to 10^{-24} cm^2.

Barnett effect The very slight magnetization produced in an iron rod when it is rotated at high speed about an axis perpendicular to its length.

barograph A barometer that produces a continuous record of atmospheric pressure on a graph.

barometer An instrument that measures the pressure of the atmosphere. Two common types are aneroid and mercury barometers.

barometric altimeter An altimeter that uses a barometer for measuring height.

barometric pressure Atmospheric pressure as measured by a barometer.

barometric switch *Baroswitch.*

baroresistor A device in which electric resistance

varies as a function of atmospheric pressure. It generally consists of a barometric element mechanically linked to a resistance element.

baroswitch A device that performs electric switching functions by mechanical actuation which results from changes in atmospheric pressure. In one form an aneroid diaphragm is mechanically linked to a contact arm that slides over a commutator whose separate segments are connected to the signal circuits used to modulate a radiosonde. Also called barometric switch.

bar pattern The pattern of repeating color bars produced by a bar generator, for adjusting color television receivers.

barrage jamming The simultaneous jamming of a number of radio or radar channels.

bar relay A relay in which a moving bar actuates several contacts simultaneously.

barrel distortion Distortion in which all four sides of a received television picture bulge outward like a barrel.

barrel-stave reflector A parabolic radar antenna reflector that has the horizontal top third and bottom third cut away, to give a high vertical beam so roll of the ship will not cause the beam to miss the target.

barretter 1. A bolometer that consists of a fine wire or metal film which has a positive temperature coefficient of resistivity, so that resistance increases with temperature. Used for making power measurements in microwave devices. It is the opposite of a thermistor, in which resistance decreases with temperature. 2. British term for *ballast resistor.*

barretter mount A waveguide mount into which a barretter can be inserted to measure electromagnetic power.

barrier *Potential barrier.*

Barrier A passive acoustic detection system for submarines, consisting of hydrophones positioned on the ocean floor and connected by undersea cable to a land-based computer center. Developed by the U.S. Navy for surveillance and tracking of enemy submarines.

barrier diode A diode formed by depositing a metal film on a high-resistivity semiconductor material. It is sometimes made as part of a bipolar transistor.

barrier grid A grid, close to or in contact with a storage surface of a charge-storage tube, that establishes an equilibrium voltage for secondary-emission charging and minimizes redistribution.

barrier layer Deprecated term for *depletion layer.*

barrier-layer capacitance *Depletion-layer capacitance.*

barrier transparency The proportion of incident electrons that penetrates a potential barrier.

barrier voltage The minimum voltage required for conduction through a PN junction.

baryon One of a class of particles that includes protons, neutrons, hyperons, and cascade particles. All free baryons heavier than the proton eventually decay into end products, one of which is the proton.

base 1. [symbol B] The region that lies between an emitter and a collector of a transistor and into which minority carriers are injected. It corresponds to the grid of an electron tube. 2. The part of an electron tube that has the pins, leads, or other terminals to which external connections are made either directly or through a socket. 3. The plastic, ceramic, or other insulating board that supports a printed-wiring pattern. 4. A plastic film that supports the magnetic powder of magnetic tape or the emulsion of photographic film. 5. An integer to which all digits are related in a positional notation system for computers. The successive digits are coefficients of successive powers of the base. The most common base values are 2, 8, and 10. Also called radix. 6. The number on which a system of logarithms is based, such as 10 or 2.718.

base address A number that appears as an address in a computer instruction but serves as the starting point for subsequent addresses to be modified.

baseband The frequency band occupied by all the transmitted signals that are used to modulate a particular carrier. The band that transmits picture and synchronizing signals in television is one example; another is the band containing all the modulated subcarriers in a carrier system.

base bias The direct voltage that is applied to the base electrode of a transistor.

base charge The charge produced by excess minority carriers in the base region of a transistor or diode.

base-diffusion isolation An integrated-circuit manufacturing technique in which the epitaxial region and substrate are opposite conductivity types, and a reverse bias is applied to an isolating ring surrounding a bipolar transistor so as to punch through the interface and create a collector pocket.

base electrode An ohmic or majority carrier contact to the base region of a transistor.

base film The plastic substrate that supports the magnetic coating of magnetic tape. The base film of most instrumentation and computer tapes is made of a polyester, such as Mylar. For less critical uses, cellulose acetate and polyvinyl chloride are used.

basegroup A group of carrier channels in a particular frequency range that forms a basic unit for further modulation to a final frequency band in a carrier communication system.

base insulator A heavy-duty insulator that supports the weight of an antenna mast and insulates the mast from the ground or some other surface.

baseline 1. A line that joins the two stations between which electric phase or time is compared in determining navigation coordinates, such as a line joining a master and a slave station in a loran system. 2. The line produced on the screen in the absence of an echo in certain types of radar displays.

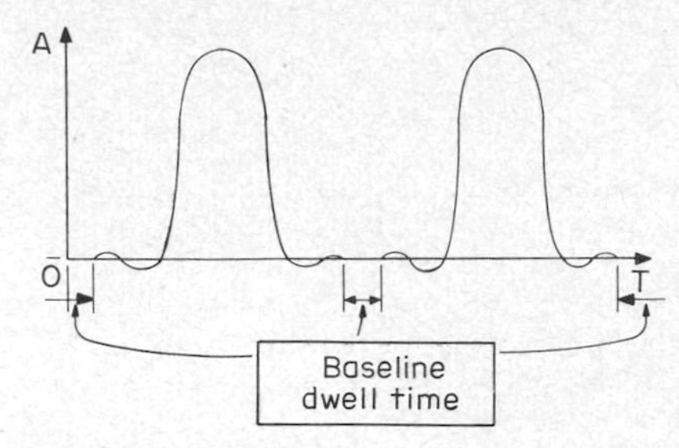

Baseline dwell time between recurring pulses.

baseline dwell time The time during which a pulse waveform coincides with its baseline.

base-loaded antenna A vertical antenna whose electrical height is increased by adding inductance in series at the base.

base modulation Amplitude modulation produced by applying the modulating voltage to the base of an amplifier transistor.

base pin *Pin.*

base point *Point.*

base region The interelectrode region of a transistor into which minority carriers are injected.

base station A fixed-location land station that provides radio communication service with mobile stations and other fixed radio stations.

BASIC [Beginners Algebraic-Symbol Interpreter Compiler] A FORTRAN-like computer programming language that can be mastered in a few days. Commonly used for business and commercial problems on time-sharing computers.

basic frequency The frequency of the sinusoidal component considered the most important, such as the fundamental frequency of an oscillator or the driving frequency of an acoustic transducer.

basic repetition rate The lowest pulse repetition rate of each of the several sets of closely spaced repetition rates in a loran system.

basic research Fundamental, theoretical, or experimental investigation to advance scientific knowledge. Immediate practical application is not a direct objective.

basket winding A crisscross coil winding in which adjacent turns are far apart except at points of crossing, giving low distributed capacitance.

bass Sounds corresponding to frequencies at the lower end of the audio range, below about 250 Hz.

bass boost A circuit that emphasizes the lower audio frequencies, generally by attenuating higher audio frequencies.

bass compensation A circuit that offsets the lowered sensitivity of the human ear to weak low frequencies by making the bass frequencies relatively stronger than the high audio frequencies as volume is lowered.

bass control A manual tone control that changes the level of bass frequencies in an audio amplifier.

bass reflex baffle A loudspeaker baffle that has an opening of such size below the loudspeaker that bass frequencies from the rear emerge to reinforce those radiated directly forward.

bass response The extent to which an AF amplifier, loudspeaker, or other device handles low audio frequencies.

bassy Pertaining to overemphasis of bass notes in sound reproduction.

batch processing A technique by which items to be processed are coded and collected into groups prior to processing.

bat-handle switch A toggle switch that has an actuating lever shaped like a baseball bat.

bathtub capacitor A paper capacitor enclosed in a metal housing that has broadly rounded corners like those on a bathtub.

bathythermograph A sensitive recording thermometer lowered into the water to determine temperatures at different levels, as required in

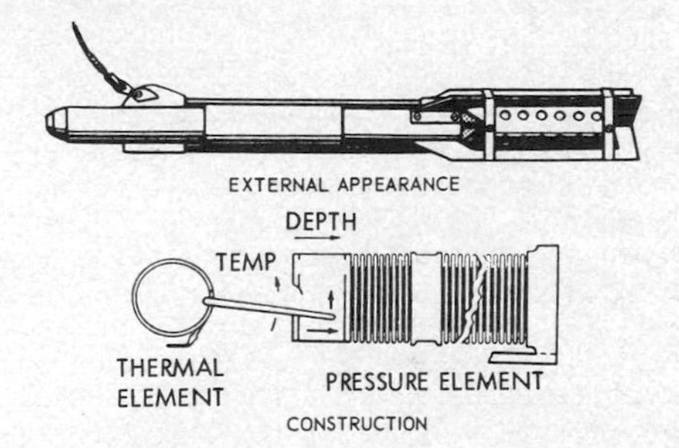

Bathythermograph in which temperature-indicating arm of thermal element moves over smoked-glass area on pressure element.

predicting sound conditions for sonar. The data may be either recorded or transmitted up the support cable as modulation on a carrier signal.

battery A DC voltage source made up of one or more cells that convert chemical, thermal, nuclear, or solar energy into electric energy.

battery analyzer A tester that automatically analyzes charge-discharge capacities of a secondary battery by using appropriate logic, timing, and switching circuits.

battery charger A rectifier unit used to change AC power to DC power for charging a storage battery. Also called charger.

battery clip A terminal that has spring jaws

Battery clip with screw terminal for connecting lead.

which can be quickly snapped on a battery terminal or other point to which a temporary wire connection is desired.

battery receiver A radio receiver that obtains its DC operating voltages from one or more batteries.

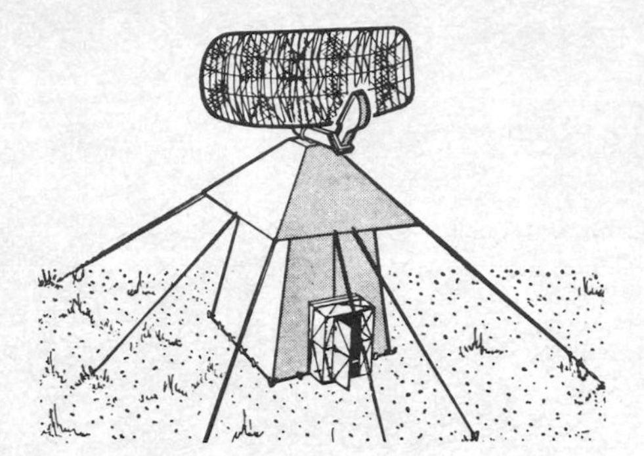

Battlefield surveillance radar installation.

battlefield surveillance radar A transportable air surveillance radar that can be set up close to front lines for monitoring friendly air cover and detecting the approach of enemy planes.

batwing antenna *Superturnstile antenna.*

baud [abbreviated Bd] A unit of signaling speed in telecommunications, equal to one element per second. Signaling speed in bauds is equal to the reciprocal of signal element length in seconds. A pulse and a space are separate elements; thus a teleprinter handling 22.5 pulses per second is operating at 45 Bd. Some modems can transmit data between computers at speeds up to 500,000 Bd (500 kBd).

Baudot code A teleprinter code that uses combinations of five or six mark and space intervals of equal duration. The five-unit code gives 32 possible characters, and the six-unit version gives 64. Used in radio and wire teleprinter operation.

bay A main division of a structure, usually containing vertically mounted electronic equipment having some common function, such as for radio communication or for the wire and cable terminations of a telephone system.

bayonet base A tube or lamp base that has two projecting pins on opposite sides of a smooth cylindrical base, to engage in corresponding slots in a bayonet socket and hold the base firmly in the socket.

bayonet coupling A connector design that uses bayonet-base pins and J-shaped slots to lock together the mating plug and receptacle of a connector.

bayonet socket A socket for bayonet-base tubes or lamps; it has J-shaped slots on opposite sides and one or more contact buttons at the bottom.

bazooka Slang term for *balun.*

B battery The battery that supplies the DC voltages required by the anode and grid electrodes of electron tubes in battery-operated equipment.

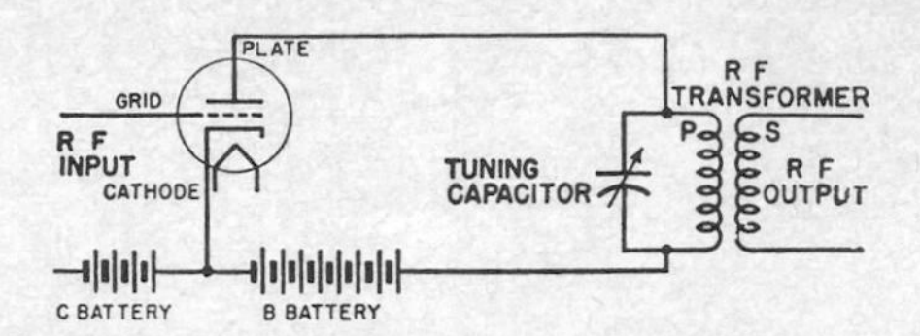

B battery connections in RF amplifier stage.

BBD Abbreviation for *bucket brigade device.*

BCD Abbreviation for *binary-coded decimal.*

BCI Abbreviation for *broadcast interference.*

Bd Abbreviation for *baud.*

BDI Abbreviation for *bearing deviation indicator.*

B display A rectangular radarscope display in which targets appear as bright spots, with target

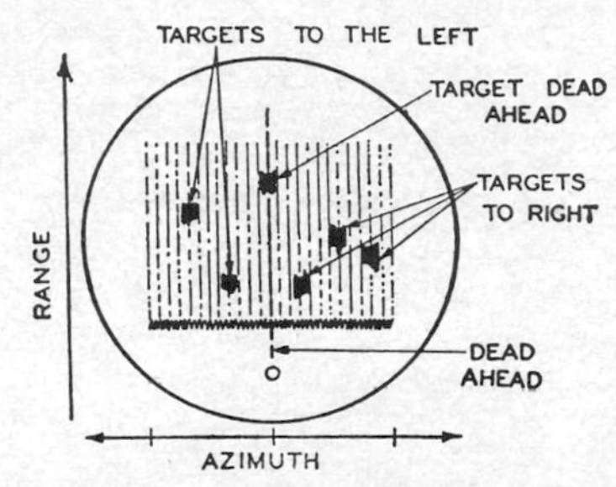

B display.

bearing indicated by the horizontal coordinate and target distance by the vertical coordinate.

beacon A navigation aid that provides a radio, radar, light, or other characteristic signal which gives bearing, course, or location guidance for vehicles.

beacon presentation The radarscope presentation that results from RF waves sent out by a radar beacon.

beacon stealing Interference between radar beacons, resulting in loss of beacon tracking by one tracking radar.

bead A glass, ceramic, or plastic insulator with a hole through its center, used to support the inner conductor of a coaxial line in the exact center of the line. Ferrite beads are placed on wire connections of high-frequency circuits to suppress parasitic oscillation by providing inductance.

bead thermistor A thermistor that consists of a small bead of semiconducting material, such as

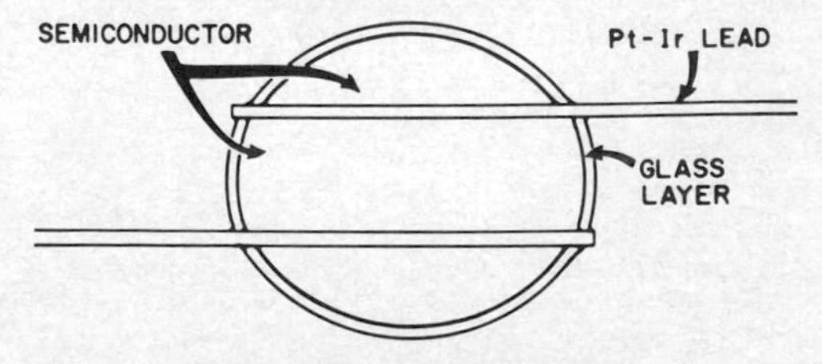

Bead thermistor using metallic oxides as semiconductor between platinum-iridium leads.

germanium, placed between two wire leads. Used for microwave power measurement, temperature measurement, and as a protective device.

beam 1. A concentrated unidirectional stream of particles, such as electrons or protons. 2. A concentrated unidirectional flow of electromagnetic waves, as from a radar antenna, a microwave relay antenna, or an A-N radio range antenna array. The beam here is a major lobe of the antenna radiation pattern and is restricted to a small solid angle in space. 3. A concentrated unidirectional flow of acoustic waves. 4. A parallel arrangement of light rays. Also called ray.

beam-accessed metal-oxide semiconductor [abbreviated BEAMOS] *Electron-beam memory.*

beam alignment An adjustment of the electron beam in a camera tube, performed on tubes employing low-velocity scanning, to cause the beam to be perpendicular to the target at the target surface.

beam antenna *Antenna array.*

beam bender *Ion-trap magnet.*

beam bending Deflection of the scanning beam by the electrostatic field of the charges stored on the target of a camera tube.

beam blanking *Blanking.*

beam capture Entry of a missile into the beam of a radar beam-rider guidance system so it can receive coded guidance signals.

beam convergence The adjustment that makes the three electron beams of a three-gun color picture tube meet or cross at a shadow-mask hole.

beam coupling The production of an alternating current in a circuit connected between two electrodes that are close to, or in the path of, a density-modulated electron beam.

beam current The electric current determined by the number and velocity of electrons in an electron beam.

beam-deflection tube An electron-beam tube in which current to an output electrode is controlled by the transverse movement of an electron beam.

beam finder A switch incorporated into a cathode-ray oscilloscope to bring the trace on the screen regardless of the settings of the horizontal, vertical, and intensity controls.

beam-indexing tube A single-beam color television picture tube in which the color phosphor strips are arranged in groups of red, green, and blue.

beam jitter A small oscillatory, angular movement of a radar antenna array and hence of the radar beam, required to develop accurate error signals for automatic tracking.

beam lead A flat thick-film lead, sometimes of gold, deposited on a semiconductor chip chemically or by evaporation, as a connecting lead for a semiconductor device or integrated circuit. Part of the semiconductor is sometimes etched away after deposition, to give a lead projecting beyond the edge of the chip so it can be bonded to a contact area on a substrate.

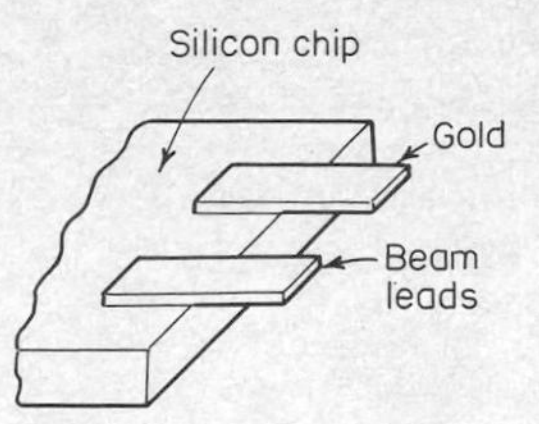

Beam leads deposited as thick gold film for making connections between silicon semiconductor chip and mounting substrate.

beam-lead bonder A thermocompression or ultrasonic bonding tool that fuses beam leads of a chip to contacts on the mounting substrate.

beam loading Absorption of power by the beam of particles in an accelerator as they gain energy.

beam-lobe switching A radio direction-finding technique that involves electric switching of antenna elements to give reception with lobes of different beam angles. Signals are compared to determine the direction to their source. A similar lobe-switching technique is used in radar.

beam magnet *Convergence magnet.*

BEAMOS Abbreviation for *beam-accessed metal-oxide semiconductor.*

beam parametric amplifier A parametric amplifier in which a modulated electron beam produces a variable reactance for current control.

beam pattern *Directivity pattern.*

beam-plasma amplifier An amplifier based on penetration of a plasma by an electron beam. It is capable of operating at millimeter and shorter wavelengths.

beam-power tube An electron-beam tube in which use is made of directed electron beams to contribute substantially to its power-handling capability. The control and screen grids are essentially aligned, and special deflecting electrodes are generally used to concentrate the electrons into beams. Also called beam tetrode.

beam rider A missile that is directed to its target by beam-rider guidance.

beam-rider guidance A missile guidance system in which a radar or other type of fixed or moving beam is directed into space to form a path along which it is desired to guide a missile to a target. Equipment in the missile detects deviations from the center of the beam and makes appropriate corrections to missile control settings.

beam separator An accelerator in which electrostatic and magnetic fields are combined to separate various particles in a beam. Because magnetic force is proportional to velocity, whereas electrostatic force is independent of velocity, the two forces cancel out for only one particle velocity. Particles with other velocities will then be deflected out of the beam.

beam signal A signal in a radio beam, or the beam itself, used as a navigation aid.

beam splitting A process for increasing radar target-locating accuracy. The azimuths at which a target first reflects data and stops reflecting data during a scan are used to define a beam that is split mathematically, to calculate mean azimuth for the target.

beam steering Changing the direction of the major lobe of a radiation pattern, usually by switching antenna elements.

beam-switching commutator An electronic commutator that uses a 10-position beam-switching tube in place of individual gate circuits for switching data points to a common output line.

beam-switching tube A vacuum tube that has 10 identical arrays of electrodes around a central cathode. These electrodes act with a ring-shaped permanent magnet surrounding the glass envelope to provide crossed electric and magnetic fields which switch the electron beam sequentially or in any desired manner from one electrode array to another. Used as a 10-position electronic switch.

beam tetrode *Beam-power tube.*

beam waveguide A waveguide based on beams whose cross-sectional field distribution is reconstituted at periodic intervals by reflective means or lens-shaped phase transformers.

beam width The angular width of a radar, radio, or other beam, measured in azimuth between points of half-power intensity.

beam-width error A radar error that occurs because the width of the scanning beam makes the target appear wider than it actually is. By covering two targets in a single sweep, beam width can also make two targets look like one on a radarscope.

bearing Angular position in a horizontal plane, expressed as the angle in degrees from true north in a clockwise direction. Also called azimuth. In navigation, azimuth and bearing have the same meaning; however, bearing is preferred for terrestrial navigation and azimuth for celestial navigation.

bearing cursor A bearing line on a rotatable transparent screen mounted in front of a PPI radar screen. The line is placed on a target, and the target bearing is read on an outer circular scale calibrated in degrees. Also used for determining the course in navigation.

bearing deviation indicator [abbreviated BDI] A sonar indicator used with a split transducer to show whether the target is to the left or right of the transducer heading. The magnitude of the aiming error is also shown.

bearing resolution The smallest angular difference in bearing at which a given radar is able to distinguish between close targets that have the same range.

bearing sensitivity The minimum signal strength required by a radio direction-finder to give repeatable bearings.

beat frequency The sum of or difference between two frequencies that are combined in a nonlinear circuit.

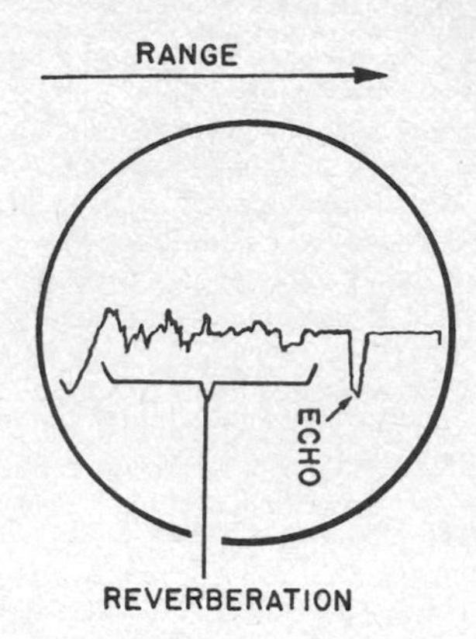

Bearing deviation indicator display. Echo pulse deflects to right when target is at right of sonar beam, and to left when target is at left of beam. Operator adjusts sonar projector until spot brightens at echo location without deflecting to right or left.

beat-frequency oscillator [abbreviated BFO] An oscillator in which a desired audio difference frequency is produced by combining two different RF signals. Used in audio signal generators for test purposes. Used in communication receivers to produce an audible signal when tuned to continuous-wave signals.

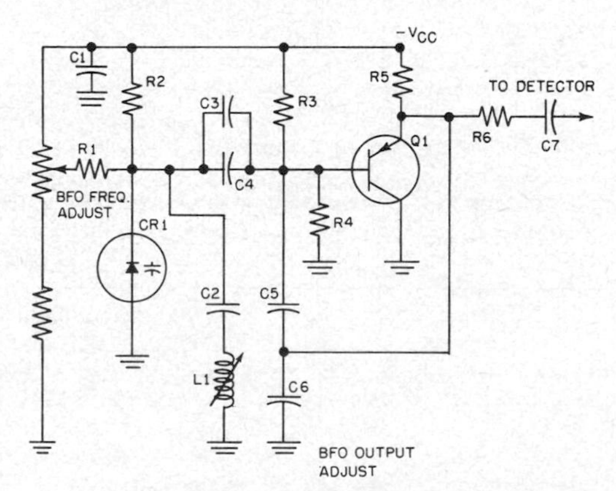

Beat-frequency oscillator using transistors.

beating A phenomenon in which two or more periodic quantities of different frequencies produce a resultant that has pulsations of amplitude.

beating-in Adjusting the frequency of one of two interconnected oscillators until no beat frequency is heard in a connected receiver. The oscillators are then operating at very nearly the same frequency.

beat note The difference frequency created when two sinusoidal waves of different frequencies are fed to a nonlinear device.

beat-note detector A detector that incorporates an oscillator or is fed by an external oscillator which has a frequency sufficiently close to the unmodulated incoming carrier frequency so an audible signal frequency is produced.

beat reception *Heterodyne reception.*

beats Beat notes, generally at a sufficiently low audio frequency to be counted.

beavertail A fan-shaped radar beam, wide in the horizontal plane and narrow in the vertical plane. The beavertail is swept up and down for height-finding.

bedspring array *Billboard array.*

beeper 1. A simple remote control system that controls target drones and other vehicles; the carrier is modulated with a different audio frequency for each desired on-off control function. 2. One who directs a pilotless aircraft or missile by remote control.

bel [abbreviated B] The fundamental unit of sound level, equal to the logarithm to the base 10 of the ratio of two amounts of power. One power value is a reference value. The decibel, a smaller unit equal to ¹⁄₁₀ B, is more commonly used.

B eliminator A separate power supply that changes the AC power-line voltage to the DC anode voltages required by electron tubes. It eliminated the need for B batteries in early radios.

bell wire Cotton or plastic-covered copper wire, usually No. 18, used chiefly for doorbell and thermostat connections.

benchmark problem A problem used to compare performances of computers.

bend 1. A smooth change in the direction of the longitudinal axis of a waveguide. 2. Departure of a defined navigation course from a straight line.

bender element A combination of two different thin strips of piezoelectric material bonded together in such a way that when voltage is applied,

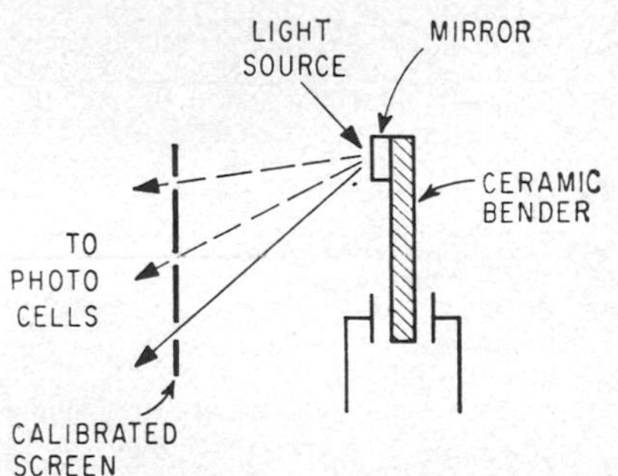

Bender element with mirror reflects light beam to different photocells, depending on voltage applied to bender strips.

one strip increases in length and the other becomes shorter. This causes the combination to bend.

bender transducer A transducer in which a voltage is generated when the bender element is bent. Conversely, application of voltage will cause bending.

bent-gun ion trap An ion trap that consists of a bend in the electron gun of a cathode-ray tube. The electrons are successfully bent by a small external permanent magnet so they pass through the gun. The ions, being heavier and hence less sensitive to the magnetic field, strike the sides of the gun harmlessly and are trapped.

berkelium [symbol Bk] A transuranic radioactive element that was initially synthesized by bombarding americium 241 with helium ions. Atomic number is 97.

beryllium [symbol Be] A metallic element that has low neutron-absorption cross section, high slowing-down power, and high thermal conductivity. Atomic number is 4.

Bessel filter An active filter that has linear phase characteristics and approximately constant time delay over a limited frequency range. Transient waveforms are passed with minimum distortion.

Bessel response A linear phase response (constant time delay) in the passband of a tunable active filter, with a shallower attenuation rate than for a Butterworth response and no overshoot.

beta [Greek β] The current gain of a transistor when connected as a grounded-emitter amplifier. It is the ratio of a small change in collector current to the resulting change in base current, collector voltage being constant.

beta-absorption gage *Beta gage.*

beta activity Radioactivity in which beta particles are emitted.

beta applicator A beta-ray source, usually strontium 90, used for treatment of superficial skin conditions.

beta cutoff frequency The frequency at which the beta of a transistor is down 3 dB from its low-frequency value.

beta decay Radioactive transformation of a nuclide in which the atomic number increases or decreases by 1 and the mass number remains unchanged. The atomic number increases when a negative beta particle (negatron) is emitted and decreases when a positive beta particle (positron) is emitted or an electron is captured.

beta disintegration The disintegration energy of a beta-decay process. For electron emission, it is equal to the sum of the kinetic energies of the beta particle, the neutrino, and the recoil atom. For positron emission, the energy equivalence of two electron rest masses must be added. For electron capture, the disintegration energy is equal to the sum of the kinetic energy of the neutrino and the electronic excitation energy of the product atom.

beta emitter A radionuclide that disintegrates by beta-particle emission.

beta gage A penetration-type thickness gage that measures the thickness or density of a sample by measuring the absorption of beta rays in the sample. Also called beta-absorption gage and gamma-absorption gage.

beta-gamma survey meter An ionization-chamber-type monitor that is sensitive primarily to beta particles and gamma rays.

beta light A self-luminous light source in which beta particles (electrons) from a radioisotope such as tritium gas strike a layer of phosphor powder and cause it to emit visible light. Useful life is about 20 years.

beta particle An electron or positron emitted from a nucleus during beta decay.

beta ray A stream of beta particles.

beta-ray spectrometer An instrument that determines the energy distribution of beta particles and secondary electrons.

beta-ray spectrum The distribution in energy or momentum of the beta particles emitted in a beta-decay process.

betatron An accelerator that consists of a horizontal doughnut-shaped vacuum enclosure, an electron gun as a source of electrons inside, and an external AC electromagnet which produces magnetic lines of force passing vertically through the enclosure. As emitted electrons travel in a circular path around the doughnut, they are continuously

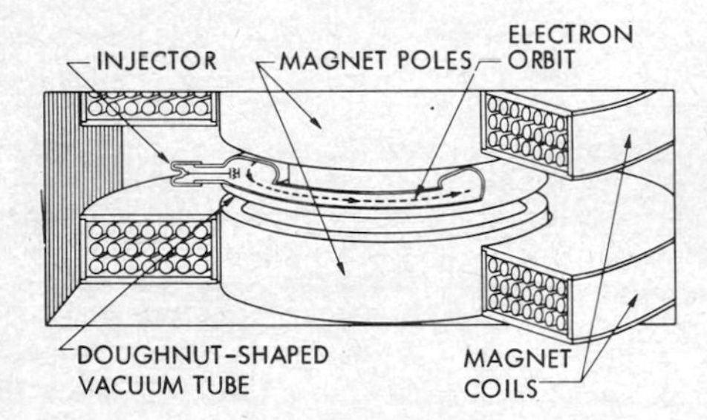

Betatron construction.

accelerated by the rapidly changing magnetic field. The resulting high-energy electron beam is then deflected out through the window for direct use as beta rays or directed against a target to produce high-energy x-rays.

beta wave A brain wave that has a frequency above 14 Hz.

Bethe-hole directional coupler A directional coupler in which the amounts of electric and magnetic coupling are balanced by rotating one waveguide with respect to the other about a coupling hole that is common to a broad face of each waveguide.

BeV Abbreviation for *billion electronvolt.*

bevatron *Proton synchrotron.*

Beverage antenna *Wave antenna.*

beyond-the-horizon communication *Scatter propagation.*

bezel A grooved rim that holds a transparent glass or plastic window or lens for a meter, tuning dial, or other indicating device.

BFO Abbreviation for *beat-frequency oscillator.*

***B-H* curve** A characteristic curve that shows the

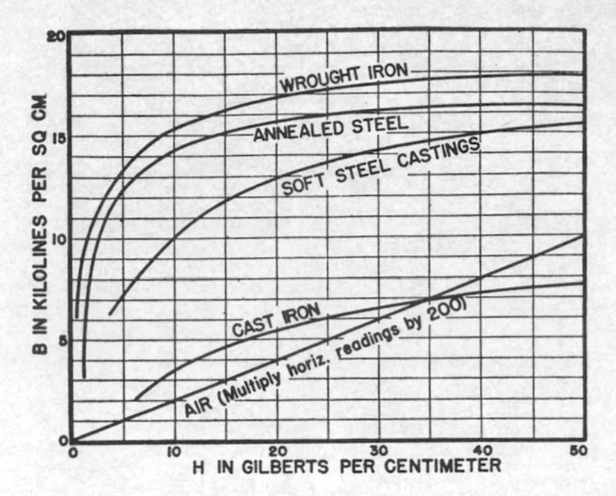

B-H curves for four different ferrous materials.

relation between magnetic induction *B* and magnetizing force *H* for a magnetic material. It shows the manner in which the permeability of a material varies with flux density. Also called magnetization curve.

***B-H* meter** A meter that measures the intrinsic hysteresis loop of a sample of magnetic material.

bialkali photocathode A dual-function photocathode used in some multiplier phototubes to achieve high quantum efficiency, high speed, low noise, and low dark current.

bias 1. The DC voltage applied to the control electrode of a transistor or electron tube to establish the desired operating point. 2. The direct current sent through the bias winding of a magnetic amplifier to establish desired operating conditions. 3. An electric, mechanical, or magnetic force applied to a device to establish a desired electric or mechanical reference level for its operation. 4. The alternating current, usually 3 to 10 times the highest audio frequency, sent through the recording head of a magnetic tape recorder to linearize the recording process.

bias current An alternating electric current above about 40 kHz, added to the audio current being recorded on magnetic tape to reduce distortion.

biased automatic gain control *Delayed automatic gain control.*

bias modulation Amplitude modulation in which the modulating voltage is superimposed on the bias voltage of an RF amplifier tube.

bias oscillator An oscillator used in a magnetic recorder to generate an AC signal that has a frequency in the range of 40 to 80 kHz, as required for magnetic biasing to give a linear recording characteristic. The bias oscillator usually also serves as the erase oscillator.

bias resistor A resistor used in the cathode or grid circuit of an electron tube to provide a voltage drop that serves as the bias voltage.

bias stabilization Use of a zener diode or other device in a transistor circuit to maintain essentially constant bias throughout the normal range of operating conditions.

bias telegraph distortion A distortion that causes telegraph mark and space pulses to be lengthened or shortened. Often caused by changes in the amplitude of incoming pulses.

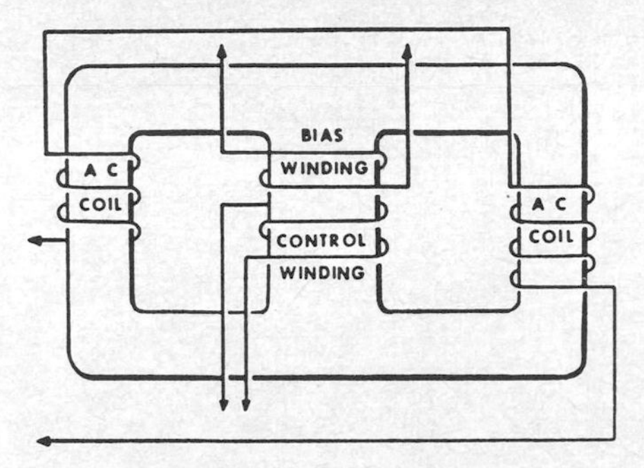

Bias winding on magnetic amplifier.

bias winding A control winding that carries a steady direct current which establishes desired operating conditions in a magnetic amplifier or other magnetic device.

biconical antenna An antenna that consists of two metal cones which have a common axis, with their vertices coinciding or adjacent and coaxial

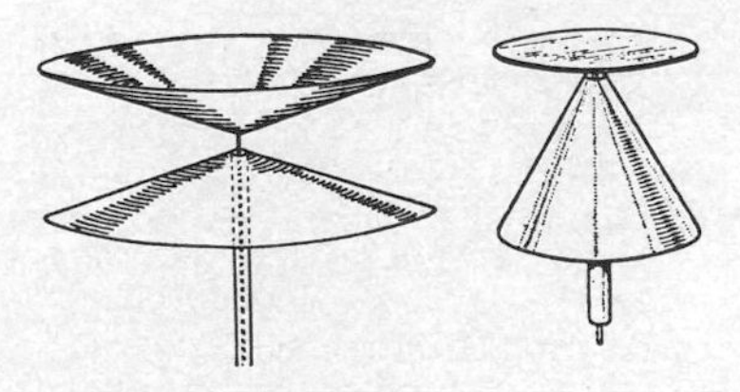

Biconical antenna and discone antenna, both fed by coaxial cable.

cable or waveguide feed to the vertices. The radiation pattern is circular in a plane perpendicular to the axis. Also called biconical horn antenna. When the vertex angle of one of the cones is 180°, the antenna is called a discone antenna.

biconical horn antenna *Biconical antenna.*

biconjugate network A linear network that has four resistances, conjugate in pairs, so a voltage in series with one resistance causes no current to flow in the other of a pair. Examples include directional couplers, hybrid junctions, and hybrid rings.

bicotar A trajectory-measuring system that uses a cotar at each of two separated ground stations. The intersection of the pointing vectors generated by the cotars gives target location.

bidirectional Responsive in two opposite directions. Also called bilateral.

bidirectional antenna An antenna that radiates or receives most of its energy in only two directions.

bidirectional breakdown diode A breakdown diode that has similar reverse characteristics for both polarities of the applied voltage.

bidirectional clamping circuit A clamping circuit which operates whenever the control pulse is applied, irrespective of the polarity of the input signal source at that time.

bidirectional counter A counter that has two or more operating modes, determined by gate control signals, to provide various combinations of addition and subtraction for pulses applied to either or both inputs.

bidirectional coupler A directional coupler that has terminals for sampling both directions of transmission.

bidirectional diode A semiconductor junction diode that has essentially the same forward, reverse, and/or breakdown characteristics for both polarities of the applied voltage.

bidirectional microphone A microphone that responds equally well to sounds coming from its front and rear, corresponding to sound incidences of 0 and 180°.

bidirectional pulse A pulse in which the variation from the normally constant value occurs in both directions.

bidirectional pulse train A pulse train, some pulses of which rise in one direction and the remainder in the other direction.

bidirectional thyristor A three-terminal semiconductor device that acts as a thyristor for either polarity of the voltage applied to its anode and cathode. It can be used as an AC switch. The triac is an example.

bidirectional transducer A transducer capable of measuring in both positive and negative directions from a reference position.

bidirectional transistor A transistor that pro-

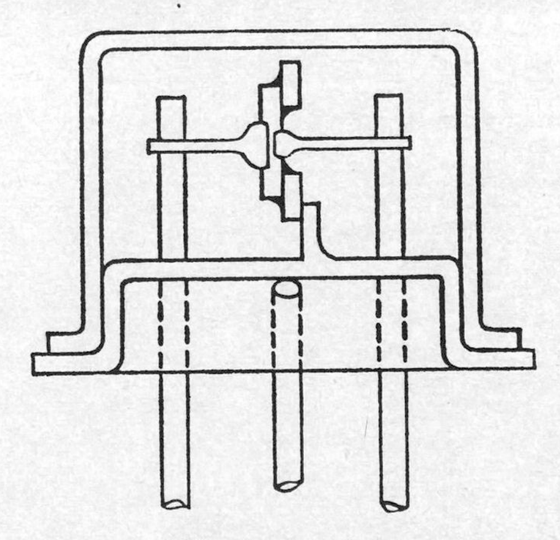

Bidirectional transistor.

vides switching action in either direction of signal flow through a circuit. Widely used in telephone switching circuits.

bidops [BI-DOPpler Scoring] A system for scoring miss distance of air-to-air missiles during target practice. A two-frequency transceiver in the airborne drone or towed target compares the phase of signals reflected from the missile and transmits this data to a ground station over a telemetry link, for conversion to miss distance.

bifilar resistor A resistor wound with a wire doubled back on itself to reduce the inductance.

bifilar suspension A meter movement that has two supporting conductors at each end of the moving element.

bifilar transformer A transformer in which wires for the two windings are wound side by side to give extremely tight coupling. When used as television IF transformers to couple stagger-tuned IF stages, the high coupling eliminates the need for a DC blocking capacitor.

bifilar winding A winding that consists of two insulated wires side by side, with currents traveling

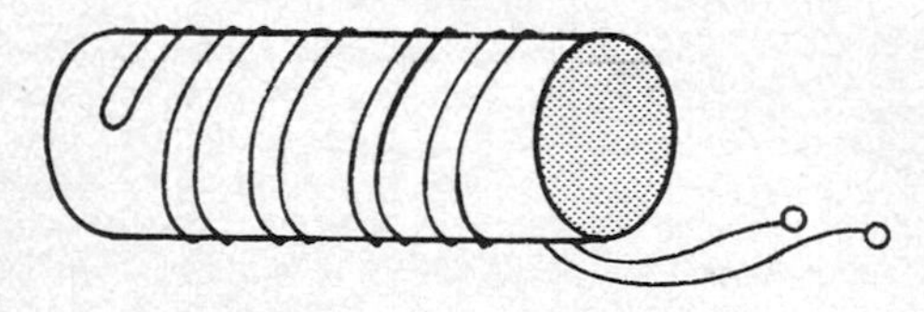

Bifilar winding as used in some resistors.

through them in opposite directions, to get maximum coupling between two circuits, minimum inductance in a resistor, or minimum pulse voltage between the two wires when they carry heater current to a tube whose cathode is driven by a pulse transformer.

Bifurcated connector.

bifurcated connector A hermaphroditic connector in which the identical mating contacts are fork-shaped.

bilateral 1. Having two sides. 2. *Bidirectional.*

bilateral amplifier An amplifier capable of receiving as well as transmitting signals. Used primarily in transceivers.

bilateral-area track A photographic sound track that has the two edges of the central area modulated according to the signal.

bilateral bearing A bearing that is either true or 180° displaced from true.

bilateral network A network in which a given current flow in either direction causes the same voltage drop.

bilateral transducer A transducer capable of transmission simultaneously in both directions between at least two terminations.

billboard array A broadside antenna array that consists of stacked dipoles spaced 1/4 to 3/4 wavelength apart in front of a large sheet-metal reflector. Also called bedspring array and mattress array.

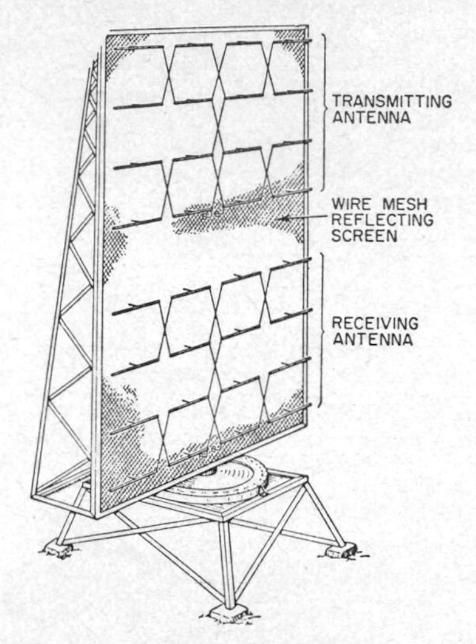

Billboard array for radar.

billion electronvolt [abbreviated BeV] Former name for *gigaelectronvolt.*

bimetallic strip A strip of two dissimilar metals welded together. The metals have different temperature coefficients of expansion, causing the strip to bend or curl when the temperature changes. Used in thermostats and thermal time-delay switches.

bimorph cell Two piezoelectric plates cemented together in such a way that an applied voltage causes one to expand and the other to contract; thus the cell bends in proportion to the applied voltage. Conversely, applied pressure will generate double the voltage of a single cell. Used in pickups and microphones.

binary Having only two possible alternatives.

binary cell An elementary unit of computer storage that can have one or the other of two stable states and can thus store one bit of information.

binary chain A series of binary circuits, each of which can affect the next circuit.

binary code A code in which each allowable position has one of two possible states. A common symbolism for binary states is 0 and 1. The binary number system is one of many binary codes.

binary-coded character An alphameric character represented by a predetermined configuration of consecutive binary digits.

binary-coded decimal [abbreviated BCD] A system of number representation in which each

Decimal	Binary	Binary Coded Decimal	Gray Code
0	0000	0000 0000	0000
1	0001	0000 0001	0001
2	0010	0000 0010	0011
3	0011	0000 0011	0010
4	0100	0000 0100	0110
5	0101	0000 0101	0111
6	0110	0000 0110	0101
7	0111	0000 0111	0100
8	1000	0000 1000	1100
9	1001	0000 1001	1101
10	1010	0001 0000	1111
11	1011	0001 0001	1110
12	1100	0001 0010	1010
13	1101	0001 0011	1011
14	1110	0001 0100	1001
15	1111	0001 0101	1000

Binary-coded decimal codes for 0 through 15, with binary and Gray codes shown for comparison.

decimal digit is represented by a group of four binary digits. Thus in the 8-4-2-1 coded decimal notation, the number 17 is 0001 0111 for 1 and 7, respectively.

binary code disk A disk that has patterns of concentric clear and opaque bars; the bars convert shaft angle directly to a nonambiguous natural binary code.

Binary code disk.

binary-coded octal An octal notation system in which each octal digit (0 through 7) is represented by a three-place binary-coded character.

binary counter *Binary scaler.*

binary digit *Bit.*

binary encoder An encoder that changes angular, linear, or other forms of input data into binary-coded output characters.

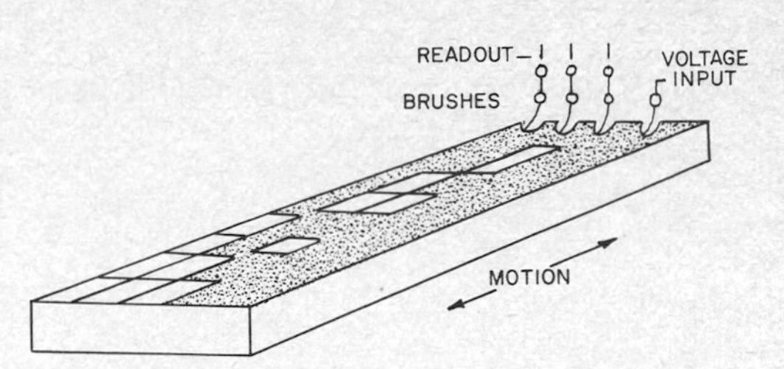

Binary encoder for linear motion. Insulated rectangles on conducting surface are sensed by brushes connected to output.

binary magnetic core A ferromagnetic core that can be made to take either of two stable magnetic states.

binary notation *Binary number system.*

binary number A numerical value expressed in binary digits as a sequence of 0s and 1s representing 1, 2, 4, 8, 16, 32, 64, 128, and other powers of 2 according to position from right to left in a group. These positional values are added to get the equivalent decimal number. Thus 010 is 2; 101 is 5; 1010 is 10; 10000 is 16; 11110 is 30.

binary number system A system of positional notation in which the successive digits are interpreted as coefficients of the successive powers of the base 2, as in binary numbers. Also called binary notation.

binary phase-shift keying [abbreviated BPSK] Keying of binary data or Morse-code dots and dashes by ±90° phase deviation of the carrier.

binary point The character, or the location of an implied symbol, that separates the integral part of a numerical expression from its fractional part in binary notation.

binary scaler A scaler that produces one output pulse for every two input pulses. Two binary scaler stages in sequence give an output pulse for every 4 input pulses; three in sequence give one for 8; four in sequence give one output pulse for every 16 input pulses. Also called binary counter and scale-of-two circuit.

binary search A search in which a set of items is divided into two parts, one of which is rejected. The process is repeated on the accepted part until the items that have the desired property are found.

binary signaling A communication system in which information is conveyed by the presence and absence, or positive and negative variations, of only one parameter of the signaling medium.

binary-to-decimal conversion The mathematical process of converting a number written in binary notation to the equivalent number written in ordinary decimal notation.

binaural Pertaining to sound that reaches the listener over two paths, to give the effect of auditory perspective.

binaural effect The ability to determine the direction from which a sound is coming by sensing the difference in arrival times of a sound wave at each ear.

binder A resin or other cementlike material used to hold particles together and provide mechanical strength. Used in phonograph records, carbon resistors, and fluorescent screens.

binder-type photoconductor A finely divided photoconductive material, such as zinc oxide, combined with a resin and used as a coating on a paper or metal substrate. Electrofax paper is an example.

binding energy 1. The net energy required to remove a particle from a system. 2. The net energy required to decompose a system into its constituent particles.

binding post A manually turned screw terminal used for making electric connections.

binomial antenna array A broadside array that has major lobes in opposite directions and no side lobes. This is achieved by spacing the antennas of the array at half-wavelength intervals and feeding them all in phase, with the relative current amplitudes of the various elements being proportional to the coefficients of successive terms in a binomial series.

bioastronautics The branch of astronautics that deals with the effect of space flight, space environment, and the environment of the moon and other planets on people, animals, and living organisms in general. Problems include weightlessness, radiation, inactivity, an artificial atmosphere, and extremes of acceleration and deceleration.

biochemical fuel cell A fuel cell in which it is hoped that small amounts of electric power can be produced continuously for many years by some form of biological system.

biocommunication Communication by mental telepathy or other parapsychical means not accounted for by presently known natural laws.

biocomputer A computer that can be programmed to show behavior characteristics comparable to those of living organisms, such as curiosity, goal seeking, memory, self-learning, and self-repair.

bioelectrogenesis The generation of electricity by living organisms.

bioelectronics The application of electronic theories and techniques to the problems of biology.

bioengineering The application of engineering knowledge to the fields of medicine and biology.

biogalvanic battery A battery that is implanted in the body and depends on interaction of metal electrodes with oxygen and fluids of the body to generate sufficient power for a heart pacemaker or other implanted medical device.

bioinstrument An instrument attached to the body for sensing and transmitting one or more forms of physiological data.

biological half-life The time required for the amount of a particular substance in a biological system to be reduced to half its original amount by natural biological elimination processes.

bionics The application of biological knowledge to electronic engineering and other fields of engineering, such as by duplicating the functions of living systems.

bionucleonics The application of ionizing radiation to living things and, in a broad sense, the study of the effects of other forms of radiant energy on people, plants, and animals.

biosatellite A satellite that carries a living being or plant for biological research in space.

biosensor A device that senses a biological function like blood pressure or heart rate, generally for recording or telemetry.

biosphere The part of the earth and its atmosphere in which animals and plants live.

biotelemetry Telemetry of biological data, as from inside the human body, without interfering with the function or process being monitored.

biotron A chamber in which pressure, temperature, and other environmental factors can be accurately controlled for biological research.

Biot-Savart's law A law that gives the intensity of the magnetic field produced by a current-carrying conductor.

biperipheral drive A cassette tape player drive in which both inside and outside edges of the flywheel are used in combination with motor-driven idlers to improve stabilization of tape speed.

bipolar Having two poles, polarities, or directions.

bipolar amplifier An amplifier capable of supplying a pair of output pulse signals corresponding to the positive or negative polarity of the input signal.

bipolar junction transistor A bipolar transistor in which excess minority carriers are injected from an emitter region into a base region and from there pass into a collector region. Large current gains can be obtained by using a common-emitter connection.

bipolar power supply A high-precision, regulated DC power supply that can be set to provide any desired voltage between positive and negative design limits, with a smooth transition from one polarity to the other. Some types can be remotely programmed with an external resistance or control voltage.

bipolar transistor A transistor that uses both positive and negative charge carriers. Both NPN and PNP types of bipolar transistors can be manufactured as discrete devices or incorporated into integrated circuits.

bipotential cathode A cathode that has two different surface potentials, to eliminate electron emission from portions of the cathode which are under the grid conductors while maintaining control of electron flow through the grid openings.

biquartic filter An active filter that uses operational amplifiers in combination with resistors and capacitors to provide infinite values of *Q* and simple adjustments for bandpass and center frequency.

biquinary notation A mixed-base notation system in which the first of each pair of digits counts 0 or 1 unit of five, and the second counts 0, 1, 2, 3, or 4 units. Thus decimal number 7 is biquinary 12;

43 is 04 03; 901 is 14 00 01; 4719 is 04 12 01 14. This is the code of the Japanese abacus and is used in binary digit form in some computers.

bird A missile, target drone, earth satellite, or other inanimate object that flies.

birdie A high-pitched whistle sometimes heard while tuning a radio receiver. It is due to beating between two carrier frequencies differing by about 10 kHz.

bismuth [symbol Bi] An element. Atomic number is 83. It is a brittle, heavy metal with low melting point and low capture cross section for neutrons. High absorption for gamma rays makes it useful as a filter or window to reduce gamma rays but transmit neutrons.

bismuth telluride [symbol Bi_2Te_3] An intermetallic compound that has high thermoelectric power and shows promise for thermoelectric refrigeration and power generation.

bistable Having two stable states.

bistable multivibrator *Flip-flop circuit.*

bistable phosphor A phosphor that has two stable states. When used in storage-oscilloscope cathode-ray tubes, each portion that has been written by the electron beam permits low-energy electrons from a second electron gun (flood gun) to hit the phosphor and keep it bright. The unwritten portions repel electrons from the flood gun and thereby stay dark.

bistable switching circuit An internally triggered switching circuit that has two operating levels or states.

bistable unit A physical element that can be made to assume either of two stable states. A binary cell is an example.

bistatic radar A radar system in which the receiver is some distance from the transmitter, with

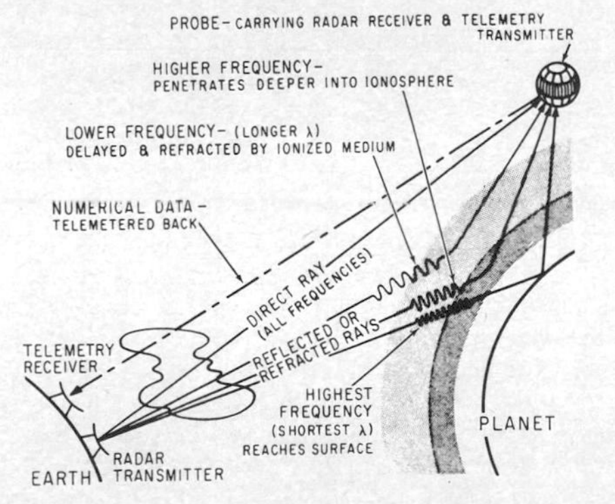

Bistatic radar, with transmitter on earth and receiver in space probe.

separate antennas for each. In conventional monostatic radar, transmitter and receiver are at the same site.

bit [abbreviated b] A single character of a language employing exactly two distinct kinds of characters. They correspond to on and off conditions in a digital computer and are usually designated as 0 and 1. Computer word size and storage capacity can be expressed in bits. Thus one computer has a storage capacity of 2048 36-bit words. Also called binary digit.

bit density The number of bits recorded per unit of length or area.

bit error rate The ratio of erroneous bits to total received bits in a data-transmission system.

bit parallel Transmission of character-forming bits simultaneously over parallel paths, as contrasted to bit serial, where the bits for a character are transmitted in sequence over a single path.

bit per second [abbreviated b/s] A rating that specifies the modulation rate of a digital transmission system.

bit rate The number of binary digits or equivalent pulses passing a given point in a data-transmission system per unit of time.

bit serial Transmission of character-forming bits in sequence, as contrasted to bit parallel, where all the bits for a character are transmitted simultaneously.

black after white A television receiver defect in which an unnatural black line follows the right-hand contour of any white object on the picture screen. The same defect also causes a white line to follow a sudden change from black to a lighter background. It is caused by receiver misalignment.

black and white television *Monochrome television.*

blackbody A perfect absorber of all incident radiant energy. It radiates energy solely as a function of its temperature.

black box A unit of electronic equipment, such as a receiver or amplifier, that can be put into or removed from an aircraft or other location as a single package. So named because the housings of such units are usually painted black.

black compression A reduction in television picture-signal gain at levels corresponding to dark areas in a picture. The effect reduces contrast in the dark areas of the picture as seen on monitors and receivers. Also called black saturation.

blacker-than-black region The portion of the standard television signal in which the electron beam of the picture tube is cut off and synchronizing signals are transmitted. These synchronizing signals have greater peak power than those for the blackest portions of the picture.

black level The level of the television picture signal corresponding to the maximum limit of black peaks. This level is generally set at 75% of the maximum signal amplitude of the synchronizing pulses.

black light Deprecated term for invisible *ultraviolet radiation.*

black negative A television picture signal in which the voltage corresponding to black is negative with respect to the voltage corresponding to the white areas of the picture.

blackout effect A temporary loss of sensitivity of

an electron tube after it handles a strong, short pulse.

black peak A peak excursion of the television picture signal in the black direction.

black positive A television picture signal in which the voltage corresponding to black is positive with respect to the voltage corresponding to the white areas of the picture.

black recording Facsimile recording in which the maximum received power corresponds to the maximum density of the record medium for amplitude modulation or to the lowest received frequency for frequency modulation.

black saturation *Black compression.*

black signal The signal at any point in a facsimile system produced by the scanning of a maximum-density area of the subject copy.

black transmission Facsimile transmission that uses black recording.

blade A flat moving conductor in a switch.

blade antenna An antenna shaped much like the tail fin of an airplane. It can be mounted on the fuselage of an airplane.

blank 1. The result of the final cutting operation on a natural crystal. 2. To cut off the electron beam of a cathode-ray tube. 3. A machine character that represents a space in the printout of a printer.

blanked picture signal The signal resulting from blanking a television picture signal. Adding the sync signal to the blanked picture signal gives the composite picture signal.

blanket area The area in the immediate vicinity of a broadcast station, where the signal of that station is so strong (above 1 V/m) that it interferes with reception of other stations.

blanketing Interference caused by a nearby transmitter whose signals are so strong that they override other signals over a wide band of frequencies.

blank groove *Unmodulated groove.*

blanking The process of cutting off the electron beam of a television picture tube, camera tube, or cathode-ray oscilloscope tube during retrace by applying a rectangular pulse voltage to the grid or cathode during each retrace interval. Also called beam blanking. The opposite action is called gating.

blanking level The level that separates picture information from synchronizing information in a composite television picture signal. It coincides with the level of the base of the synchronizing pulses. Also called pedestal and pedestal level.

blanking pulse One of the pulses that make up the blanking signal in television.

blanking signal A wave of recurrent pulses, related in time to the scanning process, to effect blanking in television. The pulses occur at both the line and field frequencies and cut off the electron beam during retrace at both transmitter and receiver.

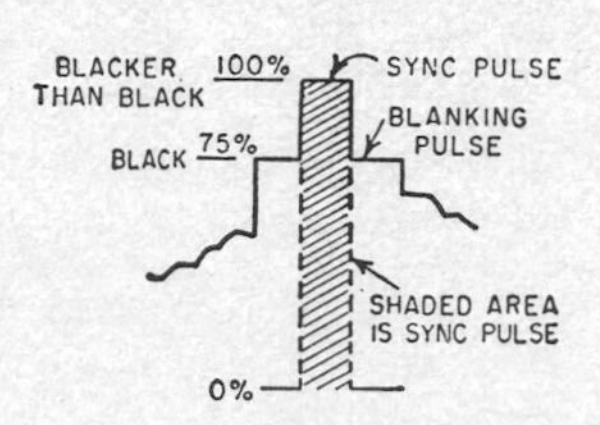

Blanking pulse at end of line, with horizontal sync pulse in its center.

blasting Distortion due to overloading of any part of a radio transmitter, receiver, or audio amplifier.

bleeder current Current drawn continuously from a voltage source to lessen the effect of load changes or provide a voltage drop across a resistor.

bleeder resistor A resistor connected across a power pack or other voltage source to improve voltage regulation by drawing a fixed current value continuously. Also used to dissipate the charge remaining in filter capacitors when equipment is turned off.

bleeding whites A condition in which white areas in a television picture appear to flow into black areas, caused by excessive signal strength at the picture tube.

bleep A short-duration high-pitched sound produced by electronic equipment.

blind approach An aircraft approach for a landing when visibility is poor, usually made with the aid of instruments and radio.

blind bombing *Bombing through overcast.*

blind flying *Instrument flying.*

blind landing An aircraft landing made with the aid of ground-controlled approach, an instrument landing system, or some other guidance system designed for use when visibility is poor.

blind navigation Navigation of an aircraft with the aid of instruments, radio, or electronic equipment when visibility is poor.

blind sector A shadow on a radarscope screen caused by a fixed obstruction near the radar antenna, like the funnels on a ship.

blind speed A radar target speed at which a moving target cannot be distinguished from a stationary target. With a radar moving-target indicator, blind speeds are those at which the radial velocity of the target is such that it traverses one half-wavelength, or multiples of, between successive pulses. With Doppler radar, blind speeds are those at which the Doppler shift in hertz is a multiple of the pulse-repetition rate of the radar.

blind takeoff An airplane takeoff under conditions of no visibility, usually made with the aid of instruments or radio.

blinking A method of providing information in pulse systems by modifying the signal at its source so the signal presentation on the display alternately appears and disappears. In loran this indicates that a station is malfunctioning. With two aircraft flying over enemy territory covered by radar, the planes can alternately spot-jam the radar to interfere with the accuracy of antiaircraft guns or missiles.

blip 1. To remove a portion of the recorded sound from a videotape of a television program, such as deleting an expletive or other undesired words. 2. *Pip.*

blip-scan ratio The number of scans necessary to produce one recognizable pip on a radarscope. The blip-scan ratio of a radar set varies with range, antenna tilt, ability of operator, set performance, target aspect, wind, and other factors.

blister 1. A radome that protrudes beyond the normal skin contours of an aircraft. 2. A protruding housing for a sonar transducer on a ship or submarine.

Bloch band *Energy band.*

Bloch wall The transition layer that separates adjacent ferromagnetic domains which are magnetized in different directions.

block A group of words considered or transported as a unit in a computer.

block diagram A diagram in which the principal divisions of an electronic system are indicated by rectangles or other geometric figures, and the signal paths are represented by lines.

blocked-grid keying A method of keying a radiotelegraph transmitter; sufficient grid bias is provided to block one or more tubes when the transmitting key is open. Closing the key removes this bias and applies full transmitter power to the antenna.

blocked impedance The impedance at the input of a transducer when the impedance of the output system is made infinite, as by blocking or clamping the mechanical system.

blocked resistance The real part of blocked impedance.

blockette A subdivision of a group of consecutive machine words transferred as a unit.

blockhouse A reinforced concrete structure, often partly or entirely underground, that protects instrumentation and personnel against blast, heat, or explosion during rocket launchings or related activities.

blocking 1. Applying a high negative bias to the grid of an electron tube to block its anode current, or producing an equivalent current-blocking effect in a transistor or other active solid-state device. 2. Overloading of a receiver by an unwanted signal so the automatic gain control reduces the response to a desired signal. 3. Combining two or more computer records into one block. 4. Preventing forward current flow in a semiconductor device.

blocking capacitor *Coupling capacitor.*

blocking layer Deprecated term for *depletion layer.*

blocking oscillator An oscillator in which the negative grid bias increases gradually during oscillation as a capacitor is charged, until a point is reached where anode current is cut off and oscillations stop. The capacitor then discharges until the grid is unblocked and oscillation is resumed. This process produces a sawtooth voltage waveform that may be used as the sweep voltage for a cathode-ray tube. Also called squegging oscillator.

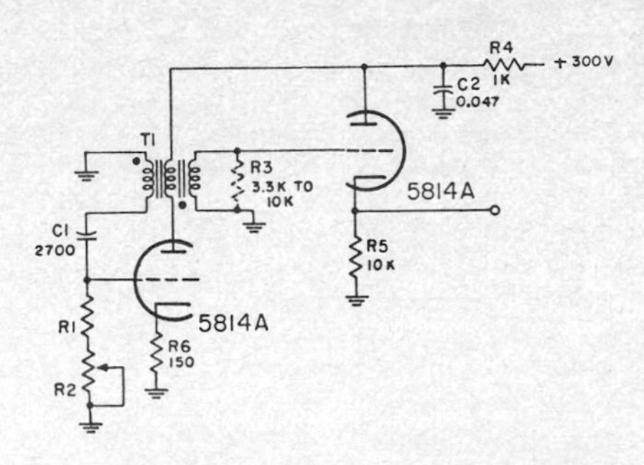

Blocking oscillator using circuit design preferred by National Bureau of Standards. Pulse repetition frequency is in range from 200 to 2000 pulses per second, depending on values of C1, R1, and R2, which are best determined experimentally.

blocking-oscillator driver A blocking oscillator that develops and shapes an essentially square pulse for driving radar modulator tubes.

blocking period The portion of the idle period in the cycle of a mercury-arc rectifier in which the anode is positive but the start of commutation is blocked by phase control.

block length The total number of records, words, or characters contained in one block.

block sort A sort in which the file is first divided according to the most significant character of the key. The separated portions are then sorted one at a time. Used chiefly in punched-card sorting of large files.

blooming 1. Defocusing of television picture areas where brightness is excessive because of enlargement of spot size and halation of the fluorescent screen. 2. An increase in radarscope spot size caused by an increase in signal intensity. Blooming may be used to convey information in navigation systems that have intensity modulation.

blooper A radio receiver that is radiating an excessively strong oscillator signal.

blower An electric fan that supplies air for cooling purposes.

blown-fuse indicator A neon warning light connected across a fuse so it lights when the fuse is blown.

blowout magnet An electromagnet or permanent magnet that deflects and extinguishes the arc formed when a high-current circuit breaker or switch is opened.

blue-beam magnet A small permanent magnet used as a convergence adjustment to change the direction of the electron beam for blue phosphor dots in a three-gun color television picture tube.

blue gain control A variable resistor in the matrix of a three-gun color television receiver, used to adjust the intensity of the blue primary signal.

blue glow A glow normally seen in electron tubes that contain mercury vapor; the glow is due to ionization of the molecules of mercury vapor. A blue glow near the electrodes of a vacuum tube means that the tube is gassy and hence defective. A soft blue fluorescent glow is normal on the glass envelopes of some vacuum tubes.

blue gun The electron gun whose beam strikes phosphor dots emitting the blue primary color in a three-gun color television picture tube.

blue restorer The DC restorer for the blue channel of a three-gun color television picture tube circuit.

blue video voltage The signal voltage output from the blue section of a color television camera, or the signal voltage between the receiver matrix and the blue gun grid of a three-gun color television picture tube.

BMEWS Abbreviation for *ballistic-missile early-warning system.*

boat A small crucible shaped somewhat like a boat, used in evaporation methods of depositing metallic films.

bobbin An insulated spool that is a support for a coil.

bobbin core A tape-wound core in which the ferromagnetic tape has been wrapped on a form or bobbin that provides mechanical support for the tape.

Bode diagram A diagram in which the phase shift or the gain of an amplifier, servomechanism, or other device is plotted against frequency to show frequency response.

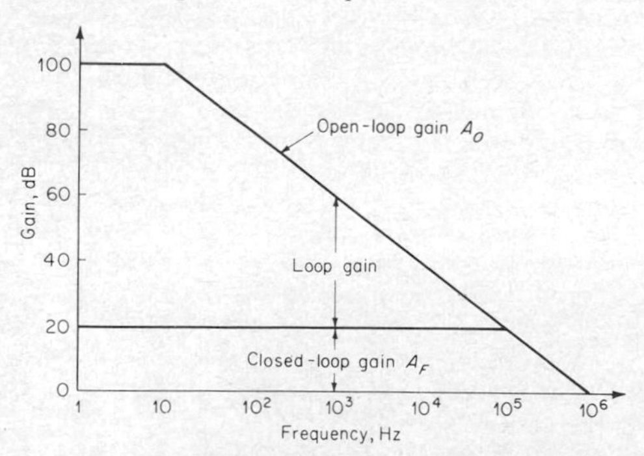

Bode diagram for typical operational amplifier.

body capacitance Capacitance that exists between the hand or body and a circuit.

body-capacitance alarm An alarm system that is triggered by the capacitance between the body of an intruder and a sensing wire or metal plate.

body-section radiography *Laminography.*

bogie An indication of an enemy or unidentified aircraft on a radar screen.

Bohr atom An atom model conceived by Bohr and Rutherford; it consists of a positive nucleus about which circulate a number of orbital electrons.

Bohr magneton A unit of magnetic moment used to specify the magnetic moment of an atomic particle or system of particles. The electronic Bohr magneton is 9.27×10^{-21} erg·G^{-1}. For the nuclear Bohr magneton or nuclear magneton, this value is divided by 1836. Also called magneton.

Bohr radius The radius of the lowest-energy electron orbit in the Bohr model of the hydrogen atom.

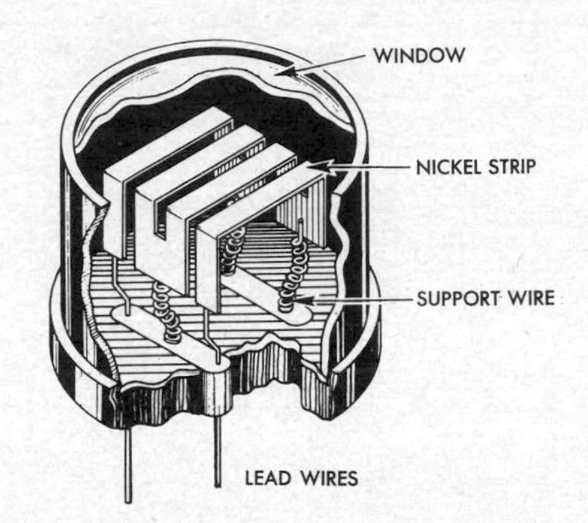

Bolometer using very thin, black, oxidized nickel strip in evacuated chamber, for infrared homing system. Also used to measure temperatures of stars.

bolometer A device that measures microwave and infrared energy. It contains a resistance element that changes in resistance when heated by the radiant energy. Used in infrared search and homing equipment for missiles, and in waveguides to measure microwave power and standing-wave ratio.

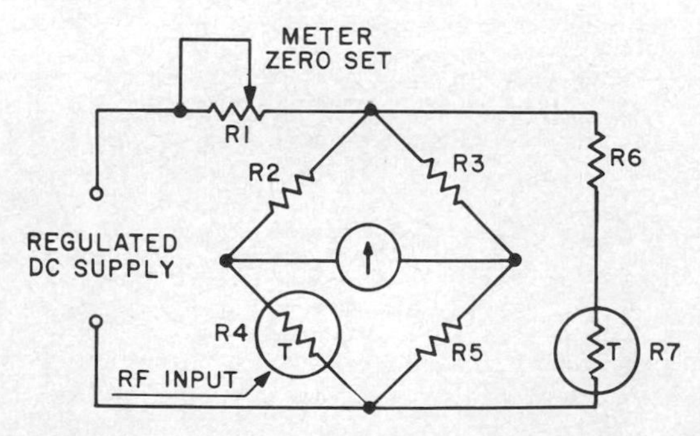

Bolometer bridge in which microwave energy to be measured is directed at bolometer R4. Thermistor R7 provides compensation for changes in ambient temperature.

bolometer bridge A bridge circuit that has a bolometer in one arm, to measure RF power.

bolometer mount A waveguide termination in

which a bolometer can be inserted to measure electromagnetic power.

Boltzmann's constant A physical constant equal to 1.380662×10^{-23} J/K or 1.380662×10^{-16} erg/K.

Boltzmann's equation The equation for particle conservation, based on the description of individual collisions.

Boltzmann's factor A number, dependent on temperature and energy difference, that gives the ratio of the number of particles with one energy to the number of particles with another energy in an atomic system.

Bomarc An Air Force surface-to-air guided missile that has supersonic speed, is launched from the ground, and seeks out and destroys enemy aircraft. Slant range is about 300 mi (480 km).

bombard To direct a stream of high-energy particles or photons against a target.

bombardment 1. The process of directing high-speed electrons at an object, causing secondary emission of electrons, heating, fluorescence, disintegration, or production of x-rays. 2. The process of directing electrons or other high-speed particles at atoms or smaller particles. 3. Use of induction heating for heating electrodes of tubes to drive out gases during evacuation.

bombardment-induced conductivity An increase in the number of charge carriers in semiconductors or insulators, caused by bombardment with ionizing particles.

bombing through overcast [abbreviated BTO] Blind bombing through clouds, using radar, infrared equipment, or other electronic equipment for guidance. Also called blind bombing.

bond A low-resistance junction of two conducting members. Semiconductor devices may use ball, die, face, stitch, thermocompression, ultrasonic, wedge, and wire bonds.

bonded-barrier transistor A transistor made by alloying the base with material that is on the end of a wire.

bonded diode A semiconductor diode made by pressing a metal wire against a semiconductor surface and producing a junction by an alloying process.

bonding Connecting together the metal parts of an aircraft, vehicle, structure, or housing to prevent interference-producing static or RF voltage buildup between adjacent metal parts.

bonding pad A metallized area on the surface of a semiconductor device, to which connections can be made.

bond strength A measure of the force required to separate a layer of material from an adjoining surface, such as printed wiring from its base. Also called adhesion.

bone conduction The process by which sound is conducted to the inner ear through the cranial bones.

bone seeker A compound or ion that migrates preferentially into living bone. Used in radiobiology.

Boolean algebra Algebra that deals with classes, propositions, ON/OFF circuit elements, and other nonnumerical elements associated with such operators as AND, NAND, NOR, NOT, and OR.

Boolean calculus Boolean algebra that has been modified to include time, to permit handling such elements as: (a) states and events; (b) operators like "after," "while," "happen," "delay," and "before"; (c) classes whose members change with time; (d) circuit elements whose ON/OFF state changes from time to time, like delay lines, flip-flops, and sequential circuits; (e) step-functions and their combinations.

Boolean function A mathematical function in Boolean algebra.

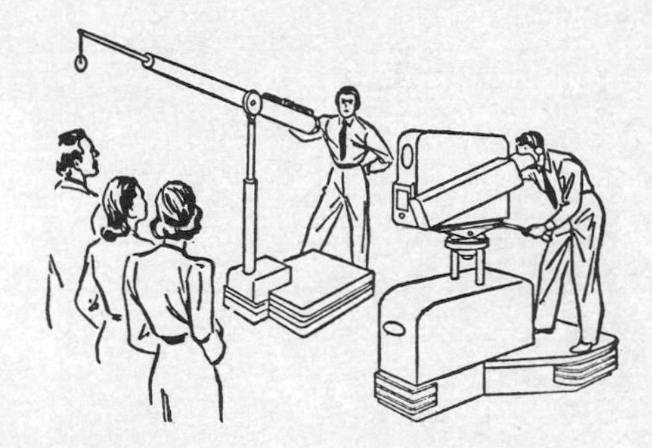

Boom for microphone in television studio.

boom A movable mechanical support that suspends a microphone within range of actors but above the field of view of television cameras.

boost To increase or amplify.

boost charge A fast partial charge of a storage battery at a high current rate.

booster 1. A separate RF amplifier connected between an antenna and a television receiver to amplify weak signals. 2. An RF amplifier that amplifies and rebroadcasts a received television or communication radio signal at higher power without change in carrier frequency, for reception by the general public.

booster amplifier An audio amplifier between mixer controls and the master volume control of a studio audio console. It compensates for mixing-circuit losses. A booster amplifier is also frequently used to increase the output voltage or current capability of an operational amplifier without polarity inversion or appreciable loss of accuracy.

booster rocket 1. A rocket engine, which uses either solid or liquid fuel, that assists the normal propulsive system of a rocket in some phase of its flight. 2. *Launch vehicle.*

booster voltage The additional voltage supplied by the damper tube to the horizontal output, horizontal oscillator, and vertical output tubes of a television receiver to give greater sawtooth sweep output.

boostglide vehicle A vehicle (half aircraft, half spacecraft) that flies to the limits of the atmosphere and is boosted by rockets into the space above and returns to earth by gliding under aerodynamic control.

boot A protective covering over any portion of a cable, wire, or connector.

bootstrap A special coded instruction at the beginning of a computer routine to make the routine assemble itself in the computer.

bootstrap circuit A single-stage amplifier in which the output load is connected between the negative end of the anode supply and the cathode while signal voltage is applied between grid and cathode. A change in grid voltage changes the input signal voltage with respect to ground by an amount equal to the output signal voltage.

bootstrap driver A vacuum-tube circuit that produces a square pulse which drives a radar modulator tube. The duration of the square pulse is determined by a pulse-forming line. The circuit is called a bootstrap driver because voltages on both sides of the pulse-forming line are raised simultaneously with voltages in the output pulse, but their relative difference (on both sides of the pulse-forming line) is not affected by the considerable voltage rise in the output pulse.

bootstrapped sawtooth generator A circuit capable of generating a highly linear, positive, sawtooth waveform by the use of bootstrapping.

bootstrapping A technique for lifting a generator circuit above ground by a voltage value derived from its own output signal.

borazon A man-made semiconductor, produced at temperatures above 1100°C and pressures of about 1×10^6 lb/in^2, that has the structure and hardness of diamond.

boresighting Initial alignment of a directional microwave or radar antenna system, using an optical procedure or a fixed target at a known location.

boresight tower A tower on which a visual target and an antenna fed from a signal generator are mounted. These targets are used for parallel alignment of the electrical axis of a receiving antenna and the optical axis of a telescope mounted on that antenna.

Born approximation A system of successive approximations for treating scattering problems in quantum mechanics. Based on the assumption that the effect of the scattering process on the wave function is small.

boron [symbol B] An element used in instruments for detecting and measuring neutrons because the isotope ^{10}B, on absorbing a neutron, breaks into two charged particles that can be detected easily. Atomic number is 5.

boron chamber An ionization chamber that is lined with boron or boron compounds and/or filled with a gaseous boron compound.

boron counter tube A counter tube that is filled with boron fluoride and/or has electrodes coated with boron or boron compounds. Used for detecting slow neutrons.

boron thermopile A thermopile in which alternate thermocouple junctions are coated with boron. Exposure to a flux of slow neutrons generates heat in these junctions, producing an output voltage proportional to neutron flux.

borrow A computer carry signal that is produced when the difference between digits being subtracted is less than zero.

Bose-Einstein condensation The concentration of bosons that occurs in the lowest energy level at low temperatures.

Bose-Einstein statistics The quantum statistics obeyed by photons and certain other quantum-mechanical systems that are in thermal equilibrium. Any number of identical particles can be in a given state at a given time.

boson A particle that obeys Bose-Einstein statistics. Its spin is an integer. Bosons include photons, pi mesons, and all nuclei that have an even number of particles.

bottoming A limiting action that occurs on positive grid-voltage peaks in a beam-power tube or pentode because of the formation of a virtual cathode under certain operating conditions. The virtual cathode limits anode current and hence flattens the corresponding anode voltage peak.

bottom metallization Metallization of the bottom surface of a semiconductor chip, for solder-bonding to a substrate.

boule A pure crystal, as of silicon, formed synthetically by rotating a small seed crystal while pulling it slowly out of molten material in a special furnace. The resulting atomic structure is that of a single crystal. For semiconductor devices, the boule is sawed into circular slices that in turn are cut into chips.

bounce A sudden variation in television picture brightness or size, independent of illumination of the original scene.

boundary An interface between P- and N-type semiconductor materials, at which donor and acceptor concentrations are equal.

boundary marker beacon A fan marker beacon near the approach end of the runway in an instrument landing system. Used at military airports but generally omitted at civil airports. Also called inner marker beacon.

bound charge The residual charge held on a conductor by the inductive action of a neighboring charge.

bound circuit A circuit that limits the excursion of an output signal to an approximate maximum value for protection purposes or to a precise maximum value required for operational amplifier applications.

bound electron An electron bound to the nucleus of an atom by electrostatic attraction.

bounding The process of limiting the magnitude of a circuit output value when the input voltage exceeds a predetermined full-scale value.

bound level An energy level in a nucleus, so close to the ground state that it can only decay by gamma emission.

bow-tie antenna A dipole antenna in which the two rods are replaced by triangular metal plates to give a bow-tie appearance. Used chiefly with a reflector for UHF television reception.

boxcar circuit A circuit used in radar for sampling voltage waveforms and storing the latest

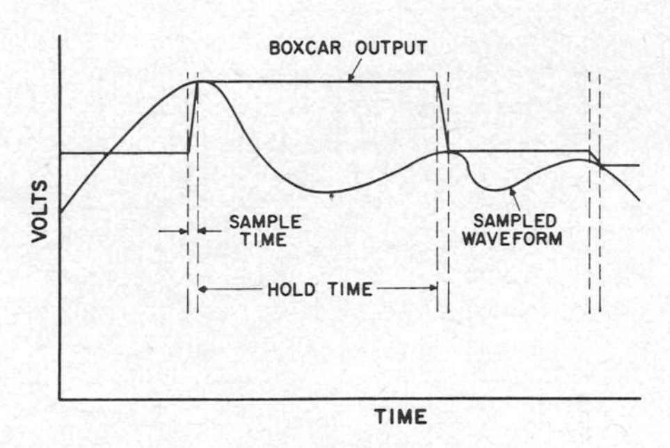

Boxcar-circuit output waveform.

value sampled. The term is derived from the flat, steplike segments of the output voltage waveform.

boxcar function A function that is zero except over a finite interval, during which it takes a constant value, often +1.

boxcar integrator A signal processor that uses a narrow filter to reduce noise without appreciably affecting signal bandwidth. Operation is based on the fact that most noise has an average value of zero. The integrator uses a gated switch or other sampling means for feeding the repetitive input signal to a low-pass filter that acts as integrator.

boxcar lengthener A pulse-lengthening circuit that lengthens a series of pulses without changing their height.

boxcars Long pulses separated by very short intervals.

BPSK Abbreviation for *binary phase-shift keying.*

Bragg angle The glancing angle for x-rays at the reflecting planes of a crystal. Used in x-ray orientation of quartz crystals for radio use.

Bragg curve 1. A curve that shows the average number of ions per unit distance along a beam of initially monoenergetic ionizing particles, usually alpha particles, passing through a gas. 2. A curve that shows the average specific ionization of an ionizing particle of a particular kind as a function of its kinetic energy, velocity, or residual range.

Bragg scattering Scattering of x-rays and neutrons by the regularly spaced atoms in a crystal, for which constructive interference occurs only at definite angles called Bragg angles.

Bragg's law A statement of the conditions under which a crystal will reflect a beam of x-rays with maximum intensity.

Bragg spectrometer An instrument for x-ray analysis of crystal structure, in which a homogeneous beam of x-rays is directed on the known face of a crystal, and the reflected beam is detected in a suitably placed ionization chamber. As the crystal is rotated, the angles at which Bragg's law is satisfied are identified as sharp peaks in the ionization current. Also called crystal spectrometer and ionization spectrometer.

Bragg's rule An empirical rule, according to which the mass stopping power of an element for alpha particles is inversely proportional to the square root of the atomic weight. The atomic stopping power is then directly proportional to the square root of the atomic weight.

braided wire A tube of fine wires woven around a conductor or cable for shielding, or used alone in flattened form as a grounding strap.

braid random-access memory A computer read-only memory constructed by selective threading of word wires through cores. A 1 is encoded in a core by threading a word wire through the core, and a 0 by passing the wire around the outside of the core.

brain A computer or other automatic device that duplicates some functions of the human brain. Used in such applications as missile guidance, process control, and navigation.

brain pacemaker An electric device that applies low-voltage stimulation either continuously or at appropriate intervals to electrodes implanted in the muscle-controlling portion of the brain, for treatment of epilepsy and other diseases.

brain wave A rhythmic fluctuation of voltage between parts of the brain, ranging from about 1 to 60 Hz and 10 to 100 μV. It is called a delta wave when the frequency is below 9 Hz, an alpha wave when the frequency is 9 to 14 Hz, and a beta wave when the frequency is above 14 Hz.

branch 1. A portion of a network that consists of one or more two-terminal elements in series. Also called arm. 2. A product that results from one mode of decay of a radioactive nuclide which has two or more modes of decay. 3. A line segment that joins two nodes, or joins one node to itself. 4. A set of computer instructions executed between two successive decision instructions. 5. *Conditional jump.*

branching The occurrence of two or more modes by which a radionuclide can undergo radioactive decay. Also called multiple decay and multiple disintegration.

branch instruction An instruction that makes the computer choose between alternative subprograms, depending on the conditions determined by the computer during the execution of the program.

branch point 1. A terminal common to two or

more branches of a network, or a terminal on a branch of a network. Also called junction point. 2. A location in a computer routine at which one of two or more choices is made.

branch transmittance The ratio of branch output signal to branch input signal.

Braun tube Former name for *cathode-ray tube.*

breadboard model An experimental arrangement of a circuit on a board or other flat surface without regard for final locations of components, to prove the feasibility of the circuit and facilitate changes when necessary.

break 1. Interruption of a radio transmission, as for sending in the opposite direction. 2. A fault in a circuit.

break-before-make contacts Contacts that interrupt one circuit before establishing another.

break contact *Back contact.*

breakdown 1. A disruptive discharge through insulation, involving a sudden and large increase in current through the insulation because of complete failure under electrostatic stress. Also called puncture. 2. Initiation of a desired discharge between two electrodes in a gas, occurring at a voltage dependent on gas density, electrode shape, electrode spacing, and polarity. 3. An undesired runaway increase in an electrode current in a gas tube. 4. Loss of blocking action in a reverse-biased semiconductor PN junction, causing a sudden current increase that is not normally destructive.

breakdown diode A semiconductor diode in which the reverse-voltage breakdown mechanism is based either on the Zener effect or the avalanche effect.

breakdown region The entire region of the semiconductor-diode voltampere characteristic beyond the initiation of breakdown for increasing magnitude of reverse current.

breakdown voltage 1. The voltage measured at a specified current in the breakdown region of a semiconductor diode. Also called zener voltage. 2. The voltage at which breakdown occurs in a dielectric or in a gas tube.

break-in keying A method of operating a radiotelegraph communication system in which the receiver is capable of receiving signals during transmission-spacing intervals.

break-in operation A method of radio communication that involves break-in keying, allowing the receiving operator to interrupt the transmission.

break-in relay A relay used for break-in operation.

breakout A joint at which one or more conductors are brought out from a multiconductor cable.

breakover In a thyristor, the transition from the forward-blocking to the forward-conducting state.

breakpoint A point in a computer program at which conditional interruption can occur to permit visual check, printing out, or other special action. Often used during debugging.

breakpoint instruction An instruction that will cause a computer to stop or transfer control to a supervisory routine for monitoring the progress of the interrupted program.

breakpoint switch A manually operated switch that controls conditional operation at breakpoints. Used primarily in debugging.

breezeway The time interval between the trailing edge of the horizontal synchronizing pulse and the start of the color burst in the standard NTSC color television signal.

Breit-Wigner formula An equation that relates the cross section of a particular nuclear reaction to the energy of the incident particle and the energy of a resonance level of the compound nucleus.

bremsstrahlung Electromagnetic radiation produced when a fast charged particle (usually an electron) is deflected by another charged particle (usually a nucleus). The radiation corresponds to a continuous spectrum of x-rays.

brevium Uranium X_2, one of the decay products in the uranium series.

Brewster angle The angle of incidence for which a wave polarized parallel to the plane of incidence is wholly transmitted, with no reflection.

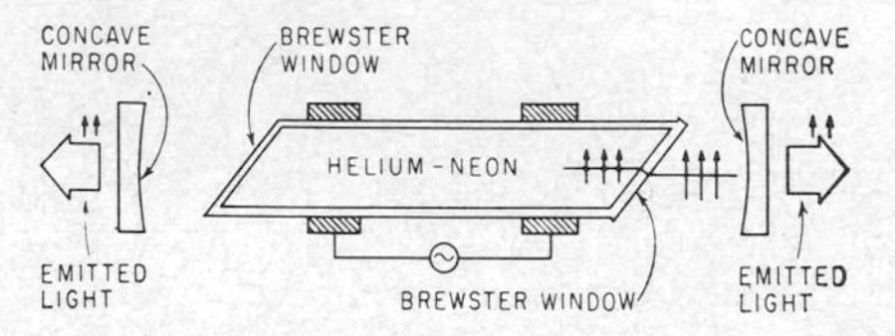

Brewster window used in gas laser.

Brewster window A special glass window at opposite ends of some gas lasers to transmit the laser output beam while reflecting other light.

bridge An instrument or circuit that has four or more arms, by means of which one or more of the electrical constants of an unknown component may be measured.

bridge amplifier An amplifier across the output of a bridge, in place of a meter. It may be an operational amplifier to which precision external resistors are added to set gain or an instrumentation amplifier that has its own gain-setting resistor network.

bridge circuit A circuit that consists basically of four sections connected in series to form a diamond. An AC voltage source is connected between one pair of opposite junctions, and an indicating instrument or output circuit is connected between the other pair of junctions. When the bridge is balanced, the output is zero.

bridged-T network A T network with a fourth branch connected across the two series arms of the T, between an input terminal and an output terminal.

bridge hybrid *Hybrid junction.*

bridge magnetic amplifier A magnetic ampli-

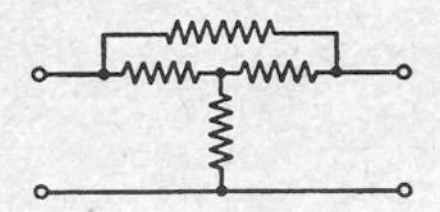

Bridged-T network.

fier in which each gate winding is connected in series with an arm of a bridge rectifier. The rectifiers provide self-saturation and DC output.

bridge rectifier A full-wave rectifier with four elements connected in series, as in a bridge circuit.

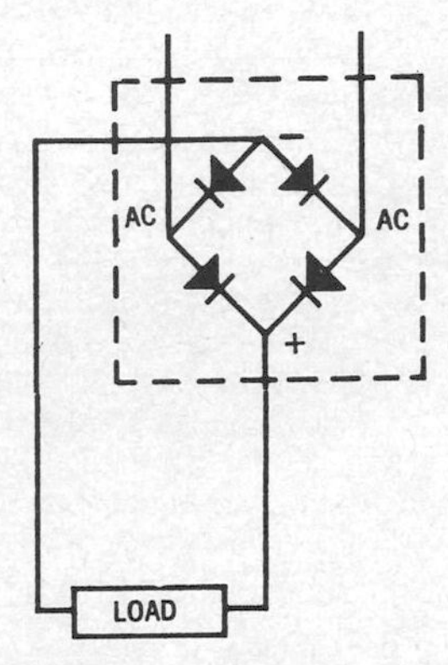

Bridge rectifier using silicon diodes. When all four diodes are good, AC terminals can be identified with ohmmeter because there will be infinite resistance between them.

Alternating voltage is applied to one pair of opposite junctions, and direct voltage is obtained from the other pair of junctions.

bridging 1. Connecting one electric circuit in parallel with another. 2. Selector-switch action in which the movable contact is wide enough to touch two adjacent contacts so that the circuit is not broken during contact transfer.

bridging amplifier An amplifier with an input impedance sufficiently high so that its input may be bridged across a circuit without substantially affecting the signal level.

bridging gain The ratio of the power a transducer delivers to a specified load impedance under specified operating conditions to the power dissipated in the reference impedance across which the input of the transducer is bridged. Usually expressed in decibels.

bridging loss The reciprocal of the bridging gain ratio. Usually expressed in decibels.

brightening pulse A pulse applied either to the grid or cathode of a radar cathode-ray tube at the beginning of the sweep, to intensify the beam during the sweep.

brightness 1. The characteristic of light that gives a visual sensation of more or less light. 2. Former name for *luminance.*

brightness control A control that varies the brightness of the fluorescent screen of a cathode-ray tube by changing the grid bias of the tube, thereby changing the beam current. Used in television receivers, radar receivers, and cathode-ray oscilloscopes. Also called brilliance control and intensity control.

brilliance The degree to which higher audio frequencies are present when a sound recording is played back.

brilliance control *Brightness control.*

Brillouin function A mathematical function that relates the magnetic moment, applied magnetic field, and temperature of a paramagnetic material to its magnetic susceptibility.

Brillouin scattering Interaction of sound waves at microwave frequencies with light waves from laser-generated coherent sources.

broadband amplifier An amplifier that has essentially flat response over a wide range of frequencies. Also called wideband amplifier.

broadband antenna An antenna that will function satisfactorily over a wide range of frequencies, such as for all 12 VHF television channels.

broadband channel A data-transmission channel that can handle frequencies higher than the normal voice-grade line limit of 3 to 4 kHz. A broadband channel can carry many voice or data channels simultaneously or can be used for high-speed single-channel data transmission.

broadband interference Interference distributed over a wider spectrum of frequencies than the tuning range of the affected receiver.

broadband klystron A klystron that has three or more resonant cavities which are externally loaded and stagger-tuned to broaden the bandwidth.

broadband noise Thermal noise that is uniformly distributed across the frequency spectrum at a wide range of energy levels.

broadcast A television or radio transmission intended for public reception.

broadcast band The band of frequencies extending from 535 to 1605 kHz, corresponding to assigned carrier frequencies that increase in multiples of 10 kHz between 540 and 1600 kHz for the United States. Also called standard broadcast band.

broadcast day The time between local sunrise and 12 midnight local standard time.

broadcasting Transmission of television and radio programs by radio waves for public reception.

broadcast interference [abbreviated BCI] Interference with standard radio broadcast reception, such as that produced by nearby amateur radio transmitters.

broadcast station A television or radio station that transmits programs to the general public. Also called station.

broadcast transmitter A transmitter for use in a commercial AM, FM, or television broadcast channel.

broad dimension The wider of the two cross section dimensions of a rectangular waveguide. It determines the critical frequency. Also called critical dimension.

broadside Perpendicular to an axis or plane.

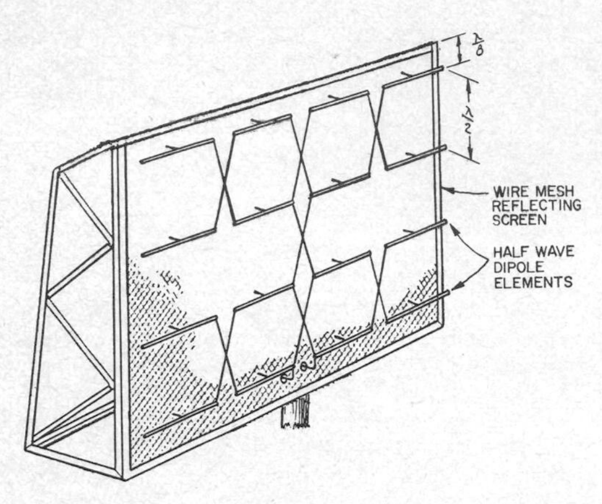

Broadside array.

broadside array An antenna array whose direction of maximum radiation is perpendicular to the line or plane of the array.

broad tuning Poor selectivity in a radio receiver, causing reception of two or more stations at a single setting of the tuning dial.

bromine [symbol Br] A nonmetallic liquid element. Atomic number is 35.

Brown and Sharpe gage [abbreviated B and S gage] *American wire gage.*

Bruce antenna *Rhombic antenna.*

brush A conductive metal or carbon block that makes sliding electric contact with a moving part.

brush discharge A luminous electric discharge that starts from a conductor when its potential exceeds a certain value but remains too low for the formation of an actual spark. Generally accompanied by a hissing sound.

brush encoder An encoder in which brushes that make contact with conductive segments on a rotating or linearly moving surface convert positional information to digitally encoded data.

brush plating Electroplating in which the plating solution is applied with a pad or brush that contains an anode, which is moved over the cathode to be plated.

brute-force filter A power-supply filter that uses large values of capacitance and inductance to smooth out ripple, instead of depending on resonant effects of active or passive tuned filters.

b/s Abbreviation for *bit per second.*

B scope A radarscope that produces a B display.

B station The loran station whose signal is always transmitted more than half a repetition period after the signal from the master or A station of the pair. Also called slave station.

B supply A power source that provides a positive voltage for the anode and other electrodes of an electron tube.

BT-cut crystal A crystal plate cut from a plane that is rotated about an X axis so the angle made with the Z axis is approximately −49°. This cut has an essentially zero temperature coefficient.

BTO Abbreviation for *bombing through overcast.*

bubble *Magnetic bubble.*

bubble chamber A chamber in which the movements and interactions of charged particles can be observed as visible tracks in a superheated liquid such as liquid hydrogen under pressure. The tracks are gas bubbles that form along the paths of the moving particles.

bubble memory A computer memory in which the presence or absence of a magnetic bubble in a localized region of a thin magnetic film designates a 1 or 0. Bubbles representing stored data can be moved by selectively exciting thin-film conductive loops placed on the surface to produce localized magnetic fields. Storage capacity can be well over 1 Mb/in^3. Also called magnetic-bubble memory.

bucket brigade device [abbreviated BBD] A semiconductor device in which majority carriers store charges that represent information, and minority carriers transfer charges from point to point in sequence, much as buckets of water were once passed along a line of volunteer firemen. Applications include shift registers and delay lines.

bucket counter A counter used in place of a binary counter in a ramp-type analog-to-digital converter to eliminate need for a stable clock frequency.

bucking coil A coil connected and positioned so that its magnetic field opposes the magnetic field of another coil. The hum-bucking coil of an excited-field loudspeaker is an example.

bucking voltage A voltage that has exactly opposite polarity to that of another voltage against which it acts.

Buckley gage An ionization gage that measures very low gas pressures.

Bucky diaphragm *Potter-Bucky grid.*

buffer 1. An isolating circuit used in a digital computer to avoid any reaction of a driven circuit on the corresponding driving circuit. 2. A storage device that compensates for differences in rates of data flow when it is transmitting information from one computer device to another. 3. *Buffer amplifier.*

buffer amplifier An amplifier used after an oscillator or other critical stage to isolate it from the effects of load impedance variations in subsequent stages. Also called buffer and buffer stage.

buffer capacitor A capacitor connected across the secondary of a vibrator transformer or between the anode and cathode of a cold-cathode rectifier tube to suppress voltage surges that might otherwise damage other parts in the circuit.

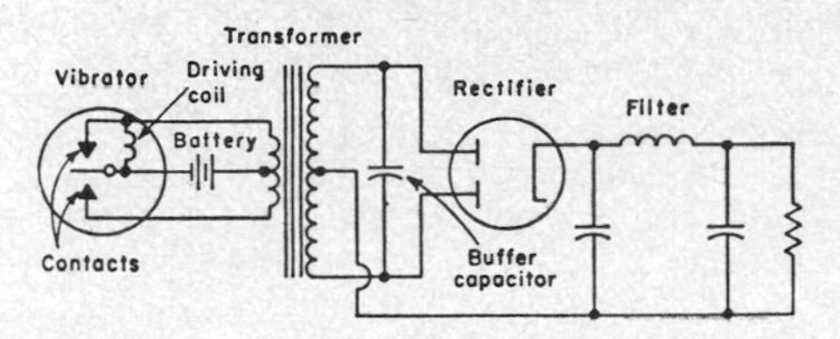

Buffer capacitor used across secondary of vibrator transformer.

buffered computer A computer that has a storage device which permits input and output data to be stored temporarily, to match the slow speed of input/output devices with the higher speeds of the computer and permit simultaneous input/output and computer operations.

buffer stage *Buffer amplifier.*

buffer storage A synchronizing element between two different forms of storage in a computer. Computation continues while transfers take place between buffer storage and the secondary or internal storage.

bug 1. A semiautomatic code-sending key in which movement of a lever to one side produces a series of correctly spaced dots, and movement to the other side produces a single dash. 2. Slang term for trouble in a piece of equipment. 3. An electronic listening device, generally concealed, used for commercial or other espionage.

bugging Use of electronic eavesdropping devices, such as high-gain directional microphones, hidden microphones wired directly to listening or recording devices or miniature radio transmitters, and inductive pickups or direct wire taps on telephones or telephone lines to monitor both sides of conversations.

building-out section A short section of transmission line, either open or short-circuited at the far end, shunted across another transmission line for tuning or matching purposes.

built-in antenna An antenna located inside the cabinet of a radio or television receiver.

built-up mica Large laminated plates of mica made by bonding thin splittings of natural mica with shellac, glyptol, or some other suitable adhesive.

bulb *Envelope.*

bulk acoustic wave An acoustic wave that travels through a piezoelectric material, as in a,quartz delay line. (With a surface acoustic wave, propagation is only on the surface.) Also called volume acoustic wave.

bulk-acoustic-wave delay line A delay line in which the delay is determined by the distance traveled by a bulk acoustic wave between input and output transducers mounted on a piezoelectric block.

bulk diode A semiconductor microwave diode that uses the bulk effect, such as Gunn diodes and

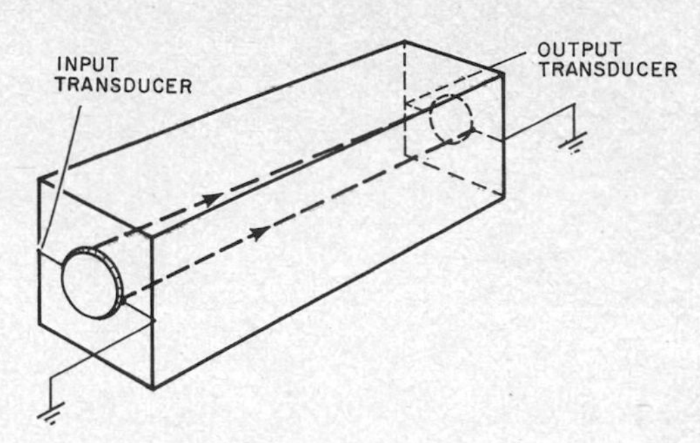

Bulk-acoustic-wave delay line.

diodes operating in limited space-charge-accumulation modes.

bulk effect An effect that occurs within the entire bulk of a semiconductor material rather than in a localized region or junction.

bulk-effect delay A semiconductor device that uses the bulk effect to provide microsecond time delays for signals ranging from audio to microwave frequencies.

bulk-effect device A semiconductor device that depends on a bulk effect, as in Gunn and avalanche devices.

bulk eraser A device that erases an entire reel of magnetic tape at once without running it through a recorder. The reel of tape is placed over a strong 60-Hz AC magnetic field, rotated a few times, and then slowly withdrawn at least 3 ft (1 m) before the power is turned off. Also called tape eraser.

bulk-ionized laser A laser that uses a dumbbell-shaped crystal which produces coherent light from its central bridge when a high-voltage electric field is applied between the ends.

bulk memory A high-capacity memory used in connection with a computer for bulk storage of large quantities of data. It may use magnetic tapes, magnetic disks, magnetic drums, shift registers, punched cards, punched paper tapes, or other nonprogrammable storage techniques.

bulk photoconductor A photoconductor that has high power-handling capability and other unique properties which depend on the semiconductor and doping materials used. Examples include cadmium selenide, germanium, indium an-

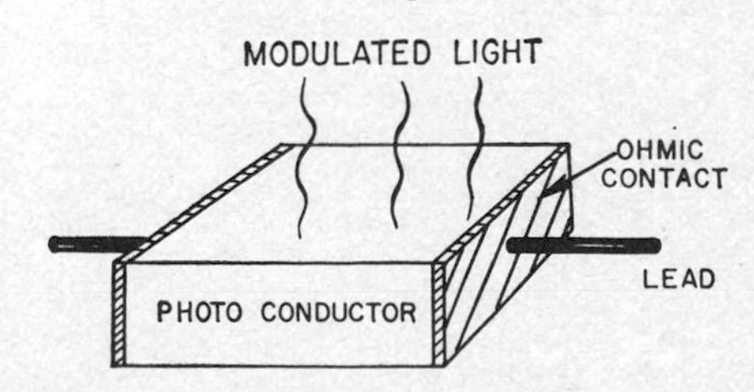

Bulk-photoconductor geometry for use as microwave demodulator.

timonide, indium arsenide, lead selenide, and silicon. With cadmium selenide semiconductors, thousands of kilowatts of power from pulsed ruby lasers can be handled.

bulk resistor An integrated-circuit resistor in which the N-type epitaxial layer of a semiconducting substrate is used as a noncritical high-value resistor. The spacing between the attached terminals and the sheet resistivity of the material together determine the resistance value.

Bull Goose A surface-to-air decoy missile that contains electronic countermeasure equipment, to confuse enemy radar systems.

bullhorn A portable loudspeaker, generally with built-in amplifier and microphone, used for voice messages to crowds or from one ship to another at sea.

Bullpup A Navy air-to-surface radio-guided missile used against enemy tanks, bridges, and other small targets.

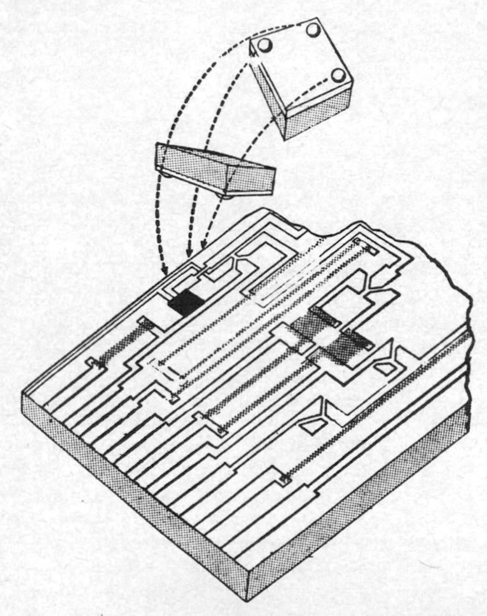

Bump contacts for base, emitter, and collector of transistor chip at top are alloyed to substrate after flipping chip, as shown by dotted lines.

bump contact A large-area contact used for alloying directly to the substrate of a transistor for mounting or interconnecting purposes.

buncher resonator The first or input cavity resonator in a velocity-modulated tube, next to the cathode. Here the faster electrons catch up with the slower ones to produce bunches of electrons. Also called input resonator.

bunching The flow of electrons from cathode to anode of a velocity-modulated tube as a succession of electron groups rather than as a continuous stream. It is a direct result of the differences of electron transit time produced by the velocity modulation.

buried diffused layer A low-resistance layer formed by impurity diffusion in semiconductor material before formation of a surface epitaxial layer.

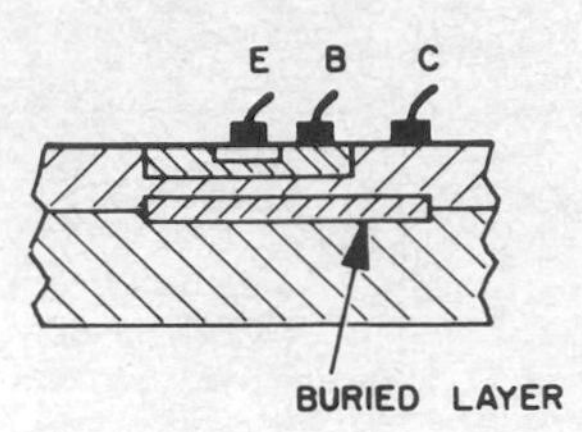

Buried diffused layer extending under collector region of transistor.

burned-in image An image that persists in a fixed position in the output signal of a television camera tube after the camera has been turned to a different scene.

burn-in Operation of a device under normal or extreme conditions to stabilize its characteristics and eliminate early failures, prior to actual use.

burnishing surface The portion of the cutting stylus, directly behind the cutting edge, that smooths the groove in mechanical recording.

burnout Failure of a device because of excessive heat produced by excessive current.

burst 1. A sudden increase in the strength of a signal being received from beyond line-of-sight range. It is believed due to meteors passing through the upper atmosphere and momentarily affecting the ionized layers that reflect radio waves back to earth. 2. An exceptionally large pulse observed in an ionization chamber, signifying the arrival of several ionizing particles simultaneously. It may be caused by a cosmic-ray shower. 3. To separate continuous-form line-printer paper into sheets. 4. *Color burst.*

burst amplifier In a color television receiver, an amplifier stage keyed into conduction and amplification by a horizontal pulse at the exact instant of each arrival of the 3.58-MHz color-burst signal. Also called chroma bandpass amplifier.

burst generator *Tone-burst generator.*

burst pedestal *Color-burst pedestal.*

burst separator The circuit in a color television receiver that separates the color burst from the composite video signal.

burst transmission Transmission in which messages are stored for a given time, then released at from 10 to 100 or more times the normal speed. The received signals are recorded and then slowed down to the normal rate for the user.

bus 1. One or more conductors used as a path for transmitting information from any of several sources to any of several destinations in an electronic computer. 2. *Busbar.*

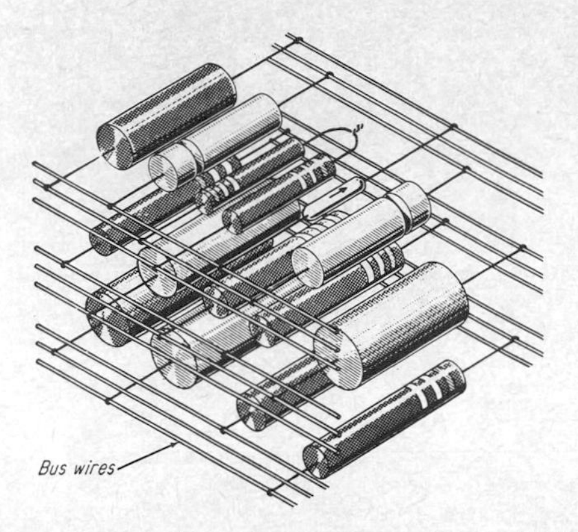

Bus wires to which leads of components are welded.

busbar A heavy, rigid, metallic conductor, usually uninsulated, that carries a large current or makes a common connection between several circuits. Also called bus.

bus driver A driver circuit that provides standard voltage levels to a large number of inverter transistor bases or diode loads.

busing Connecting a large number of data sources to a common bus.

Butler oscillator A crystal oscillator in which the crystal is connected between the cathodes of two triode sections, one connected as a cathode follower and the other as a grounded-grid amplifier.

butt contact A hemispherically shaped contact that mates against a similarly shaped contact. The contacts are usually held together by spring pressure.

butterfly capacitor A variable capacitor that has stator and rotor plates shaped like butterfly wings. Each stator plate has an outer ring that forms an inductance which varies with rotor position. Both inductance and capacitance are at a minimum when the stator and rotor plates form the four quarters of a circle and increase simultaneously to a maximum when the plates are rotated to the fully meshed position. This greatly increases the tuning range when the capacitor is used as a tuned circuit in VHF and UHF circuits.

Butterworth filter A filter that has essentially flat amplitude response in the passband and an attenuation rate beyond cutoff of 6 dB per octave for a single-pole filter. Transient response is much better than for a comparable Chebyshev filter.

Butterworth response A maximally flat amplitude response in the passband of a tunable active filter, combined with moderate settling time and moderate overshoot.

butt joint 1. A connection that gives physical contact between the ends of two waveguides to maintain electric continuity. 2. A connection formed by placing together the ends of two conductors and joining them by welding, brazing, or soldering.

button 1. A small round piece of metal that is alloyed to the base wafer of an alloy-junction transistor. Also called dot. 2. The container that holds the carbon granules of a carbon microphone. Also called carbon button.

buttonhook contact A curved hooklike contact often used on feedthrough terminals of headers to facilitate soldering or unsoldering of leads.

buzz *Dither.*

buzzer An electromagnetic device that has an armature which vibrates rapidly, producing a buzzing sound.

BW Abbreviation for *bandwidth.*

B wind Magnetic tape wound with the oxide surface facing outward. Seldom used today.

BWO Abbreviation for *backward-wave oscillator.*

BX cable Insulated wires in flexible metal tubing, used for bringing electric power to electronic equipment.

bypass A low-impedance path provided around part or all of a circuit.

bypass capacitor A capacitor connected to provide a low-impedance path for RF or AF currents around a circuit element.

bypassed mixed highs The mixed-highs signal, containing frequencies between 2 and 4 MHz, that is shunted around the chrominance-subcarrier modulator or demodulator in a color television system.

bypassed monochrome Deprecated term for *shunted monochrome.*

B − Y signal A blue-minus-luminance color-difference signal used in color television. It is combined with the luminance signal in a receiver to give the blue color-primary signal.

byte A sequence of adjacent binary digits, operated upon as a unit in a computer and usually shorter than a word. It is the smallest addressable unit. A byte is usually 8 bits, which can represent two numerals or one character.

byte per second [abbreviated byte/s] A rating used in specifying the speed of a digital transmission system.

byte/s Abbreviation for *byte per second.*

C

c 1. Abbreviation for *centi-*. 2. Abbreviation for *curie.* 3. Abbreviation for *character.*

C 1. Abbreviation for *capacitor.* 2. Symbol for the grid-bias voltage source for an electron tube. 3. Abbreviation for *centigrade* or *Celsius.* 4. Symbol for *collector.* Used on transistor circuit diagrams. 5. Abbreviation for *coulomb.* 6. Abbreviation for *capacitance.*

°C Abbreviation for degree *Celsius.*

^{14}C The artificially produced radioisotope of carbon, used as a tracer in research.

C− [C minus] The negative terminal of a C battery or other grid-bias voltage source for an electron tube, or the grid-circuit terminal to which this source terminal should be connected.

C+ [C plus] The positive terminal of a C battery or other grid-bias voltage source for an electron tube, or the grid-circuit terminal to which this source terminal should be connected.

C Mathematical symbol for *capacitance.*

CaAs Symbol for *cadmium arsenide.*

cabinet The housing for a radio receiver, television receiver, or other piece of electronic equipment.

cable A transmission line, group of transmission lines, or group of insulated conductors mechanically assembled in compact flexible form.

cable clamp A clamp that gives mechanical support to a cable at the rear of its plug or outlet.

cable television [abbreviated CATV] A television program distribution system in which signals from all local stations and usually a number of distant stations are picked up by one or more high-gain antennas at elevated locations, amplified on individual channels, then fed directly to individual receivers of subscribers by overhead or underground coaxial cable. Used to improve reception and to make more stations available in a given area. The system sometimes includes facilities for originating local programs, time and weather reports, news bulletins, and other services. Also called community antenna television.

cache memory A small but very fast supplementary computer memory between the main memory and the central processing unit to give the effect of a larger and faster memory. The cache is automatically loaded with the addressed word plus words from adjacent memory locations; since programs are usually sequential in nature, the next word to be requested is very likely to be in the cache. Reduces the size and cost of the main memory.

CAD Abbreviation for *computer-aided design.*

cadmium [symbol Cd] A metallic element widely used as a plated coating on steel hardware for electronic equipment because it improves solderability and surface conductivity and prevents corrosion. Atomic number is 48.

cadmium arsenide [symbol CaAs] A semiconductor material used with various phosphor coatings in light-emitting diodes to produce blue, green, or red light.

cadmium cell A standard cell used as a voltage reference. At 20°C its voltage is 1.0186 V.

cadmium cutoff A neutron energy level of about 0.4 eV. At energies above this, the capture section for cadmium drops rapidly.

cadmium selenide photoconductive cell A photoconductive cell that uses cadmium selenide as the semiconductor material. It has fast response time and high sensitivity to longer light wavelengths, such as those of incandescent lamps and some infrared light sources.

cadmium-silver oxide cell An alkaline-electrolyte cell that can be used without recharging in primary batteries or recharged for secondary battery use.

cadmium sulfide [symbol CdS] A semiconductor that has a forbidden band gap of 2.4 eV and a maximum operating temperature of 870°C.
cadmium sulfide photoconductive cell A photoconductive cell in which a small wafer of cadmium sulfide provides an extremely high dark-light resistance ratio. Some models can be used directly as a light-controlled switch that operates directly from a 120-VAC power line.
cadmium telluride detector A photoconductive cell capable of operating continuously at ambient temperatures up to 750°F (400°C). Used in solar cells and infrared, nuclear-radiation, and gamma-ray detectors.
Caesar Project name for a network of underwater sonar listening posts at the edge of the continental shelf, for submarine detection. The hydrophone arrays, encased in a plastic that is transparent to sound, are anchored below surface turbulence and connected by cable to shore control stations.
cage To lock the gyroscope of a gyro-controlled instrument in a fixed position with reference to its case.
cage antenna An antenna consisting of long parallel wires or rods arranged in the form of a cylinder.
CAI Abbreviation for *computer-aided instruction.*
cal Abbreviation for *calorie.*
calcium [symbol Ca] A silver-white soft metallic element used in cathode coatings for some types of phototubes. Atomic number is 20.
calcium iodide scintillator A scintillator material that consists of grown crystals of calcium iodide, which fluoresces in proportion to the strength of nuclear radiation. A newer version, europium-activated calcium iodide, gives almost twice as much light output.
calculating punch A calculator that has a card reader and card punch.
calculator A device that performs arithmetic operations based on numerical data which is entered by pressing numerical and control keys. Pocket-size electronic versions operate from batteries or a plugged-in AC power supply. Larger desk versions operate from AC power and may also have paper-tape printout facilities.
calibrate 1. To determine, by measurement or comparison with a standard, the correct value of each scale reading on a meter or other device. 2. To determine the settings of a control that correspond to particular values of voltage, current, frequency, or some other characteristic.
calibration curve A plot of calibration data that gives the correct value for each indicated reading of a meter or control dial.
calibration marker A marker line or circle that divides a radar screen into accurately known intervals for determination of range, bearing, height, or time.
californium [symbol Cf] A transuranic radioactive element produced by bombarding curium isotope ^{242}Cm with helium atoms. Atomic number is 98.
call 1. A radio transmission that identifies the transmitting station and designates the station for whom the transmission is intended. 2. To transfer control to a specified closed computer subroutine.
call letters Identifying letters, sometimes including numerals, assigned to radio and television stations by the Federal Communications Commission and other regulatory authorities throughout the world.
calorescence The production of visible light by infrared radiation. The transformation is indirect, the light being produced by heat, not by any direct change of wavelength.
calorie [abbreviated cal] The metric unit of quantity of heat. It is approximately the amount of heat required to raise the temperature of 1 g of water 1°C.
calorimeter A device that measures quantity of heat. Used to measure microwave power in terms of its heating effect.
CAM Abbreviation for *content-addressable memory.*
camera *Television camera.*
camera chain A television camera, associated amplifiers, a monitor, and the cable needed to bring the camera output signal to the control room.
camera signal The video output signal of a television camera.
camera storage tube A storage tube into which the information is introduced by electromagnetic radiation, usually light, and read at a later time as an electric signal.
camera tube An electron-beam tube in a television camera; it converts an optical image into a corresponding charge-density electric image and scans the resulting electric image in a predetermined sequence to provide an equivalent electric signal. Examples are the iconoscope, image dissector, image orthicon, and vidicon. Also called pickup tube and television camera tube.
Campbell bridge A bridge specifically designed for comparison of mutual inductances.
Campbell-Colpitts bridge An AC bridge that measures capacitance by the substitution method.
camp-on circuit A circuit in a telephone system that holds a call for a preset time if the line is busy and makes the connection automatically as soon as the busy condition is terminated.
canceled video The radar video output that remains after moving-target-indicator cancellation.
canceller The portion of a radar system in which signals reflected from fixed targets are canceled.
candela [abbreviated cd] The SI unit of luminous intensity. One candela is defined as the luminous intensity of 1/60 cm^2 of a blackbody radiator operating at the temperature of solidification of platinum. Formerly called candle.

candela per square meter [abbreviated cd/m²] The SI unit of luminance. The term nit is sometimes used for this unit.

candle *Candela.*

candlepower Luminous intensity expressed in candelas.

cannibalize To remove serviceable parts from one piece of equipment to repair another piece of equipment.

capacitance [abbreviated C; mathematical symbol *C*] 1. The electrical size of a capacitor. The basic unit is the farad, but the smaller microfarad, nanofarad, and picofarad units are commonly used. Also called capacity (deprecated) and permittance (obsolete). 2. The property that exists whenever two conductors are separated by an insulating material, permitting the storage of electricity.

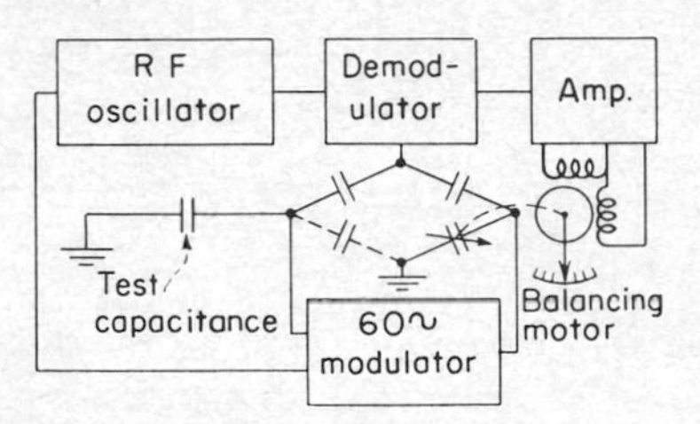

Capacitance bridge as used in capacitance level indicator. Probe in tank serves as one capacitor in bridge. Variation in capacitance unbalances bridge, producing 60-Hz modulated RF output that adds to signal from oscillator, is demodulated, then amplified to drive balancing motor in correct direction to rebalance bridge by changing variable capacitor.

capacitance bridge A bridge for comparing two capacitances, such as a Schering bridge.

capacitance level indicator A level indicator in which the material being monitored serves as the dielectric of a capacitor formed by a metal tank and an insulated electrode mounted vertically in the tank. The increase in capacitance with level can be measured accurately for depth ranges up to 200 ft (60 m).

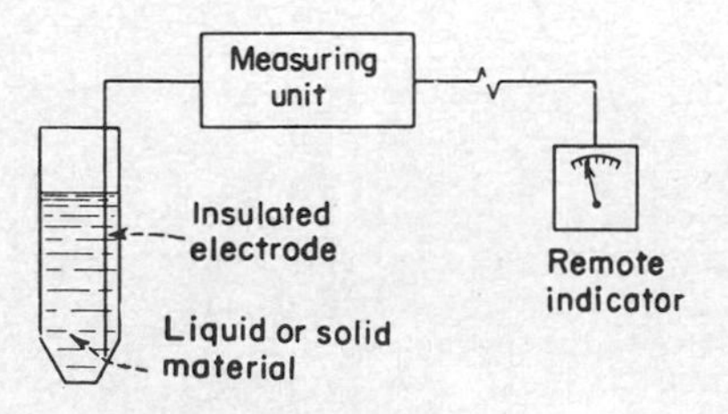

Capacitance level indicator.

capacitance-loop directional coupler A directional coupler in which a coupling link much shorter than a quarter-wavelength is placed lengthwise in a waveguide.

capacitance meter An instrument that measures capacitance values of capacitors or circuits which contain capacitance.

capacitance multiplier A circuit that uses operational amplifiers to multiply the value of an input capacitor by a fixed or adjustable factor.

capacitance-operated intrusion detector A boundary alarm system in which the approach of an intruder to an antenna wire encircling the protected area changes the antenna-ground capacitance and sets off the alarm.

capacitance relay An electronic relay that responds to a small change in capacitance, such as that created by bringing a hand near a pickup wire or plate.

capacitance standard *Standard capacitor.*

capacitive coupling Use of a capacitor to transfer energy from one circuit to another.

capacitive diaphragm A resonant window in a waveguide to provide the equivalent of capacitive reactance at the frequency being transmitted.

capacitive-discharge ignition An automotive ignition system in which energy is stored in a capacitor and discharged across the gap of a spark plug through a step-up pulse transformer and distributor each time a silicon controlled rectifier is triggered.

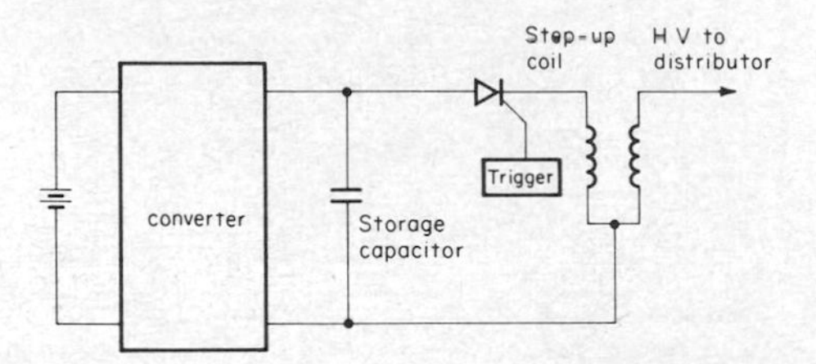

Capacitive-discharge ignition system using triggered silicon controlled rectifier with DC/DC converter.

capacitive-discharge pilot light An electronic ignition system, operating off an AC power line or battery power supply, that produces a spark for lighting a gas flame.

capacitive-discharge storage welding Resistance welding in which the energy is stored in banks of capacitors that are repeatedly discharged through the primary of the welding transformer. The current through the low-voltage secondary flows through and heats the joint that is to be welded.

capacitive feedback Feedback through a capacitor connected between the output and input of a circuit.

capacitive load A load in which the capacitive reactance exceeds the inductive reactance. Such a load draws a leading current.

capacitive post A metal post or screw that extends across a waveguide at right angles to the E field, to provide capacitive susceptance in parallel with the waveguide for tuning or matching purposes.

capacitive reactance [symbol X_c] Reactance due to the capacitance of a capacitor or circuit. Capacitive reactance is measured in ohms and is equal to 1 divided by 6.28*fC*, where *f* is in hertz and *C* is in farads.

capacitive tuning Tuning that involves use of a variable capacitor.

capacitive tuning screw An adjustable screw that projects into a waveguide or microwave cavity to provide a variable reactance, as required for tunable microwave filters.

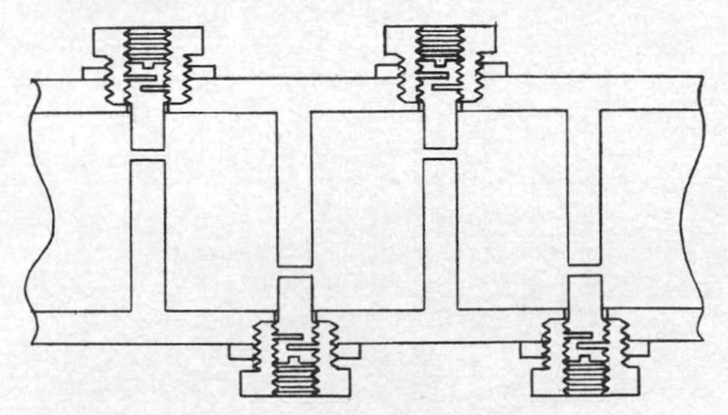

Capacitive tuning screw used in bandpass filter for waveguide.

capacitive window A conducting diaphragm that extends into a waveguide from one or both sidewalls, producing the effect of a capacitive susceptance in parallel with the waveguide.

capacitor [abbreviated C] A device that consists essentially of two conducting surfaces separated by a dielectric material like air, paper, mica, ceramic, glass, or Mylar. A capacitor stores electric energy, blocks the flow of direct current, and permits the flow of alternating current to a degree dependent on its capacitance and the frequency. Also called condenser (deprecated).

capacitor antenna An antenna in which the essential characteristic is the capacitance between two conductors or sets of conductors.

capacitor bank A number of capacitors connected together in series or in parallel.

capacitor color code A method of marking the value on a capacitor by dots or bands of colors as specified in the EIA color code.

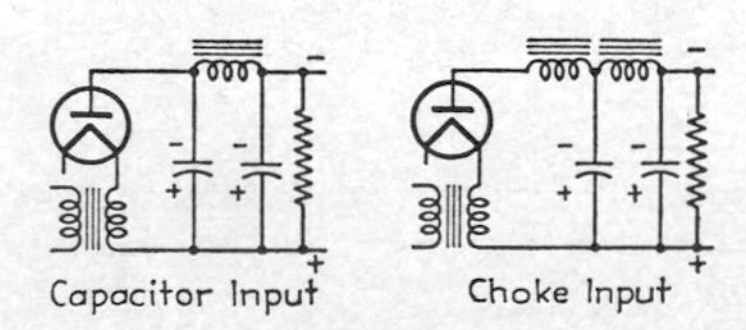

Capacitor-input filter and choke-input filter.

capacitor-input filter A power-supply filter in which a shunt capacitor is the first element after the rectifier.

capacitor integrator An integrator circuit in which a current proportional to the function to be integrated is fed into a capacitor and the capacitor voltage is read after the desired time interval of integration.

capacitor ionization chamber *Capacitor R-meter.*

capacitor loudspeaker *Electrostatic loudspeaker.*

capacitor microphone A microphone that consists essentially of a flexible metal diaphragm and a rigid metal plate that together form a two-plate air capacitor. Sound waves set the diaphragm in vibration, producing capacitance variations that are converted into AF signals by a suitable amplifier circuit. Also called condenser microphone (deprecated) and electrostatic microphone.

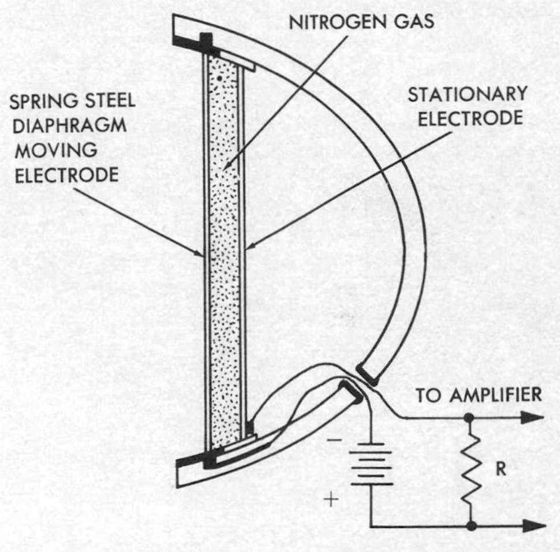

Capacitor-microphone construction.

capacitor motor A single-phase induction motor that has a main winding connected directly to a source of AC power and an auxiliary winding connected in series with a capacitor to the source of AC power.

capacitor pickup A phonograph pickup in which movements of the stylus in a record groove cause variations in the capacitance of the pickup.

capacitor R-meter An R-meter that consists of an electrometer and a detachable ionization chamber which embodies a capacitor. The chamber is attached to the electrometer and charged, then detached and partly discharged by exposure to the radiation being measured. The chamber is then returned to the electrometer for reading the decrease in charge. The instrument can be calibrated to read directly in roentgens. Also called capacitor ionization chamber.

capacitor-start motor A capacitor motor in which the capacitor is in the circuit only during the starting period. The capacitor and its auxiliary winding are disconnected automatically by a centrifugal switch or other device when the motor reaches a predetermined speed, after which the motor runs as an induction motor.

capacitor start-run motor *Permanent-split capacitor motor.*

capacitron An externally fired pool tube in which each conducting cycle is started by a high-voltage pulse applied between the mercury-pool cathode and a metal band on the outside of the glass envelope just above the pool level. Current control is achieved by varying the time at which each cycle is started.

capacity 1. The rated or maximum load or capability of a machine or device, such as the upper and lower limits of the numbers that may be processed or stored in a digital computer. 2. Deprecated term for *capacitance.*

capstan The shaft that rotates against the tape in a magnetic-tape recorder, pulling the tape past the head at a constant speed during recording and playback.

capstan idler A rubber-tired roller that holds the magnetic tape against the capstan by spring pressure.

captive fastener A screw-type fastener that does not drop out after it has been unscrewed.

capture 1. A process in which an atomic or nuclear system acquires an additional particle, such as the capture of electrons by positive ions or nuclei, or neutrons by nuclei. The gamma rays emitted as a result of capture are called capture gamma rays. 2. The act of placing a missile under control of a guidance system after flight speed has been reached.

capture cross section The cross section that is effective for radiative capture.

capture effect The effect wherein a strong FM signal in an FM receiver completely suppresses a weaker signal on the same or nearly same frequency.

capture gamma rays Gamma rays emitted during capture.

capture ratio A measure of the ability of an FM tuner to reject the weaker of two stations that are on the same frequency. The lower the ratio in decibels of desired and undesired signals, the better the performance of the tuner.

carbometer An instrument that measures the carbon content of steel by measuring magnetic properties of the steel in a known magnetic field.

carbon [symbol C] An element widely used in the construction of resistors and dry cells. Atomic number is 6.

carbon button *Button.*

carbon cycle A series of thermonuclear reactions, with release of energy, that presumably occurs in the sun and other stars. The net accomplishment is the synthesis of four hydrogen atoms into a helium atom, with attendant release of nuclear energy.

carbon dioxide laser [abbreviated CO_2 laser] A molecular gas laser in which the chief gas is carbon dioxide, with other gases such as helium and nitrogen added to increase output. Continuous operation at kilowatt output power levels makes the laser suitable for such applications as cutting, drilling, evaporating, heating, welding, optical radar, and laser communication. Pumping can be chemical, electric, or optical. Pulsed operation gives much higher average output power.

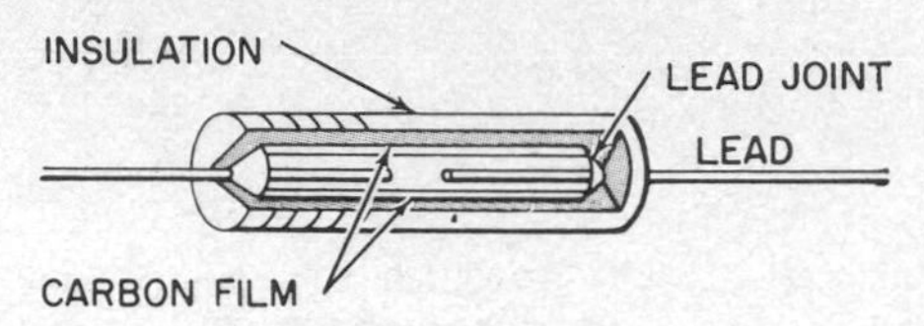

Carbon-film resistor.

carbon-film resistor A resistor made by depositing a thin carbon film on a ceramic form.

carbon-14 dating Determining the approximate age of organic material associated with archeological or fossil artifacts by measuring the rate of radiation of the carbon-14 isotope that is present in all organic matter and has a known half-life of 5568 years.

carbonized filament A thoriated tungsten filament treated with carbon to form a coating of tungsten carbide. This permits higher filament temperatures and correspondingly greater electron emission without excessive evaporation of thorium.

carbon microphone A microphone in which a flexible diaphragm moves in response to sound waves and applies a varying pressure to a container filled with carbon granules, causing the resistance of the microphone to vary correspondingly.

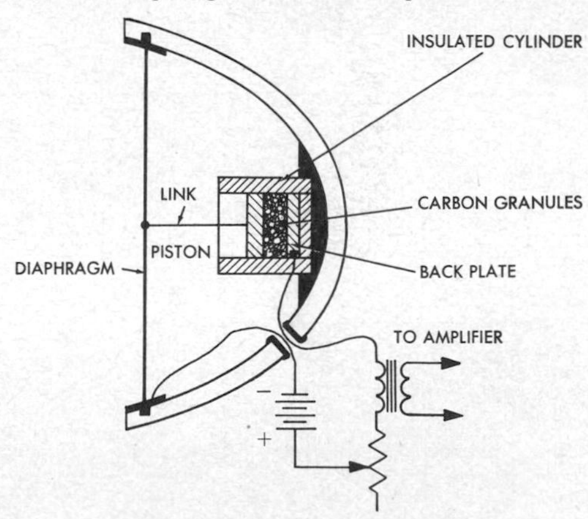

Carbon-microphone construction.

carbon monoxide laser A molecular gas laser in which the active laser molecule is carbon monoxide, and the strongest wavelengths are 4.9 to 5.7 μm. Also called CO laser.

carbon pile A variable resistor that consists of a stack of carbon disks mounted between fixed and

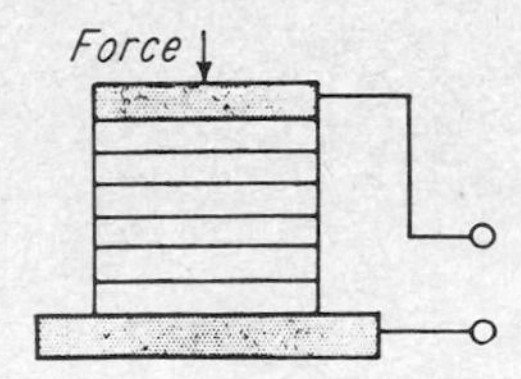

Carbon pile acting as variable resistor for high-current power circuits.

movable metal plates which serve as the terminals of the resistor. The resistance value is reduced by applying pressure to the movable metal plate.

carbon-pressure recording Electromechanical recording in which a stylus or other pressure device acts on carbon paper placed over the record sheet.

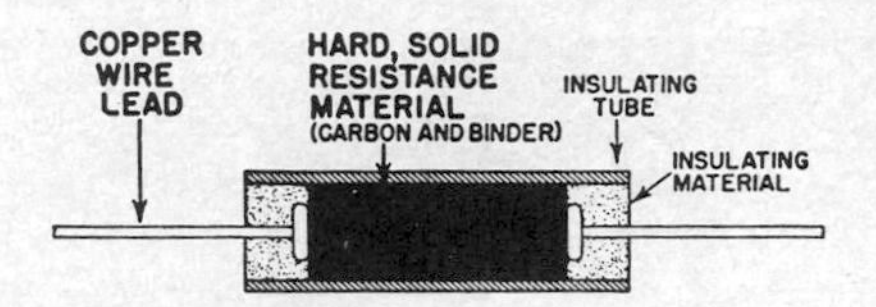

Carbon-resistor construction.

carbon resistor A resistor that consists of carbon particles mixed with a binder, molded into a cylindrical shape, and baked. Terminal leads are attached to opposite ends. The resistance of a carbon resistor decreases as temperature increases. Also called composition resistor.

carbon-zinc cell The most common type of dry cell. It has a carbon rod in the center as the positive terminal, a zinc can as the negative terminal, and a sal ammoniac paste in between. Voltage is about 1.5 V.

carborundum A compound of carbon and silicon. Used in crystal form to rectify or detect radio waves.

carborundum thermistor A thermistor in which the temperature-dependent resistor is a disk or rod formed at high temperature from a mixture of carborundum and ceramic materials, to give a wide range of negative temperature characteristics and resistance values.

carborundum varistor A varistor in which the voltage-dependent resistor is a mixture of carborundum and ceramic materials fired at high temperature. The nonlinear voltampere characteristic is expressed as $I = KE^n$, where n is typically a value between 1 and 6 and K is a constant.

carcinotron A voltage-tuned backward-wave oscillator tube that generates frequencies ranging from UHF values up to over 100 GHz. The desired output wave travels in the opposite direction to the electron beam, so the output termination is at the electron-gun end of the tube.

card *Punched card.*

card column A punched-card column into which information is entered by punches. Also called column.

card dialer A telephone in which a number can be dialed automatically and almost instantly by inserting a coded plastic card for that number in a slot on the dialer.

card-edge connector A connector that mates with printed-wiring leads running to the edge of a printed-circuit board on one or both sides. Also called edgeboard connector.

card feed A mechanism that moves cards into a machine one at a time.

card field A set of punched-card columns fixed as to number and position, into which the same item of information is regularly entered.

cardiac monitor A monitor that contains an electrocardiograph and one or more other cardiac instruments, usually combined with a cathode-ray display and circuits that trigger audible and/or visual alarms when limits of a monitored input are exceeded.

cardiac pacemaker *Pacemaker.*

card image A one-to-one representation of the contents of a punched card in the internal storage of a computer, in which a 1 represents a punch and a 0 represents the absence of a punch.

cardiogram *Electrocardiogram.*

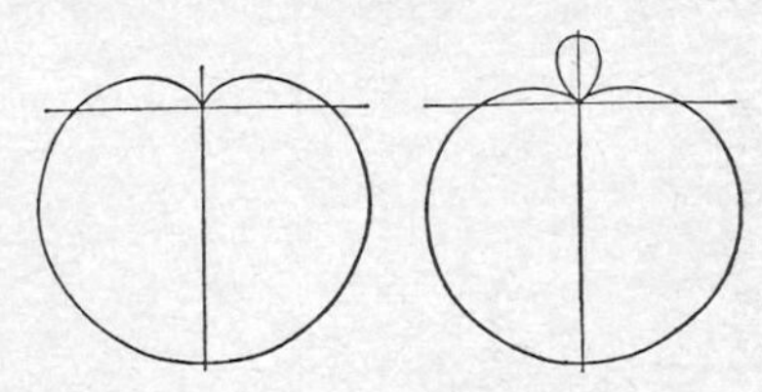

Cardioid diagram representing pickup pattern of directional microphone, and modified cardioid diagram (right) having still greater suppression of sounds arriving from sides and rear. Microphone is at intersection of horizontal and vertical lines.

cardioid diagram A polar diagram in the shape of a heart, such as the radiation pattern of a dipole antenna with reflector.

cardioid microphone A microphone that has a heart-shaped or cardioid response pattern, so it has nearly uniform response for a range of about 180° in one direction and minimum response in the opposite direction. In one form it is a combination of dynamic-microphone and ribbon-microphone elements.

cardiometer A medical electronic instrument that measures the force of the action of the heart.

cardiotachometer An electronic amplifier that times and records pulse rates of the heart.

cardiotocogram A medical electronic instrument that measures and records the instantaneous fetal heart rate and labor activity before and during childbirth.

card punch *Keypunch.*

card reader A mechanism that senses information punched on cards by using wire brushes, metal feelers, or a photoelectric system.

card reproducer A machine that makes a duplicate of a punched card.

card row A row of punch positions parallel to the long edge of a punched card. A standard card has 12 rows and 80 columns.

card sorter A machine that arranges punched cards in a desired sequence of data punched in the vertical columns.

card-to-tape converter A machine that transfers information from punched cards to punched or magnetic tape.

Carey-Foster bridge A type of Wheatstone bridge that measures the difference between nearly equal resistances. Coils connected by a slide wire act as two ratio arms.

carnauba wax A natural wax used for insulating purposes; it has a melting point of 85°C.

carpet A noise-modulated airborne jamming transmitter or other electronic device used for radar jamming.

carriage A device that moves in a predetermined path and carries some other part, such as a recorder head.

carrier 1. The radio wave produced by a transmitter when there is no modulating signal, or any other wave, recurring series of pulses, or direct current capable of being modulated. Also called carrier wave. 2. A mobile conduction electron or hole in a semiconductor. 3. A substance that, when associated with a radioactive trace of another substance, will carry the trace with it through a chemical or physical process. An isotope is often used for this purpose. Also called isotopic carrier. 4. A wave generated locally at a receiver that, when combined with the sidebands of a suppressed-carrier transmission in a suitable detector, produces the modulating wave. 5. *Carrier system.*

carrier amplifier A direct-current amplifier in which the DC input signal is filtered by a low-pass filter, then used to modulate a carrier so it can be amplified conventionally as an AC signal. The amplified DC output is obtained by rectifying and filtering the rectified carrier signal. A more common version of the carrier amplifier is the chopper amplifier, which uses either one or two choppers to convert the DC input signal into a square-wave AC signal and synchronously rectify the amplified square-wave signal.

carrier-amplitude regulation The change in amplitude of the carrier wave in an amplitude-modulated transmitter when modulation is applied symmetrically.

carrier channel The equipment and lines that make up a complete carrier-current circuit between two or more points.

carrier chrominance signal *Chrominance signal.*

carrier-controlled approach system An aircraft-carrier radar system that provides information by which aircraft approaches may be directed by radio.

carrier current A higher-frequency alternating current superimposed on ordinary telephone, telegraph, and power-line frequencies for communication and control purposes. The carrier current is modulated with voice signals to provide telephone communication between points on the power system or tone-modulated to actuate switching relays or convey data.

carrier-free radioisotope A radioisotope in which all the atoms are radioactive. The term is sometimes applied to highly radioactive material to which no inactive carrier has been intentionally added.

carrier frequency The frequency generated by an unmodulated radio, radar, carrier communication, or other transmitter, or the average frequency of the emitted wave when modulated by a symmetrical signal. Also called center frequency and resting frequency.

carrier-frequency pulse A carrier that is amplitude-modulated by a pulse. The amplitude of the modulated carrier is zero before and after the pulse.

carrier-interference ratio [abbreviated C/I ratio] An interference-specifying ratio used in microwave relays and other communication systems. It is based on measuring the desired signal, turning it off, and then measuring the undesired signal. The ratio of the two measurements is expressed in decibels.

carrier leak The carrier frequency remaining in a suppressed-carrier system.

carrier level The strength or level of an unmodulated carrier signal at a particular point in a radio system, expressed in decibels in relation to some reference level.

carrier line Any transmission line used for multiple-channel carrier communication.

carrier mobility The average drift velocity of carriers per unit electric field in a homogeneous semiconductor. The mobility of electrons is usually different from that of holes.

carrier modulation The process of varying some characteristic of a carrier in accordance with a modulating wave.

carrier noise level The noise level produced by undesired variations of an RF signal in the absence of any intended modulation. Also called residual modulation.

carrier-power transformer A transformer that supplies AC carrier power to a magnetic amplifier.

carrier repeater A one- or two-way repeater used in a carrier channel.

carrier shift Radioteletypewriter transmission in which the carrier frequency is shifted in one direction for a mark signal and in the opposite direction for a space signal.

carrier signaling Use of tone signals for ringing and other signaling functions in a carrier communication system.

carrier suppression 1. Suppression of the carrier frequency after conventional modulation at the transmitter, with reinsertion of the carrier at the receiving end before demodulation. 2. Suppression of the carrier when there is no modulation signal to be transmitted. Used on ships to reduce interference between transmitters.

carrier swing The total deviation of a frequency- or phase-modulated wave from the lowest to the highest instantaneous frequency.

carrier system A system that permits a number of independent communications over the same circuit. Also called carrier.

carrier telegraphy Telegraphy in which a single-frequency carrier wave is modulated by the transmitting apparatus for transmission over wire lines.

carrier telephony Telephony in which a single-frequency carrier wave is modulated by a voice-frequency signal, for transmission over wire lines.

carrier-to-noise ratio The ratio of the magnitude of the carrier to that of the noise after selection and before any nonlinear process such as amplitude limiting and detection.

carrier transmission Transmission in which a single-frequency carrier wave is modulated by the signal to be transmitted.

carrier wave *Carrier.*

carry A signal produced in a computer when the sum of two digits in the same column equals or exceeds the base of the number system in use or when the difference between two digits is less than zero.

Cartesian control A guided-missile control that is dependent on two sets of control surfaces, each set producing movement perpendicular to that of the other.

cartridge 1. *Phonograph pickup.* 2. *Tape cartridge.*

cartridge lamp A pilot or dial lamp that has a tubular glass envelope with metal-ferrule terminals at each end. It resembles a television or radio fuse and is mounted in the same type of socket clips.

CAS Abbreviation for *collision avoidance system.*

cascade *Avalanche effect.*

cascade amplifier An amplifier that contains two or more stages arranged in the conventional series manner, in which the output of one stage is amplified by the succeeding stage.

cascade-amplifier klystron A klystron that has three resonant cavities to provide increased power amplification and output. The extra resonator, located between the input and output resonators, is excited by the bunched beam emerging from the first resonator gap and produces further bunching of the beam.

cascade connection A series connection of amplifier stages, networks, or tuning circuits, in which the output of one feeds the input of the next.

cascade control An automatic control system in which various control units are linked in sequence, each control unit regulating the operation of the next control unit in line.

cascaded carry A carry process in which the addition of two numerals results in a sum numeral and a carry numeral that are in turn added together, this process being repeated until no new carries are generated.

cascaded image tube An image tube that has a number of sections stacked together, the output image of one section serving as the input for the next section. Used for light detection at very low levels.

cascade limiter A limiter circuit that uses two vacuum tubes in series to give improved limiter operation for both weak and strong signals in an FM receiver. Also called double limiter.

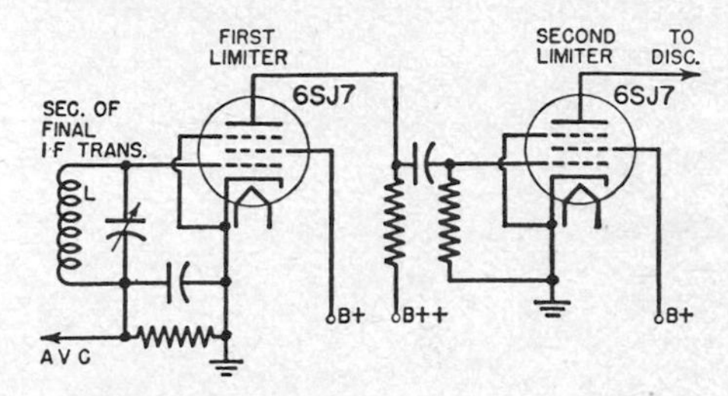

Cascade limiter for frequency-modulated receiver.

cascade particle A baryon particle that has negative or zero charge and a mass about 2600 times that of an electron. It decays into a lambda particle and a pi meson in about 10^{-10} s.

cascade shower A cosmic-ray shower that is initiated when a high-energy electron, in its passage through matter, produces one or more photons which have energies comparable with its own. These photons convert into electrons and positrons by the process of pair production. The secondary electrons in turn produce the same effects as the primary. The cascade shower of electrons and positrons builds up until the level of energy is so low that photon emission and pair production can no longer occur.

cascode amplifier A two-stage amplifier in which the collector of the input stage feeds the emitter of the output stage, or, in the electron-tube version, the anode of the input stage feeds the cathode of the grounded-grid triode output stage.

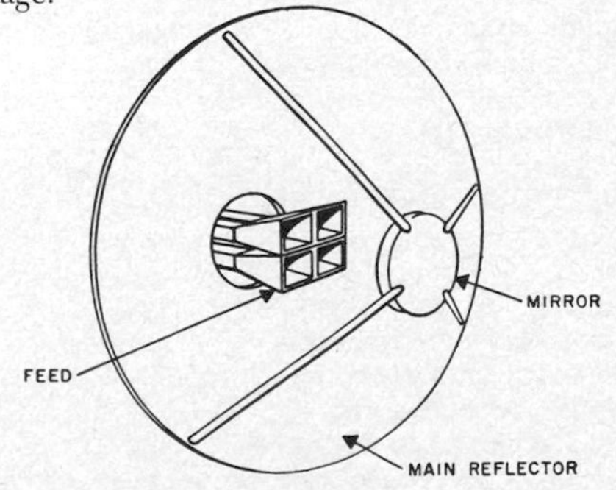

Cassegrain antenna.

Cassegrain antenna A microwave antenna in which the feed radiator is mounted at or near the surface of the main reflector and aimed at a mirror at the focus. Energy from the feed first illuminates the mirror, then spreads outward to illuminate the main reflector. This technique, adapted from optical telescope technology, eliminates the need for mounting a heavy feed radiator far in front of the main reflector.

cassette A two-reel magnetic-tape cartridge designed for easy insertion into a tape recorder or player, without threading of tape. Also called tape cassette.

cassette memory A removable magnetic-tape cassette that stores computer programs and data. The cassette is generally inserted manually into the computer as needed.

cassette player A magnetic-tape player designed for playback of prerecorded cassettes.

cassette recorder A magnetic-tape recorder designed for recording and playback of cassettes.

CAT Abbreviation for *clear-air turbulence.*

catalyst A material or condition that starts or speeds up a chemical or other reaction, such as the hardening of an epoxy cement.

cataphoresis *Electrophoresis.*

catastrophic failure The sudden and complete failure of a component or piece of equipment.

catcher *Output resonator.*

catcher resonator *Output resonator.*

catcher space The space between the output resonator grids in a velocity-modulated tube, where the density-modulated electron beam excites oscillations in the output resonator.

catching diode A diode connected to act as a short-circuit when its anode becomes positive. The diode then prevents the voltage of a circuit terminal from rising above the diode cathode voltage.

catelectrotonus Increased excitability of a nerve or muscle that is near a cathode during passage of direct current through living tissue. See also anelectrotonus.

cathode 1. [symbol K] The primary source of electrons in an electron tube. In directly heated tubes the filament is the cathode. In indirectly heated tubes a coated metal cathode surrounds a heater. Other types of cathodes emit electrons under the influence of light or high voltage. 2. The negative electrode of a battery or other electrochemical device. 3. The terminal of a semiconductor diode that is negative with respect to the other terminal when the diode is biased in the forward direction.

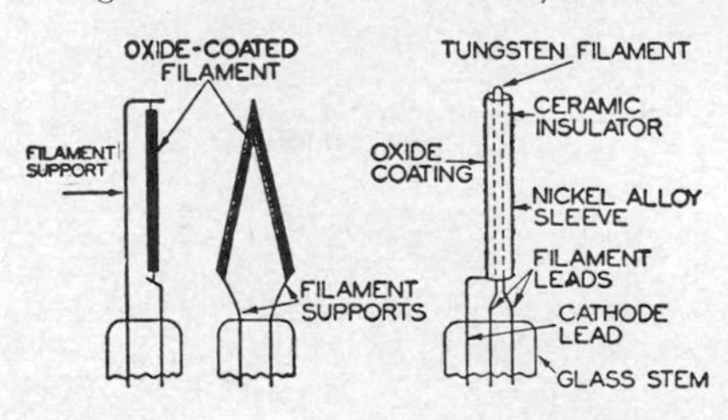

Cathodes for electron tubes. At left are two directly heated types, and at right is heater-type cathode.

cathode bias Bias obtained by placing a resistor in the common cathode return circuit, between cathode and ground. Flow of electrode currents through this resistor produces a voltage drop that makes the control grid negative with respect to the cathode of the vacuum tube.

cathode-coupled amplifier A cascade amplifier in which the coupling between two stages is provided by a common cathode resistor.

cathode dark space The relatively nonluminous region between the cathode and negative glows in a glow-discharge cold-cathode tube. Also called Crookes dark space.

cathode disintegration The destruction of the active area of a cathode by positive-ion bombardment.

cathode follower A vacuum-tube circuit in which the input signal is applied between the control grid and ground, and the load is connected between the cathode and ground. A cathode follower has a low output impedance, high input impedance, and a gain of less than unity. The anode is at ground potential at the operating frequency.

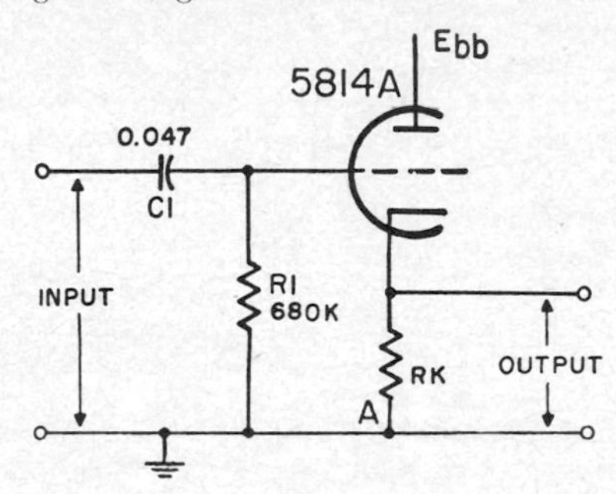

Cathode follower for pulses, using circuit design preferred by National Bureau of Standards. Value of RK is 10 kΩ when anode voltage is 150 V.

cathode glow The luminous glow that covers all or part of the cathode in a glow-discharge cold-cathode tube.

cathode keying Transmitter keying by a key in the cathode lead of the keyed vacuum-tube stage, opening the DC circuits for the grid and anode simultaneously.

cathode luminous sensitivity The photoelectric emission current divided by the luminous flux on a photocathode under specified conditions of illumination.

cathode modulation Amplitude modulation accomplished by applying the modulating voltage to the cathode circuit of an electron tube in which the carrier is present.

cathode poisoning The chemical effect of residual gases on the emissivity of the cathode of an electron tube.

cathode preheating time The minimum period of time during which heater voltage should be applied before electrode voltages are applied in an electron tube.

cathode pulse modulation Modulation produced in an amplifier or oscillator by applying externally generated pulses to the cathode circuit.

cathode radiant sensitivity The photoelectric emission current divided by the radiant flux on a photocathode at a given wavelength, under specified conditions of irradiation.

cathode ray A stream of electrons, such as that emitted by a heated filament in a tube or emitted by the cathode of a gas-discharge tube when the cathode is bombarded by positive ions.

cathode-ray charge-storage tube A charge-storage tube in which the information is written by a cathode-ray beam.

cathode-ray oscillograph A cathode-ray oscilloscope in which a photographic or other permanent record is produced by the electron beam of the cathode-ray tube.

cathode-ray oscilloscope [abbreviated CRO] A test instrument that uses a cathode-ray tube to make visible on a fluorescent screen the instantaneous values and waveforms of electrical quantities which are rapidly varying as a function of time or another quantity. Also called oscilloscope and scope.

cathode-ray storage tube A storage tube in which the information is written by a cathode-ray beam.

cathode-ray terminal *Cathode-ray-tube terminal.*

cathode-ray tube [abbreviated CRT] An electron-beam tube in which the electrons emitted by a hot cathode are formed by an electron gun into a narrow beam that can be focused to a small cross section on a fluorescent screen. The beam can be varied in position and intensity by internal electrostatic deflection plates or external electromagnetic deflection coils to produce a visible trace, pattern, or picture on the screen. Originally called Braun tube. Also called kinescope and picture tube when used in television receivers.

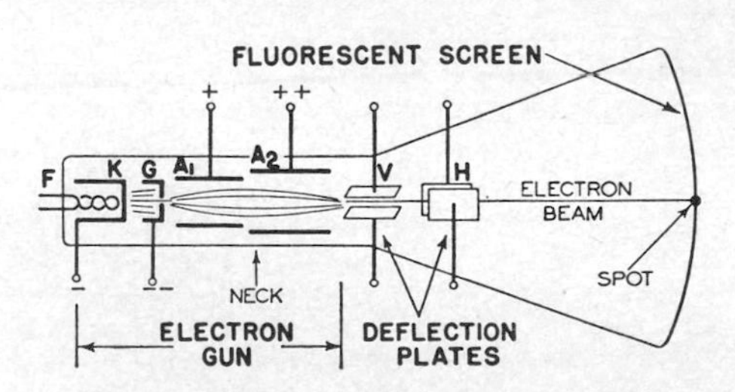

F--FILAMENT--Pins 1 & 11
K--CATHODE--Pin 11
G--CONTROL GRID--Pin 10
A_1--FIRST ANODE--Pin 4
A_2--SECOND ANODE--Pin 7
V--VERT. DEFL. PLATES--Pins 6 & 9
H--HORIZ. DEFL. PLATES--Pins 3 & 8

Cathode-ray tube having electrostatic deflection.

cathode-ray-tube display The presentation of a received signal on the screen of a cathode-ray tube.

cathode-ray-tube terminal [abbreviated CRT terminal] A computer terminal, usually interactive, in which data and queries that are fed into the computer by means of typewriterlike keys appear simultaneously in alphameric characters on the screen of a cathode-ray tube mounted above the keyboard, along with information requested from the computer and sometimes graphic displays stored or generated by the computer. Also called cathode-ray terminal, display terminal, and video terminal.

cathode-ray tuning indicator A small cathode-ray tube that has a fluorescent pattern whose size varies with the voltage applied to the grid. Used in radio receivers to indicate accuracy of tuning. Also used as a modulation indicator in some tape recorders and in place of a meter in some instruments.

cathode resistor A resistor in the cathode circuit of a vacuum tube; it has a resistance value such that the voltage drop across it due to tube current provides the correct negative grid bias for the tube.

cathode spot The small cathode area from which an arc appears to originate in a discharge tube.

cathode sputtering *Sputtering.*

cathodochromic Capable of being colored by cathode rays. Materials having this characteristic include potassium chloride and sodalite. Used in screens for dark-trace cathode-ray tubes. The coloration can be bleached out with light or heat.

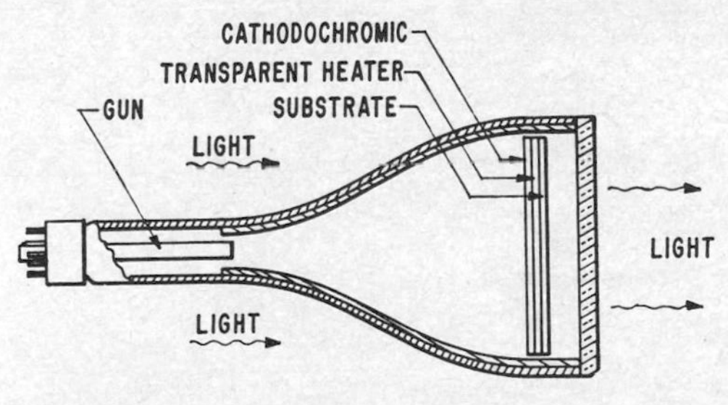

Cathodochromic cathode-ray tube using sodalite screen and heater for erasing colored image.

cathodochromic cathode-ray tube A cathode-ray tube that has a bromine sodalite or other cathodochromic screen on which information is written with an electron beam which changes the screen color from white to purple. The image is retained indefinitely without further power. It can be partially erased by visible light and completely erased by heating the screen to about 300°C with the electron beam.

cathodoluminescence Luminescence produced by high-velocity electrons. When these electrons bombard a metal in a vacuum, small amounts of the metal are vaporized in an excited state and emit radiation characteristic of the metal.

cathodophosphorescence A phosphorescence produced when high-velocity electrons bombard a metal in a vacuum.

cation A positive ion.

CATT [Controlled Avalanche Transit-time Triode] A solid-state microwave device that uses a combination of IMPATT diode and NPN bipolar transistor technologies. Carefully designed avalanche and drift zones are located between the base and collector regions.

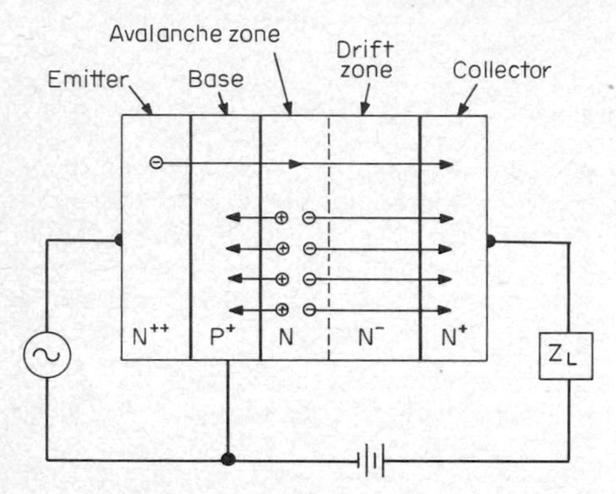

CATT construction, showing avalanche zone next to base and drift zone near collector.

CATV Abbreviation for *cable television.*

catwhisker A sharply pointed flexible wire that makes contact with the surface of a semiconductor crystal at a point which provides rectification. Also called whisker.

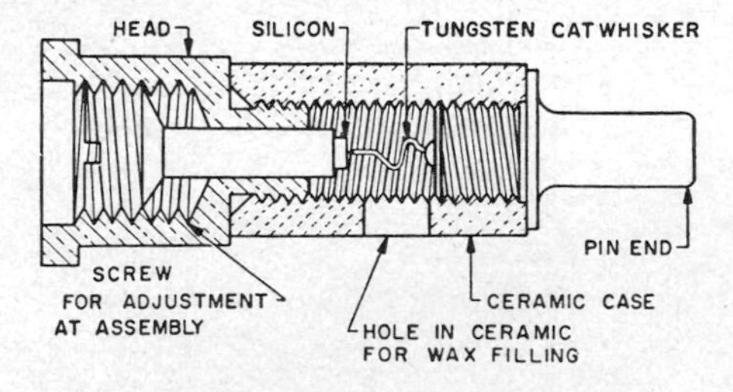

Catwhisker in microwave crystal diode.

cautery A medical electronic instrument that burns, cuts, or scars skin and tissue by precisely controlled electric current.

cave A heavily shielded compartment in which highly radioactive material can be handled, generally by remote control.

cavitation The formation of local cavities in a liquid when pressure is reduced because of movement of a body through the fluid. These bubbles are beneficial in ultrasonic cleaning.

cavity *Cavity resonator.*

cavity coupling A method of introducing or removing energy from a resonant cavity. Wire probes and loops are commonly used with coaxial lines; aperture or slot coupling is used with waveguides.

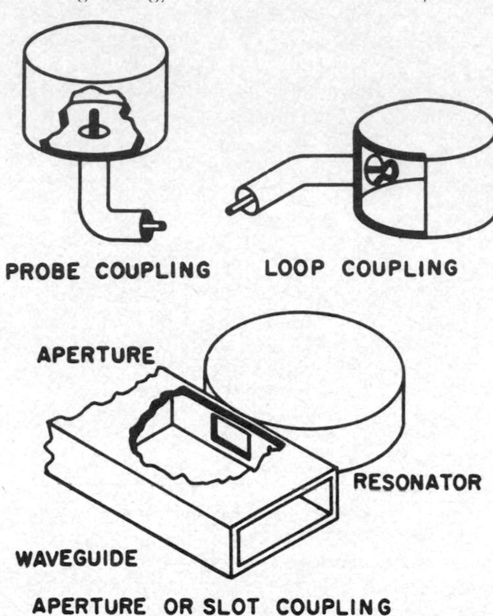

Cavity coupling methods.

cavity filter A microwave filter that uses quarter-wavelength-coupled cavities inserted in waveguides or coaxial lines to provide bandpass or other response characteristics at frequencies up into the gigahertz range.

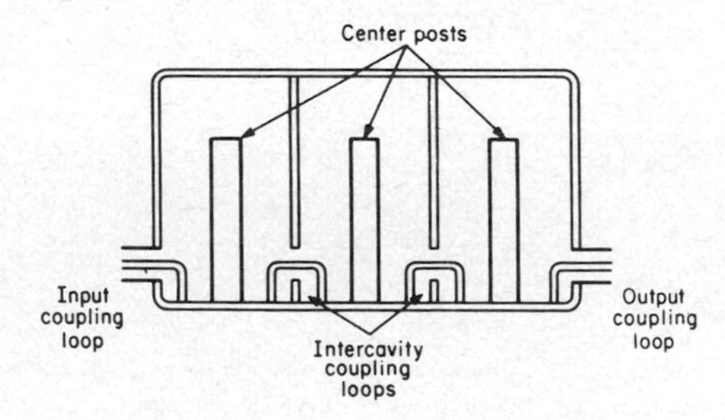

Cavity filter designed for use in coaxial line.

cavity frequency meter *Cavity-resonator frequency meter.*

cavity magnetron A magnetron that has a number of resonant cavities which form the anode, for use as a microwave oscillator. The cavities are usually radial slots in a circular anode, with the openings facing the cathode.

cavity resonance 1. The natural resonant frequency of a loudspeaker baffle. If in the audio range, it is evident as unpleasant emphasis of sounds at that frequency. 2. The resonant frequency of a cavity resonator.

cavity resonator A space totally enclosed by a metallic conductor and excited in such a way that it becomes a source of electromagnetic oscillations. The size and shape of the enclosure determine the

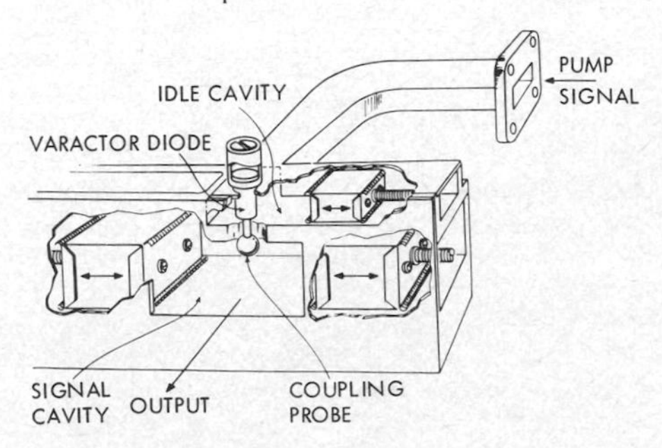

Cavity resonators as used in parametric amplifier.

resonant frequency. Used with klystrons, magnetrons, and other microwave devices. Also called cavity, resonant cavity, resonant chamber, resonant element, rhumbatron, tuned cavity, and waveguide resonator.

cavity-resonator frequency meter A variable cavity resonator that determines the frequency of

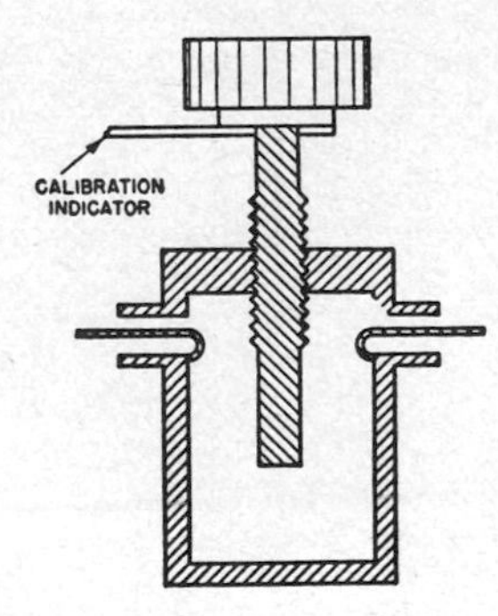

Cavity-resonator frequency meter for use with coaxial cable.

an electromagnetic wave. Also called cavity frequency meter and cavity-resonator wavemeter.

cavity-resonator wavemeter *Cavity-resonator frequency meter.*

CB Abbreviation for *citizens band.*

C band A band of radio frequencies extending from 3.9 to 6.2 GHz, corresponding to wavelengths of 7.69 to 4.84 cm.

C-band waveguide A rectangular waveguide, 3.48 by 1.58 cm, used in the dominant mode for 3.7- to 5.1-cm wavelengths.

CCD Abbreviation for *charge-coupled device.*

CCIR Abbreviation for *International Radio Consultative Committee.*

CCITT Abbreviation for *International Telegraph and Telephone Consultative Committee.*

CCM Abbreviation for *counter-countermeasure.*

C core A spirally wound magnetic core that is formed to a desired rectangular shape before it is cut into two C-shaped pieces and placed around a transformer or magnetic amplifier coil.

CCSL Abbreviation for *compatible current-sinking logic.*

CCTV Abbreviation for *closed-circuit television.*

C^3L Abbreviation for *complementary constant-current logic.*

cd Abbreviation for *candela.*

CD-4 sound Abbreviation for *compatible discrete four-channel sound.*

C display A rectangular radarscope display in which targets appear as bright spots, with target

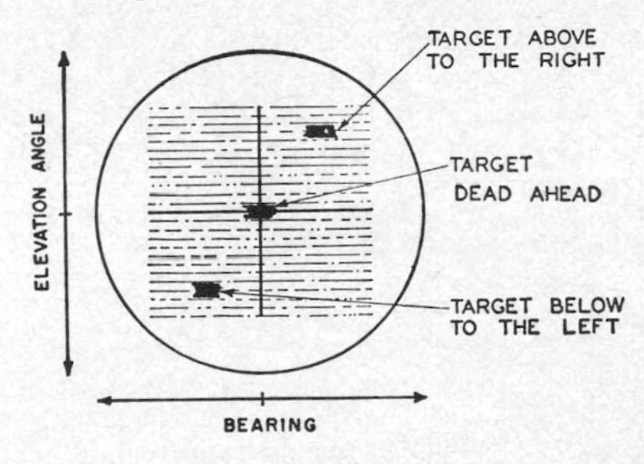

C display.

bearing indicated by the horizontal coordinate and target angle of elevation by the vertical coordinate.

CDM Abbreviation for *code-division multiplex.*

CDMA Abbreviation for *code-division multiple access.*

cd/m² Abbreviation for *candela per square meter.*

CdS Symbol for *cadmium sulfide.*

ceiling-height indicator A photoelectric instrument for measuring the height of a cloud ceiling with the aid of a vertical beam of light. Also called ceilometer and cloud-height detector.

ceilometer *Ceiling-height indicator.*

celestial guidance Guidance of a long-range missile by reference to celestial bodies. The missile is equipped with gyroscopes, optical or radio star trackers, servos, computers, and other devices that together sight stars, calculate positions, and direct the missile. Also called stellar guidance.

celestial-inertial guidance An inertial guidance system into which supplementary position and/or velocity information is fed by celestial navigation equipment that is also built into the missile.

celestial navigation Navigation by observations of celestial bodies.

celestial radio tracking A navigation technique in which the microwave emanations of the sun, moon, or certain stars are used to determine the position of a missile or other object in space.

celestial sphere An imaginary sphere of infinite radius, the center of which coincides with the center of the earth. Assumed for space navigation purposes.

cell 1. A single unit of a primary or secondary battery that converts chemical energy into electric energy. Also called electric cell. 2. A single unit of a device that converts radiant energy into electric energy, such as a nuclear cell, solar cell, or photovoltaic cell. 3. A single unit of a device whose resistance varies with radiant energy, such as a selenium cell. 4. An elementary unit of storage in a computer, such as a binary cell or decimal cell.

cellular horn *Multicellular horn.*

cellulose acetate A thermoplastic material that is widely used as the base for magnetic tape and movie film. It can be made transparent or opaque in various colors. It is tough, flexible, slow-burning, and long-lasting. Used for molding receiver cabinets. Also called acetate.

cellulose nitrate disk *Lacquer disk.*

Celsius [abbreviated C; after Anders Celsius] *Centigrade.* Actually, the two scales differ slightly because the Celsius scale uses the triple point of water, at 0.01° centigrade, in place of the ice point as a reference because of its greater convenience in theoretical computations. Although the term centigrade was officially changed to Celsius at an international conference in 1948, public acceptance has been slow in Europe and still slower in the United States. The centigrade scale was devised by Celsius.

center-feed tape Perforated paper tape that has feed holes in line with the centers of the intelligence holes.

center frequency *Carrier frequency.*

center-frequency stability The ability of a transmitter to maintain an assigned center frequency in the absence of modulation.

centering The process of adjusting the position of the trace or image on a cathode-ray-tube screen so it is centered on the screen.

centering control One of the two controls used for positioning the image on the screen of a cathode-ray tube. The horizontal centering control moves the image horizontally, and the vertical centering control moves the image vertically. Centering is achieved by adjusting the DC voltage applied to deflection plates or by adjusting the direct current flowing through deflection coils.

center line The locus of points equidistant from two reference points or lines. In loran the center line is the perpendicular bisector of the baseline connecting the master and slave stations.

center-of-mass system A frame of reference for nuclear collisions, in which the center of mass of the two colliding objects is at rest since the total momentum of the particles in this system is always zero.

center tap [symbol CT] A terminal at the electrical midpoint of a resistor, coil, or other device.

center-tap keying A method of keying a radiotelegraph transmitter by interrupting the anode return lead going to the filament transformer center tap. Used when modulating a filament-type tube.

centi- [abbreviated c] A prefix representing 10^{-2}, which is 0.01 or one-hundredth.

centigrade [abbreviated C] The metric temperature scale, in which the interval between the freezing and boiling points of water is divided into 100 equal parts or degrees, with 0°C as the freezing point and 100°C as the boiling point. Absolute zero is −273.16°C. Also called Celsius, although the Celsius scale uses 0.01°C as the cold reference. To change degrees centigrade to degrees Fahrenheit, multiply by 9/5 and add 32 to the result. Fahrenheit and centigrade scales agree at one point: −40°F = −40°C.

centimeter [abbreviated cm] A unit of length in the metric system, equal to 0.01 m or 0.394 in.

centimeter-gram-second unit [abbreviated CGS unit] An absolute unit based on the centimeter, gram, and second as fundamental units.

centimetric wave A radio wave between the wavelength limits of 1 and 10 cm, corresponding to the superhigh-frequency (SHF) range of 3 to 30 GHz.

central force A nuclear force that is a simple attraction or repulsion directed along the line joining a pair of nucleons.

central potential A nuclear potential that is spherically symmetric, so the potential energy of a particle is the same in all directions and is a function only of the distance from the center of the field.

central processing unit [abbreviated CPU] The section of a computer that contains the arithmetic, logic, and control circuits. In some systems it may also include the storage unit and the operator's console. Also called mainframe.

centrifugal separation Separation of isotopes by spinning a mixture in gas or vapor form at high speed. Heavier isotopes then concentrate near the outside of the container, and lighter isotopes concentrate near the axis of the centrifuge.

centrifugal switch A switch that is opened or closed by centrifugal force. Used on some induction motors to open the starting winding when the motor has almost reached synchronous speed.

centrifuge A large motor-driven apparatus with a long arm, at the end of which human and animal subjects or equipment can be revolved and rotated at various speeds to simulate the prolonged accelerations encountered in rockets and spacecraft.

cepstrum The Fourier cosine transform of the log-power spectrum. Also called pseudo-autovariance.

ceramet *Metal-ceramic.*

ceramic amplifier An amplifier that utilizes the

piezoelectric properties of ceramics like barium titanate in combination with the piezoresistive properties of semiconductors like silicon. An AC signal applied to electrodes on a barium titanate bar produces motion of the bar, thereby deforming the attached silicon strip and producing a corresponding variation in resistance. This resistance change varies the load current. The device is essentially a current amplifier that has extremely high input impedance and low output impedance.

ceramic capacitor A capacitor whose dielectric is a ceramic material such as steatite or barium titanate, the composition of which can be varied to give a wide range of temperature coefficients. The

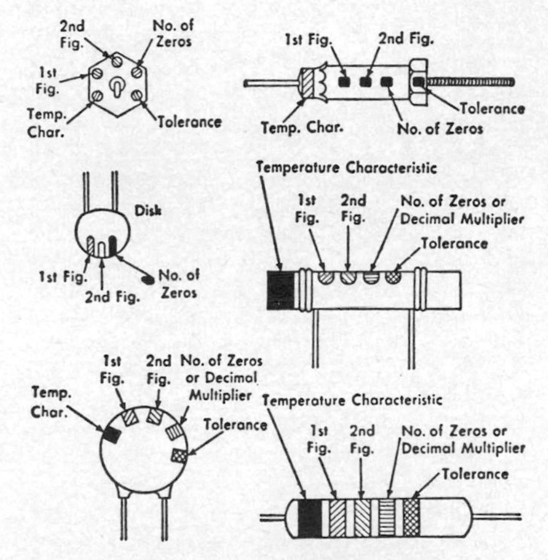

Ceramic capacitor types, and methods of applying color codes to indicate capacitance values in picofarads and other significant characteristics.

electrodes are usually silver coatings fired on opposite sides of the ceramic disk or slab or on the inside and outside of a ceramic tube. After connecting leads are soldered to the electrodes, the unit is usually given a protective insulating coating.

ceramic cartridge A device that contains a piezoelectric ceramic element, used in phonograph pickups and microphones. Ceramic cartridges deliver somewhat lower output voltage than crystal cartridges but are less affected by heat and humidity.

ceramic chip capacitor A ceramic capacitor constructed as a chip that has gold, silver, or other deposited metal terminations which facilitate automatic assembly in hybrid integrated circuits.

ceramic ladder filter A ladder filter that consists of a number of piezoelectric ceramic elements which are coupled electrically in a ladder network. The number of elements used determines the width of the bandpass response.

ceramic magnet A permanent magnet made from pressed and sintered mixtures of ceramic and magnetic powders.

ceramic microphone A microphone that uses a ceramic cartridge.

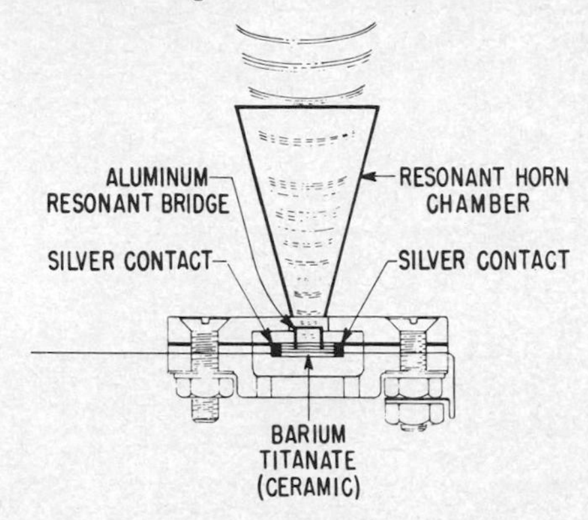

Ceramic microphone for use with 40-kHz ultrasonic remote control.

ceramic pickup A phonograph pickup that uses a ceramic cartridge.

ceramic tube An electron tube that has a ceramic envelope capable of withstanding operating

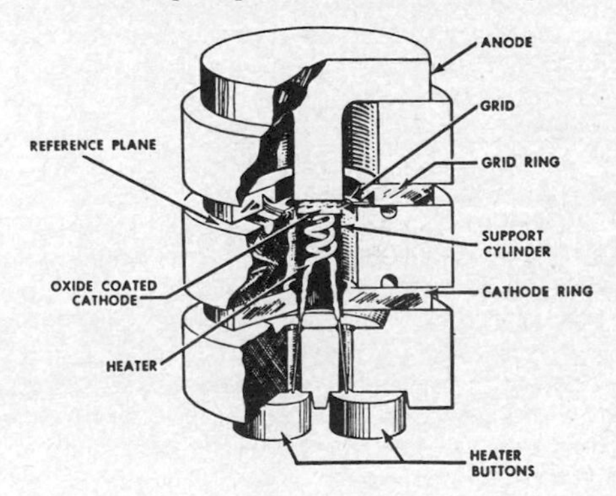

Ceramic-tube construction.

temperatures of over 500°C, as required to withstand reentry temperatures of guided missiles.

ceramoplastic A high-temperature insulating material made by bonding synthetic mica with glass.

Cerenkov counter An apparatus that detects high-energy charged particles, such as those present in cosmic rays, by observation of the Cerenkov radiation produced.

Cerenkov radiation Visible light produced when charged particles traverse a transparent medium with a velocity exceeding the velocity of light in the medium. The index of refraction of the medium must be considerably greater than unity.

cerium [symbol Ce] A rare-earth metallic element. Atomic number is 58.

cermet A combination of powdered precious-metal alloys and an inorganic material like alumina. Used in manufacturing resistors, capacitors, and other components for high-temperature applications.

cermet resistor A metal-glaze resistor that consists of a mixture of finely powdered precious metals and insulating materials fired onto a ceramic substrate.

cesium [symbol Cs] An alkali metallic element used in forming the cathodes of certain types of phototubes. Atomic number is 55. Cesium 137 is a silvery metal found as a fission product or residue of uranium 235 in an atomic pile. It has a half-life of 30 years, emits gamma rays, and is used for cancer therapy.

cesium atomic-beam resonator A device in which atoms evaporated from liquid cesium are formed into a beam that is acted on by a magnetic field, then passed into a microwave cavity where further magnetic interaction takes place. One application is frequency control of a microwave oscillator.

cesium-beam frequency standard A frequency standard that uses a precision quartz oscillator in combination with a frequency synthesizer and multiplier stages to generate standard frequencies such as 1 and 5 MHz with an accuracy of the order of 1 part in 10^{11}, by continuous comparison of the output with the 9192.631770-MHz output of a cesium atomic-beam resonator.

cesium clock An atomic clock regulated by the natural vibration frequency of atoms in a cesium atomic-beam resonator.

cesium hollow cathode A cathode in which cesium is heated at the bottom of a cylinder that serves as the cathode of an electron tube, to give current densities which can be as high as 800 A/cm^2.

cesium-ion engine An ion engine that uses a stream of cesium ions to produce a thrust for space travel.

cesium magnetometer A magnetometer that uses a cesium atomic-beam resonator as a frequency standard in a circuit that detects very small variations in magnetic fields. Applications include location of buried explosive mines, location of archeological artifacts containing traces of magnetic materials, and monitoring of magnetic fields from aircraft, spacecraft, and satellites.

cesium phototube A phototube that has a cesium-coated cathode. It has maximum sensitivity in the infrared portion of the spectrum.

cesium-vapor lamp A lamp in which light is produced by the passage of current between two electrodes in ionized cesium vapor.

cesium-vapor rectifier A gas tube in which cesium vapor serves as the conducting gas, and a condensed monatomic layer of cesium serves as the cathode coating. The tube is heated to about 180°C to give the desired vapor pressure.

CGS unit Abbreviation for *centimeter-gram-second unit.*

chad The small piece of paper removed during punching of a hole in a card or paper tape.

chadless paper tape Paper tape in which each chad is left fastened by about a quarter of the circumference of the hole, at the leading edge. Commonly used in teletypewriters, to avoid accumulation of chad. Chadless punched paper tape must be sensed by mechanical fingers because the presence of chad interferes with electrical or photoelectric reading. Corresponding letters and numerals can be printed on the tape.

chad tape Paper tape in which perforations for code characters are completely punched.

chaff A confusion reflector that consists of narrow metallic or metal-coated paper strips cut to various lengths. The strips are resonant at expected enemy radar frequencies. When dropped

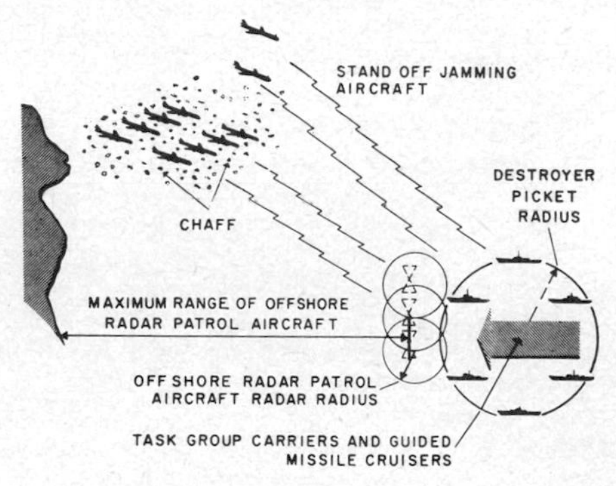

Chaff dropped by aircraft attacking destroyer-protected aircraft carriers multiplies targets, confusing antiaircraft gunners.

in clusters, chaff gives strong echo signals on enemy radarscopes. Passive chaff reflects radar signals without providing gain. Semiactive chaff derives power from the radar signal to modulate the basic skin return without providing gain. Active chaff is an expendable battery-powered jammer.

chaff rocket A rocket fired from ship or land and exploded in air to disperse enough chaff to produce a large radar return. Used to confuse enemy radar-guided missiles attacking ship targets.

chain A network of radio, television, radar, navigation, or other similar stations connected by special telephone lines, coaxial cables, or radio relay links so all can operate as a group for broadcast or communication purposes or determination of position.

chain code A cyclic sequence of n-bit words in which each word is derived from its neighbor by displacing the bits one digit position to the left or right, dropping the leading bit, and inserting a bit at the end. There is no repetition within the cyclic sequence. For 3-bit words, the example is 000 001 010 101 011 111 110 100 000.

chained list A randomly arranged computer list in which each item contains an identifier that locates the next item to be considered.

chain printer An impact printer in which the type slugs for one or more complete sets of characters are carried by the links of a chain, track, or other endless-loop arrangement for moving the characters horizontally along a line. A desired character is printed on the fly by hammer impact as it passes each print position requiring that character.

chain radar beacon A beacon with a fast recovery time, to permit simultaneous interrogation and tracking of the beacon by a number of radars.

chain radar system A number of radar stations located at various sites on a missile range to enable complete radar coverage during a missile flight. The stations are linked by data and communication lines for target acquisition, target positioning, and/or data-recording purposes.

chain reaction A reaction in which one of the agents necessary to the reaction is itself produced by the reaction, thereby causing additional reactions. In the neutron-fission chain reaction, a neutron plus a fissionable atom cause a fission, resulting in a number of neutrons that in turn cause other fissions.

chance coincidence *Accidental coincidence.*

chance failure Failure within the operational time period of a piece of equipment after all efforts have been made to eliminate design defects and unsound components, but before any foreseen wearout phenomena have time to appear.

channel 1. A band of radio frequencies allocated for a particular purpose. A standard broadcasting channel is 10 kHz wide, whereas a television channel is 6 MHz wide. 2. A path for a signal. An audio amplifier may have several input channels. A stereo amplifier has at least two complete channels. 3. A path for carrier-current signals. 4. A path for information flow in a computer. 5. The section of a storage medium that is accessible to a given reading station in a computer, such as a path parallel to the edge of a magnetic tape or drum or a path in a delay-line memory. 6. The main current path between the source and drain electrodes in a field-effect transistor or other semiconductor device.

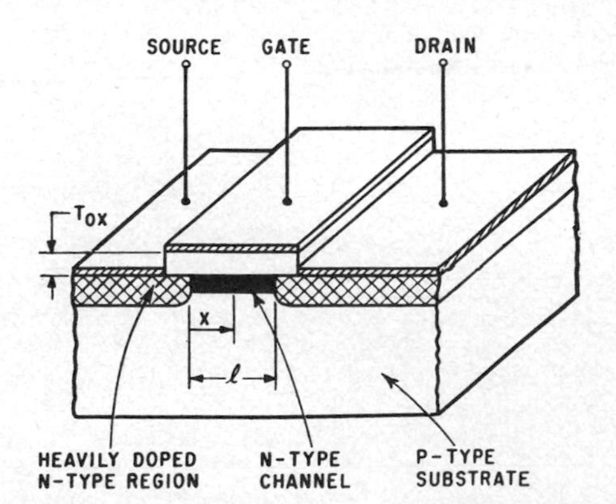

Channel location in insulated-gate MOS field-effect transistor.

channel breakdown Avalanche breakdown of a transistor channel.

channel capacity The maximum number of bits or other information elements that can be handled in a particular channel per unit time.

channel effect A leakage current that flows over a surface path between the collector and emitter in some types of transistors.

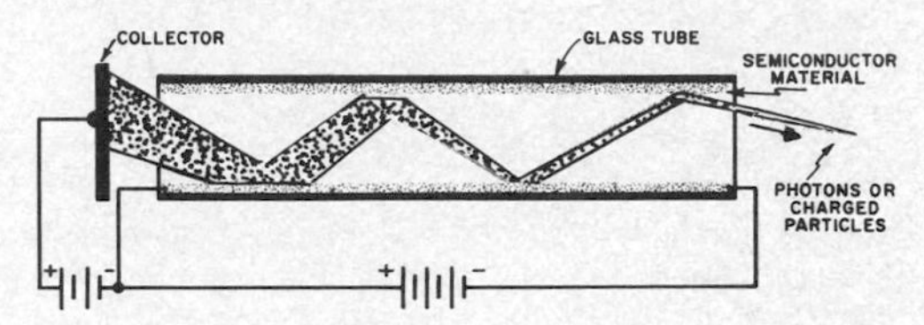

Channel electron multiplier used as radiation detector in spacecraft.

channel electron multiplier An electron multiplier in which the entire inner surface of the tubular envelope is coated with material that emits secondary electrons.

channelizing The process of subdividing a wideband transmission facility to handle a number of different circuits requiring comparatively narrow bandwidths.

channel selector A switch or other control that tunes in the desired channel in a television receiver.

channel separation The electric or acoustic difference between the left and right channels in a stereo system.

channel shifter A radiotelephone carrier circuit that shifts one or two voice-frequency channels from normal channels to higher voice-frequency channels to reduce crosstalk between channels. The channels are shifted back by a similar circuit at the receiving end.

channel stopper A ring of opposite-polarity semiconductor material diffused around each transistor in a multiple-transistor integrated circuit to provide the electrical isolation required to prevent formation of parasitic devices in the field between transistors.

channel strip An amplifier that has sufficient bandpass for one television channel. Used in cable television systems and fringe-area home locations to improve reception of a single desired station.

character [abbreviated c] A letter, digit, elementary mark, or event which may be used in various combinations to express information that a computer can read, store, or write. A group of characters in one context may become a single character in another, as in the binary-coded decimal system.

character density The number of characters recorded per unit of length or area. Character densities for magnetic tape range from 200 to over 1000 characters per linear inch (80 to over 400 characters per linear centimeter). Also called record density.

character generator A generator that creates letters, numerals, and symbols on a cathode-ray screen or other viewing surface in a desired sequence. There may also be facilities for recording the information on photographic negatives or paper.

characteristic 1. A measurable property of a device. 2. The integral part of a logarithmic value, at the left of the decimal point.

characteristic curve A curve plotted on graph paper to show the relation between two changing values.

characteristic impedance The impedance that, when connected to the output terminals of a transmission line of any length, makes the line appear to be infinitely long. There are then no standing waves on, the line, and the ratio of voltage to current is the same for each point on the line. For a waveguide, the characteristic impedance is the ratio of RMS voltage to total RMS longitudinal current at specified points on a diameter when the guide is match-terminated. For an acoustic device it is the ratio of the effective sound pressure at a point to the effective particle velocity at that point. Also called surge impedance.

characteristic radiation Radiation that originates in an atom following removal of an electron. The wavelength of the emitted radiation depends only on the element concerned and the energy levels involved.

characteristic telegraph distortion Distortion that does not affect all signal pulses alike. The effect on each transition depends on the signal previously sent, owing to remnants of previous transitions or transients that persist for one or more pulse lengths.

characteristic x-rays Electromagnetic radiation emitted as a result of rearrangements of the electrons in the inner shells of atoms. The spectrum of the radiation consists of lines that are characteristic of the element in which the x-rays are produced. The target of an x-ray tube will in general emit both continuous x-rays and characteristic x-rays.

character per inch [abbreviated c/in] A unit that specifies the number of characters which are printed or otherwise produced in 1 in.

character printer *Serial printer.*

character reader A device that scans printed or handwritten characters and delivers corresponding machine-readable code characters which can be fed to a computer or other data-processing machine. The two most common commercial versions are optical and magnetic character readers.

character recognition The technology of using a machine to sense and encode into a machine language characters that are written or printed to be read by human beings.

character set A list of characters acceptable for coding to a specific computer or input/output device.

character-writing storage tube A character-writing tube that retains its display as long as the necessary operating voltages are supplied, but can be erased by lowering one electrode voltage momentarily.

character-writing tube A cathode-ray tube that forms alphameric and symbolic characters on its screen for viewing or recording purposes.

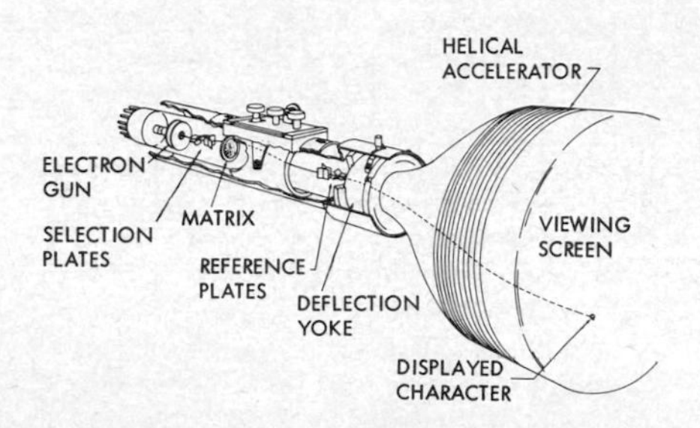

Charactron shaped-beam cathode-ray tube.

Charactron Trademark for one manufacturer's cathode-ray tube that produces a display in the form of letters or numbers.

charge 1. The quantity of electric energy stored in a capacitor, battery, elementary particle, or insulated object. Also called electric charge. 2. The material or part to be heated by induction or dielectric heating. 3. The conversion of electric energy to chemical energy in a storage battery by sending direct current through the battery in the opposite direction to that of discharge current.

charge amplifier An amplifier that converts capacitance changes of a transducer to corresponding changes in output voltage, as required for capacitor microphones and other capacitive transducers. Operational amplifiers, frequently used for this purpose, do this by converting changes in capacitor charge to changes in output voltage.

charge carrier A mobile conduction electron or mobile hole in a semiconductor.

charge-coupled device [abbreviated CCD] A semiconductor charge-transfer device that consists basically of a bottom semiconductor layer, a metal semiconductor-oxide insulation layer, and a top layer of metal electrodes. A negative voltage on an electrode creates a depletion region under the electrode in the bottom layer for storing minority carriers that represent information. When the negative voltage is shifted to an adjacent electrode,

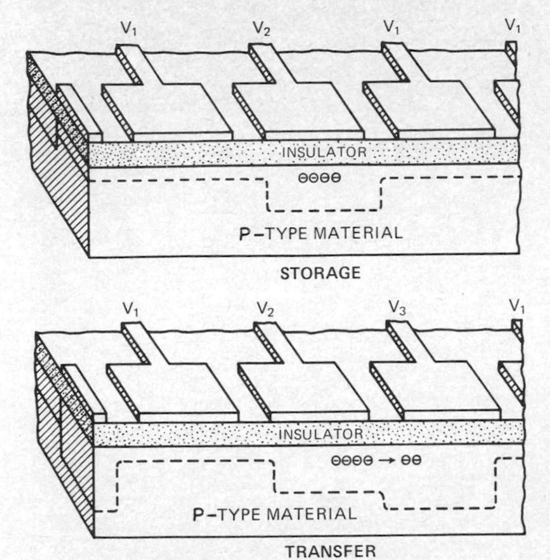

Charge-coupled device, showing how minority-carrier charge stored initially under electrode V_2 is transferred to electrode at right when voltage on V_3 is made greater than that on V_2.

the stored information moves correspondingly.

charge-coupled image sensor A charge-coupled device in which charges are introduced when light from a scene is focused on the surface of the device. The image points are accessed sequentially to produce a television-type output signal. Also called solid-state image sensor.

charge-coupled memory A computer memory that uses a large number of charge-coupled devices for storage and retrieval of data. Data can be read out nondestructively, except while the input terminal is enabled, in which case the new input bit replaces that previously stored.

charge density The charge per unit area on a surface or per unit volume in space.

charge-exchange phenomenon The phenomenon in which a positive ion that possesses sufficient kinetic energy is neutralized by colliding with a molecule and capturing an electron from it. The molecule is transformed into a positive ion.

charge-injection device [abbreviated CID] A charge-transfer device used as an image sensor in which the image points are accessed in an XY manner.

charge-mass ratio The ratio of the electric charge of a particle to its mass.

charger *Battery charger.*

charger-reader An auxiliary device used to charge and read small portable ionization chambers.

charge storage The buildup of carriers at the base of a semiconductor diode or bipolar transistor by charge-carrier flow, to give the concentration gradient required for an output current at the collector or back face.

charge-storage diode A semiconductor diode in which the turnoff time is substantially increased by charge storage.

charge-storage transistor A transistor in which the collector-base junction will charge when forward bias is applied with the base at a high level and the collector at a low level.

charge-storage tube A storage tube in which information is retained on a surface in the form of electric charges.

charge-storage varactor A varactor that uses semiconductor techniques to achieve power outputs above 50 W at UHF and microwave frequencies. Operation in the forward voltage region gives charge storage and step recovery leading to harmonic generation for frequency multiplier applications.

charge-transfer device A semiconductor device that depends upon movements of stored charges between predetermined locations, as in charge-coupled and charge-injection devices.

charging 1. The process of converting electric energy to chemical energy in a secondary battery. 2. The process of feeding electric energy to a capacitor or other device that can store electric energy.

charging current The current that flows into a capacitor when a voltage is first applied.

chart The paper or other material on which a graphic record is made by a recording instrument.

chart comparison unit A device that permits simultaneous viewing of a radar PPI display and a navigation chart in such a manner that one appears superimposed on the other.

chart recorder A recorder in which a dependent variable is plotted against an independent variable by an ink-filled pen moving on plain paper, a heated stylus on heat-sensitive paper, a light beam or electron beam on photosensitive paper, an electrode on electrosensitive paper, or other means. The plot may be linear or curvilinear on a strip-chart recorder, or polar on a circular chart recorder.

chassis The metal frame on which circuit components are mounted. The plural is the same as the singular.

chassis ground A connection to the metal chassis on which the components of a circuit are mounted, to serve as a common return path to the power source. The chassis may or may not be connected to an earth ground.

chassis punch A hand tool used to make round or square holes in sheet metal.

chatter 1. Prolonged undesirable opening and closing of electric contacts, as on a relay. Also called contact chatter. 2. Vibration of a disk recorder cutting stylus in a direction other than that in which it is driven.

cheater cord A special extension cord used to

apply AC power to a television or radio receiver when the back cover with its protective power interlock is removed for servicing.

Chebyshev Correct English transliteration of the surname of Russian mathematician Pafnuti Lvovich Chebyshev (1821–1894). Spelled Tschebyscheff in German, Tchebycheff in French, Cebisceff in Italian, and Czebyszew in Polish.

Chebyshev array An antenna array in which the elements are fed to produce an array factor that can be expressed in terms of a Chebyshev polynomial in which, for a given side-lobe level, the width of the main beam is minimized.

Chebyshev filter A constant-*k* filter that achieves sharp frequency cutoff at the expense of amplitude ripple in the passband.

check Partial or complete testing of the accuracy of a computer operation.

check bit *Check digit.*

check digit A redundant digit in a self-checking code for a computer, used to indicate a malfunction. Thus, a given check digit may be 0 if the sum of other digits in the word is odd, or 1 if the sum of other digits in the word is even. When a single error occurs in a word, the sum of the word digits no longer agrees with the check digit. Used in a parity check. Also called check bit.

checkout Testing and calibrating a piece of equipment to verify that it is ready for its intended use.

check point 1. A point in a computer routine at which processing can be halted for any reason, such as to obtain a printout of the status of processing. Also used as a program-restarting point if an error occurs later. 2. *Way point.*

check problem *Check routine.*

check routine A routine or problem designed primarily to indicate whether a fault exists in a computer, without giving detailed information on the location of the fault. Also called check problem, test program, and test routine.

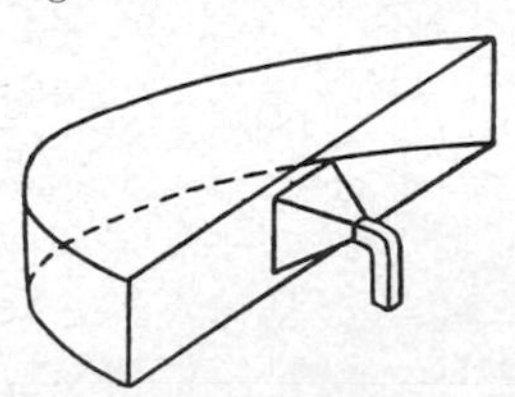

Cheese antenna fed by horn at end of rectangular waveguide.

cheese antenna An antenna that has a parabolic reflector between two metal plates, dimensioned to permit propagation of more than one mode in the desired direction of polarization. It is fed by either a dipole or flared-out waveguide facing into the reflector cavity.

chelate laser A liquid laser that uses a rare-earth chelate (a metallo-organic compound), with initial excitation taking place within the organic part of the liquid molecule and then transferring to the metallic ions to give lasing action. The normal wavelength range is 0.3 to 1.2 μm. Also called rare-earth chelate laser.

chemical binding effect The dependence of the neutron cross sections of a material on the chemical binding of the atoms composing the material.

chemical dosimeter A dosimeter in which the accumulated radiation exposure dose is indicated by color changes accompanying chemical reactions induced by the radiation.

chemical etching Removal of material from a semiconductor surface by applying a suitable acid etching solution.

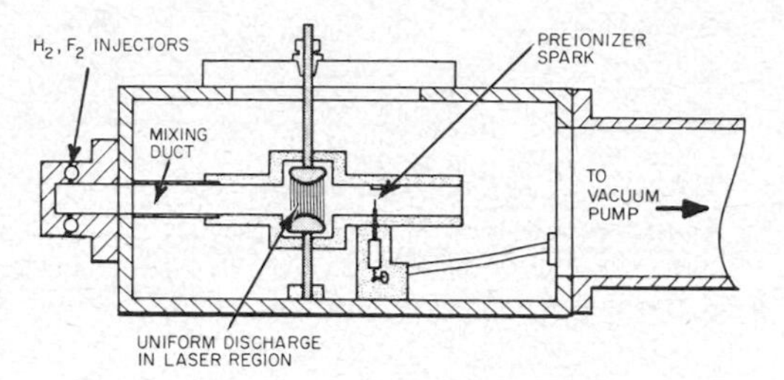

Chemical laser producing pulsed output, using spark-initiated discharge in injected mixture of hydrogen and fluorine gases.

chemical laser A gas laser in which the required population inversion for lasing is produced directly from a chemical reaction between gases like hydrogen and chlorine, deuterium and fluorine, or between either hydrogen fluoride or deuterium fluoride and carbon dioxide. The chemical reaction can be pure, with no external energy source, or it can be initiated by external energy inputs like electric discharges or light flashes. The normal wavelength range is 2 to 100 μm.

chemically deposited printed circuit A printed circuit formed on a base by the reaction of chemicals alone. Dielectric, magnetic, and conductive circuits can be applied.

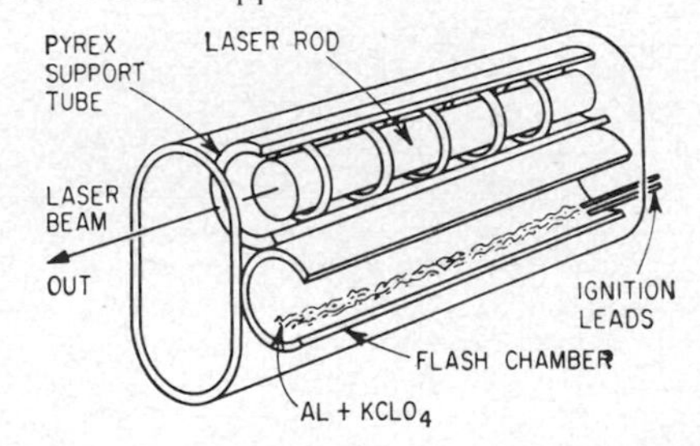

Chemically pumped laser. Chemical powders in flash chamber are ignited electrically to provide flash for pumping neodymium glass laser rod.

chemically pumped laser A laser in which pumping is achieved by using a chemical action rather than electric energy to produce the pulses of light required for pumping action. One method involves using the light given off by the chemical reaction. Another method involves using the shock wave of an explosion to generate light in a flash tube, which in turn triggers the laser. In the latter method, the shock wave from pyrotechnic squibs compresses argon gas, causing intense radiation that is rich in ultraviolet wavelengths.

chemical polishing Removal of several micrometers of semiconductor surface material with a suitable acid solution, usually after the semiconductor slice has been lapped and mechanically polished.

chemical tracer A tracer that has chemical properties similar to those of the substance with which it is mixed.

chemical vapor deposition Deposition of a thin film on a semiconductor or other surface by exposure to a vapor produced by a reaction between chemicals, either under pressure or in a vacuum.

chemiluminescence Luminescence produced by chemical action.

chemosphere The portion of the earth's atmosphere between altitudes of about 20 and 50 mi (30 and 80 km), in which photochemical activity is present. Here nitric oxide forms an ionized cloud that reflects radio signals.

Child's law An equation which states that the current in a thermionic diode varies directly with the three-halves power of anode voltage and inversely with the square of the distance between the electrodes, provided operating conditions are such that the current is limited only by the space charge.

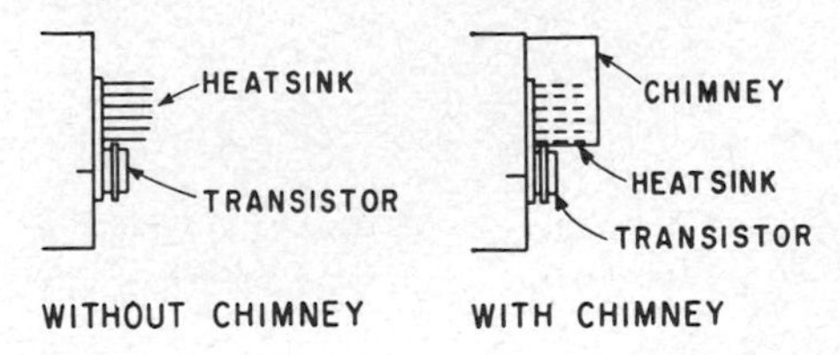

Chimney at right increases effectiveness of heatsink for transistor.

chimney A pipelike enclosure placed over a heatsink to improve natural upward convection of heat and thereby increase the dissipating ability of the sink.

chip 1. The material removed from the recording medium by the recording stylus while the groove in mechanical recording is cut. Also called thread. 2. The shaped and processed semiconductor die that is mounted on a substrate to form a transistor, diode, or other semiconductor device. 3. An integrated-circuit assembly that may contain all the circuits for a calculator, microcomputer, or microprocessor. It can have the equivalent of thousands of transistors and other components yet be no larger than a small postage stamp.

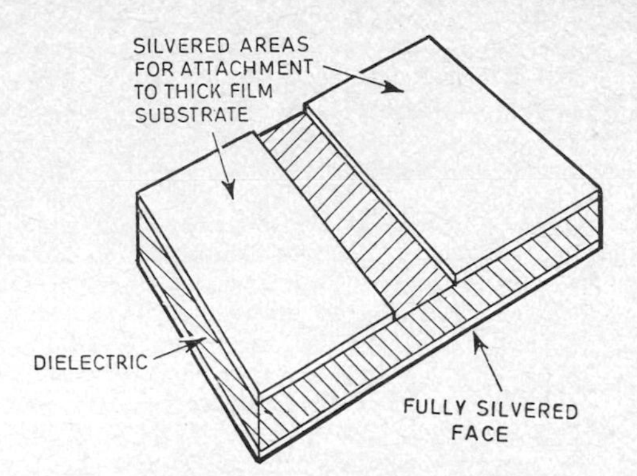

Chip capacitor using barrier-layer ceramic as dielectric.

chip capacitor A single-layer or multilayer monolithic capacitor constructed in chip form, with metallized terminations to facilitate direct bonding on hybrid integrated circuits.

chip resistor A thick-film resistor constructed in chip form, with metallized terminations to facilitate direct bonding on hybrid integrated circuits.

chirp 1. An undesirable variation in the frequency of a continuous-wave carrier when it is keyed. 2. The sound heard in a code receiver when the transmitted carrier frequency is increased linearly for the duration of a pulse code.

chirp filter A filter used for pulse compression in chirp radar, such as a surface-wave chirp filter.

chirp modulation Modulation in which the carrier signal is swept through the available band of frequencies repeatedly at a fixed rate. In the receiver, this wide pulse with high average power is compressed into a narrow pulse by delay-line techniques that enhance the desired signal while discriminating against unwanted signals. Used in radar and sonar to minimize the effects of multipath echoes, noise crosstalk, and frequency shifts. Also called swept-frequency modulation.

chirp radar Radar in which a swept-frequency signal is transmitted, received from a target, then compressed in time to give a final narrow pulse called the chirp signal. Compression is achieved with a network that introduces a delay proportional to frequency. Advantages include high immunity to jamming and inherent rejection of random noise signals.

chisel bond A thermocompression bond in which a contact wire is attached to a contact pad on a semiconductor chip by applying pressure with a chisel-shaped tool.

chlorine [symbol Cl] A gaseous element. Atomic number is 17.

choke 1. An inductance in a circuit to present a high impedance to frequencies above a specified frequency range without appreciably limiting the flow of direct current. Also called choke coil. 2. A groove or other discontinuity in a waveguide surface so shaped and dimensioned as to impede the passage of guided waves within a limited frequency range.

choke coil *Choke.*

choke coupling Coupling between two parts of a waveguide system that are not in direct mechanical contact with each other.

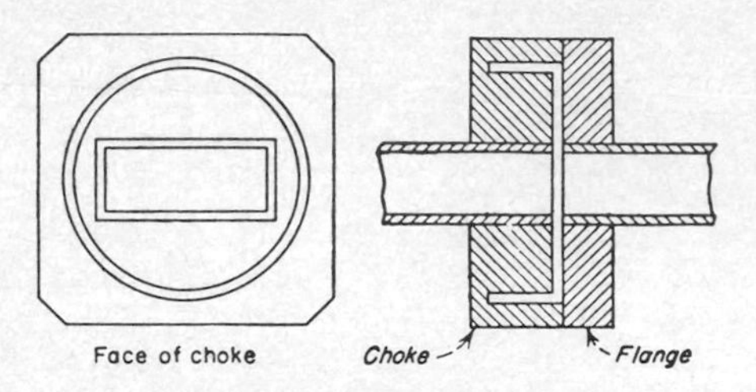

Choke-flange construction.

choke flange A waveguide flange that has in its mating surface a slot so shaped and dimensioned as to restrict leakage of microwave energy within a limited frequency range.

choke-input filter A power-supply filter in which the first filter element is a series choke.

choke joint A connection between two waveguides that uses two mating choke flanges to provide effective electric continuity without metallic continuity at the inner walls of the waveguide.

choke piston A piston in which there is no metallic contact with the walls of the waveguide at the edges of the reflecting surface. The short-circuit for high-frequency currents is achieved by a choke system. Also called noncontacting piston and noncontacting plunger.

cholesteric material A liquid crystal material in which the elongated molecules are parallel to each other within the plane of a layer, but the direction of orientation is twisted slightly from layer to layer to form a helix through the layers. Used in some types of liquid crystal displays.

chopper A device that interrupts a current or beam of light or infrared radiation at regular intervals, to permit amplification of the associated electrical quantity or signal by an AC amplifier. A chopper may also be a single-ended inverter that uses silicon controlled rectifiers for transforming DC to DC of a different voltage or DC to AC.

chopper amplifier A carrier amplifier in which the DC input is filtered by a low-pass filter, then converted into a square-wave AC signal by either one or two choppers. After amplification, the square-wave signal is synchronously rectified by chopper action to obtain the desired amplified DC output.

chopper-stabilized amplifier A direct-current amplifier in which a direct-coupled amplifier is in parallel with a chopper amplifier. The chopper amplifier provides increased stability against drift in the direct-coupled amplifier, particularly when negative feedback is used. Conversely, the direct-coupled amplifier extends the frequency range beyond that of the chopper amplifier.

chopper transistor A bipolar or field-effect transistor operated as a repetitive ON/OFF switch to produce square-wave modulation of an input signal.

chopping Removing peaks of a waveform at a predetermined amplitude level.

chorus A natural electromagnetic phenomenon in the VLF band, in which a number of different rising tones, resembling the dawn chorus of birds, in the range of 1 to 4 kHz are heard in VLF radio receivers. The phenomenon may last several hours, at any time of the day. Also called dawn chorus.

Christmas-tree pattern *Optical pattern.*

chroma The dimension of the Munsell system of color that corresponds most closely to saturation, which is the degree of vividness of a hue. Chroma is frequently used, particularly in English works, as the equivalent of saturation. Also called Munsell chroma.

chroma bandpass amplifier *Burst amplifier.*

chroma control The control that adjusts the amplitude of the carrier chrominance signal fed to the chrominance demodulators in a color television receiver, to change the saturation or vividness of the hues in the color picture. When in its zero position, the received picture becomes black and white. Also called color control and color-saturation control.

chroma oscillator A crystal oscillator used in color television receivers to generate a 3.579545-MHz signal for comparison with the incoming 3.579545-MHz chrominance-subcarrier signal being transmitted. Also called chrominance-subcarrier oscillator, color oscillator, and color-subcarrier oscillator.

chromatic Relating to color.

chromatic aberration 1. An optical lens defect that causes color fringes because the lens material brings different colors of light to a focus at different points. An achromatic lens corrects for this error. 2. An electron-gun defect that causes enlargement and blurring of the spot on the screen of a cathode-ray tube because electrons leave the cathode with different initial velocities and are hence deflected differently by the electron lenses and deflection coils.

chromaticity The color quality of light that can be defined by its chromaticity coordinates. Chromaticity depends only on hue and saturation of a color, not on its luminance (brightness). Chromaticity applies to all colors, including shades of gray, whereas chrominance applies only to colors other than grays.

chromaticity coordinate One of the two coordinates (x or y) that precisely specify the exact identity or chromaticity of a color on the CIE chromaticity diagram. Also called color coordinate and trichromatic coefficient.

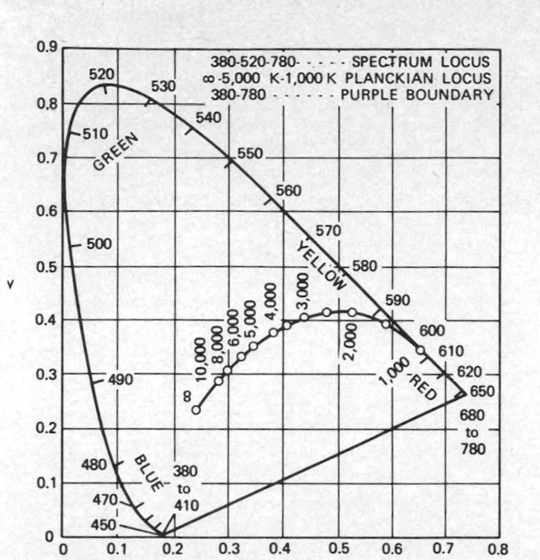

Chromaticity diagram as prepared by CIE. Temperature values on curve are in kelvins, and wavelength values are in nanometers.

chromaticity diagram A diagram in which one of the three chromaticity coordinates is plotted against another. The most common version is the CIE chromaticity diagram used in color television.

chromaticity flicker Flicker in a color television receiver caused by fluctuation of chromaticity only.

chromatron A single-gun color picture tube that has color phosphors deposited on the screen in strips instead of dots. The red, green, and blue color signals are applied in sequence to the single grid of the tube as the beam is deflected to the correct color strip by horizontal grid wires adjacent to the screen. Also called Lawrence tube.

Chromel Trademark for a nickel-chromium alloy used in thermocouples.

chrominance The difference between any color and a specified reference color of equal brightness. In color television, this reference color is white, having coordinates $x = 0.310$ and $y = 0.316$ on the chromaticity diagram.

chrominance bandwidth *Chrominance-channel bandwidth.*

chrominance carrier *Chrominance subcarrier.*

chrominance-carrier reference A continuous signal that has the same frequency as the chrominance subcarrier in a color television system and fixed phase with respect to the color burst. This signal is the reference with which the phase of a chrominance signal is compared for modulation or demodulation. In a color receiver it is generated by a crystal-controlled oscillator. Also called chrominance-subcarrier reference, color-carrier reference, and color-subcarrier reference.

chrominance channel Any path that is intended to carry the chrominance signal in a color television system.

chrominance-channel bandwidth The bandwidth of the path intended to carry the chrominance signal in a color television system. Also called chrominance bandwidth.

chrominance demodulator A demodulator in a color television receiver for deriving the I and Q components of the chrominance signal from the chrominance signal and the chrominance-subcarrier frequency. Also called chrominance-subcarrier demodulator.

chrominance modulator A modulator used in a color television transmitter to generate the I or Q components of the chrominance signal from the video-frequency chrominance components and the chrominance subcarrier. Also called chrominance-subcarrier modulator.

chrominance primary The nonphysical color represented by either the I or Q chrominance signal component in a color television system. These chrominance signals are chosen to be electrically convenient components remaining after the luminance signal is removed, from a full color signal at the transmitter.

chrominance signal One of the two components, called the I and Q signals, that add together to produce the total chrominance signal in a color television system. Also called carrier chrominance signal.

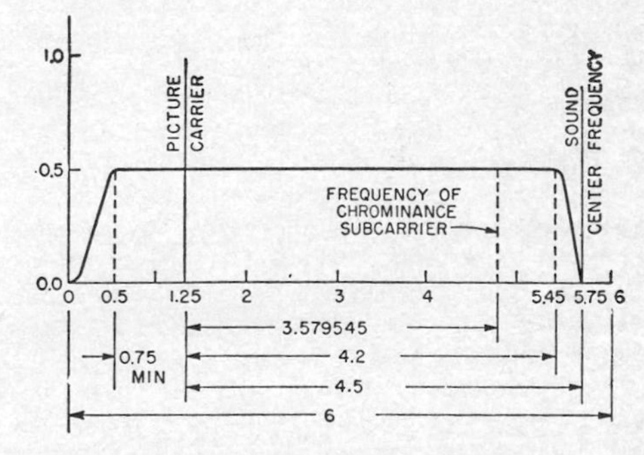

Chrominance subcarrier in standard 6-MHz television channel. Vertical scale gives relative maximum radiated field strength with respect to picture carrier. Frequency values are in megahertz.

chrominance subcarrier The 3.579545-MHz carrier whose modulation sidebands are added to the monochrome signal to convey color information in a color television receiver. The chrominance subcarrier is transmitted unmodulated in the form of color bursts that are used for synchronizing purposes in the receiver. Also called chrominance carrier, color carrier (deprecated), color subcarrier (deprecated), and subcarrier.

chrominance-subcarrier demodulator *Chrominance demodulator.*

chrominance-subcarrier modulator *Chrominance modulator.*

chrominance-subcarrier oscillator *Chroma oscillator.*

chrominance-subcarrier reference *Chrominance-carrier reference.*

chromium [symbol Cr] A metallic element. Atomic number is 24.

chromium dioxide tape A magnetic recording tape developed primarily to improve quality and brilliance of reproduction when used in cassettes operated at 1⅞ in/s (4.76 cm/s). Requires special recorders that provide high bias.

chromium-gold metallizing A metal film used on a silicon or silicon oxide surface in semiconductor devices because it is not susceptible to purple-plague deterioration. A layer of chromium is applied first for adherence to silicon, then a layer of chromium-gold mixture, and finally a layer of gold to which bonding contacts can be applied.

chromoradiometer A radiation meter that uses a substance whose color changes with x-ray dosage.

chronistor A subminiature elapsed-time indicator that uses electroplating principles to totalize operating time of equipment up to several thousand hours.

chronograph 1. An instrument that records intervals of time with a high degree of accuracy. 2. An instrument that measures and records projectile velocity by measuring the time required for the projectile to travel a known distance.

chronometer A highly accurate clock or watch used in air and surface navigation.

chronometric encoder An encoder that uses an electronic counter to time or count electrical events and deliver in digital form a number equivalent to the input magnitude.

chronoscope An electronic instrument that measures extremely short intervals of time, such as the time of passage of a rifle bullet between two points.

chronotron A device that measures millimicrosecond time intervals between pulses. In one type, interval-determining pulses are fed to a transmission line, and the spacing between the pulses on the line is measured.

chuck The clamping device that holds the stylus or needle of a phonograph pickup.

Ci Abbreviation for *curie.*

CID Abbreviation for *charge-injection device.*

CIE Abbreviation for *Commission Internationale de l'Eclairage.*

CIE chromaticity diagram A chromaticity diagram established as an international standard by the Commission Internationale de l'Eclairage. In this diagram, used in color television, the color wavelengths are plotted as coordinates of x and y.

cifax Enciphered facsimile communication in which the output of a keyed pulse generator is mixed with the output of the facsimile converter. The same key signal is subtracted from the facsimile signal at the receiving station. Unauthorized listeners cannot reconstruct the picture unless they have an identical key generator and the daily key setting.

c/in Abbreviation for *character per inch.*

cinching Creases produced in magnetic tape when the supply reel is wound at low tension and suddenly stopped during playback. Cinching may cause data-processing errors.

cineangiograph An x-ray movie camera that photographs heart action, progress of radiopaque dyes and catheters, and other medically interesting actions within the body that can be made visible on a fluoroscope screen.

cipher A transposition or substitution code for transmitting messages secretly.

ciphony equipment [CIPHer + telephONY] Any equipment attached to a radio transmitter, radio receiver, or telephone for scrambling or unscrambling voice messages. In one form speech is converted into a series of ON/OFF pulses that are mixed with pulses from a key generator. To recover the speech at the receiving terminal, an identical key generator must be set to the daily key setting and its output subtracted from the received signal. The ON/OFF pulses can then be converted to the original speech patterns.

C/I ratio Abbreviation for *carrier-interference ratio.*

circle-dot mode A mode of cathode-ray-tube storage of binary digits in which one kind of digit is represented by a small circle of excitation of the screen and the other kind by a similar circle with a concentric dot.

circuit 1. An arrangement of one or more complete paths for electron flow. 2. A complete wire, radio, or carrier communication channel.

circuit board A sheet of insulating plastic or ceramic material with or without printed wiring, on which circuit components are mounted.

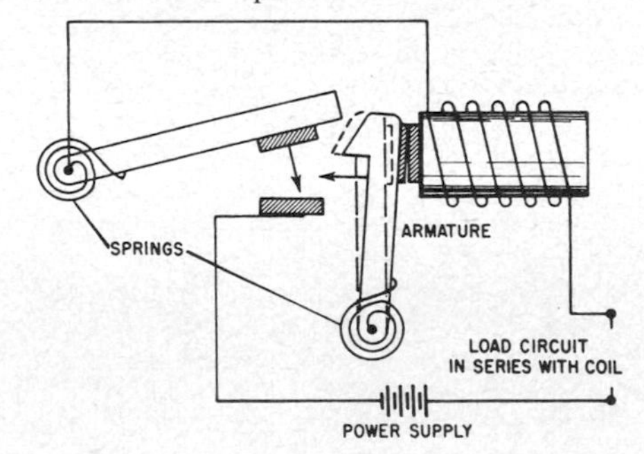

Circuit breaker in which armature normally locks contacts in closed position. Overload current through solenoid attracts armature, and spring at left opens contacts to break circuit.

circuit breaker An electromagnetic device that opens a circuit automatically when the current exceeds a predetermined value. It can be reset by operating a lever or by other means.

circuit conditioning Improving the quality of a communication circuit to meet a specific transmission requirement, such as by reducing noise, equalizing phase response, stabilizing level, improving frequency response, and correcting impedance discontinuities.

circuit diagram *Schematic circuit diagram.*

circuit element A component, integrated circuit, or other distinct part of a circuit to which internal or external connections are made.

circuit noise Undesired electric noise generated within a circuit or arriving from external sources.

circuit-noise level The ratio of the circuit noise at a point in a transmission circuit to some arbitrary amount of circuit noise chosen as a reference. Usually expressed in decibels above reference noise (dBrn) or in adjusted decibels (dBa).

circuit-noise meter An instrument that measures the electric noise level in a communication circuit, in decibels above reference noise. Also called noise-measuring set.

circuitry The complete combination of circuits in a piece of electronic equipment.

circular antenna A folded dipole that is bent into a circle so the transmission line and the abutting folded ends are at opposite ends of a diameter. When mounted horizontally, it radiates uniformly in all directions and has very little vertical radiation.

circular-beam multiplier A multiplier in which a circular beam of electrons is projected on four isolated metallic quadrants. The currents collected from the quadrants are combined to give an output current proportional to the product of the input variables that deflect the beam.

circular-chart recorder A recorder in which one or more writing pens or other recording devices are actuated by signals as a circular chart makes one complete revolution slowly under the control of a time-clock motor. After each such revolution, a new chart must be placed on the recorder.

circular electric wave A transverse electric wave for which the lines of electric force form concentric circles.

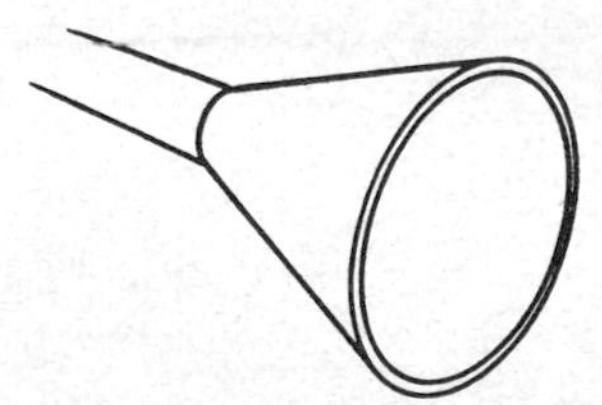

Circular horn at end of circular waveguide.

circular horn A circular-waveguide section that flares outward into the shape of a horn, to serve as a feed for a microwave reflector or lens.

circularly polarized wave An electromagnetic wave for which the electric and/or magnetic field vectors at a point describe a circle. This term is usually applied to transverse waves.

circular magnetic wave A transverse magnetic wave for which the lines of magnetic force form concentric circles.

circular mil [abbreviated cmil] A unit of area equal to the area of a circle whose diameter is 1 mil (0.001 in or 0.00254 cm). Used chiefly in specifying cross-sectional areas of round conductors.

circular orbit An orbit that makes a complete constant-altitude revolution around the earth.

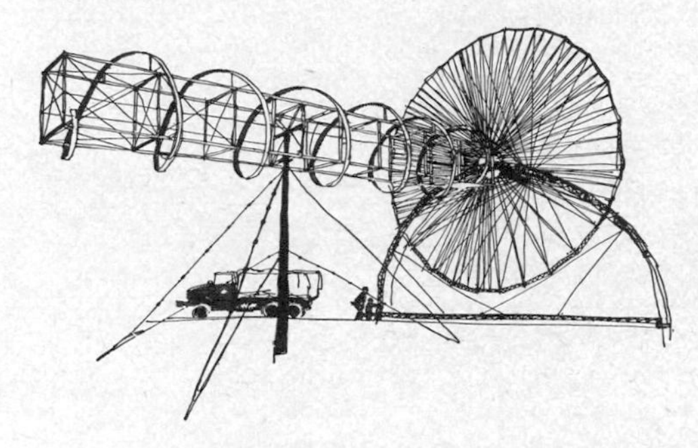

Circular polarization is produced by broadband helical antenna used for long-distance ionospheric scatter communication and radio direction-finding.

circular polarization Polarization in which the vector that represents the wave has a constant magnitude and rotates continuously about a point.

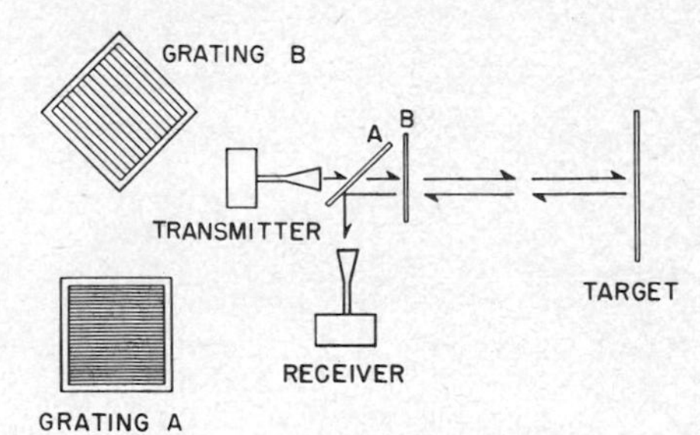

Circular-polarization duplexer as used in pulsed or continuous-wave radar system.

circular-polarization duplexer A duplexer in which simultaneous transmission and reception at millimeter wavelengths is achieved by two selective reflector gratings that consist of closely spaced metal slats. A vertically polarized wave going out from the transmitter is converted to a circularly polarized wave that has two components perpendicular to each other and 90° out of time phase by the combined action of the two gratings. A circularly polarized wave coming back through the first grating is converted to a linear horizontally polarized wave that is completely reflected by the second grating so as to go into the receiving horn.

circular probable error The radius of a circle within which half of a given number of guided missiles can be expected to fall.

circular scanning Scanning in which the radar antenna rotates in a complete circle so that the beam generates a plane or a cone whose vertex angle is close to 180°.

circular trace A time base produced by applying sine waves of the same frequency and amplitude, but 90° out of phase, to the horizontal and vertical deflection plates of a cathode-ray tube. The trace is then a circle, and signals give inward or outward radial deflections from the circle.

circular waveguide A waveguide whose cross-sectional area is circular.

circulating memory A digital computer device that uses a delay line to store information in the form of a pattern of pulses in a train. The output pulses are detected electrically, amplified, reshaped, and reinserted in the delay line at the beginning. Also called cyclic memory.

circulating register A shift register in which output data is fed back to the input, as in a closed loop.

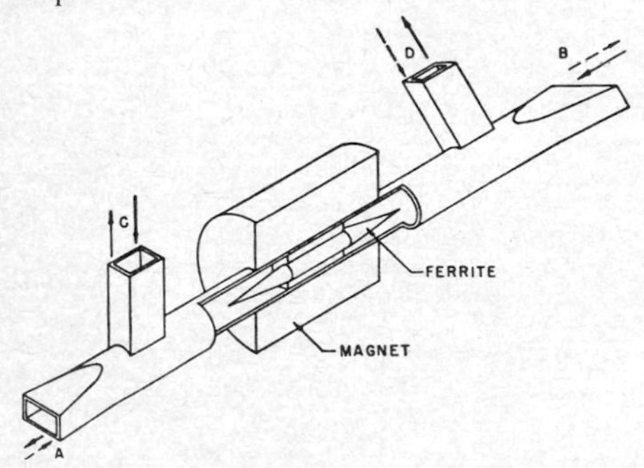

Circulator using ferrite rod positioned in longitudinal magnetic field in circular section of waveguide to rotate plane of polarization in traveling plane wave.

circulator A waveguide component that has a number of terminals so arranged that energy entering one terminal is transmitted to the next adjacent terminal in a particular direction. Also called microwave circulator.

circumlunar mission A mission in which a vehicle circles the moon and returns to earth.

cislunar Pertaining to space between the earth and the orbit of the moon. This space is a sphere centered on the earth, with a radius equal to the distance between the earth and the moon.

citizens band [abbreviated CB] One of the frequency bands allocated for citizens radio service, such as 460–470 MHz and 26.965–27.405 MHz.

citizens radio service A radio communication service for private or personal radio communication, including radio signaling, control of objects by radio, and other purposes. The frequency bands include 460–470 MHz for general use and the band around 27 MHz for ON/OFF unmodulated or tone-modulated carrier for remote control.

cladding A process of covering one metal with another, usually by bringing together the two metals and rolling, extruding, drawing, or swaging until a bond is produced.

clamp-and-hold digital voltmeter *Sample-and-hold digital voltmeter.*

clamper *DC restorer.*

clamping The introduction of a reference level that has some desired relation to a pulsed waveform, as at the negative peaks or the positive peaks. Also called DC reinsertion and DC restoration.

clamping circuit A circuit that reestablishes the DC level of a waveform. Used in the DC restorer stage of a television receiver to restore the DC component to the video signal after its loss in capacitance-coupled AC amplifiers, to reestablish the average light value of the reproduced image.

clamping diode A diode that clamps a voltage at some point in a circuit.

clamp-on A method of holding a call for a line that is in use and of signaling when it becomes free.

clamp-on ammeter *Snap-on ammeter.*

clamp-tube modulation Amplitude modulation in which the AF signal is applied to the screen grid of a tetrode or pentode operated as a modulator.

clapper A relay armature that is hinged or pivoted.

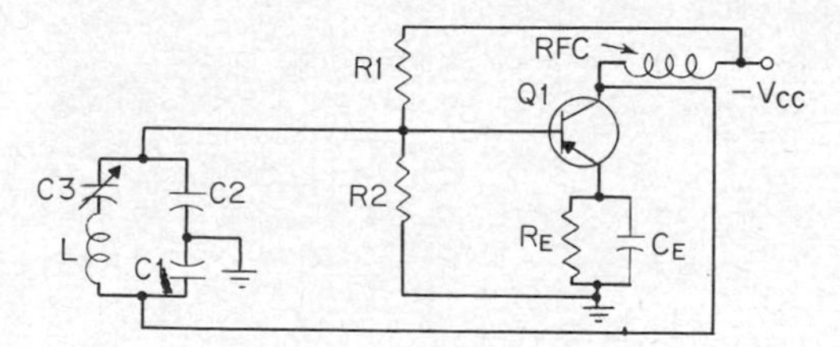

Clapp oscillator using bipolar transistor.

Clapp oscillator A series-tuned Colpitts oscillator that has low drift.

clarifier A fine-tuning control on some single-sideband citizens-band transceivers. It is adjusted for maximum naturalness of the received voice signals.

Clark cell An early form of standard cell that had a voltage of 1.433 V at 15°C. It has been largely replaced by the Weston standard cell as a voltage standard.

class A amplifier An amplifier in which the grid bias and alternating grid voltages are such that anode current in a specific tube flows at all times. The suffix 1 is added to the letter or letters of the class identification to denote that grid current does not flow during any part of the input cycle. The

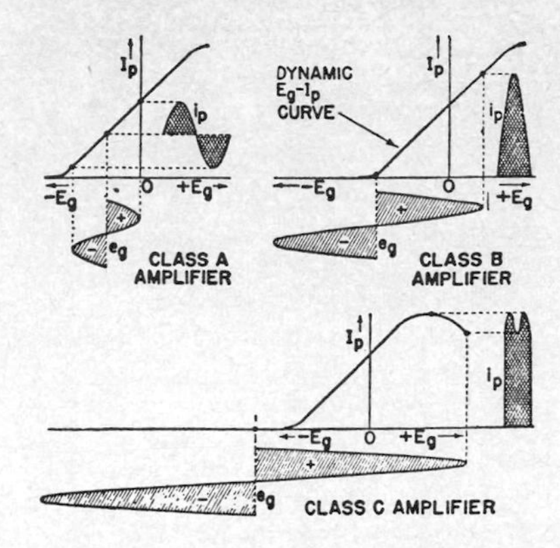

Class A, B, and C amplifier operation.

suffix 2 denotes that grid current flows during some part of the cycle. In a class A transistor amplifier, each transistor is in its active region for the entire signal cycle.

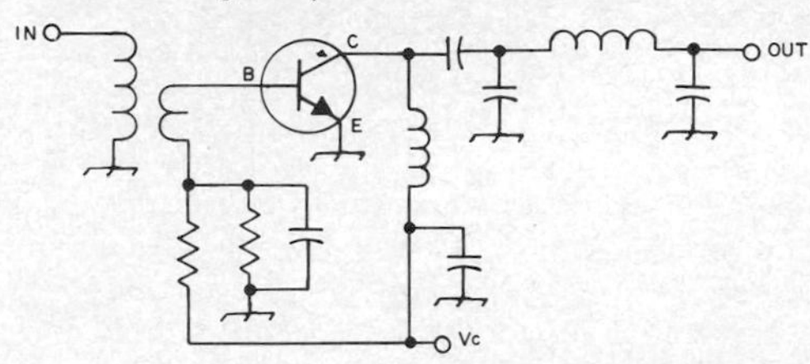

Class AB amplifier as used for linear amplification.

class AB amplifier An amplifier in which the grid bias and alternating grid voltages are such that anode current in a specific tube flows for appreciably more than half but less than the entire electric cycle. The suffix 1 denotes that grid current does not flow. The suffix 2 denotes that grid current flows during some part of the cycle. In a class AB transistor amplifier, operation is class A for small signals and class B for large signals.

class A insulation Insulation that consists of cotton, silk, paper, and other similar organic materials which are impregnated with or immersed in a liquid dielectric and can withstand temperatures up to 105°C.

class A modulator A class A amplifier that supplies the necessary signal power to modulate a carrier.

class A push-pull sound track Two single photographic sound tracks side by side; the transmission of one is 180° out of phase with the transmission of the other. Both positive and negative halves of the sound wave are linearly recorded on each track.

class B amplifier An amplifier in which the grid bias is approximately equal to the cutoff value, so that anode current is approximately zero when no exciting grid voltage is applied, and flows for ap-

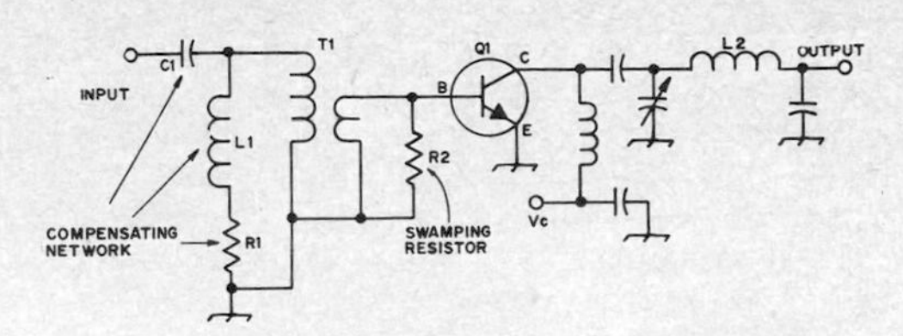

Class B amplifier.

proximately half of each cycle when an alternating grid voltage is applied. The suffix 1 denotes that grid current does not flow. The suffix 2 denotes that grid current flows during some part of the cycle. In a class B transistor amplifier, each transistor is in its active region for approximately half the signal cycle.

class B insulation Insulation that consists of asbestos, mica, glass fiber, and other similar inorganic materials combined with organic binders and capable of withstanding temperatures up to 130°C.

class B modulator A class B amplifier that supplies the necessary signal power to modulate a carrier. It is usually connected in push-pull.

class B push-pull sound track Two photographic sound tracks side by side; one track carries the positive half of the signal only, and the other carries the negative half. During the inoperative half-cycle, each track transmits little or no light.

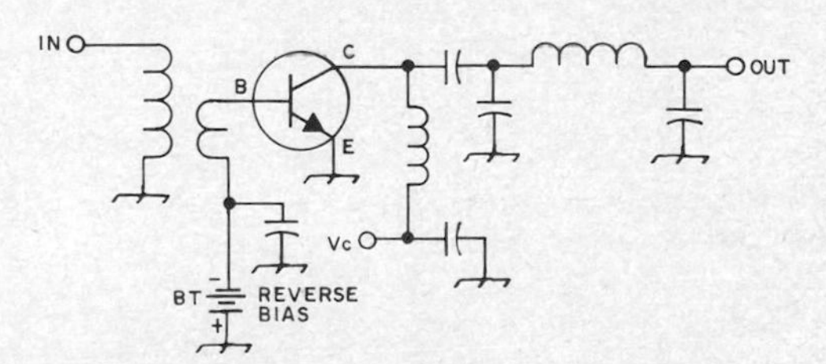

Class C amplifier using battery for reverse bias.

class C amplifier An amplifier in which the grid bias is appreciably greater than the cutoff value, so that anode current in each tube is zero when no alternating grid voltage is applied, and flows for appreciably less than half of each cycle when an alternating grid voltage is applied. The suffix 1 denotes that grid current does not flow. The suffix 2 denotes that grid current flows during some part of the cycle. In a class C transistor amplifier, each transistor is in its active region for significantly less than half the signal cycle.

class C insulation Insulation that consists of glass, mica, porcelain, quartz, and similar organic materials capable of withstanding temperatures over 220°C.

class D channel A data-transmission circuit that can transmit punched-tape data at up to 240 words per minute, or 80-column punched-card data at about 10 cards per minute.

class D stage A bipolar transistor stage operating as an ON/OFF switch in a circuit that changes suddenly between saturation and cutoff conditions. Can be used in inverters and converters.

class E channel A data-transmission circuit that can transmit at up to 1200 b/s.

class H insulation Insulation that consists of asbestos, glass fiber, mica, and similar inorganic materials combined with silicone or equivalent binders and capable of withstanding up to 180°C.

classical scattering cross section *Scattering cross section.*

classical system *Nonquantized system.*

classified Containing information whose disclosure to a prospective enemy is not in the best interests of the nation.

classify To sort into groups that have common properties. Thus intercepted electromagnetic radiations might be classified into radar, navigational, jamming, and missile-control signal groups.

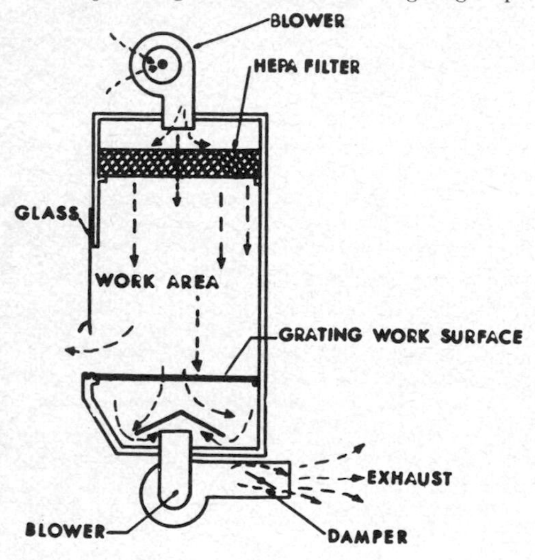

Clean-room exhaust hood for work position.

clean room A room in which elaborate precautions are used to reduce dust particles and other contaminants in the air, as required for assembly of delicate equipment. Dust-suppression techniques include pressurization of the entire room, lint-free uniforms and caps, air locks at doors, adhesive floor mats at the entrance to remove dust from the soles of shoes, and glass-enclosed assembly compartments into which workers insert only their hands for assembly work. Also called white room.

cleanup Gradual disappearance of gases from an electron tube during operation because of absorption by getter material or the tube structure.

clean weapon A nuclear warhead in which the amount of residual radioactivity is comparable to that of a normal weapon which has the same energy yield.

clear 1. To restore a memory device or binary circuit to a prescribed state, usually that representing logic 0. Also called reset. 2. A function key on calculators, to delete an entire problem or just the last keyboard entry. Usually labeled C. On some calculators, the first depression of the clear key deletes the previous entry, and the next depression clears everything. Other calculators have a separate key, labeled CE, for clearing the previous entry.

clear-air turbulence [abbreviated CAT] A meteorological phenomenon that occurs in the upper troposphere and lower stratosphere, in which high-speed aircraft are subject to violent up and down drafts.

clearance The minimum distance between two conducting objects or a conductor and ground.

clear channel A standard broadcast channel in which the dominant station or stations render service over wide areas. Stations on a clear channel are cleared of objectionable interference within their primary service areas and over all or a substantial portion of their secondary service areas.

clear text A message that is not coded and has no hidden meanings.

clear-voice override The ability of a speech scrambler to receive a clear message even when the scrambler is set for scrambler operation.

click A short-duration electric disturbance, such as that sometimes produced by a code-sending key or a switch.

click filter *Key-click filter.*

click tuner A mechanical television tuner that clicks into position for each of the 70 available UHF channels and the 12 VHF channels.

clipper *Limiter.*

clipper amplifier An amplifier that limits the instantaneous value of its output to a predetermined maximum.

clipper diode A bidirectional breakdown diode that clips signal voltage peaks of either polarity when they exceed a predetermined amplitude.

clipper-limiter A device whose output is a function of the instantaneous input amplitude for a range of values between two predetermined limits but is approximately constant, at another level, for input values above the range.

clipping 1. Cutoff of initial or final speech sounds in voice transmission, due to operation of voice-operated or other switching devices. 2. Voice distortion caused by severe overloading of amplifier circuits so that peaks of audio waveforms are cut off. 3. Any action that cuts off the peaks of a television signal. This may affect either the positive (white) or negative (black) picture-signal peaks or the synchronizing signal peaks. 4. *Limiting.*

clipping level The amplitude level at which a waveform is clipped.

clipping time The time constant of a limiter.

clock A source of accurately timed pulses, used for synchronization in a digital computer or as a time base in a transmission system.

clocked flip-flop Two flip-flops connected in a master-slave combination, to eliminate the need for capacitors or other circuit delay elements. The master flip-flop stores the input information when the clock voltage is high, and transfers it to the slave when the clock voltage is low.

clocked gate A gate circuit that is actuated by a clock pulse.

clocked logic A logic circuit in which the switching action is controlled by repetitive pulses from a clock.

clocked RS flip-flop A flip-flop that gives a 1 output if the set (S) input is enabled when the clock pulse arrives and a 0 output if the reset (R) input is enabled when clocked.

clock frequency The master frequency of the periodic pulses that schedule the operation of a digital computer.

clock rate The rate at which bits or words are transferred from one internal element of a computer to another.

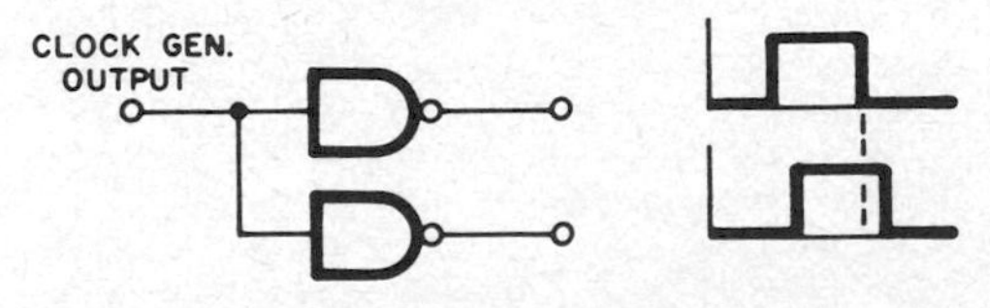

Clock skew caused by different delays in gates.

clock skew A phase shift caused by different delays in various parts of a clock-signal distribution system. Excessive skew may cause data-processing errors.

clock synchronization Any method of making independent indicators of time agree with high accuracy. When loran-C navigational transmissions are used for this purpose, clocks can be synchronized over long distances to an accuracy of better than 1 μs.

clock track A recorded track that contains a signal pattern which acts as a time reference.

clockwise capacitor A variable capacitor whose capacitance increases with clockwise rotation of its rotor, as viewed from the end of the control shaft.

clockwise polarized wave *Right-hand polarized wave.*

close control The control exercised by an air controller, who is using radio and radar, over friendly aircraft in a tactical air situation.

close-control radar Ground radar used with radio to position an aircraft over a target that is normally difficult to locate or invisible to the pilot.

close coupling The coupling obtained when the primary and secondary windings of an RF or IF transformer are close together.

closed circuit A complete path for current.

closed-circuit communication system A self-contained communication system that has no provision for interconnection with other systems. An intercom is an example.

closed-circuit signaling Signaling in which current flows in the idle condition, and a signal is initiated by increasing or decreasing the current.

closed-circuit television [abbreviated CCTV] Any application of television that does not involve broadcasting for public viewing. Theater television and industrial television are examples. The programs can be seen only on specified receivers connected to the television camera by circuits that may or may not include microwave relays and coaxial cables. Used also to permit large groups of medical students or doctors to watch surgical operations.

closed-circuit voltage The voltage at the terminals of a source when a specified load current is being drawn.

closed-cycle control system A control system in which changes in the quantity being controlled are utilized to actuate the controller directly.

closed-cycle cryogenic cooling Cryogenic cooling in which a gas such as nitrogen is liquefied by a refrigerator-type compressor, for cooling infrared detectors and other cryogenic components. The gas recovered after cooling goes back to the compressor for liquefaction.

closed loop 1. A computer program in which a cycle of instructions is executed repeatedly. 2. A signal path for feeding the output of a control system back to the input for comparison with reference values, to achieve a desired form of regulation.

closed-loop control system *Feedback control system.*

closed-loop recorder A tape recorder in which the tape is driven twice by the capstan, to reduce flutter.

closed magnetic circuit A complete circulating path for magnetic flux around a core of ferromagnetic material.

closed shell A shell that contains the maximum number of nucleons in the independent particle model of the nucleus.

closed shop A computer facility in which programming service to the user is the responsibility of a group of specialists who are not allowed in the computer room to run or oversee the running of their programs.

closed subroutine A computer subroutine that is stored away from the routine which refers to it. Such a subroutine is entered by a jump, and provision is made to jump back to the proper point in the main routine at the end of the closed subroutine.

closed waveguide A waveguide in which the entire electromagnetic field is shielded from the environment.

close-talking microphone A microphone designed for use close to the mouth, so noise from more distant points is suppressed. Also called noise-canceling microphone.

clothing monitor An instrument that monitors radioactive contamination on clothing.

cloud The nucleons that are in the nucleus of an atom but not in closed shells.

cloud absorption Attenuation of electromagnetic radiation by absorption within the water particles of a cloud.

cloud and collision warning system An aircraft radar that gives a cathode-ray display of storm clouds and high ground in the intended course, at ranges sufficient to permit course changes when necessary.

cloud attenuation Attenuation of electromagnetic radiation by scattering action in clouds.

cloud chamber An enclosure that contains air supersaturated with water vapor by sudden expansion, in which moving ionizing particles are revealed by streaks of droplets called cloud tracks. Also called expansion chamber and fog chamber.

cloud echo The echo pattern produced on a radar screen by reflection of radar signals from a cloud.

cloud-height detector *Ceiling-height indicator.*

cloud pulse The output that results from space charge effects produced by turning the electron beam on or off in a charge-storage tube.

clover-leaf antenna A nondirectional VHF transmitting antenna that consists of a number of horizontal four-element radiators stacked vertically a half-wave apart. Each horizontal unit has four loops arranged like a four-leaf clover. The units are energized to give maximum radiation in the horizontal plane.

clutter Unwanted echoes on a radar screen, such as those caused by ground, sea, and rain clutter; stationary objects; chaff; enemy jamming transmissions; and grass. Also called background return and radar clutter.

clutter gating A gating technique that gives a normal radar video display in regions which have no clutter, with moving-target-indicator video being switched in only for clutter areas.

cm Abbreviation for *centimeter.*

cmil Abbreviation for *circular mil.*

CML Abbreviation for *current-mode logic.*

CMOS Abbreviation for *complementary metal-oxide semiconductor.*

CMOS RAM A random-access memory that uses complementary metal-oxide semiconductor technology.

CMOS SOS Silicon-on-sapphire technology combined with complementary metal-oxide semiconductor technology.

CMOS SOS RAM A random-access memory that uses a combination of silicon-on-sapphire and complementary metal-oxide semiconductor technologies.

CMR Abbreviation for *common-mode rejection.*

CMRR Abbreviation for *common-mode rejection ratio.*

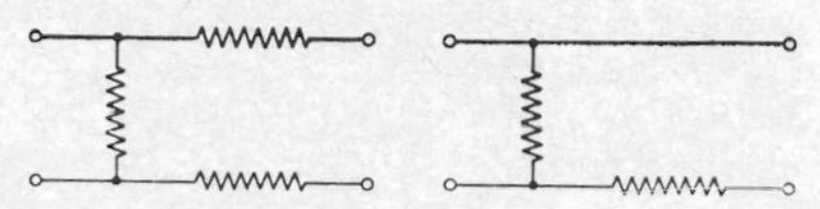

C network has three impedance branches, whereas L network at right has only two branches.

C network A network composed of three impedance branches in series; the free ends of the network are connected to one pair of terminals, and the junction points are connected to another pair of terminals.

coagulator A medical electronic instrument used during surgery to stop bleeding, by applying either electric current or light.

coal-sensing probe A nucleonic instrument that measures the thickness of coal left on the floor or ceiling of a seam by a cutter. Gamma rays from a radioactive source in the instrument are scattered by coal and rock in inverse proportion to density, so more radiation comes back to the associated Geiger counter from coal than from rock. The indicating meter can therefore be calibrated directly in coal thickness.

coarse chrominance primary The less important of the two chrominance primaries in a color television signal, called the Q signal. Because its bandwidth is limited to 0.5 MHz, this signal affects only the larger, coarser variations in the color picture. The other primary is the fine chrominance primary or I signal, going up to 1.5 MHz.

coarse control A control that makes a rough adjustment of some characteristic or quantity.

coast A radar memory feature that causes the range and/or angle systems to continue to move in the same direction and at the same speed an original target was moving. Used to prevent lock-on to a stronger target near the target being tracked.

coastal refraction The bending of the path of a direct radio wave as it crosses the coast at or near the ground due to changes in electrostatic conditions between earth and water.

coated cathode A cathode that has been coated with compounds to increase electron emission. An oxide-coated cathode is an example.

coated filament A vacuum-tube filament that has been coated with metal oxides to provide increased electron emission.

coated lens A lens whose air-glass surfaces have been coated with a thin transparent film that has an index of refraction which minimizes light loss by reflection. Coatings used include magnesium fluoride, silicon oxide, sodium fluoride, and titanium oxide.

coax *Coaxial cable.*

coaxial antenna An antenna that consists of a quarter-wave extension of the inner conductor of a coaxial line and a radiating sleeve which is formed by folding back the outer conductor of the coaxial line for a length of approximately one quarter-wavelength.

coaxial attenuator An attenuator that has a coaxial construction and terminations suitable for use with coaxial cable.

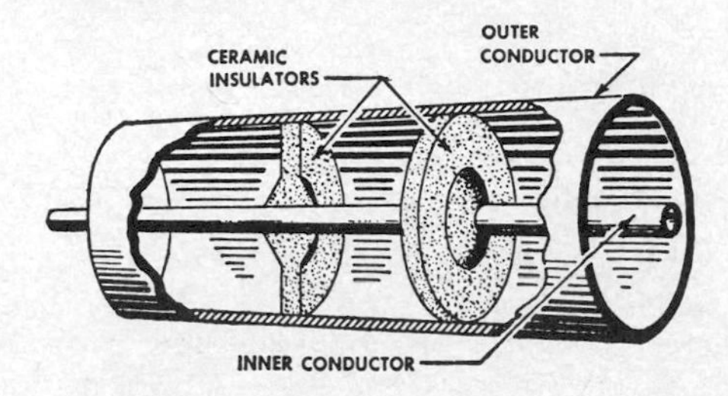

Coaxial cable using dielectric spacers and essentially rigid conductors.

coaxial cable A transmission line in which one conductor is centered inside and insulated from a metal tube that serves as the second conductor. The insulation may be a continuous solid dielectric, or there may be dielectric spacers and an insulating gas. Also called coax, coaxial line, coaxial transmission line, concentric line, and concentric transmission line.

coaxial cavity A cylindrical resonating cavity that has a central conductor in contact with its pistons or other reflecting devices. The conductor picks up a desired wave.

coaxial connector A connector that consists of a mating plug and receptacle for making a temporary or permanent connection between two lengths of coaxial cable without affecting impedance. The three most common types use bayonet, threaded, and push-pull quick-disconnect coupling methods.

coaxial-cylinder magnetron A magnetron in which the cathode and anode consist of coaxial cylinders.

coaxial diode A diode that has the same outer diameter and terminations as a coaxial cable or that is otherwise designed to be inserted into a coaxial cable.

coaxial-dipole antenna A dipole antenna that has lengths of metal tubing as the radiating elements. The twin-lead transmission line connects conventionally to the inner ends of the tubing. The outer ends of the tubing are connected by a metal rod centered in the radiating elements.

coaxial dry load A sand load for a coaxial cable.

coaxial filter A section of coaxial line that has reentrant elements which provide the inductance and capacitance of a filter section.

coaxial isolator An isolator in a coaxial cable that provides a higher loss for energy flow in one direction than in the opposite direction. All types use a permanent magnetic field in combination with ferrite and dielectric materials.

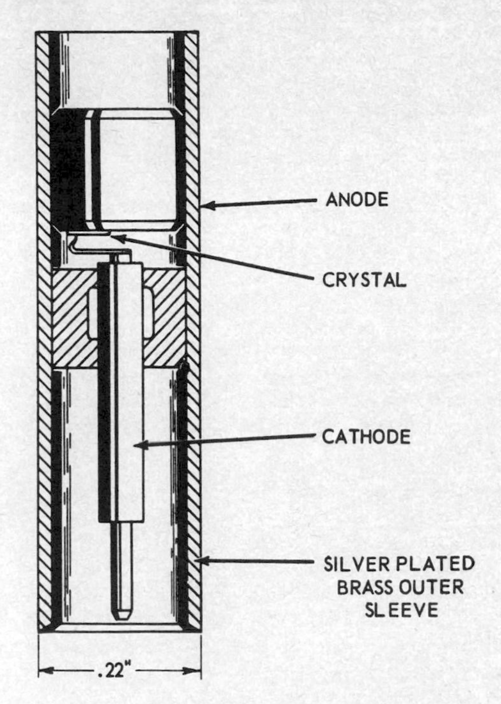

Coaxial diode using silicon crystal.

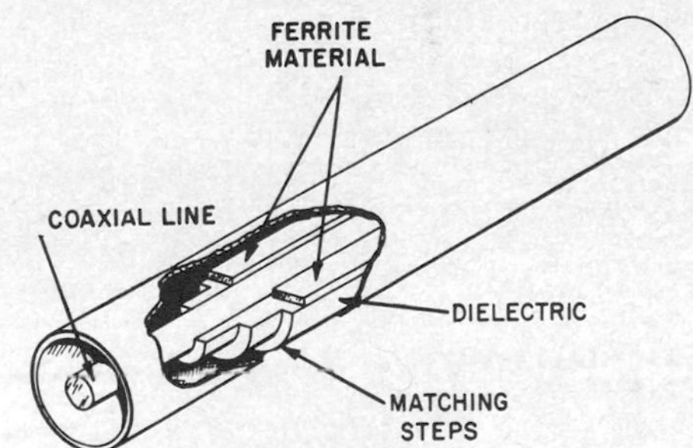

Coaxial isolator. Magnetic field may be provided by either rod-type permanent magnet inside central conductor or external horseshoe magnet with its poles near ferrite strips.

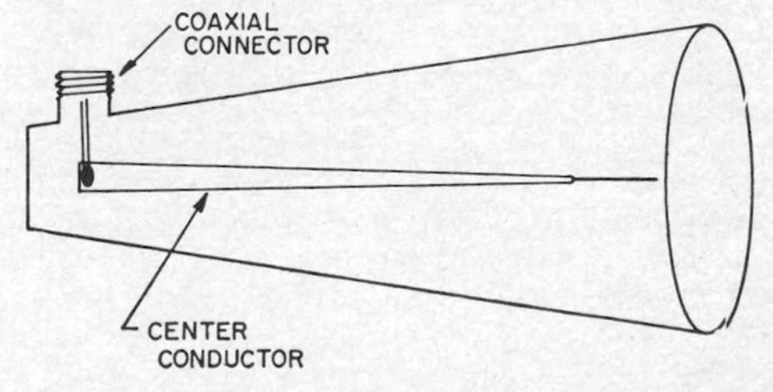

Coaxial launcher.

coaxial launcher A transducer that couples a coaxial cable to a waveguide, cavity, or other microwave device.

coaxial line *Coaxial cable.*

coaxial-line frequency meter A shorted section of coaxial line that acts as a resonant circuit and is calibrated in terms of frequency or wavelength. Also called coaxial wavemeter.

coaxial-line oscillator *Coaxial-line tube.*

coaxial-line resonator A resonator that consists of a length of coaxial line short-circuited at one or both ends.

coaxial-line tube A tube used for RF generation in the range of 4 to 6 GHz. It consists of a quarter-wave coaxial line, part of which is cut away to permit an electron beam to be fired through it. The beam thus crosses the gap between outer and inner conductors twice. The field in the first gap causes velocity modulation of the beam. Interaction of the bunched beam with the field in the second gap causes oscillation. Also called coaxial-line oscillator.

coaxial loudspeaker A loudspeaker in which a tweeter is mounted in the center of the woofer.

coaxial plasma accelerator *Plasma rocket engine.*

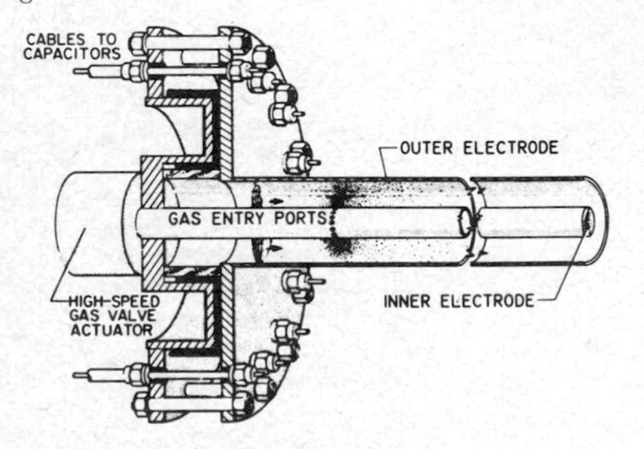

Coaxial plasma gun in which repeated discharges of capacitor bank cause plasma to be ejected at right in bursts, to propel spacecraft.

coaxial plasma gun A pulsed accelerator in which a capacitor bank is discharged between coaxial electrodes about 1000 times per second, to produce an electromagnetic acceleration force that ejects the plasma propellant in pulses for maximum thrust in spacecraft.

coaxial reed relay A relay that has terminations which correspond to those of a coaxial cable, to permit inserting the reed in the cable to interrupt the central conductor without introducing an impedance mismatch.

coaxial relay A relay that opens or closes a coaxial cable circuit without introducing a mismatch that would cause wave reflections.

coaxial sheet grating A sheet grating that consists of concentric metal cylinders each about one wavelength long, centered in a coaxial waveguide to suppress undesired modes of propagation.

coaxial stop filter A movable tuned filter placed around a conductor to limit its electrical radiating length at a given frequency.

coaxial stub A length of nondissipative cylindrical waveguide or coaxial cable branched from the side of a waveguide to produce some desired change in its characteristics.

coaxial switch A switch that changes connections between coaxial cables going to antennas, transmitters, receivers, or other high-frequency devices without introducing impedance mismatch.

coaxial transistor A point-contact transistor in which the emitter and collector are point electrodes that make pressure contact at the centers of opposite sides of a thin disk of semiconductor material serving as the base.

coaxial transmission line *Coaxial cable.*

coaxial wavemeter *Coaxial-line frequency meter.*

cobalt [symbol Co] A metallic element that has weak magnetic characteristics; it is sometimes combined with iron and steel to make special alloys used in permanent magnets. Atomic number is 27. Cobalt 60 is a radioisotope that has a half-life of 5.3 years and is widely used in place of x-ray equipment.

cobalt-beam therapy Therapy that involves the use of gamma radiation from a cobalt-60 source mounted in a cobalt bomb.

cobalt bomb A quantity of cobalt 60 mounted in a housing that has walls with up to 8 in (20 cm) of lead for protection and means for removing a lead plug to release a beam of gamma rays for use in cobalt-beam therapy.

COBOL [COmmon Business Oriented Language] A business data-processing language that can be given to a computer as a series of English statements describing a complete business operation. A master program or compiler in the computer translates the English statements into machine language, allocates storage, and produces a program for the job.

cobs Bell-shaped deflections produced on an oscilloscope by frequency-modulated continuous-wave jamming.

cochannel interference Interference between two signals of the same type in the same radio channel.

Cockcroft-Walton accelerator An early accelerator in which rectified alternating voltage is used to charge a series of capacitors to a high voltage for use in accelerating protons.

codan [Carrier-Operated Device AntiNoise] A device that silences a receiver except when a modulated carrier signal is being received.

CODAR [COrrelation, Detection, And Ranging] A method of detecting submerged submarines with sonobuoys dropped in a characteristic pattern from Navy P-3 aircraft, based on difference in time and phase of signals transmitted to the search plane by the sonobuoys.

code A system of symbols and rules for expressing information, such as the Morse and EIA color codes and the binary and other machine languages used in digital computers.

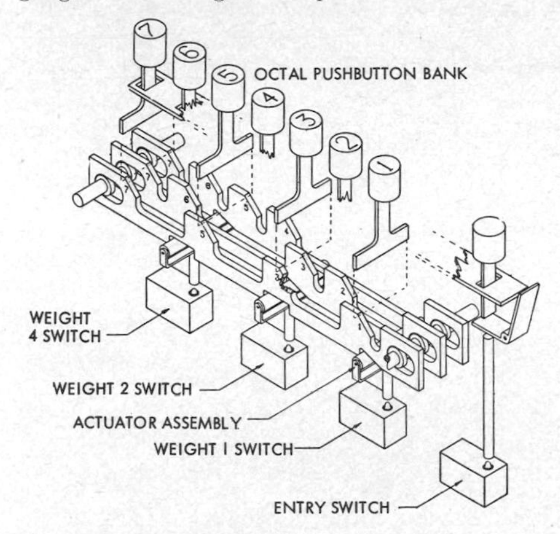

Code-bar switch for converting octal information to binary-coded information on output switches 1, 2, and 4. Button 3 closes output switches 1 and 2, and button 6 closes switches 2 and 4.

code-bar switch A switch that contains a number of pushbuttons, cams, latching members, and snap-action switches for converting octal or decimal information into binary-coded information.

code character A combination of code elements, such as dots and dashes or 0s and 1s, used to represent a character.

code converter A converter that changes coded information to a different code system. Also called decoder.

coded decimal digit A decimal digit that is expressed by a pattern of four 1s and 0s.

coded decimal notation A form of notation in which each decimal digit is converted separately into a pattern of binary 1s and 0s. For example, in the 8-4-2-1 coded decimal notation, the number 13 is represented as 0001 0011, whereas in pure binary notation it is represented as 1101.

code delay An arbitrary interval of time introduced, in addition to other time intervals, between pulsed signals sent by master and slave transmitters in loran and similar systems. The code delay is used as a security measure or to allow a navigator to resolve ambiguity in plotting a line of position. Also called coding delay (deprecated) and suppressed time delay.

coded interrogator An interrogator whose output signal forms the code required to trigger a specific radio or radar beacon. The IFF interrogator is an example.

code-division multiple access [abbreviated CDMA] A multiple-access technique used in satellite communication systems, in which neither the satellite frequency nor the time domain is divided among the earth stations. Each earth station has common usage of the full satellite bandwidth and time by transmitting with its own unique pseudonoise-coded waveform. Correlation-detection techniques are used at each receiver to recover the original information.

code-division multiplex [abbreviated CDM] Multiplex in which two or more communication links each occupy the entire transmission channel simultaneously, with code signal structures designed so a given receiver responds only to its own signals and sees the other signals as noise.

coded passive reflector A radar reflector whose reflecting properties can be varied according to a predetermined code, to produce a recognizable indication on a radarscope.

coded program A program that has been expressed in the required code for a computer.

coded stereo A stereophonic sound system used in theaters, in which a single audio channel is accompanied by a subsonic code signal that controls the volume of sound fed to loudspeakers at the left, center, and right. Thus, if drums are to seem to be at the right of the listener, the coded subsonic signal causes the system to supply relatively more power to the loudspeakers on the right during loud drum passages.

code element One of the separate elements or events constituting a code message, such as the presence or absence of a pulse, dot, dash, or space.

code group A combination of letters or numerals, or both, assigned to represent one or more words of plain text in a coded message.

code hole An information hole in punched paper tape, as distinguished from the feed holes.

code practice oscillator An oscillator used with a key and either headphones or a loudspeaker to practice sending and receiving Morse code.

coder 1. A device that generates a code by generating pulses which have varying lengths and/or spacings, as required for radio beacons and interrogators. Also called moder, pulse coder, and pulse-duration coder. 2. A person who translates a sequence of computer instructions into codes that are acceptable to the machine.

code reader A reader that responds to printed, painted, magnetic, radioactive, or other types of coded marks placed on products, shipping cartons, freight cars, or other moving objects.

code recorder An instrument that makes a permanent record of code messages received by radio or wire, as by punching holes in a tape or by making dot-and-dash marks on a tape.

code ringing Party-line telephone ringing in which the number and/or duration of rings identifies the station being called.

coding A list, in computer code, of the successive operations required to carry out a given routine or solve a given problem.

coding delay Deprecated term for *code delay.*

coefficient of coupling A numerical rating between 0 and 1 that specifies the degree of coupling between two circuits or the corresponding percentage value. Maximum coupling is 1, and no coupling is 0. Also called coupling coefficient.

coenzymometer A photoelectric instrument that measures light absorption as an indication of enzyme activity.

coercimeter An instrument that measures the magnetic intensity of a natural magnet or electromagnet.

coercive force The magnetizing force required to bring the flux density to zero in a magnetic material that has been magnetized alternately by equal and opposite magnetizing forces. It is the reverse magnetizing force needed to remove the residual magnetism.

coercivity The property of a magnetic material that is measured by the coercive force which corresponds to the saturation induction for the material.

coffin A box of heavy shielding material, usually lead, used for transporting radioactive objects. It has walls thick enough to attenuate radiation from the objects within it to an allowable level.

cogging Variations in torque and speed of an electric motor due to variations in magnetic flux as rotor poles move past stator poles.

cognitive machine A machine that has the ability to learn.

cohered video The video detector output signal in a coherent moving-target-indicator radar system.

coherence The existence of a statistical or time correlation between the phases of two or more waves.

coherence function A measured spectrum value that determines if two measurements which produce a spectral line are correlated. The function is normalized at all frequencies, so it can have values only between 0 and 1. A value of 1 means that the spectral line at the monitored point is completely coherent with the measured source, and values of 0.5 and 0 mean that 50 and 0% of the power at a given frequency at the monitored point is coherent with the measured source.

coherent Moving in unison or having some other fixed relationship, as between particles in a synchrotron or photons in a coherent laser beam.

coherent-carrier system A transponder system in which the interrogating carrier is retransmitted at a definite multiple frequency for comparison.

coherent decade frequency synthesizer A frequency synthesizer that provides a wide range of output frequencies, such as from direct current to 100 kHz, in decimal steps.

coherent detection Detection in which the received pulse-modulated signal is cross-correlated with locally generated signals that represent admissible signaling elements.

coherent detector A detector in moving-target-indicator radar that gives an output signal amplitude which depends on the phase of the echo signal instead of on its strength, as required for a display that shows only moving targets.

coherent echo A radar echo that has relatively constant phase and amplitude at a given range.

coherent frequency synthesizer A frequency synthesizer that derives its frequency from a single source, for example, from an atomic resonance device.

coherent interrupted wave An interrupted continuous wave in which the phase of the waves is maintained through successive wave trains.

coherent light Light that has essentially a single wavelength, with definite phase relationships between different points in a beam. A laser beam is an example.

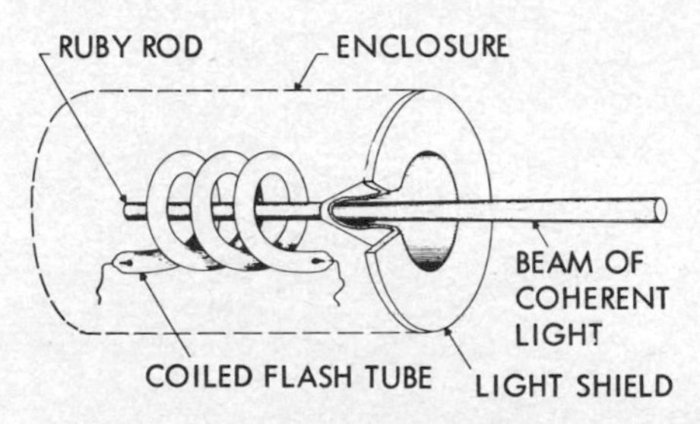

Coherent light produced by ruby laser.

coherent-light communication Communication that uses amplitude or pulse-frequency modulation of a laser beam.

coherent moving-target indicator A radar system in which the Doppler frequency of the target echo is compared to a local reference frequency generated by a coherent oscillator.

coherent noise Noise that affects all tracks across a magnetic tape equally and simultaneously. Tape speed variations develop coherent noise.

coherent oscillator [abbreviated COHO] An oscillator used in moving-target-indicator radar to serve as a reference by which changes in the RF phase of successively received pulses may be recognized. The COHO is locked to the radar transmitter frequency, so beating of this signal with the echo signal will reveal phase changes.

coherent-pulse radar A radar in which the RF oscillations of recurrent pulses bear a constant phase relation to those of a continuous oscillation.

coherent radiation Radiation in which there are definite phase relationships between different points in a cross section of the beam. In noncoherent radiation these relationships are random. Interference bands are observed only between coherent beams.

coherent reference A reference signal, usually of stable frequency, to which other signals are phase-locked to establish coherence throughout a system.

coherent scattering Scattering in which there is a definite phase relationship between incoming and scattered particles or photons.

coherent-scattering cross section The cross section for coherent scattering.

coherent system A navigation system in which the signal output is obtained by demodulating the received signal after the received signal is mixed with a local signal which has a fixed-phase relation to that of the transmitted signal, to permit use of the information carried by the phase of the received signal.

coherent transponder A transponder in which a fixed relation between frequency and phase of input and output signals is maintained.

coherent video The video signal produced in a moving-target-indicator system by combining a radar echo signal with the output of a continuous-wave oscillator. After delay, this signal is detected, amplified, and subtracted from the next pulse train to give a signal that represents only moving targets.

coherer A cell that contains a granular conductor between two electrodes. The cell becomes highly conducting when subjected to an electric field, and conduction can then be stopped only by jarring the granules. Formerly used as a detector in wireless telegraphy.

COHO Abbreviation for *coherent oscillator.*

coil [symbol L] A number of turns of wire that introduce inductance into an electric circuit, to produce magnetic flux or to react mechanically to a changing magnetic flux. In high-frequency circuits a coil may be only a fraction of a turn. The electrical size of a coil is called inductance and is expressed in henrys. The opposition that a coil offers to alternating current is called impedance and is expressed in ohms. The impedance of a coil increases with frequency. Also called inductor.

coil form The tubing or spool of insulating material on which a coil is wound.

coil loading The insertion of loading coils at regular intervals in a transmission line to improve its transmission characteristics over the required frequency band.

coil neutralization *Inductive neutralization.*

coil serving *Serving.*

coil winder A manual or motor-driven mechanism for winding coils individually or in groups.

coincidence The occurrence of counts in two or more nuclear-particle detectors simultaneously or within an assignable time interval.

coincidence circuit A circuit that produces a specified output pulse when and only when a specified number or combination of two or more input terminals receive pulses within an assigned time interval. Also called coincidence counter and coincidence gate.

coincidence correction *Dead-time correction.*

coincidence counter *Coincidence circuit.*

coincidence counting A method of distinguishing particular types of events from background events by coincidence circuits.

coincidence-current magnetic core A binary magnetic core in which information is stored only when current flows simultaneously in two or more independent windings.

coincidence gate *Coincidence circuit.*

coincidence loss Loss of counts due to the occurrence of ionizing events at intervals less than the resolution time of the counting system.

coincidence multiplier A multiplier based on the fact that the probability of a simultaneous occurrence of two independent events is the product of the probabilities of the separate events. Each event signal is converted to a pulse whose width is proportional to the amplitude of the signal. The duration of time in which the two trains of pulses are in coincidence is then proportional to the product of the probabilities for the events. Also called probability multiplier.

coincident-current selection The selection of a particular magnetic cell, for reading or writing in computer storage, by simultaneously applying two or more currents.

CO laser *Carbon monoxide laser.*

cold area An area in a plant or laboratory in which there is little or no contact with radioactive chemicals or radiations.

cold cathode A cathode whose operation does not depend on its temperature being above the ambient temperature.

cold-cathode counter tube A counter tube that has one anode and three sets of 10 cathodes. Two sets of cathodes act as guides that direct the glow discharge to each of the 10 output cathodes in correct sequence in response to driving pulses. Used for data storage, preset counting, tuning, and gating.

cold-cathode magnetron ionization gage A vacuum gage that has the structure of an inverted magnetron, with an auxiliary cathode to provide the initial field emission for starting. A cold-cathode discharge then occurs in crossed electric and magnetic fields, with ion current flow to the collector depending on the degree of vacuum.

cold-cathode rectifier A cold-cathode gas tube in which the electrodes differ greatly in size, so electron flow is much greater in one direction than in the other. An example is the 0Z4 full-wave gas rectifier tube. Also called gas-filled rectifier.

cold-cathode tube An electron tube that contains a cold cathode, such as a cold-cathode rectifier, mercury-pool rectifier, neon tube, phototube, and voltage regulator.

cold cutting stylus A stylus that has its cutting edge burnished at a plane substantially different from the cutting face, to cut and polish the groove in an acetate disk at normal room temperature.

cold junction The junction of thermocouple wires with conductors leading to the measuring

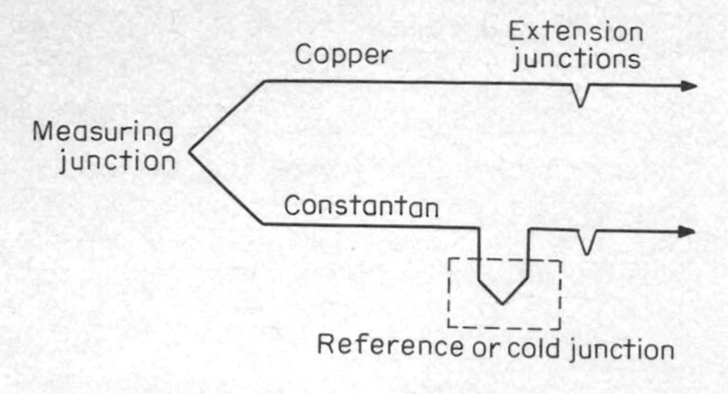

Cold junction of copper-constantan thermocouple, in which constantan alloy wire is connected to copper wire going to measuring or recording instrument.

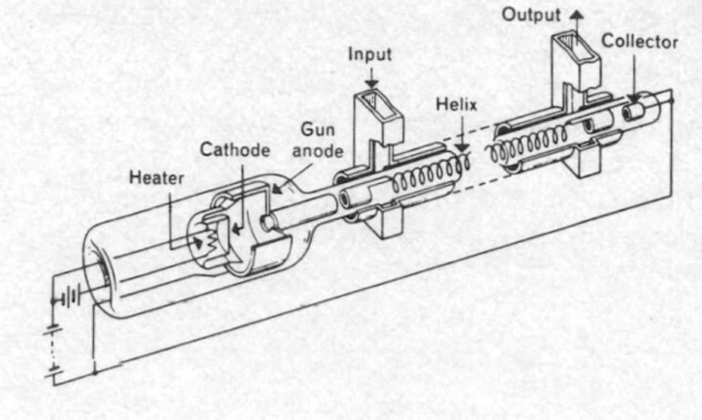

Collector at output end of traveling-wave tube serving as amplifier.

instrument. This junction is normally at room temperature.

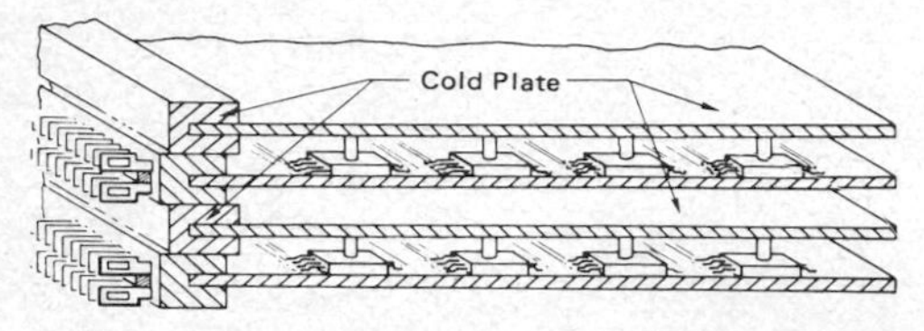

Cold plates and heat-transferring studs used with integrated-circuit packages mounted on printed-wiring boards.

cold plate An aluminum or other heat-conductive metal plate used as a heatsink for transistors and other components that are thermally coupled with or directly mounted on the plate. Heat absorption may be increased by circulating freon or some other liquid coolant through internal tubing in the plates or by blowing cooling air over the plates or the heat-transferring studs.

cold weld 1. A weld in which a molecular bond is obtained by cold flow of metal under high pressure, without heat. Used for sealing metal housings of certain components, such as some transistors. 2. A weld achieved by bringing together two metal faces in a perfect vacuum, without pressure or heat, as in outer space.

collate To combine two or more similarly ordered sets of values into one set that may or may not have the same order as the original sets. Merging is collating without changing the nature of the order.

collateral series A radioactive decay series, initiated by transmutation, that eventually joins into one of the four radioactive decay series.

collating sequence An ordering assigned to a set of items, such that any two sets in that assigned order can be collated.

collator A punched-card machine that can be used for merging two similarly sequenced sets of cards into one sequence or for matching detail cards with master cards.

collector 1. [symbol C] A semiconductive region through which a primary flow of charge carriers leaves the base of a transistor. The electrode or terminal connected to this region is also called the collector and corresponds to the anode of an electron tube. 2. An electrode that collects electrons or ions which have completed their functions within an electron tube. A collector receives electrons after they have done useful work, whereas an anode receives electrons whose useful work is to be done outside the tube. Also called electron collector.

collector capacitance The depletion-layer capacitance associated with the collector junction of a transistor.

collector characteristic curves A set of characteristic curves of collector voltage versus collector current, for a fixed value of transistor base current.

collector-current runaway The continuing increase in collector current as collector junction temperature is increased by collector-current flow.

collector efficiency The ratio of useful power output to DC power input for a transistor, usually expressed as a percentage.

collector-follower effect An effect used in constructing a transformerless single-transistor flip-flop with a conventional bipolar junction transistor. As the base of the saturated common-emitter transistor is moved rapidly toward cutoff, the collector voltage follows and remains at the extreme value for a time before rising to the collector supply voltage.

collector junction A semiconductor junction between the base and collector electrodes of a transistor. It is normally biased in the high-resistance direction, and the current through it is controlled by introducing minority carriers.

collector transition capacitance The capacitance across the collector-to-base transition region, in which the impurity concentration changes.

collimate To modify the paths of electrons in a flooding beam, or of various rays of a scanning beam, to cause them to become more nearly parallel as they approach the storage assembly of a storage tube.

collimating lens An electron lens that collimates an electron beam in a storage tube.

collimation 1. The precise alignment of the mechanical system of a radar antenna by comparison with an optical device aligned on known points in azimuth and elevation. 2. The process of making light rays or the paths of electrons or other particles in a beam parallel to each other.

collimator A device that confines the elements of a beam within an assigned solid angle.

collinear array An antenna array that consists of a number of half-wave dipoles mounted end to end and connected to operate in phase. Also called linear array.

collinear heterodyning An optical processing system in which the correlation function is developed from an ultrasonic light modulator. The output signal is derived from a reference beam in such a way that the two beams are collinear until they enter the detection aperture. Variations in optical path length then modulate the phase of both signal and reference beams simultaneously, and phase differences cancel out in the heterodyning process.

collision A close approach of two or more particles, photons, atomic systems, or nuclear systems, during which there is an interchange of quantities like energy, momentum, and charge.

collision avoidance system [abbreviated CAS] An airborne electronic system that provides automatic warning of a potential midair collision between planes. Also called airborne collision avoidance.

collision broadening The broadening or spreading of a radiation line due to interruption of the radiating process when the radiator collides with another particle. In cyclotron radiation, the collision will also change the phase of the radiation.

collision-course homing Homing in which an offset antenna is used on the missile in conjunction with built-in computers that anticipate the motion of the target and direct the missile ahead of the present position of the target on a converging course which gives a collision in minimum missile travel time.

collision density The number of neutron collisions with matter per unit volume per unit time.

collision excitation The excitation of a gas by collisions of moving charged particles.

collision frequency The average number of collisions made by a particle per unit time.

collision ionization The ionization of atoms or molecules of a gas or vapor by collision with other particles.

colloidal graphite Extremely fine flakes of graphite suspended in water, petroleum oil, castor oil, glycerine, or other liquids. Used to provide conductive shields on the inside or outside surfaces of electron tubes.

colloidal silver Finely divided particles of silver, sometimes used on terminals of components to give a larger surface area for connections.

color A characteristic of light that can be specified in terms of luminance, dominant wavelength, and purity. Luminance is the magnitude of brilliance. Wavelength determines hue and ranges from about 4000 Å for violet to 7000 Å for red (400 to 700 nm). Purity corresponds to chroma or saturation and specifies vividness of a hue.

color balance Adjustment of the circuits that feed the three electron guns of a color picture tube to compensate for differences in light-emitting efficiencies of the three color phosphors on the screen of the tube.

color-bar code A bar code that uses one or more different colors of bars in combination with black bars and white spaces, to increase the density of binary coding of data printed on merchandise tags or directly on products for inventory control and other purposes.

color-bar generator A signal generator that delivers to the input of a color television receiver the signal needed to produce a color-bar test pattern on one or more channels.

color-bar test pattern A test pattern of different colors of vertical bars, used to check the performance of a color television receiver.

color breakup Momentary separation of a color picture into its primary components as a result of a sudden change in the condition of viewing, such as fast movement of the head or blinking of the eyes.

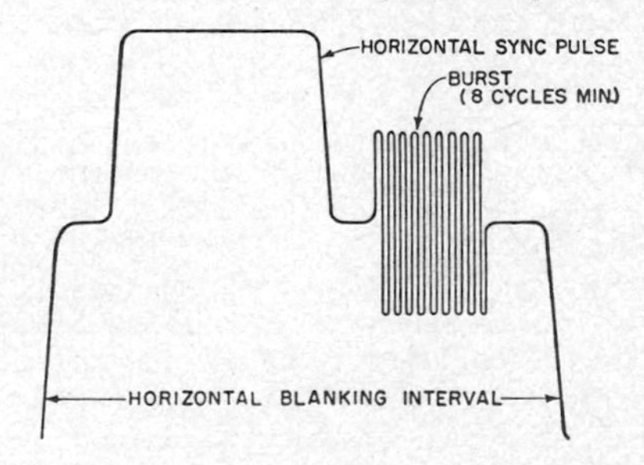

Color burst.

color burst A short series of oscillations at the chrominance subcarrier frequency of 3.579545 MHz, following each transmitted horizontal sync pulse in a color television system. This burst is a frequency reference for generating a continuous wave that is at the burst frequency and locked to it in phase. Also called burst.

color-burst pedestal The rectangular pulselike component that is part of the color burst when the axis of the color-burst oscillations does not coincide with the back porch. Also called burst pedestal.

color carrier Deprecated term for *chrominance subcarrier.*

color-carrier reference *Chrominance-carrier reference.*

color code A system of colors that indicates the electrical value of a component or identifies terminals and leads. The EIA color codes are the standards for the electronics industry.

color coder *Matrix.*

color comparator A photoelectric instrument that compares an unknown color with a standard color sample for matching purposes. The sample and unknown can be placed alternately in the measuring position, or two identical measuring systems can be used to feed a null indicator. Also called photoelectric color comparator.

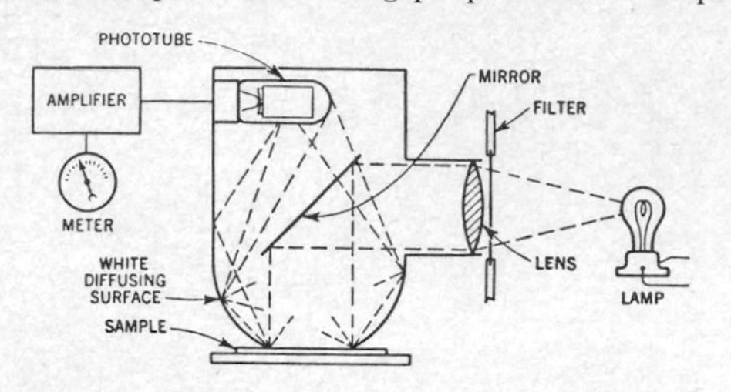

Color comparator in which sample and unknown are alternately placed in measuring position.

color contamination Poor rendition of color in a color television receiver, caused by incomplete separation of color component paths.

color control *Chroma control.*

color coordinate *Chromaticity coordinate.*

color decoder *Matrix.*

color-difference signal A signal that is added to the monochrome signal in a color television receiver to obtain a signal representative of one of the three tristimulus values needed by the color picture tube. There are three color-difference signals; the G − Y signal feeds the green channel, the R − Y signal feeds the red channel, and the B − Y signal feeds the blue channel.

color edging Spurious color at the boundaries of differently colored areas in a color television picture. Color edging includes color fringing and misregistration.

color encoder *Matrix.*

color fidelity The degree to which a color television system is capable of reproducing faithfully the colors in an original scene.

color field corrector A device used outside a color picture tube to produce an electric or magnetic field that acts on the electron beam after deflection to produce more uniform color fields.

color filter A sheet of material that absorbs certain wavelengths of light while transmitting others.

color flicker Flicker due to fluctuations of both chromaticity and luminance in a color television receiver.

color fringing Spurious chromaticity at boundaries of objects in a color television picture. Small objects may appear separated into different colors. It can be caused by a change in position of the televised object from field to field or by misregistration.

colorimeter An instrument that measures color by determining the intensities of the three primary colors which will give that color.

colorimetry The science of color measurement.

color killer circuit The circuit in a color television receiver that biases chrominance amplifier tubes or transistors to cutoff during reception of monochrome programs.

color organ An arrangement of colored lamps controlled by an audio system. Filters divide the audio spectrum into three frequency bands, each band controlling one primary-color set of lamps. The brightness of each set of lamps varies with the average loudness of the corresponding frequency band.

color oscillator *Chroma oscillator.*

color phase The difference in phase between a chrominance signal (I or Q) and the chrominance-carrier reference in a color television receiver.

color phase alternation [abbreviated CPA] The periodic changing of the color phase of one or more components of the chrominance subcarrier between two sets of assigned values after every field in a color television system.

color phase detector The color television receiver circuit that compares the frequency and phase of the incoming burst signal with those of the locally generated 3.579545-MHz chroma oscillator and delivers a correction voltage to a reactance tube to ensure that the color portions of the picture will be in exact register with the black and white portions on the screen.

color picture signal The electric signal that represents complete color picture information, excluding all synchronizing signals. In one form it consists of a monochrome component plus a subcarrier modulated with chrominance information.

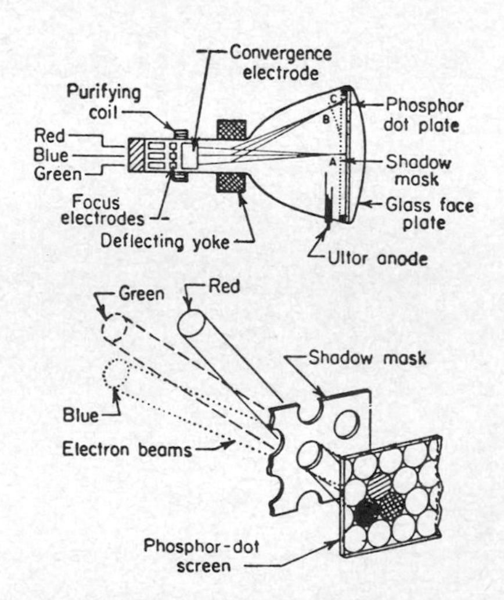

Color picture tube having three electron guns and shadow mask.

color picture tube A cathode-ray tube that has three different colors of phosphors. When these are appropriately scanned and excited in a color television receiver, a color picture is obtained.

colorplexer The section of a color television transmitter in which red, green, and blue signals are combined by matrixing and multiplexing to produce a single compatible color television signal.

color primaries The red, green, and blue primary colors that are mixed in various proportions to form all the other colors on the screen of a color television receiver.

color purity Absence of undesired colors in the spot produced on the screen by each beam of a television color picture tube.

color-purity magnet A magnet on the neck of a color picture tube, to improve color purity by changing the path of the electron beam.

color registration The accurate superimposing of the red, green, and blue images used to form a complete color picture in a color television receiver.

color response The sensitivity of a device to different wavelengths of light.

color sampling rate The number of times per second that each primary color is sampled in a color television system.

color saturation The degree to which a color is mixed with white. High saturation means little or no white, as in a deep red color. Low saturation means much white, as in light pink. The amplitudes of the I and Q chrominance signals determine color saturation in a color television receiver. Also called saturation.

color-saturation control *Chroma control.*

color sidebands The signals that extend for about 0.4 MHz above and below the 3.579545-MHz color subcarrier signal which is broadcast as part of a color television signal. The color sidebands contain picture chrominance information, which is removed from the color subcarrier by a synchronous detection process in receivers.

color signal Any signal that controls the chromaticity values of a color television picture, such as the color picture signal and the chrominance signal.

color subcarrier Deprecated term for *chrominance subcarrier.*

color-subcarrier oscillator *Chroma oscillator.*

color-subcarrier reference *Chrominance-carrier reference.*

color-sync signal A sequence of color bursts that is continuous except for a specified time interval during the vertical blanking period, each burst occurring at a fixed time with respect to horizontal sync in a color television system.

color television A television system that reproduces an image in its original colors. In the United States a compatible dot-sequential color television system is used. The bandwidth is 6 MHz, just as for monochrome television, and color synchronizing information is transmitted on a 3.579545-MHz subcarrier.

color television signal The entire signal used to transmit a full-color picture. It consists of the color picture signal and all the synchronizing signals.

color temperature The temperature of a blackbody radiator that produces the same chromaticity as the light under consideration.

color transmission The transmission of a signal wave for controlling both the luminance and chromaticity values in a picture.

color triangle A triangle drawn on a chromaticity diagram, to represent the entire range of chromaticities obtainable as additive mixtures of three prescribed primaries.

Colossus Project name for a Navy submarine detection system that has about 15 sonar heads per mile connected by submerged cables to land-based centers where the signals are analyzed and classified by comparison with submarine sound signatures.

Colpitts oscillator An oscillator in which a parallel-tuned tank circuit has two voltage-dividing capacitors in series, with their common connection

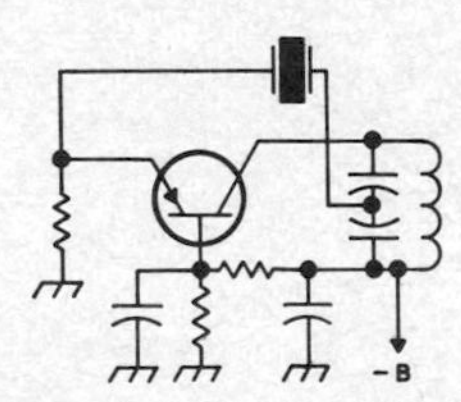

Colpitts oscillator using PNP transistor, with quartz crystal in feedback path.

going to the cathode in the electron-tube version and the emitter circuit of the transistor version.

columbium Former name for *niobium.*

column 1. *Card column.* 2. *Place.*

columnar ionization Regions of such dense ionization that even a strong external electric field cannot prevent some recombination. The term refers usually to ionization produced by alpha particles.

columnar recombination Recombination that takes place before ions have left the track. The ionization then takes place along a column, as in the dense ionization produced along the track of an alpha particle.

column-binary code A code used with punched cards, in which successive bits are represented by the presence or absence of punches on contiguous positions in successive columns as opposed to rows. Used in connection with 36-bit-word computers, where each group of three columns is used to represent a single word.

column printer A small line printer used with some calculators to provide hard-copy printout of input and output data. It typically may have 20 columns of numerals and a limited number of alphabetic or other identifying characters.

COM Abbreviation for *computer-output microfilm.*

coma A cathode-ray-tube image defect that makes the spot on the screen appear comet-shaped when the spot is away from the center of the screen.

coma lobe A side lobe that occurs in the radiation pattern of a microwave antenna when the reflector alone is tilted back and forth to sweep the beam through space. The coma lobe is produced under these conditions because the feed is no longer always at the center of the reflector. Used to eliminate the need for a rotary joint in the feed waveguide.

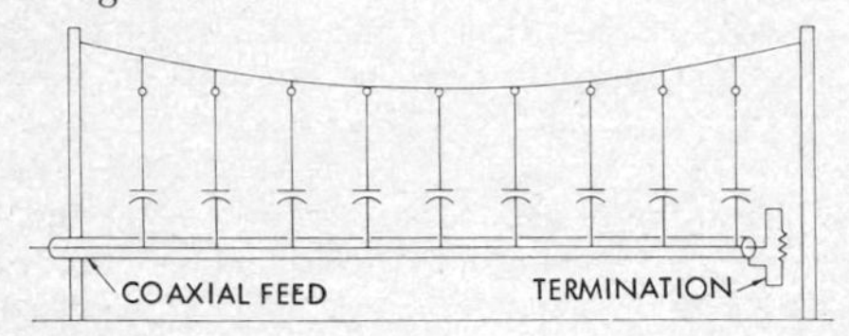

Comb antenna, supported from wire between two poles.

comb antenna A broadband antenna for vertically polarized signals, in which half of a fishbone antenna is erected vertically and fed against ground by a coaxial line.

combat information center A shipboard location at which tactical information from radar, sonar, and other equipment is received, displayed for rapid analysis, and evaluated. The results are distributed over wire lines or other communication links to appropriate points for action.

comb filter A wave filter whose frequency spectrum consists of a number of equispaced elements that resemble the teeth of a comb.

comb generator A signal generator that converts a single-frequency RF input signal into an RF output signal which has a large number of spectral lines, each line harmonically related to the input frequency.

combination microphone A microphone consisting of two or more dissimilar microphone elements.

combinatorial logic Logic in which the outputs are dependent only on input states and the delays encountered in the logic path. In contrast, sequential logic produces outputs dependent on the previous state of the logic array, on the presence of a discrete timing interval, and on the input states and delays.

combined double-T A combination of two T-shaped waveguide junctions, designed so the paths of power flow are determined by the loading of the arm and the matching between the arms.

combiner circuit The circuit that combines the luminance and chrominance signals with the synchronizing signals in a color television camera chain.

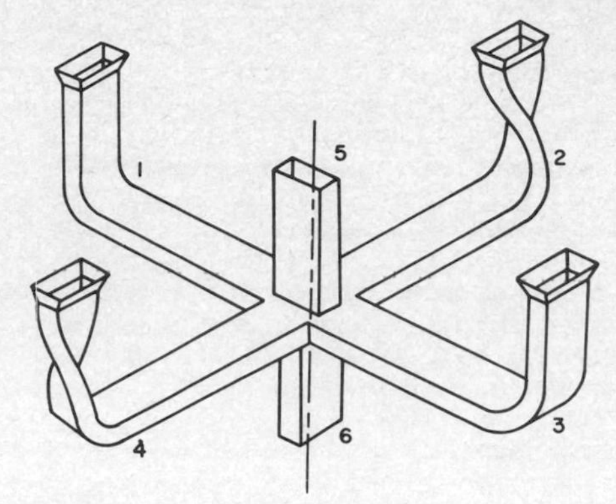

Combined double-T waveguide array for controlling power flow.

COMIT A user-oriented, general-purpose, symbol-manipulation programming language for computers.

COMLOGNET [COMbat LOGistic NETwork] A long-line data-transmission network for rapid processing and distribution of logistic information at computer-equipped automatic switching centers. The network translates input data and transmits the information to the addressee, with messages routed on a priority basis.

command 1. A signal that initiates a predetermined type of computer operation which is defined by an instruction. 2. The control of an airborne guided missile or other pilotless aircraft by electronic signals. 3. An independent signal in a feedback control system from which the dependent signals are controlled in a predetermined manner.

command-destruct signal A radio signal that detonates the destruction device of a guided missile in case of a malfunction.

command guidance A missile guidance system in which flight direction information is transmitted to the missile from a point external to the missile. The information needed about missile performance may be derived by telemetering, ground-based radar, optical tracking, or other means. The commands may be transmitted to the missile by radio, radar, or other means.

command module The spacecraft module that carries the crew, the main communication and telemetry equipment, and the reentry capsule during cruising flight. In the Apollo project, the command module orbited the moon as the lunar excursion module descended to the moon's surface, explored it, and returned to the command module.

command resolution The maximum change in command that can be made in a feedback control system without causing a change in the ultimately controlled variable.

command set A radio set that receives or gives commands, as between one aircraft and another or between an aircraft and the ground.

commentator One who edits and broadcasts news at a radio or television station, often interspersed with personal comments.

commercial An advertising message that is broadcast by television or radio.

Commission Internationale de l'Eclairage [abbreviated CIE] An international group that has set most of the basic standards of light and color now used in color television. Also abbreviated ICI (deprecated).

common Shared by two or more services, circuits, or devices.

commonality Having similar characteristics or similar operating and maintenance requirements.

common-base connection *Grounded-base connection.*

common carrier A company recognized by an appropriate regulatory agency as having a vested interest in furnishing communication services.

common cathode A cathode for two or more sections of an electrode tube, as in a duodiode pentode.

common-collector connection *Grounded-collector connection.*

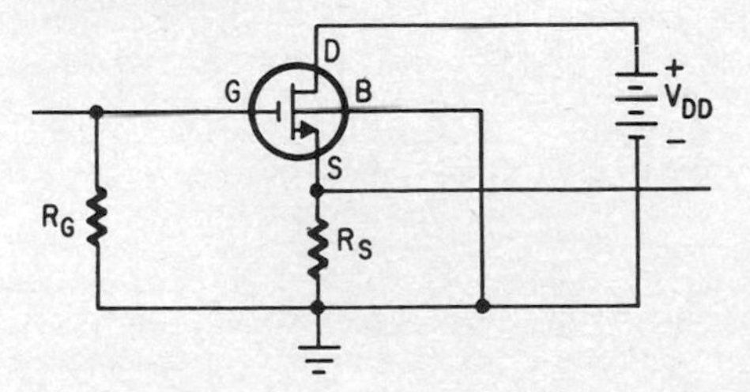

Common-drain amplifier using MOS field-effect transistor.

common-drain amplifier An amplifier which uses a field-effect transistor in such a way that the input signal is injected between gate and drain and the output is taken between the source and drain. Used chiefly in applications that require a low input capacitance or the ability to handle large input signals. Also called source-follower amplifier.

common-emitter connection *Grounded-emitter connection.*

common-gate amplifier An amplifier that uses a field-effect transistor in which the gate is common to both the input and output circuits. Used for transforming a low input impedance to a high output impedance and for high-frequency amplifier applications.

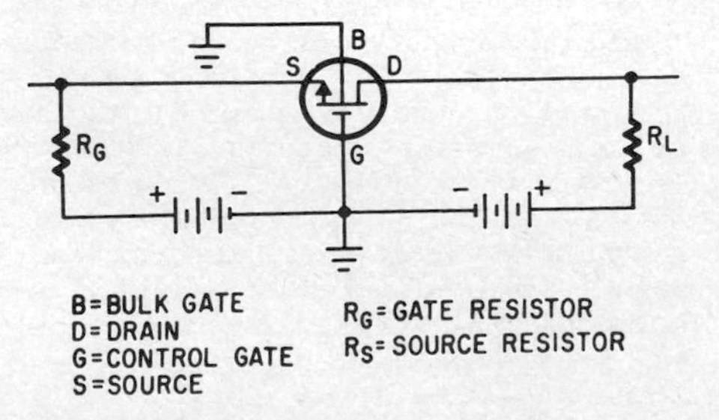

Common-gate amplifier using MOS field-effect transistor.

common language A machine-readable language that is common to a group of computers and associated equipment.

common mode Having signals that are identical in amplitude and phase at both inputs, as in a differential operational amplifier.

common-mode error The error voltage that exists at the output terminals of an operational amplifier due to the common-mode voltage at the input.

common-mode interference Interference signals that appear between both inputs of a differential circuit and a common reference like ground.

common-mode rejection [abbreviated CMR] The ability of an amplifier to cancel a common-mode signal while responding to an out-of-phase signal. Usually expressed in decibels. CMR = 20 $\log_{10}$ (CMRR); thus if the CMRR is 10^6, the corresponding CMR value is 120 dB. Also called in-phase rejection.

common-mode rejection ratio [abbreviated CMRR] The ratio of the RMS value of the common-mode interference voltage at the input terminals of an operational amplifier to the effect or error produced by that common-mode interference at the output when referred to the input (when divided by amplifier gain). A high ratio is desirable. This ratio expresses the ability of the device to reject the effect of a voltage that is applied simultaneously to both input terminals.

common-mode signal A signal applied equally to both ungrounded inputs of a balanced amplifier stage or other differential device. Also called in-phase signal.

common-mode voltage A voltage that appears in common at both input terminals of a device with respect to the output reference (usually ground).

common return A return conductor that serves two or more circuits.

common-source amplifier An amplifier stage that uses a field-effect transistor in which the input signal is applied between gate and source and the

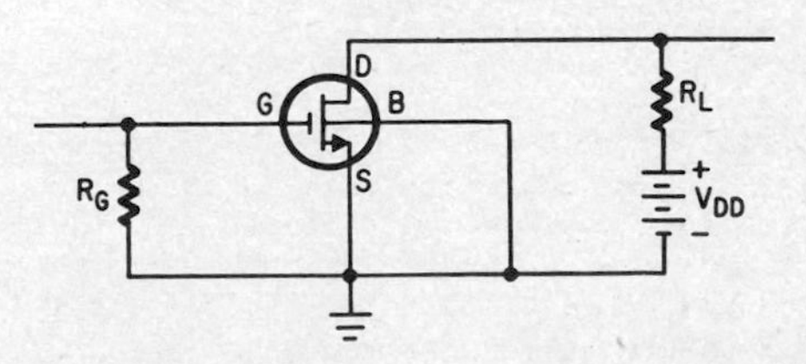

Common-source amplifier using MOS field-effect transistor.

output signal is taken between drain and source. It has high input impedance, medium to high output impedance, and voltage gain greater than unity.

common system A system of air navigation and traffic control for common use by all civil and military aviation.

communication The transmission of intelligence between two or more points over wires or by radio. The terms telecommunication and communication are often used interchangeably, but telecommunication is usually the preferred term when long distances are involved.

communication band The band of frequencies effectively occupied by a radio transmitter for the type of transmission and the speed of signaling used.

communication channel The wire or radio channel that conveys intelligence between two or more terminals.

communication countermeasure Any electronic countermeasure against communications.

communication deception Use of devices or techniques that confuse or mislead the user of a communication link or radio navigation system.

communication receiver A receiver for reception of voice or code messages transmitted by radio communication systems.

communication satellite An orbiting satellite that relays radio, television, and other signals between ground terminal stations thousands of miles apart. A satellite is placed in orbit 22,300 mi (35,-

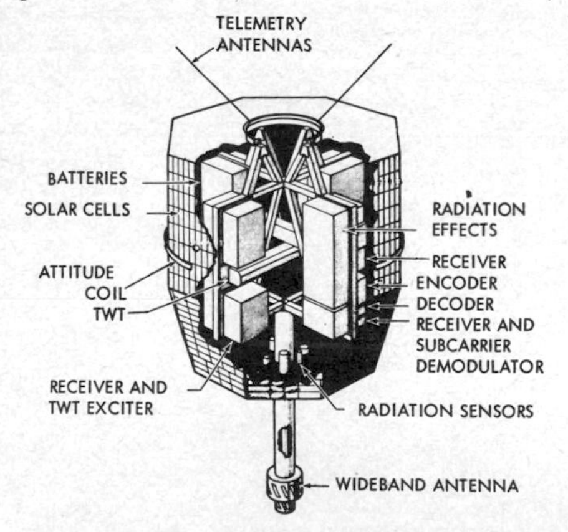

Communication satellite (early design) in which solar cells are mounted directly on housing, rather than on paddles that spread out after launch.

880 km) above the earth to give a 24-h orbital period that makes it appear stationary. A radio wave traveling at the speed of light then takes about 0.12 s to travel from an earth station to the satellite and another 0.12 s to return to another earth station.

communications engineer An engineer who specializes in the design, construction, and operation of all types of equipment used for radio, wire, or other types of communication.

communications intelligence Information derived from transmitted signals by other than the intended recipients.

Communications Satellite Corp. [abbreviated COMSAT] A common carrier created under the provisions of the Communications Satellite Act of 1962, to provide communication satellite service.

communications security Protection of transmitted information by converting it to a form that is unintelligible to an unauthorized listener yet can be reconverted to its original form at the intended receiving station, generally by using a crypto system.

community antenna television *Cable television.*

commutated-antenna direction-finding A radio receiver that is switched in sequence to one antenna after another in a circular array, to determine the direction of arrival of radio waves by sensing the commutation phase shifts.

commutating angle The fraction of an AC cycle, expressed in degrees, during which current is commutated from one rectifying element to another in an inverter or rectifier circuit.

commutating capacitor A capacitor in gas-tube rectifier circuits that prevents the anode from going highly negative immediately after extinction.

commutating diode A diode in a switching regulator circuit that provides current for the filter choke during the time the switching transistor is not conducting.

commutating reactance An inductive reactance placed in the cathode lead of a three-phase mercury-arc rectifier to ensure that tube current holds over during transfer of conduction from one anode to the next. Without this reactance the arc would go out, and the tube would have to be restarted by auxiliary means.

commutation 1. The transfer of current from one anode to another in a gas tube. 2. The sampling of various quantities in a repetitive manner, for transmission over a single channel in telemetering.

commutator 1. A circular arrangement of copper bars insulated from each other and the rotor of the DC motor or generator on which they are mounted, with current-carrying brushes bearing on the exposed surfaces to provide commutation of direct current to armature coils in sequence. 2. A device in time-division multiplexing that provides repetitive sequential switching of signals from a multiplicity of channels.

commutator switch A switch, usually rotary and mechanically driven, that performs a set of switching operations in repeated sequential order, for example, as required for telemetering many quantities. Also called sampling switch and scanning switch.

compactron A multifunction electron tube that

has a special 12-pin base. A single compactron may replace two or more tubes or transistors. Originally (up to 1962) the term was the trademark of General Electric Co. for devices of this type.

companding A process in which compression is followed by expansion. Often used for noise reduction, in which case compression is applied before noise exposure and expansion after exposure.

compandor A system that improves the signal-to-noise ratio by compressing the volume range of the signal at a transmitter or recorder by a compressor and restoring the normal range at the receiving or reproducing apparatus with an expandor.

comparator 1. A device that compares two transcriptions of the same information to verify the accuracy of transcription, storage, arithmetic operation, or some other process in a computer and delivers an output signal of some form to indicate

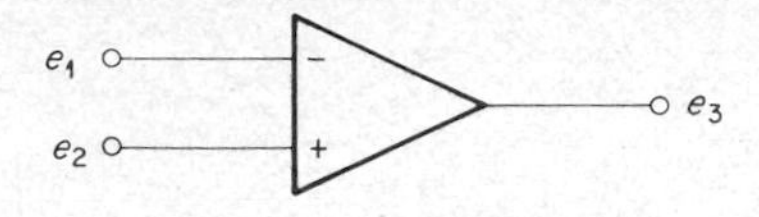

Comparator using operational amplifier without feedback. If inverting input e_1 changes while e_2 is held constant, e_3 will change in opposite direction from e_1. If noninverting input e_2 changes while e_1 is held constant, e_3 will change in same direction as e_2.

whether or not the two sources are equal or in agreement. 2. An electronic instrument that measures a quantity and compares it with a precision standard. 3. An operational amplifier used without feedback to detect changes in voltage level, as required in analog-to-digital, digital-to-analog, and other types of converters.

compare To examine the representation of two groups of characters or machine words in a computer, to determine whether they are the same or different.

comparison A computer operation in which two numbers are compared as to identity, relative magnitude, or sign.

comparison bridge 1. A bridge circuit in which any change in the output voltage with respect to a reference voltage creates a corresponding error signal that by means of negative feedback is used to correct the output voltage and thereby restore bridge balance. 2. A bridge that compares the values of two impedances and gives the result as a ratio.

compass bearing A bearing measured relative to compass north.

compatibility The ability of a new system to serve users of an old system.

compatible color television system A color television system that permits the substantially normal monochrome reception of the transmitted color picture signal on a typical unaltered monochrome receiver. This is accomplished in the U.S. color television system by dividing the color video information into a luminance signal and two chrominance signals. The luminance signal is the equivalent of a monochrome television picture signal and is utilized alone by a monochrome receiver.

compatible current-sinking logic [abbreviated CCSL] A semiconductor integrated-circuit logic family that is interchangeable with diode-transistor logic for comparable functions.

compatible discrete four-channel sound [abbreviated CD-4 sound] A sound system in which a separate channel is maintained from each of the four sets of microphones at the recording studio or other input location to the four sets of loudspeakers that serve as the output of the system. Four channels are therefore required on records and magnetic tape.

compatible integrated circuit An integrated-circuit family that has input/output logic levels and operating characteristics which make it compatible with one or more other families of integrated circuits.

compatible single-sideband system A single-sideband system that can be received by an ordinary amplitude-modulation radio receiver without distortion.

compatible stereo system A stereo system that gives satisfactory single-channel sound for people who have single-channel phonographs, tape equipment, or radio receivers.

compensated amplifier A broadband amplifier in which the frequency range is extended by choice of circuit constants.

compensated impurity A donor or acceptor impurity that is neutralized by impurities of the opposite type in a semiconductor.

compensated ionization chamber An arrangement of two ionization chambers in parallel, with potentials reversed, used as a radiation null indicator. One chamber has an adjustable source of ionizing radiation, such as uranium. This is adjusted until both chambers have the same ionization. The instrument then acts as a null indicator that shows both increases and decreases from normal ionization in the main chamber.

compensated-loop direction finder A radio direction finder that has a loop antenna whose polarization error is compensated by an additional antenna system.

compensated semiconductor A semiconductor in which one type of impurity or imperfection partially cancels the effects of the other type of impurity or imperfection. Thus a donor impurity may offset an acceptor impurity.

compensated volume control *Loudness control.*

compensation 1. Modification of the amplitude-frequency response of an amplifier to broaden the bandwidth or make the response more nearly uniform over the existing bandwidth. Also called frequency compensation. 2. The introduction of donors into a P-type semiconductor or acceptors into an N-type semiconductor.

compensation signal A signal recorded on magnetic tape, along with computer data, for use during playback to correct for the effects of tape-speed errors.

compensator A component that offsets an error or other undesired effect.

compile To prepare a machine-language program automatically from a computer program written in another programming language.

compiler A computer program that performs the item-by-item translating functions of an assembler and replaces certain items of input with series of instructions, usually called subroutines. Also called compiling routine.

compiling routine *Compiler.*

complement A number whose representation for a computer is derived from the finite positional notation of another number by subtracting each digit from one less than the base, adding 1 to the least significant digit, and executing all carries required. This is the true complement. Thus the two's complement of binary 10010 is 01110; the ten's complement of decimal 2546 is 7454. In many machines a negative number is represented as a complement of the corresponding positive number.

complementary Having PNP and NPN or P- and N-channel semiconductor elements on or within the same integrated-circuit substrate or working together in the same functional amplifier state.

complementary color A color that, when added to another given color in proper proportion, produces white.

complementary constant-current logic [abbreviated C^3L] A form of bipolar logic in which Schottky and PNP transistors are combined with Schottky barrier diodes to achieve high packing density and high resistance to radiation.

complementary direct-coupled amplifier A direct-coupled amplifier in which PNP and NPN transistor stages are in alternating sequence, to simplify the use of direct coupling and negative feedback.

Complementary constant-current logic for NAND gate.

complementary functions Two driving-point functions whose sum is a positive constant.

complementary logic switch A complementary transistor pair which has a common input and interconnections such that one transistor is on when the other is off, and vice versa. Load current flows only briefly during switching, when the voltage at the junction between the transistors changes from the level of logic 1 to the level of logic 0.

complementary metal-oxide semiconductor [abbreviated CMOS] A combination of N- and P-channel enhancement-mode devices on a single silicon chip, connected into push-pull complementary digital circuits. Advantages include low quiescent power dissipation and high operating speed. Also called complementary-symmetry metal-oxide semiconductor.

complementary-output circuit A logic circuit that has two outputs, one of which is logic 0 when the other is logic 1.

complementary symmetry A circuit that uses both PNP and NPN transistors in a symmetrical arrangement which permits push-pull operation without an input transformer or other form of phase inverter.

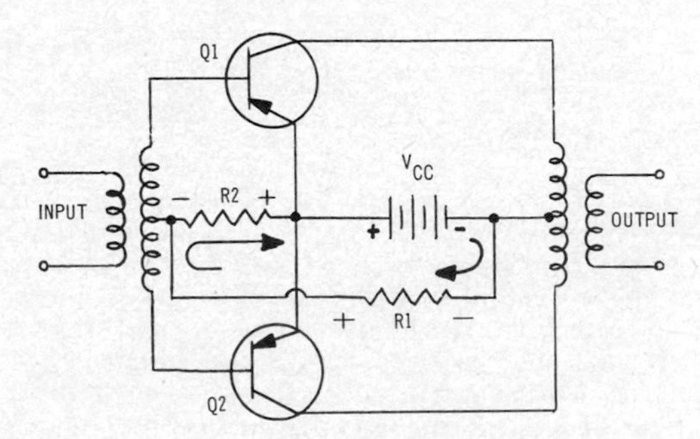

Complementary symmetry in class B push-pull amplifier having small bias voltage developed across R2.

complementary-symmetry metal-oxide semiconductor *Complementary metal-oxide semiconductor.*

complementary tracking Interconnection of

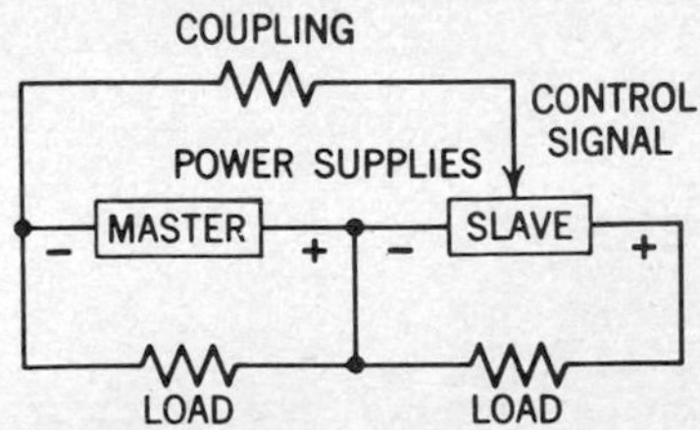

Complementary tracking of master and slave regulated power supplies.

two regulated power supplies so one is acting as master to control the other. The output voltage of the slave supply is equal or proportional to that of the master but of opposite polarity with respect to a common point.

complementary-transistor amplifier An amplifier that uses the complementary symmetry of NPN and PNP transistors.

complementary-transistor logic [abbreviated CTL] Logic that uses complementary transistors, generally in medium- and small-scale integration.

complementary transistors Two transistors whose characteristics and ratings are similar but opposite in sense, such as PNP and NPN bipolar transistors or P- and N-channel field-effect transistors, used together in a stage to give push-pull output from a single input.

complementary unijunction transistor [abbreviated CUJT] A semiconductor device with characteristics like those of a standard unijunction

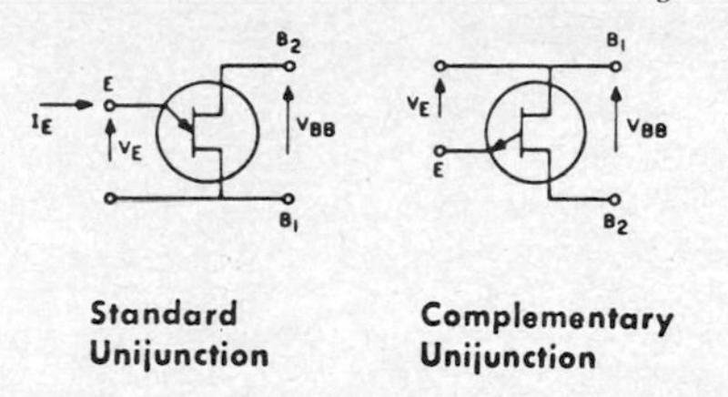

Complementary unijunction transistor symbol, with standard symbol shown for comparison.

transistor, except that the currents and voltages applied to it are of opposite polarity.

complementary wavelength The wavelength of light that, when combined with a sample color in suitable proportions, matches a reference standard light. The purples that have no dominant wavelengths, including nonspectral violet, purple, magenta, and nonspectral red colors, are specified by use of their complementary wavelengths.

complete carry The condition wherein a carry that results from the addition of carries is allowed to propagate in a computer.

complex operator The letter *j*, used to designate the reactive component of a complex impedance.

complex reflector A structure or group of structures having many radar-reflecting surfaces facing in different directions.

complex target A radar target composed of a number of reflecting surfaces that, in the aggregate, are smaller in all dimensions than the resolution capabilities of the radar.

complex tone A sound wave produced by the combination of simple sinusoidal components of different frequencies.

complex value An impedance value that consists of magnitude and phase angle presented as a function of frequency.

compliance The acoustical and mechanical equivalent of capacitance. It is the opposite of stiffness. In a phonograph, compliance is the force needed to move the stylus a given distance.

compliance voltage The output voltage range needed in a regulated DC power supply to maintain a specified constant value of current for a specified range of load resistances.

component Any electric device, such as a coil, resistor, capacitor, line, transistor, or electron tube, that has distinct electrical characteristics and terminals at which it may be connected to other components to form a circuit. Also called element.

component density The number of components per unit area or unit volume.

composite circuit A circuit used simultaneously for voice communication and telegraphy, with frequency-discriminating networks separating the two types of signals.

composite color signal The color picture signal plus all blanking and synchronizing signals. The composite color signal thus includes the luminance signal, the two chrominance signals, vertical and horizontal sync pulses, vertical and horizontal blanking pulses, and the color-burst signal.

composite color sync The signal comprising all the sync signals necessary for proper operation of a color receiver. This includes the horizontal and vertical sync and blanking pulses and the color-burst signal.

composite modulation voltage The combined output voltage of the subcarrier oscillators in a telemetering system, applied as modulation to the transmitter.

composite picture signal The complete picture signal as it leaves the television transmitter. The picture consists of picture data, blanking pulses, synchronizing pulses for monochrome, and the color subcarrier, color burst, and other information needed for transmission of color pictures. Also called composite signal and composite video signal.

composite pulse A pulse composed of a series of overlapping pulses received from the same source over several paths in a pulse navigation system.

composite signal *Composite picture signal.*

composite video signal *Composite picture signal.*

composite wave filter A combination of two or more low-pass, high-pass, bandpass, or band-elimination filters.

composition resistor *Carbon resistor.*

compound-connected transistors An arrangement of two transistors in which the base of one is connected to the emitter of the other and the two collectors are connected together. The combination may be considered a single transistor that has a high current amplification factor.

compound horn An electromagnetic horn of rectangular cross section, the four sides of which diverge to coincide with or approach four planes. The line of intersection of two opposite planes does not intersect the line of intersection of the remaining planes.

compound modulation Modulation in which one or more signals modulate their respective subcarriers, and these subcarriers in turn modulate the carrier.

compound nucleus An excited nucleus formed as an intermediate stage in an induced nuclear reaction. It is characterized by a long lifetime compared with the normal transit times of nuclear particles across the nucleus.

compound target A radar target composed of a number of randomly disposed reflecting surfaces, the aggregate extent of which exceeds any of the dimensions of the pulse packet.

compressed-air loudspeaker A loudspeaker that has an electrically actuated valve which modulates a stream of compressed air.

compression Reduction of the volume range of an AF signal. Weak signal components are made stronger so that they will not be lost in background noise, whereas loud passages are reduced in strength so they will not overload any part of the system. This action is achieved by making the effective gain vary automatically as a function of signal magnitude. The same action is used in disk recording to prevent the cutting stylus from going into adjacent grooves at high volume levels, and in facsimile to reduce the white-to-black amplitude range or frequency swing.

compressional wave A wave in an elastic medium that causes an element of the medium to change its volume without undergoing rotation. A compressional plane wave is a longitudinal wave.

compression molding Molding a record or transcription by compressing a preform of plastic.

compression ratio The ratio of the amplification at a reference signal level to that at a higher stated signal level.

compressive intercept receiver An electromagnetic surveillance receiver that instantaneously analyzes and sorts all signals within a broad RF spectrum by using pulse compression techniques which perform a complete analysis up to 10,000 times faster than a superheterodyne receiver or spectrum analyzer. One version has three identical, highly dispersive, surface acoustic-wave filters. Two filters act as pulse expansion lines in a passive sweeping local-oscillator configuration, and the third filter acts as a pulse compression line in the signal-processing portion of the receiver.

compressor The part of a compandor that compresses the intensity range of signals at the transmitting or recording end of a circuit. The compressor amplifies weak signals and attenuates strong signals, to produce a smaller amplitude range.

compromise net A network used with a hybrid junction to balance a connected communication circuit such as a subscriber's loop, other lines, or equipment. It gives a compromise between the extremes of impedance balance.

Compton absorption The absorption of an x-ray or gamma-ray photon in the Compton effect. The energy of the electromagnetic radiation is not completely absorbed since another photon of lower energy is simultaneously created.

Compton effect The elastic scattering of photons by electrons. Since the total energy and total momentum are conserved in the collisions, the wavelength of the scattered radiation is changed by an amount that depends on the angle of scattering, and part of the photon energy is transferred to electrons. Also called Compton scattering.

Compton electron An electron whose energy has been increased as a result of the Compton effect during interaction with a photon.

Compton meter An ionization chamber that has a balance chamber with a uranium source which is adjusted until it balances out normal cosmic radiation. Variations in cosmic radiation are then shown on an electrometer.

Compton recoil electron An electron that has been set in motion through interaction with a photon in the Compton effect.

Compton recoil particle Any particle which has acquired its momentum in a scattering process similar to that in the Compton effect.

Compton scattering *Compton effect.*

Compton shift The change in wavelength of scattered radiation caused by the Compton effect.

Compton wavelength A wavelength characteristic of a particle of given mass. Its value for the electron is 2.42621×10^{-10} cm, and for the proton it is 1.32140×10^{-13} cm.

computer A machine capable of accepting information, processing the information, and supplying results in a desired form. Examples include analog, digital, and hybrid computers.

computer-aided design [abbreviated CAD] Use of a computer to make the calculations required for optimizing the design of a new electronic device, circuit, or system.

computer-aided instruction [abbreviated CAI] Use of a computer with a large number of on-line student terminals to supplement or replace classroom instruction. Also called computer-assisted instruction.

computer-assisted instruction *Computer-aided instruction.*

computer code The system of binary digits or other machine codes used by a given computer.

computer control Process control in which the process variables are fed into a computer and the output of the computer is used to control the process.

computer graphics The generation of drawings, patterns, and graphs on paper, microfilm, or a cathode-ray screen by a computer.

computer-independent language A programming language that can be translated into a variety of computer languages.

computer instruction An instruction expressed

in the code required by a specific computer.

computerize To use a computer in connection with an accounting, process control, or other function previously performed without computers.

computer language A language that consists of a vocabulary of commands and a set of rules for using those commands to make a computer solve a particular problem. Commonly used languages include ALGOL, APL, BASIC, COBOL, and FORTRAN.

computer-limited A situation in which the time required for computation exceeds the time required to read inputs and write outputs.

computer network Two or more interconnected computers.

computer operation The electronic action required in a computer to give a desired computation.

computer-output microfilm [abbreviated COM] Microfilm produced in a recorder by computer-generated signals, either directly on-line or off-line from a magnetic-tape drive. Recording techniques include photographing a page at a time on the face of a cathode-ray tube, writing directly on unexposed film with an electron beam in a vacuum, and photographing one line at a time on an illuminated optical-fiber matrix.

computer-output typesetting Production of graphic arts quality printout of computer information on photographic paper or film. Some systems include page makeup and automatic insertion of computer-stored or slide-projected illustrations.

computer program A routine for solving a problem or carrying out an information-processing operation on a computer.

computer terminal A serial printer in which characters are produced one at a time in sequence on paper and/or on a cathode-ray screen as corresponding signals are fed directly or over wire lines from the terminal to the computer. The same terminal is used in reverse to print and/or show the output of the computer.

computer utility A computer that provides service on a time-sharing basis, generally over telephone lines, to subscribers who have appropriate terminals. Users may have their own programs and data stored in the computer or may use general-purpose programs that are available to all customers.

computing linkage An assembly of rigid links, pivots, and sliding members so arranged that the motion of a selected output link is a predetermined function of the motions of all the input links.

COMSAT Abbreviation for *Communications Satellite Corp.*

concentrator A switching system that lets a large number of telephone or data-processing subscribers use a lesser number of transmission lines or a narrower bandwidth, such as by storing messages until lines become available and then selecting messages for transmission on a first-come or other priority basis.

concentric groove *Locked groove.*

concentric line *Coaxial cable.*

concentric transmission line *Coaxial cable.*

concurrent Occurring or being performed simultaneously or within the same specified interval of time.

condenser 1. A system of lenses that concentrates or focuses light rays to a point. Also called condensing lens. 2. Deprecated term for *capacitor.*

condenser loudspeaker Deprecated term for *electrostatic loudspeaker.*

condenser microphone Deprecated term for *capacitor microphone.*

condensing lens *Condenser.*

conditional Subject to the result of a comparison made during computation in a computer, or subject to human intervention.

conditional breakpoint instruction A conditional jump instruction that, if some specified switch is set, will cause a computer to stop. The routine may then be continued as coded or a jump may be forced.

conditional jump A computer instruction that will cause the proper one of two or more addresses to be used in obtaining the next instruction, depending on some property of a numerical expression which may be the result of some previous instruction. Also called branch, conditional transfer, and discrimination.

conditional stability A property of a system that causes it to be stable for certain values of input signal and gain and unstable for other values. Also called limited stability.

conditional transfer *Conditional jump.*

conditioning Modifying data-transmission lines and equipment as required to match transmission levels and impedances or provide equalization between facilities.

Condor A Navy air-to-surface missile that uses optoelectronic guidance, developed for use beyond the range of antiaircraft guns which protect heavily defended ground targets. Range is about 50 mi (80 km).

conductance [symbol *G*] A measure of the ability of a material to conduct electric current. It is the reciprocal of the resistance of the material and is expressed in mhos. Conductance is the resistive or real part of admittance.

conducted interference Interfering signals arriving on power lines.

conducting polymer A plastic which has high conductivity approaching that of metals.

conduction The transmission of energy by a medium without movement of the medium itself.

conduction band An energy band in which electrons can move freely in a solid. The conduction band is normally empty or partly filled with electrons.

conduction current A current due to a flow of

conduction electrons through a body or charge carriers through a semiconductor.

conduction electron An electron in the conduction band of a solid, where it is free to move under the influence of an electric field. Also called outer-shell electron and valence electron.

conduction pump A pump in which liquid metal or some other conductive liquid is moved through a pipe by sending a current across the liquid and applying a magnetic field at right angles to current flow.

conductive coating A coating that reduces surface resistance and thus prevents the accumulation of static electric charges.

conductive elastomer A rubberlike silicone material in which suspended metal particles conduct electricity.

conductive gasket A flexible metallic gasket that reduces RF leakage at joints in shielding.

conductive pattern A design formed of any conductive or resistive material.

conductive silver paste Silver powder in a suitable vehicle for applying to ceramic or other insulating materials by silk-screening or other methods, then fixing or firing at appropriate temperatures to provide a hard conductive surface or joint.

conductivity The ability of a material to conduct electric current, as measured by the current per unit of applied voltage. It is the reciprocal of resistivity. In a semiconductor, N-type conductivity is associated with conduction electrons, and P-type conductivity is associated with holes.

conductivity modulation Variation of the conductivity of a semiconductor by variation of the charge-carrier density.

conductivity-modulation transistor A transistor in which the active properties are derived from minority carrier modulation of the bulk resistivity of a semiconductor.

conductor A wire, cable, or other body or medium that is suitable for carrying electric current.

conductor pattern A conductive pattern that has low electric resistance.

conduit Solid or flexible metal or other tubing through which insulated electric wires are run.

cone The cone-shaped paper or fiber diaphragm of a loudspeaker.

cone antenna *Conical antenna.*

cone loudspeaker A loudspeaker that uses a magnetic driving unit which is mechanically coupled to a paper or fiber cone.

cone of nulls A conical surface formed by directions of negligible radiation from an antenna.

cone of protection The volume enclosed by a cone whose apex is the highest point of a tower or other grounded structure and whose base radius is related to the height of the cone and the height of a cloud from which lightning might strike. The minimum radius at the ground is about equal to tower height for low clouds, with the radius of protection increasing to about twice this value as cloud height increases.

cone of silence A cone-shaped region, directly over the antenna of a radio-beacon transmitter, in which no signal is heard by the pilot of an aircraft.

cone resonance The frequency at which the diaphragm or cone of a loudspeaker vibrates most easily. Cone resonance must be minimized by careful design and by placing the loudspeaker in a proper enclosure, to prevent abnormally high acoustic output at the resonant frequency.

conference call A telephone or radio call that permits simultaneous conversation between three or more locations.

confidence The degree of assurance that a specified failure rate is not exceeded.

confidence curve A curve from which the maximum failure rate of a component in the first 1000 h of operation can be determined with a given percentage of confidence.

confidence interval The probability, generally expressed as a percentage, that a characteristic or performance specification will be within a specified range of values.

confidential Having such security status that unauthorized disclosure could be prejudicial to the defense interests of the nation.

configuration 1. A group of machines interconnected and programmed to operate as a system. 2. A group of components interconnected to perform a desired circuit function.

confinement Restriction of a hot plasma to a given volume as long as possible, by such means as magnetic mirrors and pinch effect.

confocal resonator A wavemeter for millimeter wavelengths; it consists of two spherical mirrors which face each other. Changing the spacing between the mirrors affects propagation of electromagnetic energy between them, permitting direct measurement of free-space wavelength. Confocal operation occurs when the center of curvature of one mirror is on the surface of the other.

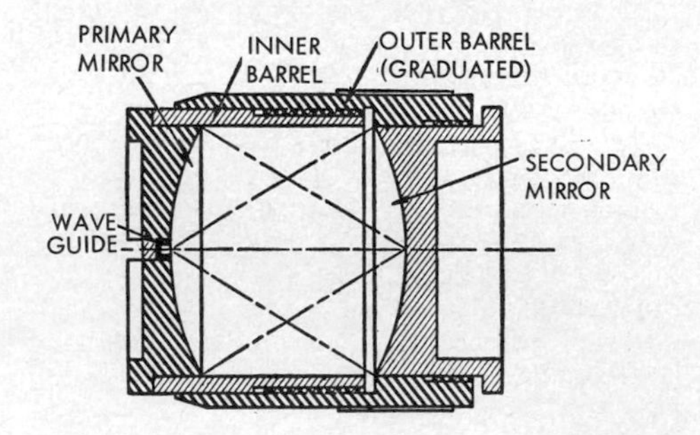

Confocal resonator mounted on end of waveguide for use as wavemeter in millimeter range.

conformal array A circular, cylindrical, hemispherical, or other shaped array of electronically switched antennas, to provide the special radiation patterns required for tacan, IFF, and other air navigation, radar, and missile control applications.

conformal coating A thin layer of epoxy, silicone, or other insulating material applied by spraying or any other method that gives a protective coating conforming to the irregular surfaces of an integrated circuit or other electronic device, generally cured by heating.

confusion jamming An electronic countermeasure technique in which the signal from an enemy tracking radar is amplified and retransmitted with distortion to create a false echo that affects accuracy of target range, azimuth, and velocity data.

confusion reflector An electromagnetic-wave reflector that is dropped from aircraft to create false signals on enemy radarscopes. It usually consists of strips of aluminum foil or metallized paper such as chaff or window, cut to lengths that are approximately resonant to the expected enemy radar frequency. The strips give strong echo signals when dropped in clusters.

conical antenna A wideband antenna in which the driven element is conical in shape. Also called cone antenna.

conical helix antenna A frequency-independent circularly polarized antenna that resembles an inverted cone and provides a unidirectional radiation pattern.

conical horn A horn that has a circular cross section and straight sides.

conical horn antenna A horn antenna that has a circular cross section and straight sides, as in a cone. It is energized by a circular waveguide that feeds the smaller end of the horn.

conical scanning Scanning in which the radar beam describes a cone whose axis coincides with the axis of the reflector.

conjugate branches Two network branches so related that a voltage applied to either branch produces no response in the other.

conjugate bridge A bridge in which the detector circuit and the supply circuits are interchanged, as compared with a normal bridge of the given type.

conjugate impedances Impedances that have resistance components which are equal and reactance components which are equal in magnitude but opposite in sign.

conjunction 1. The logic operation that uses the AND operator or logic product. 2. The condition wherein a planet or spacecraft has the same celestial longitude as the sun, when using the center of the earth as a reference.

conjunction-class trip A round trip to Mars that is made when the earth and Mars are on opposite sides of the sun. Round-trip time will be about 900 days. In contrast, an opposition-class trip is made when the earth is between Mars and the sun.

connected network A network in which there is at least one path, composed of branches of the network, between every pair of nodes of the network.

connection A direct wire path for current between two points in a circuit.

connection diagram A diagram that shows the connections needed to place in operation an electronic system which consists of one or more assemblies, power supplies, and devices being controlled.

connector A complete electric connecting device, consisting of a mating plug and receptacle for cables or of mechanically mating flanges for waveguides.

connect time 1. In a computer-based data communication system, the switching time required to set up a connection between two terminal points. 2. In a telephone system, the time duration of a connection between two points.

conoscope An optical instrument that locates the optical or Z axis of a quartz crystal.

Consol A radio navigation aid that provides a number of characteristic signal zones which rotate

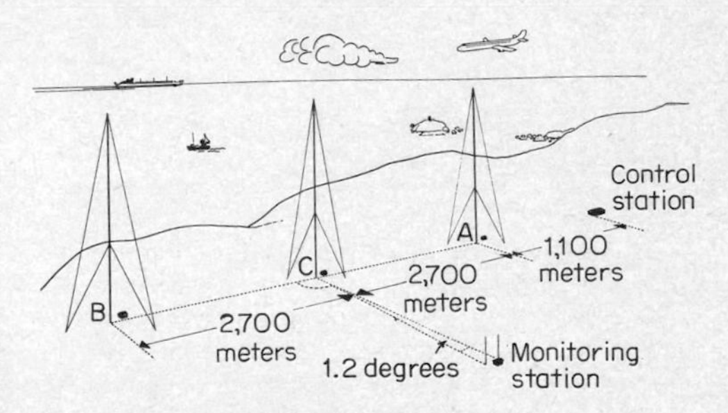

Consol installation in Norway, with antennas 2700 m apart.

in a time sequence. A bearing may be determined by observation of the instant at which transition occurs from one zone to the following zone. Also called sonne.

Consolan A long-range directional navigation system that transmits a slowly rotating keyed radio field pattern, from which a line of position can be obtained that is accurate within 20 mi (32 km) at 1500 mi (2400 km) range. Only a standard radio receiver is required in the aircraft.

console 1. A large cabinet for a radio or television receiver; it stands on the floor rather than on a table. 2. A main control desk for electronic equipment, as at a radar station, radio or television station, or airport control tower. Also called control desk. 3. The section of a computer that controls the machine manually, corrects errors, manually revises the contents of storage, and provides communication in other ways between the operator or service engineer and the central processing unit.

console receiver A television or radio receiver in a console.

constant A value that does not change during a particular process.

constant-amplitude recording A sound-recording method in which all frequencies that have the same intensity are recorded at the same amplitude. The resulting recorded amplitude is independent of frequency.

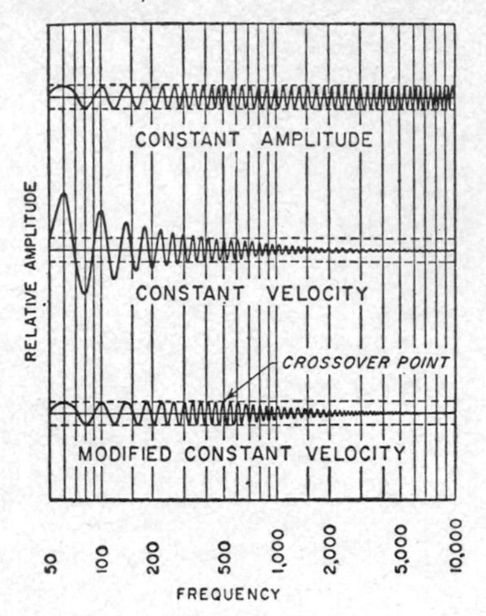

Constant-amplitude, constant-velocity, and modified constant-velocity recording having constant amplitude below crossover frequency of 500 Hz.

constantan An alloy that contains 60% copper and 40% nickel, used in making precision wire-wound resistors because of its low temperature coefficient of resistance. It is often used with iron or copper in thermocouples.

constant-current generator A generator in which output current remains essentially constant despite variations in load resistance.

constant-current modulation A system of amplitude modulation in which the output circuits of the signal amplifier and the carrier-wave generator or amplifier are connected through a common coil to a constant-current source. Changes in anode current of the signal amplifier thus produce equal and opposite changes in anode current of the RF carrier stage, thereby giving the desired modulation of the carrier. Also called Heising modulation.

constant-current transformer A transformer that automatically maintains a constant current in its secondary circuit under varying loads, when supplied from a constant-voltage source.

constant-delay discriminator *Pulse demoder.*

constant-gradient synchrotron A synchrotron in which the focusing force derived from the gradient of the magnetic field is constant along the orbit. Constant-gradient focusing is also referred to as weak focusing, in contrast to the strong focusing of the alternating-gradient synchrotron.

constant-*k* filter A filter in which the product of the series and shunt impedances is a constant that is independent of frequency.

constant-*k* lens A microwave lens constructed as a solid dielectric sphere. A plane electromagnetic wave brought to a focus at one point on the sphere emerges from the opposite side of the sphere as a parallel beam. Focusing properties are thus similar to those of a Luneberg lens.

constant-*k* network A ladder network in which the product of the series and shunt impedances is independent of frequency within the operating frequency range.

constant-luminance transmission That type of color television transmission in which the transmission primaries are a luminance primary, controlled only by the monochrome signal, and two chrominance primaries. This system is currently being used in the United States.

constant-potential accelerator An accelerator in which constant DC voltage is applied to an accelerating tube to produce high-energy ions or electrons.

constant-velocity recorder A disk recorder that has a turntable which rotates in such a manner that constant velocity is effected at the recording stylus irrespective of diameter.

constant-velocity recording A sound-recording method in which, for input signals of a given amplitude, the resulting recorded amplitude is inversely proportional to the frequency. The velocity of the cutting stylus is then constant for all input frequencies having that given amplitude.

constant-voltage power supply A regulated power supply that maintains a predetermined DC voltage across a load for a specified range of load resistance values, line voltages, temperatures, and other variables by automatically varying load current.

constraint The condition wherein a particle or group of particles has less than $3N$ degrees of freedom, where N is the number of particles in the group.

contact 1. A conducting part of a relay, connector, or switch that coacts with another such part to make or break a circuit. 2. Initial detection of an enemy aircraft, ship, submarine, or other object on a radarscope or other detecting equipment.

contact bounce The uncontrolled making and breaking of contact one or more times, but not continuously, when relay contacts are moved to the closed position.

contact chatter *Chatter.*

contact electromotive force *Contact potential.*

contact follow The distance two contacts travel together after just touching. Also called contact overtravel.

contact force The force exerted by the moving contact of a switch or relay on a stationary contact.

contact gettering The absorption of gas by contact with a dispersed getter film in an electron tube.

contact-making meter *Meter-type relay.*

contact material A metal that has high electric and thermal conductivity, low contact resistance, minimum sticking or welding tendencies, and high corrosion resistance. Commonly used contact materials include copper, silver, and gold and their alloys, platinum and palladium alloys, tungsten, molybdenum, and certain mixtures of metals.

contact microphone A microphone that picks up mechanical vibrations directly and converts them into corresponding electric currents or voltages. When used with wind, string, and percussion musical instruments, it is attached to the housing of the instrument. When used in vibration analysis of machinery, it is held against various parts of the machinery. When used as a throat microphone, it is strapped against the throat of the speaker. When used as a lip microphone, it is held against the lip of the speaker.

contact-modulated amplifier An amplifier that has a chopper at its input, to change DC and very low-frequency AC signals to a higher frequency such as 60 or 400 Hz. The resulting modulated wave is amplified in an AC amplifier to a suitable level, then demodulated, sometimes by the same contact system used to accomplish the original modulation.

contact noise The fluctuating electric resistance observed at the junction of two metals or at the junction of a metal and a semiconductor.

contactor A heavy-duty relay that controls electric power circuits.

contact overtravel *Contact follow.*

contact piston A waveguide piston that makes contact with the walls of the waveguide. Also called contact plunger.

contact plunger *Contact piston.*

contact potential 1. The voltage due to contact between two different metals, bodies in different physical states, or materials that have different chemical compositions. Also called contact electromotive force and Volta effect. 2. The voltage that exists between the control grid and cathode of an electron tube when there is no external grid bias, due to the difference in work functions of the electrode surfaces.

contact-potential barrier The potential hill at the contact surfaces of two bodies, due to formation of a barrier layer.

contact-potential difference The difference between the work functions of two materials in contact, divided by the electronic charge.

contact pressure The amount of pressure that holds together a set of contacts.

contact rectifier *Metallic rectifier.*

contact resistance The resistance in ohms between the contacts of a relay, switch, or other device when the contacts are touching each other. The value is generally a small fraction of an ohm.

contact wipe The distance that two contacts slide with respect to each other while making or breaking contact.

contaminant A material, generally in the form of airborne particles, that can enter a manufacturing or assembly process and adversely affect the quality of the product.

contaminated Made radioactive by the addition of radioactive material.

contamination The deposit of radioactive materials, such as fission fragments or radiological warfare agents, on any objective or surface.

contamination monitor A radiation counter that detects radioactive contamination of laboratory working surfaces. A thin-walled Geiger counter is commonly used for beta and gamma contamination, and a scintillation counter with a zinc sulfide screen is used for alpha particles.

content-addressable memory [abbreviated CAM] A computer memory from which words or data can be retrieved without prior knowledge of their location, by inputting a portion of the desired information. Also called associative memory.

contention A method of operating a multiterminal communication channel in which any station may transmit if the channel is free. If the channel is in use, the queue of contention requests may be maintained manually or by a computer in chronological or other predetermined sequence.

contents The information stored in any part of a computer storage.

continuity The presence of a complete path for current flow.

continuity test An electrical test that determines the presence and location of a broken connection.

continuous carrier A carrier over which information is transmitted without interrupting the carrier.

continuous control Automatic control in which the controlled quantity is measured continuously and corrections are a continuous function of the deviation.

continuous-duty rating The rating that defines the load which can be carried for an indefinite time without exceeding a specified temperature rise.

continuous film scanner A television film scanner in which the motion-picture film moves continuously while being scanned by a flying-spot kinescope.

continuous linear antenna array An antenna array that consists of an infinite number of infinitesimally spaced sources, as in some dielectric antennas.

continuous loading Loading in which the added inductance is distributed uniformly along a line by wrapping magnetic material around each conductor.

continuous-loop cartridge A magnetic-tape cartridge in which the tape is removed from the center of a roll, fed past the tape head, and returned to the outside of the roll.

continuous power The power-handling rating of an audio or other amplifier, expressed in watts RMS for a sine-wave signal.

continuous power spectrum A power spectrum

that can be represented by the indefinite integral of a suitable spectral density function. All power spectra of physical systems are continuous.

continuous recorder A recorder whose record sheet is a continuous strip or web rather than individual sheets.

continuous spectrum The spectrum of a wave whose components are continuously distributed over a frequency region, without being broken up into lines or bands.

continuous-tone squelch Squelch in which a continuous subaudible tone, generally below 200 Hz, is transmitted by FM equipment along with a desired voice signal. The tone activates a frequency-sensitive circuit that unblocks the squelch circuit of the receiver to allow reception of the desired message. Signals without the correct tone frequency or with no tone are not heard.

continuous-transmission frequency-modulated sonar [abbreviated CTFM sonar] A sonar system in which the transmitted frequency is varied continuously in linear sawtooth fashion. The frequency received by reflection from an object is

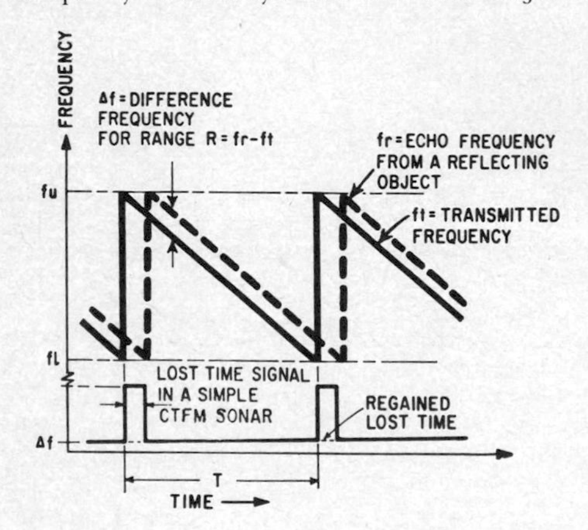

Continuous-transmission frequency-modulated sonar principles.

then proportional to the range to that object. The difference between the transmitted and received frequencies is measured with a multichannel frequency analyzer. Results are fed to a PPI cathode-ray display.

continuous tuner *Spiral tuner.*

continuous wave [abbreviated CW] A radio or radar wave that maintains a constant amplitude and a constant frequency.

continuous-wave Doppler radar *Continuous-wave radar.*

continuous-wave gas laser A laser that has a quartz envelope filled with a mixture of helium and neon at a low pressure, with Brewster-angle mirrors at opposite ends and an external optical system. An applied RF field excites the atoms in the tube, causing spontaneous emission of photons. These photons are reflected back into the gas to stimulate neon atoms, with the process repeating and building up to a self-sustained oscillation that constitutes the desired coherent laser radiation. The useful portion of this radiation passes through the 1% transmissive mirrors in an extremely narrow beam.

continuous-wave jamming Transmission of constant-amplitude constant-frequency unmodulated jamming signals as a radar countermeasure to change the gain characteristics of enemy radar receivers.

continuous-wave laser [abbreviated CW laser] A laser in which the beam of coherent light is generated continuously, as required for communication and certain other applications. The maximum average power is generally less than can be obtained with pulse operation.

continuous-wave radar A radar system in which a transmitter sends out a continuous flow of radio energy. The target reradiates a small fraction of this energy to a separate receiving antenna so located and oriented that only a small fraction of the transmitted power leaks directly into the receiver. The reflected wave is distinguished from the transmitted signal by a slight change in radio frequency called the Doppler shift. Continuous-wave radar can distinguish moving targets against a stationary reflecting background and is more conservative of bandwidth than pulse radar. Also called continuous-wave Doppler radar.

continuous x-rays The electromagnetic radiation, having a continuous spectral distribution, that is produced when high-velocity electrons strike a target. For a given x-ray tube current, the intensity associated with each wavelength is dependent on the material and thickness of the target and the voltage applied to the tube.

contouring control system A numerical positioning control system that provides simultaneous movement of the cutting tool in two or more axes.

contouring temperature recorder An ocean-water temperature recorder in which thermistors on a towed metal chain are electronically scanned in sequence from top to bottom, at time intervals adjusted to the speed of the ship so each set of readings is taken from a vertical column in the water.

contraorbit missile A missile sent backward along the calculated orbit of an approaching spacecraft, satellite, or aerospace weapon to destroy it by a head-on collision with an explosive warhead.

contrast The degree of difference in tone between the lightest and darkest areas in a television or facsimile picture. Contrast is measured in terms of gamma, a numerical indication of the degree of contrast. Pictures with high contrast have deep blacks and brilliant whites, and pictures with low contrast have an overall gray appearance.

contrast control A manual control that adjusts the range of brightness between highlights and shadows on the reproduced image in a television receiver. Usually, the contrast control varies the gain of a video amplifier tube. In a color television receiver a dual control may be used, with one section controlling the luminance signal and the other section controlling the chrominance signals; this permits adjustment of contrast without changing color.

contrast range The ratio of the brightness of the whitest portion of a picture to the blackest portion.

contrast ratio The ratio of the maximum to the minimum luminance values in a television picture.

control 1. A component that starts, stops, or adjusts a piece of equipment. 2. The section of a digital computer that carries out instructions in proper sequence, interprets each coded instruction, and applies the proper signals to the arithmetic unit and other parts in accordance with this interpretation. 3. A mathematical check used in some computer operations. 4. A test that determines the extent of error in experimental observations or measurements.

control accuracy The degree of correspondence between the ultimately controlled variable and the ideal value in a feedback control system.

control and read-only memory [abbreviated CROM] A read-only memory that also provides storage, sequencing, execution, and translation logic for a number of microinstructions.

control character A character whose occurrence in a particular context initiates, modifies, or stops a control operation in a computer or associated equipment.

control circuit 1. The circuit that feeds the control winding of a magnetic amplifier. 2. One of the circuits that responds to the instructions in the program for a digital computer. 3. A circuit that controls some function of a machine, device, or piece of equipment.

control counter A computer device that records the storage location of the instruction word to be operated upon following the instruction word in current use.

control desk *Console.*

control diagram *Flowchart.*

control electrode An electrode that initiates or varies the current between two or more electrodes in an electron tube.

control element The portion of a feedback control system that acts on the process or machine being controlled.

control field A constant location where information for control purposes is placed, such as specified columns on punched cards.

control grid A grid, ordinarily placed between the cathode and an anode, that controls the anode current of an electron tube.

controlled avalanche device A semiconductor device that has rigidly specified maximum and minimum avalanche voltage characteristics and is able to operate and absorb momentary power surges in this avalanche region indefinitely without damage.

controlled avalanche rectifier A silicon rectifier in which carefully controlled nondestructive internal avalanche breakdown across the entire junction area protects the junction surface, thereby eliminating local heating that would impair or destroy the reverse blocking ability of the rectifier.

controlled-carrier modulation A method of modulation in which the percentage modulation is held constant at all times by varying the amplitude of the carrier wave automatically to offset the variations produced by conventional amplitude modulation of the carrier wave. Also called floating-carrier modulation and variable-carrier modulation.

controlled-coupling transformer A cryogenic transformer in which the coefficient of coupling can be switched from zero to almost unity by a control current.

controlled-devices countermeasure Any electronic countermeasure against guided missiles, pilotless aircraft, proximity fuzes, or similar controlled devices.

controlled fusion The use of thermonuclear fusion reactions in a controlled manner to generate

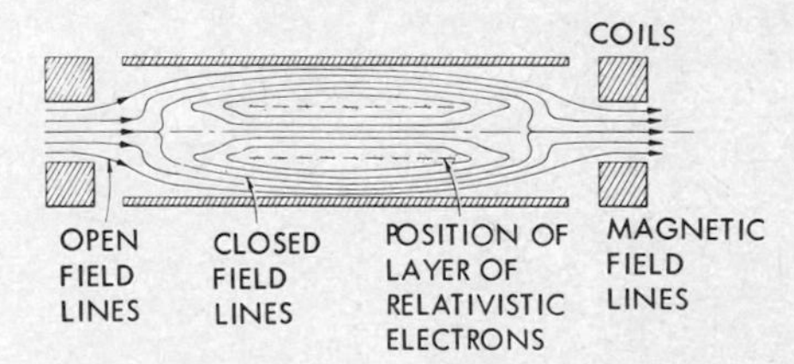

Controlled fusion using astron concept based on injection of relativistic (very high energy) electrons into evacuated cylindrical chamber having coils at each end to produce magnetic bottle for confining plasma.

power. Temperatures of at least 20 million degrees are needed.

controlled mercury-arc rectifier A mercury-arc rectifier in which one or more electrodes control the start of the discharge in each cycle and thereby control output current.

controlled rectifier A rectifier that has provisions for regulating output current, such as with thyratrons, ignitrons, or silicon controlled rectifiers.

controlled variable The quantity or condition that is measured and controlled.

controller *Automatic controller.*

control point The value of controlled variable that is maintained by an automatic control system.

control register A register that holds the identification of the instruction word to be executed next in time sequence, following the current operation in a computer.

control room A room from which engineers and production personnel control and direct a television or radio program. It is adjacent to the main studios and separated from them by large, soundproof, double-glass windows.

control signal The signal applied to the device that makes corrective changes in a controlled process or machine.

control synchro *Control transformer.*

control system 1. An arrangement of a sensing element, amplifier, and control device acting together to control some condition of a process or machine. 2. A system used in a ballistic or guided missile to maintain altitude stability during powered flight and to correct deflections caused by gusts or other disturbances. A control system uses jet vanes and other devices in common with the guidance system.

control total The sum of the numbers in a specified record field of a batch of records, determined repetitiously during computer processing so that any discrepancy from the control indicates an error. A control total may have no significance in itself, as when it is simply the sum of identification numbers of records.

control track A supplementary sound track that is usually placed on the same motion-picture film with the sound track which is carrying program material. It usually contains tone signals that control the reproduction of the sound track, such as by changing feed levels to loudspeakers in a theater to achieve stereophonic effects.

control transformer A synchro in which the electric output of the rotor is dependent on both the shaft position and the electric input to the stator. Also called control synchro.

control unit The unit of a computing system that contains the circuits which interpret and control the execution of instructions.

control winding A winding on a magnetic amplifier or saturable reactor that applies control magnetomotive forces to the core.

convection cooling Heat transfer by natural upward flow of hot air from the device being cooled.

convection current The time rate at which the electric charges of an electron stream are transported through a given surface.

convection-current modulation The time variation in the magnitude of the convection current passing through a surface, or the process of directly producing such a variation in a microwave tube.

convective discharge The movement of a visible or invisible stream of charged particles away from a body that has been charged to a sufficiently high voltage. Also called electric wind and static breeze.

convenience receptacle *Outlet.*

convergence A condition in which the electron beams of a multibeam cathode-ray tube intersect at a specified point, such as at an opening in the shadow mask of a three-gun color television picture tube. Both static and dynamic convergence are required.

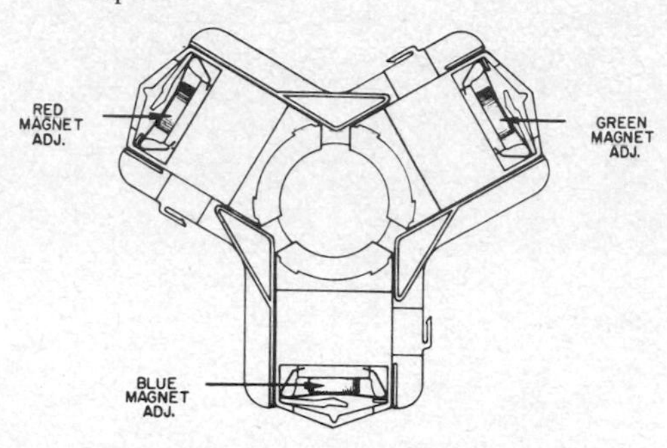

Convergence coil assembly for color television picture tube.

convergence coil One of the coils used to obtain convergence of electron beams in a three-gun color television picture tube.

convergence control A control used in a color television receiver to adjust the potential on the convergence electrode of the three-gun color picture tube to achieve convergence.

convergence electrode An electrode whose electric field converges two or more electron beams.

convergence magnet A magnet assembly whose magnetic field converges two or more electron beams. Used in three-gun color picture tubes. Also called beam magnet.

convergence plane A plane that contains the points at which the electron beams of a multibeam cathode-ray tube appear to experience a deflection applied for the purpose of obtaining convergence.

convergence surface The surface generated by the point of intersection of two or more electron beams in a multibeam cathode-ray tube during the scanning process.

convergence zone A sound transmission channel produced in sea water by a combination of pressure and temperature changes in the depth range between 2500 and 15,000 ft (760 and 4570 m). In this channel a downward sonar signal is refracted back toward the surface, to reach the surface about 30 nautical miles (55 km) away from the sonar transmitter. If the signal encounters a reflecting object along this path, the signal returns along the same route to the sonar set.

conversational mode A computer operating mode that permits queries and responses between the computer and human operators at keyboard terminals.

conversion efficiency 1. The ratio of AC output power to the DC power input to the electrodes of an electron tube. 2. The ratio of the output voltage of a converter at one frequency to the input voltage at some other frequency.

conversion electron An electron emitted by internal conversion during deexcitation of a nucleus.

conversion fraction The ratio of the number of internal conversion electrons to the total number of quanta plus the number of conversion electrons emitted in a given mode of deexcitation of a nucleus.

conversion gain ratio The ratio of signal power output to signal power input for a frequency converter or mixer.

conversion time The time of one complete measurement by an analog-to-digital converter.

conversion transducer An electric transducer in which the input and output frequencies are different. An example is the converter used in superheterodyne receivers.

convert To change the representation of data from one form to another, as from binary to decimal or from cards to tape.

converter 1. The section of a superheterodyne radio receiver that converts the desired incoming RF signal to the IF value. The converter section includes the oscillator and the mixer-first detector. Also called heterodyne conversion transducer and oscillator-mixer-first detector. 2. An auxiliary unit used with a television or radio receiver to permit reception of channels or frequencies for which the receiver was not originally designed. 3. In facsimile, a device that changes the type of modulation delivered by the scanner. 4. A computer unit that changes numerical information from one form to another, as from decimal to binary or vice versa, from fixed point to floating-point representation, or from punched cards to magnetic tape. 5. A device for converting a low DC or AC voltage to a much higher DC voltage, generally by making the low voltage drive an oscillator feeding a step-up transformer whose secondary voltage is rectified. In effect, a converter is an inverter whose output is rectified. It may also be a rotary electromechanical device. 6. *Remodulator.* 7. *Dynamotor.* 8. *Synchronous converter.*

converter tube An electron tube that combines the mixer and local-oscillator functions of a heterodyne conversion transducer.

convolver A surface-acoustic-wave device in which signal processing is performed by a nonlinear interaction between two waves traveling in opposite directions. In one version, the two different input frequencies are applied at opposite ends of the structure, and the sum-frequency signal is detected by an output transducer structure on the surface between the two inputs. Also called acoustic convolver.

cooled infrared detector An infrared detector that must be operated at cryogenic temperatures, such as at the temperature of liquid nitrogen, to obtain the desired infrared sensitivity.

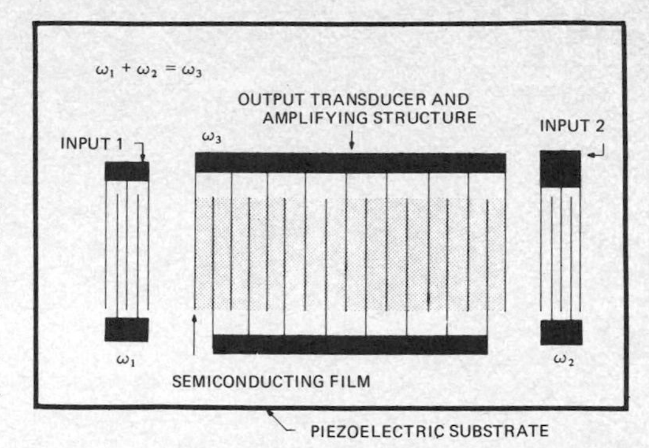

Convolver in which output transducer also provides amplification at the sum frequency.

Coolidge tube An x-ray tube in which the needed electrons are produced by a hot cathode.

cooling Setting aside a highly radioactive material until the radioactivity has diminished to a desired level.

cooperative system A missile guidance system that requires transmission of information from a remote ground station to a missile in flight, processing of the information by the missile-borne equipment, and retransmission of the processed data to the originating and/or other remote ground stations.

coordinate Any one of two or more magnitudes that determine position relative to the reference axes of a coordinate system.

coordinate data receiver A receiver specifically designed to accept the signal of a coordinate data transmitter and reconvert this signal into a form suitable for input to associated equipment such as a plotting board, computer, or radar set.

coordinate data transmitter A transmitter that accepts two or more coordinates, such as those representing a target position, and converts them into a form suitable for transmission.

coordinate indexing An indexing scheme in which equal-rank descriptors describe a document, for information retrieval by a computer or other means.

coordinate system A system for specifying the location of a point, using two coordinates if on a surface and three coordinates if in space.

coplanar electrodes Electrodes mounted in the same plane.

copper [symbol Cu] A metallic element that has good conductivity. Atomic number is 29.

copper loss Power loss in a winding due to current flow through the resistance of the copper conductors. Also called I^2R loss.

copper-oxide photovoltaic cell A photovoltaic cell in which light acting on the surface of contact between layers of copper and cuprous oxide causes a voltage to be produced.

copper-oxide rectifier A metallic rectifier in which the rectifying barrier is the junction between metallic copper and cuprous oxide. A disk

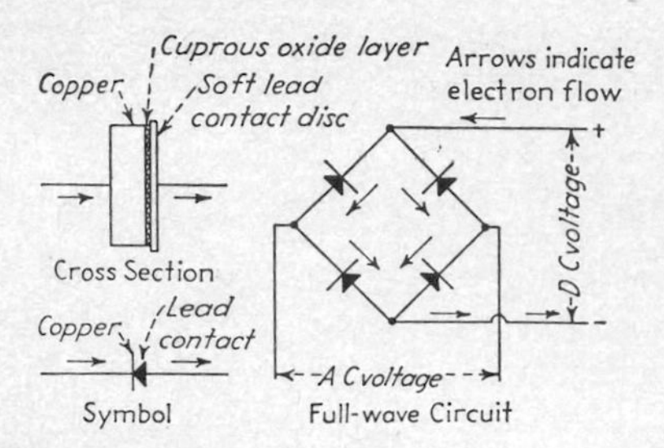

Copper-oxide rectifier construction, symbol, and typical circuit.

of copper is coated with cuprous oxide on one side, and a soft lead washer is used to make contact with the oxide layer.

copper sulfide rectifier *Magnesium-copper sulfide rectifier.*

Corbino disk A variable-resistance device that utilizes the effect of a magnetic field on the flow of carriers from the center to the circumference of a semiconductor disk.

cord A small, very flexible insulated cable.

cordless Having self-contained batteries that eliminate the need for an AC line cord.

cordwood module A module in which the individual axial-lead components are stacked like cordwood between a pair of end plates on which

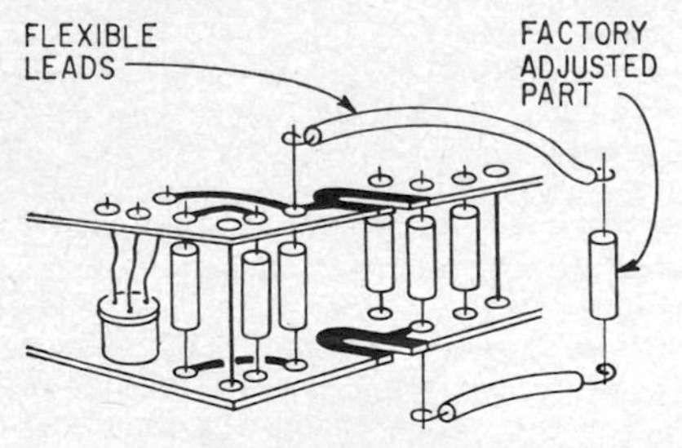

Cordwood-module packaging of tubular components.

interconnections are made. Subassemblies and integrated-circuit boards may be similarly mounted in the module.

core *Magnetic core.*

core iron A grade of soft iron suitable for cores of chokes, transformers, and relays.

coreless-type induction furnace *High-frequency furnace.*

coreless-type induction heater A device in which a charge is heated directly by induction, with no magnetic core material linking the charge to the induction coil.

core logic Logic performed in ferrite cores that serve as inputs to diode and transistor circuits.

core loss The power loss in an iron-core transformer or inductor caused by eddy currents and hysteresis effects in the iron core. Also called iron loss.

core memory *Magnetic-core storage.*

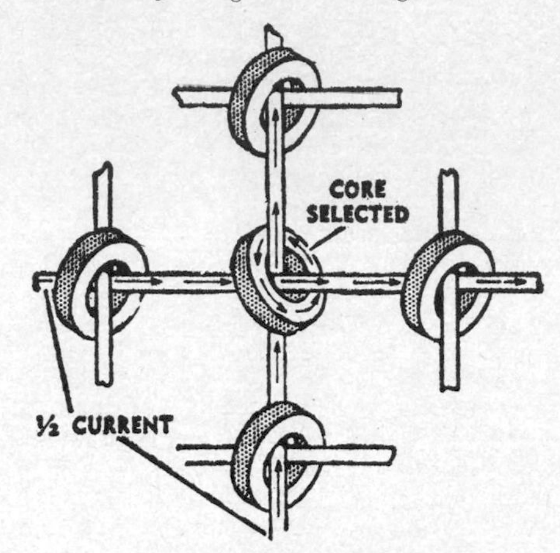

Core plane, showing how half-value currents through two wires magnetize center core in counterclockwise direction for storing 1 bit of information.

core plane A plane of magnetic cores threaded by wires, to provide one computer storage position per core.

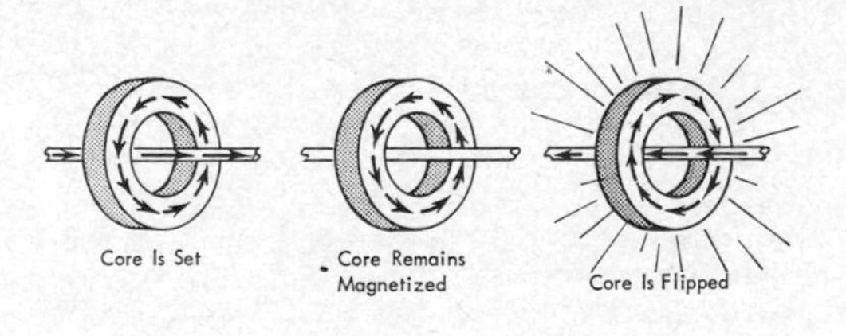

Core storage.

core storage Computer storage in which binary information is represented by the direction of magnetization of a ferromagnetic material.

core-type induction heater A device in which a charge is heated by induction, with a magnetic core linking the induction coil to the charge.

Coriolis force Deflection of a projectile during its flight across the surface of the earth, caused by the rotation of the earth.

corner A sharp bend in a waveguide. Also called elbow.

corner antenna *Corner-reflector antenna.*

corner reflector A radar reflector that consists of three conducting surfaces mutually intersecting at right angles. This reflector returns electromagnetic radiation to its source. Used to make a posi-

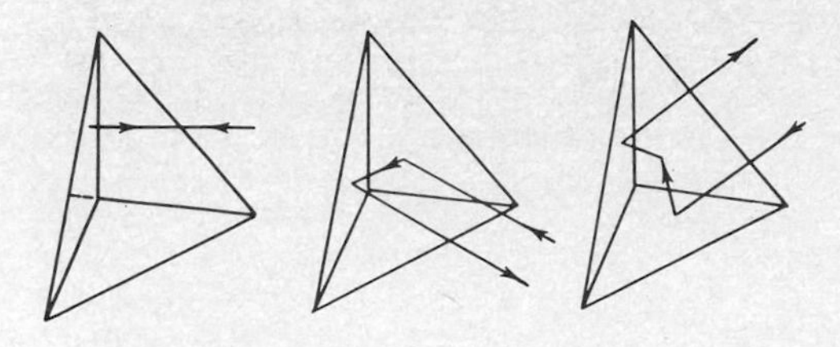

Corner reflector, showing how wave arriving from any angle is reflected back to its source.

tion more conspicuous for radar observations.

corner-reflector antenna An antenna that consists of two conducting surfaces intersecting at an angle which is usually 90°, with a dipole or other antenna located on the bisector of the angle. The surfaces are often made of wire mesh to reduce wind resistance. Maximum pickup is along the bisector of the reflector angle. Used as a UHF television and radio receiving antenna. Also called corner antenna.

corona A discharge of electricity that appears as a bluish purple glow on the surface of and adjacent to a conductor when the voltage gradient exceeds a certain critical value. It is due to ionization of the surrounding air by the high voltage.

corona shield A shield placed about a point of high potential to redistribute electrostatic lines of force.

corona tube A gas-discharge voltage-reference tube that employs a corona discharge.

corona voltmeter A voltmeter in which the crest value of a voltage is indicated by the inception of corona at a known electrode spacing.

corpuscular radiation Radiation that consists of subatomic particles, such as electrons, protons, deuterons, and neutrons, as distinguished from electromagnetic radiation.

corrected compass course *Magnetic course.*

corrected compass heading *Magnetic heading.*

correction A quantity added to a calculated or observed value to obtain the true value.

correction element The element in a fire-control system that introduces corrections based on variations of conditions from standard firing table conditions and arbitrary corrections based on observations of previous firings.

correction time The time required for the controlled variable to reach and stay within a predetermined band about the control point following any change of the independent variable or operating condition in a control system. Also called settling time.

corrective network An electric network inserted in a circuit to improve its transmission properties, its impedance properties, or both. Also called shaping network.

correed A combination of one or more reed relays with an iron core, operating coil, and sometimes a latching permanent magnet and diodes, usually packaged for mounting on a printed-wiring board.

correlation direction finder A radar direction-finding receiver that receives jamming signals. By correlating signals received at several such stations, range and azimuth of an enemy jammer can be obtained.

correlation-type receiver *Correlator.*

correlator A device that detects weak signals in noise by performing an electronic operation which approximates the computation of a correlation function. Examples include the autocorrelator and crosscorrelator. Also called correlation-type receiver.

corrugated-surface antenna A microwave antenna that consists of a waveguide feed to a mode

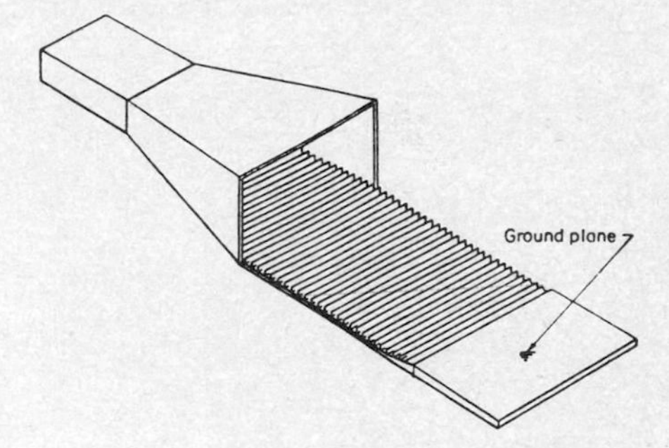

Corrugated-surface antenna.

transformer or horn launcher and a transversely corrugated metal surface that guides surface waves.

cortical stimulator An electronic instrument used in nerve and mental therapy to deliver an electric shock of prescribed strength by means of a pulsating current. A low-frequency relaxation oscillator is sometimes used to produce the pulses.

cosecant antenna An antenna that gives a beam whose amplitude varies as the cosecant of the angle of depression below the horizontal. Used in navigation radar. It may use a cheese antenna with a line source or a distorted parabolic reflector with a point source.

cosecant-squared antenna An antenna that has a cosecant-squared pattern.

cosecant-squared pattern A ground radar antenna radiation pattern that sends less power to nearby objects than to those farther away in the same sector. The field intensity varies as the square of the cosecant of the elevation angle. The pattern is achieved by either bending the top portion of the parabolic reflector forward or using a spoiler on the reflector. With this pattern, approximately equal echo signals are received from objects at the same altitude but at varying distances. Also used in airborne antennas to produce a uniform electric field along a line on the earth's surface.

cosine emission law The energy emitted by a radiating surface in any direction is proportional to the cosine of the angle which that direction makes with the normal.

cosine winding A winding used in the deflection yoke of a cathode-ray tube to prevent changes in focus as the beam is deflected over the entire area of the screen.

cosmic noise Radio static caused by a phenomenon outside the earth's atmosphere. Examples are sunspots.

cosmic radio wave A radio wave that originates in an extraterrestrial source. Examples include galactic radio noise and solar radio noise coming from the sun.

cosmic rays High-energy radiation that originates outside the atmosphere of the earth and is capable of producing ionizing events in passing through the air or other matter. The rays consist almost entirely of positively charged atomic nuclei, about two-thirds of which are protons. The balance includes mesons, alpha particles, and heavier nuclei such as those of carbon, nitrogen, oxygen, and iron. Also called primary cosmic rays.

cosmic-ray shower The simultaneous appearance of a number of downward-directed ionizing particles, with or without accompanying photons, caused by a single cosmic ray. Cosmic-ray showers reveal themselves by simultaneous actuation of separated counters. They can be roughly classified according to their properties as cascade, extensive, or penetrating shower. Also called shower.

cosmic-ray telescope An instrument that determines the direction of arrival of a cosmic ray. Two or more Geiger counters are mounted end to end on the axis of the telescope, so only particles moving along the axis can pass through all the counters to produce an output signal for a recorder.

cosmology The study of the general nature of the universe and its matter and energy, in space and time.

cosmonautics The science of travel beyond the solar system.

COS/MOS RCA abbreviation for *complementary-symmetry metal-oxide semiconductor.*

cosmotron *Proton synchrotron.*

COTAR [Correlated Orientation Tracking And Range] A passive system that tracks a vehicle in space by determining the line of direction between a remote ground-based receiving antenna and a telemetering transmitter in the missile, using phase comparison techniques.

COTAT [COrrelation Tracking And Triangulation] A trajectory-measuring system that uses several antenna baselines, each separated by large distances. Used to measure direction cosines to an object. From these measurements its space position is computed by triangulation.

Cottrell precipitator An electrostatic air-pollution eliminator in which negatively charged dust or fume particles are attracted to a positively charged wire electrode enclosed in a metal flue that acts as the negative electrode. Applications include treating sulfuric acid mist, cement-mill dust, power-plant fly ash, and metallurgical fumes.

CO_2 laser *Carbon dioxide laser.*

Coulmer antenna array A high-gain planar antenna array that consists of nonresonant elements stacked vertically and horizontally to produce both vertically and horizontally polarized waves.

coulomb [abbreviated C] The SI unit of electric charge. It is the amount of electric charge that passes through a given cross section of a conductor when a steady current of 1 A is flowing.

coulomb barrier radius The nuclear radius deduced from the rate of alpha disintegration or from cross sections of nuclear reactions that involve charged particles. Also called Gamow barrier radius.

coulomb collision The collision of two charged particles. The collision cross section is considerably larger than when one of the particles is neutral because the electric fields of the two particles can interact at much larger distances.

coulomb force The electrostatic force of attraction or repulsion exerted by one charged particle on another.

coulomb friction Friction that occurs between dry surfaces.

coulombmeter A measuring instrument that measures quantity of electricity in coulombs by integrating a stored charge in a circuit which has very high input impedance. When incorporated into an electrometer, switches provide a choice of ranges.

coulomb potential A scalar point function equal to the work per unit charge done against or by the coulomb force in transferring a particle that bears an infinitesimal positive charge from infinity to the field of a charged particle in a vacuum.

coulomb scattering Scattering that occurs when charged particles passing through matter are acted on by the electrostatic forces of other charged particles.

Coulomb's law The attraction or repulsion between two electric charges is proportional to the product of their magnitudes and inversely proportional to the square of the distance between them. Also called law of electrostatic attraction.

coulometer An electrolytic cell that measures quantity of electricity in coulombs by the chemical action produced.

coulometric titration Measurement of the integrated current passing through an electrode during the chemical process of titration.

count 1. A single response of the counting system in a radiation counter. 2. The total number of events indicated by a counter.

countdown 1. The ratio of the number of interrogation pulses not answered by a transponder to the total number received. 2. The step-by-step

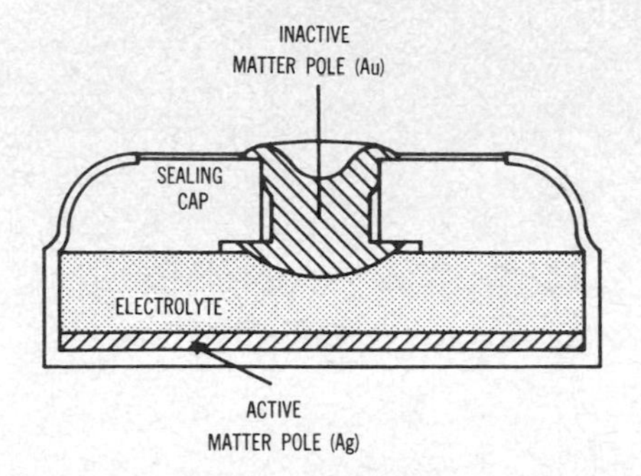

Coulometer using dry solid electrolyte to integrate electric charge over periods of from 1.5 min to 8000 h.

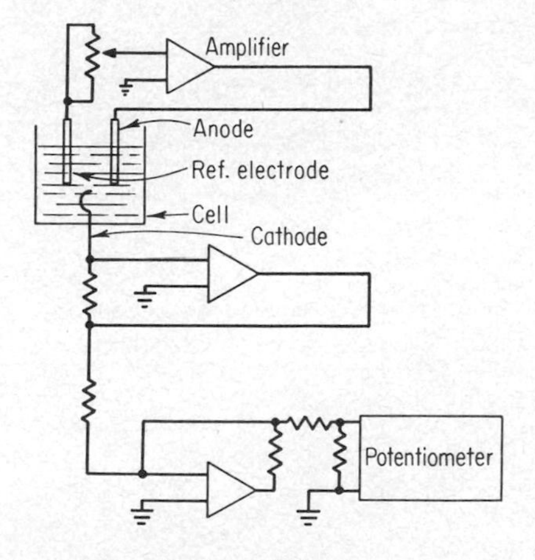

Coulometric titration.

process of preparing a missile for launching.

countdown circuits Circuits that are connected to a guided missile through its umbilical cord to actuate and check the firing controls during the final audible counting of seconds before firing.

counter 1. A complete instrument for detecting, totalizing, and indicating a sequence of events. When used to measure frequency by counting the periods of a waveform for 1 s and displaying the count as a numerical value in hertz, it is usually known as an electronic counter or digital frequency meter. 2. *Accumulator.* 3. *Radiation counter.* 4. *Scaler.*

counter circuit A circuit that receives uniform pulses which represent units to be counted and produces a voltage proportional to the total count.

counterclockwise capacitor A variable capacitor whose capacitance increases with counterclockwise rotation of its rotor, as viewed from the end of the control shaft.

counterclockwise polarized wave *Left-hand polarized wave.*

counter-controlled cloud chamber A cloud chamber whose expansion is triggered by a counter, for studying particular events.

counter-countermeasure [abbreviated CCM] Any technique, device, or system used against enemy electronic countermeasures, such as methods of extracting radar target signals from a jamming environment.

counter dead time The time interval between the start of a counted event and the earliest instant at which a new event can be counted by a radiation counter.

counter decade *Decade scaler.*

counterelectromotive force The voltage developed in an inductive circuit by a changing current. The polarity of the voltage is at each instant opposite that of the generated or applied voltage. Also called back electromotive force.

countermeasures The use of devices and/or techniques intended to impair the operational effectiveness of enemy activity.

counter plateau The region of a radiation-counter characteristic curve in which the counting rate is substantially independent of voltage. The counter is normally operated in this region.

counterpoise A system of wires or other conductors that is elevated above and insulated from the ground to form a lower system of conductors for an antenna. Used as a substitute for a ground connection. Also called antenna counterpoise.

counter tube 1. An electron tube that converts an incident particle or burst of incident radiation into a discrete electric pulse, generally by utilizing the current flow through a gas that is ionized by the radiation. Used in radiation counters. Also called radiation-counter tube. 2. An electron tube that has one signal input electrode and 10 or more output electrodes, with each input pulse transferring conduction sequentially to the next output electrode. Beam-switching tubes and cold-cathode counter tubes are examples. Used for preset counting, data storage, timing, and gating.

counting circuit A circuit that counts pulses by frequency-dividing techniques, by charging a capacitor to produce a voltage proportional to the pulse count, or by other means.

counting dial A dial that fits on the end of a control shaft and gives an accurate indication of shaft position. Some types provide digital indications of up to 1000 turns and fractions of a turn.

counting efficiency The ratio of the average number of photons or ionizing particles that produce counts to the average number incident on the sensitive area of a radiation counter.

counting ionization chamber *Pulse ionization chamber.*

counting loss The counting error due to events occurring within the dead time of a radiation detector.

counting rate The average rate of occurrence of events as observed by means of a counting system.

counting-rate curve A curve that shows how

counting rate varies with applied voltage in a radiation counter. It generally starts with a sharp rise at the threshold voltage, has a flat region known as the plateau, then ends with a sudden sharp rise. The counter is usually operated in the plateau region, where the counting rate is not appreciably affected by changes in applied voltage.

counting-rate meter An instrument that indicates the time rate of occurrence of input pulses to a radiation counter, averaged over a time interval. Also called rate meter.

couple 1. Two metals placed in contact, as in a thermocouple. 2. To connect two circuits so signals are transferred from one to the other.

coupler 1. The portion of a navigation system that receives signals of one type from a sensor and

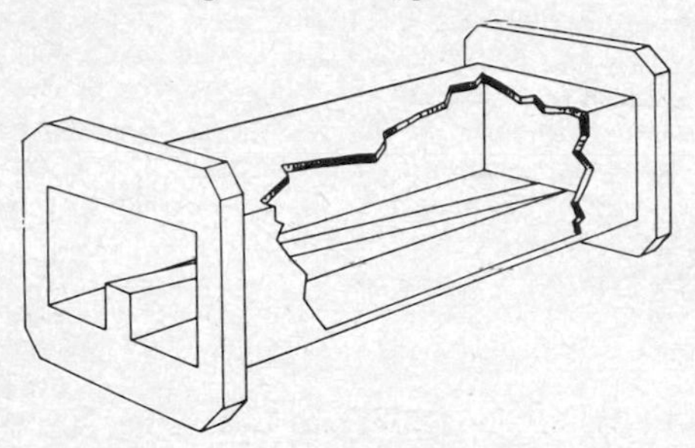

Coupler used between rectangular and ridge waveguides.

transmits signals of a different type to an actuator. 2. A component that transfers energy from one circuit to another.

coupling 1. A mutual relation between two circuits that permits energy transfer from one to the other. Coupling may be direct through a wire,

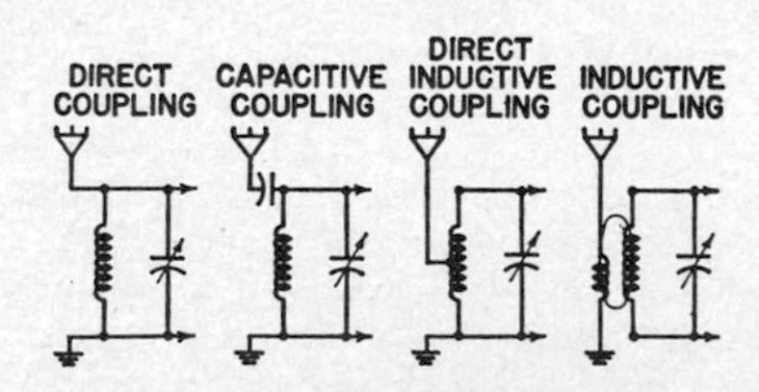

Coupling methods used with antennas.

resistive through a resistor, inductive through a transformer or choke, or capacitive through a capacitor. 2. A flexible or rigid device used to fasten together two shafts end to end.

coupling aperture An aperture in the wall of a waveguide or cavity resonator, to transfer energy to or from an external circuit. Also called coupling hole and coupling slot.

coupling capacitor A capacitor that blocks the flow of direct current while allowing alternating or signal currents to pass. Widely used for joining two circuits or stages. Also called blocking capacitor and stopping capacitor.

coupling coefficient 1. The ratio of the maximum change in energy of an electron traversing an interaction space to the product of the peak alternating gap voltage and the electronic charge. 2. *Coefficient of coupling.*

coupling hole *Coupling aperture.*

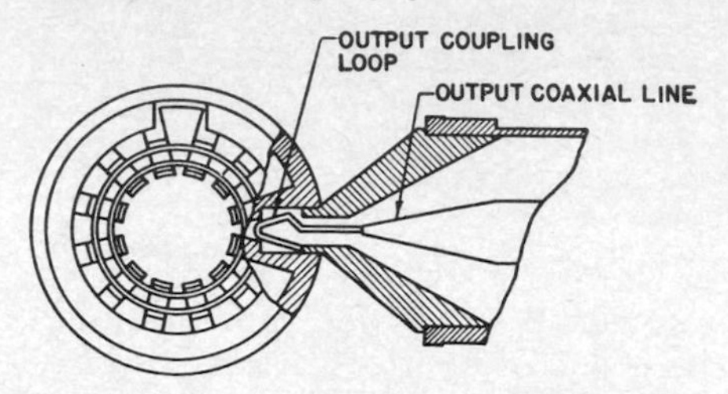

Coupling loop used between magnetron and coaxial line.

coupling loop A conducting loop that projects into a waveguide or cavity resonator, to transfer energy to or from an external circuit.

coupling probe A probe that projects into a waveguide or cavity resonator, to transfer energy to or from an external circuit.

coupling slot *Coupling aperture.*

course 1. The intended direction of travel, expressed as an angle in the horizontal plane between a reference line and the course line, usually measured clockwise from the reference line. Also called course line and desired track. 2. A radio range beam.

course error Deprecated term for *drift angle.*

course line *Course.*

course-line computer An airborne computer that continually computes an aircraft's position, in terms of its departure from course and its distance from destination or an intermediate point, by utilizing omnirange and distance-measuring-equipment transmissions. Also called offset-course computer.

course-line deviation The angular or linear difference between the actual track of a vehicle and the intended course line.

course-line deviation indicator An instrument that indicates deviation from a desired course line.

course made good The resultant direction of actual travel of a vehicle, equivalent to its bearing from the point of departure.

course pull *Course push.*

course push An erroneous deflection of the indicator of a navigation aid, due to a change in the attitude of the receiving antenna. Also called course pull.

course sector A wedge-shaped section of airspace that contains the course line and spreads with distance from the ground station of an instrument landing system.

course sensitivity The amount of displacement

of a vehicle from the course line that produces a given change of course indication.

course softening An intentional decrease in course sensitivity upon approaching a navigation aid, such that the ratio of indicator deflection to linear displacement from the course line tends to remain constant.

course width The arithmetic sum of the plus and minus lateral deviations from the course line within which the course-defining parameters do not vary by a detectable amount. Also called angular width.

covalent bond A pair of electrons shared by two neighboring atoms.

covalent crystal A crystal held together by covalent bonds.

covalent radius The effective radius of an atom in a covalent bond.

coverage *Service area.*

coverage diagram A diagram that depicts the service area of a navigation aid.

cozi [COmmunication Zone Indicator] An ionospheric sounding system for determining propagation characteristics of the ionosphere at various angles at any instant. Used to determine how well long-distance high-frequency broadcasts are reaching their intended destinations. The broadcast is interrupted momentarily to transmit a short pulse that is reflected from the ionosphere to the earth at some distant point; the backscatter signal from the earth returns to the transmitter site in a time that can be related to the height of the ionosphere and distance along the earth's surface. Also used to detect backscatter and other electromagnetic perturbations from missile plumes.

CPA Abbreviation for *color phase alternation.*

cpm Abbreviation for *cycle per minute.*

cps Abbreviation for *cycle per second,* now called hertz and abbreviated Hz.

CPU Abbreviation for *central processing unit.*

crab angle *Drift correction angle.*

crash locator beacon An automatic radio beacon carried in aircraft to guide searching forces in the event of a crash.

crater The depression formed in the surface of the earth by an explosion.

crater lamp A glow-discharge tube used as a point source of light whose brightness is proportional to the signal current sent through the tube. Used for photographic recording of facsimile signals. The glow discharge is concentrated in a crater-shaped depression at one end of the cathode.

creep Any slow change in a dimension or characteristic with time or usage.

creepage The conduction of electricity across the surface of a dielectric.

creep recovery The slow return to an original dimension or characteristic with time, after removal of the load or other condition that caused the original creep.

crest *Peak.*

crest factor The ratio of the peak value to the effective value of any periodic quantity such as a sinusoidal alternating current.

crest value *Peak value.*

crest voltmeter A voltmeter reading the peak value of the voltage applied to its terminals.

crimp contact A contact whose back portion is a hollow cylinder that will accept a wire. After a bared wire is inserted, a swedging tool is applied to crimp the contact metal firmly against the wire. Also called solderless contact.

crippled leapfrog test A variation of the leapfrog test, modified so the computer tests are repeated from a single set of storage locations rather than a changing set of locations.

critical absorption wavelength The wavelength, characteristic of a given electron energy level in an atom of a specified element, at which an absorption discontinuity occurs.

critical angle The smallest angle away from the vertical at which a radiated radio wave of a given frequency will still be reflected by the ionosphere. At smaller angles the radio waves penetrate the ionosphere and are not returned to earth.

critical coupling The degree of coupling that provides maximum transfer of signal energy from one RF resonant circuit to another when both are tuned to the same frequency. Also called optimum coupling.

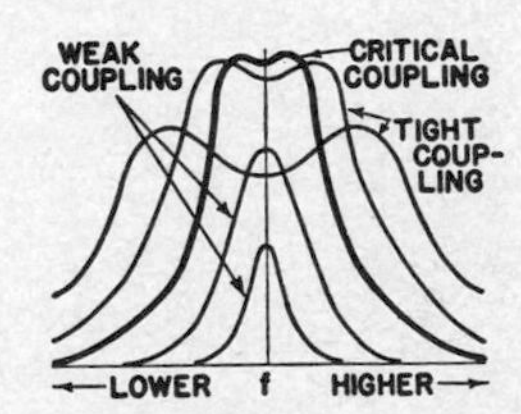

Critical coupling curve.

critical current The current in a superconductive material above which the material is normal and below which the material is superconducting, at a specified temperature and in the absence of external magnetic fields.

critical damping The degree of damping required to give the most rapid transient response without overshooting or oscillation. Thus, a critically damped meter moves its pointer to a new value without going past it.

critical dimension *Broad dimension.*

critical field The smallest theoretical value of steady magnetic flux density that would prevent an electron emitted from the cathode of a magnetron at zero velocity from reaching the anode. Also called cutoff field.

critical frequency 1. The limiting frequency below which a radio wave will be reflected by an ionospheric layer at vertical incidence at a given time. Higher frequencies will penetrate the layer. Also called penetration frequency. 2. *Cutoff frequency.*

critical inductance The minimum input choke inductance required to prevent the input choke current from going to zero during any part of the cycle in a choke-input filter for a full-wave rectifier.

critical magnetic field The field below which a superconductive material is superconducting and above which the material is normal, at a specified temperature and in the absence of current.

critical temperature The temperature below which a superconductive material is superconducting and above which the material is normal, in the absence of current and external magnetic fields.

critical wavelength The free-space wavelength that corresponds to the critical frequency.

CRO Abbreviation for *cathode-ray oscilloscope.*

CROM Abbreviation for *control and read-only memory.*

Crookes dark space *Cathode dark space.*

Crookes radiometer A radiometer which demonstrates that radiant energy from the sun can produce motion. A miniature four-vane windmill is mounted in a glass-envelope vacuum tube. Each vane is polished on one side and black on the other side. Absorption of radiant energy by the black sides warms these sides and makes adjacent residual molecules of gas rebound more rapidly than from the polished sides. The black sides then rotate away from the source of radiation.

Crookes tube An early form of discharge tube.

Crosby system The stereo FM broadcasting system used in the United States.

cross assembler An assembler program that is run on one computer to produce a program suitable for use on another type of computer.

crossband Two-way communication in which one radio frequency is used in one direction and a frequency that has different propagation characteristics is used in the opposite direction.

crossbanding Use of one interrogation frequency with several reply frequencies or one reply frequency with several interrogation frequencies.

crossband transponder A transponder that replies in a different frequency band from that of the received interrogation.

crossbar switch A switch that has a three-dimensional arrangement of contacts and a magnet system which selects individual contacts according to their coordinates in the matrix.

cross-bombardment A method for assigning the mass of a radioactive species based on its production by several nuclear bombardments that use different projectiles and/or different target materials.

cross-channel communication Two-way communication in which one radio frequency is used in one direction and a different frequency that has similar propagation characteristics is used in the opposite direction.

cross-color interference Interference produced in the chrominance channel of a color television receiver by crosstalk from the monochrome signal.

cross-control circuit A compandor circuit in which input signals to the compressor also control the operation of the expandor at the same end of the circuit.

crosscorrelation function A mathematical quantity defined as the product of two functions of time.

crosscorrelator A correlator in which a locally generated reference signal is multiplied by the incoming signal, and the result is smoothed in a low-pass filter to give an approximate computation of the crosscorrelation function. It can be used for detecting weak signals in noise in cases where the important signal characteristics are known prior to detection. Also called synchronous detector.

cross-coupling A measure of the undesired power transferred from one channel to another in a transmission medium.

crossed-field amplifier A forward-wave beam-type microwave amplifier that uses crossed-field interaction to achieve good phase stability, high efficiency, high gain, and wide bandwidth for most of the microwave spectrum.

crossed-field backward-wave oscillator One of several types of backward-wave oscillators that utilize a crossed field, such as the amplitron and carcinotron.

crossed-field multiplier phototube A multiplier phototube in which repeated secondary emission is obtained from a single active electrode by the combined effects of a strong RF electric field and a perpendicular DC magnetic field. Laser beams modulated in the gigahertz range have been detected and amplified by this tube.

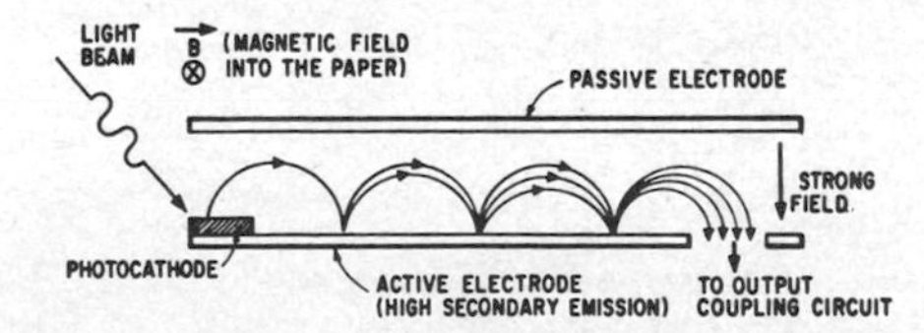

Crossed-field multiplier phototube, showing leapfrog trajectory of electrons from photocathode to output.

crossed-field tube A space-charge-wave tube in which the directions of the static magnetic field, the static electric field, and the electron beam are mutually perpendicular, as required for converting the potential energy of the electron beam into RF energy. Examples include magnetrons, crossed-field amplifiers, and some backward-wave oscillators.

crossed-pointer indicator A two-pointer indicator used with an instrument-landing system to indicate the position of an airplane with respect to the glide path.

crossed stripline cavity A cavity in which two striplines intersect at right angles, with a sphere of yttrium-iron garnet between them at the intersection to provide coupling that is maximum at low power levels and negligible at high power levels.

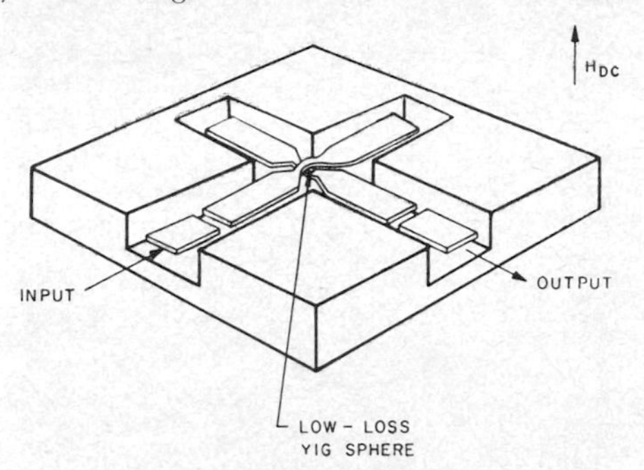

Crossed stripline cavity.

crossfoot To add numbers in several different ways in a computer, for checking purposes.

crosshatch generator A signal generator that generates a crosshatch pattern for adjusting color television receiver circuits.

cross-modulation A type of interference in which the carrier of a desired signal becomes modulated by the program of an undesired signal on a different carrier frequency. The program of the undesired station is then heard in the background of the desired program. Cross-modulation occurs because the first tube circuit in the receiver is nonlinear and hence acts as a detector for the strong undesired signal.

cross-neutralization A method of neutralization used in push-pull amplifiers, in which a portion of the AC anode-cathode voltage of each tube is applied to the grid-cathode circuit of the other tube through a neutralizing capacitor.

crossover A point at which two conductors cross, with appropriate insulation between them to prevent contact.

crossover frequency 1. The frequency at which a dividing network delivers equal power to the upper and lower frequency channels when both are terminated in specified loads. 2. *Transition frequency.*

crossover network A selective network that divides the audio-frequency output of an amplifier into two or more bands of frequencies. The band below the crossover frequency is fed to the woofer loudspeaker, and the high-frequency band is fed to the tweeter. Also called dividing network and loudspeaker dividing network.

crossover region A zone in space, close to the localizer on-course line or guide slope of an instrument approach system, in which the pointer of the indicator is in a position between the full-scale indications.

crossover spiral *Leadover groove.*

crosspoint reed relay A reed relay that has from two to five reed switches, one holding coil, and two coincident-count coils. Simultaneous energization of the two coincident-count coils closes one reed switch that is permanently wired in series with the holding coil. The holding coil then takes over and keeps all contacts in all reed switches closed, until the holding coil is externally interrupted or until a reverse-polarity pulse is applied to one of the coincident-count coils.

cross-polarization The component of the electric field vector normal to the desired polarization component.

cross-polarized operation Operation of two independent microwave digital data transmitters on the same carrier frequency and the same antenna by using cross-polarized feeds.

cross section 1. The probability, per unit flux and per unit time, that a given ionization or capture reaction will occur in a nuclear reactor. The cross section of an atom or nucleus has the dimensions of an area and is actually the effective target area presented to an incident particle or photon for a particular reaction. If the reaction cannot take place, the cross section is zero. A commonly used unit of area for cross sections is the barn, equal to 10^{-24} cm^2. 2. A section at right angles to an axis.

crosstalk 1. The sound heard in a receiver along with a desired program because of cross-modulation or other undesired coupling to another communication channel. 2. Interaction of audio and video signals in a television system, causing video modulation of the audio carrier or audio modulation of the video signal at some point. 3. Interaction of chrominance and luminance signals in a color television receiver.

crosstalk unit [abbreviated CU] A measure of the coupling between two circuits. The number of crosstalk units is 1 million times the ratio of the current or voltage at the observing point to the current or voltage at the origin of the disturbing signal, the impedances at these points being equal.

crowbar A device or circuit that monitors the output of a power supply and rapidly places a low-resistance shunt (crowbar) across the output terminals whenever a preset voltage limit is exceeded, to provide protection until slower fuses or circuit breakers can act.

CRT Abbreviation for *cathode-ray tube.*

CRT terminal Abbreviation for *cathode-ray-tube terminal.*

cruciform core A transformer core in which all windings are on one center leg, and four additional legs arranged in the form of a cross serve as return paths for magnetic flux.

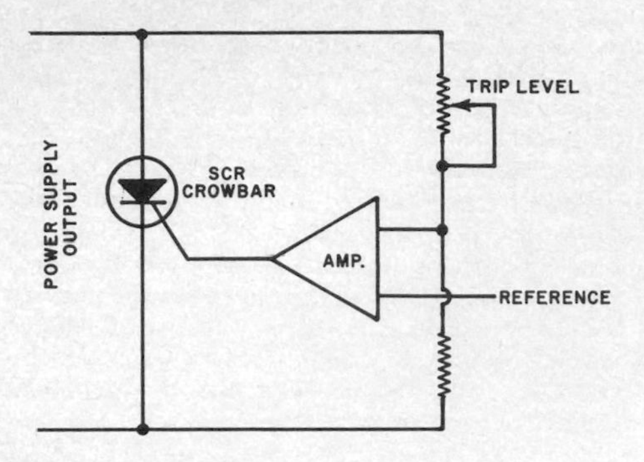

Crowbar circuit using silicon controlled rectifier to protect regulated power supply from damage by overload and to protect load from power-supply malfunction or line-voltage surges.

cruise missile A pilotless airplane that can be launched from a submarine, surface ship, ground vehicle, or another airplane. Range can be up to 1500 mi (2400 km), flying at a constant altitude that can be as low as 60 m to minimize detection by enemy radar. It may use radar terrain-contour-matching guidance that is immune to jamming and carry conventional or nuclear warheads.

cryoelectronics A branch of electronics concerned with the study and application of superconductivity and other low-temperature phenomena to electronic devices and systems.

cryogenic device A device whose operation depends on superconductivity as produced by temperatures near absolute zero.

cryogenic film A storage element that uses superconducting thin films of lead at liquid helium temperature. The application of a magnetic field forces the material into a normal state, allowing flux lines to penetrate and become trapped when the material goes superconducting again.

cryogenic gyroscope A gyroscope in which a spherical rotor of superconducting niobium spins while in levitation at cryogenic temperatures. Initially, starting coils outside a Dewar flask induce currents in rings surrounding the rotor. When the rings and rotor are cooled to superconducting temperature, the trapped currents in the rings produce magnetic fields that support the rotor. After spinning is started with a jet of helium gas, the rotor continues spinning in its magnetic suspension for as long as superconducting temperature is maintained.

cryogenic laser A laser designed for operation at cryogenic temperatures, usually below 80 K, to increase the average output.

cryogenic parametric amplifier A parametric amplifier that is cooled to about 17 K in a refrigerator to increase operating life.

cryogenics The science of physical phenomena at very low temperatures, approaching absolute zero. At such temperatures, small changes in magnetic field strength can produce large current changes in superconducting materials.

cryogenic temperature A temperature within a few degrees of absolute zero, which is −273.16°C.

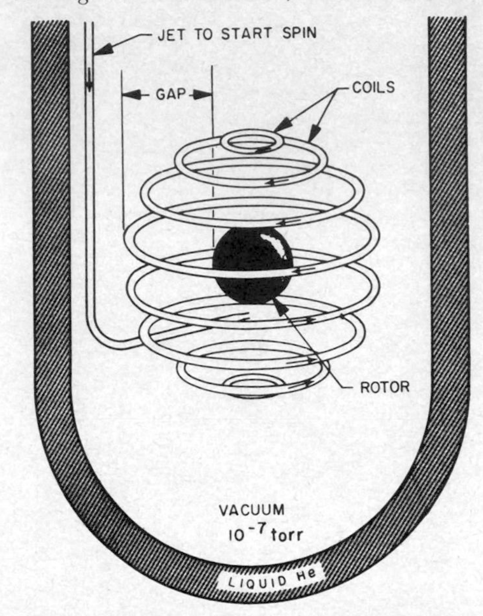

Cryogenic gyroscope using niobium sphere as rotor. Starting coil is wound around outer wall of Dewar flask (not shown).

cryogenic thermistor A thermistor designed for operation in a cryogenic liquid. Applications include liquid-level detection, for which the thermistor is heated slightly by passing a small current through it. This heat is dissipated more rapidly in the liquid than when the thermistor is above the liquid, producing a change in resistance that can be used to detect or control the level of the cryogenic liquid. Germanium is commonly used as the thermistor element because of its excellent stability at cryogenic temperatures.

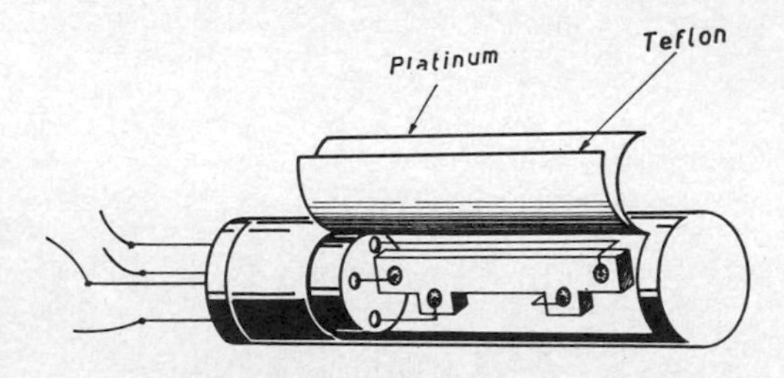

Cryogenic thermistor using germanium element cut in bridge configuration having four terminals.

cryogenic transformer A transformer that operates in digital cryogenic circuits. A controlled-coupling transformer is an example.

cryophysics Physics as restricted to phenomena occurring at very low temperatures, approaching absolute zero.

cryosar [CRYOgenic Switching by Avalanche and Recombination] A cryogenic, two-terminal, negative-resistance semiconductor device that consists essentially of two contacts on a germanium wafer which is operating in liquid helium. This bistable device switches on by avalanche or ionization and switches off by recombination of the free carriers in the single-crystal germanium. It can be used as a pulse generator, flip-flop, or other logic circuit in computers.

cryosistor A cryogenic semiconductor device in which a reverse-biased PN junction controls the ionization between two ohmic contacts. It can serve as a three-terminal switch, pulse amplifier, or oscillator after ionization.

cryostat An apparatus that maintains cryogenic temperatures.

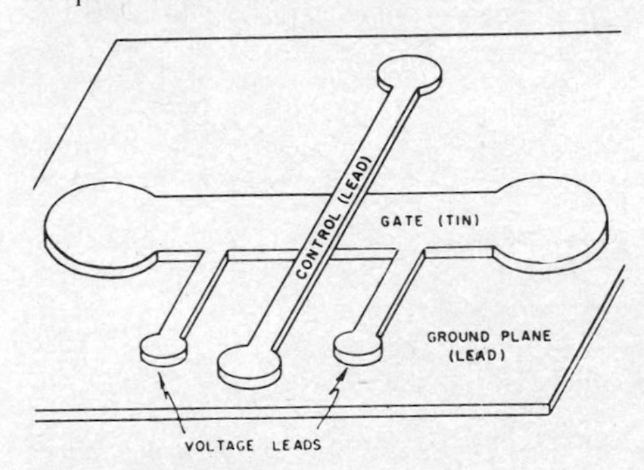

Cryotron using thin layers of superconducting tin and lead deposited on lead substrate and separated by insulating films. Current through control lead produces magnetic field that stops superconductivity in tin film.

cryotron A switch that operates in liquid helium; it consists of a short superconducting element and a control wire. When current is sent through the control wire to produce a magnetic field, the short main wire changes from a superconductive zero-resistance state to its normal resistive state. It may also be used as a computer storage element.

cryotronics The branch of electronics that deals with the design, construction, and use of cryogenic devices.

cryptanalysis The process of converting intercepted encrypted text into plain text without initial knowledge of the key used for encryption.

cryptochannel A complete system of communication that uses electronic encryption and decryption equipment and has two or more radio or wire terminals.

cryptogear Electronic equipment that uses cryptographic circuits or logic for encryption and decryption of messages so that they are meaningless when intercepted by enemy agents.

cryptography The encryption and decryption of messages for secret transmission. This term is usually shortened to crypto for both the noun and adjective (cryptographic) forms.

crystal A natural or synthetic piezoelectric or semiconductor material whose atoms are arranged with some degree of geometric regularity.

crystal activity A measure of the amplitude of vibration of a piezoelectric crystal plate under specified conditions.

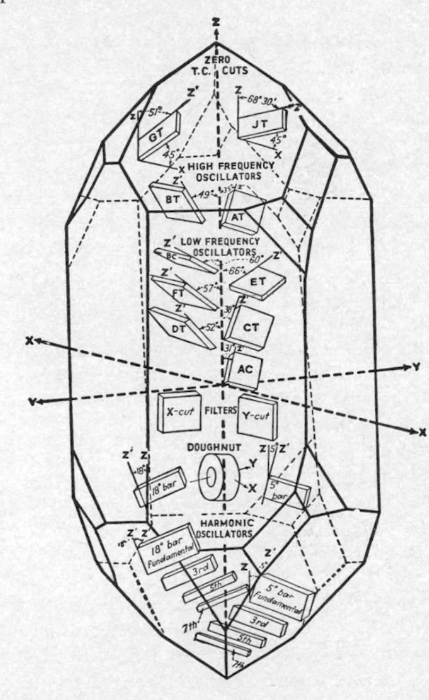

Crystal blank positions for different cuts in natural quartz. Angle of cut with respect to natural faces of crystal determines electrical characteristics of finished crystal blank.

crystal blank The result of the final cutting operation on a piezoelectric or semiconductor crystal.

crystal cartridge *Crystal pickup.*

crystal control Control of the frequency of an oscillator by a quartz crystal unit.

crystal-controlled transmitter A transmitter whose carrier frequency is directly controlled by the electromechanical characteristics of a quartz crystal unit.

crystal-controlled watch An electronic watch that uses a quartz crystal oscillator and integrated-circuit frequency dividers to drive a conventional moving-hand display or energize an electronic direct-reading digital display.

crystal counter A counter that uses one of several known crystals which are rendered momentarily conducting by ionizing events. One such crystal is a diamond.

crystal current The actual alternating current flowing through a crystal unit.

crystal cutter A cutter in which the mechanical displacements of the recording stylus are derived from the deformations of a crystal that has piezoelectric properties.

crystal detector A crystal diode or equivalent earlier crystal-catwhisker combination used to rectify a modulated RF signal to obtain the audio or video signal directly. Crystal diodes are also used in microwave receivers to combine an incoming RF signal with a local oscillator signal to produce an IF signal.

crystal diode *Semiconductor diode.*

crystal filter A highly selective tuned circuit that uses one or more quartz crystals. Sometimes used in IF amplifiers of communication receivers to improve the selectivity.

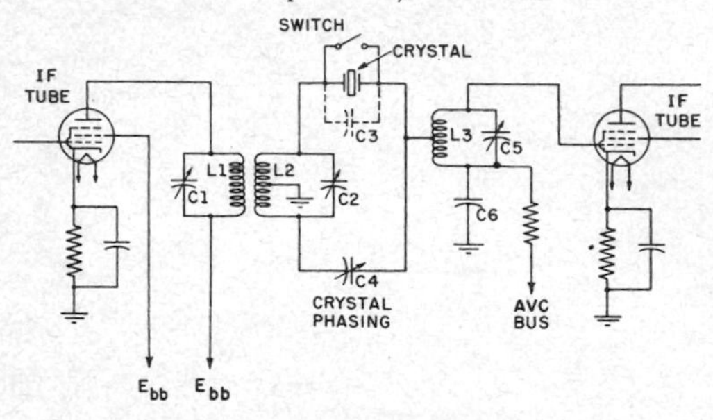

Crystal filter in IF amplifier.

crystal headphones Headphones that use Rochelle-salt or other crystal elements to convert AF signals into sound waves.

crystal holder A housing that provides proper support, mechanical protection, and connections for a quartz crystal plate. When the crystal plate is installed, the combination is called a crystal unit.

crystal hydrophone A crystal microphone that responds to waterborne sound waves.

crystal lattice filter A crystal filter that uses two matched pairs of series crystals and a higher-frequency matched pair of shunt or lattice crystals. Used chiefly in single-sideband equipment. Additional filter sections can be cascaded to improve the passband response.

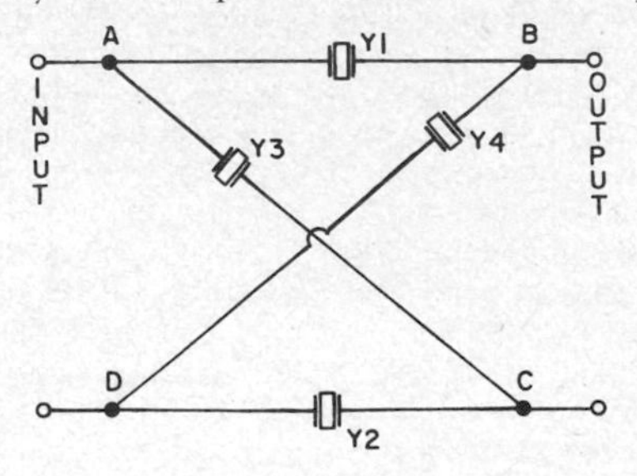

Crystal lattice filter.

crystalline laser A solid laser in which the lasing material is a pure crystal like ruby or a doped crystal like neodymium-doped ruby or neodymium-doped yttrium-aluminum garnet (YAG).

crystal loudspeaker A loudspeaker in which movements of the diaphragm are produced by a piezoelectric crystal unit that twists or bends under the influence of the applied AF signal voltage. Also called piezoelectric loudspeaker.

crystal microphone A microphone in which deformation of a piezoelectric bar by the action of sound waves or mechanical vibrations generates the output voltage between the faces of the bar. Rochelle-salt crystals or barium titanate ceramic bars are most often used. Also called piezoelectric microphone.

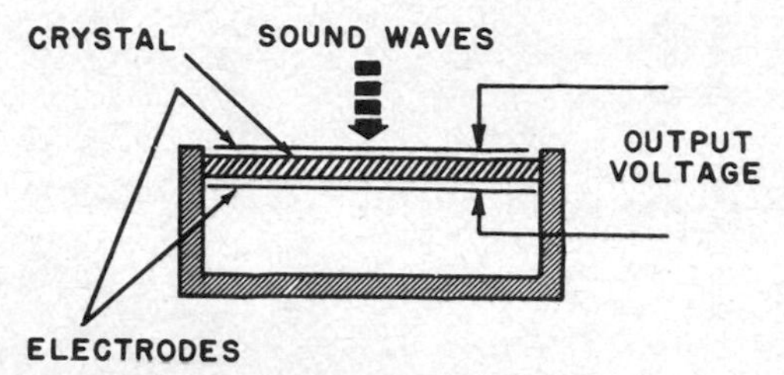

Crystal microphone in which sound waves act directly on one metallized face of piezoelectric crystal.

crystal mixer A mixer that uses the nonlinear characteristic of a crystal diode to mix two frequencies. Widely used in radar receivers to convert the received radar signal to a lower IF value by mixing it with a local oscillator signal.

crystal orientation The angle, with respect to crystal faces, at which a silicon or other semiconductor crystal is sliced.

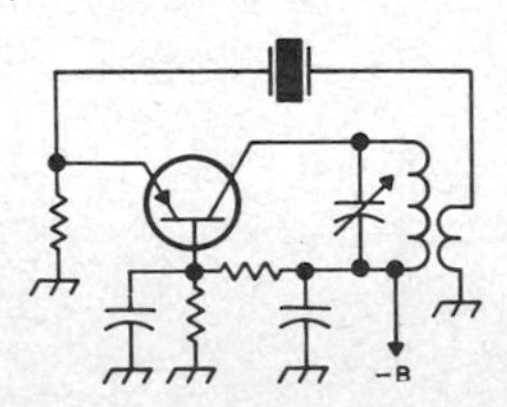

Crystal-oscillator circuit using transistor with tickler-coil feedback through crystal to emitter.

crystal oscillator An oscillator in which the frequency of the AC output is determined by the mechanical properties of a piezoelectric crystal.

crystal oven A temperature-controlled oven in which a crystal unit is operated to stabilize its temperature and thereby minimize frequency drift.

crystal pickup A phonograph pickup in which

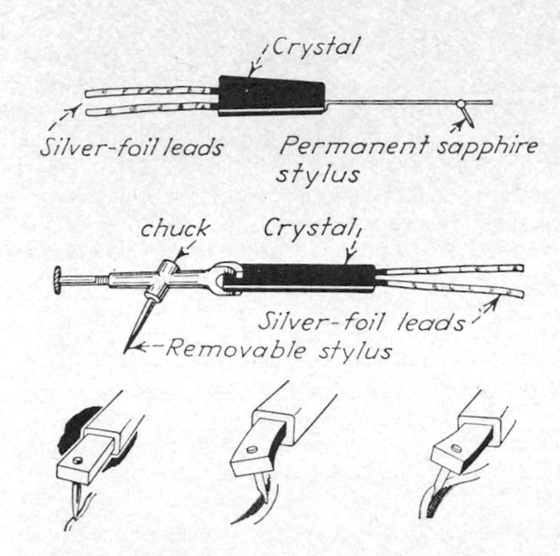

Crystal-pickup construction, and twisting action when stylus follows groove of record.

movements of the needle in the record groove cause deformation of a piezoelectric crystal, thereby generating an AF output voltage between opposite faces of the crystal. The piezoelectric material used is generally Rochelle salt or barium titanate. Also called crystal cartridge and piezoelectric pickup.

crystal plate A precisely cut slab of quartz crystal that has been lapped to final dimensions, etched to improve stability and efficiency, and coated with metal on its major surfaces for connecting purposes. Also called quartz plate.

crystal pulling A method of crystal growing in which the developed crystal is gradually withdrawn from a melt.

crystal rectifier *Semiconductor diode.*

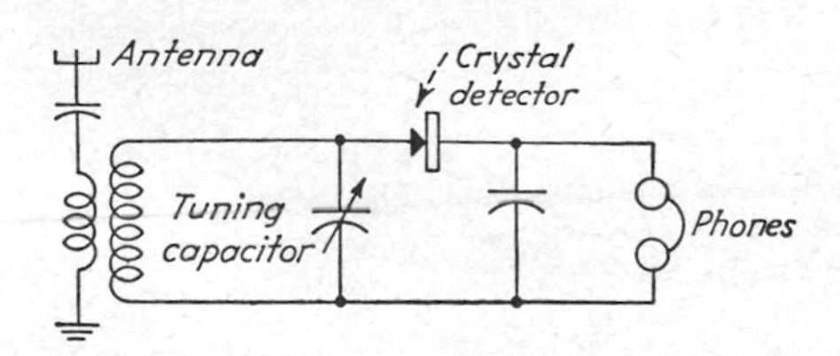

Crystal-set circuit. Long outdoor antenna is usually required.

crystal set A radio receiver that has a crystal detector stage for demodulation of the received signals, but no amplifier stages.

crystal shutter A mechanical waveguide or coaxial-cable shorting switch that, when closed, prevents undesired RF energy from reaching and damaging a crystal detector.

crystal slab A relatively thick piece of crystal from which crystal blanks are cut.

crystal spectrometer *Bragg spectrometer.*

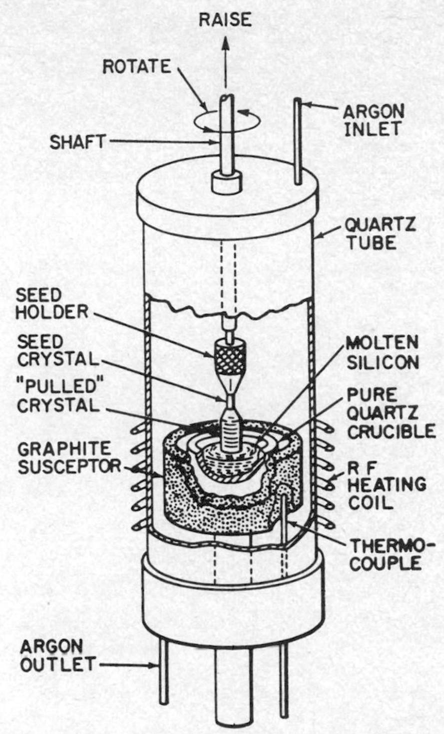

Crystal pulling as started with silicon seed crystal that is rotated as it is slowly pulled upward out of graphite crucible filled with molten polycrystalline silicon.

crystal-stabilized transmitter A transmitter which employs automatic frequency control, in which the reference frequency is that of a crystal oscillator.

crystal system One of the seven main categories to which a crystal may be assigned according to the symmetry of its external form or internal structure. The systems are cubic, tetragonal, hexagonal, trigonal, orthorhombic, monoclinic, and triclinic.

crystal transducer A transducer in which a piezoelectric crystal is the sensing element.

crystal unit A complete assembly of one or more quartz plates in a crystal holder.

crystal video receiver A broad-tuning radar or other microwave receiver that consists only of a crystal detector and a video or audio amplifier.

C-scale sound level in decibels [abbreviated dBC] The sound level in decibels as read when a standard sound-level meter is switched to weighting scale C, which weights frequencies between 70 and 4000 Hz uniformly and discriminates slightly against frequencies above and below this range.

C scan [Carrier System for Controlled Approach of Naval aircraft] An all-weather instrument landing system on aircraft carriers. Two shipborne transmitters produce a cross-shaped beam whose

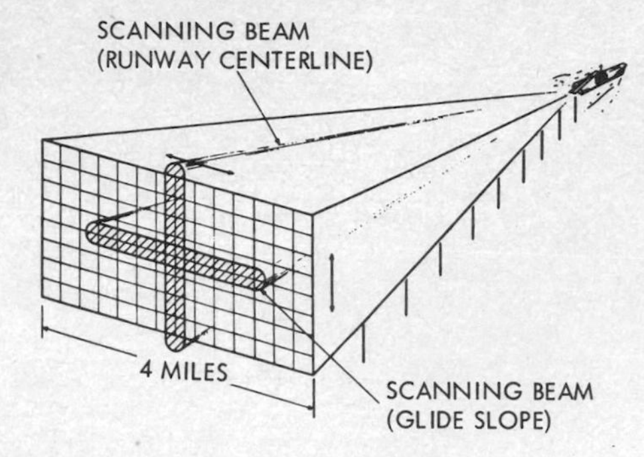

C-scan beams transmitted by aircraft carrier intersect at glide path.

intersection corresponds to the glide path. Receivers and instrumentation in the aircraft actuate a cockpit display that shows the pilot exact location with respect to the glide path.

C scope A radarscope that produces a C display.

CT Symbol for *center tap*. Used on circuit diagrams.

CT-cut crystal A quartz crystal cut at an orientation such that its resonant frequency is below 500 kHz.

CTFM sonar Abbreviation for *continuous-transmission frequency-modulated sonar*.

CTL Abbreviation for *complementary-transistor logic*.

CU Abbreviation for *crosstalk unit*.

cube tap *Multiple plug*.

cubical antenna An antenna array in which the elements form the 12 edges of a cube.

cubicle An enclosure for high-voltage equipment.

Cuccia coupler *Electron coupler*.

cue circuit A one-way communication circuit that conveys program control information.

CUJT Abbreviation for *complementary unijunction transistor*.

cumulative dose The total dose that results from repeated exposures to radiation.

cumulative ionization *Avalanche effect*.

cup core A core that encloses a coil to provide magnetic shielding. It usually has a powdered iron center post through the coil.

cuprous oxide [symbol Cu_2O] A semiconductor material formed on copper by heat. Electrons flow readily only in the direction from the metallic copper toward the oxide layer on the surface. This effect is utilized for rectification in copper-oxide rectifiers.

curie [abbreviated Ci] A unit of radioactivity, defined by international agreement in 1953 as that quantity of any radioactive nuclide which has 3.700×10^{10} disintegrations per second. Before 1953, the curie was defined as the quantity of radon that is in equilibrium with 1 g of radium.

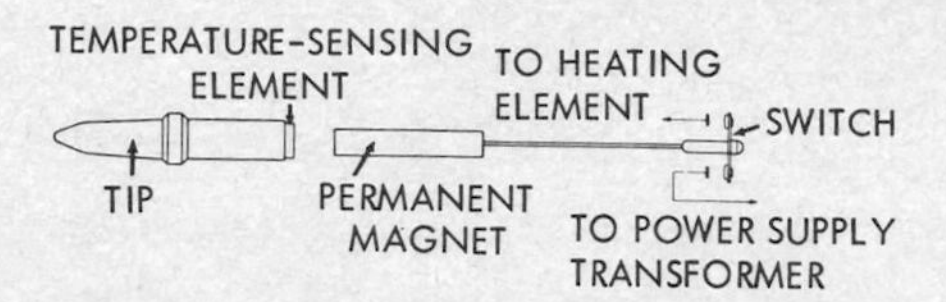

Curie-point application. When sensing element in tip of soldering iron reaches Curie point, it becomes nonmagnetic and releases permanent magnet, thereby opening switch.

Curie point The temperature above which a ferromagnetic material becomes substantially nonmagnetic.

curium [symbol Cm] A transuranic radioactive element produced artificially by bombarding plutonium with helium nuclei. Atomic number is 96.

current [symbol *I*] The rate of transfer of electricity from one point to another. Current is usually a movement of electrons but may also be a movement of positive ions, negative ions, or holes. Current is measured in amperes, milliamperes, microamperes, and nanoamperes. Also called juice (slang).

current amplification The ratio of output signal current to input signal current for an electron tube, transistor, or magnetic amplifier, the multiplier section of a multiplier phototube, or any other amplifying device. Often expressed in decibels by multiplying the common logarithm of the ratio by 20.

current amplifier An amplifier capable of delivering considerably more signal current than is fed in.

current antinode A point at which current is a maximum along a transmission line, antenna, or other circuit element that has standing waves. Also called current loop.

current attenuation The ratio of input signal circuit for a transducer to the current in a specified load impedance connected to the transducer. Often expressed in decibels.

current-balance relay A relay that operates when the magnitudes of two current inputs reach a predetermined ratio.

current calibrator A current source that provides adjustable and accurately known alternating and/or direct currents for calibrating ammeters and other current-measuring instruments.

current-carrying capacity The maximum current that can be continuously carried without causing permanent deterioration of electrical or mechanical properties of a device or conductor.

current compensation A means of compensating for stray shunt conductance across the terminals of a constant-current power supply.

current-controlled switch A semiconductor device in which the controlling bias sets the resistance at either a very high or very low value, corresponding to the off and on conditions of a switch.

current cutoff A negative-resistance circuit used

in some regulated power supplies to reduce load current automatically as load resistance is reduced and to minimize overload damage and protect sensitive loads.

current density The current per unit cross-sectional area of a conductor.

current drain The current taken from a voltage source by a load. Also called drain.

current feed Feed to a point where current is a maximum, as at the center of a half-wave antenna.

current feedback Feedback introduced in series with the input circuit of an amplifier.

current gain The ratio of output current to input current under specified conditions for a transistor or other amplifying device.

current generator A two-terminal circuit element whose terminal current is independent of the voltage between its terminals.

current hogging The condition whereby one of several parallel components or circuits takes more than its designed share of available current; usually malfunction or damage results. Applies particularly to paralleled logic gates.

current-leak detector A safety device that indicates current leakage through insulation and other undesired paths in electronic and electric equipment.

current limiter A device that restricts the flow of current to a certain amount, regardless of applied voltage.

current limiting A regulated power-supply-circuit feature that primarily protects against over-

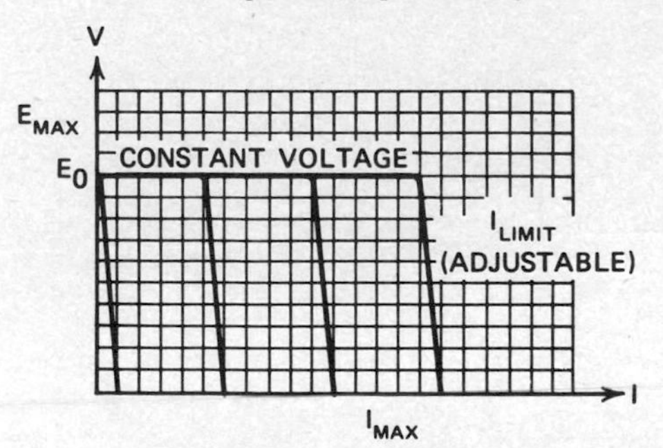

Current limiting in which circuit is adjustable to four different values of maximum current.

loads and shorts, rather than supplying constant-current operation.

current-limiting resistor A resistor inserted in an electric circuit to limit the flow of current to some predetermined value. Used chiefly to protect tubes and other components during warmup.

current loop *Current antinode.*

current-mode logic [abbreviated CML] Logic in which unsaturated transistors operate from a constant-current source that is switched at very high speed from one transistor to another.

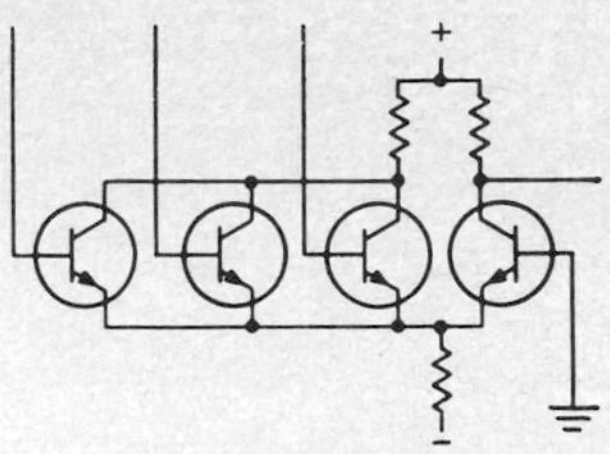

Current-mode logic.

current node A point at which current is zero along a transmission line, antenna, or other circuit element that has standing waves.

current regulator A device that maintains the output current of a voltage source at a predetermined, essentially constant, value despite changes in load impedance.

current relay A relay that operates at a specified current value rather than at a specified voltage value.

current saturation *Anode saturation.*

current-sensing resistor A resistor placed in series with a load to develop a voltage proportional to load current, as required for regulating load current.

current-sinking logic Logic in which the output current of a transistor flows back to the preceding stage through any inputs that are low. When all inputs are high, the transistor is saturated so its

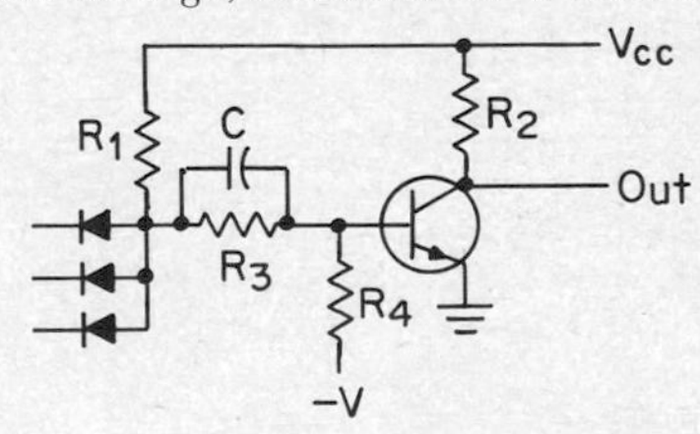

output is low and is capable of absorbing or sinking the input currents of several additional gates.

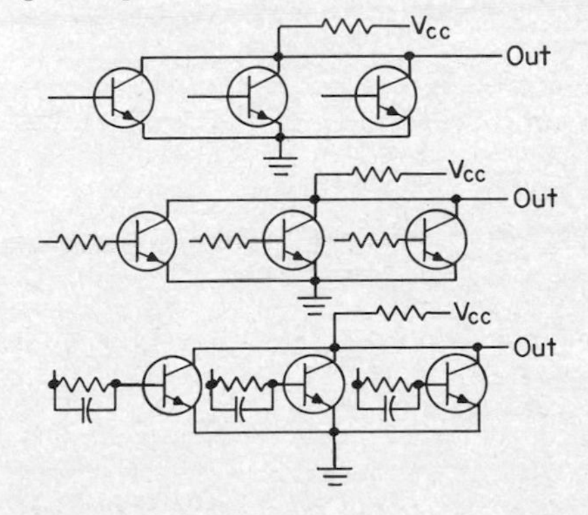

Current-sourcing logic types. Top to bottom: DCTL, RTL, RCTL.

current-sourcing logic Logic in which current flows from the output of the driving gate to the inputs of the driven gate when the output of the driving gate is high. Resistor-transistor logic is an example.

current transformer An instrument transformer whose primary winding is connected in series with a circuit that carries the current to be measured or controlled. The current is measured across the secondary winding.

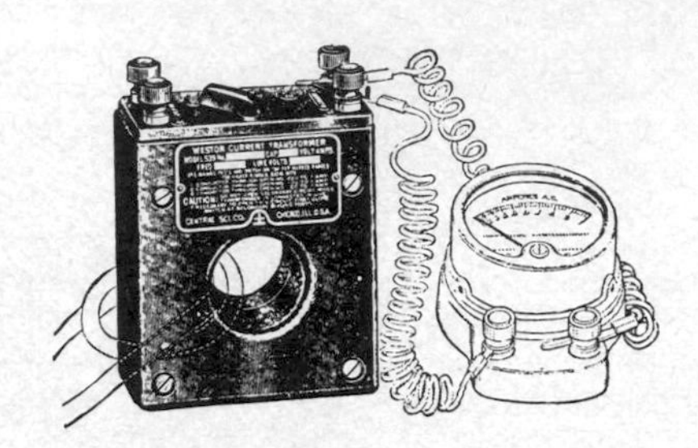

Current transformer, showing meter connections and method of looping current-carrying line through transformer.

current-type telemeter A telemeter in which the magnitude of a single current is the translating means.

cursor 1. A clear or amber-colored plastic sheet that can be placed over a radar screen and rotated until an etched diameter line passes through a target echo. The bearing from radar to target can then be read on a stationary 360° scale. Another type of cursor is used on ground radar to give the bearing of an enemy aircraft with respect to an attacking aircraft being directed from the ground by radio. 2. An indicator that moves along a line of print on a cathode-ray screen, under manual or automatic control, to show the position at which the next keyboarded character will be entered or an existing character will be changed.

cursor target bearing Target bearing as measured by a cursor on a PPI radar display.

curtain array An antenna array that consists of vertical wire elements stretched between two suspension cables. It may be backed by a second curtain that acts as reflector. The active elements are usually half-wave dipoles.

curtain rhombic antenna A multiple-wire rhombic antenna that has a constant input impedance over a wide frequency range. Two or more conductors join at the feed and terminating ends but are spaced apart vertically from 1 to 5 ft (30 to 150 cm) at the side poles.

Curtis-winding resistor A wirewound resistor in which residual inductance and capacitance are reduced by reversing the direction of alternate turns. This is achieved by passing the wire through a diametral slot in the coil form after each complete turn.

curve follower An instrument, usually photoelectric, that is capable of reading data represented by a curved line on a graph.

cusped magnetic field A magnetic field created by adjacent parallel coils that carry current in opposite directions. The magnetic strength is zero at the center of the field, and all magnetic lines of force adjacent to this point are convex. Used in fusion research, to contain a plasma of high-energy deuterium ions.

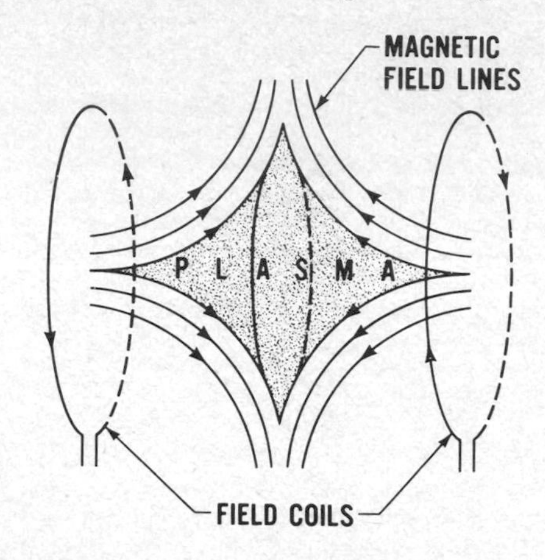

Cusped magnetic field as used to contain plasma for thermonuclear fusion.

cut 1. A section of a crystal that has two parallel major surfaces. Cuts are specified by their orientation with respect to the axes of the natural crystal, such as X cut, Y cut, BT cut, and AT cut. 2. The fraction that is removed as product or advanced to the next separative element in an isotope separation process. 3. An order to stop an action, turn off a television camera, or disconnect all microphones in a radio studio.

Cutler feed A resonant cavity that transfers RF energy from the end of a waveguide to the reflector of a radar spinner assembly.

cutoff The minimum value of bias voltage, for a given combination of supply voltages, that just stops output current in an electron tube, transistor, or other active device.

cutoff attenuator An adjustable-length waveguide that varies attenuation of signals passing through the waveguide.

cutoff bias The DC bias voltage that must be applied to the grid of an electron tube to stop the flow of anode current.

cutoff field *Critical field.*

cutoff frequency The limiting frequency beyond which the attenuation, gain, efficiency, or other performance characteristic of a device begins changing so rapidly that the output can no longer be considered useful. Also called critical frequency.

cutoff wavelength 1. The ratio of the velocity of

electromagnetic waves in free space to the cutoff frequency in a uniconductor waveguide. 2. The wavelength that corresponds to the cutoff frequency.

cutout An electric device that is operated manually or automatically to interrupt current flow, such as a circuit breaker, fuse, or switch.

cut paraboloidal reflector A paraboloidal reflector that is not symmetrical with respect to its axis.

cut-set A set of branches of a network such that the cutting of all the branches of the set increases the number of separate parts of the network, but the cutting of all the branches except one does not.

cutter An electromagnetic or piezoelectric device that converts an electric input to a mechanical output, used to drive a stylus which cuts a wavy groove in the highly polished wax surface of a disk recording. Also called cutting head, head, and recording head.

cutting angle The angle between the vertical cutting face of the stylus and the surface of the record. It ordinarily should be 90°. Deviation from this value is sometimes specified as dig-in angle or drag angle.

cutting head *Cutter.*

cutting stylus A recording stylus with a sharpened tip that removes material, to produce a groove in the recording medium.

CW Abbreviation for *continuous wave.*

CW laser Abbreviation for *continuous-wave laser.*

cybernetics A comparative study of the methods of automatic control and communication that are common to people and machines. Used in analyzing and improving the efficiency of communication systems, information-handling machines, and feedback control systems.

cycle 1. One complete sequence of values of an alternating quantity, including a rise to a maximum in one direction, a return to zero, a rise to a maximum in the opposite direction, and a return to zero. The number of cycles occurring in 1 s is called the frequency. 2. A set of operations that is repeated as a unit. 3. To run a machine through an operating cycle.

cycle-matching loran *Low-frequency loran.*

cycle per minute [abbreviated cpm] A unit of frequency of action.

cycle per second [abbreviated cps] Former unit of frequency, now called hertz and abbreviated Hz.

cycle timer A timer that opens or closes circuits according to a predetermined time schedule.

cyclically magnetized Under the influence of a magnetizing force that varies between two specific limits long enough so the magnetic induction has the same value for corresponding points in successive cycles.

cyclic binary code Any binary code that changes by only 1 bit when going from one number to the number immediately following. The gray code is one example.

cyclic function A function that assumes a given sequence of values repetitively at an arbitrarily varied rate.

cyclic memory *Circulating memory.*

cyclic shift A computer shift in which the digits dropped off at one end of a word are returned at the other end of the word.

cycling A periodic change of the controlled variable from one value to another in an automatic control system. Also called oscillation.

cycloconverter A frequency converter that changes an AC voltage of one frequency to a lower frequency without using a DC link. A single-phase version requires a separate set of silicon controlled

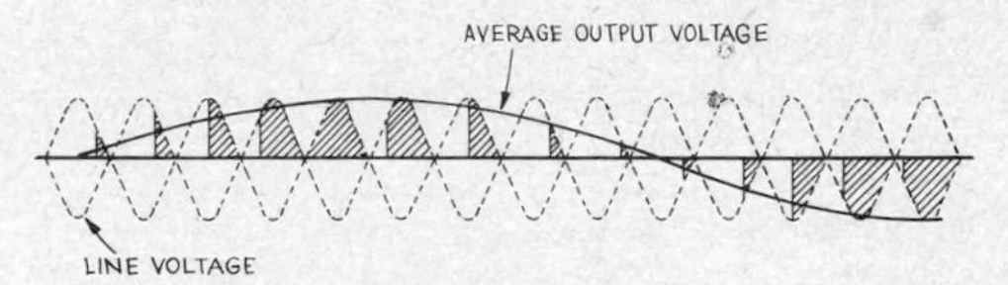

Cycloconverter waveforms, showing how SCR firing signals vary during period in which each input half-cycle conducts, so average output voltage after filtering is at desired lower frequency.

rectifiers for each polarity of the output voltage, supplied with firing signals obtained from a reference AC voltage that has the desired output frequency.

cyclograph An electronic instrument that produces on a cathode-ray screen a pattern which changes in shape according to core hardness, carbon content, case depth, and other metallurgical properties of a test sample of steel inserted in a sensing coil.

cycloinverter An inverter, operating at about 10 times the desired output frequency, coupled to a cycloconverter that is producing the desired output frequency.

cyclophon A vacuum tube in which a beam of electrons serves as a switching element.

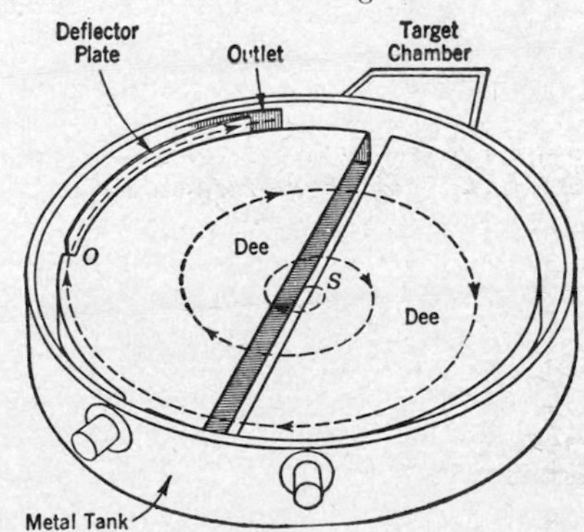

Cyclotron. Positive ions emitted by source at S in center are accelerated in steps between dee-shaped hollow electrodes and bent on half-circles by axial magnetic field. Ions leave left dee at O and hit target in chamber at top.

cyclotron An accelerator in which charged particles are successively accelerated by a constant-frequency alternating electric field that is synchronized with movement of the particles on spiral paths in a constant magnetic field normal to their path.

cyclotron frequency The frequency at which an electron beam will rotate about an axis when moving at right angles to a uniform magnetic field. This frequency depends upon only the strength of the magnetic field. When rotating in this manner, the electron beam traces the surface of a cone in space.

cyclotron-frequency magnetron A magnetron whose frequency of operation depends on synchronism between the AC electric field and the electrons oscillating in a direction parallel to this field. An example is a split-anode magnetron that has a resonator between the anodes.

cyclotron radiation The electromagnetic radiation emitted by charged particles as they orbit in a magnetic field. At low velocities the radiation is concentrated in a single spectral line, at the cyclotron frequency. At higher velocities the spectral line is spread into a band of frequencies, including harmonics of the cyclotron frequency.

cyclotron resonance A method of accelerating electrons, as for electric propulsion of spacecraft, that uses the physical principles of a cyclotron. The electrons of a plasma are trapped by crossed electric and magnetic fields. As the electrons spiral around the magnetic lines of force, they absorb energy continuously from the electric field until they collide with other particles and scatter at high speeds. The only direction of escape from the trap is through the exhaust nozzle, where they can provide a propulsion effect.

cyclotron resonance heating A modification of magnetic pumping that involves compressing and expanding plasma at a frequency which approximates the cyclotron frequency of the ions in the plasma. The goal is stellar temperatures, above several million degrees, as required for controlled fusion.

cyclotron wave A wave associated with the electron beam of a traveling-wave tube.

cylindrical antenna An antenna in which hollow cylinders serve as radiating elements.

cylindrical array An electronic scanning antenna that may consist of several hundred columns of vertical dipoles mounted in cylindrical radomes arranged in a circle. Used as a beacon interrogator for air traffic control.

cylindrical-film storage A computer storage in which each storage element consists of a short length of glass tubing that has a thin film of nickel-iron alloy on its outer surface. Wires threaded through the cylinders serve as bit and sense lines, while conducting straps at right angles to the cylinders serve as word lines.

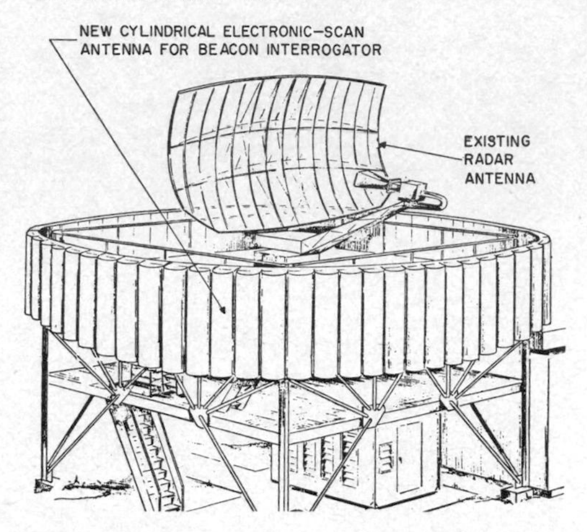

Cylindrical array having 224 columns of dipoles arranged in groups of four in cylindrical radomes to form stationary antenna that replaces conventional rotating radar antenna shown in center of array.

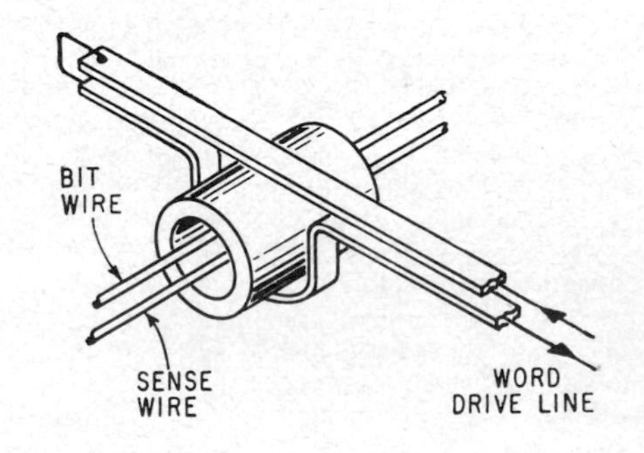

Cylindrical-film storage, in which word line produces circumferential magnetic flux in permalloy film on tubing, and bit line produces axial flux.

cylindrical reflector A reflector that is a portion of a cylinder. This cylinder is usually parabolic.

cylindrical wave A wave whose equiphase surfaces form a family of coaxial cylinders.

Czochralski process A method of producing large single crystals by inserting a small seed crystal of germanium, silicon, or other semiconductor material into a crucible filled with similar molten material, then slowly pulling the seed up from the melt while rotating it.

D

d 1. Abbreviation for *deci-*. 2. Symbol for *deuterium*.

D 1. Symbol for *drain*. 2. Symbol for *deuterium*.

D/A Abbreviation for *digital-to-analog*.

DAC Abbreviation for *digital-to-analog converter*.

D/A converter Abbreviation for *digital-to-analog converter*.

daisy-wheel printer A serial printer in which the printing element is a plastic hub that has a large number of flexible radial spokes, each spoke having one or more different raised printing characters. The wheel is rotated as it is moved horizontally step by step under computer control, and it stops when a desired character is in a desired print position so a hammer can drive that character against an inked ribbon, just as in a typewriter.

damage-risk criterion The maximum sound pressure levels of a noise, as a function of frequency, to which people can be exposed for a specified time without risk of significant hearing loss.

damp To make an oscillating indicator or other device come to rest.

damped oscillation Oscillation in which the amplitude of the oscillating quantity decreases with time.

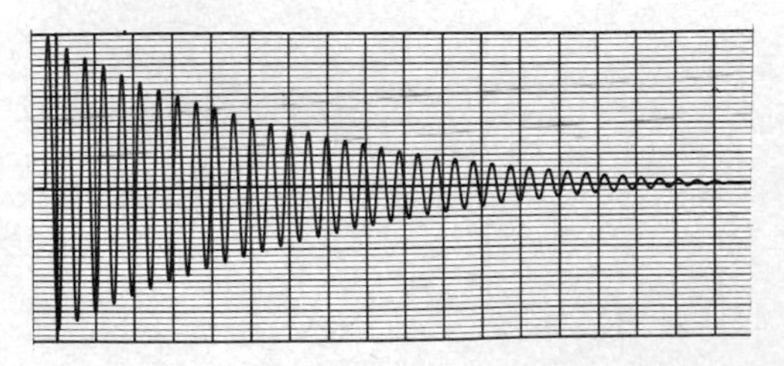

Damped wave.

damped wave A wave in which the amplitudes of successive cycles progressively diminish at the source.

damper A diode used in the horizontal deflection circuit of a television receiver to make the sawtooth deflection current decrease smoothly to zero instead of oscillating at zero. The diode conducts each time the polarity is reversed by a current swing below zero.

damper tube A diode, pentode, or other vacuum tube used in a damper circuit.

damping 1. Any action or influence that extracts energy from a vibrating system in order to suppress the vibration or oscillation. 2. Reducing or eliminating reverberation in a room by placing sound-absorbing materials on the walls and ceiling. Also called soundproofing.

damping factor The ratio of the amplitude of any one of a series of damped oscillations to that of the following one. Also called decrement.

damping magnet A permanent magnet used in conjunction with a disk or other moving conductor to produce a force that opposes motion of the conductor and thereby provides damping.

daraf [farad spelled backward] The unit of elastance, which is the reciprocal of capacitance in farads.

dark conduction Residual conduction in a photosensitive substance that is not illuminated.

dark current The current flowing through a photoelectric device in the absence of irradiation.

dark-current pulse A phototube dark-current excursion that can be resolved by the system employing the phototube.

dark discharge An invisible electric discharge in a gas.

dark resistance The resistance of a selenium cell or other photoelectric device in total darkness.

dark space A region in a glow discharge that produces little or no light.

dark-trace tube A cathode-ray tube having a bright face that does not necessarily luminesce, on which signals are displayed as dark traces or dark blips where the potassium chloride screen is hit by the electron beam. When the screen is illuminated externally by powerful lamps, the image may be enlarged greatly by optical lenses and projected onto a conventional rear-projection glass screen. In World War II this screen was usually mounted horizontally at table height, and a translucent map of the area under radar survey was placed over the screen for viewing by many controllers during naval actions. Also called skiatron.

Darlington amplifier A current amplifier consisting essentially of two separate transistors, with the collectors connected together and the emitter of one connected to the base of the other. The Darlington pair is usually mounted in a common housing.

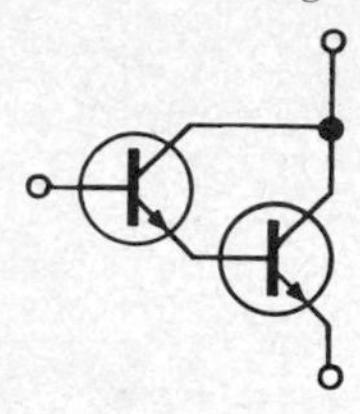

Darlington amplifier has same terminations as single transistor.

d'Arsonval movement Deprecated term for *permanent-magnet moving-coil instrument.*

dashpot A device that uses a piston moving in a gas or liquid to absorb energy and thereby delay the operating time of a relay, circuit breaker, or other electric device.

data General term for the numbers, letters, and symbols that serve as input for computer processing. Commonly treated as a collective noun, for use with a singular verb.

data acquisition The phase of data handling that begins with the sensing of variables and ends with a magnetic recording or other record of raw data. It may include a complete radio telemetering link.

data amplifier An amplifier that has low offset error and high common-mode rejection, used to condition the electric output signals of a data transducer by modifying the bandwidth, providing amplification, or reshaping the signal. The resulting output is usually converted to a digital code.

data bank One or more collections of data, generally stored on magnetic tapes, disks, or drums in a form suitable for computer processing. Also called data base.

data base *Data bank.*

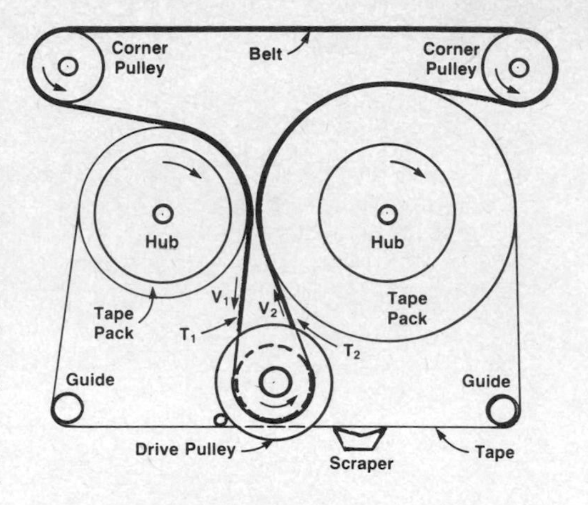

Data cartridge in which elastomer belt goes over both tape reels and drive pulley, for driving magnetic tape in either direction when cartridge is inserted into drive mechanism so its drive pulley bears against shaft of drive motor. Insertion of cartridge opens door at bottom (not shown) so tape can bear against read/write head.

data cartridge A tape cartridge used for nonvolatile and removable data storage in small digital systems.

data center An organization established primarily to acquire, analyze, process, store, retrieve, and disseminate one or more types of data. Examples include centers for processing raw data received from spacecraft and centers for collecting and compiling data on properties of materials.

data circuit A wire or radio link that allows transmission of digital data pulses with minimum distortion.

data compression Reducing the size of a collection of data by eliminating redundant data elements prior to transmission or storage.

data concentrator A device that takes data from several different teletypewriter or other slow-speed lines and feeds them to a single higher-speed line. A microprocessor can be used to perform this function.

data converter A converter that changes data from one form to another, as from punched holes in cards or paper tape to binary form on magnetic tape.

data domain The aspect of digital systems that is characterized by data flows, data formats, equipment architecture, and state-space concepts.

data element A basic unit of information, such as age, sex, payroll number, geographic location, date, or time. Also called data item.

date encryption The transformation of computer data into a form that is unreadable by nonauthorized recipients. The protection system is based on use of a unique encryption key assigned to each customer. Data is encrypted with its key at the point of transmission and decrypted at the receiving point.

data-entry terminal A portable keyboard and small numerical display designed for interactive communication with a computer. One version has 16 keys for 0 through 9 and A through F, and larger versions have all the ASCII characters.

data-handling capacity The maximum number of units of information that can be transmitted, received, processed, stored, or otherwise handled by a specific piece of equipment.

data-handling system Automatically operated equipment used to interpret data gathered by instrument installations. Also called data-reduction system.

data item *Data element.*

data link A wire, radio, or other data-transmission channel used for connecting data-processing equipment to an input terminal, output or display device, or other remotely located data-processing equipment.

data logging Recording of data in the time sequence at which related events occur.

data mile A military unit of distance, equal to 6000 ft (1828.8 m).

Data-Phone A Bell System device that permits transmission of data over telephone channels at speeds up to 2000 b/s. It uses a modem to convert binary input information into tones suitable for transmission over voice-grade lines. At the other end of the line, a receiving modem converts the tones back to binary data for digital processing equipment.

data processing Changing the form, meaning, appearance, location, or other characteristics of data. Processing includes data handling and data reduction.

data processor A machine for handling data in a sequence of operations. Data processors include desk calculators, punched-card machines, and computers.

data reduction The process of converting recorded data into desired useful forms.

data-reduction system *Data-handling system.*

data repeater A repeater that corrects highly distorted serial data signals and retransmits them.

dataset *Modem.*

data signaling rate The data-handling capacity of a set of parallel transmission channels, expressed in bits per second.

data sink A device capable of accepting signals from a data-transmission system. It may also check these signals and produce error-control signals.

data source A device capable of originating signals for a data-transmission system.

data stabilization Stabilization of the display of radar signals with respect to a selected reference regardless of changes in radar-carrying vehicle attitude, as in azimuth-stabilized PPI.

data transcription Conversion of data from one recorded form to another, as from magnetic tape to punched cards.

data-transmission system A system used to transmit data from one instrument to another.

data-transmitting element The element in a fire-control system that transmits data between elements of the system which are located some distance apart.

data under voice [abbreviated DUV] An AT&T digital data service that allows digital signals to travel on the lower portion of the frequency spectrum of existing microwave radio systems. Digital channels initially available handled speeds of 2.4, 4.8, 9.6, and 56 kb/s.

datum line A reference line from which calculations or measurements are made.

dawn chorus *Chorus.*

dB Abbreviation for *decibel.*

dBa Abbreviation for *decibels adjusted.*

dBC Abbreviation for *C-scale sound level in decibels.*

dBf Abbreviation for *decibels above 1 femtowatt.*

dBk Abbreviation for *decibels above 1 kilowatt.*

dBm Abbreviation for *decibels above 1 milliwatt,* a unit used in specifying input levels.

dB meter *Decibel meter.*

dBmp Abbreviation for *decibels above 1 milliwatt psophometrically weighted.*

dBp Abbreviation for *decibels above 1 picowatt.*

dBrn Abbreviation for *decibels above reference noise.*

dBV Abbreviation for *decibels above 1 volt.*

dBW Abbreviation for *decibels above 1 watt.*

dBx Abbreviation for *decibels above reference coupling.*

DC Abbreviation for *direct current.*

DC amplifier An amplifier capable of amplifying DC voltages and slowly varying voltages. The basic types of DC amplifiers are the direct-coupled amplifier, the carrier amplifier, and combinations of direct-coupled and carrier amplifiers. The most common type of carrier amplifier is the chopper amplifier; when the chopper amplifier is used with a direct-coupled amplifier, the combination is called a chopper-stabilized amplifier.

DC component The average value of a signal. In television it represents the average luminance of the picture being transmitted. In radar it is the level from which the transmitted and received pulses rise.

DC coupling That type of coupling in which the zero-frequency term of the Fourier series representing the input signal is transmitted.

DC/DC An abbreviation used to indicate that a converter will change one DC voltage value to a different (higher or lower) DC voltage value.

DC/DC converter An electronic circuit that converts a DC voltage to a different value by using an inverter or chopper followed by a step-up or step-down transformer and a rectifier.

DC dump The removal of all DC power from a computer system or component intentionally, accidentally, or conditionally. In some types of storage this results in loss of stored information.

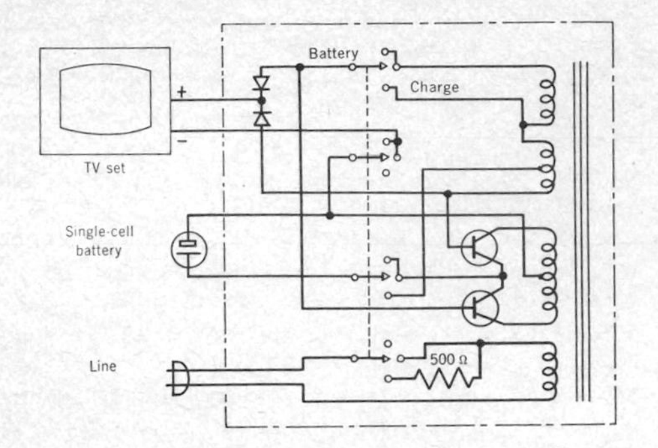

DC/DC converter that steps up 1.5-V voltage of single dry cell to 12 V for operating portable television set. Circuit also permits operation from AC line.

DC erase Use of direct current to energize an erasing head.

DC generator A rotating electric machine that converts mechanical power into DC power.

DC inserter A television transmitter stage that adds to the video signal a DC component known as the pedestal level.

DC magnetic biasing Magnetic biasing by means of direct current in magnetic recording.

DC motor A motor that operates from a DC voltage source and converts electric energy to mechanical energy.

DC motor control *Electronic motor control.*

DC offset A direct-current level that may be added to the input signal of an amplifier or other circuit.

DC picture transmission Television transmission in which the signal contains a DC component that represents the average illumination of the entire scene.

DC potentiometer A voltage-measuring instrument in which the unknown DC voltage is balanced against the precisely known voltage (1.018636 V) supplied by a standard cell. Higher unknown voltages must be stepped down with a voltage divider to obtain a balance.

DC power supply A power supply that provides one or more DC output voltages, such as a DC generator, rectifier-type power supply, converter, or dynamotor.

DC receiver A radio receiver designed to operate directly from a 120-VDC power line.

DC reinsertion *Clamping.*

DC restoration *Clamping.*

DC restorer A clamping circuit used in television receivers to restore the DC component of the video signal after AC amplification. The resulting DC voltage serves as the bias voltage for the grid of the picture tube, to make average reproduced brightness correspond to the average brightness of the scene being transmitted. Also called clamper, reinserter, and restorer.

DC self-synchronous system A system for transmitting angular position or motion, in which an arrangement of resistors serves as a transmitter that furnishes a receiver with two or more voltages which are functions of transmitter shaft position. The receiver has two or more stationary coils that set up a magnetic field which causes a rotor to take an angular position corresponding to the angular position of the transmitter shaft.

DC signaling A transmission method that uses direct current.

DC tachometer A DC generator operating with negligible load current and with constant field flux provided by a permanent magnet, so its DC output voltage is proportional to speed.

DCTL Abbreviation for *direct-coupled transistor logic.*

DC transducer A transducer that requires DC excitation and provides a DC output which varies with the parameter being sensed.

DC transmission Television transmission in which the DC component of the picture signal is still present. The true level of background illumination is thus maintained at all times.

DC voltage *Direct voltage.*

DC working volts [abbreviated DCWV] Deprecated term for *volts DC.*

DCWV Abbreviation for *DC working volts.* Use of VDC is preferred.

DDA Abbreviation for *digital differential analyzer.*

DDC Abbreviation for *direct digital control.*

DDD Abbreviation for *direct distance dialing.*

D display A radarscope C display in which targets appear as bright spots that are vertically elongated in rough proportion to range.

DDS Abbreviation for *digital data service.*

deaccentuator A circuit used in an FM receiver to offset the preemphasis of higher audio frequencies introduced at the FM transmitter.

dead 1. Having no connection to a voltage source. 2. Having no output when a signal is applied to the input.

dead band 1. The range of values over which a measured variable can change without affecting the output of a magnetic amplifier or automatic control system. Also called dead zone and neutral zone. 2. The portion of a potentiometer element that is shortened by a tap. When the wiper traverses this area, there is no change in output.

deadbeat Coming to rest without vibration or oscillation, as when the pointer of a meter moves to a new position without overshooting.

dead end 1. The end of a sound studio that has the greater sound-absorbing characteristics. 2. The portion of a tapped coil through which no current is flowing at a particular switch position.

dead-end effect Absorption of energy by unused portions of a tapped coil.

dead-end switch A switch used to short-circuit

unused portions of a tapped coil to prevent dead-end effects.

dead reckoning Determination of the approximate position of a vehicle by combining vectors for speed, direction, and other factors with the last known position. Usually allows for wind in air navigation, but excludes wind and currents in marine navigation; allowance for wind and current then gives estimated position at sea. The term comes from deduced reckoning, which was abbreviated ded. reckoning.

dead-reckoning tracer A mechanical device used to produce a continuous plot of all ships within range of a ship's radar.

dead room *Anechoic room.*

dead short A short-circuit path that has extremely low resistance.

dead spot 1. A geographic location in which signals from a radio, television, or radar transmitter are received poorly or not at all. 2. A portion of the tuning range of a receiver in which stations are heard poorly or not at all because of improper design of tuning circuits.

dead time 1. The time interval, after a response to one signal or event, during which a system is unable to respond to another. For a radiation counter it is the interval after the start of a count during which the counter is insensitive to further ionizing events. For a transponder it is the interval after the start of a pulse during which no new pulse can be received or produced. Also called insensitive time. 2. The time interval between a change in the input signal to a process-control system and the response to the signal.

dead-time correction A correction applied to an observed counting rate to allow for the probability of the occurrence of events within the dead time. Also called coincidence correction.

dead zone *Dead band.*

deathnium center An imperfection in the arrangement of atoms in a semiconductor crystal. It facilitates the generation and recombination of electron-hole pairs.

death ray A ray that can kill living cells. Ultraviolet rays can kill bacteria, radio waves of certain frequencies can kill insects, and x-rays can kill small animals.

de Broglie wavelength The wavelength ascribed by wave or quantum mechanics to a particle having a given momentum. This wavelength is equal to Planck's constant divided by the momentum of the particle.

debugging 1. The process of eliminating from a newly designed system the components and circuits that cause early failures. 2. Removing mistakes from a computer program. 3. The detection and removal of secretly installed listening devices, popularly known as bugs. The basic counterespionage instrument is a sensitive receiver that can be tuned to and homed on a concealed miniature radio transmitter.

debunching A tendency for electrons in a beam to spread out both longitudinally and transversely due to mutual repulsion. The effect is a drawback in velocity-modulation tubes.

Debye effect Selective absorption of electromagnetic waves by a dielectric, due to molecular dipoles.

Debye length The distance at which a given negative particle in a plasma is shielded by surrounding positive particles.

Debye-Sears effect The generation of acoustic waves, consisting of alternate regions of compression and refraction one half-wavelength apart, by a piezoelectric crystal vibrating in a longitudinal mode in a liquid. When a parallel beam of light is sent through the liquid in a tank having plate-glass walls, the acoustic waves act as a diffraction grating that can be used to determine the velocity of sound in the liquid.

decade 1. A group of 10. 2. The interval between any two quantities having the ratio of 10 to 1.

decade box An assembly of precision resistors, coils, or capacitors whose individual values vary in submultiples and multiples of 10. Each section contains 10 equal-value components connected in series. The total value of each section is 10 times

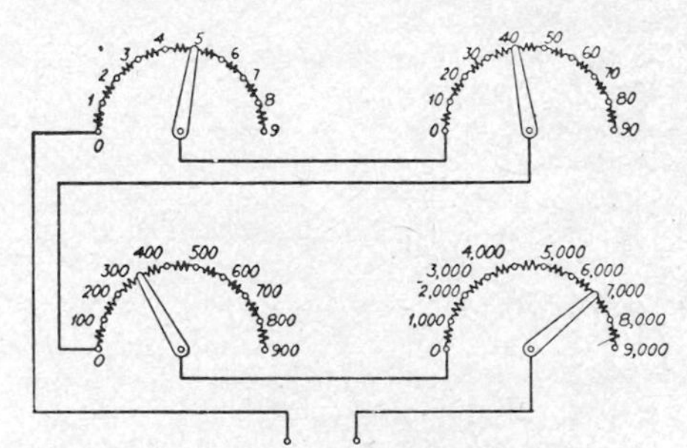

Decade box using precision resistors.

that of the preceding section. By appropriately setting the 10-position selector switch for each section, the decade box can be set to any desired value within its range.

decade counter *Decade scaler.*

decade scaler A scaler that produces 1 output pulse for every 10 input pulses. Also called counter decade, decade counter, and scale-of-ten circuit.

decametric wave A radio wave between the wavelength limits of 10 and 100 m, corresponding to the high-frequency (HF) range of 3 to 30 MHz.

decay 1. Gradual reduction in the magnitude of a quantity, as of current, magnetic flux, a stored charge, or phosphorescence. 2. *Radioactive decay.*

decay characteristic *Persistence characteristic.*

decay curve A graph showing how the activity of a radioactive sample varies with time. Alternatively, it may show the amount of radioactive material remaining at any time.

decay gammas The characteristic gamma rays emitted during the decay of most radioisotopes.

decay heat Heat produced by the decay of radioactive nuclides.

decay product A nuclide resulting from the radioactive disintegration of a radionuclide. The nuclide is formed directly or is the result of successive transformations in a radioactive series. A decay product may be either radioactive or stable.

decay rate The time rate of disintegration of radioactive material, generally accompanied by emission of particles and/or gamma radiation.

decay time The time taken by a quantity to decay to a stated fraction of its initial value. The fraction is commonly $1/e$, where e is the base of natural logarithms.

Decca A continuous-wave hyperbolic radio aid to navigation in which a receiver measures and indicates the relative phase difference between signals received from two or more synchronized ground stations. The system provides differential distance information from which position can be determined. Developed by British in World War II. Useful range is up to about 500 mi (800 km).

decelerating electrode An electrode whose potential provides an electric field to decrease the velocity of the beam electrons in an electron-beam tube.

deceleration time The time between completion of reading or writing of a magnetic-tape record and actual stopping of the tape.

deception Use of electronic countermeasures to reduce the effectiveness of enemy radar, radio, or other equipment involving electromagnetic signals.

deception jamming Transmission of jamming signals in such a manner as to mislead or confuse the data-processing system of an enemy radar.

deci- [abbreviated d] A prefix representing 10^{-1}, or one-tenth.

decibel [abbreviated dB] A unit used to express the magnitude of a change in signal or sound level. One decibel is the change in level of a pure sine-wave sound that is just barely detectable by the average human ear. The difference in decibels between two signals is 10 times the common logarithm of their ratio of powers or 20 times the common logarithm of their ratio of voltages or currents. One decibel is $1/10$ bel.

decibel meter An instrument used to measure directly the power level of a signal in decibels above or below an arbitrary reference level. Also called dB meter.

decibels above 1 femtowatt [abbreviated dBf] A power level equal to 10 times the common logarithm of the ratio of the given power in watts to 1 fW (10^{-15} W).

decibels above 1 kilowatt [abbreviated dBk] A power level equal to 10 times the common logarithm of the ratio of a given power in watts to 1 kW.

decibels above 1 milliwatt [abbreviated dBm] A power level equal to 10 times the common logarithm of the ratio of a given power in watts to 0.001 W. A negative value, such as −2.7 dBm, means decibels below 1 mW.

decibels above 1 milliwatt psophometrically weighted [abbreviated dBmp] A unit used to specify telephone-channel noise levels measured with the standard international psophometer weighting curve, which more nearly gives the actual interfering effect of line noise than does a measuring set having flat frequency response. The power of each interfering tone is compared with the power of an 800-Hz tone creating the same interference during listening tests.

decibels above 1 picowatt [abbreviated dBp] A power level equal to 10 times the common logarithm of the ratio of a given power in watts to 1 pW, or 10^{-12} W.

decibels above 1 volt [abbreviated dBV] A voltage level equal to 20 times the common logarithm of the ratio of a given voltage in volts to 1 V.

decibels above 1 watt [abbreviated dBW] A power level equal to 10 times the common logarithm of the ratio of a given power in watts to 1 W.

decibels above reference coupling [abbreviated dBx] A measure of the coupling between two circuits, expressed in relation to a reference value of coupling that gives a specified reading on a specified noise-measuring set when a test tone of 90 dBa is impressed on one circuit.

decibels above reference noise [abbreviated dBrn] A unit used to show the relationship between the interfering effect of a noise frequency, or band of noise frequencies, and a fixed amount of noise power commonly called reference noise. A 1-kHz tone that has a power level of −90 dBm was originally selected as the reference noise power. This reference level was later changed to −85 dBm, and the new unit called decibels adjusted, abbreviated dBa.

decibels adjusted [abbreviated dBa] A unit used to show the relationship between the interfering effect of a noise frequency, or band of noise frequencies, and a reference noise power level of −85 dBm. This unit replaces dBrn, which was based on a reference noise level of −90 dBm. The new unit gives less weight to low tones and thus more nearly matches the effect of sound on people. Also called adjusted decibel.

decimal-binary conversion Conversion of a number written in the decimal scale of 10 to the same number written in the binary scale of 2.

decimal-binary switch A switch that connects a single input lead to appropriate combinations of

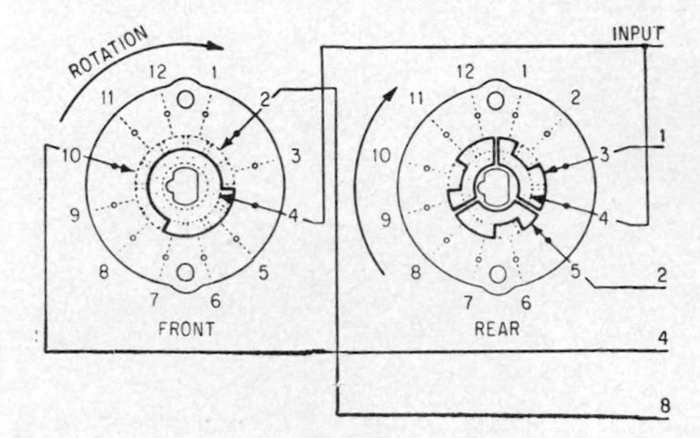

Decimal-binary switch.

four output leads (representing 1, 2, 4, and 8) for each of the decimal-numbered settings of its control knob. Thus, for position 7, output leads 1, 2, and 4 would be connected to the input.

decimal code A code in which each allowable position has 1 of 10 possible states. The conventional decimal number system is a decimal code.

decimal-coded digit One of 10 arbitrarily selected patterns of 1s and 0s used to represent the decimal digits.

decimal digit One of the 10 digits, 0, 1, 2, 3, 4, 5, 6, 7, 8, and 9, used in the scale of 10. Two of these digits, 0 and 1, also serve as binary digits in the scale of 2.

decimal notation A system of notation that uses the scale of 10.

decimal number system A system of positional notation in which the successive digits relate to successive powers of the base 10.

decimal point The point that marks the place between integral and fractional powers of 10 in a decimal number.

decimal-to-binary conversion The mathematical process of converting a number written in the scale of 10 into the same number written in the scale of 2.

decimetric wave A radio wave between the wavelength limits of 10 and 100 cm, corresponding to the ultrahigh-frequency (UHF) range of 300 to 3000 MHz.

decimillimetric wave A radio wave between the wavelength limits of 0.1 and 1 mm, corresponding to the frequency range of 300 to 3000 GHz.

decineper One-tenth of a neper.

decision A computer operation in which two numerical, alphabetic, or alphameric words are compared as to rank or size, with the result determining the next action by the computer.

decision element A circuit that performs a logic operation such as AND, OR, NOT, or EXCEPT on one or more binary digits of input information representing "yes" or "no," and expresses the result in its output.

decision table A table of all possibilities to be considered in a computer problem, along with the action to be taken for each. Sometimes used in place of a flowchart.

deck 1. A magnetic-tape transport mechanism. 2. A set of punched cards.

deck motion predictor A device that predicts the roll and fore-and-aft motion of a ship, to permit firing a missile at a desired deck angle.

deck switch *Gang switch.*

declassify To remove the security classification from a document or piece of equipment.

declination The angle between the horizontal component of the earth's magnetic field and true north.

declinometer An instrument for measuring the exact direction of a magnetic field. Used to determine magnetic declination.

decode To translate coded characters to a more understandable form, or to reverse previous encoding.

decoder 1. A matrix network in which a combination of inputs produces a single output. Used for converting digital information to an analog form. 2. A circuit that responds to a particular coded signal while rejecting others. 3. A device that decodes. 4. A device that unscrambles matrix-encoded signals for quadraphonic sound systems. 5. *Matrix.* 6. *Code converter.*

decollator A machine that separates the copies of multipart continuous forms and removes any carbon paper.

decometer An adding-type phasemeter that gives a continuous indication of phase shift between two signals, such as those received from two transmitters in the Decca navigation system.

decommutation The process of recovering a signal from the composite signal previously created by a commutation process.

decommutator The section of a telemetering system that extracts analog data from a time-serial train of samples representing a multiplicity of data sources transmitted over a single RF link.

decontamination The removal of unwanted radioactive material.

decoupling network Any combination of resistors, coils, and capacitors placed in power-supply leads or other leads that are common to two or more circuits, to prevent unwanted interstage coupling.

decoy An object that simulates the reflective or other characteristics of a target, for use as a countermeasure which protects the target from guided missiles or other enemy action.

decoy transponder A transponder that returns a strong signal when triggered directly by a radar pulse. When used for electronic countermeasures, the transponders produce large and misleading target signals on enemy radar screens.

decrement 1. The quantity by which a variable is decreased. 2. *Damping factor.*

decremeter An instrument for measuring the logarithmic decrement or damping of a wave train.

decrypt To convert a cryptogram or series of electronic pulses into plain text by electronic means.

decryption The conversion of an encrypted message to its original form by cryptogear after transmission.

Dectra A British radio navigation aid that provides coverage over a specific section of a long ocean route, using equipment and techniques similar to Decca. A master and slave station are located at each end of the route.

dedicated Reserved for a specific use or application, such as a dedicated line or dedicated microprocessor.

dedicated computer A computer used for only one special type of data processing, such as for numerical control of machine tools.

dedicated line A permanently wired line used between two points exclusively for one type of service, such as for data communication or for a radio studio-transmitter link.

dee A hollow accelerating electrode in a cyclotron, shaped like the letter D.

dee line A structural member that supports the dee of a cyclotron and acts with the dee to form the resonant circuit.

deemphasis A process for reducing the relative strength of higher audio frequencies before reproduction, to complement and thereby offset the preemphasis that was previously introduced to help these components override noise or reduce distortion. Used chiefly in frequency- and phase-modulated receivers and in phonograph amplifiers. Also called postemphasis and post-equalization.

deemphasis network An RC filter inserted into a system to restore preemphasized signals to their original form.

deenergize To disconnect from the source of power.

deep-space probe A spacecraft designed for exploring space, beyond the gravitational and magnetic fields of the earth. Examples include lunar probe, Mars probe, and solar probe.

deerhorn antenna A dipole antenna whose ends are swept back to reduce wind resistance when mounted on an airplane.

defect conduction Conduction by holes in a semiconductor.

defibrillator An electronic instrument used for stopping fibrillation during a heart attack by applying controlled electric pulses to the heart muscles.

definition 1. The fidelity with which a television or facsimile receiver forms an image. 2. The extent to which the fine-line details of a printed circuit correspond to the master drawing. 3. The fidelity with which details are reproduced in an image.

deflection The displacement of an electron beam from its straight-line path by an electrostatic or electromagnetic field.

deflection coil One of the coils in a deflection yoke.

deflection defocusing Defocusing that becomes greater as deflection is increased in a cathode-ray tube because the beam hits the screen at an increasingly greater slant, and its spot becomes increasingly more elliptical as it approaches the edges of the screen.

deflection electrode An electrode whose potential provides an electric field that deflects an electron beam. Also called deflection plate.

deflection factor The reciprocal of the deflection sensitivity in a cathode-ray tube. Deflection factor is usually expressed in amperes per inch for electromagnetic deflection and volts per inch for electrostatic deflection.

deflection plane A plane perpendicular to the cathode-ray-tube axis containing the deflection center.

deflection plate *Deflection electrode.*

deflection sensitivity The displacement of the electron beam at the target or screen of a cathode-ray tube per unit of change in the deflection field. Usually expressed in inches per volt applied between deflection electrodes or inches per ampere in a deflection coil. Deflection sensitivity is the reciprocal of deflection factor.

deflection voltage The voltage applied between a pair of deflection electrodes to produce an electric field.

deflection yoke An assembly of one or more electromagnets that is placed around the neck of an electron-beam tube to produce a magnetic field for deflection of one or more electron beams. Also called yoke.

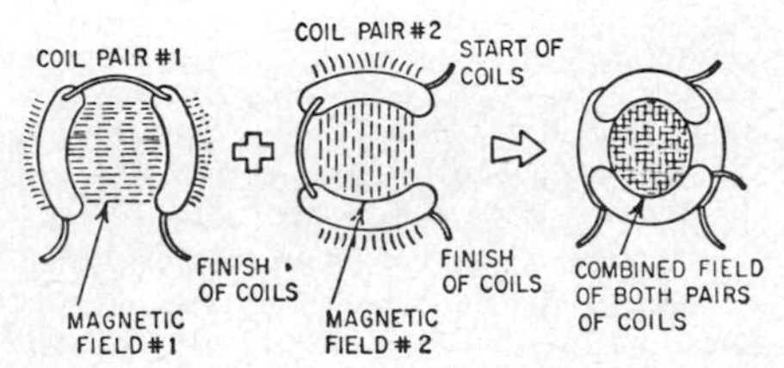

Deflection yoke for television picture tube contains four separate coils.

deflection-yoke pullback 1. In a color picture tube, the distance between the maximum possible forward position of the yoke and the position of the yoke that gives optimum color purity. 2. In a monochrome picture tube, the maximum distance the yoke can be moved back along the tube axis without producing neck shadow.

deflector A device used to distort a particle orbit so it can be removed from an accelerator to strike a target or form an external beam.

defocus To make a beam of x-rays, electrons, light, or other radiation deviate from an accurate focus at the intended viewing or working surface.

defocus-dash mode A mode of cathode-ray-tube storage of binary digits in which the writing beam is initially defocused so as to excite a small circular area on the screen. For one kind of binary digit it remains defocused. For the other kind it is suddenly focused to a concentric dot and drawn out into a dash during the time interval before the beam is cut off and moved to the next position.

defocus-focus mode A variation of the defocus-dash mode, in which the focused dot is not drawn out into a dash.

defruit To remove random asynchronous replies from the video input of a display unit in a radar beacon system by comparing the video signals on successive sweeps.

degassing The process of driving out and exhausting the gases occluded in the internal parts of an electron tube, generally by heating during evacuation.

degauss To remove, erase, or clear information from a magnetic tape, disk, drum, or core.

degaussing 1. *Demagnetizing.* 2. A method of neutralizing the magnetic field of a ship by placing a cable around its hull and sending through the cable a direct current of the correct value to neutralize the magnetic effect of the hull. The current adjustment is made at a degaussing station equipped with underwater equipment that indicates when the resultant magnetic field has been sufficiently weakened so it will not actuate a magnetic mine. Also called deperming.

degaussing coil A plastic-encased coil, about 12 in (30 cm) in diameter, that can be plugged into a 120-VAC wall outlet and moved slowly toward and away from a color television picture tube to demagnetize adjacent parts.

degaussing control A control that automatically varies the current in degaussing coils as a ship changes heading or rolls and pitches.

degeneracy 1. The condition wherein two or more modes in a resonant device have the same resonant frequency. 2. The condition wherein several states of an atom, which differ in many of their properties, nevertheless have the same value of some particular quantity, usually the total energy.

degenerate amplifier A parametric amplifier in which the pump frequency is exactly twice the signal frequency, so the idler frequency produced is equal to the input signal frequency.

degenerate conduction band A band in which exist two or more orthogonal quantum states that have the same energy, the same spin, and zero mean velocity.

degenerate electron gas An electron gas that is far below its Fermi temperature and is therefore described by the Fermi distribution. Most of the electrons completely fill the lower energy levels and are unable to take part in physical processes until excited out of these levels.

degenerate matter Matter that has been stripped of its orbital electrons, so the nuclei are packed close together.

degenerate modes A set of modes having the same resonant frequency or propagation constant.

degenerate semiconductor A semiconductor in which the number of electrons in the conduction band approaches that of a metal.

degeneration *Negative feedback.*

deglitcher A nonlinear filter or other special circuit used to limit the duration of switching transients in digital converters.

degradation *Moderation.*

°C Abbreviation for degree *Celsius.*

°F Abbreviation for degree *Fahrenheit.*

°R Abbreviation for degree *Rankine.*

deion circuit breaker An AC circuit breaker that uses a magnetic field to blow the current-interrupting arc into a stack of insulated copper plates. The resulting short arcs in series are almost instantly deionized when the current drops to zero in the AC cycle, so the arc cannot reform.

deionization The recombination of ions with electrons in a glow or arc discharge to form neutral atoms and molecules.

deionization potential The potential at which ionization of the gas in a gas-filled tube ceases and conduction stops.

deionization time The time required for a gas tube to regain its preconduction characteristics after interruption of anode current, so the grid regains control. Also called recontrol time.

deionizing grid A grid used in a gas tube to speed up deionization of the gas.

delay The amount of time by which an event is retarded.

delay circuit A circuit in which the output signal is delayed by a specified time interval with respect to the input signal.

delay coincidence circuit A coincidence circuit that is actuated by two pulses, one of which is delayed by a specified time interval with respect to the other.

delay distortion Phase distortion in which the rate of change of phase shift with frequency of a circuit or system is not constant over the frequency range required for transmission. Also called envelope delay distortion.

delay-Doppler mapping Mapping of a planet by

illuminating it with a radar beam and measuring the Doppler shift caused by rotation of the planet. Reflection from the side of the planet rotating toward the earth will be higher than the transmission frequency, and the side rotating away from the earth will reflect a lower frequency.

delayed alpha particle An alpha particle emitted by an excited nucleus that was formed an appreciable time after a beta-disintegration process.

delayed automatic gain control An automatic-gain-control system that does not operate until the signal exceeds a predetermined magnitude. Weaker signals thus receive maximum amplification. Also called biased automatic gain control, delayed automatic volume control, and quiet automatic volume control.

delayed automatic volume control *Delayed automatic gain control.*

delayed coincidence Occurrence of a count in one detector at a short but measurable time later than a count in another detector, the two counts being caused by successive events in the same nucleus.

delayed fission neutron A fission neutron emitted by fission products.

delayed neutron A neutron emitted by excited nuclei formed during beta disintegration, an appreciable time after the fission. Since neutron emission itself is prompt, the observed half-life is that of the preceding beta emitter.

delayed neutron fraction The ratio of the mean number of delayed neutrons per fission to the mean total number of neutrons (prompt plus delayed) per fission.

delayed PPI A plan-position indicator in which initiation of the time base is delayed a fixed time after each transmitted pulse, to give expansion of the range scale for distant targets so they show more clearly on the screen.

delayed-repeater satellite A satellite that stores information received from one ground terminal and retransmits it only when interrogated by a different ground terminal.

delayed sweep A sweep whose beginning is delayed for a definite time after the pulse that initiates the sweep.

delay equalizer A corrective network used to make the phase delay or envelope delay of a circuit or system substantially constant over a desired frequency range.

delay line A device utilizing the time of wave propagation to produce a time delay of a signal. A transmission line using either lumped or distributed constants gives a delay determined by the electrical length of the line. In one form, the inductance is increased by winding a helix on a flexible powdered iron core to serve as the inner conductor. High capacitance is achieved by using only a few layers of thin paper to separate the inner conductor from the stranded outer conductor. An ultrasonic delay line gives a delay determined by the length of the path taken by acoustic waves through the medium. Also called artificial delay line.

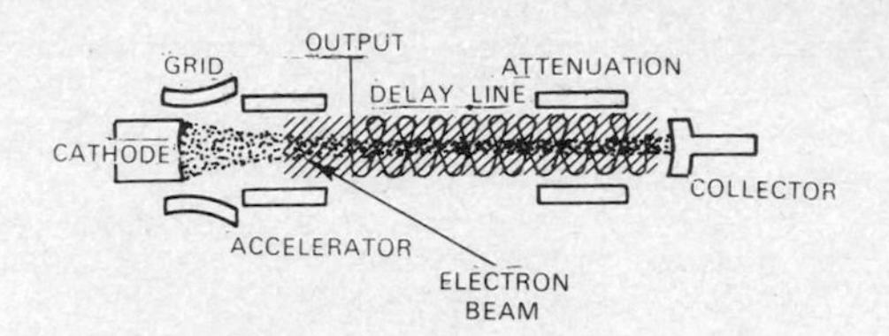

Delay line in O-type backward-wave oscillator.

delay-line cable A special cable used in the monochrome channel of a color television receiver to provide a time delay just long enough to make the monochrome and chrominance signals arrive together at the cathode-ray tube.

delay-line memory *Delay-line storage.*

delay-line storage A computer storage or memory device consisting of a delay line and means for regenerating and reinserting information into the delay line. Also called delay-line memory.

delay multivibrator A monostable multivibrator that generates an output pulse a predetermined time after it is triggered by an input pulse.

delay relay A relay having predetermined delay between energization and closing of contacts or between deenergization and dropout.

delete character A character used to erase an erroneous or unwanted computer input character. With punched tape or cards, it consists of holes in all punching positions.

delimiter A character used to separate items of data at the input of a computer. Commas or spaces are widely used for this purpose.

Dellinger effect A type of shortwave radio fadeout believed to be caused by rapid shifting of ionosphere layers during solar eruptions.

Delrac A British radio navigation system designed to provide worldwide coverage by using 21 pairs of master-slave stations, with a 3000-mi (4800-km) range for each pair of stations. Frequencies used are in the band from 10 to 14 kHz. Decca indicating equipment can be used with Delrac.

delta The difference between a partial-select output of a magnetic cell in a 1 state and a partial-select output of the same cell in a 0 state.

delta connection A combination of three components connected in series to form a triangle like the Greek letter delta. Also called mesh connection.

delta function A distribution profile large at points close to an origin and zero for all significant distances from the origin.

delta impulse function An extremely narrow pulse having such great amplitude that the product of height and duration is unity.

delta-matched antenna A single-wire antenna,

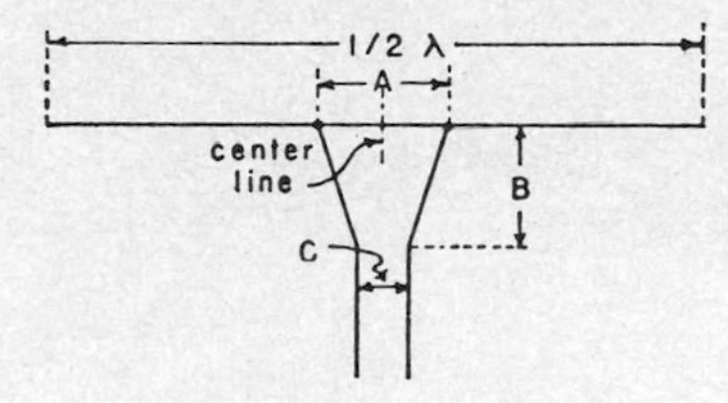

Delta-matched antenna. Dimensions A, B, and C determine matching to transmission line.

usually one half-wavelength long, to which the leads of an open-wire transmission line are connected in the shape of a Y. The flared parts of the Y match the transmission line to the antenna. Because the top of the Y is not cut, the matching section has the triangular shape of the Greek letter delta. Also called Y antenna.

delta-matching transformer The Y-shaped matching section of a delta-matched antenna.

delta modulation A pulse-modulation technique in which a continuous signal is converted into a binary pulse pattern, for transmission through low-quality channels.

delta network A set of three branches connected in series to form a mesh.

delta noise The noise signal voltage induced in the winding of a ferrite-core storage by partially selected cores.

delta pulse-code modulation A modulation system that converts audio signals into corresponding trains of digital pulses to give greater freedom from interference during transmission over wire or radio channels.

delta ray An electron or proton ejected by recoil when a rapidly moving alpha particle or other primary ionizing particle passes through matter.

delta wave A brain wave having a frequency below 9 Hz.

DELTIC [DElay-Line-TIme-Compression] A method of sampling incoming radar, sonar, seismic, speech, or other waveforms along with reference signals, compressing the samples in time, and comparing them by autocorrelation. The time compression greatly reduces the complexity and size of the equipment needed for signal analysis.

demagnetization Removal of residual magnetism.

demagnetization curve A portion of the hysteresis loop of a magnetic material, showing the peak value of residual induction and the manner in which magnetization reduces to zero when demagnetizing force is applied.

demagnetizer A device for removing undesired magnetism, as from the playback head of a tape recorder or from a recorded reel of magnetic tape that is to be erased.

demagnetizing Removing magnetism from a ferrous material. Also called degaussing.

demagnetizing force A magnetizing force applied in the direction that reduces the residual induction in a magnetized object.

demand factor The ratio of the maximum power drawn by an electronic system to the total connected load of the system. The power drawn is either the instantaneous value or the value measured over a specified period of time.

demand meter A meter that indicates or records the maximum demand for electric power during a predetermined interval of time.

demand processing Processing of data on essentially an on-line basis, to minimize storage of unprocessed data.

Dember effect The development of a DC voltage between two regions of a photoconductive semiconductor when one of the regions is illuminated, by diffusion of an optically generated hole and electron pairs away from the illuminated region.

demodulation The process of converting a modulated RF carrier signal to a form that can be heard or displayed. When the carrier is unmodulated, the process is called detection.

demodulator The stage in a receiver that removes the modulation signal from a modulated RF carrier signal.

demountable tube An electron tube that can be taken apart for repair or replacement of electrodes.

Dempster mass spectrograph A mass spectrograph in which ions are accelerated through a slit by an electric field, then deflected by a magnetic field so all ions of the same charge-mass ratio pass through a second slit.

demultiplexer A device used to separate two or more signals previously combined by a compatible multiplexer and transmitted over a single channel.

demultiplexing circuit A circuit used to separate the signals that have been combined for transmission by multiplex.

dendritic web A silicon or other crystal grown as a thin, narrow strip several yards long.

dense binary code A code in which all possible states of the binary pattern are used. A 4-bit code, with positions valued at 1, 2, 4, and 8, thus gives 16 combinations for representing numerals 0 through 15.

densimeter *Density indicator.*

densitometer 1. An instrument used to measure the optical density of a material. 2. An instrument used to measure the amount of darkening of film badges to determine the radiation dosage received by the wearer.

density 1. Weight per unit volume. 2. A measure of the light-transmitting properties of an area. It is expressed as the common logarithm of the ratio of incident light to transmitted light. 3. Amount per unit cross-sectional area, as for current, magnetic flux, or electrons in a beam.

density indicator An instrument that measures the density of a liquid or solid material. One ver-

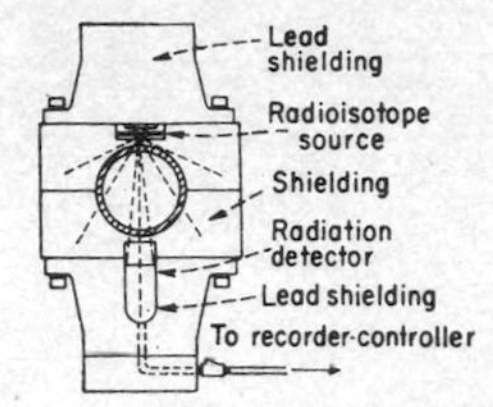

Density indicator in which gamma radiation reaching detector from radioisotope source is proportional to density of liquid flowing through pipe.

sion measures the absorption of gamma rays by the material since this absorption is proportional to density. Also called densimeter.

density logger An instrument that records formation densities in boreholes by measuring backscattered gamma radiation from a radiation source which is usually cobalt 60.

density modulation Modulation of an electron beam by making the density of the electrons in the beam vary with time.

density packing The number of units of useful information that can be stored within a given linear dimension on a single track of a magnetic tape or drum by a single head.

dependent node A node having one or more incoming branches.

deperming *Degaussing.*

depletion Reduction of the charge-carrier density in a semiconductor below the normal value for a given temperature and doping level.

depletion layer An electric double layer formed at the surface of contact between a metal and a semiconductor having different work functions. Electrons diffuse from the substance having the lower work function toward the other substance, leaving equivalent positive charges at the layer in the first substance. This action occurs because the mobile charge-carrier density is insufficient to neutralize the fixed charge density of donors and acceptors. Also called space-charge layer. Formerly called barrier layer and blocking layer.

depletion-layer capacitance The capacitance of the imaginary capacitor formed by the charges of a depletion layer. This capacitance is a function of reverse voltage. Also called barrier-layer capacitance (deprecated).

depletion-mode transistor A field-effect transistor in which charge carriers are present in the channel when the gate-source voltage is zero. Channel conductivity is increased or decreased by changing the magnitude and polarity of the gate-source voltage.

depletion region The portion of the channel in a metal-oxide field-effect transistor in which there are no charge carriers. The region acts as an insulator for isolating different sections of a semiconductor, making possible the construction of planar bipolar integrated circuits.

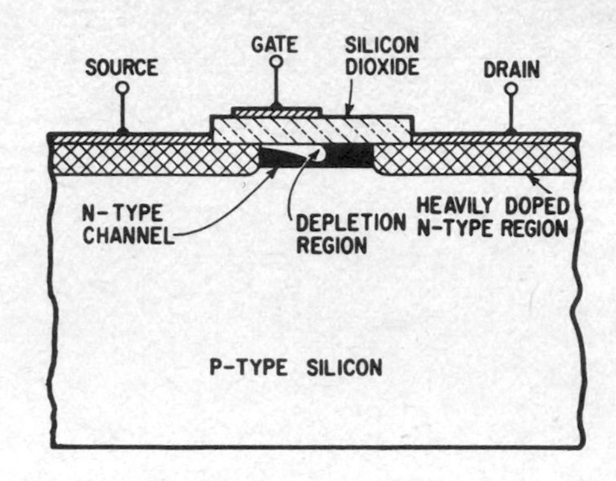

Depletion region in depletion-type MOS field-effect transistor.

depolarization Prevention of polarization in an electric cell or battery.

deposited-carbon resistor A resistor that has a thin coating of carbon deposited on a supporting insulator.

deposited oxide An oxide layer deposited on a surface by a method such as evaporation or sputtering that does not require a chemical reaction with the substrate.

deposition The process of applying a material to a base by means of vacuum, electrical, chemical, screening, or vapor methods.

depression deviation indicator A sonar depth-of-target indicator used with a tilting transducer to show whether the transducer is aimed at, above, or below the target. Depth is automatically computed from transducer angle and echo range after the transducer is aligned with the target.

depth-determining sonar Sonar in which the depth-determining transducer can be tilted to send its searchlight beam down into the water at any desired angle from a surface vessel. It contains a depth-deviation indicator that permits determination of true depth, for accurate setting of depth bombs.

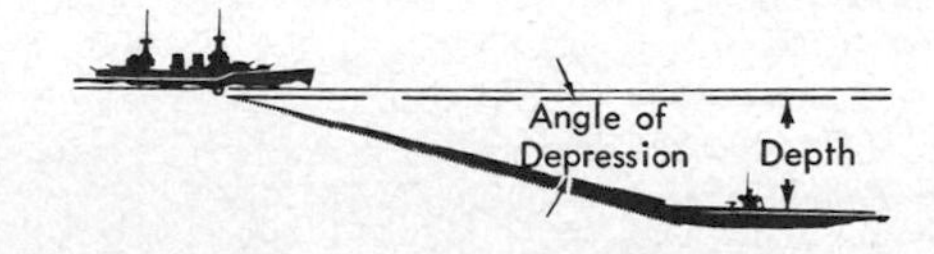

Depth-determining sonar in use.

depth-deviation indicator A device used with depth-determining sonar to tell the operator when the beam is aimed directly at the submarine target. The equipment takes into account the bending of the sonar beam because of changes in water temperature, as determined from bathythermograph records.

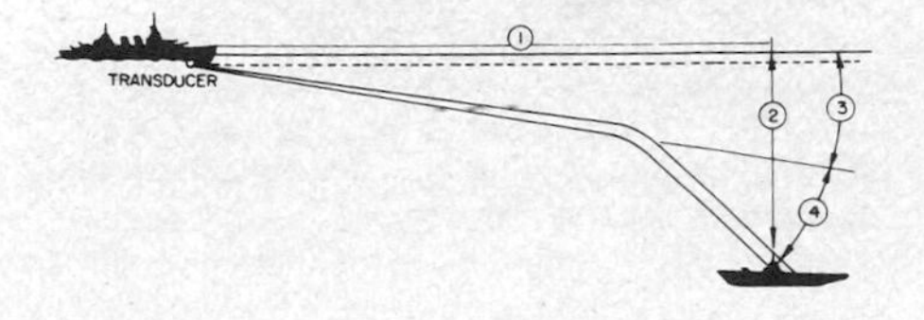

Depth-deviation indicator on depth-determining sonar compensates for bending of sonar beam by thermal changes in water. True depth of target (2) is computed from apparent target depth (3), angle of beam-direction change (4), and surface range (1).

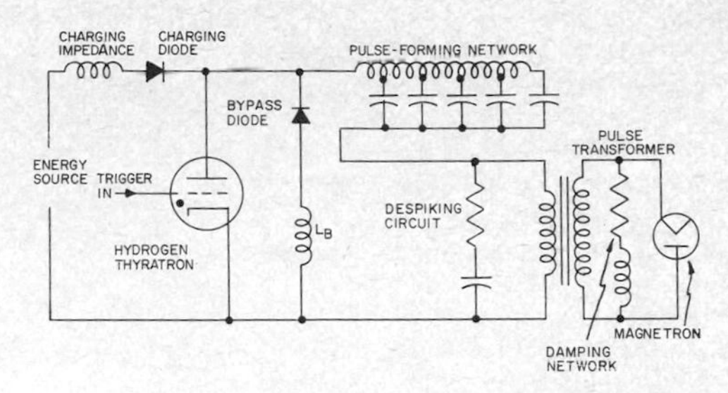

Despiking circuit for microwave radar magnetron.

depth dose The radiation dose delivered at a particular depth beneath the surface of a body. It is usually expressed as a percent of surface dose or air dose.

depth finder *Fathometer.*

depth of heating The depth below the surface of a material in which effective dielectric heating can be confined when the applicator electrodes are applied adjacent to one surface only.

depth of modulation The ratio of the difference in field strength of the two lobes of a directional antenna system to the field strength of the greater at a given point in space. Used to determine direction in a radio guidance system.

depth of penetration The thickness of a layer, extending inward from the surface of a flat conductor, that has the same resistance to direct current as the conductor as a whole has to alternating current of a given frequency in induction heating.

depth sounder *Fathometer.*

derate To reduce the rating of a device to improve reliability or to permit operation at high ambient temperatures.

derivative action Control action in which the speed at which a correction is made depends on how fast the system error is increasing. Also called rate action.

derived sound system A four-channel sound system that is artificially synthesized from conventional two-channel stereo sound by an adapter, to provide feeds to four loudspeakers for approximating quadraphonic sound.

dermohygrometer An instrument that measures skin resistance without passing direct current through the skin.

descriptor A word or phrase used to identify a document in a computer-based information storage and retrieval system.

desensitization An automatic-gain-control effect that occurs when a receiver is tuned to one channel and there is a strong signal on a nearby channel. The strength of the desired signal appears to be decreased by the presence of the nearby signal.

desired track *Course.*

despiking circuit A series resistor-capacitor circuit used in parallel with the primary of a magnetron pulse transformer to remove the spike that would otherwise appear at the leading edge of the output pulse.

despun antenna A satellite antenna that is rotated at a rate equal to and opposite from the rate at which the satellite is spinning for stabilization, so the directional antenna can be pointed continuously at the earth.

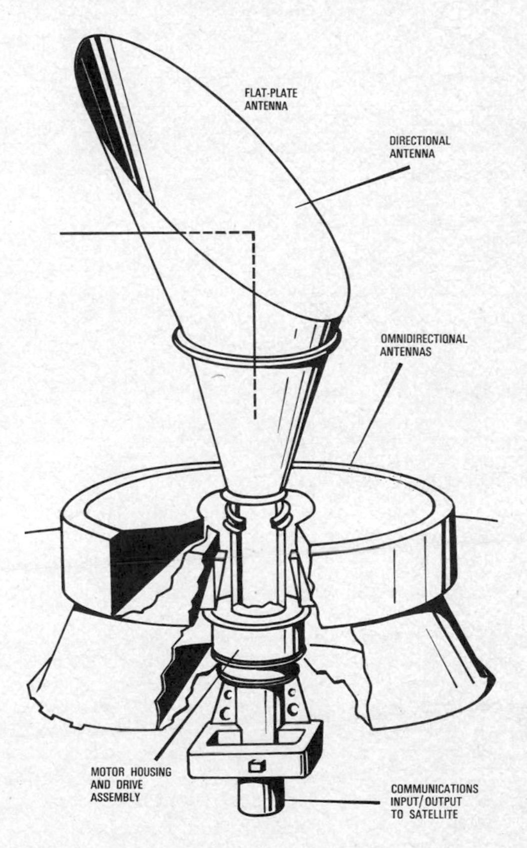

Despun antenna mounted on axis of communication satellite.

destaticization Treatment of a plastic or other material to minimize accumulation of static electricity. Used on long-playing phonograph records to reduce dust pickup.

Destriau effect Sustained emission of light by suitable phosphor powders that are embedded in an insulator and subjected only to the action of an alternating electric field.

destruct The deliberate action of exploding a missile or vehicle after it has been launched but before it has completed its course. Destructs are executed by radio when the missile gets off its plotted course or otherwise becomes a hazard.

destructive breakdown Breakdown of the barrier between the gate and channel of a field-effect transistor, causing failure of the transistor.

destructive readout [abbreviated DRO] A computer memory in which readout of data destroys it, requiring the use of circuits to restore the stored data if it will be required later.

destructive testing Intentional operation of equipment until it fails, to reveal design weaknesses.

destructor An explosive device used intentionally to destroy a missile after launching.

detail The extent to which image elements that are close together can be individually distinguished. Detail requires contrast for its recognition.

detection The process of converting an RF carrier signal to a form that can be heard or displayed. If the carrier is unmodulated, the detected result is a DC voltage that will act on a simple diode wavemeter. If the carrier is modulated, the modulation signal is obtained as output, and the process is more often called demodulation.

detection probability The probability, expressed as a percentage, that a single sonar ping will return a recognizable echo at various distances and directions from a sonar transmitter.

detection satellite A satellite designed to detect a particular type of radiation, such as radiation from a nuclear explosion.

detectivity A figure of merit representing gain over noise for low-level solid-state radiation detectors like photodiodes.

detectophone An audio system used for listening secretly to conversations. A sensitive microphone is concealed in the room and connected to an audio amplifier feeding headphones or a magnetic tape recorder in a nearby room. Alternatively, the microphone may feed a wired radio transmitter that sends the signals over power lines to a more remote listening location.

detector The stage in a receiver at which demodulation takes place. In a superheterodyne receiver this is called the second detector, even though it is actually a demodulator. The so-called first detector in a superheterodyne is simply a frequency converter that changes the incoming carrier frequency to the intermediate frequency of the receiver.

detent A mechanism used on a multiposition control to hold it firmly in each position. One common type consists of a spring-loaded ball that falls into equally spaced indentations on a plate as the control shaft is rotated.

detent tuning A television receiver tuning control in which a detent mechanism determines the correct position of the tuning shaft for receiving a desired station.

detune To change the inductance or capacitance of a tuned circuit so its resonant frequency is different from the incoming signal frequency.

detuning stub A quarter-wave stub used to match a coaxial line to a sleeve-stub antenna. The stub detunes the outside of the coaxial feed line while tuning the antenna itself.

deuterated potassium dihydrogen phosphate [abbreviated DKDP] A ferroelectric electrooptical crystal.

deuterium [symbol d, D, or H^2] The hydrogen isotope having mass number 2. It is one form of heavy hydrogen, the other being tritium.

deuterium oxide *Heavy water.*

deuteron The nucleus of a deuterium atom, consisting of a neutron and a proton.

deuteron accelerator An accelerator that produces a flux of slow neutrons by bombarding a metal-tritium target with deuterons.

Deutsche Industrie Normenausschus [abbreviated DIN] A West German institute that sets industrial standards. In the audio field, plugs and sockets having DIN geometry are used throughout the world.

deviation 1. The difference between the actual value of a controlled variable and the desired value corresponding to the set point. 2. *Frequency deviation.*

deviation distortion Distortion in an FM receiver caused by inadequate bandwidth, inadequate amplitude-modulation rejection, or inadequate discriminator linearity.

deviation ratio The ratio of the maximum possible frequency deviation to the maximum audio modulating frequency in an FM system.

deviation sensitivity 1. The rate of change of course indication with respect to the change of displacement from the course line in a navigation indicator. 2. The lowest frequency deviation that produces a specified output power in an FM receiver.

device An electronic element that cannot be divided without destroying its stated function.

device under test [abbreviated DUT] The device that is plugged into or otherwise connected to a special circuit designed for checking performance characteristics.

Dewar flask A glass container having two or more walls, with the space between the walls evacuated to prevent transfer of heat. The insides of the walls are usually silvered. Used for holding liquid gases and for cryogenic devices.

DEW line [Distant Early-Warning line] A line of

radar stations extending for about 3000 mi (4800 km) from Alaska to Greenland along the Arctic circle, to give the earliest possible warning of the approach of enemy airplanes or missiles. Automatic target-recognition circuits eliminate the need for constant attention by human observers.

dewpoint indicator An instrument that measures the dewpoint temperature at which a vapor begins to condense. In one version, a photoelectric

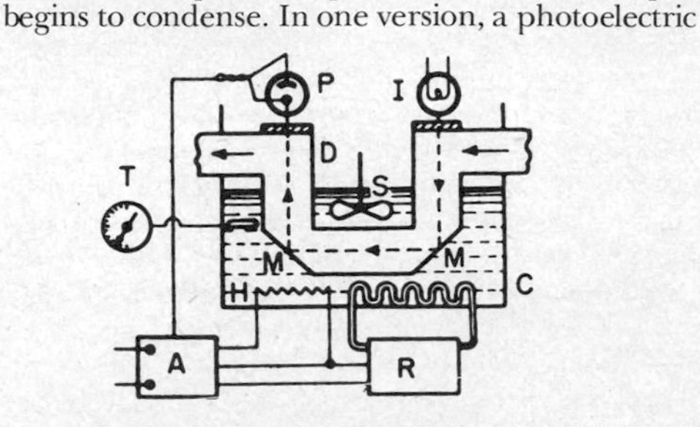

Dewpoint indicator in which vapor-bearing air or gas flows over mirrors M. When refrigerator R cools liquid bath sufficiently, vapor condenses on mirrors at M, blocking light from lamp I to photocell P. This turns off R and turns on heater at bottom of water until mirrors clear, when R is turned on again. Recorder T shows low-temperature valleys corresponding to dewpoint temperature.

system detects the instant at which the vapor flowing over a mirror begins to condense and cloud up a mirror that is being alternately heated and cooled.

DF Abbreviation for *direction finder.*

D flip-flop A delay flip-flop in which the input data is delayed by one clock pulse period. The

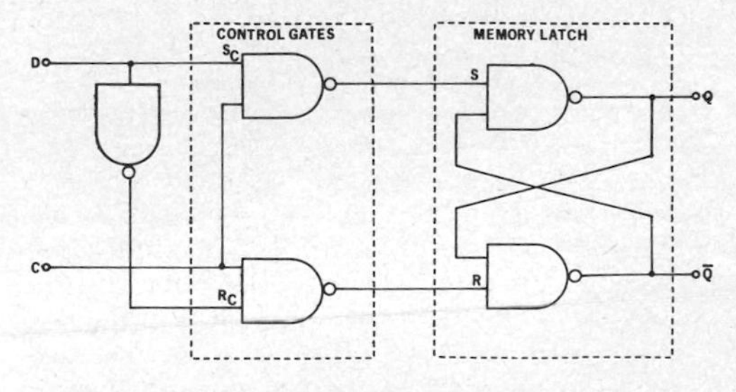

D flip-flop, made up of basic RS flip-flop serving as memory latch and preceded by control gates.

output is therefore a function of the input that appeared one pulse earlier.

DFSK Abbreviation for *double frequency-shift keying.*

diac [DIode AC switch] A bidirectional diode that has a symmetrical switching mode. It is triggered to its on state when its breakover voltage is exceeded in either direction by an applied voltage or a trigger spike. Applications include triggering of triacs. Also called trigger diode.

diagnostic routine A routine designed to locate a computer malfunction or a mistake in coding.

diagonal horn antenna A horn antenna in which all cross sections are square and the electric vector is parallel to one of the diagonals. The

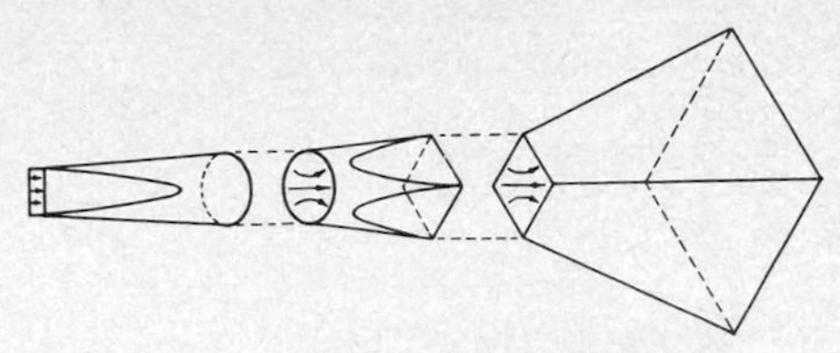

Diagonal horn antenna (right), with transition from rectangular waveguide at left.

radiation pattern in the far field has almost perfect circular symmetry.

diagram A schematic or other line drawing that explains the operation of a circuit or piece of equipment.

dial 1. A separate scale or other device for indicating the value to which a control is set. 2. A telephone calling device that generates the pulses or tones required for establishing a desired connection.

dial cable Braided cord or flexible wire cable used to make a pointer move over a dial when a separate control knob is rotated or used to couple two shafts together mechanically.

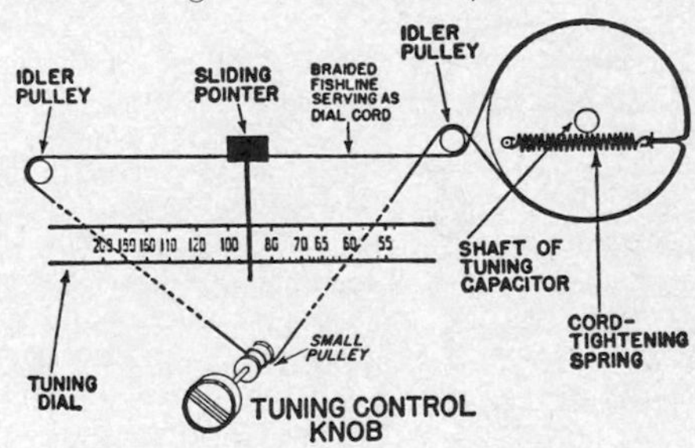

Dial-cord layout for typical slide-rule dial.

dial cord A braided cotton, silk, or glass fiber cord used as a dial cable.

dial exchange A telephone exchange area in which all subscribers originate their calls by dialing.

dial lamp A small lamp used to illuminate a dial. When used to indicate that a circuit is energized, it is called a pilot lamp.

dial-pulse interpreter A device that converts the signaling pulses of a dial telephone to a form suitable for data entry to a computer. It may provide binary-coded decimal output along with optional digital display of dialed digits and relay actuation for one or more combinations of dialed characters. Used in areas where telephone exchanges do not yet have Touch-Tone capability.

dial telephone system A telephone system in which connections between customers are established automatically by electronic and mechanical switching systems controlled by rotary pulse-generating dials or pushbutton equivalents.

dial tone A tone used in a dial telephone system to indicate that the equipment is ready for the dialing of a number.

dial-up operation Use of an ordinary voice-grade telephone line for data transmission. The telephone number of the computer is dialed, and then the handset is placed in the receptacle provided for it on the dataset or modem that converts binary data to tones. Telephone charges are the same as for regular voice calls.

diamagnetic material A material that has a magnetic permeability less than 1, such as bismuth and antimony. Diamagnetic materials are repelled by a magnet and therefore tend to position themselves at right angles to magnetic lines of force.

diamagnetic plasma An ionized gas that is bounded by a magnetic field but contains no magnetic field within it.

diameter equalization Increasing the high-frequency response in proportion to decreasing diameter of a disk recording.

diamond antenna *Rhombic antenna.*

diamond circuit A gate circuit that provides isolation between input and output terminals when in its off state, by operating transistors in their cutoff region. In the on state, the output voltage follows the input voltage as required for gating both analog and digital signals, while the transistors provide current gain to supply output current on demand.

diamond stylus A stylus having a carefully ground diamond as its point.

diaphragm 1. A thin flexible sheet that can be moved by sound waves, as in a microphone, or can produce sound waves when moved, as in a loudspeaker. 2. An adjustable opening used in television cameras to reduce the effective area of a lens to increase the depth of focus. 3. *Iris.*

diathermy Therapeutic use of RF energy to produce heat within some part of the body.

diathermy interference Television interference caused by diathermy equipment. It produces a herringbone pattern in a dark horizontal band across the picture.

diathermy machine An RF oscillator, sometimes followed by RF amplifier stages, used to generate high-frequency currents that produce heat within some part of the body, for therapeutic purposes.

dibit A pair of binary digits represented by a single modulation condition in a data-transmission system. Thus, in four-phase modulation a phase shift of 225° is 00, 315° is 01, 45° is 11, and 135° is 10.

dichotomizing search A computer search in which an ordered set of items is divided into two parts, one of which is rejected, and the process is repeated on the accepted part until the items with the desired property are found.

dichroic mirror A glass surface coated with a special metal film that reflects certain colors of light while allowing other colors to pass through. Used in some color television cameras.

dicing Sawing or otherwise machining a semiconductor wafer into small squares or dice from which transistors and diodes can be fabricated.

Dicke fix A technique used to protect a receiver from fast sweep jamming.

Dicke radiometer A radiometer-type receiver that detects weak signals in noise by modulating or switching the incoming signal before it is processed by conventional receiver circuits. After amplification and detection, the modulation frequency is recovered. The product of this signal and a reference waveform from the input modulator is smoothed in a low-pass filter, and the filter output is used to drive a display device that indicates the presence of a signal.

didymium A mixture of the rare-earth metals neodymium and praseodymium.

die [plural dice] The tiny sawed or otherwise machined piece of semiconductor material used in the construction of a transistor, diode, or other semiconductor device.

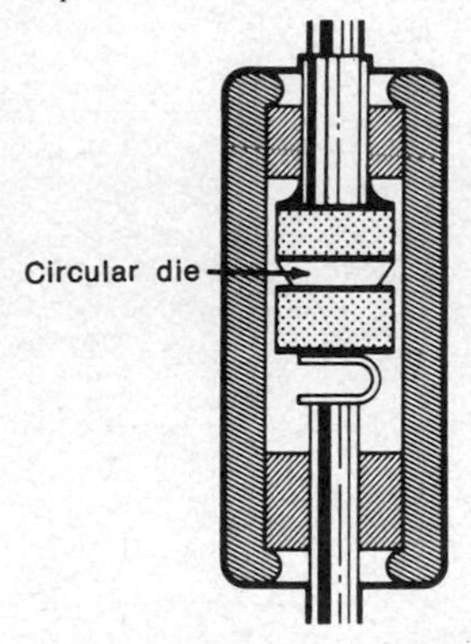

Die in silicon rectifier has truncated circular shape.

die bonding Mounting of an integrated circuit die on a substrate by brazing or other means, to provide mechanical contact along with a thermal path and sometimes an electric connection.

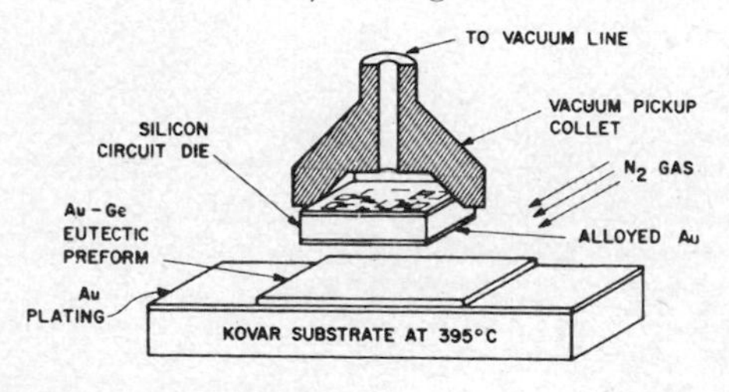

Die bonding of completed silicon integrated circuit die to substrate by brazing with eutectic alloy of gold and germanium in atmosphere of nitrogen at oven temperature of 395°C.

dielectric A material that can serve as an insulator because it has poor electric conductivity. A dielectric such as air, mica, paper, or plastic film is used between the metal-foil plates of a capacitor to separate the plates electrically and store electric energy. A dielectric undergoes electric polarization when subjected to an electric field.

dielectric absorption The persistence of electric polarization in certain dielectrics after removal of the electric field. The effect may last for several years in certain mixtures of wax that are allowed to harden in a strong electric field. Electrets are based on this effect.

dielectric amplifier An amplifier that uses a ferroelectric capacitor whose capacitance varies with applied voltage in such a way as to give signal amplification.

dielectric antenna An antenna in which a dielectric is the major component used to produce a desired radiation pattern.

dielectric constant The property of a material that determines how much electrostatic energy can be stored per unit volume when unit voltage is applied. In effect, it is the ratio of the capacitance of a capacitor filled with a given dielectric to that of the same capacitor having only a vacuum as dielectric. Also called permittivity.

dielectric current The current flowing at any instant through a surface of a dielectric that is located in a changing electric field.

dielectric diode A capacitor in which the negative electrode can emit electrons into the normally insulating region between the plates. The charge stored on the capacitor is thus continuously in transit between the electrodes, to give current flow in one direction. Cadmium sulfide crystals can serve as the dielectric.

dielectric dispersion The phenomenon in which the dielectric constant of an insulating material varies with frequency.

dielectric dissipation factor The cotangent of the dielectric phase angle of a dielectric material.

dielectric fatigue The property of some dielectrics in which resistance to breakdown decreases after a voltage has been applied for a considerable time.

dielectric gas A gas having a high dielectric constant, such as sulfur hexafluoride. Used in laser cavities and high-powered microwave equipment.

dielectric heating The heating of a nominally insulating material by placing it in a high-frequency electric field. The heat results from internal losses during the rapid reversal of polarization of molecules in the dielectric material.

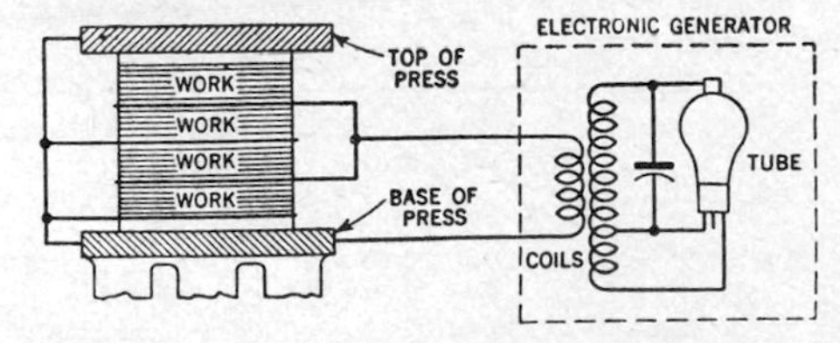

Dielectric heating setup for setting glue in plywood.

dielectric hysteresis A lagging of the electric field in a dielectric with respect to the alternating voltage applied to the dielectric. This effect causes a dielectric hysteresis loss comparable to that produced by magnetic hysteresis in a ferrous material.

dielectric isolation The separation of integrated-circuit elements by means of dielectric insulating layers created during the manufacturing process; the layers usually consist of silicon dioxide formed in a polycrystalline silicon substrate.

dielectric lens A lens made of dielectric material in such a way that it refracts radio waves in the same manner that an optical lens refracts light waves. Used with microwave antennas.

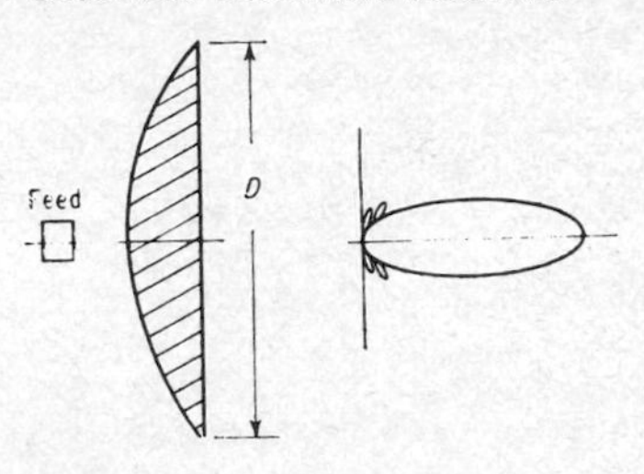

Dielectric-lens antenna and radiation pattern.

dielectric-lens antenna An aperture antenna in which the beam width is determined by the dimensions of a dielectric lens through which the beam passes.

dielectric loss The electric energy that is converted into heat in a dielectric subjected to a varying electric field.

dielectric matching plate A dielectric plate used as an impedance-matching transformer in a waveguide.

dielectric phase angle The angular difference in phase between the sinusoidal alternating voltage applied to a dielectric and the resulting alternating current.

dielectric power factor The cosine of the dielectric phase angle.

dielectric-rod antenna A surface-wave antenna in which an end-fire radiation pattern is produced by propagation of a surface wave on a tapered dielectric rod.

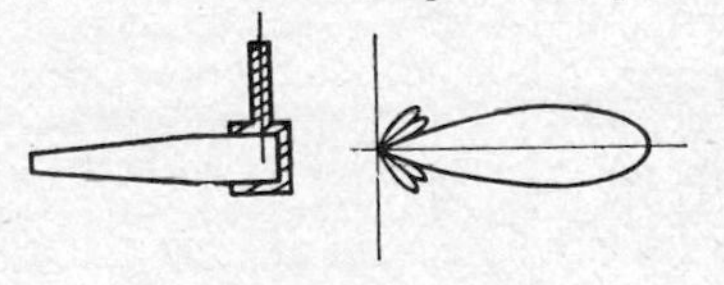

Dielectric-rod antenna.

dielectric strength The maximum potential gradient a material can withstand without rupture.

Usually specified in volts per millimeter of thickness. Also called electric strength.

dielectric test A test involving application of a voltage higher than the rated value for a specified time, to determine the margin of safety against later failure of insulating materials.

dielectric waveguide A waveguide consisting of a dielectric material surrounded by air. Electromagnetic waves travel through the solid dielectric in much the same manner that sound waves travel through a speaking tube.

dielectric wedge A wedge-shaped piece of dielectric used in one waveguide to match its impedance to that of a different waveguide.

dielectric wire A dielectric waveguide used to transmit UHF radio waves short distances between parts of a circuit.

DIFAR [DIrectional Finding And Ranging] A Navy antisubmarine warfare program in which sonobuoys are dropped when a submerged submarine is detected by a search plane, to pick up the sounds of the submarine and broadcast them up to the plane for correlation processing that gives the depth and location of the submarine.

difference amplifier *Differential amplifier.*

difference channel An audio channel that handles the difference between the signals in the left and right channels of a stereophonic sound system.

difference detector A detector circuit in which the output is a function of the difference between the amplitudes of the two input waveforms.

difference limen The increment in a stimulus that is barely noticed in half of the trials. Also called difference threshold.

difference number *Neutron excess.*

difference of potential The voltage between two points.

difference threshold *Difference limen.*

differential The difference between levels for turn-on and turn-off operation in a control system.

differential amplifier An amplifier whose output is proportional to the difference between the voltages applied to its two inputs. Some operational amplifiers can operate in a differential mode. Also called difference amplifier.

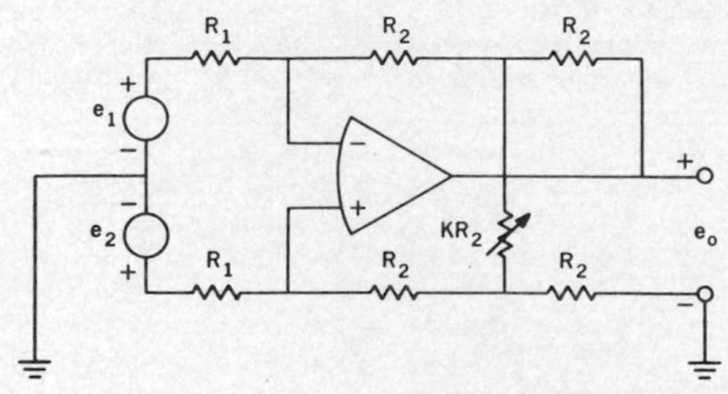

Differential amplifier with gain-adjusting control.

differential analyzer An analog computer designed and used for integrating and solving differential equations.

differential capacitance The derivative with respect to voltage of a charge characteristic, such as an alternating charge characteristic or a mean charge characteristic, at a given point on the characteristic.

differential capacitor A two-section variable capacitor having one rotor and two stators so arranged that as capacitance is reduced in one section, it is increased in the other.

differential comparator A comparator having at least two high-gain differential-amplifier stages, followed by level-shifting and buffering stages, as required for converting a differential input to single-ended output for digital logic applications.

differential cross section The cross section for scattering of particles or photons in a small solid angle in a specified direction. Integration over all angles from 0 to 180° gives the ordinary cross section for a nuclear process.

differential delay The difference between the maximum and minimum frequency delays occurring across a band.

differential discriminator A discriminator that passes only pulses whose amplitudes are between two predetermined values, neither of which is zero.

differential frequency circuit A circuit that provides a continuous output frequency equal to the absolute difference between two continuous input frequencies.

differential frequency meter A circuit that converts the absolute frequency difference between two input signals to a linearly proportional DC output voltage that can be used to drive a meter, recorder, oscilloscope, or other device.

differential gain The amount that unity is exceeded by the ratio of the output amplitudes of a small high-frequency sine-wave signal at two stated levels of a low-frequency signal on which it is superimposed in a video transmission system. Differential gain is expressed in percent by multiplying the above difference by 100 and expressed in decibels by multiplying the common logarithm of the ratio itself by 20.

differential gain control *Sensitivity-time control.*

differential input An input circuit that rejects voltages which are the same at both input terminals and amplifies the voltage difference between the two input terminals. The circuit may be either balanced or floating and may also be guarded.

differential-input impedance The impedance between the inverting and noninverting input terminals of a differential amplifier, consisting primarily of resistance and capacitance.

differential ionization chamber A two-section ionization chamber in which electrode potentials are such that output current is equal to the difference between the separate ionization currents of the two sections.

differential keying Chirp-free break-in keying of a continuous-wave transmitter, achieved by making the oscillator turn on fast before the keyed amplifier stage can pass any signal and turn off fast after the keyed amplifier stage has cut off.

differential linearity The degree of variation in size of adjacent steps in the output of a digital converter over its operating range.

differentially coherent phase-shift keying [abbreviated DPSK] Phase-shift keying in which a particular signal phase can be decoded only by comparison with the phase of the preceding bit. Used in some modems to give higher rates of data transmission over wire lines.

differential microphone *Double-button carbon microphone.*

differential-mode signal A signal applied between the two ungrounded terminals of a balanced three-terminal system.

differential modulation Modulation in which the choice of the significant condition for any signal element is dependent on the choice for the previous signal element.

differential null detector A null indicator in which a differential transformer delivers an output voltage that is proportional to the vector difference between two input voltages. One input is a 1-kHz source voltage, and the other is the unknown signal, applied through an amplifier and phase shifter that can be adjusted so the difference between the two voltages is zero.

differential operational amplifier An amplifier that has two input terminals, used with additional circuit elements to perform mathematical functions on the difference in voltage between the two input signals.

differential permeability The slope of the magnetization curve for a magnetic material.

differential phase The difference in output phase of a small high-frequency sine-wave signal at two stated levels of a low-frequency signal on which it is superimposed in a video transmission system.

differential phase-shift-keying modulation A type of differential modulation in which information is represented by changes in carrier phase in one interval with respect to the carrier phase existing for the preceding interval.

differential pressure pickup An instrument that measures the difference in pressure between two pressure sources and translates this difference into a change in inductance, resistance, voltage, or some other electrical quality.

differential pulse-height discriminator *Pulse-height selector.*

differential relay A two-winding relay that operates when the difference between the currents in the two windings reaches a predetermined value.

differential-scattering cross section The cross section for scattering of a particle from its initial velocity to a new velocity, per unit solid angle per unit speed at the new velocity.

differential synchro 1. *Synchro differential receiver.* 2. *Synchro differential transmitter.*

differential transducer A transducer that simultaneously senses two separate sources and provides an output proportional to the difference between them.

differential transformer A transformer used to join two or more sources of signals to a common transmission line.

differential-transformer transducer A transducer in which movement of the iron core of a transformer varies the output voltage across two series-opposing secondary windings.

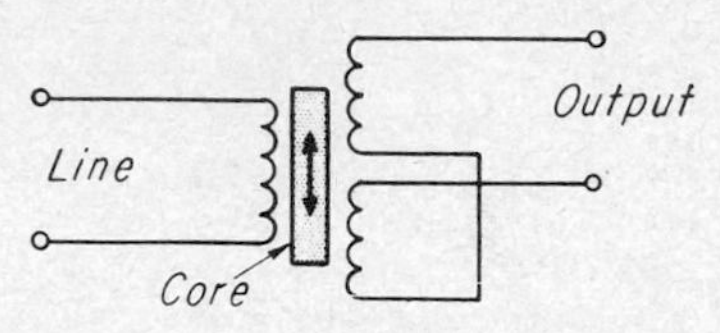

Differential-transformer transducer.

differential voltmeter A voltmeter that measures only the difference between a known and an unknown voltage.

differential winding A winding whose magnetic field opposes that of a nearby winding.

differentiating circuit A circuit whose output voltage is proportional to the rate of change of the input voltage. The output waveform is then the time derivative of the input waveform, and the phase of the output waveform leads that of the input by 90°. An RC circuit gives this differentiating action. Also called differentiating network and differentiator.

differentiating network *Differentiating circuit.*

differentiator 1. A device, usually of the analog type, whose output is proportional to the derivative of an input signal in a computer. 2. *Differentiating circuit.*

diffracted wave A wave whose front has been changed in direction by an obstacle or other nonhomogeneity in a medium, other than by reflection or refraction.

diffraction The bending of a wave as it passes the edges of an object or opening. It is caused by interference between wave components scattered by different parts of the object.

diffraction grating A polished metal or glass surface having closely spaced parallel reflecting grooves that produce a spectrum by interference between different colors of light when white light arrives at a certain angle. There may be as many as 50,000 lines per inch (20,000 lines per centimeter). Also called grating.

diffraction instrument *Diffractometer.*

diffraction pattern The pattern produced on film exposed in an x-ray diffraction camera, consisting of portions of circles having various spacings depending on the material being examined.

diffraction propagation Propagation of electromagnetic waves around objects, or over the horizon, by diffraction. The action occurs because every point in a wavefront generates a spherical front that falls off in intensity away from the forward direction. A continuous series of such actions carries radiation around objects, but with rapidly decreasing intensity.

diffraction scattering Elastic scattering that occurs when inelastic processes remove particles from a beam.

diffraction velocimeter *Laser velocimeter.*

diffractometer An instrument used to study the structure of matter or the properties of radiation by means of the diffraction of x-rays, electrons, neutrons, or other waves.

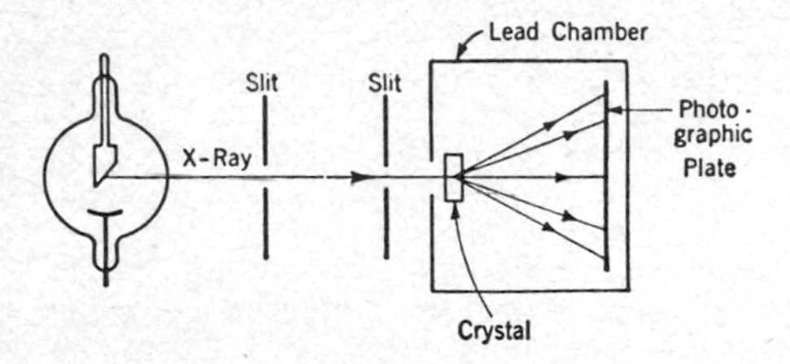

Diffractometer using x-rays to study crystals.

diffused-alloy transistor A transistor in which the semiconductor wafer is subjected to gaseous diffusion to produce a nonuniform base region, after which alloy junctions are formed in the same manner as for an alloy-junction transistor. It may also have an intrinsic region, to give a PNIP unit.

diffused-base transistor A transistor in which a nonuniform base region is produced by gaseous diffusion. The collector-base junction is also formed by gaseous diffusion, and the emitter-base junction is a conventional alloy junction.

diffused emitter-collector transistor A transistor in which both the emitter and collector are produced by diffusion.

diffused junction A semiconductor junction that has been formed by the diffusion of an impurity within a semiconductor crystal.

diffused-junction rectifier A semiconductor diode in which the PN junction is produced by diffusion.

diffused-junction transistor A transistor in which the emitter and collector electrodes have been formed by diffusion of an impurity metal into the semiconductor wafer without heating.

diffused-mesa transistor A diffused-junction transistor in which an N-type impurity is diffused into one side of a P-type wafer. A second PN junction, required for the emitter, is produced by alloying or diffusing a P-type impurity into the newly formed N-type surface. After contacts have been applied, undesired diffused areas are etched away to create a flat-topped peak called a mesa.

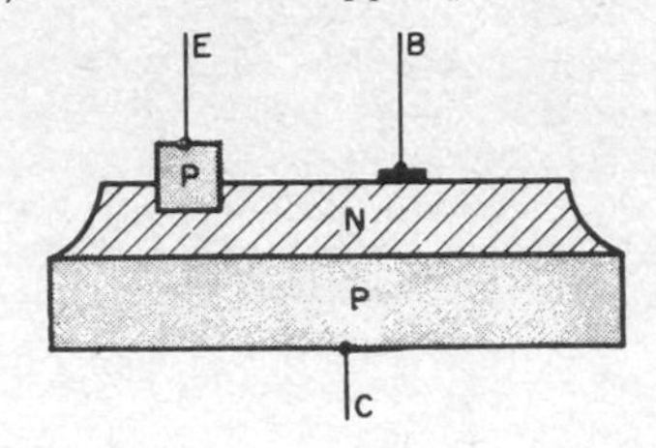

Diffused-junction transistor having N-type mesa base.

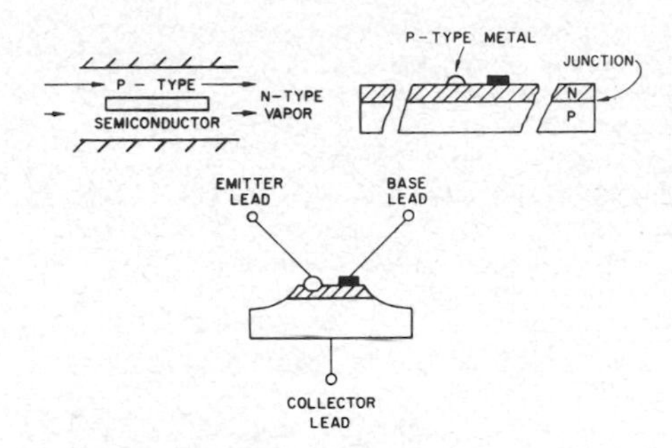

Diffused-mesa transistor production process.

diffused resistor An integrated-circuit resistor produced by a diffusion process in a semiconductor substrate.

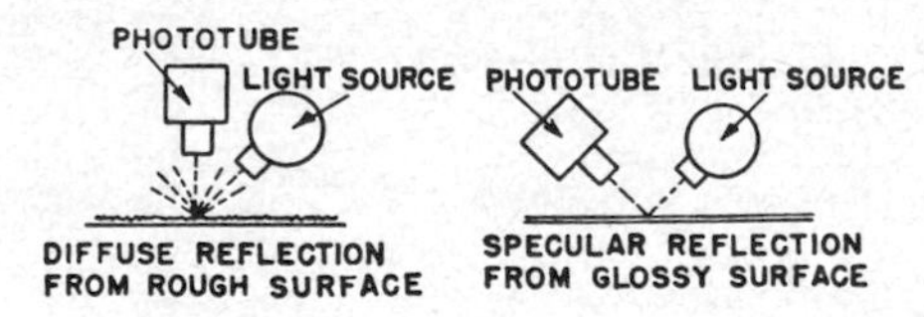

Diffuse-reflection measurement, as compared to direct or specular reflection.

diffuse reflection Reflection of light, sound, or radio waves from a surface in all directions according to the cosine law.

diffuse sound Sound that has uniform energy density in a given region and is such that all directions of energy flux at all parts of the region are equally probable.

diffuse transmission Transmission in which all the emergent radiation is observed.

diffuse-transmission density The value of the photographic transmission density obtained when light flux impinges normally on the sample and all the transmitted flux is collected and measured.

diffusion 1. The passage of particles through matter in such circumstances that the probability of scattering is large. 2. The movement of carriers, donors, or acceptors in a semiconductor. 3. The migration of atoms and vacancies of one material into another. 4. A method of producing a junction by diffusing an impurity metal into a semiconductor at a high temperature.

diffusion bonding Bonding in which the molecules of one metal diffuse into the crystalline lattice structure of another metal to form a solid solution that is equivalent to an electric connection.

diffusion capacitance The rate of change of stored minority-carrier charge with the voltage across a semiconductor junction.

diffusion cloud chamber A cloud chamber in which vapor diffuses from a source near a hot plate and condenses on a cold plate. The resulting layer of supersaturated vapor between the plates is sensitive to the passage of ionizing particles.

diffusion coefficient The constant of proportionality in Fick's law.

diffusion constant The diffusion current density in a homogeneous semiconductor divided by the charge-carrier concentration gradient. The resulting constant is also equal to the product of the drift mobility and the average thermal energy per unit charge of carriers.

diffusion length 1. The average distance to which minority carriers diffuse between generation and recombination in a homogeneous semiconductor. 2. The average distance traveled by a thermal neutron from its point of formation to its point of absorption.

diffusion process A method of producing a junction by diffusing an impurity metal into a semiconductor at a high temperature.

diffusion pump A vacuum pump in which a stream of heavy molecules, such as mercury vapor, carries gas molecules out of the volume being evacuated. Also used for separating isotopes according to weight, the lighter molecules being pumped preferentially by the vapor stream.

diffusion transistor A transistor in which current flow is a result of diffusion of carriers, donors, or acceptors, as in a junction transistor.

digicom *Digital communication.*

dig-in angle A stylus cutting angle such that the point is driving into the coating. It is the opposite of drag angle.

digit A character that stands for zero or for a positive integer smaller than the base of an ordinary number system used in a digital computer.

digital Pertaining to data in the form of digits.

digital adder An adder that accepts two numbers in digital form and gives their sum in the same digital form.

digital attenuator An absorptive microwave attenuator that uses a single current-controlled diode attenuator in combination with a digital driver which provides the correct level of control current for giving the discrete attenuation level corresponding to the binary control code. Each attenuation level is determined by a unique logic gate and resistor.

digital circuit A circuit that operates like a switch in that it has only two states: on and off.

digital clock A clock that has a large direct-reading digital display of time in the form 12:15 or 12:15:55 and sometimes a day-month-year display. A digital clock may also provide time-code signals in a variety of forms, such as series or parallel binary-coded decimal, level shift, or modulated carrier.

digital communication A communication system in which speed is transmitted in the form of trains of pulse signals, and digital information is transmitted directly from computers, radar, tape readers, teleprinters, and telemetering equipment. Delta pulse-code modulation is generally used to convert audio signals to digital signals. The system may also include automatic switching circuits that choose the fastest path from origin to destination. Also called digicom.

digital comparator A comparator that accepts two different values of input digital data and determines which is larger, or compares one digital input value against preset upper and lower digital limits and provides pass/fail information.

digital computer A computer that processes information in digital form. Electronic digital computers generally use binary or decimal notation and solve problems by repeated high-speed use of the fundamental arithmetic processes of addition, subtraction, multiplication, and division.

digital converter A converter that changes one form of digital input data to its equivalent in some other code.

digital data Data represented by a sequence of code characters.

digital data service [abbreviated DDS] An AT&T communication system developed specifically for digital data, using existing local digital lines combined with data-under-voice microwave transmission facilities.

digital delay generator A high-precision adjustable time-delay generator in which delays may be selected in increments such as 1, 10, or 100 ns by means of panel switches and sometimes by remote programming. Used in acoustic, laser, radar, sonar, and other applications involving precise measurement of very short time intervals.

digital differential analyzer [abbreviated DDA] A differential analyzer that will synthesize a waveform digitally and perform integration.

digital display A display in which the result is indicated in directly readable digital form. Alphabetic characters and special symbols are sometimes also generated. Digital displays may use incandescent lamps, light-emitting diodes, liquid crystals, Nixies and newer forms of gas-discharge displays, and a variety of other electric, electroluminescent, electronic, mechanical, and optical devices. Also called digital readout.

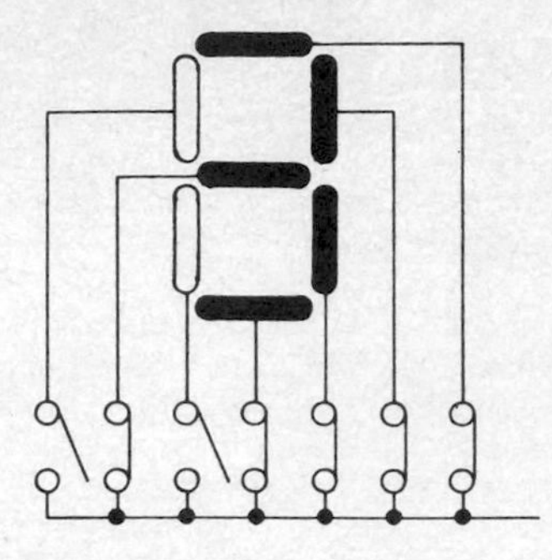

Digital display using seven segments to form numerals.

digital echo modulation A modem transmitter design technique in which the line signal is synthesized by generating signal elements in a time sequence. The signal elements may be stored in a digital memory as pulse-code modulation or delta-modulation samples.

digital filter A filter used in digital signal processing, such as for removing noise and improving quality of old acoustic phonograph records.

digital frequency meter A frequency meter in which the exact value of the frequency being measured is indicated on a digital display, after the number of cycles in the input signal is counted for a period of exactly 1 s. The time interval is determined by a master oscillator and frequency dividers.

digital gaussmeter A gaussmeter that measures magnetic flux density and indicates its value in gauss on a direct-reading digital display.

digital integrated circuit An integrated circuit that is used primarily for pulse processing, as contrasted to a linear integrated circuit which provides linear amplification of signals.

digital ionosonde An ionosonde in which digital filtering techniques are used to separate desired ionospheric echoes from undesired noise.

digitally programmed amplifier An amplifier in which the gain can be controlled by digital signals coming from a computer or other source.

digitally programmed power supply A power supply in which the value of the output voltage or current can be controlled by digital signals from a computer or other source.

digital meter A meter that provides a direct-reading digital display, eliminating the need for reading the value represented by the position of a moving pointer on a dial. The meter includes circuits for sampling a measured analog quantity, converting the instantaneous value to digital form, and presenting it as a continuously updated display that may use light-emitting diodes, liquid crystals, cold-cathode indicators, or other display devices. A three-digit display provides readings up to 999, with or without a floating decimal. A 3½-digit display has a fourth digit position, commonly limited to 1 for reading up to 1999, but some displays give limits of 2999 or 3999. Polarity is sometimes also indicated.

digital microwave radio Transmission of voice and data signals in digital form on microwave links, as in the 2-GHz common-carrier bands. Pulse-code modulation is used.

digital modulation A method of placing digital traffic on a microwave system without use of modems, by transmitting the information in the form of discrete phase or frequency states determined by the digital signal.

digital multimeter [abbreviated DMM] A multimeter in which the measured value of voltage, current, resistance, and sometimes level in decibels is shown on a direct-reading digital display. An analog-to-digital converter serves as interface between the analog input and the digital output of the multimeter.

digital multiplier A multiplier that accepts two numbers in digital form and gives their product in the same digital form, usually by making repeated additions. The multiplying process is simpler if the numbers are in binary form wherein digits are represented by a 0 or 1.

digital organ An electronic organ in which all waveforms are synthesized from a single frequency standard, which in one version is 4 MHz. Circuits provide for storage of normalized digital representations of all desired pipe-organ tones in a read-only memory.

digital output An output signal consisting of a sequence of discrete quantities coded in an appropriate manner for driving a printer or digital display.

digital panel meter [abbreviated DPM] A digital meter designed for flush mounting on a panel.

digital phase shifter A phase shifter in which a control pulse provides a predetermined amount of signal-phase shift, with the polarity of the pulse determining the direction of the shift.

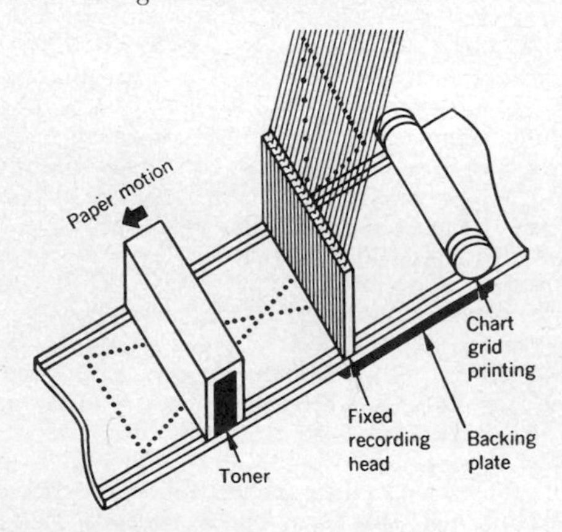

Digital plotter using electrostatic technique with up to 1400 fixed styli to place electric charges at desired positions on moving paper, after which charged areas receive carbon particles from toner and heat is used to fuse particles to paper.

digital plotter A recorder that produces permanent hard copy in the form of a graph from digital input data.

digital printer A printer that provides a permanent readable record of binary-coded decimal or other coded data in a digital form that may include some or all alphameric characters and special symbols along with numerals. The printout usually has less than 20 columns, in contrast to line printers, which usually have more than 100 columns and include all characters in all print positions. Also called digital recorder.

digital range tracker A radar range tracker in which clock pulses and counting circuits convert analog input to a digital output.

digital readout *Digital display.*

digital recorder *Digital printer.*

digital recording Magnetic recording in which the information is first coded in a digital form, generally with a binary code that uses two discrete values of residual flux. In nonreturn-to-zero recording, the two values correspond to saturation of the tape in opposite directions. In return-to-zero recording, the tape either is saturated in one direction or is in a neutral or biased state.

digital remote control A remote control in which the user selects or adjusts a desired function by punching out one or more digits on a calculatorlike keyboard. Applications include direct selection of television programs by channel number. In one version used for remote control of color television, an ultrasonic transmitter converts the 14 possible ultrasonic frequencies into digital codes that control the desired ON/OFF, channel-select, color, tint, and volume functions.

digital scrambler A scrambler used in suppressed-carrier digital modulation systems to modify the bit pattern in a nearly random manner to minimize interference errors. An inverse descrambling operation is required at the receiving end of the system.

digital signal analyzer A signal analyzer in which one or more analog inputs are sampled at regular intervals, converted to digital form, and fed to a memory. The desired function for a series of the stored inputs is computed in an arithmetic unit for use in driving a digital display. Used chiefly for analyzing random signals, signals obscured by noise, and VLF signals from about 20 kHz down to DC.

digital signal processing The conversion of analog signals to digital form for computer processing in such fields as acoustics, biomedical engineering, image enhancement, radar, seismology, and speech communication.

digital sort A sorting procedure used chiefly with punched cards, wherein all cards are first sorted according to the least significant digit, then resorted on each next higher-order digit until the items have all been sorted on the most significant digit.

digital speech communication The transmission of voice messages in binary or other digital form by wire lines or radio.

digital subtracter A subtracter that accepts two numbers in digital form and gives their difference in the same digital form.

digital synchronometer A time comparator that provides a direct-reading digital display of time with high precision by making accurate comparisons between its own digital clock and high-accuracy time transmissions from WWV or a loran-C station.

digital telemetering The telemetering of data in digital form over wire or radio links.

digital television Television in which picture redundancy is reduced or eliminated by transmitting only the data needed to define motion in the picture, as represented by changes in the areas of continuous white or continuous black. At the same time, the conventional analog signal is converted to a digital signal that can be more readily coded for secret transmission or is compressed in bandwidth for transmitting at slow speeds over ordinary telephone lines.

digital television converter A converter used to convert television programs from one system to another, such as for converting 525-line 60-field U.S. broadcasts to 625-line 50-field European PAL or SECAM standards. The video signal is digitized before conversion.

digital thermometer A thermometer in which the measured value of temperature is shown on a direct-reading digital display.

digital-to-analog [abbreviated D/A] Pertaining to systems having analog input and digital output.

digital-to-analog converter [abbreviated D/A converter or DAC] A converter that changes digital input signals to corresponding analog signal values.

digital-to-synchro converter [abbreviated D/S converter or DSC] A converter that changes BCD or other digital input data to a three-wire synchro output signal representing corresponding angular data.

digital traffic Traffic consisting of digital data that may or may not be combined with voice, video, or other forms of analog information which has been converted to a format suitable for pulse-code modulation.

digital transducer A transducer that measures physical quantities and transmits the information as coded digital signals rather than as continuously varying currents or voltages.

Digital voltmeter.

digital voltmeter [abbreviated DVM] A voltmeter in which the measured value of voltage is shown on a direct-reading digital display, usually with polarity sign as well as decimal point.

digital volt-ohm-milliammeter [abbreviated DVOM] A volt-ohm-milliammeter in which a digital display serves in place of an indicating meter.

digital watch An electronic watch in which time and sometimes calendar data is indicated by a digital display instead of by hands moving on a dial. The display may use light-emitting diodes or liquid crystal displays.

digitize To convert an analog measurement of a quantity into a numerical value.

dihedral reflector A corner reflector having two sides meeting at a line.

diheptal base A tube base having 14 pins or possible pin positions. Used chiefly on television cathode-ray tubes.

dilution Reducing the intensity of a color by adding white.

dimmer An electric or electronic control for varying the intensity of a lamp or other light source.

DIN Abbreviation for *Deutsche Industrie Normenausschus.*

diode 1. A two-electrode electron tube containing an anode and a cathode. 2. *Semiconductor diode.*

diode amplifier A microwave amplifier using an IMPATT, TRAPATT, or transferred-electron diode in a cavity, with a microwave circulator providing the input/output isolation required for amplification. Center frequencies are in the gigahertz range, from about 1 to 100 GHz, and power outputs are up to 20 W continuous wave or more than 200 W pulsed, depending on the diode used.

diode characteristic The composite electrode characteristic of an electron tube when all electrodes except the cathode are connected together.

diode-connected transistor A bipolar transistor in which two terminals are shorted to give diode action. Diodes are commonly produced this way in integrated circuits.

diode demodulator A demodulator that uses one or more crystal or electron-tube diodes to provide a rectified output whose average value is proportional to the original modulation. Also called diode detector.

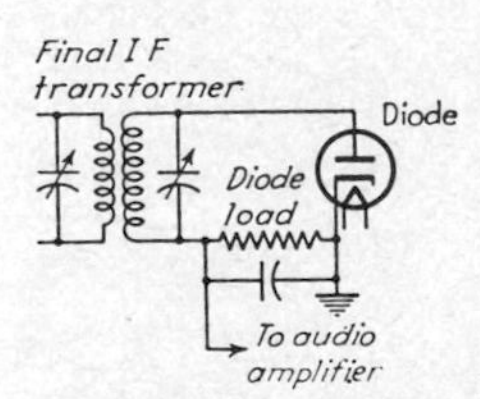

Diode-demodulator circuit.

diode detector *Diode demodulator.*

diode function generator A single-ended amplifier that has adjustable nonlinear gain characteristics. Diode-shaping networks containing potentiometers are used to vary the gain for each segment of the input voltage range.

diode gate An AND gate that uses diodes as switching elements.

diode laser *Semiconductor laser.*

diode limiter A peak-limiting circuit employing a diode that becomes conductive when signal peaks exceed a predetermined value.

diode logic Logic in which a diode is used in each input lead, with the bias voltage of each diode being varied appropriately to make the diode turn on or off.

diode matrix A matrix of diodes used between a set of input connections and a set of output connections. Usually used for code conversion.

diode mixer A mixer that uses a crystal or electron-tube diode. It is generally small enough to fit directly into an RF transmission line.

diode modulator A modulator that uses one or more diodes to combine a modulating signal with a carrier signal. Used chiefly for low-level signaling because of its inherently poor efficiency.

diode-pentode An electron tube that has a diode and a pentode in the same envelope.

diode pumping The use of an electroluminescent diode to produce excitation for population inversion and lasing action in a solid laser, as in a YAG laser.

diode ring modulator A matched set of four diodes connected cathode-to-anode as a ring for use as a modulator or demodulator.

diode-transistor logic [abbreviated DTL] Logic in which each input diode of a gate circuit performs an AND or an OR function to control the base current of a bipolar transistor that provides power gain for driving additional gates.

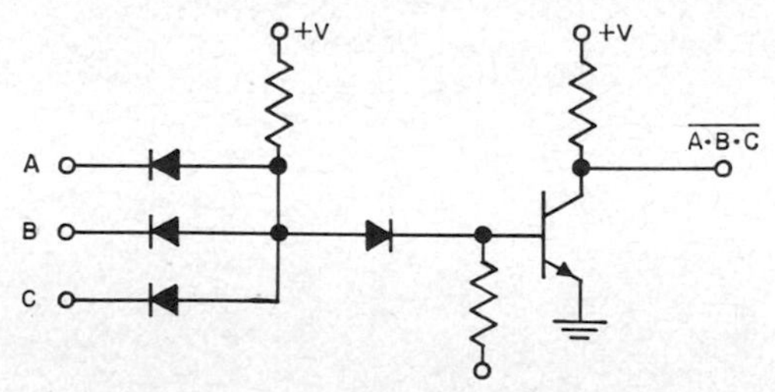

Diode-transistor logic as used for basic NAND gate.

DIP Abbreviation for *dual in-line package.*

diparaxylene *Parylene.*

dip brazing A brazing process in which the parts to be joined are clamped together with appropriate preforms of nonferrous filler metal and dipped in a molten chemical or metal bath. The filler metal is distributed in the joints by capillary attraction. A metal bath may also provide the filler metal. Used for assembling waveguide components.

dip coating Application of a protective plastic coating to a transformer or other device by dipping.

diplexer A coupling system that allows two different transmitters to operate simultaneously or separately from the same antenna.

diplex radio transmission The simultaneous transmission of two signals using a common carrier wave.

diplex reception Simultaneous reception of two signals having some feature in common, such as a single receiving antenna or a single carrier frequency.

dip meter An absorption wavemeter in which bipolar or field-effect transistors replace the electron tubes used in older grid-dip meters.

dip needle A magnetometer in which a magnetized needle is used to indicate the angle between the earth's magnetic field and the horizontal.

dipole 1. An antenna approximately one half-wavelength long, split at its electrical center for connection to a transmission line. The impedance of the antenna is about 72 Ω. The radiation pat-

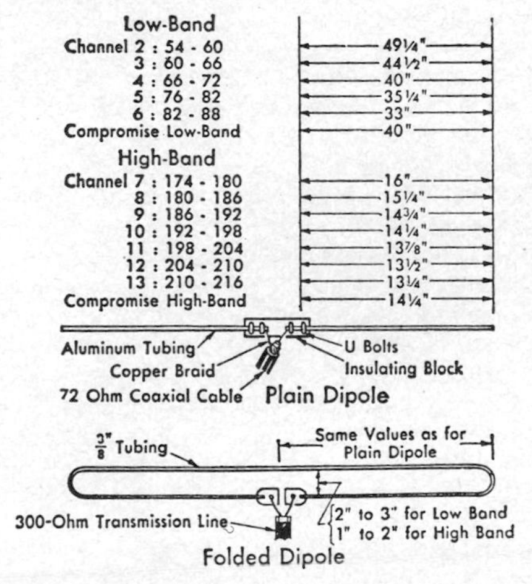

Dipole lengths for specific channels in television bands between 54 and 216 MHz. Multiply dimension values by 2.54 to convert to centimeters.

tern is a maximum at right angles to the axis of the antenna. Also called dipole antenna, doublet, doublet antenna, and half-wave dipole. 2. Two nuclear particles a very small distance apart, having opposite electric charges or opposite magnetic polarities. Also called doublet and electric doublet.

dipole antenna *Dipole.*

dipole moment A term used to specify mathematically the field caused by a given distribution of electric or magnetic charges.

dipping sonar Sonar designed to be lowered into the water by cable from a helicopter, hydrofoil, or other hovering craft. Also called dunking sonar.

dip soldering Soldering of printed-wiring boards or other assemblies by immersion in a pool of molten solder.

dipulse Pulse transmission in which the presence of one cycle of a sine-wave tone represents a binary 1 and the absence of one cycle represents binary 0.

Dirac equation A relativistic wave equation for an electron in an electromagnetic field. The solution requires that for the total angular momentum of the electron to be constant, the electron must possess its own intrinsic angular momentum, called spin.

direct-access memory *Random-access memory.*

direct-acting recorder A recorder in which the marking device is mechanically connected to, or directly operated by, the primary detector.

direct address A computer address that indicates the location where the referenced operand is to be found or stored.

direct command guidance Control of a missile or drone entirely from the launching site, by radio or by signals sent over a wire.

direct-coupled amplifier [always spell out, to avoid confusion with DC amplifier] A DC amplifier in which a resistor or a direct connection provides the coupling between stages so the DC component of a signal is preserved. The frequency response starts at zero frequency (DC) and extends to some specified upper limit.

direct-coupled transistor logic [abbreviated DCTL] Logic in which transistors in gate, flip-flop, and inverter circuits are coupled together

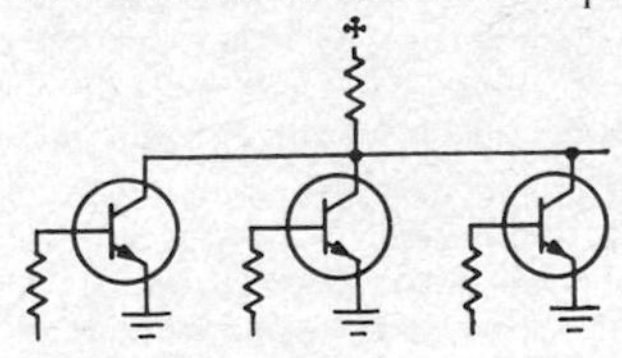

Direct-coupled transistor logic.

directly, without resistors or other coupling components.

direct coupling Coupling of two circuits by means of a nonfrequency-sensitive device such as a wire, resistor, or battery so both direct and alternating current can flow through the coupling path. In a coaxial cavity, direct coupling is achieved by connecting directly to the center conductor in the cavity.

direct current [abbreviated DC] An electric current that flows in one direction. For direct-current terms, see DC entries.

direct digital control [abbreviated DDC] Process control that uses a digital computer, usually on a time-sharing or multiplexing basis. Widely used in chemical and petroleum plants. Also called direct numerical control.

direct distance dialing [abbreviated DDD] A telephone exchange service that allows a telephone user to dial subscribers outside the local area.

directed branch A branch having an assigned direction.

direct inductive coupling Coupling by means of inductance that is common to the two circuits. One circuit may be connected directly to a tap on a coil in the other circuit.

direct inward dialing The capability for dialing individual telephone extensions in a large organization directly from outside, without going through a central switchboard.

direction The position of one point in space relative to another, without reference to the distance between them. Although direction is not an angle, it is often indicated in terms of its angular difference from a reference direction.

directional antenna An antenna that radiates or receives radio waves more effectively in some directions than others.

directional beam A radio or radar wave concentrated in a given direction.

directional characteristic The variation in the behavior of a transducer or other device with respect to direction.

directional control Control exercised over an aircraft to move it in a desired direction.

directional counter A counter more sensitive to nuclear radiation from some directions than from others.

directional coupler A device that couples a secondary system only to a wave traveling in a particular direction in a primary transmission system, while completely ignoring a wave traveling in the opposite direction. The amount of coupling is

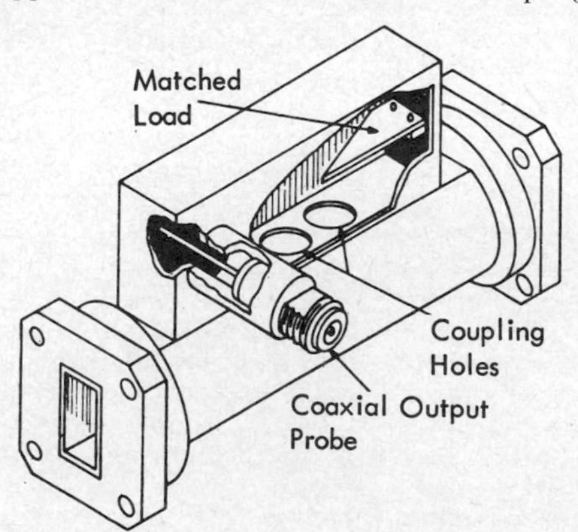

Directional coupler used between coaxial cable and waveguide.

ordinarily expressed in decibels of attenuation that the signal undergoes in passing through the coupling to the secondary line. Coaxial lines use loop-type directional couplers or two-hole directional couplers. For waveguide directional couplers, the arrangement is comparable to a two-hole coupler.

directional filter A low-pass, bandpass, or high-pass filter that separates the bands of frequencies used for transmission in opposite directions in a carrier system.

directional gain *Directivity index.*

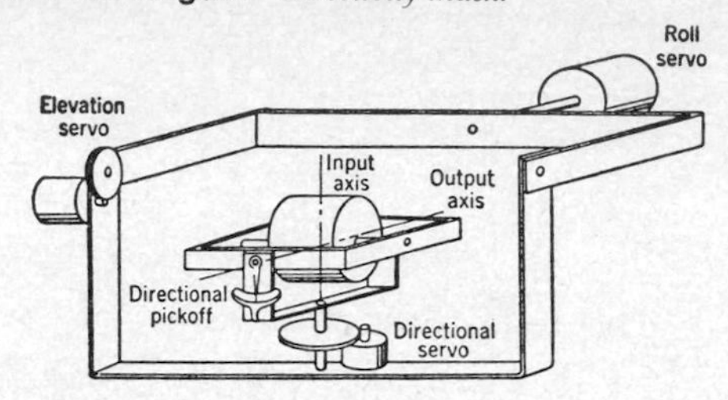

Directional gyroscope mounted on platform.

directional gyroscope A gyroscope that holds its position in azimuth and indicates deviation from a desired heading.

directional homing Homing in which the relative bearing is maintained constant.

directional hydrophone A hydrophone whose response varies significantly with the direction of sound incidence.

directional microphone A microphone whose response varies significantly with the direction of sound incidence.

directional phase shifter A passive phase shifter in which the phase change for transmission in one direction differs from that for transmission in the opposite direction.

directional response pattern *Directivity pattern.*

directional stabilizer A directional gyro used to maintain direction automatically, as in an autopilot or bombsight.

direction finder [abbreviated DF] *Radio direction finder.*

direction-finder bearing indicator An instrument used with an airborne radio direction finder to indicate the relative, magnetic, or true bearing of a station from an aircraft or the reciprocals of these bearings.

direction-finder deviation The difference between the observed radio bearing obtained with a direction finder and the true bearing of the transmitter.

direction-finding station A radio station that has equipment for determining the direction of arrival of radio waves. Two or more such stations working together can determine the location of the transmitter by triangulation.

direction-independent radar Doppler radar used for detecting the motion of intruders in a protected space.

direction of polarization The direction of electric lines of force or the electric vector in a polarized wave.

direction of propagation The direction of time-average energy flow at any point in a homogeneous isotropic medium. For a linearly polarized wave it is the direction of the electric vector. In a uniform waveguide the direction of propagation is often taken along the axis.

directive gain An antenna rating equal to 4π times the ratio of the radiation intensity in a given direction to the total power radiated by the antenna.

directivity 1. The value of the directive gain of an antenna in the direction of its maximum value. The higher the directivity value, the narrower the beam in which the radiated energy is concentrated. 2. The ratio of the power measured at the forward-wave sampling terminals of a directional coupler, with only a forward wave present in the transmission line, to the power measured at the same terminals when the direction of the forward wave in the line is reversed. The ratio is usually expressed in decibels. For a perfect coupler it would be infinitely high.

directivity factor 1. The ratio of radiated sound intensity at a remote point on the principal axis of a loudspeaker or other transducer to the average intensity of the sound transmitted through a sphere passing through the remote point and concentric with the transducer. The frequency must be stated. 2. The ratio of the square of the voltage produced by sound waves arriving parallel to the principal axis of a microphone or other receiving transducer to the mean square of the voltage that would be produced if sound waves having the same frequency and mean-square pressure were arriving simultaneously from all directions with random phase. The frequency must be stated. Directivity factor in acoustics is equivalent to directivity as applied to antennas.

directivity index The directivity factor expressed in decibels. It is 10 times the logarithm to the base 10 of the directivity factor. Also called directional gain.

directivity pattern A graphical or other description of the response of a transducer used for sound emission or reception. It is presented as a function of the direction of the transmitted or incident sound waves in a specified plane and at a specified frequency. Also called beam pattern and directional response pattern.

directivity signal A spurious signal present in the output of a coupler because the directivity of the coupler is not infinite.

directly heated cathode *Filament.*

directly ionizing particles Charged particles having sufficient kinetic energy to produce ionization by collision. Examples include electrons, protons, and alpha particles.

direct memory access [abbreviated DMA] A computer feature, set up by the central processing unit, that provides for direct data transfer from a peripheral device to the computer memory or to magnetic disk or tape storage units.

direct numerical control [abbreviated DNC] *Direct digital control.*

director A parasitic element placed a fraction of a wavelength ahead of a dipole receiving antenna to increase the gain of the array in the direction of

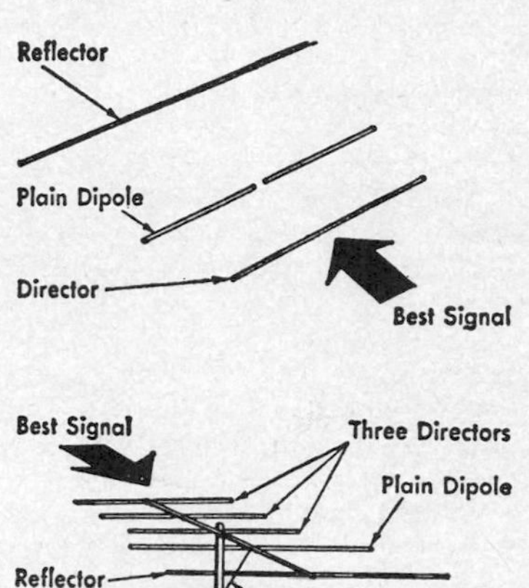

Director used with plain diode and reflector, and example of commercial plain-dipole Yagi antenna using three directors.

the major lobe. It is usually a rod slightly shorter than the receiving dipole, with no connection to the lead-in.

direct outward dialing [abbreviated DOD] A private automatic branch telephone exchange that permits all local stations to dial outside numbers.

direct pickup The transmission of television images without intermediate photographic or magnetic recording.

direct radiator loudspeaker A loudspeaker in which the radiating element acts directly on the air, without a horn.

direct recording Recording in which a record is produced immediately, without subsequent processing, in response to received signals.

direct reflection Reflection of light, sound, radar, or radio waves in accordance with the laws of optical reflection, as by a mirror. Also called mirror reflection, regular reflection, and specular reflection.

direct scanning A scanning method in which the subject is illuminated at all times and only one elemental area of the subject is viewed at a time by the television camera. Used in live television broadcasting, whereas indirect or flying-spot scanning is sometimes used in industrial television systems.

direct-view storage tube A cathode-ray tube in

which secondary emission of electrons from a storage grid is used to provide an intensely bright display for long and controllable periods of time. The tube includes one or more writing guns, a flooding gun, and selective erasing capability. The resulting image can be viewed on an aircraft instrument panel without a hood, even when in direct sunlight, for fire control and weather radar.

direct voltage A voltage that forces electrons to move through a circuit in the same direction continuously, thereby producing a direct current. Also called DC voltage.

direct wave A radio wave that is propagated directly through space from transmitter to receiver without being refracted by the ionosphere.

direct-writing galvanometer A direct-writing recorder in which the stylus or pen is attached to a moving coil positioned in the field of the permanent magnet of a galvanometer. Recording mechanisms include an ink pen, a stylus moving on carbon-coated film or paper, a heated stylus on heat-sensitive paper, and an energized stylus on electrosensitive paper.

direct-writing recorder A recorder in which the permanent record of varying electrical quantities or signals is made on paper, directly by a pen attached to the moving coil of a galvanometer or indirectly by a pen moved by some form of motor under control of the galvanometer. Pen types in common use include capillary ink, pressurized ink, and thermal acting on heat-sensitive paper.

dirty weapon An atomic weapon that produces a larger amount of radioactive residues than does a normal weapon of the same yield.

disc Alternate spelling for *disk.*

discharge 1. The passage of electricity through a gas, usually accompanied by a glow, arc, spark, or corona. 2. To remove a charge from a battery, capacitor, or other electric energy storage device. 3. The conversion of chemical energy to electric energy in a battery.

discharge lamp A lamp in which light is produced by an electric discharge between electrodes in a gas or vapor at low or high pressure. Examples include fluorescent lamps, mercury-vapor lamps, neon tubing, and sodium-vapor lamps. Also called gas discharge lamp.

discharger A silver-impregnated cotton wick encased in a flexible plastic tube with an aluminum mounting lug, used on aircraft to reduce precipitation static. The many fine high-resistance fibers provide a multitude of discharge points for static electricity.

discharge tube An evacuated enclosure containing a gas at low pressure, through which current can flow when sufficient voltage is applied between metal electrodes in the tube. The resulting current flow is chiefly due to ionization of a gas or vapor. The tube may be made conducting by triggering with a positive voltage pulse, such as for discharging a capacitor at periodic intervals to generate a sawtooth waveform.

discomposition The process in which an atom is knocked out of its position in a crystal lattice by direct nuclear impact, as by fast neutrons or fast ions that have been previously knocked out of their lattice positions. The atom so displaced eventually comes to rest at an interstitial position or at a lattice edge.

discone antenna A biconical antenna in which one of the cones is spread out to 180° to form a disk. The center conductor of the coaxial line terminates at the center of the disk, and the cable shield terminates at the vertex of the cone. Both the input impedance and radiation pattern remain essentially constant over a wide frequency range. The disk is normally parallel to the earth, giving an omnidirectional radiation pattern in a horizontal plane.

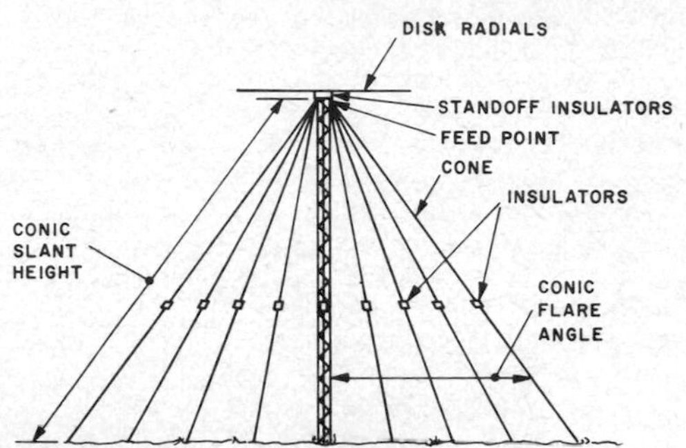

Discone antenna using wires to form cone below disk.

disconnect To open a circuit by removing wires or connections, as distinguished from opening a switch to stop current flow.

disconnector A switch intended to open a circuit only after load current has been interrupted by other means.

discontinuity A break in the continuity of a medium or material, at which a reflection of wave energy can occur.

discontinuous amplifier An amplifier in which the input waveform is reproduced on some type of averaging basis. Magnetic amplifiers and thyratron amplifiers are examples.

discotheque French word for record library. Now applied to any facility for dancing to recorded music, as played by a disk jockey or selected from a jukebox by patrons.

discrete Having an individually distinct identity.

discrete-component microcircuit A microcircuit consisting chiefly of separate active and passive components that were manufactured before installation on the microcircuit board or substrate.

discrete sampling Sampling in which the individual samples are of such long duration that the frequency response of the channel is not deteriorated by the sampling process.

discrete sentence intelligibility The percent intelligibility obtained when the speech units under consideration are sentences, usually of simple form and content.

discrete sound system A quadraphonic sound system in which the four input channels are preserved as four discrete channels during recording and playback processes. Sometimes referred to as a 4-4-4 system; a matrix sound is a 4-2-4 system because it is recorded or broadcast on only two channels. The CD-4 system as developed by RCA and Japan Victor is an example.

discrete transistor A single transistor encapsulated in a suitable package, with appropriate terminations for external connections. Two transistors are sometimes formed on the same wafer to give thermal matching.

discrete variable A quantity that may assume any one of a number of individually distinct or separate values.

discrete word intelligibility The percent intelligibility obtained when the speech units under consideration are words, usually presented so as to minimize the contextual relation between them.

discrimination 1. The degree of rejection of unwanted signals in a receiver or other equipment containing tuned circuits. 2. *Conditional jump.*

discriminator A circuit in which magnitude and polarity of the output voltage depend on how an input signal differs from a standard or another signal. Thus a frequency discriminator converts

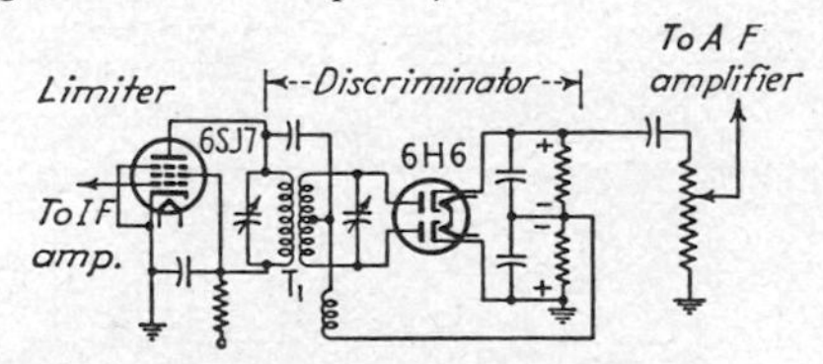

Discriminator circuit for FM receiver.

frequency deviations from a carrier frequency into corresponding amplitude variations. A pulse-height discriminator delivers an output voltage only for pulses that exceed a predetermined height. A phase discriminator converts phase variations into corresponding amplitude variations.

discriminator transformer A transformer designed to be used in a stage where FM signals are converted directly to AF signals or in a stage where frequency changes are converted to corresponding voltage changes.

disengage To break the contact between two objects.

dish A concave reflector that has a surface which is parabolic or part of a sphere, used in a microwave antenna. The periphery is usually circular.

disintegration 1. Radioactive decay in which energy is emitted from the nuclei of atoms. 2. The transformation of one nuclide into one or more different nuclides by bombardment with high-energy particles like alpha particles or helium ions, deuterons, protons, neutrons, or gamma rays.

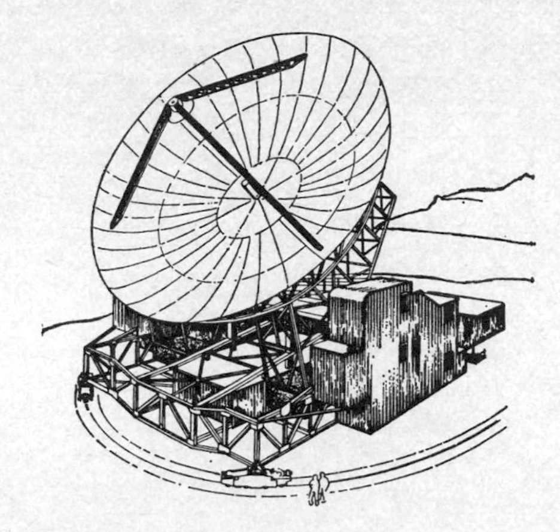

Dish used in microwave antenna for satellite communication station.

disintegration energy The energy released during radioactive decay. It is distributed among the decay products. Also called *Q* value.

disintegration rate 1. The absolute rate of decay of a radioactive substance, usually expressed in terms of disintegrations per unit of time. 2. The absolute rate of transformation of a nuclide under bombardment.

disintegration voltage The lowest anode voltage at which destructive positive-ion bombardment of the cathode occurs in a hot-cathode gas tube.

disjunctive search A search defined in terms of a logic sum.

disk A thin circular object, such as a phonograph record or a metal or plastic sheet having a magnetic or other type of coating, used for storing computer data or video information. Sometimes spelled disc.

disk capacitor A capacitor that has a disk-shaped ceramic dielectric, with leads attached to metallized opposite faces. Used for temperature compensation in tuned circuits and for bypass applications.

diskette *Floppy disk.*

disk file *Disk storage.*

disk jockey A person who plays phonograph records on a radio or television program and intersperses personal comments and announcements.

disk operating system [abbreviated DOS] A computer operating system in which source programs and sometimes incoming data are stored in a disk file. This system holds programs that can be many times the size of the computer memory, for almost instantaneous loading under control of supervisory and utility programs.

disk pack A set of magnetic disks that can be removed as a unit from a disk storage.

disk recorder A mechanical recorder that uses a disk as the recording medium.

disk resistor A resistor consisting of a stack of

washers punched from a resistance alloy, with the inner and outer edges of the washers coated with silver to provide mechanical contact with a coaxial line in which the resistor is mounted.

disk-seal tube An electron tube having disk-shaped electrodes arranged in closely spaced parallel layers, to give low interelectrode capacitance

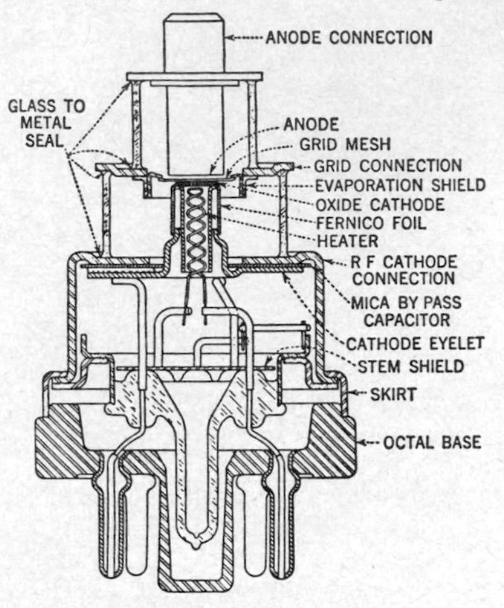

Disk-seal tube construction.

along with high-power output up to 2.5 GHz. The edges of the disk electrodes are fused into and project through the glass or ceramic envelope, to serve as external contacts. Also called lighthouse tube.

disk storage A computer storage in which bits representing data are stored in circular tracks in the magnetic coating of a rigid or flexible rotating disk. Other versions may use nonmagnetic coatings in conjunction with a laser beam or other form of read/write head. The head may be moved mechanically to a desired track, or a separate head may be provided for each track to give fastest access time and maximum reliability. Also called disk file.

dislocation An imperfection in the geometric arrangement of atoms in a crystal.

dispenser cathode An electron-tube cathode having provisions for continuously replacing evaporated electron-emitting material.

dispersal gettering Absorption of gas during dispersal of a getter in an electron tube.

disperse A data-processing operation in which grouped input items are distributed among a larger number of groups in the output.

dispersion 1. The process of separating radiation into components having different frequencies, energies, velocities, or other characteristics. A prism or diffraction grating disperses white light into its component colors. A magnetic field disperses or sorts electrons according to their velocities. 2. Scattering of microwave radiation by an obstruction. 3. A distribution of finely divided particles in a medium.

dispersive line A delay line that delays each frequency a different length of time. Both quartz and aluminum delay lines can be designed to have this property.

dispersive medium A medium in which the phase velocity of an electromagnetic wave is a function of frequency. A plasma is a dispersive medium, but free space is not because waves of all frequencies travel in space with the velocity of light.

displacement current A hypothetical current assumed to exist in the presence of time-varying electric fields. Postulated by Maxwell to explain the transfer of current through space between capacitor plates.

displacement gyroscope A gyroscope that senses, measures, and transmits angular displacement data.

displacement law *Radioactive displacement law.*

displacement transducer A transducer that converts a linear or angular movement into an electric signal. Methods of accomplishing this conversion include the use of variable-inductance, variable-resistance, variable-capacitance, and electron-tube devices.

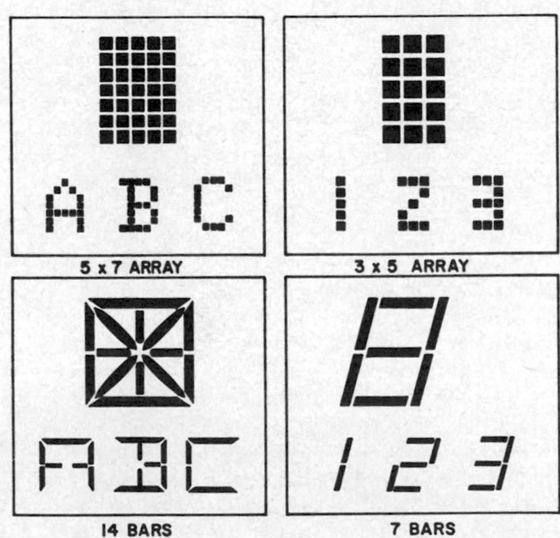

Display styles obtained with various configurations of light-emitting diodes or other light sources.

display A visual presentation of output information, as on the screen of a cathode-ray tube or in readable characters of a digital display.

display loss *Visibility factor.*

display primaries The television receiver primary colors that, when mixed in proper proportions, produce other desired colors. The three primaries usually used are red, green, and blue. Also called receiver primaries.

display storage tube A storage tube into which the information is introduced as an electric signal and read at a later time as a visible output.

display tube A cathode-ray tube used to provide a visual display. One example is a direct-view storage tube, used in the cockpit of an aircraft for navigation or fire control. Another example is a character-writing tube, used as a computer output device.

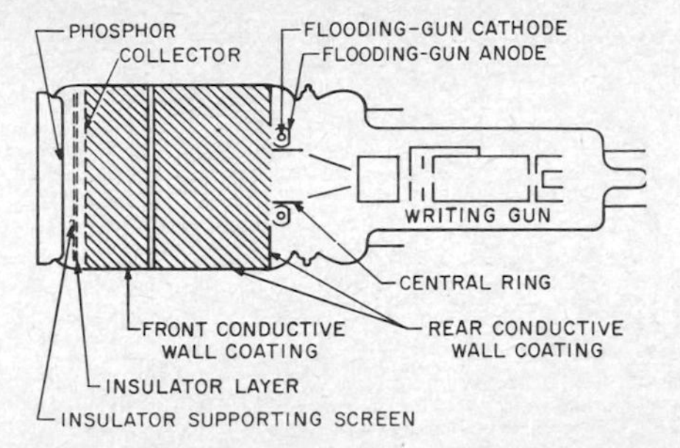

Display tube having writing gun for creating characters of display and larger coaxial flooding gun for spraying electrons in broad beam to intensify pattern created by writing gun.

disruptive discharge A sudden and large increase in current through an insulating medium due to complete failure of the medium under electrostatic stress.

dissector tube *Image dissector.*

dissipation An undesired loss of energy, generally by conversion into heat. Thus, the collector dissipation rating in watts for a transistor is the maximum amount of energy that can be lost as heat at the collector electrode without damage to the transistor.

dissipation factor The reciprocal of *Q*.

dissipation line A length of stainless steel or Nichrome wire used as a noninductive terminating impedance for a rhombic transmitting antenna when several kilowatts of power must be dissipated.

dissociation Separation of a complex molecule by collision with a second body. The energy required for this process comes from the kinetic energy of the colliding particles.

dissociative recombination The capture of an electron by a positive molecular ion. The electron combines with the ion and dissociates it into two neutral atoms.

dissolve The merging of two television camera signals in such a way that as one scene disappears, another slowly appears.

dissonance An unpleasant combination of harmonics heard when certain musical tones are played simultaneously.

distance mark A mark produced on a radar display by a special signal generator. When the mark is moved to a target position on the screen, the range to that target can be read directly on the calibrated dial of the signal generator.

distance-measuring equipment [abbreviated DME] A radio aid to navigation that provides distance information by measuring total round-trip time of transmission from an airborne interrogator to a ground-based transponder and return.

distance resolution The minimum radial distance by which targets must be separated to be separately distinguishable by a particular radar. Also called range discrimination (deprecated) and range resolution (deprecated).

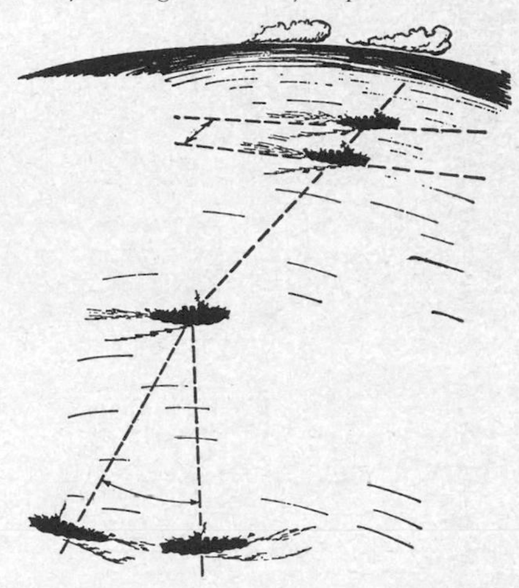

Distance resolution is illustrated by upper two ships, and bearing resolution by lower two ships, for radar on ship at center.

distortion 1. Any undesired change in the waveform of a signal. 2. Any undesired deviation of an image from proportionality with the original scene.

distortion analyzer An analyzer that measures total harmonic distortion of an audio signal by removing the fundamental frequency with a narrow-band rejection filter and measuring the amplitude of the remaining components for comparison with that of the fundamental frequency.

distortion meter An instrument that measures distortion of a sinusoidal signal by removing the fundamental-frequency component and measuring any harmonics that remain.

distress frequency A frequency allotted to distress calls, generally by international agreement. For ships at sea and aircraft over the sea, it is 500 kHz.

distress signal The international signal used when a ship, aircraft, or other vehicle is threatened by grave and imminent danger and requests immediate assistance. In radiotelegraphy, it consists of three dots, three dashes, and three dots transmitted as a single signal in which the dashes are emphasized. In radiotelephony, it is the word mayday.

distributed amplifier A wideband amplifier in which tubes are distributed along artificial delay lines made up of coils acting with the input and output capacitances of the tubes. Gain can be increased indefinitely by adding more tubes.

distributed capacitance The capacitance that exists between adjacent turns in a coil or between adjacent conductors in a cable. Also called self-capacitance.

distributed constant A circuit parameter that exists along the entire length of a transmission line. For a transverse electromagnetic wave on a two-conductor transmission line, the distributed constants are series resistance, series inductance, shunt conductance, and shunt capacitance per unit length of line.

distributed-emission photodiode A broadband photodiode proposed for detection of modulated laser beams at millimeter wavelengths. Incident

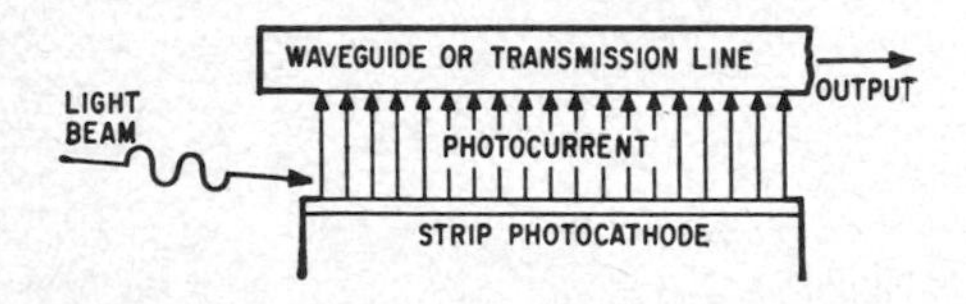

Distributed-emission photodiode, with modulated laser beam entering at left.

light falls on a photocathode strip that generates a traveling wave of photocurrent having the same wave velocity as the transmission line into which the photodiode feeds.

distributed inductance The inductance that exists along the entire length of a conductor, as distinguished from inductance concentrated in a coil.

distributed paramp A paramagnetic amplifier that consists essentially of a transmission line shunted by uniformly spaced, identical varactors. The applied pumping wave excites the varactors in sequence to give the desired traveling-wave effect.

distribution amplifier 1. An AF power amplifier used to feed a speech or music distribution system and having sufficiently low output impedance so changes in load do not appreciably affect the output voltage. 2. An RF power amplifier used to feed television or radio signals to a number of receivers, as in an apartment house or hotel.

distribution coefficients The tristimulus values of monochromatic radiations having equal power.

distribution control *Linearity control.*

disturbance 1. An undesired interference or noise signal affecting radio, television, or facsimile reception. 2. An undesired command signal in a control system.

disturbed-one output A one output of a magnetic cell to which partial-read pulses have been applied since that cell was last selected for writing.

disturbed-zero output A zero output of a magnetic cell to which partial-write pulses have been applied since that cell was last selected for reading.

ditch antenna A flush antenna in which excitation is applied to a groove or ditch formed in the flat metal ground plane.

dither A force having a controlled amplitude and frequency, applied continuously to a device driven by a servomotor so the device is constantly in small-amplitude motion and cannot stick at its null position. Used in some recorders to make the pen ready to move instantly. Also called buzz.

dither injector *Antistiction oscillator.*

dither-tuned magnetron A magnetron that has a motor-driven tuning plunger which provides rapid repetitive tuning over a narrow band as

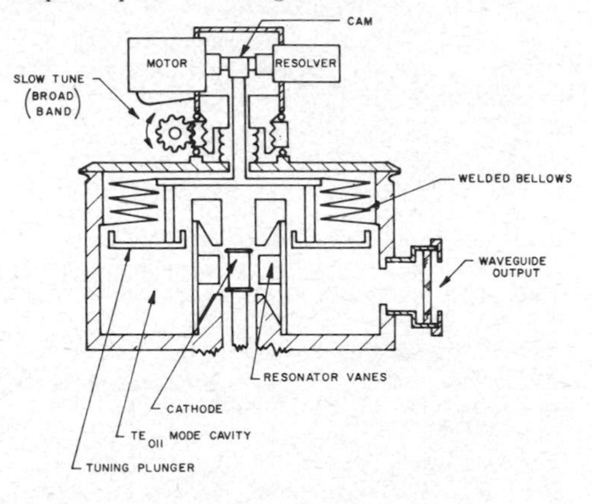

Dither-tuned magnetron in which motor-driven cam moves tuning plunger up and down rapidly in output cavity.

required for reception of frequency-agile radar signals that have a different frequency for each pulse.

divalent silver oxide cell A silver oxide primary cell in which highly active divalent silver oxide replaces the monovalent type as a depolarizing cathode, to increase power output.

divergence 1. The spreading of a cathode-ray stream due to repulsion of like charges (electrons). 2. A scalar quantity used in computations involving vectors.

divergence loss The portion of the transmission loss in an acoustic system caused by the divergence or spreading of sound rays.

diversionary missile A decoy missile used to draw enemy fire away from an attacking plane or missile.

diversity radar A radar that uses two or more transmitters and receivers, each pair operating at a slightly different frequency but sharing a common antenna and video display, to obtain greater effective range and reduce susceptibility to jamming.

diversity receiver A radio receiver designed for space or frequency diversity reception.

diversity reception Radio reception in which the effects of fading are minimized by combining two or more sources of signal energy carrying the same modulation. Space diversity takes advantage of the fact that fading does not occur simultaneously for antennas spaced several wavelengths apart. Frequency diversity takes advantage of the fact that signals differing slightly in frequency do not fade simultaneously.

divided-carrier modulation Modulation in which the carrier is divided into two components that are 90° out of phase, each modulated by a different signal. When these components are added, the frequency is unchanged, but the resulting signal varies in amplitude and phase in accordance with the modulating signals.

divider A circuit or device capable of dividing one quantity or a variable by another or by a fixed number such as 2 or 10.

dividing network *Crossover network.*

DKDP Abbreviation for *deuterated potassium dihydrogen phosphate.*

D layer The lowest layer of ionized air above the earth, occurring in the D region only in the daytime hemisphere. It reflects frequencies below about 50 kHz and partially absorbs higher-frequency waves.

DMA Abbreviation for *direct memory access.*

DME Abbreviation for *distance-measuring equipment.*

DME-COTAR A navigation system that gives complete trajectory information at a single-site ground station by combining the range indication of DME with the direction indication of cotar.

DMM Abbreviation for *digital multimeter.*

DMOS Abbreviation for *double-diffused metal-oxide semiconductor.*

DNC Abbreviation for *direct numerical control.*

docking The process of bringing two spacecraft together in space.

document The source of data for input into a computer.

documentation The flowcharts, written procedures, and operator instructions produced with a computer program, for guidance in making later modifications and for running the program on a computer.

DOD Abbreviation for *direct outward dialing.*

doghouse A small enclosure placed at the base of a transmitting antenna tower to house antenna tuning equipment.

Doherty amplifier A linear RF power amplifier that is divided into two sections whose inputs and outputs are connected by quarter-wave networks. Operating parameters are so adjusted that for all values of input signal voltage up to one-half maximum amplitude, section No. 1 delivers all the power to the load. Above this level, section No. 2 comes into operation. At maximum signal input, both sections are operating at peak efficiency, and each section is delivering half the total output power to the load.

Dolby system A noise-reduction system developed by Dr. Ray M. Dolby to reduce hiss and other high-frequency noise originating in magnetic tape. During quiet passages the level of tape noise is comparable to that of the music, whereas during loud passages the noise is masked by the music. The Dolby system provides a predetermined amount of extra amplification for low levels of the higher audio frequencies during recording, with corresponding attenuation during playback to restore the music to its correct level while reducing tape noise. In the A-type professional Dolby version, the audio spectrum is divided into four bands, with the gain in each band automatically varied with signal level. The simplified B-type version developed for home tape recorders and cassette decks uses only one frequency range, extending upward from about 600 Hz.

dolly A wheeled platform on which a television camera or other apparatus is mounted.

dolorimeter An instrument for measuring pain, based on the dol as the unit of pain intensity.

domain *Magnetic domain.*

domain-tip memory [abbreviated DOT memory] A computer memory in which the presence or absence of a magnetic domain in a localized region of a thin magnetic film designates a 1 or 0. Stored data can be changed, moved, or read out by varying external magnetic fields produced by drive currents flowing through etched copper conductors on the surface of the film. Also called magnetic-domain memory.

dome An enclosure for a sonar transducer, projector, or hydrophone and associated equipment. It is designed to have minimum effect on sound waves traveling under water.

domestic satellite [abbreviated DOMSAT] A satellite in stationary orbit 22,300 mi (35,680 km) above the equator for handling up to 12 separate color television programs, up to 14,000 private-line telephone calls, or an equivalent number of channels for other communication services within the United States.

dominant mode *Fundamental mode.*

dominant wave The electromagnetic wave that has the lowest cutoff frequency in a given uniconductor waveguide. It is the only wave that will carry energy when the excitation frequency is between the lowest cutoff frequency and the next higher cutoff frequency.

dominant wavelength The single wavelength of light that when combined in suitable proportions with a reference standard light matches the color of a given sample.

DOMSAT Abbreviation for *domestic satellite.*

donor An impurity added to a pure semiconductor material to increase the number of free electrons. Because conduction is then chiefly due to movements of electrons, which are negative charges, the resulting alloy is called an N-type semiconductor. Antimony, arsenic, and phosphorus are frequently used as donors for germanium. Also called donor impurity.

donor impurity *Donor.*

donor level An intermediate energy level close to the conduction band in the energy diagram of an extrinsic semiconductor.

donutron An all-metal tunable magnetron.

doorknob capacitor A high-voltage plastic-encased capacitor resembling a doorknob in size and shape. Sheet plastic insulation is used. A typical voltage rating is 20 kV.

doorknob tube An ultrahigh-frequency electron tube having the approximate size and shape of a doorknob. Electrodes are small and closely spaced, and leads are brought out directly through the glass envelope. The Western Electric 316A is one example.

dopant An impurity element added to a semiconductor material under precisely controlled conditions to create PN junctions required for transistors and semiconductor diodes. Also called doping agent.

doped junction A junction produced by adding an impurity to the melt during growing of a semiconductor crystal.

doped-junction transistor *Grown-junction transistor.*

doping The addition of impurities to a semiconductor to achieve a desired characteristic, such as to produce an N- or P-type material.

doping agent *Dopant.*

doping compensation The addition of donor impurities to a P-type semiconductor or of acceptor impurities to an N-type semiconductor.

Doppler broadening Frequency spreading that occurs in single-frequency radiation when the radiating atoms, molecules, or nuclei do not all have the same velocity. Each radiating particle may then give rise to a different Doppler shift.

Doppler current meter An ultrasonic instrument used to measure the speed of water. An ultrasonic beam is projected into the water and its volume reverberation signal picked up. The difference between transmitted and received frequencies, as produced by the Doppler effect, is then proportional to water speed.

Doppler effect The change in the observed frequency of a wave due to relative motion of source and observer. When the distance between source and observer is decreasing, the observed frequency is higher than the source frequency. When the distance is increasing, the observed frequency is lower. The effect occurs for sound waves as well as radio waves. Named after Christian Doppler (1803–1853), a German mathematician.

Doppler frequency *Doppler shift.*

Doppler radar A radar that makes use of the Doppler shift of an echo due to relative motion of target and radar. This shift in frequency permits differentiation between fixed and moving targets. The velocity of a moving target can be determined with high accuracy by measuring the frequency shift. A Doppler radar may be either continuous wave or pulsed.

Doppler radar altimeter A radar altimeter based on the Doppler effect. Used in lunar modules as part of the landing system.

Doppler radar guidance Missile guidance that makes use of a Doppler navigation system built into the missile to determine surface velocity and drift angle.

Doppler radar-inertial navigation An airborne navigation system in which the ground velocity signals developed by a Doppler radar navigation system are used to null accumulative errors of the inertial navigation system.

Doppler radar navigation An airborne navigation system that utilizes the Doppler effect to determine drift and ground speed. Four beams of pulsed microwave energy are beamed toward the earth along the corners of an imaginary pyramid whose peak is at the aircraft. Echoes from the front-pointing beams undergo upward Doppler shift, and echoes from the rearward beams undergo downward Doppler shift. Similarly, drift causes Doppler shift of echoes from beams on one side with respect to beams on the other side of the aircraft. Comparison of the Doppler shifts in a computer provides complete information for navigation even in zero-visibility weather, at all altitudes, without reference to ground stations.

Doppler shift The amount of the change in the observed frequency of a wave due to Doppler effect, expressed in hertz. Also called Doppler frequency.

Doppler VOR A VHF radio range that is compatible with conventional VOR but is much less critical in site requirements. The variable signal that produces azimuth information is developed by feeding an RF signal sequentially to the elements of a ring-shaped antenna array.

DORAN [DOppler RANge] A Doppler ranging system that uses phase comparison of three different modulation frequencies on the carrier wave, such as 0.01, 0.1, and 1 MHz, to obtain missile range data with high accuracy. Some aspects of DORAN are similar to DOVAP.

doroid A coil that resembles half a toroid and uses a removable core segment to simplify the winding process.

DOS Abbreviation for *disk operating system.*

dosage *Dose.*

dose The amount of ionizing radiation delivered to a specified volume, measured in rads, reps, or rems. For x-rays or gamma rays having quantum energies up to 3 MeV, the roentgen unit may be used. Also called absorbed dose and dosage.

dose equivalent The product of absorbed dose

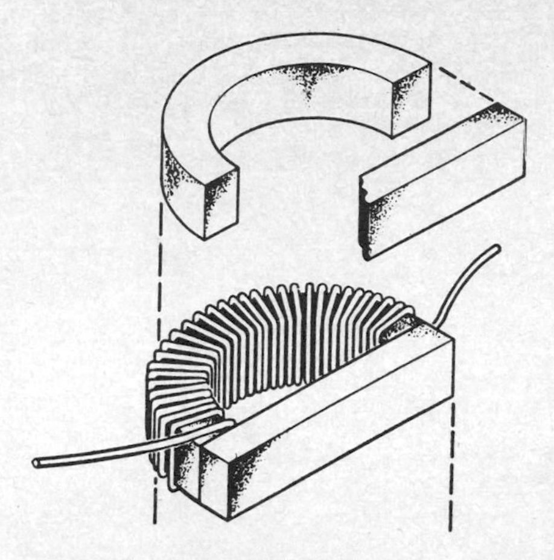

Doroid coil, showing method of assembling core.

in rads and a number of modifying factors due to nonuniform distribution of internally deposited isotopes in radiobiology. The unit of dose equivalent is the rem.

dosemeter *Dosimeter.*

dose rate The rate at which nuclear radiation is delivered.

dose-rate meter An instrument that measures radiation dose rate.

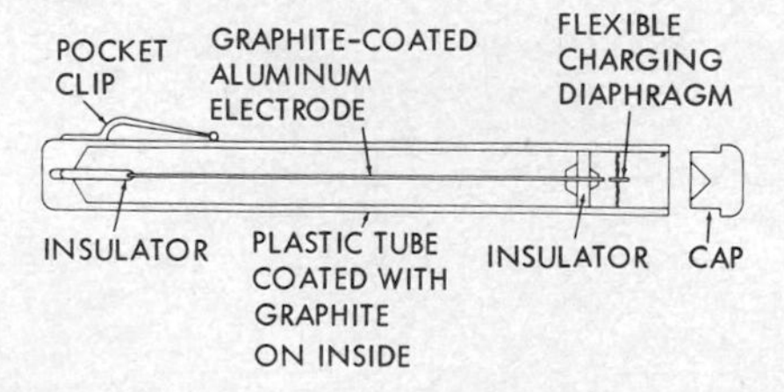

Dosimeter for wearing in pocket. Central wire and graphite-coated plastic case together form capacitor that is initially charged by DC voltage. Ionizing radiation dissipates charge. Minometer is used to measure remaining charge, to determine dosage to which wearer was exposed.

dosimeter An instrument that measures the total dose of nuclear radiation received in a given period. Also called dosemeter.

dosimetry The measurement of radiation doses.

dot *Button.*

dot angel An angel that appears on the screens of vertically pointing radars, often on clear cloudless days, as a bright dot. It is believed to be produced by a vertical column of rising air passing through air layers having different indexes of refraction. The rising air forms a hemispherical bubble at the junction of the two layers. On calm days, the top of this bubble can be a perfect reflector for radar signals. Dot angels are not seen on windy days.

dot cycle A mark pulse followed by an equal-length space element, as used in teletypewriter systems.

dot generator A signal generator that produces a dot pattern on the screen of a three-gun color television picture tube, for use in convergence adjustments. When convergence is out of adjustment, the dots occur in groups of three, one for each of the receiver primary colors. When convergence is correct, the three dots of each group converge to form a single white dot.

ABCDEFGHIJKLMNO
PQRSTUVWXYZ1234
567890+-[]•;/=?*

Dot-matrix characters and numerals produced with 5 × 7 format.

dot matrix A method of generating characters with a matrix of dots.

dot-matrix printer A printer in which a vertical row of five or seven stiff wires can be pushed against an inked ribbon under control of individual solenoids, to make dots on paper. As the row of wires is moved horizontally, the wires are appropriately energized so the resulting 5 × 7 or 7 × 9 dot-matrix pattern forms desired characters. In a thermal-matrix printer the character-forming dots are produced by heat rather than impact. Also called matrix printer.

DOT memory Abbreviation for *domain-tip memory.*

dot-sequential color television A color television system in which the red, blue, and green primary-color dots are formed in rapid succession along each scanning line.

double-amplitude-modulation multiplier A multiplier in which one variable is amplitude-modulated by a carrier, and the modulated signal is again amplitude-modulated by the other variable. The resulting double-modulated signal is applied to a balanced demodulator to obtain the product of the two variables.

double-balanced mixer A balanced mixer that uses accurately matched components, balanced transformers, and the bridge or ring configuration of diodes to minimize conversion loss and

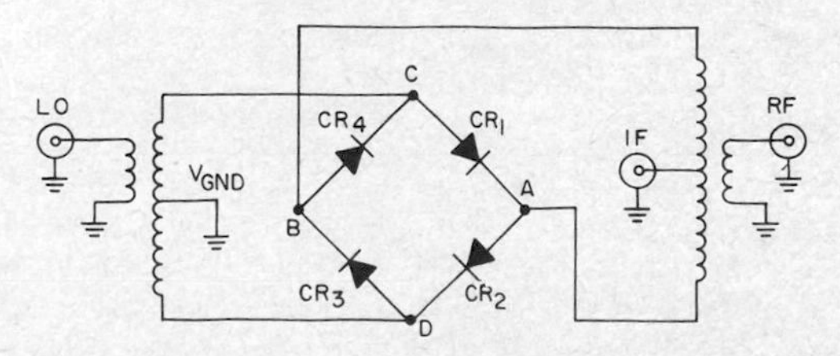

Double-balanced mixer having symmetrical configuration, with local oscillator feed at left.

reduce oscillator radiation from the antenna in critical applications using superheterodyne receivers.

double-beam cathode-ray tube A cathode-ray tube having two beams and capable of producing two independent traces that may overlap. The beams may be produced by splitting the beam of one gun or by using two guns.

double-break switch A switch that opens a conductor at two different points.

double bridge *Kelvin bridge.*

double-button carbon microphone A carbon microphone having two carbon-filled buttonlike containers, one on each side of the diaphragm, to give twice the resistance change obtainable with a single button. Also called differential microphone.

double-channel duplex A method that provides for simultaneous communication between two stations through use of two RF channels, one in each direction.

double-channel simplex A method that provides for nonsimultaneous communication between two stations through use of two RF channels, one in each direction.

double-conversion receiver *Double superheterodyne.*

double-diffused metal-oxide semiconductor [abbreviated DMOS] A metal-oxide semiconductor manufacturing process involving two-stage diffusion of impurities through a single mask

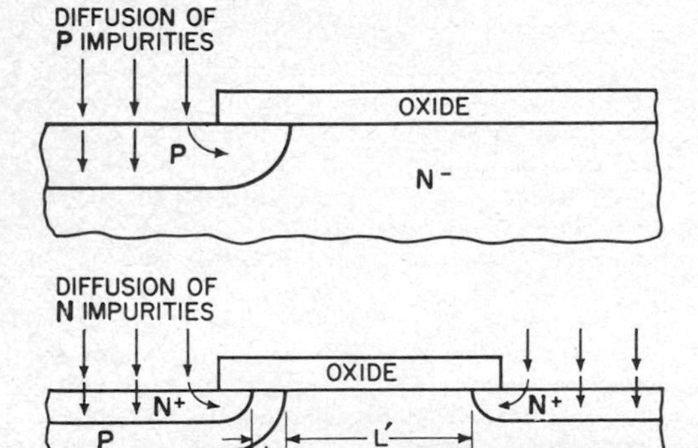

Double-diffused metal-oxide semiconductor process.

opening. Either depletion or enhancement modes can be produced. Short diffusion times keep channel lengths short, as required for production of diodes and transistors for high-speed logic and microwave applications.

double-diffused transistor A transistor in which two PN junctions are formed in the semiconductor wafer by gaseous diffusion of both P- and N-type impurities. An intrinsic region can also be formed.

double-diode *Duodiode.*

double-doped transistor The original grown-junction transistor, formed by successively adding P- and N-type impurities to the melt during growing of the crystal.

double-doublet antenna Two half-wave doublet antennas crisscrossed at their center, with one

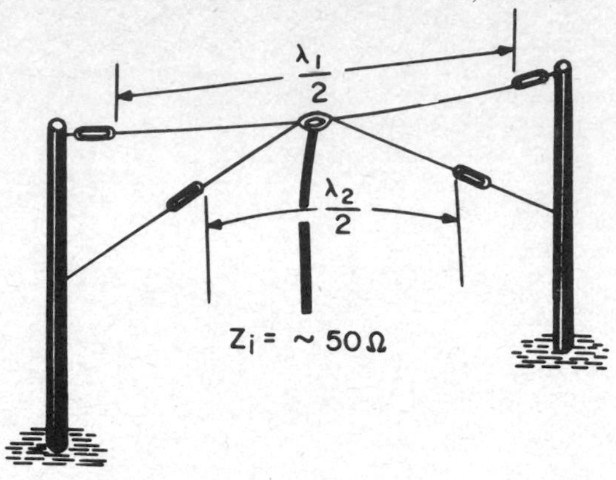

Double-doublet antenna.

shorter than the other to give broader frequency coverage.

double frequency-shift keying [abbreviated DFSK] A multiplex system in which two different telegraph signals are combined and transmitted simultaneously by using four different radio frequencies.

double-hump response *Bandpass response.*

double image A television picture consisting of two overlapping images, due to reception of the signal over two paths that differ in length so signals arrive at slightly different times. The longer path generally involves reflection of the signal by a hill, building, or large metal structure. The later-arriving reflected signal is often called a ghost because it is usually weaker than the direct signal and produces a phantomlike image to the right of the regular image.

double-length number A number having twice as many digits as are ordinarily used in a given computer. Also called double-precision number.

double limiter *Cascade limiter.*

double local oscillator A circuit that generates two RF signals which are accurately spaced several hundred hertz apart and mixes these signals to give an accurate audio frequency for use as a reference.

double-moding Undesirable shifting of a magnetron from one frequency to another at irregular intervals.

double modulation A method of modulation in which a subcarrier is first modulated with the desired intelligence, and the modulated subcarrier is then used to modulate a second carrier having a higher frequency.

double-pole double-throw [abbreviated DPDT] A six-terminal switch or relay contact arrangement that simultaneously connects one pair of terminals to either of two other pairs of terminals.

double-pole-piece magnetic head A magnetic head having two pole pieces, opposite in polarity and mounted on opposite sides of the magnetic recording medium. Either or both may have an energizing winding.

double-pole single-throw [abbreviated DPST] A four-terminal switch or relay contact arrangement that simultaneously opens or closes two separate circuits or both sides of the same circuit.

double-precision computation Use of twice as many digits as are normally handled in a digital computer, by keeping track of the numerical fragments that go beyond the computer capacity.

double-precision number *Double-length number.*

double-pulse recording Magnetic recording in which each stored bit consists of two regions magnetized in opposite polarities, with unmagnetized regions on both sides. The order of the polarity determines whether the bit is a 1 or a 0.

double-pulsing station A loran station that belongs to two pairs and emits pulses at two pulse rates.

double-quantum stimulated-emission device A laser in which the crystal contains two species of fluorescent ions: *A* and *B*. Ion species *A* can serve as the active ion in a four-level laser. The fluorescence frequencies of the two ions are so

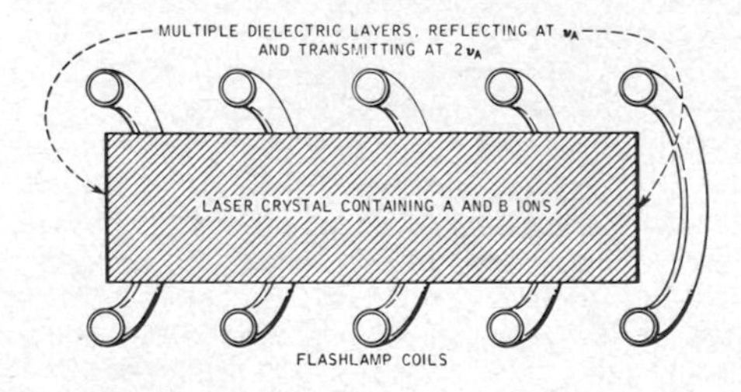

Double-quantum stimulated-emission device. Dielectric layers reflect at frequency of ions A and transmit at twice this frequency.

related that when the flashlamp coils produce pumping action, the *B* ions contribute photons to the primary *A* ion fluorescence. The effect is cumulative, triggering an avalanche that results in a giant output pulse. Films deposited on the ends of the crystal are highly reflective at the frequency of the *A* ions and highly transmissive at twice this frequency.

doubler *Frequency doubler.*

double sheath A sheath that develops when electrons and positive ions flow in opposite directions. The sheath then has a positive charge on one side and a negative charge on the other side. A double sheath develops in front of a hot cathode that emits electrons to the plasma and collects ions from the plasma.

double-sideband reduced-carrier transmission [abbreviated DSBRC transmission] Double-sideband transmission in which the useless RF power

Double-sideband reduced-carrier transmission spectrum, in which carrier power is reduced and power in sidebands increased without exceeding FCC limits for average power output.

in the carrier is reduced and the power in the intelligence-carrying sidebands correspondingly increased. Used in some citizens-band transceivers.

double-sideband transmission The transmission of a modulated carrier wave accompanied by both of the sidebands resulting from modulation.

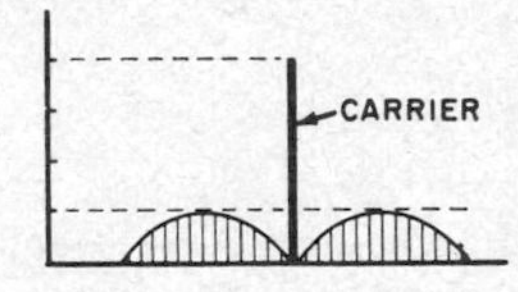

Double-sideband transmission in which carrier is 100% amplitude-modulated, giving upper and lower sidebands that carry useful intelligence.

The upper sideband corresponds to the sum of the carrier and modulation frequencies, whereas the lower sideband corresponds to the difference between the carrier and modulation frequencies.

double-sided mosaic An array of photosensitive elements insulated one from the other and mounted in a television camera tube in such a way that an image can be projected optically on one side of the mosaic. The corresponding electric signal is obtained by electronically scanning the other side of the mosaic.

double-spot tuning The reception of a given station by a superheterodyne receiver at two different dial settings, one where the local oscillator is above the station frequency by the IF value, and the other where the local oscillator is below the station frequency by the IF value. This effect can occur only if the receiver has poor selectivity. Also called repeat-point tuning.

double-stream amplifier A traveling-wave amplifier in which amplification occurs through interaction of two electron beams having different average velocities.

double-stub tuner A tuner consisting of two stubs, usually 3/8 wavelength apart, connected in parallel with a transmission line. Used for impedance-matching.

double superheterodyne A superheterodyne receiver in which the incoming carrier frequency is lowered to the first IF value by beating with one oscillator frequency in the first mixer, then reduced to the final IF value by beating with another oscillator frequency in a second mixer. Used in UHF receivers to obtain higher gain without instability, along with greater suppression of image frequencies and higher adjacent-channel selectivity. Also called dual-conversion receiver, double-conversion receiver, and triple-detection receiver.

doublet *Dipole.*

doublet antenna *Dipole.*

double-throw switch A switch having two operating positions, each providing different circuit connections.

double-track tape recording Magnetic recording in which two adjacent tracks are placed on the tape either for stereophonic reproduction or for doubling playing time by recording the second monophonic track in the opposite direction.

double triode An electron tube that has two triodes in the same envelope. Also called duotriode.

doublet trigger A trigger signal consisting of two pulses spaced a predetermined amount for coding purposes.

double-tuned amplifier An amplifier in which the stages are tuned to two different resonant frequencies to obtain wider bandwidth than is possible with single-frequency tuning.

double-tuned circuit A circuit that is resonant to two adjacent frequencies so there are two approximately equal values of peak response, with a dip between.

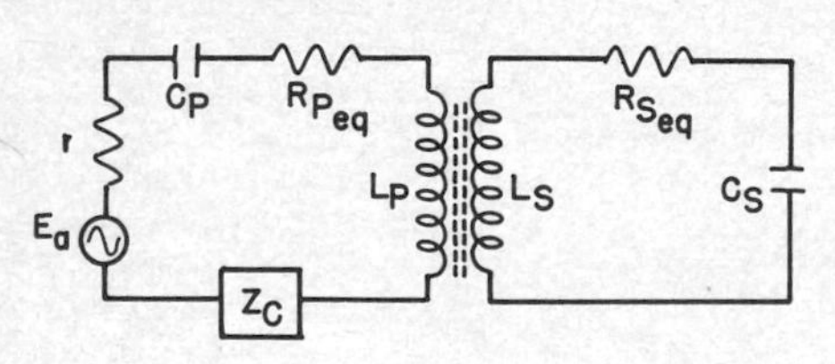

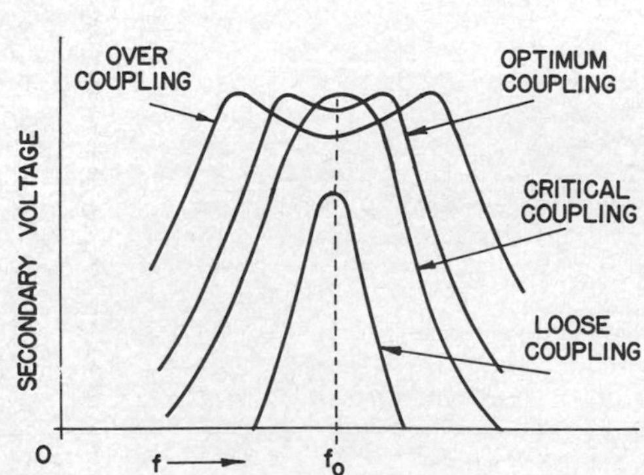

Double-tuned circuit and response curves for various degrees of coupling between transformer windings. The coupling is usually adjusted for twin-peak response giving desired bandwidth.

double-tuned detector A type of FM discriminator in which the limiter output transformer has two secondaries, one tuned above the resting frequency and the other tuned an equal amount below. Without modulation, both diodes conduct equally at the resting frequency, and the AF output is zero. Signal frequency deviation makes one diode conduct more than the other, giving AF output.

doughnut The toroidal vacuum chamber in which electrons are accelerated in a betatron or synchrotron. It is generally made of glass or ceramic material and is placed between magnet poles.

DOVAP [DOppler Velocity And Position] A phase-coherent tracking system used to determine velocity and position of missiles and space vehicles. A ground transmitter radiates a signal to a transponder on the missile and to receivers at several ground stations. The transponder doubles the frequency of the signal and retransmits it to the ground stations. The Doppler frequency shift, separated by comparing the received signal with twice the original transmitted signal frequency, is then proportional to missile velocity. Position is determined by combining the readings of two or more ground stations.

downconverter A converter that changes an incoming modulated or unmodulated carrier frequency to a lower frequency which is within the tuning range of a receiver or radio test set.

down counter A pulse counter that starts at its maximum count and decreases one count at a time to zero. Thus, a modulus 16 binary down counter decreases in 16 steps from 1111 to 0000.

down Doppler The sonar situation wherein the target is moving away from the transducer, so the frequency of the echo is less than the frequency of the reverberations received immediately after the end of the outgoing ping. Opposite of up Doppler.

down-lead *Lead-in.*

downlink The radio or optical transmission path downward from a communication satellite to the earth or an aircraft, or from an aircraft to the earth. The upward path is the uplink.

downrange Any area along the flight course of a missile. Downrange tracking stations report on missile flight behavior and receive telemetered data from the missile.

downtime The time during which a piece of equipment is not in operation because of a breakdown.

downward modulation Modulation in which the instantaneous amplitude of the modulated wave is never greater than the amplitude of the unmodulated carrier.

Dow oscillator *Electron-coupled oscillator.*

DPDT Abbreviation for *double-pole double-throw.*

DPM Abbreviation for *digital panel meter.*

DPSK Abbreviation for *differentially coherent phase-shift keying.*

DPST Abbreviation for *double-pole single-throw.*

drag angle A stylus cutting angle such that the point drags in the coating during recording on a disk. It is the opposite of dig-in angle.

drag-cup motor An induction motor having a cup-shaped rotor of conducting material, inside of which is a stationary magnetic core. It can be reversed by reversing connections to one phase of the two-phase stator winding. The lightweight rotor permits quick starts, high speed, quick stops, and sudden reversals.

drag-cup transducer A transducer in which a rotating permanent magnet induces eddy currents in a spring-mounted metal cylinder. The resulting torque rotates the cylinder and moves its attached pointer in proportion to the speed of the rotating shaft.

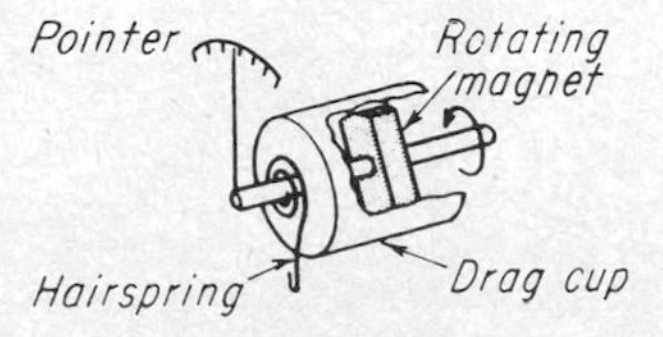

Drag-cup transducer.

drain [symbol D] 1. The region into which majority carriers flow from the channel of a field-effect transistor. The drain terminal of the transistor is connected to this region. The drain is comparable to the collector of a bipolar transistor and the anode of an electron tube. 2. *Current drain.*

drain wire A bare metallic conductor used in contact with foil shielding of a signal cable to provide a low-resistance path to ground.

drawer A frame that slides or rolls on tracks within a cabinet and has a front panel, behind which are instruments, controls, and electronic circuitry.

D region The region of the ionosphere between about 25 and 56 mi (40 and 90 km) above the earth, responsible for most of the attenuation of radio waves in the range of 1 to 100 MHz. It is below the E region.

dress The arrangement of connecting wires in a circuit to prevent undesirable coupling and feedback.

drift 1. A slow change in some characteristic of a device, such as frequency, balance current, direction (as in a gyro), or desired course of travel. Temperature variations are a common cause of frequency drift and unbalance in circuits. 2. The movement of current carriers in a semiconductor under the influence of an applied voltage. 3. Gradually developing changes in the offset voltage and both offset currents of an operational amplifier.

drift angle The angular difference between the course and the course made good. Also called course error (deprecated).

drift correction angle The angular difference between course and heading. Also called crab angle.

drift mobility The average drift velocity of carriers per unit electric field in a homogeneous semiconductor. Also called mobility.

drift space A space in an electron tube that is substantially free of externally applied alternating fields, in which repositioning of electrons takes place. In a klystron the velocity-modulated electrons form bunches in this space.

drift transistor A transistor in which an electric field dominates or assists in the transfer of charge carriers from input to output terminals, thereby permitting operation at higher frequencies.

drift tube A tubular electrode placed in the vacuum chamber of a circular accelerator, to which RF voltage is applied to accelerate the particles.

drift tunnel A piece of metal tubing, held at a fixed potential, that forms the drift space in a microwave tube or linear accelerator.

drift velocity The average velocity of an electron that is moving under the influence of an electric field. Drift velocity corresponds to the net current in an electron tube or a semiconductor device.

drip loop A downward loop formed in an antenna lead-in or other cable just before it enters a building. Water drips off at the bottom of the loop.

drive *Excitation.*

drive control *Horizontal drive control.*

driven array An antenna array consisting of a number of driven elements, usually half-wave dipoles, fed in phase or out of phase from a common source. Examples include broadside, collinear, and end-fire arrays.

driven element An antenna element that is directly connected to the transmission line.

driven sweep A cathode-ray sweep that is triggered only by an incoming signal.

drive pattern An undesired pattern of density variations caused by periodic errors in the position of the recording spot in a facsimile system. When caused by drive gears, the effect is called gear pattern.

drive pin An upward-projecting rod near the center pin of a turntable, used with a two-hole disk record to prevent slippage during recording.

drive-pin hole An off-center hole in a disk record, positioned to mate with the turntable drive pin.

drive pulse A pulsed magnetomotive force applied to a magnetic cell from one or more sources.

driver 1. The amplifier stage preceding the output stage in a receiver or transmitter. Also called driver stage. 2. The portion of a horn loudspeaker that converts electric energy into acoustic energy and feeds the acoustic energy to the small end of the horn.

driver stage *Driver.*

driving-point admittance The complex ratio of alternating current to applied alternating voltage for an electron tube, network, or other transducer.

driving-point function A response function for which the variables are measured at the same port.

driving-point impedance The complex ratio of applied alternating voltage to the resulting alternating current in an electron tube, network, or other transducer.

driving power The power supplied to the grid circuit of a tube in which the grid swings positive and draws current for a part of each cycle of the input signal.

driving signal A signal that times horizontal or vertical scanning at a television transmitter. Driving signals are usually provided by a central sync generator at the transmitter.

DRO Abbreviation for *destructive readout.*

drone A remotely controlled pilotless aircraft. It is generally controlled by radio, either from the ground or from a mother plane. Used on missions too hazardous for a human pilot, such as for target practice, probing the cloud of a nuclear explosion, taking photographs at low level over enemy territory, or carrying a television camera and transmitter over enemy territory.

drone antisubmarine helicopter A remotely controlled helicopter capable of operating from a destroyer or larger vessel and delivering a torpedo or other antisubmarine weapon to an enemy submarine up to 30 mi (48 km) away.

drop-in An unwanted character, digit, or bit formed accidentally on a magnetic recording surface.

dropout A reduction in output signal level during reproduction of recorded data, sufficient to cause a processing error.

dropout count The number of dropouts detected in a given length of magnetic tape.

dropout current The maximum current at which a relay or other magnetically operated device will release to its deenergized position.

dropout error The loss of a recorded bit or any other error occurring in recorded magnetic tape because of foreign particles on or in the magnetic coating or defects in the backing.

dropout voltage The maximum voltage at which a relay or other magnetically operated device will release to its deenergized position.

dropping resistor A resistor used in series with a load to decrease the voltage applied to the load.

dropsonde A radiosonde dropped by parachute from a high-flying aircraft to measure weather conditions and report them back to the aircraft. Used over water or other areas in which no ground station can be maintained.

drum A magnetic storage consisting of a cylinder coated with a magnetic material and driven by an electric motor. Data is recorded on, and read back from, the surface of this cylinder by individual heads for each track, staggered around the circumference of the drum. Also called drum storage.

drum parity A parity error occurring during transfer of information onto or from drums.

drum printer An impact printer in which a complete set of characters for each print position on a line is on a continuously rotating drum behind an inked ribbon, with paper in front of the ribbon. Identical characters are printed simultaneously at all required positions on a line, on the fly, by signal-controlled hammers.

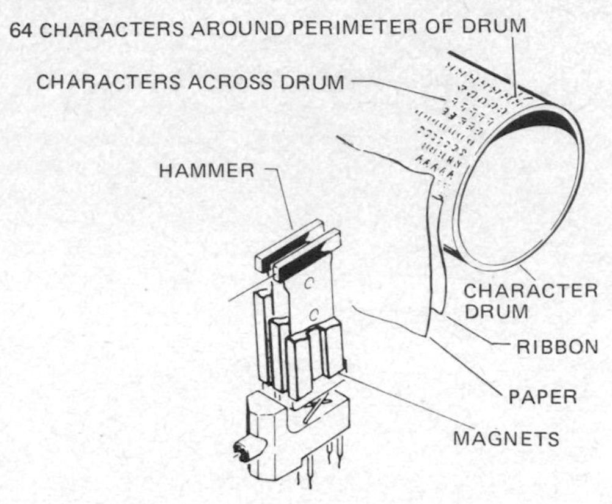

Drum printer.

drum recorder A facsimile recorder in which the record sheet is mounted on a rotating drum or cylinder.

drum storage *Drum.*

drum switch A multiposition rotary switch that has various patterns of conducting segments mounted on a cylinder of insulating material, rotating against fixed contacts surrounding the cylinder. Each switch position provides a different combination of connections between the fixed contacts.

dry battery A battery made up of a series, parallel, or series-parallel arrangement of dry cells in a single housing to provide desired voltage and current values.

dry cell A voltage-generating cell having an immobilized electrolyte. The most common form has a positive electrode of carbon and a negative electrode of zinc in an electrolyte of sal ammoniac paste.

dry-charged battery A storage battery in which the electrolyte is drained from the battery after the plates are formed. The battery can be stored for several years in this dry condition. To place in service, it is filled with electrolyte and charged for only a few minutes.

dry circuit A relay circuit in which open-circuit voltages are very low and closed-circuit currents extremely small, so there is no arcing to roughen the contacts. As a result, an insulating film can develop that prevents closing of the circuit when the contacts are brought together mechanically by the relay.

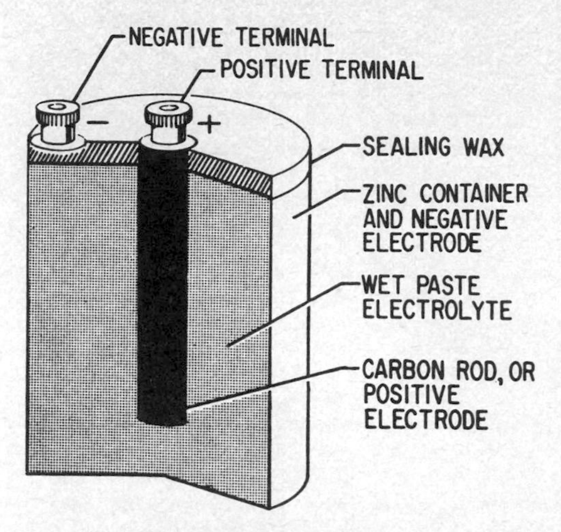

Dry-cell construction as used in large 1.5-V cell having binding-post terminals.

dry contact A contact that does not break or make current.

dry-disk rectifier *Metallic rectifier.*

dry electrolytic capacitor An electrolytic capacitor in which the electrolyte is a paste rather than a liquid. The dielectric is a thin film of gas formed on one of the plates by chemical action.

dry-reed relay A relay that has contacts mounted on magnetic reeds sealed into a length of small glass tubing. An actuating coil is wound around the tubing or wound on an auxiliary ferrite-core structure to provide the magnetic field required to operate the relay. The contacts are dry in the sense that they are not mercury-wetted.

dry-reed switch A switch that has contacts mounted on magnetic reeds in a vacuum enclosure, designed for actuation by an external permanent magnet. The contacts are dry in the sense that they are not mercury-wetted.

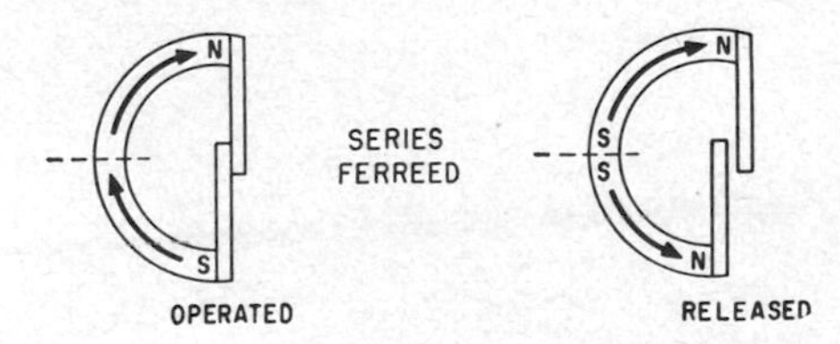

Dry-reed switch operation.

dry-tape fuel cell A fuel cell in which the fuel is in the form of a dry tape coated with fuel, oxidant, and electrolyte. The tape is fed into the cell at a rate corresponding to the demand for electric energy.

D/S Abbreviation for digital-to-synchro, usually used in connection with converters.

DSBRC transmission Abbreviation for *double-sideband reduced-carrier transmission.*

DSC Abbreviation for *digital-to-synchro converter.*

D/S converter Abbreviation for *digital-to-synchro converter.*

D scope A radarscope that produces a D display.

DT-cut crystal A quartz crystal cut to give a resonant frequency below about 500 kHz.

DTL Abbreviation for *diode-transistor logic.*

DTL/TTL An abbreviation used to indicate that a circuit will operate with inputs and/or outputs that are either diode-transistor or transistor-transistor logic.

dual automatic radio compass A combination of two automatic radio compasses feeding an azimuth indicator having two pointers, each indicating direction to one of the two radio stations being used to obtain a radio fix. Complete data for a fix is thus visible to the pilot at all times, eliminating the need for tuning first to one station and then to the other.

dual-beam oscilloscope An oscilloscope in which the cathode-ray tube has two individually controlled electron beams, for displaying two different quantities simultaneously on the same screen.

dual-channel amplifier An AF amplifier having two separate amplifiers for the two channels of a stereophonic sound system, usually operating from a common power supply mounted on the same chassis.

dual-conversion receiver *Double superheterodyne.*

dual-diversity receiver A diversity radio receiver in which the two antennas feed separate RF systems, with mixing occurring after the converter. An automatic selection system may connect the output to whichever channel is stronger at each instant.

dual-emitter transistor A passivated PNP silicon planar epitaxial transistor having two emitters, for use in low-level choppers.

dual-frequency induction heater An induction heater in which the charge receives energy by induction, simultaneously or successively, from a work coil or coils operating at two different frequencies.

dual-gate MOSFET A metal-oxide semiconductor field-effect transistor having two separate gate electrodes. The construction improves performance in mixers, amplifiers, demodulators, and other circuits.

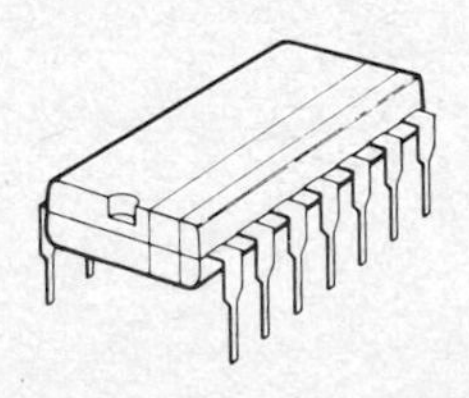

Dual in-line package for integrated circuit.

dual in-line package [abbreviated DIP] An integrated-circuit package that has two parallel rows of terminals at right angles to the body, as required for insertion in mating holes in a printed-wiring board. The other common package is the flatpack, in which the coplanar terminals project outward from the body without being bent.

dual laser A gas laser that has Brewster windows and concave mirrors at opposite ends, the mirrors having different reflectivities to produce two different visible or infrared wavelengths from a helium-neon laser beam.

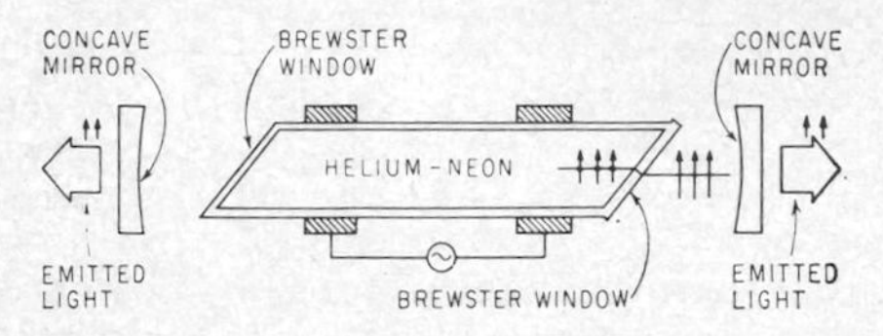

Dual laser using helium-neon gas.

dual-mode transducer A transducer inserted into a circular waveguide to select between two types of polarization.

dual modulation The process of modulating a common carrier wave or subcarrier with two different types of modulation, each conveying separate information.

dual transistors Two matched transistors mounted in a single package.

dub To transfer recorded material from one recording to another, with or without the addition of new sounds, background music, or sound effects.

dubbing Combining two or more sources of sound into a complete recording, at least one of the sources being a recording.

duct 1. An atmospheric condition that makes possible abnormally long-range radio and radar signal propagation in the troposphere. Temperature inversions cause abnormal changes in dielectric constant that make radio waves refract up and down between the two air layers forming the duct or between one air boundary and the ground. Also called tropospheric duct. 2. An enclosed runway for cables.

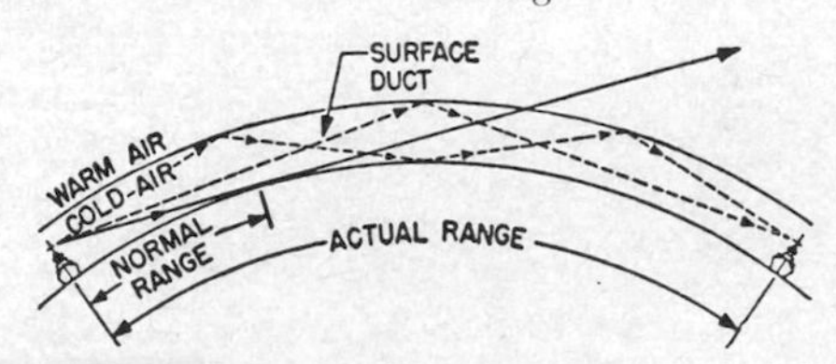

Duct of cold air under warm air makes radio waves follow curvature of earth to ship at right.

dumbbell marker An improved version of the fan marker used for air navigation, radiating a dumbbell-shaped signal pattern about 1.5 mi (2.4 km) wide and 12 mi (19 km) across.

dumbbell slot A dumbbell-shaped hole in a wall or diaphragm of a waveguide, designed to serve as a slot radiator.

dummy An artificial address, instruction, or other unit of information inserted into a digital computer solely to fulfill prescribed conditions (such as word length or block length) without affecting operations.

dummy antenna A device that has the impedance characteristic and power-handling capability of an antenna but does not radiate or receive radio waves. Used chiefly for testing transmitters. Also called artificial antenna.

dummy instruction An artificial instruction or address inserted into a computer program to serve some purpose other than execution as an instruction.

dummy load A dissipative device used at the end of a transmission line or waveguide to convert transmitted energy into heat, so essentially no energy is radiated outward or reflected back to its source.

dummy tube An electron-tube substitute that can be plugged into a socket to make normal connections between socket terminals for alignment or test purposes, without providing any of the additional functions of the tube.

dump 1. To withdraw all power from a computer accidentally or intentionally. 2. In digital computer programming, to transfer all or part of the contents of one section of computer memory into another section.

dump check A computer check that usually consists of adding all the digits during dumping and verifying the sum when retransferring.

dunking sonar *Dipping sonar.*

duobinary frequency modulation Frequency modulation using three signaling frequencies instead of two for transmitting digital data. The center frequency represents binary 0, and the two outside frequencies represent 1. The code has some error-detecting capability because the 1 frequency always relates back to the frequency of the previous 1. Used in some modems for data transmission over wire lines.

duodecimal number system A number system using the equivalent of the decimal number 12 as a base.

duodiode An electron tube that has two diodes in the same envelope, with either a common cathode or separate cathodes. Also called double diode.

duodiode-pentode An electron tube that has two diodes and a pentode in the same envelope, generally with a common cathode.

duodiode-triode An electron tube that has two diodes and a triode in the same envelope, generally with a common cathode.

duoplasmatron An ion-beam source in which

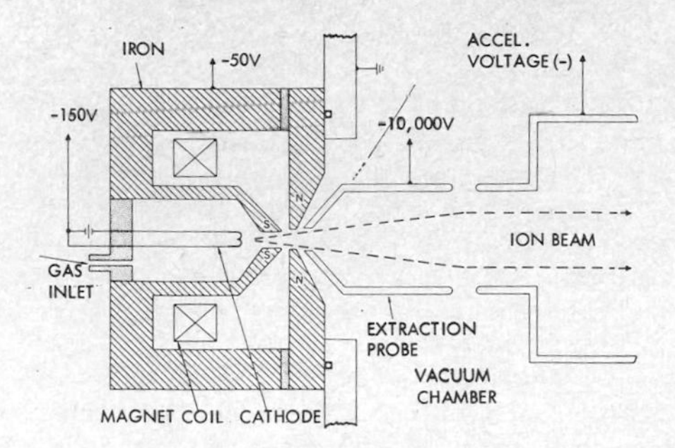

Duoplasmatron ion-beam source.

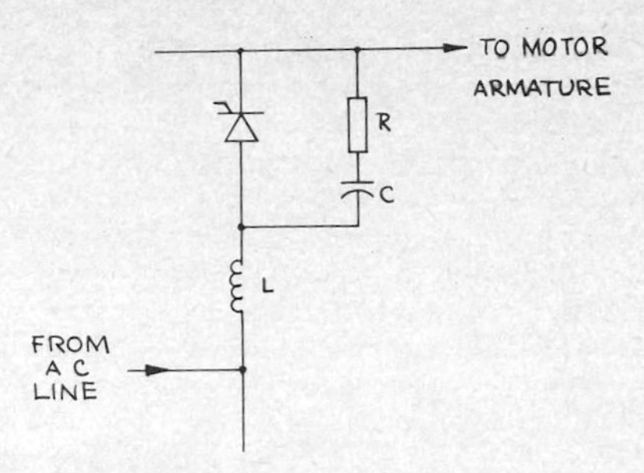

Dv/dt protection for thyristor in motor control circuit, using RLC network around thyristor.

electrons from a hot filament are accelerated sufficiently to ionize a gas by impact. The resulting positive ions are drawn out by high-voltage electrons and focused into a beam by electrostatic lens action. Used in mass spectrometers.

duotriode *Double triode.*

duplex *Duplex operation.*

duplex cable Two insulated stranded conductors twisted together. They may or may not have a common insulating covering.

duplex channel A communication channel providing simultaneous transmission in both directions.

duplexer A switching device used in radar to permit alternate use of the same antenna for both transmitting and receiving. It contains the TR tube that blocks out the receiver when the transmitter is operating. Other forms of duplexers serve for two-way radio communication using a single antenna at lower frequencies.

duplexing *Duplex operation.*

duplex operation The operation of associated transmitting and receiving apparatus concurrently, as in ordinary telephones, without manual switching between talking and listening periods. A separate frequency band is required for each direction of transmission. Also called duplex and duplexing.

duplication check A computer check which requires that the results of two independent performances (either concurrently on duplicate equipment or at a later time on the same equipment) of the same operation be identical.

dust core *Ferrite core.*

DUT Abbreviation for *device under test.*

duty cycle 1. The ratio of working time to total time for an intermittently operating device. Usually expressed as a percentage. 2. The ratio of pulse width to the interval between like portions of successive pulses. Usually expressed as a percentage.

DUV Abbreviation for *data under voice.*

***dv/dt* protection** Protection of a circuit or device from excessive voltage rate-of-change, usually achieved with a resistor-coil-capacitor network.

DVM Abbreviation for *digital voltmeter.*

DVOM Abbreviation for *digital volt-ohm-milliammeter.*

dwell A controlled time interval or delay during which a specified action occurs, such as the closing of contacts or the maximum lift position of a cam.

dwell meter An instrument for measuring the time that distributor breaker points are closed in an automobile ignition system. The reading is usually given in angular degrees through which the distributor cam rotates from the instant the contact points close until they open again.

DX Abbreviation for distance reception, used in connection with reception of, or communication with, distant radio stations.

dyadic operation Simultaneous operation on two operands in a computer.

dye laser A liquid laser in which the lasing material is an active organic fluorescent material dissolved in a solvent. One organic dye used for this purpose is anthracene. The dye can be excited by

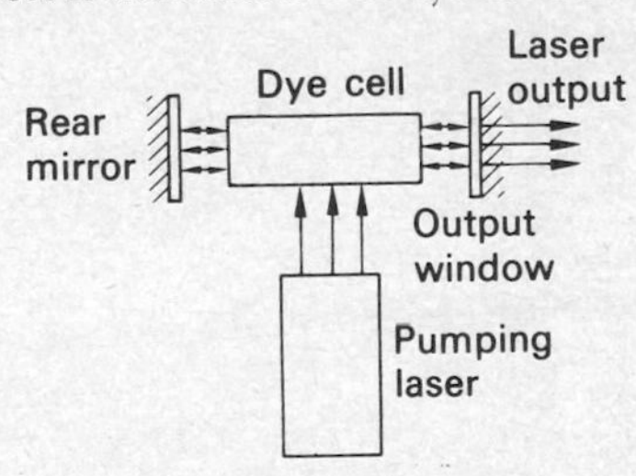

Dye laser using separate laser for pumping.

another laser, a flashlamp, or some other pulsed light source, to give either pulsed or continuous tunable output in the visible spectrum. Tuning is achieved by such means as rotating a diffraction grating or varying the optical path length of the lasing cell. The normal wavelength range is 0.3 to 1.2 μm.

dyn Abbreviation for *dyne.*

dynamic analogies Analogies that make it possible to convert the differential equations for mechanical and acoustic systems to equivalent electrical equations which can be represented by electric networks and solved by circuit theory.

dynamic braking The braking of an electric motor by using it as a generator that feeds energy to a heat-dissipating resistance load or back into the power system. Also called regenerative braking and resistance braking.

dynamic check A performance check made under operating conditions.

dynamic convergence The process whereby the locus of the point of convergence of electron beams in a color television or other multibeam cathode-ray tube is made to fall on a specified surface during scanning. Without dynamic convergence that varies with beam angle, the locus would be a spherical surface at a constant radius from the center of deflection of the beam.

dynamic demonstrator A large schematic circuit diagram that has been cemented to a board, with all components mounted near or on their symbols and connected together to give a working circuit of a radio, television receiver, or other electronic apparatus. Used for training purposes in classrooms and laboratories.

dynamic dump A computer dump that is performed periodically during the execution of a program.

dynamic focusing The process of varying the focusing electrode voltage for a color picture tube automatically so the electron-beam spots remain in focus as they sweep over the flat surface of the screen. Without dynamic focusing, part of the image would be out of focus at all times.

dynamic headphone A headphone in which a flexible miniature cone or diaphragm is attached to a voice coil positioned in the magnetic field of a

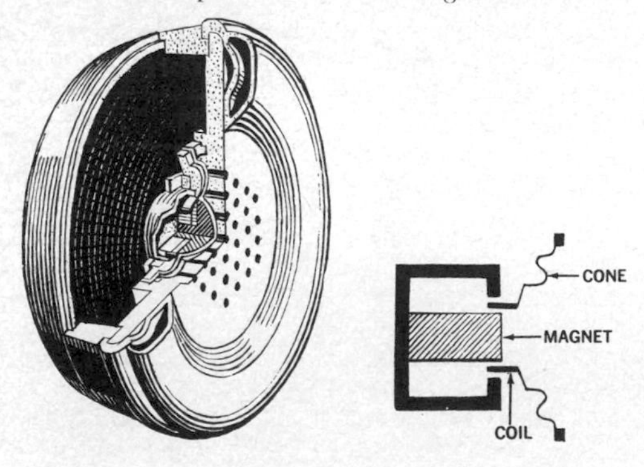

Dynamic-headphone construction.

permanent magnet. The operation is the same as for a dynamic loudspeaker.

dynamic logic A sampling-type logic in which the state of the input signal is sampled periodically by means of a clock pulse, and a corresponding output is provided in the form of a stored charge on a capacitor. Power dissipation is much lower than with conventional static logic because current is drawn only during the sampling period.

dynamic loudspeaker A loudspeaker in which the moving diaphragm is attached to a current-carrying voice coil that interacts with a constant magnetic field to give the in-and-out motion required for the production of sound waves. These waves correspond to the audio-frequency current flowing through the voice coil. In a permanent-magnet loudspeaker the constant magnetic field is produced by a permanent magnet, and in an excited-field loudspeaker it is produced by a field coil. Also called dynamic speaker and moving-coil loudspeaker.

dynamic memory *Dynamic storage.*

dynamic microphone A moving-conductor microphone in which the flexible diaphragm is attached to a coil positioned in the fixed magnetic

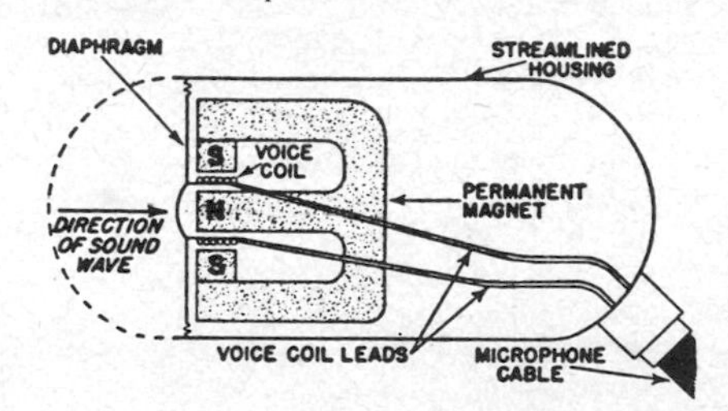

Dynamic-microphone construction.

field of a permanent magnet. When sound waves move the diaphragm back and forth, the attached voice coil cuts magnetic lines of force. The desired AF voltage is thus induced in the coil. Also called moving-coil microphone.

dynamic noise suppressor An AF filter circuit that automatically adjusts its bandpass limits according to signal level, generally by means of reactance tubes. At low signal levels, when noise becomes more noticeable, the circuit reduces the low-frequency response and sometimes also reduces the high-frequency response.

dynamic pickup A pickup in which the electric output is due to motion of a coil or conductor in a constant magnetic field. In a moving-coil phono pickup, the coil is moved by the needle that follows the grooves of a record. Also called dynamic reproducer and moving-coil pickup.

dynamic range The ratio of the specified maximum signal level capability of a system or component to its noise level. Usually expressed in decibels.

dynamic reproducer *Dynamic pickup.*

dynamic sensitivity The alternating component of phototube anode current divided by the alternating component of incident radiant flux.

dynamic speaker *Dynamic loudspeaker.*

dynamic storage Computer storage in which information at a certain position is not always available instantly because it is moving, as in an acoustic delay line or magnetic drum. Also called dynamic memory.

dynamic subroutine A subroutine that involves parameters, such as decimal-point position or item size, from which a subroutine is derived by the computer itself.

dynamo *Generator.*

dynamoelectric Pertaining to the conversion of mechanical energy to electric energy, or vice versa.

dynamoelectric amplifier A generator that serves as a power amplifier at low frequencies or direct current. The input signal is applied to the stationary field to change the excitation, and the amplified output is taken from the rotating armature.

dynamometer An electric generator or motor that measures the torque of a rotating shaft.

dynamometer multiplier A multiplier in which a fixed and a moving coil are arranged so the deflection of the moving coil is proportional to the product of the currents flowing in the coils.

dynamometer-type instrument An instrument in which current, voltage, or power is measured by the force between a fixed coil and a moving coil.

dynamotor A rotating electric machine having two or more windings on a single armature containing a commutator for DC operation and slip rings for AC operation. When one type of power is fed in for motor operation, the other type of power is delivered by generator action. Used chiefly as an inverter for converting DC to AC to operate electronic equipment from a storage battery in an automobile, boat, or other vehicle or location. Also called converter, rotary converter, and synchronous inverter.

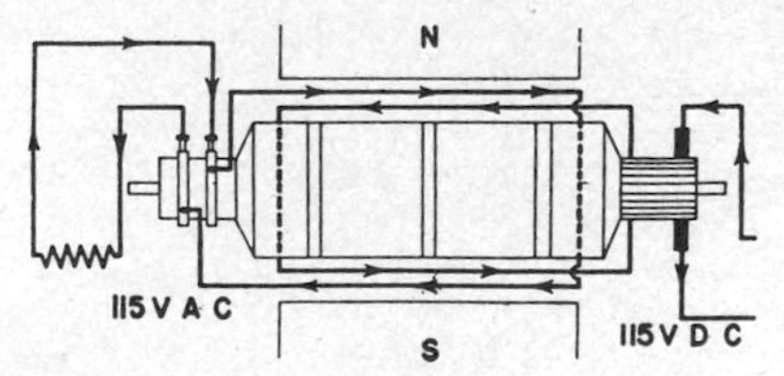

Dynamotor operating principle. Here DC input power drives motor section, and generator section feeds AC power to resistor load.

dynatron oscillator An oscillator in which secondary emission of electrons from the anode of a screen-grid tube causes the anode current to decrease as anode voltage is increased, giving the negative resistance characteristic required for oscillation. This action is achieved by making the screen grid more positive than the anode.

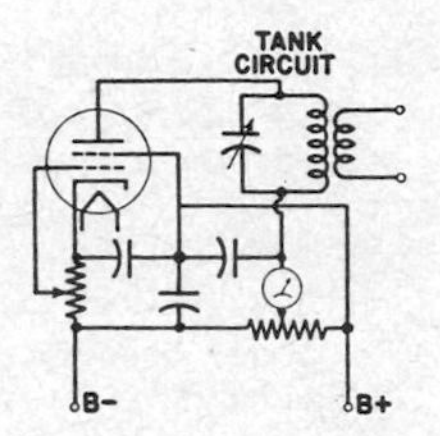

Dynatron-oscillator circuit.

dyne [abbreviated dyn] The CGS unit of force. One dyne is the force that gives an acceleration of 1 centimeter per second per second to a mass of 1 gram.

dynode An electrode whose primary function is secondary emission of electrons. Used in multiplier phototubes and some types of television camera tubes.

dysprosium [symbol Dy] A rare-earth element. Atomic number is 66.

Dzus fastener Trademark for a lock-type fastener used on airplane cowlings, in which the stud locks on a spring permanently attached to the mounting.

e 1. Symbol for the base of the system of natural or napierian logarithms, having the approximate value of 2.71828. 2. Symbol for the instantaneous value of an alternating voltage.

E 1. Symbol for *emitter.* Used on transistor circuit diagrams. 2. Abbreviation for *exa-.*

E 1. Symbol for *electric field strength.* 2. Symbol for *voltage.*

EAM Abbreviation for *electric accounting machine.*

earhanger A wire or plastic loop designed to hook over one ear and hold a single miniature earphone.

Early effect The reduction in the effective base width of a bipolar transistor when the width of the collector-base PN junction is increased by increasing the collector-base voltage. The effect was discovered by J. M. Early.

early-warning radar A long-range search radar used near the periphery of a defended area to detect approaching aircraft.

early-warning satellite A reconnaissance satellite used to detect enemy missile firings and other space activities early enough for appropriate countermeasures.

EAROM Abbreviation for *electrically alterable read-only memory.*

earphone A small, lightweight electroacoustic transducer that fits inside the ear, to function the same as a headphone or telephone receiver. Used chiefly with hearing aids.

earphone coupler A shaped cavity used for testing earphones. A microphone is mounted inside to measure pressures developed in the cavity.

earth British term for *ground.*

earth current *Telluric current.*

earthed British term for *grounded.*

earth-moon-earth [abbreviated EME] Long-distance radio communication achieved by using the moon as a reflector for radio signals. Amateur radio operators have used 144, 432, and 1296 MHz successfully for this purpose. Also called moonbounce.

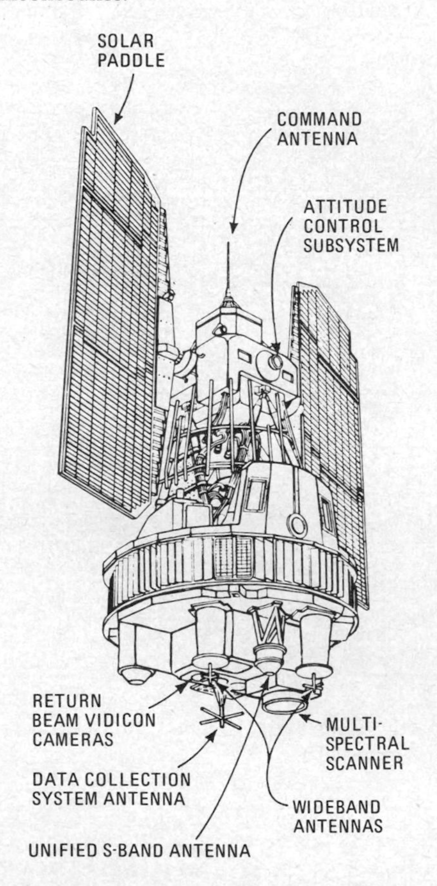

Earth resources technology satellite.

earth resources technology satellite [abbreviated ERTS] One of a series of satellites designed primarily to measure the natural resources of the earth. Functions include mapping, cataloging water resources, surveying crops and forests, tracing sources of water and air pollution, identifying soil and rock formations in search for new mineral sources, and acquiring oceanographic data for commercial fishermen.

earthshine Faint illumination of the dark side of the crescent moon, caused by reflection of sunlight from the earth.

earth-stabilized vehicle A space vehicle that is stabilized so one axis points toward the center of the earth.

EBAM Abbreviation for *electron-beam-accessed memory.*

EBCDIC [Extended Binary-Coded Decimal Interchange Code] A code that uses eight binary positions to represent a single character, giving a possible maximum of 256 characters. Used as the system code for many computers. The EBCDIC code is simply the BCD code extended to 8 binary bits.

EBICON [Electron Bombardment Induced CONductivity] A television camera tube that differs from orthicon and vidicon tubes chiefly in the construction of its target.

EBS Abbreviation for *electron-bombarded semiconductor.*

ebullator A cooling system in which a fluorocarbon cooling agent maintains a constant temperature through bubbling, despite severe changes in

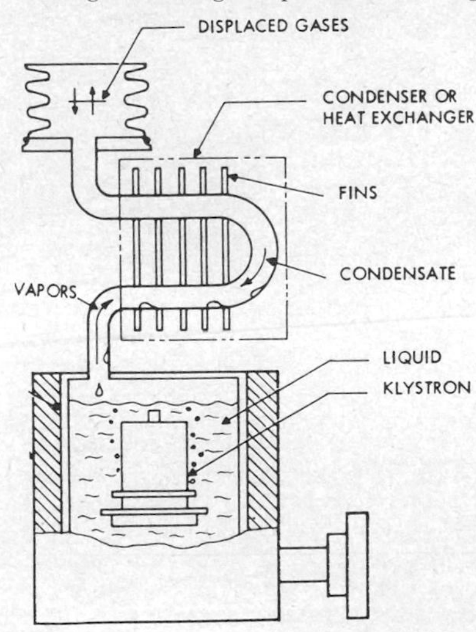

Ebullator used as enclosure for klystron of airport surveillance radar.

ambient external temperature. Used for enclosing a klystron oscillator when constant frequency is required.

eccentric circle *Eccentric groove.*

eccentric groove A locked groove whose center is different from that of a disk record, to provide an in-and-out motion of the pickup arm for actuating the trip mechanism of an automatic record changer at the end of the record. Also called eccentric circle.

Eccles-Jordan circuit *Flip-flop circuit.*

ECCM Abbreviation for *electronic counter-countermeasure.*

E cell A timing device that converts the current-time integral of an electrical function into an equivalent mass integral (or the converse operation) up to a maximum of several thousand mi-

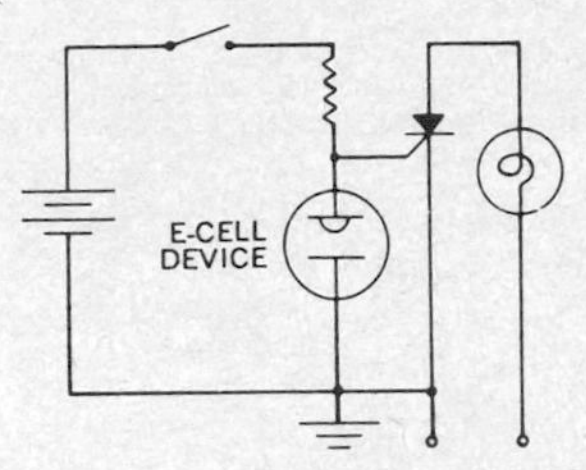

E-cell circuit.

croampere-hours. It consists essentially of a gold central electrode surrounded by an electrolyte, with the silver container serving as the second electrode. The action is that of a small integrating coulometer, in which metal is transferred through the electrolyte. Timing range is from seconds to months. Also called electrolytic cell.

ECG 1. Abbreviation for *electrocardiogram.* 2. Abbreviation for *electrocardiograph.*

echo 1. A wave that has been reflected or otherwise returned with sufficient delay and magnitude to be perceived in some manner as a wave distinct from that directly transmitted. 2. The signal reflected by a radar target, or the trace produced by this signal on the screen of the cathode-ray tube in a radar receiver. Also called radar echo and return. 3. A sound wave reflected from a hard surface. 4. *Ghost signal.*

echo area *Radar cross section.*

echo box A calibrated high-Q resonant cavity that stores part of the transmitted radar pulse

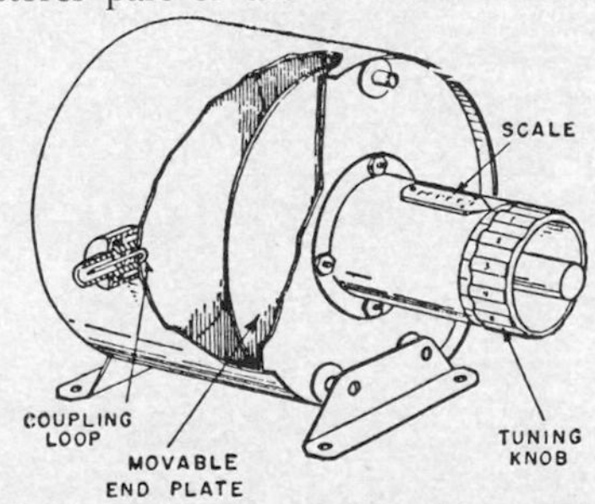

Echo box for coaxial cable feed from 10-cm radar.

power and gradually feeds this energy into the receiving system after completion of the pulse transmission. Used to provide an artificial target signal for test and tuning purposes. Also called phantom target.

echocardiograph An instrument that uses ultrasonic pulses and echo-ranging techniques to give a pictorial representation of the heart.

echo chamber A reverberant room or enclosure used in a studio to add echo effects to sounds for radio or television programs.

echo checking A checking system in which the transmitted information is reflected back to the transmitter and compared with that which was transmitted.

echo depth sounder *Fathometer.*

echoencephalograph An instrument that uses ultrasonic pulses and echo-ranging techniques to give a pictorial representation of intracranial structures. Also called sonoencephalograph (to prevent confusion with electroencephalograph).

echogram The recording or display obtained by using ultrasonic pulses and echo-ranging techniques, as in medical applications, ocean-bottom profile measurements, and geological exploration.

echoing area *Backscattering coefficient.*

echo matching Rotating a radar antenna or antenna array to a position at which the two echoes corresponding to the two directions of an echo-splitting radar are equal.

echoplex A method of checking for data-transmission errors by arranging for the computer to transmit its incoming data back to the sending terminal for printout and comparison with the originally transmitted data. Generally used only at startup or for periodic checks.

echo-ranging sonar Active sonar, in which underwater sound equipment generates bursts of ultrasonic sound and picks up echoes reflected from submarines, fish, and other objects within range, to determine both direction and distance to each target.

echorenograph An instrument that uses ultrasonic pulses and echo-ranging techniques to give a pictorial representation of the kidneys.

echo-repeater target An electronic target used to simulate a submarine for testing sonar-equipped homing torpedoes. It consists of two transducers lowered over the side of a small target boat to the desired test depth. One transducer picks up the sonar pulse from the torpedo. After amplification, this signal is fed to the other transducer to send out an echo approximating that which would be reflected from an actual submarine at that location.

echo-splitting radar Radar in which the echo is split by special circuits associated with the antenna lobe-switching mechanism, to give two echo indications on the screen of the radarscope. When the two echo indications are equal in height, the target bearing is read from a calibrated scale.

echo suppressor 1. A circuit that desensitizes electronic navigation equipment for a fixed period after the reception of one pulse, for the purpose of rejecting delayed pulses arriving from indirect reflection paths. 2. A relay or other device used on a transmission line to prevent a reflected wave from returning to the sending end of the line.

echouterograph An instrument that uses ultrasonic pulses and echo-ranging techniques to give a pictorial representation of the uterus.

ECL Abbreviation for *emitter-coupled logic.*

ECM Abbreviation for *electronic countermeasure.*

ECO Abbreviation for *electron-coupled oscillator.*

E core A transformer core made from E-shaped laminations used in conjunction with I-shaped laminations.

eddy current A circulating current induced in a conducting material by a varying magnetic field. These currents are undesirable in most instances because they represent loss of energy and cause heat. Laminations are used for the iron cores of transformers, filter chokes, and AC relays to shorten the paths for eddy currents and thus keep eddy-current losses at a minimum. At radio frequencies eddy-current paths must be broken up still more by using powdered iron cores.

eddy-current heating *Induction heating.*

eddy-current loss Energy loss caused by undesired eddy currents circulating in a magnetic core.

edgeboard connector *Card-edge connector.*

edge effect An outward-curving distortion of lines of force near the edges of two parallel metal plates that form a capacitor. A correction must be made for this effect when computing capacitance from the geometry of a structure, or a special guard ring must be used to eliminate the effect.

edge-lighted readout A digital display in which 10 or more transparent plastic plates, each with a different engraved numeral or other symbol, are arranged in depth behind the viewing window. Each engraved plate can be edge-lighted independently by its own miniature incandescent lamp. Light is reflected outward from the engraving onto the edge-lighted plate, causing that character to be visible.

edgewise bend A bend in a rectangular waveguide such that the longitudinal axis remains in a plane parallel to the wide side of the waveguide.

Edison base The standard screw-thread base used for ordinary electric lamps in the United States.

Edison distribution system A three-wire DC power distribution system that usually provides 120 V for lights and appliances and 240 V for ranges, large DC motors, and other high-power loads.

Edison effect The emission of electrons from hot bodies. The rate of emission increases rapidly with temperature. Discovered by Edison in 1883, when a current flow was obtained between the filament of an incandescent lamp and an auxiliary electrode inside the lamp. Also called Richardson effect.

Edison storage battery An alkaline storage battery that produces an open-circuit voltage of 1.2 V per cell. The active material on the negative plates is an iron alloy, whereas that on the positive plates is nickel oxide. Also called nickel-iron battery.

E display A rectangular radarscope display in which targets appear as bright spots, with target distance indicated by the horizontal coordinate and target elevation by the vertical coordinate.

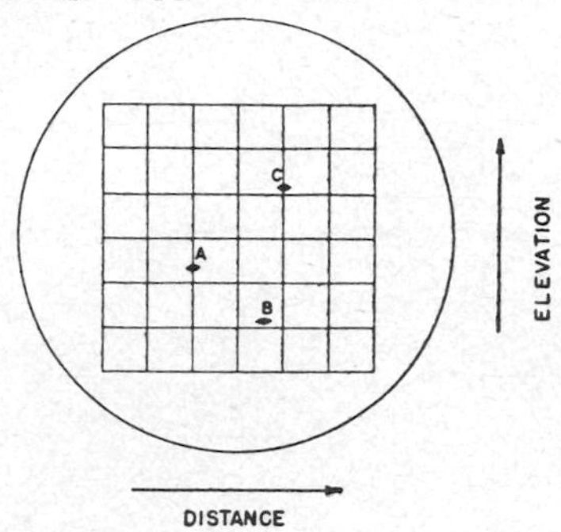

E display.

edit To arrange or rearrange digital computer output information before printing it out. Editing may involve deletion of unwanted data, selection of pertinent data, insertion of invariant symbols such as page numbers and typewriter characters, and the application of standard processes such as zero-suppression.

EDP Abbreviation for *electronic data processing.*

educational television [abbreviated ETV] Television used primarily for educational purposes, such as for broadcasting lectures from a master studio to receivers in satellite classrooms or to receivers in homes of those unable to attend schools.

EEG 1. Abbreviation for *electroencephalogram.* 2. Abbreviation for *electroencephalograph.*

EEP Abbreviation for *electroencephalophone.*

effective acoustic center The point on or near a loudspeaker or other acoustic generator from which spherically divergent sound waves appear to diverge.

effective antenna length The electrical length of an antenna, as distinguished from its physical length.

effective area A directional antenna rating, equal to the square of the wavelength multiplied by the power gain (or directive gain) of an antenna in a specified direction, and the result divided by 4π.

effective bandwidth The bandwidth of an assumed rectangular bandpass filter having the same transfer ratio at a reference frequency as a given actual bandpass filter, and passing the same mean-square value of a hypothetical current having even distribution of energy throughout that bandwidth.

effective confusion area The amount of chaff or other confusion reflector that gives a radar cross-sectional area equal to that of a particular aircraft at a particular frequency.

effective current The value of alternating current that will give the same heating effect as the corresponding value of direct current. The effective value is 0.707 times the peak value in the case of sine-wave alternating currents.

effective echoing area of target The area of a hypothetical perfect radar target, perpendicular to the incident beam, that would produce at the receiver a signal equal to that produced by the actual target. For an average aircraft it is generally from 1 to 10 m^2.

effective half-life The half-life of a radioisotope in a biological organism, resulting from a combination of radioactive decay and biological elimination.

effective height The height of the center of radiation of a transmitting antenna above the effective ground level.

effective isotropic radiated power [abbreviated EIRP] The product of the radiated RF power of a transmitter and the gain of the antenna system in a given direction relative to an isotropic (omnidirectional) radiator.

effective particle velocity The root-mean-square value of the instantaneous particle velocities at a point. Also called root-mean-square particle velocity.

effective perceived noise decibel [abbreviated epndB] A subjective noise unit that includes the effects of tone and duration. Used in Federal Aviation Administration noise-certification requirements for aircraft.

effective radiated power [abbreviated ERP] The product of antenna input power and antenna power gain, expressed in kilowatts.

effective radius of earth A radius value used in place of the geometric radius to correct for atmospheric refraction when the index of refraction in the atmosphere changes linearly with height. Under conditions of standard refraction the effective radius of the earth is 8500 km, or four-thirds the geometric radius.

effective resistance *High-frequency resistance.*

effective sound pressure The root-mean-square value of the instantaneous sound pressures at a point during a complete cycle, expressed in dynes per square centimeter. Also called pressure, root-mean-square sound pressure, and sound pressure.

effective thermal resistance The effective temperature rise per unit power dissipation of a designated junction of a semiconductor device, above the temperature of a stated external reference point, under conditions of thermal equilibrium.

effective value *Root-mean-square.*

effective wavelength The wavelength of a monochromatic x-ray that undergoes the same percentage attenuation in a specified filter as the heterogeneous x-ray beam under consideration.

efficiency 1. The ratio of useful output of a device to total input, generally expressed as a percentage. 2. The probability that a count will be produced in a counter tube by a specified particle or quantum incident.

EFL 1. Abbreviation for *emitter-follower logic.* 2. Abbreviation for *emitter-function logic.*

EFTS Abbreviation for *electronic funds transfer system.*

EHF Abbreviation for *extremely high frequency.*

EHT Abbreviation for *extra-high tension.*

E-H T junction A waveguide junction composed of a combination of E-plane and H-plane T junctions having a common point of intersection with the main waveguide.

E-H tuner An E-H T junction used for impedance transformation, having two arms terminated in adjustable plungers.

EHV Abbreviation for *extra-high voltage.*

EIA Abbreviation for *Electronic Industries Association.*

EIA color code One of the systems of color markings developed by the Electronic Industries Association for specifying electrical values and terminal connections of resistors, capacitors, and other components. Formerly called RETMA color code and RMA color code.

Eidophor system A large-screen projection television system in which a thin layer of oil is electrostatically distorted by an electron gun modulated by the television signal. Light from an external source is beamed through the oil film onto a screen that can be up to 40 ft (12 m) wide.

eight-level code A teletypewriter code that uses eight impulses, in addition to the start and stop impulses, to define a character.

Einstein–de Haas effect The rotation induced in a freely suspended ferromagnetic object when magnetization of the object is reversed.

einsteinium [symbol Es] A transuranic radioactive element produced artificially by bombardment of uranium or plutonium with nuclei. Atomic number is 99.

EIRP Abbreviation for *effective isotropic radiated power.*

EKG 1. Abbreviation for *electrocardiogram.* 2. Abbreviation for *electrocardiograph.*

elastance The reciprocal of capacitance, measured in darafs.

elastic collision A collision of nuclear particles in which the physical contents and energies of the colliding systems are unchanged, although the directions of their relative motions may be changed.

elastic scattering Scattering in which the kinetic energy of neutron plus nucleus is unchanged by the collision, and the nucleus itself is unchanged.

elastic wave *Acoustic wave.*

elastooptical effect The change that is produced in the index of refraction of a material by an internal strain resulting from either stationary or traveling elastic waves.

E layer A layer of ionized air occurring at various heights in the E region of the ionosphere, capable of bending radio waves back to earth. Average height is about 100 km. Also called Kennelly-Heaviside layer.

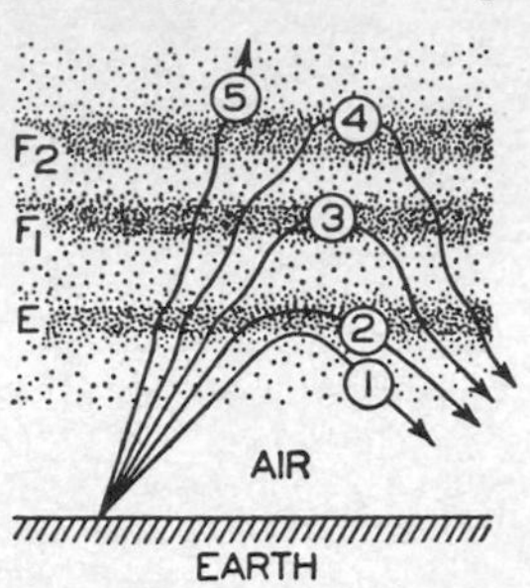

E layer, closest to earth, reflects signals along paths 1 and 2. F layers give reflection for paths 3 and 4. Signals along path 5 pass through all layers into outer space.

elbow *Corner.*

electra A continuous-wave radio navigation aid that uses special radio beacons to provide a number (usually 24) of equisignal zones. Electra is similar to sonne except that in sonne the equisignal zones as a group are periodically rotated in bearing.

electret A permanently polarized piece of dielectric material, produced by heating the material and placing it in a strong electric field during cooling. Some barium titanate ceramics, carnauba wax, and mixtures of certain other organic waxes can be polarized in this way. The electric field of an electret corresponds to the magnetic field of a permanent magnet.

electret microphone A capacitor microphone in which the diaphragm is a charged dielectric foil electret of Mylar or other plastic having a thin layer of gold on its upper surface. When sound waves move the diaphragm, the permanently

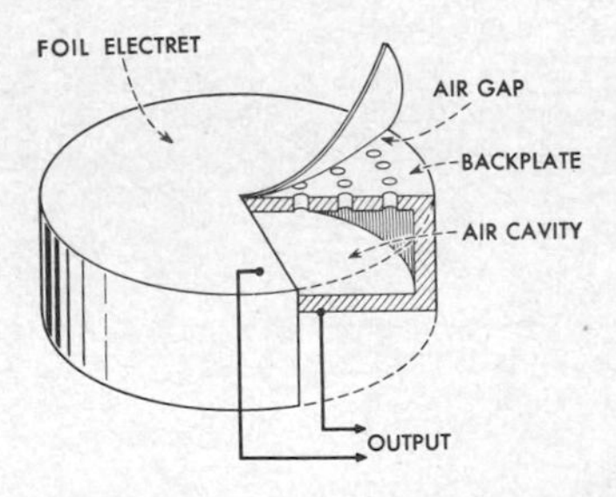

Electret-microphone construction.

stored static charges in the electret produce a correspondingly varying AF voltage between the output terminals.

electret recorder A recorder in which events lasting as little as 0.1 μs are stored as electrets on the surface of plastic foil by a puncture of a thin air gap. The foil is not damaged and can be erased after readout.

electric Containing, producing, arising from, or actuated by electricity. Examples are electric energy, electric lamp, and electric motor. Often used interchangeably with electrical.

electric accounting machine [abbreviated EAM] An electromechanical automatic data-processing machine, such as a keypunch, sorter, collator, tabulator, or other punched-card processing equipment.

electrical Related to or associated with electricity, but not containing it or having its properties or characteristics. Examples are electrical engineer, electrical handbook, and electrical rating. Often used interchangeably with electric.

electrical angle The angle that specifies a particular instant in an AC cycle. Usually expressed in degrees. One cycle is equal to 360°; hence a quarter-cycle is 90°. The phase difference between two alternating quantities is expressed as an electrical angle.

electrical axis The X axis in a quartz crystal. There are three in a crystal, each parallel to one pair of opposite sides of the hexagon. All pass through and are perpendicular to the optical or Z axis.

electrical boresight The tracking axis of a radar antenna or highly directional radio antenna, corresponding to the null of a conical-scanning antenna or the maximum of a directional antenna.

electrical center The point that divides a component into two equal electrical values.

electrical degree A unit equal to $^1\!/_{360}$ cycle of an alternating quantity.

electrical distance The distance between two points, expressed in terms of the duration of travel of an electromagnetic wave in free space between the two points. A convenient unit is the light-microsecond, which is approximately 983 ft or 300 m.

electrical engineer An engineer whose training includes a degree in electrical engineering from an accredited college or university (or comparable knowledge and experience), to prepare the person for dealing with the generation, transmission, and utilization of electric energy.

electrical engineering Engineering that deals with practical applications of electricity. Generally restricted to applications involving current flow through conductors, as in motors and generators.

electrical length The length of a conductor expressed in wavelengths, radians, or degrees. Distance in wavelengths is multiplied by 2π (6.2832) to give radians or by 360 to give degrees.

electrically alterable read-only memory [abbreviated EAROM] A read-only memory that can be reprogrammed electrically in the field any number of times, after the entire memory is erased by applying an appropriate electric field.

electrically tuned oscillator An oscillator whose frequency is determined by the value of a voltage, current, or power. Electric tuning includes electronic tuning, electrically activated thermal tuning, electromechanical tuning, and tuning methods in which the properties of the medium in a resonant cavity are changed by external electric means. An example is the tuning of a ferrite-filled cavity by changing an external magnetic field.

electrically variable coil An iron-core coil whose inductance can be varied over a wide range

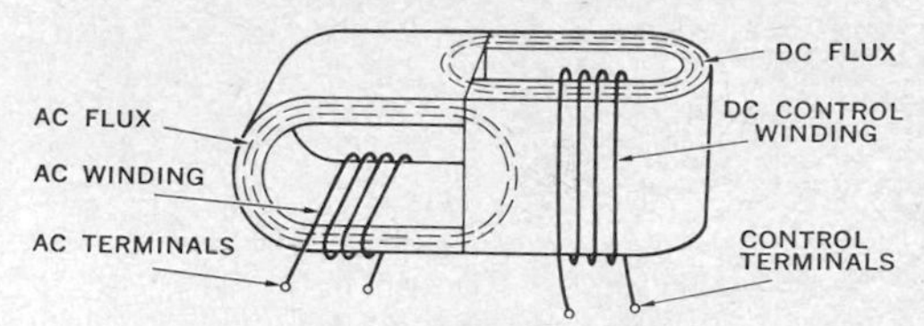

Electrically variable coil having two C cores rotated at 90° to each other, with DC control winding on one half and AC winding on other half.

by changing a small DC control current.

electrical transcription *Transcription.*

electrical zero A standard reference position from which rotor angles are measured in synchros and other rotating devices.

electric anesthesia Temporary anesthesia produced by sending an electric current through a part of the body. Also called electroanesthesia.

electric arc A discharge of electricity through a gas, normally characterized by a voltage drop approximately equal to the ionization potential of the gas. Also called arc.

electric bell A signaling device in which an electromagnet attracts an armature and causes an at-

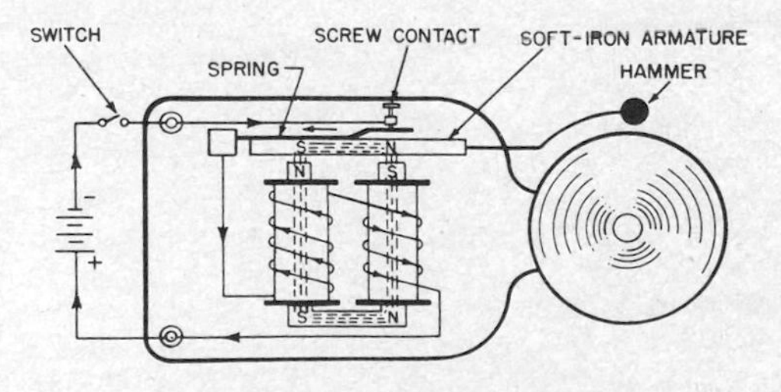

Electric-bell construction and operating circuit.

tached hammer to strike a bell. When the armature is pulled down, the contacts in the energizing circuit are opened, the armature is pulled back by a spring, and contact is reestablished for the start of the next operating cycle.

electric cell *Cell.*

electric charge *Charge.*

electric circuit A path or group of interconnected paths capable of carrying electric currents.

electric conduction The conduction of electricity by means of electrons, ionized atoms, ionized molecules, or semiconductor holes.

electric contact A physical contact that permits current flow between conducting parts.

electric control The control of a machine or device by switches, relays, or rheostats, as contrasted to electronic control by electron tubes or devices that do the work of electron tubes.

electric controller A device that governs in some predetermined manner the electric power delivered to apparatus.

electric coupling A rotating machine in which the torque is transmitted or controlled by electric or magnetic means.

electric delay line A delay line using properties of lumped or distributed capacitive and inductive elements. It can be used for storage by recirculating information-carrying wave patterns.

electric dipole A pair of equal and opposite charges an infinitesimal distance apart.

electric-discharge lamp A lamp in which light is produced by current flow through a gas or vapor in a sealed glass enclosure. Examples of such lamps include argon glow, mercury-vapor, neon glow, and sodium-vapor.

electric-discharge machining A metal-cutting process in which high-frequency discharges from a negatively charged metal tool remove metal from the work piece by electroerosion. There is no electrolyte, but the work is submerged in oil to flush away eroded particles and to delay each spark until peak energy is built up.

electric disintegration Removal of metal by an electric spark acting in air. Used chiefly where precise control is not required, such as for removing broken drills and taps.

electric displacement *Electric flux density.*

electric displacement density *Electric flux density.*

electric doublet *Dipole.*

electric dynamometer An electric generator or motor equipped with means for indicating torque.

electric energy The integral with respect to time of the instantaneous power input or power output of a circuit or device. The basic unit is the watthour.

electric eye Slang term for *photocell.*

electric field 1. The region around an electrically charged body in which other charged bodies are acted on by an attracting or repelling force. 2. The electric component of the electromagnetic field associated with radio waves and electrons in motion.

electric field strength [symbol E] The magnitude of the electric field vector.

electric field vector The force on a stationary positive charge per unit charge at a point in an electric field. Usually measured in volts per meter. Also called electric vector.

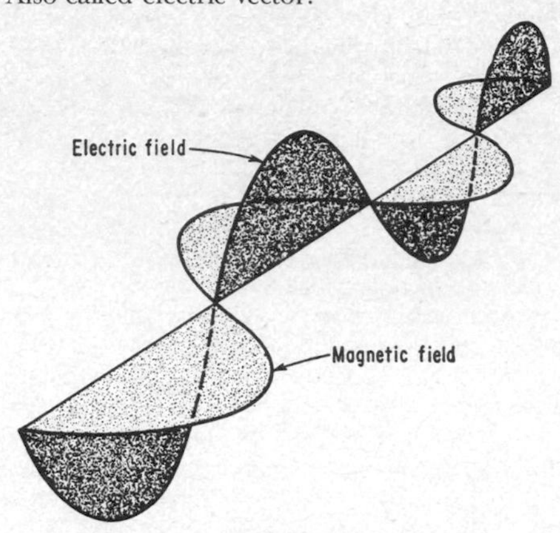

Electric field and magnetic field of electromagnetic wave.

electric fishing Fishing in which direct current is passed through water to make fish swim into nets or traps. Based on the fact that fish will swim parallel to the direction of current flow toward the anode. Also called electrofishing.

electric flux *Electric line of force.*

electric flux density A vector whose magnitude is equal to the charge per unit area that would appear on one face of a thin metal plate which is placed at a point in an electric field and oriented for maximum charge. The vector is perpendicular to the plate and directed from the negative to the positive face. Also called electric displacement and electric displacement density.

electric forming The process of applying electric energy to a semiconductor device to modify permanently its electrical characteristics.

electric guitar A guitar in which a contact microphone placed under the strings picks up the acoustic vibrations for amplification and reproduction by a loudspeaker. Volume and tone controls are usually provided.

electric hygrometer An instrument for indicating by electric means the humidity of the ambient atmosphere. Usually based on the relation between the electric conductance of a film of hygroscopic material and its moisture content.

electric image An array of electric charges, either stationary or moving, in which the density of charge is proportional at each point to the light values at corresponding points in an optical image to be reproduced.

electricity A fundamental quantity in nature, consisting of electrons and protons at rest or in motion. Electricity at rest has an electric field that possesses potential energy and can exert force. Electricity in motion (an electric current) has both electric and magnetic fields that possess potential energy and can exert force.

electric lamp A lamp in which light is produced by electricity. Examples include incandescent, arc, glow, mercury-vapor, and fluorescent lamps.
electric line of force An imaginary line, each segment of which represents the direction of the electric field at that point. Also called electric flux.
electric log The record or log of a borehole, obtained by lowering electrodes into the hole and measuring electrical properties of the rock formations traversed.
electric megaphone *Electronic megaphone.*
electric motor *Motor.*
electric noise Unwanted electric energy in a receiver or transmission system, other than crosstalk. Sources include electric appliances, electric motors, engine ignition, and power lines.
electric precipitation *Electrostatic precipitation.*
electric propulsion Propulsion of spacecraft and other vehicles by electrothermal, electrostatic, or

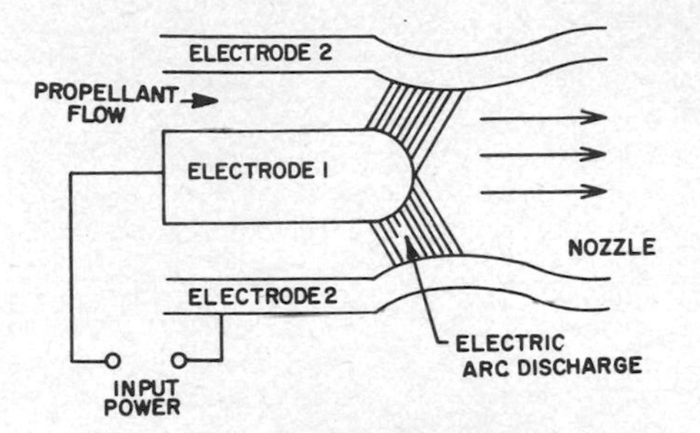

Electric propulsion using arc jet between two electrodes to heat and accelerate gaseous propellant.

plasma techniques, as contrasted to chemical propulsion, which involves direct use of fuel.
electric prospecting Prospecting for metals and minerals by measuring such fundamental electrical properties of rocks as resistivity, dielectric constant, and electrochemical activity.
electric reset relay A relay that remains in the picked-up condition after actuation until reset by applying an independent electric input.
electric rock fracturing The process of passing an electric arc through rock to lower the resistance of the puncture path, followed by high-frequency current that rapidly heats the rock and produces thermal stresses which cause fracturing. Used chiefly for secondary fragmentation in quarries.
electric scanning Scanning in which the required changes in radar-beam direction are produced by variations in phase and/or amplitude of the currents fed to the various elements of the antenna array.
electric sheet Special sheet iron or steel, generally alloyed with silicon, from which laminations are punched for transformers and other iron-core devices.
electric shield A housing, usually aluminum or copper, placed around a circuit to prevent interaction with other circuits by providing a low resistance and a reflecting path to ground for high-frequency radiation.
electric shock *Shock.*
electric shock therapy *Electroshock therapy.*
electric shock tube A gas-filled tube used in plasma physics to produce an ionized gas. A capacitor bank charged to a high voltage is discharged into the gas at one end of the tube. This ionizes and heats the gas, producing a shock wave that may be studied as it travels down the tube. The shock wave is preceded by an arc of hundreds of amperes, called a precursor arc.
electric signal A signal that makes use of electricity to convey information.
electric-signal storage tube A storage tube into which the information is introduced as an electric signal and later read as an electric signal.
electric storm 1. A meteorological disturbance in which the air is highly charged with electricity, occurring in fine weather, without clouds or rain, and often accompanied by dry, dusty winds. 2. A sudden change in the pattern of earth currents, causing interference with radio reception.
electric strength *Dielectric strength.*
electric tachometer An instrument used for measuring rotational speed by measuring the output voltage of a generator driven by the rotating shaft.
electric telegraph A ship telegraph in which the relationship of moving parts in the transmitter and receiver is maintained by self-synchronous motors or equivalent devices.
electric telemetering *Telemetering.*
electric thermometer An instrument that utilizes electric means to measure temperature, such as a thermocouple or resistance thermometer.
electric thruster A pulsed plasma generator that uses solid Teflon fuel to produce controlled thrust

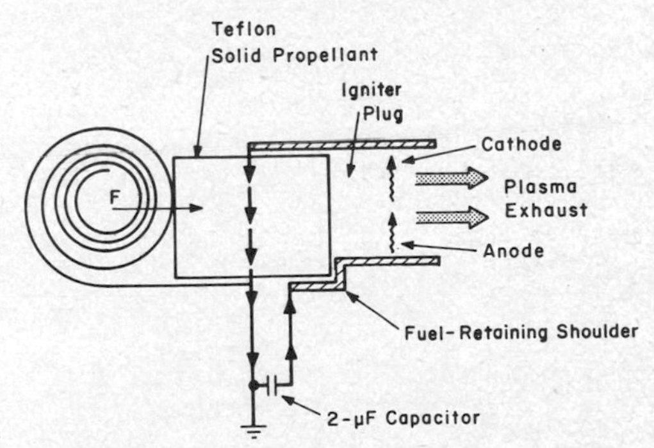

Electric thruster in which coil spring feeds fuel slug into discharge region as face of slug is consumed by electric discharge. Plasma thrust levels are typically from millionths to thousandths of a kilogram, for auxiliary propulsion of spacecraft.

as required for station keeping in some spinning synchronous communication satellites.

electric transducer A transducer in which all the waves concerned are electric.

electric tuning Tuning a receiver to a desired station by switching a set of preadjusted trimmer capacitors or coils into the tuning circuits.

electric twinning A defect occurring in natural quartz crystals, in which adjacent regions of quartz have their electric axes oppositely poled. Each type of axis is usable, but both cannot be in the same plate. During manufacture, the crystal is cut apart on the dividing line.

electric typewriter A typewriter having an electric motor that provides power for all operations initiated by the touching of keys. An electric typewriter is a serial printer, with the printing characters mounted on bars, balls, daisy wheels, or other devices. It may also be signal-controlled, as in some teletypewriters and computer terminals.

electric vector *Electric field vector.*

electric well logging Measurement of resistivity, electrochemical activity, and other characteristics of formations in the earth while coils, electrodes, or other appropriate sensors are being lowered into a borehole.

electric wind *Convective discharge.*

electrification The process of establishing a charge in an object.

electroacoustic Pertaining to a device involving both electricity and acoustics, such as a loudspeaker or microphone.

electroacoustic effect *Acoustoelectric effect.*

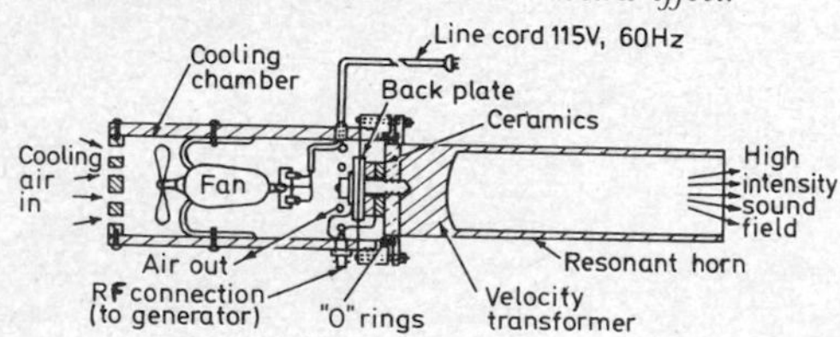

Electroacoustic transducer operating in frequency range from 10 to 100 kHz and producing sound intensity levels up to 160 dB at mouth of horn. Variable-frequency input from RF generator drives ceramic transducer.

electroacoustic transducer A transducer that receives waves from an electric system and delivers waves to an acoustic system, or vice versa.

electroanesthesia *Electric anesthesia.*

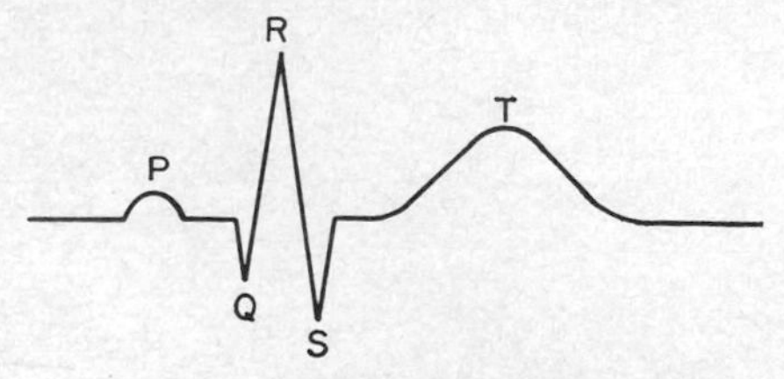

Electrocardiogram, showing typical tracing using electrodes on surface of body. Initial P wave, produced by electric activation of atrium, is about 90 ms in duration. Peaked QRS complex due to activation of ventricles lasts about 80 ms. Final slowly varying T wave is related to electric recovery of the ventricles.

electrocardiogram [abbreviated ECG or EKG] A record made by an electrocardiograph. Also called cardiogram.

electrocardiograph [abbreviated ECG or EKG] An instrument for recording the waveforms of voltages developed in the chest and lower parts of the human body in synchronism with action of the heart.

electrocautery A medical instrument used for cauterizing by using electrically generated heat to coagulate or destroy tissues.

electrochemical machining A metal-cutting process that is the reverse of electroplating. A low DC voltage is applied between the workpiece and a tool having the shape of the desired cut, and saltwater or some other electrolyte is pumped at high pressure through the gap between workpiece and tool. Electrochemical action in the gap erodes metal from the workpiece.

electrochemical recording Recording by means of a chemical reaction brought about by the passage of signal-controlled current through the sensitized portion of the record sheet.

electrochromic display A solid-state passive display that uses organic or inorganic insulating solids which change color when injected with positive or negative charges. Suitable materials include tungsten oxide or strontium titanate doped with such materials as molybdenum and iron. In most cases the image goes from colorless to dark blue and remains until intentionally erased or altered, even if power is removed. The process involves only a change in light-absorption properties, with no emission of light.

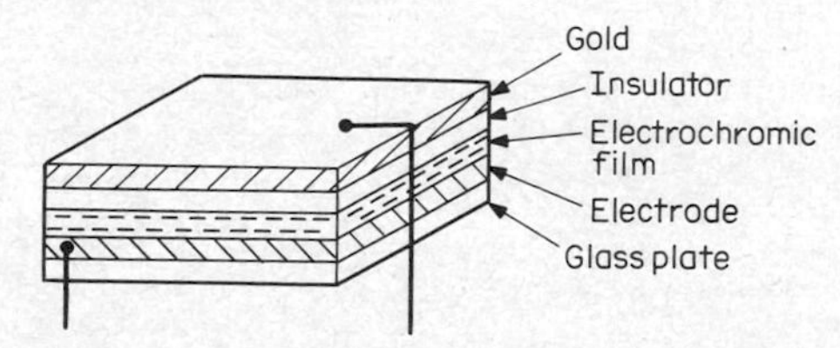

Electrochromic display in which electrochromic film turns deep blue when lower electrode is negative and becomes transparent again when polarity of DC supply voltage is reversed.

electrocoagulation The clotting, hardening, or destruction of tissue by applying high-frequency electric current.

electrocorticogram The record produced by an electrocorticograph.

electrocorticograph An instrument for recording electric activity of the brain, using electrodes placed directly on the external gray layer of the brain. This recording gives greater accuracy and more exact localization than does an electroencephalogram.

electrode 1. A conducting element that performs one or more of the functions of emitting, collecting, or controlling the movements of electrons or ions in an electron tube, or the movements of electrons or holes in a semiconductor device. 2. A terminal or surface at which electricity passes from one material or medium to another, as at the electrodes of a battery, electrolytic capacitor, or welder. 3. One of the terminals used in dielectric heating or diathermy for applying the high-frequency electric field to the material being heated.

electrodeless discharge A luminous discharge produced by a high-frequency electric field in a gas-filled glass tube having no internal electrodes.

electrodeposition The process of depositing a substance on an electrode by electroplating or electroforming. Also called electrolytic deposition.

electrode potential 1. The instantaneous voltage of an electrode with respect to the cathode of an electron tube. 2. The voltage existing between an electrode and the solution or electrolyte in which it is immersed.

electrode radiator A metal structure, often of large area, that is used as an external extension of an electrode of an electron tube to facilitate the dissipation of heat.

electrodermal reaction The change in electric resistance of the skin during emotional stress. This is one of the functional variables measured by polygraphs.

electrodesiccation The destruction of tissue by electric sparks generated at the tip of a small movable electrode.

electrode voltage The voltage between an electrode and the cathode or a specified point on a filamentary cathode of an electron tube.

electrodiagnosis Diagnosis of disease by studying electric activity of parts of the body and responses to stimulation of electrically excitable tissues.

electrodynamic instrument An instrument that depends for its operation on the reaction between the current in one or more movable coils and the current in one or more fixed coils.

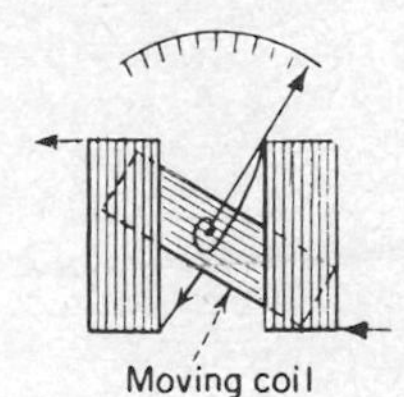

Electrodynamic instrument.

electrodynamic loudspeaker *Excited-field loudspeaker.*

electrodynamometer A two-coil measuring instrument in which the torque resulting from the reaction between the two series-connected current-carrying coils is balanced by a spiral spring.

electroencephalogram [abbreviated EEG] The record produced by an electroencephalograph.

electroencephalograph [abbreviated EEG] An instrument for recording the waveforms of voltages developed in the brain, using electrodes applied to the scalp.

electroencephalophone [abbreviated EEP] An instrument that provides an audible presentation of brain waves. Newer versions provide binaural or stereophonic effects that convey information more effectively.

electroencephaloscope An instrument for displaying on a cathode-ray screen the waveforms of voltages generated by various sections of the brain.

Electrofax Trademark of RCA for a special zinc-oxide-coated paper used in an electrostatic reproduction process capable of creating and maintaining an image on the paper.

electrofishing *Electric fishing.*

electroflor A material that changes color when electrically activated but does not radiate light.

electrofluiddynamic converter A converter that transforms the dynamic energy of a gaseous fluid into electric energy by passing through an electrostatic field a gas which contains electrically charged particles. The action is similar to that of a Van de Graaff generator, but higher power density can be obtained.

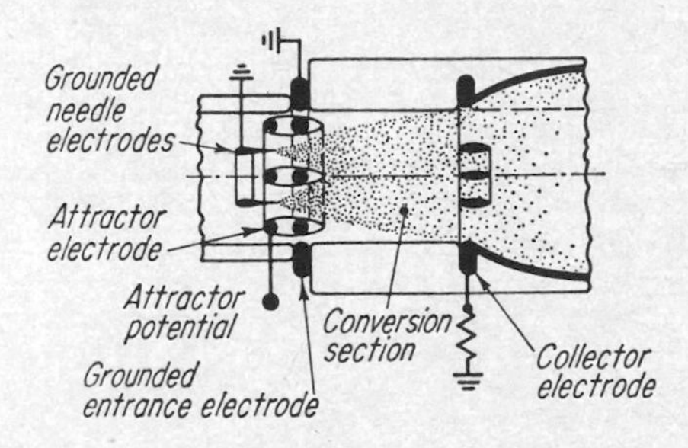

Electrofluiddynamic converter.

electroforming 1. The electrodeposition of metal on a conducting mold in sufficient thickness to make a desired metal object, such as a complex waveguide structure. The mold is often of graphite-coated wax so it can be removed by melting. 2. Production of a PN junction in a point-contact diode or transmitter by passing a large current pulse through the semiconductor material.

electrogasdynamics Conversion of the kinetic energy of a moving gas to electricity, for such applications as high-voltage electric power generation, air-pollution control, and paint spraying.

electrogasdynamic spray gun A portable electrostatic spray gun in which a convective discharge

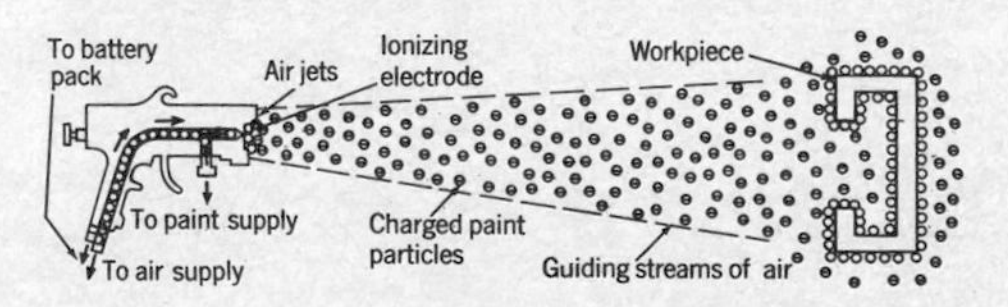

Electrogasdynamic spray gun.

or electric wind is used to charge particles of paint negatively in an electrogasdynamic channel. Guiding streams of air keep the charged paint particles in a narrow beam that minimizes overspray.

electrographic pencil A pencil used to make a conductive mark on paper, for detection by a conductive-mark sensing device.

electrographic recording Electrography in which the electrostatic image is formed by one or more rows of closely spaced parallel wires to which voltages are applied at appropriate instants to

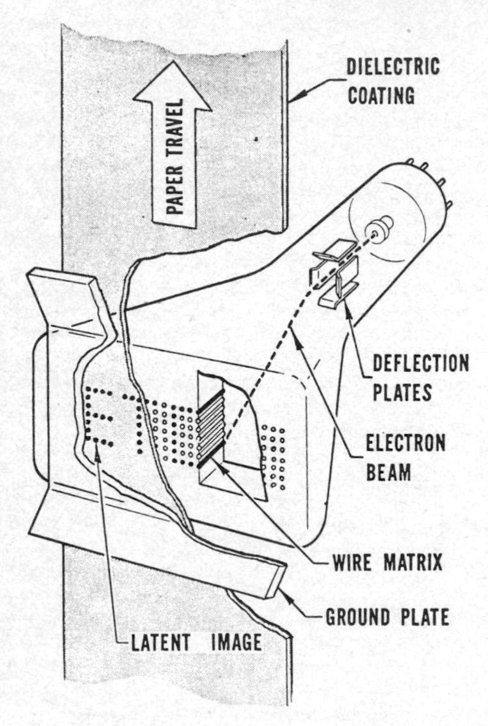

Electrographic recording in which electron beam applies voltages to fine parallel wires fused into face of cathode-ray tube.

form the desired image charge pattern. Widely used for printing magazine mailing labels at high speed on rolls of paper, as in the A. B. Dick Videograph.

electrography The branch of electrostatography in which electrostatic images are formed on an insulating medium without the aid of electromagnetic radiation. It includes xeroprinting, where the charged image is permanent, as required for repetitive printing, and electrographic recording, in which the charged image is formed by electric means.

electrojet A stream of electricity moving in the upper atmosphere around the equator and also in polar regions, where it produces auroras.

electrokymograph An instrument that provides a continuous recording of the movements of an internal organ such as the heart, generally by recording the movements or the changes in density of the shadow of the organ as presented on a fluoroscope.

electroless deposition Chemical deposition of a metal on a material, without electrolytic or electroplating action.

electroluminescence Luminescence produced in a gas by electric discharge, or in a solid by an

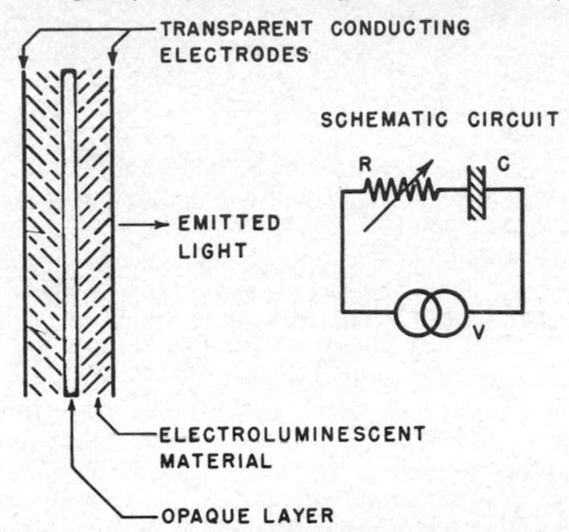

Electroluminescence principle. Incident radiation arriving from left passes through transparent electrodes to act on photoconductor (shaded area).

electric field, electron beam, x-rays, or other radiation.

electroluminescent display A digital display in which a phosphor layer is sandwiched between transparent conductive segments that form characters by electroluminescence when energized by

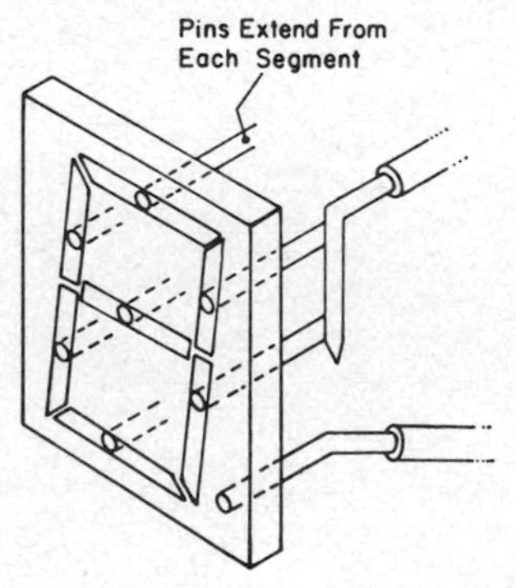

Electroluminescent display.

either AC or DC voltages. The phosphor most commonly used is zinc sulfide doped with manganese; this emits yellow light, but colors from red to green can be obtained by filtering.

electroluminescent display screen An electroluminescent device having an extra semiconductor control layer that permits storing of images

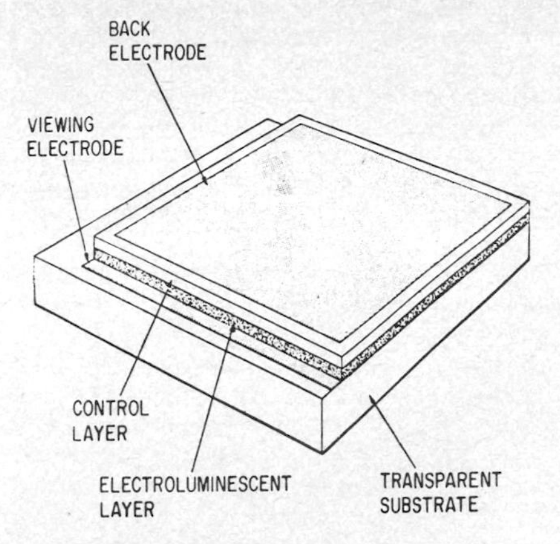

Electroluminescent display screen having additional control layer that is sensitive to ultraviolet light. When image is applied to semitransmissive back electrode with ultraviolet light, visible image lasting up to 2 h can be seen through transparent substrate on other side.

for controllable periods of at least 1 h. The desired image is projected onto the screen with ultraviolet light, and is completely erased with infrared radiation.

electroluminescent panel A light source consisting of a suitable phosphor placed between sheet-metal electrodes (one of which is essentially transparent) separated by only a few thousandths of an inch (0.001 cm), with an AC voltage applied between the electrodes.

electrolysis The production of chemical changes by passing current from an electrode to an electrolyte, or vice versa, as in electroplating, electroforming, or electropolishing. Used also in separating isotopes, as in the concentration of deuterium (heavy water) by electrolysis of ordinary water.

electrolyte A liquid, paste, or other conducting medium in which the flow of electric current takes place by migration of ions.

electrolytic capacitor A capacitor consisting of two electrodes separated by an electrolyte. A dielectric film, usually a thin layer of gas, is formed on the surface of one electrode.

electrolytic cell 1. A cell consisting of electrodes separated by an electrolyte. It can be used to store electric energy for use on demand, as in a storage cell; to generate electric energy, as in a dry cell; or to produce a desired electrochemical reaction when electric energy is applied. 2. *E cell.*

electrolytic cleaning Alkaline cleaning during which a current is passed through the cleaning solution and the metal to be cleaned.

electrolytic deposition *Electrodeposition.*

electrolytic iron A pure iron that has excellent magnetic properties, produced by an electrolytic process.

electrolytic recording Electrochemical recording in which the chemical change is made possible by the presence of an electrolyte.

electrolytic rectifier A rectifier consisting of metal electrodes in an electrolyte, in which rectification of alternating current is accompanied by electrolytic action. A polarizing film formed on one of the electrodes permits current flow in one direction but not the other.

electrolytic switch A switch having two electrodes projecting into a chamber containing a precisely measured quantity of a conductive electrolyte, leaving an air bubble of predetermined width. When the switch is tilted from true horizontal, the bubble shifts position and changes the amount of electrolyte in contact with the electrodes, thereby changing the amount of current passed by the switch. Used as a leveling switch in gyro systems.

electrolytic tank A tank in which voltages are applied to an enlarged scale model of a tube electrode system immersed in a poorly conducting liquid. The equipotential lines between electrodes are traced with measuring probes, as an aid to electron-tube design.

electromagnet A magnet consisting of a coil wound around a soft iron or steel core. The core is strongly magnetized when current flows through the coil and almost completely demagnetized when the current is interrupted. Used for attracting a movable external iron object such as the armature of a relay. In a solenoid, the iron core itself is movable.

electromagnetic Pertaining to the combined electric and magnetic fields associated with radiation or movements of electrons or other charged particles through conductors or space.

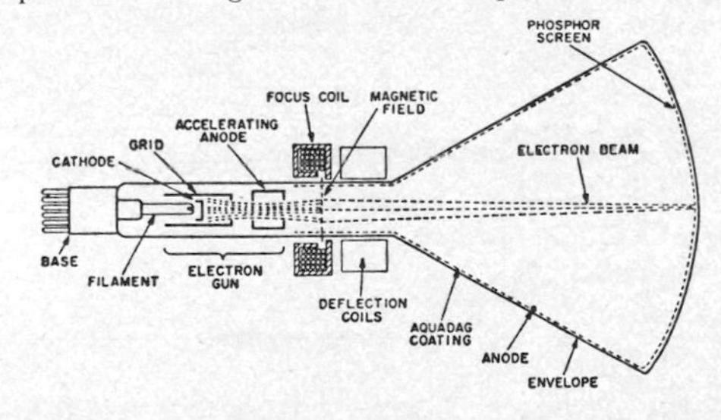

Electromagnetic cathode-ray tube with electromagnetic focus coil.

electromagnetic cathode-ray tube A cathode-ray tube in which electromagnetic deflection is used to deflect the electron beam.

electromagnetic compatibility [abbreviated EMC] The ability of electronic equipment to operate in its intended electromagnetic environment without causing or undergoing unacceptable degradation because of undesired electromagnetic radiation or response.

electromagnetic constant The speed of propagation of electromagnetic waves in a vacuum. Latest measurements, using microwave techniques, give a value of 299,793 km/s.

electromagnetic coupling Coupling that exists between circuits or conductors when they are mutually affected by the same electromagnetic field.

electromagnetic crack detector An instrument that detects cracks in iron or steel objects by applying a strong magnetizing force to the object and measuring the resulting magnetic flux through the object. When a flawed portion passes through the magnetizing coil, the magnetic flux drops.

electromagnetic deflection Deflection of an electron stream by means of a magnetic field. In a television picture tube, the magnetic fields for horizontal and vertical deflection of the electron beam are produced by sending sawtooth currents through coils in a deflection yoke that goes around the neck of the picture tube.

electromagnetic delay line A delay line consisting simply of a transmission line carrying pulse trains. The delay time generally available is not sufficient for storing a large number of pulses within a reasonable line length.

electromagnetic disturbance An electromagnetic phenomenon, usually impulsive, that is superimposed on a desired signal. The disturbance may be random or periodic.

electromagnetic energy Energy associated with radio waves, heat waves, light waves, x-rays, and other types of electromagnetic radiation.

electromagnetic environment The RF fields existing in a given area.

electromagnetic field The field associated with electromagnetic radiation, consisting of a moving electric field and moving magnetic field acting at right angles to each other and at right angles to their direction of motion.

electromagnetic flowmeter A flowmeter that offers no obstruction to liquid flow. Two coils produce an electromagnetic field in the conductive moving fluid. The current induced in the liquid, detected by two electrodes, is directly proportional to the rate of flow.

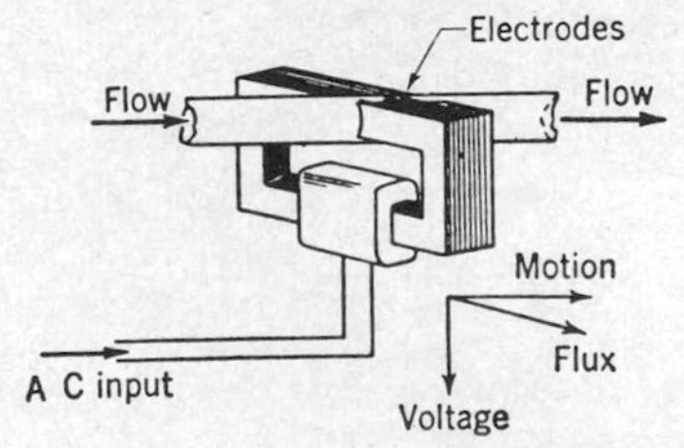

Electromagnetic flowmeter principle.

electromagnetic focusing Focusing the electron beam in a television picture tube by means of a magnetic field parallel to the beam, produced by sending an adjustable value of direct current through a focusing coil mounted on the neck of the tube.

electromagnetic forming *Magnetic forming.*

electromagnetic horn A horn-shaped antenna structure used to provide highly directional radiation characteristics. Signal power is fed to the horn by a waveguide or an exciting dipole or loop at the input end of the horn.

electromagnetic induction The production of a voltage in a coil by a change in the number of magnetic lines of force passing through the coil.

electromagnetic interference [abbreviated EMI] Any electromagnetic disturbance that interrupts, obstructs, or otherwise impairs the performance of electronic equipment. The interference may be induced intentionally, as in some forms of electronic warfare, or unintentionally by spurious emissions, intermodulation products, and other undesired effects.

electromagnetic lens An electron lens in which electron beams are focused by an electromagnetic field.

electromagnetic loudspeaker *Magnetic-armature loudspeaker.*

electromagnetic mirror A surface or region capable of reflecting radio waves, such as an ionized layer in the upper atmosphere.

electromagnetic mixing Mixing of molten alloys by exposing the melt to a strong magnetic field while passing direct current between electrodes at opposite ends of the crucible. Stirring action results from interaction of the magnetic field of the current-carrying molten alloy with the external transverse magnetic field.

electromagnetic noise An electromagnetic disturbance that is not sinusoidal.

electromagnetic plane wave A transverse electric wave, transverse electromagnetic wave, or transverse magnetic wave.

electromagnetic prospecting Prospecting for ore bodies by measuring electromagnetic waves.

electromagnetic pulse The pulse of electromagnetic radiation generated by a large thermonuclear explosion. Hardening of underground missile sites and control centers includes shielding to prevent the pulses from interfering with communication and electronic equipment.

electromagnetic pump A pump in which a conductive liquid is made to move through a pipe by sending a large current transversely through the liquid. This current reacts with a magnetic field that is at right angles to the pipe and current flow, to move the current-carrying liquid conductor just as a solid conductor is moved in an electric motor.

electromagnetic radiation Radiation associated with a periodically varying electric and magnetic field that is traveling at the speed of light, including radio waves, light waves, x-rays, and gamma radiation.

electromagnetic reconnaissance Reconnaissance for locating and identifying potentially hostile transmitters of electromagnetic radiation, including radar, communication, missile guidance, and navigation-aid equipment. Identification generally includes determination of frequency, type of modulation, pulse data, antenna characteristics, and bearing to the transmitter.

electromagnetic relay A relay in which current flow through a coil produces a magnetic field that results in contact actuation.

electromagnetic rocket engine *Plasma rocket engine.*

electromagnetic separation Separation of ions of varying mass by a combination of electric and magnetic fields. In the most common application, an isotopic mixture of ions is produced by either electron bombardment of a gas or thermionic emission. The ionized particles are accelerated and collimated into a beam by a system of electrodes, and the beam is projected into a magnetic field where the paths of the ions depend on their mass-to-charge ratio. Properly located collectors can be placed to receive ions of specified masses, as in the mass spectrograph.

electromagnetic spectrum The total range of wavelengths or frequencies of electromagnetic radiation, extending from the longest radio waves to the shortest known cosmic rays. Also called spectrum.

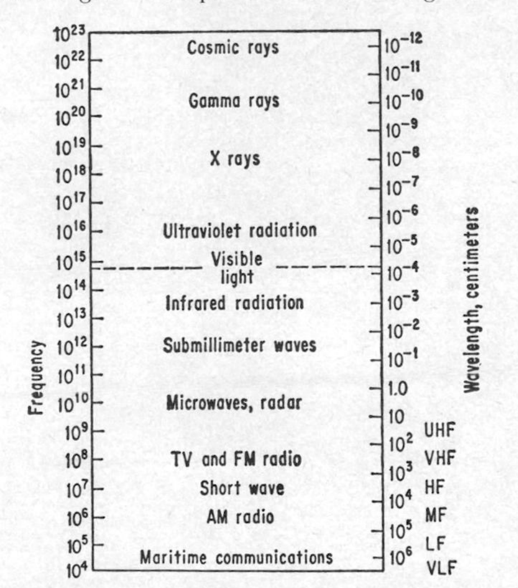

Electromagnetic spectrum. Scale at left gives frequency values in hertz.

electromagnetic survey An underground survey made by generating electromagnetic waves at the surface of the earth. The waves penetrate the earth and induce currents in conducting ore bodies, thereby generating new waves that are detected by instruments at the surface or by a receiving coil lowered into a borehole.

electromagnetic susceptibility The tolerance of circuits and components to all sources of interfering electromagnetic energy.

electromagnetic unit [abbreviated EMU] A CGS unit based on the assignment of unity to the strength of each of two like magnetic poles that repel each other with a force of 1 dyn at a distance of 1 cm in a vacuum.

electromagnetic wave A wave of electromagnetic radiation, characterized by variations of electric and magnetic fields.

electromagnetism Magnetism produced by an electric current rather than by a permanent magnet.

electromanometer An electronic instrument used for measuring pressure of gases or liquids.

electromechanical commutator A commutator that in its simplest form uses motor-driven rotating brushes to connect a number of data points in sequence to a common output line.

electromechanical recording Recording by means of a signal-actuated mechanical device.

electromechanical transducer A transducer for receiving waves from an electric system and delivering waves to a mechanical system, or vice versa.

electrometer An instrument for measuring voltage without drawing appreciable current. Older electrometer designs are based on the electrostatic force exerted between bodies that are charged with the voltage to be measured, such as suspended parallel strips of gold leaf. Modern vacuum-tube and solid-state electrometers can have an input resistance above 10^{14} Ω.

electrometer amplifier A low-noise amplifier having sufficiently low current drift and other characteristics required for measuring currents smaller than 10^{-12} A.

electrometer tube A high-vacuum electron tube that has a high input impedance (low control-electrode conductance) to facilitate measurement of extremely small direct currents or voltages.

electromigration The mass transport of metal by momentum exchange between thermally activated metal ions and conducting electrons. When electromigration occurs in transistors having aluminum or other metal interconnecting films, it may produce voids that cause failure.

electromotive force [abbreviated EMF] The force that tends to produce an electric current in a circuit. Usually called voltage.

electromotive series An arrangement of the metal elements in the order of the amount of electromotive force (voltage) set up between metal and solution when the metal is placed in a normal solution of any of its salts. Each metal is negative to those ahead of it in the list, and positive to those following it. The series for the more common metals is: sodium, magnesium, aluminum, manganese, zinc, chromium, iron, cadmium, cobalt, nickel, tin, lead, antimony, copper, silver, mercury, platinum, and gold.

electromyogram [abbreviated EMG] The record produced by an electromyograph.

electromyograph [abbreviated EMG] An instrument for measuring and recording voltages generated by muscles in the body.

electron An elementary negative particle having a mass of 9.1095×10^{-28} g and charge of 1.60219×10^{-19} C. It is the smallest electric charge that can exist. Also called negatron. An electron may also have a positive charge, in which case it is usually called a positron. The term electron was first used in an article by George J. Stoney in the July 1891 issue of *The Scientific Transactions of the Royal Dublin Society:* "On the Cause of Double Lines and of Equidistant Satellites in Spectra of Gases."

electronarcosis Unconsciousness, stupor, or arrested activity produced by sending an electric current through the body. Used for therapeutic purposes, such as anesthesia.

electron beam A narrow stream of electrons moving in the same direction, all having about the same velocity.

electron-beam-accessed memory [abbreviated EBAM] *Electron-beam memory.*

electron-beam drilling Drilling of tiny holes in a ferrite, semiconductor, or other material by using a sharply focused electron beam to melt and evaporate or sublimate the material in a vacuum. Used in making apertures in ferrite wafers or memory elements of digital computers.

Electron-beam drilling of ferrite wafers on mechanical stage platform in vacuum enclosure.

electron-beam laser A semiconductor laser in which the electron beam that provides pumping action in a thin plate of cadmium sulfide or other material is swept electrically in two dimensions by a deflection yoke, much as in a cathode-ray tube. The resulting laser output beam moves correspondingly, to provide high-speed scanning for data retrieval and imaging applications. The beam can be electronically blanked, unblanked, or modulated with analog video signals for the projection of pictures or other graphic data.

electron-beam lithography Lithography in which the radiation-sensitive film or resist is placed in the vacuum chamber of a scanning-beam electron microscope and exposed by an electron beam under digital computer control. Used to produce fine lines in very small areas, as required for the production of integrated circuits. After exposure, the film is removed from the vacuum chamber for conventional development and other production processes.

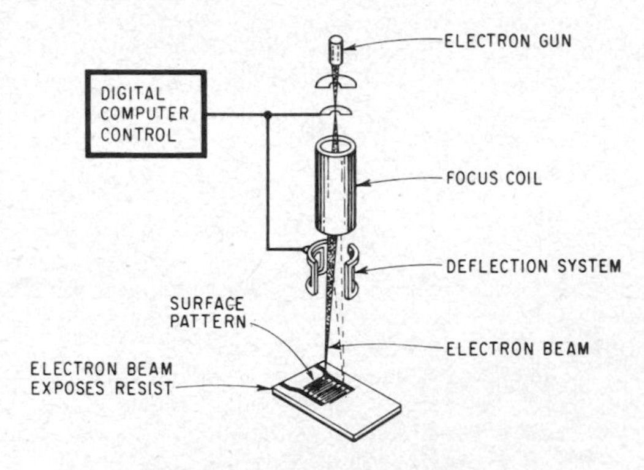

Electron-beam lithography setup.

electron-beam machining A machining process that takes place in a vacuum. Heat is produced by a focused electron beam at a sufficiently high temperature to volatilize and thereby remove metal in a desired manner. Drilling and cutting are examples of specific applications.

electron-beam magnetometer A magnetometer that depends for its operation on the change in intensity or direction of an electron beam that passes through the magnetic field to be measured.

electron-beam melting A melting process that takes place in a vacuum, heat being produced by a focused electron beam. Used principally for refining metals to a higher degree of purity than is possible with conventional vacuum-melting techniques. Chief advantage is the ability to control the temperature of the molten material and the time it remains molten, both of which affect the degree of volatilization of impurities.

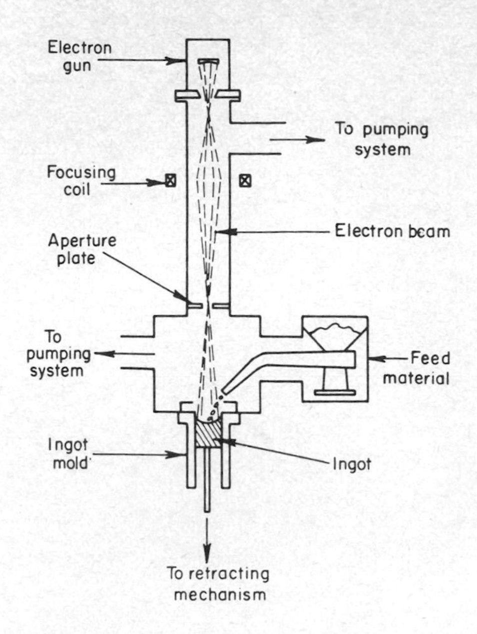

Electron-beam melting for purifying metal. During melting, impurities are volatilized and drawn off by vacuum pump.

electron-beam memory A memory that uses a high-resolution electron beam to store information on a target in a vacuum tube. The same beam is used to read out the data, which is stored in the form of electrostatic charges or some other reversible manner. Also called beam-accessed metal-oxide semiconductor (BEAMOS) and electron-beam-accessed memory (EBAM).

electron-beam parametric amplifier A parametric amplifier in which energy is pumped from an electrostatic field into a beam of electrons as the

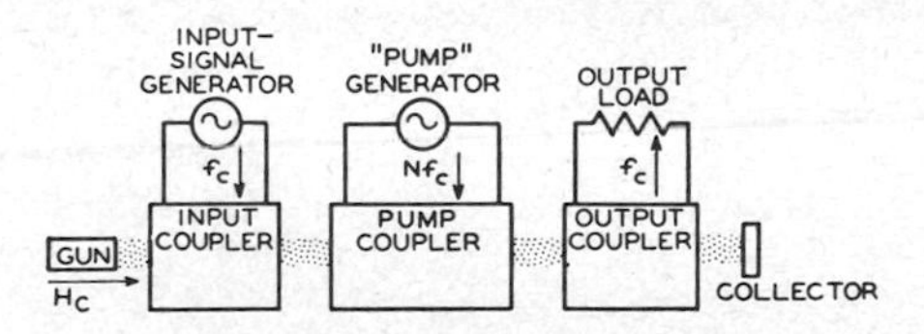

Electron-beam parametric amplifier operating diagram.

electrons travel down the length of the tube. The resulting outward-spiraling motion of the electrons is translated into an electric output by an electron coupler at the end of the tube. Another electron coupler at the beginning of the tube impresses the input signal on the beam.

electron-beam pumping The use of an electron beam to produce excitation for population inversion and lasing action in a semiconductor laser.

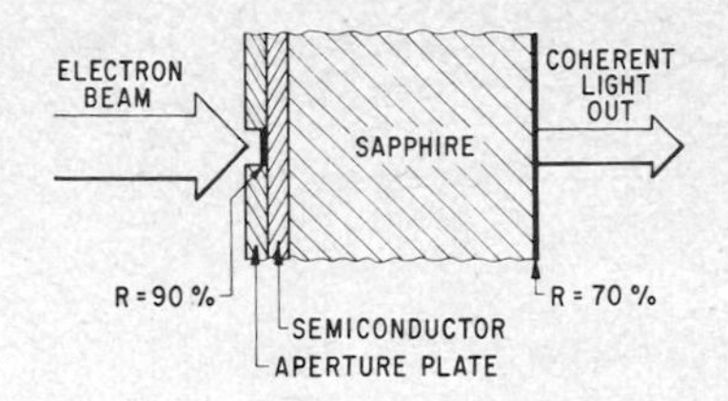

Electron-beam pumping of semiconductor laser.

electron-beam readout A technique for using an electron beam to read out data that has been stored on a scintillator-coated silver halide photographic film. The amount of light which reaches the output multiplier phototubes from a phosphor spot during a readout depends on the density of

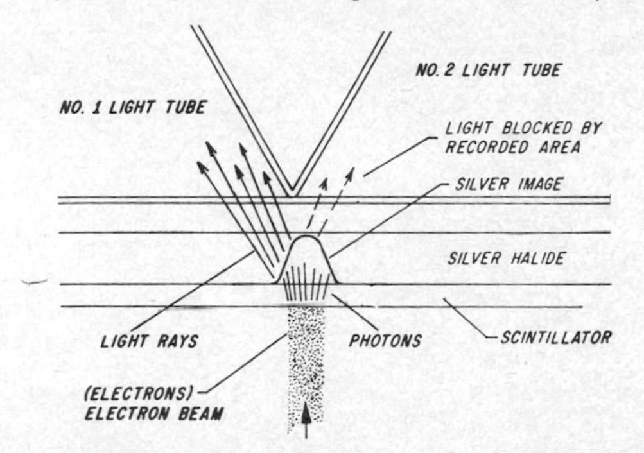

Electron-beam readout, showing how silver image blocks light rays from scintillator to No. 2 light tube.

silver deposited over that spot during recording. The electron beam is modulated by the input signal when recording on the film in a vacuum. The film is removed for conventional chemical developing, coated with a thin plastic scintillator film and an aluminum conductive layer, and then put back in the vacuum chamber for electron-beam readout.

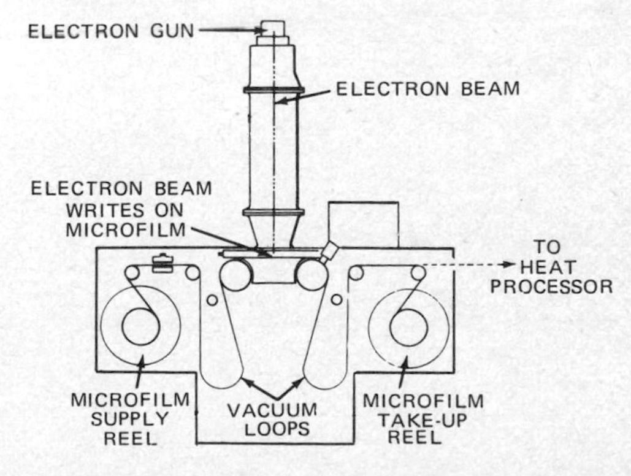

Electron-beam recorder as used to produce computer-output microfilm.

electron-beam recorder A recorder similar to early cathode-ray oscillographs, in which a moving electron beam is used to record signals or data on photographic or thermoplastic film in a vacuum chamber. Two applications are electronic video recording, and a computer-output microfilm recorder, in which an electron beam generates characters on microfilm in a vacuum chamber at speeds up to 20,000 lines per minute.

electron-beam tube An electron tube whose performance depends on the formation and control of one or more electron beams.

electron-beam welding A welding process that takes place in a vacuum. Heat is produced by a focused electron beam that can produce welds having depth-to-width ratios of up to 20 to 1. Applications include welding of thin metal foils to thicker metal without burning, sealing of metal cans containing uranium fuel elements for reactors, and direct fusion welding of ceramic objects.

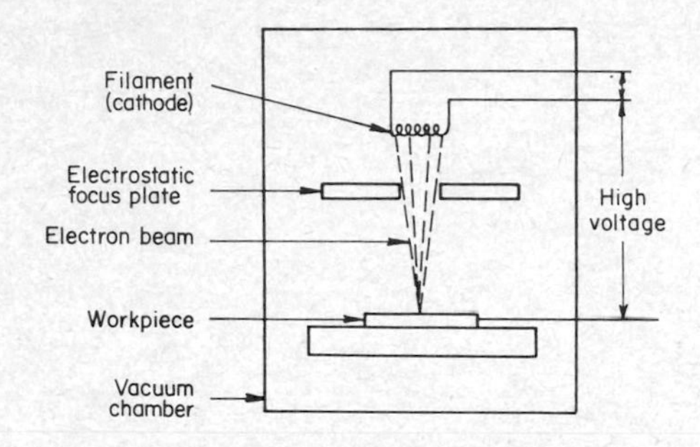

Electron-beam welding, using electrostatic focusing system.

electron binding energy *Ionization voltage.*

electron-bombarded semiconductor [abbreviated EBS] An electron-beam transistor device in which the modulated electron beam bombards a semiconductor diode serving as a target, generating hole-electron pairs in the depletion region of the diode. The resulting diode current variations develop an output pulse across an external load resistor. Current gain (the ratio of electron-beam current to diode current) can be greater than 2000. The control grid can be used for density modulation, or a traveling-wave structure can be used for deflection modulation. Applications include high-performance RF modulators and power amplifiers up to 10 GHz.

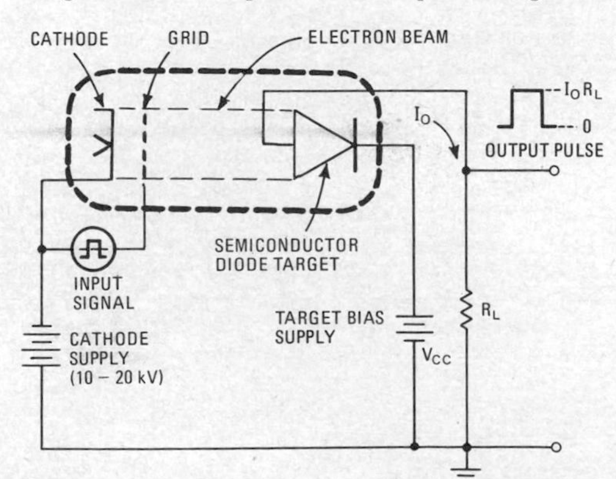

Electron-bombarded semiconductor using construction comparable to that of vacuum triode.

electron-bombardment engine An ion engine developed for deep-space flight, in which mercury propellant is vaporized in an electrically heated boiler. The resulting mercury vapor flows into an ionization chamber, where it is ionized by collisions with electrons emitted by a heated cathode. A high-voltage electric field drives the ions out at a high exhaust velocity to give propulsion. Outside the engine, the ions are neutralized by electrons from a heated tantalum filament to prevent the ion stream from being attracted back to the vehicle.

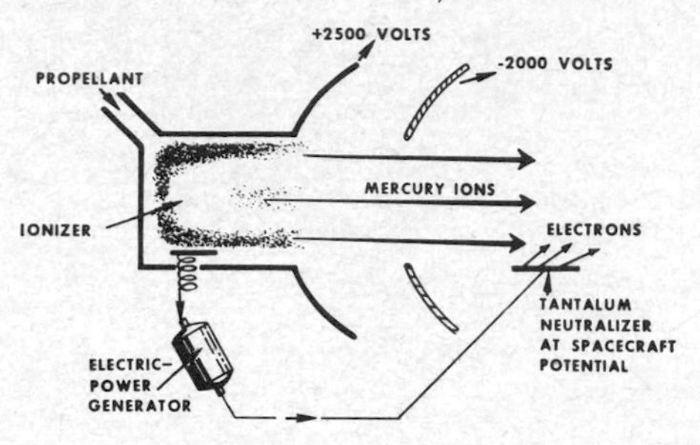

Electron-bombardment engine.

electron-bombardment-induced conductivity A method of writing and storing large numbers of information elements electrostatically on the storage tape of a television information storage tube. A dielectric-coated optical grating on the tape is bombarded with 10-keV electrons to induce momentary conductivity. This causes electrons to flow from the dielectric to the metal base of the tape. Elemental areas on the surface of the tape lose

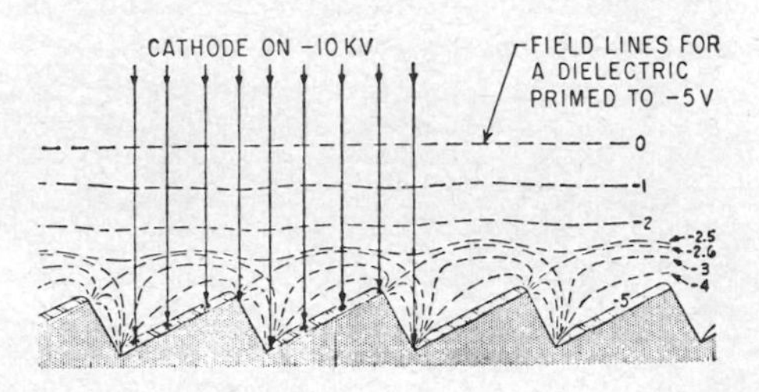

Electron-bombardment-induced conductivity as used to store television images on dielectric-coated metal optical grating having 6000 grooves per inch (2360 grooves per centimeter).

charge in proportion to light from corresponding elemental areas of the image being stored.

electron capture A radioactive transformation of a nuclide in which a bound electron merges with its nucleus. This decreases the atomic number by 1 but leaves the mass number unchanged in the new nuclide. A proton is transformed to a neutron within the nucleus, a bound electron is taken up, and a neutrino emerges. Examples are K-electron, L-electron, and M-electron capture.

electron collector *Collector.*

electron-coupled oscillator [abbreviated ECO] An oscillator employing a multigrid tube in which the cathode and two grids operate as an oscillator.

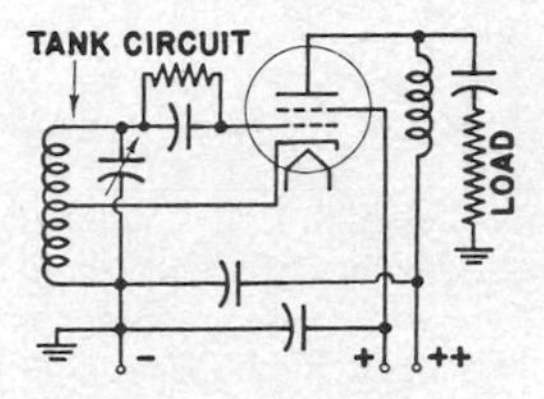

Electron-coupled oscillator circuit.

The anode-circuit load is coupled to the oscillator through the electron stream. Also called Dow oscillator.

electron coupler A microwave amplifier tube developed by C. L. Cuccia, in which electron bunching is produced by an electron beam projected parallel to a magnetic field while subjected

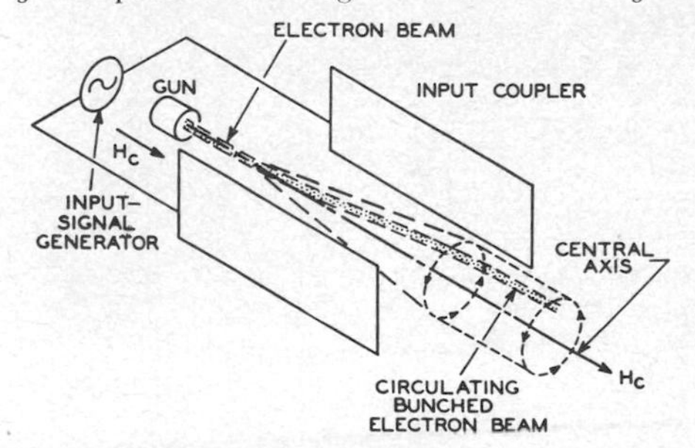

Electron coupler as used at input of electron-beam parametric amplifier.

to a transverse electric field produced by a signal generator. The electron beam traces the surface of a cone as it rotates about the central axis at the signal-generator frequency. A second coupler, known as the output coupler, is required to extract power from the spiraling motions of the electrons and deliver this power to an output load. Also called Cuccia coupler.

electron coupling A method of coupling two circuits inside an electron tube, used principally with multigrid tubes. The electron stream passing between electrodes in one circuit transfers energy to electrodes in the other circuit.

electron device A device in which conduction is principally by electrons moving through a vacuum, gas, or semiconductor, as in a crystal diode, electron tube, transistor, or selenium rectifier.

electron diffraction camera A camera used to obtain a photographic record of the position and intensity of the diffracted beams produced when a specimen is irradiated by a beam of electrons.

electronegative Having a negative electric polarity.

electron emission The liberation of electrons from an electrode into the surrounding space, usually under the influence of heat, light, or a high voltage. Also called emission.

electron flow A current produced by the movement of free electrons toward a positive terminal.

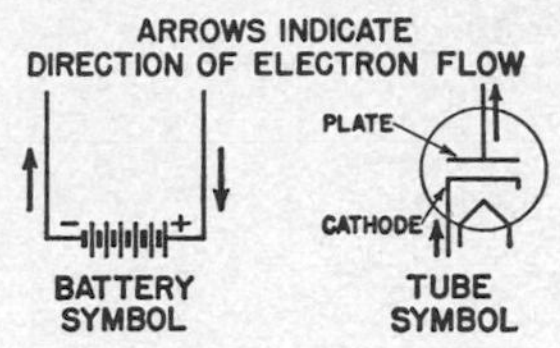

Electron flow in battery and tube.

The direction of electron flow is opposite to that of current.

electron fluence *Fluence.*

electron gun An electrode structure that produces and may control, focus, deflect, and con-

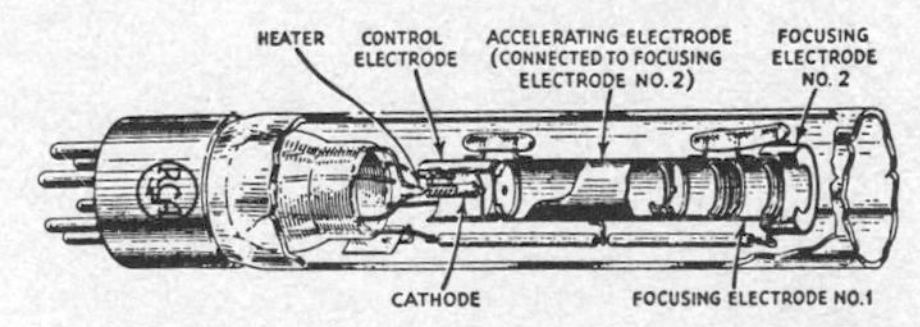

Electron gun of cathode-ray tube.

verge one or more electron beams in an electron tube. Also called gun.

electronic Pertaining to electron devices or to circuits or systems utilizing electron devices, including electron tubes, magnetic amplifiers, transistors, and other devices that do the work of electron tubes.

electronic air cleaner *Precipitator.*

electronically agile radar An airborne radar that uses a phased-array antenna which changes radar beam shapes and beam positions at electronic speeds. When combined with digital processing of the radar returns, it can simultaneously provide such functions as beacon-locating, forward-looking mapping, navigation updating, terrain avoidance, and terrain-following in manned strategic bombers and other aircraft.

electronically tuned oscillator An oscillator in which the operating frequency is changed by changing an electrode voltage or current.

electronic altimeter *Radio altimeter.*

electronic anesthesia Anesthesia produced by

sending through the brain a string of low-current pulses, usually supplemented by tranquilizers and sleep-inducing drugs. As developed in France, the 10-mA pulses are 0.003 s wide, with 0.01-s intervals between pulses.

electronic banking Banking carried out at computer terminals located in nonbank locations, such as retail stores. A customer inserts a magnetically encoded plastic card in the terminal, and then presses appropriate keys to make deposits or withdrawals, transfer money to pay bills, and even to borrow money. Also called electronic funds transfer system.

electronic Bohr magneton A unit of magnetic moment equal to 9.2732×10^{-21} erg·G^{-1}.

electronic bug A bug that automatically produces both dots and dashes having the correct lengths and spacings for international Morse code signals.

electronic calculator A calculator in which integrated circuits perform calculations and show results on a digital display. Most basic models provide all four arithmetic operations (+, −, ×, ÷), usually with floating decimal. More complex models include exponential and trigonometric functions, sometimes with a large choice of programs that can be inserted by changing a prerecorded magnetic program card. Larger desk models, operating from AC power, may provide hard copy along with the digital display. Eight-digit displays predominate in pocket-sized battery-operated models. The displays usually use either seven-segment light-emitting diodes or liquid crystals.

electronic camouflage Any method of reducing or eliminating the radar-echoing properties of a target.

electronic carillon A carillon that uses electric and electronic circuits to generate, amplify, and reproduce musical tones approximating those of bells.

electronic chimes A set of tubular chimes actuated by strikers electromagnetically controlled from a keyboard, with the resulting sounds being picked up, amplified, and reproduced by loudspeakers.

electronic circuit A circuit containing one or more electron tubes, transistors, magnetic amplifiers, or other devices providing comparable functions.

electronic commutator An electron-tube or transistor circuit that switches one circuit connection rapidly and in succession to many other circuits, without the wear and noise of mechanical switches. An example is the radial-beam tube, in which a rotating magnetic field causes an electron beam to sweep over one anode after another and produce the desired switching action.

electronic composition Typesetting in which characters are generated by electron or laser beams at speeds above about 6000 words per minute.

electronic control The control of a machine or process by circuits that use electron tubes, transistors, magnetic amplifiers, or other devices having comparable functions.

electronic controller An electric controller in which some or all of the basic functions are performed by electron devices.

electronic counter A circuit using electron tubes or equivalent devices for counting electric pulses.

electronic counter-countermeasure [abbreviated ECCM] Action taken to ensure effective use of electromagnetic radiations despite enemy countermeasures.

electronic countermeasure [abbreviated ECM] An offensive or defensive tactic using electronic and reflecting devices to reduce the military effectiveness of enemy equipment involving electromagnetic radiations.

electronic-countermeasure reconnaissance Reconnaissance by aircraft equipped with electronic devices capable of locating enemy radar stations, determining their area coverage, and making radarscope photographs for combat mission folders.

electronic curve tracer A photoelectric instrument in which a spot of light automatically traces along an inked line. It can be used to measure the area within a closed curve or to control a cutting torch for duplicating an irregular design.

electronic data processing [abbreviated EDP] Automatic data processing with a computer or other electronic equipment.

electronic deception An electronic countermeasure in which electromagnetic waves are radiated or reradiated in such a way as to mislead the enemy or decoy enemy missiles away from their targets.

electronic device A piece of equipment that uses circuits containing electron tubes, transistors, magnetic amplifiers, or other devices that do the work of electron tubes.

electronic differential analyzer A differential analyzer that integrates by means of high-gain feedback amplifiers.

electronic engineering Engineering that deals with practical applications of electronics.

electronic engraving A method of producing printing plates, wherein the original is photoelectrically scanned. The amplified scanning current controls an engraving tool that removes metal in proportion to white and dark areas of the original.

electronic fence An electronic barrier consisting of anti-intrusion devices and warning systems, installed across a demilitarized zone to detect violations.

electronic flash tube *Flash tube.*

electronic fuel injection The forced injection of fuel under pressure into an engine, under electronic control.

electronic funds transfer system [abbreviated EFTS] *Electronic banking.*

electronic fuze A fuze, such as the radio proximity fuze, that is set off by an electronic circuit incorporated in the fuze.

electronic game A self-contained version of a video game, having its own microprocessor-controlled screen or other type of display. Used in place of pinball machines in arcades and in place of slot machines and other gambling setups in casinos, as well as for entertainment in homes.

electronic generator A high-power oscillator used to generate RF energy for electronic heating. Also called electronic heater.

electronic heater *Electronic generator.*

electronic heating Heating by means of RF current produced by an electron-tube oscillator or an equivalent RF power source. The two types of electronic heating are induction heating for metals and dielectric heating for nonmetals. Also called high-frequency heating and RF heating.

electronic ignition An ignition system in which transistors or other electronic components are used instead of distributor points to control the firing of spark plugs.

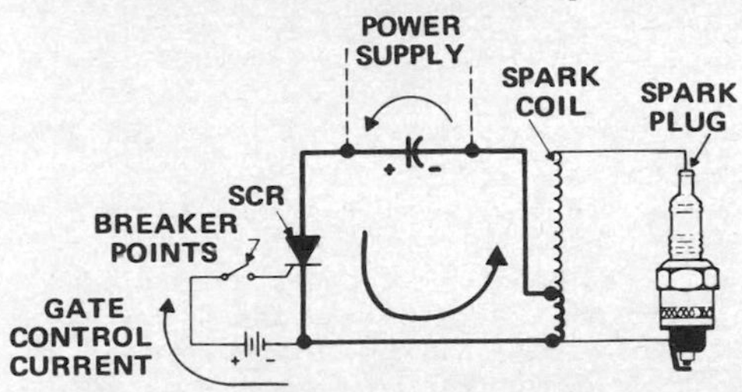

Electronic ignition using single silicon controlled rectifier (SCR) triggered by breaker points of distributor.

Electronic Industries Association [abbreviated EIA] A trade association made up chiefly of electronic component and equipment manufacturers. Its functions include standardization of sizes, specifications, and terminology for electronic products. Known as Radio Manufacturers Association (RMA) 1924–1950, Radio-Television Manufacturers Association (RTMA) 1950–1953, and Radio-Electronics-Television Manufacturers Association (RETMA) 1953–1957.

electronic intelligence [abbreviated ELINT] A worldwide Air Force network that uses fixed stations, specially equipped aircraft, and reconnaissance satellites to monitor and record enemy electromagnetic emissions. These signals are processed to give the nature and deployment of enemy warning and missile guidance radars, fire control, and countermeasures systems.

electronic interference An electric or electromagnetic disturbance that causes undesirable response in electronic equipment. Electric interference refers specifically to interference caused by the operation of electric apparatus that is not designed to radiate electromagnetic energy.

electronic jamming *Jamming.*

electronic keying Keying accomplished solely by electronic means.

electronic larynx An electronically actuated substitute for the human larynx, designed for persons who have lost the use of their vocal cords. A pulse generator feeds the entire spectrum of voice frequencies into the throat through either a tube inserted in the mouth or a small loudspeaker held against the throat. The resulting sound waves in the throat are formed into words by essentially normal movements of the jaws, lips, and tongue.

electronic locator *Metal detector.*

electronic lock A lock that uses a magnetically coded key. In one version, developed for hotels and motels, the lock code can be changed electronically from a central console as soon as a guest checks out, with simultaneous preparation of new coded keys.

electronic log The record or log of a borehole obtained by lowering a radiation counter into the hole and measuring the gamma-ray emissions of the rock formations traversed.

electronic megaphone A megaphone consisting of a microphone, audio amplifier, and horn loudspeaker built as a single unit. Also called electric megaphone.

electronic micrometer An electronic instrument for measuring and indicating small linear distances in air or across nonmetallic materials.

electronic microphone A microphone in which vibrations or sound waves act on one of the electrodes in an electron tube.

electronic missile acquisition [abbreviated EMA] A crossed-baseline interferometer system that operates on the DOVAP transponder frequency and gives azimuth and elevation angles.

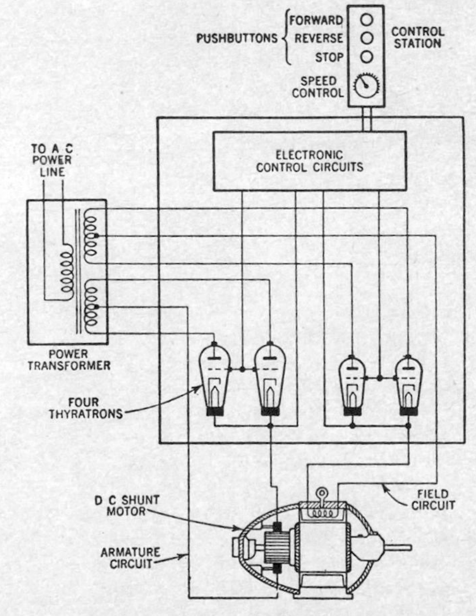

Electronic motor control using four thyratrons.

electronic motor control A control circuit used to vary the speed of a DC motor operated from an AC power line. Thyratrons or ignitrons are used to rectify the voltage and vary the field current of the motor. Also called DC motor control and motor control.

electronic multimeter A multimeter that uses semiconductor or electron-tube circuits to drive a conventional multiscale meter. When a digital display serves in place of the meter, it is called a digital multimeter.

electronic music Music consisting of tones originating in electronic sound and noise generators used alone or in conjunction with electroacoustic shaping means and sound-recording equipment. The resulting sounds may or may not resemble those of conventional musical instruments.

electronic musical instrument A musical instrument in which an audio signal is produced by a pickup or audio oscillator and amplified electronically to feed a loudspeaker, as in an electric guitar, electronic carillon, electronic organ, or electronic piano.

electronic navigation Navigation by means of electronic devices, including radio and radar equipment.

electronic organ A musical instrument which uses electronic circuits to produce music similar to that of a pipe organ.

electronic photometer *Photoelectric photometer.*

electronic piano A piano without a sounding board, in which vibrations of each string affect the capacitance of a capacitor microphone and thereby produce audio-frequency signals that are amplified and reproduced by a loudspeaker.

electronic pickup A phonograph pickup in which the output signal is produced by causing the needle to move an electrode of an electron tube.

electronic pilotage A method of pilotage in which landmark images appearing on a radarscope are used for guidance.

electronic position indicator [abbreviated EPI] A pulsed time-measuring system, similar to loran and shoran, that measures the time taken for a radio signal to travel from the ship to a shore station and return, and converts this time to distance. Arcs of distance can then be drawn on a hydrographic sheet for two or more shore stations, to determine position.

electronic potentiometer A potentiometer circuit that is continuously balanced by an electronic servosystem.

electronic profilometer An electronic instrument for measuring surface roughness. The stylus of a pickup is moved over the surface to be examined, and the resulting varying voltage is amplified, rectified, and measured with a meter calibrated to read directly in microinches of deviation from smoothness.

electronic reconnaissance The detection, location, identification, and evaluation of foreign electromagnetic radiations by means of electronic devices, as carried out by aircraft, drones, missiles, earth satellites, or fixed monitoring stations. It includes both radar reconnaissance and electronic-countermeasure reconnaissance.

electronic relay An electronic circuit that provides the function of a relay but has no moving parts. A solid-state relay is an example.

electronics 1. The branch of science and technology that deals with electron devices, including electron tubes, magnetic amplifiers, transistors, and other devices that do the work of electron tubes in controlling the flow of electricity in a vacuum, gas, liquid, semiconductor, conductor, or superconductor. 2. Pertaining to the field of electronics, as in electronics consultant, electronics course, electronics engineer, electronics laboratory, and electronics training.

electronic scanning 1. Scanning a television image with an electron beam in a cathode-ray television camera tube, as distinguished from mechanical scanning. 2. Scanning a region in space with a stationary multielement antenna so designed that the beam can be electronically aimed by changing the current feeds to the various elements.

electronic security Methods of preventing enemy personnel from obtaining useful information by intercepting radar and navigation signals.

electronics engineer An engineer whose training includes a degree in electronic engineering from an accredited college or university, a degree in electrical engineering with a major in electronics, or comparable knowledge and experience as required for working with electronic circuits and devices.

electronic sewing The use of dielectric heating for uniting thermoplastic sheet materials.

electronic sextant A sextant that uses a highly directional radio receiving system or a photoelectric system for determining position with respect to a selected member of the solar system.

electronics industry The industrial organizations engaged in the design, development, manufacture, and substantial assembly of electronic equipment, systems, assemblies, and components.

electronic skyscreen equipment [abbreviated ELSSE] Equipment used to track the telemetry-transmitted signal from a fired missile for range-safety purposes. The missile is destroyed if it deviates a predetermined amount from its intended course.

electronic speedometer A speedometer in which a transducer sends speed and distance pulses over wires to the speed and mileage indicators, eliminating the need for a mechanical link involving a flexible shaft.

electronics serviceman A serviceman who is qualified to repair and maintain electronic equipment.

electronics technician A technician who is qualified to work under the direction of an electronics engineer in assembling, testing, and repairing electronic equipment, generally in factories and laboratories.

electronic stimulator A pulse generator used to apply voltages to the body for activating muscles, identifying nerves, and other medical purposes.

electronic switch An electronic circuit used to perform the function of a high-speed switch. Applications include switching a cathode-ray oscilloscope back and forth between two inputs at such high speed that both input waveforms appear simultaneously on the screen.

electronic switching system A telephone switching system that uses a computer with a storage containing program-switching logic. The computer output actuates reed switches that set up telephone connections automatically.

electronic thermometer A thermometer in which a sensor, usually a thermistor, is placed on or near the object being measured. An oral version gives a reading in less than 20 s, as compared to 3 min or more with conventional glass-mercury thermometers.

electronic timer A timer that uses an electronic circuit to operate a relay at a predetermined interval of time after the circuit is energized, as in timing exposures for a photographic printer or in controlling an electronic generator.

electronic tonometer *Tonometer.*

electronic tube *Electron tube.*

electronic tuning Tuning of a transmitter, receiver, or other tuned equipment by changing a control voltage rather than by adjusting or switching components by hand.

electronic tuning range The frequency range of continuous tuning, between two operating points of specified minimum power output, for an electronically tuned oscillator.

electronic tuning sensitivity The rate of change in oscillator frequency with changes in electrode voltage or current for an electronically tuned oscillator.

electronic video recording [abbreviated EVR] The recording of black and white or color television visual signals on a reel of photographic film as coded black and white images that can be used for broadcasting, feeding cable television systems, or feeding individual entertainment or educational television receivers. Aural signals are recorded on magnetic stripes on the edges of the film.

electronic viewfinder A television camera viewfinder that uses a small cathode-ray picture tube to show the image being picked up.

electronic voltmeter A voltmeter that uses the rectifying and amplifying properties of electron devices and their associated circuits to secure desired characteristics, such as high input impedance, wide frequency range, and peak indications. Called a vacuum-tube voltmeter when the electron devices are vacuum tubes.

electronic warfare [abbreviated EW] Warfare directed at the electronic capabilities of the enemy, to detect and prevent hostile use of the electromagnetic spectrum. Electronic warfare includes electronic countermeasures and counter-countermeasures.

electronic watch A watch that uses a quartz crystal or a tuning fork with battery-powered electronic circuits to provide greater accuracy than is possible with conventional spring-type mechanical movements.

electronic wattmeter A wattmeter that uses two matched electronic voltmeters to give a reading proportional to the product of two voltages, on a scale calibrated to read power values directly. One voltage is that appearing across the load, and the other is obtained across a resistor in series with the line.

electron image 1. An image formed in a stream of electrons. The electron density in a cross section of the stream is at each point proportional to the brightness of the corresponding point in an optical image. 2. A pattern of electric charges on an insulating plate, with the magnitude of the charge at each point being proportional to the brightness of the corresponding point in an optical image.

electron injector The electron gun used to inject a beam of electrons into the vacuum chamber of a mass spectrometer, betatron, or other large electron accelerator.

electron lens An arrangement of electrodes, with or without magnetic focusing coils, used to control the size of a beam of electrons in an electron tube.

electron linear accelerator A linear accelerator used to accelerate electrons in a straight line up to energies in the range of 3 to 24 MeV. Used chiefly for industrial radiation and research applications.

electron microprobe An x-ray machine in which electrons emitted from a hot-filament source are accelerated electrostatically, then focused to an extremely small point on the surface of a specimen by an electromagnetic lens. Nondestructive analysis of the specimen can then be made by measuring the resulting backscattered electrons, the specimen current, the new x-ray radiation resulting from electron bombardment, the resulting fluorescence, or any other phenomenon resulting from such electron bombardment.

electron microscope A microscope in which a beam of electrons focused by an electron lens in a vacuum chamber is sent through a thin sample of the material being examined. The sample absorbs electrons in proportion to the density at each point, so the emerging beam is an electron image of the sample. This image is magnified thousands of times by another electron lens, then made visible on a fluorescent screen or recorded on photographic film.

electron mirror An electrode or other element that produces total reflection of an electron beam.

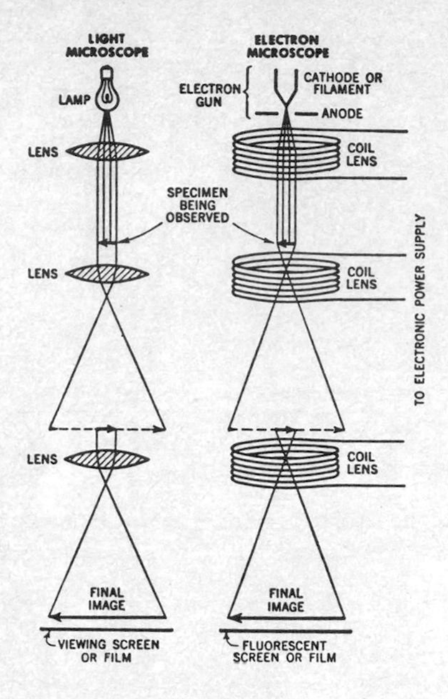

Electron microscope using electromagnetic lenses, as compared to ordinary optical microscope.

electron multiplier An electron-tube structure that employs secondary electron emission from solid reflecting electrodes (dynodes) to produce current amplification. The electron beam containing the desired signal current is reflected from each dynode surface in turn. At each reflection, an impinging electron releases two or more secondary electrons, so the beam builds up in strength. A typical arrangement of nine dynodes can give an amplification of several million. Used in multiplier phototubes and television camera tubes. Also called multiplier and secondary-electron multiplier.

electron-multiplier phototube *Multiplier phototube.*

electron-multiplier tube An electron tube that has an electron multiplier.

electron optics The branch of electronics that deals with the control of electron beams in a vacuum by means of electron lenses using electric or magnetic fields, or both.

electron paramagnetic resonance Paramagnetic resonance involving conduction electrons in a metal or semiconductor.

electron-positron pair The electron and positron simultaneously created by the process of pair production.

electron radius The classical value of 2.81777×10^{-13} cm for the radius of an electron. It is obtained by equating the rest-mass energy of the electron to its electrostatic self-energy.

electron rest mass A physical constant equal to 9.1091×10^{-28} g.

electron shell The arrangement of electrons in a given orbital outside the nucleus of an atom. All electrons in a shell have the same energy level.

electron spectrometer A spectrometer in which analysis of a substance is based on electron emission induced by x-rays.

electron spin The rotation of an electron about its own axis, contributing to the total angular momentum of the electron.

electron spin resonance Interaction of electric and magnetic fields with the spin of an electron about its own axis.

electron spin resonance spectrometer A spectrometer based on electron paramagnetic resonance.

electron synchrotron A synchrotron designed to accelerate electrons. The electron beam is allowed to strike an internal target, producing high-energy gamma rays that are used outside the machine.

electron telescope A telescope in which an infrared image of a distant object is focused on the photosensitive cathode of an image converter tube. The resulting electron image is enlarged by electron lenses and made visible by a fluorescent screen. An electron telescope can be used in complete darkness. The sniperscope and snooperscope are examples of early military versions.

electron trajectory The path of one electron in an electron tube.

electron tube An electron device in which conduction of electricity is provided by electrons moving through a vacuum or gaseous medium within a gastight envelope. A tube may provide rectification, amplification, modulation, demodulation, oscillation, limiting, and a variety of other functions. Examples include cathode-ray tubes, gas tubes,

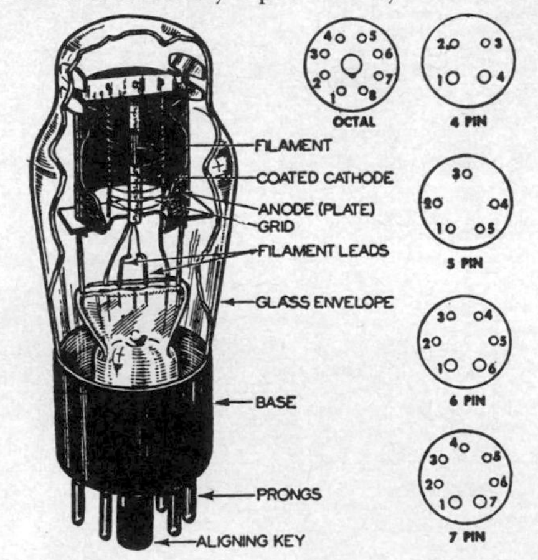

Electron-tube construction, showing typical heater-type triode with octal base, and bottom views of five different tube bases. On octal tubes, one or more base pins are sometimes omitted.

phototubes, and vacuum tubes. Also called electronic tube, radio tube, tube, and valve (British).

electron-tube amplifier An amplifier in which electron tubes provide the required increase in signal strength.

electron-tube coupler A coupler specifically designed to be inserted between an electron tube and an input or output device, as between a magnetron and a transmission line.

electron-tube generator A generator in which DC energy is converted to RF energy by an electron tube in an oscillator circuit.

electronvolt [abbreviated eV] A unit of energy equal to the energy acquired by an electron when it passes through a potential difference of 1 V in a vacuum. One electronvolt is equal to 1.60219×10^{-12} erg.

electron-wave tube An electron tube in which mutually interacting streams of electrons having different velocities cause a signal modulation to change progressively along the length of the electron streams.

electronystagmograph An instrument for recording eye movements that have been induced by electric stimulation.

electrooptical effect The effect wherein certain transparent dielectrics become doubly refracting when placed in an electric field.

electrooptical material A material that is capable of transforming electrical information into optical information or performing some optical function in response to an electric signal. One example is lead lanthanum zirconate titanate, a transparent ferroelectric ceramic whose optical properties can be changed by an electric field. In lasers, such materials can be used for beam deflection, beam modulation, and *Q* switching.

electrooptical modulator An optical modulator in which a Kerr cell, an electrooptical crystal, or other signal-controlled electrooptical device is

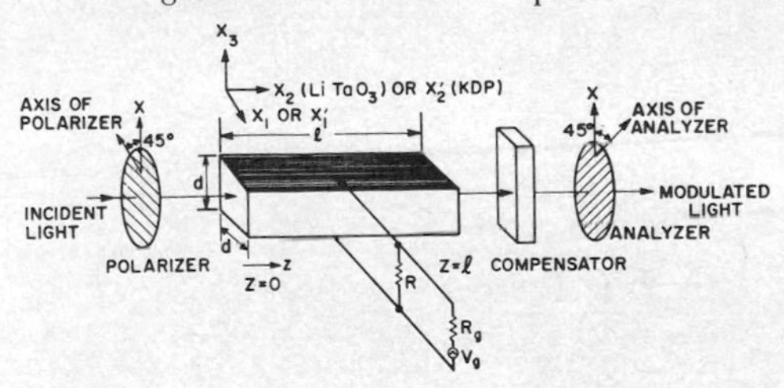

Electrooptical modulator in which intensity of light beam is ideally proportional to DC voltage applied to opposite faces of ferroelectric or piezoelectric crystal.

used to modulate the amplitude, phase, frequency, or direction of a beam of light. With a laser beam, modulating frequencies up into the gigahertz range are possible.

electrooptical shutter A shutter that uses a Kerr cell to modulate a beam of light.

electrooptics The branch of optics that deals with the effect of an electric field on light rays passing through an electrooptical material. One application is the Kerr cell, in which a signal voltage is applied to the liquid cell to modulate a light beam directed through the cell. The term electrooptics is sometimes used interchangeably with the broader term optoelectronics. For consistency and simplicity in this dictionary, most terms relating to combinations of electricity and electronics with optics will be found near the entries for optoelectronics and photoelectric.

electroosmosis The movement of fluids through diaphragms as a result of the application of an electric current.

electrophonic effect The sensation of hearing produced when an alternating current of suitable frequency and magnitude is passed through a person.

electrophoresis The movement of charged particles suspended in a fluid medium, under the influence of an electric field. Also called cataphoresis.

electrophoretic display A liquid crystal display in which a light-absorbing dye has been added to the liquid to improve both color and luminance contrast. Individual electrically charged dye parti-

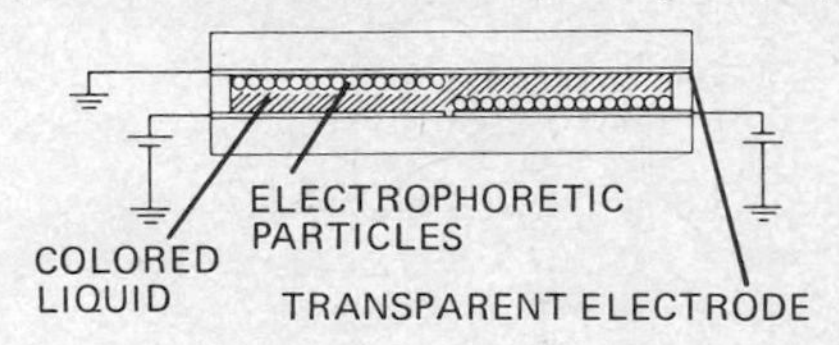

Electrophoretic display.

cles move when an electric field is applied. If white dye particles are suspended in a black fluid between transparent electrodes, a DC voltage makes the particles deposit on one electrode, and the display appears white when viewed through that electrode. When the polarity of the voltage is reversed, the particles move to the other electrode, and the display appears dark.

electrophorus A device used to produce electric charges by induction, consisting of a metal plate and a disk of resinous insulating material. In operation, the insulating disk is negatively charged by rubbing with fur. The metal plate, held by an insulating handle, is placed on the disk so it is charged by induction (bottom surface positive and top surface negative). The top surface is touched with a finger to remove the negative charge. When lifted off, the plate now has a strong positive charge all over.

electrophotography Original name for xerography, as invented by Chester F. Carlson in 1937. Electrophotography now includes both xerography and xeroradiography.

electrophrenic respiration Artificial respiration in which the nerves that control breathing are stimulated electrically through appropriately placed electrodes. The equipment needed is commercially available in portable form and is used by many rescue squads in preference to manual methods of artificial respiration.

electroplating The electrodeposition of an adherent metal coating on a conductive object for protection, decoration, or other purposes. The object to be plated is placed in an electrolyte and connected to one terminal of a DC voltage source. The metal to be deposited is similarly immersed and connected to the other terminal. Ions of the metal provide transfer of metal as they make up the current flow between the electrodes.

electropolishing The process of producing a smooth, lustrous surface on a metal by making it the anode in an electrolytic solution and preferentially dissolving the minute protuberances.

electropositive Having a positive electric polarity.

electrorefining The process of dissolving a metal from an impure anode by means of electrodeposition and redepositing it in a purer state on a cathode.

electroresistive effect The change in the resistivity of certain materials with changes in applied voltage. The varistor uses this effect.

electroretinogram A graphic record of the manner in which the voltage developed by the retina of the eye varies with time.

electroscope An instrument for detecting an electric charge by means of the mechanical forces exerted between electrically charged bodies. In one form, two narrow strips of gold leaf suspended in a glass jar spread apart when charged. The angle between the strips is then related to the charge.

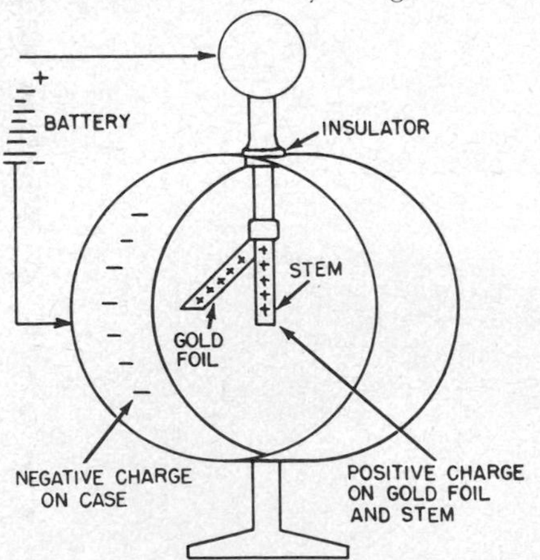

Electroscope.

electrosensitive paper A conductive paper that darkens when electric current is sent through it.

electrosensitive recording Recording in which the image is produced by passing electric current through the record sheet.

electroshock therapy [abbreviated EST] The use of an electric current, generally passed through the brain, to induce convulsions in the treatment of certain types of mental disorders. Also called electric shock therapy.

electrosleep Sleep induced by passing a low-voltage electric current through the brain.

electrostatic Pertaining to electricity at rest, such as an electric charge on an object.

electrostatic accelerator *Electrostatic generator.*

electrostatic air cleaner *Precipitator.*

electrostatically focused traveling-wave tube *Estiatron.*

electrostatic cathode-ray tube A cathode-ray tube in which electrostatic deflection is used to deflect the electron beam.

electrostatic copier A copying machine in which a photosensitive material is electrically charged in the pattern of the original being copied, and the latent image is developed by applying a finely powdered carbon toner that has been oppositely charged. Examples include Xerox and Electrofax copying processes.

electrostatic deflection The deflection of an electron beam by means of an electrostatic field produced by electrodes on opposite sides of the beam. Used chiefly in cathode-ray tubes for oscilloscopes. The electron beam is attracted to a positive electrode and repelled by a negative electrode.

electrostatic field An electric field having constant intensity, such as that produced by stationary charges.

electrostatic focusing A method of focusing an electron beam by the action of an electric field, as in the electron gun of a cathode-ray tube, so that the beam will have the required small area at a screen or other surface.

electrostatic generator A high-voltage generator in which electric charges are generated by friction or induction, then transferred mechanically to an insulated electrode to build up a voltage that may be as high as 9 MV. Examples include the Van de Graaff generator and the Wimshurst machine. Also called electrostatic accelerator and electrostatic machine.

electrostatic gyroscope A gyroscope in which a small beryllium ball is electrostatically suspended within an array of six electrodes in a vacuum inside a ceramic envelope. External coils provide a rotating magnetic field for spinning the ball at about 30,000 rpm, after which it may take up to 500 days to coast to a stop. Two photoelectric detectors sense the positions of marks on the ball and thereby obtain signals for determining and controlling the spin axis of the ball.

electrostatic headphone A headphone in which the moving diaphragm is an extremely thin membrane having a conductive coating. The diaphragm is positioned between two acoustically

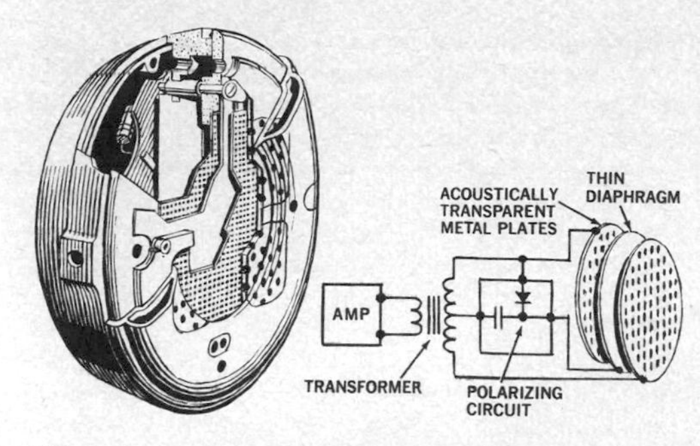

Electrostatic-headphone construction and operating circuit.

transparent metal plates to which it is alternately attracted and repelled as the audio signal varies the high DC voltage between the plates.

electrostatic induction The process of charging an object electrically by bringing it near another charged object.

electrostatic instrument A meter that depends for its operation on the forces of attraction and repulsion between electrically charged bodies.

electrostatic lens A lens consisting of coaxial metal cylinders and pierced diaphragms operated at potentials that produce electrostatic focusing of an electron beam directed along the axis of the lens.

electrostatic loudspeaker A loudspeaker in which the mechanical forces are produced by the action of electrostatic fields. In one type, the fields are produced between a thin metal diaphragm and a rigid metal plate. Also called capacitor loudspeaker.

electrostatic machine *Electrostatic generator.*

electrostatic memory *Electrostatic storage.*

electrostatic microphone *Capacitor microphone.*

electrostatic painting A painting process that uses the particle-attracting property of electrostatic charges. A DC voltage of about 100 kV is applied to a grid of wires through which the paint

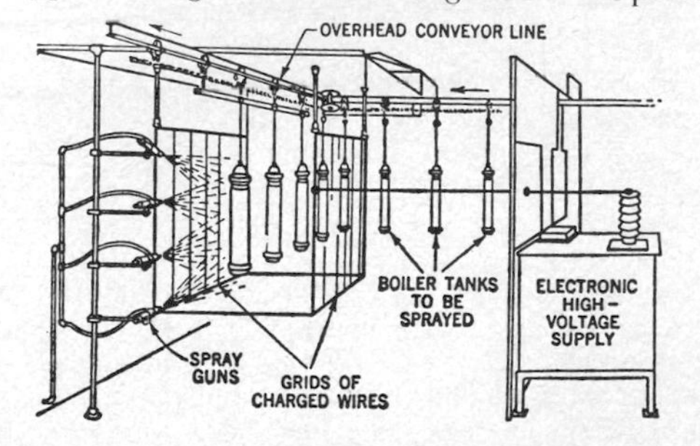

Electrostatic-painting setup for hot-water boiler tanks.

is sprayed, to charge each particle. The metal objects to be sprayed are connected to the opposite terminal of the high-voltage circuit so that they attract the particles of paint and thereby minimize waste of paint.

electrostatic photomultiplier A photomultiplier in which electrostatic fields cause the electron stream to be reflected off each dynode in turn.

electrostatic precipitation The removal of dust, smoke, or other finely divided particles from air by charging the particles with an electric field so they are attracted to oppositely charged collector electrodes. Also called electric precipitation.

electrostatic precipitator *Precipitator.*

electrostatic printer A line printer in which high-intensity lamps project images of characters onto a sensitized drum to form electrostatic patterns that attract ink powder. The images are then transferred to paper and fused just as in electrostatic copiers. More complex versions have character-forming capabilities, as in laser-beam printers, cathode-ray versions like the Videograph, and other charge-forming computer-output printers.

electrostatic propulsion Propulsion of spacecraft or other vehicles by using electric fields to accelerate charged particles in a desired direction.

electrostatic radius The nuclear radius as deduced from an analysis of nuclear binding energies, especially of mirror nuclides.

electrostatic recording Recording by means of a signal-controlled electrostatic field.

electrostatic relay A relay in which the actuator element consists of nonconducting media separating two or more conductors that change their relative positions because of the mutual attraction or repulsion of electric charges applied to the conductors.

electrostatics The science that deals with electricity at rest such as with charged objects and constant-intensity electric fields.

electrostatic scanning Scanning that involves electrostatic deflection of an electron beam.

electrostatic separator A separator in which a finely pulverized mixture falls through a powerful electric field between two electrodes. Materials having different specific inductive capacitances are deflected by varying amounts and fall into different sorting chutes.

electrostatic shield A grounded metal screen, sheet, or enclosure placed around a device or between two devices to prevent electric fields from acting through the shield. Often used to prevent interaction between the electric fields of adjacent parts on a chassis.

electrostatic storage A storage in which information is stored in the form of the presence or absence of electrostatic charges at specific spot locations, generally on the screen of a special type of cathode-ray tube known as a storage tube. Also called electrostatic memory.

electrostatic storage tube *Storage tube.*

electrostatic tape camera A camera in which images are stored electrostatically on a plastic tape. Designed for use in satellites, where the stored

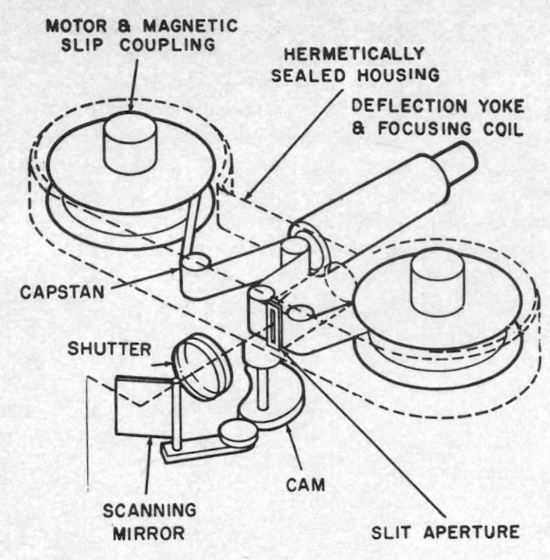

Electrostatic tape camera.

image is not damaged by Van Allen or other radiation. The four-layer tape consists of optically transparent Mylar having a conducting thin film of gold, a thin film of photoconductive material, and a protective coating of polystyrene. To record, a broad electron beam charges the tape. The optical image then discharges elemental areas in proportion to their illumination.

electrostatic tweeter A tweeter loudspeaker in which a flat metal diaphragm is driven directly by a varying high voltage applied between the diaphragm and a fixed metal electrode.

electrostatic unit [abbreviated ESU] A CGS unit based on a unit of charge that exerts a force of 1 dyn on another unit charge at a distance of 1 cm in a vacuum.

electrostatic voltmeter A voltmeter in which the voltage to be measured is applied between fixed and movable metal vanes. The resulting elec-

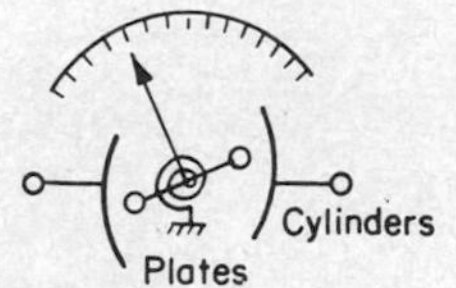

Electrostatic voltmeter.

trostatic force deflects the movable vane against the tension of a spring. An attached pointer moving over a scale indicates the voltage of the circuit. Usually used for measuring high values of DC voltage.

electrostatography A generic term covering all processes involving the forming and use of electrostatic charged patterns for recording and reproducing images. The most important application today is xerography, as used in office copying machines.

electrostethophone A stethoscope consisting of a microphone and audio amplifier feeding headphones, used for detection and study of sounds arising within the body.

electrostriction The change in dimensions that occurs in some dielectric materials when placed in an electric field. The change is independent of the polarity of the electric field. The reverse effect does not take place, whereas in piezoelectricity the effect is reversible. Also called electrostrictive effect.

electrostriction transducer A transducer that depends on production of an elastic strain in certain symmetric crystals when an electric field is applied.

electrostrictive effect *Electrostriction.*

electrostrictive relay A relay in which an electrostrictive dielectric serves as the actuator.

electrosurgery Use of electric current in surgery to cut and coagulate tissue simultaneously.

electrotape A phase-comparison method of measuring distance with radar techniques, using a 10-MHz crystal as a frequency reference.

electrotherapy The use of electricity in treating disease.

electrothermal instrument An instrument that depends for its operation on the heating effect of a current.

electrothermal propulsion Propulsion of spacecraft by using an electric arc or other electric

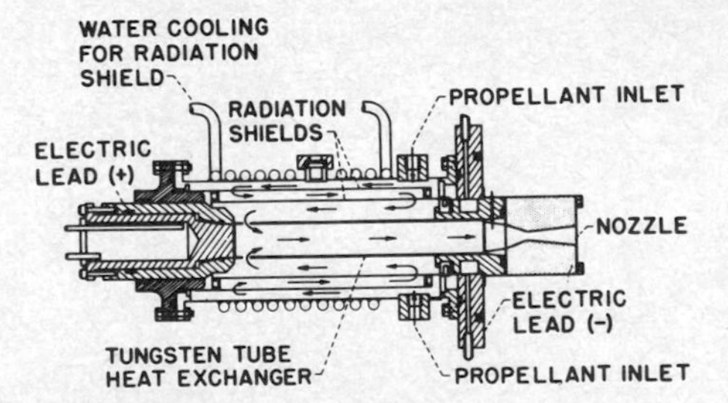

Electrothermal propulsion using electric resistance heater.

heater to bring hydrogen gas or other propellant to the high temperature required for maximum thrust. An arc-jet engine is an example.

electrothermal recording Recording in which the marking on the record sheet is produced principally by signal-controlled thermal action.

electrotonus A change in the characteristics of a nerve or muscle during or after the passage of an electric current.

element 1. A distinct functioning part of an electron tube, semiconductor device, antenna array, or other device, contributing directly to performance. 2. One of the 106 or more known substances that cannot be divided into simpler substances by chemical means. Scientists believe the total may eventually reach 110. The most recently discovered elements are generally very short-lived nuclear fission products. 3. *Component.*

elemental area *Picture element.*

elementary charge The unit charge of electricity corresponding to the charge on a single electron, and equal to about 4.80298×10^{-10} electrostatic unit.

elementary particle One of more than 60 particles from which all matter is made up. The list now includes the photon, two types of neutrino, the electron, the mu meson, two pi mesons, two K particles, the proton, the neutron, the lambda particle, three sigma particles, and two xi particles, each with its antiparticle. Also called fundamental particle.

elevated duct A tropospheric radio duct that has both its upper and lower boundaries above the ground.

elevation 1. *Altitude.* 2. *Elevation angle.*

elevation angle The angle that a radio, radar, or other beam of radiation makes with the horizontal. Also called elevation.

elevation deviation indicator An indicator that presents visually the relationship between a target and the elevation angle of a radar or radio beam.

elevation indicator A component that presents visually the angle between a fixed reference point and a target in the same vertical plane.

ELF Abbreviation for *extremely low frequency.*

ELINT Abbreviation for *electronic intelligence.*

ellipsometer An instrument for measuring thickness and other parameters in extremely thin insulating films used in semiconductor manufacturing. Based on a polarimetric technique in which the change in polarization states caused by reflection of light from a film-covered surface is measured at a fixed angle of incidence. This change in polarization is a function of thickness.

elliptically polarized wave An electromagnetic wave for which the component of the electric vector in a plane normal to the direction of propagation describes an ellipse at a given frequency.

elliptical polarization Polarization in which the magnitude of the vector representing the wave varies as the radius of an ellipse while the vector rotates about a point.

elliptical waveguide A flexible waveguide having an elliptical cross section, sometimes used in place of a rectangular waveguide.

elliptic-function filter A multistage *LC* microwave filter having extremely sharp selectivity, low insertion loss, and low delay distortion. Used in IF amplifiers of satellite communication receivers, where cascaded bridged-T networks give equalization for a bandwidth of 5 MHz at a 70-MHz IF value.

elongation The extension of the envelope of a signal caused by delayed arrival of multipath components.

ELSSE Abbreviation for *electronic skyscreen equipment.*

EMA Abbreviation for *electronic missile acquisition.*

embedding The process of molding an insulating plastic around a component or assembly to form a solid block having only the leads or terminals exposed.

embossed-foil printed circuit A printed circuit formed by indenting the desired pattern of metal foil into an insulating base, then mechanically removing the remaining unwanted raised portions.

embossed-groove recording Disk recording in which a comparatively blunt stylus pushes aside the material in the modulated groove. No material is removed from the disk.

embossing stylus A recording stylus with a rounded tip that forms a groove by displacing material in the recording medium.

EMC Abbreviation for *electromagnetic compatibility.*

EME Abbreviation for *earth-moon-earth.*

emergency broadcast system A system of broadcast stations and interconnecting facilities authorized by the Federal Communications Commission to operate in a controlled manner during a war, threat of war, state of public peril or disaster, or other national emergency.

emergency power supply A source of 60-Hz power that becomes available, usually automatically, when normal 60-Hz power-line service fails. The emergency source is usually an engine-driven alternator or a motor-generator set operating from a large storage battery.

emergency radio channel Any radio frequency reserved for emergency use, particularly for distress signals. The emergency frequencies for standard channels are 500 kHz and 8.364, 121.5, and 243 MHz.

EMF Abbreviation for *electromotive force.*

EMG 1. Abbreviation for *electromyogram.* 2. Abbreviation for *electromyograph.*

EMI Abbreviation for *electromagnetic interference.*

emission 1. Any radiation of energy by means of electromagnetic waves, as from a radio transmitter. 2. *Electron emission.*

emission bandwidth The band of frequencies comprising 99% of the total radiated power, extended to include any other discrete frequency at which the power is more than 0.25% of the total radiated power.

emission line A characteristic frequency of electromagnetic radiation emitted by atoms, molecules, or ions under certain conditions, as when electrons move from one energy level to another.

emission spectrometer A spectrometer that measures percent concentrations of preselected elements in samples of metals and other materials. When the sample is vaporized by an electric spark or arc, the characteristic wavelengths of light emitted by each element are measured with a diffrac-

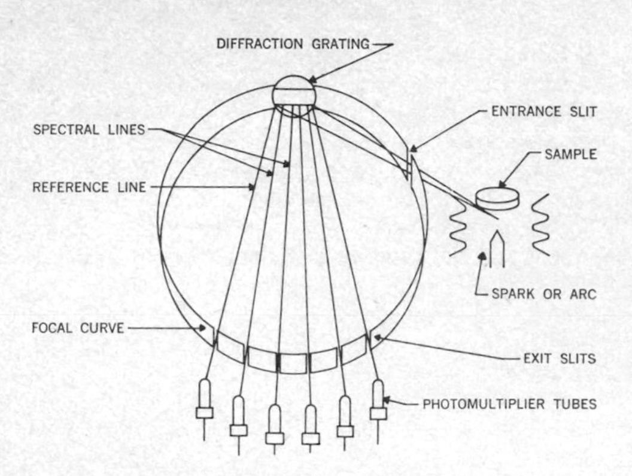

Emission spectrometer.

tion grating and an array of photodetectors.

emission spectrum The spectrum produced by radiation from any emitting source, such as the spectrum of radiation from an invisible infrared source.

emissivity The ratio of the radiation emitted by a surface to the radiation emitted by a perfect blackbody radiator at the same temperature.

emitron A television camera tube similar to an iconoscope, made in Great Britain.

emittance The power radiated per unit area of a radiating surface.

emitter [symbol E] A transistor region from which charge carriers that are minority carriers in the base are injected into the base. The emitter roughly corresponds to the cathode of an electron tube. In the symbol for a transistor, the emitter has an arrow, pointing outward for an NPN transistor and inward for a PNP transistor.

emitter bias A bias voltage applied to the emitter electrode of a transistor.

emitter channeling Formation of an undesirable low-resistance path between the emitter and one of the bases of a unijunction transistor.

emitter-coupled logic [abbreviated ECL] A form of current-mode logic in which the emitters of two transistors are connected to a single current-carrying resistor in such a way that only one transistor conducts at a time. The logic state of the output depends on which transistor is conducting.

emitter follower A gounded-collector transistor amplifier whose operation is similar to a cathode follower that uses a vacuum tube.

emitter-follower logic [abbreviated EFL] A form of logic, used in large-scale integration, in which NPN and PNP transistors are combined to give high-performance structures that are similar to complementary-transistor logic but use minimum silicon area.

emitter-function logic [abbreviated EFL] A form of large-scale integration that combines performance advantages of emitter-coupled logic with the compactness of multiemitter structures. Simplification of gate design reduces propagation delay, power dissipation, and the number of logic levels required.

emitter junction A transistor junction normally biased in the low-resistance direction to inject minority carriers into a base.

emitter stabilization Use of a zener diode or similar device to maintain essentially constant emitter voltage in a transistor circuit despite normal variations in base bias.

$E_{m,n}$ mode British term for *$TM_{m,n}$ mode.*

$E_{m,n}$ wave British term for *$TM_{m,n}$ wave.*

emphasis *Preemphasis.*

emphasizer *Preemphasis network.*

empire cloth Cotton cloth coated with insulating varnish.

empirical Based on actual measurement, observation, or experience, rather than on theory.

empty band A band of possible energy levels in an atom, none of which correspond to the energy of any electron in the given substance in the given state.

EMU Abbreviation for *electromagnetic unit.*

emulation The process by which one computer mimics the performance of another computer.

emulator A combination of hardware, microprograms, and software that can be added to one computer system to make it execute programs written for another computer system. An emulator may or may not have software, but usually must have either added hardware or added microprograms. In contrast, a simulator achieves the same results with software only.

enable To activate a circuit, such as by applying a signal or pulse of appropriate form or by removing a suppression signal.

enabling gate A circuit that initiates the start and determines the length of a generated pulse.

enabling pulse A pulse that prepares a circuit for some subsequent action.

enameled wire Wire coated with an insulating layer of baked enamel.

encapsulating The process of placing a heavy protective coating on a component or assembly by dipping it in a thick insulating plastic fluid or other insulating material.

encephalogram A radiograph made by passing x-rays through the brain to a sensitive film.

encephalography Radiography of the brain.

encipher To convert plain text into unintelligible form by means of a cipher system. Also called encode and encrypt.

enciphered facsimile communication Communication in which security is accomplished by mixing key pulses, produced by a key generator, with the output of the facsimile converter. Plain text is recovered by subtracting the identical key at the receiving terminal.

enclosure A housing for loudspeakers or other electronic equipment.

encode 1. To express given information by

means of a code. 2. To prepare a routine in machine language for a specific computer. 3. *Encipher.*

Encoder disk used to convert shaft position to corresponding digital value representing angular degrees.

encoder 1. Any device that converts analog signals, shaft positions, or other analog parameters to digital form. 2. *Matrix.*

encrypt *Encipher.*

encryption The conversion of a message by cryptogear to a meaningless form that permits secret transmission.

end-around carry A computer carry signal that is sent directly to the least significant digit place when generated in the most significant digit place.

end effect The effect of capacitance at the ends of an antenna. It requires that the actual length of a half-wave antenna be about 5% less than a half-wavelength.

end-fed vertical antenna *Series-fed vertical antenna.*

end-fire array A linear array whose direction of maximum radiation is along the axis of the array. It may be either unidirectional or bidirectional. The elements of the array are parallel and in the same plane, as in a fishbone antenna.

end-of-tape marker A metal tab, light-reflecting strip, transparent section, or special code placed at the end of the permissible recording area on magnetic tape.

endoradiosonde A miniature battery-powered radio transmitter encapsulated like a pill, designed to be swallowed for measuring and transmitting physiological data from the gastrointestinal tract.

endothermic Involving absorption of heat or other energy.

end-point control Quality control through continuous, automatic analysis of the final product. In automatic control the necessary process or machine corrections are made automatically by a controller.

end shield A shield placed at each end of the interaction space of a magnetron to prevent electrons from bombarding the end seals.

energize To apply rated voltage.

energized Electrically connected to a voltage source. Also called alive, hot, and live.

energy Ability to do work. Energy may be transferred from one form to another, but it cannot be created or destroyed.

energy band The sets of discrete but closely adjacent energy levels, equal in number to the number of atoms, that arise from each of the quantum states of the atoms of a substance when the atoms condense to a solid state from a nondegenerate gaseous condition. Also called Bloch band. For a semiconductor the highest energy level is the conduction band, containing only the excess electrons resulting from crystal impurities. The next highest level is the valence band, usually completely filled with electrons. Between these bands is the forbidden band, which is wider for an insulating material than for a semiconductor and vanishes in a conducting material.

energy beam An intense beam of light, electrons, or other nuclear particles, used to cut, drill, form, weld, or otherwise process metals, ceramics, and other materials.

energy conservation law Energy can be neither created nor destroyed, only changed from one form to another.

energy dependence The characteristic response of a radiation detector to a given range of radiation energies or wavelengths, as compared to the response of a standard open-air chamber.

energy diagram *Energy-level diagram.*

energy gap The energy range between the bottom of the conduction band and the top of the valence band in a semiconductor.

energy level A constant-energy state for particles in an atom. Only a limited number of electrons can exist at each energy level. An electron radiates energy when moving to a lower energy level, and absorbs energy when moving to a higher energy level.

energy-level diagram A diagram in which the energy levels of the particles of a quantized system are indicated by distances of horizontal lines from a zero energy level. Also called energy diagram.

energy product curve A curve obtained by plot-

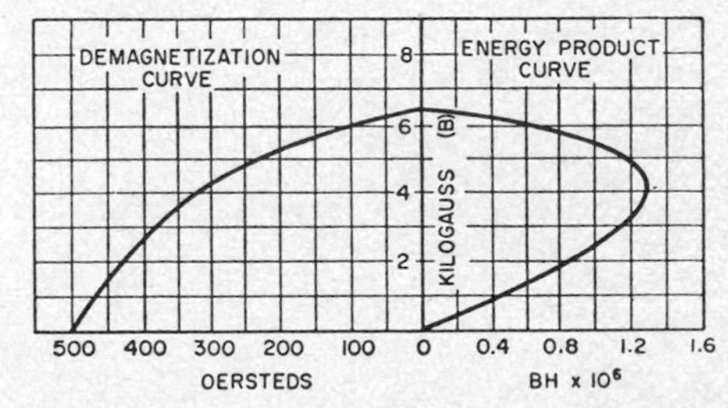

Energy product curve at right is drawn by obtaining values of *B* and *H* from demagnetization curve at left, multiplying them for each point, then plotting the *BH* products against *B*.

ting the product of the values of magnetic induction B and magnetic field strength H for each point on the demagnetization curve of a permanent-magnet material.

engineer A person having the training and experience required to perform professional duties in a branch of engineering. An engineer generally designs, constructs, operates, or supervises, whereas a scientist seeks to uncover new knowledge, new principles, or new materials.

engineering A profession in which a knowledge of the mathematical and physical sciences, gained by study, experience, and practice, is applied with judgment to the utilization of the materials and forces of nature.

enhanced-carrier demodulation An amplitude demodulation system in which a synchronized local carrier of proper phase is fed into the demodulator to reduce demodulation distortion.

enhancement An increase in the density of charged carriers in a particular region of a semiconductor. Usually produced by a built-in or applied electric field.

enhancement-mode transistor A field-effect transistor in which there are no charge carriers in the channel when the gate-source voltage is zero. Channel conductivity is increased by applying a gate-source voltage of appropriate polarity.

entropy The transmission efficiency of an information-handling system, expressed as the logarithm of the number of possible equivalent messages that can be sent by selection from a particular set of symbols.

envelope 1. The glass or metal housing of an electron tube. Also called bulb. 2. A curve drawn to pass through the peaks of a graph showing the waveform of a modulated RF carrier signal.

envelope delay The time required for the envelope of a wave to travel between two points in a system.

envelope delay distortion *Delay distortion.*

envelope demodulator A diode detector whose output is shunted by a capacitor, to make the output proportional to the peaks of the rectified amplitude-modulated carrier. Commonly used as the second detector in a television receiver.

environics The application of system concepts to all aspects of environment control required for a product or system. As applied to space electronic equipment it would include provisions for radio-interference suppression, temperature control, pressure control, and all the other environmental aspects involved in successful manned or unmanned space travel.

environment The aggregate of all the conditions and influences that affect the operation of equipment and components. Natural uncontrolled environments include temperature, humidity, rain, snow, ice, sleet, hail, fog, wind, lightning, sand, dust, fungi, radiation, pressure, sun, salt spray, and static electricity. Environments caused by operation or location include vibration, shock, noise, acceleration, aerodynamic heating, erosion in flight, electromagnetic effects, and force.

ephemeris A table that gives the calculated positions of a satellite or celestial body at regular intervals of time.

ephemeris time The fundamental standard of time, based on the orbital motion of the earth about the sun. The ephemeris second is equal to 1/31556925.9747 of the tropical year defined by the mean motion of the sun in longitude at noon on January 1, 1900.

EPI Abbreviation for *electronic position indicator.*

epitaxial diffused-junction transistor A junction transistor produced by growing a thin high-

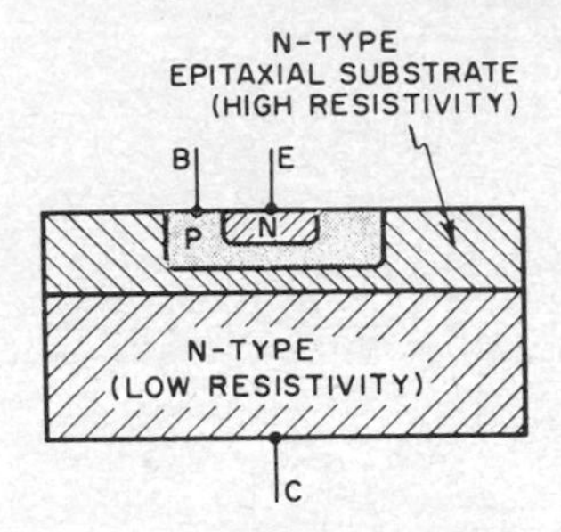

Epitaxial diffused-junction transistor cross section.

purity layer of semiconductor material on a heavily doped region of the same type.

epitaxial diffused-mesa transistor A diffused-mesa transistor in which a thin high-resistivity epi-

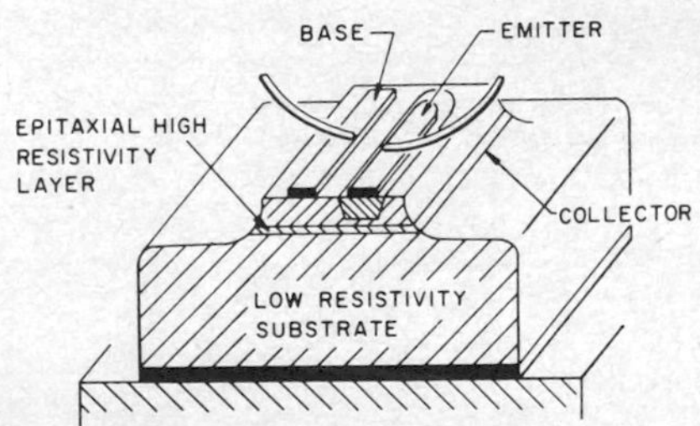

Epitaxial diffused-mesa transistor construction.

taxial layer is deposited on the substrate to serve as the collector.

epitaxial junction A doped epitaxial layer of one conductivity type, grown on an epitaxial substrate having the opposite conductivity to produce a PN junction.

epitaxial layer A semiconductor layer having the same crystalline orientation as the substrate on which it is grown. The process involves condensing silicon atoms on a silicon substrate at 1200°C; at this temperature, the atoms are mobile and able to take up the orientation of the substrate lattice.

epndB Abbreviation for *effective perceived noise decibel.*

epoxy resin A good insulating plastic having

high strength and low shrinkage during curing. Widely used in encapsulating and embedding electronic assemblies.

EPPI Abbreviation for *expanded plan-position indicator.*

EPROM Abbreviation for *erasable programmable read-only memory.*

epsilon [symbol ϵ] An elementary particle in the meson family, not isolated until recently because of its close attachment to the rho meson.

equal-energy source A light source for which the time rate of emission of energy per unit of wavelength is constant throughout the visible spectrum.

equal-energy white The light produced by a source that radiates equal energy at all visible wavelengths.

equalization The effect of all frequency-discriminative means employed in transmitting, recording, amplifying, or other signal-handling systems to obtain a desired overall frequency response. Also called frequency-response equalization.

equalization curve A curve showing the frequency response needed in a high-fidelity sound-reproducing system to compensate for preemphasis introduced at the broadcasting or recording studio.

equalizer A network designed to compensate for an undesired amplitude-frequency or phase-frequency response of a system or component. It is usually some combination of coils, capacitors, and resistors.

equalizing pulse One of the pulses occurring just before and just after the vertical synchronizing pulses in a television signal and serving to minimize the effect of line-frequency pulses on interlace. The equalizing pulses occur at twice the line frequency and make each vertical deflection start at the correct instant for proper interlace.

equally tempered scale A series of notes selected by dividing the octave into 12 equal intervals.

equatorial mounting A mounting for a radiotelescope antenna that provides for continuous motion about an axis parallel to the earth's axis of rotation. This allows the telescope to be directed constantly at any point in the sky with a simple constant-speed motor that compensates for the earth's rotation.

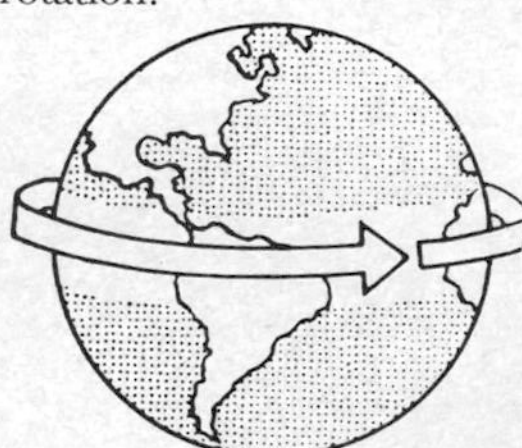

Equatorial orbit of satellite around earth, in which satellite passes only over equatorial regions despite rotation of earth.

equatorial orbit An orbit in the plane of the earth's equator.

equiangular spiral antenna A frequency-independent broadband antenna, cut from sheet metal, that radiates a very broad circularly polarized beam on both sides of its surface. This bidirectional radiation pattern is its chief limitation.

Equiangular spiral antenna, cut from sheet metal.

equilibrium orbit *Stable orbit.*

equiphase surface Any wave surface over which the field vectors at the same instant are in phase or 180° out of phase.

equiphase zone The region in space within which a difference in phase of two radio signals is indistinguishable.

equipment One or more assemblies capable of performing a complete function.

equipotential Having the same potential at all points.

equipotential cathode *Indirectly heated cathode.*

equipotential line An imaginary line in space or in a medium, having the same potential at all points.

equipotential surface An imaginary surface in space, or in a medium, on which all points have the same potential.

equisignal localizer An aircraft guidance localizer in which the localizer on-course line is centered in a zone of equal amplitude of two transmitted signals. Deviations from this zone are detectable as unbalance in the levels of the two signals. Also called equisignal radio-range beacon and tone localizer.

equisignal radio-range beacon *Equisignal localizer.*

equisignal sector *Equisignal zone.*

equisignal zone The region in space within which the difference in amplitude of two radio signals (usually emitted by a single station) is indistinguishable.

equivalent absorption The area of perfectly absorbing surface that will absorb sound energy at the same rate as the given surface or object under the same conditions. The acoustical unit of equivalent absorption is the sabin.

equivalent binary digits The number of binary digits that is equivalent to a given number of decimal digits or other characters. When a decimal number is converted into a binary number, the number of binary digits needed is in general about 3.3 times the number of decimal digits. In coded decimal notation, the number of binary digits needed is ordinarily 4 times the number of decimal digits.

equivalent circuit A circuit that is electrically equivalent to a more complex circuit or device. Used to simplify circuit analysis.

equivalent loudness level *Loudness level.*

equivalent noise temperature The absolute temperature at which a perfect resistor would generate the same noise as an actual resistor having the same resistance value.

equivalent roentgen *Roentgen equivalent physical.*

equivalent stopping power 1. *Relative stopping power.* 2. *Stopping equivalent.*

erasable programmable read-only memory [abbreviated EPROM] A read-only memory in which stored data can be erased by ultraviolet light or other means and reprogrammed bit by bit with appropriate voltage pulses.

erasable storage A storage whose data can be altered at any time.

erase 1. To remove recorded material from magnetic tape by passing the tape through a strong, constant magnetic field (DC erase) or through a high-frequency alternating magnetic field (AC erase). 2. To change all the binary digits in a digital computer storage device to binary zeros. 3. To eliminate previously stored information in a charge-storage tube by charging or discharging all storage elements.

erase oscillator The oscillator used in a magnetic recorder to provide the high-frequency signal needed to erase a recording on magnetic tape. The bias oscillator usually serves also as the erase oscillator.

erasing head A magnetic head used to obliterate material previously recorded on magnetic tape.

erbium [symbol Er] A rare-earth element. Atomic number is 68.

E region The region of the ionosphere extending from about 60 to 90 mi (95 to 160 km) above the earth, between the D and F regions.

erg The absolute CGS unit of energy and work. It is the work done when a force of 1 dyn is applied through a distance of 1 cm. One foot-pound is equal to 13.56×10^6 ergs.

ergometer An instrument for measuring the amount of work performed by specific muscles in the body under controlled conditions.

Erlang A unit of communication traffic density, equal to the average number of simultaneous calls originated during a specific hourly period.

ERP Abbreviation for *effective radiated power.*

error 1. The difference between the true value and a calculated or observed value. 2. *Malfunction.*

error code A specific character punched into a card or tape to indicate that a conscious error was made in the associated block of data. Machines reading the error code may be programmed to throw out the entire block automatically.

error-correcting code A code in which each data signal conforms to rules such that errors in received signals can be automatically detected and some or all of the errors corrected automatically.

error-detecting code *Self-checking code.*

error rate The ratio of the number of erroneous characters to the total number of characters received in a data-transmission system.

error-rate damping A type of damping in which servo control is accomplished by two voltages: one proportional to the error, and the other proportional to the rate at which the error changes. Also called proportional plus derivative control.

error signal 1. A voltage that depends on the signal received from the target in a tracking system, having a polarity and magnitude dependent on the angle between the target and the center of the scanning beam. 2. *Error voltage.*

error voltage A voltage, usually obtained from a selsyn, that is proportional to the difference between the angular positions of the input and output shafts of a servosystem. This voltage acts on the system to produce a motion that tends to reduce the error in position. Also called error signal.

ERTS Abbreviation for *earth resources technology satellite.*

Esaki diode *Tunnel diode.*

escape velocity The minimum velocity that will enable an object to escape from the surface of a planet or other body without further propulsion. The escape velocity for the earth is 25,020 mi/h (40,266 km/h); for the moon, 5364 mi/h (8633 km/h).

E scope A radarscope that produces an E display.

escutcheon An ornament used around a dial, window, control knob, or other panel-mounted part.

EST Abbreviation for *electroshock therapy.*

ESU Abbreviation for *electrostatic unit.*

etalon An optical device consisting of two spaced, parallel, and partially silvered glass plates. When used in the cavity of a laser, the etalon is tilted at an angle that provides enough reflection to prevent oscillation at all unwanted axial modes of the long cavity.

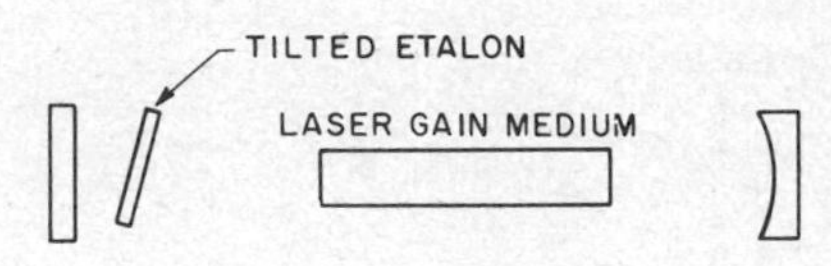

Etalon in single-frequency laser.

eta particle An uncharged elementary particle

that has 1074 times the mass of an electron and an extremely short life.

etched printed circuit A printed circuit formed by chemical etching or chemical and electrolytic removal of unwanted portions of a layer of conductive material bonded to an insulating base.

etching Removal of unwanted portions of a metal or semiconductor by chemical or electrolytic action.

Ettingshausen effect When a metal strip is placed with its plane perpendicular to a magnetic field and an electric current is sent longitudinally through the strip, corresponding points on opposite edges of the strip will have different temperatures.

ETV Abbreviation for *educational television*.

E unit A method of designating radar signal-to-noise ratio. E-1 is a 1 to 1 ratio, barely perceptible; E-2 is 2 to 1, weak; E-3 is 4 to 1, good; E-4 is 8 to 1, strong; E-5 is 16 to 1, very strong or saturating.

Eureka The ground radar beacon of the Rebecca-Eureka navigation system, operating at spot frequencies between 215 and 235 MHz. Range is about 20 to 40 mi (30 to 65 km), depending on the altitude of the aircraft carrying interrogating Rebecca equipment.

europium [symbol Eu] A rare-earth element. Atomic number is 63. Used as the red phosphor in some color television picture tubes. The red color is much brighter than in earlier tubes and does not change at higher beam currents. This permits operating all three guns at higher levels, giving brighter color pictures.

eutectic The liquid alloy composition having the lowest freezing point. For lead-tin solder, the eutectic is 63% tin and 37% solder, giving a freezing point of 361°F.

eV Abbreviation for *electronvolt*.

evacuate To remove gases and vapors from an enclosure. Also called exhaust.

evacuation The removal of gases and vapors from the envelope of an electron tube during manufacture.

evanescent mode The mode of oscillation in which the amplitude diminishes along a waveguide without change of phase.

evaporant The material that is to be deposited as a thin film by evaporation.

evaporated dielectric film A film of dielectric material deposited by vacuum evaporation for passivation of components having active and passive films. Also used as a capacitor dielectric in thin-film capacitors.

evaporation The process by which material is vaporized by heat in a vacuum to produce a thin film.

evaporative-cooled *Vaporization-cooled.*

E vector A vector that represents the electric field of an electromagnetic wave. In free space it is perpendicular to the H vector and the direction of propagation.

even harmonic A harmonic that is an even multiple of the fundamental frequency.

even parity check A method of detecting when bits are dropped by adding 1 bit whenever a character is represented by an odd number of bit patterns. All characters are then represented by an even number of bits, and failure to do so is thus a parity error.

event recorder A recorder that plots ON/OFF information against time, to indicate when events start, how long they last, and how often they recur. In one type, the writing pen is displaced a fixed

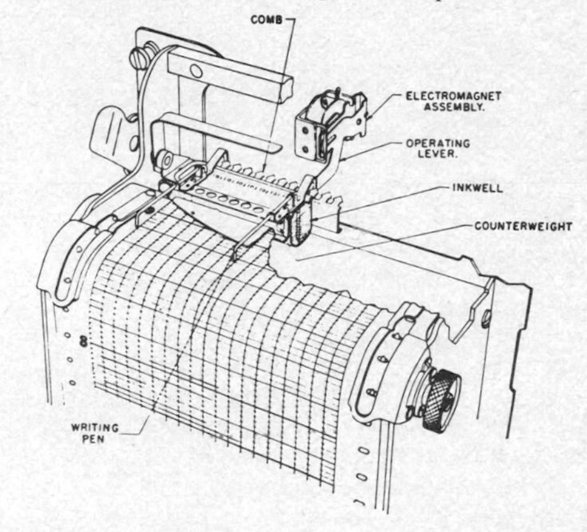

Event recorder in which writing pen is moved fixed distance to one side of electromagnet each time event occurs. Paper of strip chart is moved continuously by time-clock motor.

amount to one side when energized. In another type, the pen makes a line only when voltage is applied, so that off time is indicated by the absence of a line.

EVR Abbreviation for *electronic video recording*.

EW Abbreviation for *electronic warfare*.

E wave British term for *transverse magnetic wave* (TM wave).

exa- [abbreviated E] A prefix representing 10^{18}.

EXCEPT A logic operator which has the property that if P and Q are two statements, then the statement P EXCEPT Q is true only when P alone is true. It is false for the other three combinations (P false Q false, P false Q true, and P true Q true). The EXCEPT operator is equivalent to AND NOT.

EXCEPT gate A gate that produces an output pulse only for a pulse on one or more input lines and the absence of a pulse on one or more other lines.

exception reporting A form of programming in which only values that are outside predetermined limits, representing significant changes, are selected for printout at the output of a computer.

excess conduction Conduction by excess electrons in a semiconductor.

excess electron An electron in excess of the

number needed to complete the bond structure in a semiconductor, generally resulting from donor impurities.

excess-three code A number code in which the decimal digit n is represented by the 4-bit binary equivalent of $n + 3$, so decimal digits 0 through 9 become 0011, 0100, 0101, 0110, 0111, 1000, 1001, 1010, 1011, and 1100, respectively.

excitation 1. The signal voltage that is applied to the control electrode of an electron tube. Also called drive. 2. Application of signal power to a transmitting antenna. 3. The transfer of a nuclear system from its ground state to an excited state by adding energy. 4. The application of voltage to field coils to produce a magnetic field, as required for the operation of an excited-field loudspeaker or a generator.

excitation anode An anode used to maintain a cathode spot on a pool cathode of a gas tube when output current is zero.

excitation energy The minimum energy required to change a system from its ground state to a particular excited state.

excitation purity *Purity.*

excitation winding The magnetic amplifier winding that applies a unidirectional magnetomotive force to the core.

excited Having higher energy than the ground state of a nucleus, atom, or molecule.

excited-field loudspeaker A dynamic loudspeaker in which the steady magnetic field is produced by an electromagnet called the field coil, through which a direct current is sent. Also called electrodynamic loudspeaker.

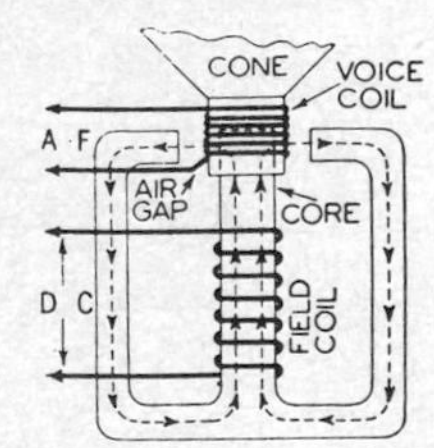

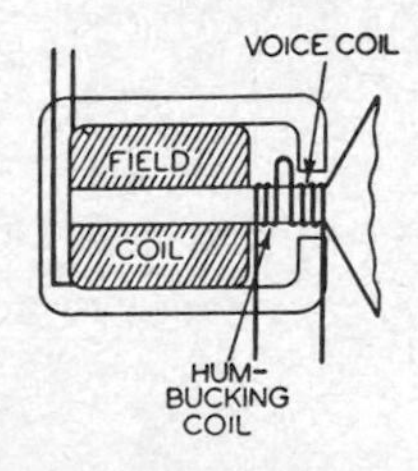

Excited-field loudspeaker, and method of using hum-bucking coil in series with voice coil.

excited state An energy level in which a nucleus or other system may exist if given sufficient energy to reach this state from a lower state. An excited nucleus will usually give off a particle in a short time and return to the ground state.

exciter 1. The portion of a directional transmitting antenna system that is directly connected to the transmitter. 2. A crystal oscillator or self-excited oscillator used to generate the carrier frequency of a transmitter. 3. A small auxiliary generator that provides field current for an AC generator. 4. A loop or probe extending into a resonant cavity or waveguide. 5. *Exciter lamp.*

exciter lamp A bright incandescent lamp having a concentrated filament. Used in variable-area sound-on-film recording, in reproducing photographic sound tracks on film, and in illuminating subject copy in facsimile transmitters. The lamp serves to excite a phototube or photocell. Also called exciter.

exciting current *Magnetizing current.*

exciton The combination of an electron and a hole in a semiconductor that is in an excited state. In attracting the electron, the hole acts as a positive charge.

excitron A single-anode mercury pool tube provided with means for maintaining a continuous cathode spot.

exclusion principle *Pauli exclusion principle.*

EXCLUSIVE-OR gate A logic gate whose output is 1 if any one but not more than one of its inputs is 1. The output is 0 if more than one input is 1 or all inputs are 0. Also called XOR gate.

execute To perform the operations in the stored program of a computer.

executive routine A digital computer routine designed to process and control other routines.

exhaust *Evacuate.*

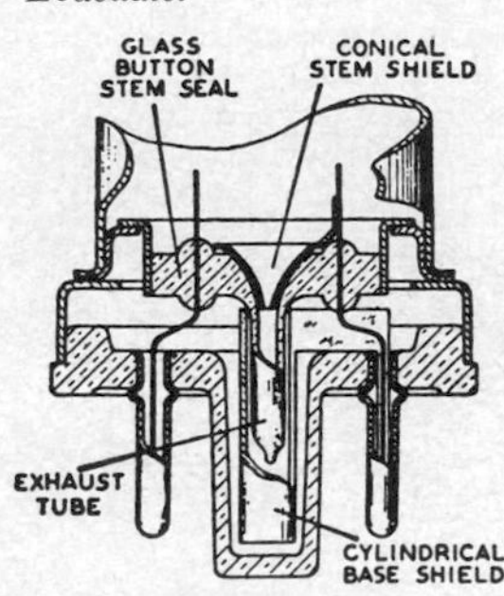

Exhaust tube of metal tube is protected by molded plastic aligning plug in base.

exhaust tube A glass or metal tube through which air is evacuated from the envelope of an electron tube.

exit A method of interrupting or leaving a repeated cycle of operations in a digital computer program.

exit dose The dose of radiation at the surface from which an irradiating beam emerges after passing through a body.

exoergic *Exothermic.*

exosphere A region of the atmosphere above the thermosphere, beginning at roughly 600 km, where the density is so low that an upward-traveling molecule makes no further collisions until it falls back to the base of the exosphere. At roughly 10,000 km the exosphere merges into the interplanetary medium. The exosphere overlaps the magnetosphere.

exothermic Involving the evolution of heat or

other energy. All radioactive processes are exothermic. Also called exoergic.

exp An abbreviation sometimes used in equations to indicate that the quantity immediately following in brackets is to be considered as the exponent of *e*, which is 2.718, the base of the natural system of logarithms.

expanded-center PPI display A modified PPI display in which each sweep starts about 1 cm out from the center of the tube screen. Bearings of nearby objects can then be determined more accurately.

expanded plan-position indicator [abbreviated EPPI] A special high-precision indicator used in some ground-controlled approach equipment to display azimuth, elevation, and range information simultaneously. Also called azel indicator.

expanded scope A magnified portion of a cathode-ray presentation.

expanded sweep A cathode-ray sweep in which the movement of the electron beam across the screen is speeded up during a selected portion of the sweep time.

expander The part of a compandor that is used at the receiving end of a circuit to return the compressed signal to its original form. It attenuates weak signals and amplifies strong signals.

expansion A process in which the effective gain of an amplifier is varied as a function of signal magnitude, the effective gain being greater for large signals than for small signals. The result is greater volume range in an audio amplifier and greater contrast range in facsimile.

expansion chamber *Cloud chamber.*

expansion orbit The final portion of the electron path in a betatron or other circular-orbit electron tube, in which electrons move outside the equilibrium orbit and impact on the target.

Explorer One of a series of 29 unmanned, highly instrumented earth satellites launched between 1958 and 1965 to obtain data on the atmosphere,

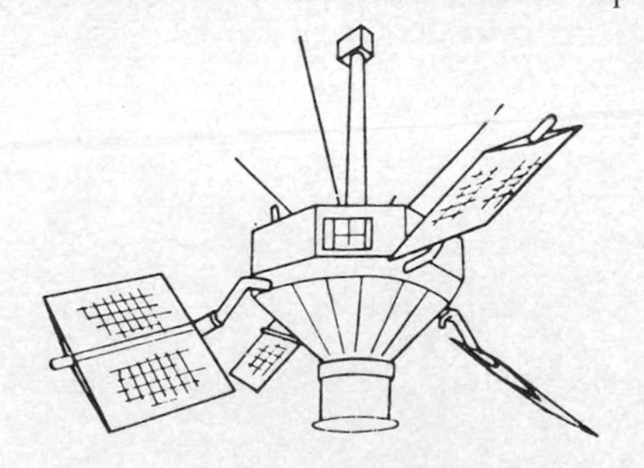

Explorer XIV satellite, showing paddle-wheel mountings for solar cells.

ionosphere, magnetosphere, the near-earth region of interplanetary space, and the earth's surface, for telemetering to receiving stations on earth.

exploring coil A small coil used to measure a magnetic field or to detect changes produced in a magnetic field by a hidden object. The coil is connected to an indicating instrument either directly or through an amplifier. Also called magnetic test coil and search coil.

explosive bonding The use of small explosive charges to make connections by pressure in microelectronic circuits.

exponential absorption The removal of particles or photons from a beam at an exponential rate as the beam passes through matter.

exponential amplifier An amplifier capable of supplying an output signal proportional to the exponential of the input signal.

exponential antenna A television receiving antenna that has a series of active elements mounted parallel to each other, with element lengths adjusted so their ends form two natural logarithmic curves. The antenna gives good gain over the VHF and UHF television bands.

exponential decay Decay of radiation, charge, signal strength, or some other quantity at an exponential rate.

exponential flareout *Altimetric flareout.*

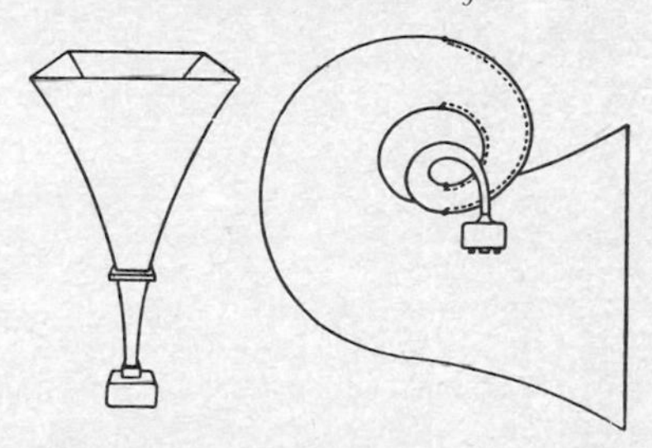

Exponential horns, straight and curled.

exponential horn A horn whose cross-sectional area increases exponentially with axial distance.

exponential transmission line A two-conductor transmission line whose characteristic impedance varies exponentially with electrical length along the line.

exponential well A potential well having an exponentially changing value.

exposure dose A measure of x-ray or gamma radiation at a point, based on its ability to produce ionization. The unit of exposure dose is the roentgen.

exposure-dose rate The exposure dose per unit time. The unit of exposure-dose rate is the roentgen per unit time.

exposure meter An instrument used to measure the intensity of light reflected from an object, for the purpose of determining proper camera exposure. Modern exposure meters consist of a photovoltaic cell connected to an indicating meter.

extended-area service A telephone exchange service, without toll charges, that extends over an area where there is a community of interest, in return for a somewhat higher exchange service rate.

extended-interaction tube A microwave tube in which a moving electron stream interacts with a traveling electric field in a long resonator. Bandwidth is between that of klystrons and traveling-wave tubes.

extended-play record A 7-in (17-cm) 45-rpm record having closely spaced grooves giving up to 8 min of playing time per side, as compared to 3 to 5 min per side with conventional 45-rpm microgroove records.

extended-range DOVAP A baseline extension of the DOVAP system for tracking missiles and spacecraft.

extension cord A line cord having a plug at one end and an outlet at the other end.

extensive shower A cosmic-ray shower occurring over an area of about 100 m in diameter, presumably initiated high in the atmosphere by a single cosmic ray having energy as high as 10^{17} eV. Also called Auger shower and giant air shower.

externally quenched counter tube A radiation-counter tube that requires an external quenching circuit to inhibit reignition.

external storage A storage that is separate from a digital computer but holds information stored in language acceptable to the machine, such as on recorded magnetic tape or punched cards.

extinction voltage The lowest anode voltage at which a discharge is sustained in a gas tube.

extract 1. To form a new computer word by extracting and putting together selected segments of given words. 2. To remove from a computer register or memory all items that meet a specified condition.

extragalactic radio source A discrete source of radio signals outside known galaxies or quasars. More than 10,000 such sources have been detected with radio telescopes, at frequencies ranging from 10 MHz to 100 GHz.

extra-high tension [abbreviated EHT] British term for the high DC voltage applied to the second anode in a cathode-ray tube, ranging from about 4 to 50 kV in various sizes of tubes.

extra-high voltage [abbreviated EHV] A voltage above 345 kV when used for power transmission.

extraordinary component The component of light that is plane-polarized and passes through a Nicol prism. The ordinary component is totally reflected by the prism.

extraordinary-wave component The magneto-ionic wave component in which the electric vector rotates in the opposite sense to that for the ordinary-wave component. Also called X wave.

extrapolate To extend the range of known values by estimating. Interpolate means to estimate missing values between those that are known.

extrapolation ionization chamber An ionization chamber so designed that volume, electrode separation, or some other factor can be varied in suitable steps for measurement purposes. The resulting measured values are plotted in appropriate form, and the desired result is obtained by extrapolation of the curve.

extraterrestrial noise Cosmic, solar, radio, and other electromagnetic noise from sources not related to the earth.

extremely high frequency [abbreviated EHF] A Federal Communications Commission designation for a frequency in the band from 30 to 300 GHz, corresponding to a millimetric wave between 1 and 10 mm.

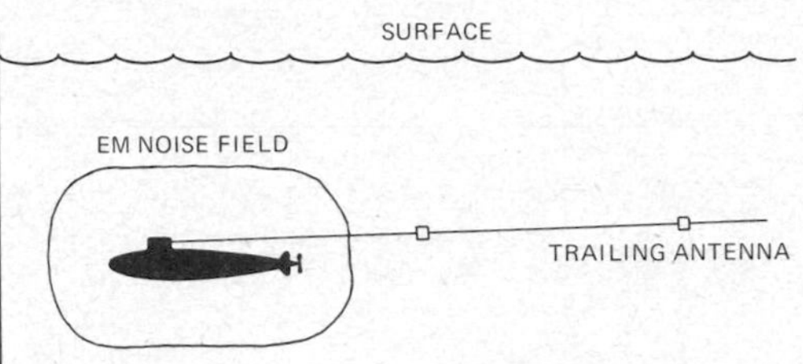

Extremely low frequency signals, in range of 30 to 100 Hz, would reach submerged nuclear submarine trailing long antenna.

extremely low frequency [abbreviated ELF] A frequency below 300 Hz in the radio spectrum, corresponding to a wavelength above 1000 km.

extrinsic properties The properties of a semiconductor as modified by impurities or imperfections within the crystal.

extrinsic semiconductor A semiconductor whose electrical properties are dependent on impurities added to the semiconductor crystal, in contrast to an intrinsic semiconductor whose properties are characteristic of an ideal pure crystal.

eyelet A short length of metal tubing sometimes used in terminal holes of printed-circuit boards to provide a through connection and a tight mechanical contact with printed wiring prior to soldering. The opposite ends of the tubing are flared out to provide the mechanical contact, one flared end being preformed, and the other being flared after insertion of the eyelet.

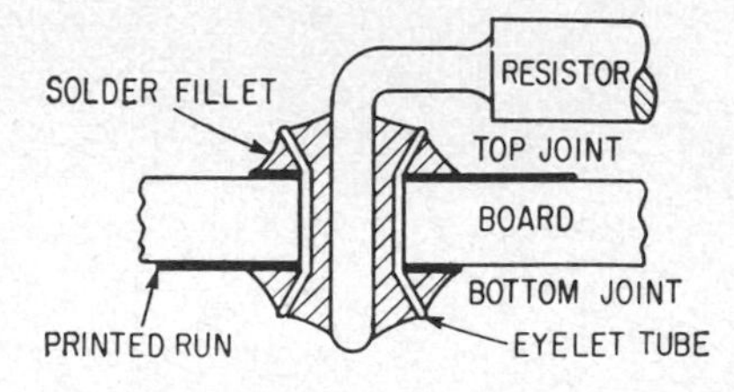

Eyelet in printed-circuit board.

F

f Abbreviation for *femto-*.
f Symbol for *frequency*.
F 1. Symbol for *filament*. 2. Symbol for *fuse*. 3. Abbreviation for *Fahrenheit*. 4. Abbreviation for *farad*.
°F Abbreviation for degree Fahrenheit.
F− *A−*.
F+ *A+*.
fA Abbreviation for *femtoampere*.
FAA Abbreviation for *Federal Aviation Agency*.
face *Faceplate*.
face bond A bond that is made directly between the bonding pad of a face-down semiconductor chip and a mating contact on a mounting substrate.
faceplate The transparent or semitransparent glass front of a cathode-ray tube, through which the image is viewed or projected. The inner surface of the face is coated with fluorescent chemicals that emit light when hit by an electron beam. Also called face.
facility A manufacturing plant, piece of equipment, or other operating entity that serves a specific purpose.
facsimile [abbreviated FAX] A system of communication in which a photograph, map, or other fixed graphic material is scanned and the information converted into signal waves for transmission by wire or radio to a facsimile receiver at a remote point. Also called phototelegraphy, radiophoto, telephoto, telephotography, and wirephoto.
facsimile converter A device that changes the type of modulation in a facsimile system.
facsimile receiver The receiver used to translate the facsimile signal from a wire or radio communication channel into a facsimile record of the subject copy.
facsimile recorder The section of a facsimile receiver that performs the final conversion of electric picture signals to an image of the subject copy on the record medium. The image may be produced by a modulated light beam focused on photographic printing paper mounted on a rotating drum, by sending electric current through special paper mounted on the drum, or by using a pressure stylus with appropriate pressure-sensitive paper on the drum.
facsimile signal The picture signal produced by scanning the subject copy in a facsimile transmitter.
facsimile transmission The transmission of signal waves produced by scanning fixed graphic material, including pictures, for reproduction in record form.
facsimile transmitter The apparatus used to translate the subject copy into facsimile signals suitable for delivery to a communication system. In most transmitters the subject copy is placed on a rotating drum driven by a synchronous motor, for scanning by a photoelectric system mounted on a carriage that moves along the length of the drum. Alternatively, the rotating drum may move axially past a stationary photoelectric system.
facsimile transmitting converter A converter that changes amplitude modulation to frequency-shift modulation in a facsimile system.
fade To change signal strength gradually.
fade in To increase signal strength gradually, as at the start of a radio or television program or when changing to a new scene. This makes sound volume and picture brightness increase gradually.
fade out To reduce signal strength gradually in a sound or television channel. Often done at the end of a program to make a picture go dark gradually, with volume of background music decreasing correspondingly.

fadeout A gradual and temporary loss of a received radio or television signal caused by magnetic storms, atmospheric disturbances, or other conditions along the transmission path. A blackout is a fadeout that may last several hours or more at a particular frequency.

fader A multiple-unit volume control used for gradual changeover from one microphone, audio channel, or television camera to another. In each case the level is held essentially constant because one section of the fader increases signal level in its channel while the other fader section reduces signal level a corresponding amount.

fading Variations in the field strength of a radio signal, usually gradual, that are caused by changes in the transmission medium.

fading margin An attenuation allowance made in radio system planning so that anticipated fading will still keep the signal above a specified minimum signal-to-noise ratio.

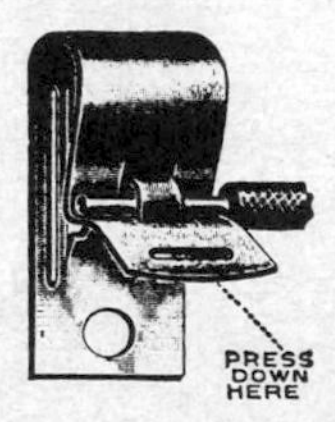

Fahnestock clip.

Fahnestock clip A spring-type terminal to which a temporary connection can be made readily.

Fahrenheit [abbreviated F] A temperature scale in which the freezing point of water is 32° and the boiling point of water is 212° at normal atmospheric pressure. Absolute zero is −459.6°F. To change degrees Fahrenheit to degrees centigrade, subtract 32 and then multiply the result by 5/9. Fahrenheit and centigrade scales agree at one point: −40°F = −40°C. Multiply degrees centigrade by 9/5 and add 32 to get degrees Fahrenheit.

fail-safe control A control so designed that control-circuit failure cannot cause a dangerous condition under any circumstances.

fairing A contoured structure mounted on a sonar dome that projects from the hull of a ship and reduces the water resistance around the dome when the ship is in motion.

fairlead A plastic, wood, or metal tube with a funnel-shaped end through which a trailing antenna is reeled on an aircraft.

Falcon An Air Force air-to-air guided missile having either radar or infrared homing guidance, a speed of about Mach 2, and a range of about 5 mi (8 km). It can be carried either internally or externally on interceptor aircraft. Some models have nuclear warheads.

fall time The time for the voltage at the trailing edge of a pulse to change from 90 to 10% of the original negative or positive amplitude.

false course A spurious additional courseline indication produced by a navigation aid, caused by undesired reflections or maladjusted equipment.

false echo A misleading extra echo indication on the plan-position indicator of a ship radar set. On rivers, shore structures may cause false echoes. A target may also be hit by a side lobe of the beam, and the echo reflected back into the main beam path by a mast or funnel.

fan beam 1. A radio beam having an elliptically shaped cross section in which the ratio of the

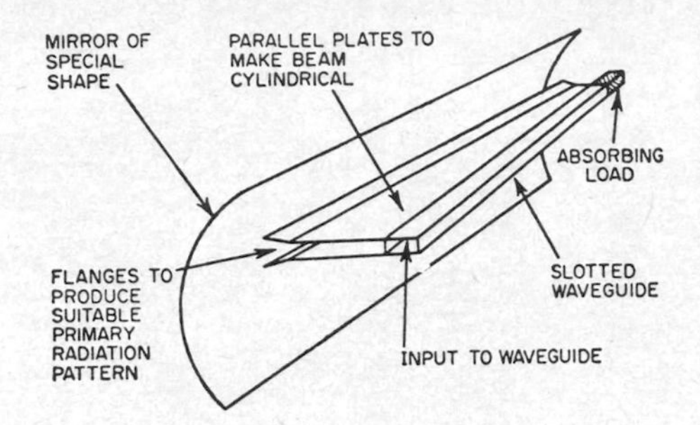

Fan beam is produced by special waveguide feed to mirror that is section of cylinder.

major to minor axes usually exceeds 3 to 1. The beam is broad in the vertical plane and narrow in the horizontal plane. 2. A radar beam having the shape of a fan.

fan dipole A dipole antenna that uses triangular sheets of metal instead of rods for reception of all

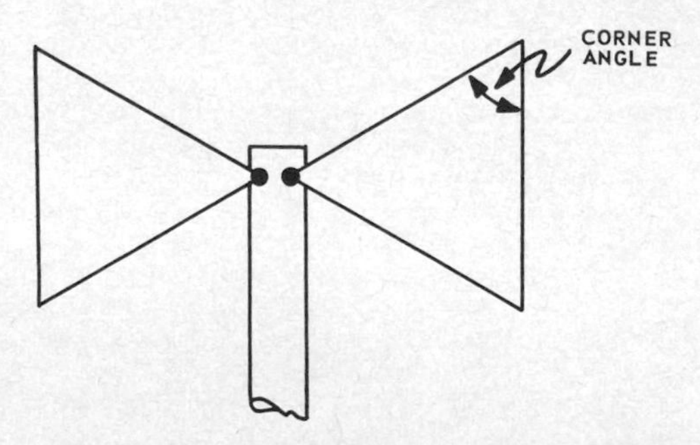

Fan dipole using triangular sheets of metal connected to 300-Ω transmission line. Total width is about 16 in (41 cm), and corner angle is 70°.

signals in the UHF television band. Input impedance is about 300 Ω. The figure-eight radiation pattern can be changed to a single directional lobe by placing a wire-screen reflector behind the dipole.

fan-in The number of inputs that can be connected to a logic circuit.

fan-in circuit A circuit having many inputs feeding to a common point.

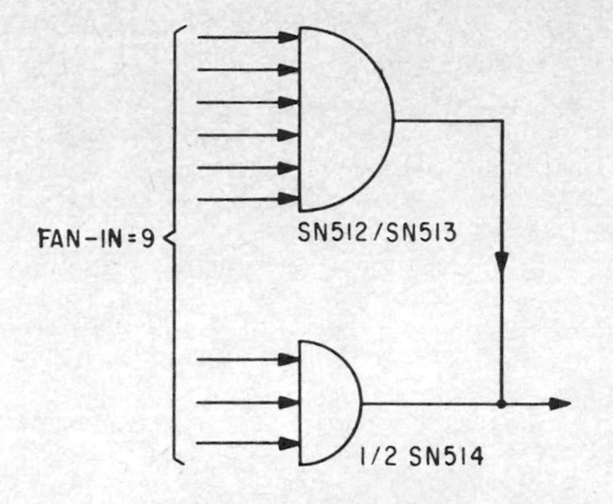

Fan-in circuit for logic gates.

fan marker *Fan marker beacon.*

fan marker beacon A VHF radio facility having a vertically directed fan beam intersecting an airway to provide a fix. When used near an airport as part of an instrument landing system, it may be called a boundary, middle, or outer marker beacon. Also called fan marker and radio fan marker beacon.

fanned-beam antenna A unidirectional antenna so designed that transverse cross sections of the major lobe are approximately elliptical. Used to produce a fan beam.

fanning beam A radio beam that is swept back and forth repeatedly over a limited arc.

fan-out The number of parallel loads that can be driven simultaneously from one state to another by a particular logic circuit.

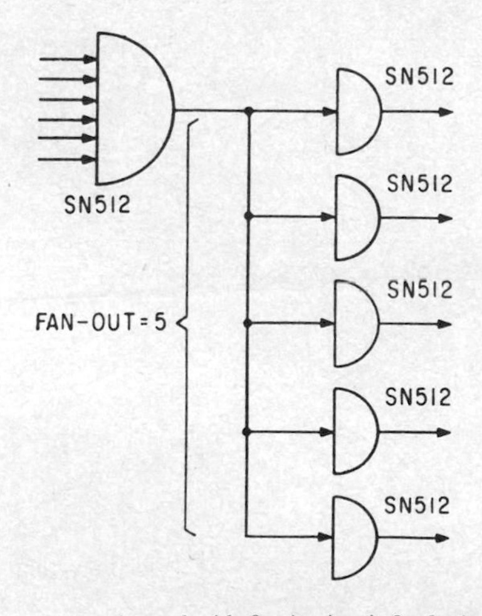

Fan-out circuit used with fan-in circuit for logic gates.

fan-out circuit A circuit having a single output point that feeds many branches.

fan-top radiator A heat dissipator designed to be pushed over the metal housing of a transistor mounted close to a printed-circuit board. The thin

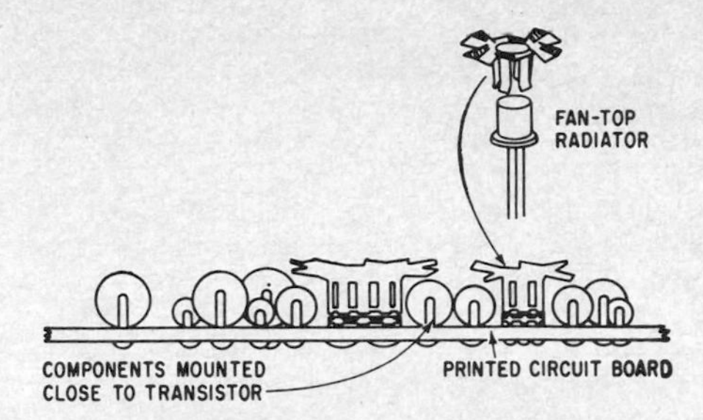

Fan-top radiator.

fans direct cooling air downward from the forced-air path above the board.

farad [abbreviated F] The SI unit of capacitance. A capacitor has a capacitance of 1 F when a voltage change of 1 V/s across it produces a current of 1 A. The farad is too large a unit for practical work, so two smaller units are generally used. The microfarad is equal to one-millionth of a farad, and the picofarad is equal to one-millionth of a microfarad.

faraday A unit of quantity of electricity equal to 96,500 C.

Faraday dark space The relatively nonluminous region that separates the negative glow from the positive column in a cold-cathode glow-discharge tube.

Faraday effect When a plane-polarized beam of light passes through certain transparent substances in a direction parallel to the lines of a strong magnetic field, the plane of polarization is rotated a certain amount. The same effect governs the action of a ferrite rotator in a waveguide.

Faraday rotation isolator An isolator used in circular waveguides to pass a wave traveling in one direction but not in the other direction. A ferrite

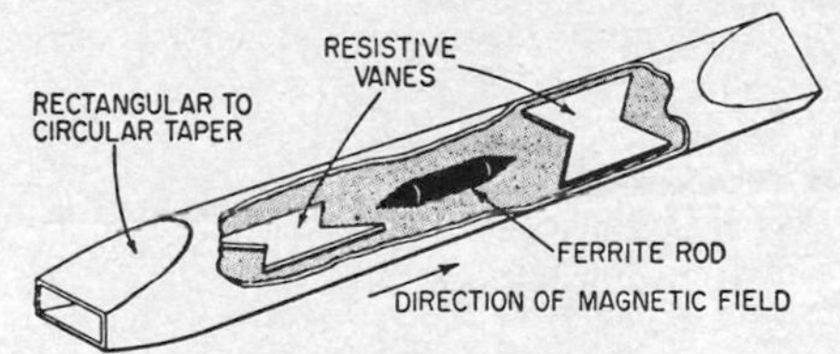

Faraday rotation isolator.

rod is positioned between resistive vanes to rotate the plane of polarization of the waves 45°. With appropriate rectangular-to-circular transitions, the isolator may be used with rectangular waveguides.

Faraday shield An electrostatic shield made of a series of parallel wires connected at one end like a comb. The common point is grounded. The shield provides electrostatic shielding while passing electromagnetic waves.

Faraday's law The voltage induced in a circuit is proportional to the rate at which the magnetic flux

Faraday's law in action—when copper disk is rotated between poles of permanent magnet, voltage is developed between center of disk and circumference.

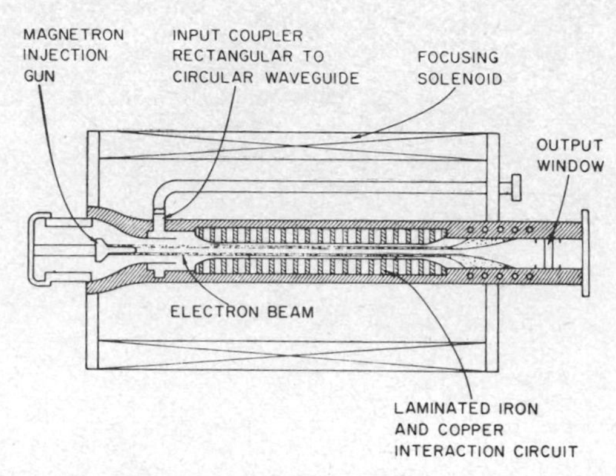

Fast-wave tube feeding circular waveguide. Peak power for low duty cycle is 150 kW at 54 GHz.

linkages of the circuit are changing. Also called law of electromagnetic induction.

faradic current An intermittent and nonsymmetrical alternating current like that obtained from the secondary winding of an induction coil. Used in electrobiology.

faradization Use of a faradic current to stimulate muscles and nerves.

far-end crosstalk Crosstalk that travels along the disturbed circuit in the same direction as desired signals in that circuit. When it occurs at carrier telephone repeater stations, the output signals of one repeater go out also over the output line for the other repeater.

far field The radiation field in the Fraunhofer region surrounding a transmitting antenna.

far region *Fraunhofer region.*

far zone *Fraunhofer region.*

fast-access storage The section of a computer storage from which data can be obtained most rapidly.

fastener A device for holding two or more parts together.

fast forward High-speed forward travel of magnetic tape, used to reach another part of the tape when intervening recorded material is not desired.

fast Fourier transform [abbreviated FFT] A Fourier transform method for calculating the frequency spectrum, in both magnitude and angle, for any function of time by means of special operating programs that speed machine computation of complex Fourier series.

fast-time-constant circuit A circuit that has a short time constant, used in radar to emphasize short-duration signals, discriminate against low-frequency components of clutter, and suppress rain and snow echoes.

fast-wave tube A high-power linear-beam electron tube in which interaction occurs between a waveguide mode with a phase velocity above the velocity of light and a spatially periodic electron beam.

fathometer A sonar-type instrument used to measure ocean depth and locate underwater objects like schools of fish. A sound pulse is transmitted vertically downward by a piezoelectric or magnetostriction transducer mounted on the hull of the ship. The time required for the pulse to return, after reflection from the bottom of the sea or

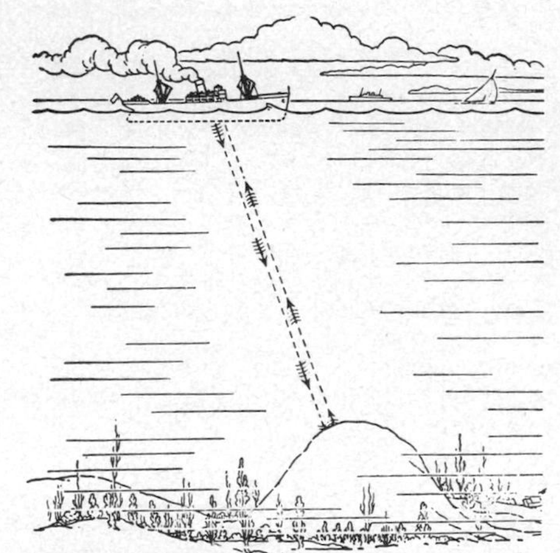

Fathometer on ship sends ultrasonic pulses downward, and echo pulses are reflected back to ship from bottom.

an intervening object, is measured electronically and converted into a depth indication on a dial or recorder chart. Most fathometers are calibrated for an assumed average velocity of sound in sea water of 4800 ft (1460 m)/s. The exact velocity varies with temperature, pressure, and salinity of the water. Also called acoustic depth finder, depth finder, depth sounder, echo depth sounder, and sonic depth finder.

fatigue The decrease of efficiency of a luminescent or light-sensitive material as a result of excitation.

fault A defect, such as an open, short-circuit, or ground in a circuit, component, or line.

fault electrode current The current to an electrode under fault conditions, such as during arcbacks and load short-circuits.

FAX Abbreviation for *facsimile*.

fc Abbreviation for *footcandle*.

FCC Abbreviation for *Federal Communications Commission*.

F display A rectangular radarscope display in which a target appears as a centralized bright spot

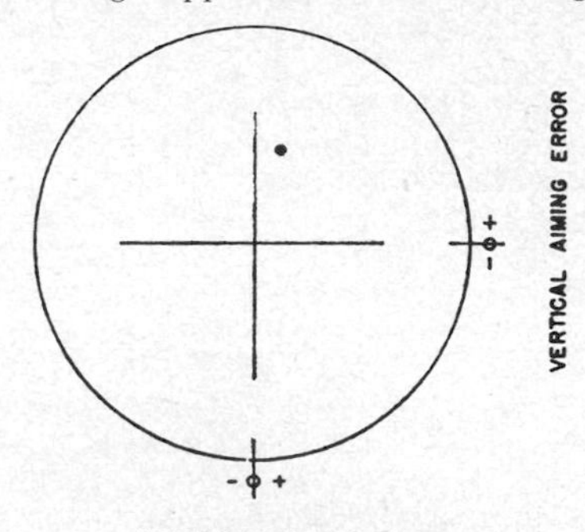

F display.

when the radar antenna is aimed at it. Horizontal and vertical aiming errors are respectively indicated by the horizontal and vertical displacement of the spot.

FDM Abbreviation for *frequency-division multiplex*.

FDMA Abbreviation for *frequency-division multiple-access*.

FDS law Abbreviation for *Fermi-Dirac-Sommerfield velocity-distribution law*.

Federal Aviation Agency [abbreviated FAA] An agency created by Congress in 1958, with full authority over both military and civilian airspace requirements.

Federal Communications Commission [abbreviated FCC] A board of seven commissioners appointed by the President under the Communications Act of 1934, having the power to regulate all electric communication systems originating in the United States, including radio, television, facsimile, telegraph, telephone, and cable systems.

feed 1. To supply a signal to the input of a circuit, transmission line, or antenna. 2. The part of a radar antenna that is connected to or mounted on the end of the transmission line and radiates RF energy to the reflector or receives energy therefrom.

feedback The return of a portion of the output of a circuit or device to its input. With positive feedback, the signal fed back is in phase with the input and increases amplification but may cause oscillation. With negative feedback, the signal fed back is 180° out of phase with the input and decreases amplification but stabilizes circuit performance and tends to minimize noise and distortion.

feedback amplifier An amplifier in which a passive network is used to return a portion of the output signal to its input in such a way as to change the performance characteristics of the amplifier.

feedback controller *Feedback control system*.

feedback control system A control system, comprising one or more feedback control loops, in which functions of the controlled signals are combined with functions of the commands to tend to maintain prescribed relationships between the commands and the controlled signals. Also called closed-loop control system and feedback controller.

feedback cutter A disk recording cutter that has an auxiliary feedback coil in the magnetic field. Signals exciting the cutter are induced into the feedback coil and fed back to the input of the cutter amplifier. This feedback gives a substantially uniform frequency response.

feedback oscillator An oscillating circuit, including an amplifier, in which the output is fed back in phase with the input. Oscillation is maintained at a frequency determined by the values of the components in the amplifier and the feedback circuits.

feedback path The transmission path from the loop output signal to the loop feedback signal in a feedback control loop.

feedback regulator A feedback control system that tends to maintain a prescribed relationship between certain system signals and other predetermined quantities. Some of the system signals in a regulator are adjustable reference signals. Under certain methods of operation, a feedback regulator may also be a servomechanism.

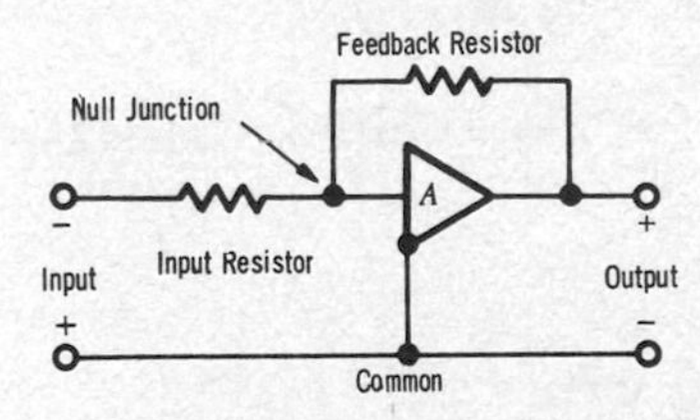

Feedback resistor as used with inverting operational amplifier A.

feedback resistor The resistor used with an operational amplifier to couple output to input and act as a voltage control.

feedback signal *Primary feedback*.

feedback winding A winding to which feedback connections are made in a magnetic amplifier.

feeder A transmission line used between a transmitter and an antenna.

feed-forward control Process control in which

changes are detected at the process input, and an anticipating correction signal is applied before process output is affected.

feed-forward network An external distortion-reducing network that can be used with a class C amplifier in communication-satellite service. Signal samples taken before and after the distorting amplifier are used to obtain an error signal. This error signal is processed and combined beyond the amplifier output, to provide an undistorted signal by cancellation of common error terms.

feed hole One of a series of holes punched in paper tape at fixed intervals to engage with the sprocket teeth of a drive wheel.

feed pitch The distance between the centers of feed holes in paper tape or sprocketed film.

feed reel The reel from which paper or magnetic tape is being fed.

feedthrough A conductor that connects patterns on opposite sides of a printed-circuit board. Also called interface connection.

feedthrough capacitor A feedthrough insulator that provides a desired value of capacitance between the feedthrough conductor and the metal chassis or panel through which the conductor is passing. Used chiefly for bypass purposes in UHF circuits.

feedthrough insulator *Feedthrough terminal.*

feedthrough terminal An insulator designed for mounting in a hole in a panel, wall, or bulkhead, with a conductor in the center of the insulator to permit feeding electricity through the partition. Also called feedthrough insulator.

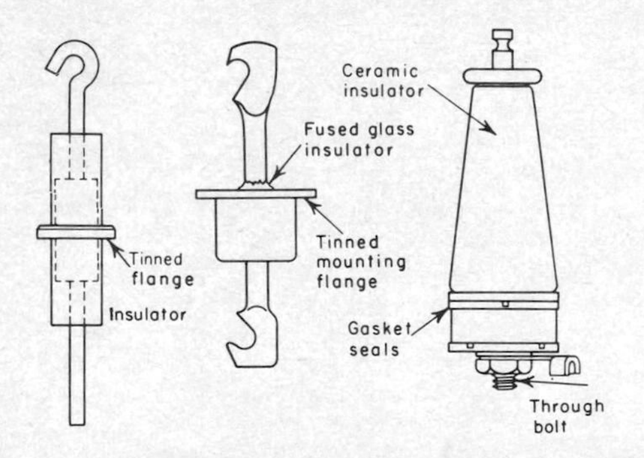

Feedthrough terminals.

female connector A connector that has one or more contacts set into recessed openings. Jacks, sockets, and wall outlets are examples of female connectors.

femto- [abbreviated f] A prefix representing 10^{-15}, which is 0.000 000 000 000 001, or one-thousandth of a millionth of a millionth.

femtoampere [abbreviated fA] A unit of current equal to 10^{-15} A.

femtosecond [abbreviated fs] A unit of time equal to 10^{-15} s.

femtovolt [abbreviated fV] A unit of voltage equal to 10^{-15} V.

femtowatt [abbreviated fW] A unit of power equal to 10^{-15} W.

fence 1. A line or network of early-warning radar stations. 2. A concentric steel fence erected around a ground radar transmitting antenna to serve as an artificial horizon and suppress ground clutter that would otherwise drown out weak signals returning at a low angle from a target.

Fermat's principle An electromagnetic wave will take a path that involves the least travel time when propagating between two points.

Fermi-age model A model used in studying the slowing down of neutrons by elastic collisions. It is assumed that the slowing down takes place by a very large number of very small energy changes.

Fermi constant A universal constant, introduced in beta-disintegration theory, that expresses the strength of the interaction between the transforming nucleon and the electron-neutrino field.

Fermi-Dirac distribution function A function that has a value in the range from zero to unity, specifying the probability that an electron in a semiconductor will occupy a certain quantum state of energy when thermal equilibrium exists. The energy for which the value of the function is 0.5 is called the Fermi level.

Fermi-Dirac-Sommerfield velocity-distribution law [abbreviated FDS law] An equation that gives the number of particles in a quantized system in equilibrium.

Fermi-Dirac statistics Quantum statistics in which no more than one of a set of identical particles may occupy a particular quantum state.

Fermi level The level of the electron energy at which the Fermi-Dirac distribution function has a value of 0.5.

fermion A particle that obeys Fermi-Dirac statistics. All known fermions have total angular moments of $(n + \frac{1}{2})\ h$, where n is zero or an integer, and h is the quantum or unit of orbital angular momentum.

Fermi plot *Kurie plot.*

Fermi selection rules A set of selection rules for allowed beta transitions.

fermium [symbol Fm] A transuranic radioactive element produced artificially by bombardment of einsteinium with alpha particles or neutrons. Atomic number is 100.

fernico An iron-nickel-cobalt alloy used for metal-to-glass seals. Not a trademark.

ferpic [FERroelectric PICture] A storage and display device in which an image can be stored as a spatially varying birefringence and viewed directly or projected onto a screen by suitably polarized light. Based on electrooptical properties of lead zirconate-lead titanate ferroelectric ceramics.

ferreed switch A switch whose contacts are mounted on magnetic blades or reeds sealed into an evacuated tubular glass housing, the contacts being operated by external electromagnets or permanent magnets. Originally, ferrite material was used for the external magnetic members. Recent ferreed switches use a cobalt-iron-vanadium alloy such as Remendur for the external magnetic structure. Here, the switch stays closed by magnetic attraction of the reeds and can be opened only by applying another pulse to reverse the polarity of the external Remendur magnet.

ferret An aircraft equipped to detect, locate, record, and analyze electromagnetic radiation.

ferric Containing a trivalent compound of iron.

ferric oxide [symbol Fe_2O_3] A magnetic iron oxide (red) used as a coating on magnetic recording tapes.

ferrimagnetic amplifier A microwave amplifier that uses ferrites.

ferrimagnetic limiter A power limiter used in microwave systems to replace TR tubes. This limiter uses ferrimagnetic material that exhibits nonlinear properties, such as a piece of ferrite or garnet.

ferrimagnetic material A ferrite material that exhibits ferrimagnetism, such as yttrium-iron garnet polycrystalline materials.

ferrimagnetism A type of magnetism in which the magnetic moments of neighboring ions tend to align antiparallel to each other. The moments are of different magnitudes, however, so there can be a large resultant magnetization. Observed in the ferrites and similar compounds.

ferristor A miniature two-winding saturable reactor that operates at a high carrier frequency and may be connected as a coincidence gate, current discriminator, free-running multivibrator, oscillator, or ring counter.

ferrite A powdered, compressed, and sintered magnetic material that has high resistivity, consisting chiefly of ferric oxide combined with one or more other metals. The high resistance makes eddy current losses extremely low at high frequencies. Examples of ferrite compositions include nickel ferrite, nickel-cobalt ferrite, manganese-magnesium ferrite, yttrium-iron garnet, and single-crystal yttrium-iron garnet. Also called ferrospinel.

ferrite bead A bead made of ferrite powder, placed on a connecting wire in a high-frequency circuit to introduce inductance for suppressing parasitic oscillations. Also used to store information in computer memories.

ferrite circulator A combination of two dual-mode transducers and a 45° ferrite rotator, used with rectangular waveguides to control and switch microwave energy.

ferrite core A magnetic core made of ferrite material. Also called dust core and powdered iron core.

ferrite-core memory A magnetic memory consisting of read-in and read-out wires threaded through a matrix of tiny toroidal cores molded from a square-loop ferrite. Some cores of this type are only 0.018 in (0.5 mm) in diameter.

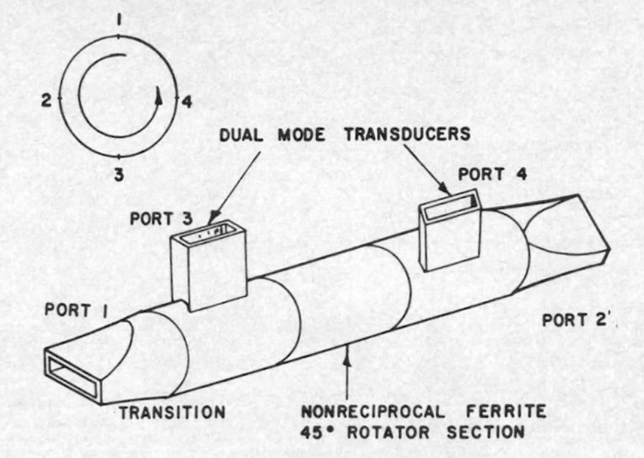

Ferrite circulator with four rectangular ports.

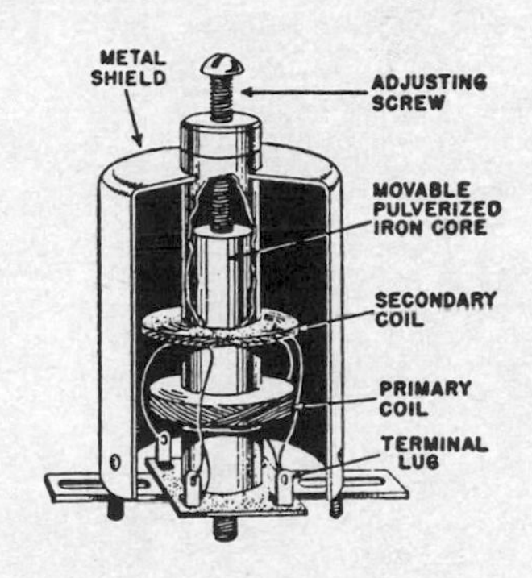

Ferrite core used to change coupling between primary and secondary of IF transformer.

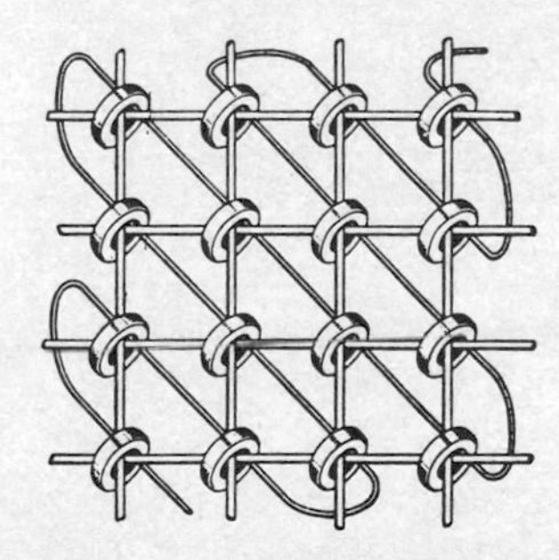

Ferrite-core memory construction. Single matrix for computer may have more than 50,000 such cores.

ferrite isolator An isolator that passes energy in one direction through a waveguide while making possible the absorption of energy from the opposite direction. A ferrite rod, centered on the axis of a short length of circular waveguide, is located between rectangular-waveguide end sections displaced 45° with respect to each other. A signal in the desired direction is rotated 45° by the ferrite rod to make it correct for the output waveguide. A backward signal is rotated 45° in the wrong direc-

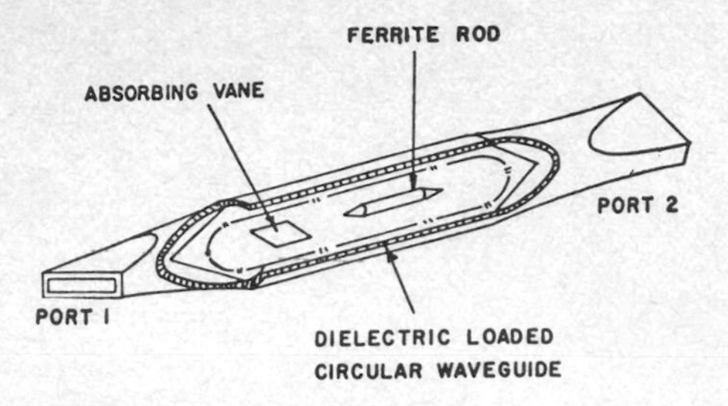

Ferrite isolator.

tion, and its energy is absorbed by resistance cards.

ferrite limiter A passive low-power microwave limiter that has an insertion loss of less than 1 dB when it is operating in its linear range, with minimum phase distortion. The input signal is coupled to a single-crystal sample of either yttrium-iron garnet or lithium ferrite, which is biased to resonance by a magnetic field. Used to protect sensitive receivers from burnout and blocking by a strong interfering signal.

ferrite phase-differential circulator A combination microwave duplexer and load isolator that functions as a switching device between a high-power radar magnetron, a radar receiver, and a radar antenna. It consists of a T junction, a ferrite section, and a short-slot hybrid coupler with a dual adapter that connects a matched load and the antenna to the hybrid. The ferrite section consists of a double waveguide unit, with a wall between the two waveguide lines. Ferrite slabs are placed in the waveguide lines to produce the desired differential phase shift.

ferrite-rod antenna An antenna that consists of a coil wound on a rod of ferrite. Used in place of a loop antenna in radio receivers. The coil generally serves as the tuning inductance for the first stage of the receiver. Also called ferrod and loopstick antenna.

ferrite rotator A gyrator consisting of a ferrite cylinder surrounded by a ring-type permanent magnet, inserted in a waveguide to rotate the

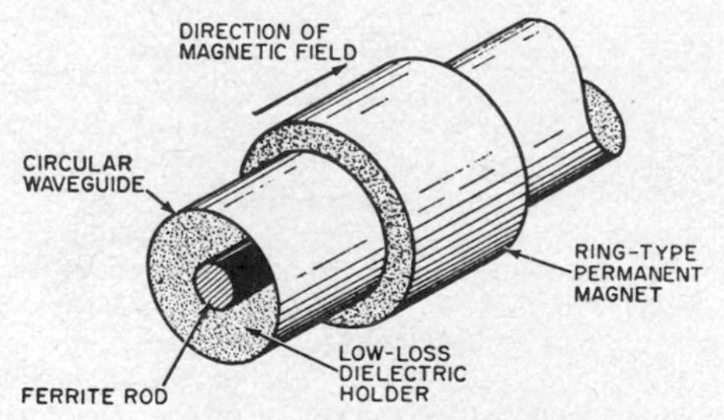

Ferrite rotator in circular waveguide.

plane of polarization of the electromagnetic wave passing through the waveguide. The dimensions of the ferrite cylinder and the magnetization of the ferrite determine the amount of rotation.

ferrite switch A ferrite device that blocks the flow of energy through a waveguide by rotating the electric field vector 90°. The switch is ener-

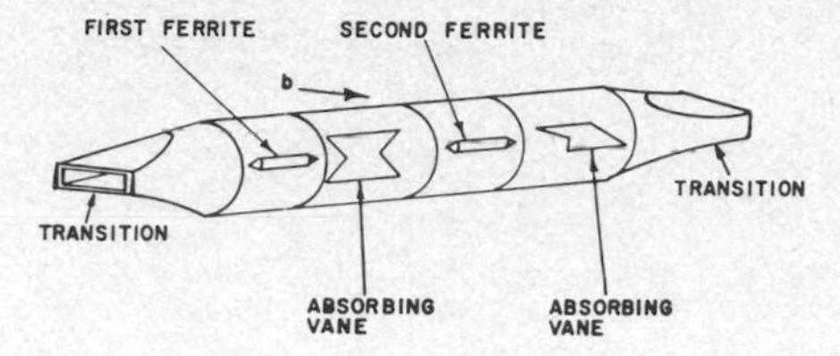

Ferrite switch.

gized by sending direct current through its magnetizing coil. The rotated energy is then reflected from a reactive mismatch or absorbed in a resistive card.

ferrite-tuned oscillator An oscillator in which the resonant characteristic of a ferrite-loaded cavity is changed by varying the ambient magnetic field, to give electronic tuning. In one example, the tuning range is 500 to 1300 MHz.

ferroacoustic storage A delay-line type of storage consisting of a thin tube of magnetostrictive material, a central conductor passing through the tube, and an ultrasonic driving transducer at one

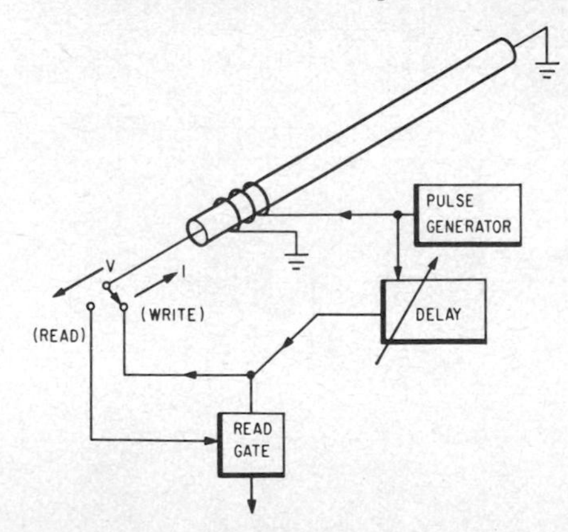

Ferroacoustic-storage principle. At top is thin tube of magnetostrictive material having central conductor wire.

end of the tube. To write, an ultrasonic pulse is first sent down the line from the transducer. After a delay corresponding to the time required for the ultrasonic pulse to reach the desired storage point on the line, a short current pulse is applied to the central conductor. This alters the magnetic state of the line at that point. For readout, an ultrasonic pulse is again sent down the line; after a delay corresponding to the location of the desired stored bit, the gate is opened momentarily to provide access to the voltage pulse produced if a bit is stored at that point when the ultrasonic pulse passes it.

ferrod *Ferrite-rod antenna.*

ferrodynamic instrument An electrodynamic instrument in which the forces are materially augmented by the presence of ferromagnetic material.

ferroelectric converter A converter that transforms thermal energy into electric energy by utilizing the change in the dielectric constant of a ferroelectric material when heated beyond its Curie temperature. A large capacitor using a ferroelectric dielectric, such as barium strontium titanate, initially at its Curie temperature, is charged by a battery through a diode, then heated beyond the Curie temperature. The dielectric constant then drops, and capacitance drops correspondingly, so that capacitor voltage goes up. After this voltage is discharged through a load, the capacitor is cooled and the cycle is repeated. Solor radiation on spacecraft could provide the required heat.

ferroelectric material A nonlinear dielectric material in which electric dipoles line up spontaneously by mutual interaction, just as magnetic dipoles line up in a magnetic material. Examples of ferroelectric materials include barium titanate, potassium dihydrogen phosphate, and Rochelle salt. Used in ceramic capacitors, acoustic transducers, and dielectric amplifiers.

ferroelectric shutter A shutter consisting of a slab of ferroelectric crystal located between polar-

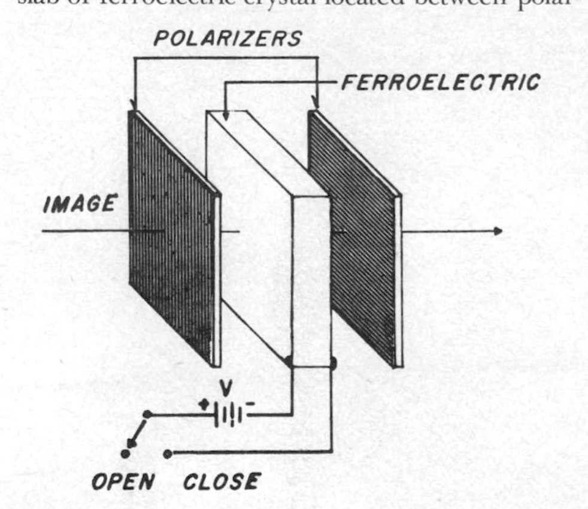

Ferroelectric shutter using single crystal of ferroelectric barium titanate.

izers whose planes are at right angles. The shutter opens to pass light when activated by a pulse of up to 100 V.

ferrofluid A colloidal suspension of ultramicroscopic magnetic particles in a carrier liquid, used as a lubricant or damping liquid. Permanent magnets or electromagnets hold the ferrofluid in position even when under pressure, as in rotary shaft seals.

ferromagnetic amplifier A parametric amplifier based on the nonlinear behavior of ferromagnetic resonance at high RF power levels. In one version, microwave pumping power is supplied to a garnet or other ferromagnetic crystal mounted in a cavity

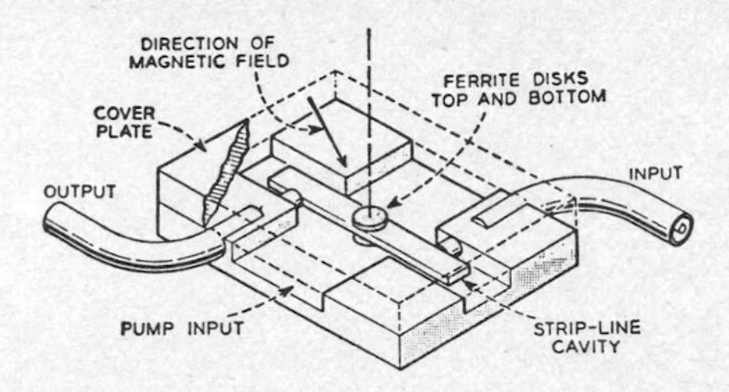

Ferromagnetic amplifier for 4.5-GHz input and output signals, using 9-GHz pump input.

containing a stripline. A permanent magnet provides sufficient field strength to produce gyromagnetic resonance in the garnet at the pumping frequency. The input signal is applied to the crystal through the stripline, and the amplified output signal is extracted from the other end of the stripline. Sometimes incorrectly called a garnet maser, but the operating principle differs from that of masers.

ferromagnetic material A magnetic material that has a permeability considerably greater than the permeability of a vacuum and which varies with the magnetizing force. The various forms of iron, steel, cobalt, nickel, and their alloys are examples.

ferromagnetic resonance Resonance at which the apparent permeability of a magnetic material at microwave frequencies reaches a sharp maximum. This resonance occurs in the presence of a steady transverse magnetic field when the microwave frequency equals the precession frequency of the electron orbits in the atoms of the magnetic material. The resonance frequency depends on the strength of the transverse field.

ferromagnetics The science that deals with storage of information and control of pulse sequences by means of the magnetic polarization properties of materials.

ferromagnetic tape A tape made of magnetic material for use in winding closed magnetic cores of toroids and transformers.

ferromagnetography A printing technique in which magnetic iron particles are applied to a latent image magnetized onto a metal sheet or drum, then transferred to ordinary paper and fixed in position by heat, pressure, or other means. The image can be formed by a row of tiny electromagnets under computer control, to give high-speed printout of characters.

ferrometer An instrument used to make permeability and hysteresis tests of iron and steel.

ferroresonant circuit A resonant circuit in

which a saturable reactor provides nonlinear characteristics, with tuning being accomplished by varying circuit voltage or current.

ferrospinel *Ferrite.*

ferrous Containing a divalent compound of iron.

ferrule terminal A cylindrical end terminal sometimes used on resistors, cartridge fuses, and other parts to permit quick insertion and removal from holders that have corresponding spring contacts.

FET Abbreviation for *field-effect transistor.*

fetal monitor An instrument that monitors fetal heart rate and electric activity during pregnancy or labor by providing phonocardiograph and electrocardiograph recordings of the fetal heart while suppressing maternal heart sounds and extraneous noises.

FET resistor A field-effect transistor in which the gate is generally tied to the drain; the resultant structure is used as a resistance load for another transistor.

FFT Abbreviation for *fast Fourier transform.*

FHP motor Abbreviation for *fractional-horsepower motor.*

fiber bundle A flexible bundle of glass or other transparent fibers, parallel to each other, used to transmit an image from one end of the bundle to the other.

Fiberglas Trademark of Owens-Corning Fiberglas Corp. for their glass fiber materials.

fiber optics The technique of transmitting light through long, thin, flexible fibers of glass, plastic, or other transparent materials. Bundles of parallel fibers can be used to transmit complete images.

fiberscope An arrangement of parallel glass fibers, with an objective lens on one end and an eyepiece at the other end. The assembly can be bent as required to view objects that are inaccessible for direct viewing.

fiber waveguide *Optical waveguide.*

Fibonacci series A series in which each number is equal to the sum of the two preceding numbers, as in 1, 2, 3, 5, 8, 13, and 21. Used in some computer sort programs to control the sequence of distribution of strings of data on work tapes.

Fick's law Diffusion current density is proportional to the negative of the gradient of neutron density.

fidelity The degree to which a system accurately reproduces at its output the essential characteristics of the signal impressed on its input.

field 1. One of the equal parts into which a frame is divided in interlaced scanning for television. A field includes one complete scanning operation from top to bottom of the picture and back again. In the present U.S. television broadcasting system there are two fields per frame, with each field taking 1/60 s and including 262.5 lines. 2. A region containing electric or magnetic lines of force, or both. 3. A set of one or more columns reserved for a particular type of information on a punched card. 4. The area covered by a lens. 5. A set of characters treated as a unit for computer processing, such as a name or amount field. 6. An operating location for equipment.

field coil A coil used to produce a constant-strength magnetic field in an electric motor, generator, loudspeaker, or other electromagnetic device.

field-displacement isolator A ferrite isolator that can be used directly in a rectangular waveguide, eliminating the rectangular-to-circular transitions needed with a Faraday rotation isolator.

field effect The change produced by an electric field on the equilibrium balance of free electrons and holes in a semiconductor material.

field-effect capacitor A capacitor in which the effective dielectric is a region of semiconductor material that has been depleted or inverted by the field effect.

field-effect device A semiconductor device whose properties are determined largely by the effect of an electric field on a region within the semiconductor.

field-effect diode A semiconductor diode in which the charge carriers are of only one polarity. It is usually constructed as an enhancement-mode MOSFET, with the gate shorted to the drain.

field-effect phototransistor A field-effect transistor that responds to modulated light as the input signal. The action is similar to that of a gate signal voltage.

field-effect transistor [abbreviated FET] A transistor in which the resistance of the current path from source to drain is modulated by applying a transverse electric field between grid or gate electrodes. The electric field varies the thickness of the depletion layer between the gates, thereby reducing the conductance. The two basic types are

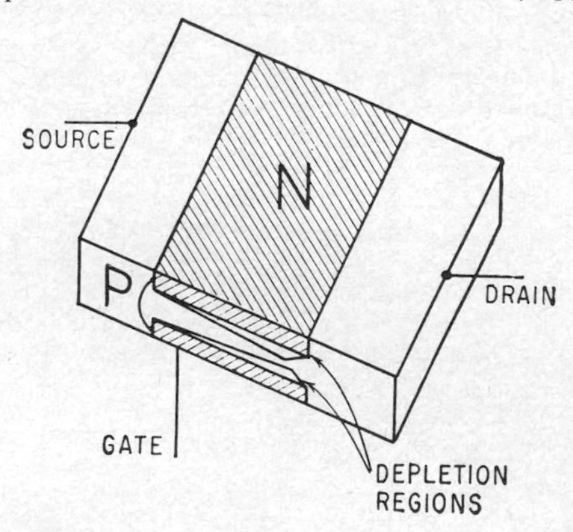

Field-effect transistor having N-type gate regions connected to gate terminal. Source and drain connections are made at opposite ends of P-type wafer.

the junction field-effect transistor (JFET) and the metal-oxide semiconductor field-effect transistor (MOSFET).

field-effect varistor A passive two-terminal nonlinear semiconductor device that maintains constant current over a wide voltage range.

field emission The liberation of electrons from an unheated solid or liquid by a strong electric field at the surface.

field-emitter cathode An unheated metal or semiconductor cathode that emits electrons when exposed to a strong electronic field.

field-enhanced photoelectric emission The increased photoelectric emission resulting from the action of a strong electric field on the emitter.

field-enhanced secondary emission The increased secondary emission resulting from the action of a strong electric field on the emitter.

field-free emission current The electron current emitted by a cathode when the electric field at the surface of the cathode is zero.

field frequency The number of fields transmitted per second in television. In the United States, it is 60 fields per second. The field frequency is equal to the frame frequency multiplied by the number of fields that make up one frame. Also called field repetition rate.

field intensity *Field strength.*

field ion emission microscope A high-magnification microscope in which an intense electric field is applied to a sharp metal point to make movements of atoms visible on the point.

field-neutralizing coil A coil used around the faceplate of a color television picture tube. Direct current is sent through this coil to produce a constant magnetic field that offsets the effect of the earth's magnetic field on the electron beams.

field-neutralizing magnet A permanent magnet mounted near the edge of the faceplate of a color picture tube to serve the same function as a field-neutralizing coil. Also called rim magnet.

field of force A region in space in which force is exerted on electric charges by other stationary or moving charges.

field pattern *Radiation pattern.*

field period The time required to transmit one television field, equal to $1/60$ s in the United States.

field pole A structure of magnetic material on which a field coil of a loudspeaker, motor, generator, or other electromagnetic device may be mounted.

field-programmable logic array [abbreviated FPLA] A programmed logic array in which the internal connections of the logic gates can be programmed once in the field by passing high current through fusible links, by using avalanche-induced migration to short base-emitter junctions at desired interconnections, or by other means. Also called programmable logic array.

field quantum The fundamental field particle that is the result of quantizing a field.

field repetition rate *Field frequency.*

field-sequential color television A color television system in which the individual red, blue, and green primary colors are associated with successive fields.

field-simultaneous system A color television system in which a complete full-color field is presented simultaneously as a unit. The eyes then see a succession of full-color images rather than a succession of primary color fields.

field strength The strength of an electric, magnetic, or electromagnetic field at a point. For electromagnetic radiation, it is generally expressed in volts, millivolts, or microvolts per meter of effective antenna height. Also called field intensity.

field-strength meter A calibrated radio receiver used to measure the field strength of radiated electromagnetic energy from a radio transmitter.

field telephone A portable telephone designed for field or combat use.

field wire An insulated flexible wire or cable used in field telephone and telegraph systems.

FIFO Abbreviation for *first-in first-out.*

figure-eight radiation pattern A radiation pattern that has equal broad lobes 180° apart, resembling the numeral 8.

figure of merit A performance rating that governs the choice of a device for a particular application. Thus the figure of merit of a magnetic amplifier is the ratio of usable power gain to the control time constant.

filament [symbol F] A cathode made of resistance wire or ribbon, through which an electric current is sent to produce the high temperature required for emission of electrons in a thermionic tube. Also called directly heated cathode, filamentary cathode, and filament-type cathode. A filament emits electrons directly, whereas a heater merely provides heat for a separate cathode.

filamentary cathode *Filament.*

filament current The current supplied to the filament of an electron tube for heating purposes.

filament emission Liberation of electrons from a heated filament wire in an electron tube.

filament resistance The resistance in ohms of the filament of an electron tube. For metal filaments, the resistance increases with temperature, so the hot resistance is many times the cold resistance.

filament saturation *Temperature saturation.*

filament transformer A small transformer used exclusively to supply filament or heater current for one or more electron tubes.

filament-type cathode *Filament.*

filament-type tube An electron tube in which electron emission is produced directly by a heated filament.

filament voltage The voltage applied to the terminals of the filament in an electron tube.

filament winding The secondary winding of a power transformer that furnishes AC heater or filament voltage for one or more electron tubes.

filament winding The secondary winding of a power transformer that furnishes AC heater or filament voltage for one or more electron tubes.

file A collection of related records on magnetic tape.

file maintenance A data-processing operation in which a master file is updated on the basis of one or more transaction files.

file protection device A ring that must be in position in the hub of a tape reel if data is to be recorded on that particular tape. A tape reel not equipped with a file protection device can be read but not written.

filled band An energy band in which each energy level is occupied by an electron.

film 1. A thin sheet or coating of material. 2. The layer adjacent to the valve metal in an electrochemical valve, in which is located the high voltage drop when current flows in the direction of high impedance.

film badge A badge containing one or more pieces of unexposed x-ray film, worn on the person as a means of measuring radiation exposure. The films may have different sensitivities, and some of the film may be shielded by a filter from certain types of radiation. The films are removed from time to time, developed, and the resulting emulsion density measured to determine exposure.

film capacitor *Plastic-film capacitor.*

film pickup A special motion-picture film projector combined with a television camera.

film reproducer Equipment in which sprocket-hole film is the medium from which a magnetic recording is reproduced.

film resistor A fixed resistor in which the resistance element is a thin layer of conductive material on an insulated form. The conductive material does not contain either binders or insulating material.

film ring A small film badge mounted on a finger ring.

film scanning The process of converting motion-picture film into corresponding electric signals that can be transmitted by a television system.

film sound recorder Equipment that uses oxide-coated sprocket-hole film as the medium for magnetic recording.

filter A selective device that transmits a desired range of matter or energy while substantially attenuating all other ranges. Thus an electric filter is a network that transmits alternating currents of desired frequencies while substantially attenuating all other frequencies. An acoustic filter transmits only desired sound frequencies. An optical filter transmits desired wavelength ranges in the visible, ultraviolet, and infrared spectrums.

filter capacitor A capacitor used in a power-supply filter system to provide a low-reactance path for alternating currents and thereby suppress ripple currents, without affecting direct currents. Electrolytic capacitors are generally used for this purpose.

filter center An information center at which all radar and other observed information concerning movements of friendly and enemy planes within a certain sector is screened and disseminated.

filter choke An iron-core coil used in a power-supply filter system to pass direct current while offering high impedance to pulsating or alternating currents.

filtration Removal of some components of a heterogeneous beam of x-ray and other radiation by inserting in the beam path a sheet of material like aluminum or copper.

final amplifier The transmitter stage that feeds the antenna.

finder An optical or electronic device that shows the field of action covered by a television camera.

fine chrominance primary The chrominance primary that is associated with the greater transmission bandwidth in the two-primary U.S. system of color television. The fine chrominance primary is the I signal, and has frequency components up to 1.5 MHz. The coarse chrominance primary is the Q signal, and has a bandwidth of only 0.5 MHz.

fine structure The occurrence of a spectral line in spectroscopy as a doublet, triplet, or other multiple, due to interaction between the spin angular momentum and the orbital angular momentum of the electrons in the emitting atoms.

fine tuning control A control used to make small changes in the frequency of the RF oscillator in a television tuner, usually by means of an adjustable capacitor, after switching to a desired channel.

finished crystal blank The finished crystal product after the completion of all processes, including application of electrodes to the faces of the crystal. Also called piezoid.

finite Having fixed and definite limits of magnitudes.

finite clipping Clipping in which the threshold level is large but below the peak input signal amplitude.

fin waveguide A waveguide containing a thin longitudinal metal fin that increases the wavelength range over which the waveguide will transmit signals efficiently. Usually used with circular waveguides.

Firebee A high-speed drone used chiefly as a target.

fire control Control of the aiming and firing of guns, rockets, or guided missiles.

fire-control radar Radar equipment used in a fire-control system.

fire-control system A system that determines the azimuth and elevation at which a gun or missile launcher must be pointed, the instant at which

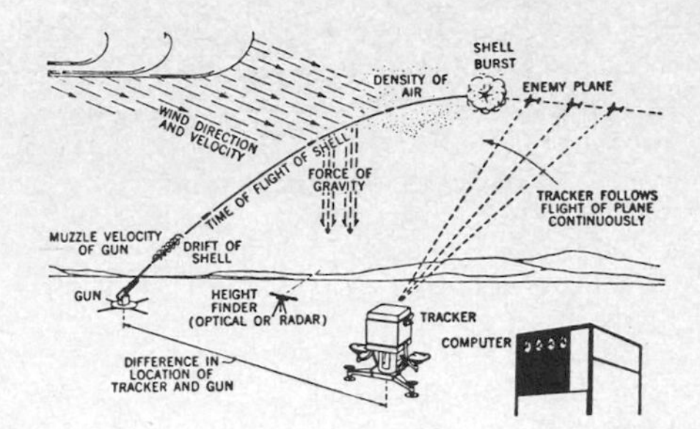

Fire control of antiaircraft gun.

the missile must be launched, and the time for which the fuze must be set if using a time-fuze missile. A typical system includes position-finding, tracking, predicting, ballistic, correcting, data-transmitting, compensating, pointing, and fuze-setting elements.

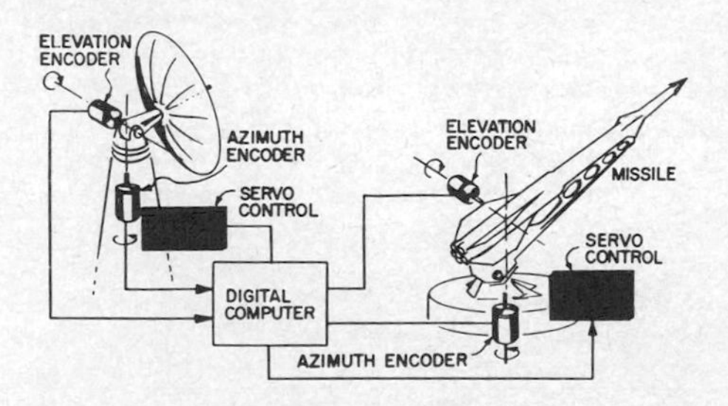

Fire-control system for guided missile.

fired state The on state of a silicon controlled rectifier or other semiconductor switching device, occurring when a suitable triggering pulse is applied to the gate. In a thyratron or other grid-controlled gas tube, the fired state corresponds to complete ionization and maximum conduction in the gas.

fired tube A TR, ATR, or pre-TR tube in which a radio-frequency glow discharge exists at the resonant gap, resonant window, or both.

firing 1. The transition from the unsaturated to the saturated state of a saturable reactor. 2. The gas ionization that initiates current flow in a gas discharge tube. 3. Excitation of a magnetron or TR tube by a pulse. 4. Exposure of a material to high temperature, to oxidize and vaporize organic binders as required for achieving desired properties. Used in thick-film production processes.

firmware A computer program or instruction used so often that it is stored in a read-only memory instead of being included in software. Often used in computers that monitor production processes. A microprogram is an example of firmware, being somewhere between hardware and software in permanence. Firmware may be marketed separately as programmed read-only memories (PROMs).

first breakdown The normal avalanche or zener breakdown of a semiconductor device, as distinct from second breakdown.

first detector *Mixer.*

first Fresnel zone The Fresnel zone that is centered on the line-of-sight path between two microwave antennas and bounded by all paths whose lengths are one half-wavelength longer than the direct path.

first generation In computers and numerical control, the period of technology associated with electron tubes, telephone relays, and stepping relays.

first harmonic *Fundamental frequency.*

first-in first-out [abbreviated FIFO] A method of establishing the order in which data items are taken from a computer memory or products are taken from inventory. In a FIFO memory, input data propagates automatically toward the output terminals. When data is removed from the output, other data in the memory moves down the line automatically to fill the vacant locations.

first-order servo A servo that has zero static error but a finite steady following error for a velocity input.

first quantum number *Main quantum number.*

fir-tree antenna A vertical array of horizontal dipoles fed by transposed two-wire line and backed by a reflector array.

fishbone antenna An end-fire array in which the elements are arranged in a plane along both sides of a transmission line, as in the skeleton of a fish.

Fishbowl A special sonar system used with the Caesar underwater-sound surveillance system.

fishpaper A type of fiber used in sheet form for insulating purposes where high mechanical strength is required, as in insulating transformer windings from the transformer core.

fishpole antenna *Whip antenna.*

fission The splitting of the nucleus of a heavy atom into two or sometimes more nuclei of lighter elements, with the release of substantial amounts of energy. The most important fissionable materials are uranium 235 and plutonium 239.

fissionable Having the property of being able to capture neutrons and thereupon split into two particles having high kinetic energy.

fission chamber An ionization chamber used to detect slow neutrons. The inside wall of the chamber has a thin coating of uranium. A slow neutron produces a fission in the uranium, and the resulting highly ionizing fission fragments produce a count in the chamber.

fix A determination of position without reference to any former position.

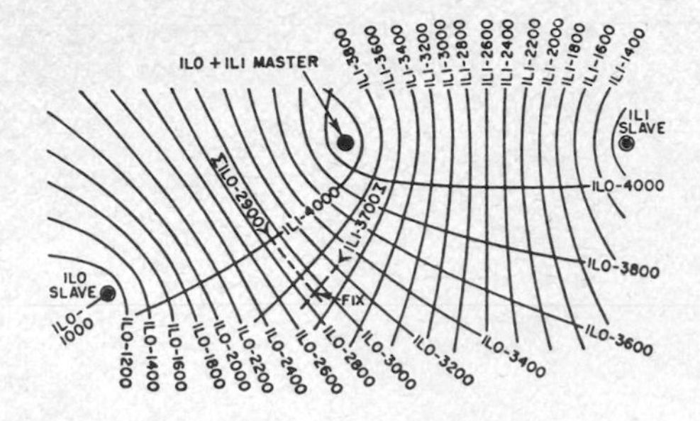

Fix obtained from one master and two slave loran stations. Pair 1L0 gave time difference of 2900 μs and pair 1L1 gave time difference of 3700 μs, so location is at intersection of loran lines 1L0-2900 and 1L1-3700.

fixed attenuator *Pad.*

fixed bias A constant value of bias voltage, independent of signal strength. For an electron-tube circuit it may be provided by a battery or other DC voltage source between cathode and ground or grid and ground, or it may be obtained from a voltage divider connected across the anode voltage source.

fixed capacitor A capacitor that has a definite capacitance value which cannot be adjusted.

fixed contact A relatively immovable contact that is engaged and disengaged by a moving contact to make and break a circuit, as in a switch or relay.

fixed-cycle operation A computer operation that is performed in a fixed amount of time, under synchronous or clock-type control.

fixed echo An echo indication that remains stationary on a radar PPI display, indicating the presence of a fixed target.

fixed-frequency IFF A type of IFF equipment that responds immediately to every interrogation, permitting display of the response on a PPI indicator.

fixed-frequency transmitter A transmitter designed for operation on a single carrier frequency.

fixed-loop radio compass An aircraft radio compass that has a loop antenna fixed in position in such a way that the pointer of the left-right indicator is centered when the aircraft is headed directly toward or away from the transmitting station to which the compass receiver is tuned.

fixed-point Pertaining to a number system in which the location of the point is fixed with respect to one end of the numerals, according to some convention.

fixed-point arithmetic 1. A method of calculation in which the computer does not consider the location of the decimal or radix point because it is given a fixed position. 2. A type of arithmetic in which the operands and results of all arithmetic operations must be properly scaled so as to have a magnitude between certain fixed values.

fixed-point calculation A computer calculation in which a fixed location of the decimal point or binary point in each number is used or assumed.

fixed-point system A point system of positional notation in which the location of the point is assumed to remain fixed with respect to one end of the numerical expressions.

fixed-program computer A computer in which the operating program is wired in or otherwise permanently stored.

fixed resistor A resistor that has no provision for varying its resistance value.

fixed service A radio communication service between specific fixed points.

fixed storage A storage that stores data not alterable by computer instructions, such as magnetic-core storage with a lockout feature.

fixed word length The length of a computer machine word that always contains the same number of characters or digits.

fixer network A network of radio or radar direction-finding stations used together to plot the ground positions of aircraft in flight.

fixture A device that holds a chassis or other part in a desired position during assembly operations but does not guide the tools performing the operations.

fL Abbreviation for *footlambert.*

flag 1. A large sheet of metal or fabric used to shield television camera lenses from light when not in use. 2. A small metal tab that holds the getter during assembly of an electron tube. 3. A special character used in data processing to indicate a change in some condition, such as the beginning or ending of a computer word. Also called tag.

flag alarm A semaphore-type indicator used in electronic navigation instruments to warn that the indications are unreliable.

flame attenuation Attenuation of radio signals by the ionized gases in the exhaust flame of a guided missile when the transmission path between the missile and the ground station passes through the flared-out flame.

flame-failure control A control that senses failure of a pilot burner in a gas-fired furnace and cuts off the fuel supply.

flame laser A molecular gas laser in which gases such as carbon disulfide and oxygen are mixed at low pressures and ignited. The flame is then self-sustaining and produces CO laser emission.

flame-sprayed conductor A conductor deposited on a board in a molten form, generally through a metal mask or stencil, by means of a spray gun that feeds wire into a gas flame and drives the molten particles against the work.

flange A fitting used at the end of a waveguide to bolt it to a microwave component or another waveguide. Also called waveguide flange.

flap attenuator A waveguide attenuator in which a contoured sheet of dissipative material is moved into the guide through a nonradiating slot to provide a desired amount of power absorption. Also called vane attenuator.

flareout That portion of the approach path of an aircraft in which the vertical component is modified to lessen the impact of landing.

flash arc A sudden increase in the emission of large thermionic vacuum tubes, probably due to irregularities in the cathode surface. It sometimes causes complete breakdown.

flashback voltage The peak inverse voltage at which ionization occurs in a gas tube.

flasher A switch that turns lamps on and off at a regular rate. It may be a motor-driven switch, a combination of a bimetallic strip and a heater element, or a solid-state switch operating continuously as an astable multivibrator at the flashing rate.

flashing Application of a high-frequency electromagnetic field to an electron tube to flash its getter during evacuation or to test the quality of the vacuum. A glow is seen inside a tube if there is enough gas left to cause ionization.

flash magnetization Magnetization of a ferromagnetic object by a current impulse of such short duration that magnetization does not penetrate beyond a shallow surface layer of the material. Sometimes used in electromagnetic crack detectors.

flashover An electric discharge around or over the surface of an insulator.

flash photolysis Photolysis in which an intense flash of light is used to decompose an unknown chemical under study. The resulting gaseous products can then be analyzed by spectroscopy.

flash radiography Radiography in which the exposure time is extremely short, such as 1 μs.

flash ranging Determining the position of the burst of a projectile or an enemy gun by observing or measuring its flash.

flash spectroscopy Spectroscopy in which the light source is a flash that is triggered at a precisely controlled time interval, to provide an instantaneous photograph of the spectrum of a chemical during its decomposition.

flash test A method of testing insulation by applying momentarily a voltage much higher than the rated working voltage.

flash tube A gas discharge tube used in a photoflash unit to produce high-intensity, short-duration flashes of light for photography. Also called electronic flash tube and photoflash tube.

flat cable A cable made of round or rectangular, parallel copper wires arranged in a plane and laminated or molded into a ribbon of flexible insulating plastic.

flat-conductor cable A cable made of wide, flat conductors arranged side by side in a plane and protected by ribbons of insulating plastic.

flat fading Fading in which all components of the received radio signal fluctuate simultaneously in the same manner.

flatpack A square or rectangular encapsulated integrated-circuit package with leads coming out from the sides of the package, in the same plane as the package.

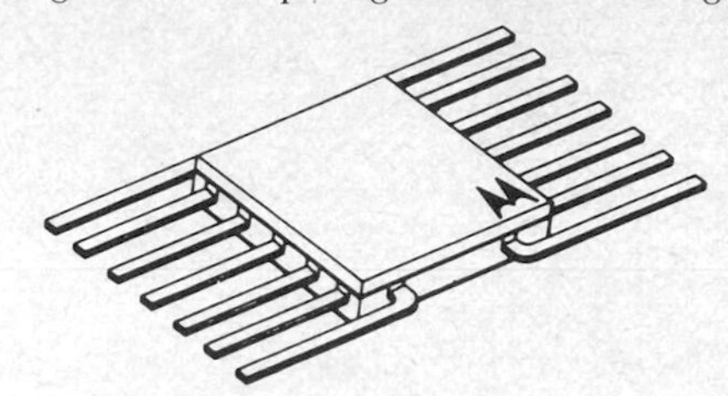

Flatpack with 14 leads.

flat response Uniform amplification or reproduction of a specified band of frequencies.

flat television receiver A television receiver in which the picture-forming device is thin enough so the entire receiver can be hung on the wall like a picture. Experimental cathode-ray picture tubes, and experimental flat panels that use thin-film integrated-circuit technology involving electroluminescence, ferroelectric ceramics, liquid crystals, or other optoelectronic techniques have approached this goal.

flat-top antenna An antenna that has two or more lengths of wire parallel to each other and in a plane parallel to the ground, each fed at or near its midpoint.

flat-top response *Bandpass response.*

flaw detector An ultrasonic, magnetic, or other type of electronic discontinuity sensor used to detect normally invisible cracks or other flaws in a solid material such as a casting, forging, steel rail, or aircraft structural member.

F layer One of the layers of ionized air occurring at various heights in the F region of the ionosphere, capable of reflecting radio waves back to earth at frequencies up to about 50 MHz. In the daytime hemisphere there are normally two F layers, called the F_1 and F_2 layers.

F_1 layer The lower of the two ionized layers normally existing in the F region in the daytime hemisphere. It is usually somewhere between 90 and 150 mi (145 and 240 km) above the earth, and it chiefly affects frequencies from about 1.5 to about 25 MHz.

F_2 layer The single ionized layer normally existing in the F region in the nighttime hemisphere. In the daytime hemisphere it is the higher of the two F layers.

Fleming's rule 1. *Left-hand rule.* 2. *Right-hand rule.*

Fletcher-Munson curves Equal-loudness curves for pure tones, plotted against frequency. They show the average sound intensity needed to produce a given loudness sensation throughout the audio-frequency range.

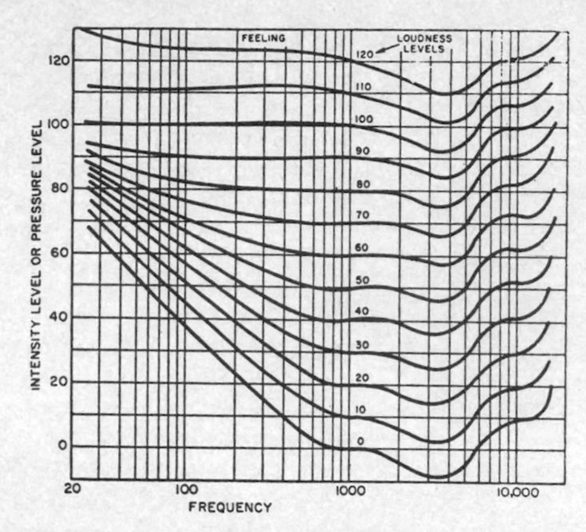

Fletcher-Munson curves of equal-loudness contours for 13 different loudness levels in phons. Frequency scale is in hertz, and intensity level is in decibels.

flexible circuit A printed circuit made on a flexible plastic sheet that is usually die-cut to fit between large components.

flexible coupling 1. A coupling designed to allow a limited angular movement between the axes of two waveguides. 2. A coupling that connects two shafts end to end and permits rotation even though the shafts are not exactly aligned.

flexible resistor A wirewound resistor that has the appearance of a flexible lead. It is made by winding Nichrome resistance wire around a length of asbestos or other heat-resistant cord, then covering the winding with asbestos and a braided insulating covering. This covering is generally color-coded to indicate the resistor value.

flexible shaft A shaft that transmits rotary motion at any angle up to about 90°. Used in electronic equipment to permit mounting adjustable controls at optimum positions with respect to other parts while still securing desirable groupings of controls on the panel.

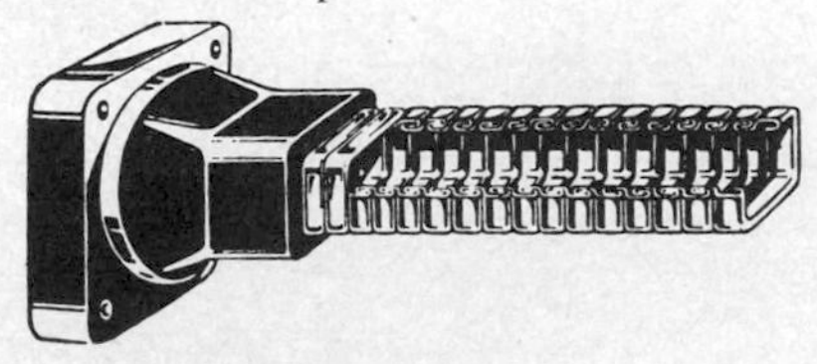

Flexible waveguide.

flexible waveguide A waveguide that can be bent or twisted without appreciably changing its electrical properties.

flicker A visual sensation produced by periodic fluctuations in light at rates ranging from a few cycles per second to a few tens of cycles per second. In U.S. television, interlaced scanning eliminates the flicker that might otherwise be noticed at 30 frames per second.

flicker effect Random variations in the output current of an electron tube that has an oxide-coated cathode, caused by random changes in cathode emission.

flicker noise Electric noise produced by the flicker effect in an electron-tube circuit. It is low-frequency noise that is in excess of shot noise. The noise also occurs in semiconductors as a result of trapping of charges at a semiconductor surface, but it is insignificant at most semiconductor operating frequencies.

flight indicator *Artificial horizon.*

flight instrument Any aircraft instrument that indicates the altitude, attitude, airspeed, drift, or direction of an aircraft.

flight path A line in space planned or used as the path for an aircraft or missile.

flight-path angle The acute angle between the horizontal and the flight path of an aircraft or missile during ascent or descent.

flight-path computer A computer that includes all the functions of a course-line computer and provides means for controlling the altitude of an aircraft in accordance with a desired plan of flight.

flight-path deviation indicator An instrument that provides a visual indication of deviation from a planned flight path.

flight-path-reference flight Stabilized flight in which control information is obtained from a navigation system capable of providing heading or altitude guidance, or both, with respect to a desired flight path.

flight-path selector An instrument used with a flight-path computer to preset the values defining the flight path to a way point.

flight simulator Equipment that simulates any or all of the conditions of actual flight. Used chiefly for training purposes. Also called flight trainer.

flight track The path in space actually traced by a vehicle. Flight track is the three-dimensional equivalent of track.

flight trainer *Flight simulator.*

flip chip A tiny semiconductor die having terminations all on one side in the form of solder pads or bump contacts. After the surface of the chip has been passivated or otherwise treated, it is flipped over for attaching to a matching substrate on which interconnecting thin films and possibly also thin-film components have previously been deposited. All connections are then made simultaneously by applying heat or a combination of ultrasonic energy and pressure.

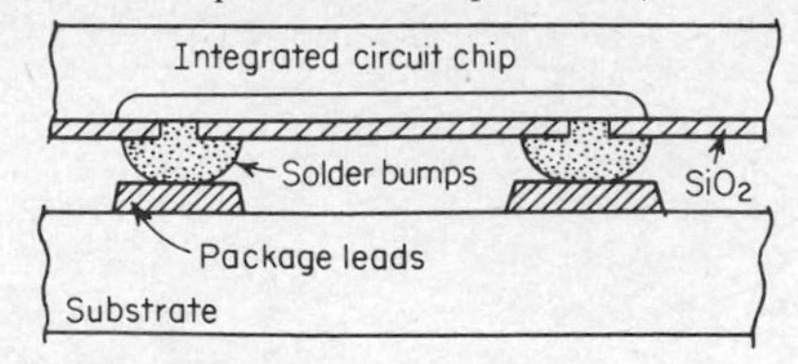

Flip-chip bonding to substrate.

flip coil A small coil used to measure the strength of a magnetic field. It is placed in the field, connected to a ballistic galvanometer or other instrument, and suddenly flipped over 180°. The resulting generated electric energy is a measure of magnetic field strength. Alternatively, the coil may be held stationary and the magnetic field reversed.

flip-flop circuit A two-stage multivibrator circuit that has two stable states. In one state, the first stage is conducting and the second is cut off. In the other state, the second stage is conducting and the first stage is cut off. A trigger signal changes the circuit from one state to the other, and the next trigger signal changes it back to the first state.

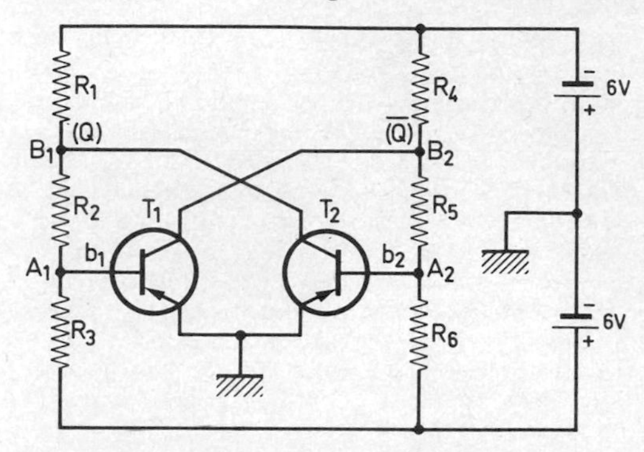

Flip-flop circuit using two PNP transistors.

For counting and scaling purposes, a flip-flop can be used to deliver one output pulse for each two input pulses. Also called bistable multivibrator, Eccles-Jordan circuit, and trigger circuit.

flipover cartridge A phonograph pickup that has two needles pointing in opposite directions. The entire cartridge can be flipped over in its pivoted mounting to place the desired needle in playing position.

floating The condition wherein a device or circuit is not grounded and not tied to an established voltage supply.

floating action Controller action in which there is a predetermined relation between the deviation and speed of a final control element. A neutral zone, in which no motion of the final control element occurs, is often used in floating action.

floating address *Symbolic address.*

floating-average-position action Floating action in which there is a predetermined relation between deviation of the controlled variable and the rate of change of the time-average position of a final control element that is moved periodically from one of two fixed positions to the other.

floating battery A storage battery connected permanently in parallel with another power source. The battery normally handles only small charging or discharging currents, but it takes over the entire load upon failure of the main supply.

floating-carrier modulation *Controlled-carrier modulation.*

floating charge Application of a constant voltage to a storage battery, sufficient to maintain an approximately constant state of charge while the battery is idle or on light duty.

floating grid An electron-tube grid that is not connected to a circuit. The grid assumes a negative potential with respect to the cathode, due to electrons hitting the grid wires, and the tube is then sensitive to external effects such as movement of a hand near the envelope. Also called free grid.

floating input An amplifier or other circuit in which no input terminal is connected to circuit ground.

floating junction A transistor junction through which the average current is zero.

floating-point Pertaining to a number system in which the location of the point does not remain fixed with respect to one end of the numerals.

floating-point arithmetic A method of calculation that automatically accounts for the location of the decimal or radix point.

floating-point calculation A computer calculation in which provisions are made for varying the location of the decimal point (if base 10) or binary point (if base 2).

floating-point routine A computer routine that permits floating-point operation for a specific problem. Usually used in computers that were not originally designed for floating-point operation.

floating-point system A point system of positional notation in which the position of the point is regularly recalculated and may be moved. A floating-point system usually locates the point by expressing a power of the base, and involves the use of two sets of digits. For floating decimal notation the base is 10, so 6,200,000 would be 6.2, 6. For floating binary notation the base is 2, so 88 would be 11, 3.

float switch A switch actuated by a float at the surface of a liquid.

flock Finely divided felt used on phonograph turntable surfaces, underneath microphone stands, and in similar locations where a nonscratching surface is desired. It is sifted over a layer of cement applied to the surface.

flood To direct a large-area flow of electrons toward a storage assembly in a charge storage tube.

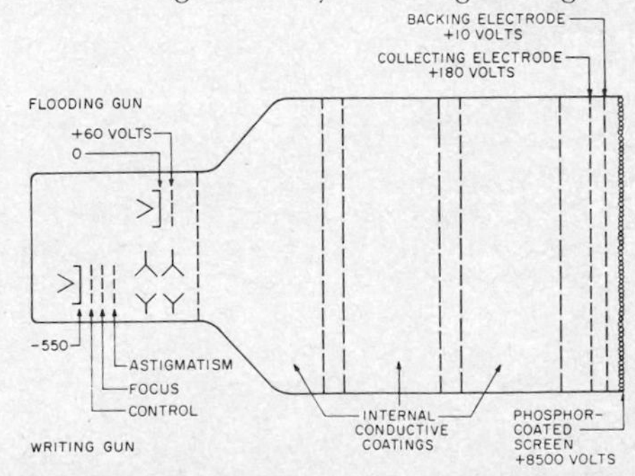

Flooding gun, located alongside writing gun of display tube to intensify image on screen.

flooding gun A large-diameter electron gun used in a cathode-ray display tube to produce an unmodulated beam of electrons that sprays outward to the walls of the tube. Here, wall coatings serve as anodes that convert the spray to a parallel beam. This intensifies the brightness of the pattern created by the writing gun on the viewing phosphor.

floodlighting Covering a wide area with radar waves.

flood projection An optical method of facsimile scanning in which the subject copy is floodlighted and the scanning spot is defined by an aperture that moves in the path of the reflected or transmitted light.

floppy disk A flexible plastic disk about 7½ in (19 cm) in diameter, coated with magnetic oxide and used for data entry to a computer. A slot in its protective envelope or housing exposes the track positions for the magnetic read/write head of the drive unit. A central opening is provided in the housing for the drive hub that goes into a 1½-in (3.8-cm) hole in the center of the disk. The housing remains stationary while the disk rotates inside. Drive speed, number of tracks, and bit capacity vary with different models; capacity can range up to 3 Gb. Also called diskette.

flow The movement of electric charges, gases, liquids, or other materials or quantities.

flowchart A graphical representation of a program or routine for a digital computer. Also called control diagram.

flowmeter An instrument that measures and indicates the rate of flow of a liquid or gas.

flow soldering Soldering of printed-circuit boards by moving them over a flowing wave of molten solder in a solder bath. The process permits precise control of the depth of immersion in the molten solder and minimizes heating of the board. Also called wave soldering.

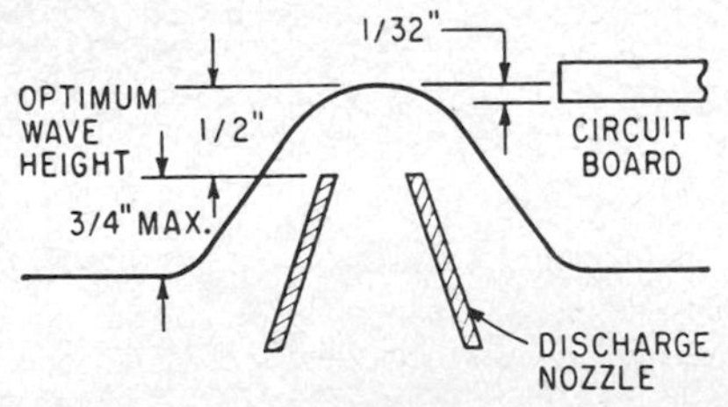

Flow soldering, in which molten solder is forced up through discharge nozzle in solder bath to create rounded wave. Multiply dimensions by 2.54 to convert to centimeters.

flow transmitter A device used to measure the flow of liquids in pipe lines and convert the results into proportional electric signals that can be transmitted to distant receivers or controllers.

fluctuation noise *Random noise.*

fluence A measure of time-integrated particle flux, expressed in particles per square centimeter. Used for electrons in electron irradiation and for neutrons in connection with effects of nuclear radiation on electronic components.

fluid amplifier An amplifier in which all amplification is achieved by interaction between jets of fluid, with no electronic circuits and usually no moving parts.

fluid computer A digital computer constructed entirely from air-powered fluid logic elements. It contains no moving parts and no electronic circuits. All logic functions are carried out by interaction between jets of air.

fluidics The technology that uses the interaction of flowing gases or liquids to perform sensing, logic, amplification, and control functions, employing devices which have no moving parts.

fluorescence Emission of light or other electromagnetic radiation by a material exposed to another type of radiation or a beam of particles, with the luminescence ceasing within about 10^{-8} s after irradiation is stopped. Certain minerals fluoresce in characteristic colors during exposure to ultraviolet radiation. A cathode-ray screen fluoresces when hit by the electron beam in the tube. Other materials give off characteristic x-rays when irradiated by higher-frequency x-rays.

fluorescence analysis Analysis based on irradiation of a sample with ultraviolet light, followed by observation or measurement of the color and intensity of the emitted fluorescent radiation.

fluorescence spectroscopy Spectroscopic study of radiation emitted by a fluorescent sample.

fluorescence yield The probability that an excited atom will emit an x-ray photon rather than an Auger electron in the first transition. The yield is a number between 0 and 1, depending on the element and its state of excitation.

fluorescent lamp A tubular discharge lamp in which ionization of mercury vapor produces radiation that activates the fluorescent coating on the inner surface of the glass. The three common types of fluorescent lamps are instant-start, preheat, and rapid-start.

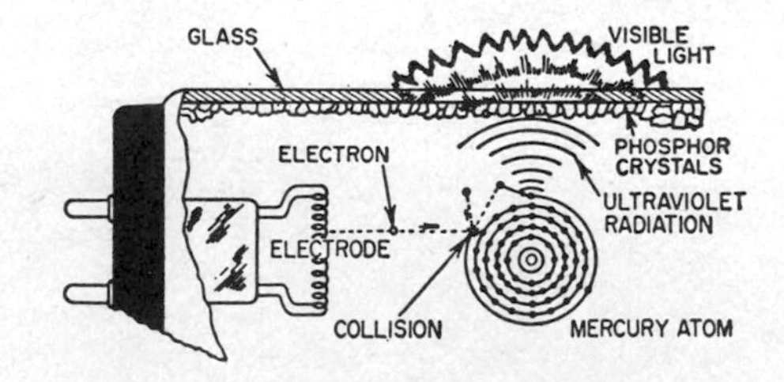

Fluorescent-lamp construction and operation. Electron emitted by electrode at one end collides with one of electrons in outer ring of mercury atom, producing ultraviolet radiation, which in turn acts on phosphor crystals inside glass wall to produce visible light.

fluorescent screen A sheet of material coated with a fluorescent substance that emits visible light when irradiated with ionizing radiation like x-rays or electron beams. In a cathode-ray tube the fluorescent screen is a coating on the inside surface of the tube face.

fluorimeter An instrument for measuring small amounts of uranium in liquids. The sample is irradiated with ultraviolet light, and the resulting fluorescence of uranium and its salts is measured photoelectrically.

fluorine [symbol F] A highly corrosive and poisonous gaseous element. Atomic number is 9.

fluorocarbon resin General term for a family of plastics that has excellent electrical insulating qualities and relatively high service temperatures. Examples include polychlorotrifluoroethylene resin, marketed as Kel-F, and polytetrafluoroethylene resin, marketed as Teflon and Fluon.

fluorod A rod made from silver-activated phosphate glass, used in solid-state dosimeters. Under irradiation, the rod absorbs ultraviolet light and emits orange fluorescent light. Measurement of the intensity of the emitted light with a photomultiplier gives a measure of the absorbed dose of radiation.

fluorography Photography of an image produced on a fluorescent screen.

fluorometer An instrument that measures the intensity of x-rays and other radiation by measuring the intensity of the fluorescence produced.

fluorometry Measurement of the intensity and color of fluorescent radiation.

fluoroscope A fluorescent screen designed for use with an x-ray tube to permit direct visual

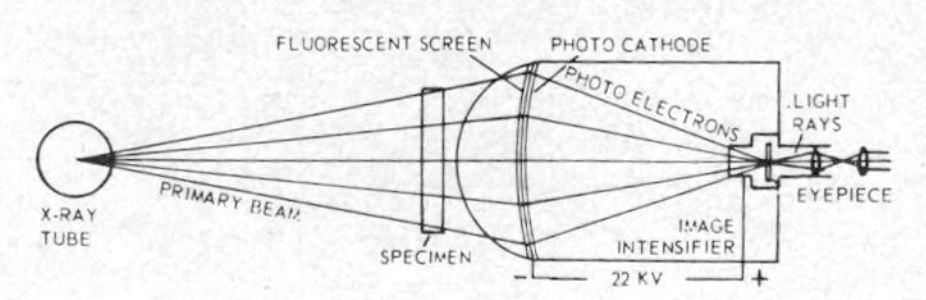

Fluoroscope used with image intensifier for inspection of welds, honeycomb assemblies, thick plastic or rubber products, and explosive charges.

observation of x-ray shadow images of objects interposed between the x-ray tube and the screen. The screen transforms invisible x-ray radiation into visible light.

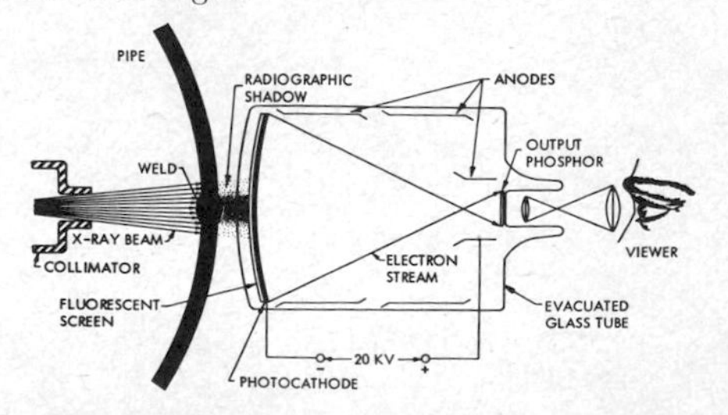

Fluoroscopic image intensifier used for inspecting weld in pipe.

fluoroscopic image intensifier An electron-beam tube that converts a relatively feeble fluoroscopic image on the fluorescent input phosphor into a much brighter image on the output phosphor. The input screen is backed by a photoemissive cathode layer that emits electrons in proportion to light at each point. High-voltage electrodes accelerate and focus this electron stream to high velocities at the output screen.

fluoroscopy Use of a fluoroscope for x-ray examination.

flush antenna An aircraft antenna that has no projections beyond the surface of the aircraft. Examples include ditch antennas and slot antennas.

flutter 1. Distortion that occurs in sound reproduction as a result of undesired speed variations during the recording, duplicating, or reproducing process. The variations in speed and hence pitch occur at a much higher rate than for wow. 2. A fast-changing variation in received signal strength such as may be caused by antenna movements in a high wind or interaction with a signal on another frequency.

flutter echo A radar echo that consists of a rapid succession of reflected pulses resulting from a single transmitted pulse.

flux 1. The electric or magnetic lines of force in a region. 2. The rate of flow of particles or photons through a unit area. 3. The product of the num-

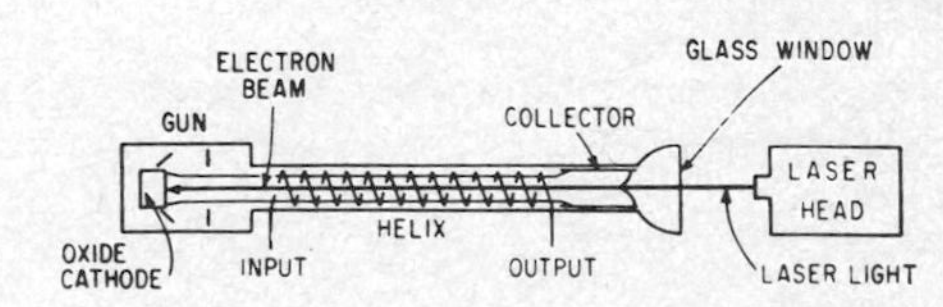

Flux of magnetic lines of force focuses electron beam in traveling-wave phototube for laser beam.

ber of particles per unit volume and their average velocity. 4. A material used to remove oxide films from the surfaces of metals in preparation for soldering, brazing, or welding. Rosin is widely used as a flux for soldering electronic circuits.

flux gate A detector that gives an electric signal whose magnitude and phase are proportional to the magnitude and direction of the external magnetic field acting along its axis. A flux gate may consist of three magnetic cores that have appropriate excitation windings and load windings to give perfect balance in the absence of external magnetic fields.

flux-gate compass *Gyro flux-gate compass.*

flux-gate magnetometer A magnetometer in which the degree of saturation of the core by an external magnetic field is used as a measure of the strength of the field.

flux guide A shaped piece of magnetic material used to guide electromagnetic flux in desired paths in induction heating. The guide may be used either to direct flux to preferred locations or to prevent the flux from spreading beyond definite regions.

flux leakage Magnetic flux that does not pass through an air gap or other part of a magnetic circuit where it is required.

flux linkage The product of the number of turns in a coil and the number of magnetic lines of force passing through the turns. Also called linkage.

fluxmeter An instrument for measuring magnetic flux. It is usually calibrated to read either in maxwells or webers.

flyback *Retrace.*

flyback power supply A high-voltage power supply used to produce the DC voltage of about 10 to 25 kV required for the second anode of a cathode-ray tube in a television receiver or oscilloscope. The sudden reversal of horizontal deflection-coil current in the horizontal output transformer during each flyback induces a voltage pulse that is increased to the required higher value by autotransformer action, then rectified and filtered. Also called kickback power supply.

flyback transformer *Horizontal output transformer.*

fly-by-wire system A flight-control system that uses electric wiring instead of mechanical or hydraulic linkages to control the actuators for the ailerons, flaps, and other control surfaces of an aircraft. Wiring makes multiple redundancy feasible, for greater combat survivability. Control signals are provided by a computer whose inputs are flight sensors, preflight programs, and manual actions by the crew. Also used in manned and unmanned spacecraft.

flying clock A cesium-controlled high-precision time standard that can be kept in continuous operation on commercial aircraft as well as in automobiles. Used for worldwide synchronization of time standards to fractional-microsecond accuracy.

flying-spot scanner A television scanner in which a simple phototube replaces a more complex iconoscope or other pickup tube at the camera. A moving spot of light, controlled either mechanically or electrically, scans the image field to be transmitted. The light reflected from or transmitted by the image field is picked up by the phototube to generate the video signal. Used today chiefly for film and slide transmission. A high-intensity cathode-ray tube is generally used as the flying-spot light source. Also used in optical character recognition and other image-converting applications.

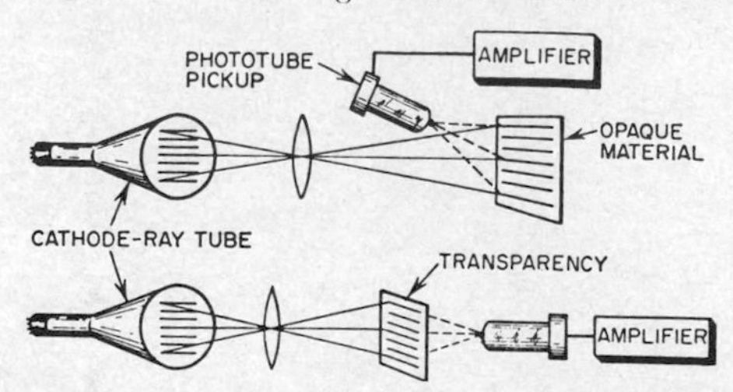

Flying-spot scanner in which light source is spot that traces raster pattern on screen of cathode-ray tube. Position of phototube depends on whether material being televised is opaque or transparent.

flywheel effect The ability of a resonant circuit to maintain oscillation at an essentially constant frequency when fed with short pulses of energy at constant frequency and phase.

flywheel synchronization Automatic frequency control of a scanning system by using the average timing of the incoming sync signals, rather than by making each pulse trigger the scanning circuit. Used in high-sensitivity television receivers designed for fringe-area reception, where noise pulses might otherwise trigger the sweep circuit prematurely.

flywheel tuning Tuning in which the control knob is associated with a heavy flywheel that provides momentum for faster coverage of the tuning dial.

FM 1. Abbreviation for *frequency-modulated.* 2. Abbreviation for *frequency modulation.*

FM/AM Amplitude modulation of a carrier by subcarriers that are frequency-modulated by information.

FM/AM multiplier A multiplier in which the frequency deviation from the central frequency of a carrier is proportional to one variable, and its amplitude is proportional to the other variable. The frequency-amplitude-modulated carrier is then consecutively demodulated for FM and for AM. The final output is proportional to the product of the two variables.

FM cyclotron *Synchrocyclotron.*

FM/FM Frequency modulation of a carrier by subcarriers that are frequency-modulated by information.

FM/FM telemetering A telemetering system used in tests of guided missiles. The subcarrier bands are frequency-modulated on subcarrier waves, and these are in turn frequency-modulated on the main RF carrier wave.

FM pickup *Variable-capacitance pickup.*

FM/PM Phase modulation of a carrier by subcarriers that are frequency-modulated by information.

FM/PM telemetering A telemetering system in which the several frequency-modulated subcarriers are used to phase-modulate the main RF carrier.

FM stereo Deprecated term for *stereo FM.*

f number A lens rating obtained by dividing the focal length of the lens by the effective maximum diameter of the lens. The lower the f number, the shorter the exposure required or the lower the illumination needed for satisfactory results with a television or ordinary camera. An f number of 3.5, for example, is usually expressed as f/3.5.

foamed plastic A resinous material that has been expanded into a multicellular structure which has low density and relatively high strength.

focal length The distance between the optical center of a lens and the television camera screen or photographic camera film when the camera is focused on a distant object.

focal spot The small area on the target of an x-ray tube that gives off x-rays when hit by the electron stream.

focus 1. The point at which rays of light or electrons of a beam converge to form a minimum-diameter spot. 2. To move a lens or adjust a voltage or current to obtain a focus.

focus control A control that adjusts spot size at the screen of a cathode-ray tube, to give the sharpest possible image. It may vary the current through a focusing coil or change the position of a permanent magnet.

focus-defocus mode A mode of storage of binary digits in which the writing beam of a cathode-ray storage tube is initially focused. For one type of binary digit it remains focused, and for the other type it is suddenly defocused to a small concentric circular area, in the time interval before the beam is cut off and moved to the next position.

focusing 1. The process of controlling convergence or divergence of the electron paths within one or more beams to obtain a desired image or current density distribution in the beam. 2. The process of moving an optical lens toward or away from a screen or film to obtain the sharpest possible image of a desired object.

focusing anode An anode used in a cathode-ray tube to change the size of the electron beam at the screen. Varying the voltage on this anode alters the paths of electrons in the beam and thus changes the position at which they cross or focus.

focusing coil A coil that produces a magnetic field parallel to an electron beam, for the purpose of focusing the beam. The coil is usually mounted on the neck of a cathode-ray picture tube, and carries a direct current whose value can be adjusted by a focus control rheostat.

focusing electrode An electrode to which a potential is applied to control the cross-sectional area of the electron beam in a cathode-ray tube.

focusing grid A focusing electrode.

focusing magnet A permanent magnet used to produce a magnetic field for focusing an electron beam.

focus rectifier A special electron tube or selenium-rectifier stage used in some color receivers to provide a separate voltage source for the color television cathode-ray tube.

fog chamber *Cloud chamber.*

fog track A line of condensation, produced in supersaturated water vapor by the passage of charged particles. Used in studying the courses and collisions of particles in cloud chambers.

foil A flexible sheet of thin aluminum, lead, or tin, widely used in fixed capacitors.

folded cavity A cavity used in some klystrons to make the incoming wave act on the electron stream from the cathode at several places to produce a cumulative effect.

folded-dipole antenna A dipole antenna whose outer ends are folded back and joined together at the center. The impedance is about 300 Ω, as compared to 70 Ω for a single-wire dipole. Widely used with television and FM receivers.

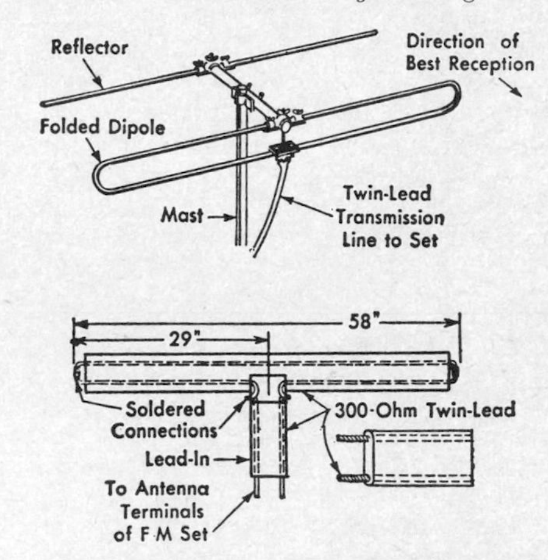

Folded-dipole antenna with reflector mounted on mast for television reception, and folded-dipole antenna made from twin-line for FM reception. Multiply dimensions by 2.54 to convert to centimeters.

folded horn An acoustic horn in which the path from throat to mouth is folded or curled to give the longest possible path in a given volume.

foldover Picture distortion seen as a white line on either side, top, or bottom of a television picture. Generally caused by nonlinear operation in either the horizontal or vertical deflection circuits of a receiver.

font A family of characters, such as those provided by a computer output printer, a computer output display, or a typewriter.

foot [abbreviated ft] A unit of length, equal to 0.3048 m. Use of the SI unit of length, the meter, is preferred.

footcandle [abbreviated fc] A former unit of illumination, replaced by the lumen per square foot. Use of the SI unit of illumination, the lux, is preferred.

footlambert [abbreviated fL] A unit of luminance. Use of the SI unit of luminance, the candela per square meter, is preferred.

foot per second [abbreviated ft/s] A unit used in specifying the speed at which sound waves travel through a medium. In air at standard sea-level conditions, the speed of sound is about 1080 ft/s (330 m/s), and 4800 ft/s (1460 m/s) in water.

forbidden band An energy band in which there

can be no electrons in a given substance.

forbidden-combination check A test for the occurrence of a nonpermissible code expression in a computer. Used to detect computer errors.

forbidden transition A transition between two states of a quantum-mechanical system for which the change in the quantum numbers involved is less probable than a competing allowed transition.

Forbush decrease The observed decrease in cosmic-ray activity on the earth about a day after a solar flare. It is now believed to be caused by the shielding effect of the magnetic fields in the plasma cloud emitted by the sun during a flare.

force To intervene manually in a digital computer program and cause the computer to execute a jump instruction.

force-balance transducer A transducer in which the output of the sensing member is amplified and fed back to a device that returns the sensing member to its rest position. The feedback signal then also serves as the output of the transducer.

forced oscillation The oscillation of some physical quantity of a system when external periodic forces determine the period of the oscillation. Also called forced vibration.

forced vibration *Forced oscillation.*

forcing The application of control impulses that are larger than warranted by the error in a system, to achieve a greater rate of correction.

forcing resistance A resistance used in series with a control winding of a magnetic amplifier or dynamoelectric amplifier to give higher response speed.

foreground processing The processing of high-priority programs by a computer, often by interrupting background processing of lower-priority programs.

foreign-body locator A device for locating foreign metallic bodies in tissue by use of suitable probes that generate a magnetic field. The presence of a magnetic body within this field is indicated by a meter or sound signal.

fork oscillator An oscillator that uses a tuning fork as the frequency-determining element.

formal logic A discipline that investigates propositions by methods which abstract the contents of the propositions and deal only with their logic forms.

format The predetermined arrangement of information on a form, page, file, or message.

form factor 1. The ratio of the effective value of an alternating quantity to the average value during a half-cycle. The form factor is about 1.11 for a pure sine wave and equal to the ratio of the readings obtained for a given AC quantity on root-mean-square and rectifier-type meters. 2. A factor that takes into account the shape of a coil when computing its inductance. Also called shape factor.

Formica Trademark for certain laminated plastic products used chiefly for surface finishes.

forming Application of voltage to an electrolytic capacitor, electrolytic rectifier, or semiconductor device to produce a desired permanent change in electrical characteristics as a part of the manufacturing process.

form-wound coil A coil that is formed or bent to an irregular shape, as for a deflection yoke.

forsterite A ceramic having low dielectric loss over a wide range of temperatures up to about 1000°C. Good at high frequencies. Used in making insulators.

FORTRAN [FORmula TRANslation] One of several specific procedure-oriented programming languages that can be used on a computer to translate into machine language a program whose steps are written in relatively simple language.

fortuitous telegraph distortion Distortion that results in departure, for one occurrence of a particular signal pulse, from the average combined effects of bias and characteristic telegraph distortion.

forty-five rpm record A 7-in-diameter (17-cm-diameter) disk recorded and reproduced at a speed of 45 rpm, having a center hole 1.5 in (3.8 cm) in diameter and grooves designed for a stylus having a point radius of 1 mil (0.0254 mm).

forward bias A bias voltage applied to a PN junction in the direction that causes a large current flow. Used in some semiconductor diode circuits.

forward coupler A directional coupler used to sample incident power.

forward current The current that flows through a rectifying junction in the conducting direction.

forward direction The direction of lesser resistance to current flow through a rectifier or semiconductor diode.

forward path The transmission path from the loop actuating signal to the loop output signal in a feedback control loop.

forward recovery time The time required for the forward current of a semiconductor diode to reach a specified value after a forward bias is instantaneously applied.

forward-scatter propagation *Scatter propagation.*

forward voltage A voltage having the correct polarity to send current through a rectifier in the forward direction.

forward wave A wave whose group velocity is in the same direction as the electron-stream motion in a traveling-wave tube.

Foster-Seeley discriminator *Phase-shift discriminator.*

Foucault current A current induced in the interior of conductors by variations of magnetic flux.

four-channel sound system *Quadraphonic sound system.*

four-course radio range A radio range that beams on-course signals in four different directions for aircraft guidance. The Adcock radio range is an example.

four-horn feed A cluster of four rectangular horn antennas used as the radiating and receiving elements of a parabolic or lens-type radar antenna, to define the four quadrants of coverage.

Fourier analysis The process of determining the amplitude, frequency, phase, coherence functions, correlation functions, power spectra, transfer functions, and other functions of each sinusoidal component in a given waveform.

Fourier analyzer A digital spectrum analyzer that provides pushbutton or other switch selection of averaging, coherence function, correlation, power spectrum, and other mathematical operations involved in calculating Fourier transforms of time-varying signal voltages for such applications as identification of underwater sounds, vibration analysis, oil prospecting, and brain-wave analysis.

Fourier series A mathematical expression by which any periodic function can be represented as a combination of sine and cosine terms that are integral multiples of a fundamental frequency.

Fourier transform A mathematical expression relating the energy in a transient to that in a continuous-energy spectrum of adjacent frequency components.

four-layer device A PNPN semiconductor device that has four layers of alternating P- and N-type material to give three PN junctions. A silicon controlled rectifier is an example.

four-layer diode A semiconductor diode that has three junctions, terminal connections being made to the two outer layers which form the junctions. A Shockley diode is an example.

four-layer transistor A junction transistor that has four conductivity regions but only three terminals. A thyristor is an example.

four-level laser A laser in which the lowest level for a laser transition is an excited state rather than the ground level. Less energy is ordinarily required to obtain the necessary population inversion in a four-level laser because the terminal level may be initially almost empty.

four-of-eight code An 8-bit code in which 4 bits are always 1 and 4 bits are always 0, to give high immunity to errors. Used in Bell System Touch-Tone units.

4PDT Abbreviation for *four-pole double-throw.*

four-phase modulation Modulation in which data is encoded on a carrier frequency as a succession of phase shifts that will be 45, 135, 225, or 315°. Each phase shift contains 2 bits of information called dibits, as follows: 225° represents 00, 315° is 01, 45° is 11, and 135° is 10. Used in some Western Electric modems for data transmission over wire lines.

four-pole double-throw [abbreviated 4PDT] A 12-terminal switch or relay contact arrangement that simultaneously connects two pairs of terminals to either of two other pairs of terminals. Used in switching sets of stereo loudspeakers.

four-pole single-throw [abbreviated 4PST] An eight-terminal switch or relay contact arrangement that simultaneously opens or closes two separate pairs of circuits.

4PST Abbreviation for *four-pole single-throw.*

four-quadrant multiplier A multiplier in which both the reference signal and the number represented by the input may be bipolar, and the multiplication rules for algebraic sign are obeyed. The output may thus be of either polarity, as called for by any combination of polarities of the two input signals. Commercial versions can sometimes be used to divide, square, and square-root. Also called quarter-square multiplier.

four-tape sort A merge-sort in which input data on two tapes is sorted into complete sequences alternately on two output tapes. The output tapes are used for input for the next run, the process being repeated until the data is in one sequence on one output tape.

four-track tape Magnetic tape on which two tracks are recorded for each direction of travel, to double the amount of stereo music that can be recorded on a given length of ¼-in (0.635-cm) tape.

four-wire circuit A circuit in which communication signals are transmitted in one direction on one path and in the other direction on the other path. Multiplexing methods such as frequency division or time division may be used to reduce the actual number of wires required.

four-wire repeater A repeater that provides amplification in opposite directions on two transmission paths.

fox message A standard message used for testing teletypewriter circuits and machines, that includes all the alphamerics: THE QUICK BROWN FOX JUMPED OVER A LAZY DOG'S BACK 1234567890.

FPLA Abbreviation for *field-programmable logic array.*

fractional-horsepower motor [abbreviated FHP motor] Any motor built into a frame smaller than that for a motor having an open construction and a continuous rating of 1 hp at 1800 rpm.

Frahm frequency meter *Vibrating-reed frequency meter.*

frame 1. One complete coverage of a television picture. In the United States a frame contains 525 horizontal scanning lines, repeated at the rate of 30 frames per second. Each frame is scanned in two interlaced fields, each covering 262.5 lines. 2. A single complete picture on motion-picture film. For 35-mm film the standard rate of projection is 24 frames per second. This means that a special projector is required to convert this to 30 frames per second for U.S. television. 3. A rectangular area representing the size of copy handled by a facsimile system. The width of a facsimile frame is the available line width, and the length is determined by the service requirements.

frame frequency The number of times per second that the frame is completely scanned in television.

frame-grid tube A tube in which the grid is wound on a rectangular frame consisting of two short rigid rods held securely by two crossbars. This construction permits use of a greater number of turns of much smaller grid wire, wound under greater tension.

frame of reference A set of lines or surfaces used as references for coordinates defining a moving or stationary point.

frame period A time interval equal to the reciprocal of the frame frequency. In the United States the frame period is 1/30 s.

framer A device for adjusting facsimile equipment so that the start and end of a recorded line are the same as on the corresponding line of the subject copy.

framing 1. Adjusting a television picture to a desired position on the screen of the picture tube. 2. Adjusting a facsimile picture to a desired position in the direction of line progression. Also called phasing.

framing control 1. A control that adjusts the centering, width, or height of the image on a television receiver screen. 2. A control that shifts a received facsimile picture horizontally.

framing signal A signal used to adjust a facsimile picture to a desired position in the direction of line progression.

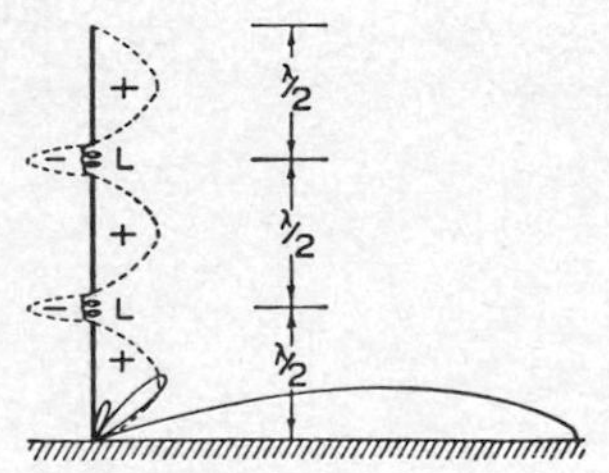

Franklin antenna, showing vertical radiation pattern for all directions (solid curves) and current distribution curves (dotted).

Franklin antenna An antenna several half-wavelengths long, having nonradiating phasing coils between half-wave sections.

Fraunhofer diffraction Diffraction that occurs when radiation passes the edges of one or more apertures.

Fraunhofer lines Dark lines in the spectrum of sunlight, as obtained with a spectroscope. They are due to absorption of certain wavelengths by gases and vapors in the solar atmosphere.

Fraunhofer region The region in which energy flow from an antenna proceeds essentially as though coming from a point source located in the vicinity of the antenna. It is beyond the Fresnel region, and begins at a distance equal to about twice the square of antenna length divided by wavelength. Also called far region and far zone.

free-air ionization chamber An air-filled ionization chamber in which a sharply defined beam of radiation passes between the electrodes without striking them or other internal parts of the equipment. The observed ionization current is then caused entirely by ions and electrons resulting from the action of radiation on the air. Used for x-ray dosimetry. Also called open-air ionization chamber.

free electron An electron that is not constrained to remain in a particular atom. It is therefore able to move freely in matter or a vacuum, when acted on by external electric or magnetic fields.

free field A field in which the effects of boundaries are negligible over the region of interest. An object in a free sound field will have the same disturbing effect as a boundary, however, unless the acoustic impedance of the object matches the acoustic impedance of the medium.

free-field emission The electron emission that occurs when the electric field at the surface of an emitter is zero.

free-field room *Anechoic room.*

free grid *Floating grid.*

free gyro A gyroscope mounted in two or more gimbal rings so that it is free to maintain a fixed orientation in space, with no external means for changing its normal precession.

free oscillation Oscillation that continues in a circuit or system after the applied force has been removed. The frequency is determined by the parameters in the system or circuit. Also called free vibration and shock-excited oscillation.

free progressive wave A wave in a medium free from boundary effects. Also called free wave.

free-running Operating without external synchronizing pulses.

free-running frequency The frequency at which a normally synchronized oscillator operates in the absence of a synchronizing signal.

free-running multivibrator *Astable multivibrator.*

free space A region high enough so that the radiation pattern of an antenna is not affected by surrounding objects like buildings, trees, hills, and the earth.

free-space field intensity The field intensity that would exist at a point in the absence of waves reflected from the earth or other reflecting objects.

free-space loss The theoretical radiation loss, depending only on frequency and distance, that would occur if all variable factors were disregarded when transmitting energy between two antennas.

free-space radiation pattern The radiation pattern that an antenna would have in free space.

free vibration *Free oscillation.*

free wave *Free progressive wave.*

F region The region of the ionosphere above about 90 mi (145 km). It includes the F_1 and F_2 layers in the daytime hemisphere, and a single F layer at night, all capable of reflecting radio waves back to earth at frequencies up to about 50 MHz.

frequency [symbol f] The number of complete cycles per unit of time for a periodic quantity such as alternating current, sound waves, or radio waves. A frequency of 1 cycle/s is 1 hertz (1 Hz). Other frequency units are kilohertz (kHz), megahertz (MHz), gigahertz (GHz), and terahertz (THz). Frequency in megahertz is equal to 300 divided by wavelength in meters.

frequency agility The ability to shift the frequency of a radar transmitter rapidly and continually, to avoid jamming by the enemy, to reduce mutual interference with friendly sources, to enhance echoes from targets, or to provide the required patterns of electronic countermeasures or electronic counter-countermeasures radiation.

frequency allocation Assignment of available frequencies in the radio spectrum to specific stations and for specific purposes, to give maximum utilization of frequencies with minimum interference between stations. Allocations in the United States are made by the Federal Communications Commission.

frequency band A continuous range of frequencies extending between two limiting frequencies. A band may include a number of channels; thus the broadcast band extends from 535 to 1605 kHz. Frequency and wavelength limits for all bands in the radio spectrum are given in the entry for *band.*

frequency bridge A bridge in which the balance varies with frequency in a known manner, such as the Wien bridge. Used to measure frequency.

frequency calibrator An instrument that generates a highly accurate signal at one or more fixed frequencies, for use in calibrating other frequency sources.

frequency changer *Frequency converter.*

frequency-change signaling A telegraph signaling method in which one or more frequencies correspond to each desired signaling condition of a telegraph code.

frequency characteristic *Amplitude-frequency response.*

frequency compensation *Compensation.*

frequency conversion The process of converting the carrier frequency of a received signal from its original value to the intermediate-frequency value in a superheterodyne receiver.

frequency converter A circuit, device, or machine that changes an alternating current from one frequency to another, with or without a change in voltage or number of phases. In a dyne receiver the oscillator and mixer-first detector stages together serve as the frequency converter. Also called frequency changer.

frequency cutoff The frequency at which the current gain of a transistor drops 3 dB below the low-frequency gain value.

frequency departure *Frequency drift.*

frequency deviation 1. The peak difference between the instantaneous frequency of a frequency-modulated wave and the carrier frequency. Also called deviation. 2. *Frequency drift.*

frequency-deviation meter An instrument that indicates the number of cycles a broadcast transmitter has drifted from its assigned carrier frequency.

frequency discriminator A discriminator circuit that delivers an output voltage which is proportional to the deviations of a signal from a predetermined frequency value. Used in frequency-modulated receivers and automatic frequency-control circuits. Examples include the double-tuned detector, locked-in oscillator, and ratio detector.

frequency distortion Nonlinear distortion in which the relative magnitudes of the different frequency components of a wave are changed during transmission or amplification. Frequency distortion occurs when an audio amplifier cannot amplify equally well all the frequencies present in the input signal. Also called amplitude distortion, amplitude-frequency distortion, and waveform-amplitude distortion.

frequency diversity Diversity reception involving the use of carrier frequencies separated 500 Hz or more and having the same modulation, to take advantage of the fact that fading does not occur simultaneously on different frequencies. The receiver minimizes the effects of fading by using at each instant the frequency that has the higher signal strength.

frequency-diversity radar A radar that uses two or more frequencies simultaneously, with means for combining the resulting echoes or selecting the strongest of the echoes from the target at each instant.

frequency divider A harmonic conversion transducer in which the frequency of the output signal is an integral submultiple of the input frequency.

frequency-division multiple-access [abbreviated FDMA] A satellite multiplex communication system that uses an analog method of modulating telephone signals. The satellite repeaters can accept several signals at different frequencies within the repeater bandwidth.

frequency-division multiplex [abbreviated FDM] A multiplex system for transmitting two or more signals over a common path by using a different frequency band for each signal.

frequency-domain oscilloscope A calibrated oscilloscope designed as a spectrum analyzer for such applications as the design of amplifiers, filters, mixers, modulators, and oscillators.

frequency-domain reflectometer A tuned reflectometer used for measuring reflection coefficients and impedance of waveguides over a wide frequency range, by sweeping a band of frequencies and analyzing the reflected returns.

frequency doubler An amplifier stage whose resonant anode circuit is tuned to the second harmonic of the input frequency. The output frequency is then twice the input frequency. Also called doubler.

frequency-doubling transponder A transponder that doubles the frequency of the signal before retransmission.

frequency drift A gradual change in the frequency of an oscillator or transmitter, due to temperature or other changes in the circuit components that determine frequency. Also called frequency departure and frequency deviation.

frequency-exchange signaling Signaling in which the change from one signaling condition to another is accompanied by decay in amplitude of one or more frequencies and buildup in amplitude of one or more other frequencies.

frequency frogging Interchanging of frequency allocations for carrier channels to prevent singing, reduce crosstalk, and reduce the need for equalization. Modulators in each repeater translate a low-frequency group to a high-frequency group, and vice versa.

frequency-hour One frequency used for 1 h, as in international shortwave broadcasting.

frequency hysteresis Failure of an oscillator to change frequency smoothly when continuously tuned because of use of a long transmission line between oscillator and load.

frequency-independent antenna An antenna whose operation is essentially independent of frequency for all frequencies above a lower limit.

frequency interlace Interlace of interfering signal frequencies with the spectrum of harmonics of scanning frequencies in television, to minimize the effect of interfering signals by altering the appearance of their pattern on successive scans.

frequency keying Keying in which the carrier frequency is shifted alternately between two predetermined values.

frequency lock A means of recovering, in a single-sideband suppressed-carrier receiver of a power-line carrier communication system, the exact modulating frequency that is applied to a single-sideband transmitter.

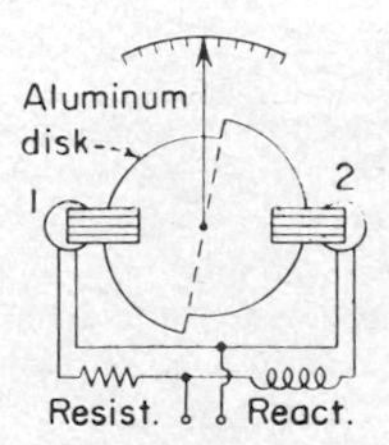

Frequency meter in which left half of off-center aluminum disk is acted on by induction voltmeter element 1 in series with resistor, giving constant torque at all frequencies. Right half of disk is acted on by element 2 in series with reactance. It has opposite torque but is affected by frequency; increases in frequency reduce current, so disk turns to right until torques are balanced.

frequency meter An instrument for measuring the frequency of an alternating current.

frequency-modulated [abbreviated FM] Pertaining to frequency modulation.

frequency-modulated jamming Jamming in which a constant-amplitude RF signal is varied above and below a center frequency to produce a signal covering a wide band of frequencies.

frequency-modulated laser A helium-neon or other laser in which an ultrasonic modulation cell is used to impress a frequency-modulated video signal on the output beam of the laser.

frequency-modulated radar A continuous-wave radar in which the carrier frequency is alternately increased and decreased at a predetermined rate. The frequency of the beat between the returning echo and the wave transmitted at the instant of echo arrival is proportional to range.

frequency-modulated radio altimeter An altimeter used in aircraft to give accurate absolute altitude indications from a few feet above the surface up to a limit of about 5000 ft (1.5 km). The frequency of the downward-radiated radio wave is varied back and forth continuously at some cyclic rate. Some of the energy from the transmitter is also fed to the input of the receiver for mixing with the energy reflected from the surface. The beat between the direct and reflected waves, equal to the change in transmitter frequency during the time of travel of the wave to the surface and back, is a direct indication of altitude. Also called low-altitude radio altimeter and terrain-clearance indicator.

frequency-modulated scanning sonar Scanning sonar in which the frequency of the transmitter is electronically decreased at a constant rate for several seconds, then suddenly increased to its original value, and the process repeated. The tone of the echo is then related to target position. The chief drawback is Doppler error when there is relative motion of transmitter and target.

frequency modulation [abbreviated FM] Modulation in which the instantaneous frequency of the modulated wave differs from the carrier

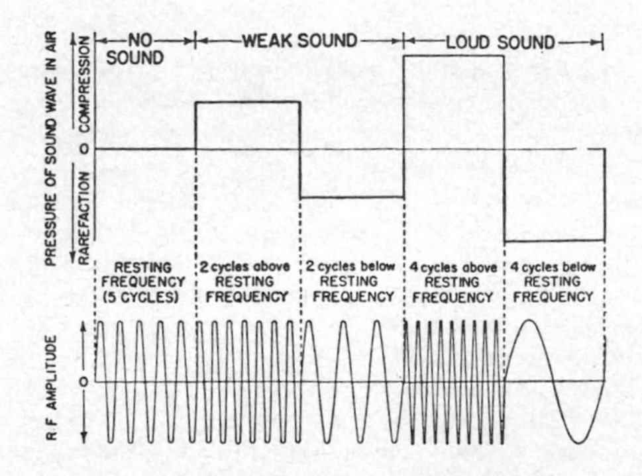

Frequency modulation, showing how carrier or resting frequency varies with strength of audio modulation.

frequency by an amount proportional to the instantaneous value of the modulating wave. The amplitude of the modulated wave is constant. A frequency-modulation broadcast system is practically immune to atmospheric and manmade interference.

frequency-modulation broadcast band The band of frequencies extending from 88 to 108 MHz, used for frequency-modulation radio broadcasting in the United States.

frequency-modulation broadcast channel A frequency band 200 kHz wide in the frequency-modulation broadcast band, designated by its center frequency. Assigned center frequencies begin at 88.1 MHz and continue in 100 successive steps of 0.2 MHz up to 107.9 MHz in the United States.

frequency-modulation receiver A radio receiver that receives frequency-modulated waves and delivers corresponding sound waves.

frequency-modulation recording Data recording in which the analog signal is converted into frequency deviations above and below a center frequency.

frequency-modulation transmitter A radio transmitter that transmits a frequency-modulated wave.

frequency-modulation tuner A tuner containing an RF amplifier, converter, IF amplifier, and demodulator for frequency-modulated signals, used to feed a low-level AF signal to a separate AF amplifier and loudspeaker.

frequency modulator A circuit or device that produces frequency modulation.

frequency monitor An instrument that indicates the amount of deviation of the carrier frequency of a transmitter from its assigned value.

frequency multiplier A harmonic conversion transducer in which the frequency of the output signal is an exact integral multiple of the input frequency. Also called multiplier.

frequency-offset transponder A transponder that changes the signal frequency by a fixed amount before retransmission.

frequency overlap The portion of a 6-MHz television channel that is common to both the monochrome and chrominance signals in a color television system. Frequency overlap is a form of bandsharing.

frequency pulling A change in the frequency of an oscillator due to a change in load impedance.

frequency pulsing Oscillator operation at two different frequencies alternately. Buildup of oscillation at one frequency creates conditions favorable for oscillation at the second frequency, and vice versa.

frequency pushing A change in the operating frequency of an oscillator, resulting from changes in bias voltage within the specified bias voltage limits for the circuit.

frequency range The range of frequencies over which a transmission system or device may be considered useful when used with different circuits under a variety of operating conditions. In contrast, bandwidth is a measure of useful frequency range with fixed circuits and fixed operating conditions.

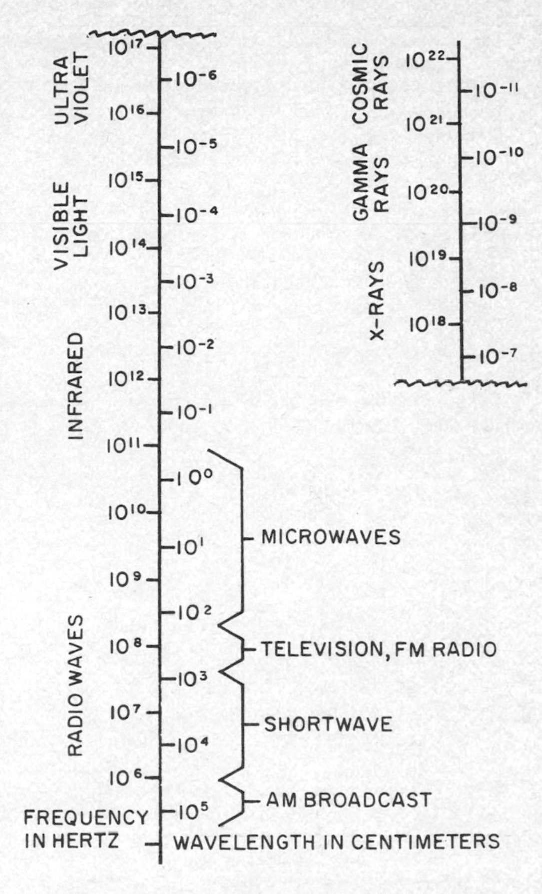

Frequency ranges for major categories of radiation in frequency spectrum. Frequencies below 10 kHz, in VLF range, are used by Navy for communication with submerged submarines.

frequency record A recording of various known frequencies at known amplitudes, usually for testing or measuring.

frequency-regulated power supply An AC power supply in which the dominant parameters are the range and regulation of the output frequency. The output voltage usually will also be regulated to some extent. A power oscillator operating from a DC power supply is usually used to generate the output frequency.

frequency regulator A device that maintains the frequency of an AC generator at a predetermined value.

frequency relay A relay that operates only at a predetermined frequency, such as a resonant-reed relay.

frequency response *Amplitude-frequency response.*

frequency-response equalization *Equalization.*

frequency run A series of tests made to determine the amplitude-frequency response characteristic of a transmission line, circuit, or device.

frequency-scan antenna A radar antenna in which scanning is produced in one dimension by frequency variation, somewhat as in a phased-array antenna.

frequency scanning Scanning of an oscillator output frequency back and forth over a desired frequency band.

frequency-selective ringing The use of two or more different frequencies of ringing current in a telephone system, in combination with ringers that are mechanically or electrically tuned so they operate only for one of the ringing frequencies.

frequency-selective surface An array of passive resonant elements that acts as a reflector over its resonant frequency band but passes RF energy essentially without attenuation at other frequencies. Used to provide dual-frequency capability for microwave tracking and radar systems.

frequency selectivity The ability of an electric circuit, apparatus, or device to differentiate between a desired signal and signals at other frequencies.

frequency-separation multiplier A multiplier in which each of the variables is split into a low-frequency and a high-frequency part. These parts are multiplied separately and the results added to give the required product. This system makes it possible to get high accuracy and broad bandwidth.

frequency separator The circuit that separates the horizontal and vertical synchronizing pulses in a monochrome or color television receiver.

frequency shift A change in the frequency of a radio transmitter or oscillator.

frequency-shift converter A device that converts a received frequency-shift signal to an amplitude-modulated or a DC signal.

frequency-shift keying [abbreviated FSK] A form of frequency modulation in which the modulating wave shifts the output frequency between predetermined values corresponding to the frequencies of correlated sources. When frequency-shift keying is used for code transmission, operation of the keyer shifts the carrier frequency back and forth between two distinct frequencies to designate mark and space. When frequency-shift keying is used for facsimile transmission, one carrier frequency represents picture black, and another, generally 800 Hz away, represents picture white.

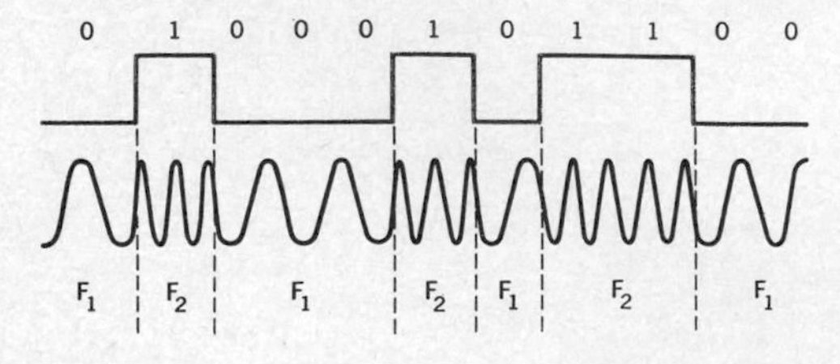

Frequency-shift keying.

frequency-slope modulation Modulation in which the carrier signal is swept periodically over the entire width of the band, much as in chirp radar. Modulation of the carrier with a voice or other communication signal changes the bandwidth of the system without affecting the uniform distribution of energy over the band. The desired information can thus be recovered from any part of the system bandwidth; this permits filtering out portions of the band that have interference without losing desired information.

frequency splitting Undesirable rapid shifting of the output frequency of a magnetron, caused by alternation between two modes of operation.

frequency stability The ability of an oscillator to maintain a desired frequency. Usually expressed as percent deviation from the assigned frequency value.

frequency stabilization The process of controlling the center frequency of an oscillator so it does not differ from that of a reference source by more than a prescribed amount.

frequency standard A stable oscillator, usually controlled by a crystal or tuning fork or an atomic clock, used primarily for frequency calibration.

frequency swing 1. The instantaneous departure of the frequency of an emitted wave from the center frequency during frequency modulation. 2. The difference between the maximum and minimum design values of the instantaneous frequency in a frequency-modulation system.

frequency synthesizer A device that provides a choice of a large number of different frequencies by setting 10-position knobs or equivalent rows of 10 pushbuttons to the exact frequency value desired. The instrument may cover part or all of the entire radio spectrum from DC up into the gigahertz range and provide a choice of waveforms that usually includes pulse, sine, square, and triangle. The output frequencies are derived from one or more precision quartz crystal oscillators, one of which is usually 1 or 5 MHz, acting with harmonic generators, dividers, multipliers, and mixers to provide direct or indirect frequency synthesis.

frequency tolerance The extent to which the carrier frequency of a transmitter may be permitted to depart from the assigned frequency.

frequency-to-voltage converter [abbreviated F/V converter] A converter that provides an analog output voltage which is proportional to the frequency or repetition rate of the input signal derived from a flowmeter, tachometer, or other AC generating device. In a process control system, this analog output voltage can be fed back to regulate the process. F/V converters are also used for stabilization and linearization of voltage-controlled oscillators.

frequency translation Moving a modulated RF carrier signal to a new location in the frequency spectrum without disturbing the relationship of the carrier to its sidebands.

frequency tripler An amplifier or other device that delivers output voltage at a frequency equal to three times the input frequency.

frequency-type telemeter A telemeter that uses the frequency of a signal as the translating means.

Fresnel diffraction Diffraction of microwaves that permits reception behind obstacles, with no focusing by the obstacles. The Fresnel diffraction region of an antenna is the near-field region in which the radiation pattern varies with distance.

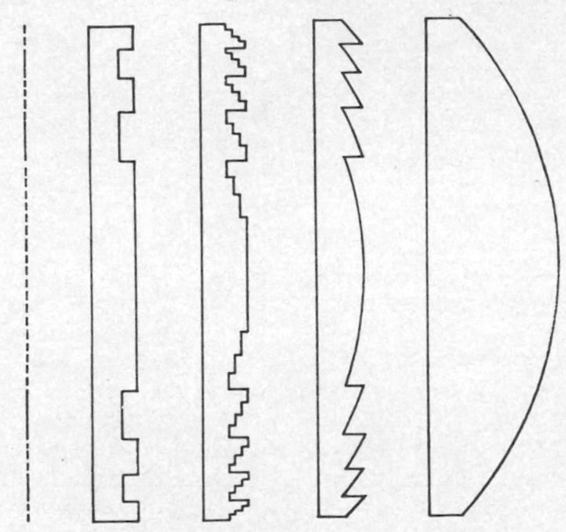

Fresnel-lens configurations. Left to right: simple Fresnel-zone plate, phase-reversing half-period zone plate, quarter-period zone plate, optical Fresnel lens, and equivalent conventional optical lens.

Fresnel lens A thin lens constructed with stepped setbacks so as to have the optical properties of a much thicker lens.

Fresnel region The region between the near field of an antenna and the Fraunhofer region. The boundary between the two is generally considered to be at a radius equal to twice the square of antenna length divided by wavelength.

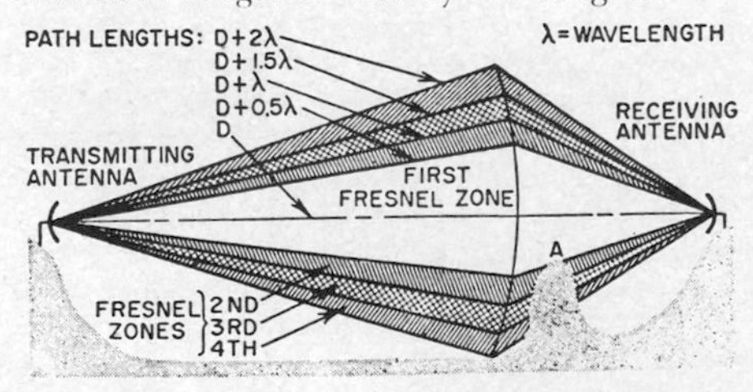

Fresnel zones between microwave antennas. Hill at A cuts off interfering waves, making received signal stronger than if hill did not exist. Hill may be higher, to give clearance of 0.6 of first Fresnel zone, without appreciably reducing signal strength.

Fresnel zone [after Augustin Jean Fresnel, 1788–1827] One of the conical zones that exist between microwave transmitting and receiving antennas due to cancellation of some portions of the wavefront by other portions that travel different distances. The boundary of the first Fresnel zone includes all paths half a wavelength longer than the line-of-sight path. The outer boundaries of the second, third, and fourth Fresnel zones are formed by paths 1, 1½, and 2 wavelengths longer than the direct path.

Fresnel-zone plate A thin plastic disk that has alternate radiopaque and radiolucent rings spaced so millimeter-wave incident radiation is brought to

Fresnel-zone plate, in which black rings are radiopaque grooves in plastic sheet.

focus at a point beyond the zone plate, just as an optical lens brings light waves to a focus. The action differs from that of a Fresnel lens in that diffraction at the angular apertures and subsequent interference of the diffracted radiation give the focusing action.

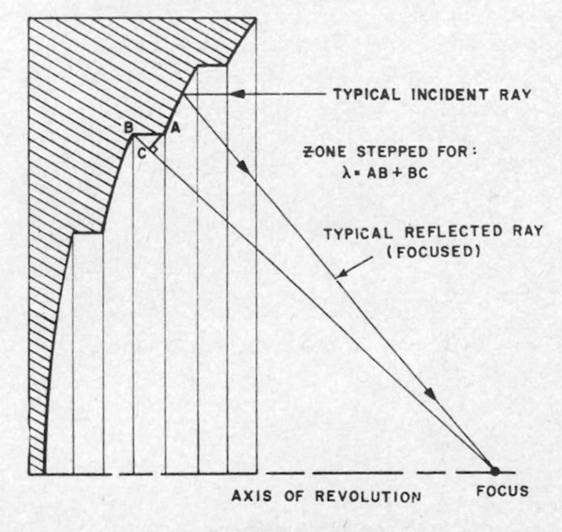

Fresnel-zone reflector.

Fresnel-zone reflector A stepped reflector that reflects incident electromagnetic radiation to a point focus. The distance between steps, in the direction of the incident radiation, is approximately half a wavelength at the receiving frequency.

friction bonding Soldering of a semiconductor chip to a substrate by vibrating the chip back and forth under pressure to create friction that breaks up oxide layers and helps alloy the mating terminals.

friction tape Cotton tape impregnated with a sticky moisture-repelling compound. Used chiefly to hold rubber-tape insulation in position over a joint or splice. Friction tape itself has relatively poor insulating qualities.

fringe area An area just beyond the limits of the reliable service area of a television transmitter, in which signals are weak and erratic. High-gain directional receiving antennas and high-sensitivity receivers are generally required for satisfactory fringe-area reception.

fringe effect The extension of the electrostatic field of an air capacitor outside the space between its plates.

fringe howl A howl or squeal heard when some circuit in a radio receiver is on the verge of oscillation.

frit seal A seal made by fusing together metallic powders with a glass binder, for such applications as hermetically sealing ceramic packages for integrated circuits.

frogging repeater A carrier repeater that has provisions for frequency frogging to permit use of a single multipair voice cable without having excessive crosstalk.

front contact The stationary contact of the normally open contacts on a relay.

front end The tuner of a television receiver, containing one or more RF amplifier stages, the local oscillator, and the mixer, along with all channel-tuning circuits.

front loading Placing material in front of a loudspeaker to change the acoustic impedance and thereby alter the radiation pattern. One method of achieving front loading is by placing an inverted exponential horn in front, to reduce the size of the opening through which acoustic energy emerges, and thereby improve the low-frequency radiation pattern.

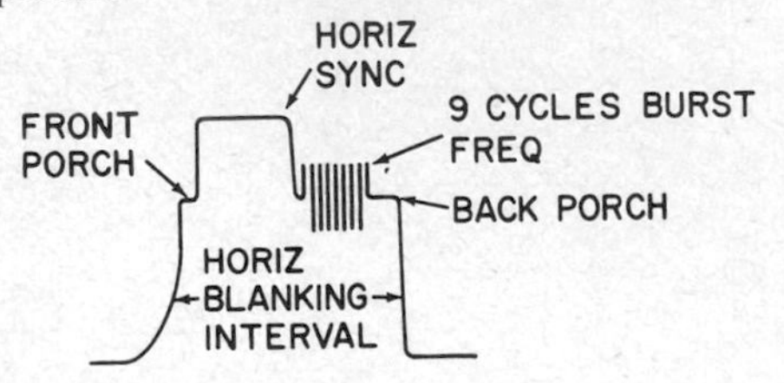

Front porch of horizontal sync pulse for color telecast, and back porch with 3.58-MHz color-burst frequency.

front porch The portion of a composite picture signal that lies between the leading edge of the horizontal blanking pulse and the leading edge of the corresponding sync pulse. The duration of the front porch is 1.27 μs in the standard U.S. television signal.

front projection A projection television system that uses a nontranslucent reflecting screen. Rear projection uses a translucent screen, with viewers and the projector on opposite sides of the screen.

front-to-back ratio *Front-to-rear ratio.*

front-to-rear ratio The ratio of the effectiveness of a directional antenna, loudspeaker, or microphone toward the front and toward the rear. Usually expressed in decibels. Also called front-to-back ratio.

frozen flux The continuous magnetic flux associated with current flowing continuously in a superconducting ring.

fruit pulse An unsynchronized pulse reply resulting from interrogation of a transponder by interrogators not associated with the responsor in question.

fs Abbreviation for *femtosecond.*

FS Abbreviation for *full scale.*

F scope A radarscope that produces an F display.

FSK Abbreviation for *frequency-shift keying.*

ft Abbreviation for *foot.*

ft/s Abbreviation for *foot per second.*

fuel cell A cell that converts chemical energy directly into electric energy, with electric power being produced as a part of a chemical reaction

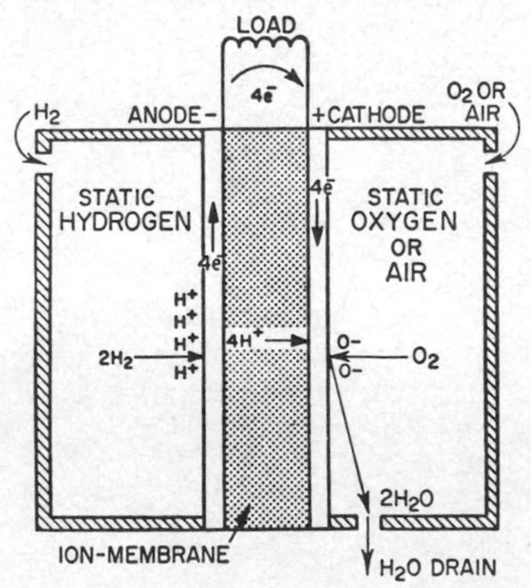

Fuel cell connected to inductive load. Hydrogen gas, serving as fuel, is fed into cell at upper left as required to meet power requirements of load. Cell shown is ion-membrane type, with no liquid electrolyte. By-product is water, which drains off at lower right.

between the electrolyte and a fuel like hydrazine, hydrogen, or even kerosene. The fuel is fed into the cell from an external tank. In one application, a fuel cell is used to charge a storage battery continuously, so power is instantly available from the battery even though the fuel cell has a start-up lag.

full adder *Adder.*

full-duplex operation Operation of telegraph or other signal systems in opposite directions simultaneously.

full load The greatest load that a circuit or piece of equipment is designed to carry under specified conditions. Any additional load is an overload.

full scale [abbreviated FS] The maximum accurate reading on a particular scale for a particular range of a measuring instrument.

full-wave bridge A bridge arrangement of four diode or tube rectifiers, providing full-wave rectification of the full secondary voltage of the power

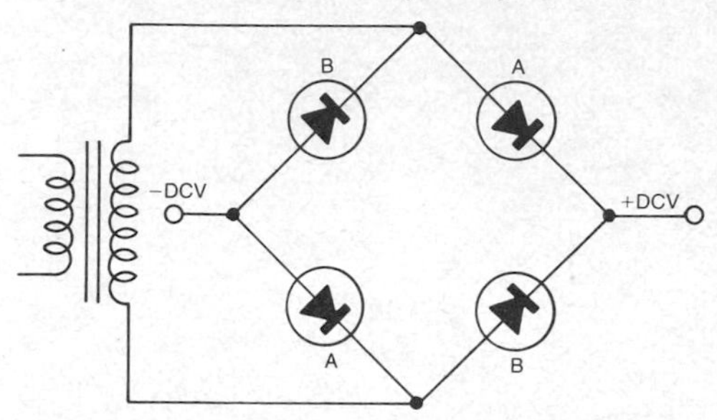

Full-wave bridge using four diodes. Polarity of DC output voltage is indicated.

transformer. A center tap is not needed on the secondary winding of the transformer. The output voltage contains both halves of each input cycle.

full-wave control Phase control that acts on both halves of each AC cycle, for varying load power over the full range from zero to the full-wave maximum value. The output voltage is usually utilized in its AC form, but a rectifying stage can be added to provide a pulsating DC output voltage. The control element may be a silicon controlled rectifier or other solid-state power control device, or a thyratron or other controllable gas tube.

full-wave gas rectifier A cold-cathode rectifier in which the cathode is much smaller than the two anodes, to give rectification.

full-wave rectification Rectification in which output current flows in the same direction during both half-cycles of the alternating input voltage.

full-wave rectifier A double-element rectifier that provides full-wave rectification. One element functions during positive half-cycles and the other during negative half-cycles.

full-wave vibrator A vibrator having an armature that moves back and forth between two fixed contacts, to change the direction of direct-current flow through a transformer at regular intervals and thereby permit voltage step-up by the transformer. Used in battery-operated power supplies for mobile and marine radio equipment.

fully active homing Homing in which the missile generates its own radar signals and carries a computer that provides guidance signals for lock-on to give a collision course.

fully ionized plasma Plasma in which all the neutral particles have lost at least one electron.

function A quantity whose value depends upon the value of one or more other quantities.

function digit A computer instruction-word digit that determines the arithmetic or logic operation to be performed.

function generator 1. An analog computer device that indicates the value of a given function as the independent variable is increased. 2. A signal generator that delivers a choice of a number of different waveforms, with provisions for varying the frequency over a wide range.

function key A special key on a keyboard to control a mechanical function, initiate a specific computer operation, or transmit a signal that would otherwise require multiple key strokes.

function multiplier An analog computer device that takes in the changing values of two functions and puts out the changing value of their product as the independent variable is changed.

function switch A network that has a number of inputs and outputs so connected that input signals expressed in a certain code will produce output signals which are a function of the input information but in a different code.

function table 1. A table that lists the values of a function for various values of the variable. 2. Sets of computer information arranged so an entry in one set selects one or more entries in the other sets. 3. A computer device that converts multiple inputs into a single output or encodes a single input into multiple outputs.

fundamental *Fundamental frequency.*

fundamental component *Fundamental frequency.*

fundamental field particle The field quantum that is the result of quantizing a field.

fundamental frequency 1. The lowest frequency component of a complex vibration, sound, or electric signal. Used as the basis for harmonic analysis of a wave. The fundamental frequency is the reciprocal of the period of a wave. Also called first harmonic, fundamental, and fundamental component. The frequency that is twice the fundamental frequency is called the second harmonic. 2. The first order or lowest frequency of an intended mode of vibration for a quartz plate or other vibrating object. Also called fundamental mode.

fundamental-frequency magnetic modulator A magnetic modulator in which the output is at the fundamental frequency of the supply.

fundamental mode 1. The waveguide mode that has the lowest critical frequency. Also called dominant mode and principal mode. 2. *Fundamental frequency.*

fundamental particle *Elementary particle.*
fundamental tone The component tone of lowest pitch in a complex tone.
fundamental unit An arbitrarily defined unit that is the basis of a system of units. All other units in the system are derived from a set of fundamental units. The dimensions of any physical quantity may be expressed as combinations of fundamental units.
fundamental wavelength The wavelength corresponding to the fundamental frequency.
fungiproofing Application of a protective chemical coating that inhibits growth of fungi on electronic equipment in humid tropical regions.
fuse [symbol F] A protective device containing a short length of special wire that melts when the

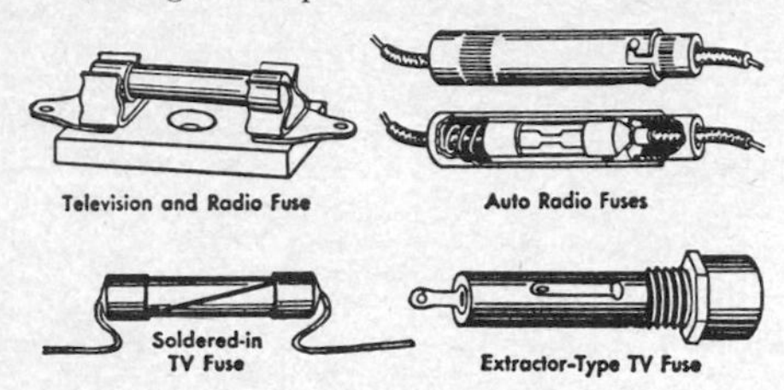

Fuses used in electronic equipment.

current through it exceeds the rated value for a definite period of time. A fuse is inserted in series with the circuit being protected, so it opens the circuit automatically during a serious overload.
fuse alarm A lamp that indicates the presence and location of a blown fuse.
fuse block An insulating base on which are mounted fuse clips or other contacts for fuses.
fuse clip A spring contact used to hold and make connection to a cartridge-type fuse.
fuse diode A diode that opens under specified current surge conditions.
fused junction *Alloy junction.*
fused quartz A glasslike insulating material made by melting crushed crystals of natural quartz or a certain type of quartz sand.
fused-quartz delay line An ultrasonic delay line made from fused quartz. Delays as high as 2.5 ms can be obtained by utilizing multiple internal reflections from a large polygon-shaped piece of fused quartz.
fuse wire Wire made from an alloy that melts at a relatively low temperature and overheats to this temperature when carrying a particular value of overload current.

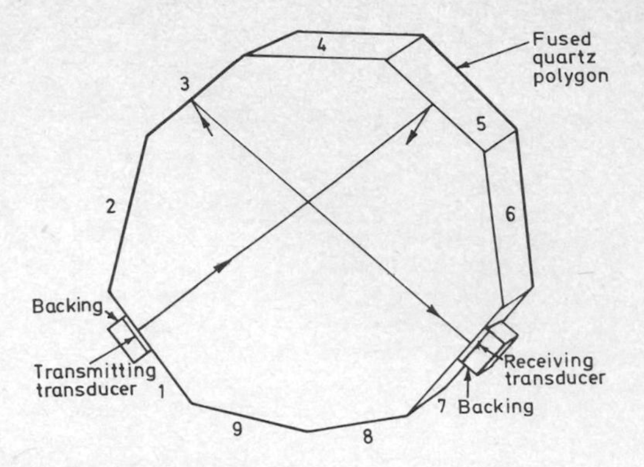

Fused-quartz delay line having nine reflecting sides. Path of beam from transmitting transducer is 1-5-9-3-6-9-2-4-6-8-2-5-8-3-7.

fusible resistor A resistor that protects a circuit against overload. Its resistance limits current flow and thereby protects against surges when power is first applied to a circuit. Its fuse characteristic opens the circuit when current drain exceeds design limits. Generally used in series with anode supply circuits.
fuze A device used to detonate an explosive charge automatically when the proper conditions of impact, elapsed time, external command, or proximity are achieved in a bomb, missile, torpedo, or mine.
fuze chronograph A chronograph that accurately measures and records the time interval between firing of a time-fuzed projectile and the air burst as detected by a photoelectric pickup aimed at the expected point of burst. The recorder may be started by the passage of a magnetized projectile through a coil mounted on the muzzle of the gun.
fuze-setting element The element in a fire-control system that sets the time fuze of the missile when such a fuze is used.
fV Abbreviation for *femtovolt.*
F/V Abbreviation for frequency-to-voltage, usually used in connection with converters.
F/V converter Abbreviation for *frequency-to-voltage converter.*
F/V-V/F An abbreviation used to indicate that a converter will convert in both directions, from frequency-to-voltage and from voltage-to-frequency.
fW Abbreviation for *femtowatt.*

G

g Abbreviation for *gram.*
G 1. Symbol for *conductance.* 2. Symbol for *gate.* 3. Abbreviation for *gauss.* 4. Symbol for *generator.* 5. Abbreviation for *giga-.* 6. Abbreviation for *gravitational force.* 7. Symbol for *grid.*
G_m Symbol for *transconductance.*
GaAlAs Symbol for *gallium aluminum arsenide.*
GaAs Symbol for *gallium arsenide.*
GaAsP Symbol for *gallium arsenide phosphide.*
gadolinium [symbol Gd] A rare-earth metallic element. Atomic number is 64.
gage A measuring device or measuring instrument.
gain The increase in signal power that is produced by an amplifier. Generally expressed in decibels.
GaInAs Symbol for *gallium indium arsenide.*
gain-bandwidth product A figure of merit for amplifiers, based on the gain and the bandwidth as measured under specified conditions.
gain control A device for adjusting the gain of a system or component.
gain margin The amount of increase in gain that would cause oscillation in a feedback control system.
GaInP Symbol for *gallium indium phosphide.*
gain-time control *Sensitivity-time control.*
galactic noise Radio-frequency noise that originates in space from all celestial bodies except the sun.
galena A crystalline form of lead sulfide, once used in crystal detectors.
gallium [symbol Ga] A metallic element. Atomic number is 31. Used as an acceptor impurity in germanium and silicon semiconductor devices.
gallium aluminum arsenide [symbol GaAlAs] A semiconductor material used in light-emitting diodes to produce a red light.
gallium arsenide [symbol GaAs] A semiconductor that has a forbidden band gap of 1.4 eV and a maximum operating temperature of 400°C. Used in infrared light-emitting diodes, varactors, and other semiconductor devices.

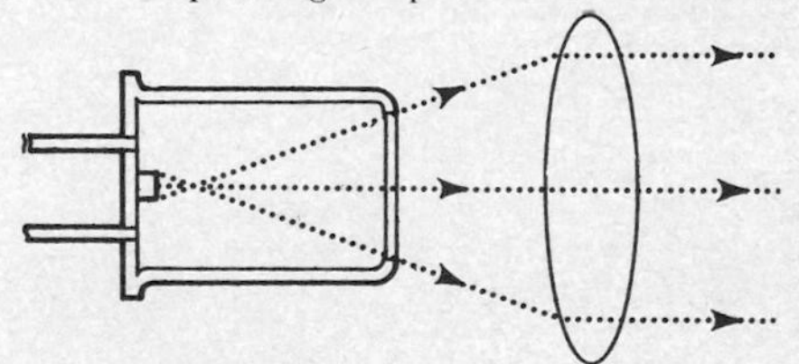

Gallium arsenide lamp used with lens to give infrared beam that can be projected over 1000 ft (300 m) for use in intrusion alarms.

gallium arsenide laser A laser that emits light at right angles to a junction region in gallium arsen-

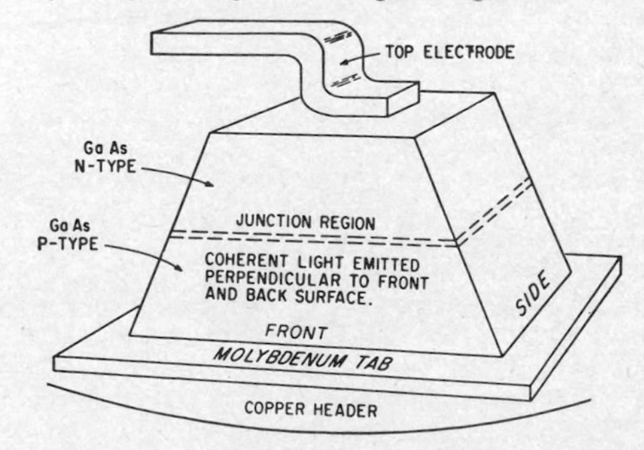

Gallium arsenide laser mounted above parabolic reflector. Greater transparency of N-type material, facing outward, increases light output.

ide, at a wavelength of 9000 Å. It can be modulated directly at microwave frequencies. Cryogenic cooling is required.

gallium arsenide phosphide [symbol GaAsP] A semiconductor material used in light-emitting diodes. The color of the visible light produced is

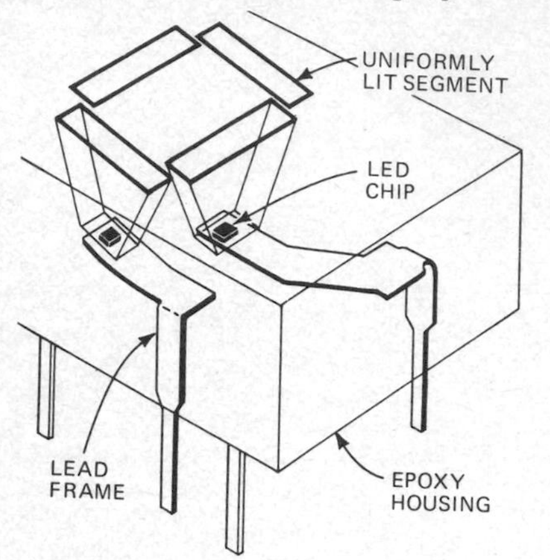

Gallium arsenide phosphide chips provide either red or yellow seven-segment display.

red or amber, depending on the proportions of arsenide and phosphide. The wavelength range is 650 to 670 nm.

gallium indium arsenide [symbol GaInAs] A semiconductor material used as a cathode surface in high-sensitivity multiplier phototubes for detecting very low light levels, as in astronomy and medicine.

gallium indium phosphide [symbol GaInP] A semiconductor material used in light-emitting diodes to produce a yellow light.

gallium phosphide [symbol GaP] A semiconductor material used in light-emitting diodes to give either green or red light in the wavelength range of 690 to 790 nm.

galvanic Pertaining to electricity flowing as a result of chemical action.

galvanic cell An electrolytic cell that is capable of producing electric energy by electrochemical action.

galvanometer An instrument for indicating or measuring a small electric current by means of a mechanical motion derived from electromagnetic or electrodynamic forces produced by the current.

galvanometer constant A number by which a particular relative reading of a galvanometer must be multiplied to obtain the current value.

galvanometer light-beam recorder A direct-writing recorder in which a galvanometer moves only the tiny mass of a mirror to deflect a light beam that writes on photosensitive film or paper.

galvanometer recorder A sound recorder in which the audio signal voltage is applied to a coil suspended in a magnetic field. The resulting movements of the coil cause a tiny attached mirror to move a reflected light beam back and forth across a slit in front of a moving photographic film. This provides a variable-area photographic record of the signal.

galvanometer shunt A resistor connected in parallel with a galvanometer to reduce its sensitivity under certain conditions. It allows only a known fraction of the current to pass through the galvanometer.

game theory A mathematical process of selecting an optimum strategy for competing with an opponent's strategy.

gamma 1. A numerical indication of the degree of contrast in a television or photographic image. It is equal to the slope of the straight-line portion of the H and D curve for the emulsion or screen. 2. A unit of magnetic field strength, equal to 10 microoersteds or 0.00001 oersted.

gamma-absorption gage *Gamma gage.*

gamma cascade Successive gamma transitions occurring in radioactive transformations.

gamma correction Correction of the effective value of gamma by introducing a nonlinear output-input characteristic.

gamma decay Emission of a gamma ray by a nucleus, with loss of energy but without change of atomic or mass number.

gamma emitter An atom whose radioactive decay process involves the emission of gamma rays.

gamma ferric oxide The ferromagnetic form of ferric oxide, used as a coating on practically all magnetic tape.

gamma gage A penetration-type thickness gage that measures the thickness or density of a sample by measuring the absorption of gamma rays in the sample. Also called gamma-absorption gage.

gamma heating Heating resulting from absorption of gamma-ray energy by a material.

gamma radiation Radiation of gamma rays.

gamma radiography Radiography by means of gamma rays.

gamma ray A quantum of electromagnetic radiation emitted by a nucleus as the result of a quantum transition between two energy levels of the nucleus. Gamma rays have energies usually

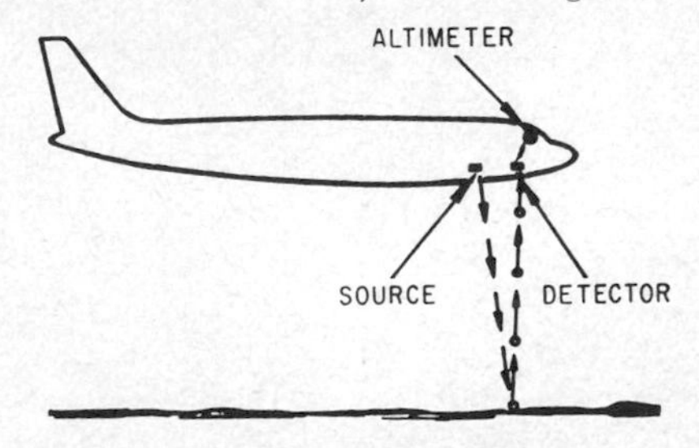

Gamma-ray altimeter on airplane coming in for landing during zero-zero visibility.

between 10 keV and 10 MeV, with shorter wavelengths than x-rays. They are more penetrating than alpha and beta particles and are not affected by magnetic fields.

gamma-ray altimeter A low-altitude altimeter that depends on transmission of photons to earth from a cobalt-60 gamma source in the belly of a plane. The resulting photon backscatter from the earth is measured by a detector, amplifier, and pulse-height discriminator. Accuracy is high in the critical range under several hundred feet, as required for all-weather aircraft landings and lunar vehicle landings.

gamma-ray camera A scintillation camera that uses a single large sodium iodide imaging crystal to convert gamma radiation from an area of the human body containing radionuclides to an equivalent light image. A matrix of multiplier phototubes detects the location of each source of light on the crystal and translates this to a corresponding point of light on the face of an oscilloscope or imaging tube.

gamma-ray counter A radiation counter designed for detecting and recording the intensity of gamma rays.

gamma-ray detector An instrument used on ships to identify and measure abnormal gamma-ray concentrations in the ocean that result from dumping or release of nuclear wastes.

gamma-ray level indicator A level indicator in which the rising level of the liquid or other mate-

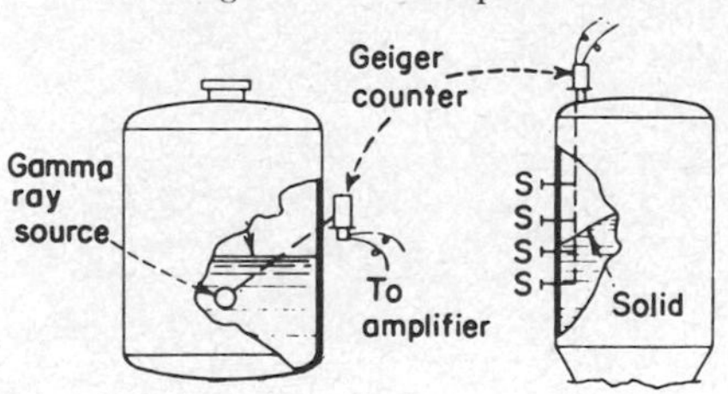

Gamma-ray level indicators using single radioactive source (left) and multiple sources S (right). Accurate measuring range is about 2 ft (0.6 m) of level change per source.

rial reduces the amount of radiation passing from a gamma-ray source through the container to a Geiger counter or other radiation detector.

gamma-ray logging Obtaining, by means of a gamma-ray probe, a record of the intensities of gamma rays emitted by the rock strata penetrated by a borehole.

gamma-ray probe A gamma-ray counter built into a watertight case small enough to be lowered into a borehole.

gamma-ray source A quantity of radioactive material that emits gamma radiation and is in a form convenient for radiology.

gamma-ray spectrometer An instrument that measures the energy distribution of gamma rays.

gamma-ray thickness gage A thickness gage

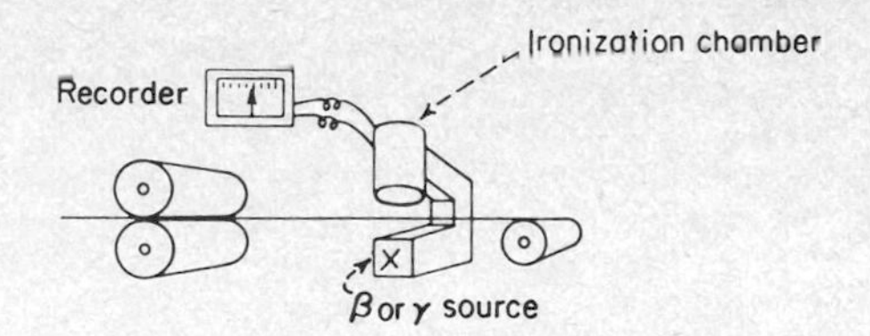

Gamma-ray thickness gage.

that uses a radioactive source and a radiation detector to measure the amount of radiation transmitted by a moving sheet of material. Absorption (and hence transmission) of radiation by the material is proportional to its thickness or weight.

gamma-ray tracking Use of three tracking stations, located at the three corners of a triangle centered on a missile about to be launched, to

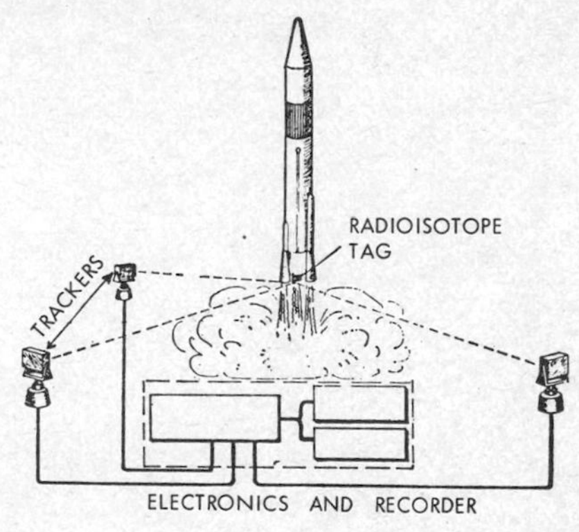

Gamma-ray tracking of missile during critical early stage of launch.

obtain accurate azimuthal tracking of a cobalt-60 gamma source in the tail. High accuracy is obtained up to about 1000 ft (300 m), despite heat shimmer, exhaust plasma, and other phenomena that affect photoelectric, radar, radio, or other conventional methods of tracking.

gammas Gamma rays.

Gamow barrier radius *Coulomb barrier radius.*

Gamow-Teller selection rules [abbreviated GT selection rules] A set of selection rules for allowed beta transitions.

gang capacitor A combination of two or more variable capacitors mounted on a common shaft to permit adjustment by a single control.

ganged tuning Simultaneous tuning of two or more circuits with a single control knob.

ganged volume control A combination of two or more volume controls, one for each channel of a stereophonic sound system, mounted on a common shaft to permit changing simultaneously the volume of all loudspeakers without changing their balance with respect to each other.

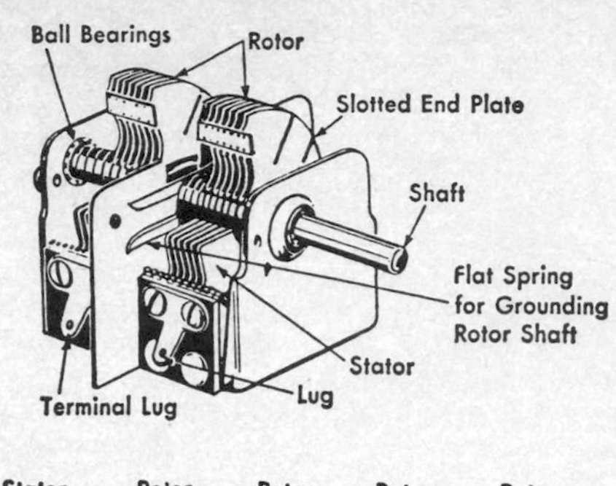

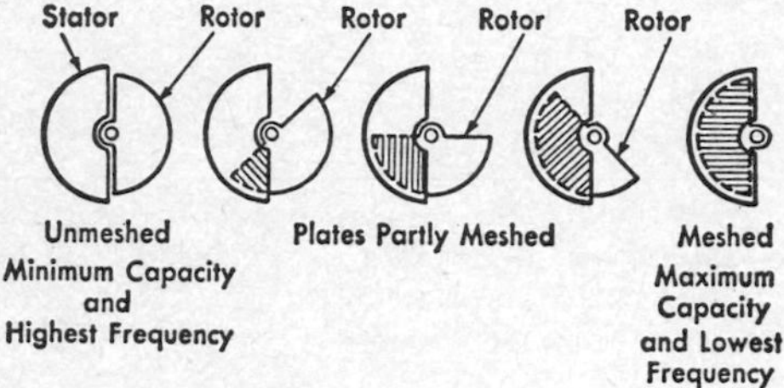

Gang-capacitor construction and tuning action.

ganging A mechanical means of operating two or more controls with one control knob.

gang punch To punch identical or constant information into all of a group of punched cards.

gang switch A combination of two or more switches mounted on a common shaft to permit operation by a single control. Also called deck switch.

gang tuning capacitor A gang capacitor used for tuning purposes in a receiver.

gantry A structure used alongside a spacecraft being prepared for launching to permit access by scientists and technicians to all levels of the vehicle and, for manned vehicles, to allow the astronauts to enter their spacecraft at the top. A gantry rolls away on tracks while remaining vertical, whereas an erector tilts away to a horizontal position before firing.

gap 1. A break in a closed magnetic circuit, containing only air or filled with a nonmagnetic material. 2. The spacing between two electric contacts. 3. A region not adequately covered by the main lobes of a radar antenna. 4. A space or a time interval used to indicate the end of a word, record, or file of data on magnetic tape or some other storage medium.

GaP Symbol for *gallium phosphide.*

gap coding A method of communicating information by interrupting the transmission of an otherwise regular signal in such a way that the interruptions form a telegraphic-type message. Used in some radar beacons.

gap-filler An auxiliary radar antenna used to cover gaps in the main radar antenna pattern.

gap length The physical distance between adjacent surfaces of the poles of a longitudinal magnetic recording head.

gap loading The electronic gap admittance that results from the movement of electrons in the gap of a microwave tube. The three types are multipactor gap loading, primary transit-angle gap loading, and secondary-electron gap loading.

gap loss The reduction in signal strength that occurs during playback of a magnetic recording as the wavelength of the recorded signal approaches the effective length of the magnetic path between the playback head and the magnetic recording film.

garbage Meaningless sequences of characters at the input or output of a computer.

garble To alter a message intentionally or unintentionally so it is difficult to understand.

garnet maser Term sometimes used incorrectly for *ferromagnetic amplifier.*

gas capacitor A capacitor that consists of two or more electrodes separated by a gas, other than air, which serves as a dielectric.

gas-cell frequency standard An atomic frequency standard in which the frequency-determining element is a gas cell containing rubidium, cesium, or sodium vapor.

gas counter A counter in which the radioactive sample is prepared in the form of a gas and introduced into the counter tube.

gas current A positive-ion current produced by collisions between electrons and residual gas molecules in an electron tube. Also called ionization current.

gas detector A gas-sensing semiconductor or other device whose resistance decreases in the presence of a deoxidizing gas like carbon monoxide, hydrogen, methane, propane, and volatile oils.

gas diode A tube that has a hot cathode and an anode in an envelope which contains a small amount of an inert gas or vapor. When the anode is made sufficiently positive, the electrons flowing to it collide with gas atoms and ionize them. As a result, anode current is much greater than that for a comparable vacuum diode.

gas discharge Conduction of electricity in a gas, due to movements of ions produced by collisions between electrons and gas molecules.

gas-discharge display A display in which seven or more cathode elements form the segments of numerical or alphameric characters when energized by about 160 VDC. The segments are vacuum-sealed in a neon-mercury gas mixture. A red glow forms on each energized cathode, much as on the individual cathode characters of older Nixie tubes.

gas-discharge lamp *Discharge lamp.*

gas doping The introduction of impurity atoms into a semiconductor material by epitaxial growth, using streams of gas that are mixed before being fed into the reactor vessel.

gasdynamic laser A gas laser that converts thermal energy directly into coherent radiation at an efficiency high enough to offer promise of wireless power transmission. The gas is usually a mixture of nitrogen and carbon dioxide with a catalyst of helium or water vapor, heated to about 2000 K by an arc jet, shock tube, or other means. The wavelength is the standard CO_2 laser transition of 10.6 μm in the infrared spectrum and can be pulsed or continuous wave.

gas etching The removal of material from a semiconductor circuit by reaction with a gas that forms a volatile compound. The etching of silicon slices with hydrogen chloride at 1000°C or higher is an example.

gas-filled cable A coaxial or other cable that contains gas under pressure and insulates and keeps out moisture.

gas-filled radiation-counter tube A gas tube used to detect radiation by means of gas ionization.

gas-filled rectifier *Cold-cathode rectifier.*

gas-flow counter tube A radiation-counter tube in which an appropriate atmosphere is maintained by a flow of gas through the tube.

gas focusing A method of concentrating an electron beam by utilizing the residual gas in a tube. Beam electrons ionize the gas molecules, forming a core of positive ions along the path of the beam. This core of ions attracts beam electrons and thereby makes the beam more compact. Also called ionic focusing.

gas-ionization readout A readout that consists of a single anode and a number of separate cathode elements inside the glass envelope of a gas-filled glow-discharge tube. The cathodes are in the shape of numerals or other characters desired for a display. Application of a negative voltage to the selected cathode makes its character visible.

gas laser A laser in which the active medium is a discharge in a gas contained in a glass or quartz tube with a Brewster-angle window at each end. The gas can be excited by a high-frequency oscillator or direct-current flow between electrodes inside the tube. The function of the discharge is to pump the medium, to obtain a population inversion. Operation can be pulsed or continuous. Basic types are: 1. atomic (neutral) gas lasers, such as helium-neon lasers; 2. ion lasers, such as metal vapor lasers; 3. molecular gas lasers, such as carbon dioxide lasers; 4. chemical lasers, in which the population inversion for lasing is produced by a chemical reaction between gases.

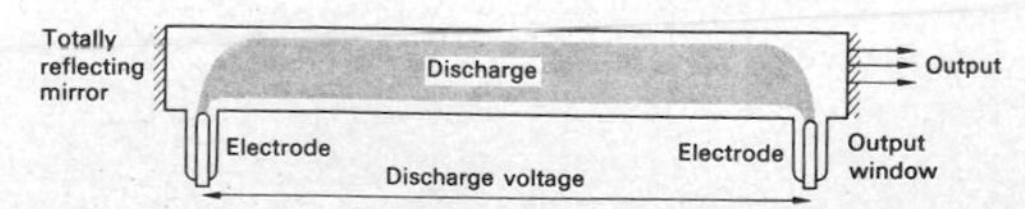

Gas laser in which ionizing discharge is axial, between end electrodes.

gas lens A lens that uses variations in the refractive indexes of gases to guide light. A long gas lens can be used to confine a laser beam to the center of a pipe, for long-distance laser communication. One method of achieving variations in refractive indexes is to place in the center of the pipe a helical heating element that concentrates the gas in the center of the pipe. Another method involves the use of two different gases, flowing in opposite directions in the pipe.

gas maser A maser in which the microwave electromagnetic radiation interacts with the molecules of a gas such as ammonia. Use is limited chiefly to highly stable oscillator applications, as in atomic clocks.

gas noise Electric noise caused by random ionization of gas molecules in gas tubes and partially evacuated vacuum tubes.

gas phototube A phototube into which a quantity of gas has been introduced after evacuation, usually to increase its sensitivity. In a vacuum phototube, no appreciable gas is present.

gas scattering The scattering of electrons or other particles in a beam by residual gas in the vacuum system.

gassy tube A vacuum tube that has not been fully evacuated or has lost part of its vacuum because of release of gas by the electrode structure during use, so that enough gas is present to impair operating characteristics appreciably. Also called soft tube.

gas thermostatic switch A thermostatic switch in which heat causes the pressure of gas in a sealed metal bellows to increase, thereby moving the bellows and closing the contacts of a switch.

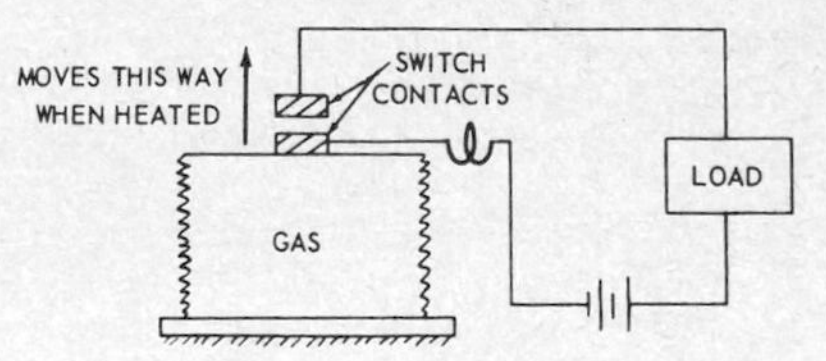

Gas thermostatic switch.

gas tube An electron tube in which the contained gas or vapor performs the primary role in the operation of the tube.

gas x-ray tube An x-ray tube in which the emission of electrons from the cold cathode is produced by positive-ion bombardment when the applied cathode-anode voltage is made sufficiently high.

gate 1. A circuit that has an output and a multiplicity of inputs so designed that the output is energized when and only when a certain combination of pulses is present at the inputs. An AND-gate delivers an output pulse only when every input is energized simultaneously in a specified manner. An OR-gate delivers an output pulse when any one or more of the input pulses meet the specified conditions. Used in digital computers. 2. A circuit in which one signal, generally a square wave, switches another signal on and off.

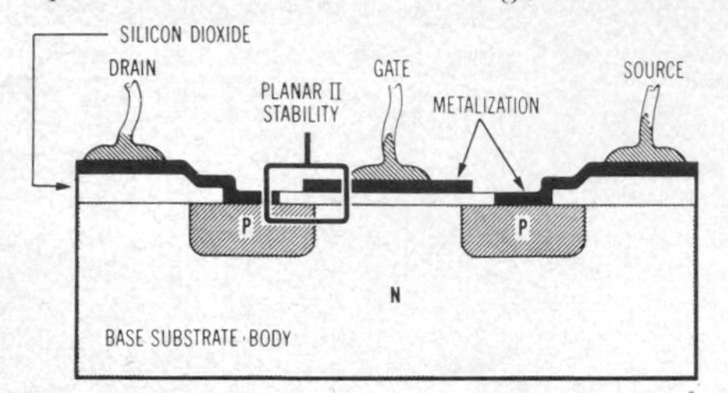

Gate in Fairchild MOS field-effect transistor. Planar process applied to area between drain and gate locks surface charge, thereby improving reliability and stability.

Used in radar to block the receiver, except for brief instants when echo signals may be returning from a target. 3. [symbol G] One of the electrodes in a field-effect transistor, silicon controlled rectifier, or other semiconductor device. 4. An output element of a cryotron. 5. To control the passage of a pulse or signal.

gate circuit A circuit that admits and amplifies or passes a signal only when a gating pulse is present.

gate-controlled rectifier A three-terminal semiconductor device in which the unidirectional current flow between the rectifier terminals is controlled by a signal applied to a third terminal called the gate. The silicon controlled rectifier is an example.

gate-controlled switch [abbreviated GCS] A semiconductor device that can be switched from its nonconducting or off state to its conducting or on state by applying a negative pulse to its gate terminal and turned off at any time by applying reverse drive to the gate. (Once a silicon controlled rectifier is turned on, it can be turned off only by interrupting its cathode-anode current.)

gate current The alternating or pulsating direct current that flows through the gate winding of a magnetic amplifier.

gated-beam tube A pentode electron tube that has special electrodes which form a sheet-shaped beam of electrons. This beam may be deflected away from the anode by a relatively small voltage applied to a control electrode, thus giving extremely sharp cutoff of anode current. Used in some FM detector circuits.

gated buffer A low-impedance inverting driver circuit that may be used as a line driver, in multivibrators, or for pulse differentiation.

gated sweep A radar sweep in which the duration and starting time are controlled, to exclude undesired echoes from the screen.

gated transistor A transistor in which a gate electrode covers the emitter and collector junc-

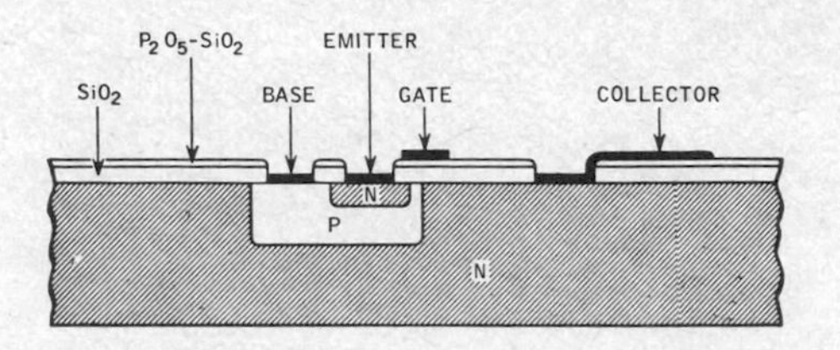

Gated transistor cross section.

tions, to permit applying an electric field at the surface of the base region.

gate equivalent circuit A unit of measure for specifying relative complexity of digital circuits, equal to the number of individual logic gates that would have to be interconnected to perform the same function as the digital circuit under evaluation.

gate generator A circuit used to generate gate pulses. In one form it consists of a multivibrator that has one stable and one unstable position.

gate pulse A pulse that triggers a gate circuit so it will pass a signal.

gate tube An electron tube that produces an output signal only when two signal voltages, from independent circuits, are applied simultaneously to two separate electrodes.

gate turnoff [abbreviated GTO] A type of silicon controlled rectifier that can be turned on by a pulse of gate current and turned off by applying a pulsed negative bias between gate and cathode terminals. Used for power-switching applications at power-line and higher frequencies.

gate-turnoff switch An all-diffused three-junction semiconductor switching device that can be turned on or off from its gate input terminal.

gate voltage The voltage across the terminals of the gate winding in a magnetic amplifier.

gate winding A winding used in a magnetic amplifier to produce ON/OFF action of load current.

gating The process of selecting those portions of a wave that exist during one or more selected time intervals or that have magnitudes between selected limits. Usually achieved by applying a pulsed voltage to a normally cut-off electron tube, transistor, or magnetic amplifier to make the device conductive (open the gate) for the duration of the pulse. The opposite action (the device is cut off for the duration of the pulse) is used to black out a television cathode-ray tube during retraces and is called blanking.

gauss [abbreviated G; plural is gauss] The CGS electromagnetic unit of magnetic induction *B*. Use of the SI unit, the tesla, is preferred.

Gaussian distribution A distribution of random variables comparable to that found in nature, characterized by a symmetrical and continuous distribution decreasing gradually to zero on either side of the most probable value. Also called normal distribution.

Gaussian noise Noise that has a frequency distribution which follows the Gaussian curve.

Gaussian noise generator A signal generator that produces a random noise signal whose frequency components have a Gaussian distribution centered on a predetermined frequency value. A gas thyratron may be used as the primary source of noise.

Gaussian well A potential well whose value varies according to a Gaussian distribution.

gaussmeter A magnetometer with a scale graduated in gauss or kilogauss. One version is used to measure the magnetic field strength between the pole faces of magnetic structures for magnetrons, ranging from 1.2 to 9.6 kG.

Gb 1. Abbreviation for *gigabit.* 2. Abbreviation for *gilbert.*

GCA Abbreviation for *ground-controlled approach.*

GCI Abbreviation for *ground-controlled interception.*

GCS Abbreviation for *gate-controlled switch.*

G display A rectangular radarscope display in which a target appears as a laterally centralized bright spot when the radar antenna is aimed at it

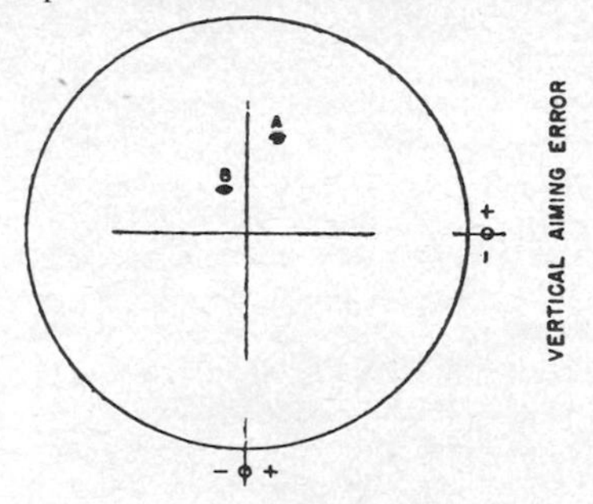

G display

in azimuth. Wings appear to grow on the spot as the distance to the target is diminished. Horizontal and vertical aiming errors are respectively indicated by horizontal and vertical displacement of the spot, just as in an F display.

gearmotor A motor combined with a set of speed-reducing gears.

Gee [Ground Electronics Engineering] A VHF navigation system in which three or more ground stations transmit synchronized pulses. Hyperbolic lines of position are determined by measuring the differences in the times of arrival of these pulses. Developed in Great Britain, and similar to loran.

Gee-H System A combination of Gee and H systems of navigation, used to give greater precision for bombing.

Geiger counter A radiation counter that uses a Geiger counter tube in appropriate circuits for detecting and counting ionizing particles, such as cosmic-ray particles. Each particle ionizes the gas in the tube in such a way that the total ionization per event is independent of the energy of the ionizing particle. Ionization results in electron flow from the tube wall to the center electrode, producing one output voltage pulse for each particle. Also called Geiger-Mueller counter. 2. *Geiger counter tube.*

Geiger counter tube A radiation-counter tube operated in the Geiger region. It usually consists of a gas-filled cylindrical metal chamber contain-

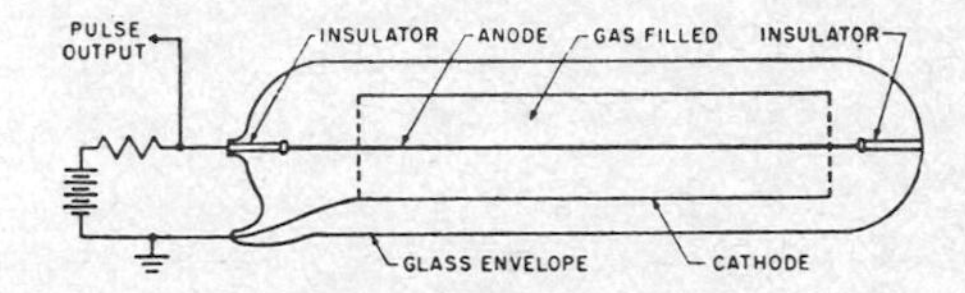

Geiger-counter-tube construction and circuit.

ing a fine-wire anode at its axis. Also called Geiger counter, Geiger-Mueller counter tube, and Geiger-Mueller tube.

Geiger-Mueller counter [abbreviated G-M counter] *Geiger counter.*

Geiger-Mueller counter tube *Geiger counter tube.*

Geiger-Mueller region *Geiger region.*

Geiger-Mueller threshold *Geiger threshold.*

Geiger-Mueller tube *Geiger counter tube.*

Geiger-Nuttall law The logarithm of the decay constant of an alpha emitter is linearly related to the logarithm of the range of the alpha particles emitted by it.

Geiger region The range of applied voltage in which the charge collected per isolated count is independent of the charge liberated by the initial ionizing event of a radiation-counter tube. Also called Geiger-Mueller region.

Geiger threshold The lowest applied voltage at which the charge collected per isolated tube count in a radiation-counter tube is substantially independent of the nature of the initial ionizing event. Also called Geiger-Mueller threshold.

Geissler tube An experimental two-electrode discharge tube used to show the luminous effects of electric discharges through various gases at low pressures. The electrodes are at opposite ends of the tube, just as in modern neon signs.

gelled cell A lead-acid cell using a nonspillable gelled electrolyte for portable use. It can be recharged many times, like an ordinary storage battery.

generalized sort A sort program that will accept the introduction of parameters at run time and does not generate a program.

general-purpose computer A computer designed to solve a wide variety of problems.

general routine A routine expressed in computer coding for solving an entire class of problems.

generate To produce, trace out, originate, or otherwise create, as for example, to generate a needed computer subroutine from parameters and skeletal coding.

generating electric field meter A device in which a flat conductor is alternately exposed to the electric field to be measured and then shielded from it. The resulting current to the conductor is rectified and used as a measure of the potential gradient at the conductor surface. Also called gradient meter.

generating magnetometer A magnetometer in which a coil is rotated in the magnetic field to be measured. The resulting generated voltage is proportional to the strength of the magnetic field.

generating voltmeter A device in which a capacitor is connected across the voltage to be measured, and its capacitance is varied cyclically. The resulting current in the capacitor is rectified and used as a measure of the voltage. Also called rotary voltmeter.

generation rate The time rate of creation of electron-hole pairs in a semiconductor.

generator [symbol G] 1. A machine that converts mechanical energy into electric energy. In its most common form, a large number of conductors are mounted on an armature that is rotated in a magnetic field produced by field coils. Also called dynamo. 2. A vacuum-tube oscillator or any other nonrotating device that generates an alternating voltage at a desired frequency when energized with DC power or low-frequency AC power. Such generators are used to produce large amounts of RF power, such as for high-frequency heating and ultrasonic cleaning. 3. A circuit that generates a desired repetitive or nonrepetitive waveform, such as a pulse generator.

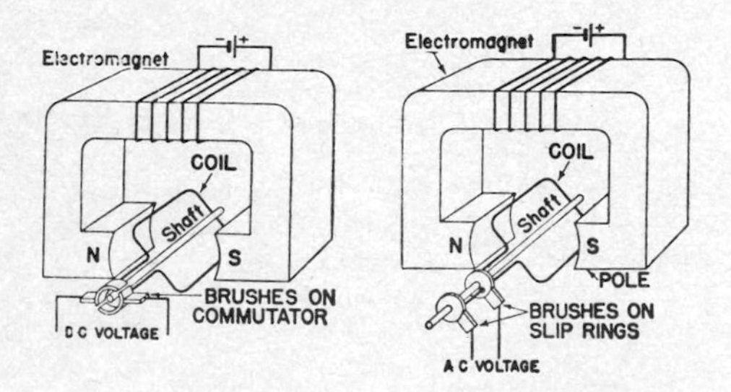

Generator principles, showing use of commutator for DC generator and slip rings for AC generator.

geodesic The shortest line between two points, as measured on any specified surface that includes the points, such as the surface of the earth.

geodesic-lens scanning antenna A scanning antenna in which scanning is accomplished by moving a small feed mechanism along the periphery of a geodesic Luneberg lens.

geodesic Luneberg lens A microwave lens that consists of a pair of conducting surfaces of revolution separated a constant distance that is less than half a wavelength. Radiation fed between the surfaces by a point source on the periphery emerges as a collimated beam diametrically opposite.

geodetic satellite A satellite that can be used to determine with high accuracy the relative locations of land masses separated by large bodies of water, such as continents and islands. The system requires three ground stations at known points and one at the unknown point. Each ground station makes accurate slant-range measurements to the satellite by measuring the delay of an electromagnetic wave.

geodimeter An optoelectronic distance-measuring instrument in which the time required for a modulated light beam to travel from the master unit to a distant mirror and return is measured and converted into distance. In one version, a Kerr cell provides 10-MHz modulation of the light beam.

geomagnetic electrokinetograph An instrument that can be suspended from the side of a ship to measure the direction and speed of ocean currents while the ship is under way. Electrodes suspended in the conductive sea water are connected to a potentiometer recorder to measure the voltage induced in the moving conductive sea water by the magnetic field of the earth. Two runs at right angles to each other are generally made to determine the direction and rate of drift caused by local ocean currents.

geomagnetism The magnetic phenomena associated with the earth and its atmosphere.

geometric distortion An aberration that causes a reproduced picture to be geometrically dissimilar to the perspective plane projection of the original scene.

geometric factor The ratio of the change in a navigation coordinate to the change in distance, taken in the direction of maximum navigation-coordinate change.

geometric horizon The locus of points on the earth at which straight lines from a point of reference in space are tangent to the earth's surface.

geometric inertial navigation Inertial navigation in which computations are based on the outputs of accelerometers that are maintained in alignment with predetermined geographic directions, such as local vertical and geographic north.

geometric mean The square root of the product of two quantities.

geometry The physical arrangement of a device or assembly, such as a neutron counter.

geometry factor The average solid angle at a radiation source that is subtended by the aperture or sensitive volume of a radiation detector, divided by the complete solid angle. Used loosely to denote counting yield or counting efficiency.

geophone A transducer, used in seismic work, that responds to motion of the ground at a location on or below the surface of the earth.

geophysical prospecting Prospecting that involves the use of acoustic, electric, and electronic equipment for measuring the physical properties of the earth as influenced by underground mineral and oil deposits.

geophysics A branch of science that deals with factors affecting the structure of the earth.

georef [GEOgraphic REFerence] Abbreviation for *world geographic reference system.*

georef grid [GEOgraphic REFerence grid] The grid system used on U.S. Air Force aeronautical charts for identifying the location of any point or area in the world. The world chart is divided into 24 parallel north-south strips 15° wide lettered from A through Z (I and O omitted), beginning at the 180th meridian, and into 12 parallel east-west strips 15° wide lettered from A through M (I omitted), beginning at the South Pole. Each quadrangle is subdivided into 15 lettered units eastward and 15 lettered units northward. These are lettered from A through Q (I and O omitted). Each 1° quadrangle is subdivided into 60 numbered minute units. Minute units may be subdivided further into decimal parts.

geostationary satellite A satellite in a stationary or synchronous orbit with respect to the earth.

germanium [symbol Ge] A brittle, grayish white metallic element that has semiconductor properties. Atomic number is 32. Now largely replaced by silicon in transistors, semiconductor diodes, and integrated circuits.

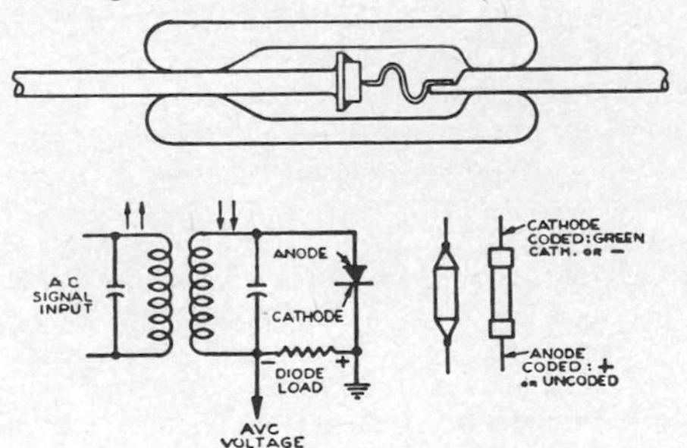

Germanium diode mounted in glass, typical circuit showing diode symbol, and polarity markings.

germanium diode A semiconductor diode that uses a germanium crystal pellet as the rectifying element.

germanium transistor A transistor in which the semiconductor material is germanium.

getter A special metal that is placed in a vacuum tube during manufacture and vaporized after the tube has been evacuated. When the vaporized metal condenses, it absorbs residual gases. Metals used as getters include barium, calcium, magnesium, potassium, sodium, and strontium. Some getters leave a silvery film on the inside of the glass envelope.

gettering Removal of residual gas from an electron tube, after evacuation, by evaporation of a getter. An ionizing electron discharge is sometimes passed through the gas during the gettering process, to accelerate absorption of the residual gas by the getter.

getter ion pump An ion pump in which the ionized gas molecules are attracted or propelled to a getter containing material that collects and holds the molecules, thereby improving the vacuum.

getter sputtering The deposition of high-purity thin films at ordinary vacuum levels by using a getter to remove contaminants remaining in the vacuum. Ionized argon is used to sputter metal

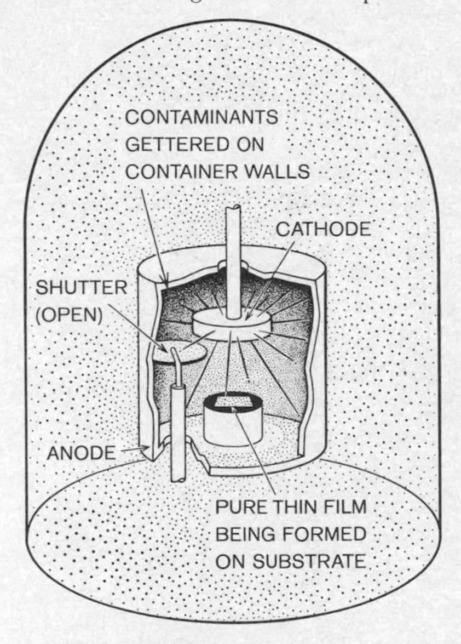

Getter sputtering equipment.

from the cathode when the high DC voltage is applied, and the resulting metal atoms carry contaminants from the vacuum to the walls of the container. A shutter is then swung aside so pure metal can travel down from the cathode to the substrate and form the desired film.

GeV Abbreviation for *gigaelectronvolt.*

ghost *Ghost image.*

ghost image 1. An undesired duplicate image at the right of the desired image on a television receiver. It is caused by multipath effect: a reflected signal traveling over a longer path arrives slightly later than the desired signal. Hills and large buildings on either side of the direct path between transmitter and receiver are the most common cause of ghost images. Also called ghost. 2. An undesired or unknown echo indication on the screen of a radar indicator. Also called ghost.

ghost mode A waveguide mode that has a trapped field associated with an imperfection in the wall of the waveguide. A ghost mode can cause trouble in a waveguide operating close to the cutoff frequency of a propagation mode.

ghost pulse *Ghost signal.*

ghost signal 1. The reflection-path signal that produces a ghost image on a television receiver. Also called echo. 2. Any signal appearing on a loran or Gee display at other than the basic repetition rate being observed. Also called ghost pulse.

GHz Abbreviation for *gigahertz.*

giant air shower *Extensive shower.*

Gibson girl Slang term for one type of portable, hand-operated radio transmitter carried on an airplane, for use in cranking out distress signals when down at sea.

giga- [abbreviated G] A prefix representing 10^9, which is 1,000,000,000, or a billion. Formerly called kilomega-.

gigabit [abbreviated Gb] One thousand megabits, or 10^9 bits.

gigacycle One thousand megacycles per second (10^9 or 1,000,000,000 cycles per second). Formerly called kilomegacycle. Now called gigahertz.

gigaelectronvolt [abbreviated GeV] A unit of energy equal to 10^9 eV. Formerly called billion electronvolt (BeV) in the United States and France.

gigahertz [abbreviated GHz] One thousand megahertz, or 10^9 Hz. Formerly called kilomegacycle.

gigawatt [abbreviated GW] One thousand megawatts, or 10^9 W.

gigohm [abbreviated GΩ] One thousand megohms, or 10^9 Ω.

gilbert [abbreviated Gb] The CGS unit of magnetomotive force. The magnetomotive force in gilberts is equal to the line integral of the magnetic field strength in oersteds around the magnetic circuit. One gilbert is equivalent to 0.7956 ampere-turn. One gilbert per centimeter is equal to 1 oersted. Use of the SI unit, the ampere-turn, is preferred.

Gill-Morrell oscillator A retarding-field-type oscillator in which the frequency of oscillation is dependent on associated circuit parameters and electron transit time within the tube.

gimbal A mounting that has two mutually perpendicular and intersecting axes of rotation. A body mounted on gimbals is free to incline in any direction.

gimbal lock Alignment and locking of gimbals that are normally at right angles to each other in a two-axis gyroscope. This catastrophic malfunction occurs when the precession angle reaches 90°, usually as a result of excessive angular motion of the aircraft or missile in which the gyroscope is mounted.

gimmick A small capacitance formed by twisting together two insulated wires.

glass-ambient diode A silicon diode that has been glass-passivated with layers of glass to give complete protection from contaminants.

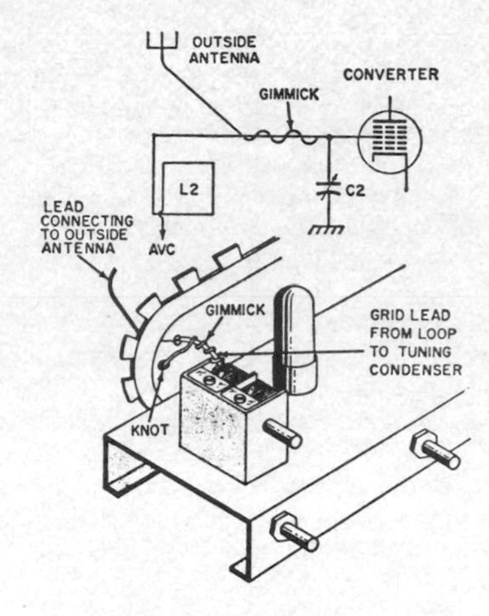

Gimmick in radio circuit.

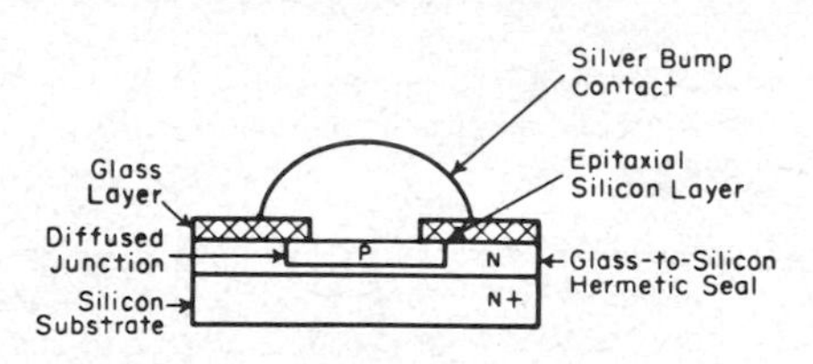

Glass-ambient diode cross section.

glass-ambient seal *Glassivated hermetic seal.*

glass-bonded mica An insulating material made by compressing a mixture of powdered glass and powdered natural or synthetic mica at high temperature.

glass capacitor A capacitor in which the glass dielectric and protective glass housing are fused to give monolithic construction with low losses and high stability.

glass dosimeter A dosimeter that uses as its radiation-sensing element a fluorod of special glass that fluoresces under ultraviolet light following gamma irradiation. The amount of fluorescence is measured with a photomultiplier circuit feeding a meter calibrated to read gamma irradiation directly in roentgens.

glass fiber A glass thread less than 0.001 in thick, used loosely or in woven form as an acoustic, electric, or thermal insulating material and as a reinforcing material in laminated plastics. Also used in fiber optics.

glassine An insulating paper used between layers of iron-core transformer windings.

glassivated hermetic seal A hermetic seal used for encapsulating semiconductor chips, involving pyrolytic deposition of glass followed by densifying to provide an encapsulating layer that will withstand extremely hostile ambient conditions.

Also called glass-ambient seal, glassivation, and glass-passivated seal.

glassivation *Glassivated hermetic seal.*

glass laser A solid laser in which glass serves as the host for laser ions of such materials as erbium, holmium, neodymium, and ytterbium. Neodymium glass lasers and neodymium-ytterbium glass lasers operate at 1.06 μm and neodymium-ytterbium-erbium glass lasers operate at 1.54 μm. Also called amorphous laser.

glass-passivated seal *Glassivated hermetic seal.*

glass-to-metal seal An airtight seal between glass and metal parts of an electron tube, made by fusing together a special glass and special metal alloy that have nearly the same temperature coefficients of expansion.

glass-type tube An electron tube that has a glass envelope.

glide bomb A bomb fitted with airfoils to provide lift. It is carried and released in the direction of a target by an airplane and may or may not be remotely controlled.

glide path The three-dimensional path used by an aircraft in airport approach procedures, as defined by the radio transmitters of an instrument landing system.

glide-path station A directional radio beacon used with an instrument landing system to provide aircraft guidance in the vertical plane during an instrument landing.

glide slope An inclined surface that includes a glide path, generated by an instrument landing system.

glide-slope angle *Slope angle.*

glide-slope deviation The difference between the actual vertical-plane path and the planned slope of an aircraft making an instrument landing, expressed in angular or linear measurements.

glide-slope facility The radio transmitter and associated equipment used at an airport to provide a glide slope.

glide-slope sector The sector that contains the glide slope in an instrument landing system. The sector is bounded above and below by radial lines from the glide-slope transmitter; along each line there is a specified difference in depth of modulation.

glide-slope transmitter A transmitter used to feed the antennas that radiate the beam for the glide slope of an instrument landing system at an airport.

G line A single dielectric-coated round wire used for transmitting microwave energy by means of surface waves.

glint A pulse-to-pulse variation in the amplitude and apparent origin of a reflected radar signal, due to reflection of the radar beam from different surfaces of a rapidly moving target or to reflections from a propeller on the target. Also called glitter.

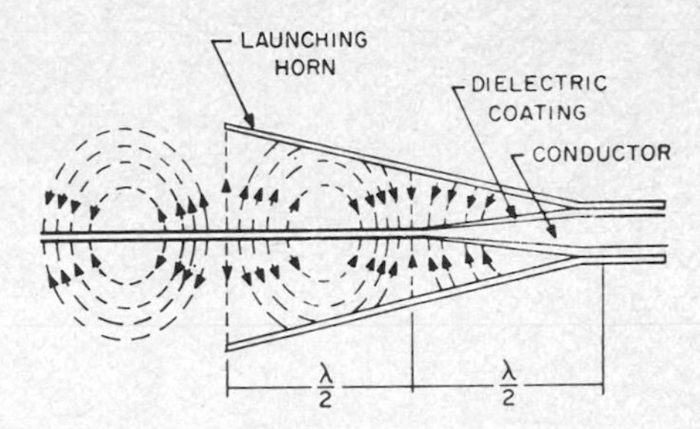

G line, showing use of horn for feed from coaxial cable at left.

glitch 1. An undesired transient voltage spike occurring on a signal being processed. In a digital-to-analog converter, a glitch can occur at a major carry, such as when switching from 0111111111 to 1000000000 because there is an interim condition in which all bits are 0. 2. A minor technical problem arising in electronic equipment.

glitter *Glint.*

global positioning system [abbreviated GPS] A satellite-based worldwide Air Force navigation system scheduled to become fully operational about 1987, to replace loran.

glomb [GLide bOMB] A glider adapted for use as a glide bomb. It is towed to the target area, released, and guided to its target by radio from a mother airplane.

glossmeter An instrument, often photoelectric, for measuring the ratio of the light reflected from a surface in a definite direction to the total light reflected in all directions.

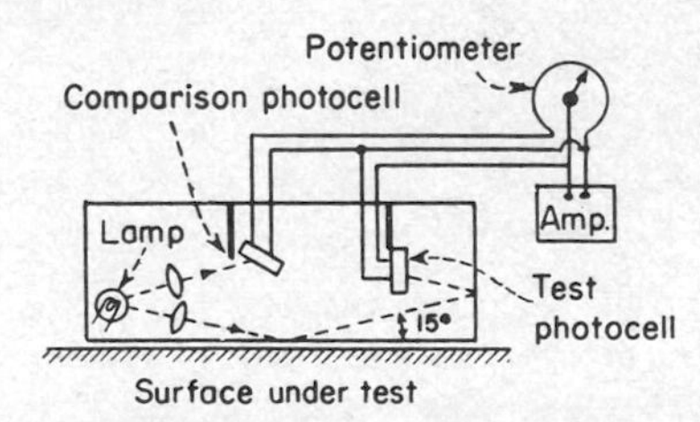

Glossmeter in which light received directly from lamp by one photocell is compared with light reflected from glossy surface to second photocell.

glove box A sealed box in which workers, using gloves attached to and passing through openings in the box, can handle certain radioactive materials safely. Also used in delicate production processes such as the assembly and encapsulation of semiconductor devices and integrated circuits. The interior of the box is sometimes maintained above atmospheric pressure, as in a clean room, to prevent contamination by foreign particles in the air of the room.

glow discharge A discharge of electricity through a gas in an electron tube, characterized by a cathode glow and a voltage drop in the vicinity of the cathode that is much higher than the ionization voltage of the gas.

glow-discharge cold-cathode tube *Glow-discharge tube.*

glow-discharge rectifier A glow-discharge tube used as a rectifier.

glow-discharge tube A gas tube that depends for its operation on the properties of a glow discharge. Also called glow-discharge cold-cathode tube and glow tube.

glow lamp A two-electrode electron tube in which light is produced by a negative glow close to the negative electrode when voltage is applied between the two electrodes. The envelope contains a small quantity of an inert gas like neon or argon. With an AC voltage, both electrodes appear to glow because each is negative half the time. Neon glow lamps, which have an orange-red glow, are the most common example. Less-used argon glow lamps have a blue-violet glow. Wattages of glow lamps range from 0.04 to 3 W.

glow switch An electron tube that contains contacts which are operated thermally by a glow discharge. Used as a starter in some fluorescent lamp circuits. The heat of the glow discharge causes a bimetallic strip to close the contacts.

glow tube *Glow-discharge tube.*

glow voltage The voltage at which a glow discharge begins in a gas tube.

glue-line heating Dielectric heating in which the electrodes give preferential heating to a thin film of glue or other relatively high-loss material located between layers of relatively low-loss material such as wood.

G-M counter Abbreviation for *Geiger-Mueller counter.*

GMT Abbreviation for *Greenwich mean time.*

gnd Abbreviation for *ground.*

gnomonic projection An aeronautical chart made by projecting the surface of a sphere onto a plane touching the sphere at one point. The projection is made by radials from the center of the sphere. Used in determining great-circle courses for aircraft and for tracking aircraft electronically.

gobo 1. A sound-absorbing shield used with a microphone to block unwanted sounds. 2. A panel used to shield a television camera lens from direct light.

GΩ Abbreviation for *gigohm.*

Golay cell A radiometer that measures the increase in pressure in a gas chamber as temperature rises when radiation is absorbed.

gold [symbol Au] A precious-metal element. Atomic number is 79. Used for electroplating electronic components that must withstand severe corrosive conditions, such as those that exist in the tropics.

gold doping A technique for controlling the lifetime of minority carriers in a diffused-mesa transistor. Gold is diffused into the base and collector regions to reduce storage time in transistor circuits.

gold-leaf electroscope An electroscope in which the two suspended strips are gold foil or leaf.

goniometer An instrument for measuring angles. Used to calculate and resolve mathematical problems or electrical functions and to establish directional phase difference between two transmitted or received signals. In one form it has two fixed windings crossed at 90° to each other, along with a rotatable third winding. In another form it is used for measuring the angles between the reflecting surfaces of a crystal or prism. An x-ray version measures the angular positions of the axes of a quartz crystal.

go/no-go test A test that is based on the measurement of one or more parameters but can have only one of two possible results, to pass or reject the device under test.

GPI Abbreviation for *ground-position indicator.*

GPS Abbreviation for *global positioning system.*

graded filter A power-supply filter in which connections to the output stage of a receiver or amplifier are made at or near the filter input to obtain maximum DC voltage. Ripple is less important at this stage because it has low gain, and there are no subsequent stages that might accentuate the ripple.

graded insulation A method of insulating high-voltage devices such as pulse transformers, wherein the insulation to ground is reduced more or less uniformly from the high-potential end of the winding to the ground or low-potential end.

graded junction *Rate-grown junction.*

graded-junction transistor *Rate-grown transistor.*

graded periodicity technique A technique for modifying the response of a surface acoustic wave filter by varying the spacing between successive electrodes of the interdigital transducer.

graded seal A seal that consists of several types of glass differing successively in expansion characteristics. Used for glass-to-metal seals of electron tubes.

gradient The rate at which a variable quantity increases or decreases. Thus voltage gradient is voltage per unit length.

gradient hydrophone A gradient microphone that responds to waterborne sound waves.

gradient meter *Generating electric field meter.*

gradient microphone A microphone whose output corresponds to a gradient of the sound pressure. A pressure microphone is a gradient microphone of zero order. A velocity microphone is a first-order gradient microphone.

gradiometer A flux-gate magnetometer whose output is proportional to the gradient of the magnetic field being measured.

grain A small particle of metallic silver remaining

in a photographic emulsion after developing and fixing. These grains together form the dark areas of a photographic image.

graininess Visible coarseness in a photographic image under specified conditions, due to silver grains.

grain-oriented iron-silicon alloy An iron-silicon alloy that has been specially rolled and heat-treated to align the grains of metal with their cube edges, parallel with the direction of rolling. This gives far better magnetic properties than hot-rolled sheet.

gram [abbreviated g] One-thousandth of a kilogram.

gramophone British term for *phonograph.*

gram-rad A unit of integral absorbed dose of radiation, equal to 100 ergs/g.

gram-roentgen A unit of energy conversion, equal to a dose of 1 R delivered to 1 g of air. One gram-roentgen is about 84 ergs. Used to describe total energy absorption by a patient undergoing radiation therapy. The quantity obtained by integrating gram-roentgens throughout a region is called the integral dose.

grand-scale integration [abbreviated GSI] Integration in which a complete major subsystem or system is fabricated on a single integrated-circuit chip that contains more than 1000 gate equivalent circuits.

graph A line drawing that shows the relation between two variable quantities.

graphechon *Storage tube.*

graphic cathode-ray-tube terminal A computer terminal that uses a cathode-ray tube to provide dynamic and manipulatable displays of charts, data, graphs, and circuit diagrams, as well as geometric figures in two or three dimensions. Any portion of the display may be changed, erased, moved, rotated, or otherwise manipulated by terminal keying, positioning of a light pen, or other means.

graphic terminal A cathode-ray-tube or other type of computer terminal that is capable of producing some form of line drawing based on data being processed by or stored in a computer.

graphite A form of carbon used in resistor elements and as a moderator in nuclear reactors.

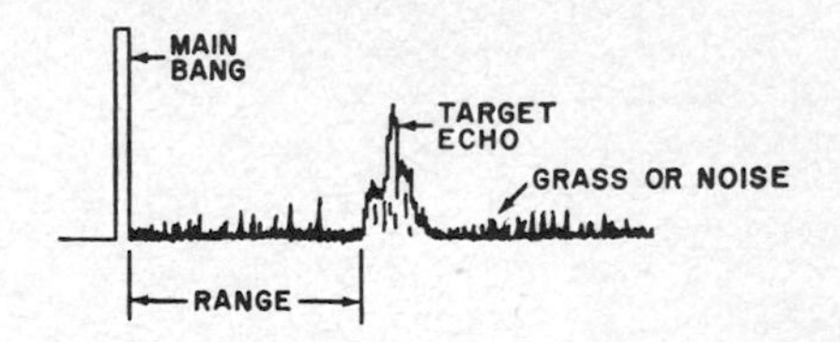

Grass showing on radar A display.

grass Clutter due to circuit noise in a radar receiver, seen on an A scope as a pattern resembling a cross section of turf. Also called hash.

grasshopper fuse A small fuse that has a spring which releases when the fuse wire melts, closing an auxiliary alarm circuit and providing a visual indication to help locate the blown fuse.

graticule A scale or network of lines on a transparent sheet, placed over a cathode-ray or other display to facilitate measurement of the quantities displayed.

grating 1. An arrangement of fine parallel wires used in waveguides to pass only a certain type of wave. 2. An arrangement of crossed metal ribs or wires that acts as a reflector for a microwave antenna and offers minimum wind resistance. 3. *Diffraction grating.*

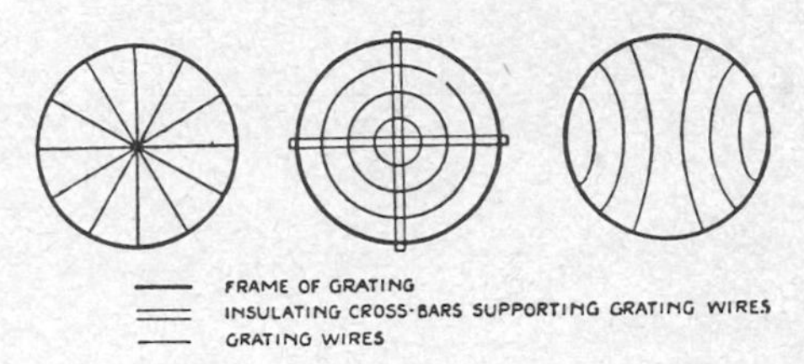

Gratings for circular waveguides. Grating of radial wires blocks all types of transverse electric waves, and other two gratings block certain types of transverse magnetic waves.

grating converter A wave converter that consists of a double grating positioned just ahead of a coaxial sheet grating in a circular waveguide. One grating conforms to the pattern of the arriving wave and the other to the pattern of the converted wave.

grating reflector An openwork metal structure that provides a good reflecting surface for microwave antennas.

Gratz rectifier A three-phase full-wave rectifying circuit that uses six rectifying elements.

gravimeter An instrument for measuring variations in the magnitude of the earth's gravitational field.

gravitational force [abbreviated G] The gravitational pull of the earth, or the comparable force required to accelerate or decelerate a freely movable body at the rate of approximately 32.16 feet per second per second (9.8 meters per second per second).

gravitational wave A hypothetical wave that travels at the speed of light and causes intense fluctuation in electron density in the F region of the earth's atmosphere. This wave propagates the effect of gravitational attraction.

graybody A body whose spectral emissivity remains constant through the spectrum and is less than that of a blackbody radiator at the same temperature.

Gray code A modified binary code in which sequential numbers are represented by expressions

Gray-code disk for converting shaft angle to cyclic or inverted Gray code.

that differ only in one bit, to minimize errors as follows:

Decimal	Binary	Gray
0	000	000
1	001	001
2	010	011
3	011	010
4	100	110
5	101	111
6	110	101
7	111	100

gray filter *Neutral-density filter.*

gray scale A series of achromatic tones having varying proportions of white and black, to give a full range of grays between white and black. A gray scale is usually divided into 10 steps.

green gain control A variable resistor used in the matrix of a three-gun color television receiver to adjust the intensity of the green primary signal.

green gun The electron gun whose beam strikes phosphor dots emitting the green primary color in a three-gun color television picture tube.

green laser A gas laser that uses mercury and argon to generate a green line at 5225 Å, corresponding to the wavelength that is most readily transmitted through sea water. Antisubmarine warfare is a possible application.

green restorer The DC restorer for the green channel of a three-gun color television picture tube circuit.

Green's function A kernel in which the integral operator is the inverse of a differential operator.

green video voltage The signal voltage output from the green section of a color television camera, or the signal voltage between the receiver matrix and the green-gun grid of a three-gun color television picture tube.

Greenwich mean time [abbreviated GMT] Former name for universal time (UT).

grenz ray An x-ray produced at the long-wavelength end of the x-ray spectrum, involving wavelengths of the order of 1 to 10 Å, by using special tubes that operate at voltages from only 5 to 15 kV. Used in skin therapy and radiography of insects and botanical specimens.

grenz tube A low-voltage x-ray tube that has a special glass window capable of transmitting x-ray wavelengths ranging from 1 to 10 Å, which are blocked by ordinary glass.

grid [symbol G] 1. An electrode located between the cathode and anode of an electron tube and having one or more openings through which electrons or ions can pass under certain conditions. A grid controls the flow of electrons from cathode to anode. 2. A network of equally spaced lines forming squares, used for determining permissible locations of holes on a printed-circuit board or a chassis. 3. *Potter-Bucky grid.*

grid bearing A bearing in which the reference line is grid north.

grid cap A top-cap terminal for the control grid of an electron tube.

grid control The control of anode current of an electron tube by varying the voltage of the control grid with respect to the cathode.

grid-controlled mercury-arc rectifier A mercury-arc rectifier in which one or more electrodes are used exclusively to control the starting of the discharge.

grid-control tube A mercury-vapor tube that has external grid control.

grid current Electron flow to a positive grid in an electron tube.

gridded tube A high-power, high-frequency, grid-controlled vacuum tube that provides wideband linear amplification.

grid detection Detection in the grid circuit of a vacuum tube, as in a grid-leak detector.

grid-dip meter An absorption wavemeter in which the resonance indicator is a vacuum-tube oscillator that has in its grid circuit a sensitive current-indicating meter. The meter dips (reads lower grid current) when energy is drawn from the oscillator by a calibrated tuned circuit that is lightly coupled to the source whose frequency is being measured. The equivalent solid-state version is called a dip meter.

grid dissipation The power lost as heat at the grid of an electron tube.

grid-drive characteristic A relation between electric or light output of an electron tube and the control-electrode voltage as measured from cutoff.

grid driving power The average product of the instantaneous value of the grid current and the alternating component of the grid voltage of an electron tube over a complete cycle.

grid emission Electron or ion emission from a grid of an electron tube.

grid-glow tube A glow-discharge tube in which one or more control electrodes initiate but do not limit the anode current except under certain operating conditions.

gridistor A field-effect transistor that has a multichannel structure of embedded grids which can be produced by integrated techniques. This structure combines the advantages of field-effect transistors with those of minority-carrier injection transistors.

grid-leak detector An electron-tube detector in which the desired AF voltage is developed across a parallel combination of a capacitor and resistor by

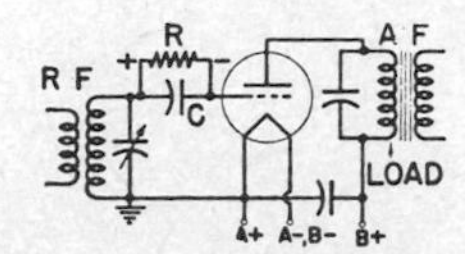

Grid-leak detector circuit.

the flow of modulated RF grid current. The circuit provides square-law detection on weak signals and linear detection on strong signals, along with amplification of the AF signal.

grid limiting Limiting action achieved by placing a high-value resistor in series with the grid of a vacuum tube. The voltage drop across this resistor increases with input signal strength, giving a varying negative grid bias that serves to level input signals which are above a certain value.

grid modulation Modulation produced by feeding the modulating signal to the control-grid circuit of any electron tube in which the carrier is present.

grid neutralization Neutralization in which a portion of the grid-cathode AC voltage of a vacuum tube is shifted 180° and applied to the anode-cathode circuit through a neutralizing capacitor.

grid north An arbitrary reference direction used in connection with the grid system of navigation. The reference direction is the top of a grid that for polar navigation consists of rectangular coordinates superimposed over the polar regions. One line on this grid coincides with the Greenwich meridian. On this grid, north is usually the direction of the North Pole from Greenwich, England. Also called polar grid.

grid-pool tank A grid-pool tube that has a heavy metal envelope resembling a tank.

grid-pool tube An electron tube that has a mercury-pool cathode, one or more anodes, and a control electrode or grid which controls the start of current flow in each cycle. The excitron and ignitron are examples.

grid pulse modulation Modulation produced in an amplifier or oscillator by applying one or more pulses to a grid circuit.

grid pulsing Pulsing of an RF oscillator by biasing the grid sufficiently negative to block oscillation, and applying a positive pulse to the grid to remove this bias.

grid resistor A resistor used to limit grid current.

grid return The portion of a grid circuit that completes the electric path from the grid to the cathode of an electron tube.

grid swing The total variation in grid-cathode voltage from the positive peak to the negative peak of the applied signal voltage.

grid voltage The voltage between a grid and the cathode of an electron tube.

grille An arrangement of wood, metal, or plastic bars across the front of a loudspeaker in a cabinet for decorative and protective purposes.

grille cloth A loosely woven cloth stretched across the front of a loudspeaker to keep out dust and provide protection without appreciably impeding sound waves.

grivation [GRId VAriaTION] The variation between grid north and magnetic north, usually expressed as an angle.

grivation computer A computer that calculates the angle between grid north and magnetic north as a function of longitude convergence in the polar regions of the earth.

Grommets made from rubber and from plastics.

grommet A circumferentially grooved self-retaining insulating washer used to cover sharp edges of a hole through which conductors are run.

groove The track inscribed in a record by the cutting or embossing stylus during mechanical recording, including undulations or modulations caused by the vibration of the stylus.

groove angle The angle between the two walls of an unmodulated disk-recording groove.

groove shape The contour of the groove in a disk recording.

groove speed The linear speed of the groove with respect to the stylus in disk recording.

ground [abbreviated gnd] 1. A conducting path, intentional or accidental, between an electric circuit or equipment and the earth, or some conducting body serving in place of the earth. Also called earth (British). 2. The lowest energy state of a nucleus, atom, or molecule. All other states are excited.

ground absorption Energy loss caused by dissipation of radio waves in the ground during transmission.

ground clamp A clamp used to connect a grounding conductor to a grounded object.

ground clutter Clutter on a ground or airborne radar caused by reflection of signals from the ground or objects on the ground. Also called ground flutter, ground return, land return, and terrain echoes.

ground control Control of an aircraft or missile in flight by a person on the ground.

ground-controlled approach [abbreviated GCA] A ground radar system used at an airport to provide information by which aircraft approaches for landings may be directed from the ground by radio during conditions of poor visibility and low cloud ceiling. It consists of an airport surveillance radar for guiding the aircraft to the start of the final approach path and a precision approach radar for showing the exact position of the aircraft on its final approach path. Also called talk-down system.

ground-controlled interception [abbreviated GCI] A radar system by means of which a controller at the ground or ship radar may direct an aircraft by radio to make an interception of another aircraft.

ground distance The mean sea-level great-circle component of distance from one point to another.

grounded Connected to earth or to some conducting body that serves in place of the earth. Also called earthed (British).

grounded-base amplifier An amplifier that uses a transistor in a grounded-base connection.

grounded-base connection A transistor circuit in which the base electrode is common to both the

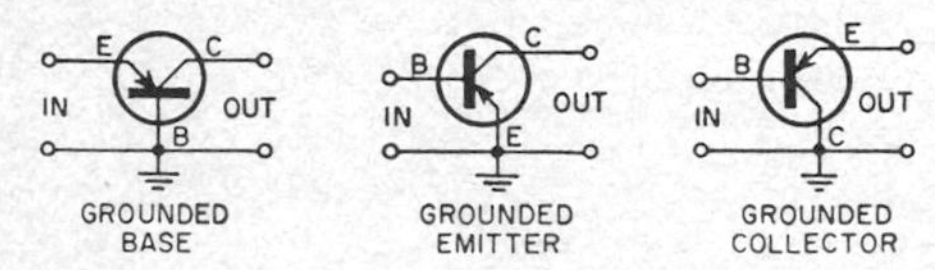

Grounded-base, grounded-emitter, and grounded-collector connections for PNP junction transistor.

input and output circuits. The base need not be directly connected to circuit ground. Also called common-base connection.

grounded-cathode amplifier A conventional electron-tube amplifier circuit. The cathode is at ground potential at the operating frequency. The input is applied between control grid and ground. The output load is connected between anode and ground.

grounded-collector amplifier An amplifier that uses a transistor in a grounded-collector connection.

grounded-collector connection A transistor circuit in which the collector electrode is common to both the input and output circuits. The collector need not be directly connected to circuit ground. Also called common-collector connection.

grounded-emitter amplifier An amplifier that uses a transistor in a grounded-emitter connection.

grounded-emitter connection A transistor circuit in which the emitter electrode is common to both the input and output circuits. The emitter need not be directly connected to circuit ground. Also called common-emitter connection.

grounded-gate amplifier An amplifier that uses thin-film transistors in which the gate electrode is

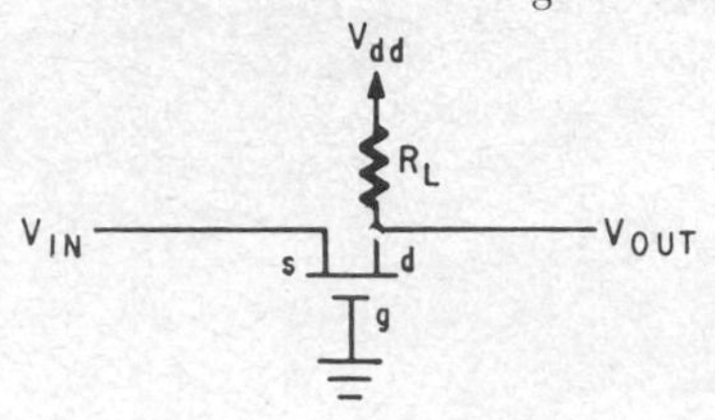

Grounded-gate amplifier using thin-film transistor.

connected to ground. The input signal is fed to the source electrode, and the output is obtained from the drain electrode.

grounded-grid amplifier An electron-tube amplifier circuit in which the control grid is at ground potential at the operating frequency. The input signal is applied between cathode and ground, and the output load is connected between anode and ground. Chief advantages are low input impedance and freedom from oscillation due to feedback.

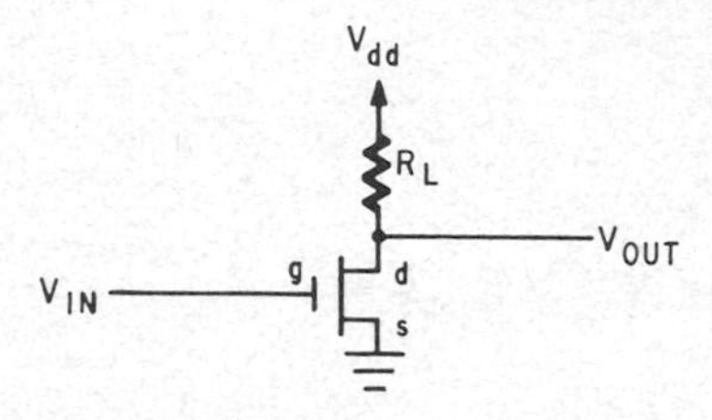

Grounded-source amplifier using thin-film transistor.

grounded-source amplifier An amplifier that uses thin-film transistors whose source electrodes are connected to ground.

ground-effect machine A vehicle that is supported by a downward-directed cushion of air produced by its own fans while being propelled over land or water by one or more aircraft engines. One application is transport of troops over mine fields.

ground environment 1. The entire complement of equipment installed on the ground to make up a communication or electronic system, facility, or station. 2. The environment that surrounds and affects a system or piece of equipment operating on the ground.

ground-equalizer coil A coil that has relatively low inductance, placed in one or more of the circuits connected to the grounding points of an antenna to distribute the current to the various points in a desired manner.

ground fault Accidental grounding of a conductor.

ground-fault interrupter A fast-acting circuit

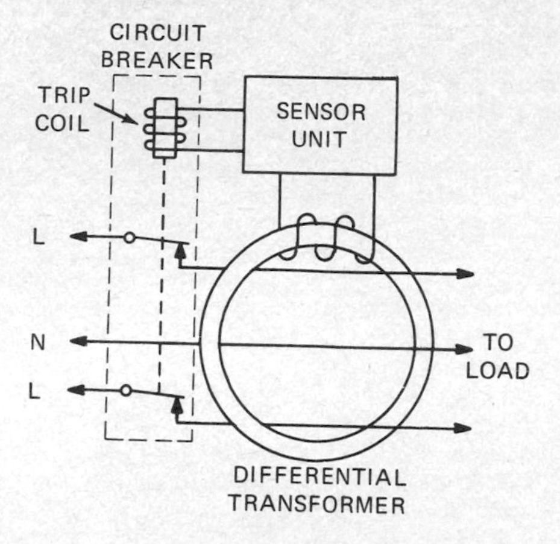

Ground-fault interrupter using differential transformer. Currents in line wires L are normally equal and opposite, so sensor output is zero. When fault path occurs from one line to grounded neutral N, sensor reacts to ground-fault current and trips circuit breaker.

breaker that also senses very small ground-fault currents such as might flow through the body of a person standing on damp ground while touching a hot AC line wire. The interrupter limits the time the current can flow through the fault by tripping the circuit breaker in as little as 0.025 s; this limits the total energy flow through the human body to a safe value. A typical trip current setting for homes is 5 mA. Used chiefly for wall-outlet circuits into which potentially dangerous appliances might be plugged. Larger versions are used in power stations.

ground flutter *Ground clutter.*

ground-guided missile A missile guided by radio control from the ground.

ground influence mine An aerial mine that rests on the ground under water and is detonated by magnetic or other influence.

grounding outlet An outlet that has, in addition to the current-carrying contacts, one grounded contact which can be used for grounding portable appliances and equipment.

grounding plate An electrically grounded metal plate on which a person stands to discharge static electricity picked up by his body, or a similar plate buried in the ground to act as a ground rod.

ground loop A circuit that has more than one ground point connected to earth ground, with the points differing enough in ground potential to produce a circulating current in the ground system. These undesirable circulating currents can be avoided by connecting the DC distribution system to earth ground with only one wire.

ground noise The residual system noise in the absence of the signal in recording and reproducing.

ground plane A grounding plate, above-ground counterpoise, or arrangement of buried radial

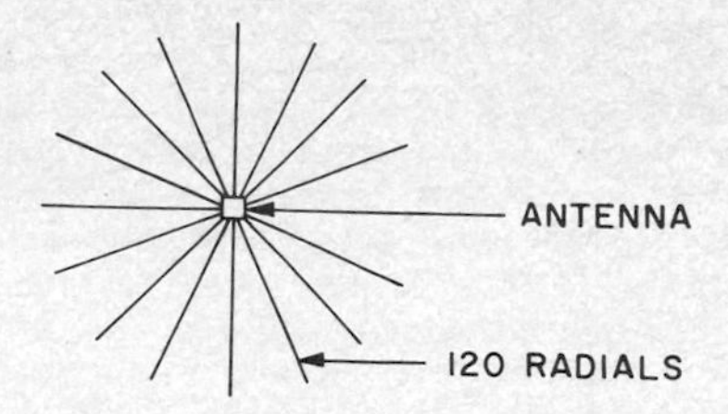

Ground plane consisting of buried radial wires at least one-quarter wavelength long at lowest frequency in use.

wires required with a ground-mounted antenna that depends on the earth as the return path for radiated RF energy.

ground-plane antenna A vertical antenna combined with turnstile or other horizontal antenna elements to lower the angle of radiation. A concentric base support and center conductor to-

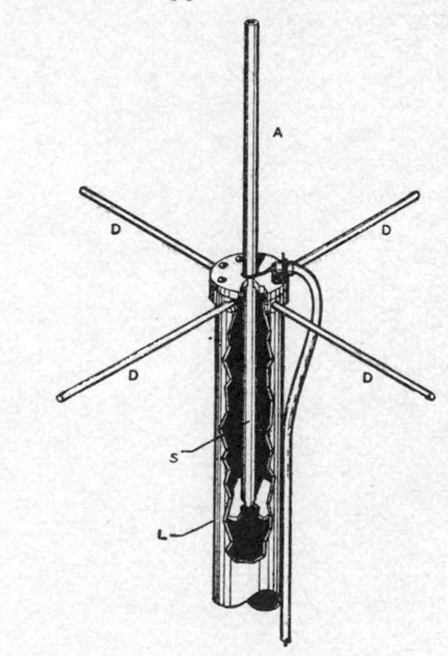

Ground-plane antenna, formed by adding turnstile elements D to high-frequency vertical antenna A, having metallic support L and center conductor S acting as inductance. Combination is fed by coaxial transmission line running outside support.

gether place the antenna at ground potential even though it may be located several wavelengths above ground. The horizontal radiation pattern is nondirectional.

ground position A position on the ground directly under an airborne aircraft.

ground-position indicator [abbreviated GPI] An air-position indicator that makes allowances for drift, to give the actual position with respect to a fixed ground point.

ground potential Zero voltage with respect to the ground or to a chassis that acts as a ground connection.

ground-reflected wave A radio wave that is reflected from the ground somewhere along the transmission path.

ground return 1. An echo received from the ground by an airborne radar set. 2. Use of the earth or a chassis as the return path for a transmission line. 3. Use of a chassis as a return path for a circuit. 4. *Ground clutter.*

ground rod A rod that is driven into the earth to provide a good ground connection.

groundscatter propagation Multihop ionospheric radio propagation along other than the great-circle path between transmitting and receiving stations. Radiation from the transmitter is first reflected back to earth from the ionosphere, then scattered in many directions from the earth's surface. Also called non-great-circle propagation.

ground speed The speed of a vehicle along its track. For an aircraft, it is the speed relative to the earth's surface.

ground-speed recorder A recorder that makes a permanent record of the horizontal speed of a moving object with respect to the earth.

ground state The lowest energy state of a nucleus, atom, or molecule. Also called normal state.

ground-state beta disintegration The total energy released in a beta transition between isobars in their ground states. It includes the energies of any gamma and associated radiations following the beta process.

ground-state disintegration The disintegration energy that is present when all reactant and product nuclei are in their ground states.

ground surveillance radar A surveillance radar operated at a fixed point for observation and control of the position of aircraft or other vehicles in the vicinity.

ground system The portion of an antenna that is closely associated with an extensive conducting surface, which may be the earth itself.

ground-to-air communication One-way radio communication from ground stations to aircraft.

ground trace The theoretical mark traced on the surface of the earth by a perpendicular line passing through a spacecraft, missile, or orbiting satellite.

ground vector A vector that represents the track and ground speed of an aircraft.

ground wave 1. A radio wave that is propagated over the earth and is ordinarily affected by the presence of the ground and the troposphere. The ground wave includes all components of a radio wave over the earth, except ionospheric and tropospheric waves. The ground wave is refracted because of variations in the dielectric constant of the troposphere, including the condition known as a surface duct. Also called surface wave. 2. One of the waves formed in the ground by an explosion. It may be a longitudinal wave (compression), transverse wave (shear), or surface wave (similar to water ripples), induced by direct ground shock of a ground or subsurface burst or by blast transmitted through the air.

ground wire A conductor used to connect electric equipment to a ground rod or other grounded object.

grouped-frequency operation Use of different frequency bands for channels in opposite directions in a two-wire carrier system.

grouping 1. Periodic error in the spacing of recorded lines in a facsimile system. 2. Nonuniform spacing between the grooves of a disk recording.

group modulation The process by which a number of channels, already separately modulated to a specific frequency range, are modulated upon a new carrier to shift the entire group to another frequency range.

group velocity The velocity of propagation of the envelope of a plane wave occupying a frequency band over which the envelope delay is approximately constant. Group velocity differs from phase velocity in a medium in which the phase velocity varies with frequency.

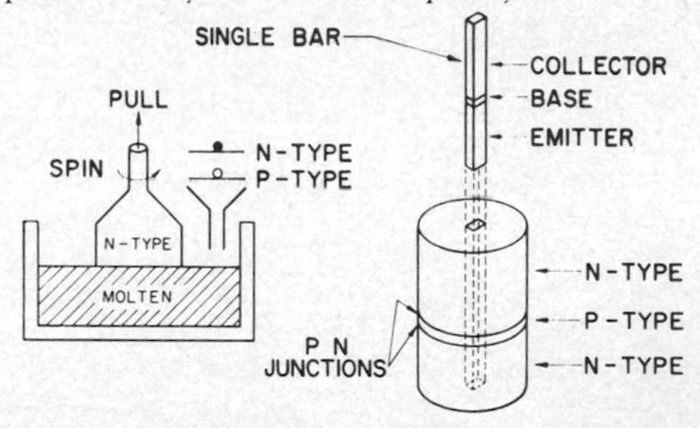

Growing of junctions in semiconductor crystal during pulling from molten material, by alternately adding N- and P-type impurities.

growing Production of semiconductor crystals by slow crystallization from a melt.

growler An electromagnetic device that consists essentially of two field poles arranged as in a motor, used for locating short-circuited coils in the armature of a generator or motor and for magnetizing or demagnetizing objects. A growling noise indicates a short-circuited coil.

grown-diffused transistor A junction transistor in which the final junctions are formed by diffusion of impurities near a grown junction.

grown junction A junction produced by changing the types and amounts of donor and acceptor impurities that are added during the growth of a semiconductor crystal from a melt.

grown-junction photocell A photodiode that consists of a bar of semiconductor material which has a PN junction at right angles to its length and an ohmic contact at each end of the bar.

grown-junction transistor A junction transistor in which different impurities are placed in the melt in sequence as the silicon or germanium seed crystal is slowly withdrawn, to produce the alternate PN and NP junctions. In another process, both impurities are introduced at the same time, and the temperature is suddenly raised and lowered to create the alternate P- and N-type layers. Also called doped-junction transistor.

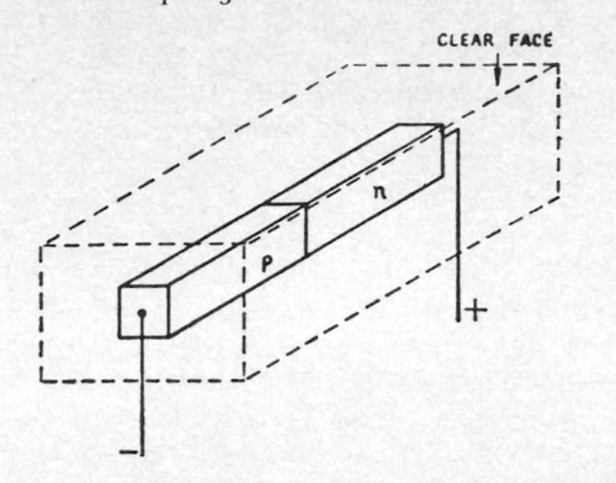

Grown-junction photocell mounted in transparent plastic block.

growth curve A curve showing how a quantity increases with time.

G scope A radarscope that produces a G display.

GSI Abbreviation for *grand-scale integration.*

GTO Abbreviation for *gate turnoff.*

GT selection rules Abbreviation for *Gamow-Teller selection rules.*

guard To monitor constantly a specific radio frequency or channel.

guard band A narrow frequency band provided between adjacent channels in certain portions of the radio spectrum to prevent interference between stations.

guard circle An inner concentric groove inscribed on disk records to prevent the pickup from being damaged by being thrown to the center of the record.

guarded input An amplifier or other circuit in which the ungrounded input terminal is electrically shielded and isolated to minimize pickup of interference.

guard ring A ring-shaped auxiliary electrode used in an electron tube or other device to modify the electric field or reduce insulator leakage. In a counter tube or ionization chamber a guard ring may also define the sensitive volume.

Gudden-Pohl effect The momentary illumination produced when an electric field is applied to a phosphor previously excited by ultraviolet radiation.

guidance Control of the path of a missile along part or all of its path.

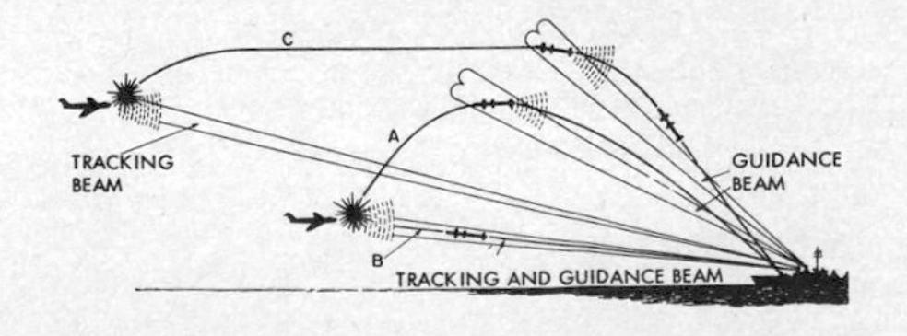

Guidance beams used to control three antiaircraft guided missiles launched from warship.

guidance beam The beam that is aimed directly at a guided missile for transmitting control instructions to the missile. In contrast, the tracking beam in antimissile warfare is aimed directly at the target at all times, rather than at the missile being sent up to intercept the target.

guidance system Any system that controls the path of a missile from launch to target. Examples are homing guidance and beam-rider guidance.

guidance tape A magnetic or punched paper tape that is placed in a missile or its computer to program desired events during flight.

guide *Waveguide.*

guided Controlled or controllable as to direction by preset mechanisms, radio commands, or built-in self-reacting devices.

guided aircraft missile A guided missile designed for launching from an aircraft in flight.

guided ballistic missile A ballistic missile that is guided only during the powered portion of its trajectory.

guided missile An unmanned missile that is guided entirely or substantially all the way to its target by internal or external means.

guided wave A wave whose energy is concentrated near a boundary or between substantially parallel boundaries separating materials of different properties and whose direction of propagation is effectively parallel to these boundaries. Waveguides transmit guided waves.

guide field The magnetic flux that holds particles in a stable circular orbit during the accelerating period in a betatron or synchrotron.

guide wavelength The wavelength in a waveguide.

Guillemin line A network or artificial line used in high-level pulse modulation to generate a nearly

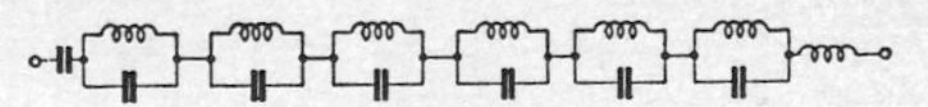

Guillemin line.

square pulse, with steep rise and fall. Used in radar sets to control pulse width.

gulp A small group of bytes, processed as a unit in a computer. The number of bytes in a gulp depends on both hardware and software for the computer. A gulp is processed in the same way as a character or word.

gun 1. *Electron gun.* 2. *Soldering gun.*

gun killer An adapter that can be inserted between the socket and base of a color picture tube, to permit killing independently any one or all of the three guns (red, green, and blue) in the tube during repair or adjustment procedures.

gun-laying radar Radar equipment specifically designed to determine range, azimuth, and elevation of a target and sometimes to automatically aim and fire antiaircraft artillery or other guns.

Gunn amplifier A microwave amplifier in which a Gunn diode functions as a negative-resistance amplifier when placed across the terminals of a microwave source. The reflected power is then greater than the incident RF power. A nonreciprocal device, such as a ferrite circulator, is used to separate the incident wave from the amplified reflected wave.

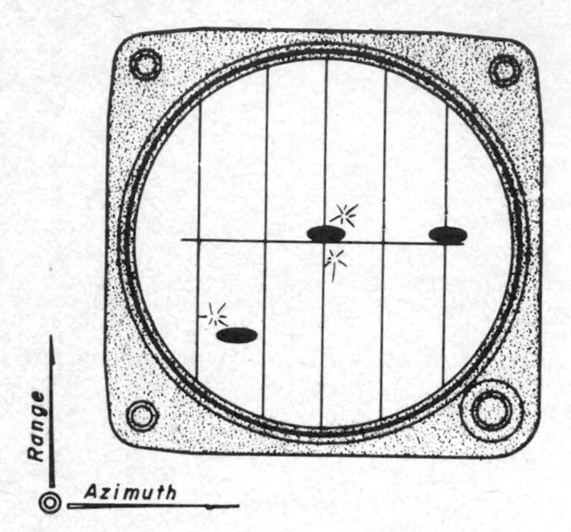

Gun-laying radar display. Solid black areas are targets, and stars indicate shell splashes.

Gunn diode A two-terminal semiconductor device that utilizes the Gunn effect to produce microwave oscillation or to amplify an applied microwave signal. The frequency of oscillation depends on domain transit time and can be well over 50 GHz. Operation is in the transit-time mode. A Gunn diode is one type of transferred-electron diode.

Gunn effect An effect discovered by J. B. Gunn in 1963; microwave oscillation occurs in a small block of N-type gallium arsenide when a constant DC voltage above a critical value is applied to contacts on opposite faces. Generated frequencies range from 500 MHz to well over 50 GHz, depending on the dimensions of the block and other factors.

Gunn oscillator An oscillator that uses a Gunn diode to generate a frequency that can be well over 50 GHz. It may have mechanical, varactor, or YIG tuning to meet the requirements of applications such as local oscillators, low-power transmitters, and microwave laboratory equipment.

gustsonde A radiosonde dropped by parachute from a high altitude for use in measuring the vertical component of turbulence aloft. An accelerometer is used to feed the telemetry transmitter.

gutta-percha A natural vegetable gum similar to rubber, used principally as insulation for wires and cables.

guy wire A wire used to hold a pole or tower in an upright position.

GW Abbreviation for *gigawatt.*

gyrator A waveguide component that uses a ferrite section to give zero phase shift for one direction of propagation and 180° phase shift for the other direction. Also called microwave gyrator.

gyrator filter A highly selective active filter that uses a gyrator which is terminated in a capacitor so as to have an inductive input impedance. The resulting synthetic inductor can be tuned with another capacitor. Gyrators for filters can be constructed in monolithic integrated-circuit form.

gyro *Gyroscope.*

gyrocompass A compass that uses a gyroscope to provide a reference direction.

gyro flux-gate compass A compass in which a flux gate, horizontally stabilized by a gyroscope, senses the horizontal component of the earth's magnetic field. Being fixed with respect to the aircraft, the compass reacts to each change in heading with a change in current. This current is amplified and used to actuate the dial of a master indicator. Also called flux-gate compass.

gyrofrequency The natural frequency of rotation of a charged particle under the influence of a constant magnetic field, such as the magnetic field of the earth.

gyro horizon *Artificial horizon.*

gyromagnetic Pertaining to the magnetic properties of rotating electric charges, such as spinning electrons moving within atoms.

gyromagnetic compass A magnetic compass in which gyroscopic stabilization is used to indicate direction.

gyromagnetic coupler A coupler in which a single-crystal YIG resonator provides coupling at the required low signal levels between two crossed stripline resonant circuits. Used as signal limiters and electronically tunable filters.

gyromagnetic effect The rotation induced in a body by a change in its magnetization, or the magnetization resulting from a rotation.

gyromagnetic ratio The ratio of the magnetic moment to the angular momentum for a charged particle moving in a closed orbit.

gyroscope A wheel mounted and driven at high speed in such a way that its spinning axis is free to

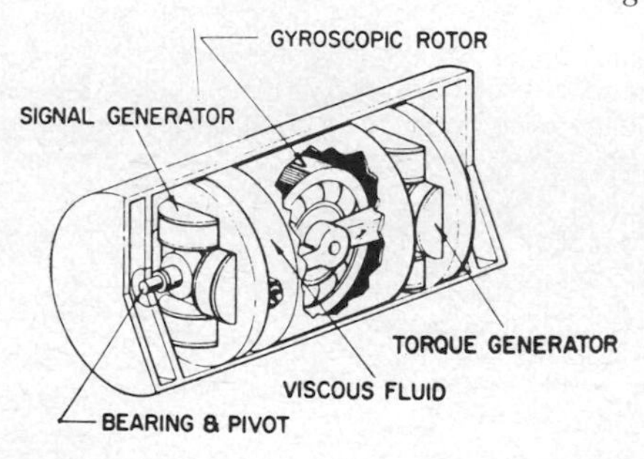

Gyroscope used in airplanes, missiles, and spacecraft. Mechanical rotor driven at high speed here provides positional reference.

rotate about either of two other axes perpendicular to itself and to each other. Used to maintain a stable equilibrium, such as is required for autopilots. Also called gyro.

gyroscopic horizon A gyroscopic instrument that indicates the lateral and longitudinal attitude of an aircraft by simulating the natural horizon.

gyro sight A gun sight that uses a gyroscope.

gyro stabilizer A stabilizer that uses a gyroscope to compensate for the roll and pitch of a ship.

gyrotron A device that detects motion of a system by measuring the phase distortion which occurs when a vibrating tuning fork is moved. The altered fork frequency is compared to a reference frequency that was originally in phase with the fork frequency.

G – Y signal The green-minus-luminance color-difference signal used in color television. It is combined with the luminance signal in a receiver to give the green color-primary signal.

h Abbreviation for *hour.*
h Symbol for *Planck's constant.*
H 1. Symbol for *heater.* 2. Abbreviation for *henry.*
H² Symbol for *deuterium.*
H³ Symbol for *tritium.*
H Symbol for *magnetic field strength.*

halation An area of glow surrounding a bright spot on a fluorescent screen, caused by phosphor scattering or multiple reflections at front and back surfaces of the glass faceplate.

half-adder A digital computer circuit that has two input and two output channels for binary digit signals, related in such a way that two half-adders can be combined to give one binary adder.

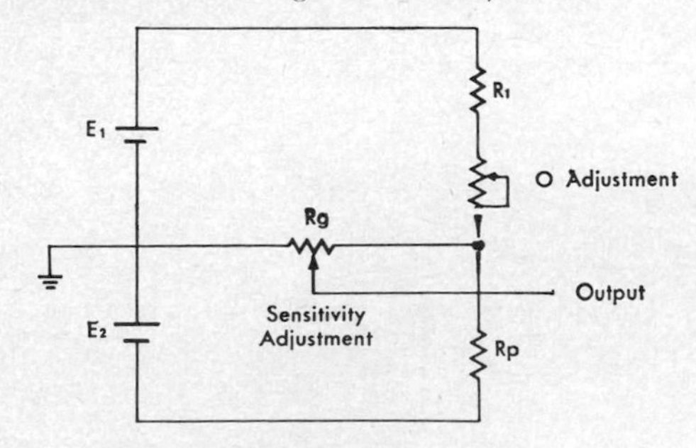

Half-bridge circuit.

half-bridge A bridge that has two power supplies, located in two of the bridge arms, to replace the single power supply of a conventional bridge.

half-cycle The time interval corresponding to half a cycle or 180° at the operating frequency of a circuit or device.

half-cycle transmission A data-transmission and control system that uses synchronized sources of 60-Hz power at the transmitting and receiving ends. Either of two receiver relays can be actuated by choosing the appropriate half-cycle polarity of the 60-Hz transmitter power supply.

half-duplex operation Operation of a telegraph system in either direction over a single channel, but not in both directions simultaneously.

half-life The average time required for half the atoms of a sample of a radioactive substance to lose their radioactivity by decaying into stable atoms. Also called half-value period.

half-power frequency One of the two values of frequency, on the sides of an amplifier response curve, at which the voltage is 70.7% of a midband or other reference value.

half-power width An angular rating for antenna beam width. In a plane containing the direction of the maximum of the radiation lobe, it is the full angle between the two directions in that plane in which the radiation intensity is half the maximum value of the lobe.

half-reflecting dielectric sheet A sheet of di-

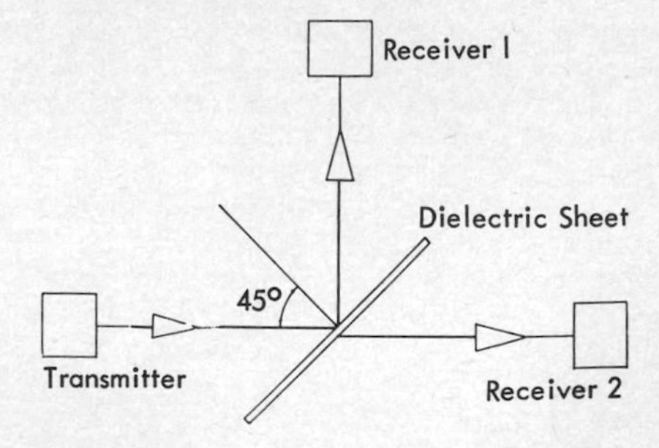

Half-reflecting dielectric sheet divides transmitter radiation equally between two receivers. Open arrows are antenna symbols.

electric material that is positioned at 45° to an incident millimeter-wave beam and is of such thickness and dielectric constant that it reflects exactly half the incident energy and transmits the other half. Used in millimeter-wave duplexers. At a wavelength of 9 mm, a sheet of ruby mica 0.5 mm thick will serve this purpose.

half-rhombic antenna A long-wire antenna array in which the lengths of the sides are several wavelengths long at the lowest frequency of operation. The radiating sides form half of a rhombic antenna, with the lines connected to opposite ends. When terminated in a resistor, the radiation pattern is unidirectional; when unterminated, it is bidirectional.

half-select current A current value that is insufficient to switch a ferrite core but will, when doubled, fully switch the core.

half-shift register A logic circuit that consists of a gated input storage element, with or without an inverter.

half-step *Semitone.*

halftone characteristic A relation between the density of the recorded copy and the density of the subject copy in a facsimile system.

half-value layer *Half-value thickness.*

half-value period *Half-life.*

half-value thickness The thickness of a given substance that, when introduced into the path of a given beam of radiation, will reduce its intensity to half the initial value. Also called half-value layer.

half-wave 1. Having an electrical length of one half-wavelength. 2. Pertaining to half of one cycle of a wave.

half-wave antenna An antenna whose electrical length is half the wavelength being transmitted or received.

half-wave control Phase control that acts on only half of each AC cycle, for varying load power from zero to half the full-wave maximum value or from half-power to full-power. The output is a pulsating DC voltage. The control element may be a silicon controlled rectifier or other solid-state power control device or a thyratron or other controllable gas tube.

half-wave dipole *Dipole.*

half-wavelength The distance corresponding to an electrical length of half a wavelength at the operating frequency of a transmission line, antenna element, or other device.

half-wave line A transmission line whose electrical length is half the wavelength of the signal being transmitted.

half-wave rectification Rectification in which current flows only during alternate half-cycles.

half-wave rectifier A rectifier that provides half-wave rectification.

half-wave vibrator A vibrator that has only one pair of contacts. It interrupts the flow of direct current through the primary of a power transformer but does not reverse the current, whereas a full-wave vibrator changes the direction of current flow twice per operating cycle.

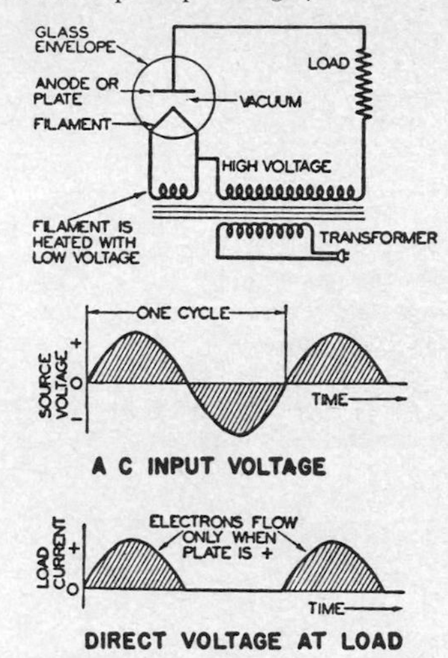

Half-wave rectifier circuit and waveforms.

Hall coefficient The constant of proportionality in the relation of the transverse electric field (Hall field) to the product of current density and magnetic flux density in an electric conductor. The sign of the majority carrier can be inferred from the sign of the Hall coefficient.

Hall constant The constant of proportionality in the equation for a current-carrying conductor in a magnetic field. The constant is equal to the transverse electric field (Hall field) divided by the product of the current density and the magnetic field strength. The sign of the majority carrier can be inferred from the sign of the Hall constant.

Hall effect The development of a voltage between the two edges of a current-carrying metal strip whose faces are perpendicular to a magnetic field.

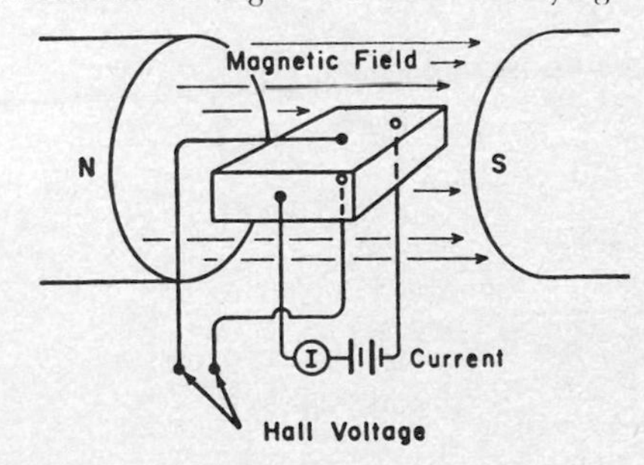

Hall effect, in which voltage is developed between top and bottom surfaces of current-carrying metal strip in magnetic field.

Hall-effect gaussmeter A gaussmeter that consists of a thin piece of silicon or other semiconductor material which is inserted between the poles of a magnet to measure the magnetic field strength by means of the Hall effect.

Hall-effect gyrator A gyrator in which the Hall effect reverses the polarity of a signal when input and output are interchanged.

Hall-effect isolator An isolator that makes use of the Hall effect in a semiconductor plate mounted in a magnetic field, to provide greater loss in one direction of signal travel through a waveguide than in the other direction.

Hall-effect modulator A Hall-effect multiplier used as a modulator to give an output voltage that is proportional to the product of two input voltages or currents. In one version, a rectangular semiconductor slab of indium arsenide is cemented to a ferrite plate. Connections are made to the four sides of the slab, two for one input variable and two for the output. The ferrite plate is then used as one flange of a ferrite bobbin for a coil energized by the other input variable to produce the magnetic field for Hall-effect action.

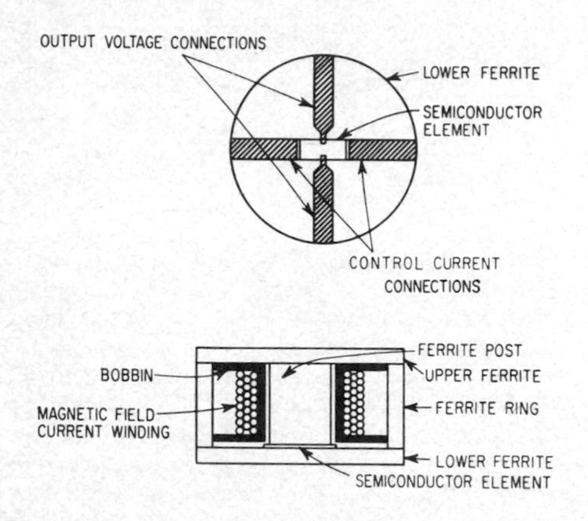

Hall-effect modulator. Indium arsenide slab on ferrite plate in upper diagram becomes lower flange of ferrite bobbin for windings.

Hall-effect multiplier A multiplier based on the Hall effect, used in analog computers to solve such problems as finding the square root of the sum of the squares of three independent variables.

Hall-effect switch A magnetically activated switch that uses a Hall generator, trigger circuit, and transistor amplifier on a silicon chip. The actuating means can be a small rod-shaped permanent magnet.

Hall generator A generator that uses the Hall effect to give an output voltage proportional to magnetic field strength.

Hall mobility The product of conductivity and the Hall constant for a conductor or semiconductor. It is a measure of the mobility of the electrons or holes in a semiconductor.

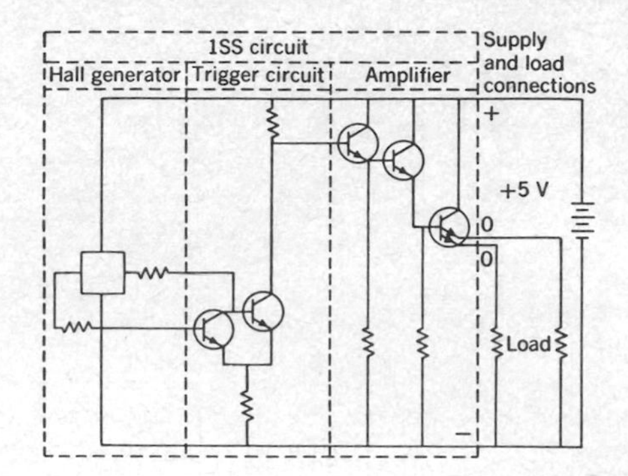

Hall-effect switch.

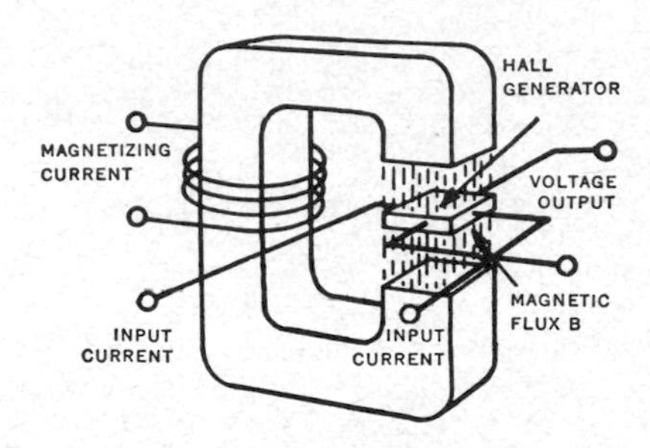

Hall generator.

Hall voltage The no-load voltage developed across a semiconductor plate due to the Hall effect, when a specified value of control current flows in the presence of a specified magnetic field.

Hallwachs effect The ability of ultraviolet radiation to discharge a negatively charged body in a vacuum.

halo An undesirable bright or dark ring surrounding an image on the fluorescent screen of a television cathode-ray tube. Generally due to overloading or maladjustment of the camera tube.

halogen counter A Geiger counter in which the self-quenching action is provided by a halogen gas like chlorine or bromine.

halt instruction A computer instruction that stops execution of the program.

ham Slang for *amateur.*

Hamming code An error-detecting and correcting code used in data transmission. When used with a 7-bit block, there would be information bits at positions 3, 5, 6, and 7 and parity check bits at 1, 2, and 4. This code detects all single and double errors in the block.

H and D curve A characteristic curve of a photographic emulsion, obtained by plotting film density against the logarithm of the exposure. Also called Hurter and Driffield curve.

Handie-Talkie Trademark of Motorola, Inc., for their line of portable two-way radio communication units.

hand-reset Requiring manual resetting after an operation, as with a circuit breaker.

handset A hand-held combination telephone-type receiver and telephone-type transmitter, as commonly used in telephone sets.

handshaking The process by which predetermined characters are exchanged by receiving and transmitting equipment to establish synchronization. A similar process is used with data transfer in computer systems.

handwheel A wheel used in place of a knob on a control shaft.

handwriting reader A pattern-recognition system that translates handwritten characters into suitable codes for computer processing. The characters, generally numbers and special control symbols, usually must be written in a predetermined manner.

hangover 1. Faulty reproduction of bass notes by a loudspeaker that is poorly damped or improperly mounted. 2. *Tailing.*

hang-up A nonprogrammed stop in a computer routine. It can be caused by improper coding of a machine instruction, attempted use of an invalid operation code, an input data error, or a machine malfunction.

hard copy Human-readable typewritten or printed characters produced on paper at the same time that information is being keyboarded in a coded machine language, as when punching cards or paper tape.

hard cosmic ray A cosmic radiation component that penetrates a moderate thickness of an absorber, such as 4 in (10 cm) of lead. The hard component consists chiefly of mesons but has some fast protons and electrons.

hard-drawn copper wire Copper wire that has been drawn to size through several dies without being annealed, to increase its hardness and tensile strength.

hardened circuit A circuit that uses components whose tolerance to radiation released by a nuclear explosion has been increased by various radiation-hardening procedures. One technique involves use of silicon dioxide to isolate components in microcircuits to prevent substrate photocurrents from affecting adjacent devices. Metallic doping of gate insulator oxides is used to reduce gamma-radiation damage in transistors.

hardened site An underground missile-launching site, control center, or other facility that can withstand in varying degrees the effects of an enemy nuclear blast. Thus Titan and Minuteman missiles are housed in underground silos from which they can be fired.

hard landing A landing made without deceleration, as by impact on the moon. A hard landing generally means complete destruction of the vehicle, as in the Ranger moon probes.

hard-limited repeater A repeater used in communication satellites to clip the composite input signal at a very low level, then amplify, filter, frequency-translate, and retransmit the resulting signal. The repeater output power level remains constant regardless of the input power level, whereas in a linear repeater it varies with input power.

hard limiter A limiter in which there are negligible variations in output once the range of limiting action is reached.

hard magnetic material A magnetic material that is not easily demagnetized. Used in permanent magnets.

hardness 1. The penetrating ability of x-rays. The shorter the wavelength, the harder the rays and the greater their penetrating ability. 2. The degree of evacuation in an x-ray or other vacuum tube. The harder the tube, the better its vacuum.

hard rubber Rubber that has been vulcanized at high temperatures and pressures, to give hardness. Once used extensively as an insulating material.

hard superconductor A superconductor that requires a strong magnetic field, more than 1000 oersteds, to destroy superconductivity. Niobium and vanadium are examples.

hard tube A vacuum tube that has been evacuated to a high degree, approaching a perfect vacuum. A tube that contains an appreciable amount of gas is called a soft tube.

hardware 1. Electronic components, subassemblies, and finished pieces of equipment, as contrasted to a drawing-board design. 2. Bolts, nuts, fasteners, brackets, handles, and similar small structural parts used in assembling electronic equipment. 3. The physical and permanent parts of a computer, as contrasted to the operating programs collectively known as software.

hard-wired Having a fixed wired program or control system built in by the manufacturer and not subject to change by programming.

hard x-ray An x-ray that has high penetrating power.

harmonic A sinusoidal component of a periodic wave, having a frequency that is an integral multiple of the fundamental frequency. The frequency of the second harmonic is twice that of the fundamental frequency or first harmonic. Also called harmonic component and harmonic frequency.

harmonic analysis 1. Any method of identifying and evaluating the harmonics that make up a

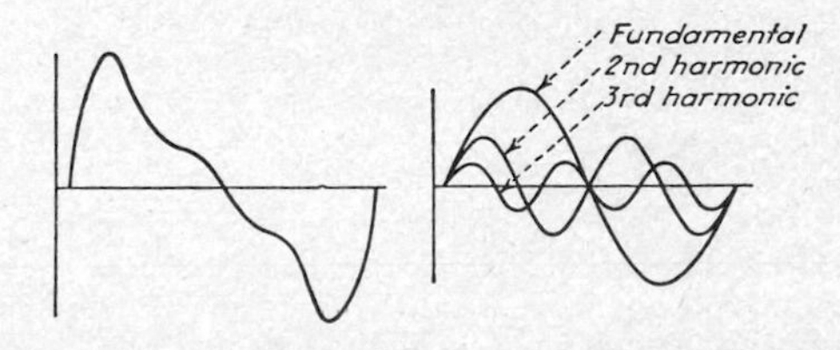

Harmonic analysis of complex wave, showing fundamental and harmonic components.

complex waveform of voltage, current, or some other varying quantity. 2. The expression of a given function as a series of sine and cosine terms that are approximately equal to the given function, such as a Fourier series.

harmonic analyzer An analyzer that measures the strength of each frequency component in a complex wave. Also called harmonic wave analyzer.

harmonic antenna An antenna whose electrical length is an integral multiple of a half-wavelength at the operating frequency of the transmitter or receiver.

harmonic attenuation Attenuation of an undesired harmonic component in the output of a transmitter, as by use of a pi network whose shunt reactances are tuned to have zero impedance at the harmonic frequency to be suppressed.

harmonic component *Harmonic.*

harmonic content The components that remain after the fundamental frequency has been removed from a complex wave.

harmonic conversion transducer A conversion transducer in which the output signal frequency is a multiple or submultiple of the input frequency, as in frequency dividers and frequency multipliers.

harmonic distortion Nonlinear distortion in which undesired harmonics of a sinusoidal input signal are generated because of circuit nonlinearity.

harmonic filter A filter that is tuned to suppress an undesired harmonic in a circuit.

harmonic frequency *Harmonic.*

harmonic generator A generator operated under conditions such that it generates strong harmonics along with the fundamental frequency.

harmonic interference Interference caused by the presence of harmonics in the output of a radio station.

harmonic-mode crystal unit *Overtone crystal unit.*

harmonic wave analyzer *Harmonic analyzer.*

harness An assembly of insulated wires of various lengths, bent to a pattern and tied together before installation in a piece of equipment.

Harpoon A Navy antiship guided missile that has a range of about 50 mi (80 km), designed for firing from either a ship or plane.

hartley A unit of information content, equal to the designation of 1 of 10 possible and equally likely values or states of anything used to store or convey information. One hartley is equal to $\log_2 10$ bits or 3.323 bits.

Hartley oscillator A vacuum-tube oscillator in which the parallel-tuned tank circuit is connected between grid and anode. The tank coil has an intermediate tap at cathode potential, so the grid-cathode portion of the coil provides the necessary feedback voltage.

Hartley's law The total number of bits of information that can be transmitted over a channel in a given time is proportional to the product of channel bandwidth and transmission time.

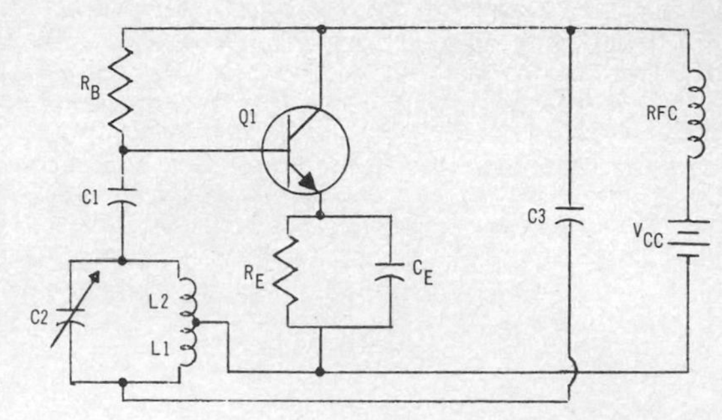

Hartley oscillator using shunt-fed transistor.

hash 1. Electric noise produced by the contacts of a vibrator or the brushes of a generator or motor. 2. *Grass.*

hash total A total obtained by adding all the digits in a group without regard to their significance, including assigned numerical values of alphabetic characters, for checking purposes in a computer.

Hawk [Homing-All-the-Way Killer] An Army surface-to-air guided missile that has a range of about 25 mi (40 km), a maximum speed of about Mach 3, and a ceiling of about 45,000 ft (14,800 m). It was originally guided by radio for attacking low-flying enemy aircraft, but newer models are radar-guided.

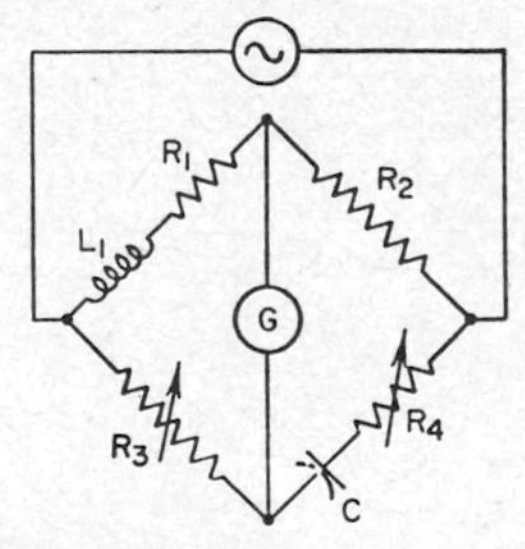

Hay-bridge circuit.

Hay bridge A four-arm bridge used to measure inductance in terms of capacitance, resistance, and frequency. Bridge balance depends on frequency.

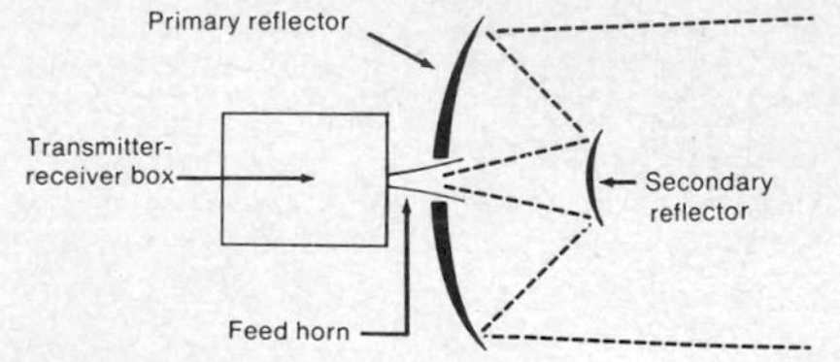

Haystack antenna, showing how secondary reflector is used for transmitting and receiving main beam.

Haystack antenna A large and powerful radar-type antenna erected by the U.S. Air Force in

Massachusetts for microwave research, including satellite relay communication, long-range tracking of spacecraft, and radio astronomy. It uses two reflectors arranged according to the Cassegrainian principle of optical telescopes, housed in a metal-frame radome about 50 m in diameter. Movement is controlled by a computer, for exact tracking of orbiting spacecraft or radio stars.

H display A radarscope B display modified to include indication of angle of elevation. The target appears as two closely spaced bright spots that

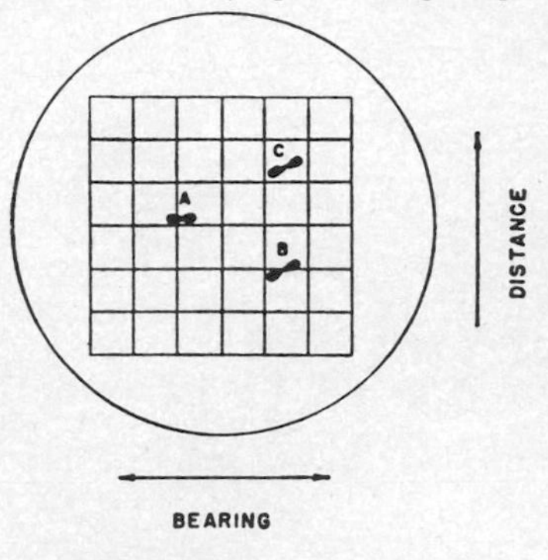

H display.

approximate a short bright line, the slope of which is proportional to the sine of the angle of target elevation.

head 1. The photoelectric unit that converts the sound track on motion-picture film into corresponding audio signals in a motion-picture projector. 2. *Cutter.* 3. *Magnetic head.*

head amplifier 1. An amplifier that is mounted close to the head which serves as its signal source, for amplifying the weak signal before it is fed through a cable to the main amplifier. 2. British term for the video amplifier that is mounted close to the pickup tube in a television camera.

head demagnetizer A device that eliminates residual magnetism built up in a magnetic-tape recording head.

header A mounting plate through which the insulated terminals or leads are brought out from a

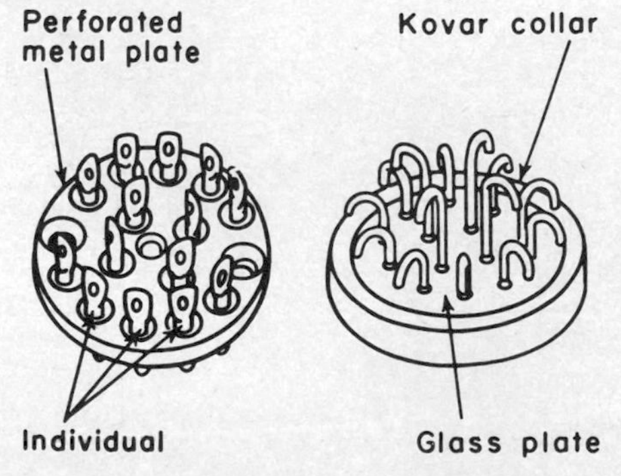

Headers with feedthrough terminals.

hermetically sealed relay, transformer, transistor, tube, or other device.

header record A computer input record that contains common, constant, or identifying information for records which follow.

heading The horizontal direction in which a vehicle is directed, expressed as an angle between a reference line and the line extending in the direction the vehicle is headed, usually measured clockwise from the reference line. Heading is the instantaneous actual direction in which an aircraft, ship, or other vehicle is pointed, whereas course is the intended direction of travel. Also called relative heading (deprecated).

heading marker A line of light produced on a marine radar PPI display at the instant when the rotating radar antenna is at the position corresponding to the ship's heading, to indicate whether objects ahead of the ship are located to port or starboard.

headlight antenna A radar antenna small enough to be housed within the thickness of an airplane wing in the manner of an automobile headlight, yet producing a beam that can be directed much like a searchlight.

headphone An electroacoustic transducer designed to be held against an ear by a clamp passing over the head, for private listening to the audio

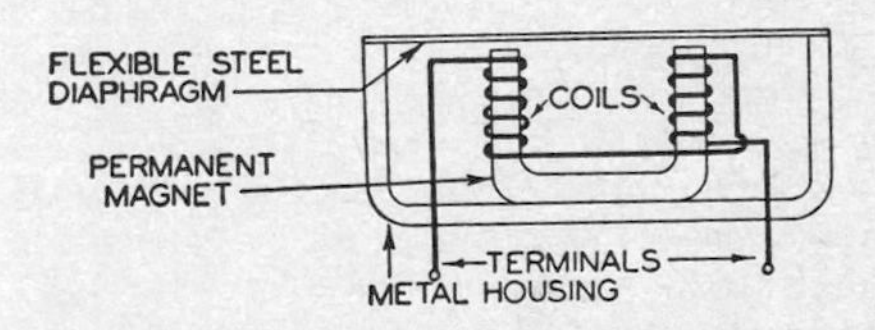

Headphone construction. Audio-frequency currents flowing through diaphragm produce magnetic field that alternately aids and opposes fixed magnetic field of permanent magnet, causing diaphragm to move and produce sound waves.

output of a communication, radio, or television receiver or other source of AF signals. Usually used in pairs, one for each ear, with the clamping strap holding both in position. Also called headset and phone.

headset *Headphone.*

health physics The protection of personnel from harmful effects of ionizing radiation by such means as routine radiation surveys, area and personnel monitoring, protective equipment, and protective procedures.

hearing aid A miniature portable sound amplifier for persons with impaired hearing, consisting of a microphone, audio amplifier, earphone, and batteries.

hearing loss The ratio of the threshold of audibility of a particular human ear to the normal threshold, expressed in decibels.

heart pacer *Pacemaker.*

heat *Thermal radiation.*

heater [symbol H] An electric heating element

for supplying heat to an indirectly heated cathode in an electron tube.

heater current The current flowing through a heater for an indirectly heated cathode in an electron tube.

heater-type cathode *Indirectly heated cathode.*

heater voltage The voltage applied to the terminals of the heater in an electron tube.

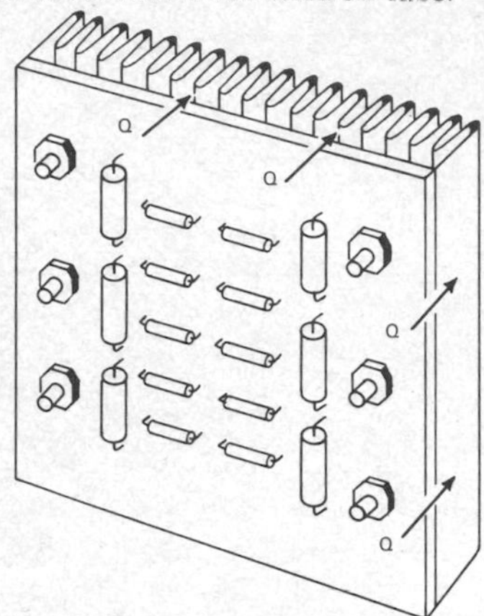

Heat-exchanger design using air as heat-transferring medium from back of printed-circuit board to corrugated metal heatsink.

heat exchanger Any device that transfers heat from a heated solid, liquid, or gas to a cooler location or to the environment.

heating element A coiled or other arrangement of resistance wire, supported by one or more ceramic insulators, through which current is sent to produce heat.

heat loss Power loss that is due to the conversion of electric energy into heat.

heat pipe A sealed metal tube that has an inner lining of wicklike capillary material and a small amount of fluid in a partial vacuum. When one end of the pipe becomes hot, the fluid boils, and the molecules of steam move along the pipe at high speed. They condense at the other end of the pipe, giving up latent heat, and the capillary lining sucks the liquid back to the hot end for a repetition of the heat-transferring cycle. Also constructed in flattened or plate-shaped form.

heat run A series of temperature measurements made on an electric device during operating tests under various conditions.

heat-seal To bond or weld a flexible plastic film or sheet of material to itself or another material by heat alone. Dielectric heating is frequently used for this purpose.

heatseeker A guided missile that incorporates an infrared device for homing on heat-radiating machines or installations, such as an aircraft engine or blast furnace.

heat shield A layer of material that gives protection from heat. Used on the front of a reentry capsule.

heat-shrinkable plastic A plastic material that can be shrink-fitted over terminals and other objects of varying size and shape, for insulating or other purposes. Commonly used in tubing form, but also available as tape, sheets, and caps. After it is placed on the object to be covered, the material is heated with a hot-air blower to a critical temperature of about 100°C, to make it shrink up to about 50% and conform to the shape of the object.

heat shunt A heatsink placed in contact with the lead of a delicate component to prevent overheating during soldering.

heatsink A mass of metal that is added to a device, to absorb and dissipate heat. Used with

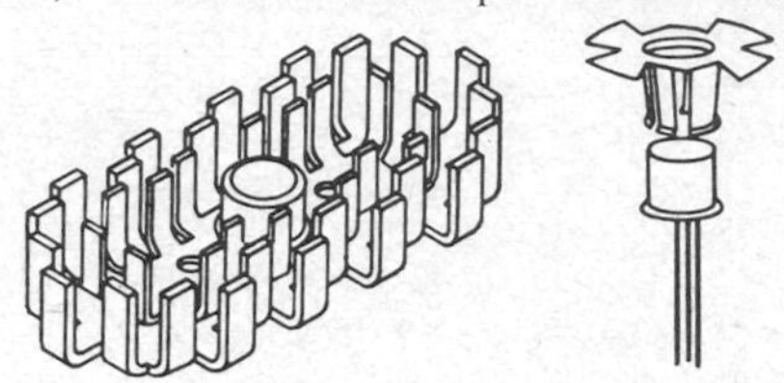

Heatsink for power transistor (left) and signal transistor (right).

power transistors and many types of metallic rectifiers.

heat transfer The transfer of heat from one location to another by conduction through fluids or solid materials, by convection involving actual movement of a heated fluid or gas, or by radiation of electromagnetic heat waves.

heat wave Infrared radiation much higher in frequency than radio waves.

Heaviside-Campbell mutual-inductance bridge A Heaviside mutual-inductance bridge in which one of the inductive arms contains a separate in-

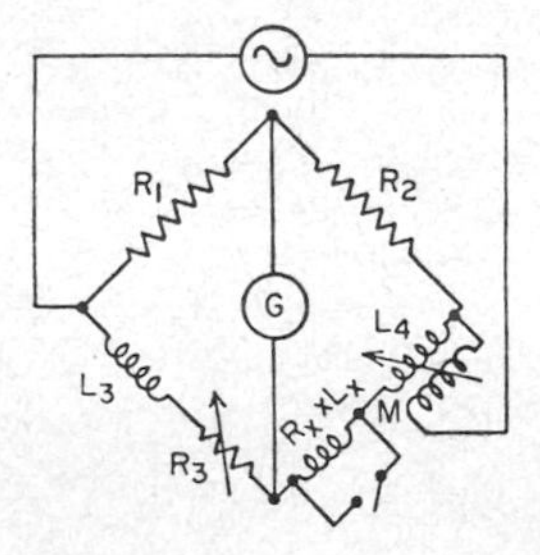

Heaviside-Campbell mutual-inductance bridge circuit.

ductor that is included in the bridge arm during the first of a pair of measurements and short-circuited during the second. Bridge balance is independent of frequency.

Heaviside mutual-inductance bridge An AC bridge used to compare self-inductances and mutual inductances. Bridge balance is independent of frequency.

heavy hydrogen An isotope of hydrogen that has a mass number of either 2 or 3. The isotope with a mass of 2 is called deuterium; the isotope

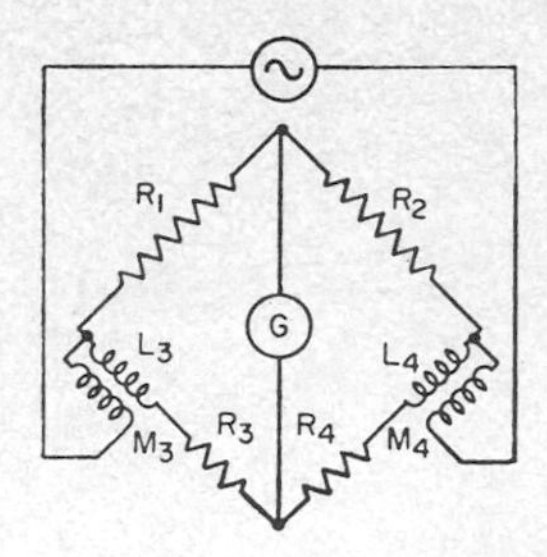

Heaviside mutual-inductance bridge circuit.

with a mass of 3 is called tritium.

heavy-ion linear accelerator A linear accelerator that has ion-beam energies up to about 10

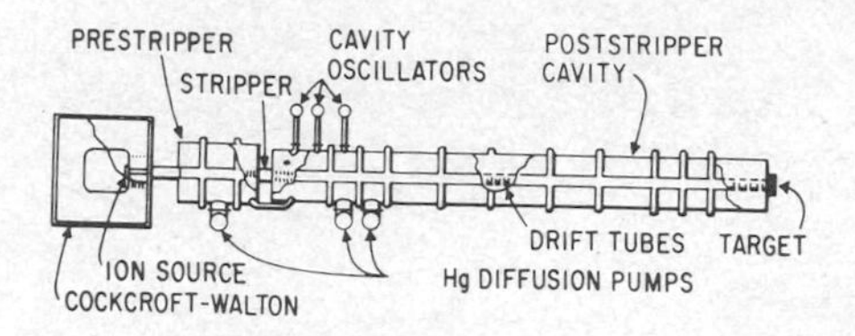

Heavy-ion linear accelerator.

MeV. Used primarily for high-energy-particle physics research.

heavy water [symbol D_2O] Water in which the hydrogen of the water molecule consists entirely of the heavy-hydrogen isotope that has a mass number of 2. Also called deuterium oxide. Density is 1.1076 at 20°C.

hecto- A prefix representing 10^2, or hundreds.

hectometric wave A radio wave between the wavelength limits of 100 and 1000 m, corresponding to the medium-frequency (MF) range of 300 to 3000 kHz.

height The vertical distance to a target from a horizontal plane passing through the location of the height computer.

height control The television receiver control that adjusts picture height.

height effect *Antenna effect.*

height finder A radar set that measures and determines the height of an airborne object.

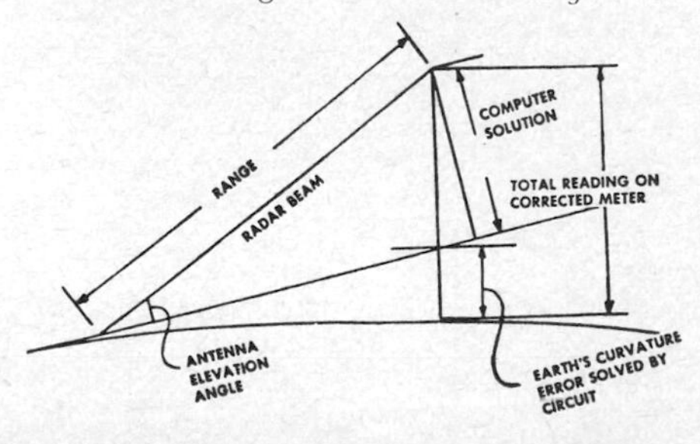

Height finder requires computer to combine earth's curvature with range and elevation data provided by radar set.

height marker A type of calibration marker used on some radar displays.

height-range indicator A radarscope that gives both height and range of a target.

Heisenberg uncertainty principle The simultaneous precise determination of velocity (or any related property) of a particle and its position is impossible. The smaller the particle, the greater the degree of uncertainty.

Heising modulation *Constant-current modulation.*

helical antenna An antenna that has the form of a helix. When the helix circumference is much smaller than one wavelength, the antenna radiates

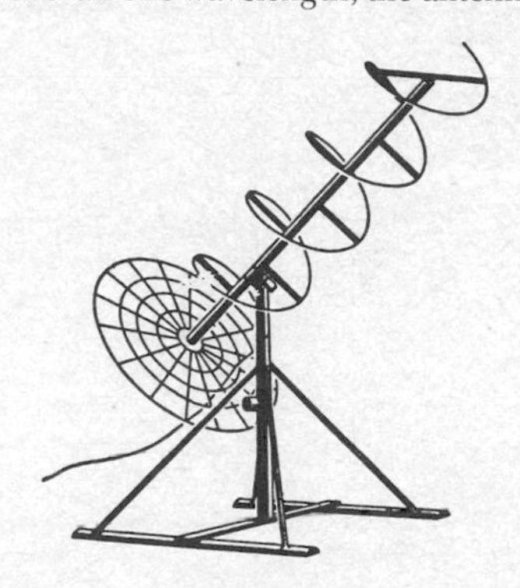
Helical antenna used with telemetering receivers.

at right angles to the axis of the helix. When the helix circumference is one wavelength, maximum radiation is along the helix axis. Also called helix antenna.

helical potentiometer A multiturn precision potentiometer in which a number of complete turns of the control knob are required to move the

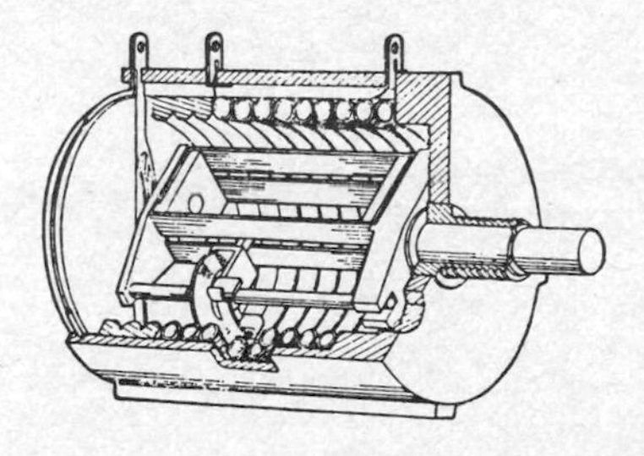
Helical potentiometer having 10-turn resistance element.

contact arm from one end of the helically wound resistance element to the other end.

helical scanning 1. A method of facsimile scanning in which a single-turn helix rotates against a stationary bar to give horizontal movement of an elemental area. 2. A method of radar scanning in which the antenna beam rotates continuously about the vertical axis while the elevation angle changes slowly from horizontal to vertical, so that

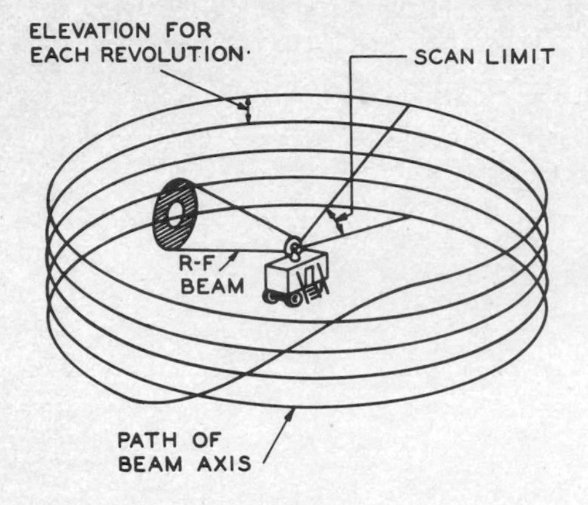

Helical scanning.

a point on the radar beam describes a distorted helix.

helicon wave An energy-carrying transverse electromagnetic wave that is supported in solid material by free carriers in the presence of a magnetic field.

helitron oscillator An electrostatically focused low-noise backward-wave oscillator. The microwave output signal frequency can rapidly be swept over a wide range by varying the voltage applied between the cathode and the associated RF circuit.

helium [symbol He] A light gaseous element that exists as a monatomic gas except at extremely low temperatures, is chemically inert, and has practically no absorption cross section for neutrons. Atomic number is 2.

helium-cadmium laser A metal vapor ion laser in which cadmium vapor, produced by heat or other means, migrates through a high-voltage glow discharge in helium, generating a continuous laser beam at wavelengths in the ultraviolet and blue parts of the spectrum (about 0.3 to 0.5 μm).

helium cryostat A cryostat in which liquefied helium-3 isotope is used with adiabatic demagnetization to produce a temperature of 0.3 K.

helium-neon laser [abbreviated HeNe laser] An atomic gas laser in which a combination of helium and neon gases is used. Applications include reading of bar-code labels in stores. The visible red beam can also replace a string for establishing levels, grades, and verticals in the construction industry. Operation is normally at three distinct wavelengths: 0.6328, 1.15, and 3.39 μm.

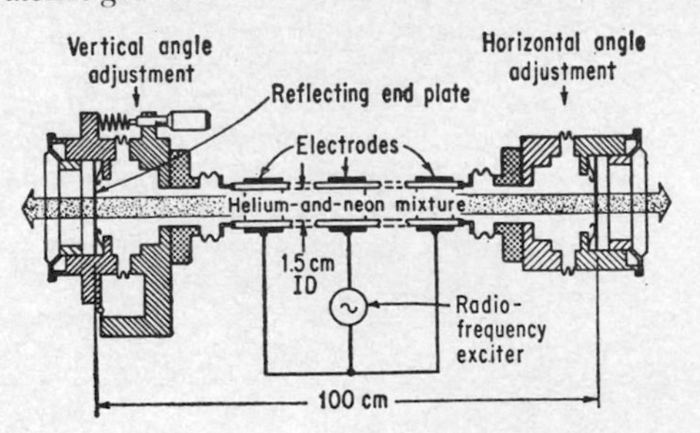

Helium-neon laser. Output beams emerge through windows at left and right.

helium spectrometer A small-mass spectrometer used to detect the presence of helium in a vacuum system. For leak detection, a jet of helium is applied to suspected leaks in the outer surface of the system, and the output indicator of the spectrometer is watched to determine the exact point at which helium enters.

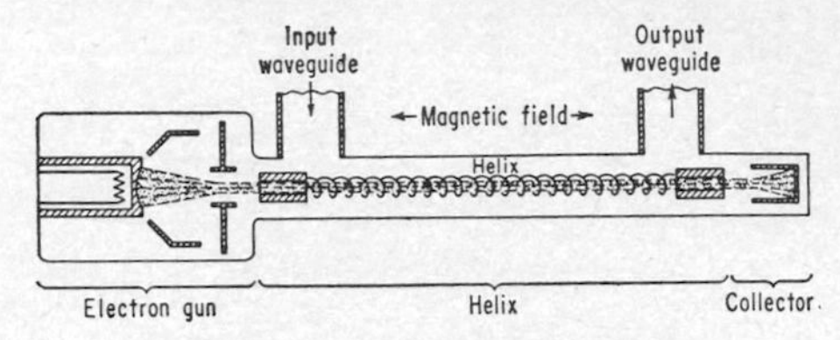

Helix in traveling-wave tube.

helix A spread-out single-layer coil of wire, either wound around a supporting cylinder or made of stiff enough wire to be self-supporting.

helix antenna *Helical antenna.*

helix recorder A recorder that uses helical scanning.

helix waveguide A waveguide that consists of closely wound turns of insulated copper wire covered with a lossy jacket.

Helmholtz resonator An acoustic enclosure that has a small opening of such dimensions that the enclosure resonates at a single frequency determined by the geometry of the resonator.

HEM wave Abbreviation for *hybrid electromagnetic wave.*

HeNe laser Abbreviation for *helium-neon laser.*

henry [abbreviated H; plural henrys] The SI unit of inductance or mutual inductance. The inductance of a circuit is 1 H when a current change of 1 A/s induces 1 V.

heptode A seven-electrode electron tube that contains an anode, a cathode, a control electrode, and four additional electrodes which are ordinarily grids.

HERALD [Harbor Echo Ranging And Listening Device] A sensitive ultrasonic underwater sound-detection system used at harbor entrances to detect submerged submarines running at silent speed. The operator is able to both listen to the sounds and determine range and bearing for their source.

hermaphroditic connector A connector in which both mating parts are exactly alike at their mating surfaces. A bifurcated connector is an example. Also called sexless connector.

hermetic Permanently sealed by fusion, soldering, or other means, to prevent the transmission of air, moisture vapor, and all other gases.

hermetically sealed crystal unit A crystal unit sealed in its glass or metal holder, usually by soldering, for protection against all external conditions except vibration and temperature.

hermetically sealed relay A relay that is permanently sealed in its metal, glass, or ceramic housing by fusion or soldering.

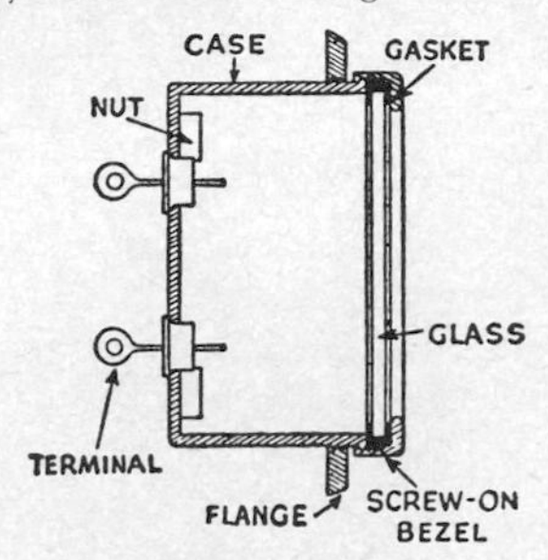

Hermetic seals around glass window and terminals of meter housing.

hermetic seal A seal that prevents passage of air, water vapor, and all other gases.

herringbone pattern An interference pattern sometimes seen on television receiver screens, consisting of a horizontal band of closely spaced V- or S-shaped lines.

hertz [abbreviated Hz] The SI unit of frequency, equal to 1 cycle per second.

Hertz antenna An ungrounded half-wave antenna.

Hertzian wave *Radio wave.*

Hertz vector A vector used to specify the electromagnetic field of a radio wave.

heterodyne To mix two AC signals of different frequencies in a nonlinear device to produce two new frequencies, corresponding respectively to the sum of and the difference between the two original frequencies. This action is the basis of all superheterodyne receivers.

heterodyne conversion transducer *Converter.*

heterodyne detector A detector in which an unmodulated carrier frequency is combined with the signal of a local oscillator that has a slightly different frequency, to provide an audio-frequency beat signal which can be heard with a loudspeaker or headphones. Used chiefly for code reception.

heterodyne frequency Either of the two new frequencies resulting from heterodyne action. One is the sum of the two input frequencies; the other is the difference between the input frequencies.

heterodyne frequency meter A frequency meter in which a known adjustable frequency is heterodyned with an unknown frequency until a zero beat is obtained. Alternatively, the unknown frequency may be heterodyned with a fixed known frequency to produce a lower-frequency signal, usually in the audio-frequency range, whose value is measured by other means. Also called heterodyne wavemeter.

heterodyne harmonic analyzer A harmonic analyzer in which a complex input voltage is mixed with the output of a variable-frequency oscillator, and the magnitude of the sum or difference frequency for each input harmonic is measured with a meter.

heterodyne interference *Heterodyne whistle.*

heterodyne oscillator A separate variable-frequency oscillator used to produce the second frequency required in a heterodyne detector for code reception.

heterodyne reception Radio reception in which the incoming RF signal is combined with a locally generated RF signal of different frequency, followed by detection. The resulting beat frequency may be audible or at a higher intermediate frequency, as in a superheterodyne receiver. Also called beat reception.

heterodyne repeater A radio repeater in which the received radio signals are converted to an intermediate frequency, amplified, and reconverted to a new frequency band for transmission over the next repeater section.

heterodyne wavemeter *Heterodyne frequency meter.*

heterodyne whistle A steady high-pitched audio tone heard in an ordinary amplitude-modulation radio receiver under certain conditions when two signals that differ slightly in carrier frequency enter the receiver and heterodyne to produce an audio beat. Also called heterodyne interference.

heterogeneous radiation Radiation that has a number of different frequencies, particles, or particle energies.

heterojunction A junction between two dissimilar semiconductor materials that have different energy gaps between their valence and conduction bands, such as between germanium and gallium arsenide. Used in some types of photodiodes, semiconductor lasers, and transistors.

heuristic Pertaining to trial-and-error exploratory methods of problem solving, in which solutions are discovered by evaluation of the progress made toward the final result.

heuristic routine A routine in which a computer attacks a problem by a trial-and-error approach, frequently involving the act of learning.

hevea rubber Rubber from the hevea brasiliensis tree, used for insulation.

hexadecimal notation A notation in the scale of 16, using decimal digits 0 to 9 and six more digits that are sometimes represented by A, B, C, D, E, and F. Also called sexadecimal notation.

hexode A six-electrode electron tube that contains an anode, a cathode, a control electrode, and three additional electrodes which are ordinarily grids.

HF Abbreviation for *high frequency.*

HgTe Symbol for *mercuric telluride.*

Decimal	Binary	Octal	Hexa-decimal
0	0	0	0
1	1	1	1
2	10	2	2
3	11	3	3
4	100	4	4
5	101	5	5
6	110	6	6
7	111	7	7
8	1000	10	8
9	1001	11	9
10	1010	12	A
11	1011	13	B
12	1100	14	C
13	1101	15	D
14	1110	16	E
5	1111	17	F

Hexadecimal notation, with decimal and binary equivalents.

hierarchy A classification scheme based on rank, order, or some other logical relationship.

hi-fi *High fidelity.*

hi-fix Decca A short-range Decca radio navigation system that can be used in a hyperbolic configuration giving lines of position, or in a range configuration based on phase comparison of continuous-wave signals from two stations.

high-altitude radio altimeter *Radar altimeter.*

high band The television band extending from 174 to 216 MHz, which includes channels 7 to 13.

high boost *High-frequency compensation.*

high-confidence countermeasure A countermeasure that is very difficult for the enemy to overcome and may therefore be expected to retain its usefulness for a number of years.

high-contrast image An image in which the contrast between black and white is great and intermediate tones are poor.

high definition The television equivalent of high fidelity, in which the reproduced image contains such a large number of accurately reproduced elements that picture details approximate those of the original scene.

high-energy physics The branch of physics that deals with the properties and behavior of charged elementary particles which are produced by collisions involving energies above 150 MeV. The collisions generally involve heavy unstable particles like antiparticles, hyperons, and mesons.

high-epithermal neutron range The neutron energy range of 1 to 100 keV.

high fidelity Fidelity of audio reproduction so perfect that listeners hear exactly what they would have heard if they had been present at the original performance. Also called hi-fi.

high-fidelity receiver A radio receiver that reproduces audio frequencies with high fidelity, to duplicate faithfully the original sound picked up by the microphone.

high frequency [abbreviated HF] A Federal Communications Commission designation for a frequency in the band from 3 to 30 MHz, corresponding to a decametric wave between 10 and 100 m.

high-frequency bias A sinusoidal signal that is mixed with a data signal being recorded on magnetic tape, to increase the linearity and dynamic range of the recorded signal. The bias frequency is usually three to four times the highest information frequency to be recorded.

high-frequency compensation Increasing the amplification at high frequencies with respect to that at low and middle frequencies in a given band, such as in a video band or an audio band. Also called high boost.

high-frequency furnace An induction furnace in which the heat is generated within the charge, within the walls of the containing crucible, or in both, by currents induced by high-frequency magnetic flux produced by a surrounding coil. Also called coreless-type induction furnace.

high-frequency heating *Electronic heating.*

high-frequency induction heater An induction heater that produces electric current flow in a charge to be heated, at a frequency higher than that of the AC line.

high-frequency resistance The total resistance offered by a device in an AC circuit, including the DC resistance and the resistance due to eddy current, hysteresis, dielectric, and corona losses. Also called AC resistance, effective resistance, and RF resistance.

high-frequency trimmer A trimmer capacitor that controls the calibration of a tuning circuit at the high-frequency end of a tuning range in a superheterodyne receiver.

high-frequency welding Welding with high-frequency current obtained from an electronic generator.

high-gamma tube 1. A television camera tube in which the voltage output increases uniformly with the intensity of the light on the image. 2. A television picture tube in which the light intensity on the screen is directly proportional to the control-grid voltage.

high level The more positive of the two logic levels or states in a binary digital logic system. When both high and low levels are negative, the opposite connotation is sometimes used, wherein the larger negative voltage (the less positive voltage) is assumed to be the high level or high state.

high-level detector A power detector or linear detector, for which the voltage-current characteristic is essentially a straight line or two intersecting straight lines extending up to high input voltage levels.

high-level language A computer programming

language in which each instruction corresponds to several machine code instructions. This allows the programmer to write in a more familiar notation that uses numerals, words, or even single alphameric characters, for automatic translation into machine language by a compiler. Examples include APL, BASIC, COBOL, and FORTRAN.

high-level modulation Modulation produced in the anode circuit of the last stage of a transmitter, where the power level approximates that at the output of the system.

highlight A bright area in a television image.

high-mu tube A vacuum tube that has a high amplification factor.

high-order digit A digit that occupies the most significant position in a positional notation system. In conventional notation it is the digit at the left.

high-pass filter A filter that transmits all frequencies above a given cutoff frequency and substantially attenuates all others.

high-potting Testing with a high voltage, generally on a production line.

high-pressure cloud chamber A cloud chamber in which the gas is maintained at high pressure to reduce the range of high-energy particles and thereby increase the probability of observing events.

high-pressure laminate A laminated plastic that is generally produced at pressures above 1000 lb/in^2 (70.3 kg/cm^2).

high-pressure mercury-vapor lamp A discharge tube that contains an inert gas and a small quantity of liquid mercury. The initial glow discharge through the gas heats and vaporizes the mercury, after which the discharge through mercury vapor produces an intensely brilliant light.

high Q A characteristic wherein a component has a high ratio of reactance to effective resistance so that its Q factor is high.

high-recombination-rate contact A semiconductor-to-semiconductor or metal-to-semiconductor contact at which thermal-equilibrium carrier densities are maintained substantially independent of current density.

high-resistance voltmeter A voltmeter that has a resistance considerably higher than 1 kΩ/V, so that it draws little current from the circuit in which a measurement is made.

high-speed carry A special adding technique sometimes used in computers, in which the carries generated in several columns are executed simultaneously to speed the operation. Also called standing-on-nines carry.

high-speed memory *Rapid memory.*

high-speed printer *Line printer.*

high-speed regulator A power-supply regulator in which the output capacitor is omitted to permit rapid programming of output voltage changes and/or quick response as a current source.

high-speed relay A relay specifically designed for short operate time, short release time, or both.

high tension 1. High voltage, of the order of thousands of volts. 2. British term for *anode voltage,* of the order of several hundred volts.

high-threshold logic [abbreviated HTL] Logic characterized by higher supply voltages (about 15 V), higher noise immunity, and higher threshold voltages than other logic families. Usually obtained by adding a zener-diode voltage drop to the normal diode voltage drop of diode-transistor logic circuits.

high vacuum A degree of vacuum at which essentially no gases or vapors are present, so that ionization cannot occur.

high-velocity scanning The scanning of a target with electrons of such velocity that the secondary-emission ratio is greater than unity, so that more than one secondary electron leaves the target for every electron hitting it.

high-voltage direct current [abbreviated HVDC] A long-distance DC power-transmission system that uses DC voltages up to about 1 MV to keep transmission losses down. At one 2-GW hydroelectric power plant at Cabora Bassa on the Zambezi River in Mozambique, the generated power is transmitted 1400 km at ±533 kV to an electronic converter station near Johannesburg.

hill-and-dale recording *Vertical recording.*

hiran [HIgh-precision shoRAN] A modification of shoran that uses special operating techniques to measure distance with accuracy comparable to first-order triangulation.

hiss Random noise in the audio-frequency range, similar to prolonged sibilant sounds.

histogram A frequency-distribution graph, in which the number of quantities having each particular range of values is represented by a vertical bar on the graph.

hit An exact match between two items of data.

hit-on-the-fly printer A line printer in which the paper, the print head, or both are in continuous motion.

$H_{m,n}$ mode British term for *$TE_{m,n}$ mode.*

$H_{m,n}$ wave British term for *$TE_{m,n}$ wave.*

H network An attenuation network composed of five branches. Two are connected in series between an input terminal and an output terminal.

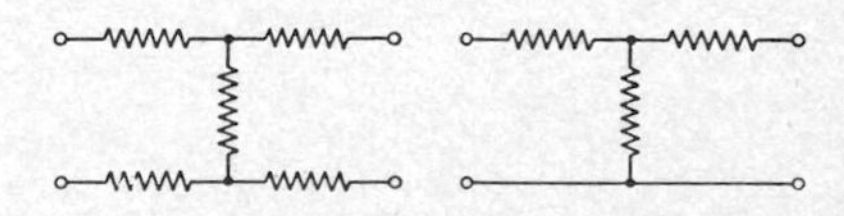

H network has five impedance branches, whereas pi network at right has only three branches.

Two are connected in series between another input terminal and output terminal. The fifth is connected from the junction point of the first two branches to the junction point of the second two branches. Also called H pad.

hodoscope An array of small Geiger counters used in tracing the paths of cosmic rays.

hold 1. To maintain storage elements at equilibrium voltages in a charge-storage tube by electron bombardment. 2. To retain information in a computer storage device for further use after it has been initially utilized. 3. A condition in which an integrator, sample-and-hold amplifier, or other charge-storing circuit maintains constant output after the input signal has been removed. 4. A designed stop in an operation or test, such as a delay in the countdown for launching a missile or in completing a telephone call.

hold control A manual control that changes the frequency of the horizontal or vertical sweep oscillator in a television receiver so that the frequency more nearly corresponds to that of the incoming synchronizing pulses. The two controls used are called the horizontal hold control and the vertical hold control. Also called speed control.

holder A device that mechanically and electrically accommodates one or more crystals, fuses, or other components in such a way that they can readily be inserted or removed.

holding anode A small auxiliary anode used in a mercury-pool rectifier to keep a cathode spot energized during the intervals when the main anode current is zero.

holding beam A diffused beam of electrons used to regenerate the charges stored on the dielectric surface of a cathode-ray storage tube.

holding coil A separate relay coil that is energized by contacts which close when a relay pulls in, to hold the relay in its energized position after the original operating circuit is opened.

holding current The minimum current required to maintain a switching device in a closed or conducting state after it is energized or triggered.

hold time The time during which pressure is applied to a welded joint after welding current ceases to flow.

hole A mobile vacancy in the electronic valence structure of a semiconductor. It is equivalent to a positive charge. A hole exists when an atom has less than its normal number of electrons.

hole-and-slot anode A magnetron anode in which the cavity resonators are circular holes connected by slots to the space between the central cathode and the anode.

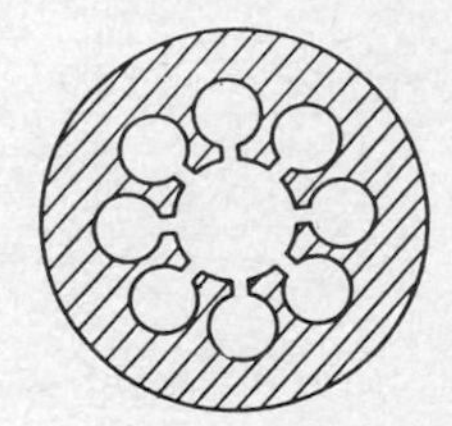

Hole-and-slot anode for magnetron.

hole conduction Conduction occurring in a semiconductor when electrons move into holes under the influence of an applied voltage and thereby create new holes. The apparent movement of such holes is toward the more negative terminal and is hence equivalent to a flow of positive charges in that direction.

hole injection The production of holes in an N-type semiconductor when voltage is applied to a sharp metal point in contact with the surface of the material.

hole mobility The ability of a hole to travel readily through a semiconductor.

Hollerith code A code used for letters, numbers, or special symbols punched in standard 80-column punched cards.

hollow-cathode lamp A glow lamp in which the glow forms only inside a small tubular cathode, to give an intense light source that has high spectral purity. Used in some spectrophotometers.

holmium [symbol Ho] A rare-earth element. Atomic number is 67.

hologram The special photographic plate used in holography. When this negative is developed and illuminated from behind by a coherent gas-laser beam, it produces a three-dimensional image in space.

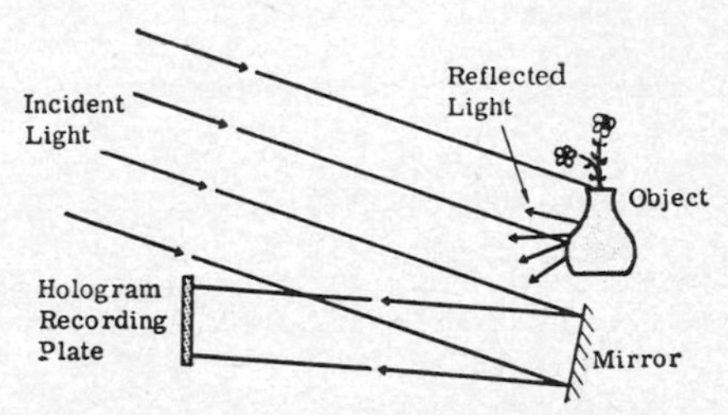

Hologram being produced. Light from laser at upper left is reflected from both object and mirror onto photographic plate at lower left.

holographic memory A memory in which information is stored in the form of holographic images on thermoplastic or other recording films. Used in optical computers.

holography Three-dimensional photography that involves the use of laser light with ordinary black and white photographic plates. The resulting images can be viewed in three dimensions without special glasses. The coherent light output of the laser beam is channeled to illuminate both the subject and the photographic plate during production of the hologram. The result is an interference pattern on the photographic plate that bears no resemblance to the original until developed and illuminated from behind by a similar coherent laser beam.

holomicrography The use of holography to produce three-dimensional images with various types of microscopes.

home 1. To fly toward a radiation-emitting

source, using the radiated waves as a guide. 2. To travel to a target by guidance of heat radiation, laser beams, radar echoes, radio waves, sound waves, or other phenomena originating in or reflected from the target. 3. The normal or starting position for a stepping relay.

homer 1. A ground-based direction-finding station that utilizes radio transmissions from aircraft to determine their bearing, then guides the aircraft toward the station by voice communication. 2. *Target seeker.*

hometaxial-based transistor A transistor in which both the emitter and connector junctions are formed by a single-diffusion process in a uniformly doped silicon slice.

homing 1. The process of approaching a desired point by maintaining constant some indicated navigation parameter other than altitude. 2. Use of radiation from a target to establish a collision course in missile guidance. The three types of homing are active, semiactive, and passive. 3. Flying toward a radio or radar transmitter by using the transmitted radiation for navigation guidance. 4. Returning to the starting position, as in a stepping relay or tuning motor.

homing adapter A device attached to an aircraft radio receiver to permit homing the aircraft on a radio beacon or other radio transmitting station. The adapter indicates direction to the transmitting station by aural or visual signals.

homing aid A system that guides an aircraft to an airport or carrier.

homing-and-busting aircraft Armed aircraft capable of homing on and destroying sources of electromagnetic radiation.

homing antenna A directional antenna array used to fly directly to a target that is emitting or reflecting radio or radar waves.

homing beacon A radio beacon, either airborne or on the ground, toward which an aircraft can fly if equipped with a radio compass or homing adapter. Also called radio homing beacon.

homing device 1. A transmitter, receiver, or adapter used for homing aircraft or by aircraft for homing purposes. 2. A device incorporated into a guided missile or the like to home it on a target. 3. A control device that automatically starts in the correct direction of motion or rotation to achieve a desired change, as in a remote-control tuning motor for a television receiver. A nonhoming device may first go to the end of its travel in the wrong direction.

homing guidance A guidance system in which a missile directs itself to a target by a self-contained mechanism that reacts to a particular characteristic of the target. Such characteristics include heat, light, sound, a reflected radar echo or other electromagnetic radiation, ionization of air, or air pollution by the target's exhaust gases. Homing guidance may be active, semiactive, or passive.

homing range The maximum distance from a target or homing beacon at which a homing device is effective.

homing relay A stepping relay that returns to a specified starting position before each operating cycle.

homing station A station at which a beacon emits signals that may be used for homing.

homing torpedo A torpedo that has homing guidance for homing on a surface vessel or a submerged submarine.

homodyne reception A system of radio reception for suppressed-carrier systems of radiotelephony, in which the receiver generates a voltage that has the original carrier frequency and combines it with the incoming signal. Also called zero-beat reception.

homogeneous Having essentially uniform characteristics or composition.

homogeneous radiation Radiation that has an extremely narrow band of frequencies, or a beam of monoenergetic particles of a single type, so that all components of the radiation are alike.

homojunction A PN junction that is formed in a single type of semiconductor material.

homologous Having a similarity in characteristics, function, position, or structure.

homopolar generator A DC generator in which the poles presented to the armature are all of the

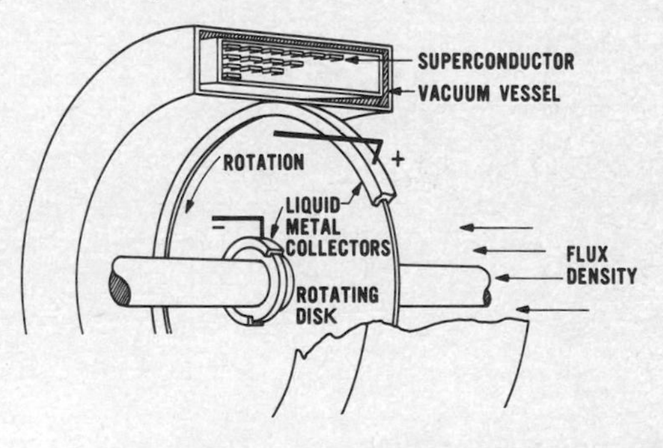

Homopolar generator using superconducting field coils immersed in liquid helium. DC output of 17 kA at 9 V is obtained from brushes of liquid sodium making contact with center and circumference of copper-disk rotor.

same polarity, so that the voltage generated in active conductors has the same polarity at all times. A pure direct current is thus produced without commutation. Also called acyclic, homopolar, or unipolar machine.

homopolar machine *Homopolar generator.*

honeycomb coil A coil wound in a crisscross manner to reduce distributed capacitance. Also called lattice-wound coil.

hood 1. An opaque shield placed above or around the screen of a cathode-ray tube or other display device to eliminate extraneous light. 2. A protective covering, usually ventilated to carry away dust, fumes, and gases, to provide a safe working position for handling dangerous chemicals or radioactive materials.

hook A circuit phenomenon occurring in four-zone transistors, wherein hole or electron conduction can occur in opposite directions to produce voltage drops that encourage other types of conduction.

hook transistor A transistor that has four alternating P- and N-type layers, with one layer floating between the base layer and the collector layer.

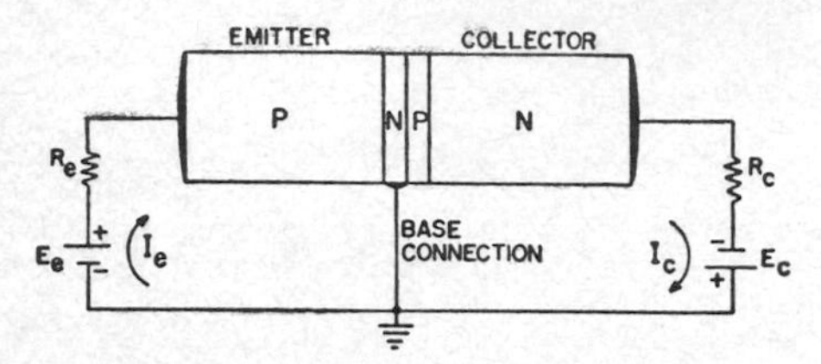

Hook-transistor construction and circuit.

This arrangement gives high emitter-input current gains. A PNPN transistor has a P-type floating layer, and an NPNP transistor has an N-type floating layer.

hookup An arrangement of circuits and apparatus for a particular purpose.

hookup wire Tinned and insulated solid or stranded soft-drawn copper wire used in making low-power circuit connections. The size is usually No. 18, 20, or 22.

hop A single reflection of a radio wave from the ionosphere back to the earth in traveling from one point to another. Usually used in such expressions as single-hop, double-hop, and multihop. The number of hops is called the order of reflection.

horizon The apparent junction of earth and sky as seen from a transmitting antenna site. The horizon bounds that part of the earth's surface which is reached by the direct wave of a radio station.

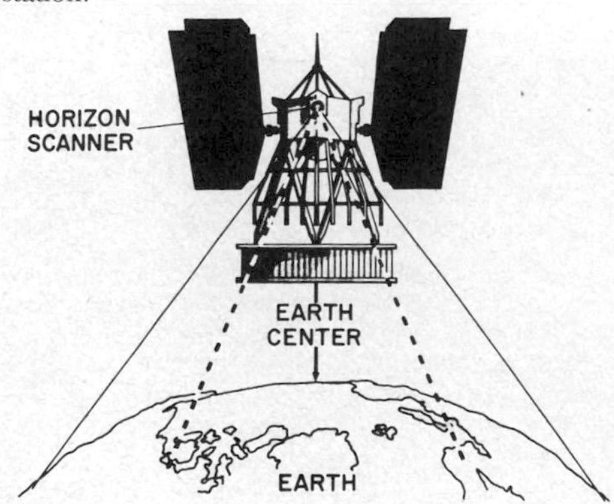

Horizon sensor on earth-orbiting weather satellite scans horizon in four directions to locate center of earth for reference.

horizon sensor A passive infrared device that detects the thermal discontinuity between the earth and space, to establish a stable vertical reference for control of the attitude or orientation of a missile or satellite in space. Thermistors serve as the infrared detectors.

horizontal amplifier An amplifier for signals used to deflect the electron beam horizontally in a cathode-ray tube.

horizontal blanking Blanking of a television picture tube during the horizontal retrace.

horizontal blanking pulse The rectangular pulse that forms the pedestal of the composite television signal between active horizontal lines. This pulse causes the beam current of the picture tube to be cut off during retrace. Also called line-frequency blanking pulse.

horizontal centering control The centering control provided in a television receiver or cathode-ray oscilloscope to shift the position of the entire image horizontally in either direction on the screen.

horizontal convergence control The control that adjusts the amplitude of the horizontal dynamic convergence voltage in a color television receiver.

horizontal definition *Horizontal resolution.*

horizontal deflection electrodes The pair of electrodes that moves the electron beam horizontally from side to side on the fluorescent screen of a cathode-ray tube that uses electrostatic deflection.

horizontal deflection oscillator The oscillator that produces, under control of the horizontal synchronizing signals, the sawtooth voltage waveform which is amplified to feed the horizontal deflection coils on the picture tube of a television receiver. Also called horizontal oscillator.

horizontal drive control The control in a television receiver, usually at the rear of the set, that adjusts the output of the horizontal oscillator. Also called drive control.

horizontal flyback Flyback in which the electron beam of a television picture tube returns from the end of one scanning line to the beginning of the next line. Also called horizontal retrace and line flyback.

horizontal frequency *Line frequency.*

horizontal hold control The hold control that changes the free-running period of the horizontal deflection oscillator in a television receiver, so that the picture remains steady in the horizontal direction.

horizontal linearity control A linearity control that permits narrowing or expanding the width of the left-hand half of a television receiver image, to give linearity in the horizontal direction so circular objects appear as true circles. Usually mounted at the rear of the receiver.

horizontal line frequency *Line frequency.*

horizontally polarized wave A linearly polarized wave whose electric field vector is horizontal.

horizontal oscillator *Horizontal deflection oscillator.*

horizontal output stage The television receiver stage that feeds the horizontal deflection coils of the picture tube through the horizontal output transformer. It may also include a part of the second-anode power supply for the picture tube.

horizontal output transformer A transformer used in a television receiver to provide the horizontal deflection voltage, the high voltage for the

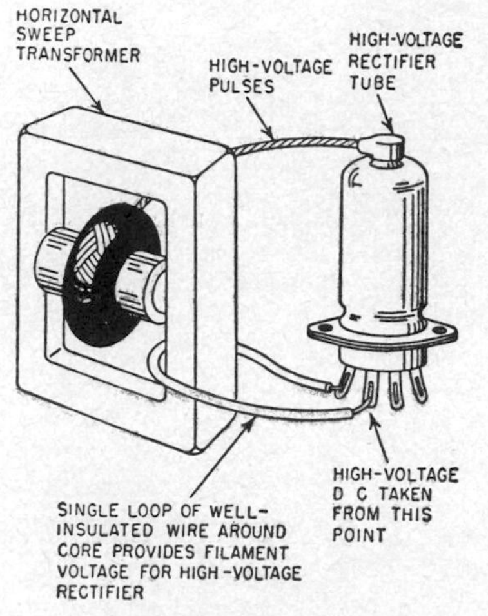

Horizontal output transformer.

second-anode power supply of the picture tube, and the filament voltage for the high-voltage rectifier. Also called flyback transformer and horizontal sweep transformer.

horizontal polarization Transmission of radio waves in such a way that the electric lines of force are horizontal and the magnetic lines of force are vertical. With this polarization, transmitting and receiving dipole antennas are placed in a horizontal plane. The U.S. television system uses horizontal polarization, whereas the British television system uses vertical polarization.

horizontal resolution The number of individual picture elements or dots that can be distinguished in a horizontal scanning line of a television or facsimile image. Usually determined by observing the wedge of fine vertical lines on a test pattern. Also called horizontal definition.

horizontal retrace *Horizontal flyback.*

horizontal-ring induction furnace An induction furnace in which a magnetic core links the output transformer primary winding with a ring-shaped horizontal trough in which the metal to be melted is placed.

horizontal scanning Rotation of a radar antenna in azimuth entirely around the horizon or in a sector of the horizontal plane.

horizontal sweep The sweep of the electron beam from left to right across the screen of a cathode-ray tube.

horizontal sweep transformer *Horizontal output transformer.*

horizontal synchronizing pulse The rectangular pulse transmitted at the end of each line in a television system, to keep the receiver in line-by-

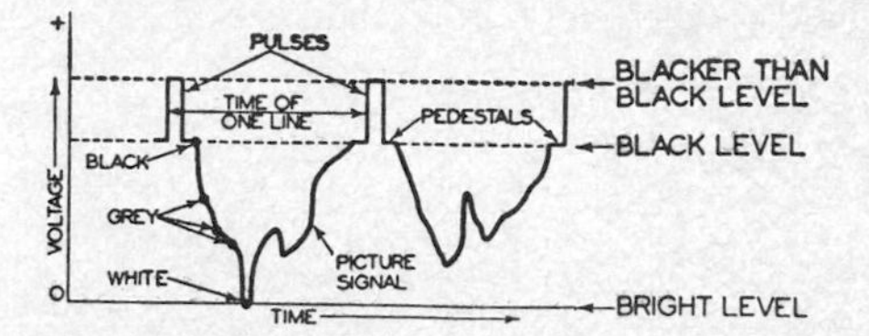

Horizontal synchronizing pulse occurs at end of each line of picture signal.

line synchronism with the transmitter. Also called line synchronizing pulse.

horizon tracker A device for establishing a vertical reference in a navigation system by precisely tracking the visible horizon.

horn 1. An acoustic device used with a loudspeaker driving element to improve the radiation of sound and achieve desired directional characteristics. The cross-sectional area increases progressively from the throat to the mouth. Also called acoustic horn. 2. An electromechanical or air-actuated signaling device.

horn antenna A microwave antenna produced essentially by flaring out the end of a circular or rectangular waveguide into the shape of a horn,

Horn antennas: rectangular, pyramidal, and conical.

for radiating radio waves directly into space. In a rectangular horn antenna, either one or both transverse dimensions increase linearly from the small end or throat to the mouth.

horn arrester A lightning arrester in which the spark gap has thick wire horns that spread outward and upward. The arc forms at the narrowest bottom part of the gap, travels upward, and extinguishes itself when it reaches the widest part of the gap.

horn feed A horn antenna used to feed a parabolic reflector in a radar antenna system.

horn loudspeaker A loudspeaker in which the radiating element is coupled to the air or another medium by means of a horn.

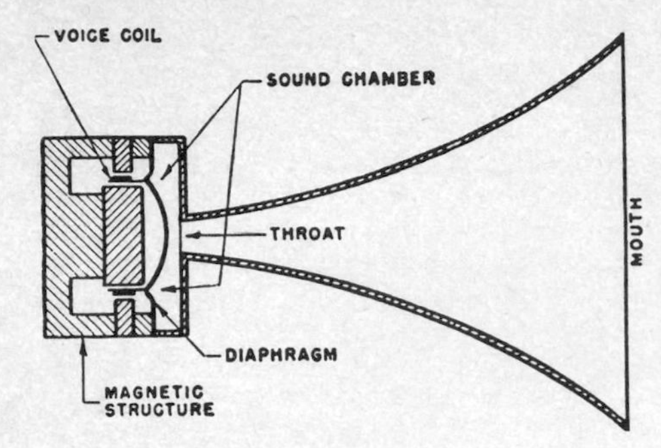

Horn loudspeaker.

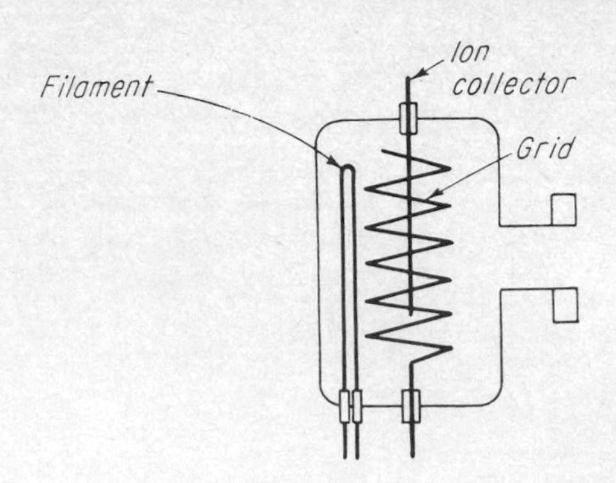

Hot-cathode ionization gage.

horn mouth The end of a horn that has the larger cross-sectional area.

horn throat The end of a horn that has the smaller cross-sectional area.

horsepower [abbreviated hp] A unit of power, equal to 746 W. Use of the SI unit of power, the watt, is preferred.

horseshoe magnet A permanent magnet or electromagnet in which the core is horseshoe-shaped or has parallel sides like a U, to bring the two poles near each other.

hot 1. Highly radioactive. 2. *Energized.*

hot-air soldering Soldering with a narrow blast of air whose temperature is closely controlled at the value required for soldering individual joints on printed-circuit boards.

hot atom An atom that has high energy as a result of a nuclear process like beta decay or neutron capture.

hot-atom chemistry The chemical reactions and properties of atoms that are in a high state of excitation or possess high kinetic energy as a result of nuclear processes.

hot carrier A carrier, which may be either an electron or a hole, that has relatively high energy with respect to the carriers normally found in majority-carrier devices like thin-film transistors. Hot carriers are injected either by emission over a potential barrier that exists at a metal-semiconductor junction or by tunneling through an extremely thin insulating layer.

hot-carrier diode *Schottky barrier diode.*

hot cathode A cathode in which electron or ion emission is produced by heat. Also called thermionic cathode.

hot-cathode ionization gage An ionization gage in which a heated filament or cathode is the source for the electrons that ionize the residual gas molecules. The resulting positive ions are collected by a negative electrode. The number of ions collected is proportional to gas pressure.

hot-cathode tube An electron tube that contains a hot cathode. Also called thermionic tube.

hot-cathode x-ray tube A high-vacuum x-ray tube in which the cathode is heated by current flow through a heater.

hot cell A heavily shielded enclosure in which radioactive materials can be handled remotely through the use of manipulators and viewed through shielded windows, with no danger to personnel.

hot electron An electron in excess of the thermal-equilibrium number and, for metals, that has an energy greater than the Fermi level. For semiconductors, the energy must be a definite amount above that of the edge of the conduction band. A hot electron (or hot hole) can be generated by photoexcitation, quantum-mechanical tunneling, minority-carrier injection across a forward-biased PN junction, high-field acceleration in nonmetallic materials, and Schottky emission over a forward-biased metal-semiconductor junction.

hot hole A hole that can move at much greater velocity than normal holes in a semiconductor. Hot holes occur when P-type silicon substrate is used with an epitaxial layer of P-type silicon.

hot laboratory A laboratory designed for research with radioactive materials which have such high strengths that special handling precautions are required.

hot line A direct wire or radio communication circuit between two points, available for immediate use without patching or switching. Communication satellite systems are used for hot lines linking the White House with the Kremlin.

hot plate A heated surface on which joints are brought to soldering temperature.

hot spot A surface area of higher than average radioactivity.

hot standby Standby equipment to which power is applied continuously, to make it ready for immediate operation.

hot-wire ammeter An ammeter in which alternating or direct current is measured by sending it through a fine wire. The resulting expansion or sag of the wire due to heat is used to deflect the meter pointer. Also called thermal ammeter.

hot-wire instrument An instrument that depends for its operation on the expansion by heat of a current-carrying wire.

hot-wire microphone A velocity microphone that depends for its operation on the change in resistance of a hot wire as the wire is cooled or heated by varying particle velocities in a sound wave.

Hound Dog An Air Force air-to-surface guided missile that has a range of about 400 km, capable of carrying a nuclear warhead, and intended for launching from long-range bombers of the Strategic Air Command.

hour [abbreviated h] A unit of time, equal to 3600 s. Use of the SI unit of time, the second, is preferred.

Housekeeper seal A vacuumtight seal made between copper and glass by bringing the copper to

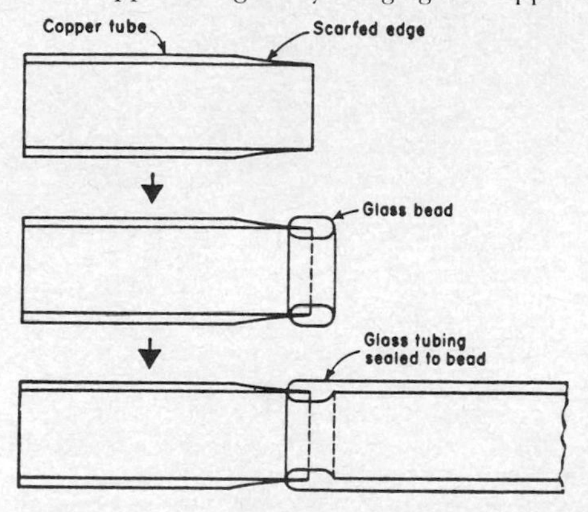

Housekeeper seal for joining copper and glass tubing.

a flexible feather edge before fusing it to the glass. The copper then flexes as the glass shrinks during cooling.

housekeeping The series of computer operations required prior to a processing run, such as the setting up of constants, variables, limits, storage locations, and any other preliminary steps that do not contribute directly to the solution of a problem.

howl An undesirable prolonged sound produced by a radio receiver or AF amplifier system, caused by electric or acoustic feedback.

howler 1. An electromechanical or other device that produces a loud attention-getting audio tone. Used in some radar and sonar stations to warn the operator that targets are appearing on the screen. 2. A signal generator used in telephone systems to feed a loud tone to the line of a subscriber whose receiver has not been correctly replaced after a call.

howling *Acoustic feedback.*

hp Abbreviation for *horsepower.*

H pad *H network.*

***h* parameter** One of a set of four transistor equivalent-circuit parameters that conveniently specify transistor performance for small voltages and currents in a particular circuit. Also called hybrid parameter.

H scope A radarscope that produces an H display.

H system A radar navigation system that uses two ground radar beacons in conjunction with airborne equipment which gives the direction and distance to each beacon. The principal H systems are Gee-H, micro-H, Rebecca-H, and shoran.

HTL Abbreviation for *high-threshold logic.*

hue The name of a color, such as red, yellow, green, blue, or purple, corresponding to the dominant wavelength. White, black, and gray are not considered hues.

hue control A control that varies the phase of the chrominance signals with respect to that of the burst signal in a color television receiver, to change the hues in the image. Also called phase control.

huff-duff [from pronunciation of first letters of High-Frequency Direction Finder] A direction finder that indicates the direction of arrival of high-frequency radio signals on the screen of a cathode-ray tube.

hum 1. An electric disturbance occurring at the power-supply frequency or its harmonics, usually 60 or 120 Hz in the United States and 50 or 100 Hz in Great Britain. An example is ripple. 2. A sound produced by an iron core of a transformer because of loose laminations or magnetostrictive effects. The frequency of the sound is twice the power-line frequency.

human engineering Engineering that involves adaptation of equipment designs to best meet the physical and psychological requirements of a human operator.

hum balancer An adjustable resistor connected between the heater leads of a tube, with the adjustable center tap connected to ground.

hum bar A dark horizontal band extending across a television picture, caused by excessive hum in the video signal applied to the input of the picture tube.

hum-bucking coil A coil wound on the field coil of an excited-field loudspeaker and connected in series opposition with the voice coil, so that hum voltage induced in the voice coil is canceled by that induced in the hum-bucking coil.

humidity detector A detector that opens or closes a switch when the amount of moisture in the atmosphere reaches a preset value. The sensing element may be moisture-absorbing paper, stretched human hairs, stretched nylon fibers, or

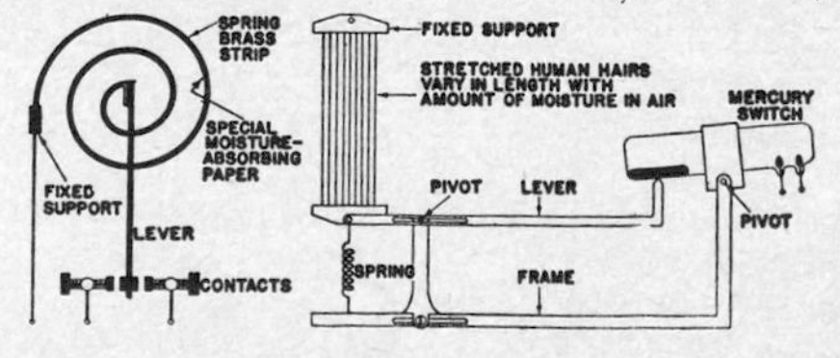

Humidity detectors.

any other material having a characteristic that changes with humidity in a known and repeatable manner.

hum modulation Modulation of an RF signal or detected AF signal by hum. This type of hum is heard in a radio receiver only when a station is tuned in.

hum slug A copper ring placed around the core of an excited-field loudspeaker adjacent to the voice coil, to serve as a single shorted turn for hum currents induced by the field coil.

hunting Undesirable oscillation of an automatic control system, wherein the controlled variable swings on both sides of the desired value.

Hurter and Driffield curve *H and D curve.*

Huygens principle Every point on an advancing wavefront acts as a source that sends out new waves. The combined effect is propagation of the wave as a whole.

HVDC Abbreviation for *high-voltage direct current.*

H vector A vector that represents the magnetic field of an electromagnetic wave. In free space it is perpendicular to the E vector and the direction of propagation.

H wave British term for *transverse electric wave* (TE wave).

hybrid 1. Having two or more different characteristics or types of structure. 2. *Hybrid junction.*

hybrid circuit A circuit in which two or more basically different types of components that perform similar functions are used together. A circuit that uses both tubes and transistors is an example.

hybrid coil *Hybrid transformer.*

hybrid computer A computer that handles both analog and digital input data. Also called analog-digital computer.

hybrid electromagnetic wave [abbreviated HEM wave] An electromagnetic wave that has components of both the electric and magnetic field vectors in the direction of propagation.

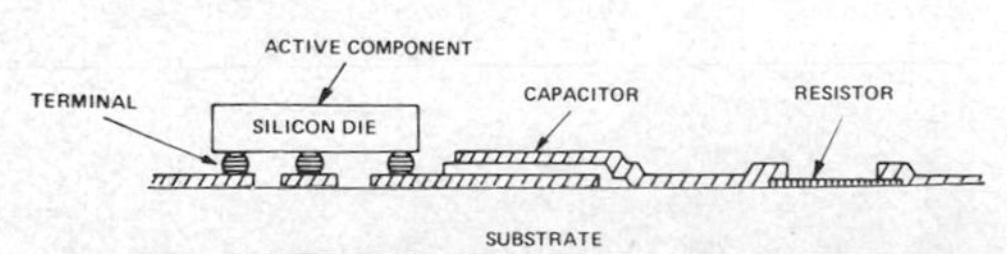

Hybrid integrated circuit constructed on insulating substrate.

hybrid integrated circuit A circuit in which one or more discrete components are used in combination with integrated-circuit construction.

hybrid junction A transformer, resistor, or waveguide circuit or device that has four pairs of terminals so arranged that a signal entering at one terminal pair will divide and emerge from the two adjacent terminal pairs but will be unable to reach the opposite terminal pair. Also called bridge hybrid and hybrid.

hybrid microcircuit A circuit in which one type of microcircuit is combined with some other type of microcircuit or with discrete components.

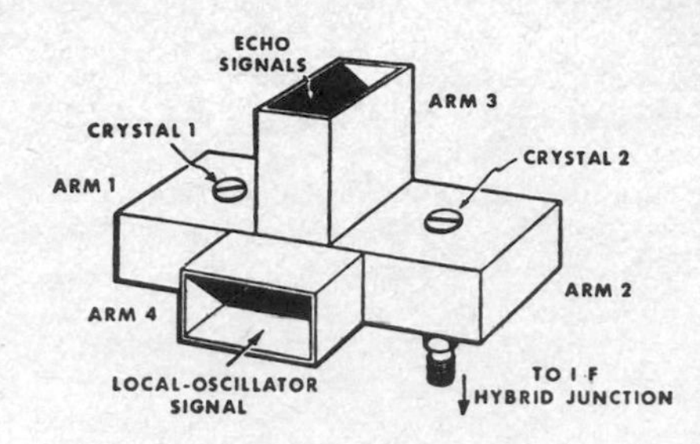

Hybrid junction for rectangular waveguides, serving as balanced mixer for radar receiver.

hybrid parameter *H parameter.*

hybrid reflectometer A reflectometer that uses highly stable rectangular-waveguide components

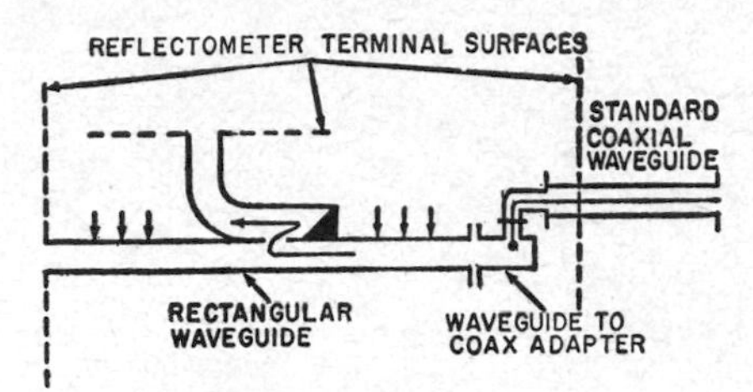

Hybrid reflectometer.

with an adapter for measuring impedance and standing-wave ratio in coaxial systems.

hybrid relay A relay in which solid-state elements are combined with moving contacts.

hybrid repeater *Hybrid transformer.*

hybrid ring A doughnut-shaped waveguide that acts as a hybrid junction for four waveguide sections. Coaxial lines are sometimes used in place of waveguides. Also called rat race.

hybrid set Two or more transformers interconnected to form a hybrid junction. Also called transformer hybrid.

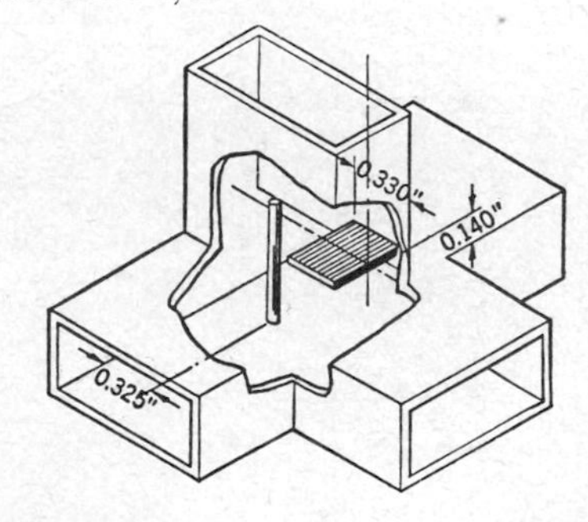

Hybrid T for waveguide, showing use of post and iris inside for matching.

hybrid T A microwave hybrid junction composed of an E-H T junction with internal matching elements. It is reflectionless for a wave propagating into the junction from any arm when the other three arms are match-terminated.

hybrid transformer A single transformer that performs the essential function of a hybrid set. Also called hybrid coil and hybrid repeater.

hydrodynamic oscillator A transducer for generating sound waves in fluids, in which a continuous flow through an orifice is modulated by a reciprocating valve system controlled by acoustic feedback.

hydrogen [symbol H] A gaseous element that has the simplest known atom, consisting of only one proton and one electron. Atomic number is 1. Its three isotopes are ordinary or light hydrogen, deuterium, and tritium.

hydrogen bomb A nuclear bomb in which heavy hydrogen nuclei fuse under intense heat and pressure to form lighter helium nuclei. An atomic bomb surrounded by lithium deuteride gives this reaction. The resulting unused mass is converted into energy and released in this nuclear fusion process.

hydrogen laser A molecular gas laser in which hydrogen is used to generate coherent wavelengths near 0.6 μm in the vacuum ultraviolet region. Peak output power above 100 kW has been obtained.

hydrogen line Monochromatic 1.42-GHz radiation from atoms of hydrogen. To date this is the only known single-frequency natural radiation in the RF portion of the electromagnetic spectrum. This line originates from cool hydrogen in the galaxy and allows radio astronomers to determine directly the velocity properties of any observed cloud of cool hydrogen by means of the Doppler effect.

hydrogen maser A gas maser in which hydrogen gas is the basis for providing an output signal that has a high degree of stability and spectral purity. The resonator frequency is 1.420 405 751 786 4 GHz.

hydrogen-oxygen fuel cell A fuel cell in which the hydrogen fuel is fed into the anode side of the cell under pressure and forced into the pores of the electrode, where it reacts with the catalytic agent and the electrolyte to release electrons to the electrode. These electrons flow through an external circuit to perform useful work and then return to the cathode side of the cell. The by-product of this reaction is water.

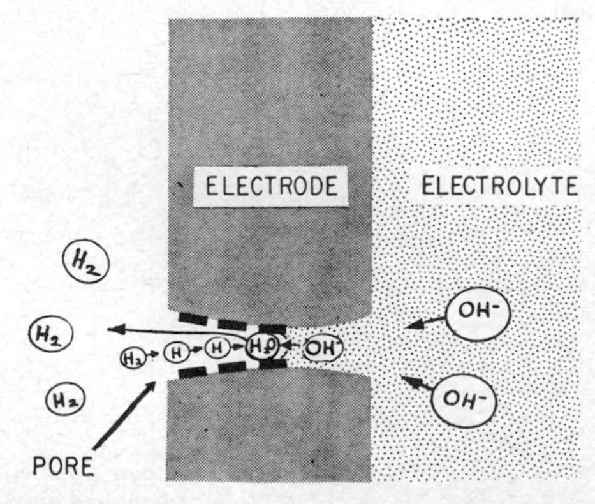

Hydrogen-oxygen fuel cell operating principle.

hydrogen thyratron A thyratron that contains hydrogen, used in radar pulse circuits to provide high peak currents at high anode voltages. Use of hydrogen instead of mercury vapor gives freedom from effects of changes in ambient temperature.

hydrometer A direct-reading instrument that has a graduated float whose position in a liquid is determined by the density or specific gravity of the liquid. Used to measure the state of charge of storage batteries.

hydronic radiation A form of underwater electromagnetic radiation used for communication under water. Voice communication has been achieved between scuba divers 250 m apart, and signals have been transmitted 50 km.

hydrophone An electroacoustic transducer that responds to waterborne sound waves and delivers essentially equivalent electric waves. Used chiefly to detect the approach of submarines and other craft.

hydrophotometer A photometer that measures light transmission through a fixed distance in sea water.

hydrostatic pressure *Static pressure.*

hygrometer An instrument that measures the humidity of the atmosphere.

hygroscopic Tending to absorb moisture.

hyperacoustic zone The region in the upper atmosphere, between 60 and 100 mi (100 and 160 km) above the earth, where the distance between the rarefied air molecules roughly equals the wavelength of sound, so that sound is transmitted with less volume than at lower levels. Above this zone sound waves cannot be propagated.

hyperbolic DOVAP A navigation system that consists of four or more DOVAP stations which have a common reference signal that is not coherent with the interrogation signal.

hyperbolic-field multiplier A multiplier in which two input variables are applied to a special cathode-ray tube that has hyperbolic deflection plates. A feedback arrangement keeps the electron beam on the vertical axis of the screen and at the same time provides a voltage proportional to the product of the two variables.

hyperbolic flareout A flareout obtained by changing the glide slope from a straight line to a hyperbolic curve at an appropriate distance from touchdown at an airport.

hyperbolic guidance Missile guidance in which the difference in the arrival times of radio signals transmitted simultaneously from two ground stations is used to control the position of the missile.

hyperbolic horn A horn whose equivalent cross-sectional radius increases according to a hyperbolic law.

hyperbolic navigation system Any system of radio navigation in which the navigating aircraft receives synchronized signals from at least three known points, as in Decca, Gee, loran, omega, and shoran. The time difference between signals received from any two stations determines a line of position in the shape of a hyperbola that has the two transmitters at its foci. Crossing this line of position with another line of position, obtained by using at least one other transmitting station, establishes a fix.

hyperfine interaction Magnetic interaction between electrons and nuclei. The effect is called hyperfine because it is really a very small interaction compared to electrostatic interaction between the charges of electrons and nuclei.

hyperfine structure A set of closely spaced lines making up a spectral line. Also called isotope structure.

hyperfragment An unstable neutral hyperon particle produced in a high-energy collision.

hyperfrequency wave Unofficial designation for microwaves that have wavelengths in the range from 1 cm to 1 m.

hyperon An unstable particle whose mass is between that of a neutron and a deuteron.

hypersonic Having a velocity greater than five times the speed of sound in air (greater than Mach 5).

hypersonics The branch of acoustics that deals with acoustic waves and vibrations at frequencies above 500 MHz. Ultrasonics covers lower frequencies down to about 20 kHz.

hypervisor A computer hardware-software technique that allows multiprogramming of two different operating systems.

hypoxia alarm An alarm that senses when the oxygen level falls below the normal level required for human breathing in aircraft and spacecraft. One version uses a small electrochemical cell as the sensor.

hysteresigraph An instrument that automatically measures and draws the hysteresis curve for a specimen of magnetic material.

hysteresis 1. An effect similar to internal friction, occurring in a material that is subjected to a varying electric field, magnetic field, or physical strain. The internal friction of the molecules causes heating of the material. With magnetic and electric hysteresis there is a lag between cause and effect, so that a measured value is different for a cause which increases to a final value than for a cause which decreases to the same final value. 2. A temporary change in the counting-rate-voltage characteristic of a radiation-counter tube, caused by its previous operation. 3. An oscillator effect wherein a given value of an operating parameter may result in multiple values of output power and/or frequency.

hysteresis curve A curve that shows the steady-state relation between the magnetic induction in a material and the steady-state alternating magnetic intensity that produces it.

hysteresis distortion Distortion that occurs in circuits which contain magnetic components, due to nonlinearity caused by hysteresis.

hysteresis error The maximum separation due to hysteresis between upscale-going and downscale-going indications of a measured variable.

hysteresis heater An induction heater in which a ferrous charge or charge container is heated principally by hysteresis losses caused by varying magnetic flux in the magnetic material. In normal induction heating, the heat is due to eddy-current losses.

hysteresis loop A curve that shows, for each value of magnetizing force, two values of the magnetic flux density in a cyclically magnetized material: one when the magnetizing force is increasing, the other when it is decreasing.

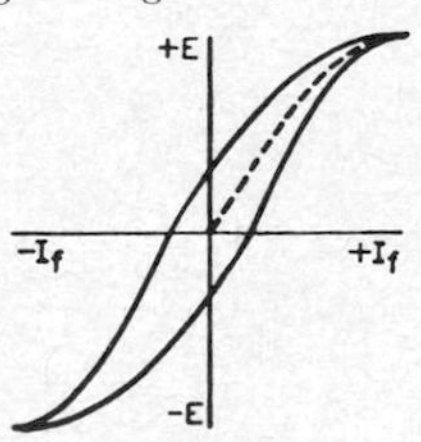

Hysteresis loop.

hysteresis loss The power loss in an iron-core transformer or other AC device because of magnetic hysteresis.

hysteresis motor A small synchronous motor sometimes used for light constant-speed duty, as for phonograph motors. It starts by virtue of the hysteresis losses induced in its hardened steel secondary member by the revolving field of the primary .

Hz Abbreviation for *hertz.* Formerly called cycle per second.

I

I [from intensity] Symbol for *current.*

IAGC Abbreviation for *instantaneous automatic gain control.*

IAT Abbreviation for *international atomic time.*

IC 1. Abbreviation for *integrated circuit.* 2. Symbol for *internal connection.* Used on tube-base diagrams.

ICAO Abbreviation for *International Civil Aviation Organization.*

ICBM Abbreviation for *intercontinental ballistic missile.*

I channel The 1.5-MHz-wide channel used in the American (NTSC) color television system for transmitting cyan-orange color information. The signals in this channel are known as I signals.

ICI Abbreviation for *International Commission on Illumination.*

iconoscope A television camera tube in which a beam of high-velocity electrons scans a photoemissive mosaic that is capable of storing an electric

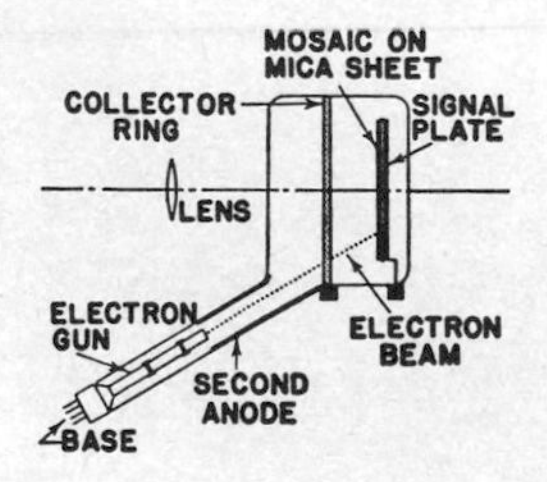

Iconoscope construction.

charge pattern corresponding to an optical image focused on the mosaic. The mosaic consists of globules of light-sensitive material on a mica sheet that has a conducting film on its back surface. A small value of capacitance thus exists between each globule and the metal film. Each globule emits electrons in proportion to the light on it, thus producing charges on the capacitances between globules. The electron beam discharges each capacitance in turn during scanning of the mosaic. The resulting variations in the current taken from the metal film form the desired output signal. Now obsolete.

ICW Abbreviation for *interrupted continuous wave.*

ideal bunching A theoretical condition in which the bunching of electrons in a velocity-modulation tube is such that all electrons in a bunch have the same velocity and phase, corresponding to an infinitely large current peak.

ideal junction A connection between two impedances or transmission lines such that the electric effects of the junction can be neglected.

ideal noise diode A diode that has an infinite internal impedance and in which the current exhibits full shot-noise fluctuations.

ideal permeability The value of permeability obtained by superimposing a large alternating magnetizing force on the desired magnetizing force, then slowly reducing the alternating component to zero.

ideal rectifier A rectifier in which the back conductance, forward resistance, and capacitance are zero.

ideal transducer A hypothetical passive transducer that transfers the maximum possible power from the source to the load, and therefore has no losses.

ideal transformer A hypothetical transformer that neither stores nor dissipates energy, has unity coefficient of coupling, and has pure inductances of infinitely great value.

I demodulator The demodulator in which the chrominance signal and the color-burst oscillator signal are combined to recover the I signal in a color television receiver.

identification The process of determining the identity of a particular displayed target or determining which blip represents a specific target in radar.

identification, friend or foe [abbreviated IFF] A beacon system used in World War II to identify a friendly aircraft when picked up by radar. A coded interrogation signal triggered the beacon carried by friendly aircraft, initiating automatic transmission of a properly coded identification signal. A modern version is now used worldwide to track aircraft and determine their altitudes for air traffic control purposes. The new system is known as an air traffic radar-beacon system, with all interrogation at 1.03 GHz and all replies at 1.09 GHz.

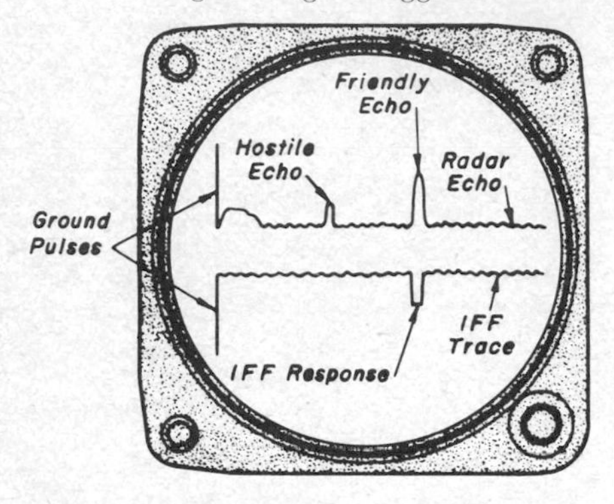

Identification, friend or foe display on radar screen.

I display A radarscope display in which a target appears as a complete circle when the radar antenna is correctly pointed at it, the radius of the circle being proportional to target distance. When the antenna is not aimed at the target, the circle reduces to a segment of a circle. The segment length is then inversely proportional to the magnitude of the pointing error, and its angular position is reciprocal to the direction of the pointing error.

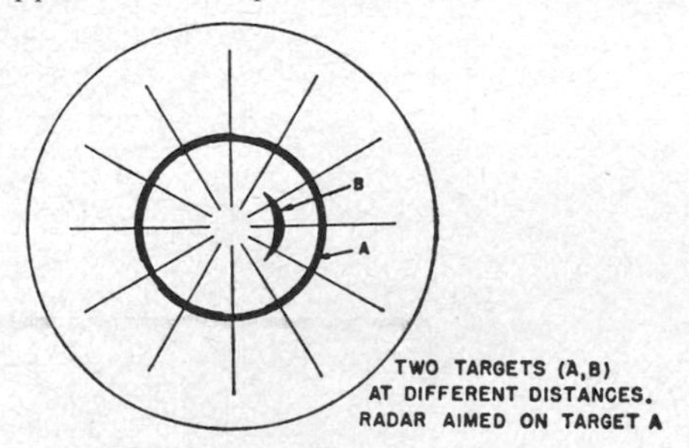

I display.

idler An intermediate rubber-tired wheel used to transmit power from a motor shaft to a driven shaft, as from a phonograph drive motor shaft to the rim of the turntable.

IDP Abbreviation for *integrated data processing.*

IEC Abbreviation for *International Electrotechnical Commission.*

IEEE Abbreviation for *Institute of Electrical and Electronics Engineers.*

IF Abbreviation for *intermediate frequency.*

IF amplifier The section of a superheterodyne receiver that amplifies signals after they have been converted to the fixed IF value by the frequency converter. It is located between the frequency converter and second detector.

IFF Abbreviation for *identification, friend or foe.*

IF harmonic interference Interference due to acceptance of harmonics of an IF signal by RF circuits in a superheterodyne receiver.

IFR Abbreviation for *instrument flight rules.*

IFR conditions Weather conditions below the minimum specified for visual flight rules, in which instrument flight rules apply.

IF rejection The ability of an FM radio receiver to reject signals of government and commercial stations operating at or near the IF value of the receiver.

IF response ratio The ratio of the field strength at a specified frequency in the IF band to the field strength at the desired frequency, each field being applied in turn, under specified conditions, to produce equal outputs.

IF signal A modulated or continuous-wave signal whose frequency is the IF value of a superheterodyne receiver and is produced by frequency conversion before demodulation. This value is usually 455 kHz for a broadcast-band radio receiver, 10.7 MHz for an FM radio receiver, approximately 45 MHz for a television receiver picture channel, and 4.5 MHz for a television receiver sound channel.

IF stage One of the stages in the IF amplifier of a superheterodyne receiver.

IF transformer The transformer used at the input and/or output of an IF amplifier stage in a superheterodyne receiver for coupling purposes and to provide selectivity.

IGFET Abbreviation for *insulated-gate field-effect transistor,* now known as *metal-oxide semiconductor field-effect transistor* [abbreviated MOSFET].

ignition interference Interference caused by the spark discharges in an automotive or other ignition system.

ignitor 1. An electrode used to initiate and sustain

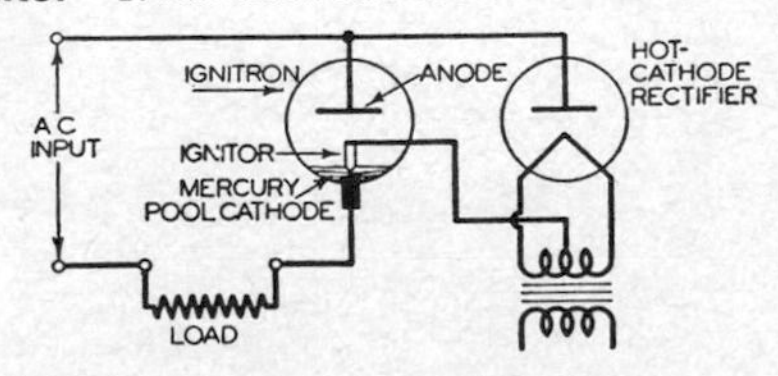

Ignitor circuit for ignitron.

the ignitor discharge in a switching tube. Also called keep-alive electrode (deprecated). 2. A pencil-shaped electrode made of carborundum or some other conducting material that is not wetted by mercury, partly immersed in the mercury-pool cathode of an ignitron and used to initiate conduction at the desired point in each AC cycle.

ignitron A single-anode pool tube in which an ignitor electrode is used to initiate the cathode

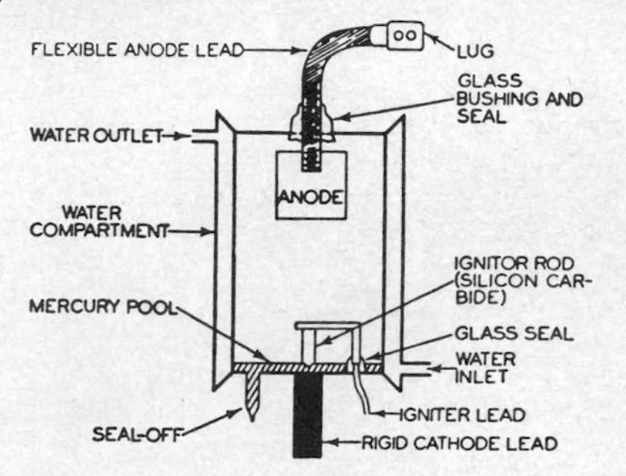

Ignitron construction, showing mercury pool that serves as cathode.

spot on the surface of the mercury pool before each conducting period.

ignitron rectifier A high-power rectifier that consists of a number of large metal-tank ignitrons, used in industrial plants like aluminum refineries and steel mills.

ignore A punched-tape code or computer instruction which indicates that no action should be taken by a computer or other device. In the Teletype and Flexowriter codes, punching of all holes is an ignore.

IHF Abbreviation for *Institute of High Fidelity.*

IHF power *Music power.*

IHF standard One of the standards adopted by the Institute of High Fidelity for specifying performance of audio amplifiers, FM tuners, and other high-fidelity equipment.

ILF Abbreviation for *infralow frequency.*

illegal character A character or combination of bits that is not accepted as a valid representation by a computer or a specific routine. Illegal characters are commonly detected and used as an indication of a machine malfunction.

illuminance Deprecated term for *illumination.*

illuminant C The reference white of color television. It closely matches average daylight.

illumination 1. The geometric distribution of power that reaches various parts of a dish reflector in an antenna system. 2. The power distribution to elements of an antenna array. 3. The density of the luminous flux falling on a surface. It is equal to the flux divided by the area of the surface when the latter is uniformly illuminated. The SI unit is the lux. Also called illuminance (deprecated) and luminous flux density.

illumination control A photoelectric control that turns on room lights when outdoor illumination decreases below a predetermined level.

illumination meter An instrument for measuring the illumination on a surface directly in lux (lumen per square meter). Also called illuminometer.

illumination sensitivity The signal output current divided by the incident illumination on a camera tube or phototube.

illuminometer *Illumination meter.*

ILS Abbreviation for *instrument landing system.*

IM Abbreviation for *intermodulation.*

image 1. An optical counterpart of an object, as a real image or virtual image. 2. A fictitious electrical counterpart of an object, as an electric image or image antenna. 3. The scene reproduced by a television or facsimile receiver.

image admittance The reciprocal of image impedance.

image antenna A fictitious electrical counterpart of an actual antenna, acting mathematically as if it

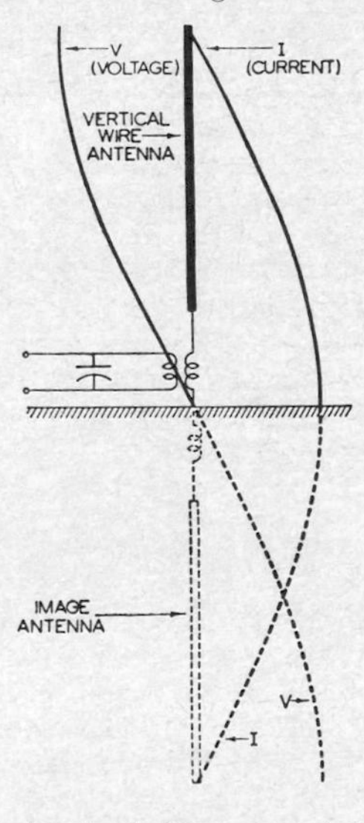

Image antenna for simple Marconi vertical-wire antenna, with voltage and current distribution curves.

existed in the ground directly under the real antenna and served as the direct source of the wave that is reflected from the ground by the actual antenna.

image converter A converter that uses optical fibers to change the form of an image, for more convenient recording and display or the coding of secret messages.

image-converter tube An electron tube that reproduces on its fluorescent screen an image of the infrared, x-ray, or other type of radiant image which is projected onto its photosensitive surface. Also called image tube.

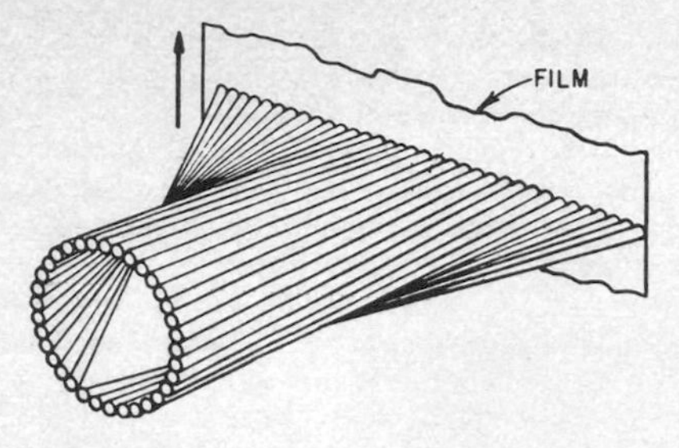

Image converter for converting circular input pattern to line pattern for recording on upward-moving film by optical fibers.

image deblurring The sharpening of photographs that have been unintentionally blurred by movement, atmospheric turbulence, improper focusing, or imperfect instruments. Optical computers involving lasers and holography may be used for this purpose.

image dissector A television camera tube in which the scene to be transmitted is focused on a light-sensitive surface, each point of which emits electrons in proportion to incident light. The resulting broad beam of electrons is drawn down the tube by a positive anode. Magnetic fields produced

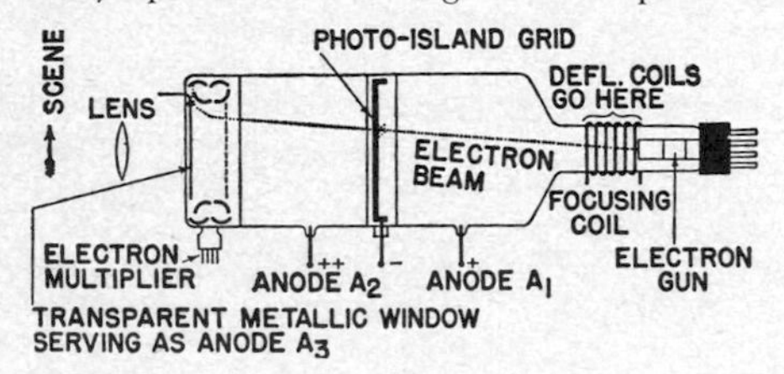

Image-dissector construction details.

by coils keep the electron image in focus as they sweep it in a scanning motion past an aperture opening into an electron multiplier. The output voltage of the electron multiplier is thus proportional at each instant to the brightness of an elemental area of the scene being scanned in orderly sequence. Also called dissector tube.

image distortion Failure of the reproduced image in a television receiver to appear the same as that scanned by the television camera.

image enhancement A method of improving color television pictures by comparing each video line, element by element, with the preceding and following lines. Any differences between vertically aligned elements are added to the middle-line element in the proper phase to enhance picture outlines and contrast.

image frequency An undesired carrier frequency that differs from the frequency to which a superheterodyne receiver is tuned by twice the intermediate frequency. Thus, if the oscillator frequency is higher than that of the desired incoming carrier, the image frequency will be the same amount lower than the oscillator frequency. If the receiver has poor selectivity, a strong image-frequency signal can get through its tuning circuits to beat with the local oscillator and produce the correct intermediate frequency.

image-frequency rejection ratio The ratio of the response of a superheterodyne receiver at the desired frequency to the response at the image frequency.

image iconoscope A camera tube in which an optical image is projected on a semitransparent photocathode, and the resulting electron image emitted from the other side of the photocathode is focused on a separate storage target. The target is scanned on the same side by a high-velocity electron beam, neutralizing the elemental charges in sequence to produce the camera output signal at the target. Sensitivity is much higher than for an iconoscope.

image impedance One of the impedances that, when connected to the input and output of a transducer, will make the impedances in both directions equal at the input terminals and at the output terminals. The load impedance and the equivalent internal impedance of the transducer are then images of each other. This condition gives maximum power transfer. When the two image impedances are equal, their value is the same as the characteristic impedance.

image intensifier 1. A solid-state optoelectronic amplifier that is capable of increasing the intensity of a radiant image. 2. An electron tube used in fluoroscopy; it contains a special screen from which electrons are released by x-rays. These electrons are accelerated and focused onto a fluores-

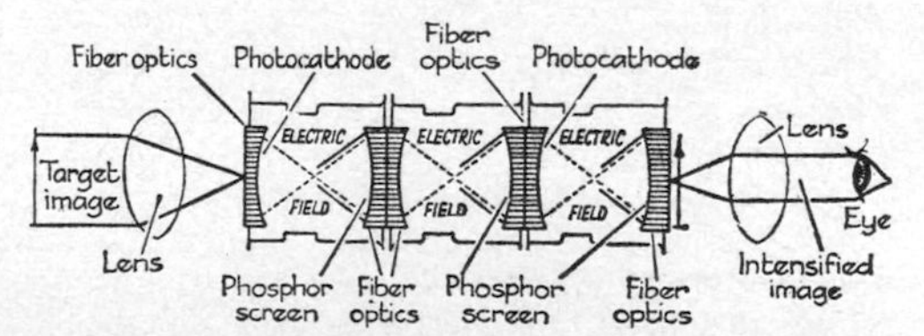

Image intensifier used in Army passive night-vision telescope for observing enemy movements by moonlight or even starlight. Faint light collected by glass lens at left passes through optical fibers to photocathode surface, casing emission of electrons. These are accelerated by electric field as they move to phosphor screen on front surface of next optical-fiber bundle. Process is repeated twice more to give intensified image at right.

cent screen, giving an image much brighter than would be produced by direct action of the x-ray beam on the fluorescent screen.

image interference Interference that occurs in a superheterodyne receiver when a station broadcasting on the image frequency is received along with the desired station. Image interference can occur when the receiver circuits do not have sufficient selectivity to reject the image-frequency signal. For a standard radio broadcast receiver, the image frequency would be 910 kHz higher than that of the desired station.

image-interference ratio A superheterodyne receiver rating that indicates the effectiveness of the preselector in rejecting signals at the image frequency.

image isocon A camera tube similar to an image orthicon but responsive to much lower light levels, including near darkness.

image orthicon A camera tube in which an electron image is produced by a photoemitting surface and focused on one side of a separate storage target that is scanned on its opposite side by an

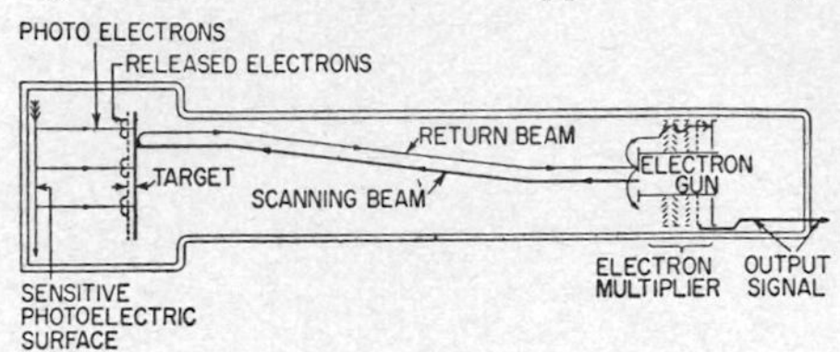

Image-orthicon construction and operation. Camera lens produces optical image shown as feathered arrow on photoelectric surface at left.

electron beam, usually consisting of low-velocity electrons. It has such high sensitivity that images can be picked up even in semidarkness. Both color and monochrome versions are widely used in television broadcasting studios.

image rejection Suppression of signals at the image frequency in a superheterodyne receiver.

image response The response of a superheterodyne receiver to an undesired signal at its image frequency.

image-storage array A solid-state panel or chip in which the image-sensing elements may be a metal-oxide semiconductor or a charge-coupled or other light-sensitive device that can be manufactured in a high-density configuration.

image-storage panel An image-storage array constructed as a thin, flat panel.

image-storage tube A storage tube into which the information is introduced by means of radiation, usually light, and read later as a visible output.

image tube *Image-converter tube.*

immersion scanning Ultrasonic scanning in which the ultrasonic transducer and the object being scanned are both immersed in water or some other liquid that provides good coupling while the transducer is being moved around the object.

immittance [IMpedance and adMITTANCE] A term to denote both impedance and admittance, as commonly applied to transmission lines, networks, and certain types of measuring instruments.

impact ionization Ionization produced by the impact of a high-energy charge carrier on an atom of semiconductor material. The effect is an increase in the number of charge carriers.

impact microphone An instrument that picks up the vibration produced when one object hits another. Used on space probes to record the impact of small meteoroids.

impact-noise analyzer An analyzer used with a sound-level meter to evaluate the characteristics of impact-type sounds and electric noise impulses that cannot be measured accurately with a noise meter alone. Applications include assessment of possible damage to hearing by punch presses, pile drivers, pneumatic drills, and office equipment.

impact predictor A computer that takes information from a missile tracking system and continuously computes the point (in real time) at which the missile will strike the earth. Based on the assumption that missile power is shut off at that instant and the remaining trajectory is ballistic in nature.

impact printer A line printer that has one or more character fonts, a ribbon or other inking device, a paper transport, and some means of impacting desired characters or character elements on the paper. Impact printers include chain and drum printers that print on-the-fly on either the front or back of the paper. Some printers use carbon paper interleaved between plain sheets to produce multiple copies; other printers use multicopy forms that have encapsulated inks on the paper itself. Operating speeds can be up to 2000 lines per minute. Dot-matrix printers can also be impact printers, with characters formed by impact of stiff wires selected from a 5 × 7 or 7 × 9 matrix.

IMPATT amplifier A diode amplifier that uses an IMPATT diode. Operating frequency range is from about 5 to 100 GHz, primarily in the C and X bands, with power output up to about 20 W continuous wave or 100 W pulsed.

IMPATT diode [IMPact Avalanche Transit-Time diode] A solid-state microwave diode that has a negative-resistance characteristic produced by a

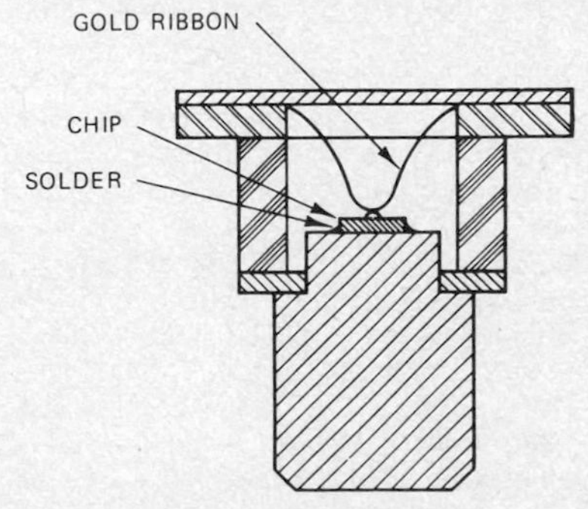

IMPATT diode chip in hermetically sealed package, with copper stud at bottom serving as terminal and heatsink. Other terminal is at top.

combination of impact avalanche breakdown and charge-carrier transit-time effects in a thin chip which is usually gallium arsenide or silicon. When suitably mounted in a tuned cavity or waveguide, the diode can be used as either an oscillator or an amplifier operating in the gigahertz range.

impedance [symbol Z] The total opposition offered by a component or circuit to the flow of an alternating or varying current. Impedance Z is expressed in ohms and is a combination of resistance R and reactance X, computed as the square root of the sum of the squares of R and X. Impedance is also computed as $Z = E/I$, where E is the applied AC voltage and I is the resulting alternating current flow in the circuit. In computations, impedance is handled as a complex ratio of voltage to current.

impedance bridge A bridge used to measure or compare impedances that may contain capacitance, inductance, and resistance.

impedance characteristic A graph in which the impedance of a circuit is plotted against frequency.

impedance coupling Coupling of two signal circuits with an impedance, usually a choke. Used in audio amplifiers when high gain and limited bandpass are required.

impedance feedback Use of a passive impedance network to provide feedback from the output terminal of an operational amplifier to the input summing junction.

impedance match The condition in which the impedance of a connected load is equal to the internal impedance of the source or the surge impedance of a transmission line, thereby giving maximum transfer of energy from source to load, minimum reflection, and minimum distortion.

impedance-matching network A network of two or more resistors, coils, and/or capacitors to couple two circuits so that the impedance of each circuit will be equal to the impedance into which it looks.

impedance-matching transformer A transformer used to obtain an impedance match between a given signal source and a load that has a different impedance than the source.

impedance triangle A diagram that consists of a right-angle triangle, with sides proportional to the resistance and reactance, respectively, of an AC circuit. The hypotenuse then represents the impedance of the circuit. The cosine of the angle between resistance and impedance lines is equal to the power factor of the circuit.

impedor A term sometimes used in place of impedance to describe a circuit element that has impedance.

imperfection Any deviation in structure from that of an ideal crystal.

implant A quantity of radioactive material in a suitable container, to be embedded in a tissue for therapeutic purposes.

implanted atom An atom introduced into semiconductor material by ion implantation.

implanted device 1. A heart pacemaker or other medical electronic device that is surgically placed in the body. 2. A resistor or other device that is fabricated within a silicon or other semiconducting substrate by ion implantation.

implode To burst inward.

implosion The inward collapse of an evacuated container, such as the glass envelope of a cathode-ray tube.

implosion weapon A nuclear weapon in which a quantity of fissionable material, less than a critical mass at ordinary pressure, has its volume suddenly reduced by compression with ordinary explosives, so it becomes supercritical and a nuclear explosion can occur.

impregnated Having spaces filled with a substance with good insulating properties.

impregnated coil A coil in which the spaces between the turns of insulated wire are filled with an insulating varnish or plastic material.

improvement threshold The value of carrier-to-noise ratio below which the signal-to-noise ratio decreases more rapidly than the carrier-to-noise ratio.

impulse 1. A pulse so short that it may be regarded mathematically as having infinitesimally small duration. 2. A pulse, generally with a waveform that rises rapidly to a sharply peaked maximum and falls rapidly to zero.

impulse bonding A variation of stitch bonding in which connections are made directly by impulse welding of the interconnecting wire to terminal pads formed on active or passive areas of the substrate or directly over plated-through holes.

impulse excitation *Pulse excitation.*

impulse generator *Surge generator.*

impulse noise Noise characterized by transient short-duration disturbances distributed essentially uniformly over the useful passband of a transmission system.

impulse relay A relay that stores the energy of a short pulse, to operate the relay after the pulse ends.

impulse solenoid A solenoid that operates on pulse power, at speeds up to several hundred strokes per second. Applications include tape drives and punches, high-speed sorting gates, and shutter drives.

impulse test *Pulse test.*

impurity An atom that is foreign to the crystal in which it exists. In a semiconductor crystal, an impurity can produce either excess electrons or holes. An acceptor impurity induces hole conduction; a donor impurity induces electron conduction.

impurity level An energy level caused by the presence of impurity atoms.

impurity scattering Scattering of electrons by impurity atoms in the crystal.

impurity semiconductor A semiconductor whose properties are caused by impurity levels produced by foreign atoms.

in Abbreviation for *inch.*

InAs Symbol for *indium arsenide.*

in-band signaling The transmission of signaling tones within the channel normally used for voice transmission.

incandescence Emission of visible radiation by a heated object, such as a lamp filament heated by electric current.

incandescent lamp An electric lamp in which light is produced by sending electric current through a filament of resistance material to heat it to incandescence.

inch [abbreviated in] A unit of length, equal to 2.54 cm. Use of the SI unit of length, the meter, is preferred.

inch per second [abbreviated in/s] A magnetic tape speed rating. Commonly used speed values include 1⅞, 3¾, 7½, 15, and 30 in/s (4.76, 9.5, 19, 38, and 76 cm/s).

inching *Jogging.*

incidence angle The angle between an approaching beam of radiation and the perpendicular (normal) to the surface that is in the path of the beam.

incidental amplitude modulation Amplitude modulation that results unintentionally from the process of frequency modulation and/or phase modulation.

incidental frequency modulation Frequency modulation that results unintentionally from the process of amplitude modulation.

incidental phase modulation Phase modulation that results unintentionally from the process of amplitude modulation.

incidental radiation device A device that radiates radio-frequency energy during normal operation, although not intentionally designed to generate such energy.

incident light The direct light that falls on a surface.

incident wave 1. A wave that impinges on a discontinuity or on a medium that has different propagation characteristics. 2. A current or voltage wave that is traveling through a transmission line in the direction from source to load.

inclination The angle that a line, surface, vector, or aircraft makes with the horizontal.

inclined synchronous orbit A nonequatorial, hence nonstationary, synchronous and circular orbit.

inclinometer 1. An instrument that measures the direction of the earth's magnetic force with relation to the plane of the horizon. 2. An instrument that measures the attitude of an aircraft with respect to the horizontal.

incoherent scattering Scattering of particles or photons, in which the scattering elements act independently of one another, so that there are no definite phase relationships among the different parts of the scattered beam.

incoherent waves Waves that have no fixed phase relationship.

increductor A variable inductance that has a saturable core, used in some high-frequency circuits.

increment A small change in the value of a variable.

incremental digital recorder A magnetic-tape recorder in which the tape advances across the recording head step by step, as in a punched-paper-tape recorder. Used for recording an irregular flow of data economically and reliably.

incremental frequency shift A method of superimposing incremental intelligence on another intelligence by shifting the center frequency of an oscillator a predetermined amount.

incremental permeability The ratio of a small cyclic change in magnetic induction to the corresponding cyclic change in magnetizing force when the average magnetic induction is greater than zero.

incremental printer A printer that prints sequentially, character by character, on each line. An example is a computer-controlled electric typewriter.

incremental sensitivity The smallest change in a quantity being measured that can be detected by a particular instrument

incremental tuner A television tuner in which the antenna, RF amplifier, and RF oscillator tuning coils are continuous or in small sections connected in series. Rotary switches make connections to the required portions of the total inductance necessary for a given channel, or short-circuit all of an inductance except that required for a given channel.

independent-particle model A nuclear model in which each proton and neutron moves independently in the field corresponding to the average positions of the other protons and neutrons. Also called individual-particle model, nuclear model, shell model, and single-particle model.

independent-sideband modulation Modulation in which the upper and lower sidebands carry entirely different information signals. The carrier may be either transmitted or suppressed.

independent variable The independent quantity or condition that, through the action of the control system of an automatic controller, directs the change in the controlled variable according to a predetermined relationship.

indexing The process of establishing memory addresses in a computer by adding the value in an address field of an instruction to a value stored in a specified index register.

index of refraction The ratio of the velocity of a wave in a vacuum to that in a specified medium.

index register A computer register whose contents are used to automatically modify addresses incorporated in instructions just prior to their execution. The original instruction remains intact and unmodified in the memory.

indicated course error An instrumental error that results in a discrepancy between the actual line of position offered by a navigation facility and the intended line of position.

indicating instrument An instrument in which the present value of the quantity being measured is visually indicated.

indicator A cathode-ray tube or other device that presents information transmitted or relayed from some other source, as from a radar receiver.

indicator gate A rectangular voltage applied to the grid or cathode circuit of an indicator cathode-ray tube to sensitize it during the desired portion of a radar operating cycle.

indicator tube An electron-beam tube in which useful information is conveyed by the variation in cross section of the electron beam at a luminescent target.

indirect address An address in a computer instruction that indicates a location where the address of the referenced operand is to be found.

indirectly controlled variable The quantity or condition that is controlled by virtue of its relation to the controlled variable and is not directly measured for control in a feedback control system.

indirectly heated cathode A cathode to which heat is supplied by an independent heater element in a thermionic tube. As a result, this cathode has the same potential on its entire surface, whereas the potential along a directly heated filament varies from one end to the other. Also called equipotential cathode, heater-type cathode, and unipotential cathode.

indirectly ionizing particles Uncharged particles that can liberate directly ionizing particles or initiate a nuclear transformation. Examples include neutrons and photons.

indirect scanning Scanning in which a narrow beam of light is moved across the area being televised. Used in flying-spot scanning of films, where the light transmitted by each illuminated elemental area in turn is picked up by one or more phototubes.

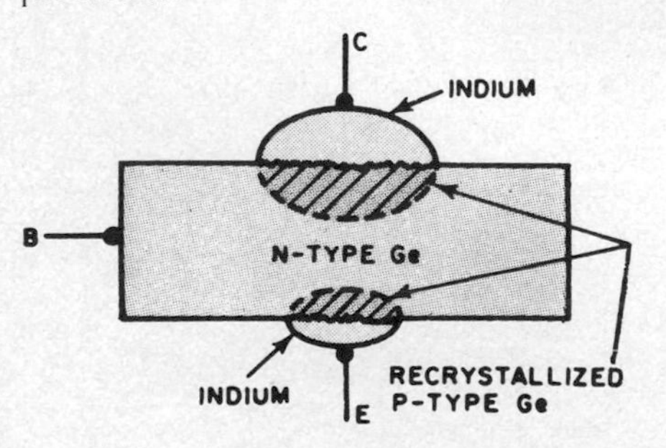

Indium, showing locations of indium pellets on alloy-junction transistor.

indium [symbol In] A metallic element. Atomic number is 49.

indium antimonide [symbol InSb] An intermetallic compound that has semiconductor properties. Used in Hall-effect devices.

indium antimonide detector A photovoltaic infrared detector that operates at liquid nitrogen temperatures and gives high sensitivity at a wavelength of about 5 μm in the infrared region. Used in search-track radar systems, missile guidance systems, target-recognition systems, early-warning radar, and for monitoring laser radiation.

indium arsenide [symbol InAs] An intermetallic compound that has semiconductor properties. Used in Hall-effect devices.

indium phosphide [symbol InP] An intermetallic compound that has semiconductor properties.

individual-particle model *Independent-particle model.*

indoor Not suitable for exposure to the weather.

indoor antenna A receiving antenna located entirely inside a building but outside the receiver.

induced charge An electrostatic charge produced on an object by an electric field.

induced current A current produced in a conductor by a time-varying electromagnetic field, as in induction heating.

induced electron emission Electron emission produced by impinging x-rays on a sample of a material being analyzed. The energy range of the resulting emitted electrons depends on the chemical composition. Used in spectrometers.

induced nuclear disintegration *Induced nuclear reaction.*

induced nuclear reaction A reaction in which a nucleus interacts with another nucleus, an elementary particle, or a photon to produce one or more other nuclei and possibly neutrons, mesons, neutrinos, and photons. The reaction occurs in about 1 ps or less. Also called induced nuclear disintegration.

induced radioactivity Radioactivity created by bombarding a substance with neutrons in a nuclear reactor or with charged particles produced by particle accelerators.

induced radionuclide A radionuclide that has a geologically short lifetime and is a product of nuclear reactions which occur currently or recently in nature. Examples are ^{14}C (natural radiocarbon), produced by cosmic-ray neutrons in the atmosphere, and ^{239}Pu, produced in uranium minerals by neutron capture.

induced voltage A voltage produced in a circuit by a change in the number of magnetic lines of force passing through a coil in the circuit.

inductance [abbreviated L; symbol L] The property of a circuit or circuit element that opposes a change in current flow. Inductance thus causes current changes to lag behind voltage changes. Inductance is measured in henrys, millihenrys, and microhenrys.

inductance bridge 1. An instrument similar to a Wheatstone bridge, used to measure an unknown inductance by comparing it with a known inductance. 2. A four-coil AC bridge circuit used for transmitting a mechanical movement to a remote location over a three-wire circuit. Half of the bridge is at each location.

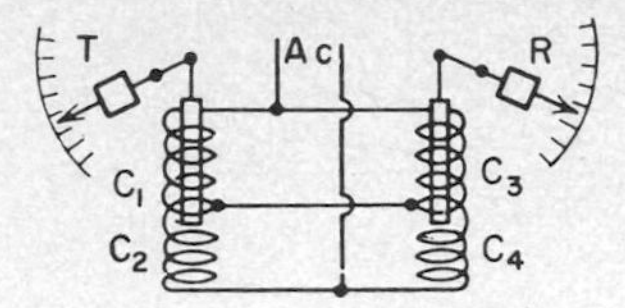

Inductance bridge used for transmitting mechanical movement at T to remote indicator at R. When soft-iron core at T is moved, resulting unbalance of bridge causes current to flow through coils at R and move its iron core until bridge is rebalanced.

inductance-capacitance [abbreviated LC] Containing both inductance and capacitance, as provided by coils and capacitors.

inductance standard *Standard inductor.*

inductance switch A cryogenic two-level variable inductor that can be switched from one level to the other by a control current.

induction The process by which a voltage, electrostatic charge, or magnetic field is produced in an object by lines of force.

induction brazing An electric brazing process in which the heat is produced by induced current flowing through the resistance of the joint being brazed.

induction coil A device for changing direct current into high-voltage alternating current. Its primary coil contains relatively few turns of heavy wire, and its secondary coil, wound over the primary, contains many turns of fine wire. Interruption of the direct current in the primary by a vibrating-contact arrangement induces a high voltage in the secondary.

induction compass A compass whose indications depend on the current generated in a coil revolving in the magnetic field of the earth.

induction-conduction heater A heating device in which electric current is conducted through a charge but restricted by induction to a preferred path.

induction coupling A coupling in which torque is transmitted by the interaction of the magnetic field produced by magnetic poles on one rotating member and induced currents in the other rotating member.

induction field 1. The portion of the electromagnetic field of a transmitting antenna that acts as if it were permanently associated with the antenna. The radiation field leaves the transmitting antenna and travels through space as radio waves. 2. The electromagnetic field of a coil carrying alternating current, responsible for the voltage induced by that coil in itself or in a nearby coil.

induction furnace A furnace in which electric energy is transformed into heat by electromagnetic induction.

induction hardening A process of hardening a ferrous alloy by using induction heating.

induction heater A generator and associated equipment used for induction heating. At low frequencies, down to power-line frequencies, rotating equipment is generally used. At higher frequencies, more suitable for surface heating, electronic generators are used.

induction heating Heating of a conducting material by placing the material in a varying electromagnetic field. The resulting eddy currents induced in the material produce heat, just as current flowing through a resistor produces heat. The frequencies used range from 60 to over 500,000 Hz, depending on the size and shape of the object to be heated. Also called eddy-current heating.

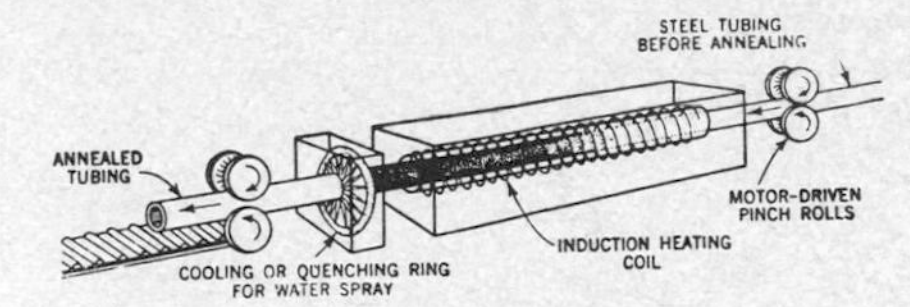

Induction-heating setup used to heat steel tubing for annealing.

induction instrument An instrument that depends for its operation on the reaction between magnetic flux set up by currents in fixed windings and other currents set up by electromagnetic induction in movable conducting parts.

induction loudspeaker A loudspeaker in which the audio-frequency current that reacts with the steady magnetic field is induced in the moving member.

induction motor An AC motor in which a primary winding on one member (usually the stator) is connected to the power source, and a polyphase secondary winding or a squirrel-cage secondary winding on the other member (usually the rotor) carries induced current.

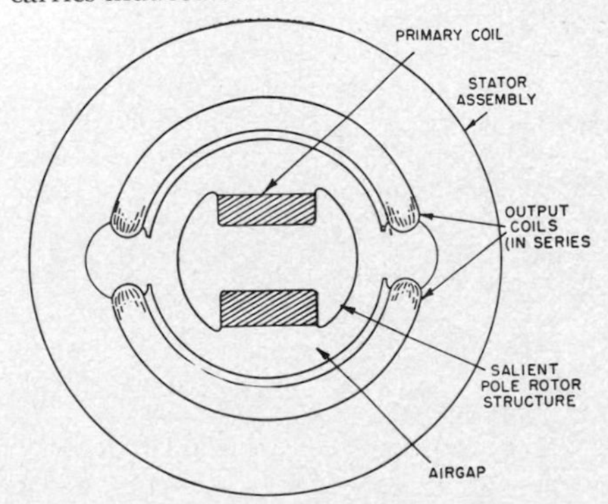

Induction potentiometer.

induction potentiometer A resolver-type synchro that delivers a polarized voltage whose magnitude is directly proportional to angular displacement from a reference position and whose phase indicates the direction of shaft rotation from the reference position. The linear range can be as great as 85°.

induction radio A carrier communication system for railroads, in which the short-range induction field of a loop antenna on a train is used to transfer signals to wayside telegraph wires connected to carrier equipment. Provides two-way communication between moving trains and wayside stations. Alternatively, the rails may be used in place of overhead wires to transmit the carrier signals.

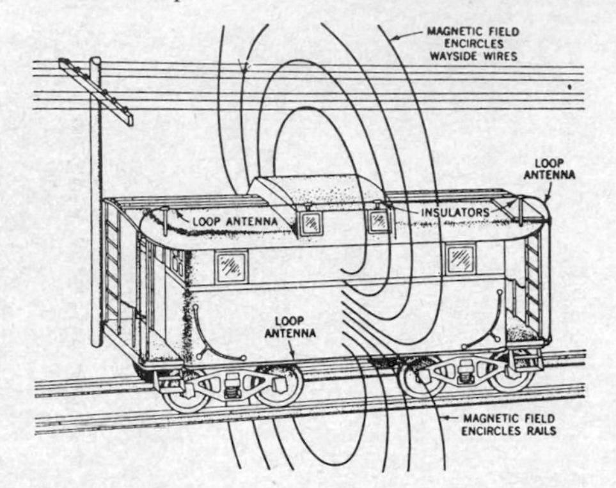

Induction radio installation on caboose.

induction ring heater A form of core-type induction heater used for heating electrically conducting charges in the form of a ring or loop, the core being open or separable to facilitate linking the charge.

induction voltage regulator A type of transformer that has a primary winding connected in parallel with a circuit and a secondary winding in series with the circuit. The relative positions of the primary and secondary windings are changed to vary the voltage or phase relations in the circuit.

inductive 1. Pertaining to inductance. 2. Pertaining to the inducing of a voltage through mutual inductance. 3. Pertaining to the inducing of an electric charge by electrostatic induction.

inductive circuit A circuit that contains a higher value of inductive reactance than capacitive reactance.

inductive coupling Coupling of two circuits by means of the mutual inductance provided by a transformer. Coupling by the self-inductance common to two circuits is called direct inductive coupling. Also called transformer coupling.

inductive diaphragm A resonant window used in a waveguide to provide the equivalent of inductive reactance at the frequency being transmitted.

inductive feedback Feedback of energy from the output circuit of an amplifier to its input through an inductance or by inductive coupling.

inductive kick The voltage induced in an iron-core coil when coil current is suddenly interrupted. The induced voltage can be many times greater than the applied voltage.

inductive load A load that is predominantly inductive, so the alternating load current lags behind the alternating voltage of the load. Also called lagging load.

inductive neutralization A method of neutralizing an amplifier, whereby the feedback susceptance due to the anode-to-grid capacitance is canceled by the equal and opposite susceptance of a coil. Also called coil neutralization and shunt neutralization.

inductive-output tube A tube in which output energy is obtained from the electron stream by electric induction between a cylindrical output electrode and the electron stream that flows through but does not touch this electrode.

inductive post A metal post or screw extending across a waveguide parallel to the E field, to add inductive susceptance in parallel with the waveguide for tuning or matching purposes.

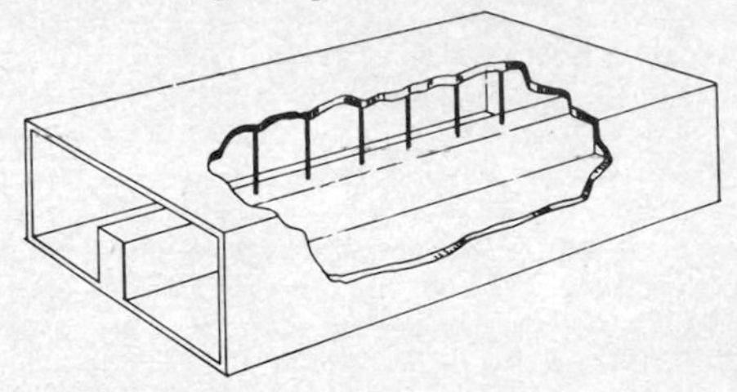

Inductive posts in filter for ridge waveguide.

inductive reactance [symbol X_L] Reactance due to the inductance of a coil or circuit. Inductive reactance is measured in ohms and is equal to $6.28fL$, where f is the frequency in hertz and L is the inductance in henrys.

inductive tuning Tuning that involves use of a variable inductance.

inductive window A conducting diaphragm extending into a waveguide from one or both sidewalls of the waveguide, to give the effect of an inductive susceptance in parallel with the waveguide.

inductometer A calibrated variable inductance, generally used with an inductance bridge.

inductor *Coil.*

inductor generator An AC generator in which all the windings are fixed and the flux linkages are varied by rotating an appropriately toothed ferromagnetic rotor. Sometimes used for generating high power at frequencies up to several thousand hertz for induction heating.

inductuner *Spiral tuner.*

industrial data processing Data processing for industrial purposes.

industrial heating equipment Equipment that uses an RF generator to produce energy for dielectric or induction heating operations in a manufacturing process.

industrial radiography Radiography of castings, welded joints, and other industrial products for quality control and flaw detection.

industrial, scientific, and medical equipment Equipment that uses radio waves for purposes other than communication. Bands authorized by the FCC for these services are at 918, 2450, 5800, and 22,500 MHz.

industrial television [abbreviated ITV] Closed-circuit television used for remote viewing of industrial processes and operations.

inelastic collision A collision in which at least one system gains internal excitation energy at the expense of the total kinetic energy of the center-of-gravity motion of the colliding systems.

inelastic gammas The gamma rays produced during inelastic scattering of neutrons.

inelastic scattering Scattering in which the kinetic energy of the neutron-plus-nucleus combination is decreased and the nucleus is left in an excited state.

inertance The acoustical equivalent of inductance.

inert gas A chemically inert gas, such as argon, helium, krypton, neon, and xenon.

inertial-celestial guidance An inertial guidance system in which position is checked automatically from time to time by means of celestial guidance.

inertial-gravitational guidance system A guidance system that is independent of all outside information except gravity.

inertial guidance A self-contained electronic guidance system that automatically follows a given course toward a ground target whose precise geographical position has been set into the computer of the system along with the starting point or a check-point position. It responds to inertial effects resulting from each change of course or speed and makes appropriate corrections under control of its built-in computer. It is independent of interference and is not subject to jamming. Used in guided missiles.

inertial navigation Inertial guidance as used in aircraft, ships, spacecraft, and submarines to provide dead-reckoning navigation based on automatic calculation of direction, distance, and speed as derived from acceleration measurements.

inertial space A frame of reference designed with respect to fixed stars.

inertia switch A switch that is actuated by an abrupt change in the velocity of the item on which it is mounted.

infinite attenuation Attenuation so great that a voltage applied to the input terminals of the filter produces no output voltage. The term is used to specify a frequency at which infinite attenuation would be produced by a filter if its coils and capacitors had zero loss.

infinite baffle A loudspeaker baffle in which no acoustic energy travels from the front of the loudspeaker diaphragm to the back. It is usually a large sealed enclosure, with the loudspeaker mounted in a hole cut into one side.

infinite clipping Clipping in which the threshold level is very small, so the output waveform is essentially rectangular.

infinite-impedance detector A detector in which the input circuit has infinite impedance.

infinite line A hypothetical transmission line whose characteristics correspond to those of an ordinary line that is infinitely long.

infinite resolution The ability to provide a stepless, continuous change in value or change in output over the entire range of a device. An example is the change in resistance of a potentiometer whose contact arm moves over a grainless ceramic-metal resistance element.

infinity [symbol ∞] 1. An indefinitely large number or amount. 2. Any number larger than the maximum number that a computer is able to store in any register. When such a number is calculated, the computer usually stops and signals an alarm indicating an overflow.

inflection point A point at which a curve takes a definite change in direction.

influence fuze *Proximity fuze.*

information A collection of facts or other data.

information center A center designed specifically for the storage, processing, and retrieval of information for dissemination at regular intervals, on demand, or on a selective basis according to expressed needs of users.

information gate A gate that passes only the information contained on one telemetering channel.

information processing The processing of data that represents information, such as index entries.

information retrieval Recovery of specific information from stored data or from a collection of documents, generally with the aid of a computer.

information science The science that deals with the properties and control of information, including its storage, retrieval, and dissemination.

information system Any means for communicating knowledge from one person to another, such as by simple verbal communication between scientists, the use of punched-card systems, optical coincidence systems based on coordinate indexing, and completely computerized methods of storing, searching, and retrieval of information. The conventional printed book is perhaps the best known and most effective of all information systems.

information theory The mathematical theory concerned with information rate, channel width, distortion, noise, and other factors affecting the transmission of information.

infradyne receiver A superheterodyne receiver in which the intermediate frequency is higher than the signal frequency, to obtain high selectivity.

infralow frequency [abbreviated ILF] A frequency in the band from 300 to 3000 Hz in the radio spectrum, corresponding to a wavelength between 100 and 1000 km.

infrared [abbreviated IR] Pertaining to infrared radiation.

infrared absorption spectrum A spectrum produced by molecular absorption of infrared radiation.

infrared beacon A source of infrared radiation used to establish a geographic reference point, the bearing of which may be determined.

infrared camera A camera that uses high-resolution scanning techniques in combination with infrared detectors to obtain thermographs that show images created by the infrared energy which is naturally radiated by all objects at all temperatures above absolute zero.

infrared communication set The components required to operate a two-way electronic system that utilizes infrared radiation to carry intelligence.

infrared detector A device used to determine the presence and/or bearing of an object by measuring infrared radiation from its surfaces. Photon-type infrared detectors have high sensitivity and fast time constants but are sensitive to only a narrow range of wavelengths in the near-infrared region and require cryogenic operating temperatures. Thermal-type infrared detectors absorb infrared energy, causing a temperature rise that changes an electrical property such as resistance.

infrared-emitting diode A light-emitting diode that has maximum emission in the near-infrared region, typically at 0.9 μm for PN gallium arsenide. Used in industrial controls, light modulators, logic circuits, optical switching, position encoders, and punched card or tape readers.

infrared fiber optics Fiber optics involving the transmission of infrared radiation, including wavelengths from about 0.8 to 100 μm.

infrared film Film that is sensitive to wavelengths in the near-infrared region.

infrared fusing Melting of metal by infrared heat. Used in fusing electrodeposited tin-lead coatings on printed-wiring boards.

infrared guidance system A missile guidance system that uses an infrared detector with amplifiers and control units as required to detect and home on a heat-emitting enemy target.

infrared heterodyne detector A heterodyne detector in which both the incoming signal and the local oscillator signal frequencies are in the infrared range and are combined in a photodetector to give an intermediate frequency in the kilohertz or megahertz range for conventional amplification.

infrared homing Homing in which the target is tracked by the infrared radiation that it emits.

infrared image converter An electron tube that converts an invisible infrared-illuminated scene into a visible image on a fluorescent screen. An infrared lens focuses the desired scene on a photocathode at the input end of the tube. The resulting stream of electrons emitted from the back of the photocathode, proportional to illumination at each point, passes through electron lenses that focus the stream on a fluorescent screen at the other end of the tube to give a visible image. Used in sniperscopes, snooperscopes, and other infrared viewing equipment.

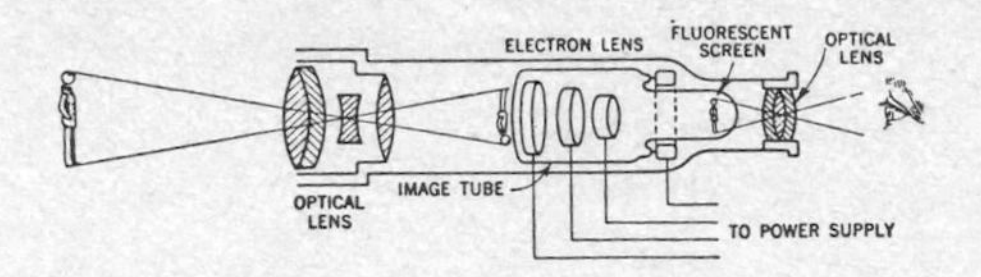

Infrared image converter.

infrared imagery Imagery produced electronically by sensing infrared electromagnetic radiations emitted or reflected from a target surface.

infrared mapping Mapping in which a sensitive infrared detector is mounted on a motor-driven scanner like that used for sector-scan radar antennas, to scan the field of view line by line, much as in television. The resulting thermal image is translated into varying shades of gray on photographic film, to give a line-pattern image similar to that seen on a television screen. Used for mapping over enemy territory at night.

infrared maser An optical maser that uses an infrared frequency as the pumping frequency, to permit signal radiation and detection at millimeter wavelengths. Also called iraser.

infrared optics Lenses, prisms, and other optical elements suitable for use with infrared radiation.

infrared photography Photography in which an infrared optical system projects an image directly on infrared film, to provide a record of point-to-point variations in temperature of a scene.

infrared polarizer A polarizer that consists of a thin film of pyrolytic graphite. It will provide up to 98% polarization of incident light at 4 μm and beyond in the infrared region.

infrared radiation Electromagnetic radiation in the infrared spectrum, ranging in wavelength from about 0.75 to 1000 μm.

infrared receiver A device that intercepts and/or demodulates infrared radiations which may carry intelligence.

infrared scanner A scanner in which mechanical or electrical techniques are used to provide line-by-line scanning of the field of view by one or more infrared detectors. In one arrangement, a motor-driven scanning mirror is simultaneously tilted by a motor-driven cam to give both horizontal and vertical scanning. Another arrangement has a fixed linear array of infrared detectors and uses the motion of the aircraft or spacecraft for vertical scanning.

infrared spectroscope A spectroscope designed to make measurements in the infrared spectrum.

infrared spectrum The portion of the electromagnetic spectrum between visible red light

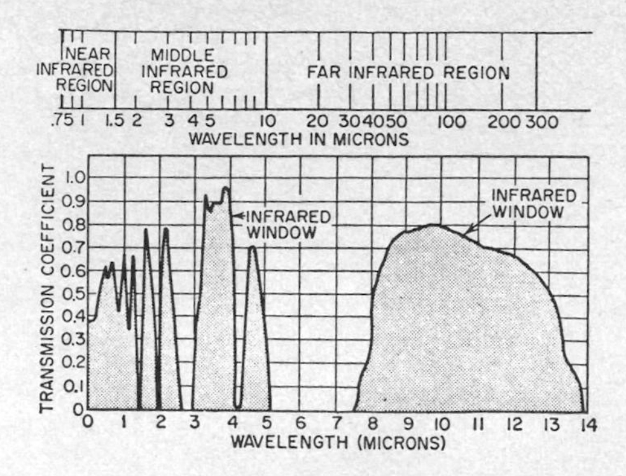

Infrared spectrum, extending between visible light at left end and microwave region starting at about 1,000 μm (1,000 microns). Below is transmission spectrum of earth's atmosphere for near- and middle-infrared regions.

(about 0.75 μm) and the shortest microwaves (about 1000 μm). All bodies that are not at absolute-zero temperature radiate in this range, making possible their detection in the dark by infrared systems. The infrared region is sometimes subdivided according to wavelength as follows: near-infrared—0.75 to 1.5 μm; middle-infrared—1.5 to 10 μm; far-infrared—10 to 1000 μm.

infrared submarine detection Detection of enemy submarines by highly sensitive infrared equipment that picks up the thermal trails in ocean water heated by submarines.

infrared transmitter A transmitter that emits energy in the infrared spectrum. It may be modulated with intelligence signals.

infrared-transparent material An optical material that transmits infrared radiation. Examples include sodium chloride (0.25 to 16 μm), cesium iodide (1 to 50 μm), and high-density polyethylene (16 to 300 μm).

infrasonic Pertaining to signals, equipment, or phenomena that involve frequencies below the range of human hearing, hence below about 15 Hz. Also called subaudio and subsonic (deprecated).

infrasonic frequency A frequency below the audible range.

infrasonic voltmeter A voltmeter that measures voltages accurately at infrasonic frequencies, generally for both sine and square waves. It may also measure higher frequencies; one commercial model has 3% accuracy at 0.05 to 30,000 Hz.

inherent filtration Filtration introduced by the wall of an x-ray tube and any permanent tube enclosure.

inherent regulation Regulation such that equilibrium is reached after a disturbance, without the aid of a control system.

inherited error An error inherited from previous steps in a computer process.

inhibit To prevent an action.

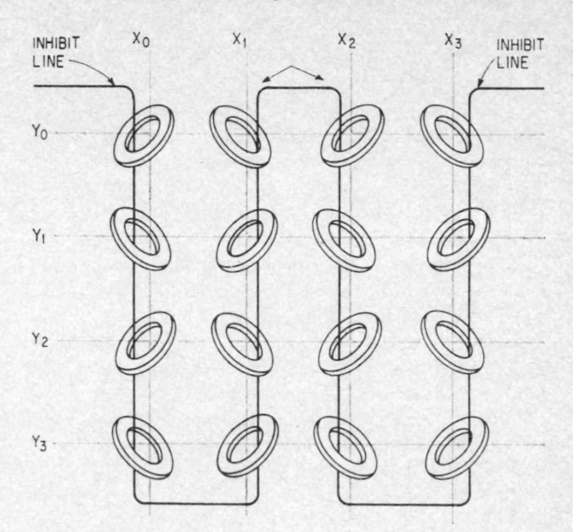

Inhibit line in ferrite-core memory prevents switching of core when writing 0, to keep core in 0 state.

inhibiting input A gate input that blocks an output which might otherwise occur in a computer.

inhibition gate A gate circuit placed in parallel with the circuit being controlled, for use as a switch.

inhibit pulse A drive pulse that tends to prevent flux reversal of a magnetic cell by certain specified drive pulses.

in-house Pertaining to an operation produced or carried on within a plant or organization, rather than done elsewhere under contract.

initial differential capacitance Differential capacitance at zero capacitor voltage.

initial ionizing event An ionizing event that initiates a count in a radiation-counter tube. Also called primary ionizing event.

initialize To set various counters, switches, and addresses in a computer to zero, or to other starting values, at the beginning of a routine or at prescribed points in a routine.

initial permeability The normal permeability that exists when both the magnetizing force and the magnetic induction approach zero.

injection 1. A process that makes the minority carrier density in a semiconductor region rise above the equilibrium value. Usually produced by a forward bias on the rectifying barrier or by irradiation with light or other penetrating radiation. 2. Placing a spacecraft in orbit.

injection grid A vacuum-tube grid used to feed the local oscillator signal into the mixer stage in some superheterodyne receivers.

injection laser A semiconductor laser in which the excitation for population inversion and lasing action is obtained by injection of electrons and holes (carriers) into the PN junction of the semiconductor.

injection laser diode A semiconductor PN junction that uses lasing to increase the light output and concentrate the light into a small area on the partially transmitting side of the semiconductor chip.

injection locking The capture or synchronization of a free-running oscillator by a weak injected signal at a frequency close to the natural oscillator frequency or to one of its subharmonics. Used for frequency stabilization in IMPATT or magnetron microwave oscillators, gas-laser oscillators, and many other types of oscillators.

ink-jet printer A nonimpact printer that uses electrostatic acceleration and deflection of ink particles emerging from nozzles to form characters on plain paper in a dot-matrix format.

ink-mist recording *Ink-vapor recording.*

ink recorder A recorder that uses an ink-filled pen or capillary tube to produce the graphic record.

ink-vapor recording A type of recording in which vaporized ink particles are directly deposited on the record sheet. Also called ink-mist recording.

in-line assembly machine An assembly machine that inserts components into a wiring board one at a time as the board is moved from station to station by a conveyor or other transport mechanism.

in-line guns An arrangement of three electron guns in a horizontal line. Used in color picture tubes that have a slot mask in front of vertical color phosphor stripes. In the Trinitron, a single gun produces three similarly positioned electron beams.

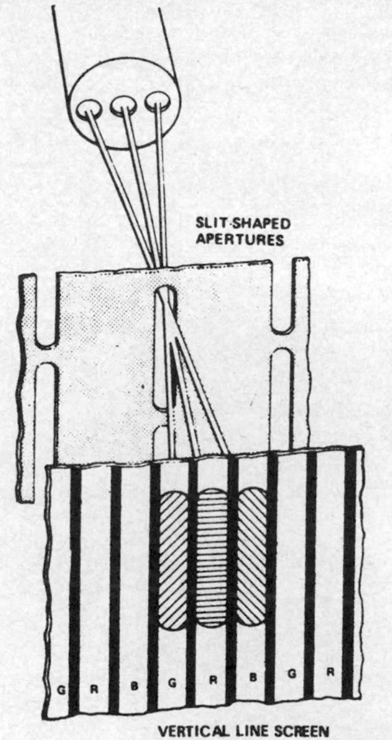

In-line guns.

in-line heads Two magnetic-tape heads mounted so that their gaps are in exact vertical alignment. This is now the standard arrangement for stereophonic tape players. Also called stacked heads.

in-line processing Processing of data by a computer without preliminary editing or sorting.

in-line stereophonic tape Magnetic stereophonic tape that has been recorded by using in-line heads. At any point on the tape the signal on one track is in line with the signal on the other. Also called stacked stereophonic tape.

inner bremsstrahlung 1. A beta-disintegration process that results in the emission of a photon whose energy is between zero and the maximum energy available in the transition. 2. The radiation resulting from the inner-bremsstrahlung process.

inner marker beacon *Boundary marker beacon.*

inorganic liquid laser A liquid laser in which an inorganic liquid such as neodymium-selenium oxychloride or neodymium-doped phosphorus chloride is used as the active material. These liquids are highly toxic and corrosive but do not require refrigeration or pulsed operation. Also called neodymium liquid laser.

InP Symbol for *indium phosphide.*

in phase Having waveforms that are of the same frequency and pass through corresponding values at the same instant.

in-phase rejection *Common-mode rejection.*

in-phase signal *Common-mode signal.*

in-port The entrance for a network.

input 1. The power or signal fed into an electronic device, or the terminals to which the power or signal is applied. 2. The data that is to be processed by a computer. 3. To feed signals or data into a computer or other piece of equipment.

input block A section of internal storage in a computer that is generally reserved for the receiving and processing of input information.

input capacitance The short-circuit transfer capacitance that exists between the input terminal and all other terminals of an electron tube (except the output terminal) connected together. This quantity is equal to the sum of the interelectrode capacitances between the input electrode and all other electrodes except the output electrode.

input equipment The equipment used for feeding data into a computer.

input extender A high-speed diode array used when increased fan-in is required, for a logic circuit.

input gap An interaction gap used to initiate a variation in an electron stream. In a velocity-modulated tube this gap is in the buncher resonator.

input impedance The impedance that exists between the input terminals of an amplifier or transmission line when the source is disconnected.

input level The ratio in decibels of audio input signal power to a reference power level of 1 mW when the signal is working into a given impedance. Commonly expressed as dBm.

input/output [abbreviated I/O] Pertaining to the equipment or procedures used for transmitting information into and out of a computer.

input/output limited Pertaining to a computer in which the speed of computer processing is limited by the speed of the input and/or output equipment.

input register A register that accepts input information for a computer at one speed and supplies the information to the central processing unit at another speed, usually much greater.

input resonator *Buncher resonator.*

input transformer A transformer used to provide a correct impedance match between a signal source and the input of a circuit or device.

inquiry station A remote terminal from which an inquiry may be sent to a computer over wire lines.

in/s Abbreviation for *inch per second.*

InSb Symbol for *indium antimonide.*

insensitive time *Dead time.*

insert earphone A small earphone that fits partially inside the ear.

insertion gain The ratio of the power delivered to a part of the system following an inserted amplifier to the power delivered to that same part before insertion of the amplifier. Usually expressed in decibels.

insertion head A mechanism used to insert a component in a printed-wiring board. The mechanism may also include automatic tools for cutting, forming, and clinching the leads of each component.

insertion loss The loss in load power due to the insertion of a component or device at some point in a transmission system. Generally expressed as the ratio in decibels of the power received at the load before insertion of the apparatus to the power received at the load after insertion.

insertion voltage gain The complex ratio of the alternating component of voltage across the output terminals of a system when an amplifier is inserted between source and output to the output voltage when the source is connected directly to the output termination.

inside spider A flexible device placed inside a voice coil to center it accurately with respect to the pole pieces of a dynamic loudspeaker.

instantaneous Having no intentionally introduced delay.

instantaneous automatic gain control [abbreviated IAGC] The portion of a radar system that automatically adjusts the gain of an amplifier for each pulse, to obtain a substantially constant output-pulse peak amplitude with different input-pulse peak amplitudes. The adjustment is fast enough to act during the time a pulse is passing through the amplifier.

instantaneous companding Companding in which the effective gain variations are made in response to instantaneous values of the signal wave.

instantaneous frequency-indicating receiver A radio receiver with a digital, cathode-ray, or other display that shows the frequency of a signal at the instant it is picked up anywhere in the band covered by the receiver. Generally used only for pulse-modulated transmission, such as from enemy radar fire-control systems. The information can be used in aircraft for tuning a jammer transmitter to the enemy frequency fast enough to protect the aircraft from artillery or missiles, even when the enemy radar may be changing frequency from pulse to pulse.

instantaneous multiplier A multiplier whose output is a continuous function of the input quantities. The Hall-effect multiplier is an example.

instantaneous particle velocity The total particle velocity at a point minus the steady velocity at that point.

instantaneous peak power An audio amplifier power rating that has little practical value because it gives a highly inflated picture of amplifier capability.

instantaneous power output The rate at which energy is delivered to a load at a particular instant.

instantaneous recording A recording intended for direct reproduction without further processing.

instantaneous sampling The process of obtaining a sequence of instantaneous values of a wave.

instantaneous sound pressure The total instantaneous pressure at a point minus the static pressure that exists at that point when no sound waves are present. The commonly used unit is the microbar.

instantaneous speech power The rate at which sound energy is being radiated by a speech source at any given instant.

instantaneous value The value of a sinusoidal or otherwise varying quantity at a particular instant.

instantaneous volume velocity The total instantaneous volume velocity at a point minus the static volume velocity at that point.

instant-on switch A switch that applies a re-

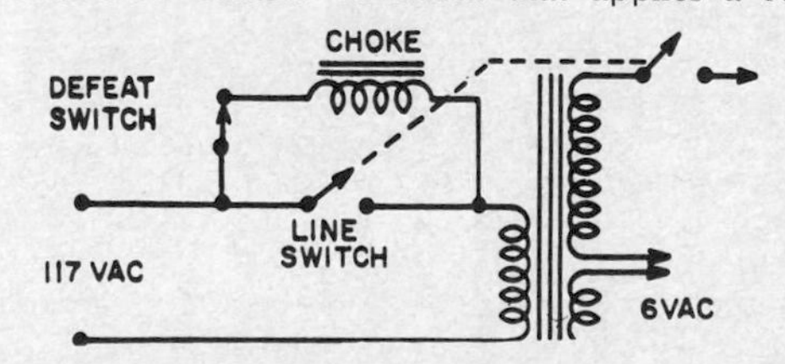

Instant-on switch in power-transformer circuit of television receiver.

duced filament voltage to all tubes in a television receiver continuously, so the picture appears almost instantaneously after the set is turned on. The switch inserts a voltage-reducing choke in series with the primary of the power transformer and opens the high-voltage secondary winding, to reduce filament voltage to about half the normal value.

instant replay A repetition of the action in a televised football game or other type of program, achieved with a slow-motion video disk recorder or ordinary video recorder. The replay may show the action previously picked up or action from one or more other cameras.

instant-start fluorescent lamp A fluorescent lamp that starts with a momentarily applied high voltage, without preheating of the electrodes.

Institute of Electrical and Electronics Engineers [abbreviated IEEE] A nonprofit professional organization of engineers and scientists, formed by combining the American Institute of Electrical Engineers and the Institute of Radio Engineers.

Institute of High Fidelity [abbreviated IHF] An organization established to develop performance specifications for audio amplifiers, FM tuners, and other high-fidelity equipment.

instruction A set of characters, with or without one or more addresses, that defines a computer operation. Also called order. The signal that initiates the operation is called a command.

instruction code An artificial language for describing or expressing the instructions that can be carried out by a digital computer. Each instruction word usually contains a part that specifies the operation to be performed and one or more addresses that identify a particular location in storage or serve some other purpose.

instruction counter A counter that indicates the location of the next computer instruction to be interpreted.

instruction register The register that temporarily stores the instruction currently being used in a computer.

instruction time The time required to carry out an instruction that has a specified number of addresses in a particular computer. A typical instruction time is 30,000 instructions per second.

instrument 1. A device for measuring and sometimes recording and controlling the value of a quantity under observation. The term "instrument" is usually applied to combinations of a meter with associated solid-state or tube circuits. 2. *Meter.*

instrumental error The error due to calibration, limited course sensitivity, and other inaccuracies introduced in any portion of a navigation system by the mechanism that translates path-length differences into navigation-coordinate information.

instrument approach An approach to a landing that is made by navigation instruments and radio guidance, with visual reference to the landing area only after the aircraft breaks through the overcast.

instrument approach system An aircraft navigation system that furnishes guidance in the vertical and horizontal planes to aircraft during descent from an initial-approach altitude to a point near the landing area. Completion of a landing requires guidance to touchdown by visual or other means.

instrumentation The use of measuring devices to determine the values of varying quantities, often for the purpose of controlling those quantities within prescribed limits.

instrumentation amplifier An amplifier that accepts a voltage signal as an input and produces a linearly scaled version of this signal at the output. It is a closed-loop fixed-gain amplifier, usually differential, and has high input impedance, low drift, and high common-mode rejection over a wide range of frequencies. Widely used for amplifying millivolt-level signals from strain-gage bridges, thermocouples, and other types of transducers.

instrumentation package The portion of a missile or artificial satellite that contains measuring and telemetering equipment, along with a power source.

instrument bombing Bombing by the use of radar or other instruments, without visual reference to the ground.

instrument conditions Weather conditions in which instrument flying is mandatory, or in which continued flight is possible only by using instruments.

instrument flight rules [abbreviated IFR] Regulations governing flying when weather conditions are below the minimum for visual flight rules.

instrument flying Flying in which navigation is carried out by the use of flight and navigation instruments, including radio and radar equipment, without visual reference to the ground. Also called blind flying.

instrument landing An aircraft landing made with the aid of an instrument landing system.

instrument landing approach An aircraft landing approach without visual reference to the ground, made with the aid of aircraft instruments and ground-based electronic devices or communication systems.

instrument landing system [abbreviated ILS] A radio system that provides in an aircraft the directional, longitudinal, and vertical guidance necessary for landing. It usually employs UHF ground transmitters and fixed directional antennas to define a beam that laterally localizes the runway extension and defines a slope plane at some angle between 2 and 5° leading to the optimum point of touchdown on the runway.

instrument multiplier A highly accurate resistor used in series with a voltmeter to extend its voltage range. Also called voltage multiplier and voltage-range multiplier.

instrument panel A panel or board that contains indicating meters.

instrument shunt A resistor connected in parallel with an ammeter to extend its current range.

instrument takeoff A takeoff that uses aircraft instruments or other aids, without visual reference to the ground.

instrument transformer A transformer that transfers primary current, voltage, or phase values to the secondary circuit with sufficient accuracy to permit connecting an instrument to the secondary rather than the primary. Used so only low currents or low voltages are brought to the instrument. With a current transformer, the primary winding is inserted into the circuit carrying the current to be measured or controlled. With a voltage transformer, the primary winding is connected across the circuit whose voltage is to be measured or controlled.

instrument-type relay *Meter-type relay.*

insulated Separated from other conducting surfaces by a nonconducting material.

insulated-gate field-effect transistor [abbreviated IGFET] Former name for *metal-oxide semiconductor field-effect transistor* [abbreviated MOSFET].

insulated wire A conductor covered with a nonconducting material.

insulating oil A mineral oil used in transformers as an insulating and cooling medium.

insulating tape Tape made or impregnated with rubber, plastic, or other insulating material, usually with an adhesive on one side. Used to cover bare joints in insulated wires or cables.

insulating varnish A varnish that has good insulating qualities. Applied to coils and windings to improve their insulation and sometimes to improve mechanical rigidity.

insulation A material that has high electric resistance and is therefore suitable for separating adjacent conductors in an electric circuit or preventing possible future contact between conductors.

insulation resistance The electric resistance between two conductors separated by an insulating material.

insulator 1. A device that has high electric resistance, for supporting or separating conductors to prevent undesired flow of current from the conductors to other objects. 2. A substance in which the normal energy band is full and separated from the first excitation band by a forbidden band that can be penetrated only by an electron having an energy of several electronvolts.

integer A whole number.

integral absorbed dose *Integral dose.*

integral action A control action in which the rate of change of the correcting force is proportional to the deviation.

integral dose The total energy imparted to an irradiated body by an ionizing radiation. Usually expressed in gram-rads or gram-roentgens. Also called integral absorbed dose and volume dose.

integrated circuit [abbreviated IC] An electronic circuit in which all the active and passive elements and the connections are made in or on a single semiconductor substrate by deposition, diffusion, etching, or other processes. Also called monolithic circuit, monolithic integrated circuit, and microcircuit. In a hybrid integrated circuit, one or more discrete components are constructed separately and mounted on the integrated-circuit substrate.

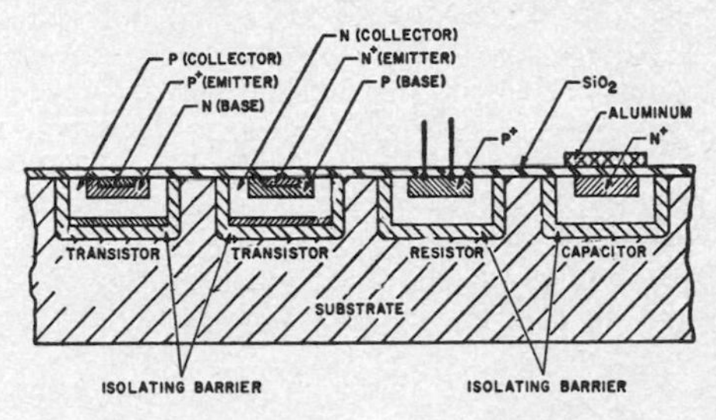

Integrated circuit using both PNP and NPN transistors.

integrated-circuit capacitor A capacitor that can be produced in a silicon substrate by conventional semiconductor production processes. A junction capacitor uses the capacitance of a reverse-biased PN junction, which can be formed at the same time as the emitter or collector junctions of transistors. For a metal-oxide semiconductor capacitor, an N^+ region is diffused into the silicon to form the bottom electrode, on which a controlled thickness of silicon oxide is formed to serve as dielectric. Maximum capacitance values are limited to a few hundred picofarads.

integrated-circuit package The arrangement used for the external leads of an integrated circuit. The common types are the dual in-line package (DIP), which has two rows of leads for insertion in punched terminal holes of a printed-wiring board, the flatpack, which has outward-spreading leads in the plane of the package, and the TO package, which resembles that used for transistors.

integrated-circuit resistor A resistor that can be produced in or on an integrated-circuit substrate as part of the manufacturing process. Film resistors are deposited on the surface of the substrate. Semiconductor resistors, which utilize the bulk resistivity of the semiconductor, can be bulk, diffused, ion-implanted, or pinched resistors.

integrated component A component incorporated into or deposited on the substrate of an integrated circuit.

integrated data processing [abbreviated IDP] Data processing in which the computer is part of an integrated system that minimizes clerical work on input and output data.

integrated electronics The branch of electronics that deals with integrated circuits, in which the interdependence of materials, circuits, and system designs is especially significant.

integrated flux The product of the number of neutrons or other nuclear particles per unit volume, the average velocity, and the exposure time.

integrated injection logic [abbreviated I^2L] Integrated-circuit logic that uses a simple and compact bipolar transistor gate structure which makes possible large-scale integration on silicon for logic arrays, memories, watch circuits, and many other analog and digital applications. The technique gives high production yields, high gate density, and low power consumption. The basic configuration uses a vertical NPN transistor, with multiple collectors for operation as an inverter, with a lateral PNP transistor serving as current source by injecting minority carriers into the emitter region of the NPN transistor. Injection can also be achieved with a light source. The transistor-pair gate can be either isolated or nonisolated. Also called merged-transistor logic.

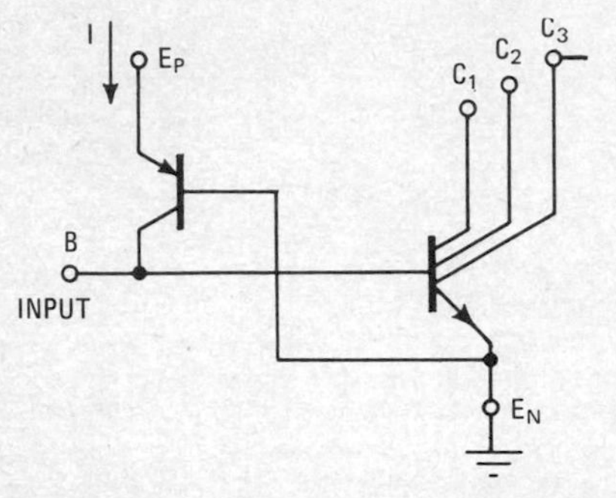

Integrated injection logic for complete gate, as shown, fits into area of one ordinary transistor on integrated-circuit substrate.

integrated optical circuit [abbreviated IOC] An integrated circuit capable of operating at optical wavelengths, using such components as microscopic lasers, modulators for laser beams, photodetectors, optical switches, and optical-fiber devices.

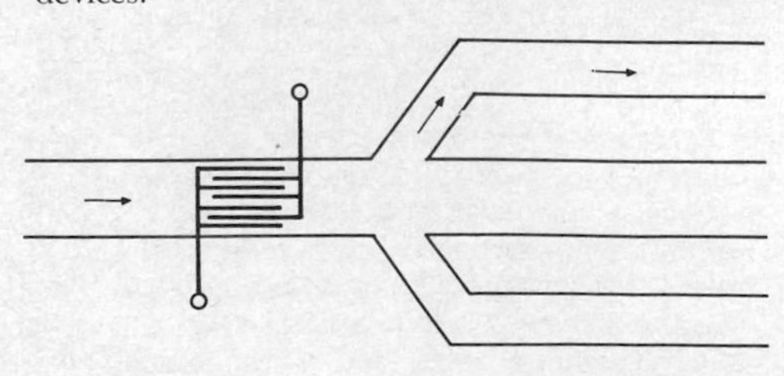

Integrated optical device serving as switch for optical communication system. Incoming signal is guided by transparent film to diffraction grating created by applying voltage to interdigital electrode structure, for routing signal beam to desired output branch.

integrated optical device A device that functions at optical wavelengths and can be constructed on a semiconducting substrate such as that used for integrated circuits.

integrated optics Optics involving the use of components that can be constructed simultaneously in a single substrate for operation as a system at optical wavelengths. Examples include optical waveguide components, surface lasers, and optical communication systems.

integrating accelerometer A transducer that measures velocity and/or distance by time integration of measured acceleration. When installed in a missile, it may be preset to cut off fuel flow when the required speed is reached.

integrating dosimeter An ionization chamber and measuring system used to measure the total radiation administered during an exposure. A device may be included to terminate the exposure when it reaches a desired value.

integrating filter A filter in which successive pulses of applied voltage cause cumulative buildup of charge and voltage on an output capacitor.

integrating gyroscope A gyroscope that senses the rate of angular displacement and measures and transmits the time integral of this rate.

integrating ionization chamber An ionization chamber whose collected charge is stored in a capacitor for subsequent measurement.

integrating meter An instrument that totalizes electric energy or some other quantity consumed over a period of time.

integrating-sphere densitometer A photoelectric densitometer in which a beam of interrupted light is directed through the specimen into a sphere that has a white inside surface. The light reflected from any part of the sphere is proportional to the total light entering the sphere. A phototube inserted in the wall of the sphere converts the reflected light into a corresponding signal that can be amplified and fed to a meter calibrated in density ratings.

integrating timer A timer that totalizes a number of small time intervals.

integrator A circuit or device whose output is the integral of its input with respect to time.

intelligence Data, information, or messages that are to be transmitted.

intelligence signal Any signal that conveys information, such as code, facsimile diagrams and photographs, music, television scenes, and spoken words.

intelligent terminal A computer terminal that has a microcomputer or microprocessor which provides text editing or other forms of processing for keyboarded information before the information is fed over wire lines to a large central computer. Also called smart terminal.

intelligibility *Articulation.*

Intelsat [INternational TELecommunications SATellite] A satellite network under international control, used for global communication by more than 80 countries. The system requires stationary satellites over the Atlantic, Pacific, and Indian Oceans and highly directional antennas at earth stations.

intensifier electrode An electrode used to increase the velocity of electrons in a beam near the end of their trajectory, after deflection of the beam.

intensifier ring A metallic ring-shaped coating on the inside of the glass envelope of a cathode-ray tube near the fluorescent screen. When a high positive voltage is applied to this ring, it increases the velocity of the electrons in the beam and thereby increases the intensity of the picture on the screen.

intensify To increase the brilliance of an image on the screen of a cathode-ray tube.

intensifying screen A thin screen coated with a substance that fluoresces readily under the influence of x-rays. It is placed next to the emulsion of x-ray film to increase the effect of x-rays on the film. In industrial radiography a sheet of thin lead is sometimes used for this purpose; here the secondary electrons and x-rays emitted by the lead produce the intensifying action.

intensitometer An instrument for determining relative x-ray intensities during radiography, to control exposure time.

intensity The strength or amount of a quantity, as of current, magnetization, radiation, or radioactivity. The symbol *I* for current comes from this word.

intensity control *Brightness control.*

intensity level A term used in acoustics to specify the relation of one sound intensity to another. The intensity level is expressed in decibels and is equal to 10 times the common logarithm of the ratio of the intensities.

intensity modulation Modulation of electron-beam intensity in a cathode-ray tube in accordance with the magnitude of the received signal. The luminance of the trace on the screen then varies with signal strength. Used in television and radar. Also called Z-axis modulation.

interaction A mutual influence, such as the strong force that holds together the nucleus of an atom, the interaction of charged particles with electromagnetic fields, the weak interaction responsible for the decay of beta and other particles, and gravitational interaction.

interaction-circuit phase velocity The phase velocity of a wave traveling through the interaction gap of a traveling-wave tube in the absence of electron flow.

interaction crosstalk Crosstalk that results from mutual coupling between two paths by means of a third path.

interaction gap An interaction space between electrodes in a microwave tube.

interaction space A region of an electron tube in which electrons interact with an alternating electromagnetic field.

interactive processing Computer processing in which the user can modify the operation appropriately while observing results at critical steps.

interactive terminal A computer terminal designed for two-way communication between operator and computer. A cathode-ray-tube terminal is one example.

intercarrier beat An interference pattern that appears on television pictures when the 4.5-MHz beat frequency of an intercarrier sound system gets through the video amplifier to the video input circuit of the picture tube.

intercarrier noise suppressor *Noise suppressor.*

intercarrier sound system A television receiver arrangement in which the television picture carrier and the associated sound carrier are amplified together by the video IF amplifier and passed through the second detector, to give the conventional video signal plus a frequency-modulated sound signal whose center frequency is the 4.5-MHz difference between the two carrier frequencies. The new 4.5-MHz sound signal is then separated from the video signal for further amplification before going to the frequency-modulation detector stage.

intercept 1. To meet or interrupt the course of a moving vessel, aircraft, or missile. 2. To tap or tune to a telephone or radio message not intended for the listener.

intercept bearing A bearing taken by electronic means on an enemy radio signal to determine the location of the station.

interceptor missile A surface-to-air or air-to-air guided missile used to intercept enemy aircraft or guided missiles.

interchannel interference Interference produced in a common channel by signals from one or more other channels.

intercom *Intercommunication system.*

intercommunication system An audio-frequency amplifier system that provides two-way voice communication between two or more locations which are usually in the same structure. Each station contains a dynamic loudspeaker that also serves as a microphone. The amplifier may be at a central station, or each station may have its own amplifier. Connections between stations may be made by wires or carrier signals traveling over electric wiring in the building. Also widely used on ships and large aircraft. Also called intercom.

interconnection diagram A diagram that shows the external connections required between two or more devices, modules, or other pieces of equipment.

intercontinental ballistic missile [abbreviated ICBM] A missile that flies a ballistic trajectory after guided powered flight, usually over ranges in excess of 4000 mi (6400 km).

intercontinental missile A missile designed for travel from one continent to another, such as an intercontinental ballistic missile.

interdigital magnetron A magnetron that has

axial anode segments around the cathode, with alternate segments connected together at one end. The remaining segments are connected together at the opposite end.

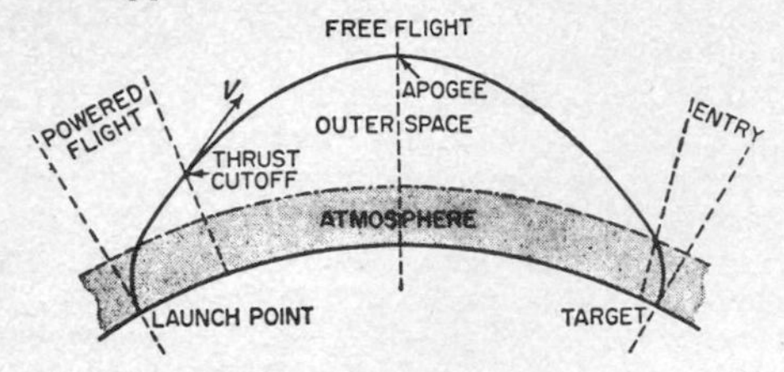

Intercontinental ballistic missile trajectory.

interdigital structure A structure in which the length of the region between two electrodes is increased by an interlocking-finger design for metallization of the electrodes. Used for transistors, capacitors, and other integrated-circuit devices. Also called interdigitated structure.

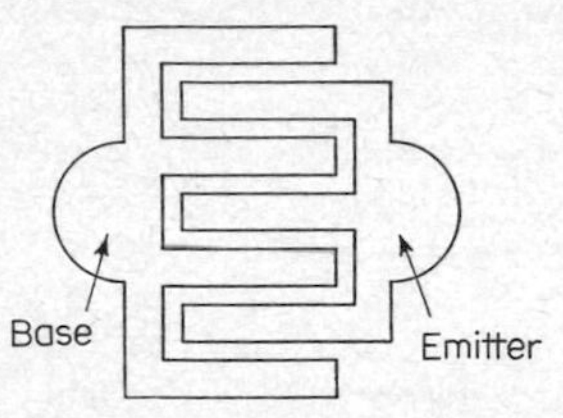

Interdigital structure for base and emitter of bipolar transistor.

interdigital transducer Two interlocking comb-shaped metallic patterns applied to a piezoelectric substrate like quartz or lithium niobate, used for converting microwave voltages to surface acoustic waves or vice versa.

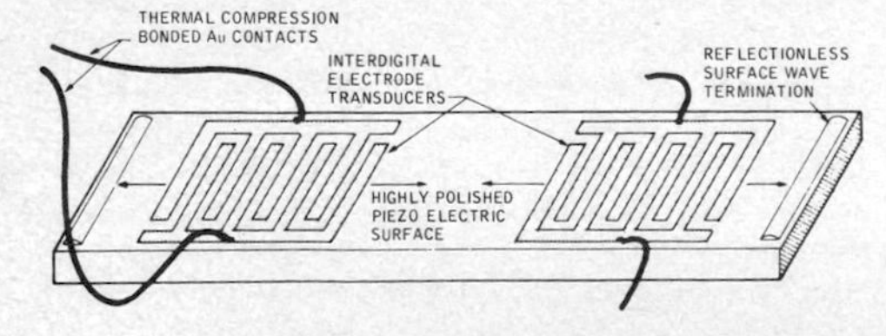

Interdigital transducers as used in delay line for frequencies from 10 MHz to 2 GHz.

interdigitated structure *Interdigital structure.*

interface A shared boundary. It may be a piece of hardware used between two pieces of equipment, a portion of computer storage accessed by two or more programs, or a surface that forms the boundary between two types of material.

interface connection *Feedthrough.*

interference 1. Any undesired energy that tends to interfere with the reception of desired signals. Man-made interference is generated by improperly operating electric devices, with the resulting interference signals either being radiated through space as electromagnetic waves or traveling over power lines. Radiated interference may also be caused by atmospheric phenomena such as lightning. Radio transmitters themselves may interfere with each other in certain locations. 2. The systematic alternate reinforcement and attenuation of two or more coherent waves when they are superimposed.

interference blanker A device that permits simultaneous operation of two or more pieces of radio or radar equipment without confusion of intelligence, or suppresses undesired signals when used with a signal receiver.

interference control Monitoring of radio frequencies assigned to missile ranges, to detect signals that might interfere with missiles.

interference eliminator *Interference filter.*

interference filter 1. A filter used to attenuate man-made interference signals entering a receiver through its power line. Also called interference eliminator. 2. A filter used to attenuate unwanted carrier-frequency signals in the tuned circuits of a receiver.

interference generator A generator designed to produce RF signals that are amplitude-modulated or frequency-modulated by random frequencies which have erratic amplitudes, to simulate atmospheric static.

interference guard band One of the two bands of frequencies that border the authorized communication band and frequency tolerance of a station, provided to minimize interference between stations on adjacent channels.

interference threshold The minimum signal-to-noise ratio required for essentially error-free message transmission and reception.

interferometer An instrument that measures very small distances by splitting a light beam into two parts directed over separate paths and then reuniting the beams. The difference in the path lengths is displayed as a light interference pattern. One application is measuring very thin films.

interferometer homing A homing guidance system in which target direction is determined by comparing the phase of the echo signal as received at two antennas precisely spaced a few wavelengths apart.

interlace 1. To assign successive memory location numbers to physically separated locations on a storage tape or magnetic drum of a computer, usually to reduce access time. 2. *Interlaced scanning.*

interlaced scanning A scanning process in which the distance from center to center of successively scanned lines is two or more times the nominal line width, so that adjacent lines belong to different fields. In U.S. television, double interlace is used, wherein 262.5 alternate lines are scanned in one field and the remaining 262.5 lines in the next field. Used to minimize flicker of the picture. Also called interlace, interlacing, and line interlace.

interlaced stacked rhombic array A combination of rhombic antennas, used for radio communication over long distances at high and very high frequencies.

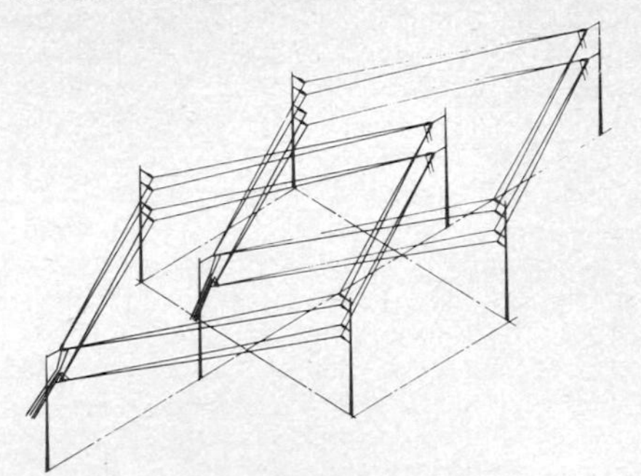

Interlaced stacked rhombic array.

interlacing *Interlaced scanning.*

interleave To alternate between parts of two or more sequences of data or operations, such as running two jobs simultaneously on a computer by switching control back and forth between two programs.

interlock A switch or other device that prevents activation of a piece of equipment when a protective door is open or some other hazard exists.

interlock circuit A circuit in which one action cannot occur until one or more other actions have first taken place. The interlocking action is generally obtained with relays.

interlock relay A relay composed of two or more coils, each with its own armature and associated contacts, so arranged that movement of one armature or the energizing of its coil is dependent on the position of the other armature.

interlock switch A switch designed for mounting on a door, drawer, or cover in such a way that it opens automatically when the door or other part is opened.

intermediate frequency [abbreviated IF] The frequency produced by combining the received signal with that of the local oscillator in a superheterodyne receiver.

intermediate horizon A hill, ridge, or building that is similar to the radar horizon but is not the most distant. An intermediate horizon may screen a valley between it and a mountain range that is the actual radar horizon.

intermediate neutron A neutron having kinetic energy in the range from about 1 to 100,000 eV.

intermediate-range ballistic missile [abbreviated IRBM] A ballistic missile that has a range between about 1700 and 3500 mi (2700 and 5600 km).

intermediate state A state of partial superconductivity that occurs when a magnetic field of appropriate strength is applied to a superconducting material below its critical temperature (the temperature below which it would be completely superconducting if no magnetic field were present).

intermediate subcarrier A carrier that may be modulated by one or more subcarriers and used as a modulating wave to modulate another carrier.

intermetallic compound A semiconductor that consists only of metallic atoms held together by metallic bonds, to give a basic crystal structure consisting of two different metallic elements. An intermetallic compound is semiconducting when the two metals together contribute just enough electrons to fill the valence band, as in bismuth telluride, gallium arsenide, gallium phosphide, indium antimonide, indium arsenide, indium phosphide, and mercuric telluride.

intermittent defect A defect that is not continuously present.

intermittent-duty rating An output rating based on operation of a device for specified intervals of time rather than continuous duty.

intermittent-duty relay A relay that must be deenergized at intervals to avoid overheating of its coil.

intermittent reception A radio receiver complaint in which the receiver operates normally for a time, then becomes defective for a time, with the process repeating itself at regular or irregular intervals.

intermittent service area The area surrounding the primary service area of a broadcast station, in which the ground wave is received but is subject to some interference and fading.

intermodulation [abbreviated IM] Modulation of the components of a complex wave by each other, producing new waves whose frequencies are equal to the sums and differences of integral multiples of the component frequencies of the original complex wave.

intermodulation distortion Nonlinear distortion characterized by the appearance of output frequencies equal to the sums and differences of integral multiples of the input frequency components. Harmonic components also present in the output are usually not included as part of the intermodulation distortion.

intermodulation interference Interference that occurs when the signals from two undesired stations differ by exactly the IF value of a superheterodyne receiver and both signals are able to pass through the preselector because of poor selectivity. The undesired signals combine in the mixer stage to give an undesired IF signal that interacts with the desired IF signal.

internal connection [symbol IC] An electron-tube connection, brought out to a base pin, that serves only for manufacturing purposes, such as for flashing the getter during evacuation.

internal contamination exposure The radiation exposure that results from the deposition of radioactive materials within body tissues. A continuous, and often long-term, hazard results from the deposit of alpha, beta, and/or gamma emitters.

internal conversion A nuclear deexcitation process in which energy is transmitted directly from an excited nucleus to an orbital electron, causing ejection of that electron from the atom. The ejected electron is called a conversion electron. Subsequent filling of the vacancy in the shell of the atom is accompanied by emission of photons, producing characteristic x-rays.

internal memory *Internal storage.*

internal radiation Nuclear radiation (alpha and beta particles and gamma radiation) that results from radioactive substances in the body. Important sources are iodine 131 in the thyroid gland and strontium 90 and plutonium 239 in bones.

internal resistance The resistance of a voltage source, acting in series with the source.

internal shield [symbol IS] A metallic coating placed on the inner surface of the glass envelope of an electron tube for shielding purposes, with a connection being brought out to a base pin.

internal storage The total memory or storage that is accessible automatically to a computer without human intervention. Also called internal memory.

international atomic time [abbreviated IAT] Time based on atomic clocks operating in conformity with the definition of the second as the SI unit of time.

international broadcast station A broadcast station that uses frequencies allocated to the broadcasting service between 5.95 and 26.1 MHz, whose transmissions are intended to be received directly by the general public in foreign countries.

International Civil Aviation Organization [abbreviated ICAO] A United Nations agency that establishes worldwide signal standards for navigation aids used by civil aviation.

International Commission on Illumination [abbreviated ICI] British version of Commission Internationale de l'Eclairage.

International Electrotechnical Commission [abbreviated IEC] An international organization formed in 1906 to facilitate the coordination and unification of national electrotechnical standards.

international Morse code The code universally used for radiotelegraphy.

International Radio Consultative Committee [abbreviated CCIR] An international organization concerned with the establishment of radio and television broadcasting standards throughout the world.

international radio silence A 3-min period of radio silence on the international distress frequency of 500 kHz only, commencing 15 and 45 min after each hour, during which radio stations may listen on that frequency for distress signals of ships and aircraft.

international standard atmosphere A standardized atmosphere, adopted internationally, used for comparing performance of aircraft and missiles. It is a pressure of 1013.2 mbar at a mean sea-level temperature of 15°C, and a lapse rate of 6.5°C/km of altitude up to 11 km, above which the temperature is assumed constant at −56.5°C.

International System of Units [abbreviated SI, for Système International d'Unités] A system of units adopted internationally in 1960, in which the ampere, candela, kelvin, kilogram, meter, mole, and second are the seven base units from which other units are derived.

International Telecommunications Union [abbreviated ITU] An international civil organization established to provide standardized communication procedures, including frequency allocations and radio regulations, on a worldwide basis.

International Telegraph and Telephone Consultative Committee [abbreviated CCITT] A United Nations advisory committee established to recommend international standards for data communication.

international unit A unit accepted internationally between 1908 and 1950 as a legal standard.

interphase transformer An autotransformer or set of mutually coupled reactors used when two or more high-power rectifiers are operated in parallel and have out-of-phase ripple voltages.

interphone An intercommunication system that uses headphones and microphones for communication between adjoining or nearby studios or offices or between crew locations on an aircraft, vessel, tank, or other vehicle. Also called talk-back circuit.

interpolate To estimate missing values between those that are known. Extrapolate means to estimate values outside the known range.

interpret To print on a punched card the information punched in that card.

interpreter 1. A computer executive routine that translates a stored program expressed in some machinelike pseudocode into machine code and performs the indicated operations by subroutines as they are translated. Also called interpretive routine. 2. A machine that senses a punched card and prints the punched information on that card.

interpretive routine *Interpreter.*

interrecord gap An unrecorded space left between records on magnetic tape, to permit tape stop-start operations without reading or recording errors.

interrogation Transmission of a signal intended to trigger a transponder, racon, or IFF system.

interrogator The transmitting section of an interrogator-responsor.

interrogator-responsor A transmitter and receiver combined, used for sending out pulses to interrogate a radar beacon and for receiving and displaying the resulting replies.

interrupt To temporarily disrupt the normal operation of a routine, as by a special signal from the computer.

interrupted continuous wave [abbreviated ICW] A continuous wave that is interrupted at a constant audio-frequency rate high enough to give several interruptions for each keyed code dot. The technique permits reception of code signals by receivers without beat oscillators but radiates more sideband frequencies than other methods.

interrupter An electric, electronic, or mechanical device that periodically interrupts the flow of a direct current to produce pulses.

interrupting capacity The highest current that a device can interrupt at its rated voltage.

interstage Between stages.

interstage transformer A transformer used to provide coupling between two stages.

interstation noise suppressor *Noise suppressor.*

interstitial implant A solid or encapsulated radiation source constructed in a form suitable for insertion directly into tissue that is to be irradiated.

interval The spacing in pitch or frequency between two sounds. The frequency interval is the ratio of the frequencies or the logarithm of this ratio.

intervalometer An electric timing device used for measuring the interval between events, the interval between an event and a reference time, or the predetermined intervals between a series of actions, such as the firing of rockets.

interval timer An instrument for signaling the expiration of a predetermined time. It may also actuate a switch at the end of the time interval. Also called timer.

Intrafax A Western Union closed-circuit facsimile system that is leased to government, military, and industrial users.

intrinsic angular momentum The angular momentum associated with axial rotation of an elementary particle.

intrinsic-barrier diode A PIN diode in which a thin region of intrinsic material separates the P- and the N-type regions.

intrinsic carrier density The equilibrium density of holes and free electrons in intrinsic semiconductor material.

intrinsic characteristic A characteristic of a material itself, independent of impurities.

intrinsic coercive force The magnetizing force required to bring to zero the intrinsic induction of a magnetic material that is in a symmetrically and cyclically magnetized condition.

intrinsic coercivity The maximum value of the intrinsic coercive force, corresponding to the saturation flux density for the material.

intrinsic condition Conduction that results from the movement of only those holes and free electrons which are present in the parent semiconductor material (not produced by impurity elements).

intrinsic flux The product of the intrinsic flux density and the cross-sectional area in a uniformly magnetized sample of magnetic material.

intrinsic flux density *Intrinsic induction.*

intrinsic hysteresis loop A curve that shows the relation between intrinsic flux density and magnetizing field strength, when the magnetizing field is cycled between equal negative and positive values. Hysteresis is indicated by the fact that the ascending and descending branches of the loop do not coincide.

intrinsic induction The additional magnetic induction that exists in a given magnetic medium, above that which would exist at the same location for the same magnetizing force if the medium were a vacuum. Also called intrinsic flux density.

intrinsic-junction transistor A four-layer transistor that has an I-type semiconductor layer between the base and collector layers, as in PNIP, NPIN, PNIN, and NPIP transistors.

intrinsic layer A layer of semiconductor material whose properties are essentially those of the pure undoped material.

intrinsic mobility The mobility of the electrons in an intrinsic semiconductor.

intrinsic noise Noise due to a device or transmission path, independent of modulation.

intrinsic permeability The ratio of intrinsic induction to the corresponding magnetizing force.

intrinsic property A property of a semiconductor that is characteristic of the ideal crystal.

intrinsic region A semiconductor region in which current flow is made up of approximately equal numbers of electrons and positive holes.

intrinsic semiconductor A semiconductor whose electrical properties are essentially characteristic of the pure ideal crystal. Current flow is made up of both electrons and positive holes in approximately equal numbers; hence the material is neither N- nor P-type. Also called I-type semiconductor.

intrinsic temperature range The temperature range in which the charge-carrier concentration of a semiconductor is substantially the same as that of an ideal crystal.

intrinsic tracer An isotope that is present naturally in a form suitable for tracing a given element through chemical and physical processes.

intrusion alarm A photoelectric, capacitance-controlled, electric, acoustic, or other system for setting off an alarm that announces the presence

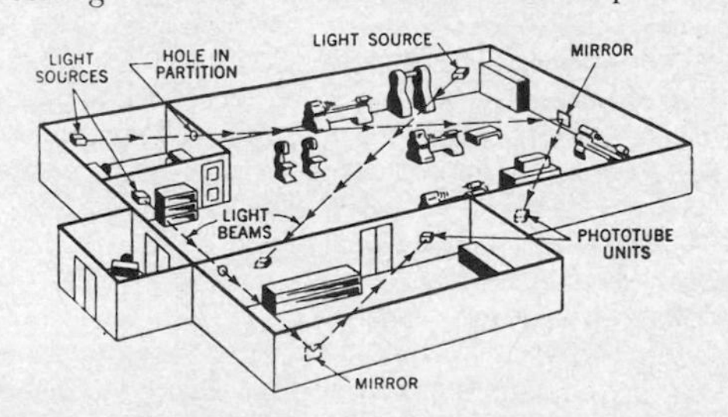

Intrusion alarm system for small factory, using three photoelectric systems.

of an intruder at the boundaries of a protected area or inside that area.

Invar Trademark for an alloy of nickel and iron, which contains about 36% nickel, that remains essentially constant in length over a wide range of temperature. Used for tuning forks and microwave cavities.

inverse conical scan jamming An electronic countermeasure technique that provides azimuth deception for an enemy tracking radar by sensing its scan-rotation frequency and retransmitting the incident signal with inverse gain modulation of the pulse amplitude.

inverse current The current that results from an inverse voltage in a contact rectifier or semiconductor device.

inverse direction The direction of greater resistance in a rectifier, going from the positive electrode to the negative electrode. It is the opposite of the conducting direction.

inverse electrode current The current that flows through an electrode of an electron tube in the direction opposite to that for which the tube is designed. Thus, for inverse anode current the electrons would flow from the anode to the cathode.

inverse feedback *Negative feedback.*

inverse-feedback filter A tuned-filter circuit used at the output of a high-selectivity amplifier that has negative feedback. The filter is adjusted so the negative-feedback output is zero at the desired resonant frequency but increases rapidly to reduce the amplification as the signal frequency departs from this value.

inverse function The function that would be obtained if the dependent and independent variables of a given function were interchanged.

inverse limiter A limiter whose output is constant for instantaneous input values within a specified range. Above and below that range it is linear or some other prescribed function of the input. Used to remove the low-level portions of signals from an output wave, such as to eliminate the annoying effects of crosstalk.

inverse logic A logic-level term whose use is deprecated because it can be construed differently for NPN devices than for PNP devices. To avoid confusion, the magnitude and polarity of the voltage levels for 1 and 0 in a specific logic circuit should always be specified.

inverse-parallel connection A connection of two rectifying elements such that the cathode of the first is connected to the anode of the second, and the anode of the first is connected to the cathode of the second.

inverse peak voltage 1. The peak value of the voltage that exists across a rectifier tube or x-ray tube during the half-cycle in which current does not flow. 2. The maximum instantaneous voltage value that a rectifier tube or x-ray tube can withstand in the inverse direction (with anode negative) without breaking down and becoming conductive.

inverse photoelectric effect The transformation of the kinetic energy of a moving electron into radiant energy at impact, as in the production of x-rays.

inverse piezoelectric effect The contraction or expansion of a piezoelectric crystal under the influence of an electric field, as in crystal headphones. The effect also occurs at PN junctions in some semiconductor materials.

inverse-square law When electromagnetic, thermal, or nuclear radiation from a point source is emitted uniformly in all directions, the amount received per unit area at any given distance from the source, assuming no absorption, is inversely proportional to the square of that distance.

inverse voltage An effective value of voltage that exists across a rectifier tube or x-ray tube during the half-cycle in which the anode is negative and current does not normally flow.

inversion 1. The process of scrambling speech for secrecy by beating the voice signal with a fixed higher audio frequency and using only the difference frequencies. The original low audio frequencies then become high audio frequencies, and vice versa. 2. A shallow layer of air in which temperature increases with altitude (temperature normally decreases with altitude). The resulting relatively sudden change in air density causes bending of radio waves.

inversion layer A surface layer of doped semiconductor material which has changed to the opposite conductivity from that of adjacent regions, as a result of surface ions, surface passivation materials, induced electric fields, or some other action.

invert To change to an opposite state.

inverted amplifier A two-tube amplifier in which the control grids are grounded and the input signal is applied between the cathodes. The grid then serves as a shield between the input and output circuits.

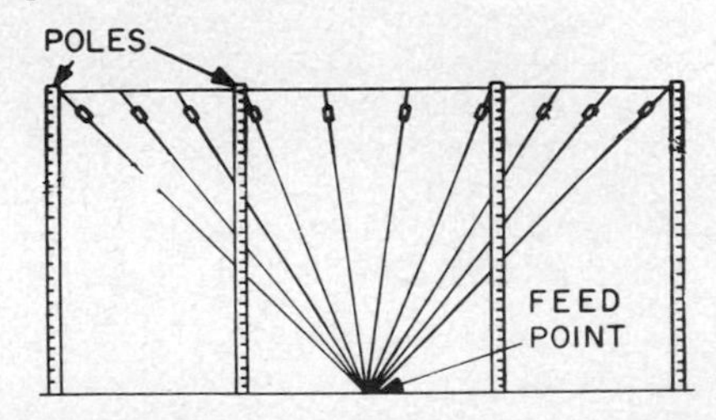

Inverted-cone antenna as used for Navy high-frequency communication.

inverted-cone antenna An antenna that consists of wires which form a cone whose apex is at ground level, with the upper ends of the wires supported by crosswires attached to a circular arrangement of poles. It is a broadband, omnidirectional, vertically polarized radiator, much like a discone antenna.

inverted exponential horn A inward-flaring horn used in front of a loudspeaker to provide

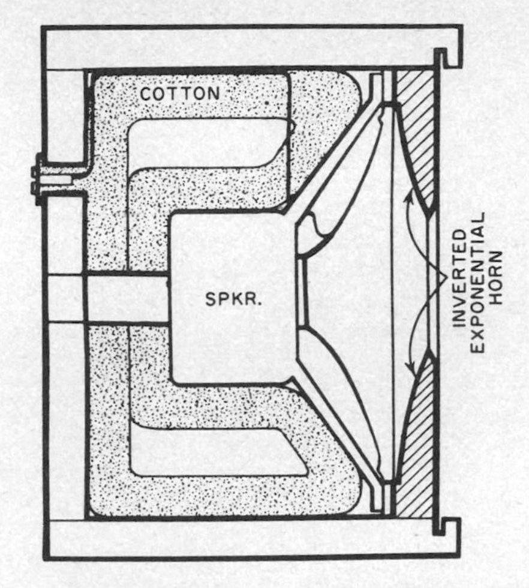

Inverted exponential horn as used in loudspeaker system.

front loading, to improve the uniformity of the radial transfer of acoustic energy into space.

inverted file A data file in which the normal sequence has been reversed.

inverted-L antenna An antenna that consists of a long horizontal wire, with the vertical lead-in wire connected to one end. Also called L antenna.

inverted speech *Scrambled speech.*

inverter. 1. A circuit or device for converting a low DC voltage to a much higher AC voltage by an oscillator or chopper, followed by a step-up transformer. When a rectifier is added to give a DC output voltage, the combination becomes a converter. 2. A circuit that takes in a positive-going pulse and delivers a negative-going pulse, or vice versa. 3. *Inverting amplifier.*

Inverter transformer A transformer (usually with four windings) used in conjunction with power transistors to convert direct current at low voltage into alternating current at higher voltages.

inverting amplifier An operational amplifier in

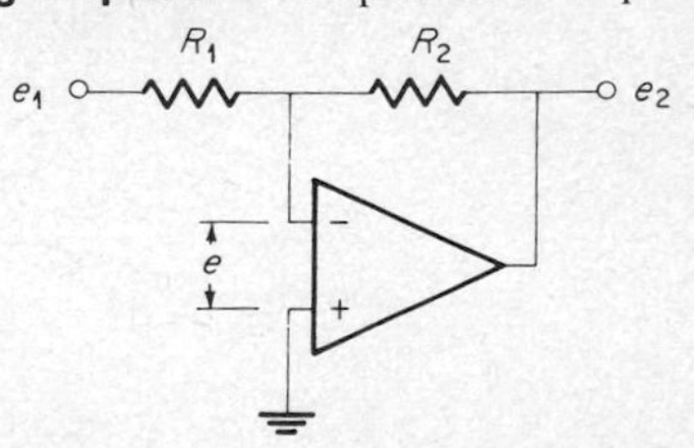

Inverting amplifier in which value of negative feedback resistor R_2 is chosen to give unity gain.

which the inverting or negative input is held virtually at ground by negative feedback regardless of the magnitudes of the two input voltages, and the output voltage change is inverted with respect to the input voltage change. Also called inverter.

inverting function A logic device that inverts the input signal, so the output is out of phase with the input. The inverting action is indicated by a circle

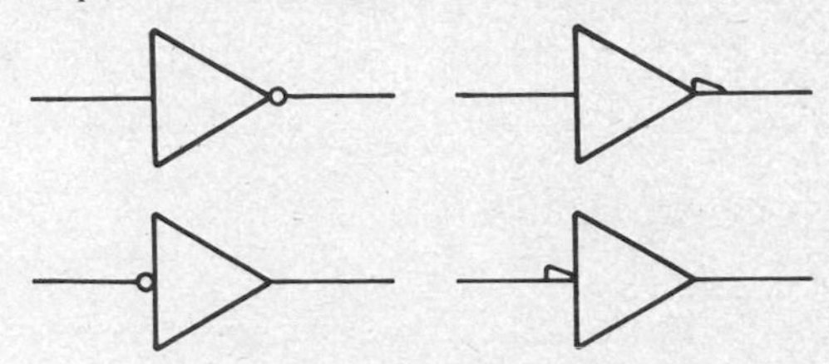

Inverting function symbols.

or small triangle placed on either the output or input terminal of the amplifier symbol.

inverting terminal The negative input terminal of an operational amplifier. A positive-going volt-

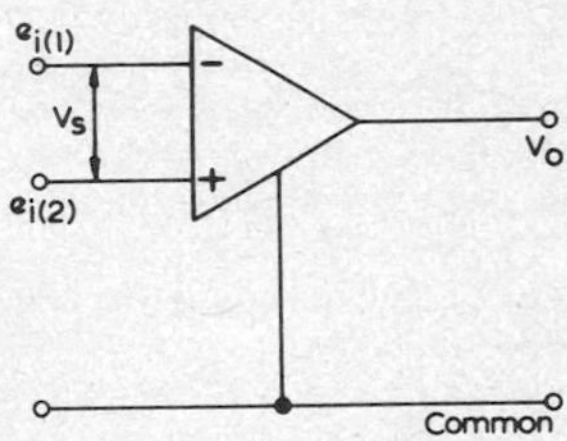

Inverting terminal is marked − on symbol of balanced-input operational amplifier.

age at the inverting terminal gives a negative-going output voltage.

I/O Abbreviation for *input/output.*

IOC Abbreviation for *integrated optical circuit.*

iodine [symbol I] A nonmetallic element. Atomic number is 53.

ion A charged atom or group of atoms. A negative ion has gained one or more extra electrons, whereas a positive ion has lost one or more electrons.

ion accelerator An accelerator that is being considered for space propulsion, in which either DC, AC, RF, or pulsed electric power is used to accelerate a plasma propellant.

ion backscattering Large-angle elastic scattering of monoenergetic ions in a beam directed at a metallized film on silicon or some other thin multilayer system. Ions are detected by a silicon surface-barrier detector that produces a pulse proportional to the energy of the backscattered ion. Used in nondestructive determination of the depth distribution of atoms after metallization.

ion beam A beam of ions drawn from a single source by a high voltage in a vacuum.

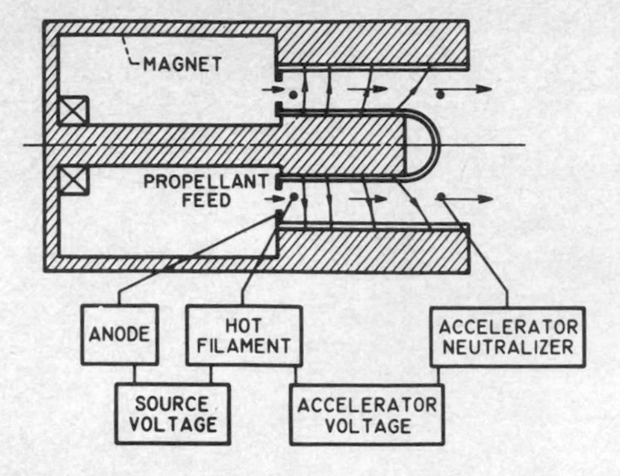

Ion accelerator using Hall effect and DC power source to force plasma out at right for spacecraft propulsion.

ion-beam scanning The process of analyzing the mass spectrum of an ion beam in a mass spectrometer either by changing the electric or magnetic fields of the mass spectrometer or by moving a probe.

ion-beam synthesis The production of chemical compounds by bombarding a source material with a beam of selected ions. Pure nitrobenzene has been produced in this way by using an electrostatic accelerator to bombard benzene molecules with a beam of nitrogen dioxide ions.

ion burn Deactivation and discoloration of a small area of phosphor at the center of the screen of a magnetically deflected cathode-ray tube, due to bombardment by heavy negative ions. An ion trap eliminates the effect by preventing the ions from leaving the electron gun.

ion chamber *Ionization chamber.*

ion charging Dynamic decay caused by ions striking the storage surface of a charge-storage tube.

ion engine A reaction engine designed for space travel, in which thrust is produced by a stream of positive ions obtained from nuclear fission or fusion. The ions are accelerated by electrostatic fields, much as in a cathode-ray tube.

ion gun *Ion source.*

ionic focusing *Gas focusing.*

ionic-heated cathode A hot cathode that is heated primarily by ionic bombardment of the emitting surface.

ion implantation A method of doping a semiconductor material by bombarding it with ions of an impurity material in a vacuum. Used to form junctions in integrated circuits. Metal or thick oxide masks are used to give desired patterns of ion-implanted regions.

ion-implanted resistor An integrated-circuit resistor produced on a semiconductor surface by ion implantation of impurities.

ionium [symbol Io] A naturally occurring radioisotope of thorium, which has an atomic weight of 230.

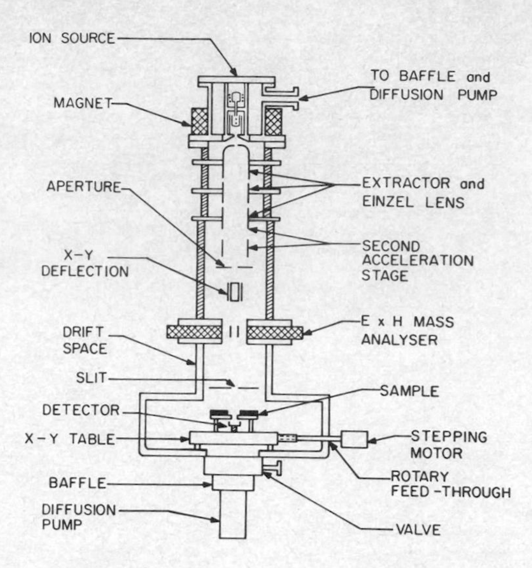

Ion-implantation equipment mounted in vacuum chamber.

ionization A process by which a neutral atom or molecule loses or gains electrons, thereby acquiring a net charge and becoming an ion. Ionization can be produced by collisions of particles, by radiation, and by other means.

ionization by collision Ionization produced by collisions of high-velocity electrons or ions with neutral atoms or molecules.

ionization chamber An enclosure that contains two oppositely charged electrodes in a gas. When the chamber is exposed to nuclear radiation, the gas is ionized and each ion is drawn to the electrode of opposite polarity. The resulting current through the chamber is proportional to the intensity of the radiation. Also called ion chamber.

ionization counter An ionization chamber in which there is no internal amplification by gas multiplication. Used for counting ionizing particles.

ionization cross section The probability that a particle or photon passing through a particular gas or other form of matter will undergo an ionizing collision.

ionization current *Gas current.*

ionization gage A vacuum gage in which electrons accelerated between a hot cathode and a nearby positive electrode cause ionization of the residual gas. The resulting positive ions are at-

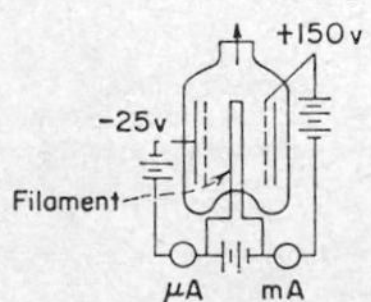

Ionization gage.

tracted to another electrode to give a measurable current whose value bears a known relation to gas pressure.

ionization instrument An ionization chamber and associated equipment used to measure the intensity of gamma rays, x-rays, and other ionizing radiation.

ionization path The trail of ion pairs produced by an ionizing particle in its passage through matter. Also called ionization track.

ionization potential The amount of energy, expressed in electronvolts, required to remove an electron from a neutral atom.

ionization spectrometer *Bragg spectrometer.*

ionization time The time interval between the initiation of conditions for conduction in a gas tube and the establishment of conduction at some stated value of tube voltage drop.

ionization track *Ionization path.*

ionization voltage The energy per unit charge, usually expressed in volts, required to remove an electron from a particular kind of atom to an infinite distance. Also called electron binding energy.

ionized layer One of the atmospheric layers, such as the E layer or F layer, that reflect radio waves back to earth under certain conditions.

ionizing energy The average energy lost by an ionizing particle when producing an ion pair in a gas. For air the ionizing energy is about 32 eV.

ionizing event An event in which one or more ions are produced.

ionizing particle A particle that produces ion pairs directly when it passes through a substance. The kinetic energy of the particle must be considerably greater than the ionizing energy of the medium.

ionizing radiation Any radiation that is capable of producing ions when it passes through a gaseous or solid material.

ion laser A gas laser in which stimulated emission takes place between two energy levels of an ion. Gases used include argon, krypton, neon, and xenon. Ionization of the gas by a glow discharge produces the excitation required for lasing action. Examples of ion lasers include helium-cadmium lasers and metal vapor lasers.

ion machining Use of a high-velocity ion beam to remove material from a surface. Used with argon ions to produce quartz crystal wafers as thin as 30 μm for oscillation at 45 MHz.

ionogram A record produced by an ionosonde, as a graph of the virtual height of the ionosphere plotted against frequency.

ionophone A high-frequency loudspeaker in which the AF signal modulates the RF supply to an arc maintained in a quartz tube, and the resulting modulated wave acts directly on ionized air to create sound waves.

ionosonde A radar system that uses a pulsed vertical beam which is swept periodically through a range of about 0.5 to 20 MHz. Photographic records of the variation of echo return time with frequency determine the vertical height at which the ionosphere reflects signals back to earth for each frequency. Also called vertical-incidence sounder.

ionosphere A region in the earth's outer atmosphere where ions and electrons are present in quantities sufficient to affect the propagation of radio waves. It begins about 30 mi (50 km) above the earth and extends above 250 mi (400 km), with

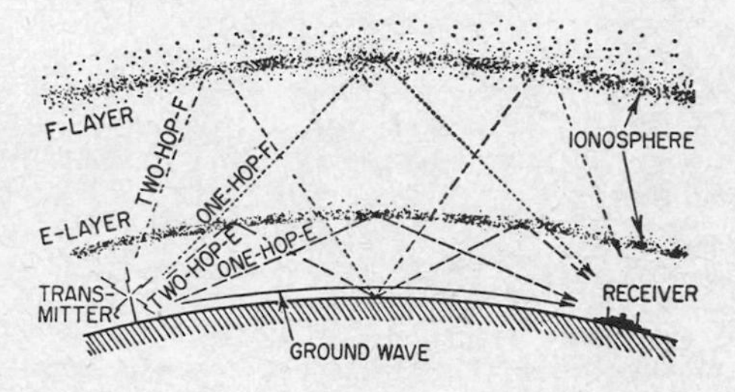

Ionosphere layers.

the height depending on the season of the year and time of day. The chief regions of the ionosphere and their approximate heights are D region: 30 to 60 mi (50 to 100 km), E region: 60 to 90 mi (100 to 150 km), F region: 90 to 250 mi (150 to 400 km).

ionospheric disturbance A disturbance that makes the heights of the ionosphere layers go beyond the normal limits for a location, date, and time of day. Ionospheric storms and radio fadeouts are examples.

ionospheric error The total systematic and random error that results from the reception of a navigation signal after ionospheric reflections. It may be due to variations in transmission paths, nonuniform height of the ionosphere, or nonuniform propagation within the ionosphere.

ionospheric scatter A form of scatter propagation in which radio waves are scattered by the lower E layer of the ionosphere to permit communication over distances of from 600 to 1400 mi (1000 to 2250 km) when using the frequency range of about 25 to 100 MHz.

ionospheric storm A turbulence in the F region of the ionosphere, usually due to a sudden burst of radiation from the sun. It is accompanied by a decrease in the density of ionization and an increase in the virtual height of the region. The higher frequencies in the band from 3 to 30 MHz are most affected by the resulting radio blackouts. Also called auroral storm.

ionospheric wave *Sky wave.*

ion pair A positive ion and an equal-charge negative ion, usually an electron, that are produced by the action of radiation on a neutral atom or molecule.

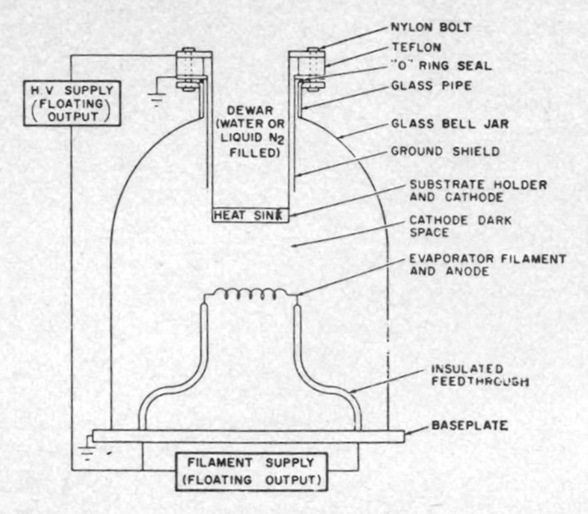

Ion plating. Most of substrate can be cooled externally as shown, to eliminate heat that might change properties of certain substrates, such as piezoelectric substrates used in transducers.

ion plating A method of depositing a thin metallic film on a ceramic or other substrate by first using ion bombardment in a vacuum to clean the surface of the substrate. The polarity is then reversed, so the substrate becomes the cathode in a high-voltage system in which the desired film material is deposited by evaporation.

ion propulsion A method of obtaining propulsion for spaceships by expelling ions and electrons from a combustion chamber. The recombination

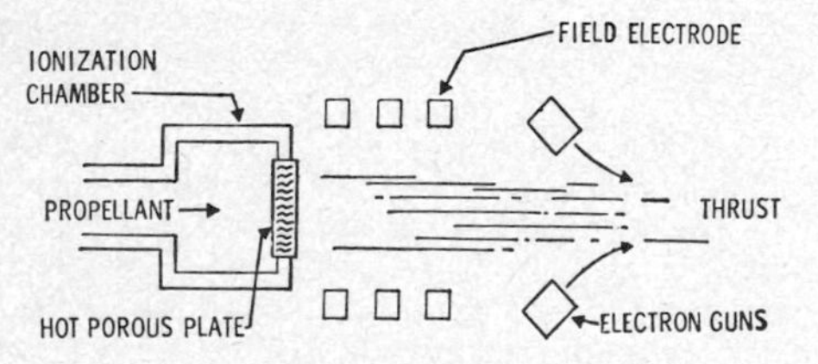

Ion-propulsion system for spacecraft.

of electrons with ions outside the chamber prevents space-charge effects that would counteract the thrust.

ion pump A vacuum pump in which the residual gas molecules are first ionized, then attracted or propelled by electric charges into an auxiliary pump or an ion trap.

ion repeller An electrode that produces a potential barrier against ions in a charge-storage tube.

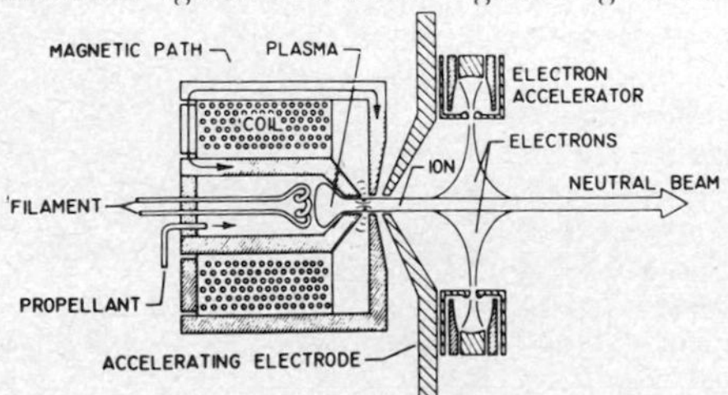

Ion-rocket power plant, in which neutral ion beam provides propulsion.

ion rocket A rocket in which a propulsion jet of ions is produced by an electric arc that is confined by an intense mirrorlike magnetic field.

ion sheath A film of positive ions that forms on or near an electrode surface in a gas tube and limits the control action.

ion source A device in which gas ions are produced, focused, accelerated, and emitted as a narrow beam. Also called ion gun.

ion spot 1. A dark spot formed near the center of the screen of a cathode-ray tube due to ion burn. 2. A spurious signal that results from bombardment of the target or photocathode of a camera tube or image tube by ions.

ion trap An arrangement whereby ions in the electron beam of a cathode-ray tube are prevented from bombarding the screen and producing an ion spot. Usually a part or all of the electron gun is tilted, and an external permanent magnet is used to bend the electron beam so it will pass through the tiny output aperture of the electron gun. The ions, being heavier and hence less affected by the magnetic field, are trapped harmlessly inside the gun.

ion-trap magnet One or more small permanent magnets with pole pieces, placed around the neck of a television picture tube to provide a magnetic

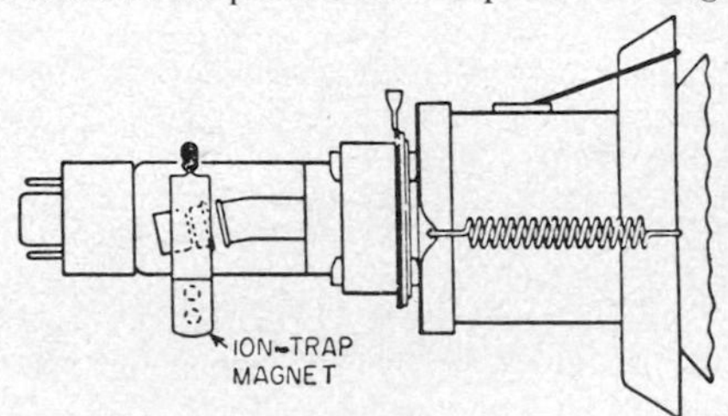

Ion-trap magnet on neck of picture tube.

field for ion-trap action in the electron gun. Also called beam bender.

ion yield The number of ion pairs produced per incident particle or quantum.

I-phase carrier A carrier phase separated by 57° from the color subcarrier in a color television receiver.

IR Abbreviation for *infrared.*

iraser [InfraRed mASER] *Infrared maser.*

IRBM Abbreviation for *intermediate-range ballistic missile.*

irdome [InfraRed DOME] A dome used to protect an infrared detector and its optical elements. Generally made from quartz, silicon, germanium, sapphire, calcium aluminate, or other material that has high transparency to infrared energy.

***IR* drop** The voltage drop produced across a resistance R by the flow of current I through the resistance. Also called resistance drop.

iridium [symbol Ir] An element. Atomic number is 77.

iris A conducting plate mounted across a wave-

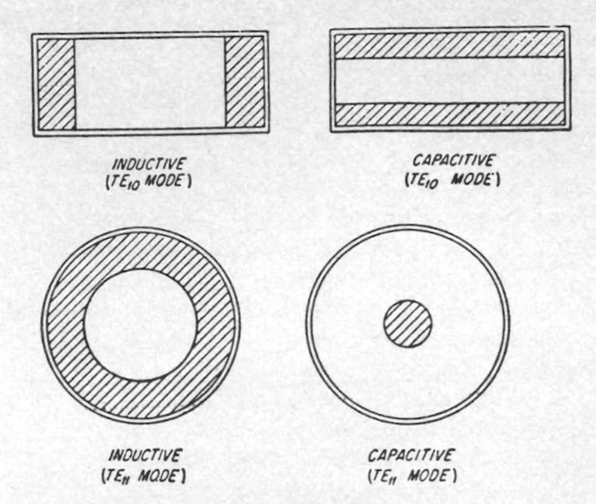

Iris designs for rectangular and circular waveguides.

guide to introduce impedance. When only a single mode can be supported, an iris acts substantially as a shunt admittance and may be used for matching the waveguide impedance to that of a load. Also called diaphragm.

iron [symbol Fe] A metallic element. Atomic number is 26.

iron-cobalt alloy An iron alloy that contains up to 65% cobalt and has high flux densities at low magnetizing currents. Chief drawbacks are high cost, relatively low permeability, and high hysteresis loss.

iron-core choke *Iron-core coil.*

iron-core coil A coil in which solid or laminated iron or other magnetic material forms part or all of the magnetic circuit linking its winding. Also called iron-core choke.

iron-core transformer A transformer in which laminations of iron or other magnetic material make up part or all of the path for magnetic lines of force that link the transformer windings.

iron-dust core A core made by mixing finely powdered magnetic material with an insulating binder and molding under pressure to form a rod-shaped core that can be moved into or out of a coil or transformer to vary the inductance or degree of coupling for tuning purposes.

iron loss *Core loss.*

iron-nickel alloy An iron alloy that contains 20 to 80% nickel. It has high permeability and low hysteresis losses at low flux densities and is more readily rolled into thin laminations than silicon steels.

iron-vane instrument A measuring instrument in which the movable element is an iron vane.

irradiance *Radiant flux density.*

irradiation The exposure of a material, object, or patient to x-rays, gamma rays, ultraviolet rays, or other ionizing radiation.

IS Symbol for *internal shield.* Used on tube-base diagrams.

I scope A radarscope that produces an I display.

I signal The in-phase component of the chrominance signal in color television; it has a bandwidth of 0 to 1.5 MHz. It consists of $+0.74(R - Y)$ and $-0.27(B - Y)$, where Y is the luminance signal, R is the red camera signal, and B is the blue camera signal.

island effect The restriction of emission from the cathode of an electron tube to certain small areas of the cathode when the grid voltage is lower than a certain value.

isobar 1. A line that connects points which have the same value of a quantity, such as a barometric pressure line on a meteorological chart. 2. One of two or more nuclides that have the same number of nucleons in their nuclei but differ in their atomic numbers and chemical properties.

isobaric spin quantum number A nuclear quantum number based on the theory that a proton and a neutron are different states of a nucleon. The nucleon is assigned an isobaric spin quantum number of ½, and its two possible orientations, +½ and −½, are assigned to the neutron and proton, respectively. Also called isotopic spin quantum number.

isochronous Having a fixed frequency or periodicity.

isochronous accelerator A circular accelerator in which particles rotate at constant frequency as they gain energy. The ordinary low-energy cyclotron is isochronous, but at high energies the relativistic increase of mass causes the particles to rotate more slowly.

isoclinic line A line passing through points of equal magnetic inclination or dip on a magnetic map of the earth. Also called aclinic line.

isodiaphere One of several nuclides that have the same difference between the numbers of neutrons and protons in their nuclei.

isodose chart A chart that shows the distribution of radiation in a medium by lines or surfaces drawn through points receiving equal doses.

isodose surface A surface on which all points receive the same dose of radiation.

isoelectronic Pertaining to atoms that have the same number of electrons outside the nucleus of the atom.

isoelectronic trap A bound state induced in a semiconductor by adding an impurity atom. The trap may bind either an electron or a hole. In some light-emitting diodes, the nature of the impurity atom used for doping determines the color of the light produced.

isogriv A line that joins points of equal magnetic variation from grid north on a map or chart.

isolated camera A television camera that is focused on an isolated activity at a sports or other event, for videotaping and replay after the overall action has been shown by the main cameras.

isolation amplifier A unity-gain amplifier that provides total isolation between input and output signal channels. Used in industrial applications in which millivolt signals must be transmitted in the presence of dangerously high voltages. Also used with medical electronic equipment to improve patient safety by interrupting ground loops and leakage current paths.

isolation diffusion Diffusion used to produce the back-to-back junctions required for isolating active devices from each other in an integrated circuit.

isolation network A network inserted into a circuit or transmission line to prevent interaction.

isolation transformer A transformer inserted into a system to separate one section of the system from undesired influences of other sections. Usually made with a 1 to 1 ratio of primary turns to secondary turns, to eliminate a direct connection without changing voltages.

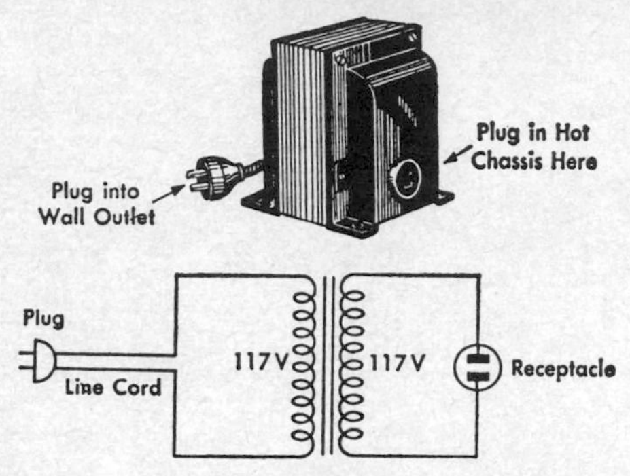

Isolation-transformer construction and circuit.

isolator A passive attenuator in which the loss in one direction is much greater than that in the opposite direction. A ferrite isolator for waveguides is an example.

isomagnetic line A line passing through points that have equal magnetic force but not necessarily the same deviation from vertical.

isomer One of two or more nuclides that have the same atomic and mass numbers but differ in other properties.

isomeric transition A radioactive transition from one nuclear isomer to another of lower energy. The deexcitation of the nuclei in the metastable state may occur by gamma emission or by internal conversion followed by emission of x-rays and/or Auger electrons. It is a type of forbidden transition.

isophotometer A direct-recording photometer that automatically scans and measures optical density of all points in a film transparency or plate and plots the measured density values in a quantitative two-dimensional isodensity tracing of the scanned areas.

isopotential path A line passing through points that have the same potential or field strength.

isopulse system A pulse-coding system in which the number of information pulses transmitted in each group is indicated by special inserted pulses.

isothermal Without temperature change.

isotone One of several nuclides that have the same number of neutrons in their nuclei but differ in the number of protons.

isotope One of two or more nuclides that have the same atomic number but differ in atomic weight and energy content because they have different numbers of neutrons. Thus natural uranium consists of three isotopes, with atomic weights of 234, 235, and 238 and designated as ^{234}U, ^{235}U, and ^{238}U, respectively.

isotope effect The effect wherein the superconducting transition temperatures of five elements—zinc, tin, mercury, thallium, and indium—are inversely proportional to the square root of the isotopic mass.

isotope shift The slight difference in wavelength for a given spectral line of one isotope as compared with that of a related isotope.

isotope structure *Hyperfine structure.*

isotopic abundance The relative number of atoms of a particular isotope in a sample of an element.

isotopic carrier *Carrier.*

isotopic enrichment The process by which the relative abundances of the isotopes of a given element are altered in a batch, thus producing a form of the element enriched in a particular isotope.

isotopic indicator *Isotopic tracer.*

isotopic mass Obsolete term for *atomic mass.*

isotopic number *Neutron excess.*

isotopic spin quantum number *Isobaric spin quantum number.*

isotopic tracer An isotope, usually radioactive, that is used as a chemical tracer for the element with which it is isotopic. Also called isotopic indicator.

isotron A device for sorting isotopes, as of uranium, by accelerating all the ions to a given energy by applying a strong electric field. Ions of different mass then have different velocities. An RF field is then applied to make the ions group themselves according to mass.

isotropic Having identical properties in all directions.

isotropic antenna *Unipole.*

isotropic dielectric A dielectric in which the dielectric constant is independent of the direction of the applied electric field.

isotropic medium A medium whose properties are the same in all directions.

isotropic radiator · A radiator that sends out energy equally in all directions.

I^2L Abbreviation for *integrated injection logic.*

***I^2R* loss** *Copper loss.*

iterated fission expectation The number of fissions produced per generation time by the daughter neutrons of a given neutron after a long period of operation of a critical assembly.

iterative array An array of a large number of interconnected identical processing modules, used with appropriate driver and control circuits in a computer to permit a large number of simultaneous parallel operations. The technique permits the use of slower, less expensive circuits for individual operations while still providing the required fast total execution time for a problem.

iterative filter A four-terminal filter that provides iterative impedance.

iterative impedance The impedance that when connected to one pair of terminals of a four-terminal transducer will cause the same impedance to appear between the other two terminals. The iterative impedance of a uniform transmission line is the same as the characteristic impedance. When a four-terminal transducer is symmetrical, the iterative impedances for the two pairs of terminals are equal and the same as the image impedances and the characteristic impedance.

iterative process A mathematical process for calculating a desired result by a repeating cycle of operations that yields a result closer and closer to the desired result.

ITU Abbreviation for *International Telecommunications Union.*

ITV Abbreviation for *industrial television.*

I-type semiconductor *Intrinsic semiconductor.*

J

j A complex operator that is mathematically equivalent to the square root of −1.

J Abbreviation for *joule.*

jack A connecting device into which a plug can be inserted to make circuit connections. The jack may also have contacts that open or close to perform switching functions when the plug is inserted or removed.

jack box A box that mounts and protects one or more jacks and sometimes one or more switches.

jacket A plastic, rubber, or other covering used over the insulation, core, or sheath of a cable for mechanical protection.

jammer A transmitter used to jam radio or radar transmissions.

jamming Radiation or reradiation of electromagnetic waves in such a way as to impair the usefulness of a specific segment of the radio spectrum that is being used by the enemy for communication or radar. Also called electronic jamming.

J antenna A half-wave antenna that is end-fed by a parallel-wire quarter-wave section, so that the radiating elements somewhat resemble the letter J.

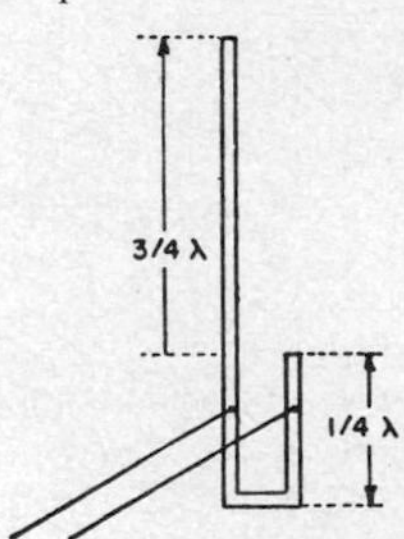

J antenna and method of connecting its quarter-wave matching section to open-wire transmission line.

Janus antenna array An array that provides both forward and backward beams, used in airborne Doppler navigation systems. Named after the Roman god of doorways, who had a face on each side of his head.

Janus technique A technique of generating a Doppler signal for an airborne navigation radar that is used to measure basic ground speed and drift angle. Microwave energy in the X band is radiated forward and backward from the aircraft toward the earth. The backscattered energy from both beams is detected, and the echo frequency of the aft beam is subtracted from that of the fore beam. The resulting low audio frequency is then a measure of ground speed.

jar An obsolete unit of capacitance used in the British Navy, equal to $1/900$ μF.

J carrier system A carrier system that provides 12 telephone channels in a bandwidth of approximately 140 kHz.

J display A modified radarscope A display in which the time base is a circle. The target signal appears as an outward radial deflection from the time base.

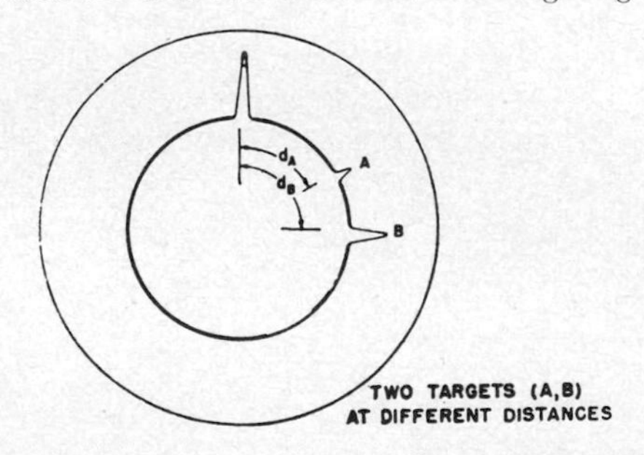

J display.

JEDEC Abbreviation for *Joint Electronic Devices Engineering Council.*

jerk A vector unit that specifies the time rate of change of an acceleration.

jewel bearing A natural or synthetic jewel, usually sapphire, that has a carefully ground conical

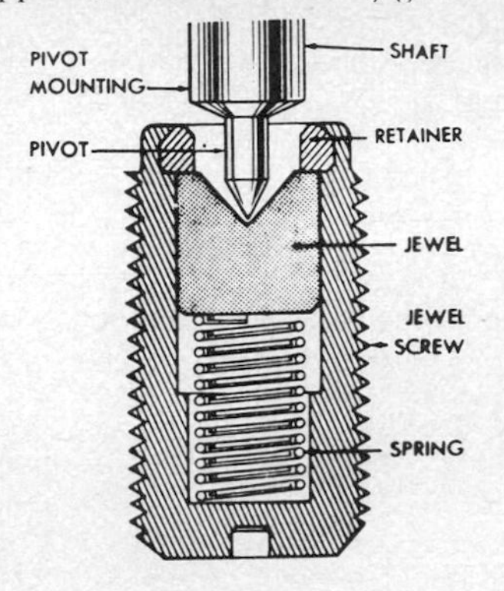

Jewel bearing in which jewel is spring-mounted for protection against excessive shock.

depression which serves as a bearing for the pivot of a meter movement or as a bearing in other delicate instruments.

Jezebel A long-life passive sonobuoy that is dropped from shore-based or carrier-based aircraft over water, to detect and transmit by radio the sounds made by a submerged enemy submarine.

JFET Abbreviation for *junction field-effect transistor.*

jitter Small, rapid variations in a waveform caused by mechanical vibrations, fluctuations in supply voltages, control-system instability, and other causes. It can cause unsteadiness of a trace or picture on a cathode-ray screen or raggedness in received facsimile copy.

JK flip-flop A flip-flop that changes the state of its output if both the J and K inputs are 1 (high) when a clock pulse arrives. If only the J input is 1,

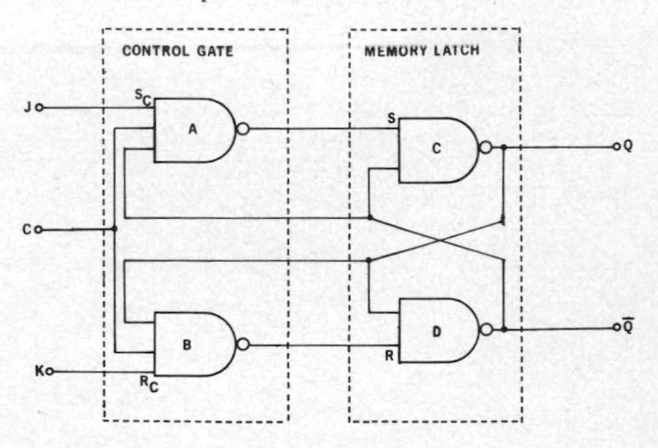

JK flip-flop arrangement.

a clock pulse drives the output to 1. If only the K input is 1, a clock pulse drives the output to 0. The choice of the letters JK for this flip-flop was purely arbitrary.

Jodrell Bank The site of a radio telescope near Manchester, England. Used also for tracking space vehicles.

jogging Quickly repeated opening and closing of a circuit to produce small movements of the driven machine. Also called inching.

Johnson noise *Thermal noise.*

joint A juncture of two wires or other conductive paths for current. Permanent joints are usually

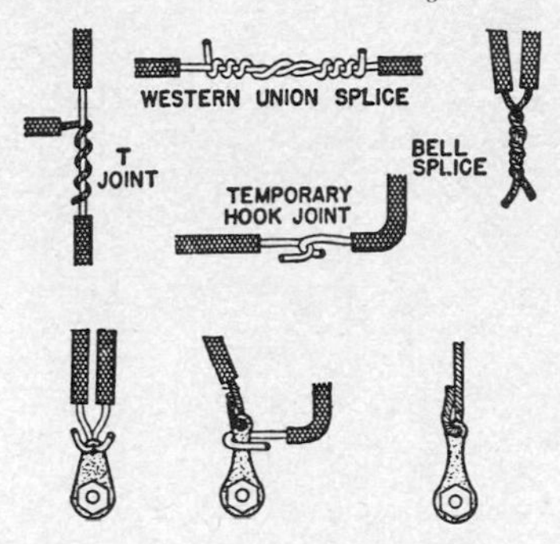

Joints used for soldered connections.

soldered, whereas temporary joints are generally held together by spring clips or screws.

Joint Electronic Devices Engineering Council [abbreviated JEDEC] A U.S. group established to deal with industrywide electronic problems.

Jones chopper A chopper that uses an autotransformer and two silicon controlled rectifiers in

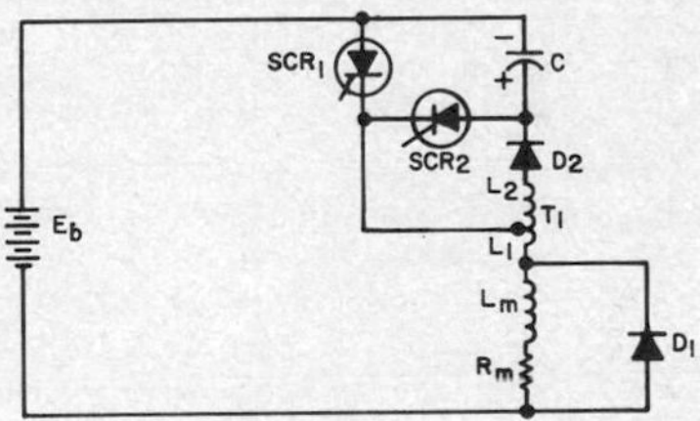

Jones chopper.

an inverter circuit which starts reliably from a DC source and provides an AC or higher DC output voltage at a desired value.

Josephson effect When a DC voltage V is applied to two superconductors separated by a very thin oxide insulating layer under cryogenic conditions (several kelvins), the frequency of the AC voltage developed across the insulating gap is equal to $2eV/h$, where e is electron charge and h is Planck's constant. Current flows through the insulators by tunneling. Applications include high-speed switching of logic circuits and memory cells (well under 100 ps), parametric amplifiers operating up to at least 300 GHz, and maintenance of the

U.S. legal volt at the National Bureau of Standards.

joule [abbreviated J] The unit of work and energy in the International System of Units (SI). One joule is the work done by a force of 1 N acting through a distance of 1 m.

Joule effect 1. The heating effect produced by the flow of current through a resistance. 2. *Magnetostriction.*

Joule's law The rate at which heat is produced in a constant-resistance electric circuit is proportional to the square of the current.

JOVIAL A computer programming language developed for computations, command and control applications, and some types of data processing.

joystick A two-axis displacement control operated by a lever or ball, for XY positioning of a device or an electron beam.

J scope A radarscope that produces a J display.

juice Slang term for *current.*

jukebox An automatic phonograph that has labeled controls which allow a choice from as many as 200 records, played by depositing a coin in a slot.

Julie An active submarine detection system in which a small charge of TNT is dropped from an antisubmarine aircraft and detonated underwater when a submerged submarine has been detected. The direct detonation sound and the echo from the target submarine are picked up by a passive Jezebel sonobuoy and transmitted via UHF to the aircraft for conversion to range and bearing data.

jump A digital-computer programming instruction that conditionally or unconditionally specifies the location of the next instruction and directs the computer to that instruction. Used to alter the normal sequence of the computer.

jumper A short length of conductor used to make a connection between two points or terminals in a circuit or to provide a path around a break in a circuit.

junction 1. A region of transition between two different semiconducting regions in a semiconductor device, such as a PN junction, or between a metal and a semiconductor. The four types of junctions are alloy, diffused, electrochemical, and grown. 2. A fitting used to join a branch waveguide at an angle to a main waveguide, as in a T junction. Also called waveguide junction.

junction battery A nuclear battery in which strontium-90 radioactive material irradiates a PN silicon junction to generate a useful voltage.

junction box An enclosure into which wires or cables are led and connected to form joints. It provides mechanical protection for the joints.

junction capacitor An integrated-circuit capacitor that uses the capacitance of a reverse-biased PN junction. This can be formed at the same time as the emitter or collector junctions of transistors. Maximum capacitance values are limited to a few hundred picofarads.

junction diode A semiconductor diode in which the rectifying characteristics occur at a junction between N- and P-type semiconducting materials.

junction field-effect transistor [abbreviated JFET] A field-effect transistor in which a bar of one type of semiconductor material has junctions of the opposite type diffused on both sides and connected together to form the gate. A terminal at one end of the bar is called the source; a terminal at the other end is called the drain. An N-channel JFET has a bar of N-type material; a P-channel JFET has a bar of P-type material. The most common type is the symmetrical JFET, in which source and drain connections can be interchanged because the geometry is symmetrical; this is indicated by a centered gate line on the symbol. In the symbol for a nonsymmetrical JFET, the gate line is off-center. The junction FET is often simply called an FET. The other basic type of FET is the MOSFET.

junction isolation Electrical isolation of a component on an integrated circuit by surrounding it with a region of a conductivity type that forms a junction, and reverse-biasing the junction so it has extremely high resistance.

junction point *Branch point.*

junction transistor A bipolar transistor in which the central base region is between the emitter and collector regions and separated from them by PN

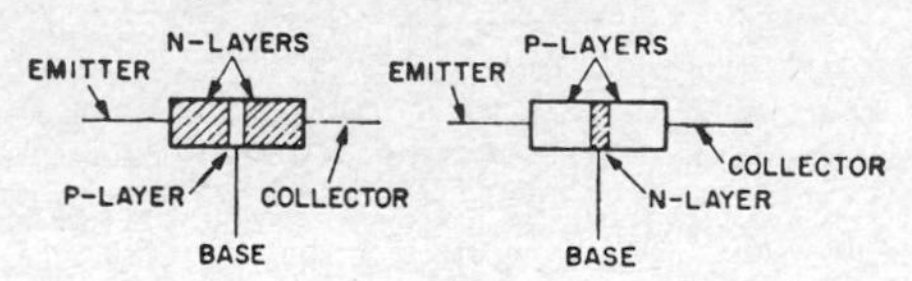

Junction-transistor construction for NPN and PNP types.

junctions. Major categories of junction transistors include grown-junction, alloy-junction, diffusion (such as mesa and planar), and epitaxial types.

justify 1. To adjust the printing positions of characters on a page so full lines have the desired length and both left and right margins are flush. 2. To shift the contents of a data storage register so either the most significant or least significant digit is at some specified position in the register.

just scale A musical scale formed by taking three consecutive triads, each having the ratio 4 to 5 to 6 or 10 to 12 to 15, with the highest note of one triad serving as the lowest note of the next.

K

k Abbreviation for *kilo-*.

K 1. Symbol for *cathode.* 2. Abbreviation for *kelvin* (plural kelvins); here K is used without the degree sign and is separated by a space from the temperature value, as in 273.16 K. 3. Unofficial abbreviation for *kilohm* (1000 Ω); here the K follows the value, with no space between, as in 10K for 10,000 Ω. Used chiefly on diagrams. 4. Symbol for *relay.*

kA Abbreviation for *kiloampere.*

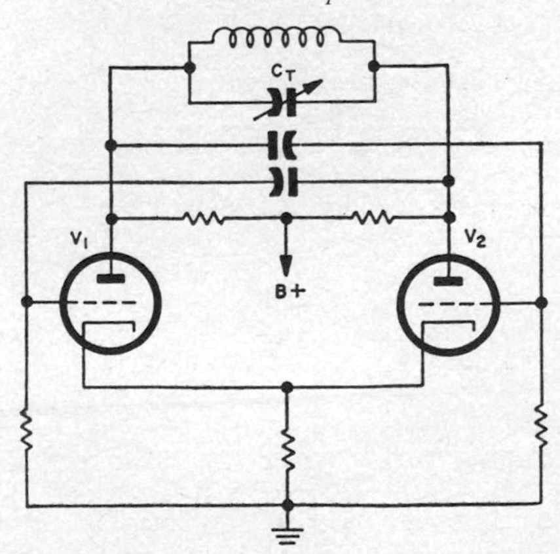

Kallitron-oscillator circuit.

kallitron oscillator A negative-resistance oscillator that uses two triodes, with the tank circuit connected between the two anodes.

Kalman filtering A computer technique used in inertial navigation systems to give a best estimate of the various navigation quantities at any instant, by taking account of the relative errors of the internal system and external measurements. The technique determines how much reliance should be placed on the inputs from the various sensors.

Kalvar Trademark for a photographic film that is developed and fixed by applying only heat. Used chiefly for microfilm applications.

Karnaugh map A truth table that has been rearranged to show a geometrical pattern of func-

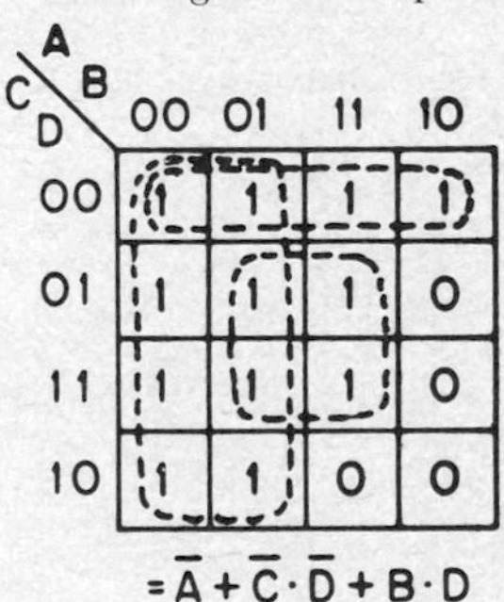

Karnaugh map, with dotted lines enclosing 1s that represent gating requirements of expression below map.

tional relationships for gating configurations. With this map, essential gating requirements can be recognized in their simplest form.

kb Abbreviation for *kilobit.*

K band A band of radio frequencies extending from 10.9 to 36 GHz, corresponding to wavelengths of 2.75 to 0.834 cm. Frequency limits for other bands are given in the entries for *band* and *X band.* Subdivisions of the K band (all values in gigahertz) are

Kp:	10.90–12.25	Kq:	20.5–24.5
Ks:	12.25–13.25	Kr:	24.5–26.5
Ke:	13.25–14.25	Km:	26.5–28.5
Kc:	14.25–15.35	Kn:	28.5–30.7
Ku:	15.35–17.25	Kl:	30.7–33.0
Kt:	17.25–20.50	Ka:	33.0–36.0

K_1 = Ku–Kq = 15.35–24.5

kBd Abbreviation for *kilobaud.*
kb/s Abbreviation for *kilobit per second.*
kbyte Abbreviation for *kilobyte.*
kbyte/s Abbreviation for *kilobyte per second.*
kc Abbreviation for *kilocycle,* now called kilohertz and abbreviated kHz.
K capture The capture by an atom's nucleus of an orbital electron from the first or K shell surrounding the nucleus.
K carrier system A carrier system that provides 12 telephone channels in a bandwidth of approximately 60 kHz.
kCi Abbreviation for *kilocurie.*
K display A modified radarscope A display in which a target appears as a pair of vertical deflections instead of as a single deflection. When the

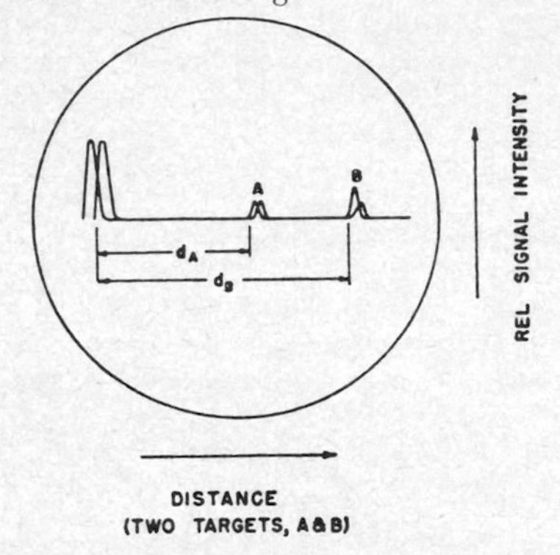

radar antenna is correctly pointed at the target in azimuth, the deflections are of equal height. When the antenna is not correctly pointed, the difference in pulse heights is an indication of direction and magnitude of azimuth pointing error.
KDP crystal Abbreviation for *potassium dihydrogen phosphate crystal.*
keep-alive circuit A circuit used with a TR, anti-TR, or other gas-discharge tube to produce residual ionization, to reduce the initiation time of the main discharge.
keep-alive electrode Deprecated term for *ignitor.*
keep-alive oscillator *Antistiction oscillator.*
keeper A bar of iron or steel placed across the poles of a permanent magnet to complete the magnetic circuit when the magnet is not in use, to avoid the self-demagnetizing effect of leakage lines. Also called magnet keeper.
K electron An electron that has an orbit in the K shell, which is the first shell of electrons surrounding the atomic nucleus, counting out from the nucleus.
K-electron capture The radioactive decay process in which an orbital electron from the K shell of an atom is captured by the nucleus of that atom. It results in the production of x-rays characteristic of the daughter atom. Other examples of such electron capture are L- and M-electron capture.
kelvin [abbreviated K, always used without the degree symbol °] The unit of temperature in the International System of Units (SI), equal to the fraction 1/273.16 of the thermodynamic temperature of the triple point of water. Formerly called degree Kelvin. The kelvin temperature scale uses Celsius degrees, but with the entire scale shifted so 0 K is at absolute zero. In this scale, water freezes at 273.16 K and boils at 373.16 K. Add 273.16 to any Celsius (formerly centigrade) value to get the corresponding value in kelvins.
Kelvin balance An ammeter in which the current to be measured is sent through two coils in series, one fixed just below the other, which is attached to one arm of a balance. The resulting force between the coils is then balanced against the force of gravity acting on a known weight at the other end of the balance arm.
Kelvin bridge A seven-arm bridge used to compare the four-terminal resistances of two four-

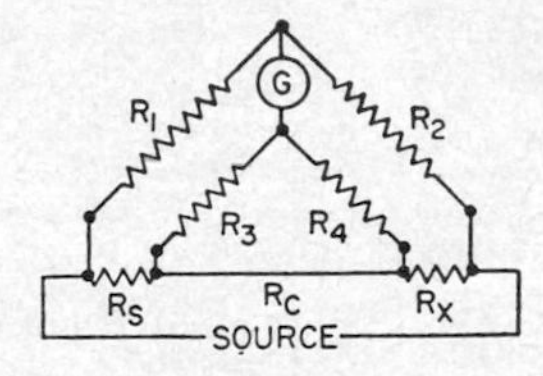

Kelvin-bridge circuit.

terminal resistors or networks. Also called double bridge and Thomson bridge.
Kelvin-Hughes projector A projector used in conjunction with a camera that photographs situation displays which appear on a character-generating tube and develops the transparencies automatically in a few seconds, for projecting onto a large screen in a combat control center.
Kennelly-Heaviside layer *E layer.*
kenotron A high-vacuum, high-voltage thermionic diode, used chiefly as a high-voltage rectifier.
Kerr cell A cell that contains a pair of electrodes in a dielectric liquid such as nitrobenzene. The dielectric becomes doubly refracting when under electric stress. If crossed Nicol prisms or Polaroid filters are put before and after the Kerr cell, no light passes through the combination when no voltage is applied to the cell. When a signal voltage

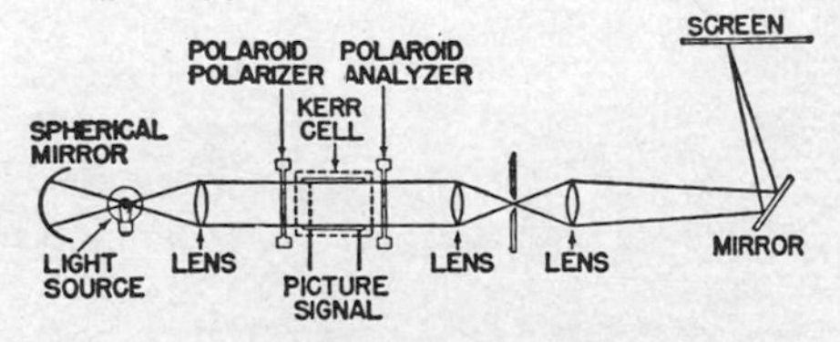

Kerr cell used in projection-type television receiver.

is applied to the cell, the plane-polarized light that enters the cell becomes elliptically polarized to an extent dependent on the voltage, and a proportional amount of light passes through the second prism. Used to modulate a beam of light or serve as a high-speed camera shutter.

Kerr magnetooptical effect Rotation of the plane of polarization of plane-polarized light when reflected from the polished pole surface of a strong magnet. The angle of rotation is proportional to the magnetizing force. Also called magnetooptical effect.

keV Abbreviation for *kiloelectronvolt.*

key 1. A hand-operated switch used for transmitting code signals. Also called signaling key. 2. A special lever-type switch used for opening or closing a circuit only as long as the handle is depressed. Also called switching key. 3. A projection that slides in a mating slot to achieve correct alignment of two parts being put together.

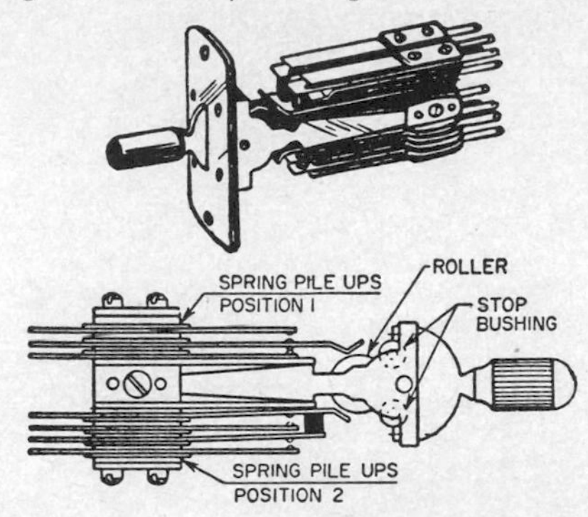

Key used for switching. Spring action returns lever handle to neutral position when it is released after being pushed up or down.

keyboard A set of keys that permits an operator to feed information manually into the central processing unit of a computer.

key click A transient signal sometimes produced when a radiotelegraph sending key is opened or closed. It is heard in a loudspeaker or headphone as a click or chirp. Also called keying chirp.

key-click filter A filter that attenuates the surges produced each time the keying circuit of a transmitter is opened or closed by the key. Also called click filter.

key-disk machine A keyboard machine used to record data directly on a magnetic disk.

keyed automatic gain control Automatic gain control in which an AGC tube in a television receiver is biased to cutoff and is unblocked only when the peaks of positive horizontal sync pulses act on its grid. This technique prevents the AGC voltage from being affected by noise pulses that occur in between sync pulses.

keyed rainbow generator A rainbow generator that has facilities for generating 3.85-MHz color-burst pulses, for making crossover adjustments and for general color television receiver troubleshooting.

keyer A device that changes the output of a transmitter from one value of amplitude or frequency to another in accordance with the intelligence to be transmitted.

keying The forming of signals, such as for telegraph transmission, by modulating a DC or other carrier between discrete values of some characteristic.

keying chirp *Key click.*

keying frequency The maximum number of times per second that a black line signal occurs when scanning the subject copy in facsimile.

keying wave *Marking wave.*

keylock switch A switch that can be operated only by inserting and turning a key like that used in ordinary locks.

keypunch A keyboard machine used to punch information manually into punched cards. Also called card punch.

key station The station at which a network radio or television program originates.

keystone distortion Camera-tube distortion such that the length of a horizontal scan line is linearly related to its vertical displacement. It occurs when the electron beam in the camera tube scans the image plate at an acute angle. A system that has keystone distortion distorts a rectangular pattern into a trapezoidal pattern. The distortion is normally corrected by special transmitter circuits.

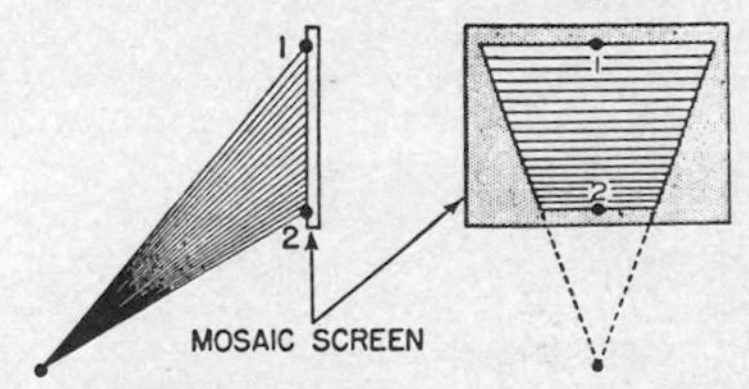

Keystone distortion in camera tube.

keyswitch A switch that is operated by depressing a key on the keyboard of a data-entry terminal. The switch may use mechanical contacts, sealed magnetic reeds, a saturable core, a change in capacitance, a Hall-effect transducer, or some other type of transducer to generate the output for encoding.

key-tape machine A keyboard machine used to record data directly onto magnetic tape.

keyword-in-context index [abbreviated KWIC index] A computer-generated listing of titles of documents, produced on a line printer, with the keywords lined up vertically in a fixed position within the title and arranged in alphabetic order.

kg Abbreviation for *kilogram.*

kG Abbreviation for *kilogauss.*

kHz Abbreviation for *kilohertz.*

kickback The voltage developed across an inductance when current flow is cut off and the magnetic field collapses.

kickback power supply *Flyback power supply.*

kickpipe A metal or plastic pipe used to protect cables from mechanical injury when brought through a floor or deck.

kick-sorter British term for *pulse-height analyzer.*

Kikuchi lines A pattern that consists of pairs of white and dark parallel lines, obtained when an electron beam is scattered (diffracted) by a crystalline solid. The pattern gives information on the structure of the crystal.

kilo- [abbreviated k] A prefix representing 10^3, or 1000.

kiloampere [abbreviated kA] One thousand amperes.

kilobaud [abbreviated kBd] One thousand bauds.

kilobit [abbreviated kb] One thousand bits.

kilobit per second [abbreviated kb/s] One thousand bits per second, used in specifying the modulation rate of a digital transmission system.

kilobyte [abbreviated kbyte] One thousand bytes.

kilobyte per second [abbreviated kbyte/s] One thousand bytes per second, a rating used in specifying the speed of a digital transmission system.

kilocurie [abbreviated kCi] One thousand curies.

kilocycle [abbreviated kc] One thousand cycles per second. Now called kilohertz and abbreviated kHz.

kiloelectronvolt [abbreviated keV] One thousand electronvolts. It is the energy acquired by an electron that has been accelerated through a voltage difference of 1000 V.

kilogauss [abbreviated kG; plural kilogauss] One thousand gauss.

kilogram [abbreviated kg] The unit of mass in the International System of Units (SI).

kilohertz [abbreviated kHz] One thousand hertz. Formerly called kilocycle.

kilohm [abbreviated kΩ: often abbreviated K on diagrams] One thousand ohms. Thus 15K on a diagram means 15,000 Ω.

kilolumen [abbreviated klm] One thousand lumens.

kilometer [abbreviated km] One thousand meters, or 3280 ft.

kilometric wave A radio wave between the wavelength limits of 1 and 10 km, corresponding to the low-frequency (LF) range of 30 to 300 kHz.

kiloton One thousand tons.

kilovolt [abbreviated kV] One thousand volts.

kilovoltage A voltage of the order of thousands of volts, such as the voltage applied to an x-ray tube.

kilovoltampere [abbreviated kVA] One thousand voltamperes.

kilovoltmeter A voltmeter whose scale is calibrated to indicate voltage in kilovolts.

kilovolt peak [abbreviated kV P] The maximum value in kilovolts of the positive peak of an applied voltage waveform.

kilovolt peak-to-peak [abbreviated kV P-P] The voltage in kilovolts as measured between the maximum positive and negative peaks of the applied voltage waveform.

kilowatt [abbreviated kW] One thousand watts.

kilowatthour [abbreviated kWh] One thousand watthours.

kiloword One thousand words, used in specifying the capacity of a computer memory.

kine [pronounced kinny] Slang term for *kinescope recording.*

kinescope *Picture tube.*

kinescope recording A motion-picture film made by photographing images on the face of the picture tube in a television monitor or receiver, to permit repeating the same television program later and at different television stations. The sound portion of the program is usually recorded separately on magnetic tape or transcription disks. Also called kine (slang) and television recording.

kinetic energy Energy associated with motion.

Kipp relay A type of bistable multivibrator circuit that can be triggered by a short-duration pulse to transfer conduction from one tube to the other and switched back by a similar pulse of opposite polarity.

Kirchhoff's laws The sum of the currents flowing to a given point in a circuit is equal to the sum of the currents flowing away from that point. The algebraic sum of the voltage drops in any closed path in a circuit is equal to the algebraic sum of the electromotive forces in that path. Also called laws of electric networks.

Kleinschmidt printer A type of page printer that is capable of receiving ASCII or Baudot code at speeds exceeding 300 b/s and providing a readable printout.

K line One of the characteristic lines in the x-ray spectrum of an atom. It is produced by excitation of the electrons of the K shell.

klm Abbreviation for *kilolumen.*

klystron An electron tube in which the electrons are periodically bunched by electric fields. The resulting velocity-modulated electron beam is fed into a cavity resonator to sustain oscillations within the cavity at a desired microwave frequency. It is

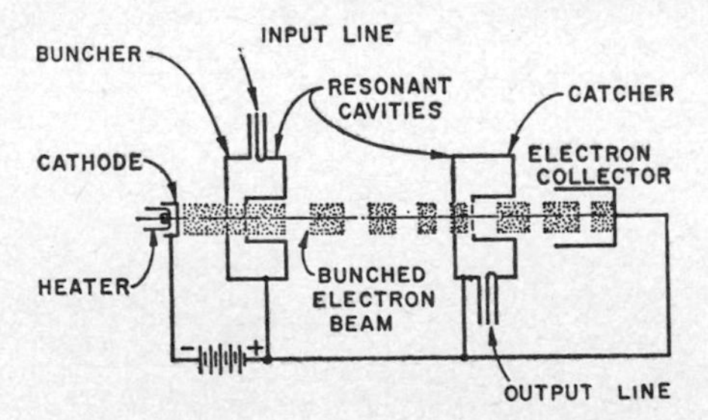

Klystron amplifier tube construction and operation.

used as an oscillator or amplifier in UHF applications such as microwave relay and radar transmitters and receivers.

klystron frequency multiplier A two-cavity klystron in which the output cavity is tuned to a multiple of the fundamental frequency.

klystron repeater A microwave repeater that consists of a klystron inserted directly into a waveguide. Incoming waves velocity-modulate the electron stream emitted by the cathode of the tube. A second cavity converts the velocity-modulated beam back into waves, but with greatly increased amplitude, and feeds them into the output waveguide.

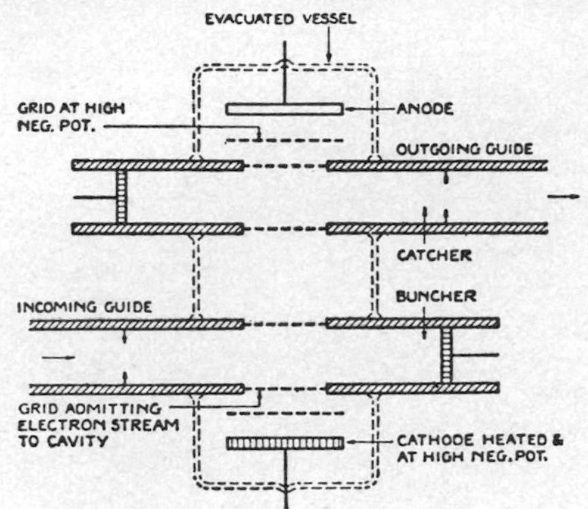

Klystron repeater in waveguide.

km Abbreviation for *kilometer.*

K meson One of the four types of K mesons, all having masses of about 970 electron masses. Formerly called tau meson.

knee The curve that joins two relatively straight portions of a characteristic curve.

knife-edge refraction A radio propagation effect wherein the normal attenuation of a signal by the atmosphere is reduced when the signal passes over and is diffracted by a sharp obstacle such as a mountain ridge.

knife switch A switch in which one or more hinged metal blades are manually pushed between spring contacts.

knob A component that is placed on a control shaft to facilitate manual rotation of the shaft. The knob sometimes has a pointer or markings to indicate shaft position.

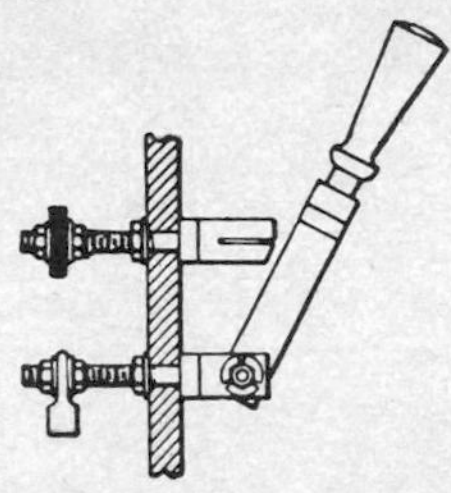

Knife switch, correctly mounted so gravity will not tend to close it.

knockout A removable portion in a metal cabinet or outlet box, readily removed with a hammer or screwdriver for installation of a cable connector or other fittings.

knot [abbreviated kt] A unit of speed equal to 1 nautical mile per hour or 1.15 mi/h (1.85 km/h).

kΩ Abbreviation for *kilohm,* equal to 1000 Ω.

Kovar Trademark for an iron-nickel-cobalt alloy used in making metal-to-glass seals.

K radiation The x-ray radiation emitted when K electrons are excited, generally by bombardment of a metal with electrons.

krypton [symbol Kr] A gaseous element. Atomic number is 36.

krypton discharge tube A cold-cathode discharge tube that contains krypton gas, used primarily as a high-voltage switching device which has a short delay time and a very fast rise time.

K scope A radarscope that produces a K display.

K shell The innermost layer of electrons surrounding the atomic nucleus; it has electrons characterized by the principal quantum number 1.

kt Abbreviation for *knot.*

Kurie plot A graph of a beta-particle spectrum. Also called Fermi plot.

kV Abbreviation for *kilovolt.*

kVA Abbreviation for *kilovoltampere.*

kV P Abbreviation for *kilovolt peak.*

kV P-P Abbreviation for *kilovolt peak-to-peak.*

kW Abbreviation for *kilowatt.*

kWh Abbreviation for *kilowatthour.*

KWIC index Abbreviation for *keyword-in-context index.*

L

L 1. Abbreviation for *lambert*. 2. Symbol for *coil*. 3. Abbreviation for *inductance*.

L Mathematical symbol for *inductance*.

label One or more characters used to identify an item, record, message, or file in a computer program.

labeled molecule A molecule that contains one or more atoms of radioactive or stable isotopes which may be followed conveniently through biological, chemical, or physical processes.

labeling The substitution, for purposes of using a tracer technique, of a radioisotope or a rare stable isotope for the naturally occurring isotope of a particular element in a chemical molecule. Also called tagging.

labile *Unstable*.

labile oscillator An oscillator whose frequency is controlled from a remote location by wire or radio.

laboratory power supply A self-contained power supply that converts an AC line voltage to a regulated DC voltage which can be varied over a specified range. Designed primarily for use with equipment being developed or tested in a laboratory.

laboratory system A frame of reference that is attached to the observer's laboratory and hence usually is at rest relative to the surface of the earth. Used in nuclear physics.

labyrinth A loudspeaker enclosure that has air chambers at the rear which absorb rearward-radiated acoustic energy, to minimize acoustic standing waves.

lacing Tying insulated wires together to support each other and form a single neat cable, with separately laced branches.

lacquer disk A mechanical recording disk made of metal, glass, or paper coated with a lacquer compound that often contains cellulose nitrate. Also called cellulose nitrate disk.

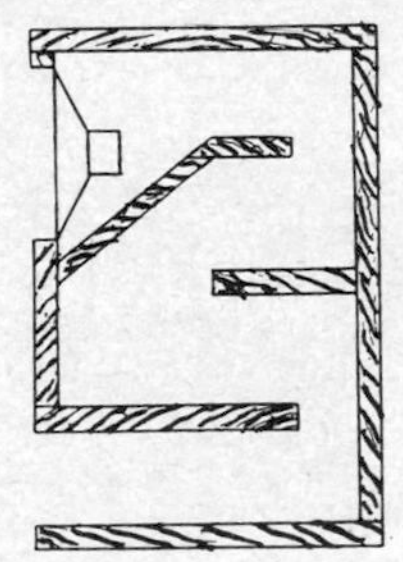

Labyrinth for single loudspeaker.

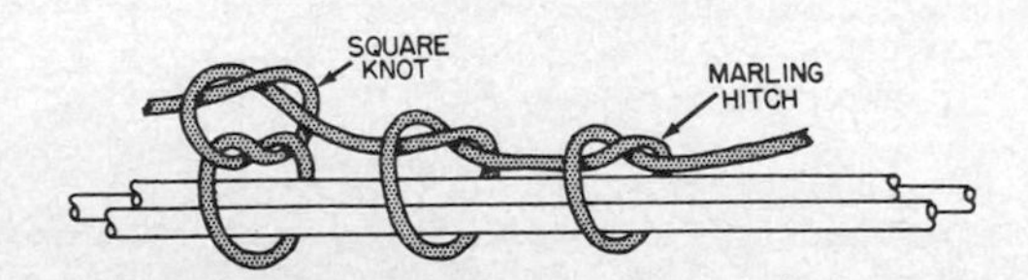

Lacing with single waxed cord.

lacquer-film capacitor A capacitor in which the dielectric is a solid lacquer film that has thin metallic coatings which act as electrodes.

lacquer master Deprecated term for *lacquer original*.

lacquer original An original recording made on a lacquer surface, for making a master. Also called lacquer master (deprecated).

lacquer recording Any recording made on a lacquer recording medium.

LADAR (LAser Detection And Ranging) *Laser radar*.

ladder adder A digital-to-analog converter that

uses a resistive ladder network and high-speed low-resistance transistor switches to produce an analog output which is proportional to the value of a binary or binary-coded-decimal digital input.

ladder attenuator An attenuator that has a series of symmetrical sections designed so the impedance remains essentially constant in both directions as the amount of attenuation is varied.

ladder network A network composed of a sequence of H, L, T, or pi networks connected in tandem. Some versions use piezoelectric ceramic

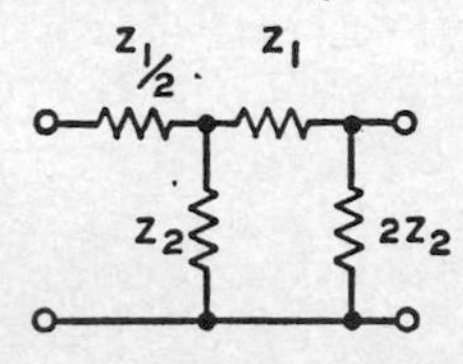

Ladder network, showing ratios of impedance values.

elements as impedances for ladder filters. Used in narrow-bandpass filters and analog-to-digital converters.

laddic A multiaperture magnetic structure that resembles a ladder, used to perform logic functions. Operation is based on a flux change in the

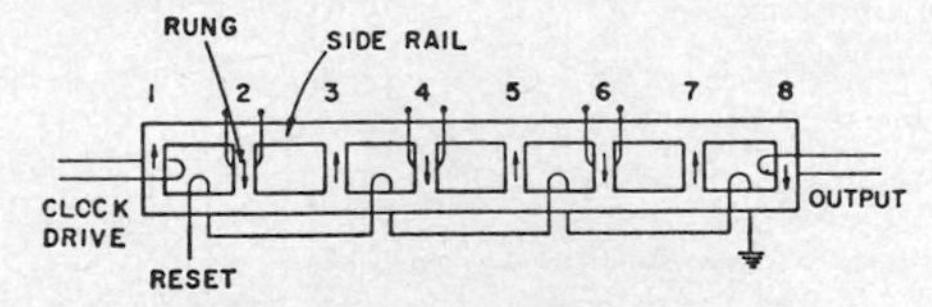

Laddic structure having eight rungs.

shortest available path when adjacent rungs of the ladder are initially magnetized with opposite polarity.

lag 1. The difference in time between two events or values considered together. Often expressed in

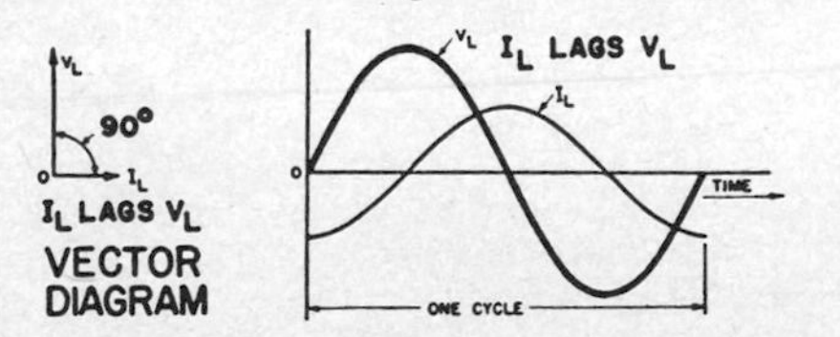

Lag of coil current behind coil voltage by 90°.

degrees when comparing alternating quantities; thus the current through a perfect coil lags the applied voltage by 90°. 2. A persistence of the electric-charge image in a camera tube for a small number of frames. 3. The time delay between an initiating action and the desired effect.

lagging current An alternating current that reaches its maximum value up to 90° behind the voltage which produces it. A lagging current flows only in a circuit that is predominantly inductive.

lagging load *Inductive load.*

lambda [symbol λ] A Greek letter that designates wavelength in meters.

lambda diode A two-terminal, negative-resistance, solid-state device that consists of a pair of complementary depletion-mode junction field-ef-

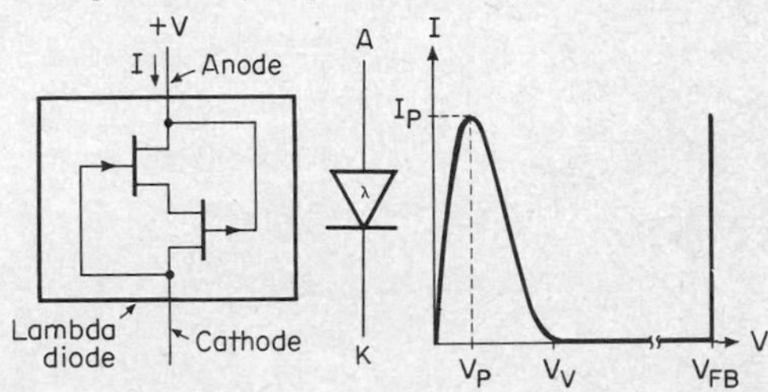

Lambda diode interconnections, symbol, and characteristic curve.

fect transistors. It can be constructed on a single chip or combined with bipolar and metal-oxide semiconductor devices in integrated circuits. Applications include amplifier, memory, oscillator, and switching circuits.

lambda particle A hyperon that has an extremely short life (about 3.7×10^{-10} s) and a mass between that of neutrons and deuterons.

lambert [abbreviated L] A CGS unit of luminance, equal to $1/\pi$ candela/ cm². It corresponds to the uniform luminance of a perfectly diffusing surface that is emitting or reflecting light at the rate of 1 lumen/cm². Use of the SI unit of luminance, the candela per square meter, is preferred.

Lamb wave An electromagnetic wave propagated over the surface of a solid whose thickness is comparable to the wavelength of the wave.

laminate A product made by bonding together two or more layers of materials.

laminated core An iron core for a coil, transformer, armature, or other electromagnetic device, built up from laminations stamped from sheet iron or steel. The laminations are more or less insulated from each other by surface oxides and sometimes by application of varnish. Laminated construction is used to minimize the effect of eddy currents.

laminated plastic A plastic material that is made by applying heat and pressure to particles or sheets of filler materials which have been impregnated with a thermosetting resin.

laminated shield A radiation shield consisting of alternate layers of materials with different shielding characteristics, such as water for neutron attenuation and steel for gamma-ray attenuation.

lamination One of the thin punchings of iron or steel used in building up a laminated core for a magnetic circuit.

laminography Radiography of a particular layer of a body or object. In one method the x-ray tube and the film are moved simultaneously in opposite directions about a pivotal point in the plane of the layer. Only the plane of the layer then produces a well-defined image. Also called body-section radiography, planigraphy, and tomography.

lamp A device that produces light, such as an electric lamp.

lamp bank A number of incandescent lamps connected together in parallel or series as a resistance load for full-load tests of electric equipment.

lamp cord Two twisted or parallel insulated wires, usually No. 18 or No. 20, used chiefly for connecting electric equipment to wall outlets.

LANAC [LAminar Navigation AntiCollision] An aircraft radio navigation system that consists of airborne interrogator and ground transponder equipments with height-coding of the airborne interrogator pulses.

Lance A surface-to-surface missile that has inertial guidance and a liquid propulsion system, to provide artillery support for infantry, armored, mechanized, and airborne divisions.

land 1. The record surface between two adjacent grooves of a mechanical recording. 2. *Terminal area.*

Landau damping Damping of a space-charge wave by electrons that move at the phase velocity of the wave.

landing aid A lamp, searchlight, radio beacon, radar device, communicating device, or any system of such devices for aiding aircraft in an approach and landing.

landing beacon A radio beacon that produces a landing beam for aircraft guidance.

landing beam A radio beam, highly directional in both elevation and azimuth, that slants upward from the landing surface of an airport. It is produced by a landing beacon and serves as the glide path in an instrument landing system for aircraft.

landline A wire connection between two ground locations.

landmine A mine that is concealed below the surface of the earth and exploded by a fuze actuated by the weight of a vehicle or person.

land mobile service Radio service between base and mobile stations or between mobile stations operating on land.

land return *Ground clutter.*

lane The surface bounded by adjacent lines of position that have the same value of the cyclic parameter for a navigation system.

Langmuir dark space A nonluminous region that surrounds a negatively charged probe inserted into the positive column of a glow discharge.

Langmuir probe A small metallic conductor inserted within a plasma to sample the plasma current.

language The characters, combining rules, and meanings used to express and process information for handling by computers and associated equipment.

L antenna *Inverted-L antenna.*

lanthanide One of the rare-earth elements; they have atomic numbers 57 to 71, inclusive. All have chemical properties similar to those of lanthanum.

lanthanum [symbol La] A rare-earth element. Atomic number is 57.

lanthanum-doped lead zirconate-lead titanate *Lead lanthanum zirconate titanate.*

lap 1. A rotating flat or curved disk, commonly of cast iron, used to grind quartz crystal plates, semiconductor blanks, flat optical objects, and other objects prior to polishing. 2. To grind a flat or curved surface with a lap.

lap dissolve Changeover from one television scene to another so that the new picture appears gradually at the same rate at which the previous picture disappears.

lapel microphone A small microphone that can be attached to a lapel or pocket on the clothing of the user, to permit free movement while speaking.

Laplace's law The strength of the magnetic field at a given point, due to an element of a current-carrying conductor, is directly proportional to the strength of the current and the projected length of the element and inversely proportional to the square of the distance of the element from the point in question.

Laplace transform A special case of a Fourier transform.

lapping Moving a quartz, semiconductor, or other crystal slab over a flat plate on which a liquid abrasive has been poured, to obtain a flat polished surface or to reduce the thickness a carefully controlled amount.

large-scale integration [abbreviated LSI] The construction of well over 100 interconnected equivalent gate circuits (or other circuits of similar complexity) on a single integrated-circuit chip to form a major system function or act as a major subsystem. Either bipolar or metal-oxide semiconductor technology may be used. Maximum packing density of circuits is achieved by interconnecting the circuits automatically as part of the chip-production process. In contrast, medium-scale integration (MSI) is defined as covering the range of 10 to 100 circuit functions.

Larmor frequency The angular frequency of precession of a charged particle rotating in a magnetic field. The frequency value is proportional to the strength of the magnetic field and the gyromagnetic ratio.

Larmor orbit The circular motion of a charged particle in a uniform magnetic field. Although the motion of the particle is unimpeded along the magnetic field, motion perpendicular to the field is always accompanied by a force perpendicular to the direction of motion and the field; the path is therefore helical.

laryngaphone A microphone held against the throat of a speaker, to pick up voice vibrations directly without responding to background noise.

LASA [Large Aperture Seismic Array] An array of 525 seismometers that covers an area 200 km in diameter in eastern Montana; each instrument is buried in a 60-m-deep hole to minimize effects of local noise. The array can be steered electronically to pick up earth disturbances in an area of interest, such as underground nuclear explosions.

LASCR Abbreviation for *light-activated silicon controlled rectifier.*

LASCS Abbreviation for *light-activated silicon controlled switch.*

lase To generate coherent electromagnetic waves, as in a laser.

laser [Light Amplification by Stimulated Emission of Radiation] An active electron device that converts input power into a very narrow, intense beam of coherent visible or infrared light. The

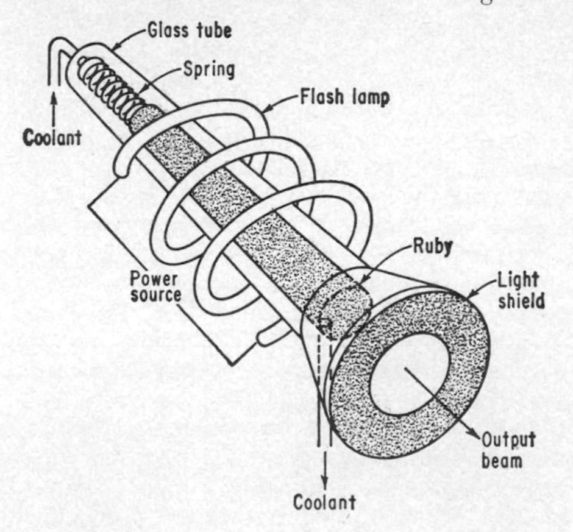

Laser, pulsed by flashes from helical flash lamp that encircles ruby rod.

input power excites the atoms of an optical resonator to a higher energy level, and the resonator forces the excited atoms to radiate in phase. The four basic types of laser are gas, liquid, semiconductor, and solid. Major applications include cutting, drilling, heating, welding, and other machining operations; distance measurement, alignment, and other surveying applications; missile guidance, ranging and tracking of moving targets, and other military applications; recording and holography; communication.

laser-acoustic delay A delay that involves interaction of a laser beam and a transparent acoustic delay line to provide a variable time delay for RF and microwave signals.

laser altimeter A continuous-wave gas laser whose output light beam is modulated by up to three radio frequencies before it is directed downward from an aircraft to scan the earth. The laser light reflected from the terrain is picked up by a telescope mirror system that is coaxial with the transmitting optics, sensed by a photomultiplier, and phase-compared with the transmitted signal to obtain round-trip propagation time and altitude.

laser anemometer An anemometer in which the wind being measured passes through two laser beams that are at right angles to each other. The wind changes the velocity of one or both of the beams, according to a principle of physics whereby

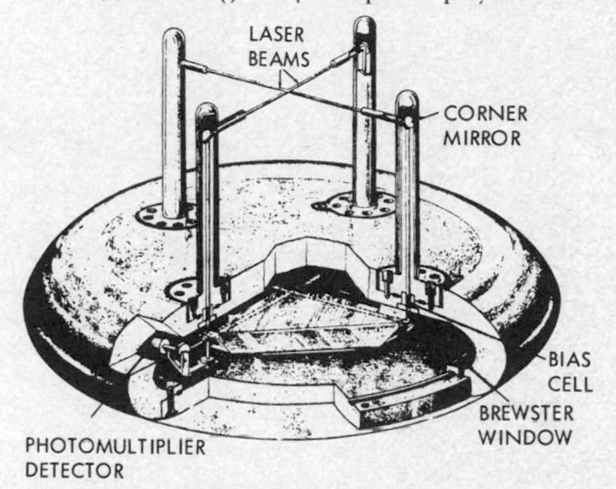

Laser anemometer.

light propagating through a moving, transparent medium undergoes a change in velocity. If one measures the changes in beam velocity with appropriate instrumentation, wind velocity can be determined with high accuracy for anything from a faint breeze to a hurricane.

laser-beam printer A nonimpact printer that operates at well over 10,000 lines per minute, using a low-power laser to produce image-forming charges a line at a time on the photoconductive surface of a drum. Dry powder that adheres only to charged areas is applied to the drum, transferred to plain paper, and fused by heat, as in Xerox copying machines. Developed to replace slower line printers for computer output. In another version, the modulated beam scans a dye-coated plastic ribbon in contact with plain paper to provide direct printing at somewhat slower speed.

laser camera An airborne camera system for night photography, in which the extremely narrow and almost invisible beam from a scanning laser is the source of illumination. The continuous-wave laser beam is split into two beams. One beam goes to a motor-driven six-sided prism scanner lens that both reflects the beam down to earth and sweeps it from side to side to give scanning action. The other beam is modulated by the output of a photomultiplier detector in a Schmidt optical system that picks up the light reflected from the ground area being scanned. The modulated beam is swept back and forth over moving

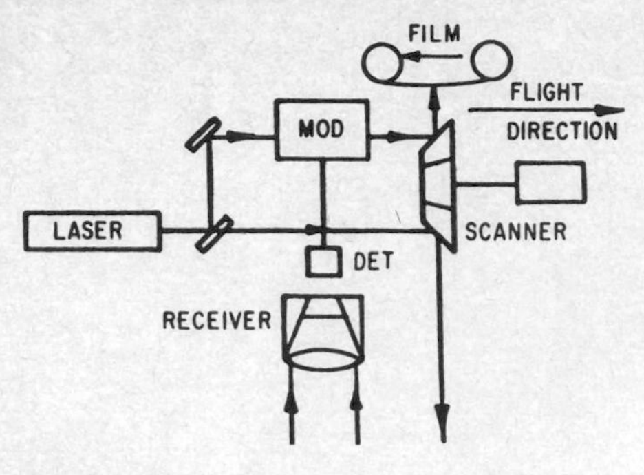

Laser camera for aerial night photography.

film by another part of the same motor-driven scanner, to give night photographs without the use of telltale magnesium flares or powerful strobe lights.

laser ceilometer A ceilometer in which the time taken by a light pulse from a ground laser to travel straight up to a cloud ceiling and be reflected back to a receiving photomultiplier is measured and converted into a cathode-ray display that indicates cloud-base height.

laser communication Optical communication in which the light source is a laser whose beam is modulated for voice, video, or data communication over information bandwidths up to 1 GHz. Developed primarily for communication between the earth and airplanes, orbiting satellites, and deep-space missions, and for satellite-to-satellite links.

laser diode *Semiconductor laser.*

laser drill A drill in which concentrated light from a ruby laser generates intense heat for drilling holes as small as 0.0001 in (2.5 μm) in diameter in gem stones, tungsten carbide, and other hard materials. Hole size is easily changed.

laser flash tube A high-power air or water-cooled xenon flash tube that produces high-intensity flashes for laser pumping applications.

laser fusion The use of an intense beam of laser light to heat a small pellet of deuterium and tritium to a temperature of about 100,000,000°C, as required for initiating a fusion reaction. One major problem is development of a laser with sufficiently high power in the visible light spectrum to achieve the required energy transfer within the pellet at the critical instant when it is hit by the beam.

laser guidance Guidance in which the target is continuously illuminated by a laser beam from an aircraft or other location so missiles, bombs, or projectiles equipped with suitable seeker heads can home in on the laser energy reflected by the target. Laser and television-guided bombs used successfully on North Vietnam targets were initially called "smart bombs."

laser-guided bomb A bomb that carries guidance for homing on a target which is illuminated by a laser beam from the bomb-carrying aircraft or a spotter aircraft.

laser gyro A gyro in which two laser beams travel in opposite directions over a ring-shaped path formed by three or more mirrors. When the position of the ring is changed in inertial space, the clockwise and counterclockwise paths have different lengths and produce a frequency difference proportional to the rate at which the gyro position changes. Rotation is thus measured without the use of a spinning mass. Also called ring laser.

laser interferometer An interferometer in which a laser is the light source for measuring displacements with a linear resolution of the order of 0.000001 in (25 nm) and angular resolution of about 0.1 arc-second, at ranges up to about 5 m. Also used to control step-and-repeat cameras for fabricating and positioning photomasks used in the production of integrated circuits.

laser intrusion-detector A photoelectric intrusion-detector in which a laser is a light source that produces an extremely narrow and essentially invisible beam around the perimeter of the area being guarded. The narrow and intense beam of a laser can be projected long distances, as required for protecting entire airports or large military areas.

laser jamming An electronic countermeasure in which a continuous-wave laser directs jamming energy back to a hostile laser receiver to prevent it from interfering with the use of laser rangefinders, radars, and tracking equipment during an attack.

laser memory A computer memory in which a controlled laser beam acts on individual and extremely small areas of a photosensitive or other type of surface, for storage and subsequent readout of digital data or other types of information.

laser photocoagulator A laser combined with an ophthalmoscope for directing bursts of coherent light through a human eye to burn selected points on a detached retina. Subsequent healing of

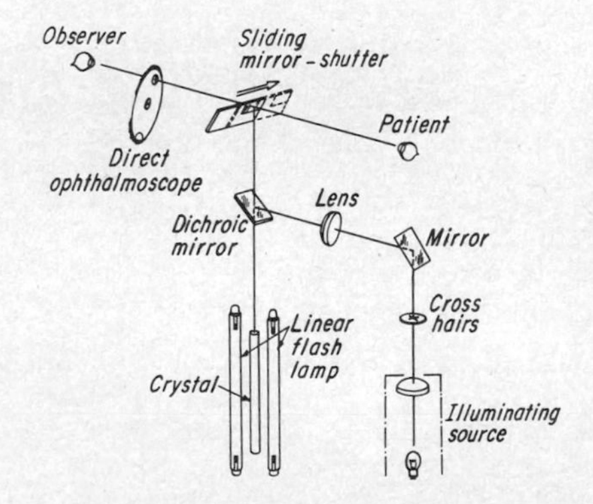

Laser photocoagulator optical paths.

the burns causes scars that weld the retina back into position.

laser radar Radar in which the coherent light beam of a laser is used in place of a microwave beam for normal radar applications and for tracking satellites, surveying the moon, and other space applications. Also called LADAR.

laser radiation detector A photodetector that responds primarily to the coherent visible, infrared, or ultraviolet light of a laser beam. One application is warning tank and other surface military vehicle operators that they are being illuminated by a laser-target designator and are therefore vulnerable to laser-guided weapons.

laser rangefinder A simplified version of laser radar, used primarily by artillery forward observers, battalion observation posts, tanks, helicopters, and other military vehicles for accurately measuring target distance. A digital counter determines elapsed time between a transmitted laser pulse and the return pulse reflected off the target. Multiplying one-way travel time by the velocity of light gives the range.

laser recorder An image reproducer that resembles a facsimile system, in which a laser beam is initially modulated by the video signal and swept over photographic film or paper to reproduce an image received over wire or radio communication systems. Applications include reproduction of news photos transmitted by wire or radio; reproduction of photos transmitted from aircraft, orbiting satellites, or spacecraft; and the recording of billions of bits of information on film for computer data storage. In a variation used for producing computer-output microfilm, the laser beam is modulated by the output of a computer to generate alphameric characters.

laserscope A pulsed high-power laser used with appropriate scanning and imaging devices to sense objects over the sea at night or in fog and provide three-dimensional images on a viewing screen. Underwater versions can extend the range of television cameras many times. Applications include ship navigation within harbors and harbor surveillance.

laser scriber A laser-cutting setup used in place of a diamond scriber for dicing thin slabs of silicon, gallium arsenide, and other semiconductor materials used in the production of semiconductor diodes, transistors, and integrated circuits. The laser beam vaporizes grooves in the upper active face of the wafer, after which a breaking operation separates the chips along the grooves. Also used for scribing sapphire and ceramic substrates.

laser seismometer A laser interferometer system that detects seismic strains in the earth by measuring changes in distance between two granite piers located at opposite ends of an evacuated pipe through which a helium-neon or other laser beam makes a round trip. Movements as small as 80 nm (one-eighth the wavelength of the 632.8-nm helium-neon laser radiation) can be detected.

laser spectrum The spectrum that includes all optical wavelengths, ranging from infrared through visible light to ultraviolet, in which coherent radiation can be produced by various types of lasers.

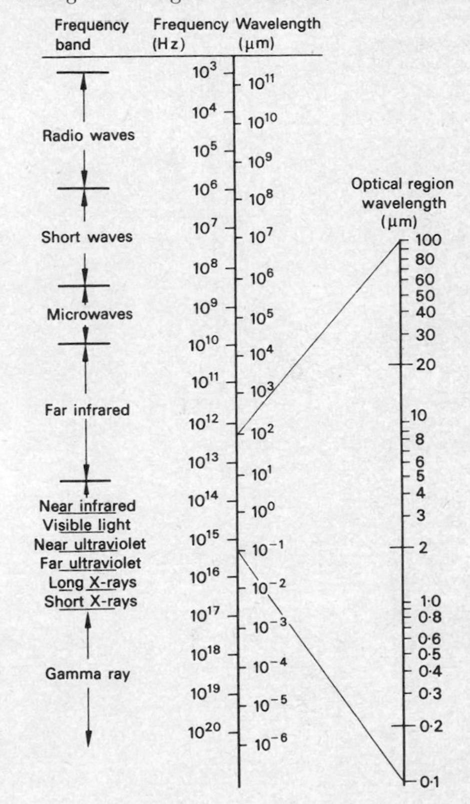

Laser spectrum covers entire optical region from wavelengths of about 0.1 μm in far ultraviolet to about 100 μm in far infrared.

laser threshold The minimum pumping energy required to initiate lasing action in a laser.

laser transit A transit in which a laser is mounted over the sighting telescope to project a clearly visible narrow beam onto a small target at the survey site. Uses include precise positioning of dredges for underwater excavations and alignment of tooling jigs for wing panels of large aircraft.

Laser transit as modified for automatic control of bulldozer. Position-sensing photodetector on machine feeds correction signal to blade through servomechanism to produce surface that is exactly parallel to laser beam.

laser-triggered switch A high-voltage high-power switch that consists of a spark gap which is triggered into conduction by a laser beam. Used for switching voltages ranging from a few kilovolts to well over 3 MV in electric power systems.

laser velocimeter A velocity-measuring instrument that uses a continuous-wave laser to send a beam of coherent light at an object moving at right angles to the beam. The needlelike diffraction lobes reflected by the moving object sweep past the optical grating in the receiver, thereby generating in a photomultiplier a series of impulses from which velocity can be determined and read out. In another version, the Doppler shift of scattered light from a moving target is measured by detecting with a photodetector the beat signal between scattered and unscattered light. The beat frequency is proportional to target velocity. Also called diffraction velocimeter and optical diffraction velocimeter.

laser welding Use of a laser beam to produce reliable spot welding of the thin foils and fine wires used in integrated circuits and other microelectronic devices.

lasing The process of generating radiation anywhere in the laser spectrum at a frequency that is characteristic of the material used in a laser by pumping or exciting electrons into higher energy states.

last-in first-out [abbreviated LIFO] A method of establishing the order in which data items are taken from a computer memory or products are taken from inventory.

latching circuit A bistable circuit that holds a switching circuit or device in its on position.

latching circulator A switchable circulator in which no holding current is required to maintain a desired microwave path after switching action is accomplished by an electromagnet or permanent magnet acting on a ferrite core. Used in striplines and waveguides.

latching phase shifter A phase shifter in which no holding current is required to maintain a desired phase shift after the device is switched magnetically to the new value. Used in striplines and waveguides.

latching reed relay A reed-type relay in which a holding coil keeps the relay contacts closed until they are released either by a reverse-polarity pulse in that coil or by a pulse applied to a separate release coil.

latching relay A relay that has contacts which lock in the energized or deenergized positions or both until reset manually or electrically. The latching action is produced by a mechanical latch. In contrast, in a lock-up relay the lock-up is accomplished magnetically or electrically.

latch-in relay A relay that maintains its contacts in the last position assumed, even without coil energization.

latency The time a digital computer takes to deliver information from its memory. It may be the time spent waiting for the desired location on a magnetic drum to appear under a reading head. In a serial storage system, latency is the access time minus the word time.

latent image A stored image, as in the form of charges on a mosaic of small capacitances.

latent period The interval between irradiation and the appearance of results.

lateral compliance The force required to move the stylus from side to side as it follows the grooves of a phonograph record.

lateral parity Parity associated with an individual character, to indicate whether the number of holes punched across paper tape is even or odd. For magnetic tape the term applies to bits recorded in parallel tracks.

lateral recording A type of disk recording in which the groove modulation is parallel to the surface of the recording medium, so that the cutting stylus moves from side to side during recording.

lateral transistor A transistor in which the emitter- and collector-base junctions are formed in separate areas, with the current flowing between the junctions in a plane parallel to the surface.

lattice 1. A pattern of identifiable intersecting lines of position laid down in fixed positions with respect to the transmitters that establish the pattern for a navigation system. 2. A structure used in a nuclear reactor, made up of discrete bodies of fissionable and nonfissionable material arranged in a geometric pattern. 3. An orderly arrangement of atoms in a crystalline material.

lattice imperfection A deviation from a perfect homogeneous lattice in a crystal.

lattice network A network composed of four branches connected in series to form a mesh. Two nonadjacent junction points serve as input terminals, and the remaining two junction points serve as output terminals.

Lattice network as drawn in two different ways.

lattice scattering Scattering of electrons by collisions with vibrating atoms in a crystal lattice, reducing the mobility of charge carriers in the crystal and thereby affecting its conductivity.

lattice-wound coil *Honeycomb coil.*

Laue pattern The characteristic photographic record obtained when x-rays from a pinhole or slit are sent through a single crystal that diffracts or bends the rays in all directions.

launcher Any device used to hold, support, and sometimes direct a missile or spacecraft during launching.

launching The process of transferring energy

from a coaxial cable or transmission line to a waveguide.

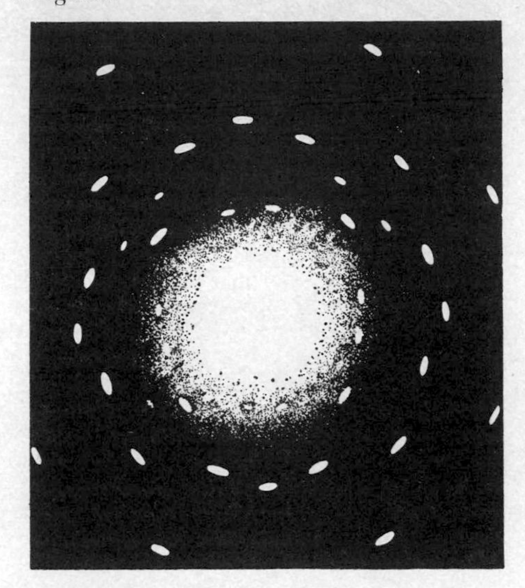

Laue pattern.

launching guidance Navigation control of a missile during launching.

launch vehicle A rocket that sets a missile or spacecraft in motion before another engine takes over. Also called booster rocket.

launch window An interval of time during which a spacecraft must be launched to reach a desired objective.

lava A natural fired stone that consists chiefly of magnesium silicate, used for insulators.

law of electric charges Like charges repel; unlike charges attract.

law of electromagnetic induction *Faraday's law.*

law of electromagnetic systems An electromagnetic system tends to change its configuration so that the flux of magnetic induction will be a maximum.

law of electrostatic attraction *Coulomb's law.*

law of induced current *Lenz's law.*

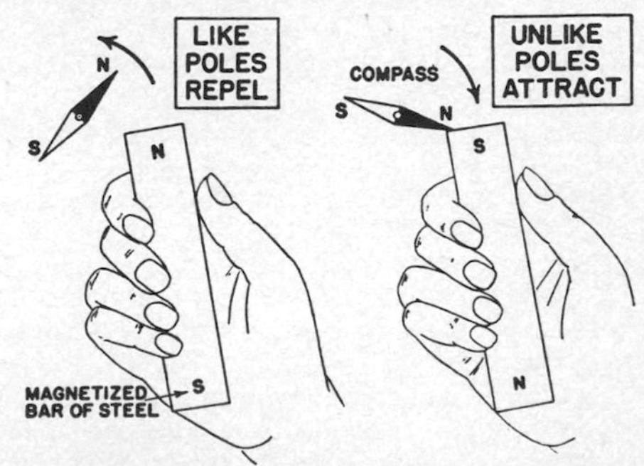

Law of magnetism.

law of magnetism Like poles repel; unlike poles attract.

Lawrence tube *Chromatron.*

lawrencium [symbol Lw] A short-lived radioactive element whose atomic number is 103, produced artificially by bombardment of californium.

laws of electric networks *Kirchhoff's laws.*

lay 1. The manner in which the turns of a coil are positioned with respect to each other. In a perfect lay, the wires are staggered with respect to the wires of the previous layer so the coil occupies minimum space. 2. The axial length of a turn of a helix in a cable that has twisted strands.

layer winding A coil-winding method in which adjacent turns are laid evenly side by side along the length of the coil form. Any number of additional layers may be wound over the first, usually with sheets of insulating material between the layers.

layout A diagram that indicates the positions of various parts on a chassis or panel.

lazy H antenna An antenna array in which two or more dipoles are stacked one above the other to obtain greater directivity.

L band A band of radio frequencies extending from 390 to 1550 MHz, corresponding to wavelengths of 76.9 to 19.37 cm. Subdivisions of this band (all values in megahertz) are

Lp:	390–465	Ls:	900–950
Lc:	465–510	Lx:	950–1150
Ll:	510–725	Lk:	1150–1350
Ly:	725–780	Lf:	1350–1450
Lt:	780–900	Lz:	1450–1550

Television UHF channels 14 through 83 cover 470 to 890 MHz in the L band. Frequency limits for other bands are given in the entries for *band.*

lb/in² Abbreviation for *pounds per square inch.*

LC Abbreviation for *inductance-capacitance.*

L-carrier system A telephone carrier system that occupies a frequency band which extends from 68 kHz to over 8 MHz, used with coaxial cable systems and microwave and tropospheric scatter radio systems.

L cathode A dispenser cathode that has a porous tungsten body covered with a layer of barium and oxygen atoms. Evaporated electron-emitting barium is replaced continuously from a compound located in a chamber behind the tungsten.

LCD Abbreviation for *liquid crystal display.*

***LC* product** The product of the inductance value L in henrys and the capacitance value C in farads.

***L/C* ratio** The ratio of inductance L in henrys to capacitance C in farads for a resonant circuit.

LDE Abbreviation for *long-delayed echo.*

L display A radarscope display in which the target appears as two horizontal pulses or blips, one extending to the right and one to the left from a central vertical time base. When the radar antenna is correctly aimed, the relative blip amplitudes indicate the pointing error. The position of the signal along the baseline indicates target distance.

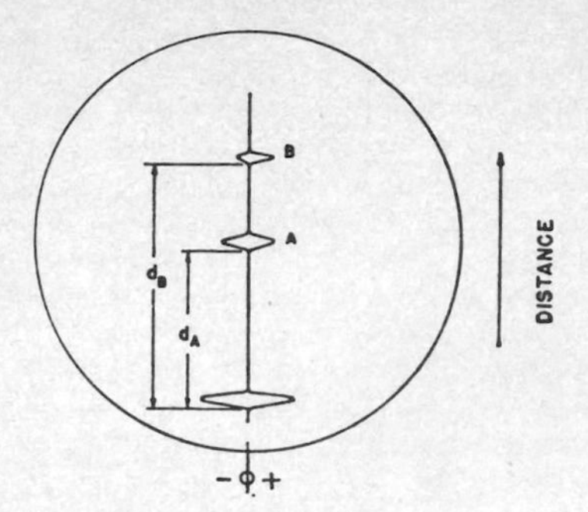

L display.

The display may be rotated 90° when used for elevation aiming instead of azimuth aiming.

LDR Abbreviation for *light-dependent resistor.*

lead [pronounced led; symbol Pb] A soft gray metallic element used as a shielding material in nuclear work and with tin in solder. Atomic number is 82.

lead [pronounced leed] 1. The angle by which one alternating quantity leads another in time, expressed in degrees or radians. Thus the current through a perfect capacitor leads the applied voltage by 90°. 2. The distance between a moving target and the point at which a gun or missile is aimed. 3. A wire that connects together two points in a circuit.

lead-acid cell The cell in an ordinary storage battery; the electrodes are grids of lead that contain lead oxides which change in composition during charging and discharging. The electrolyte is dilute sulfuric acid. Leakproof versions use a lead grid rolled like a bandage, sealed into a container with a carefully measured amount of acid.

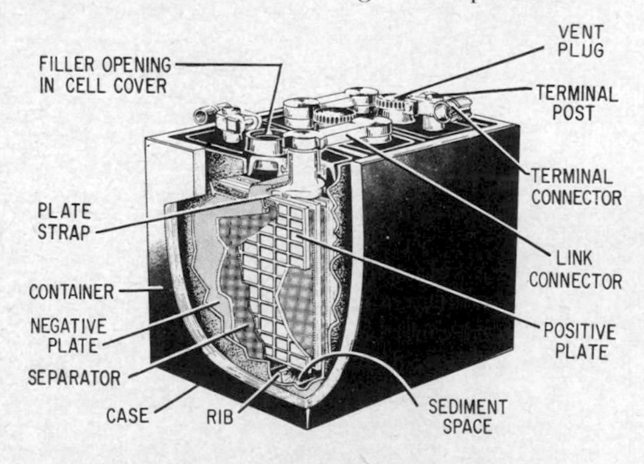

Lead-acid cell construction for 6-V storage battery having three 2-V cells in series.

lead-covered cable A cable whose conductors are protected from moisture and mechanical damage by a sheath of lead.

lead dioxide primary cell A reserve primary cell that is activated by adding liquid. It is inexpensive and withstands low temperatures but has a short life after activation.

lead equivalent The thickness of lead that gives the same reduction in radiation dose rate as the material in question.

leader An unused or blank length of tape at the beginning of a reel, used primarily for threading purposes.

leader cable A cable used as a navigation aid, in which the path to be followed is defined by the magnetic field produced by current flowing through the cable.

lead-in A single wire that connects a single-terminal outdoor antenna to a receiver or transmitter. Dipoles and other two-terminal antennas use transmission lines for this purpose. Also called down-lead.

leading current An alternating current that reaches its maximum value up to 90° ahead of the voltage which produces it. A leading current flows in any circuit that is predominantly capacitive.

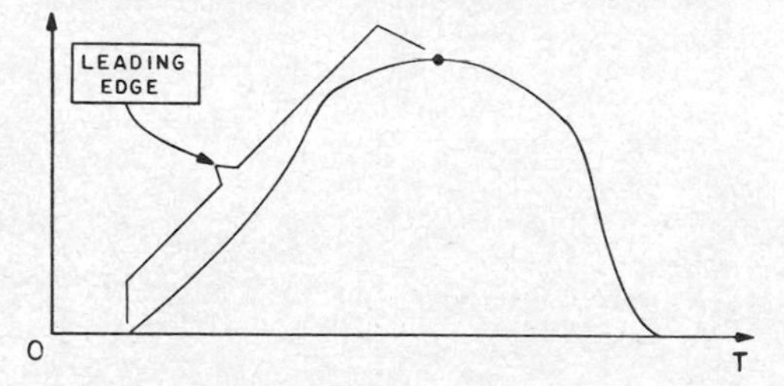

Leading edge of pulse.

leading edge The major portion of the rise of a pulse.

leading-edge pulse time The time at which the instantaneous amplitude of a pulse first reaches a stated fraction of the peak pulse amplitude.

leading ghost A ghost displaced to the left of the image on a television receiver screen.

leading load A load that is predominantly capacitive, so its current leads the alternating voltage applied to the load.

lead-in groove A blank spiral groove at the beginning of a disk recording; it generally has a pitch much greater than that of the recorded grooves, to bring the pickup stylus quickly to the first recorded groove. Also called lead-in spiral.

leading zeros Zeros preceding the first nonzero integer of a number.

lead-in insulator A tubular insulator inserted into a hole drilled through a wall, through which the lead-in wire can be brought into a building.

lead-in spiral *Lead-in groove.*

lead lanthanum zirconate titanate [abbreviated PLZT] A ferroelectric, ceramic, electrooptical material whose optical properties can be changed by an electric field or by being placed in tension or compression. Used in a variety of optoelectronic storage and display devices. Also called lanthanum-doped lead zirconate-lead titanate.

leadout groove A blank spiral groove at the end of a disk recording, generally of a pitch much greater than that of the recorded grooves, connected to either the locked or eccentric groove. It actuates the record-changing mechanism of an automatic record changer. Also called throwout spiral.

leadover groove A groove cut between separate selections or sections on a disk recording to transfer the pickup stylus rapidly from one cut to the next. Also called crossover spiral.

lead screw [pronounced leed screw] A threaded shaft that converts rotation to longitudinal motion. In a disk recorder it guides the cutter at a desired rate across the surface of an ungrooved disk during recording so the grooves will be appropriately spaced.

lead storage battery A storage battery that uses lead-acid cells.

lead sulfide [symbol PbS] The mineral galena, used as a crystal detector in the early days of radio and now used as an infrared detector.

lead sulfide cell A cell that detects infrared radiation. Either its generated voltage or change of resistance may be used as a measure of the intensity of the radiation.

lead telluride [symbol PbTe] A compound of lead used in infrared detectors.

lead time [pronounced leed time] The time allowed or required to initiate and develop a piece of equipment that must be ready for use at a given time.

lead zirconate titanate [abbreviated PZT] A ferroelectric, ceramic, electrooptical material that has lower optical transparency than PLZT but similar other properties.

leakage Undesired and gradual escape or entry of a quantity, such as loss of neutrons by diffusion from the core of a nuclear reactor, escape of electromagnetic radiation through joints in shielding, flow of electricity over or through an insulating material, and flow of magnetic lines of force beyond the region in which useful work is performed.

leakage current 1. Undesirable flow of current through or over the surface of an insulating material or insulator. 2. The flow of direct current through a poor dielectric in a capacitor. 3. The alternating current that passes through a rectifier without being rectified.

leakage flux Magnetic lines of force that go beyond their intended path and do not serve their intended purpose.

leakage inductance Self-inductance caused by leakage flux in a transformer.

leakage power The RF power transmitted through a fired TR or pre-TR tube.

leakage radiation Radiation from anything other than the intended radiating system. A common example is electromagnetic radiation that escapes through joints or defects in shielding.

leakage reactance Inductive reactance caused by leakage flux that links only the primary winding of a transformer.

leakage resistance The resistance of the path over which leakage current flows. It is normally a high value.

leak detector An instrument that finds small holes or cracks in the walls of a vessel. The helium spectrometer is an example.

leaky Having leakage. Often applied to a capacitor in which the resistance has dropped far below its normal value so that excessive leakage current flows.

leaky-wave antenna A wideband microwave antenna that radiates a narrow beam whose direction varies with frequency. It is fundamentally a perforated waveguide, thin enough to permit flush mounting for aircraft and missile radar applications. Scanning is achieved by sweeping the frequency between two limits at the desired scanning rate.

leaky waveguide A waveguide that has a narrow longitudinal slot through which energy leaks out continuously.

leapfrogging The process of phasing or delaying the ranging pulse of a tracking radar, to shift a target blip on a scope presentation past the target blip from another radar.

leapfrog test A computer test that uses a special program which performs a series of arithmetic or logic operations on one group of storage locations, transfers itself to another group, checks the correctness of the transfer, then begins the series of operations again. Eventually, all storage positions will have been tested.

learning machine A machine that is capable of improving its future actions as a result of analysis and appraisal of past actions. One simple example is character reading, in which the machine can improve its ability to recognize the various forms of a given character as more and more of these characters are shown with their correct identification, or when a human works with the machine to identify each error made by the machine.

least significant bit [abbreviated LSB] The bit that carries the lowest value or weight in binary notation for a numeral. For example, when 13 is represented by binary 1101, the 1 at the right is the least significant bit.

least significant character The character in the rightmost position in a number or word.

Lecher line *Lecher wires.*

Lecher-line oscillator A Hartley oscillator that uses Lecher wires as a tank circuit.

Lecher wires Two parallel wires that are several wavelengths long and a small fraction of a wavelength apart, used to measure the wavelength of a microwave source which is connected to one end of the wires. A sliding shorting bar is moved along the wires, the positions of standing-wave nodes are noted, and the distance between nodes is then used to determine wavelength or frequency. Also called Lecher line.

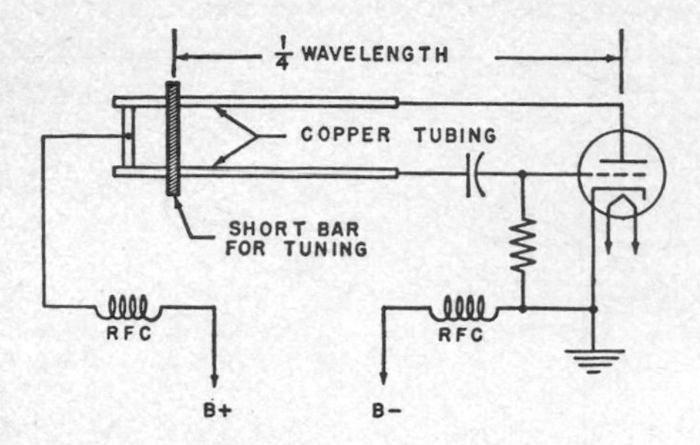

Lecher-line oscillator circuit.

Leclanche cell The common dry cell, which is a primary cell that has a carbon positive and a zinc negative electrode in an electrolyte of sal ammoniac and a depolarizer.

LED Abbreviation for *light-emitting diode.*

Leduc effect If a magnetic field is applied at right angles to the direction of a temperature gradient in a conductor, a temperature difference is produced at right angles both to the direction of the temperature gradient and the direction of the magnetic field.

left-hand polarized wave An elliptically polarized transverse electromagnetic wave in which the rotation of the electric field vector is counterclockwise for an observer looking in the direction of propagation. Also called counterclockwise polarized wave.

left-hand rule 1. For a current-carrying wire: If the fingers of the left hand are placed around the wire so that the thumb points in the direction of electron flow, the fingers will be pointing in the

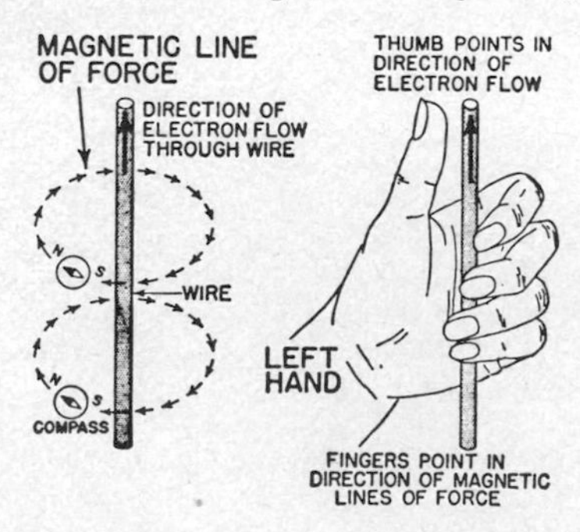

Left-hand rule for electron flow.

direction of the magnetic field produced by the wire. For conventional current flow (the opposite of electron flow), the right hand is used. 2. For a movable current-carrying wire or an electron beam in a magnetic field: If the thumb, first, and second fingers of the left hand are extended at right angles to one another, with the first finger representing the direction of magnetic lines of force and the second finger representing the direction of electron flow, the thumb will be pointing in the direction of motion of the wire or beam. Also called Fleming's rule.

left-hand taper A taper in which there is greater resistance in the counterclockwise half of the operating range of a rheostat or potentiometer (looking from the shaft end) than in the clockwise half.

left-justify 1. To adjust the printing positions of characters on a page so the left margin is lined up. 2. To shift the contents of a register so the left or most significant digit is at some specified position.

left signal The output of a microphone placed to pick up the intensity, time, and location of sounds originating predominantly to the listener's left of the center of the performing area when making a stereo recording or broadcast.

left stereo channel The left signal as electrically reproduced in stereo FM broadcasts or stereo records.

leg *Radio-range leg.*

L electron An electron that has an orbit in the L shell, which is the second shell of electrons surrounding the atomic nucleus, counting out from the nucleus.

L-electron capture A mode of electron capture similar to K-electron capture except that the electrons are initially in the L shell of the atom.

Lenard rays Cathode rays produced in air by a Lenard tube.

Lenard tube An early experimental electron-beam tube that has a thin glass or metallic foil window at the end opposite the cathode, through which the electron beam can pass into the atmosphere.

length 1. The number of subunits of data, usually digits or characters, that can be simultaneously stored or processed by a given computer device. 2. The time that data is delayed during transmission, usually expressed in microseconds. Thus the length of a delay line is rated in microseconds.

lens 1. A dielectric or metallic structure that is transparent to radio waves and can bend these waves to produce a desired radiation pattern. Used with antennas for radar and microwave relay systems. 2. One or more pieces of precisely contoured glass or other transparent material, used to focus light rays and form images by refraction. 3. An arrangement of electrodes that produces an electric field which focuses electrons into a beam. 4. An arrangement of coils or permanent magnets that produces a magnetic field which focuses electrons. Sometimes used in combination with an electric field. 5. A structure that concentrates sound waves by refraction.

lens antenna A microwave antenna in which a dielectric lens is placed in front of the dipole or horn radiator to concentrate the radiated energy into a narrow beam. Also used to focus received energy on the receiving dipole or horn.

Lenz's law The current induced in a circuit due to its motion in a magnetic field or to a change in its magnetic flux is in such a direction as to exert a mechanical force opposing the motion or to oppose the change in flux. Also called law of induced current.

lepton A particle that has a small mass, such as an electron, positron, neutrino, or antineutrino.

lethal dose A dose of ionizing radiation sufficient to cause death. Median lethal dose is the dose required (about 400 roentgens for a person) to kill half the individuals in a large group similarly exposed within a specified period of time.

lethargy A neutron energy rating. The lower the energy of a neutron, the more lethargic it is, and the higher its lethargy.

letter An alphabet character that represents sounds of a spoken language.

level 1. The difference between a quantity and an arbitrarily specified reference quantity, usually expressed as the logarithm of the ratio of the quantities. In audio and communication work, a common reference level for power is 1 mW, and power levels are expressed in dBm (decibels above 1 mW). Thus 0.01 mW is −20 dBm, 1 mW is 0 dBm, and 1 W is 30 dBm. 2. A specified position on an amplitude scale applied to a signal waveform, such as reference white level and reference black level in a standard television signal. 3. A charge value that can be stored in a given storage element of a charge-storage tube and distinguished in the output from other charge values. 4. Volume of sound. 5. A single bank of contacts, as on a stepping relay.

level above threshold The pressure level of a sound in decibels above its threshold of audibility for the individual observer. Also called sensation level.

level converter An amplifier that converts nonstandard positive or negative logic input voltages to standard DTL or other logic levels.

level indicator 1. An instrument that indicates liquid level. 2. An indicator that shows the audio voltage level at which a recording is being made. It may be a volume-unit meter, neon lamp, or cathode-ray tuning indicator.

level shifting Changing the logic level at the interface between two different semiconductor logic systems.

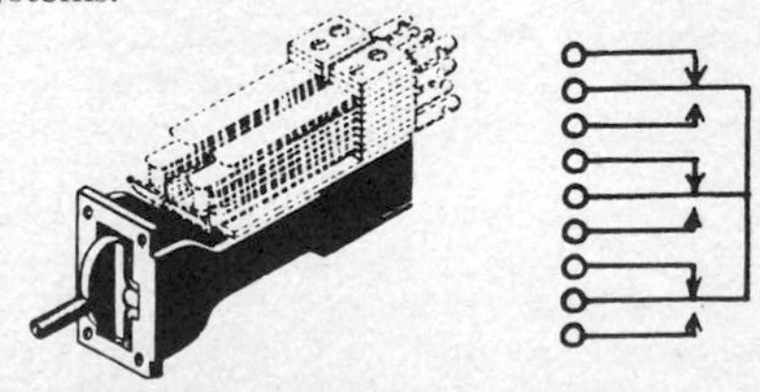

Lever pileup switch and symbol.

lever pileup switch A lever switch in which fixed and moving contacts are stacked alternately in piles, with the moving contacts being acted on by a cam when the lever-shaped handle of the switch is moved.

lever switch A switch that has a lever-shaped operating handle, such as a lever pileup switch.

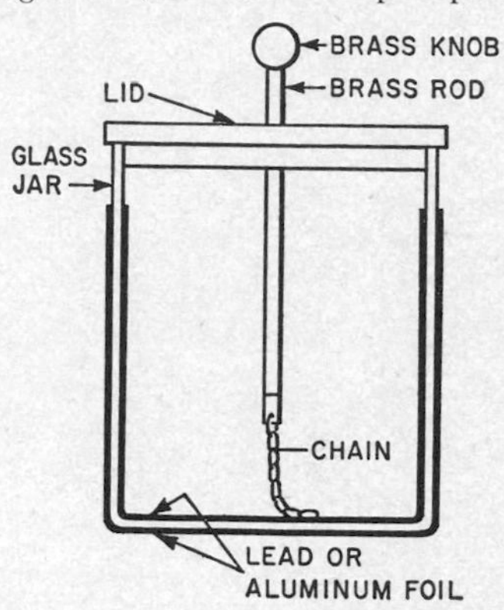

Leyden jar used as capacitor. Terminals are brass knob and outer foil.

Leyden jar An early type of capacitor that consists simply of metal foil sheets on the inner and outer surfaces of a glass jar.

LF Abbreviation for *low frequency.*

LF loran Abbreviation for *low-frequency loran.*

library A collection of standard and tested computer programs, routines, and subroutines, usually stored on magnetic tape.

libration Tilting of the moon slightly in one direction and then in the opposite direction so only three-sevenths of the surface is permanently concealed from positions on the earth's surface.

LIDAR [Laser Infrared raDAR] A meteorological instrument in which a ruby laser generates intense infrared pulses in beam widths as small as 30 seconds of arc, for measuring atmospheric conditions. Reflections and scattering effects of clouds, smog layers, and some atmospheric discontinuities are measured by radar techniques. LIDAR can also be used for tracking weather balloons, smoke puffs, and rocket trails.

lie detector *Polygraph.*

life test A test in which a device is operated under conditions that simulate a normal lifetime of use, to obtain an estimate of life expectancy.

lifetime *Mean life.*

LIFO Abbreviation for *last-in first-out.*

light Electromagnetic radiation that has wavelengths capable of causing the sensation of vision, ranging approximately from 4000 Å or 0.4 μm (far violet) to 7700 Å or 0.77 μm (far red). The velocity of light is the same as that of radio waves and is 186,282.3960 mi/s (299,792.4562 km/s). Invisible infrared radiation produced by lasers is sometimes also called light.

light-activated silicon controlled rectifier [abbreviated LASCR] A silicon controlled rectifier that has a glass window for incident light that takes the place of, or adds to the action of, an electric gate current in providing switching action. Also called photo-SCR and photothyristor.

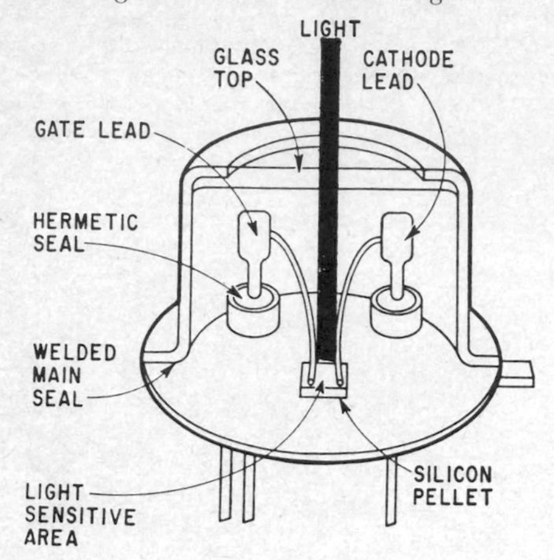

Light-activated silicon controlled rectifier.

light-activated silicon controlled switch [abbreviated LASCS] A semiconductor device that has four layers of silicon alternately doped with acceptor and donor impurities, as in a light-activated silicon controlled rectifier, but with all four of the P and N layers made accessible by terminals. When a light beam hits the active light-sensitive surface, the photons generate electron-hole pairs that make the device turn on. Removal of light does not reverse the phenomenon. The switch can be turned off only by removing or reversing its positive bias.

light amplifier An amplifier in which both the input and output signals consist of light.

light-beam pickup A phonograph pickup in which a beam of light is a coupling element of the transducer.

light carrier injection A method of introducing the carrier in a facsimile system by periodic variation of the scanner light beam, the average amplitude of which is varied by the density changes of the subject copy. Also called light modulation.

light chopper A rotating fan or other mechanical device that interrupts a light beam which is aimed at a phototube, to permit AC amplification of the phototube output and to make its output independent of strong, steady, ambient illumination.

light-controlled oscillator An oscillator whose output frequency varies with incident light.

light-dependent resistor [abbreviated LDR] *Photoconductive cell.*

light detector *Photodetector.*

light-emitting diode [abbreviated LED] A semiconductor diode that converts electric energy efficiently into spontaneous and noncoherent electromagnetic radiation at visible and near-infrared wavelengths by electroluminescence at a forward-biased PN junction. Typical semiconductor materials used include gallium phosphide (GaP) for red, green, or yellow light; gallium indium phosphide (GaInP) for yellow; gallium arsenide phosphide (GaAsP) or gallium aluminum arsenide (GaAlAs) for red; phosphor-coated cadmium arsenide (CaAs) for blue, green, or red; silicon carbide (SiC) for green or yellow; and gallium arsenide (GaAs) for infrared. The exact color is determined by both the semiconductor material used and the doping. In contrast, a semiconductor laser (laser diode) emits stimulated and coherent radiation. Also called solid-state lamp.

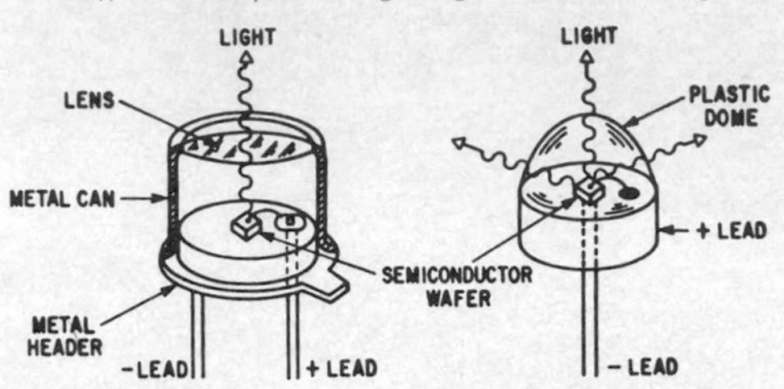

Light-emitting diodes as packaged commercially in modified transistor headers.

light flux *Luminous flux.*

light-gating cathode-ray tube A cathode-ray tube in which the electron beam varies the transmission or reflection properties of a screen that is positioned in the beam of an external light source. Used for projecting a large display on a movie-type screen. An example is the dark-trace tube.

light gun A light pen mounted in a gun-type housing.

lighthouse tube *Disk-seal tube.*

light-microsecond The distance a light wave travels in free space in 0.000001 s. Used as a unit of electrical distance, equal to approximately 983 ft (300 m).

light modulation *Light carrier injection.*

light modulator The combination of a source of light, an appropriate optical system, and a means for varying the resulting light beam to produce an optical sound track on motion-picture film.

light-negative Having negative photoconductivity, hence decreasing in conductivity (increasing in resistance) when exposed to light. Selenium sometimes exhibits this property.

lightning An electric discharge between clouds or a cloud and the earth.

lightning arrester A device that provides a discharge path to ground for lightning which strikes an antenna or transmission line. It generally contains a spark gap that has high resistance at nor-

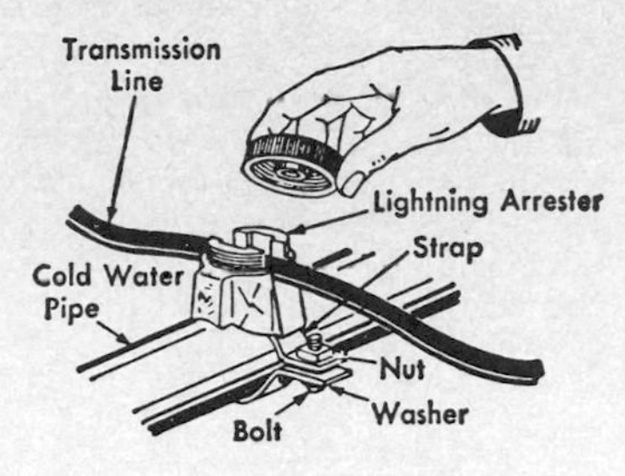

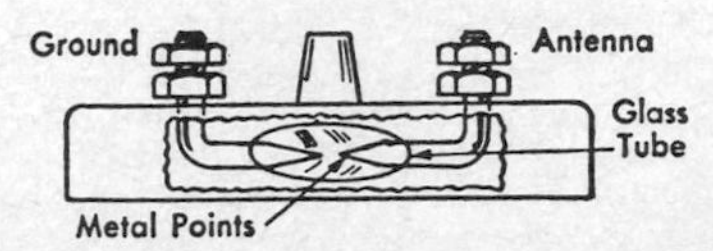

Lightning arrester designed for mounting on cold-water pipe and connecting to television twin-line, and cross section of typical lightning arrester.

mal circuit voltages. This gap breaks down to become a low resistance when acted on by the high-voltage surge of lightning. Also called arrester.

lightning generator A high-voltage power supply that generates surge voltages which resemble lightning, for testing insulators and other high-voltage components.

lightning rod A grounded metal rod that projects above the highest point on a structure; it has a sharp point at which a high charge density builds up to facilitate leakage of charged particles to ground, to minimize destructive lightning discharges during a storm.

light-operated switch A switch operated by a beam or pulse of light, such as a light-activated silicon controlled rectifier.

light pen A tiny photocell or photomultiplier, mounted with or without a fiber or plastic light pipe in a pen-shaped housing. It is held against a cathode-ray screen to detect the instant at which the electron beam goes through a particular location during its scanning sweep. Used for making measurements electronically from the screen or for changing the nature of the display.

light pipe A flexible or rigid rod made from solid transparent material or parallel transparent optical fibers, to transmit light from one end of the rod to the other with minimum loss.

light-positive Having positive photoconductivity, hence increasing in conductivity (decreasing in resistance) when exposed to light. Selenium ordinarily has this property.

light ray A beam of light that has a small cross section.

light relay *Photoelectric relay.*

light-sensitive Having photoconductive, photoemissive, or photovoltaic characteristics. Also called photosensitive.

light-sensitive cell *Photodetector.*

light-sensitive detector *Photodetector.*

light-sensitive resistor *Photoconductive cell.*

light source A lamp that supplies radiant energy, as for a photoelectric control system.

light valve A device whose light transmission can be made to vary in accordance with an externally applied electrical quantity, such as voltage, current, electric field, magnetic field, or an electron beam. Used in exposing the sound track on motion-picture film.

light water Ordinary water (H_2O), as distinguished from heavy water (D_2O).

light year The distance traveled by light through space in 1 year, equal to approximately 5.88×10^{12} mi (9.46×10^{12} km).

limen *Threshold.*

limit bridge A form of Wheatstone bridge used for rapid routine electrical tests of manufactured products. No attempt is made to balance the bridge for each test; instead, all products that produce deflections within limits corresponding to permissible tolerance are passed.

limited signal A radar signal that is intentionally limited in amplitude by the dynamic range of the radar system.

limited stability *Conditional stability.*

limiter A circuit that limits the amplitude of its output signal to some predetermined threshold level. It may act on positive or negative swings or on both. Used after the IF amplifier in some FM receivers to remove all amplitude variations from the frequency-modulated signal. For infinite limiting, the threshold level is very small, and the output has an essentially rectangular waveform. Also called amplitude limiter, automatic peak limiter, clipper, peak clipper, and peak limiter.

limiting A desired or undesired amplitude-limiting action performed on a signal by a limiter. Also called clipping.

limit switch A switch that cuts off power automatically at or near the limit of travel of a moving object which is electrically controlled.

linac [LINear ACcelerator] *Linear accelerator.*

$LiNbO_3$ Symbol for *lithium niobate.*

Lindemann glass A lithium borate-beryllium oxide glass that has no element higher in atomic number than oxygen, used as window material for low-voltage x-ray tubes because it will pass x-rays of extremely long wavelength, such as grenz rays.

line 1. A transmission or power line. 2. A production line used in mass-production assembly of electronic equipment. 3. The path covered by the electron beam of a television picture tube in one sweep from left to right across the screen. 4. One horizontal scanning element in a facsimile system. 5. *Trace.*

line amplifier An audio amplifier that feeds a program to a transmission line at a specified signal level. Also called program amplifier.

linear Having an output that varies in direct proportion to the input.

linear absorption coefficient The fractional decrease in intensity of a beam of photons or particles per unit distance traversed.

linear accelerator An accelerator that has ring-shaped electrodes arranged in a straight line. When the electrode potentials are properly varied

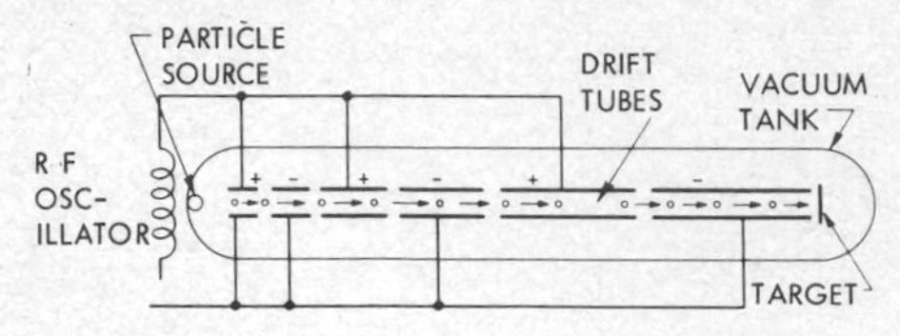

Linear-accelerator operating principle.

in amplitude at an ultrahigh frequency, particles passing through the electrodes receive successive increments of energy and are accelerated along an essentially linear path. Small versions are being used in place of cobalt therapy devices for cancer treatment. Also called linac and linear electron accelerator.

linear actuator An actuator that converts electric energy into linear mechanical motion.

linear amplifier An amplifier in which changes in output current are directly proportional to changes in applied input voltage.

linear array *Collinear array.*

linear backward-wave oscillator A backward-wave oscillator in which the required magnetic

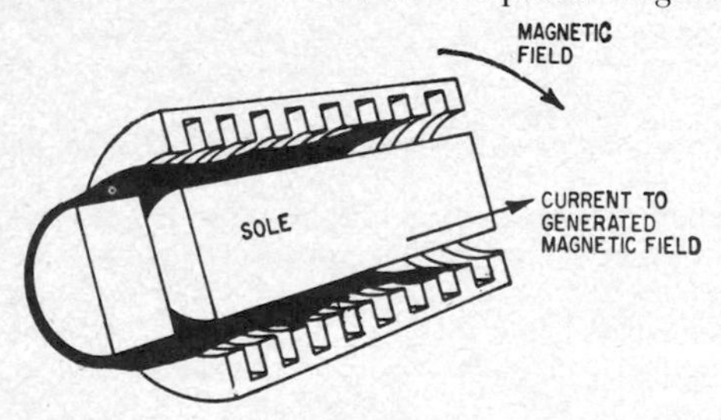

Linear backward-wave oscillator, showing current-carrying sole at right of cathode.

field is generated by current flow through an electrode called a sole, located adjacent to the cathode.

linear-beam tube A space-charge-wave tube in which the directions of the electron beam, the electric field, and the static magnetic field are parallel, as required for converting the kinetic energy of the electron beam into RF energy. Examples include traveling-wave tubes, klystrons, and some backward-wave oscillators.

linear control A rheostat or potentiometer that has uniform distribution of resistance along the entire length of its resistance element.

linear detector A detector in which the output signal voltage is directly proportional to the changes in input carrier amplitude for amplitude modulation or to the changes in input carrier frequency for frequency modulation.

linear differential transformer An electromechanical transducer that converts physical motion into an output voltage whose amplitude and phase are proportional to position. In one version, a movable iron core is positioned between two windings. Displacement of the core from its null position causes the voltage in one winding to increase, while simultaneously reducing the voltage in the other winding. The difference between the two voltages varies with linear position.

linear electric motor An electric motor that has in effect been split and unrolled into two flat sheets, so the motion between rotor and stator is linear rather than rotary. Either the stator or the

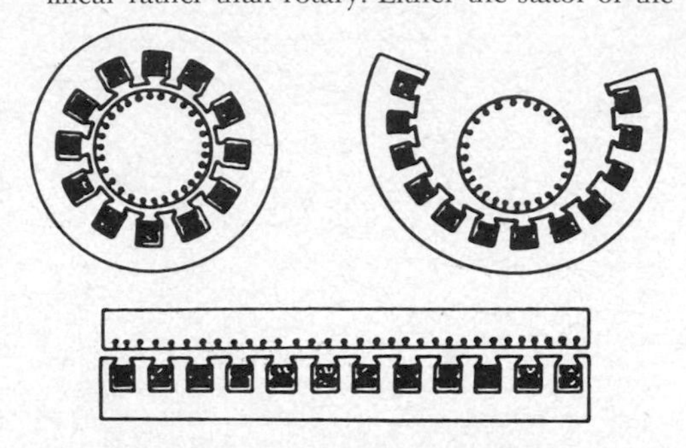

Linear electric motor is in effect ordinary induction motor that has been cut and unrolled.

rotor can be extended to form a track along which the other member moves much like a train, provided the current in the energizing coils is switched on and off in sequence so as to induce currents that produce a force acting in one direction on the moving member.

linear electron accelerator *Linear accelerator.*

linear feedback control system A feedback control system in which the relationships between the pertinent measures of the system signals are linear.

linear integrated circuit An integrated circuit that provides linear amplification of signals, as contrasted to a digital integrated circuit which is used primarily for pulse processing.

linearity 1. The condition in which the change in the value of one quantity is directly proportional to the change in the value of another quantity. 2. Uniformity of distribution of the lines and elements of an image on a television picture tube, so that straight lines in a scene are straight in the image.

linearity control A television receiver control that varies the amount of correction applied to the sawtooth scanning wave to provide the desired linear scanning of lines, so that lines appear straight, and round objects appear as true circles. Separate linearity controls, known as the horizontal linearity and the vertical linearity controls, are usually provided for the horizontal and vertical sweep oscillators. Also called distribution control.

linearly polarized wave A transverse electromagnetic wave whose electric field vector is always along a fixed line.

linear magnetron An experimental magnetron in which electrons travel in essentially linear rather than circular patterns through a transverse magnetic field.

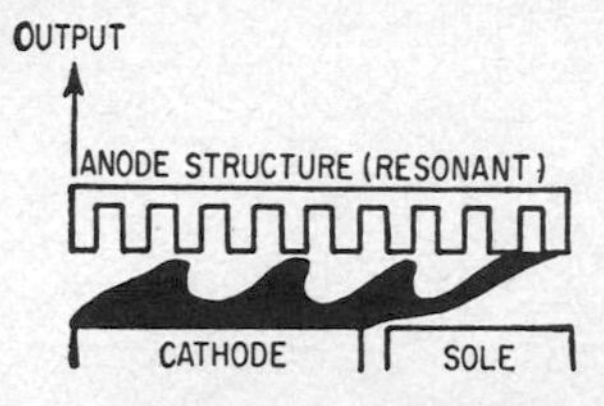

Linear magnetron.

linear modulation Modulation in which the amplitude of the modulation envelope (or the deviation from the resting frequency) is directly proportional to the amplitude of the intelligence signal at all modulation frequencies.

linear power amplifier A power amplifier in which the signal output voltage is directly proportional to the signal input voltage.

linear programming A mathematical technique for finding the optimum solution to a particular problem, in terms of lowest cost, least effort, shortest time, least equipment, or other requirements. Linear programming is an operations research technique that has no relation to computer programming, even though the linear programming operations can be performed with a computer.

linear pulse amplifier A pulse amplifier in which the peak amplitude of the output pulses is directly proportional to the peak amplitude of the corresponding input pulses.

linear recording Magnetic recording in which biasing is used to restrict operation to the linear portion of the demagnetization curve. Required for recording of analog data, sound signals, and video signals.

linear rectifier A rectifier whose output current or voltage contains a wave which has a form identical to that of the envelope of an impressed signal wave.

linear repeater A repeater used in communication satellites to amplify input signals a fixed amount, generally with traveling-wave tubes or solid-state devices operating in their linear region. The linear repeater is preceded by a soft limiter to prevent saturation. Output power varies over a wide range, whereas in a hard-limited repeater the output power is constant.

linear scan A radar scan in which the beam is oscillated back and forth over a fixed angle in a given plane, as in sector scanning.

linear stopping power The energy loss per unit distance when a charged particle passes through a medium.

linear sweep A cathode-ray sweep in which the beam moves at constant velocity from one side of the screen to the other, then suddenly snaps back to the starting side.

linear taper A taper that gives the same change in resistance with rotation over the entire range of a potentiometer.

linear time base A time base that makes the electron beam of a cathode-ray tube move at a constant speed along the horizontal time scale.

linear transducer A transducer for which the pertinent measures of all the waves concerned are linearly related.

linear variable-differential transformer [abbreviated LVDT] A transformer in which a diaphragm or other transducer sensing element moves an armature linearly inside the coils of a differential transformer, to change the output voltage by changing the inductances of the coils in equal but opposite amounts. Applications include measurement of acceleration, force, pressure, and other motion-producing parameters.

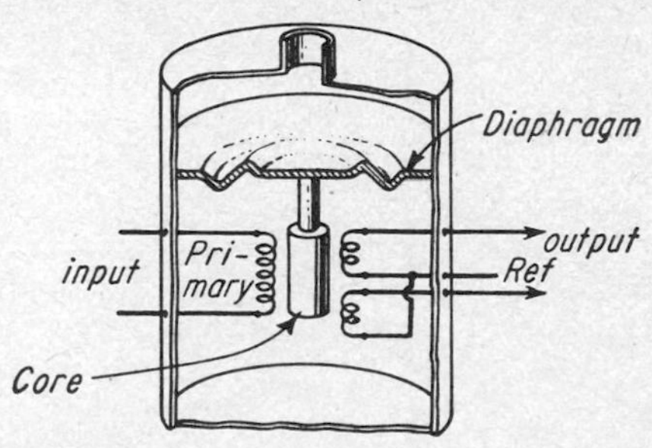

Linear variable-differential transformer in which movable core is mounted on pressure-sensing diaphragm.

line balance Balance in which the conductors of a transmission line have the same electrical characteristics with respect to each other, other conductors, and ground.

line-balance converter *Balun.*

line conditioning The addition of compensating reactances to a data-transmission line to reduce amplitude and phase delays over certain bands of frequencies.

line cord A two-wire cord that terminates in a two-prong plug at one end and is connected permanently to a radio receiver or other appliance at the other end. Used to make connections to a source of power. Also called power cord. The wire alone is commonly called lamp cord. A third wire and prong are sometimes included in a line cord to make a safety connection to ground.

line-cord resistor An asbestos-enclosed wire-wound resistor and two regular wires incorporated into a line cord.

line coupling The coupling capacitors, line-tuning circuits, and lead-in circuits that together provide a connection between power or telephone lines and the transmitter-receiver assembly of a carrier communication system.

line diffuser An oscillator used in a television monitor or receiver to produce small vertical oscillations of the spot on the screen to make the line structure of the image less noticeable at short viewing distances.

line driver An integrated circuit that acts as the interface between logic circuits and a two-wire transmission line.

line drop The voltage drop that exists between two points on a power line or transmission line, due to the impedance of the line.

line equalizer An equalizer that contains inductance and/or capacitance; it is inserted in a transmission line to modify the frequency response of the line.

line feed A signal that causes a line printer to feed paper up to the next printing line.

line filter 1. A filter inserted between a power line and a receiver, transmitter, or other unit of electric equipment to prevent passage of noise signals

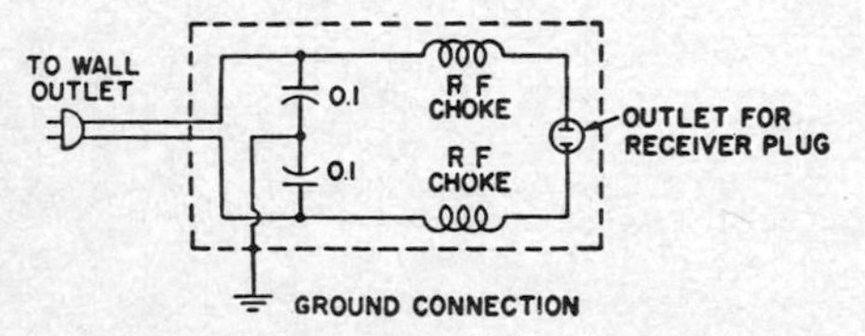

Line-filter circuit for use between radio receiver and power line. Capacitor values are in microfarads.

through the power line in either direction. Also called power-line filter. 2. A filter inserted in a transmission or high-voltage power line for carrier communication purposes.

line finder 1. A device that automatically advances the platen of a line printer or typewriter a given number of lines to the position of the next field in a printed form. 2. A switching device that automatically seeks and locates an idle telephone or telegraph circuit going to the destination being called.

line flyback *Horizontal flyback.*

line-focus tube An x-ray tube in which the focal spot is roughly a line. This gives an essentially square beam of x-rays at one angle of reflection from the target.

line frequency The number of times per second that the scanning spot sweeps across the screen in a horizontal direction in a television system. In the United States, with 525 lines and 30 complete pictures per second, it is 15,750 sweeps per second, including the horizontal scans made during the vertical return intervals. Also called horizontal frequency and horizontal line frequency.

line-frequency blanking pulse *Horizontal blanking pulse.*

line hydrophone A directional hydrophone that consists of one straight-line element, an array of suitably phased elements mounted in line, or the acoustic equivalent of such an array.

line impedance The impedance measured across the terminals of a transmission line.

line interlace *Interlaced scanning.*

line microphone A highly directional microphone that consists of a single straight-line element or an array of small parallel tubes of different lengths, with one end of each abutting a microphone element to give highly directional characteristics. Also called machine-gun microphone.

line noise Noise originating in a transmission line, from such causes as poor joints and inductive interference from power lines.

line of force An imaginary line in an electric or magnetic field, each segment of which represents the direction of the field at that point.

line of position [abbreviated LOP] The intersection of two surfaces of position, used to establish a position or fix in air navigation.

line of propagation The path taken by a radio wave through space.

line of sight The straight, unobstructed path between two points.

line-of-sight distance The distance from a transmitter to the horizon, normally representing the range limit of a radio or radar station. Usually under 200 mi (320 km). Under certain conditions, atmospheric refraction may extend the range.

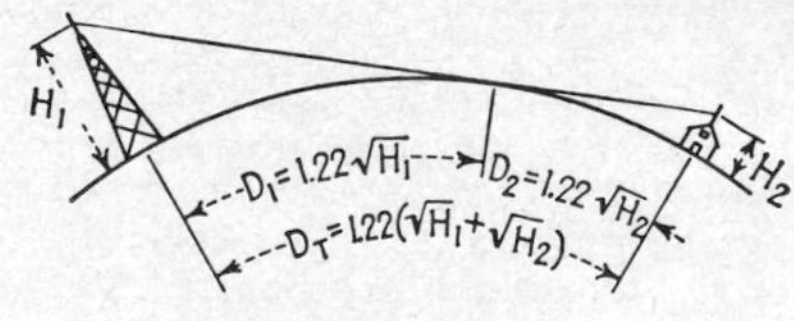

Line-of-sight path, and formula for computing path length D when heights of transmitting and receiving antennas are known. For D in kilometers and H in meters, change 1.22 to 0.674 in all equations.

line-of-sight path The direct, essentially straight, path taken by a radio wave from a transmitting antenna to a receiving antenna.

line-of-sight stabilization The stabilization of a radar antenna mounted on a ship or aircraft, to compensate for roll and pitch, by changing the elevation angle of the antenna. For horizontal scanning, the beam would then be aimed at the horizon at all times.

line pad A pad inserted between a program amplifier and a transmission line, to isolate the amplifier from impedance variations of the line.

line printer An impact or nonimpact printer that requires a full line of data and a print command signal from a computer before an entire line can be printed. The characters are not necessarily printed simultaneously. Speeds can be up to 2000 lines per minute for impact printers and more than 10,000 lines per minute for nonimpact printers. Also called high-speed printer.

line regulation The maximum change in the output voltage or current of a regulated power supply for a specified change in AC line voltage, such as from 105 to 125 V. Line regulation may be expressed as a percentage of the output or the absolute value of the change.

line-sequential color television A color television system in which an entire line is one color, with the colors changing from line to line in a red, blue, and green sequence.

line spectrum The spectrum of electromagnetic radiation emitted spontaneously from bound electrons as they jump from high to low energy levels in an atom. Each different jump in energy level has its own frequency, and these frequencies make up the line spectrum of the atom involved.

lines per minute [abbreviated LPM] A rating for the operating speed of a line printer.

line-stabilized oscillator An oscillator in which a section of high-Q transmission line is used as a frequency-controlling element.

line stretcher A section of waveguide or rigid coaxial line whose physical length is varied to change its electrical length. A telescoping mechanism is commonly used for this purpose.

line synchronizing pulse *Horizontal synchronizing pulse.*

line transformer A transformer inserted in a system for such purposes as isolation, impedance matching, or additional circuit derivation.

line trap A filter that consists of a series inductance shunted by a tuning capacitor, inserted in series with the power or telephone line for a carrier-current system to minimize the effects of variations in line attenuation and reduce carrier energy loss.

line triggering Triggering of an oscilloscope or other device by pulses derived from the power-line frequency.

line-type modulator A modulator circuit that applies a DC supply voltage to a magnetron for a predetermined period of time and repetition frequency to generate the bursts of RF power required by a radar transmitter.

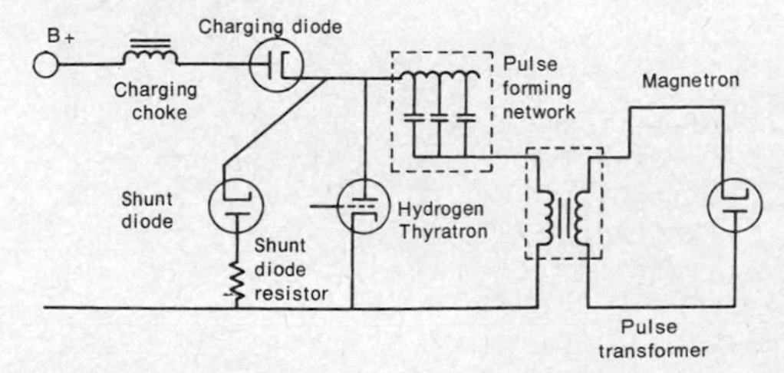

Line-type modulator circuit. Saturable reactor or combination of silicon controlled rectifiers can be used as switch in place of hydrogen thyratron.

line voltage The voltage provided by a power line at the point of use. In the United States it is usually between 110 and 125 V at outlets in homes, with 117 V as an average.

line-voltage regulator A regulator that counteracts variations in power-line voltage, to provide an essentially constant voltage for the connected load.

link 1. A radio transmitter-receiver system that connects two locations. 2. A flat strip that acts as a removable connector between two terminal screws.

linkage 1. A mechanism that transfers motion in a desired manner by using some combination of bar links, slides, pivots, and rotating members. 2. *Flux linkage.*

link coupling Coupling that consists of two coils connected together by a short length of transmission line, with each coil inductively coupled to the coil of a separate tuned circuit.

link neutralization Neutralization by link coupling between the output and input tuned circuits.

lin-log amplifier An amplifier in which the automatic-gain-control circuit operates linearly for low-amplitude input signals, but logarithmically for high-amplitude signals.

lin-log receiver A radar receiver that has a linear amplitude response for small-amplitude signals and a logarithmic response for large-amplitude signals.

Linotron A high-speed cathode-ray photocomposition machine that takes computer magnetic tapes directly and exposes negatives at speeds of 600 to more than 1000 characters per second, depending on type size. The first models were built by CBS-Mergenthaler for the U.S. Government Printing Office. An easily changed character grid gives a choice of 256 characters at a time.

lip microphone A contact microphone used against the upper lip of a person. An acoustic balancing arrangement cancels sounds originating at a distance. Used where noise level is extremely high, as in military tanks. Sound waves from the person's lips enter through only one aperture and act on the microphone, whereas sound waves from a distance enter through both apertures and act on opposite sides of the diaphragm simultaneously, so that their effects cancel.

lip-sync Synchronization of sound and picture so the facial movements of speech coincide with the sounds.

liquid cooling Use of circulating liquid to cool hermetically sealed components.

liquid crystal An organic compound that has a liquid phase and a molecular structure similar to that of a solid crystal. The liquid is normally transparent, but it becomes translucent (almost opaque) in localized areas in which the alignment of the molecules is disturbed by applying an electric field with shaped electrodes. Liquid crystals have three

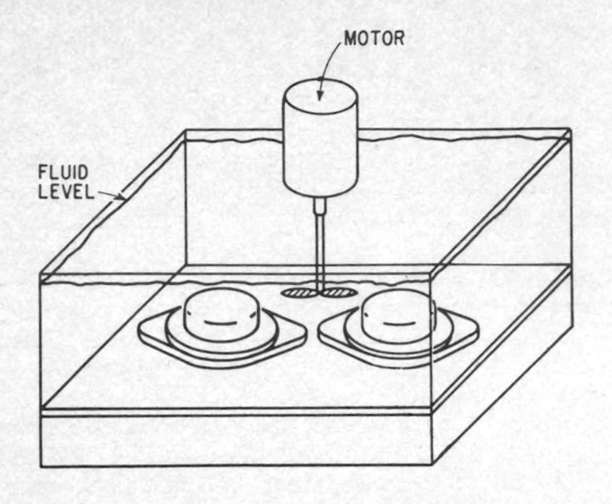

Liquid cooling of power transistors.

phases: nematic, smectic, and cholesteric; the nematic phase, in which the elongated molecules are lined up in one direction but are not in layers, is most commonly used in liquid crystal displays.

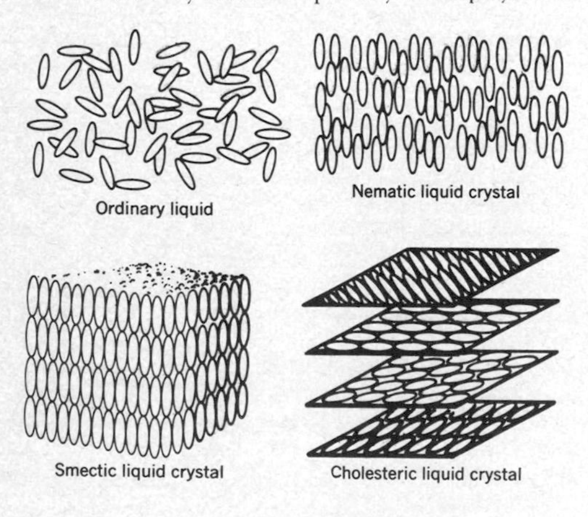

Liquid crystal phases, showing molecular arrangements in comparison with random arrangement of ordinary liquid.

liquid crystal display [abbreviated LCD] A digital display that consists essentially of two sheets of glass separated by a sealed-in liquid crystal material. The outer surface of each glass sheet has a transparent conductive coating like tin oxide or indium oxide, with the viewing-side coating etched into character-forming segments that have leads going to the edges of the display. The liquid is normally transparent. A voltage applied between front and back electrode coatings disrupts the orderly arrangement of the molecules, darkening the liquid enough to form visible characters even though no light is generated. Power drain is negligible, making liquid crystal displays ideal for digital watches. For other applications, edge or back lighting can be used to improve brightness and permit viewing in darkness.

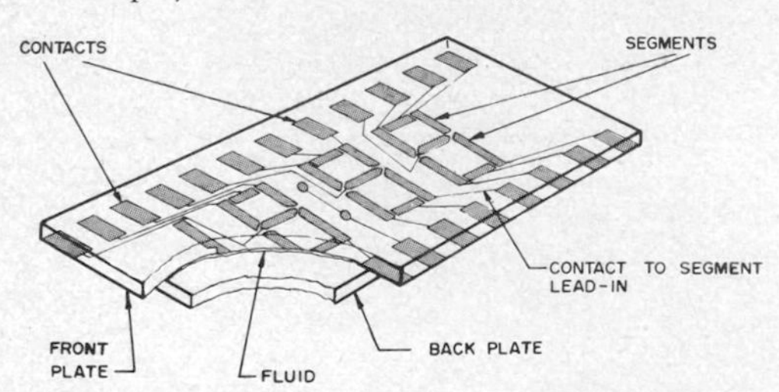

Liquid crystal display. Conductive segments are transparent, with liquid darkening behind each energized segment to form dark visible characters on bright background. The liquid layer is generally less than 0.025 mm thick. Contrast can be improved by making back plate reflective.

liquid laser A laser in which the active material is a liquid. The normal wavelength range is 0.3 to 1.2 μm. Examples include chelate, dye, and inorganic liquid lasers.

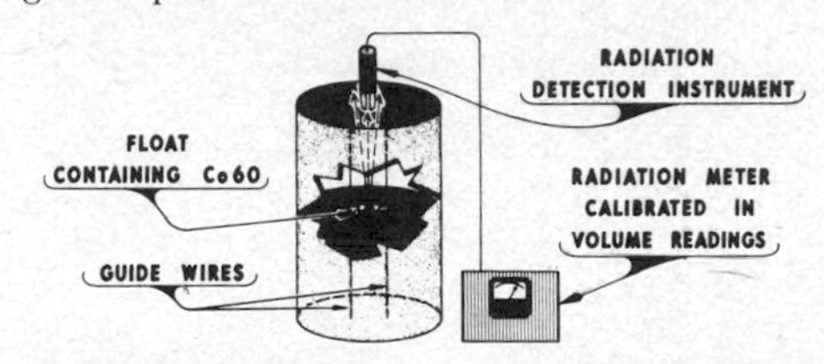

Liquid-level gage in which radiation meter measures distance of cobalt-60 float from top of tank.

liquid-level gage A gage that measures the level of liquid in a tank by sensing elements inside or outside the tank.

liquid-metal fuel cell A fuel cell that uses molten potassium and bismuth as reactants and a molten salt electrolyte. Theoretical output can be as high as 10 W/lb but relatively short life makes it suitable only for missions lasting up to 1 day, as in some types of spacecraft.

Lissajous figure The pattern that appears on an oscilloscope screen when sine waves are applied simultaneously to the horizontal and vertical deflection plates.

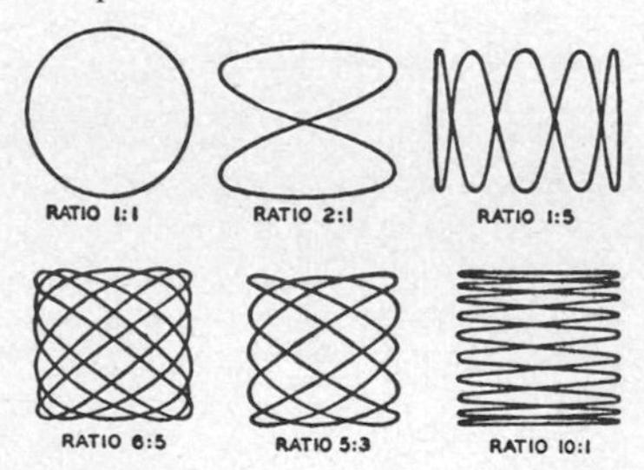

Lissajous figures for sine waves having various frequency ratios.

listening station A radio or radar receiving station that is continuously manned for various purposes, such as for radio direction-finding or for gaining information about enemy electronic devices.

$LiTaO_3$ Symbol for *lithium tantalate.*

lithium [symbol Li] An alkali metal element that has characteristics similar to those of sodium. Sometimes used on the cathodes of phototubes because it gives high response at the extreme violet end of the light spectrum. Atomic number is 3.

lithium cell A primary cell that has a nominal

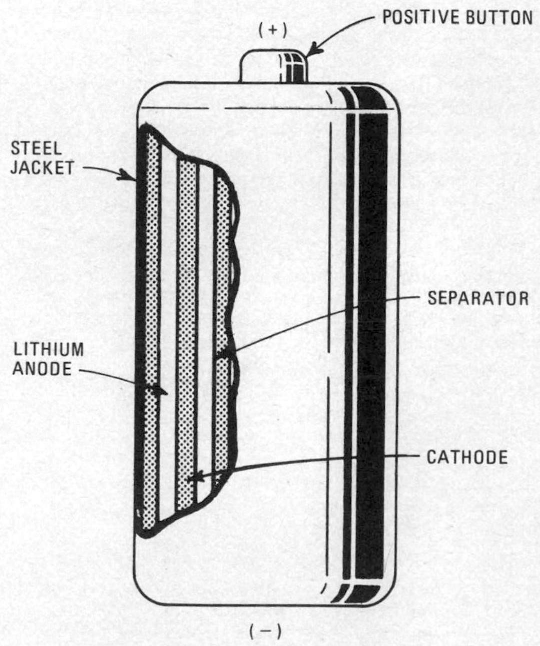

Lithium-cell construction. Venting system relieves gas pressure caused by excessive current discharge.

voltage of 2.8 V and longer shelf life than other primary cells. The lithium anode is separated from the active cathode material by an organic electrolyte that has no water. High cost limits use primarily to military and aerospace applications.

lithium-drifted detector A nuclear-particle detector in which the lithium-drifted technique provides deep depletion layers in a silicon or germanium PIN diode, to obtain excellent gamma-ray resolution and response to all other types of nuclear particles.

lithium-drifted technique A technique for increasing the sensitivity of silicon and germanium nuclear-particle detectors to gamma rays. A reverse bias is applied to an NP junction that consists of a lithium-diffused N-type region on P-type silicon, at a temperature of 100 to 150°C. This results in formation of a layer of high-resistivity material that grows from the N-type diffused layer into the P-type silicon.

lithium niobate [symbol $LiNbO_3$] A ferroelectric piezoelectric crystal that is transparent in the visible and infrared range from 0.38 to 5 μm. Applications include storage of holograms and wideband crystal filters for communication equipment.

lithium-sulfur battery A storage battery in which the cells use a molten lithium cathode and a molten sulfur anode separated by a molten-salt electrolyte that consists of lithium iodide, potassium iodide, and lithium chloride. The voltage is about 2 V per cell. Proposed for storage of reserve electric power to meet peak demands at power plants.

lithium tantalate [symbol $LiTaO_3$] A piezoelectric crystal grown at high temperature from a melt of lithium oxide and tantalum pentoxide. Used for wideband crystal filters in communication equipment.

litz wire Wire that consists of a number of separately insulated strands woven together so each strand successively takes up all possible positions in the cross section of the entire conductor, to reduce skin effect and thereby reduce RF resistance. Sometimes used for winding coils.

live 1. Broadcast directly at the time of production, instead of from recorded or filmed program material. 2. *Energized.*

live chassis A radio, television, or other chassis that has a direct chassis connection to one side of the AC line. For safety, a live chassis must be completely enclosed by an insulating cabinet.

live end The end of a radio studio that gives almost complete reflection of sound waves.

live room A room that has minimum sound-absorbing material.

L line One of the characteristic lines in the x-ray spectrum of an atom. It is produced by excitation of the electrons of the L shell.

lm Abbreviation for *lumen.*

lm/ft² Abbreviation for *lumen per square foot.*

lm/m² Abbreviation for *lumen per square meter.*

L/M ratio The ratio of the number of internal conversion electrons from the L shell to the number from the M shell, emitted in the deexcitation of a nucleus.

lm·s Abbreviation for *lumen second.*

lm/W Abbreviation for *lumen per watt.*

ln Abbreviation for *natural logarithm.*

L network A network composed of two branches in series, with the free ends connected to one pair of terminals. The junction point and one free end

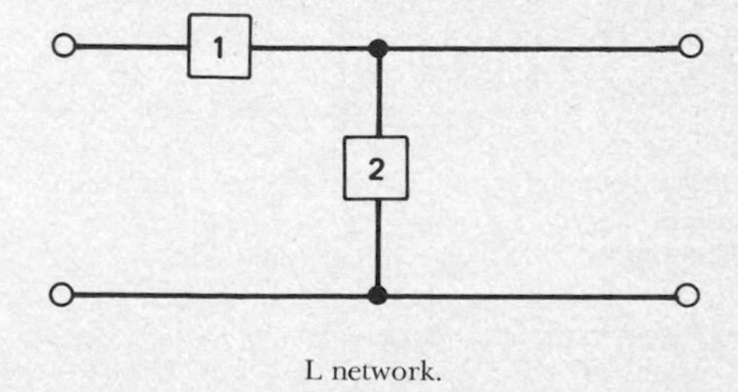

L network.

are connected to another pair of terminals.

LO Abbreviation for *local oscillator*.

load 1. The device that receives the useful signal output of an amplifier, oscillator, or other signal source. 2. A device that consumes electric power. 3. The amount of electric power that is drawn from a power line, generator, or other power source. 4. The material to be heated by an induction heater or dielectric heater. Also called work. 5. To enter data into the storage or working registers of a computer. 6. To place a reel, disk, batch of punched cards, or some other type of recording media into a machine that extracts the stored data or the audio or video content.

load cell A pressure-measuring transducer in which pressure is applied to a piezoelectric crystal, and the resulting voltage across the crystal is measured. Also used for measuring tension and other forces.

load circuit The complete circuit required to transfer power from a source to a load, such as the coupling network, leads, and load material connected to the output terminals of an induction heater.

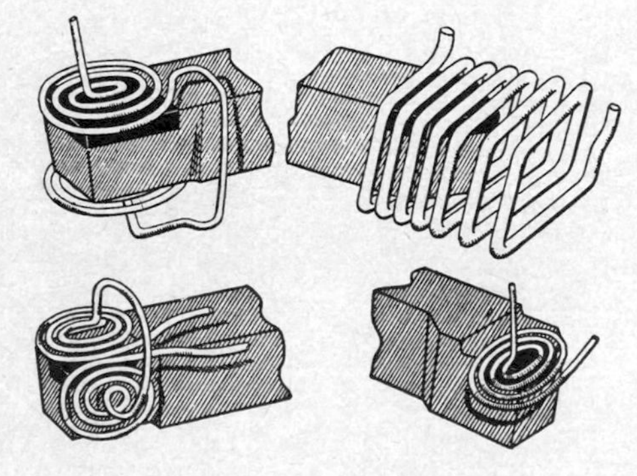

Load coils used to produce localized heat for brazing tungsten carbide tips on cutting tools.

load coil A coil that delivers AC energy by induction to a charge to be heated in induction heating. Also called work coil.

loaded antenna An antenna that has extra inductance in series, to increase its electrical length.

loaded impedance The impedance at the input of a transducer when the output is connected to its normal load.

loaded line A line that contains loading coils.

loaded Q The *Q* of a circuit or a device under working conditions.

loader A computer program that takes some other program from an input or storage device and places it in memory at some predetermined address.

load factor The ratio of average electric load to peak load, usually calculated over a 1-h period.

load impedance The complex impedance presented to a transducer by its load.

load impedance diagram A diagram that shows how the performance of an oscillator is affected by variations in load impedance.

loading The addition of inductance to a transmission line to improve its transmission characteristics throughout a given frequency band.

loading coil 1. An iron-core coil connected into a telephone line or cable at regular intervals to lessen the effect of line capacitance and reduce distortion. Also called Pupin coil and telephone loading coil. 2. A coil inserted in series with a radio antenna to increase its electrical length and thereby lower the resonant frequency.

loading disk A circular metal piece mounted at the top of a vertical antenna to increase its natural wavelength.

loading error The error introduced when more than negligible current is drawn from the output of a device. In potentiometers the loading error varies with the position of the slider and the current drawn.

load line A straight line drawn across a series of tube or transistor characteristic curves to show how output signal current will change with input signal voltage when a specified load resistance is used.

load matching Matching the load circuit impedance to that of the source to give maximum transfer of energy, as desired in induction and dielectric heating.

load-matching network A network for load matching in induction and dielectric heating.

load-matching switch A switch used in a load-matching network to compensate for a sudden change in load characteristics, such as that which occurs when passing through the Curie point of the load material in induction and dielectric heating.

load regulation The maximum change in the output voltage or current of a regulated power supply for a specified change in load conditions. Load regulation may be expressed as a percentage of the output or as the absolute value of the change.

load sharing Operation of two computers in duplex so they can share the load of the system during peak hours. At other times, one computer handles the entire load, and the other one serves as backup in case of trouble.

load transfer switch A switch that connects a generator or power source optionally to either of two load circuits.

lobe One of the three-dimensional portions of the radiation pattern of a directional antenna. The

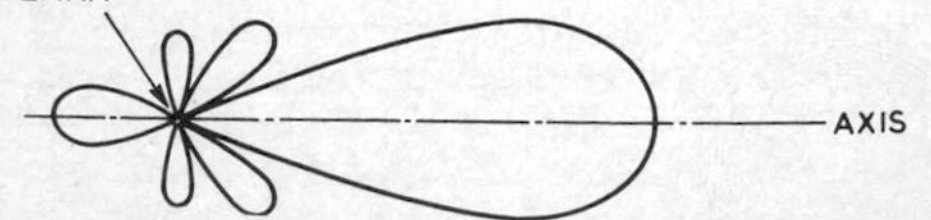

Lobe configuration for one plane of directional antenna.

direction of maximum radiation coincides with the axis of the major lobe. All other lobes in the pattern are called minor lobes.

lobe switch A switch used for systematically shifting the radiation pattern of an antenna.

lobe switching A method of determining the exact direction to a target by periodically shifting the beam of a radar antenna slightly to the left and to the right of the dead-ahead position, using electronic or mechanical means. While comparing received signal strengths, the entire antenna is turned until equal signals are received from both lobes. The antenna is then accurately aimed, without an impracticably narrow beam width being used. Used also in radio direction finding.

local channel A standard broadcast channel in which several stations may operate with powers not in excess of 250 W.

local control Control of a transmitter directly at the transmitter rather than by remote control.

localizer A radio facility that provides signals for use in lateral guidance of aircraft with respect to a runway centerline.

localizer on-course line A line in a vertical plane that passes through a localizer, on either side of which the received indications have opposite sense.

localizer sector The sector included between two radial equisignal localizer lines that have the same specified difference in depth of modulation.

local oscillator [abbreviated LO] The oscillator in a superheterodyne receiver; its output is mixed with the incoming modulated RF carrier signal in the mixer to give the frequency conversion needed to produce the IF signal.

location A place that can be uniquely specified for storage of data in a computer system.

locator A radar or other device that detects and locates airborne aircraft.

lock To fasten onto and automatically follow a target by means of a radar beam.

locked groove A blank and continuous groove placed at the end of the modulated grooves on a disk recording to prevent further travel of the pickup. Also called concentric groove.

locked-in line A telephone line that remains established after the caller has hung up. Automatic circuits for this purpose can be used at police and fire stations for tracing anonymous callers of bomb threats, false fire alarms, and other nuisance calls as well as legitimate calls for help.

locked oscillator A sine-wave oscillator whose frequency can be locked by an external signal to the control frequency divided by an integer. Used as a frequency divider.

locked-oscillator detector A type of discriminator that does not react to amplitude modulation and hence requires no limiter preceding it. The circuit uses a pentagrid tube with three tank circuits, each circuit tuned to the signal resting frequency and so located that average anode current changes only with signal frequency. Used as an FM detector.

locked-oscillator quadrature-grid FM detector An FM detector that functions as a directly driven quadrature-grid detector for strong signals and as a locked-oscillator detector for relatively weak signals. Used in some television receivers.

locked-rotor current The current drawn by a stalled electric motor.

lock-in The synchronizing of two oscillators by coupling them together or by applying sync pulses, so their frequencies are the same or the ratio of their frequencies is an integral number.

lock-in amplifier An amplifier that uses some form of automatic synchronization with an external reference signal to detect and measure very weak electromagnetic radiation at radio or optical wavelengths in the presence of very high noise levels. For optoelectronic applications the low-level light signal may be chopped for drift-free AC amplification, with the reference signal generated in synchronism with the chopping wheel. Other versions for radio frequencies use phase-sensitive detectors in combination with a signal chopper and an external reference frequency source. Locking on the very weak repetitive signal rejects noise and other interfering frequencies.

lock-in range The frequency range over which an oscillator may be synchronized by a synchronization signal.

lock-on The instant at which a radar begins to track its target automatically.

lock-up relay A relay that locks in its energized position either by permanent magnetic biasing which can be released only by applying a reverse magnetic pulse or by auxiliary contacts that keep its coil energized until the circuit is interrupted. In contrast, in a latching relay the lock-up is accomplished mechanically.

loctal base *Loktal base.*

lodar A direction finder that determines the direction of arrival of loran signals, free of night effect, by observing the separately distinguishable ground and sky-wave loran signals on a cathode-ray oscilloscope and positioning a loop antenna to obtain a null indication of the component selected to be most suitable. Also called lorad.

LOFAR A submarine detection system that uses autocorrelation techniques for long-range analysis of patterned sound picked up at the low-frequency end of the sound spectrum by underwater hydrophones of the Caesar submarine detection system.

Loftin-White circuit A direct-coupled amplifier circuit.

log 1. A written record of radio and television station operating data, required by law. 2. A record of computer operating runs, including tapes used, control settings, halts, and other pertinent data. 3. Abbreviation for *logarithm.*

logamp Abbreviation for *logarithmic amplifier.*

logarithm [abbreviated log] The power to which a number, called the base, must be raised to equal the original number. The common system of logarithms uses 10 as a base; here, the logarithm of 1000 to the base 10 is 3 because 10^3 is 1000. Another commonly used base is 2.71828, designated by *e* and known as the hyperbolic, Napierian, or natural logarithm; here the notation $\log_e N$ is often abbreviated as ln *N*.

logarithmic amplifier [abbreviated logamp] An amplifier whose output signal is a logarithmic function of the input signal.

logarithmic computer A section of a digital computer that solves problems in terms of logarithmic values or as a logarithmic function. Used in gunnery calculations.

logarithmic decrement The natural logarithm of the ratio of the amplitude of one oscillation to that of the next which has the same polarity, when no external forces are applied to maintain the oscillation.

logarithmic diode A diode that has an accurate semilogarithmic relationship between current and voltage over wide forward dynamic ranges. Used in circuit applications like dividing, multiplying, logarithmic conversions, and signal compression.

logarithmic multiplier A multiplier in which each variable is applied to a logarithmic function generator. The outputs are added together and applied to an exponential function generator, to obtain an output proportional to the product of two inputs.

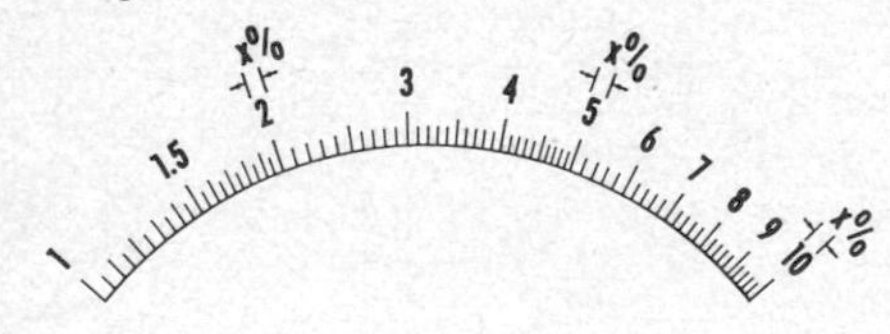

Logarithmic scale gives constant percentage accuracy of readings for all pointer positions.

logarithmic scale A scale whose graduations are spaced logarithmically rather than linearly.

logger A recorder that automatically scans measured quantities at specified times and records or logs their values on a chart.

logic 1. The basic principles and applications of truth tables, interconnections of ON/OFF circuit elements, and other factors involved in mathematical computation and the solution of problems involving step-by-step reasoning in a computer. 2. General term for the various types of gates, flip-flops, and other ON/OFF circuits used to perform problem-solving functions in a digital computer. Common forms of logic circuits include current-mode logic (CML), complementary-transistor logic (CTL), direct-coupled transistor logic (DCTL), diode-transistor logic (DTL), emitter-coupled logic (ECL), integrated injection logic (I^2L), resistor-capacitor-transistor logic (RCTL), resistor-transistor logic (RTL), and transistor-transistor logic (TTL or T^2L).

logic analysis Determination of the sequence of logic steps required during a computer run to produce the desired output files from the input files.

logic analyzer An analyzer that locates trouble in digital systems by displaying digital signal levels simultaneously for a number of locations at a predetermined instant of time which is usually determined by counting clock ticks. The logic analyzer may generate a trigger that stops data collection at the desired instant and displays the pulse signals on a cathode-ray screen for analysis of circuit or program errors, isolation of glitch sources, detection of illegal states, mapping of data flow, and diagnosis of other computer problems.

logic comparison The operation of comparing two items in a computer and producing a 1 output if they are equal or alike and a 0 output if they are not alike.

logic design The preliminary design of the complete system of logic elements required for a particular application in a digital computer.

logic diagram A diagram that represents the logic elements of a computer and their interconnections, without necessarily showing construction or engineering details.

logic element The smallest building block that can be represented by an operator in an appropriate system of symbolic logic for a computer or data-processing system. Typical logic elements are the AND gate and the flip-flop.

logic function A means of expressing a definite state or condition in magnetic amplifier, relay, and computer circuits. Examples include: (a) the AND function, where an output is produced only when a number of input signals are present and combined; (b) the OR function, where an output is obtained when any one of a number of input signals is applied; (c) the NOT function, where an output is obtained only when there is no input signal.

logic level One of the two voltages whose values have been arbitrarily chosen to represent the binary numbers 1 and 0 in a particular data-processing system. The magnitude and polarity of the voltage levels should be specified for a particular application to avoid confusion because the level used for 1 can be either higher or lower than that for 0. Terms such as positive logic, negative logic, normal logic, and inverse logic are ambiguous because they may be construed differently for NPN than for PNP devices.

logic operation A nonarithmetic operation in a computer, such as comparing, selecting, making references, matching, sorting, and merging, where yes/no decisions are involved.

logic sum A computer addition in which the result is 1 when either one or both input variables is a 1, and the result is 0 when the input variables are both 0.

logic swing The voltage difference between the logic levels used for 1 and 0. The magnitude of the swing is chosen arbitrarily for a particular system and is usually well under 10 V.

logic switch A diode matrix or other switching arrangement that is capable of directing an input signal to one of several outputs.

logic symbol A graphic symbol that represents the means of performing some specified simple computer operation, such as coincidence gating.

log-periodic antenna A broadband antenna in which the electrical lengths and spacings of the elements are chosen so the bidirectional radiation pattern, impedance, and other properties of the antenna are repeated at a number of other frequencies that are equally spaced when plotted on a logarithmic scale.

log-periodic dipole array A broadband antenna array in which dipole lengths and spacings increase with distance from a source, with the trans-

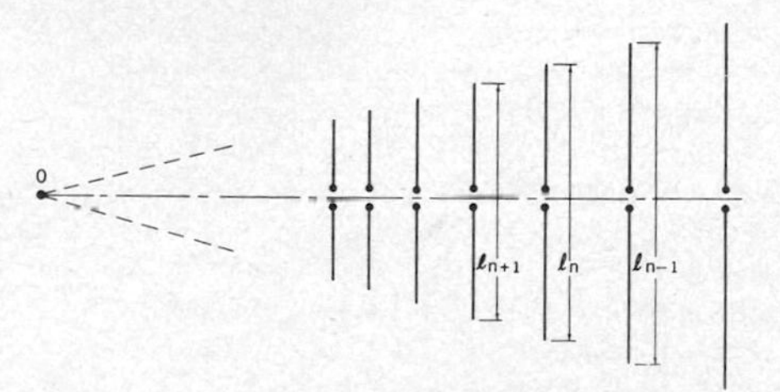

Log-periodic dipole array.

mission lines being transposed between adjacent dipole elements. The radiation pattern is unidirectional in the backfire direction, toward the source.

log-periodic folded-dipole array A unidirectional broadband antenna in which the elements are arranged much as in a log-periodic dipole array, but with all the folded dipoles connected in

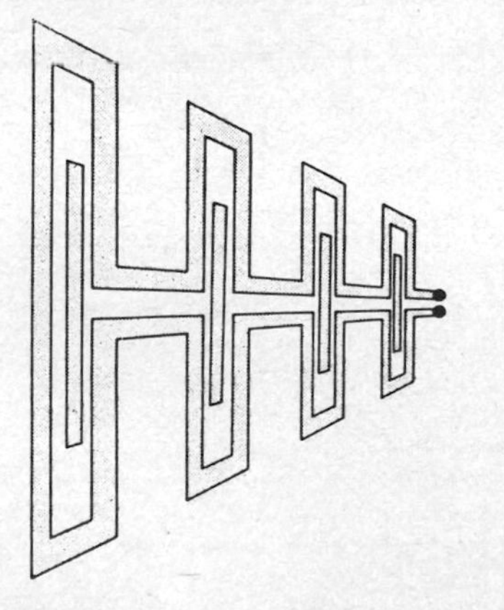

Log-periodic folded-dipole array.

series with the transmission line rather than in shunt. A phasing strip in each folded dipole is adjusted experimentally to produce a good backfire beam.

log-periodic folded-monopole array A unidirectional broadband array that is essentially half of

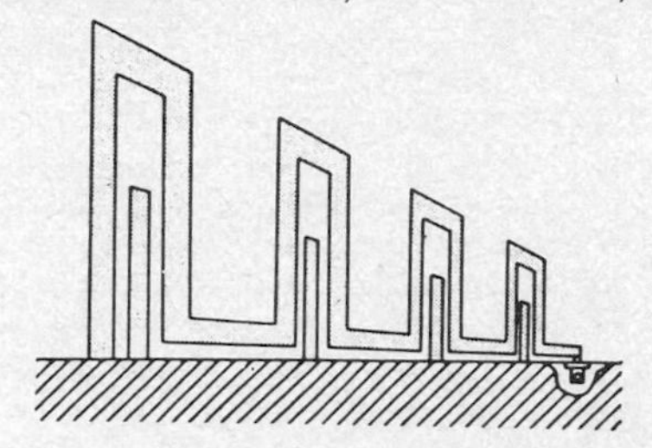

Log-periodic folded-monopole array.

a log-periodic folded-dipole array, fed against a ground plane.

log-periodic folded-slot array A unidirectional broadband antenna that consists of a single metal

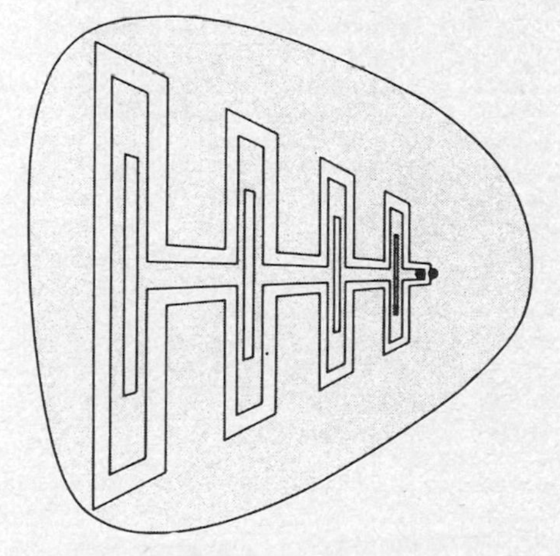

Log-periodic folded-slot array.

sheet from which slots are cut in the pattern of a log-periodic folded-dipole array.

loktal base A tube base that has a grooved metal center post which locks firmly in a corresponding eight-pin loktal socket. The tube pins are sealed directly into the glass envelope. Also called loctal base.

loktal tube An electron tube that has a loktal base.

London equation One of the partial differential equations that describe approximately the spatial and time dependence of electric and magnetic fields inside a superconductor.

lone electron An electron that is alone on an energy level.

long-delayed echo [abbreviated LDE] An echo

of a shortwave radio signal that is heard with a delay ranging from 2 to 30 s, for which no explanation can be found. Since radio waves travel at the speed of light, about 186,000 mi/s or 300,000 km/s, one complete passage around the earth takes only about 0.14 s, and even a trip to the moon and back takes less than 2 s.

long-distance xerography A facsimile system that uses a cathode-ray scanner at the microwave transmitting terminal. At the receiving terminal, a lens projects the received cathode-ray image onto the selenium-coated drum of a xerographic copying machine.

longitudinal current A current that flows in the same direction in both wires of a pair and uses the earth or other conductors for a return path.

longitudinal fuze An electronic fuze that has maximum sensitivity at right angles to the sides of the missile or other explosive, rather than in the straight-ahead direction.

longitudinal heating Dielectric heating in which the electrodes are so positioned that the electric field is parallel to the layers of a laminated material being heated.

longitudinal magnetization Magnetization of a magnetic recording medium in a direction essentially parallel to the line of travel.

longitudinal parity Parity associated with bits recorded on one track in a data block, to indicate whether the number of recorded bits in the block is even or odd.

longitudinal wave A wave in which the direction of displacement at each point of the medium is perpendicular to the wavefront.

long-line effect An effect that occurs when an oscillator is coupled to a transmission line with a bad mismatch. Two or more frequencies may then be equally suitable for oscillation, and the oscillator jumps from one of these frequencies to another as its load changes.

long-persistence screen A fluorescent screen that contains phosphorescent compounds which increase the decay time, so a pattern may be seen for several seconds after it is produced by the electron beam.

long-play record [abbreviated LP record] A 10- or 12-in (25.4- or 30.5-cm) phonograph record that operates at a speed of 33⅓ rpm and has closely spaced grooves, to give playing times up to 30 min for one 12-in side. Designed for use with a 1-mil stylus, whose point radius is 0.001 in (0.0254 mm). Also called microgroove record.

long-range alpha particle An alpha particle that is produced directly from the excited states of nuclei during beta disintegration.

long-range tracking laser In optical guidance systems, the laser that begins to track after the target has been picked up by the acquisition laser. Long-range tracking is generally done with a gallium arsenide laser that has a beam width of about 0.5°.

long-tail pair A two-tube or two-transistor circuit that has a common resistor (tail resistor) which

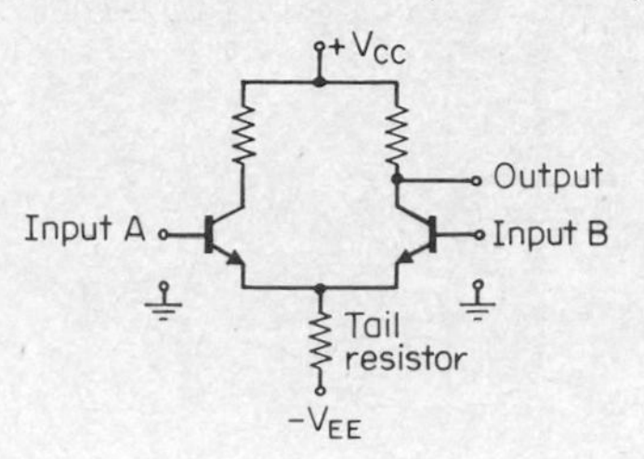

Long-tail pair using two transistors.

gives strong negative feedback. Used for differential amplification or for achieving low drift.

long wave An electromagnetic wave that has a wavelength longer than the longest broadcast band wavelength of about 545 m, corresponding to frequencies below about 550 kHz.

long-wire antenna An antenna whose length is a number of times greater than its operating wavelength, to give a directional radiation pattern.

look-through 1. Interruption of a jamming transmission for extremely short periods at irregular intervals to permit monitoring of the victim signal and determine whether jamming is still needed. 2. Observing or monitoring a desired signal that is being jammed by the enemy, during interruptions in the jamming signals.

lookup The process of matching a word or some other form of input data with similar material stored in the memory of a computer.

loop 1. A curved conductor that connects the ends of a coaxial line or other transmission line and projects into a resonant cavity for coupling purposes. 2. A closed curve on a graph, such as a hysteresis loop. 3. A closed path or circuit over which a signal can circulate, as in a feedback control system. 4. Repetition of a sequence of operations in a computer routine. 5. *Antinode.* 6. *Loop antenna.* 7. *Mesh.*

loop actuating signal The signal derived by mixing the loop input signal with the loop feedback signal of a control system.

loop antenna An antenna that consists of one or more complete turns of a conductor, usually tuned to resonance by a variable capacitor connected to the terminals of the loop. The radiation pattern is bidirectional, with maximum radiation or pickup in the plane of the loop and minimum radiation at right angles to the loop. Also called loop.

loop control *Photoelectric loop control.*

loop coupling The use of a partial or complete loop of wire to couple with the magnetic field of a tuned circuit, tuned cavity, or other tuned device, to feed in or extract RF power.

loop difference signal A type of loop actuating signal that is produced at a summing point of a feedback control loop when a particular loop input signal is applied to that summing point.

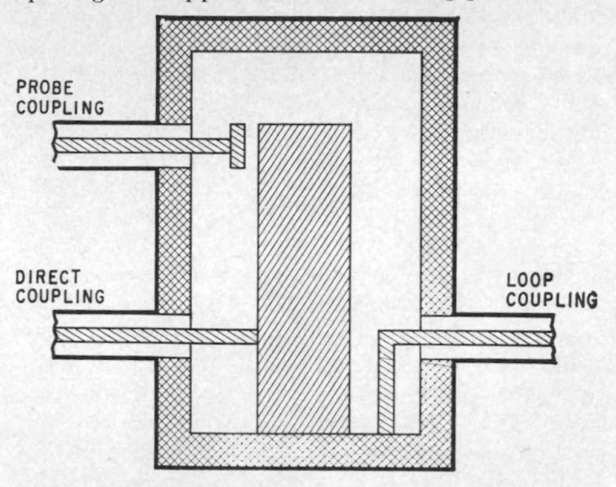

Loop coupling in coaxial cavity, with two other coupling methods shown for comparison.

loop error The desired value of the loop output signal minus the actual value in a control system.

loop error signal The loop actuating signal in those cases in which it is the loop error.

loop feedback signal The signal derived as a function of the loop output signal and fed back to the mixing point for control purposes.

loop gain The product of the gain values acting on a signal passing around a closed-loop path. In a feedback control system it is the forward gain multiplied by the feedback network gain. In a repeater, carrier terminal, or complete closed system the loop gain is the maximum gain that can be used without causing oscillation or singing, and the loop gain here may therefore be less than the product of the gain values.

loopstick antenna *Ferrite-rod antenna.*

loop test A telephone or telegraph line test that is made by connecting a faulty line to good lines, to form a loop in which measurements can be made to determine the position of the fault.

loop-type directional coupler A directional coupler for coaxial lines; it uses a loop which projects into the main line in such a way that a wave traveling in one direction transfers a part of its energy to the secondary coaxial line connected to the loop while rejecting energy traveling in the opposite direction. One end of the secondary line must be terminated in its characteristic impedance.

loop-type radio range An A-N radio range that uses two separate loop antennas fed by a single transmitter.

loose coupling Coupling that is considerably less than critical coupling, so there is very little transfer of energy. Also called weak coupling.

LOP Abbreviation for *line of position.*

LORAC [LOng-Range-ACcuracy radio system] A long-range navigation system that determines a position fix by the intersection of lines of position. Each line is defined by the phase angle between two beat-frequency waves; one wave is the beat between the continuous-wave signals from two widely spaced transmitters, and the other wave is the reference wave of the same frequency, obtained by beating the same two continuous-wave signals at a fixed location and transmitting the beat to the navigation receiver over a second RF channel. Used for surveying and ship-positioning. The operating frequency is in the band from 1.7 to 2.5 MHz.

lorad *Lodar.*

loran [LOng-RAnge Navigation] A long-distance radio navigation system for aircraft and ships; it utilizes synchronized pulses transmitted simultaneously by widely spaced transmitting stations. Hyperbolic lines of position are determined by measuring the difference in the time of arrival of these pulses. The intersection of two of these lines of position, obtained from either three or four stations, gives a position fix. Standard loran operates on frequencies between 1.8 and 2.0 MHz. Low-frequency and sky-wave-synchronized loran are two other types. The two original versions of loran, both now obsolete, were sometimes known as loran A and loran B.

loran C An extremely accurate long-range system of navigation, similar to loran, which gives accuracy comparable to that obtained by celestial observations with a sextant. Operation is in the frequency band from 90 to 110 kHz. The receiver measures pulse spacing in loran fashion to obtain a rough indication of a position, and it measures precisely the relative phase of the RF carriers in the master and slave pulse envelopes. All older loran stations are being converted to loran C.

loran chain A chain of four or more loran stations; it forms three or more pairs of stations for loran navigation.

loran chart A chart that shows loran lines of position and a limited amount of topographic detail.

loran D A tactical loran system that uses the coordinate converter of low-frequency loran C and can operate in conjunction with inertial systems on aircraft, independently of ground facilities and without radiating RF energy which could reveal the aircraft's location. The inertial system errors that build up progressively with time can be periodically corrected by loran station transmissions at randomly selected brief intervals.

loran fix A fix obtained by determining the intersection of two loran lines of position.

loran guidance Missile guidance in which a loran receiver in the missile receives signals continuously from three or more fixed ground stations. The resulting missile position data is fed into a computer that in turn produces the control signals required to guide the missile to its intended fixed ground target.

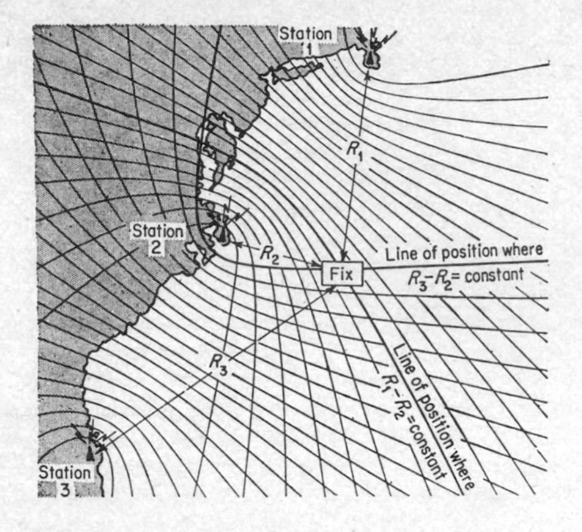

Loran fix on loran chart for East Coast of United States.

loran indicator An indicator that displays the pulse signals from two loran ground stations simultaneously and shows the time difference between reception of the two signals.

loran line A line of position on a loran chart. Each line is the locus of points whose distances from two fixed stations differ by a constant amount.

loran set A receiving set or indicator that displays the pulses from loran transmitting stations.

loran station A transmitting station in a loran system.

loran triplet A combination of three loran stations, with one of the stations forming a pair with each of the other stations.

Lorentz force equation The equation that relates the force on a charged particle to its motion in an electric and magnetic field.

Lorenz instrument landing system A continuous-wave instrument landing system used in continental Europe and Great Britain; it consists of a runway localizing beacon and two radio marker beacons.

lorhumb line A navigation course line in a lattice such that the derivative of one coordinate with respect to the other coordinate constantly equals the ratio of the difference of the coordinates at the beginning and end points of the course line.

loss 1. Power that is dissipated in a device or system without doing useful work. 2. *Transmission loss.*

losser A material or element that dissipates energy, used intentionally in a circuit to prevent oscillation and for other purposes.

Lossev effect Radiation that results from recombination of charge carriers injected in a PN or PIN junction which is biased in the forward direction.

loss factor The power factor of a material multiplied by its dielectric constant. The loss factor varies with frequency and determines the amount of heat generated in a material.

loss modulation *Absorption modulation.*

loss tangent A measure of the amount of power lost as heat when a dielectric or semiconductor material is subjected to a high-frequency electric or electromagnetic field.

lossy Having losses, generally intentional as applied to a dielectric material.

lossy line A transmission line that has intentionally high attenuation per unit length. High loss is sometimes achieved by using Nichrome for the center conductor.

loudness The intensity characteristic of an auditory sensation, in terms of which sound may be described on a scale extending from soft to loud.

loudness analyzer An analyzer that produces a cathode-ray display which shows the loudness of airborne sounds at a number of subdivisions of part or all of the audio spectrum. For continuously changing sounds, the spectrum analysis is repeated at regular intervals, such as every 25 ms. The analyzer may also have provisions for measuring peak loudness and other characteristics of sound as it affects the human ear.

loudness contour A curve that shows the related values of sound pressure level and frequency required to produce a given loudness sensation for a typical listener.

loudness control A combination volume and tone control that boosts bass frequencies when the control is set for low volume, to compensate automatically for the reduced response of the ear to low frequencies at low volume levels. Some loudness controls also provide the same automatic compensation for treble frequencies. Also called compensated volume control.

loudness level The sound pressure level, in decibels relative to 0.0002 μbar of a pure 1-kHz tone that is judged by listeners to be equivalent in loudness to the sound under consideration. Also called equivalent loudness level. One decibel of change

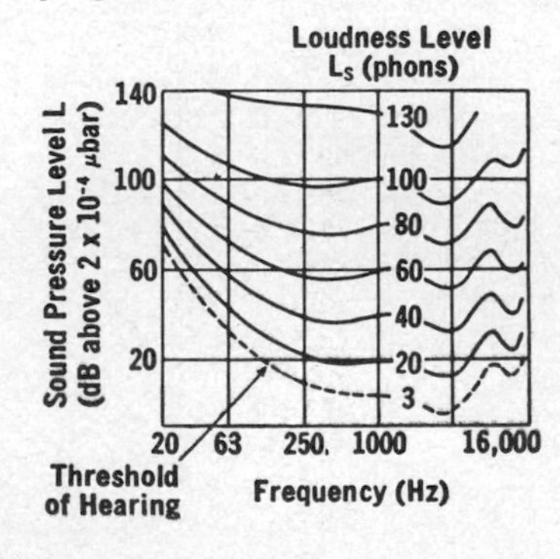

Loudness level curves for pure tones from 20 to 15,000 Hz, showing how frequency response of human ear varies with loudness.

under these conditions is called a phon.

loudspeaker [abbreviated SPKR] An electroacoustic transducer that converts audio-frequency electric power into acoustic power and radiates the acoustic power effectively at a distance in air. Also called speaker.

loudspeaker dividing network *Crossover network.*

loudspeaker impedance The impedance rating of the voice coil of a loudspeaker, corresponding to the impedance of the amplifier output terminals to which the loudspeaker should be connected to obtain its rated performance. Common impedance values are 4, 8, and 16 Ω.

loudspeaker system A combination of one or more loudspeakers with associated baffles, horns, and dividing networks, arranged to work together as a coupling means between the driving electric circuit and the air or between the circuit and another acoustic medium.

louver 1. An arrangement of concentric or parallel slats or equivalent grille members that conceals and protects a loudspeaker while allowing sound waves to pass. Often molded integrally with a plastic radio cabinet. 2. An arrangement of fixed or adjustable slotlike openings provided in a cabinet for ventilation.

low-altitude radio altimeter *Frequency-modulated radio altimeter.*

low band The band that includes television channels 2 to 6, extending from 54 to 88 MHz.

low-energy electron diffraction instrument An instrument that uses slow electron diffraction in an ultrahigh vacuum to examine the surface structure of solids. The resulting diffraction patterns are visible on a fluorescent screen.

lower sideband The sideband that contains all frequencies below the carrier-frequency value which are produced by an amplitude-modulation process.

lowest useful high frequency The lowest high frequency that is effective at a specified time for ionospheric propagation of radio waves between two specified points. The frequency value is determined by such factors as absorption, transmitter power, antenna gain, receiver characteristics, type of service, and noise conditions.

low frequency [abbreviated LF] A Federal Communications Commission designation for a frequency in the band from 30 to 300 kHz, corresponding to a kilometric wave between 1 and 10 km.

low-frequency compensation Compensation that extends the frequency range of a broadband amplifier to lower frequencies.

low-frequency induction heater An induction heater in which current flow at the commercial power-line frequency is induced in the charge to be heated.

low-frequency loran [abbreviated LF loran] A modification of standard loran, operating in the low-frequency range of approximately 100 to 200 kHz to increase range over land and during daytime. Whereas ordinary loran matches the envelopes of the RF pulses to obtain a line of position, low-frequency loran matches the cycles within the pulses to provide a much more accurate fix. Also called cycle-matching loran.

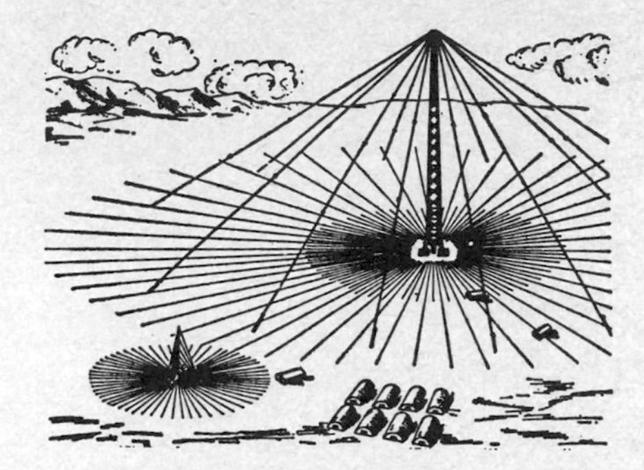

Low-frequency loran station.

low-frequency padder A trimmer capacitor connected in series with the oscillator tuning coil of a superheterodyne receiver. It is adjusted during alignment to calibrate the circuit at the low-frequency end of the tuning range.

low level The less positive of the two logic levels or states in a digital logic system. When both high and low levels are negative, the opposite connotation is sometimes used: the smaller negative voltage (the more positive voltage) is assumed to be the low level or low state.

low-level language A programming language written in the machine code of the computer in which it is to be used.

low-level modulation Modulation produced at a point in a system where the power level is low compared with the power level at the output of the system.

low-level radio-frequency signal A radio-frequency signal that has insufficient power to fire a TR, ATR, or pre-TR tube.

low-loss insulator An insulator that has negligible loss at high radio frequencies.

low-loss line A transmission line that has low power dissipation per unit length.

low-noise amplifier An amplifier in which the background noise is very weak or inaudible in the absence of desired signals.

low-order digit A digit that occupies the least significant position in a positional notation system. In conventional notation it is the digit at the right.

low-order position The rightmost position in a number or word.

low-pass filter A filter that transmits alternating currents below a given cutoff frequency and substantially attenuates all other currents.

low-pressure cloud chamber A cloud chamber in which the gas is maintained at low pressure to increase the range or decrease the scattering of particle tracks.

low tension British term for low voltage, as generally applied to filament and heater voltages of tubes.

low vacuum A degree of vacuum at which so much gas or vapor is still present that ionization can occur in an electron tube.

low-velocity scanning The scanning of a target with electrons of velocity less than the minimum velocity needed to give a secondary-emission ratio of unity. Used in the emitron, image orthicon, and vidicon types of camera tubes.

lox Liquid oxygen, a cryogenic fuel that is liquid below −183°C.

L pad A volume control that has essentially the same impedance at all settings. It consists mainly of an L network in which both elements are adjusted simultaneously.

LPM Abbreviation for *lines per minute.*

LP record Abbreviation for *long-play record.*

LSA diode [Limited Space-charge Accumulation diode] A microwave diode in which a space charge is developed in the semiconductor by the

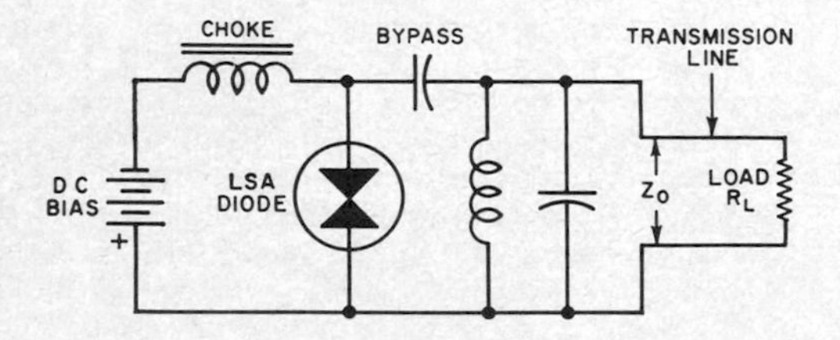

LSA diode in oscillator circuit.

applied electric field and is dissipated during each cycle before it builds up appreciably, thereby limiting transit time and increasing the maximum frequency of oscillation.

LSA mode One of the three operating modes of a transferred-electron diode, in which the entire structure becomes a negative resistance during a portion of each operating cycle. The frequency of oscillation is determined by the surrounding circuit and is independent of the transit time of the charge carriers. The other two modes are the quenched-domain and transit-time modes.

LSB Abbreviation for *least significant bit.*

L scope A radarscope that produces an L display.

L shell The second layer of electrons that surrounds the nucleus of an atom; the layer has electrons whose principal quantum number is 2.

LSI Abbreviation for *large-scale integration.*

lubber line A line placed on a compass or cathode-ray indicator parallel to the longitudinal axis of a ship or aircraft, as a reference for determining the heading.

Lucero A British interrogator-responsor used for both IFF and Rebecca-Eureka systems.

Lucite Trademark of Du Pont for their transparent acrylic resins.

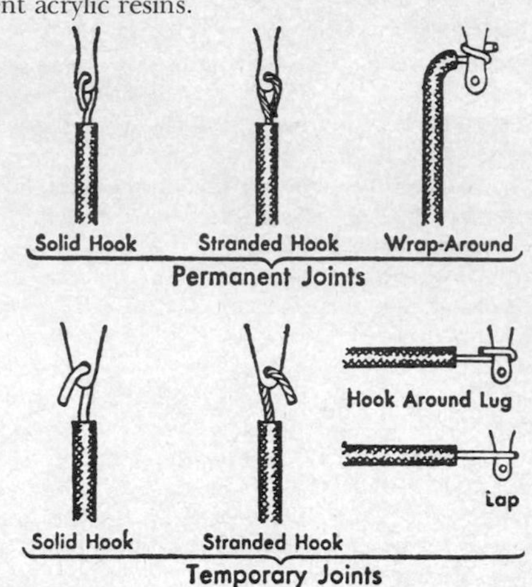

Lug connections for permanent and temporary soldered joints.

lug A stamped metal strip used as a terminal to which wires can be soldered. Also called soldering lug.

lumen [abbreviated lm] The SI unit of luminous flux equal to the flux on a unit surface, all points of which are at unit distance from a uniform point source of 1 candela.

lumen per square foot [abbreviated lm/ft²] A unit of illumination and luminous exitance. Use of the SI unit of luminous exitance, the lumen per square meter, is preferred.

lumen per square meter [abbreviated lm/m²] The SI unit of luminous exitance.

lumen per watt [abbreviated lm/W] The SI unit of luminous efficacy.

lumen second [abbreviated lm·s] The SI unit of quantity of light, equal to the quantity of light delivered in 1 s by a flux of 1 lm.

lumerg [LUMen-ERG] A unit of luminous flux, equal to 1 erg of radiant energy emitted by a source that has a luminous efficacy of 1 lm/W.

luminance The luminous intensity of any surface in a given direction per unit of projected area of the surface, as viewed from that direction. The SI unit of luminance is the candela per square meter. Formerly called brightness.

luminance carrier *Picture carrier.*

luminance channel A path intended primarily for the luminance signal in a color television system.

luminance flicker Flicker that results from fluctuation of luminance only.

luminance primary One of the three transmission primaries whose amount determines the luminance of a color in a color television system.

luminance signal The color television signal that is intended to have exclusive control of the luminance of the picture. It is made up of 0.30 red, 0.59 green, and 0.11 blue and is capable of producing a complete monochrome picture. Also called Y signal.

luminescence Emission of light by a material at lower than incandescent temperatures, as a result of chemical or electrical action, physiological processes, exposure to certain types of radiation, or other nonthermal processes.

luminescence threshold The lowest frequency of radiation that is capable of exciting a luminescent material. Also called threshold of luminescence.

luminescent Capable of exhibiting luminescence.

luminescent diode A semiconductor diode in which luminescence is produced when reversal of electric field polarity causes radiative recombination of electrons and holes at the surface of a semiconductor structure.

luminescent screen The screen in a cathode-ray tube; the screen becomes luminous when bombarded by an electron beam and maintains its luminosity for an appreciable time.

luminophor A luminescent material that converts part of the absorbed primary energy into emitted luminescent radiation.

luminosity The ratio of luminous flux to the corresponding radiant flux at a particular wavelength. Expressed in lumens per watt.

luminosity coefficients The constant multipliers for the respective tristimulus values of any color, such that the sum of the three products is the luminance of the color.

luminous efficacy The total luminous flux divided by the total radiant flux. The SI unit of luminous efficacy is the lumen per watt. Formerly called luminous efficiency.

luminous efficiency *Luminous efficacy.*

luminous emittance *Luminous exitance.*

luminous exitance The average luminous flux that leaves a surface per unit area, for which the SI unit is the lumen per square meter. Also called luminous emittance.

luminous flux The total visible energy produced by a source per unit time. It corresponds to the time rate of flow of light and is usually measured in lumens. Also called light flux.

luminous flux density The luminous flux per unit area of a surface. When it is referring to the luminous flux falling on a surface, it is called *illumination.* When it is referring to the luminous flux leaving a surface, it is called *luminous exitance.*

luminous intensity The luminous flux emitted by a source in an infinitesimal solid angle, divided by the solid angle.

luminous sensitivity The output current of a phototube or camera tube divided by the incident luminous flux.

lumped constant A single constant which is electrically equivalent to the total of that type of distributed constant existing in a coil or circuit.

lumped impedance An impedance concentrated in a single component rather than distributed throughout the length of a transmission line.

lunar module A self-powered module that transports one or more astronauts from a spacecraft in

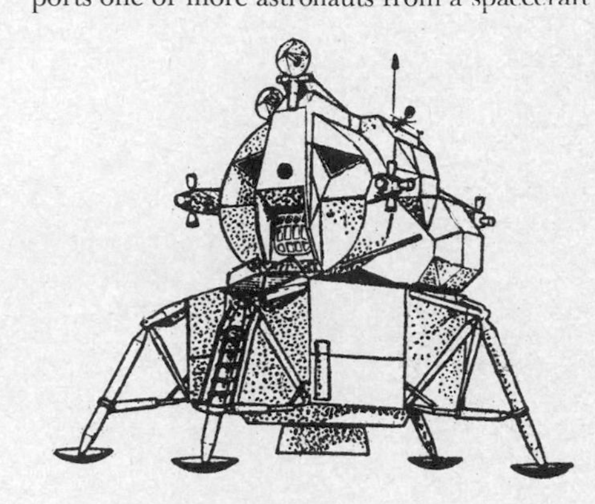

Lunar module.

lunar orbit to the moon's surface and returns them to the orbiting spacecraft.

lunar radar A special radar system that is beamed at the moon to synchronize clocks at remote locations on the earth to within at least 20 μs. Developed primarily for use by a worldwide network of stations responsible for tracking unmanned deep-space flight projects. Computers are programmed to correct the moon-reflected radar signals for transit-time delay, Doppler shifts, and other sources of errors.

lunar rocket A rocket vehicle that carries a payload to the moon. It may either circle the moon and return to earth or land on the moon.

lunar satellite A space vehicle placed in an orbit about the moon. Also called moon satellite.

lunar space The space near the moon in which the gravitational attraction of the moon is predominant.

Luneberg lens An artificial type of lens that focuses radiated electromagnetic energy at ultrahigh frequencies, to increase the gain of an antenna. The lens has a spherically symmetrical but nonuniform distribution of dielectric constant and index of refraction, designed to bend divergent rays from the feed so they are parallel as they emerge.

lutetium [symbol Lu] A rare-earth element. Atomic number is 71.

lux [abbreviated lx] The SI unit of illumination, equal to 1 lm/m^2.

Luxemburg effect Cross-modulation between two radio signals during their passage through the ionosphere, due to the nonlinearity of the propagation characteristics of free charges in space. Because of this effect, the program of a powerful station is sometimes heard when a receiver is tuned to a weaker station on a different frequency.

LVDT Abbreviation for *linear variable-differential transformer.*

Lw Symbol for *lawrencium.*

lx Abbreviation for *lux.*

Lyman alpha radiation Radiation at a wavelength of 1215 Å, associated with one of the spectral lines of hydrogen in the Lyman series.

Lyman series A group of spectral lines in the ultraviolet spectrum of hydrogen, in the wavelength range of 912 to 1215 Å.

m 1. Abbreviation for *meter* (metric unit of length). 2. Abbreviation for *milli-*.

M 1. Abbreviation for *mega-*. 2. Unofficial abbreviation for *megohm* (1,000,000 Ω); here the M follows the value, with no space between, as in 5M for 5,000,000 Ω. Used chiefly on diagrams.

M Symbol for *mutual inductance.*

MA Abbreviation for *megampere.*

Mace An Air Force surface-to-surface guided missile developed from Matador but having a self-contained navigation system. Range is about 600 mi (960 km).

machine check An automatic check in a computer, or a programmed check of machine functions.

machine error A deviation from correctness in computer-processed data, caused by equipment failure.

machine-gun microphone *Line microphone.*

machine instruction An instruction written in a machine language that a computer can recognize and execute.

machine language 1. Information in a physical form that can be handled by a computer, as on punched cards or punched paper tape. 2. Characters or instructions expressed in a form that a computer can process at once without conversion, translation, or programmed interpretation. In machine language, the value of each bit in each instruction in the program must be written out in binary, octal, or hexadecimal digital form.

machine learning The ability of a machine to improve its performance, based on past performance.

machine translation *Mechanical translation.*

machine word The standard number of characters that a computer regularly handles in each transfer. For example, a machine may regularly handle numbers or instructions in units of 36 binary digits.

Mach number [pronounced mock] The ratio of flight speed to the speed of sound in the medium in which the object moves. At sea level, Mach 1 is approximately 741 mi/h (1192 km/h) at 32°F (0°C) in dry air, but at 30,000 ft (9.14 km) altitude it is about 675 mi/h (1086 km/h).

macro *Macroinstruction.*

macroinstruction A computer instruction written in a simple higher-level language or even as a single symbol that is equivalent to one or more ordinary machine instructions. A macroinstruction is expanded into machine instructions by the assembler software of a computer, making it unnecessary for programmers to write out frequently occurring instruction sequences. Also called macro.

macrometeorite Any meteoroid particle larger than a pea.

macroprogramming The process of writing a computer program in terms of macroinstructions.

macroscopic Large enough to be read by the unaided eye.

macroscopic cross section The cross section per unit volume.

macrosonics The technology of sound at signal amplitudes so large that linear approximations are not valid, as in the use of ultrasonics for cleaning or drilling.

MAD Abbreviation for *magnetic anomaly detector.*

madistor A cryogenic semiconductor device in which injection plasma can be steered or controlled by transverse magnetic fields, to give the action of a switch.

MADRE [MAgnetic Drum Receiving Equipment] A form of over-the-horizon search radar in which a pulse of electromagnetic radiation in

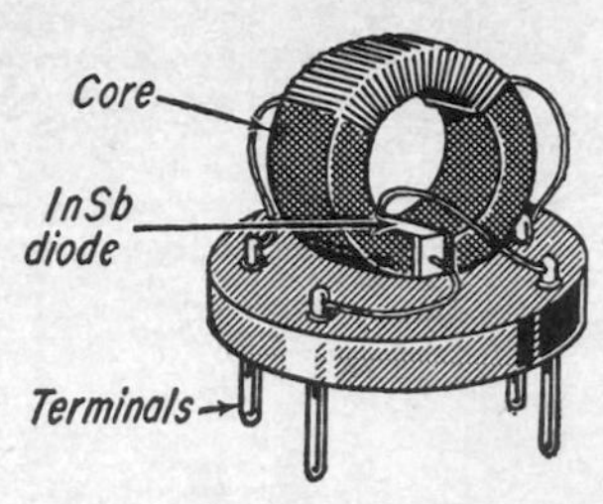

Madistor construction.

the upper portion of the high-frequency band is bounced off the ionosphere to the target. A magnetic drum stores the reflected signals reaching the receiver, to identify targets by comparison techniques. Used to detect the approach of enemy aircraft and missiles.

magamp Abbreviation for *magnetic amplifier.*

magic number The atomic number or neutron number of a nuclide that has greater-than-average stability and sometimes other exceptional properties. Some magic numbers are 2, 8, 20, 28, 50, 82, and 126.

magnal base A base that has 11 pins, used for cathode-ray tubes.

magnesium [symbol Mg] An alkaline metallic element. Its compounds are sometimes used for cathodes of phototubes when maximum response is desired in the ultraviolet region. Atomic number is 12.

magnesium cell A primary cell in which the negative electrode is made of magnesium or one of its alloys.

magnesium-copper sulfide rectifier A metallic rectifier that consists of magnesium in contact with copper sulfide. Also called copper sulfide rectifier.

magnesium-cuprous chloride cell A reserve primary cell that is activated by adding water. It is less expensive than its silver chloride counterpart but has lower capacity.

magnesium-mercuric oxide battery A reserve battery in which thin, flat anode plates of magnesium alternate with cathode plates of mercuric oxide, with thin separators between adjacent plates. The magnesium perchlorate electrolyte is stored in a separate compartment until needed, then released by rupturing a diaphragm mechanically or by firing an explosive squib.

magnesium oxide emitter A type of unheated cathode in which a nickel sleeve that serves as base is coated with a metal oxide such as magnesium oxide, about 25 μm thick. Emission of electrons is initiated by ultraviolet light, nuclear radiation, or bombarding electrons. Emission then continues after the radiation is turned off, under the influence of the anode voltage.

magnesium-silver chloride cell A reserve primary cell that is activated by adding water.

magnet An object that produces a magnetic field outside itself. It has the property of attracting other magnetic objects, such as iron, and attracting or repelling other magnets. A permanent magnet produces a permanent magnetic field, whereas an electromagnet possesses magnetic properties only when current is flowing through its windings.

magnet charger A charger that restores or establishes the field strength of a permanent magnet by applying a strong magnetic field produced by a surge of direct current through a large electromagnet.

magnetic airborne detector *Magnetic anomaly detector.*

magnetically operated solid-state switch A combination of a Hall generator, trigger circuit, and an amplifier in a single integrated circuit, opened or closed at up to 10,000 operations per second by moving a magnet or otherwise changing a magnetic field. There are no contacts to wear out and hence no contact bounce.

magnetic amplifier [abbreviated magamp] An amplifier that uses the nonlinear properties of saturable reactors, alone or in combination with other circuit elements, to provide amplification. The DC or AC input signal is applied to a control winding to change the degree of saturation of the core, thereby producing a larger change in alternating current in the output winding.

magnetic anisotropy The dependence of the magnetic properties of some materials on direction.

magnetic anomaly A variation in the expected and normal pattern of a magnetic field, as in the magnetic field of the earth. It can be caused by any large ferrous mass, such as a submerged submarine.

magnetic anomaly detector [abbreviated MAD] An airborne magnetometer that detects submerged submarines. A direct current passes through a coil wound on a high-permeability core and balances out the effect of the earth's magnetic field on the core, while an alternating current saturates the core an equal amount on both positive and negative swings. The magnetic field of a submarine makes the swings unequal, thereby producing an output signal. In Navy Orion P3 antisubmarine patrol planes, the magnetic sensor is located in a stinger-shaped housing that projects behind the tail of the plane. Also called magnetic airborne detector.

magnetic-armature loudspeaker A loudspeaker in which the diaphragm is driven by a ferromagnetic armature that is alternately attracted and repelled by interaction between the field of a permanent magnet and the field produced in the armature by a coil that carries audio-frequency currents. Also called electromagnetic loudspeaker, magnetic loudspeaker, and moving-armature loudspeaker.

magnetic bearing 1. The angle in the horizontal plane between the direction of magnetic north and the line along which an aircraft or vessel is pointing. Usually measured clockwise from magnetic north. 2. A bearing that supports a shaft in free space by magnetic fields which are generally produced by a combination of electromagnets and permanent magnets.

magnetic bias A steady magnetic field applied to the magnetic circuit of a relay or other magnetic device.

magnetic biasing Biasing of a magnetic recording medium during recording by superposing an additional magnetic field on that of the signal being recorded. Used to obtain a substantially linear relationship between the amplitude of the signal and the remanent flux density in the recording medium.

magnetic biasing coil A winding used on a saturable reactor to establish a basic magnetization of the core in both polarity and magnitude.

magnetic blowout A permanent magnet or electromagnet that produces a magnetic field which lengthens the arc between opening contacts of a switch or circuit breaker, thereby helping to extinguish the arc.

magnetic bottle A magnetic field that confines a stream of plasma to minimum volume to produce the pinch effect in an electron tube and in controlled thermonuclear fusion experiments.

magnetic brake A friction brake controlled by electromagnetic means.

magnetic bubble A cylindrical stable (nonvolatile) region of magnetization produced in a thin-film magnetic material by an external magnetic field. The direction of magnetization is perpendicular to the plane of the material. Also called bubble.

magnetic-bubble memory *Bubble memory.*

magnetic card A card that has a magnetic stripe or some other magnetizable configuration on which data can be stored by selective magnetization, much as on computer magnetic tape. Applications include credit cards for on- or off-line point-of-sale terminals, cards for obtaining cash outside banking hours at self-service money-dispensing terminals, and monthly commuter tickets in which each use changes the recorded information appropriately to represent the number of rides remaining.

magnetic cartridge *Variable-reluctance pickup.*

magnetic cell One unit of a magnetic memory, capable of storing 1 bit of information as a 0 or a 1 state.

magnetic character A character printed with magnetic ink, as on bank checks, for reading by both machines and humans.

magnetic-character reader A character reader that reads special type fonts printed in magnetic ink, such as those used on bank checks, and feeds the character data directly to a computer for processing.

magnetic circuit A complete closed path for magnetic lines of force; it has a reluctance that limits the amount of magnetic flux which can be sent through the circuit by the magnetomotive force.

magnetic clutch A clutch in which motion is transmitted from one rotating shaft to another by the attraction between magnetized poles.

magnetic contactor A contactor actuated by electromagnetic means.

magnetic core A quantity of ferrous material placed in a coil or transformer to provide a better

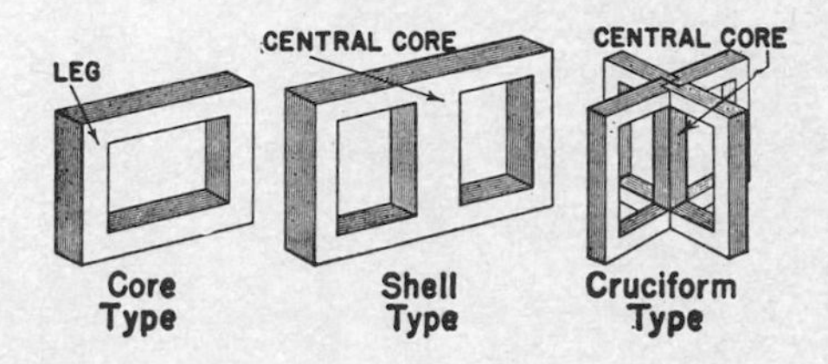

Magnetic cores for power transformers and iron-core chokes.

path than air for magnetic flux, thereby increasing the inductance of the coil and increasing the coupling between the windings of a transformer. Also called core.

magnetic-core storage A computer storage system in which thousands of tiny doughnut-shaped magnetic cores each store 1 bit of information. Current pulses are sent through wires threading through the cores to record or read out data. Also called core memory.

magnetic course A course in which the direction of the reference line is magnetic north. Also called corrected compass course.

magnetic current sheath The sheath that develops at the surface of a plasma immersed in a magnetic field. The space charge may be either positive or negative. A net current flows along the sheath surface, perpendicular to the magnetic field.

magnetic cutter A cutter in which the mechanical displacements of the recording stylus are produced by the action of magnetic fields.

magnetic damping Damping of a mechanical motion by the reaction between a magnetic field and the current generated by the motion of a coil through the magnetic field.

magnetic declination The angle between true north (geographical) and magnetic north (the direction of the compass needle). The angle is different for different places and may vary from year to year.

magnetic deflection Deflection of an electron beam by a magnetic field, as in a television picture tube.

magnetic delay line A delay line used for storage of data in a computer; it consists essentially of a metallic medium along which the velocity of propagation of magnetic energy is small compared to the speed of light. Storage is accomplished by recirculation of wave patterns that contain information, usually in binary form.

magnetic dip The angle that the magnetic field of the earth makes with the horizontal at a particular location. Also called magnetic inclination.

magnetic dipole An elementary dipole associated with nuclear particles. It consists of two equal magnetic poles which have opposite polarity, closely spaced together, and so small that directive properties are independent of size and shape. Also called magnetic doublet.

magnetic discriminator A magnetic amplifier used with a transformer and other components to sense the polarity and magnitude of coded pulses, to produce output control voltages such as might be required for making rudder corrections during the flight of a guided missile.

magnetic disk A disk that has a magnetizable coating on which binary data or other types of information can be recorded for storage and readout. Used in some types of storage units for computers.

magnetic domain A movable magnetized area in a thin-film magnetic material, the presence or absence of which can designate a 1 or 0. Either state is stable (nonvolatile), so is not affected by power failure in a computer. Used in domain-tip memories. The direction of magnetization in a domain is parallel to the plane of the material and is about 100 times as great as that in a magnetic bubble. Also called domain.

magnetic-domain memory *Domain-tip memory.*

magnetic doublet *Magnetic dipole.*

magnetic drum A rapidly rotating cylinder coated with magnetic material. Information is stored in the cylinder in the form of magnetized dipoles. The orientation or polarity of the dipoles is used to store binary information.

magnetic electron multiplier An electron multiplier in which the paths of the emitted secondary electrons are controlled by an applied magnetic field.

magnetic field Any space or region in which a magnetic force is exerted on moving electric charges. The magnetic field may be produced by a current-carrying coil or conductor, by a permanent magnet, or by the earth itself.

magnetic field strength [symbol H] The magnitude of the magnetic field vector. Expressed in amperes per meter in SI units. Formerly expressed in oersteds in CGS units. Also called magnetizing force.

magnetic film memory A magnetic memory that consists essentially of a magnetic film and associated read-in and read-out wiring, producible with packing densities of several thousand elements per square inch by vacuum deposition, electroplating, chemical etching, and/or other integrated-circuit production techniques. Rotational switching of the magnetic field of an anisotropic magnetic material such as permalloy is generally used, instead of changing the magnetic field strength of an element.

magnetic flaw detector A flaw detector in which a ferrous object is magnetized with an electromagnet or permanent magnet and sprayed with an ink that contains fine iron particles. Surface or near-surface flaws then appear as black lines that are even more visible if the surface being examined is first painted white.

magnetic flowmeter A flowmeter that depends on the presence of magnetic constituents in a liquid or slurry. It must be initially calibrated for the amount of magnetic material present per unit volume.

magnetic fluid A suspension of iron particles or colloidal ferrite particles in a carrier fluid. The colloidal suspension can be controlled in position, location, shape, specific gravity, surface, trajectory, or velocity by a magnetic field, whereas in magnetic clutch fluids the iron particles chain together and solidify when a magnetic field is applied.

magnetic fluid clutch A friction clutch that is engaged by magnetizing a liquid suspension of powdered iron located between pole-pieces mounted on the input and output shafts.

magnetic flux The magnetic lines of force produced by a magnet.

magnetic flux density [symbol B] The number of magnetic lines of force per unit area at right angles to the line. The SI unit is the tesla. The gauss was formerly used as the CGS unit of magnetic flux density. Also called magnetic induction.

magnetic focusing Focusing an electron beam by using the action of a magnetic field.

magnetic forming The forming of metal into desired shapes by using strong magnetic fields to push the metal against a forming die. The magnetic field is produced by charging a large capacitor bank, then dumping the stored energy into an induction coil in less than a millionth of a second. Also called electromagnetic forming.

magnetic friction clutch A friction clutch in which the pressure between the friction surfaces is produced by magnetic attraction.

magnetic gap A nonmagnetic section in a magnetic circuit, such as an air gap.

magnetic hardness comparator An instrument that compares the hardness of steel parts. A sample part that has the desired hardness is placed in one coil, and the parts to be tested are inserted one by one into a similar coil. If the two coils then have the same magnetic properties as displayed on a cathode-ray oscilloscope screen, the parts in the two coils have the same hardness.

magnetic head The electromagnet used for reading, recording, or erasing signals on a magnetic disk, drum, or tape. Also called head.

magnetic heading A heading in which the direction of the reference line is magnetic north. Also called corrected compass heading.

magnetic hysteresis Internal friction that occurs between the molecules of a magnetic material when subjected to a varying magnetic field. It results in heat loss and makes the magnetic induction dependent on the previous state of magnetization of the material.

magnetic inclination *Magnetic dip.*

magnetic induction 1. The process of using a magnetic field to generate or induce currents or voltages in conductors. In general, the magnetic field must be varying, or there must be relative motion between the conductor and the field. 2. *Magnetic flux density.*

magnetic-induction gyroscope A nuclear gyroscope in which DC and AC magnetic fields act on

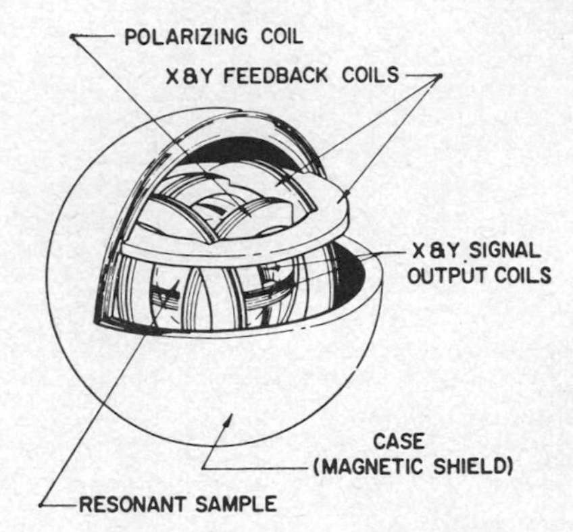

Magnetic-induction gyroscope using spin of nuclei in liquid resonant sample.

water that has been doped with paramagnetic salts. There are no moving parts.

magnetic ink Ink that contains magnetic particles to permit reading of printed characters by both a magnetic character reader and humans.

magnetic-ink character recognition [abbreviated MICR] A check-processing system in which, after conventional printing of blank checks, bank and customer identifying numerals and special characters are imprinted in magnetic ink to specifications established by the American Banking Association. After the check is cashed, up to ten numerals and four special characters are added to provide the amount of the check and any special control data required for automatic processing by computer.

magnetic integrated circuit An integrated circuit in which magnetic elements perform part or all of the intended function.

magnetic leakage Passage of magnetic flux outside the path along which it can do useful work.

magnetic lens A lens that uses an arrangement of electromagnets or permanent magnets to produce magnetic fields that focus a beam of charged particles.

magnetic levitation The use of magnetic forces for stable suspension of a ground vehicle above or below a suitable guideway. One approach is ferromagnetic attraction, in which AC or DC magnets on the vehicle ride below a ferromagnetic rail to provide attraction forces for suspension from the rail. Another approach is superconductive induction, in which suspension is provided by a repulsive-force interaction between superconducting magnet coils of the vehicle and currents induced in the guideway conductors. Magnetic levitation proposals are generally combined with linear electric motors or some other form of propulsion for high-speed ground transportation.

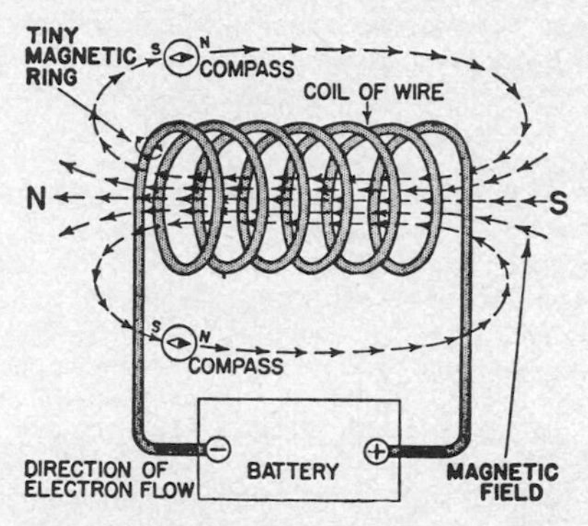

Magnetic lines of force around current-carrying coil make up magnetic field of coil.

magnetic line of force An imaginary line, each segment of which represents the direction of the magnetic flux at that point.

magnetic loudspeaker *Magnetic-armature loudspeaker.*

magnetic material A material that has a permeability considerably greater than that of air or a vacuum. Ferromagnetic materials are strongly magnetic; paramagnetic materials are feebly magnetic.

magnetic memory *Magnetic storage.*

magnetic-memory plate A magnetic memory that consists of a ferrite plate which has a grid of small holes through which the read-in and read-out wires are threaded. Printed wiring may be applied directly to the plate in place of conventionally threaded wires, permitting mass production of plates that have a high storage capacity.

magnetic meridian A horizontal line oriented, at each point on the surface of the earth, in the direction of the horizontal component of the magnetic field of the earth at that point.

magnetic microphone *Variable-reluctance microphone.*

magnetic mine An underwater mine that is detonated when the hull of a passing vessel changes the magnetic field at the mine.

magnetic mirror A magnetic field used in controlled-fusion experiments to reflect charged particles back into the central region of a magnetic bottle. Reflection occurs in the region where the

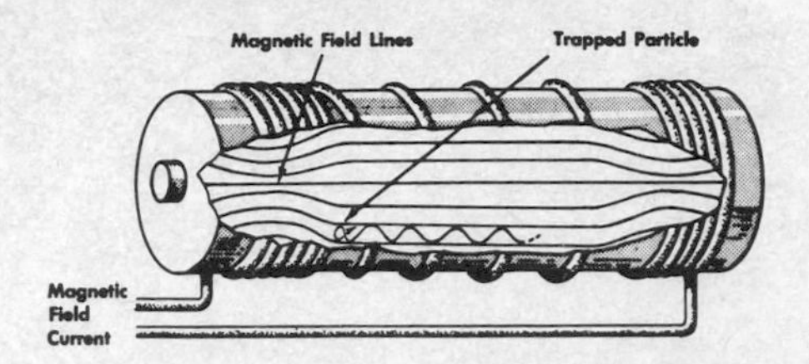

Magnetic mirrors exist where magnetic field lines are closest together.

magnetic field increases abruptly in strength.

magnetic modulator A modulator in which a magnetic amplifier is the modulating element for impressing an intelligence signal on a carrier.

magnetic moment The moment of a magnetic dipole. A magnetic moment is associated with the intrinsic spin of a particle and the orbital motion of a particle in a system.

magnetic north The direction indicated by the north-seeking element of a magnetic compass when influenced only by the earth's magnetic field. Because magnetic meridians often follow zigzag lines, the compass needle at any given place does not necessarily point to the magnetic pole.

magnetic pickup *Variable-reluctance pickup.*

magnetic plated wire A magnetic wire that has a core of nonmagnetic material and a plated surface of ferromagnetic material.

magnetic pole 1. One of the two poles of a magnet near which the magnetic intensity is greatest. These poles are known as the north and south poles. 2. Either of two locations on the surface of the earth toward which a compass needle points. The north magnetic pole is near the geographic North Pole and attracts the south pole of a compass needle.

magnetic pressure The magnetic force that confines an ionized gaseous fuel, such as plasma, in a magnetic bottle. The magnetic pressure is generally defined as being proportional to the square of the intensity of the magnetic field in gauss.

magnetic printing The permanent and usually undesired transfer of a recorded signal from one section of a magnetic recording medium to another when these sections are brought together, as on a reel of tape. Also called magnetic transfer.

magnetic probe A small pickup coil inserted in a magnetic field to measure changes in field strength.

magnetic pumping A method of heating plasma in a stellarator or other controlled-fusion device by alternately increasing and decreasing the strength of the magnetic field acting on the plasma.

magnetic reaction analyzer An analyzer in which a Hall detector measures magnetic field intensities directly with high resolution, making it possible to locate tiny defects during magnetic inspection of welds and other materials.

magnetic reading head A magnetic head that transforms magnetic variations in magnetic tape or wire into corresponding voltage or current variations.

magnetic recorder A recorder that records audio-frequency signals on magnetic tape or wire as magnetic variations in the medium. It usually also contains provisions for playback, to convert the

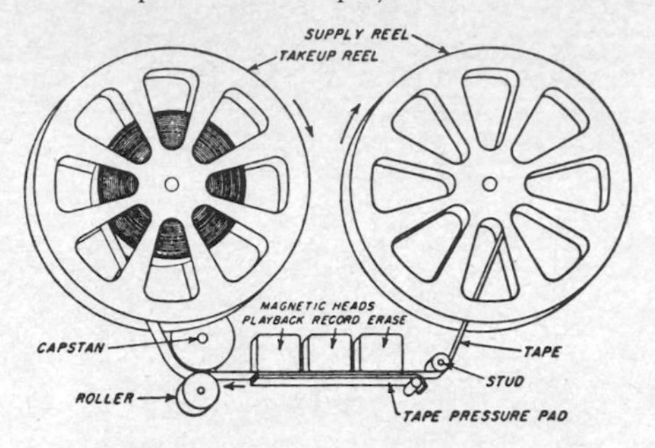

Magnetic-recorder construction.

recorded magnetic variations back into electric variations, with or without amplification and reproduction as sound waves by a loudspeaker.

magnetic recording Recording by a signal-controlled magnetic field.

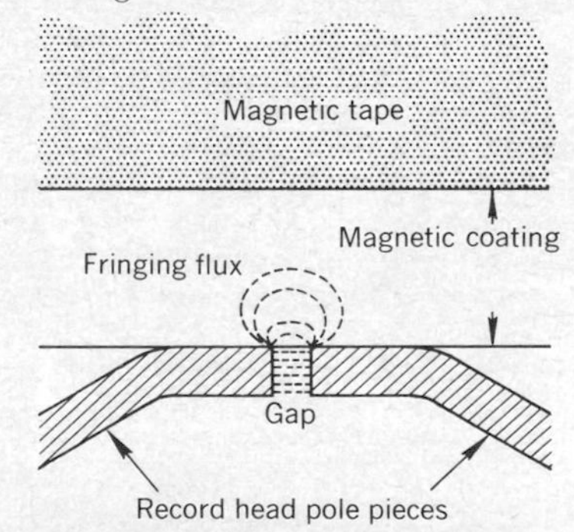

Magnetic recording head. For instrumentation applications, plastic base of tape has magnetic coating about 5 μm thick, moving over air gap of about 1 μm in poles of recording head. Playback head is similar but has even smaller gap.

magnetic recording head A magnetic head that transforms electric variations into magnetic variations for storage on magnetic media.

magnetic recording medium A magnetizable material used in a magnetic recorder to retain the magnetic variations imparted during the recording process. It may have the form of a wire, tape, cylinder, card, or disk.

magnetic recording paper A particle-oriented paper in which both machine-readable and visible traces can be produced by a magnetic recording head. The paper is reusable because the trace can be erased by a combination of AC and DC magnetic fields. Used in circular-chart, strip-chart, and XY recorders.

magnetic reed relay *Reed relay.*

magnetic reed switch *Reed switch.*

magnetic reproducing head A magnetic head that converts magnetic variations on magnetic media into electric variations.

magnetic resonance spectrum A spectrum produced by varying the RF electromagnetic field that is superimposed on a steady or slowly varying magnetic field about which the atoms of a material precess, to make molecules change their magnetic quantum numbers as they absorb or emit quanta of radio waves.

magnetic rigidity A measure of the momentum of a particle, equal to the product of the magnetic intensity perpendicular to the path of the particle and the resultant radius of curvature of the path of the particle.

magnetics The branch of science that deals with magnetic phenomena.

magnetic saturation The maximum possible magnetization of a magnetic substance. Also called saturation.

magnetic separator An apparatus for separating powdered magnetic ores from nonmagnetic ores. An electromagnet deflects magnetic materials from the path taken by nonmagnetic materials.

magnetic shield An enclosure made from high-permeability magnetic material, used to protect instruments and electronic assemblies from the effects of stray magnetic fields.

magnetic shift register A shift register in which the pattern of settings of a row of magnetic cores is shifted one step along the row by each new input pulse. Diodes in the coupling loops between cores prevent backward flow of information.

magnetic shunt A piece of magnetic material used to divert an adjustable amount of magnetic flux around an air gap, usually for calibration purposes.

magnetic sound track A magnetic stripe placed on motion-picture film to form a magnetic track on which the sound accompaniment for the film can be recorded.

magnetic spectrograph A spectrograph based on the action of a constant magnetic field on the paths of electrons or other charged particles. Used to separate particles that have different velocities.

magnetic storage Storage in which information is represented by varying degrees of magnetization of a magnetic material. The material may be in the form of magnetic cores, disks, drums, plates, or tapes. Also called magnetic memory.

magnetic storm A storm that causes rapid and erratic changes in the strength of the magnetic fields of the earth, affecting both radio and wire communications. It is believed to be caused by sun-spot activity.

magnetic-stripe credit card A credit card that has one or more magnetic stripes which contain the data required for establishing credit at an on- or off-line point-of-sale terminal. Added features include secret personal identification codes and encryption of on-line encoded data to minimize fraud and abuse of cards.

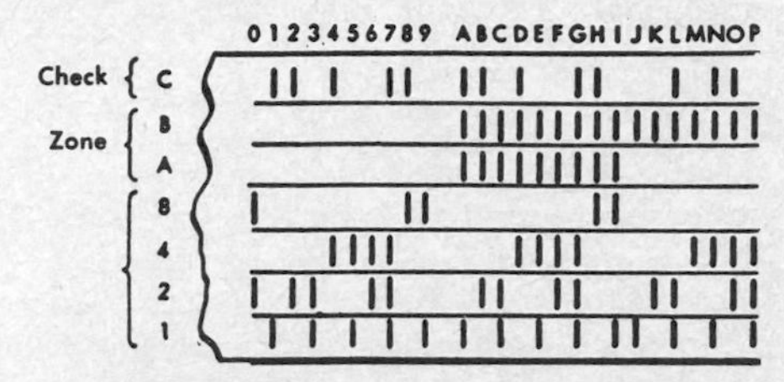

Magnetic tape as used in some digital computers. Black lines in tracks represent magnetized areas or bits that form binary codes for numerals, letters, and other characters.

magnetic tape A plastic, paper, or metal tape that is coated or impregnated with magnetizable iron oxide particles. Also called tape.

magnetic-tape core A toroidal core formed by winding a strip of thin magnetic-core material around a form. Coils may be wound around the core with a toroidal winder, or the completed core may be sawed in half to permit insertion of prewound coils.

magnetic-tape player A player that plays back prerecorded magnetic tape. It has no recording capability.

magnetic-tape reader A computer device that is capable of reading information recorded on magnetic tape and delivering the corresponding electric pulses.

magnetic-tape recorder A magnetic recorder that uses magnetic tape as the recording medium. It may also have playback facilities.

magnetic-tape unit A computer unit that usually consists of a tape transport, reading and recording heads, and associated electric and electronic equipment.

magnetic test coil *Exploring coil.*

magnetic thin film A layer of magnetic material, usually less than 1 μm thick, used chiefly for data storage and logic devices.

magnetic transfer *Magnetic printing.*

magnetic unit A unit used in measuring magnetic quantities. The SI magnetic units are the ampere-turn, ampere per meter, tesla, and weber.

magnetic-vane meter An AC meter in which the moving element is a metal vane pivoted inside a coil.

magnetic variometer An instrument that measures differences in a magnetic field with respect to space or time, as contrasted to a magnetometer, which measures the absolute value of the intensity and/or direction of a magnetic field.

magnetic wave A magnetostatic wave, one that

does not depend on the motion of magnetic fields. A surface magnetic wave is an example.

magnetic-wave device A device that depends on magnetoelastic or magnetostatic wave propagation through or on the surface of a magnetic or dielectric material. Applications are comparable to those for other types of microwave delay lines.

magnetic wire A wire made from magnetic material suitable for magnetic recording.

magnetism A property possessed by iron, steel, and certain other magnetic materials, wherein these materials can produce or conduct magnetic lines of force capable of interacting with electric fields or other magnetic fields.

magnetization 1. The degree to which a particular object is magnetized. 2. The process of magnetizing a magnetic material.

magnetization curve *B-H curve.*

magnetizing current The current that flows through the primary winding of a power transformer when no loads are connected to the secondary winding. This current establishes the magnetic field in the core and furnishes energy for the no-load power losses in the core. Also called exciting current.

magnetizing force *Magnetic field strength.*

magnet keeper *Keeper.*

magneto An AC generator that uses one or more permanent magnets to produce its magnetic field. Also called magnetoelectric generator.

magnetocardiograph [abbreviated MCG] An instrument that records the intensity of the magnetic field of the heart as produced by currents sent through the body by the voltages associated with each heart beat. The maximum strength of this magnetic field at the surface of the chest is about one-millionth of the strength of the earth's magnetic field; hence measurements must be made in a shielded room or in an isolated building where all magnetic noise sources can be turned off during record-taking.

magnetochemistry The branch of chemistry that deals with the effects of magnetism on chemicals.

magnetodiode A semiconductor diode in which an external magnetic field controls the forward current by varying the lifetime of injected carriers.

magnetoelastic energy The energy associated with the change in the dimensions of a ferromagnetic material during magnetization.

magnetoelectric Pertaining to the generation of voltages by magnetic techniques, as in an ordinary generator.

magnetoelectric effect One of the effects observed when a material carrying an electric current is placed in a transverse magnetic field. These effects include the Hall effect, Ettingshausen effect, Nernst effect, and magnetoresistance.

magnetoelectric generator *Magneto.*

magnetoelectric transducer A transducer that measures the voltage generated by the movement of a conductor in a magnetic field.

magnetograph An instrument that records variations in the magnetic field of the earth or sun.

magnetohydrodynamics [abbreviated MHD] The study of the effects of magnetic fields on superheated ionized gases and conducting fluids.

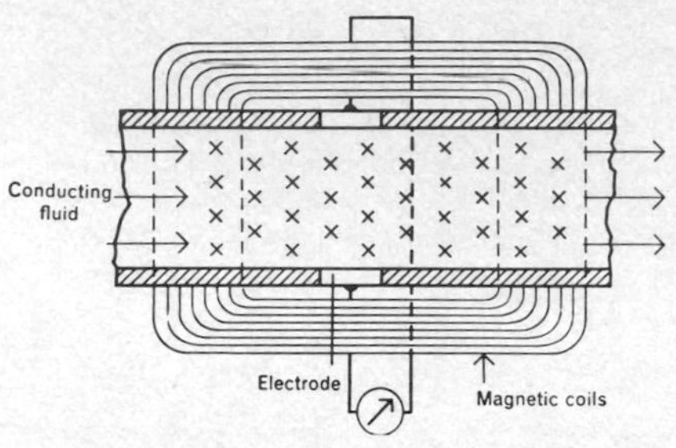

Magnetohydrodynamics as applied to electromagnetic flowmeter for measuring rate of flow of liquids.

Used in research on controlled fusion for nuclear power plants. Proposed as a method of generating electric power at higher efficiencies than steam generators.

magnetohydrodynamic shock wave A highly ionized shock wave that propagates through a magnetic field. Very large currents can be generated on the surface of the shock front, and these currents in turn produce a jump change in the magnetic field across the shock front.

magnetohydrodynamic system A propulsion system based on the reaction of a weakly ionized arc-produced plasma or on the reaction of highly conductive nuclear plasma produced by controlled nuclear reactions.

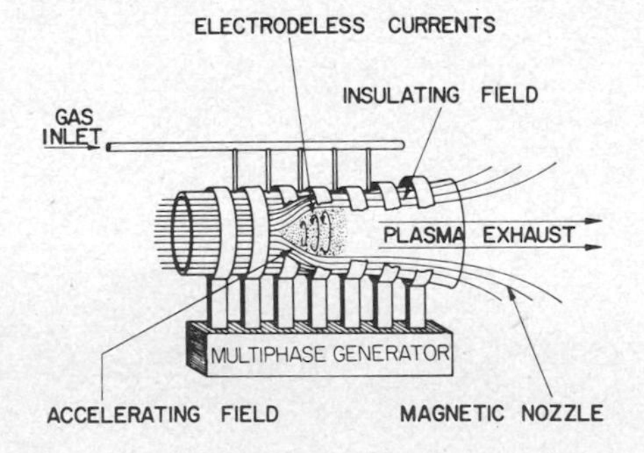

Magnetohydrodynamic system.

magnetoionic Pertaining to the combined effect of the earth's magnetic field and atmospheric ionization on the propagation of electromagnetic waves.

magnetoionic wave component Either of the two elliptically polarized wave components into which a linearly polarized wave incident on the ionosphere is separated because of the earth's magnetic field.

magnetometer An instrument that measures the magnitude and sometimes the direction of a magnetic field, such as the earth's magnetic field.

magnetomotive force The force that produces a magnetic field. It is the total magnetizing force acting around a complete closed magnetic circuit, and it corresponds to voltage (electromotive force) in an electric circuit. If due to current in a coil, magnetomotive force is proportional to ampere-turns. The CGS unit of magnetomotive force is the gilbert, equal to about 0.8 ampere-turn. Use of the corresponding SI unit of magnetomotive force, the ampere-turn, is now preferred.

magneton *Bohr magneton.*

magnetooptical effect *Kerr magnetooptical effect.*

magnetooptical laser A laser in which a continuous magnetic field contributes to the generation of coherent radiation.

magnetooptical material A material in which one or more optical properties can be changed by applying a magnetic field; for example, the Faraday effect can be used in certain transparent materials to deflect a laser beam.

magnetooptical modulator An arrangement for modulating a beam of light by passing it through a single crystal of yttrium-iron garnet,

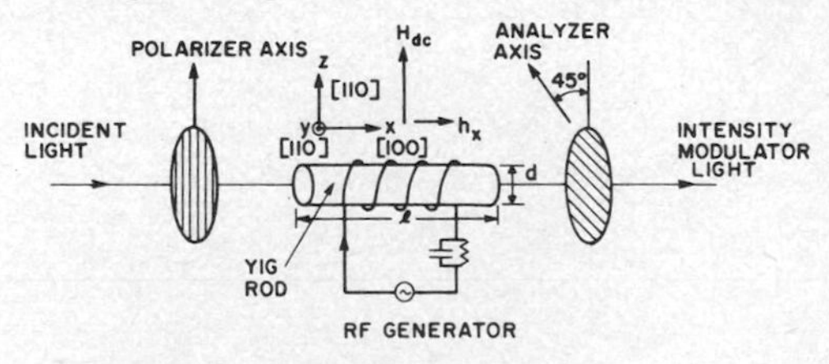

Magnetooptical modulator in which modulation of light beam is proportional to magnetization in direction of light propagation through YIG crystal.

which provides intensity modulation by using a magnetic field to produce optical rotation.

magnetooptics The branch of physics that deals with the effect of magnetic fields on light.

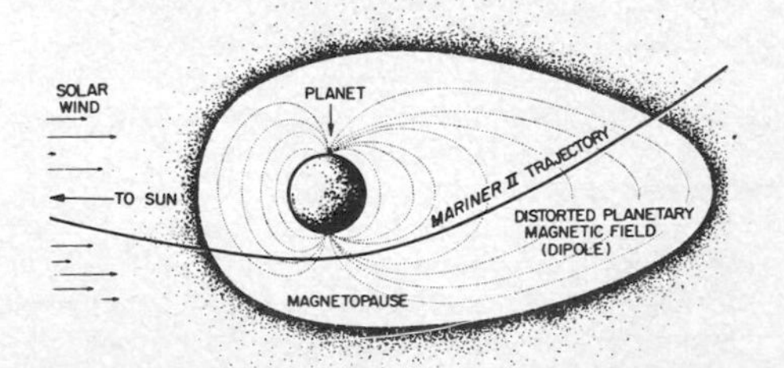

Magnetopause as believed to exist around planet Venus, and Mariner II trajectory for Venus flyby mission.

magnetopause The boundary between the magnetosphere of the earth or another planet and the interplanetary plasma or solar wind of the solar atmosphere.

magnetophone An early German version of a magnetic-tape recorder.

magnetophotophoresis Movement of dust and other particles under the combined influences of a magnetic field and radiant energy such as light.

magnetoplasmadynamics The generation of electricity by shooting a beam of ionized gas through a magnetic field, to give the same effect as moving copper bars near a magnet.

magnetoresistance The change in resistance associated with a change in magnetization in some materials.

magnetoresistance multiplier An analog multiplier in which thin-film magnetoresistors are mounted in air gaps in the magnetic core of the multiplier. An input signal applied to a coil wound on the core of the multiplier produces push-pull resistance swings in the two magnetoresistors, thereby unbalancing the bridge in which they are connected. Used in control systems and analog computers for measuring power and providing such functions as division, multiplication, squaring, and square-rooting.

magnetoresistor A resistor in which the resistance value changes with the strength of the applied magnetic field. Generally used in pairs in a bridge circuit, with a permanent magnet provid-

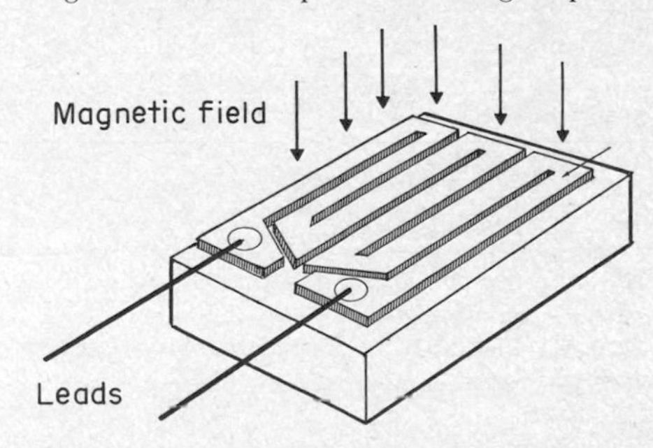

Magnetoresistor using 20-μm-thick layer of indium antimonide etched into serpentine pattern on substrate 0.1 mm thick.

ing a magnetic bias. Current flow through series-opposing coils wound on the magnet poles reduces the flux at one resistor while increasing it at the other, to unbalance the bridge and give an output current.

magnetosphere A region that extends to heights of several earth radii above the earth, existing by virtue of the earth's magnetic field. It includes the Van Allen radiation belt and consists of trapped particles, chiefly electrons and protons, that spiral about the magnetic lines from pole to pole.

magnetostatic Pertaining to magnetic properties that do not depend upon the motion of magnetic fields.

magnetostriction The change in the dimensions of a ferromagnetic object when the object is placed in a magnetic field. Also called Joule effect.

magnetostriction hydrophone A magnetostriction microphone that responds to waterborne sound waves.

magnetostriction loudspeaker A loudspeaker in which the mechanical displacement is derived from the deformation of a material that has magnetostrictive properties.

magnetostriction microphone A microphone that depends for its operation on the generation of an electromotive force by the deformation of a material that has magnetostrictive properties. Used chiefly in ultrasonic and underwater sound applications.

magnetostriction oscillator An oscillator in which the anode circuit is inductively coupled to the grid circuit through a magnetostrictive element. The frequency of oscillation is determined by the magnetomechanical characteristics of the coupling element.

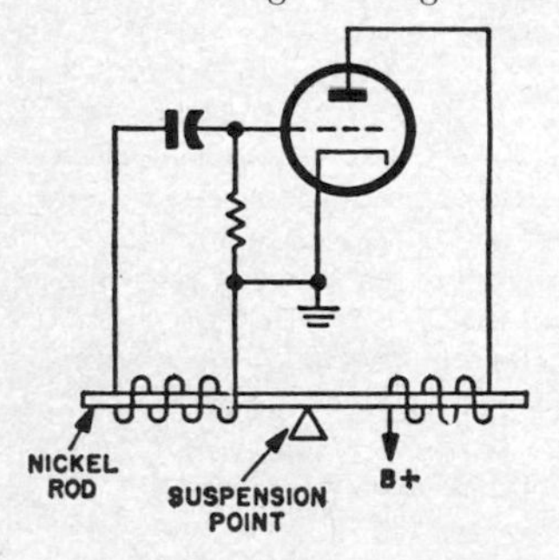

Magnetostriction-oscillator circuit.

magnetostriction transducer A transducer used with sonar equipment to change an alternating current to sound energy at the same frequency and form the sound energy into a beam. A large number of tubes and coils are usually connected in a series-parallel arrangement. One end of each tube is attached to a diaphragm that is in contact with the sea water. The transducer acts also as a microphone for returning echoes.

magnetostrictive Changing in dimensions when placed in a magnetic field.

magnetostrictive delay line A delay line made of nickel or other magnetostrictive material, in which the amount of delay is determined by a shock wave traveling through the length of the line at the speed of sound.

magnetostrictive relay A relay that functions as a result of dimensional changes occurring in a magnetic material subjected to a magnetic field.

magnetostrictive resonator A ferromagnetic rod that can be made to vibrate at one or more definite resonant frequencies by applying an alternating magnetic field.

magnetostrictor A device for converting electric oscillations to mechanical oscillations by employing the property of magnetostriction. The device consists of a bar of magnetic material, anchored at one point and surrounded by a coil that carries the oscillating current. For maximum energy, the system must be driven at or near its natural frequency. Used in ultrasonic applications.

magnetothermoelectric single crystal A crystal that can be the basis for a solid-state refrigerator. When the crystal carries an electric current in the presence of an intensely strong transverse magnetic field of over 100 kG, it is capable of cooling to temperatures of 100°C below room temperature.

magnetron A two-electrode electron tube in which the flow of electrons to the anode is controlled by a combination of crossed steady electric and magnetic fields, to produce AC power output. Used as an oscillator in microwave radio and radar transmitters. Basic types include interdigital, multicavity, multisegment, packaged, rising sun, split-anode, and traveling-wave magnetrons.

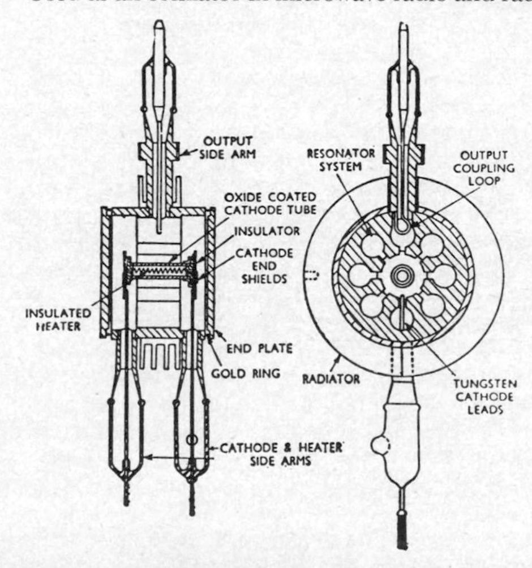

Magnetron construction, shown in side and top views. Coupling loop takes output microwave power from one cavity of hole-and-slot anode.

magnetron amplifier A magnetron used as an amplifier.

magnetron oscillator An oscillator circuit that uses a magnetron.

magnetron package A complete magnetron assembly that consists of a magnetron, its permanent magnet, and its output-matching device.

magnetron rectifier A cold-cathode gas diode rectifier in which the electron stream is controlled by an external magnetic field.

magnetron strapping Connecting together alternate segments of a multiple-cavity resonator in a magnetron by copper straps, to improve efficiency and prevent frequency jumping.

magnetron vacuum gage A vacuum gage that is essentially a magnetron operated beyond cutoff in the vacuum being measured. When used with an electron multiplier, it is sufficiently sensitive to measure the arrival of an individual ion.

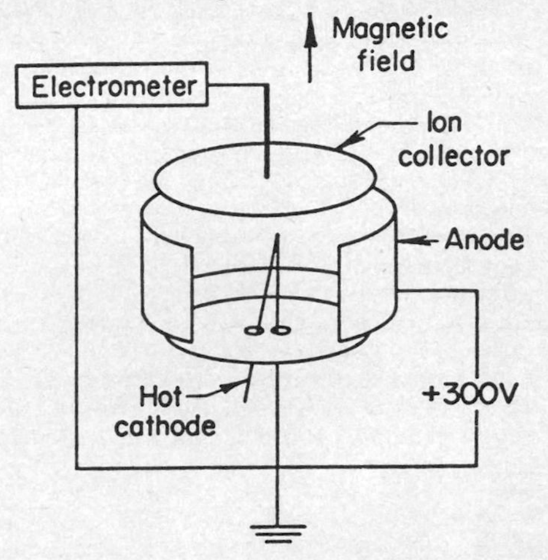

Magnetron vacuum gage.

magnet steel Steel that has high retentivity and usually contains some combination of tungsten, cobalt, chromium, and manganese with steel. Used in permanent magnets.

magnet wire Insulated copper wire in any of the sizes commonly used for winding the coils of transformers, relays, and other electromagnetic devices.

magnistor A device that utilizes the effects of magnetic fields on injection plasmas in semiconductors such as indium antimonide. A diode version can serve as an oscillator.

mag-slip British term for *synchro*.

mAh Abbreviation for *milliampere-hour*.

main A line that brings power from a generator, converter, or service-cutoff switch to the main distribution center for power lines inside a building.

main anode An anode that conducts load current in a pool-cathode tube.

main bang The transmitted pulse in a radar system.

mainframe *Central processing unit*.

main gap The conduction path between a principal cathode and a principal anode of a glow-discharge tube.

main quantum number A positive integer that specifies the size of an electron orbit. Also called first quantum number.

mains British term for *power line*.

main storage The general-purpose storage of a computer; it provides fast access to stored instructions and data being processed.

maintenance 1. Preventive or corrective measures to keep equipment in satisfactory operating condition. 2. Operating and check runs to keep computer programs and files up to date and free of errors.

main terminal In a bidirectional thyristor, the terminal marked 1. The other terminal through which the principal current flows is marked 2.

majorana particle A neutrino that corresponds to an antineutrino. Double beta decay of such a particle gives emission and absorption of a neutrino.

major apex face One of the three large sloping faces extending to the apex or pointed end of a natural quartz crystal. The other three smaller sloping faces are the minor apex faces.

major cycle The time interval between successive appearances of a given storage position in a serial-access computer storage.

majority carrier The type of carrier that constitutes more than half the total number of carriers in a semiconductor device. Majority carriers may be either holes or electrons. The other type of carrier is then known as the minority carrier. In N-type material, electrons are the majority carriers. In P-type material, holes are the majority carriers.

majority-carrier contact The electric contact across which the ratio of majority-carrier current to applied voltage is substantially independent of the polarity of the voltage applied to a semiconductor. The ratio of minority-carrier current to applied voltage is not independent of the polarity of the voltage.

majority emitter An electrode from which a flow of majority carriers enters the interelectrode region of a transistor.

majority logic Logic in which the output depends on the state of the majority of inputs.

major lobe The radiation lobe that contains the direction of maximum radiation.

make-before-break contacts Double-throw contacts arranged so that the moving contact establishes a new circuit before interrupting the old one.

make contact A normally open stationary contact on a relay. Its circuit is closed when the relay is energized.

malfunction A performance error that results from failure of a component, circuit, or system in a computer or other piece of equipment. Also called error.

management information system [abbreviated MIS] A computer-based business information system in which cost, production, sales, and other business data are inputted and processed at regular intervals. The system provides desired types of summary reports on demand as guides for management decisions, as printouts, or as a visual display in response to on-line interrogation at a computer terminal.

Manchester code A code used internally in some computers, in which a binary 1 is represented by a transition from the 1 to the 0 level in the middle of a bit, and a binary 0 is a transition from the 0 to the 1 level in the middle of a bit. In this code, one bit time is the longest interval permitted without a transition. Also called phase coding.

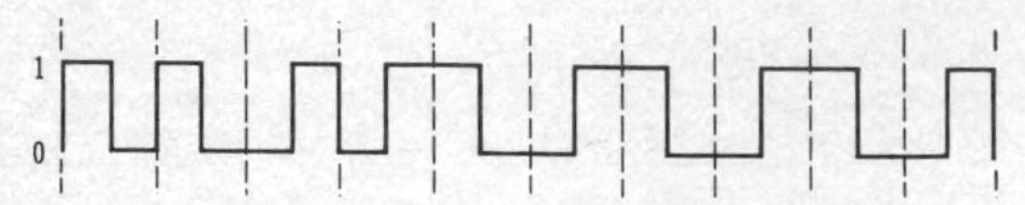

Manchester code for binary data. The space between each pair of dashed vertical lines represents one bit time.

manganese [symbol Mn] A metallic element. Atomic number is 25.

manganin An alloy that contains 84% copper, 12% manganese, and 4% nickel, used in making precision wirewound resistors because of its low temperature coefficient of resistance.

Manhattan project A project of the War Department lasting from August 1942 to August 1946, from which the atomic bomb was developed.

manipulated variable The quantity or condition that is varied by the controller, to change the value of the controlled variable in a feedback control system.

manipulator A mechanical arm that handles radioactive materials from a safe distance.

man-made interference Electromagnetic interference that results from normal or abnormal operation of electric or electronic equipment, such as harmonic or spurious signals from transmitters. Noise produced by sparking is generally called static rather than interference.

man-made static High-frequency noise signals created by sparking in an electric circuit. When picked up by radio receivers, man-made static causes buzzing and crashing sounds.

manned orbiting laboratory [abbreviated MOL] An earth-orbiting satellite that contains instrumentation and personnel for continuous measurement and surveillance of the earth, its atmosphere, and space.

manual 1. Used or operated by human action, such as with the hands. 2. A reference book, especially one that gives operating instructions.

manual direction finder A rotatable-loop radio compass that is operated manually.

manual telephone system A telephone system in which telephone connections between customers are ordinarily established manually by telephone operators according to orders given verbally by calling parties.

manual tuning Tuning in which a control knob is rotated by hand to tune in a desired station on a radio or television receiver.

many-one function switch A function switch in which a combination of several inputs must be excited at one time to produce a corresponding single output.

map-matching guidance Guidance of an aircraft or missile by a radarscope film previously obtained by a reconnaissance flight over the terrain of the route, compared with radar echoes received during the new flight.

Marconi antenna An antenna that is connected to ground at one end through the receiver or transmitter input coil and suitable tuning reactances.

marginal checking A preventive-maintenance checking procedure in which certain operating conditions, such as supply voltage or frequency, are varied about their normal values to detect and locate incipient defective units.

MARISAT [MARItime SATellite] A geostationary communication satellite equipped with a repeater operating at microwave frequencies, for ship-to-shore communication by satellite.

mark 1. The closed-circuit condition in teletypewriter systems that use neutral operation, during which the signal actuates the printer. In other than neutral operation, the mark impulse is the one that produces the same result. The other signal condition in any type of operation is called the space. 2. A designation followed by a serial number, to identify different models of a piece of military equipment.

marker 1. A radio facility used in an instrument landing system to provide a signal that designates a small area immediately above the marker location. Also called marker beacon and radio marker beacon. 2. An electronic range or bearing indication on a radar indicator.

marker beacon *Marker.*

marker generator 1. An RF generator used to inject one or more frequency-identifying pips on the pattern produced by a sweep generator on a cathode-ray oscilloscope screen. Used for adjusting response curves of tuned circuits, as when aligning FM and television receivers. 2. An RF generator that generates pulses which have precise amplitude, shape, duration, and recurrence characteristics, as required for producing reference indices on a radarscope for such quantities as target range, azimuth, and elevation.

marker pip An identifying mark on a cathode-ray display.

marking wave The emission that takes place in telegraphic communication while the active or mark portions of the code characters are being transmitted. Also called keying wave.

mark sensing A technique for marking positions on a card or sheet with an ordinary or special pencil, for automatic optical or electric sensing and conversion to magnetic tape, punched paper tape, or punched cards. When used for educational tests, partial and complete scores are usually printed directly on the sheets during mark sensing.

mark-space ratio The ratio of the duration of a single pulse to the interval between two successive recurrent pulses.

MARS [Military Affiliate Radio System] A worldwide network of radio stations operated by amateurs both on and off military installations, sponsored jointly by the U.S. Air Force, Army, and Navy to provide an auxiliary and emergency alternate communication svstem.

maser [Microwave Amplification by Stimulated Emission of Radiation] A microwave amplifier in which amplification is achieved by raising atoms or molecules of a paramagnetic material to an unstable high energy level. A microwave input signal then triggers the radiation of the excess energy at a specific wavelength, with radiated energy greatly exceeding that in the signal. Examples include gas, resonant-cavity, solid-state, and traveling-wave masers.

mask 1. A frame used in front of a television picture tube to conceal the rounded edges of the screen. 2. A thin sheet of metal or other material that contains an open pattern, to shield selected portions of a semiconductor or other surface during a deposition process.

masking 1. The amount by which the threshold of audibility of a sound is raised by the presence of another sound. The unit customarily used is the decibel. Also called audio masking and aural masking. 2. A programmed procedure for eliminating radar coverage in areas where such transmissions may be useful to the enemy for navigation purposes, by weakening the beam in such directions, or by using additional transmitters on the same frequency at suitable sites to interfere with homing. 3. Using a covering or coating on a semiconductor surface to provide a masked area for selective deposition or etching. 4. The use of tones, noise, music, or other sounds to hide or mask a clear signal for secrecy purposes. The masking signal must be available at the receiving terminal for subtraction, leaving the desired signal.

masking audiogram A graphical presentation of the masking caused by a stated noise, plotted in decibels as a function of the frequency of the masked tone.

mask microphone A microphone for use inside an oxygen mask or other type of respiratory mask.

mask-programmed read-only memory A read-only memory in which a mask produces the metallized interconnection pattern corresponding to the desired permanently stored program or data.

mass The quantity of matter in a body. Mass is a measure of inertia and determines resistance to acceleration independently of gravitational force, whereas weight is the force exerted by a body under the influence of gravity at a particular location.

mass conversion factor *Atomic mass conversion factor.*

mass defect The difference between the mass of an atom and the sum of the masses of its components if they were in a free or unbound state.

mass-energy equation The equation developed by Albert Einstein for interconversion of mass and energy, written as $E = mc^2$, where E is the energy in ergs, m is mass in grams, and c is the velocity of light in centimeters per second.

mass formula An equation that gives the atomic mass of a nuclide as a function of its atomic and mass numbers.

mass migration Metallic erosion that occurs when a high current density is passed through a conductor which has a small cross-sectional area, such as the conductors used in integrated circuits.

mass number The sum of the protons and neutrons in the nucleus of an atom or nuclide. The mass number is also equal to the sum of the atomic number and the neutron number. In the symbol for a nuclide, the mass number is usually shown as a superscript preceding the element symbol; thus, in ^{235}U, the mass number is 235. Mass number is the nearest whole number to the atomic weight of an atom. Also called nuclear number and nucleon number.

mass spectrograph A mass spectrometer that provides a permanent record of the mass spectrum lines of a material on a photographic plate.

mass spectrometer A spectrometer that analyzes a substance in terms of the ratios of mass to charge of its components. A gas or a compound in

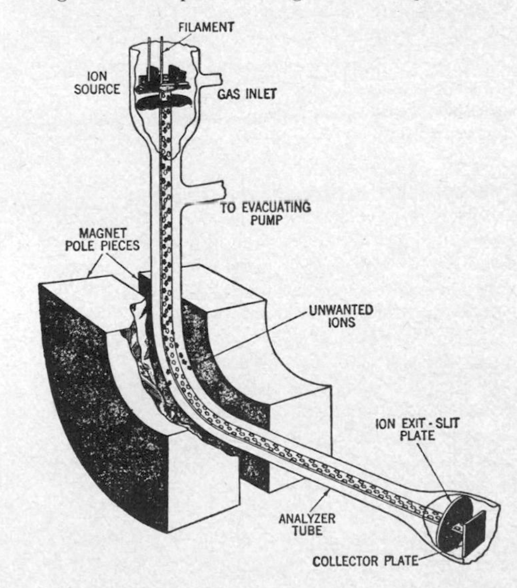

Mass spectrometer for analysis of gases by magnetic deflection.

the vapor state is bombarded by electrons, and the resulting ions are accelerated and separated according to their mass-to-charge ratios. In the most common type, combined electric and magnetic fields deflect the ions of the substance and focus each type in turn on an output electrode for detection and measurement. In another type, sorting of ions is based on the time of flight of the ions through a drift tube during acceleration by electric fields.

mass spectrum A spectrum that shows the distribution in mass or in mass-to-charge ratio of ionized atoms, molecules, or molecular fragments. The mass spectrum of an element shows the relative abundances of the isotopes of the element.

mass storage A computer device that has a large capacity for data storage in digital form, such as a disk file or magnetic drum.

mast A vertical metal pole that acts as an antenna or antenna support.

master 1. The negative metal counterpart of a disk recording, produced by electroforming as one step in the production of phonograph records. 2. *Master station.*

master-antenna television [abbreviated MATV] An antenna system that consists of an antenna array which is capable of receiving available broadcast signals and amplifying them as required for distribution over coaxial cables to a number of individual television receivers that are normally within a single home, apartment, hotel, or other building.

master brightness control A variable resistor that adjusts simultaneously the grid bias on all guns of a three-gun color picture tube.

master clock The electronic or electric source of standard timing signals, often called clock pulses, required for sequencing the operation of a computer.

master control The control console that contains the main program controls for a radio or television transmitter or network.

master file A computer file that contains relatively permanent information, usually updated periodically, such as subscriber records or payroll data other than time worked.

master gain control 1. A variable resistor or potentiometer in a stereo amplifier, to control the gain of both audio channels simultaneously. 2. A control in a radio, television, or recording studio, to change the overall audio output level without affecting the mixer controls that determine the balance of the microphones and other sound sources. Used also for fading out or fading in the sound volume.

master instruction tape A computer magnetic tape on which all programs for a system of runs are recorded.

master oscillator An oscillator that establishes the carrier frequency of the output of an amplifier or transmitter.

master oscillator–power amplifier [abbreviated MOPA] An oscillator state followed by an RF power amplifier stage that serves also as a buffer.

master routine *Routine.*

master-slave flip-flop A combination of two flip-flops, one of which (the master) receives its information on the leading edge of a clock pulse, and the other (the slave or output flip-flop) receives its information on the trailing edge of the pulse. Used to prevent false triggering when two or more gate inputs are applied almost simultaneously.

master-slave manipulator A pair of mechanical hands that handles radioactive materials by remote control from behind a protective shield.

master station The station of a synchronized group of radio stations to which the emissions of other stations of the group are referred. In a loran system it is the A station. Also called master.

master synchronization pulse A pulse distinguished from other telemetering pulses by amplitude and/or duration, to indicate the end of a sequence of pulses.

match A data-processing operation similar to a merge, except that instead of producing a sequence of items made up from the input sequences, the sequences are matched against each other on the basis of some key.

matched diodes Two diodes that have exactly the same dimensions and electrical characteristics. One may have forward polarity and the other reverse polarity, or both may have the same polarity.

matched load A load that has the impedance value which results in maximum absorption of energy from the signal source.

matched power gain The power gain obtained when the impedance of a load is matched to the effective output impedance of the amplifier to which it is connected.

matched termination A termination that produces no reflected wave at any transverse section of a waveguide or other transmission line.

matched transmission line A transmission line that has a matched termination.

matched waveguide A waveguide that has a matched termination.

matching Connecting two circuits or parts together so that their impedances are equal or are equalized by a coupling device, to give maximum transfer of energy.

matching device A device that matches unequal impedances, such as the output transformer of a radio receiver.

matching diaphragm A diaphragm that consists of a slit in a thin sheet of metal, placed transversely across a waveguide for matching purposes. The orientation of the slit with respect to the long dimension of the waveguide determines whether the diaphragm acts as a capacitive or inductive reactance.

matching stub A short length of two-wire transmission line connected at the antenna or receiver end of a regular transmission line to add inductive or capacitive reactance for matching purposes. The shorting conductor is sometimes in the form of a slider that can be moved along the stub.

matching transformer A transformer used between unequal impedances for matching purposes, to give maximum transfer of energy.

match-terminated Terminated in a load equal to the characteristic impedance of the transmission line.

mathematical check A programmed computer check of a sequence of operations; it uses the mathematical properties of that sequence.

matrix 1. A computer logic network that consists of a rectangular array of intersections of input/output leads, with diodes, magnetic cores, or other circuit elements connected at some of these intersections. The network usually functions as an encoder or decoder. 2. The section of a color television transmitter that transforms the red, green, and blue camera signals into color-difference signals and combines them with the chrominance subcarrier. Also called color coder, color encoder, and encoder. 3. The section of a color television receiver that transforms the color-difference signals into the red, green, and blue signals needed to drive the color picture tube. Also called color decoder and decoder. 4. A set of mathematical elements arranged in rows and columns, used in solving certain types of problems. 5. The precisely shaped form used as the cathode in electroforming. 6. *Translator.*

matrixing The process of performing a code conversion with a matrix, as in converting color television signal components from one form to another.

matrix printer *Dot-matrix printer.*

matrix sound system A quadraphonic sound system in which the four input channels are combined into two channels by a coding process for recording or for stereo FM broadcasting and decoded back into four channels for playback of recordings or for quadraphonic stereo reception. The matrix system is sometimes referred to as a 4-2-4 system, in contrast to discrete quadraphonic sound, which is a 4-4-4 system because four channels are used throughout. Matrix examples include the SQ system developed by CBS and the QS system developed by Sansui.

mattress array *Billboard array.*

MATV Abbreviation for *master-antenna television.*

maximum output The greatest average output power delivered to the rated load of a receiver or amplifier, regardless of distortion.

maximum permissible dose The dose of ionizing radiation that persons may receive in their lifetime without appreciable bodily injury.

maximum permissible exposure The total amount of radiation exposure to which normal persons may be subjected day by day without any harmful effects becoming evident during their lifetimes.

maximum retention time The maximum time between writing into and reading an acceptable output from a storage element of a charge-storage tube.

maximum sound pressure The maximum absolute value of the instantaneous sound pressure at a point during any given cycle. Usually expressed in microbars.

maximum undistorted output The greatest average output power into the rated load of an amplifier at which distortion does not exceed a specified limit when the input is sinusoidal. Also called maximum useful output.

maximum usable frequency [abbreviated MUF] The upper limit of the frequencies that can be used at a specified time for point-to-point radio transmission which involves propagation by reflection from the regular ionized layers of the ionosphere. Higher frequencies may be transmitted only by sporadic and scattered reflections.

maximum useful output *Maximum undistorted output.*

maxwell [abbreviated Mx] The CGS unit of magnetic flux. It is equal to 1 G/cm^2, or to one magnetic line of force. Use of the SI unit, the weber, is now preferred.

Maxwell-Boltzmann law A law that gives the distribution of velocities among the molecules of a perfect steady-state gas.

Maxwell-Boltzmann statistics Statistics which represent the distribution of particles among the various possible energy levels at such high temperatures that a large number of energy levels are excited.

Maxwell bridge A four-arm AC bridge that measures inductance (or capacitance) in terms of resistance and capacitance (or inductance). Bridge balance is independent of frequency.

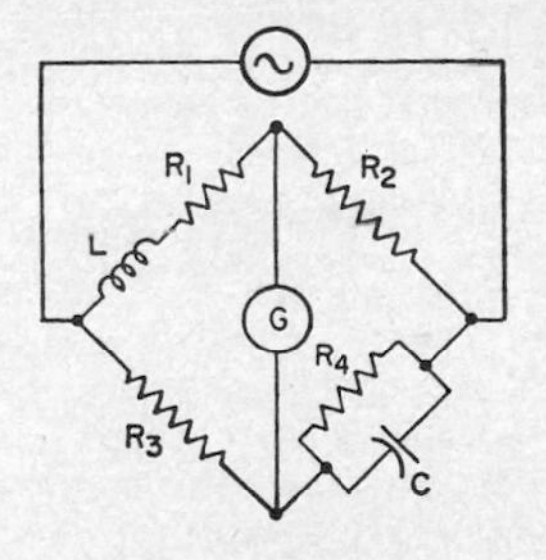

Maxwell bridge.

Maxwellian distribution The velocity distribution of the molecules of a gas in thermal equilibrium. This distribution is often assumed to hold for neutrons in thermal equilibrium with the moderator in a nuclear reactor.

Maxwell inductance bridge A four-arm AC

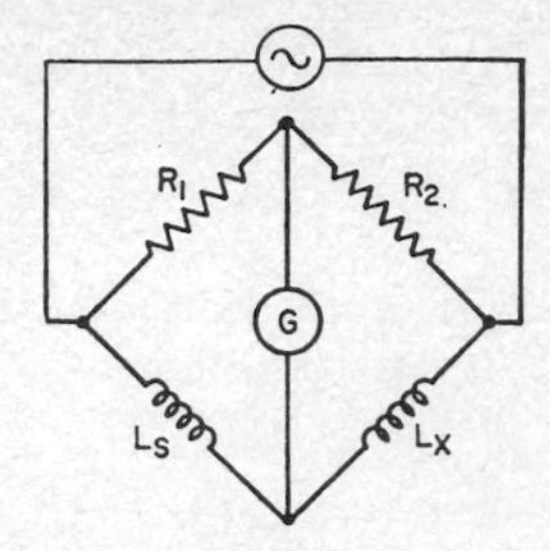

Maxwell inductance bridge circuit.

bridge that compares inductances. Bridge balance is independent of frequency.

Maxwell mutual-inductance bridge An AC bridge that measures mutual inductance in terms

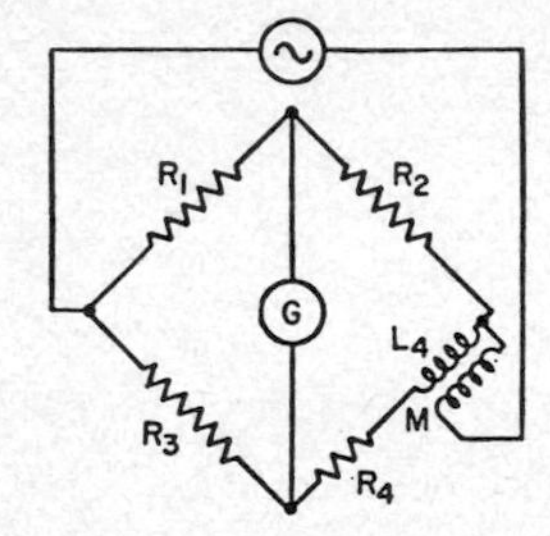

Maxwell mutual-inductance bridge circuit.

of self-inductance. Bridge balance is independent of frequency.

Maxwell triangle The equilateral-triangle form of a chromaticity diagram, in which the primary colors are represented at the corners of the triangle.

mayday [French m'aider] The international radiotelephone distress signal for ships, aircraft, and spacecraft.

mb Abbreviation for *millibarn.*

Mb Abbreviation for *megabit.*

mbar Abbreviation for *millibar.*

Mbar Abbreviation for *megabar.*

Mb/s Abbreviation for *megabit per second.*

Mbyte Abbreviation for *megabyte.*

Mbyte/s Abbreviation for *megabyte per second.*

Mc Abbreviation for *megacycle.* Now called megahertz and abbreviated MHz.

mcd Abbreviation for *millicandela.*

MCG Abbreviation for *magnetocardiograph.*

McGill fence *Mid-Canada line.*

mCi Abbreviation for *millicurie.*

MCi Abbreviation for *megacurie.*

McNally tube A single-cavity velocity-modulated microwave local oscillator tube, the frequency of which may be controlled over a wide range by electrical means.

MCU Abbreviation for *microprogram control unit.*

MCW Abbreviation for *modulated continuous wave.*

***m*-derived section** A T or pi network section designed so that when two or more sections are joined in a filter unit, their impedances are matched at all frequencies, even though the sections may have different resonant frequencies.

M display A modified radarscope A display in which target distance is determined by moving an adjustable pedestal signal along the baseline until

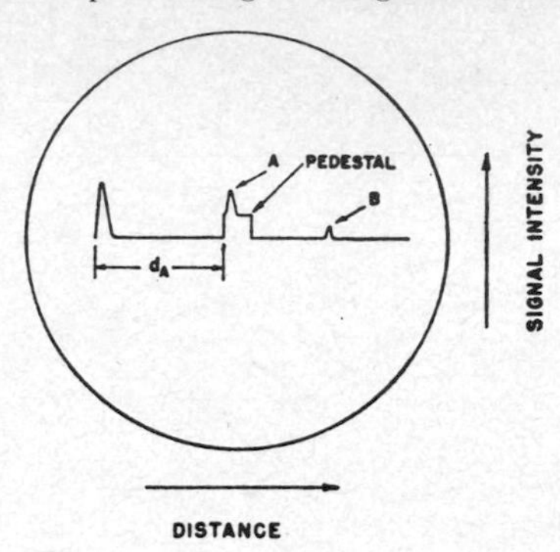

M display.

it coincides with the horizontal position of the target deflection. The control that moves the pedestal signal is calibrated in distance.

meaconing [MEAsuring and CONfusING] Measuring received radio navigation signals and broadcasting confusing signals instantly and automatically on the same frequency. Used to give enemy navigators a false indication of position.

mean The average of two or more quantities, obtained by adding the quantities and dividing by the number of quantities. Also called arithmetic mean.

mean carrier frequency The average carrier frequency of a transmitter, corresponding to the resting frequency in a frequency-modulation system.

mean free path 1. The average distance that a particle travels between successive collisions. 2. The average distance that sound waves travel between successive reflections in an enclosure.

mean free time The average time between successive collisions of a particle.

mean lethal dose *Median lethal dose.*

mean life The average time during which an atom or other system exists in a particular form. For a radionuclide, the mean life is the reciprocal of the disintegration constant. Mean life is 1.443 times the radioactive half-life. In a semiconductor, mean life is the time required for injected excess carriers to recombine with others of the opposite sign. Also called average life and lifetime.

mean pulse time The arithmetic mean of the leading-edge pulse time and the trailing-edge pulse time.

mean range The range that is exceeded by half the particles under consideration.

mean time between failures [abbreviated MTBF] The average time, usually expressed in hours, between failures in a continuously operating device, circuit, or system.

measurand The physical quantity, property, or condition that is to be measured.

measurement Determination of the magnitude, amount, or other parameter of a characteristic or quantity.

mechanical axis One of the Y axes in a quartz crystal. There are three, each perpendicular to one pair of opposite sides of the hexagon.

mechanical bandspread Bandspread in which a vernier tuning dial or other mechanical means provides greater angular rotation of a control knob for a given tuning range, to simplify tuning in crowded shortwave bands.

mechanical filter A filter that consists of shaped metal rods which act as coupled mechanical resonators when used with piezoelectric or magnetostrictive input and output transducers. Sometimes used in IF amplifiers of highly selective receivers.

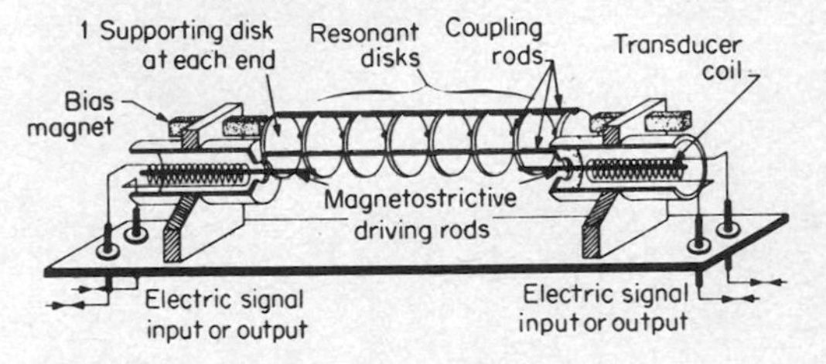

Mechanical filter for single-sideband equipment, using magnetostrictive transducer at each end.

mechanical joint A joint in which a conductor is clamped to another conductor or to a terminal mechanically, without the use of solder.

mechanical latching relay A relay in which either the armature or the contacts may be latched mechanically in either the operated or unoperated position until reset electrically or manually.

mechanical mass The portion of the mass of a particle that is considered to be an intrinsic property of that particle. It does not include the mass increment due to the interaction of the particle with itself through the medium of some field.

mechanical modulator A device that varies a carrier wave by moving some part of a circuit element. Examples include motor-driven capacitor plates and motor-driven choppers.

mechanical rectifier A rectifier in which rectification is accomplished by mechanical action, as in a synchronous vibrator.

mechanical register An electromechanical device that indicates a count of pulses.

mechanical reproducer *Acoustic pickup.*

mechanical scanning Scanning in which a beam of light controlled by a rotating scanning disk, rotating mirror, or other mechanical device is used to break up a scene or image into a rapid succession of narrow lines, as required for conversion into electric pulses.

mechanical translation Automatic translation of one language into another by a computer or other machine that contains a dictionary lookup in its memory, along with the programs needed to make logical choices from synonyms, supply missing words, and rearrange word order as required for the new language. The text to be translated must be converted into machine-readable form first, either manually by special typewriters or automatically by photoelectric character readers. Also called machine translation.

median The value at the halfway point in a series, wherein there is the same number of values above as below the median.

median energy of fission The neutron energy above (or below) which half the total number of fission-producing neutron absorptions occur.

median lethal dose The dose of radiation required to kill, within a specified period, 50% of the individuals in a large group of animals or organisms. Also called mean lethal dose.

median lethal time The time required, following administration of a specified dose of radiation, for death of 50% of the individuals in a large group of animals or organisms.

medical electronics A branch of electronics in which electronic instruments and equipment are used for such medical applications as diagnosis, therapy, research, anesthesia control, cardiac control, and surgery.

medium frequency [abbreviated MF] A Federal Communications Commission designation for a frequency in the band from 300 to 3000 kHz, corresponding to a hectometric wave between 100 and 1000 m.

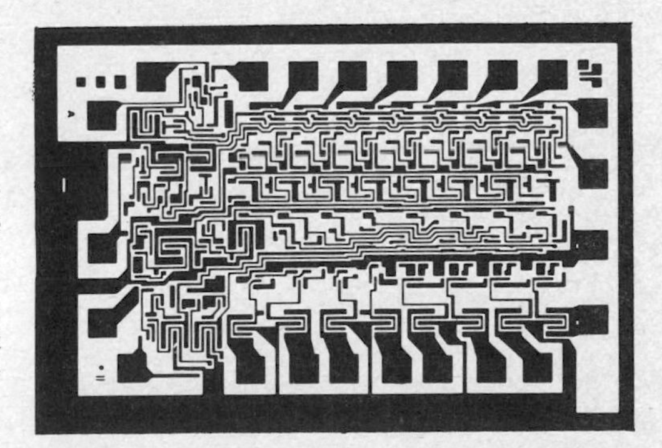

Medium-scale integration in which metallization mask has connections for about 70 equivalent gates of scaler.

medium-scale integration [abbreviated MSI] Integration in which a complete major subsystem or system is fabricated on a single integrated-circuit chip that contains more than about 12 gate equivalent circuits.

mega- [abbreviated M] A prefix representing 10^6, or 1,000,000.
megabar [abbreviated Mbar] An absolute unit of pressure equal to 1,000,000 bars. One megabar is almost exactly equal to normal atmospheric pressure.
megabit [abbreviated Mb] One million bits.
megabit per second [abbreviated Mb/s] One million bits per second, used in specifying the modulation rate of a digital transmission system.
megabyte [abbreviated Mbyte] One million bytes.
megabyte per second [abbreviated Mbyte/s] One million bytes per second, used in specifying the speed of a digital transmission system.
megacurie [abbreviated MCi] One million curies.
megacycle [abbreviated Mc] One million cycles per second. Now called megahertz and abbreviated MHz.
megaelectronvolt [abbreviated MeV] One million electronvolts.
megagauss [abbreviated MG] One million gauss.
megagauss physics The production, measurement, and application of megagauss fields, as produced by discharge of capacitor banks or explosive flux-compression techniques.
megahertz [abbreviated MHz] One million hertz. Formerly called megacycle.
megamile One million miles.
megampere [abbreviated MA] One million amperes.
megaton 1. One million tons. 2. A unit used in specifying the yield of an atom bomb, equal to the explosive power of 1 million tons of trinitrotoluene (TNT).
megavolt [abbreviated MV] One million volts.
megavoltampere [abbreviated MVA] One million voltamperes.
megawatt [abbreviated MW] One million watts.
megawatthour [abbreviated MWh] One million watthours.
Megger Trademark of James G. Biddle Co. for their high-range ohmmeter, which has a hand-

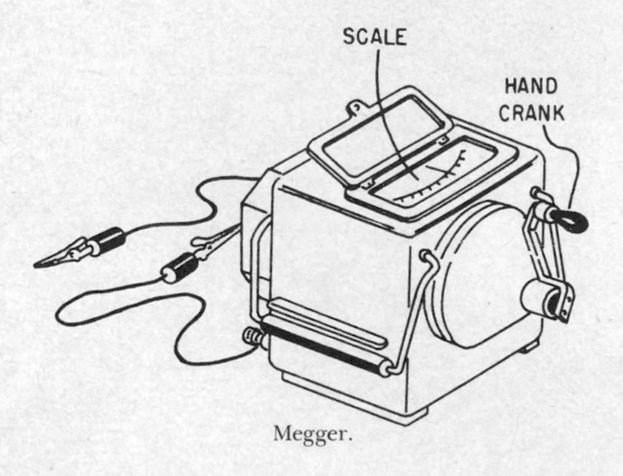

Megger.

driven DC generator as its voltage source. It is used chiefly for measuring insulation resistance values in megohms.
megohm [abbreviated MΩ] One million ohms.
megohm-farad A rating for insulation resistance of energy-storage capacitors.
megohmmeter An ohmmeter that measures resistance in megohms, gigohms, and teraohms. High test voltages are required, sometimes more than 1 kV for the highest ranges.
Meissner effect When a superconductor is cooled below the temperature required for superconductivity, the material appears to become perfectly diamagnetic. The induced magnetization opposes the applied magnetic field to such an extent that there is no magnetic field in the material.
Meissner oscillator An oscillator in which the grid and anode circuits are inductively coupled

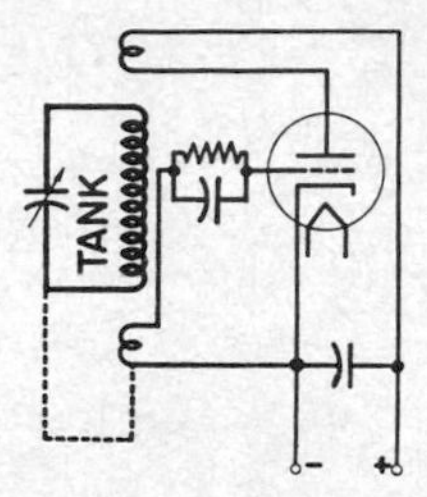

Meissner-oscillator circuit.

through an independent tank circuit that determines the frequency of oscillation.
mel A unit of pitch. By definition, a simple 1-kHz tone 40 dB above a listener's threshold produces a pitch of 1000 mels. The pitch of any sound that is judged by the listener to be *n* times that of the 1-mel tone is *n* mels.
M electron An electron that has an orbit in the M shell, which is the third shell of electrons surrounding the atomic nucleus, counting out from the nucleus.
M-electron capture A mode of electron capture similar to K-electron capture except that the electrons are initially in the M shell of the atom.
melodeon A broadband panoramic receiver used for countermeasures reception. All types of received electromagnetic radiation are presented as vertical pips on a frequency-calibrated cathode-ray indicator screen.
meltback transistor A junction transistor in which the junction is made by melting a properly doped semiconductor and letting it solidify again.
melting channel The restricted portion of the charge in a submerged horizontal-ring induction furnace, in which the induced currents are concentrated to heat and melt the charge.
melt-quench transistor A junction transistor that is made by suddenly cooling a melted-back region.
memory The portion of a digital computer that

stores information for later use. The terms memory and storage are used interchangeably today in the computer field, even though purists insist that storage is the more correct term.

memory capacity *Storage capacity.*

memory cell A single storage element of a memory, together with associated circuits for storing and reading out 1 bit of information.

memory dump *Storage dump.*

memory tube Deprecated term for *storage tube.*

memristor A circuit element that behaves like a linear resistor with memory but exhibits nonlinear characteristics.

mercuric telluride [symbol HgTe] An intermetallic compound that has characteristics similar to those of indium antimonide.

mercury [symbol Hg] A silvery white liquid metal that becomes a solid at −40°C. Used in mercury switches and electron tubes because the vapor of mercury ionizes readily and conducts electricity. The green line of mercury 198 very closely approaches pure monochromatic light. Atomic number is 80.

mercury arc An electric discharge through ionized mercury vapor; it gives off a brilliant bluish green light that contains strong ultraviolet radiation.

mercury-arc inverter An inverter that uses mercury-arc rectifiers.

mercury-arc rectifier A gas-filled rectifier tube in which the gas is mercury vapor. Small sizes use a heated cathode, and larger sizes rated up to 8 MW and higher use a mercury-pool cathode. Also called mercury-vapor rectifier.

mercury battery A battery made up of mercury cells.

mercury cell A primary dry cell that delivers an essentially constant output voltage throughout its

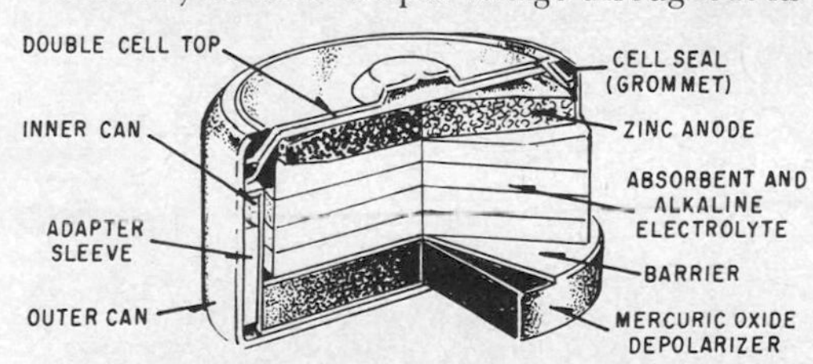

Mercury cell using flat pellet structure. Other versions use cylindrical structure.

useful life by a chemical reaction between zinc and mercury oxide. Widely used in cameras and hearing aids.

mercury delay line An acoustic delay line in which mercury is the medium for sound transmission.

mercury-hydrogen spark-gap converter A spark-gap generator that uses the oscillatory discharge of a capacitor through a coil and a spark gap as a source of RF power. The spark gap consists of a solid electrode and a pool of mercury

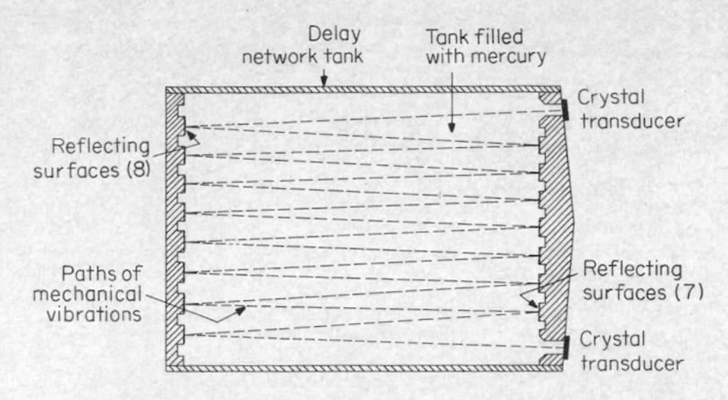

Mercury delay line providing 16 passes.

in a hydrogen atmosphere.

mercury-jet commutator A commutator in which a jet of mercury is the wiper. A motor drives a scoop that lifts mercury from a pool into a rotating reservoir, from which the mercury is ejected in a fine stream through a small nozzle by centrifugal force. As the stream rotates, it impinges in sequence on a circle of contact pins, thereby completing connections sequentially.

mercury-pool cathode A pool cathode in which the pool is liquid mercury.

mercury-pool tube *Pool tube.*

mercury relay A relay in which mercury, moved by a magnetic plunger, connects the relay contacts together when the plunger is pulled into the mercury by the energized relay coil.

mercury switch A switch that is closed by making a large globule of mercury move up to the

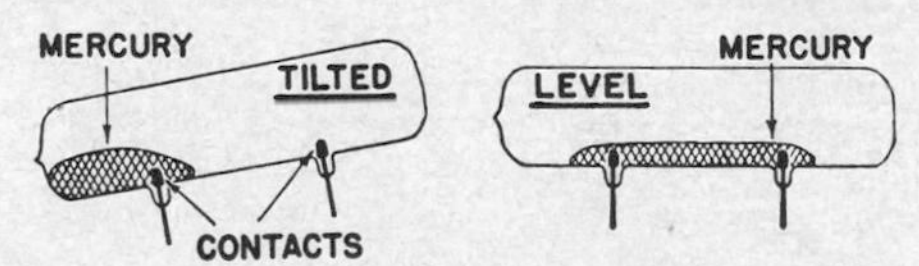

Mercury-switch construction and operation.

contacts and bridge them. The mercury is usually moved by tilting the entire switch.

mercury thermostatic switch A thermostatic switch in which heat causes mercury to expand

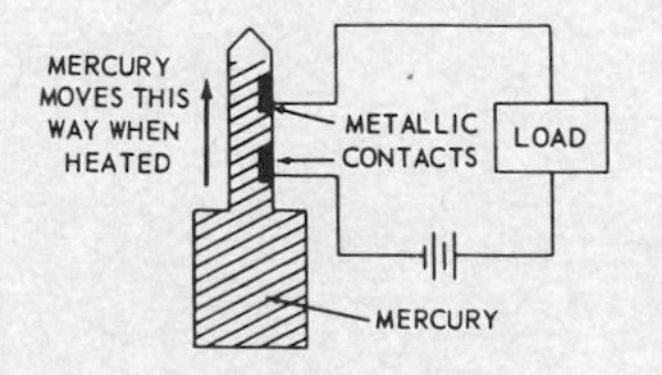

Mercury thermostatic switch.

and complete a circuit between contacts that project into the mercury column.

mercury-vapor lamp A lamp in which light is produced by an electric arc between two electrodes in an ionized mercury-vapor atmosphere. It gives off a blue-green light that is rich in ultraviolet radiation.

mercury-vapor rectifier *Mercury-arc rectifier.*

mercury-vapor tube A gas tube in which the active gas is mercury vapor.

mercury-wetted reed relay A reed relay in which the contacts are covered with a mercury film, just as in a mercury-wetted reed switch.

mercury-wetted reed switch A reed switch that contains a pool of mercury at one end and is normally operated vertically. The mercury keeps the contacts on the reeds covered with a mercury film by capillary action. Each operation of the

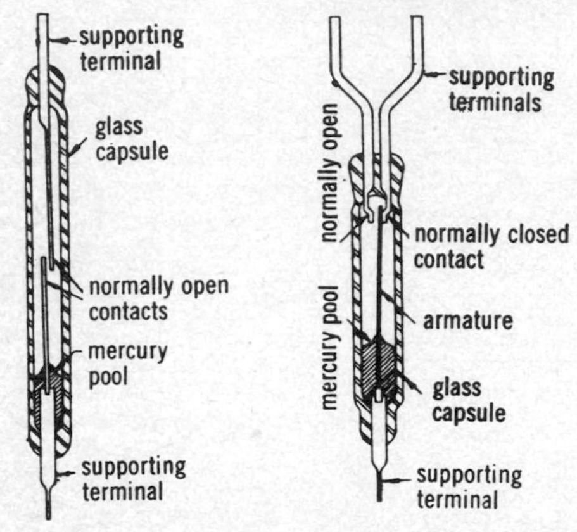

Mercury-wetted reed switch, showing single- and double-pole construction.

switch renews this mercury film contact, thereby increasing the operating life of the switch many times. In another version, mercury-wetted contacts are obtained without using a mercury pool, permitting mounting in any position.

merge To collate two or more similarly ordered sets of values into one set that has the same ordered sequence.

merged-transistor logic [abbreviated MTL] *Integrated injection logic.*

mesa The raised area that remains when semiconductor material is etched away for access to regions beneath the surface.

mesa construction A method of constructing semiconductor devices; the junctions are termi-

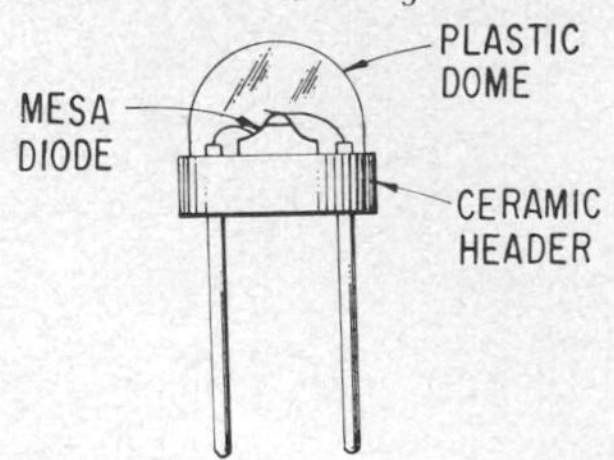

Mesa construction in light-emitting diode.

nated at the edges of the layers of semiconductor material, instead of being brought to a common surface as in planar diodes.

mesa diffusion A method of growing PN junctions by creating a single base region over the entire surface of a semiconductor slice, then etching away valleys between emitters, to leave islands or mesas of processed material for use as transistor elements.

mesa transistor A transistor in which a germanium or silicon wafer is etched down in steps so the base and emitter regions appear as physical plateaus above the collector region.

MESFET Abbreviation for *metal semiconductor field-effect transistor.*

mesh A set of branches that forms a closed path in a network in such a way that if any one branch is omitted from the set, the remaining branches of the set do not form a closed path. Sometimes called loop.

mesh connection *Delta connection.*

Mesny circuit An ultrahigh-frequency oscillator that uses a symmetrical arrangement of two vacuum tubes, with inductances in the grid and anode supply leads.

mesomorphism A state of matter intermediate between a crystalline solid and a normal isotropic liquid, in which long rod-shaped organic molecules contain dipolar and polarizable groups. The three main types of mesomorphic states are smectic, nematic, and cholesteric. The nematic state is widely used in liquid crystal displays.

meson Any elementary particle that has a rest mass intermediate in value between the mass of an electron and that of a proton. Mesons are found in cosmic rays and are produced in high-energy nuclear reactions. They may have a positive or negative charge or be neutral. Average lives are always shorter than a microsecond. They differ in masses and spins and can in some cases change from one kind of meson to another. Known mesons include mu, pi, and tau mesons.

mesosphere A region of decreasing temperature immediately above the stratosphere, extending roughly from an altitude of 60 to 90 km.

message An oral or written communication made in any plain or secret language or code; it has a definite beginning or end. In a computer, it may be a group of words transported or processed as a unit in binary or other coded form.

metal-backed screen *Aluminized screen.*

metal-ceramic A ceramic mixed with a metal oxide, carbide, or nitride to obtain greater ductility and strength. Also called ceramet.

metal-cone tube A television picture tube that has a metal rather than a glass cone between the glass faceplate and the glass neck of the tube.

metal detector An electronic device that detects concealed metal objects, such as guns, knives, or buried pipe lines, generally by radiating a high-frequency electromagnetic field and detecting the change produced in that field by the ferrous or nonferrous metal object being sought. Also called electronic locator, metal locator, and radio metal locator.

metal-film resistor A resistor in which a film of a metal, metal oxide, or alloy is deposited onto an

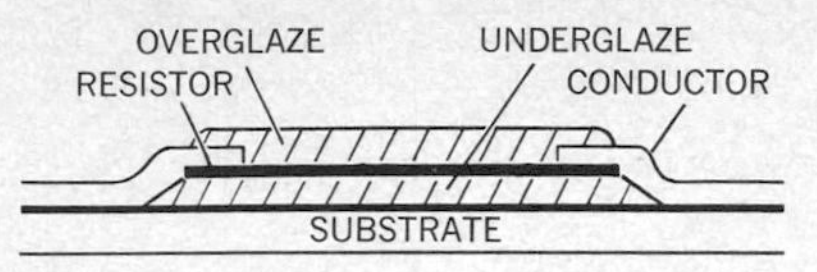

Metal-film resistor on insulating substrate.

insulating substrate for an integrated circuit or discrete resistor.

metal-glaze resistor A thick-film resistor in which a mixture of fine metal particles and powdered glass is placed on a ceramic substrate by dipping, brushing, or spraying, then fired to produce a glaze. The resistance value can be adjusted or increased by grinding a spiral groove. The metal particles are generally equal parts of palladium and silver. The cermet resistor is an example.

metal halide lamp A discharge lamp in which metal halide salts are added to the contents of a discharge tube in which there is a high-pressure arc in mercury vapor. The added metals generate different wavelengths, to give substantially white light at an efficiency approximating that of high-pressure sodium lamps. Common sizes are 400, 1000, and 1500 W.

metal-inert gas welding Welding in which an arc plasma radiates heat onto the work surface to create a weld puddle in a protective atmosphere provided by a flow of inert shielding gas. Heat travels by conduction from this puddle to melt the desired depth of metal for the weld.

metal-insulator semiconductor [abbreviated MIS] A semiconductor construction in which an insulating layer, generally a fraction of a micrometer thick, is deposited on the semiconducting substrate before the pattern of metal contacts is applied. Used to produce a field-effect region at the surface of the semiconductor material. Applications include capacitors, diodes, field-effect transistors, luminescent diodes, microstrip devices, varactors, and other semiconductor devices.

metallic antenna lens A lens that consists of contoured parallel metal surfaces placed in front of an antenna to focus the beam. The vanes change the phase velocity in proportion to the distance traveled between metal surfaces at each part of the lens.

metallic circuit A complete circuit in which the earth or ground forms no part.

metallic insulator A shorted quarter-wave section of a transmission line. Such a line acts as an extremely high impedance at a frequency corresponding to its quarter-wavelength and may therefore be used as a mechanical support.

metallic rectifier A rectifier that consists of one or more disks of metal under pressure contact with semiconductor coatings or layers, such as a

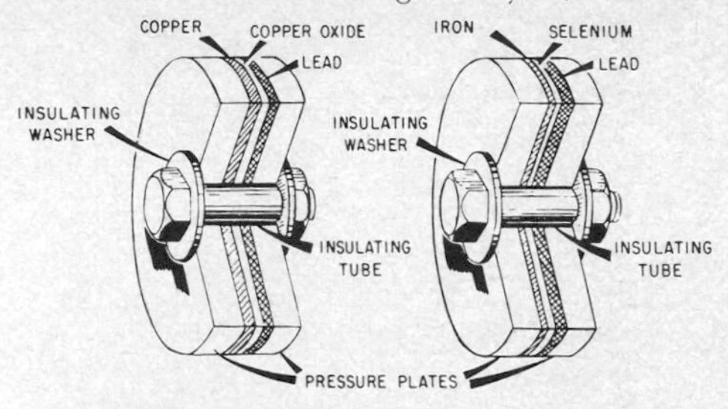

Metallic-rectifier construction for copper-oxide (left) and selenium (right) rectifying junctions.

copper-oxide, selenium, or silicon rectifier. Also called contact rectifier, dry-disk rectifier, and semiconductor rectifier.

metallic rectifier cell An elementary metallic rectifier that has one positive electrode, one negative electrode, and one rectifying junction.

metallization The deposition of one or more layers of metal onto the surface of a semiconductor material or onto a substrate, to act as contact areas and interconnections. The metal may be deposited selectively to form desired patterns, or the patterns may be etched out after deposition.

metallized capacitor A capacitor in which a film of metal is deposited directly onto the dielectric to serve in place of a separate foil strip. The capacitor has self-healing characteristics.

metallized resistor A resistor made by depositing a thin film of high-resistance metal on the surface of a glass or ceramic rod or tube.

metallized screen *Aluminized screen.*

metal locator *Metal detector.*

metal master *Original master.*

metal negative *Original master.*

metal-nitride-oxide semiconductor [abbreviated MNOS] A semiconductor structure that has a double insulating layer (instead of the usual silicon dioxide gate insulator used in metal-oxide semiconductor structures). Typically, a layer of silicon dioxide (SiO_2) is nearest the silicon substrate, with a layer of silicon nitride (Si_3N_4) over it. The ability of the double insulating layer to store charges makes it useful in memory transistor arrays, capacitors, and other semiconductor devices.

metal-oxide resistor A metal-film resistor in which an oxide of a metal such as tin is deposited as a film onto an insulating substrate.

metal-oxide semiconductor [abbreviated MOS] A metal-insulator-semiconductor structure in which the insulating layer is an oxide of the substrate material. For a silicon substrate, the insulating layer is silicon dioxide (SiO_2). Used in field-effect transistors, capacitors, resistors, and other semiconductor devices. MOS processes include CMOS, DMOS, NMOS, PMOS, and VMOS.

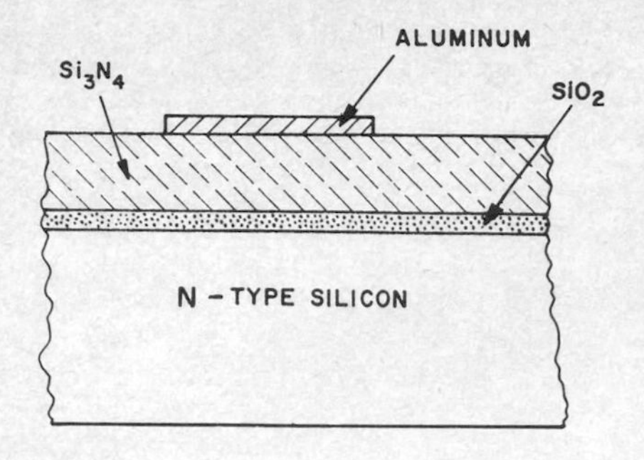

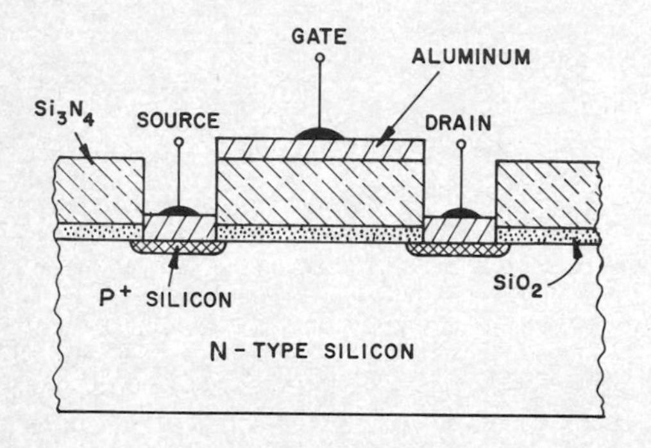

Metal-nitride-oxide semiconductor construction as used in integrated-circuit capacitor and field-effect transistor.

metal-oxide semiconductor capacitor [abbreviated MOS capacitor] An integrated-circuit capacitor in which an N^+ region is diffused into the silicon substrate to form the bottom electrode. A controlled thickness of silicon oxide is formed on this, to serve as the dielectric. The top electrode is a layer of metal that is deposited at the same time as the interconnections for the integrated circuit. Maximum capacitance values are limited to a few hundred picofarads.

metal-oxide semiconductor field-effect transistor [abbreviated MOSFET] A field-effect transistor in which the insulating layer between each gate electrode and the channel is a metal-oxide semiconductor. Typically, a silicon dioxide layer is deposited or otherwise produced on the surface of a silicon substrate, and metal electrodes are deposited on the surface of the oxide by vacuum evaporation. When operated in the normal depletion mode, a negative gate voltage depletes the charge carriers normally present in the conductive channel with zero gate bias. In the enhancement mode of operation, the gate is forward-biased, to enhance the channel charge and increase channel conductance. Either an N- or P-type substrate may be used in both modes of operation. Formerly called insulated-gate field-effect transistor. Also called metal-oxide semiconductor transistor [abbreviated MOS transistor or MOST].

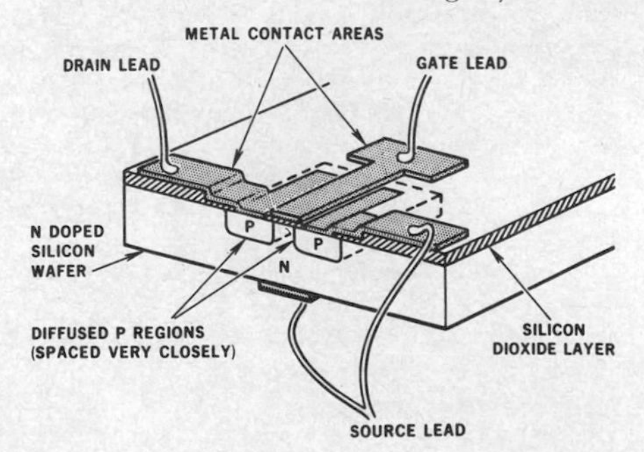

Metal-oxide semiconductor field-effect transistor.

metal-oxide semiconductor resistor [abbreviated MOS resistor] A metal-oxide semiconductor field-effect transistor used in place of a resistor in an integrated circuit.

metal-oxide semiconductor transistor [abbreviated MOS transistor or MOST] *Metal-oxide semiconductor field-effect transistor.*

metal-oxide silicon device A diode, capacitor, or other semiconductor device in which a metallic oxide such as silicon dioxide serves as an insulating layer.

metal-oxide varistor A varistor that consists of an encapsulated polycrystalline ceramic body which has metal electrodes and wire leads. In one version, zinc oxide and bismuth oxide are mixed with a ceramic powder, pressed into disks, and sintered.

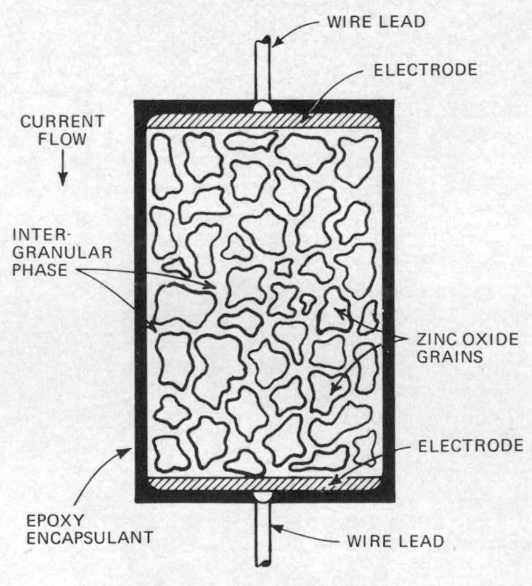

Metal-oxide varistor construction.

metal positive *Mother.*

metal semiconductor field-effect transistor [abbreviated MESFET] A field-effect transistor that uses a thin film of gallium arsenide, with a

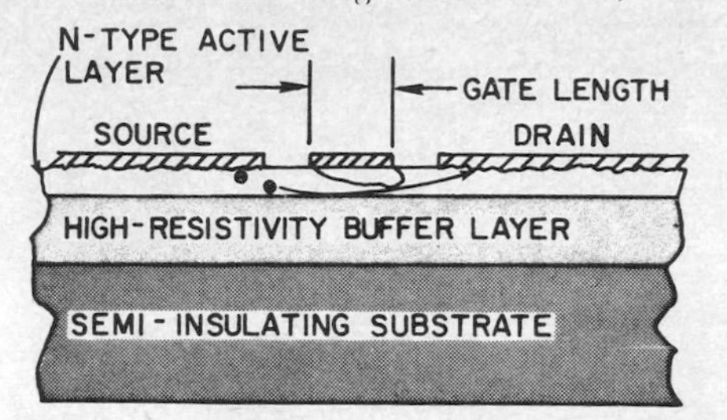

Metal semiconductor field-effect transistor construction.

Schottky barrier gate formed by depositing a layer of metal directly onto the surface of the film. It can be used as a microwave amplifier in place of traveling-wave tubes. A dual-gate version serves as a pulsed amplitude modulator operating above 20 GHz.

metal-tank mercury-arc rectifier A mercury-arc rectifier in which the anodes and the mercury cathode are enclosed in a metal container or chamber.

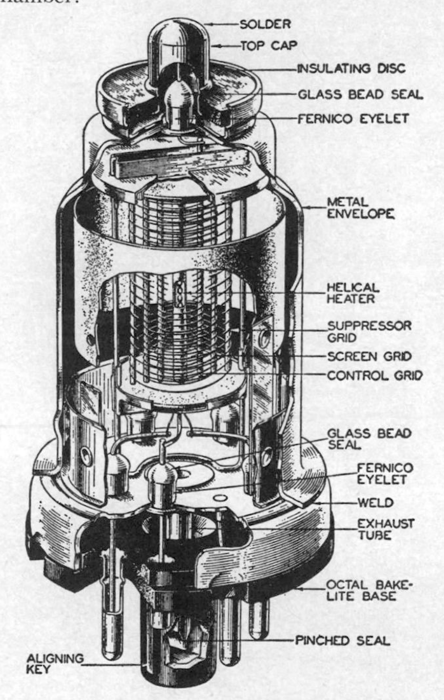

Metal tube, showing how leads pass through glass beads sealed into metal base.

metal tube A vacuum tube that has a metal envelope, with electrode leads passing through glass beads fused into the metal housing.

metal vapor laser An ion laser based on vaporization of a solid or liquid metal. Metals used include cadmium, calcium, copper, lead, manganese, selenium, strontium, and tin, vaporized with a buffer gas like helium. Operation can be either pulsed or continuous. The normal wavelength range is 0.2 to 0.6 μm.

metascope An infrared receiver that converts pulsed invisible infrared rays into visible signals for communication purposes. Also used with an infrared source for reading maps in darkness.

metastable Capable of undergoing a quantum transition to a state of lower energy, but having a relatively long lifetime as compared with the most rapid quantum transitions of similar systems.

metastable equilibrium A condition of pseudoequilibrium in which the free energy of a system is at a minimum with respect to infinitesimal changes, but not with respect to major changes. Although a condition of greater thermodynamic stability exists, the system may remain in the metastable state because the transition to the more stable condition is extremely sluggish.

metastable state An excited unstable state of a nucleus, from which all quantum transitions to lower states are generally forbidden transitions.

meteoric scatter A form of scatter propagation in which meteor trails scatter radio waves back to earth. Two radio links, working in opposite directions, are used. Any message transmitted on one link is sent back on the other, so the sender can verify satisfactory reception. The sender transmits only the first character of the message until it is received and returned as a result of scatter by a meteor. The desired message, previously recorded on magnetic tape, is then transmitted at high speed as a burst that may last from a fraction of a minute to several minutes, depending on the size and course of the meteor.

meteorograph An instrument that measures and records meteorological data such as air pressure, temperature, and humidity. When carried aloft, it is also called an aerograph. When used with a radio transmitter, it is called a radiosonde.

meter 1. A device that measures the value of a quantity under observation. Also called instrument. The term meter is usually applied to an indicating instrument alone, such as a voltmeter or an ohmmeter. 2. [abbreviated m] The basic unit of length in the metric system, equal to 39.37 in or 3.28 ft. A meter is equal to 100 cm or 1000 mm. A meter was defined in 1960 as 1,650,763.73 wavelengths of orange-red line of krypton 86.

meter-kilogram-second-ampere unit [abbreviated MKSA unit] A practical absolute electrical unit based on the meter, kilogram, second, and ampere as fundamental units.

meter-kilogram-second unit [abbreviated MKS unit] An absolute unit based on the meter, kilogram, and second as fundamental units.

meter-type relay A relay that uses a meter movement which has a contact-bearing pointer that moves toward or away from a fixed contact

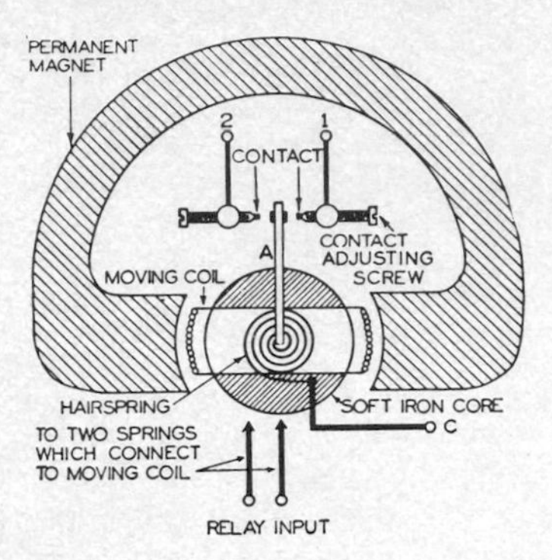

Meter-type relay.

mounted on the meter scale. Also called contact-making meter and instrument-type relay.

metric sabin A unit of sound absorption for a surface, equivalent to 1 m^2 of perfectly absorbing surface.

metric wave A radio wave between the wavelength limits of 1 and 10 m, corresponding to the very high-frequency (VHF) range of 30 to 300 MHz.

metrology The science of measurement.

MeV Abbreviation for *megaelectronvolt.*

MEW Abbreviation for *microwave early warning.*

MF Abbreviation for *medium frequency.*

mG Abbreviation for *milligauss.*

MG Abbreviation for *megagauss.*

mH Abbreviation for *millihenry.*

MHD Abbreviation for *magnetohydrodynamics.*

mho Former unit of conductance and admittance, now replaced by the siemens as the SI unit of conductance. Both are the reciprocal of the ohm. A resistance or impedance of 1 Ω is equal to a conductance or admittance of 1 siemens.

mhometer A meter used in a tube tester or other instrument to indicate conductance in mhos.

MHz Abbreviation for *megahertz.*

mi Abbreviation for *mile.*

MIC Abbreviation for *microwave integrated circuit.*

mica A transparent mineral that splits readily into thin sheets which have excellent insulating and heat-resisting qualities. Used as the dielectric in mica capacitors and as electrode spacers in electron tubes.

mica capacitor A fixed capacitor that uses mica sheets as the dielectric.

Michelson interferometer An interferometer that uses a half-reflecting dielectric sheet, a fixed sheet-metal reflector, and a movable sheet-metal reflector to give accurate frequency measurements at millimeter wavelengths. When the two reflectors are the same distances from the 45° half-reflecting sheet the path lengths are identical, and the two waves arrive at the receiver in phase to give a power maximum. As the movable reflector is moved toward or away from the half-reflecting sheet, alternate maxima and minima of detected power are observed at the receiver.

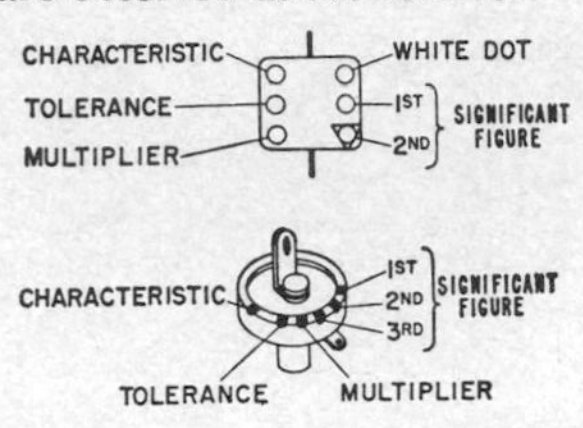

Mica capacitors—conventional molded plastic type and silvered mica button type—with two methods of using color codes to give capacitance values in picofarads and other significant characteristics.

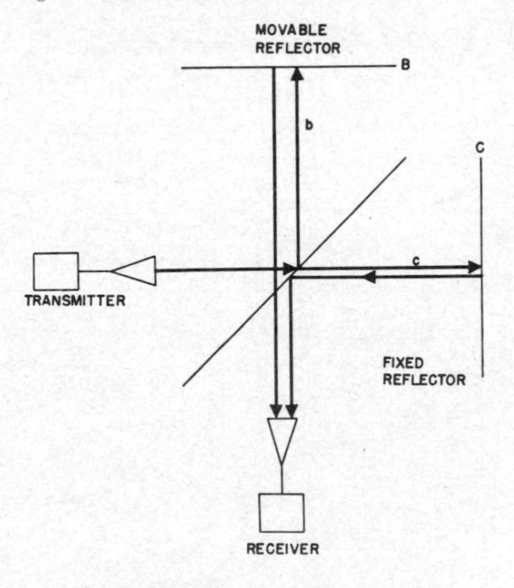

Michelson interferometer using 45° half-reflecting dielectric sheet at center and two metal-sheet reflectors, for frequency measurements at 35 GHz. Open arrows are antenna symbols.

MICR Abbreviation for *magnetic-ink character recognition.*

micro *Microcomputer.*

micro- 1. [abbreviated μ] A prefix that represents 10^{-6} or one-millionth. 2. A prefix that indicates smallness, as in microwave. 3. A prefix that indicates extreme sensitivity, as in microradiometer.

microammeter An ammeter whose scale is calibrated to indicate current values in microamperes.

microampere [abbreviated μA] One-millionth of an ampere.

microbar [abbreviated μbar] One-millionth of a bar. The microbar was the unit of pressure formerly used in acoustics. The newton per square meter is now the SI unit of pressure.

microchannel image intensifier An image intensifier in which a microchannel plate provides sufficient intensification to permit viewing at extremely low light levels, such as that of starlight.

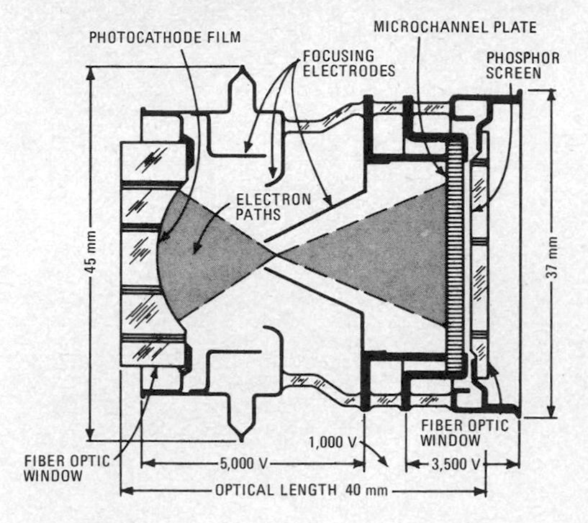

Microchannel image intensifier, with microchannel plate directly behind phosphor viewing screen.

When gated in synchronism with a pulsed laser light source, objects in the laser beam can be seen through fog, mist, or smoke screens.

microchannel plate A plate that consists of extremely small cylinder-shaped electron multipliers mounted side by side, to provide image intensification factors as high as 100,000. Applications include night-viewing binoculars, telescopes, and television camera tubes.

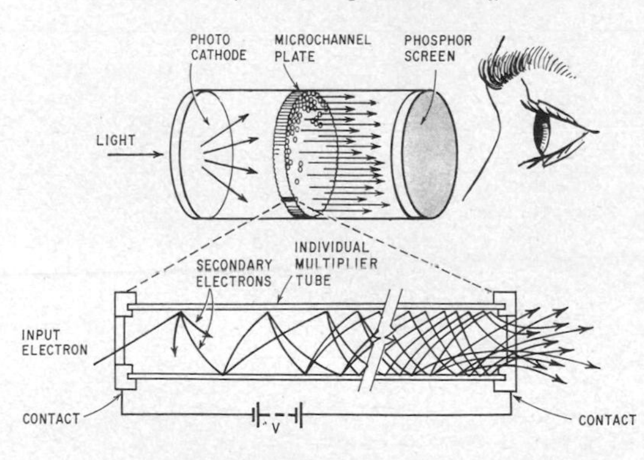

Microchannel plate and electron multiplier action in one of individual etched-out cylindrical tubes that make up plate.

microcircuit *Integrated circuit.*

microcomputer [abbreviated μC] A microprocessor combined with input/output interface devices, some type of external memory, and the other elements required to form a working computer system. A microcomputer is smaller, lower in cost, and usually slower than a minicomputer. Also called micro.

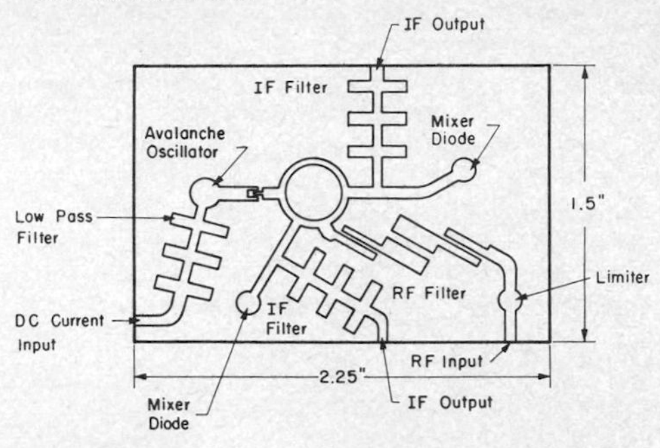

Microcircuit for complete L-band (390–1550 MHz) receiver using balanced mixer and limiter. Microwave diodes and transistors are added after passive microstrip elements have been fabricated on dielectric substrate. Multiply dimensions in inches by 2.54 to convert to centimeters.

microcontroller A microcomputer, microprocessor, or other equipment used for precise process control in data handling, communication, manufacturing, and other fields.

microcurie [abbreviated μCi] One-millionth of a curie.

microdensitometer A high-sensitivity densitometer used in spectroscopy to detect spectrum lines too faint on a negative to be seen by the human eye.

microelectronics The technology of constructing and using electronic circuits and devices in extremely small packages by using integrated-circuit or other special manufacturing techniques. Also called microminiaturization.

microenergy logic circuit A logic circuit for use where very low power drain, such as a few hundred microwatts, is required along with fast response.

microfarad [abbreviated μF] One-millionth of a farad.

microfiche A sheet of photographic film that contains a number of microimages.

microfilm A length of photographic film that contains a number of microimages.

microflash A high-intensity, short-duration light source used in photography.

microgroove record *Long-play record.*

microhenry [abbreviated μH; plural microhenrys] One-millionth of a henry.

microhertz [abbreviated μHz] One-millionth of a hertz.

microhm [abbreviated μΩ] One-millionth of an ohm.

microimage A source material reproduced on photographic film in a size too small to be readable.

microinstruction The portion of a microprogram that specifies the operation of individual computing elements and such related subunits as the main memory and the input/output interfaces. A microinstruction usually includes a next-address field that eliminates the need for a program counter.

microinstruction sequence The series of microinstructions that a microprogram control unit selects from a microprogram for the purpose of executing a single control command or macroinstruction. Microinstruction sequences can be shared by a number of macroinstructions.

microlock A phase-locking loop system for transmitting and receiving information by radio with reduced bandwidth. Used in tracking satellites and telemetering data to ground stations at line-of-sight distances as great as 3000 mi (4800 km).

microlock network A network of radio stations that uses microlock equipment to track missiles and satellites.

microlux [abbreviated μlx] One-millionth of a lux.

micrometeorite A fine dust particle, composed mainly of iron and silicates as in larger meteorites, distributed throughout the solar system. It is generally too small to do structural damage to spacecraft, but it has an erosive effect on exposed surfaces and is a potential hazard to unprotected astronauts.

micrometer [abbreviated μm] One-millionth of a meter, a unit now used for specifying wavelengths of light. Visible light is in the range from about 0.4 μm for purple to 0.75 μm for red. The angstrom was also used for specifying wavelengths of light; 1 Å is equal to 10^{-4} μm, so 1 μm is equal to 10,000 Å. Formerly called micron; use of the term micron was abrogated by international agreement in 1967.

micromho [abbreviated μmho] One-millionth of a mho.

micromicro- [abbreviated $\mu\mu$] A prefix that represents one-millionth of a millionth, or 10^{-12}. Now called pico-.

microminiaturization *Microelectronics.*

micron [abbreviated μ] One-millionth of a meter. Use of this term for specifying wavelengths of light and thin-layer thicknesses was abrogated by international agreement in 1967, and micrometer is now used.

microoptics The technology of constructing and using optical devices manufactured in extremely small packages by using integrated-circuit or other special manufacturing techniques.

microphone An electroacoustic transducer that responds to sound waves and delivers essentially equivalent electric waves. Also called mike (slang).

microphone boom An overhead extension arm that supports a microphone within range of the sound to be picked up but outside the range of a television camera.

microphone button A button-shaped telescoping container that is filled with carbon particles and serves as the resistance element of a carbon microphone.

microphone cable A special shielded cable used to connect a microphone to an audio amplifier.

microphone mixer A mixer that feeds two or more microphones into the input of an AF amplifier. Separate controls permit adjusting the output level of each microphone.

microphone preamplifier An AF amplifier that amplifies the output of a microphone before the signal is sent through a transmission line to the main AF amplifier. The preamplifier is often built into the microphone housing or stand.

microphone shield A protective covering that protects a microphone diaphragm from condensed moisture originating from the operator's breath when in use, or protects it from other environmental conditions.

microphone stand A stand that supports a microphone in a desired position above the floor or on a table.

microphone transformer An iron-core transformer that couples certain types of microphones

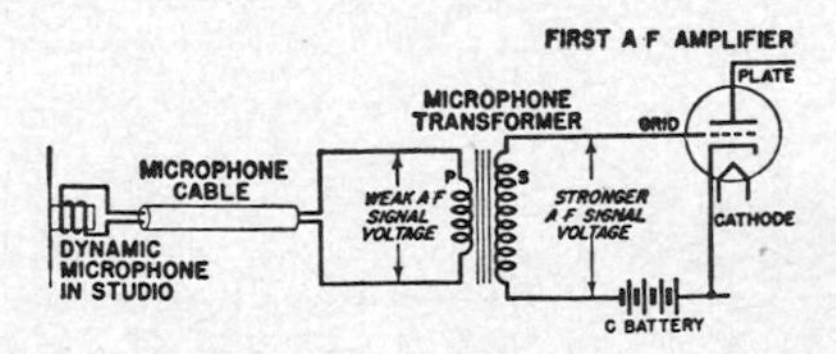

Microphone-transformer connections.

to a microphone preamplifier, to a transmission line, or to the main AF amplifier.

microphonic Vulnerable to vibration that produces microphonics. A tube in a radio receiver is microphonic if a pinging sound is heard from the loudspeaker when the side of the tube is tapped with a finger.

microphonics Noise caused by mechanical vibration of the elements of an electron tube, component, or system. The vibration causes modulation of the signal currents flowing through or controlled by the vibrating device. Heard as noise in an AF system and seen as an undesirable interference pattern in facsimile and television images. Also called microphonism.

microphonism *Microphonics.*

microphotometer A photometer that provides highly accurate illumination measurements. In one form, the changes in illumination are picked up by a phototube and converted into current variations that are amplified by vacuum tubes.

microplasma noise The portion of zener noise that is caused by the zener breakdown phenomenon. It is generally considered to be white noise.

micropower circuit A circuit in which the power consumption of individual components is down to microwatts or less, reducing heat generation so greatly that high packaging densities become feasible. Microwatt circuits and nanowatt circuits are examples.

microprocessing unit [abbreviated MPU] A microprocessor with its external memory, input/output interface devices, and buffer, clock, and driver circuits. It is usually assembled on a single printed-circuit board that is mounted in a microcomputer or in the instrumentation of an automatic control system.

microprocessor [abbreviated μP] A computer central processing unit that is manufactured on a single integrated-circuit chip (or on several chips) by utilizing large-scale integration technology. A microprocessor may be incorporated directly into the instrumentation of an automatic control system or used as the main element of a microcomputer.

microprogram A computer program that consists only of basic elemental commands which directly control the operation of each functional element in a microprocessor. Microprograms were at one time stored in read-only memories, but newer microprocessor designs permit changing part or all of a microprogram in the field without physically replacing anything. A microprogram is an example of firmware because it is somewhere between hardware and software in permanence.

microprogram control unit [abbreviated MCU] The portion of a computer that extracts from memory the next microprogram address and the associated microinstruction sequences.

microprogramming A means of controlling the execution of instructions in a computer through an orderly array of control words that are known as microinstructions.

microrad [abbreviated μrd] One-millionth of a rad.

microradiography The radiography of small objects that have details too fine to be seen by the unaided eye, with optical enlargement of the resulting negative.

microradiometer A radiometer that measures weak radiant power, in which a thermopile is supported on and connected directly to the moving coil of a galvanometer.

microreflectometer A reflectometer that measures the reflectance of very small areas of an image.

microrem [abbreviated μrem] One-millionth of a rem.

microrep [abbreviated μrep] One-millionth of a rep.

microroentgen [abbreviated μR] One-millionth of a roentgen.

microscopic cross section The cross section per atom of isotope. The dimensions are those of an area.

microsecond [abbreviated μs] One-millionth of a second.

microseism A weak and generally regularly recurring earth tremor propagated along the floor of the ocean.

microseismograph A microseismometer that has recording facilities.

microseismometer A seismometer that has sufficient sensitivity for measuring microseisms.

microsiemens [abbreviated μS] One-millionth of a siemens.

microsonics *Microwave acoustics.*

microstrip A strip transmission line that is capable of transmitting signals at microwave frequencies. It consists basically of a thin-film strip in

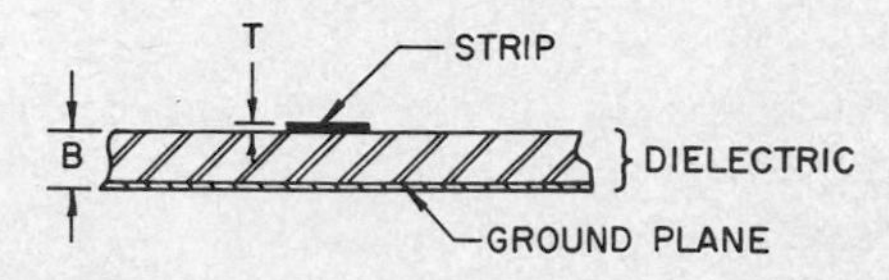

Microstrip construction.

intimate contact with one side of a flat dielectric substrate, with a similar thin-film ground-plane conductor on the other side of the substrate.

microswitch *Snap-action switch.*

Micro Switch Trademark of Micro Switch Division of Honeywell for their line of switches.

microtron A type of accelerator in which electrons are accelerated over a circular path to energies of several million electronvolts, with the time of successive particle revolutions increasing by exactly one cycle of the RF accelerating voltage, to maintain synchronism. A microtron combines the features of a cyclotron with those of a linear accelerator.

microvision An aircraft blind-landing aid in which the pilot sees on a display an outline of the airport runway and the pilot's position in relation to it, much as if the pilot were viewing runway lights at that airport at night. The system includes two beacons on the ground and two pairs of microwave horn-type antennas on the aircraft, feeding a four-channel microwave receiver and the cathode-ray display.

microvolt [abbreviated μV] One-millionth of a volt.

microvoltmeter A voltmeter whose scale is calibrated to indicate voltage values in microvolts.

microvolts per meter A measure of the intensity of the signal produced by a radio transmitter at a given point. It is equal to the signal strength in microvolts at the receiving antenna divided by the effective height of the antenna in meters. Stronger signals are expressed in millivolts per meter.

microwatt [abbreviated μW] One-millionth of a watt.

microwatt circuit A micropower circuit in which the power consumption of individual components is only a few microwatts. Construction generally involves planar diffusion technology and deposited thin-film technology on the same silicon wafer.

microwattmeter A wattmeter designed primarily to measure and indicate RF power values in microwatts. It may also have nanowatt and milliwatt scales.

microwave Pertaining to wavelengths in the microwave spectrum, ranging from about 30 to 0.3 cm and corresponding to frequencies ranging from 1 to 100 GHz.

microwave acoustics The branch of acoustics that deals with acoustic waves traveling through or on the surface of a solid material at microwave frequencies. Also called microsonics.

microwave altimeter An altimeter that operates in the microwave spectrum. A 13.9-GHz version in the Skylab satellite vehicle has detected variations of up to 110 km in the surface of the ocean, corresponding to gravity-altering changes in ocean-floor topology.

microwave antenna A combination of an open-end waveguide and a parabolic reflector or horn, to receive and transmit microwave signal beams at microwave repeater stations.

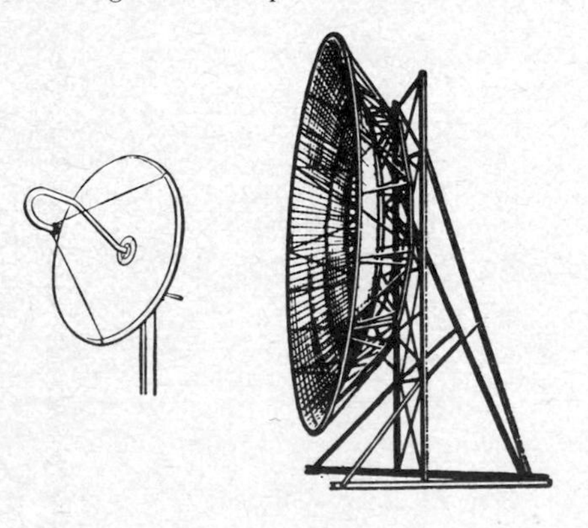

Microwave antennas using parabolic reflectors.

microwave circuit A circuit that operates at microwave frequencies.

microwave circulator *Circulator.*

microwave counter A counter that measures frequency in the microwave spectrum by counting cycles for a precise interval of time and converting the count to the corresponding frequency value in gigahertz on a digital display.

microwave delay line A small solid-state passive device that introduces controllable time delay into a microwave signal line.

microwave detector A device that reacts to the presence of microwave radiation by changing its resistance or some other electrical parameter. One basic type is the crystal detector. Another type is the bolometer, which depends on the temperature-sensitive resistance characteristic of a barretter or thermistor.

microwave diathermy Diathermy that uses frequencies in the microwave spectrum, generally at 2.45 GHz, to heat deep tissues for medical purposes.

microwave diode A diode designed primarily for operation at microwave frequencies. Examples include BARITT, Gunn, IMPATT, LSA, TRAPATT, and varactor diodes. All these microwave diodes convert DC energy directly to RF energy by utilizing negative resistance.

microwave early warning [abbreviated MEW] A high-power, long-range, early-warning radar with a number of indicators; it gives high resolution and large traffic-handling capacity.

microwave filter A filter that consists of resonant cavity sections or other elements, built into a microwave transmission line to pass desired frequencies while rejecting or absorbing other frequencies.

microwave frequencies Frequencies of 890 MHz and above, as used in FCC regulations.

microwave gyrator *Gyrator.*

microwave heating Heating of food by means of electromagnetic energy in or just below the microwave spectrum for cooking, dehydration, sterilization, thawing, and other purposes. Frequencies most often used are 915 MHz and 2.45 GHz, in the bands assigned for industrial, scientific, and medical purposes.

microwave integrated circuit [abbreviated MIC] A microwave circuit that uses integrated-circuit production techniques involving such features as thin or thick films, substrates, dielectrics, conductors, resistors, and microstrip lines, to build passive assemblies on a dielectric. Active elements such as microwave diodes and transistors are usually added after photoresist, masking, etching, and deposition processes have been completed.

microwave interferometer An instrument that measures the free-electron density of an ionized gas by measuring transmission of extremely high-frequency radio waves through the gas.

microwave landing system An instrument landing system that operates in the microwave spectrum, developed to replace lower-frequency systems at airports. The microwave system provides a high-accuracy easy-to-read cockpit display for pilot guidance in making blind landings under all weather conditions.

microwave multiplier phototube A multiplier phototube that can be used to demodulate light

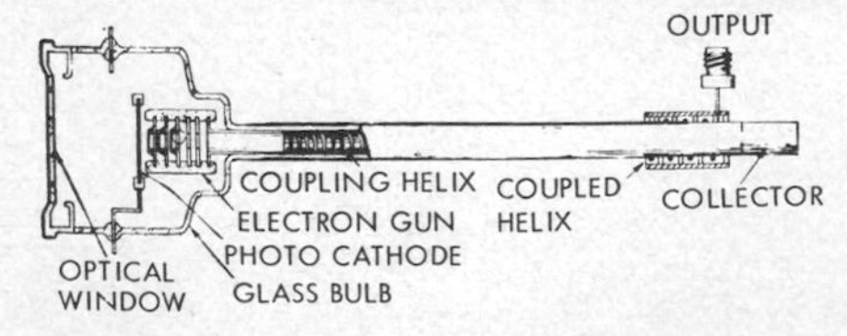

Microwave multiplier phototube using transmission secondary-emission dynodes in electron gun.

beams which are modulated at microwave frequencies. The phototube uses transmission secondary-emission dynodes to amplify the photoelectrons emitted by the photocathode. The resulting bunched photoelectrons inside the tube are focused by an axial magnetic field created by a helix. This electron beam excites on the helix an RF wave that is coupled through the glass envelope of the tube to another short length of helix connected to the output coaxial connector.

microwave oven An oven that uses microwave heating for fast cooking of meat and other foods. It is particularly useful for fast thawing and cooking of frozen foods in restaurants, airplanes, and homes.

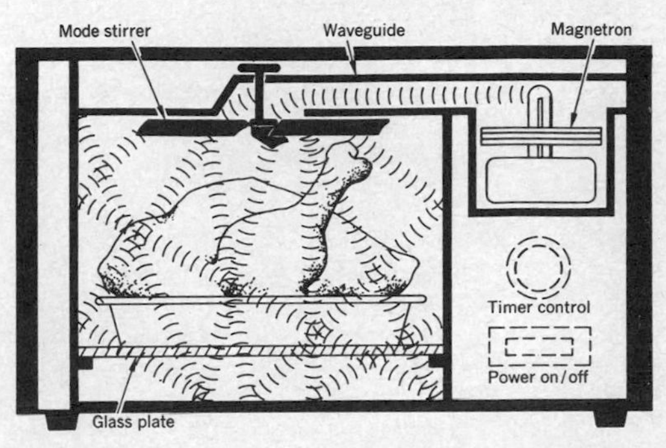

Microwave oven using RF power generated by magnetron at 2.45 GHz. Essentially all power is absorbed by food load, so cooking time must be based on weight.

microwave phototube A phototube combined with a traveling-wave tube or other microwave electron-tube structure that responds to modulated laser light beams.

microwave plasma Plasma produced by a microwave source. Present and potential applications include display devices, gas lasers, magnetohydrodynamic energy conversion, noise signal sources, and thermonuclear fusion.

microwave power transmission Transmission of power through space from a microwave transmitting antenna to a remote receiving antenna.

microwave radiometer A radiometer in which a microwave receiver detects microwave thermal radiation and similar weak wideband signals that resemble noise and are obscured by receiver noise. Most designs are based on the original Dicke radiometer, generally with a latching ferrite circulator to switch back and forth between an internal reference source and the microwave power source being measured.

microwave radiometry Radiometry that depends primarily on measurement of microwave wavelengths which emanate from beneath a surface. The measurements are thus independent of surface irregularities or variations in surface temperatures. One application involves determining the orientation of the line that joins a space vehicle with the center of mass of the earth, the moon, or the sun.

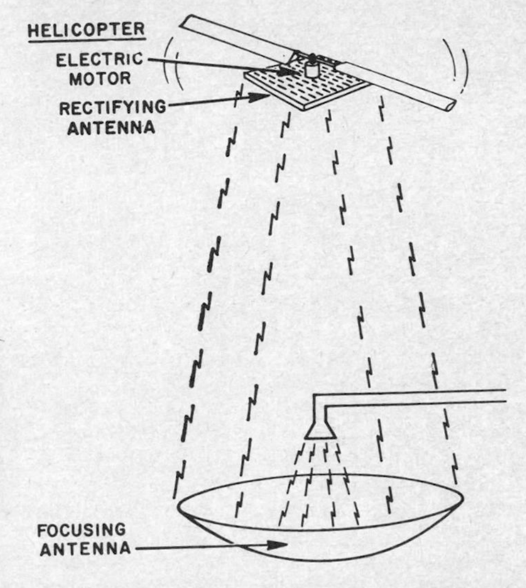

Microwave power transmission from parabolic reflector on ground supports platform in air by energizing DC motor driving helicopter rotor.

microwave refractometer An instrument that measures the refractivity of the atmosphere, which is proportional to the dielectric constant of the atmosphere. The instrument uses two precision microwave transmission cavities, one hermetically sealed to serve as a reference. Travel time of a microwave signal through each cavity is measured by associated electronic circuits and converted into the refractivity value of the air being sampled.

microwave repeater A radio repeater that uses

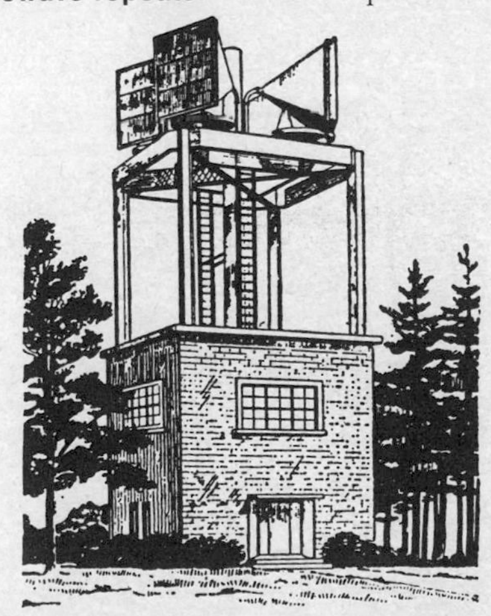

Microwave repeater used for television networks.

highly directional radio beams at microwave frequencies to link towers spaced up to 50 mi (80 km) apart. Each tower has a receiver and transmitter for each direction in which signals are picked up, amplified, and passed on.

microwaves *Microwave spectrum.*

microwave spectroscopy Spectroscopy that involves determination of the selective absorption of microwaves at various frequencies by solid or gaseous materials. Used in studying atomic, crystalline, and molecular structures.

microwave spectrum A spectrum of wavelengths between the shortwave region and the far-infrared region, commonly considered to include wavelengths from about 30 to 0.3 cm (1 to 100 GHz). Also called microwaves.

microwave system A radio system that utilizes microwaves.

microwave therapy Therapeutic use of electromagnetic energy to generate heat within the body, using frequencies in the microwave spectrum.

microwave transistor A transistor that operates at microwave frequencies. Both bipolar and field-effect transistors can be made to operate efficiently in this part of the frequency spectrum. Practically all microwave transistors are planar NPN silicon, with dimensions in the micrometer range.

microwave tube An electron tube that operates at wavelengths in the range of about 30 to 0.3 cm.

microwave turbulence Irregular fluctuations in the microwave refractive index in the atmosphere, due to either thermal turbulence or irregular distribution of water vapor.

mid-Canada line A chain of radar stations along the 55th and 56th parallels in Canada. Also called McGill fence.

midcourse guidance The guidance applied to a missile between the launching and the target approach phases.

middle marker A fan marker beacon approximately 3500 ft (1000 m) from the approach end of the runway on the localizer course of an instrument landing system.

migration 1. The uncontrolled movement of certain metals, particularly silver, from one location to another, usually with associated undesirable effects like oxidation or corrosion. It can cause serious problems on printed-circuit boards that have gold-plated silver contacts; here the silver migrates through the protective gold plating. 2. The movement of charges through a semiconductor material by diffusion or drift of charge carriers or ionized atoms. 3. The movement of crystal defects through a semiconductor crystal under the influence of high temperature, strain, or a continuously applied electric field.

migration area One-sixth the mean square distance that a neutron travels in a medium from its birth in fission until its absorption.

migration length The square root of the migration area.

mike Slang for *microphone.*

mil 1. A unit of angular measurement used in launching bombs and guided missiles. A true mil is the angle determined by an arc whose length is one-thousandth of the radius. For practical purposes, the mil is considered to be 1/6400 (instead of 1/6283) of 360°. 2. One-thousandth of an inch (0.00254 cm).

MIL Abbreviation for Military Specification, which replaces JAN specifications.

mile [abbreviated mi] A unit of length, equal to 5280 ft or 1.609344 km.

Miller bridge A bridge that measures amplification factors of vacuum tubes.

Miller code A code used internally in some computers, in which a binary 1 is represented by a transition in the middle of a bit (either up or down), and a binary 0 is represented by no transi-

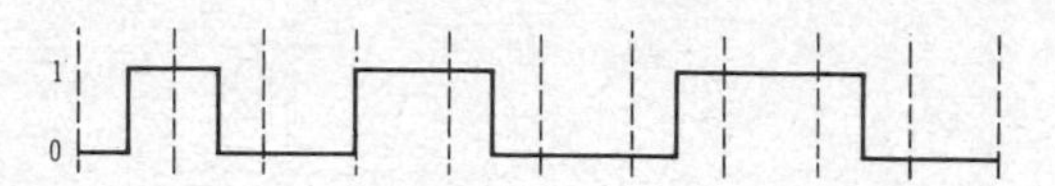

Miller code for binary data. Space between each pair of dashed vertical lines represents one bit time.

tion following a binary 1. A transition between bits represents successive 0s. In this code, the longest period possible without a transition is two bit times.

Miller effect The increase in the effective grid-cathode capacitance of a vacuum tube due to the charge induced electrostatically on the grid by the anode through the grid-anode capacitance.

Miller integrator A resistor-capacitor charging network that has a high-gain amplifier which parallels the capacitor. Used to produce a linear time-base voltage. Also called Miller time base.

Miller time base *Miller integrator.*

milli- [abbreviated m] A prefix that represents 10^{-3}, or one-thousandth.

milliammeter An ammeter whose scale is calibrated to indicate current values in milliamperes.

milliampere [abbreviated mA] One-thousandth of an ampere.

milliampere-hour [abbreviated mAh] One-thousandth of an ampere-hour.

millibar [abbreviated mbar] One-thousandth of a bar.

millibarn [abbreviated mb] One-thousandth of a barn.

millicandela [abbreviated mcd] One-thousandth of a candela.

millicurie [abbreviated mCi] One-thousandth of a curie.

milligauss [abbreviated mG; plural milligauss] One-thousandth of a gauss.

milligaussmeter A gaussmeter that measures magnetic field strength in milligauss.

millihenry [abbreviated mH; plural millihenrys] One-thousandth of a henry.

millilumen [abbreviated mlm] One-thousandth of a lumen.
millimeter [abbreviated mm] One-thousandth of a meter.
millimeter of mercury A unit of pressure for vacuum equipment, approximately equal to 1/760 atmosphere, or 1 torr.
millimeter-wave amplifier An amplifier capable of amplifying millimeter waves.

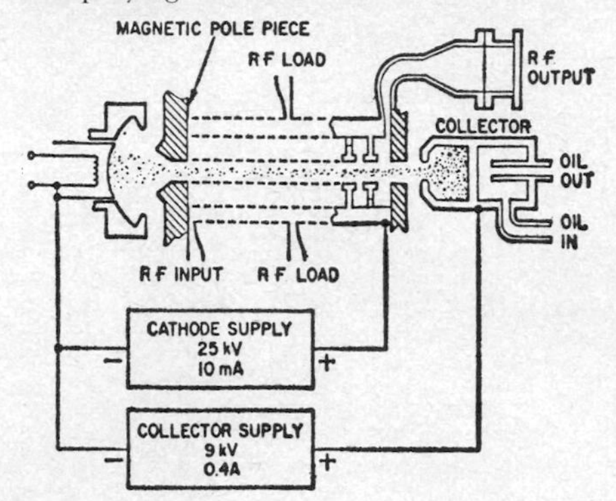

Millimeter-wave tube capable of oscillating at wavelength of 5 mm.

millimeter-wave tube A tube capable of operating as an oscillator or amplifier at wavelengths of a few millimeters.
millimetric wave A radio wave between the wavelength limits of 1 and 10 mm, corresponding to the extremely high-frequency (EHF) range of 30 to 300 GHz.
milliohm [abbreviated mΩ] One-thousandth of an ohm.
milliohmmeter An ohmmeter whose scale is calibrated to indicate resistance values in milliohms.
millirad [abbreviated mrd] One-thousandth of a rad.
milliradian [abbreviated mrad] One-thousandth of a radian, equal to 0.0572957795°.
millirem [abbreviated mrem] One-thousandth of a rem.
millirep [abbreviated mrep] One-thousandth of a rep.
milliroentgen [abbreviated mR] One-thousandth of a roentgen.
millisecond [abbreviated ms] One-thousandth of a second.
millisiemens [abbreviated mS] One-thousandth of a siemens.
millisone A unit of loudness equal to 0.001 sone.
millitorr One-thousandth of a torr.
millivolt [abbreviated mV] One-thousandth of a volt.
millivoltage A voltage whose magnitude is most conveniently expressed in millivolts.
millivoltmeter A voltmeter whose scale is calibrated to indicate voltage values in millivolts.
millivolts per meter A rating often used for signal intensities greater than 1000 μV/m.
milliwatt [abbreviated mW] One-thousandth of a watt.
min Abbreviation for *minute.*
mine detector A metal detector designed specifically to locate explosive mines that are buried or submerged. A higher-frequency version is used for locating nonmetallic antipersonnel mines.
miniature tube A small electron tube that has no base, with tube electrode leads projecting through the glass bottom in positions corresponding to those of pins for either a 7- or 9-pin tube base.

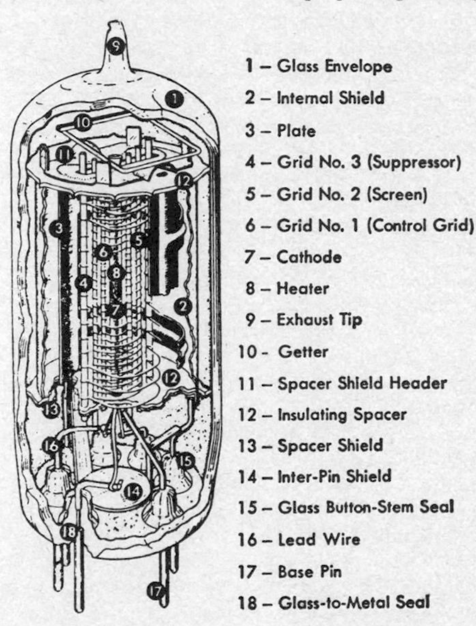

Miniature-tube construction for pentode.

miniaturization Reduction in the size and weight of a system, package, or component by using small parts arranged for maximum utilization of space.
miniaturize To redesign a component or piece of electronic equipment to fit in less space.
minicartridge A self-contained package of reel-to-reel magnetic tape that resembles a cassette or cartridge but is slightly different in design and dimensions. Used primarily for data storage.
minicomputer A small general-purpose stored-program digital computer, capable of performing essentially the same arithmetic, logic, and input/output operations as a large computer. Being smaller in size, it will have a smaller primary memory, shorter word lengths (usually a 16-bit word size as compared to up to 64 bits per word in large computers), and lower price. A minicomputer is generally usable by only a single operator or control system at a time, with no time-sharing. A minicomputer has greater performance capability than a microcomputer or programmable calculator.
minimum-access programming The program-

ming of a digital computer so that minimum waiting time is required to obtain information out of the memory.

minimum ionization The smallest possible value of the specific ionization that a charged particle can produce in passing through a particular substance. For singly charged particles in ordinary air, the minimum ionization is about 50 ion pairs per centimeter of path. In general, it is proportional to the density of the medium and to the square of the charge of the particle.

miniprinter A printer without a keyboard, driven by an external digital source such as a computer, communication line, or tape handler to print lines of alphameric data on paper. This term is generally applied to printers that have less than 80 characters per line or lower printing speeds than 120 characters per second.

miniscope A portable cathode-ray oscilloscope, generally small enough to hold in one hand as it is being used for troubleshooting.

minitrack The track of a miniature telemeter transmitter carried in an orbiting satellite, missile, or rocket.

minitrack network A network of minitrack radio stations placed at different points around the world to track the flight of an earth satellite.

minitrack radio A radio receiver that tracks an earth satellite or other object equipped with a miniature transmitter which is emitting telemeter-type signals.

minometer A capacitance-measuring instrument that measures accurately the charge remaining in a dosimeter after use, to determine the dosage of radiation to which the dosimeter was exposed.

minor apex face One of the three smaller sloping faces near but not touching the apex of a natural quartz crystal. The three larger sloping faces are the major apex faces.

minor cycle The time required for the transmission or transfer of one machine word, including the space between words, in a digital computer that uses serial transmission. Also called word time.

minor face One of the three longer sides of a natural hexagonal quartz crystal.

minority carrier The type of carrier that constitutes less than half of the total number of carriers in a semiconductor. The minority carriers are holes in an N-type semiconductor and electrons in a P-type semiconductor.

minority emitter An electrode from which a flow of minority carriers enters the interelectrode region of a transistor.

minor lobe Any lobe except the major lobe of a radiation pattern. Also called secondary lobe and side lobe.

minus zone The bit positions in a computer code that represent the algebraic minus sign.

minute [abbreviated min] A unit of time, equal to 60 s.

Minuteman An Air Force intercontinental ballistic missile that uses solid fuel to permit almost instant firing when needed. It is a three-stage missile which carries a nuclear warhead and uses inertial guidance. Range is over 6000 nautical miles (over 11,000 km).

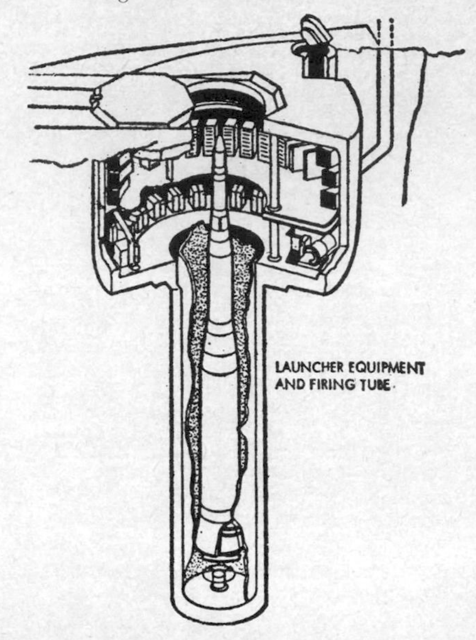

Minuteman intercontinental ballistic missile in hardened underground launch silo.

MIRAN [MIssile RANging] A microwave omnidirectional pulse-type missile-tracking system that measures range only. A master interrogator at the master ground station triggers a missile beacon. Two or more slave stations measure the transit time of the resulting pulse from the beacon, for conversion to position data at the master station.

mirror-backed screen *Aluminized screen.*

mirror galvanometer A galvanometer that has a small mirror attached to the moving element, to permit use of a beam of light as an indicating pointer.

mirror interferometer An interferometer used in radio astronomy, in which the sea surface acts as a mirror to reflect radio waves up to the single antenna, where the reflected waves interfere with the waves arriving directlv from the source.

mirror nuclides Pairs of nuclides, each member of which would be transformed into the other by exchanging all neutrons for protons, and vice versa.

mirror point A point at which a charged particle, moving in a spiral path along the lines of a magnetic field, is reflected back as it enters a stronger magnetic field region.

mirror reflection *Direct reflection.*

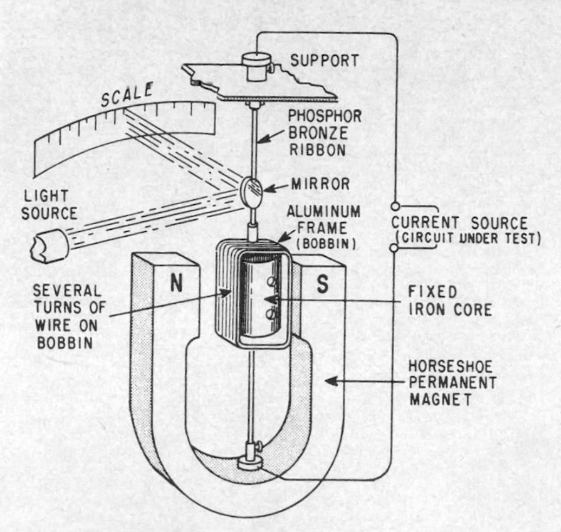

Mirror-galvanometer construction.

mirror-reflection echo A radar echo that undergoes multiple reflections, as by reflection from the side of an aircraft carrier or other large flat surface, before being reflected from a nearby target.

MIRV [Multiple Independently-targetable Reentry Vehicle] A multiple-warhead nuclear missile for underwater firing from Polaris, Poseidon, and Trident nuclear-powered submarines. Also used on Minuteman intercontinental ballistic missiles. The warheads can be programmed independently to hit widely separated targets following ejection at a prescribed altitude and velocity after the last stage of the missile has been spent.

MIS 1. Abbreviation for *management information system.* 2. Abbreviation for *metal-insulator semiconductor.*

misch metal An alloy of cerium, lanthanum, and didymium, sometimes used on the cathodes of glow tubes.

misfire Failure to establish an arc between the main anode and the cathode of an ignitron or other mercury-arc rectifier during a scheduled conducting period.

mismatch The condition in which the impedance of a source does not match or equal the impedance of the connected load or transmission line.

mismatch factor *Reflection coefficient.*

missile An object that is dropped, propelled, or otherwise projected through air or water toward a target. Although a missile is normally used to damage a target, it may also serve for photoreconnaissance, detection, meteorographic measurements, and other purposes. When the trajectory of a missile can be changed after launching, by remote control or by an internal control system, it is known as a guided missile.

missile acquisition A measuring system that provides angular data (azimuth and elevation angle) by phase comparison of radio-frequency signals received from a missile.

missile decoy A vehicle that simulates a missile in flight, attracting enemy radar and drawing fire, to increase the odds for penetration by manned bombers or other weapon systems.

missile master A system of computers, communication equipment, and data-processing equipment used to collect data on approaching aircraft from all available sources, evaluate it, select targets, and direct fire of guided missile batteries automatically.

missile plume The region of electromagnetic and other disturbances that follow a missile during reentry and make the missile more readily detectable.

missile range A marked-off course or area over which test missiles are flown under observation.

missing-pulse detector A detector that detects the absence of a pulse by comparing the energy of incoming pulses with that of preceding presumably normal pulses or by comparing the energy content of each pulse with a preset DC reference voltage.

mission The assigned task or objective of a person or group, or of a military, naval, or aerospace operation.

mistake A human action or failing that produces an unintended result. Examples include incorrect programming of computers, incorrect coding of computer inputs, or incorrect operation of manual controls by an operator.

mixed-base notation A computer number system in which a single base, such as 10 in the decimal system, is replaced by two number bases used alternately, such as 2 and 5. Biquinary notation is an example. Also called mixed-radix notation.

mixed highs The high-frequency signal components that are intended to be reproduced achromatically (without color) in a color television picture.

mixed-radix notation *Mixed-base notation.*

mixer 1. A device that has two or more inputs,

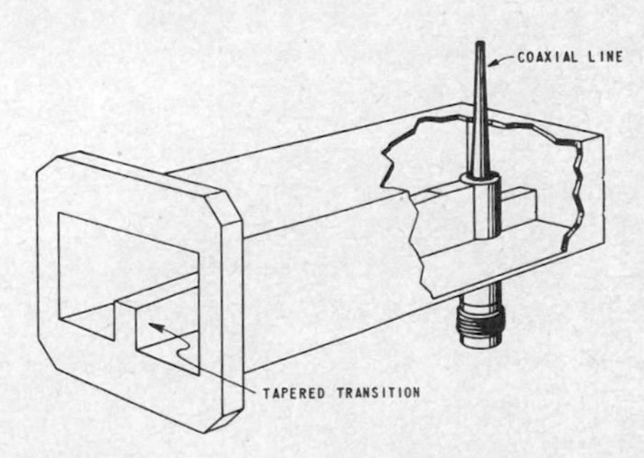

Mixer in which modulated microwave carrier signal in ridge waveguide is mixed with oscillator signal in crystal diode inside coaxial line, to provide IF output.

usually adjustable, and a common output. It combines separate audio or video signals linearly in desired proportions, to produce an output signal. 2. The stage in a superheterodyne receiver in which the incoming modulated RF signal is combined with the signal of a local RF oscillator to produce a modulated IF signal. Crystal diodes are widely used as mixers in radar and other microwave equipment. Also called first detector and mixer-first detector. The mixer and oscillator together form the converter. 3. A nonlinear device in which two light beams are combined to form new beams that have frequencies equal to the sum or the difference of the input wavelengths. The difference frequency between two light beams may be sufficiently low that it falls in the spectrum of radio waves rather than that of light waves.

mixer-first detector *Mixer.*

mixer tube An electron tube that performs only the frequency-conversion function of a converter in a superheterodyne receiver. It is supplied with voltage or power by a separate local oscillator.

mixing Combining two or more signals, such as the outputs of several microphones.

mixing amplifier An amplifier that has inputs to which two or more different signals are applied, and a common output which delivers a composite signal.

MKSA unit Abbreviation for *meter-kilogram-second-ampere unit.*

MKS unit Abbreviation for *meter-kilogram-second unit.*

M line One of the characteristic lines in the spectrum of an atom. It is produced by excitation of the electrons of the M shell.

mlm Abbreviation for *millilumen.*

mm Abbreviation for *millimeter.*

mnemonic code A programming code that is easy to remember because the codes resemble the original words, such as mpy for multiply and acc for accumulator.

MNOS Abbreviation for *metal-nitride-oxide semiconductor.*

mobile station A radio station intended to be used while in motion or during halts at unspecified points.

mobile telemetering Telemetering between points that may have relative motion, using radio in place of interconnecting wires.

mobile telephone service A service that provides radiotelephone communication from a mobile vehicle to a regular telephone or to another similarly equipped vehicle.

mobile transmitter A radio transmitter designed for installation in a vessel, vehicle, or aircraft, and normally operated while in motion.

mobile unit A truck or other vehicle equipped with television studio equipment. It is used for television pickups at remote locations. Picture and sound signals are sent back to the main transmitter by either a relay transmitter on the truck or through coaxial cables.

mobility 1. Freedom of particles to move, either in random motion or under the influence of fields or forces. 2. *Drift mobility.*

Mobius resistor A nonreactive resistor made by placing strips of aluminum or other metallic tape on opposite sides of a length of dielectric ribbon, twisting the strip assembly half a turn, joining the ends of the metallic tape, then soldering leads to opposite surfaces of the resulting loop.

mode 1. A state of a vibrating system that corresponds to a particular field pattern and one of the possible resonant frequencies of the system. The three common modes of vibration in a quartz plate are the extensional, flexural, and shear modes. 2. A form of propagation of guided waves that is characterized by a particular field pattern in a plane transverse to the direction of propagation. The field pattern is independent of position along the axis of the waveguide. For uniconductor waveguides the field pattern of a particular mode of propagation is also independent of frequency. The TE mode is the transverse electric mode, and the TM mode is the transverse magnetic mode. Also called transmission mode. 3. One of several alternative methods of operating a system or device.

mode changer *Mode transducer.*

mode filter A selective device that passes energy along a waveguide in one or more modes of propagation and substantially reduces energy carried by other modes.

mode jump A sudden and irregular change in the oscillation frequency and power output of a magnetron, due to a change in the mode of operation from one pulse to the next.

mode locking Locking of the internal cavity modes of a pulsed laser in proper phase and amplitude to divide each output pulse into a train of extremely sharp and equally spaced pulses.

modem [MOdulator-DEModulator] A combination modulator and demodulator used at each end of a telephone line to convert binary digital information to audio tone signals suitable for transmission over the line, and vice versa. With a two-wire line, transmission can be in only one direction at a time (simplex or half-duplex). With a four-wire line, simultaneous transmission is possible in both directions (full duplex), with only one modem at each end. Asynchronous modems are used for speeds up to about 1.8 kb/s. Synchronous modems that have an internal clock source can handle up to 9.6 kb/s. but require leased telephone lines which have special conditioning. An acoustic coupler permits use of a modem at any telephone handset without making direct wire connections to the line. Also called dataset and subset.

mode purity 1. The ratio of power present in the forward-traveling wave of a desired mode to the total power present in the forward-traveling waves of all modes. 2. The extent to which an ATR tube in its mount is free from undesirable mode conversion.

moder *Coder.*

moderation The slowing down of a particle, usually a neutron, as a result of collisions with nuclei. Also called degradation.

mode separation The frequency difference between resonator modes of oscillation in a microwave oscillator.

mode shift A change in the mode of magnetron operation during the interval of a pulse.

mode skip Failure of a magnetron to fire on successive pulses.

mode transducer A device for transforming an electromagnetic wave from one mode of propagation to another. Also called mode changer and mode transformer.

mode transformer *Mode transducer.*

modifier A quantity used to alter the address of an operand in a computer, such as the cycle index.

modify 1. To alter the address of the operand in a computer instruction. 2. To alter a computer subroutine according to a defined parameter.

modular construction 1. Construction that involves the use of integral multiples of a given length for the dimensions of electronic components and electronic equipment and for spacings of holes in a chassis or printed-wiring board. 2. Made from modules.

nodulate To vary the amplitude, frequency, or phase of a wave or the velocity of the electrons in an electron beam in some characteristic manner.

modulated amplifier The amplifier stage in a transmitter at which the modulating signal is introduced to modulate the carrier.

modulated-beam photoelectric system A photoelectric intrusion-detector system in which reliable beam ranges of about 1000 m are obtained by interrupting the light beam at the source with a rotating punched or slotted disk, so the output of the photodetector is an AC signal that can readily be amplified.

modulated carrier An RF carrier whose amplitude or frequency has been varied in accordance with the intelligence to be conveyed.

modulated continuous wave [abbreviated MCW] A form of emission in which the carrier is modulated by a constant AF tone. In telegraphic service, the carrier is keyed.

modulated light Light that has been made to vary in intensity in accordance with variations in an audio, facsimile, or code signal.

modulated oscillator An oscillator in which an input signal varies the output frequency.

modulated stage The RF stage to which the modulator is coupled and in which the carrier wave is modulated by an audio, video, code, or other intelligence signal.

modulated wave A carrier wave whose amplitude, frequency, or phase varies in accordance with the value of the intelligence signal being transmitted.

modulating-anode klystron A klystron that has between the cathode and the drift-tube section an electrode which can be used to turn the electron beam on and off for pulse generation.

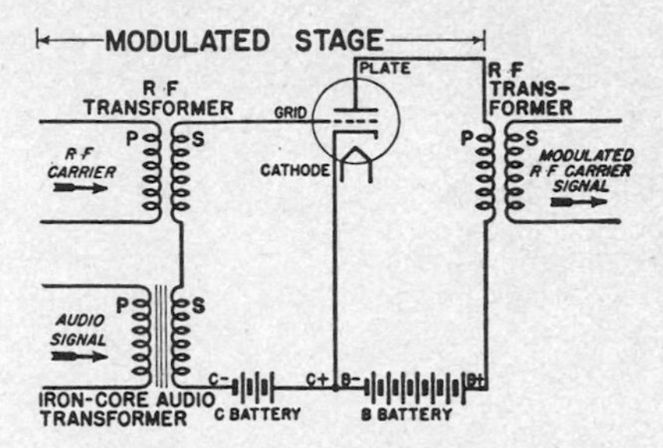

Modulated stage for radio transmitter.

modulating electrode An electrode to which a potential is applied to control the magnitude of the beam current in a cathode-ray tube.

modulating signal A signal that causes a variation of some characteristic of a carrier.

modulation The process by which some characteristic of one wave is varied in accordance with another wave. In radio broadcasting, some stations use amplitude modulation, whereas others use frequency modulation. In television, the picture portion of the program uses amplitude modulation, and the sound portion uses frequency modulation. Other types of modulation include phase, pulse-amplitude, pulse-code, pulse-duration, pulse-frequency, pulse-position, and pulse-time modulation.

modulation capability The maximum percentage modulation that is possible without objectionable distortion.

modulation-demodulation amplifier A unidirectional amplifier in which an amplitude modulator that has conversion gain is followed by a demodulator. Used in wideband microwave systems.

modulation envelope A curve drawn through the peaks of a graph that shows the waveform of a modulated signal. The modulation envelope represents the waveform of the intelligence carried by the signal.

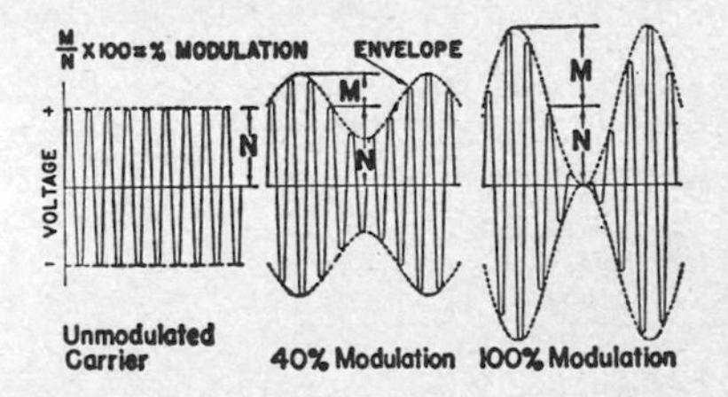

Modulation envelope and modulation percentage.

modulation factor The ratio of the peak variation in the modulation actually used in a transmitter to the maximum variation for which the transmitter was designed. Often expressed as a percentage. In amplitude modulation, the modulation factor is the ratio of half the difference between the maximum and minimum amplitudes of an amplitude-modulated wave to the average amplitude. For frequency modulation, it is the ratio of the actual frequency swing to the frequency swing required for 100% modulation, expressed in percentage.

modulation frequency The rate of variation of a frequency-modulated signal about the carrier frequency.

modulation index The ratio of the frequency deviation to the frequency of the modulating wave in a frequency-modulation system when a sinusoidal modulating wave is used.

modulation index The ratio of the frequency deviation to the frequency of the modulating wave in a frequency-modulation system when a sinusoidal modulating wave is used.

modulation meter An instrument that measures the modulation factor of a modulated wave train at a transmitter. The readings are usually expressed in percent.

modulation noise Noise that is caused by the modulating signal, making the noise level a function of the strength of the signal. Also called noise behind the signal.

modulation percentage The percentage value obtained by multiplying the modulation factor by 100.

modulator A transmitter circuit or device that varies the amplitude, frequency, phase, or other characteristic of a carrier signal in accordance with the waveform of a modulating signal which contains useful information. The carrier can also be a direct current, pulse train, light beam, laser beam, or other transmission medium.

module 1. A packaged assembly built in some multiple of a standardized size, used in combination with other modules and components to form a complete electronic system. 2. One of the self-contained building-block segments of a computer or other piece of equipment, not necessarily related in size, such as the command, service, and lunar excursion modules of the Apollo lunar spacecraft.

modulo *n* check A computer check system in which each number to be operated on is associated with a check number. This check number is equal to the remainder obtained when the number operated on is divided by *n*. Thus, in a modulo 4 check, the remainder serving as the check number will be 0, 1, 2, or 3.

mΩ Abbreviation for *milliohm.*

MΩ Abbreviation for *megohm.*

moire A spurious pattern in a reproduced television picture, resulting from interference beats between two sets of periodic structures in the image. Moires may be produced, for example, by interference between patterns in the original subject and that of the target grid in an image orthicon, or between subject patterns and the line and phosphor-dot patterns of a color picture tube.

mol Abbreviation for *mole.*

MOL Abbreviation for *manned orbiting laboratory.*

mold The metal part derived from the master by electroforming in disk recording. A mold has grooves similar to those of a recording.

molded capacitor A capacitor that has been encased in a molded plastic insulating material to keep out dust and moisture.

mole [abbreviated mol] The SI unit for amount of substance. One mole is that amount of a substance which contains as many elementary entities as there are atoms in 0.012 kg of carbon 12. When the mole is used, the type of particle (atom, molecule, ion, electron, and so forth) or group of particles must be specified.

molecular beam A unidirectional beam of neutral molecules. Measurement of the transmission of the beam through appropriate magnetic and electric fields in a vacuum can yield accurate values for such quantities as nuclear magnetic moments.

molecular-beam frequency standard A frequency standard in which millimeter-wave resonance lines of molecules determine the output frequency.

molecular circuit A circuit in which the individual components are physically indistinguishable from each other.

molecular electronics The branch of electronics that deals with the production of complex electronic circuits in microminiature form by producing semiconductor devices and circuit elements integrally while growing multizoned crystals in a furnace.

molecular gas laser A gas laser that can operate continuously at output powers in the kilowatt range, with a coherent output at practically any desired frequency in the entire gas laser spectrum from 0.15 to 773.5 μm. Laser action results from molecular transitions. Pulsed operation is also feasible. Pumping can be chemical, electric, or optical. The carbon dioxide (CO_2) laser is the most common and most important. Molecular hydrogen lasers operate in the vacuum ultraviolet region, near 0.16 μm.

molecular microwave amplifier A solid-state amplifier in which the interaction is between uncharged molecular matter and the microwave field. Examples include the maser and the parametric amplifier.

molecular pump A vacuum pump in which the molecules of the gas to be exhausted are carried away by the friction between them and a rapidly revolving disk or drum.

molecular stopping power The energy loss per molecule of a compound, per unit area normal to the motion of the particle, when a charged particle passes through a medium. It is very nearly if not exactly equal to the sum of the atomic stopping powers of the constituent atoms.

molecule The smallest particle into which an element or compound may be divided and still retain the chemical properties of the element or compound in mass.

mole fraction The number of atoms of a certain isotope of an element, expressed as a fraction of the total number of atoms of that element, which are present in an isotopic mixture.

molybdenum [symbol Mo] A metallic element, sometimes used for electrodes of electron tubes. Atomic number is 42.

molybdenum permalloy A high-permeability alloy that consists chiefly of nickel, molybdenum, and iron. A typical formulation has 4% molybdenum, 79% nickel, and 17% iron.

momentary-contact switch A switch that returns from the operated condition to its normal circuit condition when the actuating force is removed.

monatomic layer A coating that consists of a single layer of atoms.

monaural *Monophonic.*

monitor 1. An instrument that measures continuously or at intervals a condition which must be kept within prescribed limits, such as radioactivity at some point in a nuclear reactor, the image picked up by a television camera, the sound picked up by a microphone at a radio or television studio, a variable quantity in an automatic process control system, the transmissions in a communication channel or band, or the position of an aircraft in flight. 2. A person who watches a monitor.

monitor head An additional playback head on some tape recorders, to permit playing back the recorded sounds off the tape while the recording is being made.

monitoring Using a monitor.

monitoring amplifier A power amplifier used primarily for evaluation and supervision of a program.

monitoring antenna An antenna that picks up the RF output of a transmitter at the transmitter site for overall monitoring purposes.

monitor ionization chamber 1. An ionization chamber mounted in an x-ray beam and connected to a continuously reading instrument, to serve as an indicator of constancy of x-ray output. 2. An ionization chamber that detects the presence of undesirable radiation in connection with health protection.

monkey chatter *Adjacent-channel interference.*

mono *Monostable multivibrator.*

monochromatic Having only one color, corresponding to a negligibly small region of the spectrum.

monochromatic radiation Electromagnetic radiation that has a single wavelength, or photons that all have the same energy. Although no actual radiation is strictly monochromatic, it can have an extremely narrow band of wavelengths, as does sodium light and the spectrum of mercury 198.

monochromatic sensitivity The response of a device to light of a given color.

monochromator An instrument that isolates a narrow portion of the spectrum for analysis, transmission, or other purposes, as in a spectrometer.

monochrome Having only one chromaticity. This is achromatic in black and white television, involving only shades of gray between black and white.

monochrome bandwidth The video bandwidth of the monochrome channel or the monochrome signal in color television.

monochrome channel Any path intended to carry the monochrome signal in a color television system. This path may also carry other signals, such as the chrominance signal.

monochrome signal 1. A signal wave that controls luminance values in monochrome television. 2. The portion of a signal wave that has major control of the luminance values in a color television system, regardless of whether the picture is displayed in color or in monochrome.

monochrome television Television in which the final reproduced picture is monochrome, having only shades of gray between black and white. Also called black and white television.

monochrome transmission Transmission of a signal wave that controls the luminance values in a television picture but not the chromaticity values. The result is a monochrome picture.

monoenergetic radiation Radiation that consists of particles of a given type, all of which have the same energy.

monofier A complete master oscillator–power amplifier system in a single evacuated-tube envelope. Electrically it is equivalent to a stable low-noise oscillator, an isolator, and a two- or three-cavity klystron amplifier. The oscillator can be tuned electronically and manually.

monolithic Constructed from a single crystal or other single piece of material.

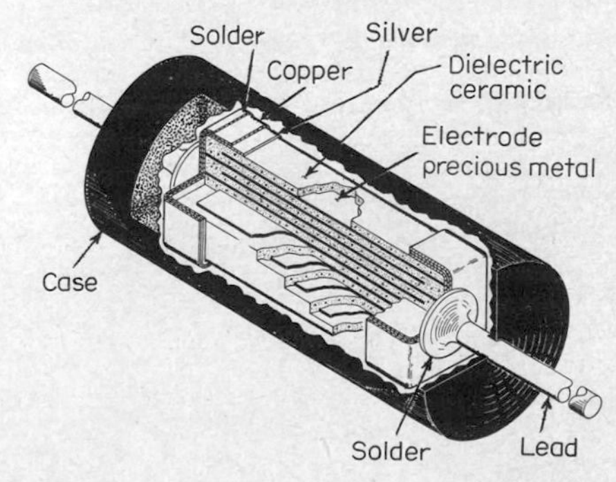

Monolithic ceramic capacitor mounted in tubular case, with axial leads.

monolithic ceramic capacitor A capacitor that consists of thin dielectric layers interleaved with staggered metal-film electrodes. After leads are connected to alternate projecting ends of the electrodes, the assembly is compressed and sintered to form a solid monolithic block.

monolithic circuit *Integrated circuit.*

monolithic crystal filter A filter in which a single quartz plate provides the functions of multiple resonators.

monolithic integrated circuit *Integrated circuit.*

monophonic Pertaining to sound that is transmitted, recorded, or heard over a single path. Binaural and stereophonic pertain to sounds arriving over two paths, to give the effect of auditory perspective. Also called monaural.

monophonic recorded tape Magnetic tape that has a single recording track, for use in a monophonic sound system.

monophonic recorder A magnetic-tape recorder that has a single-channel audio system which consists of a microphone, amplifier, and one recording head, for producing a single recorded track on magnetic tape.

monophonic sound system A sound-reproducing system in which one or more microphones feed a single channel that terminates in one loudspeaker system.

monopole An antenna mounted on an imaging ground plane, to produce a radiation pattern approximating that of a dipole.

monopulse radar Radar in which directional information is obtained with high precision by using a receiving antenna system that has two or more partially overlapping lobes in the radiation patterns. Sum and difference channels in the receiver compare the amplitudes of the antenna outputs (sum-and-difference monopulse radar) or compare the phases of the antenna outputs (phase-sensing monopulse radar).

monoscope A signal-generating electron-beam tube in which a picture signal is produced by scanning an electrode that has a predetermined pattern of secondary-emission response over its surface. This fixed image is printed on the electrode during manufacture of the tube, to give a useful test pattern for testing and adjusting television equipment.

monostable Having only one stable state.

monostable circuit A circuit that has only one stable condition, to which it returns in a predetermined time interval after being triggered.

monostable multivibrator A multivibrator with one stable state and one unstable state. A trigger signal is required to drive the unit into the unstable state, where it remains for a predetermined time before returning to the stable state. Also called mono and one-shot multivibrator.

monostable timer A monostable multivibrator that has controls for adjusting the width or time duration of the output pulse.

monostatic radar Conventional radar, in which the transmitter and receiver are at the same location and share the same antenna. In contrast, in bistatic radar the receiver and transmitter are some distance apart and use different antennas.

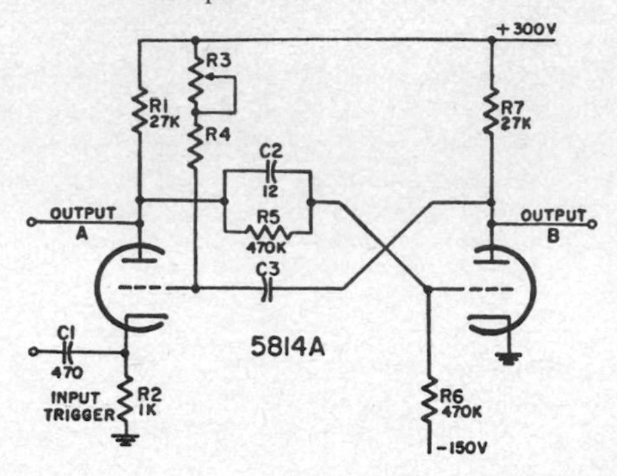

Monostable multivibrator, using circuit design preferred by National Bureau of Standards, gives positive or negative gate output pulse having 210-V peak when triggered by 25-V positive input pulse. Sum of R3 and R4 must be between 0.5 and 3 MΩ; when sum is 2 MΩ, output pulse duration in microseconds is equal to value in picofarads used for C3.

monostatic reflector A reflector that reflects energy only along the line of the incident ray. A corner reflector is an example.

monotonicity In an analog-to-digital converter, the condition wherein there is an increasing output for every increasing value of input voltage over the full operating range. The derivative of the output with respect to the input is therefore always positive.

Monte Carlo method A method of solving a group of physical problems by making a series of statistical experiments. Accuracy of results depends on the number of trials. Usually practical only when set up to be performed by a computer.

moon A natural celestial body that orbits as a satellite about the earth. Mean diameter is 2160 mi (3476 km), and mean distance from the earth is 238,857 mi (384,403 km). Surface gravity is about one-sixth of that on the earth.

moon satellite *Lunar satellite.*

Moore code An equal-length message-transmitting code in which 70 possible characters have four marking intervals and four spacing intervals. If a received character has other than eight intervals, an error signal is printed, or repetition is automatically requested. A 7-unit version that has 35 characters uses three marking and four spacing intervals.

MOPA Abbreviation for *master oscillator–power amplifier.*

Morse code A message-transmitting code that consists of dot and dash signals. The international Morse code is universally used for radiotelegraphy. The American Morse code is used only for wire telegraphy.

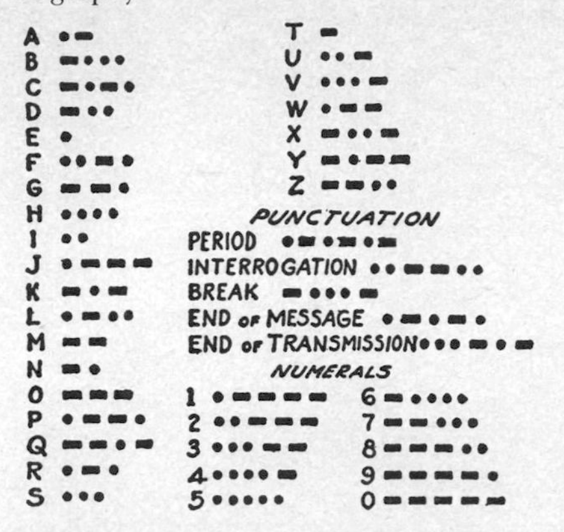

Morse code (international) used in radiotelegraphy.

Morse-code generator An automatic code generator that is operated by a typewriter keyboard. Depression of a key causes a corresponding international Morse-code character to be generated, followed by a letter space.

mortar-locating radar Radar combined with a computer to monitor the trajectory of an artillery or mortar shell or rocket and compute the location of the source of enemy fire.

MOS Abbreviation for *metal-oxide semiconductor.*

mosaic A light-sensitive surface used in television camera tubes; it consists of a thin mica sheet coated on one side with a large number of tiny photosensitive silver-cesium globules, insulated from each other. The picture is optically focused on the mosaic, and the resulting charges on the globules are scanned by the electron beam of the camera tube.

MOS capacitor Abbreviation for *metal-oxide semiconductor capacitor.*

MOS EAROM An electrically alterable read-only memory that uses metal-oxide semiconductor technology.

MOSFET Abbreviation for *metal-oxide semiconductor field-effect transistor.*

MOS LSI Large-scale integration that uses metal-oxide semiconductor technology.

MOS PROM A programmable read-only memory that uses metal-oxide semiconductor technology.

MOS RAM A random-access memory that uses metal-oxide semiconductor technology.

MOS resistor Abbreviation for *metal-oxide semiconductor resistor.*

MOS ROM A read-only memory that uses metal-oxide semiconductor technology.

Mossbauer effect The nuclear-resonance scattering effect that involves a combination of a recoil-free gamma-emission process and the inverse recoil-free absorption process. Used to detect very small changes in gamma-ray energy, such as those due to gravitational red shift or nuclear hyperfine splitting.

Mossbauer-effect analyzer An analyzer that uses the Mossbauer effect to provide information on molecular structure, quadrupole splitting of nuclear levels, and other nuclear properties.

Mossbauer-effect velocity sensor A velocity sensor that uses the Mossbauer effect to measure the Doppler shift of gamma rays emitted by a moving radioactive material.

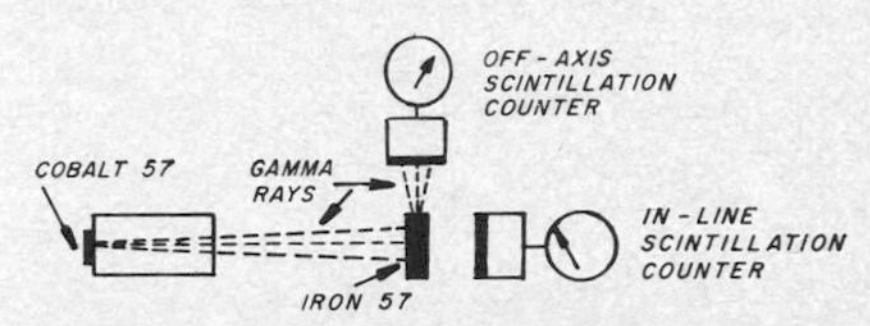

Mossbauer-effect velocity sensor. Gamma rays from cobalt-57 source are directed at iron-57 absorber. When both source and absorber are stationary, nuclear resonance occurs in the iron. Resulting gamma radiation is detected by off-axis scintillation counter. When source is in motion, resonance does not occur, and source gamma rays pass through absorber to in-line counter. To determine velocity of source, absorber is mounted on loudspeaker cone driven by variable-frequency triangular-wave source. When absorber and source are moving at same speed, nuclear resonance again occurs. Measured velocity of cone then corresponds to velocity of cobalt-57 source. Second coil on loudspeaker gives voltage proportional to this velocity.

Mossbauer spectrometer A spectrometer in which the Mossbauer effect is utilized for analysis of materials.

MOST Abbreviation for *metal-oxide semiconductor transistor.*

MOS transistor Abbreviation for *metal-oxide semiconductor transistor.*

most significant bit [abbreviated MSB] The bit that carries the highest value or weight in binary notation for a numeral.

mother A mold derived by electroforming from a master. Used to produce the stampers from which disk records are molded in large quantities. Also called metal positive.

mother aircraft An aircraft that carries the electronic equipment needed to direct a drone.

mother crystal A raw piezoelectric crystal as found in nature or grown artificially.

motion-picture pickup Use of a television camera to pick up scenes directly from motion-picture film.

motor A machine that converts electric energy into mechanical energy by utilizing forces exerted

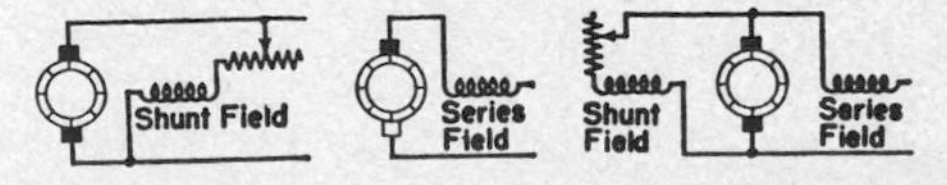

Motor connections, showing three ways in which commutator of DC motor can be connected to field: left—shunt motor; center—series motor; right—compound motor.

by magnetic fields produced by current flow through conductors. Also called electric motor.

motorboating Oscillation in a system or component, usually manifested by a succession of pulses occurring at a very low audio-frequency rate. Generally caused by excessive positive feedback, such as through a common power supply. When a loudspeaker is connected to the system, as in a radio, the pulses produce a put-put sound like that of a motorboat.

motor control *Electronic motor control.*

motor effect The repulsion force exerted between adjacent conductors carrying currents in opposite directions.

motor-generator set A motor and generator that are coupled mechanically for use in changing one power-source voltage to other desired voltages or frequencies.

Mott scattering formula The formula that gives the differential cross section for the scattering of identical particles due to a coulomb force.

mount 1. A shock or vibration isolator, usually consisting of one elastic member and one or more relatively inelastic members, used as a support for the equipment to be isolated. 2. The flange or other means by which a switching tube, or tube and cavity, is connected to a waveguide.

mountain effect The effect of rough terrain on radio-wave propagation, causing reflections that produce errors in radio direction-finder indications.

mouth The end of a horn that has the larger cross-sectional area.

movable contact The relay contact that is mechanically displaced to engage or disengage one or more stationary contacts. Also called armature contact.

movement Deprecated term for *moving element.*

moving-armature loudspeaker *Magnetic-armature loudspeaker.*

moving-coil galvanometer A galvanometer in which the current to be measured is sent through a coil suspended or pivoted in a fixed magnetic field.

moving-coil hydrophone A moving-coil microphone that responds to waterborne sound waves.

moving-coil loudspeaker *Dynamic loudspeaker.*

moving-coil meter A meter in which a pivoted coil is the moving element.

moving-coil microphone *Dynamic microphone.*

moving-coil pickup *Dynamic pickup.*

moving-conductor hydrophone A moving-conductor microphone that responds to waterborne sound waves.

moving-conductor loudspeaker A loudspeaker in which the mechanical forces result from reactions between a steady magnetic field and the magnetic field produced by current flow through a moving conductor.

moving-conductor microphone A microphone whose electric output results from motion of a conductor or coil in a magnetic field. Examples include dynamic or velocity microphones.

moving element The instrument element that moves as a direct result of a variation in the quantity which the instrument is measuring. Also called movement (deprecated).

moving-iron instrument An instrument that depends on current in one or more fixed coils acting

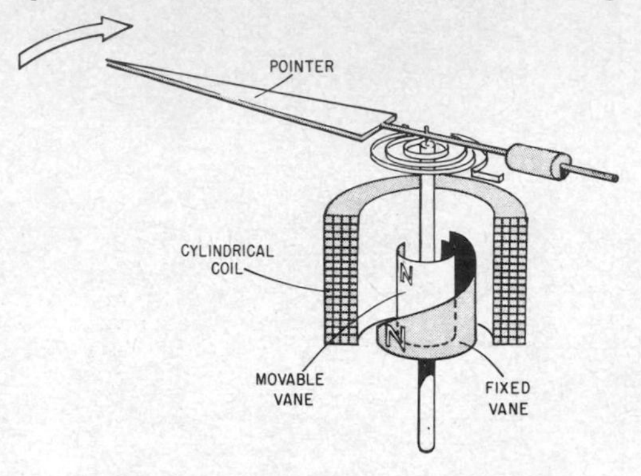

Moving-iron instrument in which fixed and rotatable soft iron vanes are similarly magnetized by coil that carries current being measured. Like poles then repel, causing meter pointer to move against force of torsion spring. Meter will measure either alternating or direct current.

on one or more pieces of soft iron, at least one of which is movable.

moving load A section of waveguide with a plunger that controls the position of a sliding, tapered, low-reflection load, to adjust the phase of residual reflection from the load.

moving-magnet instrument An instrument that depends on the action of a movable permanent magnet in aligning itself in the resultant field produced either by a fixed permanent magnet and an adjacent coil or coils carrying current or by two or more current-carrying coils whose axes are displaced by a fixed angle.

moving-magnet magnetometer A magnetometer that depends for its operation on the torques acting on a system of one or more permanent magnets which can turn in the field to be measured.

moving-target indicator [abbreviated MTI] A device that limits the display of radar information primarily to moving targets. Signals due to reflections from stationary objects are canceled by a memory circuit.

MPU Abbreviation for *microprocessing unit.*

mR Abbreviation for *milliroentgen.*

mrad Abbreviation for *milliradian.*

mrd Abbreviation for *millirad.*

mrem Abbreviation for *millirem.*

mrep Abbreviation for *millirep.*

ms Abbreviation for *millisecond.*

mS Abbreviation for *millisiemens.*

MS Designation for Military Standard.

MSB Abbreviation for *most significant bit.*
M scope A radarscope that produces an M display.
M shell The third layer of electrons about the nucleus of an atom; the layer has electrons characterized by the principal quantum number 3.
MSI Abbreviation for *medium-scale integration.*
MTBF Abbreviation for *mean time between failures.*
MTI Abbreviation for *moving-target indicator.*
MTL Abbreviation for *merged-transistor logic.*
M-type backward-wave oscillator An ordinary or noncrossed-field backward-wave oscillator.

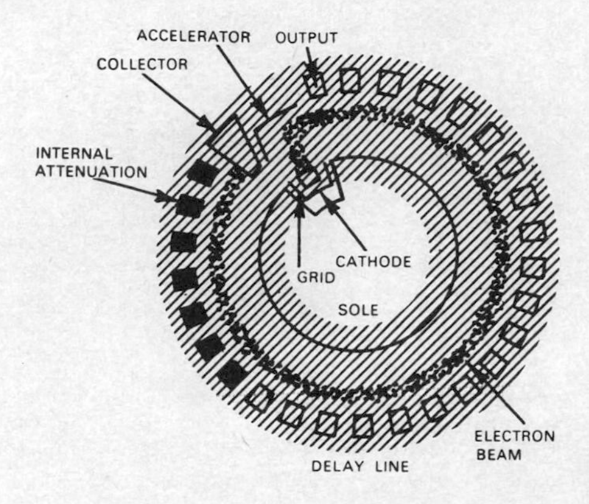

M-type backward-wave oscillator.

μ [Greek letter mu] 1. Symbol for *amplification factor.* 2. Abbreviation for *micro-.*
μA Abbreviation for *microampere.*
μbar Abbreviation for *microbar.*
μC Abbreviation for *microcomputer.*
μCi Abbreviation for *microcurie.*
Mueller bridge A bridge that measures with high precision the resistance values of three- and four-terminal resistors, particularly resistance thermometers.
μF Abbreviation for *microfarad.*
MUF Abbreviation for *maximum usable frequency.*
μ-factor The ratio of the magnitude of an infinitesimal change in the voltage at the jth electrode of an n-terminal electron tube to the magnitude of an infinitesimal change in the voltage at the lth electrode, with the current to the mth electrode remaining unchanged.
μH Abbreviation for *microhenry.*
μHz Abbreviation for *microhertz.*
multianode tube An electron tube that has two or more main anodes and a single cathode. This term is used chiefly for pool-cathode tubes.
multiaperture device A magnetic core that has two or more openings through which windings are threaded. In one arrangement, used as a variable-gain adaptive component, flux can be switched around the small apertures without disturbing the flux around the main aperture.

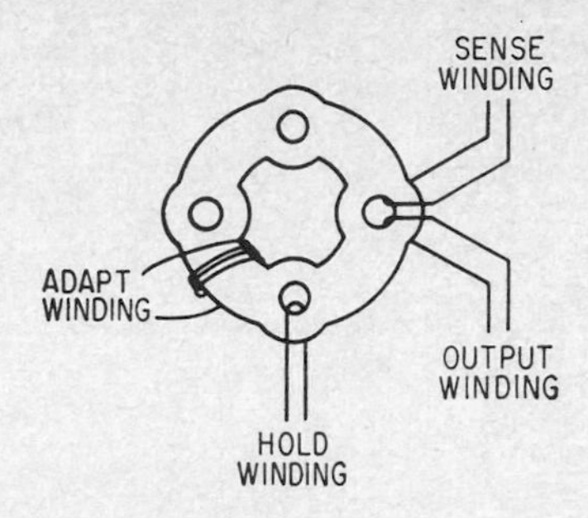

Multiaperture device.

multiband antenna An antenna that may be used satisfactorily on more than one frequency band.
multibeam cathode-ray tube A cathode-ray tube that has a number (typically seven) of electron beams that can be varied in intensity individually as well as simultaneously while being acted on by common focus and deflection coils. Applications include alphameric and graphic displays.

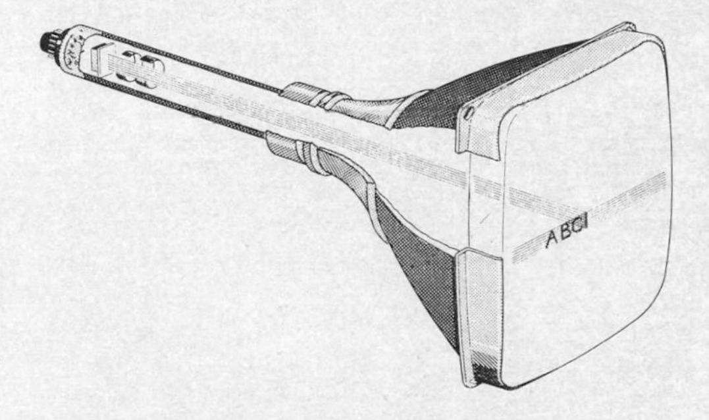

Multibeam cathode-ray tube having seven beams for generating alphameric characters while being swept horizontally across screen.

multibeam oscilloscope A cathode-ray oscilloscope that has two or more electron beams which may be controlled individually or jointly in intensity and position.

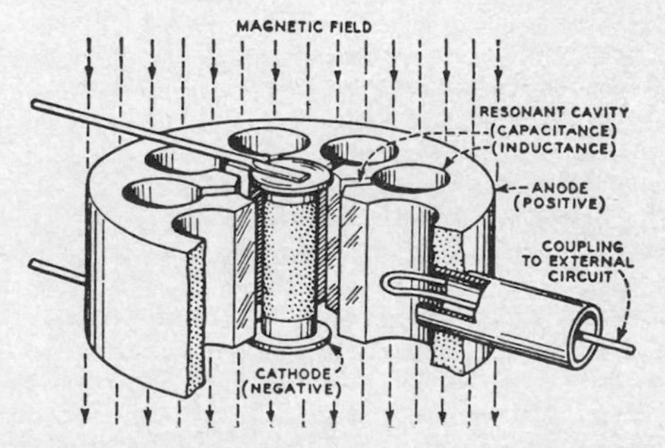

Multicavity magnetron, showing central cathode, coupling loop, and direction of magnetic field.

multicavity magnetron A magnetron in which the circuit includes a plurality of cavities, generally cut into the solid cylindrical anode in such a way that the mouths of the cavities face the central cathode.

multicellular horn 1. A cluster of horn antennas that have mouths which lie in a common surface. The horns are fed from openings spaced one wavelength apart in one face of a common waveguide, to provide a desired directional radiation pattern for the radiated energy. 2. A combination of individual horn loudspeakers that have individual driver units or are joined in groups to a common driver unit. Subdivision of a large horn into smaller horns makes it easier to control the pressure and phase of the acoustic waves across the mouths of the horns. Also called cellular horn.

multichannel analyzer *Pulse-height analyzer.*

multichannel field-effect transistor A field-effect transistor in which appropriate voltages are applied to the gate to control the space within the current flow channels. The use of more than one channel permits handling higher currents without incurring the reduction in frequency response that normally takes place when the size of a single-channel device is increased to handle higher current.

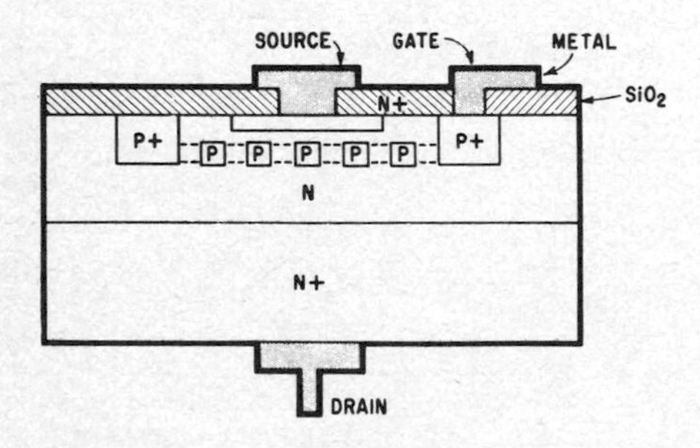

Multichannel field-effect transistor.

multichannel radio transmitter A radio transmitter that has two or more complete RF portions capable of operating on different frequencies, either individually or simultaneously.

multichip integrated circuit An integrated circuit in which two or more semiconductor chips are mounted on an insulating substrate that has terminals and interconnections.

multicolor cathode-ray tube A cathode-ray tube that provides four or more colors from a single gun, with no shadow mask, by using two different colors of phosphor layers on the screen, one activated by a low (6-kV) electron-beam voltage and the other by a high (12-kV) electron-beam voltage. Intermediate beam voltages give various mixtures of the two basic phosphor colors, making possible alphameric or graphic displays in four or more distinct colors. The phosphor color pairs are chosen from red, green, blue, and white. Also called penetration tube.

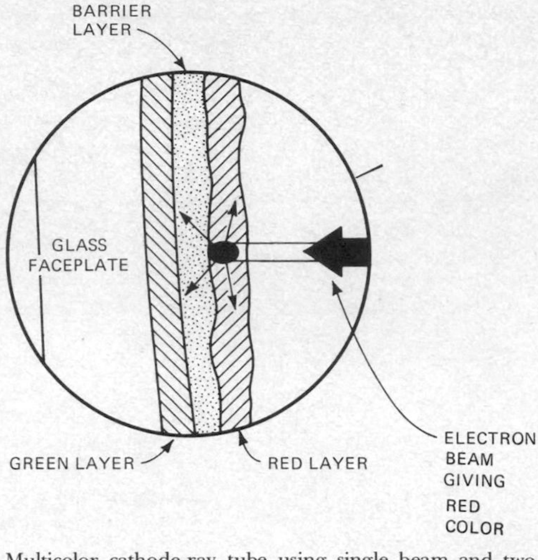

Multicolor cathode-ray tube using single beam and two phosphor layers.

multicoupler A device that connects several receivers to one antenna and properly matches the impedances of the receivers to the antenna.

multicurrent-range diode A diode that has a nearly constant coefficient over a wide range of operating terms. A silicon planar epitaxial diode is an example.

multielectrode tube An electron tube that contains more than three electrodes associated with a single electron stream.

multiemitter transistor A transistor that has one or more extra emitters. Used chiefly in logic circuits.

multifrequency transmitter A radio transmitter capable of operating on two or more selectable frequencies, one at a time, using preset adjustments.

multigrid tube An electron tube that has more than one grid electrode.

multigun tube A cathode-ray tube that has more than one electron gun.

multihole coupler A directional coupler in which two rows of graduated-diameter holes are spaced a quarter-wavelength apart on the broad waveguide faces of the coupler. Power flowing in one direction through the primary waveguide couples through the holes and excites waves in only one direction in the secondary waveguide. Because of the spacing between the holes, waves traveling in the reverse direction are out of phase and cancel.

multihop transmission Transmission in which radio waves are reflected and refracted between the earth and ionosphere several times in their path of travel to a receiver far beyond the direct transmission range. Also called multiple-hop transmission.

multilayer dielectric reflector A sequence of thin layers of transparent material of controlled thicknesses and indexes of refraction. High reflectivity is established at certain wavelengths by constructive interference of the Fresnel reflections from the dielectric interfaces. Such reflectors are often used to establish the optical cavity of a laser.

multimeter 1. A meter that has a number of measuring functions, usually with several scale ranges for each function. 2. *Volt-ohm-milliammeter.*

multimode radar Radar that has two or more functions such as simultaneous ground mapping; terrain-following maneuvers during air strikes; and identification, tracking, and rangefinding of fixed or moving targets.

multimode waveguide A waveguide that propagates more than one mode at its operating frequency.

multipactor A high-power, high-speed microwave switching device in which a thin electron cloud is driven back and forth between two parallel plane surfaces in a vacuum by an RF electric field. Applications include switching of signals in waveguides, high-speed switching of pulses in other microwave systems, and protecting front ends of radar receivers from overload by transmitter pulses.

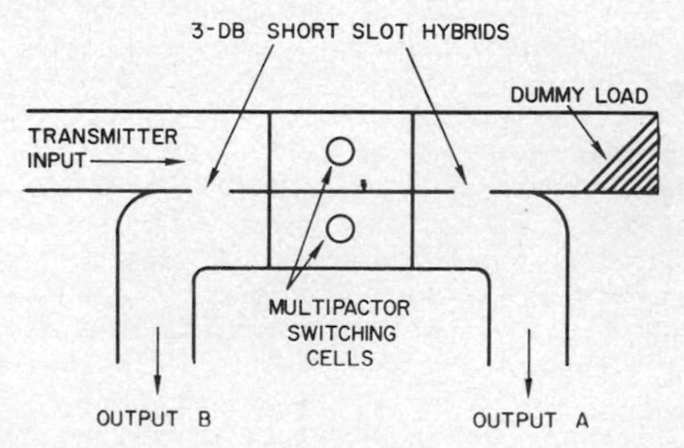

Multipactors used to switch signal in one waveguide to either of two other waveguides.

multipactor rectifier A microwave rectifier that converts microwave energy directly into DC energy by a multiple electron impact process. A thin electron cloud is driven back and forth across a gap by the applied RF field, to give repeated buildup of the discharge current.

multipass sort A sort program that sorts more data than can be contained within the internal storage of a computer. Intermediate storage, such as disk, tape, or drum, is required.

multipath *Multipath transmission.*

multipath cancellation Effectively complete cancellation of radio signals because of the relative amplitude and phase differences of components arriving over different paths.

multipath reception Reception in which the transmitter signals arrive at a receiving antenna over two or more paths, one direct and the others involving reflections from buildings or other obstacles. One result is ghosts in television pictures.

multipath transmission The propagation phenomenon that results in signals reaching a radio receiving antenna by two or more paths, causing distortion in radio and ghost images in television. At least one of the paths involves reflection from some object. Also called multipath.

multiple *Parallel.*

multiple-address code A computer instruction code in which more than one address or storage location is specified. The instruction may give the locations of the operands, the destination of the result, and the location of the next instruction.

multiple-beam klystron A klystron that has a number of electron guns, each of which produces its own beam which interacts with its own sequence of RF gaps. All beams are ultimately dissipated in a common collector. Each beam section thus acts as a conventional klystron, with the beams contributing to the power output in a rigidly phase-locked manner that cannot be achieved readily with separate klystrons.

multiple-break contacts Contacts so arranged that a circuit is interrupted in two or more places when the contacts open.

multiple circuit A circuit in which two or more identical or different circuits are connected in parallel to a common signal or power source.

multiple-contact switch *Selector switch.*

multiple course One of a family of lines of position defined by a navigation system, any one of which may be selected as a course line.

multiple decay *Branching.*

multiple disintegration *Branching.*

multiple-hop transmission *Multihop transmission.*

multiple modulation Modulation in which the modulated wave from one process becomes the modulating wave for the next. Thus, in an amplitude-modulated pulse-position-modulation system, one or more signals are used to position-modulate their respective pulse subcarriers, which are spaced in time and are used to amplitude-modulate a carrier.

multiple plug A device that is inserted into a single receptacle or wall outlet, to act as more than one receptacle. Also called cube tap.

multiple scattering Scattering in which the final displacement is the vector sum of many small displacements. Multiple scattering is greater than single or plural scattering.

multiple sound track A group of sound tracks printed adjacently on a common base, independent in character but having a common time relationship, such as for stereophonic sound recording.

multiple-speed floating action Floating action in which a final control element is moved at two or more speeds, each corresponding to a definite range of values of deviation.

multiple-spot scanning Scanning by two or more scanning spots simultaneously; each spot analyzes its fraction of the total scanned area of the subject copy.

multiple-styli recorder A direct-writing recorder that has a large number of fixed styli positioned in a line at right angles to the direction of movement of electrosensitive paper. Often used to provide readout in the form of printed alphameric characters.

multiple-track range A range system that uses two closely spaced synchronized pulse stations, similar to Gee. The indicator in the aircraft has several predetermined time difference settings by which a number of approximately parallel tracks may be flown.

multiple-trip echo An echo returned from a target so distant that the time required for the radar pulse to go out to the target and return to the set is longer than the interval between two successive pulses. These echoes show up at false ranges, identifiable because they change when the pulse repetition rate is changed. Called a second-trip echo when the echo arrives between the second and third pulses.

multiple tube counts Spurious counts induced by previous tube counts in a radiation-counter tube.

multiple-tuned antenna A low-frequency antenna that has a horizontal section connected to a multiplicity of tuned vertical sections.

multiple-unit tube An electron tube that contains within one envelope two or more groups of electrodes associated with independent electron streams, such as a duodiode, duotriode, diode-pentode, duodiode-triode, duodiode-pentode, and triode-pentode.

multiplex The simultaneous transmission of two or more programs or signals over a single RF channel, such as by time division, frequency division, or phase division. Also called multiplexing and multiplex transmission.

multiplex channel A carrier channel that provides two or more services simultaneously in the same or opposite directions.

multiplexer A device for combining two or more signals, as for multiplex, or for creating the composite color video signal from its components in color television. In a data-communication system, it is used to combine two or more data channels, such as by placing them side by side in frequency in frequency-division multiplexing, or by placing them in time slots of time-division multiplexing.

multiplexing *Multiplex.*

multiplex operation Simultaneous transmission of two or more messages in either or both directions over a multiplex channel.

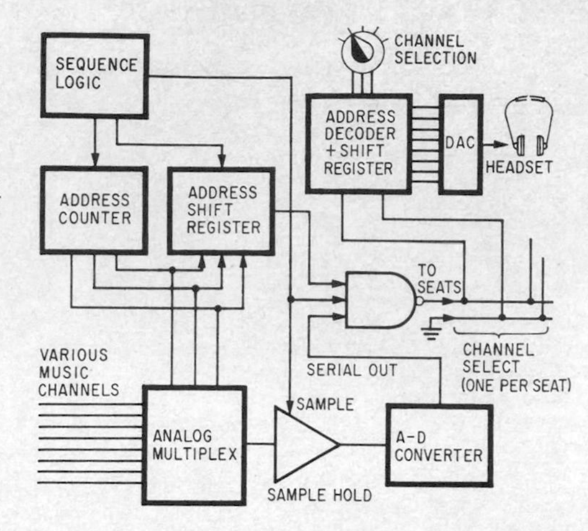

Multiplex as used for distributing eight different music channels to each seat of commercial aircraft over single pair of wires. A/D converter sends out serial words corresponding to samples of channel, in sequence. A 3-bit address code is added to 8 bits of analog information. At each seat an address decoder is linked with channel selection switch, with D/A converter operating only for digital words corresponding to selected channel.

multiplex radio transmission The simultaneous transmission of two or more signals by radio, using a common carrier wave.

multiplex transmission *Multiplex.*

multiplication 1. The ratio of neutron flux in a subcritical reactor to that supplied by a neutron source. It is the factor by which, in effect, the reactor multiplies the source strength. 2. An increase in current flow through a semiconductor due to increased carrier activity.

multiplication point A mixing point whose output is obtained by multiplication of its inputs in a feedback control system.

multiplier 1. A device that has two or more inputs and an output which is a representation of the product of the quantities represented by the input signals. Voltages are the quantities commonly multiplied. 2. A resistor used in series with a voltmeter to increase the voltage range. Scale readings are then multiplied by a factor equal to the ratio of total resistance (meter resistance plus multiplier resistor value) to meter resistance. Also called multiplier resistor. 3. *Electron multiplier.* 4. *Frequency multiplier.*

multiplier phototube A phototube with one or more dynodes between its photocathode and the output electrode. The electron stream from the photocathode is reflected off each dynode in turn, with secondary emission adding electrons to the stream at each reflection. Also called electron-multiplier phototube and photomultiplier tube.

multiplier resistor *Multiplier.*

multiplier traveling-wave photodiode A photodiode in which the construction of a traveling-wave tube is combined with that of a multiplier

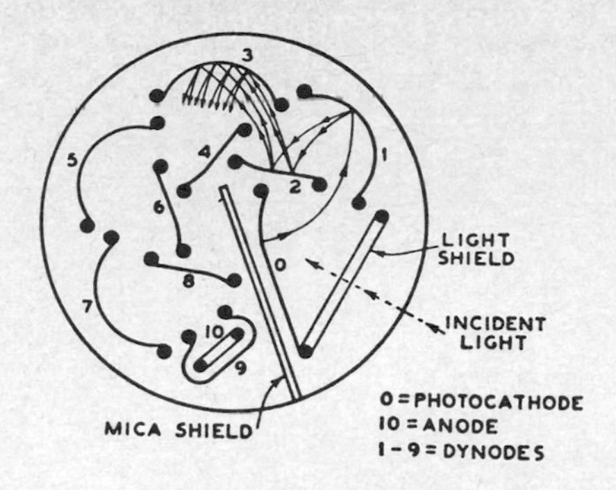

Multiplier-phototube operation.

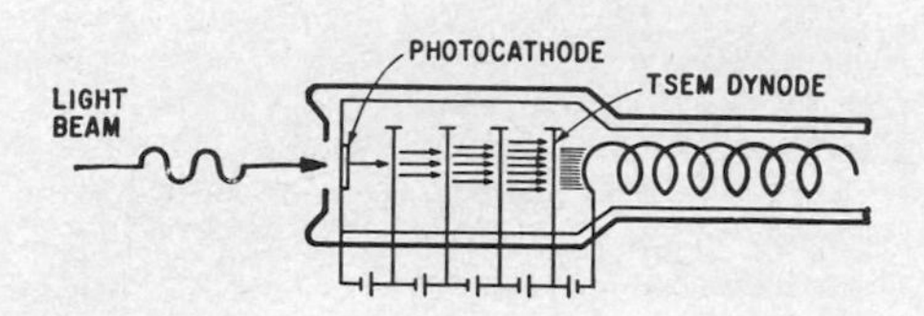

Multiplier traveling-wave photodiode, showing use of transmission-type photocathode and transmission secondary-emission dynodes.

phototube to give increased sensitivity.

multiplying factor The number by which the reading of a given meter must be multiplied to obtain the true value.

multipolar Having more than one pair of magnetic poles.

multipole moment A measure of the charge, current, and magnet distributions of a system of particles.

multiport network A network that has more than one port.

multiposition action Automatic control action in which a final control element is moved to one of three or more predetermined positions, each corresponding to a definite value or range of values of the controlled variable.

multiprocessing The simultaneous or interleaved execution of two or more programs by a computer.

multiprogramming Programming that allows two or more arithmetic or logic operations to be performed by a computer either simultaneously or on a time-sharing basis.

multisegment magnetron A magnetron with an anode divided into more than two segments, usually by slots parallel to its axis.

multitrace oscilloscope An oscilloscope in which a single beam in a cathode-ray tube is time-shared by two or more input signals.

multitrack recording system A recording system that provides two or more recording paths on a medium, carrying related or unrelated recordings that have a common time relationship.

multiturn potentiometer A precision wire-wound potentiometer in which the resistance element is formed into a helix, generally having from 2 to 10 turns. The actuating shaft must be rotated

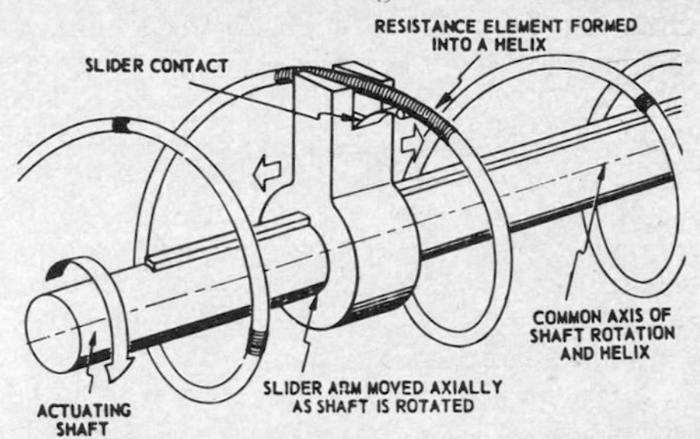

Multiturn potentiometer.

a corresponding number of turns to cover the full resistance range. The contact arm slides along the shaft as its contact moves along the helix.

multivibrator [abbreviated MVBR] A relaxation oscillator circuit that uses two electron tubes, transistors, or other active elements coupled so that one is cut off when the other is conducting. With an astable multivibrator, which is free-running,

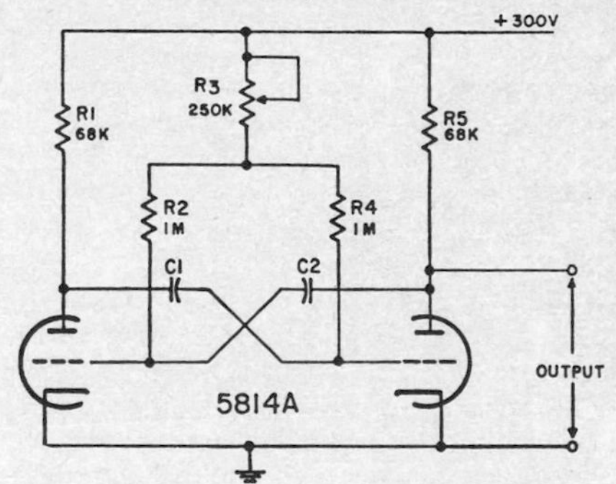

Multivibrator, using circuit design preferred by National Bureau of Standards, gives square-wave 260-V output waveform having pulse repetition frequency *f* determined by values used for capacitors; *f* in hertz is 790,000 divided by the value of Cl or C2 in picofarads.

the frequency of spontaneous transition between these two states is determined by the time constants of the coupling elements and sometimes by an external voltage. With a monostable multivibrator, a trigger signal drives the circuit into its unstable state, and the circuit constants determine the time for returning to the stable state. The third basic type of multivibrator is the flip-flop or bistable multivibrator, in which an external trigger signal is required for each transition.

μlx Abbreviation for *microlux*.

μm Abbreviation for *micrometer*.

mu meson A meson that has a mean life of 2.1 μs

and a mass 210 times that of an electron. Also called muon.

μmho Abbreviation for *micromho.*

Munsell chroma *Chroma.*

Munsell value The dimension, in the Munsell system of object-color specification, that indicates the apparent luminous transmittance or reflectance of the object on a scale which has approximately equal perceptual steps under the usual conditions of observation.

μΩ Abbreviation for *microhm.*

muon *Mu meson.*

muonium A bound system that consists of a positive mu meson and an electron.

μP Abbreviation for *microprocessor.*

μR Abbreviation for *microroentgen.*

μrd Abbreviation for *microrad.*

μrem Abbreviation for *microrem.*

μrep Abbreviation for *microrep.*

Murray loop test A method of localizing a fault in a cable by replacing two arms of a Wheatstone bridge with a loop formed by the cable under test and a good cable connected to the far end of the defective cable.

μs Abbreviation for *microsecond.*

μS Abbreviation for *microsiemens.*

MUSA [Multiple-Unit Steerable Antenna] An electrically steerable receiving antenna whose directional pattern can be rotated by varying the phases of the contributions of the individual units. These units are usually stationary rhombic antennas.

musical echo A flutter echo that is periodic and has a flutter whose frequency is in the audio range.

music-controlled lamp A lamp whose brightness is made to vary with the average loudness of the output of an audio system.

music power An output capability rating for high-fidelity audio amplifiers; it takes into account such factors as transient response, power supply regulation, and the RMS power capability. The music power rating may be as much as 20% higher than the RMS power rating of an amplifier. Also called IHF power because the measurement procedure has been standardized by the Institute of High Fidelity (IHF).

muting Silencing, or reducing in volume.

muting switch 1. A switch used in connection with automatic tuning systems to silence the receiver while tuning from one station to another. 2. A switch that grounds the output of a phonograph pickup automatically while a changer is in its change cycle.

mutual conductance *Transconductance.*

mutual-conductance meter *Transconductance meter.*

mutual inductance [symbol *M*] A measure of the amount of inductive coupling that exists between two coils. It is related to the flux linkages produced in one coil by current in the other coil. Measured in henrys, millihenrys, and microhenrys, the same as for inductance.

mutual interference Interference that affects two or more pieces of equipment because of interactions such as those caused by harmonics and other spurious emissions of transmitters and spurious responses and images of receivers.

μV Abbreviation for *microvolt.*

μW Abbreviation for *microwatt.*

mV Abbreviation for *millivolt.*

MV Abbreviation for *megavolt.*

MVA Abbreviation for *megavoltampere.*

MVBR Abbreviation for *multivibrator.*

mW Abbreviation for *milliwatt.*

MW Abbreviation for *megawatt.*

MWh Abbreviation for *megawatthour.*

Mx Abbreviation for *maxwell.*

Mylar Trademark of Du Pont for their polyester film that is widely used for insulation and as a backing for magnetic tape.

Mylar capacitor A capacitor that uses Mylar film as a dielectric between rolled strips of metal foil.

myriametric wave A radio wave between the wavelength limits of 10 and 100 km, corresponding to the very low-frequency (VLF) range of 3 to 30 kHz. (The FCC designated all frequencies below 30 kHz as VLF.)

N

n Abbreviation for *nano-*.

N 1. Abbreviation for *negative*. 2. Abbreviation for *newton*. 3. Symbol for *nitrogen*.

N 1. Symbol for *neutron number*. 2. Symbol for number of turns.

nA Abbreviation for *nanoampere*.

NAB Abbreviation for National Association of Broadcasters, known before January 1, 1958, as NARTB.

NAB curve The standard playback equalization curve adopted by the National Association of Broadcasters for disk recordings.

nail-head bond A bond or connection that involves the use of a flattened end on a wire, to obtain a larger surface area for welding, brazing, or soldering.

Nancy A system of visual blinker-signal communication that uses invisible infrared rays, for ship-to-ship signaling at night.

NAND A logic operator which has the property that if $P, Q, R, \ldots$ are statements, then the NAND of $P, Q, R, \ldots$ is true if and only if at least one statement is false; the NAND of $P, Q, R, \ldots$ is false if and only if all statements are true.

NAND gate An AND gate followed by an inverter, to give an output logic state of 0 when all input signals are 1.

nano- [abbreviated n] A prefix that represents 10^{-9}, which is 0.000000001 or one-thousandth of a millionth.

nanoampere [abbreviated nA] One-thousandth of a microampere, or 10^{-9} A.

nanocircuit A flip-flop or other logic circuit that operates at speeds above 20 MHz, in the nanosecond range.

nanofarad [abbreviated nF] One-thousandth of a microfarad, or 10^{-9} F.

nanohenry [abbreviated nH] One-thousandth of a microhenry, or 10^{-9} H.

nanometer [abbreviated nm] One-thousandth of a micrometer, or 10^{-9} m.

nanosecond [abbreviated ns] One-thousandth of a microsecond, or 10^{-9} s. Light travels approximately 1 ft (0.3 m) in 1 ns.

nanovolt [abbreviated nV] One-thousandth of a microvolt, or 10^{-9} V.

nanovoltmeter A voltmeter that is sufficiently sensitive to measure thousandths of a microvolt.

nanowatt [abbreviated nW] One-thousandth of a microwatt, or 10^{-9} W.

nanowatt circuit A circuit in which the power consumption of individual components is down to a few nanowatts, permitting extremely high packaging densities. Generally achieved by combining thin-film and planar-diffusion technology on a silicon wafer.

narrow-band axis The direction of the phasor that represents the coarse chrominance primary in color television; it has a bandwidth extending from 0 to 0.5 MHz.

narrow-band frequency modulation An FM broadcasting system used primarily for two-way voice communication. For police, fire, taxicab, and other mobile communication, 15 kHz is the maximum permissible deviation. For amateur radio, the deviation limit is 3 kHz.

***N*-ary code** A code that uses *N* distinct types of elements. As an example, with pulse-code modulation there would be *N* distinct pulse height levels.

NASA Abbreviation for *National Aeronautics and Space Administration*.

National Aeronautics and Space Administration [abbreviated NASA] A civilian agency established in 1958 to control aeronautical and space research and exploration activities sponsored by the United States, except those associated with military and defense systems.

National Electrical Code A set of regulations

that governs construction and installation of electric wiring and apparatus in the United States, established by the American National Board of Fire Underwriters for safety purposes.

National Electrical Manufacturers Association [abbreviated NEMA] An organization of manufacturers of electric products.

National Television System Committee [abbreviated NTSC] A committee organized in 1940 by representatives of U.S. companies and organizations interested in television. It formulated television standards for black and white television in 1940–1941 and for color television in 1950–1953 that were approved by the Federal Communications Commission. Countries using NTSC standards include the United States, Canada, Japan, and Mexico.

natural frequency 1. A frequency at which a body or system will oscillate freely. 2. The lowest resonant frequency of an antenna, circuit, or component.

natural logarithm [abbreviated ln] A logarithm that uses the base 2.71828.

natural period The period of a free oscillation of a body or system.

natural radioactivity Radioactivity exhibited by naturally occurring radionuclides.

natural uranium Uranium as found in nature. It is a mixture of the fertile uranium 238 isotope (almost 99.3%), the fissionable uranium 235 isotope (0.711%), and a small percentage of uranium 234. Also called normal uranium.

natural wavelength The wavelength that corresponds to the natural frequency of an antenna or circuit.

nautical mile A measure of distance equal to 1 minute of arc on the earth's surface. The United States has adopted the International Nautical Mile, equal to 1852 m or 6076.11549 ft, which is approximately 1.15 mi.

Navaglide An instrument low-approach system for aircraft that uses a single frequency on a time-sharing basis for both the localizer and glide-path beams and provides a distance indication.

Navaglobe A long-distance continuous-wave low-frequency navigation system of the amplitude-comparison type; it provides bearing information automatically in an aircraft with respect to an omnidirectional radio range.

NAVAR [NAVigation And Ranging] A coordinated series of radar air navigation and traffic-control aids that use transmissions at wavelengths of 10 and 60 cm to provide in an aircraft the distance and bearing from a given point, along with a display of other aircraft in the vicinity and commands from the ground. The system also provides on the ground a display of all aircraft in the vicinity, with their altitudes, identities, and means for transmitting certain commands.

Navarho [NAVigation Aid RHO] A long-distance continuous-wave low-frequency navigation system that provides simultaneous bearing and distance information.

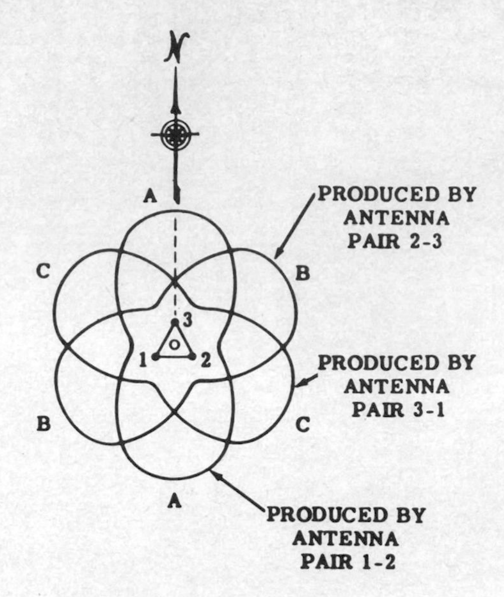

Navarho radiation pattern. Bearing to transmitter is obtained by comparing amplitudes of overlapping radiation patterns at given point, and distance is measured by phase-comparison techniques.

Navascreen A system for displaying and computing air traffic control data, using information obtained from radar and other sources.

navigation The process of directing a vehicle to reach the intended destination.

navigation aid A device or system that provides a navigator with some or all of such navigation data as present position, heading, speed, location of fixed objects and other craft, right-left steering directions or automatic steering control, and altitude. Also called navigation instrument.

navigation beacon A light, radio beacon, or radar beacon that provides navigation aid to aircraft and ships.

navigation computer A computer that uses electronic or electric circuits to compute two or more navigation factors such as altitude, direction, and velocity.

navigation coordinate A quantity whose measurement defines a surface of position that contains the vehicle or a line of position if one surface is already known.

navigation instrument *Navigation aid.*

navigation parameter A visual or aural output of a navigation aid; it has a specific relation to navigation coordinates.

navigation radar A search radar used on ships primarily for navigation purposes, to provide a visual indication of bearing and distance to any object that projects above the surface of the water within the range of the radar.

navigation satellite An earth-orbiting satellite from which radio Doppler shift measurements can be made under all weather conditions, to give the

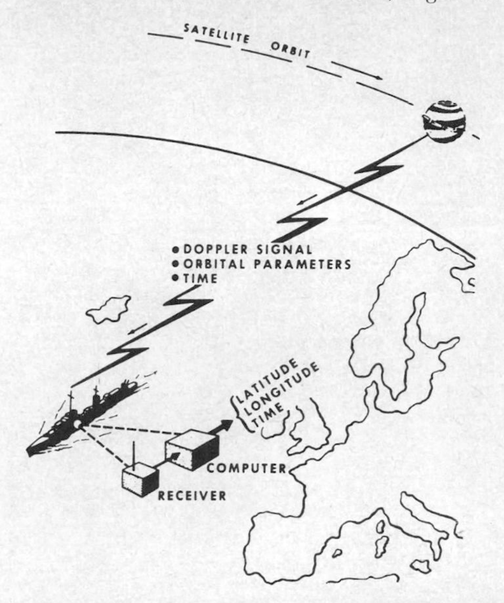

Navigation satellite, as part of Transit navigation system, transmits accurate time signals and position data to receiver and computer on ship. Central ground tracking station (not shown) transmits correction signals to satellite several times a day to maintain high accuracy.

position of a ship or aircraft anywhere on the earth with an accuracy of better than 0.5 nautical mile (0.926 km).

NAVSTAR [NAVigation System using Time And Ranging] A global system of up to 24 navigation satellites developed to provide instantaneous and highly accurate worldwide three-dimensional location by air, sea, and land vehicles equipped with suitable receivers.

NC 1. Symbol for *no connection.* Used on tube-base diagrams. 2. Symbol for *normally closed,* used with reference to relay contacts. 3. Abbreviation for *numerical control.*

N-channel A conduction channel formed by electrons in an N-type semiconductor, as in an N-type field-effect transistor.

N-channel MOS [abbreviated NMOS] A metal-oxide semiconductor process in which selective diffusion of an N-type dopant forms closely spaced source and drain regions within a silicon substrate, with the conducting channel consisting of electrons. In contrast, the channel consists of holes in the P-channel MOS process.

N display A modified radarscope K display in which the target appears as a pair of vertical deflections from the horizontal time base. Direction is indicated by the relative amplitudes of the vertical deflections. Target distance is determined by moving an adjustable pedestal signal along the baseline until it coincides with the horizontal position of the vertical deflections. The pedestal control is calibrated in distance.

NDRO Abbreviation for *nondestructive readout.*

Nd:YAG laser Abbreviation for *neodymium-doped yttrium-aluminum garnet laser.*

near-end crosstalk A type of interference that may occur at carrier telephone repeater stations when output signals of one repeater leak into the same end of the other repeater.

near field 1. The acoustic radiation field that is close to the loudspeaker or other acoustic source. 2. The electromagnetic field that exists in the near region, within a distance of one wavelength from a transmitting antenna.

near infrared That portion of the infrared spectrum that contains the shortest wavelengths, adjacent to the 0.7-μm red end of the visible spectrum.

near region The region immediately surrounding a transmitting antenna, extending out to a distance of one wavelength, in which the strength of the induction field varies inversely with the square of the distance. Also called near zone.

near zone *Near region.*

neck The small tubular part of the envelope of a cathode-ray tube, extending from the funnel to the base and housing the electron gun.

needle *Stylus.*

needle drag *Stylus drag.*

needle force *Stylus force.*

needle pressure Deprecated term for *stylus force.*

needle scratch *Surface noise.*

negative 1. A terminal or electrode that has more electrons than normal. Electrons flow out of the negative terminal of a voltage source. 2. A designation used to describe an opposite character to positive, as in negative feedback, negative image, negative resistance, and negative transmission.

negative bias A grid-bias voltage that makes the control grid of an electron tube negative with respect to its cathode.

negative charge An electric charge in which the object in question has more electrons than protons.

negative conductance A property of certain semiconductors wherein the local current density decreases whenever the local electric field exceeds a threshold level. The effect occurs in silicon transit-time devices, gallium arsenide Gunn devices, and in other semiconductor crystals that have a limited-space-charge mode of oscillation.

negative electricity A negative charge, such as that produced in a resin object by rubbing with wool.

negative electron An electron, as distinguished from a positive electron or positron. Also called negatron.

negative electron affinity A characteristic of

some semiconductor surfaces that permits their use as photoemitters, secondary emitters, and cold-cathode emitters when they are mounted in a vacuum. With this characteristic, an electron that has energy above a minimum level encounters no work function barrier at the surface of the semiconductor and can escape into the vacuum much like other emitter materials.

negative feedback Feedback in which a portion of the output of a circuit, device, or machine is fed back 180° out of phase with the input signal. Negative feedback decreases amplification, to stabilize the amplification with respect to time or frequency and to reduce distortion and noise. Also called degeneration, inverse feedback, and stabilized feedback.

negative-feedback amplifier An amplifier that uses negative feedback to improve stability and/or widen the frequency response.

negative ghost image A ghost image in which white is black and vice versa because of particular phase and amplitude relations between the direct-path and reflection-path signals.

negative glow The luminous glow in a glow-discharge cold-cathode tube; it occurs between the cathode dark space and the Faraday dark space.

negative image An image in which dark areas are bright and bright areas are dark. Also called reversed image.

negative impedance An impedance such that when the current through it increases, the voltage drop across the impedance decreases.

negative input–positive output [abbreviated NIPO] A logic circuit that accepts a negative-going input pulse and delivers a positive-going output pulse.

negative ion An atom that has more electrons than normal and therefore has a negative charge.

negative logic Digital logic in which the more negative logic level represents 0. Use of this logic level term is deprecated because it can be construed differently for NPN devices than for PNP devices. To avoid confusion, the magnitude and polarity of the voltage levels for 1 and 0 in a specific logic circuit should always be specified.

negative-matrix picture tube A color television picture tube in which the phosphor strips or dots are surrounded by a black background, to improve contrast. The electron beam is enlarged to illuminate the entire phosphor area instead of just the center, to increase brightness.

negative modulation 1. Modulation in which an increase in brightness corresponds to a decrease in amplitude-modulated transmitter power. Used in U.S. television transmitters and in some facsimile systems. 2. Modulation in which an increase in brightness corresponds to a decrease in the frequency of a frequency-modulated facsimile transmitter. Also called negative transmission.

negative picture phase The video signal phase in which the signal voltage swings in a negative direction for an increase in brilliance.

negative plate The internal plate structure that is connected to the negative terminal of a storage battery. Electrons flow from the negative terminal through the external load circuit to the positive terminal.

negative proton *Antiproton.*

negative reactance A negative inductance characterized by the fact that its reactance increases with frequency, or a negative capacitance whose reactance decreases as frequency increases. It is therefore necessary to measure at more than one frequency to identify a negative reactance.

negative resistance A resistance such that when the current through it increases, the voltage drop across the resistance decreases. This characteristic is possessed by an electric arc, some electron tubes, and some semiconductor devices.

negative-resistance device A device in which an increase in the applied voltage increases the resistance and thereby produces a proportional decrease in current. Examples include tunnel diodes, silicon unijunction transistors, and the dynatron tetrode vacuum-tube oscillator.

negative-resistance magnetron A magnetron operated in such a way that it acts as a negative resistance.

negative-resistance oscillator An oscillator in which a parallel-tuned resonant circuit is connected to a two-terminal negative-resistance device. Examples include arc converters, dynatron oscillators, transitrons, and some types of transistors and diodes.

negative-resistance region An operating region in which the current decreases when the applied voltage is increased.

negative temperature coefficient [abbreviated NTC] The condition wherein the resistance, length, or other characteristic of a material decreases when temperature increases. Carbon has a negative temperature coefficient of resistance.

negative terminal The terminal of a battery or other voltage source that has more electrons than normal. Electrons flow from the negative terminal through the external circuit to the positive terminal.

negative thermion *Thermoelectron.*

negative-transconductance oscillator An electron-tube oscillator in which the output of the tube is coupled back to the input without phase shift, the phase condition for oscillation being satisfied by the negative transconductance of the tube.

negative transmission *Negative modulation.*

negative zero The zero value reached by counting toward zero from a negative number in the binary system.

negatron 1. A four-electrode vacuum tube that has the characteristics of a negative resistance. 2. *Electron.* 3. *Negative electron.*

N electron An electron that has an orbit in the N

shell, which is the fourth shell of electrons surrounding the atomic nucleus, counting out from the nucleus.

NEMA Abbreviation for *National Electrical Manufacturers Association.*

nematic material A liquid crystal material in which the elongated molecules are parallel to each

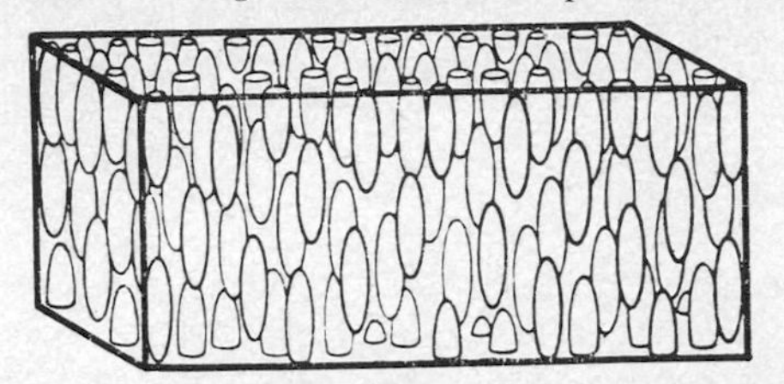

Nematic material, showing arrangement of elongated molecules.

other but not in layers. Widely used in commercial liquid crystal displays.

NEMO [Not Emanating from Main Office] A radio or television broadcast that originates outside the studio. Also called remote.

neodymium [symbol Nd] A rare-earth element. Atomic number is 60.

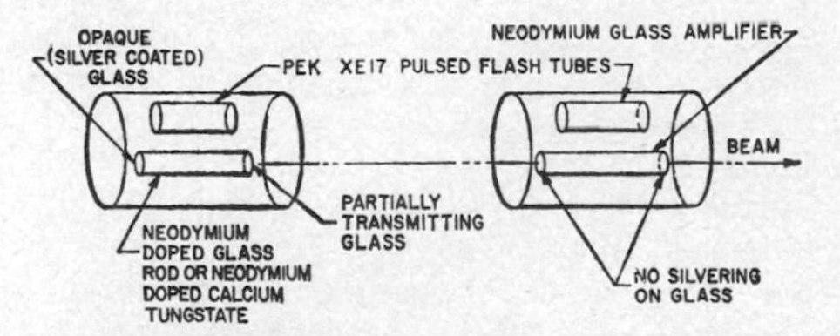

Neodymium amplifier using straight-through operation.

neodymium amplifier A light amplifier in which amplification is provided by the action of pulsed flash tubes on neodymium-doped glass rods.

neodymium-doped yttrium-aluminum garnet laser [abbreviated Nd:YAG laser] A crystalline solid laser in which the YAG crystal is doped with neodymium. Also called YAG laser.

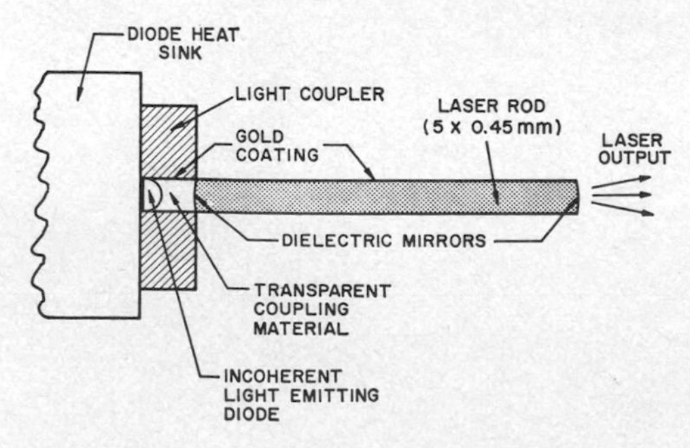

Neodymium-doped yttrium-aluminum garnet laser using end-pumping by single light-emitting diode.

neodymium glass laser An amorphous solid laser in which glass is doped with neodymium. Characteristics are comparable to those of a pulsed ruby laser, but the wavelength of radiation is outside the visible range.

neodymium liquid laser *Inorganic liquid laser.*

neon [symbol Ne] An inert gaseous element used in neon signs and some electron tubes. It produces a characteristic bright red glow when ionized. Atomic number is 10.

neon glow lamp A glow lamp that contains neon gas. It produces a characteristic red glow. Wat-

Neon glow lamp with screw base (left) and bayonet base (right).

tages of standard sizes range from 0.04 to 3 W. Also called neon lamp.

neon indicator A neon glow lamp used as a visual indicator of voltage.

neon lamp *Neon glow lamp.*

neon oscillator An oscillator circuit that consists of a neon glow lamp, a capacitor, and sometimes also a resistor.

neon tubing A glow lamp in which neon gas is ionized by the flow of electric current through

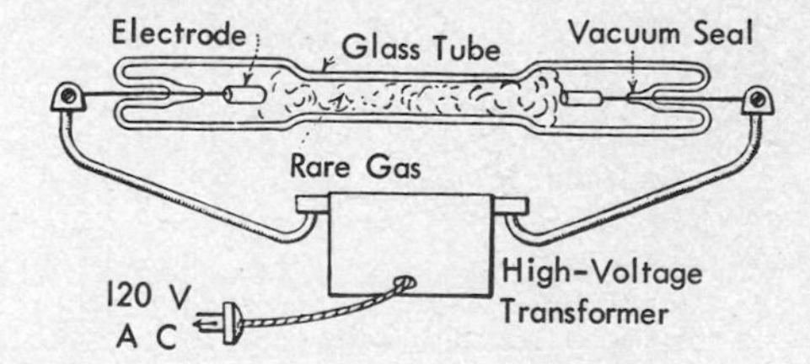

Neon tubing and transformer.

long lengths of gas tubing, to produce a luminous red glow discharge. Used chiefly in outdoor advertising signs.

Neoprene Trademark of Du Pont for polychloroprene.

NEP Abbreviation for *noise equivalent power.*

neper [abbreviated Np] A unit that expresses the ratio of two voltages, two currents, or two power values in a logarithmic manner. The number of nepers is the natural logarithm of the square root of the ratio of the two values being compared. The neper thus uses the base of 2.71828, whereas the decibel uses the common-logarithm base of 10. One neper is equal to 8.686 dB.

nephelometer An instrument that measures the size and number of dust particles suspended in a medium, based on directing a beam of light through the particles to a photomultiplier.

neptunium [symbol Np] A transuranic radioactive element formed by radioactive decay of uranium 239, which emits a beta particle (high-energy electron) to become neptunium 239. Atomic number is 93.

neptunium series The series of nuclides that result from the decay of the long-lived synthetic nuclide ^{237}Np. Mass numbers of all members are given by $4n + 1$, where n is an integer, so the sequence is also known as the $4n + 1$ series.

Nernst bridge A four-arm bridge that contains capacitors instead of resistors, for measuring capacitance values at high frequencies.

Nernst effect If heat flows through a strip of metal whose surface is perpendicular to a magnetic field, a voltage is developed between opposite edges of the strip.

Nernst-Ettinghausen effect A thermomagnetic effect that occurs in certain pure crystals, in which a temperature difference is produced in a direction perpendicular to a longitudinal electric current and an applied magnetic field.

Nernst lamp An electric lamp that consists of a short slender rod of zirconium oxide in open air, heated to brilliant white incandescence by current.

net A number of communication stations equipped for communication with each other, often on a definite time schedule and in a definite sequence.

network 1. A combination of electric elements, such as interconnected resistors, coils, and capacitors. An active network also contains a source of energy, whereas a passive network does not. 2. A number of radio or television broadcast stations connected by coaxial cable, radio, or wire lines, so all stations can broadcast the same program simultaneously.

network analyzer 1. An analog computer in which networks are set up to simulate power-line systems or other physical systems and obtain solutions to various problems before actual construction. 2. An RF analyzer used to make accurate measurements of magnitude, phase, and group delay of either of two signals with respect to a reference signal. Also used for measuring impedance and admittance over a wide frequency range.

network constant One of the values of resistance, inductance, mutual inductance, or capacitance that make up a network.

neuristor A device which behaves like a nerve fiber in that it has attenuationless propagation of signals. One goal of research is development of a complete artificial nerve cell, containing many neuristors, that could duplicate the function of the human eye and brain in recognizing characters and other visual images.

neuroelectricity A current or voltage generated in the nervous system.

neuron A living nerve cell.

neutral Having the same number of electrons as protons, hence in a normal condition wherein there is no electric charge.

neutral atom An atom in which the number of positive charges in the nucleus is equal to the number of electrons that surround the nucleus.

neutral conductor The conductor, in a power line that has more than two wires, to which the voltages from every other conductor are approximately equal in amplitude and, for an AC system, equally spaced in phase.

neutral-density faceplate A television picture-tube faceplate into which a neutral-density filter has been incorporated to increase picture contrast by attenuating external light reflected from the screen. Reflected light must pass through the filter twice and be doubly attenuated, whereas the desired light from the screen passes through only once.

neutral-density filter An optical filter that reduces the intensity of light without appreciably changing its color. Also called gray filter.

neutral ground A ground connection made to the neutral conductor or neutral point of a power line, transformer, motor, generator, or other device connected to the line. The connection may be made directly or through a grounding device that has resistance and/or impedance.

neutralization A method of nullifying oscillation-producing voltage feedback from the output to the input of an amplifier through tube interelectrode impedances. An external feedback path produces at the input a voltage that is equal in magnitude but opposite in phase to that fed back through the interelectrode capacitance.

neutralize To stop regeneration in an amplifier stage by neutralization.

neutralized stable gain The gain achieved in a neutralized transistor circuit by nullifying feedback from output to input, usually by placing a capacitor between drain and gate.

neutralizing capacitor A capacitor, usually variable, used in an anode-to-grid feedback path in an amplifier circuit for neutralization purposes.

neutralizing indicator A neon lamp or other indicator used to show when an amplifier has been neutralized. It is usually coupled to the anode tank circuit.

neutralizing tool A small screwdriver or socket wrench, partly or entirely nonmetallic, used to adjust a neutralizing capacitor.

neutralizing voltage The AC voltage that is fed from the anode circuit to the grid circuit of an amplifier, or vice versa, 180° out of phase with and equal in amplitude to the AC voltage that is transferred between these circuits over an undesired path, usually that through the grid-to-anode tube capacitance.

neutral relay A relay in which the movement of the armature is independent of the direction of current flow through the relay coil. Also called nonpolarized relay.

neutral zone *Dead band.*

neutrino An electrically neutral elementary particle with a mass so small that it is extremely difficult to detect. Existence was verified in 1956. It is produced in many nuclear reactions, including beta decay, and has high penetrating power. Neutrinos from the sun usually pass right through the earth.

neutrodyne An amplifier circuit used in early tuned-radio-frequency receivers, in which a capac-

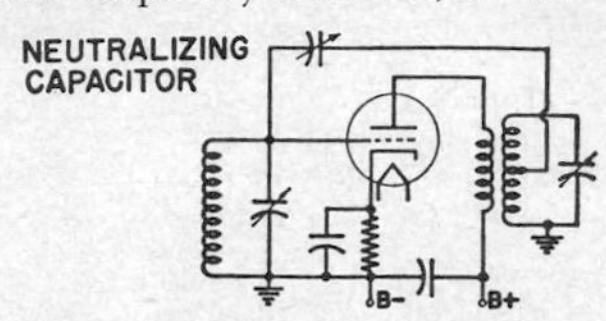

Neutrodyne circuit used as tuned RF amplifier stage.

itor was connected between the anode and grid circuits of a triode stage for neutralization purposes.

neutron An elementary nuclear particle that has zero charge and mass number 1, making its mass approximately the same as that of a proton. Ionization is produced by the products of neutron collisions.

neutron absorber A material with which neutrons interact significantly, resulting in the disappearance of the neutrons without production of other neutrons.

neutron absorption Nuclear interaction in which the incident neutron disappears. One or more neutrons may subsequently be emitted, accompanied by other particles, as in fission.

neutron activation analysis Activation analysis in which the specimen is bombarded with neutrons. Identification is made by measuring the resulting radioisotopes.

neutron albedo The probability, under specified conditions, that a neutron entering into a region through a surface will return through that surface.

neutron binding energy The energy required to remove a single neutron from a nucleus. Most neutron binding energies are in the range from 5 to 12 MeV.

neutron bomb A precision nuclear weapon for battlefields of limited size, designed to release very large quantities of radioactive neutrons for killing enemy personnel without destroying buildings, vehicles, or other structures outside a radius of about 150 m from ground zero. Residual radiation dissipates quickly. Developed primarily for delivery by artillery shells or missiles.

neutron bombardment Bombardment of atomic nuclei with neutrons, generally to produce nuclear fission.

neutron capture The reaction that occurs when an atomic nucleus absorbs or captures a neutron. The probability that a given material will absorb neutrons is proportional to its neutron capture cross section and depends on the energy of the neutrons and the nature of the material.

neutron-capture gammas The energetic gamma rays emitted by an isotope upon capture of a neutron.

neutron-capture theory Treatment of certain types of brain cancer with a beam of thermal neutrons from a nuclear reactor, after a boron-10 compound has been injected into the patient's bloodstream. The boron compound localizes briefly in the tumorous brain tissue. The neutrons are preferentially captured by the boron-10 isotope, resulting in prompt emission of an energetic alpha particle. The alpha particles are absorbed within a few millimeters, for extremely localized radiation therapy of the tumor.

neutron collision radius The nuclear radius determined by fast-neutron transmission experiments.

neutron converter A device placed in a flux of slow neutrons to produce fast neutrons.

neutron cross section A measure of the probability of reaction of a neutron with a single atomic nucleus or with one of the nuclei in a unit volume of material.

neutron crystallography The study of crystals by neutron diffraction.

neutron current density The number of neutrons that pass through a unit area of surface per unit time.

neutron curtain A thin shield, usually of cadmium, that stops slow neutrons.

neutron detector Any radiation detector that detects neutrons by measuring the charged particles or gamma rays released by a neutron-induced nuclear reaction. Neutrons cannot produce measurable ionization directly since they have no charge.

neutron diffractometer A diffractometer in which a beam of neutrons is used for diffraction analysis. The intensities of the diffracted beams at different angles are measured with an ionization chamber or radiation counter.

neutron diffusion Migration of neutrons, through successive scattering collisions, from regions of high concentration to regions of low concentration.

neutron excess The number of neutrons in a nucleus in excess of the number of protons. Also called difference number and isotopic number.

neutron flux [symbol *nv*] The intensity of neutron radiation, expressed as the number of neutrons passing through a unit area per unit time.

neutron generator A low-energy accelerator whose beam of charged particles is directed at a suitable target for producing neutrons.

neutron hardening The effect caused by the diffusion of thermal neutrons through a medium whose absorption cross section decreases as energy increases. Because the slower neutrons are preferentially absorbed, the average energy of the diffusing neutrons becomes greater.

neutron-induced activity Radioactivity produced when a neutron is captured by a nucleus.

neutron lifetime The average time a neutron spends in slowing down and diffusion before capture.

neutron logging A logging method used in boreholes to detect and identify fluid-bearing zones in the earth. A neutron source moved through the borehole induces additional gamma radiation that is measured by an ionization chamber. The measured radiation at each depth is related to the hydrogen content of the rock or soil.

neutron magnetic moment A physical constant equal to 1.9125 Bohr magnetons. The direction of the moment is opposite that of the proton.

neutron multiplication The process in which a neutron produces an average of more than one neutron in a medium that contains fissionable material.

neutron number [symbol *N*] The number of neutrons in the nucleus of an atom. It is equal to the difference between the mass number and the atomic number. In the symbol of a nuclide, the neutron number is added as a subscript following the element symbol; thus in $^{59}_{26}Fe_{33}$ the neutron number for iron is 33, the mass number is 59, and the atomic number is 26.

neutron radiative capture Capture of a slow neutron by a nucleus, with prompt emission of gamma rays.

neutron radiography Radiography in which a beam of neutrons is used in place of x-rays.

neutron rest mass A physical constant equal to 1.67482×10^{-24} g.

neutron shield A shield that protects personnel from neutron irradiation. It consists of very light neutron-moderating materials.

neutron source 1. Any material that emits neutrons, such as a mixture of radium and beryllium. 2. A source that uses a nuclear reaction to generate neutrons.

neutron spectrometer A spectrometer that determines the wavelengths of neutrons and the relative intensities of different wavelengths in a neutron beam.

neutrons per fission The total number of neutrons emitted per fission, usually including delayed neutrons.

neutron therapy Medical therapy that involves irradiation with neutrons.

neutron velocity selector An instrument that isolates and detects neutrons which have a particular range of velocities.

neutron wavelength The wavelength of the de Broglie waves associated with a neutron.

newton [abbreviated N] The SI unit of force, equal to the force that produces an acceleration of 1 m/s^2 to a mass of 1 kg. One newton is equal to 10^5 dyn.

newton per square meter [abbreviated N/m^2] The SI unit of pressure or stress. A pressure of 1 N/m^2 is equal to 1 pascal.

nF Abbreviation for *nanofarad.*

NF Abbreviation for *noise figure.*

nH Abbreviation for *nanohenry.*

nicad battery Abbreviation for *nickel-cadmium battery.*

NiCd battery Abbreviation for *nickel-cadmium battery.*

Nichrome Trademark of Driver-Harris Co. for an alloy of nickel, chromium, and sometimes iron. The alloy has high electric resistance and the ability to withstand high temperatures for long periods of time. Used in wirewound resistors and electric heating elements.

nickel [symbol Ni] A metallic element. Atomic number is 28.

nickel-cadmium battery [abbreviated NiCd battery or nicad battery] A sealed storage battery that has a nickel anode, a cadmium cathode, and

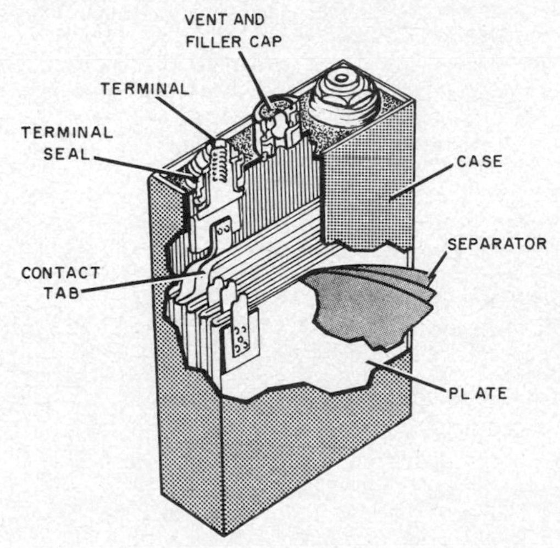

Nickel-cadmium battery, showing construction of single cell. Smaller versions use sealed cylindrical cells resembling flashlight cells.

an alkaline electrolyte. Widely used in cordless appliances. Without recharging, it can serve as a primary battery. Nominal voltage is 1.25 V per cell.

nickel-iron battery *Edison storage battery.*

Nicol prism A prism made by cementing together two pieces of transparent crystalline Iceland spar with Canada balsam. It produces plane-polarized light from ordinary unpolarized light by eliminating the ordinary component of the original light by total reflection at the cementing layer. Only the extraordinary component passes.

night effect A polarization error that occurs in radio direction finders at night.

night-vision binoculars Binoculars that are worn like eyeglasses but use a battery-powered television camera to pick up images. After being amplified, the images are viewed on tiny television picture tubes built into the binoculars.

night-vision telescope A telescope that has sufficient electronic amplification of images to be used at night without artificial illumination. It may have television, optoelectronic, or other means of providing the necessary image amplification.

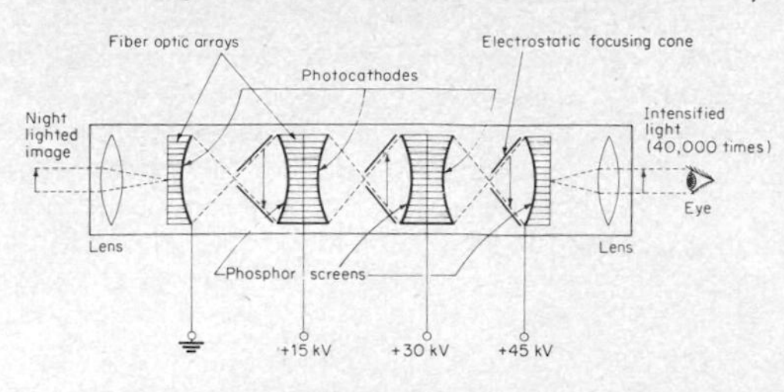

Night-vision telescope using optoelectronic components.

NI junction A junction between N-type material and intrinsic material in a semiconductor.

Nike An Army surface-to-air guided missile used in place of antiaircraft guns. The first version, Nike-Ajax, had a speed of about Mach 2, a range of about 30 mi (48 km), and was guided by radio command. The second version, Nike-Hercules, has a speed of about Mach 3, a range of about 70 mi (112 km), is guided by either radar or radio command, and can carry an atomic warhead. The third version, Nike-Zeus, has a capability of defending against ballistic-missile attacks, with guidance by phased-array radar.

Nimbus One of a series of NASA weather satellites used primarily for transmitting visual and infrared photographs of cloud formations to ground stations for meteorological applications.

niobium [symbol Nb] A metallic element. Atomic number is 41. Formerly called columbium.

Nipkow disk A disk that has one or more spirals of holes around the outer edge, with successive openings positioned so rotation of the disk provides mechanical scanning of a scene.

NIPO Abbreviation for *negative input–positive output.*

nit [abbreviated nt] A name sometimes given to the SI unit of luminance, the candela per square meter.

nitrogen [symbol N] A gaseous element. Atomic number is 7. Often used in coaxial transmission lines and other enclosed volumes to keep out moisture because it is chemically relatively inert and does not support combustion.

Nixie Trademark of Burroughs Corp. for a cold-cathode gas readout tube that has a common anode and 10 different metallic cathodes, each formed in the shape of a different numeral, alphabetic character, or special symbol. The desired character is surrounded with a brilliant glow when the corresponding cathode is energized.

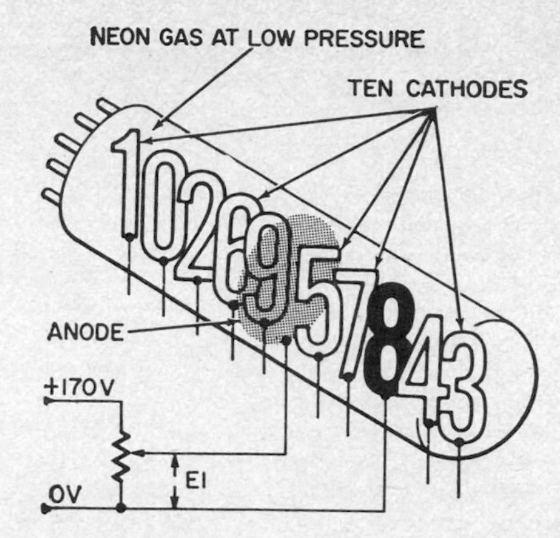

Nixie tube having 10 number-shaped cathodes, only one of which is energized at a time. One Nixie tube is needed for each character position in a display.

nm Abbreviation for *nanometer.*

NMOS Abbreviation for *N-channel MOS.*

NMOS RAM A random-access memory that uses N-channel metal-oxide semiconductor technology.

NMR Abbreviation for *nuclear magnetic resonance.*

NMRR Abbreviation for *normal-mode rejection ratio.*

N/m² Abbreviation for *newton per square meter.*

NO Symbol for *normally open.*

nobelium [symbol No] A transuranic radioactive element produced artificially by bombarding curium 244 with carbon-13 nuclei. Atomic number is 102.

noble metal A metal, such as gold, silver, or platinum, that has high resistance to corrosion and oxidation. Used in the construction of thin-film circuits, metal-film resistors, and other metal-film devices.

no connection [symbol NC] An electron-tube pin for which there is no connection internally to electrodes. This generally permits use of the corresponding tube socket terminal as an insulated standoff terminal at which two or more leads may be connected together independently of the tube.

nodal diagram A diagram that shows the order and mode of waves propagated in a waveguide.

nodal-point keying Keying of an arc transmitter at a point in the antenna circuit that is essentially at ground potential at all times.

node 1. A point, line, or surface in a standing-

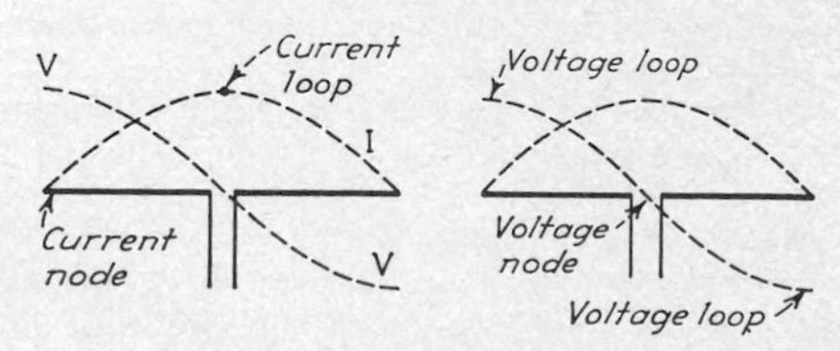

Nodes of current and voltage for half-wave dipole.

wave system where some characteristic of the wave has essentially zero amplitude. 2. A junction point in a network.

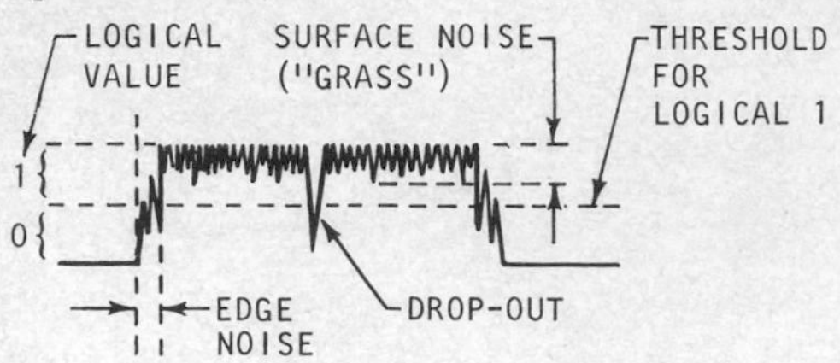

Noise as it may occur in connection with logic levels 1 and 0. Edge noise and dropouts may cause errors.

noise An undesired electric disturbance or sound that tends to interfere with the normal reception or processing of a desired signal.

noise analyzer An instrument that determines the amplitudes of the frequency components in a noise signal.

noise behind the signal *Modulation noise.*

noise-canceling microphone *Close-talking microphone.*

noise-current generator A current generator whose output is a random function of time.

noise diode A diode designed for operation at saturation in a noise-generating circuit.

noise dosimeter An instrument that determines if the total daily exposure of a person to noise is within government-established limits. One version has a detachable microphone that can be positioned anywhere on the body. It may have a digital display in addition to a sealed-in meter that reads the maximum measured level.

noise equivalent power [abbreviated NEP] The power that would be present at a given point in a circuit due to noise alone if the useful signal were removed without changing operating conditions.

noise factor The ratio of the total noise power per unit bandwidth at the output of a system to the portion of the noise power that is due to the input termination at the standard noise temperature of 290 K.

noise figure [abbreviated NF] The noise factor expressed in decibels. Noise figure is equal to 10 $\log_{10}$ (noise factor).

noise filter A filter that is inserted in an AC power line to block noise interference which would otherwise travel through the power line in either direction and affect the operation of receivers.

noise generator An instrument that generates one or more types of noise signals, covering a specified portion of the frequency spectrum. Types of noise generated may include pink or white noise and various industry-standardized noise signals. The noise may be essentially random, such as occurs in nature, or may be pseudorandom, in which noise patterns of known content and duration occur at a fixed repetition rate.

noise immunity The weakest signal that can be accepted by a circuit without getting lost in internally generated circuit noise.

noise jamming A brute-force electronic countermeasure technique that produces clutter over the entire display of an enemy tracking radar. The disclosure to the enemy of jamming action or the ability of some missiles to home on the noise jammer are two drawbacks.

noise level The level of electric or acoustic noise at a particular location. Usually expressed in decibels with respect to a specified reference level. The value of the noise is integrated over a specified frequency range.

noise limiter A limiter circuit that cuts off all noise peaks that are stronger than the highest peak in the desired signal being received, thereby reducing the effects of atmospheric or man-made interference. Also called automatic noise limiter.

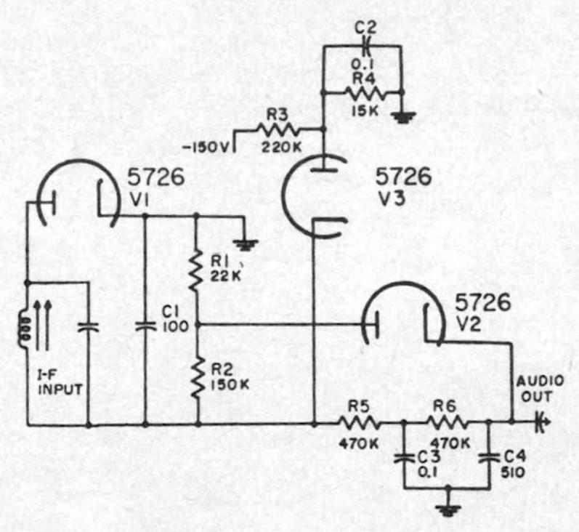

Noise limiter using circuit design preferred by National Bureau of Standards. Signal from IF amplifier of amplitude-modulated receiver is demodulated by diode detector V1. Series noise limiter V2 then clips noise peaks, while shunt noise limiter V3 acts as short-circuit across IF output during noise peaks, preventing them from operating the automatic-gain-control circuit.

noise-measuring set *Circuit-noise meter.*

noise pollution The condition wherein the loudness and other characteristics of a noise together make that noise undesirable in the human environment. Factors affecting the degree of noise pollution include the responses of different people to various sounds, exposure duration, time of day, and the amount of pain or physical damage caused by the noise.

noise power ratio The ratio in decibels of the noise level in a telephone measuring channel which has the baseband fully noise-loaded to the level in that channel when all the baseband except the measuring channel is noise-loaded.

noise quieting The ability of a receiver to reduce noise background in the presence of a desired signal. Usually expressed in decibels.

noise reduction A process whereby the average transmission of the sound track of a sound motion-picture print, averaged across the track, is decreased for signals of low level and increased for signals of high level. Since background noise introduced by the sound track is less at low transmission, this process reduces film noise during soft passages.

noise source A device that generates a random noise signal for test purposes. Common noise sources are a temperature-limited diode operated at cathode saturation, an electron multiplier, a crystal diode with positive bias, a nonoscillating reflex klystron, and a nonoscillating magnetron.

noise suppressor 1. A circuit that blocks the AF amplifier of a radio receiver automatically when no carrier is being received, to eliminate background noise. Also called intercarrier noise suppressor, interstation noise suppressor, and squelch circuit. 2. A circuit that reduces surface noise when phonograph records are played, generally by a filter that blocks out the higher frequencies where such noise predominates.

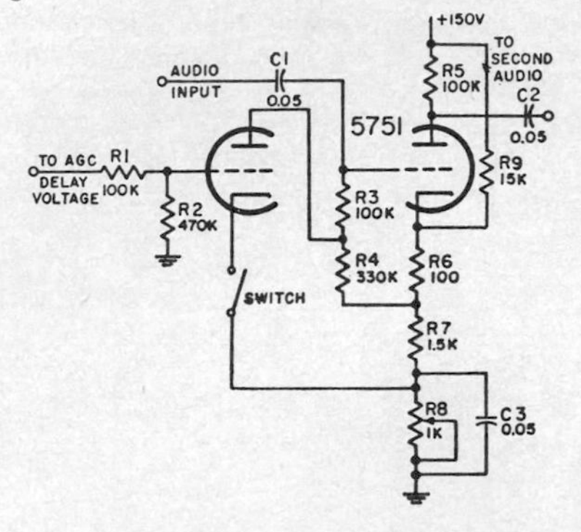

Noise suppressor using circuit design preferred by National Bureau of Standards.

noise temperature The temperature at which the thermal noise power of a passive system per unit bandwidth is equal to the noise at the actual terminals. The standard reference temperature for noise measurements is 290 K.

noise-voltage generator A voltage generator whose output is a random function of time.

no-load loss The power loss of a device that is operated at rated voltage and frequency but is not supplying power to a load.

nominal band A frequency band equal in width to that between zero frequency and the maximum modulating frequency in facsimile.

nominal frequency The specified frequency value of a crystal unit, as distinguished from the actual frequency measured during operation.

nominal line width The average separation between centers of adjacent scanning or recording lines.

nomograph A chart that has three or more scales, on which equations can be solved graphically by placing a straightedge on two known values and reading the answer where the straightedge crosses the scale for the unknown value.

nonbridging Switching action in which the movable contact leaves one fixed contact before touching the next.

noncoherent MTI A moving-target indicator system in which ground clutter is used in place of a coherent reference oscillator as the reference signal. A moving target can be detected only when ground clutter exists at the same range and azimuth as the target.

noncoherent radiation Radiation in which there are no definite phase relations between different points in a cross section of the beam.

nonconductor An insulating material.

noncontacting piston *Choke piston.*

noncontacting plunger *Choke piston.*

noncooperative system An instrumentation system in which data is transmitted from airborne missile equipment to a ground station for recording and processing, with no control of the transmission by the ground station.

noncorrosive flux Flux that is free from acid and other substances which might cause corrosion when used in soldering.

nondegenerate gas A gas formed by a system of particles whose concentration is sufficiently weak so the Maxwell-Boltzmann law applies. Examples are molecules or atoms of a body in the gaseous state, electrons emitted by a hot cathode, electrons and ions in a cloud or a plasma, electrons supplied to a conduction band by donor levels in an N-type semiconductor, and holes resulting from the passage of electrons from the normal band to an impurity band of acceptor levels in a P-type semiconductor.

nondestructive breakdown Breakdown of the barrier between the gate and channel of a field-effect transistor without causing failure of the device. In a junction field-effect transistor, avalanche breakdown occurs at the PN junction.

nondestructive readout [abbreviated NDRO] A computer memory in which the magnetic disturbance that gives readout of desired data is self-reversible, eliminating the need for special circuits to restore the stored data.

nondestructive testing Testing by techniques that do not damage or destroy the items being tested.

nondirectional *Omnidirectional.*

nondirectional antenna *Omnidirectional antenna.*

nondirectional beacon An omnidirectional beacon or other radio transmitter facility designed for use with an airborne direction finder to provide a line of position.

nondirectional microphone *Omnidirectional microphone.*

nonerasable storage A computer storage medium that cannot be erased and reused, such as punched paper tapes and punched cards.

nonferrous Not containing iron.

non-great-circle propagation *Groundscatter propagation.*

nonhoming Not returning to the starting or home position, as when the wipers of a stepping relay remain at the last-used set of contacts instead of returning to their home position.

nonhoming tuning system A motor-driven automatic tuning system in which the motor starts in the direction of previous rotation. If this is incorrect for the new station, the motor reverses after tuning to the end of the dial, then proceeds to the desired station.

nonimpact printer A line printer in which the characters are produced electrically, electronically, or optically rather than mechanically. Only a single copy can be produced. Examples include electrostatic, ink-jet, laser-beam, and thermal-matrix printers. Operating speeds are much higher than for impact printers and can be well over 10,000 lines per minute for laser-beam printers.

noninduced current A current that flows through a winding of a transformer but is not due to a voltage induced in the transformer. The filament current in a bifilar winding is an example.

noninduced voltage An alternating or direct voltage which is applied uniformly to an entire winding so that no appreciable potential difference is induced along the winding. The filament voltage supplied to a load through a bifilar winding is an example.

noninductive Having negligible or zero inductance.

noninductive capacitor A capacitor constructed so it has practically no inductance. Foil layers are staggered during winding, so an entire layer of foil projects at either end for contact-making purposes. All currents then flow laterally rather than spirally around the capacitor.

noninductive circuit A circuit that has practically no inductance.

noninductive load A load that has practically no inductance. It may consist entirely of resistance or may be capacitive.

noninductive resistor A wirewound resistor that has practically no inductance if a hairpin winding is used or if connections to adjacent sections of the winding are reversed.

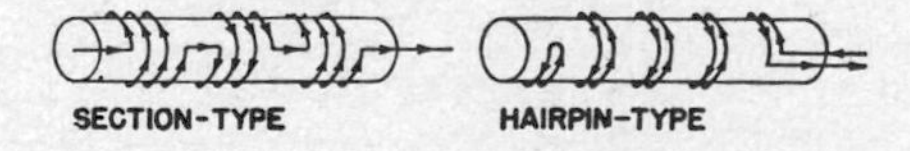

Noninductive windings.

noninductive winding A winding constructed so the magnetic field of one turn or section cancels the field of the next adjacent turn or section.

noninverting amplifier An operational amplifier in which the input signal is applied to the ungrounded positive input terminal to give a gain greater than unity and make the output voltage change in phase with the input voltage.

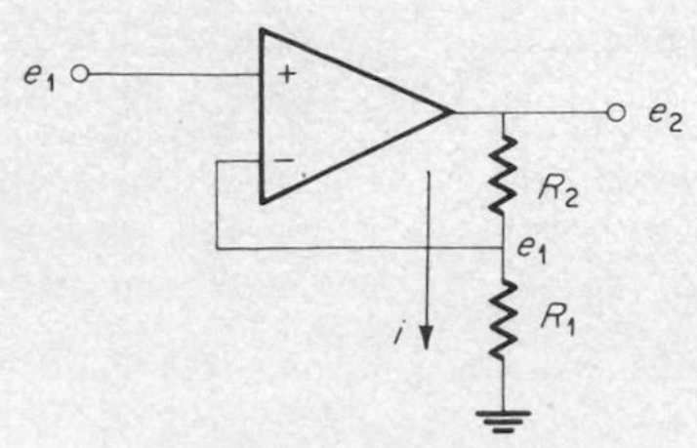

Noninverting amplifier. Value of feedback resistor R_2 is chosen to provide desired circuit gain and make output voltage e_2 increase in phase with input voltage e_1.

noninverting parametric device A parametric device in which the operation depends on a harmonic of the pump frequency and two signal frequencies, one of which is the sum of the other plus the pump harmonic. If either is moved upward in frequency, the other will also increase in frequency, corresponding to noninverting operation. There is no gain at either signal frequency.

nonlinear Not directly proportional.

nonlinear amplifier An amplifier in which the output is not directly related to the input.

nonlinear capacitor A capacitor that has a nonlinear charge characteristic or a reversible capacitance that varies with bias voltage.

nonlinear circuit A circuit not specifiable by linear differential equations in which time is the independent variable.

nonlinear detection Detection based on the curvature of a tube characteristic, such as square-law detection.

nonlinear distortion Distortion in which the output of a system or component does not have the desired linear relation to the input. Amplitude, harmonic, and intermodulation distortions are examples of nonlinear distortion.

nonlinear element An element in which an increase in applied voltage does not produce a proportional increase in current.

nonlinear feedback control system A feedback control system in which the relationships between the input and output signals are not linear.

nonlinearity The deviation of any functional relationship from direct proportionality. As an example, crowding of picture elements in one region of a television screen is due to nonlinearity of scanning action.

nonlinear network A network not specifiable by linear differential equations in which time is the independent variable.

nonlinear taper Nonuniform distribution of resistance throughout the element of a potentiometer or rheostat.

nonmagnetic Not magnetizable, and hence not affected by magnetic fields. Examples are air, glass, paper, and wood. All have a magnetic permeability of 1, the same as a vacuum.

nonmagnetic shim A nonmagnetic material attached to the armature or core of a relay to prevent iron-to-iron contact between armature and core when the relay is energized.

nonmagnetic steel A steel alloy that contains about 12% manganese and sometimes a small quantity of nickel. It is practically nonmagnetic at ordinary temperatures.

nonmicrophonic *Antimicrophonic.*

nonplanar network A network that cannot be drawn on a plane without crossing of branches.

nonpolarized electrolytic capacitor An electrolytic capacitor in which the dielectric film is formed adjacent to both metal electrodes, to give the same construction in both directions of current flow.

nonpolarized relay *Neutral relay.*

nonquantized system A system of particles whose energies are assumed to be capable of varying in a continuous manner. The number of microscopic states of the system, defined by the positions and velocities of the particles at a given instant, is then infinite. Also called classical system.

nonradiative transition A change from one energy level to another in an atom without absorption or emission of radiation. The necessary energy may be supplied or carried away by vibrations in a solid material or by motions of atoms or electrons in a plasma.

nonrelativistic particle A particle whose velocity is small with respect to that of light.

nonresonant line A transmission line in which there are no standing waves.

nonreturn-to-zero code [abbreviated NRZ code] A code used in some computers, in which a binary 1 is represented by one bit time at the 1

Nonreturn-to-zero code for binary data. Space between each pair of dashed vertical lines represents one bit time.

level and a binary 0 is represented by one bit time at the 0 level. This permits storing about twice as much data as can be stored with the return-to-zero code.

nonreturn-to-zero-inverted code [abbreviated NRZI code] A code used in some computers, in which the modulation switches from one state to the opposite state to represent a binary 0 and remains unchanged to represent a binary 1.

nonshorting switch A selector switch in which the width of the movable contact is less than the distance between contacts, so the old circuit is broken before the new circuit is completed.

nonsinusoidal wave A wave whose form differs from that of a sine wave and therefore contains harmonics.

nonspecular reflection Reflection from a rough surface, producing diffraction and scattering of waves.

nonstorage camera tube A television camera tube in which the picture signal is at each instant proportional to the intensity of the illumination on the corresponding elemental area of the scene at that instant.

nonstorage display Display of nonstored information in a storage tube without appreciably affecting the stored information.

nonsynchronous Not related in phase, frequency, or speed to other quantities in a device or circuit.

nonsynchronous transmission A data-transmission process in which a clock is not used to control the unit intervals within a block or a group of data signals.

nonsynchronous vibrator A vibrator that interrupts a DC circuit at a frequency unrelated to other circuit constants and does not rectify the resulting stepped-up alternating voltage.

nonthermal radiation Electromagnetic radiation emitted by accelerated charged particles that are not in thermal equilibrium. Aurora light and fluorescent lights are examples.

nonvolatile storage A computer storage medium that retains information in the absence of power, such as a magnetic tape, drum, or core.

NOR A logic operator which has the property that if $P, Q, R, \ldots$ are statements, then the NOR of $P, Q, R, \ldots$ is true if and only if all statements are false, false if and only if at least one statement is true.

NORAD The joint North American Air Defense Command that coordinates operations of various services in defense of continental United States and portions of the North American continent.

NOR device A logic device whose output is 1 only when all its control signals are 0.

NOR gate A gate whose output represents 1 only when all the input signals represent 0. The output is thus inverted from that of an OR gate.

normal 1. The perpendicular to a line or surface at the point of contact. 2. The expected or regular value of a quantity.

normal distribution *Gaussian distribution.*

normal induction The limiting induction, either positive or negative, in a magnetic material that is under the influence of a magnetizing force varying between two specific limits.

normalize To multiply all quantities by a constant so they fall within the operating ranges of a computer or the scale ranges of a graph.

normalized admittance The reciprocal of the normalized impedance.

normalized impedance An impedance divided by the characteristic impedance of a transmission line or waveguide.

normalized plateau slope A figure of merit for a radiation-counter tube, equal to the percentage change in counting rate divided by the percentage change in voltage. The midpoint of the plateau of the counting-rate–voltage characteristic of the tube is used as a reference.

normal logic A logic level term whose use is deprecated because it can be construed differently for NPN devices than for PNP devices. To avoid confusion, the magnitude and polarity of the voltage levels for 1 and 0 in a specific logic circuit should always be specified.

normally closed [symbol NC] A term applied to relay contacts that are connected to complete a circuit when the relay is not energized.

normally open [symbol NO] A term applied to relay contacts that are connected to break a circuit when the relay is not energized.

normal mode A characteristic distribution of vibration amplitudes among the parts of a system, each part of which is vibrating freely at the same natural frequency and phase.

normal-mode rejection ratio [abbreviated NMRR] The ability of an amplifier to reject spurious signals at the power-line frequency or at harmonics of the line frequency.

normal permeability The ratio of the normal induction B of a magnetic material to the corresponding magnetic intensity H.

normal position The deenergized position of the contacts of a relay.

normal state *Ground state.*

normal threshold of audibility The minimum sound pressure level at the entrance to the ear, at a specified frequency, that produces an auditory sensation in a large percentage of normal persons in the age group from 18 to 30 years.

normal uranium *Natural uranium.*

north magnetic pole The magnetic pole located approximately at 71°N latitude and 96°W longitude, about 1140 nautical miles south of the North Pole.

north pole The pole of a magnet at which magnetic lines of force are considered as leaving the magnet.

North Pole A geographical point on the earth that is one end of the axis about which the earth revolves. It is almost directly beneath the star Polaris.

north-stabilized PPI An azimuth-stabilized plan-position indicator in which the reference direction is magnetic north.

nose cone A hollow cone-shaped shield that fits over or serves as the nose of a guided missile. The cone is made from materials that will withstand high temperatures generated by friction with air particles during passage through the earth's atmosphere.

nose whistler A whistler whose frequency is determined primarily by the electron gyrofrequency at the top of the whistler path along a line of the earth's magnetic field.

NOT A logic operator which has the property that if P is a statement, then the NOT of P is true if P is false, false if P is true.

notation A method of representing numbers, in which a number is expressed as a sum of coefficients multiplied by successive powers of a chosen base. Examples of notation systems and their bases are binary–2; ternary–3; quaternary–4; quinary–5; decimal–10; duodecimal–12; hexadecimal–16; duotricenary–32; and biquinary–2, 5.

notch antenna A microwave antenna in which the radiation pattern is determined by the size and shape of a notch or slot in a radiating surface.

notch filter A band-rejection filter that produces a sharp notch in the frequency response curve of a system. Used in television transmitters to provide attenuation at the low-frequency end of the channel, to prevent possible interference with the sound carrier of the next lower channel.

notching relay A relay whose operation depends on successive input pulses.

note A conventional sign used to indicate the pitch, duration, or both of a tone sensation. It is also the sensation itself or the vibration causing the sensation. The word serves when no distinction is desired between the symbol, the sensation, and the physical stimulus.

NOT gate A gate whose output is energized only when its single input is not energized, and vice versa.

NOT logic A logic circuit in which the absence of input will produce an output. Conversely, if there is an input, there will be no output.

noval base An electron-tube base that has positions for nine pins which extend directly through the glass envelope. The spacing between pins 1 and 9 is greater than the spacings between the other pins, for orienting purposes. Used on miniature tubes.

novar A beam-power tube that has a nine-pin base. This permits bringing out a connection from grid No. 3 to a separate pin, to minimize snivet-type interference originating in the horizontal deflection amplifier circuit of a television receiver.

Np Abbreviation for *neper.*

NPIN transistor An intrinsic-junction transistor in which the intrinsic region is sandwiched between the P-type base and N-type collector layers.

NPIP transistor An intrinsic-junction transistor in which the intrinsic region is between P regions.

N+-type semiconductor An N-type semiconductor in which the excess conduction electron concentration is very large.

NPN Negative-positive-negative, used in referring to a semiconductor in which a layer with P-type conductivity is located between two layers that have N-type conductivity.

NPNP transistor An NPN junction transistor that has a transition or floating layer between P and N regions, to which no ohmic connection is made.

NPN transistor A junction transistor that has a

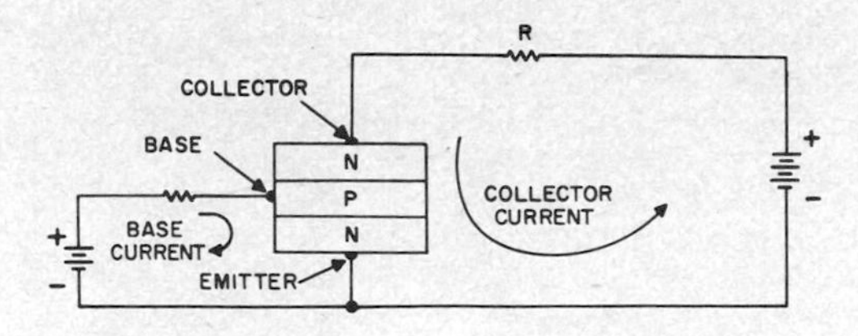

NPN transistor circuit. Input signal is applied across base resistor to produce output signal across load resistor R.

P-type base between an N-type emitter and an N-type collector. The emitter should then be negative with respect to the base, and the collector should be positive with respect to the base.

NPO material A temperature-compensating dielectric material that has an ultrastable temperature coefficient, used in ceramic capacitors. The term is derived from negative-positive-zero.

N quadrant One of the two quadrants in which the N signal of an A-N radio range is heard.

N region The region in a semiconductor where conduction electron density exceeds hole density.

NRZ code Abbreviation for *nonreturn-to-zero code.*

NRZI code Abbreviation for *nonreturn-to-zero-inverted code.*

ns Abbreviation for *nanosecond.*

N scope A radarscope that produces an N display.

N shell The fourth layer of electrons about the nucleus of an atom; it has electrons characterized by the principal quantum number 4.

N signal A dash-dot signal heard in either a bisignal zone or an N quadrant of a radio range.

nt Abbreviation for *nit.*

NTC Abbreviation for *negative temperature coefficient.*

NTSC Abbreviation for *National Television System Committee.*

N-type conductivity The conductivity associated with conduction electrons in a semiconductor.

N-type negative resistance A current-stable negative resistance, in which a given current value can correspond to only one possible voltage value. Examples include four-layer diodes, silicon unijunction transistors, and gas discharge tubes.

N-type semiconductor An extrinsic semiconductor in which the conduction electron density exceeds the hole density. The net ionized impurity concentration is donor-type.

nuclear Pertaining to the nucleus of an atom.

nuclear absorption Absorption of energy by the nucleus of an atom.

nuclear battery A primary battery in which the energy of radioactive material is converted into electric energy by solar cells or other energy converters. It may have one or more nuclear cells, each using a radioisotope such as promethium 147. Also called atomic battery.

nuclear binding energy The energy that would be required to separate an atom of atomic number Z into Z hydrogen atoms.

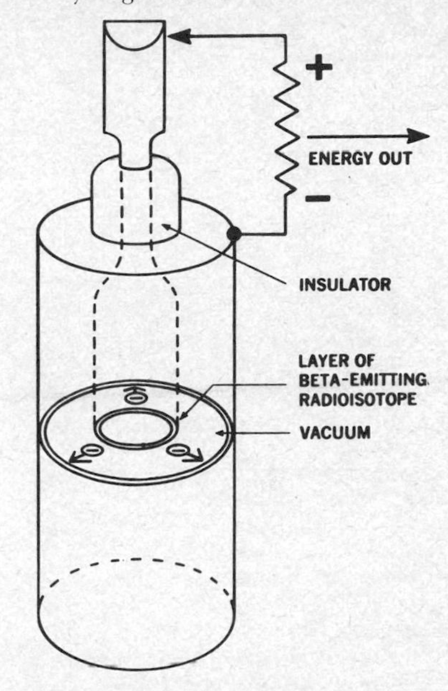

Nuclear battery in its simplest form.

nuclear Bohr magneton A unit of magnetic moment equal to $1/1836$ of the electronic Bohr magneton. Also called nuclear magneton.

nuclear cell A single cell of a nuclear battery.

nuclear charge The sum of the charges of the protons in a nucleus.

nuclear chemistry The branch of chemistry concerned with substances produced or affected by nuclear reactions.

nuclear cooling A technique of adiabatic demagnetization that uses nuclear paramagnetism in place of electron paramagnetism, for cooling down to 0.00001 K.

nuclear cross section The cross section of an atomic nucleus for a particular process.

nuclear device A nuclear explosive used for peaceful purposes, such as for experiments, tests, and commercial blasting applications. The same explosive becomes a nuclear weapon when packaged for military use.

nuclear disintegration Any transformation or change that involves nuclei, including radiative capture, inelastic scattering, beta decay, and isomeric transition. If the disintegration is spontaneous, it is said to be radioactive; if it results from a collision, it is said to be induced.

nuclear emulsion A photographic emulsion specially designed to register individual tracks of ionizing particles.

nuclear energy level The energy of an atomic nucleus. Nuclei can assume only a discrete number of energy levels.

nuclear equation An equation that shows the changes in composition, usually in terms of mass number and charge, of an atomic nucleus during a nuclear reaction. The equation shows any particles captured or radiations absorbed by the nucleus and the particles or radiations emitted.

nuclear force A nonelectromagnetic force peculiar to protons and neutrons in the atomic nucleus; the force holds the nucleus together.

nuclear fusion A thermonuclear reaction in which the nuclei of an element of low atomic weight unite under extremely high temperature and pressure to form a nucleus of a heavier atom. The associated loss in mass is released as energy. This reaction takes place in the hydrogen bomb, where hydrogen nuclei combine to form helium nuclei.

nuclear gyromagnetic ratio The ratio of the magnetic moment of the nucleus to the nuclear angular momentum quantum number.

nuclear gyroscope A gyroscope in which the conventional spinning mass is replaced by the spin of atomic nuclei and electrons. One version uses optically pumped mercury isotopes, and another uses nuclear magnetic resonance techniques.

nuclear incident An unexpected event that involves a nuclear weapon, facility, or component but is not an accidental explosion. Examples include an increase in radioactive contamination, assembly errors, malfunctioning of associated equipment, and unfavorable environmental conditions that result in damage to the weapon, facility, or component.

nuclear induction Magnetic induction originating in the magnetic moments of nuclei. The effect depends on the unequal population of energy states available when the material is placed in a magnetic field.

nuclear isobar One of two or more nuclides that have the same number of nucleons in their nuclei and therefore have about the same atomic mass.

nuclear isomer One of two or more nuclides that have the same mass number and atomic number but exist for measurable times in different quantum states with different energies and radioactive properties. The state of lowest energy is the ground state. Those of higher energies are metastable states.

nuclear isomerism The occurrence of nuclear isomers.

nuclear laser A gas laser in which the gas molecules are excited by high-energy fission particles produced by a pulsed nuclear reactor.

nuclear level control A level control in which a radioactive source and a radiation detector are mounted on opposite sides of a column of liquids, solids, or gases. Radioactive cesium 137 and a scintillation counter are used as source and detector, respectively, in one commercial version, to monitor the degree of filling of metal cans on a conveyor belt.

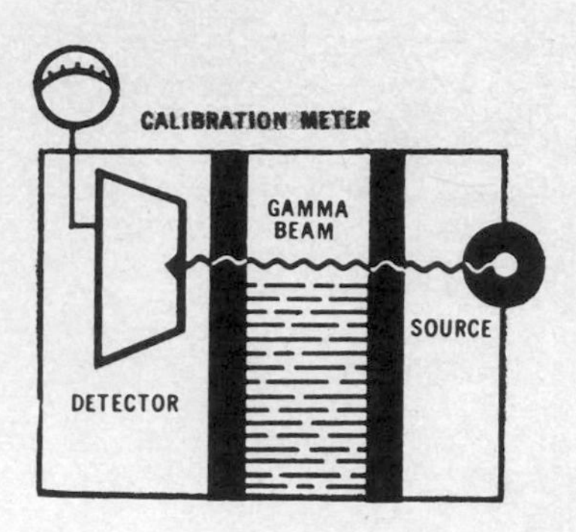

Nuclear level control.

nuclear magnetic moment The magnetic moment of an electrically charged particle that possesses angular momentum.

nuclear magnetic resonance [abbreviated NMR] Resonance encountered in energy transfers between an RF magnetic field and a nucleus placed in a constant magnetic field that is suffi-

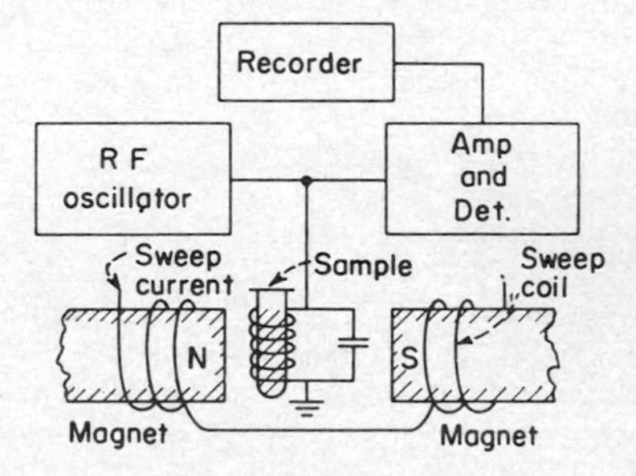

Nuclear magnetic resonance as used for chemical analysis. Sample is acted on by RF field and unidirectional magnetic field that is swept through nuclear resonant values. Recorder shows energy absorption of sample due to magnetic moments and spins of atomic nuclei, as function of magnetic field strength.

ciently strong to decouple the nucleus from its orbital electrons. The amount of energy absorbed by the atoms at resonance is a clue to identification of the atoms involved.

nuclear magnetic resonance spectrometer A spectrometer in which nuclear magnetic resonance is used for the analysis of protons and nuclei and for the study of changes in chemical and physical quantities over wide frequency ranges.

nuclear magnetometer *Nuclear resonance magnetometer.*

nuclear magneton *Nuclear Bohr magneton.*

nuclear medicine 1. The branch of medicine concerned with the problems that result from the use of nuclear energy and nuclear weapons. 2. Any medicine prepared by the use of fissionable material.

nuclear model *Independent-particle model.*

nuclear number *Mass number.*

nuclear packing The concentration of particles

within the nucleus of an atom.

nuclear paramagnetism Paramagnetism associated with nuclear magnetic moments.

nuclear particle 1. A particle in the atomic nucleus. It may be either a proton or a neutron. Also called nucleon. 2. A particle emitted by an atomic nucleus, such as an alpha or beta particle, a positron, neutrino, or meson.

nuclear photodisintegration *Nuclear reaction.*

nuclear physics The branch of physics concerned with atomic nuclei.

nuclear polarization Alignment of the spin magnetic moments of atomic nuclei in the same direction.

nuclear potential The potential energy of a nuclear particle as a function of its position in the field of a nucleus or of another nuclear particle.

nuclear potential energy The average total potential energy of all the protons and neutrons in a nucleus because of the nuclear forces between them, excluding the electrostatic potential energy.

nuclear propulsion Propulsion by nuclear energy.

nuclear quadrupole resonance spectrometer A spectrometer that obtains nuclear quadrupole resonance signals from solid organic and inorganic materials. Used as a nondestructive analyzer and for measuring bond energies, dipole moments, and polarizabilities.

nuclear radiation The radiation of neutrons, gamma particles, and other particles from an atomic nucleus as a result of nuclear fission or nuclear fusion. Also called radiation.

nuclear radius The radius of a spherical volume within which the density of protons and neutrons in a nucleus is effectively large. It is not a precisely determinable quantity.

nuclear reaction A process that involves transformation of an atomic nucleus by a photon, elementary particle, or another nucleus. Also called nuclear photodisintegration.

nuclear reaction energy The disintegration energy of a nuclear reaction. It is equal to the sum of the kinetic or radiant energies of the reactants minus the sum of the kinetic or radiant energies of the products.

nuclear reactor An apparatus in which nuclear fission may be sustained in a self-supporting chain reaction. The fissionable material used as fuel is uranium or plutonium. Most reactors use a moderating material like carbon or heavy water to slow down the neutrons and thereby control the reaction. Also called reactor.

nuclear resonance magnetometer A magnetometer in which the magnetic field of the earth is measured by nuclear magnetic resonance. Hydrogen atoms in water are generally used. Also called nuclear magnetometer.

nuclear rocket A rocket propelled by the reaction to released nuclear energy.

nuclear species *Nuclide.*

nuclear spin The total angular momentum of the atomic nucleus when considered as a single particle.

nuclear structure The internal structure of the atomic nucleus.

nuclear warhead A warhead that contains fissionable or fissionable-fusionable material.

nuclear weapon An atom or hydrogen bomb.

nuclear yield The energy released by the detonation of a nuclear weapon. It is measured in terms of megatons of trinitrotoluene (TNT) required to release the same energy.

nucleogenesis Large-scale formation of nuclei in nature.

nucleon *Nuclear particle.*

nucleonics The science and technology of nuclear energy and its applications.

nucleon number *Mass number.*

nucleus [plural nuclei] The central part of an atom; it possesses a positive charge and contains nearly all the mass of the atom. The nucleus consists of protons and neutrons, together known as nucleons, except for the hydrogen nucleus, which consists only of one proton. Also called atomic nucleus.

nuclide A species of atom characterized by the number of protons and neutrons and energy content in the nucleus, or alternatively by the atomic and mass numbers and atomic mass. To be regarded as a distinct nuclide, the atom must be capable of existing for a measurable lifetime, generally greater than 10^{-10} s. Also called nuclear species.

nuclidic mass *Atomic mass.*

NUDET [NUclear Detonation Evaluation Technique] A system for detecting nuclear explosions.

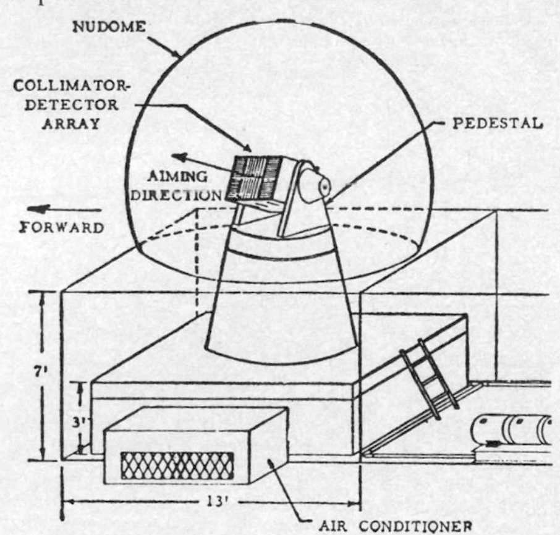

Nudome enclosing detector array used in gamma-ray tracking.

nudome A dome-shaped enclosure that is transparent to nuclear radiation. Used in gamma-ray tracking of missiles during initial stages of takeoff.

null A position of minimum or zero indication or strength, such as that between adjacent lobes of an antenna radiation pattern.

null astatic magnetometer A magnetometer in which the magnetization of a small specimen is measured by balancing its magnetic moment against that of a small current-carrying coil.

null-balance system A measuring system in which the input quantity is measured by producing a null with a corresponding calibrated balancing voltage, current, or other parameter.

null character A control character used as a filler in data processing. It may be inserted or removed from a sequence of characters without affecting the meaning of the sequence, but it may affect format or control equipment. In ASCII standards the null character is the all-zeros character, which is different from numerical zero.

null detector *Null indicator.*

null indicator A galvanometer or other device that indicates when voltage or current is zero. Used chiefly to determine when a bridge circuit is in balance. Also called null detector.

nullity The number of independent meshes that can be selected in a network.

null method A method of measurement in which the measuring circuit is balanced to bring the pointer of the indicating instrument to zero, as in a Wheatstone bridge, and the settings of the balancing controls are then read.

number A designation that represents a quantity, a position in a sequence, a direction, or some other magnitude. It may be one or more digits, a word, a sequence of pulses, or any equivalent designation. As an example, fourteen, 14, XIV, and 1110 all represent the same number. Also called numeral.

numeral 1. A conventional symbol that represents a number, such as the Arabic digits 0 to 9. 2. *Number.*

numeric 1. A number or numeral. 2. *Numerical.*

numerical Pertaining to numbers. Also called numeric.

numerical control [abbreviated NC] A control system for machine tools and industrial processes, in which numerical values corresponding to the desired positions of tools or controls are recorded on punched cards or tapes or magnetic tapes in such a way that they can provide automatic control of the operation. Generally used with a computer or microprocessor to maximize operating efficiency.

numerical display A display that gives the exact numerical value of a quantity in readable form.

nutating feed An oscillating antenna feed for producing an oscillating deflection of a tracking radar beam, in which the plane of polarization remains fixed.

nutation A periodic variation in the inclination of the axis of a spinning gyroscope to the vertical, between certain limiting angles, or a corresponding motion of a radar dipole or other object.

nutation field The time-variant three-dimensional field pattern of a directional or beam-producing antenna that has a nutating feed.

nutator A mechanical or electric device that moves a radar beam in a circular, conical, spiral, or other manner periodically to obtain greater air surveillance than could be obtained with a stationary beam.

nuvistor An electron tube in which all electrodes are cylindrical, to permit assembly by automatic machines that drop the metal electrodes into place, one inside the other, in a ceramic envelope.

nV Abbreviation for *nanovolt.*

nv Symbol for *neutron flux.*

nvt The time integral of neutron flux, equal to neutron flux nv multiplied by time t.

nW Abbreviation for *nanowatt.*

nybble A slang term used to represent some fraction of a byte. When a byte is 8 bits, a nybble is half a byte or 4 bits. A nybble is generally assumed to be 4 bits, so there are 4 nybbles in a 16-bit byte. Use of this term is deprecated unless its meaning and that of byte are precisely specified.

nylon Generic name for a family of polyamide resins, widely used for molded coil forms and as a protective coating over insulation on wires.

Nyquist criterion A parameter used in servomechanism theory; it corresponds to the open-loop harmonic response function.

Nyquist diagram A diagram from which stability of a control system may be determined. It is a closed polar plot of the loop transfer function of the system.

O

OBI Abbreviation for *omnibearing indicator.*

objective lens The first lens through which rays pass in an optical or electronic lens system.

object language A computer language that is generated by an automatic coding routine. It is usually in machine language form, but it may have some output steps in object language that acts as the source language for the next step. Also called target language.

oblique-incidence transmission The transmission of a radio wave at a slant to the ionosphere and back to earth, as in long-distance radio communication.

OBS Abbreviation for *omnibearing selector.*

obsolescence-free Designed to be as universal as possible, so the possibility of becoming outdated because of new developments is minimized.

obstacle gain The increase in signal strength obtained over a long radio-communication path when a mountain obstacle or range of hills is located about halfway between transmitting and receiving antennas. This obstacle gain offsets some of the path losses normally expected.

occluded gas Gas absorbed in a material, as in the electrodes, supports, leads, and insulation of a vacuum tube.

OCR 1. Abbreviation for *optical character reader.* 2. Abbreviation for *optical character recognition.*

octal base An electron-tube base that has a central aligning key and positions for eight equally spaced pins. Pins not needed for a particular tube are omitted without the positions of the remaining pins being changed.

octal digit One of the symbols 0, 1, 2, 3, 4, 5, 6, and 7 when used as a digit in the scale of eight.

octal notation Notation that uses the scale of eight. Here the number 235 means 5 times 1, plus 3 times 8, plus 2 times 8 squared, which is equal to 157 in decimal notation.

octal socket A socket for an octal tube.

octal tube An electron tube having an octal base.

octantal error An error that occurs in a measured bearing when spaced antennas in an electronic navigation system are used. The error varies sinusoidally through 360°, with four positive and four negative maximums.

octave The interval between any two frequencies that have a ratio of 2 to 1. Thus, going one octave higher means doubling the frequency. Going one octave lower means changing to one-half the original frequency.

octave-band noise analyzer An analyzer that contains filters which permit measurements of sound levels in a number of different one-octave bands centered on ANSI preferred frequencies. For evaluating speech interference levels, the most

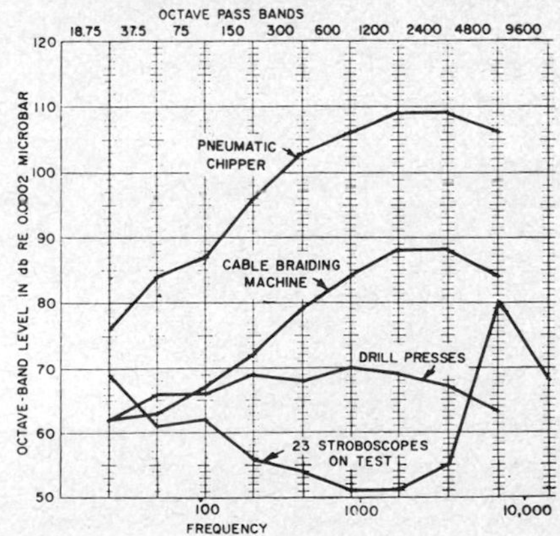

Octave-band noise analyzer measurements of pressure levels encountered in industry at frequencies up to 10,000 Hz. To prevent damage to hearing, level should not exceed 85 dB for daily exposure over a period of years in any band from 300 to 4800 Hz. Operators of cable-braiding machines and pneumatic chippers should therefore have ear protection.

important octave bands are 0.6 to 1.2, 1.2 to 2.4, and 2.4 to 4.8 kHz.

octave-band oscillator An oscillator that can be tuned over a frequency range of 2 to 1, so that its highest frequency is twice its lowest frequency.

octave-band pressure level The band pressure level for a frequency band that corresponds to a specified octave of sound. The location of an octave-band pressure level on a frequency scale is usually specified as the geometric mean of the upper and lower frequencies of the octave.

octave filter A bandpass filter in which the upper cutoff frequency is twice the lower cutoff frequency.

octet An 8-bit byte.

octode An eight-electrode electron tube that contains an anode, a cathode, a control electrode, and five additional electrodes which are ordinarily grids.

odd-even check A forbidden-combination check in which an extra digit is carried along with each word to indicate whether the total number of 1s in the word is odd or even.

odd-even nuclei Nuclei that have an odd number of protons and an even number of neutrons.

odd harmonic A harmonic that is an odd multiple of the fundamental frequency.

odd-line interlace Interlace in which each field contains an extra half-line. Thus in the standard 525-line television picture, each field contains 262.5 lines.

odd-odd nuclei Nuclei that have an odd number of protons and an odd number of neutrons.

O display An A display that has an adjustable notch for measuring distance to a radar target.

odograph An automatic electronic map-tracer that plots on a map or cross-section paper the exact course taken by a jeep or other vehicle. Phototubes and thyratrons transfer the indications of a precision magnetic compass to a plotting unit actuated by the speedometer drive cable, causing a pen to trace the course.

odoriferous homing Homing on the ionized air produced by the exhaust gases of a snorkeling submarine. A hunter-killer aircraft equipped with

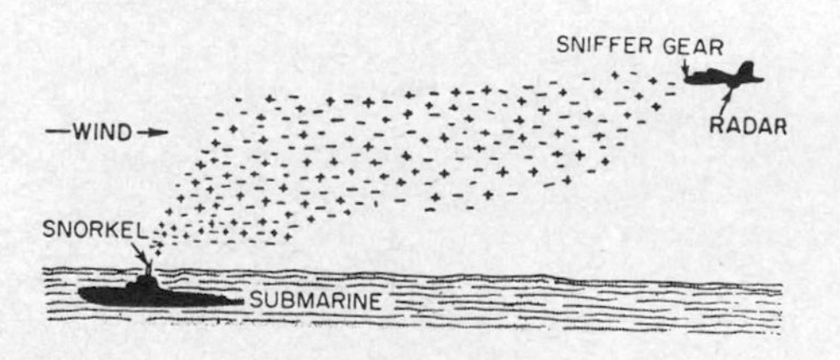

Odoriferous homing being used by hunter-killer aircraft.

sensitive electronic equipment approaches the suspected location from downwind, to pick up the scent from miles away. Also called sniffer gear.

Oe Abbreviation for *oersted.*

O electron An electron that has an orbit in the O shell, which is the fifth shell of electrons surrounding the atomic nucleus, counting out from the nucleus.

OEM Abbreviation for *original equipment manufacturer.*

oersted [abbreviated Oe] The unit of magnetic field strength H in the CGS system up to 1930, when it was replaced by the gauss. Both terms have now been replaced by ampere per meter as the SI unit.

off A term used (in connection with on) to designate the inoperative state of a device or one of two possible conditions in a circuit. Sometimes spelled OFF for emphasis or clarity.

off-center dipole A rotating dipole mounted in a parabolic reflector at an angle to the axis of rotation, to give conical scanning.

off-center PPI display A plan-position-indicator display in which the zero position of the time base is not at the center of the display. The off-center arrangement permits enlargement of the display for a selected portion of the radar service area.

off-line Pertaining to peripheral equipment or devices not in direct communication with the central processing unit of a computer.

off-line storage A storage device not under control of the central processing unit.

OFF period The portion of an operating cycle in which an electron tube, transistor, or other active device is nonconducting.

off-scale Beyond the normal indicating range, as when excessive current makes the pointer of a meter swing beyond the right-hand limit of the printed scale.

offset 1. In a process control system, the steady-state difference between the desired control point and that actually obtained. Offset is an inherent characteristic of positioning controller action. 2. In a linear amplifier, the change in input voltage that is required to produce zero output voltage in the absence of an input signal. 3. In a digital circuit, the DC voltage on which a signal is impressed. 4. In a DC amplifier, the small and relatively constant temperature-dependent voltage that exists between the two input terminals. 5. The amount of unbalance between the two halves of a symmetrical circuit.

offset angle The smaller of the two angles between the vibration axis of a phonograph pickup stylus and a line connecting the stylus point to the vertical pivot of the pickup arm. The angles are measured in the plane of the disk record.

offset-course computer *Course-line computer.*

offset current The direct current that appears as an error at either input terminal of a DC amplifier when the input current source is disconnected.

offset voltage The DC voltage that remains between the input terminals of a DC amplifier when the output voltage is zero.

off-target jamming Use of a jammer at a location

well removed from the main units of a force, to prevent utilization of the jamming signals by enemies to their own advantages.

off the shelf Available for immediate shipment.

OFHC Abbreviation for *oxygen-free high-conductivity copper.*

O guide A surface-wave transmission line that consists of a hollow cylindrical structure made of a thin dielectric sheet.

ohm [abbreviated Ω] The SI unit of resistance and impedance. A voltage of 1 V applied across a resistance of 1 Ω will produce a current of 1 A.

ohm-centimeter A unit of resistivity. The resistivity of a sample in ohm-centimeters is equal to its resistance R in ohms multiplied by its cross-sectional area A in square centimeters and the result divided by the sample length L in centimeters (resistivity = RA/L).

ohmic contact A purely resistive contact between two surfaces or materials; the contact has a linear relationship of voltage to current.

ohmic heating 1. Heating produced by sending current through a resistance. 2. Heating of plasma by using a pulsed voltage to accelerate plasma electrons and thereby cause more heat-producing collisions with plasma ions.

ohmic value The resistance in ohms.

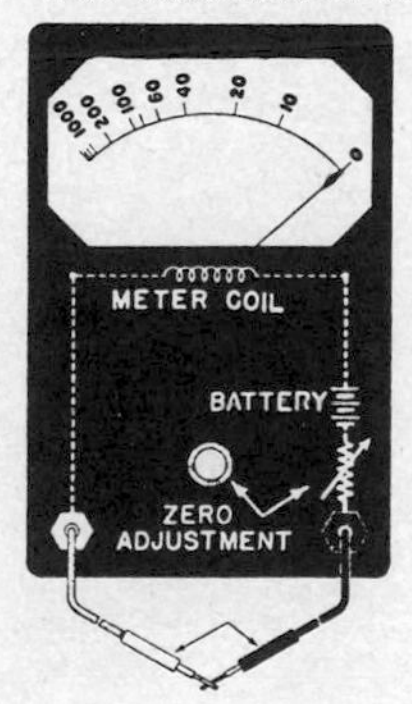

Ohmmeter circuit, showing zero adjustment and scale with zero value at right.

ohmmeter An instrument that measures electric resistance. Its scale may be graduated in ohms or megohms.

ohm per square The resistance of any square area, measured between parallel sides, of thin films of resistive materials.

ohm per volt [abbreviated Ω/V] A sensitivity rating for measuring instruments, obtained by dividing the resistance of the instrument in ohms at a particular range by the full-scale voltage value at that range. The higher the ohm-per-volt rating, the more sensitive the meter, and the less current it will draw from a circuit during a measurement.

Ohm's law The current I in a circuit is directly proportional to the total voltage E in the circuit and inversely proportional to the total resistance R

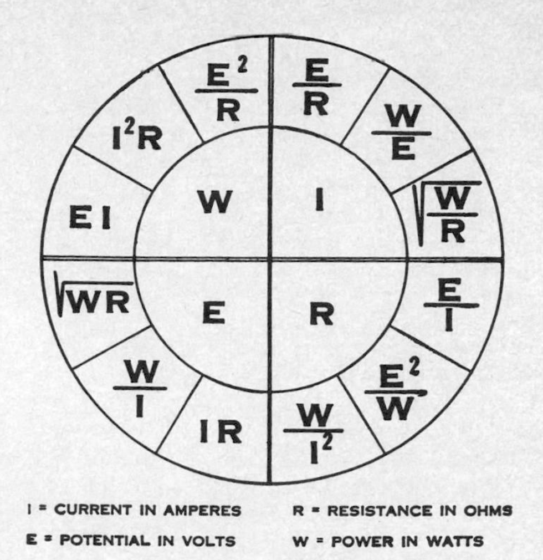

Ohm's law and related expressions for power.

of the circuit. The law may be expressed in three forms: $E = IR$, $I = E/R$, $R = E/I$.

oil-cooled tube An electron tube in which the heat produced is dissipated, directly or indirectly, by oil.

oil diffusion pump A diffusion pump that is similar to a mercury-vapor vacuum pump but uses oil instead of mercury vapor.

oil-filled device A device filled with an insulating oil.

olfactronics The science that deals with the detection and identification of odors. Highly sensitive electronic circuits are generally required for positive identification of predetermined odors, such as those given off by drugs, gunpowder, and other explosives.

omega A Greek letter, used in its capital form Ω to represent the word ohms. Used in its lowercase

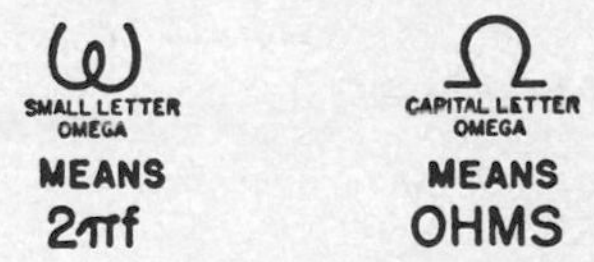

Omega uses in electronics.

form ω as a letter symbol for a value equal to 6.28 times frequency ($\omega = 2\pi f$).

Omega A long-range navigation system originally developed for totally submerged submarines, giving worldwide coverage with 6 to 10 ground stations. Now used for guidance by aircraft, ships, land vehicles, and submarines. The basic frequency is 10.2 kHz, with additional transmissions every 10 s on a time-shared basis at 11.33 and 13.6 kHz.

Ω [Greek letter omega] Abbreviation for *ohm.*

Ω/V Abbreviation for *ohm per volt.*

omni *VHF omnirange.*

omnibearing A bearing indicated by a navigation receiver, using transmissions from an omnirange.

omnibearing converter An electromechanical device that combines an omnibearing signal with vehicle-heading information to give the signals needed to drive the pointer of a radio magnetic indicator. The converter becomes an omnibearing indicator when a pointer and dial are added.

omnibearing-distance facility A radio facility that consists of an omnidirectional range in combination with distance-measuring equipment.

omnibearing-distance navigation Radio navigation that uses a system of polar coordinates as a reference in connection with omnibearing-distance facilities. It can furnish sufficiently accurate data to permit use of a computer that will provide arbitrary course lines anywhere within the coverage area of the system. Also called rho-theta navigation.

omnibearing indicator [abbreviated OBI] An instrument that presents an automatic and continuous indication of an omnibearing.

omnibearing line One of an infinite number of straight lines radiating from the geographical location of a VHF omnirange.

omnibearing selector [abbreviated OBS] An instrument that can be set manually to any desired omnibearing, to control a course-line deviation indicator.

omnidirectional Radiating or receiving equally well in all directions. Also called nondirectional.

omnidirectional antenna An antenna that has an essentially circular radiation pattern in azimuth and a directional pattern in elevation. Also called nondirectional antenna.

omnidirectional beacon A beacon that radiates radio signals equally well in all directions.

omnidirectional hydrophone A hydrophone that responds equally well in all directions to waterborne sound waves.

omnidirectional microphone A microphone whose response is essentially independent of the direction of sound incidence. Also called nondirectional microphone and astatic microphone.

omnidirectional range A radio facility that provides bearing information to or from its location at all azimuths within its service area, as directional guidance for pilots. Also called omnirange.

omnidistance The distance between a vehicle and an omnibearing-distance facility.

omnigraph An instrument that converts Morse code signals on punched tape into corresponding buzzer-produced audio signals for training purposes.

omnirange *Omnidirectional range.*

on A term used (in connection with off) to designate the operating state of a device or one of two possible conditions in a circuit. Sometimes spelled ON for emphasis or clarity.

on-course curvature The rate of change of the indicated navigation course with respect to distance along the course line or path.

on-course signal A signal indicating that the aircraft in which it is received is on course, following a guiding radio beam.

ondograph An instrument that draws the waveform of an AC voltage step by step. A capacitor is charged momentarily to the amplitude of a point on the voltage wave, then discharged into a recording galvanometer.

ondoscope A glow-discharge tube that detects high-frequency radiation, as in the vicinity of a radar transmitter. The radiation ionizes the gas in the tube and produces a visible glow.

one-address instruction A digital computer programming instruction that explicitly describes one operation and one storage location. Also called single-address instruction.

one-group model A neutron-behavior model in which neutrons of all energies are treated as having the same characteristics.

one-many function switch A function switch in which only one input is excited at a time and each input produces a combination of outputs.

one-port A self-impedance, such as a choke.

one's complement A number modified in such a way that addition of the modifying number to its original value, plus 1, will equal an even power of 2. A one's complement is obtained mathematically by subtracting the original value from a string of 1s, and electronically by inverting the states of all bits in the number. Thus, the one's complement of 010101 (representing 21) is 101010 (representing 42).

one-shot multivibrator *Monostable multivibrator.*

one state A state of a magnetic cell wherein the magnetic flux through a specified cross-sectional area has a positive value, when determined from an arbitrarily specified direction for positive flux. The opposite state wherein the magnetic flux has a negative value, when similarly determined, is called a zero state.

one-to-partial-select ratio The ratio of a 1 output to a partial-select output from a magnetic cell.

O network A network composed of four impedance branches connected in series to form a closed circuit. Two adjacent points serve as input terminals, and the remaining two junction points serve as output terminals.

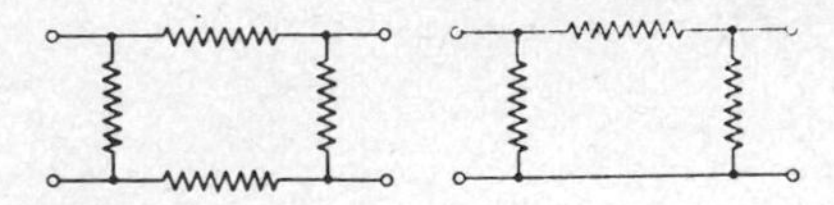

O network has four impedance branches, whereas pi network at right has only three branches.

one-way repeater A repeater that amplifies signals traveling in one direction over wire lines or through space as radio waves.

on-line Pertaining to peripheral equipment under direct control of the central processing unit of a computer, so input data is processed as it is received, and output data is transmitted on a real-time basis.

on-line data reduction The processing of information as rapidly as it is received by the computing system.

ON/OFF control A simple control system in which the device being controlled is either full on or full off, with no intermediate operating positions.

ON/OFF keying Keying in which the modulated wave is transmitted to form the mark signal and suppressed to form the space signal.

ON/OFF switch A switch that turns a receiver or other equipment on or off. Often combined with a volume control in radio and television receivers. Also called power switch.

ON period The portion of an operating cycle in which an electron tube, transistor, or other active device is conducting.

on the air Transmitting a radio signal.

on the beam Following a radio beam.

on-the-fly printer An impact printer in which the character fonts are mounted on a moving drum, chain, wheel, or other mechanism that brings each character in turn to each print position on a line at high speed. Computer-controlled hammers at each print position impact the paper when a desired character is in position. The entire line is not necessarily printed simultaneously, even though it may appear to be on impact printers operating at up to 2000 lines per minute.

opacimeter *Turbidimeter.*

opacity The ability of a substance to block the transmission of radiant energy. Opacity is the reciprocal of transmission.

opamp Abbreviation for *operational amplifier.*

opaque Preventing the passage of radiation or particles.

opaque plasma A plasma through which an electromagnetic wave cannot propagate and is therefore either absorbed or reflected. In general, a plasma is opaque for frequencies below the plasma frequency.

opcom Abbreviation for *optical communication.*

OPDAR Abbreviation for *optical radar.*

open A break in a path for electric current.

open-air ionization chamber *Free-air ionization chamber.*

open-center PPI A plan-position-indicator display in which the initiation of the time base is shown before the transmission of the radar pulse.

open circuit An electric circuit that has been broken, so there is no complete path for current flow.

open-circuit jack A jack that has no circuit-shorting contacts. The circuit can be closed only by the connections made to the plug that is inserted into the jack.

open-circuit parameter One of a set of four transistor equivalent-circuit parameters that specify transistor performance when the input and output currents are chosen as the independent variables. When these parameters are being measured, either the input or output circuit must be open for alternating current.

open-circuit voltage The voltage at the terminals of a source when no appreciable current is flowing.

open-ended Capable of being extended or expanded.

open loop A signal path that has no feedback.

open-loop gain The gain of an amplifier as measured without feedback.

open-loop system A control system that has no means for comparing the output with the input for control purposes.

open shop A computing installation at which computer programming, coding, and operating can be performed by any qualified company employee. Programmers may thus assist with or oversee the running of their programs.

open subroutine A computer subroutine that is inserted directly into the linear operational sequence, in contrast to a closed subroutine, which must be entered by a jump. An open subroutine must be recopied at each point that it is needed in a routine.

open-wire feeder *Open-wire transmission line.*

open-wire transmission line A transmission line that consists of two spaced parallel wires supported by insulators at the proper distance to give a desired value of surge impedance. Such a line acts as a pure resistance when properly terminated. Also called open-wire feeder.

operand A quantity that is to be operated on by a computer.

operate time The total elapsed time from the instant a relay coil is energized until the contacts have opened or firmly closed.

operating angle The electrical angle (portion of the grid voltage cycle) during which anode current flows in an amplifier or electron tube. Operating angles for three types of amplifiers are class A–360°, class B–180° to 360°, class C–less than 180°.

operating frequency The frequency at which a device is designed to operate.

operating point A point on the family of characteristic curves for an electron tube, transistor, or other active device that corresponds to the average electrode voltages or currents in the absence of a signal.

operating power The power that is actually supplied to the antenna of a transmitter.

operating system [abbreviated OS] An integrated set of data-processing programs and a control program that together improve the operating effectiveness of a computer. It may include software for such functions as accounting, compilation, data management, debugging, input/output control, scheduling (batch, real-time, or time-sharing), and storage assignment.

operation The action specified by a single computer instruction.

operational Immediately usable.

operational amplifier [abbreviated opamp] A DC amplifier that has high gain and excellent stability. Originally developed for mathematical operations in analog computers. Now widely used in integrated-circuit form for a variety of control, instrumentation, signal-processing, and other applications at frequencies ranging from DC to many megahertz. Negative feedback provides precise control of response characteristics. One common configuration uses two differential amplifier stages directly coupled in cascade to give balanced input terminals, feeding a single output stage.

operational power supply A regulated power supply in which the control amplifier design has been optimized for signal-processing applications. It will generally have one or more operational amplifiers and a power amplifier in addition to conventional power-supply circuits.

operational programming The process of controlling the output voltage of an operational power supply by voltage, current, or other signals that are operated on by the power supply in a predetermined fashion.

operational trigger A circuit that provides trigger action, in response to very small changes in input, by combining some of the features of a Schmitt trigger with those of an operational amplifier which has positive feedback.

operation code 1. The list of operation parts occurring in an instruction code, together with the names of the corresponding operations. 2. *Operation part.*

operation number A number that indicates the position of an operation or its equivalent subroutine in the sequence which forms a problem routine for a computer.

operation part The part of a computer instruction that usually specifies the kind of operation to be performed but not the location of the operands. Also called operation code.

operations research [abbreviated OR] The use of analytic methods, adapted from mathematics, to provide criteria for decisions concerning the actions of people, commercial or military organizations. and machines.

operation time The time after simultaneous application of all electrode voltages for a current to reach a stated fraction of its final value.

operator 1. A person whose duties include operation, adjustment, and maintenance of a piece of equipment, such as a computer or transmitter. 2. A symbol that represents a mathematical operation to be performed.

opposition The condition in which the phase difference between two periodic quantities that have the same frequency is 180°, corresponding to one half-cycle.

optic 1. Pertaining to the eye. 2. Pertaining to the lenses, prisms, and mirrors of a camera, microscope, or other conventional optical instrument.

optical Pertaining to or utilizing visible or near-visible light. The extreme limits of the optical spectrum are about 1000 Å (0.1 μm or 3×10^{15} Hz) in the far ultraviolet and 300,000 Å (30 μm or 10^{13} Hz) in the far infrared.

optical amplifier An optoelectronic amplifier in which the electric input signal is converted to light, amplified as light, then converted back to an electric signal for the output.

optical axis 1. The straight line that passes through the centers of curvature of the surfaces of a lens. Light rays passing along this direction are neither refracted nor reflected. 2. In a quartz crystal, the Z axis is the optical axis and runs lengthwise through the mother crystal from apex to apex.

optical character reader [abbreviated OCR] A character reader that scans printed or handwritten characters with some form of photoelectric system and compares combinations of light and dark areas with character data stored in a computer or other recognition unit to find a match. The corresponding machine-readable code for the identified character is then fed into a computer or other data-processing machine. Also called photoelectric character reader.

ABCDEFGH
IJKLMNOP

Optical character recognition font developed by American National Standards Institute.

optical character recognition [abbreviated OCR] Automatic recognition of printed characters by optical methods that generally involve a scanning light beam and a light-sensitive device.

optical communication [abbreviated opcom] Communication over short distances by means of beams of visible, infrared, or ultraviolet radiation, or over much longer distances with laser beams.

optical computer A computer that uses various combinations of holography, lasers, and mass-storage memories for such applications as ultrahigh-speed signal processing, image deblurring, and character recognition. Specific applications include fingerprint identification and the sharpening of images obtained with coherent side-looking synthetic-aperture radar. A typical system consists of an optical input device, image scanner, image analyzer, bandwidth compressor, and digitizer feeding a digital computer. Used in machine processing, enhancement, and analysis of aerial, radiographic, and other images.

optical computing A form of data processing that is based largely on the concepts of Fourier optics and holography. Used chiefly for processing images obtained from acoustic, electronic, microwave, optical, or other sources. Applications include analysis and enhancement of aerial and radiographic images.

optical contact bond A bond that depends on the fact that two highly polished plane metal surfaces will adhere with great force when pressed together. Used for cold welding of parts that must transfer acoustic waves at microwave frequencies, where the bond must be a small fraction of the wavelength of the energy being transferred.

optical coupler *Optoisolator.*

optical coupling Coupling between two circuits by a light beam or light pipe that has transducers at opposite ends, to isolate the circuits electrically.

optical density *Photographic transmission density.*

optical diffraction velocimeter *Laser velocimeter.*

optical encoder An encoder that converts positional information into corresponding digital data by interrupting light beams directed on photoelectric devices.

optical fiber A transparent fiber, usually of glass or plastic, used in fiber optics.

optical-fiber cable *Optical waveguide.*

optical filter A filter that consists of a pane of glass or other selectively transparent material which transmits only certain wavelength ranges in the visible, ultraviolet, and infrared spectrums.

optical heterodyning The process of combining an incoming laser signal beam with a laser local oscillator beam by making both beams fall on the

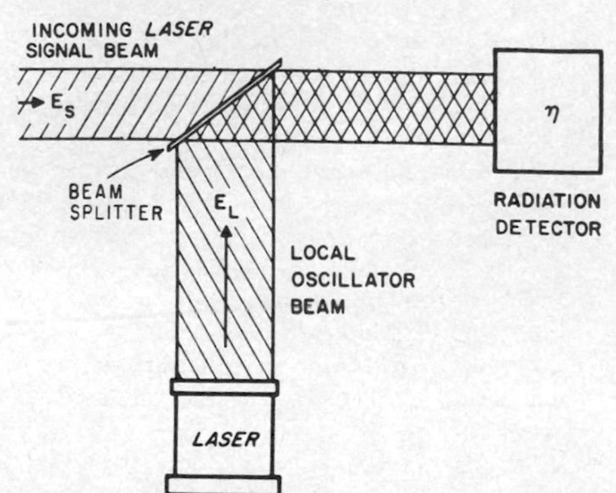

Optical heterodyning.

surface of a radiation detector. A current at the difference frequency of the two laser beams is generated by the nonlinear process of photodetection, for conventional amplification.

optical isolator *Optoisolator.*

optically coupled isolator *Optoisolator.*

optical maser *Laser.*

optical memory A computer memory that uses optical techniques which generally involve an addressable laser beam, a storage medium which responds to the beam for writing and sometimes for erasing, and a detector which reacts to the altered character of the medium when it uses the beam to read out stored data.

optical modulator A device for modulating or chopping a beam of radiant energy. Tuning fork, motor-driven, or vibrator devices are used at frequencies up to several thousand hertz, and electrooptical devices are used for frequencies up into the gigahertz range.

optical oscillograph A recorder in which a light source, a mirror galvanometer, and appropriate optical lenses focus a small spot of light on photographic paper. The signal current being re-

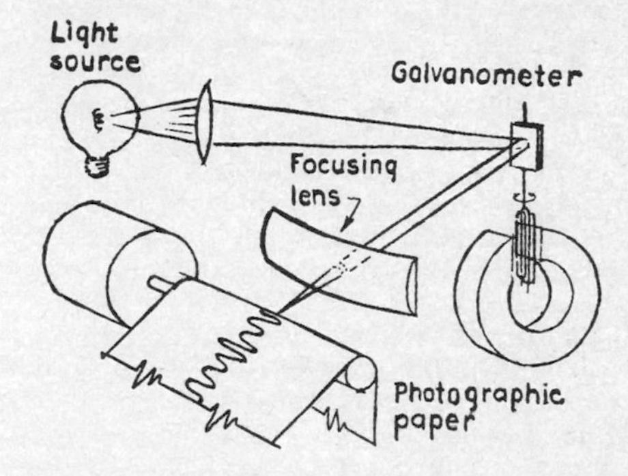

Optical oscillograph.

corded is fed to the coil of the galvanometer, thereby moving the mirror and deflecting the recording light beam in proportion to the magnitude of the current. A motor drive moves the paper during the recording process, to give a waveform of current recorded against time.

optical pattern A pattern observed when the surface of a laterally recorded phonograph record is illuminated by a light beam that is essentially parallel to the surface of the record. When bands of different constant frequencies are recorded, the width of the pattern is approximately related to the frequency response of the record. Also called Christmas-tree pattern.

optical photon A photon that has energy which corresponds to wavelengths in or near the visible spectrum, ranging from about 0.2 to 1.5 μm.

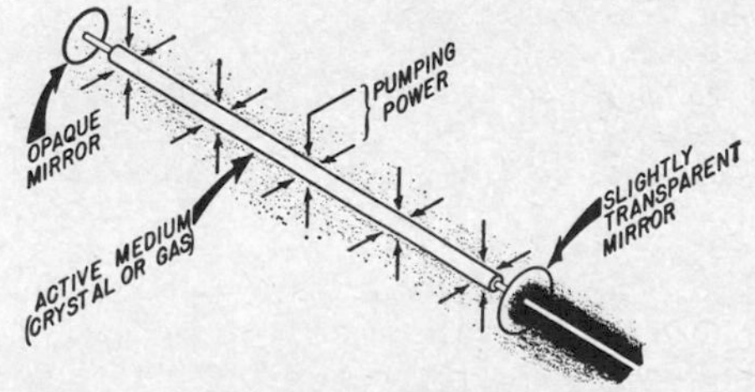

Optical pumping in laser.

optical pumping The process of altering the number of atoms or atomic systems in a set of energy levels by absorption of incident light, thereby raising the energy levels. In a semiconductor laser, the semiconductor material is irradiated with photons that have sufficient energy to produce the electron-hole pairs required for population inversion and lasing action in a semiconductor laser.

optical radar [abbreviated OPDAR] A laser system that measures elevation and azimuth angles and the slant range of a missile during its firing period.

optical reader A computer data-entry machine that converts printed characters, bar or line codes, and pencil-shaded areas into a computer-input code format. When scanning is used, the optical image is converted into an electric signal that is unique for each character, and this signal in turn is recognized and converted into the corresponding computer code. Examples include bar-code, mark-sense, and optical character readers.

optical relay An optoisolator in which the output device is a light-sensitive switch that provides the same on and off operations as the contacts of a relay.

optical resonance Luminescence in which the frequencies of the exciting and emitted radiation are essentially the same.

optical scanner A scanner that moves a beam of light across printed or written characters, symbols, or bar codes, measures the reflected light photoelectrically, and converts the resulting signals into corresponding digital representations.

optical sound recorder *Photographic sound recorder.*

optical sound reproducer *Photographic sound reproducer.*

optical sound track A sound track that consists of a variable-width or variable-density photographic recording of sound signals, placed at one side of the frames of a motion-picture film.

optical star tracker A star tracker in which optoelectronic equipment locks onto the light of a single star.

optical storage Storage of large amounts of data in permanent form on photographic film or its equivalent, for nondestructive readout by a light source and photodetector. Used for storing fixed data in such applications as airborne navigation computers and machine-tool or process-control computers. Optical storage is generally in the form of a rotating disk, with reading heads for each recorded circle of data on the disk or with heads that are moved to the desired track for readout.

optical switch A switch in which a photodetector provides ON/OFF switch action when illuminated.

optical-track command guidance Command guidance in which tracking of missile and target is done optically. The resulting data is converted into appropriate information for transmitting to the missile for correcting its flight path to the target.

optical twinning A defect that occurs in natural quartz crystals. This generally results in small regions of unusable material that are discarded when a crystal for piezoelectric use is cut.

optical waveguide A waveguide in which a light-transmitting material such as a glass or plastic fiber is used for transmitting information from point to point at wavelengths somewhere in the ultraviolet, visible-light, or infrared portions of the spectrum. Also called fiber waveguide and optical-fiber cable.

optical window The spectral region between 3000 and 20,000 Å (0.3 and 2 μm in wavelength), in which visible and near-visible radiation will pass through the atmosphere of the earth.

optics The branch of science that deals with the phenomena of light and vision, involving the portion of the electromagnetic spectrum between microwaves and x-rays. This range includes ultraviolet, visible, and infrared radiation.

optimization The process of maximizing or minimizing a desirable performance characteristic during design, construction, or operation of a device or system.

optimum bunching The bunching condition that produces maximum power at the desired frequency in an output gap of a microwave tube.

optimum coupling *Critical coupling.*

optimum load The load impedance value at which there is maximum transfer of power from source to load.

optimum programming Computer programming in which instructions and data are so stored that access time is a minimum.

optimum working frequency The most effective frequency at a specified time for ionospheric propagation of radio waves between two specified points.

optoacoustic modulator A modulator in which an acoustic wave interacts with a coherent light beam in a medium like a single crystal of lead molybdate. Used to deflect and modulate a laser beam.

optocoupler *Optoisolator.*

optoelectronic amplifier An amplifier in which the input and output signals and the method of amplification may be either electronic or optical.

optoelectronic isolator *Optoisolator.*

optoelectronics The branch of electronics that deals with solid-state and other electronic devices for generating, modulating, transmitting, and sensing electromagnetic radiation in the ultraviolet, visible-light, and infrared portions of the spectrum. Examples of optoelectronic devices include cathode-ray tubes, electroluminescent displays, holographic equipment, image converters, lasers, light-emitting diodes, light modulators, liquid crystal displays, optical memories, optical waveguides, optoisolators, photodetectors, solar cells, television camera tubes, and electrooptical devices such as Kerr cells. Although sometimes used interchangeably with electrooptics, optoelectronics is a much broader term.

optoisolator A coupling device in which a light-emitting diode, energized by the input signal, is optically coupled to a photodetector such as a light-sensitive output diode, transistor, or silicon controlled rectifier. The optical coupling through air, optical fibers, or some other form of light pipe provides essentially perfect electric isolation between input and output while transferring signals, data pulses, or control voltages. Also called optical coupler, optical isolator, optically coupled isolator, optocoupler, optoelectronic isolator, photocoupler, and photoisolator.

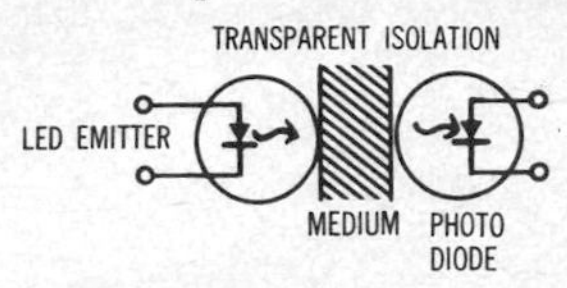

Optoisolator using light-emitting diode at input and photodiode at output.

optophone A photoelectric device that converts ordinary printed characters into characteristic sounds which can be recognized by a blind person.

OR 1. A logic operator in which the output is a true logic 1 if at least one of the inputs is a true logic 1. 2. Abbreviation for *operations research.*

orbit 1. The path described by a body in its revolution about another body, as by a man-made satellite revolving about the earth. 2. To revolve about another body.

orbital 1. Pertaining to electrons outside the nucleus of an atom. 2. An energy state or wave function from which the probability of finding an electron at a particular point can be calculated.

orbital electron An electron that is moving in an orbit around the nucleus of an atom.

orbital-electron capture Electron capture in which the electron usually comes from an orbit of the atom or molecule that contains the transforming nucleus.

orbital quantum number A number equal to the angular momentum of an electron in its orbital motion around a nucleus. This number can have all whole values from zero to $n - 1$, where n designates the main quantum number.

orbital velocity The velocity of a body in orbit. An earth satellite achieves a stationary orbit at an orbital velocity of about 1.9 mi/s (3 km/s), corresponding to an altitude of 22,300 mi (35,700 km).

orbiting astronomical observatory A satellite launched in orbit around the earth primarily for astronomical measurements. One version carries instrumentation to obtain ultraviolet absorption data of intergalactic gas clouds, for mapping the sky in the ultraviolet region.

orbiting geophysical observatory An instrumentation satellite that collects and transmits to earth data on cosmic rays, magnetic fields in space, interplanetary dust particles, and other geophysical data.

orbiting solar observatory An astronomical satellite intended for launching in a solar orbit, for telemetering to earth continuous measurements of solar phenomena.

orbit-shift coil One of the coils used to alter the orbit of beam particles in a betatron or synchrotron, to make the particles strike a target placed outside the stable orbit or enter a deflector to form an external beam.

order 1. The number of vibrations or half-period variations of a field along diameters of a circular waveguide or along the wider transverse axis of a rectangular waveguide. 2. A sequence of items or events. 3. *Instruction.*

ordinary component The component of light that is totally reflected at the cementing layer of a Nicol prism. Only the extraordinary component, which is plane-polarized, passes through the prism.

ordinary-wave component One of the two components into which a radio wave entering the ionosphere is divided under the influence of the earth's magnetic field. The ordinary-wave component has characteristics more nearly like those to be expected in the absence of a magnetic field. Also called O-wave component. The other component is the extraordinary-wave component.

ordinate The value that specifies distance in a vertical direction on a graph.

organic semiconductor An organic material that has unusually high conductivity, often enhanced by the presence of certain gases, and other properties commonly associated with semiconductors. An example is anthracene.

OR gate A multiple-input gate circuit whose output is energized when any one or more of the inputs is in a prescribed state. Used in digital computers.

orientation 1. The relationship between the length, width, and thickness directions of a quartz plate and the rectangular axes of the mother crystal. 2. The physical positioning of a directional antenna or other device that has directional characteristics.

orifice An opening in a waveguide through which energy is transmitted.

origin The intersection of the reference axes on a graph.

original equipment manufacturer [abbreviated OEM] A manufacturer who produces equipment in its final form; the manufacturer generally uses components and subassemblies obtained from other sources.

original master The master produced by electroforming from the face of a wax or lacquer recording. Also called metal master and metal negative.

origin distortion An apparent loss of deflection sensitivity in the region of the undeflected position of the spot in a gas-focused electrostatic-deflection cathode-ray tube, due to a space-charge effect.

OR logic A logic circuit in which any input will produce an output.

orthicon A camera tube in which a beam of low-velocity electrons scans a photoemissive mosaic that is capable of storing a pattern of electric

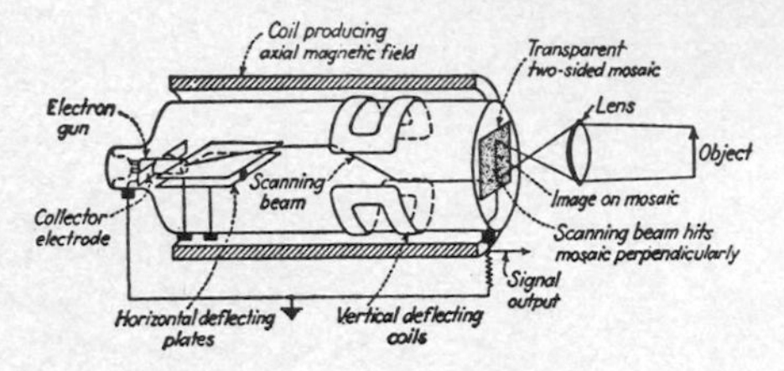

Orthicon construction.

charges. The orthicon has higher sensitivity than the iconoscope.

orthoferrite A magnetic material that contains a combination of ferrites and rare-earth elements. Used in magnetic domains of bubble memories.

orthogonal mode In a magnetic device, a mode of operation characterized by two mutually perpendicular directions of excitation.

orthohydrogen A hydrogen molecule in which the two nuclear spins are parallel, forming a triplet state.

OS Abbreviation for *operating system.*

oscillating-crystal method X-ray diffraction analysis in which the crystal is oscillated through an angle of a few degrees to simplify correlation between the diffracted beams and the crystal planes.

oscillation 1. A periodic change in a variable, as in the amplitude of an alternating current or the swing of a pendulum. 2. *Cycling.* 3. *Vibration.*

oscillator 1. A circuit that generates alternating current at a frequency determined by the values of its components. The oscillations are produced by using positive feedback with an electron tube, transistor, magnetic amplifier, or other amplifying device. Oscillations are maintained by drawing power from a battery or other source of power. For stable frequency characteristics the circuit may include a crystal, tuning fork, or other essentially unvarying source of vibrations. 2. The stage of a superheterodyne receiver that generates an RF signal of the correct frequency to mix with the incoming signal and produce the IF value of the receiver. 3. The stage of a transmitter that generates the carrier frequency of the station or some fraction of the carrier frequency.

oscillator coil The RF transformer used in the oscillator circuit of a superheterodyne receiver or in other oscillator circuits to provide the feedback required for oscillation.

oscillator harmonic interference Interference that occurs in a superheterodyne receiver when an undesired carrier signal beats with a harmonic of the local oscillator to produce the correct IF value.

oscillator-mixer-first detector *Converter.*

oscillator padder An adjustable capacitor used in series with the oscillator tank circuit of a superheterodyne receiver to permit adjusting the tracking between the oscillator and preselector at the low-frequency end of the tuning dial.

oscillator radiation The field strength produced at a distance by the local oscillator of a television or radio receiver.

oscillatory circuit A circuit in which oscillations can be generated or sustained.

oscillatory surge A surge that includes both positive and negative polarity values. A unidirectional surge is a pulse.

oscillistor A bar of semiconductor material, such as germanium, that will oscillate much like a quartz crystal when placed in a magnetic field and carrying direct current that flows parallel to the magnetic field. The current must flow between an ohmic contact and a rectifying contact so injection of plasma is possible. The frequency and waveform of oscillation are changed by exposure to various gas and liquid chemicals.

oscillogram The permanent record produced by an oscillograph, or a photograph of the trace produced by an oscilloscope.

oscillograph A recorder that produces a permanent record of the instantaneous values of one or more varying electrical quantities. In a cathode-ray oscillograph the record is produced, usually on photographic film, by the electron beam of a cathode-ray tube. In other types of oscillographs, the trace is made by a light beam, pen, heated stylus, or other means of writing directly on plain or treated chart paper.

oscilloscope *Cathode-ray oscilloscope.*

oscilloscope tube A cathode-ray tube that produces a visible pattern which is a graphic representation of electric signals.

O shell The fifth layer of electrons about the nucleus of an atom; it has electrons characterized by the principal quantum number 5. It occurs with rubidium (atomic number 37) and all elements that have higher atomic numbers.

osmium [symbol Os] A hard metallic element of the platinum group. Atomic number is 76. When alloyed with iridium, it is used in styli for disk recorders and phonographs.

outage A failure in an electric power system. Since an outage lasting only a fraction of a second can cause errors in data processing, many computers use uninterruptible power supplies as regulators for line-voltage transients and as backups for complete power-line failures.

outdiffusion Diffusion of impurity atoms out of a semiconductor surface by applying heat.

outdoor antenna A receiving antenna erected

outside a building, usually in an elevated location.

outer marker A marker approximately 5 mi (8 km) from the approach end of the runway in an instrument landing system for aircraft. It provides a fix along the localizer course line.

outer-shell electron *Conduction electron.*

outer space The space beyond the atmosphere of the earth.

outgassing Heating an electron tube during evacuation, to remove residual gases occluded in the tube elements.

outlet A power-line termination from which electric power can be obtained by inserting the plug of a line cord. Also called convenience receptacle and receptacle.

out of phase Having waveforms that are of the same frequency but do not pass through corresponding values at the same instant.

out-port The exit for a network.

output 1. The useful energy delivered by a circuit or device. 2. The information that a computer feeds to an external device.

output block A portion of the internal storage of a computer that is reserved for receiving, processing, and transmitting data to be transferred out.

output capacitance The short-circuit transfer capacitance between the output terminal and all other terminals of an electron tube, except the input terminal, connected together.

output gap An interaction gap by which usable power can be abstracted from an electron stream in a microwave tube.

output impedance The impedance presented by a source to a load. For maximum power output, the output impedance should match the load impedance.

output indicator A meter or other device that is connected to a radio receiver to indicate variations in output signal strength for alignment and other purposes, without indicating the exact value of output.

output-limited Restricted by the need to await completion of an output operation, as in process control or data processing.

output meter An AC voltmeter connected to the output of a receiver or amplifier to measure output signal strength in volume units or decibels.

output power The power delivered by a system or component to its load.

output register The computer register that holds processed data until it can be fed to an output device or line.

output resonator The resonant cavity that is excited by density modulation of the electron beam in a klystron and delivers useful energy to an external circuit. Also called catcher and catcher resonator.

output stage The final stage in electronic equipment. In a radio receiver, it feeds the loudspeaker directly or through an output transformer. In an AF amplifier, it feeds one or more loudspeakers, the cutting head of a sound recorder, a transmission line, or any other load. In a transmitter, it feeds the transmitting antenna.

output transformer The iron-core AF transformer used to match the output stage of a radio receiver or AF amplifier to its loudspeaker or other load.

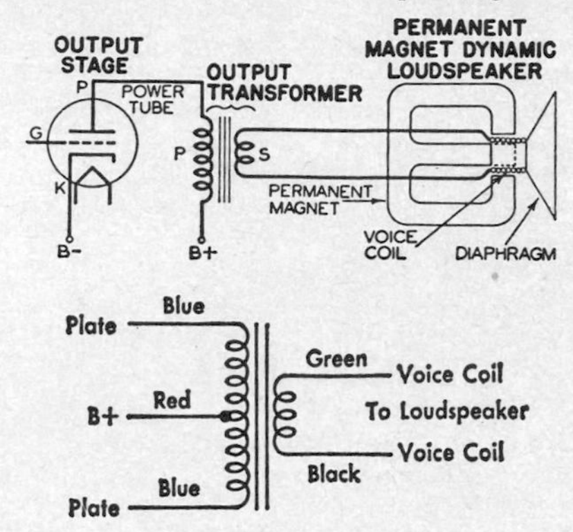

Output-transformer connections and color code.

output tube A power-amplifier tube designed for use in an output stage.

output winding A winding other than the feedback winding of a saturable reactor, through which power is delivered to the load.

oven An enclosure that contains heaters and temperature sensors for maintaining components at a desired constant or programmed variable temperature. In a microwave oven, microwave energy is the source of heat.

overall electric efficiency The ratio of the power absorbed by the load material to the total power drawn from the supply lines in induction and dielectric heating.

overall sound-pressure level The sound-pressure level measured over the entire frequency range of interest, such as from 20 to 20,000 Hz.

overbunching The bunching condition produced by continuation of the bunching process beyond optimum bunching in a velocity-modulation tube.

overcast bombing Bombing of a target through an overcast above the target, using radar or other equipment to aid in sighting through the clouds.

overcoupling The condition in which two resonant circuits are tuned to the same frequency but coupled so closely that two response peaks are obtained. Used to obtain broadband response with substantially uniform impedance.

overcurrent protection Protection against abnormally high currents in a circuit, component, power supply, or load, generally achieved by limiting the duration and magnitude of the abnormally high current.

overcutting Recording on a disk at an excessively

high signal level so the stylus cuts through into adjacent grooves.

overdamping Damping greater than that required for critical damping.

overdriven amplifier An amplifier in which the input signal waveform is intentionally distorted by driving the grid past cutoff or into anode-current saturation.

overflow 1. The condition that arises when the result of an arithmetic operation exceeds the capacity of the number representation in a digital computer. 2. The carry digit arising from this condition.

overlap The amount by which the effective height of the scanning spot in a facsimile system exceeds the nominal width of the scanning line.

overlapping contacts Contacts arranged so one set opens only after the other has closed.

overlay A technique for bringing routines into high-speed storage from some other form of storage during computer processing, so that several routines will occupy the same storage locations at different times. Used when the total storage requirement for instructions exceeds the available main storage.

overlay transistor A bipolar transistor that has many separate emitter areas interconnected by an overlay of metal film, to provide greater power amplification at high frequencies.

overload A load greater than that which a device is designed to handle. It may cause overheating of power-handling components and distortion in signal circuits.

overload capacity The current, voltage, or power level beyond which permanent damage occurs to the device in question.

overload level The level at which operation of a system ceases to be satisfactory because of signal distortion, overheating, or other effects.

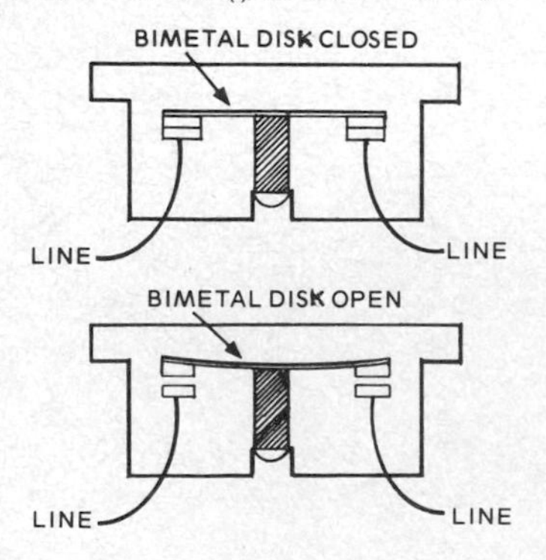

Overload protection in which metal disk is heated by overload current to open contacts.

overload protection Protection against excessive current by a device that automatically interrupts current flow when an overload occurs.

overload relay A relay which operates when current flow in a circuit exceeds the normal value for that circuit, to provide overload protection.

overmodulation Amplitude modulation greater than 100%, causing distortion because the carrier

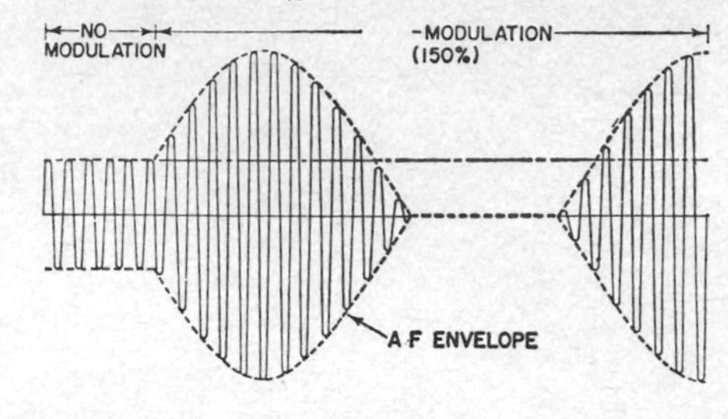

Overmodulation.

voltage is reduced to zero during portions of each cycle.

overpunch To add holes in a column of a punched card or a row on punched tape, to give a code combination that cannot be obtained with a single key stroke.

over-radiation alarm A radiation detector that trips an alarm when a predetermined level of radioactivity is reached.

override To cancel the influence of an automatic control by means of a manual control.

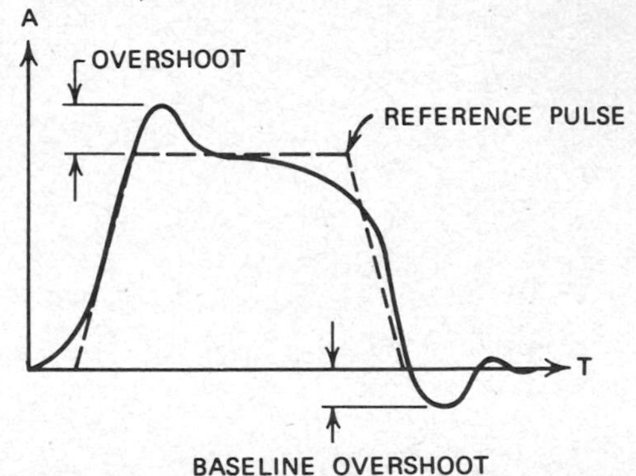

Overshoot occurring with square-wave input.

overshoot An excessive response to a change.

overshoot distortion *Overthrow distortion.*

over-the-horizon communication *Scatter propagation.*

over-the-horizon radar Long-range radar in which the transmitted and reflected beams are bounced off the ionosphere layers, to achieve ranges far beyond line of sight. It can be used to detect and identify launches of ballistic missiles because the missiles change the reflective characteristics of the ionosphere and thereby alter the transmitted radar signals.

overthrow distortion Distortion that occurs when the maximum amplitude of a signal wavefront exceeds the steady-state amplitude of the signal wave. Also called overshoot distortion.

overtone A component of a complex tone that has a pitch higher than that of the fundamental pitch. The term overtone is sometimes used in place of harmonic, the nth harmonic being called the $(n - 1)$st overtone.

overtone crystal unit A crystal unit in which the quartz plate operates at a higher order than the fundamental. Also called harmonic-mode crystal unit.

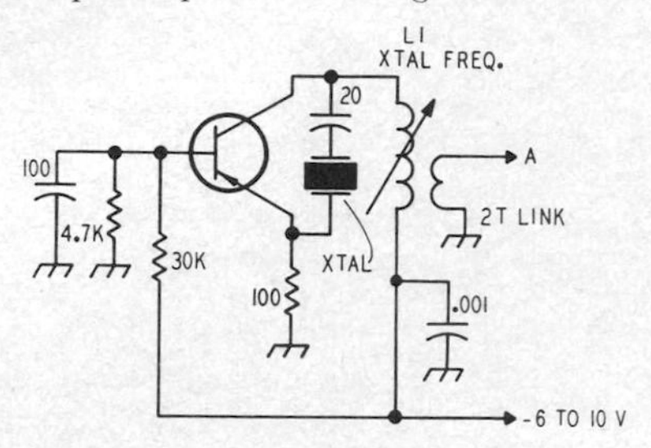

Overtone crystal unit in RF oscillator using 2N1744 or other VHF transistor. Crystal is third-overtone type.

overtravel Movement beyond the operating position in a limit switch.

overvoltage A voltage higher than the normal or predetermined limiting value.

overvoltage crowbar A circuit that monitors the output of a power supply and prevents the output voltage from exceeding a preset voltage, under any failure condition, by having a low resistance (crowbar) placed across the output terminals when an overvoltage occurs. A silicon controlled rectifier is often used as the crowbar.

overvoltage protection Interruption of power to a circuit, or reduction of voltage, by a device that responds to excessive voltage.

overvoltage relay A relay that operates when the voltage applied to its coil reaches a predetermined value.

overvoltage test A test made at a voltage well above rated operating voltage.

overwrite To place new data in a memory location in such a way that any data previously stored in that location is destroyed.

Ovshinsky effect The ability of certain amorphous or glassy semiconductor materials to serve as switches that respond to applied electric fields independently of polarity.

O-wave component *Ordinary-wave component.*

Owen bridge A four-arm AC bridge that measures self-inductance in terms of capacitance and resistance. Bridge balance is independent of frequency.

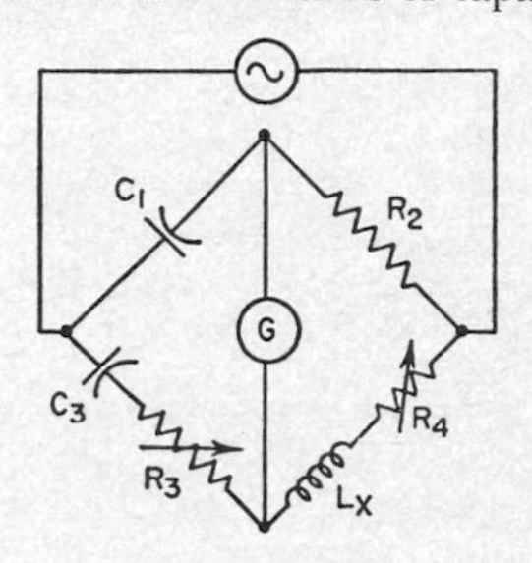

Owen-bridge circuit.

oxide-coated cathode A cathode that has been coated with oxides of alkaline-earth metals to improve electron emission at moderate temperatures. Also called Wehnelt cathode.

oxide isolation Isolation of the elements of an integrated circuit by forming a layer of silicon oxide around each element.

oxide passivation Passivation of a semiconductor surface by producing a layer of an insulating oxide on the surface.

oximeter A photoelectric instrument that measures continuously the oxygen content of the blood in a person, by photoelectric measurement of the intensity of a light beam passed through part of the ear.

oxygen [symbol O] A gaseous element. Atomic number is 8.

oxygen-free high-conductivity copper [abbreviated OFHC] Pure copper that has 100% conductivity, used for the construction of high-power electron tubes because it does not release appreciable gas when hot.

ozone-producing radiation Ultraviolet radiation shorter in wavelength than about 220 nm, at which oxygen is decomposed to produce ozone.

P

p Abbreviation for *pico-*.

P 1. Symbol for *plate* (anode) in an electron tube. 2. Symbol for *primary winding,* used on circuit diagrams to identify the primary winding of a transformer. 3. Abbreviation for *peak.* 4. Abbreviation for *peta-*. 5. Abbreviation for *positive.*

P 1. Symbol for *permeance.* 2. Symbol for *power.*

pA Abbreviation for *picoampere.*

Pa Abbreviation for *pascal.*

PABX Abbreviation for *private automatic branch exchange.*

pacemaker A pulsed battery-operated oscillator implanted in the body to deliver electric impulses to the muscles of the lower heart, either at a fixed rate or in response to a sensor that detects when the patient's pulse rate slows or ceases. Several types of power sources are used. The most common and least costly is a mercury-zinc, lithium, or other primary cell which requires replacement surgically under local anesthetic at intervals of 2 to 6 years. A nuclear battery, although more costly, can last up to 20 years. Some patients prefer a nickel-cadmium storage cell, which requires weekly recharging and is recharged by holding a recharging device against the chest for about an hour. In earlier pacemakers, the output of an external oscillator was inductively coupled to an output wire running to the heart. Also called cardiac pacemaker and heart pacer.

pack To combine several different brief fields of information into one machine word for a digital computer.

packaged magnetron An integral structure comprising a magnetron, its magnetic circuit, and its output matching device.

packaging The process of physically locating, connecting, and protecting devices or components.

packaging density The number of components per unit volume in a working system or subsystem.

packet transmission The transmission of standardized packets of data over transmission lines in a fraction of a second by networks of high-speed switching computers that have the message packets stored in fast-access core memory. The system provides automatic checking for transmission errors.

packing density The number of units of digital information per unit length or unit area of a recording or storage medium for a digital computer, such as the number of binary digits of polarized spots per inch of magnetic-tape length. Commonly used packing densities are 200, 556, and 800 b/in.

packing fraction The mass defect of an atom divided by its mass number. It is negative for most atoms that have mass numbers between 16 and 180 and positive for all atoms outside this range.

pad 1. An arrangement of fixed resistors used to reduce the strength of an RF or AF signal a desired fixed amount without introducing appreciable distortion. Also called fixed attenuator. The corresponding adjustable arrangement is called an attenuator. 2. *Terminal area.*

padder A trimmer capacitor inserted in series with the oscillator tuning circuit of a superheterodyne receiver to control calibration at the low-frequency end of a tuning range.

padding A technique used to fill out a block of information with dummy records in a computer.

paddle A large, flat, paddle-shaped support for solar cells, used on some satellites.

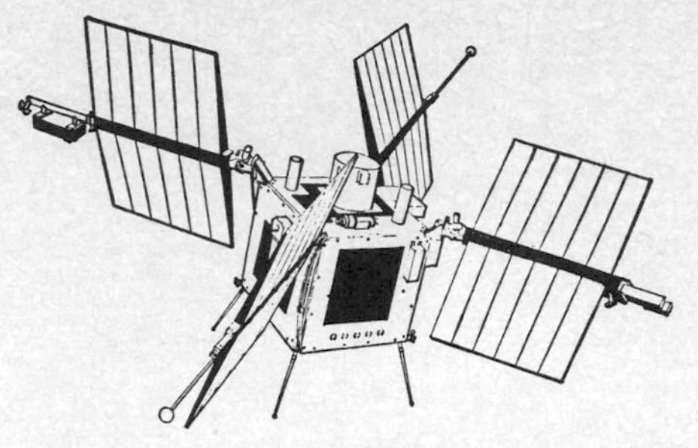

Paddles used on orbiting satellite to support solar cells that convert solar radiation into electric power.

pad electrode One of a pair of electrode plates between which a load is placed for dielectric heating.

page The contents of one section of a computer storage, generally a subdivision of a program that is moved as a block into the main computer memory. A typical page size is 4096 consecutive bytes.

page printer A printer used with a computer that composes and organizes an entire page of text, to provide printout in final page format, including allowance of space for illustrations.

paging system A radio common-carrier service that has one or more central stations which transmit beeps only or beeps plus voice on a selective calling basis to small battery-operated pagers carried by each person who may have to be paged. In a beep-only system, the person being paged hears only an audio beep and must call the dispatcher by telephone to get the message. In a beep-plus-voice system, the person hears both the alerting beep and the message.

painted printed circuit A printed circuit in which the desired conductive pattern is produced by applying a conductive liquid, using such techniques as spraying, silk screens, and offset printing.

paired cable A cable in which all conductors are twisted pairs.

pairing Faulty interlace in a television picture, wherein the lines of one field do not fall exactly between those of the preceding field. When this defect is serious, the lines of alternate fields tend to pair up and fall on one another, cutting the vertical resolution in half.

pair production The conversion of a photon into an electron and a positron when the photon traverses a strong electric field, such as that surrounding a nucleus or an electron.

pair-production absorption The absorption of gamma rays or other photons in the process of pair production.

PAL [Phase-Alternation Line] A 625-line 50-field color television system originally developed in West Germany, in which the hue and saturation information are carried by quadrature modulation, but with one of the two modulations switched 180° from line to line at the transmitter. A delay line in the receiver restores the correct phase of the two modulations by delaying one modulation for the duration of one line. Other countries now using this system include Austria, Belgium, Brazil, Denmark, Finland, the Netherlands, Norway, Sweden, Switzerland, and the United Kingdom. Also called phase-alternation line system.

palladium [symbol Pd] A metallic element. Atomic number is 46.

Palmer scan A combination of a circular or raster-type radar antenna scan with a conical scan.

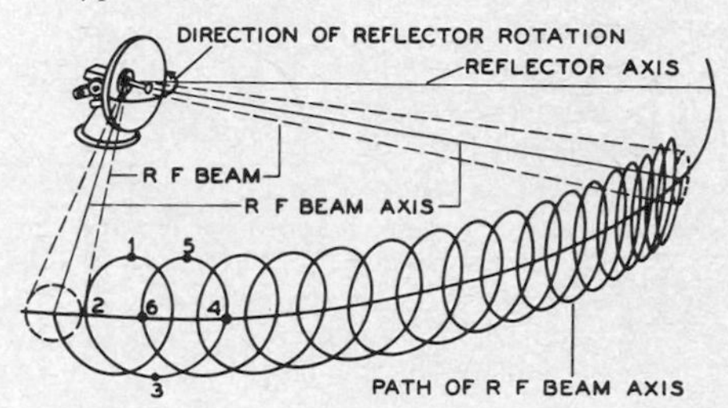

Palmer scan.

The beam is swung around the horizon concurrently with the conical scan.

palm reader An automatic personnel identification device in which employees insert encoded plastic cards into the unit and place their palms on top of the reading element. The device scans the size and shape of each hand and compares the resulting digital readout with that encoded on each card.

PAM Abbreviation for *pulse-amplitude modulation.*

PAM/FM Pulse-amplitude modulation of subcarriers, used to frequency-modulate a carrier.

pan To tilt or otherwise move a television camera vertically and horizontally to keep it trained on a moving object or secure a panoramic effect.

pancake coil A coil whose turns are arranged in a flat spiral.

pancake motor A servomotor that has a large diameter and thin cross section. Used to provide controlled torque for small angular movements, such as for driving scanning antennas. The large diameter permits mounting as many as 80 poles, to provide a desired slow speed for direct-drive requirements. Also called torquer.

panchromatic Sensitive to all wavelengths within the visible spectrum, although not uniformly so.

panel 1. A metallic or nonmetallic sheet on which the operating controls of a receiver, transmitter, or other electronic unit are mounted. 2. A group of guests participating in a television or radio forum or quiz game.

panel meter A meter designed for mounting on a panel, usually in a round or rectangular hole.

panning Moving a television camera across a field of view.

panoramic adapter An adapter used with a search receiver to provide a visual presentation, on an oscilloscope screen, of a band of frequencies extending above and below the center frequency to which the search receiver is tuned.

panoramic display A display that simultaneously shows the relative amplitudes of all signals received at different frequencies.

panoramic indicator An indicator connected to a radio receiver, to show signals received over a band of frequencies centered about the specific frequency to which the receiver is tuned.

panoramic radar A nonscanning radar that transmits signals omnidirectionally over a wide beam in the direction of interest.

panoramic receiver A radio receiver that permits continuous observation, on a cathode-ray-tube screen, of the presence and relative strength of all signals in a wide frequency band through which the receiver is periodically tuned.

paper capacitor A fixed capacitor that consists of two strips of metal foil separated by oiled or waxed paper or other insulating material and rolled together in compact tubular form. The foil strips are staggered so one projects from each end of the roll, and the connecting wires are attached to the projecting foil strips.

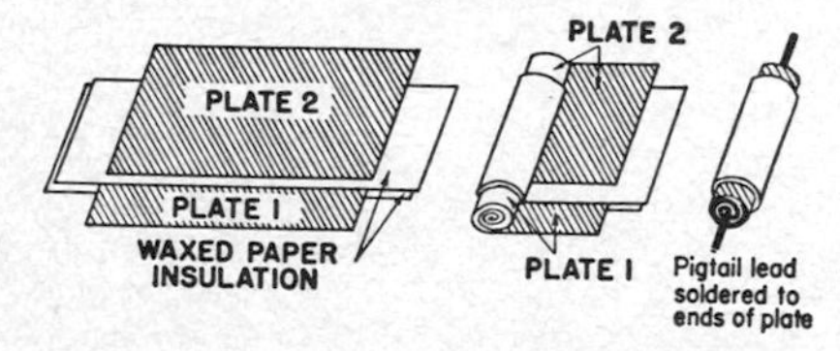

Paper-capacitor construction.

paper tape A tape in which data may be recorded by a pattern of partly or completely punched holes. Tape width varies with the maximum number of holes punched across it. Common widths are 5 holes plus feed hole—$1\frac{1}{16}$ in; 6 or 7 holes plus feed hole—$\frac{7}{8}$ in; 8 holes plus feed hole—1 in (1.75, 2.22, and 2.54 cm).

paper-tape reader A reader that senses information punched in paper tape as a series of holes. Also called tape reader.

PAR Abbreviation for *precision approach radar.*

parabolic antenna A directional microwave antenna that uses some form of parabolic reflector to give directional characteristics.

parabolic microphone A microphone used at the focal point of a parabolic sound reflector to give improved sensitivity and directivity, as required for picking up a band marching down a football field.

parabolic reflector A reflector whose inner surface is shaped by rotating a parabola about its axis. When a microwave transmitting dipole, horn, or other antenna is placed at the focal point, the reflector concentrates the radiation into a parallel beam. For reception, incoming radiation is reflected to the receiving antenna at the focal point. The reflector may be made of wire screen or sheet metal.

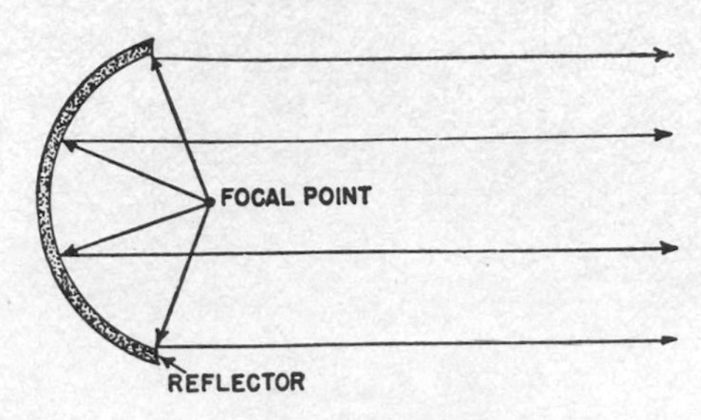

Parabolic reflector action.

paraboloid A reflecting surface formed by rotating a parabola about its axis of symmetry. Used as a reflector for sound waves and microwave radiation.

parahydrogen A hydrogen molecule in which the two nuclear spins are antiparallel, forming a singlet state.

parallax The apparent displacement of the position of an object, caused by a shift in the point of observation. Thus the pointer of a meter will appear to be at different positions on the scale, depending on the angle from which the meter is read.

parallel 1. Connected to the same pair of terminals. Also called multiple and shunt. 2. Simultaneous transmission of, storage of, or logic operations on the parts of a word, character, or other subdivision of a word in a computer, using separate facilities for the various parts. 3. Extending in the same direction and equally distant at all points, as of lines and surfaces.

parallel access A memory characteristic wherein all the bits of a byte or word are entered simultaneously at several inputs or retrieved simultaneously from several outputs.

parallel adder A computer logic circuit that adds corresponding parts of two binary numbers all at once during one execution cycle, with additional cycles only if required to handle carries.

parallel circuit A circuit in which the same voltage is applied to all components, and the current divides among the components according to their resistances or impedances.

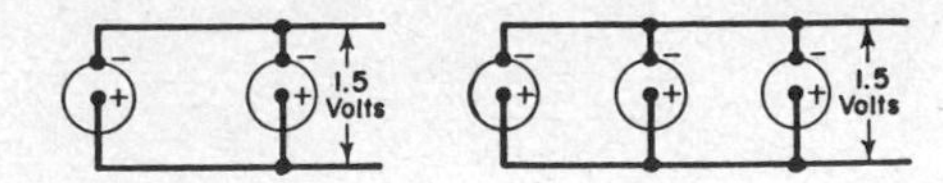

Parallel circuit for dry cells.

parallel cut A Y cut in a quartz crystal.

parallel digital computer A computer in which the digits are handled in parallel. The bits that comprise a digit may be handled either serially or in parallel.

parallel feed *Shunt feed.*

parallel operation The flow of information through a part or all of a computer over two or more lines or channels simultaneously.

parallel-plate counter chamber A radiation-counter chamber that has parallel metal plates as electrodes.

parallel-plate oscillator A push-pull ultrahigh-frequency oscillator circuit in which two parallel plates serve as the main frequency-determining elements.

parallel-plate waveguide A waveguide that consists of two metal strips whose width is large compared to the spacing between them. For the dominant transverse electromagnetic wave, this waveguide has an infinite cutoff wavelength, and guide wavelength is equal to free-space wavelength.

parallel processing Processing of two or more programs simultaneously in a computer that has more than one active processor.

parallel-resistance formula The combined resistance of resistors in parallel is less than that of the smallest resistor and is equal to the product of

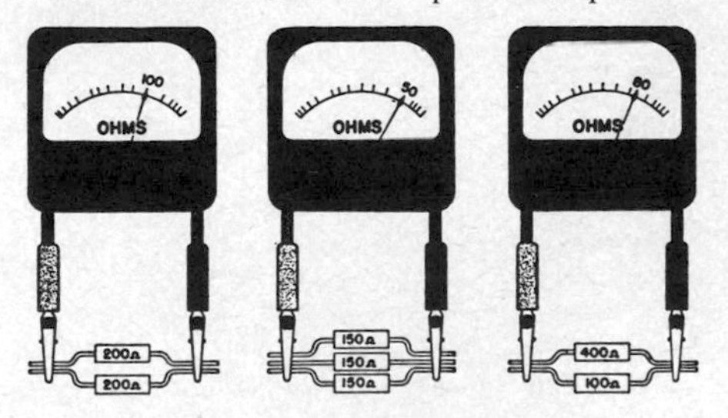

Parallel-resistance formula as illustrated by ohmmeter.

the resistor values divided by their sum. For like values, divide the ohmic value of one resistor by the number of like resistors in parallel.

parallel resonance Resonance in a parallel resonant circuit, wherein the inductive and capacitive reactances are equal at the frequency of the applied voltage. The impedance of the parallel resonant circuit is then a maximum, so maximum signal voltage is developed across it. Also called antiresonance.

parallel resonant circuit A resonant circuit in which the capacitor and coil are in parallel with the applied AC voltage. Also called antiresonant circuit.

parallel-rod oscillator An ultrahigh-frequency oscillator in which parallel rods form the tank circuits.

parallel storage Computer storage in which all bits, characters, or words are essentially equally available.

parallel-T network *Twin-T network.*

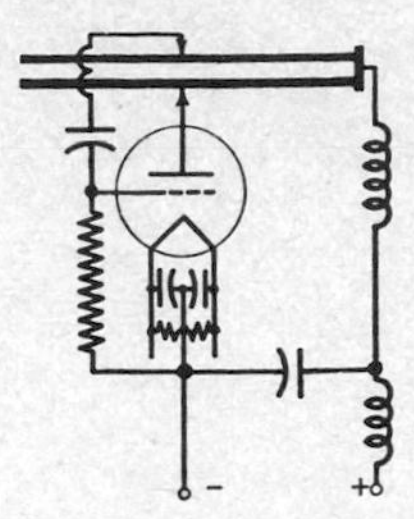

Parallel-rod oscillator in which rods are quarter-wavelength long.

parallel transfer Computer data transfer in which the characters of an element of information are transferred simultaneously over a set of parallel paths.

parallel-triggered blocking oscillator A triggered blocking oscillator in which the trigger pulse is applied to the anode of the blocking-oscillator tube rather than to the grid as in series triggering.

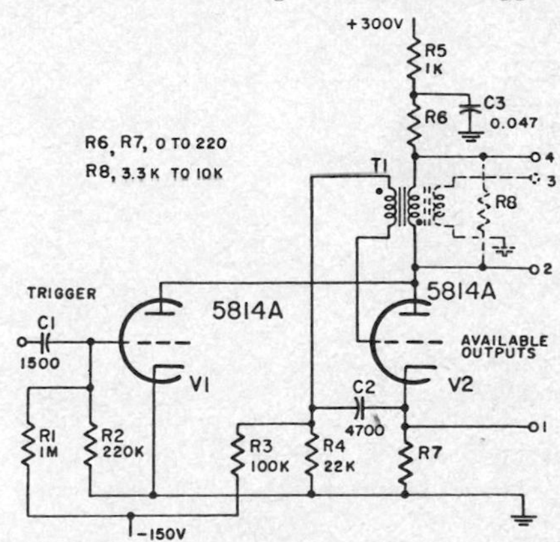

Parallel-triggered blocking oscillator using circuit design preferred by National Bureau of Standards. R8 is used only if required to prevent ringing. Outputs are nearly rectangular pulses from 1 to 5 μs wide. Triggering may be from 200 to 2000 pulses per second.

parallel-wire line A transmission line that consists of two parallel wires.

parallel-wire resonator A resonator that consists of a length of parallel-wire transmission line short-circuited at one end.

paralysis The overloading of an electron tube by a strong signal, causing the tube to charge the capacitances in the circuit to a point where they take too long to discharge and thus fail to amplify part of a succeeding signal. In radar, paralysis can obscure an echo from a target at short range because the capacitances do not have time to discharge between transmission of the pulse and reception of the echo.

paramagnetic Having a magnetic permeability greater than that of a vacuum and essentially independent of the magnetizing force. In ferromagnetic materials, the permeability varies with magnetizing force.

paramagnetic amplifier *Maser.*

paramagnetic resonance Resonance observable in a paramagnetic material as a peak in the energy absorption spectrum at a frequency related to the strength of the applied magnetic field and the gyromagnetic ratio. Used in studying the energy states of nuclei, atoms, molecules, and crystal lattices.

paramagnetism Magnetism that involves a permeability only slightly greater than unity.

parameter A quantity to which arbitrary values may be assigned, such as the value of a transistor or tube characteristic, or the value of a circuit component. A parameter is usually not changed during a given set of conditions.

parametric amplifier [abbreviated paramp] A microwave amplifier that uses an electron tube or

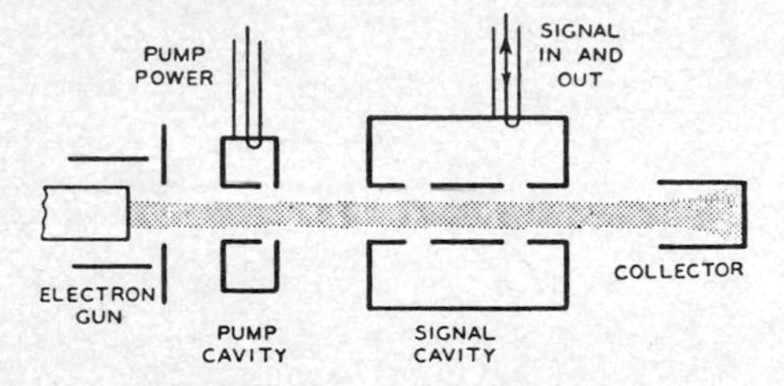

Parametric amplifier using electron-tube construction.

solid-state device whose reactance can be varied periodically by an AC voltage at a pumping frequency.

parametric device A device whose operation depends essentially on variation of some parameter with time. As an example, the parameter can be a reactance that is varied by an AC control voltage.

parametric frequency converter A frequency converter that utilizes the variation of the reactance parameter of an energy-storage element to obtain frequency conversion.

parametric modulator A modulator that utilizes the variation in the reactance parameter of an energy-storage element to produce modulation.

parametric multiplier modulator A microwave circuit in which the functions of phase modulation and frequency multiplication are combined in a single circuit, in which a varactor frequency multiplier is phase-modulated by variations in its bias voltage.

parametric oscillator An oscillator in which the reactance parameter of an energy-storage device is varied to obtain oscillation.

parametron A resonant circuit in which either the inductance or capacitance is made to vary periodically at one-half the driving frequency. Used as a digital computer element, in which the oscillation represents a binary digit.

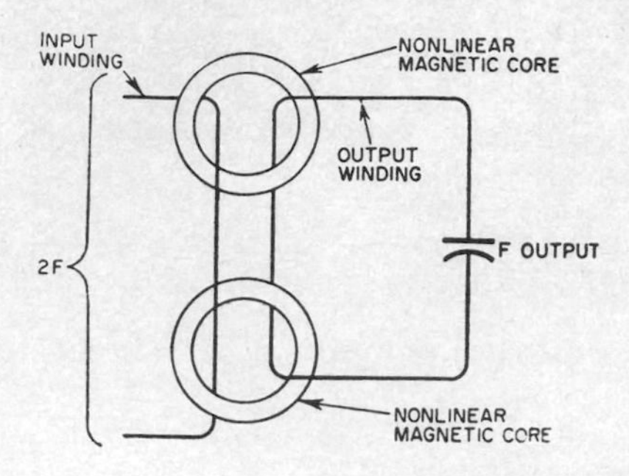

Parametron element.

paramp Abbreviation for *parametric amplifier.*

paraphase amplifier An amplifier that uses the out-of-phase relation of the signal voltages at the anode and cathode of a tube to convert a single input signal into two out-of-phase output signals, for driving a push-pull stage.

parasitic An undesired and energy-wasting signal current, capacitance, or other parameter of an electronic circuit. In an integrated circuit this parameter may result from formation of a parasitic transistor in the semiconductor substrate.

parasitic array An antenna array that contains one or more parasitic elements.

parasitic capture Any absorption of a neutron that does not result in a fission or the production of a desired element.

parasitic element An antenna element that serves as part of a directional antenna array but has no direct connection to the receiver or transmitter. A parasitic element reflects or reradiates the energy that reaches it, in such a phase relationship as to give the desired radiation pattern. Also called passive element.

parasitic oscillation An undesired self-sustaining oscillation or a self-generated transient impulse in an oscillator or amplifier circuit, generally at a frequency above or below the correct operating frequency.

parasitic suppressor A suppressor, usually in the form of a coil and resistor in parallel, inserted in a circuit to suppress parasitic high-frequency oscillations.

parasitic transistor A transistor formed unintentionally in the semiconductor substrate of an integrated circuit, resulting in signal currents that are usually undesirable.

parity A property of a wave function. The parity is 1 or even if the wave function is unchanged by an inversion of the coordinate system, and −1 or odd if the wave function is changed only in sign.

parity bit A binary digit that is added to an array of bits to make the sum of the bits always odd or always even, for checking accuracy.

parity check A type of odd-even check that is used when the total number of 1s or 0s in each permissible computer code expression is always made even or odd. A check may be made for even or odd parity.

parsec [PARallax-SECond] A unit of distance for interstellar space, equal to 3.26 light-years, 206,000 astronomical units, or 19.15 × 10^{12} mi (30.82 × 10^{12} km).

part An article that is an element of a subassembly, is not normally useful by itself, and is not amenable to further disassembly for maintenance purposes. The term is used chiefly for structural members in electronic equipment, whereas tubes, transistors, resistors, capacitors, coils, switches, relays, transformers, and similar items that have distinct electrical characteristics are usually called components.

partial A sound-sensation component that is distinguishable as a simple tone, cannot be further analyzed by the ear, and contributes to the character of the complex sound. The frequency of a partial may be higher or lower than the basic frequency and may or may not be an integral multiple or submultiple of the basic frequency.

partial carry The computer condition wherein a carry that results from the addition of carries is not allowed to propagate.

partial differential equation A differential equation that has more than one independent variable.

partial node The points, lines, or surfaces in a standing-wave system, where some characteristic of the wave field has a minimum amplitude that differs from zero.

partial-read pulse A current pulse that is applied to a magnetic memory to select a specific magnetic cell for reading.

partial-select output The voltage response produced by applying partial-read or partial-write pulses to an unselected magnetic cell.

partial-write pulse A current pulse that is applied to a magnetic memory to select a specific magnetic cell for writing.

particle Any very small part of matter, such as an atom, electron, proton, molecule, neutron, alpha particle, or beta particle.

particle accelerator *Accelerator.*

particle-oriented paper A chart paper that has a magnetic coating which is produced by combining microscopic magnetic flakes with oil to form droplets and then forming these particles into an emulsion that can be applied to the surface of ordinary bond paper or to a clear plastic substrate. The magnetic field of a small-diameter recording head rotates the magnetic flakes so they absorb or scatter incident light to give a visible dark trace that can also be read magnetically. The trace is completely erasable and reusable. Developed for use in chart, strip, and XY recorders. Electrostatic particle-oriented paper also gives visible and permanent dark traces, but at much faster writing speeds.

particle velocity The instantaneous velocity of a given infinitesimal part of a medium, with reference to the medium as a whole, due to the passage of a sound wave.

particulate Having the form of separate, very small particles.

partition noise Noise that arises in an electron tube when the electron beam is divided between two or more electrodes, as between screen grid and anode in a pentode.

parts per million [abbreviated PPM] A method of specifying the precision with which a frequency, voltage, or other parameter is generated, measured, or controlled.

party-line carrier system A single-frequency carrier telephone system in which the carrier energy is transmitted directly to all other carrier terminals of the same channel.

parylene A crystalline material, which has excellent dielectric properties, that can be vapor-deposited on almost any substrate to form extremely thin insulating films. The complete chemical name is diparaxylylene.

parylene capacitor A highly stable fixed capacitor that uses parylene film as the dielectric. It can be operated at temperatures up to 170°C as well as at cryogenic temperatures.

pascal [abbreviated Pa] The SI unit of pressure or stress. A pressure of 1 Pa is equal to 1 N/m^2.

Paschen's law The sparking potential between two parallel-plate electrodes in a gas is proportional to the product of gas pressure and electrode spacing.

pass A complete cycle of reading, processing, and writing in a computer.

passband A frequency band in which the attenuation of a filter is essentially zero.

pass element An active element, such as a transistor or tube, used in series with the load of a regulated DC power supply and controlled by an amplifier that varies its series resistance as required to maintain a constant DC output voltage.

passivate To treat the surface of a semiconductor chip with a relatively inert material like silicon dioxide, to make the surface inactive and provide protection from contamination.

passivated alloy silicon diode A temperature-compensated reference element that provides a

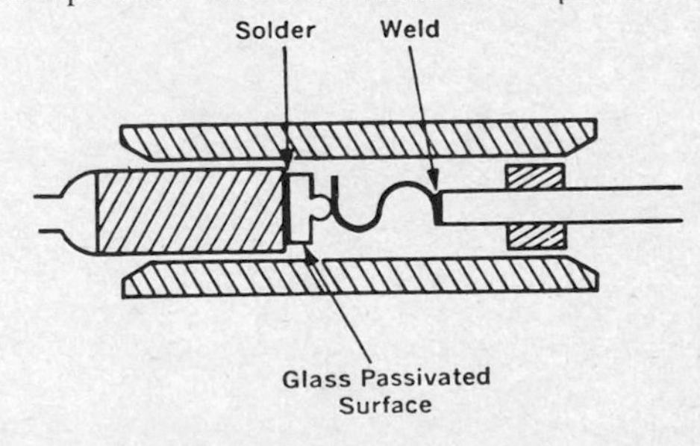

Passivated alloy silicon diode.

reference voltage for potentiometric and comparison measurements, with stability comparable to that of standard cells. Typical standard voltages for these reference diodes are 6.2 and 8.4 V.

passivated transistor A transistor that has been protected against premature failure by passivation.

passivation The process of making a semiconductor device insensitive to water, ions, and other contamination that might cause drift of parameters or premature failure. Passivation is achieved by protecting the junction area with a grown oxide layer, such as silicon dioxide, that becomes chemically bonded to the semiconductor crystal.

passive 1. Involving only energy that is radiated or reflected naturally by an object. 2. Not contributing to signal energy.

passive communication satellite A satellite that reflects communication signals between stations, without providing amplification. An example is the Echo satellite.

passive component A component that does not provide amplification, such as a resistor or capacitor.

passive corner reflector A corner reflector that is energized by a distant transmitting antenna. Used chiefly to improve the reflection of radar signals from objects that would not otherwise be good radar targets.

passive detection The detection of a target or other object by means that do not reveal the position of the detecting instrument.

passive double reflector A combination of two passive reflectors positioned to bend a microwave beam over the top of a mountain or ridge, generally without appreciably changing the general direction of the beam.

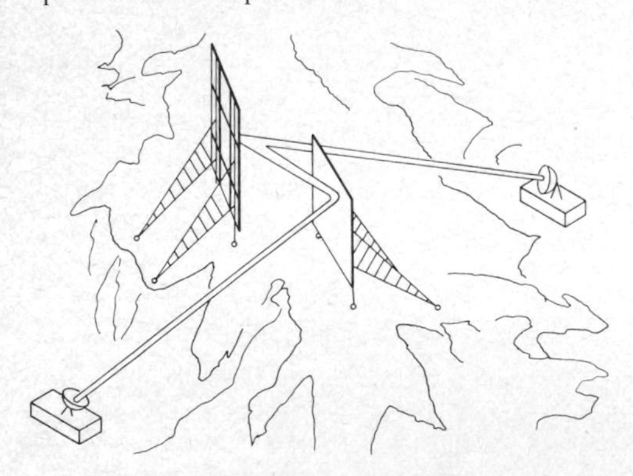

Passive double reflector mounted on high ridge.

passive electronic countermeasures Electronic countermeasures that are not detectable by the enemy, such as techniques for measuring and analyzing enemy electromagnetic radiation.

passive element *Parasitic element.*

passive filter A filter that has only passive components, with no active elements.

passive guidance Guidance of a vehicle by preset or inertial devices, without reliance on external signals or observations.

passive homing Homing that depends only on energy emanating naturally from the target, as in the form of infrared radiation, light, sound, electromagnetic radiation, ionization of air, or air pollution by exhaust gases.

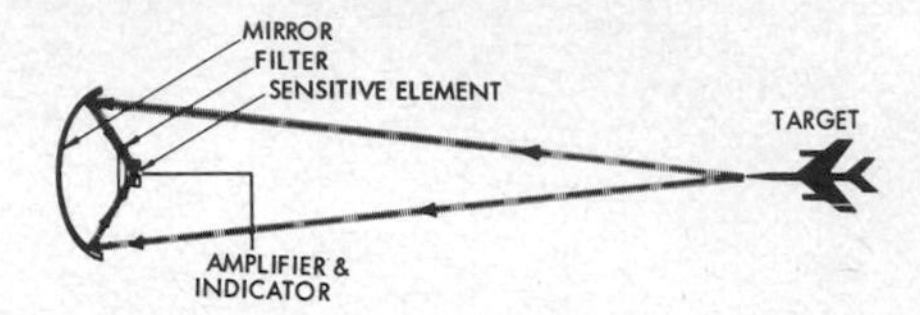

Passive infrared detection depends on normal heat given off by target and can therefore operate without revealing location to target.

passive infrared detection An infrared detection system in which only the infrared radiation normally given off by the target is detected.

passive infrared tracking Tracking by an infrared detector that responds to the normal infrared radiation emitted by all objects. No infrared light source is used, and tracking can therefore be done without conveying information to the target being tracked.

passive jamming Use of confusion reflectors to return spurious and confusing signals to enemy radars.

passive navigation countermeasures Control of transmissions from equipment that is capable of producing electromagnetic radiation, to prevent enemy use of such radiation for navigation purposes.

passive network A network that has no source of energy and hence no active gain elements.

passive radar Detection of an object at a distance by picking up the microwave electromagnetic energy that any object normally radiates when it is above absolute zero in temperature. Passive radar requires an apparent temperature difference between the object and its surroundings and radio astronomy techniques in receivers to distinguish from noise a desired signal whose level may be less that 1 pW.

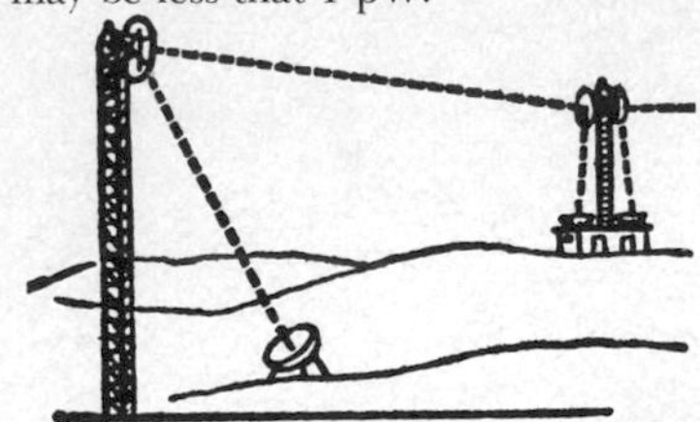

Passive reflectors on towers reflect microwave beams to antennas at convenient ground locations.

passive reflector A reflector that changes the direction of a microwave or radar beam. Often used on microwave relay towers to permit location of transmitter, repeater, and receiver equipment on the ground rather than at the tops of towers.

passive satellite A satellite that simply reflects radio signals back to earth, without amplification or other processing.

passive sonar Sonar that uses only underwater listening equipment, with no transmission of location-revealing pulses.

passive substrate A substrate that may serve as physical support and thermal sink for an integrated circuit but does not become a part of the circuit. Glass and ceramics are examples.

passive transducer A transducer that contains no internal source of power.

paste solder Finely powdered solder metal combined with a flux.

PA system Abbreviation for *public-address system.*

patch 1. A temporary connection between jacks or other terminations on a patchboard. 2. A section of coding inserted into a computer routine to correct a mistake or alter the routine.

patchboard A board or panel that has a number of jacks at which circuits are terminated. Short cables called patchcords are plugged into the jacks to connect various circuits temporarily as required in broadcast, communication, and computer work. Patchboards for computers are often designed for quick removal without disturbing the patches made on the board, to permit plugging in a patchboard already set up for the next problem.

patchcord A cord equipped with plugs at each end, to connect two jacks on a patchboard.

patent A document that confers on inventors for a term of years the exclusive right to make, use, and sell their inventions in practical form.

path A line that connects a series of points and constitutes a proposed or traveled route.

path length The length of a magnetic flux line in a core.

pattern generator A signal generator that generates a test signal which can be fed into a television receiver to produce on the screen a pattern of lines for servicing purposes.

pattern recognition Analysis of printed, photographic, or other patterns by scanning, contour tracing, or other electronic techniques, for identifying significant features in recorded medical and other patterns. Also called automatic pattern recognition.

Pauli exclusion principle Any wave function that involves several identical particles must change sign when the coordinates, including the spin coordinates, of any identical pair are interchanged. Only one particle of a given kind can occupy a particular quantum state. The principle applies to electrons, fermions, protons, and neutrons but not to bosons. Also called exclusion principle.

Pauli-Fermi principle Each level of a quantized system can include one, two, or no electrons. If there are two electrons, they must have spins in opposite directions.

pay-cable Cable television in which one or more pay channels, bringing new movies and sometimes special programs not available on ordinary television, are provided for an additional monthly fee on top of the basic cable service fee.

payload The total weight of instrumentation and/or passengers carried by a space vehicle for the direct purposes of the flight. Fuel, navigation equipment, and control equipment are not considered parts of the payload.

pay television A television system in which special programs are provided only to subscribers who make regular payments for the service. The programs may be broadcast in coded or scrambled form requiring a decoding or unscrambling device at the receiver, or they may be placed on a cable television channel that is blocked by filters in lines going to nonsubscribers.

P band A band of radio frequencies that extends from 225 to 390 MHz, corresponding to wavelengths of 133.3 to 76.9 cm.

PbS Symbol for *lead sulfide.*

PBS Abbreviation for *Public Broadcasting Service.*

PbTe Symbol for *lead telluride.*

PBX Abbreviation for *private branch exchange.*

PC Abbreviation for *printed circuit.*

P-channel A conduction channel formed by holes in a P-type semiconductor, as in a PMOS field-effect transistor.

P-channel MOS [abbreviated PMOS] A metal-oxide semiconductor process in which selective diffusion of a P-type dopant forms closely spaced source and drain regions within a silicon substrate, with the conducting channel consisting of holes. In contrast, the channel consists of electrons in the N-channel MOS process.

PCM Abbreviation for *pulse-code modulation.*

PCM/FM Pulse-code modulation on frequency modulation.

PDBM Abbreviation for *pulse-delay binary modulation.*

P display *Plan-position indicator.*

PDM Abbreviation for *pulse-duration modulation.*

PDM/FM Pulse-duration modulation on frequency modulation.

PDM/PM Pulse-duration modulation on phase modulation.

peak [abbreviated P] The maximum instantaneous value of a quantity. Also called crest.

peak amplitude The maximum amplitude of an alternating quantity, measured from its zero value.

peak clipper *Limiter.*

peak detector A detector whose output voltage approximates the true peak value of an applied signal. The detector tracks the signal in its sample mode and preserves the highest input signal in its hold mode.

peak electrode current The maximum instantaneous current that flows through an electrode.

peak energy density The maximum absolute value of the instantaneous energy density in a specified time interval.

peak envelope power [abbreviated PEP] The average RF power supplied by a transmitter during one RF cycle at the highest peak-to-peak values of the modulation envelope. It is equal to the input power indicated by an ammeter and voltmeter when the output amplifier is driven by a continuous RF signal that has the peak amplitude which the amplifier can handle within allowable distortion limits. For single-sideband transmitters, modulation must be present during measurement because the carrier is suppressed.

peaker A small fixed or adjustable inductance used to resonate with stray and distributed capacitances in a broadband amplifier to increase the gain at the higher frequencies.

peak field strength *Peak magnetizing force.*

peak flux density The maximum flux density in a magnetic material in a specified cyclically magnetized condition.

peaking Increasing the response of a circuit at a desired frequency or band of frequencies.

peaking circuit A circuit that improves the high-frequency response of a broadband amplifier. In shunt peaking, a small coil is placed in series with the anode load. In series peaking, the coil is placed in series with the grid of the following stage. Used in video amplifiers, often with both types of peaking in the same stage. In effect, the circuit converts an input signal to a more peaked waveform.

peaking coil A small coil placed in a circuit to resonate with the distributed capacitance of the circuit at a frequency for which peak response is desired, as in a video amplifier near the cutoff frequency.

peaking control An adjustable resistor-capacitor circuit that controls the wave shape of the horizontal oscillator output pulses, as required to give a linear sweep.

peaking transformer A transformer in which the number of ampere-turns in the primary is high enough to produce many times the normal flux density values in the core. The flux changes rapidly from one direction of saturation to the other twice per cycle, inducing a highly peaked voltage pulse in a secondary winding. Used to fire ignitrons and thyratrons.

peak inverse voltage [abbreviated PIV] The maximum rated value of an AC voltage acting in the direction opposite to that in which a device is designed to pass current.

peak level The maximum instantaneous level that occurs during a specified time interval, such as the peak sound pressure level in acoustics.

peak limiter *Limiter.*

peak load The maximum instantaneous load or the maximum average load over a designated interval of time.

peak magnetizing force The upper or lower limiting value of magnetizing force associated with a cyclically magnetized condition. Also called peak field strength.

peak particle velocity The maximum absolute value of the instantaneous particle velocity in a specified time interval.

peak power The maximum instantaneous power of a transmitted radar pulse. Since the resting time of a radar transmitter is long compared to its operating time, the average power is low compared to the peak power.

peak power output The output power of a radio transmitter as averaged over one carrier cycle at the maximum amplitude that can occur for any combination of transmitted signals.

peak pulse amplitude The maximum absolute peak value of a pulse, excluding spikes and other unwanted portions.

peak pulse power The power at the maximum of a pulse of power, excluding spikes.

peak response The maximum response of a device or system to an input stimulus.

peaks Momentary high volume levels during a radio program; they make the volume indicator at the studio or transmitter swing upward.

peak signal level The maximum instantaneous signal power or voltage at a specified point in a facsimile system.

peak sound pressure The maximum absolute value of the instantaneous sound pressure in a specified time interval. The SI unit of sound pressure is the newton per square meter.

peak speech power The maximum value of the instantaneous speech power within the time interval considered.

peak-to-peak [abbreviated P-P] From a positive to a negative peak in an alternating quantity.

peak-to-peak amplitude The sum of the extreme swings of an alternating quantity in positive and negative directions from its zero value. For a sinusoidal waveform, the peak amplitude in either direction is half the peak-to-peak amplitude.

peak-to-peak voltmeter A voltmeter that measures the voltage difference between the positive and negative peaks of a voltage. Two peak-reading voltmeters connected in series opposition can be used for this purpose.

peak value The maximum instantaneous value of a varying current, voltage, or power during the time interval under consideration. For a sine wave, it is equal to 1.414 times the effective value. Also called crest value.

peak voltmeter A voltmeter that reads peak values of an alternating voltage.

pedestal 1. The structure that supports a radar antenna. 2. A flat-topped pulse that elevates the base level for another pulse. 3. *Blanking level.*

pedestal level *Blanking level.*

P electron An electron that has an orbit in the P shell, which is the sixth shell of electrons surrounding the atomic nucleus, counting out from the nucleus.

pellet A small piece of semiconductor material, used in crystal diodes and transistors.

pellet film resistor A resistor in which a resistive film is deposited on a tiny alumina or other insulating pellet that has silvered end faces which serve as terminations. Used in striplines and other microwave applications that require low residual inductance and minimum skin effect.

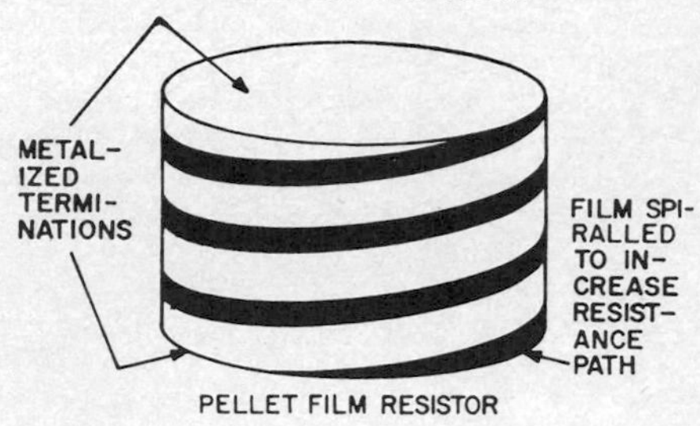

Pellet film resistor.

pellet resistor A resistor that consists of a compressed powder mixture of noble metals and oxides bonded with an organic flux. Terminations are produced on the pellet by firing platinum or gold on its ends.

Peltier effect The production or absorption of heat at the junction of two metals when a current is passed through the junction. Heat generated by current in one direction will be absorbed when the current is reversed.

pencil *Soldering pencil.*

pencil beam A narrow radar beam that has an essentially circular cross section.

pencil-beam antenna A unidirectional antenna so designed that cross sections of its major lobe are approximately circular.

pencil tube A long, thin disk-seal tube designed for use as a UHF oscillator or amplifier.

penetrating shower A cosmic-ray shower in which some or all of the particles can penetrate over about 20 cm of lead. The particles are often pi mesons.

penetration depth 1. The nominal depth below the surface of a conductor within which current is concentrated by the skin effect during induction heating. The higher the frequency, the less the penetration. 2. The depth to which an external magnetic field penetrates a superconductor.

penetration frequency *Critical frequency.*

penetration probability The probability of transmission of a particle through a potential barrier. Examples of barrier penetration are the passage of alpha particles through the barrier at the nuclear wall and the motions of electrons in a metal through the interatomic coulomb barriers. Also called transmission coefficient.

penetration tube *Multicolor cathode-ray tube.*

penetration-type thickness gage A radioactive thickness gage in which a radioactive beta or gamma source is on one side of the sheet being monitored, and a radiation counter or meter is on

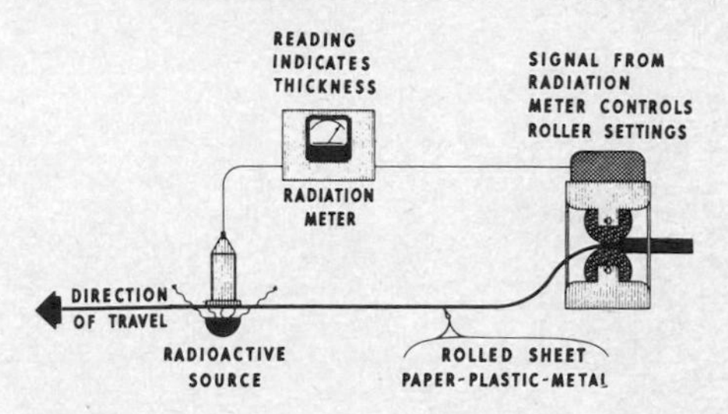

Penetration-type thickness gage.

the other side. The intensity of the beam measured by the meter is an inverse function of the mass of material in the moving sheet; hence the meter can be calibrated directly in terms of thickness for a given material. Beta and gamma gages are examples.

penetrometer An instrument that measures the penetrating power of a beam of x-rays or other penetrating radiation.

pentagrid converter A pentagrid tube used as a converter (combination oscillator, mixer, and first detector) in a superheterodyne receiver. The cathode and the first two grids are in the oscillator circuit, and the incoming RF signal is applied to the third grid.

pentagrid mixer A pentagrid tube that mixes two signals. The first and third grids are control grids to which the signals are applied. The second and fourth grids are screen grids, and the fifth grid is a suppressor grid. When used in a superheterodyne receiver, the local oscillator signal may be applied to either the first or third grid.

pentagrid tube An electron tube that has five grids.

pentode A five-electrode electron tube that contains an anode, a cathode, a control electrode, and

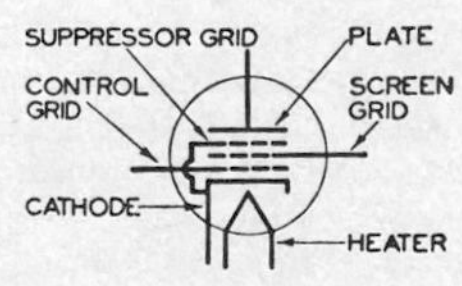

Pentode symbol.

two additional electrodes which are ordinarily grids.

pentode field-effect transistor A five-lead tran-

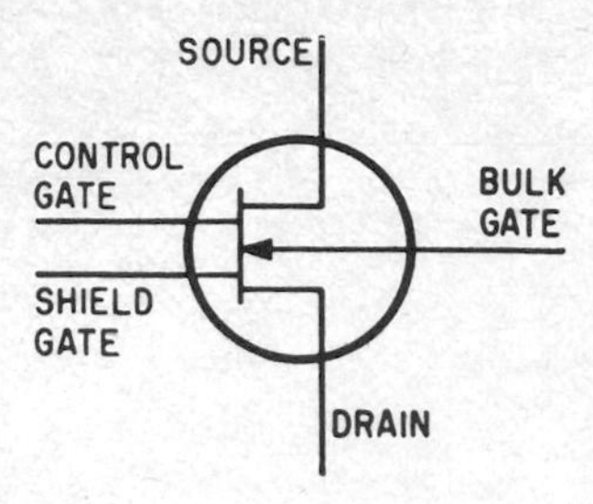

Pentode field-effect transistor.

sistor that has three gates. It can be operated as a pentode if independent bias supplies are provided for each gate.

PEP Abbreviation for *peak envelope power*.

perceived noise level The sound-pressure level of a reference sound that is judged as noisy as a given sound. The value is commonly expressed in PNdB.

percent depth dose The amount of radiation delivered at a specified depth in tissue, expressed as a percentage of the amount delivered at the skin.

percent harmonic distortion A measure of the harmonic distortion in a system or component. It is equal to 100 times the ratio of the square root of the sum of the squares of the root-mean-square voltages of each of the individual harmonic frequencies, to the root-mean-square voltage of the fundamental. Current values may be used in place of voltage values.

percent modulation The modulation factor expressed as a percentage.

percent ripple The ratio of the effective value of the ripple voltage to the average value of the total voltage, expressed as a percentage.

perceptron A network of artificial neurons that has pattern-recognizing capabilities.

percussion welding Welding in which a sudden discharge of electric current at the junction produces an arc that is extinguished by a percussive blow.

perfect dielectric A dielectric in which all the energy required to establish an electric field in the dielectric is returned to the electric system when the field is removed. A vacuum is the only known perfect dielectric.

perforator A machine that punches code signals in paper tape.

perigee The point on an elliptical orbit at which a satellite is farthest from the earth.

period The time required for one complete cycle of a regularly repeated series of events.

periodic Having a repetition rate.

periodic damping Damping in which a pointer or other moving object oscillates about a new position before coming to rest.

periodic duty Intermittent duty in which the load conditions are regularly recurrent.

periodic electromagnetic wave A wave in which the electric field vector is repeated in detail at a fixed point after the lapse of a time known as the period.

periodic law Certain properties of the elements are periodic functions of their atomic numbers. When the elements are arranged in the order of their atomic numbers, these properties recur in regular cycles.

periodic line A transmission line that has successive identical sections, with nonuniform electrical properties within each section. An example is a loaded line that has uniformly spaced loading coils.

periodic permanent magnet An assembly of axially magnetized ring-shaped permanent magnets whose adjacent faces have like polarity. Used to produce a sinusoidal or otherwise nonuniform permanent magnetic field.

periodic pulse train A pulse train made up of identical groups of pulses, the groups being repeated at regular intervals.

periodic quantity An oscillating quantity whose values recur at equal increments of time, space, or some other independent variable.

periodic rating A rating that defines the load which can be carried for specified alternate periods of load and rest.

periodic table A table in which the elements are arranged according to the periodic law, so elements with similar characteristics are logically grouped together.

periodic wave A wave that repeats itself at regular intervals, such as a sine wave.

periodic waveguide A waveguide that has discontinuities at spaced intervals.

period meter An instrument that indicates the period of a nuclear reactor in seconds. It may also operate interlocks and scram the reactor if the period is dangerously small.

peripheral equipment Equipment that works in conjunction with a computer but is not part of the computer itself, such as a card or paper-tape reader or punch, magnetic-tape handler, line printer, or cathode-ray-tube terminal.

peristaltic charge-coupled device A high-speed charge-transfer integrated circuit in which the movement of the charges is similar to the peristaltic contractions and dilations of the digestive system. Applications include imaging devices, delay lines, filters, and memories.

permalloy A magnetic alloy that has high permeability; it usually consists of iron, nickel, and small quantities of other metals. Used to shield components, tubes, and equipment from stray magnetic fields.

permanent echo A signal reflected from an object that is fixed with respect to a radar site.

permanent magnet [abbreviated PM] A piece of hardened steel or other magnetic material that has been strongly magnetized and retains its magnetism indefinitely.

permanent-magnet centering Centering of the image on the screen of a television picture tube by magnetic fields produced by permanent magnets mounted around the neck of the tube.

permanent-magnet dynamic loudspeaker *Permanent-magnet loudspeaker.*

permanent-magnet erasing head An erasing head that uses the fields of one or more permanent magnets for erasing magnetic tape.

permanent-magnet focusing Focusing of the electron beam in a television picture tube by the magnetic field produced by one or more permanent magnets mounted around the neck of the tube.

permanent-magnet loudspeaker [abbreviated PM loudspeaker] A moving-conductor loudspeaker in which the steady magnetic field is produced by a permanent magnet. Also called permanent-magnet dynamic loudspeaker.

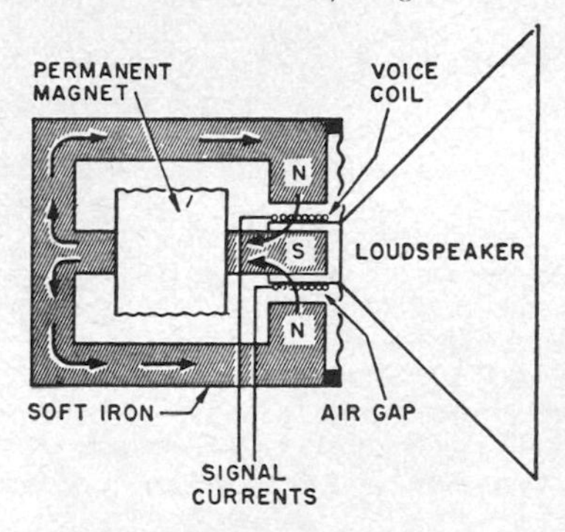

Permanent-magnet loudspeaker construction.

permanent-magnet moving-coil instrument A DC meter movement that consists of a small coil of wire supported on jeweled bearings between the poles of a permanent magnet. Spiral springs serve as connections to the coil and keep the coil and its attached pointer at the zero position on the meter scale. When the direct current to be measured is sent through the coil, its magnetic field interacts with that of the permanent magnet to produce rotation of the coil. Also called d'Arsonval movement (deprecated).

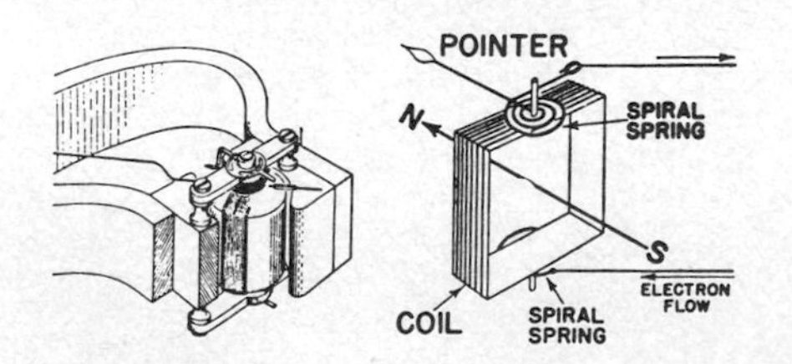

Permanent-magnet moving-coil instrument and diagram showing electron flow path through springs and coil.

permanent-magnet moving-iron instrument A meter that for its operation depends on a movable iron vane which aligns itself in the resultant field of a permanent magnet and an adjacent current-carrying coil.

permanent-magnet second-harmonic self-synchronous system A remote indicating arrangement that consists of a transmitter unit and one or more receiver units. All units have permanent-magnet rotors and toroidal stators that use saturable ferromagnetic cores and are excited with alternating current from a common external source. The coils are tapped at three or more equally spaced intervals, and the corresponding taps are connected together to transmit voltages that consist principally of the second harmonic of the excitation voltage. The rotors of the receiver units will assume the same angular position as that of the transmitter rotor.

permanent-magnet stepper motor A stepper motor in which the rotor is a powerful permanent magnet, and each stator coil is energized independently in sequence. The rotor aligns itself with the stator coil that is energized, thus giving rotation in 90° steps if there are four stator coils. For smaller steps, more stator coils or speed-reducing gears are used.

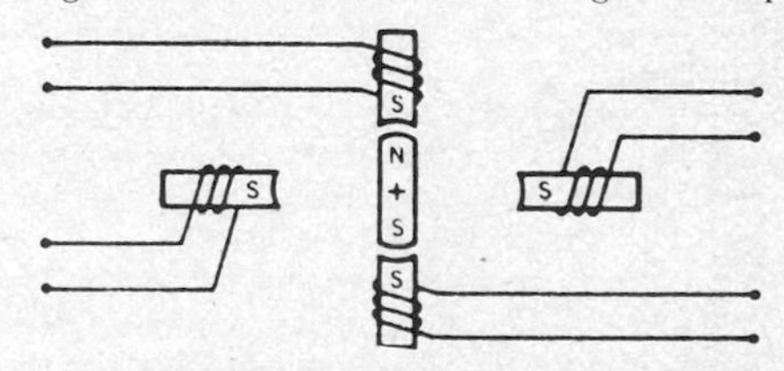

Permanent-magnet stepper motor. Energizing stators in clockwise sequence gives clockwise 90° stepping of permanent-magnet rotor.

permanent memory A computer memory that does not lose its stored data when computer power is turned off.

permanent-split capacitor motor [abbreviated PSC motor] A capacitor motor in which the starting capacitor and the auxiliary winding remain in the circuit for both starting and running. Also called capacitor start-run motor.

permatron A thermionic-cathode gas diode in which conduction is initiated by an external magnetic field instead of a grid. The resulting action is similar to that in a thyratron. Used chiefly as a controlled rectifier.

permeability A measure of how much better a given material is than air as a path for magnetic lines of force. The permeability of air is assumed as 1.

permeability tuner A television or radio tuner in which the tuning dial moves the powdered iron cores of coils in the tuning circuits.

permeability tuning Tuning of a resonant circuit by moving a ferrite core in or out of a coil, thereby changing the effective permeability of the core and the inductance of the circuit.

permeameter An instrument that measures the magnetic flux or flux density produced in a test specimen of ferromagnetic material by a given magnetic intensity, to permit computation of the magnetic permeability of the material.

permeance [symbol *P*] A characteristic of a portion of a magnetic circuit, equal to magnetic flux divided by magnetomotive force. Permeance is the reciprocal of reluctance.

permittance Obsolete term for *capacitance.*

permittivity *Dielectric constant.*

perpendicular magnetization In magnetic recording, magnetization of the recording medium in a direction perpendicular to the line of travel and parallel to the smallest cross-sectional dimension of the medium.

Pershing An Army surface-to-surface guided missile that uses a solid propellant. It is capable of carrying a nuclear warhead for ranges up to 400 nautical miles (740 km).

persistence A measure of the length of time that the screen of a cathode-ray tube remains luminescent after excitation is removed. Long-persistence screens are used for PPI radar displays. Medium-persistence screens are used chiefly in television receivers. Short-persistence screens are used in cathode-ray oscilloscopes and some types of radar displays. The last number in the type designation of a cathode-ray tube indicates its persistence, on a scale ranging from 1 for short persistence to 7 for long persistence. Also called afterglow.

persistence characteristic The relation between luminance and time after excitation of a luminescent screen. Also called decay characteristic.

persistence of vision The ability of the eye to retain the impression of an image for a short time after the image has disappeared. This characteristic enables the eye to fill in the dark intervals between successive images in movies and television and give the illusion of motion.

persistent current A magnetically induced current that flows undiminished in a superconducting material or circuit. This current in turn produces a persistent magnetic field.

persistent-image device An optoelectronic amplifier capable of retaining an image for a definite length of time.

persistor A superconducting thin-film memory element. It consists of a superconducting inductor in parallel with a switch element that is normally superconducting but becomes resistive when the current exceeds a critical value. A current pulse is applied to start a circulating current for read-in; a pulse in the opposite direction reverses the circulating current and produces an output voltage for readout.

persistron A solid-state electroluminescent and photoconductive display panel that provides amplification of light.

personal locator beacon A beacon capable of providing homing signals to help search and rescue operations after an accident.

personnel decontamination The removal of radioactive materials from human skin by appropriate mechanical and/or chemical means.

personnel monitoring Determination by standard survey meters of the degree of radioactive contamination on individuals and determination by dosimeters of the dose received.

perspective representation A radar PPI display in which the region ahead of the ship is produced practically in perspective, much as would be seen from the bridge when visibility is good.

Perspex Trademark of a British plastic similar to Plexiglas.

PERT [Program Evaluation and Review Technique] A management control tool for defining, integrating, and interrelating what must be done to accomplish a desired objective on time. A computer compares current progress against planned objectives and gives management the information needed for planning and decision making.

perturbation A change in a known system.

perturbation theory The study of the effect of small changes on the behavior of a system.

peta- [abbreviated P] A prefix that represents 10^{15}.

pF Abbreviation for *picofarad.*

PF Abbreviation for *power factor.*

PFM Abbreviation for *pulse-frequency modulation.*

PGBM Abbreviation for *pulse-gated binary modulation.*

ph Abbreviation for *phot.*

phanotron A hot-cathode gas diode used as a rectifier. The type 866 mercury-vapor rectifier tube is an example.

phantastran A solid-state phantastron.

phantastron A monostable pentode circuit that generates sharp pulses at an adjustable and accurately timed interval after receipt of a triggering signal.

phantom A volume of material that has the radiation-absorbing characteristics of tissue, used to simulate a portion of the human body. Radiation measurements made at a point in a phantom permit determination of the radiation dose delivered to a corresponding point within the body. Materials commonly used for x-rays are beeswax, Masonite (unit density), and water.

phantom circuit A communication circuit derived from two other communication circuits or from one other circuit and ground, with no additional wire lines.

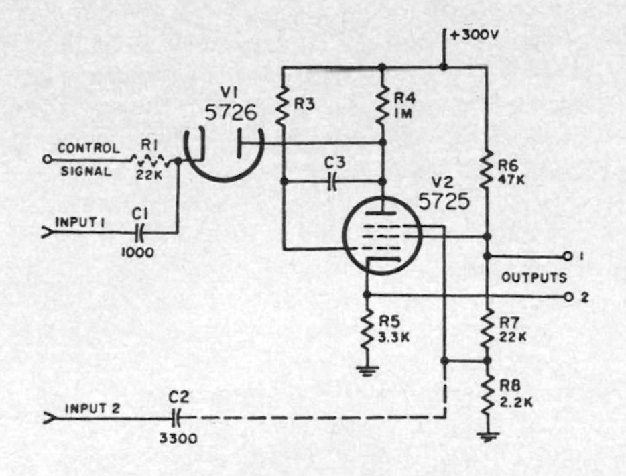

Phantastron, using circuit design preferred by National Bureau of Standards. Duration of output waveform is almost directly proportional to voltage of control signal. For durations under 1 ms, use 1.0 MΩ for R3 and 100 to 1000 pF for C3.

phantom target *Echo box.*

phase [abbreviated ϕ] The position of a point on the waveform of an alternating or other periodic quantity with respect to the start of the cycle. Usually expressed in degrees, with 360° representing one complete cycle.

phase-alternation line system *PAL.*

phase-amplitude modulation multiplier A multiplier in which the phase of a carrier is made proportional to one variable, and its amplitude is made proportional to the other variable. This signal is applied to a detector whose averaged output is proportional to the product of the two variables. The detector can be a balanced demodulator or a synchronous detector.

phase-angle meter *Phasemeter.*

phase-angle voltmeter A voltmeter that provides both a direct reading of the phase angle and the magnitude of an AC voltage.

phase comparator A comparator that accepts two RF input signals of the same frequency and provides two video outputs which are proportional, respectively, to the sine and cosine of the phase difference between the two inputs.

phase-comparison tracking system A tracking system that provides target trajectory information by continuous-wave phase-comparison techniques.

phase conjugacy A fundamental requirement for retrodirectivity in antenna arrays, wherein each element in the array must have an outgoing wave that is delayed exactly as much as the incoming wave was advanced.

phase constant A rating for a line or medium through which a plane wave of a given frequency is being transmitted. It is the imaginary part of the propagation constant and is the space rate of decrease of phase of a field component (or of the voltage or current) in the direction of propagation, in radians per unit length. The real part of the propagation constant is the attenuation constant.

phase control 1. A control that changes the phase angle at which the AC line voltage fires a thyratron, ignitron, silicon controlled rectifier, or

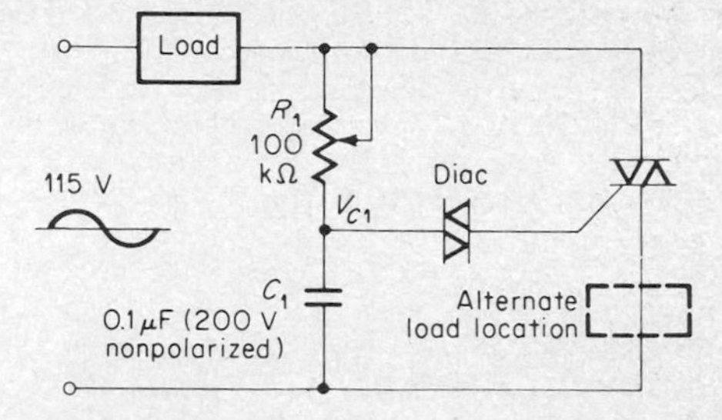

Phase-control circuit using diac-triggered triac.

other control device. Also called phase-shift control. 2. *Hue control.*

phase converter A converter that changes the number of phases in an AC power source without changing the frequency.

phased-array antenna An antenna array in which the radiation pattern of the fixed beam is determined by the phase relationships of the signals that excite the radiating elements. With adjustable phase shifters operating under computer control, the beam can be scanned in azimuth or elevation without mechanical movements.

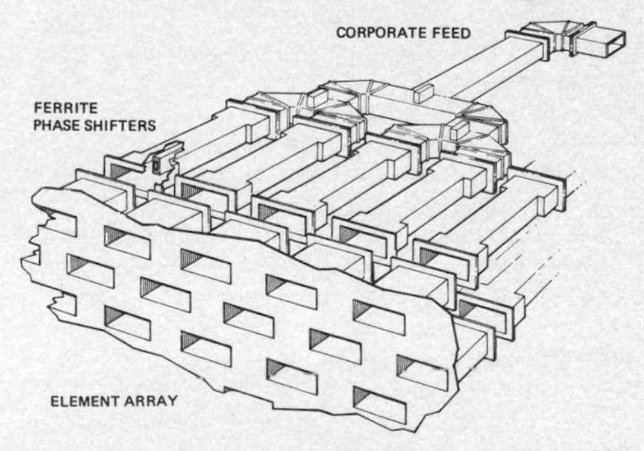

Phased-array radar.

phased-array radar *Array radar.*

phase delay The very short time delay that occurs when a single-frequency wave is transferred from one point to another in a system.

phase detector A circuit that provides a DC output voltage which is related to the phase difference between an oscillator signal and a reference signal, for use in controlling the oscillator to keep it in synchronism with the reference signal. Used in color television receivers to maintain the subcarrier oscillator in synchronization with the color-burst reference signal. Also called phase discriminator and phase-to-voltage converter.

phase deviation The peak difference between the instantaneous angle of a modulated wave and the angle of the sine-wave carrier in phase modulation.

phase difference The time in electrical degrees by which one wave leads or lags another.

phase discriminator *Phase detector.*

phase distortion *Phase-frequency distortion.*

phase equalizer A network that compensates for phase-frequency distortion within a specified frequency band.

phase focusing An automatic action that helps to keep the electrons of a multicavity magnetron in phase with the rotating field. Lagging electrons receive energy from the radial component of the gap field to reduce the phase lag, and leading electrons give up energy to the gap field to reduce the phase lead.

phase-frequency distortion Distortion occurring because phase shift is not proportional to frequency over the frequency range required for transmission. Also called phase distortion.

phase generator An instrument that accepts single-phase input signals over a given frequency range, or generates its own signal, and provides continuous shifting of the phase of this signal by one or more calibrated dials.

phase inverter A circuit or device that changes the phase of a signal by 180°, as required for feeding a push-pull amplifier stage without using a coupling transformer, or for changing the polarity of a pulse. A triode is commonly used as a phase inverter.

phase jitter Jitter that undesirably shortens or lengthens pulses intermittently during data processing or transmission.

phase localizer A localizer in which the on-course line is centered in an equiphase zone, and right-left deviations from this zone are detectable as reversals of phase of one of the two radiated signals.

phase lock The technique of making the phase of an oscillator signal follow exactly the phase of a reference signal, by comparing the phases between the two signals and using the resultant difference signal to adjust the frequency of the reference oscillator.

phase-locked loop [abbreviated PLL] A circuit that consists essentially of a phase detector which compares the frequency of a voltage-controlled oscillator with that of an incoming carrier signal or reference-frequency generator. The output of the phase detector, after passing through a loop filter, is fed back to the voltage-controlled oscillator to keep it exactly in phase with the incoming or reference frequency. Used in color television, telemetry, and many other types of receivers.

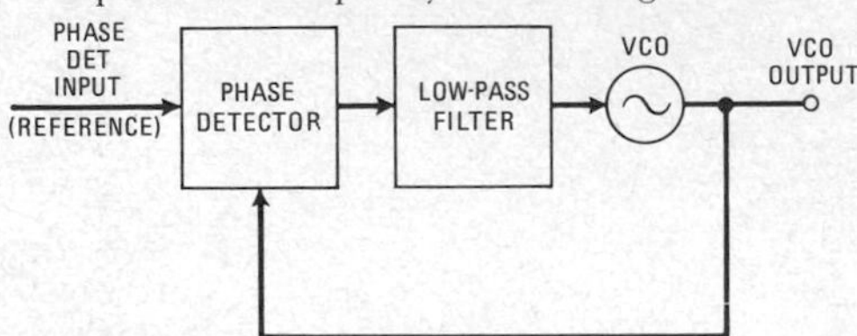

Phase-locked loop block diagram.

phasemeter An instrument that measures the difference in phase between two alternating quantities which have the same frequency. Also called phase-angle meter.

phase-modulated transmitter A transmitter that transmits a phase-modulated wave.

phase modulation [abbreviated PM or ϕM] Angle modulation in which the phase (usually expressed as an angle in degrees) of a carrier varies with the amplitude of the modulating signal wave.

phase modulator A modulator that provides phase modulation of a carrier signal.

phase multiplier A device that multiplies the frequency of signals used for phase comparison, so that phase differences may be measured to a higher degree of resolution.

phaseout The discontinuation of a major project or operation according to a gradual schedule.

phase-plane analysis Nonlinear analysis that provides a dynamic graphic time portrait of a simple circuit which has variable inductance and/or capacitance. Used with parametric-amplifier circuits.

phase-propagation ratio The propagation ratio divided by its magnitude.

phase quadrature *Quadrature.*

phaser 1. A microwave ferrite phase shifter that uses a longitudinal magnetic field along one or more rods of ferrite in a waveguide. 2. A device for adjusting facsimile equipment so the recorded elemental area bears the same relation to the record sheet as the corresponding transmitted elemental area bears to the subject copy in the direction of the scanning line.

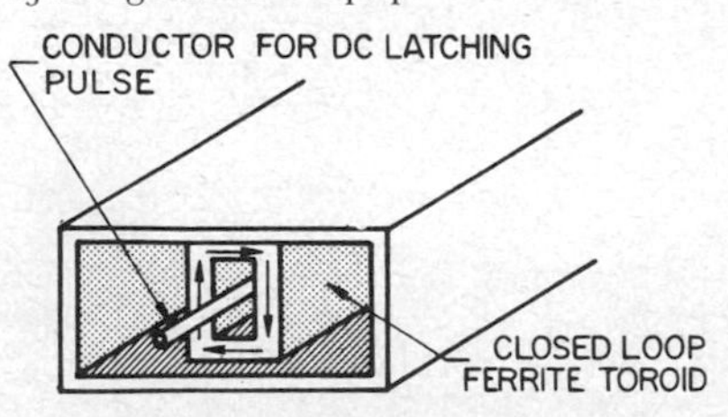

Phaser using rectangular ferrite toroid in waveguide.

phase resolution The minimum phase change that can be distinguished by a given system.

phase reversal A change of 180° or one half-cycle in phase.

phase-reversal keying [abbreviated PRK] Keying by ±90° phase deviation of the carrier.

phase-reversal modulation A form of pulse

modulation in which reversal of signal phase serves to distinguish between the two binary states used in data transmission.

phase-reversal switch A switch in a stereophonic sound system that reverses the connections to one loudspeaker so its acoustic output is reversed 180° in phase.

phase-sensing monopulse radar Monopulse radar in which the receiving antenna has two or more apertures separated by several wavelengths, each with its own feed. The apertures give identical radiation patterns. Phase comparison of the arriving signals gives the desired directional information with high precision.

phase-sensitive amplifier A servoamplifier whose output signal polarity or phase is dependent on the phase relationship between an input and a reference voltage.

phase-sequence relay A relay that responds to the order in which the voltages or currents in a polyphase system reach maximum positive values.

phase-shaped antenna *Shaped-beam antenna.*

phase shift 1. A change in the phase relationship between two alternating quantities. 2. The phase relationship between a scattered wave and the incident wave associated with a particle or photon that undergoes scattering.

phase-shift bridge A mutual-inductance bridge that measures the ratio of two voltages in both magnitude and phase.

phase-shift circuit A network that provides a voltage component which is shifted in phase with respect to a reference voltage.

phase-shift control *Phase control.*

phase-shift discriminator A discriminator that uses two similarly connected diodes and requires a limiter in its input to remove amplitude variations from the frequency- or phase-modulated input signal. The diodes are fed by a transformer that is tuned to the center frequency. When the frequency of the input signal swings away from this center frequency, one diode receives a stronger signal than the other. The net output of the diodes is then proportional to the frequency displacement. Also called Foster-Seeley discriminator.

phase shifter A device that changes the phase relation between two AC values.

phase-shift keying [abbreviated PSK] A modulation system used for data transmission in some modems. In its simplest form, the binary modulating signal produces the 0 and 180° phases of the carrier, for representing either mark and space or binary 1 and 0. For higher data rates over wire lines, four or eight different phase shifts are used. The system requires an accurate and stable reference phase at the receiver to distinguish between the various phases employed. With differentially coherent phase-shift keying, this reference requirement is eliminated by encoding data in terms of phase changes and detecting these by comparison with the phase of the preceding bit.

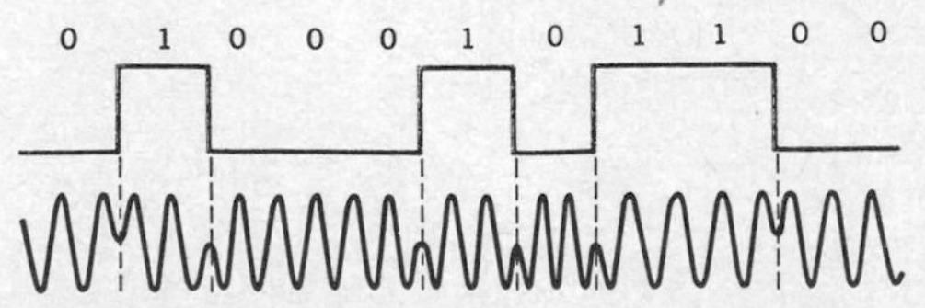

Phase-shift keying.

phase-shift microphone A microphone that uses phase-shift networks to produce directional properties.

phase-shift omnidirectional radio range An omnidirectional radio range that indicates the azimuthal position of an aircraft by two carrier waves, one of which is continuously changed in phase. The two waves are in phase only along a reference line that is usually north.

phase-shift oscillator An oscillator in which a network that has a phase shift of 180° per stage is connected between the output and input of an amplifier.

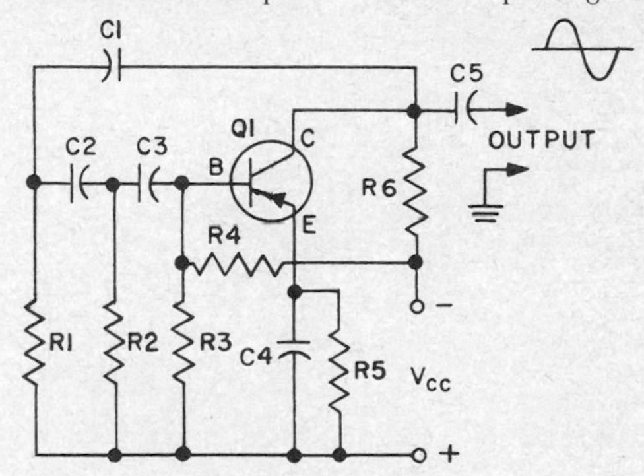

Phase-shift oscillator in which RC network provides required 180° phase shift for amplifier using transistor.

phase simulator A precision test instrument that generates reference and data signals at the same frequency but precisely separated in phase.

phase-splitter A circuit that takes a single input signal voltage and produces two output signal voltages 180° apart in phase.

phase-to-voltage converter *Phase detector.*

phase velocity The velocity with which a point that has a certain phase in an electromagnetic wave travels in the direction of propagation. In a waveguide the phase velocity may be greater than the wave velocity.

phasing *Framing.*

phasing capacitor A capacitor used in a crystal filter circuit to neutralize the capacitance of the crystal holder.

phasing line The portion of the length of a facsimile scanning line that is used for the phasing signal.

phasing link A delay line that connects together the bays of a stacked antenna, so the signals from all bays are in phase at the transmission line.

phasing signal A signal that adjusts the picture position along the scanning line in a facsimile system.

phasor A quantity expressed in complex form, with or without time variation. A phasor may be used to represent a vector, but a vector does not involve a complex plane and hence is not a phasor.

phenolic A thermosetting plastic material available in many combinations of phenol and formaldehyde, often with added fillers, to provide a broad range of physical, electrical, chemical, and molding properties.

φ [Greek letter phi] Abbreviation for *phase.*

Phillips screw A screw that has in its head a recess in the shape of an indented cross. It is

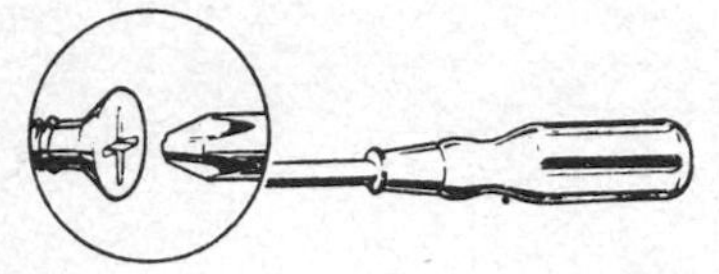

Phillips screw and Phillips screwdriver.

inserted or removed with a special Phillips screwdriver that automatically centers itself in the screw.

φM Abbreviation for *phase modulation.*

pH indicator An instrument that measures and indicates the hydrogen ion concentration of a solution on a scale of pH values from 0 to 14, where 7 indicates a neutral solution, lower numbers indicate acidity, and higher numbers indicate alkalinity.

Phoenix An air-to-air solid-propellant guided missile that has both radar and infrared acquisition, a speed of about Mach 5, and a range of about 400 nautical miles (740 km). Used on Navy planes to provide fleet air defense.

phon The unit of loudness level of a sound. It is numerically equal to the sound pressure level, in decibels relative to 0.0002 μbar, of a pure 1-kHz tone that is judged by listeners to be equivalent in loudness to the sound under consideration.

phone 1. *Headphone.* 2. *Telephone.*

phone jack A jack for standard ¼-in-diameter phone plugs. Also called telephone jack.

phone patch A device that connects an amateur or citizens band transceiver temporarily to a telephone system.

phone plug A standard plug that has a ¼-in-diameter shank, used with headphones, microphones, and other audio equipment. Usually designed for use with either two or three conductors. Also called telephone plug.

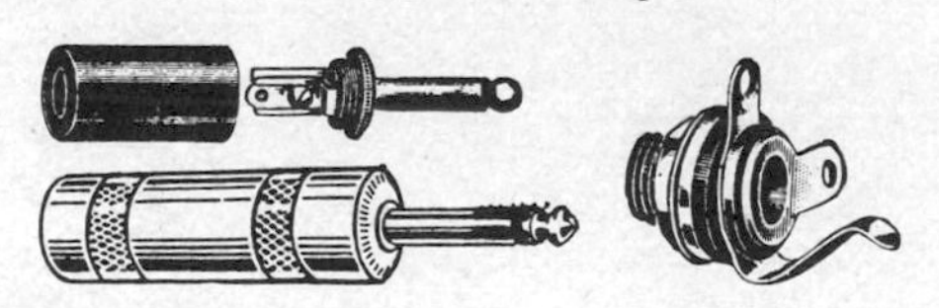

Phone plugs having standard ¼-in (0.635-cm) shank, with mating jack.

phonetic alphabet A list of standard words used for positive identification of letters in a voice message transmitted by radio or telephone. An example is

A Alfa	J Juliet	S Sierra
B Bravo	K Kilo	T Tango
C Charlie	L Lima	U Uniform
D Delta	M Mike	V Victor
E Echo	N November	W Whiskey
F Foxtrot	O Oscar	X X-ray
G Golf	P Papa	Y Yankee
H Hotel	Q Quebec	Z Zulu
I India	R Romeo	

phonocardiogram A graphic recording of the sounds of the heart.

phonocardiograph An instrument that provides a graphic record of heart murmurs and other sounds.

phono cartridge *Phonograph pickup.*

phonoelectrocardioscope An electronic medical instrument that uses a doublebeam cathode-ray oscilloscope to show simultaneously the waveforms of two different quantities related to the heart.

phonograph An instrument that converts the sound groove variations of a phonograph record into sound waves. In an electric phonograph, the needle movements in the record grooves are converted into audio-frequency currents and ampli-

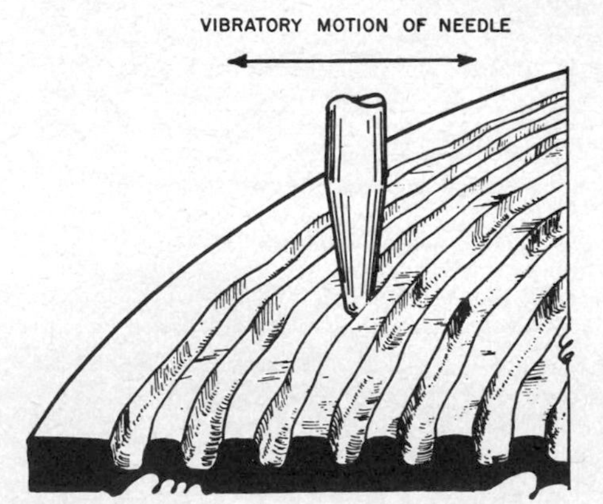

Phonograph, showing how needle is vibrated from side to side in conventional monophonic record.

fied sufficiently for reproduction by a loudspeaker. In a mechanical phonograph, the needle actuates a sound-producing diaphragm directly. The turntable on which the record is placed may be driven by an electric or a spring motor. Also called gramophone (British).

phonograph amplifier An AF amplifier that amplifies the AF output signal of a phonograph pickup.

phonograph connection Two terminals sometimes provided at the rear of a radio receiver, connected to the input of the first AF amplifier stage. When a phonograph pickup is connected to these terminals, its output is amplified by the AF amplifier and reproduced by the loudspeaker.

phonograph oscillator An RF oscillator that can be modulated by a phonograph pickup. The resulting modulated RF signal is fed through wires to the antenna and ground terminals of a radio receiver so the entire radio receiver can amplify and reproduce phonograph records. In wireless phonograph oscillators, the output is fed to a small loop antenna and broadcast through space to the radio receiver, eliminating wire connections.

phonograph pickup A pickup that converts variations in the grooves of a phonograph record into corresponding electric signals. Also called cartridge, phono cartridge, and phono pickup. An acoustic pickup converts groove variations directly into sound waves.

phonograph record A shellac-composition or vinyl plastic disk, usually 7, 10, or 12 in (17.8, 25.4, or 30.5 cm) in diameter, on which sounds have been recorded as modulations in grooves. Common speeds used are 16⅔, 33⅓, 45, and 78 rpm. Also called record.

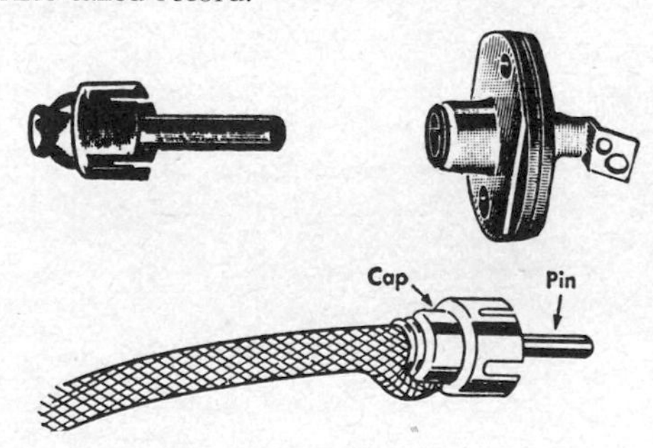

Phono jack and plug, and method mounting plug on end of shielded conductor.

phono jack A jack that accepts a phono plug and provides a ground connection for the shield of the conductor connected to the plug.

phonon A unit of thermal energy in a crystal lattice, equal in value to the product of Planck's constant and the thermal vibration frequency.

phono pickup *Phonograph pickup.*

phono plug A plug that attaches to the end of a shielded conductor, for feeding AF signals from a phonograph or other AF source to a mating phono jack on a preamplifier or amplifier.

phosphate-glass dosimeter A dosimeter in which a fluorod made from phosphate glass measures cumulative gamma-radiation doses.

phosphene A visual sensation experienced by a human subject during the passage of current through the eye.

phosphor Any material having phosphorescent, fluorescent, or luminescent properties.

phosphor bronze An alloy of copper, tin, and phosphorus, used for contact springs in switches and relays.

phosphor dot One of the tiny dots of phosphor material that are used in groups of three, one for each primary color, on the screen of a color television picture tube.

phosphor-dot faceplate The glass faceplate on which the trios of color phosphor dots are mounted in a shadow-mask three-gun color television picture tube.

phosphorescence A form of luminescence in which the emission of light continues more than 10^{-8} s after excitation by radiation that has a shorter wavelength, such as by electrons, ultraviolet light, or x-rays. When emission of light occurs only during excitation, the result is fluorescence. Also called afterglow.

phosphorogen A substance that promotes phosphorescence in another substance, as manganese does in zinc sulfide.

phosphorus [symbol P] A nonmetallic element. Atomic number is 15.

phot [abbreviated ph] The CGS unit of illumination, equal to 1 lm/cm^2. The SI unit (now preferred) is the lux, equal to 1 lm/m^2.

photalysis The use of radiant energy to produce chemical changes.

photocathode A photosensitive surface that emits electrons when exposed to light or other suitable radiation. Used in phototubes, television camera tubes, and other light-sensitive devices.

photocell A solid-state photosensitive electron device whose current-voltage characteristic is a function of incident radiation. Examples include photoconductive cells, phototransistors, and photovoltaic cells. Also called electric eye (slang) and photoelectric cell. A phototube is not a photocell.

photochemical activity Chemical changes caused by radiant energy, such as light.

photochromic compound A chemical compound that changes in color when exposed to visible or near-visible radiant energy. The effect is reversible. Used to produce very high-density microimages.

photochromic glass A glass that darkens when exposed to light but regains its original transparency a few minutes after light is removed. The rate of clearing increases with temperature.

photocoagulator A medical electronic instrument that uses an intense focused beam of light to coagulate tissue and stop bleeding during surgery. The light source is usually a laser. Another application is welding a detached retina back in position inside the eye.

photoconduction A process by which the conductance of a material is changed by incident electromagnetic radiation, such as visible, infrared, or ultraviolet light. An increase in radiation intensity increases the conductance (decreases the resistance).

photoconductive cell A photocell whose resistance varies with the illumination on the cell. The selenium cell is an example. When made from a semiconductor material such as lead sulfide, lead selenide, lead telluride, or germanium, it can give good response to infrared radiation. Cooling with liquid air or gas improves the infrared response of many photoconductive detectors. Also called light-dependent resistor, light-sensitive resistor, and photoresistor.

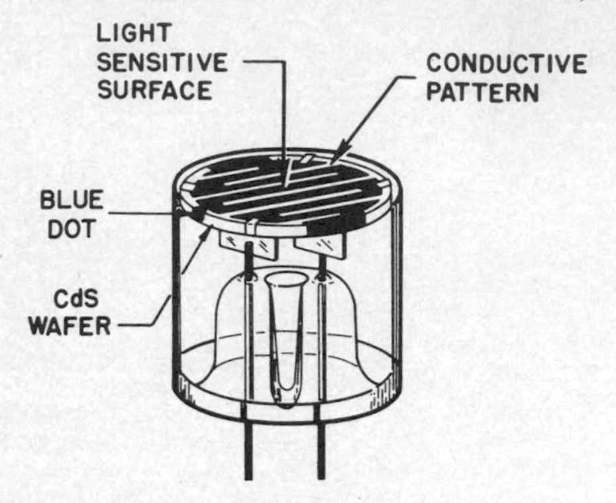

Photoconductive cell using cadmium sulfide wafer.

photoconductivity Conductivity that varies with illumination.

photoconductor A semiconductor in which conductivity varies with illumination.

photocoupler *Optoisolator.*

photocurrent An electric current that varies with illumination.

photodarlington A Darlington amplifier in which the input transistor is a phototransistor. The action is essentially the same as if a photodiode were feeding a standard Darlington amplifier in the same housing. Used in some optoisolators.

photodetector A detector that responds to radiant energy. Examples include photoconductive cells, photodiodes, photoresistors, photoswitches, phototransistors, phototubes, and photovoltaic cells. Also called light detector, light-sensitive cell, light-sensitive detector, and photosensor.

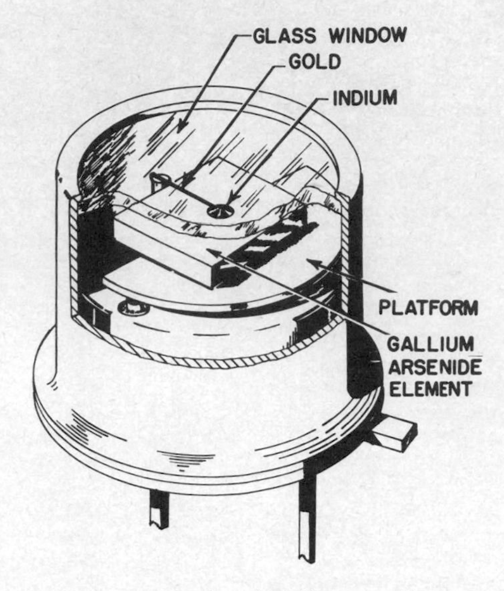

Photodetector construction with gallium arsenide element.

photodiode A semiconductor diode in which the reverse current varies with illumination. Examples include the alloy-junction and the grown-junction photocells.

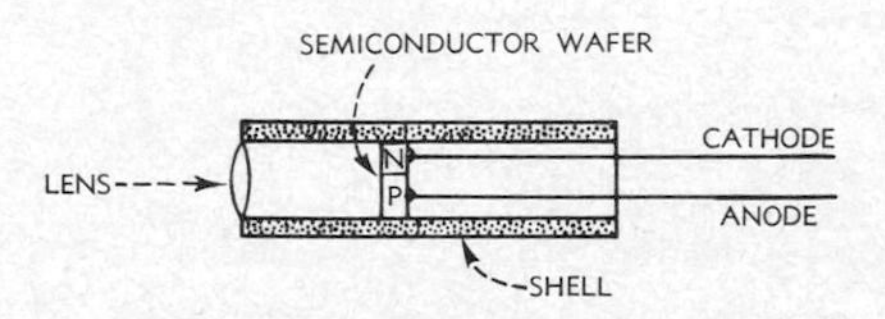

Photodiode construction.

photodiode parametric amplifier A photodetector arrangement for modulated laser beams, in which a diode serves simultaneously as a photodiode and a varactor diode, with parametric amplification being used to raise the apparent equivalent resistance of the photodetector. The pumping frequency of the parametric amplifier portion is equal to the sum of the resonant frequencies of the two tank circuits.

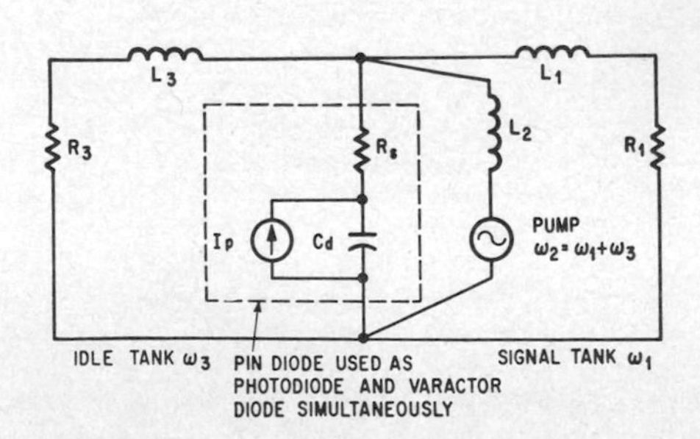

Photodiode parametric amplifier arrangement used as demodulator for laser communication or radar system.

photodisintegration The disintegration of an atomic nucleus by radiant energy.

photodissociation The removal of one or more atoms from a molecule by the absorption of a quantum of electromagnetic or photon energy.

photodosimetry Determination of the cumulative dose of ionizing radiation by use of photographic film.

photoelastic effect The change in the optical properties of a dielectric when subjected to mechanical stress.

photoelectric Pertaining to the electrical effects of light, such as the emission of electrons, generation of a voltage, or a change in resistance when exposed to light.

photoelectric abridged spectrophotometry Analysis of color by from three to eight spectral filters used in a simplified spectrophotometer to isolate spectral bands that make up color. The process is approximate since the bands employed are considerably wider than in true spectrophotometry.

photoelectric absorption The absorption of photons in the photoelectric effect.

photoelectric autocollimator An instrument that automatically produces precise electric error signals which are functions of the magnitude and direction of an angular displacement. Accuracy is achieved by using a photodetector at the location of the human eye in the basic optical instrument.

photoelectric cathode A cathode that functions primarily by photoelectric emission.

photoelectric cell *Photocell.*

photoelectric character reader *Optical character reader.*

photoelectric color comparator *Color comparator.*

photoelectric colorimeter A colorimeter that uses a phototube or photocell, a set of color filters, an amplifier, and an indicating meter for quantitative determination of color. Widely used to determine the constituents of a liquid in which the color varies with the constituents in a known manner.

photoelectric color register control A photoelectric control system used as a longitudinal position regulator for a moving material or web, to maintain a preset relationship between repetitive register marks when printing successive colors.

photoelectric constant A quantity that, when multiplied by the frequency of the radiation which is causing emission of electrons, gives the voltage absorbed by the escaping photoelectron. The constant is equal to h/e, where h is Planck's constant and e is the electronic charge.

photoelectric control Control of a circuit or piece of equipment in response to a change in incident light.

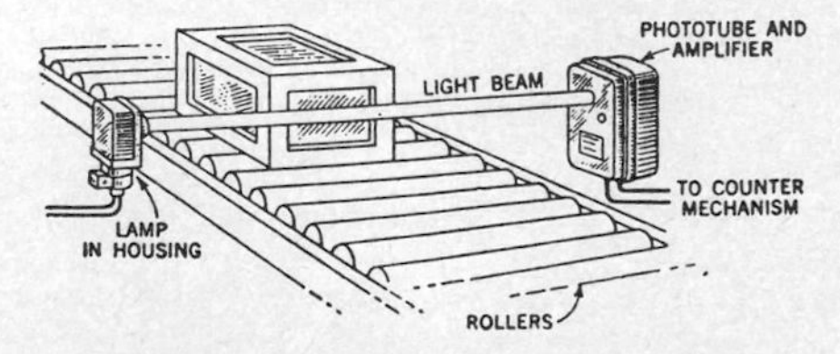

Photoelectric counter for cartons on roller conveyor.

photoelectric counter A photoelectrically actuated device that records the number of times a given light path is intercepted by an object.

photoelectric current A current of electrons emitted from the cathode of a phototube under the influence of light.

photoelectric cutoff register controller A photoelectric control system used as a longitudinal position regulator to maintain the position of the point of cutoff with respect to a repetitive pattern on a moving material.

photoelectric densitometer An electronic instrument that measures the density or opacity of a film or other material. A beam of light is directed through the material, and the amount of light transmitted is measured with a photocell and meter.

photoelectric directional counter A photoelectrically actuated device that records the number of times a given light path is intercepted by an object moving in a given direction.

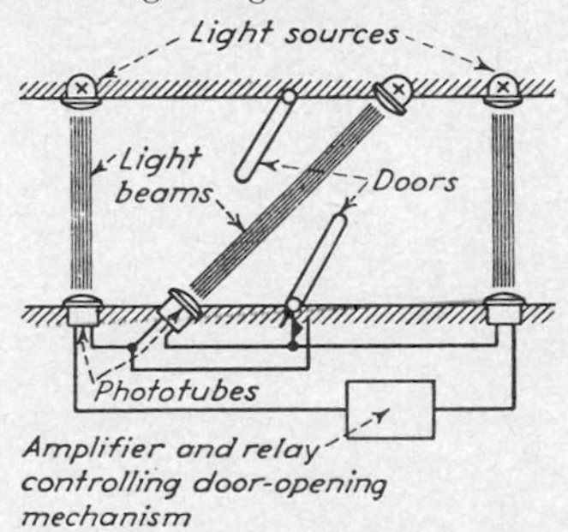

Photoelectric door opener, arranged so persons can pass in either direction and doors cannot close until person has passed through all three beams.

photoelectric door opener A photoelectric control system that opens and closes a power-operated door.

photoelectric effect The emission of electrons from a body because of visible, infrared, or ultraviolet radiant energy. The energy of a photon is absorbed for each electron emitted.

photoelectric emission The emission of electrons by certain materials upon exposure to radiation in and near the visible region of the spectrum.

photoelectric flame-failure detector A photoelectric control that cuts off fuel flow when the fuel-consuming flame is extinguished.

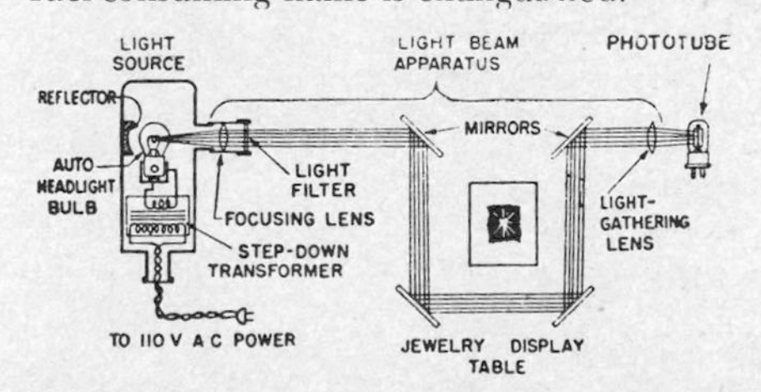

Photoelectric intrusion-detector setup for jewelry store or museum.

photoelectric intrusion-detector A burglar-alarm system in which interruption of a light beam by an intruder reduces the illumination on a phototube and thereby closes an alarm circuit.

photoelectric lighting controller A photoelectric relay actuated by a change in illumination to control the illumination in a given area or at a given point.

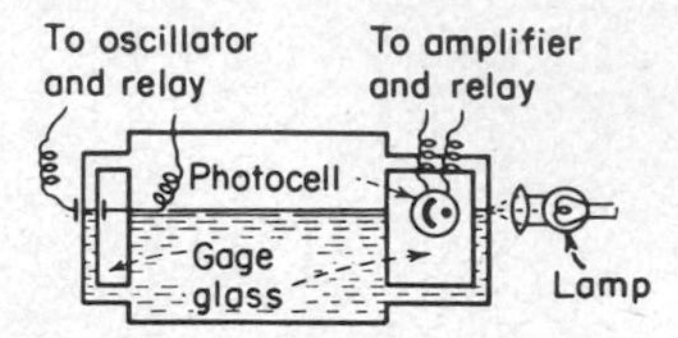

Photoelectric liquid-level indicator.

photoelectric liquid-level indicator A level indicator in which rising liquid interrupts the light beam of a photoelectric control system.

photoelectric loop control A photoelectric control system used as a position regulator for a loop of material passing from one strip-processing line to another that may travel at a different speed. Also called loop control.

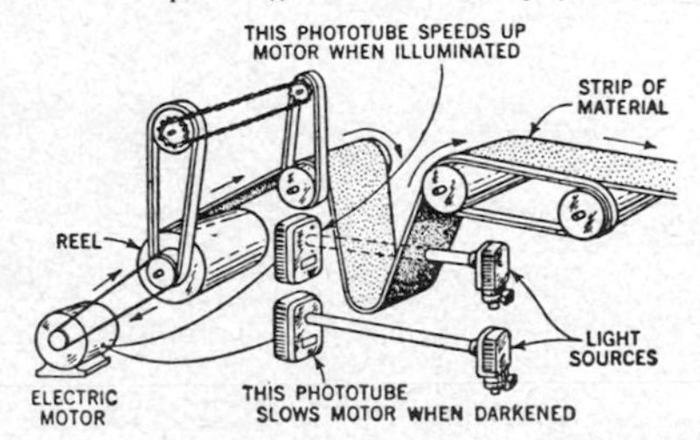

Photoelectric loop control for loop in strip of paper, metal, or fabric.

photoelectric material A material that will emit electrons when exposed to radiant energy in a vacuum. Examples are barium, cesium, lithium, potassium, rubidium, sodium, and strontium.

photoelectric opacimeter *Photoelectric turbidimeter.*

photoelectric phonograph pickup A phonograph pickup in which a tiny mirror or vane on the stylus varies the amount of light reaching two photodetectors that are providing stereo output.

photoelectric photometer A photometer that uses a photocell, phototransistor, or phototube to measure the intensity of light. Also called electronic photometer.

photoelectric pickup A pickup that converts a change in light to an electric signal.

photoelectric pinhole detector A photoelectric control system that detects minute holes in an opaque material.

photoelectric plethysmograph A photoelectric medical instrument for measuring and recording ear opacity by a tiny phototube and lamp clipped to the ear, as a measure of the state of fullness of blood vessels. Also worn by aircraft pilots during high-altitude flights as an alarm to indicate the need for more oxygen.

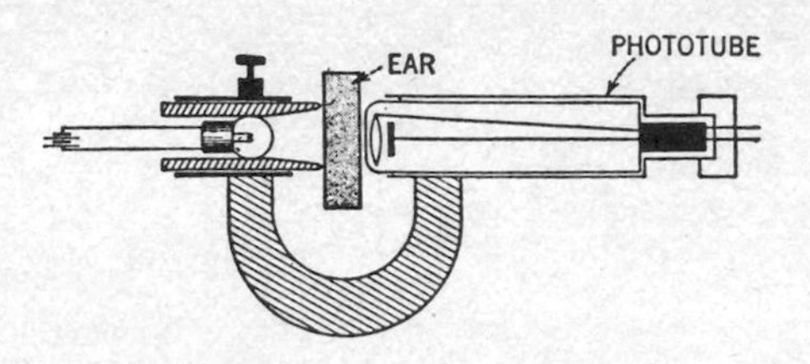

Photoelectric plethysmograph.

photoelectric pulse generator A pulse generator in which apertures on a shaft-mounted disk or drum alternately transmit and interrupt light falling on a photocell. The output pulse frequency derived from the photocell is thus directly proportional to shaft speed.

photoelectric pyrometer An instrument that measures high temperatures by using a photoelectric arrangement to measure the radiant energy given off by the heated object.

photoelectric reader A device capable of reading information, in the form of holes punched in paper tape or cards, by sensing a light passed through these holes.

photoelectric reflectometer A reflectometer that uses a photocell or phototube to measure the diffuse reflection of surfaces, powders, pastes, and opaque liquids.

photoelectric register control A register control that uses a light source, one or more phototubes, a suitable optical system, an amplifier, and a relay to actuate control equipment when a change occurs in the amount of light reflected from a moving surface due to register marks, dark areas of a design, or surface defects.

photoelectric relay A relay combined with a phototube and amplifier, arranged so changes in incident light on the phototube make the relay contacts open or close. Also called light relay.

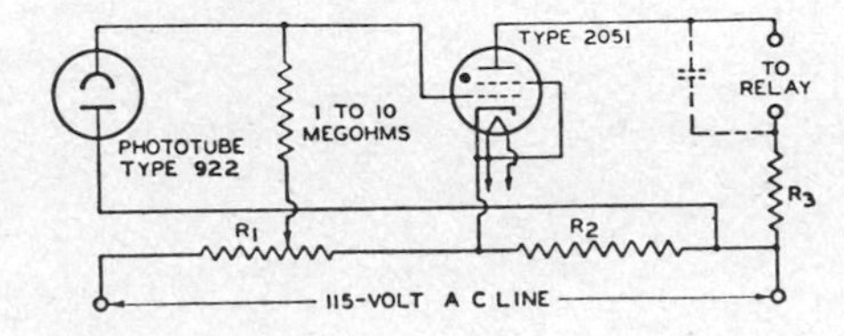

Photoelectric relay circuit.

photoelectric scleroscope A scleroscope that uses a phototube-light beam system to measure the rebound of a steel ball during hardness tests.

photoelectric sensitivity The ratio of photoelectric emission current to incident radiant energy. Also called photoelectric yield.

photoelectric side-register control A photoelectric control system used as a lateral position regulator to maintain the edge of a moving material or web at a fixed position.

photoelectric smoke-density control A photoelectric control system that measures, indicates, and controls the density of smoke in a flue or stack.

photoelectric smoke meter A photoelectric instrument that measures the density of smoke.

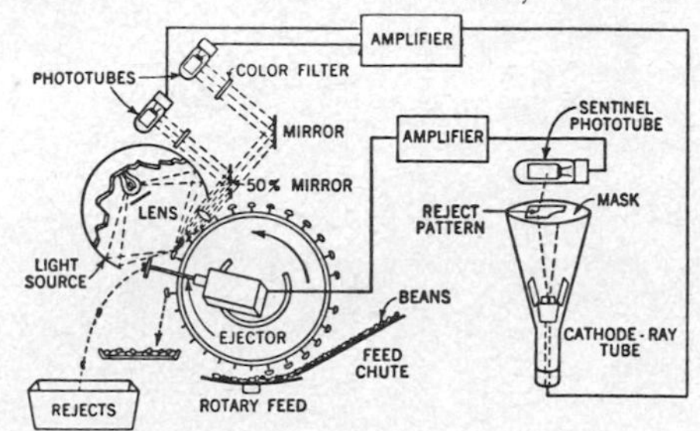

Photoelectric sorter as used to reject dark beans at high speed. Beans are held on sorting wheel by vacuum.

photoelectric sorter A photoelectric control system used to sort objects according to color, size, shape, or other light-changing characteristics.

photoelectric threshold The quantum of energy just sufficient to release electrons from a given surface by the photoelectric effect.

photoelectric timer A timer that automatically turns off an x-ray machine when the film has received the correct exposure as determined by an integrating photoelectric measuring system which monitors a fluorescent screen placed behind the film.

photoelectric transducer A transducer that converts changes in light energy to changes in electric energy.

photoelectric tristimulus colorimeter A colorimeter that uses three or more combinations of light sources, filters, and phototubes to measure colors with high accuracy.

photoelectric tube *Phototube.*

photoelectric turbidimeter A photoelectric instrument that determines the turbidity of almost clear solutions. Also called photoelectric opacimeter.

photoelectric work function The energy required to transfer electrons from a given metal to a vacuum or other adjacent medium during photoelectric emission. Usually expressed in electronvolts.

photoelectric yield *Photoelectric sensitivity.*

photoelectromagnetic effect When light falls on a flat surface of an intermetallic semiconductor located in a magnetic field that is parallel to the surface, excess hole-electron pairs are created. These carriers diffuse in the direction of the light

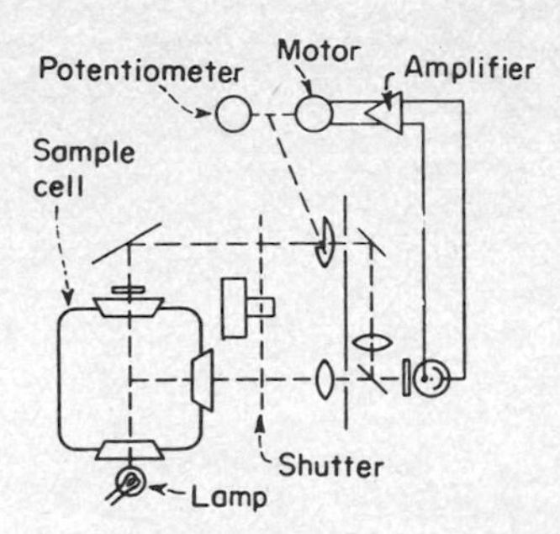

Photoelectric turbidimeter in which light from lamp reaches phototube at lower right alternately over two paths, under control of motor-driven rotary shutter. One path measures light transmitted by liquid in sample cell; other path measures light reflected at right angles by particles suspended in liquid.

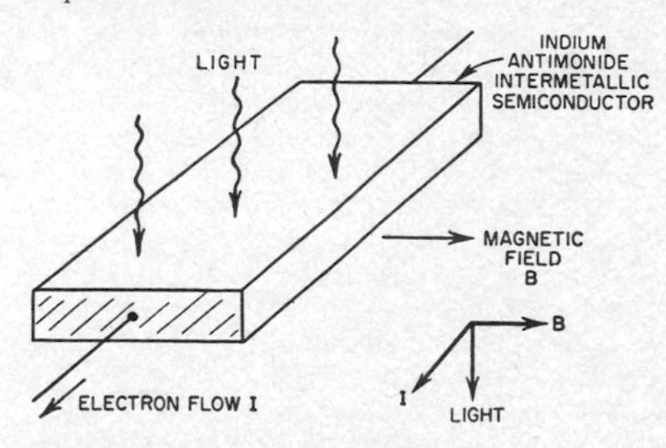

Photoelectromagnetic effect.

but are deflected by the magnetic field to give a current flow through the semiconductor that is at right angles to both the light rays and the magnetic field.

photoelectron An electron emitted by the photoelectric effect.

photoemissive Capable of emitting electrons upon exposure to radiation in and near the visible region of the spectrum.

photoemitter A material that emits electrons when it is illuminated sufficiently. With a photocathode, the emitted electrons are collected in a vacuum. With an internal photoemitter, such as a photoconductive or photovoltaic device, the electrons are detected by measuring the change they produce in the resistance or some other property of the material. In a negative-electron-affinity photoemitter, high efficiency of light detection is obtained because the electrons are generally deep within the material and are diffused to the surface.

photoemitter cathode An unheated cathode that emits electrons when it is exposed to light. Electron emission may occur at the surface that is exposed to incident light, as in conventional tubes, or may be from the opposite side, as in a transmission photocathode.

photofission Nuclear fission induced by photons.

photoflash tube *Flash tube.*

photoflash unit A portable electronic light source for photographic use; it consists of a capacitor-discharge power source, a flash tube, a battery for charging the capacitor, and sometimes a high-voltage pulse generator to trigger the flash.

photoformer A signal generator that delivers an output signal which has a waveform corresponding to that of a mask cut to the shape of the desired function. The mask is placed on the face of a cathode-ray tube, and a phototube in front of the screen is connected to the deflection circuits in such a way as to make the trace follow the contour of the mask. The vertical deflection signal is then proportional to the desired function.

photogenerator A semiconductor-junction device capable of generating or emitting light when pulsed.

photoglow tube A gas-filled phototube used as a relay by making the operating voltage sufficiently high so ionization and a glow discharge occur, with considerable current flow, when a certain illumination is reached.

photogoniometer A goniometer that uses a phototube or photocell as a sensing device for studying x-ray spectra and x-ray diffraction effects in crystals.

photographic recording Facsimile recording in which a photosensitive surface is exposed by a signal-controlled light beam or spot.

photographic sound recorder A sound recorder that has means for producing a modulated light beam and moving a light-sensitive medium relative to the beam to give a photographic recording of sound signals. Also called optical sound recorder.

photographic sound reproducer A sound reproducer in which an optical sound record on film is moved through a light beam directed at a light-sensitive device, to convert the recorded optical variations back into audio signals. Also called optical sound reproducer.

photographic transmission density The common logarithm of opacity. A film that transmits 100% of the light has a density of 0, and a film transmitting 10% has a density of 1. Also called optical density.

photoionization The removal of one or more electrons from an atom or molecule by absorption of a photon of visible or ultraviolet light. Also called atomic photoelectric effect.

photo-island grid The photosensitive surface of an image dissector tube for television cameras.

photoisolator *Optoisolator.*

photoklystron A high-frequency cavity-type photodetector. A modulated light beam hitting the photocathode causes emission of a correspondingly modulated electron beam. An anode draws these electrons into the coupling cavity, which is tuned to the modulation frequency, thereby inducing a signal in the output loop of the tube.

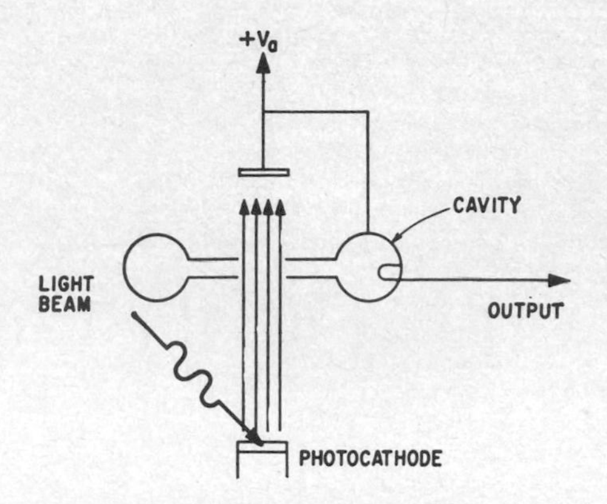

Photoklystron in which modulated light beam is converted to modulated electron current that induces signal in coupling cavity tuned to modulation frequency. Arrows from photocathode represent electron flow to anode.

photoluminescence Luminescence stimulated by visible, infrared, or ultraviolet radiation.

photomagnetic effect The direct effect of light on the magnetic susceptibility of certain substances.

photomagnetoelectric effect The generation of a voltage when a semiconductor material is positioned in a magnetic field and one face is illuminated.

photomask A film or glass negative that has many high-resolution images, used in the production of semiconductor devices and integrated circuits.

photomeson A meson, usually a pi meson, that is ejected from a nucleus by an impinging photon.

photometer An instrument that measures the intensity of a light source or the amount of illumination on a surface.

photometry Measurement of luminous flux and related quantities, such as illuminance, luminance, luminosity, and luminous intensity.

photomixer A phototube that detects optical beats in the superheterodyne receiver of an optical communication system which uses coherent laser sources.

photomultiplier counter A scintillation counter that has a built-in multiplier phototube.

photomultiplier tube *Multiplier phototube.*

photon A quantum of electromagnetic radiation, equal to Planck's constant multiplied by the frequency in hertz. Electromagnetic radiation can be considered as photons of light, x-rays, gamma rays, or radio waves.

photon coupling Coupling of two circuits by photons passing through a light pipe.

photonegative Having negative photoconductivity, hence decreasing in conductivity (increasing in resistance) under the action of light. Selenium sometimes exhibits this property.

photon emission spectrum The relative numbers of optical photons emitted by a scintillator material per unit wavelength as a function of wavelength. The emission spectrum may also be given in alternative units such as wave number, photon energy, or frequency.

photon engine A reaction engine in which thrust is obtained from a stream of light rays. Although the thrust is very small, it can be applied indefinitely in outer space to build up speeds approaching the speed of light.

photonephelometer An instrument that measures the clarity of a liquid.

photoneutron A neutron released from a nucleus in a photonuclear reaction.

photon flux The total amount of luminous flux arriving at the photocathode of a multiplier phototube per unit time, expressed in photons per second.

photonuclear reaction A nuclear reaction induced by a photon.

photoparametric amplifier A microwave amplifier for light that is amplitude-modulated in the microwave range; it consists of a photodiode which has integral parametric amplification.

photopositive Having positive photoconductivity, hence increasing in conductivity (decreasing in resistance) under the action of light. Selenium ordinarily has this property.

photoproton A proton released from a nucleus in a photonuclear reaction.

photoresist A light-sensitive coating that is applied to a substrate or board, exposed, and developed prior to chemical etching. The exposed areas serve as a mask for selective etching.

photoresistor *Photoconductive cell.*

photoscanner A scanner used to make a film record of gamma rays passing through tissue from an injected radioactive material.

photo-SCR *Light-activated silicon controlled rectifier.*

photosensitive *Light-sensitive.*

photosensitive recording Recording by the exposure of a photosensitive surface to a signal-controlled light beam or spot.

photosensor *Photodetector.*

photosphere The apparent surface of the sun or of a star from which light is seen to radiate.

phototelegraphy *Facsimile.*

photothermoelectric effect The generation of the voltage in a heat-transmitting semiconductor when it is exposed to light.

photothyristor *Light-activated silicon controlled rectifier.*

phototransistor A junction transistor that may have only collector and emitter leads or also a base lead, with the base exposed to light through a tiny lens in the housing. Collector current increases with light intensity, as a result of amplification of base current by the transistor structure.

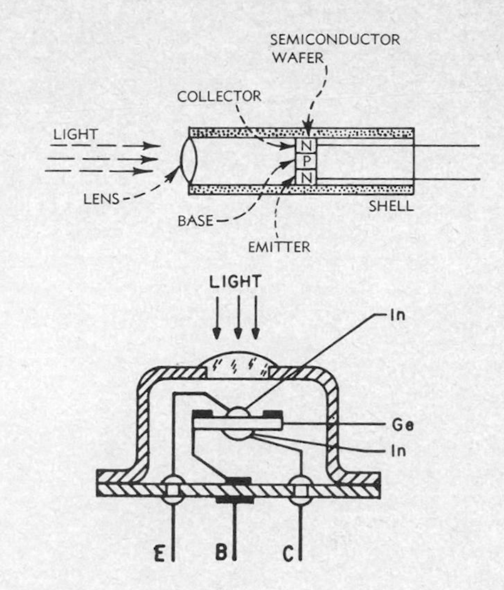

Phototransistors having two and three leads.

phototube An electron tube in which the output signal is related to the total radiation that is producing photoelectric emission from the photocathode. The photocathode surface can be chosen for maximum response to a particular part of the visible, infrared, or ultraviolet spectrum. A phototube is not a photoelectric cell. Also called electric eye (slang) and photoelectric tube.

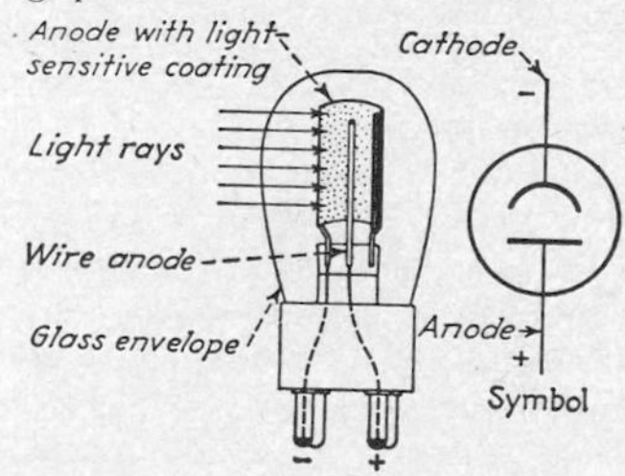

Phototube construction and symbol.

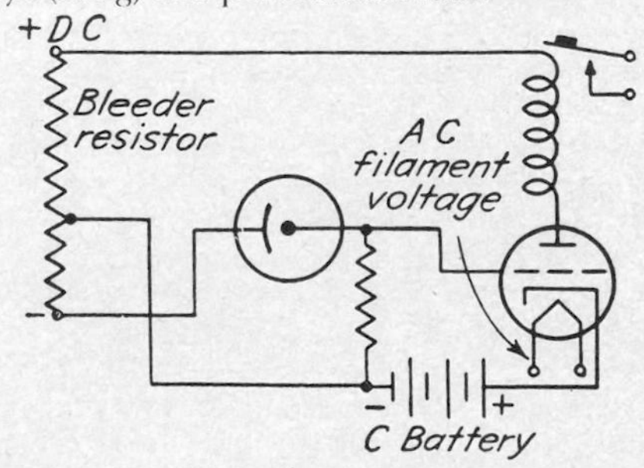

Phototube relay in which light on phototube at left makes grid of triode at right more negative, reducing anode current and thereby causing relay at upper right to drop out, opening circuit being controlled.

phototube relay A photoelectric relay in which a phototube serves as the light-sensitive device.

photovaristor A varistor in which the current-voltage relation may be modified by illumination. Cadmium sulfide and lead telluride may be used in varistors for this purpose.

photovoltaic Capable of generating a voltage as a result of exposure to radiation.

photovoltaic cell A photocell in which radiant energy causes electrons to pass through the surface of contact between a conductor and semiconductor, thereby generating a voltage. The solar cells used on satellites and space probes are photovoltaic cells in which a semiconductor such as silicon converts solar radiation into useful electric power.

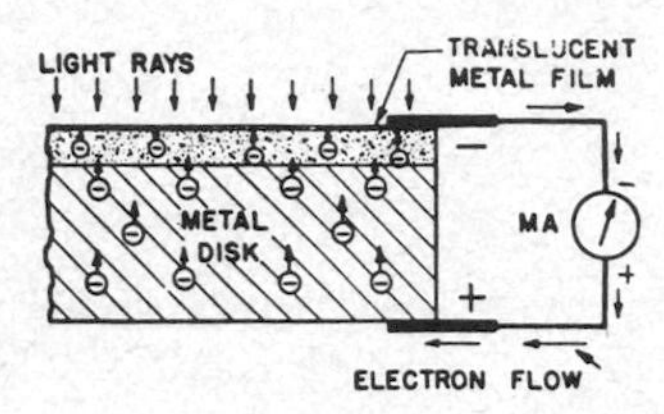

Photovoltaic cell.

photovoltaic detector A high-intensity photovoltaic cell that uses a single gallium arsenide crystal to provide high sensitivity for both visible and near-infrared radiation, as required for celestial navigation of satellites and space vehicles.

photovoltaic effect The development of a voltage across the junction of two dissimilar materials, such as a PN junction or a metal-semiconductor junction, when the junction is exposed to light or other radiant energy.

phthalocyanine Q switching Laser *Q* switching in which a solution of metal-organic compounds known as phthalocyanines is placed in a cell between an uncoated ruby laser crystal and a high-reflectivity mirror. When the incident ruby light reaches a certain level, the solution suddenly becomes almost perfectly transparent to this light, permitting the release of all the energy stored in the ruby as a giant pulse. The solution then returns to its absorbing state, in readiness for formation of another pulse.

physical optics The branch of optics that considers light as a form of wave motion, in which the energy of the light is propagated by wavefronts rather than by rays.

pi The Greek letter π, used to designate the ratio of the circumference of a circle to its diameter. A complete circle contains 2π radians. The value of pi has been determined to 500,000 places by computer, but for most purposes sufficient accuracy is obtained with 3.14159265359, 3.14159, or simply 3.1416.

picket ship A radar-equipped ship, generally anchored, that is used to extend radar-early-warning coverage seaward.

pickoff A device that converts mechanical motion into a proportional electric signal.

pickup 1. A device that converts a sound, scene, measurable quantity, or other form of intelligence into corresponding electric signals, as in a microphone, phonograph pickup, or television camera. A pickup is a transducer only when energy conversion is also involved, as in a microphone or phonograph pickup. In a telemetering system the end instrument is a pickup. 2. The minimum current, voltage, power, or other value at which a relay will complete its intended function. 3. Interference from a nearby circuit or system. 4. A type of nuclear reaction in which the incident particle takes a nucleon from the target nucleus and proceeds with this nucleon bound to itself. Pickup is the inverse of stripping. 5. A potentiometer used in an automatic pilot to detect the motion of the airplane around the gyro and initiate corrective adjustments.

pickup arm A pivoted arm that holds a phonograph pickup cartridge. Also called tone arm.

pickup cartridge A cartridge that contains the electromechanical translating elements and the reproducing stylus of a phonograph pickup.

pickup current The current at which a magnetically operated device starts to operate. Also called pull-in current.

pickup spectral characteristic The spectral response of a pickup that converts radiation to electric signals, as measured at the output terminals of the pickup tube.

pickup tube *Camera tube.*

pickup value The minimum voltage, current, or power at which the contacts of a previously deenergized relay will always assume their energized position. Also called pull-in value.

pickup voltage The voltage at which a magnetically operated device starts to operate.

pico- [abbreviated p] A prefix that represents 10^{-12}, which is 0.000000000001, or one-millionth of a millionth. Pronounced pie-ko. Formerly called micromicro-.

picoammeter An ammeter whose scale is calibrated to indicate current values in picoamperes.

picoampere [abbreviated pA] One-millionth of a microampere.

picofarad [abbreviated pF] One-millionth of a microfarad. Called puff in England.

picosecond [abbreviated ps] One-millionth of a microsecond.

picowatt [abbreviated pW] One-millionth of a microwatt, or 10^{-12} W.

pictorial wiring diagram A wiring diagram that contains actual sketches of radio parts and shows clearly all connections between the parts. Used in service manuals.

picture The image on the screen of a television receiver.

picture black The signal produced at any point in a facsimile system by the scanning of a selected area of subject copy that has maximum density.

picture brightness The brightness of the highlights of a television picture, usually expressed in candelas per square meter.

picture carrier A carrier frequency located 1.25 MHz above the lower frequency limit of a standard NTSC television signal. In color television, this carrier is used for transmitting luminance information; the chrominance subcarrier, which is 3.579545 MHz higher, transmits the color information. The sound carrier, 5.75 MHz above the picture carrier, transmits the sound information for both black and white and color television. Also called luminance carrier.

picture detail The total number of lines or elements that make up a picture on the screen of a television receiver.

picture element The smallest subdivision of a television or facsimile image. In a color television receiver it is one color phosphor dot. In a black and white television receiver it is a square segment of a scanning line whose dimension is equal to the nominal line width. Also called elemental area.

picture frequency 1. A frequency that results solely from scanning of subject copy in a facsimile system. 2. *Frame frequency.*

picture inversion Reversal of black and white shades in the recorded copy in a facsimile system.

picture line-amplifier output The junction between the television studio facility and the line feeding a relay transmitter, a visual transmitter, or a network.

picture line standard The number of horizontal lines in a complete television image. The U.S. standard is 525 lines.

picture monitor A cathode-ray tube and associated circuits, arranged to view a television picture or its signal characteristics at station facilities.

Picturephone An experimental telephone-television system built by Bell Laboratories and demonstrated at the 1964–1965 New York World's Fair. A small television screen alongside each telephone allowed parties to see each other as they carried on a telephone conversation.

picture signal The signal that results from the scanning process in television or facsimile.

picture-signal amplitude The difference between the white peak and the blanking level of a television signal.

picture-signal polarity The polarity of the signal voltage that represents a dark area of a scene, with respect to the signal voltage representing a light area. Expressed as black negative or black positive.

picture size The useful viewing area on the screen of a television receiver, in square inches.

picture synchronizing pulse *Vertical synchronizing pulse.*

picture transmission The transmission, over wires or by radio, of a picture that has a gradation of shade values.

picture transmitter *Visual transmitter.*

picture tube A cathode-ray tube used in television receivers to produce an image by varying the electron-beam intensity as the beam is deflected from side to side and up and down to scan a raster on the fluorescent screen at the large end of the tube. Also called kinescope and television picture tube.

picture-tube brightener A small step-up transformer that can be inserted between the socket and base of a picture tube to increase the heater voltage and thereby increase picture brightness to compensate for normal aging of the tube.

picture white The signal produced at any point in a facsimile system by the scanning of copy that has minimum density. Also called white.

Pierce oscillator An oscillator in which a piezoelectric crystal is connected between the anode and grid of a tube. It is basically a Colpitts oscillator, with voltage division provided by the grid-to-cathode and anode-to-cathode capacitances of the circuit.

Pie winding having three pie sections.

pie winding A coil winding that is divided into sections called pies to reduce the distributed capacitance of the coil.

piezodielectric Having the ability to change in dielectric constant when a mechanical force is applied.

piezoelectric Having the ability to generate a voltage when mechanical force is applied, or to produce a mechanical force when a voltage is applied, as in a piezoelectric crystal.

piezoelectric axis One of the directions in a crystal in which either tension or compression will generate a voltage.

piezoelectric ceramic A ceramic material that has piezoelectric properties similar to those of some natural crystals.

piezoelectric crystal A crystal that has piezoelectric properties. Used in crystal loudspeakers, crystal microphones, and crystal cartridges for phono pickups.

piezoelectric effect Generation of a voltage between opposite faces of a piezoelectric crystal as a result of strain due to pressure or twisting, and the reverse effect in which application of a voltage to opposite faces causes deformation to occur at the frequency of the applied voltage.

piezoelectric gage A pressure-measuring gage that uses a piezoelectric material to develop a voltage when subjected to pressure. Used for measuring blast pressures that result from explosions and pressures developed in guns.

piezoelectricity Electric energy resulting from the piezoelectric effect.

piezoelectric loudspeaker *Crystal loudspeaker.*

piezoelectric microphone *Crystal microphone.*

piezoelectric pickup *Crystal pickup.*

piezoelectric transducer A transducer in which the output voltage is produced by deformation of a crystal or ceramic material that has piezoelectric properties.

piezoelectric transistor A transistor in which the output current is a function of the mechanical pressure applied to the silicon or other piezoelectric semiconductor element.

piezoelectric vibrator An element cut from piezoelectric material, usually in the form of a plate, bar, or ring, with electrodes attached to or supported near the element to excite one of its resonant frequencies.

piezoid *Finished crystal blank.*

piezooptical effect The change produced in the index of refraction of a light-transmitting material by externally applied stress.

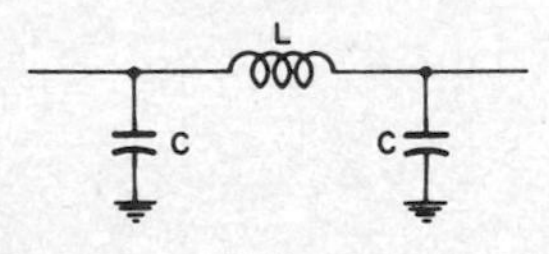

Pi filter.

pi filter A filter that has a series element and two parallel elements connected in the shape of the Greek letter pi.

pig A heavily shielded container, usually lead, used to ship or store radioisotopes and other radioactive materials.

piggyback twistor An electrically alterable nondestructive-readout storage that uses a thin, narrow tape of magnetic material wound spirally around a fine copper conductor to store information. Another similar tape is wrapped on top of the first, piggyback fashion, to sense the stored information. A binary digit or bit is stored at the intersection of a copper strap and a pair of these twistor wires.

pigment A solid that will naturally reflect some photons of light while absorbing photons of other wavelengths, without producing appreciable luminescence.

pigtail A short, flexible wire, usually stranded or braided, used between a stationary terminal and a terminal that has a limited range of motion, as in relay armatures.

pigtail splice A splice made by twisting together the bared ends of parallel conductors.

pileup A set of moving and fixed contacts, insulated from each other, formed as a unit for incorporation in a relay or switch. Also called stack.

pillbox antenna A microwave antenna that consists of a cylindrical parabolic reflector enclosed by two plates perpendicular to the cylinder, so spaced as to permit the propagation of only one mode in the desired direction of polarization. It is fed on the focal line.

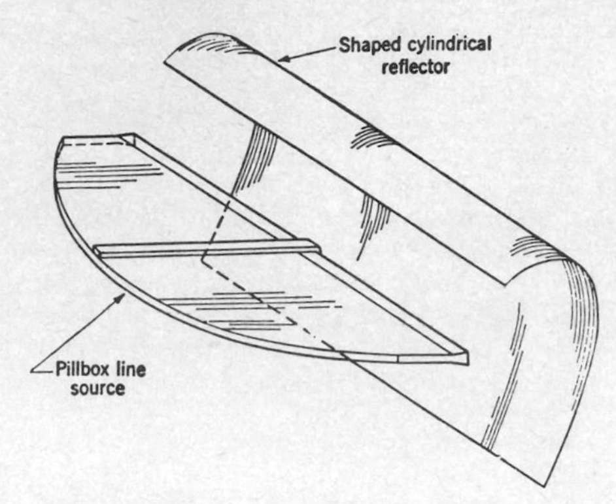

Pillbox antenna used in airborne navigation radar.

pilot A single frequency, sent over a transmission system to measure or control its characteristics.

pilotage The process of directing the movement of a vehicle by reference to recognizable landmarks or soundings, without computing positions.

pilot channel A narrow channel over which a single frequency is transmitted to operate an alarm or automatic control.

pilot circuit The portion of a control circuit that carries the controlling signal from the master switch to the controller.

pilot lamp A small lamp used to indicate that a circuit is energized. Also called pilot light. When used to illuminate a dial, it is called a dial lamp.

pilotless aircraft An aircraft equipped to function without benefit of a human pilot. It may carry such functional equipment as cameras for photographic reconnaissance, bombs for bombing, guns for air defense, counters for radiation measurements, or radar for early warning. Target aircraft and drones are special examples of pilotless aircraft.

pilot light *Pilot lamp.*

pilot regulator A regulator that maintains a constant level at the receiving end of a carrier-derived circuit despite variations in the attenuation of the transmission line. The regulator usually monitors the resistance of a pilot wire that is exposed to essentially the same temperatures as the transmission circuit being regulated. Also called pilot-wire regulator.

pilot spark A weak spark used to ionize the air in preparation for a greater spark discharge.

pilot subcarrier A subcarrier that serves as a control signal for the reception of stereo FM broadcasts.

pilot wire An auxiliary conductor used in connection with remote measuring devices or for operating apparatus at a distant point.

pilot-wire regulator *Pilot regulator.*

pi meson A meson that has a mean life of 28 ns and a mass of about 270 times that of an electron. Also called pion.

pi mode The mode of operation of a magnetron for which the phases of the fields of successive anode openings facing the interaction space differ by π radians.

pin A terminal on an electron-tube base, plug, or connector. Also called base pin and prong.

PIN [Positive-Intrinsic-Negative] A semiconductor structure that has a high-resistance intrinsic region between low-resistance P- and N-type regions. Used in many types of microwave diodes, photodiodes, switching diodes, and voltage-dependent variable resistors.

pinboard A perforated board into which special pins can be inserted to control the operation of equipment.

pinch A pressed glass stem used to support the internal leads of electron tubes. Also called press.

pinch effect 1. Constriction of ionized gas to a narrow thread in the center of a straight or doughnut-shaped electron tube through which a heavy current is passed. Also called rheostriction. 2. Constriction, and sometimes momentary rupture, of molten metal through which a heavy current is flowing. 3. Pinching of the reproducing stylus tip twice each cycle during reproduction of lateral disk recordings, due to a decrease in the groove angle cut by the recording stylus as it swings from a negative to a positive peak.

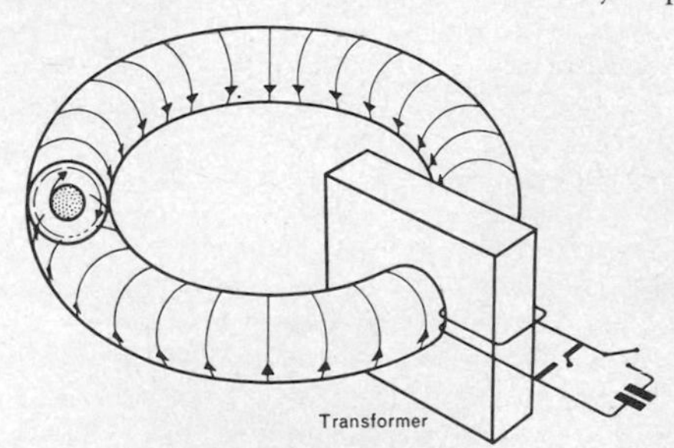

Pinch effect in toroid. External windings on toroid produce stabilizing magnetic field for current flowing through gas inside toroid when capacitor bank at right is discharged through pulse transformer.

pinchoff The equivalent of collector cutoff in a field-effect transistor. It occurs when reverse bias is increased enough that the depletion regions touch and practically cut off the drain current.

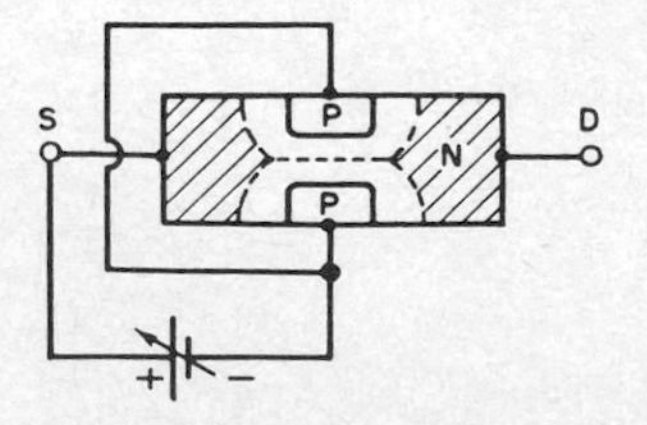

Pinchoff effect in one type of field-effect transistor, in which depletion regions around each P-type gate spread out and touch, to cut off current to drain D.

pinchoff voltage The voltage at which the current flow between source and drain in a field-effect transistor is blocked because the channel between these electrodes is completely depleted. For enhancement-type transistors that use an end-type channel, the pinchoff voltage is positive. For depletion-type metal-oxide-semiconductor transistors that use a P-type channel material, the pinchoff voltage is negative.

pinch resistor A silicon integrated-circuit resistor produced by diffusing an N-type layer over a P-type resistor. This narrows or pinches the resistive channel, thereby increasing the resistance value.

pin connection A connection made to a terminal pin at the base of an electron tube. The following abbreviations are used to identify pin connections: NC—no connection; IS—internal shield; IC—internal connection (but not an electrode connection); BS—base shield connection; P—anode; G—grid; SG—screen grid; K—cathode; H—heater; F—filament; RC—ray-control electrode; TA—target.

pincushion corrector A circuit used in television receivers to compensate for pincushion distortion. The horizontal pincushion corrector uses the vertical sawtooth voltage to vary the load on the horizontal sweep system at the vertical rate and thereby straighten the sides of the picture. The vertical corrector circuit uses a parabolic voltage at the horizontal sweep rate to make the picture straight at the top and bottom of the screen.

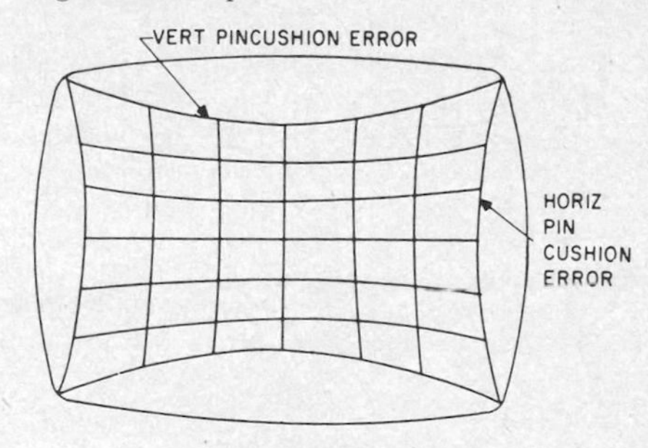

Pincushion distortion.

pincushion distortion Distortion in which all four sides of a received television picture are concave (curve inward).

PIN device A silicon semiconductor device that uses PIN construction.

PIN diode A silicon semiconductor diode that uses PIN construction.

pine-tree array An array of horizontal dipole antennas arranged to form a radiating curtain, with reflectors behind.

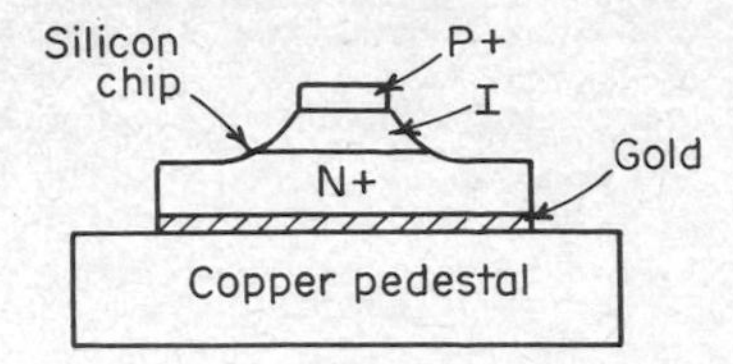

PIN diode construction, showing intrinsic layer I between P- and N-type layers.

pine-tree line A chain of radar stations built by Canada and the United States along the Canadian-American border.

pi network A network that has three impedance branches connected in series with each other to form a closed circuit, with the three junction points forming an input terminal, an output terminal, and a common input and output terminal.

ping A sonic or ultrasonic pulse sent out by an echo-ranging sonar.

ping analyzer An analyzer that stores a selected reflected sonar return or ping, analyzes the waveform, and displays the results on a storage-type cathode-ray tube for identification of the underwater target.

Ping Pong A camera-carrying photoreconnaissance missile that takes photographs automatically at a predetermined height over an enemy position, then returns to the launch area and descends by parachute for recovery of the film and reuse of the missile.

pinhole detector A photoelectric device that detects extremely small holes and other defects in moving sheets of material.

pin jack A small jack used with a plug whose thickness is comparable to that of an ordinary pin.

pink noise Noise whose intensity is inversely proportional to frequency over a specified range, to give constant energy per octave.

pion *Pi meson.*

pip The target echo indication on a radarscope. It may be a bright spot of light, as on a PPI display, or a sharply peaked pulse, as on an A display. Also called blip.

piped program A radio or television program sent over commercial transmission lines.

pip-matching display A navigation display in which the received signal appears as a pair of pips, comparison of which provides a measure of the desired quantity.

pi point The frequency at which the insertion phase shift of a device is an integral multiple of π radians or 180°.

Pirani gage A vacuum gage based on the fact that the temperature and resistance of a heated filament vary with the pressure of the surrounding gas. The less gas there is to conduct heat away from the filament, the hotter the filament becomes and the greater its resistance.

piston A sliding metal structure used in waveguides and cavities for tuning purposes or for reflecting essentially all the incident energy. Also called plunger and waveguide plunger.

piston action Movement of the entire diaphragm of a loudspeaker as a unit when driven at low audio frequencies.

piston attenuator A microwave attenuator inserted in a waveguide to introduce an amount of attenuation that can be varied by moving an output coupling device along its longitudinal axis.

pistonphone A small chamber equipped with a reciprocating piston that has a measurable displacement. Used to establish a known sound pressure in the chamber, as for testing microphones.

pitch 1. The attribute of auditory sensation that depends primarily on the frequency of the sound stimulus but also on the sound pressure and the waveform of the stimulus. Pitch determines the position of a sound on a musical scale, for which the standard pitch is 440 Hz for the tone A. With this standard, middle C is 261.6 Hz. 2. The rising and falling motion of the bow of a ship or the tail of an airplane as the structure oscillates about its transverse axis. 3. The distance between the peaks of two successive grooves on a disk recording or a screw.

pitch attitude The angle between the longitudinal axis of the vehicle and a horizontal plane.

pitchover The programmed turn from the vertical that a rocket under power takes as it describes an arc and points in a direction other than vertical.

PIV Abbreviation of *peak inverse voltage.*

PLA Abbreviation for *programmed logic array.*

place A position that corresponds to a given power of the base in positional notation. A digit located in any particular place is a coefficient of a corresponding power of the base. Places are usually numbered from right to left, zero being the place at the right if there is no decimal, binary, or other point, or the column immediately to the left of the point if there is one. Also called column.

planar array An array of ultrasonic transducers that can be mounted in a single plane or sheet, to permit closer conformation with the hull design of a sonar-carrying ship.

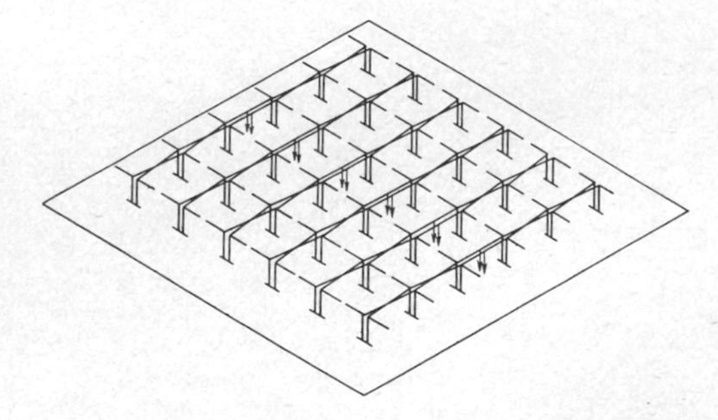

Planar-array antenna.

planar-array antenna An array antenna in

which the centers of the radiating elements are all in the same plane.

planar diode A diode that has planar electrodes in parallel planes.

planar-electrode tube An electron tube that has parallel planar electrodes.

planar epitaxial passivated diode A diode that has an oxide-passivated planar structure built into a high-resistivity epitaxial layer grown on a low-resistivity silicon substrate. This construction gives high conductance, fast recovery time, low leakage, and low capacitance.

planar junction transistor A junction transistor similar to a diffused-junction transistor but in which localized penetration of the impurity is achieved by coating portions of the wafer surface

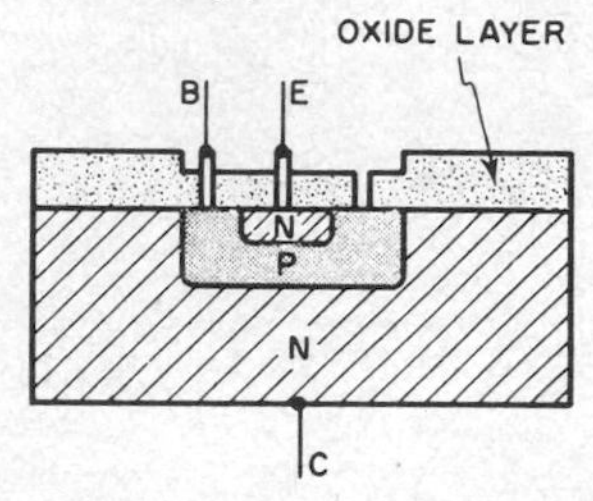

Planar junction transistor, showing base B, emitter E, and collector C.

with an oxide compound like silicon dioxide. This process is known as surface passivation.

planar network A network that can be drawn on a plane without crossing of branches.

planar photodiode A vacuum photodiode that consists simply of a photocathode and an anode. Light enters through a window sealed into the base, behind the photocathode. When mounted as

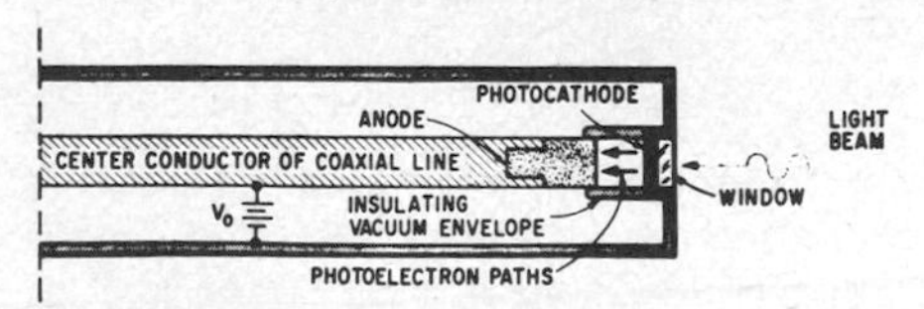

Planar photodiode installed as part of center conductor of coaxial line, for detection of light signals modulated at up to 11 GHz.

part of the central conductor of a coaxial line or mounted in a waveguide, it has detected light beams modulated at frequencies as high as the X band (5.2 to 10.9 GHz).

planar process A silicon transistor manufacturing process in which a fractional-micron-thick oxide layer is grown on a silicon substrate. A series of etching and diffusion steps is then used to produce the transistor inside the silicon substrate.

Planckian locus A line drawn through all points on a chromaticity diagram that represents light radiation from a blackbody radiator at various temperatures up to 10,000 K.

Planck's constant [symbol h] A universal constant equal to 6.626×10^{-27} erg · s. It is the proportionality factor that, when multiplied by the frequency of a photon, gives the energy of the photon.

Planck's law The fundamental law of quantum theory; it states that energy transfers associated with radiation are proportional to the frequency of the radiation. The proportionality factor is Planck's constant.

plane An assembly of magnetic-storage cores in a single plane.

plane-earth factor The ratio of the electric field strength that would result from propagation over an imperfectly conducting plane earth to that which would result from propagation over a perfectly conducting plane.

plane of polarization The plane that contains the electric field vector and the direction of propagation of a plane-polarized wave. In a horizontally polarized wave, this plane is horizontal.

plane-polarized light Light in which the electric vectors of all components of the radiation are in the same fixed plane.

plane-polarized wave An electromagnetic wave whose electric field vector at all times lies in a fixed plane that contains the direction of propagation through a homogeneous isotropic medium.

plane wave A wave whose equiphase surfaces form a family of parallel planes.

planigraphy *Laminography.*

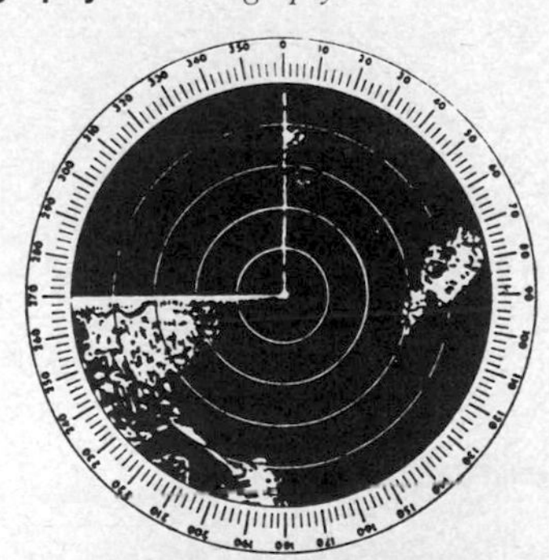

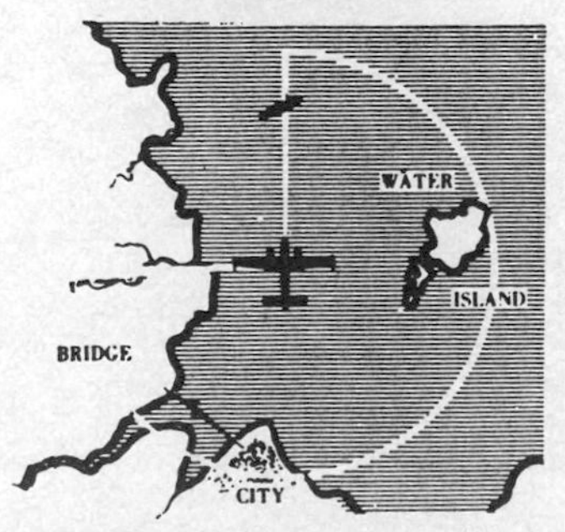

Plan-position-indicator presentation obtained with search radar in airplane, and map showing location of plane. Radar antenna is pointing at 270°.

plan-position indicator [abbreviated PPI] A radarscope display in which targets appear as bright spots at the same locations as they would on a circular map of the area being scanned, the radar antenna being at the center of the map. The sweep moves radially outward from the center of the screen, and the sweep line rotates synchronously with the radar antenna. The radial distance from the center at which an echo appears is an indication of range, and the angle measured clockwise from true north (usually at the top of the screen) is an indication of bearing. Used also in sonar. Also called P display.

Plante cell A type of lead-acid cell in which the active material is formed on the plates by electrochemical means during repeated charging and discharging, instead of being applied as a prepared paste.

plasma Any mixture of particles that contains approximately equal numbers of positive and negative particles along with neutral particles, so the mixture is electrically neutral. Generally applied to a gas that is sufficiently ionized to be conductive and affected by magnetic fields. A true plasma, which is completely ionized and has no neutral particles, is produced by temperatures above 20,000 K. In solids, plasma can exist either as electrons and positively charged donors or as holes and negatively charged acceptors. Another type of plasma occurs as holes and electrons in an intrinsic semiconductor.

plasma accelerator An accelerator that forms a high-velocity jet of plasma by using a magnetic field, an electric arc, a traveling wave, or other similar means.

plasma anodization A method of making passive thin-film circuits by using a low-pressure gas plasma of oxygen ions as the electrolyte to anodize evaporated aluminum films on a glass substrate in a vacuum. The result is an aluminum oxide film, to which aluminum electrodes can be applied.

plasma arc coating A coating of any meltable material, blasted onto practically any metal, ceramic, plastic, glass, wood, or other base by a high-velocity plasma arc gas stream. Used for applying wear-resistant coatings to metal tools and for applying heat-resisting coatings.

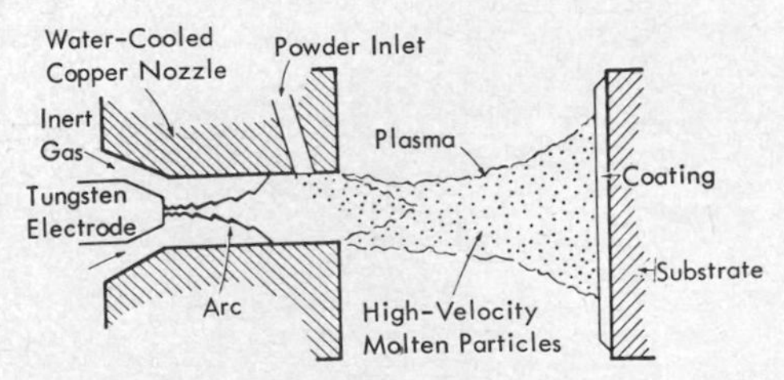

Plasma arc coating being applied to substrate at right by plasma torch at left.

plasma ball A high-intensity light source in which noble gases under high pressure are ionized between electrodes. The resulting luminous glow is concentrated into a ball about 6 mm in diameter by a surrounding toroidal magnet. Peak brightness can be as high as 500 cd/mm².

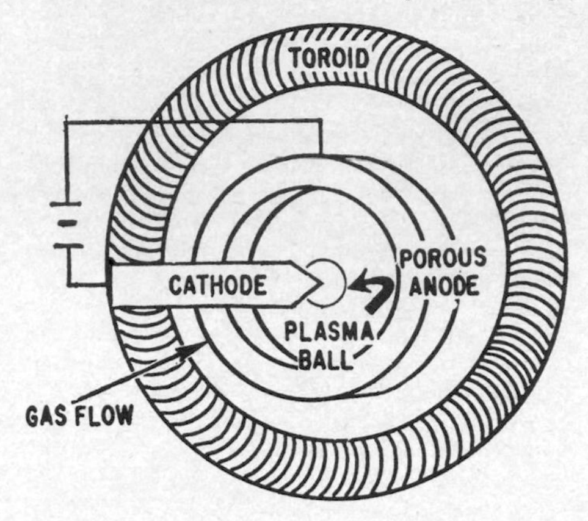

Plasma ball.

plasma cathode A cathode in which the source of electrons is a gas plasma rather than a solid. In one version, a self-sustaining plasma is maintained over a pool of mercury by initial RF bombardment combined with thermionic emission from the mercury as a result of bombardment by positive ions. In another version, a plasma of cesium ions is initiated by electron emission from the cesium-coated inner surface of an indirectly heated hollow cathode.

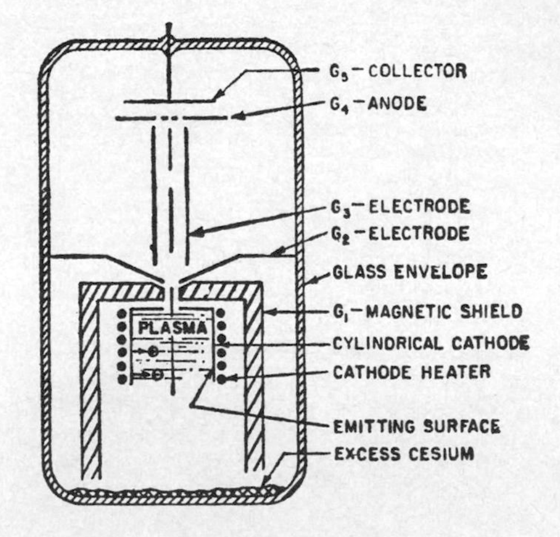

Plasma cathode in electron tube.

plasma confinement Use of a magnetic field to confine a gas plasma in the central region of an evacuated chamber long enough to develop the high temperature needed for a controlled-fusion reaction.

plasma diode A diode that converts heat directly into electricity. It consists of two closely spaced electrodes serving as cathode and anode, mounted in an envelope in which a low-pressure cesium vapor fills the interelectrode space. Heat is applied to the cathode, causing emission of electrons. These electrons are attracted to the anode, which is at a lower temperature, and travel back to the cathode through the load. The anode generally must be cooled by external means to maintain its required lower operating temperature.

plasma display A display in which sets of parallel conductors at right angles to each other are deposited on glass plates, with the very small space between the plates filled with a gas. Each intersection of two conductors defines a single cell that can be energized to produce a gas discharge forming one element of a dot-matrix display.

plasma engine An engine for space travel in which neutral plasma is accelerated and directed by external magnetic fields that interact with the magnetic field produced by current flow through the plasma. A nuclear power plant provides the necessary power.

plasma frequency A natural frequency for coherent motion of electrons in a plasma. The frequency is related to the electron mass and the restoring force of the space-charge field that arises from the displacement of the electrons. Since the space-charge field is proportional to the electron density, the plasma frequency will be proportional to the square root of the electron density.

plasma jet A magnetohydrodynamic rocket engine in which the ejection of plasma generates thrust.

plasma oscillation Electrostatic or space-charge oscillations in a plasma that are closely related to the plasma frequency. There is usually enough damping caused by electron collisions to prevent self-generation of the oscillations. They can be excited, however, by such techniques as shooting a modulated electron beam through the plasma.

plasma oscillator An oscillator in which a constant voltage is applied to a semiconductor through injection contacts in a magnetic field, to produce current oscillations in the electron-hole plasma of the semiconductor.

plasma pinch Application of the pinch effect to plasma in attempts to produce controlled nuclear fusion. A large current is passed through the stream of plasma in an electron tube. A magnetic field constricts the plasma into a smaller diameter, keeping it away from the envelope and raising the temperature of the plasma.

plasma propulsion Propulsion of spacecraft and other vehicles by using electric and/or magnetic fields to accelerate both positively and negatively charged particles (plasma) to a very high velocity.

plasma rocket engine An electric propulsion system for spacecraft, based on cyclotron resonance in a plasma that is trapped by crossed electric and magnetic fields. Also called coaxial plasma accelerator and electromagnetic rocket engine.

plasma sheath An envelope of ionized gas that surrounds a spacecraft or other body moving through an atmosphere at hypersonic velocities. The plasma sheath affects transmission, reception, and diffraction of radio waves.

plasma torch A torch in which temperatures as high as 50,000°C are achieved by injecting a plasma gas tangentially into an electric arc formed between electrodes in a chamber. The resulting

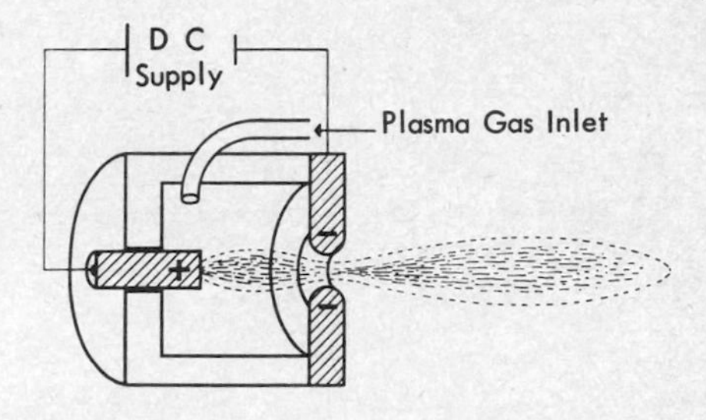

Plasma torch for producing gas stream (at right) having temperatures between 8000 and 50,000°C.

vortex of hot gases emerges at very high speed through a hole in the negative electrode, to form a jet for welding, spraying of molten metal, and cutting of hard rock or hard metals.

plasmatron A gas-discharge tube in which independently generated plasma serves as a conductor between a hot cathode and an anode. The anode current is modulated by varying either the conductivity or the effective cross section of the plasma.

plastic One of a large and varied group of organic materials that can be formed into shape by flow at some stage, usually by applying heat alone or with pressure.

plastic-film capacitor A capacitor in which al-

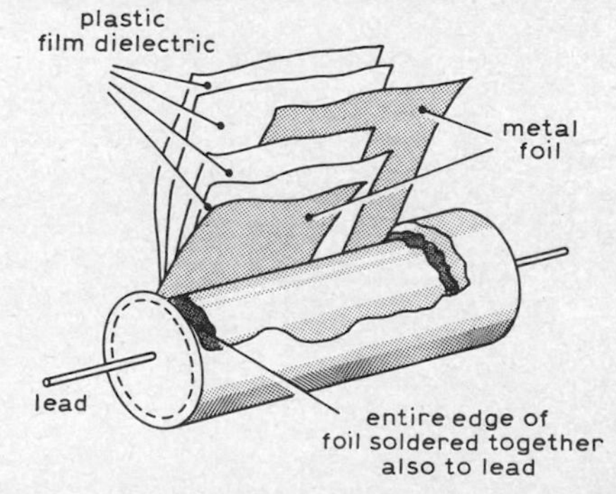

Plastic-film capacitor construction.

ternate layers of metal foil (usually aluminum) and thin films of plastic dielectric are rolled into compact tubular form, just as for a paper capacitor. The foil strips are staggered to project alternately from the row ends, and leads are soldered to the foil. Plastic films in common use include Mylar (a polyester), polycarbonate, polypropylene, polystyrene, and polysulfone. All but polystyrene can be metallized to eliminate the foil strips. Also called film capacitor.

plastisol A mixture of resins and plasticizers that can be molded, cast, or converted to continuous films by applying heat.

plate 1. One of the conducting surfaces in a capacitor. 2. One of the electrodes in a storage battery. 3. [symbol P] *Anode.*

plateau The portion of the plateau characteristic of a counter tube in which the counting rate is substantially independent of the applied voltage.

plateau characteristic The relation between counting rate and voltage for a counter tube when radiation is constant, showing a plateau after the rise from the starting voltage to the Geiger threshold.

plateau length The range of applied voltage over which the plateau of a radiation-counter tube extends.

plate bypass capacitor *Anode bypass capacitor.*

plate circuit *Anode circuit.*

plate current *Anode current.*

plated circuit A printed circuit produced by electrodeposition of a conductive pattern on an insulating base. In one method, a negative of the required pattern is printed on a metal-clad or metal-coated sheet with a plating resist. Silver or solder is electroplated on the area not protected by the resist, to give the desired wiring pattern. The resist is then removed, and the unwanted metal under it is etched away with a solution that does not attack the silver or solder plating. Also called plated printed circuit.

plated crystal unit A crystal unit in which the electrodes are metal films deposited directly on the quartz surfaces.

plate detection *Anode detection.*

plated printed circuit *Plated circuit.*

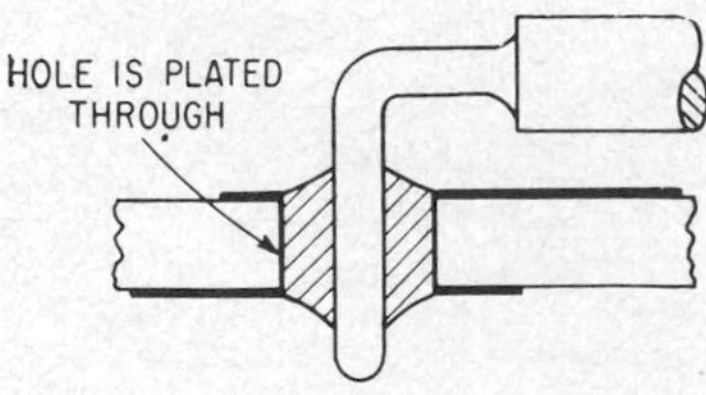

Plated-through hole in printed-wiring board.

plated-through hole An interface connection formed by electrodeposition of metal on the sides of a hole in a printed-wiring board.

plated-wire memory A memory in which copper bands or wires are electroplated with a thin magnetic alloy, usually 80% nickel and 20% iron, and combined with insulated digit and word lines to produce addressable memory cells comparable to those of core memories.

plate efficiency *Anode efficiency.*

plate input power *Anode input power.*

plate keying *Anode keying.*

plate load impedance *Anode load impedance.*

plate modulation *Anode modulation.*

plate neutralization *Anode neutralization.*

plate penetrameter A plate of material similar to that of a specimen under radiographic examination; it has a thickness of about 2% of the specimen thickness and holes of different diameters. Also called strip penetrameter.

plate power input *Anode power input.*

plate pulse modulation *Anode pulse modulation.*

plate saturation *Anode saturation.*

plate supply *Anode supply.*

plate voltage *Anode voltage.*

platform stabilization Radar antenna stabilization in which a platform is pivoted so it can be driven to a horizontal position regardless of the motion of the vehicle. Gyroscope signals actuate drive motors that maintain this horizontal position as the vehicle pitches and rolls.

platinotron A microwave tube that may be used as a high-power saturated amplifier or oscillator in pulsed radar applications. It requires a permanent magnet, just as a magnetron does.

platinum [symbol Pt] A heavy, almost white metal that resists the action of practically all acids, is capable of withstanding high temperatures, and is little affected by sparks. Used for contact points in switches and relays. Atomic number is 78.

platter Slang term for *phonograph record.*

playback Reproduction of a recording.

playback head A head that converts a changing magnetic field on magnetic tape into corresponding electric signals. Also called reproduce head.

playback loss *Translation loss.*

plethysmograph An instrument that monitors changes in the volume of blood in an organ or other part of the body.

Plexiglas Trademark of Rohm & Haas Co. for their clear acrylic molding powders and clear acrylic sheets.

pliers A two-handled tool used for holding, cutting, bending, and shaping wire and for other purposes.

pliotron General term for any hot-cathode vacuum tube that has one or more grids.

PLL Abbreviation for *phase-locked loop.*

PL/M A high-level programming language developed primarily for microcomputers and microprocessors.

plotter A visual display on which an automatically controlled pen or pencil traces the curves of one or more variables as functions of one or more other variables.

plug 1. The half of a connector that is normally movable and generally attached to a cable or removable subassembly. A plug is inserted into a jack, outlet, receptacle, or socket. 2. Radiation-blocking material used to close an aperture in the shield of a nuclear reactor. 3. A mention of a commercial product in a radio or television program.

plugboard A connection board that has a large number of jacks into which short interconnecting cords can be plugged to provide a particular pattern of connections for computers, punched-card machines, and other processing and control equipment. The boards are easily interchangeable, so a number of them can be set up in advance to form a program library.

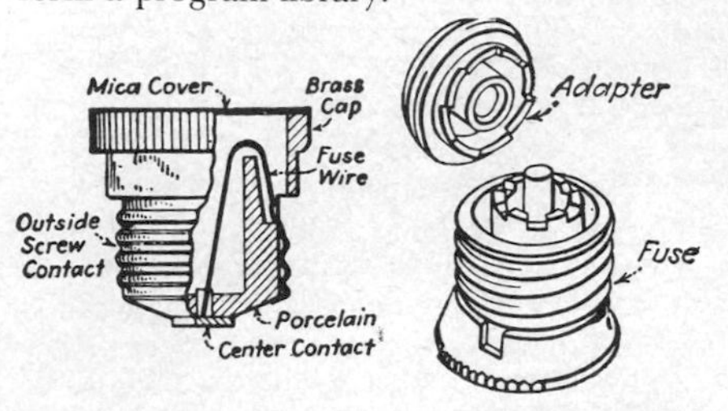

Plug fuse. Conventional version is at left. Tamper-resistant version (right) has adapter that stays in fuse socket when first inserted, to prevent use of conventional fuse. One adapter accepts only fuses up to 15 A, and a second type accepts only 20- to 30-A fuses.

plug fuse A fuse used in a standard screw-base lamp socket.

plugging Braking an electric motor by reversing its connections, so it tends to turn in the opposite direction. The circuit is opened automatically when the motor stops, so the motor does not actually reverse.

plug-in unit A component or subassembly that has plug-in terminals so all connections can be made simultaneously by pushing the unit into a suitable socket.

Plumbicon Trademark of N. V. Philips' Gloeilampenfabrieken for their line of television camera tubes.

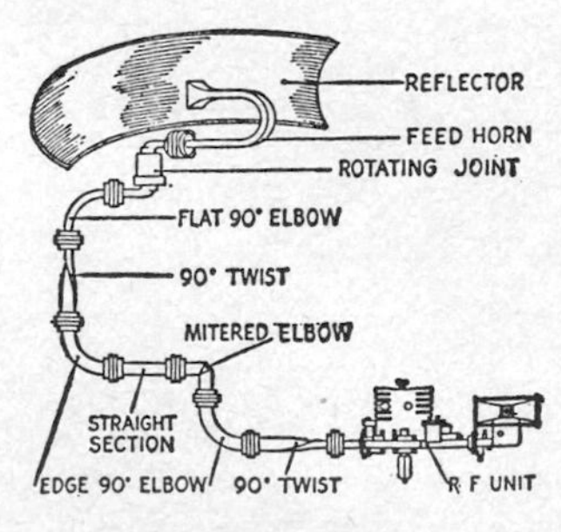

Plumbing setup used to feed microwave energy from radar transmitter to rotating antenna.

plumbing Slang term for the pipelike waveguide circuit elements used in microwave radio and radar equipment.

plunger *Piston.*

plunger relay A relay in which a plunger moves through the center of the coil by solenoid action when the relay is energized. Contacts are mounted on one or both ends of the plunger, which also serves as the core. Also called solenoid relay.

plural scattering Scattering in which the final displacement is the vector sum of a small number of displacements. Plural scattering is intermediate between single and multiple scattering.

plutonium [symbol Pu] A transuranic radioactive element formed by the decay of certain isotopes of neptunium. Plutonium does not occur in nature. Atomic number is 94.

PLZT Abbreviation for *lead lanthanum zirconate titanate.*

PM 1. Abbreviation for *permanent magnet.* 2. Abbreviation for *phase modulation.*

PM loudspeaker Abbreviation for *permanent-magnet loudspeaker.*

PMOS Abbreviation for *P-channel MOS.*

PMOS RAM A random-access memory that uses P-channel metal-oxide semiconductor technology.

PN Positive-negative, used in referring to an interface between P- and N-type conductivities in a semiconductor.

PN boundary A surface in a PN junction at which the donor and acceptor concentrations are equal.

PNdB A rating used for perceived noise level in decibels. The maximum allowable noise made by a plane taking off is specified in this way; a typical limit is 112 PNdB.

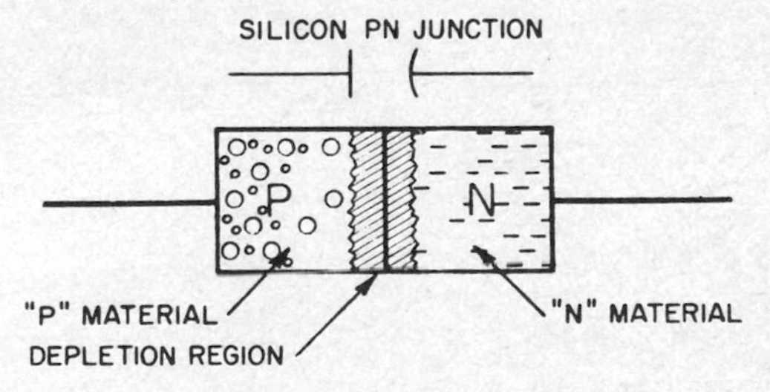

PN diode having silicon junction.

PN diode A diode, usually silicon, that has a single PN junction.

pneumatic detector An infrared detector that consists of a small gas-filled chamber which has an infrared transmitting window, a material inside the chamber to absorb radiation, and a means for converting pressure changes in the chamber into measurable signal output that is usually electric or optical.

pneumatic loudspeaker A loudspeaker in which the acoustic output results from controlled variation of an air stream.

PN hook A thin region of P-type material sandwiched between regions of N-type base and collector material in a transistor to create a potential hook at which holes are trapped in such a way as to increase the current gain.

PNIN transistor An intrinsic-junction transistor in which the intrinsic region is between N regions.

PNIP transistor An intrinsic-junction transistor in which the intrinsic region is sandwiched between the P-type base and the N-type collector.

PN junction A two-terminal device made of a semiconductor that has been treated to conduct current more easily in one direction than in the other. The treatment leaves one end of the crystal a P-type semiconductor and the other end an N-type semiconductor. Widely used in rectifiers and solar cells.

PNP Positive-negative-positive, used in referring to a semiconductor in which a layer with N-type conductivity is located between two layers that have P-type conductivity.

PNPN device A device that consists of four alternate layers of P- and N-type semiconductor material. The device always blocks in one direction only, when the positive end is positively biased. PNPN devices include four-layer diodes, gate-turnoff silicon controlled rectifiers, silicon controlled rectifiers, and silicon controlled switches. The end product depends on how many of the layers are brought out to terminals.

PNP transistor A junction transistor having an N-type base between a P-type emitter and a P-type collector. The emitter should then be positive with respect to the base, and the collector should be negative with respect to the base.

Pockel's effect The change produced in the refractive properties of certain transparent piezoelectric crystals when an electric field is applied. One application is readout of stored electrostatic patterns.

point The character or space that separates the integral and fractional parts of a numerical expression in positional notation, such as the binary point in binary notation and the decimal point in decimal notation. Also called base point and radix point.

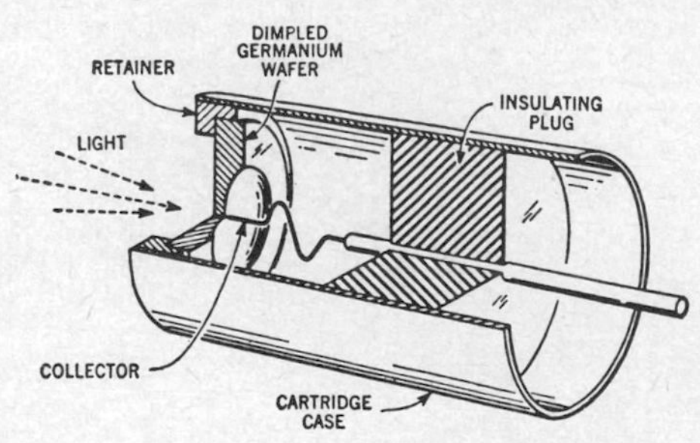

Point contact as used in phototransistor.

point contact Pressure contact between a semiconductor body and a metallic point.

point-contact diode A semiconductor diode that makes use of a point contact to provide rectifying action. Used as a detector in RF and microwave circuits.

point-contact rectifier A diode in which stray capacitance is minimized by using a point contact

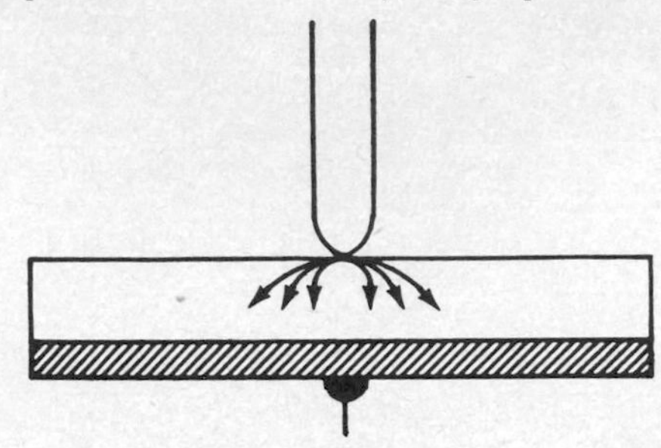

Point-contact rectifier.

as a junction, so the flow of current is essentially radial away from the junction. Used in high-frequency applications.

point-contact transistor A transistor that has a base electrode and two or more point contacts located near each other on the surface of N-type germanium. Pressure of the points creates a small volume of P-type material under each point to produce the necessary junctions for a PNP transistor.

point-contact transistor tetrode A point-contract transistor that has a collector and two emitters.

point counter tube A radiation-counter tube, which uses gas amplification, in which the central electrode is a point or a small sphere.

pointer The needle-shaped rod that moves over the scale of a meter.

pointing element The element in a fire-control system that points the missile-launching projector as directed by the other elements of the system.

point-junction transistor A transistor that has a base electrode, a point contact, and junction electrodes.

point-of-sale terminal [abbreviated POS terminal] A computer-connected terminal used in place of a cash register in a store, for customer checkout and such added functions as recording inventory data, transferring funds from the customer's bank account to the merchant's bank account, and checking credit on charged or charge-card purchases. The terminals can be modified for many nonmerchandising applications, such as checkout of books in libraries.

point source A radiation source whose dimensions are small compared with the distance from which the radiation is used.

point-to-point communication Radio communication between two fixed stations.

point-to-point wiring Wiring in which the connections between components or parts are made directly, without intermediate terminals or supports, by using the leads of the components.

poison A contaminating substance that impairs the efficiency of a material. It may reduce phosphorescence in a luminescent material or reduce emission of a cathode surface.

polar Pertaining to, measured from, or having a pole, such as the poles of the earth or of a magnet.

polar circuit A teletypewriter circuit in which current flows in one direction for a marking pulse and in the opposite direction for a spacing pulse.

polar coordinates A system of coordinates in which the location of a point is specified by its distance from a fixed point and the angle that the line from this fixed point to the given point makes with a fixed reference line.

polar diagram A diagram that uses polar coordinates to show the magnitude of a quantity in some or all directions from a point. Examples include directivity patterns and radiation patterns.

polar grid *Grid north.*

polarimeter An instrument that measures the state of polarization of polarized light.

Polaris A Navy surface-to-surface intermediate-range ballistic missile designed to be launched from submarines and surface ships for accurate bombardment of small target areas with conventional or nuclear warheads at ranges up to 2500 nautical miles (4600 km). It uses inertial guidance, with true north, the ship's position, the ship's speed, and the position of the target being fed into a computer in the missile at the instant of firing.

polarity The characteristic wherein an object exhibits opposite properties within itself, such as opposite charges (positive and negative) or opposing magnetic poles (north and south).

polarity of picture signal The polarity of the black portion of a picture signal with respect to the white portion. In a black-negative picture, the potential corresponding to the black areas of the picture is negative with respect to the potential corresponding to the white areas of the picture. In a black-positive picture the potential corresponding to the black areas of the picture is positive.

polarity-reversing switch A switch that interchanges the connections to a device.

polarization 1. The direction of the electric field as radiated from a transmitting antenna. Horizontal polarization is standard for television in the United States, and vertical polarization is standard in Great Britain. 2. Vibration of the vectors of a beam of light or other electromagnetic radiation in a particular direction or manner, as in plane-polarized light. 3. A chemical change occurring in dry cells during use, increasing the internal resistance of the cell and shortening its useful life. 4. A displacement of bound charges in a dielectric when placed in an electric field.

polarization diversity A communication technique that doubles the number of available channels by using two signals per channel, each signal having its own distinctive polarization. One signal may have vertical polarization and the other horizontal polarization, or one may be circularly polarized in a clockwise direction while the other is counterclockwise.

polarization-diversity reception Diversity reception that involves use of a horizontal dipole and a vertical dipole at the same location, with the individual receiver outputs being combined, just as in space-diversity reception. The arrangement counteracts the changes in polarization of a received radio wave during fading.

polarization error An error in a radio direction-finder indication due to changes in the polarization of the received wave as atmospheric conditions change. The error is generally greatest at night.

polarization modulation A form of modulation used in optical communication.

polarization receiving factor The ratio of the power received by an antenna from a given plane wave of arbitrary polarization to the power received by the same antenna from a plane wave of the same power density and direction of propagation, whose state of polarization has been adjusted for maximum received power.

polarized double-biased relay A polarized relay that is in addition magnetically biased or latched in either of two positions.

polarized electrolytic capacitor An electrolytic capacitor in which the dielectric film is formed adjacent to only one metal electrode. The impedance to the flow of current is then greater in one direction than in the other.

polarized light Light that vibrates in only one plane.

polarized meter A meter that has a zero-center scale, with the direction of deflection of the pointer depending on the polarity of the voltage or the direction of the current being measured.

polarized plug A plug that can be inserted into its receptacle only when in a predetermined position.

polarized receptacle A receptacle used with a polarized plug, to ensure that the grounded side of an AC line or the positive side of a DC line is always connected to the same terminal on a piece of equipment.

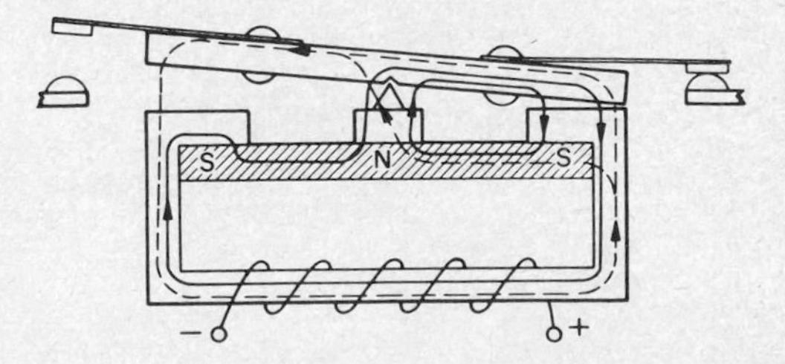

Polarized relay. When coil is momentarily energized with reverse polarity at sufficiently high level, armature moves to other position and is held there by permanent magnet.

polarized relay A relay in which the direction of movement of the armature depends on the direction of the energizing current in the relay coil.

polarizer A Nicol prism or other device for polarizing light.

polarizing current The direct current that is sent through the coil of an iron-core component to establish a reference value of magnetic flux.

polar keying A form of telegraphy in which circuit current flows in one direction for a mark signal and in the other direction for a space signal.

polar modulation Amplitude modulation in which the positive excursions of the carrier are modulated by one signal and the negative excursions by another signal.

Polaroid Trademark of Polaroid Corp. for a sheet material that produces plane-polarized light.

polar orbit A satellite orbit running north and south, so the satellite vehicle orbits over both the North and South Poles. United States polar-orbit satellites are launched from Vandenberg Air Force Base, California. A satellite in a polar orbit sees a different view on each orbit, since the earth is rotating beneath it.

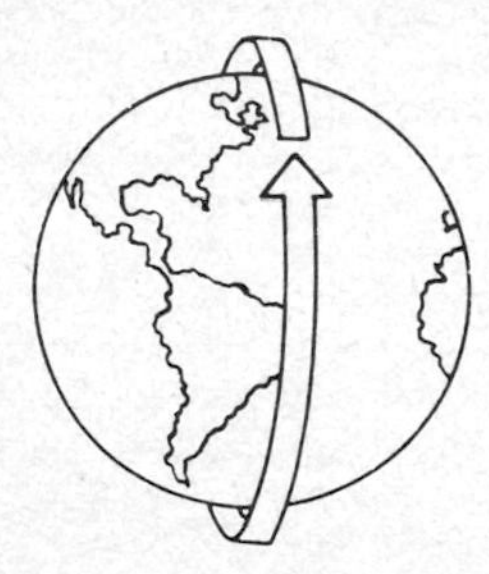

Polar orbit of earth satellite. Since earth is rotating continuously about its polar axis, each orbit traverses a different part of earth.

pole 1. A region in a magnet that has polarity, such as the north or the south pole. 2. A characteristic of a function used in circuit analysis. 3. An output terminal on a switch; a double-pole switch has two output terminals.

pole face The end of a magnetic core that faces the air gap in which the magnetic field performs useful work.

pole piece A piece of magnetic material that forms one end of an electromagnet or permanent magnet, shaped to control the distribution of the magnetic flux in the adjacent air gap.

police radio Two-way radio communication equipment installed in police cars for car-to-headquarters and sometimes car-to-car communications.

Polish notation A notation system for digital-computer logic, developed by J. Lukasiewicz, in which arithmetic and logic expressions are written without using parentheses. In Polish notation, each operator precedes its operands; thus, (a + b)c could be written as +abc. Also called prefix notation.

polling A method of calling a number of remote data-transmission terminals in sequence from a central point to allow each in turn to transmit information data on hand.

polonium [symbol Po] A radioactive element. Atomic number is 84.

polybutadiene An improved form of synthetic rubber, reduced from butadiene.

polycarbonate A thermosetting synthetic resin used as a dielectric in some metallized and foil capacitors.

polychloroprene A rubberlike compound used for jacketing on wire and cable that is subject to rough usage, moisture, oil, greases, solvents, and chemicals. Called Neoprene by Du Pont.

polychlorotrifluoroethylene resin A fluorocarbon resin that has high dielectric strength, widely used for electrical insulation.

polydirectional microphone A microphone that has provisions for changing its directional characteristics.

polyester A resin used as a base for several types of plastic.

polyester film A plastic film made from a polyester, such as Mylar. Used as a backing for magnetic tape to obtain high strength and resistance to humidity change, and used as a dielectric film in metallized and foil capacitors.

polyethylene A tough, flexible plastic compound that has excellent insulating properties at ultrahigh frequencies. Widely used as insulating material in coaxial cables and other transmission lines.

polyflop A circuit that has three, four, or more states and will remember the last of the states to which it has been set.

polygon-type delay line A delay line made from strain-free fused silica shaped in a polygon, with the surfaces designed to reflect ultrasonic beams a number of times.

polygraph An instrument that indicates or records one or more functional variables of a body that may change when a person undergoes the emotional stress associated with a lie. These variables include blood pressure, heart action, and skin resistance. Also called lie detector.

polygraphy The science of using lie detectors.

polyimide A tough high-temperature plastic used in film and sheet form for multilayer printed-circuit boards, hybrid-circuit substrates, film capacitor dielectrics, and as a base for flexible circuits and tape cable.

polyolefin A plastic insulation used on wires in either irradiated or nonirradiated form. It offers performance comparable to that of Teflon in extreme environmental conditions encountered at high altitudes in missiles and spacecraft.

polyoptic sealing Sealing of two glass parts of a vacuum tube without heat, by mating and polishing the surfaces to such a degree that they are held together by molecular attraction to give a vacuum seal.

polyphase Having or utilizing two or more phases of an AC power line.

polyplexer A radar unit that combines the functions of duplexing and lobe switching.

polypropylene A thermoplastic material made by polymerization of propylene. Applications include dielectric films for capacitors.

polyrod antenna A microwave antenna that consists of a parallel arrangement of rods made from polystyrene or other good dielectric material. When excited at one end, as from the end of a waveguide, the rods will radiate from their other ends.

polystyrene A thermoplastic material that has excellent dielectric properties. crystal clarity, and good chemical resistance. Used for insulators, coil insulation, insulation between plates in capacitors, insulating beads in coaxial cables, and optical lenses.

polystyrene capacitor A capacitor that uses film polystyrene as a dielectric between rolled strips of metal foil.

polytetrafluoroethylene resin A fluorocarbon resin that has high dielectric strength and a slippery feel, widely used for electrical insulation. The standard designation of the Society of the Plastics Industry is TFE-fluorocarbon resin. Trademark names include Teflon (Du Pont) and Fluon (Imperial Chemical Industries, England).

polyvinyl chloride [abbreviated PVC] A thermoplastic synthetic resin used for electrical insulation, as on wires.

pool cathode A cathode at which the principal source of electron emission is a cathode spot on a liquid metal electrode, usually mercury.

pool-cathode tube *Pool tube.*

pool tube A gas-discharge tube that has a mercury-pool cathode. Also called mercury-pool tube and pool-cathode tube.

poor geometry A nuclear measuring arrangement such that the angular aperture between source and detector is large, thus introducing an uncertainty in particle energy measurements.

popcorn noise Noise produced by erratic jumps of bias current between two levels at random intervals in operational amplifiers and other semiconductor devices.

population inversion The condition in which a higher energy state in an atomic system is more heavily populated with electrons than a lower energy state of the same system. This condition must be established to create lasing action in a laser.

pop-up seismometer A seismometer, generally combined with a tape recorder, that is allowed to sink freely from a ship to the bottom of the ocean and there record microseisms. After a predetermined time interval, or when the reel of tape is filled, the equipment automatically releases its anchor and floats to the surface for recovery.

porcelain A fired ceramic material used for insulators.

porcelain capacitor A fixed capacitor in which the dielectric is a high grade of porcelain, molecularly fused to alternate layers of fine silver electrodes to form a monolithic unit that requires no case or hermetic seal.

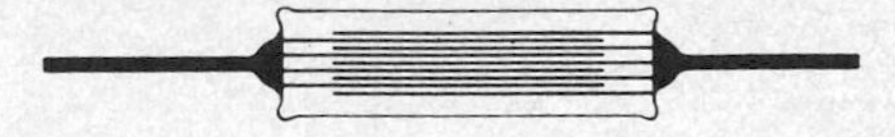

Porcelain-capacitor cross section.

port 1. An opening in a waveguide component, through which energy may be fed or withdrawn, or measurements made. 2. An opening in a base-reflex enclosure for a loudspeaker, designed and positioned to improve bass response. 3. An entrance or exit for a network. 4. An opening in a nuclear research reactor through which objects are inserted for irradiation or from which beams of radiation emerge for experimental use.

portable computer A computer about the size of a portable typewriter, designed to be taken wherever needed. It generally includes a typewriterlike keyboard, a cathode-ray or other type of display, a cartridge or other form of tape drive for loading program packages as well as for storing data, and a microprocessor that has sufficient working storage for the intended applications.

portable receiver A radio or television receiver small enough to be carried and having a carrying handle. It may or may not have batteries. The same set without a carrying handle would be called a table-model receiver.

portable storage A computer storage medium that can be easily transported without loss of stored data.

portable transmitter A transmitter that can be readily carried by a person and may or may not be operated while in motion.

port radar installation A radar installation near a harbor, used to guide ships entering or leaving the harbor during fog or darkness.

Poseidon A submarine-launched multiple-warhead nuclear missile that replaces the Polaris missile in nuclear submarines.

posistor A thermistor that has a large positive resistance-temperature characteristic. One version uses a barium titanate ceramic that has been doped to reduce its resistance.

position The location of a vehicle as determined by specific values of three navigation coordinates.

positional crosstalk The variation in the path followed by any one electron beam as the result of a change impressed on any other beam in a multibeam cathode-ray tube.

positional notation A system of notation in which the significance of each digit of a number depends on its position in the sequence. Successive digits are interpreted as coefficients of successive powers of an integer called the base.

position control system A positioning system in which the controlled motion is required only to reach a given end point, with no path control during the transition from one point to the next.

position-finding element The element in a fire-control system that locates the position of the target in space.

position fix The intersection of two plotted bearing lines on a map.

positioning action Automatic control action in which there is a predetermined relation between the value of a controlled variable and the position of a final control element.

position sensor A device that measures a position and converts this measurement into a form convenient for transmission. Also called position transducer.

position storage Storage media for machine-tool control; it contains major positions of the tool and instructions for auxiliary functions.

position telemeter A telemeter that measures and transmits angular or linear position.

position transducer *Position sensor.*

position-type telemeter Deprecated term for *ratio-type telemeter.*

positive 1. [abbreviated P] Having fewer electrons than normal, and hence having the ability to attract electrons. 2. Having the same rendition of light and shade as in the original scene.

positive bias A bias such that the control grid of an electron tube is positive with respect to the cathode.

positive charge The charge that exists in a body which has fewer electrons than normal.

positive column The luminous glow, often striated, that occurs between the Faraday dark space and the anode in a glow-discharge tube. The positive column is a plasma and has equal numbers of electrons and positive ions.

positive electricity The positive charge that is produced in a glass object by rubbing it with silk.

positive electrode The electrode that serves as the anode in a primary cell when the cell is discharging. It is connected to the positive terminal of the cell. Electrons flow through the external circuit to the positive electrode.

positive electron *Positron.*

positive feedback Feedback in which a portion of the output of a circuit or device is fed back in phase with the input, to increase the total amplification. Excessive positive feedback causes instability and distortion. When positive feedback is sufficiently high, oscillation occurs. In a sound system, excessive transfer of acoustic energy back from the loudspeaker to the microphone causes howling. Also called reaction (British), regeneration, and retroaction (British).

positive ghost A ghost image that has the same tonal variations as the primary television image.

positive-going Increasing in a positive direction.

positive-grid oscillator *Retarding-field oscillator.*

positive image A picture as normally seen on a television picture tube; it has the same rendition of light and shade as in the original scene being televised.

positive ion An atom that has less electrons than normal and therefore has a positive charge.

positive-ion emission Thermionic emission of positive particles that are ions of the metal used as the cathode of a vacuum tube or are due to some impurity in the cathode.

positive logic Digital logic in which the more positive logic level represents 1. Use of this logic level term is deprecated because it can be construed differently for NPN devices than for PNP devices. To avoid confusion, the magnitude and polarity of the voltage levels for 1 and 0 in a specific logic circuit always should be specified.

positive magnetostriction Magnetostriction in which the application of a magnetic field causes expansion of a material.

positive modulation Modulation in which an increase in brightness increases the transmitted power in an amplitude-modulation facsimile or television system or increases the transmitted frequency in a frequency-modulation facsimile system. Used in British television systems. Also called positive transmission.

positive ray A stream of positively charged atoms or molecules, produced by a suitable combination of ionizing agents, accelerating fields, and limiting apertures.

positive temperature coefficient [abbreviated PTC] The condition wherein the resistance, length, or other characteristic of a material increases when temperature increases. All metals and most metallic alloys have a positive temperature coefficient of resistance.

positive terminal The terminal of a battery or other voltage source toward which electrons flow through the external circuit.

positive transmission *Positive modulation.*

positive zero The zero value reached by counting down from a positive number in the binary system.

positron [POSItive elecTRON] A nuclear particle that has the mass of an electron and a positive charge which is exactly equal in magnitude to the negative charge of an electron. Positrons are formed in the beta decay of many radionuclides. Also called antielectron and positive electron.

positron camera A scintillation camera that uses a single large sodium iodide imaging crystal on one side of the patient. A number of scintillation counters are placed on the opposite side, in conjunction with a coincidence circuit that selects the correct scintillations from the crystal to form the desired image of positron-emitting nuclides on the screen of an oscilloscope. Used for medical diagnosis.

positronium A quasi-stable system that consists

of a positron and an electron bound together. The set of energy levels is similar to that of a hydrogen atom.

post *Waveguide post.*

postacceleration Acceleration of beam electrons after deflection in an electron-beam tube.

postacceleration cathode-ray tube An electrostatic cathode-ray tube in which the electron beam is accelerated to its final high velocity after passing through the deflection electrodes.

postdeflection focus Focusing of beam electrons after deflection, as in the chromatron.

postedit To edit the output data of a computer.

postemphasis *Deemphasis.*

postequalization *Deemphasis.*

POS terminal Abbreviation for *point-of-sale terminal.*

postforming The forming, bending, or shaping of cured laminated sheets by applying heat rapidly and pressing the sheet over a mold.

postmortem A digital computer diagnostic routine that prints out automatically or on demand the contents of the registers and storage locations of the computer when a problem has become stalled, to help locate the error in coding the problem.

postmortem dump A static dump, used for debugging purposes, that is performed at the end of a computer run.

postmortem routine A service routine used in computer troubleshooting, such as a routine that dumps out the content of a store after a failure.

post office bridge A form of Wheatstone bridge in which desired resistance values are obtained by inserting or removing plugs that fit between special terminals.

pot Slang term for *potentiometer.*

potassium [symbol K] An alkali metal that has photosensitive characteristics, used on the cathodes of phototubes when maximum response is desired to blue light. It has a moderate thermal neutron-absorption cross section. Atomic number is 19.

potassium-argon dating Dating of archeological, geological, or organic specimens by measuring the amount of argon accumulated in the matrix rock through decay of radioactive potassium.

potassium dihydrogen phosphate crystal [abbreviated KDP crystal] A piezoelectric crystal. Applications include sonar transducers.

pot core A ferrite magnetic core that has the shape of a pot, with a magnetic post in the center and a magnetic plate as a cover. The coils for a choke or transformer are wound on the center post.

potential The degree of electrification as referred to some standard, such as the earth. The potential at a point is the amount of work required to bring a unit quantity of electricity from infinity to that point. Potential and voltage are often used interchangeably.

potential barrier A semiconductor region in which the voltage is such that moving electric charges attempting to pass through it encounter opposition and may be turned back. Also called barrier and potential hill.

potential difference The voltage that exists between two points. More often called voltage.

potential energy Energy due to the position of a particle or piece of matter with respect to other particles.

potential gradient The difference in the values of the voltage per unit length along a conductor or through a dielectric.

potential hill *Potential barrier.*

potential scattering Scattering that has its origin in reflection from the nuclear surface, leaving the interior of the nucleus undisturbed, or scattering of an incident wave by reflection at a change or discontinuity.

potential transformer *Voltage transformer.*

potentiometer 1. A resistor that has a continuously adjustable sliding contact which is generally mounted on a rotating shaft. Used chiefly as a voltage divider. Also called pot (slang). 2. An instrument for measuring a voltage by balancing it against a known voltage.

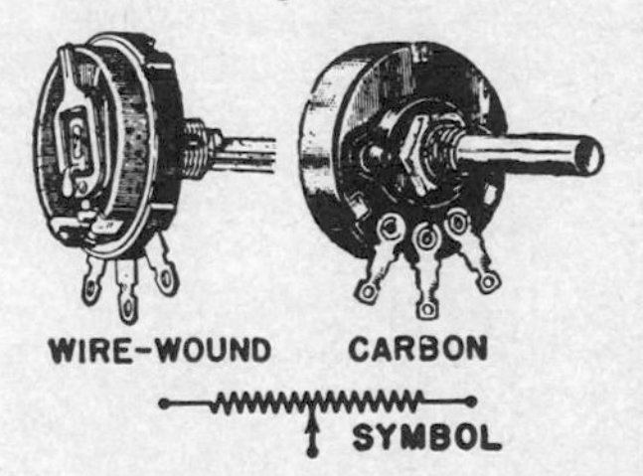

Potentiometers and potentiometer symbol.

potentiometer recorder A recorder that gives a permanent record of a varying voltage by using a potentiometer to balance the unknown voltage against a known voltage.

potentiometer-type transducer A transducer in which the pressure or other mechanical sensing

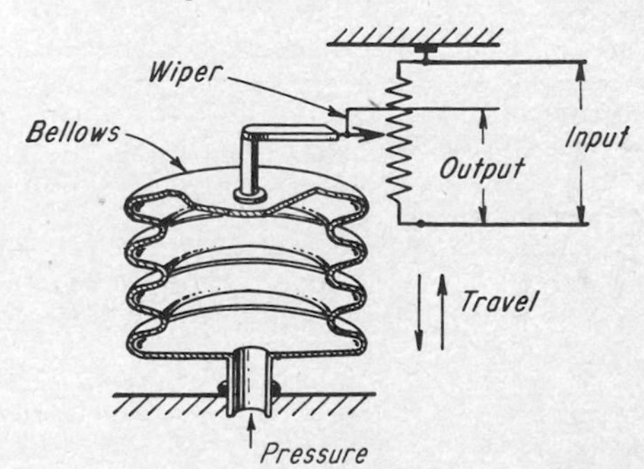

Potentiometer-type transducer attached to pressure-sensing bellows.

element moves the contact arm or wiper of a potentiometer.

potentiometric controller A controller that operates on the null balance principle, in which an error signal is produced by balancing the sensor signal against a set-point voltage in the input circuit. The error signal is amplified for use in keeping the load at a desired temperature or other parameter.

Potter-Bucky grid An assembly of lead strips resembling an open venetian blind, placed between a patient being x-rayed and the screen or film, to reduce the effects of scattered radiation. The grid is kept in motion during exposure to eliminate grid shadows. Also called Bucky diaphragm and grid.

potting A process for protecting a component or assembly by mounting the component or assembly in a can and pouring in an insulating compound.

pounds per square inch [abbreviated lb/in^2 or psi] A unit of pressure used to specify hydraulic and steam pressure and force.

powder-diffraction camera A metal cylinder that has a window through which an x-ray beam of known wavelength is sent by an x-ray tube to strike a finely ground powder sample mounted in the

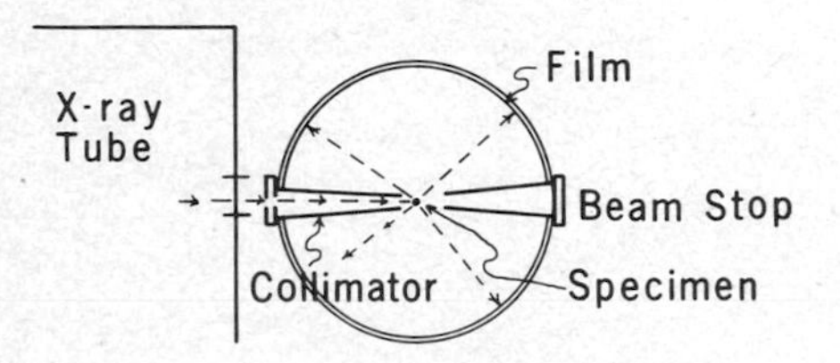

Powder-diffraction camera.

center of the cylinder. Crystal planes in this powder sample diffract the x-ray beam at different angles to expose a photographic film that lines the inside of the cylinder. A black diffraction line appears on the film at each diffraction angle.

powdered iron core *Ferrite core.*

powder metallurgy The production of magnetic cores, permanent magnets, and other molded metal objects by compressing finely powdered metals in molding dies, then heating to a temperature below the fusion point of the metal.

powder pattern A diffraction pattern obtained from a sample that consists of many small crystals oriented at random. The developed x-ray film pattern consists of a central spot surrounded by a series of concentric rings from whose diameters the Bragg angles may be computed.

power 1. [symbol P] The rate at which electric energy is fed to or taken from a device, measured in watts. Power is a definite quantity, whereas level is a relative quantity. 2. The result obtained when a number is multiplied by itself a particular number of times. Thus 125 is the third power of 5.

power amplification *Power gain.*

power amplifier An AF or RF amplifier that delivers maximum output power to a loudspeaker or other load, rather than maximum voltage gain, for a given percent distortion.

power-amplifier stage 1. An AF amplifier stage that is capable of handling considerable AF power without distortion. 2. An RF amplifier stage that serves primarily to increase the power of the carrier signal in a transmitter.

power-amplifier tube *Power tube.*

power attenuation *Power loss.*

power bandwidth The frequency range for which half the rated power of an audio amplifier is available at rated distortion. This specification is a measure of the power available at the critical high and low frequencies.

power converter A converter that changes DC power to AC power.

power cord *Line cord.*

power density 1. The power generation per unit volume of a nuclear-reactor core. 2. The amount of power per unit area in a radiated microwave or other electromagnetic field, usually expressed in watts per square centimeter. A power density of 10 mW/cm^2 is generally regarded as the maximum safe dosage for constant exposure of radar transmitter personnel.

power detector A detector circuit that will handle strong input signals without objectionable distortion.

power divider A device that produces a desired distribution of power at a branch point in a waveguide system.

power dump The removal of all power from a computer, accidentally or intentionally.

power factor [abbreviated PF] The ratio of active power to apparent power. As a percentage rating, it is equal to the resistance of a part or circuit divided by the impedance at the operating frequency, with the result multiplied by 100. A pure resistor has a power factor of 100%. A pure capacitor has a power factor of 0% leading, and a pure coil has a power factor of 0% lagging. Power factor is equal to the cosine of the phase angle between the current and voltage when both are sinusoidal.

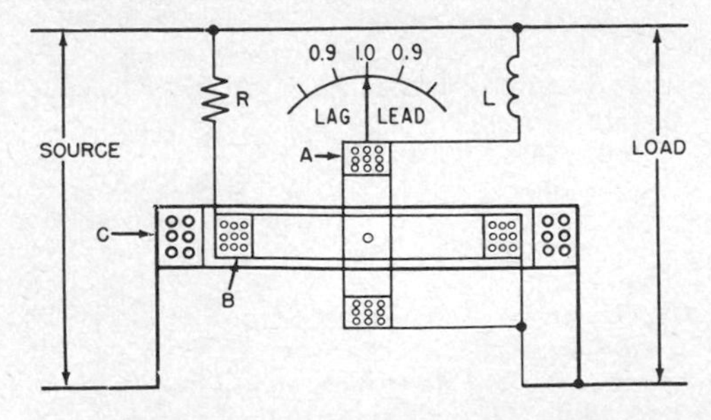

Power-factor meter using crossed-coil construction.

power-factor correction Addition of capacitors to an inductive circuit to increase the power factor. The capacitors offset part or all of the inductive reactance, making the total circuit current more nearly in phase with the applied voltage.

power-factor meter A direct-reading instrument that measures power factor.

power frequency The frequency at which electric power is generated and distributed. Most power companies in the continental United States are interconnected and operated at exactly the same frequency, which is usually maintained between 59.98 and 60.02 Hz. When the time indicated by electric clocks in this power grid differs more than 3 s from the standard time signals of WWV, steps are taken to bring back the entire power grid to correct time.

power gain 1. The ratio of the power delivered by a transducer to the power absorbed by the input circuit of the transducer. The power gain in decibels is 10 times the logarithm of the ratio of the power values. Also called power amplification. 2. An antenna rating equal to 12.56 times the ratio of the radiation intensity in a given direction to the total power delivered to the antenna.

power level The amount of power being transmitted past any point in an electric system. When expressed in decibels, it is equal to 10 times the logarithm to the base 10 of the ratio of the given power to a reference power value. Also expressed in volume units.

power-level indicator An AC voltmeter calibrated to read AF power levels directly in decibels or volume units.

power line Two or more wires conducting electric power from one location to another.

power-line carrier A carrier frequency, generally below 600 kHz, used to transmit control signals or information over power lines.

power-line filter *Line filter.*

power-line interference Interference caused by radiation from high-voltage power lines. The interference is generally noticeable only when the receiving antenna is within a few hundred feet of the line.

power loss The ratio of the power absorbed by the input circuit of a transducer to the power delivered to a specified load. Usually expressed in decibels. Also called power attenuation.

power output The AC power in watts delivered by an amplifier to a load.

power output tube *Power tube.*

power pack A power supply unit that converts the available power line or battery voltage to the voltage values required by a piece of electronic equipment.

power rating The power available at the output terminals of a component or piece of equipment that is operated according to the manufacturer's specifications.

power relay 1. A relay that functions at a predetermined value of input power. 2. The final relay in a sequence of relays controlling a load or a magnetic contactor.

power semiconductor A semiconductor device that is capable of dissipating appreciable power (generally over 1 W) in normal operation. It may handle currents of thousands of amperes or voltages up into thousands of volts, at frequencies up to 10 kHz.

power spectrum level The power level for the acoustic power in a band 1 Hz wide, centered at a specified frequency.

power supply A power line, generator, battery, power pack, or other source of power for electronic equipment.

power switch *ON/OFF switch.*

power transformer An iron-core transformer that has a primary winding which is connected to an AC power line and one or more secondary windings which provide different alternating voltage values.

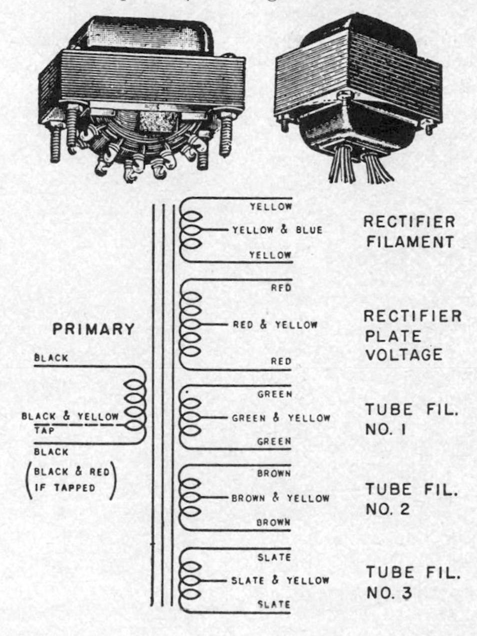

Power-transformer construction using terminals and insulated leads, with standard EIA color code for leads.

power transistor A junction transistor that handles high current and power. Used chiefly in audio and switching circuits.

power tube An electron tube capable of handling more current and power than an ordinary voltage-amplifier tube. It is used in the last stage of an AF amplifier or in high-power stages of an RF amplifier. Also called power-amplifier tube and power output tube.

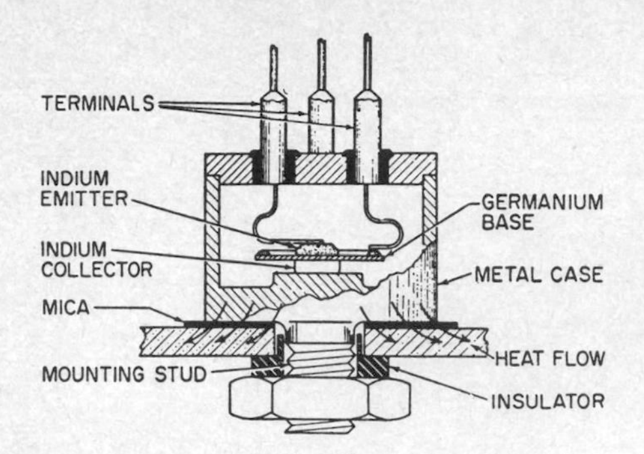

Power-transistor construction.

power winding A saturable reactor winding that receives power from a local source.

Poynting's vector A vector that represents the direction and amount of energy flow at a point in a wave at a given instant of time.

P-P Abbreviation for *peak-to-peak.*

PPBM Abbreviation for *pulse-polarization binary modulation.*

PPI Abbreviation for *plan-position indicator.*

PPI repeater A radarscope unit that duplicates at a remote location the plan-position-indicator display at the main radar installation.

PP junction A region of transition between two regions that have different properties in P-type semiconducting material.

P+-type semiconductor A P-type semiconductor in which the excess mobile hole concentration is very large.

PPM 1. Abbreviation for *parts per million.* 2. Abbreviation for *pulse-position modulation.*

PPPI Abbreviation for *precision plan-position indicator.*

pps Abbreviation for *pulses per second.*

practical system A system of electrical units in which the units are convenient multiples or submultiples of CGS units. Practical units are the ampere, coulomb, farad, henry, joule, ohm, volt, watt, and watthour.

praseodymium [symbol Pr] A rare-earth element. Atomic number is 59.

preamble The portion of a commercial radiotelegraph message that is sent first; it contains the message number, office of origin, date, and other numerical data not part of the original message.

preamp Abbreviation for *preamplifier.*

preamplifier [abbreviated preamp] An amplifier that is connected to a low-level signal source to provide gain and impedance matching so the signal can be further processed without appreciable degradation of the signal-to-noise ratio. A preamplifier may also have provisions for equalizing and mixing.

precession A change in the orientation of the axis of a gyroscope or other rotating body.

precipitation attenuation Attenuation of radio waves by passage through regions of precipitation in the atmosphere.

precipitation clutter Clutter caused by rain or other precipitation within the range of a radar station.

precipitation static Static interference due to the discharge of large charges built up on an aircraft or other object by rain, sleet, snow, or electrically charged clouds.

precipitator An electronic apparatus for removing smoke, dust, oil mist, or other small particles from air. A high direct voltage, of the order of 10 kV, is obtained from high-voltage rectifier tubes and applied to a fine wire mesh through which the

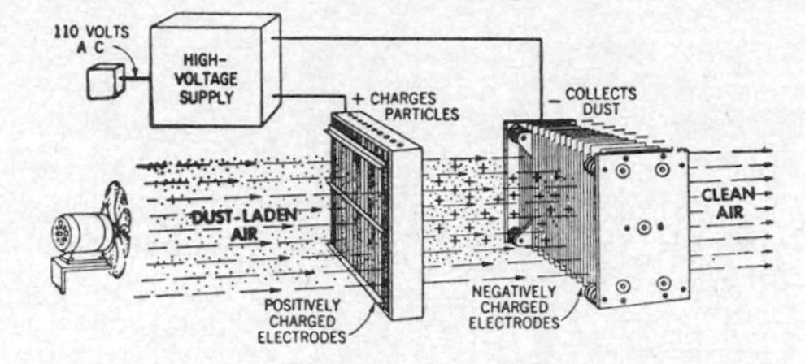

Precipitator operating principles.

air is drawn by a fan. Particles in the air are charged by this screen and are then drawn through a system of parallel charged plates that attract the particles and remove them from the air. Also called air cleaner, electronic air cleaner, electrostatic air cleaner, and electrostatic precipitator.

precision The quality of being exactly or sharply defined or stated. A six-place table has greater precision than a four-place table. The accuracy of either table would be reduced by errors in compilation or printing, however.

precision approach radar [abbreviated PAR] A radar set located on an airport, used in ground-controlled approach to provide a display that shows the distance, azimuth, and elevation of an airplane moving along the final approach path. Precision approach radar and airport surveillance radar together constitute a ground-controlled approach system.

precision-guided munitions Bombs that use television, laser, or electrooptical guidance with or without microwave remote control links between the bomb and its launching aircraft. Also called smart bomb (slang).

precision plan-position indicator [abbreviated PPPI] A plan-position indicator combined with a B display radarscope for precise measurement of the coordinates of a target. A range strobe and a mechanical azimuth marker are adjusted roughly to select the desired target on the PPI display, then adjusted more accurately while the same target is watched as it appears on the B display. Range and bearing can then be read directly on dials.

precision snap-acting switch A mechanically

operated electric switch that has predetermined and accurately controlled characteristics and contacts other than the blade and jaw, or mercury type, where the maximum separation between any butting contacts is ⅛ in (3.175 mm).

precision sweep A small portion of a normal radar sweep, usually 2000 yd (1,828.8 m), that is expanded over the entire screen to permit precise range measurements.

precision switch A switch whose output is a precise function of the position of the actuating lever.

preconduction current The low value of anode current flowing in a thyratron or other grid-controlled gas tube prior to the start of conduction.

precursor That which precedes, such as a decoy missile or a substance from which another substance is formed.

precursor arc An arc, generally several hundred amperes, that moves ahead of the shock wave down the length of an electric shock tube.

predetection Before detection. In predetection recording, the modulated carrier is recorded as received or as taken from an IF stage of the receiver.

predicted-wave signaling A communication system in which detection is optimized in the presence of severe noise by using mechanical resonator filters and other circuits in the detector to take advantage of known information on arrival and completion times of each pulse, as well as on pulse shape, pulse frequency and spectrum, and possible data content. Used in phase-shift telegraph systems for synchronous or asynchronous telegraph signals or synchronous business machine data.

predicting element The element in a fire-control system that predicts, from data provided by the position-finding, tracking, and ballistic data elements, the position of the target at the instant the missile is expected to reach it.

predissociation A process by which a molecule that has absorbed energy dissociates before it loses energy by radiation.

predistortion *Preemphasis.*

preedit To edit data before feeding it to a computer.

preemphasis The first part of a process for increasing the strength of some frequency components with respect to others, to help these components override noise or reduce distortion. Used chiefly for emphasizing the higher audio frequencies in frequency- and phase-modulated transmitters and in sound recording systems. Also called accentuation, emphasis, predistortion, and preequalization. The original relations are restored by the complementary process of deemphasis before reproduction of the sounds.

preemphasis network An *RC* filter inserted in a system to emphasize one range of frequencies with respect to another. Also called emphasizer.

preemption The act of interrupting a lower-priority transmission or process to accommodate a higher-priority user.

preequalization *Preemphasis.*

preferential recombination Recombination that takes place immediately after an ion pair is formed.

preferred numbers A series of numbers adopted by EIA and the military services for use as nominal values of resistors and capacitors, to reduce the number of different sizes that must be kept in stock for replacements.

preferred tube type A tube type recommended to designers of electronic equipment for general use, to minimize the number of tube types required for stock supply.

prefix A combining form used with a unit of measure to indicate a larger or smaller quantity as a power of 10.

prefix notation *Polish notation.*

preform A piece of material formed into a convenient shape and size for further processing, such as a pellet or slab of plastic or powdered metal prepared for a molding press.

P region The region in a semiconductor where hole density exceeds conduction electron density.

preheat fluorescent lamp A fluorescent lamp in which a manual switch or thermal starter is used to preheat the cathode for a few seconds before high voltage is applied to strike the mercury arc.

preoscillation current *Starting current.*

prerecord To record program material before it is required for broadcasting, transmission, or other use.

prerecorded tape *Recorded tape.*

preregulator A separate regulator circuit that increases the power-handling capability of a regulated power supply by minimizing power dissipation in the series regulator elements. Modern preregulators use silicon controlled rectifiers or triacs in place of the thyratrons of older tube circuits.

prescaler A scaler that extends the upper frequency limit of a counter by dividing the input frequency by a precise amount, generally 10 or 100.

preselector A tuned RF amplifier stage used ahead of the frequency converter in a superheterodyne receiver to increase the selectivity and sensitivity of the receiver.

presence The impression, as created by a recording or radio receiver, that the original program source is in the room.

presentation The form that radar echo signals are made to take on the screen of a cathode-ray tube.

preset To establish an initial value or condition, generally by setting one or more controls in advance.

preset guidance Guidance in which the path of a missile is determined by controls that are set before launching.

preset parameter A parameter whose value is not changed during the running of a given computer subroutine.

press 1. To mold a phonograph record from a stamper. 2. *Pinch.*

pressed cathode A dispenser cathode made by compacting and heating a mixture of tungsten-molybdenum alloy and barium-calcium aluminate, in a ratio such that evaporated electron-emitting surface barium is continuously replenished from within the cathode body.

pressed-glass base An electron-tube base in which heated powdered glass is pressed around the electrode leads and supports.

press fit A type of mounting in which the knurled body of a thyristor or other device is forced into an undersized mounting hole in a metal chassis that acts as a heatsink. The flow of metal during the press operation provides good thermal and electric contact between the chassis and the housing of the device.

pressing A phonograph record produced in a record-molding press from a master or stamper.

pressure *Effective sound pressure.*

pressure air-gap crystal unit A crystal unit in which the electrodes are recessed metal plates held firmly against opposite faces of the quartz plate.

pressure altimeter An altimeter that measures and indicates altitude by differences in atmospheric pressure.

pressure hydrophone A pressure microphone that responds to waterborne sound waves.

pressure microphone A microphone whose output varies with the instantaneous pressure produced by a sound wave acting on a diaphragm. Examples are capacitor, carbon, crystal, and dynamic microphones.

pressure pad A felt pad mounted on a spring-brass arm, used to hold magnetic tape in close contact with the head on some tape recorders.

pressure pickup A device that converts changes in the pressure of a gas or liquid into corresponding changes in some more readily measurable quantity such as inductance or resistance.

pressure spectrum level The effective sound pressure level for the sound energy contained within a band 1 Hz wide, centered at a specified frequency.

pressure switch A switch that is actuated by a change in pressure of a gas or liquid.

pressure transducer A transducer that uses a potentiometer, strain gage, variable-reluctance device, crystal, or other electric device for converting pressure to a proportional electric output.

pressure-type capacitor A fixed or variable capacitor mounted in a metal tank filled with nitrogen at a pressure high enough to permit a voltage rating several times that of the air rating. Used chiefly in transmitters.

pressure welding A welding process in which mechanical pressure is applied during welding. Examples are percussion, resistance, seam, and spot welding.

pressurization Use of an inert gas or dry air, at a pressure several pounds above atmospheric pressure, inside the components of a radar system or in a sealed coaxial line. Pressurization prevents corrosion by keeping out moisture and minimizes high-voltage breakdown at high altitudes.

prestore To store a quantity in an available computer location before it is required in a routine.

pretersonics The branch of electronics that involves use of acoustic waves at microwave frequencies (above 500 MHz), traveling on or in piezoelectric or other solid substrates. Also called acoustoelectronics.

pretravel The distance or angle through which the actuator of a switch moves from the free to the operating position.

pre-TR tube [pre-Transmit-Receive tube] A gas-filled RF switching tube that protects the TR tube from excessively high power and protects the radar receiver from frequencies other than the fundamental.

preventive maintenance A procedure of inspecting, testing, and reconditioning a system at regular intervals, according to specific instructions intended to prevent failures in service or retard deterioration.

PRF Abbreviation for *pulse repetition frequency.*

primaries The colors of constant chromaticity and variable amount that, when mixed in proper proportions, are used to produce or specify other colors.

primary *Primary winding.*

primary battery A battery that consists of one or more primary cells.

primary carrier flow The current flow that is responsible for the major properties of a semiconductor device. Also called primary flow.

primary cell A cell that delivers electric current as a result of an electrochemical reaction which is not efficiently reversible.

primary color A color that cannot be matched by any combination of other primary colors. In color television, the three primary colors emitted by the phosphors in the color picture tube are red, green, and blue.

primary-color unit The area within a color cell in a color picture tube that is occupied by one primary color.

primary cosmic rays *Cosmic rays.*

primary current The current flowing through the primary winding of a transformer.

primary dark space A narrow nonluminous region that appears between the cathode and the cathode glow of some gas-discharge tubes.

primary detector *Sensor.*

primary electron An electron emitted directly by a material rather than as a result of a collision.

primary element *Sensor.*

primary emission Electron emission directly

caused by the temperature of a surface, irradiation of a surface, or the application of an electric field to a surface.

primary feedback Feedback that is obtained from the controlled variable and compared with the reference input to obtain the actuating signal for a feedback control system. Also called feedback signal.

primary filter A sheet of material, usually metal, that is placed in a beam of radiation to absorb the less penetrating components.

primary flow *Primary carrier flow.*

primary frequency standard The national standard of frequency as maintained by the National Bureau of Standards, Washington, D.C. The operating frequency of a radio station is determined by comparison with multiples of this standard frequency as broadcast by station WWV.

primary fuel cell A fuel cell in which the fuel and oxidant are continuously consumed.

primary grid emission *Thermionic grid emission.*

primary ionization 1. The ionization produced by primary particles in a collision, as contrasted to the total ionization, which includes the secondary ionization produced by delta rays. 2. The total ionization produced in a counter tube by incident radiation without gas amplification.

primary ionizing event *Initial ionizing event.*

primary ion pair An ion pair produced directly by the causative primary particle or photon. An ion cluster is a group of ion pairs produced at or near the site of a primary ionizing event; it includes the primary ion pair and any secondary ion pairs formed.

primary radar Radar in which the incident beam is reflected from the target to form the return signal.

primary radiation Radiation that arrives directly from its source without interaction with matter.

primary radiator The portion of an antenna system from which energy leaves the transmission system. The distribution of the energy may be subsequently modified by other parts of the antenna system.

primary radionuclide A radionuclide that has a lifetime which exceeds several hundred million years.

primary service area The area in which the ground wave of a broadcast station is not subject to objectionable interference or fading.

primary skip zone The area around a radio transmitter that is beyond the ground-wave range but not far enough out for good skip-distance reception. Radio reception is not reliable in the primary skip zone.

primary specific ionization The number of ion clusters produced per unit track length.

primary spectrum The first-order spectrum produced by a diffraction grating.

primary standard A unit directly defined and established by some authority, against which all secondary standards are calibrated.

primary storage The main internal storage of a computer.

primary transit-angle gap loading The electronic gap admittance that results from the traversal of the gap by an initially unmodulated electron stream.

primary voltage The voltage applied to the terminals of the primary winding of a transformer.

primary winding [symbol P] The transformer winding that receives signal energy or AC power from a source. Also called primary.

prime To charge or discharge storage elements to a potential suitable for feeding data into a charge-storage tube.

prime contractor A contractor who has a direct contract for an entire project. A prime contractor may in turn assign portions of the work to subcontractors.

priming speed The rate of priming successive storage elements in a charge-storage tube.

principal axis A reference direction for angular coordinates, used in describing the directional characteristics of a transducer. It is usually an axis of structural symmetry or the direction of maximum response.

principal E plane A plane that contains the direction of maximum radiation and the electric vector.

principal H plane A plane that contains the direction of maximum radiation and the magnetic vector. The electric vector is everywhere perpendicular to the H plane.

principal mode *Fundamental mode.*

printed board *Printed-wiring board.*

printed circuit [abbreviated PC] A printed-wiring board that has components inserted into holes

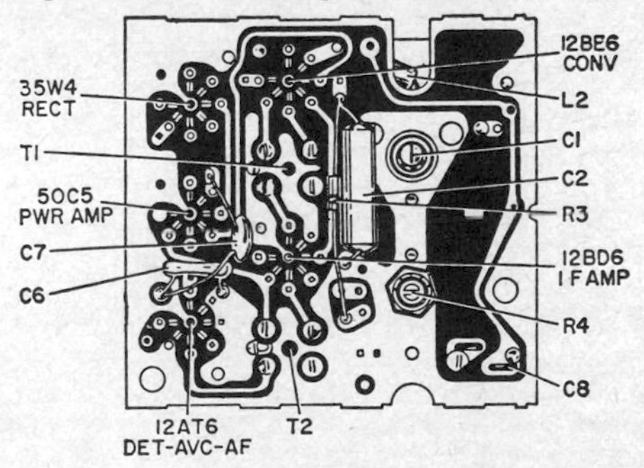

Printed circuit for table-model radio using tubes.

and soldered to form a complete circuit. It may also have printed components such as coils, formed simultaneously with the wiring.

printed component A printed coil, resistor, capacitor, switch, transmission line, or other component that provides an electric or magnetic function other than point-to-point connections or shielding. Printed components are sometimes formed on printed-wiring boards along with connections.

printed motor An electric motor in which the armature conductors are formed by printed wiring on both sides of one or more thin insulated disks, to give a low-inertia armature.

printed-wiring board A conductive pattern formed on one or both sides of an insulating base by etching, plating, or stamping. Also called printed board.

printed-wiring connector A connector that makes interconnections between printed-wiring terminations and conventional external wiring.

printer A computer output mechanism that prints characters one at a time, as in typewriters,

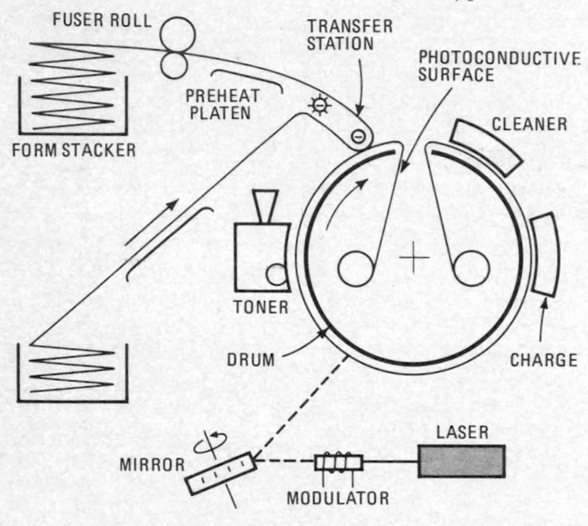

Printer in which characters are formed by modulated laser beam on photoconductive drum, for transfer by electrostatic techniques to continuous-form paper at speeds up to 14,000 lines per minute.

teleprinters, and serial printers; one line at a time, as in line printers; or one page at a time.

printer telegraph code A five- or seven-unit code used to operate telegraph-type printers. The seven-unit code requires only three marking elements out of the seven available per character. If fewer than three elements are received because of fading, or if more than three impulses arrive because of static, the receiving machine prints a special error-indicator sign rather than an improper letter.

printing calculator A desk-model electronic calculator that provides a printed record on paper tape with or without a digital display.

printout The printed output of a computer as produced by a printer.

print-through Transfer of signals from one recorded layer of magnetic tape to the next on a reel.

private automatic branch exchange [abbreviated PABX] A private branch exchange in which all telephone connections are made by remotely controlled switches.

private branch exchange [abbreviated PBX] A switching system, usually on a telephone customer's premises, that allows a customer's employee to interconnect telephone extensions, connect any of the extensions to a limited number of outgoing lines, and route incoming calls to the desired extension.

private line A communication line or channel reserved for one user.

PRK Abbreviation for *phase-reversal keying.*

probability multiplier *Coincidence multiplier.*

probability of collision The probability that an electron will collide with an atom or molecule when moving through a distance of 1 cm.

probability of ionization The ratio of the number of collisions followed by ionization to the total number of collisions in a gas during a specified time.

probability theory A measure of the likelihood of occurrence of a chance event. Used to predict behavior of a group rather than single items.

probable error The amount of error that is most likely to occur during a measurement. Half the results will have a greater error than this value, and the other half will have less error.

probe 1. A metal rod that projects into but is insulated from a waveguide or resonant cavity. Used to provide coupling to an external circuit for injection or extraction of energy. When a probe is movable in a slot, it can be used to measure the

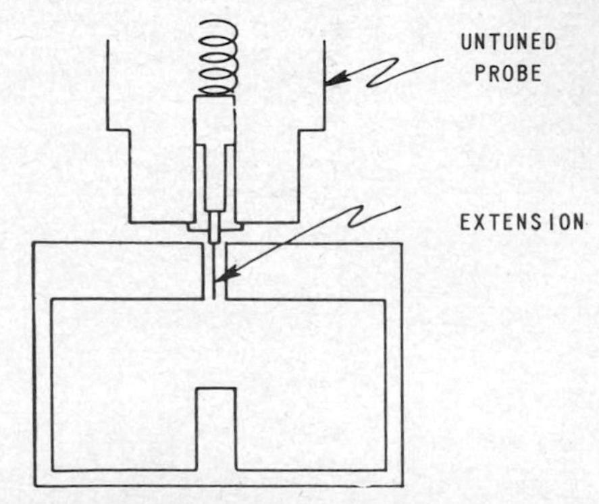

Probe used with slotted section of waveguide.

standing-wave ratio. Also called waveguide probe. 2. A large test prod that has interconnected components or circuits built into its handle. 3. An unmanned, instrumented vehicle sent into the upper atmosphere or space to gather environmental information.

probe microphone A small microphone that measures the sound pressure at a point without significantly altering the sound field in the neighborhood of the point.

problem-oriented language A computer language designed for convenience of program specification in a general problem area, rather than for easy conversion to machine instruction code.

procedure-oriented language A machine-independent language that describes how the process of solving the problem is to be carried out. Examples include ALGOL, COBOL, and FORTRAN.

process To assemble, compile, generate, interpret, compute, and otherwise act on information in a computer.

process control Automatic control of a complex industrial process.

processor The program or equipment used in a computer for assembling, compiling, translating, generating, and computing operations, simultaneously or independently.

prod *Test prod.*

product demodulator A demodulator whose output is the product of an amplitude-modulated carrier input voltage and a local oscillator signal voltage at the carrier frequency. With proper filtering, the output is proportional to the original modulation.

production control The procedure for planning, routing, scheduling, dispatching, and expediting the flow of materials, parts, subassemblies, and assemblies within a plant, from the raw state to the finished product, in an orderly and efficient manner.

production model A model in its final mechanical and electrical form, made by the production tools, jigs, fixtures, and methods to be used in turning out subsequent units.

product modulator A modulator whose output is proportional to the product of the carrier and modulating-signal voltages. The carrier is then normally suppressed.

professional engineer An engineer who is competent by virtue of fundamental education and training to apply the scientific method and outlook to the analysis and solution of engineering problems and is licensed by appropriate state or other governmental authorities.

profile chart A vertical cross-sectional drawing of a microwave path between two stations, indicating terrain, obstructions, and antenna height requirements.

program 1. A precise sequence of coded instructions used by a digital computer in solving a problem. A complete program includes plans for transcription of data and effective use of results. 2. A sequence of audio signals alone, or audio and video signals, transmitted for entertainment or information.

program amplifier *Line amplifier.*

program circuit A telephone circuit that has been equalized to handle a wider range of frequencies than is required for ordinary speech signals, for use in transmitting musical programs to the stations of a radio network.

program control A controller in which the desired value is automatically changed from time to time in accordance with a predetermined program.

program counter A register that provides the address of the next instruction to be fetched from memory in a computer. The register is normally incremented one step automatically after each instruction fetched.

program failure alarm A signal-actuated electronic relay circuit that gives a visual and aural alarm when the program fails on the line being monitored. A time delay is provided to prevent the relay from giving a false alarm during intentional short periods of silence.

program generator A program that permits a computer to write other programs automatically.

program level The level of the program signal in an audio system, expressed in volume units.

programmable calculator An electronic calculator that has some provision for changing its internal program, usually by inserting a new magnetic card on which the desired calculating program has been stored. With some calculators, new programs can be keyed in manually.

programmable controller An industrial process controller whose characteristics can be changed in the field by changing memory cards or by other means.

programmable counter A counter that divides an input frequency by a number which can be programmed into decades of synchronous down counters. These decades, with additional decoding and control logic, give the equivalent of a divide-by-N counter system.

programmable decade resistor A decade box designed so the value of its resistance can be remotely controlled by programming logic as required for the control of load, time constant, gain, and other parameters of circuits used in automatic test equipment and automatic controls.

programmable-gain amplifier An amplifier that can be operated under direct control of a digital computer or controlled by autoranging techniques, for signal-scaling under control of a digital input.

programmable logic General term that covers programmable devices which do not form a complete computing system and programmable logic systems that are actually complete computers.

programmable logic array *Field-programmable logic array.*

programmable logic device A logic device that forms only a part of a complete computing system. Examples include content-addressable memories, erasable programmable read-only memories, field-programmable logic arrays, microprocessors, programmable logic arrays, random-access memories, random logic, and read-only memories.

programmable logic system A self-contained logic system that includes all the computer elements needed to implement a specific algorithm or function. Examples include large-scale computers, microcomputers, minicomputers, and programmable calculators.

programmable power supply A power supply whose output voltage can be changed by digital control signals. There may also be provisions for manual selection of a choice of preset output voltages.

programmable read-only memory [abbreviated PROM] A read-only memory that can be programmed only once, either at the factory or in the field, by electrically fusing or otherwise removing unwanted internal links. A mistake in doing this cannot be corrected.

programmable unijunction transistor [abbreviated PUT] A PNPN device (thyristor) with an anode gate. For a constant gate voltage, the device remains off or nonconducting until the anode

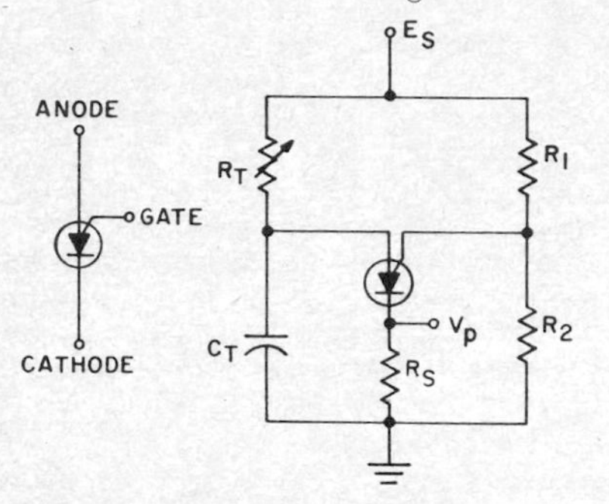

Programmable unijunction transistor symbol and example of use in relaxation oscillator.

voltage exceeds the gate voltage by a predetermined amount, at which point it turns on. The action is much like that of a silicon controlled switch.

programmed check 1. A computer check in which a sample problem with known answer, selected to have programming similar to that of the next problem to be run, is put through the computer. 2. A series of self-checking tests inserted in the computer program for a problem.

programmed electron-beam welding Electron-beam welding in which the beam is switched in sequence from connection to connection on an integrated circuit, in a vacuum, by digital control data stored in a computer or elsewhere. Similar programming can be used for removing unwanted portions of a thin film.

programmed logic array [abbreviated PLA] An array of AND/OR logic gates that provides logic functions for a given set of inputs programmed during manufacture and serves as a read-only memory. Applications include code conversion and decoding.

programmed turn The automatically controlled turn of a ballistic missile into the curved path that will lead to the correct velocity and vector for the final portion of the trajectory.

programmer 1. A person who prepares sequences of instructions for a computer, without necessarily converting them into the detailed codes. 2. A device that controls the motion of a missile in accordance with a predetermined plan.

programming Preparing a sequence of operating instructions for a computer.

programming language The language used by a programmer to write a program for a computer.

program monitor A monitor that observes the quality of a radio or television broadcast.

program parameter A parameter that may have different values during the course of a given program in a digital computer.

program register The register in the control unit of a digital computer which stores the current instruction of the program and controls the operation of the computer during the execution of that instruction.

program-sensitive error A computer error that occurs only when a unique combination of program steps is executed.

program tape A magnetic or punched-paper tape that contains the sequence of instructions required by a digital computer for solving a particular problem.

progressive scanning Scanning all lines in sequence, without interlace, so all picture elements are scanned during one vertical sweep of the scanning beam. Also called sequential scanning.

progressive wave A wave that is propagated freely in a medium.

progressive-wave antenna *Traveling-wave antenna.*

projection cathode-ray tube A television cathode-ray tube that produces an intensely bright but relatively small image which can be projected onto a large viewing screen by an optical system.

projection chamber A spark chamber in which tracks of nuclear particles that are traveling approximately parallel to the plates of the chamber are made visible by streamers along the electric field. Since the tracks are viewed parallel to the electric field, at least one plate of the chamber must be made of conducting glass or transparent metal gauze. The track is thus viewed as a projection on a plane at right angles to the field.

projection optics A system of mirrors and lenses used to project the image onto a screen in projection television. The Schmidt system is an example. Also called reflective optics.

projection television receiver A television receiver that uses a system of lenses and mirrors to project on a large-size screen an intensely bright image formed on the screen of a projection cathode-ray tube. Used chiefly for theater television.

projector 1. A horn that projects sound chiefly in one direction from a loudspeaker. 2. A machine that projects film images onto a screen. 3. *Underwater sound projector.*

projector efficiency *Transmitting efficiency.*

PROM Abbreviation for *programmable read-only memory.*

promethium [symbol Pm] A rare-earth element

that has no known stable isotopes in nature. Atomic number is 61.

promethium cell A nuclear energy cell in which beta particles from promethium 147 cause a phosphor to glow. The light output is converted to electric energy by photocells.

prompt radiation Radiation emitted within a time too short for measurement, including gamma rays, characteristic x-rays, conversion and Auger electrons, prompt neutrons, and annihilation radiation.

prong *Pin.*

propagation The travel of electromagnetic waves or sound waves through a medium, or the travel of a sudden electric disturbance along a transmission line. Also called wave propagation.

propagation anomaly A change in propagation characteristics due to a discontinuity in the medium of propagation.

propagation constant A rating for a line or medium through which a plane wave of a given frequency is being transmitted. It is a complex quantity; the real part is the attenuation constant in nepers per unit length, and the imaginary part is the phase constant in radians per unit length.

propagation delay The time required for a signal to pass through a given complete operating circuit. It is generally of the order of nanoseconds and of extreme importance in computer circuits.

propagation factor *Propagation ratio.*

propagation loss The attenuation of signals passing between two points on a transmission path.

propagation ratio The ratio, for a wave propagating from one point to another, of the complex electric field strength at the second point to that at the first point. Also called propagation factor. The field strength is a vector, the magnitude of which is less than 1 and is the attenuation ratio.

propagation time delay The time required for a wave to travel between two points on a transmission path.

propagation velocity The velocity of propagation of radio waves, within the accuracy demanded of radar equipment, is usually taken as the velocity of light, equal to 2.998×10^8 m/s or 299.8 m/μs. This corresponds to 983.6 ft/μs, 0.1863 mi/μs, or 0.1618 nautical mi/μs.

proportional band The range of values of the controlled variable that will cause a controller to operate over its full range.

proportional control Control in which the amount of corrective action is proportional to the amount of error.

proportional counter A radiation counter that consists of a proportional counter tube and its associated circuits.

proportional counter tube A radiation-counter tube operated at voltages high enough to produce ionization by collision and adjusted so the total ionization per count is proportional to the ionization produced by the initial ionizing event.

proportional ionization chamber An ionization chamber in which the initial ionization current is amplified by electron multiplication in a region of high electric field strength, as it is in a proportional counter. Used for measuring ionization currents or charges over a period of time, rather than for counting.

proportional navigation A homing guidance technique in which the missile turn rate is directly proportional to the turn rate of the missile-target line-of-sight.

proportional plus derivative control *Error-rate damping.*

proportional-position action Control action in which there is a continuous linear relation between the value of the controlled variable and the position of a final control element.

proportional region The range of applied voltage for a radiation-counter tube in which the charge collected per isolated count is proportional to the charge liberated by the initial ionizing event.

proportional response *Rate control.*

proportional-speed floating action Floating action in which the final control element is moved at a speed proportional to the deviation.

proportioning reactor A saturable-core reactor used for regulation and control. Increasing the input control current from zero to rated value makes output current increase in proportion from cutoff up to full load value.

protactinium [symbol Pa] A radioactive element that yields actinium by disintegration. Atomic number is 91. Formerly spelled protoactinium.

protection survey Evaluation of the radiation hazards incidental to the production, use, or existence of radioactive materials or other sources of radiation under a specific set of conditions.

protective resistance A resistance used in series with a gas tube or other device to limit current flow to a safe value.

protector tube A glow-discharge cold-cathode tube that becomes conductive at a predetermined voltage, to protect a circuit against overvoltage.

protium A name sometimes applied to the lightest hydrogen isotope; it has a mass number of 1 and consists of a single proton and electron. The other isotopes of hydrogen are deuterium and tritium.

proton An elementary particle that has a positive charge equal in magnitude to the negative charge of the electron. The atomic number of an element indicates the number of protons in the nucleus of each atom of that element. The rest mass of a proton is 1.67×10^{-24} g, or 1836.13 times that of an electron.

proton binding energy The energy required to remove a single proton from a nucleus. Most proton binding energies are in the range of 5 to 12 MeV.

proton magnetometer A highly sensitive magnetometer in which the spin precession principle is used for high-accuracy measurements of the total magnetic field of the earth. It can be towed through the air or underwater for making magnetic surveys.

proton microscope A microscope that is similar to the electron microscope but uses protons instead of electrons as the charged particles.

proton moment A physical constant equal to 1.41049×10^{-23} erg/G.

proton-proton chain A series of thermonuclear reactions initiated by a reaction between two protons.

proton-recoil counter A counter that measures fast neutrons.

proton rest mass A physical constant equal to 1.67252×10^{-24} g.

proton scattering microscope A microscope in which protons produced in a cold-cathode discharge are accelerated and focused on a crystal in a vacuum chamber. Protons reflected from the crystal strike a fluorescent screen to give a visual and photographable display that is related to the structure of the target crystal. Used for analysis and orientation of crystals.

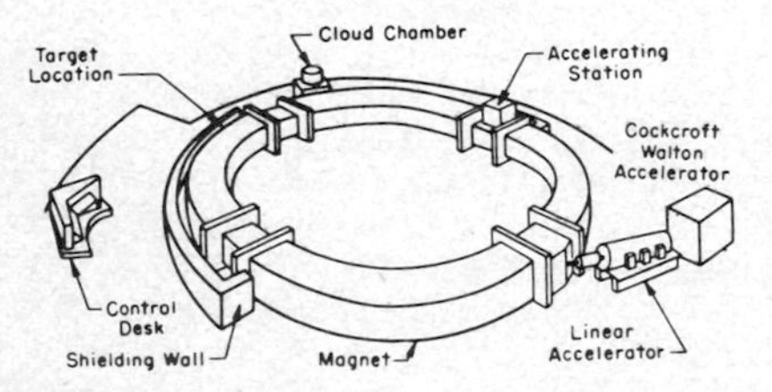

Proton synchrotron at University of California Radiation Laboratory accelerates protons to 99% of speed of light, with 6.2 GeV of energy.

proton synchrotron A synchrotron that accelerates protons. Also called bevatron and cosmotron.

prototype A model suitable for use in complete evaluation of form, design, and performance.

proximity detector A sensing device that produces an electric signal when approached by an object or when approaching an object.

proximity effect The redistribution of current in a conductor due to the presence of another current-carrying conductor.

proximity fuze A fuze that detonates a warhead when the target is within some specified region near the fuze. Radio, radar, photoelectric, or other devices may be used as activating elements. Also called influence fuze and variable-time fuze.

proximity switch A switch that is operated when the actuating device is moved near it, without physical contact. An example is a magnetic reed switch that operates when a permanent magnet is brought near it.

proximity warning indicator An airborne anticollision system that gives a warning of aircraft approaching on a possible collision course.

PRR Abbreviation for *pulse repetition rate.*

ps Abbreviation for *picosecond.*

PSC motor Abbreviation for *permanent-split capacitor motor.*

pseudoautocovariance *Cepstrum.*

pseudocode An arbitrary code, independent of the hardware of a computer, that must be translated into computer code before it can be used to direct the computer.

pseudorandom sequence A sequence of codes or numbers that appears to be random and satisfactorily meets the randomness requirements of a specific application, even though the sequence is of finite length and eventually repeats itself.

P shell The sixth layer of electrons about the nucleus of an atom; it has electrons whose principal quantum number is 6.

psi Abbreviation for *pounds per square inch.*

PSK Abbreviation for *phase-shift keying.*

psophometer An instrument that measures noise in electric circuits. When connected across a 600-Ω resistance in the circuit under study, the instrument gives a reading that by definition is equal to half the psophometric electromotive force actually existing in the circuit.

psophometric electromotive force The true noise voltage that exists in a circuit.

psophometric voltage The noise voltage as actually measured in a circuit under specified conditions.

psophometric weighting An international standard weighting curve used in specifying telephone-channel noise levels, to give more nearly the actual interfering effect of line noise than does a measuring set that has flat frequency response. The power of each interfering tone is compared with the power of an 800-Hz tone that creates the same interference during listening tests. Expressed in dBmp.

psychogalvanometer An instrument that tests mental reaction by determining how skin resistance changes when a voltage is applied to electrodes in contact with the skin.

psychosomatograph An instrument that records muscular action currents or physical movements during tests of mental-physical coordination.

PTC Abbreviation for *positive temperature coefficient.*

PTM Abbreviation for *pulse-time modulation.*

P-type conductivity The conductivity associated with holes in a semiconductor, which are equivalent to positive charges.

P-type semiconductor An extrinsic semiconductor in which the hole density exceeds the conduction electron density, so the majority carriers are holes. The net ionized impurity concentration is acceptor-type.

public-address amplifier An AF amplifier that provides sufficient output power to loudspeakers for adequate sound coverage at public gatherings.

public-address system [abbreviated PA system] A complete system for amplifying sounds and providing adequate volume for large public gatherings.

Public Broadcasting Service [abbreviated PBS] The national network of public television stations.

public television Broadcasting by noncommercial television stations, supported mainly by viewers, foundation grants, and government funds. Commercials are not allowed, but large corporations sometimes are credited for underwriting production costs of special programs.

puff British abbreviation (slang) for *picofarad.*

pull-in current *Pickup current.*

pulling An effect that forces the frequency of an oscillator to change from a desired value. Causes include undesired coupling to another frequency source or the influence of changes in the oscillator load impedance.

pulling figure The total frequency change of an oscillator when the phase angle of the reflection coefficient of the load impedance varies through 360°, the absolute value of this reflection coefficient being constant at 0.20.

pull-in value *Pickup value.*

pulsar signal A pulsed radio signal that comes from essentially a point source in interstellar space.

pulsatance Angular velocity in radians, equal to 2π times frequency in hertz.

pulsating current A direct current that increases and decreases in magnitude.

pulsation welding Resistance welding in which the current is applied in timed pulses rather than continuously, to improve the transfer of surface heat to the water-cooled welding electrodes and thereby increase electrode life.

pulse A momentary, sharp change in a current, voltage, or other quantity that is normally constant. A pulse is characterized by a rise and a decay and has a finite duration. Also called impulse.

pulse amplifier An amplifier designed specifically to amplify electric pulses without appreciably changing their waveforms.

pulse amplitude The peak, average, effective, instantaneous, or other magnitude of a pulse, usually with respect to the normally constant value. The exact meaning should be specified when giving a numerical value.

pulse-amplitude modulation [abbreviated PAM] Amplitude modulation of a pulse carrier.

pulse analyzer An instrument that measures pulse widths and repetition rates and displays on a cathode-ray screen the waveform of a pulse.

pulse-averaging discriminator A telemetering subcarrier discriminator that uses resistive and capacitive tuning components to give an output which is proportional to average pulse width.

pulse bandwidth The bandwidth outside of which the amplitude of a pulse is below a prescribed fraction of the peak amplitude.

pulse carrier A pulse train used as a carrier.

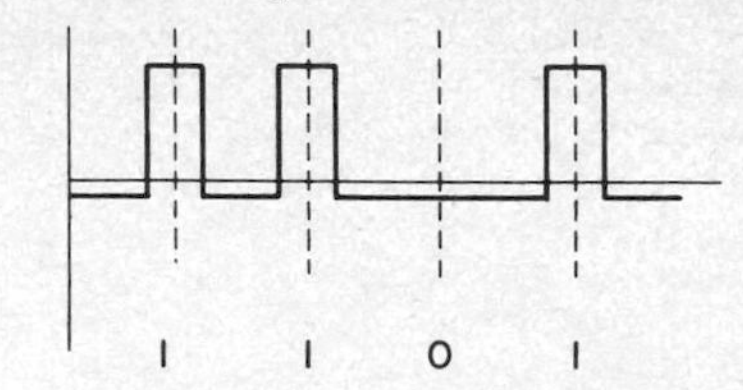

Pulse code representing binary number.

pulse code A code that consists of various combinations of pulses, such as the Morse code, Baudot code, and the binary code used in computers.

pulse-code modulation [abbreviated PCM] Modulation in which the peak-to-peak amplitude range of the signal to be transmitted is divided into a number of standard values, each value having its own three-place code. Each sample of the signal is then transmitted as the code for the nearest standard amplitude. When used in combination with frequency modulation (PCM/FM) for binary data, the carrier frequency has one deviation for a 0 and another deviation for a 1; this form is widely used in telemetry.

pulse coder *Coder.*

pulse-compression radar A radar system in which the transmitted signal is linearly frequency-modulated or otherwise spread out in time to reduce the peak power that must be handled by the transmitter. Signal amplitude is kept constant. The receiver uses a linear filter to compress the signal and thereby reconstitute a short pulse for the radar display.

pulse counter A device that indicates or records the total number of pulses received during a time interval.

pulsed accelerator An ion accelerator in which high instantaneous power levels are used to accelerate the plasma in pulses or bursts, to obtain maximum propulsion effect for spacecraft. The coaxial plasma gun is one example.

pulsed altimeter A radar altimeter that emits pulses of RF energy.

pulsed Doppler radar A radar system that uses the Doppler effect to determine target velocity.

pulse decay time The interval of time required for the trailing edge of a pulse to decay from 90 to 10% of the peak pulse amplitude.

pulse decoder A decoder that extracts useful information from a pulse-coded signal.

pulse-delay binary modulation [abbreviated PDBM] A form of pulse modulation in which a delayed pulse represents a binary 1 and an undelayed pulse represents a binary 0. Used in optical data communication.

pulse-delay network A network that consists of two or more components like resistors, coils, and capacitors, used to delay the passage of a pulse.

pulse demoder A circuit that responds only to pulse signals which have the specified spacing between pulses for which the device is adjusted. Also called constant-delay discriminator.

pulse discriminator A discriminator circuit that responds only to a pulse which has a particular duration or amplitude.

pulsed laser A laser in which a pulse of coherent light is produced at fixed time intervals, as required for ranging and tracking applications or to permit higher output power than can be obtained with continuous operation.

pulsed maser *Two-level maser.*

pulsed oscillator An oscillator that generates a carrier-frequency pulse or a train of carrier-frequency pulses as the result of self-generated or externally applied pulses.

pulse dot soldering iron A soldering iron that provides heat to the tip for a precisely controlled time interval, as required for making a good soldered joint without overheating adjacent parts.

pulsed plasma accelerator A plasma engine for space travel; a plasma current loop expands outward because of its self-induced magnetic field.

pulse droop A distortion of an otherwise essentially flat-topped rectangular pulse, characterized by a decline of the pulse top.

pulsed ruby laser A laser in which ruby is used as the active material. The extremely high pumping power required is obtained by discharging a bank of capacitors through a special high-intensity flash tube, giving a coherent beam that lasts for about 0.5 ms.

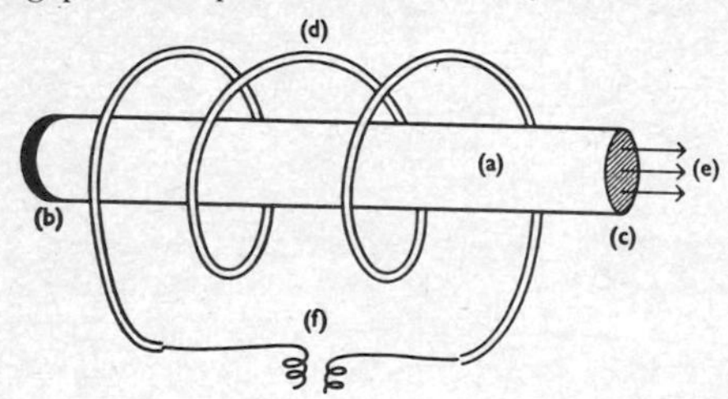

Pulsed ruby laser, showing ruby rod (a), heavily silvered flat end (b), lightly silvered flat output end (c), exciting flash tube (d), coherent output beam (e), and leads to capacitor bank (f).

pulse duration The time interval between the first and last instants at which the instantaneous amplitude reaches a stated fraction of the peak pulse amplitude. Also called pulse length (deprecated) and pulse width (deprecated).

pulse-duration coder *Coder.*

pulse-duration discriminator A circuit in which the sense and magnitude of the output are a function of the deviation of the pulse duration from a reference.

pulse-duration error An error caused by pulse duration, which makes certain targets appear longer or thicker than they actually are in the direction of the radar beam.

pulse-duration modulation [abbreviated PDM] A form of pulse-time modulation in which the duration of each pulse is varied. Also called pulse-width modulation (deprecated).

pulse duty factor The ratio of average pulse duration to average pulse spacing. This is equivalent to the product of average pulse duration and pulse repetition rate.

pulse-echo diagnosis Medical diagnosis based on the pulses of ultrasonic energy that are reflected from different structures in the human body.

pulse equalizer A circuit that produces output pulses of uniform size and shape, in response to input pulses which may vary in size and shape.

pulse excitation A method of producing oscillator current in which the duration of the impressed voltage in the circuit is relatively short compared with the duration of the current produced. Also called impulse excitation.

pulse-forming line A continuous line or ladder network that has parameters which give a specified shape to the modulator pulse in a radar modulator.

pulse-forming network A network that shapes the leading and/or trailing edge of a pulse.

pulse-frequency modulation [abbreviated PFM] A form of pulse-time modulation in which the pulse repetition rate is the characteristic varied. A more precise term for pulse-frequency modulation would be pulse repetition-rate modulation.

pulse-frequency spectrum *Pulse spectrum.*

pulse-gated binary modulation [abbreviated PGBM] A form of pulse modulation in which a binary 1 is represented by a pulse and binary 0 by the absence of a pulse. Used in optical data communication.

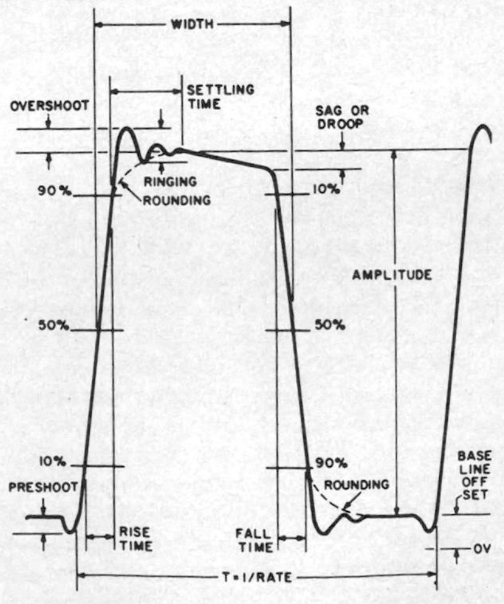

Pulse-generator output specifications.

pulse generator A generator that produces repetitive pulses or single signal-initiated pulses.

pulse-height analyzer An instrument capable of indicating the number of occurrences of pulses falling within each of one or more specified amplitude ranges. Used to obtain the energy spectrum of nuclear radiations. Also called kick-sorter (British) and multichannel analyzer.

pulse-height discriminator A circuit that produces a specified output pulse when and only when it receives an input pulse whose amplitude exceeds an assigned value. Also called amplitude discriminator.

pulse-height selector A circuit that produces a specified output pulse only when it receives an input pulse whose amplitude lies between two assigned values. Also called amplitude selector and differential pulse-height discriminator.

pulse interleaving A process in which pulses from two or more sources are combined in time-division multiplex for transmission over a common path.

pulse interrogation The triggering of a transponder by a pulse or pulse mode.

pulse interval Deprecated term for *pulse spacing.*

pulse-interval modulation Deprecated term for *pulse-spacing modulation.*

pulse ionization chamber An ionization chamber used to detect individual ionizing events. Also called counting ionization chamber.

pulse jitter A relatively small variation of the pulse spacing in a pulse train. The jitter may be random or systematic, depending on its origin, and is generally not coherent with any pulse modulation imposed.

pulse length Deprecated term for *pulse duration.*

pulse mode A finite sequence of pulses in a prearranged pattern, used for selecting and isolating a communication channel.

pulse-mode multiplex A process or device for selecting channels by means of pulse modes. This process permits two or more channels to use the same carrier frequency. Also called pulse multiplex (deprecated).

pulse moder A device for producing a pulse mode.

pulse-modulated radar Radar in which each transmission is a series of discrete pulses.

pulse-modulated waves Recurrent wave trains in which the duration of the trains is short compared with the interval between them. Used in radar.

pulse modulation Modulation in which the amplitude or time of some characteristic of a pulse carrier is varied by the modulating wave. Examples include pulse-amplitude and pulse-time modulations.

pulse modulator A device that applies pulses to the element in which modulation takes place.

pulse multiplex Deprecated term for *pulse-mode multiplex.*

pulse navigation system A navigation system that depends on the time required for a pulse of RF energy to travel a given distance. Examples include Gee, loran, radar, and shoran.

pulse-numbers modulation Modulation in which the pulse density per unit time of a pulse carrier is varied in accordance with a modulating wave by making systematic omissions without changing the phase or amplitude of the transmitted pulses. As an example, omission of every other pulse could correspond to zero modulation; reinserting some or all pulses then corresponds to positive modulation, and omission of more than every other pulse corresponds to negative modulation.

pulse packet The volume of space occupied by a single radar pulse. It is defined by the angular width of the beam, the duration of the pulse, and the distance of the pulse from the antenna.

pulse-phase modulation *Pulse-position modulation.*

pulse-polarization binary modulation [abbreviated PPBM] A form of pulse modulation in which a right circularly polarized pulse represents a binary 1 and an oppositely polarized pulse represents a binary 0. Used in optical data communication.

pulse-position modulation [abbreviated PPM] A form of pulse-time modulation in which the position in time of a pulse is varied. Also called pulse-phase modulation.

pulser A generator that produces high-voltage, short-duration pulses, as required by a pulsed microwave oscillator or a radar transmitter. In a vacuum-tube pulser, the pulse is produced by discharging a capacitor through the load. In a line-type pulser, an unterminated transmission line is charged through a high impedance and discharged through the load.

pulse radar Radar in which the transmitter sends out high-power pulses that are spaced far apart in comparison with the duration of each pulse. The receiver is active for reception of echoes in the interval following each pulse.

pulse-rate telemetering Telemetering in which the number of pulses per unit time is proportional to the magnitude of the measured quantity.

pulse recurrence interval The time, usually expressed in microseconds, between pulses in radar and loran.

pulse recurrence rate *Pulse repetition rate.*

pulse reed relay A reed relay that has one pulse coil and one holding coil for operating one or more reed switches. The holding coil alone will not operate the switch but will keep it operated after the pulse coil has received the proper signal.

pulse regeneration The process of restoring pulses to their original relative timings, forms, and magnitudes.

pulse regenerator A device or circuit capable of restoring or regenerating a pulse train to its original characteristics.

pulse repeater A device that receives signal

pulses from one circuit and transmits corresponding pulses into another circuit. It may also change the frequency and waveform of the pulses and perform other functions.

pulse repetition frequency [abbreviated PRF] *Pulse repetition rate.*

pulse repetition rate [abbreviated PRR] The number of times per second that a pulse is transmitted. In radar the pulse repetition rate is usually between 400 and 3000 pulses per second. Also called pulse recurrence rate and pulse repetition frequency.

pulse reply The transmission of a pulse or pulse mode by a transponder as the result of an interrogation.

pulse resolution The minimum time separation between input pulses that will permit a circuit or component to respond properly.

pulse rise time The interval of time required for the leading edge of a pulse to rise from 10 to 90% of the peak pulse amplitude.

pulse scaler A scaler that produces an output signal when a prescribed number of input pulses has been received.

pulse selector A circuit or device for selecting the proper pulse from a sequence of telemetering pulses.

pulse separation The time interval between the trailing edge of one pulse and the leading edge of the succeeding pulse.

pulse shaper A transducer that changes one or more characteristics of a pulse, such as a pulse regenerator or pulse stretcher.

pulse shaping Intentionally changing the shape of a pulse.

pulse spacing The interval between the corresponding pulse times of two consecutive pulses. Also called pulse interval (deprecated).

pulse-spacing analyzer An analyzer that converts the spacing between two adjacent pulses to a proportional voltage. Applications include measuring the periodicity of neural activities.

pulse-spacing modulation A form of pulse-time modulation in which the pulse spacing is varied. Also called pulse-interval modulation (deprecated).

pulse spectrum The frequency distribution of the sinusoidal components of a pulse in relative amplitude and in relative phase. Also called pulse-frequency spectrum.

pulses per second [abbreviated pps] The number of pulses per second in a varying DC quantity. For an AC quantity, use hertz.

pulse spike An unwanted pulse of relatively short duration superimposed on a main pulse.

pulse spike amplitude The peak pulse amplitude of a pulse spike.

pulse stretcher A pulse shaper that produces an output pulse whose duration is greater than that of the input pulse and whose amplitude is proportional to the peak amplitude of the input pulse.

pulse synthesizer A circuit that supplies pulses which are missing from a sequence due to interference or other causes.

pulse test An insulation test in which the applied voltage is a pulse that has a specified wave shape. Also called impulse test.

pulse tilt A distortion in an otherwise essentially flat-topped rectangular pulse, characterized by either a decline or a rise of the pulse top.

pulse-time modulation [abbreviated PTM] Modulation in which the time of occurrence of some characteristic of a pulse carrier is varied from the unmodulated value. Examples include pulse-duration, pulse-frequency, pulse-position, and pulse-spacing modulation.

pulse train A sequence of pulses that have similar characteristics.

pulse-train spectrum The frequency distribution of the sinusoidal components of a pulse train, in amplitude and in phase angle.

pulse transformer A transformer capable of operating over a wide range of frequencies, used to transfer nonsinusoidal pulses without materially changing their waveforms.

pulse transmitter A pulse-modulated transmitter whose peak power-output capabilities are usually large with respect to the average power-output rating.

pulse-type altimeter *Radar altimeter.*

pulse-type scanning sonar Scanning sonar in which a pulse of sound power is transmitted in all directions simultaneously. The volume of surrounding water is then scanned rapidly, for viewing of all echoes on a PPI screen.

pulse-type telemeter A telemeter that employs characteristics of intermittent electric signals other than their frequency as the means of conveying information.

pulse valley The portion of a pulse between two specified maxima.

pulse width Deprecated term for *pulse duration.*

pulse-width modulation Deprecated term for *pulse-duration modulation.*

pulsing circuit A circuit that provides abrupt changes of voltage or current in some characteristic pattern.

pumped tube An electron tube that is continuously connected to evacuating equipment during operation. Large pool-cathode tubes are often operated in this manner.

pumping A method of exciting electrons to a higher energy state in a laser. Pumping can be done optically with a light source, electrically with a discharge in a gas or with current in a semiconductor, or chemically by a reaction between two gases. Pumping produces the electron population inversion that is required to initiate and sustain laser action. In a maser, pumping is achieved with a microwave signal that differs in frequency from the output signal.

pumping band A group of energy states to which

ions in the ground state are initially excited when pumping radiation is applied to a laser medium. The pumping band is generally higher in energy than the levels that are to be inverted.

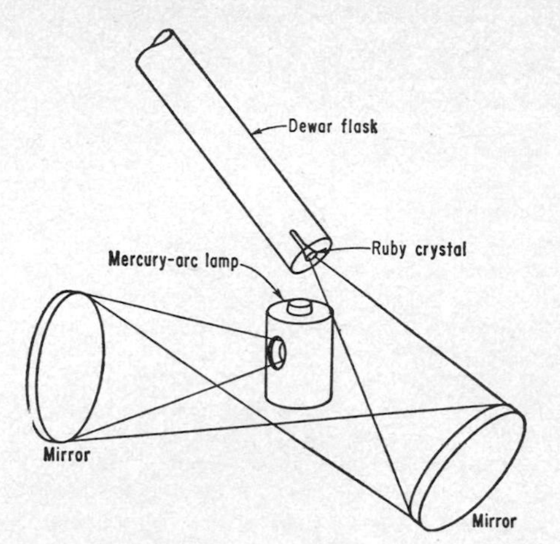

Pumping by mercury-arc lamp in continuously operating ruby laser. Liquid air in Dewar flask prevents overheating of ruby.

pumping frequency The frequency at which pumping is provided in a laser, maser, quadrupole amplifier, or other amplifier that requires high-frequency excitation.

pumping radiation Light applied to the sides or end of a laser crystal to excite the ions to the pumping band.

pumping voltage A reverse voltage delivered to the varactor of a parametric amplifier at a multiple of the RF signal frequency.

punched card A card of constant size and shape, suitable for punching in a pattern that has meaning and for being handled mechanically. The punched holes are sensed electrically by wire brushes, mechanically by metal fingers, or photoelectrically. Also called card.

punched tape Tape that is punched in coded patterns of holes which represent information. The material is usually paper, but more durable materials such as Mylar are used for tapes that must be run many times. Also called tape.

punch-through transistor A bipolar NPN transistor that has an extra diffusion step which gives a high current gain at collector-emitter voltages under about 1 V. At collector-emitter voltages above 2 or 3 V, however, the collector-base depletion layer punches through the active base region into the emitter.

punch-through varactor A high-power silicon epitaxial varactor capable of operating in the UHF band and up into the X band. Used in harmonic generators and frequency multipliers. The junction depletion-region boundary punches through the thin epitaxial layer of high-resistivity silicon material at a low reverse bias voltage.

punch-through voltage The transistor collector-to-base voltage at which the space-charge layer of the collector has widened until it touches the emitter junction.

puncture *Breakdown.*

puncture voltage The voltage at which a test specimen is electrically punctured.

Pupin coil *Loading coil.*

pup jack *Tip jack.*

pure tone *Simple tone.*

purity 1. The degree to which a primary color is pure and not mixed with the other two primary colors used in color television. 2. A ratio of distances on the CIE chromaticity diagram that compares a sample color with a reference standard light. Also called excitation purity.

purity coil A coil mounted on the neck of a color picture tube, to produce the magnetic field needed for adjusting color purity. The direct current through the coil is adjusted to a value that makes the magnetic field orient the three individual electron beams so each strikes only its assigned color of phosphor dots.

purity control A potentiometer or rheostat used to adjust the direct current through the purity coil.

purity magnet An adjustable arrangement of one or more permanent magnets used in place of a purity coil in a color television receiver.

purple boundary A straight line drawn between the ends of the spectrum locus on the CIE chromaticity diagram.

purple plague Failure of a gold-wire bond on aluminum because of the formation of porous intermetallic compounds that are mechanically weak and electrically nonconductive, occurring chiefly at temperatures above 200°C. The compound has a purple color.

pursuit-course guidance Guidance in which the missile is always moving directly toward the target. The missile thereby follows a curved path and eventually collides with the target because of its greater speed.

pushback hookup wire Tinned copper wire covered with loosely braided insulation that can be pushed back with the fingers sufficiently to expose enough bare wire for making a connection.

pushbutton control Control of machines, missiles, and other complex equipment entirely by relays and automatic electric or electronic control circuits, once a human operator has operated a pushbutton switch.

pushbutton dialing Dialing of a desired telephone number by pushing buttons in a particular sequence to initiate automatic switching in telephone exchanges, as with a Touch-Tone telephone.

pushbutton station A unit assembly of one or more pushbutton switches, with or without indicating lamps.

pushbutton switch A master switch that is operated by finger pressure on the end of an operating button.

pushbutton tuner A device that automatically tunes a radio receiver or other piece of equipment to a desired frequency when the button assigned to that frequency is pressed. The button may connect a set of preadjusted trimmer capacitors or adjustable coils into tuning circuits, control a small motor that drives the regular gang tuning capacitor, or apply force to a lever or cam system which rotates the gang tuning capacitor to the correct position.

pushdown list A computer list in which the next item to be retrieved is that most recently stored, corresponding to last-in first-out.

pushing A change in the resonant frequency of a circuit due to changes in the applied voltages.

pushing figure The amount of change in oscillator frequency produced by a specified change in oscillator electrode current, excluding thermal effects.

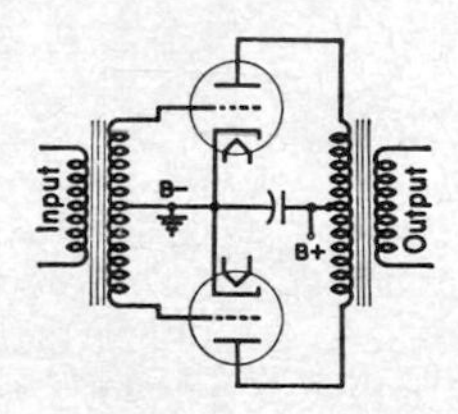

Push-pull amplifier circuit.

push-pull amplifier A balanced amplifier that uses two similar electron tubes or equivalent amplifying devices working in phase opposition.

push-pull currents *Balanced currents.*

push-pull microphone A microphone that makes use of two identical microphone elements which are actuated 180° out of phase by a sound wave as in a double-button carbon microphone.

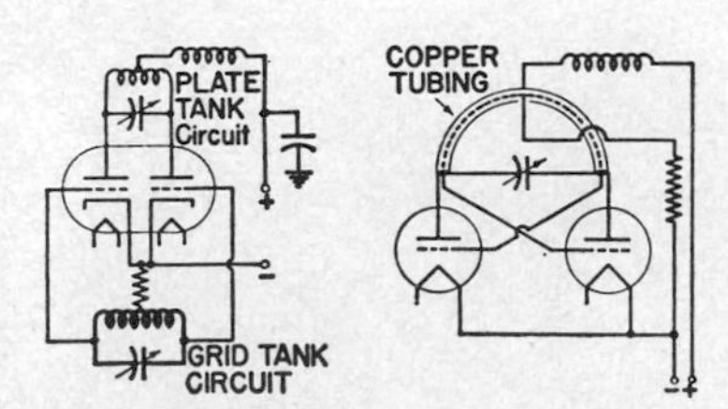

Push-pull oscillator circuit having tuned-grid and tuned-anode tank circuits, with UHF version at right.

push-pull oscillator A balanced oscillator that uses two similar electron tubes or equivalent amplifying devices in phase opposition.

push-pull-parallel amplifier A push-pull amplifier that has two or more tubes or transistors in parallel in each half of the circuit, to obtain higher output power.

push-pull track A sound track which has two recordings so arranged that the modulation in one is exactly 180° out of phase with that in the other.

push-pull transformer An AF transformer that has a center-tapped winding, for use in a push-pull amplifier.

push-pull voltages *Balanced voltages.*

push-push amplifier An amplifier that employs two similar electron tubes with grids connected in phase opposition and anodes connected in parallel to a common load. Usually used as a frequency multiplier to emphasize even-order harmonics. Transistors may be used in place of tubes.

push-push currents Currents flowing in the two conductors of a balanced line that, at every point along the line, are equal in magnitude and in the same direction.

push-push voltages Voltages (relative to ground) on the two conductors of a balanced line that, at every point along the line, are equal in magnitude and have the same polarity.

push-to-talk switch A switch mounted directly on a microphone or handset to provide a convenient means for switching two-way radiotelephone equipment, intercom equipment, or electronic dictating equipment to the talk position. The switch must be released for listening.

pushup list A computer list in which the next item to be retrieved is the oldest item placed in the list, corresponding to first-in first-out.

PUT Abbreviation for *programmable unijunction transistor.*

PVC Abbreviation for *polyvinyl chloride.*

pW Abbreviation for *picowatt.*

pylon antenna A vertical antenna constructed of one or more cylinders of sheet metal, each cylinder having a lengthwise slot.

pyramidal horn antenna An aperture antenna that has an outward-flaring rectangular horn, the

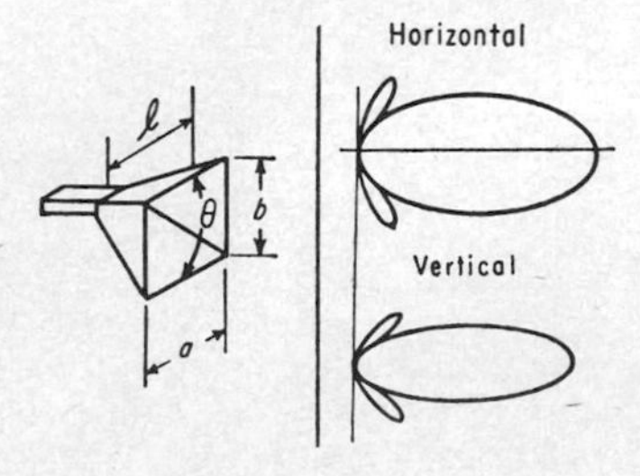

Pyramidal horn antenna, with horizontal and vertical radiation patterns.

dimensions of which determine the shape of the radiation pattern.

pyranometer An instrument that measures the intensity of the radiation received from any portion of the sky.

pyrheliometer An instrument that measures the

total intensity of solar radiation.

pyroconductivity Electric conductivity that develops in a material only at high temperature, chiefly at fusion, in solids which are practically nonconductive at atmospheric temperatures.

pyroelectric effect The development of charges in certain crystals when unequally heated or cooled.

pyromagnetic Pertaining to the interaction of heat and magnetism.

pyrometer An instrument that measures temperatures by electric means, especially temperatures beyond the range of mercury thermometers. Examples include the radiation, resistance, and thermoelectric pyrometers.

pyrotron A machine that uses magnetic mirrors in a long straight tube to reflect charged particles and prevent end leaks. Used in controlled-fusion research. The tube is surrounded by current-carrying coils, with more coils or higher currents at the ends to produce the stronger magnetic fields that act as mirrors.

Pythagorean scale A musical scale such that the frequency intervals are represented by the ratios of integral powers of the numbers 2 and 3.

PZT Abbreviation for *lead zirconate titanate.*

Q

Q A figure of merit for an energy-storing device, tuned circuit, or resonant system. It is equal to reactance divided by resistance. The *Q* of a capacitor, coil, circuit, or system thus determines the rate of decay of stored energy; the higher the *Q*, the longer it takes for the energy to be released. Also called *Q* factor and quality factor.

QAM Abbreviation for *quadrature amplitude modulation.*

Q antenna A dipole that is matched to its transmission line by stub-matching.

Q band A band of radio frequencies that extends from 36 to 46 GHz, corresponding to wavelengths of 0.834 to 0.652 cm. Subdivisions of this band (all values in gigahertz) are

Qa: 36–38
Qb: 38–40
Qc: 40–42
Qd: 42–44
Qe: 44–46

Q channel The 0.5-MHz-wide band used in the American NTSC color television system for transmitting green-magenta color information. The I channel is 1.3 MHz wide, and the luminance channel is 3.579545 MHz wide.

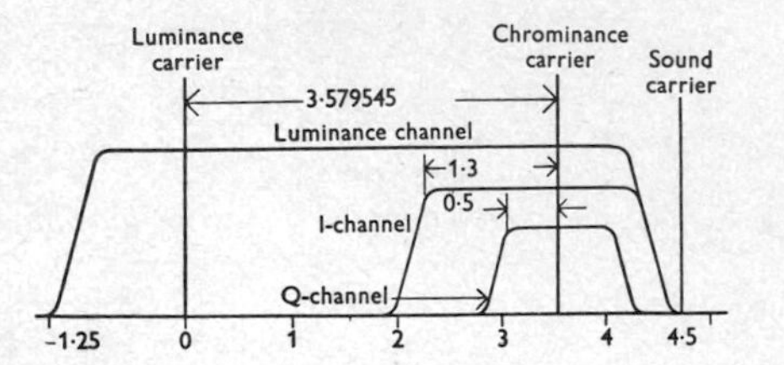

Q channel in NTSC color television system, centered on chrominance carrier. All values are in megahertz.

Q demodulator The demodulator in which the chrominance signal and 90° phase-shifted signal of the color-burst oscillator are combined to recover the Q signal in a color television receiver.

Q electron An electron that has an orbit in the Q shell, which is the seventh shell of electrons surrounding the atomic nucleus, counting out from the nucleus.

Q factor *Q.*

Q meter An instrument that measures the *Q* of a circuit or circuit element by determining the ratio of reactance to resistance. Also called quality-factor meter.

Q multiplier A filter that gives a sharp response peak or a deep rejection notch at a particular frequency, equivalent to boosting the *Q* of a tuned circuit at that frequency.

QPSK Abbreviation for *quaternary phase-shift keying.*

QRP An international radio Q signal that means "reduce your power." Also used to represent amateur radio communication that uses very low power, of the order of microwatts.

QS Trademark of Sansui Electric Co., Ltd. for their matrix-type quadraphonic sound system.

Q shell The seventh layer of electrons about the nucleus of an atom; it has electrons characterized by the principal quantum number 7.

Q signal 1. The quadrature component of the chrominance signal in color television; it has a bandwidth of 0.5 MHz. It consists of +0.48(R − Y) and +0.41(B − Y), where Y is the luminance signal, R is the red camera signal, and B is the blue camera signal. 2. A three-letter abbreviation starting with Q, used in the International List of Abbreviations for radiotelegraphy to represent complete sentences.

QSL card A card sent by one radio amateur to another to verify radio communication with each other.

Q spoiling A method of preventing laser action by quenching while a large population excess is being pumped up, so a more powerful pulse of light is obtained when the laser is triggered by *Q* switching.

Q switching A technique for keeping the *Q* of the cavity of a laser at a low value while an ion population inversion is being built up, then suddenly switching the *Q* to a high value just before instability occurs. The technique gives a very high rate of stimulated emission. Switching action can be achieved with Kerr cells, rotating reflecting prisms, or thin foils of gold inserted between the laser crystal and a high-reflectivity end plate. Phthalocyanine molecules in solution have served as repeatable *Q* switching elements for ruby lasers.

quad 1. A series of four separately insulated conductors, generally twisted together in pairs. 2. A series-parallel combination of transistors. Used to obtain increased reliability through double redundancy because failure of one transistor will not disable the entire circuit.

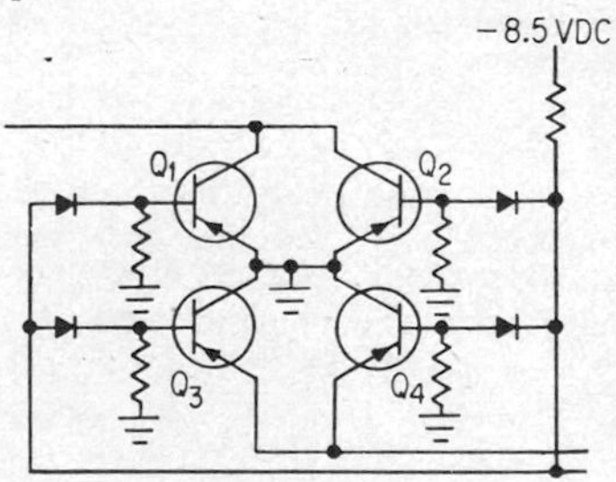

Quad configuration for four-transistor power oscillator.

quadded cable Cable in which some or all of the conductors are quads.

quadding Connecting transistors in a quad configuration, to obtain maximum reliability and efficiency.

quad in-line [abbreviated QUIL] An integrated-circuit package that has two rows of staggered pins on each side, spaced closely enough together to permit 48 or more pins per package. Used chiefly for large-scale integration.

quadraflop A four-state logic circuit that has complementary outputs. It appears as four flip-flops, with one holding a 1 and the other three holding a 0. Setting any flip-flop to 1 clears the other three.

quadrant 1. A 90° sector of a circle. 2. The fourth part of something, as one of the quadrants in a four-course radio range.

quadrantal error The angular error in a measured bearing that is due to the presence of metal in the vicinity of the direction-finding antenna, such as the metal structure and engines of an airplane or the hull of a ship.

quadrantal heading A heading to the northeast, southeast, southwest, or northwest.

quadrant electrometer An electrometer for measuring voltages and charges by means of electrostatic forces between a suspended metal plate and a surrounding metal cylinder that is divided into four insulated parts connected oppositely in pairs. The voltage to be measured is applied between the two pairs of quadrants.

quadraphonic sound system A four-channel sound system that uses speakers which are normally placed in the four corners of a room. The front pair provides stereophonic sound; the rear pair provides the sound that would normally be heard after reflection from the walls of the auditorium or other large room in which the recording was made. Listeners within the area of the four speakers then enjoy reproduction approximating that of a seat in the original concert hall. In a discrete quadraphonic system such as the CD-4, the four channels are kept acoustically separate throughout the recording and playback processes. In the QS or SQ matrix system, the four microphone channels are converted into two channels by a coding process before recording or broadcasting and decoded back into four channels for playback. In a derived quadraphonic system, the four channels are synthesized from a conventional two-channel stereo source by an adapter. Also called four-channel sound system.

quadrature Separated in phase by 90° or one quarter-cycle. Also called phase quadrature.

quadrature amplifier An amplifier that shifts the phase of a signal 90°. Used in a color television receiver to amplify the 3.58-MHz chrominance subcarrier and shift its phase 90° for use in the Q demodulator.

quadrature amplitude modulation [abbreviated QAM] Quadrature modulation in which some form of amplitude modulation is used for both digital inputs.

quadrature component The reactive component of a current or voltage, due to inductive or capacitive reactance in a circuit.

quadrature filter A filter that eliminates the quadrature components of signals in systems where information is contained in the quadrature modulation components of the carrier.

quadrature-grid FM detector An FM detector in which the sound IF signal voltage developed on the suppressor grid of a pentode lags the controlled-grid voltage by 90° (quadrature) when the incoming signal is exactly 4.5 MHz. Input frequency swings change the phase of the suppressor-grid signal with respect to that on the controlled grid, thereby changing the suppressor-grid circuit impedance and making the output voltage in the anode circuit vary with input frequency swings.

quadrature modulation Modulation in which two carrier components, differing in phase by 90°, are each modulated by a different signal.

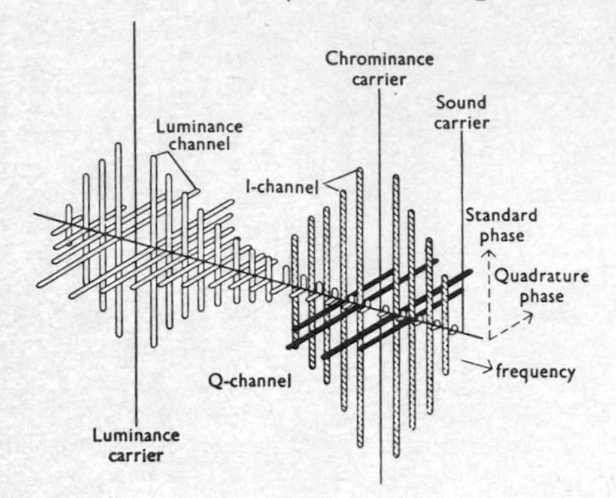

Quadrature-phase subcarrier signal and other parts of NTSC color television system.

quadrature-phase subcarrier signal The portion of the chrominance signal that leads or lags the in-phase portion by 90°.

quadricorrelator [QUADRature Information CORRELATOR] A circuit sometimes added to the automatic phase control loop in a color television receiver to obtain improved performance under severe interference conditions.

quadruple diversity Simultaneous combining of four received signals by using space, frequency, or other diversity reception techniques.

quadruplex circuit A telegraph circuit that carries two messages in each direction at the same time.

quadrupole A four-pole magnet that focuses a particle beam without bending it. Used in some alternating-gradient synchrotrons and proton linear accelerators.

quadrupole amplifier A low-noise parametric amplifier that consists of an electron-beam tube in which quadrupole fields act on the fast cyclotron wave of the electron beam to produce high amplification at frequencies in the range of 400 to 800 MHz. The cyclotron frequency is approximately

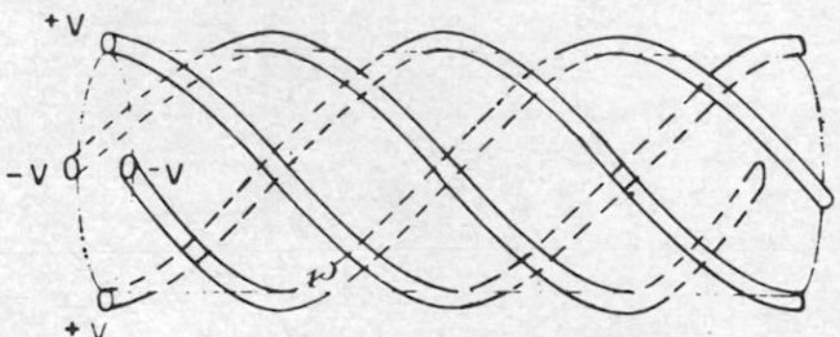

Quadrupole amplifiers may use four-wire twisted quadrupole structure to apply transverse DC electric fields to electron beam.

equal to the frequency of the signal to be amplified, and the pumping frequency is about twice this value.

quadrupole moment A term used in specifying mathematically the field caused by a given distribution of electric or magnetic charges.

Quail An air-launched decoy missile carried internally in the B-52 and used to degrade the effectiveness of enemy radar and missiles.

quality factor *Q.*

quality-factor meter *Q meter.*

quantity A positive or negative real number, used for numerical data.

quantity of electricity 1. The amount of electric charge stored in a capacitor, measured in coulombs or similar units. 2. The amount of current flowing through a circuit in a given time, measured in coulombs. One coulomb is 1 A flowing for 1 s.

quantity of radiation The total radiated energy passing through a unit area per unit of time. Expressed in ergs per square centimeter or watt-seconds per square centimeter.

quantization Division of the range of values of a wave into a finite number of smaller subranges, each subrange represented by an assigned or quantized value.

quantization distortion Inherent distortion introduced in the process of quantization. Also called quantization noise.

quantization level One of the subrange values obtained by quantization.

quantization noise *Quantization distortion.*

quantize To restrict a variable to a discrete number of possible values. Thus the age of a person is usually quantized as a whole number of years.

quantized field theory A theory in which electromagnetic fields and the fields of matter are represented by mathematical operators that describe the elementary processes of creation and annihilation of particles or photons.

quantized pulse modulation Pulse modulation that involves quantization, such as pulse-numbers and pulse-code modulation.

quantized system A system of particles whose energies can have only discrete values.

quantizer A device that measures the magnitude of a time-varying quantity in multiples of some fixed unit or quantum, at a specified instant or specified repetition rate, and delivers a proportional response which is usually in pulse code or digital form. The amplitude of the response signal is at each instant proportional to the number of quanta measured. The action of a quantizer is essentially the same as that of an analog-to-digital converter.

quantizing The process of representing any value between certain limits by the nearest of a limited number of values selected to cover the range. Used in pulse-code modulation.

quantizing encoder An encoder that converts

voltages to digital form, in which the voltage corresponding to the contents of a register is obtained from the output of a digital-voltage decoder. The error between this voltage and the input voltage is quantized to the nearest power of 2 and subtracted from the contents of the register. Each time this process is repeated, the error is diminished.

quantum [plural quanta] The smallest quantity of energy that can be associated with a given phenomenon. The quantum of electromagnetic radiation is the photon.

quantum efficiency The average number of electrons photoelectrically emitted from a photocathode per incident photon of a given wavelength in a phototube.

quantum electronics The branch of electronics associated with the various energy states of matter, such as the motions within atoms or groups of atoms and various phenomena in crystals. Examples of practical applications include the atomic hydrogen maser and the cesium atomic-beam resonator.

quantum mechanics The study of atomic structure and related problems in terms of quantities that can actually be measured.

quantum number A number assigned to one of the various values of a quantized quantity in its discrete range. As an example, the principal quantum number of an electron determines the energy level with respect to the minimum-energy level or ground state that has a quantum number of 1.

quantum state One of the states in which an atom may exist permanently or momentarily.

quantum theory A theory that atoms or molecules emit or absorb energy by a process which takes place in a series of steps, each step being the emission or absorption of an amount of energy called the quantum. For light or other radiation the quantum is the photon, the energy of which is equal to the frequency of the radiation in hertz multiplied by Planck's constant, which is 6.626×10^{-27} erg·s.

quantum voltage The voltage through which an electron must be accelerated to acquire the energy corresponding to a particular quantum.

quantum yield The number of photon-induced reactions of a specified type per photon absorbed.

quark A hypothetical particle that is believed to be a constituent of known elementary particles.

quarter-phase *Two-phase.*

quarter-square multiplier *Four-quadrant multiplier.*

quarter-wave Having an electrical length of one quarter-wavelength.

quarter-wave antenna An antenna whose electrical length is equal to one quarter-wavelength of the signal to be transmitted or received.

quarter-wave attenuator An arrangement of two wire gratings spaced an odd number of quarter-wavelengths apart in a waveguide, used to attenuate waves traveling through in one direction. The wave reflected from the first grating is canceled by that reflected from the second, so all energy reaching the attenuator is either transmitted through the gratings or absorbed by them, with no resultant reflection.

quarter-wavelength The distance that corresponds to an electrical length of a quarter of a wavelength at the operating frequency of a transmission line, antenna element, or other device.

quarter-wave line *Quarter-wave stub.*

quarter-wave plate A plate of mica or other doubly refracting crystal material of such thickness as to introduce a phase difference of one quarter-cycle between the ordinary and extraordinary components of light passing through.

quarter-wave stub A section of transmission line that is one quarter-wavelength long at the fundamental frequency being transmitted. When shorted at the far end, it has a high impedance at the fundamental frequency and all odd harmonics, and a low impedance for all even harmonics. Also called quarter-wave line and quarter-wave transmission line.

quarter-wave termination A nonreflecting waveguide termination that consists of an energy-absorbing wire grating or semiconducting film stretched across the waveguide one quarter-wavelength from a metal-plate termination. The wave reflected by the grating is canceled by the wave reflected from the plate.

quarter-wave transformer A section of transmission line approximately one quarter-wavelength long, used for matching a transmission line to an antenna or load.

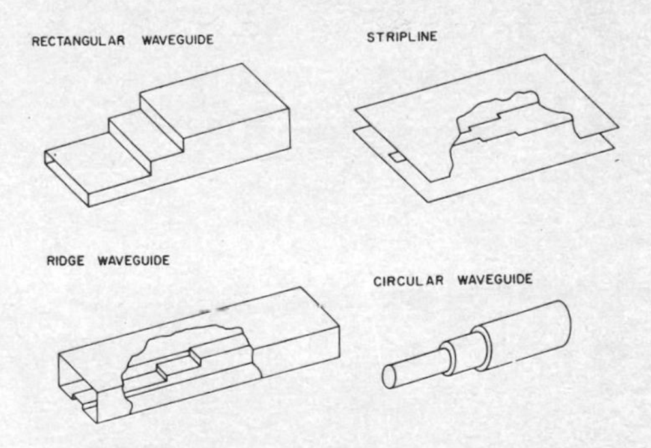

Quarter-wave transformers. Upper two are homogeneous, and lower two are nonhomogeneous.

quarter-wave transmission line *Quarter-wave stub.*

quartz A natural or artificially grown piezoelectric crystal composed of silicon dioxide, from which thin slabs or plates are carefully cut and ground to serve as a crystal plate for controlling the frequency of an oscillator.

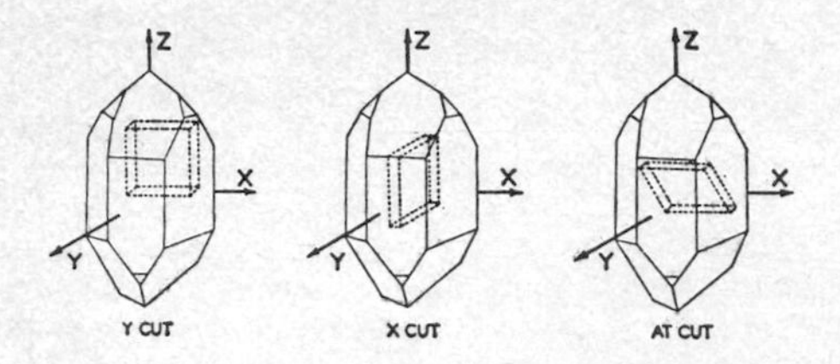

Quartz in natural form, as cut for three types of crystal plates.

quartz delay line An acoustic delay line in which quartz is used as the medium of sound transmission.

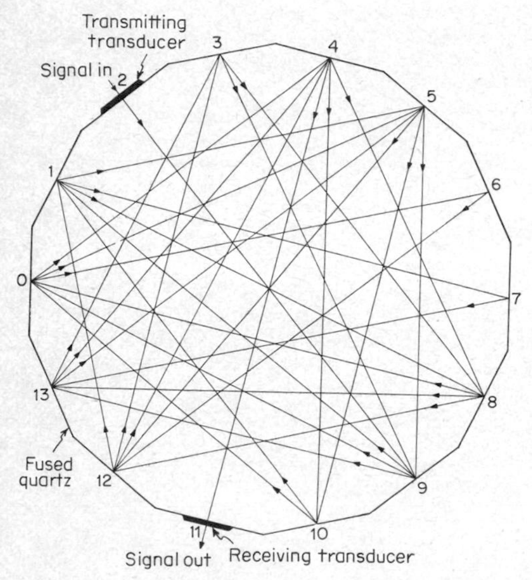

Quartz delay line providing 31 passes of transverse waves at speed of 12,300 ft/s (3750 m/s), or about 11 times the speed of sound in air.

quartz-fiber electroscope An electroscope in which a gold-plated quartz fiber serves the same function as the gold leaf of a conventional electroscope.

quartz lamp A mercury-vapor lamp that has a transparent envelope made from quartz instead of glass. Quartz resists heat, permitting higher currents, and passes ultraviolet rays that are absorbed by ordinary glass.

quartz plate *Crystal plate.*

quartz pressure gage A pressure gage that uses a highly stable quartz crystal resonator whose frequency changes directly with applied pressure.

quartz thermometer A thermometer based on the sensitivity of the resonant frequency of a quartz crystal to changes in temperature.

quartz watch An electronic watch in which the time is controlled by a quartz crystal operating in a battery-powered oscillator circuit. The time may be indicated conventionally by hands or by a digital display.

quasar A quasi-stellar source of radio signals near the limit of the observable universe, receding further into space at about half the speed of light and radiating energy beyond anything explainable by nuclear or other known phenomena.

quasi-active homing guidance Homing guidance in which the missile contains only the transmitter that illuminates the target. The receiver of the reflected energy is at some point external to the missile. Control signals must then be transmitted to the missile.

quasi-bistable circuit An astable circuit that is triggered at a rate which is high compared to its own natural frequency.

quasi-conductor A conductor that has a Q much less than unity.

quasi-dielectric A dielectric that has a Q greater than unity.

quasi-ferroelectric ceramic A ceramic whose behavior differs from that of true ferroelectric and dielectric materials. Examples include certain compositions of lead lanthanum zirconate titanate [PLZT].

quasi-linear feedback control system A feedback control system in which the relationships between the input and output signals are substantially linear despite the existence of nonlinear elements.

quasi-monostable circuit A monostable circuit that is triggered at a rate which is high compared to its own natural frequency.

quasi-optical Having properties similar to light waves, such as being limited to line-of-sight ranges.

quasi-passive satellite A passive satellite that has special reflector-type antennas which provide signal enhancement by concentrating the received signal and reflecting it back to the receiving station in a narrow beam.

quasi-single-sideband Transmitting parts of both sidebands to simulate single-sideband transmission.

quaternary phase-shift keying [abbreviated QPSK] Modulation of a microwave carrier with two parallel streams of nonreturn-to-zero data in such a way that the data is transmitted as 90° phase shifts of the carrier. This gives twice the message channel capacity of binary phase-shift keying in the same bandwidth.

quaternary signaling Signaling in which information is represented by presence and absence or plus and minus variations of four discrete levels of one parameter of the signaling medium.

quench 1. To stop abruptly. 2. To cool suddenly.

quenched-domain Gunn diode A Gunn diode in which the frequency of oscillation is increased by using a circuit that quenches the domain before it reaches the anode. This starts a new operating cycle earlier than for a free-running Gunn diode.

quenched-domain mode One of the three operating modes of a transferred-electron diode, in which the formation and extinction of space-charge domains are controlled by the surrounding circuit. The other two modes are the LSA mode and the transit-time mode.

quenched spark gap A spark gap that has provisions for rapid deionization. One form consists of many small gaps between electrodes that have relatively large mass and are good radiators of heat. The electrodes cool the gaps rapidly and thereby stop conduction.

quenched spark-gap converter A spark-gap generator that uses the oscillatory discharge of a capacitor through a coil and a quenched spark gap as a source of RF power. The spark gap usually consists of closely spaced gaps operating in series, to give good quenching action. Used in some industrial induction heating applications.

quenching 1. The process of terminating a discharge in a gas-filled radiation-counter tube by inhibiting reignition. 2. Cooling suddenly, as in heat-treating metals.

quenching circuit A circuit that diminishes, suppresses, or reverses the voltage applied to a counter tube to inhibit multiple discharges from an ionizing event.

quenching frequency The frequency at which the oscillations in a superregenerative receiver are suppressed or quenched.

queue A waiting line.

queue discipline A set of rules for selecting the next customer to receive service in a time-sharing computer system.

quick-break fuse A fuse that draws out the arc and breaks the circuit rapidly when the fuse wire melts, generally by separating the broken ends with a spring.

quick-break switch A switch that breaks a circuit rapidly, independently of the rate at which the switch handle is moved, to minimize arcing.

quiescent Without an input signal.

quiescent-carrier modulation A system of modulation in which the carrier is suppressed during intervals when no modulation is applied.

quiescent-carrier telephony A radiotelephone system in which the carrier is suppressed when there are no voice signals.

quiescent current The electrode current that corresponds to the electrode bias voltage.

quiescent push-pull amplifier A push-pull amplifier in which the control grids are biased so negatively that little anode current flows when there is no signal. There is thus no noise when tuning between stations, but tone quality is poor for weak signals.

quiescent value The voltage or current value for an electron-tube electrode when no signals are present.

quiet automatic volume control *Delayed automatic gain control.*

quieting sensitivity The least signal input for which the output signal-to-noise ratio does not exceed a specified limit in FM receivers.

quiet tuning A tuning circuit that silences the output of a radio receiver until it is accurately tuned to an incoming carrier wave.

QUIL Abbreviation for *quad in-line.*

quinary code A code based on five possible combinations for representing digits. An example is biquinary notation.

Q value *Disintegration energy.*

R

R 1. Abbreviation for *resistor*. 2. Abbreviation for *Rankine*. 3. Abbreviation for *roentgen*. 4. Abbreviation for *resistance*.

R Mathematical symbol for *resistance*.

°R Abbreviation for degree *Rankine*.

RABAL [RAdiosonde BALloon] 1. A system that involves use of a radiosonde balloon to determine atmospheric conditions at various altitudes. 2. The report obtained by radio from a radiosonde balloon.

rabbit ears A V-shaped indoor dipole television antenna whose arms are adjustable in angle and usually in length. The antenna may be attached to the television set or mounted on a base for use on top of the set.

race A transient condition in which two or more memory elements are changing state simultaneously in an asynchronous computer circuit.

race condition An ambiguous condition occurring in control counters when one flip-flop changes to its next state before a second one has had sufficient time to latch.

race track An assembly of several calutron isotope separators in the shape of a race track; it has a common magnetic field. Also called track.

raceway A channel used to hold and protect wires, cables, or busbars.

rack 1. A standardized steel cabinet that holds 19-in (48.26-cm) panels of various heights; mounted on the panels are radio receivers, amplifiers, and other units of electronic equipment. Mounting holes for the panels, usually drilled and threaded for 10-32 machine screws, are spaced apart in multiples of ½ and ⅝ in (1.27 and 1.5875 cm) to accommodate notch spacings of correspondingly standardized panels. Originally designed to hold relay panels in telephone exchanges. Also called relay rack. 2. A straight bar that has gearlike teeth designed to engage with a drive gear or pinion for producing straight-line motion.

rack-and-panel construction Construction of equipment in such a way that all chassis units and panels can be mounted on a standard rack.

rack panel A panel designed for mounting on a rack.

rad 1. [abbreviated rd] The standard unit of absorbed dose in the field of radiation dosimetry, equal to 100 ergs/g. 2. Abbreviation for *radian*.

RADAN [RAdar Doppler Automatic Navigator] A Doppler radar navigation system for aircraft, operating independently of ground-based stations.

radar [RAdio Detecting And Ranging] 1. A system that uses beamed and reflected RF energy for detecting and locating objects, measuring distance or altitude, navigating, homing, bombing, and other purposes. In detecting and ranging, the time interval between transmission of the energy and reception of the reflected energy establishes the range of an object in the beam's path. In primary radar, the return signal is produced by reflection of the transmitted energy from the target. In secondary radar, the transmitted energy triggers a responder beacon that sends an entirely new signal back to the radar set. Originally called radiolocation in Great Britain. 2. *Radar set*.

radar aid *Radar navigation aid*.

radar altimeter A high-altitude radio altimeter used in aircraft to give accurate absolute altitude indications at altitudes far above the common 5000-ft (1,524-m) limit of FM radio altimeters. Simple pulse-type radar equipment is used to send a pulse straight down and measure its total time of travel to the surface and back to the aircraft. Lowest useful altitude is about 250 ft (76.2 m). Also called high-altitude radio altimeter and pulse-type altimeter.

radar altitude Absolute altitude as determined by a radar altimeter. Also called radio altitude.

radar astronomy The study of astronomical bodies and the earth's atmosphere by radar pulse techniques, including tracking of meteors and the reflection of radar pulses from the moon and the planets.

radar band One of the frequency ranges used in radar. Early designations for these bands were P band—225 to 390 MHz; L band—390 to 1550

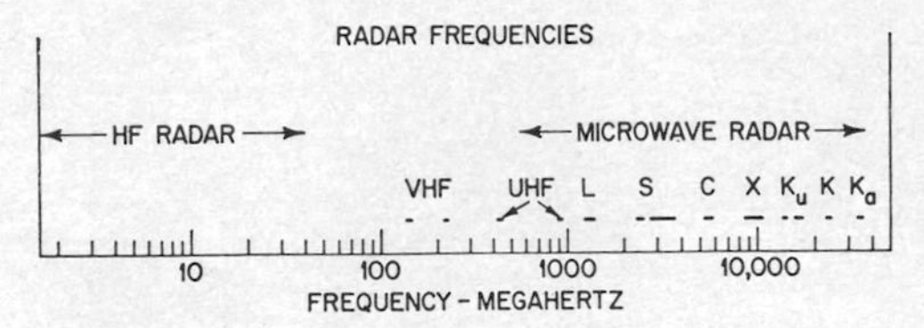

Radar bands.

MHz; S band (10-cm band)—1.55 to 5.2 GHz; X band (3-cm band)—5.2 to 10.9 GHz; K band (1-cm band)—10.9 to 36 GHz.

radar beacon A radar receiver-transmitter that transmits a strong coded radar signal whenever its radar receiver is triggered by an interrogating radar on an aircraft or ship. The coded beacon reply can be used by navigators to determine their own positions in terms of bearing and range from the beacon.

radar beam The movable beam of RF energy produced by a radar transmitting antenna. Its shape is commonly defined as the loci of all points at which the power has decreased to one-half that at the center of the beam.

radar bombardier A person trained in radar bombing.

radar bombing Bombing in which radar is used to locate the target or aiming point, to aid in positioning the bombing aircraft at the proper release point for bombing, and/or to release bombs automatically, especially under conditions of poor visibility.

radar bomb scoring A scoring system in which ground-based radar and plotting equipment are used in determining the theoretical points of bomb impact during a simulated bombing mission. Information obtained from the ground radar is combined with ballistic data and information transmitted from the aircraft regarding the wind, true heading, and true airspeed at the time of release.

radar bombsight An airborne radar set used to sight the target, solve the bombing problem, and release the bombs.

radar boresight target A target located by a survey at a known azimuth, elevation, and distance from a radar. Used for collimation and orientation of the radar antenna system.

radar calibration The process of determining the extent and accuracy of the radar coverage of a given aircraft-warning or tactical air control radar installation. Calibration includes computing the theoretical coverage, making flights to check coverage, and preparing calibration charts, diagrams, and overlays.

radar camera A special manual or automatic camera used to photograph images on a radarscope.

radar camouflage The use of special coverings or surfaces that reduce the reflection of RF energy back to a radar set, to minimize chances for detection of the object by an enemy radar set.

radar cell The volume in space that extends outward for one radar pulse length from the transmitter, with a cross section corresponding to the angular width of the radar beam as defined by points of half-power intensity.

radar chart A special map used in radar navigation. For radar-equipped ships, radar charts emphasize the coastline, hills, large buildings, and other objects that give prominent radar echoes to ships well off shore. For air navigation, radar charts similarly show outlines of cities, rivers, lakes, bridges, railroads, and other objects that appear on airborne radar screens.

radar check point A point on the surface of the earth that gives an outstanding return on an airborne radar screen for use in air navigation.

radar chronograph A radar system for measuring projectile velocities.

radar clutter *Clutter.*

radar command guidance A missile guidance system in which radar equipment at the launching site determines the positions of both target and missile continuously, computes the missile course corrections required, and transmits these by radio to the missile as commands.

radar contact Recognition and identification of an echo on a radar screen. An aircraft is said to be on radar contact when its radar echo can be seen and identified on a PPI display.

radar control Control of an aircraft, guided missile, or gun battery by radar.

radar control area The area or airspace in which radar control of aircraft or guided missiles is exercised.

radar controller 1. A person who exercises radar control over aircraft or guided missiles. 2. A radar device used in radar control.

radar countermeasure [abbreviated RCM] An electronic countermeasure used against enemy radar, such as jamming and the use of confusion reflectors.

radar coverage indicator A device that gives the maximum range at which a given aircraft can normally be tracked by a radar station, taking into account the aircraft decibel rating, the altitude of the aircraft, and characteristics of the radar.

radar cross section The area intercepting the power which, if scattered equally in all directions, would produce a received radar echo equal to that actually obtained from the target. Also called echo area.

radar data display board A board used for displaying information derived from a radar set or associated equipment.

radar display The pattern that represents the output data of a radar set, generally produced on the screen of a cathode-ray tube.

radar drift The drift of an aircraft as determined by a timed series of bearings taken on a fixed radar target.

radar echo *Echo.*

radar equation An equation that relates the transmitted and received powers and antenna gains of a primary radar system to the echo area and distance of the radar target.

radar fence A network of radar warning stations maintained as a barrier against surprise attack.

radar fire control Fire control by radar.

radar fix A determination of position by means of radar.

radar gun-layer A radar device that tracks a target and aims a gun automatically.

radar handoff Transferring radar identification and control over an aircraft from one controller to another without interrupting surveillance.

radar homing 1. Homing on the source of a radar beam. 2. Homing in which a missile-borne radar locks onto a target and guides the missile to that target.

radar homing set A complete radar set used in a missile or other vehicle to provide self-guidance to a target.

radar horizon The lowest elevation angle at which a radar can operate effectively at a particular location, taking into account the terrain in the vicinity and the curvature of the earth.

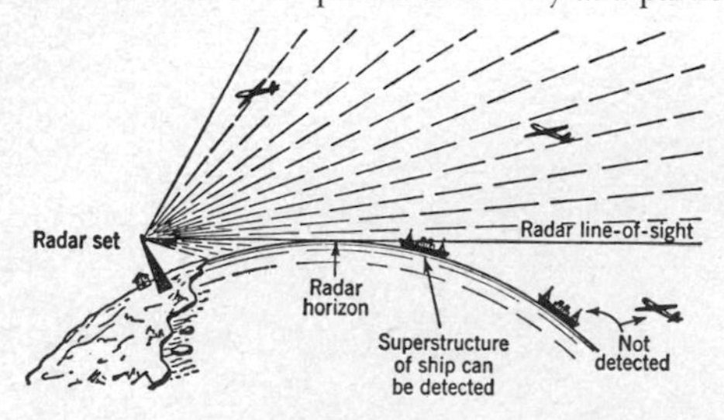

Radar horizon.

radar illumination Illumination of a target by a high-power radar external to a missile, to produce echo signals suitable for homing use by receiving equipment in the missile.

radar indicator A cathode-ray tube and associated equipment used to provide a visual indication of the echo signals picked up by a radar set.

radarman A person who operates a radar set and evaluates its indications.

radar mapping central A group of radar sets used for mapping, sometimes including facilities for automatic navigation of the mapping aircraft and retransmission of mapping data to a ground installation.

radar mile The time required for a radar pulse to travel to a target 1 statute mile (1.609 km) away and return, equal to 10.75 μs.

radar missile-tracking central A group of radar sets equipped with facilities for recording and/or indicating signals received from a transponder-equipped missile, to provide free-space position data of the missile.

radar mission An air mission that requires or uses radar.

radar modulator A modulator that varies the amplitude, phase, frequency, pulse repetition rate and/or pulse duration of a signal generated by a radar transmitter to which it is directly connected.

radar nautical mile The time interval of approximately 12.367 μs that is required for the RF energy of a radar pulse to travel 1 nautical mile (1.852 km) and return.

radar navigation Navigation by radar equipment and radar navigation aids.

radar navigation aid Any radar device designed as a navigation aid, such as a radar altimeter. Also called radar aid.

radar navigator A person trained in radar navigation.

radar net A network of radar installations set up to detect aircraft entering a defined airspace. Also called radar screen.

radar observer An aircraft observer trained in the use of airborne radar equipment for navigation, interception, search, or bombing.

radar operator 1. A person who operates a radar set. 2. An aircraft observer who operates radar equipment, such as a radar observer or radar bombardier.

radar paint A coating that absorbs radar waves.

radar performance figure The ratio of the pulse power of a radar transmitter to the power of the minimum signal detectable by the receiver.

radar picket A ship or aircraft equipped with early-warning radar and operating at a distance from the area being protected, to extend the range of radar detection.

radar pilotage Pilotage in which check points on the ground are viewed on a radarscope.

radar pilotage equipment Equipment that utilizes primary radar techniques, carried on a vehicle to determine bearing and distance of recognizable landmarks and indicate the relative positions of other vehicles.

radar plot A plot of the positions of aircraft or ships, made from data obtained by radar.

radar prediction A graphic representation of what may be expected to show on a radar screen when an actual radar scan is made.

radar prediction device A relief map of enemy territory, on which a tiny lamp is mounted at the location of an enemy radar set. The shadows cast by mountains or other features of the terrain then indicate blind spots in the enemy's radar detection system.

radar probing Obtaining data on distant objects in space by radar signals. Echoes have been received from signals sent to the planet Venus, 28,000,000 mi (45,000,000 km) away.

radar range The maximum distance at which a radar set is ordinarily effective in detecting objects. Usually assumed to be the distance at which a radar set can detect a specified object at least 50% of the time. The range in free space varies directly with receiver power sensitivity and target echo area, but it varies as the square of antenna gain and as the fourth power of transmitted power.

radar range marker A mark or line that is scribed or electronically formed on the face of a radarscope to indicate the range to the object detected. Also called range mark.

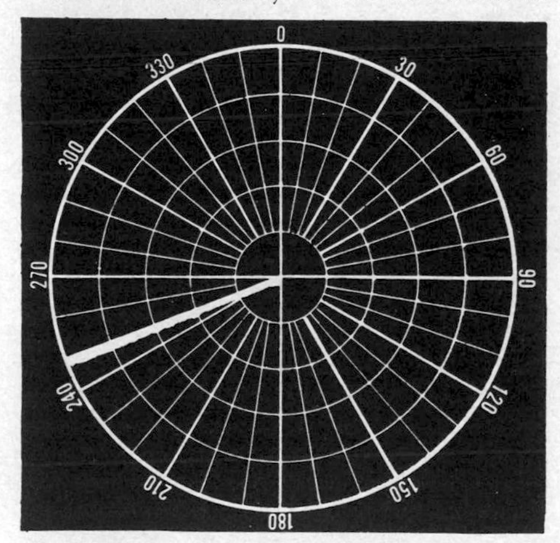

Radar range marker generates circles electronically on plan-position-indicator display, and bearing lines are usually scribed on transparent plastic sheet mounted over face of radarscope.

radar receiver A high-sensitivity radio receiver that amplifies and demodulates radar echo signals and feeds them to a radarscope or other indicator.

radar reconnaissance Use of radar to obtain information on location and strength of enemy forces and/or obtain terrain information.

radar reflector A reflector that reflects or deflects radar waves.

radar reflectoscope An arrangement of mirrors used to produce a composite image of a chart and a radarscope presentation.

radar relay Equipment used for relaying radar video and synchronizing signals to a remote location. Also called relay radar (deprecated).

radar repeat-back guidance A missile guidance system in which a search radar in the missile transmits information to the point of control.

radar repeater A cathode-ray indicator that reproduces the visible intelligence of a radar display at a remote position. When used with a selector switch, the visible intelligence of any one of several radar systems can be reproduced.

radar resolution The resolution of a radar set or system.

radar safety beacon An airborne radar beacon, used chiefly for identifying aircraft in connection with air traffic control at airports. Also called airborne beacon.

radar scan The circular, spiral, rectangular, or other motion of a radar beam as it searches for a target.

radarscope The cathode-ray oscilloscope or screen used as an indicator in a radar set. Also called scope.

radarscope display The visual presentation on a radar screen.

radar screen 1. A cathode-ray screen in a radar set. 2. *Radar net.*

radar set A complete assembly of radar equipment for detecting and ranging, consisting essentially of a transmitter, antenna, receiver, and indicator. Also called radar.

radar shadow A region shielded from radar illumination by an intervening reflecting or absorbing medium like a hill.

radar signal analysis Analysis of radar reflections from a ballistic missile or orbiting satellite to obtain as much information as possible about size, shape, and purpose.

radar signal simulator An instrument whose electric output can be applied directly to the indicator of a radar set to produce artificial radar echo indications.

radar silence A period of time during which radar transmission is stopped, generally for security reasons.

radarsonde 1. An electronic system for automatically measuring and transmitting high-altitude meteorological data from a balloon, kite, or rocket by pulse-modulated radio waves when triggered by a radar signal. 2. A system in which radar techniques are used to determine the range, elevation, and azimuth of a radar target carried aloft by a radiosonde.

radar station The ground, air, or sea location at which a radar set transmits or receives signals.

radar storm detection The detection of certain storms or stormy conditions by radar. Liquid or frozen water drops within the storm reflect radar echoes.

radar surveillance Use of one or more radar sets to locate distant targets and provide approach information. Surveillance may include facilities for identifying a target as friend or foe and means for relaying data to appropriate information and/or control centers.

radar surveying Surveying in which airborne radar is used to measure accurately the distance between two ground radio beacons positioned

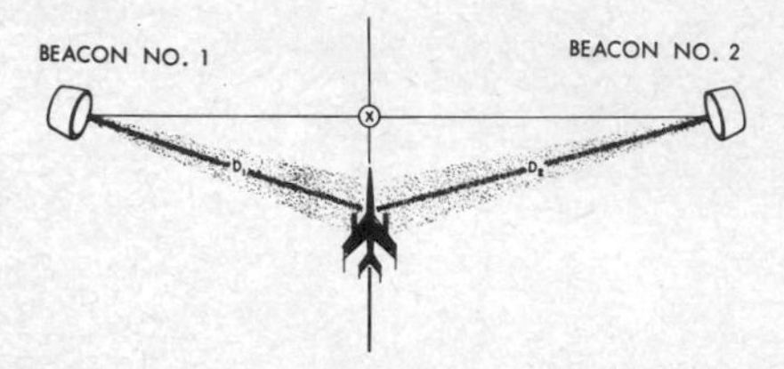

Radar surveying. When radar-measured distances D1 and D2 are minimum, aircraft is at point X. Distance between beacons can then be calculated.

along a baseline. This eliminates the need for measuring distance along the baseline in inaccessible or extremely rough terrain.

radar synchronous bombing *Synchronous radar bombing.*

radar target An object being tracked or watched on a radar screen.

radar telescope A large radar antenna and associated equipment used for radar astronomy. At Arecibo, Puerto Rico, the antenna reflector for the telescope is 1000 ft (305 m) in diameter.

radar theodolite A theodolite that uses radar to obtain azimuth, elevation, and slant range to a reflecting target, for surveying or other purposes.

radar track command guidance Command guidance that uses two radars external to a missile, one for tracking the target and the other for tracking the missile. The radar receiver outputs are fed to a computer, and the output of the computer is in turn fed into a data transmitter that transmits flight information to the missile for correcting its flight path to the target.

radar tracking Tracking a moving object by radar.

radar trainer A trainer for teaching radar techniques and operations by simulating various radar target displays.

radar transmitter The transmitter portion of a radar set.

radar video data processor A radar system that provides uncluttered displays for air traffic control by converting the input signals into digital data that can be synchronized with a system range clock. Associated circuits permit collecting and displaying weak target signals surrounded by noise.

radar warning net A radar net used for warning.

radar wave A transmitted or reflected radio wave used in radar.

radiac [RAdioactivity Detection, Identification, And Computation] 1. Detection, identification, and measurement of the intensity of nuclear radiation in an area. 2. *Radiac set.*

radiac computer A computer that scales, integrates, or counts information received from a radiac detector.

radiac data transmitting set The equipment required to detect radioactivity and transmit radioactivity data as modulation on a carrier.

radiac detector A detector that is sensitive to radioactivity or free nuclear particles and produces a reaction which can be interpreted or measured by other components.

radiac-detector charger An electrostatic generator that provides an electrostatic charge for a radiac detector.

radiac instrument *Radiac set.*

radiacmeter *Radiac set.*

radiac set A complete radioactivity detecting, identifying, and measuring system. Also called radiac, radiac instrument, and radiacmeter.

radial One of a number of radial lines of position defined by an azimuthal radio navigation facility, and identified in terms of the bearing (usually magnetic) of all points on that line from the facility.

radial-beam tube A vacuum tube in which a radial beam of electrons is rotated past circumferentially arranged anodes by an external rotating magnetic field. Used chiefly as a high-speed switching tube or commutator.

radial component A component that acts along a radius, as contrasted to a tangential component, which acts at right angles to a radius.

radial field A field of force directed toward or away from a point in space.

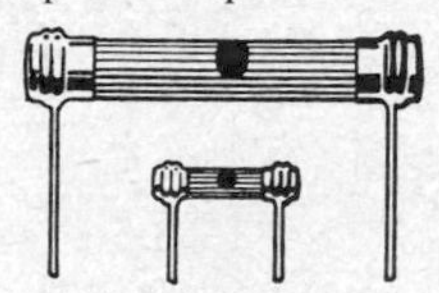

Radial leads on resistors.

radial lead A wire lead that comes from the side of a component rather than axially from the end.

radial transmission line A pair of parallel conducting planes used for propagating uniform circularly cylindrical waves that have their axes normal to the planes.

radian [abbreviated rad] The angle that intercepts an arc whose length is equal to its radius. A complete circle contains 2π rad. One radian is 57.29579°, and 1° is 0.01745 rad.

radiance The radiant flux per unit solid angle, per unit of projected area of the source. The usual unit is the watt per steradian per square meter. This is the radiant analog of luminance.

radianlength The distance between points in a sinusoidal wave that differ in phase by an angle of 1 radian. One radianlength is equal to the wavelength divided by 2π.

radian per second A unit of angular velocity.

radiansphere The boundary between the near

and far fields of a small antenna. It is a spherical surface whose radius is equal to the wavelength divided by 6.28.

radiant Emitted or transmitted along radii, as from a point source.

radiant energy Energy transmitted in the form of electromagnetic radiation, such as radio waves, heat waves, and light waves.

radiant flux The time rate of flow of radiant energy.

radiant flux density The amount of radiant power per unit area that flows across or onto a surface. Also called irradiance.

radiant gain The ratio of emitted radiant flux to incident radiant flux at specified ports in an optoelectronic device.

radiant intensity The energy emitted per unit time, per unit solid angle about the direction considered. Usually expressed in watts per steradian.

radiant sensitivity The signal output current of a camera tube or phototube divided by the incident radiant flux at a given wavelength.

radiate To send out energy, such as electromagnetic waves, into space.

radiated interference Electromagnetic interference caused by radiated noise and other undesired signals from power lines or energized electric equipment.

radiated noise Electromagnetic energy that produces undesired noise in receiving equipment.

radiated power The total power emitted by a transmitting antenna.

radiating circuit A circuit capable of sending electromagnetic waves into space, such as the antenna circuit of a radio transmitter.

radiating element A basic subdivision of an antenna that in itself is capable of radiating or receiving RF energy.

radiating guide A waveguide that radiates energy into free space through slots, gaps, or horns.

radiating microsphere A tiny ceramic particle, about the diameter of a human hair, that permanently isolates a radioisotope physically and chemically while allowing useful radiation to escape. When swallowed for medical irradiation or tracer purposes, the microspheres pass through the body intact.

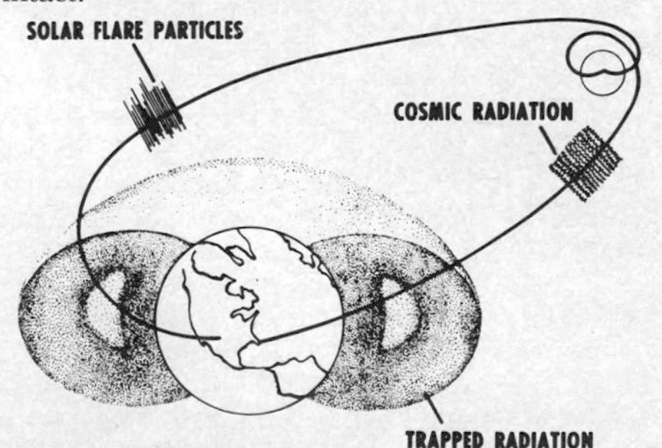

Radiation encountered on lunar landing route.

radiation 1. A stream of high-energy particles from a cyclotron or other accelerator. 2. Electromagnetic energy, such as light, sound, and radio waves, and x-rays and heat rays, traveling through a material or through space. 3. *Nuclear radiation.*

radiation absorber An insulating material in sheet form; it has a conductive backing and is used as dielectric and reflecting elements for absorbing unwanted RF energy.

radiation belt *Van Allen belt.*

radiation burn A burn caused by overexposure to radiant energy.

radiation counter An instrument that detects or measures nuclear radiation by counting the resultant ionizing events. Examples include Geiger and scintillation counters. Also called counter.

radiation-counter tube *Counter tube.*

radiation damage The effect of gamma rays, fission fragments, and neutrons on substances.

radiation danger zone A zone within which the maximum permissible constant dose rate is exceeded.

radiation detector A device for converting radiant energy to a form more suitable for observation. Used in radiac sets.

radiation dose The total amount of ionizing radiation absorbed by material or tissues. Commonly expressed in rads.

radiation efficiency The ratio of the power radiated to the total power supplied to an antenna at a given frequency.

radiation excitation *Radiation ionization.*

radiation field The electromagnetic field that breaks away from a transmitting antenna and radiates outward into space as electromagnetic waves. The other type of electromagnetic field associated with an energized antenna is the induction field.

radiation hardening Improving the ability of a device or piece of equipment to withstand nuclear or other radiation. The techniques apply chiefly to dielectric and semiconductor materials.

radiation hazard A health hazard that arises from exposure to ionizing radiation.

radiation hazards meter An instrument that detects and measures electromagnetic radiation at power density levels which are potentially hazardous to human life processes, such as those levels produced by high-energy radio sources used in transmitters, industrial processing equipment, and microwave ovens.

radiation intensity The power radiated from an antenna per unit solid angle in a given direction.

radiation ionization Ionization of the atoms or molecules of a gas or vapor by electromagnetic radiation. Also called radiation excitation.

radiation length The mean path length required to reduce the energy of relativistic charged particles by the factor $1/e$ or 0.368 as they pass through matter. The radiation length for relativistic electrons in air is 0.5 cm in lead.

radiation lobe The portion of a radiation pattern that is bounded by one or two cones of nulls.

radiation loss The portion of the transmission loss that is due to radiation of RF power from a transmission system.

radiation monitor A radiation detector that measures continuously the level of ionizing radiation and sometimes actuates an alarm when a preset danger level is exceeded.

radiation pattern A graphical representation of the radiation of an antenna as a function of direction. Also called field pattern.

radiation potential The voltage corresponding to the energy in electronvolts required to excite an atom or molecule and cause emission of one of its characteristic radiation frequencies.

radiation pressure The extremely small pressure exerted on a surface by electromagnetic radiation, or the larger pressure exerted on a surface or interface by a sound wave.

radiation pyrometer A pyrometer in which the radiant power from the object or source to be measured is focused on a thermocouple, thermopile, bolometer, or other suitable detector that provides electric output for an indicating instrument. Also called radiation thermometer.

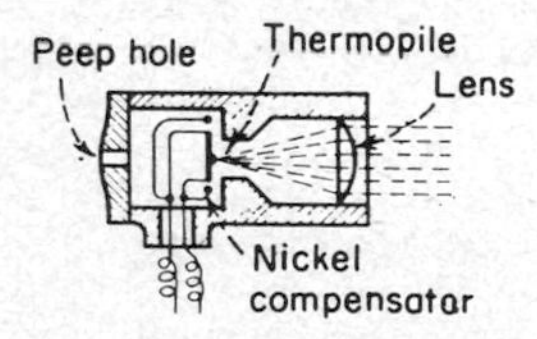

Radiation pyrometer.

radiation resistance The total radiated power of an antenna divided by the square of the effective antenna current measured at the point where power is supplied to the antenna.

radiation shield A shield or wall of lead or other material that effectively absorbs nuclear radiation.

radiation sickness The complex symptoms caused by overexposure to nuclear or other ionizing radiation, generally commencing a few hours after exposure, including nausea, vomiting, internal bleeding, and loss of white corpuscles.

radiation sterilization Use of radiation to make a plant or animal sterile (incapable of reproduction). Also used to kill all forms of life, particularly bacteria, in food, surgical sutures, space probes, and so on.

radiation therapy Treatment of disease with any type of radiation. Treatment of disease with ionizing radiation is called radiotherapy.

radiation thermometer *Radiation pyrometer.*

radiation warning symbol A standard symbol used on posters displayed in locations where radiation hazards exist. The symbol consists of a magenta trefoil printed on a yellow background.

Radiation warning symbol.

radiation window A window which is transparent to alpha, beta, gamma, and/or x-rays and protects the item that it covers from foreign matter.

radiative capture A nuclear capture process whose prompt result is the emission of electromagnetic radiation only.

radiator 1. The part of an antenna or transmission line that radiates electromagnetic waves either directly into space or against a reflector for focusing or directing. 2. A body that emits radiant energy.

radio 1. The transmission of signals through space by electromagnetic waves. Usually applied to the transmission of sound and code signals, although television and radar also depend on electromagnetic waves. 2. *Radio receiver.*

radio- 1. A prefix that denotes radioactivity or a relationship to it, as in radiocarbon. 2. A prefix that denotes the use of radiant energy, particularly radio waves.

radioacoustic position-finding A method of determining distances through water. A radio circuit is closed at the instant that a charge is exploded under water at one point. The times required for the radio signal to travel through air and the acoustic shock wave to travel through water to observing stations are measured, distances are computed from the time differences, and position is computed by triangulation.

radioacoustics The science of using sounds transmitted by radio for noncommunication purposes.

radioactinium [symbol RdAc] A thorium isotope in the actinium series, produced naturally by beta decay of actinium 227. It emits alpha particles that have a half-life of 18.8 days, and thereby changes to radium 223.

radioactive Pertaining to or exhibiting radioactivity. Also called active.

radioactive capture reaction A nuclear process in which a particle is captured and the excess energy is emitted as radiation.

radioactive contamination A condition in which radioactive material has spread to places where it may harm persons, spoil experiments, or make products or equipment unsuitable or unsafe for some specific use.

radioactive dating A technique for measuring

the age of an object or sample of material by determining the ratios of various radioisotopes in it. For example, the ratio of carbon 14 to carbon 12 reveals the approximate age of bones, pieces of wood, and other archeological specimens.

radioactive decay The spontaneous transformation of a nuclide into one or more different nuclides. The process involves (a) the emission

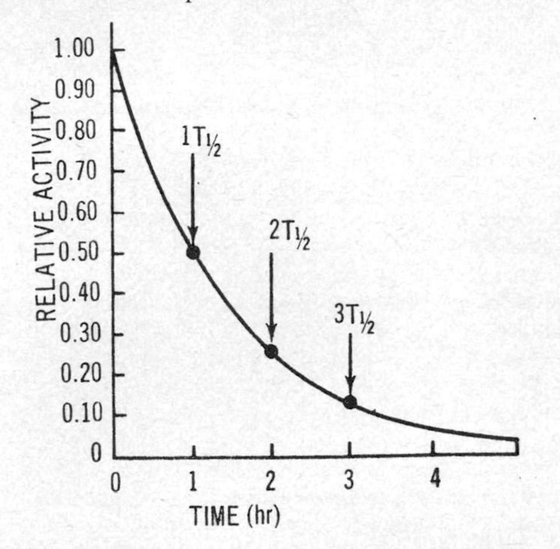

Radioactive-decay curve, showing exponential form. Rate of disintegration is proportional to number of atoms present at given time. Curve shown is for radioactive material that decays to half its original value (one half-life) in 1 h, and to half of new value (two half-lives) in next hour.

from the nucleus of alpha particles, electrons, positrons, and gamma rays, (b) the nuclear capture or ejection of orbital electrons, or (c) fission. The rate of radioactive decay is expressed in terms of the half-life of the nuclide. Also called decay.

radioactive displacement law A law governing radioactive transformations. When a nucleus emits an alpha particle, the new nucleus has an atomic number two less than the parent and a mass number of four less. When a nucleus emits a negative beta particle, the atomic number of the new nucleus is one greater than the parent and the mass number remains the same. The emission of a positron or the capture of an orbital electron decreases the atomic number by one without changing the mass number. Isomeric transition and gamma emission do not change atomic number or mass number. Also called displacement law.

radioactive element An element that disintegrates spontaneously, giving off various rays and particles. Examples include promethium, radium, thorium, and uranium.

radioactive emanation A radioactive gas given off by certain radioactive elements. Thus, radium gives off radon, thorium gives off thoron, and actinium gives off actinon.

radioactive equilibrium The condition in which the rate of decay of the atoms of a radioactive parent is equal to the rate of formation of the atoms of the radioactive descendant. This condition can exist only when no activity longer-lived than that of the parent is interposed in the decay chain.

radioactive half-life The time required for a particular radioisotope to decrease to half its initial value.

radioactive heat *Radiogenic heat.*

radioactive isotope *Radioisotope.*

radioactive material A material that has one or more constituents that exhibit significant radioactivity.

radioactive nuclide *Radionuclide.*

radioactive poisoning Medical term for illness caused by radioactive material in the human body.

radioactive series A succession of nuclides, each of which transforms by radioactive disintegration into the next until a stable nuclide results.

radioactive source Any quantity of radioactive material intended for use as a source of ionizing radiation.

radioactive standard A sample of radioactive material, usually with a long half-life, in which the number and type of radioactive atoms at a definite reference time is known. Used for calibrating radiation-measuring equipment. Also called reference source.

radioactive thickness gage An instrument that measures the thickness of the metal wall of a pipe or tank from one side, by directing a beam of

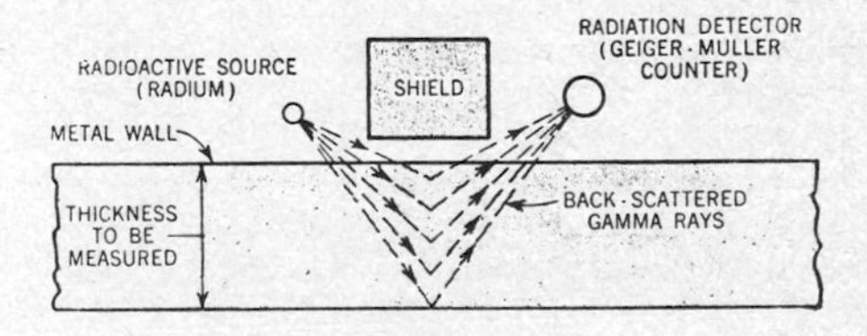

Radioactive thickness gage.

gamma rays through the wall at an angle and measuring the amount of backscattering with a radiation detector. Thicker walls give greater scattering and a correspondingly higher meter reading.

radioactive tracer A small quantity of radioisotope used to trace the progress of a biological, chemical, or other process. Also called radiotracer.

radioactivity Spontaneous nuclear disintegration, a property possessed by elements like radium, uranium, thorium, and their products. Alpha or beta particles and sometimes gamma rays are emitted by disintegration of the nuclei of atoms. Also called activity.

radioactivity log A borehole log of gamma, neutron, or other forms of radioactivity, used in prospecting for oil and minerals.

radio aid *Radio navigation aid.*

radio altimeter An absolute altimeter that depends on the reflection of radio waves from the earth for the determination of altitude, as in an FM radio altimeter and a radar altimeter. Also called electronic altimeter and reflection altimeter.

radio altitude *Radar altitude.*

radio astronomy The study of radio waves emitted by astronomical bodies.

radio aurora *Artificial radio aurora.*

radioautograph Deprecated term for *autoradiograph.*

radio-autopilot coupler A coupler that permits direct control of an automatic pilot by a radio navigation aid in an aircraft, to give automatic flight.

radio beacon A nondirectional radio transmitting station in a fixed geographic location; it emits a characteristic signal from which bearing information can be obtained by a radio direction finder on a ship or aircraft. Some types operate continuously. Other types transmit only in response to an interrogation signal and may also provide range information.

radio beam A concentrated stream of RF energy as used in radio ranges and microwave radio relays. A radar beam is a radio beam used for a particular application and in a particular manner.

radio bearing A bearing taken with respect to a radio transmitter, obtained with a radio direction finder.

radiobiology That branch of biology which deals with the effects of radiation on living tissue.

radio blackout *Radio fadeout.*

radio bomb fuze An electronic bomb fuze that is triggered by radio waves reflected from the target. It may be set to detonate at any desired distance from the target or ground by use of Doppler effect.

radio broadcast A program broadcast from a radio transmitter for general reception.

radio broadcasting Radio transmission intended for general reception.

radio broadcast station A station that transmits radio programs in the broadcast band, intended to be received by the general public.

radiocarbon Carbon 14, a weak radioisotope used in biological and agricultural tracer studies. Half-life is 5740 years.

radiocarbon age The age of a once-living material as calculated from the specific radioactivity of the carbon 14 remaining in it. Radiocarbon dating is possible because carbon 14 is produced in the atmosphere by cosmic rays and incorporated into all living objects. After death, the carbon-14 activity decays exponentially.

radiocesium Cesium 137, a radioisotope recovered from the waste of nuclear reactors. Useful for sterilizing food and as a substitute for radium in medical work. Half-life is 37 years.

radio channel A band of frequencies of a width sufficient to permit its use for radio communication. The width of a channel depends on the type of transmission and the tolerance for the frequency of emission.

radiochemistry Chemistry that involves the use of radionuclides.

radio circuit 1. An arrangement of parts and connecting wires for radio purposes. 2. *Radio communication circuit.*

radiocolloid A grouping of radioactive atoms to form colloidal aggregates.

radio command A radio control signal to which a guided missile or other remote-controlled vehicle or device responds.

radio common carrier The official designation for such miscellaneous radio services as paging and portable telephones.

radio communication Communication by radio waves, such as by radio facsimile, radiotelegraph, radiotelephone, and radioteletypewriter.

radio communication circuit A radio system for carrying out one communication at a time in either direction between two points. Also called radio circuit.

radio compass *Automatic direction finder.*

radio control The control of stationary or moving objects by signals transmitted through space by radio.

radio deception The use of radio to deceive the enemy, as by sending false dispatches or using enemy call signs.

radio detection The detection of the presence of an object by radiolocation without precise determination of its position.

radio determination The determination of position, or the obtaining of information relating to position, by the propagation properties of radio waves.

radio direction finder [abbreviated RDF] A radio aid to navigation that uses a rotatable loop or other highly directional antenna arrangement to

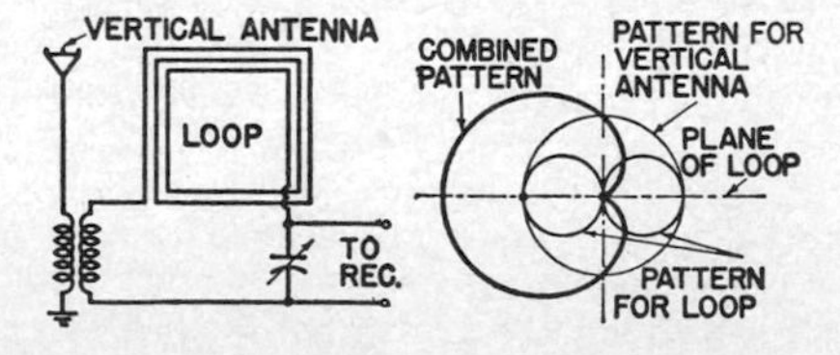

Radio direction finder antenna arrangement in which unilateral radiation pattern is obtained by combining vertical antenna with loop antenna.

determine the direction of arrival of a radio signal. Examples include aural-null and automatic direction finders. Also called direction finder.

radio distress signal The letters SOS transmitted without letter spaces in Morse code (· · · — — — · · ·) or the spoken word "mayday," transmitted on one of the international distress frequencies.

radioecology The study of the effects of radiation on animals and plants in their natural environment.

radioed Transmitted by radio.

radioelectrocardiograph A portable electrocardiograph combined with a radio transmitter. It is worn by a subject while the subject is living normally or exercising, for broadcasting the person's electrocardiographic signals to a remote receiver.

radioelement An element tagged with one or more radioisotopes.

radio engineering The field of engineering that deals with the generation, transmission, and reception of radio waves and with the design, manufacture, and testing of associated equipment.

radio facsimile Facsimile communication by radio.

radio fadeout A sudden and abnormal increase in ionization in the lower layers of the ionosphere, causing increased absorption of radio waves passing through these regions. Signals at receivers then fade out or disappear. The fadeout occurs suddenly and may last up to 1 h. Frequencies from about 3 to 10 MHz are most affected, but only where part or all of the signal path is in daylight. Transmission on frequencies below about 100 kHz is usually simultaneously improved. Also called blackout and radio blackout.

radio fan marker beacon *Fan marker beacon.*

radio field strength The effective value of the electric or magnetic field strength at a point due to the passage of radio waves of a specified frequency. Usually expressed as the electric field intensity in microvolts or millivolts per meter.

radio field-to-noise ratio The ratio of radio field strength to that of noise at a given location.

radio fix 1. Determination of the position of the source of radio signals by obtaining cross bearings on the transmitter with two or more radio direction finders in different locations, then computing the position by triangulation. 2. Determination of the position of a vessel or aircraft equipped with direction-finding equipment by obtaining radio bearings on two or more transmitting stations of known location and computing the position by triangulation. 3. Determination of position of an aircraft in flight by identification of a radio beacon or by locating the intersection of two radio beams.

radio frequency [abbreviated RF] A frequency at which coherent electromagnetic radiation of energy is useful for communication purposes. Radio frequencies are designated as follows: very low frequency, below 30 kHz; low frequency, 30–300 kHz; medium frequency, 300–3000 kHz; high frequency, 3–30 MHz; very high frequency, 30–300 MHz; ultrahigh frequency, 300–3000 MHz; superhigh frequency, 3–30 GHz; extremely high frequency, 30–300 GHz. (For entries starting with radio-frequency, see RF entries.)

radio galaxy A galaxy, consisting of billions of stars, that emits radio signals of varying strength from essentially its entire volume in the sky, thousands of light-years away.

radiogenic Produced by radioactive transformation.

radiogenic heat Heat produced within the earth by the disintegration of radioactive nuclides. Also called radioactive heat.

radioglaciology The use of short RF pulses for sounding glaciers. Similar techniques are used to measure the water content of snow fields.

radiogoniometer A goniometer used as part of a radio direction finder. In the Bellini-Tosi system, two loop antennas positioned at right angles to each other are connected to two field coils in the radiogoniometer. Bearings are obtained by a rotatable search coil that is inductively coupled to the field coils.

radiogoniometry The science of determining the direction of arrival of radio waves.

radiogram A message transmitted by radio.

radiograph A photographic image produced by a beam of penetrating ionizing radiation after the beam passes through an object, showing the variations in density or absorption in the object. When produced by x-rays, a radiograph is called an x-ray photograph.

radiographic putty A blocking medium used in radiography to reduce the effect of scattered radiation and to shield portions of the x-ray film that would otherwise be overexposed.

radiographic stereometry The process of finding the position and dimensions of details within an object by measurements made on radiographs taken from different directions.

radiography Photography that involves the use of x-rays, gamma rays, and other penetrating ionizing radiations to produce shadow images corresponding to differences in thickness, density, or absorption in the subject being examined. Widely used for medical and dental diagnosis and for nondestructive internal inspection of metal and other objects.

radio guard A military ship, aircraft, or radio station designated to listen for and record radio transmissions and handle message traffic on one or more designated frequencies.

radio guidance system A guidance system that uses radio signals to guide a flight-borne missile or other vehicle from a ground station.

radio-guided bomb An aerial bomb guided by radio control from outside the missile.

radioheliograph A radio telescope used for two-dimensional mapping of the sun.

radio homing beacon *Homing beacon.*

radio horizon The locus of points at which direct rays from a transmitter become tangential to the surface of the earth. The distance to the radio horizon is affected by atmospheric refraction.

radio inertial guidance A radio command guidance system that has an auxiliary inertial guidance system in the missile for partial guidance in case of radio guidance failure or to provide more up-to-date data for correcting radar guidance information.

radio influence Radio-frequency interference that originates on and from power lines.

radio influence field The radio noise field radiated by equipment, circuits, or conductors.

radio influence voltage The radio noise voltage induced in the conductors of equipment or circuits by nearby sources of electromagnetic radiation.

radio interference Interference with reception of a desired radio signal by an undesired radio signal or by a radio disturbance.

radio interferometer An interferometer that operates at radio frequencies. Used in radio astronomy and in tracking of satellites.

radioisotope An artificially produced isotope that is radioactive. Many elements have as many as 10 radioisotopes, produced in a cyclotron or by neutron bombardment in a nuclear reactor. Widely used in industry, medicine, and other fields as radioactive tracers and as sources of ionizing radiation. Also called radioactive isotope.

radioisotope packaging A cylindrical lead container, generally combined with fiberboard or other spacing material for centering the cylinder in a much larger container, for shipping radioactive materials without endangering the health of personnel or affecting shipments of photographic film.

radioisotope power source A low-power source in which a radioisotope such as plutonium 238 or strontium 90 generates heat by spontaneous decay, for conversion by thermoelectric or thermionic devices to electric power ranging up to several hundred watts. Used in unattended automatic weather stations, satellites, spacecraft, and other applications requiring power sources that last several years.

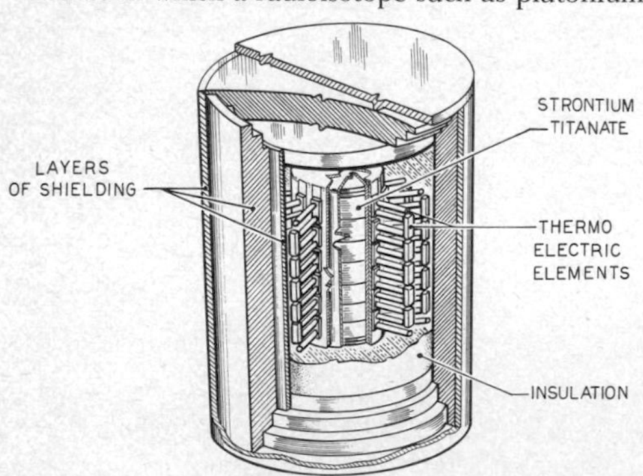

Radioisotope power source using strontium 90 locked safely in strontium titanate core to generate heat by spontaneous decay. Thermocouples surrounding core convert this heat into electric power.

radio landing beam A radio beam used for vertical guidance of aircraft during descent to a landing surface.

radiolead A radioisotope of lead.

radio line of position A line of position obtained with a radio direction finder.

radio link A radio system used to provide a communication or control channel between two specific points.

radiolocation 1. Determination of direction, position, or motion of an object by utilizing known properties of radio waves. 2. Obsolete British term for *radar*.

radio log A log of radio messages sent and received, together with other pertinent information, maintained by radio operators.

radiological Pertaining to nuclear radiation, radioactivity, and atomic weapons.

radiological defense Defense against the effects of radioactivity from atomic weapons, including detection and measurement of radioactivity, protection of persons from radioactivity, and decontamination of areas and equipment.

radiological dose The total amount of ionizing radiation absorbed by an individual exposed to any radiating source.

radiological indicator An indicator that displays the occurrence of radioactivity above a predetermined value.

radiological survey Determination of the distribution and dose rates of radiation in an area.

radiological warfare Warfare that involves weapons which produce radioactivity, such as atomic bombs and shells.

radiologist A medical specialist skilled in the use of x-rays, gamma rays, and other penetrating ionizing radiations.

radiology The branch of medicine that uses ionizing radiation for diagnosis and therapy.

radiolucent Transparent to x-rays and radio waves.

radioluminescence Luminescence produced by radiant energy, as by x-rays, radioactive emissions, alpha particles, or electrons.

radiolysis The dissociation of molecules by radiation. As an example, a small amount of water in a reactor core dissociates into hydrogen and oxygen during operation.

radio magnetic indicator [abbreviated RMI] An indicator that has a display which shows vehicle heading, relative bearing, magnetic bearing, and the omnibearing of a radio station being used for navigation purposes.

radioman *Radio operator.*

radio marker beacon *Marker.*

radio metal locator *Metal detector.*

radiometallography Examination of the crystalline structure and other characteristics of metals and alloys with x-ray equipment.

radio meteor A meteor that has been detected and tracked by radio-astronomy equipment, generally by reflection of radio waves from the ionized path of the meteor.

radiometeorograph *Radiosonde.*

radiometeorology The branch of meteorology that covers the propagation of radio energy through the atmosphere and the use of radio and radar equipment in meteorology.

radiometer An instrument that measures radiant energy. Examples include the bolometer, microradiometer, microwave radiometer, and thermopile.

radiometric analysis Quantitative chemical analysis that is based on measurement of the absolute disintegration rate of a radioactive component which has a known specific activity.

radiometry Measurement of quantities associated with radiant energy.

radio multiplexing 1. Dividing a radio channel into a number of voice or code channels through frequency division or time division. 2. Connecting two or more transmitters or receivers to the same antenna through appropriate coupling networks.

radio navigation Navigation by radio signals, using such equipment as radio direction finders, radio ranges, radio beacons, and loran.

radio navigation aid A navigation aid that uses radio signals, as contrasted to a radar navigation aid. Also called radio aid.

radio navigation guidance The guidance or control of a guided missile along a course established by external radio transmitters.

radionecrosis Destruction of living tissue by radiation. Sunburn is an example.

radio net A net of radio stations established for communication purposes.

radio noise Noise that occurs in the radio spectrum.

radionuclide A substance that exhibits radioactivity. Also called radioactive nuclide.

radio operator A person who operates radio transmitting and receiving equipment. Also called radioman.

radiopaque Not appreciably penetrable by x-rays or other forms of radiation.

radiopaque obstacle An obstacle that creates a communication blackout between a spacecraft and the earth. The sun could create such blackouts during trips to Mars.

radiophare *Radio beacon.*

radiophone *Radiotelephone.*

radiophoto 1. A photograph transmitted by radio to a facsimile receiver. 2. *Facsimile.*

radiophotoluminescence Luminescence exhibited by minerals such as fluorite and kunzite as a result of irradiation with beta and gamma rays followed by exposure to light.

radio propagation prediction A prediction of the quality or nature of radio propagation as influenced by such factors as sunspots and seasonal changes; the prediction is published periodically by the National Bureau of Standards.

radio prospecting Use of radio and electronic equipment to locate mineral or oil deposits.

radio proximity fuze A proximity fuze that contains a miniature radio transmitter and uses radio echoes from a target to trigger the fuze within predetermined limits of distance from the target.

radio range A radio transmitting facility that radiates signals which can be used by aircraft to determine bearings from the transmitting site. The A-N radio range provides four courses. Also called radio-range beacon and range.

radio-range beacon *Radio range.*

radio-range leg One of the courses or beams in a four-course radio range. Also called leg.

radio-range monitor An instrument that automatically monitors the signal from a radio-range beacon, giving a warning to attendants when the transmitter deviates a specified amount from its correct bearings and transmitting a distinctive warning to approaching planes when trouble exists at the beacon.

radio receiver A receiver that converts radio waves into intelligible sounds or other perceptible signals. Also called radio, radio set, and receiving set.

radio reception Reception of messages, programs, or other intelligence by radio.

radio relay system *Radio repeater.*

radio repeater A repeater that acts as an intermediate station in transmitting radio communication signals or radio programs from one fixed station to another. It serves to extend the reliable range of the originating station. A microwave repeater is an example. Also called radio relay system and relay system.

radioresistance The relative resistance of cells, tissues, organs, organisms, or substances to the injurious action of radiation.

radio scanner *Scanning radio.*

radio scattering *Scattering.*

radiosensitive Sensitive to damage by radiant energy.

radio serviceman A serviceman who is qualified to repair and maintain radio equipment. Also called radio technician.

radio set 1. A radio receiver and radio transmitter used together for two-way communication. 2. *Radio receiver.* 3. *Radio transmitter.*

radio signal A signal transmitted by radio.

radio silence A period during which transmissions by a radio station are stopped, such as to permit reception of signals from other stations or permit reception of weak distress signals.

radio sky The sky as it would appear if our eyes were sensitive to radio waves instead of to light.

radiosonde [pronounced radio sond] A meteorograph combined with a radio transmitter. When carried aloft by a balloon, it transmits radio signals that can be recorded at a ground station and interpreted in terms of the pressure, temperature, and humidity at regular intervals during the as-

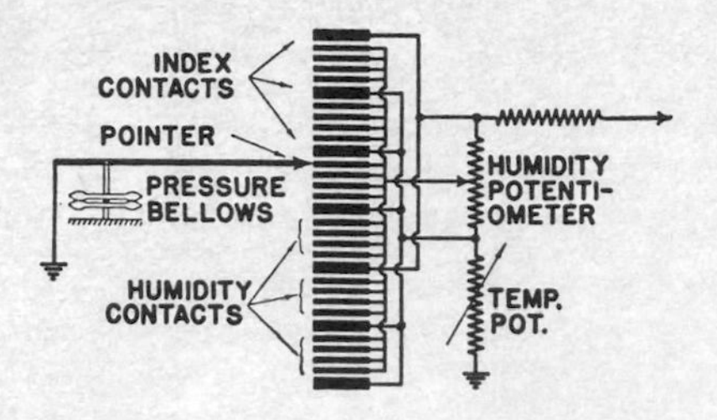

Radiosonde sensing and switching elements, used to feed amplifier connected to transmitting antenna.

cent. Also called radiometeorograph. The equipment is lowered by parachute when the balloon bursts.

radiosonde balloon A sounding balloon that carries a radiosonde.

radiosonde recorder A ground-station recorder that records the data transmitted from a radiosonde aloft.

radio sonobuoy *Sonobuoy.*

radio source A region in the sky from which radio waves are received by radio-astronomy equipment.

radio spectrum The entire range of frequencies in which useful radio waves can be produced, extending from the audio range to about 300 GHz. The radio spectrum is divided into eight bands (see band or radio frequency). Also called RF spectrum.

radio star A discrete radio source in the sky. It does not usually correspond to a known optical star and should therefore be called a radio source. The sun is the only star definitely known to emit radio waves.

radio station A station equipped to engage in radio communication or radio broadcasting.

radio-station interference Interference caused by radio stations other than that from which reception is desired.

radio strontium *Strontium 90.*

radio sun The sun as defined by its electromagnetic radiation in the radio portion of the spectrum.

radio system A complete radio equipment installation that provides multichannel communication between two points.

radio technician *Radio serviceman.*

radiotelegraph Pertaining to telegraphy over radio channels.

radiotelegraph transmitter A radio transmitter that is capable of handling code signals (type A1 and B emissions).

radiotelegraphy Telegraphy that involves the use of radio waves in place of wire lines. The international Morse code is generally used in radiotelegraphy.

radiotelephone 1. Pertaining to telephony over radio channels. 2. A radio transmitter and radio receiver used together for two-way telephone communication by radio. Also called radiophone.

radiotelephone distress call The word "mayday," corresponding to the French pronunciation of m'aider, spoken under the same conditions that the signal SOS would be transmitted in code by radiotelegraphy.

radiotelephone transmitter A radio transmitter capable of handling AF modulation, such as voice and music.

radiotelephony Two-way voice communication (telephony) carried on by radio waves, without connecting wires between stations.

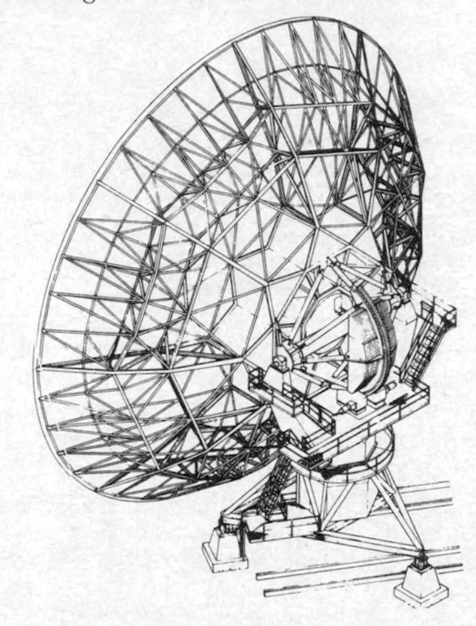

Radio telescope having dish 82 ft (25 m) in diameter.

radio telescope A sensitive radio receiver used with a large and highly directional antenna to receive signals from radio stars.

radioteletype [abbreviated RTTY] Telegraphic communication by radio, using at each station a teletypewriter connected to a radio receiver and transmitter.

radioteletypewriter A teletypewriter and the associated equipment needed for operation over a radio channel rather than over wires.

radiotherapy Treatment of disease with any ionizing radiation, including x-rays and gamma rays but not ultraviolet rays.

radiothermoluminescence Luminescence exhibited by certain vitreous and crystalline substances as a result of irradiation with beta and gamma rays followed by heating.

radiothorium [symbol RdTh] A thorium isotope that has mass number 228 and a half-life of 1.90 years. It is produced naturally by beta decay of actinium 228 and emits gamma particles to give radium 224.

radiotracer *Radioactive tracer.*

radio transmission The transmission of signals

through space at radio frequencies by radiated electromagnetic waves.

radio transmitter A transmitter that produces RF power for transmission through space in the form of radio waves. Also called radio set.

radiotransparent Permitting passage of x-rays or other forms of radiation.

radiotropism Turning or bending of a plant or other organism in response to some form of radiation.

radio tube *Electron tube.*

radio watch *Watch.*

radio wave An electromagnetic wave produced by reversal of current in a conductor at a frequency in the range from about 10 kHz to 3000 GHz. Radio waves travel through space at approximately the speed of light, which is 299,792.458 km/s or about 186,000 mi/s. Also called Hertzian wave.

radio wave propagation The transfer of energy through space by electromagnetic radiation at radio frequencies.

radio window A band of frequencies extending from about 6 MHz to 30 GHz, in which radiation from the outer universe can enter and travel through the atmosphere of the earth.

radium [symbol Ra] A highly radioactive metallic element that gives off alpha, beta, and gamma rays. Atomic number is 88.

radium age The age of a mineral as calculated from the numbers of radium atoms present originally, now, and when equilibrium is established with ionium.

radium cell A sealed thin-wall tube that contains radium.

radium needle A radium cell in the form of a needle, usually of platinum-iridium or gold alloy, designed primarily for insertion in tissue.

radium parameter The effective radius of a nucleus divided by the cube root of its mass number. The value is approximately the same for all nuclei and is about 1.4×10^{-13} cm.

radium plaque A radium container in which the radium is distributed over a surface. The shielding is usually small in one direction, to permit transmission of beta as well as gamma rays.

radium therapy Radiotherapy that uses the radiations from radium.

radix *Base.*

radix notation A positional notation in which the successive digits are interpreted as coefficients of successive integral powers of a number called the radix or base. The represented number is equal to the sum of this power series. Thus, 5762 is the sum of the power series $5 \times 10^3 + 7 \times 10^2 + 6 \times 10^1 + 2 \times 10^0$, where 10 is the base.

radix point *Point.*

RADNOS [transposition of NO RADio plus S] A radio fadeout encountered chiefly in arctic regions, considered to be caused by solar explosions, sunspots, or the aurora borealis.

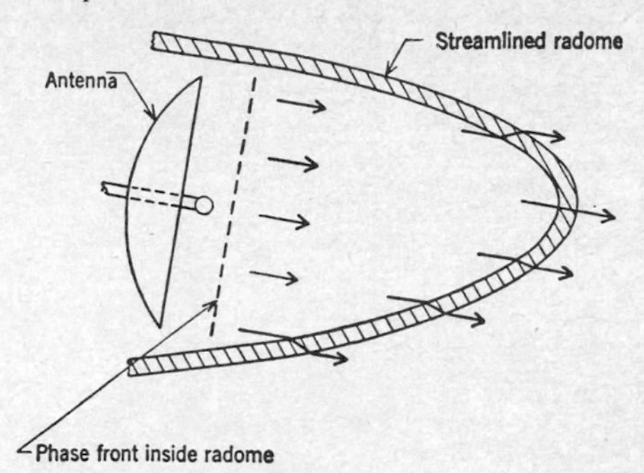

Radome and radar antenna in nose of aircraft, showing how waves passing through radome walls at an angle are refracted and distorted.

radome [RAdar DOME] A protective housing for a radar antenna, made from dielectric material that is transparent to RF energy.

radon [symbol Rn] A heavy gaseous radioactive element. Atomic number is 86 and atomic weight is 222. It is a daughter of radium in the uranium radioactive series.

radon seed A small metal or glass tube that contains radon.

rad per unit time A unit of absorbed dose rate.

radux A long-distance continuous-wave low-frequency navigation system of the phase comparison type, providing hyperbolic lines of position.

railing The jamming of radar transmissions by transmitters at a pulse rate of 50 to 150 kHz, causing images resembling fence railings to appear on radar screens.

railroad radio service A radio communication service used in connection with the operation and maintenance of a railroad common carrier.

rail voltage British term for *supply voltage.*

rain attenuation Attenuation of radio waves when passing through moisture-bearing cloud formations or areas in which rain is falling. The attenuation increases with the density of the moisture in the transmission path.

rainbow generator A signal generator that generates a signal which, when fed into a color television receiver, produces the entire color spectrum on the screen, with the colors merging together.

rain return Clutter due to rain.

RAM Abbreviation for *random-access memory.*

Raman scattering Scattering of light by the molecules of transparent gases, liquids, and solids, resulting from a change in the frequency in the incident radiation because of interaction of this radiation with the molecules. Used in studying molecular structure of materials.

Raman spectrometer A spectrometer in which a continuous-wave laser serves as the excitation source. A helium-neon laser operating at 0.6328 μm is commonly used. Also called laser Raman spectrometer.

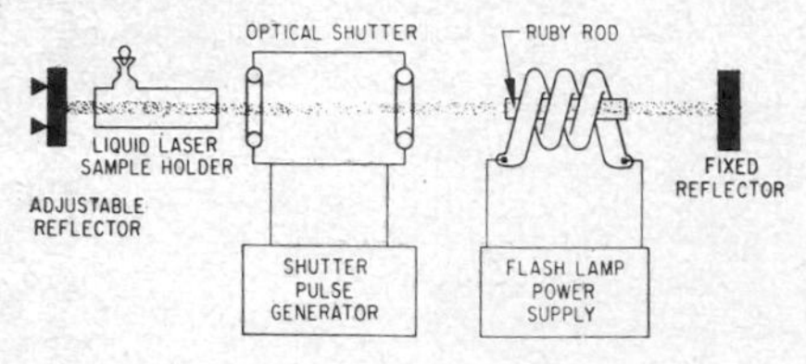

Raman scattering occurs in sample liquid along with laser action when powerful pulses of light are beamed through liquid by ruby laser.

RAMARK [RAdar MARKer] A fixed radar beacon that emits radar waves continuously to provide a bearing indication to radar-equipped ships and aircraft.

ramp generator A circuit that generates a sweep voltage which increases linearly in value during one cycle of sweep, then returns to zero suddenly to start the next cycle.

Ramsauer effect The low attenuation of slow-moving electrons by inert gases.

Rand Corporation [Research ANd Development] A nongovermental nonprofit organization engaged in research for the welfare and national security of the United States.

random access A memory or storage characteristic in which the access time is effectively independent of the location of the data.

random-access discrete-address system A radio-communication service in which a large group of users share a broad band of channels simultaneously. Voice modulation is converted into digital form, and the resulting pulses are transmitted in sequence, each at a different carrier frequency and a different assigned instant in time. Each receiver in a given service has an assigned combination of channel frequencies and time slots.

random-access memory [abbreviated RAM] A computer memory in which each storage location is accessible by X and Y coordinates, as in a core or semiconductor memory. (Magnetic tape cannot be random access.) The time required for writing in or reading out data is thus independent of location. Also called direct-access memory.

random-access programming Computer programming without regard for the time required for access to stored information.

random coincidence *Accidental coincidence.*

random error An error that can be predicted only on a statistical basis.

random interlace Interlace based on less precise timing of sweep frequencies than is required for television broadcasts. Sometimes used in industrial television.

random logic Logic that uses combinations of simple gates to provide functions which can be programmed either during manufacture or in the field.

random noise Noise characterized by a large number of overlapping transient disturbances occurring at random, such as thermal or shot noise. It is sometimes produced intentionally for test purposes or for jamming enemy transmissions. Also called fluctuation noise.

random-noise testing Testing in which a complex wave that has randomly varying frequencies and amplitudes is applied to a mechanical shake table. The signal may be obtained experimentally, as from a missile telemetering record, or may be generated electronically.

random number A number formed by a set of digits in which each successive digit is equally likely to be any of *n* digits to the base *n*. Random numbers are thus obtained entirely by chance.

random variable A discrete or continuous variable that may assume any one of a number of values, each of which has a fixed probability.

random walk The path followed by a particle as it makes random scattering collisions in a medium.

random winding A coil winding in which the turns are positioned haphazardly rather than in layers.

range 1. The distance from a radar set or weapon to a target. 2. The distance capability of an aircraft, missile, gun, radar, or radio transmitter. 3. The maximum thickness of a given medium that can be penetrated by a given ionizing particle. 4. The difference between the maximums and minimums of a variable quantity. 5. A line defined by two fixed landmarks, used for missile, vehicle, and other test purposes. 6. A line of bearing defined by a radio range. 7. *Radio range.*

range-amplitude display A radar display in which a time base provides the range scale on which echoes appear as deflections normal to the base. The base is usually a straight line, as in the A display, or a circle, as in the J display.

range circle A radar range marker in the form of a circle.

range control Control exercised over the range of a guided missile.

range deception An electronic countermeasure technique used to prevent an enemy tracking radar from obtaining accurate range information. The incident radar signal is amplified and retransmitted with suitable varying delay to create a false echo or even walk the range gate off position enough to break the lock on the target.

range discrimination Deprecated term for *distance resolution.*

range-energy relation The graphical relation between range and energy of specified particles.

range gate A gate circuit that selects radar echoes within a small range interval.

range-height indicator display A radar display that presents visually the scalar distance between a reference point and a target, along with the vertical distance between a reference plane and the target.

range mark *Radar range marker.*

range marker generator A signal generator that generates the signal required for the production of radar range markers on a radarscope. Its action is initiated by the sync pulse that starts the time base.

range rate The rate at which the distance from the measuring equipment to the signal source being tracked is changing with respect to time.

range resolution Deprecated term for *distance resolution.*

range selector A control that selects the range scale on a radar indicator.

range straggling The variation in the range of particles that have the same initial energy.

range target A reflective target at a precisely known range from a radar antenna. Used for radar range system alignment.

ranging Determining distance.

ranging crystal The crystal in a radar range unit that determines the primary range timing frequency.

rank The number of independent cut-sets that can be selected in a network. The rank is equal to the number of nodes minus the number of separate parts.

Rankine [abbreviated R] An absolute temperature scale that uses Fahrenheit degrees but with the entire scale shifted so 0°R is at absolute zero. In this scale, water freezes at 459.6°R and boils at 639.6°R. Add 427.6 to a Fahrenheit value to get the corresponding Rankine value.

RAPCON [Radar APproach CONtrol] Use of radar for direct control of aircraft in the vicinity of an airport and during the approach to the runway. Both surveillance radar and precision approach radar are required.

Raphael bridge A type of slide-wire Wheatstone bridge used for locating faults in transmission lines.

rapid memory The section of a computer from which stored information can be obtained most rapidly. Also called high-speed memory.

rapid scanning Scanning with a narrow radar beam at the rate of 10 sweeps per second or more, as required for tracking fast-moving targets.

rapid-start fluorescent lamp A fluorescent lamp used with a ballast that has a low-voltage winding which is continuously connected to the cathode heaters of the lamp, to initiate and maintain a gaseous discharge.

rare earth An element that has an atomic number in the range from 57 to 71 inclusive. The rare earths are cerium, dysprosium, erbium, europium, gadolinium, holmium, lanthanum, lutecium, neodymium, praseodymium, promethium, samarium, terbium, thulium, and ytterbium.

rare-earth chelate laser *Chelate laser.*

rare-earth magnet A permanent magnet formed from compounds of cobalt and one or two of the following: cerium, lanthanum, mischmetal, praseodymium, samarium, or yttrium. One example, the samarium-cobalt magnet, has many commercial applications.

raster A predetermined pattern of scanning lines that provides substantially uniform coverage of an

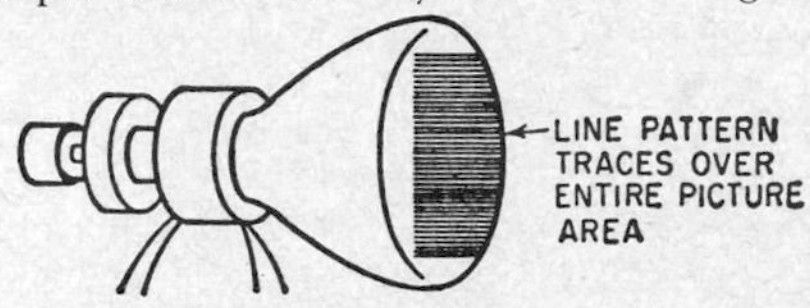

Raster is line pattern seen on screen of television picture tube when no video signal is present.

area. In television the raster is seen as closely spaced parallel lines, most evident when there is no picture.

raster burn A change in the characteristics of the scanned area on the target of a camera tube, resulting in a spurious signal when a larger or tilted raster is scanned.

ratchet relay A stepping relay actuated by an armature-driven ratchet.

rate action *Derivative action.*

rate control Control of the rate of change of the independent variable in an automatic control system. Also called proportional response and throttling control.

rated accuracy The advertised accuracy of a manufactured instrument.

rated coil current The steady-state coil current at which a relay is designed to operate.

rated coil voltage The coil voltage at which a relay is designed to operate.

rated contact current The current that contacts are designed to carry for their rated life.

rated output The output power, voltage, current, or other value at which a machine, device, or apparatus is designed to operate under specified normal conditions.

rate effect Premature triggering of some solid-state switches when the anode voltage is applied suddenly.

rate-grown junction A grown junction produced by varying the rate of semiconductor crystal growth periodically and using a melt that contains both N- and P-type impurities, so the two types of impurities alternately predominate. Also called graded junction.

rate-grown transistor A junction transistor in which both impurities (such as gallium and antimony) are placed in the melt at the same time and the temperature is suddenly raised and lowered to produce the alternate P- and N-type layers of rate-grown junctions. Also called graded-junction transistor.

rate gyroscope A spinning or vibratory gyroscope that measures the rate of change of direction of an aircraft.

rate meter *Counting-rate meter.*

rate-of-climb indicator An instrument that indicates the rate of climb or descent of an aircraft.

rate of closure The speed at which two airborne aircraft or other moving objects close the distance between them. With aircraft approaching each other, the rate of closure is the sum of their speeds.

rate of decay The time rate at which the sound pressure level, velocity level, or sound-energy density level is decreasing at a given point and at a given time. The practical unit is the decibel per second.

rate-of-turn control A gyroscopic instrument that furnishes a rate-of-turn signal to an automatic pilot system in an aircraft.

rate signal A signal proportional to the time derivative of a specified variable.

rate transmitter A transmitter in a missile being launched, used with a ground receiver to indicate the rate of speed increase.

rating A designation of an operating limit for a machine, apparatus, or device used under specified conditions.

ratio The value obtained when one quantity is divided by another of the same kind, to indicate their relative proportions.

ratio arms Two adjacent arms of a Wheatstone bridge, designed so they can be set to provide a variety of indicated resistance ratios.

ratio control Control in which a predetermined ratio between two physical quantities is maintained.

ratio detector An FM detector circuit that uses two diodes and requires no limiter at its input. The audio output is determined by the ratio of two developed IF voltages whose relative amplitudes are a function of frequency.

ratio meter A meter that measures the quotient of two electrical quantities. The deflection of the meter pointer is proportional to the ratio of the currents flowing through two coils.

rationalized unit A unit in a system of measurement that is designed to minimize occurrence of the constant 4π in equations.

ratio of transformation The ratio of the secondary voltage of a transformer to the primary voltage under no-load conditions, or the corresponding ratio of currents in a current transformer.

ratio-type telemeter A telemeter that uses the phase or magnitude relations of two or more electrical quantities as the translating means.

rat race *Hybrid ring.*

raw data Unprocessed data that may or may not be in machine-readable form.

rawin [RAdar WINd or RAdio WINd; pronounced ray win] 1. Determination of wind direction and velocity by radar or by radio direction-finding in conjunction with a radiosonde, radiosonde balloon, or a balloon carrying a radar reflector. 2. Wind information gathered by using radar tracking or radio direction-finding in connection with a specially equipped balloon.

rawin balloon A radiosonde balloon or other specially equipped balloon used in determining the movement and velocity of winds.

rawinsonde A radiosonde used in rawin.

RAWOL [RAdar WithOut Line of sight] Ground radar detection of targets below line of sight by diffraction when low hills intervene between target and radar.

ray *Beam.*

ray-control electrode [symbol RC] The electrode that controls the position of the electron beam on the screen of a cathode-ray tuning indicator.

raydist A navigation system in which a continuous-wave signal emitted from a vehicle is received at three or more ground stations. The received signals are compared in phase to determine the position of the vehicle.

Rayleigh cycle A cycle of magnetization that does not extend beyond the initial portion of the magnetization curve, between zero and the upward bend. In this region the permeability is low and there is little hysteresis.

Rayleigh disk An acoustic radiometer that measures particle velocity. A thin disk suspended by its edge from a fine fiber tends to take a position perpendicular to the horizontal component of sound particle velocity.

Rayleigh distribution A mathematical statement of a natural distribution of random variables.

Rayleigh line A spectrum line in scattered radiation that has the same frequency as the corresponding incident radiation. It arises from ordinary or Rayleigh scattering.

Rayleigh scattering Selective scattering of light by very small particles suspended in air, as by dust.

Rayleigh wave A type of wave that may be propagated near the surface of a solid and is characterized by elliptical motion of particles.

RBE Abbreviation for *relative biological effectiveness.*

RC 1. Symbol for *ray-control electrode.* Used on tube-base diagrams. 2. Abbreviation for *resistance-capacitance.*

RC amplifier *Resistance-coupled amplifier.*

RC circuit Abbreviation for *resistance-capacitance circuit.*

***RC* constant** The time constant of a resistance-capacitance circuit, equal in seconds to the resistance value in ohms multiplied by the capacitance value in farads.

RC coupling *Resistance coupling.*

RC differentiator Abbreviation for *resistance-capacitance differentiator.*

RC filter Abbreviation for *resistance-capacitance filter.*

RCG circuit Abbreviation for *reverberation-controlled gain circuit.*

RCM Abbreviation for *radar countermeasure.*

RC oscillator Abbreviation for *resistance-capacitance oscillator.*
RCTL Abbreviation for *resistor-capacitor-transistor logic.*
rd Abbreviation for *rad.*
RdAc Symbol for *radioactinium.*
RDF Abbreviation for *radio direction finder.*

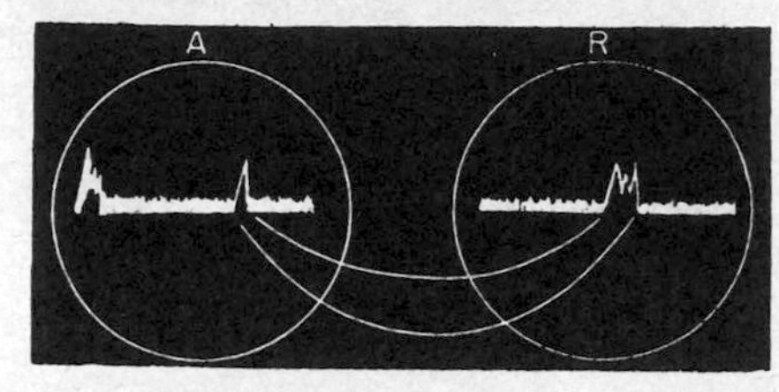

R display used as supplemental indicator on some radars, with corresponding A display for same echo shown at left.

R display A radarscope display that is essentially an expanded A display, in which an echo can be magnified for close examination.
RdTh Symbol for *radiothorium.*
reactance [symbol *X*] The opposition offered to the flow of alternating current by pure inductance or capacitance in a circuit, expressed in ohms. It is the component of impedance that is not due to resistance. Inductive reactance is due to inductance; capacitive reactance is due to capacitance.
reactance control circuit The color television receiver circuit that converts the DC correction voltage from the phase detector into a capacitive reactance change which maintains the 3.58-MHz oscillator at the correct frequency and phase.
reactance frequency divider A frequency divider that uses a nonlinear reactor to generate subharmonics of a sinusoidal source.
reactance frequency multiplier A frequency multiplier that uses a nonlinear reactor to generate harmonics of a sinusoidal source.
reactance modulator A modulator whose reactance may be varied in accordance with the instantaneous amplitude of the modulating voltage. This is normally an electron-tube circuit, used to produce phase or frequency modulation.
reactance tube An electron tube connected and operated in such a way that it acts as an inductive or capacitive reactance. The magnitude of the reactance can be changed by adjusting the grid bias voltage. Used in reactance modulators and in automatic frequency control of oscillators.
reaction 1. An action wherein one or more substances are changed into one or more new substances, as in a nuclear reaction. 2. British term for *positive feedback.*
reaction cavity A cavity that can be mounted on a side or end of a waveguide, for automatic or manual frequency-control applications. Tuning is achieved with a micrometer head that controls the position of the plunger in the cavity.
reaction motor A synchronous motor whose rotor contains salient poles but has no windings and no permanent magnets.
reaction rate The rate at which fission takes place in a nuclear reactor. This rate is fundamentally expressed as the number of nuclei undergoing fission per unit time. The reaction rate determines the reactivity of a nuclear reactor.
reactivation Application of a higher voltage than normal to the thoriated filament of an electron tube for a few seconds, to bring a fresh layer of thorium atoms to the filament surface and thereby improve electron emission.
reactive Pertaining to inductive or capacitive reactance.
reactive attenuator An attenuator that absorbs very little energy.
reactive factor The ratio of reactive power to apparent power.
reactive-factor meter A meter that measures and indicates reactive factor.
reactive load A load that has inductive or capacitive reactance.
reactive power The power value obtained by multiplying together the effective value of current in amperes, the effective value of voltage in volts, and the sine of the angular phase difference between current and voltage. Also called reactive voltamperes and wattless power. The unit of reactive power is the var.
reactive voltampere *Voltampere reactive.*
reactive voltampere-hour *Voltampere-hour reactive.*
reactive voltampere meter *Varmeter.*
reactive voltamperes *Reactive power.*
reactor 1. A device that introduces either inductive or capacitive reactance into a circuit, such as a coil or capacitor. 2. *Nuclear reactor.*
reactor-start motor A split-phase motor designed for starting with a reactor in series with the main winding. The reactor is short-circuited or otherwise made ineffective and the auxiliary circuit opened when the motor has attained a predetermined speed.
read 1. To acquire information, usually from some form of storage in a computer. 2. To generate an output corresponding to the pattern stored in a charge-storage tube. 3. To understand clearly, as in radio communication.
read-around number The number of times that priming, writing, reading, or erasing operations can be performed on storage elements adjacent to any given element in a charge-storage tube without loss of information from that element. Also called read-around ratio (deprecated).
read-around ratio Deprecated term for *read-around number.*
reader A device that converts information from one form to another, as from punched paper tape to magnetic tape.
read in To sense one form of information and transmit this information to an internal storage of a computer.

reading 1. The indication shown by an instrument. 2. To observe the readings of one or more instruments.

reading rate The number of characters, words, fields, blocks, or cards that can be sensed by an input reading device per unit of time.

reading speed The rate of reading successive storage elements in a charge-storage tube.

read number The number of times that a storage element is read without rewriting in a charge-storage tube.

read-only memory [abbreviated ROM] A random-access memory in which programming of the data pattern is fixed during manufacture and cannot be changed.

readout The presentation of output information by lights, printed or punched tape or cards, or other methods.

readout tube An electron tube that has shaped electrodes which provide a visual display of a single numeral or other character when the corresponding electrode is energized. The Nixie tube is an example.

read pulse A pulse that causes information to be acquired from a magnetic cell or cells.

read time The time interval between the instant at which information is called for from storage and the instant at which delivery is completed in a computer. Also called access time.

read/write memory A memory in which stored information can be read out or new information written in at any storage location.

ready-to-receive signal A signal sent back to a facsimile transmitter to indicate that a facsimile receiver is ready to accept the transmission.

real power The component of apparent power that represents true work. Real power is expressed in watts and is equal to voltamperes multiplied by the power factor.

real time The performance of a computation during the time of a related physical process, so the results are available for guiding the physical process.

real-time data Data presented in usable form at essentially the same time the event occurs. The delay in presenting the data must be small enough to allow a corrective action to be taken if required.

real-time delay An essentially negligible delay time, generally of the order of a few nanoseconds and controllable. Used in antenna arrays to obtain a desired radiation pattern, by connecting appropriate delay lines between the elements of the array.

rear projection A projection television system in which the picture is projected on a ground-glass screen for viewing from the opposite side of the screen.

rear-projection readout A readout in which 10 or more lamp-film-lens combinations are used as miniature rear-screen projectors all aimed at a single screen. Application of voltage to any one lamp causes its associated character image to be projected on the screen.

Rebecca The airborne interrogator-responsor of a Rebecca-Eureka system.

Rebecca-Eureka system An aircraft radar homing system in which an airborne interrogator-responsor (Rebecca) homes on a ground radar beacon (Eureka) that has been dropped or set up in advance. The system can also give the distance from the Rebecca radar to the Eureka beacon.

Rebecca-H system A British H system that uses a Rebecca radar which has been modified to display two beacon responses simultaneously.

rebroadcast Repetition of a radio or television program at a later time.

recalescent point The temperature at which there is a sudden liberation of heat as a heated metal is cooled.

receiver The complete equipment required for receiving modulated radio waves and converting them into the original intelligence, such as into sounds or pictures, or converting to desired useful information as in a radar receiver.

receiver bandwidth The frequency range between the half-power points on the frequency-response curves of a receiver.

receiver gating The application of operating voltages to one or more stages of a receiver only during that part of a cycle of operation when reception is desired.

receiver noise figure The ratio of noise in a given receiver to that in a theoretically perfect receiver.

receiver primaries *Display primaries.*

receiver radiation Radiation of interfering electromagnetic fields by the oscillator of a receiver.

receiver synchro *Synchro receiver.*

receiving antenna An antenna that converts electromagnetic waves to modulated RF currents.

receiving set *Radio receiver.*

receiving station A station used for reception of radio signals or messages.

receptacle *Outlet.*

rechargeable battery A storage battery, of which the lead-acid battery is the most common. The four other major types are nickel-iron, nickel-cadmium, silver-zinc, and silver-cadmium batteries. Some primary batteries can be recharged a limited number of times; these include zinc–manganese dioxide batteries and silver-zinc batteries.

recharger A DC power supply that recharges nickel-cadmium or other rechargeable batteries used in calculators and other battery-operated devices.

reciprocal counter A counter that measures a time interval between two events and computes the reciprocal of the measured value. Thus the reciprocal of the period of a signal is its frequency, and the reciprocal of the time for an object to travel between two points is its speed.

reciprocal-energy theorem A theorem of Rayleigh: If an electromotive force E_1 in one branch of a circuit produces a current I_2 in any other branch, and if an electromotive force E_2 inserted in this other branch produces a current I_1 in the first branch, then $I_1E_1 = I_2E_2$. This is closely related to the reciprocity theorem.

reciprocal ferrite switch A ferrite switch that can be inserted in a waveguide to switch an input signal to either of two output waveguides. Switching is done by a Faraday rotator when acted on by an external magnetic field.

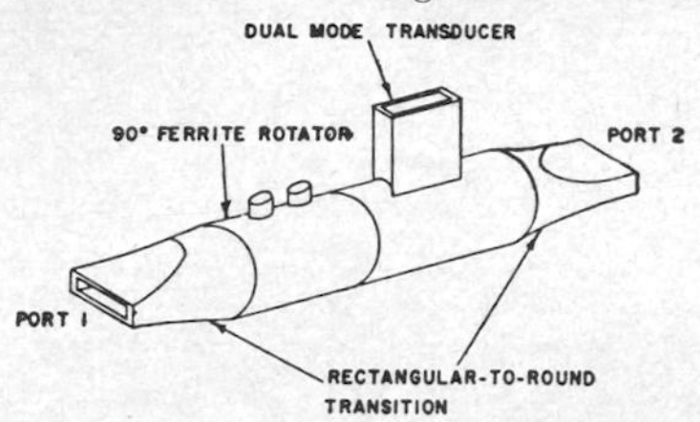

Reciprocal ferrite switch.

reciprocal impedance Two impedances Z_1 and Z_2 are said to be reciprocal impedances with respect to an impedance Z (invariably a resistance) if they satisfy the equation $Z_1Z_2 = Z^2$.

reciprocal transducer A transducer that satisfies the reciprocity principle.

reciprocal velocity region The energy region in which the capture cross section for neutrons by a given element is inversely proportional to neutron velocity.

reciprocity constant The sensitivity of a transducer used as a microphone divided by the sensitivity of the same transducer used as a source of sound.

reciprocity principle The relation between the sensitivity of a reversible electroacoustic transducer when used as a microphone and the sensitivity when used as a source of sound is independent of the type and construction of the transducer.

reciprocity theorem If a voltage located at one point in a network produces a current at any other point in the network, the same voltage acting at the second point will produce the same current at the first point.

reclosing relay A relay that functions to reclose a circuit automatically under certain conditions.

recognition differential The signal strength above noise level that gives a 50% probability of detection of an aural signal.

recoil electron An electron that has been set into motion by a collision.

recoil nucleus A nucleus that recoils when it collides with a nuclear particle or ejects a nuclear particle.

recoil particle A particle that has been set into motion by a collision or by a process involving the ejection of another particle.

recoil radiation Radiation emitted during nuclear disintegration in such a way that there is an observable recoil of the nucleus.

recombination The combination and resultant neutralization of particles or objects having unlike charges, such as a hole and an electron or a positive ion and a negative ion.

recombination radiation The radiation emitted in semiconductors when electrons in the conduction band recombine with holes in the valence band. If an actual population inversion is achieved between portions of the valence and conduction bands, or between adjacent localized states of acceptors or donors near these respective bands, stimulated emission and laser amplification or oscillation can result.

recombination rate The time rate at which free electrons and holes recombine at the surface or within the volume of a semiconductor.

recombination velocity The normal component of the electron or hole current density on a semiconductor surface divided by the excess electron or hole charge density at the surface.

recommutation Failure of load current to be completely commutated from one ignitron to another within the required time, with the result that current is commutated back to the original tube.

reconditioned-carrier receiver A receiver in which the carrier is separated from the sidebands to eliminate amplitude variations and noise, then added at an increased level to one sideband to obtain a relatively undistorted output. Generally used with single-sideband transmitters.

reconnaissance satellite An earth satellite that provides strategic information, as by television or photography.

reconstituted conductive material Conductive material formed by compressing finely divided particles.

reconstituted mica Mica sheets or shaped objects made by breaking up scrap natural mica, combining with a binder, and pressing into forms suitable for use as insulating material.

recontrol time *Deionization time.*

record 1. To preserve for later reproduction or reference. 2. A group of related facts or fields of information treated as a unit. 3. *Phonograph record.* 4. *Recording.*

record changer A record player that plays a number of records automatically in succession.

record compensator An adjustable filter used in audio systems to compensate for the differing frequency-response curves used by phonograph-record manufacturers for accentuating treble frequencies and attenuating bass frequencies during the recording process. Also called record equalizer.

record density *Character density.*

recorded program A radio program that uses phonograph records, electric transcriptions, magnetic tapes, or other means of reproduction.

recorded spot The image of the recording spot on the record sheet in a facsimile system.

recorded tape 1. A recording that is commercially available on magnetic tape. Also called prerecorded tape. 2. Any magnetic tape that has been recorded.

record equalizer *Record compensator.*

recorder An instrument that makes a permanent record of varying electrical quantities or signals. A common industrial version records one or more quantities as a function of another variable, usually time. Other types include the cathode-ray oscillograph, facsimile recorder, kinescope recorder, magnetic-tape recorder, and sound film recorder.

record gap A gap that indicates the end of a record on magnetic tape.

record head *Recording head.*

recording 1. Any process for preserving signals, sounds, data, or other information for future reference or reproduction, such as disk recording, facsimile recording, ink-vapor recording, magnetic-tape or wire recording, and photographic recording. 2. The end product of a recording process, such as the recorded magnetic tape, disk, or record sheet. Also called record.

recording blank *Recording disk.*

recording camera A camera that photographs radarscope displays and instrument readings for record purposes.

recording channel One of a number of independent recorders in a recording system or independent recording tracks on a recording medium.

recording characteristic A graph that shows the intentional attenuation of bass frequencies and accentuation of treble frequencies used in making a disk recording.

recording density The number of bits per unit length in a single linear data recording track.

recording disk An unrecorded or blank disk for recording sounds by a stylus. Also called recording blank.

recording head 1. A magnetic head used only for recording. Also called record head. 2. *Cutter.*

recording lamp A lamp whose intensity can be varied at an AF rate, for exposing variable-density sound tracks on motion-picture film and for exposing paper on film in photographic facsimile recording.

recording level The amplifier output level required to drive a particular recorder.

recording loss The difference between the amplitude recorded on a mechanical recording medium and that executed by the recording stylus.

recording noise Noise that is introduced during a recording process.

recording-playback head A magnetic head used for both recording and reproduction.

recording spot The image area formed at the record medium by a facsimile recorder.

recording storage tube A type of cathode-ray tube in which the electric equivalent of an image can be stored as an electrostatic charge pattern on

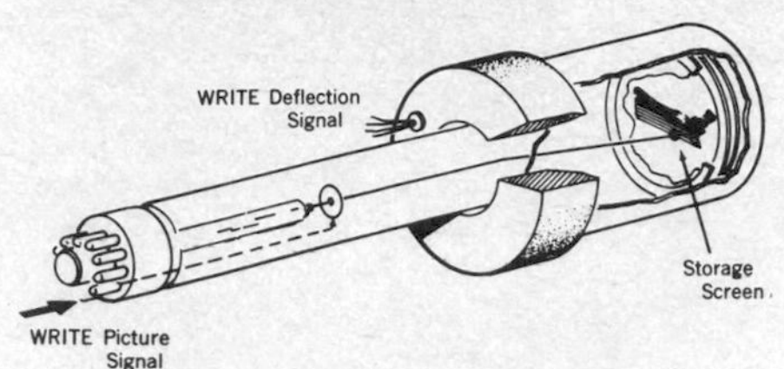

Recording storage tube.

a storage surface. There is no visual display, but the stored information can be read out at any desired later time as an electric output signal.

recording stylus A tool that inscribes the grooves in a mechanical recording medium.

record length The number of characters required for all the information in a record.

record medium The physical medium on which a facsimile recorder forms an image of the subject copy.

record player A motor-driven turntable used with a phonograph pickup to obtain AF signals from a phonograph record. These signals must be fed into an AF amplifier for additional amplification before they can be reproduced as sound waves by a loudspeaker. In an electric phonograph the amplifier and loudspeaker are combined with the record player.

record sheet The medium used to produce a visible image of the subject copy in a facsimile system.

recovery package An instrumentation package carried by a missile and designed for ejection before impact. The package usually also contains a transmitter that emits signals after ejection, to aid in recovery.

recovery time 1. The time required for the control electrode of a gas tube to regain control after anode-current interruption. 2. The time required for a fired TR or pre-TR tube to deionize to such a level that the attenuation of a low-level RF signal transmitted through the tube is decreased to a specified value. 3. The time required for a fired ATR tube to deionize to such a level that the normalized conductance and susceptance of the tube in its mount are within specified ranges. 4. The minimum time from the start of a counted pulse to the instant a succeeding pulse can attain a specific percentage of the maximum value of the counted pulse in a Geiger counter. 5. The interval required, after a sudden decrease in input signal amplitude to a system or component, to attain a specified percentage (usually 63%) of the ultimate change in amplification or attenuation due to this decrease. 6. The time required for a radar receiver to recover to half sensitivity after the end of the transmitted pulse, so it can receive a return echo.

rectangular coordinate A coordinate with respect to one of three mutually perpendicular axes, used for specifying the location of a point in space.

rectangular horn antenna A horn antenna that has a rectangular cross section, with one or both transverse dimensions increasing linearly from the small end or throat to the mouth, used at the end of a rectangular waveguide to radiate radio waves directly into space.

rectangular picture tube A television picture tube that has an essentially rectangular faceplate and screen, to show with maximum economy of screen space the total scene area scanned by a television camera.

rectangular scanning A two-dimensional sector scan in which a slow sector scan in one direction is superimposed on a rapid sector scan in a perpendicular direction.

rectangular wave A periodic wave that alternately and suddenly changes from one to the other of two fixed values. When the duty cycle is 50% (when the two fixed values have equal times), the rectangular wave becomes a square wave.

rectangular waveguide A waveguide that has a rectangular cross section.

rectenna A microwave antenna that has built-in rectifying elements, for use in converting transmitted microwave power to DC power.

rectification The process of converting an alternating current to a unidirectional current.

rectification factor The change in the average current of an electrode divided by the change in the amplitude of the alternating sinusoidal voltage applied to the same electrode, the direct voltages of this and other electrodes being maintained constant.

rectified current The direct current that results from the process of rectification.

rectifier A device that converts alternating current into a current which has a large unidirectional component, such as a gas tube, metallic rectifier, semiconductor diode, or vacuum tube.

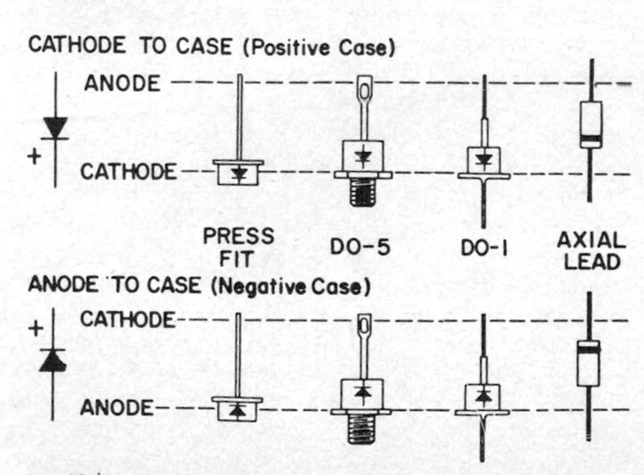

Rectifier polarity markings. Arrow points toward cathode terminal of device, which is connected to B+ of circuit; for this reason cathode of rectifier is sometimes marked +. Arrow thus points in direction of electron flow through rectifier.

rectifier instrument A DC meter used with a rectifying device, such as a copper-oxide rectifier, to measure alternating currents or voltages.

rectifier junction A semiconductor junction that has greater conductivity in one direction than in the other.

rectifier stack A stacked assembly of semiconductor rectifier disks or wafers.

rectify To convert an alternating current into a unidirectional current.

rectifying element A circuit element that provides rectification. It may be a single cell or several cells in series, parallel, or series-parallel.

rectilineal compliance The mechanical compliance that opposes a change in the applied force, such as the springiness that opposes a force acting on the diaphragm of a loudspeaker or microphone.

rectilinear Following a straight line.

rectilinear scanning The process of scanning an area in a predetermined sequence of narrow, straight parallel strips.

recurrence rate *Repetition rate.*

recycling Returning to an original condition, as to 0 or 1 in a counting circuit.

recycling detector A detector in which the capacitor across the detector output is discharged by a switching circuit just before each new carrier cycle, to give more complete elimination of the carrier and higher output.

Redeye A man-transportable shoulder-fired guided missile that has infrared guidance, used by combat troops to destroy low-flying enemy aircraft.

red gun The electron gun whose beam strikes phosphor dots emitting the red primary color in a three-gun color television picture tube.

rediffusion Wired radio in which a radio program is picked up from a broadcast station and distributed to the loudspeakers of subscribers over telephone or other wire circuits. Used in some parts of Great Britain.

redistribution The alteration of charges on an area of a storage surface by secondary electrons from any other area of the surface in a charge-storage tube or television camera tube.

redox system A chemical reduction-oxidation system, in which a potential is set up at an electrode made from an inert material such as platinum.

red restorer The DC restorer for the red channel of a three-gun color television picture tube circuit.

reduction Transformation of raw data to some usable form.

redundancy Any deliberate duplication or partial duplication of circuitry or information to decrease the probability of a system or communication failure.

redundancy check A forbidden-combination

check that uses redundant digits called check digits to detect errors made by a computer.

redundant digit A digit that is not necessary for an actual computation but reveals a malfunction in a digital computer.

red video voltage The signal voltage output from the red section of a color television camera, or the signal voltage between the matrix and the grid of the red gun in a three-gun color television picture tube.

reed frequency meter *Vibrating-reed frequency meter.*

reed relay A relay that has contacts mounted on magnetic reeds sealed into a length of small glass tubing. An actuating coil is wound around the tubing or wound on an auxiliary ferrite-core structure, to provide the magnetic field required for relay operation. The contacts may be dry or mercury-wetted. Also called magnetic reed relay.

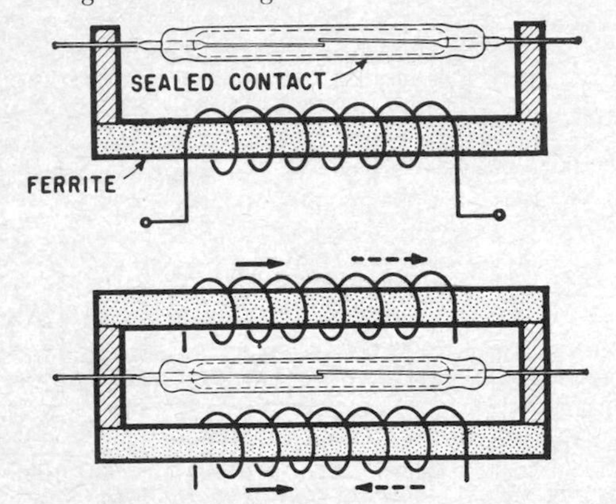

Reed relays with ferrite-core structures.

reed switch A switch that has contacts mounted on ferromagnetic reeds sealed in a glass tube, designed for actuation by an external magnetic field. The contacts may be dry or mercury-wetted. A reed switch is the contact assembly of a reed relay. Also called magnetic reed switch.

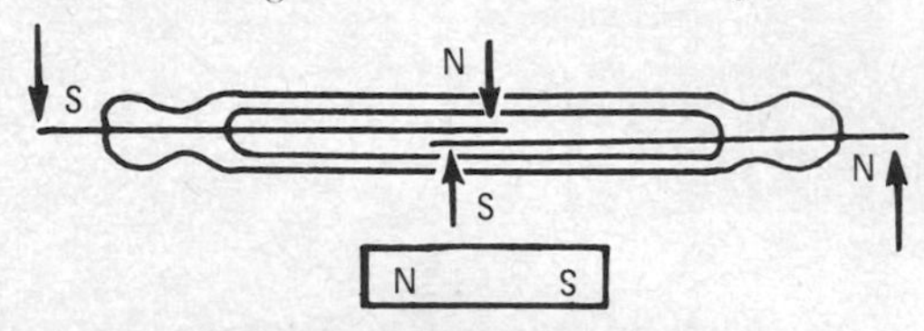

Reed switch. When a permanent magnet with polarity as shown is brought up to switch, reeds become magnetized and attract each other, closing switch.

reel A container that consists of a core and flanged ends, used for magnetic tape.

reel pack A package of small axial-lead components that have the ends of their leads affixed to a tape belt wound on a reel, for use in automatic component-inserting machines.

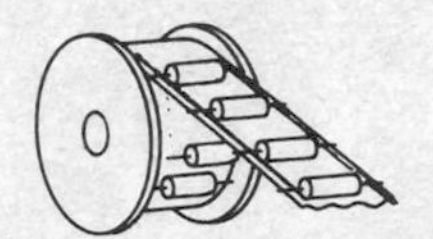
Reel pack for resistors.

reentrant Having one or more sections directed inward, as in certain types of cavity resonators.

reentrant oscillator An oscillator in which three coaxial-line resonators serve as tuning and feedback elements.

reentry The return of a missile or other object to the atmosphere of the earth after it has traveled through outer space.

reentry window The area, at the limits of the earth's atmosphere, through which a spacecraft in a given trajectory can pass to accomplish a successful reentry for a landing in a desired region.

reference black level The picture signal level that corresponds to a specified maximum limit for black peaks.

reference coupling The coupling between two circuits that gives a reading of 0 dBa (adjusted decibels) on a specified noise-measuring set connected to the disturbed circuit when a test tone of 90 dBa is impressed on the disturbing circuit. Expressed in decibels above reference coupling, abbreviated dBx.

reference dipole A straight half-wave dipole tuned and matched for a given frequency and used as a unit of comparison in antenna measurement work.

reference direction The direction used as a reference for angular measurements.

reference input An independently established signal used as a standard of comparison in a feedback control system.

reference input element A feedback control system element that establishes the relationship between the reference input and the command.

reference level The level used as a basis of comparison when designating the level of an AF signal in decibels or volume units. A common reference value in voltage, current, and power designations is 6 mW for 0 dB. For sound loudness, the reference level is usually the threshold of hearing. For communication receivers, the commonly used level is 60 μW.

reference line A line from which angular measurements are made.

reference noise The power level used as a basis of comparison when designating noise power in dBrn. The reference usually used is 1 pW (−90 dBm) at 1 kHz.

reference phase The phase of the color-burst signal voltage in a color television receiver, or the phase of the master-oscillator voltage in a color television transmitter.

reference pressure The pressure standard

used in the sealed chamber of a differential pressure gage. It is usually 1 atmosphere.

reference recording Recording a radio program for future reference.

reference source *Radioactive standard.*

reference stimulus A reference quantity that is applied to a telemetering system, such as a reference phase, pressure, or voltage, for calibrating purposes.

reference temperature A standard temperature source used as a reference for calibration of high-temperature measuring devices.

reference time An instant near the beginning of switching that is chosen as a reference for time measurements in a digital computer. It may be the first instant at which the instantaneous value of the drive pulse, the voltage response of the magnetic cell, or the integrated voltage response reaches a specified fraction of its peak pulse amplitude.

reference voltage An AC voltage used for comparison, usually to identify an in-phase or out-of-phase condition in an AC circuit.

reference volume The audio volume level that gives a reading of 0 VU on a standard volume indicator. The sensitivity of the volume indicator is adjusted so reference volume or 0 VU is read when the instrument is connected across a 600-Ω resistance to which is delivered a power of 1 mW at 1 kHz.

reference white 1. The light from a nonselective diffuse reflector that receives the normal illumination of a scene. 2. The standard white color used as a reference for specifying all other colors. The reference white used in color television approximates direct sunlight or sky light that has a color temperature of 6500 K. Primary colors are specified in units such that one unit of each primary will combine to produce reference white.

reference white level The picture signal level that corresponds to a specified maximum limit for white peaks.

reflectance *Reflection factor.*

reflected impedance The impedance value that appears to exist across the primary of a transformer when an impedance is connected as a load across the secondary.

reflected-light scanning The scanning of changes in the magnitude of the light reflected from the surface of an illuminated web.

reflected resistance The resistance value that appears to exist across the primary of a transformer when a resistive load is across the secondary.

reflected wave A wave reflected from a surface, discontinuity, or junction of two different media, such as the sky wave in radio, the echo wave from a target in radar, or the wave that travels back to the source end of a mismatched transmission line.

reflecting curtain A vertical array of half-wave reflecting antennas, generally used one quarter-wavelength behind a radiating curtain of dipoles to form a high-gain antenna.

reflecting target A target that reflects radar waves.

reflection The return or change in direction of light, sound, radar, or radio waves striking a surface or traveling from one medium into another. If the reflecting surface is smooth enough so each incident ray gives rise to a reflected ray in the same plane, the effect is known as direct, regular, specular, and mirror reflection. If the surface is so rough that reflected rays are distributed in all directions according to the cosine law, the effect is called diffuse reflection. Reflections of electromagnetic energy can occur at a mismatch in a transmission line, causing standing waves.

reflection altimeter *Radio altimeter.*

reflection coefficient The ratio of some quantity associated with a reflected wave to the corresponding quantity in the incident wave at a given point, for a given frequency and mode of transmission. Also called mismatch factor, reflection factor, and transition factor.

reflection Doppler A system that uses Doppler frequency shift to measure position and/or velocity of an object not carrying a transponder.

reflection error An error in bearing indication due to wave energy that reaches a navigation receiver by undesired reflections.

reflection factor 1. The ratio of the load current that would be delivered by a generator to a particular load without matching to the load current obtained when generator and load impedances are matched. 2. The ratio of the total luminous flux reflected by a given surface to the incident flux. Also called reflectance. 3. The ratio of electrons reflected to electrons entering a reflector space, as in a reflex klystron. 4. *Reflection coefficient.*

reflection grating A wire grating placed in a waveguide to reflect one desired wave while allowing one or more other waves to pass freely.

reflection interval The time interval between the transmission of a radar pulse or wave and the reception of the reflected wave at the point of transmission. Multiplying this time interval in microseconds by 492 gives target distance in feet; multiplying the time interval by 0.0931 gives target distance in standard miles. Multiplying microseconds by 0.15 gives target distance in kilometers.

reflection law The angle of incidence is equal to the angle of reflection.

reflection loss 1. The portion of the transition loss that is due to the reflection of power at the discontinuity. 2. The ratio in decibels of the power arriving at a discontinuity to the difference between the incident power and the reflected power.

reflection plotter An optical device used to superimpose on the face of a radar PPI tube a virtual image of a navigation chart. Used in radar piloting of ships near land. Another version is a transparent surface on which images can be marked with a grease pencil; any mark made on this plotting surface is reflected to a point directly below on the PPI display.

reflections Radio waves that have been reflected from a building, hill, or other conductive or semiconductive surface during their travel to a televi-

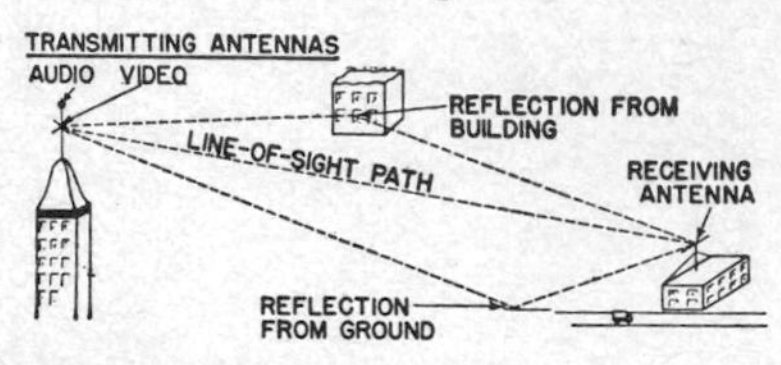

Reflections in television system can cause ghost images on receiver screen.

sion receiving antenna. The resulting longer travel time causes ghost images on the screen.

reflection seismograph A seismograph used in prospecting for underground oil deposits. A dynamite explosion near the surface produces sound waves that travel down to the oil strata and are there reflected back to the surface. Measurements of arrival times of the waves at seismographs give data for calculating the depth and extent of the underground oil pool.

reflective optics *Projection optics.*

reflectivity The fraction of incident radiant energy that is reflected from a uniformly irradiated surface.

reflectometer 1. A directional coupler that measures the power flowing in both directions through a waveguide, as a means of determining the reflection coefficient or standing-wave ratio. 2. A photoelectric instrument that measures the optical reflectance of a reflecting surface.

reflector 1. A single rod, system of rods, metal screen, or metal sheet used behind an antenna to increase its directivity. 2. A metal sheet or screen used as a mirror to change the direction of a microwave radio beam. 3. An electrode used in a reflex klystron or other electron tube to reflect a beam of electrons. 4. *Repeller.*

reflector element A single rod or other parasitic element serving as a reflector in an antenna array.

reflector space The space in a reflex klystron that follows the buncher space and is terminated by the reflector.

reflector voltage The voltage between the reflector electrode and the cathode in a reflex klystron.

reflex baffle A loudspeaker baffle in which a portion of the radiation from the rear of the diaphragm is propagated forward after controlled shift of phase or other modification, to increase the overall radiation in some portion of the AF spectrum. Also called vented baffle.

reflex bunching The bunching that occurs in an electron stream which has been made to reverse its direction in the drift space.

reflex circuit A circuit in which the signal is amplified twice by the same amplifier tube or tubes, once as an IF signal before detection and once as an AF signal after detection.

reflex klystron A single-cavity klystron in which the electron beam is reflected back through the

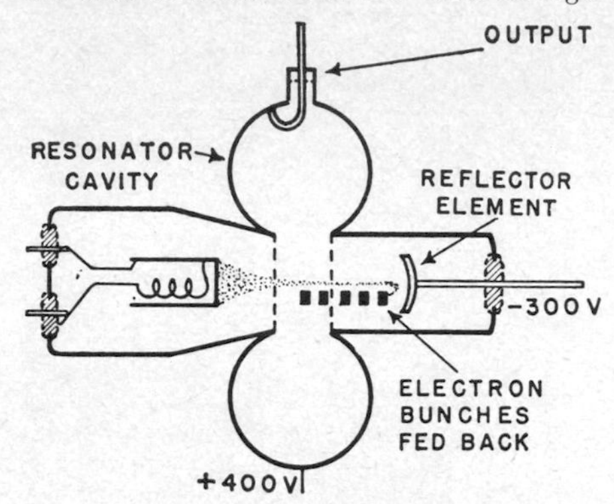

Reflex-klystron construction.

cavity resonator by a repelling electrode that has a negative voltage. Used as a microwave oscillator.

reflow soldering The application of heat to parts that have been previously coated with solder, to form a joint by reflowing the solder.

refracted wave The portion of an incident wave that travels from one medium into a second medium. Also called transmitted wave.

refraction The bending of a heat, light, radar, radio, or sound wave as it passes obliquely from

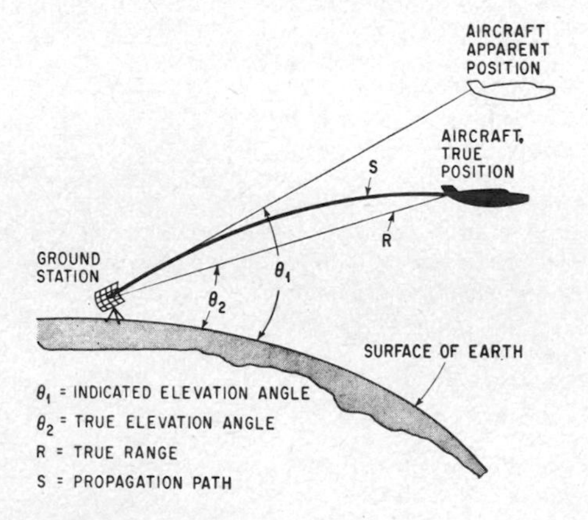

Refraction of radar waves in atmosphere. From ground station, aircraft appears to be at upper position because radar antenna is aimed there.

one medium to another in which the velocity of propagation is different.

refraction error A radio bearing error due to

bending of one or more wave paths by undesired refraction.

refraction loss The portion of the transmission loss that is due to refraction resulting from nonuniformity of the medium.

refractive index The ratio of the phase velocity of a wave in free space to that in a given medium. The refractive index of air at sea level is 1.00029.

refractive modulus The excess over unity of the modified index of refraction in the troposphere, expressed in millionths and computed as $(n + h/a - 1)10^6$, where n is the index of refraction at a height h above sea level and a is the radius of the earth.

refractivity The refractive index minus 1.

refractometer 1. An instrument that measures the refractive index of a liquid or solid, usually by measuring the critical angle at which total reflection occurs. 2. An instrument that measures the refractivity of the atmosphere, which is proportional to the dielectric constant of the air being measured. One version uses two capacitors, one hermetically sealed to serve as a reference and the other open to the air being sampled. Since capacitance varies with dielectric constant, a comparison of the two capacitances gives the difference in the dielectric constants, and this can be readily converted into the refractivity of the air being sampled.

refrangible Capable of being refracted.

regenerate 1. To restore pulses to their original shape. 2. To restore stored information to its original form in a storage tube, to counteract fading and disturbances.

regeneration 1. Replacement or restoration of charges in a charge-storage tube to overcome decay effects, including loss of charge by reading. 2. Restoration of contaminated nuclear fuel to a usable condition. 3. *Positive feedback.*

regeneration control A variable capacitor, variable inductor, potentiometer, or rheostat used in a regenerative receiver to control the amount of feedback and thereby keep regeneration within useful limits.

regeneration period The time interval in which the screen of a cathode-ray storage tube is scanned by the beam to regenerate the charge distribution that represents the stored information. Also called scan period.

regenerative amplifier An amplifier that uses positive feedback to give increased gain and selectivity.

regenerative braking *Dynamic braking.*

regenerative detector A vacuum-tube detector circuit in which RF energy is fed back from the anode circuit to the grid circuit to give positive feedback at the carrier frequency, thereby increasing the amplification and sensitivity of the circuit.

regenerative fuel cell A fuel cell in which the reaction product is processed to regenerate the reactants. This type is of particular importance for spacecraft, where electricity from a solar-cell array could be used to electrolyze the water formed during the shade part of the orbit when the fuel cells are in use.

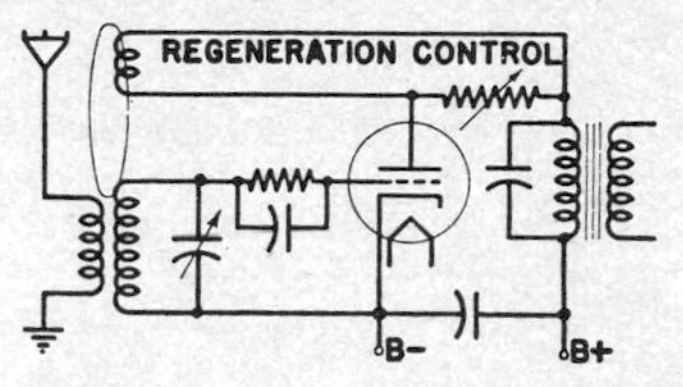

Regenerative-detector circuit.

regenerative receiver A radio receiver that uses a regenerative detector.

regenerative repeater A repeater that performs pulse regeneration to restore the original shape of a pulse signal used in teletypewriter and other code circuits. Each code element is replaced by a new code element that has specified timing, waveform, and magnitude.

regenerative storage Storage in which periodic regeneration of data is required to prevent its loss. One example is a delay line used for storage.

regional channel A standard radio broadcast channel in which several stations may operate with powers not in excess of 5 kW.

register 1. The computer hardware for holding temporarily a specific number of bits, usually equal to one or more machine words. 2. Accurate matching or superimposition of two or more patterns, such as the three images in color television or the patterns on opposite sides of a printed-circuit board. 3. To superimpose two or more designs or images with precise alignment. 4. *Registration.*

register control Automatic control of the position of a printed design with respect to reference marks or some other part of the design, as in photoelectric register control.

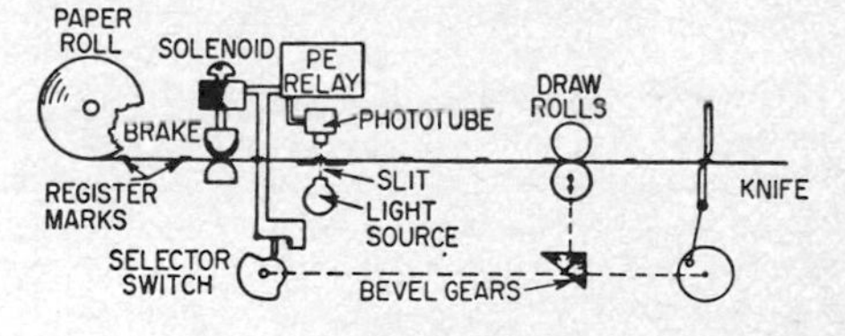

Register control used to cut paper strip at printed register marks.

register length The number of characters that can be stored by a register in a computer.

register mark A mark or line printed or otherwise impressed on a web of material for use as a reference to maintain register.

registration Exact superimposition of all three color images on the screen of a color television receiver, or superimposition of all colors of a design on a printed sheet. Also called register.

regular reflection *Direct reflection.*

regulated power supply A power supply that contains means for maintaining essentially constant output voltage or output current under changing load conditions.

regulating system *Automatic control system.*

regulation 1. The process of holding constant a quantity such as speed, temperature, voltage, or position by an electronic or other system that automatically corrects errors by feeding back into the system the condition being regulated. Regulation thus is based on feedback, whereas control is not. 2. The change in output voltage that occurs between no load and full load in a transformer, generator, or other source. Dividing this change by the rated full-load value and multiplying the result by 100 gives percent regulation. 3. The difference between the maximum and minimum tube voltage drops within a specified range of anode current in a gas tube.

regulator A device that maintains a desired quantity at a predetermined value or varies it according to a predetermined plan.

regulator tube A two-electrode gas tube that has a glow discharge with an essentially constant voltage drop. When connected in series with a resistance across a DC source, a regulator tube will maintain a constant voltage across its terminals for wide variations in the source voltage.

Regulus A Navy surface-to-surface guided missile. Regulus II uses inertial guidance, has a speed of about Mach 2, has a range of over 1000 mi (1600 km), and can carry either conventional or nuclear warheads.

reignition A process by which multiple counts are generated within a radiation-counter tube by atoms or molecules excited or ionized in the discharge accompanying a tube count.

reignition voltage The voltage that is just sufficient to reestablish conduction if applied to a gas tube during the deionization period. The value of this voltage varies inversely with time during the deionization period. Also called restriking voltage.

Reinartz crystal oscillator A crystal-controlled vacuum-tube oscillator in which the crystal current is kept low by placing in the cathode lead a resonant circuit tuned to half the crystal frequency. The resulting regeneration at the crystal frequency improves efficiency without the danger of uncontrollable oscillation at other frequencies.

reinserter *DC restorer.*

reinsertion of carrier Combining a locally generated carrier signal with an incoming suppressed-carrier signal.

rejection band The band of frequencies below the cutoff frequency in a uniconductor waveguide.

rejector *Trap.*

rejector circuit *Band-elimination filter.*

rel A unit of reluctance, equal to 1 ampere-turn per magnetic line of force.

relative abundance The fraction or percentage of the atoms of an element in a given isotope.

relative address A designation for the position of a memory location in a computer routine or subroutine. Relative addresses are translated into absolute addresses by adding a reference address, such as that at which the first word of the routine is stored.

relative aperture The ratio of the minimum vertical or horizontal clearance for particle passage to the particle orbit radius in the accelerating chamber of an accelerator.

relative bearing A bearing in which the heading of the vehicle serves as the reference line.

relative biological effectiveness [abbreviated RBE] The effectiveness of an ionizing radiation in producing a specific biological damage, as compared to the effectiveness of a reference radiation such as 200-kV x-rays.

relative coding Computer coding in which all addresses refer to an arbitrarily selected position or in which all addresses are represented symbolically.

relative damping The ratio of damping torque at a given angular velocity of a moving element in an instrument to the damping torque that would produce critical damping at that same angular velocity.

relative heading Deprecated term for *heading.*

relative humidity The ratio of the amount of water vapor present in air to the amount that would saturate it at a given temperature.

relative luminosity The ratio of the value of the luminosity at a particular wavelength to the value at the wavelength of maximum luminosity.

relative permeability *Specific permeability.*

relative plateau slope The percent change in counting rate per unit change of applied voltage near the midpoint of the plateau of a radiation-counter tube.

relative refractive index The ratio of the refractive indices of two media.

relative response The ratio, usually expressed in decibels, of the response under some particular conditions to the response under reference conditions.

relative specific ionization The specific ionization for a particle of a given medium, relative either to that for the same particle and energy in a standard medium or the same particle and medium at a specified energy.

relative stopping power The ratio of the stopping power of a given substance to that of a standard substance like aluminum, oxygen, or air. Also called equivalent stopping power.

relative target bearing The bearing of a radar target expressed relative to the heading of a ship or aircraft.

relative time delay The difference in time delay encountered by the audio signal and the composite picture signal or between the components of the picture signal traveling over a television relay system.

relative velocity The time rate of change of a position vector of a point with respect to a reference frame.

relativistic mass The mass of a particle moving at a velocity exceeding about one-tenth the velocity of light. The relativistic mass is significantly larger than the rest mass.

relativistic mass equation The equation for the relativistic mass of a particle or body that has a given rest mass and velocity.

relativistic particle A particle with a velocity so large that its relativistic mass exceeds its rest mass by a significant amount.

relativistic velocity A particle velocity sufficiently large that mass and other properties of the particle are significantly different from the at-rest values. Velocity is generally considered to be relativistic when it exceeds about one-tenth the velocity of light.

relativity A principle that postulates the interdependence of matter, space, and time in the universe, for various frames of reference.

relaxation generator *Relaxation oscillator.*

relaxation inverter An inverter that uses a relaxation oscillator circuit to convert DC power to AC power.

relaxation length The distance nuclear radiation must pass through a given material to reduce its intensity to $1/e$ of its original value, not including geometric effects.

relaxation oscillator An oscillator whose fundamental frequency is determined by the time of charging or discharging a capacitor or coil

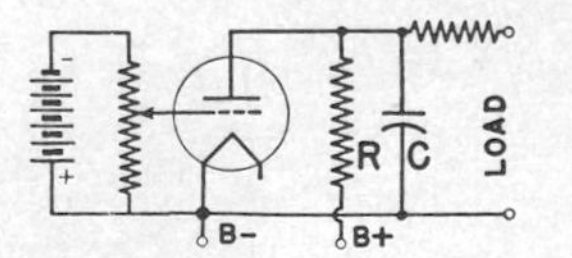

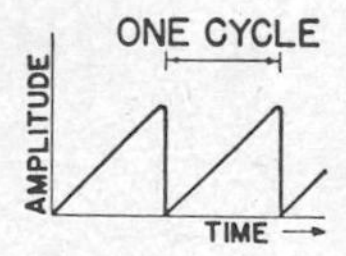

Relaxation-oscillator circuit and its sawtooth output waveform.

through a resistor, producing waveforms that may be rectangular or sawtooth. Also called relaxation generator.

relaxation time 1. The time constant required for an abrupt change of magnetizing force to make the magnetic induction reach a specified percent of its new value. 2. The travel time of an electron in a metal before it is scattered and loses its momentum.

relay [symbol K] 1. A device that is operated by a variation in the conditions in one electric circuit and makes or breaks one or more connections in the same or another electric circuit. The most common types are electromagnetic, reed, solid-state, and thermal relays. 2. A microwave or other radio system used to pass a signal from one radio communication link to another.

relay armature The movable iron part of a relay.

relay broadcast station A station licensed to transmit, from points where wire facilities are not available, programs for broadcast by one or more broadcast stations.

relay channel The band of frequencies used in transmitting a single television relay signal, including the guard bands.

relay coil One or more windings on a common form, used with an iron core to form a relay magnet.

relay computer A computer that consists chiefly of electromagnetic relays or similar electromechanical devices.

relay contact One of the pair of contacts that are closed or opened by the movement of the armature of a relay.

relay hum The low-frequency sound produced by a relay when its coil is energized by alternating current or unfiltered rectified current.

relay magnet The electromagnet that attracts the armature of a relay when energized.

relay rack *Rack.*

relay radar Deprecated term for *radar relay.*

relay receiver A receiver that accepts a television relay input signal and delivers a television relay output signal to the transmitter portion of a repeater station.

relay station *Repeater station.*

relay system 1. An assembly of relays used for switching purposes. 2. *Radio repeater.*

release A mechanical, electromagnetic, or other arrangement of parts for holding or freeing a device or mechanism as required.

release time The total elapsed time from the instant that relay coil current starts to drop until the make contacts have opened or the break contacts have closed.

reliability The probability that a device will perform its purpose adequately for the period of time intended under the operating conditions encountered.

reliability engineering A field of engineering that deals with the prevention and correction of malfunctions in equipment.

reliability index A quantitative figure of merit related to the reliability of a piece of equipment, such as the number of failures per 1000 operations or the number of failures in a specified number of operating hours.

reliability test A test designed specifically to evaluate the level and uniformity of reliability of equipment under various environmental conditions.

reluctance A measure of the opposition presented to magnetic flux in a magnetic circuit. Reluctance is the reciprocal of permeance, and is therefore equal to magnetomotive force divided by magnetic flux. The unit of reluctance is the rel, equal to 1 ampere-turn per magnetic line of force.

reluctance microphone *Variable-reluctance microphone.*

reluctance motor A synchronous motor, similar in construction to an induction motor, in which the member carrying the secondary circuit has salient poles but no DC excitation. It starts as an induction motor but operates normally at synchronous speed.

reluctance pickup *Variable-reluctance pickup.*

reluctivity The ratio of the magnetic intensity in a region to the magnetic induction in the same region. Reluctivity is the reciprocal of magnetic permeability.

rem Abbreviation for *roentgen equivalent man.*

remanence The magnetic flux density that remains in a magnetic circuit after the removal of an applied magnetomotive force. If the magnetic circuit has an air gap, the remanence will be less than the residual flux density.

remanent charge The charge that remains in a ferroelectric device when the applied voltage is removed.

remanent induction The induction that remains in a magnetic material when the magnetomotive force around the magnetic circuit is zero.

Remendur A high-performance magnetic material that can have a remanence as high as 21.5 kG. It is a malleable, ductile, cobalt-iron-vanadium alloy. Developed at Bell Laboratories.

remodulation Transferring modulation from one carrier to another.

remodulator A circuit that converts amplitude modulation to audio frequency-shift modulation for transmission of facsimile signals over a voice-frequency radio channel.

remote *NEMO.*

remote control Control of equipment from a distance over wires or by light, radio, sound, ultrasonic waves, or other means.

remote cutoff The characteristic wherein a large negative bias is required for complete cutoff of output current in an electron tube or other amplifying device.

remote indicator 1. An indicator at a distance from the data-gathering sensing element, with data being transmitted to the indicator mechanically, electrically over wires, or by light, radio, or sound waves. 2. *Repeater.*

remote line A program transmission line that runs between a remote-pickup point and a broadcast studio or transmitter site.

remotely piloted vehicle [abbreviated RPV] A robot aircraft, controlled over a two-wave radio link from a ground station or mother aircraft that can be hundreds of miles away. Electronic guidance is generally supplemented by remote-control television cameras feeding monitor receivers at the control station.

remote metering *Telemetering.*

remote pickup Picking up a radio or television program at a remote location and transmitting it to the studio or transmitter over wire lines or a shortwave or microwave radio link.

remote programming Control of the output voltage or current of a regulated power supply by a remotely varied resistance or voltage.

remote sensing Acquisition of information by sensors that are not in physical contact with the object under study, such as by measuring and recording infrared radiation from the area.

rendezvous radar Radar designed for use in orbital rendezvous and docking in space.

renormalization of mass Adding to the mechanical mass of a particle its extra mass due to self-interaction, to give a sum equal to the measured mass. Used in quantized field theory.

rep 1. Abbreviation for *roentgen equivalent physical.* 2. Abbreviation for *representative.*

repeatability 1. A measure of the variation in the readings of an instrument when identical tests are made under fixed conditions. Also called reproducibility. 2. The ability of a voltage regulator or voltage reference tube to attain the same voltage drop at a stated time after the beginning of any conducting period.

repeater 1. An amplifier or other device that receives weak signals and delivers corresponding stronger signals with or without reshaping of waveforms. It may be either a one-way or two-way repeater. Carrier, telegraph, and telephone repeaters are used in wire lines, whereas radio repeaters act directly on radio waves. 2. An indicator that shows the same information as is shown on a master indicator. Synchros are widely used for this purpose. Also called remote indicator.

repeater jammer A jammer that intercepts an enemy radar signal and reradiates the signal after modifying it to incorporate erroneous data on azimuth, range, or number of targets.

repeater station A station that contains one or more repeaters. Also called relay station.

repeating timer A timer that continues repeating its operating cycle until excitation is removed.

repeat-point tuning *Double-spot tuning.*

repeller An electrode whose primary function is to reverse the direction of an electron stream in an electron tube. Also called reflector.

reperforator *Tape reperforator.*

repetition frequency *Repetition rate.*

repetition rate The rate at which recurrent signals are produced or transmitted. Also called recurrence rate and repetition frequency.

repetitive error The maximum deviation of the controlled variable from the average value upon successive return to specified operating conditions following specified deviation in an automatic control system.

replacement tube An electron tube suitable for use in place of another tube that has the same type number but not necessarily made by the same manufacturer.

reply An RF signal or combination of signals transmitted by a transponder in response to an interrogation. Also called response.

report An output document prepared by a data-processing system.

report generation A technique for producing complete data-processing reports by giving only a description of the desired content and format, plus certain information concerning the input file.

representative [abbreviated rep] *Sales representative.*

reproduce head *Playback head.*

reproducer A punched-card machine that reads a punched card and duplicates part or all of its contents by punching another card.

reproducibility *Repeatability.*

reproducing stylus *Stylus.*

reproduction speed The area of copy recorded per unit of time by a facsimile receiver.

reprogrammable ROM [abbreviated REPROM] A programmable read-only memory that can be reprogrammed in the field as often as desired.

REPROM Abbreviation for *reprogrammable ROM.*

repulsion A mechanical force that tends to separate bodies which have electric charges or like magnetic polarity, or in the case of adjacent conductors, which have currents flowing in opposite directions.

repulsion-induction motor A repulsion motor that has a squirrel-cage winding in the rotor in addition to the repulsion-motor winding.

repulsion motor An AC motor that has stator windings connected directly to the source of AC power and rotor windings connected to a commutator. Brushes on the commutator are short-circuited and are positioned to produce the rotating magnetic field required for starting and running. This type of motor varies considerably in speed as load is changed.

repulsion-start induction motor An AC motor that starts as a repulsion motor. At a predetermined speed the commutator bars are short-circuited to give the equivalent of a squirrel-cage winding for operation as an induction motor with constant-speed characteristics.

reradiation Undesirable radiation of signals generated locally in a radio receiver, causing interference or revealing the location of the receiver.

rerecording The process of making a recording by reproducing a recorded sound source and recording this reproduction.

rerecording system A system of reproducers, mixers, amplifiers, and recorders used to combine or modify various sound recordings to provide a final sound record.

rerun To run a program or a portion of it over on a computer. Also called rollback.

rerun point A point in a computer program at which all information is available for rerunning the last-run portion of a problem when an error is detected. Several such points are usually provided for in a program, to eliminate the need for rerunning the entire problem.

rerun routine A routine used after a computer malfunction, a coding error, or an operating mistake to reconstitute a routine from the last previous rerun point. Also called rollback routine.

research Scientific investigation aimed at discovering and applying new facts, techniques, and natural laws.

research and development The conception, design, and first creation of experimental or prototype operational devices.

reserve battery A battery that has long shelf life in an unenergized state. Methods of activating the battery include heat, addition of an electrolyte or

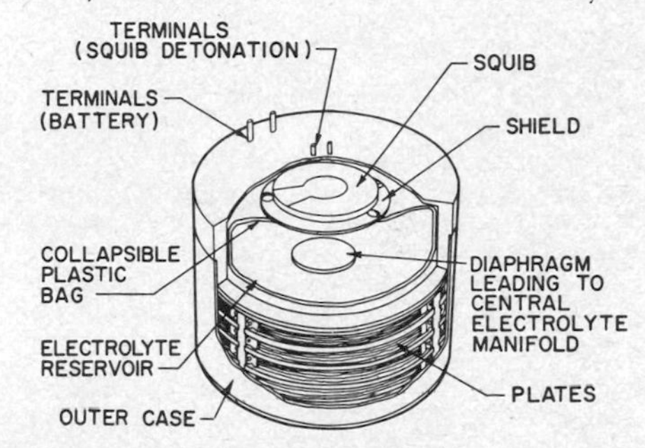

Reserve battery in which firing of explosive squib releases electrolyte from plastic bag to activate magnesium and mercuric oxide plates of battery.

water, or use of mechanical shock or other means for breaking a diaphragm or container so as to force electrolyte into the battery.

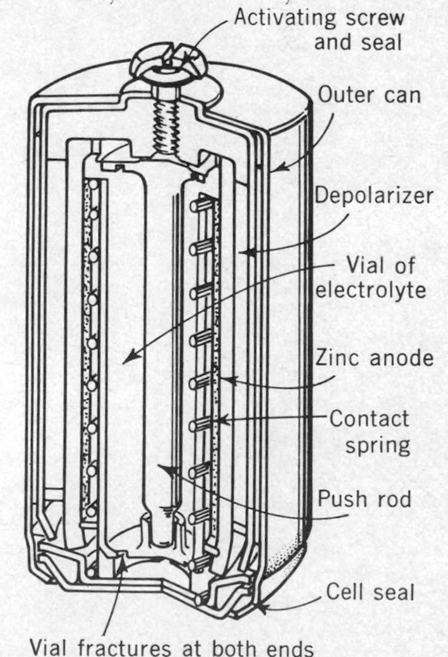

Reserve cell that is activated by turning screw at top until inner vial breaks and releases electrolyte.

reserve cell A cell used in a reserve battery.

reset *Clear.*

reset action Floating action in which the final control element is moved at a speed proportional to the extent of proportional-position action. The speed of corrective action is thus determined by both the amount and duration of the deviation from the desired value.

reset control circuit The magnetic amplifier circuit that resets the flux in the core of the saturable reactor.

reset flux level The difference between the saturation and reset levels of the flux in the core of a saturable reactor.

reset pulse 1. A drive pulse that tends to reset a magnetic cell in the storage section of a digital computer. 2. A pulse used to reset an electronic counter to zero or to some predetermined position.

reset rate The number of corrections made per minute by a control system.

reset switch A machine-operated switch that restores normal operation to a control system after a corrective action.

resettability The ability of the tuning element of an oscillator to retune the oscillator to the same operating frequency for the same set of input conditions.

resetting half-cycle The half-cycle in the AC supply voltage of a magnetic amplifier in which the flux in the core of the saturable reactor is reset.

resetting interval The portion of the resetting half-cycle in which the flux in the core of a saturable reactor is actually changing from the saturation level to the reset flux level.

reshaping circuit A circuit that changes the waveform of a current, such as a limiter.

residual activity Radioactivity that remains in a substance or system at a specified time after a period of decay.

residual charge The charge that remains on the plates of a capacitor after an initial discharge.

residual current The current that flows through a thermionic diode when there is no anode voltage, due to the velocity of the electrons emitted by the heated cathode.

residual discharge A discharge of the residual charge remaining after the initial discharge of a capacitor.

residual error The sum of random errors and uncorrected systematic errors.

residual field The magnetic field left in an iron core after excitation has been removed.

residual flux density The magnetic flux density at which the magnetizing force is zero when the material is in a symmetrically and cyclically magnetized condition. Also called residual induction, residual magnetic induction, and residual magnetism.

residual gap The length of the magnetic air gap between the armature and the center of the core face of an energized relay.

residual gas The small amount of gas that remains in a vacuum tube after the best possible exhaustion by vacuum pumps. Much of this residual gas is removed by the getter.

residual induction *Residual flux density.*

residual ionization Ionization of air or other gas in a closed chamber, not accounted for by recognizable neighboring agencies. It is now attributed to cosmic rays.

residual magnetic induction *Residual flux density.*

residual magnetism *Residual flux density.*

residual modulation *Carrier noise level.*

residual nucleus The heavy nucleus that is the end product of a nuclear transformation.

residual pin A nonmagnetic pin or screw attached to the armature or core of a relay to prevent the armature from touching and sticking to the core.

residual radiation Nuclear radiation emitted by radioactive material deposited after an atomic burst, including fission products, unfissioned nuclear material, and material in which radioactivity may have been induced by neutron bombardment.

residual radioactivity Radioactivity remaining after a reactor has been shut down.

residual range The distance in which a particle can still produce ionization after having lost some of its energy in passing through matter.

residual resistance The portion of the electric resistance of a metal that is independent of temperature.

residual resistivity The constant minimum value of electric resistivity of a metal with decreasing temperature, reached near absolute zero. The residual resistivity is not a characteristic of a particular metal, but varies with impurities in the sample. The purer the specimen, the lower its residual resistivity.

resin A natural or synthetic organic product used widely in plastics, adhesives, and surface coatings.

resist An acid-resistant nonconducting coating used to protect desired portions of a wiring pattern from the action of the etchant during manufacture of printed-wiring boards.

resistance [abbreviated R; mathematical symbol R] The opposition that a device or material offers to the flow of direct current, measured in ohms, kilohms, or megohms. In AC circuits, resistance is the real component of impedance.

resistance box A box that contains a number of precision resistors connected to panel terminals or contacts in such a way that a desired resistance value can be obtained by withdrawing plugs (as in a post office bridge) or by setting multicontact switches. Usually constructed as a decade box, wherein individual resistance values vary in submultiples and multiples of 10.

resistance braking *Dynamic braking.*

resistance bridge *Wheatstone bridge.*

resistance-bridge controller A controller that operates on the null balance principle, in which balancing of the bridge produces an error signal which is amplified for use in keeping the load at a desired temperature or other parameter.

resistance-capacitance [abbreviated RC] Containing both resistance and capacitance, as provided by resistors and capacitors.

resistance-capacitance circuit [abbreviated RC circuit] A circuit that contains resistors and capacitors which determine the time constant. The time constant in seconds is equal to the product of resistance in ohms and capacitance in farads.

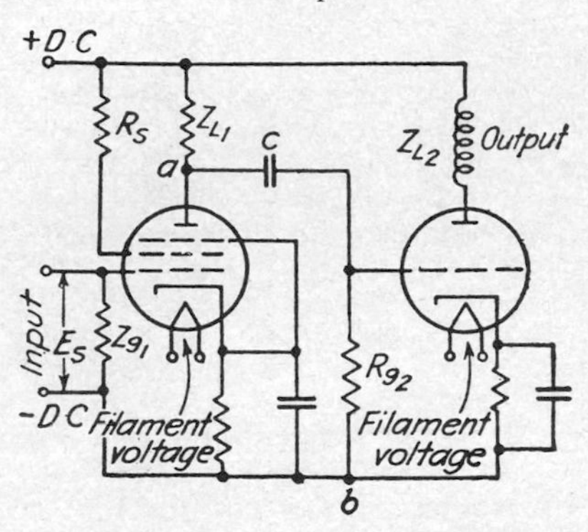

Resistance-capacitance-coupled amplifier.

resistance-capacitance-coupled amplifier An amplifier in which resistance coupling is used between stages.

resistance-capacitance coupling *Resistance coupling.*

resistance-capacitance differentiator [abbreviated RC differentiator] A resistance-capacitance circuit that produces an output voltage whose amplitude is proportional to the rate of change of the input voltage. A square-wave input thus produces sharp output voltage spikes.

resistance-capacitance filter [abbreviated RC filter] A filter that contains only resistance and capacitance elements (no inductance). Widely used in rectifier-type power supplies to reduce ripple.

resistance-capacitance oscillator [abbreviated RC oscillator] An oscillator in which the frequency is determined by resistance-capacitance elements.

resistance-coupled amplifier Commonly used term for an amplifier in which a capacitor provides a path for signal currents from one stage to the next, with resistors connected from each side of the capacitor to the power supply or to ground. It can amplify AC signals but cannot handle small changes in direct currents. Also called RC amplifier.

resistance coupling Coupling in which resistors are used as the input and output impedances of the circuits being coupled. A coupling capacitor is generally used between the resistors to transfer

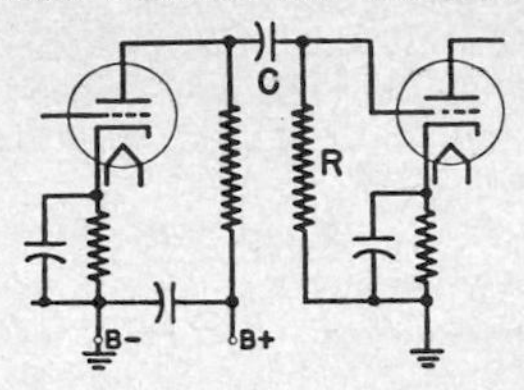

Resistance coupling between two tube stages.

the signal from one stage to the next. Also called RC coupling, resistance-capacitance coupling, and resistive coupling.

resistance decade A decade box that contains an assembly of high-accuracy resistors whose individual values are related by submultiples or multiples of 10. Each section or decade contains 10 equal-value resistors connected in series, with provisions for making a connection to any junction. The decades are similarly connected in series, with each decade having resistors 10 times larger in value than those in the next lower decade. A resistance decade can thus be set to any desired value within its range, in steps that are multiples of the smallest resistance value.

resistance drop 1. The voltage drop that occurs between two points on a conductor due to the flow of current through the resistance of the conductor. Multiplying the resistance in ohms by the current in amperes gives the voltage drop in volts. 2. *IR drop.*

resistance element An element of resistive material in the form of a grid, ribbon, or wire, used singly or built into groups to form a resistor for heating purposes, as in an electric soldering iron.

resistance furnace An electric furnace in which the heat is developed by the passage of current through a suitable internal resistance that may be the charge itself, a resistor imbedded in the charge, or a resistor surrounding the charge.

resistance hybrid A hybrid junction that consists entirely of resistors. Operation is essentially independent of frequency up to several hundred megahertz, but attenuation along desired paths is greater than with a hybrid transformer. Also called resistance junction.

resistance-inductance-capacitance circuit [abbreviated RLC circuit] A circuit that contains inductance, capacitance, and resistance.

resistance junction *Resistance hybrid.*

resistance loss Power loss due to current flowing through resistance. Its value in watts is equal to the resistance in ohms multiplied by the square of the current in amperes.

resistance magnetometer A magnetometer that depends for its operation on variations in the electric resistance of a material immersed in the magnetic field to be measured.

resistance material A material that has sufficiently high resistance per unit length or volume to permit its use in the construction of resistors.

resistance pad A pad that uses only resistances. Used to provide attenuation without altering frequency response.

resistance pyrometer A pyrometer in which the heat-sensing element is a length of wire whose resistance varies greatly with temperature.

resistance standard *Standard resistor.*

resistance-start motor A split-phase motor that has a resistance connected in series with the auxiliary winding. The auxiliary circuit is opened when the motor attains a predetermined speed.

resistance strain gage A strain gage that consists of a strip of material which is cemented to the part under test, and changes in resistance with elongation or compression.

resistance thermometer A thermometer in which the sensing element is a resistor whose resistance is an accurately known function of temperature.

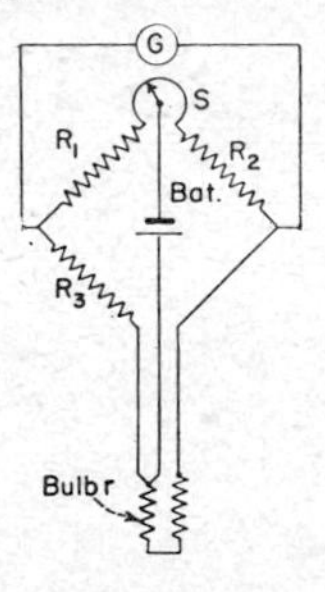

Resistance thermometer.

resistance welding Welding in which the metals to be joined are heated to melting temperature at their points of contact by a localized electric current while pressure is applied. The heating current, which may be thousands of amperes, is usually controlled electronically as to amplitude and duration. Examples of resistance welding include percussion, seam, and spot welding.

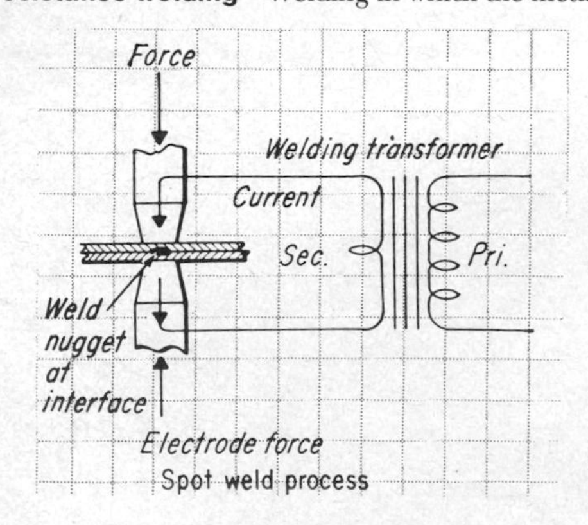

Resistance welding equipment.

resistance wire Wire made from a metal or alloy that has high resistance per unit length, such as Nichrome. Used in wirewound resistors and heating elements.

resistive conductor A conductor used primarily because it has high resistance per unit length.

resistive coupling *Resistance coupling.*

resistive-wall amplifier An electron-beam amplifier in which gain is obtained by interaction between the electron beam and a charge induced by the beam in an adjacent resistive wall. The resulting bunching action increases exponentially with distance, giving a gain comparable to that of other traveling-wave tubes.

resistivity The resistance in ohms that a unit volume of a material offers to the flow of current. Resistivity is the reciprocal of the conductivity of a material. Measured in ohm-centimeters. The resistivity of a wire sample in ohm-centimeters is equal to the resistance R in ohms multiplied by the cross-sectional area A in square centimeters and the result divided by the sample length L in centimeters (resistivity = RA/L). Resistivity is also expressed in ohms per circular mil foot, which is the resistance of a sample that is 1 circular mil in cross section and 1 ft long. Also called specific resistance.

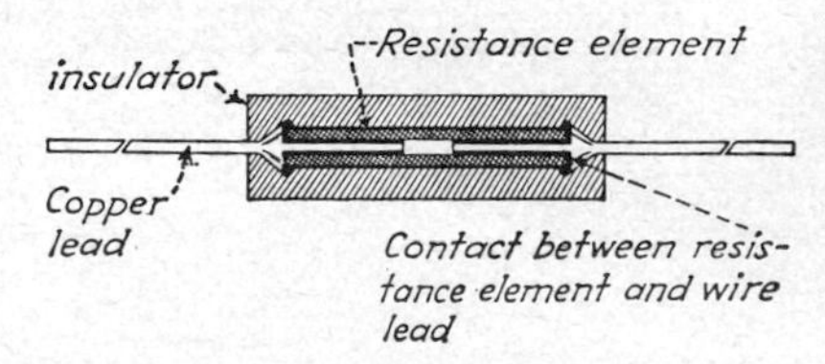

Resistor having carbon resistance element and axial leads.

resistor [abbreviated R] A device designed intentionally to have a definite amount of resistance. Used in circuits to limit current flow or provide a voltage drop.

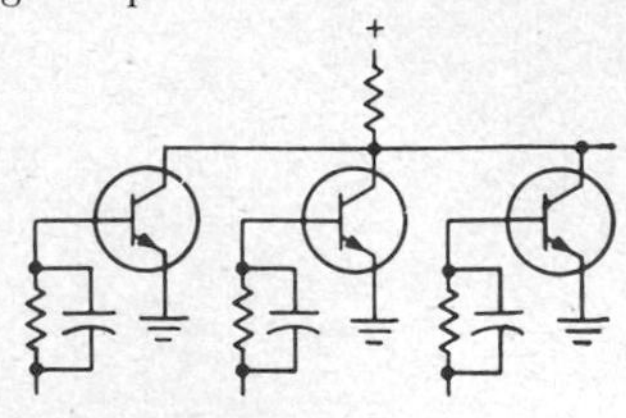

Resistor-capacitor-transistor logic.

resistor-capacitor-transistor logic [abbreviated RCTL] Resistor-transistor logic in which capacitors are added to increase switching speed.

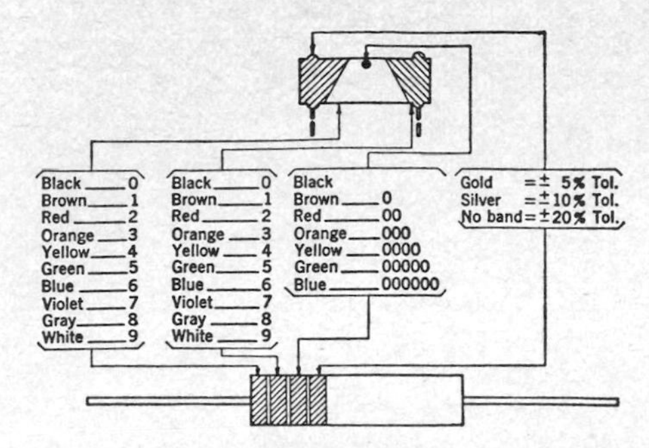

Resistor color code. On radial-lead resistors as at top, body color is read first, end color next, then dot; tolerance color on other end is easily recognized because it is always either gold or silver if present.

resistor color code A method of marking the value in ohms on a resistor by dots or bands of colors as specified in the EIA color code.

resistor core The insulating support on which a resistor element is wound or otherwise placed.

resistor element The portion of a resistor that provides resistance. It may be pure metal, an alloy, a metallic coating, a carbon-cement mixture, or a plastic that contains finely powdered metal.

resistor furnace A resistance furnace in which the heat is developed in a resistor that is not a part of the charge.

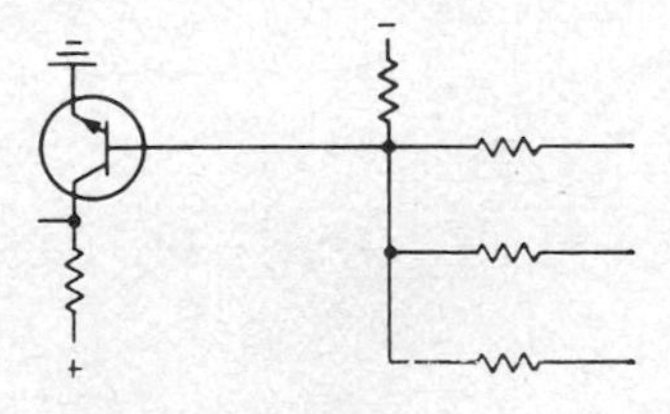

Resistor-transistor logic.

resistor-transistor logic [abbreviated RTL] Logic that uses resistors, with transistors serving only to invert the output.

resnatron [RESoNAtor-TRON] A microwave beam tetrode that contains cavity resonators, used chiefly for generating large amounts of continuous power at high frequencies. Noise-modulated 50-kW versions were used for jamming German airborne radars in World War II.

resolution A measure of ability to delineate detail or distinguish between nearly equal values of a quantity. In television, resolution is usually expressed as the maximum number of lines that can be discerned on the screen in a distance equal to tube height; this ranges from 350 to 400 lines for most receivers. In radar, resolution is the minimum separation between two targets in angle or range at which they can be distinguished on the radar screen. Also called resolving power.

resolution chart *Test pattern.*

resolution in azimuth The angle by which two targets at the same range must be separated in azimuth to be distinguishable on the screen of a particular radar set.

resolution in range The distance by which two targets at the same bearing must be separated in range to be distinguishable on the screen of a particular radar set.

resolution sensitivity The minimum change of measured variable that actuates an automatic control system.

resolution time The minimum time interval at which two successive voltage pulses can be registered by a counter.

resolution wedge A group of gradually converging lines on a test pattern, used to measure resolution in television.

resolver A synchro or other device whose rotor is mechanically driven to translate rotor angle into electric information corresponding to the sine and

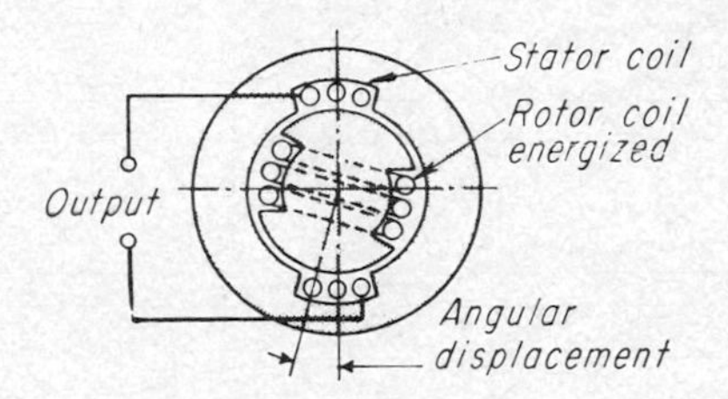

Resolver in which output voltage is exactly related to angular position of rotor.

cosine of rotor angle. Used for interchanging rectangular and polar coordinates. Also called sine-cosine generator and synchro resolver.

resolving power 1. The ability of a mass spectrometer to separate adjacent mass spectrum lines. 2. The reciprocal of the beam width of a unidirectional antenna, measured in degrees. 3. The ability of a radar set to form distinguishable images. 4. *Resolution.*

resolving time The minimum time interval between two distinct events that will permit them to be counted or otherwise detected by a particular circuit or device.

resonance [noun; also used as adjective in place of resonant] 1. The condition existing in a circuit when the inductive reactance balances out the capacitive reactance. 2. The condition existing in a body when the frequency of an applied vibration equals the natural frequency of the body. A body vibrates most readily at its resonant frequency. Also called velocity resonance. 3. The condition wherein a nuclear system in motion reacts to an external field or force applied at a natural vibration frequency, as in nuclear magnetic resonance.

resonance bridge A four-arm AC bridge that measures inductance, capacitance, or frequency.

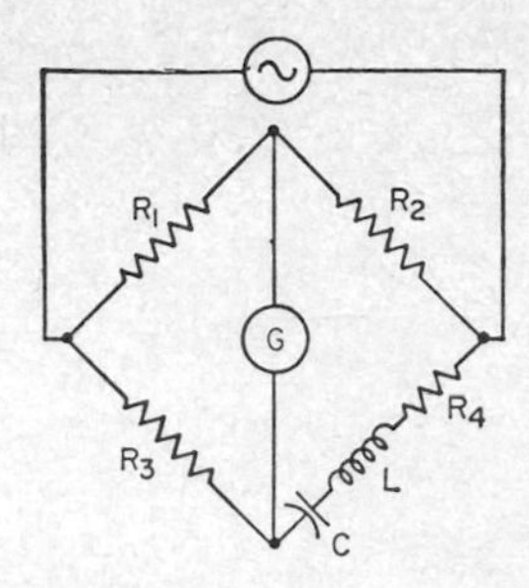

Resonance-bridge circuit.

The inductor and the capacitor, which may be either in series or in parallel, are tuned to resonance at the frequency of the source before the bridge is balanced.

resonance capture The capture of an incident particle by a nucleus in such a way that the particle enters a resonance level of the resultant compound nucleus.

resonance characteristic *Resonance curve.*

resonance current step-up The ability of a parallel resonant circuit to circulate through its coil

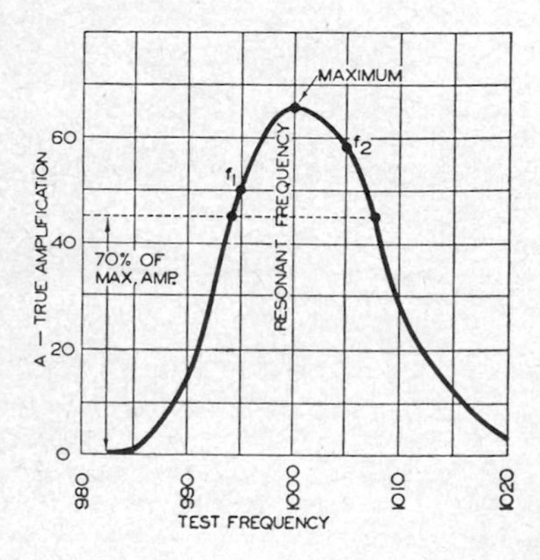

Resonance curve at 1000 kHz for typical RF amplifier.

and capacitor a current that is many times greater than the current fed into the circuit.

resonance curve An amplitude-frequency response curve that shows the current or voltage response of a tuned circuit to frequencies at and near the resonant frequency. Also called resonance characteristic.

resonance energy The kinetic energy of a particle that will be captured or scattered preferentially because of the presence of an appropriate resonance level in the compound nucleus formed by the incident particle and the target nucleus.

resonance fluorescence The emission of radiation by a gas or vapor at the same frequency as the exciting radiation.

resonance frequency 1. The frequency at which the inductive reactance of a given resonant circuit is equal to the capacitive reactance, so resonance exists. 2. The frequency at which a quartz crystal, loudspeaker diaphragm, or other object will vibrate readily.

resonance indicator A device that indicates when a circuit is tuned to resonance. It may be a meter, neon lamp, headphones, or a cathode-ray tuning indicator.

resonance lamp An evacuated quartz bulb that contains mercury, which acts as a source of radiation at the wavelength of the pure resonance line of mercury when irradiated by a mercury-arc lamp.

resonance level An energy level formed by a collision between two systems, as between a nucleon and a nucleus. The atom in question can return directly to its normal energy level by radiating the added energy.

resonance neutron A neutron that has energy in the region where the cross section of the nuclide or element is particularly large because of the occurrence of a resonance. Thus cadmium resonance neutrons have energies between 0.05 and about 0.3 eV.

resonance penetration The penetration of a nucleus by a charged particle whose energy corresponds to one of the energy levels in the nucleus.

resonance radiation The emission of radiation by a gas or vapor as a result of excitation of atoms to higher energy levels by incident photons at the resonance frequency of the gas or vapor. The radiation is characteristic of the particular gas or vapor atom but is not necessarily the same frequency as the absorbed radiation.

resonance radiometer A radiometer used for making relative measurements of weak radiation in infrared spectrometers.

resonance scattering Scattering arising from the part of the incident wave that penetrates the surface and interacts with the interior of a nucleus.

resonance spectral line One of the spectral lines emitted when a electron undergoes transfer to a lower energy state in a given atom.

resonance spectrum The spectrum of light emitted by an excited atom during its return to the ground state.

resonance transformer A high-voltage transformer in which the secondary circuit is tuned to the frequency of the power supply.

resonant Pertaining to resonance. The term resonance is often used as an adjective in place of resonant.

resonant capacitor A tubular capacitor that is intentionally wound to have inductance in series with its capacitance. Often used as a bypass capacitor in IF amplifiers, where it is made to be resonant at the IF value to give more effective bypassing of IF signals.

resonant cavity *Cavity resonator.*

resonant-cavity maser A maser in which the paramagnetic active material is placed in a cavity resonator.

resonant chamber *Cavity resonator.*

resonant-chamber switch A waveguide switch in which a tuned cavity in each waveguide branch serves the functions of switch contacts. Detuning of a cavity blocks the flow of energy in the associated waveguide.

resonant charging choke A choke that resonates with the effective capacitance of a pulse-forming network in a modulator to produce oscillation at the resonant frequency.

resonant circuit A circuit that contains inductance and capacitance of such values as to give resonance at an operating frequency. The frequency in hertz at which resonance occurs is $1/2\pi\sqrt{LC}$, where L is in henrys and C is in farads. When the capacitance and inductance are in series, the combination has a lower impedance at resonance than either alone. When in parallel, the combination has a higher impedance than either alone.

resonant diaphragm A diaphragm that has no reactance at a specified frequency.

resonant element *Cavity resonator.*

resonant gap The small internal gap in which the electric field of a TR tube is concentrated.

resonant iris A resonant window in a circular waveguide, so called because of its resemblance to an optical iris.

resonant line A transmission line that has values of distributed inductance and distributed capacitance, to make the line resonant at the frequency it is handling. Parallel resonance exists when the line is an odd number of quarter-wavelengths long and is short-circuited at the load, and series resonance exists for the same line when open at the load end.

resonant-line oscillator An oscillator in which one or more sections of transmission line serve as resonant circuits.

resonant-line tuner A television tuner in which resonant lines are used to tune the antenna, RF amplifier, and RF oscillator circuits. Tuning is achieved by moving shorting contacts that change the electrical lengths of the lines.

resonant-reed relay A reed relay in which the reed switch closes only when the required frequency is applied to the operating coil, to make one of the reeds vibrate until its amplitude is sufficient to make contact with the other reed. Used in selective paging systems.

resonant resistance The resistance of a resonant circuit or resonant line at the frequency of resonance.

resonant voltage step-up The ability of a coil and a capacitor in a series resonant circuit to deliver a voltage several times greater than the input voltage of the circuit.

resonant window A parallel combination of inductive and capacitive diaphragms, used in a waveguide structure to provide transmission at the resonant frequency and reflection at other frequencies.

resonant-window switch A waveguide switch in which a resonant window in each waveguide branch serves the function of switch contacts. Detuning of the window blocks the flow of energy in the associated waveguide.

resonate To bring to resonance, as by tuning.

resonating piezoid A piezoid (finished crystal blank) used as a resonator or oscillator rather than as a transducer.

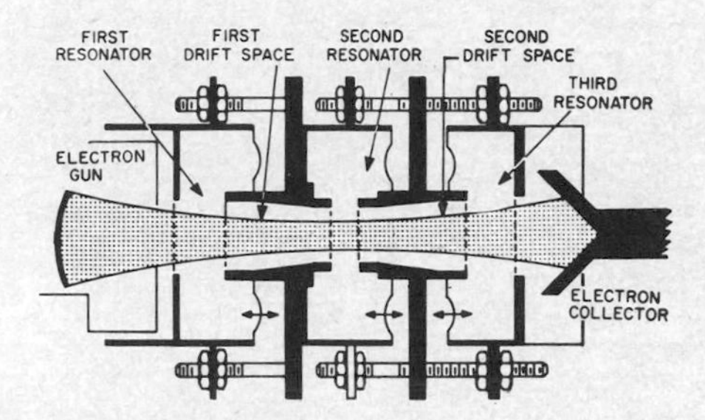

Resonators in electrostatically focused high-power traveling-wave tube.

resonator A device that exhibits resonance at a particular frequency, such as an acoustic resonator or cavity resonator.

resonator grid A grid that is attached to a cavity resonator in a velocity-modulated tube to provide coupling between the resonator and the electron beam.

resonator mode The operating mode for which an electron stream introduces a negative conductance into the coupled circuit of an oscillator.

resonator wavemeter Any resonant circuit used to determine wavelength, such as a cavity-resonator frequency meter.

responder The transmitter section of a radar beacon.

response 1. A quantitative expression of the output of a device or system as a function of the input. 2. *Amplitude-frequency response.* 3. *Reply.*

response characteristic *Amplitude-frequency response.*

responser *Responsor.*

response time The time required for the output of a control system or element to reach a specified fraction of its new value after application of a step input or disturbance. Usually given in seconds, but for magnetic amplifiers the response time is often specified in cycles of the power-line frequency. For indicating instruments the time for the pointer to come to rest at its new position is used.

responsor The receiving section of an interrogator-responsor. Also called responser.

rest energy The energy of a particle at rest,

equal to the rest mass multiplied by the square of the velocity of light.

resting frequency *Carrier frequency.*

rest mass The mass of a particle at rest or when moving with a velocity low compared with that of light. At higher velocities the particle acquires an additional relativistic mass.

restore To return a computer word to its initial value.

restorer *DC restorer.*

restoring spring The spring that moves the armature of a relay away from the core when the relay is deenergized.

restricted radiation device A device in which RF energy is intentionally generated and conducted along wires or radiated, but the total electromagnetic field does not exceed 15 μV/m at a distance in feet equal to 15,700 divided by the frequency in kilohertz (distance in meters equal to 4785 divided by the frequency in kilohertz).

restriking voltage *Reignition voltage.*

resultant A force that combines the effects of two or more forces acting on an object.

retainer A clamp or other device specifically designed to restrain the movement of an electron tube, fuse, or other removable component mounted in a socket or holder.

retarding field An electric or magnetic field that slows up electrons traveling through an interelectrode space in an electron tube.

retarding-field oscillator An oscillator that employs an electron tube in which the electrons oscillate back and forth through a grid that is maintained positive with respect to both the cathode and anode. The frequency depends on the electron transit time and may also be a function of associated circuit parameters. The field in the region of the grid exerts a retarding effect that draws electrons back after they pass through the grid in either direction. Also called positive-grid oscillator. Barkhausen-Kurz and Gill-Morell oscillators are examples.

retarding-field tube An electron tube used in a retarding-field oscillator.

retention The percentage of radioactive atoms that cannot be separated from the target materials after production of the atoms by nuclear reaction or radioactive decay.

retention time The maximum time between writing into a storage tube and obtaining an acceptable output by reading.

retentivity The property of a magnetic material that is measured by the residual flux density corresponding to the saturation induction for the material.

retina character reader A character reader that operates in the manner of the human retina in recognizing identical letters in different type fonts.

RETMA Abbreviation for Radio-Electronics-Television Manufacturers Association, now *Electronic Industries Association.*

retrace The return of the electron beam to its starting point in a cathode-ray tube after a sweep. Also called flyback.

retrace blanking Blanking a television picture tube during vertical retrace intervals to prevent retrace lines from showing on the screen. Voltage pulses for blanking are derived from vertical sweep oscillator or vertical deflection circuits and are applied to the control grid of the picture tube.

retrace ghost A ghost image produced on a television receiver screen during retrace periods. Generally caused by insufficient blanking of the camera tube at the transmitter.

retrace interval The interval of time in which the blanked scanning beam of a television picture tube or camera tube returns to the starting point of a line or field. It is about 7 μs for horizontal retrace and 500 to 750 μs for vertical retrace in U.S. television broadcasting. Also called retrace period, retrace time, return interval, return period, and return time.

retrace line The line traced by the electron beam in a cathode-ray tube in going from the end of one line or field to the start of the next line or field. Also called return line.

retrace period *Retrace interval.*

retrace time *Retrace interval.*

retransmission unit A control unit used at an intermediate station for feeding one radio receiver-transmitter unit automatically to another receiver-transmitter unit for two-way communication.

retroaction British term for *positive feedback.*

retrodirective steering An adaptive array in which transmitted energy is automatically focused back in the direction of arrival of the received signals. Examples include the corner reflector and the Luneberg lens.

retroreflector A device that reflects radiation back on a path parallel to the incident rays over a wide range of retroreflector orientation. Used in laser surveying. One version, the corner reflector, is an efficient radar target.

return 1. To go back to a planned point in a computer program and rerun a portion of the program, usually when an error is detected. Rerun points are usually not more than 5 min apart. 2. *Echo.*

return-beam mode A camera-tube operating mode in which the output current is derived from that portion of the scanning beam not accepted by the target.

return-beam vidicon A vidicon in which the electron beam that scans the target is bent back from the target to an electron multiplier and anode which surround the electron gun. Light falling on the target surface changes the resistance of the surface and thereby modulates the energy in the return beam.

return interval *Retrace interval.*

return line *Retrace line.*

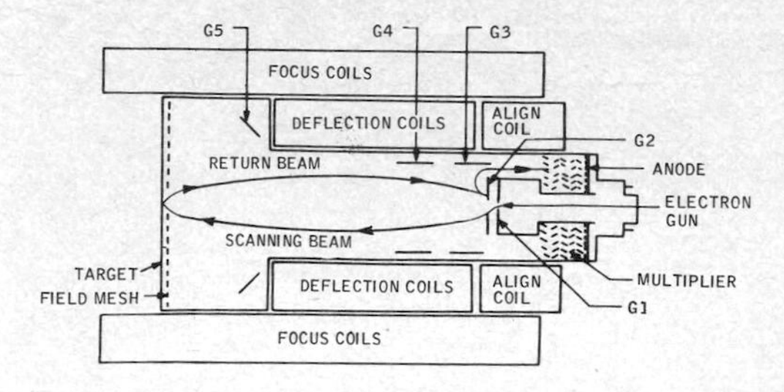

Return-beam vidicon construction.

return loss 1. The difference between the power incident upon a discontinuity in a transmission system and the power reflected from the discontinuity. 2. The ratio in decibels of the power incident upon a discontinuity to the power reflected from the discontinuity.

return period *Retrace interval.*

return time *Retrace interval.*

return to bias Magnetization of magnetic tape to saturation in a direction called minus, representing binary 0. Binary 1 signals are recorded by magnetizing the tape in the opposite direction. After each binary 1 pulse the tape returns to the minus (bias) condition. This method requires a clock to read 0s.

return-to-zero code [abbreviated RZ code] A code used in some computers, in which a binary 0 is represented by one bit time at the 0 level, and a binary 1 is pulsed in such a way that it reaches the

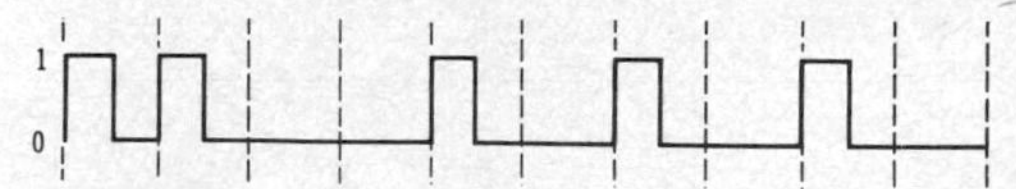

Return-to-zero code for binary data. Space between each pair of dashed vertical lines represents one bit time.

1 level for only half a bit time. The signal therefore returns to 0 (or stays at 0) after each bit. With this code, only half as much data can be stored in a given area or distance as with nonreturn-to-zero code.

return wire The ground wire, common wire, or negative wire of a DC power circuit.

reverberation The persistence of sound at a given point after direct reception from the source

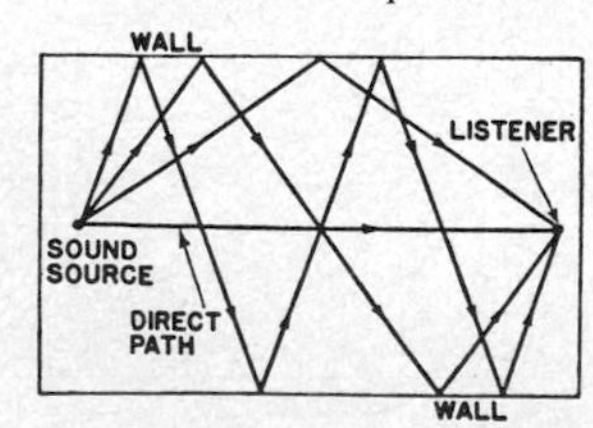

Reverberation paths for sound waves in room.

has stopped. In air it may be caused by repeated reflections from a small number of boundaries or free decay of normal modes of vibration that were excited by the sound source. In water it may be due to scattering from a large number of inhomogeneities in the medium or reflections from bounding surfaces.

reverberation absorption coefficient The term used for the sound absorption coefficient when the distribution of incident sound is completely random.

reverberation chamber An enclosure in which all surfaces have been made as sound-reflective as possible. Used for some types of acoustic measurements. Also called reverberation room.

reverberation-controlled gain circuit [abbreviated RCG circuit] A circuit used in underwater sound equipment to vary the gain of the receiving amplifier in proportion to the strength of undesired reverberations associated with the desired echo.

reverberation reflection coefficient The term used for the sound reflection coefficient when the distribution of incident sound is completely random.

reverberation room *Reverberation chamber.*

reverberation time The time in seconds required for the average sound-energy density at a given frequency to reduce to one-millionth of its initial steady-state value after the sound source has been stopped. This corresponds to a decrease of 60 dB.

reverberation-time meter An instrument that measures the reverberation time of an enclosure.

reverberation transmission coefficient The term used for the sound transmission coefficient when the distribution of incident sound is completely random.

reverberation unit A device that generates reverberation synthetically in a sound system. In one version, audio signal energy is converted into tor-

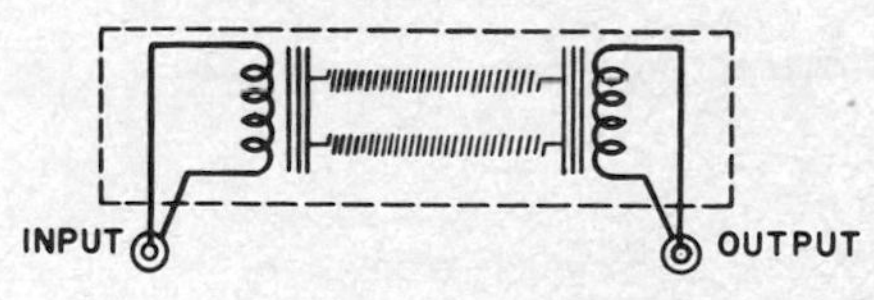

Reverberation unit using coil springs between magnetic input and output units.

sional motion of coil springs by a magnetic driver. A magnetic pickup at the other end of the springs converts the motions back into audio signals after a time delay, to give the effect of listening in a reverberant hall.

reverse bias A bias voltage applied to a diode or a semiconductor junction with polarity such that little or no current flows. It is the opposite of forward bias.

reverse-blocking thyristor A thyristor that

switches only for positive anode-to-cathode voltages. It blocks current flow for negative voltages.

reverse-conducting thyristor A thyristor that switches only for positive anode-to-cathode voltages. It conducts large currents at negative voltages comparable in magnitude to the on-state voltages.

reverse coupler A directional coupler used to sample reflected power.

reverse current The small value of direct current that flows when a metallic rectifier or semiconductor diode has reverse bias.

reverse-current relay A relay that operates whenever current flows through the relay coil in the opposite of the normal direction.

reversed image 1. A mirror image, in which the right and left sides of the picture are interchanged. 2. *Negative image.*

reverse direction The direction of higher resistance to steady DC flow through a semiconductor junction or device.

reverse emission The flow of electrons in the reverse direction (from anode to cathode) in a vacuum tube during that part of a cycle in which the anode is negative with respect to the cathode. The action is similar to arcback in gas tubes. Also called back emission.

reverse-polarity silicon diode A silicon diode in which the positions of the external tip and base are interchanged as compared with those in an ordinary silicon diode, with internal construction remaining the same. Used chiefly as crystal mixers. Thus an ordinary 1N23C silicon diode has the same electric characteristics as the 1N23CR reverse-polarity silicon diode.

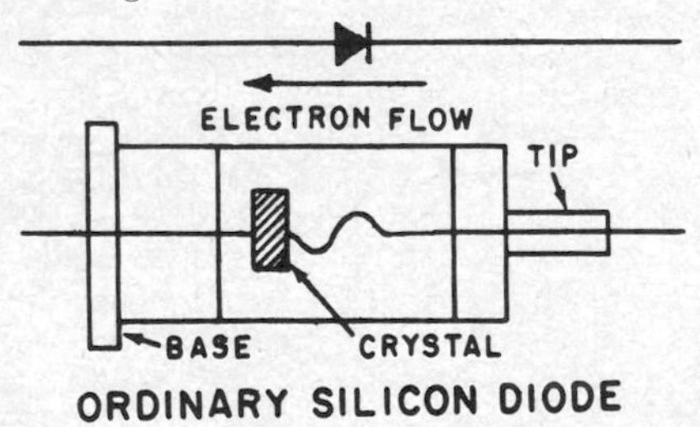

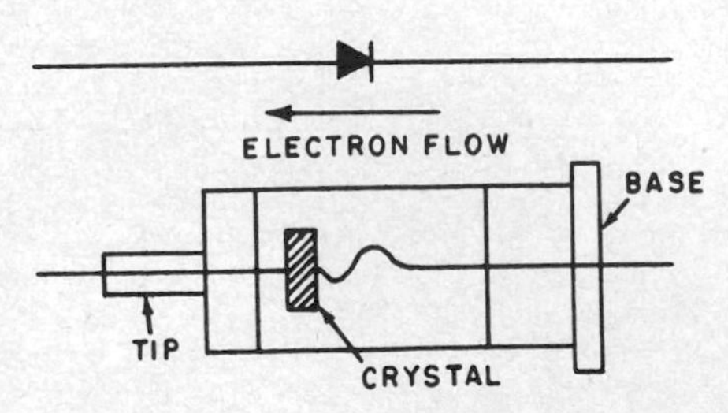

Reverse-polarity silicon diode has tip connected to crystal, whereas in ordinary silicon diode the base goes to crystal.

reverse Polish notation A type of logic used in some calculators, which permits entry of a mathematical problem from left to right exactly as it is normally written.

reverse-printout typewriter An automatic typewriter that eliminates conventional carriage return by typing one line from left to right and the next line from right to left.

reverse resistance The resistance of a metallic rectifier cell as measured at a specified value of reverse voltage or reverse current.

reverse voltage A voltage applied to a metallic rectifier with the opposite of normal polarity.

reversible counter A counter capable of counting either from left to right or right to left, usually without change of the count state at reversal.

reversible motor A motor in which the direction of rotation can be reversed by a switch that changes motor connections when the motor is stopped.

reversible permeability The term used for incremental permeability when the change in magnetic induction is vanishingly small.

reversible transducer A transducer in which the transducer loss is independent of the direction of transmission.

reversing motor A motor for which the direction of rotation can be reversed by changing electric connections or by other means while the motor is running at full speed. The motor will then come to a stop, reverse, and attain full speed in the opposite direction.

reversing switch A switch intended to reverse the connections of one part of a circuit.

rewind 1. The components on a magnetic-tape recorder that return the tape to the supply reel at high speed. 2. To return a magnetic tape to its starting position.

rewrite The process of restoring a storage device to its state prior to reading. Used when the information-storing state may be destroyed by reading.

RF Abbreviation for *radio frequency.*

RF alternator A rotating-type alternator that produces high power at frequencies above power-line values but generally lower than 100 kHz. Used today chiefly for high-frequency heating.

RF amplification Amplification of a radio signal at the same carrier frequency at which it travels through space.

RF amplifier An amplifier that increases the voltage or power of RF signals at the carrier frequency. In a superheterodyne receiver, an RF amplifier is sometimes used ahead of the converter.

RFC Symbol for *RF choke.*

RF cable A cable that has electric conductors separated from each other by a continuous homogeneous dielectric or by touching or interlocking spacer beads, designed primarily to conduct RF energy with low losses.

RF choke [symbol RFC] An RF coil designed and used specifically to block the flow of RF current while passing lower frequencies or direct current.

RF coil A coil that has one continuous untapped winding, specifically designed to furnish inductive reactance for tuning purposes in a circuit carrying RF current.

RF converter A power source for producing electric power at a frequency above about 10 kHz.

RF current An alternating current that has a frequency higher than about 10 kHz.

RF energy Alternating-current energy at any frequency in the radio spectrum between about 10 kHz and 300 GHz.

RF generator A generator capable of supplying sufficient RF energy at the required frequency for induction or dielectric heating.

RF harmonic A harmonic of a carrier frequency. The frequency of the second RF harmonic is twice the carrier frequency.

RF head A radar transmitter and part of a radar receiver, contained in a package for ready installation and removal.

RF heating *Electronic heating.*

RFI Abbreviation for *RF interference.*

RF indicator An indicator that shows the presence of RF energy at or near its own resonant frequency. It usually consists of a coil or parallel line connected to an incandescent lamp.

RF interference [abbreviated RFI] Interference from sources of energy outside a system or systems, as contrasted to electromagnetic interference (EMI) generated inside systems.

RF intermodulation distortion Intermodulation distortion that originates in the RF stages of a receiver.

RF oscillator An oscillator that generates alternating current at radio frequencies.

RF pattern A fine herringbone pattern that occurs in a television picture because of high-frequency interference.

RF pentode A pentode in which the screen grid is extended beyond the anode, to minimize electrostatic coupling between grid and anode and thereby reduce feedback at radio frequencies.

RF power probe A probe that extracts RF energy from a transmission system.

RF power supply A high-voltage power supply in which the output of an RF oscillator is stepped up by an air-core transformer to the high voltage required for the second anode of a cathode-ray tube, then rectified to provide the required high DC voltage. Used in some television receivers.

RF pulse An RF carrier that is amplitude-modulated by a pulse. The amplitude of the modulated carrier is zero before and after the pulse.

RF resistance *High-frequency resistance.*

RF response The response of a receiver to radio frequencies within and outside the channel being received.

RF signal generator A test instrument that generates the various radio frequencies required for alignment and servicing of radio, television, and electronic equipment. Also called service oscillator.

RF spectrum *Radio spectrum.*

RF stage A single stage of RF amplification.

RF transformer A transformer that has a tapped winding or two or more windings which furnish inductive reactance and/or transfer RF energy from one circuit to another by a magnetic field. It may have an air core or some form of ferrite core.

RF transmission line A transmission line designed primarily to conduct RF energy; it consists of two or more conductors supported in a fixed spatial relationship along their own length.

rhenium [symbol Re] A metallic element. Atomic number is 75.

rheo- Prefix meaning a flow of current.

rheobase The intensity of the steady current just sufficient to excite a tissue when suddenly applied.

rheoelectroencephalograph An electroencephalograph that measures differential blood flow in both sides of the brain and in any other part of the body.

rheoencephalography The measurement and recording of blood flow in the brain and other parts of the body by using two electrodes placed on opposite sides of the portion of the body being monitored.

rheography Monitoring of the impedance between two electrodes placed on the skin.

rheostat A resistor whose value may be changed readily by a control knob, to control the current in a circuit. A rheostat has one fixed terminal and one terminal that is connected to the sliding or rolling contact. Also called variable resistor. A potentiometer is a rheostat that has an additional fixed terminal at the other end of the resistance element.

rheostriction *Pinch effect.*

rheotaxial growth A chemical vapor deposition technique for producing silicon diodes and transistors on a fluid layer that has high surface mobility. The growth tends to follow the orientation of the first crystallites of silicon on top of the fluid layer. To fabricate PN structures, a rheotaxial film of silicon is deposited on an oxide-coated substrate.

Rhm Abbreviation for *roentgen-per-hour-at-one-meter.*

rhodium [symbol Rh] A metallic element. Atomic number is 45.

rhombic antenna A horizontal antenna that has four conductors which form a diamond or rhombus. Usually fed at one apex and terminated with a resistance or impedance at the opposite apex. Also called diamond antenna and Bruce antenna.

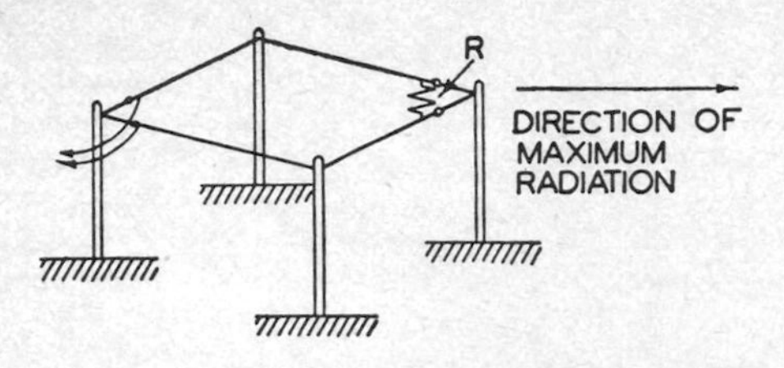

Rhombic antenna.

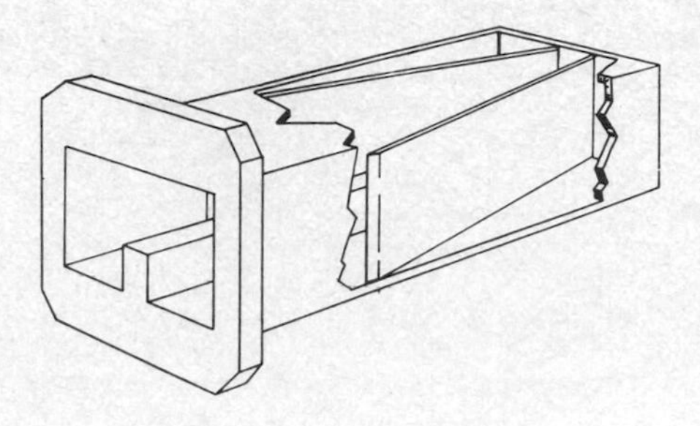

Ridge-waveguide termination.

rho meson Former name for a meson that was stopped in a nuclear emulsion without any apparent decay event or nuclear interaction. Most of the rho mesons were mu mesons, the decay event being unobserved because of insufficient emulsion sensitivity.

rhometal A high-resistivity magnetic alloy that has an initial permeability of 250 to 2000.

rho-theta navigation *Omnibearing-distance navigation.*

rhumbatron *Cavity resonator.*

RIAA curve 1. Recording Industry Association of America curve that represents standard recording characteristics for long-play records. 2. The corresponding equalization curve for playback of microgroove records.

ribbon cable A cable made of normal, round, insulated wires arranged side by side and fastened together by a cohesion process to form a flexible ribbon.

ribbon microphone A velocity microphone in which the moving element is a thin corrugated metal ribbon mounted between the poles of permanent magnets. The ribbon cuts magnetic lines of force as it is moved back and forth in proportion to the velocity of air particles in a sound wave, so an AF output voltage is induced in the ribbon.

Richardson effect *Edison effect.*

Richardson equation An equation that gives the density of thermionic emission at saturation current in terms of the absolute temperature of the filament or cathode of an electron tube.

ride gain To control the volume range of an AF circuit while watching a volume indicator.

ridge waveguide A circular or rectangular waveguide that has one or more longitudinal internal

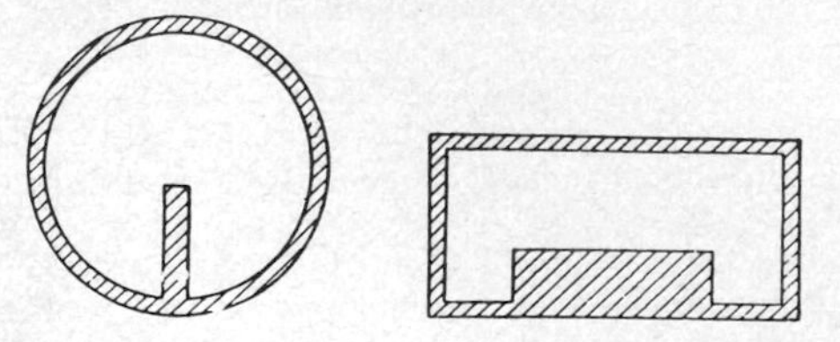

Ridge waveguides.

ridges which serve primarily to increase transmission bandwidth by lowering the cutoff frequency.

ridge-waveguide termination A component that serves as a perfect zero-impedance load when connected to the end of a ridge waveguide, so no signal is reflected back through the waveguide.

Rieke diagram A chart that shows contours of constant power output and constant frequency for a microwave oscillator, drawn on a Smith chart or

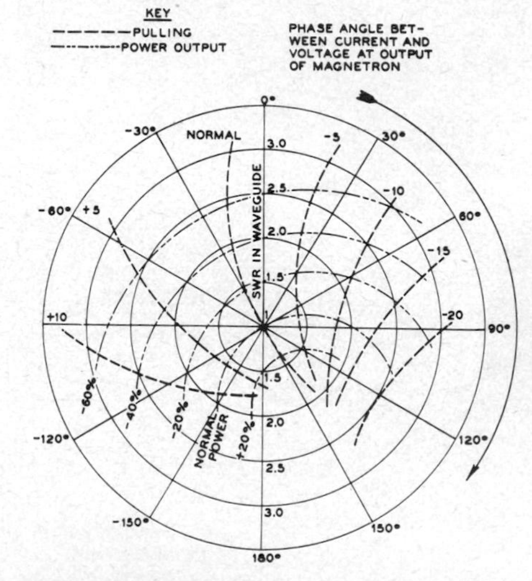

Rieke diagram showing effect of load on magnetron having normal frequency of 9 GHz. Values on dashed arcs give amount of frequency-pulling in megahertz. Ideal load is normal-power curve, passing through SWR = 1 at center of diagram. Best operating region is for load phase angle of 0 to −30° because pulling from normal 9000-MHz value is least here.

other polar diagram whose coordinates represent the components of the complex reflection coefficient at the oscillator load.

rig Slang term for a complete system of components, such as a complete amateur radio station.

Righi-Leduc effect Development of a difference in temperature between the two edges of a strip of metal in which heat is flowing longitudinally, when the plane of the strip is perpendicular to magnetic lines of force.

right-hand polarized wave An elliptically polarized transverse electromagnetic wave in which the rotation of the electric field vector is clockwise for an observer looking in the direction of propagation. Also called clockwise polarized wave.

right-hand rule 1. For a current-carrying wire: If the fingers of the right hand are placed around the wire in such a way that the thumb points in the direction of current flow, the fingers will be pointing in the direction of the magnetic field produced by the wire. For electron flow (the opposite of current flow), the left-hand rule is used. 2. For a movable current-carrying wire or an electron beam in a magnetic field: If the thumb, first, and second fingers of the right hand are extended at right angles to one another, with the first finger representing the direction of magnetic lines of force and the second finger representing the direction of current flow, the thumb will be pointing in the direction of motion of the wire or beam. Also called Fleming's rule.

right-hand taper Taper in which there is greater resistance in the clockwise half of the operating range of a potentiometer or rheostat (looking from the shaft end) than in the counterclockwise half.

right-justify 1. To adjust the printing positions of characters on a page so the right margin is lined up. 2. To shift the contents of a register so the right or least significant digit is at some specified position.

right signal The output of a microphone placed so as to pick up the intensity, time, and location of sounds originating predominantly to the listener's right of the center of the performing area, when making a stereo recording or broadcast.

right stereo channel The right signal as electrically reproduced in reception of stereo FM broadcasts or stereo records.

rim drive A phonograph or sound recorder drive in which a rubber-covered drive wheel is in contact with the inside of the rim of the turntable.

rim magnet *Field-neutralizing magnet.*

ring-around Undesired triggering of a transponder by its own transmitter, or triggering at all bearings so as to give a ring presentation on a PPI display.

ring counter A loop of binary scalers or other bistable units so connected that only one scaler is in a specified state at any given time. As input signals are counted, the position of the one specified state moves in an ordered sequence around the loop.

ring head A magnetic head in which the magnetic material forms an enclosure that has one or more air gaps. The magnetic recording medium bridges one of these gaps and is in contact with or in close proximity to the pole pieces on one side only.

ringing An oscillatory transient that occurs in the output of a system as a result of a sudden change in input. In television it may produce a series of closely spaced images or a black line immediately

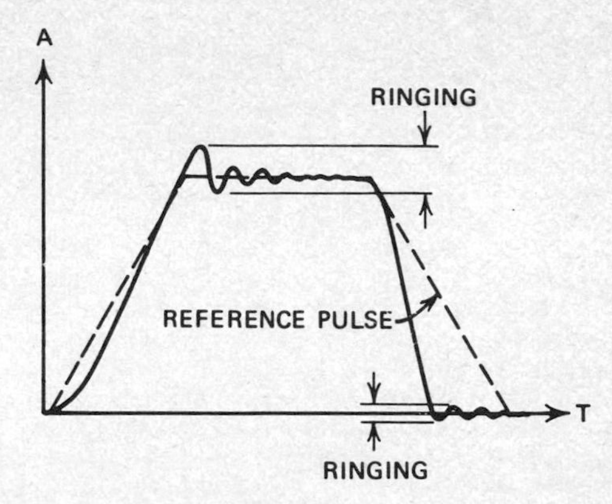

Ringing.

to the right of a white object.

ringing time 1. The time required for the output of an oscillatory circuit to decrease to a predetermined level after input power is removed. 2. The time between termination of a transmitted radar pulse and the instant at which the reradiated power from an echo box falls below the minimum required to produce an indication. Used as a measure of overall radar performance.

ring laser *Laser gyro.*

ring modulator A modulator in which four diode elements are connected in series to form a ring around which current flows readily in one direc-

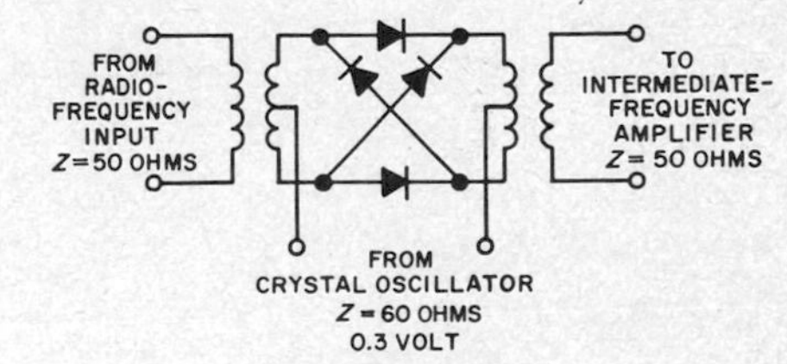

Ring modulator.

tion. Input and output connections are made to the four nodal points of the ring. Used as a balanced modulator, demodulator, or phase detector.

ring oscillator Two or more pairs of tubes operating as push-pull oscillators around a ring, usually with alternate successive pairs of grids and anodes connected to tank circuits. Adjacent tubes around the ring operate in phase opposition. The load is coupled to the anode circuits.

right-plane circuit A slow-wave structure that consists of circular rings or slotted pipes supported by one or more radial planes. Used in high-power high-frequency traveling-wave tubes.

ring scaler A scaler in which the asymmetrical condition is passed along to the next tube in line, with the last tube feeding back to the first to complete the ring.

ring-seal tube An electron tube in which the

grid and anode are radially symmetrical, with the grid connected to a metal ring sealed into the glass envelope. The construction permits insertion of the tube in a coaxial chamber, with the grid connected to the outer cylinder and the anode to the inner conductor for operation as a grounded-grid amplifier.

ring time The time during which the output of a radar echo box remains above a specified level. The interval starts when a pulse is transmitted and is usually considered ended when the energy reradiated by the echo box falls below the minimum needed for an indication on the radar screen.

riometer [Relative Ionospheric Opacity METER] An instrument that measures changes in ionospheric absorption of electromagnetic waves by determining and recording the level of extraterrestrial cosmic radio noise.

ripple The AC component in the output of a DC power supply, arising within the power supply from incomplete filtering or from commutator action in a DC generator.

ripple counter A counter that consists of flip-flops in series. When the first flip-flop changes, it affects the second, which in turn affects the third, and so on, until the last in the series is changed.

ripple factor The ratio of the effective value of the AC component of a pulsating DC voltage to the average value.

ripple filter A low-pass filter that reduces ripple while freely passing the direct current obtained from a rectifier or DC generator.

ripple frequency The frequency of the ripple present in the output of a DC source.

ripple voltage The alternating component of the unidirectional voltage from a rectifier or generator used as a source of DC power.

rise time The time required for a signal pulse to rise from 10 to 90% of its final steady value. It is a measure of the steepness of the wavefront.

rising-sun magnetron A multicavity magnetron in which resonators that have two different resonant frequencies are arranged alternately for the purpose of mode separation. The cavities appear as alternating long and short radial slots around the perimeter of the anode structure, resembling the rays of the sun.

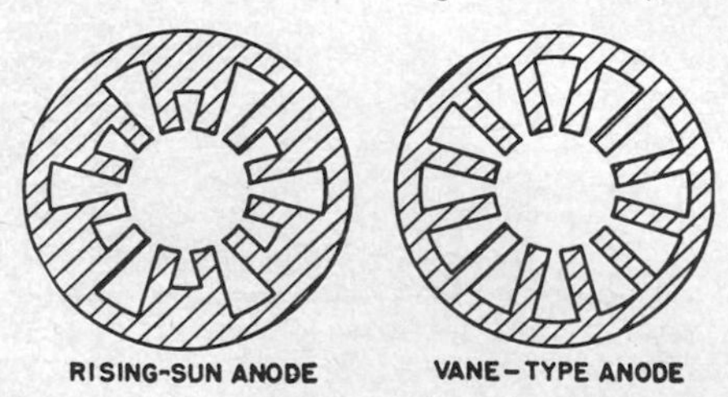

Rising-sun magnetron anode, with vane-type anode shown for comparison.

RLC circuit Abbreviation for *resistance-inductance-capacitance circuit.*

RMA Abbreviation for Radio Manufacturers Association, now *Electronic Industries Association.*

RMA color code Former name of *EIA color code.*

R-meter An ionization instrument calibrated to indicate the intensity of gamma rays, x-rays, and other ionizing radiation in roentgens.

RMI Abbreviation for *radio magnetic indicator.*

r/min SI abbreviation for revolution per minute.

R − Y signal The red-minus-luminance color-difference signal used in color television. It is combined with the luminance signal in a receiver to give the red color-primary signal.

RMS Abbreviation for *root-mean-square.*

RMS power per channel The stereo power amplifier rating now required in the United States by the Federal Trade Commission. To be meaningful, the load impedance, total harmonic distortion, bandwidth, and other amplifier parameters should also be specified.

robot A completely self-controlled electronic, electric, or mechanical device.

Rochelle-salt crystal A crystal of sodium potassium tartrate; it has a pronounced piezoelectric effect. Used in crystal microphones and crystal pickups.

rocket An unmanned self-propelled vehicle, with or without a warhead, designed to travel above the surface of the earth without control of its trajectory or course during flight.

rocket missile A missile that uses rocket propulsion.

rocking Back-and-forth rotation of the tuning control in a superheterodyne receiver while adjusting the oscillator padder near the low-frequency end of the tuning dial, to obtain more accurate alignment.

rockoon A high-altitude atmospheric sounding system that consists of a small solid-propellant research rocket launched from a large plastic balloon near the maximum altitude of the balloon flight. Used extensivelv from shipboard.

rod storage A computer storage made up of a number of cylindrical rods or wires arranged in a three-dimensional array, with each rod coated with a thin magnetic film. Currents are sent through appropriate rods to magnetize the film in predetermined directions and at predetermined locations. Readout is achieved by sensing the magnetic state of the thin film at any given location between two intersecting rods in the storage.

roentgen [abbreviated R] The international unit of exposure dose for x-rays and gamma rays. One roentgen of radiation will ionize dry air sufficiently to produce 1 electrostatic unit of electricity per 1.293 mg of air.

roentgen equivalent man [abbreviated rem] A unit of ionizing radiation, equal to the amount that produces the same damage to a person as 1 roentgen of high-voltage x-rays.

roentgen equivalent physical [abbreviated

rep] A unit of ionizing radiation, equal to the amount that causes absorption of 93 ergs of energy per gram of soft tissue. Also called equivalent roentgen.

roentgen meter A meter for measuring the cumulative quantity of x-rays or gamma rays, without reference to time.

roentgenogram Deprecated term for *x-ray photograph.*

roentgenography Radiography by x-rays.

roentgenology That branch of radiology that deals with x-rays, especially their use for diagnosis or treatment in medicine and dentistry. Radiology is a slightly more general term, covering also the use of gamma rays and other extremely high-frequency ionizing radiation.

roentgenotherapy Medical treatment by x-rays.

roentgen-per-hour-at-one-meter [abbreviated Rhm] A unit of gamma-ray source strength, corresponding to a dose rate of 1 R/h at a distance of 1 m in air.

roentgen-rate meter An electrically operated instrument that measures radioactivity and is calibrated in roentgens per unit time or any multiple thereof.

roentgen ray *X-ray.*

roger 1. A code word used in communication, meaning that a message has been received and understood. 2. An expression of agreement.

Roget spiral A helix of wire that contracts in length when a current is sent through, owing to mutual attraction between adjacent turns.

roll Slow upward or downward movement of the entire image on the screen of a television receiver, due to a lack of vertical synchronization.

roll-and-pitch control A control used with automatic pilots and remote attitude indicators; it consists of a gyroscope that provides signals for controlling an aircraft about its lateral and longitudinal axes and/or signals for providing a visual presentation of attitude to the pilot.

rollback *Rerun.*

rollback routine *Rerun routine.*

rolloff Gradually increasing attenuation as frequency is changed in either direction beyond the flat portion of the amplitude-frequency response characteristic of a system or component.

roll out To read a computer register or counter by adding a 1 to each digit column simultaneously until all have returned to 0, with a signal being generated at the instant when a column returns to 0.

ROM Abbreviation for *read-only memory.*

roof filter A low-pass filter used in carrier telephone systems to limit the frequency response of the equipment to frequencies needed for normal transmission, thereby blocking unwanted higher frequencies induced in the circuit by external sources. A roof filter improves runaround crosstalk suppression and minimizes high-frequency singing.

room tone A sound track recorded in a particular studio when unoccupied, for use between portions of dialog recorded previously in that studio to provide the quality of silence determined by the acoustics and temperature of the studio.

rooter amplifier A nonlinear amplifier in which negative feedback makes the output voltage vary as the square root or some other root of the input voltage. Used in video amplifiers of television transmitters for gamma correction to compensate for camera-tube characteristics.

root-mean-square [abbreviated RMS] 1. The square root of the average of the squares of a series of related values. 2. The effective value of an alternating current, corresponding to the DC value that will produce the same heating effect. The RMS value is computed as the square root of the average of the squares of the instantaneous amplitudes for one complete cycle. For a sine wave, the RMS value is 0.707 times the peak value. Unless otherwise specified, alternating quantities are assumed to be RMS values. Also called effective value.

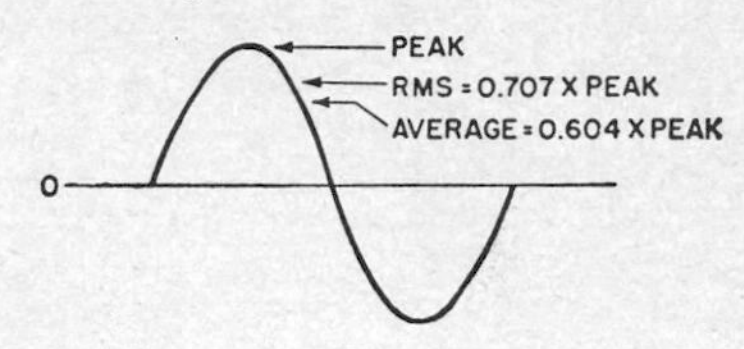

Root-mean-square (RMS) value of sine wave, as compared with peak and average values.

root-mean-square particle velocity *Effective particle velocity.*

root-mean-square sound pressure *Effective sound pressure.*

rope A long form of confusion reflector, sometimes used with tiny parachutes to reduce the rate of fall of the long strips of metal foil.

rosin-core solder Solder made up in tubular or other hollow form, with the inner space filled with rosin flux to serve as a noncorrosive flux for soldering joints.

rosin joint A soldered joint in which one of the wires is surrounded by an almost invisible film of insulating rosin, making the joint intermittently or continuously open even though it looks good. Insufficient heating of the parts before applying solder is the cause.

rotary actuator A device that converts electric energy into controlled rotary force. It usually consists of an electric motor, gear box, and limit switches.

rotary amplifier *Rotating magnetic amplifier.*

rotary beam antenna A highly directional shortwave antenna system mounted on a mast in such a manner that it can be rotated to any desired position either manually or by an electric motor drive.

rotary converter *Dynamotor.*

rotary coupler *Rotating joint.*

rotary dial A telephone calling device that generates pulses by manual rotation and release of a dial. The number of pulses is determined by how far the dial is rotated before being released.

rotary gap *Rotary spark gap.*

rotary joint *Rotating joint.*

rotary solenoid A solenoid in which the armature is rotated when actuated. The rotary stroke is usually converted to linear motion to give a longer stroke than is possible with a conventional plunger-type solenoid. The stroke may range from 25 to 95°.

rotary spark gap A spark gap in which sparks occur between one or more fixed electrodes and a number of electrodes projecting outward from the circumference of a motor-driven metal disk. Also called rotary gap.

rotary stepping relay *Stepping relay.*

rotary stepping switch *Stepping relay.*

rotary switch A switch that is operated by rotating its shaft.

rotary-tuned magnetron A magnetron that has a motor-driven slotted disk above the cylindrical cavities of its anode block. The slots in the spinning disk vary both the inductance and the capacitance of the anode, to produce a wide frequency

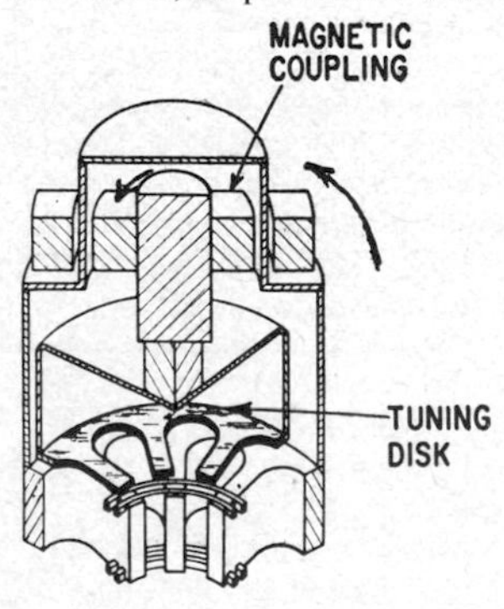

Rotary-tuned magnetron. Tuning disk and rotor of motor are on same shaft inside evacuated envelope; field coils of motor are on outside of tube.

sweep. The number of cycles of sweep is equal to the motor speed multiplied by the number of cavities in the magnetron. Frequency agility can be obtained by changing the pulse repetition rate of the radar, the speed of the tuning motor, or both, as required for electronic counter-countermeasures.

rotary-vane attenuator A device that introduces attenuation into a waveguide circuit by varying the angular position of a resistive material in the guide.

rotary voltmeter *Generating voltmeter.*

rotatable loop antenna A loop antenna that can be rotated in azimuth, for use in direction-finding.

rotatable-loop radio compass An automatic direction finder that uses a loop antenna which is rotated manually to determine the relative bearing between an aircraft or ship and a transmitter.

rotating amplifier *Rotating magnetic amplifier.*

rotating-anode x-ray tube An x-ray tube in which the anode rotates continuously to bring a

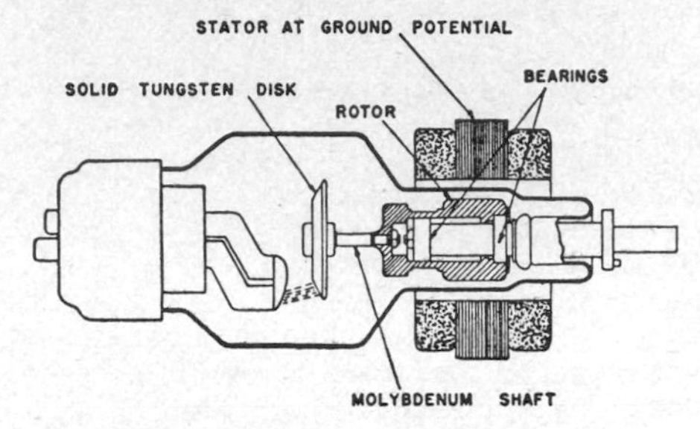

Rotating-anode x-ray tube, with rotor inside tube and stator surrounding glass neck outside tube. X-ray beam emerges downward after reflection from rotating tungsten anode disk.

fresh area of its surface into the beam of electrons, allowing greater output without melting the target.

rotating-crystal method Rotation of a crystal while it is irradiated by monochromatic x-rays and recording of the reflected beams as spots on a photographic plate, for analysis of crystal structure.

rotating joint A joint that permits one section of a transmission line or waveguide to rotate continu-

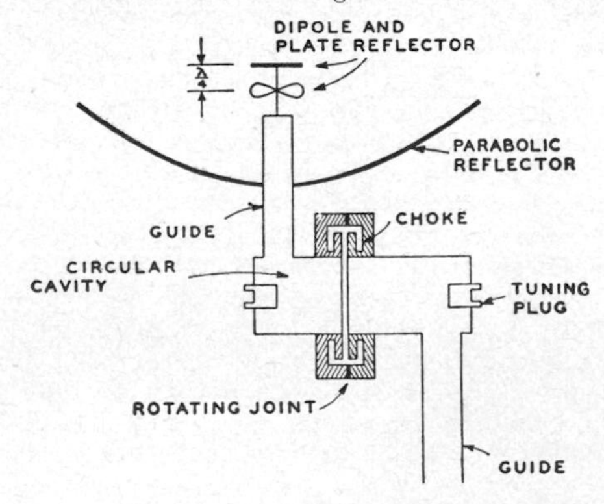

Rotating joint in waveguide feed to radar antenna.

ously with respect to another while passing RF energy. Also called rotary coupler and rotary joint.

rotating magnetic amplifier A prime-mover-driven DC generator whose power output can be controlled by small field input powers, to give power gain as high as 10,000. Used as power sources for the DC drive motors of large radar antennas as well as for other motor drives in automatic control systems. Examples include the amplidyne and metadyne. Also called rotary amplifier and rotating amplifier.

rotating radio beacon A radio transmitter arranged to radiate a concentrated beam that rotates in a horizontal plane at constant speed and transmits different signals in different directions so ships and aircraft can determine their bearings without directional receiving equipment.

rotating-type scanning sonar Scanning sonar that uses electric means to obtain a rotating receiving-beam pattern. Stationary transducer units arranged in a circle are connected in succession to the receiver by a commutator.

rotation photograph The photographic record of the diffracted beams produced when a slender beam of x-rays is directed on a rotating single crystal.

rotation therapy Radiation therapy in which either the patient or the source of radiation is rotated, to permit a larger dose at the center of rotation within the patient's body than on any area of the skin.

rotation wave *Shear wave.*

rotator A device that rotates the plane of polarization, such as a twist in a rectangular waveguide.

rotoflector An elliptically shaped rotating radar reflector that reflects a vertically directed radar beam at right angles so it radiates horizontally.

rotor The rotating member of a machine or device, such as the rotating armature of a motor or generator, or the rotating plates of a variable capacitor.

rotor plate One of the rotating plates of a variable capacitor, usually directly connected to the metal frame.

rounding error The error that results from round-off of a quantity in a computer or in calculations. Also called round-off error.

round off To change a more precise quantity to a less precise one by dropping certain less significant digits and applying some rule for changing the last significant retained digit.

round-off error *Rounding error.*

round-the-world echo A signal that occurs every 1/7 s when a radio wave repeatedly encircles the earth at its speed of 186,000 mi/s (300,000 km/s) during unusual backscatter propagation modes.

routine A set of instructions arranged in proper sequence to cause a computer to perform a desired operation, such as the solution of a mathematical problem. Also called master routine.

***r* parameter** A transistor parameter relating to resistivity.

rpm Abbreviation for revolution per minute.

rps Abbreviation for revolution per second.

RPV Abbreviation for *remotely piloted vehicle.*

r/s SI abbreviation for revolution per second.

R scope A radarscope that produces an R display.

RS flip-flop A flip-flop that has two logic inputs, designated R for reset and S for set, with only one of the inputs being at high level or 1 at a time. Operating speed is determined by the charging time of capacitors in the clock steering network connected between the R and S inputs. If the S input is at the 1 level when a clock pulse arrives, the flip-flop and its Q output go to the 1 state. A 1 at the R input gives reset to the 0 or off state. Also called set-reset flip-flop and SR flip-flop.

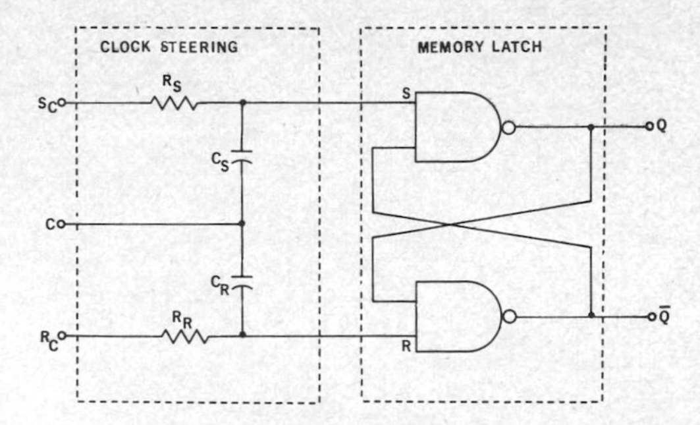

RS flip-flop arrangement, with clock steering network used in place of input control gates.

RST flip-flop An RS flip-flop that has an additional trigger input (labeled T) which can be energized to make the flip-flop change state. If the flip-flop is off, a pulse on S or T will turn it on; a pulse on the R input will cause no change. If the flip-flop is on, a pulse on the R or T input will turn it off; a pulse on the S input will cause no change. Also called set-reset-trigger flip-flop and SRT flip-flop.

RTL Abbreviation for *resistor-transistor logic.*

RTMA Abbreviation for Radio-Television Manufacturers Association, now *Electronic Industries Association.*

RTTY Abbreviation for *radioteletype.*

rubber-covered wire Wire insulated with rubber.

rubidium [symbol Rb] A photosensitive metallic element. Atomic number is 37.

rubidium magnetometer A highly sensitive magnetometer in which the spin precession principle is combined with optical pumping and monitoring for detecting and recording variations as small as 0.01 gamma (0.1 microoersted) in the total magnetic field intensity of the earth. Used for locating and mapping buried archeological sites and for airborne or oceanographic magnetic surveys.

rubidium-vapor frequency standard An atomic frequency standard in which the frequency is established by a gas cell that contains rubidium vapor and a neutral buffer gas. It is a secondary frequency standard because the rubidium gas cell is dependent on the gas mixture and pressure; hence it must be calibrated initially.

ruby A single crystal of aluminum oxide, with a small fraction of the aluminum atoms replaced by chromium atoms that serve as the source of characteristic red fluorescence under irradiation.

ruby laser A crystalline solid laser in which opti-

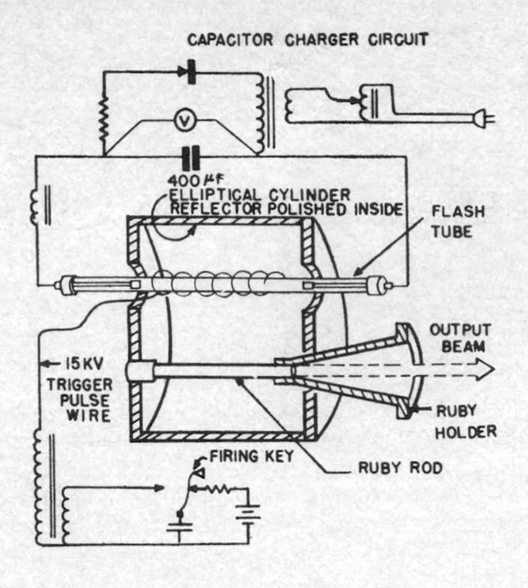

Ruby-laser circuit.

cal pumping is applied to a rod-shaped ruby crystal with a flash tube to produce an intense and extremely narrow beam of coherent red light.

ruby maser A maser that uses a ruby crystal in the cavity resonator.

ruggedization Making electronic equipment and components resistant to severe shock, temperature changes, high humidity, or other detrimental environmental influences.

ruly English A form of English developed for use in computers, in which each word has one and only one meaning. Conversely, each concept is represented by a single word.

rumble *Turntable rumble.*

run One performance of a program or routine on a computer.

runaround crosstalk Crosstalk resulting from coupling between the high-level end of one repeater and the low-level end of another repeater, as at a carrier telephone repeater station.

runaway electron An electron, in an ionized gas to which an electric field is applied, that gains energy from the field faster than it loses energy colliding with other particles in the gas.

running rabbits Random spots drifting across a radar screen, due to interference from nearby radar sets.

runway localizing beacon A small radio range used to provide accurate directional guidance

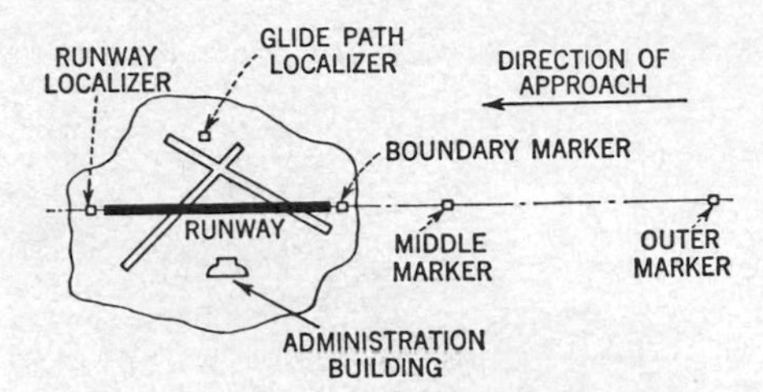

Runway localizing beacon for instrument landings, at opposite end of runway from marker beacons.

along the runway of an airport and for some distance beyond, for instrument landings.

ruthenium [symbol Ru] A metallic element. Atomic number is 44.

Rutherford scattering Scattering of moving particles at various angles as a result of interaction with atoms of a solid material.

RZ code Abbreviation for *return-to-zero code.*

S

s Abbreviation for *second.*

S 1. Symbol for *secondary winding,* used on circuit diagrams to identify the secondary winding of a transformer. 2. Abbreviation for *siemens.* 3. Symbol for *source,* used on circuit diagrams that contain field-effect transistors.

sabin A unit of sound absorption for a surface, equivalent to 1 ft^2 of perfectly absorbing surface. Also called square-foot unit of absorption. The metric sabin has an area of 1 m^2.

Sabine absorption The sound absorption defined by the equation in which reverberation time in seconds is $0.049V/A$, where V is the volume of the room in cubic feet and A is the total Sabine absorption in sabins.

Sabine coefficient The Sabine absorption of a sound-absorptive surface divided by the area of the surface.

SAC [pronounced as a word] Abbreviation for *Strategic Air Command.*

SAE Abbreviation for Society of Automotive Engineers.

safety base *Acetate base.*

safety factor The amount of load, above the normal operating rating, that a device can handle without failure.

SAGE [SemiAutomatic Ground Environment] An air defense system in which air surveillance data is processed for transmission to computers at direction centers. Here the data is further processed, evaluated, and analyzed automatically to produce weapon assignment and guidance orders.

Saint Elmo's fire A visible electric discharge sometimes seen on the mast of a ship, on metal towers, and on projecting parts of aircraft, due to concentration of the atmospheric electric field at such projecting parts.

sal ammoniac Common name for ammonium chloride, a chemical compound used as an electrolyte in some types of dry cells.

sal ammoniac cell A cell in which the electrolyte consists primarily of a solution of ammonium chloride.

sales representative A salesperson who sells the products of a number of different manufacturers in a given territory and represents those manufacturers in other ways. Also called representative.

salient pole A structure of magnetic material on which is mounted a field coil of a generator, motor, or similar device.

SAM Abbreviation for *surface-to-air missile.*

samarium [symbol Sm] A rare-earth element. Atomic number is 62.

samarium-cobalt magnet A rare-earth permanent magnet that is more efficient, has lower leakage and greater resistance to demagnetization, and can be magnetized to higher levels than conventional permanent magnets. These characteristics permit reduction of magnet size and weight for a given application. Uses include magnets for electronic watches, magnetrons, meters, and traveling-wave tubes.

SAM-D A surface-to-air missile-defense system that uses a phased-array radar which combines the functions of search, surveillance, target acquisition, and tracking with missile guidance. The missile is wingless, with maneuverability provided by four rear-mounted fins, and is designed for boost glide with very high acceleration and speed. The warhead may be either high-explosive or nuclear.

sample-and-hold A method of artificially increasing the duration of a signal pulse, or converting from analog to digital form, by sampling at fixed intervals and holding or storing the sampled

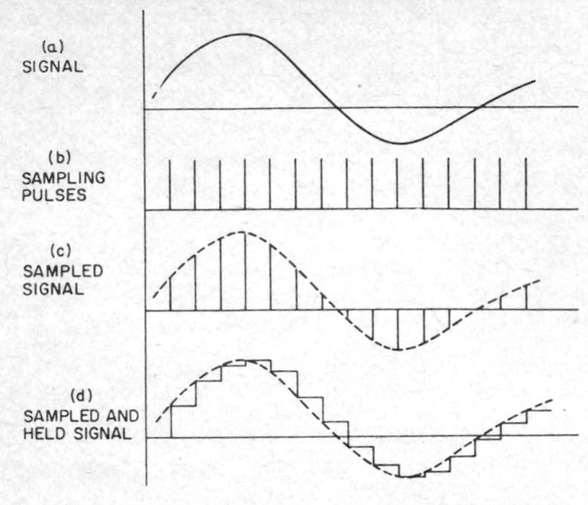

Sample-and-hold process using train of periodic sampling pulses.

amplitudes for display or other uses.

sample-and-hold digital voltmeter A digital voltmeter that measures varying voltages at a definite point in time and is unaffected by voltage changes during measurement. Also called clamp-and-hold digital voltmeter.

sampled data Data that is obtained at discrete rather than continuous intervals.

sample size The number of units in a sample.

sampling Selecting a small statistically determined portion of the total group under consideration for tests used to infer the value of one or several characteristics of the entire group.

sampling action Control action in which the difference between the set point and the value of the controlled variable is measured and correction made only at intermittent intervals.

sampling chamber A spark chamber used in the sampling mode for observing complex nuclear events that have many tracks. Sparks discharged between the parallel plates of the chamber tend to follow the direction of the track of a particle for a short distance, thereby making the track visible. Repeated discharges of sparks make the entire track visible.

sampling gate A gate circuit that extracts information from the input waveform only when activated by a selector pulse.

sampling interval The time between samples in a sampling or sample-and-hold system.

sampling multiplier *Averaging multiplier.*

sampling oscilloscope An instrument in which fast and repetitive signals are slowed down for conventional display on a cathode-ray oscilloscope. An amplitude sample of the signal is selected by a strobe pulse at an instant of time, widened, amplified, and displayed as a bright dot. The next time the signal occurs, the same process is repeated, but the strobe is automatically delayed slightly longer so it produces a sample of an adjacent portion of the signal. This process is repeated many times

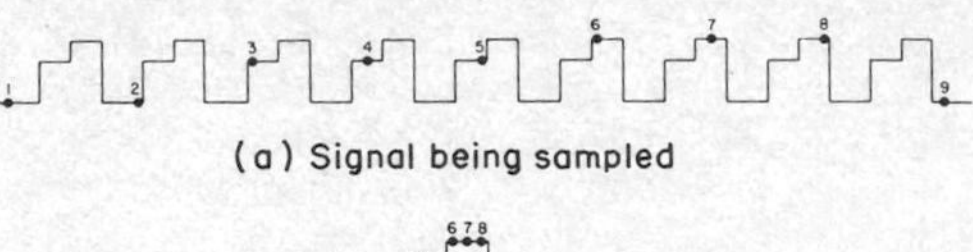

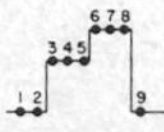

Sampling oscilloscope reconstructs waveform of signal from consecutive samples.

until a reproduction of the original signal is traced in dots.

sampling plan A plan that states sample sizes and the criteria for accepting, rejecting, or taking another sample during inspection of a group of items.

sampling rate The number of times a particular data channel is sampled in 1 s by a commutator.

sampling switch *Commutator switch.*

sampling theorem Equispaced data, in which there are two or more points per cycle of highest frequency, permits reconstruction of band-limited functions. Used in the study of information theory.

sand load An attenuator used as a power-dissipating terminating section for a coaxial line or waveguide. The dielectric space in the line is filled with a mixture of sand and graphite that acts as a matched-impedance load, preventing standing waves.

sapphire A pure variety of gem corundum that occurs in nature and is also produced artificially. Used for tips of phonograph needles because it has a hardness of 9 and takes a fine polish.

sapphire substrate Use of synthetic sapphire as a passive insulating base on which silicon can be

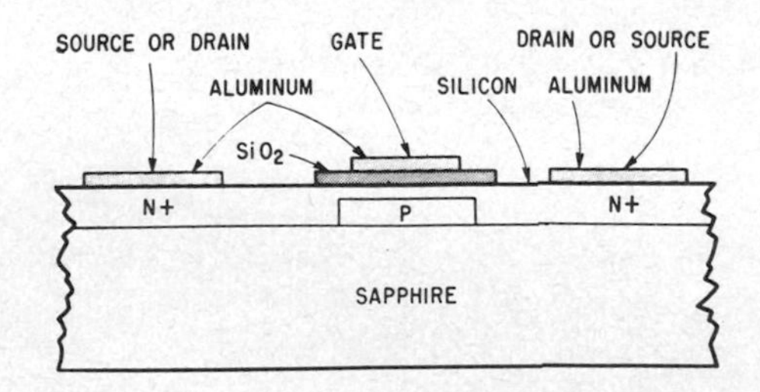

Sapphire substrate in silicon-on-sapphire field-effect transistor.

grown and then etched away selectively to form such solid-state devices as a silicon-on-sapphire field-effect transistor.

SARAH [Search And Rescue And Homing] A radio system that facilitates rescue when an aircraft goes down at sea. A small radio beacon transmitter attached to a lifebelt sends a coded pulse to a suitable receiver in search craft. Also used in spacecraft to guide rescuers after a landing on water.

satellite 1. A booster or translator operated by a television station to improve its signal strength in certain portions of its coverage area. 2. A man-made vehicle placed in orbit around the earth,

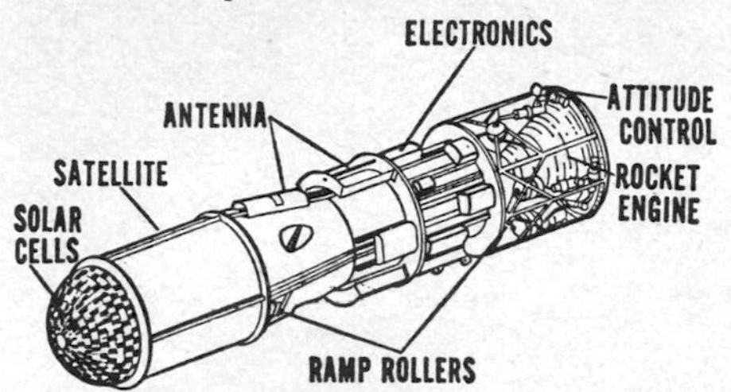

Satellite carrying instrumentation for making measurements while in orbit.

moon, or other celestial body. It usually carries measuring, recording, and transmitting equipment and may or may not be manned. Also called artificial satellite.

satellite communication Communication that involves the use of an active or passive satellite to extend the range of a radio, television, or other transmitter by returning signals to earth from an orbiting satellite. An active satellite may use a solar-cell power supply, with output power about 500 W. A passive satellite contains only reflectors.

satellite reconnaissance Strategic reconnaissance conducted by data obtained from a satellite.

saturable-core magnetometer A magnetometer that depends for its operation on the changes in permeability of a ferromagnetic core as a function of the magnetic field to be measured.

saturable reactor An iron-core reactor that has

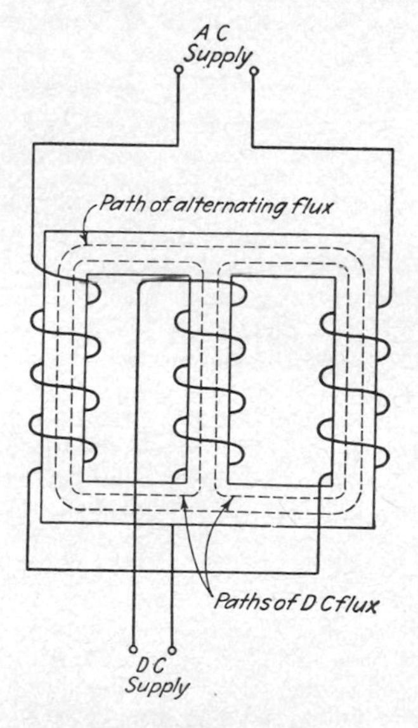

Saturable reactor.

an additional control winding which carries direct current whose value is adjusted to change the degree of saturation of the core, thereby changing the reactance that the AC winding offers to the flow of alternating current. With appropriate external circuits, a saturable reactor can serve as a magnetic amplifier.

saturable-reactor-controlled oscillator An oscillator that has a saturable reactor in its tuning circuit to control the output frequency.

saturable transformer A saturable reactor that has additional windings to provide voltage transformation or isolation from the AC supply.

saturated color A pure color, not contaminated by white.

saturated diode A diode that is passing the maximum possible current, so further increases in applied voltage have no effect on current.

saturated logic Logic in which transistors are allowed to saturate during normal operation. Examples include diode-transistor and transistor-transistor logic.

saturating signal A radar signal whose amplitude is greater than the dynamic range of the receiving system.

saturation 1. The condition in which a further increase in one variable produces no further increase in the resultant effect. 2. The condition in which the decay rate of a given radionuclide is equal to its rate of production in an induced nuclear reaction. 3. The condition in which the voltage applied to an ionization chamber is high enough to collect all the ions formed by radiation but not high enough to produce ionization by collision. 4. The condition occurring when a transistor is driven so hard that it becomes biased in the forward direction (the collector becomes positive with respect to the base, for example, in a PNP-type transistor). In a switching application, under saturation conditions the charge stored in the base region prevents the transistor from turning off quickly. 5. *Anode saturation.* 6. *Color saturation.* 7. *Magnetic saturation.* 8. *Temperature saturation.*

saturation current The maximum possible current that can be obtained as the voltage applied to a device is increased. In a gas tube it is the current at which the applied voltage is sufficient to attract all ions. In a semiconductor diode it is the portion of the steady-state reverse current that flows as a result of the transport across the junction of minority carriers thermally generated within the regions adjacent to the junction.

saturation curve A curve that shows the manner in which a quantity such as current or magnetic flux reaches saturation.

saturation flux density *Saturation induction.*

saturation induction The maximum intrinsic induction possible in a material. Also called saturation flux density.

saturation magnetostriction The value of magnetostriction that would be reached if the applied magnetizing force were increased indefinitely.

saturation reactance The reactance of the gate winding of a magnetic amplifier during the saturation interval.

saturation region An operating region in which an increase in the actuating component produces no further increase in the output effect.

saturation voltage The minimum voltage needed to produce saturation current.

Saturn A booster for launching large spacecraft, such as a manned lunar landing vehicle.

SAW Abbreviation for *surface acoustic wave.*

sawtooth current A current that has a sawtooth waveform.

sawtooth generator A generator whose output voltage has a sawtooth waveform. Used to produce sweep voltages for cathode-ray tubes.

sawtooth voltage A voltage that has a sawtooth waveform.

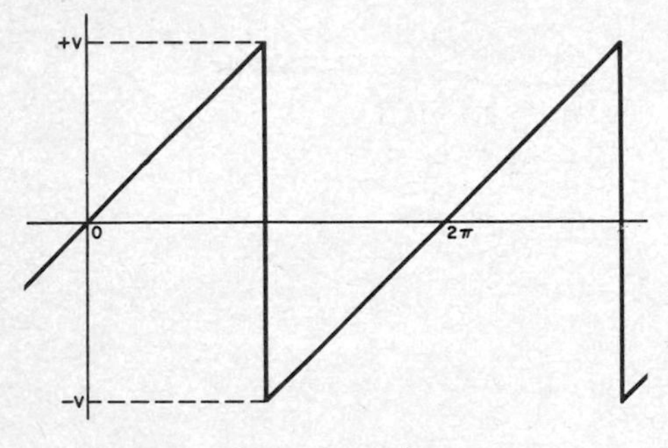

Sawtooth waveform.

sawtooth waveform A waveform characterized by a slow rise time and a sharp fall, resembling the tooth of a saw.

sb Abbreviation for *stilb.*

S band A band of radio frequencies that extends from 1.55 to 5.2 GHz, corresponding to wavelengths of 19.37 to 5.77 cm. Subdivisions of this band (all values in gigahertz) are

Se:	1.55–1.65	Ss:	2.9–3.1
Sb:	1.65–1.85	Sa:	3.1–3.4
St:	1.85–2.0	Sw:	3.4–3.7
Sc:	2.0–2.4	Sh:	3.7–3.9
Sq:	2.4–2.6	Sz:	3.9–4.2
Sy:	2.6–2.7	Sd:	4.2–5.2
Sg:	2.7–2.9		

SC Abbreviation for *suppressed carrier.*

SCA Abbreviation for *Subsidiary Communications Authorization.*

scalar Having magnitude but not direction.

scalar function A scalar quantity that has a definite value for each value of some other scalar quantity. Thus the resistance of a given conductor is a scalar function of the temperature of the conductor.

scalar quantity A quantity that has only magnitude, such as resistance, time, or temperature.

scale 1. A series of markings used for reading the value of a quantity or setting. 2. A series of musical notes arranged from low to high by a specified scheme of intervals suitable for musical purposes. 3. To change the magnitudes of the units in which a problem is expressed to bring all magnitudes within the capacity of a computer.

scale division The portion of a scale between two adjacent scale marks.

scale factor The factor by which the reading of an instrument or the solution of a problem should be multiplied to give the true final value when a corresponding scale factor is used initially to bring the magnitude within the range of the instrument or computer.

scale-of-eight circuit A counting circuit that recycles at every eighth pulse.

scale-of-ten circuit *Decade scaler.*

scale-of-two circuit *Binary scaler.*

scaler A circuit that produces an output pulse when a prescribed number of input pulses is received. A single binary scaler stage delivers an output pulse for every 2 input pulses, and a decade scaler delivers an output pulse for every 10 input pulses. Also called counter and scaling circuit.

scaling Counting pulses with a scaler when the pulses occur too fast for direct counting by conventional means.

scaling circuit *Scaler.*

scaling factor The number of input pulses per output pulse of a scaling circuit. Also called scaling ratio.

scaling ratio *Scaling factor.*

scan 1. To examine an area or a region in space point by point in an ordered sequence, as when converting a scene or image to an electric signal or when using radar to monitor an airspace for detection, navigation, or traffic control purposes. 2. One complete circular, up-and-down, or left-to-right sweep of the radar, light, or other beam or device used in making a scan.

scan axis The axis used as a reference for specifying target displacement in radar or sonar.

scan converter A cathode-ray tube that is capable of storing radar, television, and data displays for nondestructive readout over prolonged periods of time. Applications include buildup of repetitive signals submerged in noise, conversion of video displays from one scan mode to another, and special overlay and merging effects for television and radar displays.

scandium [symbol Sc] A metallic element. Atomic number is 21.

scanned array A phased-array radar antenna that uses phase shifters operating under computer control to give scanning of the beam in azimuth and elevation without mechanical movements.

scanner 1. A radar antenna and reflector assembly that oscillates back and forth about a center position during search, then moves in any required plane during tracking of a target. 2. That part of a facsimile transmitter which systematically translates the densities of the elemental areas of the subject copy into corresponding electric signals. 3. A device that automatically samples, measures, or checks a number of quantities or conditions in sequence, as in process control.

scanning The process of examining an area, a region in space, or a portion of the radio spectrum point by point in an ordered sequence.

scanning antenna A directional antenna used in radar for scanning a region.

scanning beam A beam of light, an electron beam, or a radar beam used in scanning.

scanning disk A rotating metal disk that has one or more spirals of holes near the circumference, used in early mechanical television systems to break up a scene into elemental areas at a television camera or to reconstruct a scene in a television receiver.

scanning head A light source and phototube combined as a single unit for scanning a moving strip of paper, cloth, or metal in photoelectric side-register control systems.

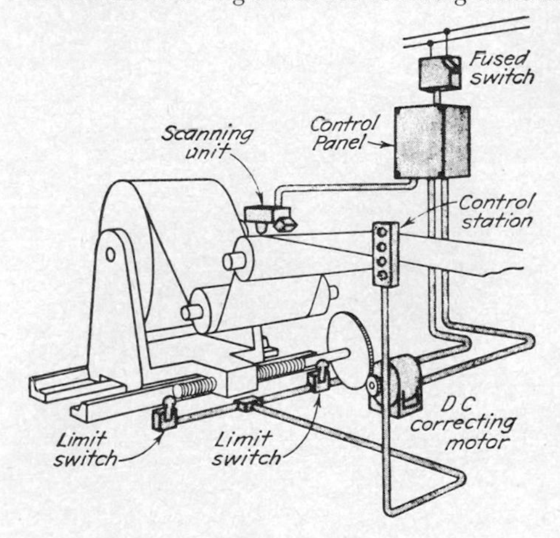

Scanning head as used for controlling position of edge of material being processed in rolls.

scanning line The single continuous narrow strip covered during one scan of a television image or one scan of the subject copy or record sheet in facsimile.

scanning loss The reduction in sensitivity when scanning across a radar target as compared with the sensitivity obtained when the radar beam is directed constantly at the target. Scanning loss is due to the change in antenna position during the interval in which the signal travels from the antenna to the target and back.

scanning radio A radio receiver that automatically scans across public service, emergency service, or other radio bands and stops at the first preselected station which is on the air. Some versions use a crystal for each channel of interest; other versions use a frequency synthesizer that permits choosing from thousands of digitally derived frequencies. Also called radio scanner.

scanning sonar Sonar in which all targets of interest are shown simultaneously, as on a radar PPI display or sector display. The sound pulse may be transmitted in all directions simultaneously and picked up by a rotating receiving transducer, or transmitted and received in only one direction at a time by a scanning transducer.

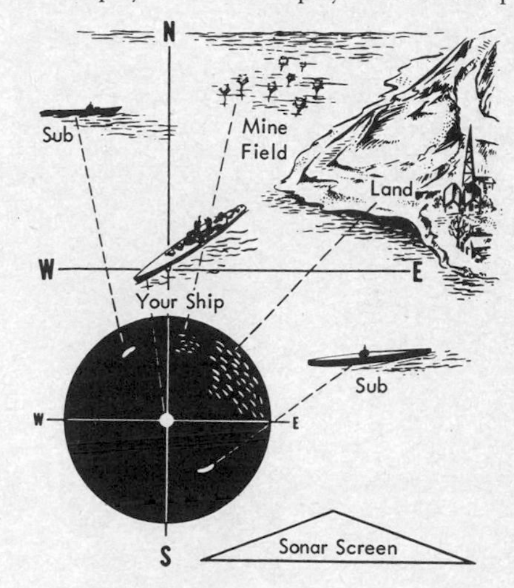

Scanning-sonar presentation on PPI screen.

scanning speed *Spot speed.*

scanning spot The area that is viewed instantaneously by the pickup system of a facsimile scanner or a television camera.

scanning switch *Commutator switch.*

scanning transducer A multielement sonar transducer in which the elements are arranged in a circle and electronically switched in sequence to give the equivalent of scanning without mechanical movement.

scan period *Regeneration period.*

scan rate The angular velocity of a radar beam. It is generally in the range of 4 to 12 rpm.

scatterband The total bandwidth occupied by the frequency spread of numerous interrogations that have the same nominal radio frequency in a pulse system.

scattered radiation Radiation that has been changed in direction during its passage through a substance. It may also be increased in wavelength.

scattering 1. The change in direction of a particle or photon because of a collision with another particle or a system. 2. Diffusion of acoustic waves because of nonuniformity of the transmitting medium. 3. Diffusion of electromagnetic waves in a random manner by air masses in the upper atmosphere, permitting long-range reception, as in scatter propagation. Also called radio scattering.

scattering amplitude A quantity closely related to the intensity of scattering of a wave by a central force field such as that of a nucleus.

scattering angle The angle between the initial and final lines of motion of a scattered particle.

scattering cross section The cross section for elastic scattering with kinetic energy conserved plus the cross section for elastic scattering with absorption followed by emission of a neutron of lower energy. It is equal to 0.657 barn per electron. Also called classical scattering cross section and Thomson cross section.

scattering loss The portion of the transmission loss that is due to scattering within the medium or roughness of the reflecting surface.

scattering parameter *S parameter.*

scatterometer A microwave sensor that is essentially a radar without ranging circuits, used to measure only the reflection or scattering coefficient while scanning the surface of the earth from an aircraft or a satellite.

scatter propagation Transmission of radio waves far beyond line-of-sight distances by using high power and a large transmitting antenna to beam the signal upward into the atmosphere and a similar large receiving antenna to pick up the small portion of the signal that is scattered by the atmosphere. In ionospheric scatter the phenomenon occurs in the lower E layer of the ionosphere. In tropospheric scatter the phenomenon is entirely in the earth's lower atmosphere, from ground level to about 30,000 ft (10 km). In meteoric scatter the trail of a passing meteor scatters the radio waves back to earth. Also called beyond-the-horizon communication, forward-scatter propagation, and over-the-horizon communication.

schematic circuit diagram A circuit diagram in which component parts are represented by simple, easily drawn symbols. Also called circuit diagram.

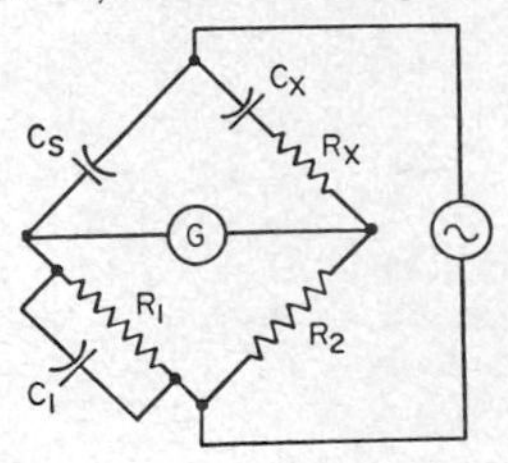

Schering-bridge circuit.

Schering bridge A four-arm AC bridge that measures capacitance and dissipation factor. Bridge balance is independent of frequency.

schlieren photography An optical system used to photograph changes in gas density due to sound waves, shock waves, or turbulence in wind tunnels. A knife edge at the focal point transmits or cuts off a light beam when the refraction of the intervening gas varies with density.

Schmidt system A projection optics system that consists of a spherical mirror, a curved image plate at the focal plane of the mirror, and a corrector

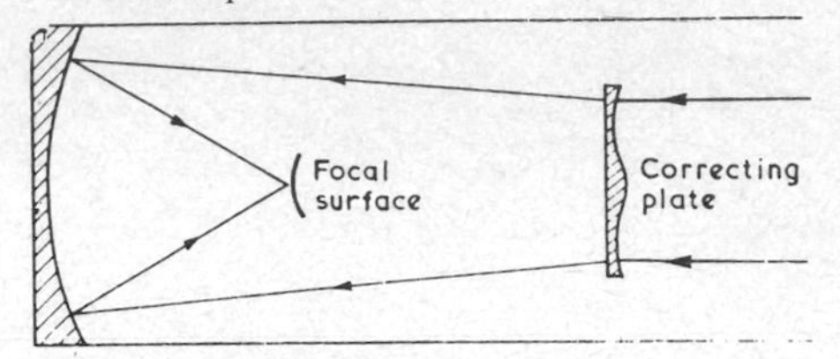

Schmidt system.

plate for correcting the spherical aberration of the mirror. Used to project a television image from a projection cathode-ray tube onto a screen.

Schmitt trigger A bistable trigger circuit that converts an AC input signal into a square-wave output signal by switching action, triggered at a

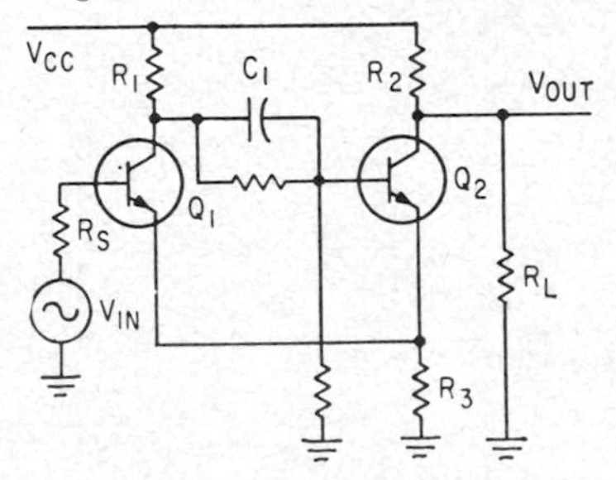

Schmitt-trigger circuit using transistors.

predetermined point in each positive and negative swing of the input signal.

Schottky barrier A transition region formed within a semiconductor surface to serve as a rectifying barrier at a junction with a layer of metal.

Schottky barrier diode A semiconductor diode formed by contact between a semiconductor layer and a metal coating; it has a nonlinear rectifying characteristic. Hot carriers (electrons for N-type material or holes for P-type material) are emitted from the Schottky barrier of the semiconductor and move to the metal coating that is the diode base. Since majority carriers predominate, there is essentially no injection or storage of minority carriers to limit switching speeds. Also called hot-carrier diode and Schottky diode.

Schottky clamped transistor A logic circuit in which a Schottky barrier diode is used to avoid saturation and thereby speed up switching action.

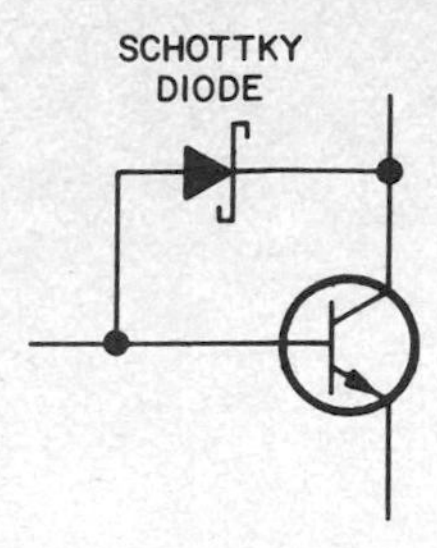

Schottky clamped transistor, with Schottky barrier diode connected between base and collector.

Used in transistor-transistor logic and integrated injection logic.

Schottky diode *Schottky barrier diode.*

Schottky effect An increase in anode current of a thermionic tube beyond that predicted by the Richardson equation, due to lowering of the work function of the cathode when an electric field is produced at the surface of the cathode by the anode. The same effect causes removal of electrons from the surface of a semiconductor material when an electric field is applied.

Schottky emission Injection of an electron over a semiconductor potential barrier into a region where allowed energy levels are available.

Schottky I²L An integrated injection logic circuit in which Schottky barrier diodes have been added for clamping purposes or built into a metal collector of the transistor.

Schottky noise *Shot noise.*

Schottky power diode A Schottky barrier diode in which heat dissipation is spread over a large area of silicon, but diode capacitance is kept low by a small metal-semiconductor contact area.

Schottky TTL gate A transistor-transistor logic gate in which each transistor has a Schottky barrier diode connected across it to reduce the number of charge carriers in the base when the transistor is on. Less time is then needed to deplete the base region of carriers so the transistor can be turned off at high switching speeds.

Schottky TTL memory A random-access memory in which each cell consists of two inverted transistors in a cross-coupled flip-flop, with emitter-base resistances serving as load.

Schroedinger equation A wave equation that gives the relation between wave function, particle mass, total energy, potential energy, and Planck's constant.

Schroedinger wave function A wave function that determines the state of a system and satisfies the Schroedinger equation. For a particle or photon, the square of the wave function is proportional to the probability that the particle will be at a particular point at a particular time.

scientific calculator An electronic calculator that has provisions for handling exponential, trigonometric, and sometimes other special functions in addition to performing arithmetic operations. In some models, the calculating program can be changed by inserting a new prerecorded magnetic program card.

scientist A person who has the training, ability, and desire to seek new knowledge, new principles, and new materials in some field of science.

scintillation 1. A flash of light (optical photons) produced in a phosphor by an ionizing particle or photon. 2. A rapid apparent displacement of a target indication from its mean position on a radar display. One cause is shifting of the effective reflection point on the target. Also called target glint, target scintillation, and wander. 3. Rapid random fading of microwave radio signals, caused by fluctuations in the refraction of the atmosphere. With optical communication links, the same fluctuations or scintillations can be produced by air turbulence.

scintillation camera A camera that provides a complete image of radionuclide distribution in a given area of the human body in one exposure, rather than by line-scanning techniques. The three main types of scintillation camera are the gamma-ray camera, positron camera, and autofluoroscope.

scintillation counter A counter in which the scintillations produced in a fluorescent material by an ionizing radiation are detected and counted by a multiplier phototube and associated circuits. Widely used in medical research, nuclear research, and prospecting for radioactive ores.

scintillation spectrometer A scintillation counter adapted to the study of energy distributions.

scintillator material A material that emits optical photons in response to ionizing radiation. The five major classes of scintillator materials are (a) inorganic crystals like sodium iodide, thallium single crystals, and zinc sulfide-silver screens; (b) organic crystals like anthracene and transstilbene; (c) liquid, plastic, or glass solution scintillators; (d) gaseous scintillators; (e) Cerenkov scintillators.

scope 1. *Cathode-ray oscilloscope.* 2. *Radarscope.*

Scophony television system An early mechanical television system that uses a rotating mirror and Kerr cell, developed in Great Britain.

scotophor A solid that exhibits reversible darkening and bleaching actions of tenebrescence under suitable irradiation.

SCR Abbreviation for *silicon controlled rectifier.*

scrambled speech Speech that has been made unintelligible by inversion, such as for privacy of transoceanic radiotelephone calls. At the receiving end, the signals are inverted again to restore the original sounds. Also called inverted speech.

scrambler A circuit that divides speech frequencies into several ranges by filters, then inverts and displaces the frequencies in each range so the resulting reproduced sounds are unintelligible. In the simplest system the entire speech frequency range is combined with the output of a fixed-frequency oscillator, and difference frequencies are used as the inverted signal. The process is reversed at the receiving apparatus to restore intelligible speech. Scramblers that provide secrecy for teletypewriter and code messages generally use cryptology techniques involving pseudorandom key generators, with similar equipment for unscrambling the received messages before or after printout. Also called secrecy system, speech inverter, and speech scrambler.

scratch filter A low-pass filter circuit inserted in the circuit of a phonograph pickup to suppress higher audio frequencies and thereby minimize needle-scratch noise.

scratchpad memory A small fast-access memory used in a computer for storage of frequently needed instructions and codes or for holding subtotals until required for final results.

screen 1. The surface on which a television, radar, x-ray, or cathode-ray oscilloscope image is made visible for viewing. It may be a fluorescent screen that has a phosphor layer which converts the energy of an electron beam to visible light, or a translucent or opaque screen on which the optical image is projected. Also called viewing screen. 2. *Screen grid.*

screen grid [symbol SG] A grid placed between a control grid and an anode of an electron tube and usually maintained at a fixed positive potential, to reduce the electrostatic influence of the anode in the space between the screen grid and the cathode. Also called screen.

screen-grid modulation Modulation produced by introducing the modulating signal into the screen-grid circuit of a multigrid tube in which the carrier is present.

screen-grid tube An electron tube that has a screen grid.

scribe projection A display system in which a servo-controlled fine-point scribe removes the coating from a metal-coated glass slide, so light passing through the scribed designs or characters is projected onto a screen.

scribing The process of scratching lines on a semiconductor wafer. Later application of pressure breaks the wafer along these lines to form desired sizes of chips for semiconductor devices and integrated circuits.

SCS Abbreviation for *silicon controlled switch.*

S/D Abbreviation for synchro-to-digital, usually used in connection with converters.

SDC Abbreviation for *synchro-to-digital converter.*

S distortion *Spiral distortion.*

sea clutter Clutter on an airborne radar due to reflection of signals from the sea. Also called sea return and wave clutter.

seal A joint between two pieces of glass, two pieces of metal, or glass and metal. For electron tubes the joint must be hermetically tight.

sealed crystal unit A crystal unit in which the quartz plate is sealed in its holder, usually by a gasket under pressure, for protection against humidity or a contaminated atmosphere.

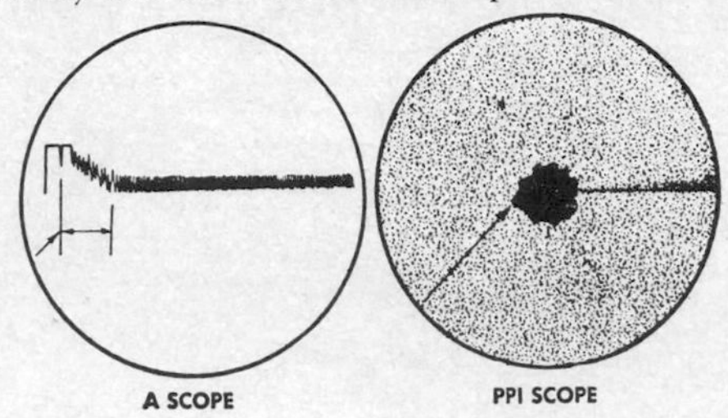

Sea clutter as seen on two types of radarscope displays.

sealed tube An electron tube that is hermetically sealed. This term is used chiefly for pool tubes.

sealing compound A compound used in dry batteries, capacitor blocks, transformers, and other components to keep out air and moisture.

sealing off Final closing of the envelope of an electron tube after evacuation.

seam welding 1. Spot welding in which the lapped metal sheets are passed between roller contacts and overlapping spot welds are made. 2. Use of dielectric heating for uniting two pieces of thermoplastic material along a prescribed line.

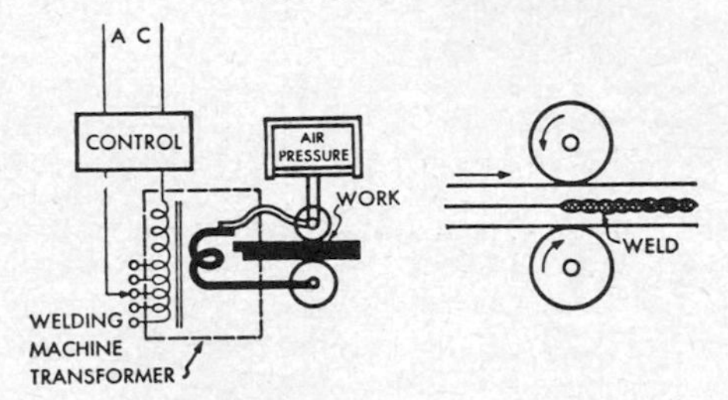

Seam-welding machine and nature of weld.

search 1. To explore a region in space with radar. 2. To examine a set of items for those which have a desired property.

search antenna A radar antenna or antenna system designed especially for search.

search coil *Exploring coil.*

searching gate A gate pulse that is made to search back and forth over a certain range.

searchlight-control radar Ground radar used to direct searchlights at aircraft.

searchlighting Projecting a radar beam continuously at a target instead of scanning the area containing the target.

searchlight sonar A sonar system in which a directional transducer concentrates the outgoing pulse of sound energy into a narrow beam and receives the echo reflected from an underwater target. The bearing of a target is determined by aiming the transducer for maximum echo strength.

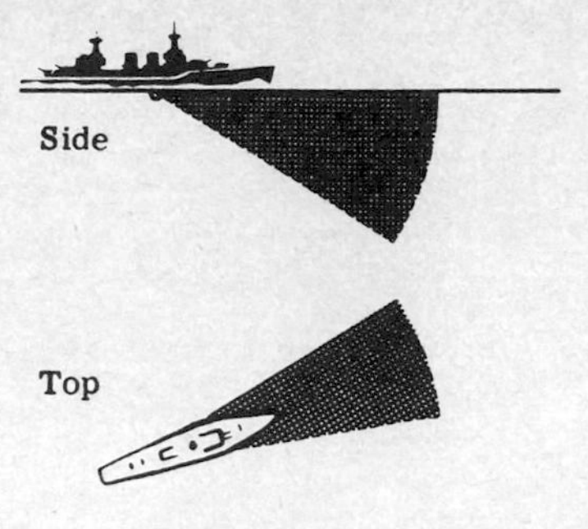

Searchlight sonar.

search radar A radar intended primarily to cover a large region of space and display targets as soon as possible after they enter the region. Used for early warning and in connection with ground-controlled approach, ground-controlled interception, and air traffic control.

search receiver A radio receiver that can be tuned over a wide frequency range for detecting and measuring RF signals transmitted by the enemy.

sea return *Sea clutter.*

SECAM [derived from French for sequential with memory] A 625-line 50-field color television system originally developed in France. In this system, one color signal is transmitted on one line and the other color signal on the next line. Delay and switching circuits in the receiver combine these signals so the three types of color video information can be displayed simultaneously. Other countries now using this system include Czechoslovakia, East Germany, Lebanon, and the U.S.S.R.

SECO [SEquential COntrol] A teletype control system that permits a primary station to control automatically the sequential transmission of data stored on perforated tape at secondary stations on the teletype circuit. Also called sequential control.

second [abbreviated s] The SI unit of time, based on the time of transition between two specific energy levels in cesium 133. This redefinition of the unit of time was approved at the Twelfth General Conference on Weights & Measures. A second was formerly $1/86{,}400$ of a mean solar day. 2. A unit of angle, equal to $1/3600°$.

secondary *Secondary winding.*

secondary battery *Storage battery.*

secondary cell *Storage cell.*

secondary cosmic rays Radiation produced when primary cosmic rays enter the atmosphere and collide with atomic nuclei. Each such collision splits one or both particles into smaller nuclear fragments like mesons, protons, neutrons, electrons, and photons, and these in turn collide with other nuclei to produce additional high-speed particles. All these results of collisions are known as secondary cosmic rays.

secondary electron 1. An electron emitted as a result of bombardment of a material by an incident electron. 2. An electron whose motion is due to a transfer of momentum from primary radiation.

secondary-electron conduction Movement of charges by free secondary electrons traveling in interparticle spaces of low-density materials under the influence of an externally applied electric field.

secondary-electron-conduction camera tube A camera tube in which an electron image, generated by a photocathode, is focused on a target that has a backplate and a secondary-electron-conduction layer which provides charge amplification and storage. Image amplification is high enough for dimly illuminated scenes.

secondary-electron multiplier *Electron multiplier.*

secondary emission The emission of electrons from a solid or liquid as a result of bombardment by electrons or other charged particles.

secondary-emission ratio The average number of electrons emitted from a surface per incident primary electron.

secondary filter A sheet of material that has a low atomic number relative to that of a primary filter used in radiology, placed in the filtered beam of radiation to remove characteristic radiation produced in the primary filter.

secondary grid emission Electron emission from a grid as a direct result of bombardment of the grid surface by electrons or other charged particles.

secondary lobe *Minor lobe.*

secondary parameter An additional rating or characteristic needed to evaluate the operation of a part beyond its normal limits, such as its temperature coefficient.

secondary radar Radar that involves transmission of a second signal when the incident signal triggers a responder beacon.

secondary radiation Particles or photons produced by the action of primary radiation on matter, such as Compton recoil electrons, delta rays, secondary cosmic rays, and secondary electrons.

secondary radionuolide A radionuclide that has a geologically short lifetime and is a decay product of natural primary radionuclides. All presently known members of this class belong to the elements from thallium to uranium.

secondary service area The area served by the sky wave of a broadcast station and not subject to objectionable interference. The signal is subject to intermittent variations in intensity.

secondary standard 1. A unit, as of length, capacitance, or weight, used as a standard of comparison in individual countries or localities, but checked against the one primary standard in existence somewhere. 2. A unit defined as a specified multiple or submultiple of a primary standard, such as the centimeter.

secondary storage Storage that is not an integral part of the computer but is directly linked to and controlled by the computer.

secondary surveillance radar [abbreviated SSR] *Air traffic control radar beacon system.*

secondary voltage The voltage across the secondary winding of a transformer.

secondary winding [symbol S] A transformer winding that receives energy by electromagnetic induction from the primary winding. A transformer may have several secondary windings, and they may provide AC voltages that are higher, lower, or the same as that applied to the primary winding. Also called secondary.

secondary x-ray Any x-ray given off by a material when irradiated by x-rays. The frequency of the secondary rays is characteristic of the material.

second breakdown Destructive breakdown in a transistor, wherein structural imperfections cause localized current concentrations and uncontrollable generation and multiplication of current carriers. Reaction occurs so suddenly that the thermal time constant of the collector regions is exceeded and the transistor is irreversibly damaged. In contrast, primary or avalanche breakdown is the normal sustaining mode of a transistor and does not cause permanent damage.

second-channel attenuation *Selectance.*

second-channel interference Interference in which the extraneous power originates from a signal of assigned type in a channel two channels removed from the desired channel.

second detector The detector that separates the intelligence signal from the IF signal in a superheterodyne receiver.

second-harmonic magnetic modulator A magnetic modulator in which the output frequency is twice the power-supply frequency.

second sound A type of heat wave that carries energy but moves at very nearly the speed of sound. The heat waves are produced by turning a heater on and off. The phenomenon has been observed in photoconducting crystals of cadmium sulfide.

second-time-around echo A radar or sonar echo received after an interval exceeding the pulse recurrence interval.

SECOR [SEquential COllation of Range] A navigation and surveying system in which an orbiting satellite acts with ground stations to give exact latitude and longitude fixes of possible targets or other points on earth. The satellite contains a transponder that receives and returns radio signals simultaneously from four ground stations, three of which are reference locations that have known latitude and longitude. Readings then give the exact values for the fourth ground station, from which map corrections can be made for intermediate targets.

secrecy system *Scrambler.*

section Each individual transmission span in a radio relay system. A system has one more section than it has repeaters.

sectionalized vertical antenna A vertical antenna that is insulated at one or more points along its length. Reactances or driving voltages are applied across the insulated points to modify the radiation pattern in the vertical plane.

sectoral horn An electromagnetic horn in which two opposite sides are parallel and the other two sides diverge.

sector display A display in which only a sector of the total service area of a radar system is shown. Usually the sector is selectable.

sector scan A radar scan through a limited angle, as distinguished from complete rotation.

secular equilibrium Radioactive equilibrium in which the parent has such a small decay constant that there has been no appreciable change in the quantity of parent present by the time the decay products have reached radioactive equilibrium. Equal numbers of atoms of all members of the series then disintegrate in unit time.

secular variation The slow variation in the strength of the magnetic field of the earth, requiring many years for a complete cycle.

secure voice Voice communication that is scrambled or coded.

security classification The classification assigned to defense information or material to denote the degree of danger to the nation that would result from unauthorized disclosure. The usual classifications are confidential, secret, and top secret.

security clearance A clearance that permits a person to have access to classified material or information up to and including a given security classification, provided the person can establish a need-to-know.

Seebeck effect Development of a voltage due to differences in temperature between two junctions of dissimilar metals in the same circuit. Discovered by J. T. Seebeck, German physicist, in 1821. Also called thermoelectric effect.

seed A small single crystal of semiconductor material, used to start the growth of a large single crystal for use in cutting semiconductor wafers.

seeding Introducing atoms that have a low ionization potential into a hot gas, to increase the electric conductivity.

Seek Bus A tactical communication system for airborne warning and control system aircraft. It is a high-capacity digital time-division multiple-access distribution system operating on a single secure broadband communication channel, to link surveillance, intelligence, and weapons systems of the Army, Navy, and Air Force.

seeker A missile or other device that finds its target by heat, light, radio waves, sound, or other radiation emitted by the target.

segment A portion of a digital-computer routine short enough to be stored entirely in the internal storage yet containing the coding necessary to call in and jump automatically to other segments of the routine.

seismic detector A microphone that detects

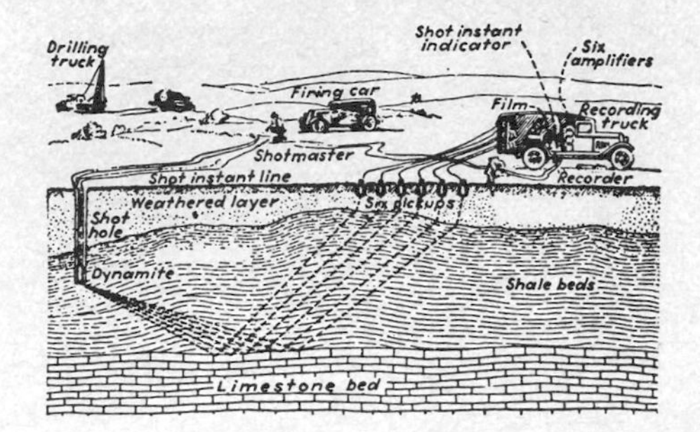

Seismic detectors being used in prospecting for petroleum.

acoustic waves transmitted through the earth.

seismic intrusion detector An acoustic sensor, connected by wire or radio to a monitor station, that detects earth noises such as those made by enemy tunnel construction or by movements of enemy troops.

seismic surveying A petroleum exploration technique based on variations in the rate of propagation of shock waves in layered subsurface media.

seismograph An instrument that records the time, direction, and intensity of earthquakes or of earth shocks produced by explosions during geophysical prospecting.

seismometer An instrument that measures earth movements or earth shocks.

selectance The reciprocal of the ratio of the sensitivity of a receiver tuned to a specified channel to its sensitivity at another channel separated by a specified number of channels from the one to which the receiver is tuned. Generally expressed as a voltage or field-strength ratio. Also called adjacent-channel attenuation and second-channel attenuation.

selection check A check made by a digital computer to verify that the correct register or other device is selected for performance of the next instruction.

selection ratio The least ratio of a magnetomotive force used to select a magnetic cell to the maximum magnetomotive force used that is not intended to select a cell, in the storage section of a digital computer.

selection rules Statements that classify transitions in terms of quantum numbers of the initial and final states of the systems involved, in such a way that transitions of a given order of inherent probability are grouped together. Transitions that have highest probability of taking place per unit time are called allowed transitions; all others are called forbidden transitions.

selective absorption Absorption of radiation as some function of frequency.

selective calling system A radio communication system in which the central station transmits a coded call that activates only the receiver to which that code is assigned.

selective diffusion Doping of isolated regions of a semiconductor material to produce individual components in a silicon integrated circuit.

selective epitaxial growth Growth of an epitaxial layer in a semiconductor material after selective masking of the surface with an oxide. Alternatively, the epitaxial layer may be formed on the entire surface and later removed from the regions where it is not required.

selective fading Fading that is different at different frequencies in a frequency band occupied by a modulated wave, causing distortion that varies in nature from instant to instant.

selective jamming Jamming in which only a single radio channel is jammed.

selective ringing Party-line telephone ringing that actuates only the bell of the desired station.

selective transmission The transmission of electromagnetic energy at wavelengths other than those reflected or absorbed in a given system.

selectivity The characteristic of a receiver that determines its ability to separate a desired signal frequency from all other signal frequencies.

selectivity control A control that adjusts the selectivity of a radio receiver.

selector An automatic or other device for making connections to any one of a number of circuits, such as a selector relay or selector switch.

selector pulse A pulse that identifies for selection one event in a series of events.

selector relay A relay capable of selecting one circuit automatically from a number of circuits.

selector switch A manually operated multiposition switch. Also called multiple-contact switch.

selenium [symbol Se] A nonmetallic element that has photosensitive properties. Its resistance varies inversely with illumination. Used also as a rectifying layer in metallic rectifiers. Atomic number is 34.

selenium cell A photoconductive cell in which a thin film of selenium is used between suitable

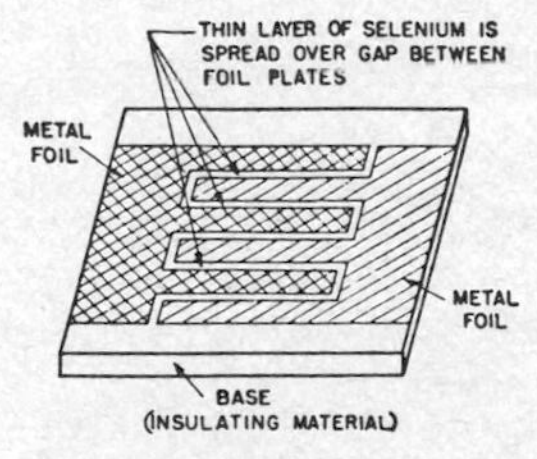

Selenium-cell construction.

electrodes. The resistance of the cell decreases when illumination is increased.

selenium rectifier A metallic rectifier in which a thin layer of selenium is deposited on one side of an aluminum plate and a conductive metal coating is deposited on the selenium. Electrons flow more freely in the direction from the metallic coating to the selenium than in the opposite direction, thus giving rectifying action.

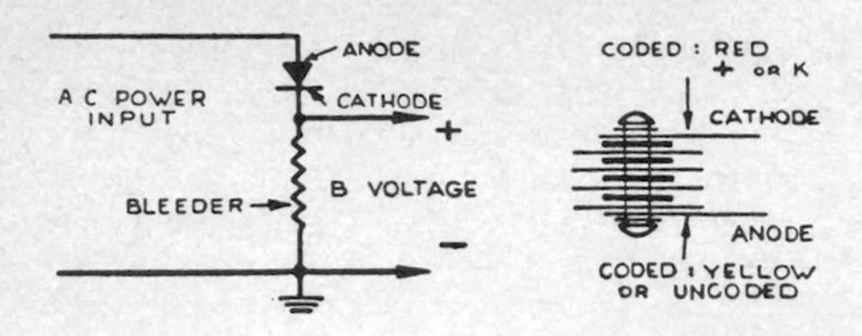

Selenium-rectifier polarity.

selenodesy The branch of applied mathematics that determines, by observation and measurement, the exact positions of points on the moon's surface and the shape and size of the moon.

selenography The geography of the moon's surface.

self-absorption Absorption of radiation by the material that emits the radiation, reducing the radiation level against which further shielding must be provided.

self-adapting Capable of changing performance characteristics automatically in response to the environment.

self-bias Grid bias provided automatically by the flow of electrode currents through a resistor in the cathode or grid circuit of an electron tube. The resulting voltage drop across the resistor serves as the grid bias.

self-capacitance *Distributed capacitance.*

self-checking code A computer code in which errors in a code expression produce a forbidden combination. Also called error-detecting code.

self-cleaning contact *Wiping contact.*

self-demagnetization The process by which a magnetized sample of magnetic material tends to demagnetize itself by virtue of the opposing fields created by its own magnetization. Self-demagnetization inhibits the successful recording of short wavelengths or sharp transitions in a recorded signal.

self-energy The energy equivalent of the rest mass of a particle. The self-energy of an electron is 511 keV.

self-excited Operating without an external source of power for excitation.

self-excited oscillator An oscillator that depends on its own resonant circuits for initiation of oscillation and frequency determination.

self-focused picture tube A television picture tube that has automatic electrostatic focus incorporated into the design of the electron gun.

self-generating transducer A transducer that does not require external electric excitation to provide specified output signals.

self-guided Directed only by built-in self-reacting devices, as in a homing missile.

self-healing capacitor A capacitor that repairs itself after breakdown caused by excessive voltage. Air capacitors, some wet electrolytic capacitors, and metallized paper capacitors have this characteristic.

self-impedance The impedance at a pair of terminals of an antenna array or of a network when all other elements or terminal pairs are open-circuited.

self-inductance Inductance that produces an induced voltage in the same circuit as a result of a change in current flow.

self-induction The production of a voltage in a circuit by a varying current in that same circuit.

self-instructed carry A carry in which information goes to succeeding locations automatically as soon as it is generated.

self-locking nut A nut that has an inherent locking action, so it cannot readily be loosened by vibration.

self-luminous light source A light source that consists of a radioactive nuclide such as tritium, firmly incorporated in solid and/or inactive materials or sealed in a protective envelope strong enough to prevent leakage of radioactive materials to the atmosphere. The nuclide incorporates or is surrounded by a phosphor that gives off light continuously in the presence of the radioactivity.

self-phased array *Adaptive array.*

self-pulse modulation Modulation by an internally generated pulse, as in a blocking oscillator.

self-quench A type of quenching used in induction heating, in which a rapidly heated surface layer of metal is cooled by the rapid conduction of heat into the cold interior.

self-quenched counter tube A radiation-counter tube in which reignition of the discharge is inhibited by gas or other internal means.

self-quenched detector A superregenerative detector in which the time constant of the grid leak and grid capacitor is sufficiently large to cause intermittent oscillation above audio frequencies. This stops regeneration just before it spills over into a squealing condition.

self-rectifying tube A hot-cathode x-ray tube in which an AC anode voltage is used, but current flows in only one direction as long as the anode is kept cool.

self-resetting Automatically returning to an original position when current is interrupted or normal conditions are restored. Applied chiefly to relays and circuit breakers.

self-saturating circuit A magnetic-amplifier circuit in which a rectifier is used in series with the output winding to give higher gain and faster response. The rectifier conducts and saturates the core during positive half-cycles of the applied AC voltage and blocks load current during negative half-cycles to permit resetting of the core by the control current.

self-scattering Scattering of radiation by the material that emits the radiation, increasing the measured activity over that expected for a weightless sample.

self-starting synchronous motor A synchronous motor provided with the equivalent of a squirrel-cage winding, to permit starting as an induction motor.

self-steering microwave array An antenna array used with electronic circuitry that senses the phase of incoming pilot signals and positions the antenna beam in their direction of arrival. Used in earth-satellite communication.

self-supporting antenna tower An antenna tower that requires no guy wires.

self-synchronous device *Synchro.*

self-wiping contact *Wiping contact.*

selsyn [SELf SYNchronous] *Synchro.* Not a trademark.

selsyn generator *Synchro transmitter.*

selsyn motor *Synchro receiver.*

selsyn receiver *Synchro receiver.*

selsyn system *Synchro system.*

selsyn transmitter *Synchro transmitter.*

semiactive chaff Chaff in which a rectifier recovers a small amount of power from the incident radar signal, for modulating the basic skin return without providing gain.

semiactive homing Homing in which the transmitter that illuminates the target is not on the missile. The missile contains only the receiver for energy reflected from the target.

semiactive tracking system A tracking system that tracks a signal source normally aboard the target for other purposes, or a system that uses a ground transmitter to illuminate the target but requires no special equipment on the missile.

semiautomatic height finder A system for automatically recording the height and azimuth of a radar target. An operator at a display scope checks the image on the scope, then operates a switch to order a computer to record the data.

semiconducting diamond An artificially made or natural diamond that has about 1% of impurities such as boron, beryllium, or aluminum, to give electric conductivity in the semiconducting range.

semiconductor A material whose resistivity is between that of insulators and conductors. The resistivity is often changed by light, heat, an electric field, or a magnetic field. Current flow is often achieved by transfer of positive holes as well as by movement of electrons. Examples include germanium, lead sulfide, lead telluride, selenium, silicon, and silicon carbide. Widely used in diodes, photocells, thermistors, and transistors.

semiconductor capacitor A reverse-biased PN junction device used as a capacitor.

semiconductor detector A nuclear-particle detector in which a semiconductor material is the sensing element. Examples include the lithium-drifted germanium and the lithium-drifted silicon detectors, both of which will detect gamma rays as well as all other types of particle.

semiconductor device An electron device in which conduction takes place within a semiconductor.

semiconductor diode A two-electrode semiconductor device that utilizes the rectifying properties of a junction between P- and N-type material in a semiconductor, as in a junction diode, or the rectifying properties of a sharp wire point in contact with a semiconductor material, as in a point-contact diode. Also called crystal diode, crystal rectifier, and diode.

semiconductor laser A laser in which stimulated emission of coherent light occurs at a PN junction when electrons and holes are driven into the junction by carrier injection, electron-beam excitation, impact ionization, optical excitation, or other means. The most common semiconductor

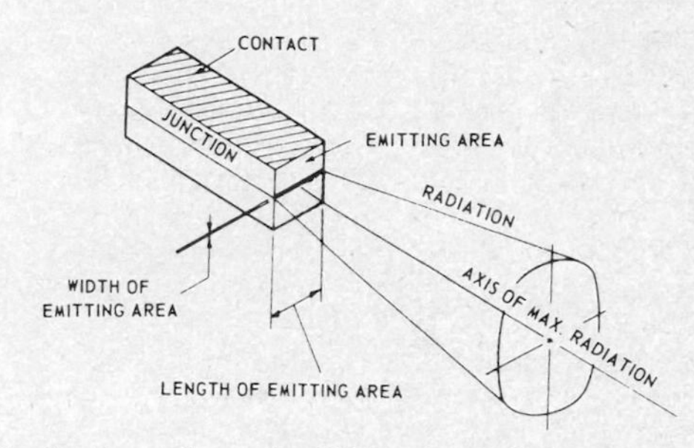

Semiconductor laser using PN junction. Pulsed power supply is required to produce high peak currents for injecting electrons across junction to produce lasing action.

material used is gallium arsenide; other materials include cadmium sulfide, lead sulfide, lead selenide, lead telluride, zinc oxide, and zinc sulfide. Output wavelengths are in the range from 0.33 to 31.2 μm. Also called diode laser and laser diode.

semiconductor memory A random-access memory in which storage cells consist of bipolar or MOS devices on a single semiconductor chip.

semiconductor microphone A microphone in

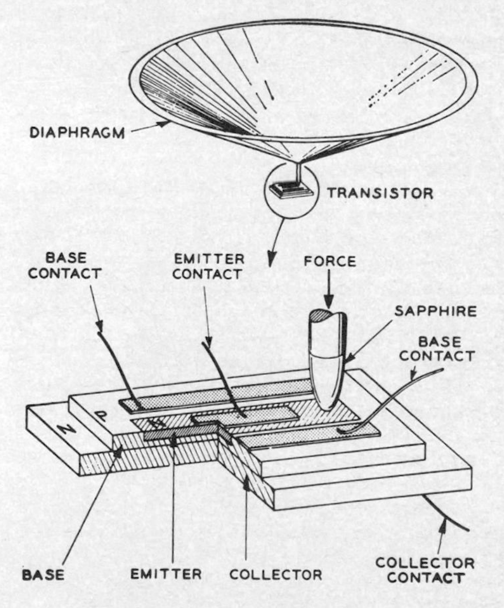

Semiconductor microphone.

which diaphragm vibrations are transmitted mechanically by a sapphire stylus to a thin semiconductor junction of a transistor. Operation is based on the piezoelectric properties of the semiconductor junction.

semiconductor rectifier *Metallic rectifier.*

semiconductor relay A semiconductor device that provides the equivalent of electromagnetic relay action, such as a silicon controlled switch.

semiconductor switch A transistor circuit that provides switching action comparable to that of a relay or switch.

semiduplex operation Operation of a communication circuit with one end duplex and the other end simplex. When used in mobile systems, the base station is usually duplex and the mobile stations are simplex.

semiremote control Remote control of a radio transmitter by devices connected to but not an integral part of the transmitter.

semitone The interval between two sounds whose basic frequency ratio is approximately equal to the twelfth root of 2. The interval, in equally tempered semitones, between any two frequencies is 12 times the logarithm to the base 2 (or 39.86 times the logarithm to the base 10) of the frequency ratio. Also called half-step.

semitransparent photocathode A camera tube or phototube photocathode in which radiant flux on one side produces photoelectric emission from the opposite side.

sending Transmitting, as Morse code.

sending-end impedance The input impedance of a transmission line.

sensation level *Level above threshold.*

sense 1. The relation of a change in the indication of a radio navigation facility to the change in the navigation parameter being indicated. 2. To resolve a 180° ambiguity in a reading. 3. To determine the arrangement or position of a device or the value of a quantity. 4. To read punched holes in tape or cards.

sense amplifier An amplifier that detects bipolar differential-input signals from a coincident-current core or semiconductor memory and is the interface between the memory and logic sections of a computer.

sense antenna An auxiliary antenna used with a directional receiving antenna to resolve a 180° ambiguity in the directional indication. Also called sensing antenna.

sense indicator A flight instrument used to determine whether an aircraft is flying toward or away from a VHF omnirange.

sense wire The wire that picks up the output signal produced by a change in the polarity of a core in a magnetic-core computer memory.

sensing The process of determining the sense of an indication.

sensing antenna *Sense antenna.*

sensing element *Sensor.*

sensitive altimeter An altimeter sufficiently sensitive for use in instrument flying.

sensitive relay A relay that will operate at small currents, usually below 10 mA.

sensitive switch *Snap-action switch.*

sensitivity A figure of merit that expresses the ability of a circuit or device to respond to an input quantity. Expressed as divisions per volt or ohms per volt for a measuring instrument, as spot displacement per volt of deflection voltage or ampere of deflection current for a cathode-ray tube, as output current per unit incident radiation density for a camera tube or other photoelectric device, and as microvolts of input signal when specifying minimum signal strength to which a receiver will respond.

sensitivity control A control that adjusts the amplification of RF amplifier stages in a receiver.

sensitivity-time control [abbreviated STC] An automatic control circuit that changes the gain of a receiver at regular intervals, to obtain desired relative output levels from two or more sequential and unequal input signals. In a loran receiver it keeps output signal amplitude essentially constant as the receiver is tuned between input signals of different strength. In a radar receiver it reduces the gain after transmission of a pulse so nearby echo signals do not overload the system, then gradually restores the gain to the maximum value required for more distant targets. Also called amplitude balance control, anti-clutter gain control, differential gain control, gain-time control, swept gain control, temporal gain control, and time-varied gain control.

sensitization *Activation.*

sensitizer *Activator.*

sensitometer An instrument that measures the sensitivity of light-sensitive materials.

sensitometry The measurement of the light-response characteristics of photographic film under specified conditions of exposure and development.

sensor Generic name for a device that senses a change in a physical quantity such as light, sound, or radio waves and converts that change into a useful input signal for an information-gathering system. A television camera is therefore a sensor. A transducer is a special type of sensor. Also called primary detector, primary element, and sensing element.

Sentinel Designation for an antiballistic missile system intended to provide area defense of the continental United States, Alaska, and Hawaii against intercontinental ballistic missiles. The system includes land-based acquisition and guidance radars for interceptor missiles.

separately excited Obtaining excitation from a source other than the machine or device itself.

separately instructed carry A carry in which information goes to the next location only on receipt of a specific signal.

separation The degree, expressed in decibels, to which left and right stereo channels are isolated from each other. A similar rating is used for front and rear quadraphonic channels. The greater the isolation in decibels, the better the separation.

separation circuit A circuit that sorts signals according to amplitude, frequency, or some other characteristic.

separation energy *Binding energy.*

separation factor The abundance ratio of two isotopes after processing, divided by their abundance ratio before processing.

separation filter A filter that separates one band of frequencies from another, as in carrier systems.

separation loss The loss in output that occurs when the surface of the coating on magnetic tape fails to make perfect contact with the surfaces of the record or reproduce head.

separator 1. A circuit that separates one type of signal from another by clipping, differentiating, or integrating action. 2. A porous insulating sheet used between the plates of a storage battery.

septate waveguide A waveguide that contains one or more septa placed across the waveguide to control the transmission of microwave power.

septinary number A number in which the quantity represented by each figure is based on a radix of 7.

septum [plural septa] A metal plate placed across a waveguide and attached to the walls by highly conducting joints. The plate usually has one or more windows or irises designed to give inductive, capacitive, or resistive characteristics.

sequence 1. An arrangement of items or events according to a specified set of rules. 2. To arrange items according to a specified set of rules.

sequence control The automatic control of a series of operations in a predetermined order.

sequencer A device that determines the order in which a number of actions occur.

sequence relay A relay that opens or closes two or more sets of contacts in a predetermined sequence.

sequence weld timer A timer that controls the sequence and duration of each portion of a complete resistance-welding cycle.

sequential-access storage Storage in which the items of information stored become available only in a one-after-the-other sequence, as in magnetic-tape storage.

sequential color television A color television system in which the primary color components of a picture are transmitted one after the other. The three basic types are the line-, dot-, and field-sequential color television systems. Also called sequential system.

sequential control 1. Control of a digital computer in such a way that instructions are fed into the computer in a given sequence during the running of a problem. 2. *SECO.*

sequential flasher A flasher circuit that turns lamps on and off in sequence to give an illusion of motion. Sometimes used as a turn-signal indicator on automobiles, with the energized lamp moving in the direction in which a turn is to be made.

sequential interlace Interlace in which the lines of one field fall directly under the corresponding lines of the preceding field.

sequential lobing A direction-finding technique that utilizes the signal derived from partially overlapped lobes which occur in sequence.

sequential logic Logic in which the outputs are dependent on the input states, delays encountered in the logic path, the presence of a discrete timing interval, and the previous state of the logic array. In contrast, combinatorial logic depends only on input states and delays.

sequential scanning *Progressive scanning.*

sequential system *Sequential color television.*

Sergeant An Army surface-to-surface guided missile that uses inertial guidance. Range is about 75 nautical miles (139 km). It can carry either conventional or nuclear warheads.

serial Pertaining to time-sequential transmission of, storage of, or logic operations on the parts of a word in a digital computer, using the same facilities for successive parts.

serial access A memory characteristic wherein all the bits of a byte or word are entered sequentially at a single input or retrieved sequentially from a single output.

serial adder A computer logic circuit that adds two binary numbers in pairs of bits, starting with the least significant bits and handling the carries while progressing step by step to the most significant bits.

serial by bit Digital-computer storage in which the individual bits that make up a computer word appear in time sequence.

serial by character Digital-computer storage in which the characters for coded-decimal or other nonbinary numbers appear in time sequence.

serial by word Digital-computer storage in which the words within a given group appear one after the other in time sequence.

serial digital computer A digital computer in which the digits are handled serially, although the bits that comprise a digit may be handled either serially or in parallel.

serial operation The flow of information through a computer in time sequence, using only one digit, word, line, or channel at a time.

serial-parallel A combination of serial and parallel, such as serial by character and parallel by bits comprising the character.

serial printer An electric typewriter that uses type bars, type balls, daisy wheels, or other mechanisms for moving a desired character into printing position and producing an image of that character on paper with a ribbon or other means. Either the paper or the printing device moves step by step to successive positions for printing one character at a time. Also called character printer.

serial programming Programming in which only one operation is scheduled at one time.

serial radiography Making a number of radiographs of the same subject in succession.

serial storage Computer storage in which time is one of the coordinates used to locate any given bit, character, or word. Access time therefore includes a variable waiting time ranging from zero to many word times.

serial transfer Transfer of the characters of an element of information in sequence over a single path in a digital computer.

series 1. An arrangement of circuit components end to end to form a single path for current. 2. The indicated sum of a set of terms in a mathematical expression, as in an alternating series or an arithmetic series.

series circuit A circuit in which all parts are connected end to end to provide a single path for current.

series coil The coil that carries the main current in a rotating machine or other device. The shunt coil is connected across the line and usually carries only a small current.

series connection A connection that forms a series circuit.

series disintegration The successive radioactive transformations in a radioactive series.

series element A two-terminal element connected to complete the only path existing between two nodes of a network. Any mesh including one series element must include all the other series elements of the mesh.

series excitation Obtaining field excitation in a motor or generator by allowing the armature current to flow through the field winding.

series-fed vertical antenna A vertical antenna that is insulated from the ground and energized at its base. Also called end-fed vertical antenna.

series feed Application of direct voltage to the anode of a vacuum tube through the load that is carrying the output signal current. In shunt feed, the direct voltage is applied to the anode through a choke, and only signal current flows through the load.

series loading Loading in which reactances are inserted in series with the conductors.

series modulation Modulation in which the modulating tube, the modulated amplifier tube, and the anode voltage supply are all in series.

series motor A commutator-type motor that has armature and field windings in series. Characteristics are high starting torque, variation of speed with load, and dangerously high speed on no-load. Also called series-wound motor.

series-parallel switch A switch that changes the connections of lamps or other devices from series to parallel, or vice versa.

series peaking Use of a peaking coil and resistor in series as the load for a video amplifier to produce peaking at some desired frequency in the passband, such as to compensate for previous loss of gain at the high-frequency end of the passband.

series regulator A power transistor or equivalent device connected in series with the load of a constant-voltage power supply in such a way that

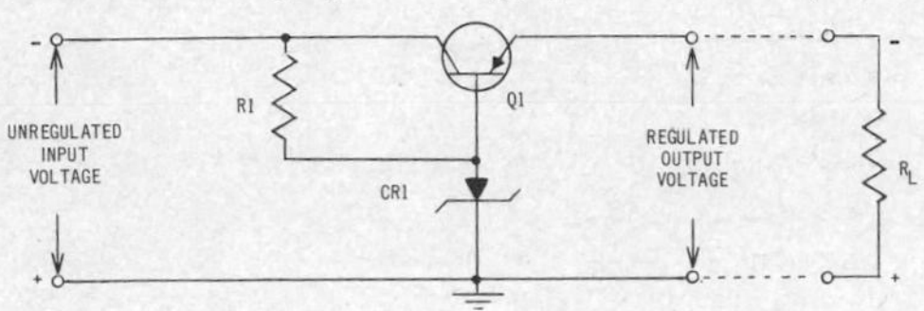

Series regulator using transistor as variable resistor in series with load to regulate load voltage.

feedback action on the series regulator changes its voltage drop as required to maintain a constant DC output voltage.

series resonance Resonance in a series resonant circuit, wherein the inductive and capacitive reactances are equal at the frequency of the applied voltage. The reactances then cancel each other, reducing the impedance of the circuit to a minimum purely resistive value. Signal current is then a maximum, and the signal voltage developed across either the coil or capacitor may be several times the voltage applied to the combination.

series resonant circuit A resonant circuit in which the capacitor and coil are in series with the applied AC voltage.

series T junction A T junction that has an equivalent circuit in which the impedance of the branch waveguide is predominantly in series with the impedance of the main waveguide at the junction.

series-wound motor *Series motor.*

serrasoid modulator A modulator that uses a crystal-controlled sawtooth oscillator in conjunction with a low-frequency modulating voltage to produce variable triggering time for a blocking oscillator.

serrated pulse A pulse that has notches or sawtooth indentations in its waveform.

serrated rotor plate *Slotted rotor plate.*

serrated vertical pulse A vertical synchronizing pulse that is broken up by five notches which extend down to the black level of a television signal, to give six component pulses, each lasting about 0.4 line and keeping the horizontal sweep circuits in step during the vertical sync pulse interval.

serrodyne A phase modulator that uses transit-time modulation of a traveling-wave tube or klystron.

serrodyning An electronic countermeasure technique used to prevent an enemy continuous-wave or pulse-Doppler tracking radar from obtaining accurate velocity information. The helix voltage of a traveling-wave tube is varied in a sawtooth manner to vary the output frequency at an audio rate over the Doppler range and cause continual loss of lock, degrading both range and velocity information.

service area The area that is effectively served by a given radio or television transmitter, navigation aid, or other type of transmitter. Also called coverage.

service band A band of frequencies allocated to a given class of radio service.

service bureau A computer installation at which processing time can be leased. It may also provide keypunching and other data-processing services.

service engineer An engineer who is qualified to maintain and repair complex electronic equipment after it has been installed.

service life The length of time that a battery or other device will provide specified performance under specified conditions of use.

serviceman A person engaged in the maintenance and repair of equipment, such as a radio serviceman.

service oscillator *RF signal generator.*

service routine A routine that assists in the operation of a computer, such as block location, correction, and tape comparison routines.

service test A test made under simulated or actual conditions of use to determine the characteristics, capabilities, and limitations of a product.

servicing Preventive maintenance and repair of equipment.

serving A covering, such as thread or tape, that protects a winding from mechanical damage. Also called coil serving.

servo *Servomotor.*

servoamplifier An amplifier used in a servomechanism.

servomechanism A feedback control system in which one or more of the system signals represent mechanical motion. Usually used to control an output position mechanically in response to input signal changes. A servomechanism may also be a regulator under certain conditions. Also called servosystem.

servometer A meter in which a tiny servomotor drives the indicating pointer.

servomotor The electric, hydraulic, or other type of motor that serves as the final control element in a servomechanism. It receives power from the amplifier element and drives the load with a linear or rotary motion. Also called servo.

servo multiplier An electromechanical multiplier in which one variable is used to position one or more ganged potentiometers across which the other variable voltages are applied.

servosystem *Servomechanism.*

sesqui-sideband transmission Transmission of a carrier modulated by one full sideband and half of the other sideband.

set 1. A radio or television receiver. 2. A combination of units, assemblies, and parts connected or otherwise used together to perform an operational function, such as a radar set. 3. To place a storage device in a prescribed state, such as to place a binary cell in the 1 state.

set analyzer An analyzer that locates trouble in a television set, radio, or other piece of electronic equipment by measuring voltages and/or currents simultaneously at a number of test points.

set point The value selected to be maintained by an automatic controller.

set-point control A digital technique for making an output device stop at a series of programmed points called set points, entered manually or automatically under control of a computer or tape reader.

set pulse A drive pulse that tends to set a magnetic cell in a computer.

set-reset flip-flop *RS flip-flop.*

set-reset-trigger flip-flop *RST flip-flop.*

setscrew A small headless machine screw, usually having a point at one end and a recessed hexagonal socket or a slot at the other end, used for such purposes as holding a knob or gear on a shaft or holding a playing needle in a phonograph pickup.

settling time 1. The time elapsed between the application of a perfect step input to an operational amplifier and the time when amplifier output has entered and remained within a specified error band that is usually symmetrical about the final value. Settling time includes the time required for the amplifier to slew from the initial value, recover from the slew-rate-limited overload, and settle within the error band. 2. *Correction time.*

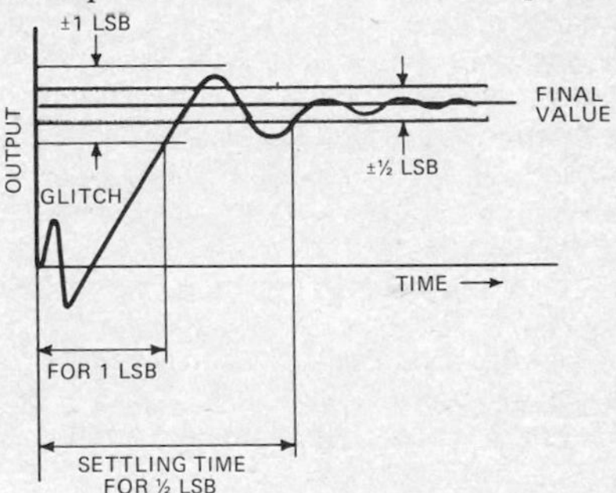

Settling time for digital-to-analog converter is often specified as time required for output to settle within range of ½ or 1 least significant bit of its final value after switching operation.

setup The ratio between reference black level and reference white level in television, both measured from blanking level. Usually expressed as a percentage.

sexadecimal notation *Hexadecimal notation.*

sexless connector *Hermaphroditic connector.*

sferics [coined from atmospherics] *Atmospheric interference.*

sferics set An electronic system that detects,

analyzes, and determines the position of electromagnetic disturbances generated by any atmospheric phenomena.

SG Symbol for *screen grid.*

shaded-pole motor A single-phase induction motor that has one or more auxiliary short-circuited windings acting on only a portion of the magnetic circuit. Generally the winding is a closed copper ring imbedded in the face of a pole. The shaded pole provides the required rotating field for starting purposes.

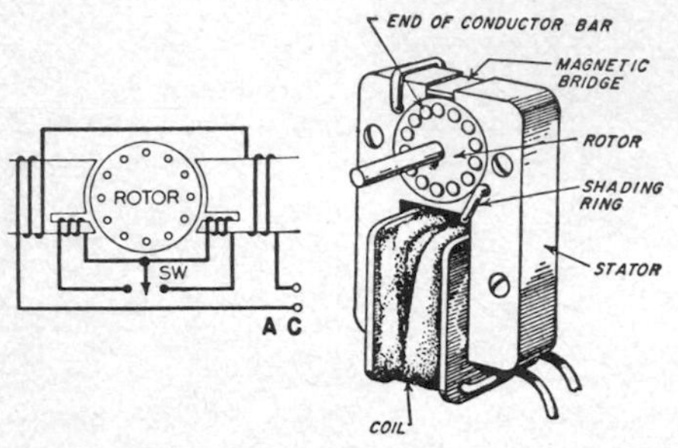

Shaded-pole motor with switch for reversing direction of rotation, and single-direction shaded-pole motor used in phonographs.

shading A variation in brightness over the area of a reproduced television picture, caused by spurious signals generated in a television camera tube during the retrace intervals. These spurious signals are generally due to redistribution of secondary electrons over the mosaic in a storage-type camera tube, and vary from scene to scene as background illumination changes.

shading coil *Shading ring.*

shading generator One of the signal generators used at a television transmitter to generate waveforms that are 180° out of phase with the undesired shading signals produced by a television camera. An operator watches the picture on a monitor and adjusts the controls of the shading generators as required to give essentially uniform scene brightness.

shading ring 1. A heavy copper ring sometimes placed around the central pole piece of an electrodynamic loudspeaker to serve as a shorted turn that suppresses the hum voltage produced by the field coil. 2. The copper ring used in a shaded-pole motor to produce a rotating magnetic field for starting purposes. Also used around part of the core of an AC relay to prevent contact chatter. Also called shading coil.

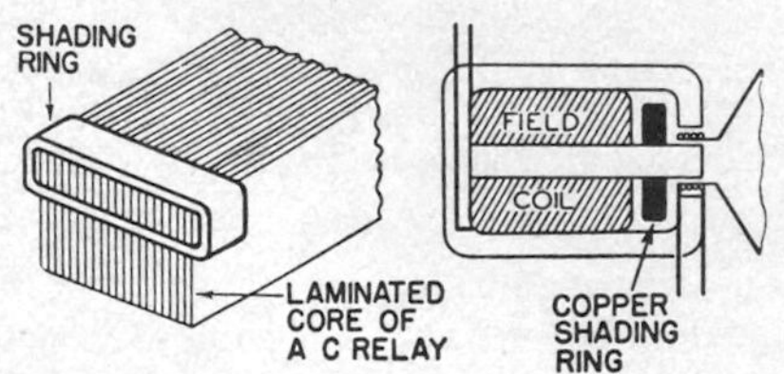

Shading ring as used on relay and on excited-field loudspeaker.

shadow factor The ratio of the electric field strength that would result from propagation of waves over a sphere to that which would result from propagation over a plane under comparable conditions.

shadow mask A thin perforated metal mask mounted just back of the phosphor-dot faceplate in a three-gun color picture tube. The holes in the mask are positioned to ensure that each of the three electron beams strikes only its intended color phosphor dot. Also called aperture mask.

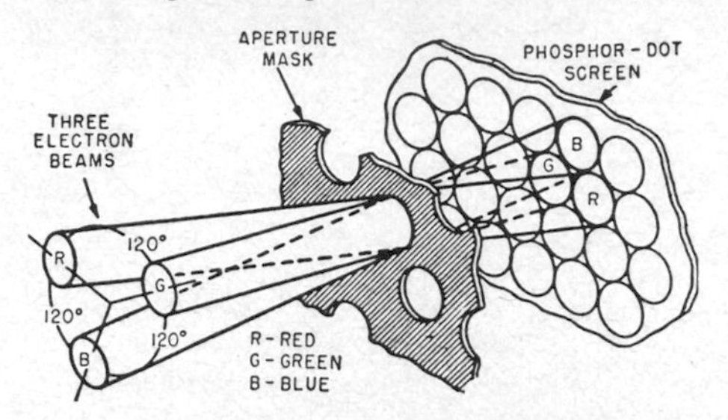

Shadow-mask details, showing positions of color phosphor dots with respect to apertures in mask.

shadow-mask color picture tube A three-gun color picture tube that uses a shadow mask.

shadow region The region in which received field strength is so reduced by some obstruction that effective reception of signals or radar detection of objects is normally improbable.

shadow tuning indicator A tuning indicator that shows by a moving shadow the accuracy with which a receiver is tuned. Older forms used the shadow of a meter pointer. Newer electron-tube types use an electron stream to produce a shadow on a small fluorescent screen, as in a cathode-ray tuning indicator.

shaft-position encoder An analog-to-digital converter in which the exact angular position of a shaft is sensed and converted to digital form. In

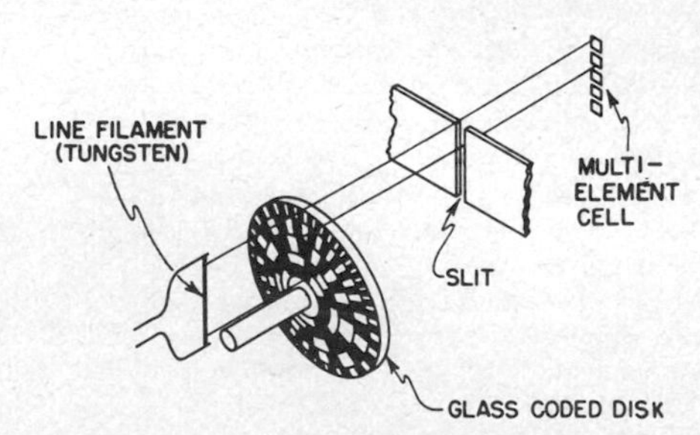

Shaft-position encoder using photoelectric readout of transparent and opaque segments produced photographically on glass disk, to give binary output directly.

one type, a continuous transducer such as a potentiometer, resolver, or synchro converts the rotary motion first to a proportional electric quantity or time interval. This in turn is converted to digital form by a voltage or time coder. In another type, conversion to digital form is accomplished directly by a disk, mounted on the shaft, that contains various combinations of segments from which a unique digital output is derived for each increment of shaft position.

shaker An electromagnetic device capable of imparting known and usually controlled vibratory acceleration to a given object. Also called shake table.

shake table *Shaker.*

shank The portion of a phonograph needle that is clamped or otherwise anchored in a phonograph pickup.

Shannon limit The best possible signal-to-noise ratio that can be obtained with the best modulation technique, based on Shannon's theorem relating channel capacity to signal-to-noise ratio.

Shape coding of control knobs.

shape coding Use of special shapes for control knobs, to permit recognition and sometimes also position monitoring by sense of touch.

shaped-beam antenna A unidirectional antenna whose major lobe differs materially from that provided by an aperture which gives uniform phase. Also called phase-shaped antenna.

shaped-beam tube A character-writing tube in which a character is formed all at once by an electron beam whose cross section corresponds to the shape of the character.

shape factor *Form factor.*

shaping network *Corrective network.*

sharp-cutoff tube An electron tube in which the control-grid openings are uniformly spaced. Anode current then decreases linearly as grid voltage is made more negative, and cuts off sharply at a particular grid voltage.

sharp tuning Having high selectivity, and therefore responding only to a desired narrow range of frequencies.

shaving Removing material from the surface of a disk recording medium to obtain a new recording surface.

shear wave A wave that causes an element of an elastic medium to change its shape without changing its volume. Also called rotation wave. A shear plane wave in an isotropic medium is a transverse wave.

sheath 1. The metal wall of a waveguide. 2. A protective outside covering on a cable. 3. A space charge formed by ions near an electrode in a gas tube.

sheath-reshaping converter A wave converter in which the change of wave pattern is achieved by gradual reshaping of the sheath of the waveguide and use of conducting metal sheets mounted longitudinally in the waveguide.

sheet grating A grating consisting of thin longitudinal metal sheets that extend inside a waveguide a distance of about one wavelength. Used to suppress undesired modes of propagation.

sheet resistivity The resistivity of a sheet of material, measured in ohms per square. The rating depends only on the resistivity and thickness of the

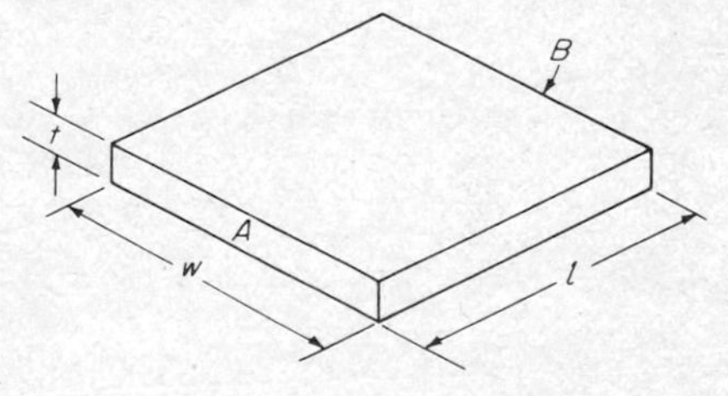

Sheet resistivity is measured between surfaces A and B of square sheet of material.

material, being independent of the size of the square. Used chiefly in connection with integrated-circuit resistor design.

shelf life The time that elapses before an unused battery or other device becomes inoperative due to age or deterioration.

shell A group of electrons that form part of the outer structure of an atom and have a common energy level.

shellac A purified lac resin, once widely used in insulating materials and commercial phonograph records.

shell model *Independent-particle model.*

shell structure The arrangement of the quantum states of nucleons (protons or neutrons) of a given kind in a nucleus in groups of approximately the same energy. Each such group is called a shell. The number of nucleons in each shell is limited by the Pauli exclusion principle. A closed shell is one that contains the maximum number. A nucleus that has all its protons and/or neutrons in closed shells has greater than average stability.

shell-type transformer A power transformer in which the primary and secondary coils are wound over each other on the center leg of the iron core

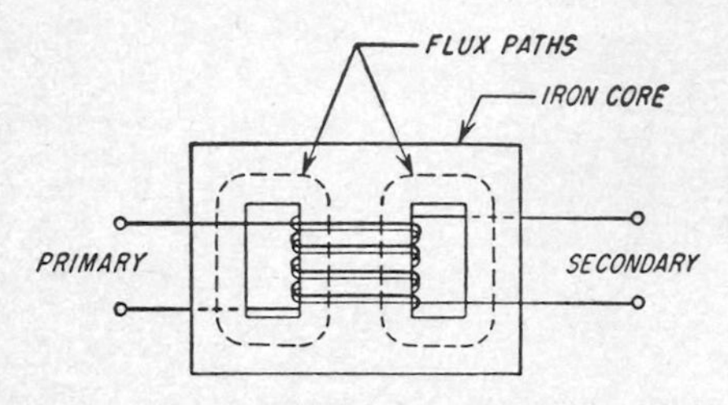

Shell-type transformer.

structure. There are no coils on the outer two legs, which serve as return paths for the magnetic circuit.

SHF Abbreviation for *superhigh frequency.*

shield A metallic housing placed around a circuit or component to suppress the effect of an electric or magnetic field within or beyond definite regions.

shielded-arc welding Arc welding in which the metal electrode is coated with a flux that produces an envelope of protective inert gas.

shielded cable A cable that has a conducting envelope around its insulated conductors. Braided wire is most often used for this purpose.

shielded line *Shielded transmission line.*

shielded nuclide A nuclide that has a charge which is one unit larger than that of a stable nuclide which has the same mass number. A shielded nuclide is generally a primary fission product.

shielded pair A two-wire transmission line surrounded by a metallic sheath.

shielded room A room that has been made free from electric interference by the application of appropriate shielding to the floor, walls, and ceiling and by the suppressing of interference which might enter over power lines.

shielded transmission line A transmission line whose elements essentially confine propagated electric energy to a finite space inside a conducting sheath, to prevent the line from radiating radio waves. Also called shielded line.

shielded wire An insulated wire covered with a metal shield, usually made of tinned braided copper wire.

shielded x-ray tube An x-ray tube enclosed in a grounded metal container except for a small window through which x-rays emerge.

shield grid A grid that shields the control grid of a gas tube from electrostatic fields, thermal radiation, and deposition of thermionic emissive material. The shield grid may also be used as an additional control electrode.

shield-grid thyratron A thyratron that has a shield grid, usually operated at cathode potential.

shielding metal A metal that has high magnetic saturation and good attenuation for electromagnetic and electrostatic fields, to protect them from stray effects. Examples include nickel alloys such as Mumetal, nickel-iron alloys, silicon iron or soft cold-rolled steel coated with ferrite powders, and sandwiches of two different shielding metals or alloys.

shift Displacement of an ordered set of characters one or more places to the left or right in a digital computer. If the characters are the digits of a numerical expression, a shift may be equivalent to multiplying by a power of the base.

shift pulse A drive pulse that initiates shifting of characters in the register of a digital computer.

shift register A computer circuit that converts a sequence of input signals into a parallel binary number or vice versa, by moving stored characters to the right or left.

shift-register memory A memory in which data is entered at one end and must be shifted stage by stage through the entire memory before becoming available again. The data can then be either recirculated or removed.

Shillelagh An Army gun-launched missile that uses a shaped-charge warhead which destroys such hard targets as tanks, bunkers, and buildings. The missile is guided to its target from its airborne or ground assault vehicle by an infrared command link.

shimming Adjustment of the strength of a magnetic field by thin spacers, shims of soft iron, or compensating coils.

ship error A radio direction finder error due to reradiation of radio waves by the metal structure of a ship.

shiran [S-band HIgh-accuracy RANging] An aerial mapping system that uses continuous-wave frequency-modulation distance-measuring equipment. Accuracy is comparable to that of first-order triangulation.

shock 1. The sudden pain, convulsion, unconsciousness, or death produced by the passage of electric current through the body. Also called electric shock. 2. *Shock motion.*

shock excitation Excitation produced by a voltage or current variation of relatively short duration. Used to initiate oscillation in the resonant circuit of an oscillator.

shock-excited oscillation *Free oscillation.*

shock heating Heating of a gas by a sudden electric discharge or by a rapidly increasing magnetic field used for plasma confinement.

Shockley diode A PNPN device that switches rapidly into its conducting state when a critical voltage is reached. Conduction continues until the anode voltage drops below a specified minimum

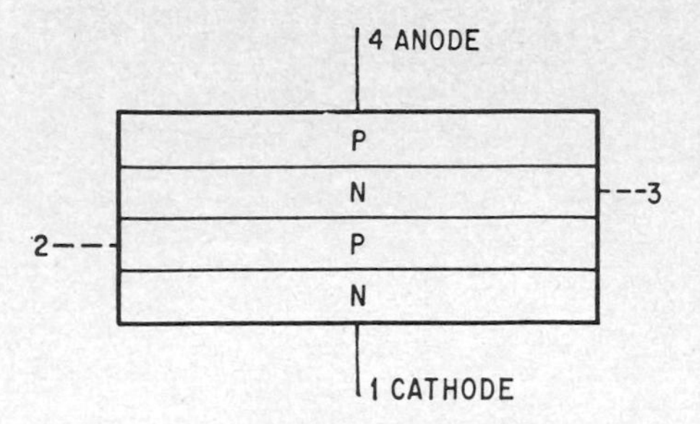

Shockley diode.

value that is called the turnoff voltage. In its blocking state, the diode impedance is very high.

shock motion Transient mechanical motion characterized by suddenness and significant relative displacements. Also called shock.

shock mount A mount used with sensitive equipment to reduce or prevent transmission of shock motion to the equipment.

shock wave A sound wave produced by a sudden change in pressure and particle velocity. It may travel faster than the velocity of sound.

SHODOP [SHOrt-range DOPpler] A short-range trajectory-measuring system based on DOVAP, used during a ballistic-missile launch.

shoran [SHOrt-RAnge Navigation] A precision position-fixing system that uses a pulse transmitter and receiver in an aircraft or other vehicle and two transponder beacons at fixed points. A receiver in the aircraft measures the round-trip times of the signals and converts these into distances to the fixed ground stations. Ordinary triangulation on a map then gives position. A high-precision version is called hiran, and an S-band hiran version is called shiran.

shoran bombing Bombing done after positioning the aircraft to the bomb-release point by shoran.

shore effect Bending of waves toward the shoreline when traveling over water near a shoreline, caused by the slightly greater velocity of radio waves over water than over land. This effect causes errors in radio direction finder indications.

shore radar station A shore radio navigation station that uses radar to determine the direction and distance of ships from the station at night or during fog.

shore-to-ship communication Radio communication between a shore station and a ship at sea.

short *Short-circuit.*

short-baseline system A trajectory-measuring system that uses a baseline whose length is very small compared with the distance to the object being tracked.

short-circuit A low-resistance connection across a voltage source or between both sides of a circuit or line, usually accidental and usually resulting in excessive current flow that may cause damage. Also called short.

short-circuit impedance The impedance of a network when a specified pair or group of its terminals is short-circuited.

short-circuit parameter One of a set of four transistor equivalent-circuit parameters that specify transistor performance when the input and output voltages are chosen as the independent variables. Two of the four measurements require short-circuiting of the input for alternating current.

short-distance navigation aid A navigation aid that is useful primarily at distances within radio line of sight, generally not more than 200 mi (320 km).

shorted out Made inactive by connecting a heavy wire or other low-resistance path around a device or portion of a circuit.

shorting-contact switch A selector switch in which the width of the movable contact is greater than the distance between fixed contacts, so the new circuit is contacted before the old one is broken. This avoids noise during switching.

shorting noise A noise that occurs in wire-wound potentiometric transducers even when no current is drawn from the device. It is due to the shorting of adjacent turns of the wire as the slider traverses the winding. The portion of the inter-turn current that flows through the slider appears as noise.

short-range attack missile [abbreviated SRAM] An air-to-surface and air-to-air solid-fuel guided missile that uses inertial guidance for complete immunity to jamming. Target coordinates are fed into the missile by a computer on the aircraft just before launch.

short-range radar Radar whose maximum line-of-sight range, for a reflecting target that has $1 m^2$ of area perpendicular to the beam, is between 50 and 150 mi (80 and 240 km).

short-time rating A rating that defines the load which a machine, apparatus, or device can carry for a specified short time.

shortwave [abbreviated SW] A general term applied to a wavelength shorter than 200 m, corresponding to frequencies higher than the highest broadcast-band frequency.

shortwave antenna An antenna that receives frequencies above the broadcast band, in the range from about 1.6 to 30 MHz.

shortwave converter A converter used between a receiver and its antenna system to convert incoming high-frequency signals to a lower carrier frequency to which the receiver can be tuned. A converter usually contains a local oscillator and a mixer, as in a superheterodyne receiver.

shortwave listener [abbreviated SWL] A person whose hobby is listening to broadcasts from domestic and foreign shortwave radio stations.

shortwave receiver A radio receiver that tunes in stations in the range from about 1.6 to 30 MHz or some portion of that range.

shortwave transmitter A radio transmitter that radiates shortwaves, generally for communication purposes or for international broadcasting.

shot effect *Shot noise.*

shot noise Noise voltage developed in a thermionic tube because of the random variations in the number and velocity of electrons emitted by the heated cathode. The effect causes sputtering or popping sounds in radio receivers and snow effects in television pictures. Also called Schottky noise and shot effect.

shower *Cosmic-ray shower.*

shower unit The mean path length required to reduce the energy of relativistic charged particles to half value as they pass through matter. One shower unit is equal to 0.693 radiation length.

Shrike An air-to-surface solid-fuel guided missile that homes in on enemy radar installations.

shunt 1. A precision low-value resistor placed across the terminals of an ammeter to increase its range by allowing a known fraction of the circuit current to go around the meter. 2. A piece of iron that provides a parallel path for magnetic flux around an air gap in a magnetic circuit. 3. To place one part in parallel with another. 4. *Parallel.*

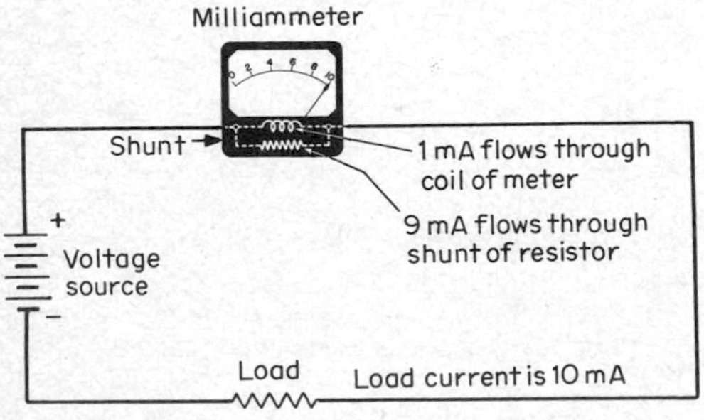

Shunt resistor across moving coil of milliammeter.

shunted monochrome A color television technique in which the luminance or monochrome signal is shunted around the chrominance modulator or chrominance demodulator. Also called bypassed monochrome (deprecated).

shunt-excited Having field windings connected across the armature terminals, as in a DC generator.

shunt-fed vertical antenna A vertical antenna connected to ground at the base and energized at a point suitably positioned above the grounding point.

shunt feed The feed of direct operating voltages to the electrodes of an electron tube through a choke coil that is parallel to and therefore separated from the signal circuit. Also called parallel feed.

shunt loading Loading in which reactances are connected between the conductors of a transmission line.

shunt neutralization *Inductive neutralization.*

shunt peaking Use of a peaking coil in a parallel circuit branch that connects the output load of one stage to the input load of the following stage, to compensate for high-frequency loss caused by the distributed capacitances of the two stages.

shunt regulator A regulator in which a transistor or other active regulating element is connected across the output and regulates the output voltage by controlling the current flowing through a resistor in series with the load.

shunt T junction A waveguide T junction that has an equivalent circuit in which the impedance of the branch guide is predominantly in parallel with the impedance of the main waveguide at the junction.

shunt-wound Having armature and field windings in parallel, as in a DC generator or motor.

shutter A device that prevents light from reaching the light-sensitive surface of a television camera except during the desired period of exposure.

SI Abbreviation for *Système International d'Unités,* the French equivalent of *International System of Units.*

SiC Symbol for *silicon carbide.*

sideband A band of frequencies on each side of the carrier frequency of a modulated radio signal, produced by the process of modulation. The upper sideband contains the frequencies that are the sums of the carrier and modulation frequencies, and the lower sideband contains the difference frequencies.

sideband attenuation Attenuation in which the transmitted relative amplitude of some sideband component of a modulated signal is smaller than that produced by the modulation process.

sideband interference *Adjacent-channel interference.*

sideband power The power contained in the sidebands. A receiver responds to sideband power rather than carrier power when it is receiving a modulated wave.

sideband-reference glide slope A modification of a null-reference instrument landing system in which the upper or sideband antenna is replaced by two antennas that are lower in height and fed out of phase, so a null is produced at the desired glide-slope angle. Used to reduce unwanted reflections of signals from rough terrain near the approach end of the runway.

sideband splash *Adjacent-channel interference.*

sideband suppression Removal of the energy of one sideband from the spectrum of a modulated carrier.

side circuit A special circuit that derives a phantom circuit from existing circuits by placing conductors effectively in parallel to form each side of the phantom circuit.

side frequency One of the frequencies of a sideband.
side lobe *Minor lobe.*
side-lobe cancellation A jamming countermeasure technique that attenuates jamming signals which are arriving through the side or back lobes of a receiving antenna.
side-lobe echo A radar echo due to a side lobe of the radar beam.
side-looking airborne radar [abbreviated SLAR] A high-resolution airborne radar that has antennas aimed to the right and left of the flight path. Used to provide high-resolution strip maps with photographlike detail. Another use is mapping unfriendly territory while flying along its perimeter at altitudes up to about 70,000 ft (21 km). Also used for detecting submarine snorkels against a background of sea clutter.
side-looking sonar Sonar in which the beam is wide vertically but narrow in the horizontal plane, to give a map of acoustic scattering from the ocean bottom on recording paper as the sonar is towed at a desired position above the bottom or mounted conventionally on a moving ship.

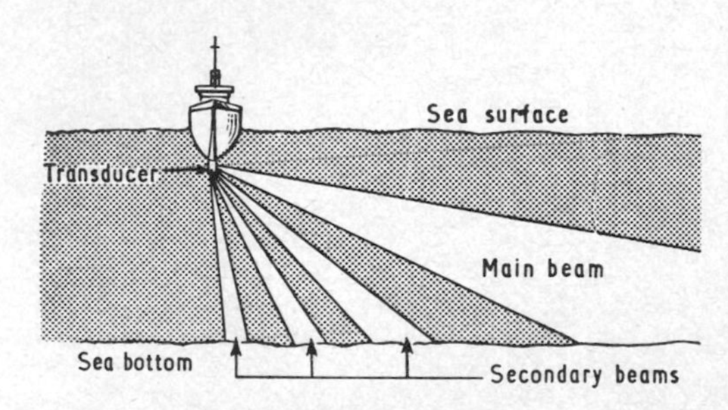

Side-looking sonar.

side thrust The radial component of the force on a pickup arm due to stylus drag.
sidetone The sound of the speaker's own voice as heard in the speaker's telephone receiver. The effect is undesirable and is usually suppressed by antisidetone circuits.
Sidewinder A Navy air-to-air guided missile that has infrared homing guidance and a speed of over Mach 2. When launched from an airplane, the missile seeks and hits an enemy bomber or fighter target by homing on the heat emitted by the target. Some versions have radar homing guidance.
siemens [abbreviated S] The SI unit of conductance, replacing the mho. A resistance or impedance of 1 Ω is equal to a conductance or admittance of 1 S.
SIGINT Abbreviation for *signal intelligence.*
sigma particle A hyperon that has an extremely short life (about 10^{-10} s), a mass between that of neutrons and deuterons, and either a positive or negative charge.
sigmatron A cyclotron and betatron operating in tandem to produce gigavolt x-rays.
sign A symbol (+ or −) used with a numerical quantity to indicate whether the quantity is above or below zero.
signal 1. A visual, aural, or other indication that conveys information. 2. The intelligence, message, or effect to be conveyed over a communication system. 3. *Signal wave.*
signal averaging Analysis of a repetitive voltage waveform, often buried in noise, by slicing it into small segments under control of a sampling circuit and storing the sampled amplitudes. Each subsequent signal slice is added to the store for its point in the cycle, to reinforce the signal of interest at an arithmetic rate while noise and other signals are added at an RMS rate. The result is continuous improvement in signal-to-noise ratio up to the limit of the instrumentation and the stability of the repetitive waveform.
signal comparator A circuit that correlates information from two or more signals.
signal conditioner A circuit that shapes or adapts a signal to the requirements of a data-transmission line.
signal contrast The ratio in decibels between the white and the black signals in facsimile.
signal data converter A circuit that converts a data-modulated signal from one form to another.
signal distance The number of digit positions in which the corresponding digits of two binary words of the same length are different.
signal frequency shift The numerical difference between the frequencies corresponding to white and black signals at any point in a frequency-shift facsimile system.
signal generator A test instrument that can be set to generate an unmodulated or tone-modulated sinusoidal RF signal voltage or an AF voltage at a known frequency and output level, as needed for aligning or servicing electronic equipment. Signal generators that are capable of producing square and triangular waveforms are generally called function generators.
signaling A procedure for indicating to the receiving end of a communication circuit that intelligence is to be transmitted.
signaling channel A tone channel used for signaling purposes.
signaling communication One-way communication from a base station to a mobile receiver to actuate a signaling device in the mobile unit or to communicate information to the desired mobile unit.
signaling key *Key.*
signal integration The summation of a succession of signals by writing them at the same location on the storage surface of a storage tube.
signal intelligence [abbreviated SIGINT] A combination of communication and electronic intelligence.

signal level The difference between the level of a signal at a point in a transmission system and the level of an arbitrarily specified reference signal. For audio signals the difference is conveniently expressed as a ratio in decibels since the reference level is then 0 dB.

signal light A light specifically designed for the transmission of code messages by visible light rays that are interrupted or deflected by electric or mechanical means.

signal output current The absolute value of the difference between output current and dark current in a camera tube or phototube.

signal plate The metal plate that backs up the mica sheet which contains the mosaic in one type of cathode-ray television camera tube. The capacitance that exists between this plate and each globule of the mosaic is acted on by the electron beam to produce the television signal.

signal-separation filter A bandpass filter that selects the desired subcarrier channel from a composite FM signal.

signal-shaping network A network inserted in a telegraph circuit, usually at the receiving end, to improve the waveform of the code signals.

signal strength The strength of the signal produced by a radio transmitter at a particular location, usually expressed as microvolts or millivolts per meter of effective receiving antenna height.

signal-strength meter A meter that is connected to the automatic volume control circuit of a communication receiver and calibrated in decibels or arbitrary S units to read the strength of a received signal. Also called S meter and S-unit meter.

signal-to-noise [abbreviated S/N] A measure of useful signal strength, generally expressed as a ratio.

signal-to-noise ratio [abbreviated SNR] The ratio of the amplitude of a desired signal at any point to the amplitude of noise signals at that same point. Often expressed in decibels. The peak value is usually used for pulse noise, and the RMS value is used for random noise.

signal tracer A test instrument that facilitates signal-tracing.

signal-tracing A servicing technique that involves tracing the progress of a signal through each stage of a receiver to locate the faulty stage.

signal voltage The effective (RMS) voltage value of a signal.

signal wave A wave whose characteristics permit some intelligence, message, or effect to be conveyed. Also called signal.

signal-wave envelope The contour of a signal wave that consists of a series of RF cycles.

signal winding The control winding to which the control signals are applied in a saturable reactor.

signature The characteristic pattern of a target as displayed by detection and classification equipment.

sign digit A character that designates the algebraic sign of a number in a digital computer. It is usually a single bit (0 or 1) in a preallocated sign position.

significant digit A digit that appears in the coefficient of a number when the number is written as a coefficient between 1.000 . . . and 9.999 . . . times a power of 10. Thus 0.009407, which is equal to 9.407×10^{-3}, has four significant digits.

sign position The position at which the sign of a number is located.

silent discharge A gradual, nondisruptive discharge of electricity from a conductor into the atmosphere. Sometimes accompanied by the production of ozone.

silent period A 3-min period twice each hour, starting at 15 and 45 min after the hour, during which International Radio Regulations require that normal transmissions must cease on all frequencies within a designated frequency band centered on 500 kHz, to permit listening for weak distress calls.

silent speed The speed at which silent motion pictures are fed through a projector, equal to 16 frames per second.

silent zone *Skip zone.*

silica *Silicon dioxide.*

silica gel A chemically inert and highly hygroscopic form of hydrated silica, used for absorbing water and vapors of solvents in coaxial lines, waveguides, and other nonevacuated enclosures.

silicide resistor A thin-film resistor that uses a silicide of molybdenum or chromium, deposited by DC sputtering in an integrated circuit when radiation hardness or high resistance values are required. Silicide films in use include $MoSi_2$, $CrSi_2$, and Si-Cr.

silicon [symbol Si] A nonmetallic element used in pure form as a semiconductor and mixed with iron or steel during smelting to improve the magnetic properties for use as transformer core material. Atomic number is 14.

silicon anodization An anodizing process that isolates the elements of an integrated circuit and improves performance by lowering the capacitance between elements and increasing the gain and speed of transistors. Applications include I^2L and T^2L circuits.

silicon capacitor A capacitor in which a pure silicon crystal slab is the dielectric. A silicon capacitor can have high Q at frequencies up to 5 GHz in low-voltage circuits. When the crystal is grown to have a P zone, a depletion zone, and an N zone, the capacitance varies with the externally applied bias voltage, as in a varactor. The higher the bias voltage of this voltage-controlled capacitor, the lower its capacitance.

silicon carbide [symbol SiC] A semiconductor material used in light-emitting diodes to produce green or yellow light. It can also be used in transistors. It will operate reliably at temperatures above 400°C.

silicon carbide lamp A light-emitting diode in which a chip of silicon carbide is the source of light.

silicon carbide rectifier A semiconductor rectifier capable of withstanding temperatures up to 1200°F (650°C).

silicon carbide varistor A voltage-dependent resistor in which silicon carbide is the resistive element.

silicon controlled rectifier [abbreviated SCR] A three-junction three-terminal PNPN thyristor that is normally an open circuit in both directions. When the proper signal is applied to the gate electrode, the device switches rapidly to a conducting state and allows current flow in the forward

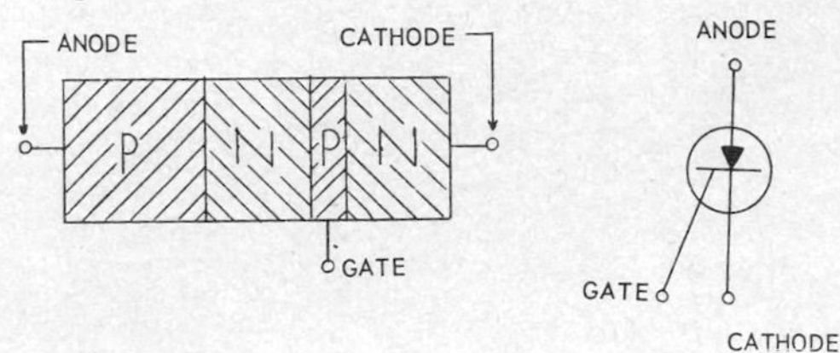

Silicon controlled rectifier construction and symbol.

direction, just as in a conventional rectifier. It remains turned on when the gate voltage is removed, and it can be turned off by removing the anode voltage, reducing it below the cathode voltage, or making the anode voltage negative as on alternate half-cycles of an AC circuit.

silicon controlled switch [abbreviated SCS] A four-terminal switching device that has four semiconductor layers, all of which are accessible. It can

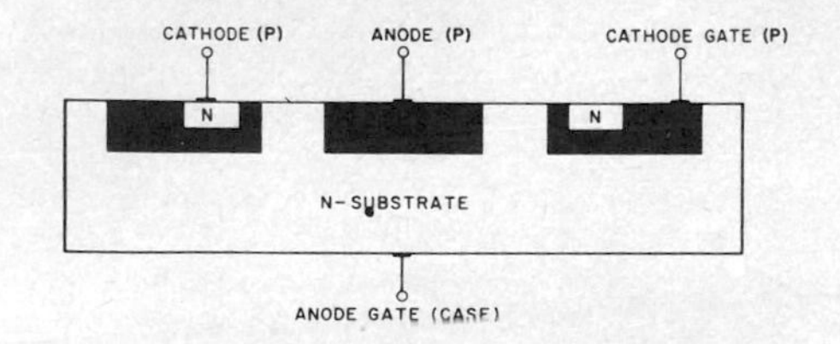

Silicon controlled switch cross section.

be used as a silicon controlled rectifier, complementary silicon controlled rectifier, gate-turnoff switch, or conventional silicon transistor.

silicon-diffused epitaxial mesa transistor A silicon transistor that has high voltage and power ratings combined with low storage time and low saturation voltage. Used in pulse amplifiers, drive amplifiers for memory cores, and data-processing equipment.

silicon diode A crystal diode that uses silicon as a semiconductor. Widely used as a detector in UHF and SHF circuits, and for zener diodes.

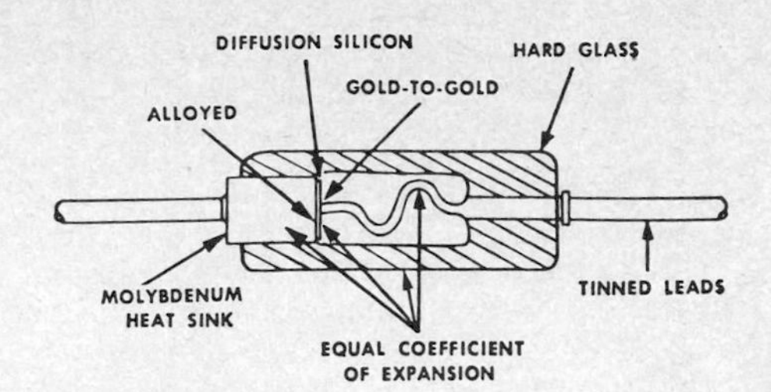

Silicon-diode construction.

silicon dioxide [symbol SiO_2[A crystalline material, of which quartz is one form, that has excellent insulating properties. Used as an insulating layer in some semiconductor devices. It can be selectively etched and doped to produce a variety of components in integrated circuits. Also called silica.

silicone A polymeric organosilicon compound that has excellent insulating, lubricating, water-resisting, and heat-resisting properties. Silicone grease is used to improve heat transfer from a power transistor or other semiconductor device to a heatsink.

silicon epitaxial planar transistor A silicon transistor in which epitaxial processing is combined with gold doping by a process that permits making thousands of transistors on one sheet at one time.

silicon gate A gate that uses polycrystalline silicon rather than a metal layer. Used in MOS technology.

silicon-gate-controlled AC switch A thyristor that can be gate-triggered from a blocking to a conducting state for either polarity of the applied voltage. Used for phase-control applications such as lamp dimming and temperature control.

silicon-gate transistor A silicon-on-sapphire field-effect transistor in which a film of silicon is the gate.

silicon image sensor A solid-state television camera in which a charge-coupled-device semiconductor chip replaces a vidicon or other camera tube. The image is focused on an array that may consist of 163,840 picture elements, charging each element in proportion to the light falling on it. The charges are removed from the elements electronically and passed on to the camera output as standard television signals.

silicon imaging device A solid-state industrial television camera that uses charge-coupled-device technology to form individual light-sensitive elements. One version has an array of 512 × 320 elements (a total of 163,840 elements) producing standard 525-line video output within a 3-MHz bandwidth.

silicon nitride [symbol Si_3N_4] A material deposited as a primary passivating layer in some semiconductor devices and integrated circuits, to protect the active devices from ionic contamination and mechanical damage.

silicon-on-sapphire [abbreviated SOS] A semiconductor manufacturing technology in which metal-oxide semiconductor devices are constructed in a thin single-crystal silicon film grown on an electrically insulating synthetic sapphire substrate. Advantages include minimizing of parasitic capacitances, low leakage, and higher operating speed. Used in large-scale integration.

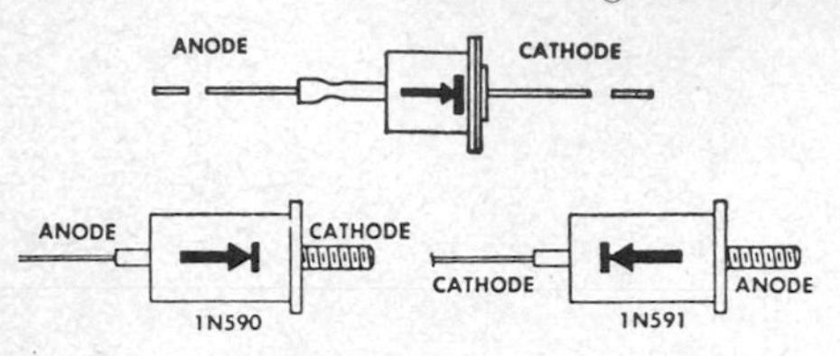

Silicon rectifiers.

silicon rectifier A semiconductor rectifier in which rectifying action is provided by an alloy junction formed in a high-purity silicon slab.

silicon resistor A resistor that uses silicon semiconductor material as a resistance element, to obtain a positive temperature coefficient of resistance which does not appreciably change with temperature. Used as a temperature-sensing element.

silicon solar cell A solar cell that consists of P and N silicon layers placed one above the other to form a PN junction at which radiant energy is converted into electricity. Theoretical maximum efficiency is 22%, and actual efficiencies better than 11% have been achieved. Used as a power source for satellite instrumentation and portable radios.

silicon steel An alloy steel that contains 3 to 5% silicon and has desirable magnetic qualities for iron cores of transformers and other AC devices.

silicon transistor A transistor in which the semiconductor material is silicon.

silo A missile shelter that consists of a hardened vertical hole in the ground, with facilities either for lifting the missile to a launch position or for direct launch from the shelter.

silo memory A first-in first-out memory.

Silsbee effect The ability of an electric current to destroy superconductivity by the magnetic field that it generates, without raising the cryogenic temperature.

silver [symbol Ag] A precious-metal element that has better electric conductivity than copper, used for contact points of relays and switches and as a plating on electronic components because it does not readily corrode. Atomic number is 47.

silver-cadmium storage battery A storage battery that combines the excellent space and weight characteristics of silver-zinc batteries with long shelf life and other desirable properties of nickel-cadmium batteries. It can be constructed as a sealed unit for use in conjunction with solar cells in spacecraft and as a power source for portable television receivers.

silvered mica capacitor A mica capacitor in which a coating of silver is deposited directly on the mica sheets to serve in place of conducting metal foil.

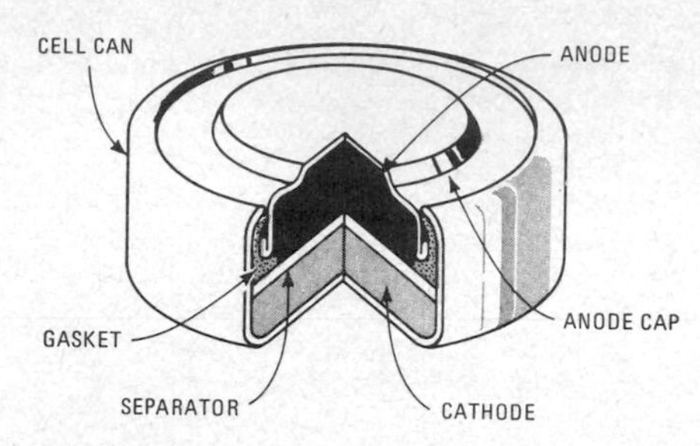

Silver oxide cell construction. Used extensively in electronic digital watches.

silver oxide cell A primary cell in which depolarization is accomplished by an oxide of silver.

silver sensitization A process of depositing a thin layer of silver on photosensitive surfaces during formation, to increase the sensitivity.

silver solder A solder composed of silver, copper, and zinc; it has a melting point lower than silver but higher than lead-tin solder.

silver storage battery An alkaline storage battery in which the positive active material is silver oxide and the negative active material contains zinc.

silver whisker A long, thin crystal of silver sometimes formed on components when solid or plated silver is used as part of the manufacturing process. The silver is transferred in its vapor phase, and it may cause failure of the device.

silver-zinc primary cell A reserve primary cell that can be activated in less than a second by adding liquid, can be discharged at high or low rates, and gives essentially constant voltage during discharge. Widely used in guided missiles and torpedoes.

silver-zinc storage battery A storage battery that gives higher current output and greater watthour capacity per unit of weight and volume than most other types, even at high discharge rates. Widely used in missiles and torpedoes, where its higher cost can be tolerated.

simple harmonic wave A wave whose amplitude at any point is a simple harmonic function of time.

simple sound source A source that radiates sound uniformly in all directions under free-field conditions.

simple target A radar target that has a reflecting surface such that the amplitude of the reflected signal does not vary with the aspect of the target.

A metal sphere is an example.

simple tone 1. A sound wave whose instantaneous sound pressure is a simple sinusoidal function of time. 2. A sound sensation characterized by singleness of pitch. Also called pure tone.

simplex *Simplex operation.*

simplex circuit A circuit derived from a pair of wires that are in effect used in parallel as one extra line, with the ground as a return path.

simplex operation A method of radio operation in which communication between two stations takes place in only one direction at a time. This includes ordinary transmit-receive operation, press-to-talk operation, voice-operated carrier, and other forms of manual or automatic switching from transmit to receive. Also called simplex.

simulation The representation of physical systems and phenomena by computers, models, or other equipment.

simulator 1. A computer or other piece of equipment that simulates a desired system or condition and shows the effects of various applied changes, such as a flight simulator. 2. Software that can be used with one computer system to make it execute programs written for another computer system. In contrast, an emulator requires added hardware and/or microprograms along with software to achieve this result.

simulcast A program broadcast simultaneously by two different types of stations, as by radio and television stations or by FM and AM broadcast stations.

simultaneous color television A color television system in which the phosphors for the three primary colors are excited at the same time, not one after another. The shadow-mask color picture tube gives a simultaneous display.

simultaneous lobing A radar direction-finding technique in which the signals received by two partly overlapping antenna lobes are compared in phase or power to obtain a measure of the angular displacement of a target from the equisignal direction.

Si_3N_4 Symbol for *silicon nitride.*

sinad ratio The ratio in decibels of signal-plus-noise-plus-distortion to noise-plus-distortion at the output of a mobile radio receiver for a modulated-signal input.

sine-cosine encoder A shaft-position encoder that has a special type of angle-reading code disk which gives an output that is a binary representation of the sine of the shaft angle.

sine-cosine generator *Resolver.*

sine potentiometer A potentiometer whose DC output voltage is proportional to the sine of the shaft angle. Used as a resolver in computer and radar systems.

sine wave A wave whose amplitude varies as the sine of a linear function of time.

sine-wave clipper A clipper circuit that cuts off the top of a sine wave so it resembles a square wave.

Sine-cosine encoder disk.

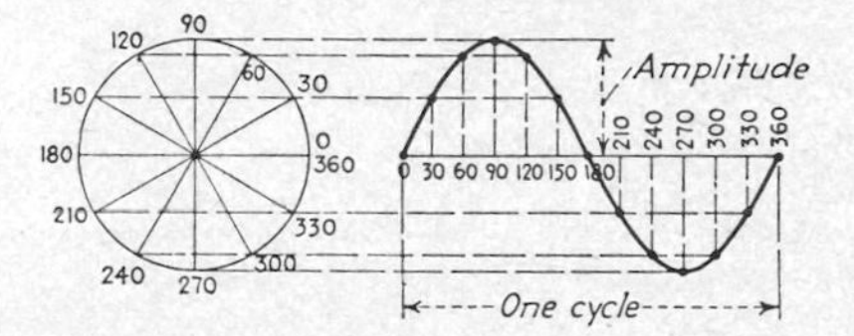

Sine-wave construction procedure.

sine-wave modulation Modulation in which the envelope of the modulated carrier signal has the waveform of a sine wave.

sine-wave response *Amplitude-frequency response.*

singing An undesired self-sustained oscillation in a system or component, at a frequency in or above the passband of the system or component. Generally caused by excessive positive feedback.

singing margin The difference in level, usually expressed in decibels, between the singing point and the operating gain of a system or component.

singing point The minimum value of gain of a system or component that will result in singing.

singing-stovepipe effect The reception and reproduction of radio signals by ordinary pieces of metal in contact with each other, such as sections of stovepipe. It occurs when rusty bolts, faulty welds, or mechanically loose connections within strong radiated fields near transmitters produce intermodulation interference. This problem is becoming increasingly severe on military ships and aircraft that contain many high-power transmitters. The mechanically poor connections serve as nonlinear diodes. Since rusting of steel is the most common cause of this trouble, the removal of steel objects from critical areas is one possible solution. Another is electroplating of steel with copper. Chief problem is locating the cause of the interference, which is one form of electromagnetic interference.

single-address instruction *One-address instruction.*

single-button carbon microphone A carbon microphone that has a carbon-filled buttonlike container on only one side of its flexible diaphragm.

single-channel simplex Simplex operation that provides nonsimultaneous radio communication between stations which use the same frequency channel.

single crystal A crystal, usually artificially grown, in which all parts have the same crystallographic orientation.

single-crystal camera An x-ray camera in which single crystals are examined. The diffracted x-ray beams are usually recorded on a cylindrical film placed coaxially with the axis of rotation of the crystal.

single defruit The process of defruiting by comparing video signals from two successive sweeps.

single-dial control Control of a number of different devices or circuits by a single control knob, as when using a gang capacitor.

single-domain particle A ferromagnetic particle so small that it can support only one permanently magnetized region in which the magnetic moments of the atoms are ordered.

single-ended Unbalanced, as when one side of a transmission line or circuit is grounded.

single-ended amplifier An amplifier in which each stage normally employs only one tube, so operation is asymmetric with respect to ground.

single-ended input An amplifier or other circuit in which one side of the input is grounded.

single-ended push-pull amplifier An amplifier that has two transmission paths designed to operate in a complementary manner and connected to provide a single unbalanced output. This circuit provides push-pull operation without the use of a transformer.

single-ended tube A tube in which all electrode connections, including the control grid, are made to base pins, so there is no top cap. The letter S after the first numerals in a receiving-tube designation indicates a single-ended tube, as in 12SK7.

single-frequency duplex Duplex carrier communication that provides communication in opposite directions, but not simultaneously, over a single-frequency carrier channel, the transfer between transmitting and receiving conditions being automatically controlled by the voices of the communicating parties.

single-frequency simplex Single-frequency carrier communication in which manual rather than automatic switching is used to change over from transmission to reception.

single-gun color tube A color television picture tube that has only one electron gun and one electron beam. The beam is sequentially deflected across phosphors for the three primary colors to form each color picture element, as in the chromatron.

single hop The range of a radio wave that is radiated at a small angle to the horizontal, so it penetrates the ionosphere only a small amount before being reflected back to the surface of the earth. The maximum range that can be spanned by single-hop transmission is about 1500 mi (2400 km) for E-layer transmissions and is obtained when the radio wave is radiated horizontally.

single in-line package [abbreviated SIP] A packaged resistor network or other assembly that has a single row of terminals or lead wires along one edge of the package.

single-particle model *Independent-particle model.*

single-phase Energized by a single alternating voltage.

single-phase circuit A circuit energized by a single alternating voltage, applied through two wires.

single-polarity pulse Deprecated term for *unidirectional pulse.*

single-pole double-throw [abbreviated SPDT] A three-terminal switch or relay contact arrangement that connects one terminal to either of two other terminals.

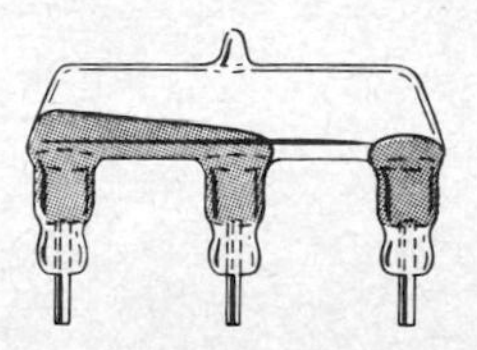

Single-pole double-throw mercury switch.

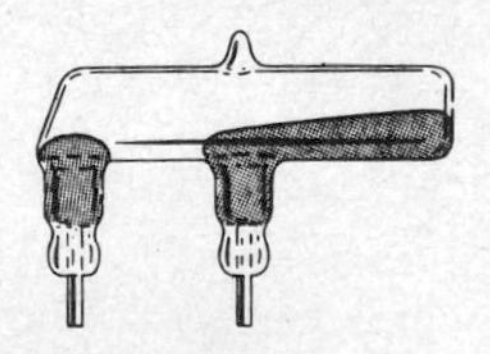

Single-pole single-throw mercury switch.

single-pole single-throw [abbreviated SPST] A two-terminal switch or relay contact arrangement that opens or closes one circuit.

single scattering The deflection of a particle from its original path owing to one encounter with a single scattering center in the material traversed. Plural scattering involves several successive encounters, and multiple scattering involves many successive encounters.

single-sideband [abbreviated SSB] Pertaining to a communication system in which one of the two sidebands for amplitude modulation is suppressed.

single-sideband communication A communication system in which the RF carrier and one sideband are suppressed. Less power is then required at the transmitter for the same effective signal at the receiver, a narrower frequency band can be used, and the signal is less affected by selective fading or man-made interference.

single-sideband converter A converter that connects to the IF amplifier output of an amplitude-modulation radio receiver and converts the receiver into a single-sideband receiver.

single-sideband filter A bandpass filter in which the slope on one side of the response curve is greater than on the other side. Used to suppress one sideband and sometimes the carrier frequency.

single-sideband modulation Modulation in which all the components of one sideband are eliminated from an amplitude-modulated wave.

single-sideband receiver A radio receiver designed for the reception of single-sideband modulation; it has provisions for restoring the carrier.

single-sideband transmission Transmission of a carrier and substantially only one sideband of modulation frequencies, as in television, where only the upper sideband is transmitted completely for the picture signal. The carrier wave may be either transmitted or suppressed.

single-sideband transmitter A transmitter in which one sideband is transmitted and the other is effectively eliminated.

single-signal receiver A highly selective superheterodyne receiver for code reception; it has a crystal filter in the IF amplifier.

single-speed floating action Floating action in which a final control element is moved at a single speed.

single-stub transformer A shorted section of coaxial line that is connected to a main coaxial line near a discontinuity to provide impedance matching at the discontinuity.

single-stub tuner A section of transmission line terminated by a movable short-circuiting plunger or bar, attached to a main transmission line for impedance-matching purposes.

single-sweep oscilloscope An oscilloscope in which the sweep must be reset and triggered for each operation, to prevent multiple displays of a trace.

single-throw switch A switch in which the same pair of contacts is always opened or closed.

single-tone keying Keying in which the modulating wave causes the carrier to be modulated with a single tone for one condition, which may be either for a mark or for a space. The carrier is unmodulated for the other condition.

single track A variable-density or variable-area sound track in which both positive and negative halves of the signal are linearly recorded. Also called standard track.

single-track recorder A magnetic-tape recorder that records only one track on the tape.

single-wire line A surface-wave transmission line that consists of a single conductor which has a dielectric coating or other treatment that confines the propagated energy close to the wire.

sink 1. A power-consuming device, such as the load in a circuit. 2. The region of a Rieke diagram where the rate of change of frequency with respect to phase of the reflection coefficient is maximum for an oscillator. Operation in this region may lead to unsatisfactory performance because of cessation or instability of oscillations.

SINS [Ship's Inertial Navigation System] A navigation system that uses a gyroscope-stabilized platform and associated inertial guidance and sonar equipment to provide automatically the ship's position, true north, and true speed of the ship with respect to the ocean bottom. Used in nuclear-powered submarines.

sintered conductor A conductor in which the metal is applied in powdered form and subjected to high temperature and pressure by heated dies to obtain intricate configurations of solid conductors.

sintering The process of bonding metal or other powders by cold-pressing into the desired shape, then heating to form a strong cohesive body.

sinusoidal Varying in proportion to the sine of an angle or time function. Ordinary alternating current is sinusoidal.

sinusoidal electromagnetic wave A wave whose electric field strength is proportional to the sine or cosine of an angle that is a linear function of time or distance.

sinusoidal quantity A quantity that varies according to a sinusoidal function of the independent variable.

SIP Abbreviation for *single in-line package.*

site error An error due to distortion of the radiated field by objects in the vicinity of a radio navigation aid.

situation-display tube A large cathode-ray tube that displays tabular and vector messages pertinent to the various functions of an air defense mission.

size control A control on a television receiver for changing the size of a picture either horizontally or vertically.

skew 1. Deviation of a received facsimile frame from rectangularity due to lack of synchronism between scanner and recorder. Expressed numerically as the tangent of the angle of this deviation. 2. The degree of nonsynchronism of supposedly parallel bits when bit-coded characters are read from magnetic tape.

skiatron *Dark-trace tube.*

skin antenna A flush-mounted aircraft antenna made by using insulating material to isolate a portion of the metal skin of the aircraft.

skin depth The depth below the surface of a conductor at which the current density has decreased 1 neper below the current density at the surface due to the action of the electromagnetic waves associated with the high-frequency current flowing through the conductor.

skin dose The dose at the center of the irradiation field on skin. It is the sum of the air dose and backscattering plus the exit dose from other parts if this is significant.

skin effect The tendency of alternating current to concentrate in the surface layer of a conductor. The effect increases with frequency and increases the effective resistance of the conductor.

skin tracking The tracking of an object by using radar reflections from its surface or skin.

skip A digital-computer instruction to proceed to the next instruction.

skip distance The minimum distance that radio waves can be transmitted between two points on the earth by reflection from the ionosphere, at a specified time and frequency. The skip distance

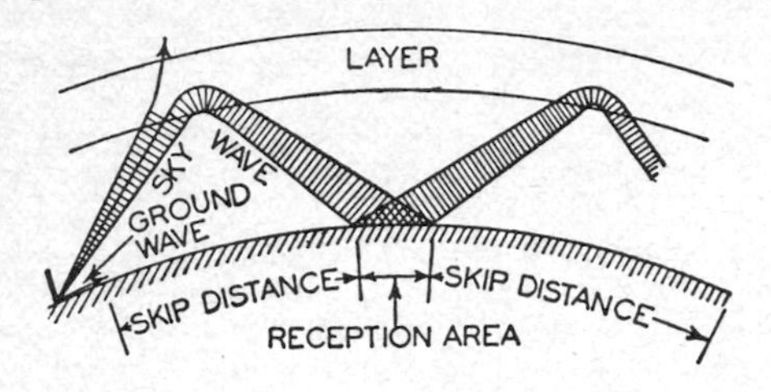

Skip distance.

thus includes the maximum ground-wave range and the width of the skip zone. In multihop transmission, the distance of each succeeding hop from earth to ionosphere and back is also the skip distance.

skip zone A ring-shaped area around a radio transmitter, surrounding the ground-wave reception region, within which no radio signals are received. The outer edge of the skip zone is at the minimum distance for reception of sky-wave signals, and the inner edge is at the maximum limit for reception of ground-wave signals. Also called silent zone and zone of silence.

sky screen Equipment that provides a positive indication to the range safety officer whenever a missile deviates from its planned trajectory. One sky screen is used to monitor flight azimuth and another to monitor vertical programming.

sky wave A radio wave that travels upward into space and may or may not be returned to earth by reflection from the ionosphere. Also called ionospheric wave.

sky-wave correction A correction for sky-wave propagation errors, applied to measured position data. The amount of the correction is based on an assumed position and an assumed ionosphere height.

sky-wave station error The error in station synchronization in sky-wave-synchronized loran that is caused by the effect of ionospheric variations on the time of transmission of the synchronizing signal from one station to the other.

sky-wave-synchronized loran [abbreviated SS loran] A type of loran in which the transmitting stations are synchronized by signals reflected from the ionosphere. Used to obtain greater range and more accurate nighttime navigation.

slab A relatively thick crystal cut from which blanks are obtained by subsequent transverse cutting.

slant distance The distance between two points not at the same elevation. Also called slant range (deprecated).

slant range Deprecated term for *slant distance.*

SLAR Abbreviation for *side-looking airborne radar.*

slave antenna A directional antenna positioned in azimuth and elevation by a servosystem. The information controlling the servosystem is supplied by a tracking or positioning system.

slaved gyromagnetic compass A compass in which the gyroscope is synchronized to a magnetic force.

slave flash A photographic flash that has a fast photosensitive switch which can be triggered by the light from a master flash. Any number of slave flashes can be used, without wire connections to the camera or master flash.

slave operation Operation of two or more stabilized power supplies or other devices in such a way that coordinated control of the assembly is achieved by controlling only the master unit.

slave station A navigation station in which some characteristic of its emission is controlled by a master station, such as the B station in loran.

slave sweep A time base that is triggered by an external waveform. Used in navigation systems to display the same information at different locations.

SLC Abbreviation for *straight-line capacitance.*

sleeve 1. The cylindrical contact that is farthest from the tip of a phone plug. 2. Insulating tubing used over wires or components. Also called sleeving.

sleeve antenna A single vertical half-wave radiator, whose lower half is a metallic sleeve through which the concentric feed line runs. The upper radiating portion, one quarter-wavelength long, connects to the center of the line.

sleeve-dipole antenna A dipole antenna surrounded in its central portion by a coaxial sleeve.

sleeve stub An antenna that consists of half a sleeve-dipole antenna projecting from a large metal surface.

sleeving *Sleeve.*

slewing Moving a radar antenna or sonar transducer rapidly in a horizontal or vertical direction, or both.

slewing motor A motor that drives a radar antenna at high speed for slewing to pick up or track a target.

slew rate The maximum rate at which the output voltage of an operational amplifier changes for a square-wave or step-signal input. The rate is usually specified in volts per microsecond.

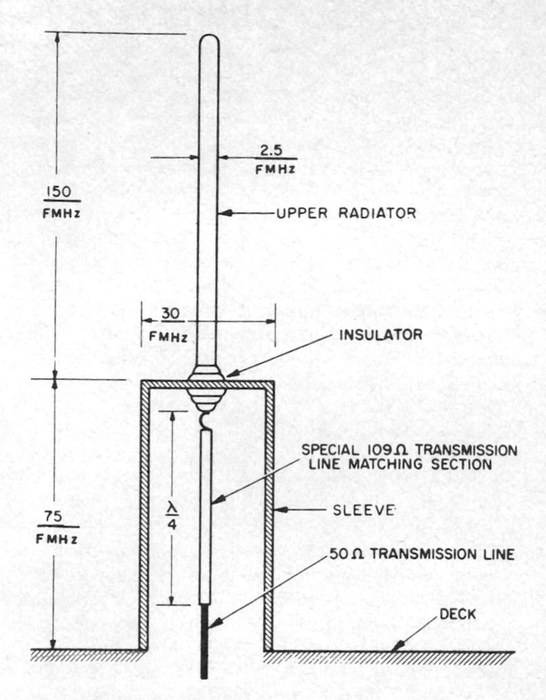

Sleeve-antenna construction, with dimensions in feet given for frequency value in megahertz. Divide numerator of each dimension by 3.28 to get dimension in meters.

SLF Abbreviation for *straight-line frequency.*

slide-back voltmeter An electronic voltmeter in which an unknown voltage is measured indirectly by adjusting a calibrated voltage source until its voltage equals the unknown voltage.

slider A sliding type of movable contact.

slide-rule dial A dial in which a pointer moves in a straight line over long straight scales that resemble the scales of a slide rule.

slide switch A switch that is actuated by sliding a button, bar, or knob.

slide-wire bridge A bridge circuit in which the resistance in one or more branches is controlled by the position of a sliding contact on a length of resistance wire stretched along a linear scale.

slide-wire rheostat A rheostat in which a sliding contact rides over a long single-layer coil of resistance wire.

sliding contact *Wiping contact.*

slip The difference between synchronous and operating speeds of an induction machine.

slip ring A conducting ring mounted on but insulated from a rotating shaft, used with a stationary brush to join fixed and moving parts of a circuit.

slope 1. The projection of a flight path in the vertical plane. 2. The degree of deviation of an essentially straight portion of a characteristic curve from the horizontal or vertical.

slope angle The angle in the vertical plane between the horizontal and the flight path of an aircraft or missile. Also called glide-slope angle.

slope detector A discriminator that uses a single tuned circuit and single diode to react to differences in frequency. Operation is on one of the slopes of the response curve for the tuned circuit. Seldom used in FM receivers because the linear portion of the curve is too short for large-signal operation.

slope deviation The difference between the projection in the vertical plane of the actual path of movement of an aircraft and the planned slope for the aircraft. Expressed in terms of angular or linear measurement.

slope equalizer An equalizer used with an amplifier to make the attenuation of a section of transmission line constant over the frequency band being transmitted.

slope filter A filter that has a response which rises or falls with frequency over a given frequency range.

slot antenna An antenna formed by cutting one or more narrow slots in a large metal surface fed by a coaxial line or waveguide. For unidirectional radiation, the rear of the metal sheet is boxed in, or the slot is energized directly by a waveguide. In another version, diagonal slots are cut into a length of waveguide at precisely spaced intervals.

slot array An antenna array that consists of a number of slot antennas, energized separately to give a desired radiation pattern.

slot coupling Coupling between a coaxial cable and a waveguide by two coincident narrow slots, one in a waveguide wall and the other in the sheath of the coaxial cable.

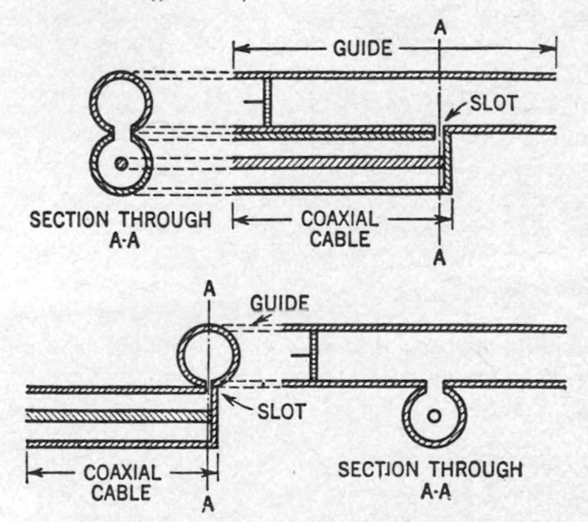

Slot coupling as used between coaxial cable and circular waveguide.

slot mask A color picture tube mask that has vertical slots positioned in front of a screen on which color phosphors are arranged in vertical stripes. Used in the Trinitron and other picture

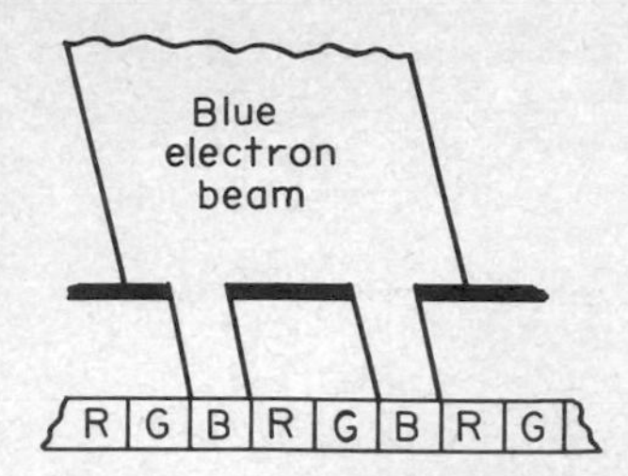

Slot-mask operation for blue electron beam. Beam angle is wide enough to pass through two slots, but it is at such an angle that it hits only two blue vertical phosphor stripes on screen. Red and green beams arrive at different angles, as required for hitting two beams of their own colors. All three beams sweep line by line across screen to form color picture.

tubes that have in-line electron guns. Also called aperture grille.

slotted line *Slotted section.*

slotted rotor plate A rotor plate that has radial slots to permit bending different sections of the plate either inward or outward to adjust the total capacitance of a variable capacitor section during alignment. Also called serrated rotor plate and split rotor plate.

slotted section A section of waveguide or shielded transmission line in which the shield is

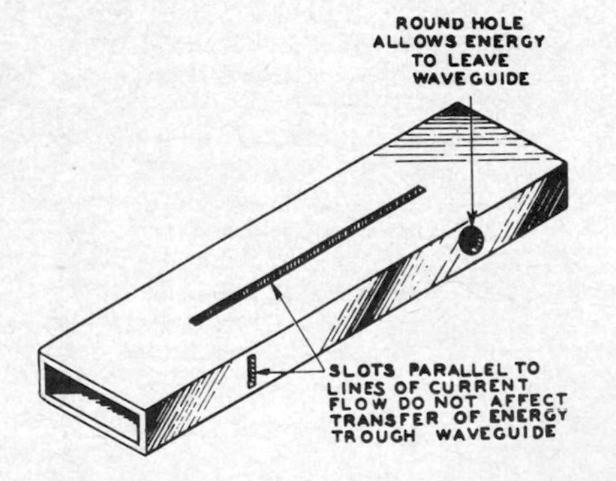

Slotted section. Probe inserted into one of slots will remove small amount of energy without affecting waveguide performance.

slotted to permit the use of a traveling probe for examination of standing waves. Also called slotted line and slotted waveguide.

slotted waveguide *Slotted section.*

slow-acting relay A time-delay relay in which an interval of several seconds may exist between energizing of the coil and pulling up of the armature. The delay can be obtained electrically by placing a solid copper ring on the core of the relay. Also called slow-operate relay.

slow-blow fuse A fuse that can withstand up to 10 times its normal operating current for a brief period, as required for circuits and devices which draw a very heavy starting current.

slow-down video A method of transmitting radar data over narrow-bandwidth circuits by storing the radar video signal over the time required for the antenna to move through one beam width. The stored signal is then sampled at a periodic rate such that each range interval of interest is sampled at least once per beam width or per azimuth quantum, and the resulting data is quantized for transmission.

slowing-down A decrease in the energy of a particle as a result of collisions with nuclei.

slowing-down area An area defined as one-sixth the mean square distance between a neutron source and the point where the neutron reaches a given energy in an infinite homogeneous medium.

slowing-down density The number of neutrons slowing down per unit volume per unit time.

slowing-down kernel The probability that a neutron will go from one position to another while slowing down through a specified energy range.

slowing-down length The square root of the slowing-down area.

slowing-down power The average decrease in the value of the natural logarithm of energy of a neutron per unit distance traveled by the neutron in a material.

slow memory A section of a computer memory from which information can be obtained automatically but not at the fastest rate.

slow-motion video disk recorder A magnetic-disk recorder that stores one field of video information per revolution, for instant replay at normal speed or any degree of slow motion down to complete stopping of action. Used extensively for instant replays during broadcasts of football games.

slow neutron A neutron that has low kinetic energy, up to about 100 eV.

slow-operate relay *Slow-acting relay.*

slow-release relay A time-delay relay in which there is an appreciable delay between deenergizing of the coil and release of the armature.

slow-scan television [abbreviated SSTV] Television in which each complete scan of the scene takes about 8 s. This reduces transmission channel requirements to that for an audio signal (usually less than 5 kHz), permitting use of ordinary telephone lines for transmission and reception of sequences of still pictures. First developed for use by amateur radio operators. A high-persistence cathode-ray tube is required, to hold the picture for 8 s. The picture may be cut to 120 lines, giving a much coarser image than with standard 525-line television.

slow wave A wave that has a phase velocity less than the velocity of light, as in a ridge waveguide.

slow-wave circuit A microwave circuit in which the phase velocity is much slower than the velocity of light.

slug 1. A heavy copper ring placed on the core of a relay to delay operation of the relay. 2. A movable iron core for a coil. 3. A movable piece of metal or dielectric material used in a waveguide for tuning or impedance-matching purposes.

slug tuner A waveguide tuner that contains one or more longitudinally adjustable pieces of metal or dielectric.

slug tuning A means for varying the frequency of a resonant circuit by introducing a slug of material into either the electric or magnetic fields or both, as in permeability tuning.

SLW Abbreviation for *straight-line wavelength.*

small-scale integration [abbreviated SSI] Integration in which a complete major subsystem or system is fabricated on a single integrated-circuit chip that contains integrated circuits which have appreciably less complexity than for medium-scale integration.

small-signal analysis Circuit analysis based on such small excursions of current and voltage from their quiescent operating points that linear operation may be assumed.

smart bomb Slang term for *precision-guided munitions.*

smart terminal *Intelligent terminal.*

smear A television picture defect in which objects appear to be extended horizontally beyond their normal boundaries in a blurred or smeared manner. One cause is excessive attenuation of high video frequencies in the television receiver.

smectic material A liquid crystal material in which the elongated molecules are arranged in layers and all molecules are parallel.

S meter *Signal-strength meter.*

Smith chart A special polar diagram that contains constant-resistance circles, constant-reactance circles, circles of constant standing-wave ratio, and radius lines which represent constant line-angle loci. Used in solving transmission-line and waveguide problems.

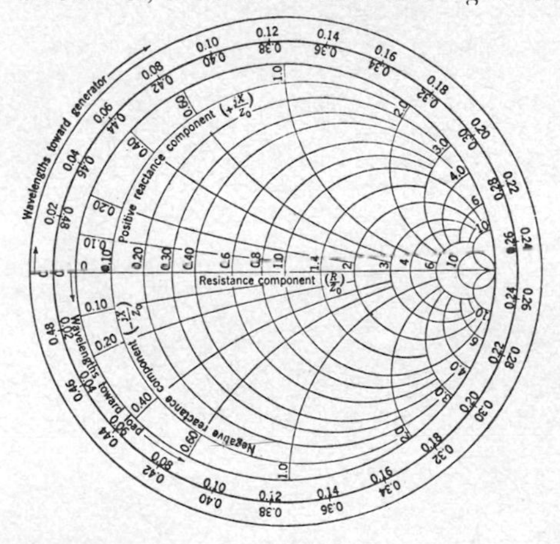

Smith chart.

smoke detector A photoelectric system for actuating an alarm when smoke in a chimney or other location exceeds a predetermined density.

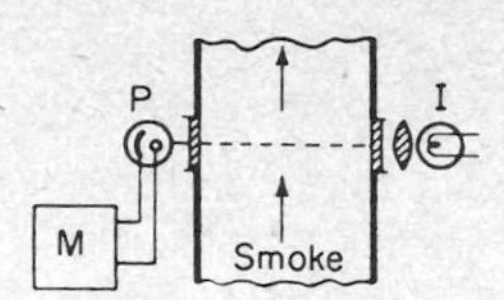

Smoke detector installed on chimney. Lamp I sends beam of light through smoke to photocell P.

smoke-puff decoy A countermeasure against enemy infrared devices.

smooth To decrease or eliminate rapid fluctuations in measured data.

smoothing choke An iron-core choke used in a power-supply filter circuit to remove ripple.

smoothing factor A factor that expresses the effectiveness of a filter in smoothing ripple-voltage variations.

SMPTE Abbreviation for *Society of Motion Picture and Television Engineers.*

S/N Abbreviation for *signal-to-noise.*

SNAP [Systems for Nuclear Auxiliary Power] A small nuclear power plant in which heat from radioisotope decay in a fuel such as strontium 90 is converted into electric energy, to provide power for spacecraft instrumentation, telemetry, and other applications.

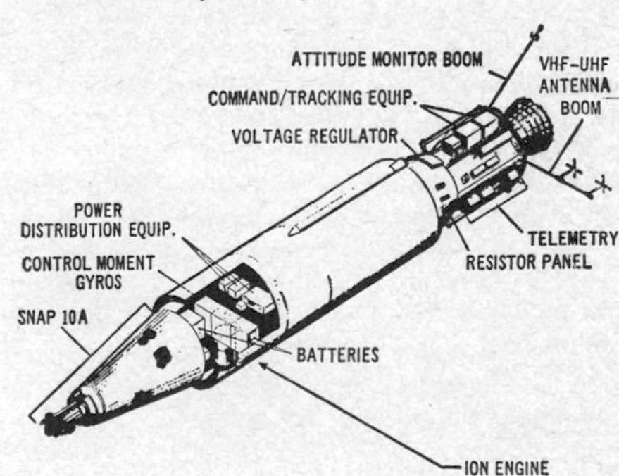

SNAP 600-W nuclear reactor installed in nose cone of satellite.

snap-action switch A switch that responds to very small movements of its actuating button or lever and changes rapidly and positively from one

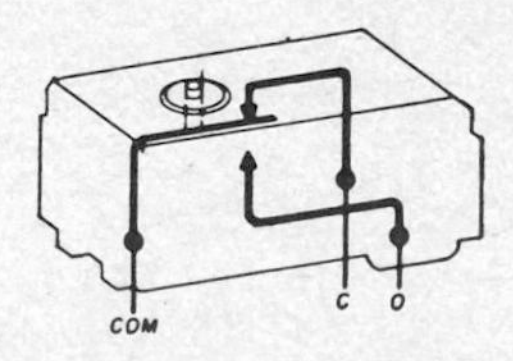

Snap-action switch providing single-pole double-throw action.

contact position to the other. Also called microswitch.

snap-off diode A planar epitaxial passivated silicon diode that is processed so a charge is stored close to the junction when the diode is conducting. When reverse voltage is applied, the stored charge then forces the diode to snap off or switch rapidly to its blocking state.

snap-on ammeter An AC ammeter that has a magnetic core in the form of hinged jaws which can be snapped around the current-carrying wire. Also called clamp-on ammeter.

snap switch A switch in which the contacts are separated or brought together suddenly by the action of a spring placed under tension or compression by the operating knob or lever.

snap varactor A varactor that stores a charge while conducting in the forward direction. When the polarity of the applied voltage is reversed, the varactor conducts for a short time in the reverse direction until the stored charge is gone, and conduction then stops suddenly. Applications include frequency multipliers, harmonic generators, oscillators, and pulse circuits. Closely related devices include punch-through varactors, snap-off diodes, and step-recovery diodes.

sneak path An undesired path formed in a computer logic circuit, generally at a particular configuration of gate and clock transistors, as a result of poor circuit design.

Snell's law When a wave travels from one medium into another that has a different index of refraction, the product of the sine of the angle of refraction and the refractive index of the refracting medium is equal to the product of the sine of the angle of incidence and the index of refraction of the medium which contains the incident beam.

sniffer gear *Odoriferous homing.*

sniperscope A snooperscope used in place of a telescopic sight on a rifle.

snivet A straight, jagged, or broken vertical black line that appears near the right-hand edge of a television receiver screen because of a horizontal amplifier tube fault that produces a discontinuity in the zero-bias anode current curve near its knee. It can be eliminated by changing the setting of almost any picture control or changing horizontal amplifier tubes.

snooperscope An infrared source, an infrared image converter, and a battery-operated high-voltage DC source constructed in portable form to permit a foot soldier or other user to see objects in total darkness. Infrared radiation sent out by the infrared source is reflected back to the snooperscope and converted into a visible image on the fluorescent screen of the image tube.

snow Small, random white spots produced on a television or radar screen by inherent noise signals that originate in the receiver. Visible on television screens only when received signal strength is inadequate, because strong incoming signals will override the noise signals.

snow static Precipitation static caused by falling snow.

SNR Abbreviation for *signal-to-noise ratio.*

Society of Motion Picture and Television Engineers [abbreviated SMPTE] A nonprofit organization established for the advancement of theory and practice related to the production and utilization of motion pictures and television programs.

socket A device that provides electric connections and mechanical support for an electronic or electric component which requires convenient replacement.

socket adapter An adapter used between a tube socket and a tube to permit use of the tube in a socket designed for some other type of tube or to permit making tube current or voltage measurements while the tube is in use.

sodalite A silicate material that has cathodochromic properties when appropriately treated. Used for screens of dark-trace cathode-ray tubes.

sodium [Symbol Na] A metallic alkali element that has a melting point of 97.5°C and moderate thermal neutron cross section. Used in liquid form as a coolant for some types of nuclear reactors. Used on cathodes of phototubes when maximum response is desired at the violet end of the visible spectrum. Atomic number is 11.

sodium iodide scintillation counter A scintillation counter in which a sodium iodide crystal serves as a scintillator for determining the energy of incident gamma rays.

sodium-vapor lamp A discharge lamp that contains sodium vapor, used chiefly for highway illumination.

sofar [SOund Fixing And Ranging] A system for fixing a position at sea by exploding a charge under water, measuring the times for the shock waves to travel through water to three widely separated shore stations, and calculating the position of the explosion by triangulation. The explosive can be dropped from a lifeboat by survivors of air or sea disasters.

soft base A missile-launching base that is not protected against atomic attack.

soft landing A landing made by decelerating a vehicle with parachutes, retrorockets, or other decelerating means, as required for piloted spacecraft landings on the moon and safe return to earth.

soft limiting Limiting in which there is still an appreciable increase in output for increases in input signal strength up into the range at which limiting action occurs.

soft magnetic material Magnetic material that is easily demagnetized.

soft radiation Less penetrating radiation.

soft superconductor A superconductor in which superconductivity can be destroyed by a weak magnetic field, as low as 25 oersteds. Zirconium and cadmium are examples. A hard superconductor may require over 1000 oersteds, or even as much as 100,000 oersteds for certain alloys.

soft tube 1. An x-ray tube that has a vacuum of about 0.000002 atmosphere, the remaining gas being left in intentionally to give less penetrating rays than those of a more completely evacuated tube. 2. *Gassy tube.*

software The programs and routines required to utilize the capabilities of computers.

soft x-ray An x-ray that has a comparatively long wavelength and poor penetrating power.

solar burst A sudden increase in the RF energy radiated by the sun, generally associated with visible solar flares.

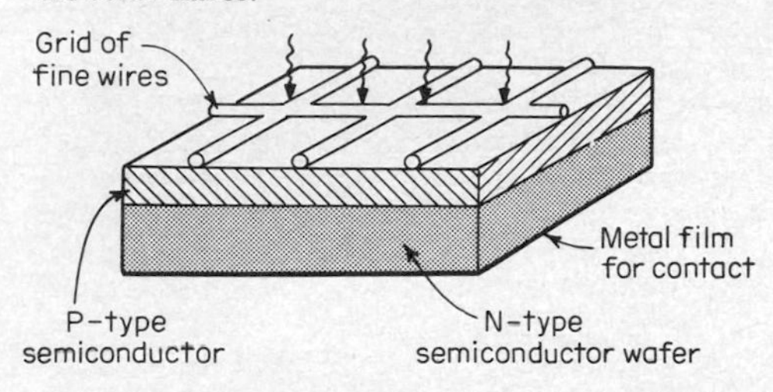

Solar-cell construction.

solar cell A silicon photovoltaic cell that converts sunlight directly into electric energy. Used in satellites to provide power for transmitting equipment.

solar-cell array A battery of solar cells mounted for oriented exposure to solar radiation. Used as a power source for satellites and spacecraft.

solar corpuscle A particle, usually a proton, sprayed out into the solar system by disturbances on the sun. The particles react with the earth's magnetic field to produce ionospheric disturbances.

solar cycle *Sunspot cycle.*

solar flare A violent eruption of the surface of the sun, sending a tremendous cloud of cosmic rays out into space. The magnetic field associated with this cloud is believed to shield the earth from cosmic rays coming from other interstellar sources, thus accounting for the sudden decrease in cosmic-ray count on the earth following a solar flare.

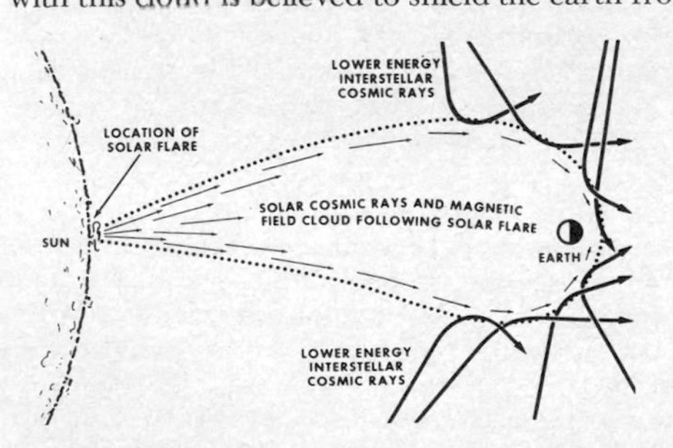

Solar flare and associated cosmic-ray and magnetic-field clouds.

solar-flare proton A proton ejected from the sun during solar flares, with energies ranging from below 10 MeV up to several gigaelectronvolts. Cosmic-ray particles from solar flares consist primarily of these protons.

solar generator An electric generator powered by radiation from the sun and used in some satellites.

solarimeter An instrument that makes direct readings of solar radiation intensity from sun and sky.

solar magnetograph A magnetograph designed primarily to study the powerful magnetic fields of sunspots.

solar probe A space vehicle that launches toward the sun, for measuring and transmitting radiation values until destroyed by the sun's heat.

solar pumping The use of sunlight focused directly into a laser rod for pumping to induce lasing action.

solar radio noise Radio noise that originates at the sun and increases greatly in intensity during sunspots and flares. Heard as a hissing noise on shortwave radio receivers.

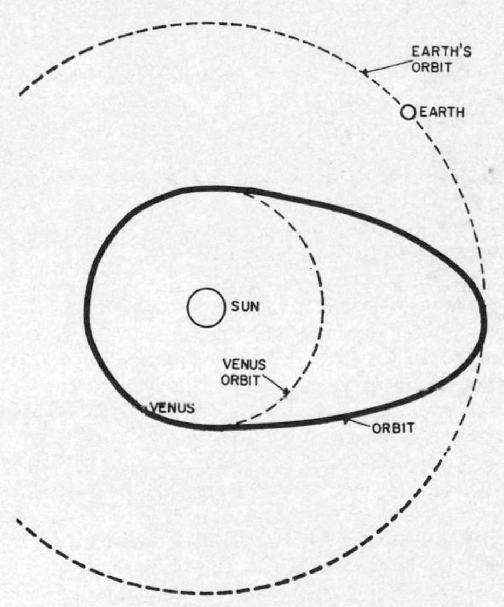

Solar satellite on elliptical orbit.

solar satellite A satellite that orbits the sun.

solar wind A stream of protons moving constantly outward from the sun.

solder 1. An alloy that can be melted at a fairly low temperature, for use in joining metals which have much higher melting points. An alloy of lead and tin in approximately equal proportions is the solder most often used for making permanent joints in circuits. 2. To join two metals with solder.

solderability The ability of a printed circuit or other metal surface to be wet by molten solder.

soldering The process of joining metals by fusion and solidification of an adherent alloy that has a melting point below about 800°F (425°C).

soldering flux A material that dissolves oxides from surfaces being soldered. Rosin is widely used for this purpose when soldering electronic circuits because of its noncorrosive qualities.

soldering gun A soldering tool that has as its tip a fast-heating resistance element which serves as a short-circuit across a high-current, low-voltage secondary winding of a step-down transformer built into the unit. A trigger-type switch in a pistol-grip handle permits intermittent use. Also called gun.

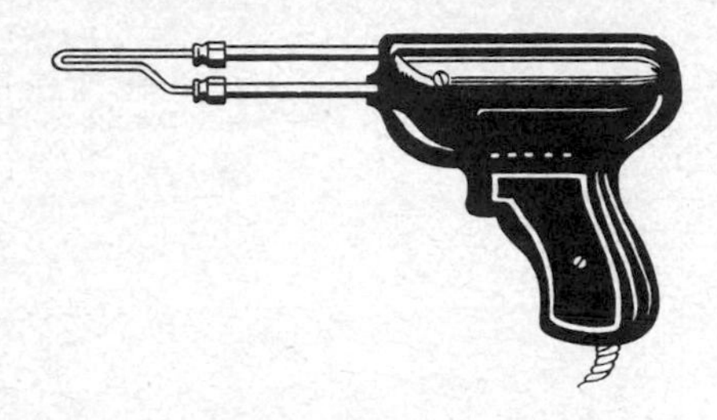

Soldering gun.

soldering iron A tool used to apply heat to a joint preparatory to soldering. The source of heat is either a gas torch or an internal electric heating element.

soldering lug *Lug.*

soldering paste A soldering flux prepared in the form of a paste.

soldering pencil A small soldering iron, about the size and weight of a standard lead pencil, used for soldering or unsoldering joints on printed-wiring boards. Also called pencil.

soldering pliers A soldering tool that sends electric heating current directly through the terminal, joint, or other part being heated for soldering. The tool resembles a pair of pliers that have carbon-electrode jaws, each connected to one terminal of a low-voltage, high-current secondary winding on a transformer.

soldering tool A tool that produces the heat needed in soldering, such as a soldering gun, soldering iron, or soldering pliers.

solderless connector A connector that holds wire ends firmly together to provide a good connection without solder. A common form is a cap with tapered internal threads, twisted over the exposed ends of the wires.

solderless contact *Crimp contact.*

solderless wrapped connection *Wire-wrap connection.*

solenoid A coil that surrounds a movable iron core. When the coil is energized by sending alternating or direct current through it, the core is pulled to a central position with respect to the coil. Used to convert electric energy into mechanical energy. Other forms of solenoids produce rotary rather than axial movement of the core. The core may be stationary and the coil movable.

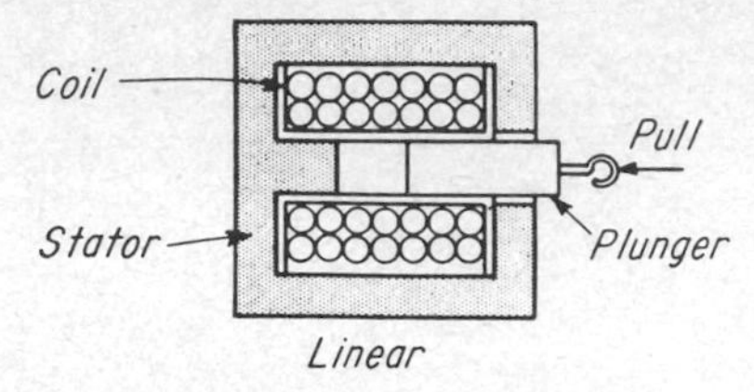

Solenoid having movable core providing linear motion.

solenoid relay *Plunger relay.*

solenoid valve A valve actuated by a solenoid, for controlling the flow of gases or liquids in pipes.

solid conductor A conductor that consists of a single wire rather than strands.

solid electrolytic capacitor An electrolytic capacitor in which the dielectric is an anodized coating on one electrode, with a solid semiconductor material filling the rest of the space between the electrodes.

solid laser A laser in which either a crystalline or amorphous solid material, usually in the form of a rod, is excited by optical pumping. The most common crystalline materials are ruby, neodymium-doped ruby, and neodymium-doped yttrium-aluminum garnet. Solid lasers are usually pulsed. Continuous operation is possible, but it is more difficult in glass than in crystals because of the lower thermal conductivity of glass. Solid lasers may have either three or four excitation levels. Ruby lasers use three-level operation. Four-level operation is usually obtained with active atoms or ions of an actinide, rare-earth, or transition metal. The normal wavelength range is 0.6 to 3 μm.

solid-state Pertaining to a circuit, device, or system that depends on some combination of electrical, magnetic, and optical phenomena within a solid that is usually a crystalline semiconductor material.

solid-state amplifier An amplifier in which the active devices are transistors constructed individually or as parts of an integrated circuit.

solid-state analog panel meter An analog panel meter in which a bar of light formed by light-emitting diodes moves on a calibrated scale.

solid-state circuit A circuit formed entirely in or on a single wafer of semiconductor material.

solid-state circuit breaker A circuit breaker in which a zener diode, silicon controlled rectifier, or solid-state device is connected to sense when load terminal voltage exceeds a safe value. This solid-state device, acting in microseconds, can be used to trip a slower-acting electromechanical circuit breaker that disconnects the load from the power line.

solid-state component A component whose operation depends on the control of electric or magnetic phenomena in solids, such as a transistor, crystal diode, or ferrite device.

solid-state DC motor A DC motor in which brushes are replaced by two Hall generators located 90° apart on the stator winding. During rotation of the cylindrical permanent-magnet rotor, the Hall units generate sinusoidal currents that control stator-winding switching by separate solid-state circuits, for precise control of motor speed.

solid-state device A device other than a conductor that uses magnetic, electrical, and other properties of solid materials, as opposed to vacuum or gaseous devices.

solid-state dosimeter A dosimeter that depends on radiophotoluminescence, thermoluminescence, or electric conductivity changes for measurement of radiation dose and dose rate. One example is the fluorod.

solid-state image sensor *Charge-coupled image sensor.*

solid-state lamp *Light-emitting diode.*

solid-state maser A maser in which a semiconductor material produces the coherent output beam. Two input waves are required; one wave, called the pumping source, induces upward energy transitions in the active material. The second wave, of lower frequency, causes downward transitions and undergoes amplification as it absorbs photons from the active material. Generally used as an amplifier, but it can also serve as an oscillator.

solid-state memory A computer memory that has metal-oxide semiconductor technology instead of conventional magnetic cores.

solid-state physics The branch of physics that deals with the structure and properties of solids, including semiconductors.

solid-state relay [abbreviated SSR] A relay that uses only solid-state components, with no moving parts.

solid-state static alternator An alternator that converts a square-wave input from a pulse generator into a single- or three-phase AC output voltage by using silicon controlled rectifiers as current-switching elements. It has no moving parts.

solid-state switch A switch in which a semiconductor material is the switching element. The switch is triggered by a voltage pulse applied to the anode or gate terminal. Examples include silicon controlled rectifiers, diacs, and triacs.

solid-state television camera A television camera that uses solid-state techniques for the image sensor as well as all circuits.

solid-state thyratron A semiconductor device, such as a silicon controlled rectifier, that approximates the extremely fast switching speed and power-handling capability of a gaseous thyratron tube.

solid-state tuner A tuner in which the energy absorption of a semiconductor such as yttrium-iron garnet is changed by varying the strength of the magnetic field applied to the material, to provide electric tuning.

solid-state voltmeter A digital voltmeter that uses transistors and diodes instead of vacuum tubes.

solid-state watch A watch that uses integrated circuits, usually driving a digital display.

solid tantalum capacitor An electrolytic capacitor in which the anode is a porous pellet of tantalum. The dielectric is an extremely thin layer of tantalum pentoxide formed by anodization of the exterior and interior surfaces of the pellet. The

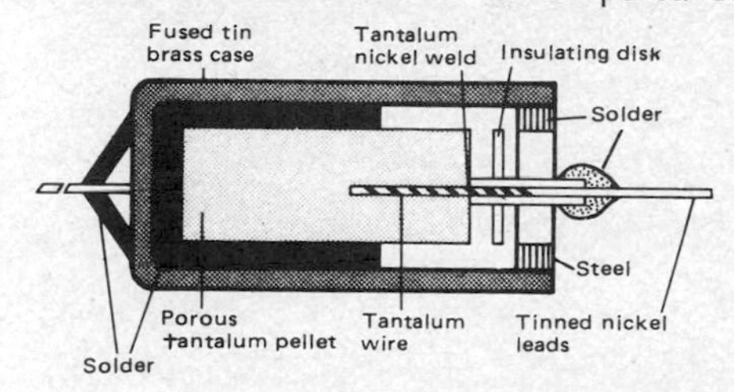

Solid tantalum capacitor.

cathode is a layer of semiconducting manganese dioxide that fills the pores of the anode over the dielectric. Polarity of connections must be observed, just as with other electrolytic capacitors.

solion [SOLution ION] An electrochemical device in which amplification is obtained by controlling and monitoring a reversible electrochemical reaction. In the tetrode solion, four inert elec-

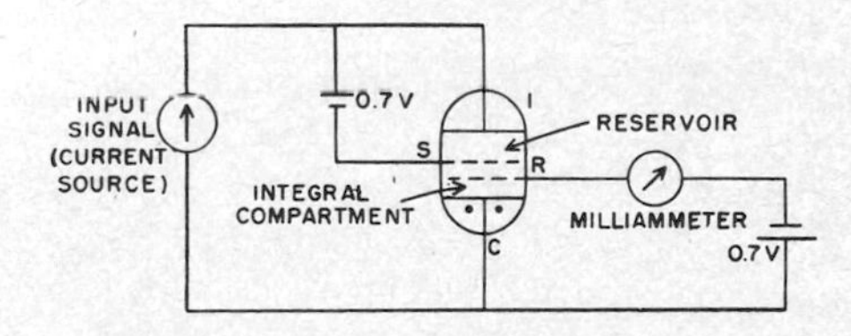

Solion connected as integrator, in which output current is proportional to integral of input current.

trodes are immersed in an electrolyte. By controlling the charge transferred between the two input electrodes, the conductivity between the two output electrodes is made to vary in proportion to the input current.

solution ceramic A nonbrittle inorganic ceramic insulating coating that can be applied to wires at a low temperature. Examples include ceria, chromia, titania, and zirconia.

Sommerfeld's equation An equation for ground-wave propagation that relates field strength at the surface of the earth at any distance from a transmitting antenna to the field strength at unit distance for given ground losses.

sonar [SOund Navigation And Ranging] Equipment used for sonic and ultrasonic underwater detection, ranging, sounding, and communication. The most common type is echo-ranging sonar, an active sonar in which a sonic or ultrasonic pulse is transmitted, reflected from an object, and received back at the transmitter location. The elapsed time for the pulse gives target distance, and the directional characteristics of the transmitting-receiving transducer give target range. For underwater communication the pulses may be spaced according to international Morse code. Other versions are passive, scanning, and searchlight sonar. British term is asdic. Also called sonar set.

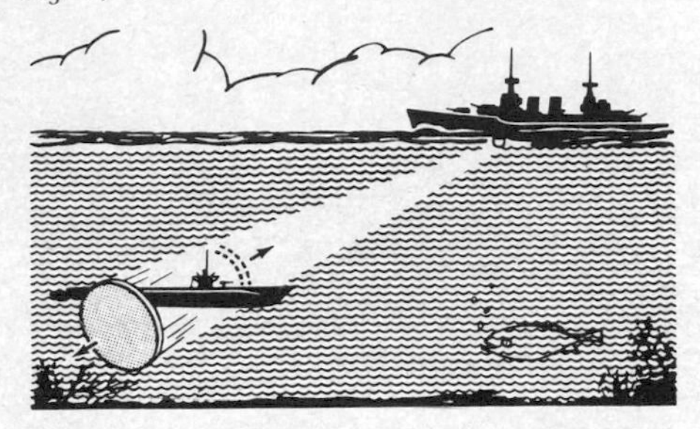

Sonar used on ship for determining range and bearing to submarine.

sonaramic indicator An indicator connected to a sonar receiver to present visually the signals received over a band of frequencies centered about the specific frequency to which the sonar receiver is tuned.

sonar attack plotter A system that coordinates information from a sonar installation, a ship's gyrocompass, and related devices, and presents graphically the information needed to plan an antisubmarine attack.

sonar beacon An underwater beacon that transmits sonic or ultrasonic signals, to provide bearing information. It may have receiving facilities that permit triggering by an external source.

sonar data computer A computer that calculates two or more factors of sonar data, such as range, bearing, depth, and sound velocity.

sonar dome A streamlined watertight enclosure that provides protection for a sonar transducer, sonar projector, or hydrophone and associated equipment, and offers minimum interference to sound transmission and reception.

sonar modulator A modulator that varies the frequency of a signal generated by a sonar transmitter.

sonar projector An electromechanical device used under water to convert electric energy to sound energy. A crystal or magnetostriction transducer is usually used for this purpose.

sonar receiver A receiver that intercepts and amplifies the sound signals reflected by an underwater target and displays the accompanying intelligence in useful form. It may also pick up other underwater sounds.

sonar receiver-transmitter A single piece of equipment that combines the functions of generating energy for sonic or ultrasonic underwater transmission and receiving the resulting echo signals. Used primarily for detection and ranging. Secondary functions may include underwater communication.

sonar resolver A resolver used with echo-ranging and depth-determining sonar to calculate and record the horizontal range of a sonar target, as required for depth-bombing.

sonar set *Sonar.*

sonar signal simulator A signal generator that feeds synthetic signals to a sonar receiver or directly to its indicator to give a presentation similar to that observed under actual operating conditions.

sonar sounding set A sonar set that determines depth of water from the point of installation on a vessel to the ocean bottom or to the surface of the ocean.

sonar trainer A trainer designed as a teaching aid to develop skills in sonar techniques.

sonar train mechanism A mechanism that rotates a sonar projector or transducer. It may also have facilities for tilting the transducer or projector.

sonar transducer A transducer used under water to convert electric energy to sound energy and sound energy to electric energy.

sonar transducer scanner A circuit or switch that permits sampling the output signals of individual elements or groups of elements in certain types of sonar transducers, to determine the direction of arriving signals.

sonar transmitter A transmitter that generates electric signals of the proper frequency and form for application to a sonar transducer or sonar projector, to produce sound waves of the same frequency in water. The sound waves may carry intelligence.

sonar window The portion of a sonar dome or sonar transducer that passes sound waves at sonar frequencies with little attenuation while providing mechanical protection for the transducer.

sonde An instrument that obtains weather data during ascent and descent through the atmosphere, in a form suitable for telemetering to a ground station by radio as in a radiosonde.

sone A unit of loudness. By definition, a simple 1-kHz tone that is 40 dB above a listener's threshold of hearing produces a loudness of 1 sone. A sound that is judged by the listener to be *n* times as loud as that of the 1-sone tone has a loudness of *n* sones. A millisone is equal to 0.001 sone.

sonic Pertaining to sound waves and the speed of sound.

sonic altimeter An instrument that determines the height of an aircraft above the earth by measuring the time taken for sound waves to travel from the aircraft to the surface of the earth and back to the aircraft again. The method is based on sound that has a velocity of 1,080 ft/s (329 m/s) through dry air at 32°F (0°C).

sonic applicator An electromechanical transducer for local application of sound for therapeutic purposes, such as for treatment of muscular ailments.

sonic barrier The turbulence encountered by an aircraft as its speed approaches the speed of sound. Also called sound barrier and transonic barrier.

sonic boom An explosionlike sound heard when a shock wave, generated by an aircraft flying at supersonic speed, reaches the ear.

sonic cleaning Cleaning of contaminated materials by the action of intense sound in the liquid in which the material is immersed.

sonic delay line *Acoustic delay line.*

sonic depth finder *Fathometer.*

sonic drilling The process of cutting or shaping materials with an abrasive slurry driven by a reciprocating tool attached to an AF electromechanical transducer.

sonic flaw detection The process of locating imperfections in solid materials by observing internal reflections or a variation in transmission through the materials as a function of sound-path location.

sonic frequency *Audio frequency.*

sonic mine *Acoustic mine.*

sonics The use of sound in any noncommunication process.

sonic speed *Speed of sound.*

sonic surgery The use of focused energy to produce precisely circumscribed alterations at predetermined sites within tissue.

sonic viscometry Determination of the coefficients of viscosity of liquids or slurries by measurement of the acoustic properties of a transmitted wave or by the reaction of such a medium on a transducer.

sonne *Consol.*

sonobuoy An acoustic receiver and radio transmitter mounted in a buoy that can be dropped from an aircraft by parachute to pick up underwater sounds of a submarine and transmit them to the aircraft. To track a submarine, several buoys are dropped in a pattern that includes the known or suspected location of the submarine, with each buoy transmitting an identifiable signal. An electronic computer then determines the location of the submarine by comparison of the received signals and triangulation of the resulting time-delay data. Also called radio sonobuoy.

sonobuoy trainer A trainer that gives aircraft crews practical experience in locating submarines by simulating the tactics necessary to obtain a fix.

sonoencephalograph *Echoencephalograph.*

sonography The use of sonic or ultrasonic energy for imaging of biological tissues, as in studying the progress of the fetus in pregnancy, determining the size of tumors, and diagnosis of other pathological conditions.

sonoluminescence Creation of light in liquids by sonically induced cavitation.

sort To arrange items of information in a computer according to specified rules.

sorter A machine that sorts punched cards according to the punches in a specified column of the card.

SOS 1. The distress signal in radiotelegraphy, consisting of the run-together letters S, O, and S of the international Morse code. 2. Abbreviation for *silicon-on-sapphire.*

sound 1. The sensation of hearing, produced when sound waves act on the brain through the nerves of the ears. The extreme frequency limits for human hearing are from about 15 to 20,000 Hz, but animals can hear much higher frequencies. Also called audio (slang) and sound sensation. 2. A vibration in an elastic medium at any frequency that produces the sensation of hearing. This vibration is propagated by the medium as a sound wave. Frequencies that produce sound are called sound or audio frequencies. Frequencies below the audio range are infrasonic, and frequencies above the audio range are ultrasonic. Sound travels at about 1100 ft/s (335 m/s) in air at sea level and about 4800 ft/s (1463 m/s) in water.

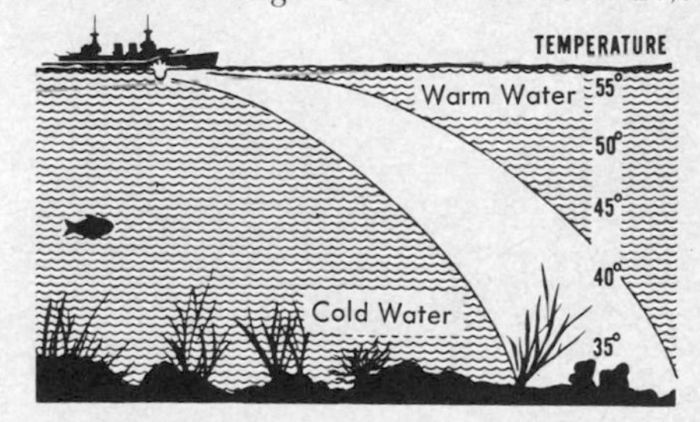

Sound beam from surface vessel is bent downward when water temperature decreases with depth because speed of sound increases with water temperature.

sound absorber A material that absorbs a high percentage of incident sound energy.

sound absorption The process by which sound energy is diminished in passing through a medium or striking a surface.

sound absorption coefficient The ratio of sound energy absorbed to that arriving at a surface or medium. Also called acoustic absorption coefficient and acoustic absorptivity.

sound analyzer A device that measures the levels of the components of a complex sound as a function of frequency.

sound articulation The percent articulation obtained when speech units are fundamental sounds, usually combined into meaningless syllables.

sound bar One of the two or more alternate dark and bright horizontal bars that appear in a television picture when AF voltage reaches the video input circuit of the picture tube.

sound barrier *Sonic barrier.*

sound box *Acoustic pickup.*

sound carrier The television carrier that is frequency-modulated by the sound portion of a television program. The unmodulated center frequency of the sound carrier is 4.5 MHz higher than the video carrier frequency for the same channel.

sound channel 1. The series of stages that handles only the sound signal in a television receiver. 2. A layer of sea water that extends from about 700 m down to about 1500 m, in which sound travels at about 1500 m/s, the slowest it can travel in sea water. The density of sea water, which depends upon temperature, pressure, and salinity, is a maximum in the sound channel because salinity and temperature decrease with depth, whereas pressure increases with depth.

sound effects Mechanical devices or recordings that provide lifelike imitations of various sounds.

sound-effects filter A filter, usually adjustable, that reduces the passband of a system at low and/ or high audio frequencies to produce special effects.

sound energy Energy existing in a medium due to sound waves.

sound-energy density Sound energy per unit volume, generally expressed in ergs per cubic centimeter.

sound-energy flux The average rate of flow of sound energy for one period through any specified area, generally expressed in ergs per second.

sounder *Telegraph sounder.*

sound field A region that contains sound waves.

sound film Motion-picture film that has a sound track along one side for reproduction of the sounds which are to accompany the film.

sound-film recorder A device that converts sound signals to a modulated light beam (variable-area or variable-density) to expose film and produce sound images.

sound filmstrip A filmstrip that has accompanying sound on a separate disk or tape which is manually or automatically synchronized with projection of the pictures in the strip.

sound frequency *Audio frequency.*

sound gate The gate through which film passes in a sound-film projector for conversion of the sound track into AF signals that can be amplified and reproduced.

sound head The section of a sound motion-picture projector that converts the photographic or magnetic sound track to audible sound signals.

sound image The photographic image of a sound, as on a film sound track.

sounding 1. Any penetration of the natural environment, under water or up into the atmosphere, for scientific observation and measurement. 2. Determining depth of water.

sounding balloon A small free balloon used for carrying radiosonde equipment aloft.

sounding electrode A probe that makes measurements in a gas discharge.

sounding rocket A research rocket that obtains upper-atmosphere data.

sound intensity The sound energy transmitted per unit of time through a unit area, expressed in ergs per second per square centimeter or in watts per square centimeter.

sound level The weighted sound pressure level at a point in a sound field, determined in the manner specified by the American Standards Association. The meter reading in decibels corresponds to a value of the sound pressure integrated over the audible frequency range with a specified frequency weighting and integration time.

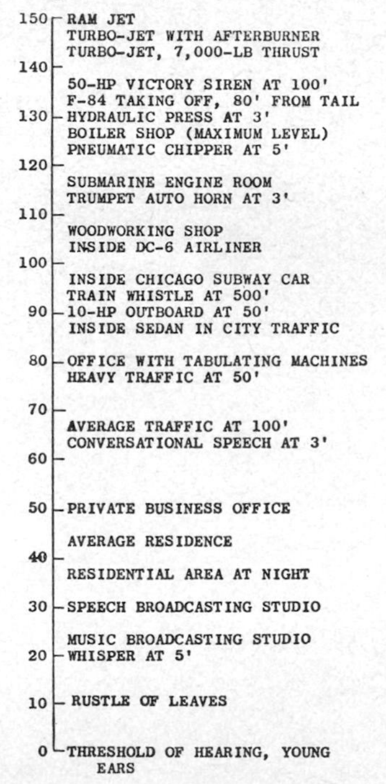

Sound levels in decibels with respect to threshold of hearing for young ears at 1000 Hz. Divide values in feet by 3.28 to change to meters.

sound-level meter An instrument that measures noise and sound levels in a specified manner. The meter may be calibrated in decibels or volume units. It includes a microphone, an amplifier, an output meter, and frequency-weighting networks. A volume-unit meter is an example.

sound power The total sound energy radiated by a source per unit time, generally expressed in ergs per second or watts.

sound-powered telephone A telephone operating entirely on current generated by the speaker's voice, with no external power supply. Sound waves cause a diaphragm to move a coil back and forth between the poles of a powerful but small permanent magnet, generating the required AF voltage in the coil.

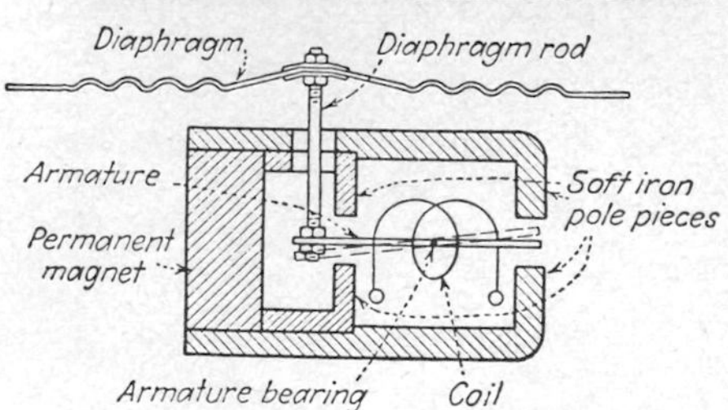

Sound-powered telephones connect together all battle stations on combat vessels, and operate even though ship's electric power system is out of commission. Diagram shows operating principle.

sound power level A value in decibels equal to 10 times the logarithm to the base 10 of the ratio of radiated sound power to a reference power.

sound pressure *Effective sound pressure.*

sound pressure level [abbreviated SPL] A value in decibels equal to 20 times the logarithm to the base 10 of the ratio of the pressure of this sound to a reference pressure. Reference pressures in common use are 0.0002 and 1 μbar.

sound probe A probe that explores a sound field without significantly disturbing the field in the region being explored. It may be a small microphone or a small tubular attachment added to a conventional microphone.

soundproofing *Damping.*

sound ranging Determining the location of a gun or other sound source by measuring travel times of the sound wave to microphones at three or more different known positions.

sound recorder A recorder that provides a permanent record of sounds, as on phonograph records, optical sound tracks, magnetic tape, or magnetic wire.

sound recording system A combination of transducing devices and associated equipment suitable for storing sound in a form capable of subsequent reproduction.

sound reflection coefficient The ratio of sound energy reflected from a surface to that reaching the surface. Also called acoustic reflection coefficient and acoustic reflectivity.

sound-reinforcing system A public-address system used in a theater or auditorium to make the sound-energy density as nearly uniform as possible throughout the audience.

sound-reproducing system A combination of transducing devices and associated equipment for picking up sound at one location and reproducing it at either the same location or some other location, at the same time or at some later time.

sound screen A motion-picture screen that has small perforations, to permit transmission of sound from loudspeakers placed behind it.

sound sensation *Sound.*

sound spectrograph An instrument that records and analyzes the spectral composition of audible sound. Applications include identification of speakers by their voice patterns.

sound spectrum A representation of the amplitudes of the components of a complex sound, arranged as a function of frequency.

sound speed The speed of sound motion-picture film, standardized at 24 frames per second. For 16-mm film this is 36 ft/min (11 m/min), and for 35-mm film it is 90 ft/min (27.4 m/min).

sound stripe A longitudinal stripe of magnetic material placed on some motion-picture films for recording a magnetic sound track.

sound takeoff The point at which the sound signal is separated from the video signal in a television receiver for separate IF amplification, demodulation, and AF amplification. Also called takeoff.

sound track A narrow band, usually along the margin of a sound film, that carries the sound record. It may be a variable-width or variable-density optical track or a magnetic track.

sound transmission coefficient The ratio of transmitted to incident sound energy at an interface in a sound medium. The value depends on the angle of incidence of the sound. Also called acoustic transmission coefficient and acoustic transmittivity.

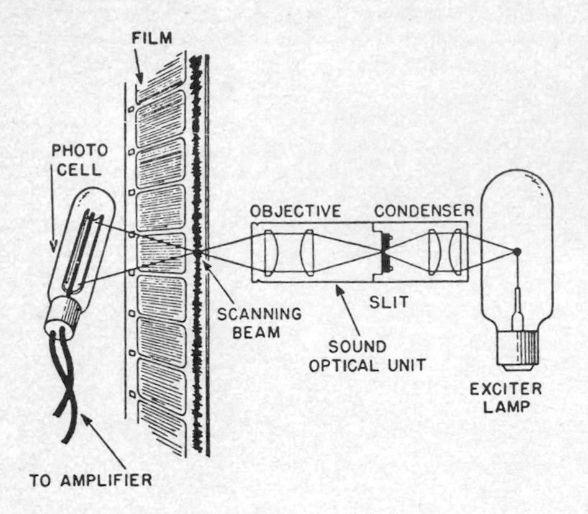

Sound track (variable-area) on motion-picture film, with lamp, optical system, and photocell for converting track to corresponding audio signal.

sound wave The traveling wave produced in an elastic medium by vibrations in the frequency range of sound. Approximate velocities of sound waves are 1100 ft/s (335 m/s) in air at 0°C, 4800 ft/s (1463 m/s) in water, and 16,400 ft/s (5000 m/s) in steel.

source 1. [symbol S] The terminal in a field-effect transistor from which majority carriers flow into the conducting channel in the semiconductor material. The source is comparable to the emitter of a bipolar transistor. 2. The circuit or device that supplies signal power or electric energy to a transducer or load circuit. 3. A radioactive material packaged so as to produce radiation for experimental or industrial use.

source-follower amplifier *Common-drain amplifier.*

source impedance The impedance presented by a source to a transducer or load circuit.

source language The language in which a problem is programmed for a computer. It must be translated into an object program in machine language by a computer routine.

source program A program that is written in a source language.

south magnetic pole The magnetic pole located approximately at 73°S latitude and 156°E longitude, about 1020 nautical miles (1900 km) north of the South Pole.

south pole The pole of a magnet at which magnetic lines of force are assumed to enter.

South Pole A geographical point on the earth that is one end of the axis about which the earth revolves.

space 1. The open-circuit condition or the signal that causes the open-circuit condition in telegraphic or teletypewriter communication. The closed-circuit condition is the mark. 2. The universe, extending from the earth's atmosphere without limit. 3. A nonprinting action in a typewriter or printer.

space attenuation The power loss of a signal traveling in free space, caused by such factors as absorption, reflection, and scattering. Usually expressed in decibels.

space charge The net electric charge distributed throughout a volume or space, such as the cloud of electrons in the space near the cathode of a thermionic vacuum tube or phototube.

space-charge-controlled tube A microwave tube that generates RF power by using a grid electrode to control a space-charge-limited current. A ceramic planar triode tube is an example.

space-charge debunching A process in which the mutual interactions between electrons in a stream spread out the electrons of a bunch.

space-charge density The net electric charge per unit volume.

space-charge effect The repulsion between the electrons emitted by the cathode of a thermionic vacuum tube and the electrons accumulated in the space charge near the cathode, resulting in a reduction in anode current.

space-charge grid A grid, usually positive, that controls the position, area, and magnitude of a potential minimum or of a virtual cathode in a region adjacent to the grid of an electron tube.

space-charge layer *Depletion layer.*

space-charge region A semiconductor region in which the net charge density is significantly different from zero.

space-charge wave An electrostatic wave in a plasma, produced by oscillating motions of charges. The waves are longitudinal in that the electric field is in the direction of propagation.

space-charge-wave tube A microwave tube in which space charges play an important role in the conversion of DC power to RF power. The two basic tube types are linear-beam and crossed-field.

spacecraft A vehicle that travels beyond the atmosphere of the earth, through outer space. A spacecraft may or may not be a satellite, and may be piloted or unpiloted.

space current The total current flowing between the cathode and all the other electrodes in a tube, including the anode current and the currents to all other electrodes.

space-diversity reception Radio reception that involves the use of two or more antennas located several wavelengths apart, feeding individual receivers whose outputs are combined. The system gives an essentially constant output signal despite fading caused by variable propagation characteristics because fading affects the spaced-out antennas at different instants of time.

space environment The environment encountered by vehicles and living creatures in space, characterized by absence of atmosphere.

space factor 1. The ratio of the space occupied by the conductors in a winding to the total cubic content or volume of the winding, or the similar ratio of cross sections. 2. The ratio of the space occupied by iron to the total cubic content of an iron core.

space pattern A geometric pattern on a test chart, used to measure geometric distortion in television equipment.

space permeability The factor that expresses the ratio of magnetic induction to magnetizing force in a vacuum. In the CGS electromagnetic system of units, the permeability of a vacuum is arbitrarily taken as unity.

space phase Reaching corresponding peak values at the same point in space.

space probe An unpiloted instrumented spacecraft that makes measurements of lunar and planetary surfaces and atmospheres, solar winds, and other phenomena in space.

space quadrature A difference of a quarter-wavelength in the position of corresponding points of a wave in space.

space research Research that involves studies of all aspects of environmental conditions beyond the atmosphere of the earth.

space service A radio communication service between space stations.

space simulator A chamber that can be hermetically sealed to simulate the environment and other conditions of space (except weightlessness and true cosmic radiation), for studying reactions of human beings and animals under conditions as close as possible to those found in spacecraft.

space station A piloted orbiting facility from which space travel, exploration, or communication may be further effected.

space suit A pressure suit with transparent airtight visor, designed to provide protection against cold and heat while maintaining a suitable atmosphere for breathing, for use if cabin pressure is lost in a spacecraft and to permit travel outside the spacecraft. Also required for walking on the moon.

Spacetrack A global system of optical, radar, and radiometric sensors tied into a computer facility, used for detecting, tracking, and cataloging all man-made objects that are orbiting the earth.

space wave The component of a ground wave that travels more or less directly through space from the transmitting to the receiving antenna. One part of the space wave goes directly from one antenna to the other; another part is reflected off the earth between the antennas.

spacing wave The signal emitted by a radiotelegraph transmitter during spacing portions of the code characters. Also called back wave.

spacistor A multiple-terminal solid-state device, similar to a transistor, that achieves high-frequency operation up to about 10 GHz by injecting electrons or holes into a space-charge layer which forces these carriers rapidly to a collecting electrode. Input and output impedances can be as high as 30 MΩ.

SPADATS [SPAce Detection And Tracking System] A powerful early-warning-radar system used for long-range detection of intercontinental enemy missiles and tracking orbiting spacecraft.

spade bolt A bolt that has a spade-shaped flattened head with a transverse hole, used to fasten shielded coils, capacitors, and other components to a chassis.

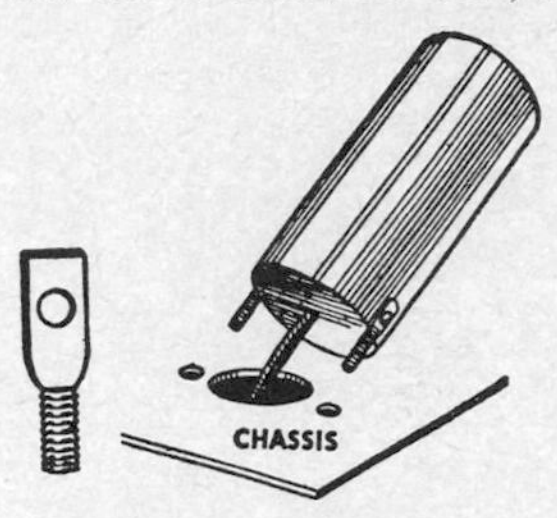

Spade bolt and typical use.

spade lug An open-ended flat termination for a wire lead, easily slipped under a terminal nut.

spaghetti Insulating tubing used over bare wires or as a sleeve for holding together two or more insulated wires. The tubing is usually made of varnished cloth or a plastic.

spallation A nuclear reaction in which the energy of each incident particle is so high that more than two or three particles are ejected from the target nucleus, and both its mass number and atomic number are changed.

S parameter One of a set of four reflection and transmission coefficients used in designing high-frequency transistor amplifiers and oscillators. Also called scattering parameter.

spark A short-duration electric discharge caused by a sudden breakdown of air or some other dielectric material that separates two terminals, accompanied by a momentary flash of light. Also called sparkover.

spark capacitor A capacitor connected between contact points to reduce sparking when the contacts open. Alternatively, the capacitor may be connected across the inductance that causes the spark. When a resistor is used in series with the spark capacitor, the network is known as a spark suppressor.

spark chamber A device that makes the tracks of nuclear particles visible during experiments. In its most common form, the trajectory of the particle is made visible by a series of short sparks that are triggered by a particle as it passes through an array of narrow spark gaps. A spark chamber may be used with or without a magnetic field.

spark coil An induction coil that produces spark discharges.

spark gap An arrangement of two electrodes between which a spark may occur. The insulation (usually air) between the electrodes is self-restoring after passage of the spark. Used as a switching device, as for protecting equipment against lightning or for switching a radar antenna from receiver to transmitter, and vice versa.

spark-gap generator A high-frequency generator in which a capacitor is repeatedly charged to a high voltage and allowed to discharge through a spark gap into an oscillatory circuit, generating successive trains of damped high-frequency oscillations.

spark-gap modulation A modulation process that produces one or more pulses of energy by a controlled spark-gap breakdown, for application to the element in which modulation takes place.

spark-gap modulator A modulator that uses a controlled spark gap to modulate a carrier.

sparking Intentional or accidental spark discharges, as between the brushes and commutator of a rotating machine, between contacts of a relay or switch, or at any other point at which a current-carrying circuit is broken.

sparking voltage The minimum voltage at which a spark discharge occurs between electrodes of given shape at a given distance apart under given conditions.

sparkover *Spark.*

spark plate A metal plate insulated from the chassis of an auto radio by a thin sheet of mica and connected to the battery lead to bypass noise signals picked up by battery wiring in the engine compartment.

spark spectrum The spectrum produced by a spark discharging through a gas or vapor. With metal electrodes, a spectrum of the metallic vapor is obtained.

spark suppressor A resistor and capacitor in series, connected between a pair of contacts to suppress sparking when the contacts open.

spark transmitter A radio transmitter that utilizes the oscillatory discharge of a capacitor through an inductor and a spark gap as the source of RF power.

Sparrow One of a series of Navy air-to-air and ship-to-air solid-fuel missiles that have various types of homing guidance.

spatial filter An optical filter that consists of a very small aperture, such as a pinhole. Used in laser cavities to change the beam shape or mode structure, and used outside lasers to change only beam shape.

SPDT Abbreviation for *single-pole double-throw.*

speaker *Loudspeaker.*

special-purpose computer A computer that solves a specific type of problem.

specific address *Absolute address.*

specific coding Digital-computer coding in which all addresses refer to specific registers and locations.

specific conductivity The conducting characteristic of a material in siemens per cubic centimeter (formerly mhos per cubic centimeter). It is the reciprocal of resistivity.

specific dielectric strength The dielectric strength per millimeter of thickness of an insulating material.

specific electronic charge The ratio of the electronic charge to the rest mass of the electron.

specific emission The rate of emission per unit area.

specific gamma-ray emission The exposure dose rate produced by the unfiltered gamma rays from a point source of a defined quantity of a radioactive nuclide at a defined distance. The unit of specific gamma-ray emission is the roentgen per millicurie hour at 1 cm.

specific ionization The number of ion pairs formed per unit distance along the track of an ion passing through matter.

specific ionization coefficient The average number of pairs of ions with opposite charges that are produced by electrons which have a specified kinetic energy when traveling a unit distance in a gas at a specified pressure and temperature.

specific permeability The permeability of a substance divided by the permeability of a vacuum. Also called relative permeability.

specific resistance *Resistivity.*

specific routine A routine expressed in computer coding to solve a particular problem, with addresses of registers and locations specifically stated.

spectral characteristic The relation between wavelength and some other variable, such as between wavelength and emitted radiant power of a luminescent screen per unit wavelength interval.

spectral color A color that appears in the spectrum of white light. The basic spectral colors are violet, blue, green, yellow, orange, and red.

spectral purity Having a single wavelength.

spectral quantum yield The average number of electrons photoelectrically emitted from a photocathode per incident photon of a given wavelength.

spectral radiant intensity The radiant intensity per unit wavelength interval, such as watts per steradian per micrometer.

spectral response *Spectral sensitivity characteristic.*

spectral selectivity The effect of radiation wavelength on the output current of a photoelectric device.

spectral sensitivity characteristic The relation between the radiant sensitivity and the wavelength of the incident radiation of a camera tube or phototube, under specified conditions of irradiation. Also called spectral response.

spectrograph A spectrometer that provides a permanent record of a spectrum of radiation.

spectrometer An instrument that disperses radiation into its component wavelengths and measures the magnitude of each component.

spectrometry The use of spectrographic techniques for deriving the physical constants of materials.

spectrophotometer An instrument that measures transmission or apparent reflectance of visible light as a function of wavelength, permitting accurate analysis of color or accurate comparison of luminous intensities of two sources at specific wavelengths.

Spectrophotometer in which motor-driven cam moves diffraction grating to make spectrum of sample sweep across phototube at known rate, for recording purposes.

spectrophotometric analysis A method of quantitative analysis based on spectral energy distribution in the absorption spectrum of a substance in solution.

spectroradiometer An instrument that measures the spectral energy distribution of any type of radiation, such as infrared radiation.

spectroscope An instrument that spreads individual wavelengths in a radiation to permit observation of the resulting spectrum.

spectroscopy The branch of physical science that deals with the measurement and analysis of visible, infrared, and ultraviolet spectra.

spectrum [plural is spectra] 1. All the frequencies used for a particular purpose. Thus the radio spectrum extends from about 10 kHz to about 300 GHz. 2. The result of dispersing an emission (such as light) in accordance with some progressive property, usually its frequency. 3. *Electromagnetic spectrum.*

spectrum analyzer An instrument that measures the amplitudes of the components of a complex waveform throughout the frequency range of the waveform. One version gives the energy at each frequency in the output of a pulsed magnetron. The instruments generally provide a cathode-ray display of amplitude versus frequency.

spectrum line A line recorded by a spectrograph to represent a specific wavelength, atomic mass, or other spectral quantity.

spectrum locus The locus of points representing the chromaticities of spectrally pure stimuli in a chromaticity diagram.

spectrum signature The spectral characteristics of the transmitter, receiver, and antenna of an electronic system, including emission spectra, antenna patterns, and other characteristics. Used in studies of electromagnetic interference created by transmitting and receiving equipment.

spectrum stripping A method of simplifying the analysis of complex pulse-height spectra by electronically subtracting spectrum components in groups.

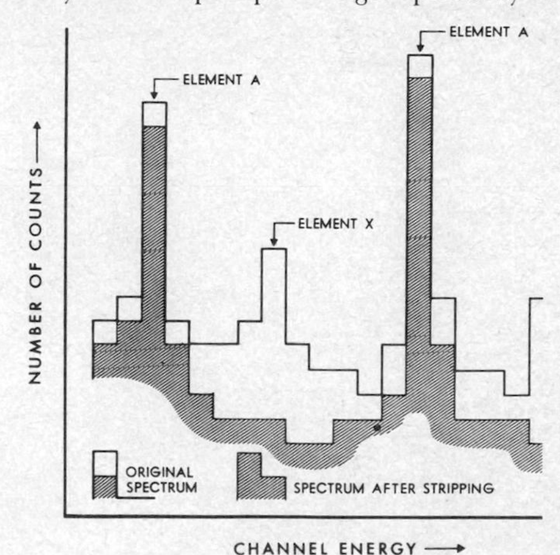

Spectrum stripping, showing how known characteristics of element X are subtracted from gamma-ray count of differential analyzer, thereby clarifying graph for element A.

specular reflection *Direct reflection.*

specular transmission Transmission in which only the emergent radiation parallel to the entrant beam is observed.

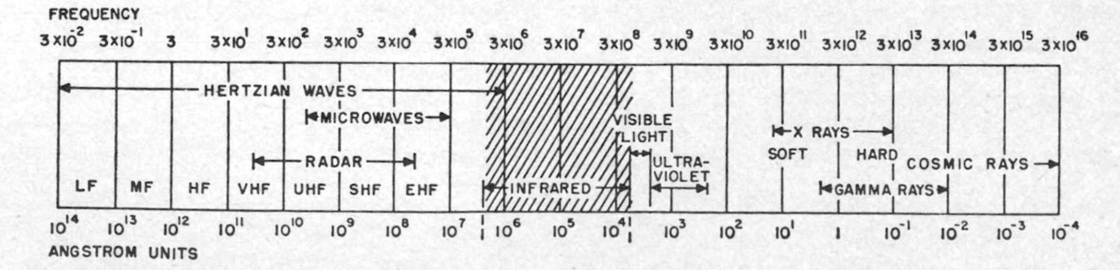

Spectrum of electromagnetic radiation, extending from 0.03 MHz at left to cosmic rays at right. Radio or Hertzian waves overlap longer infrared wavelengths. Frequency scale is in megahertz.

specular transmission density The value of the photographic density obtained when the light flux impinges normally on the sample and only the normal component of the transmitted flux is collected and measured.
speech amplifier An AF amplifier designed specifically for amplification of speech frequencies, as for public-address equipment and radiotelephone systems.
speech clipper A clipper that limits the peaks of speech-frequency signals, as required for increasing the average modulation percentage of a radiotelephone or amateur radio transmitter.
speech compression A method of eliminating certain frequency components of a speech signal to permit transmission over a narrower frequency band without appreciably affecting intelligibility.
speech digitization The conversion of analog speech waveforms to digital form.
speech frequency *Voice frequency.*
speech inverter *Scrambler.*
speech power The rate at which sound energy is being radiated by a speech source at a given instant, or the average of the instantaneous values over a given time interval.
speech processing Changing the characteristics of a voice signal to improve communication intelligibility. Methods include audio compression, audio clipping, RF compression, and RF clipping.
speech recognition system A system that recognizes spoken syllables, words, and phrases. It can be used for direct entry of voice data to a computer or control system.
speech scrambler *Scrambler.*
speech synthesizer *Voice synthesizer.*
speed 1. A scalar quantity equal to the magnitude of velocity. Speed thus specifies no direction. 2. The angular velocity of a rotating shaft or device, generally expressed in revolutions per minute. 3. The rate of performance of an act. 4. The aperture of a lens. 5. The exposure time of a shutter. 6. The frequency of a relaxation oscillator.
speed control 1. A control that changes the speed of a motor or other drive mechanism, as for a phonograph or magnetic-tape recorder. 2. *Hold control.*
speed of light A physical constant equal to 2.997925×10^{10} cm/s. This corresponds to 186,280 statute mi/s, 161,870 nautical mi/s, and 328 yd/μs. All electromagnetic radiation travels at this same speed in free space. The velocity of light is usually considered to be a vector quantity representing the speed of light in a particular direction.
speed of sound The speed at which sound waves travel through a given medium. In air under standard sea-level conditions, sound travels at 1100 ft/s (335 m/s). In water it is about 4800 ft/s (1463 m/s). Also called sonic speed.
speed regulator A device that maintains the speed of a motor or other device at a predetermined value or varies it in accordance with a predetermined plan.
speedup capacitor A capacitor connected to speed up the switching action of a circuit or device.
spheredop A modified DOVAP missile-tracking system in which a stable oscillator serves in place of a transponder in the missile.
sphere gap A spark gap between two equal-diameter spherical electrodes.
sphere-gap voltmeter A high-voltage voltmeter that consists of an adjustable sphere gap. The electrodes are moved together until the voltage being measured produces a spark, and the voltage is calculated from gap spacing and electrode diameter.
spherical aberration An image defect caused by the spherical form of an optical or electron lens or mirror, resulting in blurred focus and image distortion.
spherical antenna An antenna that has the shape of a sphere, used chiefly in theoretical studies.
spherical-earth factor The ratio of the electric field strength that would result from propagation over an imperfectly conducting spherical earth to that which would result from propagation over a perfectly conducting plane.
spherical faceplate A television picture tube faceplate that is a portion of a spherical surface.
spherical hyperbola The locus of the points on the surface of a sphere that has a specified constant difference in great-circle distances from two fixed points on the sphere.
spherical wave A wave whose equiphase surfaces form a family of concentric spheres. The direction of travel is always perpendicular to the surfaces of the spheres.
sphygmogram A graphic recording of human pulse waves.
sphygmograph An instrument that records the waveforms of the pulse of a patient.
sphygmomanometer An instrument that measures arterial blood pressure.
sphygmophone A microphone attached to the wrist to pick up the sounds of the pulse.
spider A highly flexible perforated or corrugated disk that centers the voice coil of a dynamic loudspeaker with respect to the pole piece without appreciably hindering in-and-out motion of the voice coil and its attached diaphragm.
spike A short-duration transient whose amplitude considerably exceeds the average amplitude of the associated pulse or signal.
spike train A regular succession of pulses, used in electrobiology.
spin The total angular momentum of a nucleus when considered as a single particle.
spindle A shaft, such as the upward-projecting shaft on a phonograph turntable, used for positioning the record.
spinel A hard crystalline oxide of magnesium

and aluminum ($MgAl_2O_4$); it has characteristics which make it suitable for use in place of sapphire as a substrate for semiconductor material.

spin-flip laser A semiconductor laser in which the output wavelength is continuously tunable by a magnetic field. Indium antimonide crystals can be used. Operation is based on exciting conduction-band electrons to a higher energy level by reversing the direction of the electrons as they spin about an axis in the direction of the magnetic field.

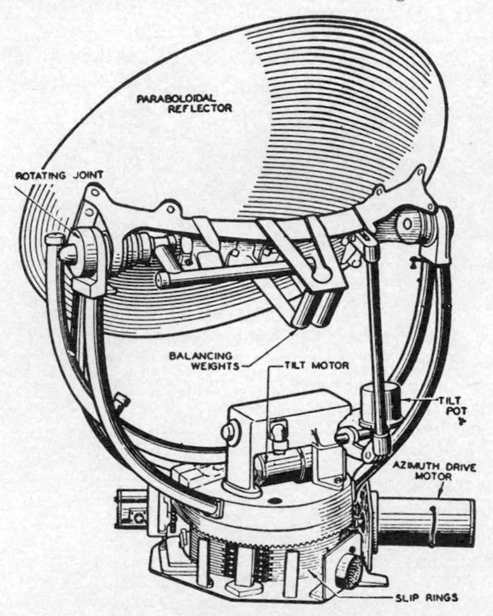

Spinner assembly for ground radar.

spinner A rotating radar antenna and reflector assembly.

spin-orbit coupling The interaction between intrinsic and orbital angular momentum of a particle.

spin precession magnetometer A magnetometer in which the proton-free precession principle is used in quantum electronic instrumentation for measuring and recording geomagnetic phenomena. Examples include the proton magnetometer and the rubidium magnetometer.

spin quantum number A number that gives the angular momentum of the electron considered as a small charged sphere rotating around an axis.

spin stabilization Rotation of a satellite on an axis that is oriented to the earth's axis, to give a stabilizing flywheel effect. If solar cells are mounted all around the periphery of the satellite, a large number of the cells are exposed to the sun for power generation. In some satellites, an inner communication package with its antennas is spun at the same speed in the opposite direction to give a platform that is stationary with respect to the earth; this permits use of high-gain directional antennas aimed at a desired receiving station on the earth.

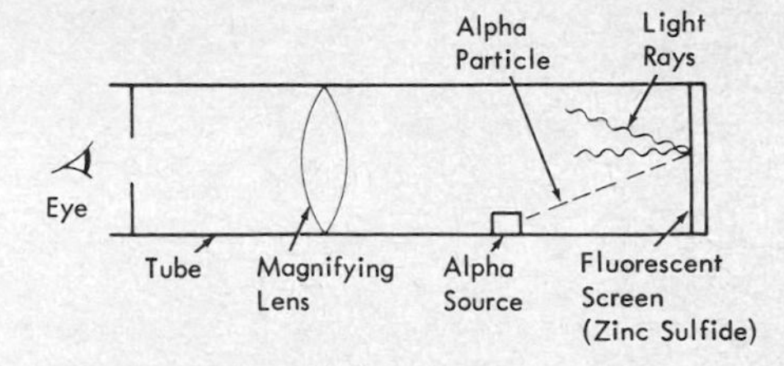

Spinthariscope cross section.

spinthariscope An instrument for viewing the scintillations of alpha particles on a luminescent screen, usually with the aid of a microscope.

spin-tuned magnetron A magnetron in which the output frequency is varied by using an external stator to rotate inside the tuning cavity a metal cylinder that has holes precisely drilled around its circumference. Rotation of the cylinder provides frequency agility for radar.

spin wave The type of wave that exists within a ferrimagnetic material when all the magnetic moments are precessing uniformly but not in phase.

spiral distortion A distortion in which image rotation varies with distance from the axis of symmetry of the electron optical system of a camera or image tube that uses magnetic focusing. Also called S distortion.

spiral-four cable *Spiral quad.*

spiral quad Four separately insulated telephone or telegraph conductors wound spirally around a supporting core. Also called spiral-four cable.

spiral scanning Scanning in which the direction of maximum radiation describes a portion of a spiral. The rotation is always in one direction. Used with some types of radar antennas.

spiral tuner A tuner that has spiral coils and a tuning mechanism which slides a contact along the spiral of each coil. The sliding contacts change the inductance of the coils and thus change the frequency. Also called continuous tuner and inductuner.

spirometer An instrument that measures the amount of air inhaled and exhaled. It may provide direct digital readouts along with analog-type recordings.

SPKR Abbreviation for *loudspeaker.*

SPL Abbreviation for *sound pressure level.*

splash baffle *Arc baffle.*

splash ring A metal shield used in an ignitron or other pool tube to prevent mercury from splashing on other electrodes.

splat cooling A method of cooling a liquid alloy by using a jet of gas to blast it against a copper surface. The liquid metal spreads out into a thin foil and cools rapidly by conduction to the copper. The atoms do not have time to align themselves into normal patterns, and the resulting new patterns provide new chemical and electrical properties. When silver and germanium are combined in this way, the new compound is superconducting.

splatter Distortion due to overmodulation of a transmitter by peak signals of short duration, particularly sounds that contain high-frequency harmonics. It is a form of adjacent-channel interference.

splice A joint that connects two lengths of conductor with good mechanical strength and good conductivity.

splicing tape A pressure-sensitive nonmagnetic tape used for splicing magnetic tape. It has a hard adhesive that will not ooze and gum up the recording head or cause adjacent layers of tape on the reel to stick together.

split-anode magnetron A magnetron that has an anode divided into two equal segments, usually by slots parallel to its axis.

split hydrophone A directional hydrophone in which each transducer or group of transducers produces a separate output voltage.

split integrator A digital-differential-analyzer integrator that can multiply its output or input by a constant.

split-phase current One of two different phases of current obtained from a single-phase AC circuit by reactances.

split-phase motor A single-phase induction motor that has an auxiliary winding connected in parallel with the main winding, but displaced in magnetic position from the main winding, to produce the required rotating magnetic field for starting. The auxiliary circuit is generally opened when the motor has attained a predetermined speed.

split projector A directional projector in which electroacoustic transducing elements are so divided and arranged that each division may be energized separately through its own electric terminals.

split rotor plate *Slotted rotor plate.*

split-series motor A DC series-connected motor that has one series field winding for each direction of rotation. Used in servomechanisms that have ON/OFF control.

split-stator variable capacitor A variable capacitor that has a rotor section which is common to two separate stator sections. Used in grid and anode tank circuits of transmitters for balancing purposes.

spoiler A rod grating mounted on a parabolic reflector to change the pencil-beam pattern of the reflector to a cosecant-squared pattern. Rotating the reflector and grating 90° with respect to the feed antenna changes from one pattern to the other.

spoiling A passive navigation countermeasure used to prevent use of a radio or radar transmission by the enemy for navigation purposes.

spoking A radar malfunction in which luminous spots continue on the screen for an abnormal length of time and form radial lines, interfering with presentations.

sponsor The advertiser who pays part or all of the cost of a television or radio program.

spontaneous emission The spontaneous decay of an excited atom into its ground-state energy level, with the energy carried off by radiation.

spontaneous fission Nuclear fission in which no particles or photons enter the nucleus from the outside. It can occur only in the heaviest elements, and it gives very long half-lives for decay.

spoofing Deceiving or misleading the enemy in electronic operations, as by continuing transmission on a frequency after it has been effectively jammed by the enemy, using decoy radar transmitters to lead the enemy into useless jamming effort, or transmitting radio messages that contain false information for intentional interception by the enemy.

sporadic E layer A layer of intense ionization that occurs sporadically within the E layer. It is variable in time of occurrence, height, geographical distribution, penetration frequency, and ionization density.

sporadic reflection *Abnormal reflection.*

spot 1. The luminous area produced on the viewing screen of a cathode-ray tube by the electron beam. 2. A commercial announcement of short duration, inserted in programs or broadcast between programs.

spot gluing Applying heat to a glued assembly by dielectric heating to make the glue set in spots that are more or less regularly distributed.

spot jamming Jamming of a specific channel or frequency.

spot-optimizer magnet A permanent magnet that resembles that of an ion trap, used on the neck of some color picture tubes to provide an adjustment for optimum picture detail.

spot projection An optical method of scanning or recording in which the scanning or recording spot is defined in the path of the reflected or transmitted light.

spot size The cross section of an electron beam at the screen of a cathode-ray tube.

spot-size error An error in interpreting a radarscope presentation; it occurs when the spot on the screen is so large that two or more objects appear as one.

spot speed The speed of a scanning or recording spot within the available line. Also called scanning speed.

spot welding 1. Resistance welding in which the fusion is limited to a small area. 2. Use of dielectric heating to join together sheets of thermoplastic material at a number of spots.

spot wobble An oscillating movement of an electron beam and its resultant spot on the screen, produced intentionally to suppress the pattern of horizontal lines across the picture.

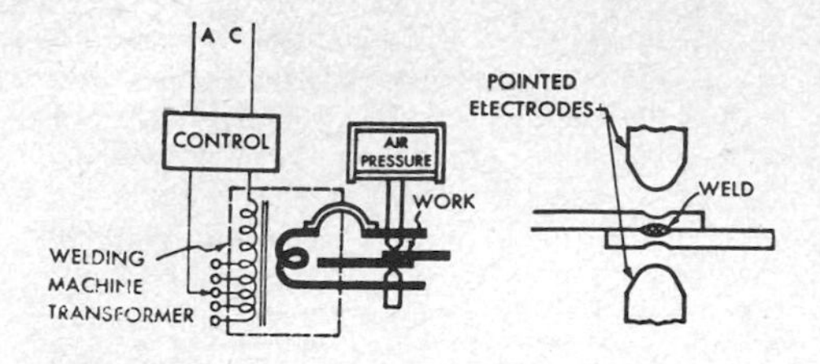

Spot-welding machine and nature of weld.

sprayed printed circuit A printed circuit formed by spraying particles of metal in molten or gaseous form onto an insulating base.

spray point One of the sharp points arranged in a row and charged to a high DC potential, used to charge and discharge the conveyor belt in a Van de Graaff generator.

spread The range within which the values of a variable quantity occur.

spreader An insulating crossarm that holds apart the wires of a transmission line or multiple-wire antenna.

spreading resistance The portion of the resistance of a point-contact rectifier that is due to the semiconducting material alone, not including the barrier-layer resistance.

spread-spectrum system A radio or radar system in which the bandwidth occupied is intentionally spread out many times that of the information content (more than 10 times as wide) for such reasons as security and privacy, antijamming security, and utilization of signals buried in noise. Spreading may take the form of frequency-hopping, time-dodging, or frequency-time dodging in a predetermined manner.

spring contact A relay or switch contact mounted on a flat spring, usually of phosphor bronze.

spring-return switch A switch in which the contacts return to their original positions when the operating lever is released.

Sprint A surface-to-air high-acceleration antimissile missile that has nuclear warhead capability.

sprocket pulse 1. The pulse generated by the magnetized spot that accompanies every character recorded on magnetic tape. This pulse is used during read operations to regulate timing of read circuits and provide a count of the number of characters read from tape. 2. A pulse generated by the feed hole in paper tape, to serve as the timing pulse for reading or punching the paper tape.

SPST Abbreviation for *single-pole single-throw.*

spurious count A count from a radiation counter other than background counts and those directly caused by ionizing radiation. Also called spurious tube count.

spurious emission *Spurious radiation.*

spurious modulation Undesired modulation that occurs in an oscillator, such as frequency modulation caused by mechanical vibration.

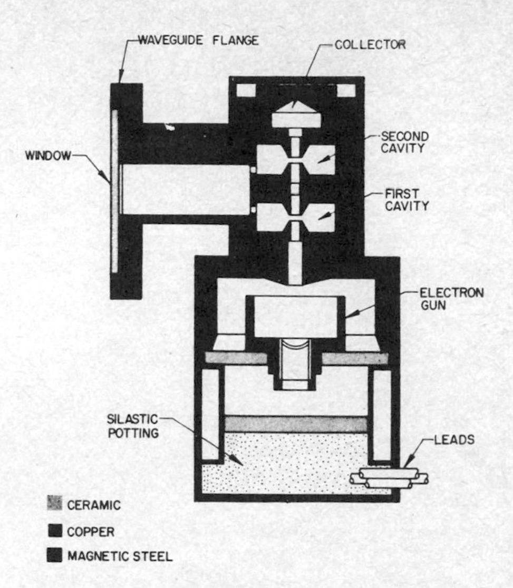

Spurious modulation is minimized by using heavy steel and copper walls in this nongridded two-cavity klystron to suppress mechanical vibration.

spurious pulse A pulse in a radiation counter other than one purposely generated or directly caused by ionizing radiation.

spurious pulse mode An unwanted pulse mode formed by the chance combination of two or more pulse modes, and indistinguishable from a pulse interrogation or a pulse reply by a transponder.

spurious radiation Any emission from a radio transmitter at frequencies outside its frequency band. Also called spurious emission.

spurious response Response of a radio receiver to a frequency different from that to which the receiver is tuned.

spurious response ratio The ratio of the field strength at the frequency that produces a spurious response to the field strength at the desired frequency, each field being applied in turn to the receiver under specified conditions to produce equal outputs. Image ratio and IF response ratio are special forms of spurious response ratio.

spurious signal An unwanted signal generated in the equipment itself, such as spurious radiation or undesired shading signals generated in a television camera tube.

spurious tube count *Spurious count.*

sputtering A method of depositing a thin layer of metal on a glass, plastic, metal, or other surface in a vacuum. The object to be coated is placed in a large demountable vacuum tube that has a cathode made of the metal to be sputtered. The tube is operated under conditions that promote cathode bombardment by positive ions. As a result, extremely small particles of molten metal fall uni-

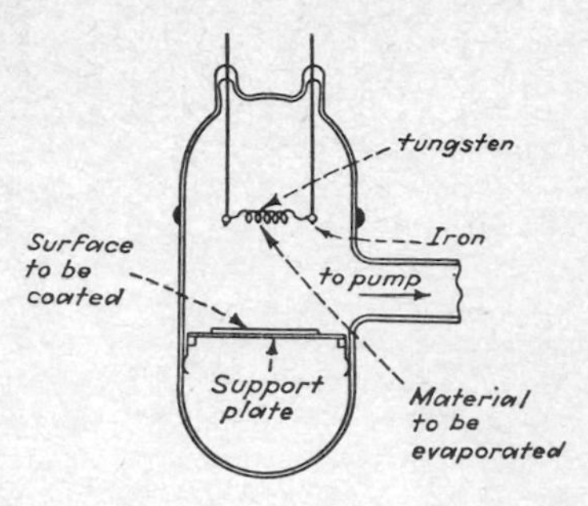

Sputtering setup.

formly on the object and produce on it a thin, conductive metal coating. The action occurs to some extent in ordinary electron tubes but is undesirable there because the positive ions knock small particles of the coating off the cathode. Also called cathode sputtering.

SQ Trademark of CBS Laboratory for their matrix-type quadraphonic sound system.

square-core oscillator An oscillator in which the core of the feedback transformer has essentially a square-loop characteristic. Used to generate square-wave pulses.

square-foot unit of absorption *Sabin.*

square-law demodulator *Square-law detector.*

square-law detection Detection in which sinusoidal input gives an output proportional to the square of the input.

square-law detector A demodulator whose output voltage is proportional to the square of the amplitude-modulated input voltage. Also called square-law demodulator.

square-loop ferrite A ferrite that has an approximately rectangular hysteresis loop.

squareness ratio The ratio of the flux density at zero magnetizing force to the maximum flux density for a material in a symmetrically cyclically magnetized condition. The ratio alternatively may be based on a magnetizing force halfway between zero and its negative limiting value.

squarer *Squaring circuit.*

square-rooting circuit A circuit that produces an output voltage which is proportional to the square root of the input voltage.

square wave A wave that alternately assumes two different fixed values for equal lengths of

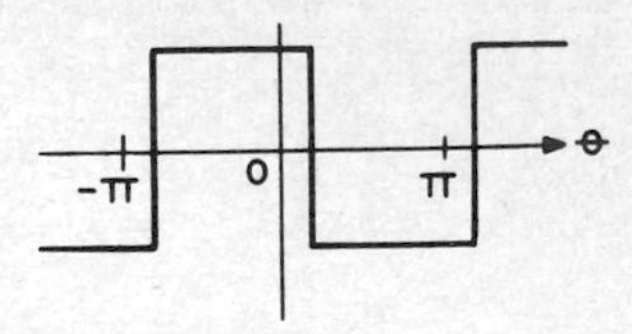

Square wave.

time, the time of transition being negligible in comparison with the duration of each fixed value. A square wave has no even harmonics. A wave with different times for the two fixed values is known as a rectangular wave.

square-wave generator A signal generator that generates a square-wave output voltage.

square-wave modulator A modulator that delivers a square-wave AF output voltage, generally at a frequency of 1 kHz, for modulating RF signal sources such as klystrons.

square-wave testing Use of a series of related step functions to determine the performance characteristics of a device or system.

squaring circuit 1. A circuit that reshapes a sine or other wave into a square wave. 2. A circuit that contains nonlinear elements and produces an output voltage proportional to the square of the input voltage. Also called squarer.

squashing A method of applying a single-crystal semiconductor film on a substrate by passing the two-layer sandwich between two heated rollers, to

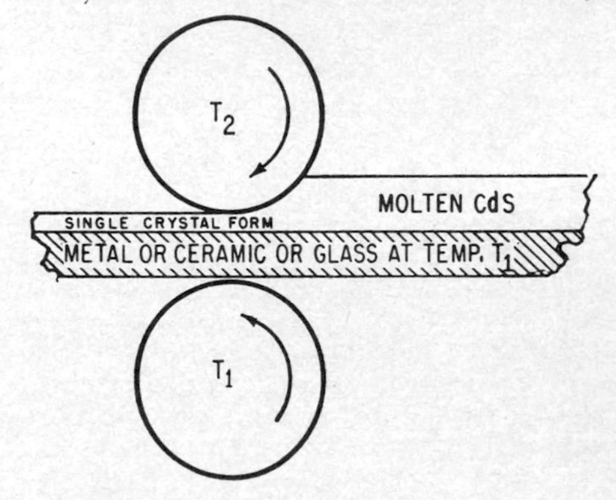

Squashing technique for producing single crystal of cadmium sulfide as thin film on substrate.

produce the controlled temperature gradient needed for single-crystal growth.

squealing A condition in which a radio receiver produces a high-pitched note or squeal along with the desired radio program, due to interference between stations or oscillation in some receiver circuit.

squeezable waveguide A section of waveguide whose width can be altered periodically to shift the phase of the RF wave traveling through it. Also called squeeze box.

squeeze box *Squeezable waveguide.*

squeeze time The time interval between the initial application of electrode force to the work and the first application of current in spot welding.

squeeze track A variable-density sound track whose width is varied by the recording operator to provide an overriding control on the amplitude of the reproduced signal for improving the signal-to-noise ratio.

squegging The condition wherein the start and

stop of oscillation is determined by the charging and discharging action of a capacitor-resistor combination in the grid circuit of a vacuum-tube oscillator.

squegging oscillator *Blocking oscillator.*

squelch To quiet a receiver automatically by reducing its gain in response to a specified characteristic of the input, as by reducing gain to suppress background noise when there is no input signal.

squelch circuit *Noise suppressor.*

squib A small, electrically activated pyrotechnic device that fires the igniter in a rocket or is used for some similar purpose. A squib does not explode as does a detonator.

SQUID [Superconducting QUantum Interference Device] A superconducting ring that couples with one or two junctions. Applications include high-sensitivity magnetometers, near-magnetic-field antennas, and measurement of very small currents or voltages.

squint 1. The angle between the two major lobe axes in a radar lobe-switching antenna. 2. The angular difference between the axis of radar antenna radiation and a selected geometric axis, such as the axis of the reflector. 3. The angle between the full-right and full-left positions of the beam of a conical-scan radar antenna.

squirrel-cage antenna A four-bay stacked array of vertical dipoles mounted on a vertical column. Each bay is balun-fed at two points to obtain omnidirectional radiation in the horizontal plane.

squirrel-cage induction motor An induction motor in which the secondary circuit consists of a squirrel-cage winding arranged in slots in the iron core.

squirrel-cage magnetron A magnetron in which the anode consists of spaced bars concentric with and parallel to the axis of the cathode.

squirrel-cage winding A permanently short-circuited winding, usually uninsulated, that has its conductors uniformly distributed around the periphery of the rotor and joined by continuous end rings.

squitter Random intentional or unintentional triggering of a transponder transmitter in the absence of interrogation, generally by noise signals.

sr Abbreviation for *steradian.*

SRAM Abbreviation for *short-range attack missile.*

SR flip-flop *RS flip-flop.*

SRT flip-flop *RST flip-flop.*

SSB Abbreviation for *single-sideband.*

SSB/AM Single-sideband operation with amplitude modulation. Also called AM/SSB.

SSI Abbreviation for *small-scale integration.*

SS loran Abbreviation for *sky-wave-synchronized loran.*

SSR 1. Abbreviation for *secondary surveillance radar.* 2. Abbreviation for *solid-state relay.*

SSTV Abbreviation for *slow-scan television.*

stability 1. Freedom from undesired variations. 2. Ability to develop restoring forces that are equal to or greater than the disturbing forces in a control system, so equilibrium is restored. 3. Freedom from undesired oscillation.

stabilization 1. Maintenance of a desired orientation independent of the roll and pitch of a ship or aircraft. 2. Treatment of a magnetic material to improve the stability of its magnetic properties.

stabilized feedback *Negative feedback.*

stabilized flight Flight in which control information is obtained from inertia-stabilized references such as gyroscopes.

stabilized platform A platform whose attitude is maintained by two or more gyroscopes and associated servosystems, despite the pitch and roll of a vehicle in space, on water, or in water. Used to support radar antennas, sonar transducers, and inertial guidance systems.

stabistor A diode component that has closely controlled conductance, controlled storage charge, and low leakage, as required for clippers, clamping circuits, bias regulators, and other logic circuits which require tight voltage-level tolerances. In one form, a stabistor consists of one to three planar, epitaxial, passivated, diode pellets mounted in a special heatsink package.

stable Incapable of spontaneous changes in atomic or nuclear systems. A stable nuclide is one that is not radioactive.

stable element A navigation instrument or device that maintains a desired orientation independently of the motion of the vehicle.

stable isotope 1. A nonradioactive isotope of an element that also has radioactive isotopes. 2. A mixture of isotopic nonradioactive nuclides different in composition from any natural mixture. 3. Any stable nuclide.

stable orbit The constant-radius circular path of accelerated particles in a betatron or synchrotron. Also called equilibrium orbit.

stack 1. To assign different altitudes by radio to aircraft awaiting their turns to land at an airport. 2. A sonar equipment assembly in the sound room of a ship. 3. A portion of a computer memory that is allocated for storage of temporary information which will be needed at some subsequent point in the program. 4. *Pileup.*

stacked array An array in which the antenna elements are stacked one above the other and connected in phase to increase the gain.

stacked dipoles Two or more dipole antennas arranged above each other on a vertical supporting structure and connected in phase to increase the gain.

stacked heads *In-line heads.*

stacked stereophonic tape *In-line stereophonic tape.*

stage A circuit that contains a single section of an electron tube or equivalent device, or two or more similar sections connected in parallel, push-pull, or push-push. It includes all parts connected between the control-grid input terminal of the device and the input terminal of the next adjacent stage.

stage-by-stage elimination method A method of locating trouble in receivers by checking performance of one stage after another with a test signal introduced by a signal generator.

staggered circuits Adjacent circuits that are alternately tuned to two different frequencies to obtain broadband response, as in a video IF amplifier.

staggered tuning Alignment of successive tuned circuits to slightly different frequencies, to widen the overall amplitude-frequency response curve.

staggering Adjusting tuned circuits to give staggered tuning.

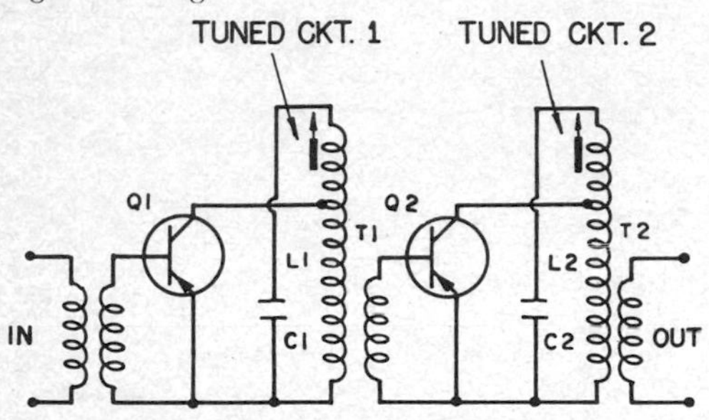

Stagger-tuned amplifier for 450-kHz IF value, with first stage tuned to 443 kHz and second stage to 457 kHz.

stagger-tuned amplifier An amplifier that uses staggered tuning to give a wide bandwidth.

staircase generator A signal generator whose output voltage increases in steps, making its output waveform on an oscilloscope screen have the appearance of a staircase.

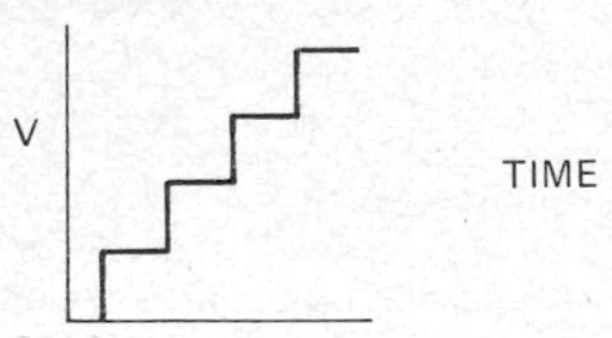

Staircase-generator waveform.

STALO [STAble Local Oscillator] A highly stable local RF oscillator used for heterodyning signals to produce an intermediate frequency in radar moving-target indicators. Only echoes that have changed slightly in frequency due to reflection from a moving target produce an output signal.

STALO cavity A cavity resonator used with a klystron oscillator to stabilize the output frequency.

stamped printed circuit A type of printed circuit formed by die-stamping a foil or film to embed the conductive pattern in an insulating base.

stamper A negative, generally made of metal by electroforming, used for molding phonograph records.

stamping A transformer lamination that has been cut out of a strip or sheet of metal by a punch press.

standard 1. A reference used as a basis for comparison or calibration. 2. A concept that has been established by authority, custom, or agreement, to serve as a model or rule in the measurement of a quantity or the establishment of a practice or procedure.

standard atmosphere An arbitrary atmosphere used in comparing performance of aircraft. For official U.S. government use, the standard atmosphere is based on the assumptions that air is a dry perfect gas, the ground temperature is 59°F (15°C), the temperature gradient in the troposphere is 0.003566°F/ft (6.5°C/km), and the temperature in the stratosphere is −67°F (−55°C). The atmospheric pressure at sea level is then 29.92 inHg (760 mmHg).

standard broadcast band *Broadcast band.*

standard broadcast channel The band of frequencies occupied by the carrier and two sidebands of a broadcast signal, with the carrier frequency at the center. Carrier frequencies are spaced 10 kHz apart, starting at 540 kHz and going up to 1,600 kHz.

standard cable A cable of particular size and construction, used as a reference for specifying transmission line losses. The standard cable has a conductor that weighs 20 lb/mi (5.7 kg/km), a loop resistance of 88 Ω/mi (55 Ω/km), a capacitance of 54 nF/mi (34 nF/km), an inductance of 1 mH/mi (0.6 mH/km), and an attenuation constant of 0.103.

standard candle The unit of candlepower, equal to a specified fraction of the visible light radiated by a group of 45 carbon-filament lamps preserved at the National Bureau of Standards, when the lamps are operated at a specified voltage. The standard candle was originally the amount of light radiated by a tallow candle of specified composition and shape.

standard capacitor A capacitor whose capacitance value is not likely to vary with temperature and is known to a high degree of accuracy. Also called capacitance standard.

standard cell A primary cell whose voltage is accurately known and remains sufficiently constant for instrument-calibration purposes. The Weston standard cell has a voltage of 1.018636 V at 20°C.

standard deviation The RMS value of the deviations of a series of like quantities from their mean.

standard-frequency service A radio communication service that involves the transmission of standard and specified frequencies of known high accuracy, intended for general reception.

standard-frequency signal One of the highly

accurate signals broadcast by the National Bureau of Standards radio station WWV on 2.5, 5, 10, 15, 20, 25, 30, and 35 MHz at various scheduled times. Used for testing and calibrating radio equipment throughout the world.

standard inductor An inductor (coil) that has high stability of inductance value, with little variation of inductance with current or frequency and with a low temperature coefficient. It may have an air core or an iron core. Used as a primary standard in laboratories and as a precise working standard for impedance measurements. Also called inductance standard.

standardize To adjust the exponent and coefficient of a floating-point result in a digital computer so the coefficient lies in the normal range of the computer.

standard loran Basic loran, in which medium-frequency pulses from transmitting stations are synchronized by ground waves rather than sky waves, as in sky-wave-synchronized loran.

standard microphone A microphone whose response is accurately known for the condition under which it is to be used.

standard noise temperature The temperature of 290 K (27°C) that is used in evaluating the noise factor of a signal-transmission system.

standard observer A hypothetical observer who requires standard amounts of primaries in a color mixture to match every color. The present standard primaries and the standard amounts required to match various wavelengths of the spectrum were established in 1931 by the International Commission on Illumination.

standard pitch A musical pitch based on 440 Hz for tone A. With this standard the frequency of middle C is 261.6 Hz.

standard propagation The propagation of radio waves over a smooth spherical earth of uniform dielectric constant and conductivity, under conditions of standard refraction in the atmosphere.

standard refraction The refraction that would occur in an idealized atmosphere in which the index of refraction decreases uniformly with height at the rate of 39×10^{-6} per kilometer.

standard resistor A resistor that is adjusted with high accuracy to a specified value and is but slightly affected by variations in temperature. Also called resistance standard.

standard sea-water conditions Sea water at a static pressure of 1 atmosphere, a temperature of 15°C, and a salinity such that the velocity of propagation is exactly 1500 m/s. Density is then 1.02338 g/cm^3, characteristic acoustic impedance is 153.507 CGS units, and pressure spectrum level of thermal noise is 82.17 dB below 1 μbar. The velocity of sound increases 0.018 m/s per meter of depth.

standard source A light source that consists of a segment of fused thoria immersed in a chamber of molten platinum. When the platinum is at its melting point, the light emitted from the chamber approximates blackbody radiation.

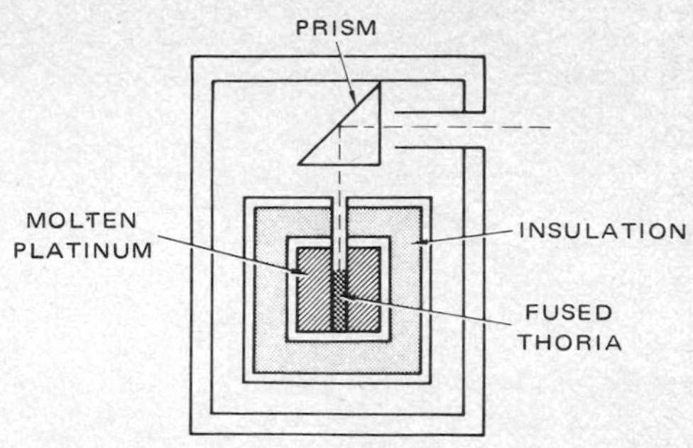

Standard source for radiation energy, as adopted by international agreement.

standard temperature and pressure A temperature of 0°C and a pressure of 760 mmHg.

standard time Mean solar time, based on the transit of the sun over a specified meridian called the time meridian, and adopted for use over an area called a time zone.

standard track *Single track.*

standard tuning frequency The frequency of 440 Hz, corresponding to the note A_4.

standard volume indicator A volume indicator that has the characteristics specified by American National Standards Association.

stand by A request to wait for additional messages to be transmitted a short time later.

standby battery A storage battery held in reserve as an emergency power source in case of failure of regular power facilities at a radio station or other location.

standby transmitter A transmitter installed and maintained for use during periods when a main transmitter is out of service for maintenance or repair.

standing-on-nines carry *High-speed carry.*

standing wave A wave in which the ratio of an instantaneous value at one point to that at any other point does not vary with time. A standing wave is produced by two waves traveling in opposite directions and having the same frequency, such as a wave and its reflection from a discontinuity.

standing-wave antenna An antenna or antenna system in which the current distributions are produced by standing waves of charges on the conductors.

standing-wave detector A detector that can be

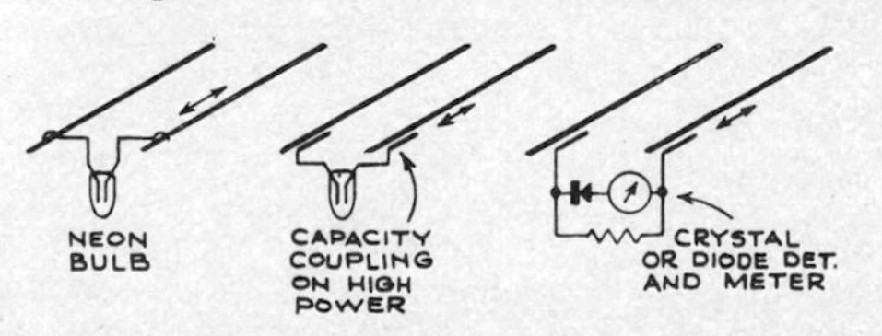

Standing-wave detectors.

moved along the length of a transmission line or waveguide to locate the nodes or antinodes of a standing wave.

standing-wave loss factor The ratio of the transmission loss in an unmatched waveguide to that in the same waveguide when matched.

standing-wave meter An indicating instrument for measuring the standing-wave ratio in a transmission line or waveguide. It may include means for finding the locations of nodes and antinodes. The detecting device is generally a bolometer, crystal diode, or thermocouple. Also called standing-wave-ratio indicator.

standing-wave ratio [abbreviated SWR] The ratio of the maximum to the minimum amplitudes of corresponding components of a field, voltage, or current along a transmission line or waveguide in the direction of propagation and at a given frequency; or, alternatively, the reciprocal of this ratio.

standing-wave-ratio bridge A bridge that measures the standing-wave ratio in a transmission line, generally to check the impedance match.

standing-wave-ratio indicator *Standing-wave meter.*

standing-wave system *Stationary-wave system.*

standing-wave voltage ratio [abbreviated SWVR] The ratio of the maximum to the minimum voltage values along a transmission line.

standoff insulator An insulator that supports a conductor at a distance from the surface on which the insulator is mounted.

star A star-shaped group of tracks made by ionizing particles originating at a common point, either in a nuclear emulsion or a cloud chamber. Some stars are produced by successive disintegrations of an atom in a radioactive series; others are produced by nuclear reactions of the spallation type, such as those due to cosmic-ray particles.

star chain A radio navigation transmitting system that comprises a master station about which three or more slave stations are symmetrically located.

star connection *Star network.*

Stark broadening Broadening or spreading of a single spectral line by random electric fields of charged particles in the plasma. These electric fields will slightly change the energy level of a radiating atom and hence cause a change in the frequency of radiation. Used to determine ion density in a high-density plasma.

star network A set of three or more branches, with one terminal of each connected at a common node to give the form of a star. Also called star connection.

starter 1. An auxiliary electrode used in a gas tube to initiate conduction. 2. A device used with one type of fluorescent lamp to preheat the cathode and then apply starting voltage to initiate conduction. 3. A controller that starts a motor and brings it up to normal speed.

starting anode An anode that establishes the initial arc in a mercury-arc rectifier.

starting current The value of electron-stream current through an oscillator at which self-sustaining oscillations will start under specified conditions of loading. Also called preoscillation current.

starting electrode An electrode that establishes a cathode spot in a pool tube.

starting rheostat A rheostat that controls the current taken by a motor during starting and acceleration.

starting voltage The minimum voltage that must be applied to a radiation-counter tube to obtain counts with a particular circuit.

star tracker A telescopic instrument, on a missile or other flight-borne object, that locks onto a celestial body and gives guidance to the missile or other object during flight. A star tracker may be optical or radiometric.

starved amplifier A direct-coupled pentode amplifier operated at unusually low screen-grid voltage; it has a high anode circuit resistance, to give high stage gain with fewer components and low current drain.

state-variable filter An active bandpass filter that consists of a summing amplifier followed by two integrators, all of which are operational amplifiers.

static 1. A hissing, crackling, or other sudden sharp sound that tends to interfere with the reception, utilization, or enjoyment of desired signals or sounds. When heard in an ordinary radio receiver it may be caused by natural electric storms or improperly operating electric devices in the vicinity. Crackling sounds heard when listening to long-playing plastic phonograph records are caused by dust particles attracted to the record by surface electric charges built up by friction on dry days. Static appears as small white specks or flashes, called snow, on a television picture. 2. Without motion or change.

static breeze *Convective discharge.*

static characteristic A relation between a pair of variables, such as electrode voltage and electrode current, with all other operating voltages for an electron tube, transistor, or other amplifying device maintained constant.

static charge An electric charge accumulated on an object.

static convergence Convergence of the three electron beams at an opening in the center of the shadow mask in a color picture tube. This is called static convergence because the beams must meet at this point when there are no scanning forces.

static converter A frequency or voltage converter that uses static switching devices (having no moving parts) like electron tubes, solid-state devices, and magnetic amplifiers.

static dump A dump that is performed at a particular time in a computer run, generally at the end of a run.

static electricity The transfer of a static charge from one object to another by actual contact or by a spark that bridges an air gap between the objects.

static eliminator A device that reduces the effect of atmospheric static interference in a radio receiver.

static focus The focus of the undeflected electron beam in a cathode-ray tube.

static frequency converter A frequency converter that has no moving parts, such as a solid-state static alternator.

static machine A machine that generates electric charges, usually by electric induction, to build up high voltages for research purposes.

static pressure The pressure that would exist at a point in a medium with no sound waves present. In acoustics, the commonly used unit is the microbar. Also called hydrostatic pressure.

static RAM A random-access memory in which data is stored in a conventional bistable flip-flop and need not be refreshed.

static storage Computer storage such that information is fixed in space and available at any time, as in flip-flop circuits, electrostatic memories, and coincident-current magnetic-core storage.

static stylus force *Stylus force.*

static subroutine A computer subroutine that involves no parameters other than the addresses of the operands.

static switching Switching of circuits by magnetic amplifiers, semiconductors, and other devices that have no moving parts.

static test A measurement taken under conditions where neither the stimulus nor the environmental conditions fluctuate.

station 1. An assembly line or assembly machine location at which a wiring board or chassis is stopped for insertion of one or more parts. 2. A location at which radio, television, radar, or other electronic equipment is installed. 3. *Broadcast station.*

stationary-anode tube An x-ray tube that has a stationary anode.

stationary battery A storage battery used only in a permanent location.

stationary contact A contact that is rigidly fastened to the frame of a switch or relay and does not move during operation.

stationary orbit A circular, equatorial, and synchronous orbit, in which the satellite appears stationary with respect to any point on the earth's surface because satellite altitude of 22,300 mi (35,-890 km) and orbital velocity of about 1.9 mi/s (3 km/s) keep it in a fixed relation to points on the earth. A stationary orbit must be synchronous, but a synchronous orbit is not necessarily stationary.

stationary satellite A satellite that has been placed in a stationary orbit.

stationary state A discrete energy state in which an electron, atom, or other quantized particle or system may exist.

stationary wave A standing wave in which amplitudes of the wave components are equal, so the energy flux is zero at all points.

stationary-wave system An interference pattern characterized by stationary nodes and antinodes. Also called standing-wave system.

statistical error An error in counting caused by random time-distributions of the disintegrations of nuclear particles.

statistical straggling A variation in range, ionization, or direction of particles that is due to fluctuations in the distance between collisions in a medium and in the energy loss and deflection angle per collision.

statistical test A procedure that determines whether observed values or quantities fit a hypothesis well enough so the hypothesis can be accepted.

stator 1. The portion of a rotating machine that contains the stationary parts of the magnetic circuit and their associated windings. 2. The stationary set of plates in a variable capacitor.

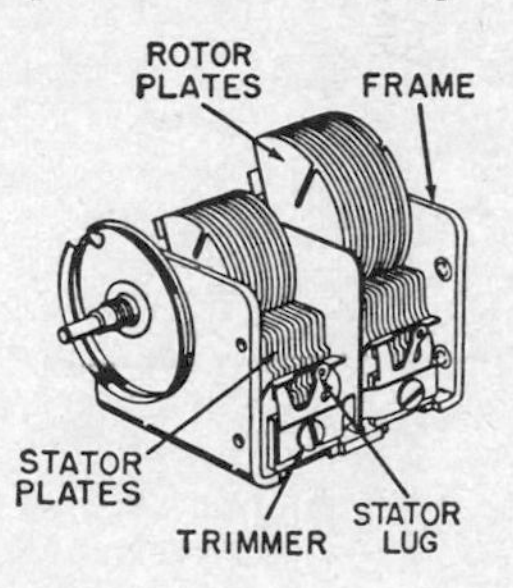

Stator plates in gang capacitor.

stator plate One of the fixed plates in a variable capacitor. Stator plates are generally insulated from the frame of the capacitor.

statute mile A unit of distance equal to 5280 ft (1.609344 km).

STC Abbreviation for *sensitivity-time control.*

steady state The condition in which circuit values remain essentially constant, after initial transients or fluctuating conditions have disappeared.

steady-state error The error that remains after transient conditions have disappeared in a control system.

steatite A dense, nonporous heat-resisting ceramic that consists chiefly of a silicate of magnesium; it has excellent insulating properties, even at high frequencies. It is molded and fired in various shapes for tube sockets and insulators.

steel-tank rectifier A mercury-arc rectifier that has a steel tank as an envelope.

steerable antenna A directional antenna whose major lobe can be readily shifted in direction.

Stefan-Boltzmann law The total emitted radiant energy of a blackbody is proportional to the fourth power of its absolute temperature.

Steinmetz formula An empirical formula for the magnetic hysteresis loss per unit volume of material per magnetization cycle, specifying that the energy loss in ergs is proportional to the 1.6th power of the maximum flux density.

stellarator [STELLAR generATOR] An instrument that studies controlled thermonuclear reactions, in which ionized gases in a glass tube sur-

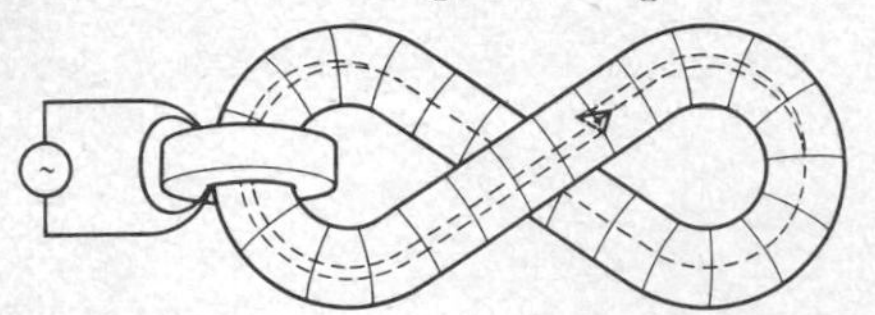

Stellarator in which transformer at right produces axial current for heating plasma in twisted tube. External windings on tube carry direct current for producing current-confining magnetic field.

rounded by magnetizing coils are heated to stellar temperatures of several million degrees by means of plasma pinch.

stellar guidance *Celestial guidance.*

stellar map matching Guidance in which a map of the stars is matched with the position of the stars as observed through a telescope, to provide guidance for a missile, spacecraft, or other vehicle.

stem The inward-projecting portion of the glass envelope of an electron tube, through which the heavy leads pass that support and make connections to the electrodes.

stenode circuit An IF amplifier circuit in which a crystal filter passes only signals at the exact IF value, giving high selectivity.

step 1. One operation in a computer routine. 2. A portion of a step-function waveform, consisting of a single sudden change in amplitude and a period of time at the new amplitude value.

step attenuator An attenuator in which the attenuation can be varied in precisely known steps by switches.

step-by-step excitation The successive transitions of an atom to higher levels of excitation.

step-by-step system 1. A control system in which the drive motor moves in discrete steps when the input element is moved continuously. 2. *Strowger system.*

step-down transformer A transformer in which the AC voltages of the secondary windings are lower than that applied to the primary winding.

step function A signal that has zero value before a certain instant of time and a constant nonzero value immediately after that instant.

step-function generator A function generator whose output waveform increases and decreases suddenly in steps that may or may not be equal in amplitude.

step-function response The time variation of an output signal when a specified step-function input signal or disturbance is applied.

step multiplier A multiplier in which a feedback network of an amplifier is controlled in a stepwise manner by one variable to make the gain of the

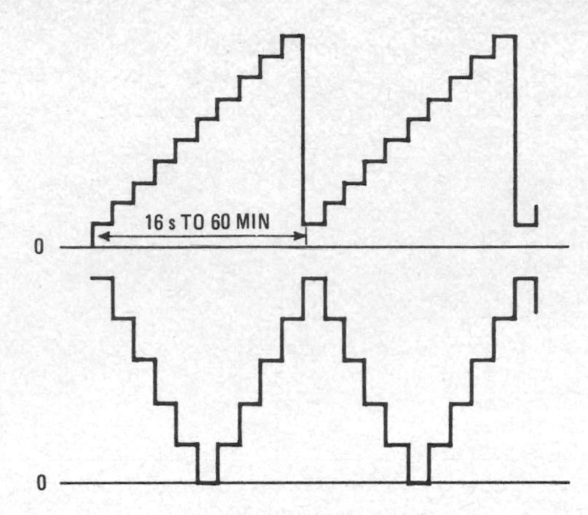

Step functions.

amplifier proportional to it. The other variable is fed into the amplifier, and the output is the required product.

stepper motor A motor that rotates in short and essentially uniform angular movements rather than continuously. Typical steps are 30, 45, and 90°. The angular steps are obtained electromagnetically rather than by ratchet and pawl mechanisms, as in stepping relays. The two basic types are permanent-magnet and variable-reluctance stepper motors.

stepping *Zoning.*

stepping register A register in which an appropriate AC waveform controls the locations of stored data in computers.

stepping relay A relay whose contact arm may rotate through 360° but not in one operation. Also

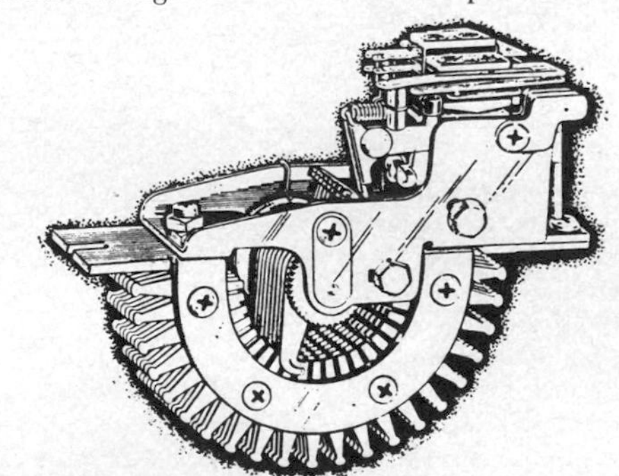

Stepping relay.

called rotary stepping relay, rotary stepping switch, and stepping switch.

stepping switch *Stepping relay.*

step-recovery diode A diode that stores a charge while conducting in the forward direction. When the applied voltage is reversed, the diode conducts for a brief time, up to about 300 ns, until the stored charge is removed, and then abruptly stops conducting. Used in harmonic-generating and pulse-sharpening circuits.

step up To increase the value of electrical quantity.

step-up 1. An increase in the value of an electrical quantity. 2. Pertaining to an increase in the value of an electrical quantity.

step-up transformer A transformer in which the AC voltages of the secondary windings are higher than that applied to the primary winding.

step wedge 1. A step-shaped block of material that absorbs x-rays, used to compare the radiographic effects of progressively weaker x-rays. 2. An optical negative whose density varies in steps, used for test purposes.

step-wedge penetrameter A penetrameter made from material similar to the specimen under radiographic examination; it has steps ranging usually from 1 to 5% of the specimen thickness. Each step may contain one or more drilled holes for assessment of definition.

steradian [abbreviated sr] The SI unit of solid angle, subtending a spherical surface whose area is equal to the square of the radius. The total solid angle about a point in space is 4π sr.

Sterba-curtain array A stacked array with a curtain reflector, suspended from messenger cables running between two steel towers. The curtain may be parasitic or excited. Used with a high-power transmitter for highly directional long-range communications.

stereo 1. Pertaining to three-dimensional pickup or reproduction of sound, as achieved by using two or more separate audio channels. Also called stereophonic. 2. *Stereo sound system.*

stereo- A prefix that designates a three-dimensional characteristic.

stereo amplifier An audio-frequency amplifier that has two or more channels, as required for use in a stereo sound system.

stereo broadcasting Broadcasting two sound channels, for reproduction by a stereo sound system that has a stereo tuner at its input.

stereocephaloid microphone A microphone arrangement that simulates normal human hearing.

stereo effect Reproduction of sound in such a manner that the listener receives the sensation that individual sounds are coming from different locations, just as did the original sounds reaching the stereo microphone system.

stereofluoroscopy A fluoroscopic technique that gives three-dimensional images.

stereo FM The stereo broadcasting system used in the United States. The main FM carrier is modulated by the sum of the left and right stereo channels, and a subcarrier at 38 kHz is suppressed after being amplitude-modulated by the difference between the left and right channel signals. A 19-kHz pilot tone is added to indicate at receivers

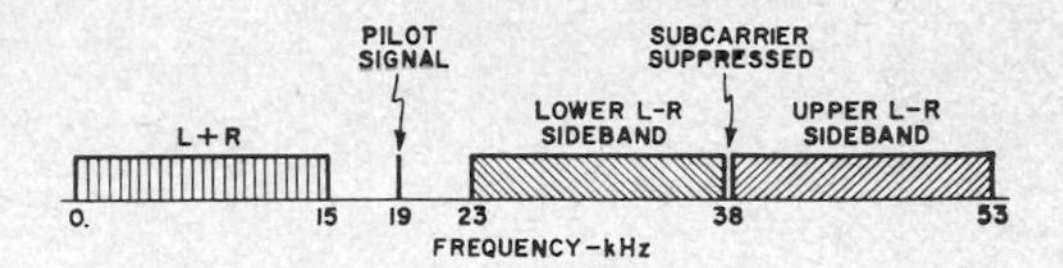

Stereo FM spectrum for one 200-kHz FM channel. Assigned carrier frequency of station would be at 0 kHz on frequency scale.

the presence of a stereo transmission. Additional bandwidth remaining in the channel assigned to the FM station is often used for SCA background music service. At receivers, the 19-kHz tone is doubled and used as a reference for demodulating the subcarrier sidebands, after which the pairs of modulations are added to get the left channel and subtracted to get the right channel. Also called FM stereo (deprecated).

stereo microphone system An arrangement of two or more microphones spaced far enough apart to give two different output signals, as required for making a stereo recording or feeding a stereo sound system directly or by radio.

stereophonic *Stereo.*

stereo pickup A phonograph pickup used with standard single-groove two-channel stereo records. The pickup cartridge has a single stylus that actuates two elements, one responding to stylus

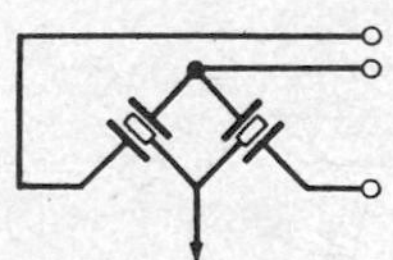

Stereo-pickup symbol.

motion at 45° to the right of vertical and the other responding to stylus motion at 45° to the left of vertical. Each cartridge element feeds one of the channels in a stereo preamplifier or stereo amplifier.

stereo preamplifier An audio-frequency preamplifier that has two channels, for use in a stereo sound system.

stereo record A single-groove disk record that has V-shaped grooves at 45° to the vertical. Each groove wall has one of the two recorded channels. Also called stereo recording.

stereo recorded tape Recorded magnetic tape that has two separate recordings, one for each channel of a stereo sound system. Also called stereo recording.

stereo recording 1. *Stereo record.* 2. *Stereo recorded tape.*

stereoscopic Pertaining to a three-dimensional visual image.

stereoscopic television Television in which the viewed images have a three-dimensional appearance.

stereo separation The ratio of the electric signal in the right stereo channel to the signal in the left stereo channel when only a right signal is transmitted, or vice versa.

stereo sound system A sound-reproducing system in which a stereo pickup, stereo tape recorder, stereo tuner, or stereo microphone system feeds two independent audio channels, each of which terminates in one or more loudspeakers arranged to give listeners the same audio perspective as they would get at the original sound source. Also called stereo.

stereo subcarrier A subcarrier whose frequency is the second harmonic of the pilot subcarrier frequency used in stereo FM broadcasting.

stereo subchannel The band of frequencies from 23 to 53 kHz, containing the stereo FM subcarrier and its associated sidebands.

stereo tape recorder A magnetic-tape recorder that has two stacked playback heads, for reproduction of stereo recorded tape.

stereo tuner A tuner that has provisions for receiving both channels of a stereo broadcast.

stick circuit A circuit used with a relay or similar device to keep it energized through its own holding contacts once the device is actuated.

sticking The tendency of a flip-flop or other bistable circuit to stay in a particular one of its two stable states or switch spontaneously to that state.

stiction [STatic frICTION] Friction that tends to prevent relative motion between two movable parts at their null position.

stilb [abbreviated sb] A CGS unit of luminance, equal to 1 cd/cm^2. The SI unit of luminance, the candela per square meter, is preferred.

stimulated emission A downward transition of electron energy levels under conditions such that a photon of the correct frequency will stimulate another transition and make two photons available, with the process repeating to give rapid cumulative buildup of coherent light output, as in lasers. Stimulated emission, population inversion, and light amplification together form the basis for laser operation.

stimulator A neurosurgical device that supplies a controlled AC voltage to two electrodes that are applied to a patient.

stimulus A signal that affects the controlled variable in a control system.

stirring effect The circulation produced in a molten charge of metal because of the combined forces of motor and pinch effects.

stitch bonding A method of making wire connections between two or more points on an integrated circuit by using impulse welding or heat and pressure while feeding the connecting wire through a hole in the center of the welding electrode.

stochastic process A random sampling process in which each and every member of the population has the same opportunity of being selected.

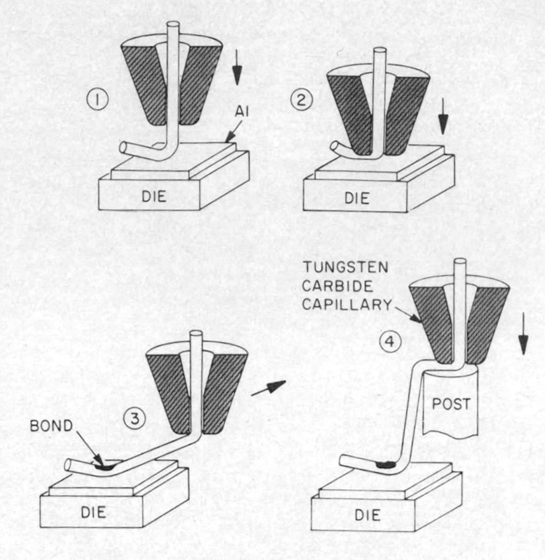

Stitch bonding as used for connecting point on die to adjacent terminal post.

stochastic variable A variable that is dependent on the random variable and is usually measured experimentally.

stoichiometric impurity A crystalline imperfection in a semiconductor.

Stokes' law The wavelength of luminescence excited by radiation is always greater than that of the exciting radiation.

STOL aircraft [Short TakeOff and Landing aircraft] An aircraft that can take off and land on a runway shorter than 1500 ft (457.2 m).

stop The aperture or useful opening of a lens, usually adjustable by a diaphragm.

stop band The band of frequencies in which a filter or other frequency-sensitive device has high attenuation.

stopping The decrease in kinetic energy of an ionizing particle as a result of energy losses along its path through matter.

stopping capacitor *Coupling capacitor.*

stopping cross section *Atomic stopping power.*

stopping equivalent The thickness of a standard substance that is capable of producing the same energy loss as does a given thickness of the substance under consideration, when a charged particle passes through. Also called equivalent stopping power.

stopping off The application of a resist to any part of a cathode or other surface prior to electroplating or etching.

stopping power The effect of a substance on the kinetic energy of a charged particle passing through it.

stopping voltage The voltage required to stop an electron emitted by photoelectric or thermionic action.

storage The portion of a digital computer that stores information for later use. The terms storage and memory are used interchangeably today in the computer field, even though purists insist that storage is the more correct term.

storage access time The time required to transfer information from a storage location in a computer to the local storage register or other location where the information becomes available for processing.

storage battery A connected group of two or more storage cells. Common usage permits application of this term to a single cell used independently. Also called accumulator (British) and secondary battery.

storage capacity The information-storing capacity of a storage device, expressed in bits, bytes, words, or other units of storage capacity. Also called memory capacity.

storage cell 1. An electrolytic cell for generating electric energy, in which the cell after being discharged may be restored to a charged condition by sending an electric current through it in a direction opposite that of the discharging current. Also called secondary cell. 2. An elementary unit of storage, such as a binary cell.

storage cycle time The total time required to transfer information from storage to a local storage register, plus the time required to return the information to storage after it is no longer needed for processing.

storage device A device into which data can be inserted, in which it can be retained, and from which it can be retrieved.

storage dump A printout of the contents of all or part of a computer storage. Also called memory dump.

storage effect Temporary storage of injected excess minority carriers in the higher-resistivity side of a semiconductor junction.

storage element 1. An area of a storage surface which retains information distinguishable from that of adjacent areas in a charge-storage tube. 2. The smallest part of a digital-computer storage, used for storing a single bit.

storage location A digital-computer storage position that holds one machine word and usually has a specific address.

storage medium Any device or recording medium into which data can be copied and held until some later time and from which the entire original data can be obtained.

storage oscilloscope An oscilloscope that can retain an image for a period of time ranging from minutes to days, or until deliberately erased to make room for a new image. Storage techniques range from long-persistence and bistable phosphors to a variety of digital techniques.

storage register A register in the main internal memory of a digital computer storing one computer word.

storage tube An electron tube that uses cathode-ray beam scanning and charge storage for the introduction, storage, and removal of information.

store To transfer an element of information to storage in a computer, for later extraction.

storecasting A Subsidiary Communications Authorization service in which a standard stereo FM broadcast carrier is modulated by an additional signal, usually 60 to 74 kHz above the carrier frequency, for transmission of background music to stores and public buildings.

stored base charge The phenomenon associated with the storage of minority charge carriers in the base region of alloy-junction-type transistors under conditions of saturation.

stored energy The potential energy of atomic displacements that is retained by a solid after irradiation.

stored-energy welding Welding in which electric energy is accumulated electrostatically, electromagnetically, or electrochemically at a relatively slow rate and released at the required rate for welding.

stored-program computer A digital computer that can translate or otherwise alter an input program by using internally stored instructions and then executing the rewritten program.

straggling Random variations of some property of particles as a result of their passage through matter.

straight dipole A dipole that consists of a single straight conductor, usually broken at its center for connection to a transmission line.

straight-line capacitance [abbreviated SLC] A variable-capacitor characteristic obtained when the rotor plates are shaped so capacitance varies directly with the angle of rotation.

straight-line frequency [abbreviated SLF] A variable-capacitor characteristic obtained when the rotor plates are shaped so the resonant frequency of the tuned circuit containing the capacitor varies directly with the angle of rotation.

straight-line path The axis of the Fresnel-zone family of paths between two microwave antennas.

straight-line wavelength [abbreviated SLW] A variable-capacitor characteristic obtained when the rotor plates are shaped so the wavelength for resonance in the tuned circuit that contains the capacitor varies directly with the angle of rotation.

strain gage A strain-sensitive element designed to be attached to a member in which strain is to be measured. It is usually connected into a bridge circuit that feeds a recorder directly or through an amplifier. The most common type is the resistance strain gage.

strain-gage bridge A bridge arrangement of four strain gages, cemented to a stressed part in such a way that two gages show increases in resistance and two show decreases when the part is stressed. The change in output voltage under stress is thus much higher than that for a single gage.

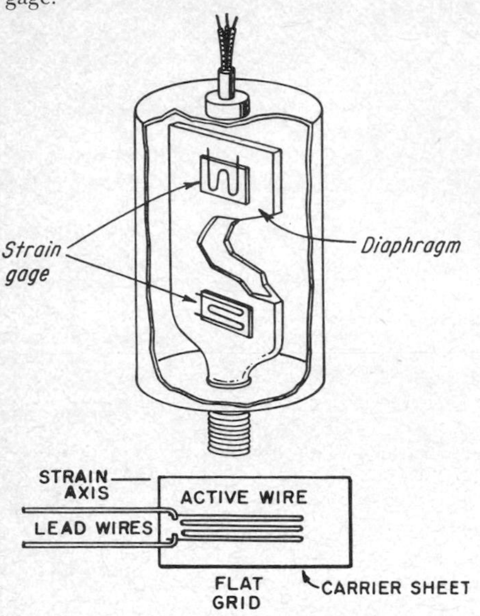

Strain gage mounted on flexible sealed diaphragm for use as air-pressure transducer, and strain gage designed for cementing on beam to sense deflection.

strain-gage multiplier A time-varying resistance multiplier that uses a strain gage as its time-varying element. One variable controls the strain in the gage, and the other variable controls the current through it. The voltage across the strain gage is proportional to the required product.

strain-gage transducer A transducer in which fine resistance wire is fixed to a member whose displacement is to be sensed. Displacement strains the wire, changing its resistance.

strain insulator An insulator used between sections of a stretched wire or antenna to break up the wire into insulated sections while withstanding the total pull of the wire.

strand One of the wires or groups of wires in a stranded wire.

stranded conductor *Stranded wire.*

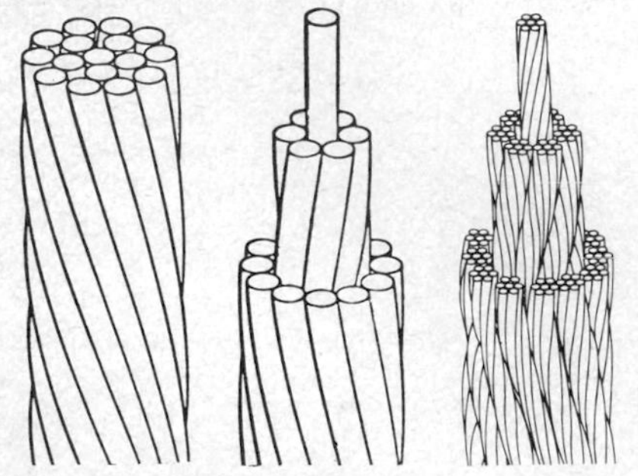

Stranded wire.

stranded wire A conductor composed of a group of wires or a combination of groups of wires, usually twisted together. Also called stranded conductor.

strange particles A class of elementary particles for which the details of production and decay cannot be fully explained by existing theories. Hyperons and mesons are in this class.

strap A conductive link used between alternate resonator segments of a magnetron.

strapping Connecting together resonator segments that have the same polarity in a multicavity magnetron, to suppress undesired modes of oscillation.

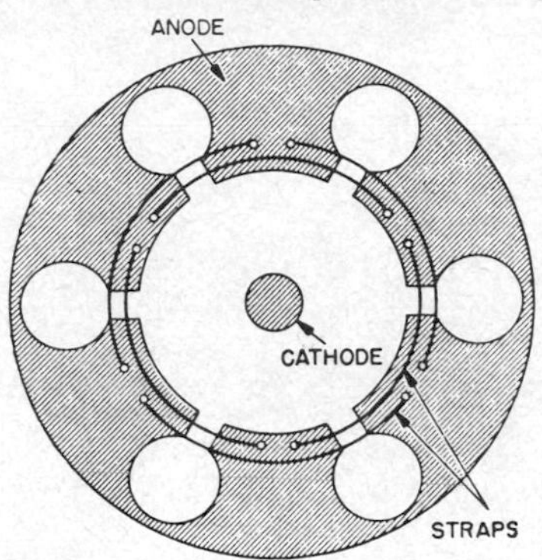

Strapping in multicavity magnetron.

Strategic Air Command [abbreviated SAC] A major air command of the U.S. Air Force, charged with carrying out strategic air operations and controlling all long-range bomber aircraft.

strategic missile A guided missile used for long-range action against an enemy's war-making capacity.

stratoscope A balloon-borne astronomical telescope for taking solar photographs and transmitting them to a ground receiving station.

stratosphere A stratum of the earth's atmosphere above the troposphere, extending from about 7 mi (11.25 km) up to about 50 mi (80 km) above the earth. The temperature is essentially constant in the stratosphere.

stray capacitance Undesirable capacitance between circuit wires, wires and the chassis, or components and the chassis of electronic equipment. Also called strays.

stray field Leakage magnetic flux that spreads outward from a coil and does no useful work.

stray radiation Radiation that serves no useful purpose.

strays *Stray capacitance.*

streaking A television picture condition in which white or black horizontal streaks or smudges appear to follow images across the screen. The effect is more apparent at vertical edges of objects where there is an abrupt transition from black to white or white to black, and may be caused by excessive low-frequency response.

streamer An indefinite wavy band that occurs when the gas pressure in a discharge tube is reduced below the value required for a glow discharge through the tube.

striated discharge An electric discharge characterized by alternate light and dark bands in the positive column adjacent to the anode of a glow-discharge tube.

striation technique A technique for making sound waves visible by using their individual abilities to refract light waves.

striking 1. Starting an electric arc by touching the electrodes together momentarily. 2. Electrodeposition of a thin initial film of metal, usually at a high current density.

striking voltage The grid-cathode voltage required to start the flow of anode current in a gas tube.

string A group of data items that are in an ascending or descending sequence according to alphabetic, numerical, or other rules.

string electrometer An electrometer in which a conducting fiber is stretched midway between two oppositely charged metal plates. The electrostatic field between the plates displaces the fiber laterally in proportion to the voltage between the plates.

strip To remove insulation from a wire.

strip-chart recorder A recorder in which one or more writing pens or other recording devices trace changes in a measured variable on the surface of a strip chart that is moved at constant speed by a time-clock motor.

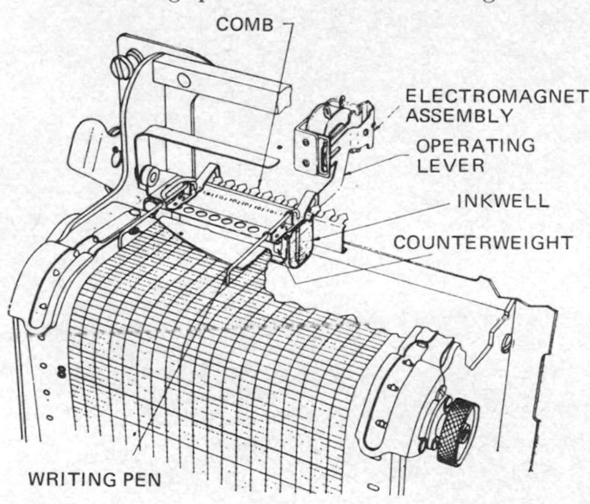

Strip-chart recorder.

stripline A strip transmission line that consists of a flat metal-strip center conductor which is separated from flat metal-strip outer conductors by dielectric strips.

stripline circuit A circuit in which one or more strip transmission lines serve as filters or other circuit components.

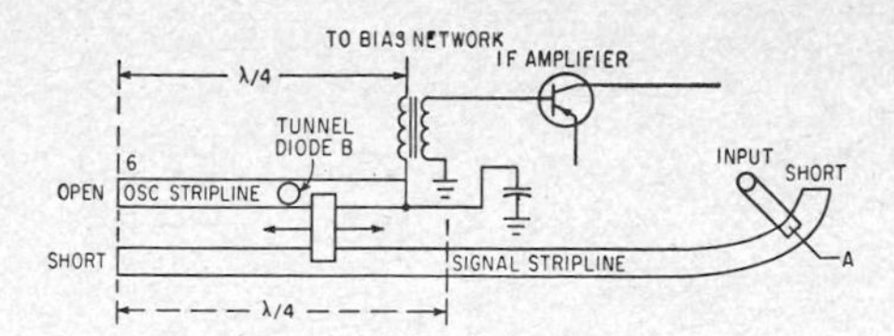

Stripline circuit for tunnel-diode L-band autodyne converter.

strip penetrameter *Plate penetrameter.*

stripper A hand or motorized tool used to remove insulation from wires.

stripping 1. Removal of a metal coating, as when etching away undesired portions of a printed circuit. 2. An effect observed in bombardment with deuterons or heavier nuclei, whereby only part of the incident particle merges with the target nucleus. The remainder proceeds with most of its original momentum in practically its original direction.

strip printer A printer that prints exactly one continuous line of alphameric characters on a thin paper tape.

strip transmission line A microwave transmission line that consists of a thin, narrow rectangular strip which is separated from a wide ground-plane conductor or mounted between two wide ground-plane conductors. Separation is usually achieved with a low-loss dielectric material on which the conductors are formed by printed-circuit techniques.

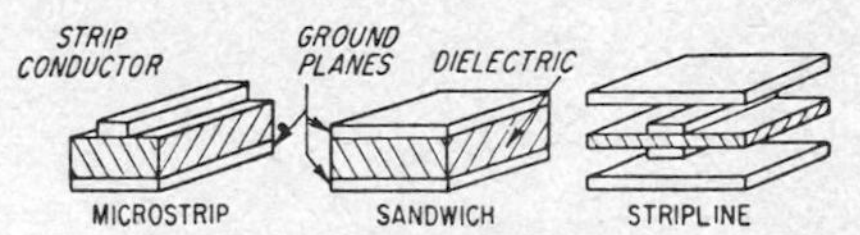

Strip transmission lines.

strobe 1. A pulse used to gate the output of a counter, shift register, or other computer circuit, to produce a desired action. 2. A pulse superimposed on a radar image to serve as a marker from which the range of the target can be measured. 3. A line or wedge produced on a radar screen by a jamming signal. 4. *Stroboscope.*

strobe marker A small bright spot or a short gap or other discontinuity produced on the trace of a radar display to indicate the part of the time base that is receiving attention.

stroboscope A controllable intermittent source of intense light, used to create the illusion of slowing down or stopping vibrating or rotating objects. The flashing frequency is adjusted until it corresponds to some multiple of the speed of vibration or rotation of the object under study. Also called strobe.

stroboscopic disk A printed disk that has a number of concentric rings, each containing a different number of dark and light segments. When the disk is placed on a phonograph turnta-

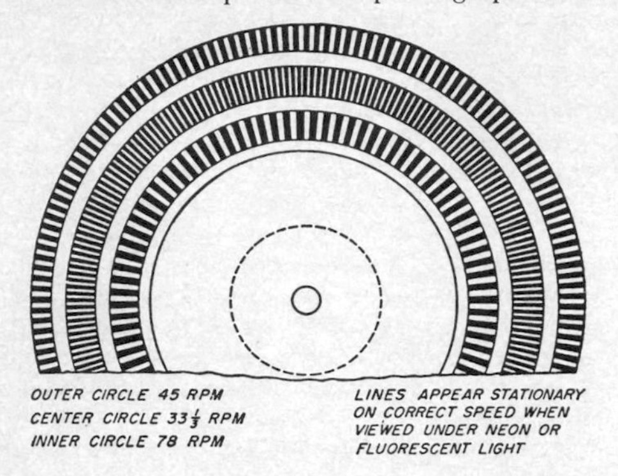

Stroboscopic disk for checking speed of phonographs with 120-Hz flashes from glow lamp operating on 60-Hz power.

ble or rotating shaft and illuminated at a known frequency by a flashing discharge tube, speed can be determined by noting which pattern appears to stand still or rotate slowly.

stroboscopic tachometer A stroboscope that has a scale which reads in flashes per minute or revolutions per minute. The speed of a rotating device is measured by directing the stroboscopic lamp on the device, adjusting the flashing rate until the device appears to be stationary, then reading the speed directly on the scale of the instrument.

stroboscopic tube A cold-cathode gas-filled arc-discharge tube that produces intensely bright flashes of light for use with a stroboscope. Also called strobotron.

strobotron *Stroboscopic tube.*

strontium [symbol Sr] A metallic element sometimes used on cathodes of phototubes to obtain maximum response to ultraviolet radiation. Atomic number is 38.

strontium 90 A radioisotope that has a half-life of about 25 years. Also called radio strontium.

Strowger system An automatic telephone switching system that uses successive step-by-step selector switches actuated by current pulses produced by rotation of a telephone dial. The selectors are electromagnetically operated and contain a number of tiers of fixed contacts, each arranged in a semicircle. A moving contact arm first rises to the height of the desired tier, then swings around horizontally and stops over the required contact. Also called step-by-step system.

stub 1. A short section of transmission line, open or shorted at the far end, connected in parallel with a transmission line to match the impedance of the line to that of an antenna or transmitter. 2. A solid projection one quarter-wavelength long,

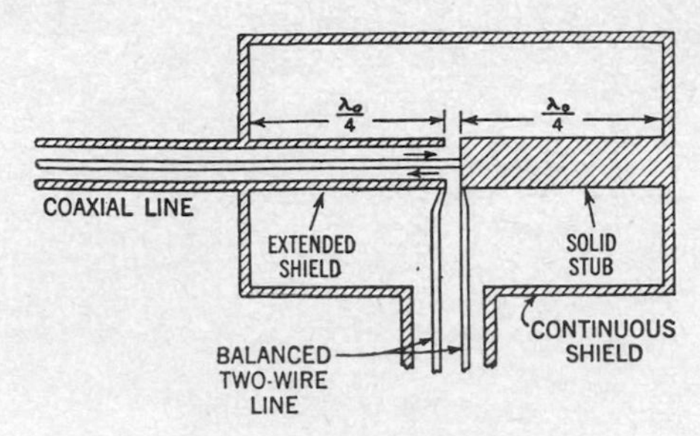

Stub used in cavity to support conductor during conversion from coaxial line to balanced two-wire line.

used as an insulating support in a waveguide or cavity.

stub-matching Use of a stub to match a transmission line to an antenna or load. Matching depends on the spacing between the two wires of the stub, the position of the shorting bar, and the point at which the transmission line is connected to the stub.

stub-supported line A transmission line that is supported by short-circuited quarter-wave sections of coaxial line. A stub exactly a quarter-wavelength long acts as an insulator because it has infinite reactance.

stub-tuned filter A microwave stopband filter that consists of a number of T junctions of different sizes, inserted in a waveguide to produce high

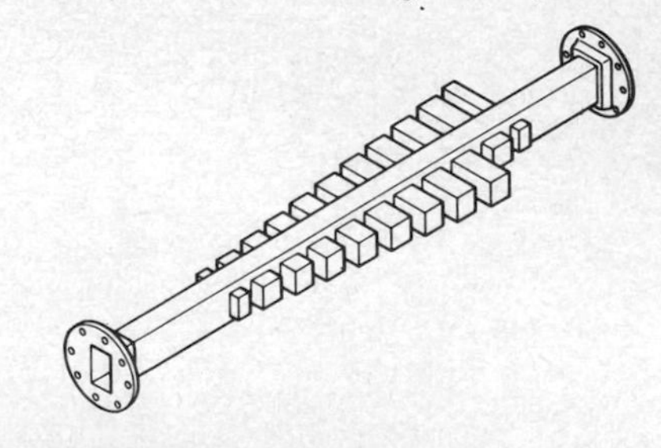

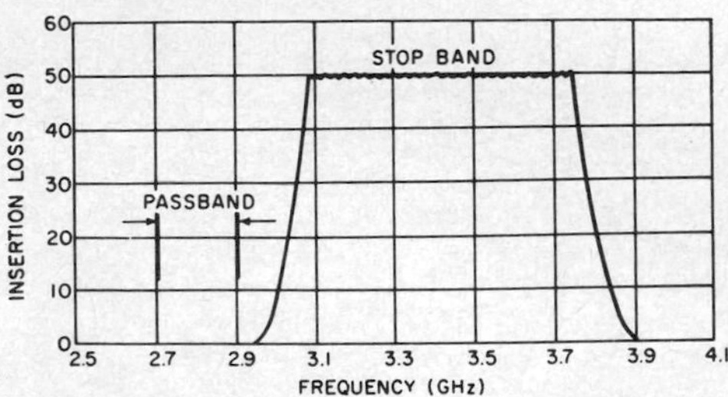

Stub-tuned filter for 2.7–2.9 GHz waveguide, providing 50-dB attenuation between 3.1 and 3.7 GHz.

attenuation over a band of frequencies. Generally used to suppress undesired frequencies that can be up to the fourth harmonic.

stub tuner An adjustable shorted stub used to

adjust a transmission line for maximum power transfer.

studio A room in which television or radio programs are produced.

stunt box A device that controls the nonprinting functions of a teletypewriter terminal.

stylus [plural is styli] The portion of a phonograph pickup that follows the modulations of a record groove and transmits the resulting me-

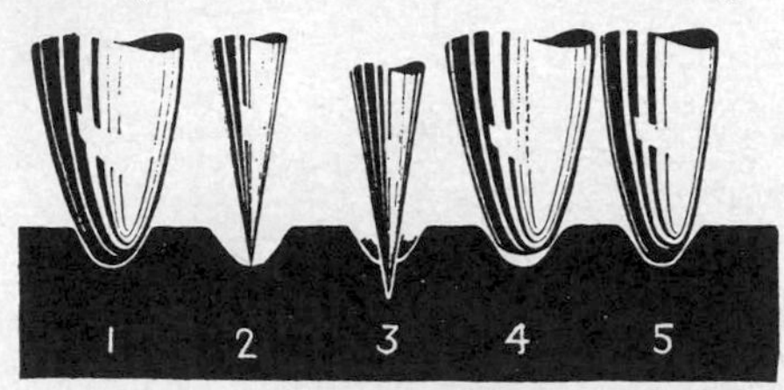

Stylus points: (1) theoretically ideal but impractical because angle changes; (2) too sharp, allowing free movement; (3) too sharp, causing gouging; (4) too blunt, scoring groove walls; (5) best practical shape.

chanical motions to the transducer element of the pickup for conversion to corresponding AF signals. Also called needle and reproducing stylus.

stylus drag The force that results from friction between the surface of a recording medium and the reproducing stylus. Also called needle drag.

stylus force The vertical force exerted on a stationary recording medium by the stylus when in its operating position. Also called needle force, needle pressure (deprecated), static stylus force, stylus pressure (deprecated), and vertical stylus force.

stylus pressure Deprecated term for *stylus force.*

s-type negative resistance A voltage-stable negative resistance, in which a given current in the operating range may have three different possible values of terminal voltage. Examples include tunnel diodes, vacuum-tube tetrodes used as dynatron oscillators, and the common-emitter input of a point-contact transistor.

subassembly Two or more components combined into a unit for convenience in assembling or servicing equipment.

subatomic Pertaining to particles smaller than atoms, such as electrons, protons, and neutrons.

subaudio *Infrasonic.*

subcadmium neutron A neutron that has less kinetic energy than the cadmium cutoff energy.

subcarrier 1. A carrier that is applied as a modulating wave to modulate another carrier. 2. *Chrominance subcarrier.*

subcarrier band A band associated with a given subcarrier and specified in terms of maximum subcarrier deviation.

subcarrier discriminator A discriminator that demodulates a telemetering subcarrier frequency.

subcarrier oscillator 1. The crystal oscillator that operates at the chrominance subcarrier or burst frequency of 3.579545 MHz in a color television receiver. This oscillator, synchronized in frequency and phase with the transmitter master oscillator, furnishes the continuous subcarrier frequency required for demodulators in the receiver. 2. An oscillator used in a telemetering system to translate variations in an electrical quantity into variations of a frequency-modulated signal at a subcarrier frequency.

subcontractor A manufacturer or other organization that receives a contract from a prime contractor for a portion of the work on a project.

subharmonic A sinusoidal quantity that has a frequency which is an integral submultiple of the frequency of some other sinusoidal quantity to which it is referred. A third subharmonic would be one-third the fundamental or reference frequency.

subliminal Below the threshold of conscious responsiveness to a stimulus. Applications include behavior modification that involves audio or video motivational stimuli.

submarine detecting set A complete airborne electronic detection system designed primarily to indicate the presence of submerged submarines and mines.

submillimeter wavelength A wavelength shorter than 1 mm, corresponding to frequencies above 300 GHz.

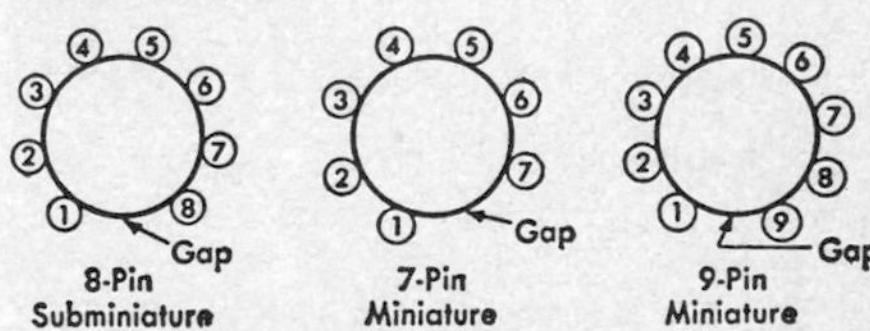

Subminiature-tube pin connections, with miniature tube connections shown for comparison. Diagrams represent bottom views of tubes and sockets. No. 1 pin is always next to gap in clockwise direction. Pins are never omitted.

subminiature tube An extremely small electron tube used in hearing aids and other miniaturized equipment. A typical subminiature tube is about 1½ in (3.8 cm) long and 0.4 in (1 cm) in diameter, with the pins emerging through the glass base.

subminiaturization Reduction of size and weight of electronic equipment, generally achieved through use of printed circuits, modules, and special heat-dissipating features.

subnanosecond Less than 1 ns, or less than one-billionth of a second.

subnanosecond radar A radar that operates with very short pulses; it has a duration of less than 1 ns. Effective range resolution is less than 3 in (7.5 cm), permitting accurate measurements by radar of ice and snow thickness, surface reflection coefficients, and signal attenuation of sea water. Also used for periscope detection in antisubmarine warfare and other conventional military applications that require a high degree of range resolution.

subprogram A portion of a computer program.

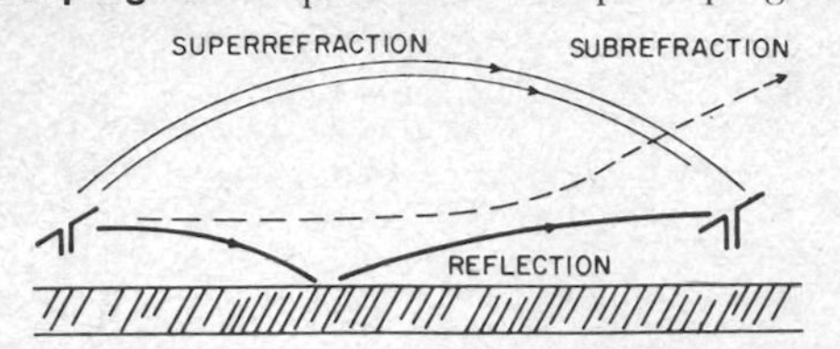

Subrefraction, superrefraction, and reflection of radio waves traveling over water are all causes of deep fading.

subrefraction Refraction in which radar waves are bent upward by the atmosphere because of an excessive vertical temperature gradient or humidity gradient, reducing the range of a radar set.

SUBROC [SUBmarine ROCket] A Navy missile that is fired underwater from a conventional torpedo tube but is rocket-powered and most of its course is through the air. The missile reenters the water at supersonic speed and detonates at the preset depth at which an enemy submarine has been located. Range is about 50 mi (80 km).

subroutine A portion of a routine that causes a computer to carry out a well-defined mathematical or logic operation. At its conclusion, control reverts to the master routine.

subset 1. A telephone or other subscriber equipment connected to a communication system. 2. *Modem.*

Subsidiary Communications Authorization [abbreviated SCA] An FCC 1955 regulation that authorized multiplexing on FM broadcasts to provide such background music services as storecasting. The control channels for this purpose could originally be located anywhere between 25 and 75 kHz. With the stereo FM passband now occupying 23 to 53 kHz, FM stations changing to stereo must move their SCA frequency above the stereo passband, generally to 67 kHz.

subsonic 1. Less than the velocity of sound in air, hence less than Mach 1, which is about 738 mi/h (1188 km/h). Used chiefly in connection with airplanes and missiles. 2. Deprecated term for *infrasonic.*

substrate The physical material on which a microcircuit is fabricated. Used primarily for mechanical support and insulating purposes, as with ceramic, plastic, and glass substrates, but semiconductor and ferrite substrates may also provide useful electrical functions.

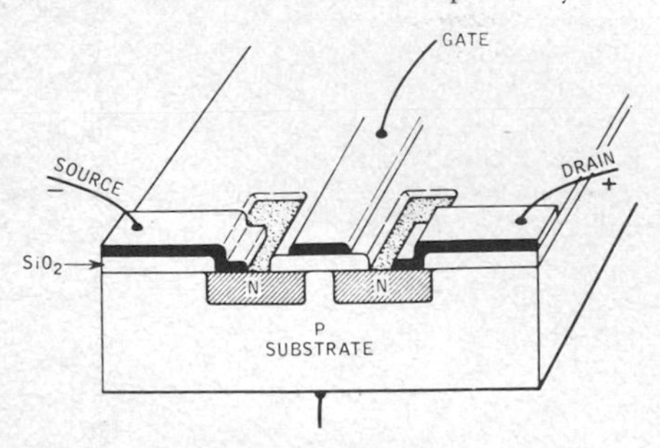

Substrate of field-effect transistor.

subsurface wave An electromagnetic wave that has an underwater or underground propagation path. Operating frequencies for communication must generally be below about 35 kHz because of attenuation of higher frequencies by water or earth.

subsynchronous Operating at a frequency or speed that is a submultiple of the source frequency.

subsystem A system subdivision that itself has the characteristics of a system.

subtracter 1. A circuit in which the analog output signal is proportional to the difference between the amplitudes of two analog input signals. 2. A computer circuit that provides a digital output equal to the difference between two digital inputs.

subtraction-type radiometer A radio receiver in which a DC voltage equal to the smoothed second detector output, when only noise is present at the receiver input, is subtracted from the smoothed second detector output. The resulting difference in voltage is applied to a galvanometer as an indication of an input signal.

sudden ionospheric disturbance A sudden increase of ionization density in low parts of the ionosphere caused by a bright solar eruption. It usually lasts a few minutes, causing a sudden increase in absorption of radio waves and sometimes simultaneous disturbances in earth currents and the earth's magnetic field.

Suhl effect When a strong transverse magnetic field is applied to an N-type semiconducting filament, holes injected into the filament are deflected to the surface, where they may recombine rapidly with electrons or be withdrawn by a probe. The overall effect is an increase in conductance.

sulfating Formation of lead sulfate on the plates of lead-acid storage batteries, reducing the energy-storing ability of the battery and eventually causing failure.

sulfur [symbol S] An element. Atomic number is 16.

sulfur hexafluoride A dielectric gas used to sup-

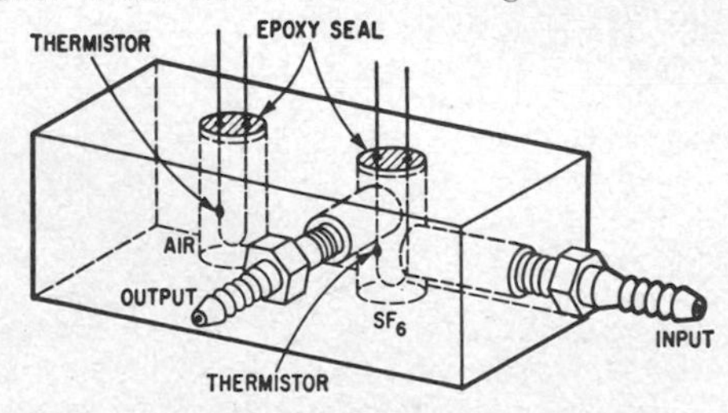

Sulfur hexameter in waveguide uses two thermistors in bridge circuit.

press arcing in high-power radar waveguides.

sulfur hexameter An instrument that measures or continuously monitors the amount of sulfur hexafluoride present in a waveguide or other device in which this gas is used as a dielectric.

sulfuric acid A compound of sulfur, hydrogen, and oxygen; it has the chemical formula H_2SO_4. Used as the electrolyte in lead-acid storage batteries.

sum-and-difference monopulse radar Monopulse radar in which the receiving antenna has one aperture and two or more closely spaced feeds, each of which produces a radiation pattern that is displaced from the antenna boresight axis. Signals arriving off the axis give unequal amplitudes in the two channels, and signals on the axis give equal amplitudes that produce a sharp null in the difference channel and a peak in the sum channel.

sum channel A combination of two stereophonic channels to give a program that can be recorded or transmitted by a single channel.

summation bridge A bridge that measures such values as temperature, frequency, speed of rotation, time, resistance, and capacitance by adding the original bridge current to the current needed for balance and presenting the result on an indicator or scale.

summation check A redundancy check in which groups of digits are summed, usually without regard for overflow, and that sum checked against a previously computed sum to verify the accuracy of a digital computer.

summation network *Summing network.*

summing amplifier An amplifier that delivers an output voltage which is proportional to the sum of two or more input voltages or currents.

summing network A passive electric network whose output voltage is proportional to the sum of two or more input voltages. Also called summation network.

summing point A mixing point whose output is obtained by addition, with prescribed signs, of its inputs in a feedback control system. In an operational amplifier circuit, the junction of the input with the feedback network is commonly called the summing point.

sun follower A photoelectric pickup and an associated servomechanism used to maintain a sun-facing orientation, as for a space vehicle. Also called sunseeker.

S unit An arbitrary unit of signal strength, sometimes used along with decibels on the scale of a signal-strength meter.

S-unit meter *Signal-strength meter.*

sunlight recorder A recorder that uses a phototube, capacitor-charging circuit, and thyratron-operated counter to record the integrated value of solar radiation over a period of time.

sun-pumped laser A continuous-wave laser in which pumping is achieved by concentrating the energy of the sun on the laser crystal with a parabolic mirror. One version uses an yttrium-aluminum garnet crystal rod, which does not require cryogenic cooling temperatures.

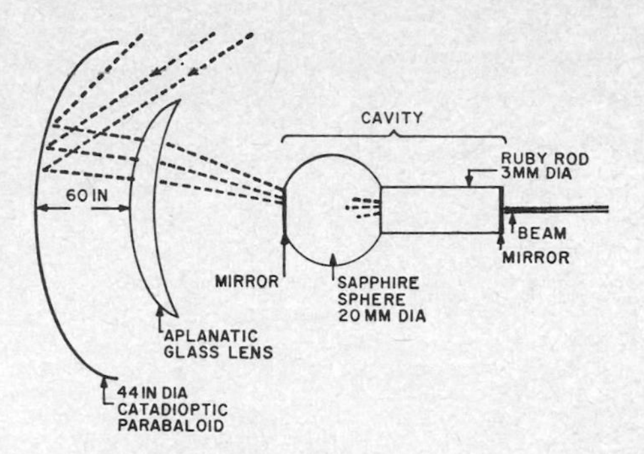

Sun-pumped laser using parabolic mirror 110 cm in diameter to send sunlight through lens and sapphire sphere to ruby rod. Mirror is 150 cm behind lens.

sun satellite A spacecraft that has gone into orbit around the sun.

sunseeker *Sun follower.*

sunspot A dark spot on the sun, usually associated with the magnetic storms on the earth that affect radio communication at the lower frequencies.

sunspot cycle A period of about 11 years in which the number and duration of sunspots and solar flares pass through a cycle of buildup to a maximum value and then drop back to a minimum value. Also called solar cycle.

sunspot number The predicted number of sunspots for a given month.

sun strobe The signal display seen on a radar PPI screen when the radar antenna is aimed at the sun. The pattern resembles that produced by continuous-wave interference and is due to RF energy radiated by the sun. Used for checking radar antenna azimuth and elevation calibration, by comparing with the known position of the sun.

SUP Symbol for *suppressor grid.* Used on circuit diagrams.

superconducting The state of a superconductor in which it exhibits superconductivity, below a critical temperature. Also called superconductive.

superconducting generator A DC generator that in one form consists of a series of flat plates of superconducting lead or niobium arranged in a circle, wired together by superconducting wire, and cooled near absolute zero by an appropriate liquid gas.

superconducting memory A memory made up of a number of cryotrons, thin-film cryotrons, superconducting thin films, or other superconducting storage devices. These operate only under cryogenic conditions and dissipate power only during the read or write operation, which permits construction of large, dense memories.

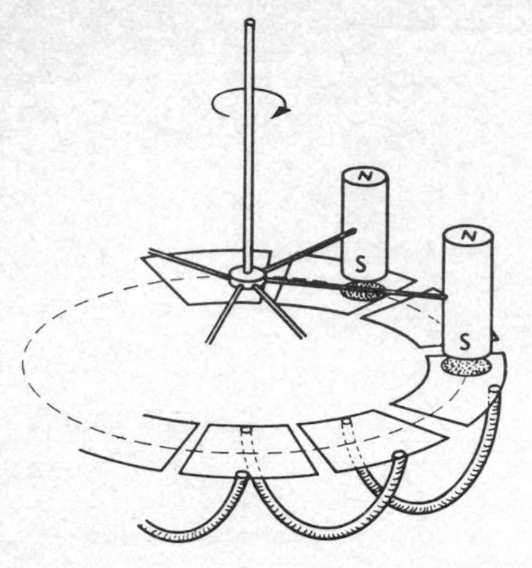

Superconducting generator using rotating permanent magnets.

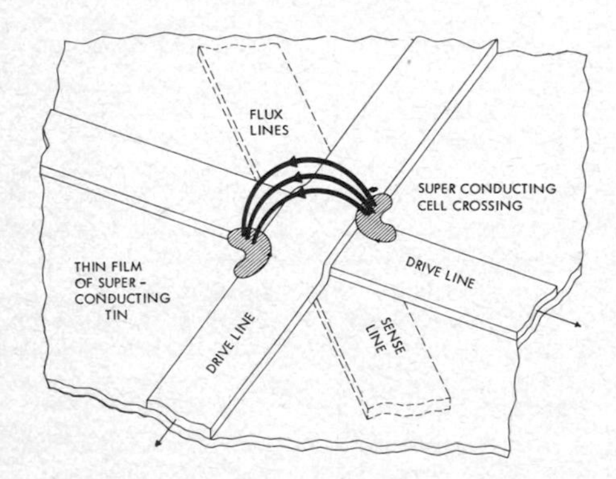

Superconducting memory in which drive lines are on one side of superconducting thin film of tin and sense line is on other side.

superconducting solenoid A solenoid that uses superconducting wires to generate strong magnetic fields.

superconducting thin film A thin film of indium, tin, or other superconducting element, used as a cryogenic switching or storage device, as in a thin-film cryotron.

superconducting transition A transition from a normal state to a superconducting state, occurring at a temperature that depends on the magnetic field as well as on the nature of the material.

superconductive *Superconducting.*

superconductivity A low-temperature phenomenon in which the resistance of certain metals, alloys, and compounds drops essentially to zero at a critical temperature near absolute zero.

superconductor Any material capable of exhibiting superconductivity. Examples include iridium, lead, mercury, niobium, tin, tantalum, and vanadium, and many alloys.

superhet *Superheterodyne receiver.*

superheterodyne receiver A receiver in which all incoming modulated RF carrier signals are converted to a common IF carrier value for additional

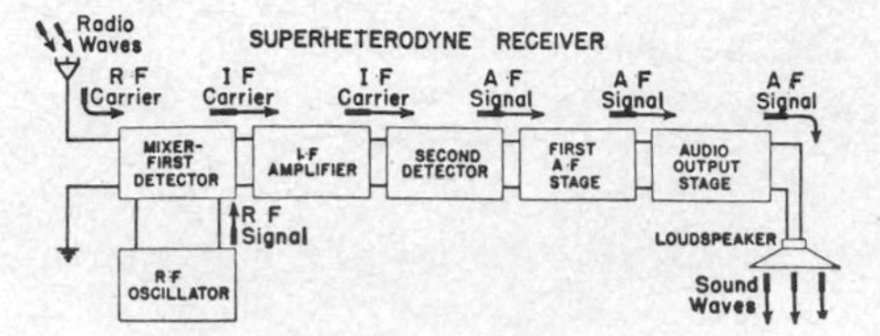

Superheterodyne receiver block diagram.

amplification and selectivity prior to demodulation, using heterodyne action. The output of the IF amplifier is then demodulated in the second detector to give the desired AF signal. Also called superhet.

superhigh frequency [abbreviated SHF] A Federal Communications Commission designation for a frequency in the band from 3 to 30 GHz, corresponding to a centimetric wave between 1 and 10 cm.

superlattice A lattice in which the atoms of one element occupy regular positions in the lattice of another element without forming a compound.

superposed circuit An additional channel obtained from one or more circuits normally provided for other channels, in such a manner that all the channels can be used simultaneously without mutual interference.

superposition theorem The current that flows at a point in a linear network during simultaneous application of a number of voltages throughout the network is the sum of the component currents at the point which would be produced by the individual voltages acting separately. Similarly, the voltage between any two points under such conditions is the sum of the voltages that would be produced between these two points by the individual voltages acting separately.

superpower station A broadcast station that uses extremely high power, generally more than 1 MW.

superrefraction Refraction in which radar waves are bent downward by the earth's atmosphere, generally over sea water that is at least 3°C colder

Superrefraction of radar waves can give detection of surface targets at ranges of more than 1400 mi (2250 km).

than the air. The effect is comparable to that of an atmospheric duct. The range of a radar set is greatly increased under this condition.

superregeneration Regeneration in which the

oscillation is broken up or quenched at a frequency slightly above the upper audibility limit of the human ear by a separate oscillator circuit connected between the grid and anode of the amplifier tube, to prevent the regeneration from exceeding the maximum useful amount.

superregenerative detector A detector in which superregeneration is used to obtain extremely high sensitivity with a minimum number of amplifier stages.

superregenerative paramagnetic amplifier A paramagnetic amplifier that gives a much higher gain-bandwidth product than conventional designs but with a higher noise figure. It may be self-quenched or use separate quenching.

superregenerative receiver A tuned-radio-frequency receiver that uses a superregenerative detector.

supersensitive relay A relay that operates on extremely small currents, generally below 250 μA.

supersonic 1. Faster than the velocity of sound in air, hence faster than Mach 1, which is about 738 mi/h (1188 km/h). Used chiefly in connection with airplanes and missiles. 2. Deprecated term for *ultrasonic.*

supersonics The study of phenomena associated with aircraft and missile speeds higher than the speed of sound.

supersync signal A combination horizontal and vertical sync signal transmitted at the end of each television scanning line to synchronize the operation of a television receiver with that of the transmitter.

superturnstile antenna A modified turnstile antenna that has wing-shaped dipole elements used in pairs mounted at right angles about a common vertical axis. The dipole pairs are fed in quadrature to give substantially omnidirectional radiation over a wide band for FM and television transmitters. Also called batwing antenna.

supervisory control system A control system that provides both indication and control for remotely located equipment by electrical means, as by carrier-current channels on power lines.

supervoltage A voltage in the range of 500 to 2000 kV, used for some x-ray tubes.

supply voltage The voltage obtained from a power source for operation of a circuit or device. Also called rail voltage (British).

suppressed carrier [abbreviated SC] A carrier that is suppressed at the transmitter. The chrominance subcarrier in a color television transmitter is an example.

suppressed-carrier transmission Transmission in which the carrier component of the modulated wave is suppressed, leaving only the sidebands to be transmitted.

suppressed time delay *Code delay.*

suppressed-zero instrument An indicating or recording instrument in which the zero position is below the lower end of the scale markings.

suppression Elimination of any component of an emission, as a particular frequency or group of frequencies in an AF or RF signal.

suppressor 1. A resistor used in series with a spark plug or distributor of an automobile engine or other internal-combustion engine to suppress spark interference that might otherwise interfere with radio reception. 2. *Suppressor grid.*

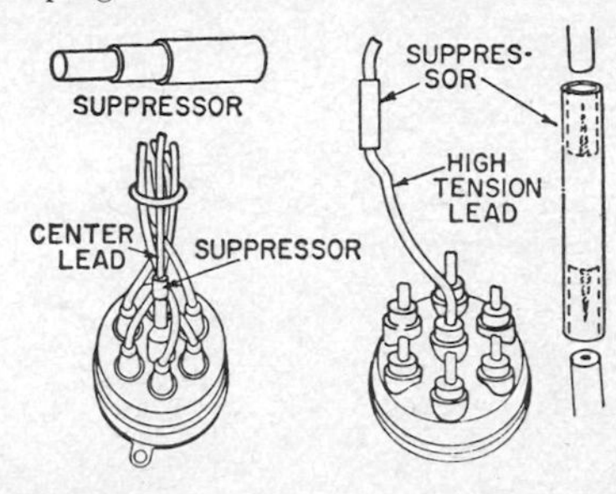

Suppressor installation in high-voltage center lead of distributor in automobile.

suppressor grid [symbol SUP] A grid placed between two positive electrodes in an electron tube primarily to reduce the flow of secondary electrons from one electrode to the other. It is usually used between the screen grid and the anode. Also called suppressor.

suppressor-grid modulation Amplitude modulation in which the modulating signal is applied to the suppressor grid of a pentode that is amplifying the carrier signal.

surface acoustic wave [abbreviated SAW] An acoustic wave that has frequencies as high as several gigahertz, traveling on the optically polished surface of a piezoelectric substrate at a velocity which is only about 10^{-5} times that of electromagnetic waves. A surface acoustic wave thus has the

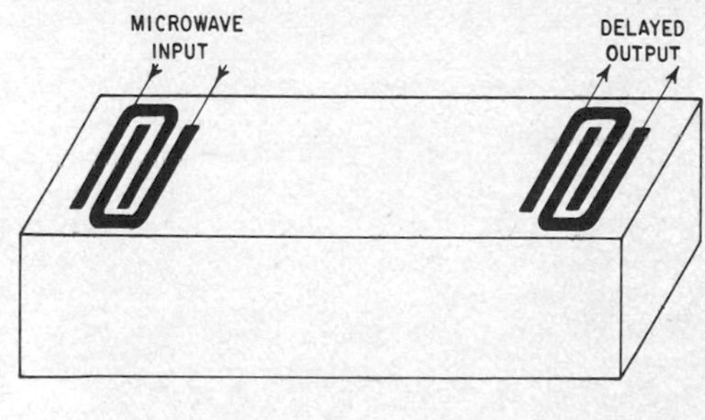

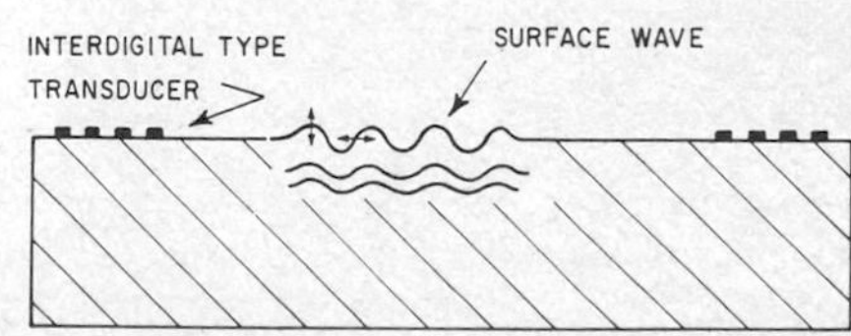

Surface acoustic wave produced by interdigital transducers of delay line.

slow-travel property of sound while retaining the microwave frequency of its source. Used in a variety of delay lines, filters, pulse processors, and other microwave devices and circuits. Piezoelectric substrates in common use include bismuth germanium oxide ($Bi_{12}GeO_{20}$), lithium niobate ($LiNbO_3$), and quartz. Also called acoustic surface wave.

surface-acoustic-wave delay line A delay line in which the delay is determined by the distance

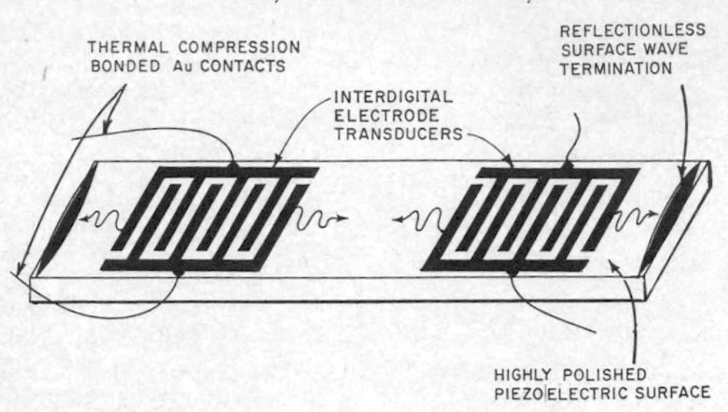

Surface-acoustic-wave delay line.

traveled by a surface acoustic wave on a piezoelectric surface.

surface-acoustic-wave filter An electric filter that consists of a piezoelectric bar which has a polished surface along which surface acoustic

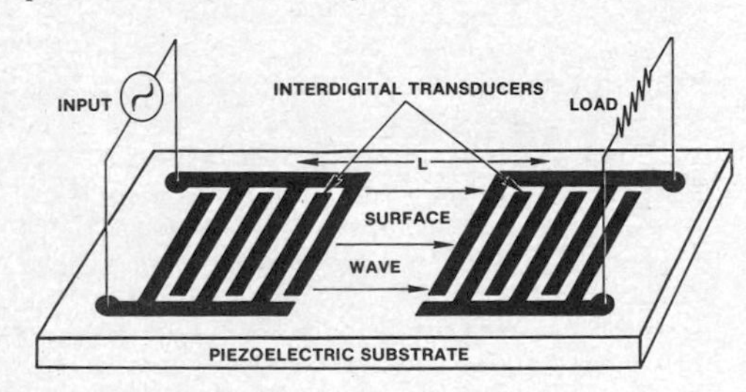

Surface-acoustic-wave filter.

waves can propagate, with metallic input and output transducer configurations deposited at opposite ends of one face of the bar.

surface analyzer An instrument that measures or records irregularities in a surface by moving the stylus of a crystal pickup or similar device over the surface, amplifying the resulting voltage, and feeding the output voltage to an indicator or recorder which shows the surface irregularities magnified as much as 50,000 times.

surface barrier A potential barrier formed at a surface of a semiconductor by the trapping of carriers at the surface. The effective area of the barrier is appreciably larger than for a point-contact transistor.

surface-barrier detector A semiconductor nuclear-particle detector in which the rectifying junction occurs between an evaporated gold layer and high-resistivity N-type silicon. Performance is much like that of a PN junction.

surface-barrier diode A diode that uses thin surface layers, formed either by deposition of metal films or surface diffusion, to serve as a rectifying junction.

surface-barrier transistor A transistor triode in which surface barriers are formed on opposite sides of a thin wafer of N-type germanium by

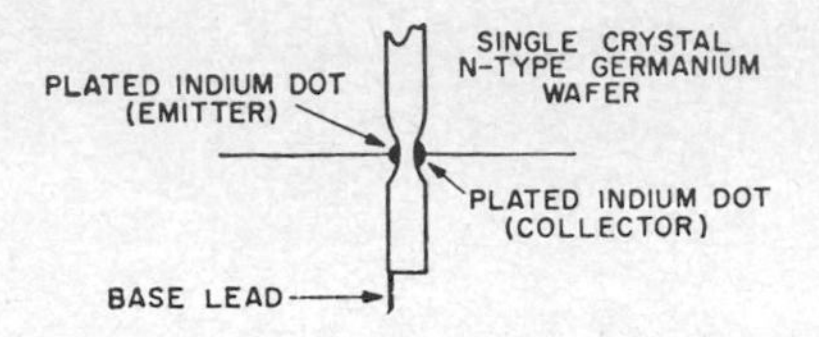

Surface-barrier transistor construction.

etching depressions, then electroplating the collector and emitter dots that serve as rectifying contacts.

surface-charge transistor An integrated-circuit transistor element based on controlling the trans-

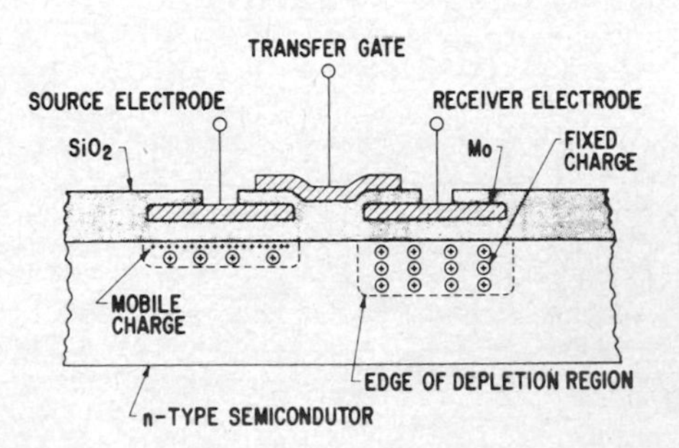

Surface-charge transistor cross section.

fer of stored electric charges along the surface of a semiconductor.

surface-contact rectifier A rectifier in which a surface barrier serves as the rectifying contact.

surface density The quantity per unit area of anything distributed over a surface. In nuclear physics, mass per unit area is used to indicate absorber, source, target, and support thicknesses for thin radioactive sources.

surface duct An atmospheric duct for which the lower boundary is the surface of the earth.

surface hardening Hardening of a metallic surface by rapid induction heating and rapid quenching.

surface leakage The passage of current over the surface of an insulator.

surface magnetic wave A magnetostatic wave that can be propagated on the surface of a magnetic material, as on a slab of yttrium-iron garnet (YIG).

surface noise The noise component in the electric output of a phonograph pickup due to irregularities in the contact surface of the groove. Also called needle scratch.

surface of position A surface defined by a constant value of some navigation coordinate.

surface-passivated diode A diode in which the semiconductor assembly on one side of a silicon substrate is hermetically sealed by a production process that combines surface passivation with glass-to-silicon seals.

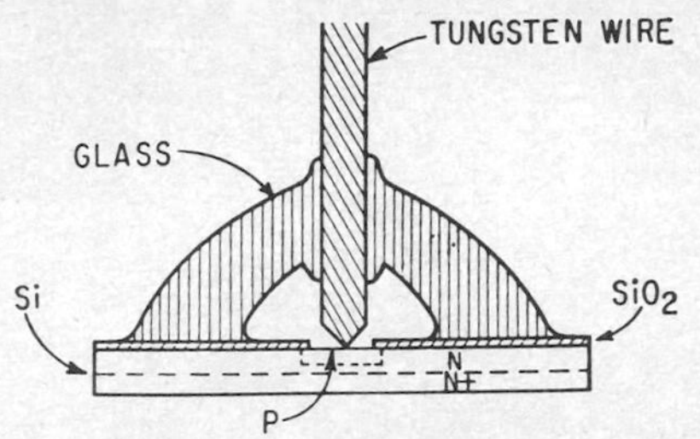

Surface-passivated diode, with silicon dioxide providing surface passivation of silicon substrate for point-contact diode.

surface-passivated transistor A transistor whose semiconductor surfaces have been protected against water, ions, and other environmental conditions by passivation, in which a protective compound is chemically bonded to the surface of the semiconductor crystal.

surface passivation A method of coating the surface of a P-type wafer for a diffused-junction transistor with an oxide compound like silicon oxide, to prevent penetration of the impurity in undesired regions.

surface photoelectric effect Ejection of an electron from the surface of a solid or liquid by an incident photon whose total energy is absorbed by the material.

surface recombination rate The time rate at which free electrons and holes recombine at the surface of a semiconductor.

surface recombination velocity The velocity with which electrons and holes drift to the surface of a semiconductor and recombine.

surface resistivity The electric resistance of the surface of an insulator, measured between the opposite sides of a square on the surface. The value in ohms is independent of the size of the square and the thickness of the surface film.

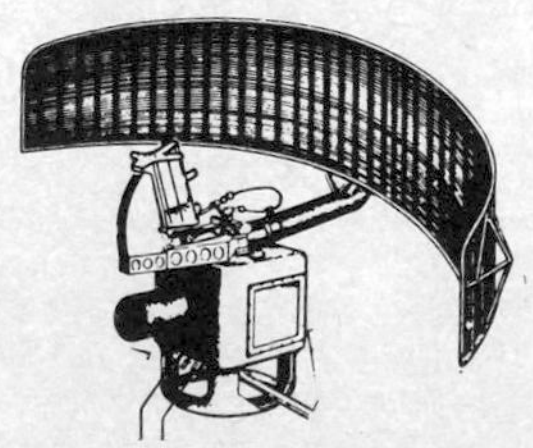

Surface-search radar used on ships for detecting low-flying aircraft and other ships within line-of-sight range.

surface-search radar Radar used on a ship to detect surface vessels and low-flying aircraft within line-of-sight distance; it gives range and bearing of each target while maintaining complete 360° azimuth search.

surface-to-air missile [abbreviated SAM] A guided missile that is fired at an airborne target from the ground or the deck of a a surface ship. Examples include Bomarc, Hawk, SAM-D, Talos, and Tartar.

surface-to-surface guided missile A guided missile that is fired at a surface target from a surface position on land or water. Examples include Mace, Pershing, Polaris, Regulus, Sergeant, and Titan.

surface wave 1. A wave that can travel along an interface between two different media without radiation. The interface must be essentially straight in the direction of propagation. The most common interface used is that between air and the surface of a metal conductor. 2. *Ground wave.*

surface-wave antenna An antenna energized in such a way that a surface wave is propagated along its structure. The dielectric-rod antenna is an example.

surface-wave chirp filter A surface-acoustic-wave filter in which the spacing of the interdigital electrodes is graded from one end of the array to the other. Used for pulse compression in chirp radar.

surface waveguide A waveguide in which the electromagnetic field is essentially on the outside of the guiding structure.

surface-wave transmission line A single-conductor transmission line energized in such a way that a surface wave is propagated along the line with satisfactorily low attenuation.

surge A momentary large increase in the current or voltage in an electric circuit.

surge admittance The reciprocal of characteristic impedance.

surge-crest ammeter A magnetometer used with magnetizable links to measure the peak value of transient electric currents.

surge generator A device that produces high-voltage pulses, usually by charging capacitors in parallel and discharging them in series. Also called impulse generator.

surge impedance *Characteristic impedance.*

surge suppressor A suppressor that responds to the rate of change of a current or voltage to prevent a rise above a predetermined value. It may include resistors, capacitors, coils, gas tubes, and semiconducting disks.

surveillance Systematic observation of air, surface, or subsurface areas or volumes by visual, electronic, photographic, or other means for intelligence or other purposes.

surveillance camera A simplified television camera, usually black and white, used with closed-circuit video equipment for security systems.

surveillance radar A radio navigation aid that

uses primary radar to display on a maplike circular screen the positions of all aircraft within its range. Examples are airport surveillance radar and ground surveillance radar.

survey instrument A portable instrument that detects and measures radiation.

susceptance The imaginary component of admittance.

susceptibility The ratio of the magnetization of a material to the magnetizing field.

susceptometer An instrument that measures paramagnetic, diamagnetic, or ferromagnetic susceptibility. In one superconducting version, the temperature may be held constant or varied over a range of 3 to 300 K.

suspension A fine wire or coil spring that supports the moving element of a meter.

sustained oscillation Continuous oscillation at a frequency essentially equal to the resonant frequency of the system.

sustaining program A program that has no commercial sponsor.

SW 1. Abbreviation for *shortwave.* 2. Symbol for *switch.* Used on circuit diagrams.

swamping resistor A resistor placed in the emitter lead of a transistor circuit to minimize the effects of temperature on the emitter-base junction resistance.

sweep 1. The steady movement of the electron beam across the screen of a cathode-ray tube, producing a steady bright line when no signal is present. The line is straight for a linear sweep and circular for a circular sweep. 2. The steady change in the output frequency of a signal generator from one limit of its range to the other.

sweep amplifier An amplifier used in a television receiver to amplify the sawtooth output voltage of the sweep oscillator and shape the waveform as required for the deflection circuits.

sweep-balance recorder A direct-writing recorder in which the chart is marked at the instant when the amplitude of a sweep voltage matches that of the input signal being recorded.

sweep circuit The sweep oscillator, sweep amplifier, and any other stages used to produce the deflection voltage or current for a cathode-ray tube.

sweep frequency The rate at which an electron beam is swept back and forth across the screen of a cathode-ray tube.

sweep-frequency reflectometer A reflectometer that measures standing-wave ratio and insertion loss in decibels over a wide range of frequencies, in either single- or sweep-frequency operation.

sweep generator A test instrument that generates an RF voltage whose frequency varies back and forth through a given frequency range at a rapid constant rate. Used to produce an input signal for circuits or devices whose frequency response is to be observed on an oscilloscope. Also called sweep oscillator.

sweep jammer A jammer that sweeps a narrow band of frequencies over a broad bandwidth occupied by enemy transmissions.

sweep jamming Jamming an enemy radarscope by sweeping the region of radar beam coverage with electromagnetic waves that have the same frequency as those received by the radarscope.

sweep oscillator 1. An oscillator that generates a sawtooth voltage which can be amplified to deflect the electron beam of a cathode-ray tube. Also called time-base generator and timing-axis oscillator. 2. *Sweep generator.*

sweep-through jamming Jamming in which a transmitter is swept through an RF band in short steps to jam each frequency briefly, producing a sound like that of an aircraft engine.

sweep voltage The periodically varying voltage applied to the deflection plates of a cathode-ray tube to give a beam displacement that is a function of time. Also called time-base voltage.

swept-frequency measurement Measurement of magnitude and phase parameters of a device, component, or system as a function of frequency.

swept-frequency modulation *Chirp modulation.*

swept gain control *Sensitivity-time control.*

swing The total variation in the frequency or amplitude of a quantity.

swinging Momentary variation in the frequency of a received radio wave.

swinging choke An iron-core choke that has a core which can be operated almost at magnetic saturation. The inductance is then a maximum for small currents and swings to a lower value as current increases. Used as the input choke in a power supply filter to provide improved voltage regulation.

switch [symbol SW] A manual or mechanically actuated device for making, breaking, or changing the connections in an electric circuit.

switchboard A single large panel or assembly of panels on which are mounted switches, circuit breakers, meters, fuses, and terminals essential to the operation of electric equipment.

switch detent A detent used on a switch to establish predetermined switching positions.

Switched Telecommunications Network A specialized common-carrier communication service owned by telephone companies; it provides both analog and digital information transmission over cable and microwave radio links.

switching diode A crystal diode that provides essentially the same function as a switch. Below a specified applied voltage it has high resistance corresponding to an open switch, and above that voltage it suddenly changes to the low resistance of a closed switch.

switching key *Key.*

switching reactor A saturable-core reactor that has several input control windings and one or more output windings which essentially duplicate the functions of a relay.

switching regulated power supply A regulated

power supply in which the line voltage is rectified, converted to a pulse-width-modulated high-frequency AC voltage such as 20 kHz, stepped up or down to a desired voltage by a transformer, then rectified and filtered.

switching time 1. The time interval between the reference time and the last instant at which the instantaneous voltage response of a magnetic cell reaches a stated fraction of its peak value. 2. The time interval between the reference time and the first instant at which the instantaneous integrated voltage response of a magnetic cell reaches a stated fraction of its peak value.

switching transistor A transistor designed for ON/OFF switching operation. A planar transistor doped with gold to reduce excess minority-carrier density in the base is one example.

switching tube A gas tube that switches high-power RF energy in the antenna circuits of radar and other pulsed RF systems. Examples are ATR, pre-TR, and TR tubes.

SWL Abbreviation for *shortwave listener.*

SWR Abbreviation for *standing-wave ratio.*

SWVR Abbreviation for *standing-wave voltage ratio.*

syllabic companding Companding in which the effective gain variations are made at speeds that allow response to the syllables of speech but not to individual cycles of the signal wave.

symbol 1. A design used on diagrams to represent a component or identify specific characteristics, quantities, or objects. 2. A letter or abbreviation used on diagrams or in equations or text to represent a quantity or unit of measure, to identify an object.

symbolic address A label that identifies a particular word, function, or other information in a digital-computer programming routine, independent of the location of the information within the routine. Also called floating address.

symbolic coding Coding that uses symbols other than actual computer addresses, to make computer programming easier.

symbolic logic Logic in which symbols suitable for calculation are used to express nonnumerical relations for a computer. Boolean algebra is an example.

symbolic programming A method of programming in which arbitrary symbols are used instead of explicit numerical codes and addresses. The computer is then used to translate the symbols into machine language.

symmetrical avalanche rectifier An avalanche rectifier than can be triggered in either direction, after which it has a low impedance in the triggered direction.

symmetrical cyclically magnetized condition A condition of a magnetic material when it is in a cyclically magnetized condition and the limits of the applied magnetizing forces are equal and of opposite sign, so the limits of flux density are equal and of opposite sign.

symmetrical relay A relay that has two identical coils which may be used interchangeably as operate and reset coils.

symmetrical transducer A transducer in which all possible pairs of specified terminations may be interchanged without affecting transmission because the input and output image impedances are all equal.

symmetrical transistor A junction transistor in which the emitter and collector electrodes are identical and their terminals interchangeable.

sync 1. *Synchronization.* 2. *Synchronize.*

sync compression The reduction in gain applied to the sync signal over any part of its amplitude range with respect to the gain at a specified reference level.

sync generator An electronic generator that supplies synchronizing pulses to television studio and transmitter equipment. Also called sync-signal generator.

synchro [SYNCHROnous] Any of several devices used for transmitting and receiving angular position or angular motion over wires, such as a synchro transmitter or synchro receiver. Also called mag-slip (British), self-synchronous device, and selsyn.

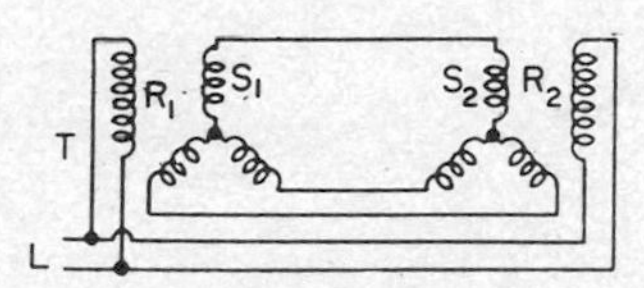

Synchro arrangement in which five wire lines connect synchro R1–S1 to R2–S2. When rotor R1 is turned, rotor R2 will follow closely.

synchro angle The angular displacement of a synchro rotor from its electrical zero position.

synchro control transformer A transformer that has its secondary winding on a rotor. When its three input leads are excited by angle-defining voltages, the two output leads deliver an AC voltage that is proportional to the sine of the difference between the electrical input angle and the mechanical rotor angle. The output voltage thus varies sinusoidally with rotor position, being essentially zero when mechanical and electrical angles are the same, and can be used for control purposes.

synchro control transmitter A high-accuracy synchro transmitter that has high-impedance windings.

synchrocyclotron A cyclotron in which the radio frequency of the electric field is frequency-modulated. Also called FM cyclotron.

synchro differential receiver A synchro receiver that subtracts one electrical angle from another and delivers the difference as a mechanical angle. One set of three input leads is excited by one set of angle-defining voltages. The other set of three input leads is excited by the other set of angle-defining voltages. The rotor rotates to the difference angle, with a torque proportional to the sine of the difference between the angles. Also called differential synchro.

synchro differential transmitter A synchro transmitter that adds a mechanical angle to an electrical angle and delivers the sum as an electrical angle. When its three input leads are excited by the electrical angle-defining voltages, the three output leads deliver voltages that define an angle which is the sum of the electrical input and the mechanical rotor angles. Also called differential synchro.

synchro generator *Synchro transmitter.*

synchro motor *Synchro receiver.*

synchronism The condition in which two or more varying quantities have the same speed or reach their peaks at the same instant of time.

synchronization The maintenance of one operation in step with another, as in keeping the electron beam of a television picture tube in step with the electron beam of the television camera tube at the transmitter. Also called sync.

synchronization error The error caused by imperfect timing of two operations in a navigation system.

synchronization indicator An indicator that presents visually the relationship between two varying quantities or moving objects.

synchronize To produce synchronization. Also called sync.

synchronized sweep A sweep voltage that is controlled by an AC voltage in such a way that the forward and return traces on a cathode-ray oscilloscope are exactly superimposed and appear as a single trace.

synchronizing Maintaining a fixed speed or phase relationship between two varying quantities or moving objects, as between two scanning processes.

synchronizing signal *Sync signal.*

synchronometer An instrument that counts the number of cycles produced by a signal source in a given time interval. It can serve as a master clock when driven by a frequency standard.

synchronous In step or in phase, as applied to two or more circuits, devices, or machines.

synchronous capacitor A synchronous motor that runs without mechanical load and draws a large leading current like a capacitor. Used to improve the power factor and voltage regulation of an AC power system.

synchronous clock An electric clock driven by a synchronous motor, for operation from an AC power system in which the frequency is accurately controlled.

synchronous computer A digital computer in which all operations are controlled by signals from a master clock.

synchronous converter A converter in which motor and generator windings are combined on one armature and excited by one magnetic field. Normally used to change AC power to DC power.

synchronous coupling An electric coupling in which torque is transmitted by attraction between magnetic poles on both rotating members.

synchronous data transmission Data transmission in which a clock defines transmission times for data. Since start and stop bits for each character are not needed, more of the transmission bandwidth is available for message bits.

synchronous demodulator *Synchronous detector.*

synchronous detector 1. A detector that inserts a missing carrier signal in exact synchronism with the original carrier at the transmitter. When the input to the detector consists of two suppressed-carrier signals in phase quadrature, as in the chrominance signal of a color television receiver, the phase of the reinserted carrier can be adjusted to recover either one of the signals. Two synchronous detectors, using carriers differing in phase by 90°, can thus extract the I and Q signals separately from the chrominance signal. Also called synchronous demodulator. 2. *Crosscorrelator.*

synchronous gate A time gate in which the output intervals are synchronized with an incoming signal.

synchronous generator *Alternator.*

synchronous inverter *Dynamotor.*

synchronous machine An AC machine whose average speed is proportional to the frequency of the applied or generated voltage.

synchronous motor A synchronous machine that transforms AC electric power into mechanical power, using field magnets excited with direct current.

synchronous orbit An orbit with the same period as that of the earth revolving about its axis, so

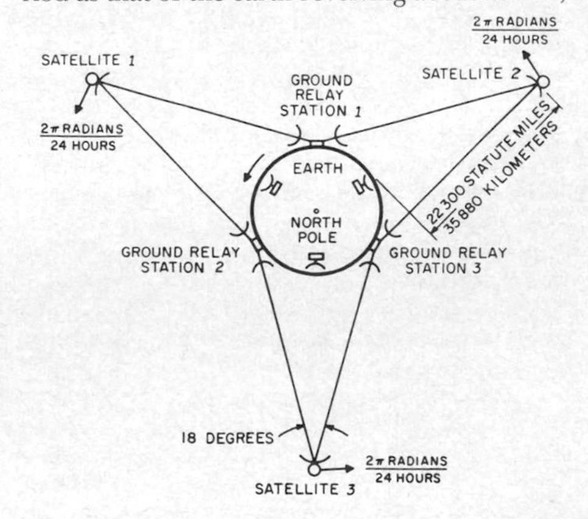

Synchronous orbit of communication satellites, at equator.

satellite rotational speed and the earth's speed of rotation are in synchronism.

synchronous radar bombing Radar bombing in which special airborne radar equipment that contains rate and steering mechanisms is used to control the direction of flight of the bombing aircraft, solve the bombing problem, and automatically drop bombs at the proper release point. Also called radar synchronous bombing.

synchronous rectifier A rectifier in which contacts are opened and closed at correct instants of time for rectification by a synchronous vibrator or a commutator driven by a synchronous motor.

synchronous satellite A satellite that is placed in orbit in a west-to-east direction 22,300 mi (35,880 km) above and parallel to the equator, where it completes an orbit of the earth in 24 h and therefore appears to be stationary. When used as a communication satellite, it can carry relay transmitters that greatly increase the range of television and radio transmission.

synchronous speed A speed value related to the frequency of an AC power line and the number of poles in a synchronous machine. Synchronous speed in revolutions per minute is equal to the frequency in hertz divided by the number of poles, with the result multiplied by 120.

synchronous switch A thyratron circuit that controls the operation of ignitrons in such applications as resistance welding.

synchronous vibrator An electromagnetic vibrator that simultaneously converts a low direct voltage to a low alternating voltage and rectifies a

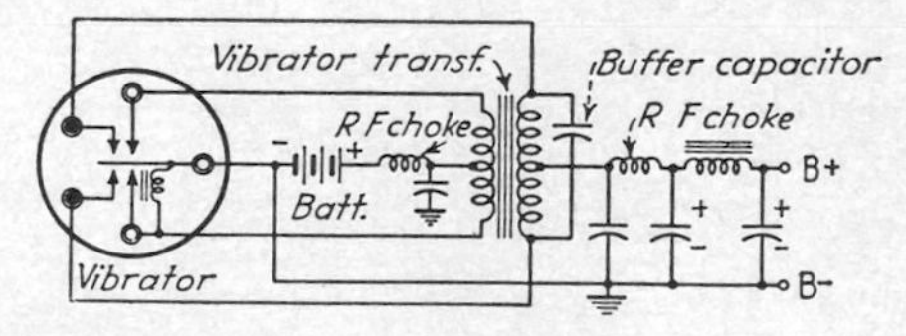

Synchronous-vibrator circuit.

high alternating voltage obtained from a power transformer to which the low alternating voltage is applied. In power packs, it eliminates the need for a rectifier tube.

synchronous voltage The voltage required to accelerate electrons from rest to a velocity equal to the phase velocity of a wave in the absence of electron flow in a traveling-wave tube.

synchro receiver A synchro that provides an angular position related to the applied angle-defining voltages. When two of its input leads are excited by an AC voltage and the other three input leads are excited by the angle-defining voltages, the rotor rotates to the corresponding angular position. The torque of rotation is proportional to the sine of the difference between the mechanical and electrical angles. Also called receiver synchro, selsyn motor, selsyn receiver, and synchro motor.

synchro resolver *Resolver.*

synchroscope 1. An instrument that indicates whether two periodic quantities are synchronous. The indicator may be a rotating-pointer device or a cathode-ray oscilloscope that provides a rotating pattern. The position of the rotating pointer is a measure of the instantaneous phase difference between the quantities. 2. A cathode-ray oscilloscope that shows a short-duration pulse by using a fast sweep which is synchronized with the pulse signal to be observed.

synchro system An electric system for transmitting angular position or motion. In the simplest form it consists of a synchro transmitter connected by wires to a synchro receiver. More complex systems include synchro control transformers and synchro differential transmitters and receivers. Also called selsyn system.

synchro-to-digital converter [abbreviated SDC] A converter that converts synchro or resolver output voltages into parallel binary data that represent angular position.

synchro transmitter A synchro that provides voltages related to the angular position of its rotor. When its two input leads are excited by an AC

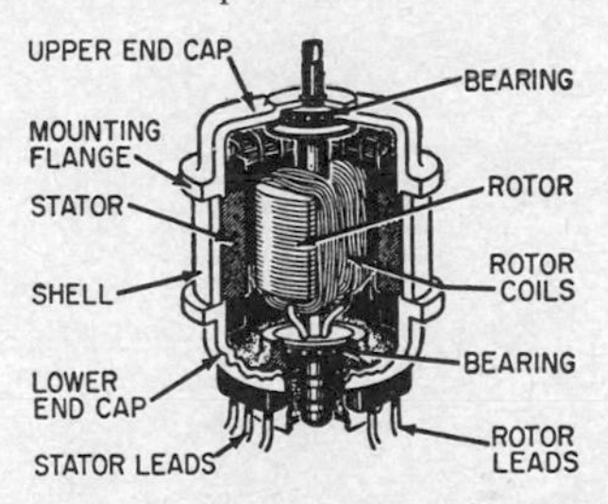

Synchro-transmitter construction.

voltage, the magnitudes and polarities of the voltages at the three output leads define the rotor position. Also called selsyn generator, selsyn transmitter, synchro generator, transmitter, and transmitter synchro.

synchrotron An accelerator similar to a betatron but also having a higher-frequency magnetic field that is applied in synchronism with the orbiting and rapidly accelerating charged particles, to give much higher beam energy than in a betatron. The two types are the electron and proton synchrotrons.

synchrotron radiation Electromagnetic radiation generated by the acceleration of charged relativistic particles, usually electrons, in a magnetic field. First encountered in a synchrotron.

synchro zeroing Lining up the zero positions of a synchro system with the zero position of the associated indicator or mechanism being controlled.

sync level The level of the peaks of the synchronizing signal in a television system.

sync limiter A limiter circuit used in television to prevent sync pulses from exceeding a predetermined amplitude.

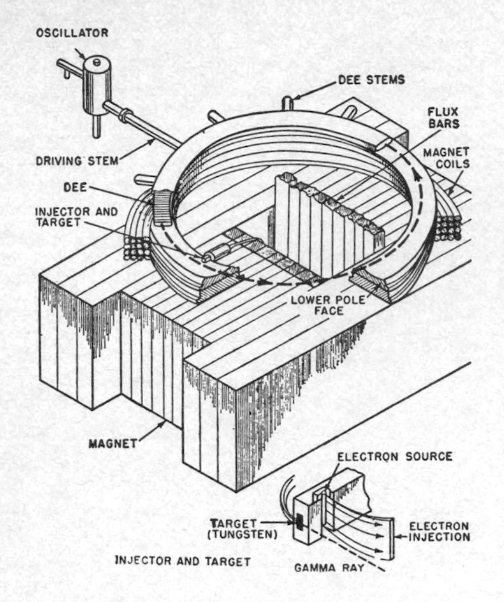

Synchrotron construction, with top half of magnet removed.

sync pulse One of the pulses that make up a sync signal.

sync separator A circuit that separates synchronizing pulses from the video signal in a television receiver. The signal for the sync separator is usually taken from the anode circuit of the video amplifier.

sync signal A signal transmitted after each line and field to synchronize the scanning process in a television or facsimile receiver with that of the transmitter. The picture, blanking, and sync signals together make up the composite picture signal in a television system. Also called synchronizing signal.

sync-signal generator *Sync generator.*

synergic curve A curve plotted for the ascent of a rocket, space-air vehicle, or space vehicle, calculated to give optimum fuel economy and optimum velocity. To minimize air resistance, the curve starts off vertically and bends toward the horizontal starting at an altitude of about 40 km.

synthetic-aperture radar A radar system in which an aircraft moving along a very straight path emits microwave pulses continuously at a frequency constant enough to be coherent for a period during which the aircraft may have traveled about 1 km. All echoes returned during this period can then be processed just as if a single antenna as long as the flight path had been used. Such a long "synthetic" aperture has high resolving power, giving displays with extremely fine detail. Used in terrain mapping.

synthetic mica A fluor-phlogopite mica made artificially by heating a large batch of raw material in an electric resistance furnace and letting the mica crystallize from the melt during controlled slow cooling.

synthetic quartz A quartz crystal that is grown commercially at high temperature and pressure around a seed of quartz suspended in a solution which contains scraps of natural quartz crystals.

syntony The condition wherein two or more oscillators have exactly the same resonant frequency.

system A combination of several pieces of equipment integrated to perform a specific function. Thus a fire control system may include a tracking radar, computer, and gun.

systematic errors Errors that have an orderly character and can be corrected by calibration.

system deviation The value of the ultimately controlled variable minus the ideal value in a feedback control system.

Système International d'Unités [abbreviated SI] French equivalent of *International System of Units.*

system engineering An engineering approach that takes into consideration all the elements related in any way to the equipment under development, including utilization of manpower and the characteristics of each component of the system.

system error The ideal value minus the value of the ultimately controlled variable in a feedback control system.

system noise The output of a system when operating with zero input signal.

systems analysis The analysis of an activity, procedure, method, technique, or business to determine what must be accomplished and how the necessary operations may best be accomplished.

T

t Abbreviation for *tonne.*

T 1. Abbreviation for *tera-*. 2. Abbreviation for *tesla.* 3. Symbol for *transformer.* Used on circuit diagrams. 4. Symbol for *tritium.*

TA Symbol for *target.* Used on tube-base diagrams.

table A collection of data, each item uniquely identified by some label or its relative position in a computer location.

table lookup Obtaining a function value that corresponds to an argument, stated or implied, from a table of function values stored in a computer.

table-model receiver A radio or television receiver that has a cabinet of suitably small size for use on a table. The same set with a carrying handle is called a portable receiver.

tabulating equipment Equipment that uses punched cards.

tabulator A machine that reads information from punched cards and produces lists, tables, and totals on separate forms or continuous paper.

tacan [TACtical Air Navigation] An air navigation system in which a single UHF transmitter sends out signals that actuate airborne equipment to provide range and bearing indications with respect to the transmitter location, when interrogated by a transmitter in the aircraft. Each tacan station broadcasts a location-identifying Morse code signal at regular intervals. Also called tactical air navigation.

tachometer An instrument that measures angular speed in revolutions per minute. An electric tachometer delivers an output voltage that is proportional to speed.

tachometer generator A tachometer that uses a construction similar to that of a drag-cup motor. A drag cup rotates in an alternating magnetic field created by an excitation winding, producing in the output winding a voltage proportional to the speed at which the tachometer generator is driven.

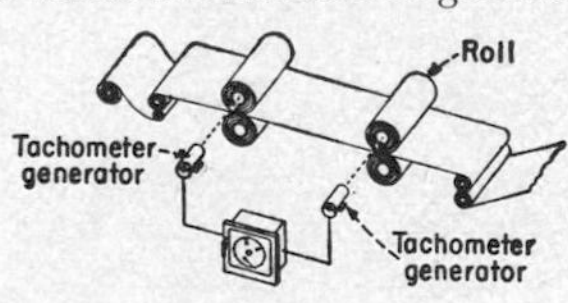

Tachometers used with control equipment for synchronizing speeds of two rolls.

tactical air navigation *Tacan.*

tactical missile A guided missile used in tactical operations against or in the presence of a hostile force.

tactical radar A radar set used in operations against or in the presence of a hostile force.

tactile feel A sudden change in pressure or a click that indicates when a key has been depressed sufficiently on a calculator or other keyboard machine.

tag *Flag.*

tagged atom An atom of an isotopic tracer.

tagging *Labeling.*

tail 1. A small pulse that follows the main pulse of a radar set and rises in the same direction. 2. The trailing edge of a pulse.

tail-end radar detector A wide-open receiver used in the tail of a military airplane to warn the crew that their craft is under radar surveillance from the ground or from another plane at their rear.

tailing Excessive prolongation of the decay of a

signal. Also called hangover.

tail warning radar Radar installed in the tail of an aircraft to warn the pilot that an aircraft is approaching from the rear.

takeoff *Sound takeoff.*

takeup reel The reel that accumulates magnetic tape after the tape is recorded or played by a tape recorder.

talk-back circuit *Interphone.*

talk-down system *Ground-controlled approach.*

talking Rebecca-Eureka system A modified version of Rebecca-Eureka that allows two-way voice communication between an aircraft and a ground station.

talk-listen switch A switch on intercommunication units to permit using the loudspeaker as a microphone when desired.

Talos A Navy surface-to-air guided missile that has a speed of about Mach 3, a range of about 25 mi (120 km), and beam-rider guidance. One version can carry nuclear warheads and can also be used against ships and shore bombardment targets.

tangent galvanometer A galvanometer in which a small compass is mounted horizontally in the center of a large vertical coil of wire. The current through the coil is proportional to the tangent of the angle of deflection of the compass needle from its normal position parallel to the magnetic field of the earth.

tangential component A component that acts at right angles to a radius.

tangential wave path The path of propagation of a direct wave that is tangential to the surface of the earth. The path is curved by atmospheric refraction.

tank 1. A unit of acoustic delay-line storage that contains a set of channels, each channel forming a separate recirculation path. 2. The heavy metal envelope of a large mercury-arc rectifier or other gas tube that has a mercury-pool cathode. 3. *Tank circuit.*

tank circuit A circuit capable of storing electric energy over a band of frequencies continuously distributed about the resonant frequency, such as a coil and capacitor in parallel. The selectivity of the circuit is proportional to the *Q* factor, which is the ratio of the energy stored in the circuit to the energy dissipated. Also called tank.

tantalum [symbol Ta] A metallic element used for the electrodes of electron tubes because of its high absorption rate for residual gases. Also used for the plates of electrolytic capacitors. Atomic number is 73.

tantalum capacitor An electrolytic capacitor in which the anode is some form of tantalum. Examples include solid tantalum, tantalum-foil electrolytic, and tantalum-plug electrolytic capacitors.

tantalum chip capacitor A miniaturized solid tantalum capacitor for installation on printed circuits and hybrid integrated circuits. A copper coating process is generally used so the capacitor can withstand the high temperature of solder reflow during mounting.

tantalum-foil electrolytic capacitor An electrolytic capacitor that uses plain or etched tantalum foil for both electrodes, with a weak acid electrolyte.

tantalum nitride resistor A thin-film resistor that consists of tantalum nitride deposited on a substrate such as industrial sapphire.

tantalum oxide capacitor A capacitor that uses as a dielectric a film of tantalum oxide which is grown or deposited on substrate material in thick-film or thin-film circuits.

tantalum-slug electrolytic capacitor An electrolvtic capacitor that uses a sintered slug of tantalum as the anode, in a highly conductive acid electrolyte. Some types may be used at operating temperatures as high as 200°C.

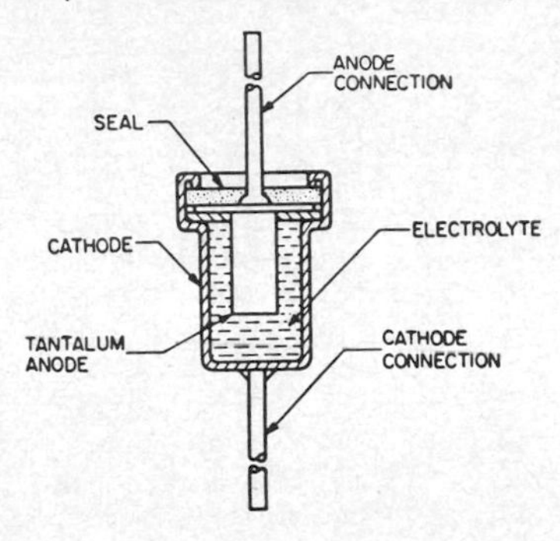

Tantalum-slug electrolytic capacitor construction.

tantalum thin-film circuit A thin-film circuit in which tantalum film is used directly for resistors, and anodically oxidized tantalum film is used for capacitors. The pieces of film are mounted on a silicon wafer along with transistors and diodes. Interconnections are produced by evaporation of molten metal through appropriate masks.

T antenna An antenna that consists of one or more horizontal wires, with a lead-in connection being made at the approximate center of each wire.

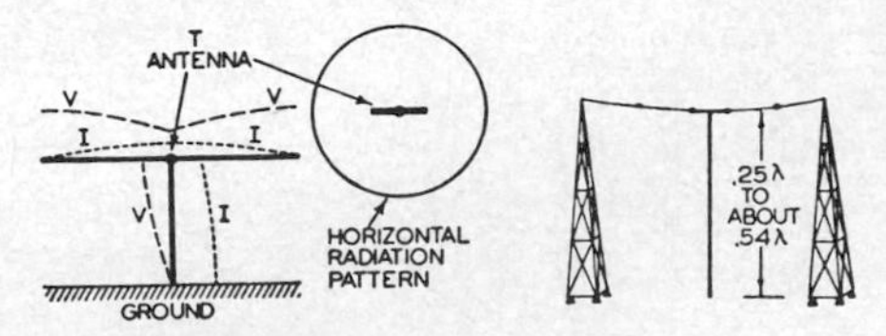

T-antenna voltage and current distribution curves, radiation pattern, and construction when supported by two towers.

tap A connection made at some point other than the ends of a resistor or coil.

tape 1. *Magnetic tape.* 2. *Punched tape.*

tape cartridge A cartridge which holds a length of magnetic tape in such a way that the cartridge can be slipped into a tape recorder and played without threading the tape. The tape may be an endless loop feeding out from the center of a roll and back onto the roll on the outside, or it may be an ordinary length of tape that runs back and forth between two reels inside the cartridge. Also called cartridge.

tape cassette *Cassette.*

tape comparator *Tape verifier.*

tape deck A tape-recorder mechanism mounted on a motor board; it includes the tape transport and the bias and erase oscillators but no preamplifier, power amplifier, loudspeaker, or cabinet.

tape drive *Tape transport.*

tape eraser *Bulk eraser.*

tape guides Grooved pins of nonmagnetic material mounted at either side of a magnetic recording head assembly to position the magnetic tape on the heads as the tape is being recorded or played.

tape handler *Tape transport.*

tape-limited Pertaining to a computer operation in which the time required to read and write tapes exceeds the time required for computation.

tape loop A length of magnetic tape that has the ends spliced together to form an endless loop. Used in message repeater units and some types of tape cartridges, to eliminate the need for rewinding the tape.

tape operating system [abbreviated TOS] A computer operating system in which source programs and sometimes incoming data are stored on magnetic tape, rather than in the computer memory.

tape player A machine only for playback of recorded magnetic tapes. Also called tape reproducer.

tape punch A machine that punches code and feed holes in paper tape.

taper The manner in which resistance is distributed throughout the element of a potentiometer or rheostat. Uniform distribution, having the same resistance per unit length throughout the element, is called linear taper. Nonuniform distribution is called nonlinear, left-hand, or right-hand taper.

Taper of potentiometer.

tape reader *Paper-tape reader.*

tape recorder A recorder that records audio signals and other information on magnetic tape by selective magnetization of iron oxide particles which form a thin film on the tape. A recorder usually also includes provisions for playing back the recorded material.

tapered transmission line *Tapered waveguide.*

tapered waveguide A waveguide in which a physical or electrical characteristic changes continuously with distance along the axis of the waveguide. Also called tapered transmission line.

tape reperforator A machine that automatically punches a paper tape from received signals. Also called reperforator.

tape reproducer *Tape player.*

tape speed The speed at which magnetic tape moves past the recording head in a tape recorder. Standard speeds are $\frac{15}{16}$, $1\frac{7}{8}$, $3\frac{3}{4}$, $7\frac{1}{2}$, 15, and 30 in/s (2.38, 4.76, 9.5, 19, 38, and 76 cm/s). Faster speeds give improved high-frequency response under given conditions.

tape splicer A device that splices magnetic tape automatically or semiautomatically by using splicing tape or by fusing with heat.

tape-to-card converter A machine that converts information directly from punched or magnetic tape to cards.

tape transmitter 1. A code-transmitting machine actuated by previously punched paper tape. Used for high-speed sending because the tape can be fed through the machine much faster than it was originally punched. 2. A facsimile transmitter that transmits subject copy printed on narrow tape.

tape transport The mechanism on the deck of a tape recorder that holds the tape reels, drives the

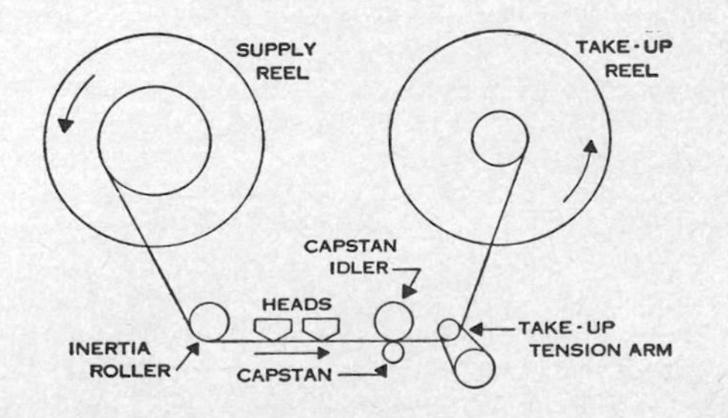

Tape transport.

tape past the recording heads, and controls various modes of operation. Also called tape drive and tape handler.

tape verifier A verifier for checking the accuracy of a punched paper tape by comparing it with a second manual punch of the same data. The machine stops whenever a character being punched the second time differs from that on the first tape. Also called tape comparator.

tape-wound core A length of ferromagnetic material in tape form, wound in such a way that each turn falls directly over the preceding turn.

tapped control A rheostat or potentiometer that has one or more fixed taps along the resistance element, usually to provide a fixed grid bias or for automatic bass compensation.

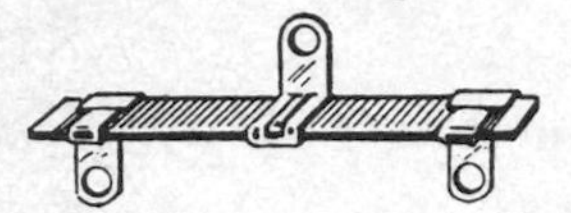
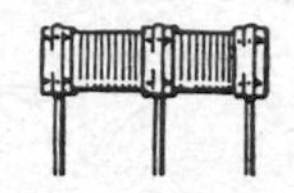

Tapped resistors.

tapped resistor A wirewound fixed resistor that has one or more additional terminals along its length, generally for voltage-divider applications.

tap switch A multicontact switch used chiefly for connecting one circuit lead to any one of a number of taps on a resistor or coil.

target [symbol TA] A substance or object exposed to bombardment or irradiation by nuclear particles, electrons, or electromagnetic radiation. In an x-ray tube the target is the anode or anticathode, from which x-rays are emitted as a result of electron bombardment. In a nuclear reaction it is the initially stationary atom or nucleus. In radar

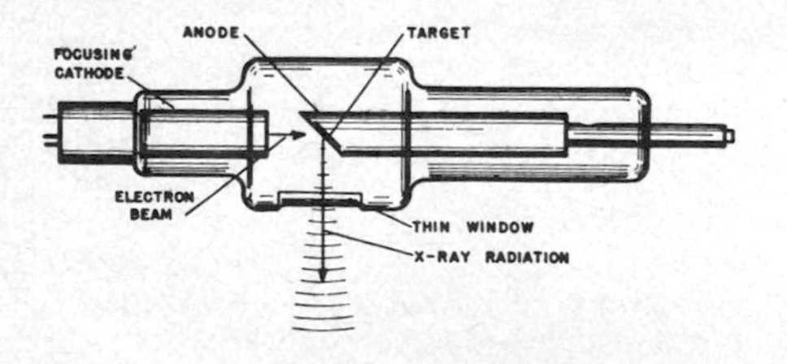

Target in conventional x-ray tube.

and sonar it is any object capable of reflecting the transmitted beam. In a television camera tube it is the storage surface that is scanned by an electron beam to generate an output signal current which corresponds to the charge-density pattern stored there. In a cathode-ray tuning indicator tube it is one of the electrodes that is coated with a material that fluoresces under electron bombardment.

target acquisition The first appearance of a recognizable and useful echo signal from a new target in radar and sonar.

target discrimination The ability of a detection or guidance system to distinguish a target from its background or to discriminate between two or more targets that are close together.

target drone A pilotless aircraft controlled by radio from the ground or from a mother ship and used exclusively as a target for antiaircraft weapons.

target fade A momentary reduction in the strength of an echo signal from a radar or sonar target, due to interference or other phenomena. Tracking radar usually includes memory circuits that maintain tracking during this period, to prevent loss of the target.

target glint *Scintillation.*

target identification Identification of a target as to its nature and whether it is friend or foe.

target language *Object language.*

target noise Statistical variations in a radar echo signal due to the presence on the target of a number of reflecting elements randomly oriented in space. Target noise can cause scintillation.

target scintillation *Scintillation.*

target seeker A missile that has a self-contained system which provides homing guidance to the target. Also called homer.

target timing A radar technique for determining wind velocity by correlating the distance that a radar target travels across a radar screen with other known speed data. Used particularly in arctic regions.

Tartar A Navy surface-to-air guided missile intended primarily for use on destroyers.

TASI [Time Assignment Speech Interpolation] A method of increasing or even doubling the capacity of a transatlantic telephone-cable circuit by using intervals of silence on each circuit for transmitting information from other channels. The signals for a large group of telephone conversations are split into time segments, and only the segments carrying information are transmitted, after having been labeled in such a way that they can be directed to the correct receiver. The cost of the complex switching circuits required is small in comparison with the cost of laying another cable.

tau meson Former name for *K meson.*

taut-band meter An analog panel meter in which the moving element is supported between two extremely fine metal bands under tension. The bands twist when the moving element of the meter rotates under magnetic action and thereby supply return torque while providing electric connections to the moving element.

taxi radar *Airport surface detection equipment.*

Tb Abbreviation for *terabit.*

TC Abbreviation for *temperature coefficient.*

T carrier A time-division multiplex system that uses digital transmission of pulse-code-modulation-encoded information over cables which contain twisted pairs or over microwave radio links.

Tchebycheff French spelling of *Chebyshev.*

T circulator A circulator in which three identical rectangular waveguides are joined asymmetrically to form a T-shaped structure, with a ferrite post or wedge at its center. Each port must be separately matched. Power entering any waveguide will emerge from only one adjacent waveguide.

TCXO Abbreviation for *temperature-compensated crystal oscillator.*

TDM Abbreviation for *time-division multiplex.*

TDMA Abbreviation for *time-division multiple-access.*

TDR 1. Abbreviation for *temperature-dependent resistor.* 2. Abbreviation for *time-domain reflectometer.*

TE Abbreviation for *transferred electron.*

TEA Abbreviation for *transferred-electron amplifier.*

TEA laser Abbreviation for *transversely excited atmospheric-pressure laser.*

tearing A television picture defect in which groups of horizontal lines are displaced in an irregular manner, caused by inadequate horizontal synchronization.

technetium [symbol Tc] An element produced by bombarding molybdenum with deuterons and neutrons. It also occurs among the fission products of uranium. Atomic number is 43.

technical representative [abbreviated tech rep] A person who represents one or more manufacturers in an area and gives technical advice on the application, installation, operation, and maintenance of their products in addition to selling the products.

technician An engineering assistant.

tech rep Abbreviation for *technical representative.*

Teflon Trademark of Du Pont for their polytetrafluoroethylene resin. Used as a dielectric in some capacitors.

telautograph A writing telegraph instrument in which manual movement of a pen at the transmitting position varies the current in two circuits in such a way as to cause corresponding movements of a pen at the remote receiving instrument. Ordinary handwriting can thus be transmitted over wires. Also called telewriter.

tele- Prefix meaning from a distance.

teleammeter A telemeter that measures and transmits current values to a remote point.

telecamera *Television camera.*

telecast [TELEvision broadCAST] The transmission of a television program intended for reception by the general public.

telecasting Broadcasting a television program.

telecommunication Any transmission, emission, or reception of signals, writing, images, sounds, or intelligence of any nature by wire, radio, visual, or other electromagnetic systems. The terms telecommunication and communication are often used interchangeably.

teleconference A conference in which the participants are some distance apart but are able to talk to each other and sometimes also see each other by telecommunication.

telegraph channel A path suitable for the transmission of telegraph signals between two telegraph stations, either over wires or by radio. It may be one of several channels on a single radio or wire circuit, providing simultaneous transmission in the same frequency range, simultaneous transmission in different frequency ranges, or successive transmission.

telegraph circuit The complete wire or radio circuit over which signal currents flow between transmitting and receiving apparatus in a telegraph system.

telegraph distributor A device that effectively associates one DC or carrier telegraph channel in rapid succession with the elements of one or more sending or receiving devices.

telegraph key A hand-operated telegraph transmitter, used to form telegraph signals.

telegraph level The signal power at a specified point in a telegraph circuit when one telegraph channel is in the marking or continuous-tone condition and all other channels are silenced.

telegraph modem The complete equipment for modulating and demodulating one or more separate telegraph circuits, each circuit containing one or more telegraph channels.

telegraph-modulated wave A wave obtained by varying the amplitude or frequency of a continuous wave by telegraphic keying.

telegraph repeater A repeater inserted at intervals in long telegraph lines to amplify weak code signals, with or without reshaping of pulses, and retransmit them automatically over the next section of the line.

telegraph sounder A telegraph receiving instrument in which an electromagnet attracts an armature each time a pulse arrives. The armature makes an audible sound as it hits against its stops at the beginning and end of each current pulse. The intervals between these sounds are varied in accordance with the telegraph code. Also called sounder.

telegraph transmitter A device that controls a source for electric power, to form telegraph signals for radio or wire transmission. It may be a telegraph key or a motor-driven sender that uses previously punched tape.

telegraphy Communication at a distance by code signals that consist of current pulses sent over wires or by radio.

telemeter 1. The complete measuring, transmitting, and receiving apparatus for indicating or recording the value of a quantity at a distance. Also called telemetering system. 2. To transmit the value of a measured quantity to a remote point over wires or by radio.

telemeter band One of up to 18 subcarrier bands used to modulate a carrier in the standard FM/FM telemetering system.

telemeter channel A complete circuit for transmission of one telemetered function, including pickup, commutator, modulator, transmitter, receiver, detector, decoder, and recorder.

telemetering Transmitting the readings of instruments to a remote location by wires, radio waves, or other means. Also called electric telemetering, remote metering, and telemetry.

telemetering antenna A highly directional antenna, generally mounted on a servo-controlled mount for tracking purposes, used at ground stations to receive telemetering signals from a guided missile or spacecraft.

Telemetering antenna.

telemetering system *Telemeter.*

telemeter pickup A device that measures and converts data to be telemetered into a form suitable for modulation of a telemetering channel.

telemetric data analyzer An analyzer that amplifies telemetered data and separates the individual values.

telemetric data monitor A monitor that provides a continuous visual display of telemetered signals.

telemetric data receiving set A complete electronic system that intercepts, demodulates, displays, and records data propagated by the transmitter of a telemeter.

telemetry *Telemetering.*

telephone A system for converting sound waves into electric variations that can be sent over wires and reproduced at a distant point. Used primarily for voice communication. It consists essentially of a telephone transmitter and receiver at each station, interconnecting wires, signaling devices, a central power supply, and switching facilities. Also called phone.

telephone capacitor A small fixed capacitor connected in parallel with a telephone receiver to bypass higher audio frequencies and thereby reduce noise.

telephone carrier current A carrier current used for telephone communication over power lines or to obtain more than one channel on a single pair of wires.

telephone channel A one- or two-way path suitable for the transmission of audio signals between two stations.

telephone circuit The complete circuit over which audio and signaling currents travel in a telephone system between the two telephone subscribers in communication with each other. The circuit usually consists of insulated conductors because ground returns are now rarely used in telephony.

telephone current An electric current produced or controlled by the operation of a telephone transmitter.

telephone dial A switch operated by a finger wheel, used to make and break a pair of contacts the required number of times for setting up a telephone circuit to the party being called.

telephone induction coil A coil used in a telephone circuit to match the impedance of the line to that of a telephone transmitter or receiver.

telephone jack *Phone jack.*

telephone line The conductors extending between telephone subscriber stations and central offices.

telephone loading coil *Loading coil.*

telephone modem A piece of equipment that modulates and demodulates one or more separate telephone circuits, each circuit containing one or more telephone channels. It may include multiplexing and demultiplexing circuits, individual amplifiers, and carrier-frequency sources.

telephone pickup A large flat coil placed under a telephone set to pick up both voices during a telephone conversation for recording purposes.

telephone plug *Phone plug.*

telephone receiver The portion of a telephone set that converts the AF current variations of a telephone line into sound waves.

telephone relay A relay that has a multiplicity of contacts on long spring strips mounted parallel to the coil, actuated by a lever arm or other projection of the hinged armature. Used chiefly for switching in telephone circuits.

Telephone-relay construction.

telephone repeater A repeater inserted at one or more intermediate points in a long telephone line to amplify telephone signals, to maintain the required current strength.

telephone repeating coil A coil used in a telephone circuit for inductively coupling two sections of a line when a direct connection is undesirable.

telephone retardation coil A coil used in a telephone circuit to pass direct current while offering appreciable impedance to alternating current.

telephone ringer An electromagnetic device that actuates a clapper which strikes one or more gongs to produce a ringing sound. Used with a telephone set to signal a called party.

telephone set An assembly that includes a telephone transmitter, a telephone receiver, and associated switching and signaling devices.

telephone switchboard A switchboard for interconnecting telephone lines and associated circuits.

telephone transmitter The microphone used in a telephone set to convert speech into AF signals.

telephony The transmission of speech and sounds to a distant point for communication purposes.

telephoto *Facsimile.*

telephotography *Facsimile.*

telephoto lens A lens that has a long focal length, used in television cameras to secure large images of distant objects.

teleprinter An electric typewriter that can be actuated by signals received over wire lines and can similarly generate output signals while producing hard copy. Used as a data-communication terminal. It may also have a paper-tape punch, paper-tape reader, and magnetic-tape transport.

teleran [TELEvision Radar Air Navigation] A radar navigation system in which the positions of aircraft are determined by ground radar. The resulting PPI display is then superimposed on a map and transmitted to the aircraft by television, so each of the pilots can see the position of their aircraft in relation to others in the vicinity.

telestimulator A system that stimulates the brain and does not use wire connections to the control unit.

Teletype Trademark of Teletype Corp. for their teletypewriters.

Teletypesetter Trademark of Fairchild Camera & Instruments Corp. for apparatus used to make a linotype machine set and cast type automatically in response to punched paper tape or equivalent electric signals.

teletypewriter [abbreviated TTY] A special electric typewriter that produces coded electric signals which correspond to manually typed characters and automatically types messages when fed with similarly coded signals produced by another machine. The signals may be transmitted directly over wires that connect the machines or used to drive a teletypewriter perforator which produces punched paper tape for later transmission.

teletypewriter exchange service [abbreviated TWX] A direct-dialing point-to-point service that uses teleprinter equipment. The service also permits interfacing with computers. Also called telex.

teletypewriter perforator An electromechanical perforator that punches teletypewriter code combinations on a paper tape when connected to a manually operated teletypewriter.

televise To pick up a scene with a television camera and convert it into corresponding electric signals for transmission by a television station.

television [abbreviated TV] A system that converts a succession of visual images into corresponding electric signals and transmits these signals by radio or over wires to distant receivers at which the signals can be used to reproduce the original images.

television broadcast band The band extending from 54 to 890 MHz, in which are the 6-MHz channels assignable to television broadcast stations in the United States. The frequencies are 54 to 72 MHz (channels 2 through 4), 76 to 88 MHz (channels 5 and 6), 174 to 216 MHz (channels 7 through 13), and 470 to 890 MHz (channels 14 through 83).

television broadcast station A station in the television broadcast band that transmits simultaneous visual and aural signals intended to be received by the general public.

television cable Coaxial cable capable of transmitting the bandwidth required for television signals, which is about 4.5 MHz.

television camera The pickup unit that converts a scene into corresponding electric signals. Optical lenses focus the scene to be televised on the photosensitive surface of a camera tube. This tube breaks down the visual image into small picture elements and converts the light intensity of each element in turn into a corresponding electric signal. Also called camera and telecamera.

television camera tube *Camera tube.*

television channel A band of frequencies 6 MHz wide in the television broadcast band, available for assignment to a television broadcast station.

television chart A test chart used for checking the resolution of a television system.

television engineering The field of engineering that deals with the design, manufacture, and testing of equipment required for the transmission and reception of television programs.

television film scanner A motion-picture projector adapted for use with a television camera tube to televise 24 frame per second motion-picture film at the 30 frame per second rate required for television.

television game *Video game.*

television-guided bomb A bomb that carries a small television camera in its nose for guidance. The camera system can be locked on the target before the bomb is dropped, for self-guidance, or the pilot can monitor the camera picture over a microwave relay link and adjust the course of the bomb by remote control after dropping the bomb.

television interference [abbreviated TVI] Interference produced in television receivers by amateur radio and other transmitters.

television picture photography To obtain photographs of the screen of a television receiver, use a speed of 1/30 s (1/25 is closest on most cameras) to get one complete frame (two fields). Set the aperture according to the reading of an exposure meter, using the daylight exposure index. Very fast pan film, such as Tri-X with an ASA daylight index of 200, is needed. A tripod is essential. Turn out room lights, and use the maximum brilliance that still gives a clear picture.

television picture tube *Picture tube.*

television receiver A receiver that converts incoming television signals into the original scenes along with the associated sounds. Also called television set.

television reconnaissance Reconnaissance in which television is used to transmit a scene from the reconnoitering point to another location on the surface or in the air.

television recording *Kinescope recording.*

television relay system *Television repeater.*

television repeat-back guidance Command guidance in which a television camera and transmitter are mounted in a guided missile or pilotless vehicle to provide a view ahead for the operator at the remote-control location.

television repeater A repeater that transmits television signals from point to point by using radio waves in free space as a medium, such transmission not being intended for direct reception by the public. Also called television relay system.

television screen The fluorescent screen of the picture tube in a television receiver.

television set *Television receiver.*

television signal General term for the aural and visual signals that are broadcast together to provide the sound and picture portions of a television program.

television studio-transmitter link A fixed station that transmits television program material and related communications from a studio to the transmitter of a television broadcast station.

television transmission standards The standards that specify the characteristics of a U.S. television signal. Channel width is 6 MHz, the visual carrier is 4.5 MHz lower than the aural center frequency, the aural center frequency is 0.25 MHz lower than the upper frequency limit of the channel, there are 525 scanning lines per frame period and they are interlaced two to one, the frame frequency is 30 per second and the field frequency 60 per second, the aspect ratio of the transmitted television picture is four units horizontally to three units vertically, the scene is scanned from left to right horizontally and from top to bottom vertically at uniform velocities, and a decrease in initial light intensity increases the radiated power.

television transmitter A visual and an aural transmitter used together for transmitting a complete television signal.

televoltmeter A telemeter that measures voltage.

telewattmeter A telemeter that measures power.

telewriter *Telautograph.*

telex *Teletypewriter exchange service.*

telluric current A natural electric current flowing through the earth. The direction and intensity of this current varies with the earth's magnetic field, auroral and solar activity, and other cosmic phenomena. Also called earth current.

tellurium [symbol Te] An element. Atomic number is 52.

tellurometer A microwave instrument used in surveying to measure distance. The time for a radio wave to travel from one observation point to the other and return is measured and converted into distance by phase comparison, much as in radar.

Telpak Trademark of the Bell System for a variety of leased wideband communication channels.

Telstar An experimental active communication satellite.

TE$_{m,n}$ mode A mode in which a particular transverse electric wave is propagated in a waveguide. Also called H$_{m,n}$ mode (British).

TE$_{m,n,p}$ mode A mode of wave propagation in a cavity that consists of a hollow metal cylinder closed at its ends, for which the transverse field pattern is similar to that of the TE$_{m,n}$ mode in a corresponding cylindrical waveguide and for which p is the number of half-period field variations along the axis. Also applicable to closed rectangular cavities.

tempco Abbreviation for *temperature coefficient.*

temperature characteristic The performance of a device as temperature is varied through specified limits.

temperature coefficient [abbreviated TC and tempco] The amount of change in the value of a performance characteristic per degree change in temperature.

temperature coefficient of frequency The rate of change of frequency with temperature.

temperature coefficient of resistance The rate of change in resistance value per degree change in temperature, usually expressed as ohms per ohm per degree Celsius.

temperature-compensated crystal oscillator [abbreviated TCXO] A crystal oscillator in which temperature-compensation networks have been adjusted to match the crystal characteristics, to give frequency stability over a wide temperature range.

temperature-compensated reference element A special temperature-compensated diode which has stability comparable to that of standard cells. An example is the passivated alloy silicon diode.

temperature-compensated zener diode A zener-diode package that consists of one reverse-biased zener PN junction which has a positive temperature coefficient, connected in series with one or more forward-biased diodes that have negative temperature coefficients. The result is a zener voltage that remains essentially constant over a wide temperature range.

temperature-compensating alloy An alloy whose magnetic properties change with temperature. The most common examples are nickel-iron alloys, in which magnetic permeability decreases at a controlled rate with increases in temperature. Widely used for shunts in watthour meters, speedometers, tachometers, and voltage regulators, to compensate for temperature changes.

temperature-compensating capacitor A capacitor whose capacitance varies with temperature in a known and predictable manner. Used in resonant circuits to compensate for changes in the values of other parts with temperature.

temperature-compensating network A network whose components are so chosen that the network characteristics change with temperature in a predetermined manner.

temperature compensation The process of making some characteristic of a circuit or device independent of changes in ambient temperature.

temperature control A control that maintains the temperature of an oven, furnace, or other enclosed space within desired limits.

temperature-controlled crystal unit A crystal unit that contains, in addition to the quartz plate or plates, a heater device that maintains the temperature of the quartz plate within specified limits.

temperature correction A correction applied to a measured value to compensate for changes that are due to a temperature that is higher or lower than some standard temperature value.

temperature-dependent resistor [abbreviated TDR] A nonlinear resistor in which the resistance value is some function of the ambient tem-

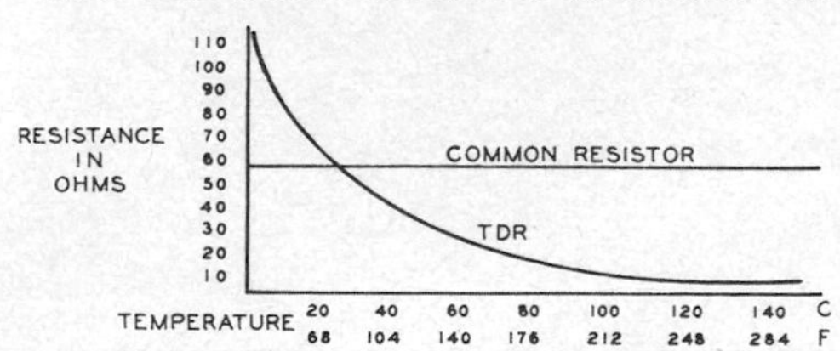

Temperature-dependent resistor characteristic, compared to that for common resistor.

perature or the temperature rise produced internally by current flow. Examples include barretters, bolometers, and thermistors.

temperature derating Lowering the rating of a device when it is to be used at elevated temperatures.

temperature element The sensing element of a temperature-measuring device or direct-reading temperature indicator.

temperature inversion 1. The ocean condition in which surface water is colder than the water below. Since the speed of sound increases with water temperature, a temperature inversion causes a sonar beam to bend in the direction of the colder water. 2. A region in the troposphere at which temperature increases rather than decreases with altitude.

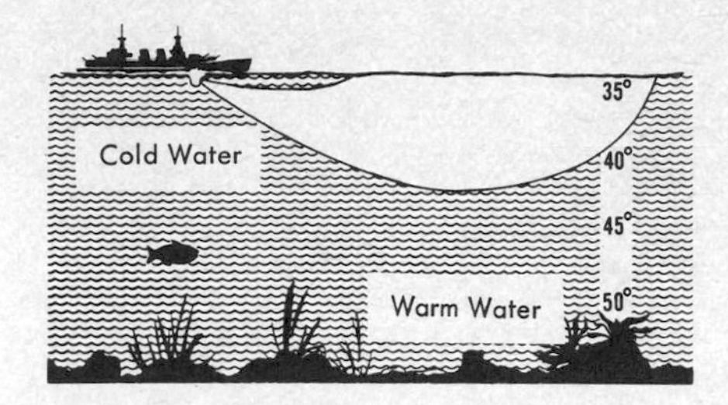

Temperature inversion causes downward-aimed searchlight sonar beam to bend toward surface because sound travels faster in warm water.

temperature range The total variation in ambient temperature for a given application, expressed in degrees Celsius.

temperature saturation The condition in which the anode current of a thermionic vacuum tube cannot be further increased by increasing the cathode temperature at a given value of anode voltage. The effect is due to the space charge formed near the cathode. Also called filament saturation and saturation.

temporal gain control *Sensitivity-time control.*

temporary storage An internal computer storage location reserved for intermediate and partial results in a digital computer.

TEM wave Abbreviation for *transverse electromagnetic wave.*

tenebrescence Darkening and bleaching under suitable irradiation. Materials that have this property are called scotophors. Darkening may be produced by primary x-rays or cathode rays, and bleaching may be produced by heat or by photons of appropriate wavelengths.

tera- [abbreviated T] A prefix that represents 10^{12}, which is 1,000,000,000,000 or a million million. Pronounced terra, rhyming with Sarah.

terabit [abbreviated Tb] One million megabits, equal to 10^{12} bits.

teracycle One million megacycles, or 10^{12} cycles per second. Now called terahertz and abbreviated THz.

teraelectronvolt [abbreviated TeV] A unit of energy equal to 10^{12} eV.

terahertz [abbreviated THz] One million megahertz, or 10^{12} Hz. Formerly called teracycle.

teraohm [abbreviated TΩ] One million megohms, equal to 10^{12} Ω.

teraohmmeter An ohmmeter that has a teraohm range for measuring extremely high insulation resistance values.

terbium [symbol Tb] A rare-earth element. Atomic number is 65.

terminal 1. A screw, soldering lug, or other point to which electric connections can be made. 2. The equipment at the end of a microwave relay system or other communication channel. 3. One of the electric input or output points of a circuit or component. 4. A device that provides access to a computer, such as a teleprinter.

terminal area The enlarged portion of conductor material that surrounds a hole for a lead on a printed circuit. Also called land and pad.

terminal board An insulating mounting for terminal connections. Also called terminal strip.

terminal equipment The equipment at a terminal of a communication channel.

terminal guidance Navigation control of a missile as it approaches its target.

terminal lug A soldering lug placed on a terminal board or at the end of a wire.

terminal phase The path of a missile as it approaches its target. For a ballistic missile, the terminal phase is that part of the trajectory between reentry and impact.

terminal strip *Terminal board.*

terminated line A transmission line terminated in a resistance equal to the characteristic impedance of the line, so there are no reflections and no standing waves.

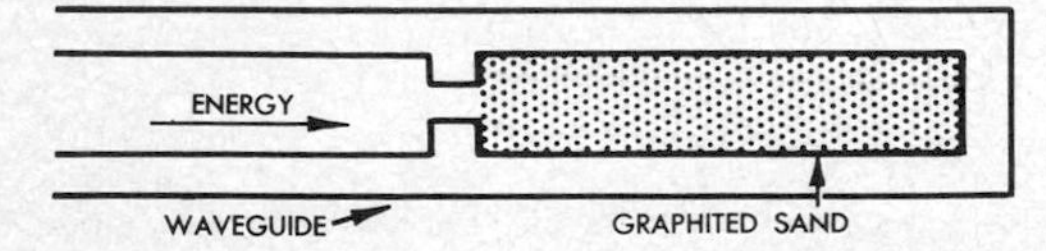

Termination used for minimum reflection in waveguide.

termination The load connected to the output end of a circuit, device, or transmission line.

ternary alloy An alloy that contains iron, silicon, and aluminum. It has high resistivity and good magnetic permeability, with magnetic properties approaching those of iron-nickel alloys.

ternary notation A system of notation that uses the base of 3 and the characters 0, 1, and 2.

terrain-avoidance radar An airborne radar that provides a display of the terrain ahead of a low-flying airplane, to permit avoidance of obstacles.

terrain-clearance indicator *Frequency-modulated radio altimeter.*

terrain-clearance warning indicator A terrain-clearance indicator that gives a warning signal when the clearance between the aircraft and the earth immediately below reaches a predetermined minimum value.

terrain echoes *Ground clutter.*

terrain error The navigation error due to distortion of a radiated field by the nonhomogeneous characteristic of the terrain over which the radiation has propagated.

terrain-following radar Airborne radar that provides autopilot control signals as required to maintain a low and constant altitude above the earth. It may also provide a display of the terrain ahead of the plane as required for manual terrain-following by the pilot at night.

terrestrial guidance *Terrestrial-reference guidance.*

terrestrial-magnetic guidance Terrestrial-reference guidance in which the control system of the missile reacts to the magnetic field of the earth.

terrestrial magnetism Magnetism produced by the earth.

terrestrial-reference flight Stabilized flight in which control information is obtained from terrestrial phenomena such as the earth's magnetic field, atmospheric pressure, or gravity.

terrestrial-reference guidance Long-range missile guidance in which the control system of the missile reacts to magnetic, gravitational, or other properties of the earth. Also called terrestrial guidance.

tesla [abbreviated T] The SI unit of magnetic flux density (magnetic induction). It is equal to 1 Wb/m^2.

Tesla coil An air-core transformer used with a spark gap and capacitor to produce a high voltage at a high frequency.

test To check the operation or performance characteristics of a component, piece of equipment, system, or computer program under controlled conditions.

test clip A spring clip at the end of an insulated wire lead used to make temporary connections quickly for test purposes.

test lead A flexible insulated lead, usually with a test prod at one end, used for making tests, connecting instruments to a circuit temporarily, or making other temporary connections.

test pattern A chart that has various combinations of lines, squares, circles, and graduated shading, transmitted from time to time by a television station to check definition, linearity, and contrast for the complete system from camera to receiver. Also called resolution chart.

test prod A metal point attached to an insulating handle and connected to a test lead for conveniently making a temporary connection to a terminal while tests are being made. Also called prod.

test program *Check routine.*

test record A phonograph record that has recorded frequencies suitable for checking and adjusting audio systems.

test routine *Check routine.*

test set A combination of instruments needed for servicing a particular type of equipment.

tetrad A group of four pulses used to express a digit in the scale of 10 or 16.

tetrafluoroethylene resin A fluorocarbon used as a base for polytetrafluoroethylene resin, marketed as Teflon.

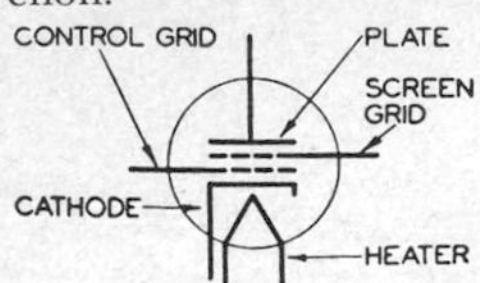

Tetrode symbol. Plate electrode is more properly called anode.

tetrode A four-electrode electron tube that contains an anode, a cathode, a control electrode, and one additional electrode which is ordinarily a grid.

tetrode transistor A four-electrode transistor.

TeV Abbreviation for *teraelectronvolt.*

TE wave Abbreviation for *transverse electric wave.*

$TE_{m,n}$ wave 1. In a circular waveguide, the transverse electric wave for which m is the number of axial planes along which the normal component of the electric vector vanishes, and n is the number of coaxial cylinders (including the boundary of the waveguide) along which the tangential component of the electric vector vanishes. The $TE_{0,1}$ wave is the circular electric wave that has the lowest cutoff frequency, and the $TE_{1,1}$ wave is the dominant wave and has electric lines of force approximately parallel to a diameter of the waveguide. 2. In a rectangular waveguide, the transverse electric wave for which m is the number of half-period variations of the electric field along the longer transverse dimension, and n is the number of half-period variations of the electric field along the shorter transverse dimension. Also called $H_{m,n}$ wave (British), for both circular and rectangular waveguides.

Texas tower A radar tower built in the sea offshore, to serve as part of an early-warning radar network. It resembles the offshore oil derricks in the Gulf of Mexico.

TFE-fluorocarbon resin Standard term of Society of the Plastics Industry for polytetrafluoroethylene resin, marketed as Teflon by Du Pont and as Fluon by Imperial Chemical Industries in Great Britain.

thallium [symbol Tl] An element. Atomic number is 81. Several thallium isotopes are members of the uranium, actinium, thorium, and neptunium radioactive series.

thallium-activated sodium iodide detector A gamma-ray detector that uses a single crystal of thallium-activated sodium iodide.

thallium oxysulfide A compound of thallium, oxygen, and sulfur that has photoconductive properties.

thalofide cell A photoconductive cell in which the active light-sensitive material is thallium oxysulfide in a vacuum. It has maximum response at the red end of the visible spectrum and in the near-infrared.

THD Abbreviation for *total harmonic distortion.*

theater television A large projection-type television receiver used in theaters, generally for closed-circuit showing of important sport events.

therapy tube An x-ray tube used in x-ray therapy.

thermal agitation Random movements of the free electrons in a conductor; they produce noise signals that may become noticeable when they occur at the input of a high-gain amplifier. Also called thermal effect.

thermal-agitation voltage The voltage produced in a circuit by thermal agitation.

thermal ammeter *Hot-wire ammeter.*

thermal battery 1. A combination of thermal cells. 2. A voltage source that consists of a number of bimetallic junctions connected to produce a voltage when heated by a flame.

thermal cell A reserve cell that is activated by applying heat to melt a solidified electrolyte.

thermal compression bonding *Thermocompression bonding.*

thermal conductivity The quantity of heat that passes through a unit volume of a material in unit time when the difference in temperature of the two faces is 1°C.

thermal converter A converter that consists of one or more thermojunctions in thermal contact with an electric heater. The voltage developed at the output terminals by thermoelectric action is then a measure of the input current to its heater. Also called thermocouple converter, thermoelectric generator, and thermoelement.

thermal cross section The cross section as measured with thermal neutrons.

thermal cutout A heat-sensitive switch that automatically opens the circuit of an electric motor or other device when the operating temperature exceeds a safe value.

thermal detector A heat detector, such as a bolometer or thermocouple.

thermal drift Drift caused by internal heating of equipment during normal operation or by changes in external ambient temperature.

thermal effect *Thermal agitation.*

thermal flasher An electric device that opens and closes a circuit automatically at regular intervals because of alternate heating and cooling of a bimetallic strip which is heated by a resistance element in series with the circuit being controlled.

thermal imager A camera or other infrared mapping system that gives an infrared image of a scene in which various objects or areas differ in temperature. In arctic regions, ice crevasses concealed by snow bridges are revealed by a temperature differential of as little as 0.1°C over a crevasse.

thermal inertia The reciprocal of thermal response.

thermal instrument An instrument that depends on the heating effect of an electric current, such as a thermocouple or hot-wire instrument.

thermal ionization The ionization of atoms or molecules by heat, as in a flame.

thermal noise Electric noise produced by thermal agitation of electrons in conductors and semiconductors. This random motion of free electrons increases with temperature. Also called Johnson noise.

thermal noise generator A generator that uses the inherent thermal agitation of an electron tube to provide a calibrated noise source.

thermal photograph A photograph made of an image tube or similar device that shows objects on the earth differentiated by their radiations of heat or infrared waves.

thermal photography *Thermography.*

thermal printer A nonimpact printer in which characters are formed by heating selected elements of a 5 × 7 or 7 × 9 dot matrix that is in contact with heat-sensitive paper.

thermal protector A temperature-sensing element that is built into a motor or other equipment to interrupt the power when overheating reaches a dangerous level. The protector may reapply power automatically when the machine has cooled, or it may require manual resetting.

thermal radiation Radiation in the form of heat, emitted by all bodies that are not at absolute zero in temperature. The wavelength range extends from the shortest ultraviolet through visible light to the longest infrared wavelengths. Also called heat.

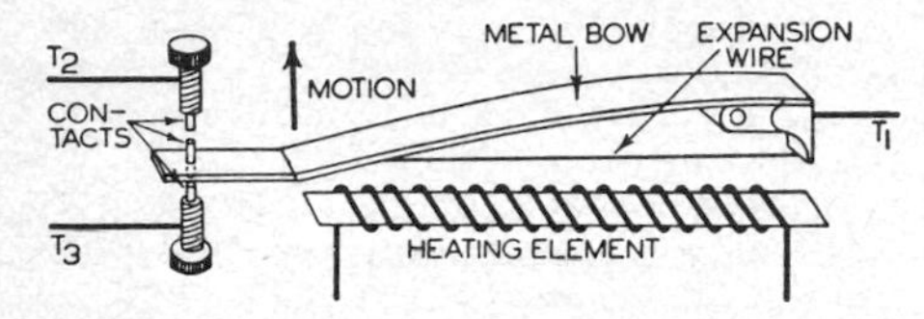

Thermal relay.

thermal relay A relay operated by the heat produced by current flow.

thermal runaway A condition that may occur in a power transistor when collector current increases collector junction temperature, reducing collector resistance and allowing a greater current to flow. The increased current increases the heating effect still more. The action may continue until the transistor is destroyed, particularly when the ambient temperature is high.

thermal switch A temperature-controlled switch.

thermal tuner A microwave tuner that uses thermal tuning of a cavity resonator.

thermal tuning The process of changing the operating frequency of a system by using controlled thermal expansion to alter the geometry of the system.

thermel A thermoelectric device that measures temperature, such as a thermocouple or thermopile.

thermion A charged particle emitted by a heated body, as by the hot cathode of a thermionic tube.

thermionic Pertaining to the emission of electrons as a result of heat.

thermionic arc An electric arc in which the cathode is heated by the arc current itself.

thermionic cathode *Hot cathode.*

thermionic converter A converter that converts heat energy directly into electric energy. In one version, two metal electrodes are separated by a gas at low pressure. When one electrode is heated to about 1100°C, electrons boiled out of it travel through the gas to the other electrode to give an electric current. Also called thermionic generator.

thermionic current A current caused by directed movements of electrons or other thermions, such as the flow of emitted electrons from the cathode to the anode in a thermionic tube.

thermionic detector A detector that uses a hot-cathode tube.

thermionic diode A diode electron tube that has a heated cathode.

thermionic emission The liberation of electrons or ions from a solid or liquid as a result of heat.

thermionic generator *Thermionic converter.*

thermionic grid emission The current produced by electrons thermionically emitted from a grid. Also called primary grid emission.

thermionic tube *Hot-cathode tube.*

thermionic work function The energy required to transfer electrons from a given metal to an adjacent medium during thermionic emission, as from a heated filament to a vacuum.

thermistor [THERMal resISTOR] A bolometer that makes use of the change in resistivity of a semiconductor with change in temperature. A thermistor has a high negative temperature coefficient of resistance, so its resistance decreases as temperature rises. Used in critical circuits to compensate for opposite temperature variations in other components. Used as a bolometer to measure temperatures and microwave energy. Used also as a nonlinear circuit element.

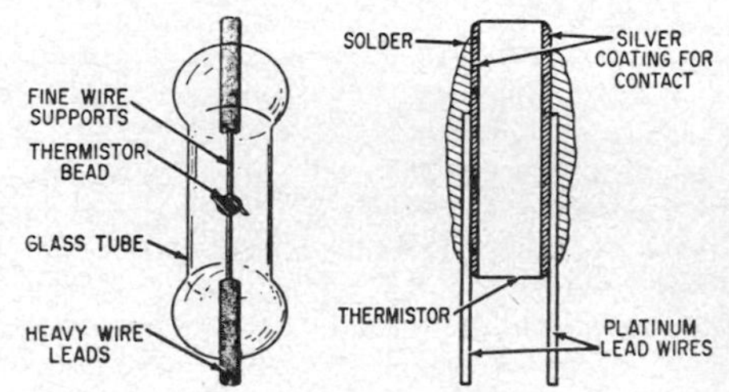

Thermistor construction.

thermistor mount A waveguide mount into which a thermistor can be inserted to measure electromagnetic power.

thermoammeter An ammeter that is actuated by the voltage generated in a thermocouple through which the current to be measured is sent. Used

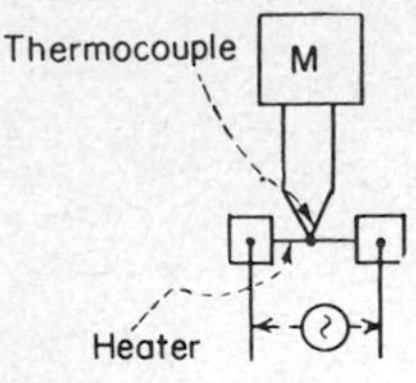

Thermoammeter in which RF input current flows through heater. Resulting heat is measured by thermocouple connected to meter M.

chiefly for measuring RF currents. Also called thermocouple ammeter, thermocouple instrument, and thermocouple meter.

thermocline An interface between warmer and colder water in the ocean, at which sound or sonar waves are so badly bent that enemy submarines can easily escape detection by hiding under the interface.

thermocompression bonding Bonding produced by a combination of heat and pressure only, without solder. Sometimes used in attaching wires to metal terminal pads of integrated circuits. Also called thermal compression bonding.

thermocouple A device that consists of two dissimilar conductors welded together at their ends to form a junction. When this junction is heated, the voltage developed across it is proportional to

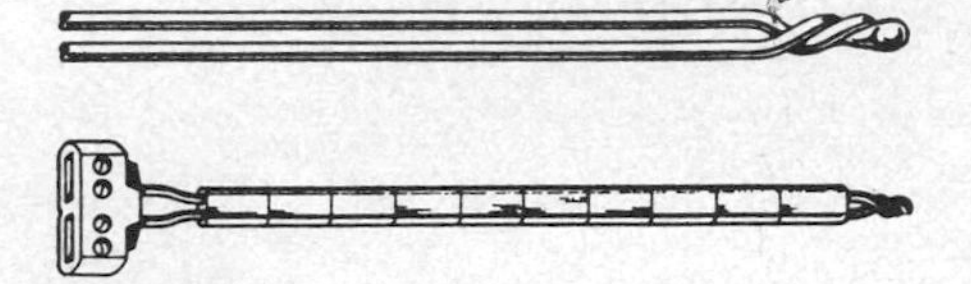

Thermocouples, uninsulated and with two-hole ceramic-bead insulators.

the temperature rise. Used for measuring temperatures, as in a thermoelectric pyrometer, or for converting radiant energy into electric energy.

thermocouple ammeter *Thermoammeter.*

thermocouple converter *Thermal converter.*

thermocouple instrument *Thermoammeter.*

thermocouple meter *Thermoammeter.*

thermocouple thermometer *Thermoelectric thermometer.*

thermocouple vacuum gage A vacuum gage that depends for its operation on the thermal conduction of the gas present. Pressure is measured as a function of the voltage of a thermocou-

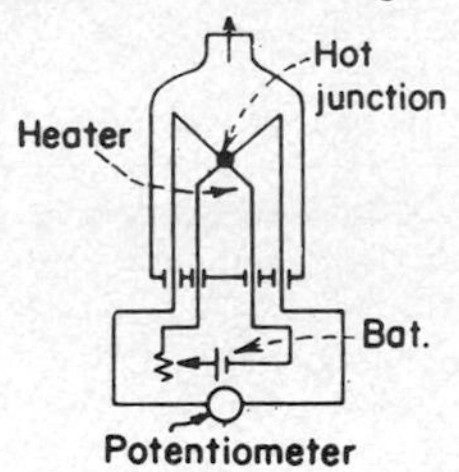

Thermocouple vacuum gage.

ple whose measuring junction is in thermal contact with a heater that carries a constant current. Ordinarily used over a pressure range of 10^{-1} to 10^{-3} mmHg.

thermoelectric converter A converter that changes solar or other heat energy to electric energy. Used as a power source on spacecraft.

thermoelectric cooler An electronic heat pump based on the Peltier effect; it involves the absorption of heat when current is sent through a junction of two dissimilar metals. It can be mounted within the housing of a device to prevent overheating or to maintain a constant temperature.

thermoelectric effect *Seebeck effect.*

thermoelectric generator *Thermal converter.*

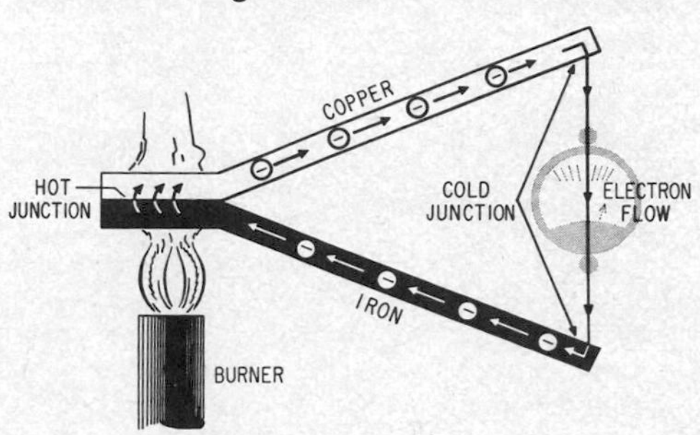

Thermoelectricity principle.

thermoelectricity Electricity produced by direct action of heat, as by unequal heating of two thermojunctions in the same circuit.

thermoelectric junction *Thermojunction.*

thermoelectric material A material that can be used to convert thermal energy into electric energy or provide refrigeration directly from electric energy. Good thermoelectric materials include lead telluride, germanium telluride, bismuth telluride, and cesium sulfide.

thermoelectric microrefrigerator A refrigeration device that uses the Peltier effect for cooling small electronic components such as infrared detectors.

thermoelectric module A device that utilizes the Peltier effect to provide spot cooling of transistors, infrared detectors, and other components when energized by direct current. Also used for precise temperature control of liquids, solids, and gases.

thermoelectric pyrometer A pyrometer in which the sensing element is a thermocouple.

thermoelectric series A series of metals arranged in the order of their thermoelectric voltage-generating ratings with respect to some reference metal such as lead.

thermoelectric solar cell A solar cell in which the sun's energy is first converted into heat by a sheet of metal, and the heat is converted into electricity by a semiconductor material sandwiched between the first metal sheet and a metal collector sheet.

thermoelectric thermometer A thermometer in which the measuring junction of a thermocouple is in thermal contact with the body of the patient. Also called thermocouple thermometer.

thermoelectron An electron liberated by heat, as from a heated filament. Also called negative thermion.

thermoelement *Thermal converter.*

thermograph 1. A far-infrared image-forming device that provides a thermal photograph by scanning a far-infrared image of an object or scene. 2. An instrument that senses, measures, and records the temperature of the atmosphere.

thermography Photography that uses radiation in the long-wavelength far-infrared region, emitted by objects at temperatures ranging from −170 to over 300°F (−112 to 150°C). Also called thermal photography.

thermojunction One of the surfaces of contact between the two conductors of a thermocouple. Also called thermoelectric junction.

thermoluminescence Luminescence produced in a material by moderate heat.

thermoluminescent dosimeter A dosimeter based on the principle that when certain irradiated solids are heated, trapped electrons or holes are restored to the ground state, with emission of light. The amount of this light is measured with a multiplier phototube.

thermomagnetic Pertaining to the effect of temperature on the magnetic properties of a substance, or to the effect of a magnetic field on the temperature distribution in a conductor.

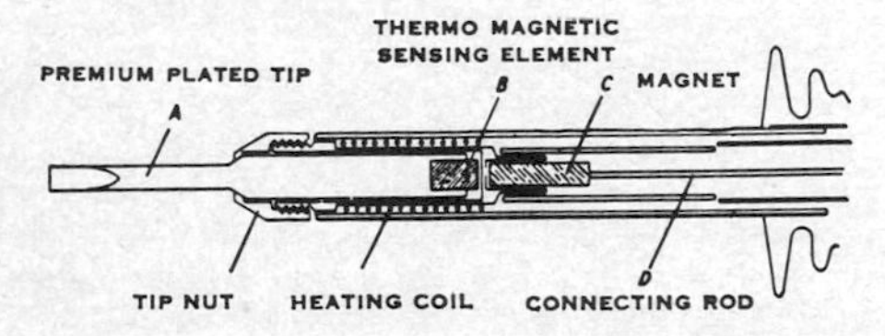

Thermomagnetic sensing element used to control tip temperature of soldering iron. Permanent magnet C is attracted to sensing element B when iron is cold. This pulls power switch on in handle, through rod D. As tip reaches its Curie point, which is selected operating temperature, element B is no longer able to hold magnet, and its release under spring action opens switch. When tip cools slightly, magnet is again attracted and tip is heated again.

thermometer An instrument that measures and indicates temperature.

thermomilliammeter A low-range thermoammeter.

thermophone An electroacoustic transducer in which sound waves that have an accurately known strength are produced by the expansion and contraction of the air adjacent to a conductor whose temperature varies in response to a current input. Used chiefly for calibrating microphones.

thermopile A group of thermocouples connected in series to give higher voltage output or in parallel to give higher current output, for measuring temperature or radiant energy or converting radiant energy into electric power.

thermoplastic A plastic that can be softened by heat and rehardened into a solid state by cooling. It may be remelted and remolded many times. Examples are cellulose acetate, cellulose nitrate, methyl methacrylate, polyethylene, polystyrene, and vinyls.

thermoplastic recording A recording process in which information is stored on a plastic tape by the action of a modulated electron beam, to give high storage density and playback within a few milliseconds. The tape consists of a thermoplastic film on a transparent conducting film supported by an ordinary plastic tape base. The electron beam deposits charges on the thermoplastic film in accordance with beam modulation. Application of heat by RF heating electrodes softens the film enough to produce deformation that is proportional to the stored electrostatic charges. Hardening preserves the deformation. An optical system is used for playback.

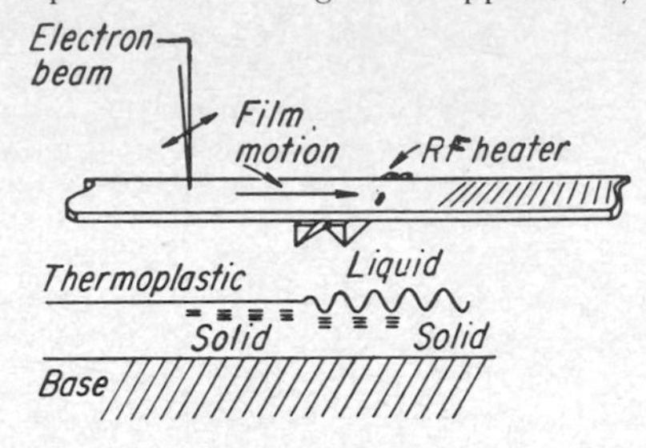

Thermoplastic recording.

thermoregulator A high-accuracy or high-sensitivity thermostat. One type consists of a mercury-in-glass thermometer with sealed-in electrodes, in which the rising and falling column of mercury makes and breaks an electric circuit.

thermorelay *Thermostat.*

thermosetting material A plastic that solidifies when first heated under pressure and cannot be remelted or remolded without having its original characteristics destroyed. Examples are epoxies, melamines, phenolics, and ureas.

thermosphere The region of the atmosphere, above the mesosphere, in which there is strong heating and increasing temperature, resulting from photodissociation and photoionization of nitrogen and oxygen atoms. The region extends roughly from 50 to 375 mi (80 to 600 km) altitude.

thermostat A device that opens or closes a circuit when the temperature deviates from a preset value or range of values, to actuate the controls of a heating element and thereby produce the required corrective action. Also called thermorelay.

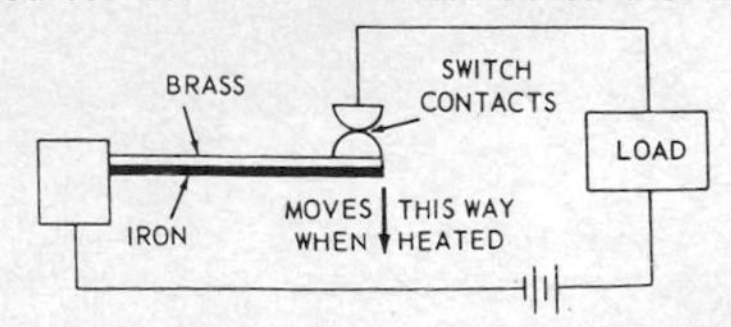

Thermostatic switch using bimetallic strip (brass and iron).

thermostatic switch A temperature-operated switch that receives its operating energy by thermal conduction or convection from the device being controlled or operated.

thermostat materials Pairs of metals that have widely different coefficients of expansion. When they are joined together, a temperature change makes one material expand more than the other,

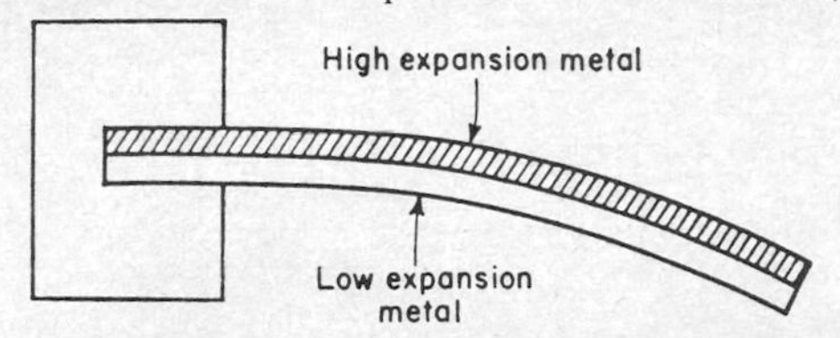

Thermostat materials, showing how heat makes combination of metals bend downward.

causing a change in shape of the combination that can be used to make or break a circuit as temperature changes. The principal combinations used are nickel and iron, chromium and iron, and pairs of various alloys.

Thevenin's theorem If an impedance is connected between two points at which there exist a voltage and an impedance, the current through the added impedance will be equal to the voltage value divided by the sum of the impedance values.

thick film A film of conductive, dielectric, or resistive material that is generally well over 10 μm thick. The film is usually applied by silk-screening, followed by drying and firing.

thick-film capacitor A capacitor in which two overlapping thick-film layers of conducting material are separated by a deposited dielectric film.

thick-film resistor A fixed resistor in which the resistance element is a thick film made from particles of noble metals, metal oxides, and glass powders suspended in organic vehicles or binders.

thickness gage A gage that measures the thickness of a sheet of material, an object, or a coating.

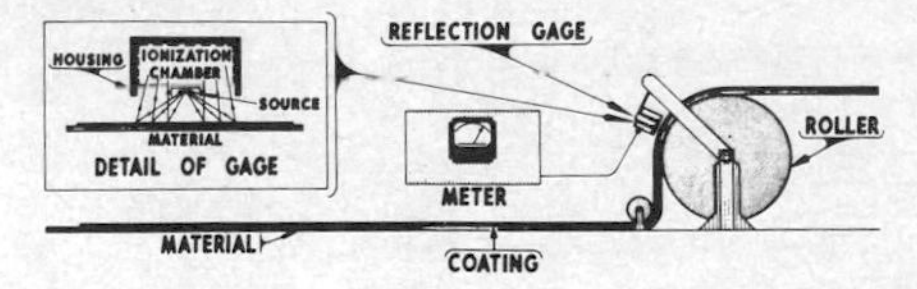

Thickness gage, backscattering type, for monitoring thickness of coating on sheet material.

Examples include penetration-type and backscattering radioactive thickness gages and ultrasonic thickness gages.

thickness vibration Vibration of a piezoelectric crystal in the direction of its thickness.

thimble ionization chamber A small cylindrical or spherical ionization chamber, usually with walls of organic material or air-filled walls.

thin film A molecularly thin film deposited on a glass, ceramic, or semiconductor substrate by sputtering, evaporation, or chemical vapor deposition through a mask. Resistors, capacitors, and active elements such as transistors can be produced in this manner.

thin-film capacitor A capacitor that can be constructed by evaporation of conductor and dielectric films in sequence on a substrate. Silicon monoxide is generally used as the dielectric.

thin-film circuit A circuit in which all active and passive elements and their interconnections are made from thin films, usually less than a few micrometers thick, deposited by sputtering, vacuum evaporation, or other means on an insulating substrate.

thin-film component A component that is deposited on the substrate of an integrated circuit as one or more thin-film layers.

thin-film cryotron A cryotron in which the transition from superconducting to normal resistivity

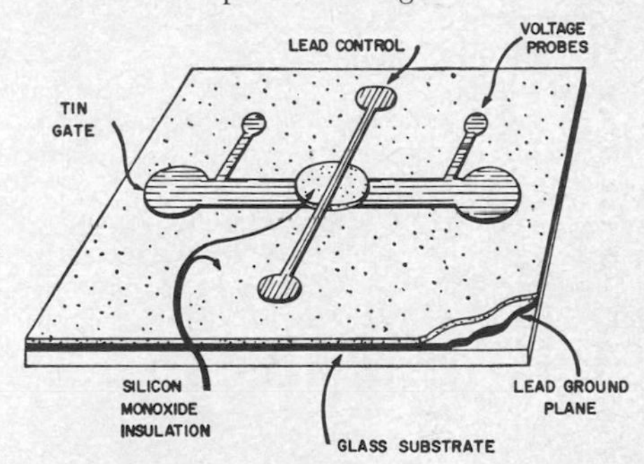

Thin-film cryotron construction.

of a thin film of tin or indium, serving as a gate, is controlled by current in a film of lead that crosses and is insulated from the gate.

thin-film ferrite coil An inductor made by depositing a thin flat spiral of gold or other conducting

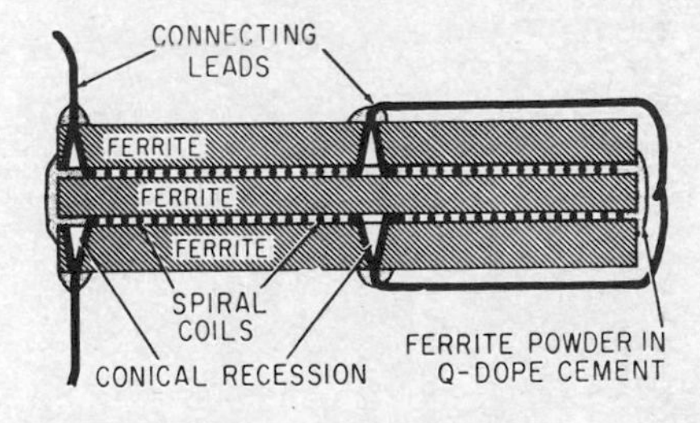

Thin-film ferrite coil sandwich, consisting of two thin-film gold spirals between three ferrite substrates.

metal on a ferrite substrate. Higher inductance values can be obtained by sandwiching two such conductor spirals between three ferrite substrates and connecting the spirals in series.

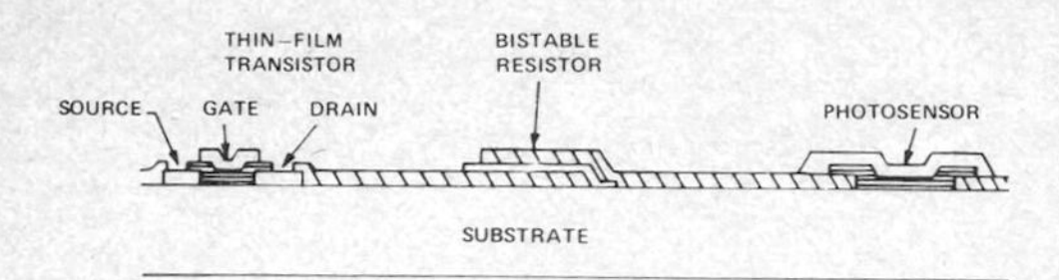

Thin-film integrated circuit constructed on insulating substrate.

thin-film integrated circuit An integrated circuit that consists entirely of thin films deposited in a patterned relationship on a substrate.

thin-film magnetoresistor A thin-film resistor whose value can be changed by applying a magnetic field.

thin-film material A material that can be deposited in a desired pattern by a variety of chemical, mechanical, or high-vacuum evaporation techniques. Thin-film resistors are often made from gold-platinum or nickel-chromium alloys as well as from tin oxide deposited on a ceramic substrate. Thin-film capacitors are made by depositing alternate layers of a dielectric and a metal, aluminum oxide and silicon oxide being typical dielectrics. Coils are made by depositing pure-metal films on ferrite.

thin-film memory A computer memory that consists of a thin film of magnetic material evaporated on a heated glass base in the presence of a DC magnetic field parallel to the surface of the base. Large magnetic memory arrays with thousands of elements can be made in one operation.

thin-film microcircuit A microcircuit that has a nonsemiconductor substrate on which thin-film passive components are produced by vapor deposition or other electrochemical processes. Active components may be discrete or deposited types.

thin-film resistor A fixed resistor whose resistance element is a metal, alloy, carbon, or other thin-film material.

thin-film solar cell A solar cell in which a thin film of gallium arsenide, cadmium sulfide, or other semiconductor material is evaporated on a thin, flexible metal or plastic substrate.

thin-film transistor A field-effect transistor constructed entirely by thin-film techniques, for use in thin-film circuits. A thin metal gate electrode is

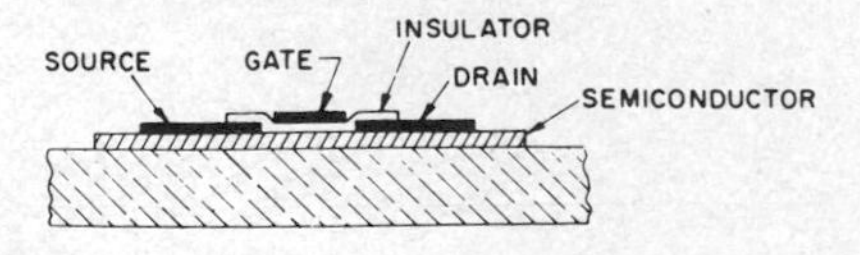

Thin-film transistor having coplanar electrodes.

separated by a thin insulating film from a semiconductor layer that is usually cadmium sulfide. Current flows through a channel in the semiconductor layer, between two electrodes called the source and drain. The amount of current is controlled by the voltage applied to the insulated gate.

thin-window counter tube A counter tube in which a portion of the enclosure has low absorption to permit the entry of short-range radiation.

third harmonic A sine-wave component that has three times the fundamental frequency of a complex wave.

Thomson bridge *Kelvin bridge.*

Thomson coefficient The ratio of the voltage existing between two points on a metallic conductor to the difference in temperature of those points.

Thomson cross section *Scattering cross section.*

Thomson effect When a current flows from a warmer to a cooler portion of a conductor, or vice versa, heat is liberated or absorbed, depending on the material the conductor is made of.

Thomson scattering Scattering of electromagnetic radiation by electrons. The scattering cross section for an electron is 0.657 barn.

Thomson voltage The voltage that exists between two points which are at different temperatures in a conductor.

Thoraeus filter A primary radiological filter of tin, combined with a secondary filter of copper to absorb the characteristic radiation of the tin and a third filter of aluminum to absorb the characteristic radiation of the copper. In the range of 200 to 400 kV such a filter hardens x-rays more efficiently than the usual combination of copper and aluminum.

thoriated tungsten filament A vacuum-tube filament that consists of tungsten mixed with a small quantity of thorium oxide to give improved electron emission.

thorium [symbol Th] A metal that emits electrons liberally when heated. Sometimes incorporated in tungsten filaments of vacuum tubes.

thread *Chip.*

three-dimensional radar Radar capable of producing position data in three dimensions for a number of targets simultaneously.

three-gun color picture tube A color television picture tube in which three electron guns emit three electron beams, one for each primary color. Each beam is directed onto phosphor dots that emit only the corresponding primary color. Each gun is controlled by its appropriate primary color signal. The shadow-mask color picture tube is an example.

three-level laser A laser that involves three energy levels, one of which is the ground state. Laser action usually occurs between the intermediate and ground states.

three-phase circuit A circuit energized by AC voltages that differ in phase by one-third of a cycle, or 120°.

three-pole switch An arrangement of three single-pole single-throw switches coupled together to make or break three circuits simultaneously.

three-way system A three-unit loudspeaker system that consists of a woofer to handle the lowest frequencies, a midrange unit, and a tweeter for the high frequencies.

threshold 1. The least value of a current, voltage, or other quantity that produces the minimum detectable response. Also called limen. 2. The level of pumping at which a laser can go into self-excited oscillation.

threshold current The minimum current value at which a nonself-sustained gas discharge changes to a self-sustained discharge.

threshold effect The inherent suppression of noise in a phase- or frequency-modulated receiver by a carrier whose peak value is only slightly greater than that of the noise.

threshold energy The energy limit, for an incident particle or photon, below which a particular endothermic reaction will not occur or a particular nuclear reaction cannot be observed.

threshold frequency The frequency of incident radiant energy below which there is no photoemissive effect.

threshold gate A computer logic element that has an output of 1 for a minimum sum of input weights, and an output of 0 for a maximum sum of input weights. Threshold gates may be used in various combinations to form flip-flops, memory cells, accumulators, and other logic circuits.

threshold of audibility The minimum effective sound pressure of a specified signal that is capable of evoking an auditory sensation in a specified fraction of the trials. The threshold may be expressed in decibels relative to 0.0002 μbar or 1 μbar.

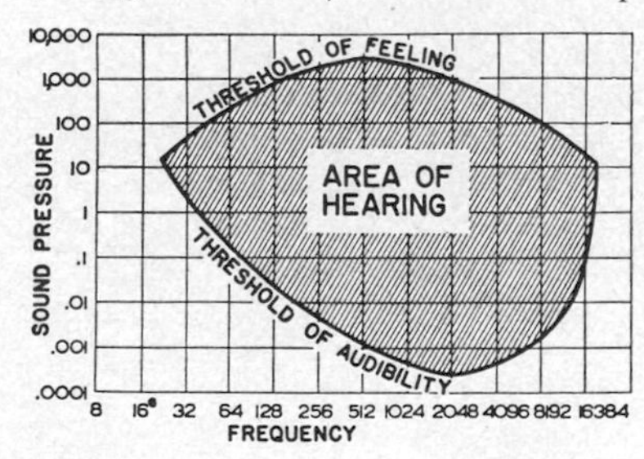

Threshold of audibility, shown as sound pressure in microbars plotted against frequency in hertz. At 1000 Hz, threshold of hearing is 0.0002 μbar, equal to 0 dB when 0.0002 μbar is reference level.

threshold of feeling The minimum effective sound pressure of a specified signal that, in a specified fraction of trials, will stimulate the ear to a point at which there is the sensation of feeling, discomfort, tickle, or pain. Customarily expressed in decibels relative to 0.0002 μbar or 1 μbar.

threshold of luminescence *Luminescence threshold.*

threshold sensitivity The smallest amount of a quantity that can be detected by a measuring instrument or automatic control system.

threshold signal The smallest signal that gives a recognizable change in positional information in a navigation system.

threshold value The minimum input that produces a corrective action in an automatic control system.

threshold voltage The lowest voltage at which a specified change in performance occurs.

threshold wavelength The wavelength of the incident radiant energy above which there is no photoemissive effect.

throat The smaller end of a horn or tapered waveguide.

throat microphone A contact microphone that is strapped to the throat of a speaker and reacts to throat vibrations directly rather than to the sound waves they produce.

throttling Control by intermediate steps between full on and full off.

throttling control *Rate control.*

throughput The maximum output of a system, such as lines per minute or pages per minute of a computer line printer or computer-output microfilm system.

throwout spiral *Leadout groove.*

thulium [symbol Tm] A rare-earth element. Atomic number is 69.

thumbwheel switch A compact multiposition switch in which finger pressure on the actuating member advances the switch and its indicator to the next contact position.

thump A low-frequency transient disturbance in an audio system.

thyratron A hot-cathode gas tube in which one or more control electrodes initiate but do not limit the anode current except under certain operating conditions.

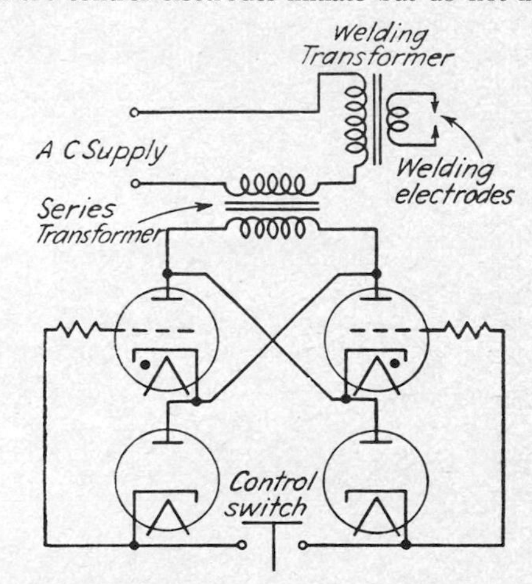

Thyratrons (gas triodes) control current for welding transformer.

thyratron inverter An inverter circuit that uses

thyratrons to convert DC power to AC power.

thyristor A semiconductor switching device in which bistable action depends on PNPN regenerative feedback. A thyristor can be unidirectional or bidirectional, have from two to four terminals, and be triggered from its blocking to its conducting state at a desired point within a single 90° quad-

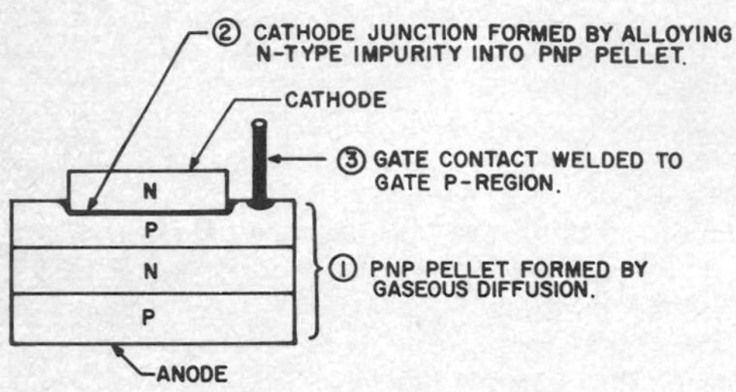

Thyristor PNPN pellet formed by alloy diffusion.

rant of the applied AC voltage. The silicon controlled rectifier is the most common unidirectional thyristor; others include the gate-turnoff switch and light-activated silicon controlled rectifier. The triac and silicon bilateral switch are examples of bidirectional thyristors, which can conduct in either direction.

THz Abbreviation for *terahertz.*

ticker A printer designed specifically for stock quotations and news, generally with printout in a single line that runs parallel to the edges of narrow paper tape.

tickler A small coil connected in series with the anode circuit of an electron tube and inductively coupled to a grid-circuit coil to provide feedback. Used chiefly in regenerative detector and oscillator circuits.

tie-down point One of the frequencies at which a radio receiver is aligned. For the broadcast band, the tie-down points are usually 600 and 1400 kHz.

tie-line A leased communication channel or circuit.

tie point An insulated terminal to which two or more wires are connected.

tie wire A wire that connects a number of terminals together.

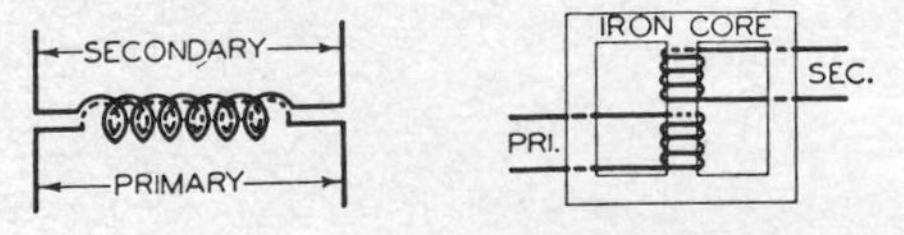

Tight coupling in air and iron-core transformers.

tight coupling Inductive coupling in which practically all the magnetic flux of one coil links another coil.

tilt The angle that an antenna axis forms with the horizontal.

tilt error The component of ionospheric height error in navigation that is caused by nonuniform height of the ionospheric layer.

tilting Forward inclination of the wavefront of radio waves traveling along the ground. The amount depends on the electrical constants of the ground.

tilt stabilization Stabilization of a radar antenna by using an additional servomotor to tilt the antenna up or down during scanning, as required to correct for pitch and roll of the ship or aircraft.

timbre That attribute of auditory sensation in terms of which a listener can judge that two sounds similarly presented and having the same loudness and pitch are dissimilar. Timbre depends primarily on the spectrum of the stimulus, but it also depends on the waveform, sound pressure, and frequency location of the spectrum of the stimulus.

time A measure of duration of an event. The fundamental unit of time is the second.

time base The line formed by sweep-circuit action on the screen of a cathode-ray tube.

time-base generator *Sweep oscillator.*

time-base voltage *Sweep voltage.*

time code generator A crystal-controlled pulse generator that produces a train of pulses with various predetermined widths and spacings, from which the time of day and sometimes day of year can be determined. Used in telemetry and other data-acquisition systems, to provide the precise time of each event.

time constant The time required for a voltage or current in a circuit to rise to approximately 63% of its steady final value or to fall to approximately 37% of its initial value. The time constant of a coil that has an inductance L in henrys and resistance R in ohms is L/R. The time constant of a capacitor that has a capacitance C in farads in series with a resistance R in ohms is RC.

time delay The time required for a signal to travel between two points in a circuit or for a wave to travel between two points in space.

time-delay circuit A circuit that delays a signal or action a definite desired period of time.

time-delay fuse A fuse in which the burnout action depends on the time it takes for the overcurrent heat to build up in the fuse and melt the fuse element.

time-delay relay A relay in which there is an appreciable interval of time between energizing or deenergizing of the coil and movement of the armature, such as a slow-acting relay and a slow-release relay.

time discriminator A circuit in which the sense and magnitude of the output is a function of the time difference between two pulses and their relative time sequence.

time-distribution analyzer An instrument that indicates the number or rate of occurrence of time

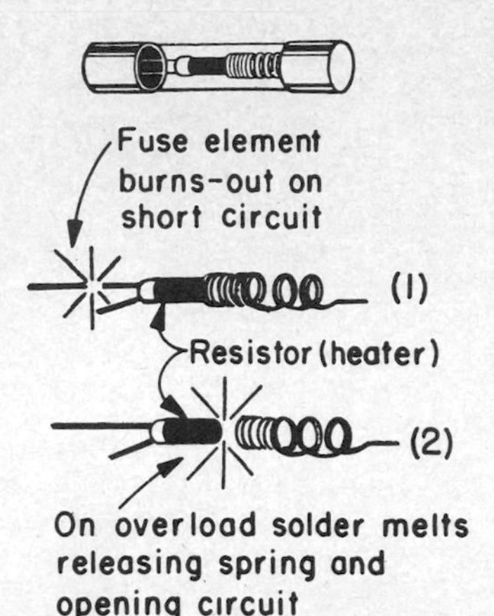

Time-delay fuse action. For high-current fault (1), fuse opens quickly. For slight overload (2), fuse opens only after solder link has been heated enough to melt.

intervals falling within one or more specified time-interval ranges. The time interval is delineated by the separation between pulses of a pulse pair. Also called time sorter.

time-division multiple-access [abbreviated TDMA] A digital method of modulating telephone signals, usually combined with pulse-code modulation and phase-shift keying.

time-division multiplex [abbreviated TDM] The transmission of two or more signals over a common path by using successive time intervals for different signals.

time-division multiplier An electronic multiplier in which the output is the average of a train of pulses that have width controlled by one variable and amplitude controlled by the other variable.

time-domain reflectometer [abbreviated TDR] An instrument that measures the electrical characteristics of wideband transmission systems, subas-

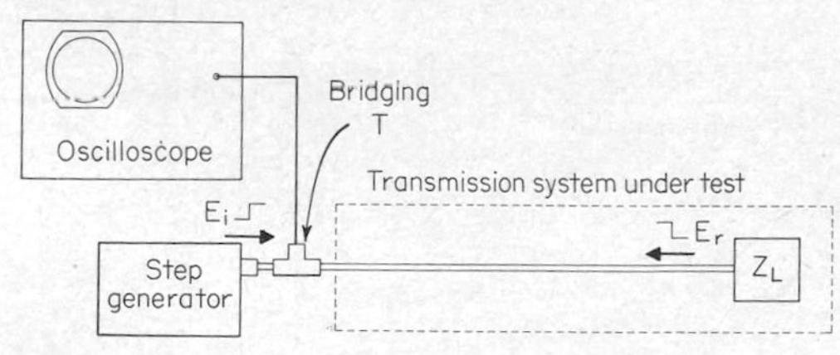

Time-domain reflectometer setup for testing transmission system.

semblies, components, and lines by feeding in a voltage step and displaying the superimposed reflected signals on an oscilloscope equipped with a suitable time-base sweep. The display gives the nature and location of each pulse-reflecting discontinuity.

time flutter *Time jitter.*

time gate A circuit that gives an output only during chosen time intervals.

time-interval counter An electronic counter that measures a time interval by counting the number of pulses received from an RF signal generator in that time interval.

time-interval selector A circuit that produces a specified output pulse when and only when the time interval between two pulses lies between specified limits.

time jitter Variations in the synchronization of the components of a radar system, causing variations in the position of the observed pulse along the time base and reducing the accuracy with which the time of arrival of a pulse can be determined. Also called time flutter.

time lag The time between an event and a resultant effect, as between occurrence of a primary ionizing event and its count by a counter.

time-mark generator A signal generator that produces highly accurate clock pulses which can be superimposed as pips on a cathode-ray screen for timing the events shown on the display.

time modulation Modulation in which the time of occurrence of a definite portion of a waveform is varied in accordance with a modulating signal.

time-of-flight mass spectrometer A mass spectrometer in which all the positive ions of the material being analyzed are ejected into the drift region of the spectrometer tube with essentially the same energies. Since the velocities of these ions will vary with mass, the ions spread out in accordance with their masses as they reach the cathode of a magnetic electron multiplier at the other end of the tube. The time of flight of each group of ions is indicated horizontally on the associated oscilloscope screen, and ion abundance is indicated by the peak height of each pulse on the screen.

time phase Reaching corresponding peak values at the same instants of time though not necessarily at the same points in space.

time quadrature Differing by a time interval corresponding to one-fourth the time of one cycle of the frequency in question.

timer 1. A circuit used in radar and electronic navigation systems to start pulse transmission and synchronize it with other actions, such as the start of a cathode-ray sweep. 2. *Interval timer.*

time response The time required for the output of a control system to show the effect of application of a prescribed input signal.

time-shared amplifier An amplifier used with a synchronous switch to amplify signals from different sources one after another.

time-sharing The use of a device or system for two or more purposes during the same overall time interval by allocating small divisions of the total time to each purpose on a fixed schedule or according to demand.

time signal A radio signal broadcast at accurately

known times each day on a number of different frequencies by WWV and other stations, for use in setting clocks.

time sorter *Time-distribution analyzer.*

time switch A clock-controlled switch that opens or closes a circuit at one or more predetermined times.

time tick An accurately controlled pulsed radio signal used for setting timepieces.

time-varied gain control *Sensitivity-time control.*

timing-axis oscillator *Sweep oscillator.*

timing signal Any signal recorded simultaneously with data on magnetic tape, for identifying the exact time of each recorded event.

tin [symbol Sn] A metallic element. Atomic number is 50.

tinned wire Copper wire that has been coated during manufacture with a layer of tin or solder to prevent corrosion and simplify soldering of connections.

tin oxide resistor A metal-film resistor that consists of tin oxide fused to the surface of a glass or ceramic substrate. The resistance value is adjusted or increased by cutting or grinding a spiral groove into the film surface.

tinsel A type of confusion reflector.

tinsel cord A highly flexible cord used for headphone leads and test leads, in which the conductors are strips of thin metal foil or tinsel wound around a strong but flexible central cord.

tip 1. The contacting part at the end of a phone plug. 2. A small protuberance on the envelope of an electron tube, resulting from the closing of the envelope after evacuation.

tip jack A small single-hole jack for a single-pin contact plug. Also called pup jack.

Tiros [Television InfraRed Observation Satellite] One of a series of meteorological satellites

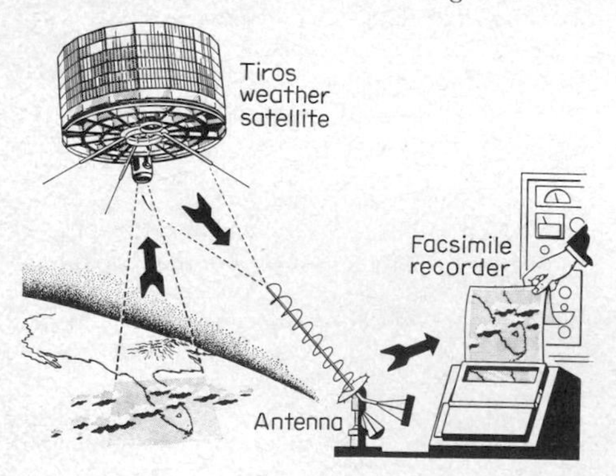

Tiros weather satellite and ground receiving system.

carrying infrared equipment and television cameras. Used for transmitting pictures of cloud cover, locations of ice floes, and other weather data.

tissue dose The dose received by a tissue in the region of interest, expressed in roentgens for x-rays and gamma rays.

tissue-equivalent material Material having the same elements in the same proportions as they occur in some particular biological tissue.

Titan An Air Force surface-to-surface intercontinental ballistic missile that has a range of over 6000 mi (9700 km) and uses inertial guidance in a nuclear warhead.

titanium [symbol Ti] A metallic element that has high strength and corrosion resistance. Atomic number is 22.

titration control An electronic control used in chemical processes to regulate acidity or alkalinity.

T junction A waveguide junction in which the branch guide intersects the main guide at right angles.

$TM_{m,n}$ mode A mode in which a particular transverse magnetic wave is propagated in a waveguide. Also called $E_{m,n}$ mode (British).

$TM_{m,n,p}$ mode A mode of wave propagation in a cavity that consists of a hollow metal cylinder closed at its ends, for which the transverse field pattern is similar to that of the $TM_{m,n}$ mode in a corresponding cylindrical waveguide and for which p is the number of half-period field variations along the axis. Also applicable to closed rectangular cavities.

TM wave Abbreviation for *transverse magnetic wave.*

$TM_{m,n}$ wave 1. In a circular waveguide, the transverse magnetic wave for which m is the number of axial planes along which the perpendicular component of the magnetic vector vanishes, and n is the number of coaxial cylinders to which the electric vector is perpendicular. The $TM_{0,1}$ wave is the circular magnetic wave that has the lowest cutoff frequency. 2. In a rectangular waveguide, the transverse magnetic wave for which m is the number of half-period variations of the magnetic field along the longer transverse dimension, and n is the number of half-period variations of magnetic field along the shorter transverse dimension. Also called $E_{m,n}$ wave (British) for both circular and rectangular waveguides.

T network A network composed of three branches, with one end of each branch connected to a common junction point, and the three remaining ends connected to an input terminal, an output terminal, and a common input and output terminal, respectively.

toe and shoulder The nonlinear portions of the H and D curve, located below and above the straight portion of this curve.

to-from indicator A sensing device used in aircraft to show whether the numerical reading of an omnibearing selector represents a bearing toward or away from an omnidirectional range.

toggle To switch over to an alternate state, as in a flip-flop.

toggle switch A small switch that is operated by

manipulation of a projecting lever which is combined with a spring to provide a snap action for opening or closing a circuit quickly.

tolerance A permissible deviation from a specified value, expressed in actual values or more often as a percentage of the nominal value.

toll call A long-distance telephone call for which an individual extra charge is made, based on such factors as distance, length of call, and time of day.

toll-free number An inward WATS service of the Bell System that allows anyone to call a business firm from any part of the country or from specified regions, free of charge, by dialing a specially assigned 10-digit number starting with 800. The monthly charge to the firm is based on the total area covered and the total time for all calls made to that number during the month.

toll line A telephone line or channel that connects different telephone exchanges.

Tomahawk A solid-fuel cruise missile designed for firing from the torpedo tube of a submarine, climbing in air to a predetermined low altitude such as 300 m, cruising horizontally to within striking range of its target, then homing on the target.

TΩ Abbreviation for *teraohm*.

tomography *Laminography*.

tone 1. A sound wave capable of exciting an auditory sensation that has pitch, or a sound sensation that has pitch. 2. The equality of reproduction of a sound program.

tone arm *Pickup arm*.

tone-burst generator An instrument that produces pulses or bursts of an input frequency provided by an external oscillator. Panel controls are provided for adjusting the number of cycles in the burst and the time interval between bursts. Also called burst generator.

tone control A control used in an AF amplifier to change the frequency response so as to secure the most pleasing proportion of bass to treble. Individual bass and treble controls are provided in some amplifiers.

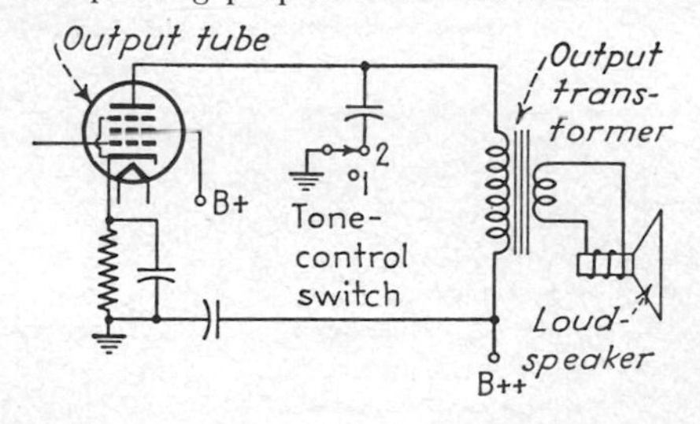

Tone-control circuit bypasses higher audio frequencies to ground when switch is in position 2, to emphasize bass notes.

tone generator A signal generator that generates an AF signal suitable for signaling purposes or for testing AF equipment.

tone localizer *Equisignal localizer*.

tone-modulated wave A continuous wave that is modulated by a single audio frequency.

toner The fine black resinous powder used in electrostatic imaging processes to make an electrostatic image readable. The toner is either deposited directly on coated paper or transferred from a charged surface to ordinary paper and then fused to the paper by heating.

tonne [abbreviated t] The metric ton, equal to 1000 kg, 2204.623 lb, or 1.1 tons.

tonometer An electronic instrument that measures hydrostatic pressure within the eye. When placed in position, a tiny movable plate is pressed against the eye, flattening a circular section of the cornea. No eyeball anesthesia is required. A current is then sent through a small electromagnet. The current is of such value that it will just pull the plate away from the eye. The value of the current is then proportional to eye pressure. Used in diagnosis of glaucoma. A measurement can be made in about 1 s. Also called electronic tonometer.

TO package A type of package used for some transistors, integrated circuits, and other semiconductor devices, in which the leads are arranged in a circle and project from the base parallel to the axis of the device. The housing is usually a cylindrical metal can made in standardized sizes, with the leads passing out of the housing through glass or other insulating eyelets in its base.

top cap A metal cap positioned at the top of an electron tube and connected to one of the electrodes, usually the control grid.

top-loaded vertical antenna A vertical antenna that is wider at the top, to modify the current distribution and give a more desirable radiation pattern in the vertical plane. A coil may be connected between the enlarged portion of the antenna and the remaining structure.

topside fathometer A fathometer mounted on top of a submarine to send sound waves upward. Used to measure the distance to the surface or to the bottom and top of surface ice.

topside sounder A sounder that sends short radio pulses from a satellite down toward the ionosphere and listens for echoes while sweeping the frequency or changing to different fixed frequencies, to obtain a profile of echo delay versus frequency. This profile is called an ionogram. Applications include measurement of ion concentration and electron density in the ionosphere.

torn-tape operation A teletypewriter operating method in which messages received on punched paper tape are separated by tearing the tape, and individual pieces of tape are fed to other circuits for retransmission.

toroid A coil or transformer wound on a doughnut-shaped core. The toroidal core gives a maximum magnetic field within itself, with minimum magnetic flux leakage externally.

toroidal klystron A klystron in which the interaction processes are paralleled by using interdigital structures to provide reentrancy and increase power output at centimeter wavelengths.

torpedo nose assembly An assembly located in front of a torpedo warhead; it contains the hydrophones and associated amplifier and control circuits required for acoustic homing on sounds made by the propellers of a ship.

torque amplifier An analog computer device that has input and output shafts and supplies work to rotate the output shaft in positional correspondence with the input shaft without imposing any significant torque on the input shaft.

torque-coil magnetometer A magnetometer that depends for its operation on the torque developed by a known current in a coil which can turn in the field to be measured.

torque gradient The amount of torque developed by a synchro per degree of angular difference between transmitter and receiver rotors.

torque motor A motor designed primarily to exert torque while stalled or rotating slowly.

torquer *Pancake motor.*

torr The new international standard unit of atmospheric pressure or vacuum. One torr is defined as $\frac{1}{760}$ of a standard atmosphere. The torr differs from the earlier millimeter-of-mercury unit by only 1 part in 7 million.

torsion galvanometer A galvanometer in which the force between the fixed and moving systems is measured by the angle through which the supporting head of the moving system must be rotated to bring the moving system back to its zero position.

torsionmeter An instrument that measures the amount of power transmitted by a rotating shaft, as by measuring the twisting of the shaft under load or measuring the twisting of components mounted on a coupling device inserted between sections of the shaft.

torsion-string galvanometer A sensitive galvanometer in which the moving system is suspended by two parallel fibers that tend to twist around each other.

TOS Abbreviation for *tape operating system.*

total electrode capacitance The capacitance between one electrode and all other electrodes connected together.

total electron binding energy The energy required to remove all the electrons of an atom to infinite distance from the nucleus and from each other, leaving only the bare nucleus.

total harmonic distortion [abbreviated THD] The ratio, expressed in percent, of the RMS voltage value for all harmonics present in the output of an audio system to the total RMS voltage at the output, for a pure sine-wave input. The lower the percentage figure, the better the audio system.

total ionization 1. The total electric charge on the ions of one sign when the energetic particle that has produced these ions has lost all its kinetic energy. 2. The total number of ion pairs produced by the ionizing particle along its entire path.

total nuclear binding energy The energy required to break up a nucleus into its constituent nucleons.

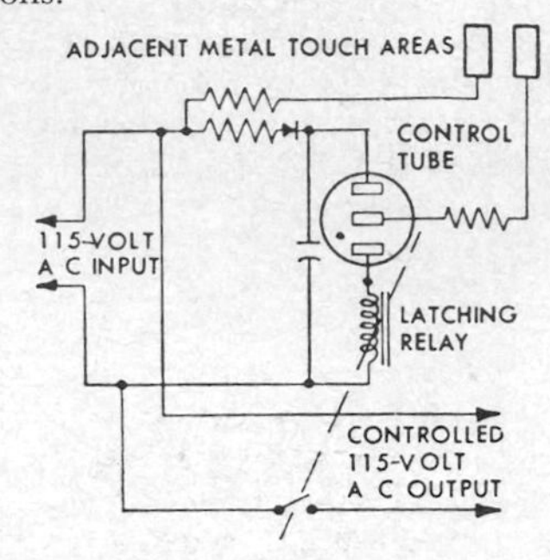

Touch control using neon gas triode.

touch control A circuit that closes a relay when two metal areas are bridged by a finger or hand.

Touch-Tone telephone A Bell System telephone that has 12 pushbuttons, each of which produces a

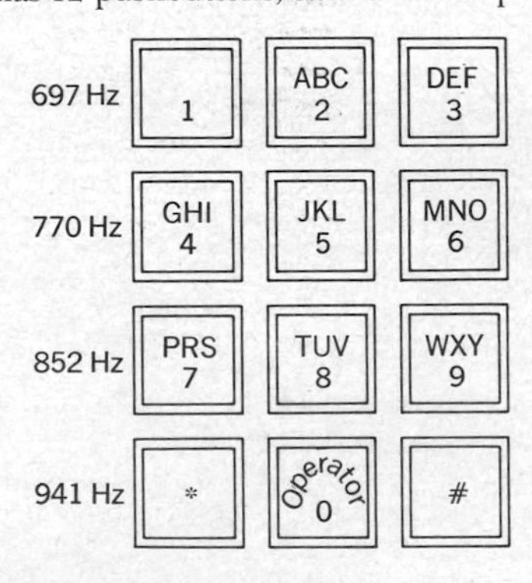

Touch-Tone telephone pushbutton arrangement. Each button produces two tone frequencies, having values indicated at left and below button pressed.

distinctive two-frequency musical tone which initiates the proper automatic switching in telephone exchanges, just as the DC pulsing of a rotating dial does.

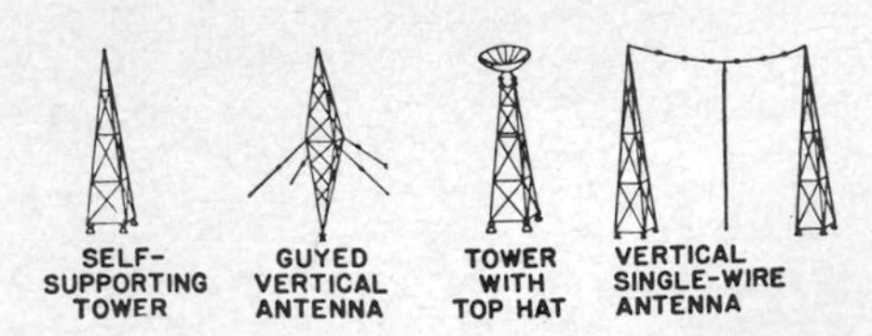

Tower designs for broadcast stations.

tourmaline A strongly piezoelectric natural crystal.

tower A tall metal structure used as a transmitting antenna, or used with another such structure to support a transmitting antenna wire.

tower radiator A tall metal structure used as a transmitting antenna.

Townsend avalanche *Avalanche effect.*

Townsend characteristic The current-voltage characteristic curve for a phototube at constant illumination and at voltages below that at which a glow discharge occurs.

Townsend coefficient The number of ionizing collisions by an electron per centimeter of path length in the direction of the applied electric field in a radiation counter.

Townsend discharge A discharge in a gas at moderate pressure (above about 0.1 mmHg), corresponding to corona. It is free of space charges.

Townsend ionization *Avalanche effect.*

T pad A pad made up of resistors arranged as a T network.

TPI Abbreviation for *tracks per inch.*

trace 1. The visible path of a moving spot on the screen of a cathode-ray tube. Also called line. 2. An extremely small quantity of a substance. 3. An interpretive diagnostic technique that provides an analysis of each executed instruction and writes it on an output device as each instruction is executed by a computer.

trace concentration A concentration of a substance below the usual limits of chemical detection. Radionuclides are often observable in trace concentration by their radioactivity.

trace interval The time interval in which a sweep traces a desired pattern on the screen of a cathode-ray tube.

tracer 1. A foreign substance, usually radioactive, that is mixed with or attached to a given substance so the distribution or location of the latter can later be determined. 2. A thread of contrasting color woven into the insulation of a wire for identification purposes.

tracing distortion The nonlinear distortion introduced in the reproduction of a mechanical recording because the curve traced by the motion of the reproducing stylus is not an exact replica of the modulated groove.

track 1. A path that records one channel of information on a magnetic tape, drum, or other magnetic recording medium. The location of the track is determined by the recording equipment rather than by the medium. 2. The horizontal component of the path actually followed by a vehicle, or (marine usage) the intended path. 3. The visible path of an ionizing particle in a cloud chamber of nuclear photographic emulsion. 4. The trace of a moving target on a PPI radar screen or an equivalent plot. 5. To follow the progress of a missile, aircraft, hurricane, or other moving object or action, generally by radar, radio direction finders, infrared, or optical equipment. 6. *Race track.*

track-command guidance Missile guidance in which the target and missile are tracked by separate radars, and corrective commands are sent to the missile by radio.

track homing The process of following a line of position known to pass through an objective.

tracking 1. The condition in which all tuned circuits in a receiver accurately follow the frequency indicated by the tuning dial over the entire tuning range. 2. A motion given to the major lobe of a radar or radio antenna such that some preassigned moving target in space is always within the major lobe. 3. The following of a groove by a phonograph needle. 4. Maintaining the same ratio of loudness in the two channels of a stereophonic sound system at all settings of the ganged volume control.

tracking beam The beam that is aimed directly at the target at all times in antimissile warfare. Data obtained from this beam is transmitted to the counter-attacking guided missile over what is known as the guidance beam.

tracking element The element in a fire-control system that receives data from the position-finding element and computes the speed and direction of movement of the target and sometimes the rates of change in speed and direction.

tracking error Deviation of the vibration axis of a phonograph pickup from tangency with a groove. True tangency is possible for only one groove when the pickup arm is pivoted. The longer the pickup arm, the less the tracking error.

tracking filter A bandpass filter whose center frequency follows the average frequency of the input signal.

tracking station A radio, radar, or other station set up to track an object moving through the atmosphere or space.

track in range To adjust the gate of a radar set so it opens at the correct instant to accept the signal from a target that is changing in range.

track made good The resultant track of an aircraft, represented as a straight line between the departure point and the last point of fix on the surface.

tracks per inch [abbreviated TPI] The number of recording tracks produced per inch of radial movement of a magnetic tape recording head on a magnetic disk or other type of recording medium.

track while scan An electronic system that detects a radar target, computes its velocity, and predicts its future position without interfering with continuous radar scanning.

tradeoff Balancing one desirable performance parameter against others for optimizing the overall performance of a system.

traffic The messages transmitted and received over a communication channel.

trailer A bright streak at the right of a dark area or dark line in a television picture, or a dark area or streak at the right of a bright part. Usually caused by insufficient gain at low video frequencies.

trailer record A data-processing record that follows a group of records and contains data related to the entire group of records.

trailing antenna An aircraft radio antenna that has one end weighted and trailing free from the aircraft when in flight.

trailing edge The major portion of the decay of a pulse.

trailing-edge pulse time The time at which the instantaneous amplitude of a pulse last reaches a stated fraction of the peak pulse amplitude.

train To aim or direct a radar antenna in azimuth.

trainer A piece of equipment used for training operators of radar, sonar, and other electronic equipment by simulating signals received under operating conditions in the field.

trajectory The path traced through space by a missile or space vehicle.

trajectory-controlled Guided or directed so its trajectory will follow a predetermined curve, as for a missile.

transaction data Random and unpredictable new input data for a data-processing system, such as hours worked, quantities shipped, and amounts invoiced.

transadmittance A specific measure of transfer admittance under a given set of conditions.

transceiver A radio transmitter and receiver combined in one unit and having switching arrangements to permit use of one or more stages for both transmitting and receiving.

transconductance [symbol G_m] An electron-tube rating, equal to the change in anode current divided by the change in control-grid voltage. The unit of transconductance is the mho. Less strictly, transconductance is the amplification factor of the tube divided by its anode resistance. In a field-effect transistor, transconductance is the ratio of a change in output current to the initiating change in input voltage. Also called mutual conductance.

transconductance meter An instrument that indicates the transconductance of a grid-controlled electron tube. Also called mutual-conductance meter.

transcontinental ballistic missile A ballistic missile that has a range of at least 12,500 mi (20,100 km), so it can be fired from any point on the earth's surface and reach any surface target.

transcribe 1. To record, as to record a radio program by electrical transcriptions or magnetic tape for future rebroadcasting. 2. To copy, with or without translating, from one external computer storage medium to another.

transcriber The equipment used to convert information from one form to another, as for converting computer input data to the medium and language used by the computer.

transcription A 16-in (40.64-cm) diameter, 33⅓-rpm disk recording of a complete program, made especially for radio broadcast purposes. Also called electrical transcription.

transducer General term for any device that converts energy from one form to another, as from acoustic energy to electric or mechanical energy. Loudspeakers, mircophones, phonograph pickups, and strain gages are examples of transducers.

transducer scanner A device that provides a means of sampling directional signals from individual magnetostriction transducers in a sonar transmitter array. Capacitor plates arranged radially on a disk rotate with respect to a stationary circular disk that contains matching plates which are connected to the transducer elements, to give scanning of all elements once per revolution of the rotor disk.

transducing piezoid A piezoid used in a transducer.

transfer To transmit or copy information from one computer device to another without changing its form.

transfer check A check on the accuracy of transfer of a word in a digital computer, usually made automatically.

transfer function The mathematical relationship between the output of a control system and its input.

transfer impedance The ratio of the voltage applied at one pair of terminals of a network to the resultant current at another pair of terminals, all terminals being terminated in a specified manner.

transfer instruction A digital-computer instruction or signal that specifies the location of the next operation to be performed.

transfer oscillator An oscillator that extends the upper frequency limits of an electronic counter. The transfer oscillator mixes the unknown signal with a harmonic of a signal derived internally from a variable-frequency oscillator that is tuned for zero beat. The counter measures the frequency of the variable-frequency oscillator signal, and the counter reading multiplied by the harmonic number gives the unknown frequency.

transferred charge The net electric charge transferred from one terminal of a capacitor to another via an external circuit.

transferred electron [abbreviated TE] A free electron that has been transferred from one minimum to another within a zone. The effective mass of the electron changes to that associated with the new minimum, but the electron does not change its location.

transferred-electron amplifier [abbreviated TEA] A diode amplifier, which generally uses a transferred-electron diode made from doped N-type gallium arsenide, that provides amplification in the gigahertz range to well over 50 GHz at power outputs typically below 1 W continuous wave. A Gunn amplifier is an example.

transferred-electron device A semiconductor device, usually a diode, that depends on internal negative resistance caused by transferred electrons in gallium arsenide or indium phosphide at high electric fields. Transit time is minimized, permitting oscillation at frequencies up to several hundred megahertz. Operation may be in the transit-time mode, as in Gunn diodes; in the quenched-domain mode; or in the limited space-charge accumulation (LSA) mode.

transferred-electron diode A transferred-electron device, generally made from gallium arsenide, that can produce microwave energy directly

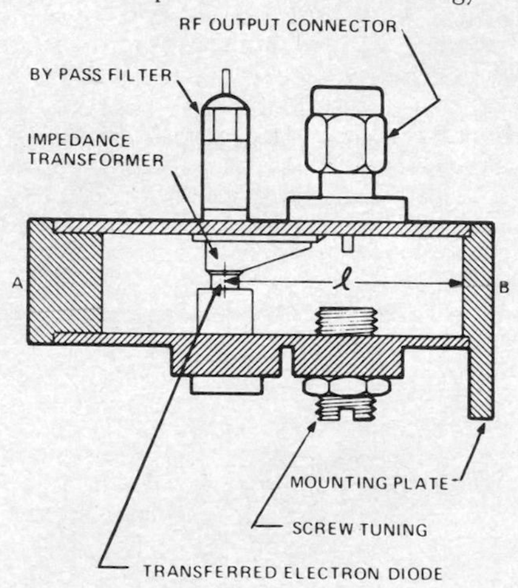

Transferred-electron diode mounted in tunable waveguide cavity of microwave oscillator.

from a DC input when combined with an appropriate cavity or a microstrip integrated circuit.

transferred-electron effect The variation in the effective drift mobility of charge carriers in a semiconductor when significant numbers of electrons are transferred from a low-mobility valley of the conduction band in a zone to a high-mobility valley, or vice versa.

transfer switch A switch for transferring one or more conductor connections from one circuit to another.

transfer time The total elapsed time between the breaking of one set of contacts on a relay and the making of another set of contacts, after all contact bounce has ceased.

transform To change the form of digital-computer information without significantly altering its meaning.

transformer [symbol T] A component that consists of two or more coils which are coupled together by magnetic induction. Used to transfer electric energy from one or more circuits to one or more other circuits without change in frequency but usually with changed values of voltage and current.

transformer bridge A network that consists of a transformer and two impedances, in which the input signal is applied to the transformer primary and the output is taken between the secondary center-tap and the junction of the impedances that

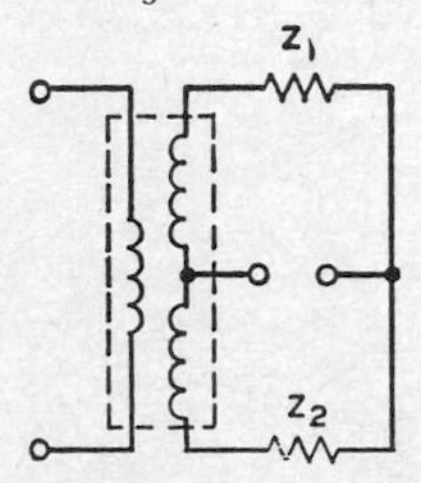

Transformer bridge, with input terminals at left.

connect to the outer leads of the secondary. When used as a crystal filter, a capacitor is used as one impedance to balance the static capacitance of the crystal that serves as the other impedance, so there is no transmission except in the vicinity of crystal resonance.

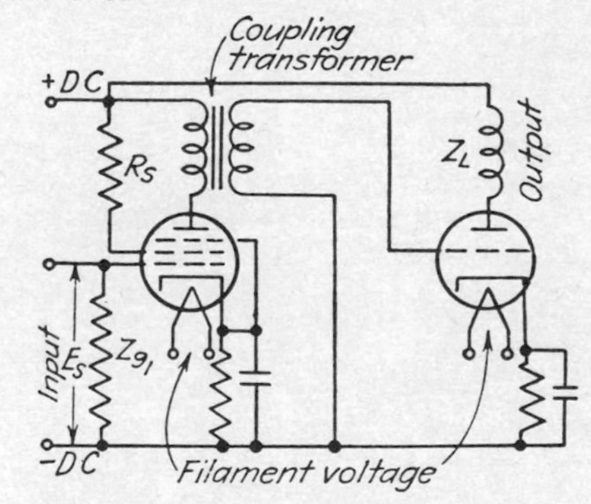

Transformer-coupled amplifier.

transformer-coupled amplifier An AF amplifier that uses untuned iron-core transformers to provide coupling between stages.

transformer coupling *Inductive coupling.*

transformer hybrid *Hybrid set.*

transformerless soldering iron A soldering iron that uses a silicon controlled rectifier in place of a transformer for lowering the line voltage.

transformer loss The ratio of the power delivered by an ideal transformer to the power delivered by an actual transformer under specified conditions. Usually expressed in decibels.

transformer oil A high-quality insulating oil in which windings of large power transformers are sometimes immersed to provide high dielectric strength, high insulation resistance, high flash point, and freedom from moisture and oxidization.

transformer read-only store A read-only store in which the presence or absence of mutual inductance between two circuits determines whether a binary 1 or 0 is stored. It contains one word line

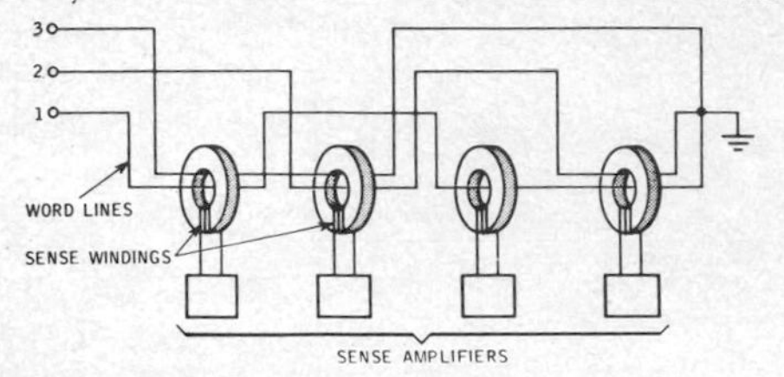

Transformer read-only store, showing storage of 1011 on word line 1, 0101 on word line 2, and 1100 on word line 3.

for each word store, and one transformer for each digit in the output word. A word line threaded through a core gives a 1, and a line bypassing a core gives a 0. For a readout, a current pulse is sent through a word line, causing output pulses to be produced in the transformers through whose cores that word line is threaded.

transforming section A length of waveguide or transmission line that has a varying cross section, used for impedance transformation.

transient A pulse, damped oscillation, or other temporary phenomenon occurring in a system prior to reaching a steady-state condition.

transient analyzer An analyzer that generates transients in the form of a succession of equal electric surges of small amplitude and adjustable waveform, applies these transients to a circuit or device under test, and shows the resulting output waveforms on the screen of an oscilloscope.

transient distortion Distortion due to inability to amplify transients linearly.

transient equilibrium Radioactive equilibrium in which the parent has such a large decay constant that the quantity of parent present decreases before radioactive equilibrium is reached.

transient motion An oscillatory or other irregular motion that occurs while a quantity is changing to a new steady-state value.

transient oscillation A momentary oscillation that occurs in a circuit during switching.

transient overshoot The maximum value of the overshoot of a quantity as a result of a sudden change in conditions.

transient phenomena Rapidly changing actions that occur in a circuit during the interval between closing of a switch and settling to steady-state conditions, or any other temporary actions which occur after some change in a circuit.

transient response The response of a circuit to a sudden change in an input quantity, such as to a step function.

transient suppressor A device used specifically to protect a circuit from destructive voltage surges. Examples include gas-discharge (spark-gap), metal-oxide, selenium, and silicon carbide devices; resistor-capacitor networks; varistors; and zener diodes.

transistance The characteristic that makes possible the control of voltages or currents so as to accomplish gain or switching action in a circuit. Examples of transistance occur in transistors, diodes, and saturable reactors.

transistor [TRANSfer resISTOR] An active semiconductor device that has three or more electrodes. The three main electrodes used are the emitter, collector, and base. Conduction is by electrons and carriers or holes. Germanium and silicon are the materials most often used as the semiconductor material. Transistors can perform practically all the functions of tubes, including amplification and rectification.

transistor amplifier An amplifier in which one or more transistors provide amplification comparable to that of electron tubes. In a class A transistor amplifier, operation is in the linear region of the collector characteristic. For class B, amplification occurs only during half of each input signal cycle. For class AB, the collector current or voltage is zero for less than half of each input cycle. For class C, collector current or voltage is zero for more than half of each input cycle.

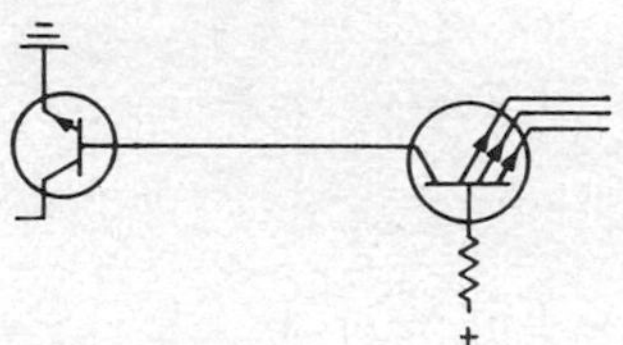

Transistor-coupled logic.

transistor-coupled logic A form of integrated-circuit logic that may be used with common bases and collectors, for multiple-emitter coupling.

transistorized Constructed with transistors being used in place of electron tubes.

transistorized DC motor 1. A conventional AC motor driven by a transistorized DC/AC converter. 2. A DC motor in which transistors replace the conventional commutator for commutating the current.

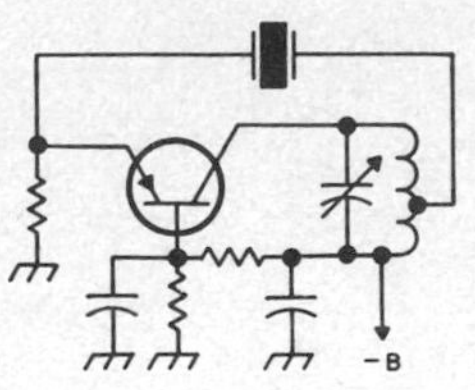

Transistor oscillator using quartz crystal in Hartley oscillator circuit.

transistor oscillator An oscillator in which a

transistor is used in place of an electron tube.

transistor radio A radio receiver in which transistors are used in place of electron tubes.

transistor symbol A schematic symbol that represents a transistor in circuit diagrams. The base is represented by a straight line at right angles to its lead. The collector line intersects the base at an angle and has no arrow. The emitter line has an arrow, pointing toward the base for a PNP transistor and pointing away from the base for an NPN transistor.

transistor-transistor logic [abbreviated TTL and T^2L] A logic circuit that contains two or more

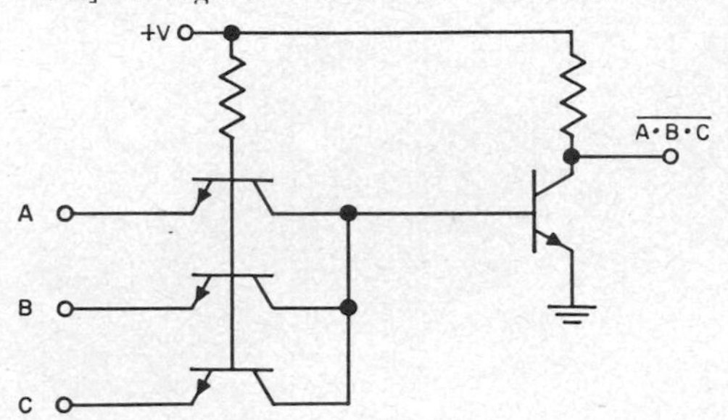

Transistor-transistor logic circuit.

transistors, for driving large output capacitances at high speed.

Transit A navigation satellite designed as an all-weather global system by which the positions of surface craft, submarines, and aircraft can be accurately fixed.

transitional coupling The amount of inductive coupling between two coils that gives the widest passband and flattest response curve without double peaks.

transition coding Color-bar coding of binary data in which the color of the previous bar determines whether the bar being read is a binary 0 or 1. Each consecutive bar is a different color; in one example, a green bar following a white space is a binary 0 bit, and a black to green transition is a binary 1. Transition coding gives higher density of coding and improves accuracy of readout when printing quality is poor.

transition effect A change in the intensity of the secondary radiation associated with a beam of primary radiation as the latter passes from a vacuum into a material medium or from one medium into another.

transition element An element that couples one type of transmission system to another, as for coupling a coaxial line to a waveguide.

transition factor *Reflection coefficient.*

transition frequency The frequency that corresponds to the intersection of the asymptotes to the constant-amplitude and constant-velocity portions of the frequency response curve for a disk recording. This curve is plotted with output voltage ratio

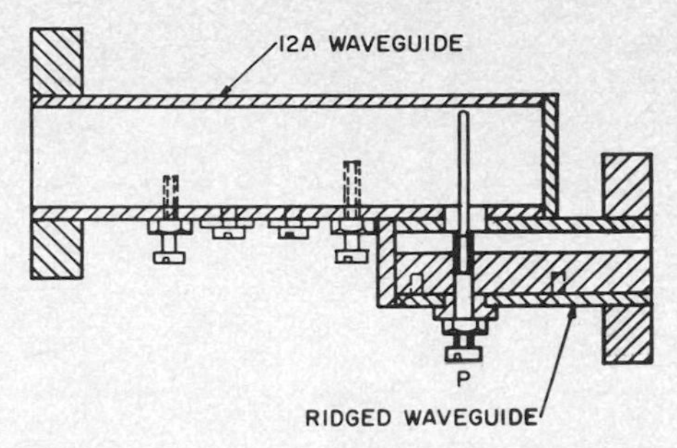

Transition element for coupling rectangular 12A waveguide to ridged waveguide is probe P, soldered to center of ridge.

in decibels as the ordinate, and the logarithm of the frequency as the abscissa. Below the transition frequency, the level is progressively reduced when cutting a record to prevent loud bass notes from overcutting the groove walls. One standard transition frequency value is 500 Hz. Also called crossover frequency and turnover frequency.

transition loss 1. The difference between the power incident upon a transition or discontinuity between two media in a wave propagation system and the power transmitted beyond the discontinuity that would be observed if the medium beyond the discontinuity were match-terminated. 2. The ratio in decibels of the power incident upon a discontinuity to the power transmitted beyond the discontinuity that would be observed if the medium beyond the discontinuity were match-terminated.

transition point A point at which the constants of a circuit change in such a way as to cause reflection of a wave being propagated along the circuit.

transition region The region between two homogeneous semiconductors in which the impurity concentration changes.

transitron oscillator A negative-resistance oscillator in which the screen grid is more positive than the anode, and a capacitor is connected between

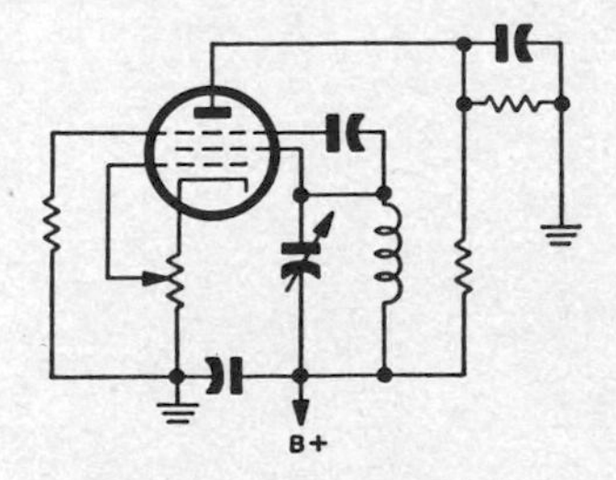

Transitron-oscillator circuit.

the screen grid and the suppressor grid. The suppressor grid periodically divides the current between the screen grid and the anode, thereby producing oscillation.

transit time The time required for an electron or other charge carrier to travel between two electrodes in an electron tube or transistor.

transit-time microwave diode A solid-state microwave diode in which the transit time of charge carriers is short enough to permit operation in microwave bands. Bulk diodes (such as Gunn and LSA) and junction diodes (such as BARITT, IMPATT, and TRAPATT) are two major types.

transit-time mode One of the three operating modes of a transferred-electron diode, in which space-charge domains are formed at the cathode and travel across the drift region to the anode. The frequency of oscillation is influenced by the dimensions of the drift region. This mode is used in Gunn diodes. The other two modes are the LSA and the quenched-domain modes.

translate To change computer information from one language to another without significantly affecting the meaning.

translation loss The amount by which the amplitude of motion of a stylus differs from the recorded amplitude in a disk record. Also called playback loss.

translator 1. A computer network or system that has a number of inputs and outputs, so connected that when signals representing information expressed in a certain code are applied to the inputs, the output signals will represent the same information in a different code. Also called matrix. 2. A combination television receiver and low-power television transmitter, used to pick up television signals on one frequency and retransmit them on another frequency to provide reception in areas not served directly by television stations. A translator usually broadcasts on a UHF channel from No. 70 to No. 83.

transliterate To convert the characters of one alphabet to another alphabet.

translunar Beyond the orbit of the moon.

transmission 1. The process of transferring a signal, message, picture, or other form of intelligence from one location to one or more other locations by wire lines, radio, light or infrared beams, or other communication systems. 2. A message, signal, or other form of intelligence that is being transmitted. 3. The ratio of the light flux transmitted by a medium to the light flux incident upon it. Transmission may be either diffuse or specular. Also called transmittance.

transmission band The frequency range above the cutoff frequency in a waveguide, or the comparable useful frequency range for any other transmission line, system, or device.

transmission coefficient 1. The ratio of transmitted to incident energy or some other quantity at a discontinuity in a transmission medium. For sound waves, it is called the sound transmission coefficient. 2. *Penetration probability.*

transmission grating A diffraction grating produced on a transparent base so radiation is transmitted through the grating instead of being reflected from it.

transmission level The ratio of the signal power at any point in a transmission system to the signal power at some point in the system chosen as a reference point. Usually expressed in decibels.

transmission limit A limiting wavelength or frequency above or below which a given type of radiation is not appreciably transmitted by a given medium.

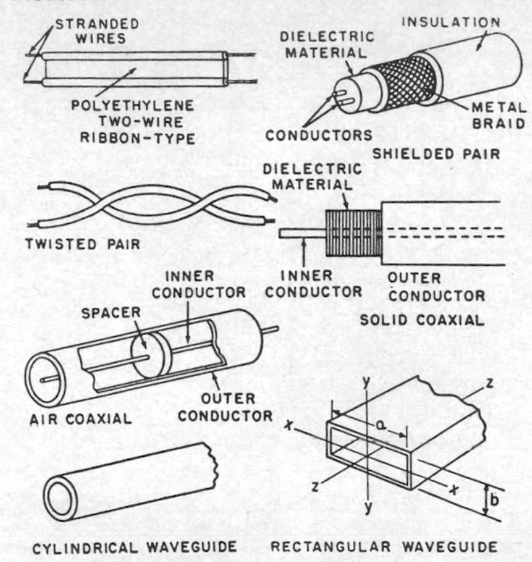

Transmission lines.

transmission line A waveguide, coaxial line, or other system of conductors used to transfer signal energy efficiently from one location to another.

transmission-line coupler A coupler that permits the passage of electric energy in either direction between balanced and unbalanced transmission lines.

transmission-line trap An interference trap that can be used with television receivers to minimize FM and other types of interference picked

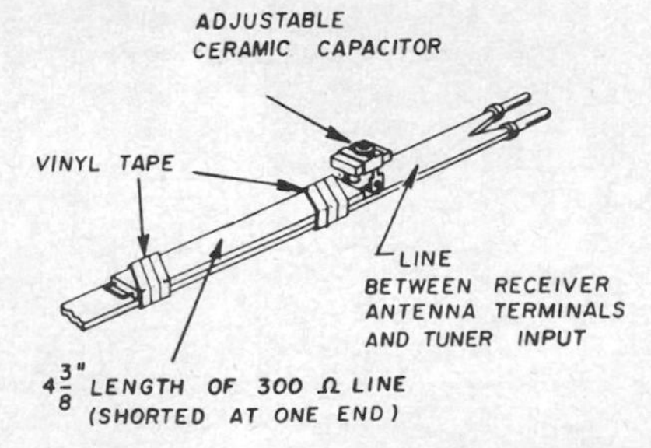

Transmission-line trap taped to television receiver twin-line. Trap is 11.1 cm long, with 2.5- to 13-pF adjustable capacitor connected to unshorted ends.

up by the television antenna in the range of 40 to 170 MHz. It consists of a 4⅜-in (11-cm) length of twin-line that has a short-circuit at one end and an adjustable ceramic capacitor at the other end, taped against the receiver twin-line.

transmission loss 1. The ratio of the power at one point in a transmission system to the power at a point farther along the line. Usually expressed in decibels. 2. The actual power that is lost in transmitting a signal from one point to another through a medium or along a line. Also called loss.

transmission measuring set A measuring instrument that consists of a signal source and receiver which have known impedances, to measure the insertion loss or gain of a network or transmission path connected between those impedances.

transmission mode *Mode.*

transmission modulation Amplitude modulation of the reading-beam current in a charge-storage tube as the beam passes through apertures in the storage surface. The degree of modulation is controlled by the stored charge pattern.

transmission plane The plane of vibration of polarized light that will pass through a Nicol prism or other polarizer.

transmission primaries The set of three color primaries that correspond to the three independent signals contained in the color television picture signal. The three receiver primaries in the color picture tube form one set. The luminance primary and the two chrominance primaries, known as the Y, I, and Q primaries, form another possible set of transmission primaries.

transmission secondary-emission multiplication Electron multiplication in which electrons hitting one side of a dynode cause emission of

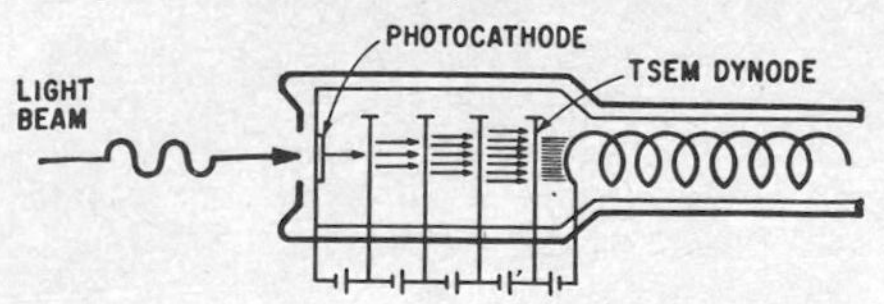

Transmission secondary-emission multiplication as used in traveling-wave phototube.

many more electrons from the opposite side of that dynode, with the process building up as the electron stream passes through a series of dynodes.

transmission security The aspect of communication security that is concerned with the transmission of messages over wires or by radio.

transmission speed The number of information elements sent per unit time. Usually expressed as bits, characters, word groups, or records per second or per minute.

transmission target An x-ray target in which the useful x-ray beam emerges from the surface remote from that on which the electron stream is incident.

transmission time The absolute time interval from transmission to reception of a signal.

transmission-type photocathode A photocathode that emits electrons from one side in proportion to the intensity of light that reaches the other side through an optical window in the phototube.

transmission unit An early signal-level unit now known as the decibel.

transmissivity *Transmittivity.*

transmissometer A photoelectric instrument that measures the visibility of the atmosphere.

transmit 1. To send a message, program, or other information to a person or place by wire, radio, or other means. 2. To reproduce information in a new location in a digital computer, replacing whatever was previously stored and clearing or erasing the source of the information.

transmit negative The transmission of facsimile signals intended for reception as a negative.

transmit positive The transmission of facsimile signals intended for reception as a positive.

transmittance *Transmission.*

transmitted-carrier operation Amplitude modulation in which the carrier wave is transmitted.

transmitted wave *Refracted wave.*

transmitter 1. The equipment used for generating and amplifying an RF carrier signal, modulating the carrier signal with intelligence, and feed-

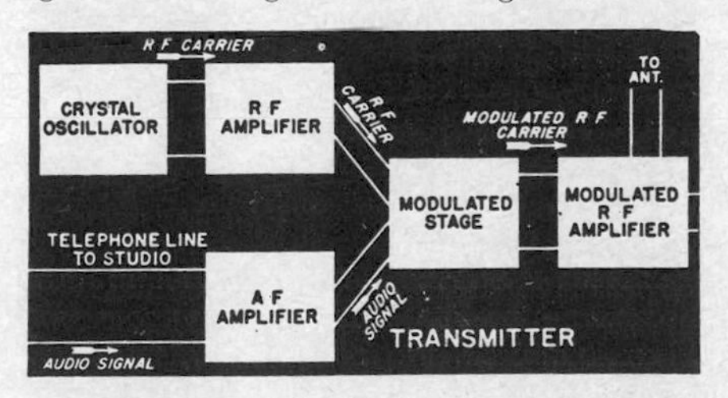

Transmitter block diagram for radio station.

ing the modulated carrier to an antenna for radiation into space as electromagnetic waves. 2. In telephony, the microphone that converts sound waves into AF signals. 3. *Synchro transmitter.*

transmitter input polarity The polarity of the portion of a television picture signal that represents a dark area of a scene, relative to the potential of a portion of the signal which represents a light area.

transmitter synchro *Synchro transmitter.*

transmitting efficiency The ratio of total acoustic power output to electric power input for an electroacoustic transducer. Also called projector efficiency.

transmitting station The location at which a transmitter, transmitting antenna, and associated transmitting equipment of a radio or television station are grouped.

transmittivity The ratio of the transmitted radiation to radiation arriving perpendicular to the boundary between two media. Also called transmissivity.

transmutation A nuclear process in which a nuclide is transformed into the nuclide of a different element.

transolver A synchro that has a two-phase cylindrical rotor within a three-phase stator, for use as a transmitter or a control transformer with no degradation of accuracy or nulls.

transonic Pertaining to transonic speed.

transonic barrier *Sonic barrier.*

transonic speed A speed in the range of about Mach 0.8 to Mach 1.2, corresponding to 600 to 900 mi/h (about 960 to 1450 km/h), at which one or more local points on the body of an aircraft or missile are moving at subsonic speed at the same time that one or more other points move at sonic or supersonic speed.

transparent Permitting passage of radiation or particles.

transparent plasma A plasma through which an electromagnetic wave can propagate. In general, a plasma is transparent at frequencies above the plasma frequency.

transpolarizer An electrostatically controlled circuit impedance that can have about 30 discrete and reproducible impedance values. Two capacitors, each having a crystalline ferroelectric dielectric with a nearly rectangular hysteresis loop, are connected in series and act as a single low impedance to an AC sensing signal when both capacitors are polarized in the same direction. Application of 1-μs pulses of appropriate polarity increases the impedance in steps.

transponder A radio device that receives an interrogating or challenging radio signal and automatically transmits a response on the same or a different frequency.

transponder dead time The time interval between the start of a pulse and the earliest instant at which a new pulse can be received or produced by a transponder.

transport To convey as a whole from one storage device to another in a digital computer.

transportable transmitter A transmitter designed to be readily carried or transported from place to place but not normally operated while in motion.

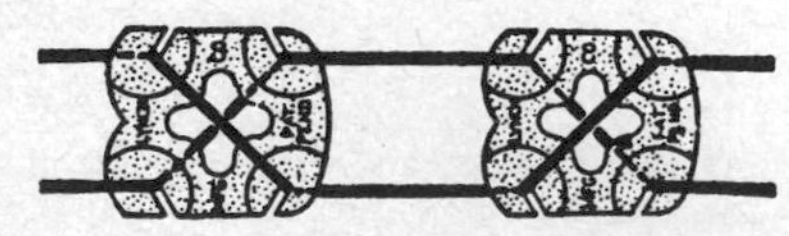

Transposition of conductors by ceramic insulators.

transposition Interchanging the relative positions of conductors at regular intervals along a transmission line to reduce crosstalk.

transradar A bandwidth compression system developed for long-range narrow-band transmission of radio signals from a radar receiver to a remote location.

transrectification Rectification that occurs in one circuit when an alternating voltage is applied to another circuit.

transuranium element An element that has an atomic number greater than uranium (which is 92), such as neptunium, plutonium, americium, curium, berkelium, californium, einsteinium, fermium, mendelevium, and nobelium.

transversal filter A filter whose frequency transmission properties exhibit a periodic symmetry.

transverse-beam traveling-wave tube A traveling-wave tube in which the direction of motion of the electron beam is transverse to the average direction in which the signal wave moves.

transverse electric wave [abbreviated TE wave] An electromagnetic wave in which the electric field vector is everywhere perpendicular to the direction of propagation. Also called H wave (British).

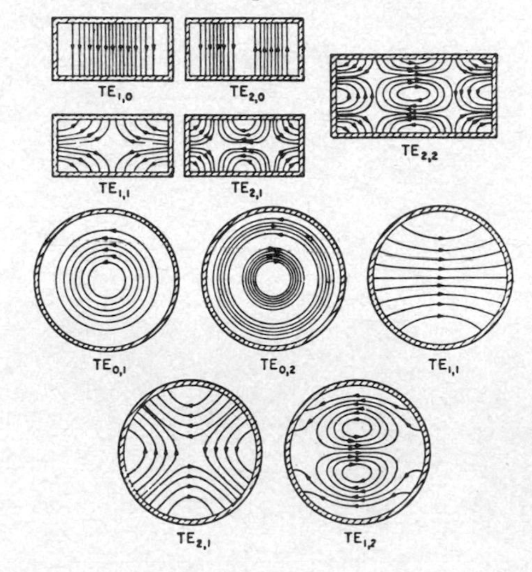

Transverse electric wave modes in rectangular and circular waveguides, with electric field configuration for each.

transverse electromagnetic wave [abbreviated TEM wave] An electromagnetic wave in which both the electric and magnetic field vectors are everywhere perpendicular to the direction of propagation.

transverse-field traveling-wave tube A traveling-wave tube in which the traveling electric fields that interact with the electrons are essentially perpendicular to the average motion of the electrons.

transverse-film attenuator An attenuator in which a conducting film is placed across a waveguide.

transversely excited atmospheric-pressure laser [abbreviated TEA laser] A gasdynamic laser in which the discharge path between electrodes is made extremely short by placing electrodes at opposite sides of the discharge tube, with many pairs of such electrodes positioned along the length of the tube to provide full excitation. Peak output powers exceeding 1 MW have been obtained.

transverse magnetic wave [abbreviated TM wave] An electromagnetic wave in which the magnetic field vector is everywhere perpendicular to the direction of propagation. Also called E wave (British).

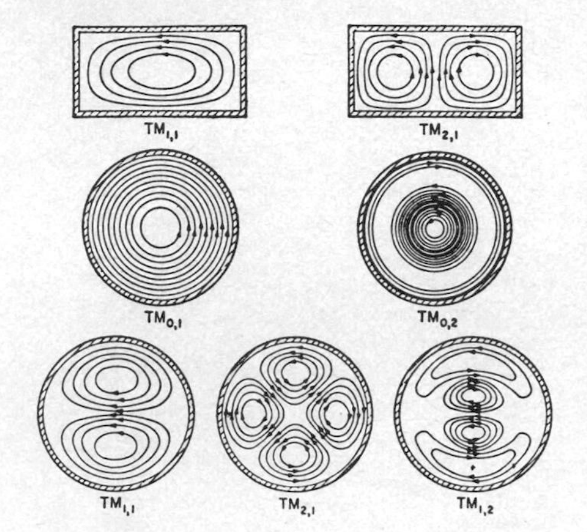

Transverse magnetic wave modes for rectangular and circular waveguides, with magnetic field configuration for each.

transverse magnetization Magnetization of a magnetic recording medium in a direction perpendicular to the line of travel and parallel to the greatest cross-sectional dimension.

transverse plate A plate of metal or highly resistant material used to close the end of a waveguide or as an adjustable piston inside the waveguide.

transverse wave A wave in which the direction of displacement at each point of the medium is parallel to the wavefront.

trap 1. A tuned circuit used in the RF or IF section of a receiver to reject undesired frequencies. Traps in television receiver video circuits keep the sound signal out of the picture channel. Also called rejector. 2. A semiconductor imperfection that prevents carriers from moving through the material. 3. *Wave trap.*

TRAPATT diode [TRApped Plasma Avalanche Transit-Time diode] A solid-state microwave diode in which the operating frequency as an oscillator is approximately determined by the thickness of the active layer. It is a transit-time device like the IMPATT diode, but it operates in a different mode; the avalanche zone moves through the drift region, creating a trapped space-charge plasma within the PN junction region.

trapezoidal wave A square wave on which is superimposed one ramp of a sawtooth wave.

trapped flux Magnetic flux that links with a closed superconducting loop.

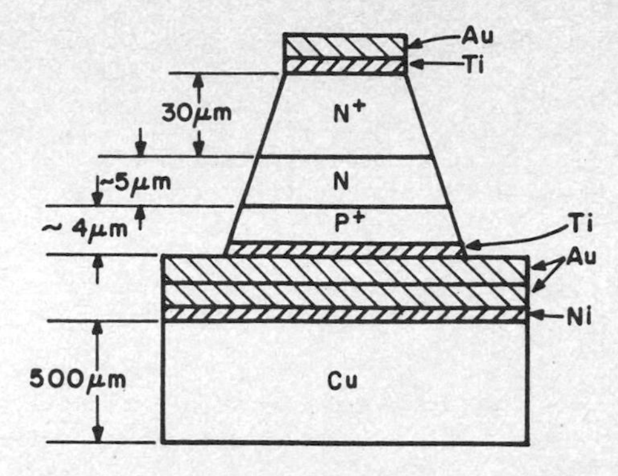

TRAPATT diode with integral copper heatsink. Dimensions are in micrometers.

trapped mode Propagation in which the energy radiated in the troposphere is almost entirely confined within a duct.

trapped radiation Radiation from space that has become trapped in the magnetic field of the earth, as in the Van Allen belt.

trapping A process wherein electrons are held at an irregularity in the crystal lattice of a semiconductor until released by thermal agitation.

traveling detector An RF probe mounted in a slotted-line section of waveguide and used with a detector to measure standing-wave ratios.

traveling wave A wave formed by translation of energy along a conductor. The energy is equally divided between current and voltage forms.

traveling-wave accelerator A plasma engine for space travel in which plasma is accelerated through a tube by a series of coils spaced along the tube and excited by polyphase RF energy.

traveling-wave acoustic amplifier An acoustic amplifier in which energy is transferred from an electron beam to an acoustic wave in such a way as to achieve mechanical amplification.

traveling-wave amplifier [abbreviated TWA] A broadband microwave amplifier that uses one or more traveling-wave tubes to provide power output which can be well over 200 W in the frequency range of 1 to 40 GHz.

traveling-wave antenna An antenna in which the current distributions are produced by waves of charges propagated in only one direction in the conductors. Also called progressive-wave antenna.

traveling-wave interaction The interaction between an electron stream and a slow wave moving through a circuit in approximate synchronism with the velocity of the electrons.

traveling-wave light modulator A two-conductor transmission line in which a portion of the dielectric is an electrooptical material capable of modulating a laser beam. The index of refraction of this material, and hence the velocity of light through it, varies with the applied electric field. Dielectric materials used include cubic crystals and

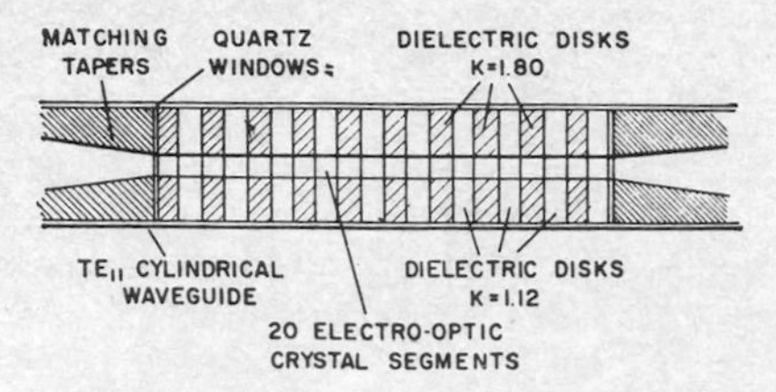

Traveling-wave light modulator using 20 crystal segments to modulate laser beam.

cuprous chloride and zinc sulfide. The amount of mismatch between the velocity of propagation of light through the crystal and the velocity of the microwave signal being transmitted at a given frequency determines the bandwidth of the modulator.

traveling-wave magnetron A traveling-wave tube in which the electrons move in crossed static electric and magnetic fields that are substantially normal to the direction of wave propagation, as in practically all modern magnetrons.

traveling-wave magnetron oscillation Oscillation sustained by the interaction between the space-charge cloud of a magnetron and a traveling electromagnetic field whose phase velocity is approximately the same as the mean velocity of the cloud.

traveling-wave maser A ruby maser used with a comblike slow-wave structure and a number of yttrium-iron garnet isolators to give amplification in the frequency range from about 400 MHz to well over 10 GHz. Operation is at the temperature of liquid helium (4.2 K).

traveling-wave parametric amplifier A parametric amplifier in which the signal, pump, and difference-frequency waves are propagated along a continuous structure that contains nonlinear reactors.

traveling-wave phototube A traveling-wave tube that has a photocathode and an appropriate window to admit a modulated laser beam. The modulated laser beam causes emission of a current-modulated photoelectron beam, which in turn is accelerated by an electron gun and directed into the helical slow-wave structure of the tube. An alternative arrangement uses a transmission-type photocathode and transmission-type dynodes in an electron multiplier that feeds the helical structure. Use of traveling-wave phototubes is restricted essentially to visible wavelengths of light because infrared photocathodes that have sufficient sensitivity for laser beam detection are not yet available.

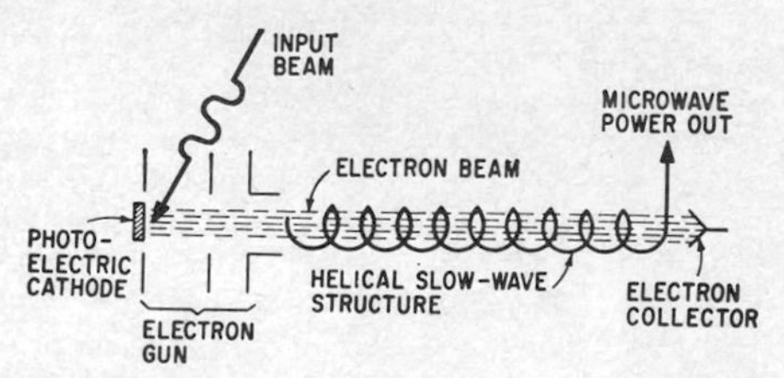

Traveling-wave phototube.

traveling-wave tube [abbreviated TWT] An electron tube in which a stream of electrons interacts continuously or repeatedly with a guided electromagnetic wave moving substantially in synchronism with it, in such a way that there is a net transfer of energy from the stream to the wave. The tube is used as an amplifier or oscillator at frequencies in the microwave region.

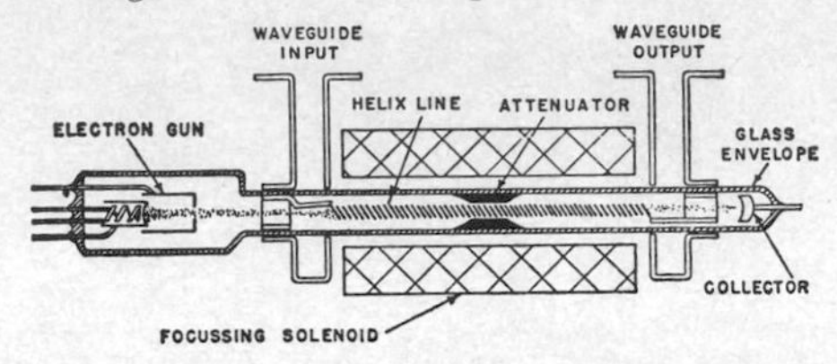

Traveling-wave tube uses solenoid to provide required longitudinal magnetic field.

TR box *TR tube.*

TR cavity The resonant portion of a radar TR tube.

treble High audio frequencies, such as those handled by a tweeter in a sound system.

treble boost Adjustment of the amplitude-frequency response of a system or component to accentuate the higher audio frequencies.

TRF Abbreviation for *tuned radio frequency.*

triac [TRIode AC semiconductor switch] A bidirectional gate-controlled thyristor that provides full-wave control of AC power. With phase control of the gate signal, load current can be varied from about 5 to 95% of full power.

Triac symbol.

triad 1. A triangular group of three small phosphor dots, each dot emitting one of the three primary colors, on the screen of a three-gun color picture tube. 2. *Triplet.*

triangle generator A signal generator whose

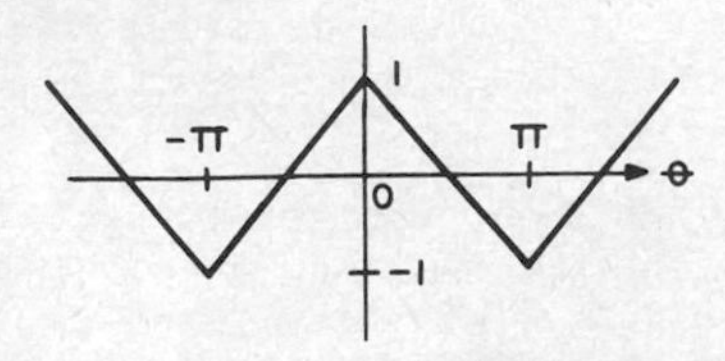

Triangle-generator output waveform.

output is a repeating ramp function that has equal positive and negative rates of change with time. The resulting triangle wave resembles a series of equilateral triangles.

triangulation Determination of the position of a ship or aircraft by obtaining bearings of the moving object with reference to two fixed radio stations a known distance apart. This gives the values of one side and all angles of a triangle, from which the position can be computed.

tribo- A prefix meaning pertaining to or resulting from friction.

triboelectricity Electric charges generated by friction.

triboelectric series A list of materials that produce an electrostatic charge when rubbed together, arranged in such an order that a material has a positive charge when rubbed with a material substance below it in the list and a negative charge when rubbed with a material above it in the list.

triboelectroemanescence Electron emission that occurs when a metal is fractured or abraded.

triboluminescence Luminescence produced by friction between two materials.

trichromatic coefficient *Chromaticity coordinate.*

trickle charge A continuous charge of a storage battery at a low rate to maintain the battery in a fully charged condition.

trickle charger A device that charges a storage battery at a low rate continuously to keep it fully charged.

triclinic Pertaining to a crystal structure that has three unequal axes intersecting at angles, not more than two of which are equal and not more than one of which is 90°.

tricon A radio navigation system in which the airborne receiver accepts pulses from a triplet or chain of three stations pulsed in variable time sequence. The time sequences vary, so pulses arrive at the same time along paths of various lengths.

tridipole antenna A horizontally polarized antenna that has three curved dipoles mounted in a horizontal plane to form a circle.

tridop A Doppler missile-tracking system that consists of a master station and three additional receiving stations on the ground. The master station triggers a continuous-wave transponder in the missile, and this in turn radiates signals to the ground station for comparison with a timing signal from the master station.

triductor An arrangement of iron-core transformers and capacitors used to triple a power-line frequency. A DC voltage is applied to some of the windings to provide enough premagnetization so the applied AC voltage can saturate the cores and thereby generate the desired third-harmonic output.

trifluorochloroethylene resin A fluorocarbon used as a base for polychlorotrifluoroethylene resin, marketed as Kel-F.

trigatron An electronic switch in which conduction is initiated by the breakdown of an auxiliary gap in a gas-filled envelope. The gap between the two main electrodes is normally nonconducting but breaks down when a pulse is applied to a trigger electrode. Used in some radar modulators.

trigger 1. To initiate a sudden action, as by applying a pulse to a trigger circuit. 2. The pulse that initiates the action of a trigger circuit. 3. *Trigger circuit.*

trigger action Use of a weak input pulse to initiate main current flow suddenly in a circuit or device.

trigger circuit 1. A circuit or network in which the output changes abruptly with an infinitesimal change in input at a predetermined operating point. Also called trigger. 2. A circuit in which an action is initiated by an input pulse, as in a radar modulator. 3. *Flip-flop circuit.*

trigger control Control of thyratrons, ignitrons, and other gas tubes in such a way that current flow may be started or stopped, but not regulated as to rate.

trigger diode *Diac.*

triggered blocking oscillator A blocking oscillator that can be reset to its starting condition by a trigger voltage. A parallel-triggered blocking oscillator has less effect on the trigger source than a series-triggered blocking oscillator, but the series-triggered type has less delay.

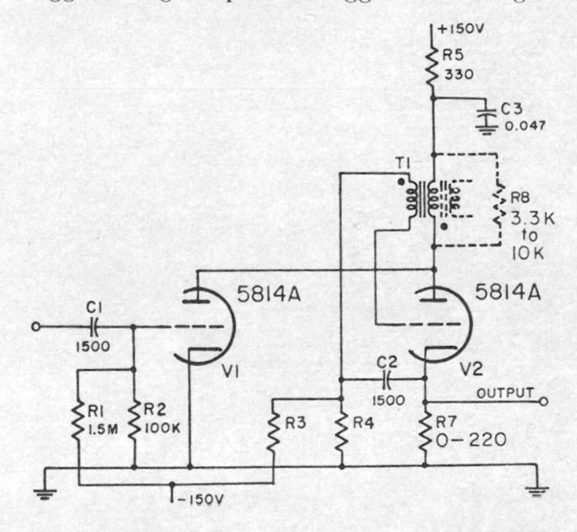

Triggered blocking oscillator using circuit design preferred by National Bureau of Standards. For pulse trigger spacings up to 60 μs, R3 is 47 kΩ and R4 is 4.7 kΩ; for spacings from 60 to 500 μs, R3 is 100 kΩ, and R4 is 10 kΩ.

triggered spark gap A fixed spark gap in which the discharge passes between two electrodes but is initiated by an auxiliary trigger electrode to which low-power pulses are applied at regular intervals by a pulse amplifier.

triggering Initiation of an action in a circuit, which then functions for a predetermined time, as for the duration of one sweep in a cathode-ray tube.

trigger pulse A pulse that starts a cycle of operation. Also called tripping pulse.

trigger switch A switch that is actuated by pulling a trigger and is usually mounted in a pistol-grip handle.

trigger tube A cold-cathode gas-filled tube in which one or more auxiliary electrodes initiate the anode current but do not control it.

trigger winding A winding added to a pulse transformer to supply a low-voltage pulse to an external load, usually for synchronizing purposes.

trihedral reflector A corner reflector that has three square or triangular sides which meet at a point. Used as an artificial radar reflector and for other applications where signals must be reflected back toward the transmitter over a greater angle than is practical with a plane sheet reflector.

trimmer A small variable or semiadjustable capacitor or variable inductance used in tuning circuits to adjust capacitance values for alignment purposes so all circuits can be tuned accurately by a single control.

trimmer capacitor A variable or semiadjustable capacitor used as a trimmer.

trimmer resistor A miniature rheostat used in place of a fixed resistor to permit convenient fingertip, screwdriver, or wrench adjustment of resistance values in a circuit.

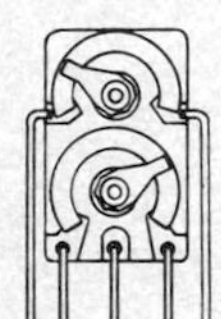
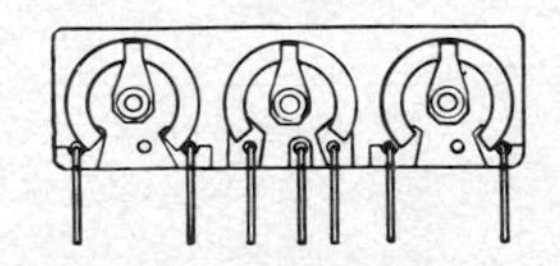
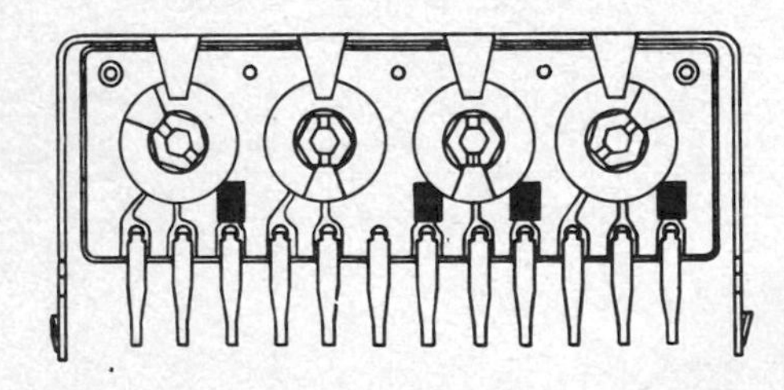

Trimmer resistors and trimmer potentiometers having conventional wire leads and plug-in printed-circuit leads.

trimming Fine adjustment of capacitance, inductance, or resistance of a component during manufacture or after installation in a circuit.

trimorph A vibrator that consists of three plates of piezoelectric material. When clamped at one end and driven by an AC voltage applied between the center plate and the paralleled outer plates, it can serve as a synchronous motor for driving a clock.

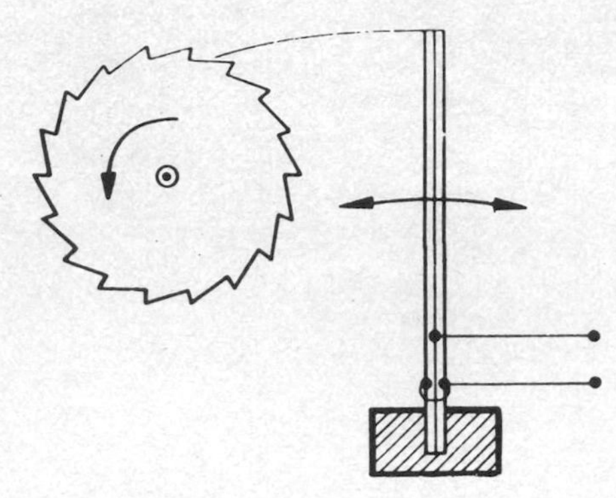

Trimorph arrangement for driving clock mechanism through sawtooth wheel and pawl attached to free end of piezoelectric vibrator.

Trinitron A color picture tube in which the phosphor screen is deposited in narrow vertical stripes instead of dots. The mask in front of the screen is a grille of vertical slots instead of a shadow mask with round holes. A single electron gun emits three beams, one for each primary color, in a horizontal line. The phosphor stripes are very narrow compared to the beams, so each beam spreads across two slots in the mask. The angle at which a beam enters a slot determines which color of phosphor stripe it hits.

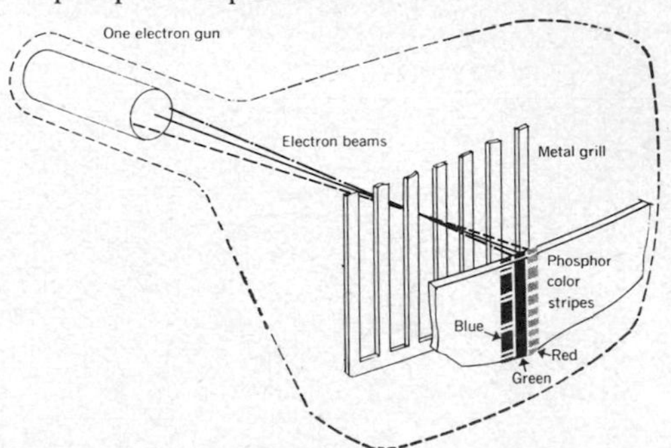

Trinitron operating principle. Beams are actually wide enough to cover two slots and hit two identical colors of phosphor stripes at the same time.

trinoscope An arrangement of three picture tubes with color filters and projection lenses, used in theater television to project the superimposed red, blue, and green images required for full-color pictures.

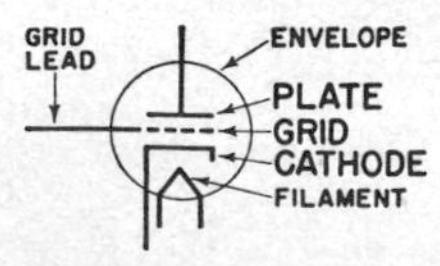

Triode symbol. Plate electrode is more properly called anode.

triode A three-electrode electron tube that contains an anode, a cathode, and a control electrode.

triode amplifier An amplifier that uses only triodes.

triode-hexode converter A triode oscillator and

a multigrid mixer in the same tube envelope.

trip action Instability that occurs in a magnetic amplifier because of excessive feedback.

trip coil A coil that opens a circuit breaker or other protective device when coil current exceeds a predetermined value.

triple-conversion receiver A communication receiver that has three different intermediate frequencies, to give higher adjacent-channel selectivity and greater image-frequency suppression.

triple-detection receiver *Double superheterodyne.*

triple-diode triode An electron tube that has three diodes and one triode served by a common cathode.

triple-stub transformer A transformer in which three stubs are placed a quarter-wavelength apart on a coaxial line and adjusted in length to compensate for impedance mismatch.

triplet Three radio navigation stations operated as a group for the determination of positions. Also called triad.

triplexer A dual-duplexer that permits use of two receivers simultaneously and independently in a radar system by disconnecting the receivers during the transmitted pulse.

tripping pulse *Trigger pulse.*

trip value The voltage, current, or power at which a polarized relay will transfer from one contact to another.

tri-state logic A form of TTL logic in which the output stages or input and output stages can assume three states; two are the normal low-impedance 1 and 0 states, and the third is a high-impedance state that allows many tri-state devices to time-share bus lines.

tristimulus colorimeter A colorimeter that measures a color stimulus in terms of tristimulus values.

tristimulus values The amounts of each of the three primary colors that must be combined to establish a match with a given sample color.

tri-tet oscillator A crystal-controlled tetrode oscillator in which the crystal circuit is isolated from the output circuit by using the screen grid as the oscillator anode. Used for multiband operation because it generates strong harmonics of the crystal frequency.

tritium [symbol T or H^3] The hydrogen isotope that has mass number 3. It is one form of heavy hydrogen, the other being deuterium.

tritium light source A self-luminous lamp that uses radioactive tritium which emits high-energy beta particles during its decay (electrons having a maximum energy of 18.6 keV). These electrons provide excitation for a phosphor coating inside a glass envelope, to give a light source that requires no power supply. Used for aircraft emergency exit signs, life-preserving equipment, and compasses. The luminous half-life is about 8 years.

triton The nucleus of tritium.

trombone A U-shaped length of waveguide that is adjustable in length.

tropicalize To prepare for use in a tropical climate by applying a coating that resists moisture and fungi.

tropopause The discontinuity that separates the stratosphere from the troposphere.

troposcatter *Tropospheric scatter.*

troposphere The portion of the earth's atmosphere that extends from the surface up to about 6 mi (10 km), in which temperature generally decreases with altitude, clouds form, and convection exists.

tropospheric bending Refraction of radio waves by adjacent layers of air masses that have different temperature and humidity characteristics in the troposphere, making possible long-distance transmission of VHF radio waves.

tropospheric duct *Duct.*

tropospheric scatter A form of scatter propagation in which radio waves are scattered by the troposphere. The phenomenon is essentially independent of frequency, hence is useful for communication at distances of several hundred kilometers over the entire RF spectrum. Also called troposcatter.

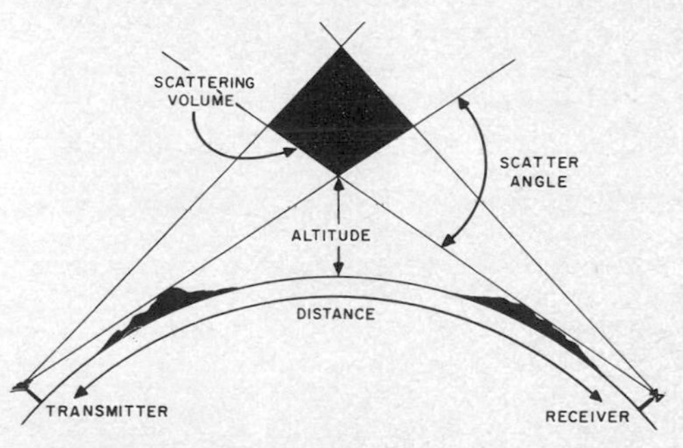

Tropospheric scatter as used for point-to-point beyond-the-horizon microwave communication using frequencies in range of 350 MHz to 8 GHz.

tropospheric superrefraction A condition in the troposphere whereby radio waves are bent sufficiently to be returned to the earth.

tropospheric wave A radio wave that is propagated by reflection from a region of abrupt change in dielectric constant or its gradient in the troposphere.

trouble-locating problem A computer test problem whose incorrect solution supplies information as to the location of a fault. Used after a check problem shows that a fault exists.

troubleshooting Locating and repairing faults in equipment after they have occurred.

TR switch *TR tube.*

TR tube [Transmit-Receive tube] A gas-filled RF switching tube used to disconnect a receiver from its antenna during the interval for pulse transmission in radar and other pulsed RF systems. Also called TR box and TR switch.

true altitude The altitude above mean sea level.

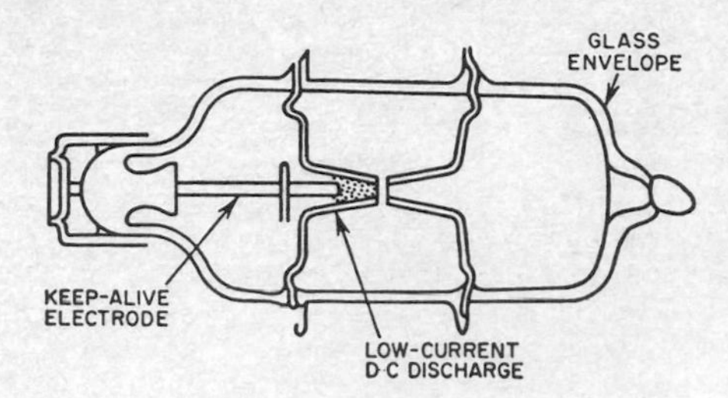

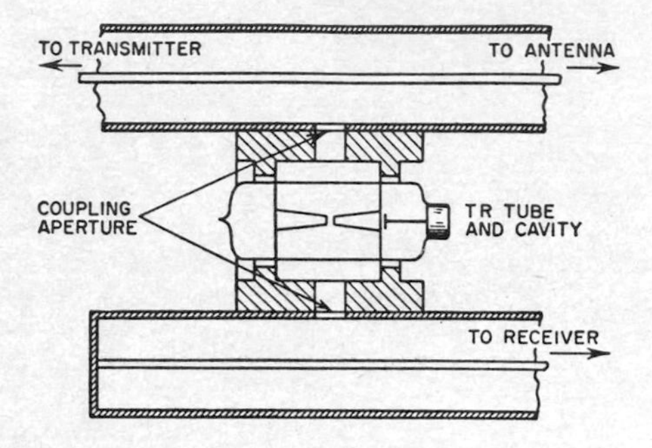

TR tube construction and method of use.

true bearing A bearing given in relation to true geographic north. A magnetic bearing is given in relation to magnetic north, and a relative bearing is given in relation to the lubber line or other axis of an aircraft or vessel.

true-bearing rate The rate of change of true bearing.

true-bearing unit A unit added to a radar set to rotate the PPI display so true north is always at the top of the screen.

true course A course indicated by an angle measured clockwise from true north.

true heading A heading measured with respect to true north.

true north The direction of the North Pole from the observer, or a line showing this direction.

truncate To drop digits at the end of a numerical value. The number 3.14159265 is truncated to five figures in 3.1415, whereas it would be 3.1416 if rounded off to five figures.

truncated paraboloid A radar parabolic reflector in which a portion of the top and bottom have been cut away to broaden the radar beam in the vertical plane.

truncation error The computation error that results from use of only a finite number of terms of an infinite series.

trunk 1. A path over which information is transferred in a computer. 2. A telephone line connecting two central offices.

truth table A table that describes a logic function by listing all possible combinations of input values and indicating for each such combination the true output values.

AND

B	A	C
0	0	0
0	1	0
1	0	0
1	1	1

OR

B	A	C
0	0	0
0	1	1
1	0	1
1	1	1

NAND

B	A	C
0	0	1
0	1	1
1	0	1
1	1	0

Truth tables for three gating functions.

Tschebycheff German spelling of *Chebyshev.*

T²L Abbreviation for *transistor-transistor logic.*

TTL Abbreviation for *transistor-transistor logic.*

TTY Abbreviation for *teletypewriter.*

tube coefficients The constants that describe the characteristics of an electron tube, such as amplification factor and transconductance.

tube complement The number of electron tubes required in a piece of electronic equipment.

tube noise Noise originating in an electron tube, such as that due to shot noise and thermal agitation.

tube shield A shield to be placed around an electron tube.

tube socket A socket that accommodates electrically and mechanically the terminals of an electron tube.

tube tester A test instrument that measures and indicates the condition of electron tubes used in electronic equipment.

tubular capacitor A paper or electrolytic capacitor that has the form of a cylinder, with leads usually projecting axially from the ends. The capacitor plates are long strips of metal foil separated by insulating strips, rolled into a compact tubular shape.

tumbling Loss of control in a two-frame free gyroscope, occurring when both frames of reference become coplanar.

tunable-cavity filter A microwave filter that can be tuned by adjusting one or more tuning screws which project into the cavity or by adjusting the positions of one or more rectangular or circular irises in the cavity or waveguide.

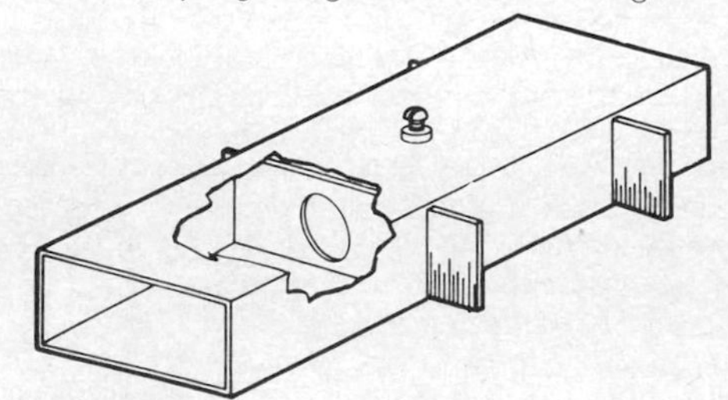

Tunable-cavity filter using two adjustable circular irises and one tuning screw.

tunable echo box An echo box that consists of an adjustable cavity operating in a single mode. It can be calibrated so the setting of the plunger at resonance indicates wavelength.

tunable laser A laser in which the frequency of the output radiation can be tuned over part or all of the ultraviolet, visible, and infrared regions of the spectrum. One commercial version uses a flashlamp-excited dye laser with a birefringent tuning element and frequency-doubling quartz crystals. Other versions use KDP crystal frequency-doubling stages and mirror-angle tuning. The choice of dye solution affects the tuning range. Applications include research in plasma physics, spectroscopy, chemistry, photoconductivity, biology, and holography.

tunable magnetron A magnetron that can be tuned over a range of frequencies by electronic or mechanical means. Generally the capacitance or inductance of the resonant structure is varied mechanically to achieve tuning.

tune To adjust for resonance at a desired frequency.

tuned amplifier An amplifier in which the load is a tuned circuit. Load impedance and amplifier gain then vary with frequency.

tuned-anode oscillator A vacuum-tube oscillator whose frequency is determined by a tank circuit in the anode circuit, coupled to the grid to provide the required feedback.

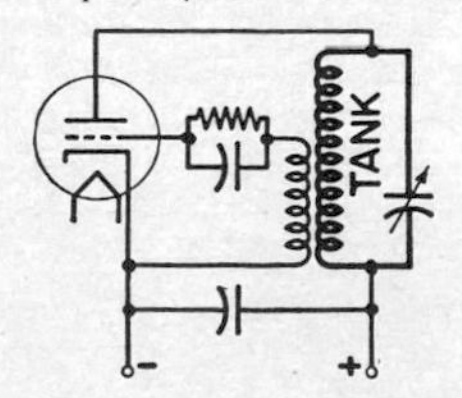

Tuned-anode oscillator circuit.

tuned antenna An antenna whose inductance and capacitance values provide resonance at the desired operating frequency.

tuned-base oscillator A transistor oscillator in which the frequency-determining resonant circuit is located in the base circuit. This is comparable to a tuned-grid electron-tube oscillator.

tuned cavity *Cavity resonator.*

tuned circuit A circuit whose components can be adjusted to make the circuit responsive to a particular frequency in a tuning range. Also called tuning circuit.

tuned-circuit oven An electrically heated compartment that accommodates and maintains tuned-circuit elements at an essentially constant temperature to prevent drifting in frequency with changes in temperature.

tuned-collector oscillator A transistor oscillator in which the frequency-determining resonant circuit is located in the collector circuit. This is comparable to a tuned-anode electron-tube oscillator.

tuned dipole A dipole that provides resonance at its operating frequency.

tuned filter A filter that uses one or more tuned circuits to attenuate or pass signals at the resonant frequency.

tuned-grid oscillator A vacuum-tube oscillator whose frequency is determined by a tank circuit in the grid circuit, coupled to the anode to provide the required feedback.

tuned-grid tuned-anode oscillator A vacuum-tube oscillator that has parallel-tuned tanks in both anode and grid circuits, with feedback being obtained through the anode-to-grid interelectrode capacitance of the tube.

tuned radio frequency [abbreviated TRF] Pertaining to a receiver in which all RF amplification is carried out at the frequency of the transmitted carrier signal.

tuned radio-frequency amplifier An RF amplifier in which all tuned circuits are adjusted to the frequency of the desired transmitted carrier signal.

tuned radio-frequency receiver A radio receiver that consists of a number of amplifier stages which are tuned to resonance at the carrier frequency of the desired signal by a gang capacitor. The amplified signals at the original carrier frequency are fed directly into the detector for demodulation, and the resulting AF signals are amplified by an AF amplifier and reproduced by a loudspeaker. Now largely replaced by superheterodyne receivers having greatly improved selectivity.

tuned radio-frequency stage A stage of amplification that is tunable to the carrier frequency of the signal being received.

tuned-reed frequency meter *Vibrating-reed frequency meter.*

tuned relay A relay that has mechanical or other resonating arrangements which limit response to currents at one particular frequency.

tuned rope Long lengths of chaff cut to the various lengths required for tuning to a number of different frequencies.

tuned transformer A transformer whose associated circuit elements are adjusted to be resonant at the frequency of the alternating current supplied to the primary, thereby causing the secondary voltage to build up to higher values than would otherwise be obtained.

tuner The portion of a receiver that contains circuits which can be tuned to accept the carrier frequency of a desired transmitter while rejecting the carrier frequencies of all other stations on the air at that time. A television tuner commonly contains only the RF amplifier, local oscillator and mixer stages, whereas a radio tuner also contains the IF amplifier and second-detector stages.

tungar tube A gas tube that has a heated thoriated tungsten filament which acts as a cathode and a graphite disk that acts as an anode in an argon-filled bulb at a low pressure. Used chiefly as a rectifier in battery chargers.

tungsten [symbol W] A hard metallic element that has a melting point of 3370°C. Sometimes used for filaments and other electrodes of electron tubes. Atomic number is 74. Usually called wolfram outside the United States.

tungsten-halogen lamp A tungsten-filament lamp in which halogen gases (related to iodine) are added to the normal gas mixture in the envelope to combine with the tungsten evaporated from the filament and prevent blackening. Used chiefly for high-intensity photographic, theater, and television lighting.

tungsten inert-gas welding Welding in which an arc plasma from a tungsten electrode radiates heat onto the work surface, to create a weld puddle in a protective atmosphere provided by a flow of inert shielding gas. Heat must then travel by conduction from this puddle to melt the desired depth of weld.

tuning Adjusting circuits for optimum performance at a desired frequency.

tuning capacitor A variable capacitor used for tuning purposes.

tuning circuit *Tuned circuit.*

tuning coil A variable inductance used for tuning purposes.

tuning control The control knob that adjusts all tuning circuits simultaneously to a desired frequency.

tuning core A ferrite core that is moved in and out of a coil or transformer to vary the inductance.

tuning fork A U-shaped bar of hard steel, fused quartz, or other elastic material that vibrates at a definite natural frequency when struck or when set in motion by electromagnetic means. Used as a frequency standard.

tuning-fork drive Use of a tuning fork to control the frequency of an oscillator. A high harmonic of the fork frequency is picked up by a coil and amplified to control the frequency of the main oscillator in a transmitter or other piece of equipment.

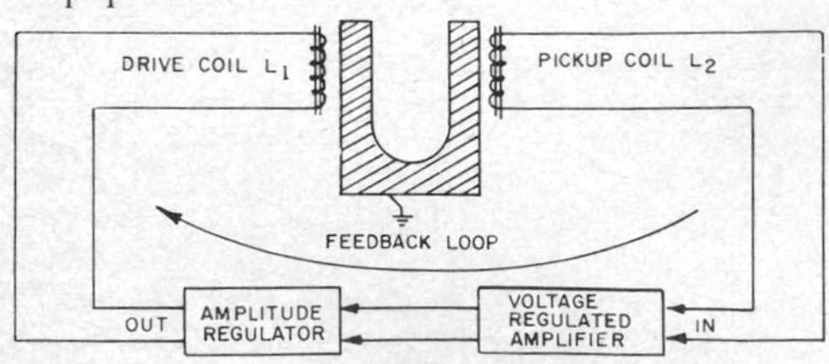

Tuning-fork resonator using feedback arrangement for continuous oscillation.

tuning-fork resonator A tuning fork and associated coils that generate an AC voltage related to the natural vibrating frequency of the fork.

tuning indicator A device that indicates when a radio receiver is tuned accurately to a radio station, such as a meter or a cathode-ray tuning indicator. It is connected to a circuit that has a DC voltage which varies with the strength of the incoming carrier signal.

tuning meter A DC voltmeter or ammeter used as a tuning indicator.

tuning probe An essentially lossless probe that can be extended through the wall of a waveguide or cavity resonator to project an adjustable distance inside.

tuning range The frequency range over which a receiver or other piece of equipment can be adjusted by a tuning control.

tuning screw A screw that is inserted into the top or bottom wall of a waveguide and adjusted as to depth of penetration inside for tuning or impedance-matching purposes.

tuning sensitivity The rate of change of frequency of an oscillator with the position of a mechanical tuner, electric tuning voltage, or other control parameter, at a given operating point.

tuning stub A short length of transmission line, usually shorted at its free end, that is connected to a transmission line for impedance-matching purposes.

tuning wand A rod of insulating material that has a brass plug at one end and a ferrite core at the other end. Used for checking receiver alignment.

tunnel cathode An unheated cathode in which a thin metal film that serves as the cathode surface is deposited on an insulating film which is on a metal substrate. When a voltage is applied between the metal film and the substrate, electrons tunnel from the substrate and reach the metal film with sufficient energy to pass through the cathode surface into the vacuum of the tube.

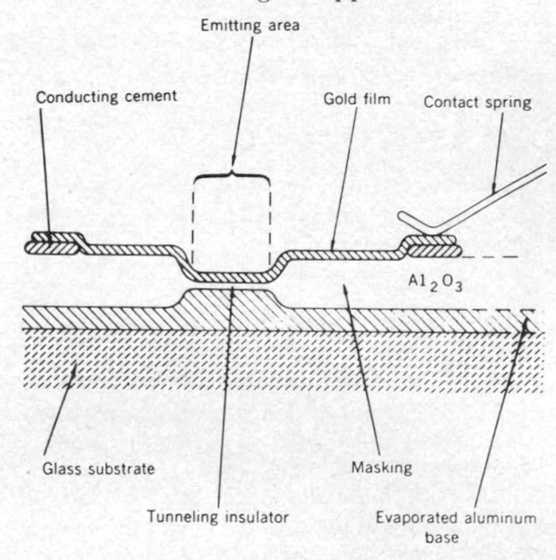

Tunnel-cathode cross section.

tunnel diode A heavily doped junction diode that has negative resistance in the forward direction over a portion of its operating range, due to quantum mechanical tunneling. It can be made from a variety of semiconductor materials, including germanium, silicon, gallium arsenide, and indium antimonide, for use as an oscillator or amplifier operating well up into microwave frequencies. Also called Esaki diode (after L. Esaki of Japan, who proposed the design).

tunnel-diode amplifier A low-noise microwave

amplifier that uses a tunnel diode as the active element. Applications include radar and telemetry receivers. Frequency range is from about 200 MHz to well over 10 GHz.

tunnel effect The piercing of a rectangular potential barrier in a semiconductor by a particle that does not have sufficient energy to go over the barrier. The wave associated with the particle is almost totally reflected on the first slope of the barrier, but a small fraction passes through the barrier.

tunneling Penetration of a potential barrier in a semiconductor by electrons whose energy is theoretically insufficient to overcome the barrier.

turbidimeter An instrument that measures the turbidity of a liquid. In a photoelectric turbidimeter, the amount of light that passes through the liquid is measured. Also called opacimeter.

turbulence Variations in refractive index that accompany microthermal fluctuations in the atmosphere, causing distortion in optical communication systems.

Turing machine A hypothetical computer that is not limited in use by storage capacity.

turn One complete loop of wire.

turnkey system A numerical control or computer system for which the manufacturer has total responsibility for building, installing, and testing the system, including hardware and software.

turnover cartridge A phonograph pickup that has two styli and a pivoted mounting which places in playing position the correct stylus for a particular record. There is usually a stylus with 3-mil radius for 78-rpm records and a stylus with 1-mil radius for 45- and 33⅓-rpm records.

turnover frequency *Transition frequency.*

turns ratio The ratio of the number of turns in a secondary winding of a transformer to the number of turns in the primary winding.

turnstile antenna An antenna that consists of one or more layers of crossed horizontal dipoles on a mast, usually energized so the currents in the two dipoles of a pair are equal and in quadrature. Used with television, FM, and other VHF or UHF transmitters to obtain an essentially omnidirectional radiation pattern. The superturnstile antenna is a more elaborate version in which the dipole elements are wing-shaped.

turntable The rotating platform on which a disk record is placed for recording or playback.

turntable rumble Low-frequency vibration that is mechanically transmitted to a recording or reproducing turntable and superimposed on the reproduction. Also called rumble.

turret tuner A television tuner that has one set of pretuned circuits for each channel, mounted on a drum which is rotated by the channel selector. Rotation of the drum connects each set of tuned circuits in turn to the receiver antenna circuit, RF amplifier, and RF oscillator.

TV Abbreviation for *television.*

TVI Abbreviation for *television interference.*

TWA Abbreviation for *traveling-wave amplifier.*

tweaking Making a small individual adjustment on a device after installing it in a system, such as making a zero correction for an operational amplifier.

tweeter A loudspeaker that handles only the higher audio frequencies, usually those well above 3 kHz. Generally used in conjunction with a crossover network and a woofer.

twin check A continuous check of computer operation, achieved by duplication of equipment and automatic comparison of results.

twin-lead *Twin-line.*

twin-line A transmission line that has two parallel conductors separated by insulating material. Line impedance is determined by the diameter and

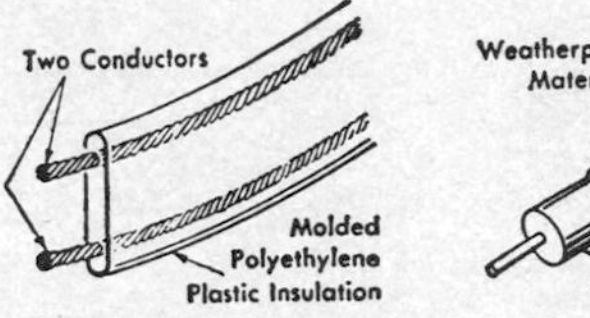

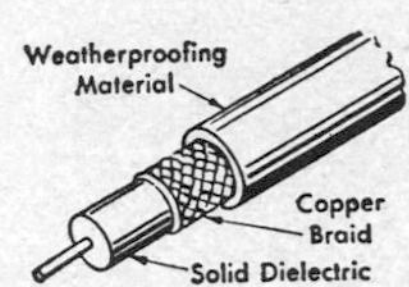

Twin-line (300-Ω impedance) and coaxial cable (72-Ω impedance) used as transmission lines between television receiving antennas and television receivers.

spacing of the conductors and the insulating material and is usually 300 Ω for television receiving antennas. Also called balanced transmission line and twin-lead.

twinning A defect that occurs in quartz crystals, resulting from structural misgrowth of otherwise perfect crystals. The two forms of twinning are electric and optical twinning.

twin-T network A network that consists of two T networks in parallel, with the capacitor and resis-

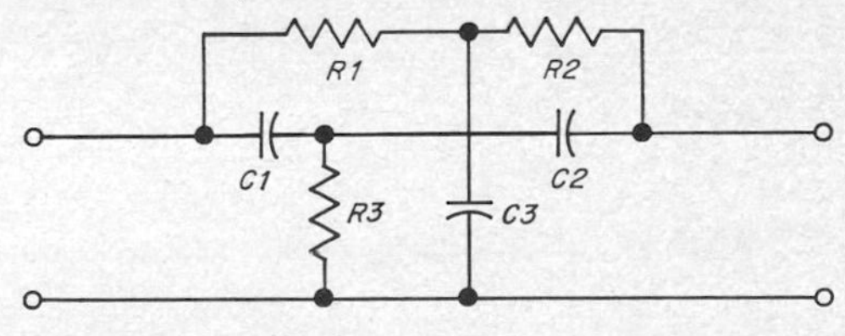

Twin-T network.

tor positions interchanged in one of them. Also called parallel-T network.

twin-triode Two triode vacuum tubes in a single envelope.

twist A waveguide section in which there is a progressive rotation of the cross section about the longitudinal axis of the waveguide.

twisted pair A cable composed of two small insulated conductors twisted together without a common covering.

twisted-ring counter A feedback shift register in which the next input digit is the complement of the outgoing digit, giving 2N states for N stages. Thus only five stages are needed for a scale-of-10 counter.

twister A piezoelectric crystal that generates a voltage when twisted.

twistor A computer memory element that consists of a helix of magnetic wire wound under tension at an angle of 45° on a short piece of nonmagnetic wire, with a fine-wire solenoid wound over the helix. Signal currents through the straight wire and the solenoid combine to produce a magnetic flux parallel to the helical wire. The direction of this magnetic flux is easily reversed by reversal of coil current. Readout is obtained by sending through a line of twistors a current strong enough to switch over all of them, and noting which ones deliver readout signals.

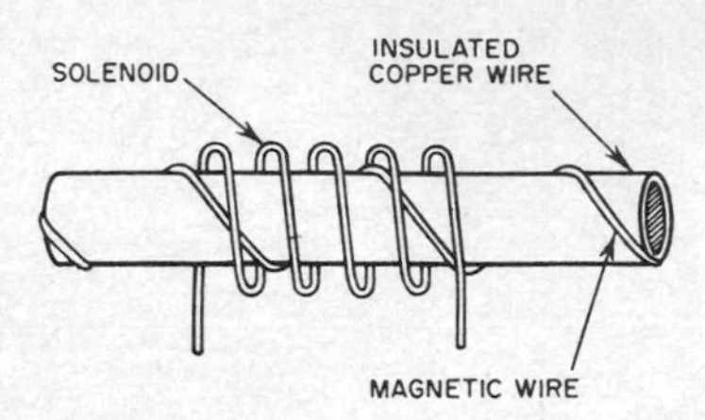

Twistor using barber-pole construction.

two-address programming Digital-computer programming in which each complete instruction includes an operation and specifies the location of two registers, usually one containing an operand and the other containing the result of the operation.

two-course radio range A radio range that beams on-course signals in opposite directions.

two-frequency duplex Carrier communication that provides simultaneous communication in both directions between two stations by using different carrier frequencies in the two directions.

two-hole directional coupler A directional coupler that consists of two parallel coaxial lines in contact, with holes or slots through their contacting walls at two points a quarter-wave apart. These permit extraction of a portion of the RF energy traveling in one direction through the main line while rejecting energy traveling in the opposite direction. One end of the secondary line must be terminated in its characteristic impedance.

two-level maser A solid-state maser in which all the molecules are excited simultaneously, permitting only pulsed operation. Each useful interval of oscillation or amplification must be followed by an interval of excitation. A paramagnetic solid material such as cerium ethyl sulfate is used. Also called pulsed maser.

two-out-of-five code An error-detecting code in which each decimal digit from 0 through 9 is represented by a different combination of five binary digits, two of which are always one kind and three the other kind, such as two 0s and three 1s.

two-phase Having a phase difference of one quarter-cycle or 90°. Also called quarter-phase.

two-position action Automatic control action in which a final control element is moved from one of two fixed positions to the other, with no intermediate positions.

two-receiver radiometer A radiometer-type receiver that uses two independent receiver channels, the IF outputs of which are multiplied and the product smoothed in a low-pass filter. A high signal-to-noise ratio is obtained at the output, as required for detection of extremely weak signals.

two's complement A binary number formed by interchanging all 1s and 0s of the original binary number and adding 1 to the result. Adding the original number and its two's complement gives a sum that is a power of 2. Thus for 10110 the two's complement is 01001 + 1 = 01010 and the sum of 10110, 01001, and 1 is 22 + 9 + 1 or 32 (binary 100000), which is the fifth power of 2.

two-source frequency keying Keying in which the modulating wave switches the output frequency between predetermined values that correspond to the frequencies of independent sources.

two-tone keying Keying in which the modulating wave causes the carrier to be modulated with one frequency for the marking condition and with a different frequency for the spacing condition.

two-value capacitor motor A capacitor motor that uses different values of effective capacitance for starting and running.

two-way communication Communication between radio stations that have both transmitting and receiving equipment.

two-way correction Register control that produces a correction in register in either direction.

two-way repeater A repeater that amplifies signals coming from either direction.

two-wire repeater A telephone repeater that provides for transmission in both directions over a two-wire telephone circuit.

TWT Abbreviation for *traveling-wave tube.*

TWX Abbreviation for *teletypewriter exchange service.*

type-bar printer A serial printer in which two characters are mounted on a type bar, as in some electric typewriters and early teletypewriters. Desired characters are printed one at a time by the action of a signal-controlled electric motor, manually depressed keys, or other means.

typewriter terminal An electric typewriter combined with a ASCII or other code generator that provides code output for feeding a computer, calculator, or other digital equipment. The terminal also produces hard copy when driven by incoming code signals.

Typhon A surface-to-air missile of advanced design for installation on carriers, cruisers, frigates, and destroyers, for use against high-performance aircraft and short-range tactical missiles. It may have a nuclear warhead.

U

u Abbreviation for *atomic mass unit.*
UHF Abbreviation for *ultrahigh frequency.*
UHV Abbreviation for *ultrahigh voltage.*
UJT Abbreviation for *unijunction transistor.*
ultor The electrode or set of electrodes to which the highest DC voltage is applied in a picture tube, to accelerate the electron beam before deflection. Also called second anode.
ultra-audion oscillator A Colpitts oscillator in which the two voltage-dividing capacitances of the tank circuit are the anode-to-cathode and grid-to-cathode capacitances of the tube.
Ultrafax Trademark of RCA for a system that combines radio, television, facsimile, and film recording techniques for transmitting printed information at high speed.
ultrafine wire Wire ranging in size from No. 47 through No. 60 American wire gage.
ultrahigh frequency [abbreviated UHF] A Federal Communications Commission designation for a frequency in the band from 300 to 3000 MHz, corresponding to a decimetric wave between 10 and 100 cm.
ultrahigh-frequency converter An electronic circuit that converts UHF signals to a lower frequency to permit reception on a VHF receiver. Used to convert UHF television signals to VHF signals for reception on VHF television receivers.
ultrahigh-frequency translator A television broadcast translator station that operates on a UHF television broadcast channel.
ultrahigh voltage [abbreviated UHV] A voltage above about 1 MV.
ultramicrometer An instrument that measures very small displacements by electrical means, such as by the variation in capacitance which results from the movement being measured.
ultrashort waves Radio waves shorter than about 10 m in wavelength (above 30 MHz in frequency). Waves shorter than 30 cm (above 1 GHz) are usually called microwaves.
ultrasonic Pertaining to signals, equipment, or phenomena involving frequencies just above the range of human hearing, hence above about 20 kHz.
ultrasonic bonding Bonding of two identical or dissimilar metals by mechanical pressure combined with a wiping motion produced by ultrasonic vibration. Can be used for attaching gold or aluminum wire leads to semiconductor devices.

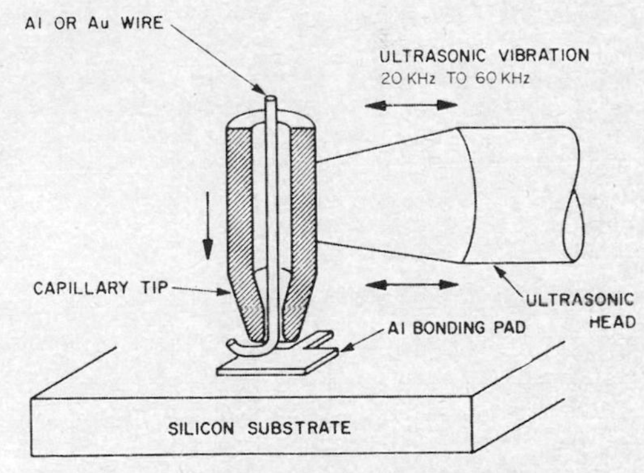

Ultrasonic bonding of wire lead to aluminum bonding pad on silicon semiconductor material. Ultrasonic head provides wiping motion that removes oxide films, to give strong intermolecular bond.

ultrasonic cleaning The cleaning of objects by immersion in a liquid that is acted on by ultrasonic waves.
ultrasonic coagulation The bonding of small particles into large aggregates by the action of ultrasonic waves.

ultrasonic communication Communication through water by keying the sound output of echo-ranging sonar on ships or submarines.

ultrasonic cross grating An ultrasonic space grating produced by crossing beams of ultrasonic waves.

ultrasonic delay line A delay line in which use is made of the propagation time of sound through a medium such as fused quartz, barium titanate, or mercury to obtain a time delay of a signal.

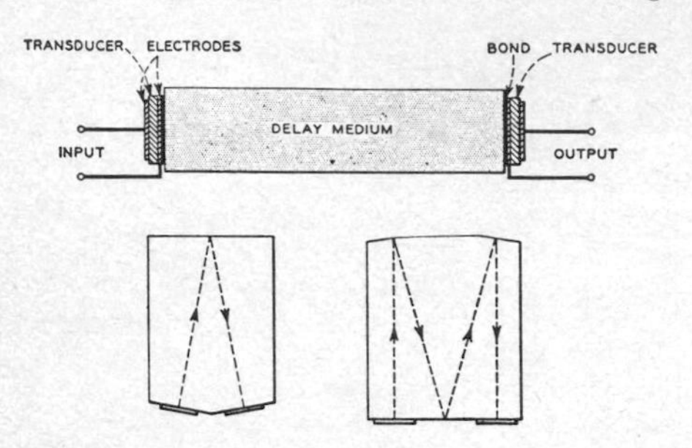

Ultrasonic delay line, and methods of placing transducers to increase path length and delay time by reflecting signal beam.

ultrasonic detector A mechanical, electric, thermal, or optical device for detecting and measuring ultrasonic waves.

ultrasonic diagnosis The use of ultrasonic echo techniques to obtain a visual image of the interior of a body for medical diagnostic purposes.

ultrasonic drill A drill in which a magnetostrictive transducer is attached to a tapered cone that serves as a velocity transformer. With an appropriate tool at the end of the transformer, practically any shape of hole can be drilled in hard, brittle materials like tungsten carbide and gems.

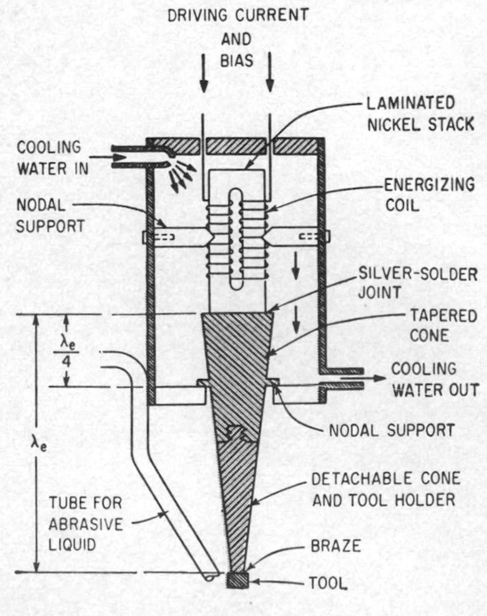

Ultrasonic drill.

ultrasonic equipment Equipment that generates AC energy at frequencies above about 20 kHz and uses that energy to excite or drive an electromechanical transducer for the production and transmission of ultrasonic energy for industrial, scientific, medical, or other purposes.

ultrasonic flaw detector An ultrasonic generator and detector used together much as in radar, to determine the distance to a wave-reflecting internal crack or other flaw in a solid object.

ultrasonic frequency A frequency lying above the audio-frequency range. The term is commonly applied to elastic waves propagated in gases, liquids, or solids.

ultrasonic generator A generator that consists of an oscillator which drives an electroacoustic transducer, used to produce acoustic waves above about 20 kHz.

ultrasonic holography The use of an interference pattern between two ultrasonic waves to reconstruct an image of the interior of an opaque object for photography. Applications include nondestructive testing.

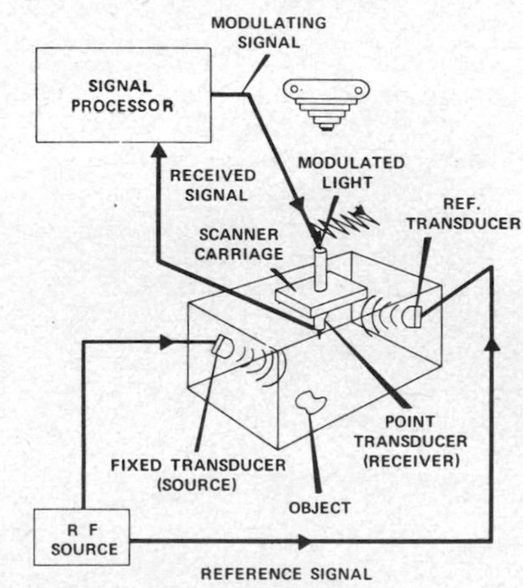

Ultrasonic-holography setup for object immersed in liquid.

ultrasonic image converter A device that makes acoustic field configurations optically visible.

ultrasonic imaging *Acoustic imaging.*

ultrasonic inspection Nondestructive inspection of the interior of an object by using ultrasonic techniques to obtain echoes from inner flaws.

ultrasonic level detector A level detector in which an ultrasonic transmitter and receiver are set into one wall of a container or tank. When the level is below that of the ultrasonic beam, reflection occurs at the opposite wall. When the level of the liquid or other material reaches the beam, reflection occurs in the material, and the elapsed time of pulse travel is shorter.

ultrasonic light diffraction The formation of optical diffraction spectra when a beam of light is

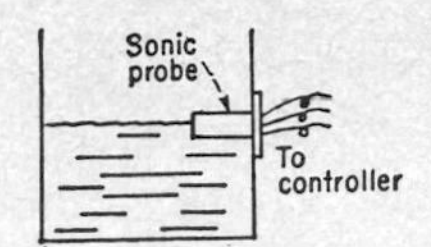

Ultrasonic level detector.

passed through a longitudinal ultrasonic wave field. The diffraction results from the periodic variation of the light refraction in the ultrasonic field.

ultrasonic light modulator A light modulator that utilizes the action of ultrasonic waves on a light beam which is passing through a fluid.

ultrasonic material dispersion The production of suspensions or emulsions of one material in another due to the action of high-intensity ultrasonic waves.

ultrasonic modulation cell A cell in which a laser beam is frequency-modulated by the action of a longitudinal ultrasonic wave derived from a frequency-modulated video signal. The ultrasonic wave scatters the incident light into various orders that are frequency-shifted and can therefore be used for transmitting information to a photodetector which is feeding a television monitor.

ultrasonics The branch of acoustics that deals with vibrations and acoustic waves above about 20 kHz. Also called ultrasound.

ultrasonic scanner A scanner that uses ultrasonic pulse-echo techniques for diagnostic imaging of tissue/organ interfaces.

ultrasonic sealing The sealing of thermoplastic packaging films by applying vibratory mechanical pressure at ultrasonic frequencies to develop localized heat that melts and fuses the mating plastic surfaces.

ultrasonic soldering iron A soldering iron in which heat is combined with high-frequency vibration to induce cavitation in molten solder. This removes oxides from the joint, eliminating the need for soldering flux.

ultrasonic sounding Determining ocean depth with a fathometer by measuring the time interval between the transmission of an ultrasonic wave downward from the surface of the water and the arrival of the echo reflected from the bottom of the ocean.

ultrasonic space grating A periodic spatial variation of the index of refraction of a medium due to the presence of acoustic waves.

ultrasonic stroboscope A light interrupter whose action is based on the modulation of a light beam by an ultrasonic field.

ultrasonic therapy Use of ultrasonic vibrations for therapeutic purposes.

ultrasonic thickness gage A thickness gage in which the time of travel of an ultrasonic beam through a sheet of material is used as a measure of the thickness of the material.

ultrasonic trainer A radar trainer in which ultrasonic waves are directed against a molded relief map under water, to simulate responses that occur in actual radar operation during flight over the terrain represented by the map.

ultrasonic transducer A transducer that converts AC energy above 20 kHz to mechanical vibrations of the same frequency. It is generally either magnetostrictive or piezoelectric.

ultrasonic waves Elastic waves that have a frequency above about 20 kHz.

ultrasonic welding The use of ultrasonic energy to produce fusion of two mating pieces of metal without heat.

ultrasound *Ultrasonics.*

ultraviolet altimeter An altimeter for manned atmospheric and space vehicles; it uses the ultraviolet spectrum to obtain accurate altitude readings up to 125,000 ft (38 km) from the surface of the earth.

ultraviolet lamp A lamp that provides a high proportion of ultraviolet radiation, such as various forms of mercury-vapor lamps.

ultraviolet radiation The portion of the electromagnetic spectrum between visible violet light (about 380 nm, 0.38 μm, or 3800 Å) and the x-ray region at about 10 nm, 0.01 μm, or 100 Å. Also called black light (deprecated).

umbilical connector The quick-disconnect connector that terminates an umbilical cable and mates with the umbilical connector receptacle.

umbilical cord A cable capable of being disconnected quickly, used to test a missile up to the instant of launching and sometimes to feed in last-minute target data. The cord may be severed by such means as a trigger-action spring, explosive charge, or guillotine blade.

umbra The region of total shadow behind an object in a beam of radiation. A straight line drawn from any point in this region to any point in the source passes through the object.

umbrella antenna An antenna in which the radiating wires run downward at an angle in some or all directions from a central tower or from a wire running between two towers, somewhat like the ribs of an open umbrella.

unbalanced circuit A circuit whose two sides are electrically unlike.

unbalanced line A transmission line in which the voltages on the two conductors are not equal with respect to ground. A coaxial line is an example.

unbalanced output An output in which one of the two input terminals is substantially at ground potential.

unblanking pulse A pulse applied to the grid or cathode of a cathode-ray tube to turn on the beam for a particular length of time, normally the time of one sweep.

unbundling The pricing of certain types of computer software and services separately from the hardware.

uncage To disconnect the erection circuit of a

displacement gyroscope system.

uncertainty The estimated amount by which an observed or calculated value may depart from the true value.

uncertainty principle A principle of quantum mechanics, stating that it is impossible to obtain simultaneously and with high precision both the position and the momentum of a particle.

uncharged Having a normal number of electrons, and hence having no electric charge.

unclassified Not having a security classification.

unconditional jump A digital-computer instruction that interrupts the normal process of obtaining instructions in an ordered sequence and specifies the address from which the next instruction must be taken. Also called unconditional transfer.

unconditional transfer *Unconditional jump.*

undamped wave A continuous wave produced by oscillations that have constant amplitude.

underbunching A condition that represents less than optimum bunching in a velocity-modulation tube.

undercurrent relay A relay that operates when its coil current falls below a predetermined value.

underdamping A circuit condition in which one or more output oscillations occur after a sudden change in input, before the output settles to its new value.

underflow The generation of a quantity smaller than the smallest nonzero quantity that a computer is capable of storing.

undermodulation Insufficient modulation at a transmitter, caused by electrical limitations or improper adjustment of the modulator.

undershoot The initial transient response to a unidirectional change in input. It precedes the main transition and is opposite in sense.

undervoltage alarm An alarm that gives an aural and/or visual indication when the voltage in an equipment falls below a specified value.

undervoltage relay A relay that operates when its coil voltage falls below a predetermined value.

underwater ambient noise The portion of the underwater background noise that is due to disturbances in the water other than the signal and man-made noise.

underwater antenna An antenna placed and used under water.

underwater background noise Underwater sound other than the desired signal, including underwater ambient noise and that noise inherent in the hydrophone and its associated equipment.

underwater burst The explosion of a nuclear weapon beneath the surface of the water.

underwater mine A mine located under water and exploded by propeller vibration, magnetic attraction, contact, or remote control.

underwater-mine coil A coil and associated equipment that detects changes in the magnetic field at an underwater mine caused by a passing ship.

underwater-mine depth compensator A hydrostatically actuated device that increases the sensitivity of the firing mechanism when an underwater mine exceeds a predetermined depth.

underwater signal An underwater disturbance that is to be detected, such as sonar transmissions from a projector, machinery noise, and propeller noise.

underwater sound communication set The components and items required to provide underwater communication by utilizing sonic or ultrasonic waves in water.

underwater sound projector A transducer that produces sound waves in water. Also called projector.

unfurlable antenna An antenna that can be folded or rolled into a size small enough to fit inside a spacecraft before launch and can be unfurled in space after launch.

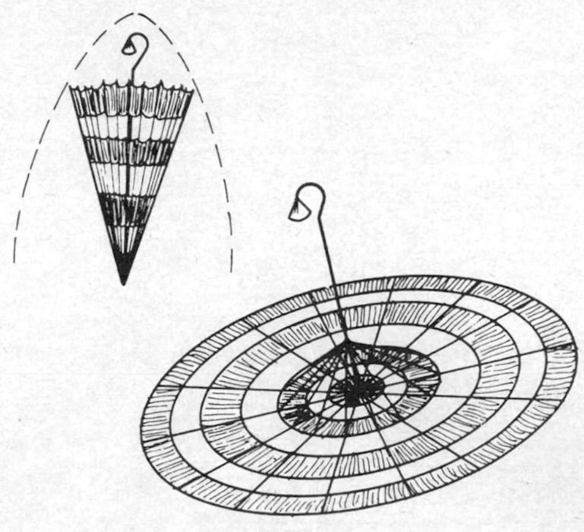

Unfurlable antenna for microwave communication, using waveguide feed for parabolic reflector.

ungrounded Without an intentional connection to ground except through voltmeters or other high-impedance devices.

unguided Not subject to guidance or control during flight.

uniconductor waveguide A waveguide that consists of a cylindrical or rectangular metallic surface which surrounds a homogeneous dielectric medium.

unidirectional 1. Flowing in only one direction, such as direct current. 2. Radiating in only one direction.

unidirectional antenna An antenna that has a single well-defined direction of maximum gain.

unidirectional coupler A directional coupler that samples only one direction of transmission.

unidirectional current A current that flows in the same direction at all times.

unidirectional hydrophone A unidirectional microphone that responds to waterborne sound waves.

unidirectional log-periodic antenna A broadband antenna in which the cut-out portions of a log-periodic antenna are mounted at an angle to each other, to give a unidirectional radiation pat-

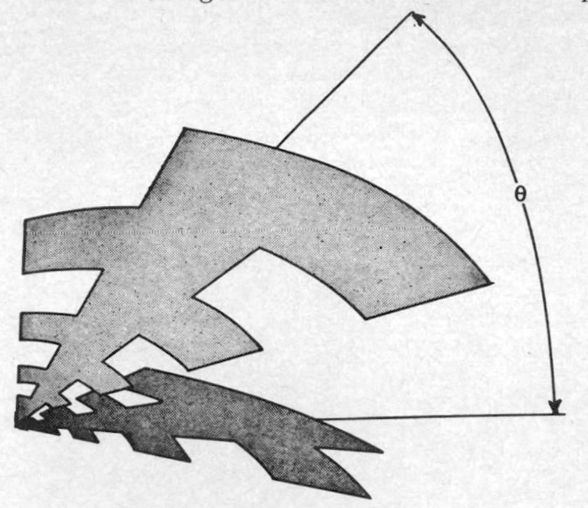

Unidirectional log-periodic antenna cut from sheet metal.

tern in which the major radiation is in the backward direction, off the apex of the antenna. Impedance is essentially constant for all frequencies, as is the radiation pattern.

unidirectional microphone A microphone that is responsive predominantly to sound incident from one hemisphere, without picking up sounds from the sides or rear.

unidirectional pulse A pulse in which pertinent departures from the normally constant value occur in one direction only. Also called single-polarity pulse (deprecated).

unidirectional transducer A transducer that measures stimuli in only one direction from a reference zero or rest position.

unidirectional voltage A voltage whose polarity, but not necessarily magnitude, is constant.

unifilar Having or using only one fiber, wire, or thread.

uniform line A line that has the same electrical properties along its entire length.

uniform plane wave A plane wave in which the electric and magnetic field vectors have constant amplitude over the equiphase surfaces. Such a wave can be found only in free space at an infinite distance from the source.

uniform transmission line A transmission line in which the physical and electrical characteristics do not change with distance along the direction of propagation.

uniform waveguide A waveguide in which the physical and electrical characteristics do not change with distance along the axis of the guide.

unijunction transistor [abbreviated UJT] A PN junction device that has the emitter connected to the PN junction on one side of a silicon bar and connections for its two bases at opposite ends of the bar. It may also be constructed by the planar process, with all three electrodes on one face of the silicon chip. The transistor has a stable negative

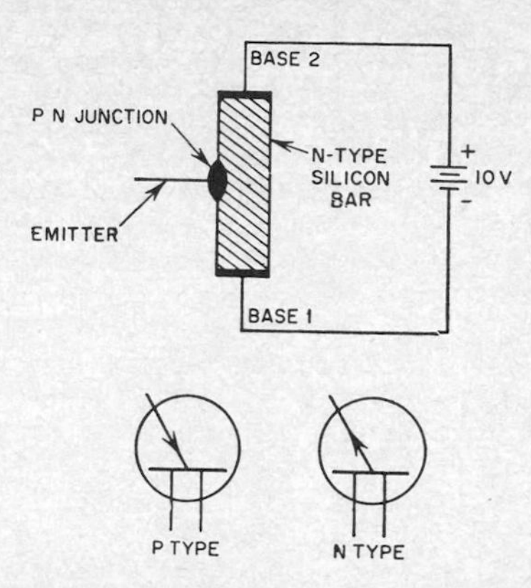

Unijunction-transistor construction for N-type unit, and symbols for both types.

resistance characteristic over a wide temperature range. Used primarily as a switching device.

unilateral-area track A sound track in which only one edge of the opaque area is modulated in accordance with the recorded signal. A second edge may be modulated by a noise-reduction device.

unilateral bearing A bearing obtained with a radio direction finder that has unilateral response, eliminating the chance of a 180° error.

unilateral conductivity Conductivity in only one direction, as in a perfect rectifier.

unilateral device A device that transmits energy in one direction only.

unilateralization Use of an external feedback circuit in a high-frequency transistor amplifier to prevent undesired oscillation by canceling both the resistive and reactive changes produced in the input circuit by internal voltage feedback. With neutralization, only the reactive changes are canceled.

unilateral transducer A transducer in which the waves at its outputs cannot affect its inputs.

unimpeded harmonic operation Operation of a magnetic amplifier in such a way that the impedance of the control circuit is substantially zero, permitting essentially unrestricted flow of all harmonic currents in the control circuit.

uninterruptible power system [abbreviated UPS] A system that provides protection against primary AC power failure and variations in power-line frequency and voltage. The most common system used to protect a computer installation includes a storage battery, battery charger, solid-state inverter, and solid-state switching circuit. The system may be used on-line between power line and load to provide voltage regulation and suppress transients, or off-line and switched in only when utility power fails.

unipolar Having only one pole, polarity, or direction.

unipolar machine *Homopolar generator.*

unipole A hypothetical antenna that radiates or receives signals equally well in all directions. Also called isotropic antenna.

unipotential cathode *Indirectly heated cathode.*

unit 1. An assembly or device capable of independent operation, such as a radio receiver, cathode-ray oscilloscope, or computer subassembly that performs some inclusive operation or function. 2. A quantity adopted as a standard of measurement.

unit charge The electric charge that will exert a repelling force of 1 dyn on an equal and like charge 1 cm away in a vacuum, assuming that each charge is concentrated at a point.

unitized construction Construction of equipment in subassemblies that can be manufactured and tested separately and are readily replaceable as individual units within the equipment of which they are a part.

unit magnetic pole A magnetic pole that will repel an equal magnetic pole of the same sign with a force of 1 dyn if the two poles are placed 1 cm apart in a vacuum.

unit operator A symbolic operator that leaves every other operator unchanged.

unity coupling Perfect magnetic coupling between two coils so all the magnetic flux produced by the primary winding passes through the entire secondary winding.

unity power factor A power factor of 1.0, obtained when current and voltage are in phase, as in a circuit that contains only resistance.

universal joint A coupling that connects two shafts whose axes intersect at an angle.

universal motor A motor that may be operated at approximately the same speed and output on either direct current or single-phase alternating current.

universal output transformer An output transformer that has a number of taps on its winding, to permit its use between the AF output stage and the loudspeaker of practically any radio receiver by proper choice of connections.

universal product code [abbreviated UPC] A 10-digit bar code that can be printed in ink on the outside of a package for laser or other electronic scanning at supermarket checkout counters. Each digit is represented by the ratio of the widths of adjacent stripes and white areas. The code is readable regardless of the orientation in which it passes through the scanner system. The term also covers the corresponding combinations of binary digits into which the scanned bars are converted for computer processing that provides continuously updated inventory data and printout of the register tape at the checkout counter. The first five digits of the code identify the manufacturer; the second five digits identify the specific product. Current price and tax data for each product and descriptive terms for the register tape are stored in the computer.

universal product code scanner A scanner ca-

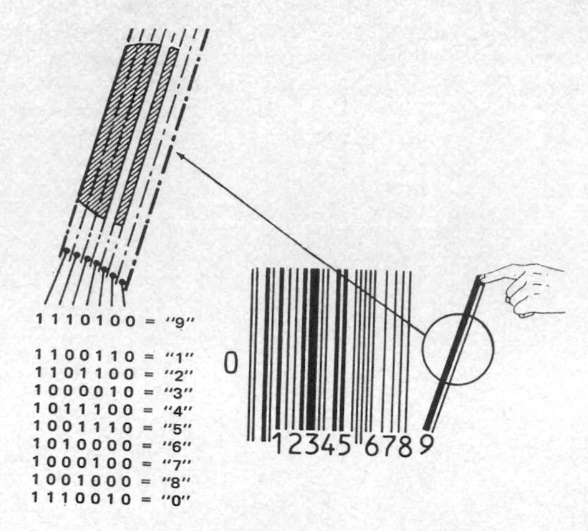

Universal product code. Each digit is allocated seven slots or bits, with black representing a 1 and white a 0, to give binary representation as shown for even-parity characters. Each digit has two runs of 1s separated by 0s.

pable of reading the standard bar codes printed on packaged supermarket products. For annual theft-loss inventories of products on shelves, the scanner may be in the form of a manually moved electronic pencil that feeds a magnetic tape recorder which later serves as computer input.

universal receiver *AC/DC receiver.*

universal shunt *Ayrton shunt.*

universal time [abbreviated UT] Mean solar time at Greenwich, England, as reckoned from midnight. Formerly called Greenwich mean time (GMT).

universal time coordinated [abbreviated UTC] A time scale coordinated by the International Bureau of Time to international atomic time (IAT) as derived from atomic clocks. It is also coordinated to universal time as corrected for polar motion.

univibrator *Monostable multivibrator.*

unloaded antenna An antenna that has no added inductance or capacitance.

unmodified scatter Radiation that is scattered

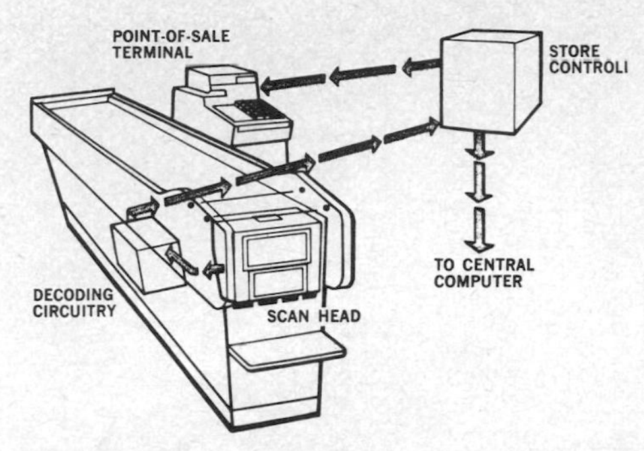

Universal product code scanner for supermarket checkout counter may use low-power helium-neon laser in scan head. Clerk must position each package for readout.

without a change in photon energy.

unmodulated Having no modulation, as during moments of silence in a radio program or a disk recording.

unmodulated groove A groove made in a mechanical recording medium when no signal is applied to the cutter. Also called blank groove.

unpack To separate packed computer information into a sequence of separate words or elements.

unpolarized light A beam of light in which the planes of vibration of the photons are oriented at random about the axis of the beam.

unquenched spark gap A spark gap that has no special means for deionization.

unstabilized antenna An antenna mounted directly on the structure of a vessel or aircraft, with no means for offsetting roll and pitch.

unstable Capable of undergoing spontaneous change, as in a radioactive nuclide or an excited nuclear system. Also called labile.

unstable servo A servo in which the output drifts away from the input without limit.

untuned Not resonant at any of the frequencies being handled.

untuned antenna *Aperiodic antenna.*

unwind To code explicitly, at length and in full, all the operations of a cycle, thus eliminating all red-tape operations in the final problem coding. Unwinding may be performed automatically by the computer during assembly, generation, or compilation.

UPC Abbreviation for *universal product code.*

upconverter 1. A converter that changes an incoming modulated or unmodulated carrier frequency to a higher frequency which is within the range of a receiver or radio test set. 2. A light-emitting diode in which excitation by infrared radiation produces visible luminescence.

up counter A pulse counter that starts at zero and increases one count at a time to its design limit. Thus a modulus 16 binary up counter increases in 16 steps from 0000 to 1111.

update 1. To put into a computer master file the changes required by current information or transactions. 2. To modify an instruction so that the address numbers it contains are increased by a stated amount each time the instruction is performed.

up Doppler The sonar situation wherein the target is moving toward the transducer, so the frequency of the echo is greater than the frequency of the reverberations received immediately after the end of the outgoing ping. Opposite of down Doppler.

up/down counter A counter that has a control input which switches the direction of the counting mode without affecting the contents of the counter.

uplink The radio or optical transmission path upward from the earth to a communication satellite, or from the earth to aircraft. The return path is the downlink.

upper sideband The higher of two frequencies or groups of frequencies produced by a modulation process.

UPS Abbreviation for *uninterruptible power system.*

uranium [symbol U] A radioactive element. Atomic number is 92. Naturally occurring uranium is a mixture of 99.28% ^{238}U, 0.71% ^{235}U, and 0.00580% ^{234}U. The nucleus of ^{235}U is capable of absorbing a neutron and thereupon undergoing fission into two highly radioactive fragments that fly apart with great energy, releasing neutrons. A chain reaction is possible because fission is induced by one neutron but releases more than one neutron.

uranium age The age of a mineral as calculated from the numbers of ionium atoms present originally, now, and when equilibrium is established with uranium.

urea plastic A thermosetting plastic that has good dielectric qualities and good strength. Used for radio receiver cabinets and instrument housings.

usable sensitivity The signal strength required to produce a program level that is 30 dB greater in amplitude than the combined amplitudes of noise and distortion, in Institute of High Fidelity standards for FM tuners.

USAF Abbreviation for United States Air Force.

useful beam The part of the primary radiation that passes through the aperture, cone, or other collimator used in radiology.

UT Abbreviation for *universal time.*

UTC Abbreviation for *universal time coordinated.*

utility routine A standard routine used to assist in the operation of a computer, such as a conversion, printout, or tracing routine.

uvicon A television camera tube that has a conventional vidicon scanning section preceded by an ultraviolet-sensitive photocathode, an electron-accelerating section, and a special target.

V

V 1. Abbreviation for *volt.* 2. Symbol used on diagrams to designate a voltmeter or an electron tube.

VA Abbreviation for *voltampere.*

VAC Abbreviation for *volts AC.*

vacancy A defect in the form of an unoccupied lattice position in a crystal.

vacuum An enclosed space from which practically all air has been removed.

vacuum capacitor A capacitor that has separated metal plates or cylinders mounted in an evacuated glass envelope to obtain a high breakdown voltage rating.

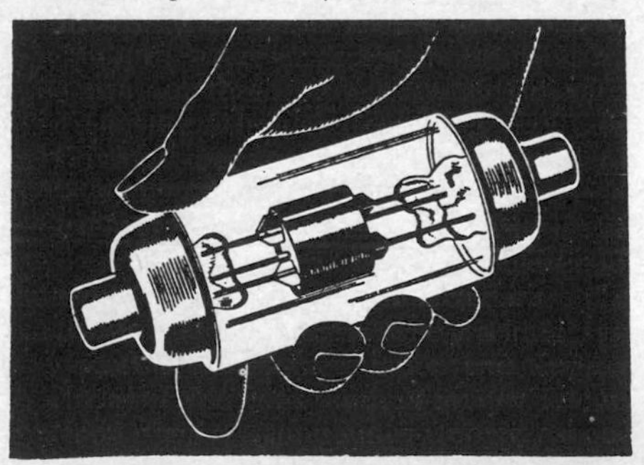

Vacuum capacitor used in radio transmitters.

vacuum diffusion Diffusion of impurities into a semiconductor material in a continuously pumped hard vacuum.

vacuum evaporation Deposition of thin films of metal or other materials on a substrate, usually through openings in a mask, by evaporation from a boiling source in a hard vacuum.

vacuum fluorescent lamp An evacuated display tube in which the anodes are coated with a phosphor that glows when electrons from the cathode strike it, to create a display. The phosphor is viewed from the side that electrons hit. Applications include displays for desktop calculators and digital scales.

vacuum forepump A vacuum pump capable of lowering the pressure down to about 0.001 mmHg, which is low enough for operation of a diffusion pump.

vacuum gage A device that indicates the absolute gas pressure in a vacuum system, in millimeters of mercury or micrometers of mercury. One micrometer is the pressure that will support a column of mercury 0.001 mm high.

vacuum-impregnated Impregnated with an insulating compound while in a vacuum, to ensure penetration of the compound into the layers of a capacitor or between the turns of a coil.

vacuum-leak detector An instrument that detects and locates leaks in a high-vacuum system. A mass spectrometer is frequently used for this purpose.

vacuum metallizing Deposition of a metal coating on a plastic or other object by evaporating the metal in a vacuum chamber that contains the object.

vacuum pencil A pencillike length of tubing connected to a small vacuum pump, for picking up semiconductor slices or chips during fabrication of solid-state devices.

vacuum phototube A phototube which is evacuated to such a degree that its electrical characteristics are essentially unaffected by gaseous ionization. In a gas phototube, some gas is intentionally introduced.

vacuum relay A relay that has its contacts mounted in an evacuated glass housing, to permit handling RF voltages as high as 20 kV without flashover between fixed and movable contacts.

vacuum seal An airtight junction.

vacuum spectrograph A spectrograph in which the optical path is in a vacuum and a reflection grating is usually used in place of a dispersive prism. Used for measurements in the extreme infrared and ultraviolet ranges, where lenses and air would absorb the radiation.

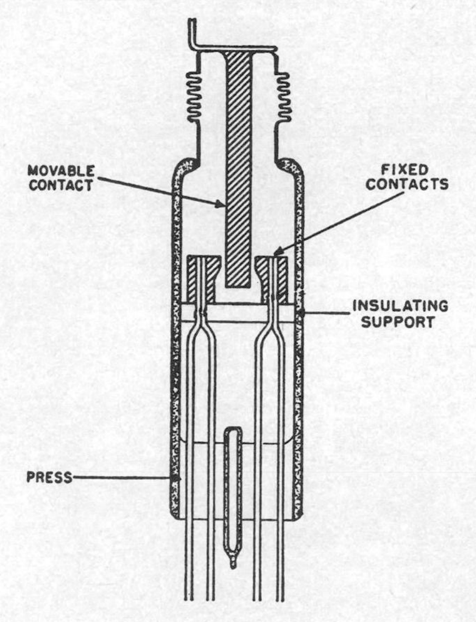

Vacuum switch.

vacuum switch A switch that has its contacts in an evacuated envelope to minimize sparking.

vacuum tank The airtight metal chamber that contains the electrodes of a mercury-arc rectifier or similar tube and in which the rectifying action takes place.

vacuum tube An electron tube evacuated to such a degree that its electrical characteristics are essentially unaffected by the presence of residual gas or vapor.

vacuum-tube amplifier An amplifier that uses one or more vacuum tubes to control the power obtained from a local source.

vacuum-tube electrometer An electrometer in which the ionization current in an ionization chamber is amplified by a special vacuum triode that has an input resistance above 10 GΩ.

vacuum-tube keying A code-transmitter keying system in which a vacuum tube is connected in series with the anode supply lead of the final stage. The grid of the keying tube is connected to its filament through the transmitting key. When the key is open, the tube blocks, interrupting the anode supply to the output stage. Closing the key lets anode current flow through the keying tube and the output tubes.

vacuum-tube modulator A modulator that uses a vacuum tube as a modulating element for impressing an intelligence signal on a carrier.

vacuum-tube oscillator A circuit that uses a vacuum tube to convert DC power into AC power at a desired frequency.

vacuum-tube rectifier A rectifier in which rectification is accomplished by the unidirectional passage of electrons from a heated electrode to one or more other electrodes within an evacuated space.

vacuum-tube transmitter A transmitter in which electron tubes are utilized to convert the applied electric power into RF power.

vacuum-tube voltmeter [abbreviated VTVM] An electronic voltmeter that uses vacuum tubes along with or without semiconductor devices.

valence A number that represents the proportion in which an atom is able to combine directly with other atoms. It generally depends on the number and arrangement of electrons in the outermost shell of each type of atom.

valence band An energy band, occurring in the spectrum of a solid crystal, in which lie the energies of the valence electrons that bind the crystal together.

valence bond The bond formed between the electrons of two or more atoms.

valence electron *Conduction electron.*

valence shell The electrons that form the outermost shell of an atom.

validity check A computer check of input data, based on known limits for variables in given fields. Thus a week may not have more than 168 working hours.

value The magnitude of a quantity.

valve British term for *electron tube.*

vanadium [symbol V] A metallic element. Atomic number is 23.

Van Allen belt One of the belts of ionizing radiation surrounding the earth, one centered about 2000 mi (3200 km) above the earth, one at about 10,000 mi (16,000 km), and one at about 20,000 mi (32,000 km). The radiation consists of protons and electrons that come chiefly from the sun as solar wind and are trapped by the earth's magnetic field. During massive solar storms, high-energy particles reach the belt directly from the sun and overload it, creating auroras as excess particles are dumped from the belt in the earth's polar regions. Also called radiation belt.

Van Atta array An antenna array in which pairs of corner reflectors or other elements equidistant from the center of the array are connected together by low-loss transmission line in such a way that the received signal is reflected back to its source in a narrow beam, to give signal enhancement without amplification.

Van de Graaff generator An electrostatic generator widely used as an accelerator. An endless

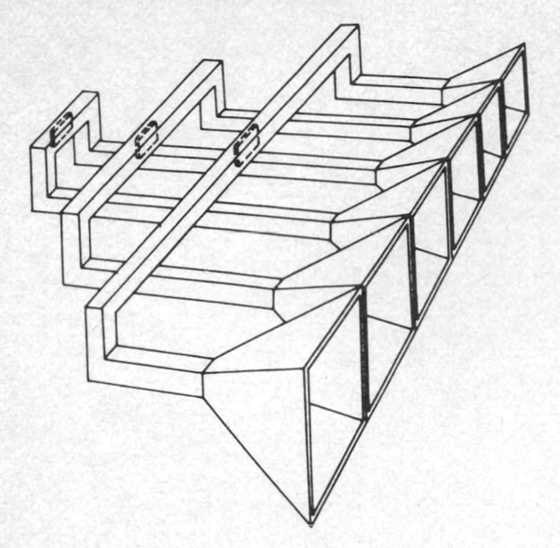

Van Atta array.

moving belt of insulating material collects electric charges by induction and discharges them inside a large hollow spherical electrode to produce voltages as high as 9 MV. Used for accelerating electrons, protons, and other nuclear particles.

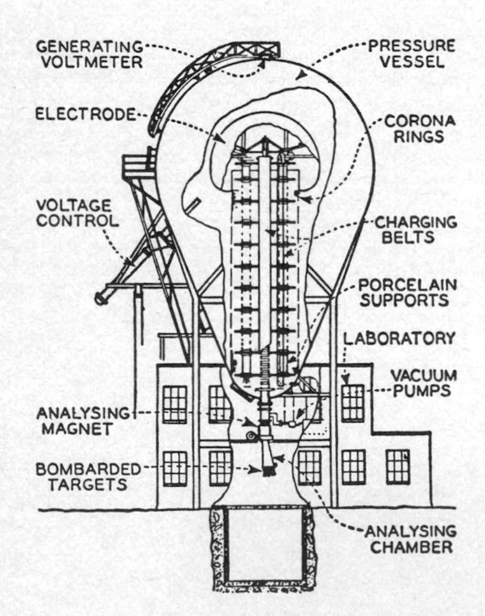

Van de Graaff generator 47 ft (14.3 m) high, capable of generating up to 3 MVDC.

vane attenuator *Flap attenuator.*

vane-type anode A magnetron anode which resembles that used in rising-sun magnetrons except that all vanes are the same size and shape.

vane-type instrument A measuring instrument that uses the force of repulsion between fixed and movable magnetized iron vanes, or the force existing between a coil and a pivoted vane-shaped piece of soft iron, to move the indicating pointer.

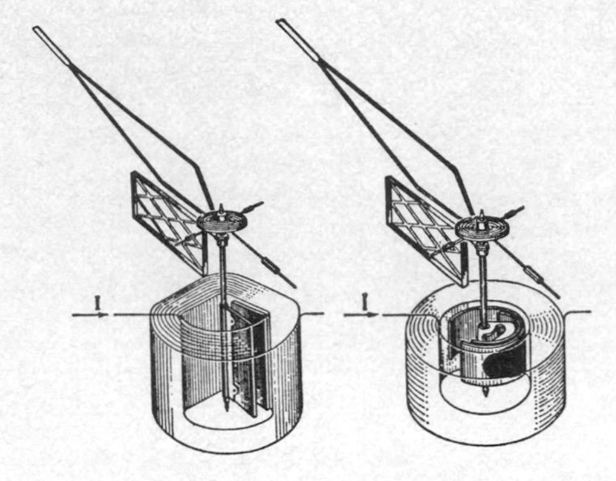

Vane-type instrument construction.

V antenna An antenna that has a V-shaped arrangement of conductors fed by a balanced line at the apex. The included angle, length, and elevation of the conductors are proportioned to give the desired directivity.

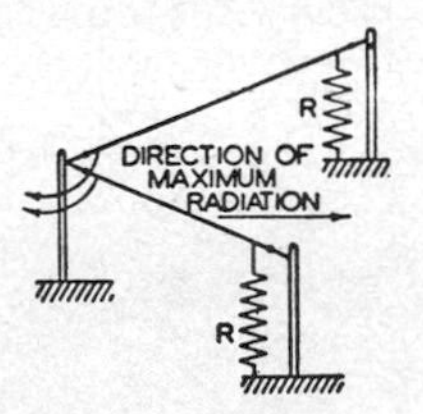

V antenna.

vapor-deposited printed circuit A printed circuit formed by condensation of a material from its gaseous state, using vacuum deposition methods or other vaporized metal techniques in conjunction with masks or other pattern-forming means.

vapor deposition The process of depositing a thin film of metal by condensation of molten metal vapor in a vacuum.

vaporization-cooled Cooled by vaporization of a nonflammable liquid that has a low boiling point and high dielectric strength. The liquid is flowed or sprayed on hot electronic equipment in an enclosure, where it vaporizes, carrying the heat to the enclosure walls, radiators, or a heat exchanger. Also called evaporative-cooled.

vapor pressure The pressure of the vapor of a liquid that is kept in confinement, as in a mercury-vapor rectifier tube.

var The name and symbol used in the International System of Units (SI) for voltampere reactive, which is the unit of reactive power.

VAR Abbreviation for *visual-aural range.*

varactor A PN semiconductor diode whose capacitance varies with the applied voltage. Used as a variable-reactance tuning element in oscillator and amplifier circuits, including parametric amplifiers. Also called varactor diode.

varactor diode *Varactor.*

varh Abbreviation for *voltampere-hour reactive.*

variable A quantity that may assume a number of distinct values.

variable-area track A sound track divided laterally into opaque and transparent areas. A sharp line of demarcation between these areas corresponds to the waveform of the recorded signal.

variable attenuator An attenuator that reduces the strength of an AC signal either continuously or in steps, without causing appreciable signal distortion, by maintaining a substantially constant impedance match.

variable-capacitance pickup A phonograph pickup in which the stylus produces a variation in capacitance that is used to frequency-modulate an oscillator. The output signal of the oscillator is converted to an audio voltage by a detector. Also called FM pickup.

variable-capacitance transducer A transducer that measures a parameter or change in a parameter by a change in capacitance.

variable capacitor A capacitor whose capacitance can be varied by moving one set of metal plates with respect to another.

variable-carrier modulation *Controlled-carrier modulation.*

variable coupling Inductive coupling that can be varied by moving one coil with respect to another.

variable-cycle operation A computer operation in which the cycles of action may be of different lengths, as in an asynchronous computer.

variable-density sound track A constant-width sound track in which the average light transmission varies along the longitudinal axis in proportion to some characteristic of the applied signal.

variable-depth sonar [abbreviated VDS] Sonar in which the projector and receiving transducer are mounted in a watertight pod that can be lowered below a vessel to an optimum depth for minimizing thermal effects when detecting underwater targets.

variable-erase recording Recording on magnetic tape by selective erasure of a prerecorded signal.

variable-focal-length lens A television camera lens system whose focal length can be changed continuously during use while maintaining sharp focusing and a constant aperture, to give the effect of gradually moving the camera toward or away from the subject. The Zoomar lens is an example.

variable-frequency oscillator [abbreviated VFO] An oscillator whose frequency can be varied over a given range.

variable-gain multiplier A multiplier in which one variable is passed through an amplifier whose gain is controlled by the other variable. The output of such an amplifier is proportional to the required product.

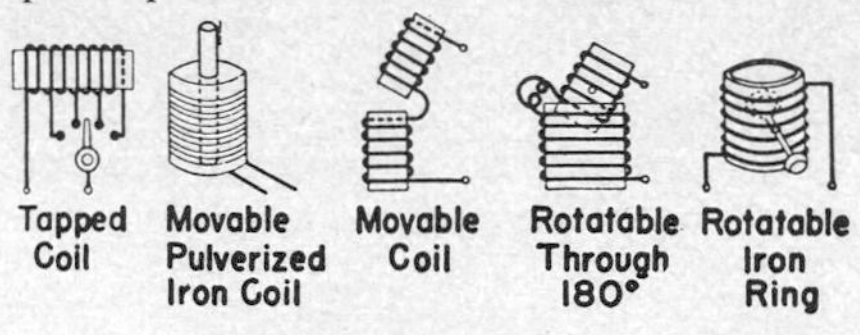

Variable inductances.

variable inductance A coil whose inductance value can be varied.

variable-inductance pickup A phonograph pickup that depends for its operation on the variation of its inductance.

variable-inductance transducer A transducer in which the output voltage is a function of the change in a variable-inductance element.

variable-iris waveguide coupler A microwave component used to couple a waveguide to the external input or output cavity of a klystron. It permits simple matching adjustments over a wide tuning range without use of stub tuners or matching sections.

variable-length record A computer input record for which the number of words, characters, bits, or fields is not fixed.

variable-mu tube An electron tube in which the amplification factor varies in a predetermined manner with control-grid voltage. This characteristic is achieved by making the spacing of the grid wires vary regularly along the length of the grid, in such a manner that a very large negative grid bias is required to block anode current completely.

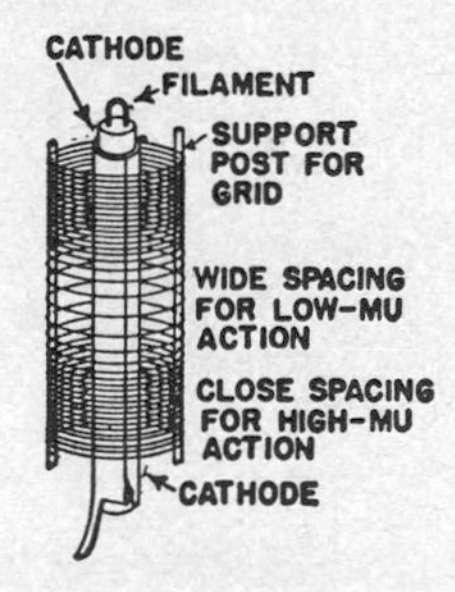

Variable-mu tube construction.

variable-reluctance microphone A microphone that depends for its operation on variations in the reluctance of a magnetic circuit. Also called magnetic microphone and reluctance microphone.

variable-reluctance pickup A phonograph pickup that depends for its operation on variations in the reluctance of a magnetic circuit due to the movements of an iron stylus assembly which is a part of the magnetic circuit. The reluctance variations alternately increase and decrease the flux through two series-connected coils, inducing in them the desired AF output voltage. Also called magnetic cartridge, magnetic pickup, and reluctance pickup.

variable-reluctance stepper motor A stepper motor that has a soft iron rotor with teeth or poles so positioned that they cannot simultaneously align with all the stator poles. In one example, four of the rotor poles align with four stator poles, and the four other rotor poles fall between stator poles. Each energization of a set of stator coils thus makes the rotor move half the angular distance between adjacent stator poles, to give stepping action. A wide range of speeds, torques, and stepping angles is possible by choosing the proper combination of stepper motor and gearbox designs.

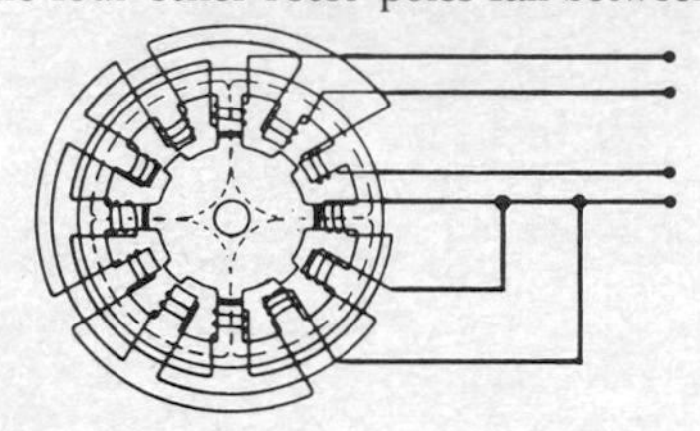

Variable-reluctance stepper motor with 12 stator poles and 8 rotor poles gives 15° stepping of rotor.

variable-reluctance transducer A transducer in which a slug of magnetic material is moved between two coils by the displacement being monitored. This changes the reluctance of the coils, thereby changing their impedance.

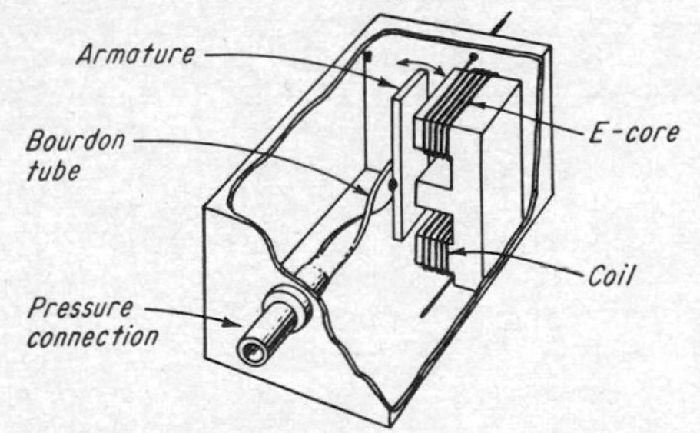

Variable-reluctance transducer used for sensing pressure. Increase in air pressure twists Bourdon tube, making attached armature move toward one pole of core and away from other pole.

variable-resistance pickup A phonograph pickup that depends for its operation on the variation of a resistance.

variable-resistance transducer A transducer in which the signal output depends on the change in a resistance element.

variable resistor *Rheostat.*

variable speech control [abbreviated VSC] A method of removing small portions of speech from a tape recording at regular intervals and stretching the remaining sounds to fill the gaps, so recorded speech can be played back at twice or even 2½ times the original speed without changing pitch and without significant loss of intelligibility. Developed for use by the blind and any others who want to listen to speech at speeds approximating normal reading speed.

variable-time fuze [abbreviated VT fuze] *Proximity fuze.*

variable transformer An iron-core transformer that has provisions for varying its output voltage over a limited range or continuously from zero to maximum output voltage, generally by means of a contact arm moving along exposed turns of the secondary winding. It may be an autotransformer.

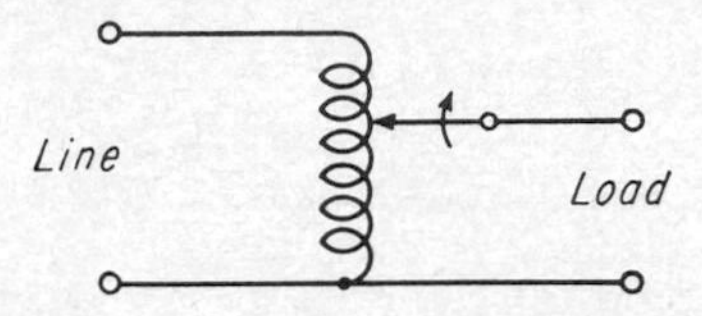

Variable transformer using single autotransformer winding.

Variac Trademark of General Radio Co. for their line of variable transformers.

varicap *Varactor.*

varindor A variable inductance in which a change in current varies the inductance value.

variocoupler An RF transformer that has provisions for varying the coupling between the two windings. Its construction is like that of a variometer, but its coils are not connected together.

variolosser A variable-loss circuit that improves the signal-to-noise ratio of a communication channel. At the transmitting end the variolosser is connected so its loss increases as input signal strength increases. At the receiving terminal the variolosser is in a circuit that restores the original dynamic range of the signal.

variometer A variable inductance that has two coils in series, one mounted inside the other, with provisions for rotating the inner coil to vary the total inductance of the unit over a wide range.

varistor A two-electrode semiconductor device that has a voltage-dependent nonlinear resistance. Its resistance drops as the applied voltage is increased. The resistance also varies in a predictable manner with temperature. Also called voltage-dependent resistor.

Varley loop test A method of using a Wheatstone bridge to determine the distance from the test point to a fault in a telephone or telegraph line or cable.

varmeter An instrument for measuring reactive power in vars. Also called reactive voltampere meter.

varnished cambric Linen or cotton fabric that has been impregnated with varnish or insulating oil and baked. Used for insulating purposes, especially for between-layer insulation in transformers.

V band A band of frequencies in the millimeter region, extending from 46 to 56 GHz, corresponding to wavelengths of 0.652 to 0.536 cm. Subdivisions of this band (all values in gigahertz) are

Va: 46–48
Vb: 48–50
Vc: 50–52
Vd: 52–54
Ve: 54–56

V-beam radar A volumetric radar system that uses two fan beams to determine the distance, bearing, and height of a target. One beam is vertical and the other inclined. The beams intersect at

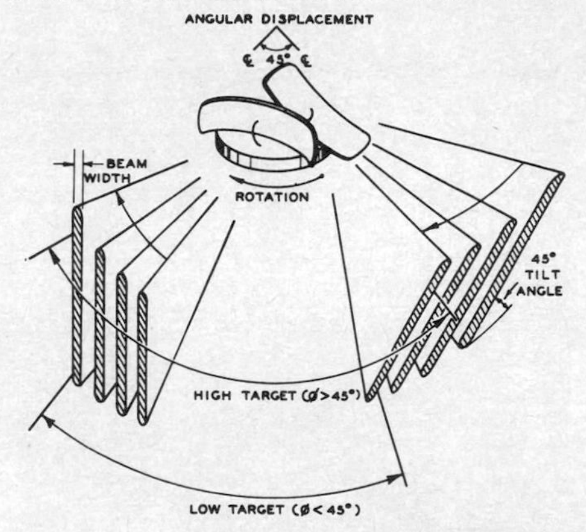

V-beam radar. Antenna at left, producing thin vertical beams at left, is followed in azimuth by right-hand antenna, which produces thin beam tilted at 45°. Angle between interceptions of target by two rotating beams is measure of elevation angle of target.

ground level and rotate continuously about a vertical axis. The time difference between the arrivals of the echoes of the two beams is a measure of target elevation.

VCO Abbreviation for *voltage-controlled oscillator.*

VCXO Abbreviation for *voltage-controlled crystal oscillator.*

VDC Abbreviation for *volts DC.*

VDS Abbreviation for *variable-depth sonar.*

vector 1. A quantity that has both magnitude and direction, represented as a line terminated by an arrowhead. Often used to portray the amplitude and phase of a sinusoidal signal. Also called vector quantity. 2. To guide a pilot, navigator, aircraft, or missile from one point to another within a given time by a direction communicated to the craft.

vectorcardiogram *Vector electrocardiogram.*

vector diagram An arrangement of vectors that shows the magnitude and phase relations between two or more alternating quantities which have the same frequency.

vector electrocardiogram The two or three-dimensional presentation of cardiac electric activity that results from displaying lead pairs against each other rather than against time. Also called vectorcardiogram.

vector quantity *Vector.*

vectorscope A cathode-ray oscilloscope that displays both the phase and amplitude of an applied signal with respect to a reference signal.

Veitch diagram A diagram used for simplifying the handling of logic associated with the design of counter circuits.

velocimeter A Doppler system that uses the Doppler shift of a continuous-wave carrier transmitted to and reflected from a moving target to measure radial velocity.

velocity 1. A vector quantity that denotes both the direction and speed of a linear motion or denotes the direction of rotation and the angular speed in the case of rotation. 2. The time rate of change of a position vector of a point with respect to an inertial frame, the axes of which are usually fixed with respect to the earth.

velocity antiresonance The condition wherein a small change in the frequency of a sinusoidal force applied to a body or system causes an increase in velocity at the driving point, or the frequency is such that the absolute value of the driving-point impedance is a maximum.

velocity correction A method of register control in which the velocity of the web is changed gradually.

velocity factor The ratio of the velocity of propagation in any medium to the velocity of propagation in free space. The velocity of an RF current is slightly less in a conductor than it would be in free space.

velocity filter A radar storage-tube device that blanks all targets which do not move more than one radar cell within a predetermined number of antenna scans.

velocity-fluctuation noise The noise component developed in a traveling-wave tube or a weak-signal photodetector because of the wide thermal distribution of electron velocities in the beam.

velocity-focusing mass spectrograph *Velocity spectrograph.*

velocity hydrophone A velocity microphone that responds to waterborne sound waves.

velocity level A sound rating in decibels, equal to 20 times the logarithm to the base 10 of the ratio of the particle velocity of the sound to a specified reference particle velocity.

velocity-limiting servo A servomechanism in which the maximum velocity of the servo is the chief limit on performance.

velocity microphone A microphone whose electric output depends on the velocity of the air particles that form a sound wave. Examples are hot-wire and ribbon microphones.

velocity-modulated oscillator An electron-tube structure in which the velocity of an electron stream is varied as the stream passes through a resonant cavity called a buncher. Energy is extracted from the bunched electron stream at a higher energy level in passing through a second cavity resonator called the catcher. Oscillations are sustained by coupling energy from the catcher cavity back to the buncher cavity.

velocity-modulated tube An electron-beam tube in which the velocity of the electron stream is alternately increased and decreased within a period comparable to the local transit time.

velocity modulation Modulation in which a time variation in velocity is impressed on the electrons of a stream.

velocity of light A physical constant equal to 2.997925×10^{10} cm/s. This corresponds to 186,280 statute mi/s, 161,870 nautical mi/s, and 328 yd/μs. All electromagnetic radiation travels at this same speed in free space. The velocity of light is usually considered to be a vector quantity that represents the speed of light in a particular direction.

velocity of sound The approximate velocities of sound waves are 1100 ft/s (335 m/s) in air at 0°C, 4800 ft/s (1463 m/s) in water, and 16,400 ft/s (5000 m/s) in steel.

velocity resonance *Resonance.*

velocity selector A mechanical monochromator for slow neutrons; it consists of several cadmium disks mounted at intervals along a rotating shaft. Slots in the disks are offset in angle for each successive disk, so only neutrons with a small range of velocities can pass through all the slits at a given speed of rotation.

velocity sorting Any process of selecting electrons according to their velocities.

velocity spectrograph A mass spectrograph in which only positive ions that have a certain velocity pass through all three slits and enter a chamber where they are deflected by a magnetic field in proportion to their charge-to-mass ratio. Also called velocity-focusing mass spectrograph.

velocity transducer A transducer that generates an output which is proportional to its velocity.

venetian-blind antenna A microwave antenna array that consists of tiltable parabolic cylinders arranged like a venetian blind. The first pair of elements is switched against the remainder of the array to give beam scanning without mechanical movement.

Venetian-blind antenna.

vented baffle *Reflex baffle.*

vented pressure pickup A pressure pickup in which a vent reestablishes a reference pressure just before each use.

Venus probe A probe for exploring and reporting on conditions on or about the planet Venus, such as Pioneer and Mariner U.S. probes.

Verdet's constant A Faraday-effect constant that determines the angle of rotation of plane-polarized light when passing through certain materials in a magnetic field.

verification Automatic comparison of one data transcription with another transcription of the same data, to reveal errors.

verifier A punched-card machine or auxiliary computer device that automatically provides verification.

vernier Any device, control, or scale used to obtain a fine adjustment or to increase the precision of a measurement.

vernier capacitor A small variable capacitor placed in parallel with a larger tuning capacitor to provide a finer adjustment after a larger unit has been set approximately to the desired position.

vernier dial A tuning dial in which each complete rotation of the control knob causes only a fraction of a revolution of the main shaft, permitting fine and accurate adjustment.

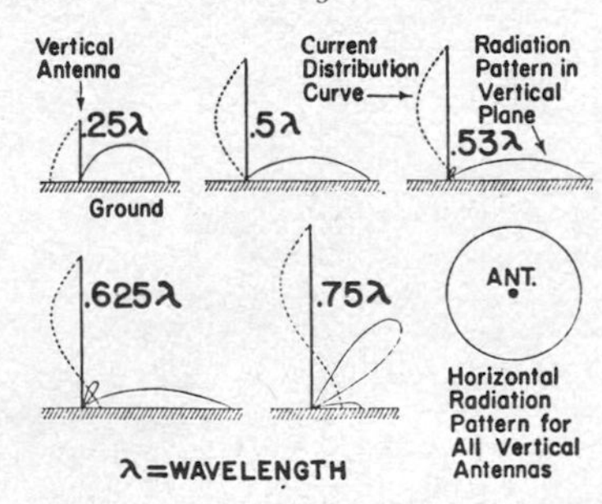

Vertical-antenna radiation patterns in vertical plane for various heights (thin solid curves) and current distribution curves (dotted).

vertical antenna A vertical metal tower, rod, or suspended wire used as an antenna.

vertical blanking Blanking of a television picture tube during the vertical retrace.

vertical blanking pulse The rectangular pulse that is transmitted at the end of each field of a television signal to cut off the beam current of the picture tube while the beam is returning to the top of the screen for the start of the next field.

vertical centering control The centering control provided in a television receiver or cathode-ray oscilloscope to shift the position of the entire image vertically in either direction on the screen.

vertical compliance The ability of a stylus to move freely in a vertical direction while in the groove of a phonograph record.

vertical convergence control The control that adjusts the amplitude of the vertical dynamic convergence voltage in a color television receiver.

vertical definition *Vertical resolution.*

vertical deflection electrodes The pair of electrodes that moves the electron beam up and down on the fluorescent screen of a cathode-ray tube which employs electrostatic deflection.

vertical deflection oscillator The oscillator that produces, under control of the vertical synchronizing signals, the sawtooth voltage waveform that is amplified to feed the vertical deflection coils on the picture tube of a television receiver. Also called vertical oscillator.

vertical field-effect transistor [abbreviated VFET] A field-effect transistor in which the gate is constructed in the form of a vertical grating between source and drain in the semiconductor material, to allow higher current flow between source and drain. Advantages include fast pulse response, as required for high-fidelity audio power amplifiers.

vertical hold control The hold control that changes the free-running period of the vertical deflection oscillator in a television receiver, so the picture remains steady in the vertical direction.

vertical-incidence sounder *Ionosonde.*

vertical interval reference [abbreviated VIR] A reference signal inserted into a television program signal every 1/60 s, in line 19 of the vertical blanking period between television frames, to provide references for luminance amplitude, black-level amplitude, sync amplitude, chrominance amplitude, and color-burst amplitude and phase. Used in television transmitters and video recorders to provide a reference for maintaining color, brightness, and contrast at exact specifications. It can also be used in receivers to provide automatic adjustment of chroma and tint.

vertical linearity control A linearity control that permits narrowing or expanding the height of the image on the upper half of the screen of a television picture tube, to give linearity in the vertical direction so circular objects appear as true circles. Usually mounted at the rear of the receiver.

vertically polarized wave A linearly polarized wave in which the electric field vector is vertical.

vertical oscillator *Vertical deflection oscillator.*

vertical polarization Transmission of radio waves in such a way that the electric lines of force are vertical, while the magnetic lines of force are horizontal. With this polarization, transmitting and receiving dipole antennas are placed in a vertical plane.

vertical quarter-wave stub An antenna that has a vertical portion which is electrically a quarter-wavelength long. Generally used with a ground plane at the base of the stub.

vertical radiator A transmitting antenna that is perpendicular to the earth.

vertical recording A type of disk recording in which the groove modulation is perpendicular to the surface of the recording medium, so the cutting stylus moves up and down rather than from side to side during recording. Also called hill-and-dale recording.

vertical resolution The number of distinct horizontal lines, alternately black and white, that can be seen in the reproduced image of a television or facsimile test pattern. Vertical resolution is primarily fixed by the number of horizontal lines used in scanning. Also called vertical definition.

vertical retrace The return of the electron beam to the top of the screen at the end of each field in television.

vertical stylus force *Stylus force.*

vertical sweep The downward movement of the scanning beam from top to bottom of the picture being televised.

vertical synchronizing pulse One of the six pulses that are transmitted at the end of each field in a television system to keep the receiver in field-by-field synchronism with the transmitter. Also called picture synchronizing pulse.

very high frequency [abbreviated VHF] A Federal Communications Commission designation for a frequency in the band from 30 to 300 MHz, corresponding to a metric wave between 1 and 10 m.

very long-range radar Radar whose maximum line-of-sight range is greater than 800 mi (1290 km) for a target that has an area of 1 m^2 perpendicular to the radar beam.

very low frequency [abbreviated VLF] A Federal Communications Commission designation for a frequency in the band from 3 to 30 kHz, corresponding to a myriametric wave between 10 and 100 km.

very short-range radar Radar whose maximum line-of-sight range is less than 50 mi (80 km) for a target that has an area of 1 m^2 perpendicular to the radar beam.

vesicular film A film that is sensitive to ultraviolet light and is developed by heat, without chemicals.

vestigial sideband [abbreviated VSB] The transmitted portion of an amplitude-modulated sideband that has been largely suppressed by a filter which has a gradual cutoff in the neighborhood of the carrier frequency. The other sideband is transmitted without much suppression.

vestigial-sideband filter A filter that is inserted between a transmitter and its antenna to suppress part of one of the sidebands.

vestigial-sideband transmission A type of radio signal transmission for amplitude modulation in which the normal complete sideband on one side of the carrier is transmitted, but only a part of the other sideband is transmitted. Used for the visual transmitter in television, where the lower sideband extends only 0.75 MHz below the carrier and the upper sideband extends 4 MHz. Also called asymmetrical-sideband transmission.

vestigial-sideband transmitter A transmitter in which one sideband and a portion of the other are intentionally transmitted.

V/F Abbreviation for voltage-to-frequency, usually used in connection with converters.

V/F converter Abbreviation for *voltage-to-frequency converter.*

VFET Abbreviation for *vertical field-effect transistor.*

VFO Abbreviation for *variable-frequency oscillator.*

VFR Abbreviation for *visual flight rules.*

VFR conditions Weather conditions equal to or better than the minimum prescribed for flights under visual flight rules.

VHF Abbreviation for *very high frequency.*

VHF antenna Any antenna designed for operation in the VHF band from 30 to 300 MHz, such

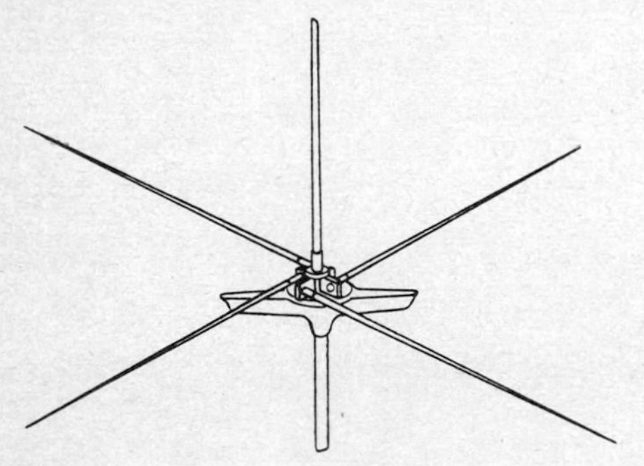

VHF antenna for shipboard radio communication, consisting of vertical quarter-wave stub and four-element ground plane.

as for VHF television reception, FM radio reception, radio communication, and other purposes.

VHF channel One of the 6-MHz television channels designated by the numbers 2 through 13. Channels 2–4 cover 54–72 MHz; 5–6 cover 76–88 MHz; 7–13 cover 174–216 MHz.

VHF homing adapter A homing adapter that operates in the VHF range.

VHF omnirange [abbreviated VOR] An omnirange that operates in the band from 112 to 118 MHz to provide bearing information for aircraft. It emits a nondirectional reference modulation and a signal whose character varies with rotation through 360° in a horizontal plane. The receiving equipment interprets the two signals in terms of the bearing between the receiver and the signal source. Also called omni.

VHF/UHF direction finder A ground-based radio direction finder capable of being used alone or in conjunction with airport surveillance radar.

vibrating capacitor A capacitor whose capacitance is varied in a cyclic manner to produce an alternating voltage proportional to the charge on the capacitor. Used in a vibrating-reed electrometer.

vibrating-reed electrometer An instrument that uses a vibrating capacitor to measure a small charge, often in combination with an ionization chamber.

vibrating-reed frequency meter A frequency meter that consists of steel reeds which have different and known natural frequencies, all excited by an electromagnet carrying the alternating cur-

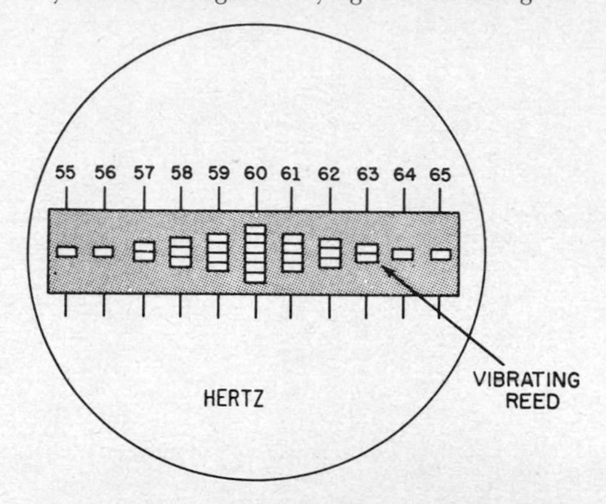

Vibrating-reed frequency meter display, showing ends of reeds below scale calibrated in hertz. Reed having natural frequency closest to that of energizing current vibrates through largest amplitude, here 60 Hz.

rent whose frequency is to be measured. Also called Frahm frequency meter, reed frequency meter, and tuned-reed frequency meter.

vibrating-reed magnetometer An instrument that measures magnetic fields by noting their ef-

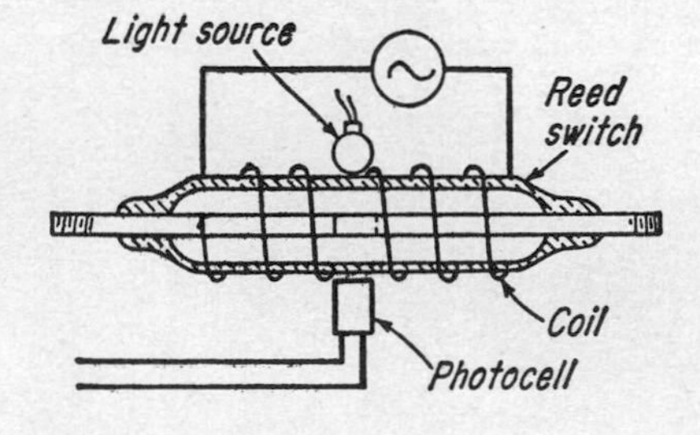

Vibrating-reed magnetometer.

fect on the vibration of reeds excited by an alternating magnetic field.

vibrating-reed rectifier An electromagnetic device that rectifies an alternating current by reversing the connections between the power line and load each time the alternating current reverses in direction. The reversing contacts are on a vibrating reed of magnetic material that is acted on by a coil carrying the alternating current, so the reed moves in synchronism with the current.

vibrating-sample magnetometer A device that determines the magnetic properties of a sample of magnetic material by vibrating it in a magnetic field and measuring the voltage induced in search coils near the sample.

vibration A periodic change in the position of an object, such as a pendulum or a diaphragm of a loudspeaker energized by a sinusoidal current. Also called oscillation.

vibration galvanometer An AC galvanometer in which the natural oscillation frequency of the moving element is equal to the frequency of the current being measured.

vibration meter An instrument that measures the displacement, velocity, and acceleration associated with mechanical vibration. In one form it consists of a piezoelectric vibration pickup that has uniform response from 2 to 1000 Hz, feeding an amplifier which has an indicating meter at its output. Also called vibrometer.

vibration pickup A pickup that responds to mechanical vibrations rather than to sound waves. In one type, twisting or bending of a Rochelle-salt crystal generates a voltage that varies in accordance with the vibration being analyzed.

vibrato A musical embellishment that depends primarily on periodic variations of frequency, which are often accompanied by variations in amplitude and waveform.

vibrator An electromagnetic device for converting a direct voltage into an alternating voltage. It contains a vibrating armature, on which are con-

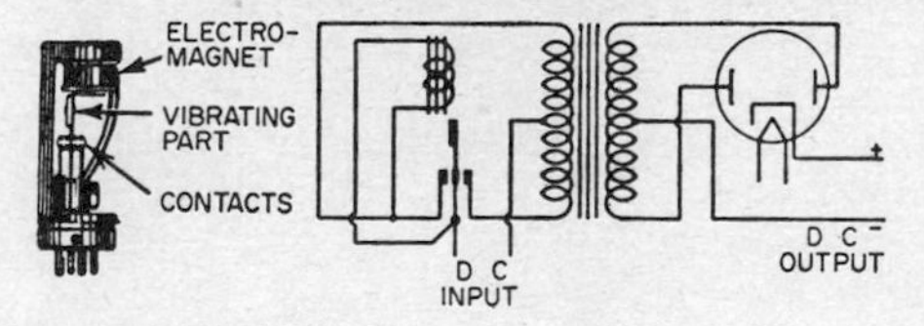

Vibrator construction and basic circuit.

tacts that reverse the direction of current flow during each vibration. Used to convert the voltage of a storage battery into a low alternating voltage that can be stepped up by a power transformer and rectified.

vibrator-type inverter A device that uses a vibrator and an associated transformer or other inductive device to change DC input power to AC output power.

vibratory gyroscope A instrument that utilizes a vibrating rod or tuning fork to perform certain equilibrium or directional functions of the spinning gyroscope.

vibrometer *Vibration meter.*

video [Latin for "I see"] 1. Pertaining to picture signals or to the sections of a television system that carry these signals in either unmodulated or modulated form. 2. Pertaining to the demodulated radar receiver output that is applied to a radar indicator.

video amplifier A wideband amplifier capable of amplifying video frequencies in radar and television.

video cable Coaxial cable used for signal transmission in cable television systems. Basic types include 75-Ω unbalanced indoor cable and outdoor cable that has a single conductor centered in a shield and 124-Ω balanced indoor cable and outdoor cable which has two parallel or twisted insulated conductors centered in a shield.

video circuit A broadband circuit that carries intelligence which could become visible.

video converter A converter that consists of a television camera tube and a cathode-ray tube between which may be inserted a transparent overlay. The image on the overlay is illuminated by the radial scan of the cathode-ray tube and converted to video signals by the camera tube, for superimposition on a radar PPI display or other radar applications.

video correlator A radar circuit that enhances automatic target-detection capability, provides data for digital target plotting, and gives improved immunity to noise, interference, and jamming.

video data digital processing Digital processing of video signals for pictures transmitted over a television link, to improve picture quality by reducing the effects of noise and distortion. The computer compares each scanned line with adjacent lines and eliminates extreme changes caused by electromagnetic interference.

video detector The detector that demodulates video IF signals in a television receiver.

video disk recorder A video recorder that records television visual signals and sometimes aural signals on a magnetic, optical, or other type of disk

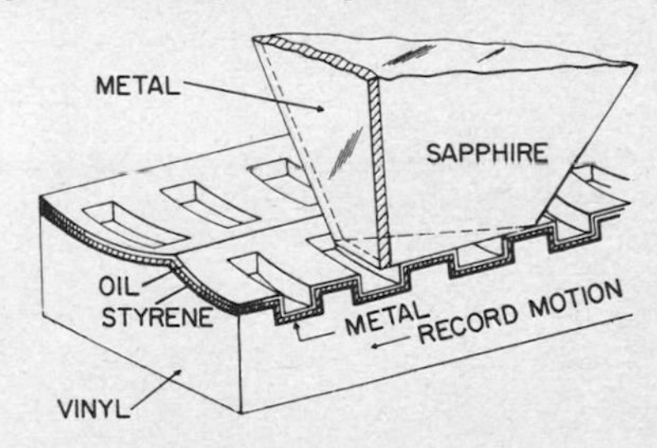

Video disk recorder in which metallized sapphire stylus detects variations in capacitance caused by changes in width and spacing of metallized grooves in each track on vinyl disk.

which is usually about the size of a long-playing phonograph record.

video frequency One of the frequencies that exist in the output of a television camera when an image is scanned. It may be any value from almost zero to well over 4 MHz.

video-frequency amplifier An amplifier capable of handling the entire range of frequencies that comprise a periodic visual presentation in television, facsimile, or radar.

video gain control A control that adjusts the gain of a video amplifier, as for varying the intensity of the echoes on a radar PPI screen to get maximum contrast between desired echoes and undesired clutter.

video game An electronic game that can be connected to or built into a television receiver, to use the television screen as a playing field or display showing player and ball movements, scores, or other actions called for by the game, race, or other type of activity which is controlled remotely by one or more players. Microprocessor-based models may also provide sound effects and interchangeable program cards for changing the rules of a game or changing the entire game. Also called television game.

Videograph Trademark of A. B. Dick Co. for their high-speed cathode-ray character generator and electrostatic printer, used for printing magazine address labels under control of magnetic-tape files of computer-processed addresses. The moving electron beam applies a charge on a dielectric-coated paper to form electrostatic images of characters. Powder is then attracted to the image areas and fused to give readable addresses.

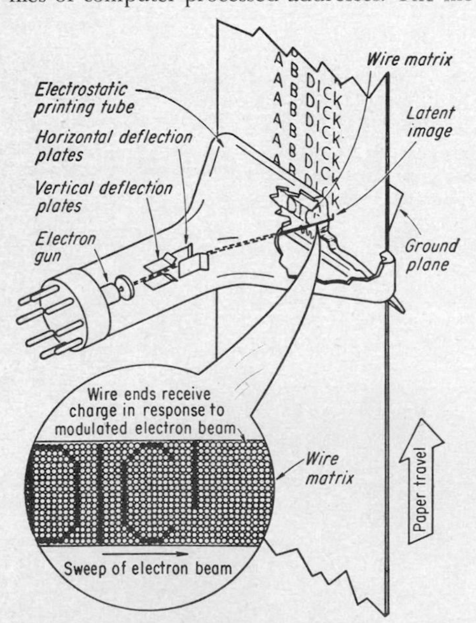

Videograph as used for printing address labels electronically at high speed on strip of dielectric-coated paper.

video integration A method of utilizing the redundancy of repetitive signals to improve the output signal-to-noise ratio, by summing the successive video signals.

video mapping A mapping procedure in which a chart of an area is electronically superimposed on a radar display.

video mixer A mixer that combines the output signals of two or more television cameras.

videophone *Video telephone.*

video player A player that converts a video disk, videotape, or other type of recorded television program into signals suitable for driving a home television receiver.

video recorder A recorder capable of storing the video signals for a television program and feeding them back later to a television transmitter or directly to a receiver. Examples include electron-beam video, holographic video, video disk, and videotape recorders.

video recording Recording of information that has a bandwidth in excess of about 500 kHz, such as television or radar signals.

video signal A signal that contains periodic visual information together with blanking and synchronizing pulses, as in a radar or television system.

video stretching A procedure whereby the duration of a video pulse is increased in a navigation system.

videotape A heavy-duty magnetic tape designed primarily for recording the video signals of television programs.

videotape recording [abbreviated VTR] A method of recording television video signals on magnetic tape for later rebroadcasting of television programs.

videotape replay A videotape recorder that uses a relatively short endless loop of magnetic tape, to permit the repetition of a televised sports scene within seconds after the original action.

video telephone A combined telephone and video receiver that allows each party to see the other party while talking. Also called videophone.

video terminal *Cathode-ray-tube terminal.*

video waveform The portion of the television signal waveform that corresponds to visual information. Synchronizing pulses are not included.

vidicon A camera tube in which a charge-density pattern is formed by photoconduction and stored on a photoconductor surface that is scanned by an electron beam, usually of low-velocity electrons. Used chiefly in industrial and other closed-circuit television cameras.

viewfinder An auxiliary optical or electronic device attached to a television camera so the operator can see the scene as the camera sees it.

viewing screen *Screen.*

viewing time The time during which a storage tube is presenting a visible output that corresponds to the stored information.

Villari effect The change in magnetic induction that occurs when a magnetostrictive material is mechanically stressed.

vinyl resin A soft plastic material used in making

long-playing phonograph records.

VIR Abbreviation for *vertical interval reference.*

virtual cathode The locus of a space-charge-potential minimum such that only some of the electrons approaching it are transmitted, the remainder being reflected back to the electron-emitting cathode.

virtual height The apparent height of an ionized layer, as determined from the time interval between the transmitted signal and the ionospheric echo at vertical incidence.

virtual memory A combination of primary and secondary memories that can be treated as a single memory by programmers because the computer itself translates a program or virtual address to the actual hardware address.

viscometer An instrument that measures the degree to which a liquid resists a change in shape. Also called viscosimeter.

viscosimeter *Viscometer.*

viscous-damped arm A phonograph pickup arm which is mechanically damped by a highly viscous liquid in such a way that the arm floats gently down to a record when dropped.

visibility factor The ratio of the minimum signal input power detectable by ideal instruments connected to the output of a receiver, to the minimum signal power detectable by a human operator through a display connected to the same receiver. Also called display loss.

visible radiation Radiation that has wavelengths ranging from about 3800 to 7800 Å (0.38 to 0.78 μm), corresponding to the visible spectrum of light.

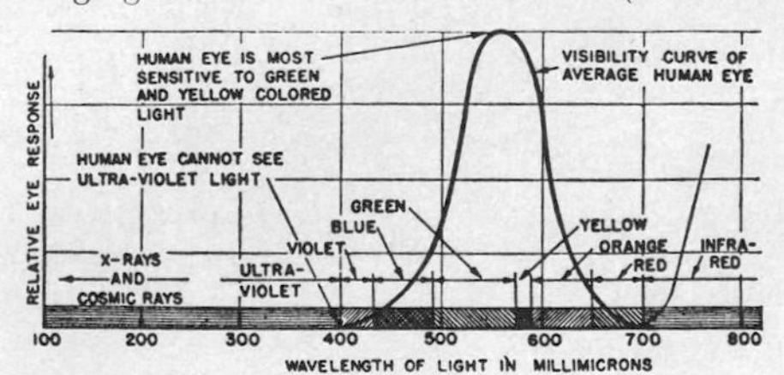

Visible radiation. To convert wavelengths in millimicrons to micrometers, divide by 1000; thus, 500 millimicrons is 0.5 μm. One angstrom is 0.0001 μm, so the upper visible limit of 7800 Å is 0.78 μm.

visual-aural range [abbreviated VAR] A VHF radio range that provides one course for display to the pilot on a zero-center left-right indicator and another course, at right angles to the first, in the form of aural A-N radio range signals. Either indication may be used to resolve the ambiguity of the other.

visual broadcast service A service rendered by stations broadcasting images for general public reception, such as by television or facsimile broadcast stations.

visual carrier frequency The frequency of the television carrier that is modulated by picture information.

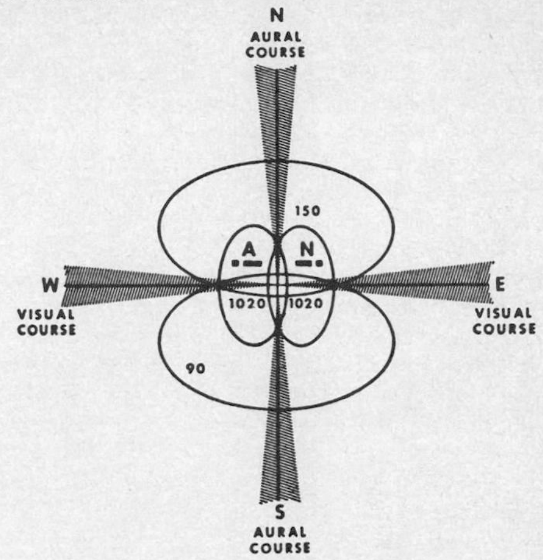

Visual-aural range radiation patterns, courses, and signals. Frequency values of tones are in hertz.

visual flight rules [abbreviated VFR] Regulations that govern flying of aircraft when visibility is good up to a specified altitude.

visual radio range Any range facility whose course is flown by visual instrumentation not associated with aural reception.

visual range The range of beta particles in an absorber, usually aluminum, as estimated by visual inspection of breaks in the absorption curve of the material.

visual signal The picture portion of a television signal.

visual transmitter The radio equipment used to transmit the video part of a television program. Also called picture transmitter. The visual and aural transmitters together are called a television transmitter.

visual transmitter power The peak power output when transmitting a standard television signal.

VLF Abbreviation for *very low frequency.*

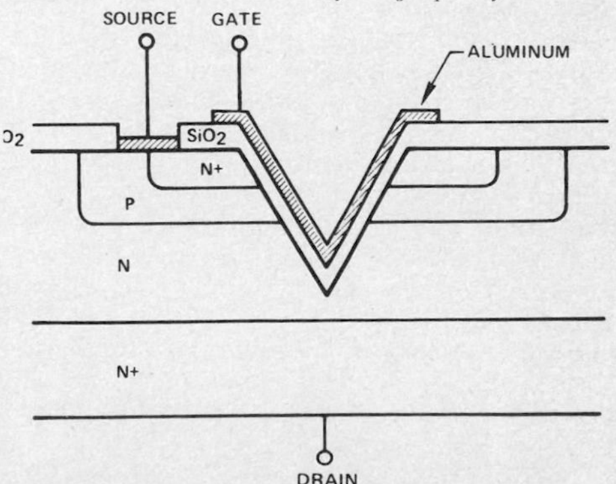

VMOS structure in N-channel MOSFET.

VMOS [Vertical Metal-Oxide Semiconductor] An MOS technology that involves essentially the formation of four diffused layers in silicon and etching of a V-shaped groove to a precisely controlled depth in the layers, followed by deposition of metal over silicon dioxide in the groove to form the gate electrode. Advantages include higher current density and higher packing density for transistors and other semiconductor devices.

vocoder A device that produces synthetic speech.

VODAS [Voice-Operated Device Anti-Singing] A voice-operated switching device used in transoceanic radiotelephone circuits to suppress echoes and singing sounds automatically. It connects a subscriber's line automatically to the transmitting station as soon as the subscriber starts speaking and simultaneously disconnects it from the receiving station, thereby permitting the use of one radio channel for both transmitting and receiving without appreciable switching delay as the parties alternately talk.

VODER [Voice Operation DEmonstratoR] An electronic system that uses electron tubes and filters, controlled through a keyboard, to produce voice sounds artificially.

VOGAD [Voice-Operated Gain-Adjusting Device] An automatic-gain-control circuit that maintains a constant speech output level in long-distance radiotelephony.

voice channel A communication channel that has sufficient bandwidth to carry voice frequencies intelligibly. The minimum bandwidth for a voice channel is about 3 kHz.

voice coder A coder that converts speech input into a digital form that can be enciphered for secure transmission. The transmitted digital signals are deciphered and converted back into speech at the receiver.

voice coil The coil that is attached to the diaphragm of a moving-coil loudspeaker and moves through the air gap between the pole pieces due to interaction of the fixed magnetic field with that associated with the AF current flowing through the voice coil.

voice digitization The conversion of analog voice signals to digital signals. Advantages include improved communication because digital signals are relatively immune to noise, crosstalk, and distortion. Faded signals can be regenerated without loss of quality. Techniques include use of various types of pulse-code modulation.

voice frequency An audio frequency in the range essential for transmission of speech of commercial quality, from about 300 to 3400 Hz. Also called speech frequency.

voice-frequency carrier telegraphy Carrier telegraphy in which the modulated currents are transmitted over a voice-frequency telephone channel.

voice-grade line An ordinary telephone line that has a bandwidth which extends from about 300 to 3400 Hz, as required for handling speech or digital data at speeds up to about 2400 b/s.

voice multiplexing The compression of two or more (typically four) simultaneous telephone conversations into one voice-grade channel in a manner that is undetectable to listeners, by using some form of voice digitization.

voice-operated transmission [abbreviated VOX] A method of radio communication in which the carrier is radiated only when a voice signal is present.

voiceprint A record of the distinctive patterns formed by a person's voice, as obtained with a sound spectrograph. It can be used for identification in place of fingerprints.

voice recognition unit A computer peripheral device that recognizes a limited number of spoken words and converts them into equivalent digital signals which can serve as computer input or initiate other desired actions. One version, which recognizes spoken digits 0 through 9 and about 40 additional preselected words, uses a sine-cosine analog transform to sum the outputs of six audio filters. The resulting sine-cosine patterns are automatically normalized to intensity and duration before they are acted on by recognition software. A display can be added to show each recognized word or digit.

voice response A computer-controlled recording system in which basic sounds, numerals, words, and/or phrases are individually stored for playback under computer control as the reply to a keyboarded query. Applications include stock market quotations and giving checking account balance when an account number is punched into a computer remote terminal.

voice synthesizer A synthesizer that simulates speech in any language by assembling a language's elements or phonemes under digital control, each with the correct inflection, duration, pause, and other speech characteristics. Applications include computer-generated voice replies to queries by cathode-ray-tube terminals and Touch-Tone telephones. When combined with voice recognition, the system can provide oral replies to a limited variety of spoken queries. Also called speech synthesizer.

voice warning system An aircraft cockpit instrument that has a digitized or prerecorded vocabulary of warning words used for such purposes as calling attention to possible trouble on the aircraft or warning of the possibility of an airborne collision.

volatile storage Computer storage in which information cannot be retained without continuous power dissipation. If power is turned off, the stored information vanishes. Delay-line memories and electrostatic storage tubes are examples.

volt [abbreviated V] The SI unit of voltage or

potential difference. One volt will send a current of 1 A through a resistance of 1 Ω. Named after the Italian physicist Alessandro Volta, 1745–1829.

Volta effect *Contact potential.*

voltage [symbol E] The term most often used to designate electric pressure that exists between two points and is capable of producing a flow of current when a closed circuit is connected between the two points. Voltage is measured in volts, millivolts, microvolts, and kilovolts. The terms electromotive force, potential, potential difference, and voltage drop are all often called voltage.

voltage amplification The ratio of the magnitude of the voltage across a specified load impedance to the magnitude of the input voltage of the amplifier or other transducer feeding that load. Often expressed in decibels by multiplying the common logarithm of the ratio by 20.

voltage amplifier An amplifier designed primarily to build up the voltage of a signal, without supplying appreciable power.

voltage-amplifier tube A tube designed primarily for use in a voltage amplifier; hence it has high gain but delivers very little output power.

voltage attenuation The ratio of the magnitude of the voltage across the input of a transducer to the magnitude of the voltage delivered to a specified load impedance connected to the transducer. Often expressed in decibels by multiplying the common logarithm of the ratio by 20.

voltage calibrator A voltage source that provides an adjustable high-accuracy calibration voltage for use in calibrating measuring instruments.

voltage-controlled capacitor A capacitor whose capacitance value can be changed by varying an externally applied bias voltage, as in a silicon capacitor. Used for automatic frequency control, remote control of tuning, and frequency modulation.

voltage-controlled crystal oscillator [abbreviated VCXO] A crystal oscillator circuit in which the oscillator output frequency can be varied or swept over a range of frequencies by varying a DC modulating voltage.

voltage-controlled magnetic amplifier A magnetic amplifier in which the flux change during the resetting interval is related to the input signal voltage and is essentially independent of control-circuit current.

voltage-controlled oscillator [abbreviated VCO] An oscillator whose frequency of oscillation can be varied by changing an applied voltage.

voltage cutoff The electrode voltage that reduces the anode current, beam current, or some other electron-tube characteristic to a specified low value.

voltage-dependent resistor *Varistor.*

voltage divider A tapped resistor, adjustable resistor, potentiometer, or a series arrangement of two or more fixed resistors connected across a voltage source. A desired fraction of the total voltage is obtained from the intermediate tap, movable contact, or resistor junction.

voltage doubler A transformerless rectifier circuit that gives approximately double the output voltage of a conventional half-wave rectifier by charging a capacitor during the normally wasted half-cycle and discharging it in series with the output voltage during the next half-cycle.

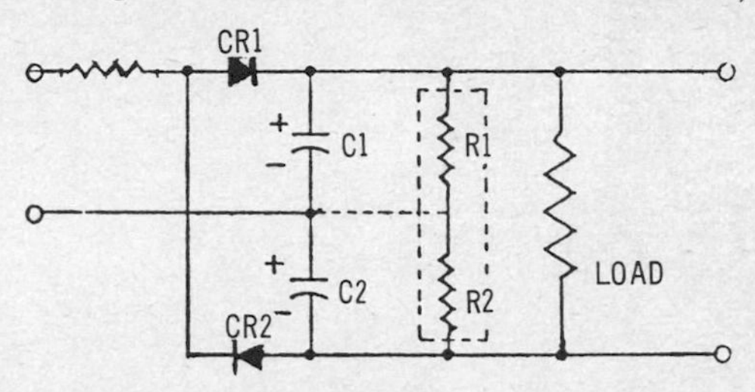

Voltage-doubler circuit.

voltage drop The voltage developed across a component or conductor by the flow of current through the resistance or impedance of that component or conductor.

voltage feed Excitation of a transmitting antenna by applying voltage at a voltage loop or antinode.

voltage feedback Feedback in which the voltage drop across part of the load impedance acts in series with the input signal voltage.

voltage follower An operational amplifier used without feedback elements but with a direct feedback connection from the output to the inverting input to give unity gain so the output voltage follows the noninverting input voltage. A voltage follower has a very high input impedance and a very low output impedance.

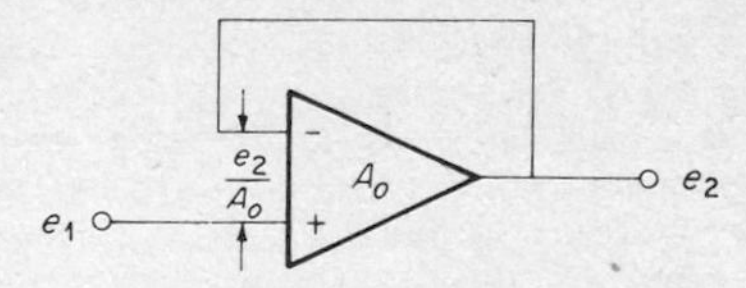

Voltage follower using direct feedback connection from output to inverting input.

voltage gain The difference between the output signal voltage level in decibels and the input signal voltage level in decibels. This value is equal to 20 times the common logarithm of the ratio of the output voltage to the input voltage. The voltage gain is equal to the amplification factor of the tube or transistor only for a matched load.

voltage generator A two-terminal circuit element in which the terminal voltage is independent of the current through the element.

voltage gradient The voltage per unit length along a resistor or other conductive path.

voltage jump An abrupt change or discontinuity in tube voltage drop during operation of a glow-discharge tube.

voltage loop An antinode at which voltage is a maximum.

voltage multiplier *Instrument multiplier.*

voltage node A point that has zero voltage in a stationary wave system, as in an antenna or transmission line. A voltage node exists at the center of a half-wave antenna.

voltage-range multiplier *Instrument multiplier.*

voltage rating The maximum sustained voltage that can safely be applied to an electric device without risking the possibility of electric breakdown. Also called working voltage.

voltage-reference diode A PN junction diode which has a sufficiently stable breakdown voltage that it can be used to develop a reference voltage.

voltage-reference tube A gas tube in which the tube voltage drop is approximately constant over the operating range of current and is also relatively stable with time at fixed values of current and temperature.

voltage reflection coefficient The ratio of the complex electric field strength or voltage of a reflected wave to that of the incident wave.

voltage-regulated AC power supply A power supply that operates from an AC line and delivers a regulated AC output voltage, usually adjustable, at the same frequency or at some other frequency.

voltage-regulating transformer A power transformer that delivers an essentially constant output voltage over a wide range of input voltage values.

voltage regulation The ratio of the difference between no-load and full-load output voltage of a device to the full-load output voltage, expressed as a percentage.

voltage regulator A device that maintains the terminal voltage of a generator or other voltage source within required limits despite variations in input voltage or load. Also called automatic voltage regulator.

voltage-regulator diode A diode that maintains an essentially constant direct voltage in a circuit despite changes in line voltage or load. A zener diode provides this regulation, but other voltage-regulator diodes do not use the Zener effect.

voltage-regulator tube A glow-discharge tube in which the tube voltage drop is approximately constant over the operating range of current. Used to maintain an essentially constant direct voltage in a circuit despite changes in line voltage or load. Also called VR tube.

voltage relay A relay that functions at a predetermined value of voltage.

voltage saturation *Anode saturation.*

voltage-sensitive resistor A resistor whose value varies markedly with applied voltage over at least a portion of its voltage range. It may consist of one or more mineral crystals or two or more metallic oxide disks, but does not have rectifying properties.

voltage stabilizer A zener diode or other device that suppresses variations in a DC voltage. It is often used in place of a capacitor across a cathode biasing resistor.

voltage-stabilizing tube A gas-filled tube normally working with a glow discharge in the part of the characteristic where the voltage is practically independent of current.

voltage standard A voltage source whose value is known to a high degree of accuracy. A standard cell is an example.

voltage standing-wave ratio [abbreviated VSWR] The ratio of the amplitude of the electric field or voltage at a voltage minimum to that at an adjacent maximum in a stationary-wave system, as in a waveguide, coaxial cable, or other transmission line.

voltage standing-wave ratio meter [abbreviated VSWR meter] An electrically operated instrument that indicates voltage standing-wave ratios and is calibrated in voltage ratios.

voltage-to-frequency converter [abbreviated V/F converter] A converter in which the output frequency is a function of some reference or control signal. One version accepts an analog input from a sensor, transducer, or other analog device and generates a train of digital output pulses at a rate directly proportional to the instantaneous amplitude of the input signal. This digital output can be fed into a computer for process control or other applications.

voltage transformer An instrument transformer whose primary winding is connected in parallel with a circuit in which the voltage is to be measured or controlled. Also called potential transformer.

voltage-tunable tube An oscillator tube whose operating frequency can be varied by changing one or more of the electrode voltages, as in a backward-wave magnetron.

voltage-type telemeter A telemeter that employs the magnitude of a single voltage as the translating means.

voltage-variable capacitor *Varactor.*

voltaic cell A primary cell that consists of two dissimilar metal electrodes in a solution which acts chemically on one or both of them to produce a voltage.

voltaic pile An early form of primary battery; it consists of a pile of alternate pairs of dissimilar metal disks, with moistened pads between pairs.

voltammeter An instrument that may be used as either a voltmeter or ammeter.

voltampere [abbreviated VA] The unit of apparent power in an AC circuit that contains reactance. Apparent power is equal to the voltage in volts multiplied by the current in amperes, without taking phase into consideration.

voltampere-hour reactive [abbreviated varh] The unit of the integral of reactive power over time in the International System of Units (SI), equal to a reactive power of 1 var integrated over 1 h. Also called reactive voltampere-hour.

voltampere meter An instrument that measures the apparent power in an AC circuit.

voltampere reactive [symbol var] The unit of reactive power in the International System of Units (SI), as adopted by the International Electrotechnical Commission in 1930. Also called reactive voltampere.

Volta's law The contact voltage developed between two dissimilar conductors is the same whether the contact is direct or through one or more intermediate conductors.

voltmeter An instrument that measures voltage. Its scale may be calibrated in volts or related smaller or larger units. A voltmeter that indicates millivolt values is often called a millivoltmeter. Similarly, a microvoltmeter indicates microvolt values, and a kilovoltmeter indicates kilovolt values.

voltmeter-ammeter A voltmeter and an ammeter combined in a single case but with separate terminals.

voltmeter sensitivity The ratio of the total resistance of a voltmeter to its full-scale reading in volts, expressed in ohms per volt.

volt-ohm-milliammeter [abbreviated VOM] A test instrument that has a number of different ranges for measuring voltage, current, and resistance. Also called multimeter.

volts AC [abbreviated VAC] The AC voltage in volts. Usage and common practice determine the exact meaning of the rating.

volts DC [abbreviated VDC] The DC voltage in volts. Usage and common practice determine the exact meaning of the rating. Also called DC working volts (deprecated) and working volts DC (deprecated).

volts root-mean-square [abbreviated VRMS] The root-mean-square voltage rating in volts.

volume 1. The magnitude of a complex AF current as measured in volume units (VU) on a standard volume indicator. 2. The intensity of a sound.

volume acoustic wave *Bulk acoustic wave.*

volume compressor An AF control circuit that limits the volume range of a radio program at the transmitter, to permit using a higher average percent modulation without risk of overmodulation. Also used when making disk recordings, to permit a closer groove spacing without overcutting. Also called automatic volume compressor.

volume control A potentiometer that varies the loudness of a reproduced sound by varying the AF signal voltage at the input of the audio amplifier.

volume dose *Integral dose.*

volume expander An AF control circuit sometimes used to increase the volume range of a radio program or recording by making weak sounds weaker and loud sounds louder. The expander counteracts volume compression at the transmitter or recording studio. Also called automatic volume expander.

volume indicator A standardized instrument for indicating the volume of a complex electric wave such as that corresponding to speech or music. The reading in volume units (VU) is equal to the number of decibels above a reference level. The sensitivity is adjusted so the reference level of 0 VU is indicated when the instrument is connected across a 600-Ω resistor that is dissipating a power of 1 mW at 1 kHz.

volume ionization The average ionization density in a given volume of ionizing particles.

volume lifetime The average time interval between the generation and recombination of minority carriers in a homogeneous semiconductor.

volume-limiting amplifier An amplifier that contains an automatic device which functions only when the input volume exceeds a predetermined level and then reduces the gain in such a manner that the output volume stays substantially constant despite further increases in input volume. The normal gain of the amplifier is restored when the input volume returns below the predetermined limiting level.

volume magnetostriction The change in the volume of a magnetostrictive material when subjected to a magnetic field.

volume recombination Recombination of positive and negative ions at low energies throughout the volume of an ionization chamber, or recombination of free electrons and holes in the volume of a semiconductor.

volume recombination rate The time rate at which free electrons and holes recombine within the volume of a semiconductor.

volumetric radar A radar capable of providing three-dimensional position data for a multiplicity of targets.

volume unit [abbreviated VU] A unit used to specify the AF power level in decibels above a reference level of 1 mW (0.001 W), as measured with a standard volume indicator. A volume unit is equal to a decibel only when changes in power are involved or when the decibel value has this same reference level. It is unnecessary to specify the reference level when dealing in volume units because the level is a part of the definition.

volume-unit meter [abbreviated VU meter] A meter calibrated to read AF power levels directly in volume units.

volume velocity The rate of flow of a medium through a specified area due to a sound wave.

VOM Abbreviation for *volt-ohm-milliammeter.*

VOR Abbreviation for *VHF omnirange.*

VORDAC [VHF Omnidirectional Range and Distance-measuring equipment for Area Coverage] A precision air navigation system for high-density air traffic routes; it consists of standard distance-measuring equipment and a high-accuracy VHF omnidirectional range.

VOR receiver An aircraft radio receiver that receives signals from a VHF omnirange and interprets them for the aircraft crew.

vortac [VHF Omnidirectional Range TACan] An air navigation system in which civilian VHF omnirange facilities are used for directional guidance of aircraft, and military UHF tacan is used for measuring distance.

vowel articulation The percent articulation obtained for vowels, usually combined with consonants into meaningless syllables.

VOX Abbreviation for *voice-operated transmission.*

VRMS Abbreviation for *volts root-mean-square.*

VR tube *Voltage-regulator tube.*

VSB Abbreviation for *vestigial sideband.*

VSC Abbreviation for *variable speech control.*

VSTOL aircraft [Vertical and Short TakeOff and Landing aircraft] An aircraft that needs little or no runway length.

VSWR Abbreviation for *voltage standing-wave ratio.*

VSWR meter Abbreviation for *voltage standing-wave ratio meter.*

VT fuze Abbreviation for *variable-time fuze.*

VTOL aircraft [Vertical TakeOff and Landing aircraft] An aircraft capable of taking off and landing vertically, then changing to high-speed level flight.

VTR Abbreviation for *videotape recording.*

VTVM Abbreviation for *vacuum-tube voltmeter.*

VU Abbreviation for *volume unit.*

vulcanized fibre A laminated plastic made by chemically treating layers of 100% rag-content paper to gelatinize the paper and fuse the layers into a solid mass. When dried under pressure, it forms a hard, tough material that has good electrical properties along with mechanical strength and dimensional stability. The British spelling "fibre" is commonly used for this product because of its British origin.

VU meter Abbreviation for *volume-unit meter.*

W Abbreviation for *watt.*

wafer A thin polished slice of a semiconductor crystal on which integrated circuits or individual semiconductor devices can be fabricated, often in duplicate, for subsequent cutting into individual dice.

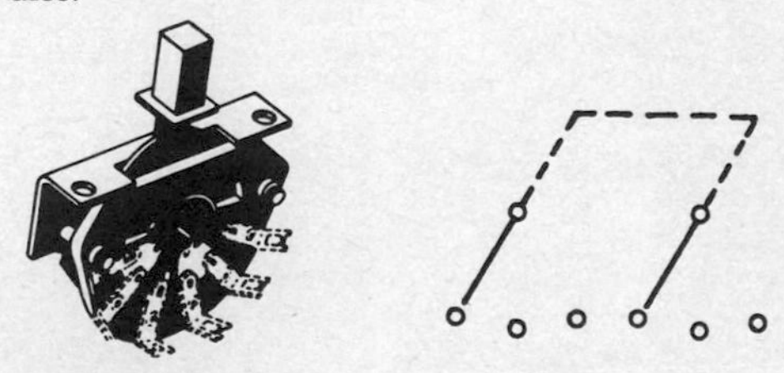

Wafer lever switch and symbol.

wafer lever switch A lever switch in which a number of contacts are arranged on one or both sides of one or more wafers, for engaging with one or more contacts on a movable wafer segment actuated by the operating lever.

wafer socket An electron-tube socket that consists of one or two wafers of insulating material; it has holes in which are spring metal clips that grip the terminal pins of a tube.

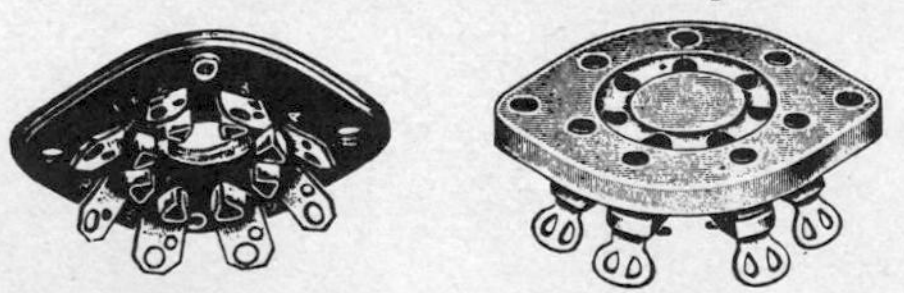

Wafer sockets made from laminated plastic (left) and molded steatite (right).

waffle-iron store A thin-film magnetic memory in which the driving and sense conductors are placed in grooves in a ferrite plate. The flux paths are completed by placing over the assembly a thin square-loop magnetic film electroplated on a polished copper substrate. Storage occurs in the film at each crossing where word and digit currents flow simultaneously. Readout is obtained by reversing word current polarity.

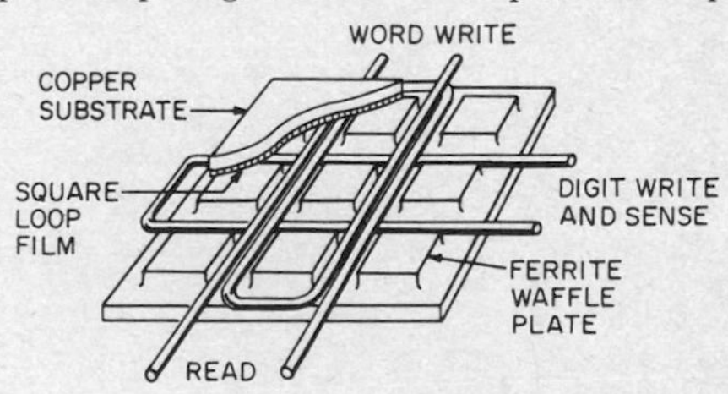

Waffle-iron store.

Wagner ground A ground connection used with an AC bridge to minimize stray capacitance errors when measuring high impedances. A potentiometer is connected across the bridge supply oscillator, with its movable tap grounded.

walkie-talkie A compact portable combination radio transmitter and receiver that can be carried by one person, usually strapped over the back, and used for communication over medium distances.

wall absorption The decrease in beta- or gamma-ray output due to absorption in the radioactive material itself.

wall effect The contribution to the ionization in an ionization chamber by electrons liberated from the walls.

wall outlet An outlet mounted on a wall, from which electric power can be obtained by inserting the plug of a line cord.

wander *Scintillation.*

warble-tone generator An AF signal generator whose frequency is varied cyclically at a subaudio rate over a fixed range.

warhead The section of a bomb, guided missile, ballistic missile, torpedo, or other missile that contains the explosive, chemical, or other charge intended to damage the enemy.

warla [Wide Aperture Radio Location Array] A directional antenna array that uses frequency-independent log-periodic elements for long-distance radio transmission and reception. Frequency range is 2 to 32 MHz.

watch The service performed by a qualified operator when on duty in the radio room of a vessel. Also called radio watch.

water-activated battery A primary battery that contains the electrolyte but requires the addition of or immersion in water before it is usable.

water calorimeter A calorimeter that measures RF power in terms of the rise in temperature of water in which the RF energy is absorbed.

water-cooled tube An electron tube that is cooled by circulating water through or around the anode structure.

water load A matched waveguide termination in which the electromagnetic energy is absorbed in

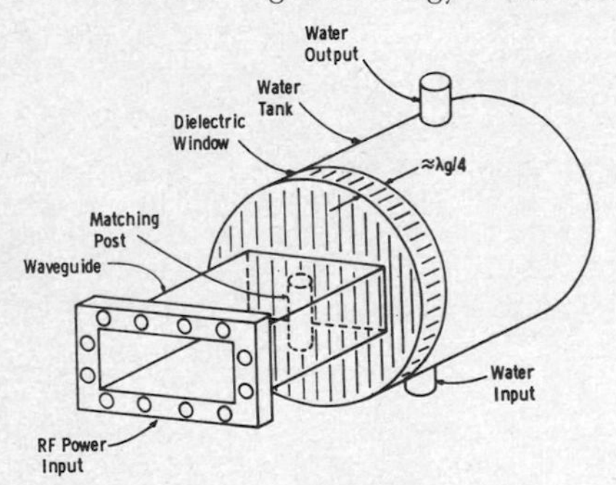

Water load serving as waveguide termination. Also used to dissipate unwanted power reaching end of waveguide.

water. The resulting rise in the temperature of the water is a measure of the output power.

water monitor A monitoring device that detects and measures waterborne radioactivity.

WATS [Wide-Area Telephone Service] A special telephone service that allows a customer to call anyone in one or more regions into which the continental United States has been divided, on a direct dialing basis, for a flat monthly charge related to the number of regions to be called.

watt [abbreviated W] The SI unit of electric power. In a DC circuit, the power in watts is equal to volts multiplied by amperes. In an AC circuit, the true power in watts is effective volts multiplied by effective amperes, then multiplied by the circuit power factor. There are 746 W in 1 horsepower. The term is named after James Watt, Scottish inventor.

wattage rating A rating that expresses the maximum power which a device can safely handle continuously.

watthour [abbreviated Wh] The practical unit of electric energy, equal to a power of 1 W absorbed continuously for 1 h. One kilowatthour is equal to 1000 Wh.

watthour meter A meter that measures and registers the integral, with respect to time, of the

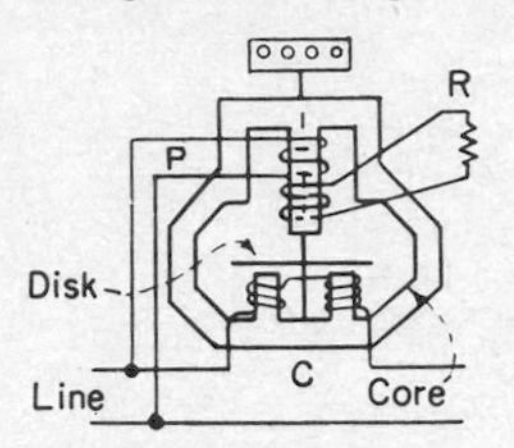

Watthour meter consisting of potential winding P on iron core, with current windings C. Armature is aluminum disk with shaft behind core. Interaction of flux produced by three poles with induced eddy currents in disk causes rotation proportional to power. Revolutions counted by register at top represent watthours.

active power of the circuit in which it is connected. The unit of measurement is usually the kilowatthour.

wattless power *Reactive power.*

wattmeter A meter that measures electric power in watts.

watt per steradian [abbreviated W/sr] The SI unit of radiant intensity.

wattsecond [abbreviated Ws] The amount of electric energy corresponding to 1 W acting for 1 s. One wattsecond is equal to 1 J.

wave A propagated disturbance for which the intensity at any point in the medium is a function of time, and the intensity at a given instant is a function of the position of the point. A wave may be electric, electromagnetic, acoustic, mechanical, or any other form.

wave amplitude The magnitude of the maximum change from zero of a characteristic of a wave.

wave analyzer An instrument that measures the amplitude and frequency of the various components of a complex current or voltage wave.

wave angle The angle in azimuth and elevation at which a radio wave arrives at a receiving antenna or leaves a transmitting antenna.

wave antenna A directional antenna composed of a system of parallel horizontal conductors from one half-wavelength to several wavelengths long, with the far ends terminated to ground by the characteristic impedance of the antenna. Also called Beverage antenna.

waveband A band of frequencies, such as that assigned to a particular radio communication service.

waveband switch A multiposition switch that changes the tuning range of a receiver or transmitter from one waveband to another.

wave clutter *Sea clutter.*

wave converter A converter that changes one type of wave to another in a waveguide.

wave equation An equation that describes a particular wave motion through a medium.

wave filter A transducer that separates waves on the basis of their frequency. It introduces relatively small insertion loss to waves in one or more frequency bands and relatively large insertion loss to waves of other frequencies.

waveform The shape of a wave, as obtained by plotting a characteristic of the wave with respect to time.

waveform-amplitude distortion *Frequency distortion.*

waveform analyzer A frequency-selective voltmeter that measures the amplitude and frequency of each component of a complex waveform.

waveform monitor A cathode-ray oscilloscope that has a time base suitable for viewing the waveform of the video signal in a television system. Also called A scope.

waveform synthesizer A signal generator whose output is variable as to frequency, phase, harmonic content, and harmonic amplitude.

wavefront The portion of a wave envelope that is between the virtual zero point and the point at which the wave reaches its crest value, as measured in either time or distance.

wave function A set of solutions to Maxwell's equations for wave propagation in a homogeneous isotropic region.

wave group The resultant of two or more wave trains of different frequency traversing the same path.

waveguide A rectangular or circular metal pipe that has a predetermined cross section, specifically designed to guide or conduct high-frequency electromagnetic waves through its interior, or any other equivalent system of material boundaries capable of guiding waves. Also called guide.

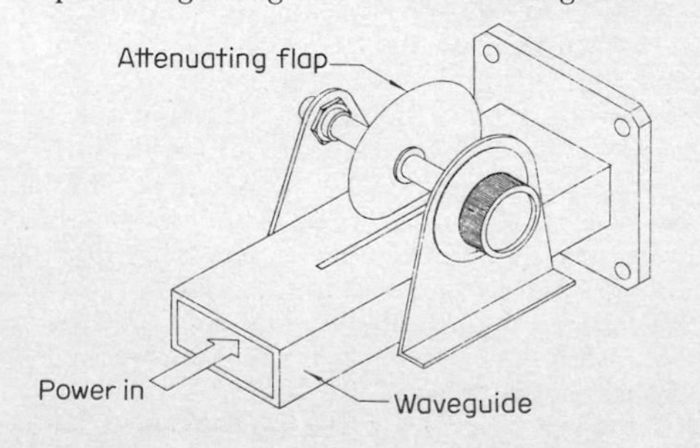

Waveguide attenuator.

waveguide attenuator An attenuator used in a waveguide to produce attenuation by any means, including absorption and reflection.

waveguide bend A section of waveguide in which the direction of the longitudinal axis is changed. An E-plane bend in a rectangular waveguide is bent along the narrow dimension, and an H-plane bend is bent along the wide dimension. Also called waveguide elbow.

waveguide component A device connected at specified ports in a waveguide system.

waveguide connector A mechanical device for electrically joining and locking together separable mating parts of a waveguide system. Also called waveguide coupling.

waveguide coupling *Waveguide connector.*

waveguide cutoff wavelength The wavelength that corresponds to the cutoff frequency of a waveguide. Below this frequency the attenuation rises rapidly.

waveguide directional coupler A directional coupler that consists of two parallel waveguides which have a common wall into which two slots are cut. These permit extraction of a portion of the RF energy traveling in one direction through the main waveguide while rejecting energy traveling in the opposite direction.

waveguide elbow *Waveguide bend.*

waveguide filter A filter made up of waveguide components, used to change the amplitude-frequency response characteristic of a waveguide system.

waveguide flange *Flange.*

waveguide gasket A gasket that maintains electric continuity between two mating sections of waveguide.

waveguide junction *Junction.*

waveguide lens A microwave lens in which the required phase changes result from transmission through suitable waveguide elements.

waveguide mode suppressor A waveguide filter that suppresses undesired modes of propagation in a waveguide.

waveguide phase shifter A device that adjusts the phase of the output signal of a waveguide system with respect to the phase of the input signal.

waveguide plunger *Piston.*

waveguide post A cylindrical rod placed in a transverse plane of a waveguide to serve substantially as a shunt susceptance. Also called post.

waveguide probe *Probe.*

waveguide radiator An open-ended waveguide, with or without a flaring horn, that radiates electromagnetic energy to a reflector or out into space.

waveguide resonator *Cavity resonator.*

waveguide seal A seal over the end of a waveguide to prevent entrance of moisture without appreciably attenuating radio frequencies.

waveguide shim A thin, resilient, metal sheet inserted between waveguide components to ensure electric continuity.

waveguide shutter A waveguide section that contains an adjustable mechanical barrier which can be set to block or divert RF energy.

waveguide slug tuner A quarter-wavelength dielectric slug that projects into a waveguide for tuning purposes. It is usually adjustable as to position and depth of penetration.

waveguide stub An auxiliary section of waveguide that has an essentially nondissipative termination, joined at some angle with a main section of waveguide.

waveguide stub tuner A waveguide tuning or detuning device that consists of an adjustable piston mounted in a waveguide stub.

waveguide switch A switch for mechanically positioning a waveguide section, to couple it to one of several other sections in a waveguide system.

waveguide taper A section of tapered waveguide.

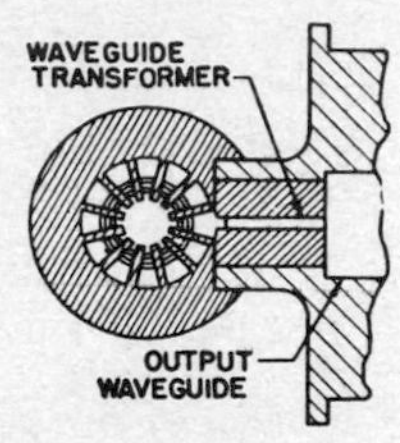

Waveguide transformer used between magnetron and output waveguide.

waveguide transformer A waveguide component that provides impedance transformation.

waveguide tuner An adjustable tuner that is used in a waveguide system to provide impedance transformation.

waveguide twist A waveguide section in which there is progressive rotation of the cross section about the longitudinal axis.

waveguide wavelength The distance along a uniform waveguide, at a given frequency and for a given mode, between similar points at which a signal component differs in phase by 2π radians.

waveguide window A thin conducting metal window placed transversely in a waveguide for impedance matching. For an inductive window, the edges of the slit in the window are parallel to the electric field in the lowest mode in the waveguide. For a capacitive window, the edges of the slit are perpendicular to the electric field.

wave heating Heating of a material by energy absorption from a traveling electromagnetic wave.

wave impedance The ratio of the transverse electric field to the transverse magnetic field in a waveguide.

wave interference The variation of wave amplitude with distance or time, caused by the superposition of two or more waves that have the same or nearly the same frequency.

wave-interference microphone A highly directional microphone that is coupled to the sound field over an area and responds to the sum of the pressures over this area. Chief drawbacks are variation of polar response with frequency and its large size. Examples include line microphones and microphones that use parabolic reflectors.

wavelength The distance between points that have corresponding phase in two consecutive cycles of a periodic wave. The wavelength in meters is approximately equal to 300 divided by the frequency in megahertz. Wavelengths of light are specified in micrometers [1 μm (formerly 1 μ) = 10^{-6} m = 1000 nm = 10,000 Å]. Multiply angstroms by 0.1 to get nanometers, or by 0.0001 to get micrometers.

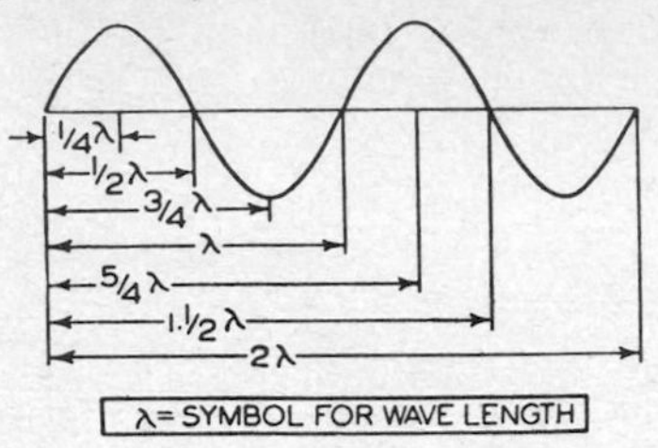

Wavelength values for sinusoidal wave.

wavelength shifter A photofluorescent compound used with a scintillator material to increase the wavelengths of the optical photons, thereby permitting more efficient use of the photons by the phototube or photocell.

wave mechanics A theory that assigns wave characteristics to the components of atomic structure and seeks to interpret physical phenomena in terms of hypothetical waveforms. Introduced by Schroedinger in 1926.

wavemeter An instrument that measures the wavelength of an RF wave. Since wavelength is related to frequency, a wavemeter also serves as a frequency meter.

wave normal A unit vector that is perpendicular to an equiphase surface, with its positive direction taken on the same side of the surface as the direction of propagation. In isotropic media, the wave normal is in the direction of propagation.

wave packet A wave function that describes a localized particle whose position is known within fairly narrow limits.

wave propagation *Propagation.*

wave soldering *Flow soldering.*

wave tail The portion of a wave envelope that is between the crest and the end of the envelope.

wave tilt The forward inclination of the waveform of radio waves arriving along the ground. Its value depends on the electric constants of the ground.

wave train A series of wave cycles produced by the same disturbance.

wave trap A resonant circuit connected to the

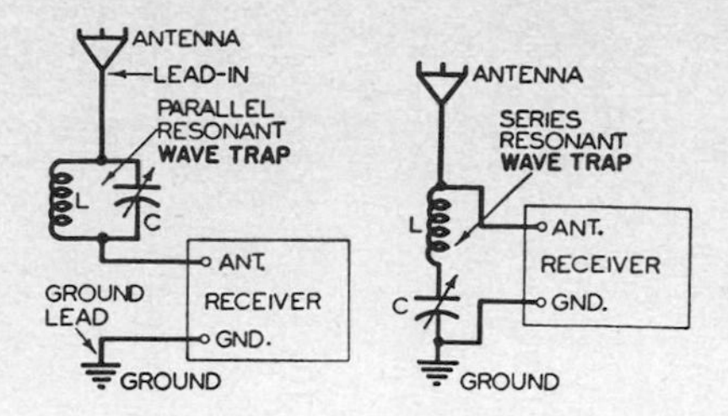

Wave-trap connections.

antenna system of a receiver to suppress signals at a particular frequency, such as that of a powerful local station which is interfering with reception of other stations. Also called trap.

wave trough The minimum value of the envelope of a progressive wave.

wave-type microphone A microphone that depends on wave interference for its directivity.

wave velocity The velocity of propagation of an electromagnetic wave, equal to 3×10^{10} cm/s in free space. In a waveguide, the rate of energy transfer is called the group velocity and is less than the wave velocity, whereas the velocity of the electric wave is called the phase velocity and may be greater than the wave velocity.

wax A blend of waxes with metallic soaps, used in mechanical recording.

wax master Deprecated term for *wax original.*

wax original An original recording made on a wax surface and used to make a master. Also called wax master (deprecated).

way point A selected point on a radio navigation course line that has some particular significance. Also called check point.

Wb Abbreviation for *weber.*

weak coupling *Loose coupling.*

weak interaction The interaction responsible for beta, pi-meson, and K-meson decays and the decays of the lambda, sigma, and xi particles. These interactions are many orders of magnitude weaker than strong or electromagnetic interactions.

weapon system The complete system required to deliver a weapon to its target, including production, storage, transport, launchers, aircraft, and guidance equipment.

weather radar Radar capable of detecting echoes from clouds or rain.

weather transmitting set The components and items required to operate a complete electronic set for automatically measuring and transmitting weather data by radio.

weber [abbreviated Wb] The SI unit of magnetic flux. It is the amount of flux that, when linked with a single turn of wire, will induce 1 V in the turn as it decreases uniformly to zero in 1 s. One weber is equal to 10^8 maxwells.

wedge 1. A waveguide termination that consists of a tapered length of dissipative material introduced into the guide, such as carbon. 2. A convergent pattern of equally spaced black and white lines, used in a television test pattern to indicate resolution. 3. An optical filter in which the transmission decreases continuously or in steps from one end to the other.

wedge-base lamp A small indicator lamp that has wire leads folded back on opposite sides of a

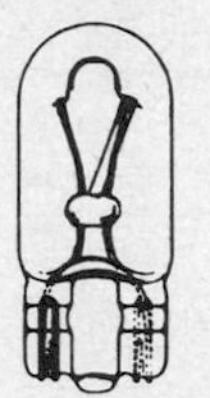

Wedge-base lamps.

flat glass base. Designed for pushing into a socket that has wedge-shaped spring contacts.

wedge bonding A type of thermocompression bonding in which a wedge-shaped tool is used to press a small section of the lead wire onto the bonding pad of an integrated circuit. The bond may be a cold, ultrasonic, or thermocompression bond.

wedge filter A radiation filter so constructed that its thickness or transmission characteristics vary continuously or in steps from one edge to the other. Used to increase the uniformity of radiation in certain types of treatment.

wedge spectrograph A spectrograph in which the density of the radiation passing through the entrance slit is varied by moving an optical wedge.

Wehnelt cathode *Oxide-coated cathode.*

Wehnelt cylinder The metal tube that encloses the cathode of a cathode-ray tube and concentrates the electrons emitted in all directions from the cathode.

weighted noise level The noise level weighted in accordance with the 70-dB equal-loudness contour of the human ear, expressed in decibels above 1 mW.

weighting The artificial adjustment of measurements to account for factors that, in the normal use of the device, would otherwise be different from conditions during the measurements. As an example, background noise measurements may be weighted by applying factors or by introducing networks to reduce measured values in inverse ratio to their interfering effects.

weighting function A measure of the relative effect on reactivity of localized changes in nuclear properties as a function of position and property change.

weightlessness A condition in which no acceleration, whether of gravity or other force, can be detected by an observer within the system in question. An unaccelerated satellite orbiting the earth is "weightless," even though gravity affects its orbit. Weightlessness can be produced within the atmosphere in aircraft that is flying a parabolic flight path.

weightlessness switch *Zero-gravity switch.*

Weissenberg method An x-ray crystal analysis method in which the crystal is rotated in the beam of x-rays and the film is moved parallel to the axis of rotation. The crystal is surrounded by a sleeve that has a slot which passes a line-shaped beam of x-rays, to give positive identification of each spot or line on the pattern.

welding A process of joining metals by the application of heat, pressure, or both.

welding current The current that is sent through a joint to produce the heat needed to make a weld.

welding cycle The complete series of events involved in making a weld.

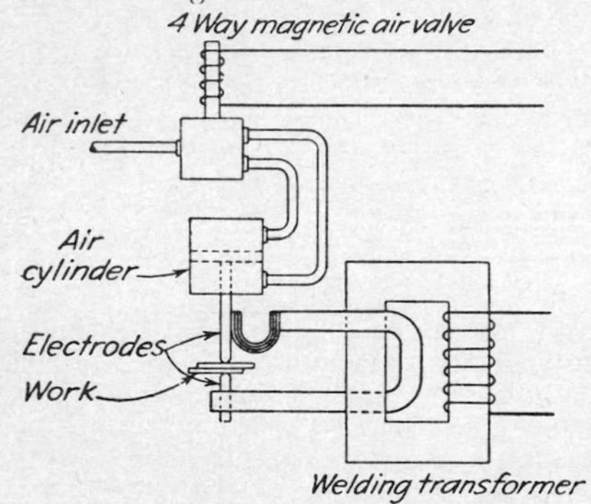

Welding transformer as used in resistance welder.

welding transformer A high-current, low-voltage power transformer used to supply current for welding.

weld interval The total of all heating and cooling times when making a single multiple-impulse weld by resistance welding.

weld-interval timer A timer that controls the heating and cooling intervals when multiple-impulse welds are being made.

weld time The time that welding current flows through the work when a weld is being made.

weld timer A timer that controls only the weld time.

well counter A radiation counter that has a heavy tubular lead shield closed at one end, in which the radiation detector and the radioactive sample are inserted to reduce the effect of background radiation. Generally used with scintillation counters that have large crystals, for counting beta particles.

Wertheim effect When a ferromagnetic wire is twisted in a longitudinal magnetic field, a voltage is produced between the ends of the wire.

Western Union joint A joint or splice that has good mechanical strength as well as good conductivity, made by crossing the cleaned ends of two

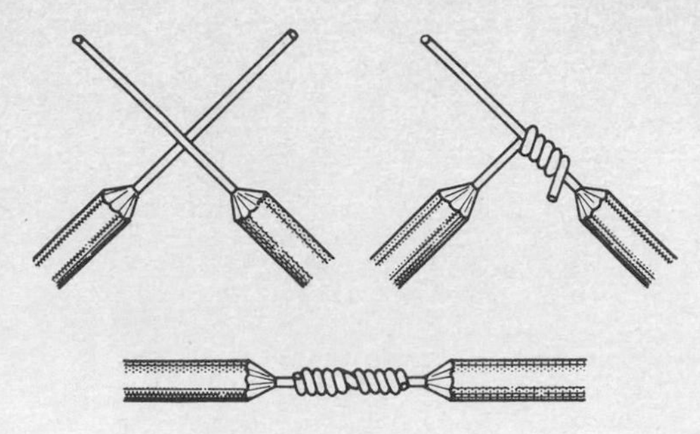

Western Union joint.

wires and then winding the end of each wire around the other wire and soldering the joint.

Weston standard cell A standard cell used as a highly accurate voltage source for calibrating purposes. The positive electrode is mercury, the negative electrode is cadmium, and the electrolyte is a saturated cadmium sulfate solution. The Weston standard cell has a voltage of 1.018636 V at 20°C.

wet cell A cell whose electrolyte is in liquid form.

wet electrolytic capacitor An electrolytic capacitor that uses a liquid electrolyte.

wetting The coating of a contact surface with an adherent film of mercury.

wetting agent A substance that decreases the surface tension of a liquid, to make the liquid spread and adhere better.

Wh Abbreviation for *watthour.*

Wheatstone bridge A four-arm bridge, all arms of which are predominantly resistive. Used for

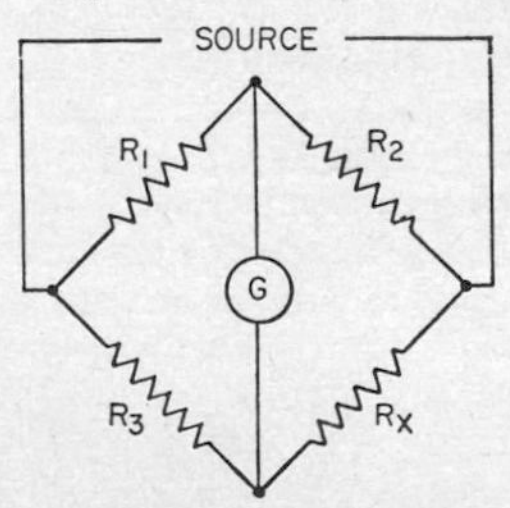

Wheatstone-bridge circuit.

measuring resistance. Also called resistance bridge.

wheel printer A line printer that prints its characters from the rim of a wheel around which is the type for the alphabet, numerals, and other characters.

wheel static Interference encountered in auto-radio installations because of static electricity developed by friction between the tires and the street.

whip antenna A flexible vertical rod antenna, used chiefly on vehicles. Also called fishpole antenna.

whisker 1. An extremely fine single-crystal filament of a metal or inorganic compound. The two main production methods are reduction of metal halides in hydrogen at about 700°C to initiate crystal growth and reaction of molten metal vapor with an oxidizing, nitriding, or other atmosphere conducive to crystal growth. Undesired whisker growth in hot and humid environments causes shorting between conductors in electronic equipment. Whiskers have much greater tensile strength and higher elasticity than the corresponding bulk form of the metal. 2. *Catwhisker.*

whistler An effect caused when an electric disturbance produced by a lightning discharge travels out along lines of magnetic force of the earth's field and is reflected back to its origin from a magnetically conjugate point on the earth's surface. The characteristic drawn-out descending pitch of the whistler is a dispersion effect caused by the greater velocity of the higher-frequency components of the disturbance. Radio signals can be transmitted along whistler paths from the northern to the southern hemisphere.

white *Picture white.*

White Alice *Alice.*

white compression The reduction in picture-signal gain at levels corresponding to light areas, with respect to the gain at the level for midrange light values. The overall effect of white compression is to reduce contrast in the highlights of the picture. Also called white saturation.

white level The carrier signal level that corresponds to maximum picture brightness in television and facsimile.

white light Any radiation that produces the same color sensation as average noon sunlight.

white noise Random acoustic or electric noise that has equal energy per cycle over a specified total frequency band. The electric disturbance caused by random movements of free electrons in a conductor or semiconductor is one example; another is the frequency spectrum of white light.

white-noise record A phonograph record used to test the frequency response of audio reproduction systems. A recording of acoustic white noise extending over the entire band of audio frequencies is reduced in bandwidth progressively in steps as the top limit of the bandwidth at each step is announced.

white object An object that reflects all wavelengths of light with substantially equal high efficiencies and considerable diffusion.

white peak A peak excursion of the picture signal in the white direction.

white recording A form of amplitude-modulation recording in which the maximum received power corresponds to the minimum density of the record medium in a facsimile system. In a frequency-modulated white recording the lowest received frequency corresponds to the minimum density of the record medium.

white room *Clean room.*

white saturation *White compression.*

white signal The signal produced at any point in a facsimile system by scanning a minimum-density area of the subject copy.

white-to-black amplitude range The ratio of signal voltage for picture white to signal voltage for picture black at a given point in a facsimile system that uses positive amplitude modulation. Generally expressed in decibels. For negative amplitude modulation the reverse ratio, of black to white, is used.

white-to-black frequency swing The difference between the signal frequencies corresponding to picture white and picture black at any point in a facsimile system that uses frequency modulation.

white transmission Amplitude-modulated transmission in which maximum transmitted power corresponds to minimum density of the subject copy, or frequency-modulated transmission in which the lowest transmitted frequency corresponds to the minimum density of the subject copy in a facsimile system.

whole-body counter A radiation counter that directly measures radioactivity in the entire human body.

whole-body scanner A device that uses electronic sensors instead of photographic film to detect x-rays beamed through the entire body of a patient. The sensor output signals are converted by a computer to a cathode-ray image or a printout of the organs scanned.

whole step *Whole tone.*

whole tone The interval between two sounds whose basic frequency ratio is approximately equal to the sixth root of 2. Also called whole step.

wicking The flow of solder up under the insulation of covered wire.

wide-angle lens An optical lens that has a large angular field, generally greater than 80°.

wideband amplifier *Broadband amplifier.*

wideband axis The direction of the phasor that represents the fine chrominance primary (the I signal) in color television.

wideband dipole A dipole that has a low ratio of length to diameter, to give resonance over a relatively wide frequency band.

wideband ratio The ratio of the occupied frequency bandwidth to the intelligence bandwidth in a multiplex system.

wide-open receiver A receiver that has essentially no tuned circuits, to receive all frequencies simultaneously in the band of coverage. An example is the tail-end radar detector used in some military aircraft. When a wide-open receiver is used for electromagnetic reconnaissance, the receiver output is recorded on magnetic tape together with a baseband reference signal. The tape must then be analyzed to isolate each frequency component recorded.

width 1. The horizontal dimension of a television or facsimile picture. 2. The time duration of a pulse.

width control The control that adjusts the width of the pattern on the screen of a cathode-ray tube in a television receiver or oscilloscope.

Wiedemann effect The twist produced in a current-carrying wire when the wire is placed in a longitudinal magnetic field.

Wiedemann-Franz law The ratio of the thermal conductivity to the electric conductivity is proportional to the absolute temperature for all metals.

Wiegand effect The ability of a mechanically stressed ferromagnetic wire to recognize rapid switching of magnetization when subjected to a DC magnetic field.

Wien bridge A four-arm AC bridge used to measure capacitance or inductance in terms of resistance and frequency. Bridge balance depends on frequency.

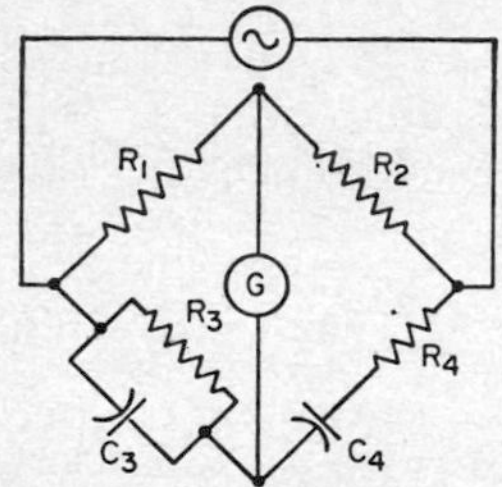

Wien bridge for measuring capacitance.

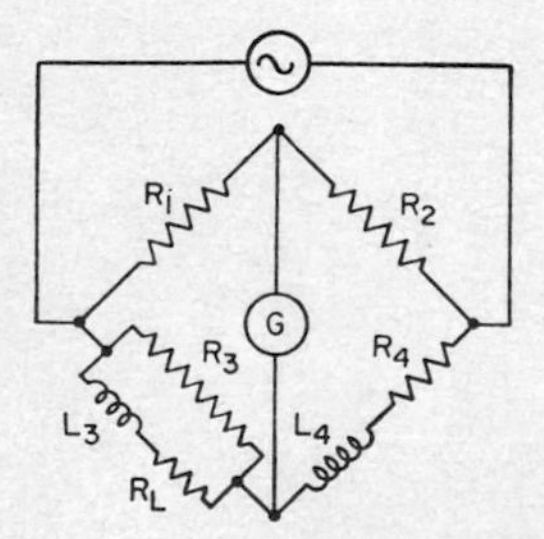

Wien bridge for measuring inductance.

Wien-bridge oscillator A phase-shift feedback oscillator that uses a Wien bridge as the frequency-determining element.

Wien's displacement law The wavelength of the peak radiation is inversely proportional to the absolute temperature of a blackbody. As temperature rises, the peak of the spectral energy distribution curve is shifted toward the short-wavelength end of the spectrum.

willemite A natural fluorescent mineral that consists chiefly of zinc orthosilicate, used for screens of cathode-ray tubes.

Williams tube A cathode-ray storage tube in which information is stored as a pattern of electric charges produced, maintained, read, and erased by suitably controlled scanning of the screen by the electron beam.

Wilson chamber A cloud chamber that contains air supersaturated with water vapor by sudden expansion, in which rapidly moving nuclear particles such as alpha or beta rays produce ionization tracks by condensation of vapor on the ions produced by the rays. These tracks may be observed or photographed through a suitable window.

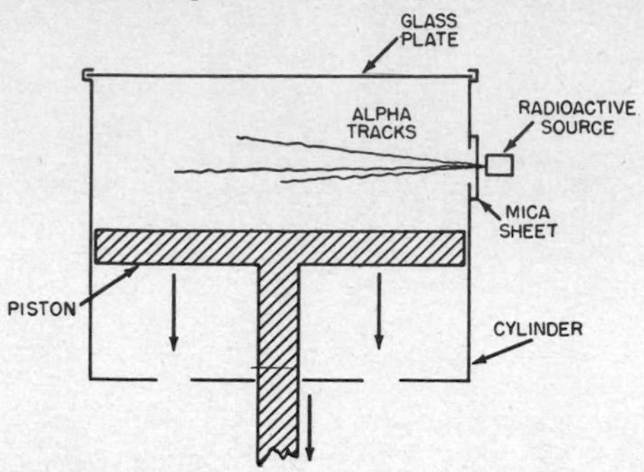

Wilson chamber, showing wiggly straight tracks of alpha rays.

Wimshurst machine An electrostatic generator that consists of two glass disks rotating in opposite directions and having sectors of tinfoil and collecting combs so arranged that static electricity is produced for charging Leyden jars or discharging across a gap.

wind The manner in which magnetic tape is wound onto a reel. In an *A* wind, the coated surface faces the hub. In a *B* wind, the coated surface faces away from the hub.

windcharger A wind-driven DC generator used for charging storage batteries.

wind-driven generator A generator that derives its power from wind acting on its own propeller.

winding 1. One or more turns of wire forming a continuous coil for a transformer, relay, rotating machine, or other electric device. 2. A conductive path, usually of wire, that is inductively coupled to a magnetic storage core or cell.

Windom antenna A multiband transmitting antenna that provides satisfactory operation on even harmonics of its fundamental frequency, such as for operation on several harmonically related amateur bands. One version is a wire one half-wavelength long at the fundamental frequency, with a 300-Ω twin-line feeder connected about 35% off center.

window 1. A confusion reflector that consists of strips of chaff, wire, or bars cut to give resonance at expected enemy radar frequencies and dropped in clusters from aircraft or expelled from shells or rockets as a radar countermeasure. Called window because the first pieces used were the size of small panes of glass. Later it was found that strips worked just as well and took less foil. 2. A hole in a partition between two cavities or waveguides, used for coupling. 3. An aperture for the passage of particles or radiation in a nuclear reactor. 4. An energy range of relatively high transparency in the total neutron cross section of a material. Such windows arise from interference between potential and resonance scattering in elements of intermediate atomic weight and can be of importance in neutron shielding. 5. A material that has minimum absorption and reflection of radiant energy, sealed into the vacuum envelope of a microwave or other electron tube to permit passage of the desired radiation through the envelope to the output device. Alumina, sapphire, beryllia, and quartz are examples of windows currently being used for microwave energy, and still other materials are being used for infrared energy output. 6. Atmospheric radio window.

window comparator A comparator that detects signal voltages at two different levels by comparing them to fixed references.

wing spot generator A circuit that produces on the target of a radar G display a pair of wings whose size is inversely proportional to target range.

wiper A wiping contact that moves in sequence over a number of different stationary contacts, as in a selector switch or stepping relay.

wiping contact A switch or relay contact that moves laterally with a wiping motion after it touches a mating contact. Also called self-cleaning contact and sliding contact.

wire A single bare or insulated metallic conductor that has solid, stranded, or tinsel construction, to carry current in an electric circuit.

wire communication Communication over a wire, as distinguished from radio communication.

wired-program computer A computer in which the sequence of instructions that form the operating program is created by interconnection of wires on a removable control panel. If the program is permanently wired, without the interchangeable panel, the arrangement becomes a fixed-program computer.

wired radio The transmission of modulated RF carrier signals as currents flowing through wires, rather than as radio waves traveling through space. Telephone wires are sometimes used for this purpose, as in the British system of rediffusion.

wire gage A system of numerical designations of wire sizes. The American wire gage starts with 0000 as the largest size, going to 000, 00, 0, 1, 2, and up to 40 and beyond for the smallest sizes.

wire-grid lens antenna A high-frequency lens antenna that consists of two circular grids suspended one over the other, with the combination surrounded by a radial wire horn. Used for radio communication in the range of 3 to 30 MHz.

wire guidance Control of a guided missile by sending control signals over wires pulled by the missile. Used chiefly in simple short-range antitank missiles.

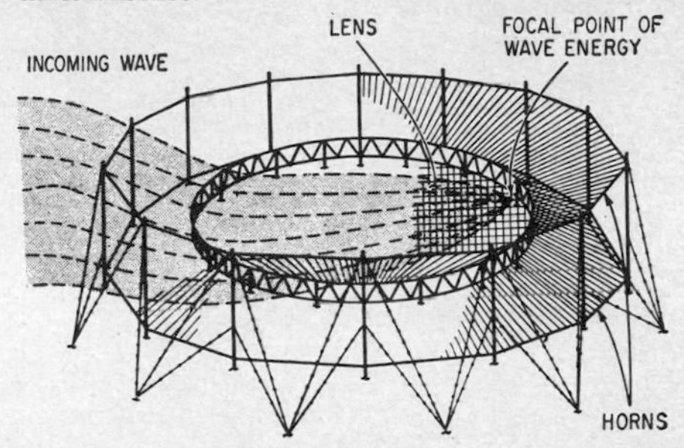

Wire-grid lens antenna, showing how radial wire horn concentrates waves in vertical direction, and two circular grids serve as lens for concentrating wave to focal point at coupler.

wireless British term for radio. Used in this country chiefly when the word radio might be misinterpreted, as in the term wireless record player.

wireless record player An electric phonograph connected to modulate an RF oscillator that feeds a small built-in antenna, used to broadcast a phonograph recording across a room or into another room of a home for reception and reproduction by a radio receiver which is tuned to the frequency at which the oscillator is operating.

wirephoto 1. A photograph transmitted over wires to a facsimile receiver. 2. *Facsimile.*

wire printer A high-speed printer that prints characterlike configurations of dots through the proper selection of wire ends from a matrix of wire ends.

wire recorder A magnetic recorder that utilizes a round stainless steel wire about 0.004 in (0.1 mm) in diameter instead of magnetic tape.

wiresonde An apparatus for gathering meteorological data at low altitudes, in which meteorological data such as temperature and humidity are transmitted over wire to ground-recording devices by sensing and sending apparatus carried aloft by a captive balloon.

wire stripper A hand-operated tool or special machine designed to cut and remove the insulation for a predetermined distance from the end of an insulated wire without damaging the solid or stranded wire inside.

wiretap A secretly made and concealed connection to a telephone line, office intercommunication line, or other wiring system, for monitoring conversations and activities in a room from a remote location without knowledge of the participants, legally or illegally.

wirewound resistor [symbol WW] A resistor that uses as the resistance element a length of high-resistance wire or ribbon, usually Nichrome, wound on an insulating form.

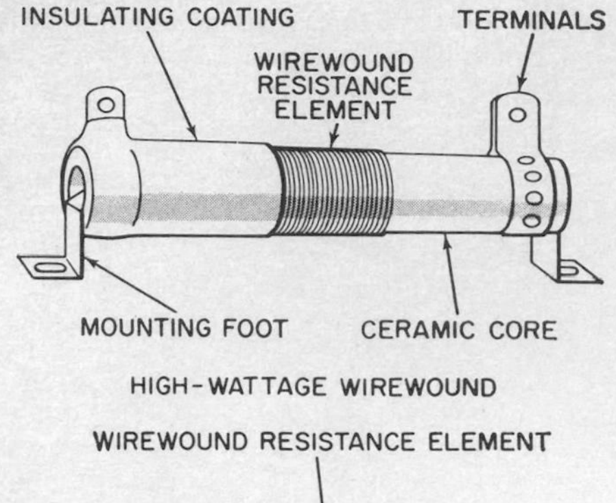

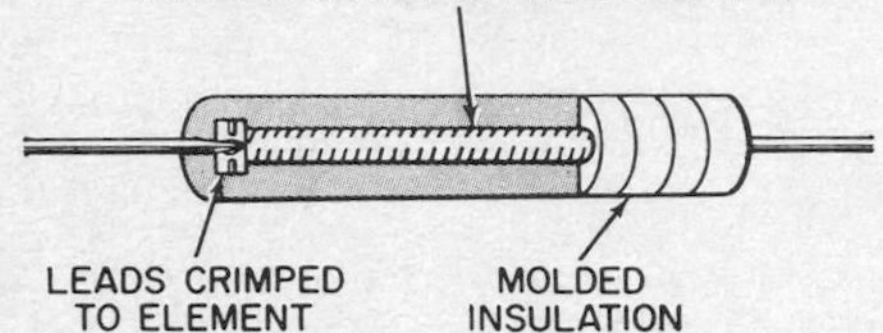

Wirewound resistors.

wirewound rheostat A rheostat in which a sliding or rolling contact moves over resistance wire that has been wound on an insulating core.

wire-wrap connection A solderless connection made by wrapping several turns of bare wire around a sharp-corner rectangular terminal un-

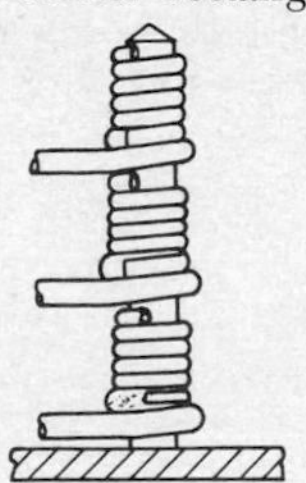

Wire-wrap connection.

der tension, using either a power tool or hand tool. Also called solderless wrapped connection and wrapped connection.

wire-wrap pin A rectangular sharp-cornered terminal, to which a connection can be made with a wire-wrap tool that winds the bare end of a wire around the terminal several times, under tension.

wiring harness An array of insulated conductors bound together by lacing cord, metal bands, or other binding, in an arrangement suitable for use only in specific equipment for which the harness was designed. It may include terminations.

wobble joint A radar waveguide joint in which the section coming from the transmitter is stationary and that going to the antenna can be wobbled in a desired manner by using a flexible section of waveguide, an airgap joint, or some other means of permitting movement between two waveguides

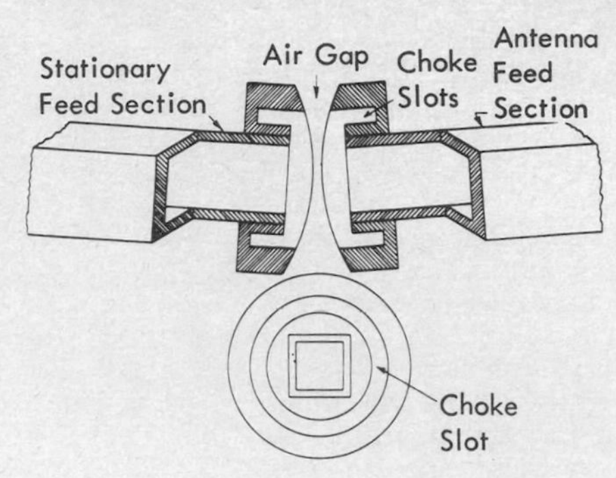

Wobble joint using choke flanges designed to permit frictionless movement of waveguides at air gap.

without introducing a mismatch.

wobbulator A signal generator in which a motor-driven variable capacitor varies the output frequency periodically between two known limits, as required for displaying a frequency-response curve on the screen of a cathode-ray oscilloscope.

wolfram Term usually used for tungsten outside the United States.

womp A sudden increase in brightness of a television screen, generally caused by a sudden increase in signal strength.

woofer A large loudspeaker that reproduces low audio frequencies at relatively high power levels.

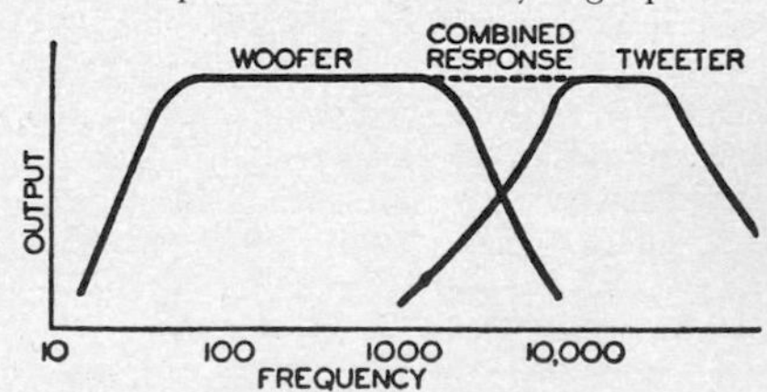

Woofer response curve, shown with tweeter curve and combined flat response of loudspeaker system.

Usually used in combination with a crossover network and a high-frequency loudspeaker called a tweeter.

word An ordered set of characters that is treated, stored, and transported by computer circuits as a unit. Word lengths may be either fixed or variable, depending on the computer.

word generator A special pulse generator that generates a programmable train of pulses (called a word) for testing data-processing and telemetry equipment.

word line The conductor that links all the bits for a given word in a magnetic memory.

word processing The creation, dissemination, storage, and retrieval of the written word by typewriter terminals that use magnetic tape for storage, automatic control, editing, and retyping. More elaborate systems use time-sharing computers.

word time *Minor cycle.*

work *Load.*

work coil *Load coil.*

work function The minimum energy needed to remove an electron from the Fermi level of a metal to infinity. Usually expressed in electronvolts.

working Carrying on radio communication with a station by telegraphy, telephony, or facsimile for a purpose other than calling.

working storage A portion of the internal storage of a computer that is reserved for the data upon which operations are being performed, including partial results.

working voltage *Voltage rating.* Working voltage is generally used in connection with DC voltage ratings for capacitors. The exact meaning of this rating in a particular usage should always be specified. The preferred abbreviation for the value of the voltage rating in volts is VDC; other abbreviations are DCWV (deprecated) and WVDC (deprecated).

working volts DC [abbreviated WVDC] Deprecated term for *volts DC.*

world geographic reference system [abbreviated georef] A geographic reference system used by the U.S. Air Force for aircraft position reports, target designations, and other tactical air operations.

world numbering plan The assignment, to each telephone main station in the world, of a unique number that has a maximum of 12 digits which represent a country code plus a national number, to be used for international direct dialing. The first digit is 1 for North America; Africa is 2; 3 and 4 are Europe; 5 is South America and Cuba; 6 is the South Pacific (Australasia); 7 is Union of Soviet Socialist Republics; 8 is the North Pacific (Eastern Asia); 9 is the Far East and Middle East; 0 is a spare.

woven-screen storage A digital storage plane made by weaving wires coated with thin magnetic films. When currents are sent through a selected pair of wires that are at right angles in the screen, storage and readout occur at the intersection of the two wires.

wow A low-frequency flutter. When caused by an off-center hole in a disk record, it occurs once per revolution of the turntable.

WPM Abbreviation for words per minute.

wrapped connection *Wire-wrap connection.*

Wratten filter A gelatin or glass filter that has specific light-transmission characteristics.

wrinkle finish A lacquer or varnish finish that may be applied with a brush or spray and dries with an attractive wrinkled surface. Often used on panels and cabinets of electronic equipment.

write To introduce information into some form of storage in a computer.

write pulse A pulse that causes information to be introduced into a magnetic cell or cells for storage purposes in a computer.

write time The time interval between the instant at which information is ready for storage and the instant at which storage is completed in a computer. Also called access time.

writing gun The gun used in a display tube to create the pattern being viewed. In a radar display tube, where the pattern must be viewed despite a high level of light coming in through the windshield of the cockpit in a fighter plane, the writing gun is used in conjunction with a flooding gun that intensifies the image.

writing speed The rate of writing on successive storage elements in a charge-storage tube.

Ws Abbreviation for *wattsecond.*

W/sr Abbreviation for *watt per steradian.*

Wullenweber antenna An antenna array that consists of two concentric circles of masts, connected to be electronically steerable. Used for ground-to-air communication at SAC bases.

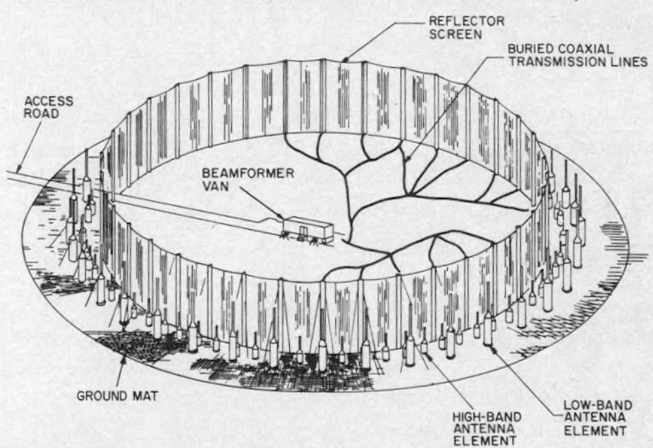

Wullenweber antenna array.

WVDC Abbreviation for *working volts DC.* Use of VDC is preferred.

WW Symbol used on diagrams to designate a *wire-wound resistor.*

WWV The call letters of a radio station maintained by the National Bureau of Standards to provide standard radio and audio frequencies and other technical services such as precision time signals and radio-propagation disturbance warnings. The station broadcasts from Boulder, Colorado, on 2.5, 5, 10, 15, 20, and 25 MHz.

WWVH The National Bureau of Standards radio station on Kauai, Hawaii, broadcasting services similar to those of WWV on 5, 10, and 15 MHz.

X Symbol for *reactance*. Inductive reactance is designated as X_L and capacitive reactance as X_C.

X axis 1. A reference axis in a quartz crystal. 2. The horizontal axis on a cathode-ray oscilloscope screen or on a graph. The corresponding vertical axis is the Y axis.

X band A radio-frequency band that extends from 5.2 to 10.9 GHz, corresponding to wavelengths of 5.77 to 2.75 cm. Frequency limits for other bands are given in the entries for *band* and *K band.* Subdivisions of the X band (all values in gigahertz) are

Xa:	5.2 –5.5	Xc:	7.0 –8.5
Zq:	5.5 –5.75	Xl:	8.5 –9.0
Xy:	5.75–6.2	Xs:	9.0 –9.6
Xd:	6.2 –6.25	Xx:	9.6 –10.0
Xb:	6.25–6.9	Xf:	10.0 –10.25
Xr:	6.9 –7.0	Xk:	10.25–10.90

X cut A quartz-crystal cut made in such a manner that the X axis is perpendicular to the faces of the resulting slab.

xenon [symbol Xe] A gaseous element. It is one of the rare gases used in some thyratrons and other gas-discharge tubes. Atomic number is 54.

xenon flash tube A flash tube that contains xenon gas, which produces an intense peak of radiant energy at a wavelength of 0.57 μm (white light) when a high DC pulsed voltage is applied between electrodes at opposite ends of the tube.

xerographic copying machine An office copying machine that uses a selenium-coated drum and xerography principles to produce a copy on paper.

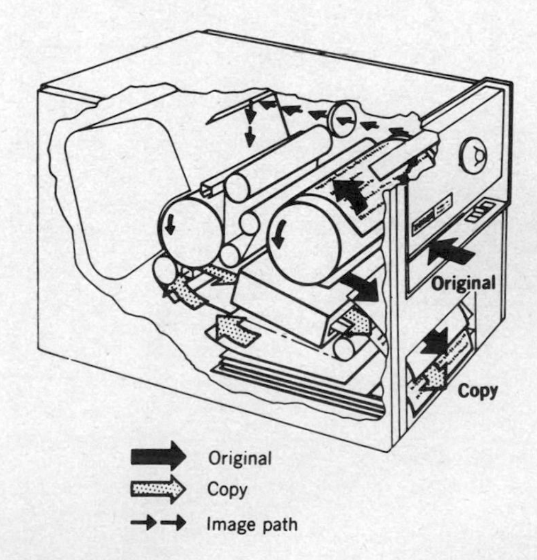

Xerographic copying machine (Xerox Corp. desktop model 813).

xerography The original form of electrophotography, in which an electrostatic image is formed on a light-sensitive selenium-coated surface when exposed to an optical image. The charged image areas attract and hold a fine black or colored resinous powder called a toner. The powder image is transferred to a sheet of paper and fused by heat to make it permanent. Invented by Chester F. Carlson in 1937. Battelle Institute undertook development of the process in 1944, and Xerox Corp. obtained commercial rights in 1947.

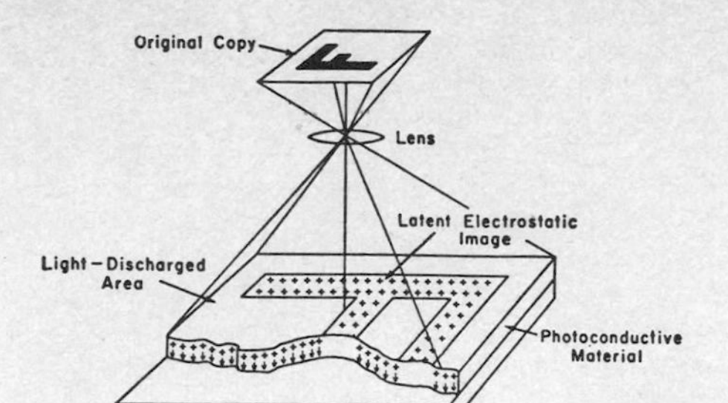

Xerography principle. Charges leak down through photoconductive selenium to grounded metal plate wherever light strikes.

xeroprinting An electrostatic image-forming process in which the printing plate is a metal substrate that has a permanent electrically insulating image. When the plate is moved under corona-charging wires, the image areas retain a static electric charge, and nonimage areas dissipate the charge. The charged areas attract and hold the black or colored powder toner. Another corona-charging process is then used to transfer the powder image to a sheet of paper, after which the image is fixed by applying heat to fuse the powder to the paper.

xeroradiography An electrostatic image-forming process in which x-rays or gamma rays form an electrostatic image on a photoconductive insulating medium. The charged image areas attract

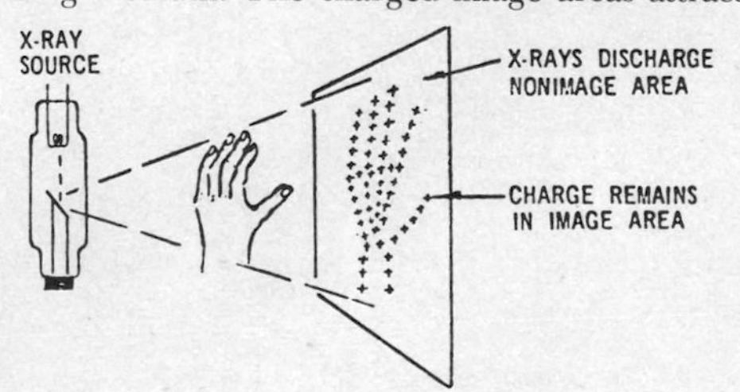

Xeroradiography gives permanent black image of hand on paper, in contrast to radiography, where image on photographic x-ray film requires chemical developing.

and hold a fine powder called a toner. The powder image is then transferred to paper and fused there by heat. Xeroradiography is one branch of electrophotography; the other is xerography, in which the images are formed by infrared, visible, or ultraviolet radiation.

X guide A surface-wave transmission line that consists of a dielectric structure which has an X-shaped cross section.

xi-zero One of the 34 elementary particles of matter. Discovered in 1959.

XOR gate *EXCLUSIVE-OR gate.*

X plate One of the two deflection electrodes used to deflect the electron beam horizontally in an electrostatic cathode-ray tube.

x-radiation Radiation of x-rays.

x-ray 1. A penetrating electromagnetic radiation similar to light but having much shorter wavelengths (from about 10^{-7} to 10^{-10} cm or 0.1 to 100 Å), between ultraviolet and gamma rays. Usually generated by accelerating electrons to high velocity and suddenly stopping them by collision with a metal target. The resulting bombardment of the atoms in the target causes the atoms to lose energy, and this energy is radiated as x-rays of definite wavelength. X-rays are also produced by transitions of atoms from higher to lower energy states. Properties of x-rays include ionization of a gas through which they pass, penetration of all solids in varying degrees, production of secondary x-rays when stopped by material bodies, and action on fluorescent screens and photographic film. Photons that originate in the nucleus of an atom are generally called gamma rays, and photons originating outside the nucleus are called x-rays. Also called roentgen ray. 2. To photograph with x-rays. 3. *X-ray photograph.*

x-ray analysis Determination of the internal structure of crystalline solids by the diffraction pattern produced when x-rays are passed through the material.

x-ray diffraction Diffraction of a beam of x-rays by the regular atomic lattice of a crystal. A characteristic diffraction pattern is obtained for each crystalline material.

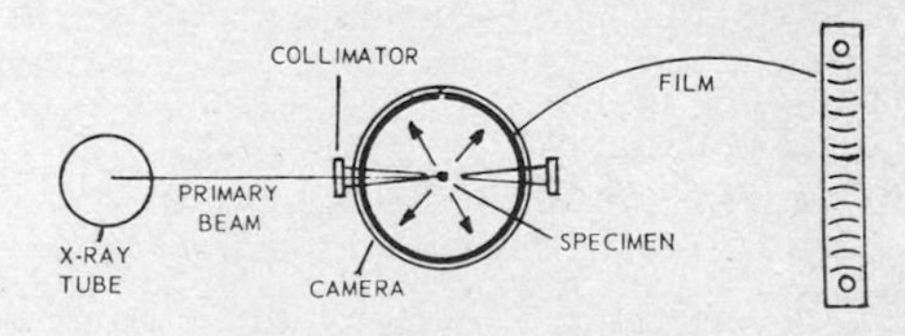

X-ray diffraction, in which atoms of specimen produce characteristic curved lines on film.

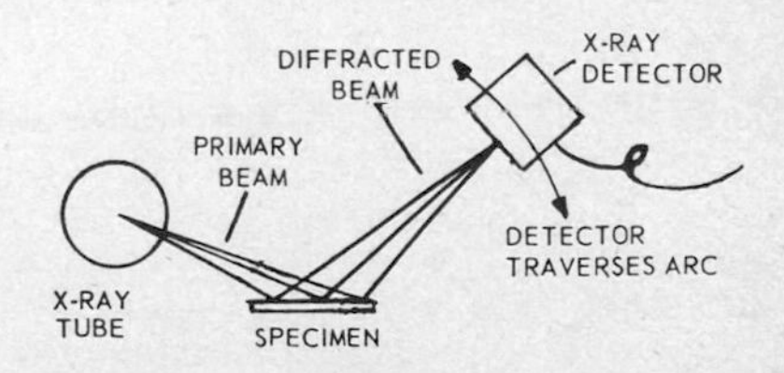

X-ray diffractometer. Output of detector may be fed to strip-chart recorder to give permanent record.

x-ray diffractometer An instrument used in x-ray analysis to measure the intensities of the diffracted beams at different angles.

x-ray film A film base coated, usually on both sides, with an emulsion designed for use with x-rays.

x-ray fluorescence absorptiometer An ab-

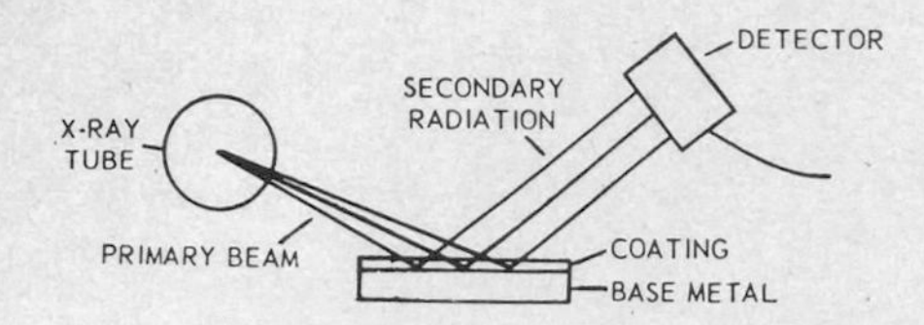

X-ray fluorescence absorptiometer.

sorptiometer that measures the thickness of a coating, such as tin on steel plate. The primary x-ray beam is directed at the coating, causing the base metal to fluoresce. The resulting secondary radiation from the fluorescence is partially absorbed by the coating, according to its thickness. The measured intensity of the secondary radiation reaching the detector is therefore a measure of the thickness of the coating.

x-ray goniometer An instrument that determines the positions of the electrical axes of a quartz crystal by reflecting x-rays from the atomic planes of the crystal.

x-ray hardness The penetrating ability of x-rays. It is an inverse function of the wavelength.

x-ray machine The x-ray tube, power supply, and associated equipment required for producing x-ray photographs.

x-ray microscope A modification of an electron microscope, in which an ultrafine-focus x-ray tube or electron gun produces an electron beam focused to an extremely small image on a transmission-type x-ray target that serves as a vacuum seal. Specimens being examined can thus be in air, as can the photographic film that records the magnified image.

x-ray photograph A radiograph made with x-rays. Also called roentgenogram (deprecated) and x-ray.

x-ray spectrogram A record of an x-ray diffraction pattern.

x-ray spectrograph An x-ray spectrometer equipped with photographic or other recording apparatus.

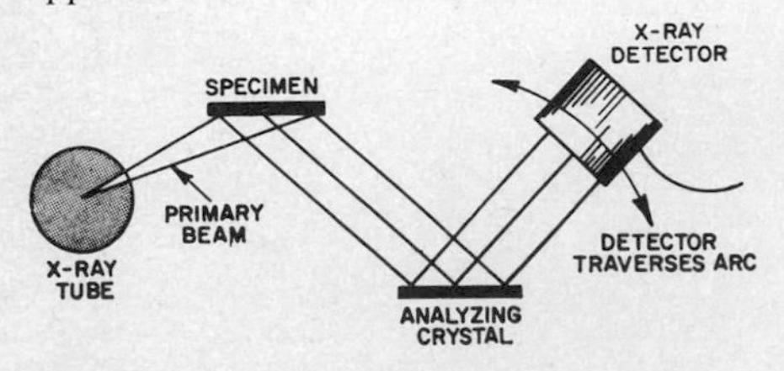

X-ray spectrometer. For x-ray spectrograph, output of detector is fed to recorder or printout mechanism.

x-ray spectrometer An instrument that produces the x-ray spectrum of a material and measures the wavelengths of the various components.

x-ray spectrum The spectrum of x-rays arranged according to wavelength, produced by electron bombardment of a target, as in an x-ray tube. It consists of a continuous spectrum on which are superimposed certain groups of much sharper lines characteristic of the element used as target. These lines, such as the K, L, and M lines, correspond to transitions between the inner energy levels of the atom.

x-ray structure The atomic structure of a substance as determined by x-ray diffraction patterns.

x-ray television Use of a closed-circuit television system in place of photographic film during x-ray inspection of welded joints and other industrial x-ray applications. The technique gives images im-

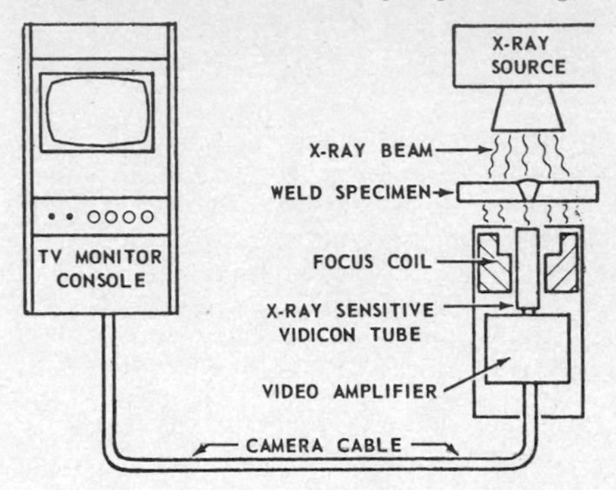

X-ray television as used for inspecting welding joints.

mediately, without the time and cost of developing film. It can show enlargements as much as 50×, for detection of small defects. For permanent records, a videotape recorder can be used, or the images on the television screen can be photographed selectively. Since the television monitor can be located remotely from the inspection area, personnel are protected from harmful x-ray radiation.

x-ray therapy Medical treatment by controlled application of x-rays. It is one type of radiotherapy.

x-ray thickness gage A thickness gage that measures and indicates the thickness of moving

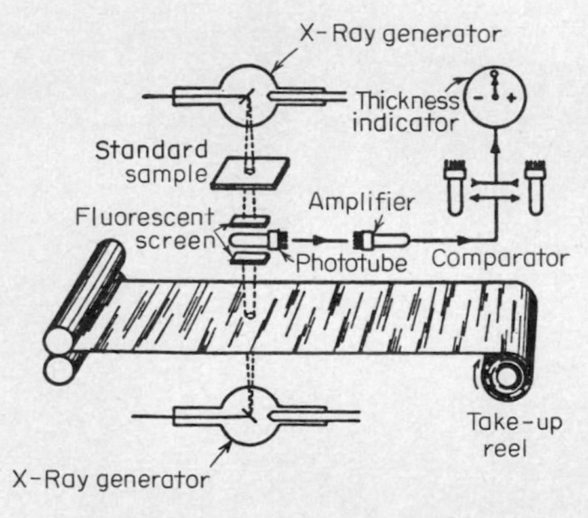

X-ray thickness gage for moving strip of metal.

cold-rolled sheet steel during the rolling process without making contact with the sheet. An x-ray beam directed through the sheet is absorbed in proportion to the thickness of the material and its atomic number.

x-ray tube A vacuum tube that produces x-rays by accelerating electrons to a high velocity by an electrostatic field, then suddenly stopping them by collision with a target.

x unit A unit of wavelength equal to 0.001 Å or 10^{-11} cm. Used for specifying wavelengths of x-rays and other highly penetrating radiations.

X wave *Extraordinary-wave component.*

XY recorder A recorder that traces on a chart the relation of two variables, neither of which is time. Sometimes the chart is moved in proportion to time, and one of the variables is so controlled that it increases in proportion to time.

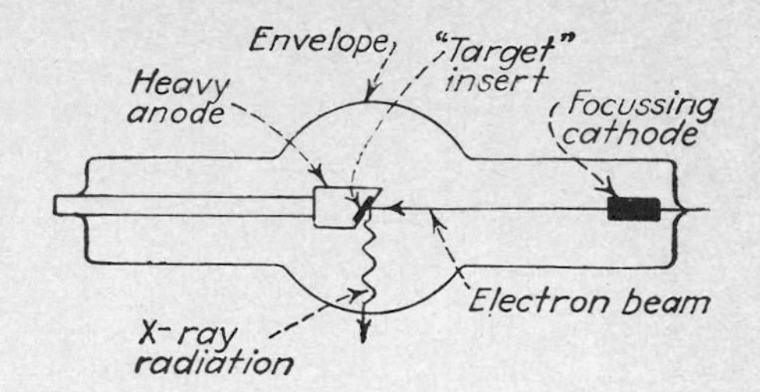

X-ray tube having heavy nonrotating anode.

XY switch A remotely controlled wiper switch in which each wiper moves in a horizontal plane, first in one direction and then at right angles to this direction.

Y

Y Symbol for *admittance.*

YAG Abbreviation for *yttrium-aluminum garnet.*

Yagi antenna An end-fire antenna array that has maximum radiation in the direction of the array line. It has one dipole connected to the transmission line and a number of equally spaced unconnected dipoles mounted parallel to the first in the same horizontal plane to serve as directors and reflectors.

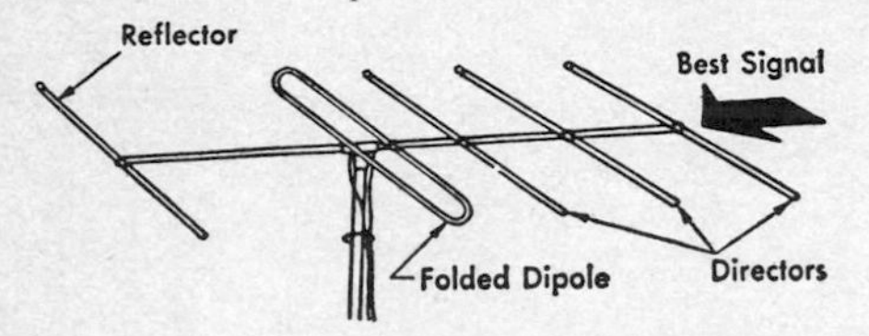

Yagi antenna for television reception.

Yag-laser *Neodymium-doped yttrium-aluminum garnet laser.*

Y antenna *Delta-matched antenna.*

Y axis 1. A reference axis in a quartz crystal. 2. The vertical axis on a cathode-ray oscilloscope screen or on a graph.

Y circulator A circulator in which three identical rectangular waveguides are joined to form a symmetrical Y-shaped configuration, with a ferrite post or wedge at its center. Power entering any waveguide will emerge from only one adjacent waveguide.

Y connection *Y network.*

Y cut A quartz-crystal cut such that the Y axis is perpendicular to the faces of the resulting slab.

Y factor A noise measurement factor used in specifying the noise figure of a receiver. Based on having known cold and hot reference temperatures.

YIG Abbreviation for *yttrium-iron garnet.*

YIG device A filter, oscillator, parametric amplifier, or other device that uses an yttrium-iron garnet crystal in combination with a variable magnetic field to achieve wideband tuning in microwave circuits.

YIG filter A filter that consists of an yttrium-iron garnet crystal positioned in a magnetic field provided by a permanent magnet and a solenoid. Tuning is achieved by varying the amount of direct current through the solenoid. The bias magnet tunes the filter to the center of the band, thus minimizing the solenoid power required to tune over wide bandwidths.

YIG-tuned parametric amplifier A parametric amplifier in which tuning is achieved by varying the amount of direct current flowing through the solenoid of a YIG filter.

YIG-tuned tunnel-diode oscillator A microwave oscillator in which precisely controlled wideband tuning is achieved by varying the current through a tuning solenoid that acts on a YIG filter in the tunnel-diode oscillator circuit.

Y junction A waveguide in which the longitudinal axes of the waveguide form a Y.

Y network A star network that has three branches. Also called Y connection.

yoke *Deflection yoke.*

Y plate One of the two deflection electrodes used to deflect the electron beam vertically in an electrostatic cathode-ray tube.

Y signal *Luminance signal.*

ytterbium [symbol Yb] A rare-earth metallic element. Atomic number is 70.

yttrium [symbol Y] A rare-earth metallic element. Atomic number is 39.

yttrium-aluminum garnet [abbreviated YAG] A crystalline material used in some solid lasers.

yttrium-iron garnet [abbreviated YIG] A crystalline material used in microwave devices.

Yukawa potential A nuclear potential used in the meson theory of nuclear forces to specify the interaction between two nucleons.

Z

Z 1. Symbol for *atomic number*. 2. Symbol for *impedance*.

Z axis The optical axis of a quartz crystal. It is perpendicular to both the X and Y axes.

Z-axis modulation *Intensity modulation.*

Z-cut crystal A quartz crystal cut in such a way that the Z axis is perpendicular to the face of the resulting slab.

Zebra time Mean time at the Greenwich meridian. Used in communication and for synchronized reckonings. The hour 2400 Zebra time is 1900 EST, 1800 CST, 1700 MST, 1600 PST, 1400 Hawaiian standard time, 1000 Sydney standard time, 0900 Tokyo standard time, 0800 Manila standard time, 0300 Moscow standard time, and 0100 Berlin standard time. Also called Z time.

Zeeman effect The increase in the number of spectrum lines produced by a light source when in a strong magnetic field.

Zeeman splitting constant A physical constant equal to 4.66858×10^{-5} cm^{-1} G^{-1}.

zener breakdown Nondestructive breakdown in a semiconductor that occurs when the electric field across the barrier region becomes high enough to produce a form of field emission which suddenly increases the number of carriers in this region.

zener diode A semiconductor breakdown diode, usually constructed of silicon, in which reverse-voltage breakdown is based on the Zener effect. Some avalanche diodes, in which breakdown is based on the avalanche effect, are often called zener diodes because the Zener effect was discovered before the theory of avalanche breakdown was formulated.

Zener effect The effect that is responsible for zener breakdown in a semiconductor.

zener noise Noise generated as a result of the breakdown phenomenon in a zener diode. It is largely white noise.

zener voltage *Breakdown voltage.*

zeppelin antenna A horizontal antenna that is some multiple of a half-wavelength long and is fed at one end by one lead of a two-wire transmission line which is also some multiple of a half-wavelength long.

zero Nothing. Most computers use both plus and minus zero. A positive binary zero is usually indicated by the absence of digits or pulses in a word, and negative binary zero (in a computer operating on one's complements) is indicated by a pulse in every pulse position in a word. In a coded decimal machine, decimal zero and binary zero may not have the same representation.

zero-access storage A computer storage for which the waiting time is negligible.

zero-address instruction A computer instruction that specifies an operation in which the locations of the operands are defined by the computer code, so no address is needed.

zero adjuster A device that adjusts the pointer position of an instrument or meter to read zero when the electrical quantity is zero.

zero beat The condition in which a circuit is oscillating at the exact frequency of an input signal, so no beat tone is produced or heard.

zero-beat reception *Homodyne reception.*

zero bias The condition in which the control grid and cathode of an electron tube are at the same DC voltage.

zero compression Any of a number of techniques used to eliminate the storage of nonsignificant leading zeros during data processing in a computer.

zero-crossing detector A comparator that determines whether the input signal is greater than or less than zero.

zero-cut crystal A quartz crystal which has been cut in such a way that its temperature coefficient with respect to frequency is essentially zero.

zero defects A program for improving product quality to the point of perfection, so there will be no failures due to defects in construction. Initiated for military electronics production after failure of million-dollar missiles because of faulty soldered joints or careless inspection of components that cost only pennies.

zero-frequency component The DC component of a complex waveform.

zero gravity The complete absence of gravitational effects; it exists when gravitational attraction is exactly nullified or is counterbalanced by centrifugal force, as in an orbiting satellite.

zero-gravity switch A switch that closes as weightlessness or zero gravity is approached. In

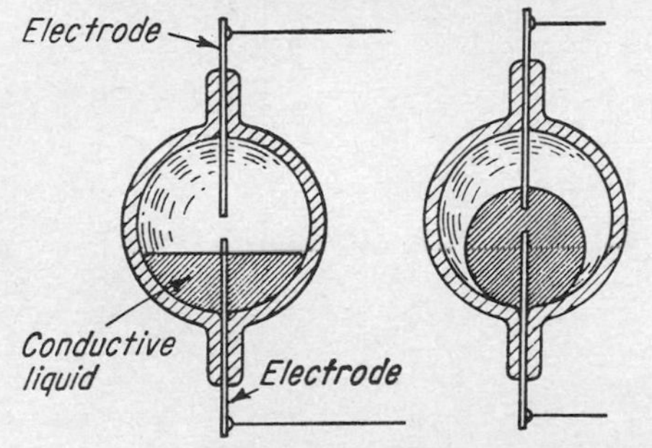

Zero-gravity switch is in normal earth-gravity position at left. In zero gravity, surface tension of mercury makes it form sphere that touches both electrodes and thereby closes switch.

one version, a conductive sphere of mercury encompasses two contacts at zero gravity but flattens away from the upper contact under the influence of gravity. Also called weightlessness switch.

zero level The reference level used for comparing sound or signal intensities. In AF work, a power of 6 mW is generally used as zero level. In sound, the threshold of hearing is generally assumed as the zero level.

zero-point energy The kinetic energy remaining in a substance at a temperature of absolute zero.

zero potential An expression usually applied to the potential of the earth, as a convenient reference for comparison.

zero shift The output of a balanced magnetic amplifier for zero control signal, due to drift.

zero stability The maximum zero shift that occurs over a given period of time during given changes in operating conditions of a balanced magnetic amplifier.

zero state A state of a magnetic cell wherein the magnetic flux through a specified cross-sectional area has a negative value, when determined from an arbitrarily specified direction for negative flux. The opposite state wherein the magnetic flux has a positive value, when similarly determined, is called a one state.

zero-subcarrier chromaticity The chromaticity that is intended to be displayed when the subcarrier amplitude is zero in a color television system.

zero suppression The elimination of nonsignificant zeros to the left of the integral part of a quantity before the results of a computer operation are printed.

zero time reference The time reference for the schedule of events in one cycle of radar operation.

zero-voltage switch A circuit which applies voltage to a load at very nearly the exact moment that an AC supply voltage reaches zero. Advantages include minimizing of RF interference.

zinc [symbol Zn] A bluish white metallic element. Atomic number is 30.

zinc-carbon cell A dry cell that has a positive electrode of carbon and a negative electrode of zinc, in an electrolyte of sal ammoniac paste.

zinc chloride cell A primary cell that uses carbon and zinc electrodes with an electrolyte which contains only zinc chloride. Nominal voltage is 1.5

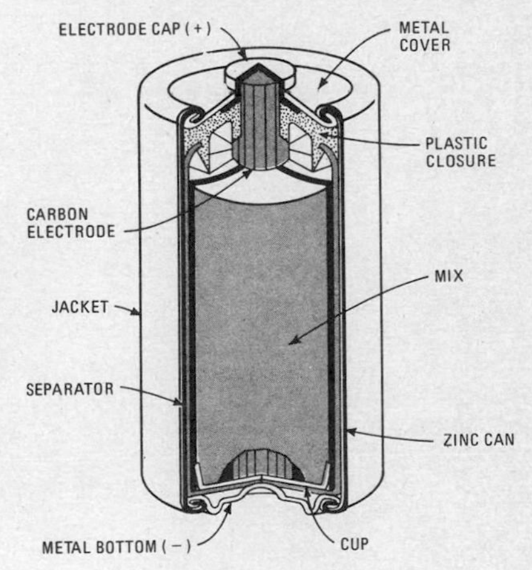

Zinc chloride cell construction.

V, just as for common Leclanche zinc-carbon cells, but current output is much higher even at low temperatures. Life is about twice that of a zinc-carbon cell under comparable conditions.

zinc–manganese dioxide primary cell A rechargeable alkaline primary cell that can have a shelf life of up to 2 years.

zinc orthosilicate The mineral willemite, used in making fluorescent screens for cathode-ray tubes. When bombarded by an electron beam, it glows with a green tint.

zinc–silver chloride primary cell A reserve primary cell that is activated by adding water. It can have a high capacity, up to 40 Wh/lb, and long life after activation.

zinc–silver oxide cell An alkaline electrolyte cell

that may be used without recharging in primary batteries or can be recharged for secondary battery use.

zinc telluride [symbol ZnTe] A semiconductor that has a forbidden-band gap of 2.2 eV and a maximum operating temperature of 780°C when used in a transistor.

zirconium [symbol Zr] A metallic element. Atomic number is 40.

zirconium lamp A high-intensity point-source lamp that has a zirconium oxide cathode in an argon-filled bulb. It is used because of its low emanation of long-wavelength light and its concentrated source.

Z marker beacon *Zone marker.*

ZnTe Symbol for *zinc telluride.*

zone leveling The passage of one or more molten zones along a semiconductor body for distributing impurities uniformly throughout the material.

zone marker A radio station that radiates signals vertically in a cone-shaped pattern to define a zone above a radio range station. Also called Z marker beacon.

zone of silence *Skip zone.*

zone purification The passage of one or more molten zones along a semiconductor for reducing the impurity concentration of part of the ingot. The semiconductor crystal is slowly moved through zones of intense heat. The crystal melts a portion at a time. As the molten region moves from one end of the crystal to the other, the impurities move with it and congregate at one end of the crystal, where they can be sawed off.

zoning The displacement of various portions of the lens or surface of a microwave reflector so the resulting phase front in the near field remains unchanged. Also called stepping.

zoom To enlarge a portion of a television picture rapidly at the transmitter, optically or electronically.

Z time *Zebra time.*

Electronics Style Manual

Contents

Contents (Continued)

TABLES

Electronics Style Manual

Time and money can be saved by following a consistent policy of spelling, hyphenating, capitalizing, and abbreviating when writing technical articles and reports. The cost of correcting style errors or inconsistencies increases exponentially from the typing process through galley and page proofs and the hand-correcting of individual finished reports. Still more costly is the time engineers spend in determining or arguing about style while they are writing their manuscripts.

This style manual presents concisely and conveniently the rules used in the fourth edition of *Electronics Dictionary*. The rules were formulated by first tabulating the practices followed in current electronics publications, books, catalogs, and ads. This picture of current usage showed such a tremendous variation in styles, often within a single issue or even within the same article, that a matrix chart was prepared to show at a glance how each of the 29 leading publications in the electronics industry handled the 60 most common classes of style problems derived from the initial tabulation. Style choices were then weighted in various ways, such as by consistency of use within a publication and by checking trends for many years back. In this way the variations for each style form were narrowed to logical and consistent rules that represent majority practices. For units of measure, the rules conform to SI units of measure adopted by the American National Standards Institute.

The final presentation in this style manual permits changing individual style forms to match strong personal or company preferences, without impairing the overall usefulness of the manual.

Many examples are given to illustrate how the rules can be applied to new problems that may arise in the future. However, when in doubt as to the handling of an unlisted word or abbreviation, try looking it up first in the body of this dictionary. For nontechnical words, use a general dictionary; the latest edition of *Webster's New Collegiate Dictionary* (published by G. & C. Merriam Co.) is highly recommended. For secretarial use when definitions are not needed, McGraw-Hill's *20,000 Words* presents the preferred spelling styles of *Webster's Collegiate* in easy-to-use form.

RULES FOR ABBREVIATIONS

Advantages and Drawbacks of Abbreviations. Familiar abbreviations contribute to faster reading and better understanding. Abbreviations save money by reducing typing, editing, composition, and lettering costs and space requirements for text, tables, and diagrams. But abbreviations can cause many headaches and errors if they are used indiscriminately without consideration for all levels of readers. Therefore, the first rule for abbreviations is to spell them out whenever there is the slightest possibility that they may not be familiar to all readers. If the unfamiliar term is used so many times that abbreviating will significantly save space, spell out the meaning of the abbreviation in parentheses after its first use.

Practically all the abbreviations in this style manual are uppercase (capital) letters in accordance with current usage. The major exceptions are units of measure, which follow the forms established by national and international agreements.

Abbreviations of Words and Phrases. Use all capital letters, run together, for abbreviations formed from one or more first letters of key words in a phrase. This rule not only eliminates worrying about hyphens within an abbreviation when the phrase is used as a compound adjective, but it also means that these abbreviations can be used as they are to begin a sentence, anywhere in a heading or article title, and on diagrams. This gives a pleasing uniformity of abbreviating style, in contrast to the awkwardness of using ac or a-c in text, AC or A-C on diagrams, and Ac, AC, A-c, or A-C in headings or at the beginning of a sentence.

Use a hyphen after such an abbreviation when the abbreviation is combined with another word to form a compound adjective, as in "three DC-coupled stages." Use a period only when an abbreviation falls at the end of a sentence.

Never insert a hyphen in an abbreviation of a compound adjective, even though the hyphen is used in the spelled-out form.

Abbreviations of phrases are generally pronounced letter by letter, but this is not a requirement for capitalizing. Thus LED can be pronounced either letter by letter or as a word. Long abbreviations such as IMPATT, JFET, MOSFET, and PMOS are usually pronounced partly or entirely as words. (Acronyms like radar and modem are treated separately in this manual because they are pronounced and written as words.)

Use a numerical exponent to indicate repetition of a letter in abbreviations for logic terms when this helps clarity and is in common use.

The diagonal (/) can mean *and, or, -to-,* or *with* when it is used in abbreviations of phrases.

The hyphen is seldom used in abbreviations. When used, the hyphen will generally have one of the meanings represented by the diagonal. Thus in A/D-D/A the hyphen stands for *or,* but in P-P it stands for *-to-*.

Do not italicize R, L, and C where these letters represent resistance or resistor, inductance or coil, and capacitance or capacitor, respectively. When R, L, and C are used in combinations, run the letters together without hyphens, as in RC coupling and RLC circuit. Italicize as *R, L,* and *C* only in mathematical equations where these letters represent values of resistance, inductance, and capacitance, respectively, and in expressions like *LC* product, *L/C* ratio, and *RC* constant.

Table 1 groups by subject representative examples of commonly used word and phrase abbreviations, making it easy to extend the abbreviating style to other terms in each category. Table 2 lists the same abbreviations alphabetically, along with their spelled-out equivalents, for quick reference when writing, typing, and proofreading. Other abbreviations, and equivalent spelled-out forms, can be found in the body of the fourth edition of *Electronics Dictionary.*

Plurals of Abbreviations. For plurals of capital-letter abbreviations, regardless of whether in text or headings, add a lowercase s. The same rule applies to plurals of numbers. Examples:

ADCs	FETs	MICs	12SK7s	0s
DACs	ICs	RAMs	1980s	1s
DPMs	LEDs	SCRs		

Use an apostrophe for the possessive form of a capital-letter abbreviation, as in "the SCR's heatsink." Use an apostrophe with plurals of lowercase letters, as in "programming has two m's."

Rules for Acronyms. An acronym is a pronounceable abbreviation coined from the initial letter or letters of each successive or major part of a compound term. The most common acronyms are normally written in lowercase letters the same as ordinary words, are pronounced as words, and form plurals the same as words (generally by adding a lowercase s, as in "modems"). These common acronyms are either initially capitalized or set in all caps in titles and tabulation headings. Table 3 gives examples of acronyms that are generally lowercased because they have become a part of the language of electronics.

New acronyms are being coined continuously, particularly for military and commercial electronic systems. At first these may be written entirely in capital letters, or the first letter may be capitalized, in text as well as in titles and headings. A few of these acronyms, such as laser, are so widely used that they are entirely lowercased in text; a few may remain capitalized or in all caps in defiance of all the logical rules of good writing. However, a great majority of these acronyms are dropped from our language as the systems become obsolete or fail to make the transition from research to production. In this edition of *Electronics Dictionary,* despite intensive research on current usage, it has proved impossible to present a consistent capitalization pattern for both older and newer acronyms because industry itself has not achieved consistency.

Computer Programming Languages. Although the names of programming languages are acronyms and therefore are pronounced as words, it is now common usage to type or print them in capital letters. Examples:

ALGOL	COBOL
BASIC	FORTRAN

SI Units of Measure. Standard letter symbols (abbreviations) for words that represent units of measure or quantity, known as SI units, were established by international agreement in 1960 for worldwide use. Additional letter symbols have been established by the American National Standards Institute. The combined standards, as presented in this section, have been adopted for mandatory use by the Department of Defense and are currently being used in almost all electronics industry publications.

When a letter symbol in the SI system of units is derived from the name of a pioneer in the field, capitalize the letter that represents the first letter of the surname. Table 4 lists the names of these pioneers to help you remember the symbols that must be capitalized.

Standard prefixes are used as multipliers with the SI letter symbols. In Table 5, these prefixes are tabulated in descending sequence of magnitude for convenience in forming letter symbols that require their use. Note that only the five largest prefixes have capital letters for their symbol.

Only two Greek letters are used in SI letter symbols: μ for the prefix micro- and Ω for ohm as the unit of resistance. These usually must be added by hand when typing. Since the only correct way to avoid using these Greek letters is to spell them out, the unit of resistance is often written out as microhms, ohms, kilohms, megohms, and so on, particularly when it is used only a few times in a report. It is also permissible to spell out any other units of measure so they can be capitalized in titles or headings. Spell out symbols whenever there is a possibility that an unfamiliar symbol may confuse some reader.

Letter symbols always represent both singular and plural forms of the units of measure. Never add a lowercase s for plural because it represents seconds in the SI system of units.

Use a period after a letter symbol only at the end of a sentence. Do not use periods after letter symbols in tabulations or diagrams. Use the diagonal (/) with letter symbols only to represent *per.*

Use an exponent after a length symbol to change it to an area or volume symbol. (The abbreviations sq for square and cu for cubic are not used with SI units.) Pronunciation is unchanged; thus in^2 is pronounced "square inches" and in^3 is pronounced "cubic inches."

Use a space between a number and its abbreviated unit of measure in noun phrases. Use a hyphen between a number and its abbreviated unit of measure in compound adjectives.

Do not use a letter symbol without a numerical value. Spell out units of measure whenever they are used with words that represent approximations. Examples:

a few kilohertz higher
several volts lower
approximately a watt
tens of kilovolts
hundreds of amperes
over a million ohms

Table 6 lists alphabetically the units of measure commonly used in electronics writing, with the approved letter symbol or abbreviation for each, and shows the correct methods of using prefix symbols and combining two or more letter symbols.

Importance of Following Style. Letter symbols must always be written *exactly as shown in Table 6,* even in headings, because capitalizing a lowercase letter in a symbol changes its meaning in many cases. Thus mW and MW represent entirely different units. Furthermore, seven of the letter symbols used in the field of electronics have distinctly different meanings in lowercase than in uppercase; Table 7 lists such symbols.

Voltage, Current, and Power Ratings. When a value is followed by one or more modifying abbreviations indicating that the value is AC, DC, peak, peak-to-peak, root-mean-square, and so forth, the styles shown in Table 8 are suitable for text, tables, headings, and diagrams. Use a space between the value and its unit of measure. For clarity, use a space after a unit of measure that is followed by a modifier such as P, PEP, PIV, and P-P; however, the space is not needed with the modifiers AC, DC, or RMS.

Plurals of Spelled-Out Units. With a few exceptions, plurals of spelled-out units are formed conventionally. Use plurals only with numbers above 1; use the singular form for values equal to or less than 1. With hyphenated symbols, make the last word plural. With symbols including *per,* make the word ahead of *per* plural. When units have modifiers, as with temperature units, make the unit itself plural. Table 9 gives examples of correct forms for plurals.

Data-Processing Units. Letter symbols and abbreviations for data units of measure are presented separately because current usage is often ambiguous and badly in need of standardization. The original SI units did not cover these units, but American Standard letter symbols have been adopted for bit and baud. Table 10 gives the forms recommended for these special units of measure.

Handling Resistance Values. Since the Greek symbol Ω is a requirement for all SI units of resistance, Table 11 summarizes examples of its correct use *in text,* for convenient reference.

Abbreviation Rules for Diagrams and Tables. Letter symbols for units of measure are used with values of components on circuit diagrams and in tables in the same manner as in text. Use lowercase and capital letters exactly as used in text. Use a space between a value and its letter symbol. Use a zero to the left of a decimal point for values less than 1.

A number of shortcuts are commonly used *on diagrams* even though they are not provided for in the SI system of letter symbols. For ohm values alongside resistor symbols, the letter symbol Ω is often omitted because the meaning of the value is obvious. For values in thousands of ohms, the capital letter K is sometimes used without a space or omega after the number. Similarly, M is used without a space or omega to represent millions of ohms. Thus a 100-ohm resistor could be labeled 100 Ω or simply 100; a 5000-ohm resistor could be labeled 5 kΩ, 5K, 5000 Ω, or simply 5000; a 2.5-megohm resistor could be labeled 2.5 MΩ, 2,500,000, 2500K, or 2.5M (2.5 MEG is equally clear but is no longer an accepted form). However, when specifying resistance values of coils and other nonresistive components, the complete letter symbol with Ω must be used for clarity. Note that a lowercase k means only *thousand* in the SI system and requires a letter symbol after it.

RULES FOR WORDS

Hyphenating of Compound Terms. More than 75 percent of the spelling questions arising in electronics writing and editing are concerned with compound terms (two words used together)—whether they should be solid, hyphenated, or open. Even general dictionaries do not always agree with each other or with the *Government Printing Office Style Manual* about the handling of compound terms.

Generally, when a compound term is used

as an adjective, separate the words by a hyphen to indicate that the entire compound term modifies the following noun. If there is no following noun, do not use the hyphen. Never use a hyphen ahead of the word *percent,* even when it is part of a compound adjective, as in "7.6 percent rejection rate." Always use a hyphen between a numerical value and a spelled-out or abbreviated unit of measure when they together modify a noun, as in "120-volt line" and "120-V line." When the first word of a compound adjective is an adverb ending in *ly,* never use the hyphen. Examples:

Switch to the broadcast band.
The broadcast-band antenna is a loop.
The set requires a 120-VAC power source.
The circuit uses up-to-date transistors.
The normally open contacts are not used.
A heavily loaded power supply may overheat.
The antenna is a 10-foot dish.
Use the 100-kilometer range.
A 10 percent increase in cost.
The vice-president is an electronics engineer.

Use of Diagonal with Compound Terms. The diagonal (/) can mean *and* or *or* when used between two words, as in "and/or," "read/write," and "up/down." (The diagonal has the same significance when used between capital-letter terms like AND/OR and ON/OFF.)

Hyphenating after Prefixes. Prefixes such as anti-, co-, de-, electro-, infra-, multi-, non-re-, semi-, sub-, super-, ultra-, and so forth, are combining forms and are normally not hyphenated. (Note, though, that the prefixes self- and quasi- are always hyphenated in electronics.) Use a hyphen after a prefix when essential for sense, as in "re-cover," meaning "to cover again"; when the prefix precedes the same prefix, as in "super-superpile"; when the letter i or a is repeated, as in "semi-infinite" or "intra-aural"; and when the prefix precedes a capitalized word or abbreviation, as in "pre-TR tube."

Plurals of Electronics Terms. Plurals are regularly formed by adding s or es to the singular, as in "card—cards" and "flux—fluxes." Irregular plurals are given in general dictionaries like *Webster's New Collegiate.* (Plurals of spelled-out units of measure are in the Rules for Abbreviations section of this Style Manual and Table 9.)

Many words have two plural forms, one of which usually can be traced back to Latin or British forms. Table 12 gives the currently preferred American plural forms for some of these words.

Problem Words. Table 13 gives electronics terms that are often misspelled and preferred spellings of terms for which there are two or more correct forms. Only the most-used form is listed. When grammatical usage determines spelling, abbreviations in parentheses identify the usage for each form of the word: adjective (adj); noun (n); verb (v).

Certain words derived from names of persons (in addition to those used as units of measure) are not capitalized because they have acquired special meanings in common usage: coulomb force; roentgenology; zener; zeppelin.

Sound-Alike Words. Table 14 lists pairs of words that frequently give trouble in electronics dictation, with brief definitions to help a secretary choose the correct dictated form from the context in which it is used.

Problems with Adjectives. When the word *electronics* is used as an adjective, two rules apply: (1) drop the s if the following noun is a circuit, device, system, or some aspect of electron flow; (2) retain the s if the following noun does not actually involve electron flow. The distinction is admittedly confusing, and the words are often used interchangeably. Thus The Institute of Electrical and Electronics Engineers follows this rule, but Electronic Industries Association and the magazines *Electronic Engineer* and *Electronic Technician/Dealer* do not.

Since the rules for adjectives are often difficult to apply, Table 15 gives representative examples of correct usage (as defined by purists) for the electronics adjectives and several others that present similar problems. Other combinations of these problem adjectives are in alphabetic sequence in this edition of *Electronics Dictionary*.

Do not waste time puzzling over an adjective that could go either way because it will probably be found both ways in current writing. For noncritical correspondence, just use the form that sounds best or comes closest to matching some other word in the appropriate list. For formal writing, where consistency of style is important, use the combinations in Table 15.

Capitalization in Titles. Always capitalize the first and last words of a title or heading. Capitalize all other words, except conjunctions (and, or, and so forth), prepositions (in, of, to, within, and so forth), articles (a, an, the), and *to* in infinitives.

Both words of a hyphenated compound adjective are normally capitalized. This is the safest rule to follow when in doubt, even though purists observe two exceptions:

1. If the second element of a compound adjective modifies the first element, do not capitalize the second word:

   Medium-sized Technical Manual

2. If both elements of a compound adjective constitute a single word, do not capitalize the second word:

   Self-sustaining Reaction
   Warm-up Time

In titles, you must spell out or capitalize letter symbols and abbreviations for units of measure exactly as you do in text. Capitalize lowercase acronyms, such as laser, modem, opamp, radar, and sonar, the same as other single words in titles. Examples:

Computing the Maximum Range of a VHF Radar Set
Rayleigh Distribution and Its Generalizations
On the Role of the Process of Reflection in Radio Wave Propagation
An Introduction to the Theory of Finite-State Machines
A Comparison of Line-of-Sight Signals at 9.6 and 34.5 GHz
How to Test FETs
Design of 455-kHz IF Transformers

Shape-Designating Letters. When a term includes a description of a shape that resembles some letter of the alphabet, use the actual capital letter instead of spelled-out versions like tee, vee, and wye. An exception is "dee," often used in place of the letter D. Examples:

dee line	J antenna	T junction
H network	L network	Y circulator

RULES FOR NUMBERS

Consistent handling of numbers is one of the most difficult aspects of technical writing because the rules vary with context, personal preference, and the nature of the writing. The following rules are examples of current good usage that can be followed safely by an engineer to achieve consistency in writing.

When to Spell Out Numbers. Use numbers as much as possible in technical writing because they are easier to read than words. But spell out a large number that starts a sentence if it is not practical to rephrase the sentence.

Numbers under 10, and sometimes 10, are often spelled out. Personal preference at the moment of writing seems to govern what is done.

Many words that are not units of measure, such as bit, byte, channel, digit, input, output, phase, pulse, quadrant, and stage, are frequently preceded by small numbers and used as either noun phrases or compound adjectives. It is common practice to spell out these small numbers if the expression is used only a few times, but numbers are generally preferable if they occur many times or with similar expressions that have numbers above 10. Use a hyphen after the number for a compound adjective, but use a space after the number for a noun phrase. Examples:

four-input gate   8-bit word

4-input gate	three-phase power
gate has four inputs	Sixty-Hz power is
gate has 4 inputs	Use 60-Hz power

Use Numbers with Units of Measure. Always use numbers with abbreviated or spelled-out units of measure. Examples:

120 V	15 years	20°C
550 kHz	200 meters	1.5 mA

Use a space between a number and its unit of measure. Omit the space only when you are deviating from SI standards by using K to represent thousands of ohms on diagrams, in text, or in tabulations. Examples:

15 ohms 1.4K 100K
273.16 K (K represents kelvins)

Handling Percentages. For occasional usage it is permissible to spell out *percent* when the word is used with figures or spelled-out values. Never use a hyphen. Always use numbers with the percent sign (%), without a space between. Examples:

five percent rate	1.5%
5 percent rate	25%

Using Zero with Decimal. Use a 0 to the left of the decimal in numbers less than 1 because sometimes the decimal point will not print clearly. Examples:

0.15 V 0.007 gram

Using Commas in Large Numbers. Use a comma between every group of three figures in large numbers, counting from the right. A comma is optional in a 4-digit number, but all such numbers should be treated alike. Two styles are equally acceptable here: use commas above 999, or use commas above 9999. Use of a space in place of a comma is also acceptable and is now preferred in the SI metric system. When space is used to separate groups of three, digits are counted from the decimal point to the left and/or right. Do not use commas in page numbers, year numbers, addresses, decimal fractions, or binary numbers. Examples:

page 1372
1976
1072 Main Street
101101 (binary)
3,000 *or* 3000 *or* 3 000
30,000 *or* 30 000
11,753.914732 *or* 11 753.914 732

Typing Fractions. Use figures separated by a diagonal for constructing fractions not provided on the typewriter. When several fractions are used in the same context and some are not on the typewriter, construct all the fractions with diagonals. Type fractions in the simplest form, even though pronounced otherwise. Examples:

5/8	1/4	1/8 inch
3-11/16	1/2	7/8 inch
15/24	5/8	3/10,000 inch

Hyphenating Spelled-Out Numbers. When spelling out compound numbers between 21 and 99, use a hyphen between the two words. When larger numbers are spelled out, as at the start of a sentence, use no other hyphens. Use figures for round numbers in hundreds, thousands, or millions. For very large numbers, the words *million* and *billion* may be used with figures for clarity. When two numbers are adjacent, spell out one. Use a hyphen between the spelled-out numerator and denominator of a fraction. Examples:

17 million
2.5 billion
forty-five
one hundred
two hundred fifty-seven
five thousand four hundred twenty-two
forty 1-volt divisions
16 two-cent stamps
two-thirds voltage
two-thirds of the space
nine-tenths of the ICs
one-millionth of a farad

Using the Degree Sign. Position the degree sign, typed as a lowercase raised letter o, between the last figure of a Celsius (centigrade), Fahrenheit, or Rankine temperature value and the capital letter for the temperature scale, without spaces. Replace the degree sign by a space in kelvin temperature values. Use the degree sign by itself after a number to designate angular degrees. Examples:

22°C	639.6°R	22.5°
70°F	273.16 K	135°

Using Apostrophes and Quotes as Symbols. The apostrophe may be used to represent feet and the quotation mark may be used for inches when many dimensions are given. In all other cases, use the abbreviations ft and in or their spelled-out forms. Use apostrophes and quotation marks for minutes and seconds of arc or angle only in combinations, as when specifying latitude or longitude. Examples:

43′ in diameter	7′ 10″ high
30″ wide	44°10′30″ longitude
a 5-ft 6-in board	8½″ × 19¼″

Logic States. Use figures for numerical logic states. Use all-capital letters for words that designate logic states if needed for emphasis or clarity. Examples:

0	HIGH	ON
1	LOW	OFF

Ranges of Values. When a hyphen (en dash in printed material) or the word *to* is used between two numbers that are the upper and lower limits of a range, the meaning is "up to and including." Use the word *to* in place of the hyphen chiefly to avoid awkwardness when the two limit values have different units of measure or when one or both values have polarity signs. Do not repeat units of measure unless they are different. Examples:

550–1500 kHz band	0 to 64°C
90–135 V range	−5 to +5 V
range of 90 to 135 V	0.1–200 mA
range of 90–135 V	25–100,000 Hz
pages 73–97	25 Hz to 100 kHz
−20 to +50°C	0 to 2.5 MHz

Use hyphens after numbers in compound adjectives when a choice or range is separated by *and, or,* or *to.* When giving an option of two values, separate them with the word *or* rather than a diagonal or hyphen to avoid ambiguity. Examples:

single- and three-phase power
120- or 240-volt line
120 or 240 VAC
25- to 20,000-Hz response
90- to 135-V range
50 or 60 Hz

TABLE 1 Subject Listing of Abbreviations for Common Words and Phrases

Circuits	ABC AFC AGC AVC C FFT L LC PLL R RC RLC TRF TWA
Computers	ADP ALU BCD CAD CAI CAM COM CPU CROM DDC DMA DOS EAROM EBCDIC EDP EPROM FPLA I/O MCU MICR MPU μC μP NC NRZ NRZI OCR OS PROM RAM ROM RZ
Construction	DIP IC MIC PC SIP
Converters	A/D ADC A/D-D/A D/A DAC D/S DSC F/V F/V-V/F S/D SDC V/F
Devices	MHD NC NMR NO PM SAW TR TWT
Displays	CRO CRT LCD LED MTI PPI
Equipment	ADF BMEWS DF DME DOMSAT DUT ECG EEG ERTS EVR GCA IFF ILS LP MARISAT OEM PA PABX PAR PBX RC RDF RPV SAM SRAM VOR VTR
Frequency	AF EHF ELF HF IF IR LF MF MUF RF SHF SW UHF UV VHF VLF
Interference	ECCM ECM EMC EMI EW RFI TVI
Logic	C^3L CML DCTL DTL DTL/TTL ECL HTL I^2L RTL T^2L TTL
Materials	KDP PLZT PVC YAG YIG
Measuring	DMM DPM DVM FS VOM VTVM
Modulation	AM AM/FM CW FDM FM FSK PAM PCM PCM/FM PM PPM PSK PTM QAM QPSK SC SCA SSB SSB/AM TDM VSB
Oscillators	BFO BWO COHO ECO LO MOPA MVBR STALO VCO VFO
Parameters	AWG BW CGS CMRR EMF EMU ERP ESU G LPM LSB MKS MSB P PEP PF PIV P-P PPM PRF PRR RMS S/N SNR SWR TC THD TPI VSWR VU WPM
Power	AC AC/DC DC DC/DC EHV HVDC UHV UPS
Production	AQL FIFO LIFO MTBF
Services	CATV CB CCTV EME ETV ITV MATV NTSC PAL SECAM SSTV TV TWX
Solid-State	BARITT CATT CCD CMOS DMOS FET IMPATT JFET LDR LSA LSI MNOS MOS MOSFET MOST MSI N NMOS NPN NPNP P PIN PMOS PN PNP PNPN SCR SOS SSI TRAPATT UJT
Switches	DPDT DPST 4PDT 4PST SPDT SPST

TABLE 2 Alphabetic Listing of Abbreviations for Common Words and Phrases

Abbreviation	Meaning
ABC	automatic bass compensation; automatic brightness control
AC	alternating current
AC/DC	AC or DC
A/D	analog-to-digital
ADC	analog-to-digital converter
A/D-D/A	analog-to-digital or digital-to-analog
ADF	automatic direction finder
ADP	automatic data processing
AF	audio frequency
AFC	automatic frequency control
AGC	automatic gain control
ALU	arithmetic-logic unit
AM	amplitude modulation
AM/FM	AM or FM
AQL	acceptable quality level
AVC	automatic volume control
AWG	American wire gage
BARITT	barrier injection transit time
BCD	binary-coded decimal
BFO	beat-frequency oscillator
BMEWS	ballistic-missile early-warning system
BW	bandwidth
BWO	backward-wave oscillator
C	capacitance; capacitor
CAD	computer-aided design
CAI	computer-aided instruction
CAM	content-addressable memory
CATT	controlled avalanche transit-time triode
CATV	cable TV
CB	citizens band
CCD	charge-coupled device
CCTV	closed-circuit TV
C^3L	complementary constant-current logic
CGS	centimeter-gram-second
CML	current-mode logic
CMOS	complementary MOS
CMRR	common-mode rejection ratio
COHO	coherent oscillator
COM	computer-output microfilm
CPU	central processing unit
CRO	cathode-ray oscilloscope
CROM	control and read-only memory
CRT	cathode-ray tube
CW	continuous wave
D/A	digital-to-analog
DAC	digital-to-analog converter
DC	direct current
DC/DC	DC to DC
DCTL	direct-coupled transistor logic
DDC	direct digital control
DF	direction finder
DIP	dual in-line package
DMA	direct memory access
DME	distance-measuring equipment
DMM	digital multimeter
DMOS	double-diffused MOS
DOMSAT	domestic satellite
DOS	disk operating system
DPDT	double-pole double-throw
DPM	digital panel meter
DPST	double-pole single-throw
D/S	digital-to-synchro
DSC	digital-to-synchro converter
DTL	diode-transistor logic
DTL/TTL	DTL or TTL
DUT	device under test
DVM	digital voltmeter
EAROM	electrically alterable ROM
EBCDIC	extended binary-coded decimal interchange code
ECCM	electronic counter-countermeasure
ECG	electrocardiogram
ECL	emitter-coupled logic
ECM	electronic countermeasure
ECO	electron-coupled oscillator
EDP	electronic data processing
EEG	electroencephalogram
EHF	extremely high frequency
EHV	extra-high voltage
ELF	extremely low frequency
EMC	electromagnetic compatibility
EME	earth-moon-earth
EMF	electromotive force
EMI	electromagnetic interference
EMU	electromagnetic unit
EPROM	erasable PROM
ERP	effective radiated power
ERTS	earth resources technology satellite
ESU	electrostatic unit
ETV	educational TV
EVR	electronic video recording
EW	electronic warfare
FDM	frequency-division multiplex
FET	field-effect transistor
FFT	fast Fourier transform
FIFO	first-in first-out
FM	frequency modulation
4PDT	four-pole double-throw
4PST	four-pole single-throw
FPLA	field-programmable logic array
FS	full scale

FSK	frequency-shift keying
F/V	frequency-to-voltage
F/V-V/F	frequency-to-voltage or voltage-to-frequency
G	gravitational force
GCA	ground-controlled approach
HF	high frequency
HTL	high-threshold logic
HVDC	high-voltage DC
IC	integrated circuit
IF	intermediate frequency
IFF	identification, friend or foe
ILS	instrument landing system
IMPATT	impact avalanche transit time
I/O	input/output
IR	infrared
I^2L	integrated injection logic
ITV	industrial TV
JFET	junction FET
KDP	potassium dihydrogen phosphate
L	coil; inductance
LC	inductance-capacitance
LCD	liquid crystal display
LDR	light-dependent resistor
LED	light-emitting diode
LF	low frequency
LIFO	last-in first-out
LO	local oscillator
LP	long-play
LPM	lines per minute
LSA	limited space-charge accumulation
LSB	least significant bit
LSI	large-scale integration
MARISAT	maritime satellite
MATV	master-antenna TV
MCU	microprogram control unit
MF	medium frequency
MHD	magnetohydrodynamics
MIC	microwave IC
MICR	magnetic-ink character recognition
MKS	meter-kilogram-second
MNOS	metal-nitride-oxide semiconductor
MOPA	master oscillator–power amplifier
MOS	metal-oxide semiconductor
MOSFET	metal-oxide semiconductor FET
MOST	metal-oxide semiconductor transistor
MPU	microprocessing unit
MSB	most significant bit
MSI	medium-scale integration
MTBF	mean time between failures
MTI	moving-target indicator
μC	microcomputer
MUF	maximum usable frequency
μP	microprocessor
MVBR	multivibrator
N	negative
NC	normally closed; numerical control
NMOS	N-channel MOS
NMR	nuclear magnetic resonance
NO	normally open
NPN	negative-positive-negative
NPNP	negative-positive-negative-positive
NRZ	nonreturn-to-zero
NRZI	nonreturn-to-zero-inverted
NTSC	National Television System Committee
OCR	optical character reader
OEM	original equipment manufacturer
OS	operating system
P	peak; positive
PA	public address
PABX	private automatic branch exchange
PAL	phase-alternation line
PAM	pulse-amplitude modulation
PAR	precision approach radar
PBX	private branch exchange
PC	printed circuit
PCM	pulse-code modulation
PCM/FM	PCM on FM
PEP	peak envelope power
PF	power factor
PIN	positive-intrinsic-negative
PIV	peak inverse voltage
PLL	phase-locked loop
PLZT	lead lanthanum zirconate titanate
PM	permanent magnet; phase modulation
PMOS	P-channel MOS
PN	positive-negative
PNP	positive-negative-positive
PNPN	positive-negative-positive-negative
P-P	peak-to-peak
PPI	plan-position indicator
PPM	parts per million; pulse-position modulation
PRF	pulse repetition frequency
PROM	programmable ROM
PRR	pulse repetition rate
PSK	phase-shift keying
PTM	pulse-time modulation
PVC	polyvinyl chloride
QAM	quadrature amplitude modulation
QPSK	quaternary phase-shift keying

TABLE 2 *(Continued)*

R	resistance; resistor
RAM	random-access memory
RC	resistance-capacitance
RDF	radio DF
RF	radio frequency
RFI	RF interference
RLC	resistance-inductance-capacitance
RMS	root-mean-square
ROM	read-only memory
RPV	remotely piloted vehicle
RTL	resistor-transistor logic
RZ	return-to-zero
SAM	surface-to-air missile
SAW	surface acoustic wave
SC	suppressed carrier
SCA	Subsidiary Communications Authorization
SCR	silicon controlled rectifier
S/D	synchro-to-digital
SDC	synchro-to-digital converter
SECAM	sequential with memory (French)
SHF	superhigh frequency
SIP	single in-line package
S/N	signal-to-noise
SNR	signal-to-noise ratio
SOS	silicon-on-sapphire
SPDT	single-pole double-throw
SPST	single-pole single-throw
SRAM	short-range attack missile
SSB	single-sideband
SSB/AM	SSB with AM
SSI	small-scale integration
SSTV	slow-scan TV
STALO	stable local oscillator
SW	shortwave
SWR	standing-wave ratio
TC	temperature coefficient
TDM	time-division multiplex
THD	total harmonic distortion
TPI	tracks per inch
TR	transmit-receive
TRAPATT	trapped plasma avalanche transit time
TRF	tuned radio frequency
T^2L	transistor-transistor logic
TTL	transistor-transistor logic
TV	television
TVI	TV interference
TWA	traveling-wave amplifier
TWT	traveling-wave tube
TWX	teletypewriter exchange service
UHF	ultrahigh frequency
UHV	ultrahigh voltage
UJT	unijunction transistor
UPS	uninterruptible power system
UV	ultraviolet
VCO	voltage-controlled oscillator
V/F	voltage-to-frequency
VFO	variable-frequency oscillator
VHF	very high frequency
VLF	very low frequency
VOM	volt-ohm-milliammeter
VOR	VHF omnirange
VSB	vestigial sideband
VSWR	voltage standing-wave ratio
VTR	videotape recording
VTVM	vacuum-tube voltmeter
VU	volume unit
WPM	words per minute
YAG	yttrium-aluminum garnet
YIG	yttrium-iron garnet

TABLE 3 Acronyms Handled Like Words

laser	opamp	sonar
logamp	paramp	tacan
loran	radar	teleran
magamp	radome	telex
maser	selsyn	vortac
modem	shoran	

TABLE 4 SI Letter Symbols Derived from Pioneers

Unit	Symbol	Name
ampere	A	Ampere
bel	B	Bell
coulomb	C	Coulomb
curie	Ci	Curie
farad	F	Faraday
gauss	G	Gauss
gilbert	Gb	Gilbert
henry	H	Henry
hertz	Hz	Hertz
joule	J	Joule
kelvin	K	Kelvin
lambert	L	Lambert
maxwell	Mx	Maxwell
neper	Np	Neper
oersted	Oe	Oersted
ohm	Ω	Ohm
roentgen	R	Roentgen
siemens	S	Siemens
tesla	T	Tesla
volt	V	Volta
watt	W	Watt
weber	Wb	Weber

TABLE 5 Magnitude Prefixes Used with Letter Symbols for SI Units

Multiplier	Symbol	Prefix
10^{18}	E	exa-
10^{15}	P	peta-
10^{12}	T	tera-
10^{9}	G	giga-
10^{6}	M	mega-
10^{3}	k	kilo-
10^{-1}	d	deci-
10^{-2}	c	centi-
10^{-3}	m	milli-
10^{-6}	μ	micro-
10^{-9}	n	nano-
10^{-12}	p	pico-
10^{-15}	f	femto-
10^{-18}	a	atto-

TABLE 6 Letter Symbols for SI and Other Units of Measure

Unit	Symbol	Unit	Symbol
ampere	A	gigahertz	GHz
ampere-hour	Ah	gigawatt	GW
ampere per meter	A/m	gigohm	GΩ
angstrom	Å	gilbert	Gb
attoampere	aA	gram	g
bar	bar	henry	H
barn	b	hertz	Hz
baud	Bd	horsepower	hp
bel	B	hour	h
bit	b	inch	in
bit per second	b/s	inch per second	in/s
calorie	cal	joule	J
candela	cd	kelvin	K
candela per square meter	cd/m^2	kiloampere	kA
centimeter	cm	kilobaud	kBd
circular mil	cmil	kilobit	kb
coulomb	C	kiloelectronvolt	keV
cubic centimeter	cm^3	kilogauss	kG
cubic foot per minute	ft^3/min	kilogram	kg
cubic meter	m^3	kilohertz	kHz
cubic meter per second	m^3/s	kilohm	kΩ
curie	Ci	kilometer	km
decibel	dB	kilometer per hour	km/h
degree (plane angle)	°	kilovolt	kV
degree Celsius	°C	kilovoltampere	kVA
degree Fahrenheit	°F	kilowatt	kW
degree Rankine	°R	kilowatthour	kWh
dyne	dyn	lambert	L
electronvolt	eV	liter	L*
erg	erg	lumen	lm
farad	F	lumen per square foot	lm/ft^2
femtoampere	fA	lumen per square meter	lm/m^2
femtovolt	fV	lumen per watt	lm/W
femtowatt	fW	lumen second	lm·s
foot	ft	lux	lx
foot per minute	ft/min	maxwell	Mx
foot per second	ft/s	megabar	Mbar
gauss	G	megabit	Mb
gigabit	Gb	megaelectronvolt	MeV
gigaelectronvolt	GeV	megahertz	MHz

*In 1976, the capital L as the symbol for liter (and mL as the symbol for milliliter) was adopted by the National Bureau of Standards and the American National Metric Council.

Unit	Symbol
megampere	MA
megavolt	MV
megawatt	MW
megawatthour	MWh
megohm	MΩ
meter	m
microampere	μA
microbar	μbar
microfarad	μF
microhenry	μH
microhm	$\mu\Omega$
micrometer	μm
microsecond	μs
microsiemens	μS
microvolt	μV
microwatt	μW
mil	mil
mile	mi
mile per hour	mi/h
milliampere	mA
millibar	mbar
milligauss	mG
milligram	mg
millihenry	mH
milliliter	mL
millimeter	mm
milliohm	mΩ
millirad	mrd
milliradian	mrad
milliroentgen	mR
millisecond	ms
millisiemens	mS
millivolt	mV
milliwatt	mW
minute (time)	min
nanoampere	nA
nanofarad	nF
nanohenry	nH
nanometer	nm
nanosecond	ns
nanovolt	nV
nanowatt	nW
nautical mile	nmi
neper	Np
oersted	Oe
ohm	Ω
ohm per volt	Ω/V
picoampere	pA
picofarad	pF
picosecond	ps
picowatt	pW
pound	lb
rad	rd
radian	rad
revolution per minute	r/min
revolution per second	r/s
roentgen	R
second (time)	s
siemens	S
square foot	ft^2
square inch	in^2
square meter	m^2
square mile	mi^2
square yard	yd^2
steradian	sr
teraelectronvolt	TeV
terahertz	THz
teraohm	TΩ
tesla	T
var	var
volt	V
voltampere	VA
watt	W
watthour	Wh
watt per steradian	W/sr
weber	Wb
yard	yd

TABLE 7 Look-Alike Letter Symbols

B	bel	b	barn
C	coulomb	c	cycle
G	gauss	g	gram
H	henry	h	hour
R	roentgen	r	revolution
S	siemens	s	second

TABLE 8 Abbreviation Style for Voltage, Current, and Power Ratings

120 V	mVAC	mV P	mVRMS
17 μV	VAC	V P	VRMS
20-A fuse	kVAC	kV P	kVRMS
50-mA meter	mVDC	μA P-P	V PIV
15-W amplifier	VDC	mA P-P	kV PIV
80-mW output	kVDC	A P-P	kW PEP

TABLE 9 Examples of Plurals for Spelled-Out Units of Measure

Singular	Plural
ampere	amperes
ampere-hour	ampere-hours
ampere per meter	amperes per meter
circular mil	circular mils
cubic meter	cubic meters
degree Fahrenheit	degrees Fahrenheit
electronvolt	electronvolts
foot	feet
gauss	gauss
henry	henrys
hertz	hertz
horsepower	horsepower
inch	inches
kilohm	kilohms
lux	lux
megohm	megohms
mho	mhos
mil	mils
siemens	siemens
voltampere	voltamperes
watthour	watthours
0.95 volt	1.05 volts

TABLE 10 Symbols and Abbreviations for Data-Processing Units

Unit	Symbol	Notes
baud	Bd	The unit of telegraph and teleprinter signaling speed. It is expressed in code elements per second, in which a code element is equal to one pulse plus one space. Thus a teleprinter handling 50 pulses per second is operating at 50 bauds. Do not use "per second" or "/s" with baud because the word itself includes this. Multiples of baud are not needed or used.
bit	b	A bit is a single character of two-valued (binary) computer language that represents the smallest unit of information being handled. A bit is a choice of two alternatives, such as ON or OFF, 0 or 1, positive or negative, magnetized or unmagnetized, hole or no hole, pulse or no pulse, etc.
gigabit	Gb	
kilobit	kb	
megabit	Mb	
bit per second	b/s	
kilobit per second	kb/s	
megabit per second	Mb/s	
bit per inch	b/in	
byte	None	A byte is the smallest number of bits processed as a unit by a given computer. Always spell out to avoid confusion with bit. This is an ambiguous term because its meaning depends on the computer or processor it is applied to; see definition of byte in this edition of *Electronics Dictionary.* Define meaning when possible: 8-bit byte; 16-bit byte; 32-bit byte; etc. Use of the letter symbol B for byte is common but is deprecated, as is use of KB and MB. Use of K alone for "thousands of bytes" is ambiguous and also deprecated.
kilobyte	kbyte	
megabyte	Mbyte	
byte per second	byte/s	
kilobyte per second	kbyte/s	
megabyte per second	Mbyte/s	
word	None	A word is a set of bits or characters that occupies one storage location in a computer and is treated as a unit. A word may have a fixed or variable length, depending on the computer. Abbreviating is not recommended.
kiloword	None	
character	c	A character is a letter, number, or symbol, such as ≠, /, =, +, etc., that is represented by a group of bits.
character per inch	c/in	
character per second	c/s	
line	None	
line per inch	line/in	
thousands	K	Deprecated, though widely used by itself to mean roughly "thousands of bits," "thousands of bytes," or "thousands of words"; the reader has to determine the intended meaning by context or by guessing. Actually, when K is used in this way, it is a binary value representing the power of 2 just above the indicated thousands; thus K may be used alone to mean 1024 bits. A "4K RAM" is actually a "4096-bit RAM."
inch per second	in/s	Used for tape search speed.

TABLE 11 Correct Handling of SI Symbol for Ohms

Noun	Spelled-Out Form	Adjective	Spelled-Out Form
0.5 μΩ	0.5 microhm	0.5-μΩ	0.5-microhm
5 μΩ	5 microhms	5-μΩ	5-microhm
5 mΩ	5 milliohms	5-mΩ	5-milliohm
5 Ω	5 ohms	5-Ω	5-ohm
5 kΩ	5 kilohms	5-kΩ	5-kilohm
5 MΩ	5 megohms	5-MΩ	5-megohm
5 GΩ	5 gigohms	5-GΩ	5-gigohm
5 TΩ	5 teraohms	5-TΩ	5-teraohm

TABLE 12 Preferred Plural Forms

Singular	Plural	Singular	Plural
antenna	antennas	locus	loci
axis	axes	matrix	matrices
chassis	chassis	maximum	maxima
criterion	criteria	nucleus	nuclei
data*	data	phenomenon	phenomena
echo	echoes	quantum	quanta
focus	foci	radius	radii
formula	formulas	spectrum	spectra
helix	helices	vacuum	vacuums
index	indexes	zero	zeros

**Data* is used with a singular verb when thinking of a group as a whole. Example: Data is lost when power fails. *Data* is used with a plural verb when thinking of individual members of a group. Example: Data are acquired from multiple sources.

TABLE 13 Preferred Spellings

acoustooptical
acoustooptics
adapter
alphameric
antiaircraft
antihunt
antijamming
antireflection
arithmetic (adj, n)
arrester
back lobe
back porch (n)
back-porch (adj)
back up (v)
backup (adj, n)
band edge (n)
band-edge (adj)
band-elimination (adj)
bandpass (adj, n)
band-rejection (adj)
band selector (n)
band-selector (adj)
band spectrum (n)
bandspread
bandstop
bandwidth
bidirectional
bipolar
bistable
blackbody
broadband
bypass
cancel, -able, -ed, -ing
cancellation
cassette
catalog
Chebyshev
citizens band
class A
coaxial
cochannel
coercive
coincidence
control, controlled, controller, controlling
cooperate
coordinate
coulomb force (*but* Coulomb's law)
couterclockwise
countermeasure
cut off (v)
cutoff (adj, n)
deemphasis
deenergize
diagramed
discrete
disk
disk file
Doppler
downconverter
down counter
down Doppler
downlink
drop-in (adj, n)
dropout
electrooptical
electrooptics
Fahnestock
fan-in
fan-out
floppy disk
fluorescent
focused
follow up (v)
followup (adj, n)
footcandle
footlambert
gage
gases
Gaussian
gray
gyrocompass
half-life
heatsink
infrared
in-house
in-line
install, -ed, -ing
installment
Kirchhoff
label, -ed, -ing
lead-in (adj, n)
logic (adj, n)
mainframe
measurable
metallization
metallize
metallizing
microwave
movable
noninductive
nonlinear
numerical
on-line
optical
optics
optoelectronics
optoisolator
overall
parallel, -ed, -ing
passband
percent
permissible
plug-in
programmable
programmed, programmer, programming
reentrant
reentry
roentgenology
sawtooth
self-excited
self-starting
self-synchronous
semiactive
semiduplex
servoamplifier
short-circuit (adj, n)
shortwave (adj, n)
sideband
signaled, signaling
stand by (v)
standby (adj, n)
standoff
step down (v)
step-down (adj, n)
step up (v)
step-up (adj, n)
subassembly
subcarrier
submillimeter
subtracter
sulfur
totaled, totaling
trademark
trade name
transferred
traveled, traveling
tunable
ultrasonic
ultraviolet
upconverter
up counter
up Doppler
up/down counter
uplink
usable
usage
vendor
videotape
warm up (v)
warm-up (adj, n)
waveguide
wavelength
wideband
x-ray (adj, n, v)
XY recorder
zener (*but* Zener effect)
zeppelin

TABLE 14 Sound-Alike Words

accept	to receive or endure
except	exclusion of
affect	to act upon
effect	the result of an action
aural	pertaining to hearing
oral	pertaining to speaking
bases	plural of base and basis
basis	foundation
complement	mutually enhance; total requirement
compliment	flattering remark
discreet	tactful
discrete	having distinct parts
extant	existing
extent	range
fuse	electric device
fuze	explosive device
igniter	a heating device
ignitor	a switching-tube electrode
its	possessive pronoun
it's	contraction of "it is" or "it has"
lightening	lessening; becoming lighter
lightning	discharge of electricity from a cloud
material	consisting of matter; important
materiel	supplies, especially military
scalar	having magnitude but not direction
scaler	pulse counter
sign	an information-bearing mark
sine	trigonometric function

TABLE 15 Matching Adjectives with Nouns

acoustic absorber, absorptivity, calibrator, clarifier, compensator, compliance, component, constant, damping, delay line, device, dispersion, duct, element, energy, environment, feedback, filter, flaw detection, generator, homing, horn, impedance, intensity, interferometer, labyrinth, lens, level, load, mass, material, measurement, medium, memory, mine, network, output, pickup, plaster, power, propagation, property, radiation, reactance, reflectivity, refraction, resistance, resonator, scattering, signal, source, stiffness, structure, system, transducer, transmittivity, transparency, wave	**acoustical** analogy, constant, engineer, equivalent, length, measurement, ohm, phase constant, phenomenon, problem, propagation constant, reciprocity theorem, theorem, unit
communication antenna, band, channel, circuit, countermeasure, receiver, satellite, system, transmitter	**communications** consultant, corporation, course, department, diagram, dictionary, draftsman, engineer, formula, industry, laboratory, manual, publication, rating, report, serviceman, standard, symbol, technician, test, theorem, training
electric anesthesia, apparatus, appliance, arc, attraction, cable, charge, circuit, clock, component, conductance, conduction, conductivity, conductor, connection, contact, continuity, controller, current, cycle, device, dipole, discharge, disturbance, doublet, elements, energy, equipment, field, flux, generator, heating, image, impedance, impulse, inertia, input, instrument, intensity, interference, lamp, layer, line of force, machine, meter, motor, network, noise, output, polarization, power, propulsion, pulse, repulsion, resistance, resistivity, shock, signal, spark, storm, strength, system, transducer, twinning, typewriter, vector, wave, wind, wire	**electrical** analog, analogy, angle, axis, center, characteristic, constant, degree, department, diagram, distance, effect, engineer, engineering, equation, formula, height, industry, insulation, knowledge, law, length, measurement, midpoint, phase, phenomenon, property, publication, quantity, rating, resistance, size, standard, symbol, tape, term, test, theorem, transcription, unit, value, zero
electronic apparatus, charge, circuit, clock, component, control, counter, countermeasure, device, discharge, efficiency, engineering, equipment, heating, ignition, instrument, interference, music, organ, power supply, system, test equipment, tester, tuning	**electronics** consultant, corporation, course, department, diagram, dictionary, draftsman, engineer, formula, industry, laboratory, manual, publication, rating, report, serviceman, standard, symbol, technician, test, theorem, training

NOTES

NOTES

NOTES

NOTES

NOTES

NOTES

NOTES

NOTES

NOTES

NOTES